[redacted]

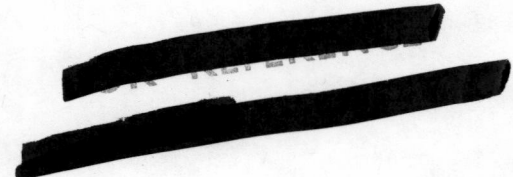

Handbook of CONCRETE ENGINEERING

Handbook of CONCRETE ENGINEERING

Second Edition

edited by **Mark Fintel**

VNR VAN NOSTRAND REINHOLD COMPANY
New York

Library of Congress Catalog Card Number: 84-7359
ISBN: 0-442-22623-3

Manufactured in the United States of America

Published by Van Nostrand Reinhold Company Inc.
135 West 50th Street
New York, New York 10020

Van Nostrand Reinhold Company Limited
Molly Millars Lane
Wokingham, Berkshire RG11 2PY, England

Van Nostrand Reinhold
480 Latrobe Street
Melbourne, Victoria 3000, Australia

Macmillan of Canada
Division of Gage Publishing Limited
164 Commander Boulevard
Agincourt, Ontario M1S 3C7 Canada

15 14 13 12 11 10 9 8 7 6 5 4 3 2

Library of Congress Cataloging in Publication Data
Main entry under title:

Handbook of concrete engineering.

Includes index.
1. Concrete construction–Handbooks, manuals, etc.
I. Fintel, Mark.
TA682.H36 1984 624.1'834 84-7359
ISBN 0-442-22623-3

Preface to Second Edition

The significant changes that have taken place in the state-of-the-art of concrete engineering and construction during the last ten years with the extreme success and popularity in the United States and abroad of the first edition of this handbook published in 1974 has made it imperative to issue this second edition. The following were the main contributors to the changes: (1) many innovations in design techniques have been introduced in the interim, particularly as a result of the wide use of computers in design offices thus allowing extreme sophistication of analysis; (2) the ACI code that serves as the basis for concrete design has undergone two significant revisions—in 1977, again in 1983. This latest edition of the handbook is based on the 1983 version of the ACI Code; (3) progress in construction equipment and methods; (4) higher strength materials.

The charts and tables of the chapter on the proportioning of sections are based on the metric system, making it possible for designers in Canada and overseas to apply the metric version of the 1983 Code to their own designs in the metric system.

Several chapters in the first edition have been deleted and replaced by new ones on post-tensioned slab systems (which are widely used in buildings), parking structures (due to their popularity), and structural plain concrete (has now been introduced in the ACI Code). All other chapters (except that covering tubular structures) have been thoroughly revised to reflect the changes brought about by the 1977 and 1983 ACI Codes and to incorporate the changes in the state-of-the-art since publication of the first edition of the handbook ten years ago.

MARK FINTEL

Preface to First Edition

This handbook contains up-to-date information on planning, design, analysis, and construction of engineered concrete structures. Its intention is to provide engineers, architects, contractors, and students of civil engineering and architecture with authoritative practical design information.

The tremendous progress and changes in all the areas of concrete engineering in the last two decades seemed to indicate a need for a new *Handbook of Concrete Engineering*. In addition, the many inquiries on subjects of concrete engineering received daily by the editor in the course of his professional activity, heightened his enthusiasm in accepting the proposal from Van Nostrand Reinhold to assemble this handbook.

Much of the information contained in this book has evolved during the last 15 to 20 years. The subjects of a number of chapters are so recent that the material has never before been published in book form.

The following traditional engineering subjects are covered in chapters on proportioning of members (ultimate strength design); deflections; flat plates and flat slabs; foundations; properties of materials for reinforced concrete (geared particularly to design engineers); joints in buildings; thin shells; prestressed concrete; chimneys; silos and bunkers; concrete masonry; sanitary structures; pipes, marine structures; paving; construction methods and equipment; and structural analysis by force, displacement and finite element methods.

A number of chapters are of particular interest because they represent areas of recent technological advances: fire resistance, multistory structures; tubular structures; ductility of reinforced concrete members; earthquake resistance; large panel structures; and finally, computer applications and computer software.

Because of space limitations, a careful selection of topics has been made. The 27 chapters have been written by 28 experts from industry and universities, recognized as outstanding authorities in their respective fields. Their wide experience has resulted in concise chapters geared toward practical application in planning, design and construction of engineered concrete structures. Each chapter contains the general philosophy, the basic concepts, and applications elaborated by design examples. While ready-to-use formulas and design approaches are presented in the chapters, theoretical development of formulas has been omitted, and references for additional source material have been appended to most chapters.

The editor gratefully acknowledges the efforts of the authors in preparing high quality manuscripts, and their cooperation and patience with the editor and the publishers in all the stages of producing the handbook. Thanks to their unstinting cooperation, this book has been produced in a relatively short number of years.

MARK FINTEL

September, 1974

Contents

Handbook of
CONCRETE ENGINEERING

1

Proportioning of Sections—Strength Design Method

MURAT SAATCIOGLU, Ph.D.*

1.1 INTRODUCTION

This chapter contains up-to-date design information on proportioning reinforced concrete sections. It is intended to provide practical design information and design aids to structural engineers and designers who are involved in various phases of the design process. Readers are expected to be familiar with the fundamentals of reinforced concrete design, although the basic concepts are reviewed in a systematic manner for bridging the gap between theory and practice. Each subject category covered in the chapter contains a brief explanation of the relevant design concepts, illustrative examples, and the commonly used design aids. The material contained in this chapter includes: (1) fundamentals of reinforced concrete behavior, (2) description of the strength design method, (3) design of members for flexure, (4) design for flexure and axial load, (5) proportioning sections for shear and torsion, and (6) properties and use of reinforcement in concrete.

1.2 SI UNITS IN STRUCTURAL DESIGN

The International System of units, abbreviated SI (from the French, Système International d'Unités), has been accepted throughout the world. In this chapter SI units are followed exclusively. All other chapters of the *Handbook* are in conventional U.S. units. While the basic ACI 318-83 continues to be written in conventional U.S. units, a metric version of the code, ACI 318M-83, has also been issued for those users who make their designs in SI units.

The SI system is an absolute system of units based on the quantities of length, time, and mass. In structural engineering the primary concern is on the units of length and force, with mass involved only when gravitational forces are being computed.

A distinct feature of SI appears in the use of the term "weight." In the SI system, "mass" is used to express the amount of material in kilograms (kg). It should not be confused with weight, since weight is a force applied by gravity on the mass. Kilogram is *not* a force unit and only indicates the amount of matter in an object. The kilogram is defined as the mass of a certain platinum–iridium cylinder that is kept at the International Bureau of Weights and Measures near Paris, France. An accurate copy of this cylinder is kept at the National Bureau of Standards in the United States. Mass of an object remains the same everywhere on earth, whereas weight differs going from sea level to the top of a mountain because of the change in gravitational acceleration. In application of SI units to structural engineering, the conventional term "unit weight" is replaced by "mass density." For example, mass density of normal density concrete is 2400 kg/m^3. Again, mass density is *not* the same as what is conventionally known as weight of a unit volume of

*Assistant Professor, Department of Civil Engineering, University of Toronto, Toronto, Canada.

material. Therefore, in computing dead loads, mass density should be multiplied by the gravitational acceleration, $g = 9.8\ m/s^2$, to obtain force.

In SI units the unit of force is the newton (N). One newton is defined as the force required to give a one-kilogram mass an acceleration of one meter per second square. Or, simply:

$$F = ma$$

$$\text{newton} = \text{kilogram} \times \text{meter/second square}$$

Since the load is a force term, loads and forces are expressed in newtons, kilonewtons, or meganewtons for concentrated loads. Distributed loads are expressed as kilonewton per square meter or kilonewton per meter. It should be noted, however, that kN/m^2 is a load unit and is not used for stress. For strength and stress, pascal, kilopascal, and megapascal can be used.

$$\text{pascal} = \text{newton/square meter}$$

Concrete strength and reinforcement yield stress are expressed in units of megapascal (MPa).

Length in structural engineering is generally expressed as millimeters and meters. The meter, originally defined as one ten-millionth of the distance from the pole to the equator along the meridian through Paris, was later defined as the length of a certain platinum–iridium bar kept at the International Bureau of Weights and Measures. A more accurate definition of a meter is 1,650,763.73 wavelengths of a certain radiation of the krypton-86 atom.

Generally, in structural design, meter is used to express span lengths and column heights. For cross-sectional characteristics such as dimensions, areas, or moments of inertia, as well as the area of reinforcement, units are expressed in millimeters.

The other unit of interest to engineers is a unit to express moment. Moments are generally expressed in kilonewton meters (kN · m).

Table 1-1 provides conversion factors from the conventional U.S. units (British system of units) to the SI units for those quantities that are commonly referred to in a structural design process. The metric bar sizes (consistent with the SI units) are discussed in Section 1.8.

1.3 BEHAVIOR OF REINFORCED CONCRETE

In designing reinforced concrete members it is important to understand the basic material and member behavior. In this

TABLE 1-1 Metric Conversion Factors

To Convert From	*To*	*Multiply By*
LENGTH		
inch (in.)	millimeter (mm)	25.4
inch (in.)	meter (m)	0.0254
foot (ft)	meter (m)	0.3048
yard (yd)	meter (m)	0.9144
AREA		
square foot (sq ft)	square meter (m^2)	0.09290
square inch (sq in.)	square millimeter (mm^2)	645.2
square yard (sq yd)	square meter (m^2)	0.8361
VOLUME		
cubic inch (cu in.)	cubic millimeter (mm^3)	16387.1
cubic foot (cu ft)	cubic meter (m^3)	0.02832
cubic yard (cu yd)	cubic meter (m^3)	0.7646
gallon (gal) Canadian	liter	4.546
gallon (gal) Canadian	cubic meter (m^3)	0.004546
gallon (gal) U.S.	liter	3.785
FORCE		
kip	newton (N)	4448.2
kip	kilonewton (kN)	4.4482
pound (lb)	newton (N)	4.4482
FORCE PER UNIT LENGTH		
pound per linear foot (plf)	newton per meter (N/m)	14.5939
FORCE PER UNIT AREA		
pound per square foot (psf)	newton per square meter (N/m^2)	47.8803
STRESS, MODULUS OF ELASTICITY		
pound per square inch (psi)	kilopascal (kPa)	6.895
pound per square foot (psf)	pascal (Pa)	47.8803
kip per square inch (ksi)	megapascal (MPa)	6.895
MOMENT		
foot kip	kilonewton meter (kN · m)	1.3558
MASS, DENSITY		
pound (lb)	kilogram (kg)	0.4536
pound per linear foot (plf)	kilogram per meter (kg/m)	1.488
pound per square foot (psf)	kilogram per square meter (kg/m^2)	4.882
pound per cubic foot (pcf)	kilogram per cubic meter (kg/m^3)	16.02

section, discussions are devoted to behavior of reinforced concrete members subjected to flexural, axial, shear, and torsional stresses. Background information on the *ACI Code* design provisions is presented. Derivations of some of the ACI design equations are given.

1.3.1 Flexure

A simple reinforced concrete beam loaded by gravity loads exhibits the basic characteristics of flexural behavior. Depending on the magnitude of bending moments, the beam deforms either in the elastic range or in the inelastic range. If the bending moment is smaller than the cracking moment, then the beam is in the elastic range. Figure 1-1(a) illustrates strain and stress distributions at the midspan section prior to flexural cracking. At this stage, stresses in concrete are proportional to strains. The internal force couple formed by tension and compression in concrete provides the resistance to the externally applied moment. The reinforement in concrete is basically inactive at this stage, mainly because it has not strained enough to develop sizable stresses. Principles of elastic theory can be employed for computing stresses in concrete. The presence of reinforcement in concrete makes the beam nonhomogeneous; therefore, the transformed area concept is used to account for the reinforcement by an equivalent concrete area. Moment of inertia of the transformed concrete section can then be used to compute stresses in concrete. Extreme fiber tension in concrete is:

$$f_t = \frac{M(h - c)}{I_{tr}} \tag{1-1}$$

As the applied load is increased, the maximum tension in concrete approaches the modulus of rupture. At this point, tension cracks start forming in concrete. The "cracking moment" can be computed using eq. (1-1) with modulus of rupture, f_r, substituted for f_t:

$$M_{cr} = \frac{f_r I_{tr}}{h - c} \tag{1-2}$$

This equation is the same as eq. (9-8) of ACI 318M-83 with the exception that I_{tr} is replaced by I_g for simplicity.

Once the cracks start forming, they propagate quickly toward the neutral axis under increasing loads. The neutral axis shifts upward with progressive cracking. The strain distribution assumes a new shape, resulting in a larger curvature. Since the concrete cannot resist tensile stresses, the reinforcing steel is called upon to resist the entire tension. At moderate loads, concrete stresses and strains continue to be proportional. Up to about 50% of f'_c the linear relationship between stresses and strains produces reasonably accurate results. Figure 1-1(b) shows strain and stress distribu-

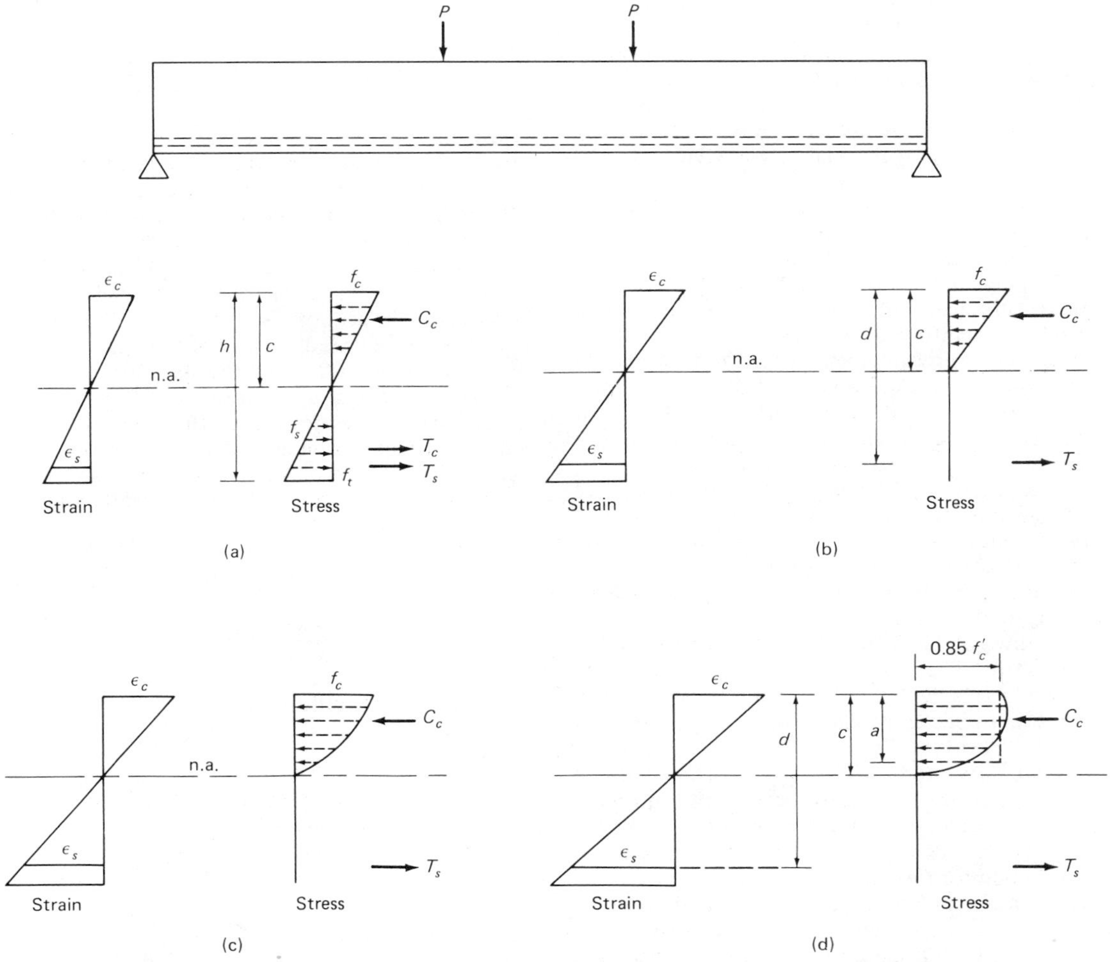

Fig. 1-1 Flexural behavior of a simple beam at different load stages.

tions at a moderate load level beyond cracking. The internal flexural resistance at the section is provided by a force couple consisting of concrete compression C_c and steel tension T_s. For a given neutral axis location c, the moment M can be found using eq. (1-3):

$$M = T_s \left(d - \frac{c}{3}\right) \tag{1-3}$$

where

$$T_s = A_s f_s.$$

To compute stresses in concrete, the device of the transformed section can still be used. Here the cracked concrete is assumed to make no contribution. The transformed section consists of compression concrete and the equivalent concrete area of the reinforcing steel.

When the applied load is further increased, stresses in concrete and steel increase correspondingly. At higher levels of stress, the proportionality of stresses and strains ceases, and the material nonlinearity is observed. Although a linear strain distribution can still be used to be consistent with the assumption that "plane sections before bending remain plane after bending," the stress distribution is of the same shape as the concrete stress–strain curve.

Figure 1-1(c) shows strain and stress distributions at higher loads, close to the ultimate load level. The internal resisting couple continues to be formed by the concrete compression and steel tension; the appropriate stress–strain relationships for concrete and steel can be used to find the coupling forces. A common practice is to utilize a bilinear idealization for the stress–strain relationship for steel. This idealization implies that the stresses and strains in steel are proportional, as before, up to the yield stress. Beyond this stress level, stresses in steel are taken equal to the yield stress, irrespective of the magnitude of strain. For concrete, one has to find the area under the stress distribution curve that has the same shape as the stress–strain relationship of concrete.

In properly proportioned members, the reinforcement does not yield under service loads. However, the concrete may crack in flexure even under low service loads. This is typical behavior for reinforced concrete and is not an undesirable occurrence. Reinforcement would not be needed if the concrete did not crack. Flexural cracks, in well-proportioned members, are very small hairline cracks. Their presence is hardly noticeable, and they do not present any undesirable appearance or corrosion problems.

The ultimate capacity of the beam is eventually reached as the applied load continues to increase. Failure is governed by one of two modes, depending on the amount of reinforcement. If relatively low percentage of steel is used in a section, the steel on the tension side starts yielding. This triggers widening of the cracks and continued shifting of the neutral axis upward. The cracks become visible with the increase in curvature. The deflection also increases to excessive levels. When this happens, the concrete is subjected to higher strains and stresses to maintain compatibility and equilibrium at the section. Eventually, concrete crushing occurs as the secondary mode of failure. Until the collapse mechanism forms, however, excessive deflections and wide cracks provide warning of an imminent failure. This type of ductile failure initiated by yielding of reinforcement is a desirable failure mode for flexural members.

Figure 1-1(d) shows the strain and stress distributions at the ultimate load condition when the failure is governed by steel yielding. This is the ultimate load condition considered in design. Although the strain distribution is linear, concrete stress distribution shows a parabolic variation. In order to find the internal resisting couple at the section, one must integrate concrete stress distribution to determine the compression force. ACI 318M-83 permits the simplification of representing concrete stress distribution by an equivalent rectangular stress block. Accordingly, concrete stress of $0.85\ f'_c$ is assumed uniformly distributed over an equivalent compression zone bounded by the edges of the cross section. The height of the rectangular stress block is defined as the neutral axis location c multiplied by the factor β_1. This factor is equal to 0.85 for concrete strengths up to and including 30 MPa. For strengths above 30 MPa, β_1 reduces at a rate of 0.08 for each 10 MPa, provided that it is not below 0.65. The *Code* recommends the maximum usable strain of 0.003 at the extreme concrete compression fiber. This value represents a conservative maximum strain for concrete. The concrete compression, C_c, the reinforcement tension, T_s, and the internal resisting moment, M_n, can be found as shown below:

$$C_c = 0.85\ f'_c\, ab \tag{1-4}$$

$$T_s = A_s f_y \tag{1-5}$$

Equilibrium requires $C_c = T_s$ and hence:

$$a = \frac{A_s f_y}{0.85\ f'_c b} \tag{1-6}$$

$$M_n = A_s f_y \left(d - \frac{a}{2}\right) \tag{1-7}$$

$$M_n = A_s f_y \left(d - \frac{f_y A_s}{1.7\ f'_c b}\right) \tag{1-8}$$

In heavily reinforced sections, relatively high tension capacity is provided, and the capacity of the concrete in compression is exhausted prior to yielding of the reinforcement. Concrete crushing occurs in a sudden and brittle manner. Because of the nonductile behavior of concrete crushing, this type of proportioning should be avoided.

If concrete crushing and steel yielding occur simultaneously, the section is said to be "balanced." A section reaches its balanced point when the maximum compression fiber strain is 0.003, and the steel strain is equal to its yield strain. Equations (1-4) through (1-8) are equally applicable to a balanced section, since this condition is a special case of an ultimate moment condition at which the limits of the two materials are reached simultaneously. In Fig. 1-1(d), if ϵ_y is substituted for ϵ_s, the linear strain condition gives:

$$\frac{c}{d} = \frac{\epsilon_c}{\epsilon_c + \epsilon_y} = \frac{0.003}{0.003 + f_y/200{,}000}$$

$$= \frac{600}{600 + f_y} \tag{1-9}$$

From force equilibrium:

$$C_c = T_s$$

$$0.08\ f'_c \beta_1\, cb = A_s f_y$$

Substituting eq. (1-9) for $\frac{c}{d}$ and $\rho_b = \frac{A_s}{bd}$:

$$\rho_b = \frac{0.85\ f'_c \beta_1}{f_y} \times \frac{600}{600 + f_y} \tag{1-10}$$

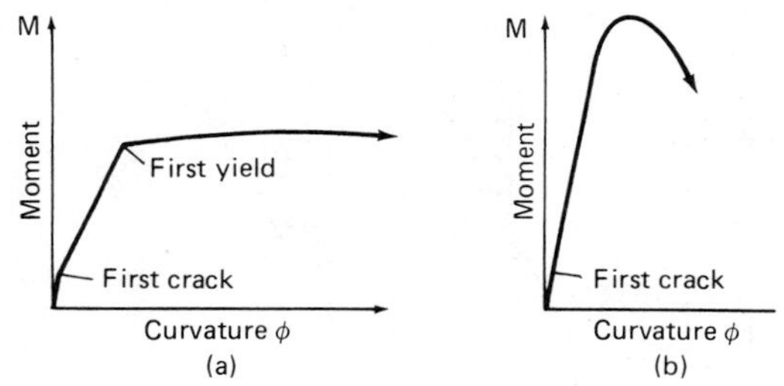

Fig. 1-2 Moment-curvature relationships.

Equation (1-10) gives the balanced section reinforcement ratio. Reinforcement ratios lower than this value produce ductile failure with reinforcement yielding occurring before crushing of concrete.

The load stages discussed above can best be summarized by a moment-curvature relationship. Each of the above load stages represents a characteristic point on the moment-curvature curve. Curvatures at different load stages can be determined from the corresponding strain diagrams as:

$$\phi = \frac{\epsilon_c + \epsilon_s}{d} \qquad (1\text{-}11)$$

Figure 1-2(a) shows a typical M–ϕ curve for a singly reinforced section failing in tension. The first characteristic point on the curve is the point at which concrete cracks. Up to this point, the section is elastic. At cracking, there is a slight increase in curvature under constant moment. The curve then continues with a lower slope (reduced stiffness due to cracking) until the first yielding in steel takes place. At this point there is a substantial reduction in the slope of the curve. The curvature increases with very little increase in moment until concrete crushing occurs. The moment-curvature relationship for a heavily reinforced section failing in compression is shown in Fig. 1-2(b). Comparison of the two curves indicates that the beam failing in tension is capable of developing larger inelastic curvatures and hence shows a ductile behavior. Moment-curvature curves for a singly reinforced rectangular section corresponding to three different areas of steel are shown in Fig. 1-3.

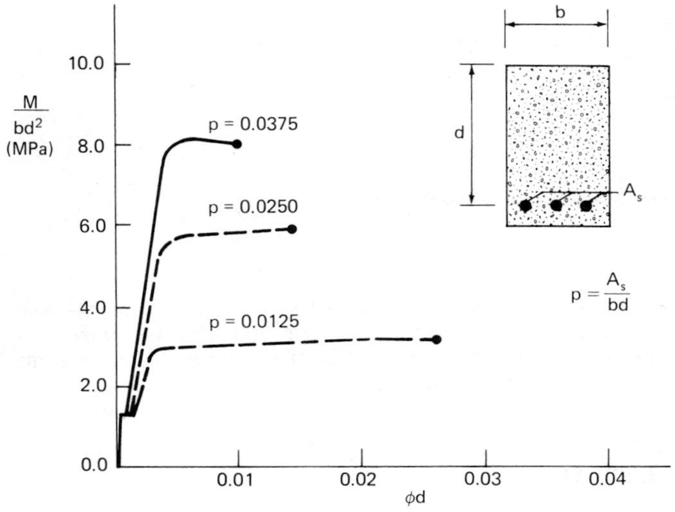

Fig. 1-3 Moment-curvature curves as affected by percentage reinforcement.

1.3.2 Flexure and Axial Load

Generally, concrete members subjected to axial loads are simultaneously subjected to bending moments. Concentrically compressed members rarely occur in practice. Even if a member appears to be resisting a concentric load, there is always some eccentricity due to construction imperfections. Furthermore, bending moments due to continuity are always present in monolithic structures. Lateral loads also impose bending moments on columns. Consequently, it is safe to assume that some eccentricity always exists in compression members. The provisions of ACI 318M-83 allow for some moment capacity in a section even if the results of analysis indicate a concentric load. This is an indirect procedure to assure a minimum design moment capacity. Earlier editions of the *Code* explicitly spelled out the minimum eccentricity requirement for compression members.

Reinforced concrete sections under combined axial force and bending moment behave in much the same manner as flexural members. The only difference between the two is the presence of additional uniform compression. A good example is a column, which behaves elastically under a relatively moderate eccentric load, the same as a beam. Figure 1-4(a) snows stress and strain distributions at a column section. By using elastic theory and the transformed area concept, we can find stresses in the concrete. The maximum and the minimum fiber stresses are computed as shown below:

$$f_{\substack{\max \\ \min}} = \frac{P}{A_{tr}} \pm \frac{M\,h}{2\,I_{tr}} \qquad (1\text{-}12)$$

Axial force and moment resistance of a column section at this load stage can be computed by applying the principles of static equilibrium.

When a column is subjected to higher eccentricity, tensile stresses can exceed the modulus of rupture, and concrete cracking occurs. Figure 1-4(b) illustrates this stress condition. If the maximum compression is less than about 50% of f_c', elastic theory is still applicable with the appropriate transformed section. However, as the applied load is increased, material nonlinearity becomes more pronounced. Concrete stress distribution forms a parabolic profile following the concrete stress–strain relationship. Steel yielding may take place with increasing load. The ultimate load capacity is reached either by crushing of the concrete or yielding of the steel. This load stage is of interest to designers. Figure 1-4(c) shows a rectangular column section eccentrically compressed by a force P_n. If a rectangular stress block is used to express concrete stress distribution at ultimate, the nominal strength for axial force and flexure can be computed using the equilibrium equations:

$$P_n = 0.85\, f_c' ab + A_s' f_s' - A_s f_s \qquad (1\text{-}13)$$

$$M_n = P_n e = 0.85\, f_c' ab \left(\frac{h}{2} - \frac{a}{2}\right) + A_s' f_s' \left(\frac{h}{2} - d'\right)$$

$$+ A_s f_s \left(d - \frac{h}{2}\right) \qquad (1\text{-}14)$$

Columns with large eccentricities fail because of the yielding of tension reinforcement. Concrete crushing takes place shortly after the steel yields. At failure, $\epsilon_{cc} = 0.003$, $f_s = f_y$ and f_s' is usually equal to f_y. Conversely, in columns with small eccentricities, concrete crushing occurs prior to tension steel yielding. In this case, $\epsilon_{cc} = 0.003$, f_s' is usually

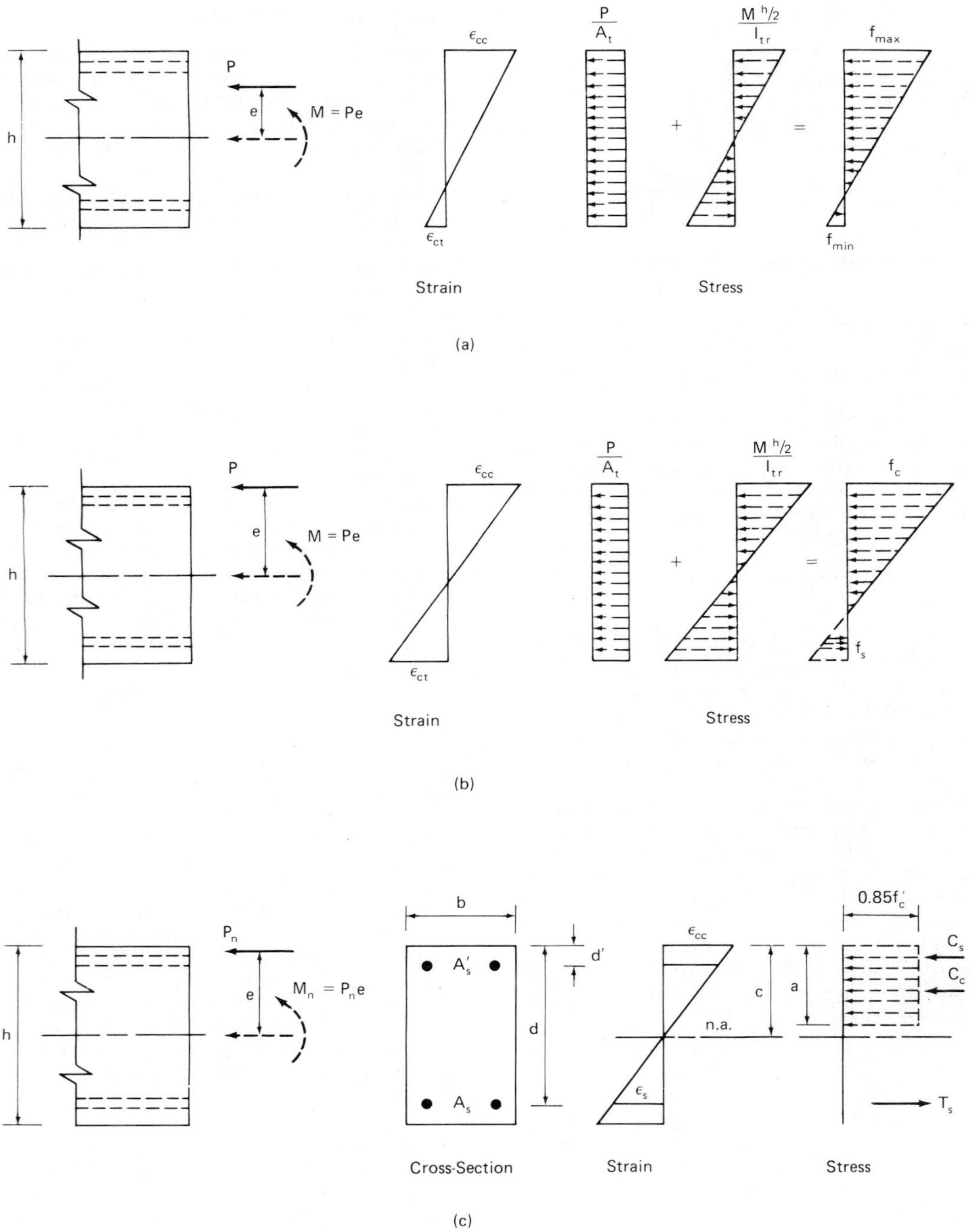

Fig. 1-4 Reinforced concrete under combined axial force and flexure.

equal to f_y, and f_s must be determined from a strain-compatibility analysis. Strain-compatibility analysis involves an iterative procedure. The neutral axis location is first assumed, and the corresponding stress in steel is found. This value is then compared with the stress computed to satisfy the static equilibrium. The iteration continues until a reasonable agreement is achieved.

Equations (1-13) and (1-14) give the axial load and moment combination that produces failure in a given column section. Depending on the eccentricity, e, different P_n and M_n combinations may govern the design. These combinations are expressed in terms of well-known moment–axial force interaction diagrams. These diagrams are extremely useful design tools. The details of the interaction diagrams are discussed in Section 1.6.

1.3.3 Shear

Shear behavior of reinforced concrete is still a subject of interest for researchers. Most structural members are subjected

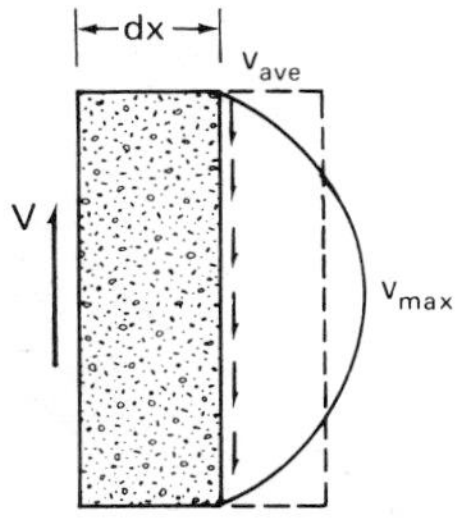

Fig. 1-5 Shear stress across concrete section.

to shear stresses that are generally combined with axial forces, flexure, and torsion. The interaction of these stresses with shear, especially in cracked concrete, becomes a complex problem. Extensive experimental work is still being conducted to clarify shear behavior of concrete, especially beyond the elastic range.

The shear force across any concrete cross section can be derived using the principles of equilibrium. Variation of shear along a member is conveniently expressed by a shear force diagram. If a differential length of a beam is isolated as a free body diagram, it can be seen that the vertical shear force, V, must be balanced by the shear stress across the section. Figure 1-5 shows the distribution of shear stress in a homogeneous beam cross section. The distribution is parabolic with the maximum shear stress equaling $1\frac{1}{2}$ times the average value. The average stress is equal to the shear force, V, divided by the cross-sectional area. In the 1971 *ACI Code*, this average stress was used to express both the applied shear and the resisting shear capacity. In the 1977 and 1983 *ACI Codes*, shear forces, rather than the stresses, are used for design purposes. Therefore, the ultimate shear force, V_u (factored load), is used to proportion sections for shear. Shear capacity V_n, is also expressed in terms of force.

The significance of shear can best be visualized by examining a simple beam under gravity loads. Stress distribution given in Fig. 1-5 clearly indicates that the shear is maximum at the neutral axis and zero at the extreme fibers. This observation is also confirmed for cross sections of other shapes. It may also be recalled that flexural stresses are exactly the opposite, having zero stress at the neutral axis and the maxima at the extreme fibers. If a small square element taken from the neutral axis of the beam shown in Fig. 1-6(a) is examined, it can be seen that the vertical shear stresses v are acting on the two vertical faces of the element. For reasons of equilibrium, equal-intensity shear stresses also act on the two horizontal faces. Figure 1-6(b) illustrates the element and the shear stresses acting on all four faces. If this element is rotated 45°, as shown in Fig. 1-6(c), an equivalent set of stresses is obtained. The components of the shear stresses v add up to normal stress t_1 and c_1, tension in one direction and compression in the other direction. The two elements shown in Fig. 1-6(b and c) are equivalent to each other, and they both show the same stress condition. This indicates that the beam is subjected to tension and compression due to shear stresses at a plane 45° with the beam axis. This tension stress is referred to as "diagonal tension" and can lead to cracking in concrete. These cracks, if due to shear stresses alone, initiate at the neutral axis.

As mentioned earlier, shear stresses usually occur in combination with other stresses. In beams, the interaction of shear and flexural stresses depends upon their location in the beam. If a small element is taken at the extreme fiber, there is no contribution of shear, and the response is governed by flexure. However, if the element is taken at location 2 in Fig. 1-6(a), both shear and flexural stresses act on the element. This is shown in Fig. 1-6(d). The element can be rotated an angle α such that the resultant stresses are tension and compression stresses t_2 and c_2, as shown in Fig. 1-6(e). These stresses are known as "principal stresses." Their values and orientation are given by:

$$t_2 = \frac{f}{2} + \frac{1}{2}\sqrt{f^2 + 4v^2} \quad (1\text{-}15)$$

$$c_2 = \frac{f}{2} - \frac{1}{2}\sqrt{f^2 + 4v^2} \quad (1\text{-}16)$$

$$\tan \alpha = \frac{v}{t_2} \quad (1\text{-}17)$$

Principal stresses in a concrete member vary in both magnitude and orientation. Depending on the relative magnitudes of bending moments and shear forces, different types of cracks may develop. Flexural cracks are dominant in those

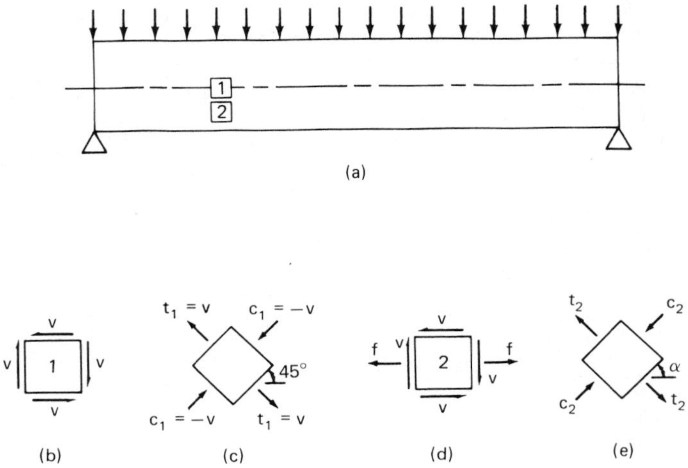

Fig. 1-6 Diagonal tension due to shear. (Adopted from Ref. 1.3)

areas where bending moments are significantly high as compared to shear forces. When high shear forces and low bending moments are present, web shear cracks may develop, propagating from the neutral axis. The combination of the two types of cracks can be seen in those areas where both shear and flexure are equally significant.

The nominal shear strength of concrete has been determined by tests. Web shear cracks usually occur at approximately $0.3\sqrt{f_c'}$ where f_c' is in MPa. When shear cracks occur in the presence of large moments, the nominal shear strength of concrete is about $0.17\sqrt{f_c'}$. It is evident from the preceding discussion that the nominal concrete shear strength greatly depends on the moment-to-shear ratio.

The following empirical equation has been developed for concrete members subjected to combined shear and flexure:

$$v_c = \left(0.16\sqrt{f_c'} + 17.2\frac{\rho_w Vd}{M}\right) \leqslant 0.3\sqrt{f_c'} \qquad (1\text{-}18)$$

This equation was adopted by the 1977 and 1983 *ACI Codes* as eq. (11-6), in the form of a nominal shear force rather than the shear stress provided by concrete.

Shear resistance of concrete also varies with axial loads. Axial compression improves the shear capacity. Axial tension, on the other hand, drastically reduces the shear resisting capacity of concrete. The relationship between the axial force and shear has been incorporated into the *Code* using empirical equations derived from test results.

The shear capacity of plane concrete may or may not be enough to resist the applied design shear force, V_u. Once a crack is formed in concrete, its shear resistance is significantly reduced. If the crack is narrow, the "aggregate interlock" between the two cracked surfaces can continue resisting shear at a somewhat lower level. As in the case of flexure, the reinforcement is called upon to take a portion of the applied shear force. Specially provided vertical shear reinforcement can resist significantly high shear forces. The shear-resisting mechanism beyond cracking has four components. Figure 1-7 illustrates the components of the total shear resistance provided by a reinforced concrete section. The equilibrium of forces requires:

$$V_u = V_e + V_a + V_d + V_s \qquad (1\text{-}19)$$

Where V_e is the resistance of uncracked concrete, V_a is the resistance by aggregate interlock, V_d is the dowel action provided by longitudinal steel, and V_s is the force in the vertical stirrups. Figure 1-8 shows the distribution of internal shear forces in a beam with stirrups. Among the four shear force components shown above, only V_s can be computed with accuracy. The other three terms cannot be determined separately, so their lumped effect is expressed empirically using the available test data. This combined shear resistance is denoted as the nominal shear strength of concrete and can be computed using eq. (1-18). If the horizontal projection of the crack is assumed to be equal to effective depth d, then:

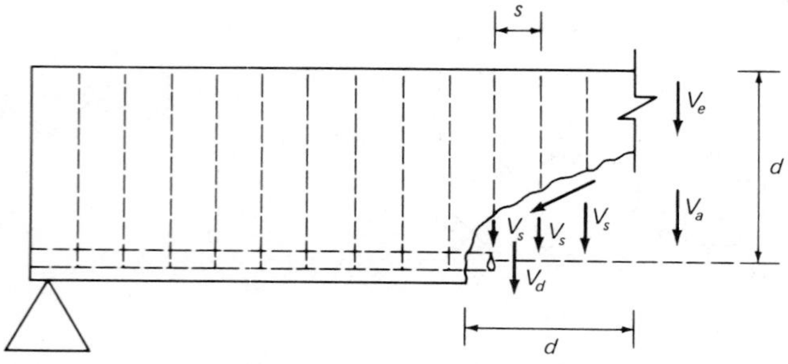

Fig. 1-7 Shear-resisting mechanism in a reinforced concrete beam.

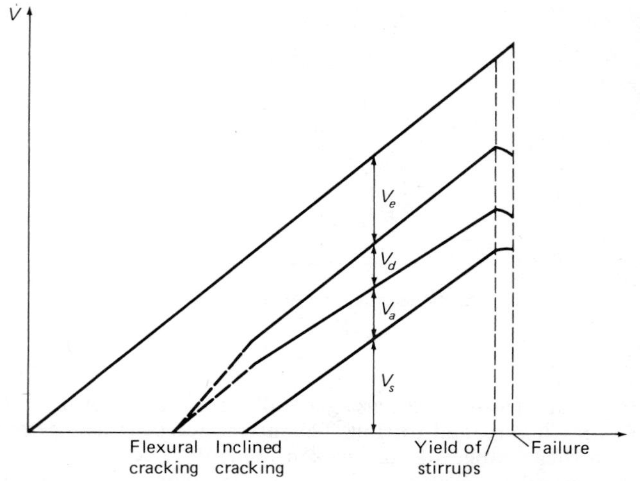

Fig. 1-8 Contribution of each factor involved in shear resistance. (Adopted from Ref. 1.3.)

$$V_s = \frac{A_v f_y d}{s} \qquad (1\text{-}20)$$

Sometimes inclined bars are used as shear reinforcement; longitudinal tension bars bent in the support region and inclined stirrups fall into this category. The vertical shear resistance, V_s, for inclined bars can be derived as shown below with reference to Fig. 1-9:

$$V_s = A_v f_y n \sin\alpha \qquad (1\text{-}21)$$

where:

n = number of inclined bar locations spaced at distance s. For uniform spacing of s crossing a diagonal crack, $n = d'/s'$.

A_v = area of shear reinforcement provided within distance s. It is the area of inclined bars (perpendicular to the line of inclination) in the case of inclined shear reinforcement.

α = angle of the inclined bars with respect to the beam axis.

$d' = d/\sin 45$

$s' = s \sin\alpha/\sin(135\text{-}\alpha)$

Substituting the expression for n into eq. (1-21) and rearranging terms:

$$V_s = A_v f_y \frac{d}{s} \frac{\sin(135-\alpha)}{\sin(45)\sin\alpha} \sin\alpha \qquad (1\text{-}22)$$

$$V_s = A_v f_y \frac{d}{s} \frac{\sin(135)\cos\alpha - \cos(135)\sin\alpha}{\sin 45} \qquad (1\text{-}23)$$

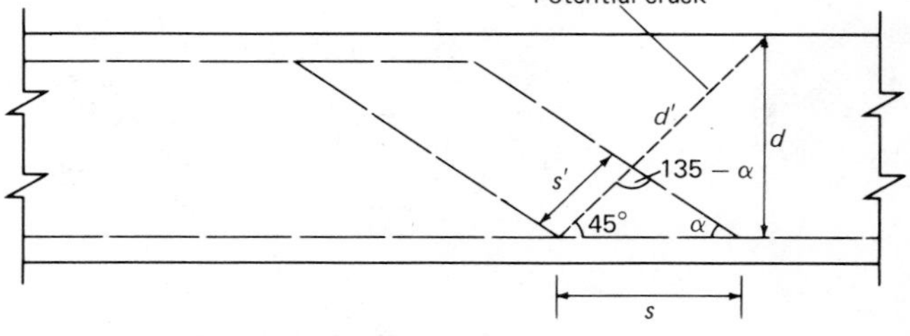

Fig. 1-9 Bent longitudinal bars or inclined stirrups.

Since sin 135 = sin 45 = - cos 135:

$$V_s = A_v f_y \frac{d}{s} (\cos \alpha + \sin \alpha) \quad (1\text{-}24)$$

Equations (1-20) and (1-24) are identical to those given in ACI 318M-83. Details of the shear design procedure are discussed in Section 1.7.

1.3.4 Torsion

Torsion in reinforced concrete members usually occurs in combination with other actions. A typical example is a spandrel beam of a slab system or an edge beam of a frame system whose torsional rigidity provides flexural restraint to the framing elements in the transverse direction. These beams, undergoing deformation due to flexure and shear, are also subjected to torsion.

Behavior of a reinforced concrete member subjected to torsion can best be evaluated by examining the elastic and inelastic response of the member. A member subjected to torsion behaves in the elastic mode prior to the formation of diagonal tension cracks. At this stage the contribution of reinforcement is negligible, and the member response depends on the elastic material and sectional properties. St. Venant's classical torsion theory can be applied to elastic rectangular concrete sections, from which the maximum torsional shearing stress, v_t, can be derived:

$$v_t = \alpha_1 \frac{T_c}{x^2 y} \quad (1\text{-}25)$$

where T_c is torsional moment applied to a rectangular section, x and y are the shorter and longer sides of the section, and α_1 is a shape factor that varies between 3 and 5 for the various ratios of y/x. The maximum stress v_t occurs at the middle of the long face of the rectangular section as shown in Fig. 1-10(a). An element taken on the face of the member, is subjected to torsional shear stresses along its four sides. As in the case of the transverse shear discussed in the previous section and illustrated in Fig. 1-6, the element is subjected to tension and compression stresses that are equal to torsional shear stresses when rotated 45° as shown in Fig. 1-10(b). However, torsional shear stresses have the opposite sign in the other half of the member. Therefore, in the case of torsion, diagonal tension stresses on one face of the member are perpendicular to tension stresses on the opposite face. As the applied torque is increased, torsional shear stresses also increase, until the tensile capacity of concrete is exceeded. At this load stage, diagonal tension cracks form with approximately 45° inclination.

If the lowest value of 3 for the shape factor α_1 is assigned to eq. (1-25), and the value of diagonal tension stress at cracking is substituted for v_t, an expression for the torque T_{cr} is obtained for design purposes. The value of v_{cr} at cracking is found to vary between $0.33\sqrt{f'_c}$ MPa and $0.58\sqrt{f'_c}$ MPa. The *ACI Code* uses $0.50\sqrt{f'_c}$ MPa for diagonal tension stress that produces cracking in concrete:

$$v_{cr} = 3 \frac{T_{cr}}{x^2 y} \quad (1\text{-}26)$$

$$0.50\sqrt{f'_c} = 3 \frac{T_{cr}}{x^2 y} \quad (1\text{-}27)$$

$$T_{cr} = \frac{\sqrt{f'_c}\, x^2 y}{6} \quad (1\text{-}28)$$

where T_{cr} is the torque at which diagonal tension cracking occurs in a reinforced concrete member.

If the member cross section is of T or L shape, an approximate but sufficiently accurate expression can be developed by dividing the section into constituent rectangles. Figure 1-11 illustrates the division of various shape cross sections for torsional analysis. Once a suitable division of the section is done, then it is assumed that each rectangle resists a portion of the external torque in proportion to its torsional rigidity. This leads to the following approximation:

$$v_{cr} = 3 \frac{T_{cr}}{\Sigma x^2 y} \quad (1\text{-}29)$$

When a member is subjected to combined flexure and torsion, the response depends on the predominance of one or the other. However, generally speaking, a portion of the diagonal crack falls into the compression zone created by flexure. Therefore, the cracked section continues to resist some torsion even without the contribution of web reinforcement. Mattock found that the torsional strength of a cracked concrete section without web reinforcement is about 50% of the uncracked strength. The ACI 318M-83 conservatively assumes that the torsional resistance of cracked concrete is 40% of cracking stress $0.5\sqrt{f'_c}$ MPa. If eq. (1-25) is used with $\alpha_1 = 3$ and $v_t = 0.5\sqrt{f'_c}$ MPa, torsional resistance of concrete after cracking can be expressed as follows:

$$0.4\, v_t = 3 \frac{T_c}{x^2 y} \quad (1\text{-}30)$$

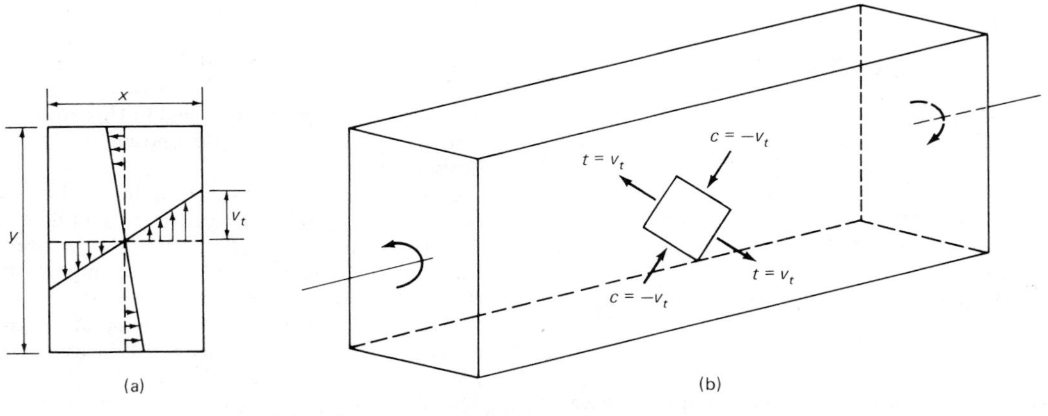

Fig. 1-10 Behavior of a concrete beam in pure torsion.

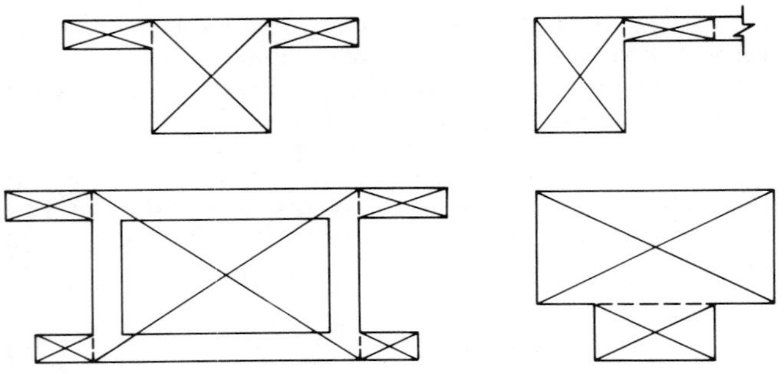

Fig. 1-11 Torsional resistance of different concrete sections and constituent rectangles.

$$(0.4)(0.5\sqrt{f'_c}) = 3\,\frac{T_c}{x^2y} \qquad (1\text{-}31)$$

$$T_c = 0.066\sqrt{f'_c}\,x^2y \qquad (1\text{-}32)$$

It was mentioned earlier that torsion generally occurs in combination with shear and flexure. Because of the lack of a satisfactory theory for the interaction effects of shear, flexure, and torsion, experimental results are utilized for the purpose of developing design equations. These experimental results indicate a circular relationship between the normalized values of flexural shear and flexural torsion. Figure 1-12 shows the relationship between T_u/T_0 and V_u/V_0, which gives the lower bound to test data. V_u and T_u are the ultimate shear and torsion carried by concrete when a member is subjected to *combined* shear, flexure, and torsion. V_0 and T_0 are the concrete capacities for shear and torsion when combined only with flexure. $V_0 = 0.17\sqrt{f'_c}\,bd$ as previously discussed in Section 1.3.3, and $T_0 = 0.066\sqrt{f'_c}\,x^2y$ as derived by eq. (1-32). If the equation of the curve in Fig. 1-12 is written as shown below, the design expressions for shear and torsion capacities of concrete can be obtained.

$$\left(\frac{T_u}{T_0}\right)^2 + \left(\frac{V_u}{V_0}\right)^2 = 1 \qquad (1\text{-}33)$$

$$\frac{T_u^2}{(0.066\sqrt{f'_c}\,x^2y)^2} + \frac{V_u^2}{(0.17\sqrt{f'_c}\,bd)^2} = 1 \qquad (1\text{-}34)$$

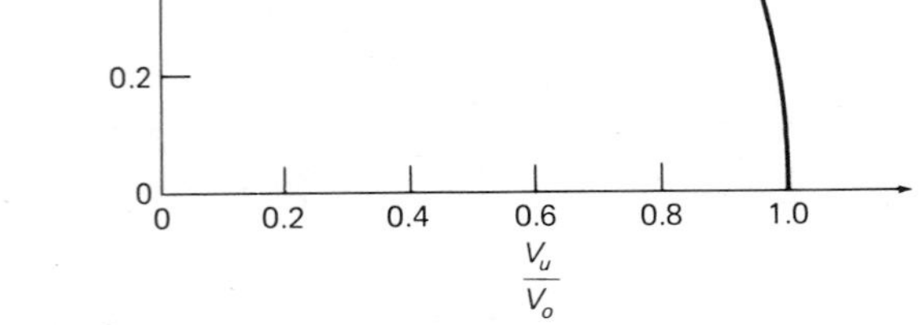

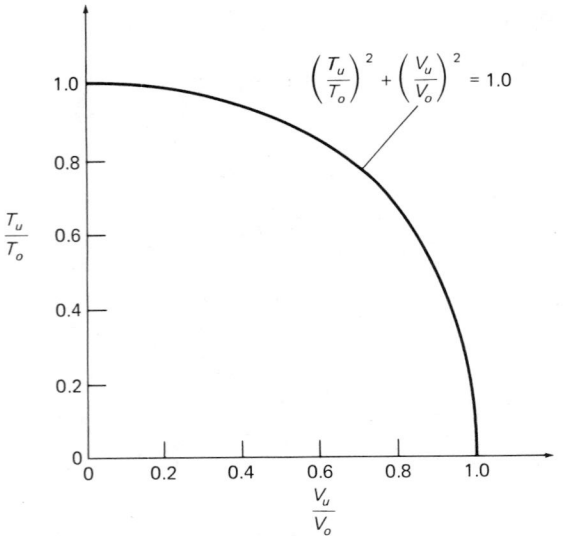

Fig. 1-12 Interaction of shear and torsion in the presence of flexure.

Multiplying all the terms by $(0.066\sqrt{f'_c}\,x^2y)^2$ and rearranging terms:

$$T_u^2\left[1 + \frac{(0.066\sqrt{f'_c}\,x^2y)^2}{(0.17\sqrt{f'_c}\,bd)^2}\left(\frac{V_u}{T_u}\right)^2\right] = (0.066\sqrt{f'_c}\,x^2y)^2 \qquad (1\text{-}35)$$

$$T_u = \frac{0.066\sqrt{f'_c}\,x^2y}{\sqrt{1 + \left(\dfrac{0.4\,V_u}{C_t\,T_u}\right)^2}} \qquad (1\text{-}36)$$

where $C_t = bd/x^2y$.

Similarly for shear capacity of concrete:

$$V_u = \frac{0.17\sqrt{f'_c}\,bd}{\sqrt{1 + \left(2.5\,C_t\,\dfrac{T_u}{V_u}\right)^2}} \qquad (1\text{-}37)$$

Equations (1-36) and (1-37) give the nominal concrete capacity for torsion and shear respectively when the member is subjected to combined shear, torsion, and flexure. These equations are identical to those given in ACI 318M-83.

If the applied torsion is more than concrete can resist, then the reinforcement is called upon to provide additional torsion capacity. Torsional reinforcement consists of closed stirrups and longitudinal bars. A space truss model is conventionally used to derive expressions for the required web reinforcement. The application of the space truss model involves tension members consisting of stirrups, and compression members consisting of concrete struts. If a potential diagonal crack forming 45° with one wide face of a rectangular beam and any angle θ with the narrow faces is considered, the number of stirrups crossing the failure plane can be determined. The total torque resisted by horizontal and vertical legs of the stirrups can then be derived as:

$$T_s = \frac{A_t\alpha_t x_1 y_1 f_y}{s} \qquad (1\text{-}38)$$

where A_t is the area of one leg of a closed stirrup spaced at distance s, and x_1 and y_1 are the shorter and longer dimensions of the rectangular stirrups. Coefficient α_t is empirically determined and specified by the *ACI Code* as:

$$\alpha_t = \frac{1}{3}\left[2 + \frac{y_1}{x_1}\right] \leqslant 1.50 \qquad (1\text{-}39)$$

Generally about the same volume of longitudinal reinforcement as stirrup steel is required. The contribution of longitudinal bars to torsional resistance should not be underestimated. They provide crack control by resisting widening of torsional cracks and also help in anchoring the stirrups.

1.4 DESIGN PROCEDURE

The design procedure followed in this chapter is the strength design method, commonly known as the ultimate strength design procedure.

Development of the strength design method can be traced back to the nineteenth century. Thullie's flexural theory, developed in 1897, and Ritter's parabolic distribution theory, developed in 1899, are examples of early attempts to use the strength method of design. In 1909, in spite of the early efforts, the elastic theory of Coignet and Tedesco gained acceptance and was adopted by the *Code*. The simple straight-line stress distribution of the elastic theory appealed to many engineers who were familiar with the application of the theory to other materials. The elastic theory formed the basis of the working stress design procedure which dominated reinforced concrete design codes for many years.

After more than a half century of experience with the working stress design, deficiencies of this procedure became evident. Laboratory tests conducted at the University of Illinois, Lehigh University, and the Portland Cement Association formed the basis of the strength design method, which appeared as an appendix to the *ACI Code* for the first time in 1956 and as a part of the *Code* proper in 1963. The two methods were covered equally in the 1963 edition of the *Code*.

The 1971 *ACI Code* was in large part based on the strength design method, allowing the working stress design as an alternate procedure. The use of the true working stress method was allowed only for design of flexural members. All other applications of working stress were in the form of a factored-down strength design, using increased safety factors.

The strength design method gained full acceptance in the 1977 and 1983 *ACI Codes*. In these editions, working stress is relegated to an appendix.

The strength design procedure is based on proportioning sections for the ultimate state of stress under ultimate load conditions. Both concrete and reinforcing steel exhibit material nonlinearities prior to reaching their ultimate strengths. Therefore, inelastic behavior of both materials must be considered and must be expressed in mathematical terms for ultimate strength formulation. A number of concrete stress distributions have been proposed, including parabolic trapezoidal and rectangular shapes. The idealization of post-yield stresses for steel is simpler. Generally, a constant value of steel stress is assigned to reinforcing bars within the post-yield region.

Figure 1-13 shows a rectangular reinforced concrete section and the corresponding stress and strain distributions. The strain distribution is assumed to vary linearly with the maximum compression strain of 0.003. Tests of reinforced concrete members confirm that the distribution of strain is essentially linear across the section. However, the maximum concrete compressive strain at crushing varies from 0.003 to 0.008. In most practical cases the variation is limited to 0.003–0.004. In current North American design practice, the 0.003 value is accepted as a conservative value for use in design.

The parabolic concrete stress distribution shown in Fig. 1-13 can be represented by an equivalent rectangular stress block. The *ACI Code* permits this simplification provided the appropriate factors are used, as specified in the *Code*, for determining the height and the width of the rectangular block. Another conservative simplification permitted by the *Code* is the assumption that the concrete does not provide any tensile resistance. Tests indicate that tensile capacity of concrete in flexure is in the order of 10% of its compressive strength.

Once the strain and stress distributions are determined, the ultimate capacity of a section can be computed using the principles of static equilibrium. This capacity is then compared with the ultimate applied moment to start the iterative design process.

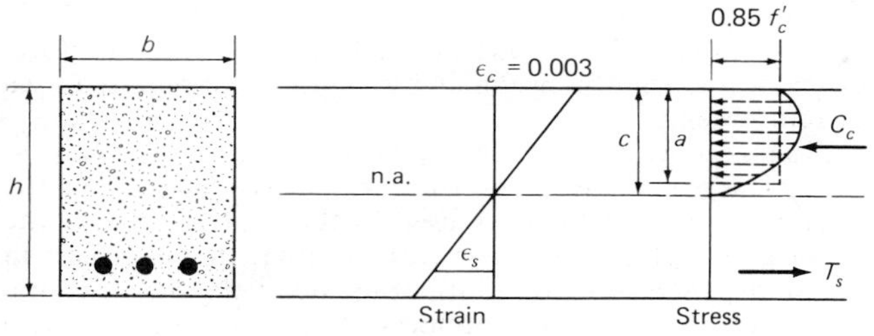

Fig. 1-13 Reinforced concrete section at ultimate load condition.

As stated earlier, strength design is based on proportioning sections for the ultimate state of stress under ultimate load conditions. The use of the term "ultimate" to express both the capacity and the applied design loads in the 1971 *ACI Code* led to some confusion among designers. Therefore, ACI 318-77 and ACI 318-83M adopted different terminology to differentiate between the required and provided ultimate strengths. Accordingly, the term "required strength" applies to service loads multiplied by the appropriate load factors, whereas the term "design strength" is used to express the nominal capacity of a section reduced by a strength reduction factor. The basic criterion for strength design can then be expressed as follows:

$$\text{Required strength} \leqslant \text{Design strength}$$

The above criterion provides for the margin of safety in two ways: (1) the required strength is computed by increasing service loads by load factors; (2) the design strength is computed by reducing the nominal strength by a strength reduction factor.

The design criteria presented above applies to all possible states of stress. The general design criteria can then be rewritten using the ACI terminology for flexure, axial load, shear, and torsion as shown below:

$$M_u \leqslant \phi M_n$$

$$P_u \leqslant \phi P_n$$

$$V_u \leqslant \phi V_n$$

$$T_u \leqslant \phi T_n$$

Each member in a structure is required to satisfy the design criteria under the most critical load combination. The load combinations are specified in section 9.2 of ACI 318M-83 and are discussed below along with the load factors.

1.4.1 Load Factors

The required strength, U, specified in the *ACI Code* is expressed in terms of combinations of factored loads. Each type of load is assigned a load factor that reflects reliability of information used in determining the actual magnitude of the load. For example, the factor 1.4 is assigned to dead loads, which are likely to be in the order of the calculated weight of the material. Live loads, on the other hand, are assigned a factor of 1.7, mainly because this type of loading is based on less reliable information. Similarly, different load factors are applied to other types of loads.

A common combination of design loads is that of dead (D) and live (L) loads. In such cases the required strength, U, must be at least equal to:

$$U = 1.4D + 1.7L$$

When wind (W) and earthquake (E) loads are present in design, the following combinations apply:

$$U = 0.75\ (1.4D + 1.7L + 1.7W)$$

$$U = 0.75\ (1.4D + 1.7L + 1.87E)$$

$$U = 0.9D + 1.3W$$

$$U = 0.9D + 1.43E$$

The first two of the above conditions include combinations of either wind or earthquake with dead and live loads. Because each of these loads is already increased by load factors and the simultaneous occurrence of all three loads at factored levels is not likely, the *Code* imposes a combination factor of 0.75 to reduce the required strength to a more realistic load level.

The last two combinations are intended to safeguard structures against overturning due to lateral loads which is resisted by gravity loads. An example of this condition is the increase in axial tension in vertical members caused by the overturning effect of lateral loads where dead and live loads are insufficient. In such cases increasing gravity loads by load factors would reduce safety against overturning.

The *Code* specifies combinations of other loads that may govern the design. If resistance to lateral earth pressure, H, or lateral liquid pressure, F, is included in design, or when structural effects, T, due to differential settlement, creep, shrinkage, or temperature change are significant in design, required strength, U, is governed by one of the following load combinations:

$$U = 1.4D + 1.7L + 1.7H$$

$$U = 0.9D + 1.7H$$

$$U = 1.4D + 1.7L + 1.4F$$

$$U = 0.75\,(1.4D + 1.4T + 1.7L)$$

$$U = 1.4\,(D + T)$$

1.4.2 Strength Reduction Factors

The design strength of a member is determined as the nominal strength multiplied by a strength reduction factor, ϕ. This factor has a value less than 1.0 and varies with the importance of a member, the consequence of the failure mode, and the degree of confidence in computing the nominal strength.

The flexural theory used for reinforced concrete design has been verified quite successfully by tests over the years, and has gained the confidence of researchers and designers. Because of the accuracy in predicting the nominal flexural strength, and the ductile nature of the flexural failure, a higher ϕ-factor of 0.9 is assigned to flexural members.

Member behavior under shear and torsion cannot be predicted with the same degree of accuracy. Furthermore, failure under shear is more abrupt and brittle without giving any warning. Therefore, a lower ϕ-factor of 0.85 is assigned to these members.

Compression members (columns and walls) generally carry stacks of floor loads. Consequence of a failure in a column is more severe than for a local beam failure. For this reason, as well as the brittle and sudden nature of a compression failure, a low ϕ-factor of 0.7 is assigned to tied columns. Because spiral reinforcement in columns improves ductility, a somewhat higher value of ϕ is assigned to spiral columns.

For compression members subjected to significantly low axial compression, the failure mode is governed by flexure. These members exhibit behavior similar to that of beams. Therefore, the *Code* permits a linear increase in ϕ-factors from 0.7 (for columns) to 0.9 (for beams) as the axial design strength is reduced from $0.10\,f_c'\,A_g$ to zero. Table 1-2 gives the capacity reduction factors for different stress conditions.

1.4.3 Design for Serviceability

The strength design method ensures structural integrity and safety against structural collapse. However, although adequate safety is provided, cracking and excessive deflections may lead to serviceability problems. Cracking may be excessive if the reinforcement is not distributed properly. Deflections may be too high if the selected member does not have adequate stiffness. These problems may occur even if the structure meets all the requirements of strength design. Therefore, for a satisfactory design, the crack widths and deflections at service load conditions should be checked. These serviceability checks require the use of elastic theory.

TABLE 1-2 Strength Reduction Factors

Action	ϕ
Flexure, with or without axial tension	0.90
Axial tension	0.90
Axial compression, with or without flexure:	
Members with spiral reinforcement conforming to section 10.9.3	0.75*
Other reinforced members	0.70*
Shear and torsion	0.85
Bearing on concrete	0.70
Flexure in plain concrete	0.65

**May be increased linearly to 0.90 as ϕP_n decreases from $0.10 f_c' A_g$ or ϕP_b, whichever is smaller, to zero.*

The *ACI Code* provides limits for maximum permissible deflections. For members that meet certain minimum thickness requirements, the deflection check is not a requirement. Table 1-3 gives the minimum thickness of nonprestressed beams or one-way slabs for which deflection checks are not required.

If necessary, an elastic analysis can be employed for deflection calculations. The flexural rigidity required in these calculations includes the effect of concrete cracking. The effective moment of inertia of concrete can be found using eq. (9-7) and the provisions of section 9.5 of ACI 318M-83.

Crack control in structural members is another serviceability requirement that has significant implications on appearance and weather resistance of concrete members. These requirements are especially stringent for tanks containing liquids, marine structures, and those structures that are faced with corrosion and freeze-thaw cycles. Crack widths are directly related to steel stress in the reinforcement. Distribution of reinforcement plays a significant role in controlling crack widths. ACI 318M-83 defines the quantity z as a function of steel stress and the distribution of reinforcement. This quantity can be computed using the provisions of ACI 318M-83 section 10.6.4. Limitations are imposed on the quantity z for different exposure conditions. Spacing of reinforcement as governed by crack control is discussed in Section 1.8.1.

1.5 FLEXURAL DESIGN

Member sizes in structural design are often dictated by architectural, aesthetic, and mechanical requirements. The deflection limits may also play a role in determining cross-sectional dimensions. Structural efficiency and strength considerations are not always the deciding factors. Once cross-sectional dimensions are determined, the structural efficiency is then considered by finding the required area of reinforcement.

Design moments are found by means of an elastic analysis. Preliminary member sizes form the basis for finding the elastic rigidities, which are used in the analysis. Under the usual gravity loads, positive design moments are largest in the midspan region. In continuous members, negative moments occur over the supports. Generally, the negative moment at the first interior support is larger than the other

TABLE 1-3 Minimum Thickness *h* of Nonprestressed Beams or One-Way Slabs Unless Deflections Are Computed—for members not supporting or attached to partitions or other construction likely to be damaged by deflections (Ref. 1.1)

Conc. Mass Dens.	*Steel Yield Str.*	*Simply Supported*		*One End Continuous*		*Both Ends Continuous*		*Cantilever*	
w_c kg/m^3	f_y MPa	*Solid One-Way Slabs*	*Beams or Ribbed One-Way Slabs*	*Solid One-Way Slabs*	*Beams or Ribbed One-Way Slabs*	*Solid One-Way Slabs*	*Beams or Ribbed One-Way Slabs*	*Solid One-Way Slabs*	*Beams or Ribbed One-Way Slabs*
1500	300	l/20.0	l/16.0	l/24.0	l/18.5	l/28.0	l/21.0	l/10.0	l/8.0
	350	18.4	14.7	22.0	17.0	25.7	19.3	9.2	7.3
	400	16.7	13.3	20.0	15.4	23.3	17.5	8.3	6.7
	450	15.8	12.7	19.0	14.7	22.2	16.6	7.9	6.3
	500	14.8	11.9	17.8	13.7	20.7	15.6	7.4	5.9
1600	300	l/20.5	l/16.4	l/24.6	l/18.9	l/28.7	l/21.5	l/10.2	l/8.2
	350	18.8	15.1	22.6	17.4	26.4	19.8	9.4	7.5
	400	17.1	13.7	20.5	15.8	23.9	17.9	8.5	6.8
	450	16.2	13.0	19.5	15.0	22.7	17.1	8.1	6.5
	500	15.2	12.2	18.2	14.1	21.3	16.0	7.6	6.1
1700	300	l/21.0	l/16.8	l/25.2	l/19.4	l/29.4	l/22.1	l/10.5	l/8.4
	350	19.3	15.5	23.2	17.9	27.1	20.3	9.7	7.7
	400	17.5	14.0	21.1	16.2	24.6	18.4	8.8	7.0
	450	16.7	13.3	20.0	15.4	23.3	17.5	8.3	6.7
	500	15.6	12.5	18.7	14.4	21.8	16.4	7.8	6.2
1800	300	l/21.6	l/17.3	l/25.9	l/20.0	l/30.2	l/22.7	l/10.8	l/8.6
	350	19.9	15.9	23.8	18.4	27.8	20.9	9.9	7.9
	400	18.0	14.4	21.6	16.7	25.2	18.9	9.0	7.2
	450	17.1	13.7	20.5	15.8	24.0	18.0	8.6	6.8
	500	16.0	12.8	19.2	14.8	22.4	16.8	8.0	6.4
1900 to 2000	300	l/22.0	l/17.6	l/26.4	l/20.3	l/30.8	l/23.1	l/11.0	l/8.8
	350	20.2	16.2	24.3	18.7	28.3	21.2	10.1	8.1
	400	18.4	14.7	22.0	17.0	25.7	19.3	9.2	7.3
	450	17.4	14.0	20.9	16.1	24.4	18.3	8.7	7.0
	500	16.3	13.1	19.6	15.1	22.8	17.1	8.2	6.5
2400	300	l/24.0	l/19.2	l/28.8	l/22.2	l/33.5	l/25.2	l/12.0	l/9.6
	350	22.0	17.6	26.5	20.4	30.9	23.1	11.0	8.8
	400	20.0	16.0	24.0	18.5	28.0	21.0	10.0	8.0
	450	19.0	15.2	22.8	17.6	26.6	20.0	9.5	7.6
	500	17.8	14.2	21.3	16.4	24.9	18.7	8.9	7.1

negative moments. For a continuous beam with equal spans, loaded with uniform loads, a typical negative design moment for an interior span is equal to the fixed end moment. In these interior spans, negative moments are about twice as large as positive moments. Generally, the negative design moment is larger than the positive design moment. It is common practice to keep the cross sections uniform along the member and adjust the area of reinforcement at critical sections according to the respective design moments.

Sections of flexural members are proportioned for tension steel. In a typical one-way continuous member, bottom reinforcement is provided in the span, whereas top reinforcement is provided over the supports. Under certain conditions, especially when member depth is limited, compression reinforcement is used along with increased tension steel for a higher moment capacity. Design of tension and compression reinforcement is discussed below with the related design aids and example problems.

1.5.1 Rectangular Beams with Tension Reinforcement

Nominal moment capacity, M_n, for a singly reinforced rectangular section is given by eq. (1-8). This equation is rewritten below in terms of the reinforcement ratio, ρ:

$$M_n = \rho b d f_y \left(d - \frac{\rho d f_y}{1.7 f'_c}\right) \tag{1-40}$$

or:

$$M_n = b d^2 q_n \tag{1-41}$$

where:

$$q_n = \rho f_y \left(1 - \frac{\rho f_y}{1.7 f'_c}\right) \tag{1-42}$$

$$\rho = \frac{A_s}{bd} \tag{1-43}$$

The factor q_n is only a function of the reinforcement ratio, ρ, and the material properties, f_y and f'_c. If the value of q_n is computed for different combinations of ρ, f_y, and f'_c, then the nominal design strength, M_n, can be determined for any rectangular cross section. Table 1-4 is generated for this purpose.

The significance of different failure modes for flexural members is discussed in Section 1.3.1. As stated earlier a

TABLE 1-4 Nominal Flexural Strength of Rectangular Sections

$$M_n = bd^2 q_n$$

f_y (MPa)	300			350			400		
f'_c (MPa)	20	30	40	20	30	40	20	30	40
ρ_b	.0321	.0482	.0582	.0261	.0391	.0472	.0217	.0325	.0393
ρ	q_n (MPA)								
.0035	1.02	1.03	1.03	1.18	1.20	1.20	1.34	1.36	1.37
.0040	1.16	1.17	1.18	1.34	1.36	1.37	1.52	1.55	1.56
.0045	1.30	1.31	1.32	1.50	1.53	1.54	1.70	1.74	1.75
.0050	1.43	1.46	1.47	1.66	1.69	1.70	1.88	1.92	1.94
.0055	1.57	1.60	1.61	1.82	1.85	1.87	2.06	2.11	2.13
.0060	1.70	1.74	1.75	1.97	2.01	2.04	2.23	2.29	2.32
.0065	1.84	1.88	1.89	2.12	2.17	2.20	2.40	2.47	2.50
.0070	1.97	2.01	2.04	2.27	2.33	2.36	2.57	2.65	2.68
.0075	2.10	2.15	2.18	2.42	2.49	2.52	2.74	2.82	2.87
.0080	2.23	2.29	2.32	2.57	2.65	2.68	2.90	3.00	3.05
.0085	2.36	2.42	2.45	2.71	2.80	2.84	3.06	3.17	3.23
.0090	2.49	2.56	2.59	2.86	2.96	3.00	3.22	3.35	3.41
.0095	2.61	2.69	2.73	3.00	3.11	3.16	3.38	3.52	3.59
.0100	2.74	2.82	2.87	3.14	3.26	3.32	3.53	3.69	3.76
.0105	2.86	2.96	3.00	3.28	3.41	3.48	3.68	3.85	3.94
.0110	2.98	3.09	3.14	3.41	3.56	3.63	3.83	4.02	4.12
.0115	3.10	3.22	3.27	3.55	3.71	3.79	3.98	4.19	4.29
.0120	3.22	3.35	3.41	3.68	3.85	3.94	4.12	4.35	4.46
.0125	3.34	3.47	3.54	3.81	4.00	4.09	4.26	4.51	4.63
.0130	3.45	3.60	3.68	3.94	4.14	4.25	4.40	4.67	4.80
.0135	3.57	3.73	3.81	4.07	4.29	4.40	4.54	4.83	4.97
.0140	3.68	3.85	3.94	4.19	4.43	4.55	4.68	4.99	5.14
.0145	3.79	3.98	4.07	4.32	4.57	4.70	4.81	5.14	5.31
.0150	3.90	4.10	4.20	4.44	4.71	4.84	4.94	5.29	5.47
.0155	4.01	4.23	4.33	4.56	4.85	4.99	5.07	5.45	5.63
.0160	4.12	4.35	4.46	4.68	4.99	5.14	5.20	5.60	5.80
.0165	4.23	4.47	4.59	4.79	5.12	5.28	5.32	5.75	5.96
.0170	4.33	4.59	4.72	4.91	5.26	5.43	5.44	5.89	6.12
.0175	4.44	4.71	4.84	5.02	5.39	5.57	5.56	6.04	6.28
.0180	4.54	4.83	4.97	5.13	5.52	5.72	5.68	6.18	6.44
.0185	4.64	4.95	5.10	5.24	5.65	5.86	5.79	6.33	6.59
.0190	4.74	5.06	5.22	5.35	5.78	6.00	5.90	6.47	6.75
.0195	4.84	5.18	5.35	5.45	5.91	6.14	6.01	6.61	6.91
.0200	4.94	5.29	5.47	5.56	6.04	6.28	6.12	6.75	7.06
.0205	5.04	5.41	5.59	5.66	6.17	6.42	6.22	6.88	7.21
.0210	5.13	5.52	5.72	5.76	6.29	6.56	6.32	7.02	7.36
.0215	5.23	5.63	5.84	5.86	6.41	6.69	6.42	7.15	7.51
.0220	5.32	5.75	5.96	5.96	6.54	6.83	6.52	7.28	7.66
.0225	5.41	5.86	6.08	6.05	6.66	6.96	6.62	7.41	7.81
.0230	5.50	5.97	6.20	6.14	6.78	7.10	6.71	7.54	7.96
.0235	5.59	6.08	6.32	6.24	6.90	7.23	6.80	7.57	8.10
.0240	5.68	6.18	6.44	6.32	7.02	7.36	6.89	7.79	8.24
.0245	5.76	6.29	6.56	6.41	7.13	7.49	6.98	7.92	8.39
.0250	5.85	6.40	6.67	6.50	7.25	7.62	7.06	8.04	8.53
.0255	5.93	6.50	6.79	6.58	7.36	7.75	7.14	8.16	8.67
.0260	6.01	6.61	6.91	6.66	7.48	7.88	7.22	8.28	8.81
.0265	6.09	6.71	7.02	6.74	7.59	8.01	7.30	8.40	8.95
.0270	6.17	6.81	7.14	6.82	7.70	8.14	7.37	8.51	9.08
.0275	6.25	6.92	7.25	6.90	7.81	8.26	7.44	8.63	9.22
.0280	6.32	7.02	7.36	6.98	7.92	8.39	7.51	8.74	9.36
.0285	6.40	7.12	7.47	7.05	8.02	8.51	7.58	8.85	9.49

TABLE 1-4 *(Continued)*

.0290	6.47	7.22	7.59	7.12	8.13	8.63	7.64	8.96	9.62
.0295	6.55	7.31	7.70	7.19	8.23	8.76	7.70	9.07	9.75
.0300	6.62	7.41	7.81	7.26	8.34	8.88	7.76	9.18	9.88
.0305	6.69	7.51	7.92	7.32	8.44	9.00	7.82	9.28	10.01
.0310	6.76	7.60	8.03	7.39	8.54	9.12	7.88	9.39	10.14
.0315	6.82	7.70	8.14	7.45	8.64	9.24	7.93	9.49	10.27
.0320	6.89	7.79	8.24	7.51	8.74	9.36	7.98	9.59	10.39
.0325	6.95	7.89	8.35	7.57	8.84	9.47	8.03	9.69	10.51
.0330	7.02	7.98	8.46	7.63	8.93	9.59	8.08	9.78	10.64
.0335	7.08	8.07	8.56	7.68	9.03	9.70	8.12	9.88	10.76
.0340	7.14	8.16	8.67	7.73	9.12	9.82	8.16	9.97	10.88
.0345	7.20	8.25	8.77	7.79	9.22	9.93	8.20	10.07	11.00
.0350	7.26	8.34	8.88	7.84	9.31	10.04	8.24	10.16	11.12
.0355	7.31	8.43	8.98	7.88	9.40	10.15	8.27	10.25	11.23
.0360	7.37	8.51	9.08	7.93	9.49	10.27	8.30	10.33	11.35
.0365	7.42	8.60	9.19	7.97	9.57	10.37	8.33	10.42	11.47
.0370	7.48	8.68	9.29	8.02	9.66	10.48	8.36	10.51	11.58
.0375	7.53	8.77	9.39	8.06	9.75	10.59	8.38	10.59	11.69
.0380	7.58	8.85	9.49	8.10	9.83	10.70	8.40	10.67	11.80
.0385	7.63	8.93	9.59	8.13	9.91	10.80	8.42	10.75	11.91
.0390	7.67	9.02	9.69	8.17	10.00	10.91	8.44	10.83	12.02
.0395	7.72	9.10	9.78	8.20	10.08	11.01	8.46	10.91	12.13
.0400	7.76	9.18	9.88	8.24	10.16	11.12	8.47	10.98	12.24
.0405	7.81	9.26	9.98	8.27	10.24	11.22	8.48	11.05	12.34
.0410	7.85	9.33	10.08	8.29	10.31	11.32	8.49	11.13	12.44
.0415	7.89	9.41	10.17	8.32	10.39	11.42	8.50	11.20	12.55
.0420	7.93	9.49	10.27	8.34	10.46	11.52	8.50	11.57	12.65
.0425	7.97	9.56	10.36	8.37	10.54	11.62	8.50	11.33	12.75
.0430	8.01	9.64	10.45	8.39	10.61	11.72	8.50	11.40	12.85
.0435	8.04	9.71	10.55	8.41	10.68	11.82	8.50	11.46	12.95
.0440	8.08	9.78	10.64	8.42	10.75	11.91	8.49	11.53	13.04
.0445	8.11	9.86	10.73	8.44	10.82	12.01	8.48	11.59	13.14
.0450	8.14	9.93	10.82	8.45	10.89	12.10	8.47	11.65	13.24
.0455	8.17	10.00	10.91	8.47	10.95	12.20	8.46	11.71	13.33

Notes:
(1) For beams with compression reinforcement $\rho = (A_s - A'_s)/bd$.
(2) For T beams $\rho = \rho_w = (A_s - A_{sf})/b_w d$.
(3) Solid lines indicate minimum and maximum permissible values for rectangular sections with tension reinforcement only.

ductile failure is the desirable failure mode. This type of failure occurs if the reinforcement ratio, ρ, is less than the balanced section ratio, ρ_b. The expression for ρ_b is developed in Section 1.3.1 and is given by eq. (1-10).

ACI 318M-83 provides restrictions to the amount of steel that can be placed in a flexural member. To ensure a ductile failure, the maximum reinforcement ratio is limited to 75% of ρ_b. Reinforcement ratios in the order of 30 to 40% of ρ_b are considered to be good design practice. Table 1-4 contains values of q_n corresponding to different levels of ρ. Relative magnitudes of ρ with respect to ρ_b can be found in this table for selecting ductile sections. Table 1-4 covers the commonly used material strengths and reinforcement ratios between the minimum and the maximum limits.

EXAMPLE 1-1: Proportion the rectangular beam shown in Fig. 1-14 for tension reinforcement. The applied ultimate design moment $M_u = 200$ kN · m, and the concrete strength is 30 MPa. Use grade 400 steel for reinforcement with a minimum net cover of 50 mm.

$$M_u \leqslant \phi M_n$$

$$M_n = 200/0.9 = 222 \text{ kN} \cdot \text{m}$$

Assuming No. 20 bars will be used:

$$d = 500 - 50 - 10 = 440 \text{ mm}$$

$$M_n = bd^2 q_n \qquad \text{(from eq. 1-41)}$$

$$q_n = \frac{222 \times 10^6}{(300)(440)^2} = 3.82 \text{ MPa}$$

From Table 1-4, $\rho = 0.0104$.

$$A_s = \rho bd$$

$$A_s = 0.0104\,(300)\,440 = 1373 \text{ mm}^2$$

$$\frac{1373 \text{ mm}^2}{300 \text{ mm}^2} = 4.58 \quad \text{Use 5 No. 20 bars.}$$

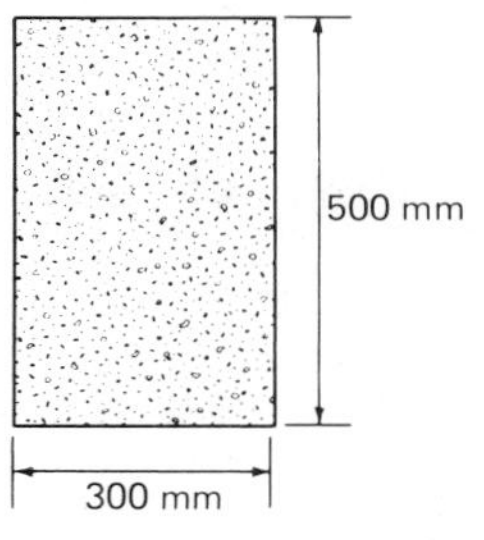

Fig. 1-14 Example 1-1.

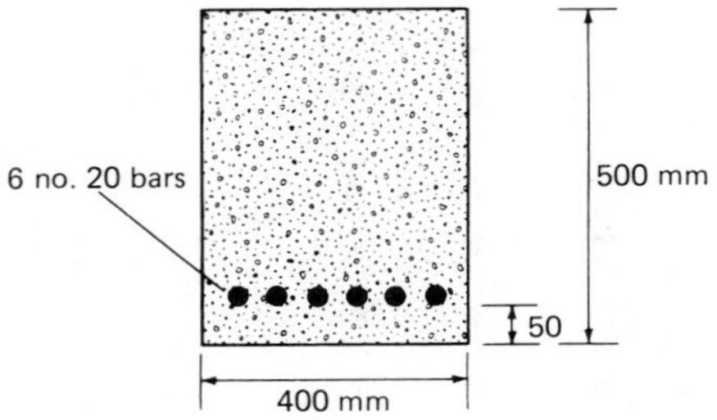

Fig. 1-15 Example 1-2.

EXAMPLE 1-2: Find the moment capacity, M_n, of the rectangular section shown in Fig. 1-15. Material properties are: $f_c' = 25$ MPa and $f_y = 300$ MPa.

$$M_n = bd^2 q_n \qquad \text{(from eq. 1-41)}$$

$$\rho = \frac{A_s}{bd} = \frac{(6)(300)}{(400)(450)} = 0.010$$

For $\rho = 0.010$, $f_c' = 25$ MPa, and $f_y = 300$ MPa:

$$q_n = 2.78 \text{ MPa} \qquad \text{(from Table 1-4)}$$

$$M_n = (0.40 \text{ m})(0.45)^2(2.78 \text{ MPa}) \times 1000$$

$$M_n = 225 \text{ kN} \cdot \text{m}$$

1.5.2 Rectangular Beams with Tension and Compression Reinforcement

Tension reinforcement is generally sufficient to provide the required capacity in rectangular beams. Under certain conditions, however, the required flexural capacity cannot be provided with tension reinforcement alone. This usually happens if the member depth is limited by architectural considerations. In such cases the increased moment capacity of a section is obtained by providing compression reinforcement.

Figure 1-16 illustrates a rectangular section with compression steel. The ultimate moment capacity consists of two components. One component (M_1) is the capacity of the concrete section reinforced with tension steel ($A_s - A_s'$). The other component (M_2) is the moment capacity of a steel section consisting of compression steel A_s' and an equal area of tension steel.

Compression steel is generally located close to the extreme compression fiber of a section. Therefore, it is sufficiently accurate to assume that the compression steel yields when the section reaches its ultimate capacity. Based on this assumption, moment component M_2 can be written as follows:

$$M_2 = A_s' f_y (d - d') \tag{1-44}$$

or:

$$M_2 = bd^2 q_n' \tag{1-45}$$

where:

$$q_n' = \rho' f_y \left(1 - \frac{d'}{d}\right) \tag{1-46}$$

$$\rho' = \frac{A_s'}{bd} \tag{1-47}$$

The factor q_n' is a function of compression reinforcement ratio ρ', ratio d'/d, and the reinforcement yield stress, f_y. If the values of q_n' are tabulated for different combinations of ρ', d'/d, and f_y, the additional moment capacities of sections due to compression reinforcement can be determined. Table 1-5 is generated for this purpose.

Moment capacity of a rectangular section with tension reinforcement was previously derived in eq. (1-8). Substituting ($A_s - A_s'$) for A_s yields:

$$M_1 = (A_s - A_s') f_y \left(d - \frac{f_y (A_s - A_s')}{1.7 f_c' b}\right) \tag{1-48}$$

or:

$$M_1 = bd^2 q_n \tag{1-49}$$

where:

$$q_n = \rho f_y \left(1 - \frac{\rho f_y}{1.7 f_c'}\right) \tag{1-50}$$

$$\rho = \frac{(A_s - A_s')}{bd} \tag{1-51}$$

Equation (1-50) is identical to eq. (1-42) derived for beams with only tension reinforcement. Therefore, the values of q_n tabulated in Table 1-4 can be used to compute moment component M_1 with reinforcement ratio ρ as defined by eq. (1-51). Total moment capacity of the section can then be expressed as:

$$M_n = M_1 + M_2 \tag{1-52}$$

$$M_n = bd^2 (q_n + q_n') \tag{1-53}$$

ACI 318M-83 places restrictions on the amount of reinforcement that can be used in a flexural member. A maximum reinforcement ratio is specified to ensure a ductile failure. For rectangular sections with compression reinforcement, the maximum tension steel is governed by the following condition:

$$\rho_{\max} = 0.75\, \bar{\rho}_b + \rho' \frac{f_{sb}'}{f_y} \tag{1-54}$$

where:

$\bar{\rho}_b$ = balanced reinforcement ratio for a rectangular section with tension reinforcement only

f_{sb}' = stress in compression reinforcement at balanced strain conditions

$$f_{sb}' = 600 - \frac{d'}{d}(600 + f_y) \leqslant f_y \tag{1-55}$$

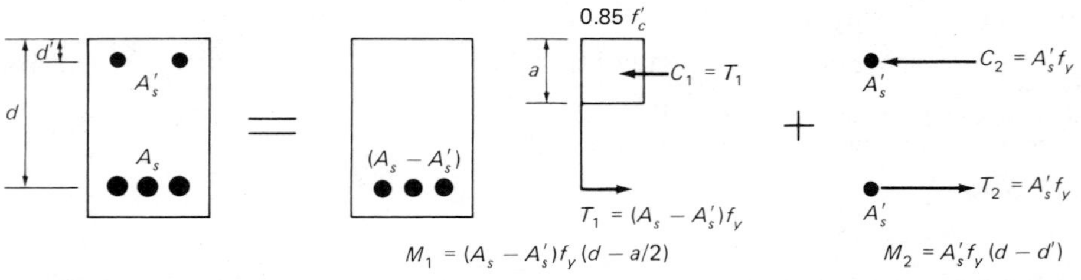

Fig. 1-16 Components of ultimate moment in a rectangular section with compression reinforcement.

TABLE 1-5 Moment Couple Due to Compression Steel

$$M_2 = bd^2 q'_n$$

f_y (MPa)	300			350			400		
d'/d	0.10	0.15	0.20	0.10	0.15	0.20	0.10	0.15	0.20
ρ'	q'_n (MPa)								
0.0005	0.13	0.13	0.12	0.16	0.15	0.14	0.18	0.17	0.16
0.0010	0.27	0.25	0.24	0.31	0.30	0.28	0.36	0.34	0.32
0.0015	0.41	0.38	0.36	0.47	0.45	0.42	0.54	0.51	0.48
0.0020	0.54	0.51	0.48	0.63	0.60	0.56	0.72	0.68	0.64
0.0025	0.68	0.64	0.60	0.79	0.74	0.70	0.90	0.85	0.80
0.0030	0.81	0.77	0.72	0.94	0.89	0.84	1.08	1.02	0.96
0.0035	0.94	0.89	0.84	1.10	1.04	0.98	1.26	1.19	1.12
0.0040	1.08	1.02	0.96	1.26	1.19	1.12	1.44	1.36	1.28
0.0045	1.21	1.15	1.08	1.42	1.34	1.26	1.62	1.53	1.44
0.0050	1.35	1.27	1.20	1.57	1.49	1.40	1.80	1.70	1.60
0.0055	1.48	1.40	1.32	1.73	1.64	1.54	1.98	1.87	1.76
0.0060	1.62	1.53	1.44	1.89	1.78	1.68	2.16	2.04	1.92
0.0065	1.75	1.66	1.56	2.05	1.93	1.82	2.34	2.21	2.08
0.0070	1.89	1.78	1.68	2.20	2.08	1.96	2.52	2.38	2.24
0.0075	2.02	1.91	1.80	2.36	2.23	2.10	2.70	2.55	2.40
0.0080	2.16	2.04	1.92	2.52	2.38	2.24	2.88	2.72	2.56
0.0085	2.29	2.17	2.04	2.68	2.53	2.38	3.06	2.89	2.72
0.0090	2.43	2.29	2.16	2.83	2.68	2.52	3.24	3.06	2.88
0.0095	2.56	2.42	2.28	2.99	2.83	2.66	3.42	3.23	3.04
0.0100	2.70	2.55	2.40	3.15	2.97	2.80	3.60	3.40	3.20
0.0105	2.83	2.68	2.52	3.31	3.12	2.94	3.78	3.57	3.36
0.0110	2.97	2.80	2.64	3.46	3.27	3.08	3.96	3.74	3.52
0.0115	3.10	2.93	2.76	3.62	3.42	3.22	4.14	3.91	3.68
0.0120	3.24	3.06	2.88	3.78	3.57	3.36	4.32	4.08	3.84
0.0125	3.37	3.19	3.00	3.94	3.72	3.50	4.50	4.25	4.00
0.0130	3.51	3.31	3.12	4.09	3.87	3.64	4.68	4.42	4.16
0.0135	3.64	3.44	3.24	4.25	4.02	3.78	4.86	4.59	4.32
0.0140	3.78	3.57	3.36	4.41	4.16	3.92	5.04	4.76	4.48
0.0145	3.91	3.70	3.48	4.57	4.31	4.06	5.22	4.93	4.64
0.0150	4.06	3.82	3.60	4.72	4.46	4.20	5.40	5.10	4.80
0.0155	4.18	3.95	3.72	4.88	4.61	4.34	5.58	5.27	4.96
0.0160	4.32	4.08	3.84	5.04	4.76	4.48	5.76	5.44	5.12
0.0165	4.45	4.21	3.96	5.20	4.91	4.62	5.94	5.61	5.28
0.0170	4.59	4.33	4.08	5.35	5.08	4.78	6.12	5.78	5.44
0.0175	4.72	4.46	4.20	5.51	5.21	4.90	6.30	5.95	5.60
0.0180	4.86	4.59	4.32	5.67	5.35	5.04	6.48	6.12	5.76
0.0185	4.99	4.72	4.44	5.83	5.50	5.18	6.66	6.29	5.92
0.0190	5.13	4.84	4.58	5.98	5.65	5.32	6.84	6.46	6.08
0.0195	5.26	4.97	4.68	6.14	5.80	5.46	7.02	6.63	6.24
0.0200	5.40	5.10	4.80	6.30	5.95	5.60	7.20	6.80	6.40
0.0205	5.53	5.23	4.92	6.46	6.10	5.74	7.38	6.97	6.56
0.0210	5.67	5.35	5.04	6.61	6.25	5.88	7.56	7.14	6.72
0.0215	5.80	5.48	5.16	6.77	6.40	6.02	7.74	7.31	6.88
0.0220	5.94	5.61	5.28	6.93	6.54	6.16	7.92	7.48	7.04
0.0225	6.07	5.74	5.40	7.09	6.69	6.30	8.10	7.65	7.20
0.0230	6.21	5.86	5.52	7.24	6.84	6.44	8.28	7.82	7.36
0.0235	6.34	5.99	5.64	7.40	6.99	6.58	8.46	7.99	7.52
0.0240	6.48	6.12	5.76	7.56	7.14	6.72	8.64	8.16	7.68
0.0245	6.61	6.25	5.88	7.72	7.29	6.86	8.82	8.33	7.84
0.0250	6.75	6.37	6.00	7.87	7.44	7.00	9.00	8.50	8.00

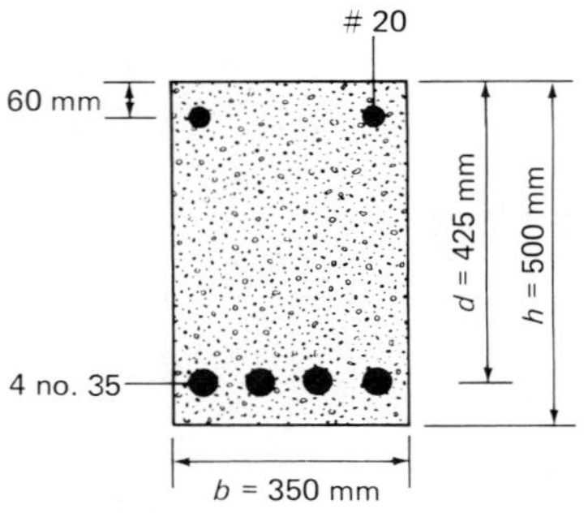

Fig. 1-17 Example 1-3.

Reinforcement yield stress, f_y, in eq. (1-55) is in MPa. If the compression steel yields at balanced strain conditions, as in the case of most beams, then $f'_{sb} = f_y$ and $\rho_{max} = 0.75\,\overline{\rho}_b + \rho'$.

EXAMPLE 1-3: Design a rectangular beam for the ultimate design moment $M_u = 500$ kN · m. The beam dimensions as shown in Fig. 1-17 cannot exceed $b = 350$ mm and $h = 500$ mm owing to architectural limitations. Use 400 MPa steel for reinforcement and 30 MPa concrete.

$$M_u \leqslant \phi M_n$$

Try a section with tension reinforcement only:

$$M_n = bd^2 q_n \qquad \text{(from eq. 1-41)}$$

$$\rho_b = 0.0325 \qquad \text{(from Table 1-4)}$$

Good design practice requires 30 to 40% of ρ_b. Use 40% of ρ_b. Assume $b = 350$ mm and $d = 425$ mm.

$$\rho = 0.40\ (0.0325) = 0.0130$$

$$q_n = 4.67 \text{ MPa} \qquad \text{(from Table 1.4)}$$

$$M_n = bd^2 q_n = (0.350)(0.425)^2 \times 4.67 \times 1000$$

$$M_n = 295 \text{ kN} \cdot \text{m}$$

$$\phi M_n = (0.9)(295) = 265 \text{ kN} \cdot \text{m} < 500 \text{ kN} \cdot \text{m} \qquad \text{N.G.}$$

Try using the maximum reinforcement ratio:

$$\rho_{max} = 0.75\ \rho_b$$

Use slightly less than the maximum value, since the actual provided area of steel may be higher than the required area of steel. Use $\rho = 0.70\ \rho_b = 0.0227$:

$$q_n = 7.47 \text{ MPa} \qquad \text{(from Table 1-4)}$$

$$M_n = (0.350)(0.425)^2 \times 7.47 \times 1000$$

$$M_n = 472 \text{ kN} \cdot \text{m}$$

$$M_n = (0.9)(472) = 425 \text{ kN} \cdot \text{m} < 500 \text{ kN} \cdot \text{m} \qquad \text{N.G.}$$

Use compression reinforcement to increase the flexural capacity of the section. Use $d' = 60$ mm. Calculating the excess moment that is to be resisted by compression reinforcement:

$$M_2 = \frac{M_u}{\phi} - M_1$$

$$M_2 = \frac{500}{(0.9)} - 472$$

$$M_2 = 555 - 472 = 83 \text{ kN} \cdot \text{m}$$

$$M_2 = bd^2 q'_n$$

$$q'_n = \frac{83}{(0.350)(0.425)^2}\ \frac{1}{1000} = 1.31 \text{ MPa}$$

$$\rho' = 0.0038 \qquad \text{(from Table 1-5)}$$

$$A'_s = (0.0038)(350)(425) = 565 \text{ mm}^2$$

Use two #20 bars as compression reinforcement:
Provided area of compression steel = 2(300) = 600 mm^2
Required area of tension reinforcement:

$$A_s = (\rho + \rho')\, bd = (0.0227 + 0.0038)(350)(425)$$

$$A_s = 3942 \text{ mm}^2$$

Use four #35 bars as tension reinforcement:
Provided area of tension steel = 4(1000) = 4000 mm^2
Check for maximum reinforcement ratio:

$$\rho' = \frac{600}{(350)(425)} = 0.004$$

$$\rho = \frac{4000}{(350)(425)} = 0.0269$$

$$\rho_{max} = 0.75\,\overline{\rho}_b + \rho' \frac{f'_{sb}}{f_y}$$

$$\overline{\rho}_b = 0.0325 \qquad \text{(from Table 1-4)}$$

$$f'_{sb} = 600 - \frac{60}{425}(600 + 400) = 459 \qquad \text{(from eq. 1-55)}$$

$f'_{sb} > f_y$; therefore since yielding occurs $f'_{sb} = f_y$

$$\rho_{max} = 0.75\ (0.0325) + 0.004 = 0.0284$$

$$\rho = 0.0269 < 0.0284 \qquad \text{O.K.}$$

EXAMPLE 1-4: Determine the nominal moment capacity of the section shown in Fig. 1-18. Concrete strength and reinforcement yield stress used are $f'_c = 25$ MPa and $f_y = 300$ MPa, respectively. Two No. 20 and three No. 30 bars are used for compression and tension reinforcement, respectively. Net concrete cover used is 50 mm.

$$d = 500 - 50 - \frac{30}{2} = 435 \text{ mm}$$

$$d' = 50 + \frac{20}{2} = 60$$

$$A_s = (3)(700) = 2100 \text{ mm}^2$$

$$A'_s = (2)(300) = 600 \text{ mm}^2$$

$$M_n = bd^2\ (q_n + q'_n) \qquad \text{(from eq. 1-53)}$$

$$q_n = 2.41 \text{ MPa for } \rho = \frac{A_s - A'_s}{bd} = \frac{1500}{(400)(435)}$$

$$= 0.0086 \qquad \text{(from Table 1-4)}$$

$$\rho' = \frac{A'_s}{bd} = \frac{600}{(400)(435)} = 0.00345$$

$$q'_n = 0.89 \text{ MPa} \qquad \text{(from Table 1-5)}$$

$$M_n = bd^2\ (q_n + q'_n) = (0.400)(0.435)^2\ (2.41 + 0.89)(1000)$$

$$M_n = 250 \text{ kN} \cdot \text{m}$$

1.5.3 "T" Beams

Rectangular beams are generally cast monolithically with concrete slabs, forming full composite action between the

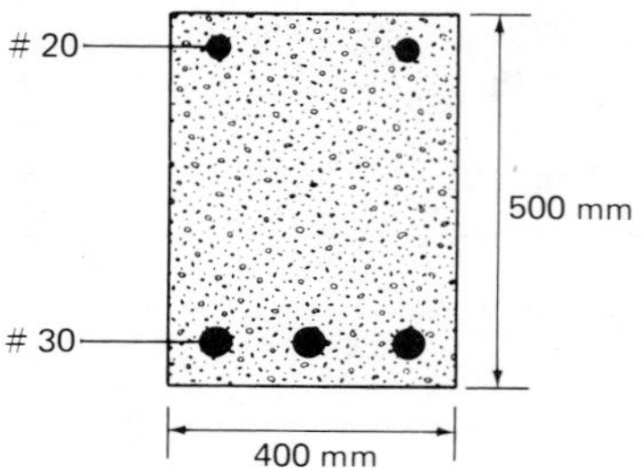

Fig. 1-18 Example 1-4.

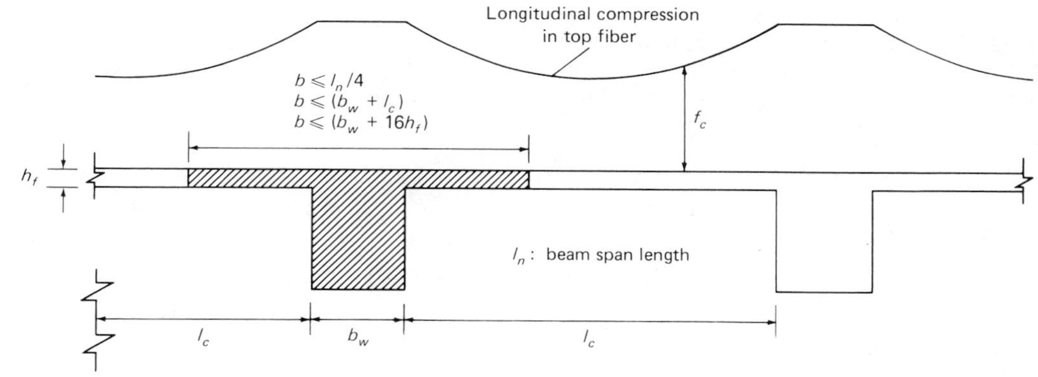

Fig. 1-19 Distribution of longitudinal compression stresses in the top fiber of a slab-beam assembly and the concept of effective width in T-sections.

slab and the beams. In the positive moment region of a slab–beam assembly, compression stresses develop in the slab, while the bottom beam reinforcement is subjected to tensile stresses. In such cases, the effective cross section of the beam has a T-shape consisting of the rectangular beam as the web and a portion of the slab as the flange. If the beams are closely spaced, the flange width can be taken as the distance between the slab centerlines. However, if the spacing is large, the flange is more highly stressed in the vicinity of the web as shown in Fig. 1-19, and the stress distribution is not uniform across the flange. In such cases it is convenient to define an effective flange width over which the stress distribution is assumed to be uniform. ACI 318M-83 recommends that the flange width for T beams should not exceed one fourth of the beam span length. Furthermore, the effective overhanging slab width on each side should exceed neither 8 times the slab thickness nor one half the clear distance to the next web. Figure 1-19 illustrates a typical T-beam and the limiting dimensions.

T-beams generally possess high compression capacity because of the large area of concrete in the flange. The neutral axis of a typical T-beam usually lies in the flange. This is generally the case unless a large area of tension steel is provided, or the flange thickness is too small relative to the web. When the neutral axis is within the flange, the section behaves as a rectangular section with a width equal to the flange width. If the tension in reinforcement exceeds the compression in flange concrete, the neutral axis falls below the flange and the section behaves as a T-section. Thus, for T-beam action the following condition must be satisfied:

Tension in steel > Compression in flange concrete

$$A_s f_y > 0.85 f'_c\, b\, h_f \qquad (1\text{-}56)$$

$$A_s > \frac{0.85 f'_c\, b\, h_f}{f_y} \qquad (1\text{-}57)$$

The ultimate moment capacity of a T-beam consists of two components, each of which can be computed easily. Figure 1-20 illustrates the two moment components (M_f and M_w) and the associated effective concrete areas in each section. According to this figure, the ultimate capacity of a T-beam is made up of the capacities of two rectangular sections. The first section consists of a concrete beam of width $(b - b_w)$ and a steel area of A_{sf} when the rectangular compression stress block has a depth equal to the flange thickness. A_{sf} is resisted by compression in the overhangs and forms only a portion of the total tension steel A_s.

$$M_f = A_{sf} f_y \left(d - \frac{h_f}{2}\right) \qquad (1\text{-}58)$$

where:

$$A_{sf} = \frac{0.85 f'_c\, h_f\, (b - b_w)}{f_y} \qquad (1\text{-}59)$$

The second section that contributes to the ultimate strength of a T-beam is a rectangular section formed by width b_w and height h and reinforced with $(A_s - A_{sf})$. The nominal moment capacity of a rectangular section with tension reinforcement was previously derived and expressed in eq. (1-41). Using the same notations, the second component, M_w, can be written as:

$$M_w = b_w d^2 q_n \qquad (1\text{-}60)$$

where q_n can be obtained from Table 1-4 using:

$$\rho_w = \frac{(A_s - A_{sf})}{b_w d} \qquad (1\text{-}61)$$

Total nominal moment capacity of a T-beam can then be found by summing the two components involved:

$$M_n = M_f + M_w \qquad (1\text{-}62)$$

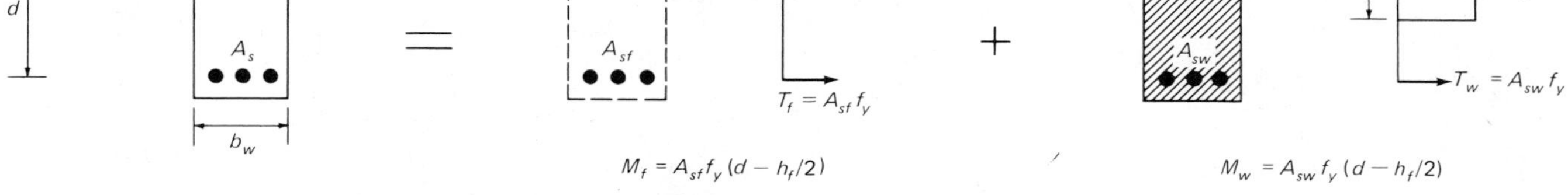

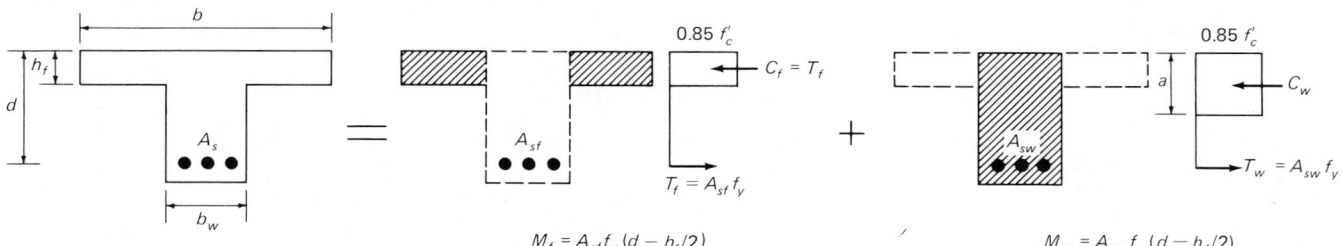

Fig. 1-20 Components of ultimate moment in a T-section.

The maximum area of tension reinforcement in T-beams is limited by the ACI 318M-83 to ensure ductile behavior. The same limit that applies to rectangular beams also applies to T-beams. The maximum allowable reinforcement ratio is limited to 75% of the balanced value. Balanced section reinforcement ratio ρ_b and the maximum reinforcement ratio ρ_{max} for T-beams are given below.

$$\rho_b = \frac{b_w}{b} (\overline{\rho}_b + \rho_f) \qquad (1\text{-}63)$$

where:

$\rho_f = \dfrac{A_{sf}}{b_w d}$ and A_{sf} is expressed by eq. (1-59)

$\overline{\rho}_b$ = balanced reinforcement ratio for a rectangular section (b_w by h) with tension reinforcement ($A_{sw} = A_s - A_{sf}$)

$$\rho_{max} = 0.75 \left[\frac{b_w}{b} (\overline{\rho}_b + \rho_f) \right] \qquad (1\text{-}64)$$

EXAMPLE 1.5: The T-beam shown in Fig. 1-21 is to be proportioned for the positive moment region. If the concrete strength is 20 MPa and the reinforcement grade is 400, determine the area of tension steel required for: (i) M_u = 500 kN · m and (ii) M_u = 700 kN · m.

i. For M_u = 500 kN · m assume the beam behaves as a rectangular beam.

$$M_u \leqslant \phi M_n$$

$$M_n = \frac{500}{0.9} = 555 \text{ kN} \cdot \text{m}$$

$$M_n = bd^2 q_n \qquad \text{(from eq. 1-41)}$$

$$q_n = \frac{555}{(1.20)(0.43)^2} = 2500 \text{ kPa}$$

$$q_n = 2.5 \text{ MPa}$$

$$\rho = 0.0068 \qquad \text{(from Table 1-4)}$$

$$A_s = \rho bd = 0.0068\ (1200)(430) = 3500 \text{ mm}^2$$

Condition for T-beam behavior (from eq. 1-57):

$$A_s > \frac{0.85 f'_c b h_f}{f_y} \frac{0.85(20)(1200)(80)}{400} = 4080 \text{ mm}^2$$

$A_s = 3500 \text{ mm}^2 < 4080 \text{ mm}^2$ The assumption of rectangular beam behavior is correct.

Therefore, the required area of tension steel is 3500 mm².

ii. For M_u = 700 kN · m first assume the beam behaves as a rectangular beam.

$$M_u \leqslant \phi M_n$$

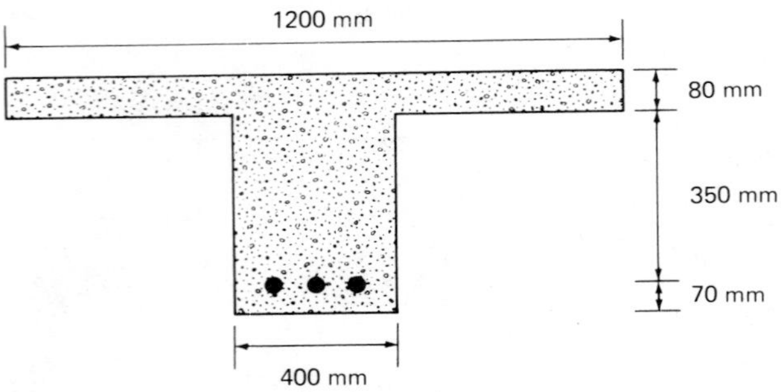

Fig. 1-21 Example 1-5.

$$M_n = \frac{700}{0.9} = 778 \text{ kN} \cdot \text{m}$$

$$M_n = bd^2 q_n \qquad \text{(from eq. 1-41)}$$

$$q_n = \frac{778}{(1.20)(0.43)^2} = 3500 \text{ kPa}$$

$$q_n = 3.50 \text{ MPa}$$

$$\rho = 0.0099 \qquad \text{(from Table 1-4)}$$

$$A_s = \rho bd = 0.0099\ (1200)(430) = 5100 \text{ mm}^2$$

Condition for T-beam behavior (from eq. 1-57):

$$A_s > \frac{0.85 f'_c b h_f}{f_y} = 4080 \text{ mm}^2$$

$$A_s = 5100 \text{ mm}^2 > 4080 \text{ mm}^2$$

Therefore, the beam behaves as a T-beam.

Find the area of steel, A_{sf}, and the moment component, M_f, that is resisted by the overhangs, using eqs. (1-59) and (1-58), respectively:

$$A_{sf} = \frac{0.85\ (20)(80)(800)}{400} = 2720 \text{ mm}^2$$

$$M_f = (2720)(400)(430 - 40) \times 10^{-6} = 424 \text{ kN} \cdot \text{m}$$

$$M_w = M_n - M_f = 778 - 424 = 354 \text{ kN} \cdot \text{m}$$

Use eqs. (1-60) and (1-61) and Table 1-4 to determine A_{sw}:

$$q_n = \frac{354 \times 10^{-3}}{(0.40)(0.43)^2} = 4.79 \text{ MPa}$$

$$\rho_w = 0.0144 \qquad \text{(from Table 1-4)}$$

$$A_{sw} = 0.0144\ (400)(430) = 2477 \text{ mm}^2$$

Total area of tension steel:

$$A_s = A_{sf} + A_{sw}$$

$$A_s = 2720 + 2477 = 5197 \text{ mm}^2$$

EXAMPLE 1.6: A T-beam has the following characteristics: b = 900 mm, b_w = 300 mm, d = 400 mm, h_f = 65 mm, f_c = 20 MPa, f_y = 400 MPa. If six No. 25 bars are used in two layers as tension reinforcement, determine the nominal moment capacity of the section. Verify if this section meets the maximum reinforcement ratio requirement.

$$A_s = 6\ (500) = 3000 \text{ mm}^2$$

Determine if the beam behaves as a T-beam. Condition for T-beam behavior is:

$$A_s > \frac{0.85 f'_c b h_f}{f_y} = 2486 \text{ mm}^2 \qquad \text{(from eq. 1-57)}$$

$3000 \text{ mm}^2 > 2486 \text{ mm}^2$ Therefore T-beam theory applies.

Calculate the nominal moment capacity:

$$A_{sf} = \frac{0.85\ (20)(65)(600)}{400} \qquad \text{(from eq. 1-59)}$$

$$A_{sf} = 1657 \text{ mm}^2$$

$$M_f = (1657)(400) \left(400 - \frac{65}{2}\right) \times 10^{-6} \qquad \text{(from eq. 1-58)}$$

$$M_f = 243 \text{ kN} \cdot \text{m}$$

$$A_{sw} = A_s - A_{sf} = 3000 - 1657 = 1343 \text{ mm}^2$$

$$\rho_w = \frac{A_{sw}}{b_w d} = \frac{1343}{(300)(400)} = 0.0112$$

$$q_n = 3.90 \text{ MPa} \qquad \text{(from Table 1-4)}$$

$$M_w = b_w d^2 q_n = (300)(400)^2\ (3.90) \times 10^{-6} = 187 \text{ kN} \cdot \text{m}$$

$$M_n = M_f + M_w = 243 + 187 = 430 \text{ kN} \cdot \text{m}$$

The maximum reinforcement ratio for T-beams is:

$$\rho_{max} = 0.75 \left[\frac{b_w}{b}(\bar{\rho}_b + \rho_f)\right] \qquad \text{(from eq. 1-64)}$$

$$\rho_f = \frac{A_{sf}}{b_w d} = \frac{1657}{(300)(400)} = 0.0138$$

The value of $\bar{\rho}_b$ can be obtained from Table 1-4, which tabulates ρ_b for rectangular sections with tension reinforcement. For $f_y = 400$ MPa and $f_c' = 20$ MPa, $\bar{\rho}_b = 0.0217$.

$$\rho_{max} = 0.75 \left[\frac{300}{900}(0.0217 + 0.0138)\right]$$

$$\rho_{max} = 0.0089$$

$$\rho = \frac{As}{bd} = \frac{3000}{(900)(400)} = 0.0083 < 0.0089 \qquad \text{O.K.}$$

1.5.4 One-Way Slabs

Reinforced concrete slab systems are classified on the basis of their structural behavior. One-way slabs are those slab panels that have length-to-width ratios equal to or greater than 2.0. These slabs deflect essentially in one direction, similarly to continuous beams. Therefore, they are usually designed as one-way members with additional minimum reinforcement in the transverse direction. The basic principles of flexural beam design also apply to one-way slabs. Tables 1-6, 1-7, and 1-8 provide nominal moment capacity, M_n, for 1.0-meter-wide slab sections, corresponding to reinforcement ratio ρ and effective depth d. Two-way slab systems, on the other hand, deflect in more than one direction and are reinforced accordingly. The discussion of two-way slab systems is beyond the scope of this chapter.

One-way slab thickness is usually governed by the deflection limitations. The ACI 318M-83 *Code* provides minimum slab thicknesses for slabs with different support conditions. These values are tabulated in Table 1-3 for one-way members.

Minimum reinforcement ratio for slabs is dictated by crack control owing to shrinkage and temperature stresses. According to the ACI 318M-83 *Code*, the minimum area of steel is 0.20% of the gross concrete area (bh) for steel grades of 300 and 350. This requirement is 0.18% of the concrete area for grade 400 steel, and $0.18 \times 400/f_y$ percent for reinforcements with higher grades.

The maximum reinforcement ratio for slabs is the same as that specified for beams. This ratio is set at 75% ρ_b to ensure ductile behavior.

EXAMPLE 1-7: Design the simply supported one-way slab panel shown in Fig. 1-22 for a live load of 10.0 kN/m^2. Use normal-density concrete with $f_c' = 30$ MPa and reinforcement with $f_y = 400$ MPa.

Determine the minimum slab thickness as governed by the deflection requirements:

$$h_{min} = l/20 = 4000/20 = 200 \text{ mm} \qquad \text{(from Table 1-3)}$$

Compute dead and live loads for a 1.0-meter slab width:

$$DL = 1.0 \times 0.2 \times 2400 \times 9.81 \times 10^{-3} = 4.71 \text{ kN/m}$$

$$LL = 1.0 \times 10.0 = 10.0 \text{ kN/m}$$

Determine the design load:

$$W_u = 1.4D + 1.7L \qquad \text{(see Section 1.4.1)}$$

$$W_u = 1.4(4.71) + 1.7(10.0) = 23.6 \text{ kN/m}$$

Calculate design moment:

$$M_u = \frac{W_u l^2}{8} = \frac{23.6(4.0)^2}{8} = 47.2 \text{ kN}\cdot\text{m}$$

(positive moment of a simply supported one-way member)

Use Table 1-7 to determine the required area of steel for 1.0-meter-wide slab strip. Allow 30 mm for the concrete cover plus the bar radius.

$$d = 200 - 30 = 170 \text{ mm}$$

$$M_n = \frac{M_u}{\phi} = \frac{47.2}{0.9} = 52.44 \text{ kN}\cdot\text{m}$$

$\rho = 0.005$ gives $M_n = 55.75$ when $d = 170$ mm (from Table 1-7)

$$A_s = 0.005 \times 1000 \times 170 = 850 \text{ mm}^2\text{/m}$$

Use No. 10 bars @ 120 mm.

1.6 DESIGN FOR FLEXURE AND AXIAL LOAD

Columns and walls are commonly used compression members that are subjected to varying degrees of bending stresses. Compression members can be classified as either short or long (slender) members, depending on their behavior. If the second-order stresses due to transverse deformations affect the ultimate strength of a member, then the member is classified as a long member. If the transverse deformations are negligibly small, the ultimate strength is not affected by secondary moments, and the member is classified as a short member.

1.6.1 Short Columns and Column Interaction Diagrams

A short column is one in which the column capacity is governed by the sectional capacity. In a short column, slenderness effects and the associated secondary moments are insignificant. Therefore, the column capacity can be found directly from the critical section capacity.

Purely concentric columns rarely occur in practice. Some bending is always present, due to either external forces or column imperfections. Column sections can resist less axial loads in the presence of moments. The *ACI Code* provides a limit to the design axial load strength so that a column is designed to carry at most 80 to 85% of its concentric load capacity. This ensures some reserved moment capacity in columns even if the design moment is zero.

At the other extreme, columns behave like beams if the applied moment is very high in comparison to the axial load. The *ACI Code* provides a limit to design axial load strength, ϕP_n, below which the ductility requirements of beams apply. This means that the maximum reinforcement ratio, in this range, is limited to $0.75\rho_b$ as in the case of the beams.

ACI 318M-83 specifies minimum and maximum limits of longitudinal reinforcement for compression members as 1% and 8% of the gross concrete area, respectively. In practice, however, more than approximately 4% reinforcement creates clearance problems during splicing. Therefore, $\rho_g = 0.04$ can be regarded as a practical upper limit. Usually a high percentage of reinforcement is used in the lower-story columns. Column capacity is reduced in the upper stories by reducing the area of reinforcement while keeping the column section constant until the reduction in applied load warrants a change in column size. It is generally more economical to keep column sections unchanged so that the same formwork can be reused.

Strength of sections under combined bending and axial force can best be determined with the help of the interaction diagrams. An interaction diagram gives a relationship between the ultimate axial load capacity and the ultimate moment capacity of a given section. If these diagrams are available for different column sections with different rein-

TABLE 1-6 Nominal Moment Capacity of 1.0-Meter-Wide Slab Strip

$f'_c = 20$ MPa

$f_y = 300$ MPa

ρ	Nominal Moment Capacity, M_n (kN · m/m)																	
	Effective Depth, d (mm)																	
	60	80	100	120	140	160	180	200	220	240	260	280	300	320	340	360	380	400
0.002	2.1	3.8	5.9	8.5	11.6	15.1	19.1	23.6	28.5	34.0	39.8	46.2	53.0	60.4	68.1	76.4	85.1	94.3
0.003	3.2	5.6	8.8	12.6	17.2	22.4	28.4	35.0	42.4	50.5	59.2	68.7	78.9	89.7	101.3	113.6	126.5	140.2
0.004	4.2	7.4	11.6	16.7	22.7	29.6	37.5	46.3	56.0	66.7	78.3	90.8	104.2	118.5	133.8	150.0	167.2	185.2
0.005	5.2	9.2	14.3	20.6	28.1	36.7	46.5	57.4	69.4	82.6	96.9	112.4	129.0	146.8	165.7	185.8	207.0	229.4
0.006	6.1	10.9	17.0	24.5	33.4	43.6	55.2	68.2	82.5	98.2	115.2	133.6	153.4	174.6	197.1	220.9	246.2	272.8
0.007	7.1	12.6	19.7	28.4	38.6	50.4	63.8	78.8	95.4	113.5	133.2	154.5	177.3	201.8	227.8	255.3	284.5	315.2
0.008	8.0	14.3	22.3	32.1	43.7	57.1	72.3	89.2	108.0	128.5	150.8	174.9	200.8	228.4	257.9	289.1	322.1	356.9
0.009	8.9	15.9	24.9	35.8	48.7	63.6	80.5	99.4	120.3	143.2	168.0	194.9	223.7	254.5	287.3	322.1	358.9	397.7
0.010	9.8	17.5	27.4	39.4	53.6	70.0	88.6	109.4	132.4	157.6	184.9	214.4	246.2	280.1	316.2	354.5	395.0	437.6
0.011	10.7	19.1	29.8	42.9	58.4	76.3	96.5	119.2	144.2	171.6	201.4	233.6	268.2	305.1	344.5	386.2	430.3	476.8
0.012	11.6	20.6	32.2	46.4	63.1	82.4	104.3	128.8	155.8	185.4	217.6	252.4	289.7	329.6	372.1	417.2	464.8	515.0
0.013	12.4	22.1	34.5	49.7	67.7	88.4	111.9	138.1	167.1	198.9	233.4	270.7	310.7	353.5	399.1	447.5	498.6	552.4
0.014	13.3	23.6	36.8	53.0	72.2	94.2	119.3	147.2	178.2	212.0	248.8	288.6	331.3	377.0	425.5	477.1	531.6	589.0
0.015	14.1	25.0	39.0	56.2	76.5	100.0	126.5	156.2	189.0	224.9	263.9	306.1	351.4	399.8	451.3	506.0	563.8	624.7
0.016	14.8	26.4	41.2	59.4	80.8	105.5	133.6	164.9	199.5	237.4	278.7	323.2	371.0	422.1	476.5	534.3	595.3	659.6
0.017	15.6	27.7	43.3	62.4	85.0	111.0	140.5	173.4	209.8	249.7	293.0	339.9	390.1	443.9	501.1	561.8	626.0	693.6
0.018	16.4	29.1	45.4	65.4	89.0	116.3	147.2	181.7	219.8	261.6	307.1	356.1	408.8	465.1	525.1	588.7	655.9	726.8
0.019	17.1	30.4	47.4	68.3	93.0	121.5	153.7	189.8	229.6	273.3	320.7	372.0	427.0	485.8	548.5	614.9	685.1	759.1
0.020	17.8	31.6	49.4	71.2	96.8	126.5	160.1	197.6	239.2	284.6	334.0	387.4	444.7	506.0	571.2	640.4	713.5	790.6
0.021	18.5	32.8	51.3	73.9	100.6	131.4	166.3	205.3	248.4	295.6	347.0	402.4	461.9	525.6	593.3	665.2	741.2	821.2
0.022	19.1	34.0	53.2	76.6	104.2	136.2	172.3	212.8	257.4	306.4	359.6	417.0	478.7	544.6	614.9	689.3	768.0	851.0
0.023	19.8	35.2	55.0	79.2	107.8	140.8	178.2	220.0	266.2	316.8	371.8	431.2	495.0	563.2	635.8	712.8	794.2	880.0
0.024	20.4	36.3	56.8	81.7	111.2	145.3	183.9	227.0	274.7	326.9	383.6	444.9	510.8	581.1	656.1	735.5	819.5	908.0
0.024	20.5	36.4	56.9	81.9	111.5	145.6	184.3	227.6	275.4	327.7	384.6	446.0	512.0	582.6	657.7	737.3	821.5	910.3

f_y = 350 MPa	0.002	2.5	4.4	6.9	9.9	13.4	17.6	22.2	27.4	33.2	39.5	46.3	53.7	61.7	70.2	79.3	88.9	99.0	109.7
	0.003	3.7	6.5	10.2	14.7	19.9	26.0	33.0	40.7	49.3	58.6	68.8	79.8	91.6	104.2	117.6	131.9	146.9	162.8
	0.004	4.8	8.6	13.4	19.3	26.3	34.4	43.5	53.7	65.0	77.3	90.7	105.2	120.8	137.5	155.2	174.0	193.8	214.8
	0.005	6.0	10.6	16.6	23.9	32.5	42.5	53.8	66.4	80.3	95.6	112.2	120.1	149.4	170.0	191.9	215.1	239.7	265.6
	0.006	7.1	12.6	19.7	28.4	38.6	50.4	63.8	78.8	95.4	113.5	133.2	154.5	177.3	201.8	227.8	255.3	284.5	315.2
	0.007	8.2	14.6	22.7	32.7	44.6	58.2	73.7	90.3	110.0	131.0	153.7	178.2	204.6	232.8	262.8	294.6	328.3	363.8
	0.008	9.2	16.4	25.7	37.0	50.4	65.8	83.2	102.8	124.4	148.0	173.7	201.4	231.2	263.1	297.0	333.0	371.0	411.1
	0.009	10.3	18.3	28.6	41.2	56.0	73.2	92.6	114.3	138.3	164.6	193.2	224.1	257.2	292.7	330.4	370.4	412.7	457.3
	0.010	11.3	20.1	31.4	45.2	61.5	80.4	101.7	125.6	152.0	180.8	212.2	246.2	282.6	321.5	362.9	406.9	453.4	502.4
	0.011	12.3	21.8	34.1	49.2	66.9	87.4	110.6	136.6	165.2	196.6	230.8	267.7	307.3	349.6	394.7	442.5	493.0	546.2
	0.012	13.3	23.6	36.8	53.0	72.2	94.2	119.3	147.2	178.2	212.0	248.8	288.6	331.3	377.0	425.5	477.1	531.6	589.0
	0.013	14.2	25.2	39.4	56.8	77.2	100.9	127.7	157.6	190.7	227.0	266.4	309.0	354.7	403.6	455.6	510.8	569.1	630.6
	0.014	15.1	26.8	41.9	60.4	82.2	107.4	135.9	167.8	203.0	241.6	283.5	328.8	377.4	429.4	484.8	543.5	605.6	671.0
	0.015	16.0	28.4	44.4	63.9	87.0	113.6	143.8	177.6	214.9	255.7	300.1	348.0	399.5	454.6	513.2	575.3	641.0	710.3
	0.016	16.8	29.9	46.8	67.4	91.7	119.7	151.6	187.1	226.4	269.4	316.2	366.7	421.0	479.0	540.7	606.2	675.5	748.4
	0.017	17.7	31.4	49.1	70.7	96.2	125.7	159.0	196.3	237.6	282.7	331.8	384.8	441.8	502.7	567.4	636.2	708.8	785.4
	0.018	18.5	32.8	51.3	73.9	100.6	131.4	166.3	205.3	248.4	295.6	347.0	402.4	461.9	525.6	593.3	665.2	741.2	821.2
	0.019	19.3	34.2	53.5	77.0	104.8	136.9	173.3	214.0	258.9	308.1	361.6	419.4	481.4	547.8	618.4	693.3	772.4	855.9
	0.020	19.7	35.0	54.7	78.8	107.2	140.1	177.3	218.9	264.8	315.2	369.9	429.0	492.5	560.3	632.5	709.1	790.1	875.5
f_y = 400 MPa	0.002	2.8	5.0	7.8	11.2	15.3	20.0	25.3	31.2	37.8	45.0	52.8	61.2	70.3	80.0	90.3	101.2	112.8	125.0
	0.003	4.2	7.4	11.6	16.7	22.7	29.6	37.5	46.3	56.0	66.7	78.3	90.8	104.2	118.5	133.8	150.0	167.2	185.2
	0.004	5.5	9.8	15.2	22.0	29.9	39.0	49.4	61.0	73.8	87.8	103.1	119.5	137.2	156.1	176.3	197.6	220.2	244.0
	0.005	6.8	12.0	18.8	27.1	36.9	48.2	61.0	75.3	91.1	108.4	127.2	147.6	169.4	192.8	217.6	244.0	271.8	301.2
	0.006	8.0	14.3	22.3	32.1	43.7	57.1	72.3	89.2	108.0	128.5	150.8	174.9	200.8	228.4	257.9	289.1	322.1	356.9
	0.007	9.2	16.4	25.7	37.0	50.4	65.8	83.2	102.8	124.4	148.0	173.7	201.4	231.2	263.1	297.0	333.0	371.0	411.1
	0.008	10.4	18.6	29.0	41.7	56.8	74.2	93.9	116.0	140.3	167.0	196.0	227.3	260.9	296.8	335.1	375.7	418.6	463.8
	0.009	11.6	20.6	32.2	46.4	63.1	82.4	104.3	128.8	155.8	185.4	217.6	252.4	289.7	329.6	372.1	417.2	464.8	515.0
	0.010	12.7	22.6	35.3	50.8	69.2	90.4	114.4	141.2	170.8	203.3	238.6	276.7	317.6	361.4	408.0	457.4	509.6	564.7
	0.011	13.8	24.5	38.3	55.2	75.1	98.1	124.1	153.2	185.4	220.6	258.9	300.3	344.8	392.3	442.8	496.4	553.1	612.9
	0.012	14.8	26.4	41.2	59.4	80.8	105.5	133.6	164.9	199.5	237.4	278.7	323.2	371.0	422.1	476.5	534.3	595.3	659.6
	0.013	15.9	28.2	44.0	63.4	86.3	112.8	142.7	176.2	213.2	253.7	297.8	345.3	396.4	451.0	509.2	570.8	636.0	704.8
	0.014	16.8	29.9	46.8	67.4	91.7	119.7	151.6	187.1	226.4	269.4	316.2	368.7	421.0	419.0	540.7	606.2	675.5	748.4
	0.015	17.8	31.6	49.4	71.2	96.8	126.5	160.1	197.6	239.2	284.6	334.0	387.4	444.7	506.0	571.2	640.4	713.5	790.6
	0.016	18.7	33.2	52.0	74.8	101.8	133.0	168.3	207.8	251.5	299.2	351.2	407.3	467.6	532.0	600.6	673.3	750.2	831.2
	0.016	19.0	33.7	52.6	75.8	103.2	134.8	170.6	210.6	254.8	303.2	355.9	412.8	473.8	539.1	608.6	682.3	760.2	842.4

TABLE 1-7 Nominal Moment Capacity of 1.0-Meter-Wide Slab Strip

$f'_c = 30$ MPa

$f_y = 300$ MPa

Nominal Moment Capacity, M_n (kN · m/m)

ρ	Effective Depth, *d* (mm) 60	80	100	120	140	160	180	200	220	240	260	280	300	320	340	360	380	400
0.002	2.1	3.8	5.9	8.5	11.6	15.2	19.2	23.7	28.7	34.2	40.1	46.5	53.4	60.7	68.5	76.8	85.6	94.9
0.003	3.2	5.7	8.8	12.7	17.3	22.6	28.6	35.4	42.8	50.9	59.8	69.3	79.6	90.5	102.2	114.6	127.7	141.5
0.004	4.2	7.5	11.7	16.9	23.0	30.0	38.0	46.9	56.7	67.5	79.2	91.9	105.5	120.0	135.5	151.9	169.2	187.5
0.005	5.2	9.3	14.6	21.0	28.5	37.3	47.2	58.2	70.5	83.9	98.4	114.1	131.0	149.1	168.3	188.7	210.2	232.9
0.006	6.3	11.1	17.4	25.0	34.0	44.5	56.3	69.5	84.0	100.0	117.4	136.1	156.3	177.8	200.7	225.0	250.7	277.8
0.007	7.2	12.9	20.1	29.0	39.5	51.5	65.2	80.5	97.5	116.0	136.1	157.9	181.2	206.2	232.8	261.0	290.8	322.2
0.008	8.2	14.6	22.9	32.9	44.8	58.5	74.1	91.5	110.7	131.7	154.6	179.3	205.8	234.2	264.4	296.4	330.3	365.9
0.009	9.2	16.4	25.6	36.8	50.1	65.5	82.8	102.3	123.8	147.3	172.9	200.5	230.1	261.8	295.6	331.4	369.2	409.1
0.010	10.2	18.1	28.2	40.7	55.3	72.3	91.5	112.9	136.7	162.6	190.9	221.4	254.1	289.1	326.4	365.9	407.7	451.8
0.011	11.1	19.8	30.9	44.4	60.5	79.0	100.0	123.5	149.4	177.8	208.6	242.0	277.8	316.1	356.8	400.0	445.7	493.8
0.012	12.0	21.4	33.5	48.2	65.6	85.7	108.4	133.8	161.9	192.7	226.2	262.3	301.1	342.6	386.8	433.6	483.1	535.3
0.013	13.0	23.1	36.0	51.9	70.6	92.2	116.7	144.1	174.3	207.5	243.5	282.4	324.2	368.8	416.4	466.8	520.1	576.3
0.014	13.9	24.7	38.5	55.5	75.5	98.7	124.9	154.2	186.5	222.0	260.5	302.2	346.9	394.7	445.5	499.5	556.5	616.7
0.015	14.8	26.3	41.0	59.1	80.4	105.0	132.9	164.1	198.6	236.3	277.4	321.7	369.3	420.1	474.3	531.7	592.5	656.5
0.016	15.7	27.8	43.5	62.6	85.2	111.3	140.9	173.9	210.5	250.5	293.9	340.9	391.3	445.3	502.7	563.5	627.9	695.7
0.017	16.5	29.4	45.9	66.1	90.0	117.5	148.7	183.6	222.2	264.4	310.3	359.9	413.1	470.0	530.6	594.9	662.8	734.4
0.018	17.4	30.9	48.3	69.5	94.6	123.6	156.4	193.1	233.7	278.1	326.4	378.5	434.5	494.4	558.1	625.7	697.2	772.5
0.019	18.2	32.4	50.6	72.9	99.2	129.6	164.0	202.5	245.0	291.6	342.3	396.9	455.7	518.4	585.3	656.2	731.1	810.1
0.020	19.1	33.9	52.9	76.2	103.8	135.5	171.5	211.8	256.2	304.9	357.9	415.1	476.5	542.1	612.0	686.1	764.5	847.1
0.021	19.9	35.3	55.2	79.5	108.2	141.4	178.9	220.9	267.3	318.1	373.3	432.9	497.0	565.4	638.3	715.6	797.3	883.5
0.022	20.7	36.8	57.5	82.7	112.6	147.1	186.2	229.8	278.1	331.0	388.4	450.5	517.1	588.4	664.2	744.7	829.7	919.3
0.023	21.5	38.2	59.7	85.9	116.9	152.7	193.3	238.7	288.8	343.7	403.3	467.8	537.0	611.0	689.7	773.3	861.6	954.6
0.024	22.3	39.6	61.8	89.0	121.2	158.3	200.3	247.3	299.3	356.2	418.0	484.8	556.5	633.2	714.8	801.4	892.9	989.4
0.025	23.0	40.9	64.0	92.1	125.4	163.8	207.3	255.9	309.6	368.5	432.4	501.5	575.7	655.1	739.5	829.1	923.7	1023.5
0.026	23.8	42.3	66.1	95.1	129.5	169.1	214.1	264.3	319.8	380.6	446.6	518.0	594.6	676.6	763.8	856.3	954.1	1057.1
0.027	24.5	43.6	68.1	98.1	133.5	174.4	220.8	272.5	329.8	392.5	460.6	534.2	613.2	697.7	787.6	883.0	983.9	1090.2
0.028	25.3	44.9	70.2	101.0	137.5	179.6	227.3	280.7	339.6	404.1	474.3	550.1	631.5	718.5	811.1	909.3	1013.2	1122.6
0.029	26.0	46.2	72.2	103.9	141.4	184.7	233.8	288.6	349.2	415.6	487.8	565.7	649.4	738.9	834.2	935.2	1042.0	1154.5
0.030	26.7	47.4	74.1	106.7	145.3	189.7	240.1	296.5	358.7	426.9	501.0	581.1	667.1	759.0	856.8	960.6	1070.3	1185.9
0.031	27.4	48.7	76.0	109.5	149.0	194.7	246.4	304.2	368.0	438.0	514.0	596.2	684.4	778.7	879.0	985.5	1098.0	1216.7
0.032	28.1	49.9	77.9	112.2	152.7	199.5	252.5	311.7	377.2	448.9	526.8	611.0	701.4	798.0	900.9	1010.0	1125.3	1246.9
0.033	28.7	51.1	79.8	114.9	156.4	204.2	258.5	319.1	386.1	459.5	539.3	625.5	718.0	817.0	922.3	1034.0	1152.1	1276.5
0.034	29.4	52.2	81.6	117.5	159.9	208.9	264.4	326.4	394.9	470.0	551.6	639.7	734.4	835.6	943.3	1057.5	1178.3	1305.6
0.035	30.0	53.4	83.4	120.1	163.4	213.5	270.2	333.5	403.6	480.3	563.7	653.7	750.4	853.8	963.9	1080.6	1204.0	1334.1
0.036	30.6	54.5	85.1	122.6	166.9	217.9	275.8	340.5	412.0	490.3	575.5	667.4	766.2	871.7	984.1	1103.3	1229.3	1362.1
0.036	30.7	54.6	85.4	123.0	167.4	218.6	276.7	341.6	413.3	491.8	577.2	669.4	768.5	874.4	987.1	1106.6	1233.0	1366.2

$f_y = 350$ MPa	0.002	2.5	4.4	6.9	9.9	13.5	17.7	22.4	27.6	33.4	39.8	46.7	54.1	62.1	70.7	79.8	89.5	99.7	110.5
	0.003	3.7	6.6	10.3	14.8	20.2	26.3	33.3	41.1	49.8	59.2	69.5	80.6	92.6	105.3	118.9	133.3	148.5	164.5
	0.004	4.9	8.7	13.6	19.6	26.7	34.9	44.1	54.5	65.9	78.4	92.0	106.7	122.5	139.4	157.4	176.5	196.6	217.9
	0.005	6.1	10.8	16.9	24.3	33.1	43.3	54.8	67.6	81.8	97.3	114.2	132.5	152.1	173.1	195.4	219.0	244.0	270.4
	0.006	7.2	12.9	20.1	29.0	39.5	51.5	65.2	80.5	97.5	116.0	136.1	157.9	181.2	206.2	232.8	261.0	290.8	322.2
	0.007	8.4	14.9	23.3	33.6	45.7	59.7	75.6	93.3	112.9	134.3	157.7	182.9	209.9	238.8	269.6	302.3	336.8	373.2
	0.008	9.5	16.9	26.5	38.1	51.9	67.7	85.7	105.9	128.1	152.4	178.9	207.5	238.2	271.0	305.9	343.0	382.1	423.4
	0.009	10.6	18.9	29.6	42.6	57.9	75.7	95.8	118.2	143.0	170.2	199.8	231.7	266.0	302.6	341.6	383.0	426.8	472.9
	0.010	11.7	20.9	32.6	46.9	63.9	83.5	105.6	130.4	157.8	187.8	220.4	255.6	293.4	333.8	376.8	422.5	470.7	521.6
	0.011	12.8	22.8	35.6	51.3	69.8	91.1	115.3	142.4	172.3	205.0	240.6	279.1	320.3	364.5	411.5	461.3	514.0	569.5
	0.012	13.9	24.7	38.5	55.5	75.5	98.7	124.9	154.2	186.5	222.0	260.5	302.2	346.9	394.7	445.5	499.5	556.5	616.7
	0.013	14.9	26.5	41.4	59.7	81.2	106.1	134.3	165.8	200.6	238.7	280.1	324.9	373.0	424.4	479.1	537.1	598.4	663.0
	0.014	15.9	28.3	44.3	63.8	86.8	113.4	143.5	177.2	214.4	255.1	299.4	347.2	398.6	453.6	512.0	574.0	639.6	708.7
	0.015	17.0	30.1	47.1	67.8	92.3	120.6	152.6	188.4	227.9	271.3	318.4	369.2	423.9	482.3	544.4	610.1	680.1	753.5
	0.016	17.9	31.9	49.9	71.8	97.7	127.6	161.5	199.4	241.3	287.1	337.0	390.8	448.7	510.5	576.3	646.1	719.8	797.6
	0.017	18.9	33.6	52.6	75.7	103.0	134.5	170.3	210.2	254.4	302.7	355.3	412.1	473.0	538.2	607.6	681.2	758.9	840.9
	0.018	19.9	35.3	55.2	79.5	108.2	141.4	178.9	220.9	267.3	318.1	373.3	432.9	497.0	565.4	638.3	715.6	797.3	883.5
	0.019	20.8	37.0	57.8	83.3	113.3	148.0	187.4	231.3	279.9	333.1	390.9	453.4	520.5	592.2	668.5	749.5	835.0	925.3
	0.020	21.7	38.7	60.4	87.0	118.4	154.6	195.7	241.6	292.3	347.9	408.2	473.5	543.5	618.4	698.1	782.7	872.1	966.3
	0.021	22.6	40.3	62.9	90.6	123.3	161.0	203.8	251.6	304.5	362.3	425.3	493.2	566.2	644.2	727.2	815.3	908.4	1006.5
	0.022	23.5	41.8	65.4	94.1	128.1	167.4	211.8	261.5	316.4	376.6	441.9	512.5	588.4	669.4	755.7	847.3	944.0	1046.0
	0.023	24.4	43.4	67.8	97.6	132.9	173.6	219.7	271.2	328.1	390.5	458.3	531.5	610.1	694.2	783.7	878.6	978.9	1084.7
	0.024	25.3	44.9	70.2	101.0	137.5	179.6	227.3	280.7	339.6	404.1	474.3	550.1	631.5	718.5	811.1	909.3	1013.2	1122.6
	0.025	26.1	46.4	72.5	104.4	142.1	185.6	234.9	289.9	350.8	417.5	490.0	568.3	652.4	742.3	838.0	939.4	1046.7	1159.8
	0.026	26.9	47.8	74.8	107.7	146.5	191.4	242.2	299.1	361.9	430.6	505.4	586.1	672.9	765.6	864.3	968.9	1079.6	1196.2
	0.027	27.7	49.3	77.0	110.9	150.9	197.1	249.4	308.0	372.6	443.5	520.4	603.6	692.9	788.4	890.0	997.8	1111.7	1231.8
	0.028	28.5	50.7	79.2	114.0	155.2	202.7	256.5	316.7	383.2	456.0	535.2	620.7	712.5	810.7	915.2	1026.0	1143.2	1266.7
	0.029	29.3	52.0	81.3	117.1	159.3	208.1	263.4	325.2	393.5	468.3	549.6	637.4	731.7	832.5	939.8	1053.6	1174.0	1300.8
	0.029	29.5	52.5	82.0	118.1	160.7	209.9	265.7	328.0	396.8	472.3	554.3	642.8	737.9	839.6	947.8	1062.6	1184.0	1311.9
$f_y = 400$ MPa	0.002	2.8	5.0	7.9	11.3	15.4	20.2	25.5	31.5	38.1	45.4	53.2	61.7	70.9	80.6	91.0	102.1	113.7	126.0
	0.003	4.2	7.5	11.7	16.9	23.0	30.0	38.0	46.9	56.7	67.5	79.2	91.9	105.5	120.0	135.5	151.9	169.2	187.5
	0.004	5.6	9.9	15.5	22.3	30.4	39.7	50.2	62.0	75.0	89.3	104.8	121.5	139.5	158.7	179.2	200.9	223.8	248.0
	0.005	6.9	12.3	19.2	27.7	37.7	49.2	62.3	76.9	93.0	110.7	129.9	150.7	172.9	196.8	222.1	249.0	277.5	307.4
	0.006	8.2	14.6	22.9	32.9	44.8	58.5	74.1	91.5	110.7	131.7	154.6	179.3	205.8	234.2	264.4	296.4	330.3	365.9
	0.007	9.5	16.9	26.5	38.1	51.9	67.7	85.7	105.9	128.1	152.4	178.9	207.5	238.2	271.0	305.9	343.0	382.1	423.4
	0.008	10.8	19.2	30.0	43.2	58.8	76.8	97.2	120.0	145.2	172.8	202.7	235.1	269.9	307.1	346.7	388.7	433.1	479.9
	0.009	12.0	21.4	33.5	48.2	65.6	85.7	108.4	133.8	161.9	192.7	226.2	262.3	301.1	342.6	386.8	433.6	483.1	535.3
	0.010	13.3	23.6	36.9	53.1	72.3	94.4	119.4	147.5	178.4	212.3	249.2	289.0	331.8	377.5	426.1	477.7	532.3	589.8
	0.011	14.5	25.7	40.2	57.9	78.8	102.9	130.3	160.8	194.6	231.6	271.8	315.2	361.8	411.7	464.8	521.0	580.5	643.3
	0.012	15.7	27.8	43.5	62.6	85.2	111.3	140.9	173.9	210.5	250.5	293.9	340.9	391.3	445.3	502.7	563.5	627.9	695.7
	0.013	16.8	29.9	46.7	67.2	91.5	119.5	151.3	186.8	226.0	269.0	315.7	366.1	420.3	478.2	539.8	605.2	674.3	747.2
	0.014	17.9	31.9	49.9	71.8	97.7	127.6	161.5	199.4	241.3	287.1	337.0	390.8	448.7	510.5	576.3	646.1	719.8	797.6
	0.015	19.1	33.9	52.9	76.2	103.8	135.5	171.5	211.8	256.2	304.9	357.9	415.1	476.5	542.1	612.0	686.1	764.5	847.1
	0.016	20.1	35.8	56.0	80.6	109.7	143.3	181.3	223.9	270.9	322.4	378.3	438.8	503.7	573.1	647.0	725.4	808.2	895.5
	0.017	21.2	37.7	58.9	84.9	115.5	150.9	190.9	235.7	285.2	339.5	398.4	462.0	530.4	603.5	681.3	763.8	851.0	942.9
	0.018	22.3	39.6	61.8	89.0	121.2	158.3	200.3	247.3	299.3	356.2	418.0	484.8	556.5	633.2	714.8	801.4	892.9	989.4
	0.019	23.3	41.4	64.7	93.1	126.8	165.6	209.5	258.7	313.0	372.5	437.2	507.0	582.1	662.3	747.6	838.2	933.9	1034.8
	0.020	24.3	43.2	67.5	97.1	132.2	172.7	218.5	269.8	326.5	388.5	456.0	528.8	607.1	690.7	779.7	874.2	974.0	1079.2
	0.021	25.3	44.9	70.2	101.0	137.5	179.6	227.3	280.7	339.6	404.1	474.3	550.1	631.5	718.5	811.1	909.3	1013.2	1122.6
	0.022	26.2	46.6	72.8	104.9	142.7	186.4	235.9	291.3	352.4	419.4	492.2	570.9	655.3	745.6	841.7	943.7	1051.5	1165.0
	0.023	27.1	48.3	75.4	108.6	147.8	193.0	244.3	301.6	365.0	434.3	509.7	591.2	678.6	772.1	871.7	977.2	1088.8	1206.5
	0.024	28.1	49.9	77.9	112.2	152.7	199.5	252.5	311.7	377.2	448.9	526.8	611.0	701.4	798.0	900.9	1010.0	1125.3	1246.9
	0.024	28.4	50.5	78.9	113.6	154.6	201.9	255.5	315.5	381.7	454.3	533.2	618.4	709.9	807.7	911.8	1022.2	1138.9	1262.0

TABLE 1-8 Nominal Moment Capacity of 1.0-Meter-Wide Slab Strip

$f'_c = 40$ MPa

$f_y = 300$ MPa

	Nominal Moment Capacity, M_n (kN · m/m)																	
	Effective Depth, d (mm)																	
ρ	60	80	100	120	140	160	180	200	220	240	260	280	300	320	340	360	380	400
0.002	2.1	3.8	5.9	8.6	11.7	15.2	19.3	23.8	28.8	34.3	40.2	46.6	53.5	60.9	68.7	77.1	85.9	95.2
0.003	3.2	5.7	8.9	12.8	17.4	22.7	28.8	35.5	43.0	51.2	60.0	69.9	79.9	90.9	102.7	115.1	128.2	142.1
0.004	4.2	7.5	11.8	17.0	23.1	30.2	38.2	47.2	57.1	67.9	79.7	92.4	106.1	120.7	136.3	152.8	170.2	188.6
0.005	5.3	9.4	14.7	21.1	28.8	37.6	47.5	58.7	71.0	84.5	99.2	115.0	132.0	150.2	169.6	190.1	211.8	234.7
0.006	6.3	11.2	17.5	25.2	34.3	44.9	56.8	70.1	84.8	100.9	118.5	137.4	157.7	179.4	202.6	227.1	253.0	280.4
0.007	7.3	13.0	20.4	29.3	39.9	52.1	65.9	81.4	98.5	117.2	137.6	159.6	183.2	208.4	235.3	263.8	293.9	325.6
0.008	8.3	14.8	23.2	33.3	45.4	59.3	75.0	92.6	112.1	133.4	156.5	181.5	208.4	237.1	267.6	300.1	334.3	370.4
0.009	9.3	16.6	25.9	37.3	50.8	66.4	84.0	103.7	125.5	149.3	175.3	203.3	233.4	265.5	299.7	336.0	374.4	414.8
0.010	10.3	18.4	28.7	41.3	56.2	73.4	92.9	114.7	138.8	165.2	193.9	224.8	258.1	293.6	331.5	371.6	414.1	458.8
0.011	11.3	20.1	31.4	45.2	61.5	80.4	101.7	125.6	152.0	180.9	212.3	246.2	282.6	321.5	363.0	406.9	453.4	502.4
0.012	12.3	21.8	34.1	49.1	66.8	87.3	110.5	136.4	165.0	196.4	230.5	267.3	306.8	349.1	349.1	441.9	492.3	545.5
0.013	13.2	23.5	36.8	52.9	72.1	94.1	119.1	147.1	177.9	211.8	248.5	288.2	330.9	376.5	425.0	476.4	530.9	588.2
0.014	14.2	25.2	39.4	56.7	77.2	100.9	127.7	157.6	190.7	227.0	266.4	308.9	354.7	403.5	455.5	510.7	569.0	630.5
0.015	15.1	26.9	42.0	60.5	82.4	107.6	136.2	168.1	203.4	242.0	284.1	329.5	378.2	430.3	485.8	544.6	606.8	672.4
0.016	16.1	28.6	44.6	64.2	87.4	114.2	144.5	178.4	215.9	257.0	301.6	349.8	401.5	456.8	515.7	578.2	644.2	713.8
0.017	17.0	30.2	47.2	67.9	92.5	120.8	152.8	188.7	228.3	271.7	318.9	369.9	424.6	483.1	545.3	611.4	681.2	754.8
0.018	17.9	31.8	49.7	71.6	97.4	127.3	161.1	198.8	240.6	286.3	336.1	389.7	447.4	509.0	574.7	644.3	717.8	795.4
0.019	18.8	33.4	52.2	75.2	102.4	133.7	169.2	208.9	252.8	300.8	353.0	409.4	470.0	534.8	603.7	676.8	754.1	835.6
0.020	19.7	35.0	54.7	78.8	107.2	140.0	177.2	218.8	264.8	315.1	369.8	428.9	492.4	560.2	632.4	709.0	790.0	875.3
0.021	20.6	36.6	57.2	82.3	112.0	146.3	185.2	228.7	276.7	329.3	386.4	448.2	514.5	585.3	660.8	740.8	825.4	914.6
0.022	21.5	38.1	59.6	85.8	116.8	152.6	193.1	238.4	288.4	343.3	402.9	467.2	536.3	610.2	688.9	772.3	860.5	953.5
0.023	22.3	39.7	62.0	89.3	121.5	158.7	200.9	248.0	300.1	357.1	419.1	486.1	558.0	634.9	716.7	803.5	895.3	992.0
0.024	23.2	41.2	64.4	92.7	126.2	164.8	208.6	257.5	311.6	370.8	435.2	504.7	579.4	659.2	744.2	834.3	929.6	1030.0
0.025	24.0	42.7	66.7	96.1	130.8	170.8	216.2	266.9	323.0	384.4	451.1	523.1	600.5	683.3	771.4	864.8	963.5	1067.6
0.026	24.9	44.2	69.1	99.4	135.3	176.8	223.7	276.2	334.2	397.7	466.8	541.4	621.5	707.1	798.2	894.9	997.1	1104.8
0.027	25.7	45.7	71.4	102.7	139.8	182.7	231.2	285.4	345.3	411.0	482.3	559.4	642.2	730.6	824.8	924.7	1030.3	1141.6
0.028	26.5	47.1	73.6	106.0	144.3	188.5	238.5	294.5	356.3	424.1	497.7	577.2	662.6	753.9	851.1	954.2	1063.1	1178.0
0.029	27.3	48.6	75.9	109.3	148.7	194.2	245.8	303.5	367.2	437.0	512.9	594.8	682.8	776.9	877.0	983.3	1095.5	1213.9
0.030	28.1	50.0	78.1	112.4	153.1	199.9	253.0	312.4	377.9	449.8	527.9	612.2	702.8	799.6	902.7	1012.0	1127.6	1249.4
0.031	28.9	51.4	80.3	115.6	157.4	205.5	260.1	321.1	388.6	462.4	542.7	629.4	722.5	822.1	928.0	1040.4	1159.3	1284.5
0.032	29.7	52.8	82.4	118.7	161.6	211.1	267.1	329.8	399.0	474.9	557.3	646.4	742.0	844.3	953.1	1068.5	1190.5	1319.1
0.033	30.5	54.1	84.6	121.8	165.8	216.5	274.1	338.3	409.4	487.2	571.8	663.2	761.3	866.2	977.8	1096.2	1221.4	1353.4
0.034	31.2	55.5	86.7	124.8	169.9	222.0	280.9	346.8	419.6	499.4	586.1	679.7	780.3	887.8	1002.2	1123.6	1251.9	1387.2
0.035	32.0	56.8	88.8	127.9	174.0	227.3	287.7	355.1	429.7	511.4	600.2	696.1	799.1	909.2	1026.4	1150.7	1282.1	1420.6
0.036	32.7	58.1	90.8	130.8	178.1	232.6	294.3	363.4	439.7	523.3	614.1	712.2	817.6	930.3	1050.2	1177.4	1311.8	1453.5
0.037	33.4	59.4	92.9	133.7	182.0	237.8	300.9	371.5	449.5	535.0	627.9	728.2	835.9	951.1	1073.7	1203.7	1341.2	1486.1
0.038	34.2	60.7	94.9	136.6	186.0	242.9	307.4	379.6	459.3	546.6	641.4	743.9	854.0	971.7	1096.9	1229.7	1370.2	1518.2
0.039	34.9	62.0	96.9	139.5	189.9	248.0	313.9	387.5	468.8	558.0	654.8	759.5	871.8	991.9	1119.8	1255.4	1398.8	1549.9
0.040	35.6	63.2	98.8	142.3	193.7	253.0	320.2	395.3	478.3	569.2	668.0	774.8	889.4	1011.9	1142.4	1280.7	1427.0	1581.2
0.041	36.3	64.5	100.8	145.1	197.5	257.9	326.4	403.0	487.6	580.3	681.1	789.9	906.8	1031.7	1164.7	1305.7	1454.8	1612.0
0.042	37.0	65.7	102.7	147.8	201.2	262.8	332.6	410.6	496.8	591.3	693.9	804.8	923.9	1051.2	1186.7	1330.4	1482.3	1642.4
0.043	37.6	66.9	104.5	150.5	204.9	267.6	338.7	418.1	505.9	602.1	706.6	819.5	940.7	1070.4	1208.3	1354.7	1509.4	1672.4
0.044	38.1	67.7	105.7	152.3	207.2	270.7	342.6	422.9	511.7	609.0	714.8	828.9	951.6	1082.7	1222.3	1370.3	1526.8	1691.7

f_y = 350 MPa	0.002	2.5	4.4	6.9	10.0	13.6	17.7	22.4	27.7	33.5	39.9	46.8	54.3	62.4	70.9	80.1	89.8	100.0	110.8
	0.003	3.7	6.6	10.3	14.9	20.3	26.5	33.5	41.4	50.0	59.5	69.9	81.0	93.0	105.9	119.5	134.0	149.3	165.4
	0.004	4.9	8.8	13.7	19.7	26.9	35.1	44.4	54.8	66.4	79.0	92.7	107.5	123.4	140.4	158.5	177.7	198.0	219.4
	0.005	6.1	10.9	17.0	24.6	33.4	43.6	55.2	68.2	82.5	98.2	115.3	133.7	153.4	174.6	197.1	221.0	246.0	272.8
	0.006	7.3	13.0	20.4	29.3	39.9	52.1	65.9	81.4	98.5	117.2	137.6	159.6	183.2	208.4	235.3	263.8	293.9	325.6
	0.007	8.5	15.1	23.6	34.0	46.3	60.5	76.5	94.5	114.3	136.0	159.7	185.2	212.6	241.8	273.0	306.1	341.0	377.9
	0.008	9.7	17.2	26.8	38.7	52.6	68.7	87.0	107.4	129.9	154.6	181.5	210.5	241.6	274.9	310.4	347.9	387.7	429.6
	0.009	10.8	19.2	30.0	43.3	58.9	76.9	97.3	120.2	145.4	173.0	203.1	235.5	270.4	307.6	347.3	389.3	433.8	480.7
	0.010	12.0	21.2	33.2	47.8	65.1	85.0	107.6	132.8	160.7	191.2	224.4	260.3	298.8	340.0	383.8	430.3	479.4	531.2
	0.011	13.1	23.2	36.3	52.3	71.2	93.0	117.7	145.3	175.8	209.2	245.5	284.7	326.9	371.9	419.9	470.7	524.5	581.1
	0.012	14.2	25.2	39.4	56.7	77.2	100.9	127.7	157.6	190.7	227.0	266.4	308.9	354.7	403.5	455.5	510.7	569.0	630.5
	0.013	15.3	27.2	42.5	61.1	83.2	108.7	137.6	169.8	205.5	244.5	287.0	332.8	382.1	434.7	490.8	550.2	613.1	679.3
	0.014	16.4	29.1	45.5	65.5	89.1	116.4	147.3	181.9	220.1	261.9	307.4	356.5	409.2	465.6	525.6	589.3	656.6	727.5
	0.015	17.4	31.0	48.4	69.8	95.0	124.0	157.0	193.8	234.5	279.1	327.5	379.8	436.0	496.1	560.0	627.9	699.6	775.1
	0.016	18.5	32.9	51.4	74.0	100.7	131.6	166.5	205.6	248.7	296.0	347.5	402.9	462.5	526.2	594.0	666.0	742.0	822.2
	0.017	19.5	34.7	54.3	78.2	106.4	139.0	175.9	217.2	262.8	312.7	367.0	425.7	488.6	556.0	627.6	703.6	784.0	868.7
	0.018	20.6	36.6	57.2	82.3	112.0	146.3	185.2	228.7	276.7	329.3	386.4	448.2	514.5	585.3	660.8	740.8	825.4	914.6
	0.019	21.6	38.4	60.0	86.4	117.6	153.6	194.4	240.0	290.4	345.6	405.6	470.4	540.0	614.4	693.6	777.6	866.3	959.9
	0.020	22.6	40.2	62.8	90.4	123.1	160.8	203.5	251.2	303.9	361.7	424.5	492.3	565.1	643.0	725.9	813.8	906.7	1004.7
	0.021	23.6	42.0	65.6	94.4	128.5	167.8	212.4	262.2	317.3	377.6	443.2	514.0	590.0	671.3	757.8	849.6	946.6	1048.9
	0.022	24.6	43.7	68.3	98.3	133.8	174.8	221.2	273.1	330.5	393.3	461.6	535.3	614.5	699.2	789.3	884.9	986.0	1092.5
	0.023	25.5	45.4	71.0	102.2	139.1	181.7	229.9	283.9	343.5	408.8	479.8	556.4	638.7	726.7	820.4	919.8	1024.8	1135.5
	0.024	26.5	47.1	73.6	106.0	144.3	188.5	238.5	294.5	356.3	424.1	497.7	577.2	662.6	753.9	851.1	954.2	1063.1	1178.0
	0.025	27.4	48.8	76.2	109.8	149.4	195.2	247.0	305.0	369.0	439.1	515.4	597.7	686.2	780.7	881.3	988.1	1100.9	1219.8
	0.026	28.4	50.4	78.8	113.5	154.5	201.8	255.4	315.3	381.5	454.0	532.8	618.0	709.4	807.1	911.2	1021.5	1138.2	1261.1
	0.027	29.3	52.1	81.4	117.2	159.5	208.3	263.6	325.5	393.8	468.7	550.0	637.9	732.3	833.2	940.6	1054.5	1174.9	1301.9
	0.028	30.2	53.7	83.9	120.8	164.4	214.7	271.8	335.5	406.0	483.1	567.0	657.6	754.9	858.9	969.6	1087.0	1211.2	1342.0
	0.029	31.1	55.3	86.3	124.3	169.2	221.1	279.8	345.4	417.9	497.4	583.7	677.0	777.1	884.2	998.2	1119.1	1246.9	1381.6
	0.030	32.0	56.8	88.8	127.9	174.0	227.3	287.7	355.1	429.7	511.4	600.2	696.1	799.1	909.2	1026.4	1150.7	1282.1	1420.6
	0.031	32.8	58.4	91.2	131.3	178.7	233.4	295.4	364.8	441.3	525.2	616.4	714.9	820.7	933.8	1054.1	1181.8	1316.7	1459.0
	0.032	33.7	59.9	93.6	134.7	183.4	239.5	303.1	374.2	452.8	538.9	632.4	733.5	842.0	958.0	1081.5	1212.4	1350.9	1496.8
	0.033	34.5	61.4	95.9	138.1	187.9	245.5	310.7	383.5	464.1	552.3	648.2	751.7	862.9	981.8	1108.4	1242.6	1384.5	1534.1
	0.034	35.3	62.8	98.2	141.4	192.4	251.3	318.1	392.7	475.2	565.5	663.7	769.7	883.6	1005.3	1134.9	1272.3	1417.6	1570.8
	0.035	36.2	64.3	100.4	144.6	196.8	257.1	325.4	401.7	486.1	578.5	678.9	787.4	903.9	1028.4	1161.0	1301.6	1450.2	1606.9
	0.035	36.5	64.8	101.3	145.9	198.6	259.4	328.3	405.3	490.4	583.6	685.0	794.4	911.9	1037.6	1171.3	1313.2	1463.1	1621.2

TABLE 1-8 *(Continued)*

f'_c = 40 MPa

		Nominal Moment Capacity, M_n (kN · m/m)																	
		Effective Depth, d (mm)																	
	ρ	60	80	100	120	140	160	180	200	220	240	260	280	300	320	340	360	380	400
f_y = 400 MPa	0.002	2.8	5.1	7.9	11.4	15.5	20.2	25.6	31.6	38.3	45.5	53.4	62.0	71.2	81.0	91.4	102.5	114.2	126.5
	0.003	4.2	7.5	11.8	17.0	23.1	30.2	38.2	47.2	57.1	67.9	79.7	92.4	106.1	120.7	136.3	152.8	170.2	188.6
	0.004	5.6	10.0	15.6	22.5	30.6	40.0	50.6	62.5	75.6	90.0	105.6	122.5	140.6	160.0	180.6	202.5	225.6	250.0
	0.005	7.0	12.4	19.4	28.0	38.0	49.7	62.9	77.6	94.0	111.8	131.2	152.2	174.7	198.8	224.4	251.6	280.3	310.6
	0.006	8.3	14.8	23.2	33.3	45.4	59.3	75.0	92.6	112.1	133.4	156.5	181.5	208.4	237.1	267.6	300.1	334.3	370.4
	0.007	9.7	17.2	26.8	38.7	52.6	68.7	87.0	107.4	129.9	154.6	181.5	210.5	241.6	274.9	310.4	347.9	387.7	429.6
	0.008	11.0	19.5	30.5	43.9	59.8	78.1	98.8	122.0	147.6	175.6	206.1	239.1	274.4	312.3	352.5	395.2	440.3	487.9
	0.009	12.3	21.8	34.1	49.1	66.8	87.3	110.5	136.4	165.0	196.4	230.5	267.3	306.8	349.1	394.1	441.9	492.3	545.5
	0.010	13.6	24.1	37.6	54.2	73.8	96.4	122.0	150.6	182.2	216.8	254.5	295.2	338.8	385.5	435.2	487.9	543.6	602.4
	0.011	14.8	26.3	41.2	59.3	80.7	105.4	133.3	164.6	199.2	237.0	278.2	322.6	370.4	421.4	475.7	533.3	594.2	658.4
	0.012	16.1	28.6	44.6	64.2	87.4	114.2	144.5	178.4	215.9	257.0	301.6	349.8	401.5	456.8	515.7	578.2	644.2	713.8
	0.013	17.3	30.7	48.0	69.2	94.1	122.9	155.6	192.1	232.4	276.6	324.6	376.5	432.2	491.8	555.2	622.4	693.5	768.4
	0.014	18.5	32.9	51.4	74.0	100.7	131.6	166.5	205.6	248.7	296.0	347.4	402.9	462.5	526.2	594.0	666.0	742.0	822.2
	0.015	19.7	35.0	54.7	78.8	107.2	140.0	177.2	218.8	264.8	315.1	369.8	428.9	492.4	560.2	632.4	709.0	790.0	875.3
	0.016	20.9	37.1	58.0	83.5	113.6	148.4	187.8	231.9	280.6	333.9	391.9	454.5	521.8	593.7	670.2	751.4	837.2	927.6
	0.017	22.0	39.2	61.2	88.1	120.0	156.7	198.3	244.8	296.2	352.5	413.7	479.8	550.8	626.7	707.5	793.1	883.7	979.2
	0.018	23.2	41.2	64.4	92.7	126.2	164.8	208.6	257.5	311.6	370.8	435.2	504.7	579.4	659.2	744.2	834.3	929.6	1030.0
	0.019	24.3	43.2	67.5	97.2	132.3	172.8	218.7	270.0	326.7	388.8	456.3	529.2	607.6	691.3	780.4	874.9	974.8	1080.1
	0.020	25.4	45.2	70.6	101.6	138.4	180.7	228.7	282.4	341.6	406.6	477.2	553.4	635.3	722.8	816.0	914.8	1019.3	1129.4
	0.021	26.5	47.1	73.6	106.0	144.3	188.5	238.5	294.5	356.3	424.1	497.7	577.2	662.6	753.9	851.1	954.2	1063.1	1178.0
	0.022	27.6	49.0	76.6	110.3	150.2	196.1	248.2	306.4	370.8	441.3	517.9	600.6	689.5	784.5	885.6	992.9	1106.3	1225.8
	0.023	28.6	50.9	79.6	114.6	155.9	203.7	257.8	318.2	385.0	458.2	537.8	623.7	716.0	814.6	919.6	1031.0	1148.7	1272.8
	0.024	29.7	52.8	82.4	118.7	161.6	211.1	267.1	329.8	399.0	474.9	557.3	646.4	742.0	844.3	953.1	1068.5	1190.5	1319.1
	0.025	30.7	54.6	85.3	122.8	167.2	218.4	276.4	341.2	412.8	491.3	576.6	668.7	767.6	873.4	986.0	1105.4	1231.6	1364.7
	0.026	31.7	56.4	88.1	126.9	172.7	225.5	285.4	352.4	426.4	507.4	595.5	690.7	792.8	902.1	1018.4	1141.7	1272.1	1409.5
	0.027	32.7	58.1	90.8	130.8	178.1	232.6	294.3	363.4	439.7	523.3	614.1	712.2	817.6	930.3	1050.2	1177.4	1311.8	1453.5
	0.028	33.7	59.9	93.6	134.7	183.4	239.5	303.1	374.2	452.8	538.9	632.4	733.5	842.0	958.0	1081.5	1212.4	1350.9	1496.8
	0.029	34.6	61.6	96.2	138.5	188.6	246.3	311.7	384.8	465.7	554.2	650.4	754.3	865.9	985.2	1112.2	1246.0	1389.3	1539.4
	0.029	35.1	62.4	97.5	140.4	191.0	249.5	315.8	389.9	471.8	561.4	658.9	764.2	877.2	998.1	1126.8	1263.2	1407.5	1559.5

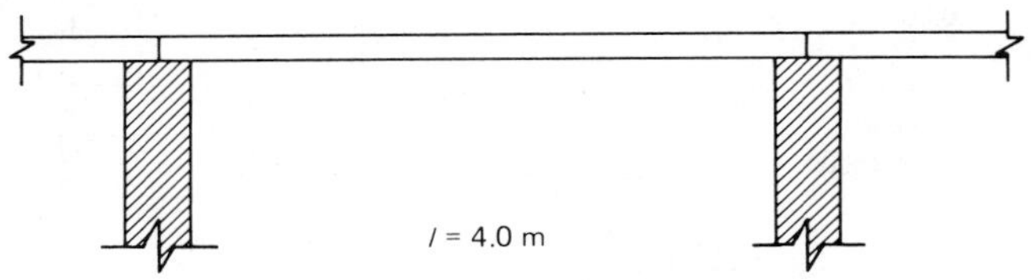

Fig. 1-22 Example 1-7.

forcement percentages and arrangements, the design process can be simplified considerably.

The nominal strengths for an axial load, P_n, and a moment, M_n, are derived in Section 1.3.2 for a rectangular cross section and are given by eqs. (1-13) and (1-14), respectively. P_n and M_n can be computed for an assumed position of the neutral axis. If a sufficient number of neutral axis positions are considered, ranging from a compression failure under almost uniform concrete compression to a pure flexural failure, enough points for the P_n versus M_n relationship can be obtained. These points are plotted to produce the interaction diagrams.

Equations (1-13) and (1-14) are derived for rectangular sections, using a rectangular concrete stress block. The interaction diagrams given in Figs. 1-23 through 1-32 are developed using these equations. The concrete area replaced by reinforcing steel is substracted from the gross concrete area for more accurate prediction of the sectional capacity. Two different reinforcement arrangements are included for rectangular column interaction diagrams. These are: equal area of steel at two end faces, and equal area of steel at all four faces. Reinforcement along the side faces is assumed to be distributed in ten equal layers in the latter case, for the purpose of compiling the interaction diagrams.

Figures 1-33 through 1-37 give the interaction diagrams for circular columns with a circular reinforcement pattern. A parabolic stress–strain relationship was used for concrete in developing the interaction diagrams for circular columns. This relationship is given below and is generally used in the design aids developed by the Portland Cement Association.

$$f_c = 0.85 f_c' \left[\left(\frac{2\epsilon}{\epsilon_0} \right) - \left(\frac{\epsilon}{\epsilon_0} \right)^2 \right] \quad \text{if} \quad 0 < \epsilon < \epsilon_0$$

$$f_c = 0.85 f_c' \qquad \epsilon_0 < \epsilon < 0.003$$

where $\epsilon_0 = 2.0 \times 0.85 \times f_c'/E_c$.

The interaction diagrams given in this chapter do *not* include the capacity reduction factor, ϕ. They are based on nominal strengths and therefore should be multiplied by the appropriate ϕ factors to obtain the design strengths. Maximum allowable axial loads as specified by the *ACI Code* are indicated on the diagrams for tied rectangular columns by dashed horizontal lines.

EXAMPLE 1-8: An elastic frame analysis indicates that a column is subjected to dead and live load moments of 140 kN · m and 60 kN · m, respectively, in the critical direction. Axial loads due to dead and live loads are 1400 kN and 600 kN, respectively. If the column can be considered as a short column, find a suitable square column section and a reinforcement pattern. Use $f_c' = 30$ MPa and $f_y = 400$ MPa.

First determine the factored loads:

$$P_u = 1.4P_D + 1.7P_L$$

$$P_u = 1.4(1400) + 1.7(600) = 2980 \text{ kN}$$

$$M_u = 1.4M_D + 1.7M_L$$

$$M_u = 1.4(140) + 1.7(60) = 298 \text{ kN} \cdot \text{m}$$

Try 450 mm square tied column with equal area of reinforcement along four faces:

$$\frac{P_n}{A_g} = \frac{P_u}{\phi A_g} = \frac{2980 \times 1000}{(0.7)(450)^2} = 21.02 \text{ MPa}$$

$$\frac{M_n}{A_g h} = \frac{M_u}{\phi A_g h} = \frac{298 \times 10^6}{(0.7)(450)^3} = 4.67 \text{ MPa}$$

Assume $\gamma = 0.8$ and use Fig. 1-31(b):

$$\rho = 0.027$$

$$A_s = 0.027(450)^2 = 5467 \text{ mm}^2$$

Use eight No. 30 bars equally distributed along four faces.

1.6.2 Slender Columns

Slender columns are those members whose ultimate load capacities are affected by the slenderness effect which produces additional bending stresses and/or instability of columns. Therefore, evaluation of a slender column involves consideration of the column length in addition to its cross section. The degree of slenderness is generally expressed in terms of "slenderness ratio" kl_u/r, where k is the effective length parameter and r is the radius of gyration ($r = \sqrt{I/A}$). Figure 1-38 gives values of k for various support conditions.

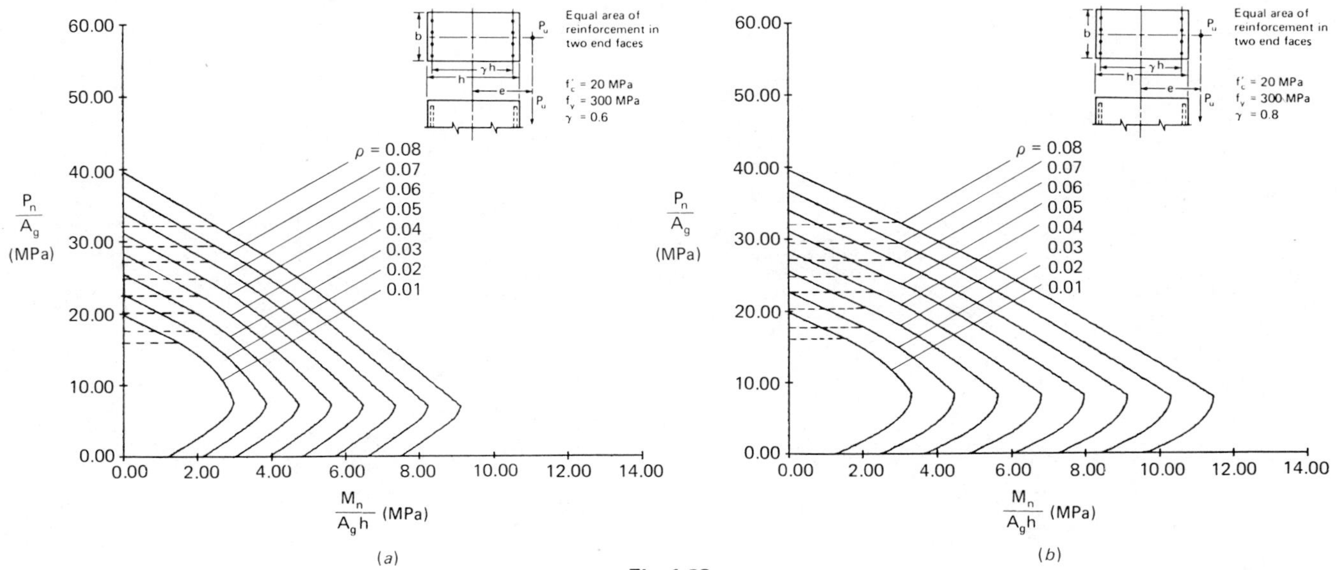

Fig. 1-23

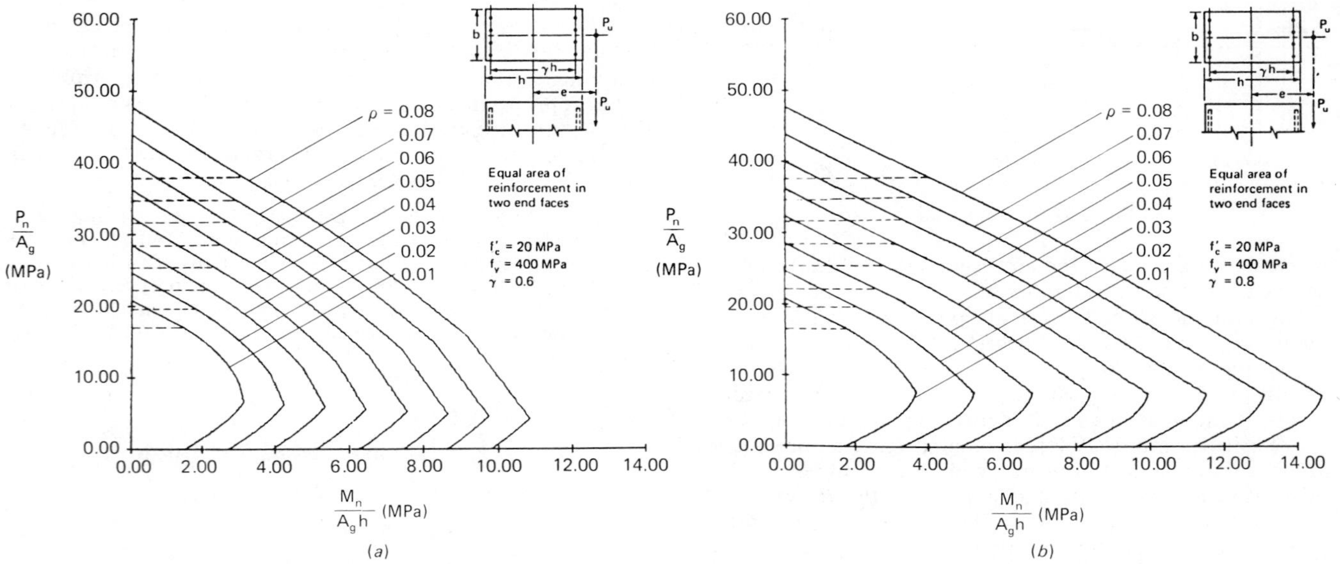

Fig. 1-24

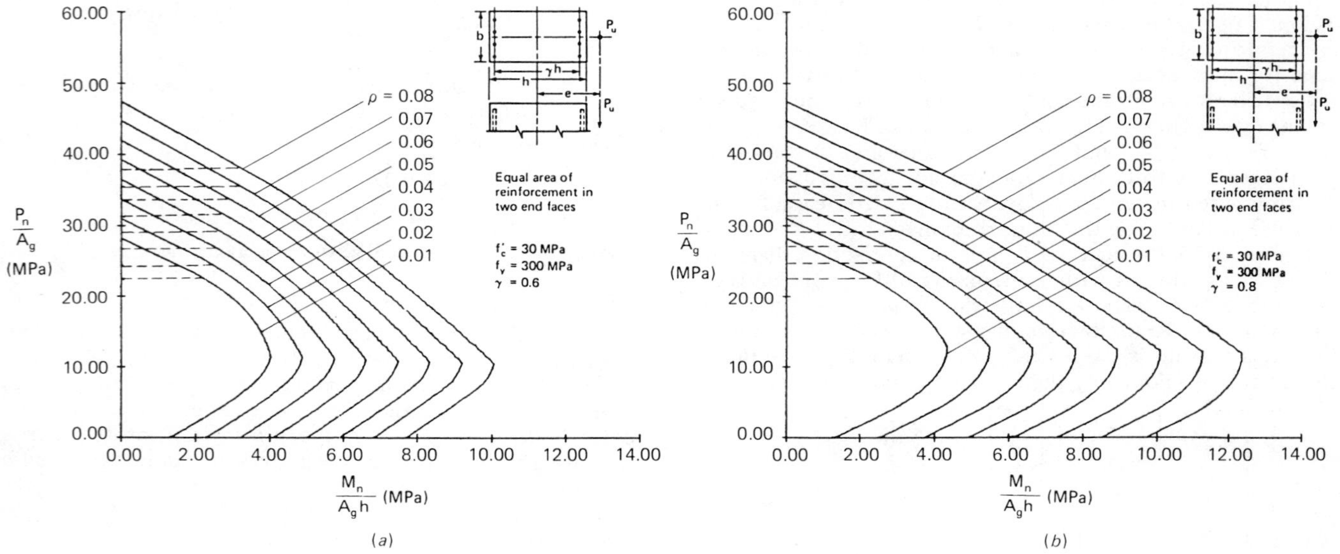

Fig. 1-25

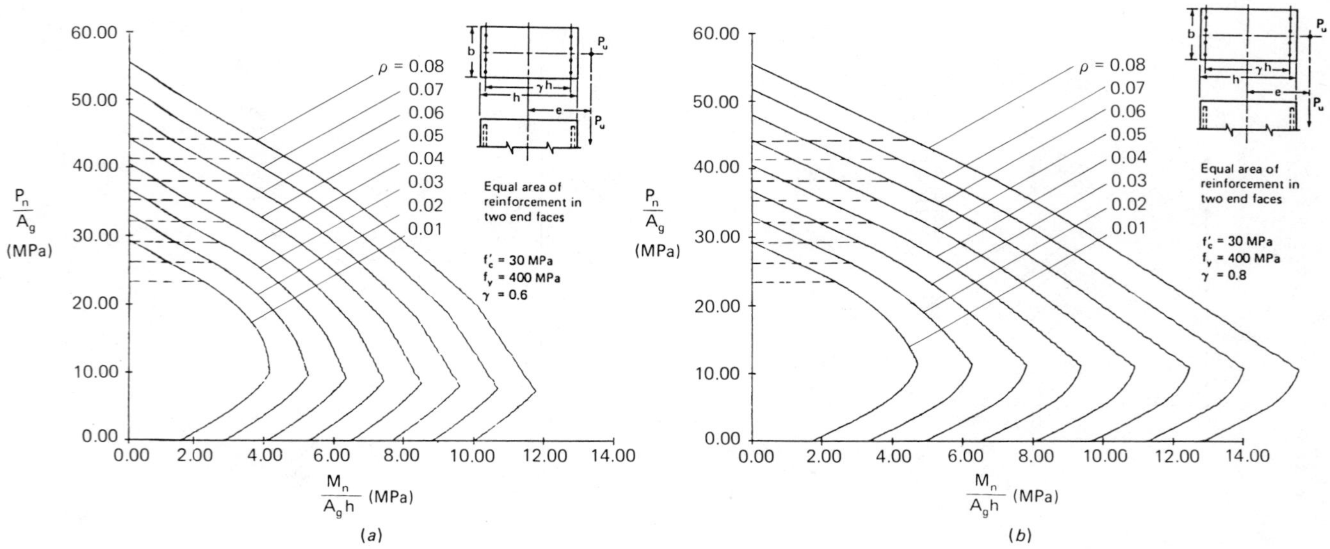

Fig. 1-26

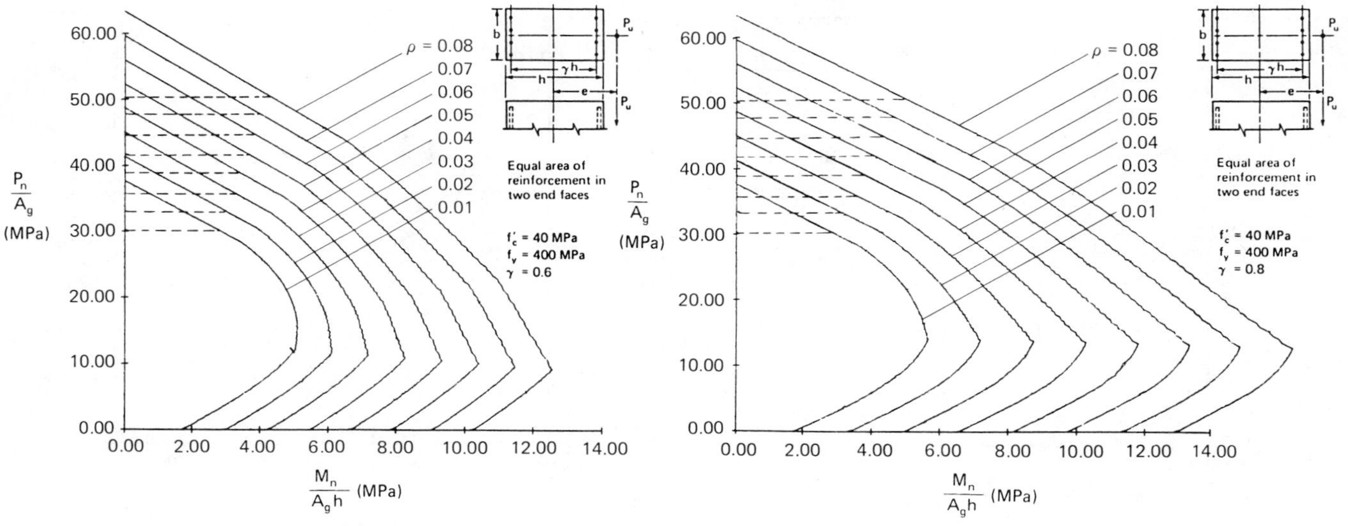

(a) (b)

Fig. 1-27

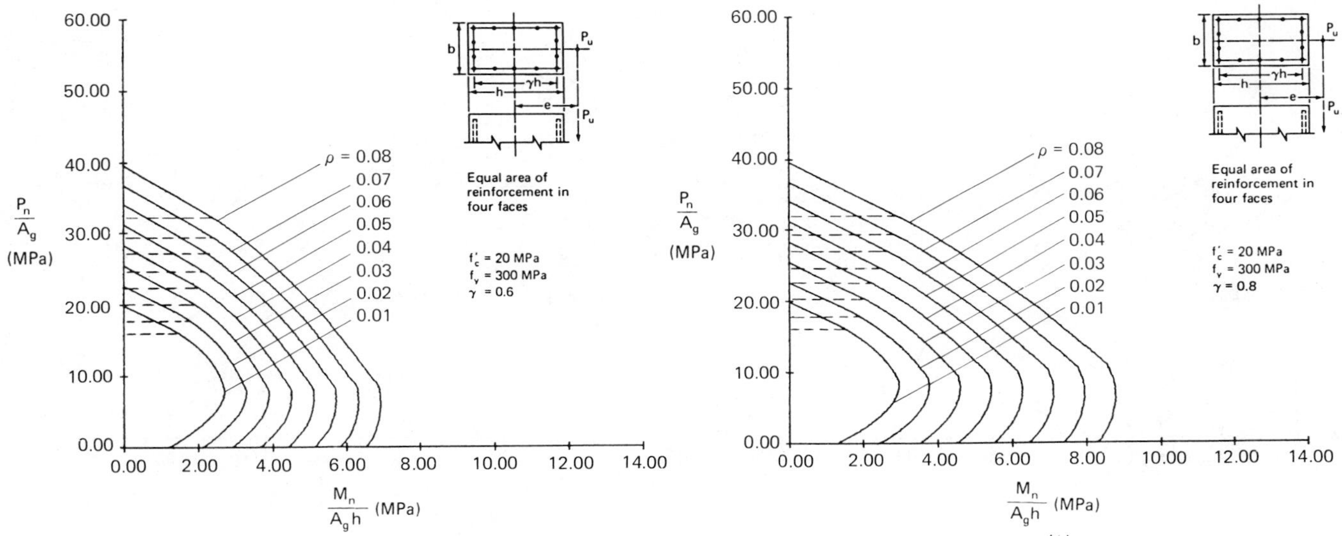

(a) (b)

Fig. 1-28

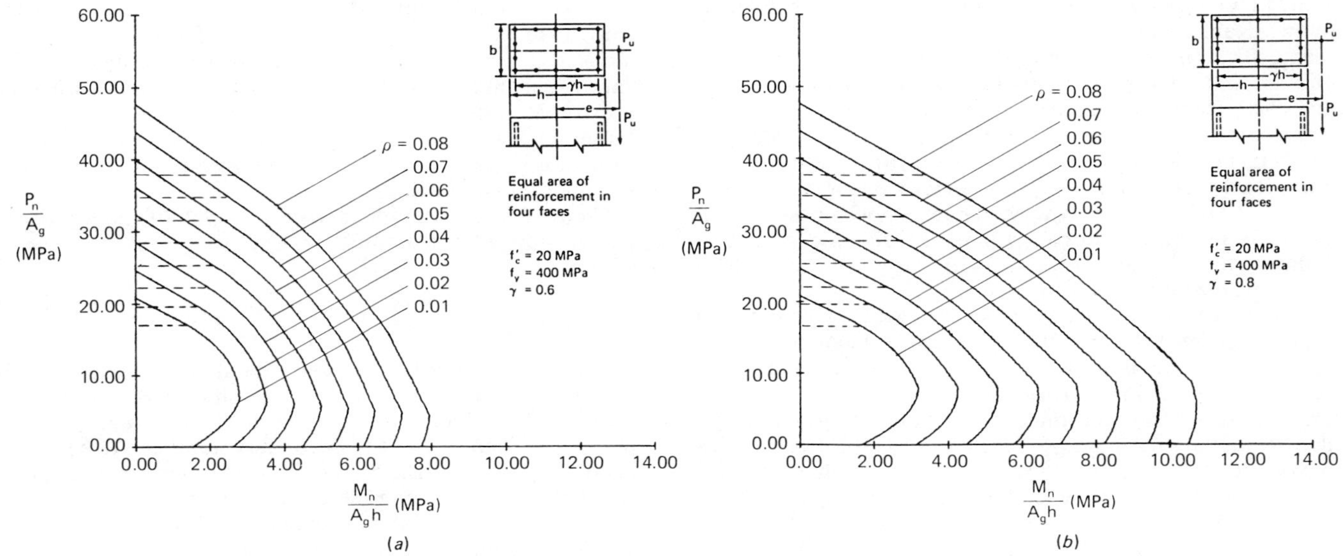

(a) (b)

Fig. 1-29

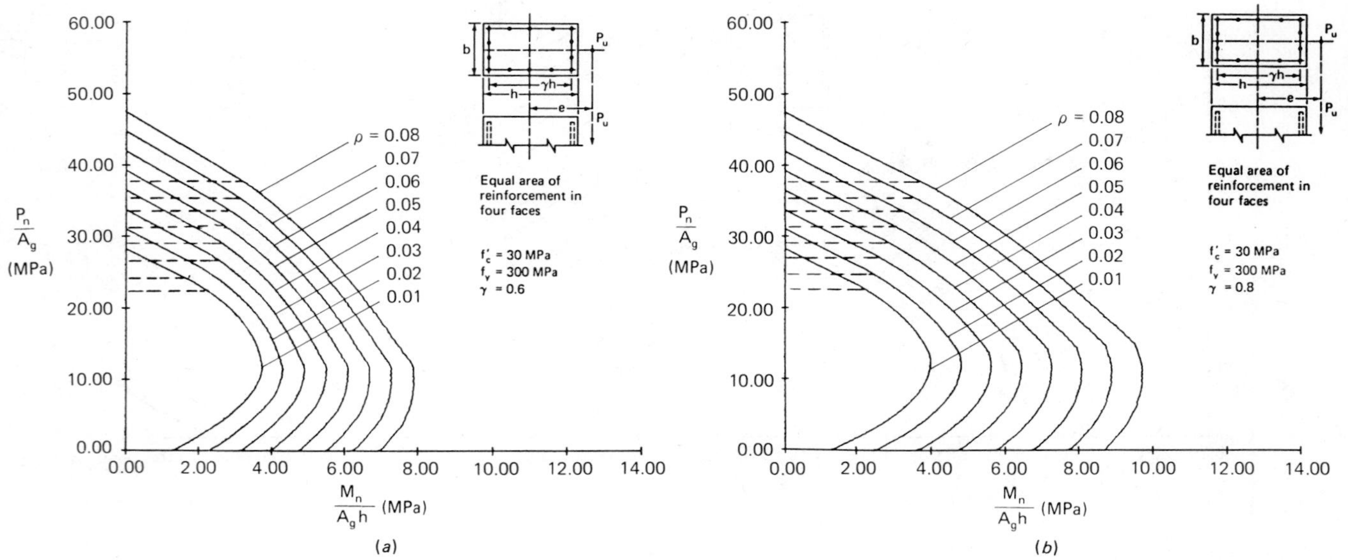

Fig. 1-30

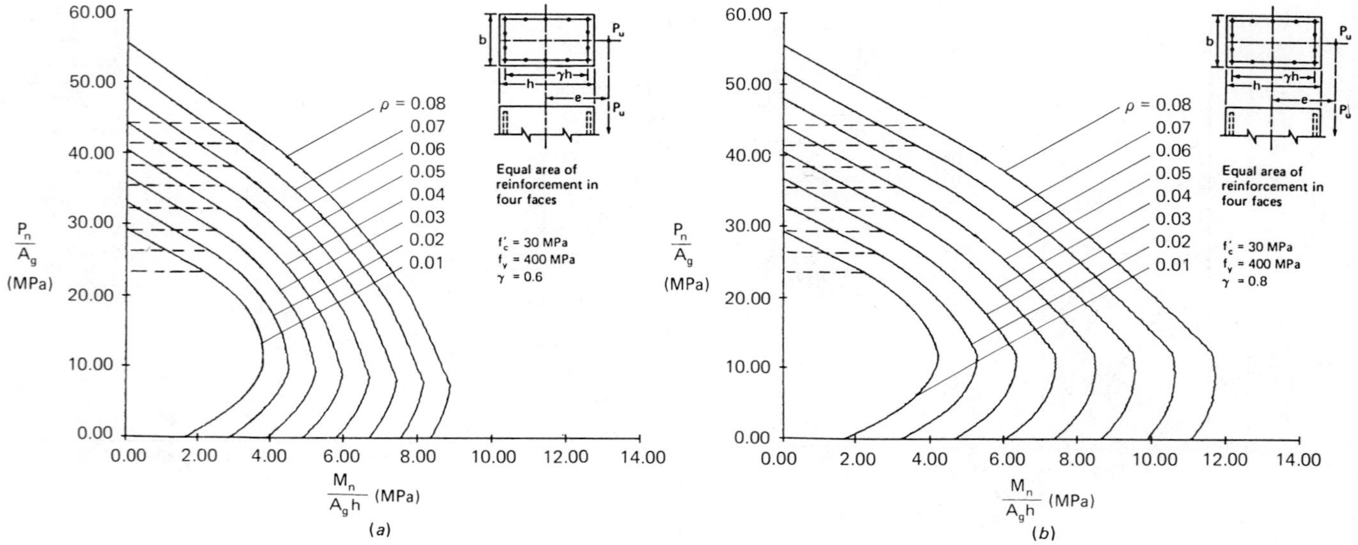

Fig. 1-31

Slender columns, when subjected to eccentric loading, show sizable deflections. These deflections produce additional flexural stresses due to the increase in eccentricity by the amount of transverse deflection (Δ). This is known as the P-Δ effect. The additional moment, $P\Delta$, is sometimes referred to as the "secondary moment." If the total moment including the secondary moment reaches the ultimate capacity of a section, the column fails owing to material failure. If the column is very slender, it becomes unstable prior to reaching material failure. In this case instability failure occurs. Figure 1-39 illustrates the types of failures for reinforced concrete columns.

Slenderness effects are more pronounced in columns of unbraced frames. Frames that do not have adequate bracing against lateral loads show excessive sway which jeopardizes stability of columns. Adequate bracing in frames helps to stabilize secondary deformations at column ends and produces more stable columns. Because of the difference in behavior between a braced and an unbraced frame, columns are treated differently depending on the bracing conditions of their frames. Generally a column is assumed to be braced if total stiffness of the bracing elements such as shear walls, diagonal members, infilled panels, and other bracing mechanisms is greater than 6.0 times the stiffness of all the columns at a given story. In many cases engineering judgment and visual inspection are sufficient to determine whether horizontal displacements can have a significant effect on the moments of the structure. Structures that are subjected to high seismic forces require special evaluation for the adequacy of lateral bracings.

The *ACI Code* commentary defines stability index Q for a given story of a structure:

$$Q = \frac{\Sigma P_u \Delta_u}{H_u h_s} \tag{1-65}$$

where H_u is the total factored lateral force acting within the story, Δ_u is the elastic deflection due to H_u (neglecting $P\Delta$ effects), and h_s is the story height. If the stability index, Q, is less than 0.04, the secondary moment ($P\Delta$) is expected to be less than 5% of the first-order moments, and the structure can be considered to be braced.

As can be seen from the foregoing discussions, design of slender columns must be based on a second-order analysis,

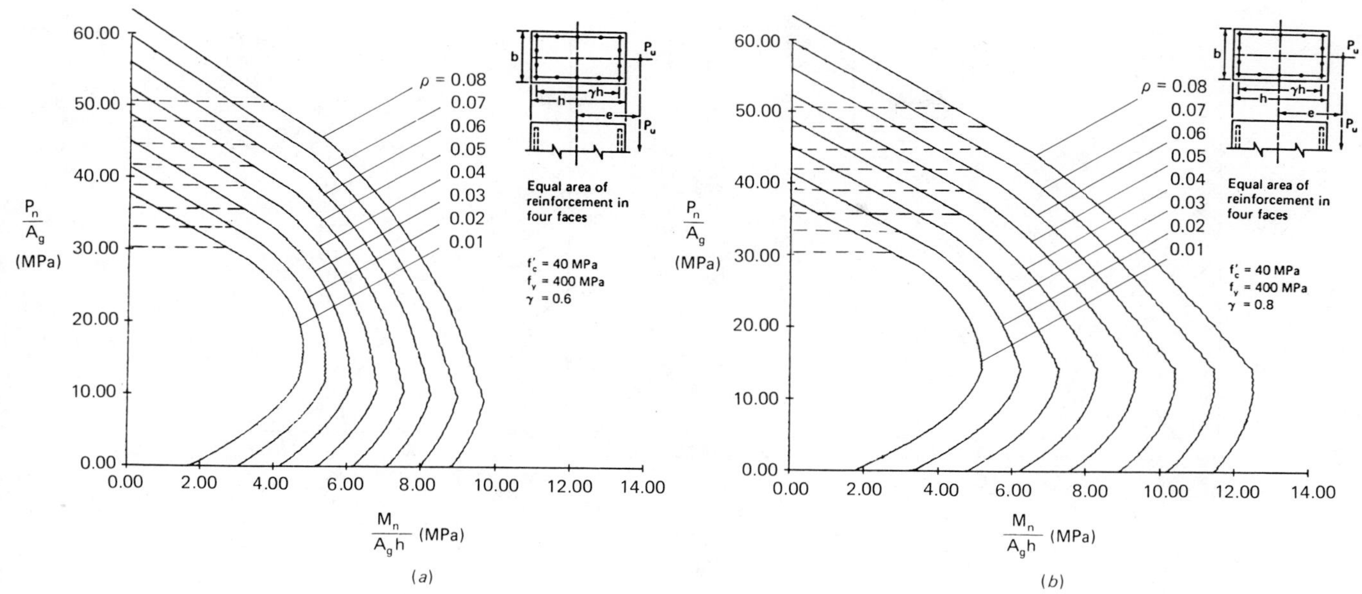

Fig. 1-32

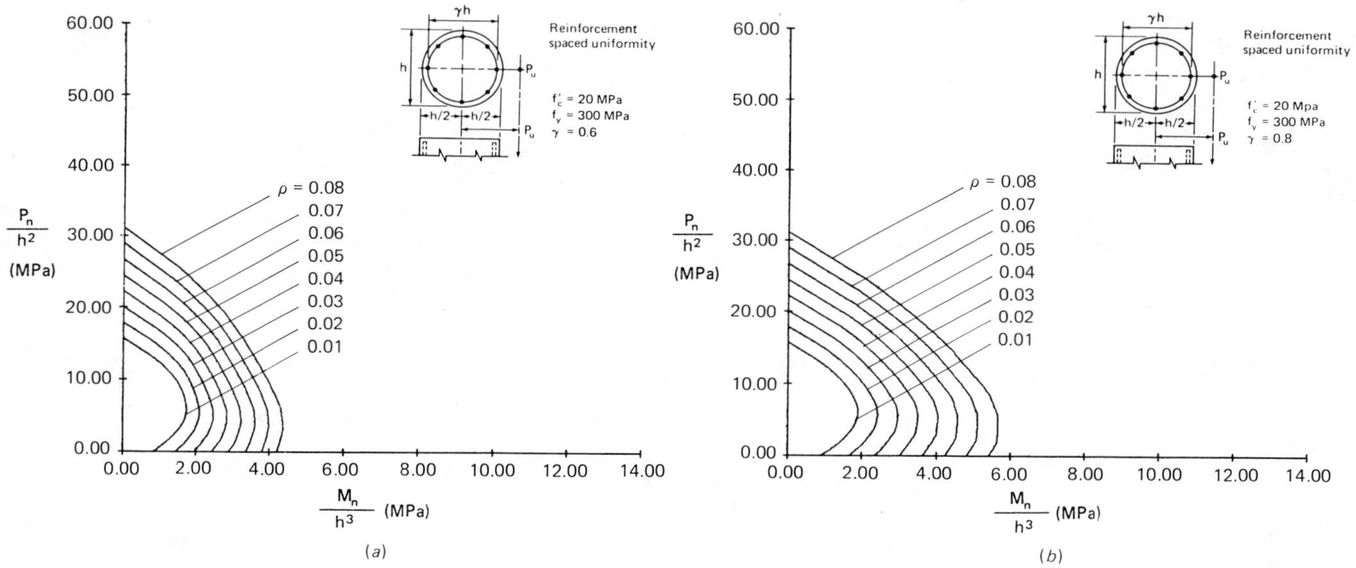

Fig. 1-33

taking into account the effects of deflection, change in stiffness, sustained load effects, and stability of columns. This kind of incremental analysis can usually be done with the aid of computers. The ACI 318M-83 *Code* recognizes such a detailed analysis for design of compression members. Because the analytical tools for a second-order analysis may not be conveniently available in a design office environment, the *Code* also outlines a procedure for approximate evaluation of slenderness effects. This procedure is known as the "Moment Magnification Method." The method is applicable to compression members with slenderness ratios that are *not* greater than 100. For compression members braced against sidesway, the slenderness effect can be neglected if the slenderness ratio is less than $34 - 12M_{1b}/M_{2b}$. The ratio M_{1b}/M_{2b} is the ratio of smaller column end moment to larger end moment. This ratio is positive if the column is bent in single curvature, and negative if it is bent in double curvature. For compression members that are not braced against sidesway, effects of slenderness may be neglected when the slenderness ratio is less than 22.

Moment Magnification Method. This method is outlined in the *ACI Code* for evaluation of slenderness effects without conducting a second-order analysis. Accordingly, compression members are designed using the factored axial load P_u and the factored moments M_{2b} and M_{2s} magnified by δ_b and δ_s to account for slenderness. The magnified moment M_c can be found by using the following equation:

$$M_c = \delta_b M_{2b} + \delta_s M_{2s} \tag{1-66}$$

where:

$$\delta_b = \frac{C_m}{1 - (P_u/\phi P_c)} \geqslant 1.0 \tag{1-67a}$$

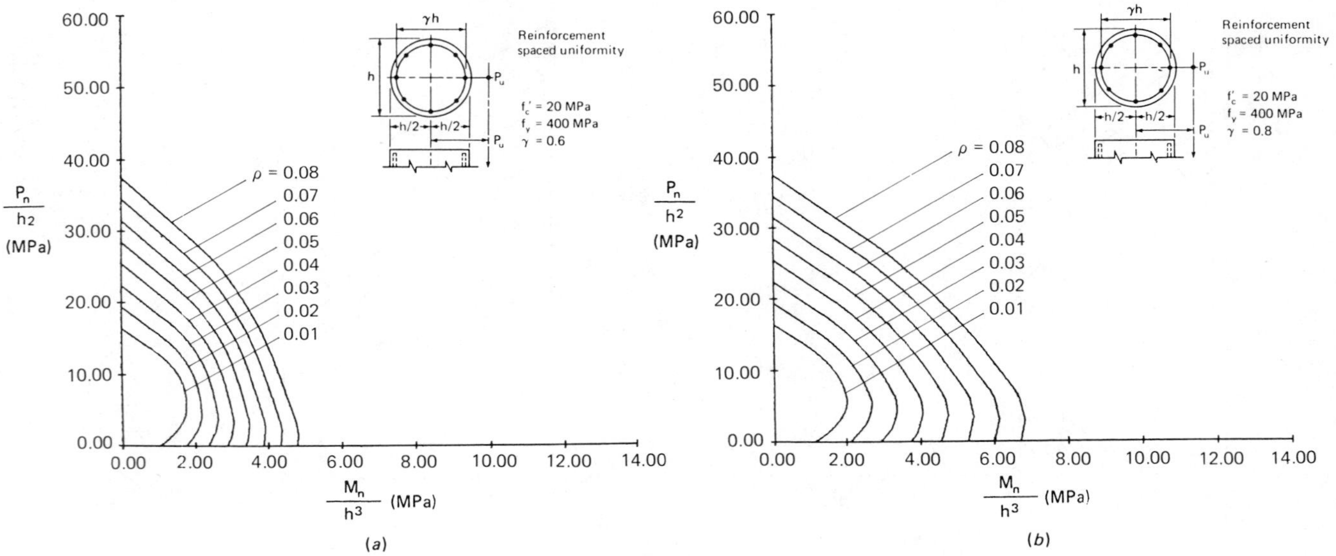

Fig. 1-34

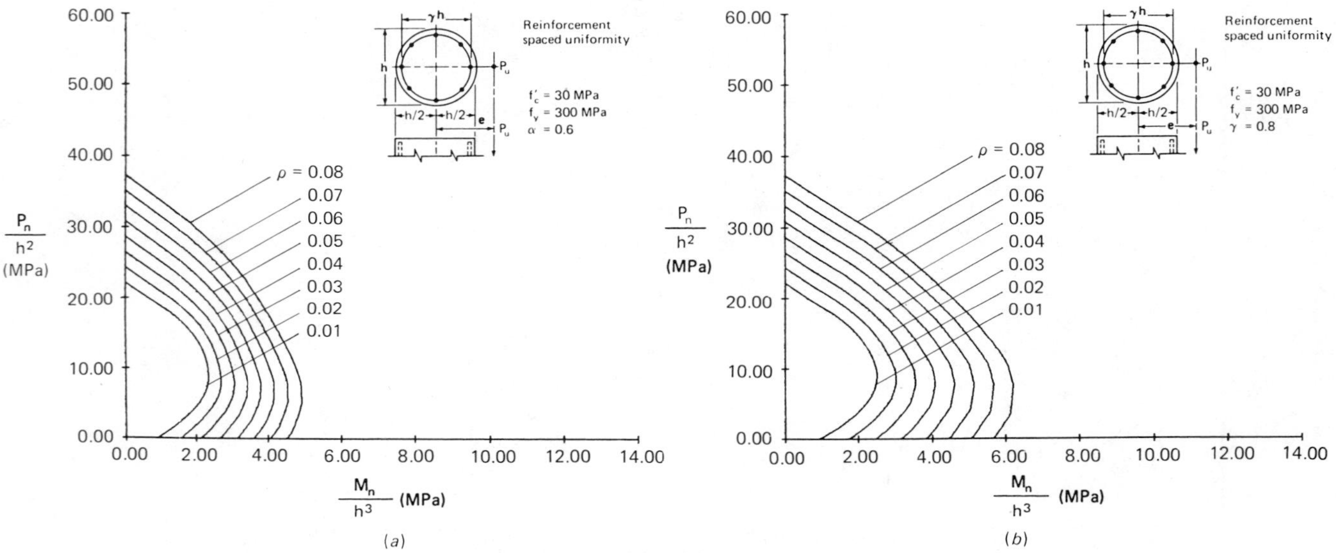

Fig. 1-35

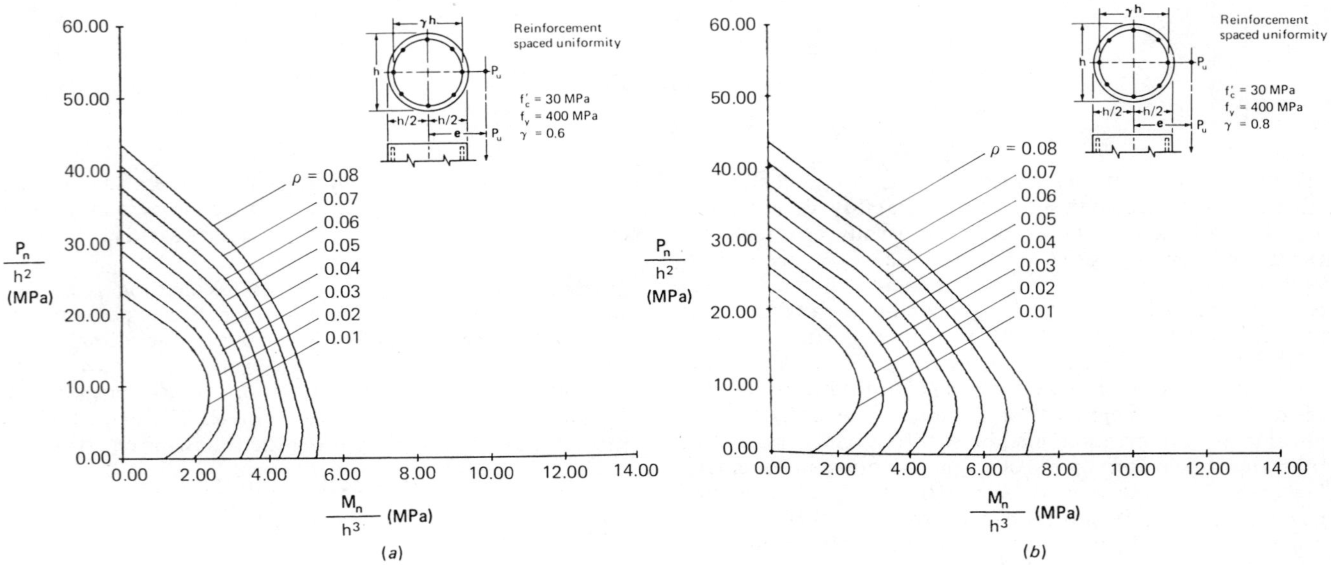

Fig. 1-36

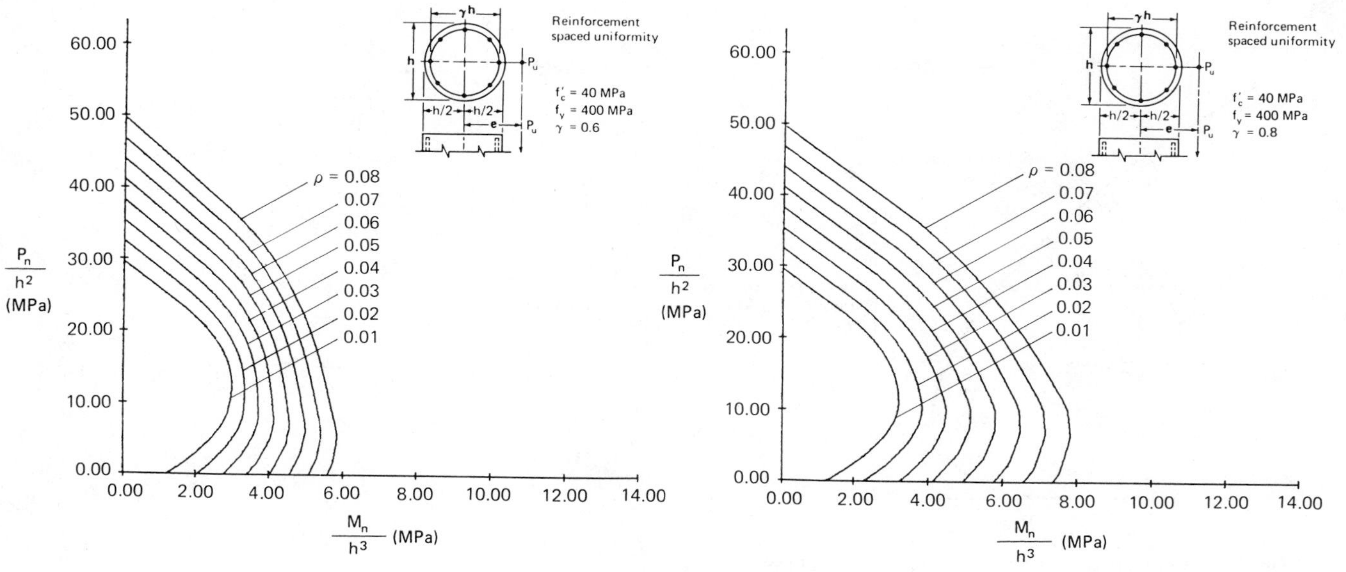

Fig. 1-37

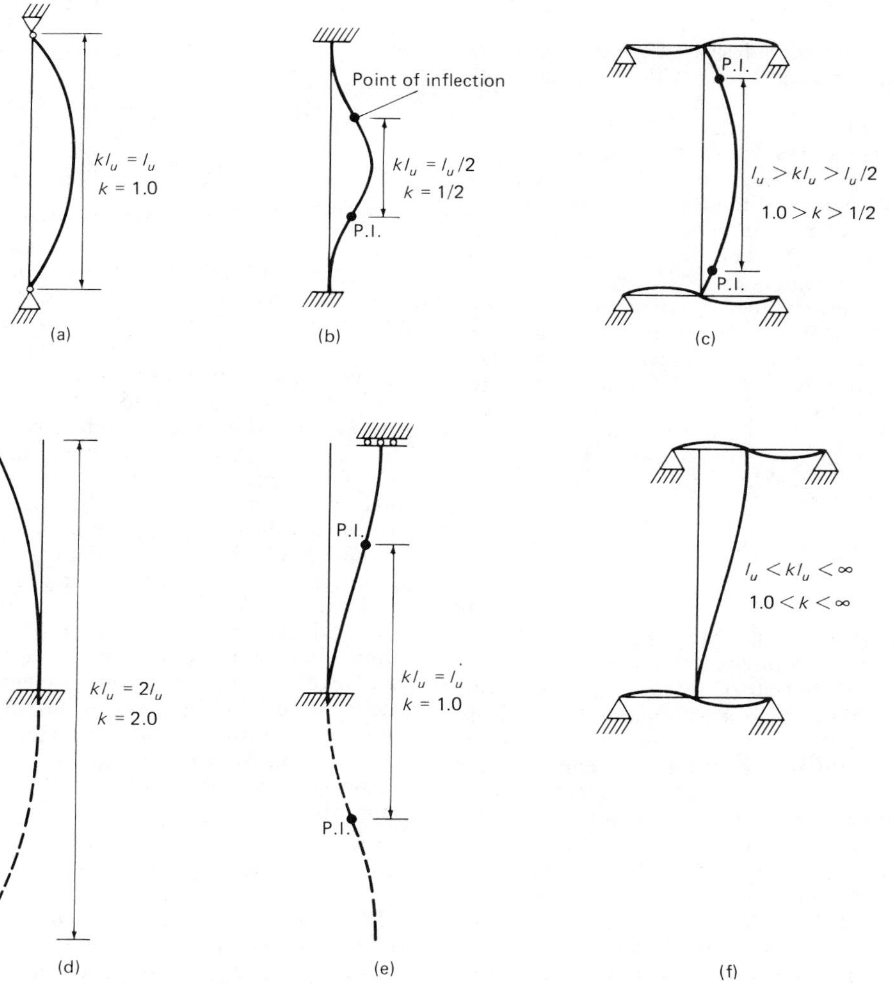

Fig. 1-38 Effective length of columns with different end restraints.

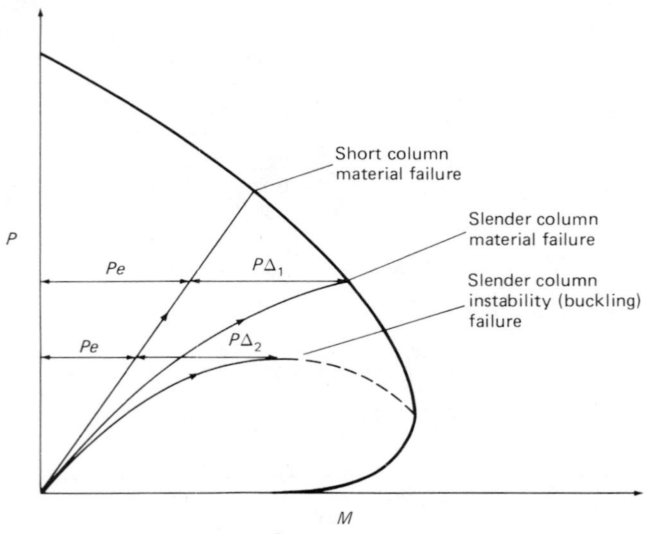

Fig. 1-39 Comparison of short and long column failures.

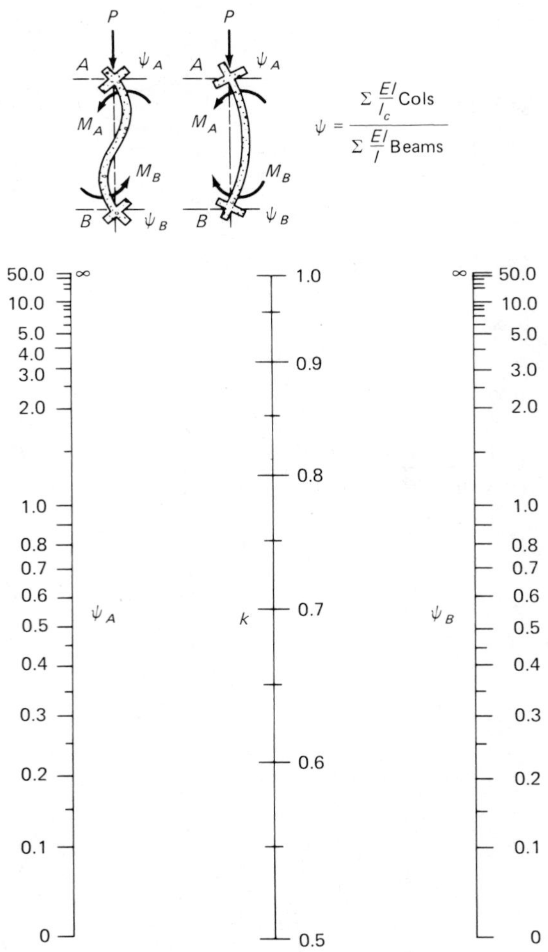

Fig. 1-40 Effective length factors for braced frames.

$$\delta_s = \frac{1}{1 - (\Sigma P_u/\phi\Sigma P_c)} \geqslant 1.0 \qquad (1\text{-}67b)$$

and:

$$P_c = \frac{\pi^2 EI}{(kl_u)^2} \qquad (1\text{-}68)$$

For frames not braced against sidesway, both δ_b and δ_s are computed. For braced frames, δ_s is equal to 1.0 and only δ_b is computed.

If a column is part of a frame that is not braced against sidesway, then the frame as a whole may fail in a lateral instability mode. Therefore, the stability of a column is related to story displacement of the frame. Moment magnification factors for a column of an unbraced frame are computed for the entire story (δ_s) and for the individual column assuming a braced frame (δ_b).

Determination of the critical load P_c, given by eq. (1-68), involves the choice of a stiffness parameter EI that includes the effects of cracking creep and nonlinearity of the concrete stress–strain relationship. In lieu of a more accurate prediction of EI, the *ACI Code* provides the following two equations:

$$EI = \frac{(E_c I_g/5) + E_s I_{se}}{1 + \beta_d} \qquad (1\text{-}69)$$

$$EI = \frac{E_c I_g/2.5}{1 + \beta_d} \qquad (1\text{-}70)$$

where β_d is the ratio of maximum factored dead load moment to maximum factored total load moment.

In eq. (1-69) the contribution of reinforcing steel is accounted for by the term $E_s I_{se}$. Gross concrete stiffness is divided by 5.0 to account for concrete cracking. Equation (1-70) is easier to evaluate and produces more conservative results. The effect of creep due to sustained loading is taken into account by dividing the EI term by $(1 + \beta_d)$.

Determination of a proper effective length factor k is also required to compute the critical load P_c. Figure 1-38 provides k factors for certain idealized cases. The majority of columns have end conditions that are neither fully fixed nor fully hinged but instead partially restrained by adjoining members. Therefore, the factor k varies with the ratio of column stiffness to flexural member stiffness that provides restraint at column ends. This variation is shown in Figs. 1-40 and 1-41 for columns of braced and unbraced frames, respectively. The stiffness parameter EI for a flexural member can be computed by considering the moment of inertia of a cracked transformed section. For columns, eq. (1-69) can be used with $\beta_d = 0$. According to the *ACI Code* commentary, the use of $0.5I_g$ for flexural members and I_g for compression members results in reasonably accurate k factors for those columns that have kl_u/r less than 60. The effective length factor k for columns that are braced against sidesway can be conservatively taken as 1.0. In calculation of P_c to determine δ_b, k factor for braced frames should be used. Similarly, calculation of P_c for δ_s should include k factor for unbraced frames.

The factor C_m in eq. (1-67) is an equivalent moment correction factor. In the derivation of this equation the maximum moment is assumed to occur at or near the column midheight. If the maximum moment occurs at one end of the column, this moment is modified by C_m to find an equivalent uniform column moment. The following equation is used to compute C_m for members braced against sidesway and not exposed to transverse loads between supports:

$$C_m = 0.6 + 0.4\,\frac{M_{1b}}{M_{2b}} \geqslant 0.4 \qquad (1\text{-}71)$$

For all other cases $C_m = 1.0$. If the analysis indicates that there is essentially no moment at both ends of a column, the ratio M_{1b}/M_{2b} is taken equal to 1.0.

Columns that are essentially concentric may be subjected to very little or no end moments. If the end eccentricity is

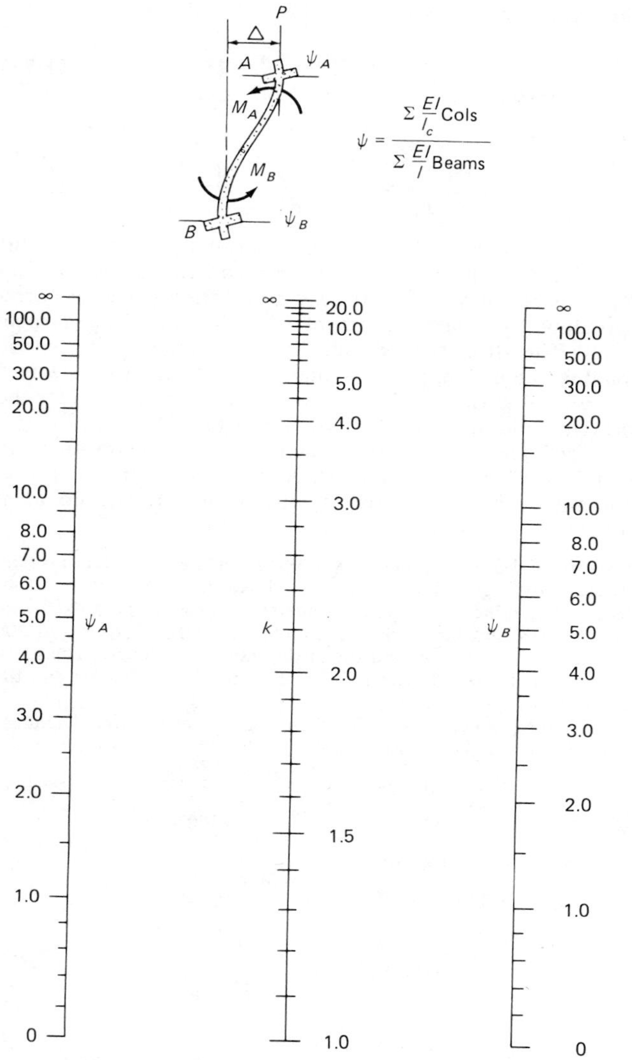

Fig. 1-41 Effective length factors for unbraced frames.

less than the minimum eccentricity, $(15 + 0.03h)$ mm, then M_{2b} in eq. (1-66) is computed using $(15 + 0.03h)$ mm. Similarly, for columns that are *not* braced against sidesway, if the end eccentricity is less than the minimum eccentricity, $(15 + 0.03h)$ mm is used in computing M_{2s}.

EXAMPLE 1-9: Design a tied column for a factored axial load P_u = 2500 kN and a factored moment M_u = 200 kN · m. The column cross section is limited to 400 mm square, and kl_u = 3500 mm. The frame is braced against sidesway, and M_{1b}/M_{2b} = 1.0. Use f_c' = 30 MPa, f_y = 400 MPa, and β_d = 0.25.

Check whether the slenderness effect can be neglected. Assume a 400 mm square section.

$$r = \sqrt{I/A} = 0.3h \quad \text{for square sections}$$

$$\frac{kl_u}{r} = \frac{3500}{0.3(400)} = 29.17 > 34 - 12\frac{M_{1b}}{M_{2b}} = 22.$$

Therefore, the slenderness effect should be considered.

$$EI = \frac{E_c I_g/2.5}{1 + \beta_d} \quad \text{(from eq. 1-70)}$$

$$E = 5000\sqrt{f_c'} = 27{,}386 \text{ MPa}$$

$$EI = \frac{27{,}386 \times 10^3 (0.4)^4/(12)(2.5)}{1 + 0.25} = 18{,}695 \text{ kN} \cdot \text{m}^2$$

Note that eq. (1-69) could be used for an assumed area of steel for more accurate prediction of EI.

$$P_c = \frac{\pi^2 EI}{(kl_u)^2} \quad \text{(from eq. 1-68)}$$

$$P_c = \frac{\pi^2(18{,}695)}{(3.5)^2} = 15{,}062 \text{ kN}$$

$$C_m = 0.6 + 0.4\frac{M_{1b}}{M_{2b}} \quad \text{(from eq. 1-71)}$$

$$C_m = 1.0$$

Determine moment magnification factor δ_b by using eq. (1-67a):

$$\delta_b = \frac{1.0}{1 - (2500/0.7 \times 15{,}062)} = 1.31$$

For columns that are effectively braced, eq. (1-66) reduces to;

$$M_c = \delta_b M_{2b}$$

$$M_c = 1.31 \times 200 = 262 \text{ kN} \cdot \text{m}$$

Proportion the 400-mm-square section for P_u = 2500 kN and M_c = 262 kN · m using the appropriate moment interaction diagram. Apply ϕ = 0.7.

$$\frac{P_n}{A_g} = \frac{2500 \times 10^{-3}}{(0.7)(0.4)^2} = 22.3 \text{ MPa}$$

$$\frac{M_n}{A_g h} = \frac{262 \times 10^{-3}}{(0.7)(0.4)^3} = 5.8 \text{ MPa}$$

Assuming γ = 0.7 and interpolating the values between Fig. 1-31(a) and Fig. 1-31(b):

$$\rho = 0.048$$

$$A_s = 0.048(400)^2 = 7680 \text{ mm}^2$$

Use eight No. 35 bars equally distributed along four faces.

1.6.3 Biaxial Bending

Columns are often subjected to simultaneous bending moments about two orthogonal axes. Generally the critical moment about one axis is sufficient to consider for design. Sometimes, however, biaxial bending effects can be significant. Corner columns of a framed building constitute a typical example of a case where biaxial bending may have to be considered in design. Circular columns can be designed for biaxial bending by using the interaction diagrams developed for uniaxial bending with a design moment $M_u = \sqrt{(M_{ux})^2 + (M_{uy})^2}$. Columns of other cross sections require three-dimensional interaction diagrams.

Figure 1-42 illustrates a three-dimensional interaction

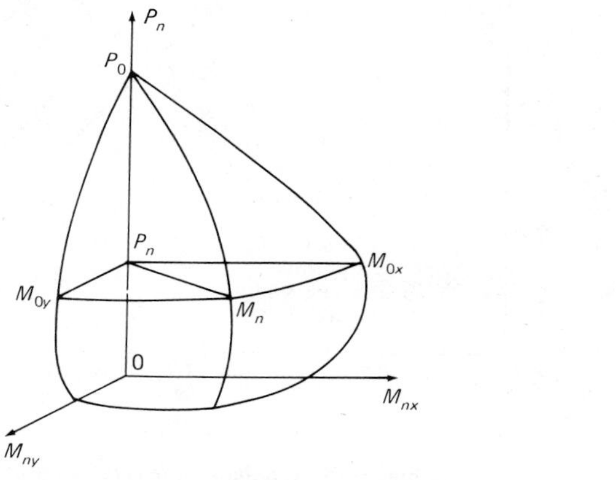

Fig. 1-42 Three-dimensional M-P interaction diagram.

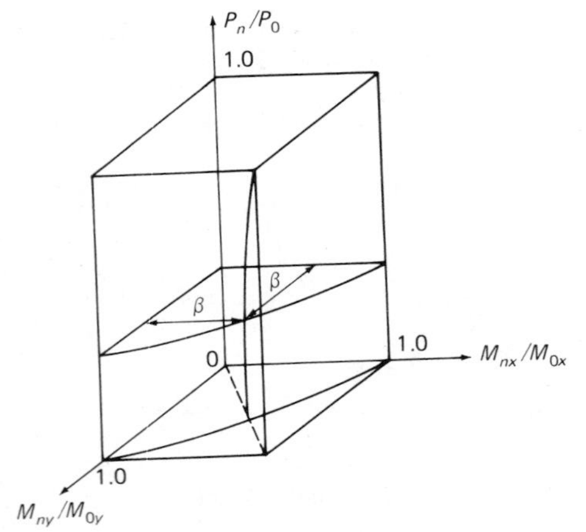

Fig. 1-43 Nondimensionalized ultimate capacity surface.

diagram, typically required for the design of a rectangular column section. Each horizontal plane corresponding to a specific value of P_n provides a failure contour for biaxial moments. When the sectional capacities are plotted in terms of the nondimensional parameters P_n/P_0, M_{nx}/M_{0x}, and M_{ny}/M_{0y}, the ultimate capacity surface assumes the shape shown in Fig. 1-43. Bresler suggested an expression for failure contours corresponding to a specific level of P_n/P_0. This expression is given below in terms of parameter β:

$$(M_{nx}/M_{0x})^n + (M_{ny}/M_{0y})^n = 1 \qquad (1\text{-}72)$$

where $n = \log 0.5/\log \beta$.

The parameter β depends primarily on the ratio P_n/P_0 and to a lesser degree on the reinforcement ratio, arrangement, and strength. Figure 1-44 gives the ultimate moment capacity of a column section for different values of β. For lightly loaded columns (i.e., when P_n/P_0 is small), variation of β is in the order of 0.5 to 0.7. As the ratio P_n/P_0 increases, the value of β also increases. A lower value of β can be used for the purpose of a conservative design.

Other approximate procedures are available for biaxial design. Bresler developed an approximate design procedure that has been satisfactorily verified by tests. Accordingly, the following expression is used for biaxial design:

$$\frac{P_n}{P_{0x}} + \frac{P_n}{P_{0y}} - \frac{P_n}{P_0} = 1 \qquad (1\text{-}73)$$

or:

$$\frac{1}{P_n} = \frac{1}{P_{0x}} + \frac{1}{P_{0y}} - \frac{1}{P_0} \qquad (1\text{-}74)$$

For a given or assumed column section and reinforcement arrangement, the nominal uniaxial load capacities P_{0x} and P_{0y} can be found from uniaxial interaction diagrams corresponding to uniaxial moments of M_{0x} and M_{0y}, respectively. The ultimate load capacity, P_0, for purely concentric loading can be computed. Equation (1-74) can then be used to determine the nominal axial load capacity, P_n, under the biaxial bending condition. Equation (1-74) does not produce reliable results if the axial load level is below approximately 10% of the concentric load capacity. In this range, columns behave like beams with very little or no axial forces.

EXAMPLE 1-10: A 300 mm × 500 mm rectangular column is reinforced with eight No. 25 bars equally distributed along the four faces. Determine if the design is adequate for a factored axial load of 2250 kN and factored design moments of 160 kN · m and 300 kN · m about the weak and the strong axes, respectively. The specified material strengths are $f_c' = 30$ MPa and $f_y = 400$ MPa, and the concrete cover is 40 mm.

First, determine P_{0x} and P_{0y} from uniaxial interaction diagrams:
Bending about x axis (Fig. 1-45):

$$\gamma = \frac{300 - 80 - 25}{300} = 0.65, \qquad \rho = \frac{8(500)}{(300)(500)} = 0.027$$

$$\frac{M_{0x}}{A_g h} = \frac{160 \times 10^6}{(300)(500)(300)} = 3.56 \text{ MPa}$$

Bending about y axis (Fig. 1-45):

$$\gamma = \frac{500 - 80 - 25}{500} = 0.79, \qquad \rho = 0.027$$

$$\frac{M_{0y}}{A_g h} = \frac{300 \times 10^6}{(300)(500)(500)} = 4.0 \text{ PMa}$$

Using Fig. 1-31 for moments about the x and y axes separately:

$$\frac{P_{0x}}{A_g} = 31.8 \text{ MPa} \quad \text{(interpolating between Figs. 1-31a and 1-31b for } \gamma = 0.65)$$

$$P_{0x} = 31.8(300)(500) = 4770 \times 10^3 \text{N} = 4770 \text{ kN}$$

$$\frac{P_{0y}}{A_g} = 24.3 \text{ MPa} \quad (\text{for } \gamma = 0.79)$$

$$P_{0y} = 24.3(300)(500) = 3645 \times 10^3 \text{N} = 3645 \text{ kN}$$

$$\frac{P_0}{A_g} = 35.5 \text{ MPa}$$

$$P_0 = 35.5(300)(500) = 5325 \times 10^3 \text{N} = 5325 \text{ kN}$$

Use eq. (1-74) to determine the adequacy of the biaxially applied axial load, P_n:

$$\frac{1}{P_n} = \frac{1}{4770} + \frac{1}{3645} - \frac{1}{5325}$$

$$P_n = 3376 \text{ kN}$$

$$\phi P_n \geqslant P_u$$

$$(0.7)(3376) = 2363 \text{ kN} > 2250 \text{ kN} \qquad \text{O.K.}$$

Therefore the design is adequate.

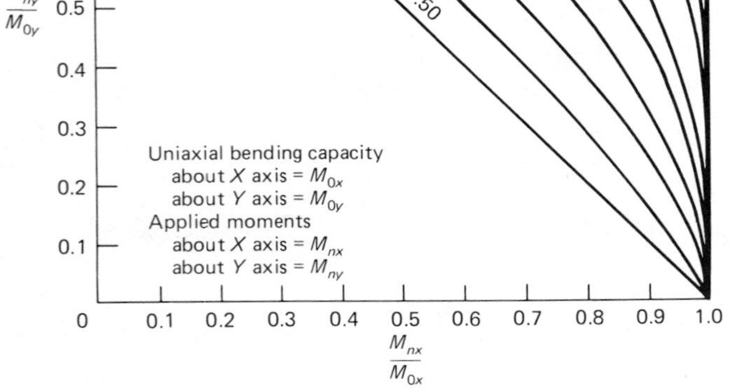

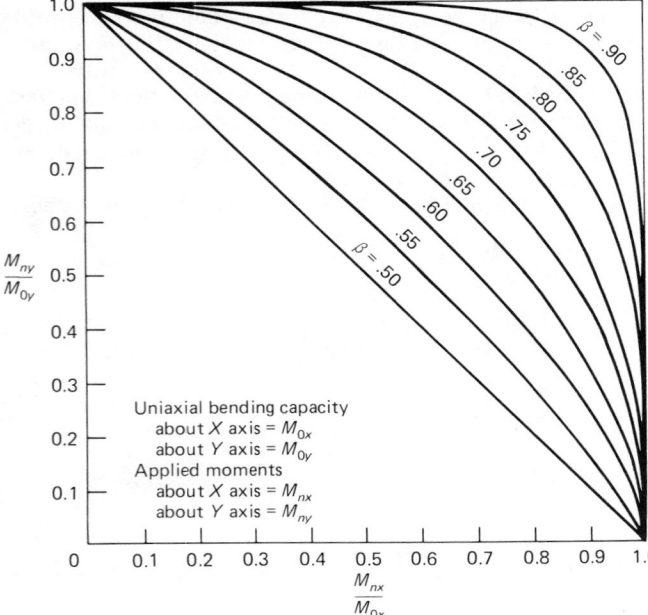

Fig. 1-44 Biaxial moment relationship.

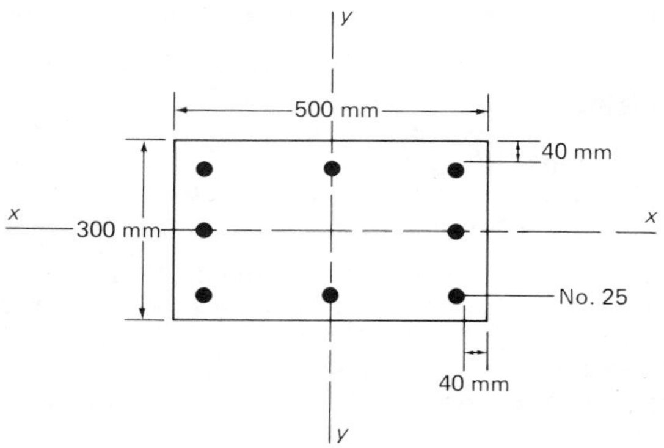

Fig. 1-45 Example 1-10.

TABLE 1-9 Maximum Tie Spacing

Longitudinal Bar Size	S^*_{max} (mm) No. 10 Tie	No. 15 Tie
10	181	181
15	256	256
20	312	312
25	403	403
30	478	478
35	542	571
45	542	699
55	542	768

**Maximum tie spacing is equal to S_{max} or the least column dimension, whichever is smaller.*

1.6.4 Lateral Reinforcement for Columns

Lateral reinforcement for columns normally consists of individual ties or closely spaced continuous spirals. Typical column tie and spiral arrangements are shown in Fig. 1-46. These reinforcements are essential components of the reinforcement cage. Lateral ties and spirals are primarily used to hold longitudinal bars in place. They provide lateral bracing to longitudinal bars to prevent local bar buckling under heavy compression loads. The structural performance of columns can be significantly affected by the amount and detailing of lateral reinforcement. Columns that are subjected to relatively high shear or torsion require additional lateral reinforcement. Closely spaced ties or spirals provide confinement of the concrete core area which improves column strength. Confinement of concrete in columns that are subjected to load reversals has significant implications on their strength and ductility.

Limiting values for size and spacing of lateral reinforcement are specified in the ACI 318M-83 *Code*. Accordingly, lateral ties for columns consist of at least No. 10 bars. Deformed wire or welded wire fabric of equivalent area may be used. Vertical spacing of ties is limited to the smaller of 16 longitudinal bar diameter, 48 tie bar diameter, or least dimension of the column. Table 1-9 summarizes the maximum allowable spacings for lateral ties for various combinations of bars. While the spacing must be sufficiently small for reinforcement to be effective, too closely spaced ties can create concrete placement problems.

Axially loaded columns reinforced with ties generally fail in a sudden and brittle manner when the column capacity is reached. At the failure load, concrete crushes and bulges outward between the ties; buckling of longitudinal bars takes place, and the column cannot withstand the applied load. If the column is reinforced with spiral lateral reinforcement, concrete within the core area is confined. When the same failure load is applied, the core concrete, because it is confined by spirals, cannot shear outward, and the longitudinal bars are not allowed to buckle. At this load level, the unconfined shell concrete spalls off, but the column core continues carrying the applied load. Depending on the amount of spiral reinforcement, the column continues carrying loads at higher inelastic deformation levels. This illustrates the ductile nature of spirally reinforced columns.

The *ACI Code* provides a minimum spiral reinforcement that can develop a load capacity equivalent to that of the shell concrete. The intention here is to build some ductility into columns so that they can continue resisting the same applied load even after the shell concrete is spalled off. The expression for the minimum ratio of spiral reinforcement that satisfies this condition is derived below.

$$P_{\text{due to shell}} = 0.85 f'_c (A_g - A_c) \qquad (1\text{-}75)$$

The spiral reinforcement contributes to column axial load capacity at least twice the yield strength of the spiral reinforcement.

$$P_{\text{due to spiral}} = 2\rho_s A_c f_y \qquad (1\text{-}76)$$

$$P_{\text{due to shell}} = P_{\text{due to spiral}} \qquad (1\text{-}77)$$

$$\rho_s = 0.425 \frac{(A_g - A_c) f'_c}{A_c f_y} \qquad (1\text{-}78)$$

or as it appears in the *ACI Code* in a slightly increased form:

$$\rho_s = 0.45 \left(\frac{A_g}{A_c} - 1 \right) \frac{f'_c}{f_y} \qquad (1\text{-}79)$$

where A_g and A_c are the gross and core (out-to-out of spirals) concrete areas, respectively. The quantity f_y is the yield strength of the spiral reinforcement, and ρ_s is the ratio of volume of spiral reinforcement to total volume of core.

For cast-in-place construction, the minimum spiral bar size is No. 10. The minimum and maximum clear spacings between the spirals are 25 mm and 80 mm, respectively.

1.6.5 Load Bearing Walls

Concrete walls are commonly used as structural members to carry vertical loads. If the primary function of a wall is to support vertical loads, it can be defined as a bearing wall.

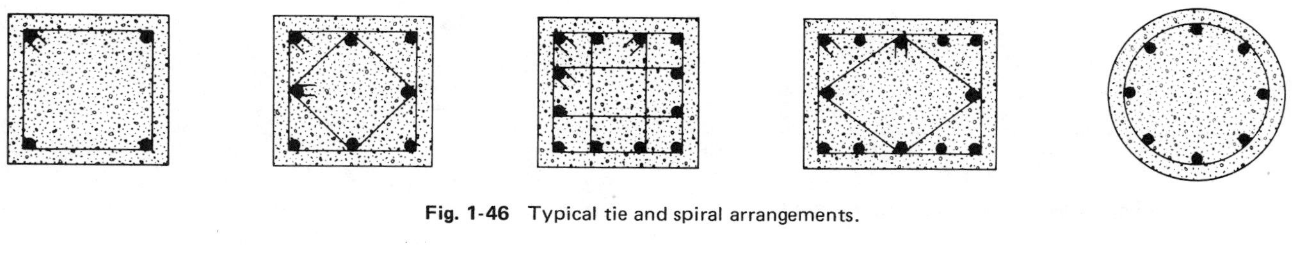

Fig. 1-46 Typical tie and spiral arrangements.

Bearing walls are also subject to horizontal loads. Horizontal forces acting parallel to the plane of the wall are usually not critical. This is mainly due to the relatively high bending capacity of the wall about its strong axis. However, bending stresses caused by either a horizontal wind load in a direction perpendicular to the plane of the wall or the eccentricity of a vertical load causing bending about the weak axis can be significant. Therefore, design of bearing walls should include the effects of axial load as well as bending about the weak axis. If the horizontal force due to high wind and seismic action is large in magnitude, then walls are positioned so that they provide the required stiffness and strength in the direction of the lateral load. These structural walls are not treated like bearing walls, and their design requirements are discussed in section 2.1.6.6.

The thickness of bearing walls varies with the height. The ACI 318M-83 *Code* limits the minimum wall thickness to $1/25$ of the unsupported height or width, whichever is shorter. Furthermore, the minimum thickness is limited to 150 mm for walls up to 5 meters high. This limit increases by 25 mm for each 8-meter increment below the top 5 meters of the wall.

The *ACI Code* permits bearing walls to be designed either by using the empirical design method specified in the *Code*, or by treating wall strips as columns. In the latter case, certain special provisions are specified for walls that make them different from columns. The empirical design method is used if the design axial load is applied within the middle third of the wall thickness. This implies that if the eccentricity of an axial load is $h/6$ or less, the following empirical design axial load strength can be used:

$$\phi P_{nw} = 0.55\phi f_c' A_g \left[1 - \left(\frac{kl_c}{32h}\right)^2\right] \tag{1-80}$$

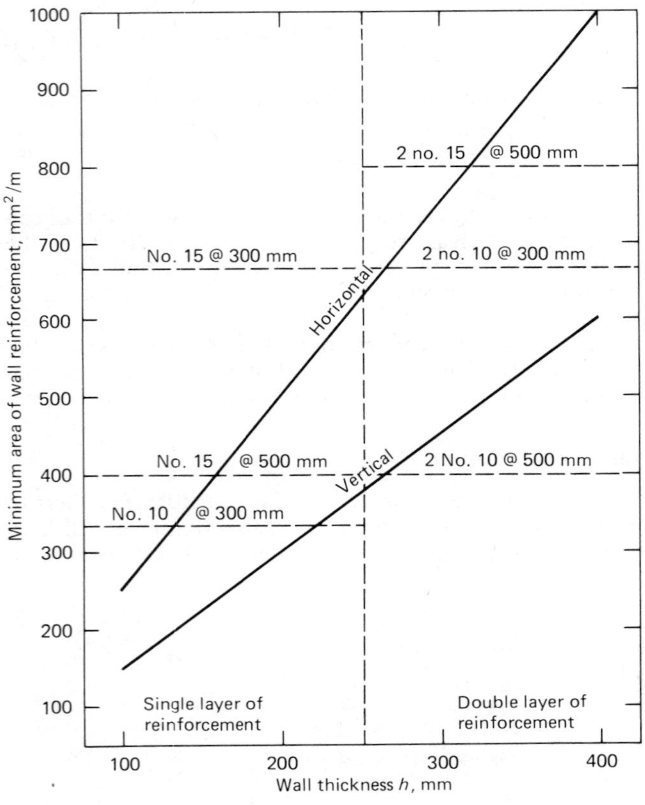

Fig. 1-47 Minimum wall reinforcement for $f_y < 400$ MPa.

where $\phi = 0.70$ and the effective length factor k is as follows:

(a) For walls braced top and bottom against lateral translation and
   i. restrained against rotation at one or both ends (top and/or bottom): 0.8
   ii. unrestrained against rotation at both ends: 1.0
(b) For walls not braced against lateral translation: 2.0

Figures 1-47 and 1-48 give the minimum area of vertical and horizontal reinforcement required in walls. In the case of a concentrated axial load or a reaction, the length of wall considered as effective shall not exceed the center-to-center distance between loads, nor the width of bearing plus four times the wall thickness.

EXAMPLE 1-11: Design a 3.75-meter-high reinforced concrete bearing wall that supports 200-mm-wide beams spaced at 1.5 meters on centers. Each beam produces a factored design load of 800 kN on the wall with an eccentricity of 20 mm. Use $f_c' = 30$ MPa, and assume that bearing stresses do not govern the design. The wall is fixed at the foundation level and is restrained against lateral translation. Assume $f_y = 350$ MPa for reinforcement.

Minimum thickness:

$$l_c/25 = 3750/25 = 150 \text{ mm}$$

Try:

$$h = 150 \text{ mm}$$

$$e = 20 < (150/6) = 25 \text{ mm}$$

Therefore the empirical design method can be used.
Effective length of bearing:

$$b' + 4h = 200 + 4(150)$$

$$= 800 \text{ mm} < 1500 \text{ mm} \quad \text{O.K.}$$

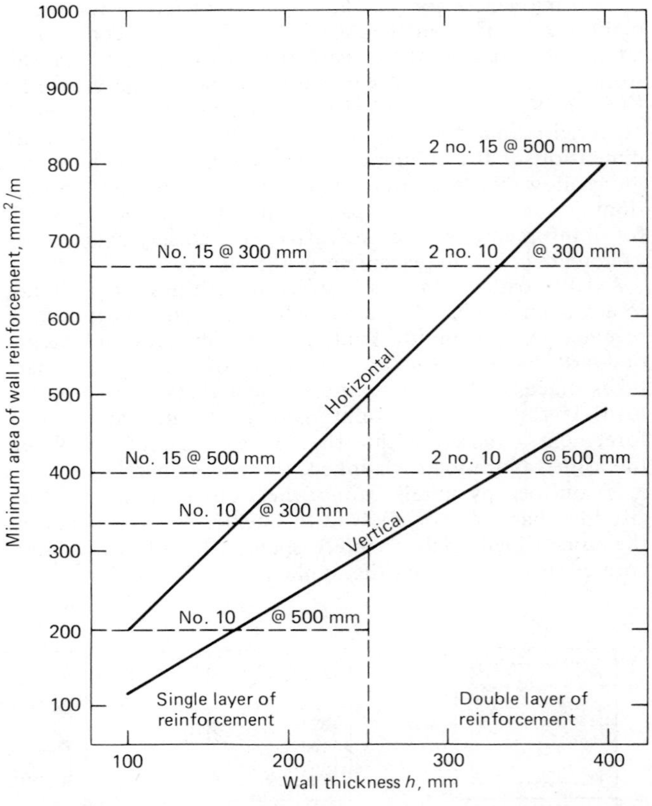

Fig. 1-48 Minimum wall reinforcement for $f_y \geqslant 400$ MPa and bars less than No. 20.

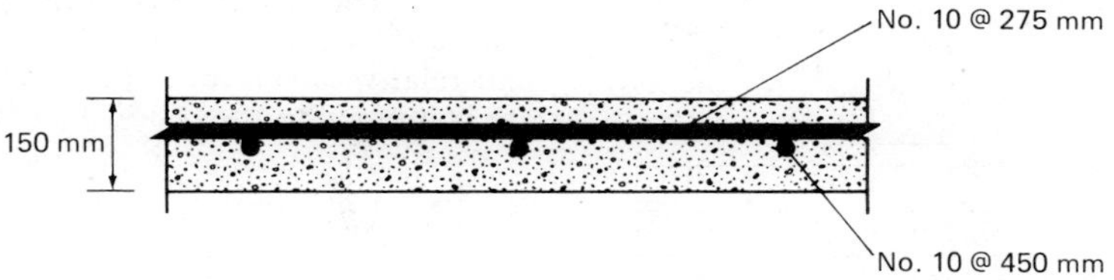

Fig. 1-49 Example 1-11.

Wall capacity:

$$\phi P_{nw} = 0.55\phi f'_c A_g \left[1 - \left(\frac{kl_c}{32h}\right)^2\right]$$

$k = 0.8$ (restrained against lateral translation, and rotation at one end)

$$\phi P_{nw} = 0.55(0.7)(30)(800)(150)\left[1 - \left(\frac{(0.8)(3750)}{(32)(150)}\right)^2\right]$$

$$\phi P_{nw} = 844590 \text{ N}$$

$$\phi P_{nw} = 845 \text{ kN} > 800 \text{ kN} \qquad \text{O.K.}$$

Use 150 mm wall.

Vertical steel:

$$0.0015A_g = 0.0015(150)(1000) = 225 \text{ mm}^2/\text{m}$$

Use No. 10 @ 450 mm.

Horizontal steel:

$$0.0025A_g = 0.0025(150)(1000) = 375 \text{ mm}^2/\text{m}$$

Use No. 10 @ 275 mm.

Bearing walls can also be designed using the strength method provisions for columns under combined bending and axial force. This procedure is usually followed for bearing walls that do not meet the limitations of the empirical design method. In such cases, all the provisions for column design, including the slenderness effects, should be considered along with the special design provisions for walls.

1.6.6 Interaction Diagrams for Structural Walls

Structural walls (or shear walls) are used primarily to stiffen multistory structures against lateral loads. In areas that are not affected by strong earthquakes, flexural strength and ductility requirements in walls are moderate. In these regions it has been a common practice to design walls with about 0.25% reinforcement, uniformly distributed in both horizontal and vertical directions.

As the strength requirement for walls resisting overturning moments increases, more area of steel is required at the ends of a wall section. Reinforcement near the neutral axis remains ineffective in resisting tensile stresses due to overturning. Furthermore, deformation characteristics of the two types of reinforcement arrangements, namely, uniform steel and steel concentrated at the ends, reveal that the latter shows a more ductile behavior. This is a characteristic much desired in a seismic-resistant wall. Recognizing these facts, the *ACI Code* recommends special vertical reinforcement concentrated near the ends of seismic-resistant structural walls.

Determination of the flexural capacity of a wall section with different reinforcement arrangements and different percentages of steel can be rather cumbersome. Therefore, interaction diagrams have been developed for wall sections, and are presented in Figs. 1-50 and 1-51. Figure 1-50 gives the moment axial force interaction diagrams for walls reinforced with uniformly distributed vertical steel. The effective depth, d, is assumed to be equal to $0.9h$.

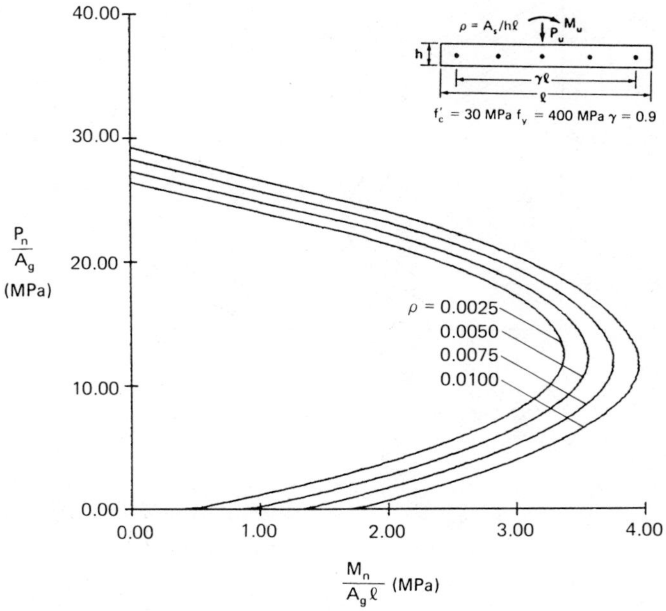

Fig. 1-50 Interaction diagrams for rectangular walls with uniformly distributed reinforcement.

Figures 1-51(a) and 1-51(b) give the interaction diagrams for ductile structural walls that include concentrated end reinforcement in addition to 0.25% of web reinforcement distributed between the end regions. Generally, each of these end regions (sometimes known as boundary elements) constitutes 10% of the cross-sectional area. In some cases, depending on the size of the section, these end regions can be greater than 10% of the cross section. Therefore, two sets of charts, one with 10% end region and the other with 20% end region, are provided. In developing these diagrams, all the end reinforcement is assumed to be concentrated at the centroid of the end region, and the web steel is assumed to be uniformly distributed in ten layers across the width of the web. Rectangular stress block is used with due consideration to the area of concrete replaced by the compression steel.

1.7 PROPORTIONING SECTIONS FOR SHEAR AND TORSION

Shear capacity of concrete members depends on the tensile strength of concrete. Members that are subjected to shear and/or torsion may develop diagonal cracks that can result in brittle shear failures. Shear reinforcement is provided in most concrete members to avoid such sudden failures. It is desirable to provide enough shear capacity in a member so that a sudden shear failure does not occur prior to a ductile failure in flexure.

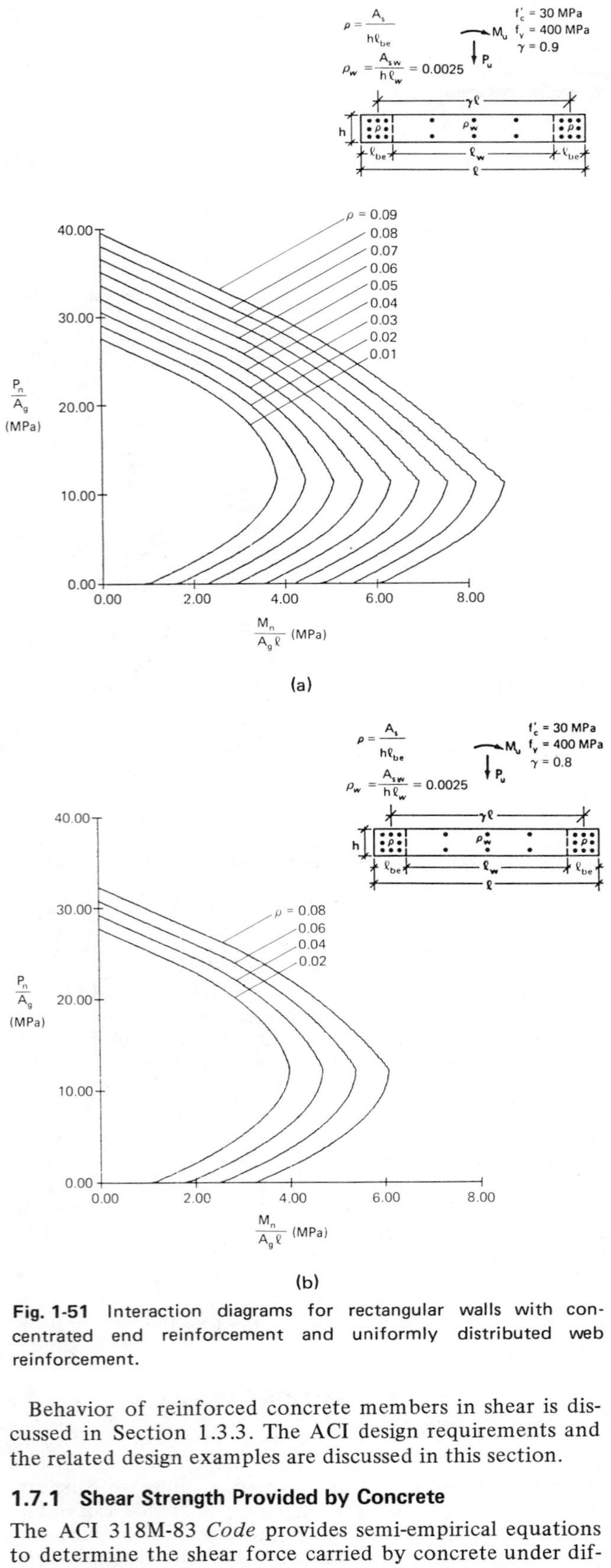

Fig. 1-51 Interaction diagrams for rectangular walls with concentrated end reinforcement and uniformly distributed web reinforcement.

Behavior of reinforced concrete members in shear is discussed in Section 1.3.3. The ACI design requirements and the related design examples are discussed in this section.

1.7.1 Shear Strength Provided by Concrete

The ACI 318M-83 *Code* provides semi-empirical equations to determine the shear force carried by concrete under different combinations of stresses. These equations are based on the results of tests and can predict conservatively the shear resistance of concrete.

Shear resistance of concrete in flexural members depends on a number of variables. Primary variables are the concrete tensile strength, crack width as controlled by flexural steel, and relative magnitudes of shear and flexure. These factors are combined in the following expression:

$$V_c = \left(\frac{1}{7}\sqrt{f_c'} + 17.0\rho_w \frac{V_u d}{M_u}\right) b_w d \leqslant 0.3\sqrt{f_c'}\, b_w d \tag{1-81}$$

where f_c' is in MPa and all other quantities are in N and mm. The ratio $(V_u d/M_u)$ in eq. (1-81) should not be taken greater than 1.0 at any section. Often the contribution of the second term in the equation is small and the following simpler expression produces satisfactory designs:

$$V_c = \tfrac{1}{6}\sqrt{f_c'}\, b_w d \tag{1-82}$$

For members subjected to axial compression, eq. (1-81) can be used with the moment M_u replaced by an equivalent moment M_m which accounts for the axial force N_u. The moment M_m can be expressed approximately as:

$$M_m = M_u - N_u \frac{(4h - d)}{8} \tag{1-83}$$

If M_m is computed to be negative, the following equation for V_c is applicable:

$$V_c = 0.3\sqrt{f_c'}\, b_w d \sqrt{1 + 0.3N_u/A_g} \tag{1-84}$$

If M_m is *not* negative, eq. (1-81) is used, but V_c need not be greater than the value found by using eq. (1-84). Furthermore, when eq. (1-81) is used for members under axial compression, the ratio $(V_u d/M_m)$ is *not* limited to 1.0.

Because of the apparent complexities involved in applying the above equations, ACI 318M-83 permits the use of the following simplified but more conservative equation for members subjected to axial compression:

$$V_c = \frac{1}{6}\left(1 + \frac{1}{14}\frac{N_u}{A_g}\right)\sqrt{f_c'}\, b_w d \tag{1-85}$$

The equations derived for axial compression have been used on members subjected to axial tension with negative N_u for tension. The results were satisfactory in most of the cases. However, some test results showed lower diagonal cracking loads than predicted with the use of the above equations. For this reason ACI 318M-83 adopted a more conservative approach for members subjected to axial tension. Accordingly, a linear interpolation is to be made between eq. (1-82) when there is no applied tension and $V_c = 0$ when axial tension is 3.33 MPa. This leads to the following expression:

$$V_c = \frac{1}{6}\left(1 + 0.3\frac{N_u}{A_g}\right)\sqrt{f_c'}\, b_w d \tag{1-86}$$

where N_u is negative for tension. The *ACI Code* also provides for a more conservative option of neglecting the shear resistance of concrete for members subjected to axial tension. In this case, the shear reinforcement is to be provided such that the total applied design shear force will be resisted only by the reinforcement.

Shear capacity of concrete for members subjected to combined shear and torsion is derived in Section 1.3.4 and is expressed by eq. (1-37). Equation (1-37) is derived for a concrete section without the web reinforcement and hence represents the ultimate capacity of the section. For a reinforced concrete section with web reinforcement, this equation provides the shear resistance contributed by the concrete:

$$V_c = \frac{\frac{1}{6}\sqrt{f_c'}\, b_w d}{\sqrt{1 + \left(2.5C_t \frac{T_u}{V_u}\right)^2}} \tag{1-87}$$

The expressions given in the foregoing discussion apply to normal density concrete. The shear capacity of low density concrete is somewhat lower. Therefore, the shear capacity computed on the basis of normal density concrete should be adjusted to obtain the capacity for low density concrete. According to the *ACI Code*, if the average splitting tensile strength of low density concrete, f_{ct}, is known (in MPa), the value of $1.8 f_{ct}$ should be substituted for $\sqrt{f_c'}$ in the above expressions, provided that this value does not exceed $\sqrt{f_c'}$. If f_{ct} is not specified, then $\sqrt{f_c'}$ shall be multiplied by 0.75 for all-low density concrete, and 0.85 for sand-low density concrete.

1.7.2 Shear Strength Provided by Shear Reinforcement

When the required shear strength, V_u, cannot be provided by the concrete alone, web reinforcement is provided for additional shear resistance. The web reinforcement can be in the form of bent longitudinal bars, stirrups, or a combination of the two. The *ACI Code* requires that the bent bars should make an angle of at least 30° with longitudinal bars and the stirrups an angle of at least 45° to provide resistance to diagonal tensile stresses.

Total nominal shear strength of a section consists of the concrete and the steel contributions:

$$V_n = V_c + V_s \tag{1-88}$$

The nominal shear resisted by concrete, V_c, and the associated design equations are discussed in Section 1.7.1. The remaining shear resistance, $V_s = V_n - V_c$, can be computed as shown below:

For vertical stirrups:

$$V_s = A_v f_y \frac{d}{s} \tag{1-89}$$

For inclined bars:

$$V_s = A_v f_y \frac{d}{s} (\sin \alpha + \cos \alpha) \tag{1-90}$$

where A_v is the area of shear reinforcement provided within distance s. In the case of conventional U stirrups, A_v is twice the area of one leg of stirrup. For inclined bars, A_v is the area of inclined bar measured perpendicular to the line of inclination. Angle α is the angle of inclination with respect to the beam axis. Derivations of eqs. (1-89) and (1-90) are presented in Section 1.3.3. The same equations are used to determine the spacing of shear reinforcement for a given bar size. One should be careful, however, to avoid widely spaced shear reinforcement. If the distance between two adjacent reinforcing bars is excessive, the concrete in between may not be able to transfer the shear. ACI 318M-83 sets a maximum spacing limit of $d/2$ for vertical stirrups. Inclined stirrups and bent longitudinal bars are required to be spaced such that at least one bar crosses every line drawn between the middepth of the section ($d/2$) and the longitudinal tension reinforcement with 45° inclination. This is shown in Fig. 1-52. Furthermore, in the case of inclined longitudinal bars, only the center three-fourths of the inclined portion is assumed to be effective. These provisions result in maximum reinforcement spacings of $0.5\, d\, (1 + \cot \alpha)$ and $0.375\, d\, (1 + \cot \alpha)$ for inclined stirrups and inclined longitudinal bars, respectively. If the shear resistance of reinforcement, V_s, exceeds $\frac{1}{3} \sqrt{f_c'}\, b_w d$, the specified maximum spacing limits are reduced by one-half.

In addition to the spacing limitations, the *Code* provides maximum and minimum limits to the area of shear reinforcement. If too much shear reinforcement is provided, concrete crushing can occur prior to reinforcement yielding. For this reason, the contribution of reinforcing steel to shear resistance, V_s, is limited to $\frac{2}{3} \sqrt{f_c'}\, b_w d$. If the design calls for a higher V_s than the specified limit, a larger concrete area requirement is indicated for increased contribution of concrete shear resistance.

It is good practice to provide a minimum area of shear reinforcement even if the computed concrete capacity appears to be sufficient. This avoids a possible brittle failure that may follow formation of premature diagonal cracking. If V_u exceeds $\frac{1}{2} \phi V_c$, ACI 318M-83 requires the following minimum shear reinforcement:

$$A_v = \frac{1}{3} \frac{b_w s}{f_y} \tag{1-91}$$

where b_w and s are in mm and f_y is in MPa.

Yield strength of shear reinforcement is also limited by the *Code*. Yield strength of 400 MPa is specified by ACI 318M-83 to control crack widths for continued availability of aggregate interlock.

EXAMPLE 1-12: The rectangular beam shown in Fig. 1-53 is subjected to factored design shear force and moment of 400 kN and

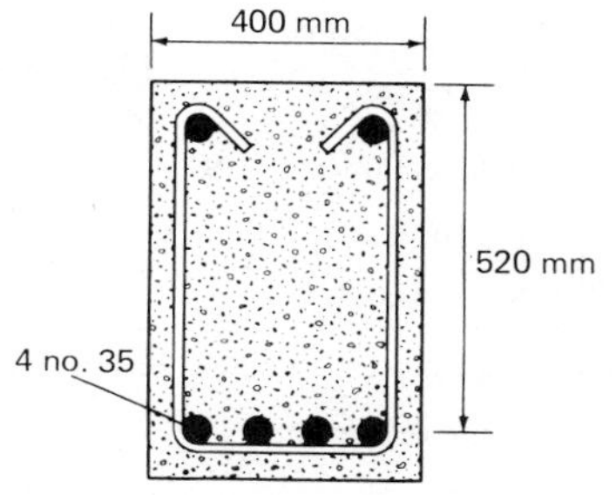

Fig. 1-53 Example 1-12.

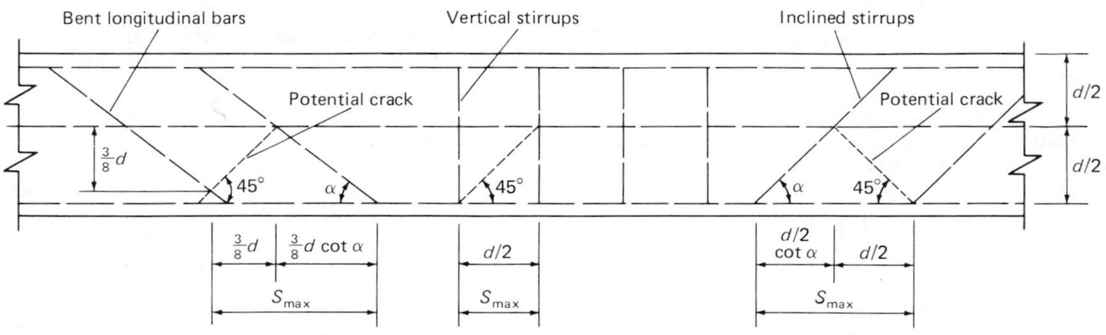

Fig. 1-52 Maximum spacing of shear reinforcement for bent longitudinal bars, vertical stirrups, and inclined stirrups.

200 kN · m, respectively. If the material properties are $f'_c = 30$ MPa and $f_y = 400$ MPa, determine the required spacing of vertical U-stirrups.

Shear resistance provided by concrete:

i. Using the simplified expression given by eq. (1-82):

$$V_c = \frac{1}{6}\sqrt{f'_c}\, b_w d = \frac{1}{6}\sqrt{30}\,(400)(520) = 190{,}000 \text{ N}$$

$$V_c = 190 \text{ kN}$$

ii. Using eq. (1-81);

$$V_c = \left(\frac{1}{7}\sqrt{f'_c} + 17.0\rho_w \frac{V_u d}{M_u}\right) b_w d \leq 0.3\sqrt{f'_c}\, b_w d$$

$$\rho_w = \frac{A_s}{b_w d} = \frac{4 \times 1000}{(400)(520)} = 0.0192$$

$$\frac{V_u d}{M_u} = \frac{(400)(0.520)}{(200)} = 1.04 > 1.00 \quad \text{Therefore use 1.0.}$$

$$V_c = \left[\tfrac{1}{7}\sqrt{30} + 17.0(0.0192)(1.0)\right](400)(520)$$

$$V_c = 230{,}000 \text{ N} < 0.3\sqrt{30}\,(400)(520) = 342{,}000 \text{ N}$$

$$V_c = 230 \text{ kN}$$

Use $V_c = 230$ kN and determine if shear reinforcement is required:

$$V_u = 400 \text{ kN} > \phi V_c = 0.85(230) = 195 \text{ kN}$$

Therefore, shear reinforcement is required.

$$V_s = V_n - V_c = \frac{V_u}{\phi} - V_c$$

$$V_s = \frac{400}{0.85} - 230 = 240 \text{ kN}$$

$$(V_s)_{max} = \tfrac{2}{3}\sqrt{f'_c}\, b_w d = 759{,}500 \text{ N} = 759 \text{ kN} > 240 \text{ kN} \quad \text{O.K.}$$

Use No. 10 stirrups and determine the spacing. Equation (1-89) can be rewritten as:

$$s = \frac{A_v f_y d}{V_s}$$

$$s = \frac{2(100)(400)(520)}{240 \times 10^3} = 173 \text{ mm}$$

Check minimum reinforcement (eq. 1-91):

$$(A_v)_{min} = \frac{1}{3}\,\frac{(400)(173)}{400} = 57.67 \text{ mm}^2 < 200 \text{ mm}^2$$

More than minimum shear reinforcement has been provided.

Check maximum spacing requirement:

$$V_s = 240 \times 10^3 \text{ N} < \tfrac{1}{3}\sqrt{30}\,(400)(520) = 380 \times 10^3 \text{ N}$$

Therefore:

$$s_{max} = d/2 = 520/2 = 260 \text{ mm}$$

$$s = 173 \text{ mm} < s_{max} = 260 \text{ mm} \quad \text{O.K.}$$

Use $s = 170$ mm.

EXAMPLE 1-13: Find the shear reinforcement arrangement required in a continuous beam subjected to shear force distribution shown in Fig. 1-54. Beam cross-sectional dimensions and the material properties are given below:

$$b_w = 300 \text{ mm} \qquad f'_c = 30 \text{ MPa}$$

$$d = 450 \text{ mm} \qquad f_y = 300 \text{ MPa}$$

Shear resistance provided by concrete, using eq. (1-82):

$$V_c = \tfrac{1}{6}\sqrt{f'_c}\, b_w d = \tfrac{1}{6}\sqrt{30}\,(300)(450) = 123 \times 10^3 \text{ N}$$

$$V_c = 123 \text{ kN}$$

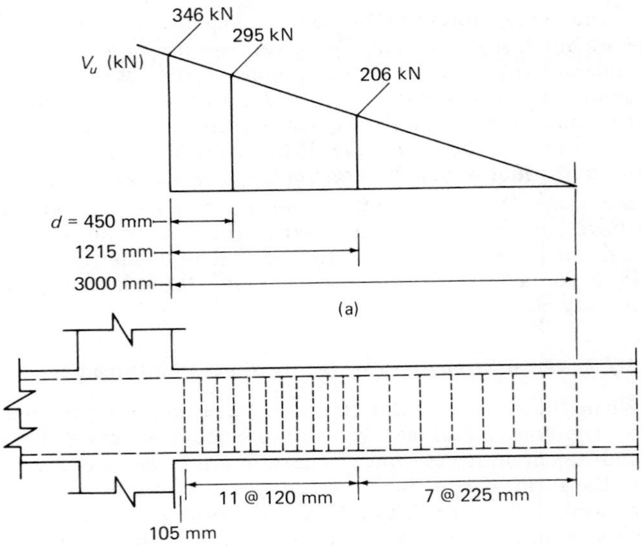

Fig. 1-54 Example 1-13.

Factored design shear force $V_u = 295$ kN at d distance from column face:

$$V_u = \phi V_n$$

$$V_n = 295/0.85 = 347 \text{ kN} > V_c = 123 \text{ kN}$$

Therefore, shear reinforcement is required.

$$V_n = V_c + V_s$$

$$V_s = 347 - 123 = 224 \text{ kN}$$

$$(V_s)_{max} = \tfrac{2}{3}\sqrt{f'_c}\, b_w d = 493 \times 10^3 \text{ N}$$

$$(V_s)_{max} = 493 \text{ kN} > 224 \text{ kN} \quad \text{O.K.}$$

Assuming No. 10 U-stirrups are used, determine the required spacing from eq. (1-89):

$$s = \frac{A_v f_y d}{V_s} = \frac{2(100)(300)(450)}{224 \times 10^3} = 120 \text{ mm}$$

Check minimum steel requirement:

$$(A_v)_{min} = \frac{1}{3}\,\frac{b_w s}{f_y} = \frac{1}{3}\,\frac{(300)(120)}{300} = 40 \text{ mm}^2 < 200 \quad \text{O.K.}$$

Check maximum spacing requirement:

$$\tfrac{1}{3}\sqrt{f'_c}\, b_w d = 246 \times 10^3 \text{ N} > V_s = 224 \times 10^3 \text{ N}$$

Therefore:

$$s_{max} = d/2 = 450/2 = 225 \text{ mm}$$

$$s = 120 \text{ mm} < s_{max} = 225 \text{ mm} \quad \text{O.K.}$$

Find the distance to the section that requires minimum stirrup steel:

$$s_{max} = 225 \text{ mm}$$

$$V_s = \frac{A_v f_y d}{s} = \frac{2(100)(300)(450)}{(225)} = 120 \times 10^3 \text{ N}$$

$$V_s = 120 \text{ kN}$$

$$V_n = V_c + V_s = 123 + 120 = 243 \text{ kN}$$

$$V_u = \phi V_n = 0.85(243) = 206 \text{ kN}$$

The section that is subjected to $V_u = 206$ kN is located at 1215 mm from the column face. Therefore, use No. 10 U-stirrups at 120 mm within the critical region and use No. 10 stirrups at 225 mm between the critical regions as shown in Fig. 1-54(b).

1.7.3 Design for Torsion

Design strength of reinforced concrete members subjected to torsion is based on the nominal torsional moment strengths of concrete, T_c, and reinforcement, T_s. ACI 318M-83 requires that torsion be considered in design if factored torsional moment, T_u, exceeds $0.04 \sqrt{f'_c}\ \Sigma x^2 y$. The *Code* requires that the combination of torsion with flexure and shear be considered. Torsional resistance of concrete and reinforcement are expressed using the equations based on the combined effects of shear flexure and torsion.

The expression for concrete contribution to torsion is derived in Section 1.3.4, and is given below:

$$T_c = \frac{\frac{1}{15}\sqrt{f'_c}\sum x^2 y}{\sqrt{1 + \left(\frac{0.4 V_u}{C_t T_u}\right)^2}} \qquad (1\text{-}92)$$

where:

$$C_t = \frac{b_w d}{\sum x^2 y}$$

x, y = shorter and longer dimensions of rectangular part of cross section

Equation (1-92) includes the effects of shear and flexure on torsional resistance. If the shear force is zero, the equation reduces to $T_c = 1/15 \sqrt{f'_c}\ \Sigma x^2 y$, which is the same as eq. (1-32) and gives concrete resistance to torsion in the presence of flexure.

For members subject to significant axial tension, N_u, torsion reinforcement is designed to carry the total torsional moment. If, however, a detailed calculation is made, the *Code* allows the use of eqs. (1-92) and (1-87), provided the values of T_c and V_c are multiplied by $(1 + 0.3N_u/A_g)$, where N_u/A_g is negative for tension.

When the required torsional strength at the factored load level is more than that resisted by concrete, torsion reinforcement is provided. The excess torque, T_s, to be resisted by reinforcement is computed using the following equation:

$$T_s = \frac{T_u}{\phi} - T_c \qquad (1\text{-}93)$$

where:

$$T_s = \frac{A_t \alpha_t x_1 y_1 f_y}{s} \qquad (1\text{-}94)$$

A_t in eq. (1-94) is the area of one leg of a closed stirrup within a distance s, and x_1 and y_1 are the shorter and longer center-to-center dimensions of closed rectangular stirrups. The factor α_t is specified by the *ACI Code* as being equal to $1/3\ (2 + y_1/x_1) \leqslant 1.5$.

The *Code* specifies a minimum area of closed stirrups if the factored torsional moment T_u exceeds $\phi(0.04 \sqrt{f'_c}\ \Sigma x^2 y)$. For members that require both shear and torsion reinforcement, the minimum area of closed stirrups is determined using the following expression:

$$(A_v + 2A_t) = \frac{1}{3}\frac{b_w s}{f_y} \qquad (1\text{-}95)$$

Torsion reinforcement, as in the case of shear reinforcement, should be placed within certain spacing limitations to be effective. The maximum allowable stirrup spacing specified by the *Code* is given below:

$$s_{max} = (x_1 + y_1)/4 < 300 \text{ mm} \qquad (1\text{-}96)$$

Once the area of stirrup reinforcement is determined, the required area of longitudinal steel for torsional resistance can be computed. ACI 318M-83 provides the following two equations for the area of longitudinal steel, A_l, to be placed around the perimeter of the closed stirrups:

$$A_l = 2A_t\left(\frac{x_1 + y_1}{s}\right) \qquad (1\text{-}97)$$

$$A_l = \left[\frac{(2.75)\,xs}{f_y}\left(\frac{T_u}{T_u + \frac{V_u}{3C_t}}\right) - 2A_t\right]\left(\frac{x_1 + y_1}{s}\right) \qquad (1\text{-}98)$$

where:

$$C_t = \frac{b_w d}{\sum x^2 y}$$

The required area of longitudinal steel, A_l, will be taken as the greater value obtained from eqs. (1-97) and (1-98). The value of A_l computed using eq. (1-98) need not be greater than the value obtained using the same equation after substituting $b_w s/3f_y$ for $2A_t$. Longitudinal bars consisting of at least No. 10 bars should be placed such that the spacing does not exceed 300 mm. At least one longitudinal bar is placed in each corner of the closed stirrups.

The amount of torsion reinforcement is limited by the *Code* for reasons of ductility. Accordingly, the maximum torsion reinforcement is one that produces torsional resistance of reinforcement equal to four times the concrete resistance ($T_s \leqslant 4T_c$). Furthermore, the reinforcement yield strength is limited to 400 MPa, as in the case of the shear reinforcement.

EXAMPLE 1-15: A 150-mm concrete slab cantilevers 2.0 meters from the centerline of a beam. The slab is designed for 2.40 kN/m^2 live load. The beam cross section is 400 mm by 600 mm as shown in Fig. 1-55, and the clear span is 10 meters. No. 25 bars are used with 35 mm net cover over stirrups as flexural tension reinforcement. Determine the required shear and torsion reinforcement at the critical section. Material properties are $f'_c = 30$ MPa, $f_y = 400$ MPa, and $W_c = 2400$ kg/m^3.

Design loads:

Slab:

$$1.4D = 1.4(2400 \times 9.81 \times 10^{-3} \times 1.8 \times 0.15) = 8.90 \text{ kN/m}$$

$$1.7L = 1.7(2.40 \times 1.8) = 7.34 \text{ kN/m}$$

Beam:

$$1.4D = 1.4(2400 \times 9.81 \times 10^{-3} \times 0.4 \times 0.6) = 7.91 \text{ kN/m}$$

$$1.7L = 1.7(2.40 \times 0.4) = 1.63 \text{ kN/m}$$

$$W_{slab} = 8.90 + 7.34 = 16.24 \text{ kN/m}$$

$$W_{beam} = 7.91 + 1.63 = 9.54 \text{ kN/m}$$

The spandrel beam is subjected to uniformly distributed factored load of 16.24 + 9.54 = 25.78 kN/m and uniformly distributed torsional moment of 16.24 × 1.1 = 17.86 kN · m/m.

Distribution of shear and torsion along the beam span is shown in Fig. 1-55 ($V_u = 25.78 \times 5.0 = 129$ kN and $T_u = 17.86 \times 5.0 = 89$ kN · m at the column face). At a distance $d = 542$ mm from the column face, $V_u = 115$ kN and $T_u = 80$ kN · m. Check whether torsion may be neglected in design.

$$0.04\sqrt{f'_c}\sum x^2 y = 0.04\sqrt{30}\,(400^2 \times 600 + 150^2 \times 450) \times 10^{-6} = 23.25 \text{ kN}\cdot\text{m}$$

$$T_u = 80 \text{ kN}\cdot\text{m} > 23.25 \text{ kN}\cdot\text{m}$$

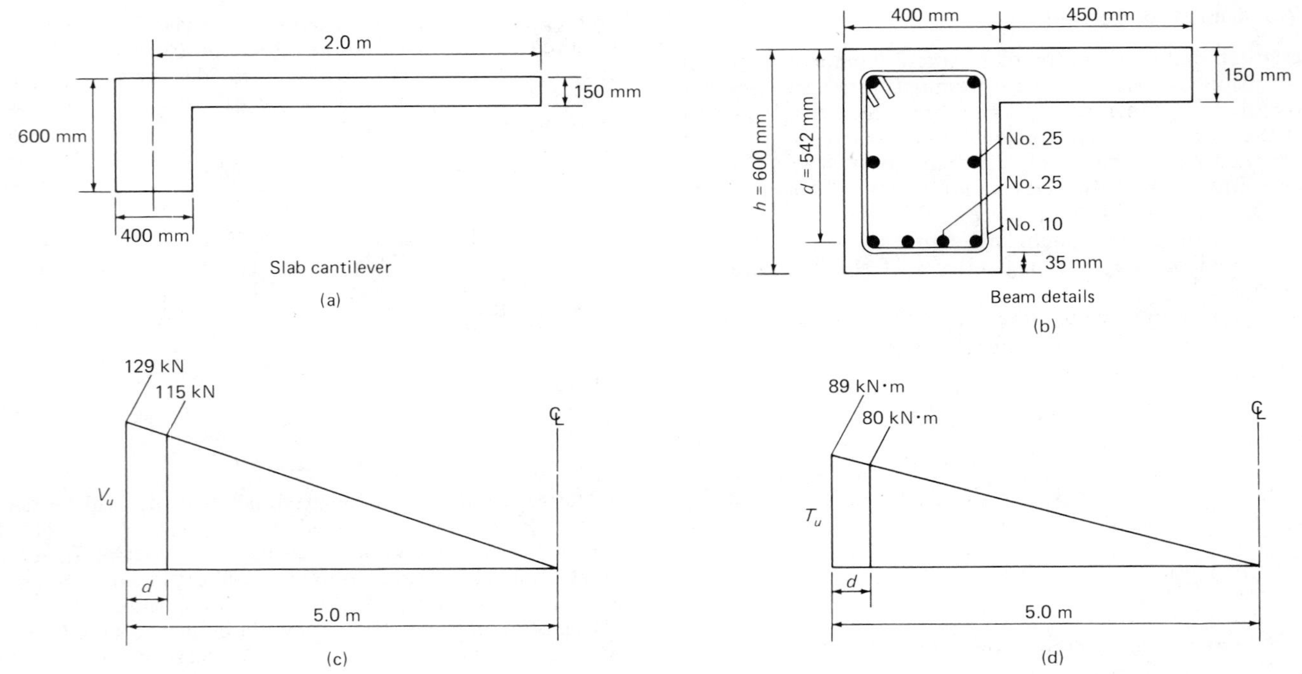

Fig. 1-55 Example 1-15.

Therefore torsion can *not* be neglected. From eqs. (1-87) and (1-92), the concrete contribution to shear resistance and concrete contribution to torsion are:

$$V_c = \frac{\frac{1}{6}\sqrt{f'_c}\, b_w d}{\sqrt{1 + \left(2.5C_t \dfrac{T_u}{V_u}\right)^2}}$$

$$C_t = \frac{b_w d}{\sum x^2 y} = \frac{400 \times 542}{400^2 \times 600 + 150^2 \times 450} = 0.0020$$

$$V_c = \frac{\frac{1}{6}\sqrt{30}\,(400)(542)}{\sqrt{1 + \left(2.5 \times 0.002\, \dfrac{80 \times 10^3}{115}\right)^2}} = 55 \times 10^3 \text{ N}$$

$$V_c = 55 \text{ kN}$$

$$T_c = \frac{\frac{1}{15}\sqrt{f'_c} \sum x^2 y}{\sqrt{1 + \left(\dfrac{0.4 V_u}{C_t T_u}\right)^2}}$$

$$T_c = \frac{\frac{1}{15}\sqrt{30}\,[(400)^2 (600) + (150)^2 (450)]}{\sqrt{1 + \left[\dfrac{(0.4)(115)}{(0.002)(80)\,10^3}\right]^2}} = 37 \times 10^6 \text{ N} \cdot \text{mm}$$

$$T_c = 37 \text{ kN} \cdot \text{m}$$

Area of shear reinforcement for transverse shear:

$$V_s = \frac{V_u}{\phi} - V_c$$

$$V_s = \frac{115}{0.85} - 55 = 80 \text{ kN}$$

Check for maximum shear to be resisted by shear reinforcement:

$$(V_s)_{\max} = \tfrac{2}{3}\sqrt{f'_c}\, b_w d = 792 \text{ kN} > 80 \text{ kN} \qquad \text{O.K.}$$

$$A_v = \frac{V_s s}{f_y d} = \frac{80{,}000 s}{(400)(542)} = 0.369s$$

Area of one leg of closed stirrup for torsion:

$$T_s = \frac{T_u}{\phi} - T_c$$

$$T_s = \frac{80}{0.85} - 37 = 57 \text{ kN}$$

Check for maximum torsion that can be resisted by torsion reinforcement:

$$(T_s)_{\max} = 4T_c = 148 \text{ kN} > 57 \text{ kN} \qquad \text{O.K.}$$

$$A_c = \frac{T_s s}{\alpha_t f_y x_1 y_1}$$

If No. 10 bars are used for stirrups, $x_1 = 400 - 70 - 10 = 320$ mm and $y_1 = 600 - 70 - 10 = 520$ mm, and using eq. (1-94):

$$\alpha_t = \frac{1}{3}\left[2 + \frac{520}{320}\right] = 1.21 < 1.50$$

$$A_t = \frac{(57 \times 10^6)\, s}{(1.21)(400)(320)(520)} = 0.708s$$

Total area of steel for two legs of closed stirrups required for combined shear and torsion:

$$2A_t + A_v = 2 \times 0.708s + 0.369s = 1.785s$$

Area of two legs of a No. 10 stirrup = 2(100) = 200 mm^2.

$$s = \frac{200}{1.785} = 112 \text{ mm}$$

Check for maximum stirrup spacing for torsion reinforcement:

$$s_{\max} = \frac{x_1 + y_1}{4} = \frac{320 + 520}{4} = 210 \text{ mm} < 300 \text{ mm}$$

$$s = 112 \text{ mm} < s_{\max} = 210 \text{ mm} \qquad \text{O.K.}$$

Check for maximum stirrup spacing for shear reinforcement:

$$s_{max} = d/2 = 542/2 = 271 \text{ mm}$$

$$s = 112 \text{ mm} < s_{max} = 271 \text{ mm} \quad \text{O.K.}$$

Minimum area of steel required by the *Code* (eq. 1-95):

$$\frac{1}{3}\frac{b_w s}{f_y} = \frac{1}{3}\frac{(400)(112)}{400} = 37 \text{ mm}^2 < 1.785(112) = 200 \text{ mm}^2$$

The required closed vertical stirrups at the critical section consists of No. 10 stirrups at 112 mm spacing. Longitudinal steel required at the critical section using eqs. (1-97) and (1-98) is:

$$A_{l_1} = 2.0(100)\,\frac{320 + 520}{112} = 1500 \text{ mm}^2$$

$$A_{l_2} = \left[\frac{(2.75)(400)(112)}{400}\left(\frac{80 \times 10^6}{80 \times 10^6 + \dfrac{115 \times 10^3}{3(0.002)}}\right) - 2(100)\right] \cdot \left(\frac{320 + 520}{112}\right)$$

$$A_{l_2} = 364 \text{ mm}^2 < 1500 \text{ mm}^2$$

Therefore, the required area of longitudinal reinforcement is 1500 mm^2. Since the maximum spacing is 300 mm, and there should be at least one bar at each corner of a closed stirrup, six bars are required, two at the middepth, two at the top, and two at the bottom. If two No. 25 bars are used at the middepth, the flexural reinforcement at each corner should be increased by $[1500 - 2(500)]/4 = 125 \text{ mm}^2$.

1.8 REINFORCEMENT PROPERTIES AND DESIGN

Reinforcing steel is classified on the basis of its minimum specified yield strength for structural purposes. The specified yield strength is generally referred to as the grade of steel. Steel grade represents a minimum guaranteed yield strength (in MPa). Grade 400 is commonly used in reinforced concrete construction.

The stress–strain relationships of three steel grades are shown in Fig. 1-56. Each curve in the figure shows a linear elastic portion followed by a flat yield plateau and a strain-hardening range. For bars lacking a well-defined yield point, the stress corresponding to a strain of 0.002 can be used as the yield strength. The modulus of elasticity for steel is the slope of the elastic portion of the stress–strain curve and is typically equal to 200,000 MPa.

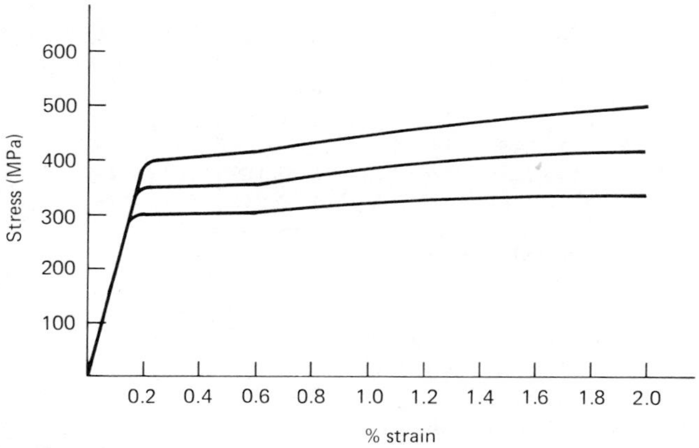

Fig. 1-56 Stress–strain relationship of reinforcing steel bars, grades 300, 350, and 400 MPa.

TABLE 1-10 Metric Reinforcing Steel Bars

	Nominal Dimensions			
Bar No.	*Cross-Sectional Areas* (mm)^2	*Diameter* (mm)	*Perimeter* (mm)	*Unit Mass* (kg/m)
10	100	11.3	35.5	0.785
15	200	16.0	50.1	1.570
20	300	19.5	61.3	2.355
25	500	25.2	79.2	3.925
30	700	29.9	93.9	5.495
35	1000	35.7	112.2	7.850
45	1500	43.7	137.3	11.775
55	2500	56.4	177.2	19.625

Eight different sizes of metric reinforcing bars are produced for use in reinforced concrete design. Each bar size is designated by its approximate bar diameter in millimeters. For example, a No. 20 bar means a steel bar of approximately 20 mm diameter. Table 1-10 gives the properties of metric reinforcing steel bars. A comparison of imperial and metric bars is presented in Table 1-11. In this chapter, metric bars consistent with the SI system of units are used exclusively.

For reinforcement to be effective in concrete, there must be a sufficient bond between the two materials. The force in reinforcement must be transmitted to the concrete by bond stresses. Deformed bars are produced on the basis of ASTM specifications for improved bond characteristics. Furthermore, reinforcing bars must be continued beyond critical sections to develop sufficient bond stresses through which forces are transmitted into the reinforcing bars. The requirements of development length and reinforcement anchorage are expressed in the ACI 318M-83 *Code*.

1.8.1 Spacing of Reinforcement

Placement of reinforcement is an important component of reinforced concrete design. The required area of steel in concrete members must be provided so that the spacing re-

TABLE 1-11 Comparison of Imperial and Metric Sizes (Ref. 1.1)

Imperial Bar			*Metric Bar*			
Size	*Area* (in.^{2})	*Area* (mm^2)	*Size*	*Area* (in.^{2})	*Area* (mm^2)	*Metric Bar Is**
# 3	0.11	71	10	0.16	100	45% L
# 4	0.20	129	10	0.16	100	20% S
# 4	0.20	129	15	0.31	200	55% L
# 5	0.31	200	15	0.31	200	Same
# 6	0.44	284	20	0.47	300	6.8% L
# 7	0.60	387	20	0.47	300	22% S
# 7	0.60	387	25	0.78	500	30% L
# 8	0.79	510	25	0.78	500	1.3% S
# 9	1.00	645	30	1.09	700	9% L
#10	1.27	819	30	1.09	700	14% S
#10	1.27	819	35	1.55	1000	22% L
#11	1.56	1006	35	1.55	1000	0.6% S
#14	2.25	1452	45	2.33	1500	3.5% L
#18	4.00	2581	55	3.88	2500	3.0% S

**Percent difference is based on cross-sectional area.*
L = *larger.*
S = *smaller.*

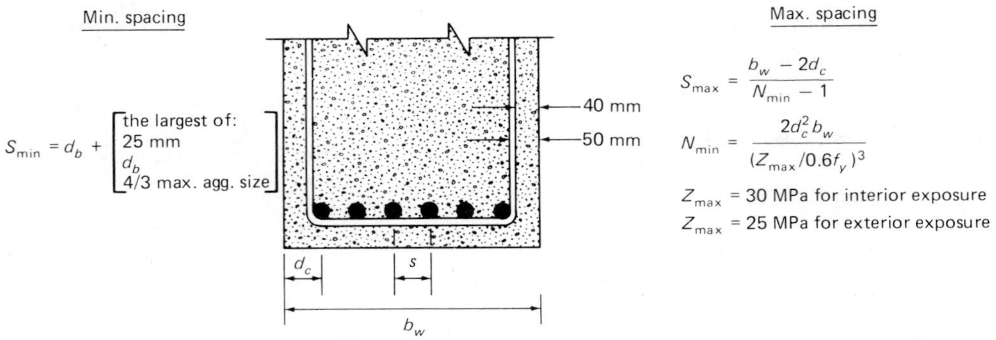

Fig. 1-57 Spacing and cover requirements for beams.

quirements are satisfied. While densely reinforced members are subject to concrete placement problems, too widely spaced bars cannot provide sufficient crack control. Spacing of web reinforcement for shear and torsion is discussed in Sections 1.7.2 and 1.7.3. Spacing of lateral reinforcement for columns is discussed in Section 1.6.4. Spacing and cover requirements for longitudinal reinforcement in beams and columns are illustrated in Figs. 1-57 and 1-58.

1.8.2 Development of Deformed Bars

Design forces in reinforcing bars must be developed on each side of every critical section in a reinforced concrete member. This can be achieved by providing sufficient length, or end anchorage, or a combination of the two.

Tension reinforcement in flexural members can be developed by bending across the web either to be anchored or made continuous along the opposite face. If a reinforcement is to be terminated, it should be extended by a distance equal to the effective depth d or 12 times the bar diameter, whichever is greater, beyond the point at which it is no longer required. Whether a bar is bent or terminated, it must have an embedment length not less than the development length l_d. The development length is given in Table 1-12 and is discussed later in this section. The *ACI Code* does not allow flexural reinforcement to be terminated in a tension zone unless one of the following conditions is satisfied:

i. The shear strength at the point of termination does not exceed two-thirds of the total shear capacity at the section;
ii. Stirrup area in excess of that required for shear and torsion is provided along each terminated bar a distance from the termination point equal to three-fourths the effective depth of the member. Excess stirrup area A_v shall not be less than $0.4 b_w s/f_y$. Spacing, s, shall not exceed $d/8\beta_b$ where β_b is the ratio of area of reinforcement cut off to total area of tension reinforcement at the section; or
iii. For No. 35 bar and smaller, the continuing bars provide double the area required for flexure at the cutoff point, and the shear does not exceed three-fourths of the total shear capacity of the cross section.

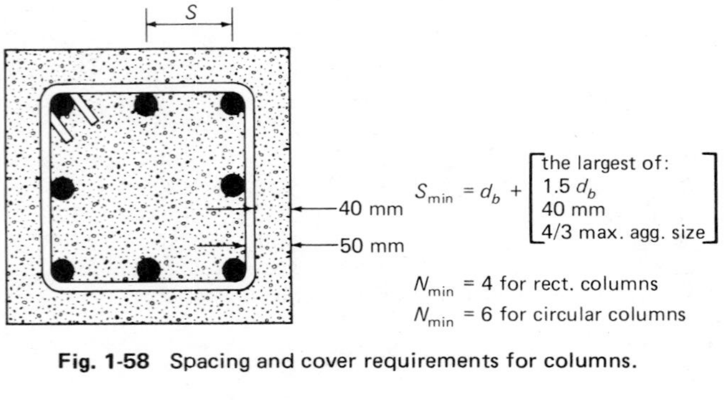

Fig. 1-58 Spacing and cover requirements for columns.

The ACI 318M-83 *Code* provides further limitations to termination of bars, depending on whether the bar is a positive or a negative moment reinforcement. Accordingly, at least one-third the positive moment reinforcement in simple beams and one-fourth the positive moment reinforcement in continuous members shall extend along the same face of the member into the support. For beams, this reinforcement will extend 150 mm or more into the support. If a flexural member is part of the primary lateral load resisting system, the positive moment reinforcement extended into the support shall be anchored so that the specified yield strength, f_y, in tension will be developed at the face of the support.

Special requirements for development length are applicable to positive moment reinforcement at simple supports and at points of inflection. In these regions, the bar sizes should be selected so that the following requirement for development length, l_d, is satisfied:

$$l_d \leqslant \frac{M_n}{V_u} + l_a \tag{1-99}$$

TABLE 1-12 Development Length l_d in Tension

$l_d = l_{bd} \times$ Modification factors $\geqslant$ 300 mm*

f'_c (MPa)	20			30			40		
f_y (MPa)	300	350	400	300	350	400	300	350	400
Bar Size	*Basic Tension Development Length l_{bd}* (mm)								
10	197	229	262	197	229	262	197	229	262
15	278	325	371	278	325	371	278	325	371
20	382	446	510	339	396	452	339	396	452
25	637	743	850	520	607	694	451	526	601
30	892	1041	1190	728	850	971	631	736	841
35	1275	1487	1699	1041	1214	1388	901	1051	1202
45	1744	2035	2326	1424	1661	1899	1233	1439	1644
55	2281	2661	3041	1862	2173	2483	1613	1882	2150

**The 300 mm limit does not apply when computing lap splices and web reinforcement development length.*

	Modification factors
Top reinforcement (more than 300 mm of concrete cast below the bars)	*1.4*
All-low-density concrete	*1.33*
Sand-low-density concrete	*1.18*
Bars spaced 140 mm or more on center	*0.8*
Excess area of steel	*(A_s required)/(A_s provided)*
Bars enclosed by spirals not less than 6 mm diameter and not more than 100 mm pitch	*0.75*

TABLE 1-13 Development length l_d in Compression

$l_d = l_{bd} \times$ Modification factors $\geqslant$ 200 mm

f_c' (MPa)	20			⩾30		
f_y (MPa)	300	350	400	300	350	400
Bar Size	*Basic Compression Development Length l_{bd}* (mm)					
10	182	212	243	149	174	199
15	259	300	343	211	246	282
20	314	366	419	257	300	343
25	406	473	541	333	388	444
30	481	562	642	395	460	526
35	575	671	766	471	550	628
45	704	821	938	577	673	769
55	908	1059	1211	744	869	993

	Modification factors
Excess area of steel	*(A_s required)/(A_s provided)*
Bars enclosed by spirals not less than 6 mm diameter and not more than 100 mm pitch	*0.75*

where M_n is nominal moment strength when all the reinforcement at the section yields, and V_u is the factored shear force at the section. The quantity l_a at a support is equal to the summation of the embedment length beyond the center of support and the equivalent embedment length of any hook or mechanical anchorage provided. At a point of inflection, l_a is limited to the effective depth of the member or 12 bar diameters, whichever is greater. The *Code* allows a 30% increase in the value of M_n/V_u if the ends of reinforcement are in a compression zone.

In the case of negative moment reinforcement, at least one-third the total tension steel should have an embedment length beyond the point of inflection. The embedment length for this case will not be less than the effective depth of the member, 12 bar diameters, or $1/16$ the clear span, whichever is greater.

The tension development length, l_d, is computed by first finding the basic development length, l_{bd}, and then multiplying this value by an appropriate modification factor to account for the factors that are applicable to the specific design case. The basic development length is a function of bar size, specified yield strength, and concrete strength. Table 1-12 tabulates the values of l_{bd} and the related modification factors. Table 1-13 provides the development length l_d for those bars that are in compression. Development of flexural reinforcement in a typical continuous beam is shown in Fig. 1-59.

Web reinforcement should also be designed to develop the design forces. The *ACI Code* requires that the ends of single leg, simple U, or multiple U stirrups be anchored by one of the following three means:

i. A standard hook plus an embedment of $0.5l_d$. The l_d embedment of a stirrup leg shall be taken as the distance between the middepth of the member, $d/2$, and the start of the hook (point of tangency).

ii. Embedment $d/2$ above or below the middepth of the compression side of the member for a full development length, l_d, but not less than 24 bar diameters or 300 mm.

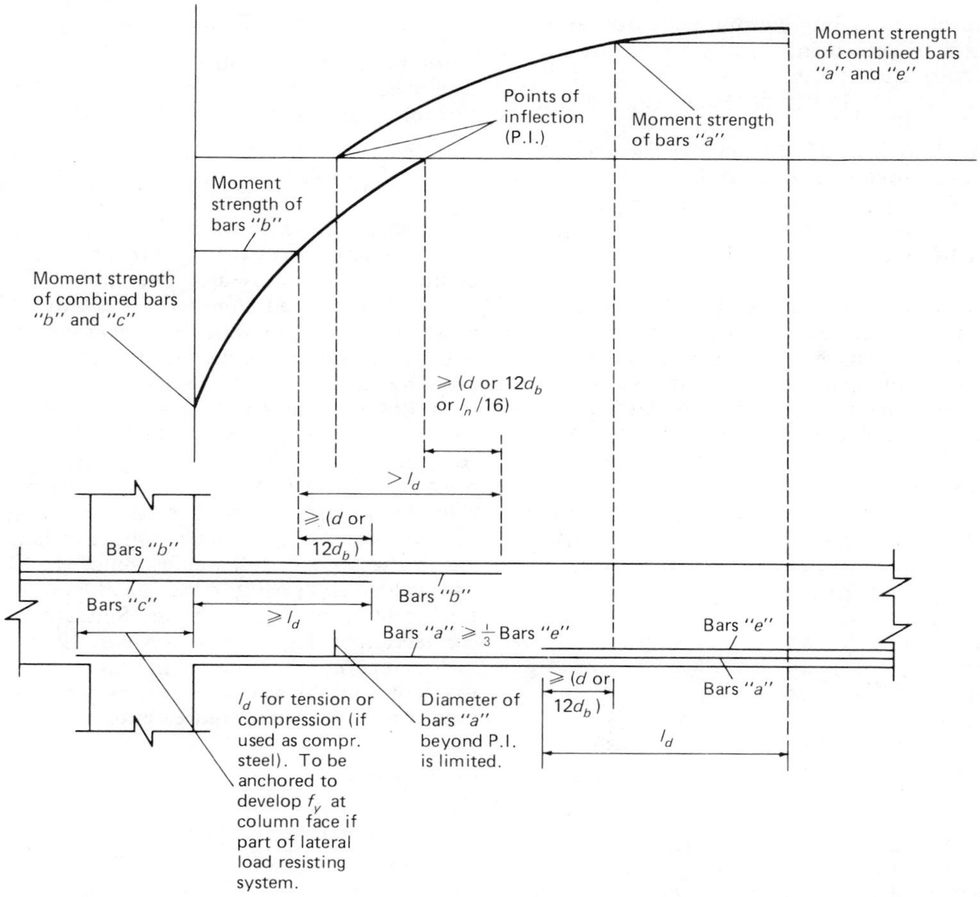

Fig. 1-59 Development of flexural reinforcement in a typical continuous beam. (Ref. 1.4)

iii. For No. 10 and No. 15 bars, bending around the longitudinal reinforcement through at least 135°. For stirrups with design stress exceeding 300 MPa, an embedment of $l_d/3$ is to be provided in addition to the 135° bent. The $l_d/3$ embedment of a stirrup leg shall be taken as the distance between the middepth of the member, $d/2$, and the start of the hook (point of tangency).

Pairs of U-stirrups or ties so placed as to form a closed unit shall be considered properly spliced when the laps are $1.7l_d$. In members that are at least 450 mm deep, such splices with $A_b f_y$ not more than 40 kN per leg may be considered adequate if the legs extend the full available depth of the member.

1.8.3 Standard Hooks in Tension

Hooks are not considered effective in developing reinforcement in compression. Standard hooks in tension are considered to develop an equivalent embedment length, l_e, where:

$$l_e = l_{bd} f_h / f_y \qquad (1\text{-}100)$$

and:

$$f_h = \xi \sqrt{f'_c} \qquad (1\text{-}101)$$

Values of the basic tension development length, l_{bd}, can be obtained from Table 1-12, and the values of ξ should not be greater than those listed in Table 1-14. The value of ξ can be increased 30% where enclosure is provided to the plane of the hook.

1.8.4 Splices of Deformed Bars

Reinforcing bars are often cut into shorter pieces for fabrication, transportation, or placement reasons. Frequently, it becomes necessary to splice them in the field. Two types of tension splices are common: (1) lap splices and (2) positive connections.

Lap splices are used for No. 35 and smaller bars except for splicing to footing dowels. Figure 1-60 shows two types of lap splices. Bars to be spliced may or may not be in contact. When they are not in contact, the clear space between the bars should not be more than one-fifth the required lap length or 150 mm.

Tension splices are classified as Class A, B, or C depending on the proportion of total steel to be spliced, and the ratio of steel area provided to that required by analysis. Table 1-15 gives the definition of each class of lap splice. The required length of lap for tension is established by tests and is specified in terms of tension development length, l_d. ACI 318M-83 specifies the following minimum lap length for each class of lap splice:

TABLE 1-14 ξ-Values*

	f_y = 400 MPa		f_y = 300 MPa
Bar Size	*Top Bars*	*Other Bars*	*All Bars*
10	45	45	30
15	45	45	30
20	37	45	30
25	30	45	30
30	30	40	30
35	30	35	30
45	27	27	27
55	18	18	18

**Values of ξ may be increased 30% where enclosure is provided perpendicular to the plane of the hook. The enclosure may consist of external concrete or internal closed ties, spirals, or stirrups.*

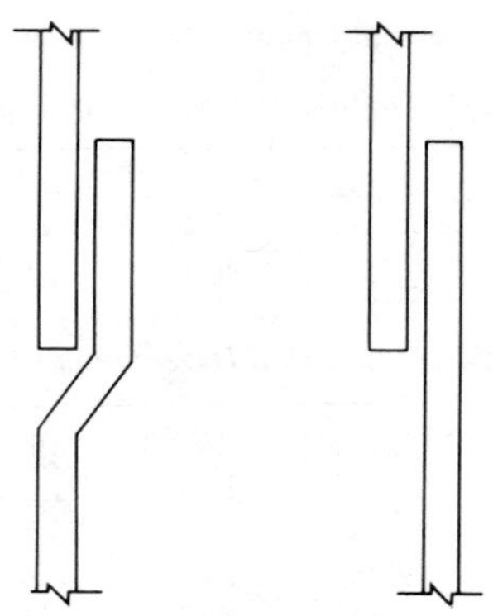

Fig. 1-60 Lap splices.

TABLE 1-15 Tension Lap Splices

	Maximum Percent of A_s Spliced Within Required Lap Length		
$\dfrac{A_s \text{ provided}}{A_s \text{ required}}$	*50*	*75*	*100*
≥2	Class A	Class A	Class B
<2	Class B	Class C	Class C

Class A: $l_s = 1.0 l_d$
Class B: $l_s = 1.3 l_d$
Class C: $l_s = 1.7 l_d$

Compression lap splices should have a minimum lap length equal to the compression development length given in Table 1-13 but should not be less than:

$$0.073 f_y d_b \qquad \text{for } f_y \leq 400 \text{ MPa}$$
$$(0.13 f_y - 24)\, d_b \qquad \text{for } f_y > 400 \text{ MPa}$$

In no case may a splice be less than 300 mm. If the concrete strength is less than 20 MPa, the lap length for bars in compression is increased by one-third.

In tied reinforced compression members, where ties have an effective area of $0.0015hs$ or more throughout the lap length, the values found for the lap length may be multiplied by 0.83, provided that the resulting value is not less than 300 mm. In spirally reinforced compression members, the lap length may be multiplied by 0.75, provided that the resulting value is not less than 300 mm.

An alternative to lap splices is the use of positive connections by welding or mechanical devices. This type of connection is generally used for splices of larger bars. According to the *Code*, a full welded splice shall have bars butted and welded to develop in tension at least 125% of the specified yield strength of the bar. Similarly a full mechanical connection is required to develop 125% of the specified yield strength, f_y, of the bar in tension or compression. Positive connections not meeting this requirement may still be used, provided other requirements of the *Code* are satisfied.

NOTATION

a = depth of equivalent stress block
A_c = area of concrete core measured out-to-out of lateral reinforcement
A_g = gross area of section

A_l = area of longitudinal steel resisting torsion
A_s = area of tension steel
A'_s = area of compression steel
A_{sf} = area of tension steel balanced by the flange overhangs
A_{sw} = area of web steel in tension
A_t = area of one leg of a closed stirrup resisting torsion
A_{tr} = transformed area of a section
A_v = area of shear reinforcement within a distance s
b = width of compression face of member
b' = width of bearing
b_w = web width
c = distance from extreme compression fiber to neutral axis
c, c_1, c_2 = compressive stress
C_c = concrete compression force
C_t = factor relating shear and torsional stress properties ($= b_w d/\Sigma x^2 y$)
d = distance from extreme compression fiber to centroid of longitudinal tension reinforcement
d' = distance from extreme compression fiber to centroid of longitudinal compression reinforcement
d_b = reinforcing bar diameter
e = eccentricity
E_c = modulus of elasticity of concrete
E_s = modulus of elasticity of steel
f'_c = concrete compressive strength
f_{ct} = average splitting tensile strength of concrete
f_h = tensile stress developed by a standard hook
f_r = modulus of rupture of concrete
f_s = stress in tension reinforcement
f'_s = stress in compression reinforcement
f_y = specified yield strength for reinforcement
h = overall member thickness
h_f = flange thickness
I = moment of inertia
I_g = moment of inertia of a gross section
I_{tr} = moment of inertia of a transformed section
k = effective length factor for compression members
l = span length
l_a = additional embedment length at support or at point of inflection
l_{bd} = basic development length
l_c = clear span length
l_d = development length
l_e = equivalent embedment length of a hook
l_n = length of clear span in long direction of two-way construction
l_s = length of splice
l_u = unsupported length of compression member
M = bending moment
M_{1b} = smaller factored end moment on a compression member due to the loads that result in *no* appreciable sidesway
M_{2b} = larger factored end moment on a compression member due to loads that result in *no* appreciable sidesway
M_{2s} = larger factored end moment on a compression member due to loads that result in appreciable sidesway
M_c = factored moment to be used for design of compression member
M_{cr} = bending moment at concrete cracking
M_f = internal moment couple due to flange overhangs
M_n = nominal moment capacity of a section
M_u = factored design moment
M_w = internal moment couple due to the web of a T-section
M_1 = smaller end moment on a compression member
M_1 = moment component of a section reinforced with ($A_s - A'_s$) as tension reinforcement
M_2 = larger end moment on a compression member
M_2 = moment couple due to compression reinforcement
N_u = factored axial force
P = axial force
P_c = critical load
P_n = nominal axial load strength
P_{nw} = nominal axial load strength of a wall
P_0 = nominal axial load strength at zero eccentricity
r = radius of gyration
s = spacing of shear or torsion reinforcement measured parallel to longitudinal reinforcement
t, t_1, t_2 = tensile stress
T_c = nominal torsional moment strength provided by concrete
T_{cr} = torsional moment at concrete cracking
T_s = tension force in longitudinal reinforcement
T_s = nominal torsional moment strength provided by torsion reinforcement
T_u = factored torsional moment at section
T_0 = torsion capacity of concrete when combined with flexure
U = required strength to resist factored loads
v = shear stress
v_c = shear strength of concrete
v_t = stress due to torsional moment
V_a = shear force resisted by aggregate interlock
V_c = nominal shear strength provided by concrete
V_d = shear force resisted by dowel action
V_e = shear force resisted by uncracked concrete
V_n = nominal shear strength
V_s = nominal shear strength provided by shear reinforcement
V_u = factored shear force at section
V_0 = shear capacity of concrete when combined with flexure
x = shorter overall dimension of rectangular part of cross section
x_1 = shorter center-to-center dimension of closed rectangular stirrup
y = longer overall dimension of rectangular part of cross section
y_1 = longer center-to-center dimension of closed rectangular stirrup
α = angle between the principal stress plane and the horizontal plane
α = angle between inclined stirrups and longitudinal axis of member
α_1 = shape factor to be used with elastic torsional stress
β = coefficient used in biaxial bending design
β_b = ratio of area of bars cut off to total area of bars at the section
β_d = ratio of maximum design dead load moment to maximum design total load moment, always positive
β_1 = ratio of depth of rectangular stress block to the distance from the compression face to the neutral axis
γ = ratio of distance between the centroid of bars to total thickness of column in the direction of bending
δ_b = moment magnification factor for frames braced against sidesway
δ_s = moment magnification factor for frames not braced against sidesway
Δ = transverse deflection

ξ = constant for standard hooks
ϵ = strain
ϵ_c = strain in concrete
ϵ_{cc} = compressive strain in concrete
ϵ_s = strain in steel
ϵ_y = strain in steel at yield
ϵ_0 = strain in concrete corresponding to f_c'
$\rho = A_s/bd$
$\rho' = A_s'/bd$
ρ_b = reinforcement ratio at balanced section
ρ_s = ratio of volume of spiral reinforcement to total volume of core (out-to-out of spirals) of a spirally reinforced concrete
ϕ = curvature
ϕ = capacity reduction factor

REFERENCES

1.1 "Metric Design Handbook for Reinforced Concrete Elements," Edited by Murat Saatcioglu, Printed by Canadian Portland Cement Association, Ottawa, Canada, 1980.

1.2 Park, R. and Paulay, T., "Reinforced Concrete Structures," John Wiley & Sons, 1975.

1.3 Winter, G. and Nilson, A. H., "Design of Concrete Structures," McGraw-Hill Book Company, 1979.

1.4 "Building Code Requirements for Reinforced Concrete," ACI 318M-83, American Concrete Institute, Box 19150, Redford Station, Detroit, Michigan.

2

Deflections

DAN E. BRANSON, Ph.D., P.E.*

2.1 INTRODUCTION

With the present-day use of higher-strength concrete and reinforcing steel, the strength of load-factor method of design, and the resulting shallower sections, the problem of predicting and controlling deflections of concrete flexural members under service loads has become increasingly important.

The objective of this chapter is to present a concise and practical treatment of the subject of initial and time-dependent deflection of nonprestressed concrete one-way and two-way flexural systems. The principal effects taken into account are those due to elastic, cracking, creep, and shrinkage deformation.

The American Concrete Institute *Building Code* and *Commentary*,[2-1] papers by ACI Committee 435 on deflections[2-2–2-7] and ACI Committee 209 on creep and shrinkage,[2-8] the AASHTO highway bridge specifications,[2-9] and several books and chapters by the author[2-10–2-13] are used as the principal guides. Additional background references on material properties,[2-14–2-17] beam deflections,[2-18–2-36] and slab deflections[2-37–2-44] are also cited. Similar information may be found in the literature[2-10, 2-11, 2-13, 2-36, 2-45–2-47] on the deflection of prestressed concrete members.

This material is concerned with deflections that occur at service load levels under static conditions and may not apply for loads with strong dynamic characteristics such as those due to earthquakes, transient winds, and machinery. The calculation procedures used are approximate in nature. Because of the variability of concrete structural deformation, in most cases it is essential that relatively simple procedures be used, so that engineers will guard against placing undue reliance on computed or predicted results.

Both English and SI units are used, with conversion factors and typical basic values and units used in computing deflections summarized in Table 2-1. Numerical examples that are similar to those herein, but entirely in SI units, may be found in the PCA *Metric Design Handbook*.[2-12]

2.2 MATERIAL PROPERTIES

The principal material parameters required for most concrete deformation calculations are based on compressive strength, elastic, cracking, creep, and shrinkage properties. A detailed and practical summary of procedures for computing these properties (recommended by the *ACI Code*, AASHTO, ACI Committees 209 and 435, the PCI *Design Handbook*, and the author) is presented in Appendix A, Chapter 5 of Ref. 2-13. This information includes the following effects: different concrete ages on the compressive strength (and hence the moduli of rupture and elasticity); moist and steam curing; different-weight concrete; and different loading ages, load durations, relative humidity conditions, and member sizes on average creep and shrinkage values. Simple equations that readily lend themselves to computer programming are presented in Ref. 2-13 for these properties. Values are also tabulated for hand calculations. In addition, both equations and graphs for these properties are presented in Refs. 2-10 and 2-47. Extensive comparisons of these procedures with experimental results are shown in Ref. 2-10.

*Professor of Civil Engineering, University of Iowa, Iowa City, Iowa.

TABLE 2-1 Conversion Factors and Typical Basic Values and Units Used in Computing Deflections in English and SI Units

	Conversion Factors	*Commonly Used Values and Units*
Stress, strength	1 ksi = 6.895 N/mm^2 = 6.895 MPa	3000 psi = 3 ksi = 20.7 N/mm^2
Unit weight	1 pcf = 16.02 kg/m^3	145 pcf = 2323 kg/m^3
Concentrated load	1 kip = 4.448 kN	10,000 lb = 10 kips = 44.5 kN
Distributed load	1 kip/ft = 14.59 kN/m	1000 lb/ft = 1 kip/ft = 14.59 kN/m
Distributed load	1 psf = 47.88 N/m^2	100 psf = 4.79 kN/m^2
Bending moment	1 ft-kip = 1.356 kN-m	100 ft-kips = 135.6 kN-m
Curvature	1/1 in. = 1/25.40 mm	100×10^{-6} 1/in. = 3.94×10^{-6} 1/mm
Deflection	1 in. = 25.40 mm	1 in. = 25 mm (usually rounded to nearest mm)

Concrete Strength

f'_c (psi)	N/mm^2	f'_c (N/mm^2)	psi
f'_c = 3000 psi	= 20.7 N/mm^2,	f'_c = 20 N/mm^2	= 2901 psi
4000	27.6	25	3626
5000	34.5	30	4351
6000	41.4	35	5076
7000	48.3	40	5801
8000	55.2	45	6526
		50	7252

Unit Weight and Distributed Dead Load

Property	*Normal Weight*	*Sand-Lightweight*	*All-Lightweight*
Unit wt. of plain conc.,	w_c = 145 pcf = 2323 kg/m^3	115 pcf = 1842 kg/m^3	100 pcf = 1602 kg/m^3
Unit wt. of R/C or P/C (including steel)	150 pcf = 2403 kg/m^3	120 pcf = 1922 kg/m^3	104 pcf = 1666 kg/m^3

For cross-sectional area of 30 cm by 60 cm–distributed dead load for Nor. Wt. Conc. = (2403 kg/m^3)(.009807 kN/m^3 per kg/m^3)(.30 m × .60 m) = 4.24 kN/m. This corresponds to an area of 11.81 in. by 23.62 in.–distributed dead load = (150 pcf)(11.81 in. × 23.62 in.)/144 $in.^2/ft^2$ = 290.6 lb/ft. × 0.01459 kN/m per lb/ft = 4.24 kN/m. Checks.

Summarized in this section are the simplified calculation procedures recommended by the *ACI Code* and *Commentary*,[2-1] ACI Committees 435[2-7] and 209,[2-8], AASHTO,[2-9] Pauw,[2-14] Christiason,[2-17] and the author.[2-10, 2-13, 2-17, 2-22]

For a general discussion of the effect of concrete material properties on deflections, see a chapter by Fling[2-28], and Ref. 2-10.

2.2.1 Concrete Modulus of Rupture

The *ACI Code*[2-1] recommends eqs. (2-1) for computing the modulus of rupture of different-weight concrete:

Normal weight concrete:

$$f_r(\text{psi}) = 7.5\sqrt{f'_c(\text{psi})}$$

$$f_r(\text{N/mm}^2) = 0.623\sqrt{f'_c(\text{N/mm}^2)}$$

Sand-lightweight concrete:

$$f_r(\text{psi}) = \overset{6.38}{(0.85)(7.5)}\sqrt{f'_c(\text{psi})}$$

$$f_r(\text{N/mm}^2) = \overset{0.530}{(0.85)(0.623)}\sqrt{f'_c(\text{N/mm}^2)}$$

All-lightweight concrete:

$$f_r(\text{psi}) = \overset{5.63}{(0.75)(7.5)}\sqrt{f'_c(\text{psi})}$$

$$f_r(\text{N/mm}^2) = \overset{0.467}{(0.75)(0.623)}\sqrt{f'_c(\text{N/mm}^2)} \qquad (2\text{-}1)$$

EXAMPLE:

When f'_c = 4000 psi for Nor. Wt. Conc.,

$$f_r = 7.5\sqrt{4000} = 474 \text{ psi } (3.27 \text{ N/mm}^2)$$

When f'_c = 25 N/mm^2 for Nor. Wt. Conc.,

$$f_r = 0.623\sqrt{25} = 3.12 \text{ N/mm}^2 \text{ (452 psi)}$$

When f'_c = 4000 psi for All-Lt. Wt. Conc.,

$$f_r = (0.75)(7.5)\sqrt{4000} = 356 \text{ psi } (2.45 \text{ N/mm}^2)$$

Equations (2-1) are to be used when the tensile splitting strength, f_{ct}, is not specified. Otherwise, for lightweight aggregate concretes eq. (2-1) in English units for normal weight concrete shall be modified by substituting $f_{ct}/6.7$ for $\sqrt{f'_c}$, but the value of $f_{ct}/6.7$ used shall not exceed $\sqrt{f'_c}$.

AASHTO[2-9] recommends eqs. (2-2) for computing the modulus of rupture of different-weight concrete:

Normal weight concrete:

$$f_r(\text{psi}) = 7.5\sqrt{f'_c(\text{psi})}$$

$$f_r(\text{N/mm}^2) = 0.623\sqrt{f'_c(\text{N/mm}^2)}$$

Sand-lightweight concrete:

$$f_r(\text{psi}) = 6.3\sqrt{f'_c(\text{psi})}$$

$$f_r(\text{N/mm}^2) = 0.523\sqrt{f'_c(\text{N/mm}^2)}$$

All-lightweight concrete:

$$f_r(\text{psi}) = 5.5\sqrt{f'_c(\text{psi})}$$

$$f_r(\text{N/mm}^2) = 0.456\sqrt{f'_c(\text{N/mm}^2)} \qquad (2\text{-}2)$$

Based on some 332 data points for different-weight concrete from Ref. 2-10, ACI Committee 435[2-7] recommends the statistically improved and less conservative (higher values than by eqs. 2-1 and 2-2) eqs. (2-3) for computing the modulus of rupture of different weight concrete in the range 90 pcf $\leqslant w_c \leqslant$ 145 pcf, or 1440 N/mm^2 $\leqslant$ $w_c \leqslant$ 2320 N/mm^2:

$$f_r(\text{psi}) = 0.65\sqrt{w_c(\text{pcf})f_c'(\text{psi})}$$

$$f_r(\text{N/mm}^2) = 0.0135\sqrt{w_c(\text{kg/m}^3)f_c'(\text{N/mm}^2)} \qquad (2\text{-}3)$$

EXAMPLE:

When $f_c' = 4000$ psi and $w_c = 145$ pcf (Nor. Wt. Conc.),

$$f_r = 0.65\sqrt{(145)(4000)}$$

$$= 495 \text{ psi versus 474 psi in eq. (2-1)}$$

When $f_c' = 25$ N/mm^2 and $w_c = 2323$ kg/m^3 (Nor. Wt. Conc.),

$$f_r = 0.0135\sqrt{(2323)(25)}$$

$$= 3.25 \text{ N/mm}^2 \text{ versus } 3.12 \text{ N/mm}^2 \text{ in eq (2-1)}$$

When $f_c' = 4000$ psi and $w_c = 100$ pcf (All-Lt. Wt. Conc.),

$$f_r = 0.65\sqrt{(100)(4000)}$$

$$= 411 \text{ psi versus 356 psi in eq. (2-1)}$$

2.2.2 Concrete Modulus of Elasticity

The *ACI Code*,[2-1] ACI Committee 435,[2-7] and AASHTO[2-9] recommend eq. (2-4) from Ref. 2-14 for computing the modulus of elasticity of different-weight concrete in the range 90 pcf $\leqslant w_c \leqslant$ 155 pcf, or 1440 N/mm^2 $\leqslant w_c \leqslant$ 2480 N/mm^2.

$$E_c(\text{psi}) = 33\sqrt{[w_c(\text{pcf})]^3 f_c'(\text{psi})}$$

$$E_c(\text{N/mm}^2) = 0.0427\sqrt{[w_c(\text{kg/m}^3)]^3 f_c'(\text{N/mm}^2)} \qquad (2\text{-}4)$$

EXAMPLE:

When $f_c' = 6000$ psi and $w_c = 145$ pcf (Nor. Wt. Conc.),

$$E_c = 33\sqrt{(145)^3(6000)}$$

$$= 4.46 \times 10^6 \text{ psi } (30.8 \text{ kN/mm}^2)$$

When $f_c' = 40$ N/mm^2 and $w_c = 2323$ kg/m^3 (Nor. Wt. Conc.),

$$E_c = 0.0427\sqrt{(2323)^3(40)}$$

$$= 30{,}200 \text{ N/mm}^2 \ (4.38 \times 10^6 \text{ psi})$$

When $f_c' = 6000$ psi and $w_c = 100$ pcf (All-Lt. Wt. Conc.),

$$E_c = 33\sqrt{(100)^3(6000)}$$

$$= 2.56 \times 10^6 \text{ psi } (17.7 \text{ kN/mm}^2)$$

Based on some 274 data points for different-weight concrete, the author[2-10] recommends the statistically improved (especially for higher strengths in the 5000 to 8000 psi range) and more conservative (lower values than by eq. 2-4 when $f_c' > 4000$ psi) eq. (2-5) for computing the modulus of elasticity of different-weight concrete:

$$E_c(\text{psi}) = a\sqrt{[w_c(\text{pcf})]^3 f_c'(\text{psi})}, \quad \text{where} \quad a = 39.0 - 0.0015 f_c'(\text{psi}) \qquad (2\text{-}5)$$

EXAMPLE:

When $f_c' = 6000$ psi and $w_c = 145$ pcf, $a = 30.00$ and

$$E_c = 30.00\sqrt{(145)^3(6000)}$$

$$= 4.06 \times 10^6 \text{ psi versus } 4.46 \times 10^6 \text{ psi in eq. (2-4)}$$

When $f_c' = 40$ N/mm^2 = 5800 psi and $w_c = 2323$ kg/m^3 = 145 pcf,

$a = 3.30$ and

$$E_c = 30.30\sqrt{(145)^3(5800)}$$

$$= 4.029 \times 10^6 \text{ psi} = 27{,}800 \text{ N/mm}^2 \text{ versus } 30{,}200 \text{ N/mm}^2 \text{ in eq. (2-4)}$$

When $f_c' = 6000$ psi and $w_c = 100$ pcf, $a = 30.00$ and

$$E_c = 30.00\sqrt{(100)^3(6000)}$$

$$= 2.32 \times 10^6 \text{ psi versus } 2.56 \times 10^6 \text{ psi in eq. (2-4)}$$

2.2.3 Steel Modulus of Elasticity

The *ACI Code*[2-1] and AASHTO[2-9] recommend eq. (2-6) for the modulus of elasticity of nonprestressed (reinforcing) steel:

$$E_s = 29 \times 10^6 \text{ psi } (200 \text{ kN/mm}^2) \qquad (2\text{-}6)$$

2.2.4 Concrete Creep and Shrinkage

The *ACI Code* does not recommend separate values for concrete creep and shrinkage. AASHTO[2-9] recommends no separate creep factors for deflections, but does suggest a value of $(\epsilon_{sh})_u = 200 \times 10^{-6}$ in./in. (mm/mm) for normal weight concrete. This appears to be a relatively low value. For average conditions, ACI Committee 435[2-7] recommends eqs. (2-7) and (2-8) for ultimate values of the creep coefficient, C_u, and shrinkage strain, $(\epsilon_{sh})_u$:

$$C_u = 1.60 \qquad (2\text{-}7)$$

$$(\epsilon_{sh})_u = 400 \times 10^{-6} \text{ in./in. (mm/mm)} \qquad (2\text{-}8)$$

Equations (2-7) and (2-8) represent average values based on some 470 data points for creep and 356 data points for shrinkage of different-weight concrete in Ref. 2-10. As can be seen in Table 1-10 of Ref. 2-10 or Table A-8, Chapter 5 of Ref. 2-13, these values correspond approximately to the following conditions: 70% average relative humidity, loading ages for creep and shrinkage from about age 20 days for both moist and steam cured concrete, and minimum thicknesses of about 6 in. (15 cm). For these conditions the upper and lower limits of the experimental data from Ref. 2-10 are $C_u = 0.90$ to 2.80 and $(\epsilon_{sh})_u = 200 \times 10^{-6}$ to 600×10^{-6} in./in. (mm/mm).

For different load durations and shrinkage periods Table 2-2 may be used as follows:

EXAMPLE: Compute C_u and $(\epsilon_{sh})_u$ at a concrete age of 6 months. This corresponds to a sustained loading period and a shrinkage period from age 20 days to age 6 months. Consider moist cured concrete.

$$C_{6\text{Mths}} = (0.68)(1.60) = 1.09$$

$$(\epsilon_{sh})_{6\text{Mths}} = (0.77)(400 \times 10^{-6}) = 308 \times 10^{-6} \text{ in./in. (mm/mm)}$$

In studying the above-mentioned data from Ref. 2-10, no consistent distinction is found between normal weight and lightweight concrete creep and shrinkage. That is, the cement gel seems to be doing most of the shrinking and creeping, without major influences from the type of aggregate. However, it is noted that higher shrinkage has been found in some cases for lightweight concrete as compared to normal weight concrete of comparable strength and quality.

2.2.5 Multipliers for Time-Dependent Deflections

Multipliers for the combined effects of creep and shrinkage on deflections, T or T_u (assumed to be the combined time-dependent multipliers for "additional long time deflections"), of nonprestressed concrete flexural systems are recommended by the *ACI Code* and *Commentary*[2-1] for both normal and lightweight concrete in Table 2-3 or Fig. 2-1. These multipliers are based on results from Ref. 2-19.

Other recommendations for the time-dependent deflection multipliers for both normal and lightweight concrete are as follows:

TABLE 2-2 Creep and Shrinkage Ratios from Age 20 Days to the Indicated Concrete Age (to Be Used with eqs. 2-7 and 2-8)

Creep, Shrinkage Ratios	*Concrete Age*					
	2 Mths	*3 Mths*	*6 Mths*	*1 Yr*	*2 Yrs*	*≥[d]5 Yrs*
C_t/C_u	[a]0.48	0.56	0.68	0.77	0.84	1.00
$(\epsilon_{sh})_t/(\epsilon_{sh})_u$–M.C.	[b]0.46	0.60	0.77	0.88	0.94	1.00
$(\epsilon_{sh})_t/(\epsilon_{sh})_u$–S.C.	[c]0.36	0.49	0.69	0.82	0.91	1.00

Creep of Moist and Steam Cured Concrete:

$$^{a}C_t/C_u = \frac{t^{0.60}}{10 + t^{0.60}} = \frac{40.8^{0.60}}{10 + 40.8^{0.60}} = 0.48$$

where t = time after loading in days = 365/6 – 20 = 40.8 days.

Shrinkage of Moist Cured Concrete:

$$^{b}(\epsilon_{sh})_t/(\epsilon_{sh})_u = \frac{\left[\frac{t}{35 + t}\right]_t - \frac{*13}{35 - 13}}{**1 - \frac{13}{35 - 13}} = \frac{0.606 - 0.271}{1 - 0.271} = 0.46$$

where t = concrete age in days – standard age of 7 days = 365/6 – 7 = 53.8 days, *20 – 7 = 13 days, and **$[t/(35 + t)]_u = 1$.

Shrinkage of Steam Cured Concrete:

$$^{c}(\epsilon_{sh})_t/(\epsilon_{sh})_u = \frac{\left[\frac{t}{55 + t}\right]_t - \frac{*19}{55 - 19}}{**1 - \frac{19}{55 - 19}} = \frac{0.521 - 0.257}{1 - 0.257} = 0.36$$

where t = concrete age in days – standard age of 1 day = 365/6 – 1 = 59.8 days, *20 – 1 = 19 days, and **$[t/(55 + t)]_u = 1$.

These creep and shrinkage equations by the author and Christiason[2-17] have also been recommended by ACI Committee 209.[2-8] The above procedure may also be applied to other than 20-day loading ages for creep or starting ages for shrinkage (by adjusting the number 20), and to other ultimate values of C_u and $(\epsilon_{sh})_u$ as well.

[d]*Considered to be unity for practical purposes.*

AASHTO[2-9] (same as 1963, 1971, and 1977 *ACI Codes*, but not the 1983 *ACI Code*):

$$\lambda = k_r T_u = [2 - 1.2(A_s'/A_s)] \geqslant 0.6 \qquad (2\text{-}9)$$

In eq. (2-9), $T_u = 2$ and $k_r = [1 - 0.6(A_s'/A_s)]$ = a reduction factor that takes into account principally the time-dependent effects of compression steel and downward movement of the neutral axis.

TABLE 2-3 Time-Dependent Multiplier, T or T_u, for Computing Additional Long-Time Deflections for Both Normal Weight and Lightweight Concrete, Unless a More Comprehensive Analysis Is Used (*ACI Code*[2-1])[a]

Load Duration	*Multiplier*	*Load Duration*	*Multiplier*
5 years or more	$T_u = 2.0$	6 months	$T = 1.2$
12 months	$T = 1.4$	3 months	$T = 1.0$

[a]*These values were obtained from Fig. 2-1 and Ref. 2-19.*

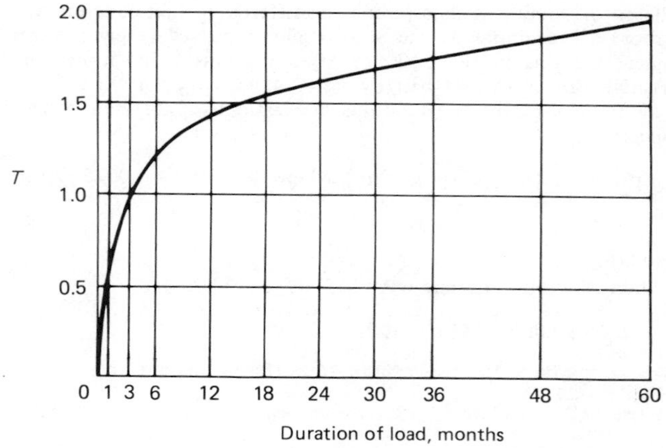

Fig. 2-1 Time-dependent multiplier, T, for computing additional long-time deflections for both normal weight and lightweight concrete, unless a more comprehensive analysis is used (*ACI Code*,[2-1] based on Ref. 2-19).

TABLE 2-4 Time-Dependent Multiplier, $\lambda = k_r T_u$, for Computing Additional Long-Time Deflections (over a Period of About 3 Years), Unless a More Comprehensive Analysis Is Used (1973 AASHTO[2-9])

Humidity Conditions	$A_s' = 0$	$A_s' = 0.5 A_s$	$A_s' = A_s$
Climate of high humidity	2.5	1.8	1.5
Climate of average humidity	3.0	2.2	1.8
Climate of low humidity	3.5	2.5	2.0

AASHTO in the 1973 edition (information still useful): Table 2-4.

ACI Committee 435[2-7]: Under normal conditions, use eq. (2-10).

Where creep and/or shrinkage are expected to be abnormally high, a larger value of T_u may be required. The use of Fig. 2-1 for different loading periods but multiplied by 2.5/2.0 is also recommended.

$$T_u = 2.5 \tag{2-10}$$

The value of 2.5, instead of the *ACI Code* and AASHTO recommendation of 2.0, is thought to be more realistic. Other codes, for example, use values of 3 (Australian), 3.5 (1973 AASHTO for low humidity), and so on.

The author[2-22]: Table 2-5.

The approximate deflection formulas using these factors are presented later in the chapter.

2.3 CONTROL OF DEFLECTIONS

Deflections of one-way and two-way nonprestressed concrete flexural systems are controlled by means of steel ratio limitations, and/or minimum thicknesses, and/or directly limiting computed deflections, as summarized in this section. A general discussion of the subject of deflection control may be found in a report by ACI Committee 435,[2-3] a chapter by Fling,[2-28] and a chapter by the author.[2-10]

2.3.1 Tension Steel Ratio Limitations for Nonprestressed One-Way Members

One method used to minimize deflection problems is to use a relatively small tension steel ratio in the flexural design. This results in a relatively deep beam because the required internal moment arm is thus relatively large. In this connection, limiting values of ρ ranging from 25% ρ_b to 40% ρ_b are recommended by ACI Committee 435[2-7] in Table 2-6.

2.3.2 Minimum Thicknesses for Nonprestressed One-Way Members

Deflections of beams and one-way slabs supporting loads commonly found in buildings will normally be satisfactory when the minimum thicknesses in Table 2-7 are met or exceeded. This table from the ACI Code, Table 9.5(a), applies only for members *not* supporting or *not* attached to partitions or other construction likely to be damaged by large deflections.

In the case of shored composite members, the minimum thicknesses apply as for monolithic T-beams. In the case of unshored construction, if the thickness of a nonprestressed precast member meets the *Code* requirements, deflections need not be computed. In addition, if the thickness of an unshored nonprestressed composite member meets the minimum thickness requirements of the *Code*, deflections occurring after the member becomes composite need not be calculated. However, the long-time deflection of the precast member should be investigated for the magnitude and duration of load prior to the beginning of effective composite action.

Values in Table 2-7 have been modified by ACI committee 435[2-7] and extended in Table 2-8 to include members that *are* supporting or attached to nonstructural elements likely to be damaged by large deflections. The committee states that deflections under uniform loads commonly encountered in buildings will normally be satisfactory when the minimum thicknesses specified in Table 2-8 for one-way members are met or exceeded. Lesser thicknesses may be used when computed deflections are shown to be satisfactory. A discussion of this information is presented in the "Commentary" of Ref. 2-7.

The method developed principally by Grossman[2-35] for the ACI Committee 318 Task Group for Code Simplification recommends eq. (2-11) and Table 2-9, as follows: Unless obtained by a more comprehensive analysis, or computation of deflection indicates a lesser thickness can be used without adverse effects, minimum thickness of beams and one-way slabs with loads uniformly distributed and not requiring compression reinforcement for strength ($\rho \leqslant 0.75\ \rho_b$) shall be governed by eq. (2-11):

$$h_{\min} = \left(\frac{l}{24}\right)\left[\frac{(l/a)_{\min}}{180}\right]\left[\frac{\lambda D + L}{D + L}\right] \geqslant \frac{l}{24} \tag{2-11}$$

TABLE 2-5 Time-Dependent Multiplier, T_u, for Computing Additional Long-Time Deflections for Both Normal Weight and Lightweight Concrete Members of Common Types and Sizes (Ref. 2-22)[a, b]

	Average Relative Humidity, Age When Loaded								
	100%			70%			[b]50%		
Concrete Strength, f_c' at age 28 Days	⩽7d	14d	⩾28d	⩽7d	14d	⩾28d	⩽7d	14d	⩾28d
2500 to 4000 psi (17 to 28 N/mm^2)	2.0	1.5	1.0	3.0	2.0	1.5	4.0	3.0	2.0
>4000 psi (>28 N/mm^2)	1.5	1.0	0.7	2.5	1.8	1.2	3.5	2.5	1.5

[a] *For loading periods: 1 Mth or Less — 3 Mths — 1 Year — 5 Years or More*
Use: 25% — 50% — 75% — 100% of Table Values

[b] *The 50% humidity values may normally be used for lower average relative humidities, such as in heated buildings.*

TABLE 2-6 Recommended Tension Reinforcement Ratios for Nonprestressed One-Way Members So That Deflections Will Normally Be within Acceptable Limits (ACI Committee 435[2-7])[a]

Members	*Cross Section*	*Normal Weight Concrete*	*Lightweight Concrete*
Not supporting or not attached to nonstructural elements likely to be damaged by large deflections	Rectangular "T" or box	$\rho \leqslant 35\% \rho_b$ $\rho_w \leqslant 40\% \rho_b$	$\rho \leqslant 30\% \rho_b$ $\rho_w \leqslant 35\% \rho_b$
Supporting or attached to nonstructural elements likely to be damaged by large deflections	Rectangular "T" or box	$\rho \leqslant 25\% \rho_b$ $\rho_w \leqslant 30\% \rho_b$	$\rho \leqslant 20\% \rho_b$ $\rho_w \leqslant 25\% \rho_b$

[a]*For continuous members, the positive region steel ratios only may be used.* ρ_b: *Refers to the balanced steel ratio based on ultimate strength.*

TABLE 2-7 Minimum Thickness of Beams or One-Way Slabs Not Supporting or Attached to Partitions or Other Construction Likley to Be Damaged by Large Deflections, Unless Deflections Are Computed (*ACI Code*)

	Minimum Thickness			
Member	*Simply Supported*	*One End Continuous*	*Both Ends Continuous*	*Cantilever*
Solid one-way slabs	$l/20$	$l/24$	$l/28$	$l/10$
Beams or ribbed one-way slabs	$l/16$	$l/18.5$	$l/21$	$l/8$

[a]*The values given in this table shall be used directly for members made with normal weight concrete (w_c = 145 pcf = 2320 kg/m^3) and Grade 60 steel (f_y = 60 ksi = 414 N/mm^2). For other conditions, the values shall be modified as follows:*

(a) For structural lightweight concrete having unit weights in the range of 90 to 120 pcf (1440 to 1920 kg/m^3), the values in the table shall be multiplied by 1.65 − 0.005 w_c in English units and 1.65 − 0.000312 w_c in SI units, but not less than 1.09 in either units. The value of w_c = 112 pcf = 1867 kg/m^3 corresponds to 1.09.

(b) For nonprestressed reinforcement with values of f_y other than 60,000 psi (414 N/mm^2), the values in the table shall be multiplied by 0.4 + f_y/100,000 in English units and 0.4 + f_y/689.5 in SI units.

Examples of these correction factors are as follows:

f_y, ksi = 40	50	60	75	w_c, pcf = 90	100	110	120
Cor. factor = 0.80	0.90	1.00	1.15	Cor. factor = 1.20	1.15	1.10	1.09
f_y, N/mm^2 = 275	345	414	517	w_c, kg/m^3 = 1440	1600	1760	1920
Cor. factor = 0.80	0.90	1.00	1.15	Cor. factor = 1.20	1.15	1.10	1.09

TABLE 2-8 Minimum Thickness of Beams or One-Way Slabs Used in Roof and Floor Construction (ACI Committee 435[2-7])[a]

	Members not *supporting or* not *attached to nonstructural elements likely to be damaged by large deflections*				*Members supporting or attached to nonstructural elements likely to be damaged by large deflections*			
Member	*Simply Supported*	*One End Cont.*	*Both Ends Cont.*	*Cantilever*	*Simply Supported*	*One End Cont.*	*Both Ends Cont.*	*Cantilever*
Roof slab[b]	$l/22$	$l/28$	$l/35$	$l/9$	$l/14$	$l/18$	$l/22$	$l/5.5$
Floor slab, and roof beam[b] or ribbed roof slab[b]	$l/18$	$l/23$	$l/28$	$l/7$	$l/12$	$l/15$	$l/19$	$l/5$
Floor beam or ribbed floor slab	$l/14$	$l/18$	$l/21$	$l/5.5$	$l/10$	$l/13$	$l/16$	$l/4$

[a]*The footnotes in Table 2-7 apply to Table 2-8 in toto.*
[b]*Refers to roofs subjected to normal snow or construction live loads only, and with minimal ponding problems.*

where $\lambda = T/(1 + 50\rho')$ and T is given in Table 2-3. Also, λ may be taken as the net difference between λ-values computed at the various time periods considered. In eq. (2-11), $(l/a)_{min}$ shall be the applicable limit stipulated in Table 2-11. The dead load D shall include any sustained portion of the live load. Equation (2-11), with its $l/24$ lower limit, shall apply for simply supported members with normal weight concrete (w_c = 145 pcf) and Grade 60 steel. For other conditions, h_{min} obtained from eq. (2-11) shall be multiplied by the applicable factor or factors for:

(a) Members with one end continuous: 0.80

TABLE 2-9 Permissible Reduction in h_{min} by eq. (2-11) for Members with $\rho < 0.75\ \rho_b$ ($\omega < 0.75\ \rho_b\ f_y/f_c'$) (Ref. 2-35)[a]

Percent Reduction =		10	20	30	40
Type of Member	f_y ksi		$\omega = \rho\ f_y/f_c'$		
Rectangular section	40	0.17	0.12	0.08	0.06
($b_w/b = 1.0$)	60	0.14	0.10	0.08	0.06
Flanged section[b]					
($b_w/b = 0.1$)	40	0.13	0.08	0.04	0.02
($h_f/h \geqslant 0.2$)	60	0.10	0.05	0.03	0.01
Flanged section[b]					
($b_w/b = 0.1$)	40	0.08	0.06	0.03	0.02
($h_f/h = 0.1$)	60	0.07	0.04	0.02	0.01

[a]*Interpolation may be used.*
[b]*For flanged sections with $b_w/b > 0.1$, linear interpolation may be used between ω values at $b_w/b = 0.1$ and $b_w/b = 1.0$ (rectangular section). ρ for flanged sections computed as $\rho = A_s/bd$.*

(b) Members with both ends continuous: 0.65
(c) Cantilevers: 2.40
(d) Members with d less than $0.9h$: $0.9\ h/d$
(e) Structural lightweight concrete having a unit weight w_c in excess of 90 lb per cu ft: $\sqrt{145/w_c}$
(f) f_y other than 60,000 psi: $0.4 + f_y/100{,}000$

Equation (2-11) was derived[2-35] to provide minimum thicknesses for members proportioned for $\rho = 0.75\ \rho_b$, and to be compatible with the permissible deflections (span–deflection ratios) in Table 2-11 or other such limitations. Table 2-9 may be used to reduce h_{min} in eq. (2-11) when ρ is much less than $0.75\ \rho_b$ ($\omega \leqslant 0.17$). A study of this information is presented in Ref. 2-35 and the discussions of the paper.

2.3.3 Minimum Thicknesses for Nonprestressed Two-Way Slab Systems

Deflections of two-way slab systems with and without beams, drop panels, and column capitals need not be computed when the thickness requirements of the *ACI Code*,[2-1] Section 9.5.3, are met, according to the *Commentary*. This procedure includes the effect of panel locations (interior, side, or corner); panel shape; span ratios; supporting columns and capitals; drop panels; and the design yield strength of the reinforcing steel. The equations in the *Code* provide for a transition from slabs on stiff beams to slabs without beams.

Typical cases from the *ACI Code* minimum thickness procedure (see Example 6, below, for equations) are summarized in Table 2-10 from Ref. 2-37. The same information is provided in Ref. 2-12 for f_y in rounded SI units. It may be noted in Table 2-10 that the difference between the controlling minimum thickness for square panels and 2 to 1 rectangular panels in each case is not very large.

2.3.4 Allowable Computed Deflections

The allowable computed deflections (minimum permissible span–deflection ratios) specified in the *ACI Code*[2-1] for both one-way and two-way systems are given in Table 2-11.

Where excessive deflections may cause damage to nonstructural or other structural elements, only that part of the deflection occurring after the construction of the element needs to be considered. The most stringent span–deflection limit of 480 in Table 2-11 is an example of such a case.

Where excessive deflections may result in either aesthetic or functional problems, such as objectionable visual sagging, ponding of water, vibration, and improper operation of machinery or sliding doors, the total deflection should be considered. Such examples are not included in Table 2-11 and must be dealt with by the engineer on a case-by-case basis.

The span–deflection ratios (allowable computed deflections) in Table 2-11 provide for a simple set of allowable deflections, instead of a more extensive set of limitations that would be required to cover all possible types of construction and conditions of loading. For more information on allowable deflections, see a report by ACI Committee 435[2-3] and Ref. 2-10.

2.4 DEFLECTION OF NONPRESTRESSED ONE-WAY MEMBERS

The problem of predicting deflections of concrete flexural members is complicated by the effect of cracking along the span (Fig. 2-2), including that due to: uncertain construction loads at an early age; continuity; repeated loading; uncertain creep and shrinkage magnitudes; time of form removal and installation of nonstructural elements; and so on. A discussion of these effects may be found in Refs. 2-10, 2-27, and 2-28. The simplified procedures used herein for computing short-time, creep and shrinkage deflections were developed by the author[2-10, 2-20, 2-25, 2-36] and are recommended by ACI Committee 435[2-7] and the *ACI Code* and *Commentary*.[2-1] The basic method is also recommended by AASHTO.[2-9]

2.4.1 Initial or Short-Time Deflection

The effective moment of inertia[2-20] for cantilevers, simple beams, or between inflection points of continuous beams, is given by eq. (2-12):

$$I_e = (M_{cr}/M_a)^3 I_g + [1 - (M_{cr}/M_a)^3] I_{cr} \leqslant I_g \qquad (2\text{-}12)$$

where:

$$M_{cr} = f_r I_g / y_t \qquad (2\text{-}13)$$

M_a is the maximum service load moment (unfactored moment) at the stage for which deflections are being considered, and f_r is computed by eq. (2-1), (2-2), or (2-3). Values of I_g and I_{cr} may be computed using the equations in Fig. 2-3 or by various charts in the literature.[2-10, 2-12, 2-26, 2-28]

For prismatic members (including T-beams with different positive and negative region cracked sections, for example) in eq. (2-12), I_e (and thus M_a) may be determined at the support section for cantilevers and the midspan section for simple and continuous spans. Alternatively for continuous prismatic and nonprismatic beams, the *ACI Code*[2-1] suggests using the average I_e of critical positive and negative moment sections, and the *ACI Commentary*,[2-1] as well as ACI Committee 435,[2-7] recommends eqs. (2-14) and (2-15) for somewhat improved results[2-10] over the other two methods:

Beams with one end continuous:

$$\text{Avg. } I_e = 0.85\ I_m + 0.15(I_e\text{-Cont. End}) \qquad (2\text{-}14)$$

Beams with both ends continuous:

$$\text{Avg. } I_e = 0.70\ I_m + 0.15(I_{e1} + I_{e2}) \qquad (2\text{-}15)$$

TABLE 2-10 *ACI Code*,[2-1] **Section 9.5.3: Minimum Thicknesses for Two-Way Nonprestressed Concrete Construction for f_y = 60 ksi,[a] Unless Deflections Are Computed (from Ref. 2-37). Expressed as a fraction of the longer clear span, l_n**

Two-Way Construction	Minimum h	But Not Less Than	Need Not Exceed
	[b]*Eq. 9-11*	[b]*Eq. 9-12*	[b]*Eq. 9-13*
Solid flat plate:			
Square interior panel (Min. h = 5 in.)	$l_n/32.7$[a,c]	$l_n/41.8$	$l_n/32.7$[c]
Square side panel (Min. h = 5 in.)	$l_n/28.7$	$l_n/37.0$	$l_n/29.7$[c]
Square corner panel (Min. h = 5 in.)	$l_n/27.7$	$l_n/35.9$	$l_n/29.7$[c]
Solid flat plate and edge beams with stiffnesses such that α = 0.8			
Square side panel	$l_n/35.2$[c]	$l_n/40.7$	$l_n/32.7$
Square corner panel	$l_n/35.0$[c]	$l_n/39.$	$l_n/32.7$
Solid flat slab with drop panels of length $\geqslant l/3$ and depth $\geqslant 1.25\,h$:			
Square interior panel (Min. h = 4 in.)	$l_n/36.4$[c]	$l_n/46.6$	$l_n/36.4$[c]
Square side panel (Min. h = 4 in.)	$l_n/31.6$	$l_n/40.7$	$l_n/32.7$[c]
Square corner panel (Min. h = 4 in.)	$l_n/30.4$	$l_n/39.5$	$l_n/32.7$[c]
Solid flat slab with drop panels of length $\geqslant l/3$ and depth $\geqslant 1.25\,h$, and edge beams with stiffnesses such that α = 0.8:			
Square side panel	$l_n/39.1$[c]	$l_n/45.2$	$l_n/36.4$
Square corner panel	$l_n/37.9$[c]	$l_n/43.9$	$l_n/36.4$
Solid flat plate:			
Rectangular interior panel ($\beta = l_{n1}/l_{n2} = 2$, Min. h = 5 in.)	$l_n/32.7$[c]	$l_n/50.9$	$l_n/32.7$[c]
Rectangular side panel ($\beta = l_{n1}/l_{n2} = 2$, Min. h = 5 in.)	$l_n/28.2$	$l_n/44.2$	$l_n/29.7$[c]
Rectangular corner panel ($\beta = l_{n1}/l_{n2} = 2$, Min. h = 5 in.)	$l_n/26.6$	$l_n/42.1$	$l_n/29.7$[c]
Solid flat plate and edge beams with stiffnesses such that α = 0.8:			
Rectangular side panel ($\beta = l_{n1}/l_{n2} = 2$)	$l_n/38.3$[c]	$l_n/48.6$	$l_n/32.7$
Rectangular corner panel ($\beta = l_{n1}/l_{n2} = 2$)	$l_n/36.6$[c]	$l_n/46.4$	$l_n/32.7$
Solid flat slab with drop panel of length $\geqslant l/3$ and depth $\geqslant 1.25\,h$:			
Rectangular interior panel ($\beta = l_{n1}/l_{n2} = 2$, Min. h = 4 in.)	$l_n/36.4$[c]	$l_n/56.6$	$l_n/36.4$[c]
Rectangular side panel ($\beta = l_{n1}/l_{n2} = 2$, Min. h = 4 in.)	$l_n/31.0$	$l_n/48.6$	$l_n/32.7$[c]
Rectangular corner panel ($\beta = l_{n1}/l_{n2} = 2$, Min. h = 4 in.)	$l_n/29.3$	$l_n/46.4$	$l_n/32.7$[c]
Solid flat slab with drop panels of length $\geqslant l/3$ and depth $\geqslant 1.25\,h$, and edge beams with stiffnesses such that α = 0.8:			
Rectangular side panel ($\beta = l_{n1}/l_{n2} = 2$)	$l_n/42.5$[c]	$l_n/54.0$	$l_n/36.4$
Rectangular corner panel ($\beta = l_{n1}/l_{n2} = 2$)	$l_n/40.7$[c]	$l_n/51.5$	$l_n/36.4$

[a]*For f_y other than 60 ksi, multiply all numbers in the table by $[800 + (0.005)(60{,}000)]/[800 + 0.005\,f_y]$ = 1.048 for f_y = 50,000 psi and 1.100 for f_y = 40,000 psi.*
Examples:

$$(32.7)(1.048) = 34.3 \text{ or } l_n/34.3 \text{ for } f_y = 50{,}000 \text{ psi}$$
$$(32.7)(1.100) = 36.0 \text{ or } l_n/36.0 \text{ for } f_y = 40{,}000 \text{ psi}$$

[b]*Equations from ACI Code,*[2-1] *Section 9.5.3.*
[c]*Controlling limit.*

where I_m refers to the midspan section I_e, and I_{e1} and I_{e2} refer to I_e at the respective beam ends. Moment envelopes should be used in computing both positive and negative values of I_e. For a single heavy concentrated load, only the midspan I_e should be used.

The use of the midspan section properties for continuous prismatic members is considered satisfactory in approximate calculations primarily because the midspan rigidity (including the effect of cracking) has the dominant effect on deflections, as shown by ACI Committee 435.[2-5] For numerical integration solutions,[2-10, 2-31] including for nonprismatic members, the value of I_e at a particular section is given by eq. (2-12), with M_a changed to the moment at the section, and the power 3 changed to 4.

I_e in eq. (2-12) provides a transition between the upper and lower bounds of I_g and I_{cr}, as a function of the level of cracking in the form of M_{cr}/M_a. The equation empirically includes the effect of tension stiffening, or tensile concrete between cracks and in low tensile stress regions. A refinement that may be desirable for heavily reinforced members, or when using lightweight concrete (low E_c and hence higher modular ratio n and higher transformed steel area), is to replace I_g and I_{ucr} in eqs. (2-12) and (2-13)—the procedure used in the original development.[2-20]

For different load levels, the deflection should be computed in each case using eq. (2-12) for the total load level being considered, such as dead load or dead-plus-live load. The incremental deflection, such as for live load, is then

TABLE 2-11 Minimum Permissible Ratios of Span (*l*) to Deflection (*a*)—for Determining Allowable Computed Deflections (*ACI Code*[2-1])

Type of Member	*Deflection (a) to be Considered*	$(l/a)_{min}$
Flat roofs not supporting and not attached to nonstructural elements *likely* to be damaged by large deflections	Immediate deflection due to live load	180[a]
Floors not supporting and not attached to nonstructural elements *likely* to be damaged by large deflections	Immediate deflection due to live load	360
Roof or floor construction supporting or attached to nonstructural elements *likely* to be damaged by large defls.	That part of total deflection occurring after attachment of nonstructural elements. Sum of long-time duflection due to all sustained loads (dead load plus any sustained portion of live load) and immediate deflection due to any additonal live load[c]	480[b]
Roof or floor construction supporting or attached to nonstructural elements *not likely* to be damaged by large defls.		240[b]

[a] *Limit not intended to safeguard against ponding. Ponding should be checked by suitable calculations of deflection, including added deflections due to ponded water, and considering long-time effects of all sustained loads, camber, construction tolerances, and reliability of provisions for drainage.*
[b] *Ratio limit may be lower if adequate measures are taken to prevent damage to supported or attached elements, but shall not be lower than tolerance of nonstructural elements.*
[c] *Long-time deflection shall be determined in accordance with Section 9.5.2.5 or 9.5.4.2 of the Code but may be reduced by amount of deflection calculated to occur before attachment of nonstructural elements. This amount shall be determined on basis of accepted engineering data relating to time–deflection characteristics of members similar to those being considered.*

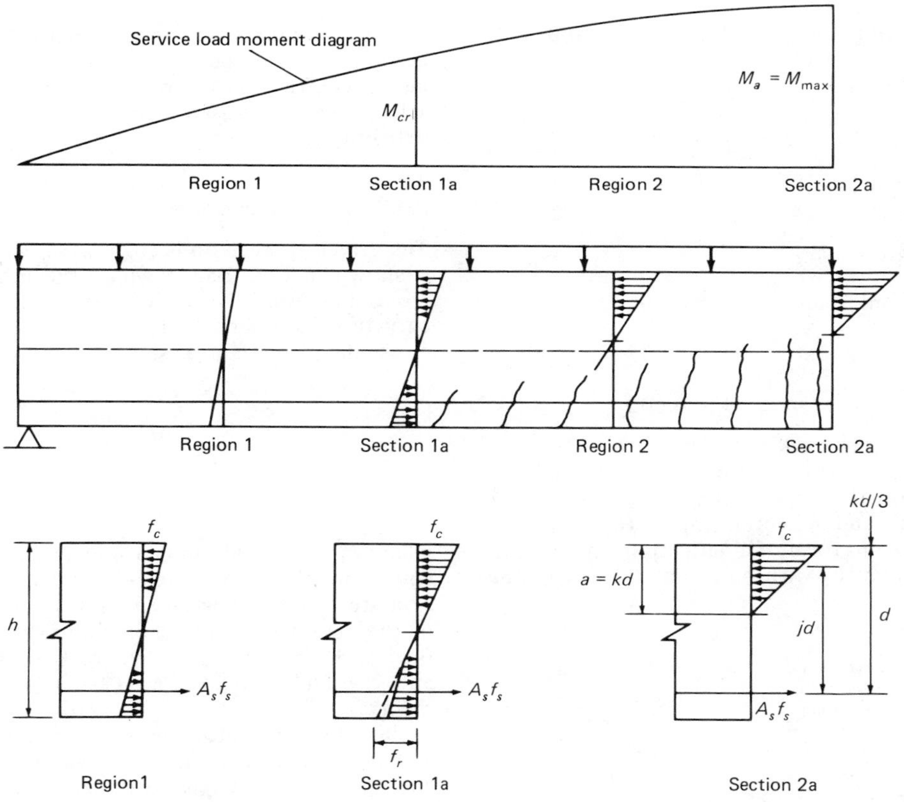

Fig. 2-2 Various regions of stress and cracking in a nonprestressed concrete beam carrying service loads.

computed as the difference in these values, as shown in idealized form in Fig. 2-4.

With regard to the effect of repeated loading, since the upper envelope of repeated load-deflection curves approximates the single loading curve,[2-10, 2-21, 2-36] even though increasing residual deflections (due primarily to creep and cracking effects) occur, it appears reasonable to compute short-time deflections using I_e and the residual deflections by an appropriate factor in terms of the total time-dependent or separate creep and shrinkage deflections.

The initial or short-time deflection, a_i, may be computed using the basic elastic eq. (2-16) for cantilever, simple, and continuous beams. In the case of continuous beams, the midspan deflection may normally be used as an approximation of the maximum deflection:

$$a_i = \frac{K(5/48)M_a l^2}{E_c I_e} \tag{2-16}$$

where M_a is the support moment for cantilevers and the midspan moment for simple and continuous beams, l is the span length (usually the clear span), and E_c is computed in eq. (2-4) or (2-5). For uniform loading, w, the theoretical values of the deflection coefficient, K, are as follows:

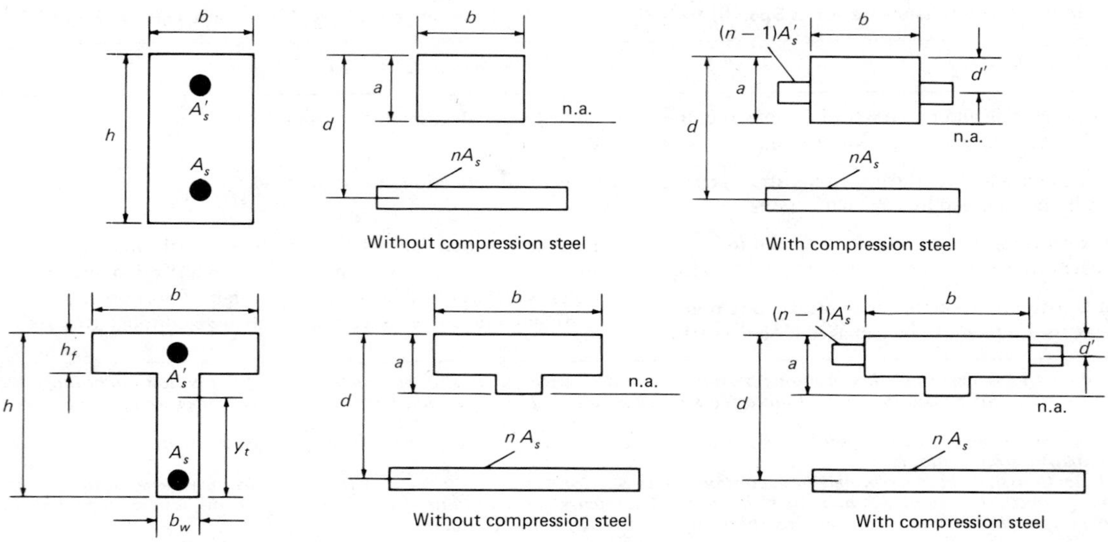

Fig. 2-3 Moment of inertia of gross section, I_g, and cracked transformed section, I_{cr}.

Cantilevers (the deflection due to rotation at the supports must be determined in addition)	$K = 12/5 = 2.400$
Simple beams	$K = 1.000$
Continuous beams (derived in Refs. 2-10, 2-12) where $M_0 = wl^2/8$ and M_a is the net midspan moment.	$K = 1.20 - 0.20M_0/M_a$
Continuous beams (with ideal end conditions):	
Fixed-hinged beam, midspan deflection	$K = 0.800$
Fixed-hinged beam, max. defl. using max. mom.	$K = 0.738$
Fixed-fixed beam	$K = 0.600$

For other types of loading, see Refs. 2-10 and 2-28.

Since deflections are logically computed for a given continuous span based on the same loading pattern for maximum positive moment and deflection, eq. (2-16) is thought to be the most convenient form of a deflection equation. In addition, in using eq. (2-16) with only the midspan I_e, the negative moments are not required in the deflection calculation for continuous beams, and the use of the *ACI Code* design coefficients for maximum positive moment is considered to be satisfactory in most cases for computing deflections.

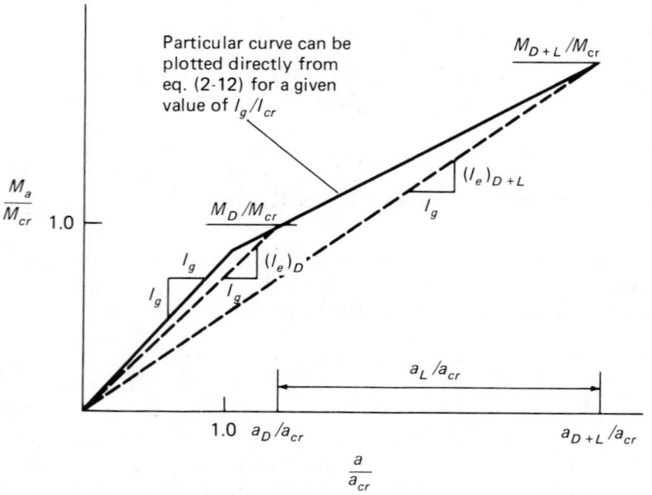

Fig. 2-4 Idealized moment versus short-time deflection diagram using eq. (2-12).[2-10]

2.4.2 Long-Time Deflection

The effect of creep on beam stresses, strains, and curvatures is shown in Fig. 2-5, in which the neutral axis is seen to move downward. As a result, the relative increase in creep curvature (and hence deflection) is less than the relative increase in creep strain, as shown in eqs. (2-17) and (2-18):

$$\frac{\phi_{cp}}{\phi_i} = k_r \frac{\epsilon_{cp}}{\epsilon_i} \tag{2-17}$$

$$\frac{a_{cp}}{a_i} = k_r \frac{\epsilon_{cp}}{\epsilon_i} \tag{2-18}$$

where k_r is a reduction factor less than 1 that takes into account principally the time-dependent effects of compression steel in reducing creep, and the downward movement of the neutral axis. Based on empirical results,[2-10, 2-25] $k_r = 0.85/(1 + 50\rho')$ in eq. (2-19) for the effect of creep alone, and $k_r = 1/(1 + 50\rho')$ in eq. (2-26) for combined creep and shrinkage effects.

The effect of reinforcing steel in reducing shrinkage strain, and thus inducing shrinkage curvature, is shown in Fig. 2-6 and eq. (2-21). The relative effect of shrinkage on curvature

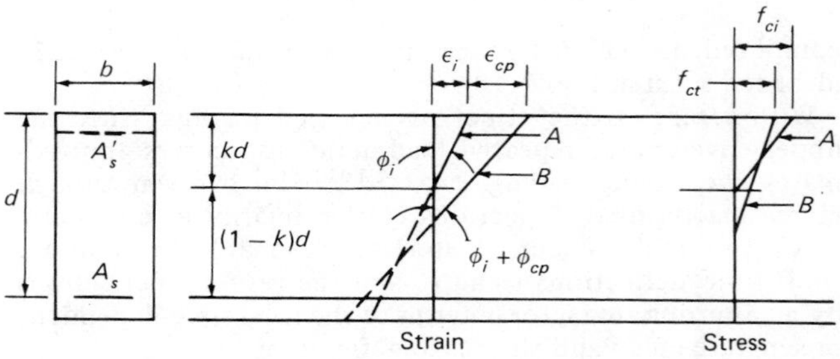

Fig. 2-5 Stress and strain distribution, and curvature in a nonprestressed concrete beam due to initial and creep deformation.

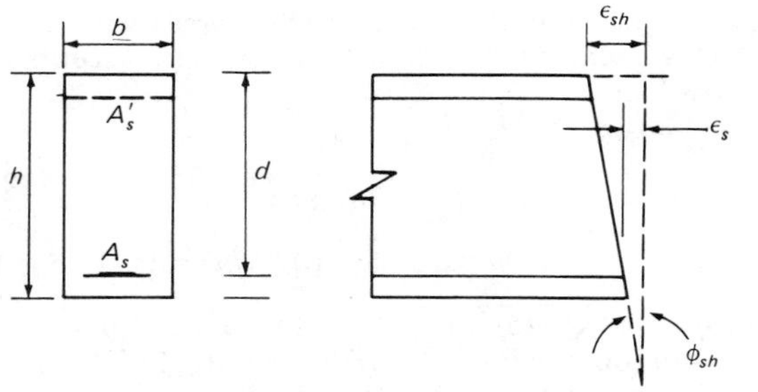

Fig. 2-6 Shrinkage curvature of a nonprestressed concrete beam.

depends on the steel content (ρ or the difference in ρ and ρ'). For example, when $\rho = \rho'$, presumably the shrinkage curvature is zero. Shrinkage curvature is computed by eq. (2-21), in which the relative effects of the tension and compression steel percentages (A_{sh} in eqs. 2-23 through 2-25, and Fig. 2-7) were determined empirically.[2-20]

For considering creep and shrinkage deflections separately, eqs. (2-19) through (2-25) are recommended by ACI Committee 435:[2-7]

$$\lambda_{cp} = k_r C_t = \frac{0.85 C_t}{1 + 50\rho'} \tag{2-19}$$

$$a_{cp} = \lambda_{cp} (a_i)_{sus} \tag{2-20}$$

$$\phi_{sh} = (\epsilon_{sh} - \epsilon_s)/d = (\epsilon_{sh}/d)\,[1 - (\epsilon_s/\epsilon_{sh})] = A_{sh}\epsilon_{sh}/h \tag{2-21}$$

$$a_{sh} = K_{sh}\phi_{sh} l^2 \tag{2-22}$$

where C_t and ϵ_{sh} may be determined by eqs. (2-7) and (2-8) and Table 2-2, and A_{sh} is obtained in Fig. 2-7, or by eq. (2-23), (2-24), or (2-25):

$$A_{sh} = 0.7(p - p')^{1/3} \left[\frac{p - p'}{p}\right]^{1/2} \quad \text{for } p - p' \leqslant 3.0\% \tag{2-23}$$

$$= 0.7p^{1/3} \quad \text{when } p' = 0 \tag{2-24}$$

$$A_{sh} = 1.0 \quad \text{for } p - p' > 3.0\% \tag{2-25}$$

The compression steel ratio, ρ', in eq. (2-19), and the steel percentages, p and p', used in determining A_{sh}, refer to the support section of cantilevers and the midspan section of simple and continuous beams. For T-beams, use $p = 100(\rho + \rho_w)/2$ in determining A_{sh}.

Based on the assumption of equal positive and negative shrinkage curvatures with an inflection point at the quarter-point of continuous spans (generally satisfactory for this purpose), the following values have been derived[2-10] for the shrinkage deflection coefficient, K_{sh}, in eq. (2-22):

Cantilevers	$K_{sh} = 1/2 = 0.500$
Simple beams	$K_{sh} = 1/8 = 0.125$
Spans with one end continuous—multi-span beams	$K_{sh} = 0.090$
Spans with one end continuous—two-span beams	$K_{sh} = 0.084$
Spans with both ends continuous	$K_{sh} = 0.065$

According to the *ACI Code*,[2-1] additional long-time deflections due to combined creep and shrinkage may be estimated by eqs. (2-26) and (2-27) for sustained loads:

$$\lambda = k_r T = \frac{T}{1 + 50\rho'} \tag{2-26}$$

$$a_{cp+sh} = \lambda (a_i)_{sus} \tag{2-27}$$

where ρ' is determined at the support section for cantilevers and the midspan section for simple and continuous spans. Various values of λ and T are given in eqs. (2-9) and (2-10), Tables 2-3, 2-4, and 2-5, and Fig. 2-1 by the *ACI Code*, AASHTO, ACI/435, and the author.

λ_{cp} in eq. (2-29) and λ in eq. (2-26) are presented in the form of two factors—k_r (a section property), and C_t or T (a material property), similar to EI. In using a weighted average of end and midspan values of I_e for continuous spans in eqs. (2-14) and (2-15), it is recommended that the same weighted averages for end and midspan values of k_r (and hence λ_{cp} and λ) be used, since both I_e and k_r relate to beam curvatures.

With regard to the choice of computing creep and shrinkage deflections separately in eqs. (2-20) and (2-22), versus combined creep and shrinkage deflections in eq. (2-27), the latter is simpler but provides only a rough approximation, since shrinkage deflections are only indirectly related to the loading (primarily by means of the steel content). One case in which the separate calculation of creep and shrinkage deflections may be preferable is that in which part of the live load is considered as a sustained load, as illustrated in an example by ACI Committee 435.[2-7]

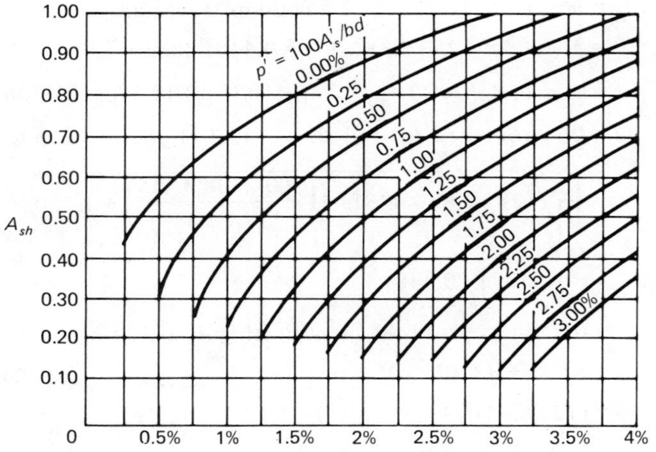

Fig. 2-7 Values of A_{sh} for calculating shrinkage curvature in eq. (2-21).[2-10]

2.4.3 Example 1—Simple-Span Rectangular Beam, Sand-Lightweight Concrete

Given: beam in Fig. 2-8, designed by the strength method:

$$f'_c = 4000 \text{ psi } (27.6 \text{ N/mm}^2),$$

$$f_y = 40{,}000 \text{ psi } (276 \text{ N/mm}^2)$$

Live load = 400 lb/ft (5.84 kN/m), 20% sustained

Required: analysis of short-time deflections, and long-time deflections at age 6 months and 5 years (ultimate values).

Loads and moments:

$$w_D = (120 \text{ pcf, T. 2-1})(12)(20)/144$$

$$= 200 \text{ lb/ft } (2.92 \text{ kN/m})$$

In eq. (2-11):

$$D = w_{sus} = 200 + (0.20)(400)$$

$$= 280 \text{ lb/ft}, L = 320 \text{ lb/ft}$$

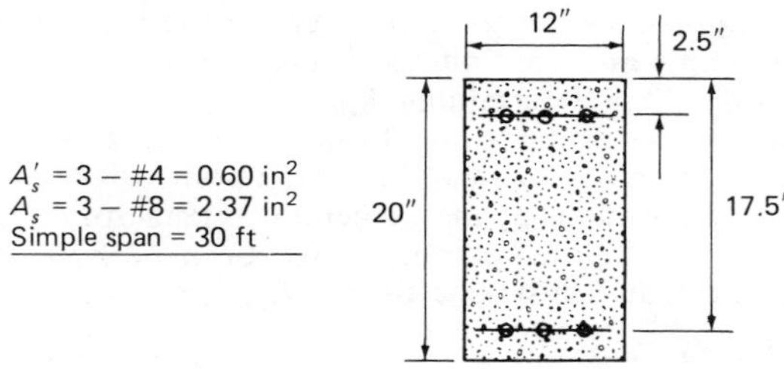

Fig. 2-8 Example 1—Simple-span rectangular beam, sand-lightweight concrete.

$$M_D = w_D l^2/8 = (0.200)(30)^2/8$$
$$= 22.50 \text{ ft-kips } (30.5 \text{ kN-m})$$
$$M_{sus} = (0.280)(30)^2/8 = 31.50 \text{ ft-kips}$$
$$M_L = (0.400)(30)^2/8 = 45.00 \text{ ft-kips},$$
$$M_{D+L} = 67.50 \text{ ft-kips}$$

Strength design, based on tension steel only (A'_s not required for strength):

$$M_u = 1.4(22.50) + 1.7(45.00) = 108.0 \text{ ft-kips } (146 \text{ kN-m})$$
$$a = \frac{A_s f_y}{(0.85 f'_c b)} = \frac{(2.37)(40)}{(0.85)(4)(12)} = 2.32 \text{ in. } (5.89 \text{ cm})$$
$$\phi M_n = \phi A_s f_y \left(d - \frac{a}{2}\right) = (0.90)(2.37)(40)(17.5 - 1.16)/12$$
$$= 116 \text{ ft-kips} \quad \text{OK}$$

Steel ratios:

$$\rho = \frac{A_s}{bd} = \frac{2.37}{(12)(17.5)} = 0.01129,$$
$$\rho' = \frac{A'_s}{bd} = \frac{0.60}{(12)(17.5)} = 0.00286$$

ρ_b (from handbook for f'_c = 4000 psi and f_y = 40,000 psi) = 0.0495

Tension steel ratio limitations (Table 2-6):

$$(\rho/\rho_b)\,100 = (0.01129/0.0495)\,100 = 22.5\%$$

versus 30% and 20% for lightweight concrete rectangular beams. Thus, the first condition for nonstructural elements in Table 2-6 is satisfied, but not the second. This agrees with the deflection check.

Minimum thickness:
1977 *ACI Code* (T. 2-7):

$$h_{min} = (l/16)(0.80 \text{ for } f_y)\,(1.09 \text{ for } w_c)$$
$$= (360/16)(0.80)(1.09)$$
$$= 19.6 \text{ in.} < h = 20 \text{ in. } (50.8 \text{ cm})$$

Thus, the condition for partitions and other construction in Table 2-7 is satisfied. This agrees with the deflection check.
ACI/435 (T. 2-8):

$$h_{min} = (l/18 \text{ or } l/14 \text{ or } l/12 \text{ or } l/10)(0.80)(1.09)$$
$$= 17.4 \text{ in. or } 22.4 \text{ in. or } 26.2 \text{ in. or } 31.4 \text{ in.}$$
$$(44.2 \text{ cm to } 79.8 \text{ cm})$$

Thus, the first condition for nonstructural elements in Table 2-8 for roof beams, $l/18$ (not for floor beams, $l/14$), is satisfied, but not the second. This agrees with the deflection check, except that floor beams are also acceptable.
Grossman:

$$\lambda = \frac{T_u}{1 + 50\rho'} = \frac{(2.0, \text{ T. 2-3})}{1 + (50)(0.00286)} = 1.75 \qquad \text{(eq. 2-26)}$$

d = 17.5 in. < 0.9h = 18 in., $(0.9h/d)$ = 18/17.5 = 1.029

$\sqrt{145/w_c} = \sqrt{145/(115, \text{ T. 2-1})}$ = 1.123; 0.4 + f_y/100 = 0.4 + 40/100 = 0.800; $\omega = \rho f_y/f'_c$ = 0.01129(40/4) = 0.113; from Table 2-9, (1 − 0.218) = 0.782.

$$h_{min} = \left(\frac{l}{24}\right)\left[\frac{(l/a)_{min}}{180}\right]\left[\frac{\lambda D + L}{D + L}\right]$$

$$(1.029)(1.123)(0.800)(0.782) \geqslant \frac{l}{24} \qquad \text{(eq. 2-11)}$$

$$= \left(\frac{360}{24}\right)\left[\frac{240 \text{ or } 480}{180}\right]\left[\frac{(1.75)(280) + 320}{280 + 320}\right](0.723)$$

$$> \frac{360}{24} = 15.0 \text{ in.}$$

$$= 19.5 \text{ in. } (49.5 \text{ cm}) \qquad \text{for } (l/a)_{min} = 240$$
$$= 39.0 \text{ in. } (99.1 \text{ cm}) \qquad \text{for } (l/a)_{min} = 480$$

Thus, the condition for nonstructural elements in Table 2-11 is satisfied for $(l/a)_{min}$ = 240 but not 480. This agrees with the deflection check. Equation (2-11) is not checked for $(l/a)_{min}$ = 180 and 360, since these values refer to live load only in Table 2-11.

Modulus of rupture, modulus of elasticity, modular ratio:

$$f_r = 6.38\sqrt{f'_c} = 6.38\sqrt{4000} = 404 \text{ psi } (2.79 \text{ N/mm}^2) \qquad \text{(eq. 2-1)}$$

$$E_c = 33\sqrt{w_c^3 f'_c} = 33\sqrt{(115)^3\,(4000)}$$
$$= 2.57 \times 10^6 \text{ psi } (17.7 \text{ kN/mm}^2) \qquad \text{(eq. 2-4)}$$

$$n = \frac{E_s}{E_c} = \frac{29}{2.57} = 11.3$$

Gross and cracked section moments of inertia (using Fig. 2-3):

$$I_g = bh^3/12 = (12)(20)^3/12 = 8000 \text{ in.}^4 (333{,}000 \text{ cm}^4)$$
$$B = b/(nA_s) = 12/(11.3)(2.37) = 0.4481 \text{ 1/in.}$$
$$r = (n - 1)A'_s/(nA_s) = (10.3)(0.60)/(11.3)(2.37) = 0.2308$$
$$a = [\sqrt{2dB(1 + rd'/d) + (1 + r)^2} - (1 + r)]/B \qquad \text{(eq. C)}$$
$$= \left[\sqrt{(2)(17.5)(0.4481)\left(1 + \frac{0.2308 \times 2.5}{17.5}\right) + 1.2308^2} - 1.2308\right]\Big/0.4481$$

= 6.65 in. (versus 6.88 in. neglecting A'_s and using eq. A)

$$I_{cr} = ba^3/3 + nA_s(d - a)^2 + (n - 1)A'_s(a - d')^2 \qquad \text{(eq. D)}$$
$$= (12)(6.65)^3/3 + (11.3)(2.37)(17.5 - 6.65)^2 + (10.3)(0.60)(6.65 - 2.5)^2$$

= 4435 in.^{4} (184,600 cm^4) (versus 4323 in.^{4} neglecting A'_s and using eq. B)

Effective moment of inertia (using eqs. 2-12 and 2-13):

$$M_{cr} = f_r I_g / y_t = (404)(8000)/(10)(12{,}000)$$
$$= 26.93 \text{ ft-kips } (36.5 \text{ kN-m})$$
$$M_{cr}/M_D = 26.93/22.50 > 1; \text{ hence } (I_e)_D$$
$$= I_g = 8000 \text{ in.}^4$$
$$(M_{cr}/M_{sus})^3 = (26.93/31.50)^3 = 0.625$$
$$(I_e)_{sus} = (M_{cr}/M_a)^3 I_g + [1 - (M_{cr}/M_a)^3]\, I_{cr} \leqslant I_g$$
$$= (0.625)(8000) + (1 - 0.625)(4435)$$
$$= 6663 \text{ in.}^4$$
$$(M_{cr}/M_{D+L})^3 = (26.93/67.50)^3 = 0.064$$
$$(I_e)_{D+L} = (0.064)(8000) + (1 - 0.064)(4435)$$
$$= 4663 \text{ in.}^4$$

Initial or short-time deflection (using eq. 2-16):

$$(a_i)_D = \frac{K(5/48)\, M_D l^2}{E_c (I_e)_D}$$
$$= \frac{(1)(5/48)(22.50)(30)^2(12)^3}{(2570)(8000)}$$
$$= 0.177 \text{ in. } (4.5 \text{ mm})$$
$$(a_i)_{sus} = \frac{K(5/48)\, M_{sus} l^2}{E_c (I_e)_{sus}}$$
$$= \frac{(1)(5/48)(31.50)(30)^2(12)^3}{(2570)(6663)} = 0.298 \text{ in.}$$
$$(a_i)_{D+L} = \frac{K(5/48)\, M_{D+L} l^2}{E_c (I_e)_{D+L}}$$
$$= \frac{(1)(5/48)(67.50)(30)^2(12)^3}{(2570)(4663)} = 0.912 \text{ in.}$$

$(a_i)_L$ from Fig. 2-4 $= (a_i)_{D+L} - (a_i)_D = 0.912 - 0.117 =$ 0.795 in. (20.2 mm) versus the following allowable deflections from Table 2-11:

Flat roofs not supporting and not attached to nonstructural elements likely to be damaged by large deflections:

$$(a_i)_L \leqslant l/180 = 360/180 = 2.00 \text{ in.} \qquad \text{OK}$$

Floors not supporting and not attached to nonstructural elements likely to be damaged by large deflections:

$$(a_i)_L \leqslant l/360 = 360/360 = 1.00 \text{ in.} \qquad \text{OK}$$

Long-time deflection (using eqs. 2-19 to 2-27):

Separate creep and shrinkage deflections (ACI/435), eqs. (2-19) to (2-23):

$$\lambda_{cp} = \frac{0.85 C_u}{1 + 50\rho'} = \frac{(0.85)(1.60, \text{ eq. 2-7})}{1 + (50)(0.00286)} = 1.19$$
$$a_{cp} = \lambda_{cp} (a_i)_{sus} = (1.19)(0.298) = 0.355 \text{ in. } (9.0 \text{ mm})$$

From Fig. 2-7, or eq. (2-23):

$$A_{sh} = 0.7(1.129 - 0.286)^{1/3} \left[\frac{1.129 - 0.286}{1.129}\right]^{1/2} = 0.571$$
$$\phi_{sh} = A_{sh} (\epsilon_{sh})_u / h = (0.571)(400 \times 10^{-6}, \text{ eq. 2-8})/20$$
$$= 11.42 \times 10^{-6} \text{ 1/in.}$$
$$a_{sh} = K_{sh} \phi_{sh} l^2 = (1/8)(11.42 \times 10^{-6})(30)^2(12)^2$$
$$= 0.185 \text{ in.}$$
$$a_{cp} + a_{sh} + (a_i)_L = 0.355 + 0.185 + 0.795$$
$$= \underline{1.34 \text{ in. } (34 \text{ mm})}$$

versus 1.18 in. at age 6 months using $C_t = 1.09$ (instead of 1.60) and $\epsilon_{sh} = 308 \times 10^{-6}$ in./in. (instead of 400×10^{-6}) from the example following eqs. (2-7) and (2-8).

Combined creep and shrinkage deflections (*ACI Code*), eqs. (2-26) and (2-27):

$$\lambda = \frac{T_u}{1 + 50\rho'} = \frac{(2.0, \text{ T. 2-3})}{1 + (50)(0.00286)} = 1.75$$
$$a_{cp+sh} = \lambda (a_i)_{sus} = (1.75)(0.298) = 0.522 \text{ in.}$$
$$a_{cp+sh} + (a_i)_L = 0.522 + 0.795 = \underline{1.32 \text{ in. } (34 \text{ mm})}$$

versus 1.11 in. at age 6 months using $T = 1.2$ (instead of 2.0) from Table 2-3.

These computed ultimate deflections, 1.34 in. and 1.32 in., are compared with the allowable deflections in Table 2-11 as follows:

Roof or floor construction supporting or attached to nonstructural elements *likely* to be damaged by large deflections (very stringent limitation):

$$a_{cp} + a_{sh} + (a_i)_L \leqslant \frac{l}{480} = \frac{360}{480} = 0.75 \text{ in.}$$

Not OK by both methods

Roof or floor construction supporting or attached to nonstructural elements *not likely* to be damaged by large deflections:

$$a_{cp} + a_{sh} + (a_i)_L \leqslant \frac{l}{240} = \frac{360}{240} = 1.50 \text{ in.}$$

OK by both methods

2.4.4 Example 2—Continuous T-Beam, Normal Weight Concrete

Given: beam in Fig. 2-9, designed by the strength method (A'_s not required for strength):

$$f'_c = 3000 \text{ psi } (20.7 \text{ N/mm}^2),$$
$$f_y = 50{,}000 \text{ psi } (345 \text{ N/mm}^2)$$
$$\text{Live load} = 80 \text{ psf } (3.83 \text{ kN/m}^2), \quad 30\% \text{ sustained}$$

The beam will be assumed to be continuous at one end only for $h_{\min}$ in Tables 2-7 and 2-8 and eq. (2-11), for Avg. I_e (and $k_r - \lambda_{cp}$ and λ) in eq. (2-14) and for K_{sh} in eq. (2-22), since the exterior end is supported by a spandrel beam. The end span might be assumed to be continuous at both ends when supported by an exterior column.

Required: analysis of short-time and long-time deflections of the end span.

Loads and moments:

$$w_D = (150 \text{ pcf, T. 2-1})(14 \times 19 + 120 \times 5)/144$$
$$= 902 \text{ lb/ft } (13.2 \text{ kN/m})$$
$$w_L = (80)(10) = 800 \text{ lb/ft}$$

In eq. (2-11):

$$D = w_{sus} = 902 + (0.30)(800) = 1142 \text{ lb/ft}, \quad L = 560 \text{ lb/ft}$$

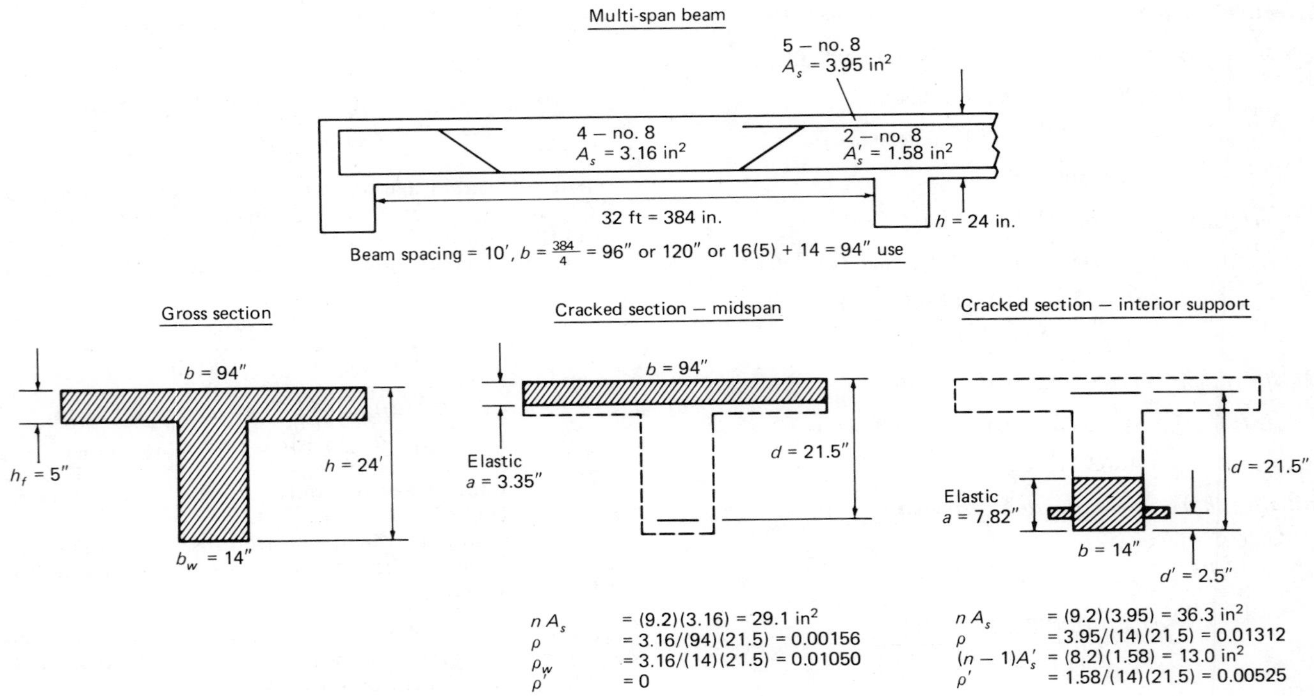

Fig. 2-9 Example 2—Continuous T-beam, normal weight concrete.

In lieu of a moment analysis, the *ACI Code* design moment coefficients may be used as follows: Pos. $M = wl_n^2/14$ for positive I_e and maximum deflection, and Neg. $M = wl_n^2/10$ for negative I_e.

$$\text{Pos. } M_D = w_D l_n^2/14 = (0.902)(32)^2/14$$
$$= 66.0 \text{ ft-kips (89.5 kN-m)}$$
$$\text{Pos. } M_{sus} = (1.142)(32)^2/14 = 83.5 \text{ ft-kips}$$
$$\text{Pos. } M_L = (0.800)(32)^2/14 = 58.5 \text{ ft-kips},$$
$$\text{Pos. } M_{D+L} = 124.5 \text{ ft-kips}$$
$$\text{Neg. } M_D = (0.902)(32)^2/10 = 92.4 \text{ ft-kips}$$
$$\text{Neg. } M_{sus} = (1.142)(32)^2/10 = 116.9 \text{ ft-kips}$$
$$\text{Neg. } M_L = (0.800)(32)^2/10 = 81.9 \text{ ft-kips},$$
$$\text{Neg. } M_{D+L} = 174.3 \text{ ft-kips}$$

Tension steel ratio limitations (Table 2-6, positive region):

ρ_b (from handbook for

$$f_c' = 3000 \text{ psi}, \ f_y = 50{,}000 \text{ psi, rect. comp. area)}$$
$$= 0.0275; \ (\rho_w/\rho_b)\,100$$
$$= [(0.01050, \text{Fig. 2-9})/0.0275]\,100 = 38.2\%$$

versus 40% and 30% for normal weight concrete T-beams. Thus, the first condition for nonstructural elements in Table 2-6 is satisfied, but not the second. Both conditions in Table 2-6 are satisfied in the deflection check.

Minimum thickness:

1977 *ACI Code* (T. 2-7):

$$h_{min} = (l/18.5)(0.90 \text{ for } f_y)$$
$$= (384/18.5)(0.90)$$
$$= 18.7 \text{ in.} < h = 24 \text{ in. (61.0 cm)}$$

Thus, the condition for partitions and other construction in Table 2-7 is satisfied. This agrees with the deflection check.

ACI/435 (T. 2-8):

$$h_{min} = (l/23) \text{ or } l/18 \text{ or } l/15 \text{ or } l/13)(0.90 \text{ for } f_y)$$
$$= 15.0 \text{ in or } 19.2 \text{ in. or } 23.0 \text{ in. or } 26.6 \text{ in.}$$
$$(38.1 \text{ cm to } 67.6 \text{ cm})$$

Thus, both conditions for nonstructural elements in Table 2-8 are satisfied, except the second condition for floor beams, $l/13$. This agrees with the deflection check, except that the latter floor-beam case is also acceptable.

Grossman:

$$\lambda = \frac{T_u}{1 + 50\rho'} = \frac{(2.0, \text{T. 2-3})}{1 + 0} = 2.00 \quad \text{(eq. 2-26)}$$

One end continuous–0.800; $0.4 + f_y/100 = 0.4 + 50/100 = 0.900$; $d = 21.5$ in. $< 0.9h = 0.9(24) = 21.6$ in.; $(0.9h/d) = 21.6/21.5 = 1.005$; $b_w/b = 14/94 = 0.149$; $h_f/h = 5/24 = 0.208$; $\omega = \rho f_y/f_c' = 0.00156(50/3) = 0.026$; from Table 2-9, $(0.149 - 0.100/0.900)(4.5) = 0.25$, $(3.75 - 2.60)/(3.75 - 1.75) = 0.58$, $(1 - 0.358) = 0.642$.

$$h_{min} = \left(\frac{l}{24}\right)\left[\frac{(l/a)_{min}}{180}\right]\left[\frac{\lambda D + L}{D + L}\right]$$
$$\cdot (0.800)(0.900)(1.005)(0.642) \geqslant \frac{l}{24} \quad \text{(eq. 2-11)}$$
$$= \left(\frac{384}{24}\right)\left[\frac{240 \text{ or } 480}{180}\right]\left[\frac{(2.00)(1142) + 560}{1142 + 560}\right]$$
$$\cdot (0.465) > \frac{384}{24} = 16.0 \text{ in.}$$
$$= 16.6 \text{ in. (42.2 cm)} \quad \text{for } (l/a)_{min} = 240$$
$$= 33.2 \text{ in. (84.3 cm)} \quad \text{for } (l/a)_{min} = 480$$

Thus, the condition for nonstructural elements in Table 2-11 is satisfied for $(l/a)_{min} = 240$ but not 480. Both of these conditions are satisfied in the deflection check. Equation (2-11) is not checked for $(l/a)_{min}$ = 180 and 360, since these values refer to live load only in Table 2-11.

Modulus of rupture, modulus of elasticity, modular ratio:

$$f_r = 7.5\sqrt{f_c'} = 7.5\sqrt{3000} = 411 \text{ psi } (2.83 \text{ N/mm}^2) \quad \text{(eq. 2-1)}$$

$$E_c = 33\sqrt{w_c^3 f_c'} = 33\sqrt{(145)^3(3000)}$$

$$= 3.16 \times 10^6 \text{ psi } (21.8 \text{ kN/mm}^2) \quad \text{(eq. 2-4)}$$

$$n = \frac{E_s}{E_c} = \frac{29}{3.16} = 9.2$$

Gross and cracked section moments of inertia (using Fig. 2-3):

Pos. sec.:

$$y_t = h - (1/2)\,[(b - b_w)\,h_f^2 + b_w h^2]/[(b - b_w)\,h_f + b_w h]$$

$$= 24 - (1/2)\,[(80)(5)^2 + (14)(24)^2]/[(80)(5) + (14)(24)] = 17.16 \text{ in.}$$

$$I_g = (b - b_w)\,h_f^3/12 + b_w h^3/12 + (b - b_w)\,h_f(h - h_f/2 - y_t)^2 + b_w h(y_t - h/2)^2$$

$$= (80)(5)^3/12 + (14)(24)^3/12 + (80)(5)(24 - 2.5 - 17.16)^2 + (14)(24)(17.16 - 12)^2$$

$$= 33{,}440 \text{ in.}^4 \ (1{,}392{,}000 \text{ cm}^4)$$

$$B = b/(nA_s) = 94/(9.2)(3.16) = 3.233 \text{ 1/in.}$$

$$a = (\sqrt{2dB + 1} - 1)/B$$

$$= [\sqrt{(2)(21.5)(3.233) + 1} - 1]/3.233 = 3.35 \text{ in.}$$

$< h_f = 5$ in. (hence, treat as rectangular compression area)

$$I_{cr} = ba^3/3 + nA_s(d - a)^2 = (94)(3.35)^3/3 + (9.2)(3.16)(21.5 - 3.35)^2 = 10{,}750 \text{ in.}^4$$

Neg. sec:

$I_g = 33{,}440$ in.^{4} (same as Pos. Sec.);

$I_c = 9399$ in.^{4}

(similar to Example 1, for $b = 14$ in., $d = 21.5$ in., $d' = 2.5$ in., $A_s = 3.95$ in.^{2}, $A_s' = 1.58$ in.^{2}).

Effective moment of inertia (using eqs. 2-12, 2-13, and 2-14):

Pos. sec.:

$$M_{cr} = f_r I_g/y_t = (411)(33{,}440)/(17.16)(12{,}000) = 66.7 \text{ ft-kips}$$

$$M_{cr}/M_D = 66.7/66.0 > 1; \text{ hence } (I_e)_D = I_g = 33{,}440 \text{ in.}^4$$

$$(M_{cr}/M_{sus})^3 = (66.7/83.5)^3 = 0.510$$

$$(I_e)_{sus} = (M_{cr}/M_a)^3\, I_g + [1 - (M_{cr}/M_a)^2]\, I_{cr} \leqslant I_g$$

$$= (0.510)(33{,}440) + (1 - 0.510)(10{,}750) = 22{,}320 \text{ in.}^4$$

$$(M_{cr}/M_{D+L})^3 = (66.7/124.5)^3 = 0.154$$

$$(I_e)_{D+L} = (0.154)(33{,}440) + (1 - 0.154)(10{,}750) = 14{,}240 \text{ in.}^4$$

Neg. sec.:

$$M_{cr} = (411)(33{,}440)/(24 - 17.16)(12{,}000) = 167.4 \text{ ft-kips}$$

$$M_{cr}/M_D = 167.4/92.4 > 1; \text{ hence } (I_e)_D = I_g = 33{,}440 \text{ in.}^4$$

$$M_{cr}/M_{sus} = 167.4/116.9 > 1; \text{ hence } (I_e)_{sus} = I_g = 33{,}440 \text{ in.}^4$$

$$(M_{cr}/M_{D+L})^3 = (167.4/174.3)^3 = 0.886$$

$$(I_e)_{D+L} = (0.886)(33{,}440) + (1 - 0.886)(9399) = 30{,}700 \text{ in.}^4$$

Average values:

$$\text{Avg. } (I_e)_D = I_g = 33{,}440 \text{ in.}^4 \ (1{,}392{,}000 \text{ cm}^4)$$

$$\text{Avg. } (I_e)_{sus} = 0.85 I_m + 0.15(I_{e\text{-Cont. End}}) = 0.85(22{,}320) + 0.15(33{,}440) = 23{,}990 \text{ in.}^4 \ (998{,}500 \text{ cm}^4)$$

$$\text{Avg. } (Ie)_{D+L} = 0.85(14{,}240) + 0.15(30{,}700) = 16{,}710 \text{ in.}^4 \ (695{,}500 \text{ cm}^4)$$

Initial or short-time deflection (using Eq. 2-16):

$$K = 1.20 - 0.20 M_0/M_a = 1.20 - (0.20)(wl_n^2/8)/(wl_n^2/14) = 0.850$$

$$(a_i)_D = \frac{K(5/48)\,M_D l^2}{E_c(I_e)_D} = \frac{(0.850)(5/48)(66.0)(32)^2(12)^3}{(3160)(33{,}440)} = 0.098 \text{ in. } (2.5 \text{ mm})$$

The same result is obtained using midspan $(I_e)_D$ and Avg. $(I_e)_D$.

$$(a_i)_{sus} = \frac{K(5/48)\,M_{sus} l^2}{E_c(I_e)_{sus}} = \frac{(0.850)(5/48)(83.5)(32)^2(12)^3}{(3160)(22{,}320 \text{ or } 23{,}990)}$$

= 0.185 in. using midspan $(I_e)_{sus}$ and 0.173 in. using Avg. $(I_e)_{sus}$

$$(a_i)_{D+L} = \frac{K(5/48)\,M_{D+L} l^2}{E_c(I_e)_{D+L}} = \frac{(0.850)(5/48)(124.5)(32)^2(12)^3}{(3160)(14{,}240 \text{ or } 16{,}710)}$$

= 0.433 in. using midspan $(I_e)_{D+L}$ and 0.369 in. using Avg. $(I_e)_{D+L}$

$(a_i)_L$ from Fig. 2-4

$= (a_i)_{D+L} - (a_i)_D = (0.433 \text{ or } 0.369) - 0.098$

= <u>0.335 in. (8.5 mm)</u> using midspan I_e and <u>0.271 in.</u> using Avg. I_e

versus the following allowable deflections from Table 2-11: *Flat roofs* not supporting and not attached to nonstructural elements likely to be damaged by large deflections:

$$(a_i)_L \leqslant \frac{l}{180} = \frac{384}{180} = 2.13 \text{ in.} \quad \text{OK}$$

Floors not supporting and not attached to nonstructural elements likely to be damaged by large deflections:

$$(a_i)_L \leqslant \frac{l}{360} = \frac{384}{360} = 1.07 \text{ in.} \qquad \text{OK}$$

Long-time deflection (*using Eqs. 2-19 to 2-27*):

Separate creep and shrinkage deflections (ACI/435), eqs. (2-19) to (2-24):

$$\text{Pos. } \lambda_{cp} = \frac{0.85C_u}{1+50\rho'} = \frac{(0.85)(1.60, \text{Eq. 2-7})}{1+0} = 1.36$$

$$\text{Neg. } \lambda_{cp} = \frac{(0.85)(1.60)}{1+(50)(0.00525)} = 1.08;$$

$$\text{Avg. } \lambda_{cp} = 0.85(1.36) + 0.15(1.08) = 1.32$$

$$a_{cp} = \lambda_{cp}(a_i)_{sus} = (1.36 \text{ or } 1.32)(0.185 \text{ or } 0.173)$$

$$= 0.252 \text{ in. using midspan values and}$$

$$0.228 \text{ in. using average values}$$

$$p = 100(\rho + \rho_w)/2 = 100(0.00156 + 0.01050)/2$$

$$= 0.603 \text{ for positive section}$$

From Fig. 2-7 or eq. (2-24):

$$A_{sh} = 0.7p^{1/3} = 0.7(0.603)^{1/3} = 0.591$$

$$\phi_{sh} = A_{sh}(\epsilon_{sh})_u/h = (0.591)(400 \times 10^{-6}, \text{eq. 2-8})/24$$

$$= 9.85 \times 10^{-6} \text{ 1/in.}$$

$$a_{sh} = K_{sh}\phi_{sh}l^2 = (0.090)(9.85 \times 10^{-6})(32)^2(12)^2$$

$$= 0.131 \text{ in.}$$

$$a_{cp} + a_{sh} + (a_i)_L$$

$$= (0.252 \text{ or } 0.228) + 0.131 + (0.335 \text{ or } 0.271)$$

$$= \underline{0.72 \text{ in. (18 mm)}} \text{ using midspan values and}$$

$$\underline{0.63 \text{ in.}} \text{ using average values}$$

Combined creep and shrinkage deflections (*ACI Code*), eqs. (2-26) and (2-27):

$$\text{Pos. } \lambda = \frac{T_u}{1+50\rho'} = \frac{(2.0, \text{T. 2-3})}{1+0} = 2.00$$

$$\text{Neg. } \lambda = \frac{2.0}{1+(50)(0.00525)} = 1.58;$$

$$\text{Avg. } \lambda = 0.85(2.00) + 0.15(1.58) = 1.94$$

$$a_{cp+sh} = \lambda(a_i)_{sus} = (2.00 \text{ or } 1.94)(0.185 \text{ or } 0.173)$$

$$= 0.370 \text{ in. using midspan values and}$$

$$0.336 \text{ in. using average values}$$

$$a_{cp+sh} + (a_i)_L$$

$$= (0.370 \text{ or } 0.336) + (0.335 \text{ or } 0.271)$$

$$= \underline{0.71 \text{ in. (18 mm)}} \text{ using midspan values and}$$

$$= \underline{0.61 \text{ in.}} \text{ using average values}$$

These computed ultimate deflections ranging from 0.61 in. to 0.72 in. are compared with the allowable deflections in Table 2-11 as follows:

Roof or floor construction supporting or attached to nonstructural elements *likely* to be damaged by large deflections (very stringent limitation):

$$a_{cp} + a_{sh} + (a_i)_L \leqslant \frac{l}{480} = \frac{384}{480} = 0.80 \text{ in.} \qquad \text{All OK}$$

Roof or floor construction supporting or attached to nonstructural elements *not likely* to be damaged by large deflections:

$$a_{cp} + a_{sh} + (a_i)_L \leqslant \frac{l}{240} = \frac{384}{240} = 1.60 \text{ in.} \qquad \text{All OK}$$

In the case of continuous rectangular beams and one-way slabs, the positive I_e tends to be larger than the negative I_e because of the smaller positive moment, and hence computed deflections using midspan values only tend to be *smaller* than those using average values. As Example 2 shows in the case of continuous T-beams, the positive I_e tends to be smaller (not larger) than the negative I_e, primarily because of the much smaller y_t and much larger M_{cr} in the negative region, and hence computed deflections using midspan values only tend to be *larger* than those using average values.

2.5 DEFLECTION OF NONPRESTRESSED COMPOSITE MEMBERS

The ultimate (in time) deflection of unshored and shored composite members is computed by eqs. (2-28) through (2-31). These equations are derived in detail in Refs. 2-10 and 2-13. Subscripts 1 and 2 are used to refer to the slab (or effect of the slab such as under slab dead load) and precast beam, respectively.

These procedures are described for a composite beam in which both unshored and shored construction are assumed. Examples 3 and 4 demonstrate the beneficial effect of shoring in reducing deflections.

2.5.1 Unshored Composite Members

$$a_u = \overbrace{(a_i)_2}^{(1)} + \overbrace{0.77k_r(a_i)_2}^{(2)} + \overbrace{0.83k_r(a_i)_2\frac{I_2}{I_c}}^{(3)} + \overbrace{0.36a_{sh}}^{(4)}$$

$$+ \overbrace{0.64a_{sh}\frac{I_2}{I_c}}^{(5)} + \overbrace{(a_i)_1}^{(6)} + \overbrace{1.22k_r(a_i)_1\frac{I_2}{I_c}}^{(7)}$$

$$+ \overbrace{a_{DS}}^{(8)} + \overbrace{(a_i)_L}^{(9)} + \overbrace{(a_{cp})_L}^{(10)} \qquad (2\text{-}28)$$

When $k_r = 0.85$ in eq. (2-19) (no compression steel in the precast beam) and a_{DS} is assumed to be equal to $0.50(a_i)_1$, eq. (2-28) reduces to eq. (2-29):

$$a_u = \overbrace{\left(1.65 + 0.71\frac{I_2}{I_c}\right)}^{(1+2+3)}(a_i)_2 + \overbrace{\left(0.36 + 0.64\frac{I_2}{I_c}\right)}^{(4+5)}a_{sh}$$

$$+ \overbrace{\left(1.50 + 1.04\frac{I_2}{I_c}\right)}^{(6+7+8)}(a_i)_1 + \overbrace{(a_i)_L}^{(9)} + \overbrace{(a_{cp})_L}^{(10)} \qquad (2\text{-}29)$$

In eqs. (2-28) and (2-29), the part of the total creep and shrinkage occurring before and after slab casting is based on the assumption of a precast beam age of 20 days when its dead load is applied and 2 months when the composite slab is cast.

Term (1) is the *initial or short-time dead load deflection of the precast beam* using eq. (2-16), in which $M_a = M_2$ = midspan moment due to the precast beam dead load. For computing $(I_e)_2$ in eq. (2-12), M_a refers to the precast beam dead load, and M_{cr}, I_g, and I_{cr} to the precast beam section at midspan.

Term (2) is the *dead load creep deflection of the precast beam up to the time of slab casting* using eq. (2-20), in which C_t = (0.48 at age 2 months in Table 2-2) (1.60 in eq. 2-7) = 0.77, and ρ' refers to the compression steel in the precast beam at midspan when computing k_r in eq. (2-19).

Term (3) is the *creep deflection of the composite beam following slab casting due to the precast beam dead load* using eq. (2-20), in which C_u = 1.60 − 0.77 = 0.83. The quantity k_r is the same as in Term (2). The ratio I_2/I_c modifies the initial stress (strain) and accounts for the effect of the composite section in restraining additional creep curvature (strain) after the composite section becomes effective. As a simple approximation, $I_2/I_c = [(I_2/I_c)_g + (I_2/I_c)_{cr}]/2$ may be used.

Term (4) is the *deflection due to shrinkage warping of the precast beam up to the time of slab casting* using eq. (2-22) to compute a_{sh}, in which 0.36 for steam cured concrete (assumed to be the usual case for precast beams) in Table 2-2 at age 2 months refers to 36% of the total shrinkage. From eq. (2-8), $(\epsilon_{sh})_u = 400 \times 10^{-6}$ in./in.

Term (5) is the *shrinkage deflection of the composite beam following slab casting due to the shrinkage of the precast beam concrete* using eq. (2-22), in which 1.00 − 0.36 = 0.64 or 64% of the total shrinkage. This term does not include the effect of differential shrinkage and creep, which is given by Term (8). The ratio I_2/I_c is the same as in Term (3).

Term (6) is the *initial or short-time deflection of the precast beam under slab dead load* using eq. (2-16), in which the incremental deflection (as shown in Fig. 2-4) is computed as follows: $(a_i)_1 = (a_i)_{1+2} - (a_i)_2$, where $(a_i)_2$ is the same as in Term (1). For computing $(I_e)_{1+2}$ and $(a_i)_{1+2}$ in eqs. (2-12) and (2-16), $M_a = M_1 + M_2$ due to the precast beam plus slab dead load at midspan, and M_{cr}, I_g, and I_{cr} refer to the precast beam section at midspan. When partitions, roofing, and so on, are placed at the same time as the slab, or soon thereafter, their dead load should be included in M_1 and M_a.

Term (7) is the *creep deflection of the composite beam due to slab dead load* using eq. (2-20), in which (0.76—loading age correction factor at age 2 months from Ref. 2-10) (1.60) = 1.22. In this term the initial strains, curvatures, and deflections under slab dead load were based on the precast section only. Hence the creep curvatures and deflections refer to the precast beam concrete, although the composite section is restraining the creep curvatures and deflections, as mentioned in Term (3). The quantity k_r is the same as in Term (2), and ratio I_2/I_c is the same as in Term (3).

Term (8) is the *deflection due to differential shrinkage and creep.* As an approximation, $a_{DS} = 0.50(a_i)_1$ may be used. For more details, see Refs. 2-10 and 2-13.

Term (9) is the *initial or short-time live load (plus other loads applied to the composite beam and not included in Term 6) deflection of the composite beam* using eq. (2-16), in which the incremental deflection (as shown in Fig. 2-4) is estimated as follows: $(a_i)_L = (a_i)_{D+L} - (a_i)_D$, based on the composite section only. This is thought to be the best and conservative [since the computed $(a_i)_D$ is on the low side, and thus the computed $(a_i)_L$ is on the high side] approximation, even though the incremental loads are actually resisted by different sections (members). This method is the same as Term (5) of eq. (2-30)—the same as for a monolithic beam. Alternatively, eq. (2-16) may be used with $M_a = M_L$ and $I_e = (I_c)_{cr}$ as a simpler rough approximation. The first method is illustrated in Example 4, and the alternative method in Example 3.

Term (10) is the *partial live load creep deflection for any sustained live load (and other loads) applied to the composite beam* using eq. (2-20), in which C_u = 1.60 from eq. (2-7), and ρ' refers to any compression steel in the slab at midspan when computing k_r in eq. (2-19).

2.5.2 Shored Composite Members

It is assumed in eqs. (2-30) and (2-31) that the composite beam supports all of the dead and live load. The calculation of deflections for shored composite beams is essentially the same as for monolithic beams, except for the deflection due to shrinkage warping of the precast beam which is resisted by the composite section after the slab has hardened, and the deflection due to differential shrinkage and creep of the composite beam. These effects are represented by Terms (3) and (4) in eq. (2-30):

$$a_u = \overbrace{(a_i)_{1+2}}^{(1)} + \overbrace{1.80k_r(a_i)_{1+2}}^{(2)} + \overbrace{a_{sh}\frac{I_2}{I_c}}^{(3)} + \overbrace{a_{DS}}^{(4)} + \overbrace{(a_i)_L}^{(5)} + \overbrace{(a_{cp})_L}^{(6)} \qquad (2\text{-}30)$$

When k_r = 0.85 in eq. (2-19) (neglecting any effect of slab compression steel), and a_{DS} is assumed to be equal to $(a_i)_{1+2}$, eq. (2-30) reduces to eq. (2-31):

$$a_u = \overbrace{3.53(a_i)_{1+2}}^{(1+2+4)} + \overbrace{a_{sh}\frac{I_2}{I_c}}^{(3)} + \overbrace{(a_i)_L}^{(5)} + \overbrace{(a_{cp})_L}^{(6)} \qquad (2\text{-}31)$$

Term (1) is the *initial or short-time deflection of the composite beam due to slab plus precast beam dead load (plus partitions, roofing, etc.)* using eq. (2-16), in which $M_a = M_1 + M_2$ = midspan moment due to slab plus precast beam (etc.) dead load. For computing $(I_e)_{1+2}$ in eq. (2-12), M_a refers to the moment, $M_1 + M_2$, and M_{cr}, I_g, and I_{cr} to the composite beam section at midspan.

Term (2) is the *creep deflection of the composite beam due to the dead load* in Term (1), using eq. (2-20). C_u = 1.80 (based on the shores being removed at about age 10 days for a moist cured slab, from Ref. 2-10), and ρ' refers to any compression steel in the slab at midspan when computing k_r in eq. (2-19).

Term (3) is the *shrinkage deflection of the composite beam after the shores are removed due to the shrinkage of the precast beam concrete*, but not including the effect of differential shrinkage and creep, which is given by Term (4). Equation (2-22) may be used to compute a_{sh}. Assuming the slab is cast at a precast beam concrete (steam-cured) age of 2 months and shores are removed about 10 days later, from Table 2-2 and eq. (2-8), $(\epsilon_{sh})_u$ = (1 − say 0.37) $(400 \times 10^{-6}) = 252 \times 10^{-6}$ in./in.

Term (4) is the *deflection due to differential shrinkage and creep.* As an approximation, $a_{DS} = (a_i)_{1+2}$ may be used. For more details, see Refs. 2-10 and 2-13.

Term (5) is the *initial or short-time live load deflection of the composite beam* using eq. (2-16). The calculation of the incremental live load deflection follows the same procedure as that of a monolithic beam. This is the same as the first method described in Term (9) of eq. (2-28).

Term (6) is the *partial live load creep deflection for any sustained live load* using eq. (2-20). This is the same as Term (10) of eq. (2-28).

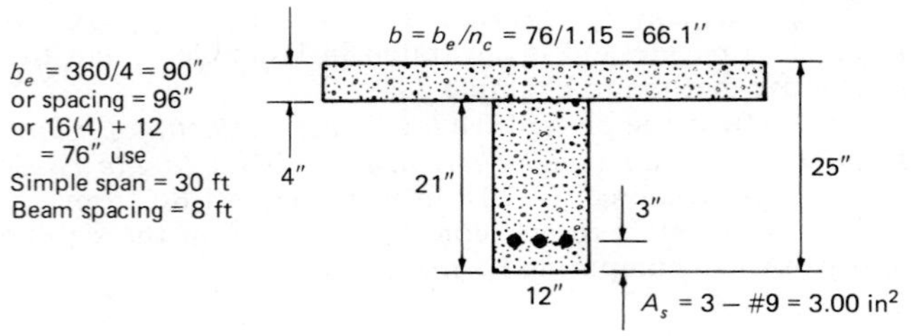

Fig. 2-10 Example 3—Unshored composite beam, normal weight concrete. Example 4—Shored composite beam, normal weight concrete.

These procedures suggest using midspan values only, which may normally be satisfactory for both simple composite beams and those with a continuous slab as well. For an example of a continuous slab in composite construction, see Ref. 2-13.

2.5.3 Example 3—Unshored Composite Beam, Normal Weight Concrete

Given: beam in Fig. 2-10, designed by the strength method:

$$\text{Slab } f'_c = 3000 \text{ psi } (20.7 \text{ N/mm}^2),$$

$$\text{precast beam } f'_c = 4000 \text{ psi}, \quad f_y = 60{,}000 \text{ psi}$$

$$\text{Live load} = 70 \text{ psf } (3.35 \text{ kN/m}^2), \quad 40\% \text{ sustained}$$

Required: analysis of short-time and long-time deflections.

Loads and moments:

$$w_1 = (150 \text{ pcf, T. 2-1})(96)(4)/144$$
$$= 400 \text{ lb/ft } (5.84 \text{ kN/m})$$
$$w_2 = (150)(12)(21)/144 = 263 \text{ lb/ft},$$
$$w_L = (70)(8) = 560 \text{ lb/ft}$$
$$M_1 = w_1 l^2/8 = (0.400)(30)^2/8$$
$$= 45.0 \text{ ft-kips } (61.0 \text{ kN-m})$$
$$M_2 = (0.263)(30)^2/8 = 29.6 \text{ ft-kips},$$
$$M_L = (0.560)(30)^2/8 = 63.0 \text{ ft-kips}$$

Minimum thickness:

1977 *ACI Code* (T. 2-7):

$$h_{\min} = l/16 = 360/16 = 22.5 \text{ in. } (57.2 \text{ cm})$$

ACI/435 (T. 2-8):

$$h_{\min} = (l/18 \quad \text{or} \quad l/14 \quad \text{or} \quad l/12 \quad \text{or} \quad l/10)$$
$$= 20.0 \text{ in.} \quad \text{or} \quad 25.7 \text{ in.} \quad \text{or} \quad 30.0 \text{ in.} \quad \text{or} \quad 36.0 \text{ in.}$$
$$(50.8 \text{ cm to } 91.4 \text{ cm})$$

The composite beam (h = 25 in.) satisfies two of these conditions, and the precast beam (h = 21 in.) only one. For the two cases that are satisfied by the composite section, only the long-time deflection of the precast beam before composite action is effective need be investigated. However, a general deflection analysis will be made and compared with the allowable deflections in Table 2-11.

Modulus of rupture, modulus of elasticity, modular ratio:

$$(E_c)_1 = 33\sqrt{w_c^3 f'_c} = 33\sqrt{(145)^3(3000)}$$
$$= 3.16 \times 10^6 \text{ psi } (21.8 \text{ kN/mm}^2)$$
$$(f_r)_2 = 7.5\sqrt{f'_c} = 7.5\sqrt{4000} = 474 \text{ psi } (3.27 \text{ N/mm}^2)$$
$$(E_c)_2 = 33\sqrt{(145)^3(4000)} = 3.64 \times 10^6 \text{ psi}$$
$$n_c = (E_c)_2/(E_c)_1 = 3.64/3.16 = 1.15,$$
$$n = E_s/(E_c)_2 = 29/3.64 = 8.0$$

Gross and cracked section moments of inertia (using Fig. 2-3):

Precast sec.:

$$I_g = bh^3/12 = (12)(21)^3/12 = 9261 \text{ in.}^4 \ (385{,}000 \text{ cm}^4)$$
$$B = b/(nA_s) = 12/(8.0)(3.00) = 0.5000 \text{ 1/in.}$$
$$a = (\sqrt{2dB + 1} - 1)/B$$
$$= [\sqrt{(2)(18)(0.5000) + 1} - 1]/0.5000 = 6.72 \text{ in.}$$
$$I_{cr} = ba^3/3 + nA_s(d - a)^2 = (12)(6.72)^3/3$$
$$+ (8.0)(3.00)(18 - 6.72)^2 = 4268 \text{ in.}^4$$

Composite sec.:

$$y_t = h - (1/2)\,[(b - b_w)\,h_f^2$$
$$+ b_w h^2]/[(b - b_w)\,h_f + b_w h]$$
$$= 25 - (1/2)\,[(54.1)(4)^2 + (12)(25)^2]/[(54.1)(4)$$
$$+ (12)(25)] = 16.90 \text{ in.}$$
$$I_g = (b - b_w)\,h_f^3/12 + b_w h^3/12$$
$$+ (b - b_w)\,h_f(h - h_f/2 - y_t)^2 + b_w h(y_t - h/2)^2$$
$$= (54.1)(4)^3/12 + (12)(25)^3/12$$
$$+ (54.1)(4)(25 - 2 - 16.90)^2$$
$$+ (12)(25)(16.90 - 12.5)^2 = 29{,}770 \text{ in.}^4$$
$$B = b/(nA_s) = 66.1/(8.0)(3.00) = 2.754$$
$$a = (\sqrt{2dB + 1} - 1)/B$$
$$= [\sqrt{(2)(22)(2.754) + 1} - 1]/2.754 = 3.65 \text{ in.}$$

$< h_f = 4$ in. (hence, treat as rectangular compression area)

$$I_{cr} = ba^3/3 + nA_s(d - a)^2 = (66.1)(3.65)^3/3$$
$$+ (8.0)(3.00)(22 - 3.65)^2 = 9153 \text{ in.}^4$$
$$I_2/I_c = [(I_2/I_c)_g + (I_2/I_c)_{cr}]/2$$
$$= [(9261/29{,}770) + (4268/9153)]/2 = 0.389$$

Effective moment of inertia (using eqs. 2-12 and 2-13):

In Term (1), eq. (2-28)—Precast sec.:

$$M_{cr} = f_r I_g/y_t = (474)(9261)/(10.5)(12{,}000)$$
$$= 34.8 \text{ ft-kips } (47.2 \text{ kN-m})$$
$$M_{cr}/M_2 = 34.8/29.6 > 1; \text{ hence } (I_e)_2$$
$$= I_g = 9261 \text{ in.}^4 \ (385{,}000 \text{ cm}^4)$$

In Term (6), eq. (2-28)—Precast sec.:

$$[M_{cr}/(M_1 + M_2)]^3 = [34.8/(45.0 + 29.6)]^3 = 0.102$$
$$(I_e)_{1+2} = (M_{cr}/M_a)^3\, I_g$$
$$+ [1 - (M_{cr}/M_a)^3]\, I_{cr} \leqslant I_g$$
$$= (0.102)(9261) + (1 - 0.102)(4268)$$
$$= 4777 \text{ in.}^4 \ (199{,}000 \text{ cm}^4)$$

Deflections (*using eqs. 2-28 and 2-29*):
Term (1):

$$(a_i)_2 = \frac{K(5/48)\,M_2 l^2}{(E_c)_2\,(I_e)_2} = \frac{(1)(5/48)(29.6)(30)^2(12)^3}{(3640)(9261)}$$

$= 0.142$ in.

Term (2):

$k_r = 0.85$ (no precast beam compression steel)

$0.77k_r(a_i)_2 = (0.77)(0.85)(0.142) = 0.093$ in.

Term (3):

$$0.83k_r(a_i)_2\,\frac{I_2}{I_c} = (0.83)(0.85)(0.142)(0.389) = 0.039 \text{ in.}$$

Term (4)–$K_{sh} = 1/8$:
Precast sec.:

$p = (100)(3.00)/(12)(18) = 1.39\%$

From Fig. 2-7 or eq. (2-24):

$$A_{sh} = 0.7p^{1/3} = 0.7(1.39)^{1/3} = 0.781$$

$$\phi_{sh} = A_{sh}(\epsilon_{sh})_u/h = (0.781)(400 \times 10^{-6})/21$$

$$= 14.88 \times 10^{-6}\ 1/\text{in.}$$

$$a_{sh} = K_{sh}\phi_{sh}l^2 = (1/8)(14.88 \times 10^{-6})(30)^2(12)^2$$

$= 0.241$ in.

$0.36a_{sh} = (0.36)(0.241) = 0.087$ in.

Term (5):

$$0.64a_{sh}\,\frac{I_2}{I_c} = (0.64)(0.241)(0.389) = 0.060 \text{ in.}$$

Term (6):

$$(a_i)_1 = \frac{K(5/48)(M_1 + M_2)\,l^2}{(E_c)_2\,(I_e)_{1+2}} - (a_i)_2$$

$$= \frac{(1)(5/48)(45.0 + 29.6)(30)^2(12)^3}{(3640)(4777)} - 0.142$$

$= 0.553$ in.

Term (7):

$$1.22k_r(a_i)_1\,\frac{I_2}{I_c} = (1.22)(0.85)(0.553)(0.389) = 0.223 \text{ in.}$$

Term (8):

$$a_{DS} = 0.50(a_i)_1 = (0.50)(0.553)$$

$= 0.277$ in. (rough estimate)

Term (9)–Using the alternative method:

$$(a_i)_L = \frac{K(5/48)\,M_L l^2}{(E_c)_2\,(I_c)_{cr}} = \frac{(1)(5/48)(63.0)(30)^2(12)^3}{(3640)(9153)}$$

$= 0.306$ in.

Term (10):

$k_r = 0.85$ (neglecting the effect of any slab compression steel)

$$(a_{cp})_L = k_r C_u[0.40(a_i)_L]$$

$$= (0.85)(1.60)(0.40 \times 0.306) = 0.166 \text{ in.}$$

In eq. (2-28):

$$a_u = 0.142 + 0.093 + 0.039 + 0.087 + 0.060$$
$$+ 0.553 + 0.223$$
$$+ 0.277 + 0.306 + 0.166 = \underline{1.95 \text{ in. } (50 \text{ mm})}$$

Checking eq. (2-29) (should get same answer):

$$a_u = \left(1.65 + 0.71\,\frac{I_2}{I_c}\right)(a_i)_2 + \left(0.36 + 0.64\,\frac{I_2}{I_c}\right)a_{sh}$$

$$+ \left(1.50 + 1.04\,\frac{I_2}{I_c}\right)(a_i)_1$$

$$+ (a_i)_L + (a_{cp})_L = (1.65 + 0.71 \times 0.389)(0.142)$$

$$+ (0.36 + 0.64 \times 0.389)(0.241)$$

$$+ (1.50 + 1.04 \times 0.389)(0.553)$$

$$+ 0.306 + 0.166 = \underline{1.95 \text{ in.}} \quad \text{Same}$$

Assuming nonstructural elements are installed after the composite slab has hardened, $a_{cp} + a_{sh} + (a_i)_L + (a_{cp})_L =$ Terms (3) + (5) + (7) + (8) + (9) + (10)–excluding Terms (1), (2), (4), and (6)– = 0.039 + 0.060 + 0.223 + 0.277 + 0.306 + 0.166 = <u>1.07 in (27 mm).</u>

The comparisons with allowable deflections are shown at the end of Example 4.

2.5.4 Example 4–Shored Composite Beam, Normal Weight Concrete

Given: same as Example 3, except using shored construction.

Required: analysis of short-time and long-time deflections, to show the beneficial effect of shoring in reducing the total deflection.

Effective moment of inertia (*using eqs. 2-12 and 2-13*):
In Term (1), eq. (2-30)–Composite sec.:

$$M_{cr} = f_r I_g / y_t$$

$$= (474)(29{,}770)/(16.90)(12{,}000)$$

$= 69.6$ ft-kips (94.4 kN-m)

$$[M_{cr}/(M_1 + M_2)]^3 = [69.6/(45.0 + 29.6)]^3 = 0.812$$

$$(I_e)_{1+2} = (M_{cr}/M_a)^3\,I_g$$

$$+ [1 - (M_{cr}/M_a)^3]\,I_{cr} \leqslant I_g$$

$$= (0.812)(29{,}770) + (1 - 0.812)(9153)$$

$= 25{,}890$ in.^{4} (1,078,000 cm^4)

In Term (5), eq. (2-30)–Composite sec.:

$$[M_{cr}/(M_1 + M_2 + M_L)]^3$$

$$= [69.6/(45.0 + 29.6 + 63.0)]^3 = 0.129$$

$$(I_e)_{D+L}$$

$$= (0.129)(29{,}770) + (1 - 0.129)(9153)$$

$= 11{,}810$ in.^{4} (492,000 cm^4)

versus the alternative method of Example 3 using $I_e = (I_c)_{cr} = 9153$ in.^{4} with the live load moment directly.

Deflections (*using eqs. 2-30 and 2-31*):

Term (1):

$$(a_i)_{1+2} = \frac{K(5/48)(M_1 + M_2)\, l^2}{(E_c)_2\,(I_e)_{1+2}}$$

$$= \frac{(1)(5/48)(45.0 + 29.6)(30)^2(12)^3}{(3640)(25{,}890)}$$

$$= 0.128 \text{ in.}$$

Term (2):

$$k_r = 0.85 \text{ (neglecting the effect of any slab compression steel)}$$

$$1.80k_r(a_i)_{1+2} = (1.80)(0.85)(0.128) = 0.196 \text{ in.}$$

Term (3)—From Term (4) of Example 3, and using $(\epsilon_{sh})_u = 252 \times 10^{-6}$ in./in.:

$$a_{sh}\frac{I_2}{I_c} = (252/400)(0.241)(0.389) = 0.059$$

Term (4):

$$a_{DS} = (a_i)_{1+2} = 0.128 \text{ in. (rough estimate)}$$

Term (5):

$$(a_i)_L = \frac{K(5/48)(M_1 + M_2 + M_L)\, l^2}{(E_c)_2\,(I_e)_{D+L}} - (a_i)_{1+2}$$

$$= \frac{(1)(5/48)(45.0 + 29.6 + 63.0)(30)^2(12)^3}{(3640)(11{,}810)}$$

$$- 0.128 = 0.391 \text{ in.}$$

Term (6):

$$k_r = 0.85 \text{ (neglecting the effect of any slab compression steel)}$$

$$(a_{cp})_L = k_r C_u[0.40(a_i)_L]$$

$$= (0.85)(1.60)(0.40 \times 0.391) = 0.213 \text{ in.}$$

In eq. (2-30):

$$a_u = 0.128 + 0.196 + 0.059 + 0.128 + 0.391 + 0.213$$

$$= \underline{1.12 \text{ in. (28 mm)}} \text{ versus 1.95 in. in the unshored case}$$

This shows the beneficial effect of shoring in reducing deflections.

Checking eq. (2-31) (should get same answer):

$$a_u = 3.53(a_i)_{1+2} + a_{sh}\frac{I_2}{I_c} + (a_i)_L + (a_{cp})_L$$

$$= (3.53)(0.128) + 0.059 + 0.391 + 0.213$$

$$= \underline{1.12 \text{ in.}} \quad \text{Same}$$

Assuming nonstructural elements are installed after the shores are removed:

$$a_{cp} + a_{sh} + (a_i)_L + (a_{cp})_L$$

$$= a_u - (a_i)_{1+2} = 1.12 - 0.13 = \underline{0.99 \text{ in. (25 mm)}}$$

Comparisons with allowable deflections in Table 2-11:

Flat roofs not supporting and not attached to nonstructural elements likely to be damaged by large deflections:

$$(a_i)_L \leqslant l/180 = 360/180 = 2.00 \text{ in.} \quad \text{OK}$$

Floors not supporting and not attached to nonstructural elements likely to be damaged by large deflections:

$$(a_i)_L \leqslant l/360 = 360/360 = 1.00 \text{ in.} \quad \text{OK}$$

versus 1.07 in. and 0.99 in. in Examples 3 and 4: one conditions OK.

Roof or floor construction supporting or attached to nonstructural elements *likely* to be damaged by large deflections (very stringent limitation):

$$a_{cp} + a_{sh} + (a_i)_L + (a_{cp})_L \leqslant l/480$$

$$= 360/480 = 0.75 \text{ in.} \quad \text{Not OK}$$

Roof or floor construction supporting or attached to nonstructural elements *not likely* to be damaged by large deflections:

$$a_{cp} + a_{sh} + (a_i)_L + (a_{cp})_L \leqslant l/240$$

$$= 360/240 = 1.50 \text{ in.} \quad \text{OK}$$

versus 1.07 in. and 0.99 in. in Examples 3 and 4: one condition OK, one not OK.

2.6 DEFLECTION OF NONPRESTRESSED TWO-WAY SLAB SYSTEMS

2.6.1 Initial or Short-Time Deflection

An approximate procedure by Nilson and Walters[2-42] that is compatible with the direct design and equivalent frame methods of the *ACI Code* is used to compute the initial or short-time deflection of two-way slab systems.[2-10, 2-42–2-44] The method is essentially the same for flat plates, flat slabs, and two-way slabs, once the appropriate stiffnesses are computed. In this procedure, the midpanel deflection is computed at the sum of the midspan column strip deflection in one direction, such as a_{cx}, and the midspan middle strip deflection in the other direction, such as a_{my}, as shown in Fig. 2-11.

Under vertical loads, the midspan deflection of an equivalent frame can be considered as the sum of three parts: that of a panel assumed to be fixed at both ends of its span, plus that due to the rotation at each of the two support lines.

The midspan fixed-end deflection of the equivalent frame under uniform loading is given by eq. (2-32):

$$\text{Fixed } a_{\text{frame}} = wl^4/384E_c I_{\text{frame}} \tag{2-32}$$

where w is the distributed load per width of frame, and l is the span center-to-center of columns. To include the effect of different positive and negative region I's (primarily when using drop panels and/or I_e in eq. 2-12), an average may be used, as given by eqs. (2-14) and (2-15).

The calculation of the midspan fixed-end deflection of the column and middle strips is then based on the M/EI ratio of the strips to the frame:

$$\text{Fixed } a_{c,m} = (LDF)_{c,m}\,(\text{Fixed } a_{\text{frame}})\,\frac{(EI)_{\text{frame}}}{(EI)_{c,m}} \tag{2-33}$$

where $(LDF)_{c,m} = M_{c,m}/M_{\text{frame}}$. Typical values of the lateral distribution factor, LDF, are shown in Table 2-12.

If the ends of the columns at the floor above and below are assumed to be fixed (usual case in the equivalent frame analysis), or ideally pinned, the rotation of the column at the floor in question is equal to the net applied moment divided by the stiffness of the equivalent column. This is

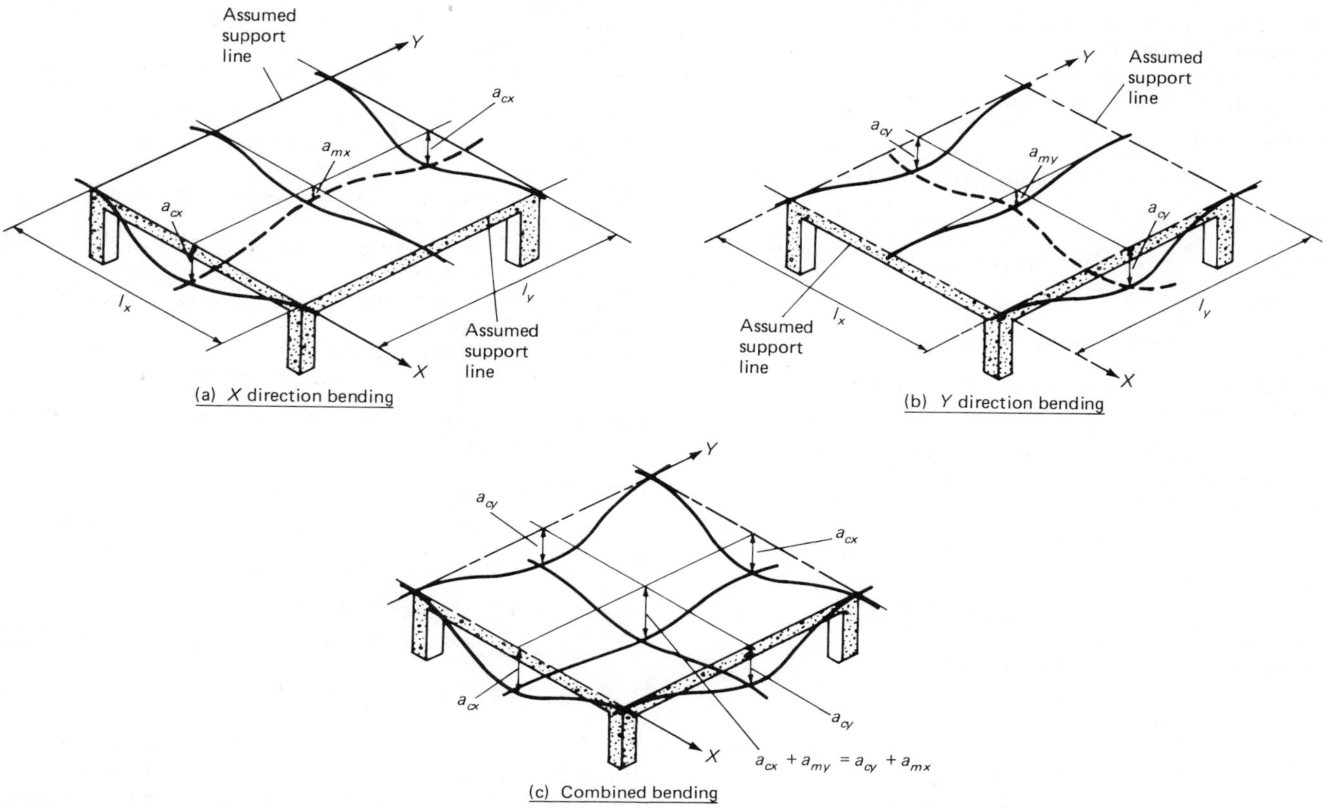

Fig. 2-11 Basis of equivalent frame method for deflection analysis of two-way slab systems, with or without beams.[2-42]

given by eq. (2-34) for the frame, column strip, and middle strip:

$$\text{End } \theta_c = \text{End } \theta_m = \text{End } \theta_{\text{frame}} = \text{End } \theta$$
$$= (M_{\text{net}})_{\text{frame}}/K_{ec} \qquad (2\text{-}34)$$

where K_{ec} is the gross-section flexural stiffness used in design.

In the direct design method, moments are based on the clear span and should theoretically be adjusted to obtain moments and rotations at the column centerlines. However, the use of moments in eq. (2-34) at the face of the columns should cause little error. Particularly in the case of flat plates and flat slabs, the span center-to-center of columns is thought to be more appropriate in deflection calculations than the clear span, as used herein.

As a practical matter, only the exterior column rotation need be considered in most cases when using the direct design moment coefficients with equal spans. Particularly when the live load is large compared with the dead load (frequently not the case), the end rotations may be computed by a simple moment–area procedure in which the effect of pattern loading may be included, as shown in a book by Nilson, page 386.[2-47]

The midspan deflection of a member having an end rotation of θ radians, with the far end fixed, is computed by eq. (2-35)—as can be easily shown in the moment–area method:

$$a_\theta = (\text{End } \theta)\, l/8 \qquad (2\text{-}35)$$

Because (End θ) is based on the gross-section properties in eq. (2-34), when the deflection calculations are based on I_e, eq. (2-36) may be used instead of eq. (2-35) for consistency:

$$a_\theta = (\text{End } \theta)(l/8)(I_g/I_e)_{\text{frame}} \qquad (2\text{-}36)$$

The combined midspan deflection of a column or middle strip is the sum of the three parts, as computed in eq. (2-37):

$$a_{c,m} = \text{Fixed } a_{c,m} + (a_{\theta 1})_{c,m} + (a_{\theta 2})_{c,m} \qquad (2\text{-}37)$$

where $a_{\theta 1}$ and $a_{\theta 2}$ refer to the midspan deflections due to rotations at both ends. As shown in Fig. 2-11, the total midpanel deflection is given by eq. (2-38):

$$a = a_{cx} + a_{my} = a_{cy} + a_{mx} \qquad (2\text{-}38)$$

For other than square symmetrical panels, eq. (2-39) may be used:

$$a = [(a_{cx} + a_{my}) + (a_{cy} + a_{mx})]/2 \qquad (2\text{-}39)$$

A review of the available information for simply estimating slab deflections by using deflection coefficients is presented in two books by the author.[2-10, 2-11]

Effective Moment of Inertia. The effective moment of inertia in eq. (2-12) is recommended by the *ACI Code*[2-1] and ACI Committee 435[2-7] for computing deflections of partially cracked two-way construction. An average I_e of the positive and negative regions in eqs. (2-14) and (2-15) may also be used. The following typical cracking locations have been found empirically in Refs. 2-38 through 2-41, and the corresponding values of I_e have been shown to apply[2-10, 2-43] in most cases (also recommended by ACI/435[2-7]):

Slabs without beams (flat plates, flat slabs):
- All dead load deflections—I_g
- Dead-plus-live load deflections:
  - For the column strips in both directions—I_e
  - For the middle strips in both directions—I_g

TABLE 2-12 Lateral Distribution Factors (*LDF*) for Column and Middle Strips

From the *ACI Code*, the column strip percentages are:

Exterior negative:

$$100 - 10\beta_t + 12\beta_t(\alpha_1 l_2/l_1)(1 - l_2/l_1)$$

Interior negative:

$$75 + 30(\alpha_1 l_2/l_1)(1 - l_2/l_1)$$

Positive:

$$60 + 30(\alpha_1 l_2/l_1)(1.5 - l_2/l_1)$$

except when $\alpha_1 l_2/l_1 > 1$ (typical two-way slab case), use $\alpha_1 l_2/l_1 = 1$.

Example: $\alpha_1 = \beta_t = 0$ assumed for a flat plate with no beams. Use an average of the positive and negative region values.

Interior panel, both directions:

$$(LDF)_c = \frac{60\ (\text{Pos}) + 75\ (\text{Neg})}{2} = 67.5\%$$

$$(LDF)_m = 100 - 67.5 = 32.5\%$$

Side panel, one end continuous direction:

$$(LDF)_c = \frac{60 + (100 + 75)/2}{2} = 73.8\%$$

$$(LDF)_m = 100 - 73.8 = 26.2\%$$

Side panel, both ends continuous direction:

$$(LDF)_c = 67.5\%, \quad (LDF)_m = 32.5\%$$

Corner panel, both directions:

$$(LDF)_c = 73.8\%, \quad (LDF)_m = 26.2\%$$

These conditions are demonstrated in Example 5.

Slabs with beams (two-way slabs):
All dead load deflections—I_g
Dead-plus-live load deflections:
For the column strips in both directions—I_g
For the middle strips in both directions—I_e

These conditions are demonstrated in an example in Ref. 2-10.

The I_e of the equivalent frame in each direction is then taken as the sum of the column and middle strip I_e's.

2.6.2 Long-Time Deflection

Since the available data on long-time deflections of two-way construction are too limited to justify more elaborate procedures, the *ACI Code*[2-1] and ACI Committee 435[2-7] recommend using the same procedures as those used for one-way members. The *ACI Code* method of eqs. (2-26) and (2-27), together with $T_u = 2.5$ as recommended by ACI/435 in eq. (2-10), is preferred by the author herein.

2.6.3 Example 5—Slab System Without Beams (Flat Plate), Normal Weight Concrete

Given: flat plate with no spandrel beams, designed by the *ACI Code* direct design method (strength method):

Slab steel f_y = 40,000 psi (276 N/mm²),
Slab f_c' = 3000 psi (20.7 N/mm²)
Square panels—16 ft by 16 ft (4.88 m by 4.88 m) c.c. of columns
Square columns—15 in. by 15 in. (38.1 cm by 38.1 cm), col. f_c' = 5000 psi
Clear span—l_n = 16 - 1.25 = 14.75 in. (4.50 m)
Story height—10 ft (3.05 m), slab thickness—h = 5.5 in. (14.0 cm)
Column strip pos. d = 5.5 - 0.75 - 0.31 (#5 Rad.) = 4.44 in. (11.3 cm)
Column strip neg. d = 5.5 - 0.75 - 0.63 (#5 Dia.) = 4.12 in. (10.5 cm)
Middle strip d's not required (slab remains uncracked in middle strips)
Live load = 70 psf (3.35 kN/m²);
check for 0% and 40% sustained live load

Required: analysis of short-time and long-time deflections of a corner panel.

Shear design:

$$w_D = (150\ \text{pcf, T. 2-1})(5.5)/12 = 68.8\ \text{psf}\ (3.30\ \text{kN/m}^2$$

$$w_u = 1.4(68.8) + 1.7(70) = 215.3\ \text{psf},$$

$$V_u = 0.2153(16)^2 = 55.1\ \text{kips}\ (245\ \text{kN})$$

$$\phi V_c = \phi 4\sqrt{f_c'}\, b_o d = (0.85)(4)\sqrt{3000}\,(4)(15 + 4.12)(4.12)/1000 = 58.7\ \text{kips} \quad \text{OK}$$

One-way shear is also satisfactory.

Minimum thickness:
From Table 2-10 for either square or rectangular panels, and using the factor 1.100 in the footnote for f_y = 40,000 psi:

Interior panel h_{min}

$$= (l_n = 14.75 \times 12)/(32.7 \times 1.100)$$

$$= 4.92\ \text{in.}\ (12.5\ \text{cm})$$

Side, corner panel h_{min}

$$= (l_n = 14.75 \times 12)/(29.7 \times 1.100) = 5.42\ \text{in.}$$

To verify by the governing *ACI Code* equation:

$$h_{min} = \frac{l_n(800 + 0.005 f_y)}{36{,}000}$$

$$\cdot\ (1.10\ \text{when no spandrel beams are used})$$

$$= \frac{(14.75 \times 12)\,[800 + (0.005)(40{,}000)]}{36{,}000}(1.10)$$

$$= 5.41\ \text{in.} \quad \text{Checks}$$

The design thickness of 5.5 in. is marginally satisfactory for both shear and deflections. However, as an illustration deflections will be checked for a corner panel, where it is found that the most stringent deflection limit of $l/480$ is not satisfied.

Modulus of rupture, modulus of elasticity, modular ratio:

$$f_r = 7.5\sqrt{f_c'} = 7.5\sqrt{3000} = 411\ \text{psi}\ (2.83\ \text{N/mm}^2) \qquad \text{(eq. 2-1)}$$

$$E_{sc} = 33\sqrt{w_c^3 f_c'} = 33\sqrt{(145)^3(3000)}$$

$$= 3.16 \times 10^6\ \text{psi}\ (21.8\ \text{kN/mm}^2) \qquad \text{(eq. 2-4)}$$

$$E_{cc} = 33\sqrt{(145)^3(5000)} = 4.07 \times 10^6\ \text{psi},$$

$$n = E_s/E_{cs} = 29/3.16 = 9.2$$

Flexural stiffness (K_{ec} and α_{ec}) of exterior equivalent column:

$K_b = 0$ (no beams)

$I_s = (I_g)_{frame} = l_2 h^3/12 = (16 \times 12)(5.5)^3/12$

$= 2662$ in.^{4} (110,800 cm^4)

$K_s = 4E_{cs}I_s/l_1 = 4E_{cs}(2662)/(16)(12) = 55.46E_{cs}$

For ext. frame, $K_s = 55.46E_{cs}/2 = 27.73E_{cs}$.

$K_c = 4E_{cc}I_c/l_c = 4E_{cc}(15)^4/(12)(10)(12) = 140.63E_{cc}$

$\sum K_c = 2K_c = 2(140.63E_{cc}) = 281.26E_{cc}$

$C = (1 - 0.63x/y)(x^3y/3)$

$$= \left(1 - 0.63\,\frac{5.5}{15}\right)\left(\frac{5.5^3 \times 15}{3}\right) = 639.7 \text{ in.}^4$$

$$K_t = \frac{9E_{cs}C}{l_2(1 - c_2/l_2)^3} = \frac{(2)(9)\,E_{cs}(639.7)}{(16)(12)\left(1 - \dfrac{15}{16 \times 12}\right)^3}$$

$= 76.54E_{cs}$

For ext. frame,

$K_t = 76.54E_{cs}/2 = 38.27E_{cs}$,

$E_{cc} = (4.07/3.16)\,E_{cs} = 1.288E_{cs}$.

$$K_{ec} = \frac{1}{1/\sum K_c + 1/K_t} = \frac{E_{cs}}{[1/(281.26)(1.288)] + (1/76.54)}$$

$= 63.19E_{cs}$

For ext. frame,

$$K_{ec} = \frac{E_{cs}}{[1/(281.26)(1.288)] + (1/38.27)} = 34.61E_{cs}.$$

To use in eq. (2-34):

Avg. $K_{ec} = (63.19 + 34.61)(3.16 \times 10^6)/(2)(12{,}000)$

$= 12{,}880$ ft-kips (17,470 kN-m)

$\alpha_{ec} = K_{ec} \Big/ \sum (K_s + K_b)$

$= 63.19E_{cs}/55.46E_{cs} = 1.139$, $1/\alpha_{ec} = 0.878$

For ext. frame,

$\alpha_{ec} = 34.61E_{cs}/27.73E_{cs} = 1.248$, $1/\alpha_{ec} = 0.801$.

To compute M_{net} for use in eq. (2-34):

Avg. $\alpha_{ec} = 1.194$, 1/Avg. $\alpha_{ec} = 0.838$

Service load moments and cracking moment:

$(M_0)_D = w_D l_2 l_n^2/8 = (68.8)(16)(14.75)^2/8{,}000$

$= 29.94$ ft-kips (40.6 kN-m)

$(M_0)_{D+L} = w_{D+L} l_2 l_n^2/8 = (68.8 + 70)(16)(14.75)^2/8{,}000$

$= 60.40$ ft-kips

See Table 2-13 for the half-column strip and half-middle strip moments:

TABLE 2-13 Moments (ft-kips) and Moments of Inertia (in.^{4}) for Example 5[a]

	Ext. Eq. Fr. $1/\alpha_{ec} = 0.801$		Int. Eq. Fr. $1/\alpha_{ec} = 0.878$	
Description	*Pos.*	*Int. Neg.*	*Pos.*	*Int. Neg.*
1. Moment ratios[b]	0.475	0.694	0.481	0.697
2. Panel M_D = (Line 1)$(M_0)_D$	14.22	20.78	14.40	20.87
3. Panel M_{D+L} = (Line 1)$(M_0)_{D+L}$	28.69	41.92	29.05	42.10
4. $(LDF)_c$–From T. 2-12 with $\alpha_1 = \beta_t = 0$	0.60	0.75	0.60	0.75
5. $(M_D)_{c/2}$ = (Line 2)(Line 4)/2	4.27	7.79	4.32	7.83
6. $(M_{D+L})_{c/2}$ = (Line 3)(Line 4)/2	8.61	15.72	8.72	15.79
7. $(M_L)_{c/2}$ = Line 6 – Line 5	4.34	7.93	4.40	7.96
8. $(M_u)_{c/2}$ = 1.4(Line 5) + 1.7(Line 7)	13.36	24.39	13.53	24.49
9. $(M_D)_{m/2}$ = (Line 2)/2 – Line 5	2.84	2.60	2.88	2.61
10. $(M_{D+L})_{m/2}$ = (Line 3)/2 – Line 6	5.74	5.24	5.81	5.26
11. $[(M_{cr})_{c/2}/(\text{Line 6})]^3$	0.893	0.147	0.859	0.145
For Half-Column Strips–Steel Design and I_e Calculation				
12. [c]Req. R_u = (Line 8)(12,000)/ϕ (48) d^2	188	399	191	401
13. Req. ρ (From Design Aid, Ref. 2-33)	0.0050	0.0120	0.0050	0.0120
14. Req. A_s = (Line 13)(48) d (in.^{2})	1.07	2.37	1.07	2.37
15. No. of #5 Used	4	8	4	8
16. A_s = (Line 15)(0.31) (in.^{2})	1.24	2.48	1.24	2.48
17. I_{cr} (separate calculation) (in.^{4})	147.3	210.3	147.3	210.3
18. [d]$(I_e)_{D+L}$ = (Ln 11) I_g + [1 – (Ln 11)] I_{cr}	610.1	277.2	592.4	276.3

[a]*Ext. neg. moments and ext. neg. I_e are not required in eq. (2-14).*

[b]*Pos. mom. ratio* $= 0.63 - \dfrac{0.28}{1 + 1/\alpha_{ec}}$, *neg. mom. ratio* $= 0.75 - \dfrac{0.10}{1 + 1/\alpha_{ec}}$.

[c]*Req.* $R_u = \dfrac{(\text{Line 8})(12{,}000)}{(0.9)(48)(\text{Pos. } d = 4.44)^2} = 188$ *psi*, $= \dfrac{(\text{Line 8})(12{,}000)}{(0.9)(48)(\text{Neg. } d = 4.12)^2} = 399$ *psi.*

[d]$(I_e)_{D+L} = 0.893(665.5) + (1 - 0.893)(147.3) = 610.1$ *in.*^{4} *(eq. 2-12), where* $I_g = (48)(5.5)^3/12 = 665.5$ *in.*^{4}.

$$(M_{cr})_{c/2} = (M_{cr})_{m/2} = f_r I_g / y_t$$
$$= (411)(16 \times 12)(5.5)^3/(4)(12)(2.75)(12{,}000)$$
$$= 8.29 \text{ ft-kips } (11.24 \text{ kN-m})$$
$$> \text{All } (M_D)_{c/2}, (M_D)_{m/2}, \text{ and } (M_{D+L})_{m/2}$$

on Lines 5, 9, and 10 of Table 2-13. Hence $I_e = I_g$ for *all* dead load and all middle-strip dead-plus-live load deflections, as suggested in the text for typical cases. The steel design and $(I_e)_{D+L}$ calculations for half-column strips are also shown in Table 2-13.

Summary of effective moments of inertia:
Middle strips:

$$(I_e)_D = (I_e)_{D+L} = I_g = (8)(12)(5.5)^3/12$$
$$= 1331 \text{ in.}^4 \ (55{,}400 \text{ cm}^4)$$

Column strips:

$$(I_e)_D = I_g = 1331 \text{ in.}^4$$

Using the half-column strip values of $(I_e)_{D+L}$ from Line 18 of Table 2-13:

$$\text{Avg. } (I_e)_{D+L} = 0.85 I_m + 0.15(I_{e\text{-Cont. End}}) \quad \text{(eq. 2-14)}$$
$$= (0.85)(610.1 + 592.4)$$
$$+ (0.15)(277.2 + 276.3) = 1105 \text{ in.}^4$$

Equivalent frame:

$$(I_e)_D = (I_g)_c + (I_g)_m = (I_g)_{\text{frame}}$$
$$= (2)(1331) = 2662 \text{ in.}^4$$
$$(I_e)_{D+L} = (I_e)_c + (I_g)_m = 1105 + 1331 = 2436 \text{ in.}^4$$

Deflections (using eqs. 2-32 through 2-38):

Fixed a_{frame}

$$= wl^4/384 E_{cs} I_{\text{frame}} \quad \text{(eq. 2-32)}$$

(Fixed $a_{\text{frame}})_{D, D+L}$

$$= \frac{(68.8 \text{ or } 138.8)(16)^5(12)^3}{(384)(3.16 \times 10^6)(2662 \text{ or } 2436)}$$
$$= 0.039 \text{ in., } 0.085 \text{ in.}$$

Fixed $a_{c,m}$

$$= (LDF)_{c,m} \text{ (Fixed } a_{\text{frame}})(I_{\text{frame}}/I_{c,m}) \quad \text{(eq. 2-33)}$$

Pos and neg. Avg. $(LDF)_c = 0.738$, $(LDF)_m = 0.262$ (Table 2-12)

$$(\text{Fixed } a_c)_D = (0.738)(0.039)(2) = 0.058 \text{ in.}$$
$$(\text{Fixed } a_c)_{D+L} = (0.738)(0.085)(2436/1105) = 0.138 \text{ in.}$$
$$(\text{Fixed } a_c)_L = 0.138 - 0.058 = 0.080 \text{ in.}$$
$$(\text{Fixed } a_m)_D = (0.262)(0.039)(2) = 0.020 \text{ in.}$$
$$(\text{Fixed } a_m)_{D+L} = (0.262)(0.085)(2436/1331) = 0.041 \text{ in.}$$
$$(\text{Fixed } a_m)_L = 0.041 - 0.020 = 0.021 \text{ in.}$$
$$(M_{\text{net}})_{D+L} = 0.65(M_0)_{D+L}/(1 + /\text{Avg. } \alpha_{ec}) \quad (\textit{ACI Code})$$
$$= (0.65)(60.40)/(1 + 0.838)$$
$$= 21.36 \text{ ft-kips } (28.96 \text{ kN-m})$$
$$(M_{\text{net}})_D = (29.94/60.40)(21.36) = 10.59 \text{ ft-kips}$$

For both column and middle strips:

$$\text{End } \theta_D = (M_{\text{net}})_D/\text{Avg. } K_{ec} = 10.59/12{,}880$$
$$= 0.000822 \text{ rad} \quad \text{(eq. 2-34)}$$
$$\text{End } \theta_{D+L} = 21.36/12{,}880 = 0.001658 \text{ rad}$$
$$a_\theta = (\text{End } \theta)(l/8)(I_g/I_e)_{\text{frame}} \quad \text{(eq. 2-36)}$$
$$(a_\theta)_D = (0.000822)(16)(12)(1)/8 = 0.020 \text{ in.}$$
$$(a_\theta)_{D+L} = (0.001658)(16)(12)(2662/2436)/8 = 0.043 \text{ in.}$$
$$(a_\theta)_L = 0.043 - 0.020 = 0.023 \text{ in.}$$
$$a_{c,m} = \text{Fixed } a_{c,m} + (a_{\theta 1})_{c,m} + (a_{\theta 2})_{c,m} \quad \text{(eq. 2-37)}$$
$$(a_c)_D = 0.058 + 0.020 + 0 = 0.078 \text{ in.}$$
$$(a_m)_D = 0.020 + 0.020 + 0 = 0.040 \text{ in.}$$
$$(a_c)_L = 0.080 + 0.023 + 0 = 0.103 \text{ in.}$$
$$(a_m)_L = 0.021 + 0.023 + 0 = 0.044 \text{ in.}$$
$$a = a_{cx} + a_{my} \quad \text{(eq. 2-38)}$$
$$(a_i)_D = 0.078 + 0.040 = 0.118 \text{ in. } (3.0 \text{ mm})$$
$$(a_i)_L = 0.103 + 0.044 = \underline{0.147 \text{ in. } (3.7 \text{ mm})}$$

Using eqs. (2-10), (2-26), and (2-27):

$$a_{cp+sh} = \lambda(a_i)_{sus}$$
$$= \frac{2.5}{1 + (50)(\rho' = 0)} [(a_i)_D + 0.40(a_i)_L]$$
$$= 2.5[0.118 + (0.40)(0.147)]$$
$$= 0.442 \text{ in. (with 40\% sustained LL)}$$
$$a_{cp+sh} = 2.5(0.118)$$
$$= 0.295 \text{ in. (with 0\% sustained LL)}$$
$$a_{cp+sh} + (a_i)_L = 0.442 + 0.147$$
$$= \underline{0.59 \text{ in. } (15 \text{ mm})} \text{ (with 40\% sustained LL)}$$
$$= 0.295 + 0.147$$
$$= \underline{0.44 \text{ in. } (11 \text{ mm})} \text{ (with 0\% sustained LL)}$$

Comparisons with allowable deflections in Table 2-11:
Flat roofs not supporting and not attached to nonstructural elements likely to be damaged by large deflections:

$$(a_i)_L \leqslant (l_n \text{ or } l)/180$$
$$= (14.75 \text{ or } 16)(12)/180$$
$$= 0.98 \text{ in. or } 1.07 \text{ in., versus } \underline{0.15 \text{ in.}} \quad \text{OK}$$

Floors not supporting and not attached to nonstructural elements likely to be damaged by large deflections:

$$(a_i)_L \leqslant (l_n \text{ or } l)/360$$
$$= 0.49 \text{ in. or } 0.53 \text{ in., versus } \underline{0.15 \text{ in.}} \quad \text{OK}$$

Roof or floor construction supporting or attached to nonstructural elements *likely* to be damaged by large deflections:

$$a_{cp+sh} + (a_i)_L \leqslant (l_n \text{ or } l)/480$$
$$= 0.37 \text{ in. or } 0.40 \text{ in., versus } \underline{0.44 \text{ in., } 0.59 \text{ in.}}$$

Not OK

Roof or floor construction supporting or attached to nonstructural elements *not likely* to be damaged by large deflections:

$$a_{cp+sh} + (a_i)_L \leqslant (l_n \text{ or } l)/240$$
$$= 0.74 \text{ in. or } 0.80 \text{ in., versus } \underline{0.44 \text{ in., } 0.59 \text{ in.}}$$

OK

A second flat plate example (same as Example 5 except using slab $f_c' = 4000$ psi, $h = 6$ in., and superimposed dead load = 20 psf in addition to the live load of 70 psf) revealed the following differences:

1. Only the negative column strip was cracked under dead-plus-live load, as compared to both positive and negative column strips in Example 5 (positive column strip only marginally cracked in Example—Line 11, Table 2-13). All parts of the slab under dead load and the middle strips under dead-plus-live load remained uncracked in both examples.
2. The deflection limit of $(l_n$ or $l)/480 = 0.37$ in. or 0.40 in. was satisfied for the 0% sustained live load—$a_{cp+sh} + (a_i)_L = \underline{0.35 \text{ in.}}$, but not the 40% sustained live load—$a_{cp+sh} + (a_i)_L = \underline{0.44 \text{ in.}}$ In Example 5 this limit was not satisfied for both 0% and 40% sustained live load.

The following typical observations can be seen in these examples and other similar examples in Refs. 2-10 and 2-12:

Analogous to 1 above: In most cases when computing deflections of flat plates and flat slabs, major cracking occurs only in the negative column strip under dead-plus-live loading. That is, all regions under dead load, and the middle strips and positive region of the column strips under dead-plus-live load tend to be uncracked or only marginally cracked according to this procedure. In addition, since the negative I_e in eqs. (2-14) and (2-15) is weighted only 15% for one end continuous cases and 30% for both ends continuous, the use of essentially I_g for all middle strip locations and the positive column strip locations means that the overall I_e of the equivalent frame is relatively close to I_g. In Example 5, the effective $(I_{D+L})_{\text{frame}} = 2436$ in.^{4} compared to $(I_g)_{\text{frame}} = 2662$ in.^{4} In a similar example in Ref. 2-12, the effective $(I_{D+L})_{\text{frame}} = 1379 \times 10^6$ mm^4 compared to $(I_g)_{\text{frame}} = 1547 \times 10^6$ mm^4. In a similar example in Ref. 2-10, the deflections were as follows: $a_{cp+sh} + (a_i)_L = 0.36$ in. using I_g and 0.39 in. using I_e for an interior panel, and 0.50 in. using I_g and 0.60 in. using I_e for a corner panel.

Analogous to 2 above: Such comparisons of computed versus allowable deflections demonstrate the difficulty in meeting the stringent *ACI Code* limitation of $(l_n$ or $l)/480$, even when the minimum thickness provisions of the *Code* are satisfied.

2.6.4 Example 6—Slab System With Beams (Two-Way Slab), Normal Weight Concrete

Given: slab system with beams in Fig. 2-12:

$f_y = 50{,}000$ psi (345 N/mm^2), slab thickness $h_f = 6$ in. (15.2 cm)
Square panels—20 ft by 20 ft (6.1 m by 6.1 m) c.c. of columns
All beams—$b_w = 12$ in. (30.5 cm), $h = 25$ in. (63.5 cm)
Clear span—$l_n = 20 - 1 = 19$ ft (5.8 m)

It is noted that the concrete strength and loading are not required in this analysis.

Required: minimum thickness analysis for deflection control (*ACI Code* and Commentary)[2-1].

Effective width, b, and section properties (Figs. 2-3, 2-12):
Interior beam:

$$I_s = (20)(12)(6)^3/12 = 4320 \text{ in.}^4 \ (179{,}800 \text{ cm}^4)$$

$$h - h_f = 25 - 6 = 19 \text{ in.} \leqslant 4h_f = (4)(6) = 24 \text{ in.} \qquad \text{OK}$$

Hence, $b = 12 + (2)(19) = 50$ in. (127 cm).

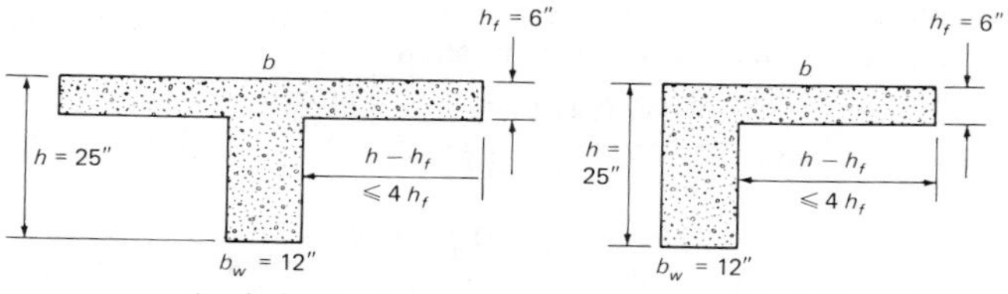

Fig. 2-12 Example 6—Slab system with beams (two-way slabs), normal weight concrete.

$$y_t = h - (1/2)[(b - b_w)h_f^2 + b_w h^2]/[(b - b_w)h_f + b_w h]$$
$$= 25 - (1/2)[(38)(6)^2 + (12)(25)^2]/[(38)(6) + (12)(25)] = 16.60 \text{ in.}$$

$$I_b = (b - b_w)h_f^3/12 + b_w h^3/12 + (b - b_w)h_f(h - h_f/2 - y_t)^2 + b_w h(y_t - h/2)^2$$
$$= (38)(6)^3/12 + (12)(25)^3/12 + (38)(6)(25 - 3 - 16.60)^2 + (12)(25)(16.60 - 12.5)^2$$
$$= 28{,}000 \text{ in.}^4 \ (1{,}165{,}000 \text{ cm}^4)$$

$$\alpha = E_{cb}I_b/E_{cs}I_s = I_b/I_s = 28{,}000/4320 = 6.48$$

Spandrel beam:

$$I_s = (10)(12)(6)^3/12 = 2160 \text{ in.}^4 \ (89{,}900 \text{ cm}^4)$$

$$b = 12 + (25 - 6) = 31 \text{ in.} \ (78.7 \text{ cm})$$

$$y_t = 25 - (1/2)[(19)(6)^2 + (12)(25)^2]/[(19)(6) + (12)(25)] = 15.12 \text{ in.}$$

$$I_b = (19)(6)^3/12 + (12)(25)^3/12 + (19)(6)(25 - 3 - 15.12)^2 + (12)(25)(15.12 - 12.5)^2$$
$$= 23{,}420 \text{ in.}^4 \ (975{,}000 \text{ cm}^4)$$

$$\alpha = I_b/I_s = 23{,}420/2160 = 10.84$$

α_m, β_s, β:
α_m (average value of α for all beams on the edges of a panel), and β_s (ratio of length of continuous edges to total perimeter of a slab panel):

Interior panel: $\alpha_m = 6.48$, $\beta_s = 1$
Side panel: $\alpha_m = [(3)(6.48) + 10.84]/4 = 7.57$, $\beta_s = 3/4$
Corner panel: $\alpha_m = [(2)(6.48) + (2)(10.84)]/4 = 8.66$, $\beta_s = 1/2$

For square panels, β = ratio of clear spans in the two directions = 1.

Minimum thickness:

$$h_{\min} = \frac{l_n(800 + 0.005f_y)}{36{,}000 + 5000\beta[\alpha_m - 0.5(1 - \beta_s)(1 + 1/\beta)]}$$
$$= \frac{(19)(12)[800 + (0.005)(50{,}000)]}{36{,}000 + (5000)(1)[6.48 - (0.5)(1 - 1)(1 + 1)]}$$
$$= 3.50 \text{ in. } (8.9 \text{ cm})\text{—interior panel}$$

Also $h_{\min} = 3.30$ in.—side panel, and $h_{\min} = 3.12$ in.—corner panel. But it is not less than:

$$h_{\min} = \frac{l_n(800 + 0.005f_y)}{36{,}000 + 5000\beta(1 + \beta_s)}$$
$$= \frac{(19)(12)[800 + (0.005)(50{,}000)]}{36{,}000 + (5000)(1)(1 + 1)}$$
$$= 5.20 \text{ in. } (13.2 \text{ cm})\text{—interior panel}$$

Also h_{min} = 5.35 in.—side panel, and h_{min} = 5.50 in.—corner panel. It need not be more than:

$$h_{min} = \frac{l_n(800 + 0.005f_y)}{36,000}$$

$$= \frac{(19)(12)\,[800 + (0.005)(50,000)]}{36,000}$$

= 6.65 in. (16.9 cm) for all panels

Hence the slab thickness of 6 in. (15.2 cm) > 5.50 in. (14.0 cm) is satisfactory for all panels, and deflections need not be checked according to the *ACI Commentary.*[2-1]

Deflections: For a similar two-way slab example in which deflections are computed in Ref. 2-10, the results were as follows: $a_{cp+sh} + (a_i)_L$ = 0.26 in. using I_g and 0.27 in. using I_e for an interior panel, and 0.31 in. using I_g and 0.36 in. using I_e for a corner panel.

NOTATION

a = deflection. Also depth of compression area (elastic), and depth of compression block (ultimate strength)
a_θ = deflection due to end rotation in two-way slab systems
A_s = area of nonprestressed tension steel
A'_s = area of compression steel
A_{sh} = shrinkage curvature coefficient
b = width of compression face of a member
b_e, b_w = effective flange width, web width
c_2 = width of column face perpendicular to slab strip being considered in a two-way slab system
C_t = creep coefficient defined as the ratio of creep strain to initial strain, at any time t
C_u = ultimate (in time) creep coefficient
d = effective depth of a section = distance from extreme compression face to centroid of tension steel
d' = distance from extreme compression face to centroid of compression steel
D = dead load plus any sustained portion of the live load in eq. (2-11)
E_c = modulus of elasticity of concrete
E_{cc}, E_{cs} = modulus of elasticity of column concrete, slab concrete
E_s = modulus of elasticity of steel
f_c, f_s = concrete flexural stress, steel stress
f = modulus of rupture of concrete
h = overall thickness of member
h_f = flange thickness
h_{min} = minimum thickness
I = moment of inertia (second moment of the area of a section)
I_b, I_c, I_s = moment of inertia of gross section of beams, columns, slabs
I_c = moment of inertia of composite section with transformed slab
I_{cr} = moment of inertia of cracked transformed section
I_e = effective moment of inertia
I_g = moment of inertia of gross concrete section, neglecting the steel
I_m = moment of inertia at midspan
I_2 = moment of inertia of precast beam of a composite member
k_r = reduction factor defined in eqs. (2-17), (2-18)
K = deflection coefficient
K_b, K_c, K_{ec}, K_s = flexural stiffness of beams, columns, effective columns, slabs
K_{sh} = shrinkage warping deflection coefficient
K_t = torsional stiffness of torsional element in two-way slab systems
l, l_n = span length, clear span
L = live load in eq. (2-11)
M = bending moment
M_a = maximum service load moment (unfactored moment) at the stage for which deflections are being considered
M_{cr} = cracking moment
M_m = midspan moment
M_0 = beam statical moment. Also used for statical moment of a slab panel
M_n, M_u = nominal moment strength, ultimate moment strength
n = modular ratio = E_s/E_c
n_c = $E_{\text{precast beam}}/E_{\text{slab}}$ for a composite member
p = tension steel percentage = $100A_s/bd$
p' = compression steel percentage = $100A'_s/bd$
T = multiplier for additional long-time deflection due to time-dependent effects, principally creep and shrinkage
T_u = multiplier for the additional ultimate long-time deflection
w = uniformly distributed load
w_c = unit weight of concrete
y_t = distance from the centroidal axis of the gross section to the extreme face in tension
ϵ = unit strain
$\epsilon_i, \epsilon_{cp}$ = initial strain, creep strain
$\epsilon_s, \epsilon_{sh}$ = steel strain, shrinkage strain
$(\epsilon_{sh})_t, (\epsilon_{sh})_u$ = shrinkage strain at any time t, ultimate shrinkage strain
$\theta, \theta_c, \theta_m, \theta_{\text{frame}}$ = end rotation in general, of column strips, of equivalent frames
λ = multiplier for additional long-time deflections due to combined creep and shrinkage
λ_{cp} = multiplier for additional long-time deflections due to creep
ρ = tension steel ratio = A_s/bd
ρ' = compression steel ratio = A'_s/bd
ρ_b = steel ratio producing a balanced ultimate strength condition
ρ_w = A_s/b_wd
ϕ = curvature. Also used to denote capacity reduction factor
ϕ_{sh} = shrinkage warping curvature

Subscripts:

1 = cast-in-place slab of a composite beam or the effect of the slab, such as due to slab dead load. Also used to denote direction being considered in a two-way slab system
2 = precast beam section of a composite member. Also used to denote perpendicular dimension to direction being considered in a two-way slab system
c = composite section. Also concrete. Also column strip
cp = concrete creep
cr = cracking
D = dead load
DS = differential shrinkage and creep

g = gross section
i = initial value (short-time value)
L = live load
m = midspan. Also used to denote middle strip
s = steel. Also used to denote slab
sh = shrinkage
sus = sustained
t = time-dependent
u = ultimate (in time) value

REFERENCES

2-1 American Concrete Institute, *Building Code Requirements for Reinforced Concrete*, and *Commentary on Building Code Requirements for Reinforced Concrete*, ACI 318-83, Detroit, Michigan, 1983.

2-2 ACI Committee 435, D. E. Branson, Chairman, "Deflections of Reinforced Concrete Flexural Members," *ACI Journal*, *Proceedings* V. 63, No. 6, June 1966, pp. 637–674. Also *ACI Manual of Concrete Practice*.

2-3 Subcommittee 1, R. S. Fling, Chairman, ACI Committee 435, D. E. Branson, Chairman, "Allowable Deflections," *ACI Journal*, *Proceedings* V. 65, No. 6, June 1968, pp. 433–444. Also *ACI Manual of Concrete Practice*.

2-4 Subcommittee 2, J. R. Benjamin, Chairman, ACI Committee 435, B. L. Meyers, Chairman, "Variability of Deflections of Simply Supported Reinforced Concrete Beams," *ACI Journal*, *Proceedings* V. 69, No. 1, Jan. 1972, pp. 29–35. Discussion, *ACI Journal*, July 1972. Also *ACI Manual of Concrete Practice*.

2-5 Subcommittee 7, R. S. Fling and A. F. Shaikh, Chairmen, ACI Committee 435, B. L. Meyers and G. M. Sabnis, Chairmen, "Deflections of Continuous Concrete Beams," *ACI Journal*, *Proceedings* V. 70, No. 12, Dec. 1973, pp. 781–787. Also *ACI Manual of Concrete Practice*.

2-6 Subcommittee 5, A. H. Nilson, Chairman, ACI Committee 435, G. M. Sabnis, Chairman, "State-of-the-Art Report on Deflection of Two-Way Reinforced Concrete Floor Systems," *Deflection of Concrete Structures*, SP 43-3, American Concrete Institute, Detroit, 1974, pp. 55–81. Also *ACI Manual of Concrete Practice*.

2-7 Building Code Subcommittee, D. E. Branson, Chairman, ACI Committee 435, J. R. Libby and G. M. Sabnis, Chairmen, "Proposed Revisions by Committee 435 to ACI Building Code and Commentary Provisions on Deflections," *ACI Journal*, *Proceedings* V. 75, No. 6, June 1978, pp. 229–238. Discussion, *ACI Journal*, Dec. 1978.

2-8 Subcommittee 2, D. E. Branson, Chairman, ACI Committee 209, J. R. Keeton, Chairman, "Prediction of Creep, Shrinkage, and Temperature Effects in Concrete Structures," *Designing for Effects of Creep, Shrinkage, and Temperature Effects in Concrete Structures*, SP 27-3, American Concrete Institute, Detroit, 1971, pp. 51–93.

2-9 American Association of State Highway and Transportation Officials, *Standard Specifications for Highway Bridges*, Washington, D.C., 1977, 496 pp. Also 1973 Ed.

2-10 Branson, D. E., *Deformation of Concrete Structures*, Mc-Graw Hill Book Co., Advanced Book Program, New York, 1977, 546 pp.

2-11 Branson, D. E., *Deflexiones de Estructuras de Concreto Reforzado y Presforzada*, Serie Concreto Estructural, CE-1, Instituto Mexicano del Cemento y del Concreto, Mexico City, 1978, 130 pp.

2-12 Branson, D. E., Chapter 5—"Designing for Deflections," pp. 5-1 to 5-61, *Metric Design Handbook for Reinforced Concrete Elements*, Canadian Portland Cement Association, Ottawa, Canada, Editor, M. Saatcioglu, 1978, 7–59 pp.

2-13 Branson, D. E., Chapter 4—"Reinforced Concrete Composite Flexural Members," pp. 97–147, and Chapter 5—"Prestressed Concrete Composite Flexural Members," pp. 148–210, *Handbook of Composite Construction Engineering*, Van Nostrand Reinhold Co., New York, Editor, G. M. Sabnis, 1979, 380 pp.

2-14 Pauw, A., "Static Modulus of Elasticity of Concrete as Affected by Density," *ACI Journal*, *Proceedings* V. 57, No. 6, Dec. 1960, pp. 679–687.

2-15 ACI Committee 213, J. A. Hanson, Chairman, "Guide for Structural Lightweight Aggregate Concrete," *ACI Journal*, *Proceedings* V. 64, No. 8, Aug. 1967, pp. 433–469. Also *ACI Manual of Concrete Practice*.

2-16 American Concrete Institute, *Designing for Effects of Creep, Shrinkage, and Temperature in Concrete Structures*, SP-27, Detroit, 1971, 430 pp.

2-17 Branson, D. E., and Christiason, M. L., "Time-Dependent Concrete Properties Related to Design—Strength and Elastic Properties, Creep and Shrinkage," *Designing for Effects of Creep, Shrinkage, and Temperature in Concrete Structures*, SP 27-13, 1971, pp. 257–277.

2-18 Washa, G. W., and Fluck, P. G., "The Effect of Compressive Reinforcement on the Plastic Flow of Reinforced Concrete Beams," *ACI Journal*, *Proceedings* V. 49, No. 2, Oct. 1952, pp. 89–108.

2-19 Yu, W. W., and Winter, G., "Instantaneous and Long-Time Deflections of Reinforced Concrete Beams Under Working Loads," *ACI Journal*, *Proceedings* V. 57, No. 1, July 1960, pp. 29–50.

2-20 Branson, D. E., "Instantaneous and Time-Dependent Deflection of Simple and Continuous Reinforced Concrete Beams," *HPR Report No. 7*, Part 1, Alabama Highway Department, Bureau of Public Roads, Aug. 1963, 78 pp.

2-21 Burns, N. H., and Siess, C. P., "Repeated and Reverse Loading in Reinforced Concrete," *Proceedings*, *ASCE*, V. 92, ST5, Oct. 1966, pp. 65–78.

2-22 Branson, D. E., "Design Procedures for Computing Deflections," *ACI Journal*, *Proceedings* V. 65, No. 9, Sept. 1968, pp. 730–743.

2-23 Beeby, A. W., "Short-Term Deformations of Reinforced Concrete Members," *Technical Report No. TRA 409*, Cement and Concrete Association, London, Mar. 1968, 32 pp.

2-24 Hollington, M. R., "A Series of Long-Term Tests to Investigate the Deflection of a Representative Precast Concrete Floor Component," *Technical Report TRA 442*, Cement and Concrete Association, London, Apr. 1970, 43 pp.

2-25 Branson, D. E., "Compression Steel Effect on Long-Time Deflections," *ACI Journal*, *Proceedings* V. 68, No. 8, Aug. 1971, pp. 555–559.

2-26 Lutz, L. A., "Graphical Evaluation of the Effective Moment of Inertia for Deflection," *ACI Journal*, *Proceedings* V. 70, No. 3, Mar. 1973, pp. 207–213.

2-27 Ferguson, P. M., *Reinforced Concrete Fundamentals*,

3rd Ed., 1973 and 4th Ed., 1979, John Wiley & Sons, Inc., New York, 724 pp.

2-28 Fling, R. S., Chapter 2—"Deflections," pp. 44-54, *Handbook of Concrete Engineering*, Van Nostrand Reinhold Co., New York, Editor, M. Fintel, 1974, 801 pp.

2-29 American Concrete Institute, *Deflections of Concrete Structures*, SP-43, Detroit, 1974, 637 pp.

2-30 Fling, R. S., "Simplified Deflection Computations," *Deflections of Concrete Structures*, SP 43-7, 1974, pp. 205-223.

2-31 Zuraski, P. D., Salmon, C. G., and Shaikh, A. F., "Calculation of Instantaneous Deflections for Continuous Reinforced Concrete Beams," *Deflections of Concrete Structures*, SP 43-12, 1974, pp. 315-331.

2-32 Kripanarayanan, K. M., and Branson, D. E., "Some Experimental Studies of Time-Dependent Deflections of Noncomposite and Composite Reinforced Concrete Beams," *Deflections of Concrete Structures*, SP 43-16, 1974, pp. 409-419.

2-33 Wang, C. K., and Salmon, C. G., *Reinforced Concrete Design*, 3rd Ed., Harper & Row Publishers, New York, 1979, 918 pp.

2-34 Winter, G., and Nilson, A. H., *Design of Concrete Structures*, 9th Ed., McGraw-Hill Book Co., New York, 1979, 647 pp.

2-35 Grossman, J. S., "Simplified Computations for Effective Moment of Inertia Ie and Minimum Thickness to Avoid Deflection Computations," *ACI Journal*, *Proceedings* V. 78, No. 6, Now.-Dec. 1981, pp. 423-439. Discussion, *ACI Journal*, Sept.-Oct. 1982.

2-36 Branson, D. E., and Trost, H., "Unified Procedures for Predicting the Deflection and Centroidal Axis Location of Partially Cracked Nonprestressed and Prestressed Members," *ACI Journal*, *Proceedings* V. 79, No. 2, Mar.-Apr. 1982, pp. 119-130.

2-37 Rice, P. F., Editor, *CRSI Handbook*, Concrete Reinforcing Steel Institute, Chicago, 2nd Ed., 1975, pp. 15-53.

2-38 Guralnick, S. A., and La Fraugh, R. W., "Laboratory Study of a 45-Foot Square Flat Plate Structure," *ACI Journal*, *Proceedings* V. 60, No. 9, Sept. 1963, pp. 1107-1185.

2-39 Vanderbilt, M. D., Sozen, M. A., and Siess, C. P., "Deflections of Multi-Panel Reinforced Concrete Floor Slabs," *Proceedings*, *ASCE*, V. 91, ST4, Part 1, Aug. 1965, pp. 77-101.

2-40 Hatcher, D. S., Sozen, M. A., and Siess, C. P., "Test of a Reinforced Concrete Flat Slab," *Proceedings*, *ASCE*, V. 95, ST6, June 1969, pp. 1051-1072.

2-41 Gamble, W. L, Sozen, M. A., and Siess, C. P., "Tests of a Two-Way Reinforced Concrete Floor Slab," *Proceedings*, *ASCE*, V. 95, ST6, June 1969, pp. 1073-1096.

2-42 Nilson, A. H., and Walters, D. B., "Deflection of Two-Way Floor Systems by the Equivalent Frame Method," *ACI Journal*, *Proceedings* V. 72, No. 5, May 1975, pp. 210-218.

2-43 Kripanarayanan, K. M., and Branson, D. E., "Short-Time Deflections of Flat Plates, Flat Slabs, and Two-Way Slabs," *ACI Journal*, *Proceedings* V. 73, No. 12, Dec. 1976, pp. 686-690.

2-44 Scanlon, A., and Murray, D. W., "Practical Calculation of Two-Way Slab Deflections," Concrete International, American Concrete Institute, V. 4, No. 11, Nov. 1982, pp. 43-50.

2-45 Subcommittee 5, A. C. Scordelis, Chairman, ACI Committee 435, D. E. Branson, Chairman, "Deflections of Prestressed Concrete Members," *ACI Journal*, *Proceedings* V. 60, No. 12, Dec. 1963, pp. 1697-1728. Also *ACI Manual of Concrete Practice*.

2-46 Prestressed Concrete Institute, *PCI Design Handbook—Precast and Prestressed Concrete*, 2nd Ed., Chicago, 1978, 8-28 pp.

2-47 Nilson, A. H., *Design of Prestressed Concrete*, John Wilay & Sons, New York, 1978, 526 pp.

3

One- and Two-Way Slabs

SIDNEY H. SIMMONDS, Ph.D.*

3.1 INTRODUCTION

Reinforced concrete slabs are one of the most widely used structural elements. In many structures, in addition to providing a versatile and economical method of supporting gravity loads, the slab also forms an integral portion of the structural frame to resist lateral forces.

In spite of their widespread use, there has never been a universally accepted method of proportioning all slab systems, owing to the complexity of formulating a theoretical analysis for the support conditions commonly used in buildings. Although "exact" analyses can be obtained for arbitrary geometry and material response using numerical techniques such as finite difference[3-3] and finite element[3-4, 3-5] methods, the time and cost of preparing input and interpreting output make these techniques unsuitable for design office use except for very exceptional cases. Fortunately most slab systems encountered in practice do not require a rigorous analysis, and simplified design procedures may be used. It is such simplified design procedures for gravity loading that are discussed in this chapter.

Structural reinforced concrete slabs are classified by the way they are supported. Slabs supported such that they can bend essentially in one direction only are referred to as one-way slabs; that is, the load is basically carried in one direction. Slabs supported by isolated supports (columns) arranged in more or less regular rows that permit the slab to deflect in two orthogonal directions are known as two-way slabs. Slabs with supports that are random in size and location are classified as irregular slabs and may exhibit both one-way and two-way behavior. Such slabs require special attention.

Slabs that carry load by two-way action but without the use of beams are one of the most efficient structural systems. Such slabs are economical, since they can be constructed with minimum field labor resulting from the use of simple formwork and reinforcing steel arrangements. In addition these slabs provide high flexibility in column layout, and require the least story height for a specified clear headroom; and in many cases the ceiling finish, if required, can be applied directly to the slab soffit, thereby eliminating costly hung ceilings. For slabs with longer spans or heavy industrial loading, drop panels and/or column capitals (see Section 3.2) may be used.

In certain cases it is advantageous to stiffen the slab by providing beams between columns. Such beams will decrease slab deflections and so permit longer spans with thinner slab sections. The use of beams greatly reduces the problems of shear and moment transfer between columns and slabs. In addition, beams frequently provide an economical solution for unusual loading conditions such as a large stationary concentrated load or line loads.

As the span length of slabs increases, the self weight of the solid slab becomes significant. The influence of self weight may be reduced considerably by forming cavities in the slab soffit using standard pan forms. When the cavities form continuous ribs in one direction only, the slab is referred to as a joist slab (see Fig. 3-4). When the cavities form continuous ribs in two directions the slab is referred to as a waffle slab (see Example 3-4).

Solid one-way slabs are generally economical for spans below 14 ft. On the other hand, one-way joist slabs have

*Professor of Civil Engineering, The University of Alberta, Edmonton, Alberta, Canada.

been used competitively for spans exceeding 40 ft with the thickness of slab between ribs varying from 3 to $4^1/_2$ in., depending on the required fire rating. Solid two-way slabs without beams are generally most efficient for spans up to 25 ft, which can be increased to 28–30 ft with the use of drop panels. Waffle slabs are used frequently in the 30–40 ft span range. The effects of self weight and deflection problems associated with longer spans can also be overcome by the use of post-tensioning techniques, as discussed in Chapter 9. A discussion on the economics of various slab systems is given by Fintel and Ghosh.[3-6]

The steps in the design of any slab system may be summarized as follows:

1. Choose the slab thickness and the dimensions of auxiliary stiffening elements, if any, based on experience and code limitations.
2. Obtain a moment field that is in equilibrium with the loading and compatible with the supports.
3. Check the shear, torsion, and shear–moment transfer requirements.
4. Select the reinforcement.

The many different design procedures that have been proposed differ from each other essentially only in the second of these steps.

To compensate for the lack of a ready means of analyzing two-way slab systems, most building codes* contain detailed rules for carrying out the above design steps for the simple case of a building slab supported only on columns forming essentially rectangular panels of approximately equal spans and supporting only uniformly distributed gravity loading. The rules for such slabs as specified by ACI 318-83[3-1] are examined in detail in Sections 3.4 and 3.6 through 3.9. As the slab geometry and loading deviate from these idealized conditions, these rules become less applicable, and considerable judgment may be required by the designer to obtain a satisfactory design. Some techniques for special problems are given in Section 3.11.

Reinforced concrete slabs are generally underreinforced and possess considerable ductility, allowing considerable moment redistribution within the slab. This permits using inelastic design procedures for determining moment fields for irregular geometry where code procedures based on approximating elastic behavior cannot be used. Such techniques are discussed in Section 3.12.

Owing to the inherent ductility of reinforced concrete slabs, all design procedures will provide sufficient strength if the equations of equilibrium are satisfied. However, even regular slabs may have serviceability problems due to excessive deflections and cracking unless sufficient care is given to these items. It is these two serviceability factors that provide the criteria on which most slab designs are ultimately judged, and their importance in the design procedure cannot be overemphasized. Serviceability is discussed in Section 3.4.

3.2 SLAB TERMINOLOGY

In the concrete construction industry, as in other fields, terms are coined that have meaning primarily to those working within the industry. Sometimes the terms coined are precise and become part of the technical vocabulary; at other times the terms are used loosely, and the meaning may change with time. Such was the case with several terms used in reinforced concrete slab design.

Prior to 1971, for reasons given in Section 3.3, the code referred to those slabs having beams between all columns as "two-way" and those without beams as "flat." These terms were neither precise nor descriptive, since both slab systems carried load by two-way action, were flat on the top surface, and could have projections below the slab soffit. Since 1971 the code has referred to such slabs as "slabs with beams" and "slabs without beams" (see Fig. 3-1). The term "two-way" is used only to refer to slab action, not slab type, and the term "flat" has disappeared.

The new terms, although superior to the former designations, are not entirely precise. In interpreting the provisions of the code, a slab without beams also includes those slabs that have beams only along discontinuous edges. For a slab to qualify as a slab with beams, there must be beams between all columns, and these beams must have stiffnesses related to their span lengths. No other slab systems are considered by the code for two-way action.

The special case of a slab without beams having neither drop panels nor column capitals is frequently referred to as a "flat plate." Although not used by the code, this is a descriptive term, and it has continued to be used in the construction industry.

The term "column capital" refers to the flaring of the column cross section immediately below the slab. The purpose of this is to increase the shear resistance periphery and to reduce the bending moments in the slab. When

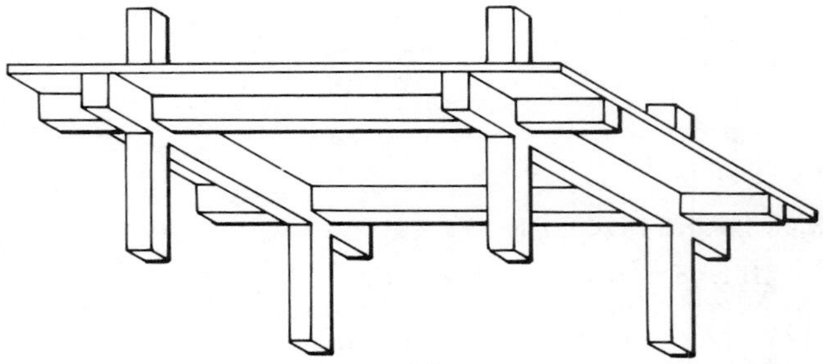

(a)

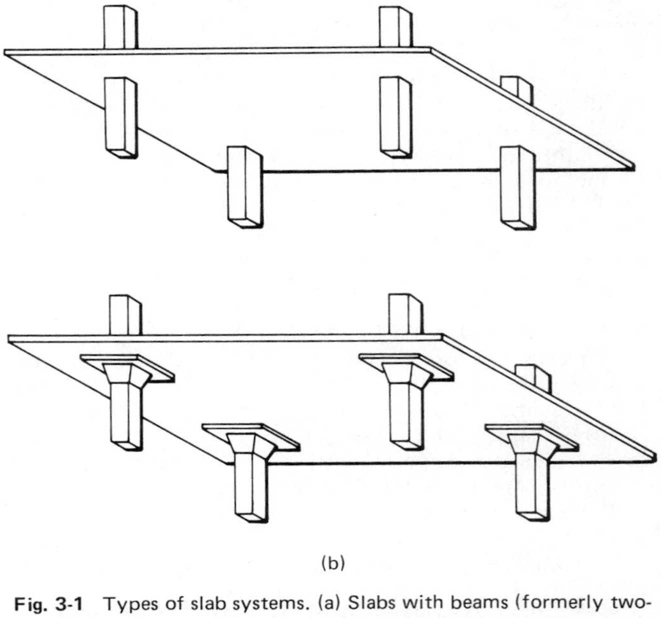

(b)

Fig. 3-1 Types of slab systems. (a) Slabs with beams (formerly two-way); and (b) slabs without beams (formerly flat slab and flat plate).

*Reference to a building code or to a specified code value or section in this chapter will be to ACI Standard 318-83 (Ref. 3-1) unless noted otherwise. Similarly, reference to the 1971 code will mean ACI Standard 318-71.

column capitals are formed, no portion of the capital shall be considered for structural purposes that lies outside the largest right circular cone or pyramid that can be formed in the concrete with planes oriented no greater than 45° to the column.

Drop panels refer to a uniform thickening of the slab in the vicinity of the column and are used to increase the effective depth of the slab when computing negative steel requirements. To qualify as a drop panel the thickened portion must extend in each direction not less than one-sixth of the span length measured center-to-center of supports in that direction, and must project below the slab at least one-fourth of the slab thickness beyond the drop.

3.3 HISTORICAL DEVELOPMENT OF SLAB DESIGN

Since the procedures for slab design given in the code have changed considerably since their introduction and are continuing to change, there is some benefit in reviewing briefly their development.

Reinforced concrete slabs were not a logical development of past construction practices. They were an invention. Patents were issued for "reinforced concrete slabs" as early as 1854 based on the concept of the concrete forming an arch with the reinforcement acting as the tie, and in 1867 based on the reinforcement acting as a catenary with the concrete used as a filler. Both systems were limited to a gridwork of closely spaced beams and hardly conform to our current definition of slab behavior.

The first patent for a recognizable reinforced concrete slab was given to Turner in 1903. He described a "mushroom" slab supported directly by columns with flared tops and reinforced both parallel to the column lines and along the diagonals. Two years later he built the first slab in Minneapolis, but as there was no precedent for the design, a performance load test was required. The slab readily carried the required loading without exceeding the specified deflection limits. So successful was his slab that others copied the idea, and by 1913 over 1000 "flat" slabs had been built.

Each builder had to develop his own design procedures and then verify the design by conducting a performance load test or by posting a performance bond. Depending on the builder, there was considerable variation in the reinforcement required. This was demonstrated in 1910 by McMillan,[3-7] who compared the quantity of reinforcement required by six design methods for a 20 ft × 20 ft interior panel carrying 200 psf live load, and found that they varied by a factor of 4.

In 1914 Nichols[3-8] established a simple criterion for the minimum total moment that must exist across the critical sections of a panel to satisfy equilibrium. His paper was not well received because he indicated total moments considerably greater than those used in many of the "successful" slab designs.

Other designers turned to classical plate theory as a basis of analysis, since the governing differential equation for elastic plate bending had been formulated by Lagrange in 1811. Solutions of this equation for rectangular panels bounded by combinations of simply supported and fixed edges had been developed. However, because these solutions were based on nondeflecting panel boundaries, the slab bending moments obtained are valid only when stiff beams are present on all four sides of each panel.

It was during this period of controversy that a committee was established to draft code provisions for the design of reinforced concrete slabs. This committee borrowed heavily from the work of Westergaard and Slater,[3-9] who had made an extensive examination of all previous work. The first slab provisions appeared in 1921 and were in two parts. The first part was placed in the body of the code and presented design coefficients for the slab obtained from solutions based on classical theory. As a result they were applicable only for "two-way" slabs with stiff beams between all columns. The second part was placed in an appendix to the code and covered "flat" slabs. These provisions were based on the formula derived by Nichols, but, even though Westergaard and Slater had reconciled the apparent differences between the theoretical and measured moments, the code, undoubtedly influenced by the success of the early builders, included a coefficient that resulted in total panel moments as low as 72% of that required by equilibrium. With only minor modifications these two procedures remained unchanged for the next half century.

It was long recognized that neither procedure was satisfactory. There was no means of designing a slab that fell between the slab types considered, that is, a slab with shallow beams. But a more serious concern was the difference in factor of safety based on ultimate load capacity that resulted from their origins. To resolve these problems a comprehensive study was initiated in the late 1950s primarily at the University of Illinois.[3-10] This study consisted of tests to failure of scaled slab systems[3-11–3-13] and analytical studies[3-14–3-16] which indicated that the behavior of slabs with or without beams was essentially the same and that the factor of safety based on live load using the existing code procedures was approximately three times greater for "two-way" slabs compared to "flat" slabs. Rather than modify existing code procedures this study was used to develop the direct design method (DDM) and the equivalent frame method (EFM) for the design of slabs with or without beams for uniformly distributed gravity loading. These procedures were incorporated in the 1971 code and resulted in a reasonably uniform factor of safety for all slab systems to which they apply.

The EFM replaces the three dimensional column-slab structure with orthogonal two-dimensional frames that are analyzed elastically. To approximate more closely actual behavior of the structure for pattern gravity loads, the stiffnesses of the members of the two-dimensional frame are modified from the stiffnesses obtained assuming prismatic members. The calculations required to make these modifications are complex and difficult to perform manually even with design aids. The procedure can be used in conjunction with a standard elastic frame computer program if a preprocessor is written to obtain equivalent stiffnesses and a postprocessor to obtain design moments.

The DDM was simple in concept and consisted of distributing the total panel static moment to design sections in accordance with a set of rules. The simplicity of the method required that certain restrictions of geometry be placed on its use. Unfortunately in the interim period between first publishing of the code in 1970 and its adoption in 1971, the basis of distributing moments in exterior panels was made a function of a stiffness ratio defined by the EFM so that much of the simplicity was lost.

Although the DDM and EFM introduced in the 1971 code were superior to previous procedures, they still were not totally satisfactory, and changes were made for the 1983 code. The DDM was returned to its simple concepts by removing the need to compute equivalent column stiffness ratios. Although the EFM is essentially unchanged, much of the explanation as to how to modify member stiffnesses has been removed from the code and placed in the *Commentary*.[3-2] Also, for the first time, the code addresses directly the role of the slab in resisting lateral loading by indicating that moments obtained from a lateral load analysis of an unbraced frame, in which stiffnesses of the members are obtained taking into account the effects of cracking and reinforcement, may be combined with the results of a gravity load analysis obtained using either the DDM or the EFM.

No building code is ever static, and future changes in the code slab sections are inevitable; hence no history is ever complete. More general approaches based on plastic design theory as used in other parts of the world may be incorporated by including permissible assumptions used by these methods. More complex and supposedly more "accurate" frame models have been proposed for combined gravity and lateral loading. And still others advocate the removal of all design procedures from the code and include only a performance specification.

3.4 SELECTING THE SLAB THICKNESS

A well-designed slab is judged on the basis of its performance under service conditions. The selection of slab thickness and the decision to use auxiliary stiffening elements such as column capitals and beams are of paramount importance in the design of slabs, since, to a great extent, these items determine the magnitude of slab deflections and the degree of cracking. As a general rule, ecomony of construction dictates that the slab be as thin as practical (minimum dead load) and have as few auxiliary stiffening elements as possible. On the other hand, the slab must provide sufficient shear capacity and flexural rigidity to keep deflections and cracking within acceptable limits. In many instances it is both practical and economical to obtain the required shear capacity and also decrease the slab deflections by providing column capitals, increasing slab thickness, or providing beams between the columns. We shall first examine the choice of slab thickness and then the use of auxiliary stiffening elements.

There are two procedures for selecting an initial slab thickness to satisfy deflection requirements. The first computes deflections for specific loading situations, using numerical techniques such as the finite element method or the methodology given in Chapter 2 of this handbook, and compares these values to permissible maximum values given in Chapter 9 of the code. The simpler and more commonly used procedure is to provide a minimum thickness that experience has indicated will result in a serviceable slab. The latter procedure is discussed in this section.

For one-way construction the minimum slab thicknesses expressed as fractions of the span length, l, are given in Table 3-1. The span length for slabs built integrally with the supports is defined as the clear span and for slabs not built integrally with the supports as the clear span plus the depth of member but not greater than the distance between centers of the supports.

For two-way slabs, minimum thicknesses for slabs consisting of regular rectangular panels on column supports are specified by the code both as fractions of the clear span, l_n, between supports and as absolute minimum values. For slabs without beams* the minimum thickness, h, is specified as:

$$h = \frac{l_n(800 + 0.005f_y)}{36{,}000} \tag{3-1}$$

which for $f_y = 40{,}000$ psi reduces to $h = l_n/36$.

*The phrase "slabs without beams" includes those slabs that have beams only along discontinuous edges (see Section 3.2).

TABLE 3-1* Minimum Thickness of Nonprestressed Beams or One-Way Slabs Unless Deflections Are Computed**

	Minimum thickness, h			
	Simply Supported	*One End Continuous*	*Both Ends Continuous*	*Cantilever*
Member	Members not supporting or attached to partitions or other construction likely to be damaged by large deflections.			
Solid one-way slabs	$l/20$	$l/24$	$l/28$	$l/10$
Beams or ribbed one-way slabs	$l/16$	$l/18.5$	$l/21$	$l/8$

**Table 9.5(a) from Ref. 3-1.*
***Span length l is in inches.*
Values given shall be used directly for members with normal weight concrete ($w_c = 145$ pcf) and Grade 60 reinforcement. For other conditions, the values shall be modified as follows:
(a) For structural lightweight concrete having unit weights in the range 90–120 lb per cu ft, the values shall be multiplied by ($1.65 - 0.005w_c$) but not less than 1.09, where w_c is the unit weight in lb per cu ft.
(b) For f_y other than 60,000 psi, the values shall be multiplied by ($0.4 + f_y/100{,}000$).

By specifying the minimum thickness in terms of the clear span, l_n, advantage is taken of the beneficial effect of large column sections or column capitals in reducing slab deflections. The term in the numerator in parentheses provides the greater slab thickness required to compensate for the decrease in stiffness caused by the greater degree of cracking associated with the higher allowable steel stresses. In ACI 318-63 this increase in thickness was 10% for each 10,000 psi increase in f_y above 40,000 psi, an increase that is retained for one-way slabs. However, in ACI 318-71 for two-way slabs this increase was reduced to 5%, since under service load conditions this type of slab is less likely to be cracked over a large portion of the area and thus is seemingly less sensitive to changes in steel stresses.

For slabs with beams the minimum thickness is:

$$h = \frac{l_n(800 + 0.005f_y)}{36{,}000 + 5000\beta\,[\alpha_m - 0.5(1 - \beta_s)(1 + 1/\beta)]} \tag{3-2}$$

The influence of the shape and position of the panel on the slab deflections is considered by the terms β and β_s, respectively. β is defined as the ratio of long to short clear spans and β_s as the ratio of length of continuous edges to total perimeter of a slab panel. The effect of the presence of stiffening beams is given by α_m, which is defined as the average value of α for the beams on all sides of the panel. The means of computing α for a beam is given later in this section. For very flexible beams, or no beams, the latter term in the denominator becomes negative, and the thickness by the above equation is greater than that required for slabs without beams. For this reason it is also specified that the thickness need not be greater than that given by eq. (3-1).

On the other hand, for very stiff beams, the required thickness can become less than desirable; so a limiting minimum thickness of:

$$h = \frac{l_n(800 + 0.005f_y)}{36{,}000 + 5000\beta(1 + \beta_s)} \tag{3-3}$$

is specified. For square interior panels, this corresponds to beams having an average flexural stiffness ratio of 2.0.

In addition, for panels having discontinuous edges, either an edge beam having a minimum stiffness of $\alpha = 0.8$ must be provided, or the minimum thicknesses must be increased by 10%. However, if a drop panel having dimensions at least one-third of the clear spans and a projection below the slab of at least $h/4$ is provided, the required minimum thickness may be reduced by 10%.

Notwithstanding the computed thicknesses from the above expressions, the thickness shall not be less than the following values:

For slabs without beams or drop panels	5 in.
For slabs without beams but with drop panels	4 in.
For slabs with beams on all four edges with a value of α_m at least equal to 2.0	$3\frac{1}{2}$ in.

When the slab thickness chosen is such that the shear capacity of the slab for the given column size is not sufficient, it is frequently good practice to provide a column capital. (See Section 3.2 for definition.)

Under certain conditions it is desirable to stiffen the slab by providing beams spanning between the columns. This happens, for example, when the vibrational properties of the slab are important or when heavy partitions or equipment are placed near column lines. The most common example is the use of beams along the outside edges to support the exterior walls directly and to provide the additional flexural rigidity required to reduce the slab thickness.

The dimensionless ratio α is defined as the ratio of flexural stiffness of beam section to the flexural stiffness of a width of slab bounded laterally by the centerline of the adjacent panel, if any, on each side of the beam. In this definition, for monolithic or fully composite construction, the beam section as shown in Fig. 3-2 is taken to mean not only the beam stem but also that portion of the slab on each side of the beam extending a distance equal to the projection of the beam above or below the slab, whichever is greater, but not greater than four times the slab thickness. The value of α may be obtained directly from the slab geometry using Fig. 3-3. (See Example 3-3.)

3.5 ONE-WAY SLABS

One-way slabs, as the name implies, resist loading by deflecting in one direction and hence have bending moments primarily in just one direction. While this type of behavior is obviously associated with isolated slabs that are supported only along opposite edges of a panel (for example, a stairway slab supported only at the top and bottom treads), it also occurs in slab panels that may be supported on all sides and may even be continuous in both directions, where owing to the geometry of the supports the curvatures in one direction are very much greater than in the other. A common example of this latter case is found in buildings with either structural steel or reinforced concrete framing

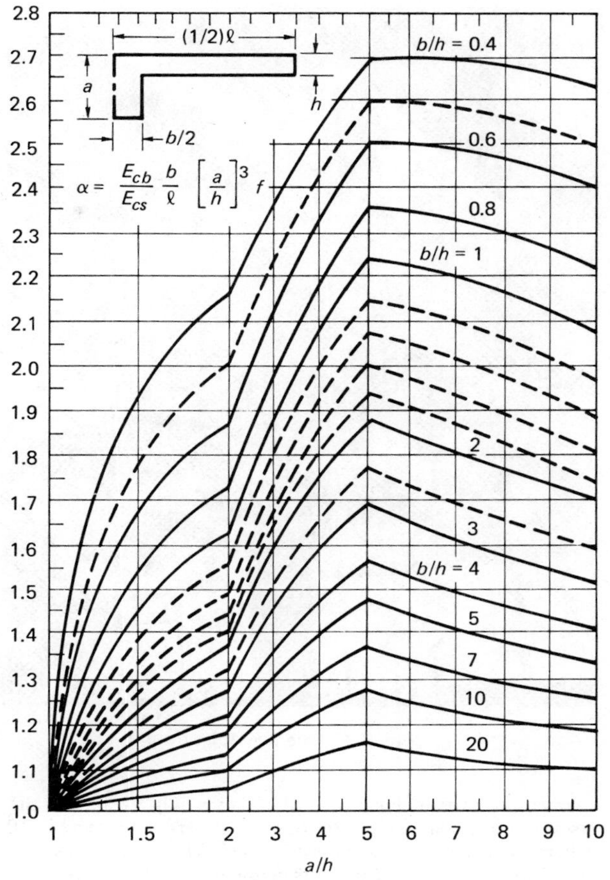

Fig. 3-3 Beam to slab ratio, α.

which make use of a grid of beams and girders between columns. When the ratio of the long to short dimension of the slab panel is large, say greater than 2, the slab may be designed as a one-way slab. An example of this type of slab system is given later in this section.

One-way behavior will also occur when a slab has substantially greater stiffness in one direction than in the other, as results when the slab is ribbed in one direction as shown in Fig. 3-4. These slabs are designed as T-beams in which the nominal shrinkage reinforcement in the transverse direction becomes the flexural reinforcement for the slab.

EXAMPLE 3-1: Design of a Typical Floor System for a Retail Sales Store

The architect has recommended a column spacing of 24 ft in the transverse direction and 30 ft in the longitudinal direction. For floors used for general merchandise the basic live loading is frequently specified as 100 psf. In addition to the dead weight of the structural slab, consideration must also be given to permanent loads such as slab topping and mechanical services such as sprinkling systems, ventilation ducts, and so on. For these loads and column spacing the use of a one-way slab supported on a grid of beams and girders offers an economical solution.

The design follows the steps given in Section 3.1. The slab thickness was selected on the basis of Table 3-1 so that deflection calculations are not required. Design moments and shears are obtained by considering a strip of slab, of unit width, spanning in the primary direction and analyzing it as a continuous beam. As the geometry meets the specified limitations, the approximate analysis given in section 8.3.3* is used to determine design shears and moments.

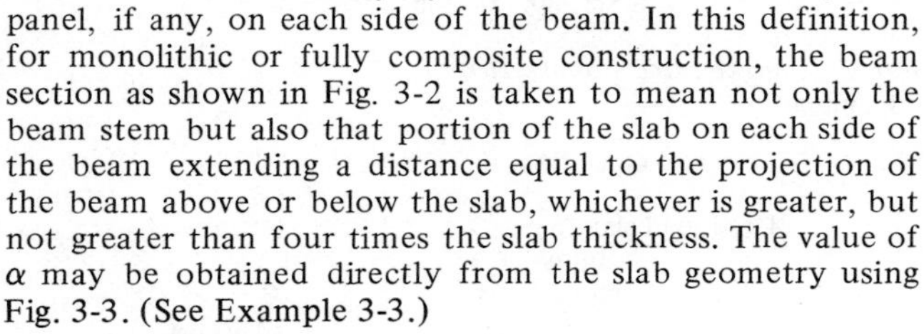

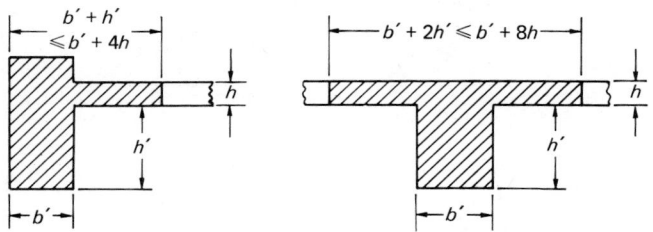

Fig. 3-2 Beam sections used in the definition of α.

*Reference in design examples to a section refer to the appropriate section in ACI 318-83.

Fig. 3-4 One-way joist floor.

Since the overall thickness of the slab is only 5.5 in. and the shear stresses are very low, a system of straight bars was selected, although a bent-up system could have been used. The maximum spacing and the minimum area requirements of the principal reinforcement must be computed because, as is the case in this example, these quantities may control the selection of reinforcement at design sections where the bending moment is small. Calculations are also included to indicate that the maximum steel ratio provided will result in a ductile failure and that the distribution of reinforcement used will result in satisfactory crack control. These latter calculations are generally not required, since for one-way slabs it would be unusual to select a thickness that would result in an overreinforced section, and for bar sizes smaller than #9, the maximum spacing is not governed by the crack control provisions. Development lengths and cut-off locations of the primary reinforcement are determined in the same manner as for other flexural members and therefore are not presented.

There will be local one-way behavior in the direction perpendicular to the design strip due to the presence of the stiffer girder. Transverse reinforcement is required to resist the resulting moments and is computed on the basis of considering the slab acting as a cantilever with span being the smaller of either eight times the slab thickness or one-half the clear distance to the web of the next girder. Since this is flexural reinforcement, it must be fully anchored and accurately placed. For this reason it is recommended that it be detailed independently of the transverse shrinkage reinforcement unless great care is taken to ensure that it will be properly placed.

The beams are an integral part of the slab system and would be designed as T-beams. It must be noted that in designing the slab strip a negative moment was assigned to the discontinuous edge. The exterior beam must be designed to resist these moments as a torsional moment applied to the beam. For this reason, although the exterior beam has a smaller tributary load area than interior beams, it may require a greater width to resist these torsional moments. The design of the beams and girders follows the examples in Chapter 1.

Materials:
Concrete–normal weight $f_c' = 3{,}000$ psi
Reinforcing–intermediate $f_y = 60{,}000$ psi

Design:
Panel aspect ratio $= l_2/l_1 = 30/12 = 2.5 > 2.0$
Design as one-way slab

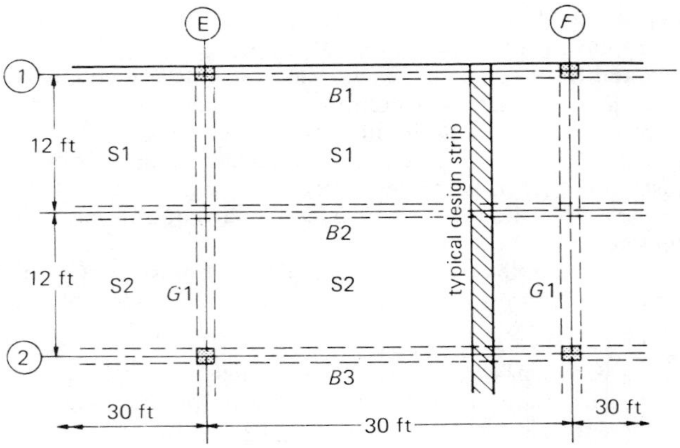

Step 1–Select slab thickness. Assume width of beam stem, b_w = 12 in., and slab and beams to be monolithic; therefore l = clear span = 12 × 12 – 12 = 132 in.
From Table 3-1:

$$\text{for exterior spans } h = \frac{l}{24} = \frac{132}{24} = 5.5 \text{ in.}$$

$$\text{for interior spans } h = \frac{l}{28} = \frac{132}{28} = 4.7 \text{ in.}$$

Try h = 5.5 in. Using #4 bars:

$$d = h - \text{cover} - \tfrac{1}{2} \text{ bar diameter}$$

$$= 5.5 - 0.75 - 0.5/2 = 4.5 \text{ in.}$$

Step 2–Analyze typical design strip:
Loading:

Live load (specified) $w_l = 100 \times 1.7 = 170$ psf

Dead load slab ($5\frac{1}{2}$") 69
topping (2") 25
ceiling + mech. 20
$w_d = 114 \times 1.4 = 160$ psf
$w_u = 330$ psf
$= 0.330$ ksf

Obtain moment and shear coefficients from *Code* section 8.3.3.

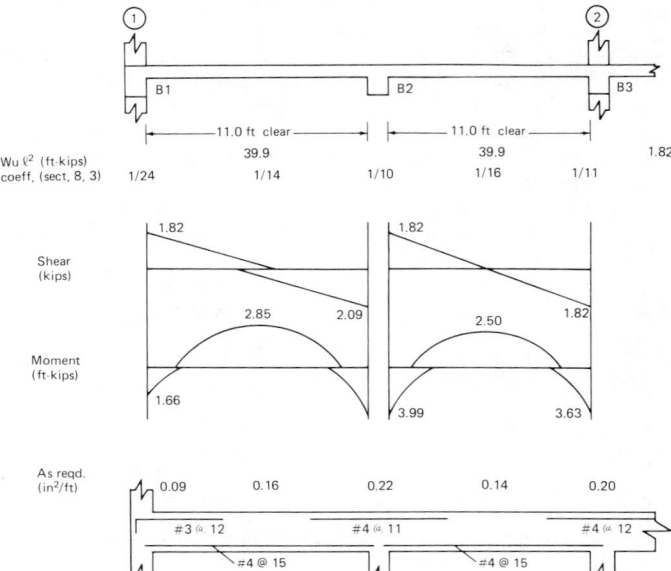

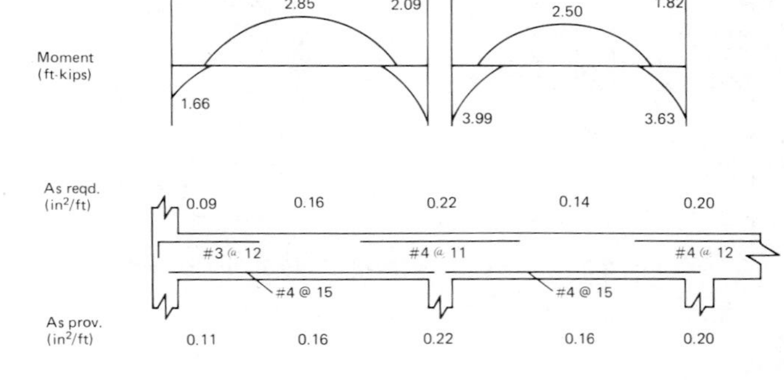

Step 3–Check shear:

$$\phi V_c = \phi v_c bd \geqslant V_u \max$$

$$\phi V_c = 0.85 \times 2\sqrt{3000} \times 12 \times 4.5/1000$$

$$= 5.03 \text{ kips} > 2.09 \text{ kips} \quad \text{O.K.}$$

Step 4–Choose reinforcement:

1. Principal reinforcement (for required area of reinforcement, A_s, see sketch above):
   (a) Maximum spacing (section 7.6.5):

$$s_{max} = 3h = 3 \times 5.5 = 16.5 \text{ in.} < 18 \text{ in.}$$

   (b) Minimum reinforcement (section 7.12):

$$A_s \min = 0.0018\, hb = 0.0018 \times 5.5 \times 12 = 0.12 \text{ in.}^2/\text{ft}$$

   (c) Maximum reinforcement (section 10.3.3):

$$\rho_b = \frac{0.85\, \beta_1 f_c'}{f_y}\left(\frac{87}{87 + f_y}\right) = \frac{0.85 \times 0.85 \times 3}{60}\left(\frac{87}{87 + 60}\right)$$

$$= 0.0214$$

$$\rho_{\max} = \frac{A_s \max}{bd} = \frac{0.22}{12 \times 4.5} = 0.0041$$

$$= 0.19\rho_b < 0.75\rho_b \quad \text{OK}$$

   (d) Distribution of reinforcement (section 10.6):

for interior exposure $z_{max} = 175$

for $f_s = 0.6 f_y$; $d_c = (0.75 + 0.25) = 1.0$;

$$A = 2 d_c s_{max}$$

$$z = f_s \sqrt[3]{d_c A} = 0.6 \times 60 \sqrt[3]{1 \times 2 \times 16.5}$$

$$= 116 < 175 \quad \text{OK}$$

For choice of primary reinforcement see sketch above.

2. Transverse reinforcement across girder (section 8.10): Effective overhanging slab width (section 8.10.2) is the smaller of 8 × 5.5 = 44 in. or ½ × 29 × 12 = 174 in. Design as cantilever with span 44 in.

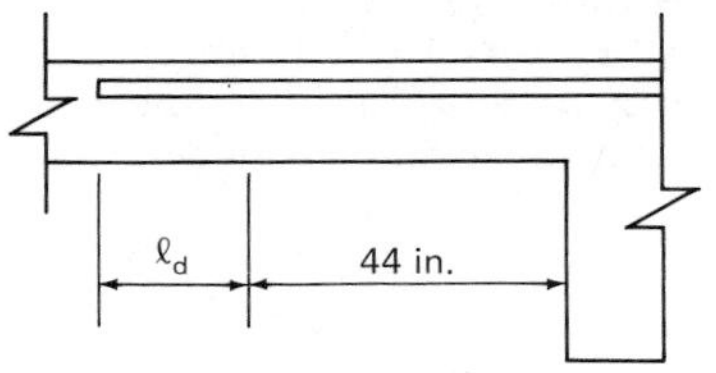

$$M_u = \frac{1}{2} w_u l^2 = \frac{1}{2} \times 0.33\left(\frac{44}{12}\right)^2 = 2.22 \text{ kip-ft/ft}$$

$$d_{\text{eff}} = h - \text{cover} - d_b \text{ (primary bar)} - \frac{1}{2} d_b \text{ (transverse bar)}$$

$$= 5.5 - 0.75 - 0.5 - 0.25 = 4.0 \text{ in.}$$

$$A_s = \frac{2.22 \times 12}{0.9 \times 60 \times 0.9 \times 4.0} = 0.14 \text{ in.}^2/\text{ft}$$

   (a) Maximum spacing (section 8.10.5.2):

$$s_{max} = 5 \times 5.5 = 27.5 > 18 \text{ in.}$$

   (b) Minimum reinforcement (section 7.12):

$$A_s \min = 0.0018 hb = 0.12 \text{ in.}^2/\text{ft}$$

Use #4 @ 17-in. spacing for transverse reinforcement across girders.

3. Temperature and shrinkage reinforcement (section 7.12): Use #4 @ 18-in. spacing.

3.6 DIRECT DESIGN METHOD

The DDM is a simple straightforward procedure for obtaining design moments in continuous two-way slabs. Owing to its simplicity, it was deemed necessary to impose the following limitations on its use. These limitations are to ensure two-way behavior with no localized irregularities.

1. Three shall be a minimum of three continuous spans in each direction.
2. The panels shall be rectangular with the ratio of longer to shorter span center-to-center of supports within a panel not greater than 2.0.
3. The successive span length center-to-center of supports in each direction shall not differ by more than one-third of the longer span.
4. Columns may be offset a maximum of 10% of the span in the direction of the offset, from either axis between center lines of successive columns.
5. The live load shall not exceed three times the dead load.
6. If a panel is supported by beams on all sides, the relative stiffness of the beams in the two perpendicular directions $\alpha_1 l_2^2/\alpha_2 l_1^2$ shall not be less than 0.2 or greater than 5.0.

During the compilation of the results of the slab program at Illinois[3-14] it became apparent that it would be impractical to incorporate all possible variables in a simplified design procedure. Two dimensionless ratios, namely $\alpha_1 l_2^2/\alpha_2 l_1^2$ and $\alpha_1 l_2/l_1$, were defined for use in determining slab moments. The first ratio is equal to unity when the beam flexural stiffness in each direction is proportional to the span length, a condition that was assumed throughout the slab program and is approximately satisfied by most slabs in practice. Limits are placed on this ratio to prevent the use of this design procedure when very stiff beams are provided in one direction compared to the stiffness of the beams in the other direction, since, for this condition, the slab tends to behave in one-way action for which this method is not applicable. The second ratio, $\alpha_1 l_2/l_1$, is the parameter used in defining effective beam flexural stiffness.

The procedure begins by considering design strips along each column line that are bounded laterally by the centerlines of panels on each side, as shown in Fig. 3-5. For each span the total factored static moment, M_0, defined as the sum of positive and average negative factored moments, is computed as:

$$M_0 = \frac{w_u l_2 l_n^2}{8} \tag{3-4}$$

where l_2 is defined as the average of adjacent transverse spans for interior design strips, and distance from edge to

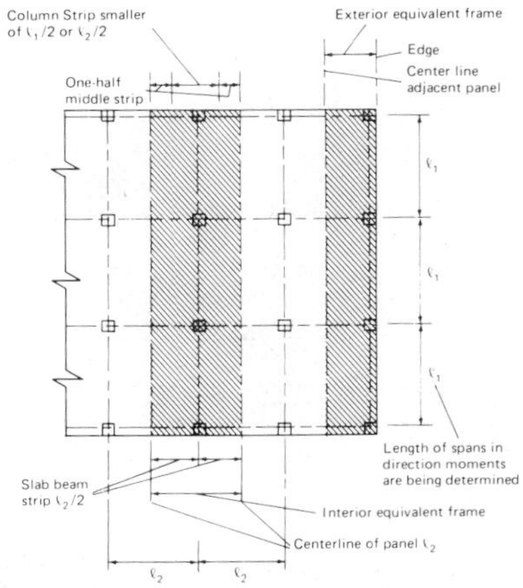

Fig. 3-5 Design strips for direct design method.

TABLE 3-2 Coefficients for Negative and Positive Factored Moments*

	(1)	(2)	(3)	(4)	(5)
			Slab Without Beams Between Interior Supports		
	Exterior Edge Unrestrained	*Slab With Beams Between All Supports*	*Without Edge Beam*	*With Edge Beam*	*Exterior Edge Fully Restrained*
Interior negative factored moment	0.75	0.70	0.70	0.70	0.65
Positive factored moment	0.63	0.57	0.52	0.50	0.35
Exterior negative factored moment	0	0.16	0.26	0.30	0.65

**From Ref. 3-1.*

panel centerline for exterior design strips and l_n is the clear span extending from face to face of columns, capitals, brackets, or walls but not less than 0.65 the center-to-center span, l_1. For purposes of computing M_0 or locating the critical design section, circular or regular polygon-shaped supports are treated as square supports with the same area.

The total static moment, M_0, is distributed to the negative design moments located at the face of rectangular supports and to the positive design moments located at midspan in much the same manner as for uniformly loaded continuous beams. For interior spans, $0.65M_0$ is assigned to the negative moment sections and $0.35M_0$ to the positive moment section. For exterior spans the distribution is dependent on the degree of restraint provided at the discontinuous edge, and suggested coefficients for various edge supports are given in Table 3-2. The coefficients for the unrestrained edge are used when the slab edge can be considered simply supported, say on a masonry wall, where no appreciable moment is transferred. The coefficients for an exterior edge with a fully restrained edge are the same as for interior panels and are used if the exterior edge of the slab is rigidly attached to a concrete wall that has a flexural stiffness sufficiently large compared to that of the slab that little rotation of the edge of the slab is permitted. Obviously some judgment is permitted for intermediate conditions. In addition, the DDM specifies that negative and positive factored design moments may be modified by 10%, provided the total static moment for each span in the direction considered is not less than that required by eq. (3-4).

It is now only necessary to distribute the factored design moments across the critical sections. This is accomplished by defining column and middle strips as indicated in Fig. 3-6. A column strip is the width of slab on each side of a column center line equal to $0.25l_2$ or $0.25l_1$, whichever is less, and includes the beam, if any. The middle strip is formed by the balance of the slab and hence consists of two portions on either side of the column strip. The code specifies only the portion of the factored design moment that is assigned to the column strip. The remaining portion is assigned to the half middle strips in proportion to their widths.

The portion of design moment assigned to the column strip is given in Fig. 3-7. It is seen that for slabs without beams, ($\alpha_1 = 0$), 60% of the positive moment and 75% of the negative moment at interior supports is assigned to the column strip. For slabs with stiff beams ($\alpha_1 l_2/l_1$ equal to or greater than 1.0) these portions are dependent on the panel aspect ratio l_2/l_1. Linear interpolation is used for values of $\alpha_1 l_2/l_1$ between zero and 1.0. For the exterior negative moment the amount assigned to the column strip is depen-

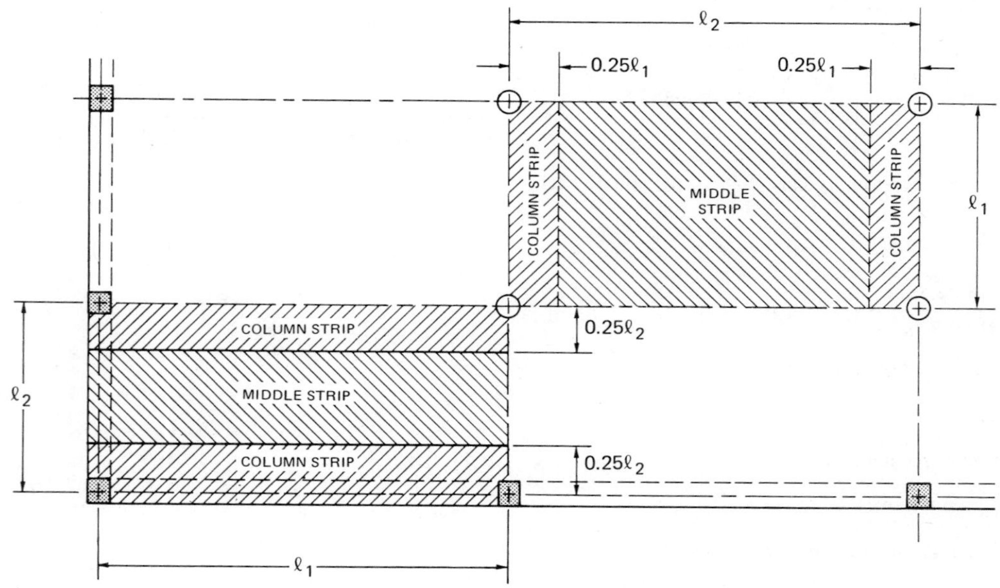

Fig. 3-6 Definition of column and middle strips.

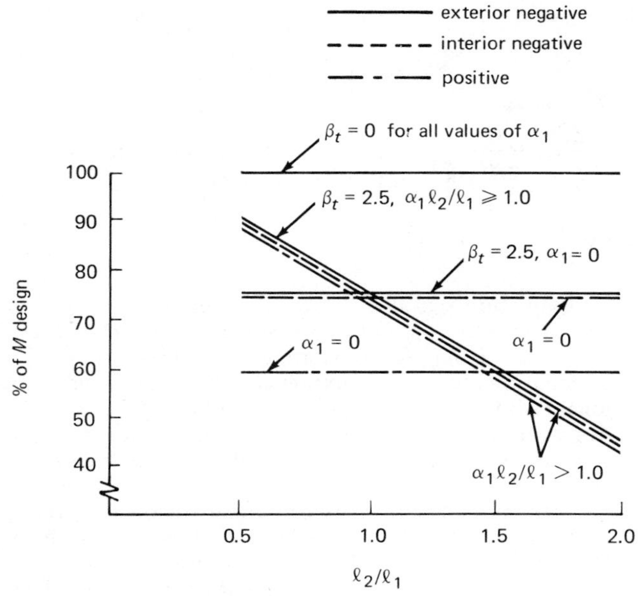

Fig. 3-7 Percentage of M_{design} assigned to column strip.

Table 3-3 α_{min}*

β_a	*Aspect ratio* l_2/l_1	*Relative beam stiffness, α* 0	0.5	1.0	2.0	4.0
2.0	0.5–2.0	0	0	0	0	0
1.0	0.5	0.6	0	0	0	0
	0.8	0.7	0	0	0	0
	1.0	0.7	0.1	0	0	0
	1.25	0.8	0.4	0	0	0
	2.0	1.2	0.5	0.2	0	0
0.5	0.5	1.3	0.3	0	0	0
	0.8	1.5	0.5	0.2	0	0
	1.0	1.6	0.6	0.2	0	0
	1.25	1.9	1.0	0.5	0	0
	2.0	4.9	1.6	0.8	0.3	0
0.33	0.5	1.8	0.5	0.1	0	0
	0.8	2.0	0.9	0.3	0	0
	1.0	2.3	0.9	0.4	0	0
	1.25	2.8	1.5	0.8	0.2	0
	2.0	13.0	2.6	1.2	0.5	0.3

**From Ref. 3-1.*

dent on the torsional stiffness of the beam along the discontinuous edge, β_t, as well as the effective beam stiffness $\alpha_1 l_2/l_1$. When no edge beam is used, 100% of the exterior negative moment is assigned to the column strip.

For slabs with beams, a portion of the moment assigned to the column strip is assumed to be resisted by the beam in proportion to its effective stiffness. This portion is 85% for beams with $\alpha_1 l_2/l_1 \geqslant 1.0$. For values of this ratio between 1.0 and 0.0 the portion resisted by the beam is obtained by linear interpolation between 85% and 0%. The DDM also includes rules for proportioning the shear force between the beam and slab. These are described in Section 3.8. The beam must also be designed to carry any loading applied directly to the beam including the self weight of the stem.

For the special case in which the exterior support consists of a wall or column extending for a distance equal to or greater than three-quarters of the transverse span, l_2, the exterior negative moment is considered to be uniformly distributed across l_2. In addition, the middle strip adjacent to and parallel with an edge supported by a wall is proportioned to resist twice the moment assigned to the adjacent half middle strip corresponding to the first row of interior supports.

A complete design for moments must include the moments assigned to the supports and the effects of pattern loading. Conditions of equilibrium require that all supporting columns and walls built integrally with the slab must resist the moments caused by the factored loads acting on the slab as computed above. However, for interior supports of equal spans and loading the unbalanced moment would be zero, and the DDM specifies a minimum support moment, obtained from eqn. (3-5), that is approximately the unbalanced moment when half the live load is applied to the longer span:

$$M = 0.07[(w_d + 0.5w_l)l_2 l_n^2 - w_d' l_2' (l_n')^2] \qquad (3\text{-}5)$$

If supports are present both below and above the slab, this moment is distributed in direct proportion to their flexural stiffnesses.

For exterior supports, since the coefficients contained in Table 3-2 are based on using the larger positive and interior negative moments occurring over the range of permissible geometry, the exterior negative moments will be the smallest possible values. For this reason the minimum moment transferred between slab and edge support must not be less than the nominal moment strength of the column strip framing into the support.

Pattern loads need be considered only when the ratio of unfactored dead load to unfactored live load, β_a, is less than 2. For such cases the designer must either ensure that the flexural stiffnesses of the columns above and below, α_c, is greater than the value of α_{min} given in Table 3-3, or multiply the positive factored moments in the panels supported by such columns by the coefficient, δ_s:

$$\delta_s = 1 + \frac{2 - \beta_a}{4 + \beta_a}\left(1 - \frac{\alpha_c}{\alpha_{min}}\right) \qquad (3\text{-}6)$$

Use of the direct design method is illustrated in Examples 3-2, 3-3, and 3-4.

3.7 EQUIVALENT FRAME METHOD

The EFM is an alternate method to the DDM for computing moments and shears for gravity loads in slabs supported on columns or walls forming orthogonal frames. The structure is considered to consist of design frames through the building along both longitudinal and transverse column lines which are bounded laterally by the centerline of the panel on each side. Frames adjacent and parallel to an edge are bounded by that edge and the centerline of the adjacent panels. It is usual to assume that each floor is unaffected by the loading on other floors so that the far ends of the columns above and below are assumed fixed. The frame to be analyzed elastically then corresponds exactly to the design strips used in the DDM with the addition of columns above and below (Fig. 3-8).

Since the interaction of the columns and slab is considered, the restrictions on geometry imposed by limitations 1, 3, and 5 for use of the DDM method (see Section 3.6) are removed. The restrictions imposed by limitations 2, 4, and 6 are still required to ensure two-way behavior, and the use of the EFM for situations outside these limitations should be done with caution, since considerable judgment may be required to obtain reasonable results. It should also be noted that the EFM does not explicitly contain guidance for the dis-

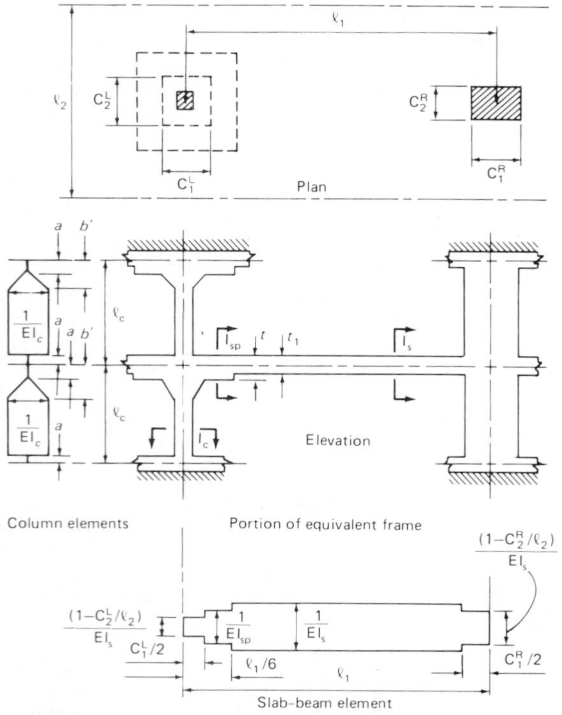

Fig. 3-8 Column and slab-beam elements for the equivalent frame method.

tribution of the design moments across the critical sections except for the case of slabs with beams satisfying limitation 6 for which the rules of distribution contained in the DDM may be used. These rules may also be used for slabs without beams if limitations 2 and 4 are satisfied.

In order to obtain moments from the two-dimensional frames that correspond to those in the three-dimensional slab–column structure particularly for pattern loading, equivalent stiffnesses must be computed for each member. If calculations are being done by hand using Cross moment distribution, the parameters of fixed end moment coefficient, stiffness factor, and carry-over factor are required.

In lieu of using moment distribution, some designers prefer to use an elastic frame analysis program based on the direct stiffness method. Unless the program has provision for considering nonprismatic beams, it is necessary when using the EFM to compute K_s and K_{ec} as in example 3-5 and then obtain an imaginary value of I such that the expression $4EI/l$ will give the same stiffness values. This computed imaginary I is then entered into the program.

A typical equivalent slab-beam element is shown in Fig. 3-8. The moment of inertia for this element between the faces of the columns, brackets, or capitals is based on the uncracked cross section of the concrete including beam or drop panel, if any. The moment of inertia between the center of the column and the face of the column, bracket, or capital is considered finite and is dependent on the transverse dimensions of the panel and support, and is indicated in Fig. 3-8. Coefficients for determining the fixed end moments, stiffness, and carry-over factors for slab-beam elements are given in Tables 3-4 to 3-6.[3-17]

The equivalent column consists of both the physical column and a transverse torsional member, as shown in Fig. 3-9. The stiffness of the equivalent column, K_{ec}, is determined from the expression:

$$\frac{1}{K_{ec}} = \frac{1}{\sum K_c} + \frac{1}{K_t} \qquad (3\text{-}7)$$

where:

$$K_c = \frac{k_c E_{cc} I_c}{l_c} = \text{stiffness of attached column}$$

$$K_t = \frac{\sum 9 E_{cs} C}{l_2(1 - c_2/l_2)^3} = \text{torsional stiffness of attached transverse members}$$

$$C = \sum \left[1 - 0.63\left(\frac{x}{y}\right)\right] \frac{x^3 y}{3} \qquad (3\text{-}8)$$

The summation sign in front of K_c in eq. (3-7) indicates that if columns are present above and below the slab, their stiffnesses are computed separately and added. Similarly, the summation sign in the expression K_t implies that the torsional stiffness of the transverse member in each side of interior columns is computed separately and added. For exterior columns, of course, there is only the one transverse member.

The torsional constant, C, is computed for the transverse member by dividing the cross section of this member into rectangular components with small dimension x and large dimension y and using either eq. (3-8) or Table 3-7. The correct division into possible rectangles is the one that results in the largest value of C. The cross section of the transverse torsional members is assumed constant throughout their length and consists of the larger of:

(a) Portion of slab having a width equal to that of the column, bracket, or capital in the direction for which moments are being determined.

(b) For monolithic or fully composite construction:
   i. Portion of slab specified in (a) plus that part of the transverse beam above and below the slab.
   ii. The beam plus that portion of the slab on each side of the beam extending a distance equal to the projection of the beam above or below the slab, whichever is greater, but not greater than four times the slab thickness.

If the design strip contains a beam parallel to the design strip, the value of K_t computed above must be increased by the ratio I_{sb}/I_s, where I_{sb} is the moment of inertia of the slab section of width l_2 and that portion of the beam stem extending above or below the slab, and I_s is the moment of inertia of the slab section of width l_2, i.e., $I_s = l_2 h^3/12$.

In determining the stiffness coefficient, k_c, for the column, variations in the moment of inertia along the column shall be taken into account. The length of the column, l_c, is defined as the distance from the midsurface of the slab to midsurface of the adjacent slab. The moment of inertia of the column between the midsurface and the soffit of the slab, drop panel, or beam may be considered infinite. Values of the stiffness coefficient k_c, are tabulated in Tables 3-8 and 3-9.

The equivalent stiffnesses for each slab-beam and column element having been evaluated, the equivalent frame is analyzed by elastic analysis. Where the live load does not exceed three quarters of the dead load or the nature of the live load is such that all panels must be loaded simultaneously, the design moments are obtained assuming full live load on all panels; otherwise maximum moments are obtained at each critical section by loading appropriate panels with three-quarters live load with no live load on the other panels.

Since the analysis of the frame is based on center-to-center lengths, the negative moments of the column centerlines may be reduced to the critical design section assumed to be at the face of the rectilinear column or capital for interior columns but not more than 0.175 l_1 from the column

TABLE 3-4 Moment Distribution Factors for Slab-Beam Elements without Drop Panels*

c_1/l_1 \ c_2/l_2		0.00	0.05	0.10	0.15	0.20	0.25	0.30	0.35	0.40	0.45	0.50
0.00	M	0.083	0.083	0.083	0.083	0.083	0.083	0.083	0.083	0.083	0.083	0.083
	k	4.000	4.000	4.000	4.000	4.000	4.000	4.000	4.000	4.000	4.000	4.000
	C	0.500	0.500	0.500	0.500	0.500	0.500	0.500	0.500	0.500	0.500	0.500
0.05	M	0.083	0.084	0.084	0.084	0.085	0.085	0.085	0.086	0.086	0.086	0.086
	k	4.000	4.047	4.093	4.138	4.181	4.222	4.261	4.299	4.334	4.368	4.398
	C	0.500	0.503	0.507	0.510	0.513	0.516	0.518	0.521	0.523	0.526	0.528
0.10	M	0.083	0.084	0.085	0.085	0.086	0.087	0.087	0.088	0.088	0.089	0.089
	k	4.000	4.091	4.182	4.272	4.362	4.449	4.535	4.618	4.698	4.774	4.846
	C	0.500	0.506	0.513	0.519	0.524	0.530	0.535	0.540	0.545	0.550	0.554
0.15	M	0.083	0.084	0.085	0.086	0.087	0.088	0.089	0.090	0.090	0.091	0.092
	k	4.000	4.132	4.267	4.403	4.541	4.680	4.818	4.955	5.090	5.222	5.349
	C	0.500	0.509	0.517	0.526	0.534	0.543	0.550	0.558	0.565	0.572	0.579
0.20	M	0.083	0.085	0.086	0.087	0.088	0.089	0.090	0.091	0.092	0.093	0.094
	k	4.000	4.170	4.346	4.529	4.717	4.910	5.108	5.308	5.509	5.710	5.908
	C	0.500	0.511	0.522	0.532	0.543	0.554	0.564	0.574	0.584	0.593	0.602
0.25	M	0.083	0.085	0.086	0.087	0.089	0.090	0.091	0.093	0.094	0.095	0.096
	k	4.000	4.204	4.420	4.648	4.887	5.138	5.401	5.672	5.952	6.238	6.527
	C	0.500	0.512	0.525	0.538	0.550	0.563	0.576	0.588	0.600	0.612	0.623
0.30	M	0.083	0.085	0.086	0.088	0.089	0.091	0.092	0.094	0.095	0.096	0.098
	k	4.000	4.235	4.488	4.760	5.050	5.361	5.692	6.044	6.414	6.802	7.205
	C	0.500	0.514	0.527	0.542	0.556	0.571	0.585	0.600	0.614	0.628	0.642
0.35	M	0.083	0.085	0.087	0.088	0.090	0.091	0.093	0.095	0.096	0.098	0.099
	k	4.000	4.264	4.551	4.864	5.204	5.575	5.979	6.416	6.888	7.395	7.935
	C	0.500	0.514	0.529	0.545	0.560	0.576	0.593	0.609	0.626	0.642	0.658
0.40	M	0.083	0.085	0.087	0.088	0.090	0.092	0.094	0.095	0.097	0.099	0.100
	k	4.000	4.289	4.607	4.959	5.348	5.778	6.255	6.782	7.365	8.007	8.710
	C	0.500	0.515	0.530	0.546	0.563	0.580	0.598	0.617	0.635	0.654	0.672
0.45	M	0.083	0.085	0.087	0.088	0.090	0.092	0.094	0.096	0.098	0.100	0.101
	k	4.000	4.311	4.658	5.046	5.480	5.967	6.517	7.136	7.336	8.625	9.514
	C	0.500	0.515	0.530	0.547	0.564	0.583	0.602	0.621	0.642	0.662	0.683
0.50	M	0.083	0.085	0.087	0.088	0.090	0.092	0.094	0.096	0.098	0.100	0.102
	k	4.000	4.331	4.703	5.123	5.599	6.141	6.760	7.470	8.289	9.234	10.329
	C	0.500	0.515	0.530	0.547	0.564	0.583	0.603	0.624	0.645	0.667	0.690
$X = (1 - c_2/l_2)^3$		1.000	0.856	0.729	0.613	0.512	0.421	0.343	0.274	0.216	0.166	0.125

**From Ref. 3-17*
FEM (uniform load w) $= Mwl_2l_1^2$
K (stiffness factor) $= kEI_2h^3/12l_1$
Carry-over factor $= C$

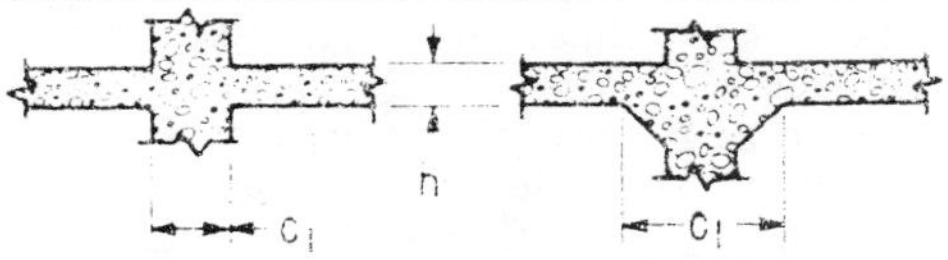

TABLE 3-5 Moment Distribution Factors for Slab-Beam Elements with Drop Panels*

$h_{dp} = 1.25h$

c_1/l_1 \ c_2/l_2		0.00	0.05	0.10	0.15	0.20	0.25	0.30
0.00	M	0.088	0.088	0.088	0.088	0.088	0.088	0.088
	k	4.795	4.795	4.795	4.795	4.795	4.795	4.797
	C	0.542	0.542	0.542	0.542	0.542	0.542	0.542
0.05	M	0.088	0.088	0.089	0.089	0.089	0.089	0.090
	k	4.795	4.846	4.896	4.944	4.990	5.035	5.077
	C	0.542	0.545	0.548	0.551	0.553	0.556	0.558
0.10	M	0.088	0.088	0.089	0.090	0.090	0.091	0.091
	k	4.795	4.894	4.992	5.039	5.184	5.278	5.368
	C	0.542	0.548	0.553	0.559	0.564	0.569	0.573
0.15	M	0.088	0.089	0.090	0.090	0.091	0.092	0.092
	k	4.795	4.938	5.082	5.228	5.374	5.520	5.665
	C	0.542	0.550	0.558	0.565	0.573	0.580	0.587
0.20	M	0.088	0.000	0.090	0.091	0.092	0.093	0.094
	k	4.795	4.978	5.167	5.361	5.558	5.760	5.962
	C	0.542	0.552	0.562	0.571	0.581	0.590	0.590
0.25	M	0.088	0.089	0.090	0.091	0.092	0.094	0.095
	k	4.795	5.015	5.245	5.485	5.735	5.994	0.261
	C	0.542	0.553	0.565	0.576	0.587	0.598	0.600
0.30	M	0.088	0.089	0.090	0.092	0.093	0.094	0.095
	k	4.795	5.048	5.317	5.601	5.902	0.219	0.550
	C	0.542	0.554	0.567	0.580	0.593	0.605	0.618

**From Ref. 3-17*
FEM (uniform load w) $= Mwl_2l_1^2$
K (stiffness factor) $= kEI_2h^3/12l_1$
Carry-over factor $= C$

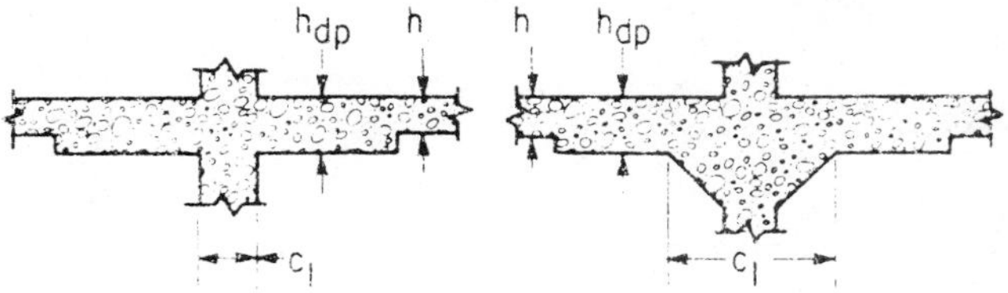

TABLE 3-6 Moment Distribution Factors for Slab-Beam Elements with Drop Panels*

$h_{dp} = 1.5h$

c_2/l_2 / c_1/l_1		0.00	0.05	0.10	0.15	0.20	0.25	0.30
0.00	M	0.093	0.093	0.093	0.093	0.093	0.093	0.093
	k	5.837	5.837	5.837	5.837	5.837	5.837	5.837
	C	0.589	0.589	0.589	0.589	0.589	0.589	0.589
0.05	M	0.093	0.093	0.093	0.093	0.094	0.094	0.094
	k	5.837	5.890	5.942	5.993	6.041	6.087	6.131
	C	0.589	0.591	0.594	0.596	0.598	0.600	0.602
0.10	M	0.093	0.093	0.094	0.094	0.094	0.095	0.095
	k	5.837	5.940	6.042	6.142	6.240	6.335	6.427
	C	0.589	0.593	0.598	0.602	0.607	0.611	0.615
0.15	M	0.093	0.093	0.094	0.095	0.095	0.096	0.096
	k	5.837	5.986	6.135	6.284	6.432	6.579	6.723
	C	0.589	0.595	0.602	0.608	0.614	0.620	0.626
0.20	M	0.093	0.093	0.094	0.095	0.006	0.096	0.097
	k	5.837	6.027	6.221	6.418	6.616	6.816	7.015
	C	0.589	0.597	0.605	0.613	0.621	0.628	0.635
0.25	M	0.093	0.094	0.094	0.095	0.096	0.097	0.098
	k	5.837	6.065	6.300	6.543	6.790	7.043	7.298
	C	0.589	0.598	0.608	0.617	0.626	0.635	0.644
0.30	M	0.093	0.094	0.095	0.096	0.097	0.098	0.090
	k	5.837	0.099	6.372	6.657	6.953	7.258	7.571
	C	0.589	0.599	0.610	0.620	0.631	0.641	0.651

*From Ref. 3-17.

FEM (uniform load w) $= Mwl_2l_1^2$

K (stiffness factor) $= kEl_2h^3/12l_1$

Carry-over factor $= C$

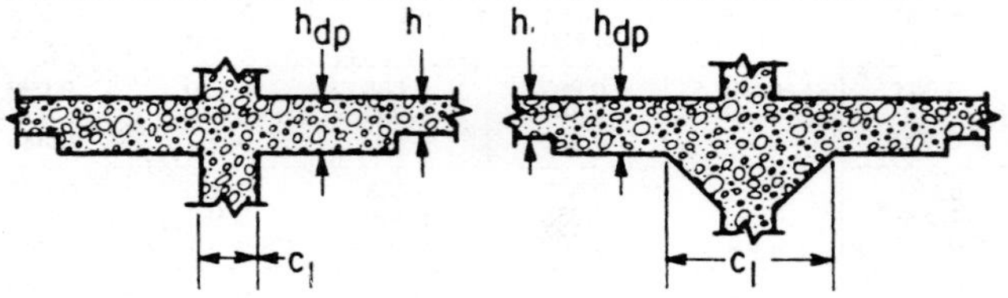

center. At exterior columns the critical section for negative moment is at the face of the column, or if there is a bracket or capital, at a distance not greater than one-half the projection beyond the column face.

The code gives no guidance as to how the reduction in negative moment to the face of the column is to be accomplished. Several authors have used free body diagrams of the joint connection to perform this reduction, but the results can differ significantly, depending on the assumption of the location of the column reaction for the span being considered relative to the column centerline. A study[3-18] in which several of the better-known procedures were compared with elastic solutions concluded that the closest agreement was obtained using the simplest method proposed by PCA.[3-19] With this procedure the negative moment at the critical section, M_u, is:

$$M_u = M_{\text{℄}} - \frac{Vc_1}{3} \tag{3-9}$$

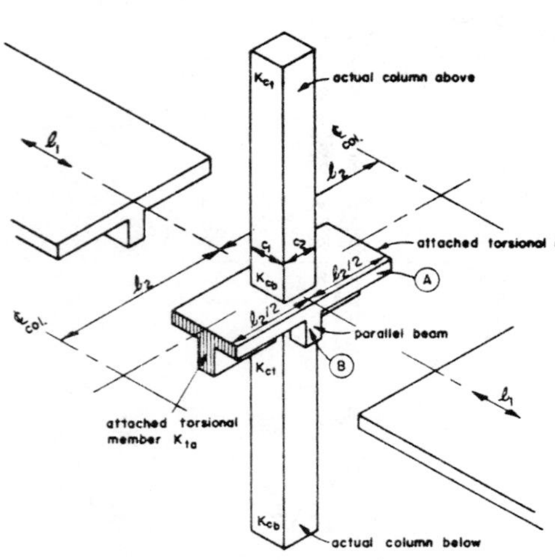

Fig. 3-9 Equivalent column.

where V is the shear force for the span considered as a simply supported beam. Where the slab geometry is within the limits specified for the use of the DDM but the design moments were computed using the EFM, the absolute sum of the positive and average negative design moments at the critical sections in any panel need not exceed the total static moment given by eq. (3-4).

Where this occurs and the negative moments have already been reduced at the critical sections, it is recommended that any additional reduction be proportioned between the positive and negative moment regions.

The unbalanced moment to be transferred between slab and column at the critical section for shear (Section 3.8) is the sum of the design moments computed for the columns above and below the slab.

Use of the EFM is illustrated in Example 3-5.

3.8 SHEAR AND SHEAR-MOMENT REQUIREMENTS

Two types of shear failure can be defined in slabs: "beam" shear, in which the slab is considered as a wide beam with a potential diagonal tension crack extending across the full width, and "perimeter" shear, where around a column the tension crack may form along the surface of a truncated cone or pyramid.

Beam shear can occur in one-way slabs as demonstrated in Example 3-1, or in two-way slabs with stiff beams where the slab is considered to carry load in the short panel direction to the beams (see Fig. 3-10). As the name implies, beam shear in slabs is considered in the same manner as shear in beams with an allowable concrete shear stress of $2\sqrt{f'_c}$.

For slabs without beams the failure mode is perimeter shear, sometimes referred to as punching shear. The critical section is assumed to be perpendicular to the plane of the slab and to have a perimeter, b_0, that is a minimum but approaches no closer than $d/2$ to the periphery of the support or concentrated load. Owing to the confining effect

TABLE 3-7 Values of Torsion Constant, C*

y \ x	4	5	6	7	8	9	10	12	14	16
12	202	369	592	868	1,188	1,538	1,900	2,557	—	—
14	245	452	736	1,096	1,529	2,024	2,566	3,709	4,738	—
16	388	534	880	1,325	1,871	2,510	3,233	4,861	6,567	8,083
18	330	619	1,024	1,554	2,212	2,996	3,900	6,013	8,397	10,813
20	373	702	1,167	1,782	2,553	3,482	4,567	7,165	10,226	13,544
22	416	785	1,312	2,011	2,895	3,968	5,233	8,317	12,055	16,275
24	548	869	1,456	2,240	3,236	4,454	5,900	9,469	13,885	19,005
27	522	994	1,672	2,583	3,748	5,183	6,900	11,197	16,628	23,101
30	586	1,119	1,888	2,926	4,260	5,912	7,900	12,925	19,373	27,197
33	650	1,243	2,104	3,269	4,772	6,641	8,900	14,653	22,117	31,293
36	714	1,369	2,320	3,612	5,284	7,370	9,900	16,381	24,860	35,389
42	842	1,619	2,752	4,298	6,308	8,828	11,900	19,837	30,349	43,581
48	970	1,869	3,184	4,984	7,332	10,286	13,900	23,293	35,836	51,773
54	1,098	2,119	3,616	5,670	8,356	11,744	15,900	26,749	41,325	59,965
60	1,226	2,369	4,048	6,356	9,380	13,202	17,900	30,205	46,813	68,157

$$*C = (1 - 0.63x/y)\,\frac{x^3 y}{3}$$

x is smaller dimension of rectangular cross section.
**From Ref. 3-17.*

on the concrete caused by the two-way action, the allowable concrete shear stress is increased to $4\sqrt{f'_c}$.

In addition to the shear force, the unbalanced moment transferred from the slab to the column must be resisted at this critical section (Fig. 3-11). When design moments are obtained using the DDM, this unbalanced moment at an interior column is either the difference in negative design moments or the moment from eq. (3-5), whichever is larger, and for exterior columns is the nominal capacity of the negative reinforcement placed in the column strip. When the EFM is used, the unbalanced moment results from the appropriate pattern loading case, from unequal spans or from unequal loading in adjacent spans.

A portion, γ_v, of the unbalanced moment, M is considered to be transferred by eccentricity of shear and the balance by flexure, where:

$$\gamma_v = 1 - \frac{1}{1 + \frac{2}{3}\sqrt{\frac{c_1 + d}{c_2 + d}}} \qquad (3\text{-}10)$$

If the distribution of the shear stresses due to the portion of unbalanced moment resisted by shear eccentricity is assumed linear (Fig. 3-11), the maximum shear stress at the critical section can be computed in a manner analogous to the maximum compressive stress in an eccentrically loaded column:

$$v_u = \frac{V}{\phi A} + \frac{\gamma_{v_1} M_1 c_1}{\phi J_1} + \frac{\gamma_{v_2} M_2 c_2}{\phi J_2} \qquad (3\text{-}11)$$

TABLE 3-8 Column Stiffness Coefficients, k_c, for Columns without Tapered Capitals*

a/h \ b/h	0.00	0.02	0.04	0.06	0.08	0.10	0.12	0.14	0.16	0.18	0.20	0.22	0.24
0.00	4.000	4.082	4.167	4.255	4.348	4.444	4.545	4.651	4.762	4.878	5.000	4.128	5.263
0.02	4.337	4.433	4.533	4.638	4.747	4.862	4.983	5.110	5.244	5.384	5.533	5.690	5.856
0.04	4.709	4.882	4.940	5.063	5.193	5.330	5.475	5.627	5.787	5.958	6.138	6.329	6.533
0.06	5.122	5.252	5.393	5.539	5.693	5.855	6.027	6.209	6.403	6.608	6.827	7.060	7.310
0.08	5.581	5.735	5.898	6.070	6.252	6.445	6.650	6.868	7.100	7.348	7.613	7.897	8.203
0.10	6.091	6.271	6.462	6.665	6.880	7.109	7.353	7.614	7.893	8.192	8.513	8.859	9.233
0.12	6.659	6.870	7.094	7.333	7.587	7.859	8.150	8.461	8.796	9.157	9.546	9.967	10.430
0.14	7.292	7.540	7.803	8.084	8.385	8.708	9.054	9.426	9.829	10.260	10.740	11.250	11.810
0.16	8.001	8.291	8.600	8.931	9.287	9.670	10.080	10.530	11.010	11.540	12.110	12.740	13.420
0.18	8.796	9.134	9.498	9.888	10.310	10.760	11.260	11.790	12.370	13.010	13.700	14.470	15.310
0.20	9.687	10.080	10.510	10.970	11.470	12.010	12.600	13.240	13.940	14.710	15.560	16.490	17.530
0.22	10.690	11.160	11.660	12.200	12.800	13.440	14.140	14.910	15.760	16.690	17.210	18.870	20.150
0.24	11.820	12.370	12.960	13.610	14.310	15.080	15.920	16.840	17.870	19.000	20.260	21.650	23.260

**From Ref. 3-17.*

$$Kc = \frac{k_c E_c I_c}{l_c}$$

Note:

a = length of rigid column section at near end
b = length at rigid column section at far end

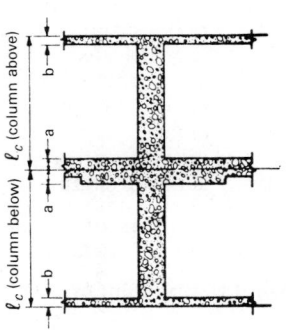

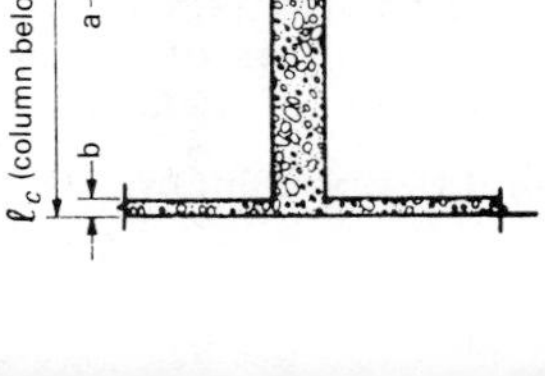

TABLE 3-9 Column Stiffness Coefficients, k_c^* for Columns with 45° Tapered Capitals

	a/l_c										
b'/l_c	0.00	0.005	0.010	0.015	0.020	0.025	0.030	0.035	0.040	0.045	0.050
0.0	4.000	4.102	4.208	4.318	4.433	4.552	4.676	4.805	4.940	5.080	5.226
	4.000	4.102	4.208	4.318	4.433	4.552	4.676	4.8051	4.940	5.080	5.226
0.02	4.013	4.115	4.222	4.332	4.448	4.567	4.692	4.822	4.957	5.098	5.245
	4.003	4.105	4.211	4.322	4.436	4.558	4.680	4.810	4.944	5.085	5.232
0.04	4.050	4.154	4.261	4.373	4.490	4.611	4.738	4.869	5.006	5.149	5.298
	4.012	4.115	4.221	4.332	4.447	4.567	4.692	4.822	4.958	5.099	5.246
0.06	4.108	4.214	4.323	4.438	4.557	4.680	4.809	4.943	5.083	5.229	5.382
	4.025	4.129	4.236	4.347	4.463	4.584	4.170	4.841	4.978	5.120	5.269
0.08	4.185	4.293	4.406	4.523	4.645	4.772	4.905	5.042	4.186	5.336	5.492
	4.042	4.147	4.225	4.367	4.484	4.606	4.733	4.866	5.003	5.147	5.297
0.10	4.280	4.392	4.508	4.629	4.755	4.886	5.022	5.164	5.313	5.467	5.629
	4.063	4.168	4.277	4.391	4.590	4.633	4.761	4.895	5.034	5.180	5.331
0.12	4.393	4.509	4.628	4.754	4.884	5.020	5.161	5.308	5.462	5.623	5.790
	4.086	4.192	4.303	4.418	4.538	4.662	4.792	4.928	5.069	5.216	5.370
0.14	4.522	4.642	4.767	4.897	4.032	5.174	5.321	5.474	5.634	5.801	5.975
	4.112	4.219	4.331	4.448	4.569	4.695	4.827	4.964	5.107	5.257	5.413
0.16	4.667	4.793	4.923	5.059	5.200	5.347	5.301	5.661	5.828	6.002	6.184
	4.139	4.248	4.362	4.480	4.603	4.731	4.864	5.004	5.149	5.301	5.459
0.18	4.830	4.961	5.097	5.239	5.387	5.541	5.701	5.869	6.044	6.266	6.417
	4.169	4.279	4.394	4.514	4.639	4.769	4.904	5.046	5.193	5.347	5.508
0.20	5.009	5.146	5.289	5.438	4.493	5.754	5.923	6.098	6.282	6.473	6.674
	4.200	4.312	4.429	4.550	4.677	4.809	4.947	5.090	5.240	5.397	5.560
0.22	5.205	5.349	5.499	5.656	5.818	5.988	6.165	6.360	6.543	6.744	6.955
	4.233	4.346	4.465	4.588	4.717	4.851	4.991	5.137	5.289	5.449	5.615
0.24	5.419	5.571	5.729	5.893	6.604	6.243	6.430	6.624	6.827	7.040	7.262
	4.266	4.382	4.502	4.628	4.758	4.805	5.037	5.185	5.340	5.502	5.672

**Upper value is for capital end; lower value is for base end of column.*

$$K_c = \frac{k_c E_{cc} I_c}{l_c}$$

a = length of column section considered rigid
b' = depth of 45 deg capital below soffit of slab

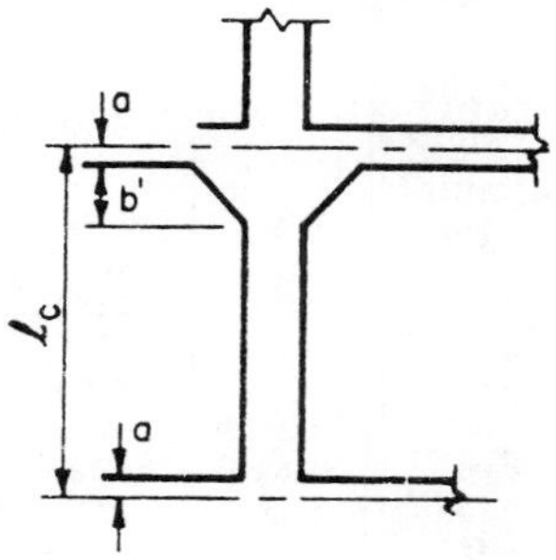

where:

- A is the area of the critical section.
- M is the moment at the centroid of the critical section transferred between the slab and column.
- c is the distance from the centroid line to a point on the critical section.
- J is the equivalent of polar moment of inertia of the critical section about the centroidal line.
- 1, 2 are subscripts denoting the two directions in which moments are applied.
- ϕ is the strength reduction factor for shear = 0.85.

The parameters of the critical section vary depending on the location of the column, as summarized below:

(a) Interior column:

$$A = 2d(c_1 + c_2 + 2d)$$

$$c = (c_1 + d)/2$$

$$J = (c_1 + d)\,d^3/6 + (c_1 + d)^3\,d/6 + d(c_2 + d)(c_1 + d)^2/2$$

(b) Corner column:

$$A = d(c_1 + c_2 + d)$$

$$c = (c_1 + d/2)^2/2(c_1 + c_2 + d)$$

$$J = [(c_1 + d/2)\,d^3 + (c_1 + d/2)^3 d]/12 + (c_2 + d/2)\,dc^2 + (c_1 + d/2)\,d[(c_1 + d/2)/2 - c]^2$$

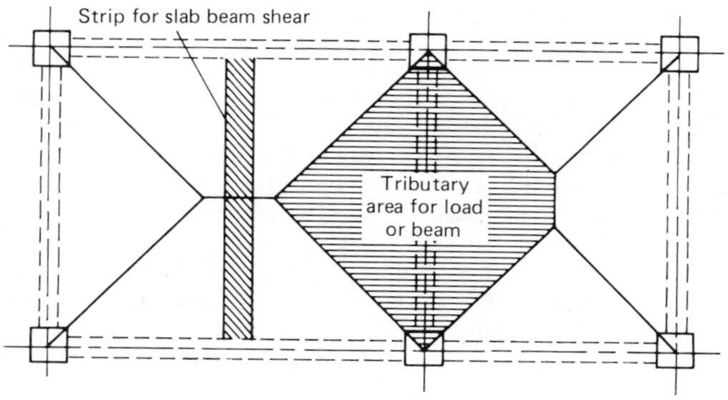

Fig. 3-10 Loaded area for determining "beam" shear.

(c) Edge column:

$$A = d(2c_1 + c_2 + 2d)$$

$$(c)_1 = (c_1 + d/2)^2/(2c_1 + c_2 + 2d)$$

$$(c)_2 = (c_2 + d)/2$$

$$J_1 = [(c_1 + d/2)\,d^3 + (c_1 + d/2)^3\,d]/6 + (c_2 + d)\,d[(c)_1]^2 + 2(c_1 + d/2)\,d[(c_1 + d/2)/2 - (c)_1]^2$$

$$J_2 = [(c_2 + d)\,d^3 + (c_2 + d)^3\,d]/12 + 2(c_1 + d/2)\,d[(c)_2]^2$$

It will be noted that no subscripts are used for the section parameters for interior and corner columns. Defining c_1 as the column dimension in the direction the moment is acting, the same expressions can be used for the other direction by interchanging the values of c_1 and c_2. For the edge column, subscripts are used, since the geometry of the critical section is different in the two directions. In the above expressions for the edge column the length c_1 is perpendicular to the discontinuous edge, and the subscript 1 is for the design strip perpendicular to the discontinuous edge. An example of shear-moment transfer calculations is contained in Example 3-2.

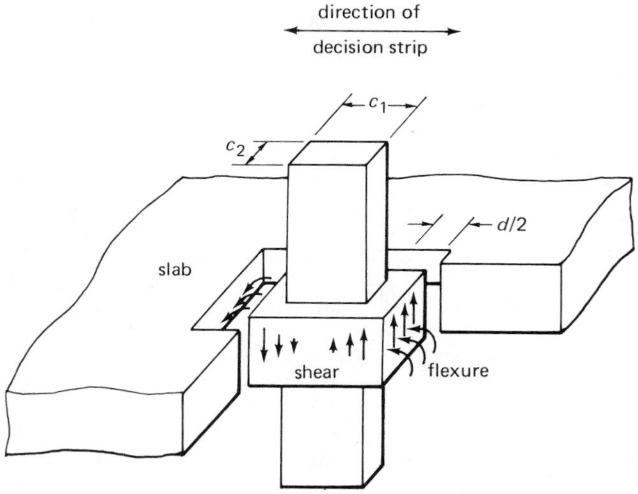

Fig. 3-11 Moment transfer between slab and column, used in conjunction with perimeter shear.

For slabs with beams some decision needs to be made as to how much of the shear transferred to the column is through the beams or through the slab directly. Guidance for this distribution is given in the direct design method. The tributary area for loads transferred to a beam is bounded by 45° lines drawn from the corners of the panels and the centerline of the panels parallel to the long sides as shown in Fig. 3-10. When the effective stiffness of the beam, $\alpha_1 l_2/l_1$, is equal to or greater than 1.0, the beam is designed to carry 100% of the shear. For beams with $\alpha_1 l_2/l_1$ less than 1.0, the portion of shear attributed to the beam decreases linearly to zero when $\alpha_1 = 0$. Of course beams must also be designed to resist the shear caused by loads applied directly to the beam, including the weight of the beam stem. The slab is designed to resist that portion of the total shear not assigned to the beams. Although the code does not state so explicitly, it is recommended that the above rule for slabs with beams be used when moments and shears are obtained using the equivalent frame method.

Generally, when shear is critical in the design, it is most economical to add column capitals, increase thickness of slab, or increase the strength of the concrete in that order. There are instances where it may be necessary to increase the shear capacity by the addition of shear reinforcement in the slab. Such reinforcement is discussed in Section 3.11.

3.9 SLAB REINFORCEMENT

The area of reinforcement required to resist factored moments at all critical sections is computed as for flexural members (Chapter 1). This reinforcement in each direction must be equal to or exceed the minimum area required for shrinkage and temperature.

For one-way slabs other than joist slabs, the primary flexural reinforcement must not be spaced farther apart than three times the slab thickness, or 18 in. Cut-off points and development lengths are computed in the same manner as for beams.

For solid two-way slabs the spacing of bars at critical sections shall not exceed two times the slab thickness. Where drop panels are used, their presence does not affect the distribution of moments. However, to be considered effective in reducing the amount of negative reinforcement over the column of a slab without beams, the drop must extend in each direction a distance of at least one-sixth of the span length in that direction and have a projection below the slab of at least one-quarter of the slab thickness beyond the drop. For determining reinforcement, the thickness of the drop panel below the slab shall not be considered to be greater than one-fourth of the distance from the edge of the drop panel to the edge of the column capital.

The portion of the unbalanced moment transferred to the column by flexure shall be resisted by the slab with an effective width between lines that are one and one-half times the slab or drop panel thickness outside opposite faces of the column or capital. The required reinforcement across this effective width can be obtained by concentration of the column strip reinforcement or by additional reinforcement. The portion of the unbalanced moment transferred by flexure is specified as:

$$\gamma_f = \frac{1}{1 + \frac{2}{3}\sqrt{\frac{c_1 + d}{c_2 + d}}} \qquad (3\text{-}12)$$

Negative moment reinforcement perpendicular to a discontinuous edge shall be bent, hooked, or otherwise anchored to develop the calculated tension in the bars. Positive reinforcing perpendicular to a discontinuous edge shall extend to the edge of the slab and have an embedment, straight or hooked, of at least 6 in. in spandrel beams, walls, or columns.

When transverse spans vary, the moment intensity calculated for two adjacent half middle strips may not be equal. It is recommended that the total reinforcement required for the two adjacent half middle strips be distributed uniformly across the width of the middle strip.

For slabs with beams along discontinuous edges having a value of α greater than 1.0, additional reinforcing proportioned to resist a moment equal to the maximum positive moment per foot of width in the slab is required at exterior corners in both the top and the bottom of the slab to resist the twisting moments not otherwise accounted for. The direction of this moment is parallel to the diagonal from the corner in the top of the slab and perpendicular to the diagonal in the bottom of the slab: hence the required reinforcement in both the top and bottom may be placed either in a single layer or in two layers parallel to the sides of the slab. This reinforcement should extend in each direction from the corner equal to one-fifth of the longer span.

In addition to the above requirements, reinforcement in slabs without beams shall have minimum bend point locations or extensions, as given in Fig. 3-12. Where adjacent spans are unequal, the extension in each span of the negative reinforcement beyond the face of the support is based on the requirements of the longer span. Bent bars may be used only when the depth–span ratio permits use of bends of 45° or less. For slabs in frames not braced against sidesway and for slabs resisting lateral loads, the lengths of reinforcement must be determined by analysis but shall not be less than those prescribed in Fig. 3-12.

Considerable economy in placing slab reinforcing can be achieved if no bent bars are used and the negative reinforcing is placed as a preformed mat. To achieve maximum economy it has been the practice of some designers to place no negative steel in the middle strips and concentrate all of the panel negative steel in the column strips, a placing sequence that is in violation of the requirements of past and present ACI codes. To investigate the effects of this practice, a floor of an apartment building was load-tested by Cardenas and Kaar;[3-20] one portion of the slab was reinforced according to ACI requirements, and an identical portion had all of the panel negative steel concentrated in the column strips. Under the standard test load ($0.3w_d + 1.7w_l$) there was no significant difference in the deflections of the two portions, and all crack widths were small with no major effects on stiffness or serviceability.

The writer is familiar with several buildings in which the slab's negative steel mat was extended into the middle strips such that although the required negative steel area in the middle strip was provided, it was concentrated in the portion adjacent to the column strips rather than uniformly distributed over the middle strip. These slabs have performed well. Thus it would appear that the negative steel can be concentrated in the column strips without problems in serviceability. However, the reader is cautioned that the above slabs were all proportioned using ACI 318-63, in which the column strip was defined as a quarter of the panel width on each side of the column line. For rectangular panels the definition of column strip in the short direction by ACI 318-83 will be less than this width. For this reason, when the panel aspect ratio is appreciably greater or less than 1.0, it would be prudent in the short direction to extend the negative steel mat into adjacent portions of the middle strip. This will not increase steel costs but will ensure a more satisfactory distribution of cracks.

3.10 TWO-WAY SLAB DESIGN EXAMPLES

The design examples contained in this section are included primarily to illustrate the design requirements for two-way slabs as specified by ACI 318-83.[3-1] For this reason certain simplifications of geometry were introduced, and calculations common to other examples are not repeated. The four examples were chosen to include most of the slab problems that are encountered in practice, including selection of slab thickness, determination of design moments at critical sections using both the DDM and EFM, shear-moment capacity checks, and some factors affecting selection of reinforcement

The following paragraphs contain a general description of each example with comments on the design procedures. These are followed by the examples, which are formatted to correspond to actual design sheets for those portions of the design considered. Calculations on the design sheets are separated with headings and are given in sufficient detail to make them self-explanatory.

Example 3-2 illustrates a typical high-rise apartment structure utilizing flat plate and shear wall construction. Such structures are characterized by light loading and relatively short spans. Reinforcement having a yield stress of 60,000 psi was used in all the design examples because of the increasing difficulty in obtaining reinforcement with a lower yield stress. The use of reinforcement with a lower yield stress would have permitted use of a thinner slab and reduced the number of locations where minimum reinforcement controlled the design. The use of edge beams would have also permitted a decrease in the slab thickness, but at the price of increased forming and construction costs. Edge beams would also have a significantly reduced the shear-moment transfer at edge columns although no difficulty was encountered here. The DDM was used in preference to the EFM because of its simplicity.

Owing to problems with moment transfer in the vicinity of columns, the *Commentary*[3-2] recommends that all flexural reinforcement in this region be detailed carefully on the construction drawings. Calculations showing the reinforcement requirements in this region are included as part of the shear-moment capacity calculations for a typical interior and edge column.

Example 3-3 illustrates the selection of slab thickness and design moments for a slab with beams. A complete design would, of course, include the flexural and shear reinforcement for the beams.

Example 3-4 is the design of a typical interior panel of a waffle-type slab. Such slabs are economically formed using generally square pans that are supported on standard scaffolding. Standard pans are widely available on a 24 in. and 36 in. module. Standard pans for the 24 in. module have plan dimensions at the base of 19 X 19 in. and are available in heights of 6, 8, 10, and 12 in. Pans for the 36 in. module measure 30 X 30 in. at the base and are available in heights of 8, 10, 12, 14, 16, and 20 in. To facilitate stripping, all standard pans have sloping sides forming a bevel of $1\frac{1}{2}$ in./ft from the vertical. In many localities other pan sizes are available for purchase or rent. In recent years, pans on a 60 in. module have also been available in some locations.

The pans forming a waffle slab may be arranged so that no cavities are found along the lines joining the columns, so that in effect a beam is formed. These are analyzed as slabs with beams with the beam stiffness computed as the ratio of solid portion to panel width of slab considering cavities. An alternate arrangement is to arrange the pans so that there are no cavities around the column, forming in effect a drop panel. This arrangement is the one indicated in the example,

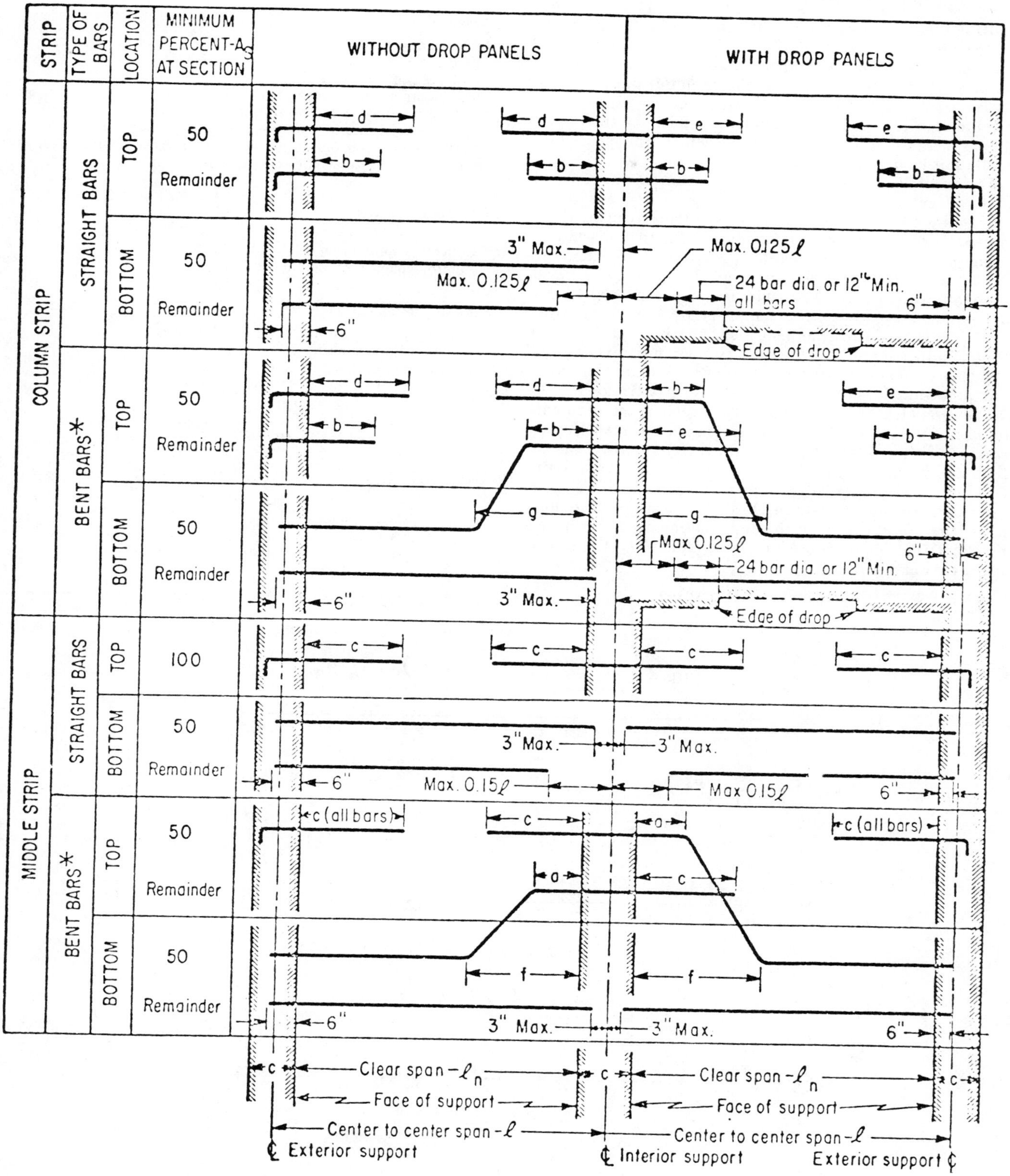

* Bent bars at exterior supports may be used if a general analysis is made

	BAR LENGTH FROM FACE OF SUPPORT						
	MINIMUM LENGTH					MAXIMUM LENGTH	
MARK	a	b	c	d	e	f	g
LENGTH	$0.14\ell_n$	$0.20\ell_n$	$0.22\ell_n$	$0.30\ell_n$	$0.33\ell_n$	$0.20\ell_n$	$0.24\ell_n$

Fig. 3-12 Minimum length of slab reinforcement, slabs without beams.[3-1]

and the computations follow the procedure for a slab without beams.

Example 3-5 illustrates the requirements of the EFM. Because of the complexity of the method, only a single design strip is considered. The variations in forming the column–slab junctions are intended to illustrate the use of the tabulated design aids, without which the procedure would not be practical.

The example computes only the design moments reduced to the face of the supports. Since this is a slab with no beams and the basic geometry is essentially regular, these moments could be proportioned to column and middle strips using the recommendations for the DDM. Since the EFM considers pattern loading explicitly, the unbalanced moments to be used in shear-moment capacity calculations are obtained directly from the appropriate loading patterns. Standard Hardy Cross moment distribution was used to obtain design moments; however, similar moments would have been obtained using the two-cycle moment distribution procedure described in Ref. 3-19 for pattern loads.

The geometry of Example 3-5 just meets the requirements for use of the DDM with respect to relative span lengths. To compare the results and to indicate the comparison of calculations required, moments were also obtained using the DDM. For more regular geometry the agreement between the two procedures would be closer. The example also illustrates the modification of the design moments as permitted by the DDM, a procedure recommended when the design spans are not equal.

EXAMPLE 3-2: Slab without Beams (Flat Plate)

Design data:

Live loading = 40 lb/ft^2
Partition & mech. allow. 25 lb/ft^2
Exterior wall 200 lb/ft
No column capitals or edge beams.
Slab designed for gravity loads only.
Story height, center-to-center = 9 ft
NOTE: slab extended 3 in. beyond external column face to support insulated panels for exterior wall

Material properties:

Concrete (normal weight) $f_c' = 4000$ psi
Reinforcement $f_y = 60{,}000$ psi

Choose slab thickness:

Assume constant slab thickness for all panels.
For slabs without beams eq. (3-1) governs.
Maximum l_n occurs in N–S direction in exterior panels:

$$l_n = 19 \times 12 - \frac{16}{2} - \frac{20}{2} = 228 - 18 = 210 \text{ in.}$$

From eq. (3-1):

$$h = \frac{210(800 + 0.005 \times 60{,}000)}{36{,}000} = 6.42 \text{ in.}$$

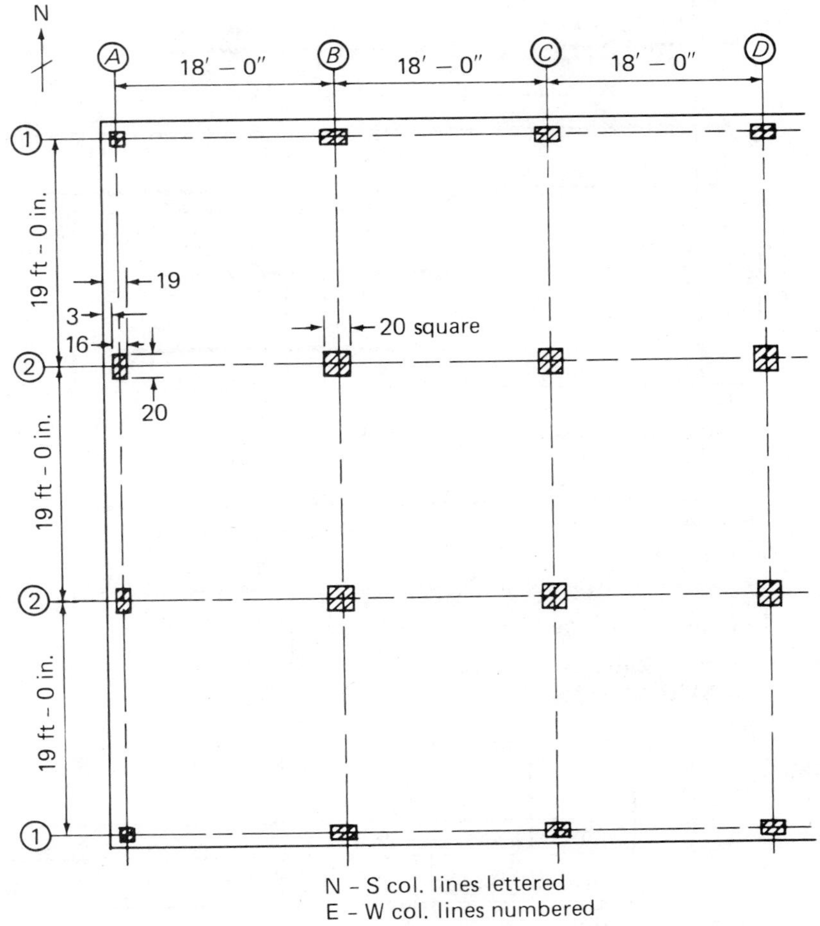

Since there is no beam provided at discontinuous edge, must increase thickness by 10% (section 9.5.3.3):

$$h = 1.1 \times 6.42 = 7.06 \text{ in.; } \underline{\text{use } 7.0 \text{ in.}}$$

Factored design loads:

$$w_d = 1.4\left(\frac{7}{12} \times 150 + 25\right) = 158 \text{ psf}$$

$$w_l = 1.7 \times 40 \qquad = \underline{\ 68\ }$$

$$w_u = \qquad 226 \text{ psf}$$

Exterior wall, $w_u = 1.4 \times 200 = 280$ lb/ft

Obtain design moments:

Check limitations for use of DDM:
1. three spans OK
2. $l_l/l_s = 19/18 = 1.06 < 2.0$ OK
3. Spans in same direction are equal OK
4. Columns located on column lines OK
5. For loading $68/158 < 3.0$ OK
6. Does not apply to slabs without beams

Use DDM.

Transverse (N–S) strips:

(a) Interior column lines Ⓑ, Ⓒ, etc:

	①			②		SYMM	
l_1		19.0			19.0		ft
l_2		18/2 + 18/2 = 18.0			18.0		ft
l_n		$19 - \frac{1}{2}(16/12 + 20/12) = 17.5$			$19 - \frac{1}{2}(20/12 + 20/12) = 17.33$		ft
M_0 (eq. 3-4)		155.7			152.8		ft-kips
$\% M_0$ (Table 3-2)	0.26	0.52	0.70	0.65		0.35	
M_{des} slab	−40.5	+81.0	−109.0	−99.3		+53.5	ft-kips

(b) Exterior column line Ⓐ:

l_2		$18/2 + \frac{1}{2}(16/12) + 3/12 = 9.92$			9.92		ft
M_0 (eq. 3-4)		85.8			84.2		ft-kips
$\% M_0$ (Table 3-2)	0.26	0.52	0.70	0.65		0.35	
M_{des} slab	−22.3	+44.6	−60.1	−54.7		+29.5	ft-kips

Moments due to exterior wall:

$M_0 = 0.28 l_n/8$		10.7			10.5		ft-kips
M_{des} wall	−2.8	+5.6	−7.5	−6.8		+3.7	ft-kips

Longitudinal (E–W) strips:

(a) Interior column line 2:

	Ⓐ			Ⓑ		Ⓒ	
l_1		18.0			18.0		ft
l_2		19/2 + 19/2 = 19.0			19.0		ft
l_n		$18 - \frac{1}{2}(16/12 + 20/12) = 16.5$			$18 - \frac{1}{2}(20/12 + 20/12) = 16.33$		ft
M_0 (eq. 3-4)		146.1			143.2		ft-kips
$\% M_0$ (Table 3-2)	.26	.52	.70	.65	.35	.65	
M_{des} slab	−38.0	+76.0	−102.3	−93.1	+50.1	−93.1	ft-kips

(b) Exterior column line 1:

l_2		$19/2 + \frac{1}{2}(16/12) + 3/12 = 10.42$			10.42		
M_0 (eq. 3-4)		80.1			78.5		ft
$\% M_0$ (Table 3-2)	.26	.52	.70	.65	.35	.65	
M_{des} slab	−20.8	+41.7	−56.1	−51.0	+27.5	−51.0	ft-kips

Moments due to exterior wall:

$M_0 = 0.28 l_n^2/8$		9.6			9.4		ft-kips
M_{des} wall	−2.5	+5.0	−6.7	−6.1	+3.3	−6.1	ft-kips

Pattern loads (section 13.3.6.1):

$$\beta_a = \frac{w_d}{w_l} \text{ (unfactored)} = \frac{112.5}{40} = 2.81 > 2.0$$

Therefore no modification of moments is required for pattern loading.

Assign moments to column and middle strips:

Moments assigned to column and half middle strips using Fig. 3-7:

$$\text{N–S direction: } \beta_t = \frac{C}{2I_s} = \frac{1325}{2 \times 6174} = 0.11$$

$$\text{E–W direction: } \beta_t = \frac{C}{2I_s} = \frac{1325}{2 \times 6517} = 0.10$$

Since β_t is small, assign all negative moment at discontinuous edges to column strips.

N–S design moments (ft-kip):

	Ⓐ		Ⓑ		Ⓒ
①	□ -2.8 -22.3 (1.03)	0 0 / 0 ( )*	□ -40.5 (1.67)	0 0 / 0 ( )*	□ -40.5 (1.67)
	5.6 26.8 (1.33)	17.8 16.2 / 34.0 (1.40)	48.6 (2.00)	16.2 16.2 / 32.4 (1.33)	48.6 (2.00)
②	□ -7.5 -45.1 (2.16)	-15.0 -13.6 / -28.6 (1.18)*	□ -81.8 (3.37)	-13.6 -13.6 / -27.2 (1.12)*	□ -81.8 (3.37)
	3.7 17.7 (0.88)	11.8 10.7 / 22.5 (0.93)*	32.1 (1.32)	10.7 10.7 / 21.4 (0.88)*	32.1 (1.32)
	ext. col. strip	middle strip	col. strip	middle strip	col. strip
	5'-5"	9'-0"	9'-0"	9'-0"	9'-0"

E–W design moments (ft-kip):

SYMM	②		①	
0 0 / 0 ( )*	□ -38.0 (1.71)	0 0 / 0 ( )*	□ -20.8 -2.5 (1.05)	Ⓐ
15.2 15.2 / 30.4 (1.36)*	+45.6 (2.05)	15.2 16.7 / 31.9 (1.43)*	25.0 5.0 (1.35)	
-12.8 -12.8 / -25.6 (1.15)*	□ -76.7 (3.44)	-12.8 -14.0 / -26.8 (1.20)*	□ -42.1 -6.7 (2.19)	Ⓑ
10.0 10.0 / 20.0 (6.90)*	30.1 (1.35)	10.0 11.0 / 21.0 (0.94)*	16.5 3.3 (0.89)	
-11.6 -11.6 / -23.2 (1.04)*	□ -69.8 (3.13)	-11.6 -12.8 / -24.4 (1.10)*	□ -38.3 -6.1 (1.99)	Ⓒ
middle strip	col. strip	middle strip	ext. col. strip	
10'-0"	9'-0"	10'-0"	5'-5"	

Selection of reinforcement:

Assume #4 bars; $d_b = 0.5$ in.; cover = 0.75 in.

Effective depths—N–S bars in layers closer to surfaces:

N–S direction: $d = 7.0 - 0.75 - 0.5/2 = 6.0$ in.

E–W direction: $d = 7.0 - 0.75 - 0.5 - 0.5/2 = 5.5$ in.

Minimum reinforcement (section 7.12):

$$A_s \min = 0.0018 \times 7 \times 12 = 0.15 \text{ in.}^2/\text{ft}$$

Maximum bar spacing (section 13.4.2):

$$s_{max} = 2h = 2 \times 7 = 14 \text{ in.}$$

Total area of reinforcement required at each critical section is each design strip is shown below the design moment in parentheses. Locations where minimum reinforcement governs are marked with an asterisk.

Positive flexural reinforcement would normally be distributed uniformly across the design strip. The distribution of negative flexural reinforcement is dependent on the moment to be transferred to the support and is considered in conjunction with the shear-moment capacity.

Column B2:

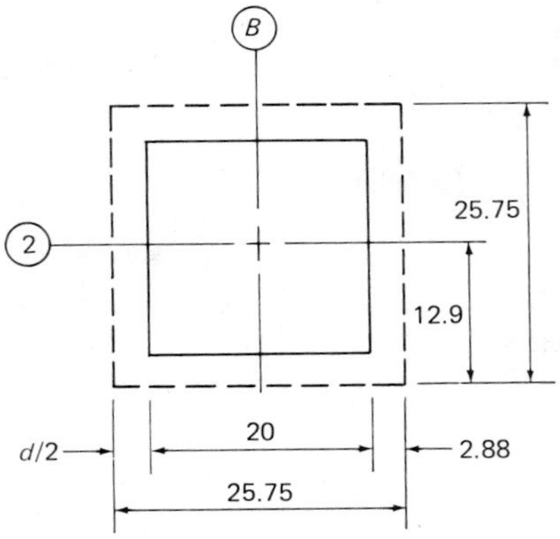

Flexural reinforcement (top mat):

N–S $M = 81.8$ ft-kips $A_s = 3.37$ in.^{2}

Try 17 #4 bars $A_s = 3.4$ in.^{2} $s = \dfrac{9 \times 12}{17} = 6.4$ in.

E–W $M = 76.7$ ft-kips $A_s = 3.44$ in.^{2}

Try 17 #4 bars $A_s = 3.4$ in.^{2} $s = \dfrac{9 \times 12}{17} = 6.4$ in.

Unbalanced moment (ft-kips):

(a) From design strips:

N–S $109.0 - 99.3 = 9.7$

E–W $102.3 - 93.1 = 9.2$

(b) From eq. (3-5):

$$0.07[(.158 + .5 \times .068)(18)(17.5)^2 - (.158)(18)(17.33)^2] = 14.3$$

$$0.07[(.158 + .5 \times .068)(19)(16.5)^2 - (.158)(19)(16.33)^2] = 13.5$$

Moment transferred by flexure (section 13.3.4.2):

From Eq. (3-10): $\gamma_f = 0.6$

Effective width $= c_2 + 3h = 20 + 3 \times 7 = 41$ in.

Moment by flexure:

N–S $M_f = 0.6 \times 14.3 = 8.6$ ft-kips

$A_s = 0.35$ in^2

for #4 bars $n = 0.35/0.2 = 2$

$s = 41/2 = 20.5 > 6.4$ in.

E–W $M_f = 0.6 \times 13.5 = 8.1$ ft-kips

$A_s = 0.36$ in^2

for #4 bars $n = 0.36/0.2 = 2$

$s = 41/2 = 20.5 > 6.4$ in.

Place flexural reinforcement uniformly across column strip in both directions.

Shear force at critical section:

$V = w_u A = 0.226[19.0 \times 18.0 - (25.75 \times 25.75)/144] = 76.3$ kips

Moment transferred by shear (section 11.12.2):

From eq. (3-10): $\gamma_v = 0.4$

N–S $M_1 = 0.4 \times 14.3 = 5.7$ ft. kips

E–W $M_2 = 0.4 \times 13.5 = 5.4$ ft. kips

Section properties at critical shear section:

$d = (6.0 + 5.5)/2 = 5.75$ in.

$A = 5.75[4(20 + 5.75)] = 592.3$ in.^{2}

$c = (20 + 5.75)/2 = 12.9$ in.

$J = (25.75)(5.75)^3/6 + (25.75)^3(5.75)/6$

$+ (5.75 \times 25.75)(25.75)^2/2 = 66{,}266$ in.^{4}

From eq. (3-11):

$$v_u = \frac{76{,}300}{0.85 \times 592.3} + \frac{5.7 \times 12{,}000 \times 12.9}{0.85 \times 66{,}266} + \frac{5.4 \times 12{,}000 \times 12.9}{0.85 \times 66{,}266} = 182 \text{ psi}$$

$v_{all} = 4\sqrt{4000} = 253$ psi

Since $v_u < v_{all}$, adequate in moment-shear transfer.

Column B1:

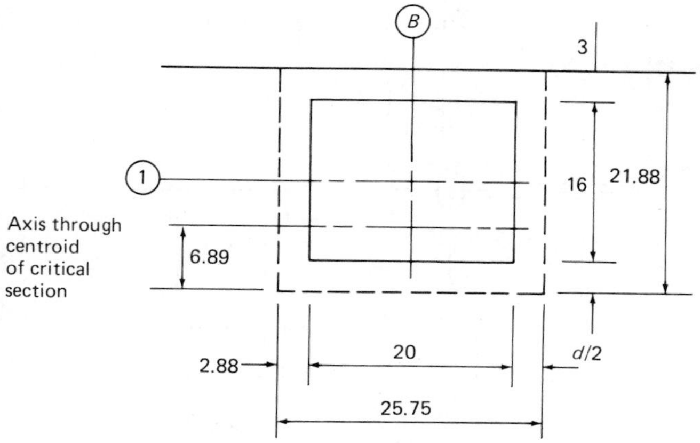

Flexural reinforcement (top mat):

N–S $M = 40.5$ ft-kips $A_s = 1.67$ in.^{2}

Try 9 #4 bars, $A_s = 1.80$ in.^{2} $s = \dfrac{9 \times 12}{9} = 12.0$ in.

(see sketch for final selection)

E–W Slab $M = 42.1$ ft-kips $A_s = 1.89$ in.^{2}

Try 10 #4 bars $A_s = 2.00$ in.^{2}

Wall $M = 6.7$ ft-kips $A_s = 0.30$ in.^{2}

Try 1 #5 $A_s = 0.31$ in.^{2}

Unbalanced moment (ft-kips):
N–S (section 13.6.3.6)

$M = A_s f_s(d - a/2)$

$$= \frac{1.8 \times 60}{12}\left(6 - \frac{1.8 \times 60}{2 \times 0.85(9 \times 12)4}\right) = 52.7$$

E–W:
(a) From design strip, $62.8 - 57.1 = 5.7$
(b) Eq. (3-5) slab loading only, wall moments considered balanced:

$$0.07[(.158 + 5 \times 0.068)(10.42)(16.5)^2 - (0.158)(10.42)(16.33)^2] = 7.4$$

Moment transferred by flexure (section 13.3.4.2):

$$\gamma_f = \frac{1}{1 + \frac{2}{3}\sqrt{\frac{21.88}{25.75}}} = 0.62$$

$M_f = 0.62 \times 52.7 = 32.6$ ft-kips $A_s = 1.34$ in.^{2}
Effective width $= c_2 + 3h = 20 + 3 \times 7 = 41$ in.
For #4 bars, $n = 1.34/.2 = 6.7$
Require equivalent of 6.7 #4 bars across effective width of 41 in. (see sketch)

E–W:

$$\gamma_f = 1/\left(1 + \frac{2}{3}\sqrt{\frac{25.75}{21.88}}\right) = 0.58$$

Effective width $= 16 + 1.5 \times 7 + 3 = 29.5$ in.
$M_f = 0.58 \times 7.4 = 4.3$ ft-kips $A_s = 0.19$ in.^{2}
For #4 bars, $n = 0.19/0.2 = 1$ bar
Can space slab flexural reinforcement uniformly across slab strip

Shear force at critical section:

$V_{slab} = 0.226[18 \times 10.42 - (25.75 \times 21.88)/144] = 41.5$ kips

$V_{wall} = 2 \times 0.28 \times 18.0/2 = 5.1$ kips

Moment transferred by shear (section 11.12.2):

N–S $\gamma_v = 1 - \gamma_f = 1 - 0\ 62 = 0.38$

$M_1 = 0.38 \times 52.7 = 20.0$ ft-kips

E–W $\gamma_v = 1 - \gamma_f = 1 - 0.58 = 0.42$

$M_2 = 0.42 \times 7.4 = 3.1$ ft-kips

Section properties for critical shear section:

$A = 5.75(2 \times 21.88 + 25.75) = 399.7$ in.^{2}

$(c)_1 = (21.88)^2/(2 \times 21.88 + 25.75) = 6.89$ in.

$(c)_2 = 25.75/2 = 12.88$ in.

$J_1 = [(21.88)(5.75)^3 + (21.88)^3(5.75)]/6$

$+ (25.75)(5.75)(6.89)^2$

$+ 2(21.88)(5.75)(21.88)/2 - 6.89)^2 = 21{,}888$ in.^{4}

$J_2 = [(25.75)(5.75)^3 + (25.75)^3(5.75)]/12$

$+ 2(21.88)(5.75)(12.88)^2 = 50{,}332$ in.^{4}

$$v_u = \frac{41{,}500 + 5{,}100}{0.85 \times 399.7} + \frac{20 \times 12{,}000 \times 6.89}{0.85 \times 218{,}888} + \frac{3.1 \times 12{,}000 \times 12.88}{0.85 \times 50{,}332}$$

$= 137.2 + 88.9 + 11.2 = 237$ psi $< v_{all} = 253$ psi

Therefore adequate in moment-shear transfer.

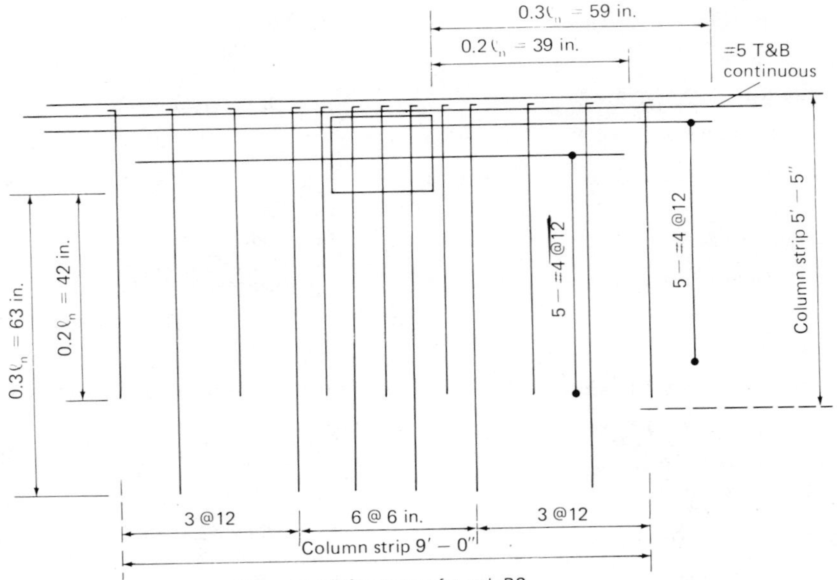

Top mat reinforcement for col. B2.
(bars anchored at discont. edge.)

EXAMPLE 3-3: Slab with beams

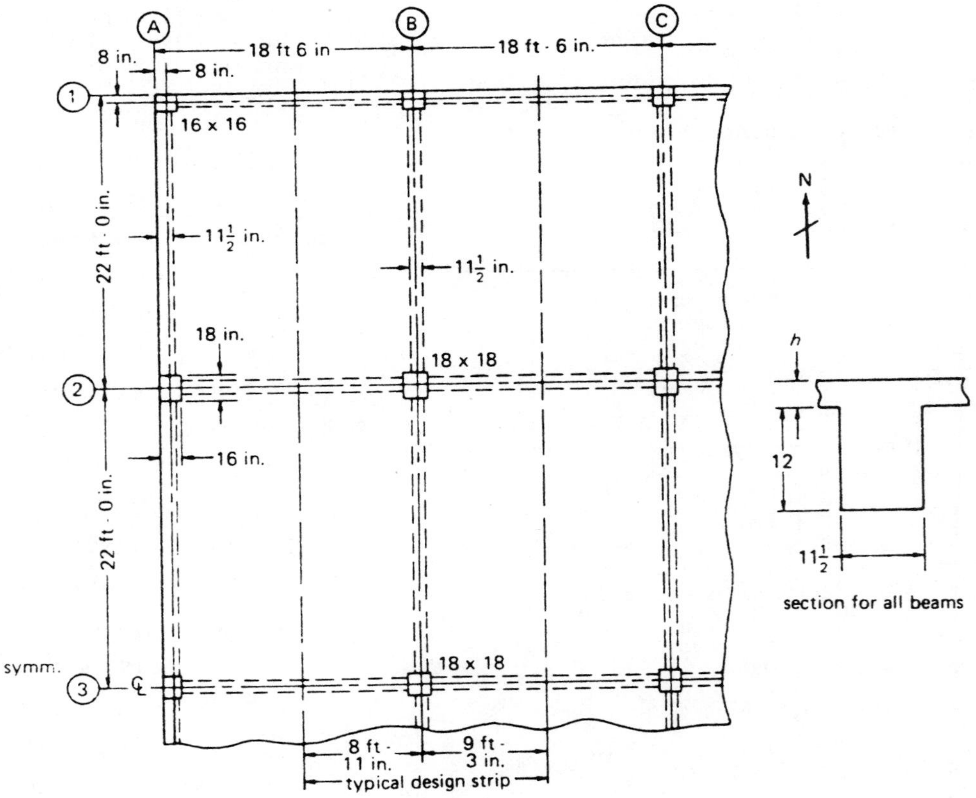

Floor Plan

FLOOR PLAN

Design data:

nominal live load	150 lb/ft^2
mechanical services	10 lb/ft^2
exterior wall	500 lb/ft
story height, center to center	14 ft

Material:

f'_c = 4000 psi γ = 150 lb/ft^3

f_y = 60,000 psi

beam depth – 12 in. below slab soffit

Choose slab thickness:

Consider limits for interior panel:

$$\text{N–S } l_n = (22 \times 12) - (9 + 9) = 246 \text{ in.}$$

$$\text{E–W } l_n = (18.5 \times 12) - (9 + 9) = 204 \text{ in.}$$

$$\beta = \frac{246}{204} = 1.21; \quad \beta_s = 1.0$$

$$h_{min} = \frac{246(800 + 0.005 \times 60{,}000)}{36{,}000 + 5000 \times 1.21(1 + 1)} = 5.64 > 5 \text{ in.}$$

$$h_{max} = \frac{246(800 + 0.005 \times 60{,}000)}{36{,}000} = 7.5 \text{ in.}$$

Try $h = 6.0$ in.

Beam sections:

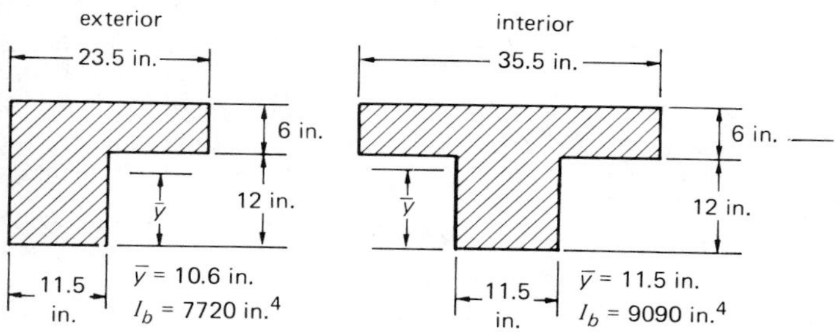

$\alpha = I_b/I_s$, for different slab strips.

$$\text{Column line (1)} \quad \alpha = \frac{7720 \times 12}{(11.33 \times 12) \times 6^3} = 3.15,$$

$$\text{Column line (B)} \quad \alpha = \frac{9090 \times 12}{(18.16 \times 12) \times 6^3} = 2.32$$

$$\text{Column line (A)} \quad \alpha = \frac{7720 \times 12}{(9.58 \times 12) \times 6^3} = 3.72,$$

$$\text{Column line (2)} \quad \alpha = \frac{9090 \times 12}{(21.63 \times 12) \times 6^3} = 1.94$$

$$\alpha_m = \frac{2(2.32 + 1.94)}{4} = 2.13$$

From eq. (3-2), interior panel:

$$h = \frac{246(800 + 0.005 \times 60{,}000)}{36{,}000 + 5000 \times 1.21[2.13 - 0.5(1 - 1)(1 + 1/1.21)]}$$

$$= 5.54 < h_{min}$$

Check corner panel:

$$\text{N–S } l_n = (22 \times 12) - (16 + 9) = 239 \text{ in.}$$

$$\text{E–W } l_n = (18.5 \times 12) - 16 + 9) - 197 \text{ in.}$$

$$\beta_s = 0.5, \quad \beta = 239/197 = 1.22, \quad \alpha_m = 2.78$$

$$h = \frac{239(800 + 0.005 \times 60{,}000)}{36{,}000 + 5000 \times 1.22[2.78 - 0.5(1 - 0.5)(1 + 1/1.22)]}$$

$$= 5.3 \text{ in.}$$

Therefore h_{min} governs. Use $h = 6$ in. for all panels.

Factored design loads:
Slab

$$w_d = 1.4(6/12 \times 150 + 10) = 119 \text{ psf}$$
$$w_l = 1.7(150) \qquad = 255$$
$$w_u = \qquad = 374 \text{ psf}$$

Beam stem:

$$w_d = 1.4\left(\frac{11.5 \times 12}{144}\right) 150 = 202 \text{ lb/ft}$$

Design moments:

First five limitations for use of DDM satisfied.
6. $\alpha_1 l_2^2/\alpha_2 l_1^2 = 2.32(18.16)^2/1.94(22)^2 = 0.815$ OK
Use DDM.

Pattern loads: (section 13.3.6.1):

$$\beta_a = w_d/w_l \text{ (unfactored)} = 85/150 = 0.57 < 2.0$$

Therefore must consider effects of pattern loading.

$$\alpha_c = \sum K_c / \sum (K_s + K_b)$$

$$\sum K_c = 2(4E(18)^4/12)/(14 \times 12) = 417E$$

$$\sum (K_s + K_b) = 2(1 + \alpha)4El_2h^3/12 \times l_1$$

$$= 2(1 + 2.32)\,4E\; 218(6)^3/12 \times 22 \times 12 = 395E$$

$$\alpha_c = 1.06 > \alpha_{min} \text{ (Table 3-3)}$$

Therefore no modification of design moments required.

Consider typical interior design strip col. line (B):

	(1)			(2)			(3)
l_1		256 in. = 21.33 ft			264 in. = 22.0 ft		
l_n		239 in. = 19.9 ft			246 in. = 20.5 ft		
l_2 (8′-11″ + 9′-3″)		218 in. = 18.17 ft			218 in. = 18.17 ft		
M_0(eq. 3-4)		336			357		ft-kips
$\%M_0$ (Table 3-2)	0.16	0.57	0.70	0.65	0.35	0.65	
M_{des} slab	−54	+192	−235	−232	+125	−232	ft-kips
Moments due to beam stem:							
M_0 $.202l^2/8$		10			11		
M beam stem	−2	+6	−7	−7	+4	−7	ft-kips

Assign moments to column and middle strips:

For design strip on column line B, widths of column strip and half middle strips as shown. Since $\alpha_1 l_2/l_1 > 1$, assign 85% of column strip moment to beams. Moments to half middle strips in proportion to their widths.

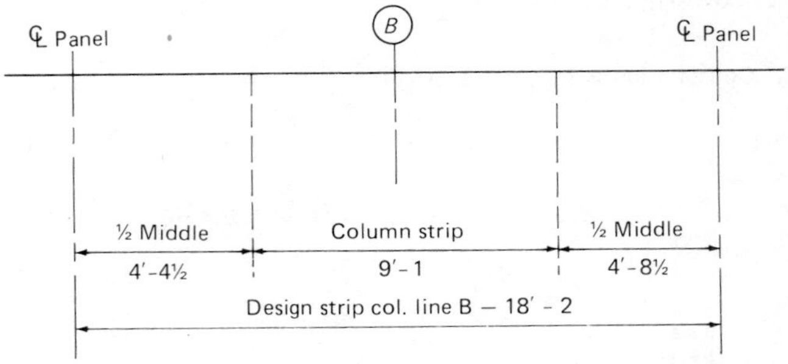

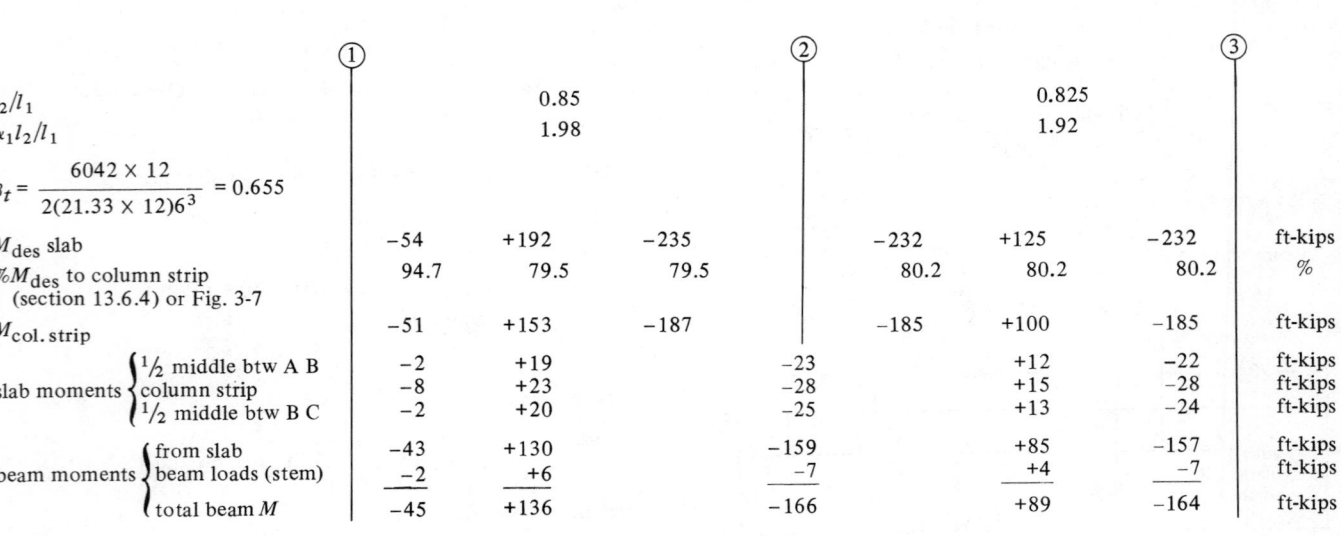

	①			②			③	
l_2/l_1		0.85				0.825		
$\alpha_1 l_2/l_1$		1.98				1.92		
$\beta_t = \dfrac{6042 \times 12}{2(21.33 \times 12)6^3} = 0.655$								
M_{des} slab	−54	+192	−235		−232	+125	−232	ft-kips
$\%M_{des}$ to column strip (section 13.6.4) or Fig. 3-7	94.7	79.5	79.5		80.2	80.2	80.2	%
$M_{col.\,strip}$	−51	+153	−187		−185	+100	−185	ft-kips
slab moments: 1/2 middle btw A B	−2	+19		−23		+12	−22	ft-kips
slab moments: column strip	−8	+23		−28		+15	−28	ft-kips
slab moments: 1/2 middle btw B C	−2	+20		−25		+13	−24	ft-kips
beam moments: from slab	−43	+130		−159		+85	−157	ft-kips
beam moments: beam loads (stem)	−2	+6		−7		+4	−7	ft-kips
beam moments: total beam M	−45	+136		−166		+89	−164	ft-kips

Check shear capacity:

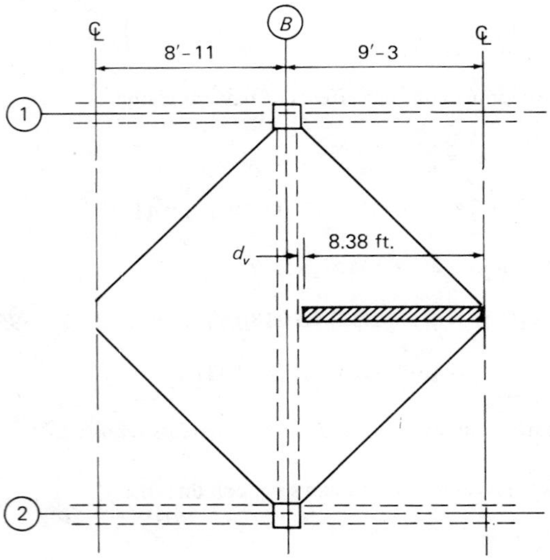

(a) "Beam" shear in slab $d_v = 4.75$
Consider shaded strip of unit width

$$V_u = w_u A = 0.374 \times 8.38 = 3.13 \text{ kips}$$

$$\phi V_n = \phi v_{all} bd = 0.85 \times 2\sqrt{4000} \times 12 \times 4.75/1000 = 6.13 \text{ kips}$$

Since $\phi V_n > V_u$ satisfactory in slab shear

(b) Shear in beams:

	①		②	③	
V critical sections	40	40	42	42	kips
V due to M end span	−6	6	—	—	
V beam stem	2	2	2	2	2
design shear V_u	36	48	44	44	kips

EXAMPLE 3-4: Slab without Beams (Waffle Slab)

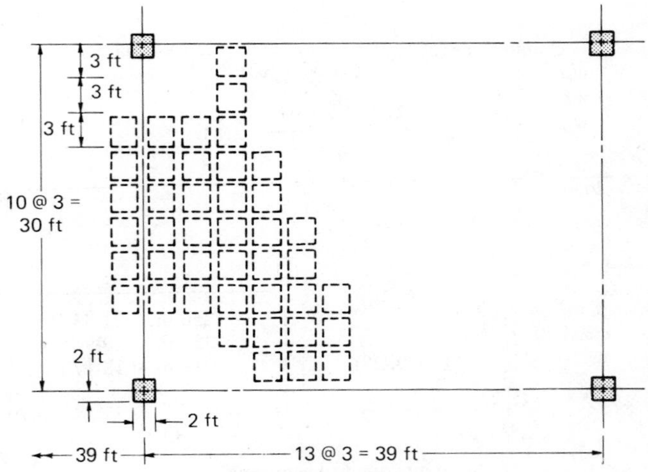

Typical Interior Panel

Design data:
Live loading, passenger car parking garage
Distributed loading 50 psf
Concentrated load 2500 lb on 2.5 ft × 2.5 ft area
Slab and columns designed for gravity load only
Select 30 in. pans on 3 ft module

Material properties:
Concrete (normal weight) $f'_c = 4000$ psi
Reinforcement $f_y = 60{,}000$ psi

Choose slab thickness:
Thickness required if solid slab, eq. (3-1):

$$l_n = 39 - 2 = 37 \text{ ft} = 444 \text{ in.}$$

$$h = \frac{444(800 + 0.005 \times 60{,}000)}{36{,}000} = 13.57 \text{ in.}$$

$$I_g/(3 \text{ ft module}) = \frac{36(13.57)^3}{12} = 7491 \text{ in.}^4/(3 \text{ ft module})$$

Select waffle section to give at least this I_g.
Try 16 in. deep pans with $3\frac{1}{2}$ in. slab.

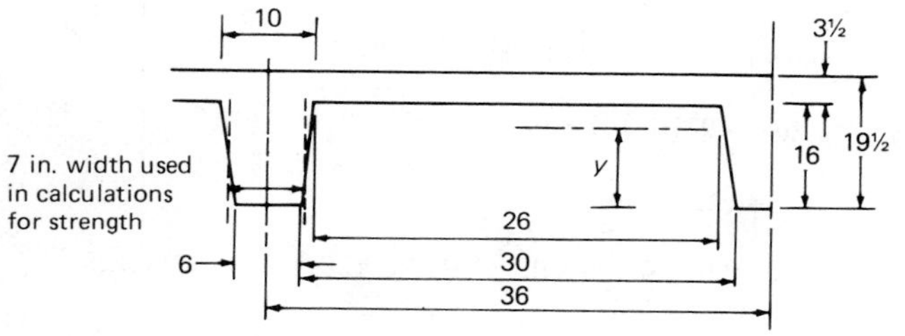

$\bar{y} = 13.17$ in. $\quad I_g = 8041$ in.^{4}

Use of 16 in. pans with $3\frac{1}{2}$ in. slab satisfactory for deflection control.

Factored design loads:
Self wt. of slab:

$$\text{Waffle section } w_w = \frac{1}{9}\left(\frac{36 \times 36}{144} \times \frac{19.5}{12} - \frac{28 \times 28}{144} \times \frac{16}{12}\right) 150$$
$$= 128 \text{ psf}$$

$$\text{Solid section } w_s = \frac{19.5}{12} \times 150 = 244 \text{ psf}$$

	Waffle	Solid
w_d	1.4 × 123 = 172 psf	1.4 × 244 = 342 psf
w_l	1.7 × 50 = 85	1.7 × 50 = 85
w_u	257	427 psf

For moment calculations use a uniform load obtained by weighting loads in proportion to areas:

$$w_u = \frac{12 \times 12}{30 \times 39} \times 427 + \frac{(30 \times 39 - 12 \times 12)}{30 \times 39} \times 257 = 278 \text{ psf}$$

Design moments:
Since all limitations for use of DDM are satisfied, use DDM.
Consider typical interior design strip in long direction (see sketch).

$$l_n = 39 - 2 = 37 \text{ ft}; \quad l_2 = 30 \text{ ft}$$

$$M_0 = \frac{0.278 \times 30 \times 37^2}{8} = 1427 \text{ ft-kips}$$

Moments at critical sections (section 13.6.3.2):

$$M^-_{des} = -0.65 \times 1427 = -928 \text{ ft-kips}$$

$$M^+_{des} = 0.35 \times 1427 = +499 \text{ ft-kips}$$

Also require negative moment at face of solid drop panel.

$$x = 13.5 - 3.5/12 = 13.21 \text{ ft}$$

$$M^-_{face} = M^+_{des} - w_u l_2 (2x)^2/8$$

$$= 499 - 0.257 \times 30(2 \times 13.21)^2 = -174 \text{ ft-kips}$$

Distribute moments to column and middle strips (section 13.6.4 or Fig. 3-7):

Location	Column strip	Middle strip
face of support	0.75(−928) = −696	−232 ft-kips
face of drop	0.75(−174) = −131	−42 ft-kips
midspan	0.6(499) = 299	200 ft-kips

Select reinforcement:
At face of support:

(a) Column strip-resistance provided by solid slab 12.58 ft wide:

$$d = 19.5 - 0.75 - 0.5 = 18.25 \text{ in.}$$

$$A_s = 696 \times 12/\phi f_y jd = 9.42 \text{ in.}$$

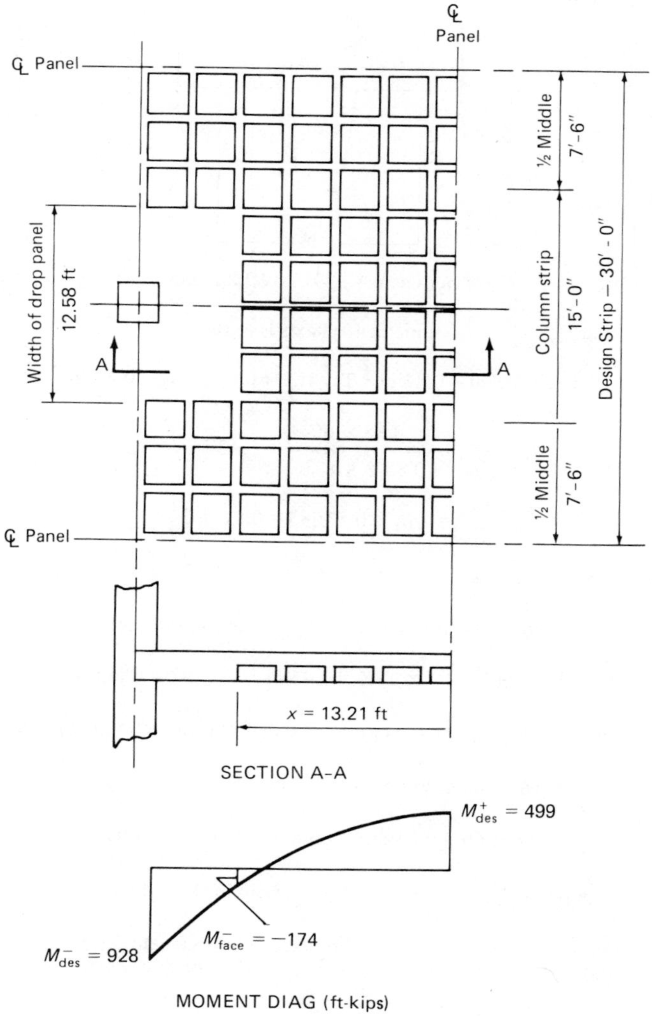

(b) Middle strip-resistance provided by five joists:

M per joist = −232/5 = −46.4 ft-kips

A_s per joist = 0.63 in.^{2}

At face of drop panel:

(a) Column strip-resistance provided by five joists:

M per joist = −131/5 = 26.2 ft-kips

A_s per joist = 0.36 in.^{2}

(b) Middle strip-resistance provided by five joists:

M per joist = −43/5 = −8.6 ft-kips

A_s per joist = 0.12 in.^{2}

NOTE: Above A_s can be provided by a number of bar arrangements. Care should be taken to ensure that all bars are properly anchored in the slab by extending them sufficiently beyond the point where they are no longer required to resist negative moment.

At midspan:

(a) Column strip-resistance provided by five joists acting as T-beams (see sketch):

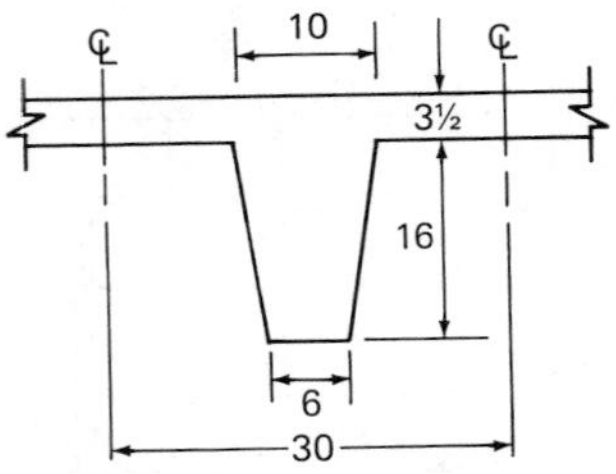

M per beam = 299/5 = 59.8 ft-kips

Assume neutral axis is in flange

A_s per beam = 0.81 in.^{2} Try two #6 bars A_s = 0.88 in.^{2}

$$a = \frac{0.88 \times 60}{0.85 \times 4 \times 36} = 0.43 \text{ in.}$$

$$c = a/\beta_1 = 0.43/.85 = 0.51 < 3\tfrac{1}{2}$$

Neutral axis in flange, design as rectangular beam.

(section 10.5.1) $\rho_{min} = 200/f_y = 200/80{,}000 = 0.0033$

A_s min = 0.0033 × 7 × 19.5 = 0.45 in.^{2} Use two #6 in each web.

(b) Middle strip-resistance provided by five joists acting as T-beams

M per beam = 200/5 = 40 ft-kips

A_s per beam = 0.54 in.^{2} Could use two #5 bars per web.

Similar analysis required for design strip in short direction.

NOTE: Temperature and shrinkage reinforcement in amount of 0.0018 × 3.5 × 12 = 0.076 in.^{2}/ft required both ways in slab.

Check shear capacity:
Perimeter shear at column:

$$d_v = 19.5 - 0.75 - 0.75 = 18.0 \text{ in.}; \quad d_v/2 = 9 \text{ in.}$$

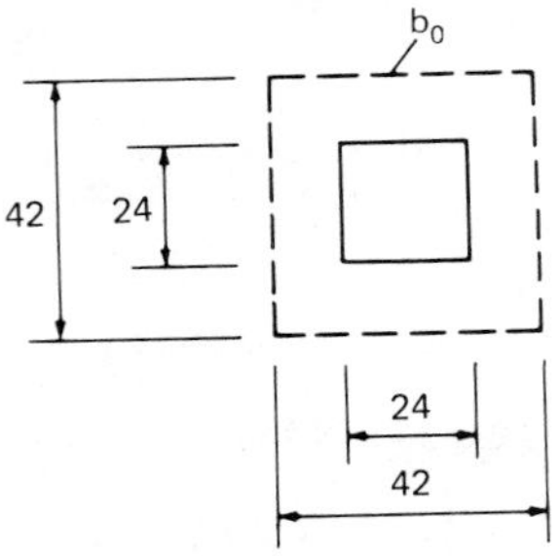

Area of solid slab = 12 × 12 = 144 ft^2

Area of waffle slab = (30 × 39) − 144 = 1026 ft^2

Area enclosed by b_0 = 3.5 × 3.5 = 12.25 ft^2

$$V_u = \sum w_u A = 0.257(1026) + 0.427(144 - 12.25) = 320 \text{ kips}$$

From eq. (3-5) and w_d weighted = 193 psf

$$M_{E\text{-}W} = 0.07[(0.193 + 0.5 \times 0.85)30(37)^2 - (0.193)30(37)^2]$$

= 122 ft-kips

$$M_{N\text{-}S} = 0.07[(0.193 + 0.5 \times .085)39(28)^2 - (0.193)39(28)^2]$$

= 91 ft-kips

Properties of critical section:

$A = db_0 = 19(4 \times 42) = 3024$ in.^{2}

$\gamma_v = 0.4$; $c = 42/2 = 21$ in.

$J = 42(18)^3/6 + 6(42)^3/6 + 18(42)^3/2 = 781794$ in.^{4}

$$v_u = \frac{320{,}000}{0.85 \times 3024} + \frac{0.4(122)12{,}000}{0.85 \times 781{,}704} + \frac{0.4(91)12{,}000}{0.85 \times 781{,}704}$$

= 126 psi

(v_c) allowable = $4\sqrt{f'_c}$ = 253 psi

Satisfactory for perimeter shear.

Beam shear:

V_u at perimeter of drop panel = 0.257 × 1026 = 264 kips

No. of webs resisting shear = 4 × 5 = 20

Shear force per web = 264/20 = 13.2 kips

NOTE: This value can be reduced to d from face of drop panel, if required.

$$\phi V_n = 0.85 \times 2\sqrt{4000} \times 7 \times 18/1000 = 13.6 \text{ kips}$$

Satisfactory in beam shear.

EXAMPLE 3-5: Design Strip Using EFM

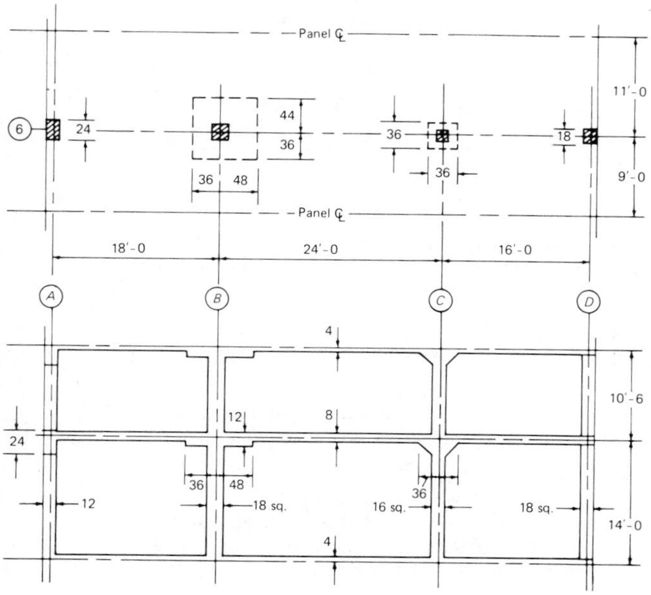

Design data:
Geometry for design strip centered on column line 6 as shown
Service live load = 125 psf
Frame designed for gravity load only
Same quality of concrete in both columns and slab

Factored design loads:

$$w_d = 1.4(8/12 \times 150) = 140 \text{ psf}$$

$$w_l = 1.7(125) = 213 \text{ psf}$$

$$w_d + w_l = 353 \text{ psf}$$

$$w_d + 0.75\, w_l = 299 \text{ psf}$$

Design moments at critical sections:

NOTE: Although the design strip meets all the limitations imposed on the use of the DDM, the EFM is used for illustration. Analysis is to be performed using moment distribution.

(a) Compute stiffness factors:

The geometry for the equivalent members in the simplified frame are shown schematically. Corresponding to the geometry of each column, the values of k_c are obtained from Tables 3-8 and 3-9. Similarly values of k_s at each end of the equivalent slab-beam elements are obtained from Tables 3-4 to 3-6.

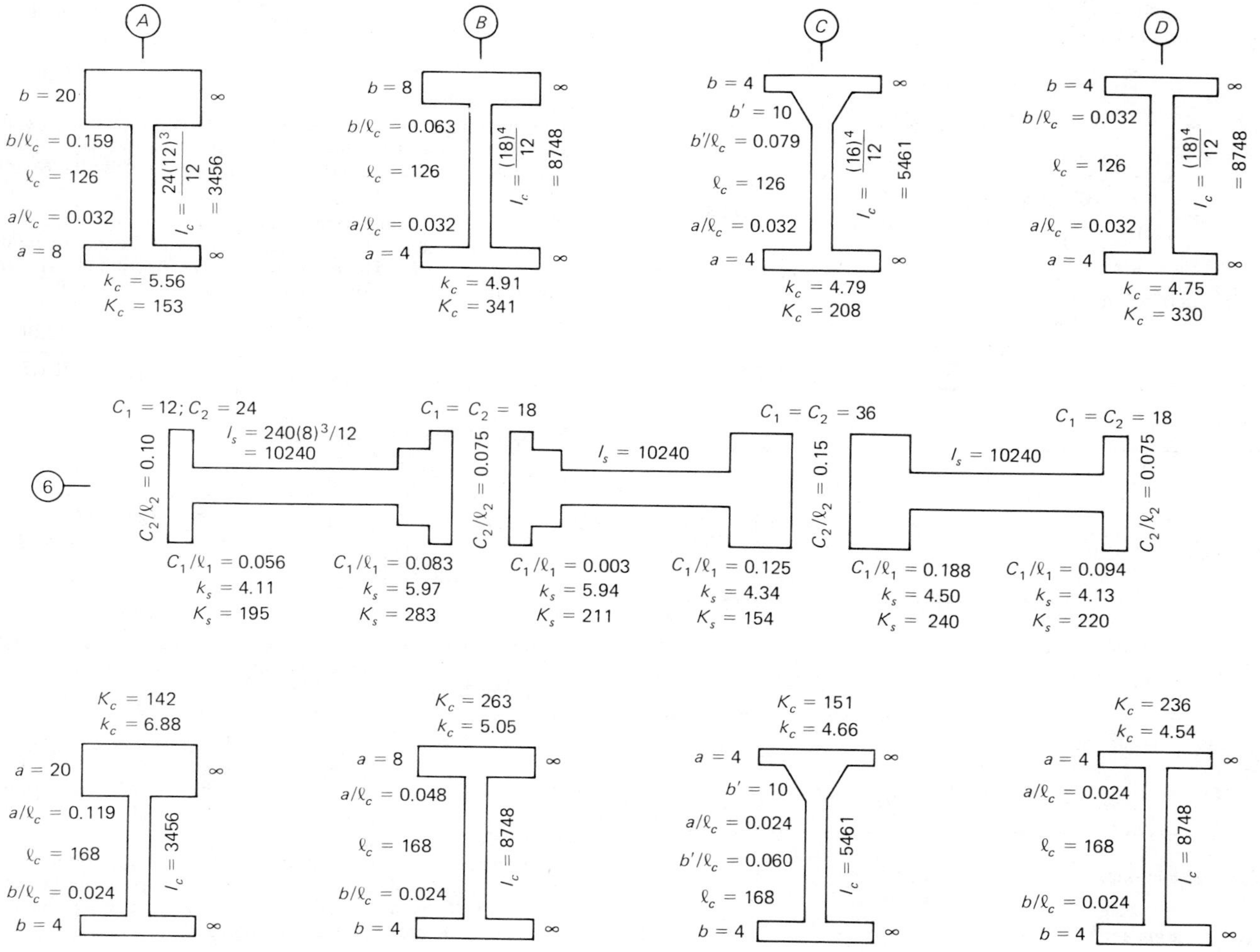

Since the same concrete is used for both slab and columns, the stiffness factors are computed for a unit value of E_c.

$$K_c = k_c I_c / l_c; \quad K_s = k_s I_s / l_s$$

Compute K_{ec} (eq. 3-7):

(i) Column Line A:

Compute values of torsional constant C (Table 3-7):

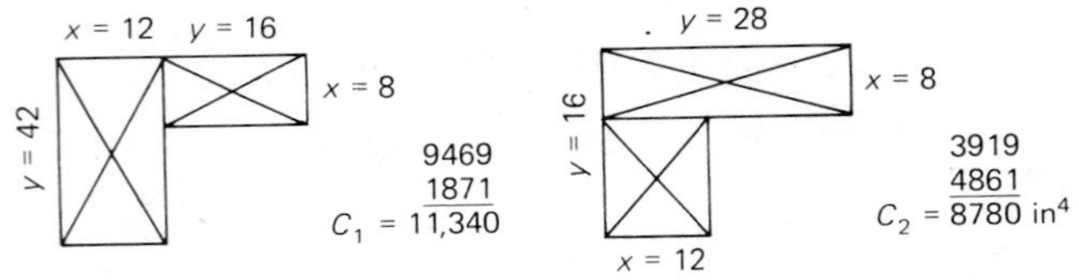

$C_1 > C_2$; therefore $C = 11{,}340$ in.^{4}

Since values of l_2 on each side of the design strip are nearly equal, an acceptable value for K_t can be obtained using the average value for l_2 in eq. (3-7) and multiplying by 2 for the summation:

$$K_t = 2\left(\frac{9 \times 11{,}340}{240(1 - 0.1)^3}\right) = 1167$$

This compares with $K_t = 1187$ using actual values of l_2. From sketch, $\Sigma K_c = 153 + 142 = 295$

$$\frac{1}{K_{ec}} = \frac{1}{295} + \frac{1}{1167}; \quad K_{ec} = 235 \text{ in.-kips}$$

Distribution factors:

$$\text{for slab-beam } DF = \frac{K_s}{K_s + K_{ec}} = \frac{195}{195 + 235} = 0.453$$

for columns $DF = 1 - 0.453 = 0.547$

(ii) Column Line B:

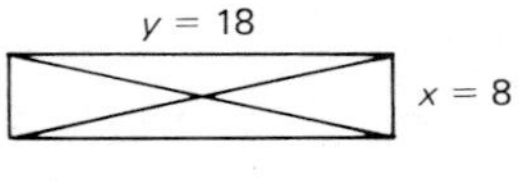

$C = 2212$ in.^{4}

$$K_t = 2\left(\frac{9 \times 2212}{240(1 - 0.075)^3}\right) = 210; \qquad K_c = 341 + 263 = 604$$

$K_{ec} = 156$ in.-kips

(iii) Column Line C:

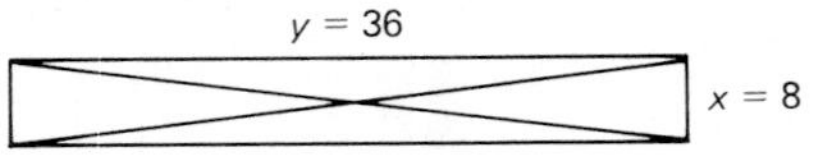

$C = 5284$ in.^{4}

$$K_t = 2\left(\frac{9 \times 5284}{240(1 - 0.15)^3}\right) = 645; \qquad K_c = 208 + 151 = 359$$

$K_{ec} = 231$ in.-kips

(iv) Column Line D:

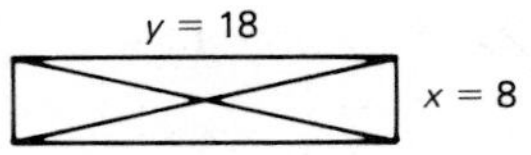

K_t same as for col. line B

$$\sum K_c = 330 + 236 = 566$$

$$K_{ec} = 153 \text{ in.-kips}$$

(b) Compute carry-over factors:
Carry-over factors obtained directly from Tables 3-4 to 3-6.

(c) Compute fixed-end moments:
Moment coefficients obtained from Tables 3-4 to 3-6.
Note moments required for each specified loading conditions.

Moment distributions:

Specified load cases (section 13.7.6):

(a) Full live load on all spans
(b) 3/4 live load on middle span, dead load all spans
(c) 3/4 live load end spans, dead load all spans

Positive moments at midspan:

Compute total panel moment $M_p = w_u l_2 l_1^2 / 8$
$+M = M_p - \frac{1}{2}$ (sum of $-M$'s in panel after distribution). NOTE: For pattern loadings design positive moment can occur only in panel with portion of live load.

Example—span BC for load case (b):

$$M_p = (0.140 + 0.75 \times .213) \times 20 \times 24^2/8 = 432 \text{ ft-kips}$$

$$+M = 432 - \tfrac{1}{2}(268 + 272) = +162 \text{ ft-kips}$$

Reduction of negative moments to critical sections:

V = shear on end of member computed as simple beam

c_1 = width of support at end of member

reduction = $Vc_1/3$

Design moments:

Slab design moments are the maximum moment at each section due to any loading case. These are then distributed to column and middle strips using the same procedure described for the DDM.

Total column moments are the unbalanced moments of the column centerlines, i.e., before negative moments are reduced (section 13.7.7.6). These moments are proportioned to columns above and below in ratio of stiffnesses.

Example—column on col. line B:

$$\text{Portion to column above} = \frac{341}{341 + 263}(64) = 36 \text{ ft-kips}$$

Summary of design moments using EFM:

	Ⓐ				Ⓑ				Ⓒ				Ⓓ	
Slab M_{des}	97	−79	+101	−279	64	−307	+169	−245	102	−191	+87	−32	56	ft-kips
Column above	50				36				59				33	ft-kips
Column below	47				28				43				23	ft-kips

Since the geometry satisfies the limitations placed on use of DDM, slab moments can be further reduced according to section 13.7.7.4. It is recommended that both positive and negative moments be reduced in proportion to ratio of M_0/M_T.

	Ⓐ				Ⓑ				Ⓒ				Ⓓ	
Total Panel Mom. M_T			278				445				198			ft-kips
(eq. 3-4) M_0			248				417				167			ft-kips
M_0/M_T Slab M_{des}		−70	+90	−249		−288	+158	−230		−161	+73	−27		ft-kips

Alternate solution using DDM:

Section 13.6.7 permits modification of design moments obtained using DDM by 10%, provided total panel moments not less than M_0. Owing to ratio of spans, modify to reduce unbalanced moments at interior columns.

Moment Distributions, Example 3-5

	A	B (left)	B (right)	C (left)	C (right)	D
Mom. Coeff. M	0.084	0.094	0.093	0.085	0.087	0.084
C.O.F. C	0.51 →	← 0.59	0.59 →	← 0.52	0.53 →	← 0.51
D.F. slab	0.453	0.435	0.325	0.246	0.384	0.590
D.F. column	0.547	0.240		0.370		0.410

Load case (a) $(w_D + w_L)$ on all spans Section 13.7.6.4

	A				B				C				D	
$w_u l_2 l_1^2/8$ $FEM = M w_u l_2 l_1^2$ Jt. bal.		−192 +87	286	−215 −71		−378 +53	508	−346 +47		−157 −72	226	−152 +89		ft-kips ft-kips
C.O. Jt. bal.		+42 −19		−44 +9		−24 −6		−31 −4		−45 +5		+38 −22		
C.O.		−5 +2		+10 −3		+2 +3		+3 +2		+11 −3		−3 +2		
$M_{℄} \sum$ $V; c_1$ reduction $Vc_1/3$	86 1.0	−85 50.9 +17	+87	−314 76.2 +38	36 1.5	−350 85.6 +43	+169	−329 83.8 +84	68 3.0	−261 69.8 +70	+72	−48 43.2 +22	48 1.5	ft-kips
reduced M		−68	+87	−276		−307	+169	−245		−191	+72	−26		ft-kips

Load Case (b) (w_D) on end spans, $(w_D + 0.75w_L)$ on middle span Section 13.7.6.3

	A				B				C				D	
$w_u l_2 l_1^2/8$ $FEM = M w_u l_2 l_1^2$ bal.		−76 +34		−85 −102		−320 +76	432	−293 +57		−62 −88		−60 +35		ft-kips ft-kips
C.O. bal.		+60 −27		−17 −6		−30 +4		−45 +7		−18 −10		+47 −27		
C.O. bal.		+4 −2		+14 −8		−4 +6		−2 +4		+14 −6		+5 −3		
$M_{℄} \sum$		−7		−204	64	−268	+162	−272	102	−170		−3		ft-kips

Load Case (c) $(w_D + 0.75w_L)$ on end spans, (w_D) on middle span Section 13.7.6.3

	A				B				C				D	
$w_u l_2 l_1^2/8$ FEM bal.		−162 +73	243	−181 −14		−150 −10		−137 +1		−132 −2	192	−128 +76		ft-kips
C.O. bal.		−8 +4		−37 +16		−1 −12		+6 −11		−39 +17		−1 +1		
C.O. bal.		−9 +4		−2 +3		+6 −3		+7 −2		−1 +3		−9 +5		
$M_{℄} \sum$	97	−97	+101	−187	17	−170		−136	18	−154	+87	−56	56	
$V; c_1$ reduction $Vc_1/3$	1.0	54 +18										48 +24	1.5	
reduced M	97	−79										−32	56	ft-kips

	A				B				C				D	
(eq. 3-4) M_0			248				417				167			ft-kips
(Table-3-2) Coef.		0.3	0.5	0.7		0.65	0.35	0.65		0.7	0.5	0.3		
M_{des}		−74	+124	−174		−271	+146	−271		−117	+84	−50		ft-kips
Mod. (13.6.7)			−9	−18			+13	+26		−10	−5			ft-kips
Modif. M_{des}		−74	+115	−192		−271	+159	−245		−127	+79	−50		ft-kips
Col. Mom. (eq. 3-5)	74				108				126				50	ft-kips

3.11 SPECIAL SLAB PROBLEMS

3.11.1 Lateral Load Analysis

Resistance to lateral loading on unbraced reinforced concrete structures is provided by frame action involving the floor system and the supporting columns. The usual method of evaluating this action is to consider the structure made up of longitudinal and transverse frames located along column lines and bounded laterally by the centerline of the panel on each side. These frames extend vertically over the entire height of the structure and are analyzed as a unit for both moments and drift (lateral desplacements). Moments obtained from the analysis may be added directly to those obtained from the gravity load analysis.

The actual frames are three-dimensional, having "beam" widths consisting of a slab of width l_2 and columns having a width c_2. To analyze these frames as two-dimensional frames requires some modeling of member stiffnesses in order to match the response of the actual structure to the idealized frame being analyzed. The problem is made more complex if the reduced stiffness caused by cracking of the concrete is considered. Many authors have studied the problem and proposed solutions, but the almost complete lack of reliable experimental data makes evaluation of the different procedures most difficult. A comprehensive review of proposed procedures is given by Vanderbilt.[3-21]

The various methods can be categorized as either the effective beam width model, in which the actual transverse width, l_2, is replaced by the effective width, αl_2, or the transverse torsional member model, in which the connection between column and beam is considered to be provided by imaginary transverse torsional members.

Using the effective width method, the members of the two-dimensional frame are assumed prismatic. Column stiffnesses are computed using gross dimensions and beam stiffnesses using slabs of width αl_2. Studies indicate that α is sensitive to variations in the ratios l_2/l_1 and c_1/l_1, but relatively insensitive to c_2/c_1, and the absolute value of l_1. Typical values for α as obtained by Pecknold[3-22] are given in Fig. 3-13. A comparison of values obtained by several investigations for square panels and square columns is given in Fig. 3-14. In this figure the group of curves in the "rigid" region are based on an upper bound estimate of the column stiffness whereas the curves in the "flexible" region are based on a lower bound.

The transverse member model uses a procedure similar to the EFM to compute the torsional stiffness of the transverse member, K_t. However, this member is associated with the slab element to obtain an equivalent stiffness for this element, K_{es}, given by:

$$\frac{1}{K_{es}} = \frac{1}{K_t} + \frac{1}{K_s}$$

In computing the slab stiffness, K_s, the effective moment of inertia is taken as βI_g where I_g is the gross moment of inertia. Studies indicate that to account for cracking of the slab a value of β in the range of 0.33 is recommended.

Since both methods satisfy equilibrium, moments obtained for a reasonable range of α or β values will, in general, be satisfactory. However, a major reason for a lateral drift analysis is to predict lateral deflection or drift of the structure. A commonly specified limit for drift in building codes is 1/500 of the height applied on a story-by-story basis. Unfortunately the magnitude of drift is sensitive to the values of α or β used in the analysis, and care is required in interpreting the results. Fintel[3-23] in a study of drift of high-rise concrete buildings, concluded: "All the shortcomings in computing deflections . . . cast serious doubt on the reliability of deflection computations for high-rise buildings. Actually, differences of several hundred percent between deflections computed by exact and simplified methods are not uncommon. Therefore, a drift limitation should be used with a deflection computation spelled out in detail. A drift limitation not corresponding to a specific method of computing deflections is almost meaningless."

3.11.2 Concentrated Loads

Concentrated loads can result from the construction procedure. Commonly occurring examples of concentrated

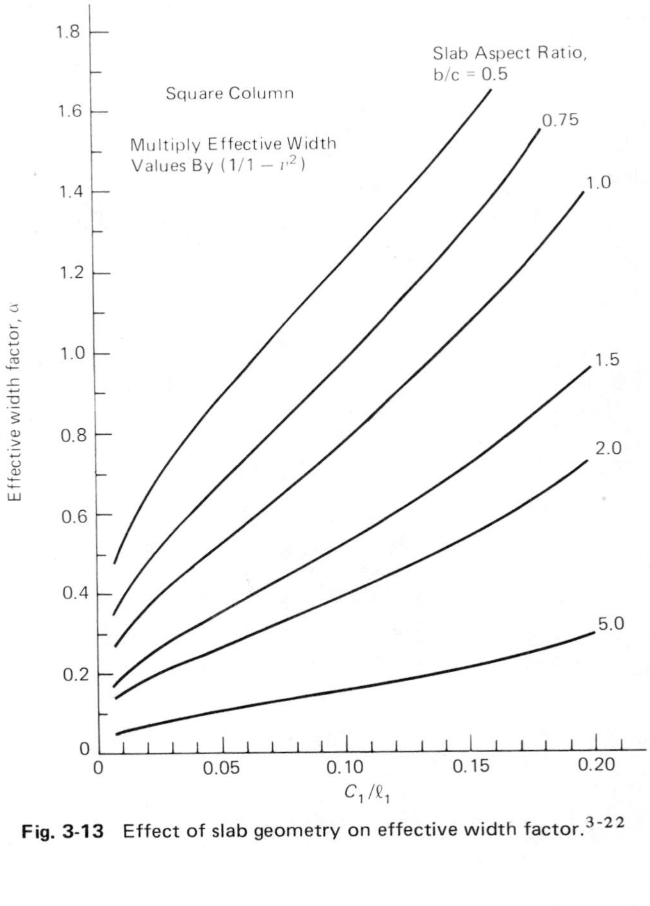

Fig. 3-13 Effect of slab geometry on effective width factor.[3-22]

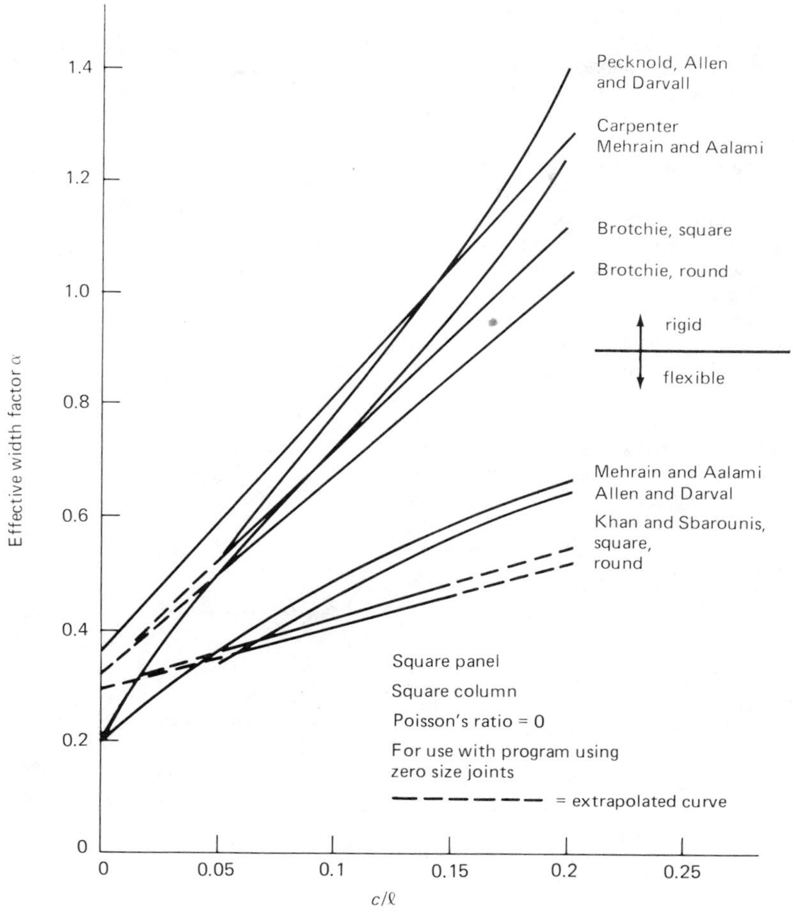

Fig. 3-14 Effect of column stiffness on effective width factor.[3-21]

loadings on completed slab systems include wheel loads in parking structures and in those warehouses where forklift trucks are used. Since in these cases the loads can move about, there is the additional problem of where to place the load to obtain the critical design moments and shears. Generally such loads are considered to act over a finite area using the recommendations for distributing wheel loads as given in highway specifications for bridge decks. This practice is satisfactory for determining the adequacy of the slab against failure by punching shear.

To determine the most unfavorable location for the load for bending moments in the slab, use is made of influence surfaces. The most complete study of the moments caused by concentrated loads in the interior panels of continuous slabs with and without beams was made by Woodring and Siess.[3-24] In keeping with the concept of dividing the slab into middle and column design strips, they obtained influence surfaces for the positive and negative design moments in each strip for concentrated loads and for area loads extending to one-third of the span length. They concluded that the effect on the design moments of spreading a concentrated load over a larger area is small unless the load is in close proximity to the design moment locations. However, the position of the concentrated load with respect to the moment location was the most important variable. They found that the critical moments in the middle strip could be reversed in sign from the moments usually associated with these positions. For example, when a concentrated load is placed on a column line midway between two columns, the moment in the middle strip directly under this load normally referred to as the negative middle strip moment is positive. Moreover the magnitude of this positive moment is more than twice as great as the maximum negative moment produced at this section by the same concentrated load located anywhere in the panel even if Poisson's ratio is taken as zero. This explains why certain warehouse slabs reinforced only for a uniformly distributed load have shown excessive cracking due to moment reversals caused by moving loads for which they are not reinforced. In an attempt to simplify the calculations for concentrated loads, Woodring proposed an equivalent load factor, C, which when multiplied by the concentrated load gave an equivalent uniform panel load. Plots giving this load factor were prepared for each design moment location for different concentrated load positions and beam stiffnesses, but the number of plots required was extensive.

The common building code requirement of considering a load of 2500 lb distributed over a 2.5 × 2.5 ft square area and located to produce maximum effect will generally not govern design of floor slabs of usual dimensions. Where the magnitude of the concentrated load is large compared to the dead load, consideration should be given when detailing reinforcing to changing locations of points of inflection and possible reversal of moment signs in the middle strips.

3.11.3 Openings in Slabs

Almost all slabs have openings, but generally they do not constitute any difficulty in the design. For large openings it may be necessary to stiffen the slab with the use of beams around the opening. Discontinuities in the moment path may require some redistribution of the moments. It is necessary that provision be made to carry the total panel moment in each direction without excessive deflection.

Most openings are sufficiently small that the moments may be obtained on the basis that the opening is absent. The reinforcement that is interrupted by the opening is added to the sides of the opening. ACI 318-83 permits this procedure for any size of opening that is located in areas common to intersecting middle strips, and for openings not greater than one-eighth the width of either column strip in areas common to intersecting column strips. In areas common to one column strip and one middle strip, not more than one-fourth of the reinforcement in either strip may be interrupted by the opening and so placed on either side of the opening.

When openings are located within a distance less than ten times the slab thickness, or when, for slabs without beams, openings are located in the column strips, the effective periphery, b_0, for shear is to be reduced by the portion that is enclosed by radial projections from the centroid of the column or loaded area to the outer edges of the openings, as shown in Fig. 3-15. Where shearheads are provided, only one-half of this reduction is required.

3.11.4 Isolated Beams

The design procedures outlined in ACI 318-83 are applicable to slab systems with beams only when the ratio $\alpha_1 l_2^2/\alpha_2 l_1^2$ is not less than 0.2 or greater than 5.0. Clearly then, where a beam is placed only along one edge of an interior panel, the code procedures will not apply, and the design must be modified in the region of this discontinuity. A suggested procedure[3-18] for this modification follows. Consider the slab system, illustrated in Fig. 3-16, in which a single line of beams is placed in the y-direction and all panels are similar in shape, although the panel aspect ratio can vary from 0.5 to 2.0. Strips are considered in each direction as in the direct design method. The total static moment for the strip in each panel can be expressed as $Cwl_2 l_1^2$ (the value l_n can be substituted for l_1 without invalidating the procedure, provided values of c/l are not excessive).

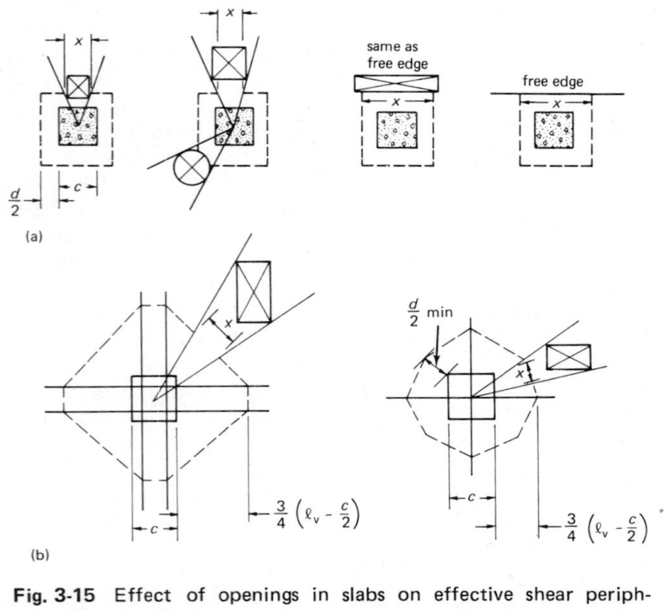

Fig. 3-15 Effect of openings in slabs on effective shear periphery. (a) Slabs without shear reinforcements, effective periphery = $b_0 - x$; (b) slabs with shear reinforcement, effective periphery = $b_0 - x/2$.

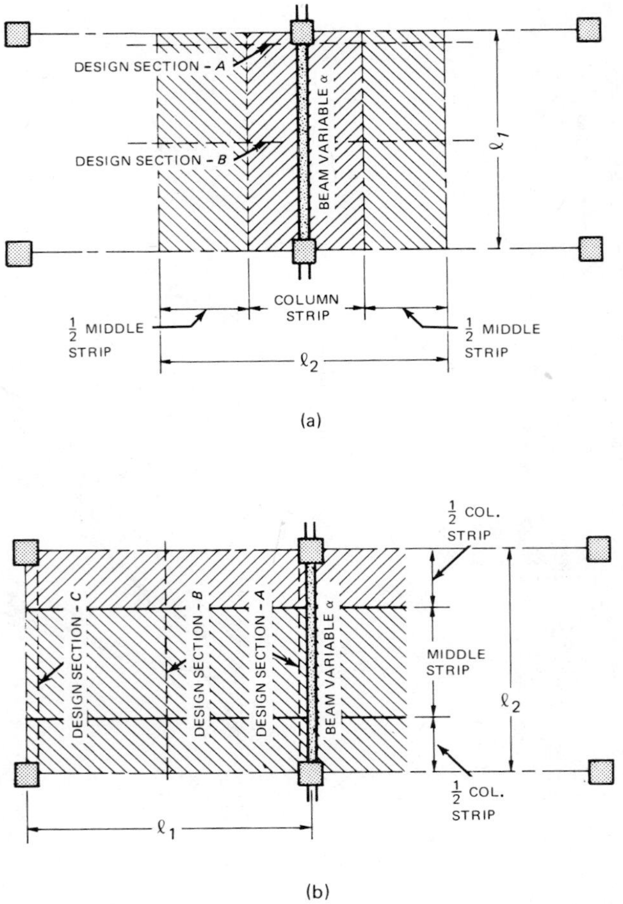

Fig. 3-16 Design sections and strip definitions for isolated beam. (a) Design sections. Table 3-10; (b) design sections, Table 3-11.

First consider the strip containing the beam. As the flexural stiffness of the beam is increased, the portion of the load carried by this strip increases with a corresponding decrease in the parallel adjacent strips. This increase in load and hence panel static moment M_o depends on the panel shape as seen in Fig. 3-17, where values of the coefficient C are given. It is recommended that the corresponding decrease in M_o for the adjacent strips be neglected and the moments in these strips be designed using code values.

The total panel moment must now be proportioned between negative and positive sections and between column and middle strips. The percentages of M_o assigned to each slab and beam section are given in Table 3-10.

Consider now the design strips perpendicular to the beam. In all cases the total panel static moment from elastic analysis is $0.125\ wl_2/l_1^2$. The distribution of this moment to the critical design sections is given in Table 3-11.

By examining the distribution of moments in Tables 3-10 and 3-11, it is seen that, if an isolated beam line is placed in an otherwise regular slab system without beams, it will attract moment in the direction in which the beam spans so that the total static moment of a strip bounded by the centerline of the panel adjacent on each side will exceed the static moment caused by the load acting on this area. In addition, the moments in the vicinity of the beam will be altered as the beam stiffness increases to approach those of a one-way slab spanning perpendicular to the beam.

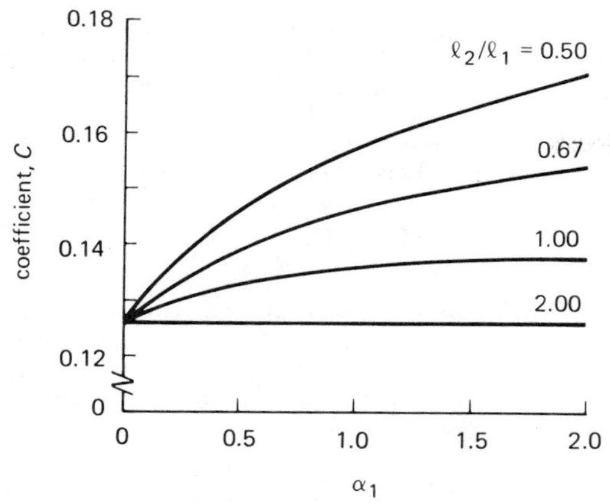

Fig. 3-17 Coefficient, *C*, for panel static moment parallel to beam.[3-18]

3.11.5 Slab Shear Reinforcement (Shearheads)

When the shear strength (punching shear) of a two-way slab without beams is not sufficient, we have seen that the designer may increase the slab thickness, use column capitals, and/or drop panels, or increase the strength of the concrete to improve the shear capacity of the slab. An alternative permitted by ACI 318-83 is to use shear reinforcement consisting of reinforcing bars, when the actual shear stress does not exceed the allowable concrete shear stress by more than 50%; or structural steel shearheads, when the actual shear stress does not exceed the allowable by more than 75%. The procedure for designing such reinforcement is illustrated in Example 3-6.

EXAMPLE 3-6: Check the adequacy in shear of an interior panel of a slab without beams (flat plate), and if it is not sufficient provide suitable shear reinforcement. You are given the following data:

$$f'_c = 3000 \text{ psi (normal weight concrete)}$$

$$f_y \text{ (reinforcing bars)} = 60{,}000 \text{ psi}$$

$$F_y \text{ (structural steel)} = 36{,}000 \text{ psi}$$

$$d = 8 \text{ in.},\quad c_1 = c_2 = 12 \text{ in.},\quad V_u = 170 \text{ kips}$$

NOTE: It is assumed that the capacity of the slab in shear as a wide beam has been found satisfactory and that two-way (punching) shear governs.

SOLUTION:

(a) Determine adequacy of slab in shear without shear reinforcement (Section 11.11.2)

$$b_o = 4(12 + 8) = 80 \text{ in.}$$

$$\varphi V_n = \varphi 4\sqrt{f'_c}\, b_o d$$

$$= 0.85 \times 4 \times \sqrt{3000} \times 80 \times 8 = 119{,}200 \text{ lbs}$$

Since $V_u > \varphi V_n$, capacity of slab in shear is not adequate without shear reinforcement.

(b) Shear reinforcement consisting of reinforcing bars (Section 11.11.3).

Maximum allowable φV_n for using reinforcing bars as shear reinforcement $= \varphi 6\sqrt{f'_c}\, b_o d = 0.85 \times 6 \times \sqrt{3000} \times 80 \times 8 = 178{,}800$ lbs. Since $\varphi V_n > V_u$ may use reinforcing bars for shear reinforcement.

$$V_c = 2\sqrt{f'_c}\, b_o d$$

TABLE 3-10 Distribution of Moments in Strip Parallel to Beam as Percentages of M_o*[3-18]

α_1	0.0		0.5		1.0		2.0	
*Section***	*A*	*B*	*A*	*B*	*A*	*B*	*A*	*B*
Beam Moment								
l_2/l_1 = 0.5	–	–	27	12	37	17	45	23
0.67	–	–	30	12	39	17	47	23
1.0	–	–	35	14	45	19	52	24
2.0	–	–	46	20	54	25	60	28
Slab Moment, Column Strip								
0.5	42	17	22	11	16	9	10	6
0.67	45	18	22	11	15	9	9	6
1.0	51	20	21	12	14	8	8	6
2.0	50	20	14	8	8	5	5	3
Slab Moment, Middle Strip								
0.5	25	16	16	12	12	9	9	7
0.67	21	16	14	11	11	9	8	7
1.0	16	13	9	9	7	7	5	5
2.0	16	14	6	6	4	4	2	2

*$M_o = Cwl_2 l_1^2$; *values of C given in Fig. 3-17.*
***For location of sections see Fig. 3-16.*

TABLE 3-11 Distribution of Moments in Strip Perpendicular to Beam as Percentages of M_o*[3-18]

α_1	0.0			0.5			1.0			2.0		
*Section***	*A*	*B*	*C*	*A*	*B*	*C*	*A*	*B*	*C*	*A*	*B*	*C*
Slab Moment, Column Strip												
l_2/l_1 = 2.0	50	20	50	41	17	49	36	15	48	31	14	47
1.5	50	20	50	42	17	50	38	16	49	34	15	48
1.0	51	20	51	46	19	50	43	19	50	41	19	50
0.5	42	17	42	37	17	42	36	17	42	35	17	42
Slab Moment, Middle Strip												
2.0	16	14	16	44	13	10	59	12	7	76	11	4
1.5	16	14	16	36	14	13	44	14	11	53	14	10
1.0	16	13	16	26	14	14	30	14	14	33	14	14
0.5	25	16	25	30	17	25	32	17	25	33	17	25

*$M_o = 0.125\, wl_2 l_1^2$.
***For location of sections see Fig. 3-16.*

$$V_s \text{ (for one face of critical section)} = \frac{1}{4}\left(\frac{V_u}{\varphi} - V_c\right)$$

$$= \frac{1}{4}\left(\frac{170{,}000}{0.85} - 2 \times \sqrt{3000} \times 80 \times 8 = 32{,}470 \text{ lbs.}\right.$$

Try #3 stirrups

$$A_v = 0.22 \text{ in.}^2 \text{ (2 legs)}$$

$$s = \frac{A_v f_y d}{V_s} = \frac{0.22 \times 60{,}000 \times 8}{32{,}470} = 3.25 \text{ in.}$$

$$s_{max} = \frac{d}{2} = 4.0 \text{ in.}$$

Now calculate length over which stirrups are required (Section 11.11.3.3). Stirrups are required until φV_c at section b_o indicated by dashed line is equal to or greater than V_u.

$$b_o = 4(c + a\sqrt{2})$$

Equating $\varphi V_c = V_u$

$$0.85 \times 2 \times \sqrt{3000} \times 4(12 + a\sqrt{2}) \times 8 = 170{,}000$$

$$a = 32 \text{ in.}$$

length required $= a + d = 32 + 8 = 40$ in.

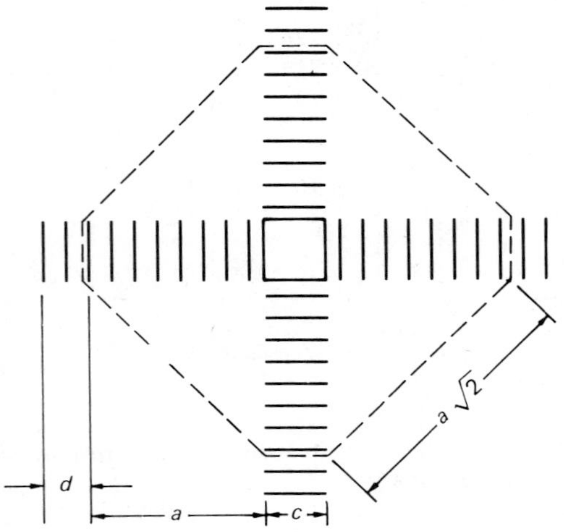

(c) Shear reinforcing consisting of structural steel section (Section 11.11.4)

Length of shearhead arm, l_v, selected such that at critical section b_o, located at ¾ distance $[l_v - (c_1/2)]$ from column face, the concrete shear stress does not exceed $4\sqrt{f_c'}$ (see sketch).

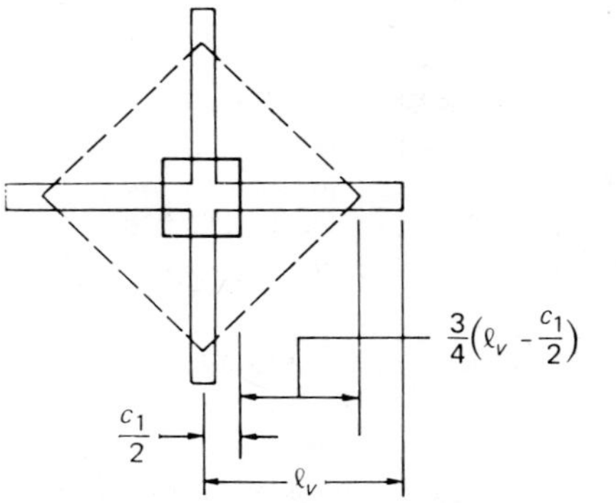

$$b_{o,\min} = \frac{V_u}{4\varphi d\sqrt{f_c'}} = \frac{170{,}000}{4 \times .85 \times 8 \times \sqrt{3000}} = 114 \text{ in.}$$

$$b_o = 4\sqrt{2}\left[\frac{c_1}{2} + \frac{3}{4}(l_v - c_1/2)\right]$$

$$\therefore l_v = \frac{1}{3}\left[\frac{b_o}{\sqrt{2}} - \frac{c_1}{2}\right] = \frac{1}{3}\left[\frac{114}{\sqrt{2}} - \frac{12}{2}\right]$$

$$= \frac{1}{3}(80.5\text{-}6) = 24.5 \text{ in.}$$

Maximum allowable φV_n for using structural steel section as shear reinforcement $= \varphi 7\sqrt{f_c'}\, b_o d = 0.85 \times 7 \times \sqrt{3000} \times 114 \times 8 = 297{,}200$ lbs $> V_u$, ∴May use structural steel section.

To ensure that the sheared reinforcing does not fail in flexure before the shear strength of the slab is reached, each arm must be capable of resisting the full plastic moment, M_p, given in the code as:

$$M_p = \frac{V_u}{\varphi 2\eta}[h_v + \alpha_v(l_v - c_1/2)]$$

where η = number of arms in shearhead; h_v = depth of steel section; and α_v = ratio of stiffness of single shearhead arm to composite cracked section of width $(c_2 + d)$, and α_v must be not less than 0.15.

Assume $h_v = 5$ in., $\alpha_v = 0.2$, then with $\eta = 4$

$$M_p = \frac{170}{0.9 \times 2 \times 4}[5 + 0.2(24.5 - 12/2)] = 205 \text{ in.-kips}$$

Try S5 × 10.0 $Z_x = 5.67$ in.^{3}

$$M_p = Z_x F_y = 5.67 \times 36{,}000 = 204 \text{ in.-kips.}$$

Although 204 < 205, it is sufficiently close if assumed value of α_v is satisfactory.

$$EI(\text{S5} \times 10.0) = 12.3 \times 29{,}000 = 356{,}700 \text{ kip-in.}^2$$

EI (cracked section) assuming five #5 bars over width of $c_2 + d$)

$$= 1{,}715{,}000 \text{ kip-in.}^2$$

$$\alpha_v = \frac{356{,}700}{1{,}715{,}000} = 0.208 > 0.15$$

$$\text{revised } M_p = \frac{170}{0.9 \times 2 \times 4}[5 + 0.208(24.5 - 12/2)) = 209 \text{ in.-kips}$$

About 2% underdesign, use anyway since V_u not reduced to critical section.

Maximum depth of steel section permitted is (70 × web thickness) = 70 × 0.210 = 14.7 > 5 in. OK

Allowable distance from compressive face of concrete to compression flange section $= 0.3d = 2.4$ in. $> (9 - 5)/2 = 2.0$ OK

Use S5 × 10; length = 24.5 in. + 5 in.; say, 30 in.

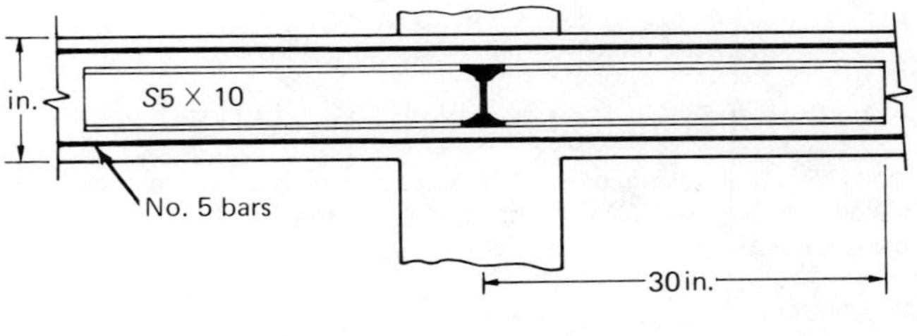

$$2(\tfrac{5}{8} + \tfrac{5}{8} + \tfrac{3}{4}) + 5 = 9 \text{ in.}$$

∴no bars need be cut to fit S section.

3.12 PLASTIC DESIGN PROCEDURES

The distribution of design moments and shear stresses used in the code methods[3-1] were selected to approximate elastic behavior. An alternate approach, widely used outside of North America, is to provide a slab with sufficient strength capacity in moment and shear to resist the factored loading. Serviceability requirements are provided by using maximum span-to-depth ratios and minimum reinforcement arrangements as outlined in earlier sections. Since reinforced concrete slabs will undergo significant plastic deformation before failure, design methods based on predicting failure loads are sometimes referred to as plastic design methods. They have the major advantage of being readily applied to slabs with very irregular geometry.

For flexural strength capacity the essential assumption is that the moment–curvature relationship at each section can be considered elastic-plastic; that is, as the reinforcement yields at any section, the plastic moment will be developed and maintained as the section rotates, permitting a redistribution of moments. Since reinforced concrete slabs are lightly reinforced, the actual moment–curvature response will closely approximate this idealized behavior.

Two procedures, corresponding to the two theorems of plasticity, are possible for slabs. The upper bound approach to the flexural strength capacity requires obtaining the load that will cause sufficient plastic yielding to occur to form a preselected mechanism. This load will be an upper limit, since it is sufficient to cause failure, but failure at a lower load corresponding to a different mechanism may occur unless all possible mechanisms are considered. The lower bound approach is to obtain a moment field that is everywhere in equilibrium with the factored loading and at no point exceeds the yield moment provided. Such a slab will certainly be strong enough to carry the loading but, unless a good moment field is selected, may not be the most economical. The most widely used upper bound method is the yield-line method proposed by Johanson,[3-25] although an ingenious procedure for flat plate structures was presented by Wiesinger.[3-26] The major lower bound procedures are attributed to Hillerborg.[3-27, 3-28]

Plastic procedures have also been proposed for the moment-shear transfer problem. Again the basic assumption is that the various mechanisms that transfer shear and moment to the column are capable of developing and maintaining their capabilities in spite of plastic deformations. A procedure that predicts shear-moment capacity for a number of varying conditions is the beam analogy proposed by Hawkins,[3-29] but it is not suitable for routine design use. A simpler procedure was prepared by Kanoh and Yoshizaki,[3-30] Because of space limitations, only a brief introduction to the concepts behind the procedures proposed by Johansen, Hillerborg, and Kanoh and Yoshizaki is presented.

3.12.1 Yield-Line Method

This is an upper bound approach to obtain the flexural capacity. Under increasing load, yielding of the reinforcement will begin and progress to form lines of yielding until the slab is subdivided into segments that form a collapse mechanism. Since the angle changes across the lines of yielding are much greater than the elastic curvatures within the segments, the segments are assumed to remain plane, which means that the lines of yielding at their intersection may be considered to be straight lines.

For many slab geometries the exact configuration of the yield lines may not be known, and a series of possible patterns must be examined by trial and error The correct pattern is that corresponding to the lowest value of load required to produce a mechanism. Generally yield lines will occur along lines of fixed support, lines of symmetry, and, for slabs supported directly by columns, an axis passing through the column. Typical yield-line patterns are shown in Fig. 3-18.

Slabs are generally reinforced in two perpendicular directions, say x and y, with corresponding plastic moment capacities m_x and m_y. The bending moment, m, and twisting

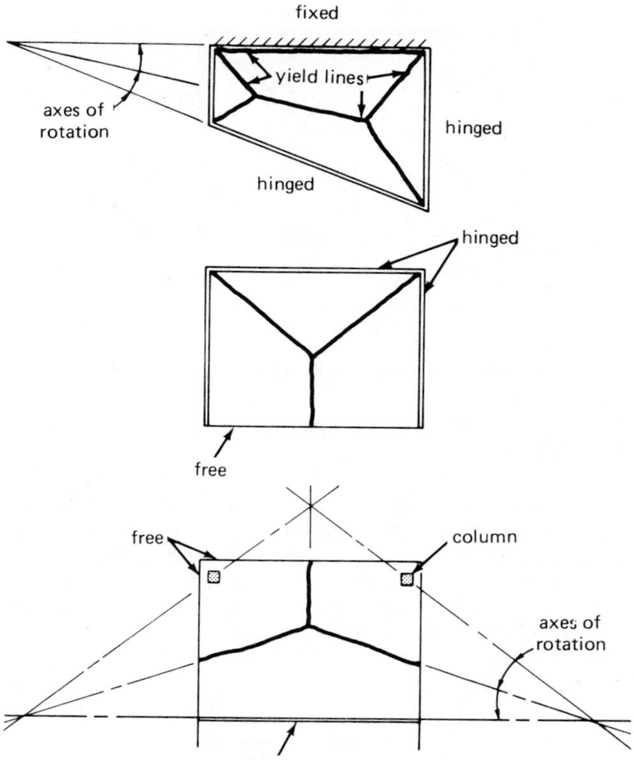

Fig. 3-18 Yield-line patterns.

moment, m_t, acting along a yield line making an angle θ with the x-axis, as shown in Fig. 3-19, are:

$$m = m_y \cos^2 \theta + m_x \sin^2 \theta$$

$$m_t = (m_y - m_x) \sin \theta \cos \theta$$

For the special case of an isotropic slab (that is, where $m_x = m_y$), $m = m_x = m_y$ and $m_t = 0$ for all values of θ.

To obtain the load required to form a selected yield pattern, two procedures are possible, equilibrium and energy. These are best presented in an example using the yield-line method.

Consider the simply supported rectangular slab in Fig. 3-20. For simplicity let us assume that the reinforcement provided gives a positive moment capacity of $m_x = m_y =$ 8000 ft.lb/ft at all locations. The problem is to determine the magnitude of the uniformly distributed load that will cause a collapse mechanism to form. A likely pattern of yield lines is shown, but the position of the intersection of the yield lines, designated at point 0, is not known precisely.

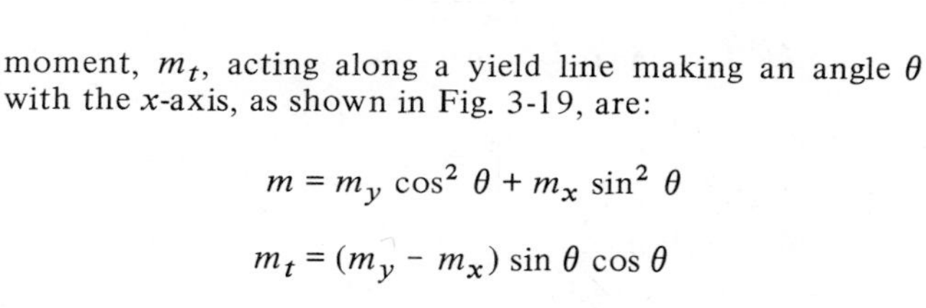

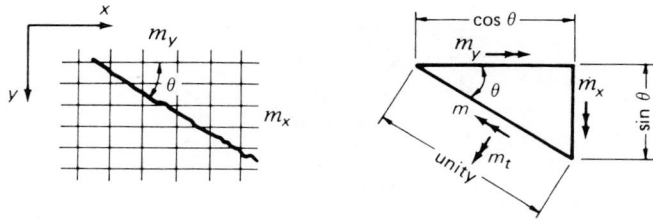

Fig. 3-19 Moment transformation.

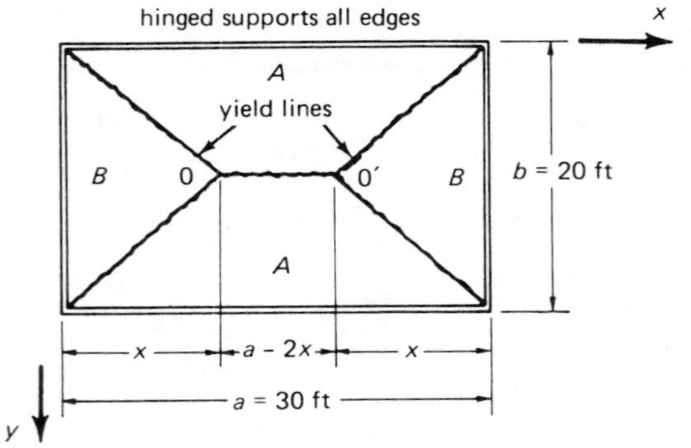

Fig. 3-20 Rectangular slab used in yield-line example.

From the symmetry the position of 0 can be defined in terms of the length x.

Equilibrium Procedure. This procedure involves computing the load required to obtain equilibrium for each segment for an assumed pattern of yield lines. If the load required for each segment is not the same, the pattern is altered until this condition is obtained. For our example the correct yield pattern can be obtained by assuming values of x until the load required for equilibrium is the same for each segment.

Trial 1: Assume $x = 10$ ft.

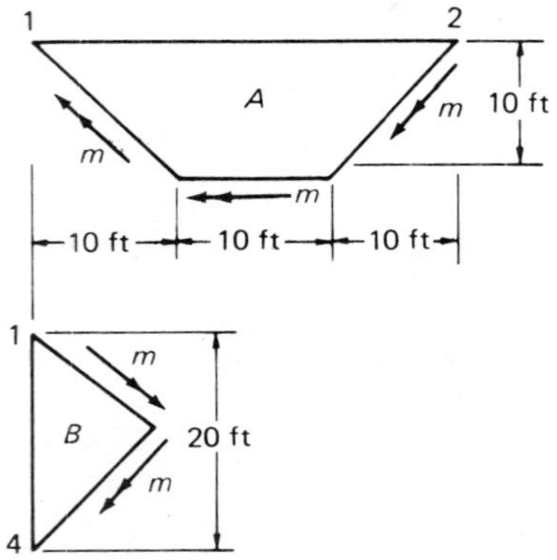

Equilibrium about line 1-2:

$$30 \times 8000 - 10 \times 10w_A \times \tfrac{10}{2} - 2(\tfrac{1}{2} \times 10 \times 10w_A \times \tfrac{10}{3}) = 0$$

$$w_A = 288 \text{ psf}$$

Equilibrium about line 1-4:

$$20 \times 8000 - \tfrac{1}{2} \times 20 \times 10w_B \times \tfrac{10}{3} = 0$$

$$w_B = 480 \text{ psf}$$

It is seen that the load, w_A, on segment A, required to satisfy equilibrium of this segment is not equal to the corresponding load on segment B. Since w_B is greater, we must increase the size of segment B.

Trial 2: Assume $x = 12$ ft

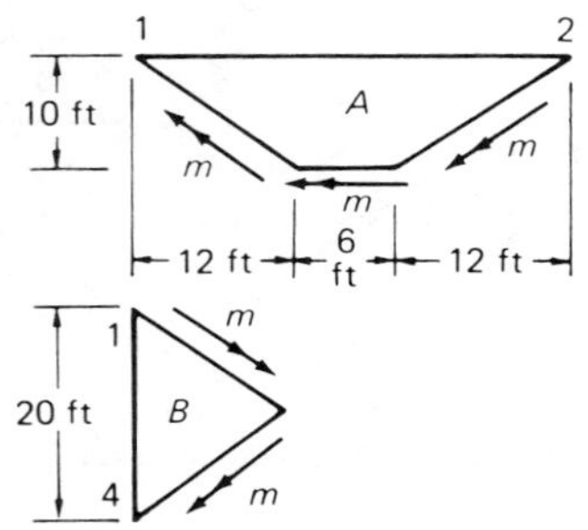

$$30 \times 8000 - 6 \times 10w_A \times \tfrac{10}{2} - 2(\tfrac{1}{2} \times 12 \times 10 \times w_A \times \tfrac{10}{3}) = 0$$

$$w_A = 343 \text{ psf}$$

$$20 \times 8000 - \tfrac{1}{2} \times 20 \times 12w_B \times \tfrac{12}{3} = 0$$

$$w_B = 324 \text{ psf}$$

The agreement between w_A and w_B in the second trial is seen to be much closer. A third trial can be made if closer agreement is desired. With $x = 11.9$ ft, $w = 340$ psf.

Energy Procedure. From the principle of virtual work, the work done by the external loading during an assumed displacement of the mechanism is equal to the work done by the resisting moments along the yield lines acting through the corresponding rotations. Although it is usual in practice to write the equation in terms of a trial numerical value of x and a unit displacement, the equation for our example will be first written algebraically to indicate the origin of the terms. δ is the displacement of line 0–0′.

Rotation of segments $A = \phi_A = \delta/b/2 = 2\delta/b$

Rotation of segments $B = \phi_B = \delta/x$

External work = Internal work

$$2\left[w(a - 2x)\,\frac{b}{2}\,\frac{\delta}{2} + 2w\left(\frac{xb}{4}\right)\frac{\delta}{3} + w\left(\frac{xb}{2}\right)\frac{\delta}{3}\right]$$

$$= 2[ma\phi_A + mb\phi_B]$$

Substituting numerical values $\delta = 1$, $a = 30$ ft, $b = 20$ ft, $m = 8000$ ft/lb/ft:

for $x = 10$, $116.67w = 5m$, $w = 342.9$ psf

for $x = 12$, $110w = 4.67m$, $w = 339.39$ psf

for $x = 11.9$, $110.33w = 4.68m$, $w = 339.38$ psf

It is immediately apparent that the energy procedure is much less sensitive than the equilibrium procedure to the precise locations of the yield lines. However, unlike the case of the equilibrium procedure, for a single trial you do not know whether you have obtained the lowest value of the load and in which direction you should alter the pattern to obtain a lower load. For this reason it is recommended that the first trial for any pattern type be done using the equilibrium method. Based on the result, the yield-line locations can be moved appropriately. For most problems, using the energy procedure for the revised pattern will give a loading with sufficient accuracy that further trials are not required.

The yield-line method can also be used successfully for concentrated loads or supports where the yield pattern will be in the form of a conical fan.[3-31] The application of the yield-line method for the design of flat plates is described by Simmonds and Ghali.[3-32]

3.12.2 Strip Method

The strip method is a design procedure in which the designer provides reinforcement for a bending moment field that is everywhere in equilibrium with the applied loading. The equilibrium equation for slabs[3-33] is:

$$\frac{\partial^2 m_x}{\partial_x^2} + \frac{\partial^2 m_y}{\partial_y^2} - 2\,\frac{\partial^2 m_{xy}}{\partial_x \partial_y} = -w$$

If the twisting moment, m_{xy}, is neglected, this equation uncouples to:

$$\frac{\partial^2 m_x}{\partial x^2} = -w_x; \qquad \frac{\partial^2 m_y}{\partial y^2} = -w_y$$

where $w_x + w_y = w$. These equations are recognized as those pertaining to beams so that in essence the slab moment field corresponds to the superposition of the moment fields for continuous beams or strips in the orthogonal directions. This simplification is justified and leads to reasonable designs for slab systems support such that each strip will be supported in a stable manner.

The designer must make two decisions; the first is the definition of the load carrying strips, and the second is the proportion of total load to be carried by each strip. Although these choices are arbitrary and all will lead to safe designs, the best choice is the one that leads to the most economical reinforcement requirements.

One selects the strips on the basis of how the load is to be transferred to the supports, keeping in mind that each strip must have sufficient supports to be stable, although strips can be supported by other strips. Load will be carried primarily in the direction of greatest stiffness. Since the two choices are closely related, the writer prefers to refer to the boundaries of the strips as load dispersion lines, which implies that the essential decision is to decide arbitrarily how the load will be carried.

Consider the simply supported rectangular slab in Fig. 3-20. Load near the edges is likely to be carried by that edge. Similarly, near the corners a portion of the load will be carried in each direction. Near midspan the manner in which the load is carried will depend on the rectangularity of the panel, i.e., equally in each direction for a square panel and increasing amounts in the short direction as the rectangularity increases.

Consider the load dispersion diagram in Fig. 3-21, where the load dispersion lines defining the strips are shown as dotted. Because of the rectangularity, load in the central portion is considered to be carried solely in the short direction. Moment diagrams are then shown for each strip corresponding to the load assigned to that strip.

Obviously other distributions of load could have been selected that would have resulted in other moment fields, all satisfying equilibrium but requiring varying amounts of reinforcement. The pattern used in the example results in a reinforcing pattern that is easy to place, and since the moment diagrams, except for strip d–d, are flat top, the reinforcement provided will be fully stressed over most of its length, a condition that leads to economy.

Hillerborg has extended the concepts of lower bound solutions to point supported slabs; the reader is directed to Ref. 3-28 for details.

3.12.3 Plastic Shear Design

The code[3-1] procedure for considering shear-moment problems is to assume a distribution of shear stresses and then limit the maximum stress to some specified value. An alternate approach is to evaluate the capacities of the various resisting mechanisms and to determine whether their sum is sufficient to carry the combined factored shear force and bending moment.

The method described here is attributed to Kanoh and Yoshizaki[3-30] as a simplified version of the beam analogies proposed by Hawkins. The price of the simplification is that although the failure load is predicted accurately, no indication of the likely failure mode is obtained.

This approach uses a critical section located $d/2$ from the column faces (Fig. 3-22) and assumes that the effects of shear and moment interact as follows:

$$\frac{V}{V_0} + \frac{M}{M_0} < 1.0$$

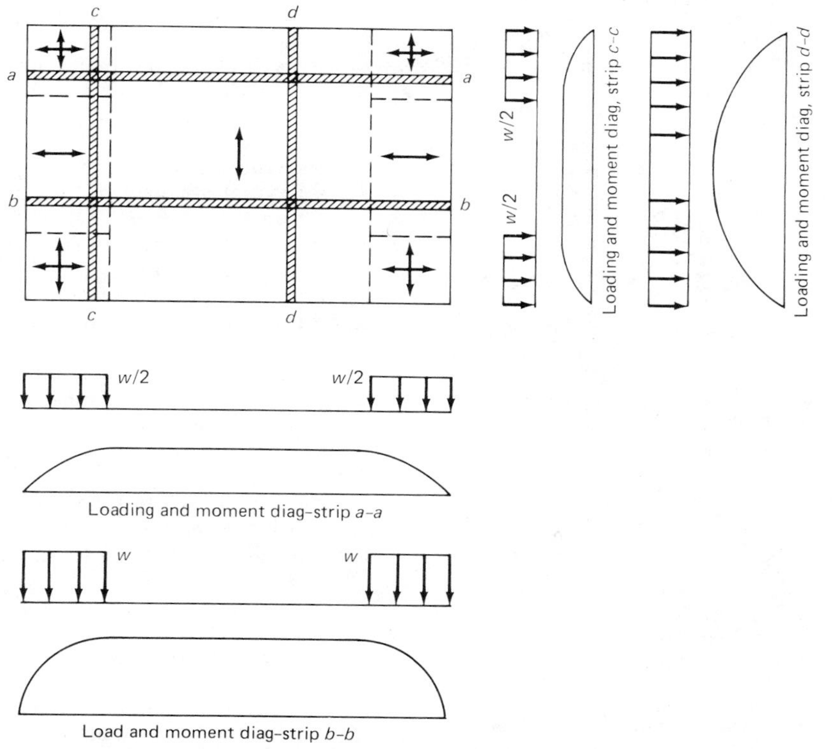

Fig. 3-21 Load dispersion diagram for strip method showing typical design strips.

where:

V = factored shear force transferred to column
M = factored moment transferred to column
V_0 = connection capacity in pure shear
M_0 = connection capacity in pure moment

The shear capacity, V_0, is obtained in the same manner as given in the code (see Fig. 3-22b), namely:

$$V_0 = \phi v_c A_0$$

where $\phi = 0.85$, $v_c = 2(1 + 2\beta_c)\sqrt{f_c'} \leqslant 4\sqrt{f_c'}$, and for an interior column $A = 2d(c_1 + c_2 + 2d)$.

The ultimate moment capacity, M_0, is given as:

$$M_0 = M_s + M_f + M_t$$

where:

M_s = portion transferred by eccentricity of shear, Fig. 3-22(c)
= $\phi v_c G_1$; ϕ and v_c as defined for V_0,
$G_1 = d(c_1 + d)(c_2 + d)$ for interior column

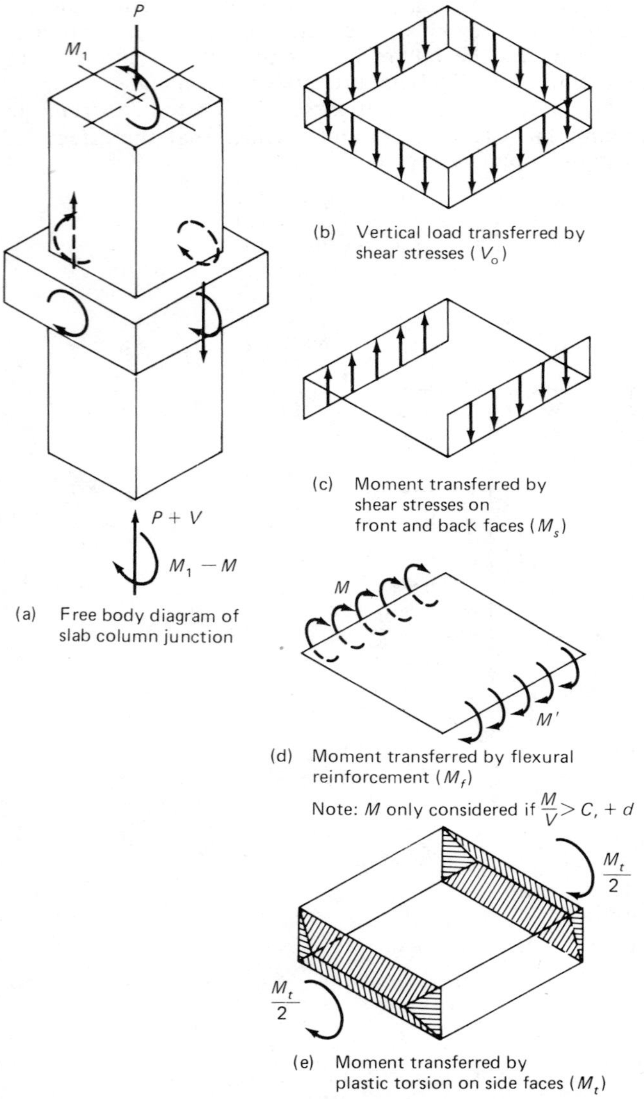

Fig. 3-22 Load transfer mechanisms for plastic shear-moment interaction equation.

M_f = portion transferred by flexural reinforcement, Fig. 3-22(d)
= $\phi A_s f_y$ (0.9d); $\phi = 0.9$ and A_s is for top steel in front face only, (i.e., M' in Fig. 3-22d) unless $M/V < c_1 + d$

M_t = portion transferred by twisting moments, Fig. 3-22(e)
= $\phi v_{tu} d^2 (c_1 + 2d/3)$; $\phi = 0.85$, $v_{tu} = 24\sqrt{f_c'}$

NOTATION

b_o = periphery of critical section for shear
b_w = width of beam web, in.
c = distance from centroid of support to critical section
c_1 = size of rectangular or equivalent rectangular column, capital, or bracket measured in the direction in which moments are being determined
c_2 = size of rectangular or equivalent rectangular column, capital, or bracket measured transverse to the direction in which moments are being determined
C = cross-sectional property to define torsional stiffness
C = coefficient for determining panel moment (Section 3.11.4)
d = distance from extreme compression fibre to centroid of tension reinforcement, in.
E_{cb} = modulus of elasticity for beam concrete
E_{cc} = modulus of elasticity for column concrete
E_{cs} = modulus of elasticity for slab concrete
f_c' = specified compressive strength of concrete, psi
f_y = specified yield strength of nonprestressed reinforcement
F_y = specified yield strength of structural steel (Section 3.11.5)
h = slab thickness, in.
h_v = total depth of shearhead cross section (Section 3.11.5)
I_b = moment of inertia about centroidal axis of gross beam section as defined in Fig. 3-2
I_c = moment of inertia of gross cross section of columns
I_s = moment of inertia about centroidal axis of gross section of slab
J = sum of the second centroidal moments of the shear area
K_b = flexural stiffness of beam; moment per unit rotation
K_c = flexural stiffness of column; moment per unit rotation
K_{ec} = flexural stiffness of an equivalent column; moment per unit rotation
K_s = flexural stiffness of slab; moment per unit rotation
K_t = torsional stiffness of torsion member; moment per unit rotation
l_n = length of clear span in long direction of two-way construction measured face to face of columns in slabs without beams and face to face of beams or other supports in other cases when determining minimum slab thickness
l_n = length of clear span, in the direction moments are being determined, measured face to face of supports
l_v = length of shearhead arm from centroid of reaction (section 3.11.3)
l_1 = length of span in the direction moments are being determined measured center to center of supports
l_2 = length of span transverse to l_1 measured center to center of supports
M = unbalanced moment transferred from slab to column
M_o = total static design moment in a panel
T_u = torsional design moment
v_c = nominal permissible shear stress carried by concrete

v_{tu} = nominal total design torsion stress
v_u = nominal total design shear stress
V_u = total applied design shear force at critical section
w_d = factored design dead load per unit area of slab
w_l = factored design live load per unit area of slab
w_u = factored design load per unit area of slab
x = shorter overall dimension of rectangular part of a cross section
y = longer overall dimension of rectangular part of a cross section
α = effective width factor (Section 3.11.1)
α = ratio of flexural stiffness of beam section (Fig. 3-2) to the flexural stiffness of a width of slab bounded laterally by the centerline of the adjacent panel, if any on each side of the beam
α_c = ratio of flexural stiffness of the columns above and below the slab to the combined flexural stiffness of the slabs and beams at a joint taken in the direction moments are being determined
α_m = average value of α for all beams on the edges of a panel
α_v = ratio of stiffness of shearhead arm to surrounding composite slab section (Section 3.11.5)
α_1 = α in the direction of l_1
α_2 = α in the direction of l_2
β = ratio of clear spans in long to short direction
β = effective stiffness factor (Section 3.11.1)
β_a = ratio of dead load per unit area to live load per unit area (in each case without load factors)
β_s = ratio of length of continuous edges to total perimeter of a slab panel
β_t = ratio of torsional stiffness of edge beam section to the flexural stiffness of a width of slab equal to the span length of the beam, center to center of supports
δ_s = factor by which the positive moments in a panel are increased to account for pattern loading
ϕ = capacity reduction factor
= 0.85 for shear
= 0.90 for flexure

REFERENCES

3-1 American Concrete Institute, ACI Standard 381-83, *Building Code Requirements for Reinforced Concrete*, ACI 381-83, American Concrete Institute, Detroit, Michigan, 1983.

3-2 American Concrete Institute, *Commentary on Building Code Requirements for Reinforced Concrete*, ACI 318-83, Detroit, Michigan, 1983.

3-3 Ang. A. H. S., and Prescott, W. S., "Equations for Plate-Beam Systems in Transverse Bending," *Proceedings, ASCE*, V. 87 (EM6), December 1961.

3-4 Zienkiewicz, O. C., *The Finite Element Method*, 3rd Ed., McGraw-Hill, New York, 1977, pp. 226–267.

3-5 Cook, R. D., *Concepts and Applications of Finite Element Analysis*, 2nd Ed., John Wiley, New York, 1981, pp. 244–270.

3-6 Fintel, M., and Ghosh, S. K., *Economics of Long Span Concrete Slab Systems for Office Buildings—A Survey*, Portland Cement Association, Skokie, Illinois, 1981.

3-7 McMillan, A. B., "A Comparison of Methods for Computing the Strength of Flat Reinforced Plates," *Engineering News*, V. 63 (13), Mar. 1910, pp. 364–367.

3-8 Nichols, J. B., "Statical Limitations Upon the Steel Requirement in Reinforced Concrete Flat Slab Floors," *Transactions, ASCE*, V. 77, 1914, pp 1670–1736.

3-9 Westergaard, H. M., and Slater, W. A., "Moments and Stresses in Slabs," *Proceedings, ACI*, V. 17, 1921, p. 415.

3-10 Sozen, M. A., and Siess, C. P., "Investigation of Multiple-Panel Reinforced Concrete Floor Slabs," *Proceedings, ACI*, V. 60 (8), Aug. 1963, pp. 999–1027.

3-11 Gamble, W. L., Sozen, M. A., and Siess, C. P., "Test of a Two-Way Reinforced Floor Slab," *Proceedings, ASCE*, V. 95 (ST6), June 1969, pp 1073–1096.

3-12 Jirsa, J. O., Sozen, M. A. and Siess, C. P. "Pattern Loadings on Reinforced Concrete Floor Slabs," *Proceedings ASCE*, V. 95 (ST6), June 1969, pp. 1117–1137.

3-13 Hatcher, D. S. Sozen, M. A., and Siess, C. P., "Test of a Reinforced Concrete Flat Slab," *Proceedings, ASCE*, V. 95 (ST6), June 1969, pp. 1051–1072.

3-14 Gamble, W. L., Sozen, M. A., and Siess, C. P., "Measured and Theoretical Bending Moments in Reinforced Concrete Floor Slabs," Civil Engineering Studies, *Structural Research Series* No. 246, University of Illinois, June 1962.

3-15 Simmonds, S. H., and Siess, C. P., "Effects of Column Stiffness on the Moments in Two Way Floor Slabs," Civil Engineering Studies, *Structural Research Series* No. 253, University of Illinois, July 1962.

3-16 Corley, W. G., and Jirsa, J. O., "Equivalent Frame Analysis for Slab Design," *Proceedings, ACI*, V. 67 (11), Nov. 1970, pp. 875–884.

3-17 Simmonds, S. H., and Misic, J., "Design Factors for Equivalent Frame Method," *Proceedings, ACI*, V. 68 (11), Nov. 1971, pp. 825–831.

3-18 Simmonds, S. H. "Effects of Supports on Slab Behavior," presented at ASCE National Structureal Engineering Meeting, Cleveland, April 1972 (meeting preprint 1697).

3-19 *Continuity in Building Frames*, Portland Cement Association, Skokie, Illinois, 1959.

3-20 Cardenas, A. E., and Kaar, P. H., "Field Test on a Flat Plate Structure," *Proceedings, ACI*, V. 68 (1), Jan. 1971, pp. 50–59.

3-21 Vanderbilt, M. D., "Equivalent Frame Analysis of Unbraced Reinforced Concrete Buildings for Static Lateral Loads," Structural Report No. 36, Dept. of Civil Engineering, Colorado State University, Fort Collins, July 1981.

3-22 Pecknold, D. A., "Slab Effective Width for Equivalent Frame Analysis," *Proceedings, ACI*, V. 72 (5) April 1975, pp. 135–137. Discussion by Allen Darvall, Robert Glover, and Author, October 1975, pp. 583–586.

3-23 Fintel, M. "Deflections of High-Rise Concrete Buildings," *Proceedings, ACI*, V. 75 (7), July 1975, pp. 324–328.

3-24 Woodring, R. E., and Siess, C. P., "An Analytical Study of the Moments in Continuous Slabs Subjected to Concentrated Loads," Civil Engineering Studies, *Structural Research Series* No 264, University of Illinois, May 1963.

3-25 Johanson, K. W., *Pladeformler* (in Danish), Polyteck-nisk Forening Copenhagen, 1949. See also *Yield-Line Theory*, English translation, Cement and Concrete Association, London, 1962.

3-26 Wiesinger, F. P., "Design of Flat Plates with Irregular Column Layout," *Proceedings, ACI*, V. 70 (z), Feb. 1973, pp. 117–123.

3-27 Hillerborg, A., "Jamviksteori for armerade betong-platter" *Betong*, V. 41 (4) 1956, pp. 171–182. See also Building Research Station, Translation LC1082.

3-28 Hillerborg, *A Strip Method of Design*, Viewpoint Publications, Cement and Concrete Association, London, 1975, 256 pp.

3-29 Hawkins, N. M. "Shear and Moment Transfer Between Concrete Flat Plates and Columns," Progress Report on NSF Grant No. GK-16375, Dept of Civil Engineering, University of Washington, Seattle, 1971.

3-30 Kanoh, Y., and Yoshizaki, S., "Strength of Slab-Column Connections Transferring Shear and Moment," *Proceedings, ACI*, V. 76 (3), Mar. 1979, pp. 461–478.

3-31 Gesund, H., "Design for Punching Strength of Interior Columns," *Advances in Concrete Slab Technology*, Pergamon Press, New York, 1979.

3-32 Simmonds, S. H., and Ghali, A., "Yield-Line Design of Slabs," *Journal of the Structural Division, ASCE*, V. 102 (STI), Jan. 1976, pp. 109–123.

3-33 Timoshenko, S., and Woinowsky-Krieger, S., *Theory of Plates and Shells*, McGraw-Hill, New York, 1959.

4

Joints in Buildings

MARK FINTEL*

4.1 INTRODUCTION

Concrete is subject to changes in length, plane, and volume caused by changes in its temperature and moisture content, reaction with atmospheric carbon dioxide, or the imposition or maintenance of loads. The effects may be permanent contractions, due, for example, to initial drying shrinkage, carbonation, and irreversible creep. Other effects are transient and depend on environmental fluctuations in humidity and temperature, or the application of loads, and may result in either expansions or contractions.

The results of these changes are movements, both permanent and transient, of the extremities of concrete elements. If contraction movements are restrained, then cracking may occur within the unit. The restraint of expansion movement may result in crushing within the unit, or crushing of its ends in the transmission of unanticipated forces to abutting units. In most concrete structures these effects are objectionable from a structural or an appearance viewpoint; one of the means of handling them is to provide joints at which movement can be accommodated without loss of integrity of the structure.

The occurrence of cracks in concrete construction due to volume changes has long been a problem. It is generally accepted that some crack formation in slabs and exterior walls is unavoidable, except in very low structures. However, much can be done to reduce or control cracking. Steel reinforcement can be used to distribute cracks. Whether or not these distributed cracks can be seen readily depends upon the magnitude of the volume change, the percent of reinforcement, and the size and shape of the structure.

The designer usually wants to encourage as many narrow, closely spaced cracks as possible in reinforced flexural members. Wide cracks are objectionable for aesthetic reasons, and because they permit the entrance of water or aggressive solutions that might corrode the reinforcement. Since the sum of all the crack widths is more or less constant and determinate, the width of a crack is inversely proportional to the number of cracks that can be encouraged to form in a certain length.

The problem of cracking cannot be ignored. It is handled either by hiding the cracks in preformed grooves, or by using reinforcement to ensure a large number of hairline cracks.

The use of joints, particularly in large buildings, is, therefore, inevitable, and rarely do we find a concrete structure built without the inclusion of either construction joints, control joints, expansion joints, shrinkage strips, isolation joints, or a combination of these. Although joints are placed in concrete so that cracks do not occur elsewhere, it is seemingly almost impossible to prevent occasional cracks between joints. The provision of joints to take up the movement occurring in concrete is a subject that does not always receive the consideration its importance should demand. Often the decision of whether or not to provide the joint becomes a matter of opinion, while in most cases it can be the result of a logical consideration of effects. Because of their great influence on correct detailing of the job, on the progress schedule, and on the appearance of the finished building, the location of joints must not be left to chance.

4.2 MOVEMENTS IN STRUCTURES

To understand the action of the various types of joints and the demands put on them by the structure, it is necessary to have some knowledge of the movements to which concrete structures are subjected during their life.

The movements in hardened concrete that can cause cracking can originate from:

*Consulting Engineer, formerly Director, Advanced Engineering Services Department, Portland Cement Association, Skokie, Illinois.

(a) The properties inherent in concrete as a material that are independent of the type of structure; these properties include shrinkage.
(b) Movements depending on the type of structure and consisting of effects of all imposed loads such as self-weight and lateral loads of wind and earthquakes. Such movements may be deflections, elastic strains, and strains due to creep caused by permanent loads or by the applied prestressing forces.
(c) Movements depending on the location of the structure caused by changes in temperature and humidity. The severity of the environment indicates the magnitude of the movement to be expected which is influenced by the effects of relative exposure to sun and prevailing winds.

Only some of the foregoing qualitative considerations of movements can be considered quantitatively by designers. If the individual movements are known, the final effects can be considered by adding them together. However, the accuracy of such assessment may be very poor. Some of the basic parameters needed to compute the individual movements are still poorly defined; for example, the modulus of elasticity of concrete, under conditions of stress close to the state of rupture. The magnitude of movements may frequently be rendered more uncertain by the presence of restraints such as friction and the interaction of structural elements which are sometimes not measurable. This is more true of complex structures such as large buildings, than of more simple structures such as bridges. Fortunately, however, limited quantitative accuracy can be accepted, since the overall knowledge of the movement in complex structures is of more interest to the designer than its actual magnitude. Knowledge of overall movements allows provision of details to accommodate the movements and relieve a buildup of stresses that may otherwise cause distress. Only rarely is it possible to predict accurately the amount of movement in a complex structure, and a precise prediction of the magnitude of a number of superimposed movements is imprudent. However, an exact analysis is sometimes possible, and in a simple structure may be of practical value. But in a complex structure, calculations should be relied upon only to a limited extent, because while theoretically the magnitude of the movements can be obtained, those movements are sure to differ from the reality due to fluctuations of environmental conditions, the various effects of restraint of the structure that are not considered in the analysis, and the variation of the shrinkage and creep properties of the concrete. It is of value, however, to consider qualitatively the effects of the various movements and to assess their magnitude for comparison of different construction processes.

Only movements due to shrinkage and creep are specific to concrete as a material; all other movements, be they due to loads or in response to temperature variations, are characteristic of structural steel as well.

4.3 TENSILE STRESSES LEADING TO CRACKING AND THEIR ORIGIN

Cracking of a concrete section occurs when the tensile stress acting on the section is larger than its tensile strength, or when tensile strain exceeds the tensile strain capacity. Since the tensile strength of concrete may be only about one-tenth of its compression strength, its tendency is to crack under relatively low tensile stresses. It should be appreciated that strains due to shrinkage, creep, and temperature variations do not result in stressing of the material in the structural sense unless restraint against free movement is provided. In floor construction, for example, external means of restraint against volume changes may be caused by columns and walls, while internal restraint is caused by the reinforcement, particularly against shrinkage and creep strains, and, in case of slabs on the ground, by subgrade restraint.

The actual mechanism by which cracks form may be quite complicated, but the basic causes involved are straightforward. If a structure were freely supported in space, if all its parts had the same rate of volume change due to shrinkage and temperature, and if, furthermore, all parts of the structure were exposed to the same atmospheric conditions, no differential volume changes and therefore no cracks would result. In an actual structure, however, restraint may be applied to beams or slabs in lower stories of buildings by rigid columns and by the foundation, since volume changes in the foundation are minimized by the insulating effect of the surrounding soil. In higher stories, as the restraint is diminished, cracking is reduced.

Observations show that in buildings with basements, or large heavy foundations under walls, the most numerous cracks are in the first-story walls, due to the restraining action of the portion of the building below ground where shrinkage is the least. Buildings with freestanding columns in the first story will crack less because the freestanding columns will accommodate the movements due to differential shrinkage between the basement and the stories above.

The use of different materials in conjunction with each other also may lead to cracking, since different coefficients of expansion and/or shrinkage result in movements of one material relative to the other.

Buildup of tensile stresses resulting from restrained shortening may be due to shrinkage and temperature drop. It has long been accepted that within reasonable limits contraction stresses are relieved by the beneficial effects of creep. This is true in many cases, but in others it occurs by chance, and not by good judgment.

4.3.1 Shrinkage

Shrinkage during the curing and drying of concrete is unavoidable unless the concrete is submerged in water. The amount of shrinkage varies with the mix and with atmospheric conditions. A typical coefficient of shrinkage for average thickness structural slabs may be up to about 600×10^{-6} in./in. Thus, for a 100 ft long slab, the contraction could be about $^3/_4$ in. The rate at which shrinkage occurs depends on the rate of loss of moisture. This rate is important, since a slow loss will allow time for the concrete to gain strength, and also to undergo a certain amount of plastic flow. More rapid drying will greatly reduce these beneficial effects. Thus, for any particular slab, a more rapid moisture loss (i.e., drying at higher air temperature and lower atmospheric humidity) will result in an increase in the width and number of shrinkage cracks.

Cracks may also result from differential moisture and temperature gradients. Tensile stresses are thus induced in the outer skin of the concrete mass when the surface is shrinking and cooling more rapidly than the interior. This effect is not too critical for thin structures such as slabs and walls, but it can be a major consideration in massive concrete work or thick pavements. Since the rate of advancing shrinkage diminishes as a square from the depth of the drying surface, the thick element may never dry out and reach its ultimate shrinkage because during the changes of seasons it will be absorbing moisture, while the thin element will dry out and crack before the change in season arrives. However, the thick element drying from one face only may crack at the outer face because of internal restraint, but the cracks most likely will be shallow. A building with exterior columns and 6 in. thick walls with openings will have more "leaky" cracks than a building with 12 in. thick bearing walls with similar openings.

4.3.2 Temperature Changes

The stresses introduced in concrete by temperature variations can lead to serious cracking. A reduction in temperature is more serious than an increase because the stresses induced are tensile and because they combine with shrinkage stresses. Also, the maximum temperature of the concrete during the first day controls the amount of subsequent expansion and contraction of the concrete. The higher the temperature during the first day, the greater the contraction during cold weather. Consequently, from this standpoint alone, it is preferable to place concrete in cool weather. Under these conditions, the difference between the temperature of the concrete after curing and drying, and the lowest temperature to which it will be subjected, will be reduced.

The amount of thermal expansion and contraction of concrete varies with factors such as type and amount of aggregate, richness of mix, water–cement ratio, temperature range, concrete age, and degree of saturation of concrete. Of these, aggregate type has the greatest influence. A typical value is on the order of 5 to 6×10^{-6} in./in. per °F, i.e., very nearly the same as for steel. If an unrestrained slab or wall 100 ft long has a temperature variation from summer to winter of 100°F, the total thermal movement might be about 0.6–0.7 in. Movements occur at the exposed surface of the concrete, which cools off more quickly, before they occur in the interior of the section, leading frequently to additional warping or curling effects. Observations of buildings in service indicate the total movement is usually less than half of that which might be anticipated by combining the contraction due to temperature drop with the shrinkage. This is due to restraining effects of the reinforcing steel and restraining effects of columns, walls, and foundation.

4.4 REINFORCING STEEL

The presence of reinforcement produces internal restraint and, thereby, reduced movement of the concrete but not necessarily an elimination of cracking. The attempt of the concrete to reduce in size places the steel in compression and concrete in tension. A highly reinforced section approaches the condition of full restraint and can cause cracking in the concrete. Average reinforced sections (under 2% steel) usually have an apparent shrinkage potential of 0.02 to 0.04%, compared to 0.03 to 0.08% for plain concrete.

The presence of properly distributed reinforcement causes numerous small cracks to occur, rather than a few wide cracks. The reinforcement, whether as a means of controlling cracks, or provided for structural reasons, can partially resist some of the shortening. Prestressing which can prevent cracks from occurring can play an important part in the control of shrinkage. It must be noted, however, that reinforcement cannot suppress entirely length changes of concrete from whatever causes.

The ACI 318-83 *Building Code* recommends that the following minimum ratios of shrinkage and temperature reinforcement be provided perpendicular to the main reinforcement in structural floors and roof slabs:

1. Where grade 40 and 50 deformed bars are used — 0.0020
2. Where grade 60 deformed bars or welded wire fabric, deformed or plain, are used — 0.0018

4.5 TYPES OF JOINTS

It is convenient to categorize into two groups the many types of joints in reinforced and plain concrete structures that have been devised to accommodate construction needs and the various movements.

4.5.1 Construction Joints

These joints are installed to break up the structure into smaller units in accordance with the production capacity of the construction site. True construction joints are not designed to provide for any movements, but are merely separations between consecutive concreting operations.

4.5.2 Movement (Functional) Joints

These joints are installed to accommodate volume changes:

1. Expansion joints
2. Control (contraction) joints
3. Shrinkage strips

Movement joints serve to prevent restraints that would otherwise occur as a result of differences in deformation of the adjacent parts. In some cases, such joints are interposed between the structure and its foundations, since the superstructure deforms owing to external loads, internal forces, and temperature variations while the foundation normally remains immovably secured in the ground.

Since properly functioning joints are usually expensive to build, it is desirable to install one joint to serve a dual purpose whenever possible. By the very nature of its construction, an expansion joint acts also as a contraction joint; and obviously expansion joints can also function as construction joints. A further saving in the number of joints required may be effected by arranging to have a contraction joint coincide with a construction joint, thus eliminating one joint.

4.6 JOINT SPACING

Spacing of a functional (contraction and expansion) joints depends upon a great number of factors: shrinkage properties of the concrete, type of exposure to temperature and humidity, resistance to movement (restraint), thickness of members, amount of reinforcement, structural function of the member, external loads, soil conditions, structural configurations, and other conditions. Many of these factors are elusive variables, sometimes difficult to establish. As a consequence, both experience and opinion on joint spacing vary greatly.

In reinforced concrete elements, joint spacing and reinforcement are interrelated variables, and the choice of one should be related to the other. As yet, however, a reliable relationship between the two quantities does not appear to have been established. Sufficient steel must be included to control cracking between the joints. If the joint spacing is increased, the reinforcement must be increased correspondingly to control cracking over the longer distance.

The shape of a building has a definite effect on joint locations. Any change in direction in such buildings shaped as T, L, and Y may require a close examination of the necessity of joints at the junctions that usually create stress con-

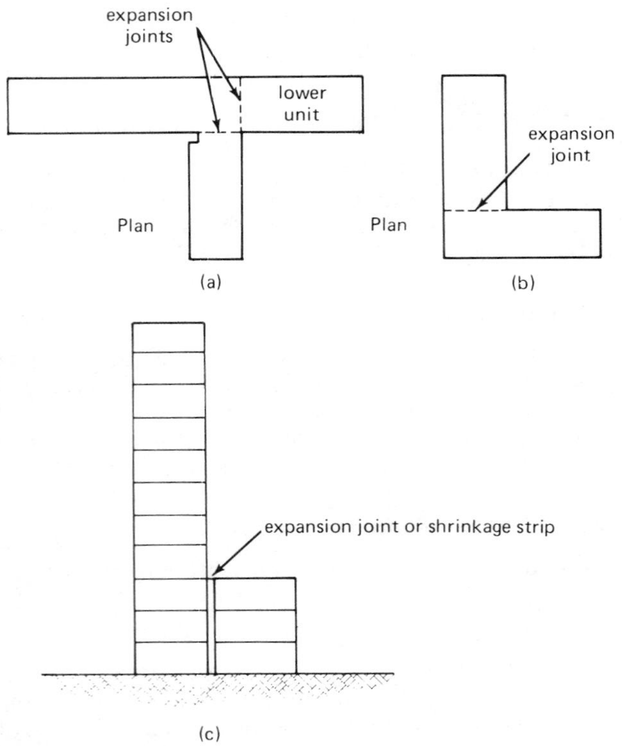

Fig. 4-1 Joints related to the shape of buildings. (a) and (b) Joints are placed at major changes of building run; in (c) joint is placed at junction of tall and low buildings.

centrations. Also, at the junction between tall and low buildings, the differential settlements require a stress relief mechanism, in the form of either an expansion joint or a shrinkage strip (Fig. 4-1).

Joints may also be needed at any location at which stress concentrations may occur, such as large openings in walls or slabs, changes in thickness of walls, slabs, etc. In each of these cases a sufficient amount of reinforcement (in lieu of joint) across the potential crack may prevent cracking.

The type of construction and the climate are also important in determining the spacing of movement joints. For example, a building that has uninsulated walls or is unheated must have joints at more frequent intervals than a heated building. Thermal movements are reduced when an insulation layer is placed on the exterior face of the structure.

There is a considerable divergence of opinion on spacing of movement joints (expansion and contraction) with recommendations for expansion joints varying from 100 to 200 ft, while for contraction joints, they vary from a few feet up to 80 ft.

The distance between construction joints depends on the production capacity of the construction site, being limited either by the formwork or casting capacity. When the distance between construction joints becomes too large, intermediate control joints are introduced.

4.7 NONSTRUCTURAL ELEMENTS

The need for movement joints must be investigated in the building as a whole and not in the structure alone. For the owner and user of the building it is not enough that the structural frame performs to the selected criteria when the doors do not close, the windows leak, the partitions crack, and the external cladding buckles away from the structure. The completed building, and not the structural frame alone, is the responsibility of the professionals involved in the design process.

To prevent development of distress in the nonstructural elements, it is often necessary to limit the extent of their movement independently of overall expansions, contractions, and deflections occurring in the concrete frame.

Brittle partitions and finishes are particularly sensitive to deflections of beams supporting them. Since the deflections due to creep and shrinkage are dependent on time, it is advisable to give consideration to controlling the time of installation of fragile partitions and finishes connected to concrete beams. For instance, a certain concrete beam deflects 0.25 in. immediately after removal of the shores. Eventually it may deflect an additional 0.50 in. due to creep and shrinkage. However, half of the above added deflection will take place during a period of about two months after removal of the shores. Thus, only 0.25 in. of the deflections will be left to affect the nonstructural fragile elements, if they are installed two months after removal of the shores.

If brittle partitions are used in flexible frames, or if the frames have exposed columns that move up and down in response to temperature variations, the partitions must either be separated from the frame or have the same flexibility as the frame. Such flexibility would allow the partitions to follow the distortions of the frame without being distressed and cracked. Unless intentionally separated from the frame, the partitions will distort with the building and contribute to the rigidity of the structure in resisting any movements of the frame.

To leave the partitions unaffected from the distortion of the structural frame, details around the edges of partitions should be provided to allow vertical as well as horizontal slippage. One of the simplest ways to achieve this is to provide a channel enclosure for partition walls, where the partitions meet the columns and ceiling. The partition details shown in Fig. 4-2, originally suggested by the dry wall partition manufacturers, have been extensively used in the design of many projects. This detail allows a partition to float and provides the necessary restraint against lateral loads acting on the partition.

4.8 JOINTS IN CLADDING

A large number of high-rise reinforced concrete buildings are being built with the exterior columns and end shearwalls clad with clay masonry or natural stone, such as travertine or granite. Although in the traditional buildings of the early twentieth century such cladding was detailed by the architects themselves, most of these buildings were made of structural steel, and, therefore, did not have to consider shrinkage and creep of the frame after construction. Also the ability of the heavy cladding to carry substantial loads reduced stresses and movements considerably.

External cladding supported by a multistory structural frame is a typical example of the need to examine vertical as well as horizontal movements. Exterior clay masonry cladding as well as thin stone cladding has been known to suffer from movement distress and cause buckling failures. Such failures in the mid-fifties were initially attributed to the shrinkage and creep shortening of the lightweight concrete columns and walls of the structural frame. Further study, however, showed that the expansion of the brickwork was a factor additive to the shortening of the concrete. Expansive movement in clay brickwork is caused mainly by gain in moisture content over a period of years after the bricks leave the kiln. Whereas clay brick expands

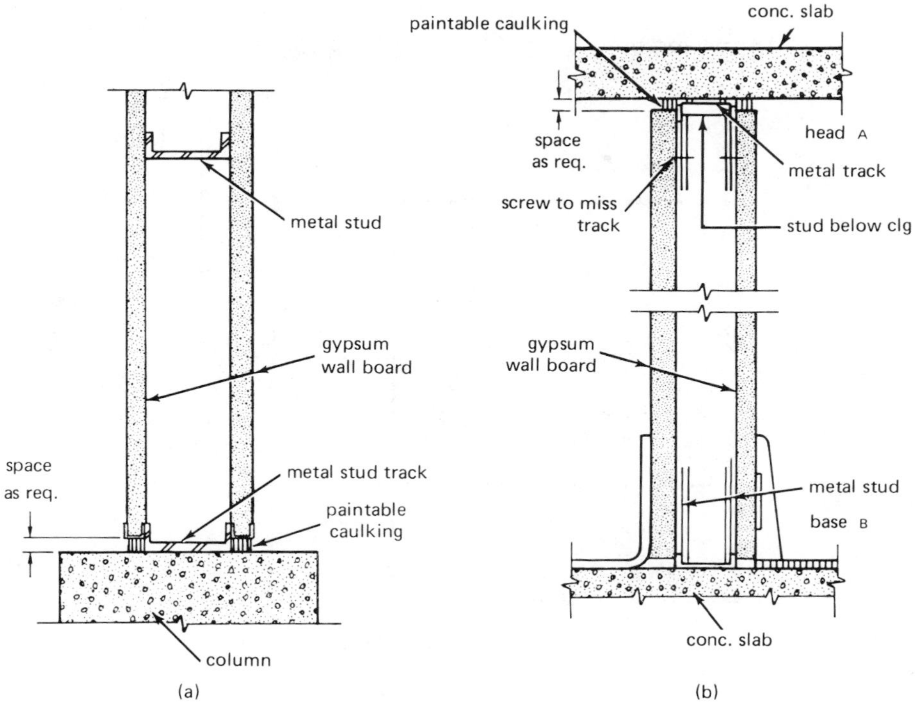

Fig. 4-2 Details of floating partitions. (a) Joint at exterior column, plan view; and (b) joint at slab and ceiling, section view.

because of a gain in moisture after leaving the kiln, concrete and concrete block lose moisture and conseqeuntly shrink in volume when exposed to the atmosphere. Clay bricks start to expand as soon as they are removed from the kiln, and their rate of expansion decreases with time. Clay bricks that go directly from the kiln into walls will, therefore, have a much larger expansion potential than similar bricks that have aged a few months. Thus, the amount of expansion to be provided for, i.e., width and spacing of expansion joints, depends on the time that has elapsed since the bricks were removed from the kiln.

Brickwork may expand as much as $^1/_2$ to 3 in. in 100 ft. Therefore, vertical expansion joints are also needed to accommodate horizontal expansion.

No single recommendation on the positioning and spacing of expansion joints can be applicable to all structures. Each building design should be analyzed to determine the potential movements, and provisions should be made to relieve excessive stress that might be expected to result from such movement.

Total unrestrained temperature expansion of clay masonry walls may be estimated from the formula:

$$w = [0.0002 + 0.000004(T_{max} - T_{min})]\ L$$

where:

L = length of wall in in.
T_{max} = maximum mean wall temperature in °F
T_{min} = minimum mean wall temperature in °F
w = total expansion of wall in in.

However, this will be reduced by indeterminate compensating factors such as restraint, shrinkage, and plastic flow of mortar, and variations in workmanship. The recommended spacing of expansion joints in masonry walls seems to range between 20 and 75 ft and is affected by the severity of the environment and by the tensile and shear strength of the walls. The means provided for resisting differential movement also affect the joint spacing.

When cladding attached to a multistory frame (columns or walls) expands and/or the frame shortens owing to shrinkage and creep, the vertical load is transferred from the columns or walls to the cladding until crushing or buckling of the cladding occurs. This problem can be eliminated by providing horizontal expansion joints at every floor level or every alternate floor level at shelf angles as shown in Fig. 4-3. This joint should consist of a compressible material or a clearance immediately below the angle. The joint should be sealed with a mortar-colored elastic sealant. The shelf angles should be secured against any rotation and against deflections over $^1/_{16}$ in. A small space ($^1/_2$ in.) should be left between length of angles to allow for thermal movement. A joint width of $^1/_4$ in. per story should be adequate for most cases to accommodate the elastic shortening of the column caused by progress of construction after placing the cladding, creep, and shrinkage effects of the column, and the expansion of the cladding due to extreme summer temperature and possible moisture absorption. The individual contribution of each of these effects is as follows, assuming a 12 ft (144 in.) story height:

$$\begin{aligned} \text{elastic strain} &= 200 \times 10^{-6} \times 144 &&= .0288 \text{ in.} \\ \text{creep} &= 300 \times 10^{-6} \times 144 &&= .0432 \text{ in.} \\ \text{shrinkage} &= 450 \times 10^{-6} \times 144 &&= .0650 \text{ in.} \\ \text{temperature} &= 70^\circ\text{F} \times 6 \times 10^{-6} \times 144 &&= \underline{.0605 \text{ in.}} \\ & \text{total} &&= .1974 = 3/16 \text{ in.} \end{aligned}$$

Vertical joints between individual stone or precast cladding units should be provided to accommodate thermal expansion of the cladding and the distortions of the frame due to lateral forces.

4.9 CONSTRUCTION JOINTS

Construction joints are stopping places in the process of placing concrete, and are required because it is impractical to place concrete in a continuous operation, except for very

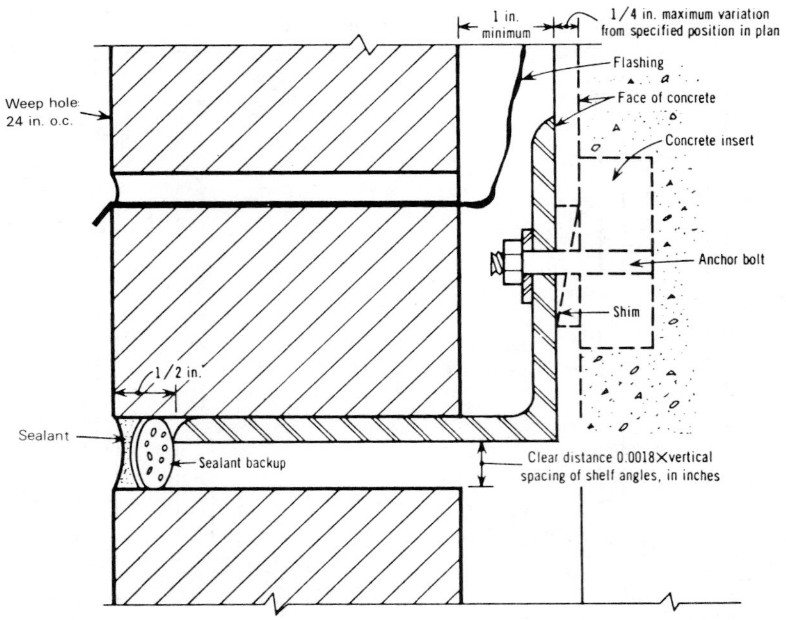

Fig. 4-3 Horizontal expansion joint at shelf angle.

small structures or special types of structures built with slip forms. Construction joints should not be confused with expansion joints, which, if considered necessary, will allow for free movement of parts of a building and should be designed for complete separation.

The main problem in the formation of a good construction joint is that of obtaining a well-bonded watertight joint between the hardened and the fresh concrete. For a sound joint, the reinforcing should be cleaned, and the aggregate of the hardened concrete should be exposed by brushing, waterblasting, or sandblasting before placing the new concrete.

If there should be any doubt as to the adequacy of the bond between the old and new concrete, the reinforcement crossing the construction joints should be supplemented by dowels.

The simplest type of construction joint is a butt type formed by the usual bulkhead board, as in Fig. 4-4(a). This joint is suitable for thin slabs.

Slabs can use a type of joint that resembles tongue and groove lumber construction. The keyway may be formed by fastening metal, wood, or premolded key material to a wood bulkhead. Concrete above the joint should be hand-tooled or saved to match a control joint in appearance. The second placing of concrete later enters the groove to form the tongue and thus allow for shear forces to be transmitted through the joint, as in Fig. 4-4(b). In plain slabs on ground this ensures that future slabs will remain level with previously cast concrete.

Joints should be made straight, exactly horizontal or vertical, and should be placed at suitable locations. In walls, horizontal construction joints can be made straight by nailing a 1 in. wood strip to the inside face of the form (Fig. 4-4c). Concrete is then placed to a level about $\frac{1}{2}$ in. above the bottom of the strip. After the concrete has settled and just before it becomes hard, any laitance which has formed on the top surface is removed. The strip is then removed, and irregularities in the joint are leveled off.

The forms are usually removed at construction joints and then re-erected for the next lift of concrete, as illustrated in Fig. 4-4(c). A variation of this procedure is to use a rustication strip instead of the 1 in. wood strip and to form a groove in the concrete for architectural effect. Rustication strips may be V-shaped or rectangular with a slight bevel. If V-shaped, the joint should be made at the point of the V. If a rectangular rustication strip is used, the joint should be made at the top edge of the inner face of the strip.

Continuous or intermittent keyways in either vertical or horizontal construction joints of reinforced slabs or walls are of questionable value. While they seemingly contribute little added resistance to the joints, they may contribute to spalling, and they interfere with getting the best quality of concrete and maximum strength at the joint. When proper concreting procedures are followed, the bond between the old and new concrete plus the doweling effect of the reinforcement can be made good enough to provide shear resistance equivalent to that of concrete placed monolithically.

Joints should be perpendicular to the main reinforcement. All reinforcement should be continued across construction joints. In providing construction joints it is essential to minimize the leakage of grout from under stop-end boards. If grout does escape, forming a thin wedge, it should be removed before subsequent concreting commences to avoid weakening the structure. For watertightness a continuous waterstop of plastic, copper, or rubber is essential. In wall construction and other reinforced concrete work, it may not be convenient to sandblast or to use water jets for cleaning joint surfaces. Good results have been obtained by constructing the form to the level of the joint, overfilling the forms an inch or two, and then removing the excess concrete just before setting occurs. The concrete then can be finished with stiff brushes.

Hardened concrete should be moistened thoroughly before new concrete is placed on it. Where the concrete has dried out it may be necessary to saturate it for a day or more. No pools of water should be left standing on the wetted surface when the new concrete is placed.

Where concrete is to be placed on hardened concrete or on rock, a layer of mortar on the hard surface is needed to provide a cushion against which the new concrete can be placed. The fresh mortar prevents stone-pockets and assists in securing a tight joint. The mortar should have a slump of less than 6 in. and should be made of the same materials as the concrete, but without the coarse aggregate. It should

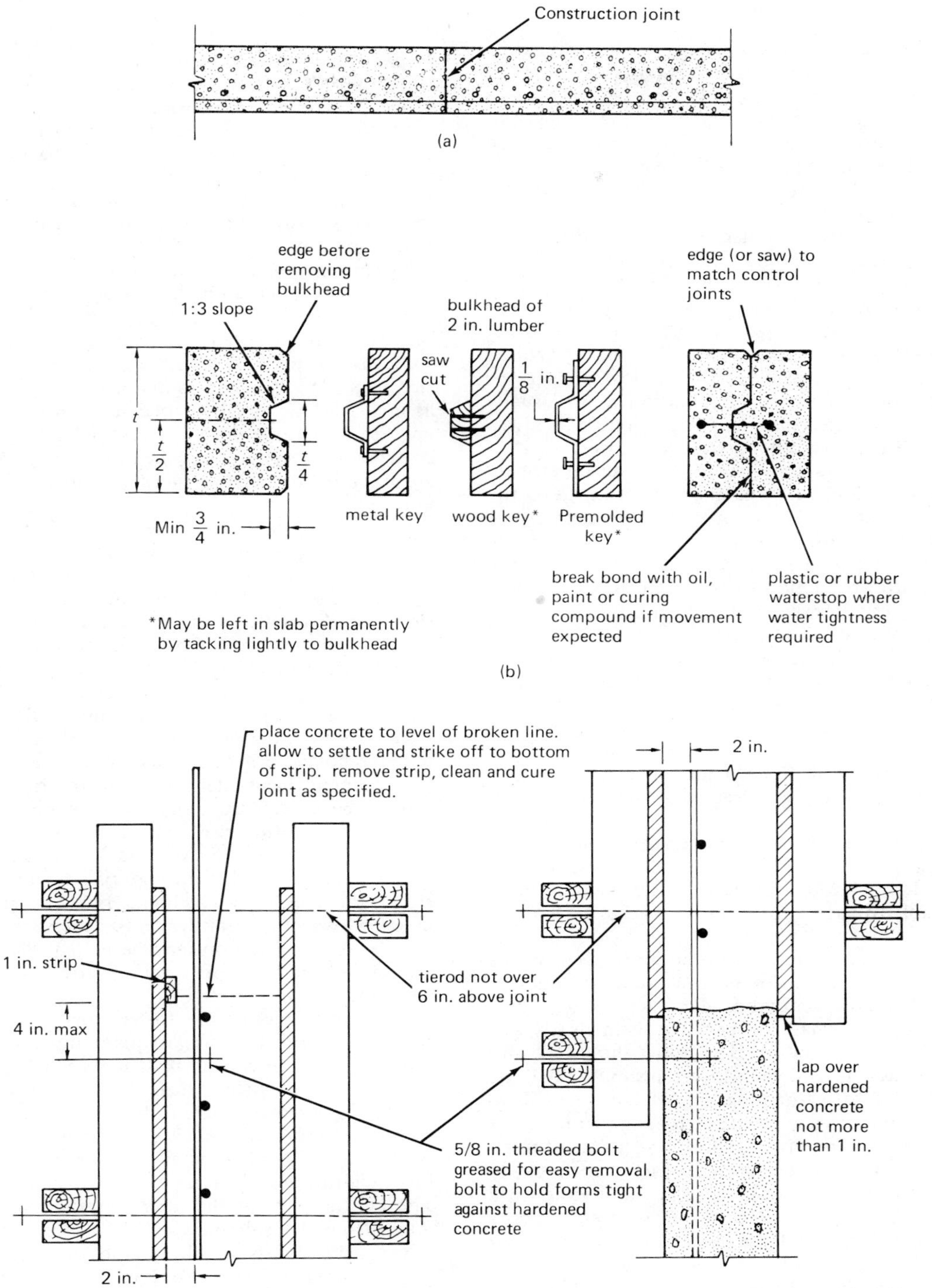

Fig. 4-4 Construction joints. (a) Butt joint in structural slabs; (b) tongue and groove joint in slabs; and (c) straight horizontal joint.

be placed to a thickness of $\frac{1}{2}$ to 1 in. and should be worked well into the irregularities of the hard surface.

Depending on the structural design, construction joints may be required later to function as expansion or contraction joints, or they may be required to be monolithic; that is, the second placement must be soundly bonded to the first so as to prevent movement and to be essentially as strong as the section without a joint.

Construction joints at which the concrete of the second placement is intentionally separated from that of the preceding placement by a bond-breaking membrane, but without space to accommodate expansion of the abutting units, function as contraction joints. Similarly, construction joints in which a filter is placed, or a gap is otherwise formed by bulkheading or the positioning of precast units, function as expansion joints. Construction joints may run horizontally or vertically, depending on the placing sequence prescribed by the design of the structure.

4.9.1 Location of Construction Joints

Location of construction joints is usually predetermined by agreement between the architect-engineer and the contractor, so as to limit the work that can be done at one time to a convenient size, with least impairment of the strength of the finished structure, though it may also be necessitated by unforeseen interruptions in concreting operations. In walls a horizontal length of placement in excess of 40 ft is not normally recommended.

Generally it is impractical to place concrete in lifts higher than one story. Designers should recognize this when locating horizontal construction joints. For buildings in which the concrete walls are to be exposed, joints may be located at bends of ornamentation, ledges, rustications, or other architectural details. It is convenient to locate horizontal joints at the floor line or in line with window sills. In the design of hydraulic structures, construction joints usually are spaced at shorter intervals than in nonhydraulic structures to reduce shrinkage and temperature stresses.

As construction joints are provided to accommodate progress of construction, their spacing is determined by the type of work, site conditions, and production capacity of the construction site. It is also important that the spacing of construction joints be planned such that the joints, while in accordance with production capacity, occur only where they may be properly constructed and do not occur where they may create stress concentrations.

Construction joints should be located by the designer to provide logical separation between segments of the structure. As a rule, construction joints are allowed only where shown on the drawings. If the placing of concrete is involuntarily stopped for a time longer than the initial setting time of the concrete used, the old surface is to be considered as a construction joint, and treated as such before casting is resumed.

It is best to avoid locating vertical construction joints at or near the corners of the building, since their presence may make it difficult to tie the corners together securely; it is, however, desirable to have a control joint within 10 to 15 ft of a corner if possible.

The appearance of a structure can be influenced by the location of the construction joints, and the aim should be to install them in a position that renders them as inconspicuous as possible; the alternative is to make them clearly visible as a feature of the structure. The joints should fit into the architectural design, and their location should facilitate the construction of forms and placing of concrete. However, from the point of view of strength of the structure, it is desirable to position construction joints at points of minimum shear. For slabs and beams it is, therefore, usual to have construction joints at midspan or in the middle third of the span. These rules are based on the assumption that a construction joint may result in less than 100% of shear capacity in the interface. If it were practicable to have such joints at the supports for slabs and beams, it would improve appearance and result in a considerable saving on the cost of the formwork.

Joints in bearing walls and columns should be located on the underside of floor slabs, beams, or girders, and at the tops of footings or floor slabs. Columns should be filled to a level preferably a few inches below the junction of a beam or haunch before making a construction joint. To avoid cracking due to settlement, concrete in columns and walls should be allowed to stand for at least two hours, and preferably overnight, before concrete is placed in slabs, beams, or girders framing into them. Haunches, drop panels, and column capitals are considered as part of the floor, or roof, and should be placed integrally with them.

4.10 EXPANSION JOINTS

Expansion joints are used to allow for expansion and contraction of concrete during the curing period and during service; to permit dimensional changes in concrete due to load; to separate, or isolate, areas or members that could be affected by any such dimensional changes; and to allow relative movements or displacements due to expansion, contraction, differential foundation movement, or applied loads. Obviously, expansion joints can also function as construction joints.

Expansion joints are frequently used to isolate walls from floors or roofs; columns from floors or cladding; pavement slabs and decks from bridge abutments or piers; and in other locations where restraint or transmission of secondary forces is not desired. Many designers consider it good practice to place expansion joints where walls change direction as in L-, T-, Y-, and U-shaped structures, and where different cross sections develop (Fig. 4-1). Expansion joints in structures are sometimes called isolation joints because they are intended to isolate structural units that behave in different ways.

Expansion joints are made by providing a space for the full cross section between abutting cast-in-place structural units by the use of filler strips of the required thickness or by leaving a gap when precast units are positioned. Expansion joints usually start about the foundation level, and continue throughout the height of the structure.

Expansion joints are installed mainly to control the effects of temperature increase, and to a lesser degree, to allow independent structural action of adjacent units.

The term "expansion joint" is a misnomer so far as cast-in-place concrete is concerned, since the shrinkage a structural element has undergone from the time it was cast will usually be greater than any possible subsequent expansion due to temperature rise, increase in humidity, or other factors. The term "expansion joint" is more appropriate for structures assembled from materials such as brick, stone, or structural steel, with which material the term possibly originated. However, in sanitary structures designed to contain liquids, the contraction may be halted when the structures are placed in use, and even reversed during hot humid weather.

If expansion joints are to be provided, they must be introduced in the preliminary planning stage. It is generally difficult to form breaks in structures when the design has reached an advanced stage without provision for such breaks. Experience and intuitive understanding of how the completed structure behaves may help in the positioning of such joints. Calculations should support the selection of a joint width, and be related to the anticipated movements. The elastic range of the filler material employed in a joint must also be considered.

Expansion joints should not be provided unless they are clearly necessary, since they can be an embarrassment to the structural and architectural designer, as they are often incompletely detailed and frequently badly constructed. There is no doubt that the proper course to adopt, where practicable, is to control a movement without permanent joints; for example, with shrinkage strips, described later in this chapter. When, however, a designer considers that danger of unacceptable cracking exists, the structure should be divided into controllable units by providing expansion joints at suitable places.

Expansion joints are designed for relative movement of adjacent sections and should be located so that:

1. They act as stress relief planes.

2. The concrete between the joints is not subjected to substantial volume change stresses.
3. Other elements supported by the concrete, such as partitions, exterior cladding, window frames, and others in the building, are not subjected to movement distress.
4. The shape, size, and type of joint will function correctly for all conditions of movement.

Factors that should be considered in the design and detailing of expansion joints are: shrinkage, creep, thermal movements, foundation settlements, and elastic deformations of adjacent structural units.

Thermal expansion of concrete roofs caused by solar radiation is a common cause of distress to buildings. Such distress can be minimized by applying thermal insulation on top of the roof to reduce the temperature differentials. Otherwise, either expansion joints at required intervals should be arranged, or the roof should be able to slide on top of the supporting walls (see Fig. 4-15) with suitable separation of the plaster finish at the junction between roof and walls.

Other than in long buildings, expansion joints may be necessary at the junction of tall and short buildings (Fig. 4-1) to avoid distress due to differential settlements. If, for example, a 40-story tower has a 2-story base, the base will have fully settled elastically when the second story is completed, and the construction progress on the tower will create differential settlements at the junction as the tower will continue to settle. An expansion joint at the junction will allow each of the two different adjacent building units to settle individually without distressing their connection. Incidentally, a shrinkage strip may fulfill the same function. Also, where a new building unit is attached to an existing building, an expansion joint may be desirable.

4.10.1 Spacing of Expansion Joints

Spacing of expansion joints in buildings is a controversial issue. There is a great divergence of opinion concerning the importance of expansion joints in concrete construction. Some experts recommend joint spacings as low as 30 ft while others consider expansion joints entirely unnecessary. Joint spacings of roughly 150 to 200 ft for concrete structures seem to be typical ranges recommended by various authorities. Divergent viewpoints are reflected both in private practice and in building codes. The existence of such opposing opinions, which, obviously, cannot be equally valid as a consideration in a single structure, is nonetheless understandable, since it is based on divergence of previous experience.

Those who advocate the complete omission of expansion joints in concrete construction state that drying shrinkage is greater than the expansion caused by a 100°F increase in temperature; therefore, any temperature increase will tend to close up shrinkage cracks, and there will be practically no compressive stress in the concrete due to thermal expansion.

In a 1940 report of a joint committee (AIA, ASCE, ACI, AREA, PCA) it was suggested that, in localities with large temperature ranges, expansion joints should be provided every 200 ft. In milder climates, 300 ft was suggested.

In the 1940s a distinct trend started toward the elimination of expansion joints in long buildings. This trend is continuing into the present time. Even in locations with large temperature ranges, buildings up to 400 and 500 ft have been constructed without expansion joints, and seemingly the performance has been satisfactory. The following are examples of such buildings:

The *General Accounting Office Building*, built in 1951,[1] Washington, D.C. is an eight-story, flat slab building 638 × 389 ft in plan, with 25 × 25-ft bays. Control joints were used on 50-ft centers. No evidence of distress resulting from the omission of expansion joints was found in this building in 1956.

Military Personnel Records Center, St. Louis, Missouri, is a six-story, flat slab building built in 1956, 728 × 282 ft in plan with 22 × 22-ft bays. Control joints were spaced every second bay (44-ft centers) in each direction. The concrete was placed between control joints in a checkerboard fashion with a 48-hour interval between adjacent sections. The slab reinforcement is continuous through the joints with a #4 dowel 3 ft long at 12 in. center added at the top of the slab. A technical paper with background information on this building and expansion joints in general appeared in the *ACI Journal*.[2]

The *Los Angeles Union Terminal* has a seven-story, flat slab warehouse 550 × 100 ft in plan, with 20 × 20-ft bays, as well as a four-story building 440 × 100 ft in plan. According to an inspection report of 1958, both buildings were in excellent condition after 40 to 50 years, showing no distress due to lack of expansion joints.

The 1972 "Minimum Property Standards–Manual of Acceptable Practices" by the FHA recommends that spacing of expansion joints for buildings not exceed the following values:

Type of Building	*Outside Temperature Variations*	*Maximum Joint Spacing (ft)*
Heated	Up to 70°F	600
	above 70°F	400–500
Unheated	up to 70°F	300
	above 70°F	200

Figure 4-5 shows joint spacing as recommended by the Federal Construction Council, and is adapted from *Expansion Joints in Buildings*, Technical Report No. 65, prepared by the Standing Committee on Structural Engineering of the Federal Construction Council, Building Research Advisory Board, Division of Engineering, National Research Council, National Academy of Sciences, 1974. Note that the spacings obtained from the graph in Fig. 4-5 should be modified for various conditions as shown in the notes beneath the graph. Values for the design temperature change can be obtained from Fig. 4-6.

When expansion joints are required in nonrectangular structures, they should always be located at places where the plan or elevation dimensions change radically.

4.10.2 Expansion Joint Details

To be effective, an expansion joint should separate the two adjacent units into completely independent structures. Joints should extend through foundation walls, but column footings need not be cut at a joint unless the columns are short and rigid. No reinforcement should pass through these joints; it should terminate 2 in. from the face of the joint. Dowels with bond breaker may be used to maintain plane.

[1] Reynolds, W. E., "Ways to Cut Building Costs Shown by General Accounting Office Building, Wash. D.C.," *Civil Engineering*, v. 22, June 1952, pp. 50–54.

[2] Cohn, Earl B., and Wahl, W. A., "Military Personnel Records Center Built Without Expansion Joints," *ACI Journal*, v. 54, June 1958.

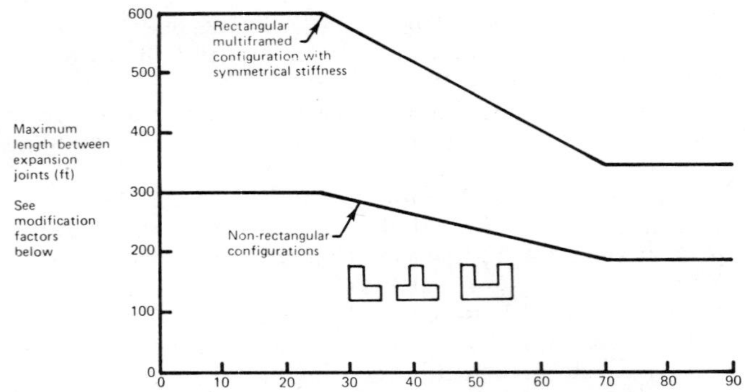

These curves are directly applicable to buildings of beam-and-column construction, hinged at the base, and with heated interiors. When other conditions prevail, the following rules are applicable:

(a) If the building will be heated only and will have hinged-column bases, use the allowable length as specified;
(b) If the building will be air conditioned as well as heated, increase the allowable length by 15% (provided the environmental control system will run continuously);
(c) If the building will be unheated, decrease the allowable length by 33%.
(d) If the building will have fixed-column bases, decrease the allowable length by 15%;
(e) If the building will have substantially greater stiffness against lateral displacement at one end of the plan dimension, decrease the allowable length by 25%.

When more than one of these design conditions prevail in a building, the percentile factor to be applied should be the algebraic sum of the adjustment factors of all the various applicable conditions.

Source: *Expansion Joints in Buildings*, Technical Report No. 65, National Research Council, National Academy of Sciences, 1974.

Fig. 4-5 Maximum building length without use of expansion joints.

Joint location is generally dictated by structural considerations, with architectural treatment of walls at the joints developed accordingly.

The simplest expansion joint is one on a column line with double columns (see Fig. 4-7(a)). This may be costly and complicated as it may involve the construction of two independent sets of columns and beams at the joints, separated by a 1 in. space. Such costs should, however, be considered against the risk and cost of extensive trouble which might occur year after year if the joints were omitted.

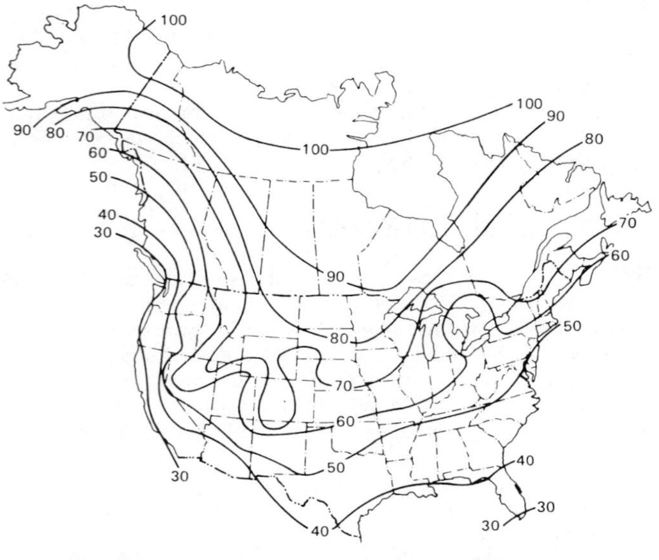

Fig. 4-6 Maximum seasonal climatic temperature change, deg F.

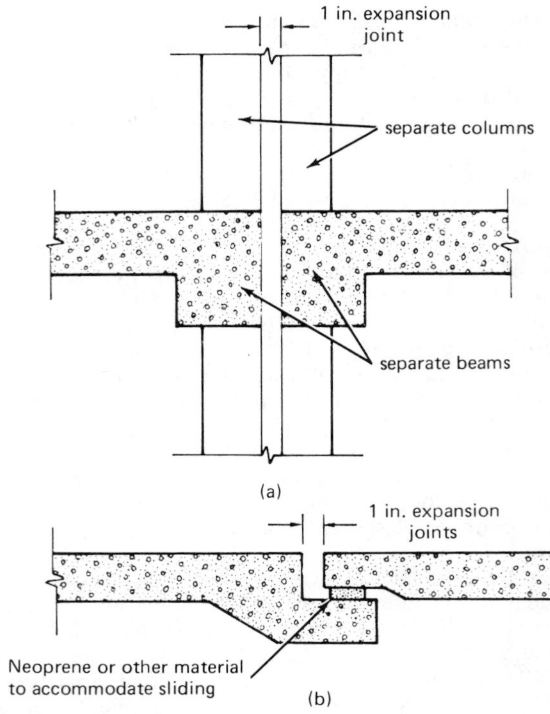

Fig. 4-7 Expansion joints. (a) Joint with separate columns and beams; and (b) joints in the slab at one-third or one-quarter of the span.

Expansion joints without a double column may be used by introducing them in the third or quarter point in the slab in a manner indicated in Fig. 4-7(b). The introduction of such a hinge near the point of inflection should, of course, be taken into account in the structural analysis. Watertightness of such joints may be difficult to achieve.

Expansion joints in buildings are usually $^3/_4$ in. to 1 in. wide. Judging from past experience, it appears safe to consider that the maximum movement at joints spaced 200–300 ft apart will not exceed 1 in. under the most unfavorable conditions. Joints that are too wide may result in loss of filler material.

Expansion joints filled with a compressible elastic, non-extruding material that is intended to accommodate the movements occurring, and at the same time provide an adequate seal against water and foreign matter. Generally, no single material has been found yet that will completely satisfy both conditions mentioned. Essentially three types of jointing materials are used: joint fillers (such as strips of asphalt-impregnated fiberboard), sealers, and waterstops. Sealers (sealing compounds) and waterstops (rubber, plastic, or metal) are used where a joint has to be sealed against the passage or pressure of water.

Expansion joints must be carefully detailed and carried through the finishes of the floor and ceiling as shown in Fig. 4-8(a) and (b).

4.10.3 Seismic Joints

Seismic joints are wide expansion joints provided to separate portions of buildings dissimilar in mass and in stiffness. The joints are provided to allow the adjacent buildings or

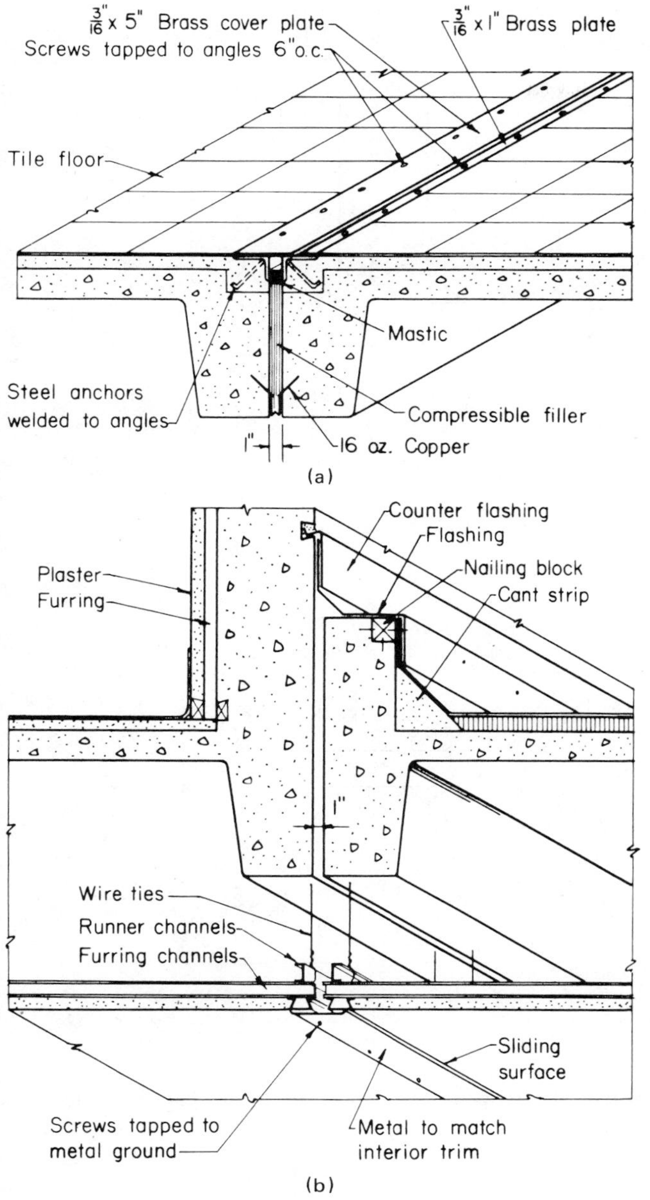

Fig. 4-8 Details of expansion joints through finishes. (a) Covering of expansion joint in floor; and (b) expansion joint in ceiling and on the roof.

building units that have various vibration characteristics to oscillate during an earthquake without hammering at each other. The alternative to a seismic joint is to tie the buildings or units together so that they can vibrate as a single system.

The width of a seismic joint should be equal to the sum of the total deflections at the level involved from the base of the two buildings, but not less than the arbitrary rule of 1 in. for the first 20 ft of height above the ground, plus $\frac{1}{2}$ in. for each 10 ft additional height. The determination of these deflections will be the summation of the story drift in addition to the building's flexural deflection (column lengthening and shortening) to the level involved. Shearwall buildings, being much stiffer, need a seismic joint only, say, half as wide, since the earthquake oscillations of shearwall buildings will be much smaller than those of framed buildings.

Figure 4-9 shows seismic separation joints at various locations in buildings. As shown in the figure, the seismic joint coverages must allow movement, be waterproof, and be architecturally acceptable. The details of the joint coverages should allow at least a doubling of the joint opening when subjected to an earthquake.

4.11 CONTROL JOINTS

The tensile stresses caused by shrinkage and temperature drop in the concrete that is not free to contract can be relieved or reduced to tolerable limits by control joints.

A control joint is a purposely made plane of weakness of a section so that cracking and contraction will occur along these preselected straight lines which are inconspicuous and less objectionable. Control joints are inexpensive and can be formed easily by tacking wooden, rubber, plastic, or metal strips to the inside of the forms as shown in Fig. 4-10(a) and (b). After removal of the form and the tacked-on strips, a narrow vertical groove is left in the concrete on the inside and/or outside surfaces.

Control joints can also serve as partial expansion joints, limited to the extent of contraction that occurred in the joint. The terms "contraction joints" and "dummy contraction joints" are often used interchangeably for control joints, causing some confusion.

Numerous ways have been devised for forming control joints. In all cases it is recommended that the combined depth of inside and outside grooves be not less than about $\frac{1}{5}$ to $\frac{1}{4}$ of the thickness of the slab or wall to make the joint effective. In slabs on grade and pavements, the necessary weakening of the section can be made by sawing a groove in the concrete soon after it has hardened (Fig. 4-10d). The groove may also be formed with a hand tool before hardening.

Quite often wooden or metal strips forming control joints are left permanently in the concrete, as shown in Fig. 4-10(a).

Control joints are appropriate only where the net result of the contraction and any subsequent expansion during service is such that the units abutting are always shorter than at the time the concrete was placed. They are frequently used to divide large, relatively thin units; for example, pavements, floors, retaining, and other walls into smaller panels. They are called control joints because they are intended to control crack location.

Control joints may form a complete break, dividing the original concrete unit into separate units. Where the joint does not open wide, some shear transfer may be maintained by aggregate interlock. Where greater shear transfer is required without restricting freedom to open and close, coated dowels may be used. If restriction of the joint opening is required for structural stability, appropriate tie bars or continuation of all or part of the reinforcing steel across the joint may be provided. Reinforcement across control joints is usually partially or completely cut.

Where small amounts of reinforcing steel or wire mesh are provided in an element, it is desirable to interrupt them at joints at least 3 in. away from the joint on either side.

If a construction joint serves as a control joint, an insert in the form provides a better finish of the casting operation which at the same time forms a straight control joint. Obviously, the surface of the joint should be treated with a bond breaker to assure free movement (Fig. 4-11).

4.11.1 Spacing of Control Joints

No exact rules for the location of control joints can be made. Each job must be evaluated individually to determine

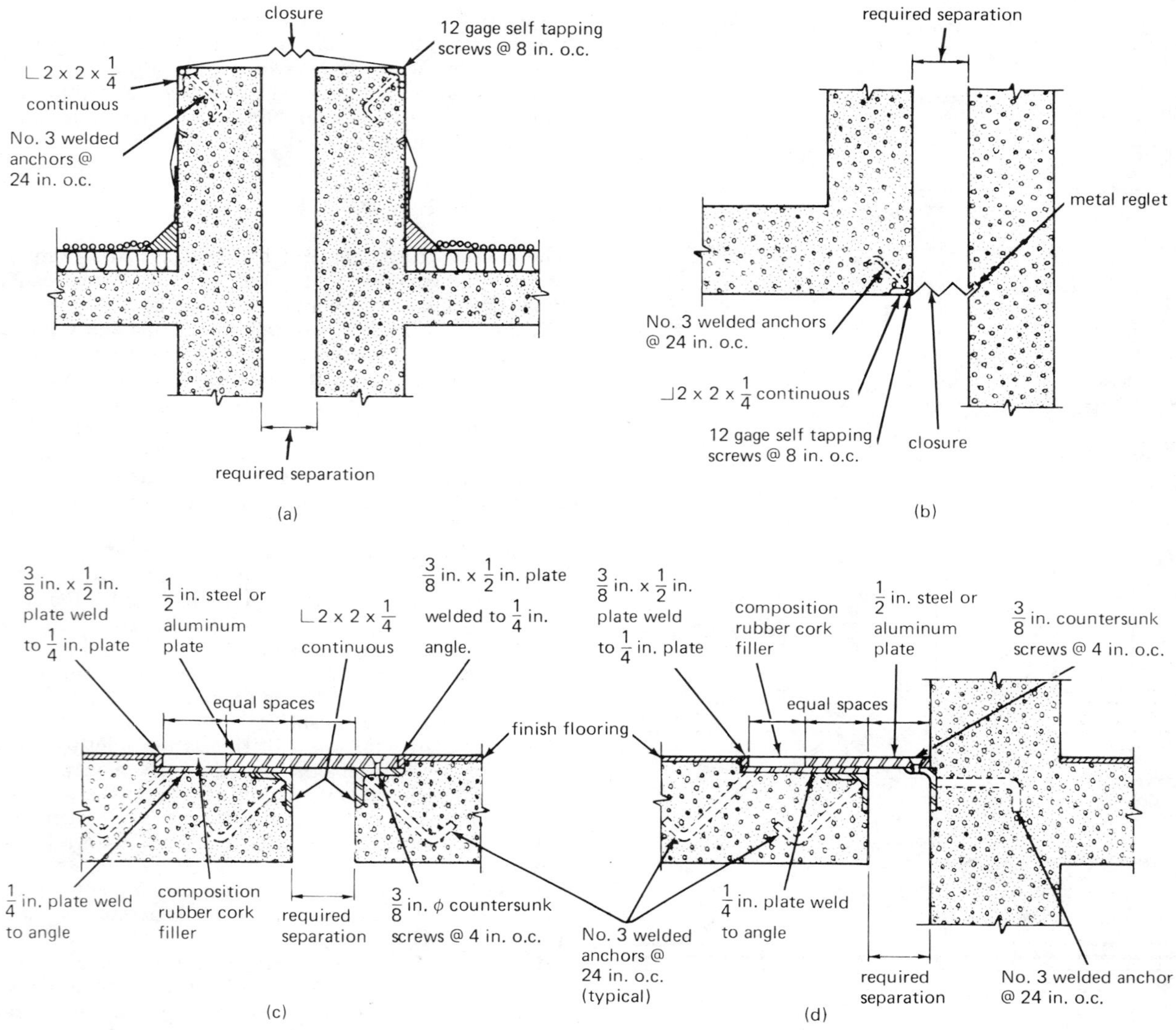

Fig. 4-9 Seismic separation joint details. (a) Roof parapet separation; (b) plan at exterior vertical closure; (c) typical floor plate closure at building separation; and (d) typical building separation floor plate closure at walls.

where and how far apart joints should be placed, taking into account the specific structural design. Joint spacings commonly used show considerable variation ranging from a few feet up to 80 ft. The spacing of joints depends on the exposure and severity of climatic conditions, and on the internal restraint (reinforcing) and external restraint (columns, walls, friction with the ground) provided by other parts of the structure. Obviously, the shrinkage of the concrete and temperature range to be expected have a major influence on the spacing. Joints should be made where abrupt changes in thickness or height occur. To some degree there is a choice between joints at short intervals with normal, or no, reinforcement and larger distances between joints with increased reinforcement. Many of the recommendations favor a 15- to 25-ft spacing between control joints in walls and slabs on grade. First joints should occur 10 to 15 ft from a corner. In walls and slabs, the optimum ratio of panel dimensions enclosed by joints is 1 to 1, with a maximum aspect ratio of $1\frac{1}{2}$ to 1.

In walls having frequent openings, spacing control joints 20 ft apart is considered maximum. The spacing in walls without windows should not be more than 25 ft, and a joint within 5 to 15 ft of each corner is desirable.

Observations show that building fronts having about 60% or more of openings crack least, but solid blank walls and walls with relatively small openings (say 25% of gross area) present problems. Cracks develop at about 10 to 15 ft on centers; the greater the spacing of the cracks, the wider the cracks.

Walls with frequent openings should have smaller joint spacings than solid walls. Cracks usually form at windows and doors where the concrete sections are weakest. Additional reinforcing should, therefore, be placed at the corners of these openings to intercept potential cracks.

Many of the references and specifications available seem to favor a 20- to 25-ft spacing between control joints in walls. If a wall is less than 10 ft high, however, and restrained at the foundation by anchorage to more massive concrete or a rock foundation, it may even crack near the center of a 20-ft section. Under such conditions, joints at 10-ft intervals may be desirable. For walls 10 to 20 ft in height, joint spacing should be approximately equal to the wall height.

Long slabs that form external balconies are prone to develop cracks originating from their edges. The cracks are caused by shrinkage and thermal differentials between the

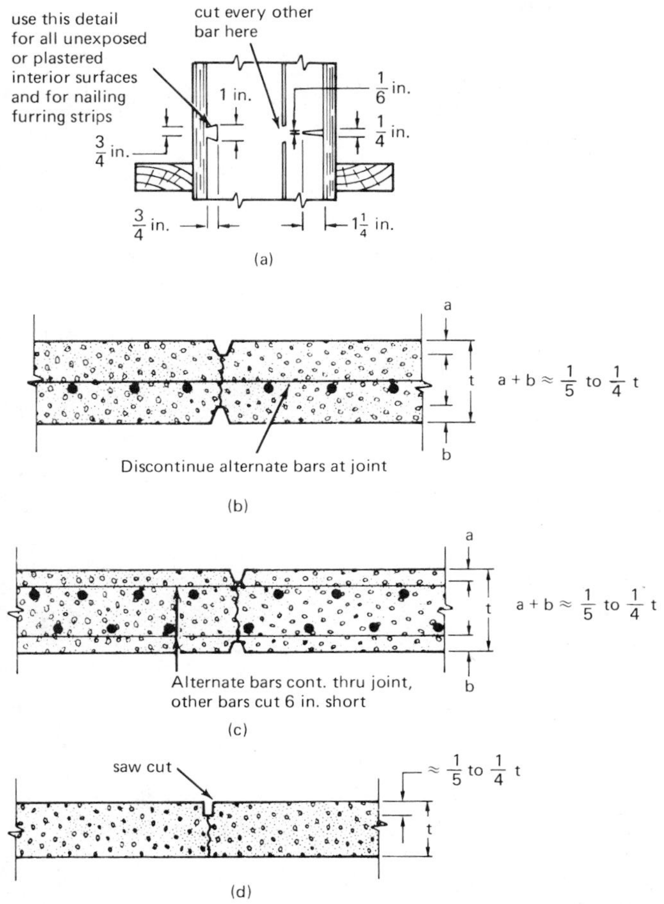

Fig. 4-10 Control joints. (a) Control joints should reduce the effective wall thickness by at least one-fourth; (b) control joints in walls 8 in. thick, or less; (c) control joints in walls over 8 in. thick; and (d) saw cut joint in a slab or pavement.

slab on the inside and the concrete near the free edge of the balcony. Control joints at sufficiently close intervals (less than 10 ft) will usually attract all the shrinkage and temperature cracking. If such joints are extended only partly across the balcony slab, the concrete near the end of the joint should be locally reinforced to prevent the prolongation of the joint as a crack.

In cases of continuous forming (slip forming or extrusion) it may be advisable to make larger units between joints (if it is impractical to form units as one would wish) as a justifiable risk, and to rely on the ability of the unit to resist the forces due to frustrated movements caused by volumetric changes. In such cases a certain amount of shrinkage and temperature reinforcement is required to improve the resistance. It is at times more prudent to trust an unjointed concrete to resist its own movement than to provide an incomplete joint which the structure will probably complete by itself in due course and in a most unsightly manner.

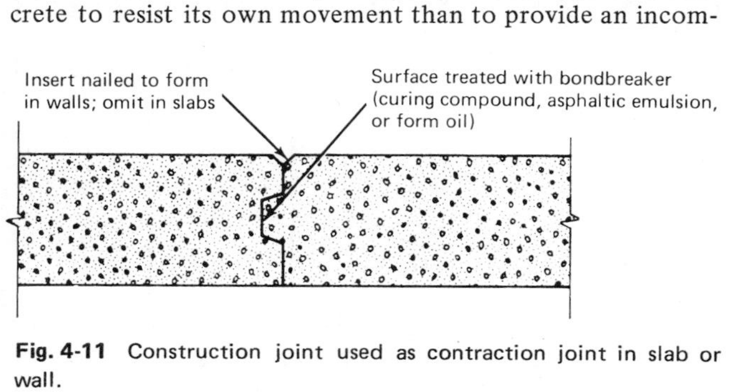

Fig. 4-11 Construction joint used as contraction joint in slab or wall.

4.12 SHRINKAGE STRIPS

Shrinkage strips are temporary joints that are left open for a certain time during construction to allow a significant part of the shrinkage to take place without inducing stresses. Such joints have been used to a considerable extent both in massive structures and in thin walls and slabs, for the purpose of reducing shrinkage stresses and minimizing shrinkage cracks. Such strips divide the structure into parts that shrink independently until they are connected by casting the strip.

In recent years shrinkage strips have replaced expansion joints to a great extent in long multistory buildings. In enclosed multistory buildings the slabs are subjected only to shrinkage, while the roof slab is subjected to temperature length changes in addition to shrinkage. Therefore, the roof slab, if not insulated sufficiently, may need an expansion or sliding joint (see Figs. 4-14 and 4-15) in addition to the shrinkage strips. Experience has shown that in long multistory buildings, shrinkage strips spaced about 100 to 150 ft apart perform successfully. The distance between horizontal shrinkage strips should be less in slabs with stiff supports (large columns or reinforced concrete walls parallel to the direction of shrinkage) or if the expected shrinkage coefficient is unusually high.

The shrinkage strip is usually 2 to 3 ft wide across the entire building and is cast two to four weeks later than the adjacent units to reestablish continuity. During the period the strip remains open, a significant part of the shrinkage takes place (about 40% of the total shrinkage may occur in the first month), while in the meantime the concrete will gain tensile strength to resist the remaining shrinkage with little or no cracking.

The flexural reinforcement crossing the shrinkage strip is generally made to act continuously from section to section after the joint is closed. However, to run the flexural reinforcement in one piece continuously from one unit through the strip into the other unit would impede unrestrained shrinkage of the concrete units on either side of the strip, thus defeating the objective of the shrinkage strip. Therefore, it is necessary to lap the reinforcing bars within the strip as shown in Fig. 4-12(a). An alternate method of detailing the reinforcement to avoid excessive stresses and possible slippage of the bars on either side of the strip while the strip is still open is special expansion bends within the strip, as shown in Fig. 4-12(b), so as to allow for a change in the length of the strip. The plane of the bends must be parallel to the face of the slab, and the bends placed alternately left hand and right hand, so as to avoid accumulation of stresses when they tend to straighten out due to the loads superimposed upon the completed structure.

Shrinkage strips can also be used in walls. In such cases the walls are constructed in lengths not exceeding 25 ft with internal gaps of 2 ft left between each length. In order to allow unrestrained contraction to take place in each length of wall, the gap should be left open for three to four weeks after the casting of the adjacent length of wall. As in slabs the reinforcement should be lapped within the gap. Another method of reducing shrinkage stresses in walls is to construct them on an alternate panel system (similar to checkerboard casting of slabs on ground), the length of each panel not exceeding about 25 ft. Both examples, or variations of them, are in common use, and they form an acceptable practical approach to limiting the amount of shrinkage cracking.

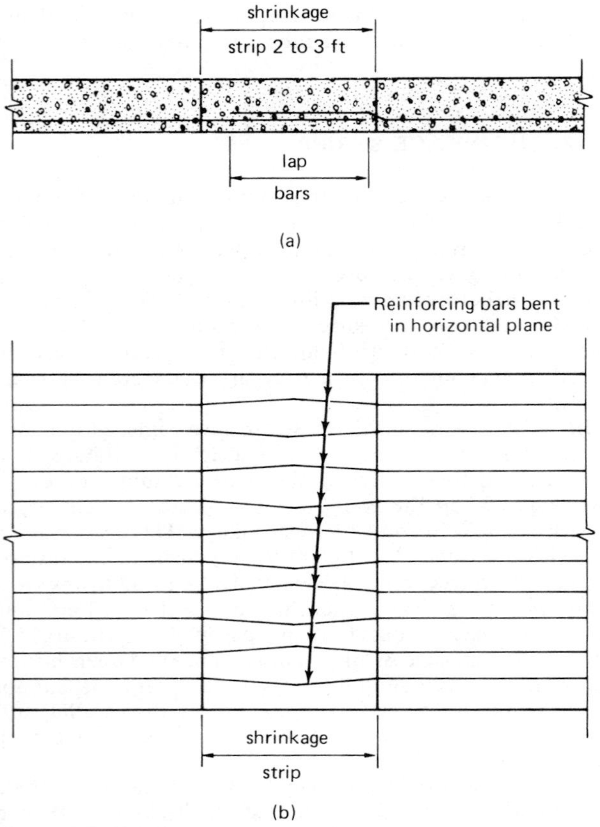

Fig. 4-12 Reinforcement details in shrinkage strips. (a) Section—lapped reinforcement; and (b) plan—bent reinforcement.

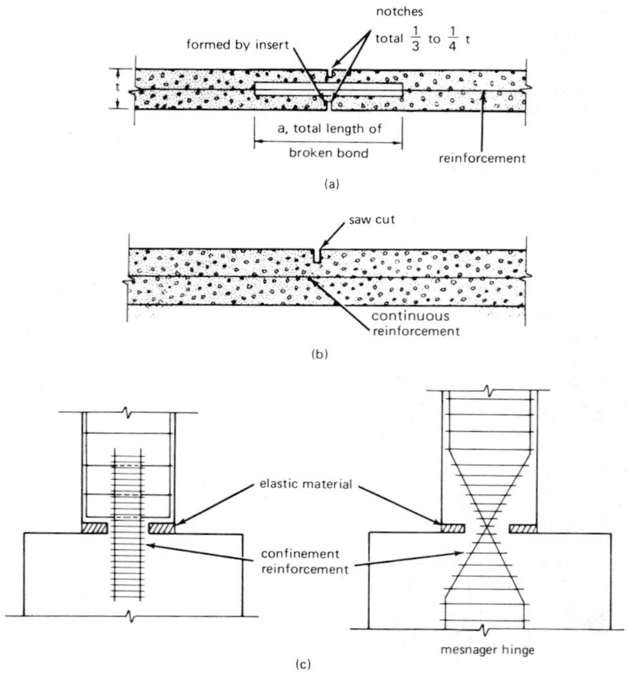

Fig. 4-13 Hinged joints. (a) Elastic control joint; (b) hinged joint in pavement; and (c) hinged joints in columns.

4.13 SPECIAL PURPOSE JOINTS

4.13.1 Elastic Control Joints

To improve the performance of control joints, the bond between reinforcement passing through the joint and the concrete can be eliminated for an appropriate length across the joint as shown in Fig. 4-13(a). This prevents yielding of the reinforcement in the vicinity of the joint (which might otherwise occur), and as a result the joint always remains elastic. The breaking of the bond can be achieved either by a paper wrapping, asphalt coating, plastic tubing, or other means. The length of bar over which bond should be eliminated is determined with due regard to the maximum joint opening, the maximum concrete temperature drop, and the yield strength of the bars. At the maximum joint opening, the bars should be stressed below their elastic limit. The maximum joint opening varies with joint spacing, other conditions being similar. The concrete temperature drop is the difference between the concrete temperature at time of set and the lowest temperature to which the concrete will be subjected.

Elastic control joints were first used in an experimental highway project in Indiana more than 40 years ago. In recent years these joints have been used experimentally in pavements in Sweden and Germany.

Table 4-1 illustrates the approximate relationship between concrete temperature drop, t, length of broken bond to joint spacing ratio, c/a, reinforcement ratio ρ, and steel stress at a joint, f_s.

4.13.2 Hinge Joints

Hinge joints permit rotation within a section, however, separation of the abutting units is limited by tie bars or the continuation of reinforcing steel across the joint. This type has wide usage in, but is not restricted to, pavements where longitudinal joints (Fig. 4-13(b)) function in this manner to overcome warping effects by forcing the same deflections on both sides of the joint due to wheel loads or settlement of the subgrade. The reinforcement usually yields in the vicinity of the joint if shrinkage and temperature drop forces are present.

In structures, hinge joints in columns are often referred to as articulated joints and are arranged as shown in Fig. 4-13(c). Such joints are arranged to preclude a moment transfer between two elements, mostly between the foundation and column above. It is usually achieved when the section is substantially reduced by blocking out notches that are filled with an elastic, easily compressible material. The reinforcement through the joint can consist of parallel bars,

TABLE 4-1 Stresses in Reinforcement of Elastic Joints

Temperature Drop t (°F)	*Length of Broken Bond to Joint Spacing* (c/a)	*Reinforcing Ratio* ρ	*Steel Stress at Joint* f_s *in ksi*
25	0.50	0.01	9
		0.02	8
	0.25	0.01	15
		0.02	13
50	0.50	0.01	17
		0.02	16
	0.25	0.01	30
		0.02	26
100	0.50	0.01	34
		0.02	32
	0.25	0.01	60
		0.02	52

or the bars can cross each other in an X-shape (Mesnager hinge).

In both cases a sufficient amount of confinement reinforcement (usually closed hoops or spirals) is necessary to increase the strain and strength capacity of the concrete in the hinge. Obviously, the hinge section must be sufficient to carry all the vertical loads and the shear forces.

4.13.3 Sliding Joints

Sliding joints may be required where one unit of a structure must move perpendicularly to another unit. The joints are usually made with a layer of a bond breaking material such as a bituminous compound, paper, neoprene felt, or other materials that facilitate sliding. Figure 4-14 shows a sliding joint utilizing a neoprene pad. Occasionally steel plates are embedded in both sliding surfaces. In such cases a lubricant is used between the plates to prevent the plates from "freezing" together. Sliding joints permit creep, shrinkage, and temperature drop shortening of elements without buildup of stresses. A distance, *d*, should be left (Fig. 4-14) for the joint to accommodate elongation of the element due to temperature rise, thus acting like an expansion joint. With neoprene as a connecting material, this joint can also facilitate rotation due to deflection of the connected element without significantly shifting the support reaction toward the edge. The required thickness of the neoprene depends on the amount of sliding to be provided for the joint and the durometer of the neoprene. Particular attention should be paid to the sealing of such joints where watertightness is required.

4.13.4 Slip Joints Between Masonry Walls and Concrete

Concrete slabs and foundations have considerably different moisture and thermal movements from those of the masonry walls with which they must work. Slabs and foundations also are usually under different states of stress from the walls, owing to different temperature and humidity environments. Therefore, it is important to break bond and provide a slip joint between these elements by positive means. With horizontal layers of material facilitating sliding between slabs and walls, each element will be able to move somewhat independently while still providing the necessary support to the other. Among materials that can be used as slip joints in structures are two sheets of building paper, bearing pads of polytetrafluoroethylene, or two layers of galvanized steel coated with grease.

Horizontal cracks have been observed in numerous loadbearing masonry walls supporting reinforced concrete slabs. Such cracking is generally attributed to thermal expansion, curling, or horizontal shrinkage of the slab. If the movement of the slab pulls the top of the wall sideways, the wall usually cracks several courses below the underside of the slab. A typical method of breaking bond between the walls and slabs to prevent horizontal cracks is shown in Fig. 4-15.

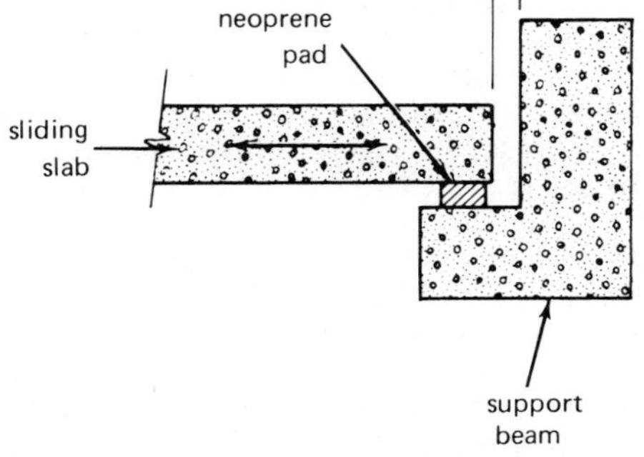

Fig. 4-14 Sliding joint on a beam support.

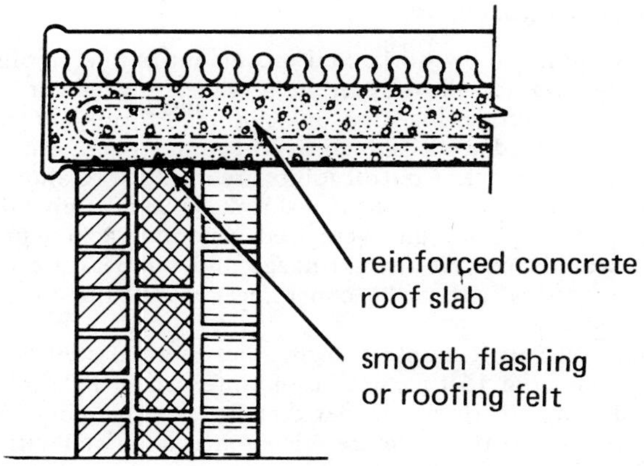

Fig. 4-15 Slip joint between masonry walls and concrete roof slab.

4.14 JOINTS FOR SLABS ON GRADE IN BUILDINGS

4.14.1 Types of Joints

Three types of joints will provide crack control in slabs on grade. They are defined below by the character and direction of the movement that will occur in the joint:

1. Control joints allow differential movement only in the plane of the floor.
2. Isolation joints allow differential movement in all directions.
3. Construction joints allow no movement in the completed floor. As this is difficult to achieve, construction joints should be made to act as control joints.

Figure 4-16 shows a plan of a slab on ground in which all the above joints are indicated.

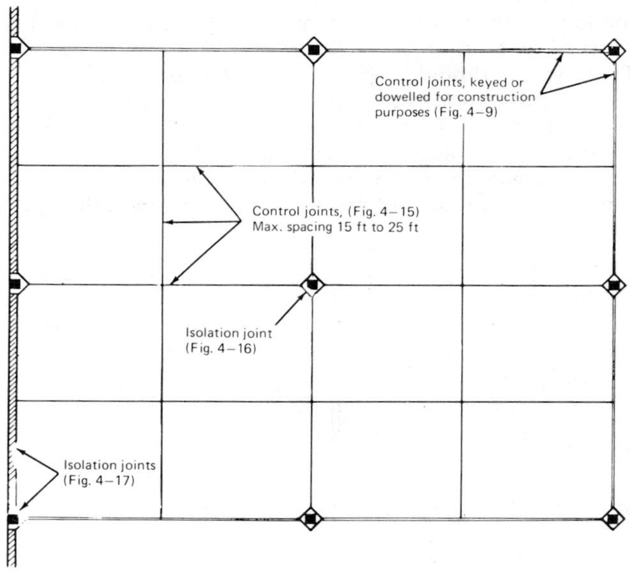

Fig. 4-16 Plan for a typical floor on ground. The double line around the perimeter indicates a key or doweled joint.

4.14.2 Control Joints

Control joints allow horizontal movement of the adjoining slabs, but do not allow differential vertical movement. This means that vertical loads on one side of the joint are transferred to the adjacent slab across the joint, usually through aggregate interlock. Control joints are installed to allow for contraction caused by drying shrinkage and temperature drop. If no control joints were used, random cracking in the floor would occur if the shrinkage and temperature drop tensile stresses within the concrete exceeded the concrete tensile strength.

Joint spacing to control drying and thermal contraction should be from 15 to 25 ft in unreinforced or lightly reinforced concrete floors. Under certain favorable conditions joints up to 30 ft, and more, have performed well without cracking. It is usual practice to place control joints on the column lines and between column lines as shown in Fig. 4-16. Variation in spacing is the result of differences in local conditions such as the concrete itself, climate, construction practices, and the type of soil. A thin slab in a dry environment, placed over a dry soil, will experience much more shrinkage, and therefore, will require a much closer spacing of joints (even less than 10 ft) than a thick slab in an environment of high humidity, placed over wet soil. Also, the joint spacing should be chosen so that the panels are approximately square. Panels with excessive length-to-width ratios (more than $1\frac{1}{2}$ to 1) are likely to crack.

Occasionally, joint spacing at much shorter distances than suggested above may be necessary to satisfy the particular geometry of a specific job. Joints may be necessary at abrupt changes in slab thicknesses. Reentrant corners are particularly sensitive to cracking and require either joints or reinforced bars placed diagonally. Proper joint location is too important to be left to chance; therefore, to ensure adequate follow-through on the job, joints should be planned in advance and detailed on the plans.

Control joints are made by purposely creating a vertical plane of weakness in the slab. The cracking then occurs at this weakened plane rather than at random locations in a slab. All control joints should be continuous, not staggered or offset. One of the most economical ways to make a control joint is to saw a slot in the top of the finished slab, as shown in Fig. 4-17. When a jagged crack forms then under the joint, it allows vertical load to be transferred to the adjoining slab without differential vertical movement. Because the joint is made after the concrete is placed and finished, control joints do not interfere with concreting operations. The saw cut should be made as early as possible, prior to the buildup of shrinkage stresses, and preferably during rising temperature. Therefore, it is generally done in the early afternoon for morning construction, or the following morning for concrete that was placed in the afternoon. The width of the cut should not exceed $^3/_{16}$ in., and the depth should be $^1/_5$ to $^1/_4$ of the slab depth, but not less than the maximum size of the aggregate. Immediately after sawing, the joint should be flushed with water or air under pressure to remove the sawing residue. It is good practice to seal the joints to avoid infiltration of foreign material which may cause spalling at the joints. If hard-wheeled traffic is expected, the joint should be sealed with lead.

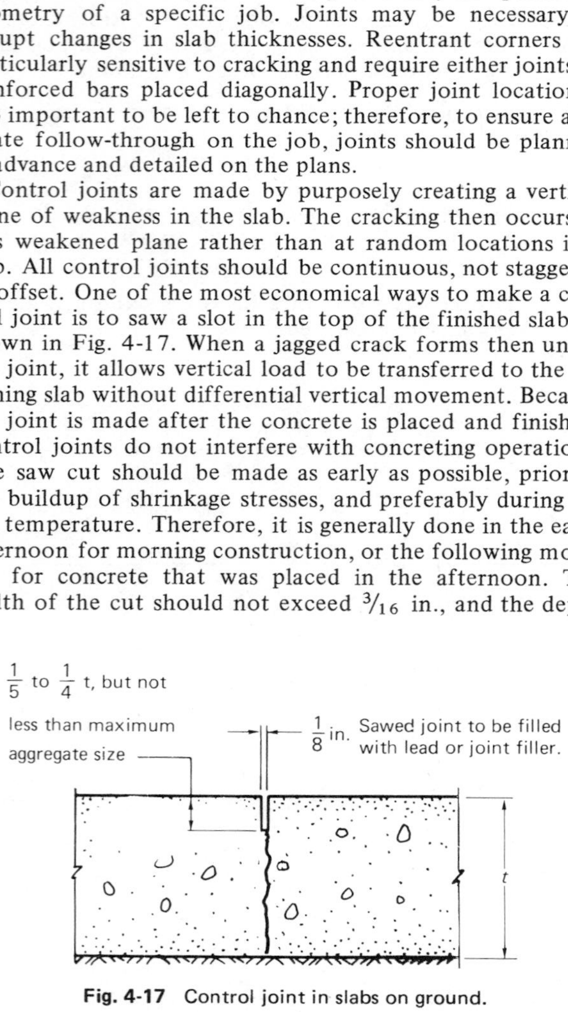

Fig. 4-17 Control joint in slabs on ground.

Another method of forming a control joint is to create a joint completely through the slab. Such joints are made by following a checkerboard pattern. In checkerboard bay placing, concrete is placed in alternate squares, like the squares of one color on a checkerboard. In this way partial shrinkage of half the squares occurs prior to placement of the alternate ones. The resulting joints between the previously placed concrete and the freshly placed concrete could be bonded as in a construction joint and later saved to form the control joint. However, it is usually simpler to use the nonbonded keyed control joint with or without dowels, as shown in Fig. 4-4(b). Breaking the bond between the new and previously placed slabs by spraying or painting with a curing compound, asphaltic emulsion, or form oil allows for the freedom of horizontal movement necessary for crack control.

The key provides the same type of vertical load transfer achieved by the aggregate interlock in the sawed control joint. Experience has shown that the tongue and groove need not be as deep as given in Fig. 4-4(b).

If wire mesh or reinforcing steel is used in the slab, it should be reduced or eliminated at a control joint, regardless of the joint construction method. Wire mesh is best reduced by cutting out alternate wires where the joint will be. If reinforcing bars are used, half the number of bars are eliminated across the joint. Reinforcing steel is sometimes used to increase the distance between joints. However, the greater the joint spacing, the more the joint will open. Where joints open wide, jolting by vehicles may result, and increased joint maintenance will be required. The easiest joint to maintain is one that is narrow and straight.

Welded wire fabric is often placed in the slab to control cracking. However, unless the steel is prestressed, the cracks will not be reduced because there is no stress in the steel to counter the tension in the slab until after the cracks occur. The steel fabric may prevent shrinkage cracks from opening too wide, but the amount of steel normally used has little, if any, effect on cracking.

4.14.3 Isolation Joints

Isolation joints separate or isolate concrete slabs from columns, footings, or walls. In addition to horizontal movements of the slab caused by shrinkage, they also permit vertical movement that occurs due to differences in unit soil pressure under floors, walls, columns, and machinery footings. They should also be used at other points of restraint such as drain pipes, sumps, or stairways. Because the isolation joint is used to allow freedom of movement, there should be no connection across the joint by reinforcement, keyways, or bond.

Interior and exterior columns should be boxed out as shown in Fig. 4-18. Circular fiberboard forms can be used in place of the square wood forms. The rectangular form of the boxout is placed so that its corners point at the control joints along the column lines, making it easier to saw up to the column. The isolation joint eliminates cracks radiating from the corners.

An isolation joint between a floor and the wall may be made by attaching an asphalt-impregnated sheet not more than $^1/_4$ in. thick to the wall prior to placing the concrete, as shown in Fig. 4-19. The joint shown in Figure 4-19(b) is

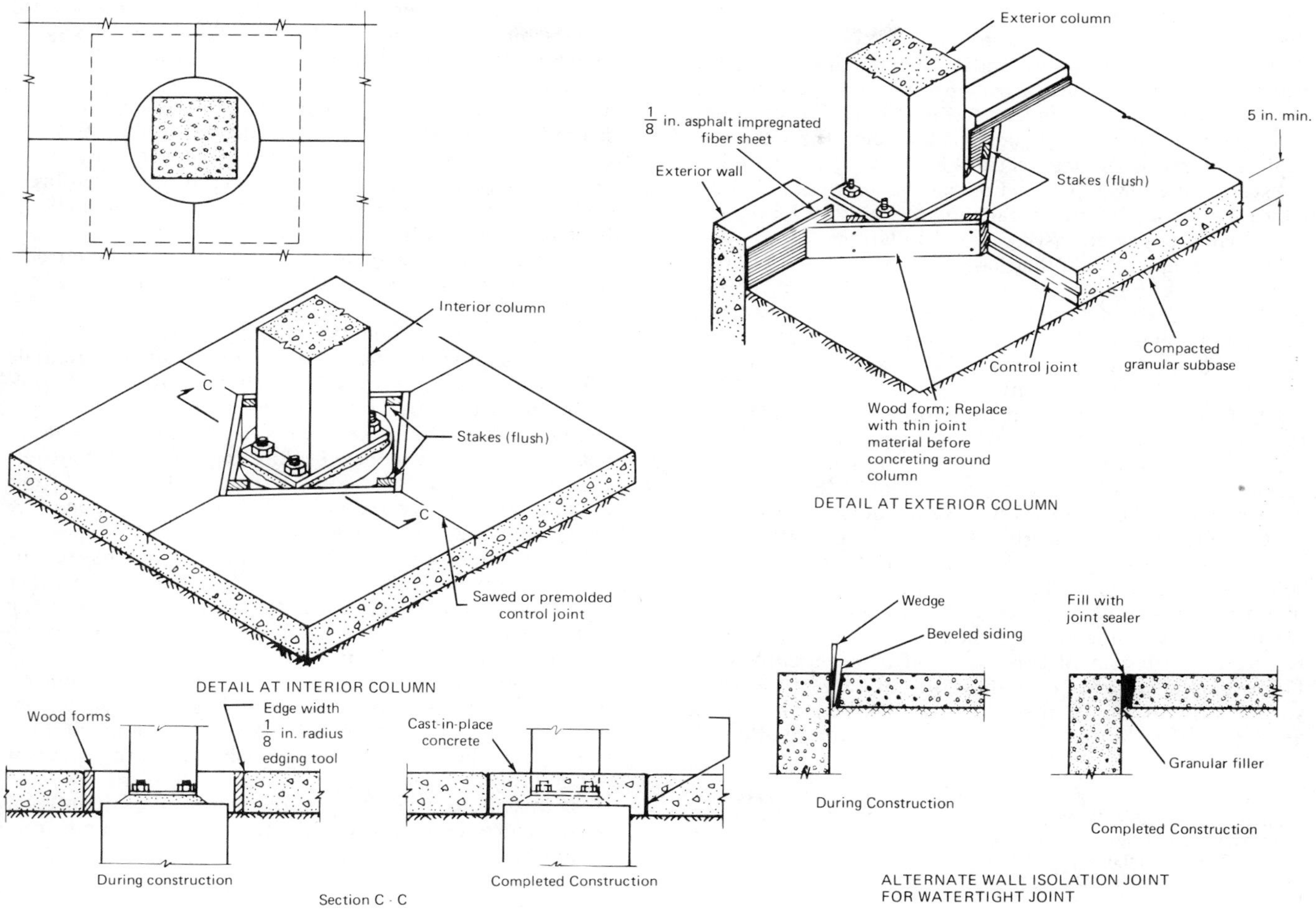

Fig. 4-18 Isolation joints at interior and exterior columns.

not recommended unless the unit load on the ground under the footing is greater than under the slab and special care is taken to consolidate the fill for some distance from the wall.

4.14.4 Construction Joints

Construction joints are temporary stopping places in concreting. They are necessary in large projects because there is a limit to how much concrete can be placed and finished in one work shift.

A true construction joint should constitute neither a plane of weakness nor an interruption in the homogeneity of the concrete; therefore, every effort must be made to ensure bonding of the newly placed concrete to that previously placed. To assure a bond, the old concrete should be thoroughly cleaned and dampened before the new concrete is placed. To avoid the possibility of breaking the bond after the floor is in use, tie bars are frequently used to assist in vertical load transfer across the joint. Tie bars have the added advantage of also being able to resist shearing stresses along the joint caused by differential shrinkage of the old and new concrete. However, since it is difficult to achieve monolithic joints in floors on grade, construction joints should be spaced and made to act as control joints.

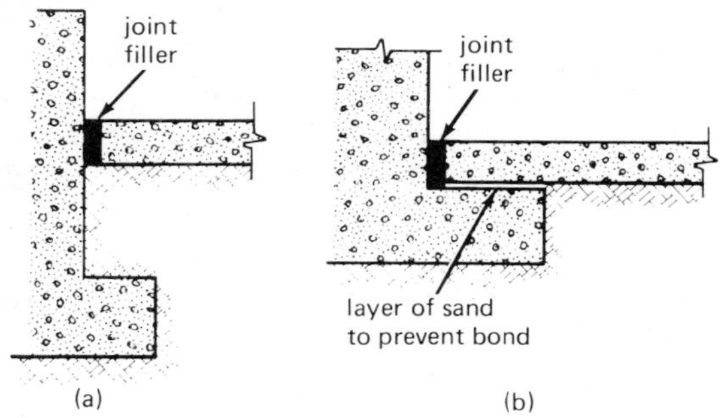

Fig. 4-19 Isolation joints at exterior wall. (a) Correct, and (b) incorrect.

4.15 MASSIVE STRUCTURAL CONCRETE WITHOUT JOINTS

In large sections the heat of hydration, which continues for several days, does not dissipate quickly and causes an increase in the temperature of the concrete. As the surfaces are subjected to ambient temperatures, they cool off, and a thermal gradient across the section is created. When the temperature differential between the cooled surface and the hot interior of the section exceeds 35°F, cracking occurs, which progresses from the surface into the interior. Traditionally, cracking of mass concrete was avoided by limiting the amount of cement, using special low-heat cement, cooling the aggregates, artificially cooling the freshly placed concrete for a number of days, limiting lift heights, and using frequent construction joints.

The extent of heat of hydration generated depends primarily upon the amount of cement per cubic yard of concrete, as well as its chemical composition. For structural concrete requiring a high cement content, the extent of heat of hydration of large sections is therefore magnified.

A technique developed in England[1] makes possible large continuous pours of structural concrete without joints. It is particularly helpful in casting mat foundations.

The technique of eliminating joints is based on controlling the temperature gradients and thus limiting internal thermal strains and avoiding externally applied restraints. Internal thermal strains are controlled by ensuring that no part of the hydrating concrete mass is allowed to become cooler than the hottest part by more than 35°F (20°C). Contrary to the traditional method of cooling the concrete, the solution here is to keep the concrete warm; in other words, to avoid rapid cooling so that the temperature gradient across the section is always less than 35°F.

To estimate the temperature rise in the concrete due to heat of hydration, the British report suggests replacing the complex formulas by a single calculation that gives the extreme temperature rise. The method indicates that the maximum temperature possible for hydrating concrete will be the concrete's temperature at the time of placing plus 23°F for each 170 lb cement content of the mix, per cubic yard. For example, if concrete with 700 lb of cement per cubic yard is cast at 50°F, the maximum temperature will be $50°\text{F} + 23°\text{F} \times 700/170 = 145°\text{F}$. This determination of the maximum temperature has been found usable for all practical purposes between 500 and 1000 lb per cubic yard cement-content mixes and for any sections with all dimensions greater than 6 ft.

In addition to controlling temperature gradients to minimize thermal cracking, it is important to avoid casting the fresh concrete against adjacent sides or between two opposite, massive sections of hardened concrete while reinforcement is continuous across each interface. The thermal expansion of the fresh concrete and subsequent inevitable cooling with its accompanying contraction cause tension cracks to form completely through the concrete, no matter how thick it is.

To employ this "no-joint" technique successfully, it is necessary to monitor the critical temperature differentials within the in-place concrete between the hottest and coolest points in the mass from the moment it is cast, for a number of days. It is sometimes necessary to measure more than two places to guard against missing the extreme differential. Usually thermo-couples are placed about 1 in. from the most exposed face and at the middepth of the pour.

When the gradient beings to increase and approaches 30°F, it is necessary at once to protect the exposed surfaces from further cooling. This can be done by a number of methods, such as: (1) insulating boards, (2) insulating blankets, (3) tenting, or (4) sand over and under polyethylene sheets. Also, the formwork of framed elements may need thermal protection to slow cooling. As the temperature rise due to heat of hydration does not usually continue beyond three days, it will take an additional number of days for the concrete to cool slowly to the ambient temperature.

This large-pour technique makes it possible to construct massive concrete sections without joints faster and more economically.

[1] FitzGibbon, Michael E., BA, MICE, "Large Pours for Reinforced Concrete Structures," Current Practice Sheets No. 28, 36, *Concrete*, a magazine published by the Concrete Society, 52 Grosvenor Gardens, London, England, SW1W0AQ.

5

Footings

FRITZ KRAMRISCH, D.Eng.*

5.1 GENERAL

The purpose of a footing is to transfer safely to the ground the dead load of the superstructure (weight), and all other external forces acting upon it. The latter consist not only of general live load (as in dwellings, warehouses, industrial buildings, and so on) or fill (as in silos, bunkers, tank supporting structures, and the like), but also the ground effects of various lateral forces such as wind, blast, or earthquake. In case any of these forces are of dynamic character, an allowance for impact should be included in the estimate of their magnitude unless a more exact evaluation is required. Footings also include foundation elements that are designed to resist uplift and overturning, and comprise in general all structural elements that will provide a stable base for the superstructure and safe transfer of all applied forces down to the ground.

The ways and means by which the transfer of these forces into the subsoil can be obtained are many. They depend primarily on the type and magnitude of the loading, the stiffness and structural behavior of the superstructure, the bearing capacity of the soil in general and that of a certain stratum in particular, the depth below ground surface where this soil stratum is encountered, and the depth of the groundwater level below grade. The type of foundation is also influenced, though to a lesser degree, by the geographical location and climatic condition of the site; the time of the year when the construction is to be executed; the speed of construction that is desired; the availability of material, equipment, and manpower; the prevailing economic conditions; and many other circumstances. The importance of these apparently secondary considerations can grow with the magnitude of the job and may, in certain cases, influence the successful outcome of an entire project. It is, therefore, not enough to approach the selection of the type of foundation with a cool, calm engineering mind; very often, experienced judgment that can envision advantages and foresee difficulties deserves serious consideration.

The type and magnitude of the loading will usually be furnished by the engineer designing the superstructure. It is up to the foundation engineer to collect all information regarding the purpose of the superstructure, the materials that will be used in its construction, its sensitivity to settlements in general and to differential settlements in particular, and all other pertinent information that may influence the successful selection and execution of the foundation design.

The evaluation of the maximum bearing capacity of the soil is to be made on the basis of soil mechanical considerations. The engineer designing the foundation should select the soil stratum that is most suitable for the support of the superstructure; must assume the appropriate safety factor to arrive at the allowable bearing pressure; and, finally, must decide on the most economical type of foundation to be used. For this reason it is essential for a foundation engineer to possess a good knowledge of the problems that are involved in the design and behavior of the superstructure, a certain familiarity with the basic principles of soil mechanics, and a good understanding of the interaction between both.

A detailed treatment of above topics does not fall within the scope of this handbook; however, a short discussion of

*Chief Civil Engineer ret., Albert Kahn Associates, Inc., Architects and Engineers, Detroit, Michigan. Lecturer at Arizona State University, Tempe, Arizona.

the basic considerations affecting the evaluation and distribution of the bearing pressures under footing bases is given below.

5.2 EVALUATION OF BEARING PRESSURES AT FOOTING BASES

5.2.1 General Principles

The distribution of the bearing pressures under a concentrically loaded, infinitely stiff footing, with frictionless base, resting on an ideal, cohesionless or cohesive subsoil,[5-1, 5-2] is generally known, and is shown in Fig. 5-1. Under ordinary conditions few soils will exhibit such behavior; no footing could be considered to be infinitely stiff. The distribution of the bearing pressure under somewhat flexible footings and ordinary soil conditions will be similar to those shown in Fig. 5-2; or it may assume any intermediate distribution. The assumption of a uniform bearing pressure over the entire base area of a concentrically loaded footing, as shown in Fig. 5-3, seems to be justified, therefore, for reasons of simplicity, and is common design practice. This assumption not only represents an average condition, but is usually on the safe side because most of the common soil types will produce bearing pressure distributions similar to that shown in Fig. 5-2(a). The foundation designer, however, shall keep in mind that the assumption of a uniform bearing pressure distribution was primarily made for reasons of simplicity and may, in special cases, require adjustment.

Any footing that is held in static equilibrium solely by bearing pressures acting against its base has to satisfy the following basic requirements regardless of whether it is an isolated or a combined footing:

1. The resultant of all bearing pressures, acting against the footing base (reaction), must be of equal intensity and opposite direction to the resultant of all loads and/or vertical effects due to moments and lateral forces, acting on the footing element (action).
2. The location where the resultant vector of the reaction intersects the footing base must coincide with the location where the resultant vector of the action is applied. Action and reaction are as defined under (1) above.
3. The maximum intensity of the bearing pressures under the most severe combination of service loads must be smaller than, or equal to, the maximum bearing pressure allowed for this kind of loading and type of soil, as determined by principles of soil mechanics.
4. The resultant vector of the least favorable combination of vertical loads, horizontal shears, and bending moments that may occur under service load conditions, including wind or earthquake, must intersect the footing base within a maximum eccentricity that will provide safety against overturning.

The method most commonly used for the design of footings and related elements for ordinary building construction, is the one in which static equilibrium is obtained by bearing pressures against the footing base only. This method is also the standard method that has been included in the *Building Code Requirements for Reinforced Concrete*, ACI 318-83.

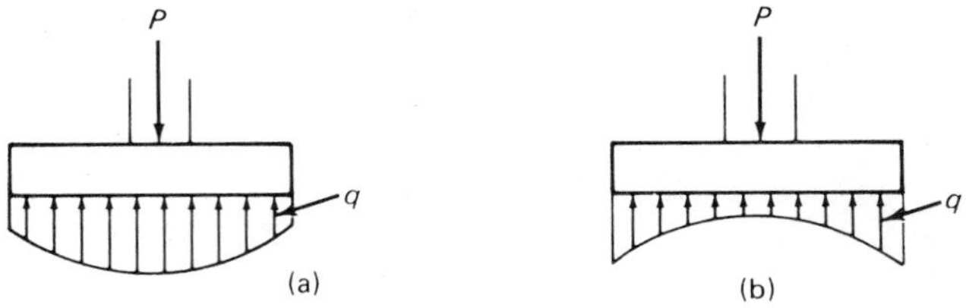

Fig. 5-2 Bearing pressure distribution for a flexible footing on ordinary soil. (a) On granular soil; and (b) on clayey soil.

For zero eccentricities, the bearing pressures will be uniformly distributed over the entire base area of the footing as shown in Fig. 5-3 and will have the intensity of $q = P/A_F$.

If the footing shall restrain the column base, i.e., if a bending moment has to be resisted by the subsoil alongside a concentric load, or if the column load is applied outside of the centroid of the base area of the footing, the bearing pressure distribution will vary depending on the magnitude of the eccentricity and its relationship to the kern distance, c_k. The kern distance can generally be evaluated as shown in Fig. 5-4.

When the eccentricity is equal to, or smaller than, the kern distance, c_k, the extreme (maximum or minimum) bearing pressures q_{min}^{max} can be found by superposing the flexural bearing pressures over the axial bearing pressures (see Fig. 5-5a).

When the eccentricity becomes greater than the kern distance, superposition cannot be applied anymore because it would result in tensile stresses between the soil and footing near the lifted edge of the base. Equilibrium can, however, be attained by resisting the load resultant by a bearing pressure resultant of equal magnitude and location. In this case the extreme bearing pressures at the edge of the base can be evaluated as shown in Fig. 5-5(b). The maximum edge pressure, q_{max}, must, under all conditions, be smaller than or equal to the maximum allowable soil pressure, q_a.

This condition applies until the eccentricity, e, of the load, P, reaches the edge of the footing base. Any greater eccentricity will result in overturning. Such a condition, however, can only occur on rock or on very hard, stiff soils. For most practical cases, edge-yielding can make a footing unusable and produce a condition that is equivalent to overturning. Edge-yielding will occur when the extreme bearing pressure at the pressed edge will cause failure in the bearing capacity of the subsoil. The eccentricity causing this condition will, therefore, limit the maximum useful eccentricity. Unless actual test results are available, the failure condition in the bearing capacity of the soil, q_f, can be assumed with about 2.5 times the allowable bearing capacity, q_a; the mininum safety factor against overturning is usually specified as 1.50, although somewhat greater safety factors are sometimes desirable. Introducing these requirements, we arrive

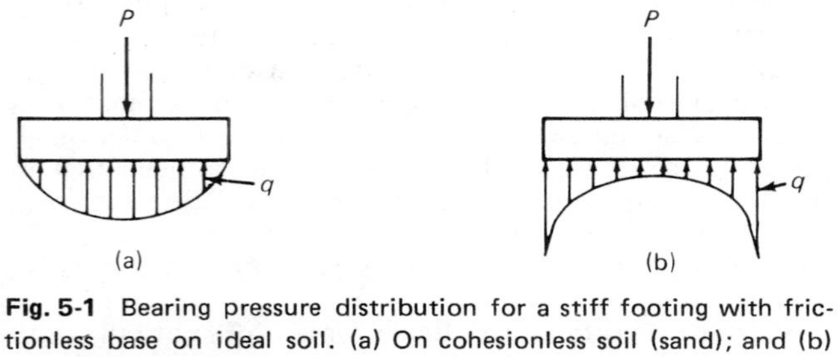

Fig. 5-1 Bearing pressure distribution for a stiff footing with frictionless base on ideal soil. (a) On cohesionless soil (sand); and (b) on cohesive soil (clay).

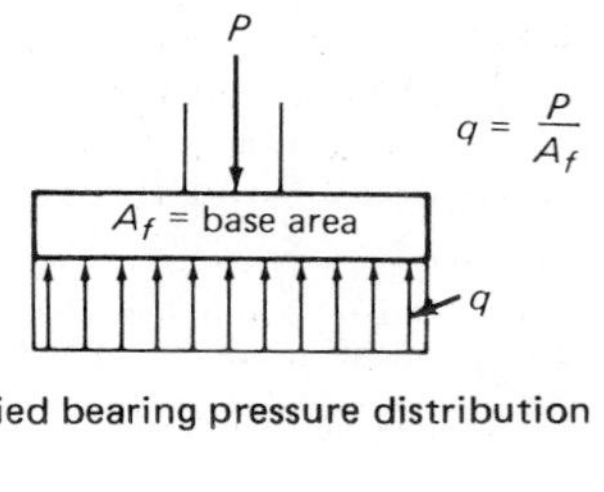

Fig. 5-3 Simplified bearing pressure distribution (commonly used).

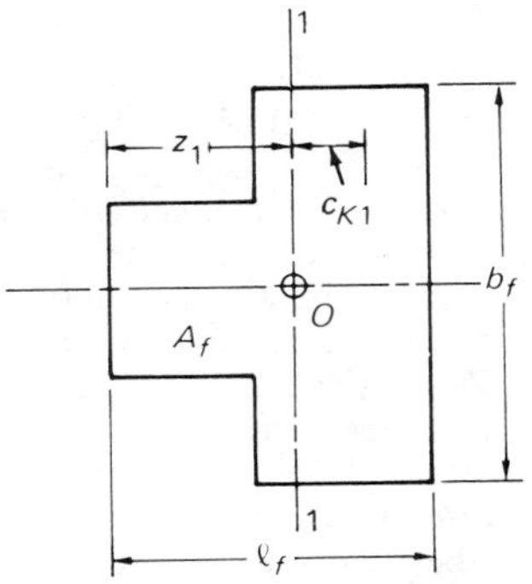

Fig. 5-4 Kern distance. (a) $c_{k1} = I_{F1}/A_{F1}z_1$; (b) I_{F1} = moment of inertia of footing base about neutral axis 1-1; (c) z_1 = distance of extreme fiber at opposite side of desired kern distance; (d) for strips, $c_k = l_f/6$.

in Fig. 5-5(c) at a maximum eccentricity, e_{max}, that can safely be utilized. The extent to which the design engineer will take advantage of this condition will depend on his judgment of the soil and on the sensitivity of the superstructure to lateral tilting that may occur if a loading, causing such an eccentricity, is applied for a longer period.[5-3]

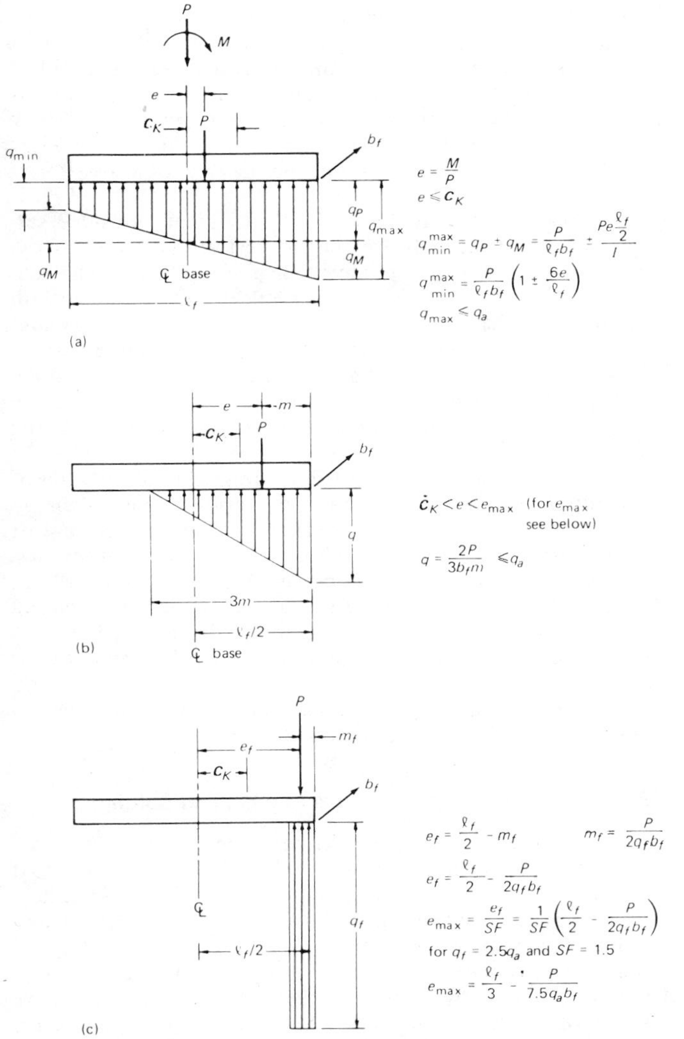

Fig. 5-5 Bearing pressure distribution under eccentric loading.

Moments occurring alongside concentric loads may be uniaxial or biaxial. If they occur in oblique directions, it is most practical to have their influence divided into two perpendicular components, each parallel to the main axes of the footing mat, and to superpose the resulting bearing pressures. Such conditions occur not only with isolated spread footings, but also with strip footings of limited length, as in the case of shear walls and the like.

Combined footings, (i.e., footings supporting more than one column load, such as exterior double-column footings), strip footings supporting spaced column loads, rafts, or mats can be designed as described above, as long as the entire foundation can be considered as infinitely stiff. In this case the resultant of all bearing pressures must be equal to the resultant of all loads, and the location of the bearing pressure resultant must coincide with the eccentricity of the load resultant. This approach is statically correct, but not necessarily close to the actual condition.

In certain cases it may be advisable to consider the footing as a beam on an elastic foundation and utilize the elasticity of the footing mat as well as that of the soil in the evaluation of the bearing pressures. The bearing pressures obtained by this method no longer follow a straight-line distribution across the contact area. They show maximum accumulations immediately below, and in the vicinity of, concentrated loads and greatly reduced intensities between them, as shown in Fig. 5-6. Such a pressure distribution can reduce the maximum design moments of a foundation considerably and is therefore in many cases quite economical. This method is, in addition, intriguing in its setup and appealing especially to the mathematically inclined engineer.[5-4]

In the analysis of a beam on elastic foundation, the soil is introduced as an elastic medium and the intensity of its reaction (bearing pressure) is assumed to be proportional to the deformation (settlement) of the footing, or soil, under the load.

Soils behave like elastic materials only to a limited degree, and only small portions of the settlements can be recovered in case of unloading the superstructure. Settlements of the subsoil very seldom occur immediately or shortly after the load application. Most soils are rather insensitive to sudden load changes and react more to average loadings applied over an extended period. Variable configurations of the line

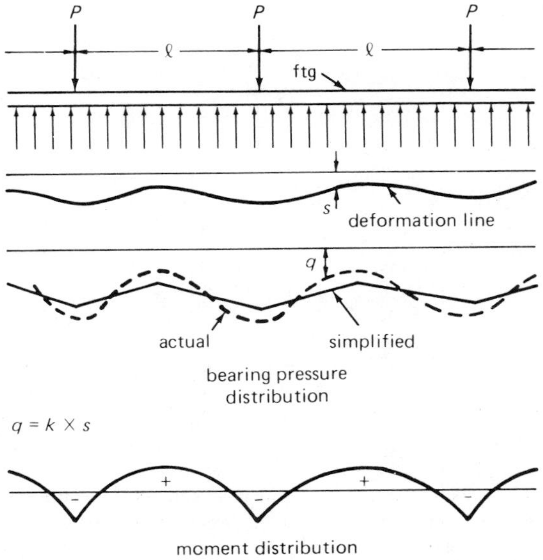

Fig. 5-6 General conditions for a beam on an elastic foundation.

of deformations during this period will cause a variable distribution of corresponding bearing pressures and internal stresses of the footing or foundation mat. In addition, the factor relating the magnitude of the bearing pressure to that of the deformation is not a constant but varies with the magnitude of the settlements.

Some of these characteristics have been overcome by refinements that have been introduced into this method, making it much more complex. If, however, foundations under combined footings are not too stiff and they follow the deformations of the subsoil, the beam-on-elastic-foundation-method will furnish bearing pressure distributions that are more closely related to the actual condition and more economical than the straight-line distribution of the bearing pressures.

Owing to the basic deficiencies of a foundation design executed with the help of elasticity relationships, as described in this chapter, it is advisable that results obtained by this method not be taken as exact solutions but more as a guide to arrive at a reasonable distribution of the bearing pressure; such foundations shall be designed with ample reserve capacity to sustain deviations from the theoretical findings. Under such considerations, the method can provide an excellent tool and a valuable aid in the design of sometimes rather difficult foundation problems.[5-5]

Figure 5-6 shows the basic relationships between loading, deformation, and bearing pressure as well as the resulting moment distribution. Formulas giving the bearing pressures and bending moments for a strip designed as a beam on elastic foundation are given later, in Fig. 5-22.

So far, bearing pressures against the footing base have been considered as the only means to resist external forces. Static equilibrium of a footing element subjected to lateral forces and/or moments in addition to vertical loads can also be attained with the help of passive earth pressure, as shown in Fig. 5-7. In this case lateral (passive) pressure of the soil is developed while it resists either lateral movements or tilting tendencies of the footing element. This method is sometimes used in the design of footings supporting tall, light superstructures such as transmission towers, light poles, and so on. One of the drawbacks of this method is that foundations are usually surrounded by backfill of questionable lateral resistance; but even where the footing is cast tightly against the original subsoil, passive earth pressure will only develop (unless the soil is very stiff or hard) after a certain movement or tilting of the superstructure has taken place.

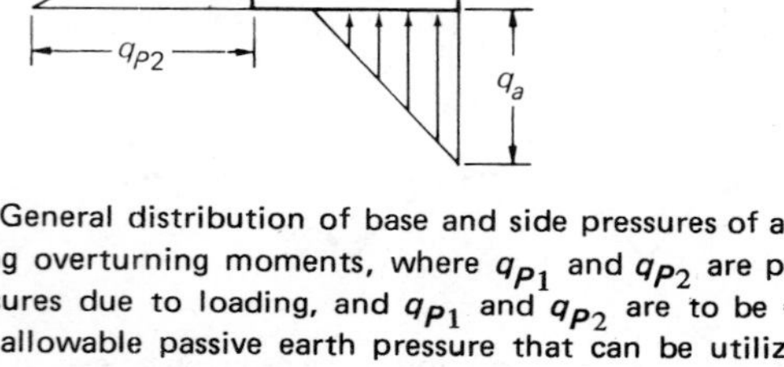

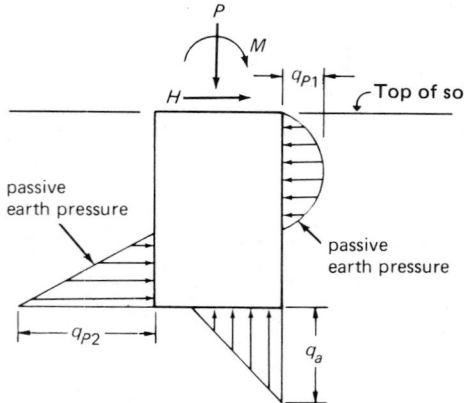

Fig. 5-7 "General distribution of base and side pressures of a footing resisting overturning moments, where q_{P_1} and q_{P_2} are passive earth pressures due to loading, and q_{P_1} and q_{P_2} are to be $\leqslant$ the maximum allowable passive earth pressure that can be utilized at each depth.

5.2.2 Strip Footings

1. Concentrically Loaded Strip Footings. The loadings of strip footings considered in the following section consist primarily of continuous line loads such as walls, or closely spaced, concentrated loads arranged in one direction. In many cases, the element causing the load, e.g., the wall itself, may provide sufficient stiffness to permit the assumption of a continuous line load, even if the intensity of the loading varies slightly (Fig. 5-8). Concentrated loads at close spacings sometimes require a stiffening member to transform the effects of the concentrated loads into a continuous loading. For small variations in the load intensity, the stiffening member may be the footing itself; or a foundation wall resting on top of the footings; or a beam formed by composite action of both. In either case the stifffening member has to be designed as an inverted continuous beam, upwardly loaded by a line load made up of the bearing pressure, q, and intermittently supported where the loads are applied, as shown in Fig. 5-9.

2. Eccentrically Loaded Strip Footings (or Strip Footings with Concentric Load and Moment at Base). For stability considerations, a free-standing wall is equivalent to one that is fully restrained at its base. Any shear and bending moment caused by lateral forces such as wind, fill, earth-pressure, simulated earthquake loadings, and so on, as well as all moments and shears caused by eccentrically applied vertical loads acting on projecting piers or brackets, must be transferred into the subsoil (see Fig. 5-10).

In cases where walls or columns are not free-standing but form a part of a frame work, such as in monolithic floor-wall or skeleton constructions, the wall or column may transfer a certain amount of moment into the ground. How successful such a design assumption may be, deserves a brief discussion.

It can be proved by statics that only small, lopsided settlements are needed to reduce the restraining effect at the base of a frame work to such a degree that it may be rendered useless. Such a rotation would produce a condition at the base that is in effect similar to that of a hinge and would increase the moments in the upper frame by the amount lost at the base. Cautious designers assume and design the base of such frameworks as hinges and make the structure this way independent of the deformation of the subsoil.

Bending moments or eccentric loadings caused by wind or earthquake simulating forces may very well be taken by restraint at the base of the footing mat; such forces are of short duration, and stable soils usually respond rather elastically to loads of short duration. It is, however, recommended that bending moments and eccentric forces caused by dead loads, fill, or average service live load conditions not be transferred into the soil unless the soil is very stiff or well consolidated.

For intermediate conditions, partial restraint in an amount selected by judgment may be utilized.

5.2.3 Isolated Spread Footings (square and oblong)

1. Concentrically Loaded Isolated Spread Footings. Isolated spread footings are used to resist column loads that are concentric with the centroid of the footing base. Theoretically such footings may have any shape. Square and oblong bases are most common, but round and polygonal shapes are being used under certain conditions. Footings may be solid or without a center portion, like a ring. We speak of a concentric loading if the resultant of all acting forces coincides with the centroid of the footing mat or bearing area, regardless of its shape. In this case the bearing

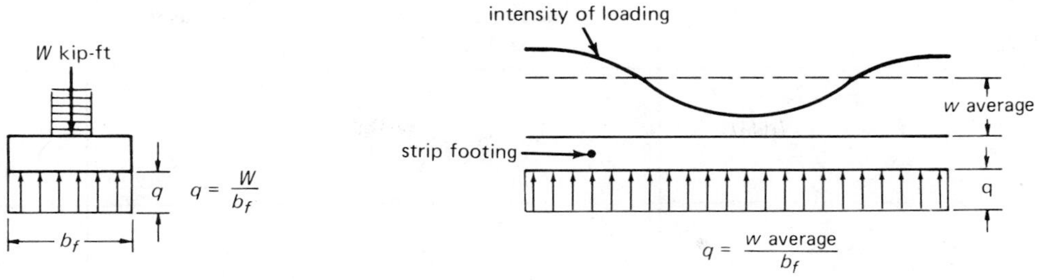

Fig. 5-8 Equalizing strip loadings of variable intensity.

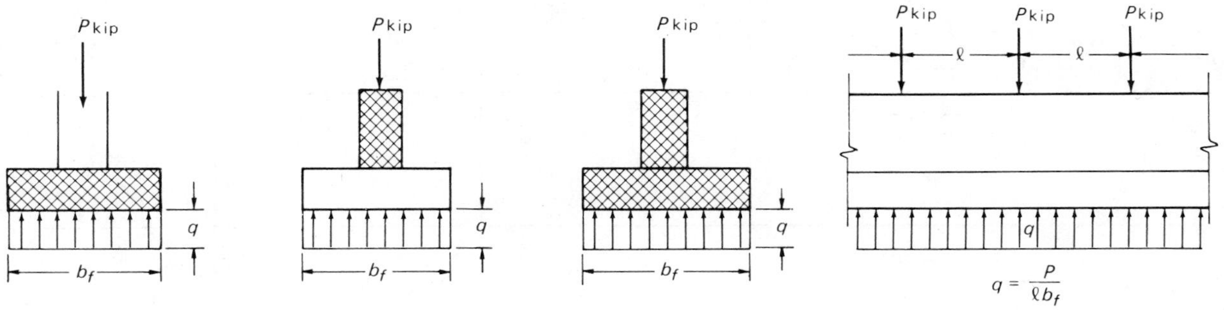

Fig. 5-9 Transforming concentrated loads into equal bearing pressures.

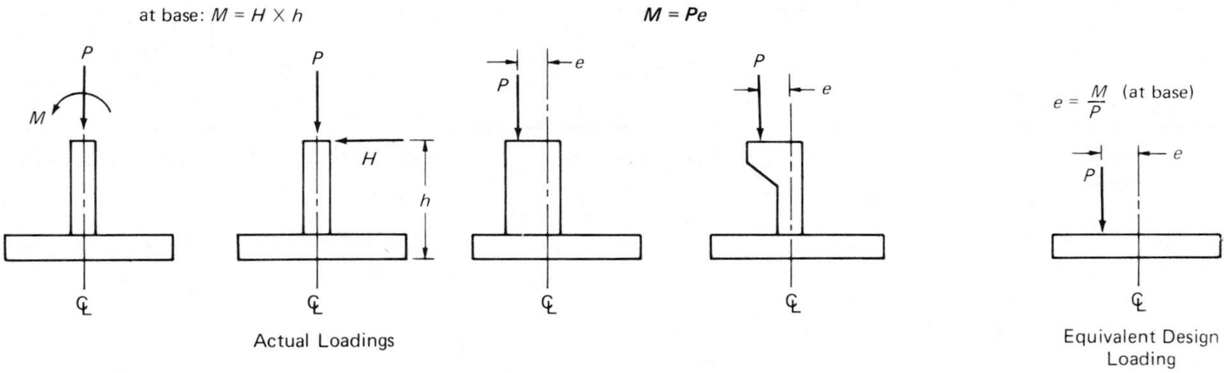

Fig. 5-10 Analogy between moments and eccentric loading conditions.

pressure distribution to be used for the design is usually assumed to be of uniform intensity over the entire area (see Fig. 5-3).

2. *Eccentrically Loaded Isolated Spread Footings or Isolated Footings with Concentric Load and Moment at Base.* In case of eccentric loadings caused by a moment at the column base, or by an eccentrically applied load, or by an unsymmetrical footing base, the bearing pressures will deviate from the uniform distribution and gradually vary across the footing base. The distribution of the bearing pressures will follow the rules as described above and shown in Fig. 5-5.[5-3] Figure 5-11 gives the base area, moment of inertia, and kern distance for various shapes of footing bases.[5-6] For irregular shapes, kern distance can be found as indicated in Fig. 5-4.

Round and polygonal footings, whether solid or ring-shaped, can only have a uniaxial eccentricity under the resultant of such a loading. Square and rectangular footings may have eccentricities in the direction of either one or both main axes, if the eccentricity falls in an oblique direction.

Figures 5-12 and 5-16 give the intensity of the extreme bearing pressure and the location of the zero pressure for various shapes of footings under eccentric loading.[5-6, 5-7]

Moments affecting the bearing pressures shall be determined about the elevation of the footing base. For evaluating the bearing pressures by means of Figs. 5-12 through 5-16, moments shall be converted first into eccentric loadings having the magnitude of the resultant loading, P, acting at the eccentricity $e = M_B/P$. In the evaluation of the resultant P it is important to investigate all conditions of loading that may occur simultaneously with the moment under consideration, i.e., the maximum as well as the minimum. In the calculation of the minimum loading it is advisable to reduce dead loads arbitrarily, as required similarly in the *ACI Building Code* 318-83, Chap. 9, to ascertain that the bearing pressure and/or eccentricity resulting from this loading condition is a maximum. See also Section 5.3.1.

5.2.4 Combined Footings

Independent, isolated footings for each column load usually provide the most economical method to design a foundation. Support of two or more column loads on one combined footing should be attempted only when required by one of the following conditions:

(a) Proximity of the building line or any other space limitation adjacent to a building column.

Shape of Base Area		*Base Area* A_F	*Moment of Inertia* I_x	*Kern Distance* c_k
CIRCLE	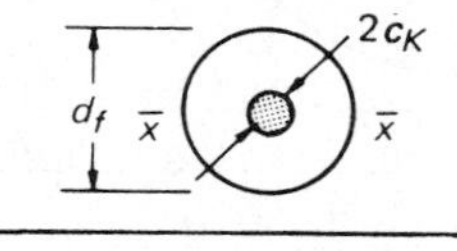	$0.785d_f^2$	$0.049d_f^4$	$0.125d_f$
OCTAGON	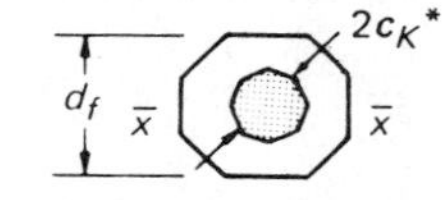	$0.828d_f^2$	$0.055d_f^4$	$0.122d_f$
HEXAGON	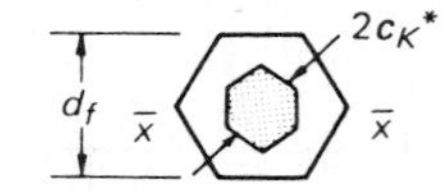	$0.866d_f^2$	$0.060d_f^4$	$0.120d_f$
SQUARE	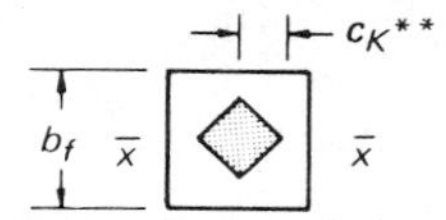	b_f^2	$0.083b_f^4$	$0.167d_f$
RECTANGLE	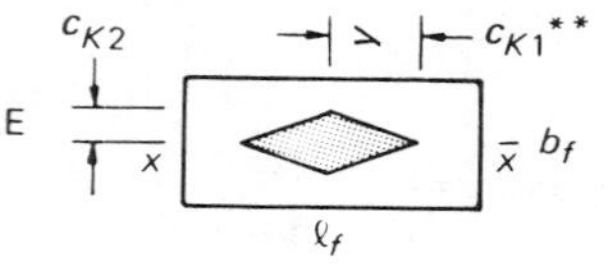	$l_f b_f$	$I_x = 0.083 l_f b_f^3$ $I_y = 0.083 b_f l_f^3$	$c_{k1} = 0.167 l_f$ $c_{k2} = 0.167 b_f$
RING	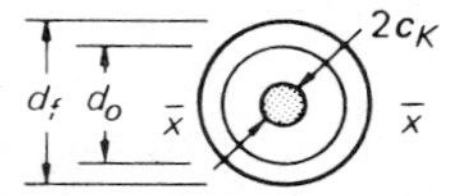	$0.785(d_f^2 - d_0^2)$	$0.049(d_f^4 - d_0^4)$	$\frac{d_f}{8}\left(1 + \frac{d_0^2}{d_f^2}\right)$

Fig. 5-11 Area properties of various cross sections. $*c_k$ = radius of circle inscribed in polygonal kern area. $**c_k$ = kern distance on main axis. Shaded area in diagrams indicates kern.

Values of C_1 and C_2 for Various Values of e/l_f

e/l_f	C_1	C_2
0.000	–	1.000
0.025	–	1.150
0.050	–	1.300
0.075	–	1.450
0.100	–	1.600
0.125	–	1.750
0.150	–	1.900
0.167	1.000	2.000
0.175	0.975	2.051
0.200	0.900	2.222
0.225	0.825	2.424
0.250	0.750	2.667
0.275	0.675	2.962
0.300	0.600	3.333
0.325	0.525	3.809
0.333	0.500	4.000
0.350	0.450	4.444
0.375	0.375	5.333
0.400	0.300	6.667
0.425	0.225	8.889
0.450	0.150	13.333
0.475	0.075	26.667
0.500	0.000	∞

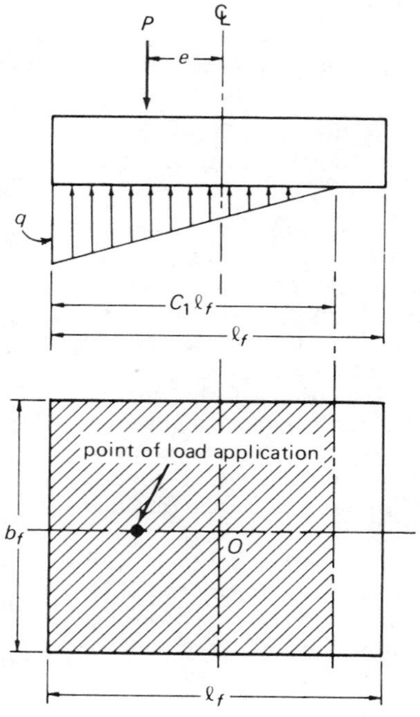

Fig. 5-12 Bearing pressure under square and oblong bases. Notation used includes: $A_F = l_f b_f$; $q_p = P/A_F$; $e = M/P$; $q = C_2 q_p$; and q_p = bearing pressure under concentric loading.

Values of C_1 and C_2 for Various Values of $e/b_f\sqrt{2}$

$e/b_f\sqrt{2}$	C_1	C_2
0.000	–	1.000
0.025	–	1.30
0.050	–	1.60
0.075	–	1.90
0.083	1.000	2.00
0.100	0.915	2.21
0.110	0.870	2.34
0.120	0.832	2.48
0.130	0.796	2.63
0.140	0.766	2.80
0.150	0.736	2.97
0.160	0.707	3.16
0.170	0.680	3.37
0.180	0.654	3.61
0.190	0.628	3.85
0.200	0.604	4.14
0.210	0.582	4.44
0.220	0.561	4.77
0.230	0.540	5.14
0.240	0.521	5.54
0.250	0.500	6.00
0.300	0.400	9.38
0.350	0.300	16.67
0.400	0.200	37.50
0.450	0.100	150.00
0.500	0.000	∞

Values of C_1 and C_2 for Various Values of e/d_f

e/d_f	C_1	C_2
0.000	–	1.00
0.025	–	1.20
0.050	–	1.40
0.075	–	1.60
0.100	–	1.80
0.125	1.000	2.00
0.150	0.910	2.23
0.175	0.830	2.48
0.200	0.755	2.76
0.225	0.685	3.11
0.250	0.615	3.55
0.275	0.550	4.15
0.294	0.500	4.69
0.300	0.485	4.96
0.325	0.420	6.00
0.350	0.360	7.48
0.375	0.295	9.93
0.400	0.235	13.87
0.425	0.175	21.08
0.450	0.120	38.25
0.475	0.060	96.10
0.500	0.000	∞

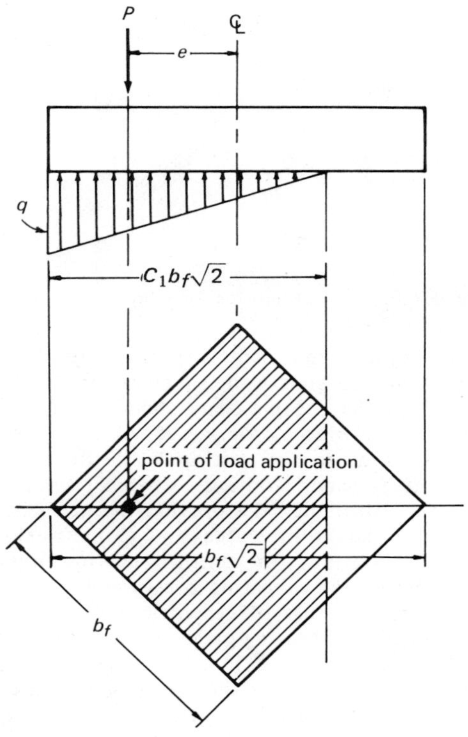

Fig. 5-13 Bearing pressure under square bases (about diagonal axis). Notation same as in Fig. 5-12 except for $A_F = b_f^2$.

$$q = C_2 q_p$$

$$q_p = \frac{P}{A_F}$$

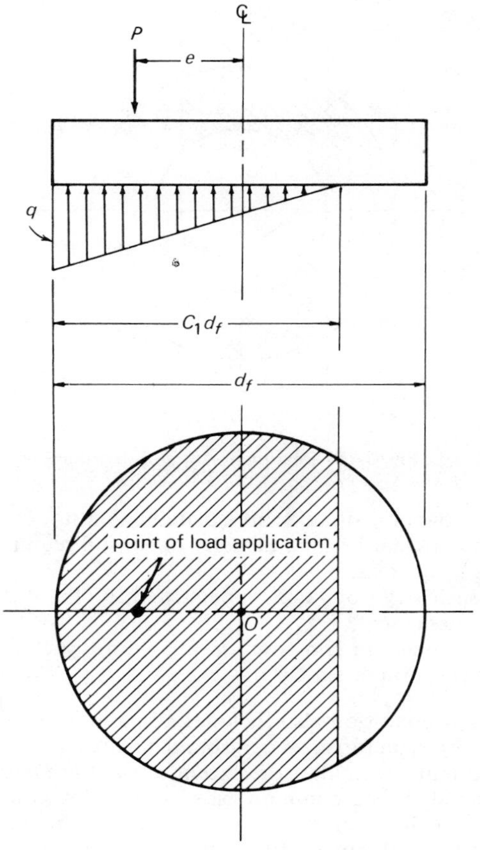

Fig. 5-14 Bearing pressure under circular and polygonal bases. Notation same as in Fig. 5-12 except for $A_F = d_f^2\pi/4$.
NOTE: For hexagonal bases use $d_f = 1.077\ d_{fo}$; for octagonal bases use $d_f = 1.041\ d_{fo}$, where d_{fo} = diameter of inscribed circle.

Values of C_1 and C_2 for Various Values of e/d_f and d_o/d_f

	C_1 – Values d_o/d_f						
e/d_f	0.0	0.5	0.6	0.7	0.8	0.9	1.0
0.125	2.00	–	–	–	–	–	–
0.150	1.82	–	–	–	–	–	–
0.175	1.66	1.89	1.98	–	–	–	–
0.200	1.51	1.75	1.84	1.93	–	–	–
0.225	1.37	1.61	1.71	1.81	1.90	–	–
0.250	1.23	1.46	1.56	1.66	1.78	1.89	2.00
0.275	1.10	1.29	1.39	1.50	1.62	1.74	1.87
0.300	0.97	1.12	1.21	1.32	1.45	1.58	1.71
0.325	0.84	0.94	1.02	1.13	1.25	1.40	1.54
0.350	0.72	0.75	0.82	0.93	1.05	1.20	1.35
0.375	0.59	0.60	0.64	0.72	0.85	0.99	1.15
0.400	0.47	0.47	0.48	0.52	0.61	0.77	0.94
0.425	0.35	0.35	0.35	0.36	0.42	0.55	0.72
0.450	0.24	0.24	0.24	0.24	0.24	0.32	0.49
0.475	0.12	0.12	0.12	0.12	0.12	0.12	0.25
0.500	0.00	0.00	0.00	0.00	0.00	0.00	0.00

	C_2 – Values d_o/d_f						
e/d_f	0.0	0.5	0.6	0.7	0.8	0.9	1.0
0.000	1.00	1.00	1.00	1.00	1.00	1.00	1.00
0.025	1.20	1.16	1.15	1.13	1.12	1.11	1.10
0.050	1.40	1.32	1.29	1.27	1.24	1.22	1.20
0.075	1.60	1.48	1.44	1.40	1.37	1.33	1.30
0.100	1.80	1.64	1.59	1.54	1.49	1.44	1.40
0.125	2.00	1.80	1.73	1.67	1.61	1.55	1.50
0.150	2.23	1.96	1.88	1.81	1.73	1.66	1.60
0.175	2.48	2.12	2.04	1.94	1.85	1.77	1.70
0.200	2.76	2.29	2.20	2.07	1.98	1.88	1.80
0.225	3.11	2.51	2.39	2.23	2.10	1.99	1.90
0.250	3.55	2.80	2.61	2.42	2.26	2.10	2.00
0.275	4.15	3.14	2.89	2.67	2.42	2.26	2.17
0.300	4.96	3.58	3.24	2.92	2.64	2.42	2.26
0.325	6.00	4.34	3.80	3.30	2.92	2.64	2.42
0.350	7.48	5.40	4.65	3.86	3.33	2.95	2.64
0.375	9.93	7.26	5.97	4.81	3.93	3.33	2.89
0.400	13.87	10.05	8.80	6.53	4.93	3.96	3.27
0.425	21.08	15.55	13.32	10.43	7.16	4.50	3.77
0.450	38.25	30.80	25.80	19.85	14.60	7.13	4.71
0.475	96.10	72.20	62.20	50.20	34.60	19.80	6.72
0.500	∞	∞	∞	∞	∞	∞	∞

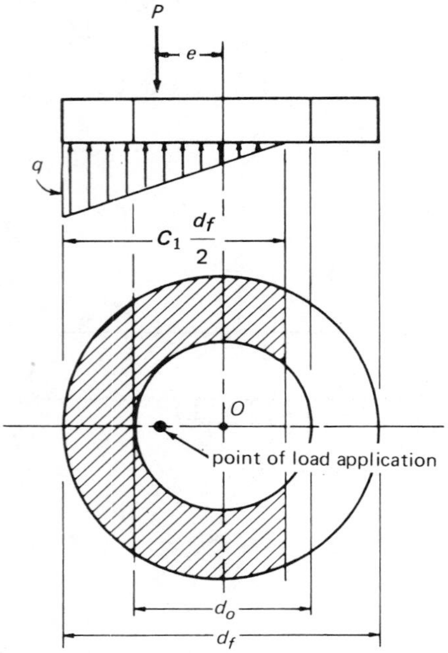

$$q = C_2 q_p$$

$$q_p = \frac{P}{A_F}$$

Fig. 5-15 Bearing pressure under ring-shaped bases. Notation same as in Fig. 5-12 except for $A_F = (d_f^2 - d_o^2)\,\pi/4$.

(b) Overlapping of adjacent isolated column footings.
(c) Insufficient bearing capacity of subsoil requiring large bearing areas.
(d) Sensitivity of the superstructure to differential settlements.
(e) Advantage in construction procedure, such as trench excavation or similar.

The first condition also includes all cases where elements of existing structures, underground facilities, excavations, embankments, etc., limit the extent of the foundation to be constructed; it is a common case for exterior building columns. If such a condition does not permit the design of a symmetrical column footing, the footing can be combined with the adjacent footing, located on the axis perpendicular to the limiting line, to balance its eccentricity. Depending on the difference in the intensity of the two combined column loads, as well as the existence of any simultaneous moments or shears, the eccentricity of the resultant load and its magnitude will govern the shape and size of the required base area for the footing. Any shape of base area can be used for the footing as long as it will satisfy the basic requirements discussed in Section 5.2.1. Rectangular, trapezoidal, and T-shaped footings are most commonly used. Figures 5-17 through 5-19 provide the necessary formulas to determine the required areas and geometrical relations to design such footings for uniform bearing pressure; or to evaluate the extreme bearing pressures for a given footing shape and size in case of eccentric load applications. It is recommended that bearing pressures be maintained over the entire base area of a combined footing, i.e., that it be designed in such a way that the eccentricity does not exceed the kern distance. Under extreme conditions, however, combined footings may also be designed for partial bearing with exclusion of tensile stresses at the footing base, as indicated.

Values of C_3 for Various Values of e_l/l_f; e_b/b_f

e_l/l_f	e_b/b_f													
	0.50	0.40	0.30	0.20	0.18	0.16	0.14	0.12	0.10	0.08	0.06	0.04	0.02	0.00
0.50	∞	∞	∞	∞	∞	∞	∞	∞	∞	∞	∞	∞	∞	∞
0.40	∞	37.5	18.8	12.5	11.6	10.9	10.2	9.6	9.0	8.5	8.0	7.5	7.1	6.7
0.30	∞	18.8	9.4	6.2	5.8	5.4	5.1	4.8	4.5	4.2	4.0	3.8	3.5	3.3
0.20	∞	12.5	6.2	4.1	3.9	3.6	3.4	3.2	3.0	2.8	2.7	2.5	2.4	2.2
0.18	∞	11.6	5.8	3.9	3.6	3.4	3.2	3.0	2.8	2.6	2.5	2.3	2.2	2.1
0.16	∞	10.9	5.4	3.6	3.4	3.2	3.0	2.8	2.6	2.5	2.3	2.2	2.1	2.0
0.14	∞	10.2	5.1	3.4	3.2	3.0	2.8	2.6	2.5	2.3	2.2	2.1	2.0	1.8
0.12	∞	9.6	4.8	3.2	3.0	2.8	2.6	2.5	2.3	2.2	2.1	2.0	1.8	1.7
0.10	∞	9.0	4.5	3.0	2.8	2.6	2.5	2.3	2.2	2.1	2.0	1.8	1.7	1.6
0.08	∞	8.5	4.2	2.8	2.6	2.5	2.3	2.2	2.1	2.0	1.8	1.7	1.6	1.5
0.06	∞	8.0	4.0	2.7	2.5	2.3	2.2	2.1	2.0	1.8	1.7	1.6	1.5	1.4
0.04	∞	7.5	3.8	2.5	2.3	2.2	2.1	2.0	1.8	1.7	1.6	1.5	1.4	1.2
0.02	∞	7.1	3.5	2.4	2.2	2.1	2.0	1.8	1.7	1.6	1.5	1.4	1.2	1.1
0.00	∞	6.7	3.3	2.2	2.1	2.0	1.8	1.7	1.6	1.5	1.4	1.2	1.1	1.0

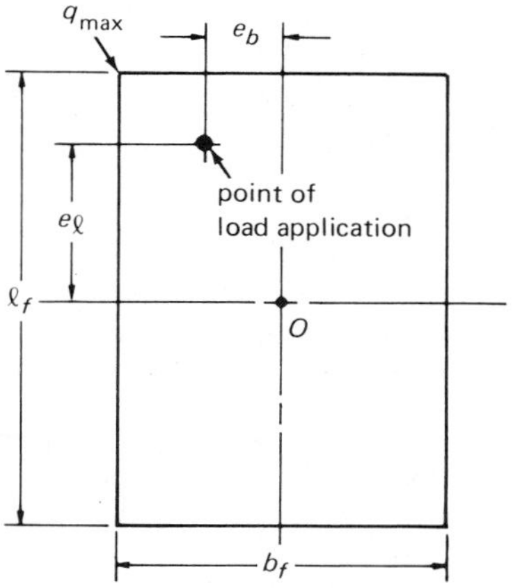

Fig. 5-16 Maximum corner pressure for biaxial eccentricities. Notation same as in Fig. 5-12 except for $q_{max} = C_3 q_p$, $q_p = P/A_F$.

If the nearest column that can be utilized for the design of a combined footing is too far away to permit a combined footing to be built economically, counterweights, or deadmen, can be provided to balance the eccentric loading of a footing, as shown in Fig. 5-20. In such a case it is advantageous to design the footing to establish concentric loading for each element; otherwise, edge pressure conditions have to be investigated carefully. In the evaluation of the available weight on a deadman, it is recommended that one rely primarily on the actual developed weight, and disregard load spreading or internal friction as far as possible. The safety factor to be used in the design of such foundations shall be at least the same as used for overturning.

Referring to condition (b), if overlapping of isolated single footings occurs, combination of the two footings into one will permit greater lateral spreading in maintaining a symmetrical base area (see Fig. 5-21).

In (c) and (d) above, it can easily be seen that a combination of two or more column loads by means of strips, rafts, or mats will not only spread the load over a bigger area, but will also give the foundation a monolithic quality that will help to bridge over soft spots in the subsoil. This will reduce the risk of differential settlements. Such a design can be structurally desirable and may also be economical on soft or irregular subsoils. If the foundation is stiff enough (either by means of its own thickness or by means of well-placed, monolithic basement walls), then the foundation may be considered as one unit. The shape of its base and the resulting bearing pressures can be evaluated as a combined footing. The stiffness of a strip, raft, or mat foundation alone is often not great enough to produce sufficient rigidity; in this case, the footing is best treated as a beam on elastic foundation. Figure 5-22 gives the basic equations required to design and evaluate the bearing pressures and moments using the simplified method.[5-3, 5-5]

As far as the last condition goes, the advantage that may be gained is primarily economic.

5.3 BASIC DESIGN PROCEDURE—REINFORCED CONCRETE FOOTINGS WITH EQUALLY DISTRIBUTED BEARING PRESSURES

5.3.1 Footing Size

It is important to keep the determination of the footing size completely separate from the design of the footing strength. The determination of the footing size, or of its width, as in the case of a strip footing or foundation raft, depends on the following criteria:

(a) The evaluation of the service loads. (The term "service" is used here to differentiate the actual quantities from the factored loads to be used in the strength design.)
(b) The evaluation of the bearing pressures at the footing base under the service loads, and the maximum allowable soil pressure.

In the evaluation of the service loads, the entire dead load (weight) of the superstructure, including the weight of the footing with surcharge and all floor live loads, should be used in their actual intensity. Floor live loads of multistory buildings, having an intensity of 100 lb/ft^2 or less may, according to the requirements of many local building codes, be reduced on the assumption that the full live load will hardly ever occur simultaneously on all floors. Reductions for live loads exceeding 100 lb/ft^2 are seldom permitted by building codes because such loads usually apply to warehouses or storage facilities.

Floors of industrial buildings are often designed for heavy loads in anticipation of machines that may have to be moved across the floor or will have to be set up at certain unforeseen locations. It is up to the judgment of the design-

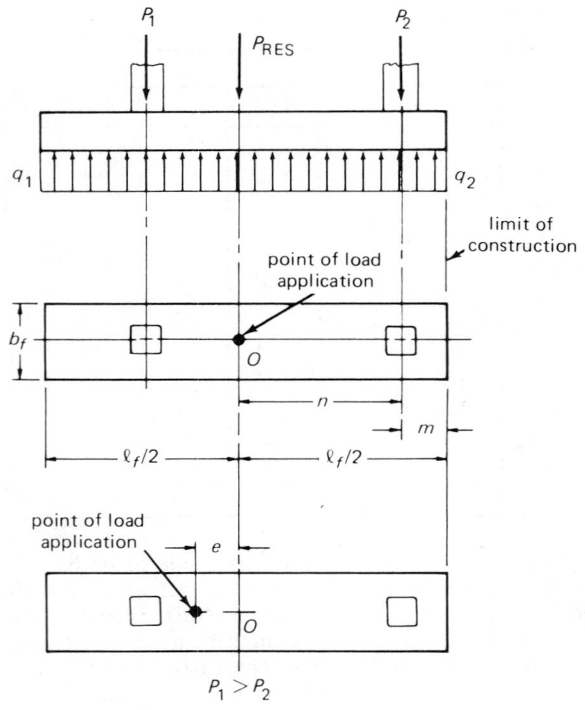

1) For uniform bearing pressure

$$l_f = 2(m+n), \quad P_{RES} = P_1 + P_2$$

$$b_f = \frac{P_{RES}}{q\,l_f}, \quad q_1 = q_2 = q$$

2) For eccentric loading e on either side of O; q_1 = bearing pressure at loaded edge

for $e \leqslant \frac{l_f}{6}$ $\quad e = \frac{M_{RES}}{P_{RES}}$

$$\frac{q_1}{q_2} = \frac{P_{RES}}{b_f\,l_f}\left(1 \pm \frac{6e}{l_f}\right)$$

for $e > \frac{l_f}{6} < e_{max}$

$$q_1 = \frac{2P_{RES}}{3\left(\frac{l_f}{2} - e\right) b_f}$$

$q = 0$ at a distance $3\left(\frac{l_f}{2} - e\right)$ from loaded edge (for e_{max} see Fig. 5-5c).

Fig. 5-17 Strap footing.

ing engineer and his personal knowledge of the manufacturing processes, to suggest a justified live load reduction to be used for the design of columns and footings and have it approved by the respective authority.

Crane loads, hoist loads, equipment loads, and the like have to be included in full, even if they are only of short duration.

Impact caused by the occasional passing of a crane need not be considered in the design load of a footing; however, impact caused by continuously operating hammers or reciprocating machines shall be considered in the loading. Footings resting on loose, granular subsoil may require special provisions to prevent the transfer of impact or vibrations to the subsoil.

Loadings caused directly or indirectly by fill, lateral earth pressure, or water pressure have to be considered in full.

Lateral loads due to wind and their related effects, such as vertical forces and moments caused by them, have to be determined and considered in the design.

In certain geographical areas, earthquake-simulating lateral forces have to be applied at all mass centers of the structure or as otherwise required by the applicable building code, and have to be transmitted through the foundation into the subsoil. These forces can act in any direction, but do not have to be considered simultaneously with the wind forces. Whichever force will have the greater effect on the element under consideration will govern.

Building codes usually permit an increase in the allowable stresses by 33% where wind or seismic forces are included. The load combination resulting in the largest required base area shall be used.

In general, footings shall be designed to be at least as strong as the design loads of the columns they support; all possible combinations of forces that can act simultaneously on the footing under consideration have to be considered. For the design of footings for equal bearing pressures under average service load conditions, see Section 5.4.2.

The *Building Code Requirements for Reinforced Concrete*, ACI 318-83, stipulates the following combinations of load effects as generally applicable:

1. $(D + L)$ The resulting maximum bearing pressure must be smaller than or equal to the maximum allowable soil pressure, $q \leqslant q_a$.

2. $(D + L + W)$, or $(D + L + E)$ The resulting maximum caused by either loading (W or E), must be smaller than $1.33q_a$. This increase of 33% above the allowable bearing pressure is usually accepted but ought to be checked with the local building code requirements.

In case of uplift or overturning, the most critical combinations shall be investigated. Live loads shall only be considered with uplift forces where they contribute to the overturning; wherever dead load counteracts the overturning, it shall be introduced with only 0.9 of its actual value, to be on the safe side. The required safety factor against overturning shall not be less than 1.50, but shall be checked with the local building code requirements.

The evaluation of the intensity and distribution of the bearing pressures shall be done along the lines discussed in Section 5.2.

The maximum allowable soil pressure shall be determined by principles of soil mechanics. Unless the engineer designing the foundation has determined the maximum allowable soil pressure himself, it is important for him to understand the basis on which it was determined and the safety factor that was used in its evaluation. It is also important for him to find out whether the weight of the overlying soil surcharge was included in the evaluation of the allowable bearing pressure, and what the minimum depth below ground surface is at which this bearing pressure can be developed. He should also know the influence that raising or lowering of a footing may have on the magnitude of the allowable bearing pressure, and the elevation and possible fluctuations of the groundwater level. All these factors can be of important influence on the design of the foundation.

5.3.2 Footing Strength

1. *General Principles.* The strength design, also called ultimate strength design (USD), of practically all types of footings can proceed along the lines described below.

For uniform bearing pressure (q) (general relationship)

$$\frac{b_f}{b_{f1}} = \frac{3(n+m) - l_f}{2l_f - 3(n+m)}$$

$$(b_f + b_{f1}) = \frac{2P_{\mathrm{RES}}}{ql_f}$$

$$C_1 = \frac{l_f(2b_f + b_{f1})}{3(b_f + b_{f1})}; \quad C_2 = \frac{l_f(b_f + 2b_{f1})}{3(b_f + b_{f1})}$$

for uniform bearing pressure of known magnitude (q).*

for $P_1 > P_2$ (shown dotted)

$$b_f = \frac{6P_{\mathrm{RES}}}{l_f^2 q}\left(n_1 + m - \frac{l_f}{3}\right)$$

$$b_{f1} = \frac{2P_{\mathrm{RES}}}{ql_f} - b_f$$

for $P_1 < P_2$ (as shown)

$$b_f = \frac{6P_{\mathrm{RES}}}{l_f^2 q}\left[\frac{2}{3}l_f - n_2 - m\right]$$

$$b_{f1} = \frac{2P_{\mathrm{RES}}}{ql_f} - b_f$$

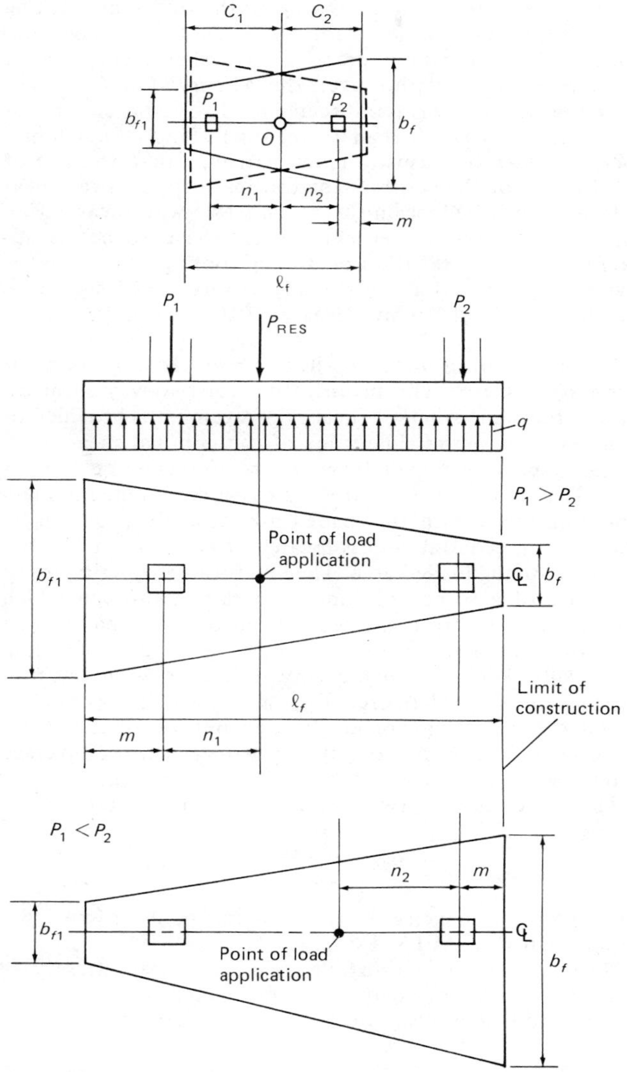

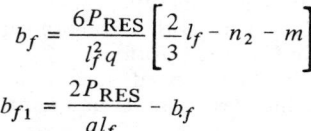

Fig. 5-18 Trapezoidal footing. "For non-uniform bearing pressure under eccentric loading see: A. Zweig, "Eccentrically loaded trapezoidal or round footings," *Proceedings, ASCE*, ST-1, Feb. 1966, pp. 161–168.

$$P_1 < P_2$$

for uniform bearing pressure $q_1 = q_2 = q$

$$b_1 = \frac{P_{\mathrm{RES}}}{q}\left[\frac{2(n+m) - l_2}{l_1(l_1 + l_2)}\right]$$

$$b_2 = \frac{P_{\mathrm{RES}}}{l_2 q} - \frac{l_1 b_1}{l_2}$$

area $A_F = l_1 b_1 + l_2 b_2$

$$n + m = \frac{l_1^2 b_1 + 2l_1 b_1 l_2 + l_2^2 b_2}{2(l_1 b_1 + l_2 b_2)}$$

$$c_{K1} = \frac{I}{(n+m)A_F}, \quad c_{K2} = \frac{I}{[l_f - (n+m)]\,A_F}$$

for eccentric load application:

$$e \leqslant c_k$$

$$q_1 = \frac{P_{\mathrm{RES}}}{A_F} \mp \frac{P_{\mathrm{RES}}\,e[l_f - (n+m)]}{I}$$

$$q_2 = \frac{P_{\mathrm{RES}}}{A_F} \pm \frac{P_{\mathrm{RES}}\,e(n+m)}{I}$$

The upper sign applies if e is towards wide side. The lower sign applies if e is towards narrow side.

$$e > c_K$$

if e is towards wide side

$$q_2 = \frac{2P_{\mathrm{RES}}}{3b_2[(n+m) - e]}$$

if e is towards narrow side

$$q_1 = \frac{2P_{\mathrm{RES}}}{3b_1[l_f - (n+m+e)]}$$

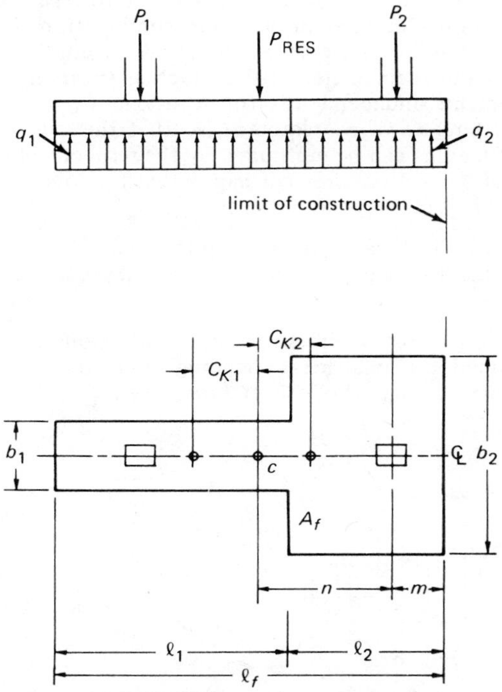

Fig. 5-19 T-shaped footings.

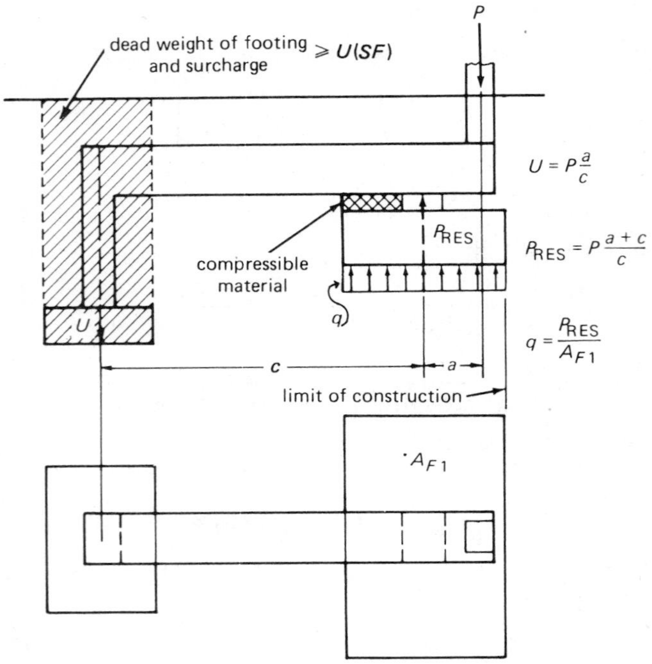

Fig. 5-20 Footing with deadman.

The size of the footing or foundation, and the resulting bearing pressures, are determined from the service loads (actual loads). In order to perform the strength design of a footing, the various types of loads have to be multiplied by the respective load factors; and the resulting bearing pressures have to be reevaluated for these factored loads. It has to be kept in mind, however, that these newly evaluated bearing pressures are of purely mathematical nature and have no soil-mechanical significance. They are calculated reactions to an imaginary factored loading condition resisted by the strength capacity of the foundation element.

The strength design requires that the minimum capacity of every structural element be sufficient to resist the factored loadings in their most severe combination. For this purpose it is advisable to assemble all different types of loadings and their related effects, such as shears and bending moments, independently for dead load, live load, wind, and earthquake (if so required), so that the sum of each type of loading can be multiplied by the respective load factor. ACI 318-83 requires in Chap. 9 that the following load factors be used:

1.4, for dead loads, fills and liquid loads
1.7, for live loads, wind loads and earth pressures
1.87, for earthquake loads

After the most severe strength combinations have been determined for axial loads, bending moments, and shears according to Chap. 9 of the *ACI Building Code*, the footing

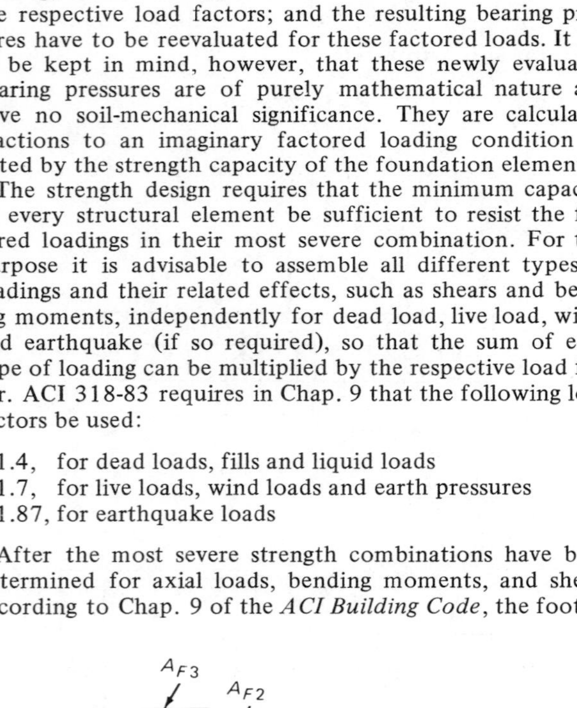

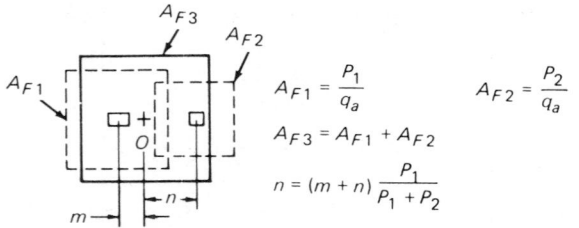

Fig. 5-21 Combined footing areas.

has to be designed strong enough to resist these factored load combinations and their resulting bearing pressures.

It has to be kept in mind, however, that factored loadings do not always furnish the same eccentricities as service loads. This is true where lateral loads, or other loads causing eccentricities or overturning, are of a different origin from the loads or load combinations causing the concentric loading. Consequently, the various loadings will have to be multiplied by different load factors, and will thus result in bearing pressure distributions that are, in principle, different from those obtained for the unfactored service load conditions. Since this is not the intent of the design, it is advisable to determine the resulting bearing pressure distribution from the service loads and then multiply it by an appropriate load factor for the strength design. In such a case, either the maximum load factor of 1.7 may be used, or an approximate load factor between 1.7 and 1.4 may be selected.

A footing, depending on its type, can be compared with a heavily loaded bracket, or with a column capital, or with a beam or slab, all in inverted position.

Because of the heavy (inverted) loading due to the bearing pressure, the thickness (depth) of a reinforced concrete footing is, with the exception of long-span rafts and mats, usually governed by shear. Owing to the different working conditions under which footings often have to be constructed, it is common practice to design them without shear reinforcement; however, the *ACI Building Code* permits the use of shear reinforcement, if so required, except for mats or slabs less than 10 in. thick. Since the column, pier, or pedestal is usually much narrower than the size of the footing or the width of the footing strip, one-way shear action (also called beam-shear), *and* two-way shear action (also called slab- or perimeter-shear), have to be investigated. With the exception of oblong footings or long rafts, two-way shear action will usually govern the design. It is therefore advisable to investigate it first.

2. Investigation for Two-Way Shear Action (Slab- or Perimeter-Shear). The investigation for two-way shear action is the same whether shear reinforcement is provided or not; only the permissible value of the nominal shear stress varies, depending on whether shear reinforcement is used or not. The nominal shear stress is to be determined along a line concentric with the loaded area, usually provided by the pier or pedestal, and is located at a distance $d/2$ from the face of the loaded area. The line forms the perimeter of the base of a truncated cone or pyramid, through which bearing stresses are considered to be transformed straight into the supporting subsoil. Considering the case of a concentrically loaded, isolated spread footing with uniform bearing pressure, the force V_{2u}, as shown in Fig. 5-23, is pressing the footing portion located outside of ΔA_{v2} upward. This force is resisted by the two-way shear of average intensity, v_{2u}, acting along the perimeter area $b_0 d$.

The shear stress for two-way shear action is therefore:

$$v_{2u} = \frac{V_{2u}}{\phi b_0 d} 1000$$

where ϕ is the capacity reduction factor for shear ($\phi = 0.85$) from Chap. 9 of ACI 318-83.

The maximum permissible two-way shear stress $v_2 = 4\phi\sqrt{f'_c}$. However, where the β_c-ratio exceeds 2, the maximum permissible shear stress shall be reduced to

$$v_2 = \left(2 + \frac{4}{\beta_c}\right)\phi\sqrt{f'_c}.$$

This shear stress is applicable to all concrete footings with-

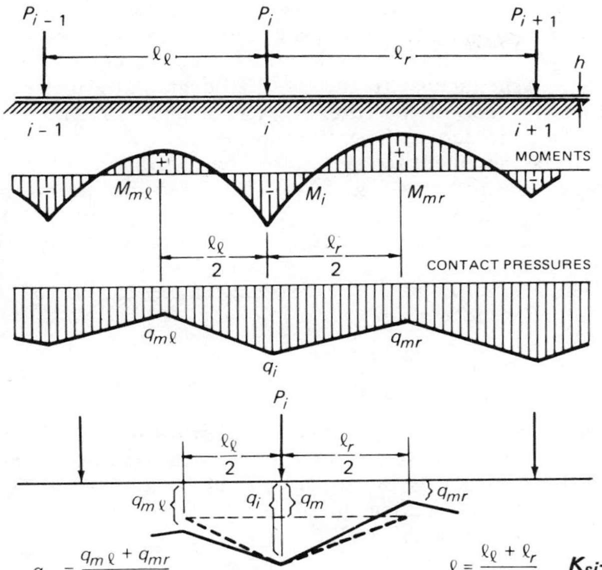

$q_m = \frac{q_{mℓ} + q_{mr}}{2}$ $\qquad ℓ = \frac{ℓ_ℓ + ℓ_r}{2}$

$$l = \frac{l_l + l_r}{2}, \qquad \frac{1.75}{\lambda} < l < \frac{3.50}{\lambda}$$

minimum total length: $\frac{5.25}{\lambda}$ or 3 bays

$$\lambda = 2.9\sqrt[4]{\frac{K_{si}F}{f'_c h_3}} : \left(\frac{1}{\text{ft}}\right)$$

Form Factors, F 1) b = average size in ft.

Sand	b 1)	5	10	15	LARGE
	F	0.36	0.30	0.275	0.25
Clay	l/b 2)	1	2	3	STRIP
	F	1	0.83	0.77	0.67

2) 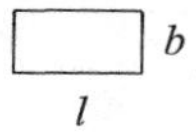

***K_{si}*-Values**

	Sand			*Clay*		
N	LOOSE 4-10	MEDIUM 10-30	DENSE 30-50	STIFF 8-15	VERY STIFF 15-30	HARD >30
K_{si}	20-60	60-300	300-1000	50-100	100-200	>200

at column (i): $\quad M_i = -\frac{P_i}{4\lambda}(0.24\lambda l + 0.16) \leqslant -\frac{P_i l}{12}$

for about equal spans

$$q_1 = \frac{5P_i}{l} + \frac{48M_i}{l^2}$$

at midspan left: $\quad q_{ml} = 2P_i\frac{l_r}{l_l l} - q_i\frac{l}{l_l}$

for equal spans $q_m = \frac{2P_i}{l} - q_i$

at midspan right: $\quad q_{mr} = 2P_i\frac{l_l}{l_r l} - q_i\frac{l}{l_r}$

at midspan left: $\quad M_{ml} = M_{ol} + \frac{M_l + M_i}{2}$

$$M_{ol} = \frac{l_l^2}{48}(q_l + 4q_{ml} + q_i)$$

at midspan right: $\quad M_{mr} = M_{or} + \frac{M_r + M_i}{2}$

$$M_{or} = \frac{l_r^2}{48}(q_r + 4q_{mr} + q_i)$$

<u>End Condition:</u> (The index e in P_e, M_e, and q_e refers to the end column.)

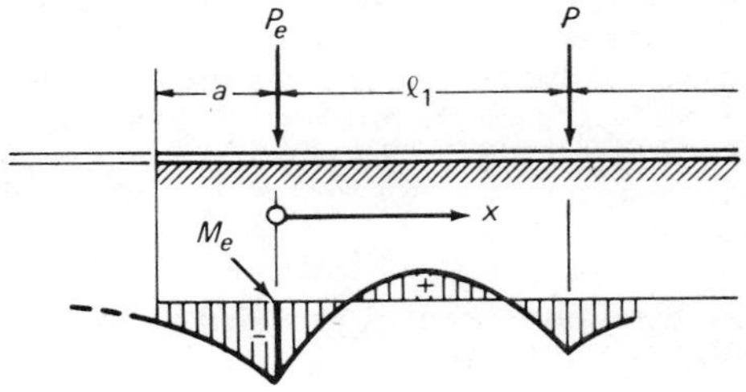

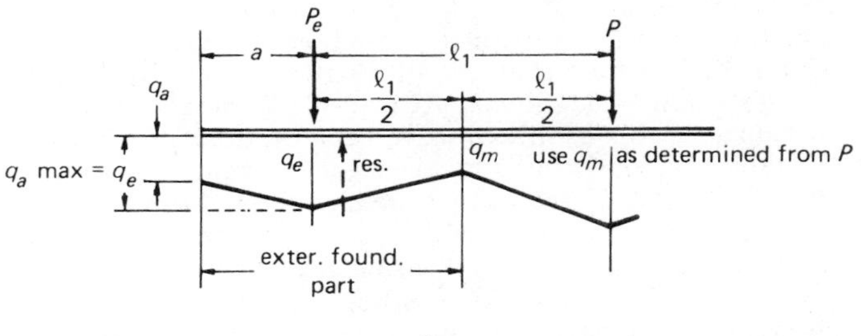

at end column: 1) $M_e = -\frac{P_e}{4\lambda}(0.13\lambda l_1 + 1.06\lambda a - 0.50)$

$$q_e = \frac{4P_e + (6M_e/a) - q_m l_1}{a + l_1} \qquad q_a = -\frac{3M_e}{a^2} - \frac{q_e}{2}$$

2) $M_e = -\left(\frac{4P_e - q_m l_1}{4a + l_1}\right)\frac{a^2}{2}$

$$q_e = q_a = \frac{4P_e - q_m l_1}{4a + l_1}$$

The smaller M_e governs.

Fig. 5-22 Simplified design of combined footings.[5-5]

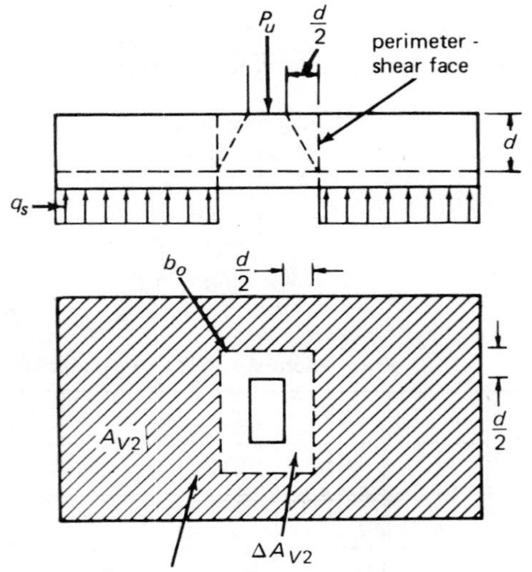

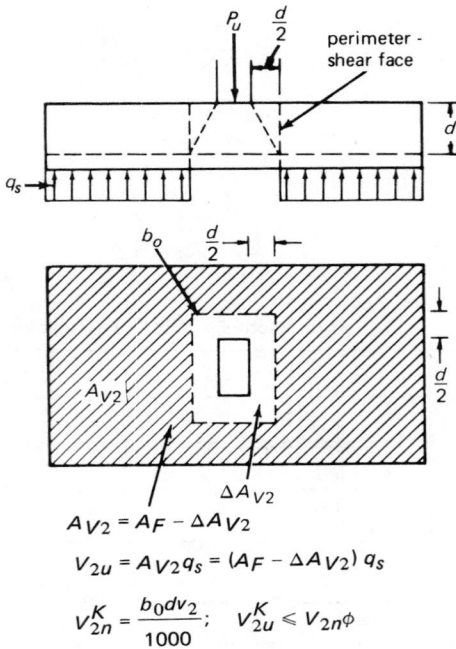

$$A_{V2} = A_F - \Delta A_{V2}$$

$$V_{2u} = A_{V2} q_s = (A_F - \Delta A_{V2})\, q_s$$

$$V_{2n}^K = \frac{b_o d v_2}{1000}; \quad V_{2u}^K \leqslant V_{2n}\phi$$

Fig. 5-23 Two-way (perimeter) shear action.

out shear reinforcement. Here, as well as in the following sections and examples, all concrete is assumed to be normal density concrete. Where sand-lightweight, or all-low-density concrete is used (which is rather unusual in connection with footings), the permissible shear values have to be reduced in accordance with the requirements of Chap. 11 of ACI 318-83. If shear reinforcement is provided in accordance with Chap. 11 of ACI 318-83, the maximum permissible shear stress is:

$$v_{2u} = 6\sqrt{f_c'}$$

If shearhead reinforcement is provided in accordance with Chap. 11 of ACI 318-83:

$$v_{2u} = 7\sqrt{f_c'}$$

However, for practical reasons, footings ought to be designed whenever possible without the use of shear reinforcement.

In the following, reference is frequently made to tables contained in other chapters of this handbook, or in the *Strength Design Handbook*, ACI SP-17. Numbers of tables contained in this handbook have been omitted herein, and the tables are referred to by their general name only, with ACI SP-17 added. Tables that are similar to those referred to in ACI SP-17 can also be found in other design aids and textbooks, and can be used in a similar manner.

Tables to determine the footing depth as required by perimeter shear have been provided in ACI SP-17 to simplify the calculation. These tables are prepared for square footings, concentrically loaded by square, round, or polygonal pedestals or piers. If the loading pressure, q_c, at the base of the column, pier, or pedestal, and the bearing pressure at the footing base, q_s, due to the factored load are known, then the effective depth of the reinforced concrete footing can be determined from the ratio d/h, which can be taken from the tables.

Similar tables are also provided in ACI SP-17 for rectangular footings loaded by square or rectangular columns, piers, or pedestals. These tables can also be used for round and polygonal columns, piers, and pedestals, if their loading areas are transformed into squares of equal cross-sectional area.

3. Investigation for One-Way Shear Action. The nominal shear stress due to one-way shear action shall be calculated as in an ordinary reinforced concrete beam, along a plane perpendicular to the footing, located at a distance d from the face of the column, pier, or pedestal. This plane shall extend across the entire footing, and the shear stresses shall be assumed to be uniformly distributed over this plane, as shown in Fig. 5-24. The governing shear force, V_{1u}, for one-way shear action consists, therefore, of the sum of all bearing pressures, q_s, acting outside of the critical section.

The average nominal shear strength along the critical section in Kips is then:

$$V_{1n}^K = \frac{(12 b_f)\, d v_1}{1000} \quad \text{and} \quad V_{1u} \leqslant V_{1n}\phi$$

where ϕ is the capacity reduction factor for shear ($\phi = 0.85$) from Chap. 9 of ACI 318-83. If no shear reinforcement is provided, which is the usual case, and the concrete is of normal density taking all of the shear stresses, then the maximum permissible stress is:

$$v_1 = v_c = 2\sqrt{f_c'}$$

In the exceptional cases, where shear reinforcement is provided in footings, it needs to be designed only for the excessive shear stress $(v_{1u} - v_c)$ in accordance with the *ACI Code.* In no case shall $(v_{1u} - v_c)$ exceed $8\sqrt{f_c'}$.

4. Anchorage Development of Column Dowels. After the minimum footing thickness that will satisfy both shear requirements has been determined, it should be checked to see whether it provides sufficient depth for the development of the column dowels. Anchorage development in compression is the most common condition for column bars; however where bending moments or uplift forces have to be transmitted to the footing, the bars must also satisfy the anchorage requirements for tension bars. Right-angle, or 90°, hooks at the bottom end of column dowels (common practice), are of no help in the development of anchorage

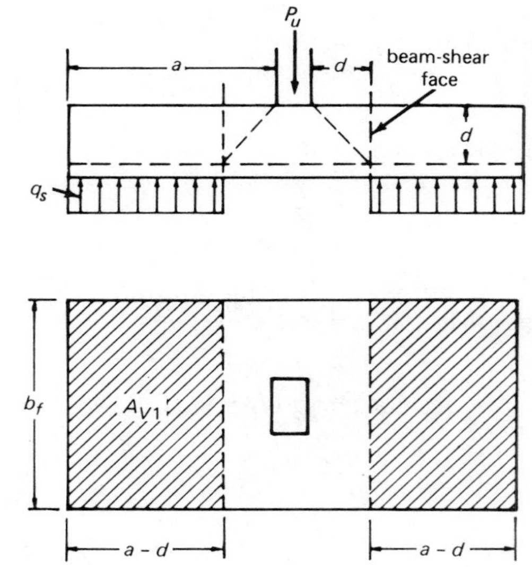

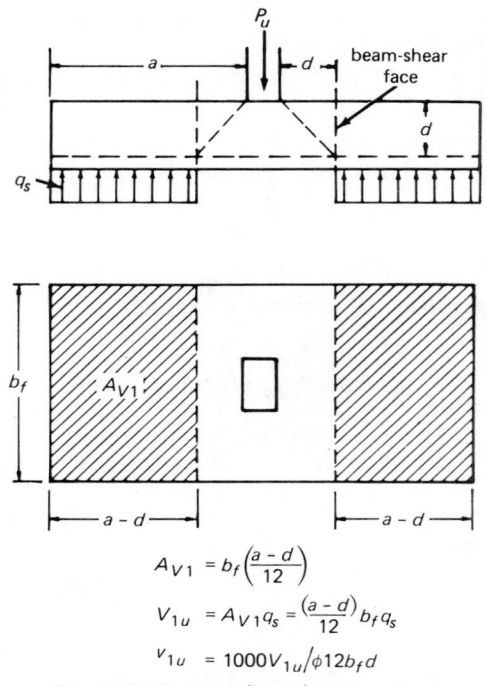

$$A_{V1} = b_f\left(\frac{a-d}{12}\right)$$

$$V_{1u} = A_{V1} q_s = \frac{(a-d)}{12} b_f q_s$$

$$v_{1u} = 1000 V_{1u} / \phi 12 b_f d$$

Fig. 5-24 One-way (beam) shear action.

for compression bars, see Fig. 5-25(a). The development length in compression shall be calculated in accordance with the requirements of Sec. 12.3 of ACI 318-83. Where column dowels have to be developed in tension, as in the case of uplift, the anchorages may be designed as shown in Fig. 5-25(b). In this case the straight development length shall be calculated in accordance with the requirements of Sec. 12.2 and the embedment length of a bar terminating in a standard hook in accordance with the requirements of Sec. 12.5 of ACI 318-83. A 90° hook with a minimum horizontal extension of 12 d_b may be substituted for a standard hook. Dowels of smaller diameter than the column bars can be used to reduce the required anchorage length, and, therefore, thickness of the footing. Dowels of bigger diameter than the column bars may be used also if desirable; the *Code*, however, does not permit the diameter of the dowel to exceed that of the column bar by more than 0.15 in. ACI 318-83 provides that under certain conditions not all longitudinal column reinforcement needs to be extended into the footings, but only enough to cover the excess beyond the permissible bearing stress of the supporting or the supported member, whichever is smaller. In this connection, column steel, which has to be counted on at or above the contact area, has to be extended or developed in the column above and in the footing below. This also applies to areas where high edge pressures are caused by eccentric loadings or moments.

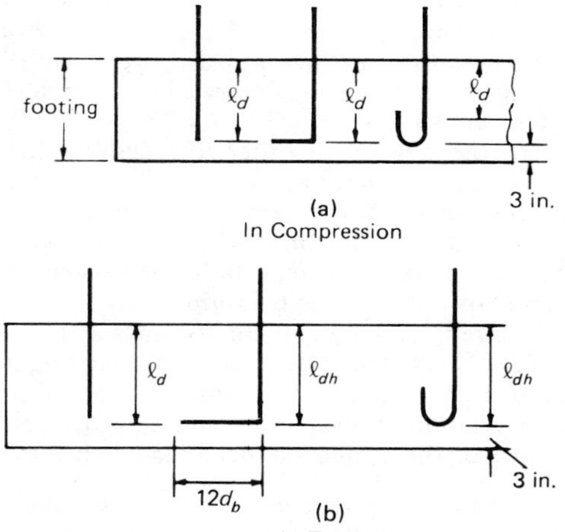

Fig. 5-25 Development (anchorage) of column bars or dowels. (a) In compression; and (b) in tension.

5. *Investigation for Flexure.* The flexural strength of a footing can be determined in a similar manner to that of an ordinary beam or cantilever member. The condition is shown in Fig. 5-26.

The reference lines about which the bending moments are to be determined shall extend all the way across the footing and shall be located as follows:

(a) At the face of the supported element, for footings supporting columns, piers, or pedestals.
(b) Halfway between the center and the edge of the supported masonry, for footings supporting masonry construction.
(c) Halfway between the face of the column and the edge of the base plate, for footings supporting steel base plates and steel columns.

The flexural moment, M_u, to be used in the calculation is therefore:

$$M_u = \left(\frac{a_L}{12}\, b_f q_s\right)\left(\frac{a_L}{2 \times 12}\right)\frac{1}{\phi} \text{ in. kip-ft}$$

For rectangular footings or for square footings supporting rectangular columns, pedestals, or piers, the larger moment, i.e., the moment in the direction of the longer projection, may influence the selection of the footing depth; to determine the reinforcement, however, the moments have to be calculated for each direction. The capacity reduction factor, ϕ, in the above equation is 0.9 for flexure according to Chap. 9 of ACI 318-83 and $M_u = \phi M_n$. If Tables of Graphs of the Design Handbook ACI 318-77 are used, a ϕ-factor of 0.9 has been incorporated in the Table values.

The flexural strength shall also be investigated at all changes in the cross section of the footing element. Such checks are important in the design of stepped or sloped footing mats.

The flexural Tables 1-4 through 1-8 provided in Chapter 1 of this handbook, for the design of ordinary beam and slab problems, can be used just as well for the flexural design of a footing. If ACI SP-17 is used, the F-factor can be determined from the selected size of the footing as:

$$F = (12 b_f)\, d^2/12000 \; (b_f \text{ in. ft, } d \text{ in in.})$$

or can be read from the F-table. With the help of $K_u = M_u/F$, the corresponding a_u or ρ can be taken from the tables and the cross-sectional area of the reinforcement be determined from:

$$A_s = M_u/a_u d \quad \text{or} \quad A_s = \rho 12 b_f d$$

If the required percentage exceeds the maximum permissible percentage $\rho_{max} = 0.75\rho_b$ (preferably $0.6\rho_b$) or if the size and spacing of the reinforcing bars appears to be impractical, the footing thickness has to be increased to satisfy the requirements.

If the flexural graphs of ACI SP-17 are used, the moments, M_u, must be determined either for a 1-ft-wide strip if the slab graphs are used; or for a 10-in.-wide strip if the beam graphs are used; and the obtained reinforcement must be multiplied by b_f for slabs or $1.2 b_f$ for beams to find the total amount required in each direction. Attention must be

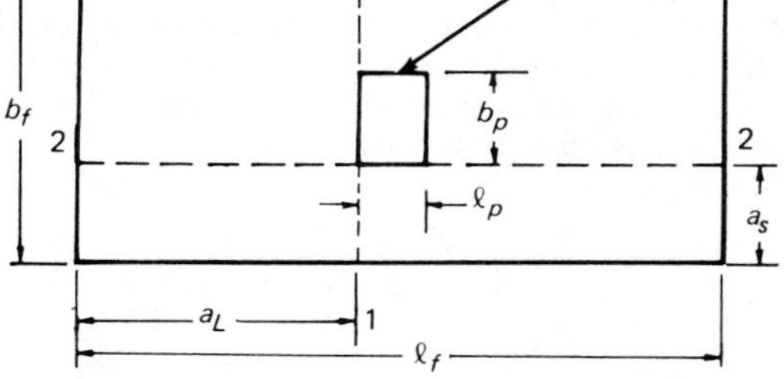

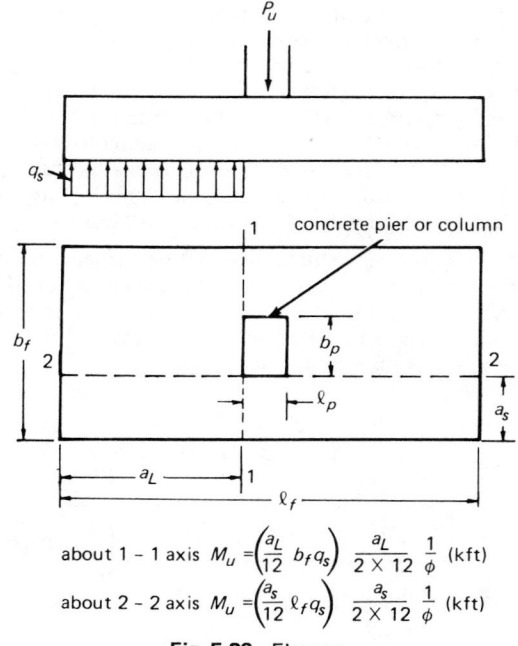

about 1 - 1 axis $M_u = \left(\frac{a_L}{12}\, b_f q_s\right) \frac{a_L}{2 \times 12} \frac{1}{\phi}$ (kft)

about 2 - 2 axis $M_u = \left(\frac{a_s}{12}\, \ell_f q_s\right) \frac{a_s}{2 \times 12} \frac{1}{\phi}$ (kft)

Fig. 5-26 Flexure.

drawn to the selection of the bar diameter to be used for the reinforcement because it must, in addition to providing the required cross-sectional area, also satisfy the anchorage requirements (see paragraph 6, below).

ACI SP-17 also contains a special set of footing tables that are rather practical to use. These tables cover a wide range of bearing pressures due to factored loadings from $q_s = 3.33$ kip/ft^2 to $q_s = 26.67$ kip/ft^2 and can be used for both structural plain and reinforced concrete footings. Since the base area (size) of the footing must be known before the strength design is attempted, its greatest projection beyond the critical line is also known. The effective footing depth, d, can be selected so that the permissible perimeter shear and beam shear values are satisfied for the given projection. After these checks have been made, the required amount of reinforcement can be read from the table.

6. *Development of Footing Reinforcement.* Most of the footings or portions thereof consist of short, heavily loaded cantilever sections. In the design of such elements it is important to see that the reinforcing bars are sufficiently anchored on either side of the critical section, as described above, to develop the full tension required. Since the projection of the footing beyond the critical section is a given length, and to satisfy the anchorage requirements, the diameters of the reinforcement selected should be small enough to provide sufficient embedment length on either side of the critical section. Most footing tables prepared for strength design, including those in ACI SP-17, contain the maximum diameters of bars that may be used to satisfy whatever projection the footing has.

5.4 SPECIAL CONDITIONS

5.4.1 Stepped Footings

In the design of isolated spread footings, the calculated thickness is required at the various critical locations near the center of the footing, but not at the edge of it. It has been common practice to give footings a tapered or stepped cross section in order to save the concrete in areas where it is not structurally needed. This method is structurally sound but getting out of practice for the following reasons:

(a) The extra cost of the formwork is often greater than the saving in the amount of concrete used.
(b) The monolithic action between the upper (cap) and lower (mat) portion of the footing, which is structurally essential, is in practice difficult to obtain if cap and mat are not cast simultaneously (see Fig. 5-27). Unless special provisions are made, this method of construction will develop a "cold" joint, and a cleavage plane will separate the two pieces.

Steps, however, are still in use in cases where the mat is getting excessively thick (usually more than 3 ft). If footings are not cast monolithically, key ways or shear-friction reinforcement have to be provided to transfer the horizontal shear and obtain monolithic action.

The size and thickness of the caps have to be designed in such a way that at each step (change in cross section) all shear stresses and flexural requirements are satisfied.

5.4.2 Footings Designed for Equal Bearing Pressures

Since settlements are practically independent of short-time fluctuations in the loading, foundations for apartment houses, office buildings, institutional buildings, and the like are often designed for equal bearing pressures under average service load conditions, with the intent to obtain equal settlements over the entire building area. Full dead load plus one-half of the live load are often considered to represent an "average service load"; however, other ratios may be substituted depending on the judgment of the designing engineer. Such an approach will, at its best, reduce the amount of differential settlements to some degree because mutual influence, dishing, and, especially, variations in footing sizes will influence the settlement of each footing in a different way, regardless of the equal bearing pressure.

If such a design is desired, proceed as follows:

(a) Determine the live load to dead load ratio for each column footing.
(b) Determine the average (reduced) service load for all column footings (usually assumed with full dead load plus one-half live load).
(c) Select the column with the greatest ratio of live load to dead load (from item a) and design its footing for the maximum allowable soil pressure under full load.
(d) Determine the bearing pressure, q_{av}, for the same footing under the average service load (as determined under item b).
(e) Design the size of all other footings for the average service load (as determined under b) and the average bearing pressure (as determined under d).
(f) Ascertain for all footings, that the bearing pressure under the maximum load does not exceed the maximum allowable bearing pressure.
(g) Make strength design of all footings at least for the factored maximum load and the bearing pressure caused by it or preferably for the q_s determined from the maximum allowable bearing pressure factored according to the applicable dead load to live load ratio.

The final result will be a somewhat overdesigned foundation, having equal bearing pressures under average service load conditions.

5.4.3 Footings of Structural Plain Concrete

Structural plain concrete footings on soil are permitted by the code. For more detailed information and example see Chapter 27, "Structural Plain Concrete," in this handbook.

5.4.4 Design Examples

The following design examples for reinforced concrete footings were prepared with and without the help of tables and

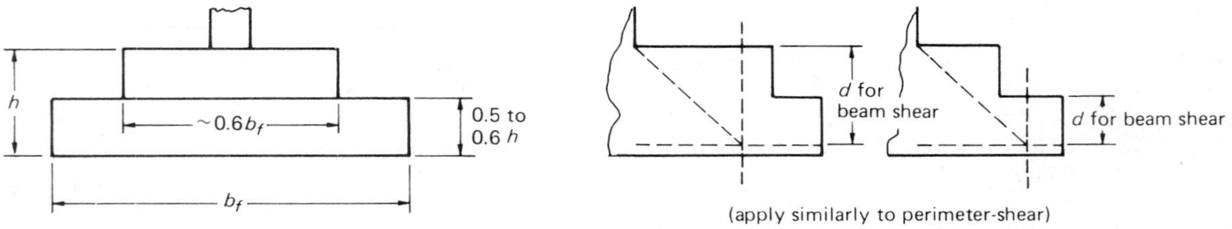

Fig. 5-27 Stepped footing.

graphs. In the first part of the design examples, the procedure was presented in easy-to-follow steps. A considerable amount of explanatory text was also added to assist in the understanding of the basic requirements. It cannot be stressed enough that a full understanding of these basic approaches makes them applicable to footings of any type or character. Although this may make the procedure appear rather lengthy, a great part of the extra steps will become unnecessary in the solution of an actual design problem; however, they were included here for illustrative purposes.

EXAMPLE 5-1: Without using footing tables and graphs, design a *concentrically loaded, square, spread footing* for the following conditions:

column load

$$P_D = 350.0 \text{ kips}, \quad P_L = 275.0 \text{ kips}$$

pier

$$b_p = 18 \text{ in.}, \qquad l_p = 24 \text{ in.}$$

$$q_a = 4.50 \text{ k/ft}^2, \quad h_s = 5 \text{ ft}$$

$$w_L = 100 \text{ lb/ft}^2,$$

$$f'_c = 3000 \text{ psi, normal density concrete}, \quad f_y = 40{,}000 \text{ psi}$$

Design footing by strength design method.

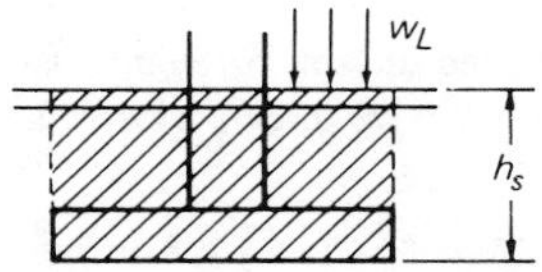

Fig. 5-28 Evaluation of surcharge.

Step 1: Determine footing size from service loads. The approximate weight of the surcharge is $\Delta q = \gamma_s h_s + w_L$ where γ_s is the assumed average unit weight for all material above the footing base. Since it consists of concrete and soil it can be estimated close enough between 150 and 100 lb/ft^3. In this example it is assumed to be 130 lb/ft^3 (see Fig. 5-28).

$$\Delta q = \gamma_s h_s + w_L = 0.13 \times 5 + 0.1 = 0.75 \text{ kip/ft}^2$$

Determine effective soil pressure (allowable soil pressure that can be utilized):

$$q_e = q_a - \Delta q = 4.50 - 0.75 = 3.75 \text{ kip/ft}^2$$

to find the minimum required base area of footing:

$$A_{F\,\min} = \frac{P_D + P_L}{q_e} = \frac{350 + 275}{3.75} = 167 \text{ ft}^2$$

Selected footing size:

$$\text{Use 13-ft square}, \quad A_F = 169 \text{ ft}^2$$

Step 2: Determine bearing pressure to be used for strength design.

$$q_s = \frac{P_{Du} + P_{Lu}}{A_F} = \frac{350 \times 1.4 + 275 \times 1.7}{169} = 5.70 \text{ kip/ft}^2$$

Step 3: Determine thickness of footing mat. In the case of a square spread footing, concentrically loaded by a square column, the mat is always governed by two-way (slab or perimeter) shear action.[5-8] It is, therefore, advisable to investigate such a footing first for this condition. Only if the footing is oblong, or if the column or pier has a rectangular cross section, does the footing thickness have to be checked also for one-way (beam) shear action (see also Step 3 of Example 5-2), and may be governed by it. The tentatively selected mat thickness must also be checked for dowel embedment length and flexural requirements.

(a) When one is investigating for two-way shear action, the most practical design approach is to assume a footing thickness and check

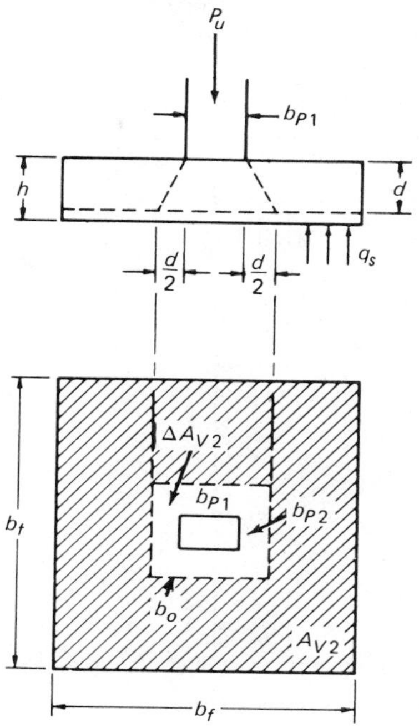

assume: $d = 26''$

ΔA_{v2} = Base Area of truncated shear pyramid (ACI 318-71, Sec. 11.10.1 & 2)

$$\Delta A_{v2} = \left(b_{P1} + 2\frac{d}{2}\right) \times \left(b_{P2} + 2\frac{d}{2}\right)$$

$$= (24 + 26)(18 + 26)$$

$$= 2200 \text{ sq in}$$

$$= 15.3 \text{ sq ft}$$

$$A_{v2} = A_F - \Delta A_{v2} = 169.0 - 15.3 = 153.7 \text{ sq ft}$$

$$V_{2u} = A_{v2} q_s = 153.7 \times 5.7 = 875.0 \text{ k}$$

$$v_{2u} = \frac{V_{2u}}{\phi b_o d} = \frac{875.0 \times 1000}{0.85\,[2(50 + 44)]\,26} = 212 \text{ psi}$$

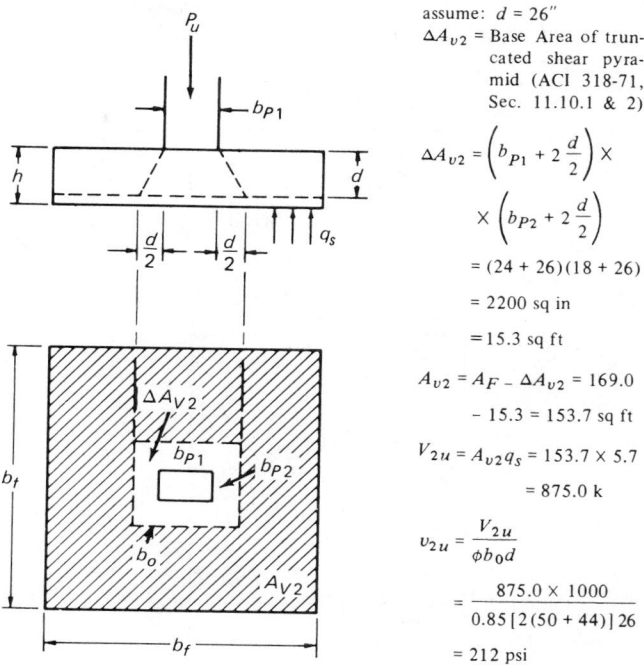

Fig. 5-29 Two-way shear action.

it for its required strength; if incorrectly assumed, it can easily be adjusted to the correct thickness. The location of the base of the truncated shear cone or pyramid is concentric with that of the pier and located at a distance $d/2$ outside of it, as shown in Fig. 5-29.

The permissible two-way shear stress (ACI 318-83, Chap. 11) is:

$$v_{2u} = 4\sqrt{f'_c} = 220 \text{ psi}$$

since $\beta_c = 50/44 = 1.14 < 2$.

NOTE: It is advisable to design a footing thick enough to satisfy the permissible shear stresses for unreinforced concrete. When, under extreme conditions, the footing cannot be made thick enough, the shear capacity of the footing has to be strengthened by reinforcement the same way as in a column-slab intersection. If shear reinforcement (usually bent-up bars) is used, the permissible average shear stress may be increased by 50%; if a steel shearhead reinforcement is provided, the permissible average shear stress may be increased by 75%. In each case the concrete section can only be stressed to the permissible stress value of $4\sqrt{f'_c}$, and the remainder has to be carried by the reinforcement.

(b) Investigation for one-way shear action begins with investigation of the footing mat in the direction in which the distance between footing edge and critical section is largest. The critical section for one-way shear runs parallel to each pier face, and at the distance d away from it, across the entire footing mat (see Fig. 5-30).

$$a = \frac{b_f - b_p}{2} = \frac{13.0 - 1.5}{2} = 5.75 \text{ ft} = 69 \text{ in.}$$

$$A_{v1} = \frac{(a - d)}{12} b_f = \frac{69 - 26}{12} 13.0 = 46.6 \text{ ft}^2$$

$$V_{1u} = A_{v1} q_s = 46.6 \times 5.7 = 268 \text{ kips}$$

$$V^K_{1n} = \frac{b_f d v_{1u}}{1000} = \frac{(13 \times 12)\, 26 \times 110}{1000} = 446^K$$

$$V^K_{1u} \leqslant \phi V_{1n}$$

$$268 \leqslant 0.85 \times 446 = 379^K$$

The permissible shear stress for one-way action is:

$$v_{1u} = 2\sqrt{f'_c} = 110 \text{ psi}$$

NOTE: In essence, the note provided at the end of Step 3(a) applies here also. (ACI 318-83, Chap. 11).

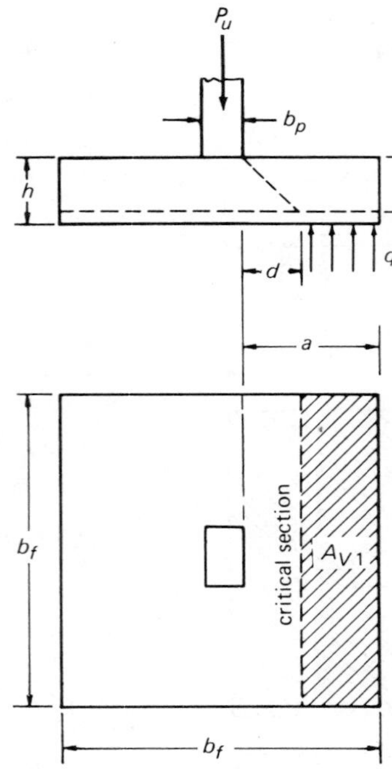

Fig. 5-30 One-way shear investigation.

Step 4: Investigation for flexure. If the projection of the footing beyond the critical section varies the flexural investigation has to be made for each direction. In each case the critical section extends across the entire footing.

For concrete piers or columns, the critical section is located at the face of the pier (ACI 318-83, Chap. 15). (See also Fig. 5-26, $l_f = b_f$.)

(a) Flexural computation in the direction of the longer footing dimension:

$$a_L = \frac{l_f - l_p}{2} = \frac{13.0 - 1.5}{2} = 5.75 \text{ ft} = 69 \text{ in.}$$

$$M_u = b_f a_L q_s \frac{a_L}{2} = 13.0 \times 5.75 \times 5.7 \frac{5.75}{2} = 1220 \text{ kip-ft}$$

$$F = \frac{b_f d^2}{12{,}000} = \frac{(13 \times 12)\, 26^2}{12{,}000} = 8.75, \quad K_u = M_u/F = \frac{1220}{8.75} = 139$$

$\rho = 0.004.$ $A_s = \rho b_f d = 0.004(13 \times 12)\, 26 = 16.2 \text{ in.}^2$

Minimum reinforcement required (ACI 318-83, Chap. 7):

$$\rho_{min} = 0.0018$$

$$A_{s\,min} = \rho_{min} b_f h = 0.0018(13 \times 12)\, 31 = 8.8 \text{ in.}^2$$

For evaluation of h see below.

$$A_s = 16.2 \text{ in.}^2 \text{ governs.}$$

(b) Flexural computation in the direction of the shorter footing dimension:

$$a_s = \frac{b_f - b_p}{2} = \frac{13.0 - 2.0}{2} = 5.5 \text{ ft} = 66 \text{ in.}$$

$$M_u = l_f a_s q_s \frac{a_s}{2} = 13.0 \times 5.5 \times 5.7 \frac{5.5}{2} = 1120 \text{ kip-ft}$$

$F = 8.75$ as above

$$K_u = \frac{M_u}{F} = \frac{1120}{8.75} = 128, \quad \rho = 0.0037$$

Minimum reinforcement required (ACI 318-83, Chap. 7):

$$\rho_{min} = 0.0018$$

$$A_s = \rho b_f d = 0.0037(13 \times 12)\, 26 = 15.0 \text{ in.}^2 \text{ governs.}$$

NOTE: It would be theoretically correct to use a different d-value for the flexural computation in each direction because the reinforcement is placed in two layers. The requirement of placing a certain layer below the other one is, economically, only seldom worth the effort, and practically difficult to control, unless the character of the two layers is drastically different; e.g., in an oblong footing where the longer bars are usually specified to be placed in the first layer from the bottom. For square footings it is advisable to stay on the safe side, and design the reinforcement in both directions for the shorter effective depth d. Care has to be exercised in the selection of the bar size that can be used for the reinforcement because bar development (anchorage) is always critical in members that are highly stressed by shear. Since every bar has to be fully anchored at either side of the critical section, the shorter length, which is (a - 3 in.) will govern the design, (see Fig. 5-31).

Any bar whose development length in tension is smaller than a - 3" can be used. For the Example under consideration, any deformed bar size up to #11 is acceptable. Bar sizes greater than #11 are commonly not used for footings, although there is no *Code* restriction in this respect. 17-#9 ($A_s = 17 \text{ in.}^2 > 16.2 \text{ in.}^2$) will be provided in each direction.

The total thickness of the footing can be found, under consideration of the above, as shown in Fig. 5-32.

$$h = d + 1.5 d_b + c_c = 26 + 1.5 \times 1.13 + 3 = 30.7 \text{ in.}$$

$$\cong 31.0 \text{ in.}$$

EXAMPLE 5-2: Below, Example 5-1 is solved with the help of footing tables contained in the *Strength Design Handbook*, ACI SP-17. Step 1 and Step 2 have to be performed as before.

Step 3:

(a) Use footing tables of ACI SP-17, furnishing the Effective Depth required by Two-Way (Perimeter) Shear action: $f'_c = 3000$ psi.

For a column aspect ratio of $\beta_c = 1.14 < 2$, as determined in Step 3(c) of Example 5-1, $K_{V6} = 1$ and $K_{V6} \times q_s = 1 \times 5.7 = 5.7$ ksf. Find the ratio of Area Footing/Area Column = $A_F/A_C = (13^2/1.5 \times 2) \cong 56$. Enter Table with $q_s = 5.7$ and $A_F/A_C = 56$ to find the ratio $d/h_c = 1.25$. h_c is size of the equivalent square pier. $h_c = \sqrt{18 \times 24} = 20.7$, $d = 1.25 \times 20.7 = 26$ in.

NOTE: Table was prepared for square piers but it can also be used for slightly rectangular piers if the rectangle is transformed into a square of equal area.

(b) From general footing tables of ACI SP-17, select the table that is closest to the required q_s, or interpolate if necessary. The footing table for $q_s = 6.67$ will be on the safe side. By entering this table with the selected depth $d = 26$ in. (interpolate between 24 and 28), we find the appropriate a_b-value equal to 78.25 in. This value represents the maximum projection that can be used for one-way shear action. Since the maximum projection, a, in the example is 69 in. (which is smaller than 78.25), the selected depth of 26 in. is satisfactory.

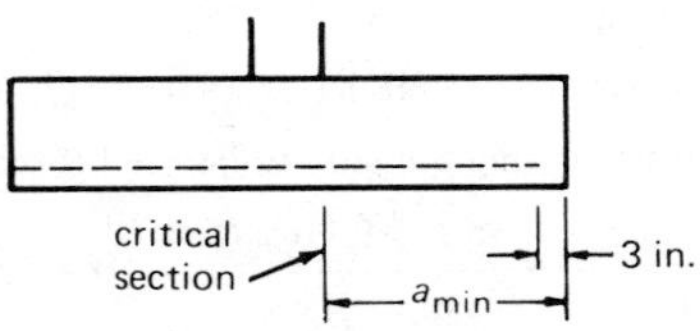

a_{min} - 3 in. = 65 - 3 = 62 in.

Fig. 5-31 Bar development (anchorage).

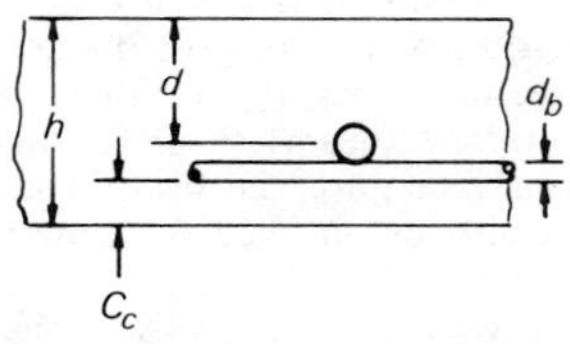

Fig. 5-32 Effective depth of spread footings.

Step 4: The footing must also be deep enough to develop the column dowels, or the column dowels must be selected so that their necessary development length is satisfied by the depth of the footing. In case the dowels have to be anchored for compression only, the l_d must be smaller than or equal to 26 in., which is satisfied by #11 bars. In case of tension, the anchorage can be increased through hooks or bends.

Step 5: The same footing table can be used to determine the reinforcement. Enter the table with the value of $d = 26$ and $a = 69$ in. Double interpolation furnishes a value of 1.22 in.²/ft for the reinforcement in each direction. The total required reinforcement is therefore

$$A_s = A_s/\text{ft (from table)} \times b_f = 1.22 \times 13 = 15.9 \text{ in.}^2$$

against 16.2 in.² found by calculation in Step 4a of Example 5-1. The column at the right hand side of the Table indicates the maximum bar size that can be used corresponding to each projection a.

EXAMPLE 5-3: Without the help of footing tables, design an oblong, concentrically loaded, spread footing for the following conditions: all data identical with those of Example 5-1 except that the width of the footing is restricted to 8 ft, and the long side of the pier is for architectural reasons, perpendicular to the long side of the footing.

Step 1: Determine footing size from service loads. Evaluation of the minimum base area remains unchanged:

$$A_F = 167 \text{ ft}^2$$

Selecting footing size:

$$b \times l_f = 8 \times 21 \text{ ft}, \quad A_F = 168 \text{ ft}^2$$

Step 2: Determine bearing pressure to be used in strength design. Unchanged from Example 5-1:

$$q_s = 5.7 \text{ kip/ft}^2$$

Step 3: Determine the thickness of the footing. In the case of an oblong, concentrically loaded, spread footing, the footing thickness is often governed by one-way (beam) shear action, depending on the length-to-width ratio of the footing. It is therefore necessary to investigate the footing for this condition first, and check the tentatively selected thickness afterward for two-way (slab) shear action (see also Step 3 of Example 5-1). The tentatively selected footing thickness must also be checked for dowel anchorage and flexural requirements.

(a) When investigating for one-way shear action, consider that the critical section for one-way shear extends across the entire footing parallel to each pier face at the distance d away from it. Similarly to Step 3 of Example 5-1, we assume the footing depth, in this case, to be 36 in. (see Fig. 5-33).

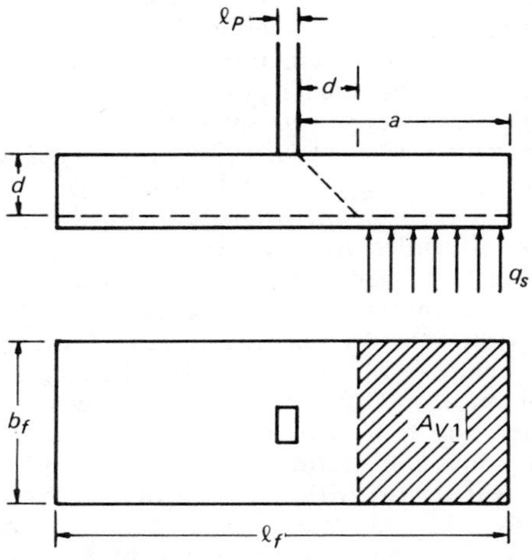

Fig. 5-33 One-way shear in oblong footings.

$$a = \frac{l_f - l_p}{2} = \frac{21.0 - 1.0}{2} = 10.0 \text{ ft} = 120 \text{ in.}$$

$$A_{V1} = \frac{(a-d)}{12} b_f = \frac{120 - 36}{12} 8.0 = 56.0 \text{ ft}^2$$

$$V_{1u} = A_{V1} q_s = 56.0 \times 5.7 = 312.0 \text{ kip}$$

$$V_{1n}^K = \frac{(12 b_f)\, d v_1}{1000} = \frac{(12 \times 8)\, 36 \times 110}{1000} = 380^K$$

$$V_{1u}^K \leq \phi V_n = 0.85 \times 380 = 323^K$$

$$312 < 323$$

The permissible shear stress for one-way action is:

$$2\sqrt{f_c'} = 2\sqrt{3000} = 110 \text{ psi}$$

(b) Investigation for two-way shear:

$$\Delta A_{V2} = (l_p + 2d/2)(b_p + 2d/2) = (18 + 36)(24 + 36)$$
$$= 3240 \text{ in.}^2 = 22.4 \text{ ft}^2$$

$$V_{2u} = A_{V2} q_s = (l_f \times b - \Delta A_{V2})\, q_s$$
$$= (8.0 \times 21.0 - 22.4)\, 5.7 = 830.0 \text{ kip}$$

$$v_{2u} = V_{2u}/\phi b_o d = 830 \times 1000/0.85[2(54 + 60)]\; 36 = 120 \text{ psi}$$

which is well below the permissible value of $4\sqrt{f_c'} = 220$ psi. The correctness of this result is questionable, if we look at the plan of the footing in Fig. 5-34, which illustrates the condition. An even distribution of the two-way shear along the perimeter is hardly probable if we consider the narrow width of the influence area parallel to the short sides of the pier.

A more reasonable result is obtained if we divide the influence area in four portions, and investigate the greatest shear stress caused by each part separately.

$$A'_{V2} = \frac{l_c - (l_p + d)}{2} = \frac{21.0 \times 12 - (18 + 36)}{2 \times 12} \times 8 - 2 \times \frac{18^2}{2 \times 144} = 63.75 \text{ ft}^2$$

$$V'_{2u} = A'_{V2} q_s = 63.75 \times 5.7 = 365 \text{ kip}$$

$$v'_{2u} = \frac{V'_{2u}}{\phi(b_p + d)\, d} = \frac{365 \times 1000}{0.85(24 + 36)\, 36} = 200 \text{ psi}$$

which is considerably greater and more realistic than the first value of v_{2u} found above, but still within the permissible limit of 220 pis.

Step 4: The depth of 36 in. will permit the use of any bar size for dowels, regarding their development in compression.

Step 5: Investigation for flexure. The oblong shape of the footing requires the design to be made independently for both directions. It is good practice, and reasonable, to assume that the reinforcement in the long direction will be placed at the bottom in order to utilize a greater depth. It is safer, however, and recommended, to design both layers for the shorter effective depth to stay independent of the field inspection.

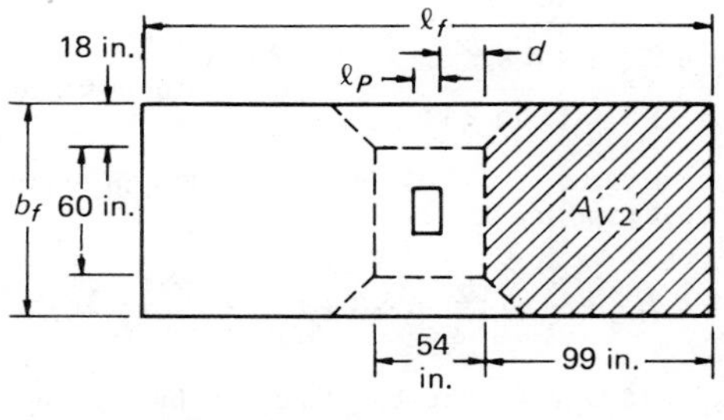

Fig. 5-34 Two-way shear in oblong footings.

(a) In the long direction of the mat:

$$a_L = \frac{21.0 - 1.0}{2} = 10.0$$

$$M_u = a_L b_f q_s \frac{a_L}{2} = 10.0 \times 8.0 \times 5.7 \times \frac{10}{2} = 2280 \text{ kip-ft}$$

$$F = b_f d^2/12{,}000 = (8 \times 12) \times 36^2/12{,}000 = 10.4,$$

$$K_u = 2280/10.4 = 218$$

$a_u = 2.85$ from Flexure Table 1.1 of the Design Handbook ACI SP-17 for the 318-83 Code. a_u in previous editions of ACI SP-17 = a_n in ACI SP-17 for the 318-83 Code.

$$A_s = \frac{M_u}{a_u d} = \frac{2280}{2.85 \times 36} = 22.2 \text{ in.}^2$$

The available anchorage length in this direction is $a - 3'' = (10 \times 12) - 3 = 117$ in. Since the development in tension for any bar up to #11 is smaller than this length, any bar can be used.

$$A_s = 15 - \#11, \quad A_s = 23.4 \text{ in.}^2$$

(b) In the short direction of the mat:

$$a_s = \frac{8.0 - 1.5}{2} = 3.25 \text{ ft}$$

$$M_u = a_s l_f q_s \frac{a_s}{2} = 3.25 \times 21 \times 5.7 \times \frac{3.25}{2} = 630 \text{ kip-ft}$$

$$F = \frac{l_f d^2}{12{,}000} = \frac{(21 \times 12) \times 36^2}{12{,}000} = 27.2, \quad K_u = \frac{540}{27.2} = 20$$

$$a_u = 2.96$$

Only minimum reinforcement will be required.

$$A_s = \frac{1.33 M_u}{a_u d} = \frac{1.33 \times 540}{2.96 \times 36} = 6.8 \text{ in.}^2$$

or:

$$A_{s\,min} = \rho_{min} l_f h = 0.0018(21 \times 12)\,41 = 18.7 \text{ in.2}$$

For evaluation of h see below. The smaller value can be used.

The available anchorage length in this direction is a $-3'' = (3 \times 12) - 3 = 33$ in. The maximum bar size which requires a development length of 33″ or less is #9. Because of the large footing length a greater number of bars is desirable, and we select #7 bars.

In an oblong footing, section 15.4.4 of ACI 318-71 requires that a portion of the reinforcement in the short direction A_{s1}, be distributed over a width b_f and the balance of the reinforcement be spread evenly over the rest of the footing length, eq. (15-1).

If A_{sT} = total required reinforcement in the short direction, then:

$$A_{s1} = \frac{A_{sT} \times 2}{(S + 1)} = \frac{6.8 \times 2}{(2.65 + 1)} = 3.7 \text{ in.}^2$$

$$S = l_f/b_f = 21/8 = 2.65$$

To avoid unequal spacings of reinforcement, the total reinforcement A_{sT} may be increased to A_{s2} in order to be spread evenly over the entire length of the footing.

$$A_{s2} = \frac{2A_{sT}S}{(S + 1)} = \frac{2 \times 6.8 \times 2.65}{(2.65 + 1)} = 9.8 \text{ in.}^2, \quad \text{still less than } 18.7 \text{ in.}^2$$

Reinforcement used is 31 #7 bars, $A_s = 18.6$ in.^{2} and the total thickness of the footing is:

$$h = d + 3 + d_{bL} + \frac{1}{2} d_{bs} = 36 + 3 + 1.41 + \frac{0.88}{2} = 40.85 \cong 41 \text{ in.}$$

where d_{bL} and d_{bs} are the bar diameters in the long and short direction.

EXAMPLE 5-4: Stepped footing. If the thickness of the footing investigated in Example 5-3 is considered to be uneconomical or otherwise excessive, it may be designed as a stepped footing and required to be cast monolithically (see Fig. 5-27).

It is common practice to make the cap (the upper portion) about $0.6b_f$ long; however, this length should be checked as described below. The cap thickness is usually assumed with a fraction of the total footing thickness, depending on the number of steps to be used. In this case the shear investigations of Example 5-3, for one-way and for two-way shear, are applicable only as long as the critical sections fall within the extent of the cap; otherwise, the depth occurring at the critical section has to be used. The shear-cone or pyramid must not intersect the step, or the size of the cap has to be adjusted as needed. The entire shear investigation has to be repeated along the perimeter of the cap.

In the flexural investigation, care has to be exercised so that only that portion of the cross section that is in compression is utilized for determining the F-value, $bd^2/12{,}000$, and consequently the amount of reinforcement. In addition to the maximum moment occurring at the critical section, the amount at the edge of each step has to be investigated, and the amount of reinforcement at these locations checked for the reduced available depth.

5.5 REINFORCED CONCRETE FOOTINGS WITH CONCENTRATED REACTIONS (PILE CAPS)

5.5.1 General Principles

Where soil conditions do not favor the design or construction of shallow foundations (spread footings), but a firm soil stratum can be found at greater depth, piles can be used to transfer the loads from the superstructure down to the soil stratum, where the required resistance is available. The piles may develop this resistance by end bearing (bearing piles) on the firm stratum; or by skin friction (friction piles) developed by driving the piles into the firm stratum. Pier-foundations or caissons can also be used for similar purposes but do not form a part of this discussion.

Similar to the action of a spread footing, a footing on piles (commonly called pile cap) has to distribute the column load to the piles in each group, which in turn will transmit it to the subsoil. The main difference between the two types of footings lies in the application of the base reactions which, in the case of a footing on piles, consists of a number of concentrated loads. If we divide the sum of all pile reactions in a group, just for reasons of comparison, by the base area of the pile cap, we obtain an equivalent bearing pressure caused by the bearing capacities of the individual piles. Such an average bearing pressure would be quite high because of the large bearing capacities of the individual piles. These large pile capacities were brought about by great progress made in the theoretical understanding of the soil resistance; by improvement in the quality of the materials used; and by the higher power and reliability of modern driving procedures and equipment.

The allowable bearing capacity that can be expected from a pile is usually based on the information gained from exploratory soil borings, and evaluated with the help of soil-mechanical principles; it should be confirmed, however, by performance tests made on the site to ascertain the actual conditions. Depending on the availability of rock, hardpan, or other firm soil stratum and on their distance below grade, the engineer will decide whether bearing piles can be used economically. Otherwise, he has to resort to friction piles of some sort to utilize the available soil condition.

Lack of a firm soil stratum at reasonable depth can sometimes be treated also with the help of floating (boatlike) foundations, which do not form a part of this discussion. The structural design of a pile cap is, in principle, not affected by the type of pile to be used, because it is primarily dependent on the magnitude of the pile reaction; however, a few explanations are necessary for a better understanding in the evaluation of the basic design approach.

5.5.2 Number of Piles Required

In the case of a spread footing, the size of the footing is determined from the total load on the footing and the allowable bearing pressure; hence, the size of the footing is rather made to order. In the case of a pile cap, however, the number of piles in the group is determined from the total load and the allowable load bearing capacity of each individual pile. Since the addition of a pile will raise the capacity of the whole group by a considerable amount, some of the pile groups may have, in order to be on the safe side, a capacity that exceeds that of the column load by a substantial amount. Furthermore, it is common practice to use, for reasons of stability, a minimum of three piles in a free-standing pile group; a minimum of two piles if a foundation beam or similar provides lateral support; and a single pile only if lateral support can be provided in two directions. These minimum requirements have to be satisfied even if the capacity provided by the pile group far exceeds the amount of the load to be supported. It is good practice to design the pile caps in any case for the full allowable capacity of the group. This is done whether required by the column load or not, and in spite of the waste that may be connected with it, to permit full utilization of the pile capacity under any circumstances.

In the case of bearing piles, every pile in a group may be considered to act as an independent pier down to the bearing stratum and to share equally in the carrying of the load. In the case of friction piles, the number of piles in a group affects their carrying capacity, especially that of the interior piles. Although this deficiency is usually averaged over the entire pile group, as far as the capacity of the group is concerned, the variation in the capacity of each individual pile requires, sometimes, consideration in the design of the pile cap.

In either case, whether we are dealing with bearing piles or friction piles, there is always a chance that some piles in the group may develop a smaller (or greater) resistance than others; a pile cap ought to be stiff enough to equalize this condition. It is therefore advisable not to keep the effective depth of a pile cap down to the minimum required, but to increase it somewhat wherever possible.

The design of a pile cap follows in general the same rules and regulations as that of a spread footing, except that the base reactions (pile reactions) are applied as concentrated loads in the center of each pile. Attention is drawn to Chap. 15 of ACI 318-83, which states that "in computing the external shear on any section through a footing supported on piles, the entire reaction from any pile whose center is located $d_p/2$ (d_p is the pile diameter at the upper end) or more outside the section shall be assumed as producing shear on the section. The reaction from any pile whose center is located $d_p/2$ or more inside the section shall be assumed as producing no shear on the section. For intermediate positions of the pile center, the portion of the pile reaction to be assumed as producing shear on the section shall be based on straight line interpolation between full value at $d_p/2$ outside the section and zero value at $d_p/2$ inside the section."

For evaluation of pile reactions under various loading conditions see the following section.

The considerable intensity of the concentrated pile reactions requires that more than usual attention be given to the design for shear in the concrete cap and the development (anchorage) of the reinforcement in the section. Owing to the importance of a crack-free entity of a pile cap in the distribution of the column load to the supporting pile group, the use of plain concrete is not permitted for pile caps.

As already indicated in Section 5.5.1 of this handbook, the present trend in the art of designing pile foundations is to use piles with large bearing capacities. Where such piles are used, the number of the piles in each group will, for obvious reasons, be smaller than if ordinary (usually up to 70 t) pile capacities are used. Since the plan (layout) of the piles within a group is, in general, not affected by the magnitude of the bearing capacity, pile caps for large-capacity piles will have to be much deeper than equivalent pile caps for ordinary piles—this primarily in order to develop the necessary shear strength. When the ratio of overall-depth/clear-span of a simple supported beam exceeds 0.8, as stipulated in Chap. 10 of ACI 318-83, then the member has to be designed for flexure and for shear as a "deep flexural member" according to the requirements of Chaps. 10 and 11 of ACI 318-83. Above provisions may be construed to be applicable to the design of two-pile caps, which behave in essence like simple supported beams. No provisions, however, exist in the code so far for dealing with two-way shear action in deep flexural members, as it occurs in connection with three-, four-, and perhaps even five-pile groups. Since such conditions are rather common, the design of pile caps for large-capacity piles may still be considered to be somewhat controversial, requiring the individual attention and judgment of the design engineer.[5-16, 5-17] He shall also be aware of the fact that it is not only good practice but rather essential to keep the design of pile caps somewhat on the safe side because neither will the actual (field) capacity of all piles in a group be the same, nor will the locations of the driven piles agree with the layout on which the design was based. Variations in both respects have to be expected, and the design of the cap must be sufficiently strong to cope with both possibilities.

Where the necessary shear strength requires excessively deep cross sections, shear reinforcement may be used to reduce the section to a reasonable size. The design of the shear reinforcement shall follow the requirements for deep flexural members and for deep (double-sided) brackets or corbels as given in Chap. 11 of ACI 318-83. In this respect engineers will find that the use of horizontal bars for this purpose is most effective, whereas vertical bars are primarily useful for corner bars, spacers, and the like.

Because of the great magnitude of the concentrated loads, it is also advisable to check the thickness of the pile cap for perimeter shear around each individual pile to ascertain that it is within the permissible limit ($4\sqrt{f_c'}$).

5.5.3 Evaluation of Pile Reactions

1. Concentric Loading Conditions. After the allowable pile reaction, R_{pa} (often incorrectly called "allowable pile capacity"), has been determined or evaluated by principles of soil mechanics,* the minimum number of piles for each column load can be determined as follows:

The effective pile reaction, R_{pe} (kips), consists of the allowable pile reaction, R_{pa} (tons), less the weight of the pile cap per pile, W_p. Any eventual surcharge shall be added to the weight of the pile cap.

*Verification of the validity of this "allowable pile reaction" is usually established by one or more pile loading tests performed at the site under actual driving conditions and at the beginning of construction. A safe assumption of the allowable pile reaction, however, has to be made, at a much earlier date to enable the engineer to design the foundation ahead of the actual construction.

$$R_{pe} \text{ (in kips)} = 2R_{pa} - W_p$$

The number of piles, n_p, required to support the unfactored total column load P is then $n_p = P/R_{pe}$, where n_p is to be rounded off to the next whole number.

Unless special conditions require a spreading of the piles, they are assembled in tight patterns to arrive at the most economical design for the pile caps. An often recommended spacing, c_p is about three times the butt diameter of the pile, usually not less than $2\frac{1}{2}$ ft. The most common spacing for piles of an average pile reaction ranging from 30 to 70 tons is 3 ft.

2. *Eccentric Loading Condition or Concentric Loading with Moment at Base.* To transform eccentric loading conditions into concentric loadings with moment at base proceed as follows:

(a) Find pile reaction R_p for concentric loading condition.
(b) Find pile reaction R_{pM} for moment at base.
(c) Superpose 1 and 2.

$$R_p + R_{pM} \leqslant 2R_{pa}$$

Where wind or earthquake is included, the R_{pa} can be increased by 33% if so allowed by the local building code. The extreme pile reaction due to a moment M is:

$$R_{pM} = \frac{M}{I_{pG}/z_{pG}}$$

To calculate the moment of inertia of a pile group I_{pG}, first find the centroid of the pile group and moment of inertia of all units in the group about the centroidal axis.

$$I_{pG} = \sum_{1}^{n} y^2$$

where y is the distance of each pile in the group from the centroidal axis.

Where a pile group consists of m equal, parallel rows of piles, the moment of inertia of the entire group is:

$$I_{pG} = mI_p/\text{Row} = m\,\frac{n_{pr}(n_{pr}^2 - 1)}{12}\,c_p^2$$

and the section modulus for the extreme piles in the group is:

$$S_{pG} = m\,\frac{n_{pr}(n_{pr} + 1)}{6}\,c_p$$

However, if the parallel rows are not of the same configuration, sum up the moments of inertia for the various rows and find the section modulus of the extreme pile by dividing the moment of inertia of the entire group by the distance of the extreme pile from the centroid, as:

$$S_{pG} = I_{pG}/z_{pG}$$

EXAMPLE 5-5: As discussed in Section 5.3.2 for ordinary spread footings, the number of piles or their arrangement in the pile group depends only on the unfactored loading conditions, as shown in Fig. 5-35, and the strength design of the pile cap has to be done by converting all loads and reactions to the factored conditions.

Column load:

$$\begin{aligned} D &= 400 \text{ kip} \\ L &= 520 \qquad R_{pa} = 50 \text{ tons} \\ \text{total} &= 920 \text{ kip} \end{aligned}$$

$$R_{pe} = 2R_{pa} - W_p = 2 \times 50 - 6.5 = 93.5 \text{ kip}$$

$$n_p = \frac{920}{93.5} = 9.8 \cong 10 \text{ piles}$$

$$W_p = A_p(\Delta q) = 3^2(50 + 75 + 150 + 450)$$

$$= 6525 \text{ lb} \cong 6.5 \text{ kip}$$

where $\Delta q = [w_L + \text{slab} + \text{fill} + \text{cap}]$. See Fig. 5-35.

EXAMPLE 5-6: Investigate Example 5-5 for an additional wind moment of 450 kip-ft in the long direction of the pile group.

The moment of inertia of the entire pile group can be considered as the sum of the moments of inertia of each row of piles:

$$I_{pG} = \sum \frac{n_{pr}(n_{pr}^2 - 1)}{12}\,c_p^2$$

$$= \left[2 \times \frac{3(3^2 - 1)}{12} + 1 \times \frac{4(4^2 - 1)}{12}\right] 3^2 = 81 \text{ ft}^3$$

The section modulus of the extreme pile in longitudinal direction is then:

$$S_{pG} = I_{PG}/1.5c_p = 81/1.5 \times 3 = 18 \text{ ft}^2$$

and the reaction on this pile due to the wind moment is:

$$R_{pM} = \frac{M}{S_{PG}} = \frac{450}{18} = 25 \text{ kip}$$

Summing up, we obtain a total maximum pile reaction under wind of:

$$R_p + R_{pM} = 98.5 + 25.0 = 123.5 \text{ kip}$$

Since the maximum allowable pile reaction under wind is:

$$R_{pa(w)} = 1.33 \times R_{pa} = 1.33 \times 100 = 133 > 123.5 \text{ kip}$$

no increase in the number of piles is required due to wind.

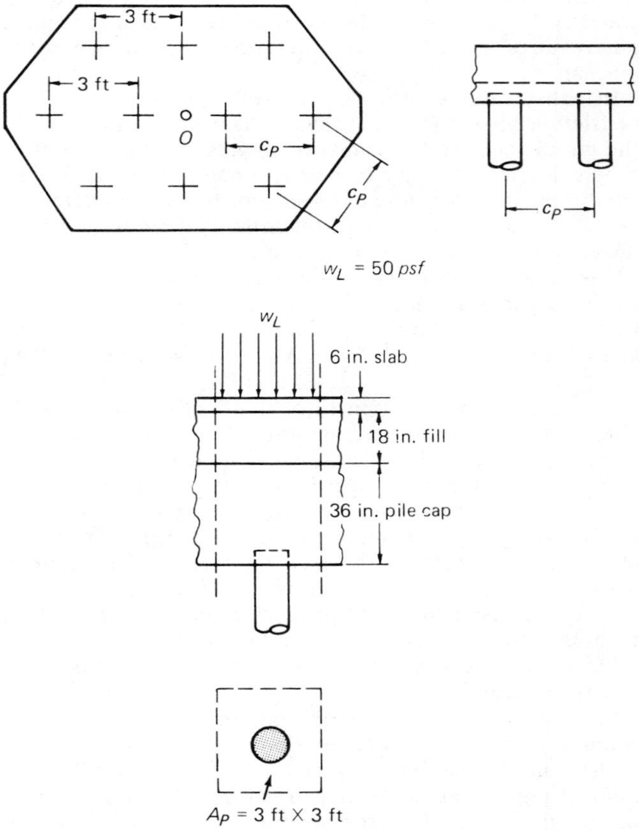

Fig. 5-35 Pile cap. Conventional pile arrangement in ten-pile cap.

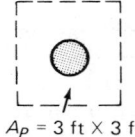

EXAMPLE 5-7: Strength design of pile cap. The column load and allowable pile capacity is the same as in Example 5-5.

$$f_c' = 3000 \text{ psi}, \quad \text{and} \quad f_y = 60{,}000 \text{ psi}$$

Pier size is 22 × 22 in.; the butt diameter of the piles is 14 in. Determine the thickness and reinforcement of the pile cap.

The strength design of the pile cap is based on the R_{pu}, which is determined from the factored loading, similar to the q_s for the spread footings, and has also here no other significance.

From dead load:

$$\frac{400 \times 1.4}{10} = 56.0 \text{ kip}$$

From live load:

$$\frac{520 \times 1.7}{10} = 88.4$$

The factored pile reaction is then 56.0 + 88.4 = 144.4 kips; it is, however, recommended to design the pile cap for the maximum factored pile reaction based on the average load factor.

Average load factor:

$$\frac{400 \times 1.4 + 520 \times 1.7}{920} \cong 1.6$$

Maximum factored pile reaction due to column load is:

$$93.5 \times 1.6 \cong 150 \text{ kip}$$

Figure 5-36 shows the layout for a ten-pile cap and the various approaches that need to be followed in the evaluation of its strength design.

Step 1: For two-way shear section (a), as indicated in the lower left quadrant of Fig. 5-36, let us assume that the necessary depth has been evaluated with 30 in. and is checked herewith: the critical base size of the truncated pyramid is 22 + 2 × 30/2 = 52 in. The critical shear force V_{2u} is then:

$$V_{2u} = 6 \times 150.0 + 2 \times 134 = 1168 \text{ kip}$$

where the contribution of the outer piles located on the y–y axis is

$$150 \times \frac{12.5}{14} = 134 \text{ kip}, \quad d_p = 14 \text{ in.}, \quad \frac{d_p}{2} + 5.5 = 12.5 \text{ in.}$$

$$v_{2u} = \frac{V_{2u}}{b_o d\phi} = \frac{1168 \times 1000}{(4 \times 52) \times 30 \times 0.85} = 220 \text{ psi}$$

which equals the permissible $4\sqrt{f_c'} = 220$ psi.

Two-way shear action (b), is indicated in the upper left quadrant of Figure 5-36. It can be realized, by inspection of Fig. 5-36, that the actual shear distribution is unequal and will be much greater in the long direction. If we take an approach similar to Example 5-3(b) we require a much greater cap thickness, as evaluated in the approach (a) described above. In this respect we divide the shear action again into two portions separated by a 45° line placed at the corner of the truncated pyramid base. Let us assume again that the necessary thickness of 34 in. has been evaluated before and is checked below.

The critical base size of the truncated pyramid is here 22 + 2 × 34/2 = 56 in. and the shear force for the most stressed quadrant becomes:

$$V_{2u}' = 1 \times 150 + 2 \times 107 = 364.0 \text{ kip}$$

where the contribution of the outer piles is:

$$150 \times 10/14 = 107.0 \text{ kip}, \quad d_p/2 + 3 = 10 \text{ in.}$$

$$v_{2u} = \frac{V_{2u}'}{b_o' d\phi} = \frac{364 \times 1000}{56 \times 34 \times 0.85} = 225 > 220 \text{ psi}$$

but acceptable. The greater depth of 34 in. is, therefore, selected.

Step 2: One-way shear action is indicated in the upper right quadrant of Fig. 5-36. The critical line is 22/2 + 34 = 45 in. away from the y–y axis. The critical shear force is then $V_{1u} = 150.0$ kip,

$$v_{1u} = \frac{V_{1u}}{b_f d\phi} = \frac{150 \times 1000}{92 \times 34 \times 0.85} = 56 \text{ psi}$$

which is smaller than $2\sqrt{f_c'} = 110$ psi.

Step 3: The critical sections for flexure, as indicated in the lower right quadrant of Figure 5-36, are at the face of the pier; the moments and reinforcements are determined for these sections.

Critical section 1:

$$M_u = 150.0\left(\frac{7 + 2 \times 25 + 43}{12}\right) = 1250 \text{ kip-ft}$$

$$F = \frac{99 \times 34^2}{12{,}000} = 9.6, \quad K_u = 1250/9.6 = 130, \quad a_u = 4.37$$

$$A_s = \frac{M_u}{a_u d} = \frac{1250}{4.37 \times 34} = 8.4 \text{ in.}^2, \quad \rho_{min} = \frac{200}{60{,}000} = 0.0033$$

$$A_{s\,min} = 0.0033 \times 99 \times 34 = 11.3 \text{ in.}^2, \quad \text{or} \quad 1.33 \times 8.4 = 11.2 \text{ in.}^2$$

which governs

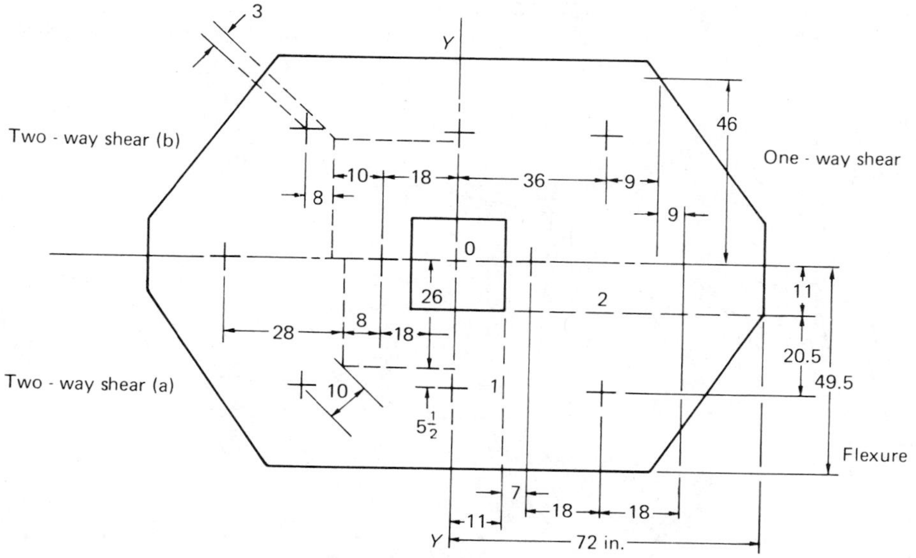

Fig. 5-36 Stress evaluation in pile caps.

Critical section 2:

$$M_u = 3 \times 150 \times 20.5/12 = 770 \text{ kip-ft}$$

$$F = \frac{144 \times 34^2}{12{,}000} = 13.7, \quad K_u = \frac{770}{13.7} = 56, \quad a_u = 4.45$$

$$A_s = \frac{M_u}{a_u d} = \frac{770}{4.45 \times 34} = 5.1 \text{ in.}^2$$

$$A_{s\,min} = \rho_{min} bd = 0.0033 \times 144 \times 34 = 16.1 \text{ in.}^2, \quad \text{or}$$

$$1.33 \times 5.1 = 6.8 \text{ in.}^2 \text{ which governs}$$

The selection and distribution of the bars is done as described in Example 5-1, Step 4(b). The maximum bar size that may be used has to be selected in such a way that the development (anchor) length of the bar is smaller than or equal to the shortest available embedment length of the bar at either side of the critical section.

5.6 RETAINING WALLS

5.6.1 General

A retaining wall is a structure designed for the purpose of providing one-sided lateral confinement of soil or fill.

All retaining walls, with the exception of true cantilever walls anchored to rock, are in principle gravity walls, i.e., their action depends primarily on their developed weight. In common practice, however, only those retaining walls are called gravity walls where the dead weight required to make the resultant vector intersect the base within safe allowable limits is made up solely of the dead weight of the concrete. Such walls are usually designed unreinforced, Fig. 5-37(a). The commonly called cantilever walls are in principle gravity walls where reinforcement is used to reduce and modify the cross section of the concrete in such a way that portions of the soil or fill are utilized for developing the necessary rightening moment, Fig. 5-37(b).

In every retaining wall design, regardless of the type used, three resultant forces, namely, the lateral confinement pressure, Q, the total developed weight, P, and the soil reaction or bearing resistance, R, have to be brought into equilibrium (Fig. 5-38a); in addition, all internal stresses in the structure and all external soil reactions have to be within the permissible limits.

Retaining walls of the commonly called cantilever type can be subdivided into two main groups:

1. Continuous walls of constant cross section, where every foot of wall length is providing its own equilibrium, Fig. 5-37(b).
2. Sectional walls, where crosswalls introduced at certain spacings, provide all stability requirements and the walls between them act only as intermediate elements, Fig. 5-37(c).

Where the crosswalls are visible in front, they are called buttresses; where they are behind the wall and inside the soil, they are called counterforts. Some retaining walls are designed to have both.

The portion of a continuous cantilever wall or crosswall that is pressed downward into the soil is called the toe, and the portion that is lifted upward is called the heel. The vertical portion is called the stem, Fig. 5-37(b).

The footing of a continuous retaining wall or crosswall must be large enough:

1. To resist the resultant vector due to confinement pressures, dead weight of the concrete and developed weight of the soil by means of safe bearing pressures.
2. To keep the wall safely from overturning.

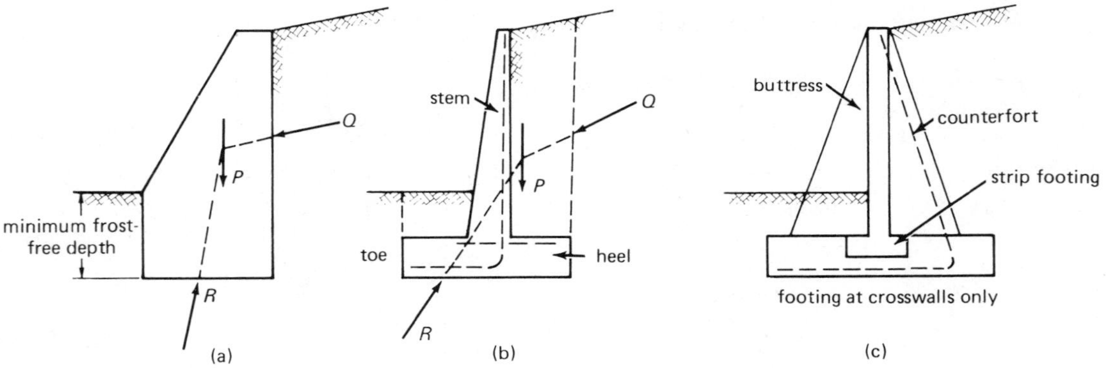

Fig. 5-37 Types of retaining walls.

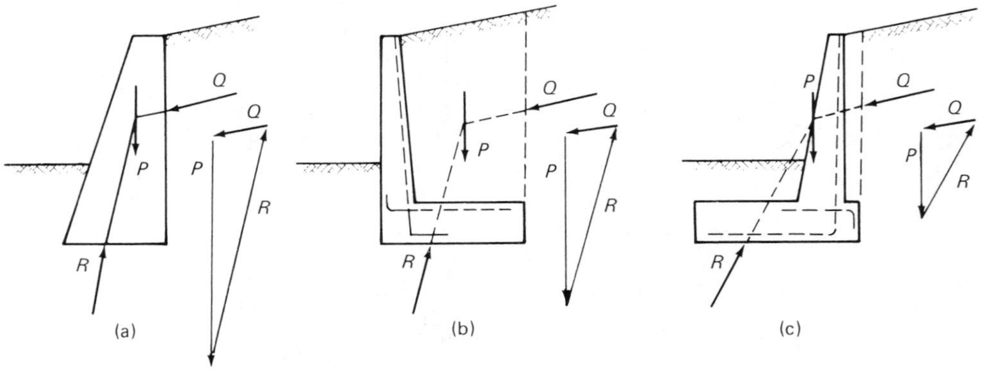

Fig. 5-38 Conditions of equilibrium for retaining walls.

5.6.2 Confinement Pressure

The magnitude and distribution of the lateral pressures exerted by the confined soil or backfill depends on the kind of material, its moisture content, existence and depth of the groundwater, slope of backfill, and eventual surcharge due to live loads, storage, building loads, etc., applied close enough to be of influence. These pressures and their distribution have to be evaluated by principles of soil mechanics and do not form a part of this discussion. Most textbooks on soil mechanics contain detailed information on this subject.[5-1, 5-2]

The resultant of these pressures is located at the centroid of the pressure wedge. The angle of inclination between the resultant and a line perpendicular to the back of the wall (ϕ') indicates the wall friction and is usually expressed as a fraction of the angle of internal friction, ϕ, of the fill material; it is often assumed with $\phi/2$. It is, however, important to keep in mind that this angle is also influenced by the slope, material, and compaction of the backfill, by the surface texture of the concrete (at the back of the wall), by the existence of groundwater behind the wall, or merely by moisture in the soil, which can act like a lubricant and reduce the friction angle to a minimal value. It is common practice with many designers to disregard the wall friction and to apply the resultant perpendicular to the back of the wall, in order to be on the safe side.

Spaced weep holes or continuous backdrains are often provided to alleviate the heavy pressure condition that can be caused by groundwater accumulating behind the back of the wall. Since weep holes and other drainage provisions may be clogged, it is recommended to investigate a retaining wall for a condition of full, or at least increased, water pressure. Such loading, however, represents an emergency condition, and it is up to the engineer's judgment to reduce the safety factor for such a design as he sees fit.

5.6.3 Bearing Pressure

The pressure distribution under the footing of a retaining wall follows the same rules and is determined by the same methods as the pressure distribution under an eccentrically loaded footing. It is of course desirable, and to be attempted wherever possible, to keep the intersection of the resultant of all active forces (confinement pressures and developed weight) within the kern of the footing base. (In the case of a strip footing under a continuous retaining wall, the kern distance is $1/6$ of the footing size in the direction of the loading.) However, in many cases this is not economically feasible, and greater eccentricities have to be accepted with a pressure distribution extending only over a part of the footing. The maximum edge pressure is then determined as discussed in Sections 5.2.2 and 5.2.3 and shown in Fig. 5-5.

5.6.4 Overturning

Overturning can also be treated as in regular column or wall footings. Where the base of the retaining wall is resting on rock or very hard soil, overturning may be calculated about the pressed edge and the safety factor can be expressed by the ratio $SF = M_R/M_O$, where M_O is the overturning moment caused by the confinement pressures acting about the pressed edge, and M_R is the resisting moment consisting of the dead weight of the retaining structure plus the developed weight of the fill material and any other frictional or passive resistances in the soil that may be mobilized during the overturning. The safety factor may also be expressed according to the "Suggested Design Procedures for Combined Footings and Mats"[5-3] as the ratio of the distance of the pressed edge from the base centroid to the eccentricity, e. In this ratio, $e = M_B/P$, where M_B is the sum of all moments about the base centroid, and P is the sum of all forces acting perpendicular to the base. The safety factor against overturning should customarily be not less than 1.5.

Where the retaining wall rests on soil, the investigation regarding overturning may proceed along similar lines, except that the critical line about which the overturning and resisting moments are to be calculated is not at the pressed edge but somewhat inside, at the distance e_f from the centroid which is the center of gravity of the pressure block evaluated for an ultimate soil bearing condition. The evaluation of this condition is similar to the one described in Section 5.2.1 and shown in Fig. 5-5(c). Where the ultimate soil pressure is not known, it may be assumed to be 2.5 times the allowable soil pressure.[5-3]

5.6.5 Sliding Resistance

Lateral resistance against the horizontal component of the confinement pressure, commonly called resistance against sliding, has to be supplied by static friction at the footing base and by passive earth pressure against the embedded front portion of the retaining wall. Where this resistance is insufficient, the passive pressure can be increased by extending a key or lug into the soil below the footing base. Since the preference for the location of the lug can be argued about, it is probably located best where it is most practical with regard to construction and placement of reinforcement (Fig. 5-39).

5.6.6 Type Selection

The quality of the available subsoil and its allowable bearing pressure, as well as the height of the wall itself and the available construction space in front of the slope, can influence the selection of an economical type of wall to a great degree.

Where the construction space is ample, the allowable soil pressure is reasonable, and the height of the structure is not excessive, gravity walls of structural plain concrete can be used with advantage. Their weight is usually large enough to develop all the necessary base friction; however, the provision of keys or lugs projecting into the firm subsoil is rather common.

Where the available bearing pressure is large and construction space is available, the designer can utilize a long heel

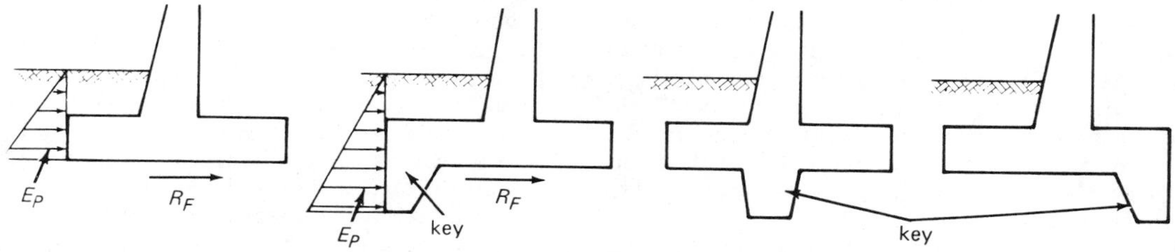

Fig. 5-39 Lateral resistance at base for retaining walls.

and develop much back fill weight to deflect the horizontal pressure resultant sharply down to the ground (Fig. 5-38b). However, where the available bearing pressure is small, too much of it would have to be used to carry the developed weight of the backfill. In this case, the designer will have to be satisfied with a smaller total dead weight, which will decrease the inclination of the resultant. Consequently he will have to increase the toe to keep the resultant sufficiently within the base to satisfy the allowable bearing pressure, and to keep the wall from overturning (Fig. 5-38c). For ordinary conditions it may be considered good practice to keep the sizes of heel and toe about the same. The base of the toe has to be placed at frost-free level.

Butresses and/or counterforts are used presently only for retaining walls of greater height (more than 20 to 25 ft, unless buttresses are architecturally desired) because of the great expense usually involved in the formwork. This type of design may see a kind of revival in combination with the use of precast wall panels.

For the enclosure of underground storage or extra large basements areas, free-standing retaining walls are sometimes used as basement walls.

In certain cases retaining walls are designed to act as such only temporarily until a full tie-in with the rest of the structure is achieved. In other cases such walls are designed only for the final loading condition, but not for the loading during construction, and will therefore require temporary shoring to maintain their stability. Some retaining walls may during construction change from a fully independent free-standing structure to a top- and bottom-supported basement wall, and in such a case every condition of loading ought to be considered for their stability as well as for their strength.

The use of precast sections may also here see a considerable field of application in the future. Such sections may, in the form of sheetings or in the form of entire wall panels, be driven or inserted into the ground before the main excavation has taken place. In such a case the same unit may serve successively as protective sheeting, retaining wall, and final basement enclosure.[5-9]

Free-standing retaining walls may also be constructed entirely of precast units. Such designs may simulate either the action of a sectional cantilever wall, as shown in Fig. 5-37(c), or that of a gravity wall, in which case they are called cribbings.

5.6.7 Cribbings[5-10]

Cribbings consist in general of two types of units, namely, face and anchor units, also called stretchers and headers.

The structural design of cribbings is usually based on an empirical evaluation. The open-faced units provide excellent drainage; in other cases drainage has to be provided to prevent groundwater from backing up and exerting pressure and unsightly leakage.

The satisfactory behavior of such precast cribbings depends to a considerable degree on the quality of the compacted backfill, which shall be installed in close coordination with the placement of the sections.

The face units are straight precast members of various cross sections, often with protruding lugs at their ends to connect them to the adjacent face and/or anchor members. Face members can be designed in open, closed, and flush-type manner, depending on the architectural requirements. They are usually set with a batter of about 1 : 6 at the front and can also be placed so as to form a curved face, up to 20°, without special units.

The anchor units usually come in two types of design, fishtail units or continuous back units. A fishtail unit is a T-shaped, precast section, placed perpendicular to the face of the wall with the purpose of tying two adjacent face units together and anchoring them back to the fill material. Continuous back units are used together with cross wall units to form boxlike openings that are filled with compacted soil. There exist also combined units, where face and anchor units are cast together, simplifying erection where transportation permits.

5.6.8 Pile-Supported Retaining Walls

Retaining walls may also be supported on piles, which is often the case along waterfronts, or where the subsoil does not have a sufficient bearing capacity or lateral stability. In such designs the vertical component of the pressure resultant is to be taken by the piles as for a strip footing and the horizontal component is to be taken by either batter piles, tie backs, or similar devices.

Where the subsoil is also capable of providing uplift anchorage for the piles, the necessary developed weight may be reduced accordingly.

5.6.9 Design Procedure

The static design and the stability investigation of a retaining wall shall be based on service load conditions; the size of the footing and its location with regard to the wall itself is, therefore, entirely governed by the actual fill and/or liquid pressures under consideration of the allowable soil bearing pressures and allowable values for soil friction and passive resistance. The structural design of the stem and footing sections and their reinforcement, however, shall be based on the strength design method. For this purpose, the actual loads and active confinement pressures have to be multiplied by the appropriate load factors, and the resulting bearing pressures and other resistances have to be evaluated from these factored loading conditions. The procedure is not always easily executed because of variations in eccentricities that may be caused by it (see Example 5-8). It is important to keep in mind, just as in the case of the ordinary footing, that these reactions are only caused by assumed factored loading conditions, but have no relationship to ultimate soil bearing values and the like.

After the factored loadings and corresponding reactions have been determined, every part of the retaining wall (stem, toe, or heel) has to be designed independently as a fully restrained cantilever section protruding from a mass center and carrying all applied loads.

The design to be performed for the service load conditions consists of a trial-and-error approach. The result is not too sensitive to slightly incorrect assumptions and can usually be adjusted in a second trial, if so desired. Because of the multitude of possible combinations, it is rather difficult to arrive at reliable recommendations in this respect; however, if the conditions are not too much out of the ordinary, the footing size perpendicular to the wall directions can be assumed to be about $^2/_5$ to $^2/_3$ of the total wall height, the footing thickness about $^1/_8$ to $^1/_{12}$ of the footing size, and the base of the stem about $^1/_{10}$ to $^1/_{12}$ of its height. The following example has not been cut down to the minimum size–i.e., it could be adjusted and recalculated for further savings; such a procedure would be mostly repetitive and more confusing than helpful. The top of the stem should not be made less than 8 in. to be able to place into it two layers of reinforcement, if so required. It is customary to place a second layer of reinforcement inside the exposed face for shrinkage and cracking control. It is practical to coordinate the reinforcements of stem and toe in such a way that the reinforcing bars of the stem can be bent right into the toe. For greater wall heights it is economical to run only a part of the vertical wall reinforcement to the top and stop the remainder at lower elevations.

EXAMPLE 5-8: Design a continuous retaining wall (cantilever wall) for the following conditions:

(a) The difference between upper and lower level is 12 ft.
(b) The upper level shall sustain a surcharge of 200 lb/ft^2 (medium heavy parking).
(c) Soil of satisfactory bearing capacity is encountered at a depth of 4 ft below lower level (must be equal or greater than frost-free depth).
(d) Provide weep holes to relieve water pressure; maximum expected ground water level 3 ft above lower level.
(e) Use concrete with minimum $f_c' = 3000$ psi, and reinforcing steel having a minimum yield point of 60,000 psi.
(f) Soil mechanical considerations provided the following values: The bearing soil is a medium dense, silty sand with an allowable bearing pressure of 3000 lb/ft^2. The weight of the moderately dry soil can be assumed with 100 lb/ft^3 and its angle of internal friction $\phi = 32°$.

1. Assumption of Size and Evaluation of Loadings. Based on the requirements and other informations given above, the designer makes an assumption for the shape, sizes, and other relationships, and arrives at an arbitrary cross section as given in Fig. 5-40. Using these assumptions, he arrives at the following values:

$$P_1 = \frac{(8 + 12)}{12 \times 2} \times 14.75 \times 0.15 \qquad = \quad 1.85 \text{ kip}$$

$$P_2 = 8.50 \times 15/12 \times 0.15 \qquad = \quad 1.60 \text{ kip}$$

$$P_3 = 3.50\,(14.75 \times 0.1 + 0.20) \qquad = \quad 5.90 \text{ kip}$$

$$P_4 = 2.75 \times 4.00 \times 0.1 \qquad = \quad 1.10 \text{ kip}$$

total dead load $P = P_1 + P_2 + P_3 + P_4 = 10.45$ kip/foot of wall

Eccentricity of total dead load P from centroid of base area, o:

$$e_P = \frac{P_1 \times 0.05 - P_3 \times 2.50 + P_4 \times 2.25}{P} = -1.15 \text{ ft}$$

The active earth pressure:

$$p_a = \gamma h \tan^2 (45° - \phi/2)$$

Introducing above values the pressure increment can be evaluated with $p_a = 0.03$ kip/ft^2 per foot of depth. Transforming the given surcharge into an equivalent height of additional soil, we obtain, for the active earth pressure at the upper level, $p_{a1} = 2 \times 0.03 = 0.06$ kip/ft^2, and at the lower level $p_{a2} = 14 \times 0.03 = 0.42$ kip/ft^2.
Below this elevation the active earth pressure remains constant.

$$Q_1 = \frac{0.06 + 0.42}{2} \times 12 = 2.88 \text{ kips}$$

$$Q_2 = 0.42 \times 4.0 \qquad = 1.68 \text{ kips}$$

Total active earth pressure is:

$$Q = Q_1 + Q_2 = 4.56 \text{ kip/lin-ft of wall.}$$

The elevations at which they are applied are, for Q_1:

$$h_{Q1} = 4.0 + \frac{12}{3}\left(\frac{2 \times 0.06 + 0.42}{0.06 + 0.42}\right) = 4.0 + 4.5 = 8.5 \text{ ft}$$

for Q_2:

$$h_{Q2} = 2.0 \text{ ft}$$

and for the resultant active earth pressure, Q:

$$h_Q = \frac{2.88 \times 8.5 + 1.68 \times 2.0}{4.56} = 6.1 \text{ ft}$$

2. Calculation of Bearing Pressures. The resultant eccentricity at the base of the retaining wall can be found from

$$e = \frac{M_Q \pm M_P}{P} = \frac{Q \times h_Q \pm P \times e_P}{P}$$

The (±) sign depends on the location of e_P with regard to the base centroid, o.

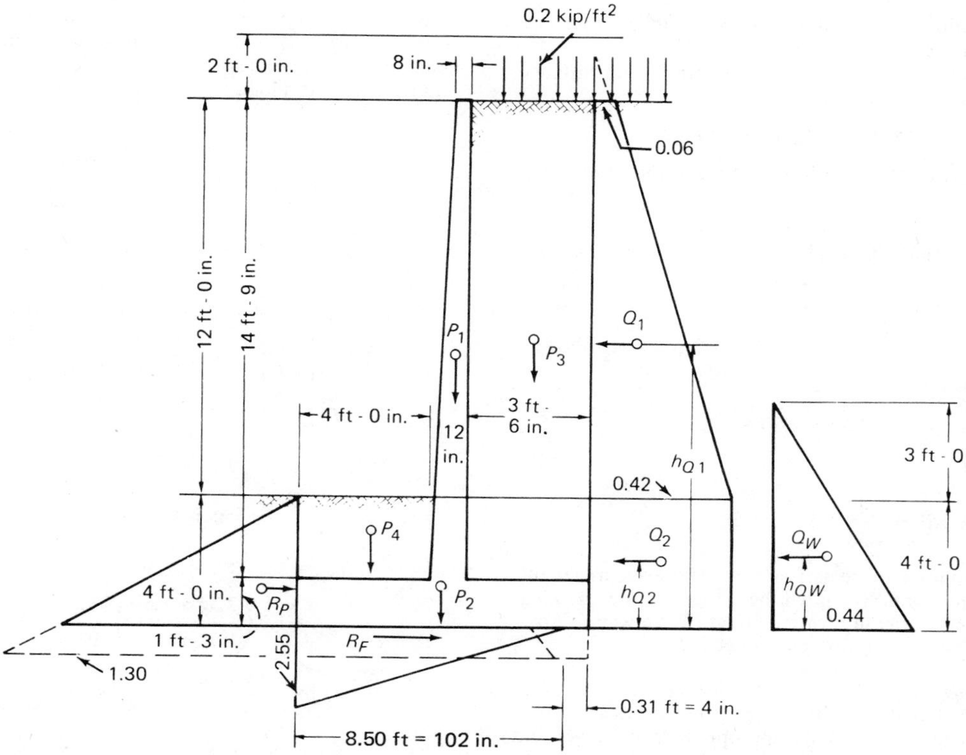

Fig. 5-40 Loadings and reactions for retaining wall, Example 5-8.

$$e = \frac{4.56 \times 6.1 - 10.45 \times 1.15}{10.45} = \frac{15.9}{10.45} = 1.52 \text{ ft}$$

which is slightly outside the kern distance $c_k = \frac{8.5}{6} = 1.43$ ft.

The maximum toe pressure can be calculated according to Fig. 5-5 as:

$$q = \frac{2P}{3mb} = \frac{2 \times 10.45}{3(4.25 - 1.52)\,1} = 2.55 \text{ kip/ft}^2 < 3.0 \text{ kip/ft}^2$$

3. Overturning. Overturning, according to Fig. 5-5(c), can be computed as follows:

$$e_f = \frac{l_f}{2} - \frac{P}{2q_f b_f} = \frac{8.5}{2} - \frac{10.45}{2(2.5 \times 3)\,1} = 3.55 \text{ ft}$$

$$SF = \frac{e_f}{e} = \frac{3.55}{1.52} = 2.30 > 1.5$$

Overturning under full water pressure is computed using: the total water pressure above base elevation:

$$Q_W = (7.0 \times 0.0624)\frac{7.0}{2} = 1.55 \text{ kip}$$

and the additional moment due to the waterpressure about the base:

$$M_W = 1.55 \times \frac{7.0}{3} = 3.6 \text{ kip-ft}$$

The eccentricity under this extreme condition would be:

$$e_W = \frac{M_Q \pm M_P + M_W}{P_{\text{RED}}} = \frac{28.0 - 12.1 + 3.6}{9.57} = 2.0 \text{ ft}$$

The P in this equation was reduced to account for the buoyancy of the submerged portions.

$$SF = \frac{e_f}{e_W} = \frac{3.55}{2.0} = 1.77 > 1.5$$

4. Sliding. A friction coefficient of 0.45 was assumed from soil-mechanical considerations.

The total horizontal force is $Q = 4.56$ kip; the frictional resistance that can be developed at the base is:

$$R_F = 0.45P = 0.45 \times 10.45 = 4.75 \text{ kip}$$

The passive earth pressure that can be developed at the front of the retaining wall is:

$$p_{p1} = \gamma \tan^2 (45° + \phi/2)\, h = 0.1 \tan^2\left(45° + \frac{32°}{2}\right) 4.0 = 1.3 \text{ kip/ft}^2$$

$$R_p = p_{p1}h/2 = 1.3 \times \frac{4.0}{2} = 2.6 \text{ kip}$$

The total resistance is therefore:

$$R = R_F + R_P = 4.75 + 2.60 = 7.35 \text{ kip}$$

and the safety factor:

$$SF = \frac{7.35}{4.56} = 1.6 > 1.5$$

This safety factor appears to be satisfactory. However, under full water pressure the horizontal force will be larger and the frictional resistance smaller due to the lubricating effect of the water. Provision of a key at the base is therefore recommended.

5. Design of Concrete Thickness and Reinforcement. In order to proceed with the strength design of the elements the actual service loads have to be multipleid by the appropriate load factors, and the respective strengths of the elements be evaluated under consideration of the appropriate ϕ-factors.

6. Stem Design. Moment about base of stem:

$$M_u = 1.7\left[Q_1(h_{Q1} - 1.25) + p_{a2}\frac{(2.75)^2}{2}\right]$$

$$= 1.7\left[2.88(8.50 - 1.25) + 0.42 \times \frac{(2.75)^2}{2}\right]$$

$$= 38.2 \text{ kip-ft}$$

For 2 in. concrete protection:

$$d_{\text{eff}} = 12 - 2.5 = 9.5 \text{ in.}$$

$$F = \frac{bd^2}{12{,}000} = \frac{12 \times 9.5^2}{12{,}000} = 0.09, \quad K_u = \frac{M_u}{F} = \frac{38.2}{0.09} = 425$$

a_u as taken from Flexure Table 1.1 of the Design Handbook ACI SP-17, 1983, $a_u = 4.00$.

$$A_s = M_u/a_u d = 38.2/4.03 \times 9.5 = 1.01 \text{ in.}^2 \ \#7 @ 7$$

Check for full water pressure:

$$M_u = 38.2 + 1.4\left[\frac{(7 - 1.25)^2}{2}\, 0.0624 \times \frac{(7 - 1.25)}{3}\right]$$

$$= 38.2 + 1.4[1.04 \times 1.9] = 40.95 \text{ kip-ft}$$

$$K_u = \frac{40.95}{0.09} = 455, \quad a_u = 4.0, \quad A_s = \frac{40.95}{4.0 \times 9.5} = 1.07$$

$$A_s \text{ provided} = \#7 @ 7 = 1.03 \text{ in.}^2; \quad \frac{1.07}{1.03} = 1.04, \quad \text{OK}$$

7. Toe Design. It is advisable to check the shear condition first. In connection with bearing pressures, it is sometimes diffiuclt to apply the corresponding load factors properly if the bearing pressures were caused by loadings with different load factors. Such a procedure may sometimes cause relocation of the resulting eccentricities and lead to pressure distributions that are in principle different from those obtained under service load conditions. Since this is not considered to be the intent of the design, it is recommended to select in such cases either the highest load factor, 1.7, in order to be on the safe side, or to select by judgment an approximate load factor between 1.4 and 1.7, to apply to the bearing pressures obtained from the investigation of the service load conditions.

In this example a load factor of 1.7 was used.

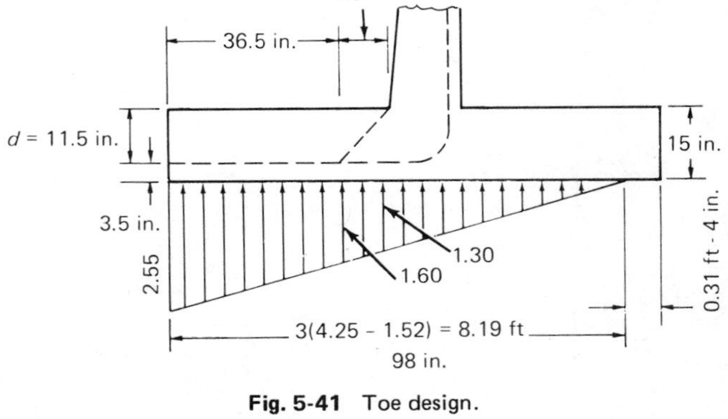

Fig. 5-41 Toe design.

$$d_{\text{eff}} = 15 - 3.5 = 11.5 \text{ in.}$$

$$V_{1u} = 1.7\left(\frac{2.55 + 1.60}{2}\right)\frac{36.5}{12} = 10.8 \text{ kips}$$

$$v_{1u} = \frac{1000V_{1u}}{\phi bd} = \frac{1000 \times 10.8}{0.85 \times 12 \times 11.5} = 92 \text{ psi}$$

$$v_{c\,\text{all}} = 2\sqrt{f_c'} = 110 \text{ psi}$$

$$92 < 110$$

For flexure, the bearing pressure outside the critical line (face of stem) is:

$$1.7\left(\frac{2.55 + 1.30}{2}\right)4 = 13.0 \text{ kip}$$

The centroid of the trapezoidal pressure distribution is 2.25 ft away from the face of the stem. Hence:

$$M_u = 13.0 \times 2.25 = 29.5 \text{ kip-ft}, \quad F = \frac{12 \times 11.5^2}{12{,}000} = 0.132$$

$$K_u = \frac{29.5}{0.132} = 220, \quad a_u = 4.27, \quad A_s = \frac{29.5}{4.27 \times 11.5} = 0.62 \text{ in.}^2$$

Since it is practical to bend the stem reinforcement right into the toe, the provided stem reinforcement, $A_s = 1.03$ in.^{2}, is compared with the required one and found to be ample.

8. Heel design.

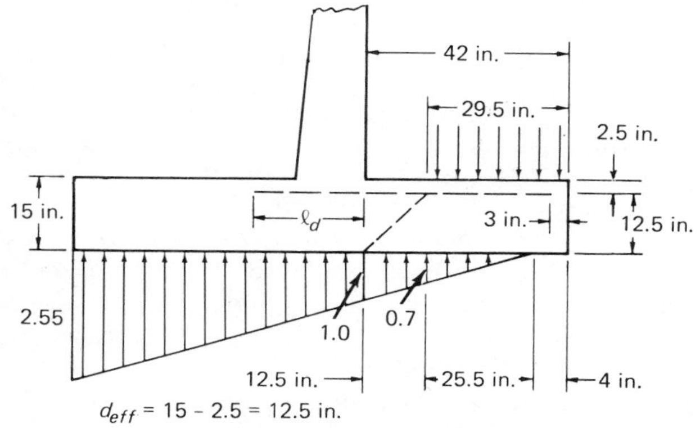

Fig. 5-42 Heel design.

The shear force is:

$$V_{1u} = 1.7\left(\frac{P_3 \times 29.5}{42} - \frac{0.7}{2} \times \frac{25.5}{12}\right) = 6.0 \text{ kip}$$

$$v_{1u} = \frac{1000V_{1u}}{\phi bd} = \frac{1000 \times 6.0}{0.85 \times 12 \times 12.5} = 47 \text{ psi} < 110 \text{ psi}$$

The flexural moment to be resisted is:

$$M_u = 1.4 \times P_3 \times 1.75 - 1.7\left[\frac{1.0}{2}\left(\frac{38}{12}\right)^2 \frac{1}{3}\right] = 14.5 - 2.8$$

$$= 11.7 \text{ kip-ft}$$

$$F = \frac{12 \times 12.5^2}{12{,}000} = 0.155, \quad K_u = \frac{11.7}{0.155} = 76 \text{ less than min.}$$

$$A_s = \frac{1.33M_u}{a_u d}, \text{ or } \rho_{min}\, bd$$

$$A_s = \frac{1.33 \times 11.7}{4.42 \times 12.5} = 0.28 \text{ in.}^2, \text{ or } 0.0033 \times 12 \times 12.5 = 0.49 \text{ in.}^2$$

$A_s = 0.28$ in.^{2} governs

A bar size has to be selected for which the development length is smaller than the available embedment 42 - 3 = 39 in. Use #5 bars @ 12" o.c., because #5 bars have a development length which is suitable and their number will satisfy the least required cross section.

NOTATION

a = footing projection in general, ft
a_L = projection of footing beyond critical face in long direction, in.
$a_n = a_u$ (as used in earlier editions of Design Handbook ACI SP-17)
a_s = projection of footing beyond critical face in short direction, in.
a_u = factor used in determining $A_s = M_u/a_u d$ (see also a_n)
A_F = base area of footing, ft^2
A_P = cross-sectional area of pier, in.^{2} (also A_C, ft^2)
A_{ip} = average influence area per pile, ft^2
A_{V1} = influence area of bearing pressure for one-way shear, ft^2
A_{V2} = influence area of bearing pressure for two-way shear, ft^2
ΔA_{V2} = base area of truncated cone or pyramid for two-way shear, ft^2
b = width of combined footing, ft
b_f = side of square footing, ft
b_f = short side of oblong footing, ft
b_o = base perimeter of truncated cone or pyramid for two-way shear, in.
b_p = dimension of pier parallel to footing side b_f, in.
β_c = longside/shortside of concentrated load or reaction area (also aspect ratio A_F/A_C)
c_c = concrete protection in.
c_k = kern distance, ft
c_p = pile spacing, ft
d = effective depth of section, in.
d_b = bar diameter, in.
d_f = diameter of round or polygonal footing, ft
d_o = diameter of opening in circular or polygonal footings, ft
d_P = pile diameter, in.
D = dead loads or their related internal moments and forces
e = eccentricity of load resultant from footing centroid, ft
e_f = eccentricity, as under e, causing failure pressure q_f at edge, ft
e_{max} = maximum permissible eccentricity to prevent overturning, ft
E = earthquake loads or their related moments and forces
$F = bd^2/12{,}000$
γ_s = average unit weight of footing + surcharge, kip/ft^3
h = lever arm of horizontal force, ft
h_f = thickness of footing, in.
h_Q = lever arm of force Q, ft
h_s = depth of footing base below floor, ft
H = horizontal force, kip
I_c = moment of inertia of concrete cross section, in.^{4}
I_F = moment of inertia of footing base area, ft^4
I_{PG} = moment of inertia of pile group, ft^2
$K_n = K_u$ (as used in earlier edition of Design Handbook ACI SP-17)
K_u = factor used in determining $F = M_u/K_u$ (see also K_n)
K_{si} = coefficient of vertical subgrade reaction for a 1 ft^2 area, tons/ft^2/ft
l = long dimension of combined footing, ft
l = distance between column centers, ft
l_d = development length (anchorage) of bar, in.
l_f = long side of rectangular footing, ft
l_P = dimension of pier parallel to footing side l_f, in.
L = live loads and their related internal moments and forces

m = distance of eccentric load from pressed edge of footing, ft
m = general footing dimension, ft
m = number of pile rows of equal configuration
$M, M_u, \phi M_n$ = flexural moment, kip-ft
M_B = moment about base, kip-ft
n = general footing dimension, ft
n_P = number of piles in group
n_{P_r} = number of piles in row
N = blow count of standard penetration test
o = centroid of footing or pile group
p_a = active earth pressure, kip/ft^2
p_p = passive earth pressure, kip/ft^2
P, P_u = total column load at base, kip
P_D, P_{Du} = column dead load, kip
P_L, P_{Lu} = column live load, kip
P_{RES} = resultant load, kip
q = bearing pressure, kip/ft^2
Δq = weight of footing + surcharge, kip/ft^2
q_a = allowable bearing pressure, kip/ft^2
q_c = bearing pressure at base of column or pier, kip/in.^{2}
q_e = effective bearing pressure $(q - \Delta q)$, kip/ft^2
q_f = bearing pressure at failure, kip/ft^2
q_m = bearing pressure due to moment, kip/ft^2
q_P = bearing pressure under concentric loading, kip/ft^2
q_s = bearing pressure at footing base due to factored loading, kip/ft^2
Q = lateral force against retaining wall, kip
R = lateral resistance, kip
R_F = frictional resistance at base, kip
R_M, R_{Mu} = pile reaction due to moment, kip
R_P, R_{Pu} = pile reaction due to concentric loading, kip
R_{P_a} = allowable pile reaction, tons
R_{P_e} = effective pile reaction $(2R_{P_a} - W_P)$, kip
ρ = percentage of reinforcement
s = soil deformation
S = ratio of long side to short side of footing
S_{PG} = section modulus of pile group, ft
S_{PR} = section modulus of pile row, ft
SF = safety factor
U = uplift force, kip
v_{1u} = average shear stress for one-way shear, psi
v_{2u} = average shear stress for two-way shear, psi
V_{1u} = shear force for one-way shear, kip
V_{2u} = shear force for two-way shear, kip
w = line load, kip/ft
w_L = live load on floor, kip/ft^2
W = wind loads or their related moments and forces
W_P = weight of pile footing + surcharge, per pile, kip
z = distance of extreme fiber from centroid, ft
z_{PG} = distance of extreme pile from centroid of pile group, ft
z_t = distance of extreme tensile fiber from centroid, in.
z_x = distance of reference line from centroid, ft

REFERENCES

5-1 Terzaghi, K., *Theoretical Soil Mechanics*, John Wiley and Sons, Inc., New York, 1942.

5-2 Terzaghi, K., and Peck, R. B., *Soil Mechanics in Engineering Practice*, 2nd Ed., John Wiley and Sons, Inc., New York, 1967.

5-3 ACI Committee 336 (formerly 436), "Suggested Design Procedures for Combined Footings and Mats," (ACI 336.2R-66) (Reaffirmed 1980) *ACI Journal, Proceedings*, V. 63, (10), pp. 1041–1057, Oct. 1966. Also *ACI Manual of Concrete Practice*, Part 4. American Concrete Institute, Detroit.

5-4 Hetenyi, M., *Beams on Elastic Foundations*, University of Michigan Press, Ann Arbor, Michigan, 1946.

5-5 Kramrisch, F., and Rogers, P., "Simplified Design of Combined Footings," *Proceedings ASCE*, Paper No. 2959, **87** (SM5), Oct. 1961.

5-6 Molitor, D. A., *A Practical Treatise on Chimney Design*, Peters Company, Detroit, 1938.

5-7 *Beton Kalender*, Wilhelm Ernst and Sohn, Berlin, 1939.

5-8 Furlong, R., "Design Aids for Square Footings," *ACI Proceedings*, **62**, March 1965, pp. 363–371.

5-9 Kramrisch, F., "Structural Walls of Precast Concrete Sheet Piling," *Building Construction*, **5** (11), Nov. 1964, pp. 64–67.

5-10 "Concrete Crib Retaining Walls," Portland Cement Association, Concrete Information No. ST-46.

5-11 Rice, P. F., and Hoffman, E. S., "Pile Caps—Theory, Code, and Practice Gaps," *CRSI Bull.* No. 2, 1978.

5-12 Gogate, A. B., and Sabhis, G. M., "Design of Thick Pile Caps," *ACI Journal*, Jan–Feb. 1980.

6

Properties of Materials for Reinforced Concrete

SIDNEY FREEDMAN*

6.1 INTRODUCTION

The properties of hardened concrete have considerable significance to designers of concrete structures or products. The physical properties of concrete depend upon a number of factors including mix proportions, aggregates, type of cement, curing conditions, and age. They are also affected by environmental conditions such as temperature and relative humidity. The durability of concrete is related to these same factors with particular emphasis on the cement content and amount of entrained air.

Tests to determine strengths are undoubtedly the most common tests made to evaluate the properties of the hardened concrete. There are three reasons for this: (1) the strength of concrete in compression or tension has in most cases a direct influence on the load carrying capacity of both plain and reinforced structures; (2) of all the properties of hardened concrete, strength can usually be determined most easily; and (3) the results of strength tests can be used as a qualitative indication of other important qualities of hardened concrete.

6.2 COMPRESSIVE STRENGTH

6.2.1 Quality of Cement Paste and Bond

The compressive strength of concrete made with aggregate of adequate strength is governed, in general, by the strength either of the cement paste or of the bond between the paste and the aggregate particles. At early ages the bond strength is lower than the paste strength; at later ages the reverse may be the case.

The water-cement ratio and the degree to which hydration has progressed to a large extent determine the quality of the portland cement paste. The water-cement ratio rule is as follows: for a given cement and acceptable aggregates the strength that may be developed by a workable, properly placed mixture of cement, aggregate, and water (under the same mixing, curing, and testing conditions) is influenced by the (a) ratio of mixing water to cement, (b) ratio of cement to aggregate, (c) grading, surface texture, shape, strength, and stiffness of aggregate particles, and (d) maximum size of the aggregate. Mix factors partially or totally independent of water-cement ratio that affect the strength are (1) type and brand of cement, (2) amount and type of admixture or pozzolan, and (3) mineralogic makeup of the aggregate.

Figure 6-1 presents a series of band curves that were developed from a large number of compressive strength tests on 6 × 12 in. cylinders made by many laboratories using a variety of materials. Note that strengths increase as the water-cement ratios decrease, and that strengths increase with age. These saturated specimens show 20 to 30% lower strength than companion specimens tested dry.

Table 6-1 shows the relative strengths of concretes made with different cements; Type I portland cement concrete is used as the basis for comparison. These values are characteristic for concretes that are moist-cured until tested.

*Director, Producer Member Services, Prestressed Concrete Institute, Chicago, Illinois; formerly, Manager, Concrete Technology Section, Portland Cement Association, Skokie, Illinois.

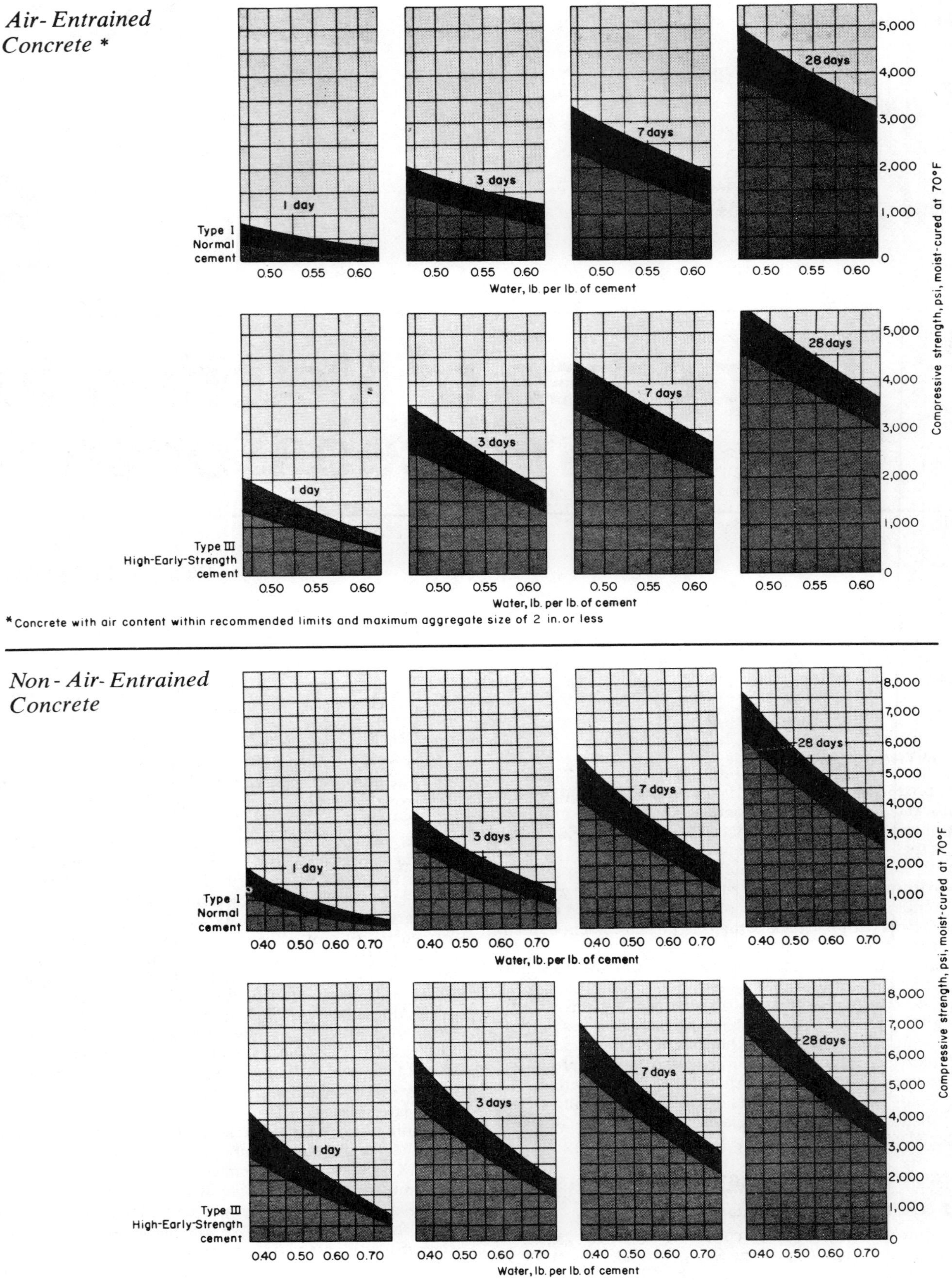

Fig. 6-1 Relationship between water–cement ratio and compressive strength for portland cements at different ages. These relationships are approximate and should be used only as a guide in lieu of data on job materials. (From: *Design and Control of Concrete Mixtures* 11th ed., Portland Cement Association, Skokie, Ill., 1968.)

TABLE 6-1 Approximate Relative Strength of Concrete As Affected by Type of Cement*

Type of Portland Cement	*Compressive Strength–percent of strength of Type I Portland Cement concrete*			
ASTM	1 day	7 days	28 days	3 months
I	100	100	100	100
II	75	85	90	100
III	190	120	110	100
IV	55	55	75	100
V	65	75	85	100

**From: "Design and Control of Concrete Mixtures," Portland Cement Association, Skokie, Ill., 1968.*

High-strength concrete shows a higher rate of strength gain at early ages as compared to normal-strength concrete, but at later ages these differences are not significant. (High-strength concrete is considered to be concrete with compressive strength higher than 6000 psi for normal-weight aggregates and 4000 psi for lightweight aggregates.) The average rate of development of strength for normal, medium, and high-strength limestone concrete cylinders is shown in Fig. 6-2. The higher rate of strength development of high-strength concrete at early ages is caused by an increase in the internal curing temperature in the concrete cylinders due to a higher heat of hydration, because of the high cement content in high strength concrete. Typical ratios of 7-day to 95-day strength were 0.60 for normal strength, 0.65 for medium strength, and 0.73 for high strength concrete. High-strength concrete shows a larger reduction in compressive strength than normal-strength concrete when allowed to dry before completion of curing.

Water-reducing admixtures in portland cement concrete have been used for three decades. Until very recently, the best of the conventional water reducers were able to yield reductions in total net mixing water of at most 10%. Although a 10% reduction helped improve many properties of concrete, greater water reductions were not possible because of excessive set retardations from high dosages of these admixtures. A new class of admixtures commonly called

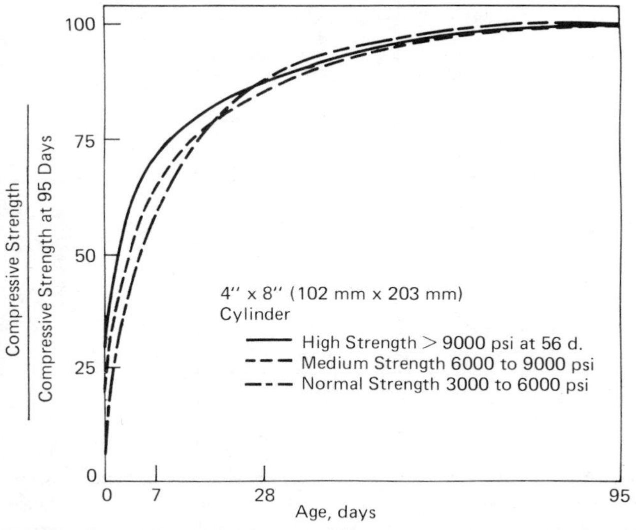

Fig. 6-2 Strength gain with age for limestone concretes moist cured until testing (From: Carrasquillo, R. L., Nilson, A. H., and Slate, F. O., "Microcracking and Behavior of High-Strength Concrete," *ACI Journal*, May–June 1981, pp. 179–186.)

superplasticizers (high-range water reducers) was introduced in Japan and Germany in the mid-1960s. These admixtures meet the requirements of ASTM C494 types F or G, and must reduce the water content of the concrete mix by a minimum of 12%.

The currently available superplasticizers in North America may be classified as: (1) sulfonated melamine formaldehyde condensates; (2) sulfonated naphthalene formaldehyde condensates; or (3) modfied lignosulfonates.

There are three ways in which superplasticizers may be used in concrete:

(a) To produce concrete with a very low water to cement ratio. To achieve high-strength concrete, water content of the mix is reduced while maintaining the same cement content. The reduced workability is compensated for by incorporating superplasticizers. By this method, water reductions of up to 30% can be achieved and concrete with water to cement ratios as low as 0.28 can be successfully placed. The efficiency of water reduction increases with an increase in cement content.

(b) To produce concretes with a reduced cement content. Superplasticizers can be used to produce concretes with reduced cement contents while the water to cement ratio is maintained constant.

(c) To produce flowing concretes. Superplasticizers have been used to produce self-compacting, self-leveling flowing concretes. In this application no attempt is made to reduce either the water to cement ratio or the cement content. Instead, the aim is to increase the slump from, say, 3 to 8 in. without causing any segregation so that the concrete can be placed in heavily steel-reinforced sections. The actual dosage needed for a particular job will be a function of the particular cement used, mix design (especially cement factor), water–cement ratio desired, temperature, type of mixer, addition time of admixture, and other variables. A combination of superplasticizers and conventional water reducers, set-modifying, or air-entraining admixtures may be used to achieve a desired performance.

The use of superplasticizers as water reducers does not require any significant change in the mix proportioning (other than reduction in water–cement ratio). A small increase in sand content (approximately 5%) may be useful to avoid a "rocky" mix. If desired, total paste content may be increased to compensate for the volume of water removed from the batch. This may be done by use of additional cement, fly ash, or other finely ground materials. The water–cement ratio should be maintained, however.

The lower water content of mixes containing high-range water reducers results in dramatically increased strengths at early ages over control mixes prepared at equal consistencies. A good rule of thumb is that with these materials, 7-day strengths can be obtained in 3 days, and 28-day strengths can be obtained in 7 days. The 28-day strength is typically increased by 20 to 30%. The increase in mechanical properties, i.e., compressive and flexural strength and modulus of elasticity, is generally commensurate with reductions in water to cement ratio, although some exceptions have been noted. Ultimate strengths over 10,000 psi at ages of 90 days and beyond are obtainable when high-range water reducers are used with moderately rich concrete mixtures.

Although these materials significantly reduce bleeding and appear to have only minor effects on setting time, they adversely affect the behavior of workability with time, resulting in high rates of what is commonly called slump loss.

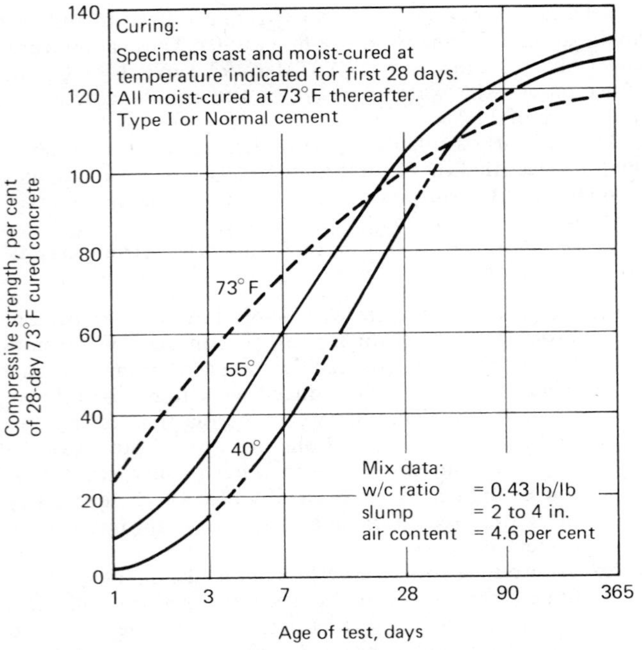

Fig. 6-3 Effect of low temperatures on concrete compressive strength at various ages. (From: *Design and Control of Concrete Mixtures*, 12th ed., PCA, 1979.)

6.2.2 Curing Temperatures

Figure 6-3 shows the age–compressive strength relationship for concrete that has been mixed, placed, and cured at temperatures between 40 and 73°F. At temperatures below 73°F, strengths are lower at early ages but higher at later periods.

Higher early strengths may be achieved through use of Type III or High-Early-Strength cement. Principal advantages occur prior to 7 days. At 40°F curing temperatures, the early advantages of this type of mixture are more pronounced and persist longer than at higher temperatures.

Figure 6-4 shows the effect of high concrete curing temperatures on compressive strength. These tests, using identical concretes of the same water-cement ratio, show that while higher concrete temperatures increase early strength, at later ages the reverse is true. If the water content had been increased to maintain the same slump (without changing the cement content), the reduction in strength would have been even greater.

Strength gain practically stops when moisture required for curing is no longer available. Concrete that is placed at low temperatures (but above freezing) may develop higher strengths than concrete placed at high temperatures, but curing must be continued for a longer period. It is not safe to expose concrete to freezing temperatures at early periods. If freezing is permitted within 24 hours, much lower strength will result.

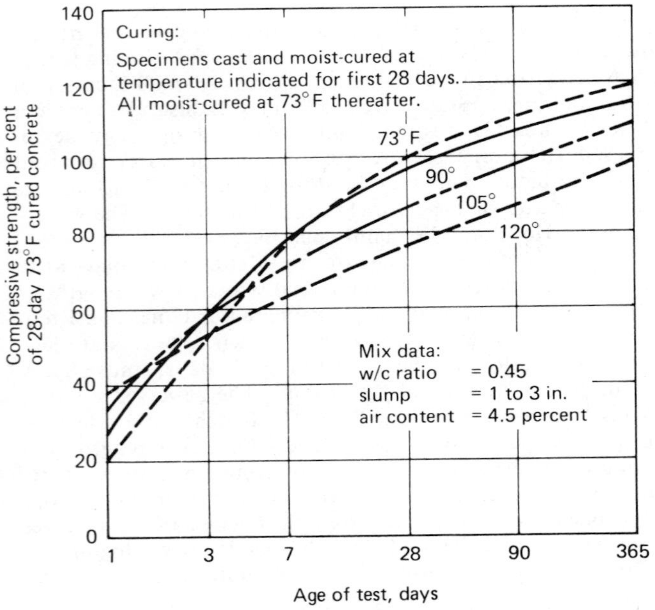

Fig. 6-4 Effect of high temperatures on concrete compressive strength at various ages. (From: *Design and Control of Concrete Mixtures*, 12th ed., PCA, 1979.)

6.2.3 Air-Entrained Concrete

Strength of air-entrained concrete depends principally upon the voids–cement ratio. For this ratio, "voids" is defined as

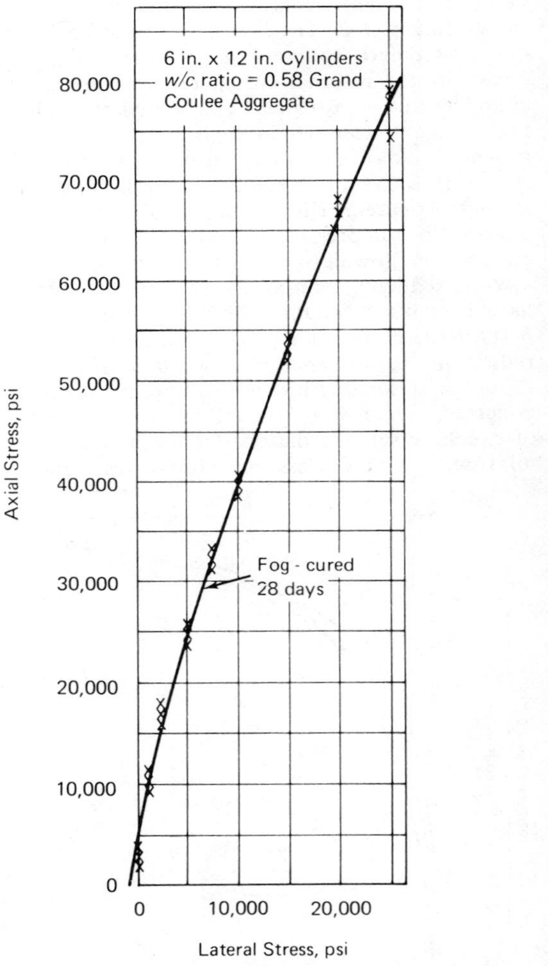

Fig. 6-5 Relations of axial stress to lateral stress at failure in triaxial compression tests of concrete. (From: Balmer, G. A., "Shearing Strength of Concrete under High Triaxial Stress—Computation of Mohr's Envelope as a Curve," *Structural Research Laboratory Report No. SP-23*, U. S. Department of the Interior, Bureau of Reclamation, Oct. 28, 1949.)

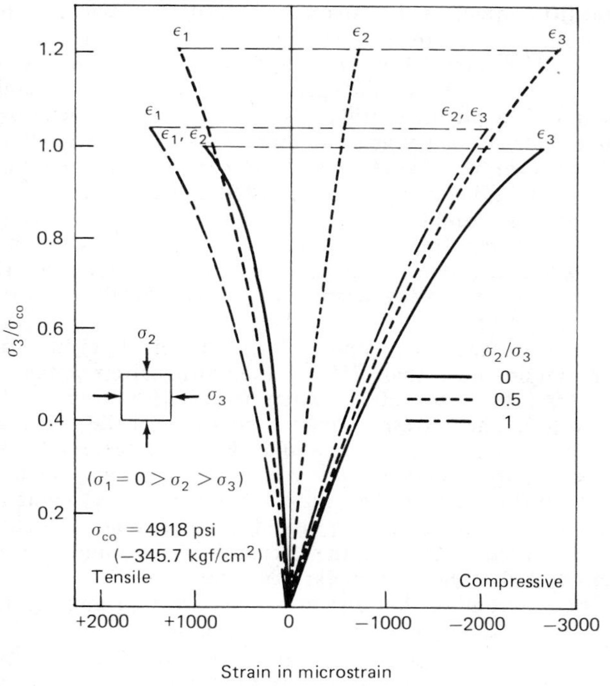

Fig. 6-6(a) Stress–strain relationships of concrete under biaxial compression. From: Tasiyi, M. E., Slate, F. O., and Nilson, A. H., "Stress–Strain Response and Fracture of Concrete in Biaxial Loading," *ACI Journal*, July 1978, pp. 306–312.)

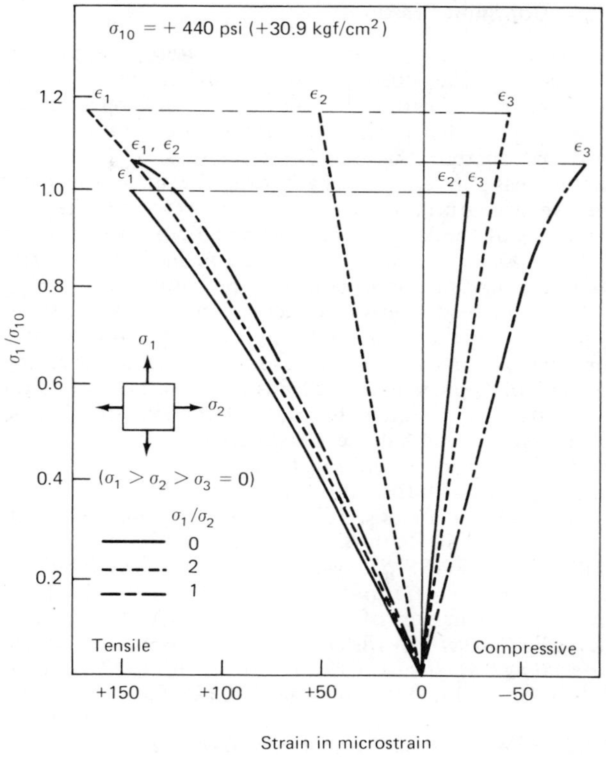

Fig. 6-6(c) Stress–strain relationships of concrete under biaxial tension. (From: Tasiyi, M. E., Slate, F. O., and Nilson, A. H., "Stress–Strain Response and Fracture of Concrete in Biaxial Loading," *ACI Journal*, July 1978, pp. 306–312.)

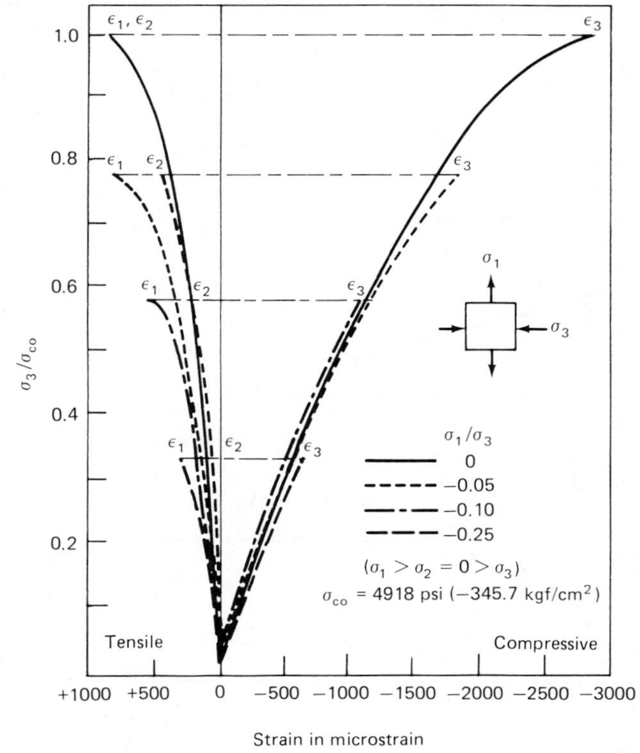

Fig. 6-6(b) Stress–strain relationships of concrete under biaxial compression–tension. (From: Tasiyi, M. E., Slate, F. O., and Nilson, A. H., "Stress–Strain Response and Fracture of Concrete in Biaxial Loading," *ACI Journal*, July 1978, pp. 306–312.)

the total volume of water plus air (entrained and entrapped). Hence, when air content is maintained constant, strength varies inversely with the water–cement ratio. As air content is increased, a given strength generally may be maintained by holding to a constant voids–cement ratio. Some increase in cement content may be necessary in richer mixes in order to accomplish this.

At a given water–cement ratio, the compressive strength of air-entrained concrete is reduced by approximately 5% for every 1% of entrained air. Flexural strength is reduced considerably less by air entrainment. Air-entrained as well as non-air-entrained concrete can usually be proportioned to provide any desired strength. Air-entrained concretes have lower water–cement ratios than non-air-entrained concretes; therefore, reductions of strength that generally accompany entrainment of air are minimized. However, some reductions in strength may be tolerable in view of other benefits. These reductions become significant only in mixes containing more than about 565 lb of cement per cubic yard. In lean or harsh mixes, strengths are generally increased by entrainment of air in proper amounts.

Attainment of high strength with air-entrained concretes may be difficult at times. Both air-entrained and non-air-entrained concretes have increased water demands when slumps are maintained constant as concrete temperatures rise. Even though a reduction in mixing water is associated with air entrainment, mixtures with high cement contents require more mixing water; hence, the increase in strength expected from the additional cement is offset somewhat by the additional water added. Also, with some aggregates it is not possible to secure extremely high strengths with air-entrained concrete.

6.2.4 Confining Pressures

The axial compressive strength of concrete is appreciably increased by the existence of lateral confining pressures such as those exerted by the spiral reinforcement in a column or by that portion of column concrete surrounded by floor concrete, Fig. 6-5. (Also see Figs. 6-20 and 6-21.) The increase in axial strength over the unconfined compressive strength ranges from as much as four to five times the value of the lateral stress for small confining pressures (0 to 1500 psi) to about two and one-half or three times the lateral stress for large confining pressures.

The ultimate strength of concrete under biaxial compression is greater than that under uniaxial compression and is dependent on the principal stress ratio. A maximum strength increase of approximately 22% is achieved at a stress ratio of $\sigma_2/\sigma_3 = 0.5$. Under biaxial compression–tension, the compressive strength decreases almost linearly as the applied tensile stress is increased, Fig. 6-6.

For additional information on the properties of concrete under multiaxial stress states, refer to two papers by Gerstle, H. K., et al., "Strength of Concrete under Multiaxial Stress States," ACI SP-55, *Douglas McHenry International Symposium on Concrete and Concrete Structures*, 1978, 684 pp.; and "Behavior of Concrete under Multiaxial Stress States," *Journal of the Engineering Mechanics Division, Proceedings of the American Society of Civil Engineers*, V. 106, No. EM6, Dec. 1980, pp. 1383–1403.

6.2.5 Effect of High Temperature Exposure

The influence of heat exposure per se on the structural properties of concrete is critically dependent on the moisture content of the concrete at the time of heating and is quite different for concrete in which the moisture is free to evaporate as compared to concrete sealed against moisture loss. Deterioration of structural properties also can be expected from any test variable that produces large temperature differentials in the concrete on heating (rapid heating or cooling or use of very large specimens). Partial loss of chemically combined water (dehydration of cement paste) occurs above 250°F and primarily affects the flexural strength. For practical situations in which the free moisture can evaporate as heating commences, it can be expected that little or no change or a slight beneficial change will take place in the compressive strength on heating to 500°F, provided the conditions for thermal shock are not present, Fig. 6-7. When free moisture is retained in the concrete during heating, deterioration in structural properties can be expected in increasing degrees of severity at all levels of heating up to 500°F.

Carbonate aggregate concrete or sanded lightweight concrete retain more than 75% of their original strengths at temperatures up to 1200°F when heated without load and tested hot. The corresponding temperature for the siliceous aggregate concrete is about 800°F, Fig. 7-7.* Strengths of specimens stressed in compression during heating generally are 5 to 25% higher than those of companion specimens that are not stressed during heating, and residual strengths are somewhat lower than the strengths of companion specimens tested at high temperatures. Strengths of specimens stressed in compression during heating are not significantly affected by the applied stress level ranging from 25 to 55% of the original strength. Original strength of the concrete also has little effect on the percentage of strength retained after heating concretes to high temperatures and then cooling the concrete to room temperature. (See Chapter 7, "Fire Resistance," for practical implications.)

6.2.6 Effect of Low Temperature Exposure

The strength of concrete at a given low temperature will depend on the cement content and the water–cement ratio

*Chapter 7, "Fire Resistance."

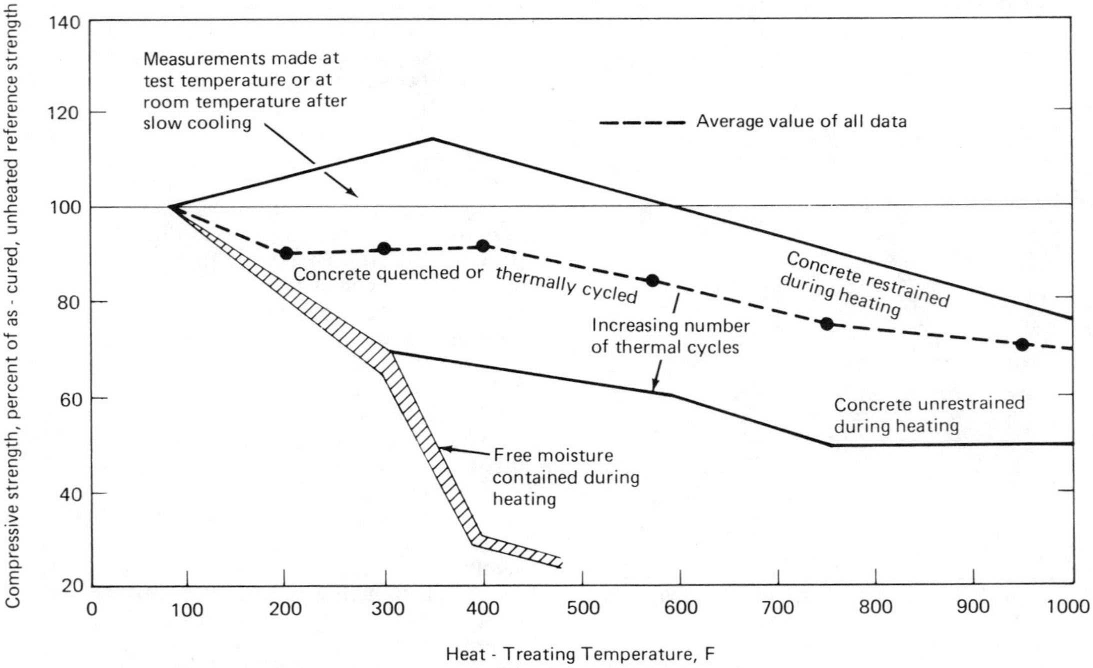

Fig. 6-7 Extremes of the influence of heat exposure on the compressive strength of concrete as determined by various investigators. (From: Lankard et al., "Effects of Moisture Content on the Structural Properties of Portland Cement Concrete Exposed to Temperatures up to 500°F," ACI SP-25, *Temperature and Concrete*, 1971.)

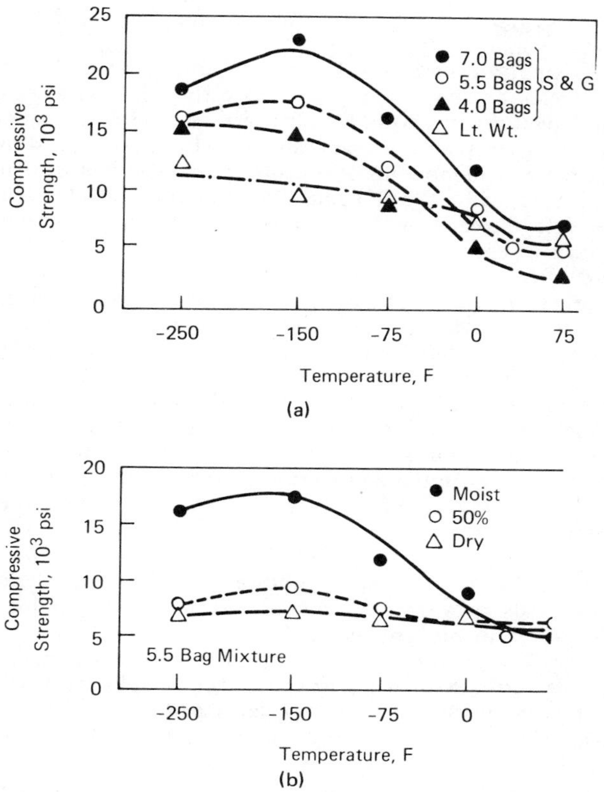

Fig. 6-8 (a) Effect of low temperatures on strength of moist concrete; and (b) effect of moisture condition on strengths at low temperatures. (From: Monfore, G. E., and Lentz, A. E., "Physical Properties of Concrete at Very Low Temperatures," *PCA Research Bulletin 145*, 1962.)

of the mix, the aggregate, the age, and especially the moisture condition.

The results of strength determinations at low temperatures are shown in Fig. 6-8(a) for various concrete mixtures in a moist condition. Although there is some scatter of points, the trends are similar for the different mixes. Tests on the 5.5-bag mix indicate very little difference in strengths between 75°F and 35°F, and it is felt that the same would be true for the other concretes. In general, strengths increase as the temperature is lowered from 75°F, reach maximum values in the range from −75°F to −150°F, then decrease as the temperature approaches −250°F.

The increases in compressive strength are related to the evaporable water contents, and Fig. 6-8(b) shows that the compressive strength of oven-dry concrete increases by only about 20% as the temperature is lowered from 75°F to −150°F, whereas the increase in comparable moist concrete is 240%. The increase for concrete equilibrated in an atmosphere of 50% relative humidity is only slightly greater than that for oven-dry concrete. These increases in strength of moist concrete with decrease in temperature below freezing must be largely due to the formation of ice, which increases progressively as the temperature is lowered.

6.2.7 Impact Loading

Loads that are applied at rates considerably higher than 10,000 psi per second are referred to as impact loads. The dynamic compressive strength of a given type of concrete is higher than the static strength, becoming relatively greater as the duration of impact decreases, Fig. 6-9. There are significant increases in compressive strengths, elastic moduli, and ability to absorb strain energy under impact loading.

6.3 TENSILE STRENGTH

Tensile strength is one of the fundamental properties of concrete. Although reinforced concrete structures are not normally designed to resist direct tension, a knowledge of the tensile strength of their members helps in understanding the behavior of these structures and designers use this property to resist loads (flexural tension), shear, shrinkage, and temperature stresses. Significant principal tension stresses may be associated with multiaxial states of stress in walls,

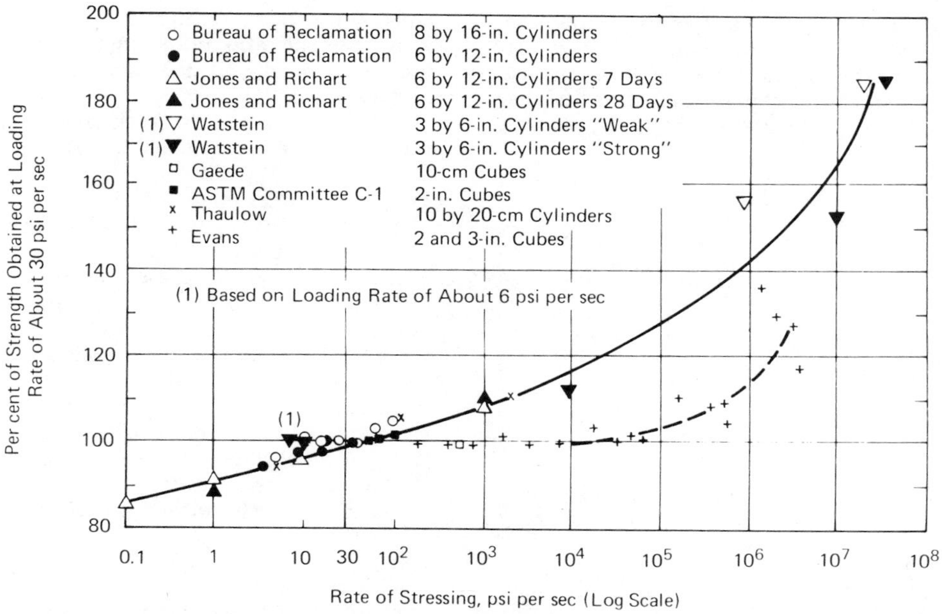

Fig. 6-9 Effect of rate of stressing on the compressive strength of concrete. (From: McHenry, D., and Shideler, J. J., "Review of Data on Effect of Speed in Mechanical Testing of Concrete," *PCA Development Bulletin D9*, 1956.)

shells, or deep beams. Under biaxial tension, the strength of concrete increases as compared to its uniaxial tensile strength. A maximum strength increase in the order of 10 to 20% is achieved at a stress ratio of $\sigma_1/\sigma_2 = 2.0$. In determining the tensile strength of concrete three types of tests have been used: (1) direct, (2) flexural, and (3) splitting tension.

6.3.1 Direct Tension

A direct application of a pure tension force free from eccentricity is difficult and is further complicated by secondary stresses induced by the grips or embedded studs. No standard test involving the application of direct tension exists, and it is usual to determine the tensile strength of concrete in flexure or by the application of splitting tension. However, the available data indicates that direct tension is about 12 to 7% of the compressive strength when strengths vary from 2500 to 7000 psi; as compressive strength increases, the ratio of direct tension to compressive strength decreases.

6.3.2 Flexural Tension

The flexural test does not measure the true tensile strength of concrete but determines what is known as the modulus of rupture. The tensile strength as determined by the modulus of rupture of a given concrete depends to a considerable extent on the distribution of load, size of the specimen, moisture condition, and the rate of loading.

The modulus of rupture may be found by cantilever, centerpoint, or third point loading. The third point method is a better estimate of flexural strength and is normally used for design.

The modulus of rupture of a 6 in. deep beam is about 20% greater than that of a 10-in. beam. This phenomenon is thought to be caused mainly by difference in strain gradient; the smaller the depth, the steeper the gradient and the higher the modulus of rupture.

Modulus of rupture tests are sensitive to shrinkage, temperature, and minor surface defects, such as large pieces of aggregate near the surface. The outer fibers of concretes undergoing drying are extremely sensitive to the transient moisture content, and under these conditions may not furnish data that are satisfactorily reproducible. Flexural and direct tension specimens register a higher strength when completely dry than when surface-wet; however, partially

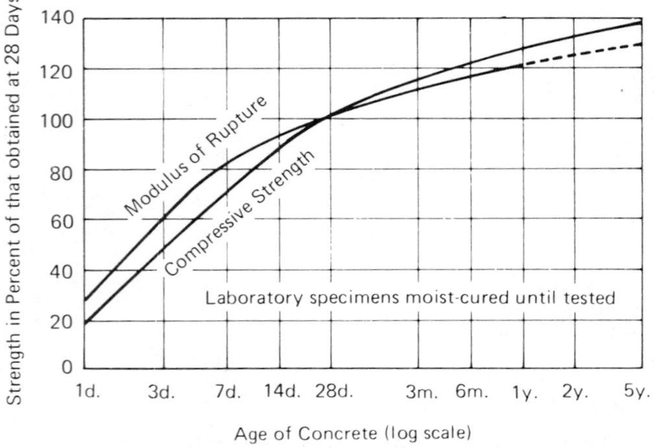

Fig. 6-10 Age-strength relationships for moist-cured concrete.

dry flexural specimens may show a reduction of 40% over wet specimens.

Flexural strengths of lightweight and sand and gravel concretes do not differ greatly at early ages, but after 28 days lightweight shows less strength gain with continuous moist curing than does sand and gravel. Figure 6-10 shows the increase in modulus of rupture for moist-cured sand and gravel concrete. Until about 28 days the modulus gains strength faster than the compressive strength. Lightweight aggregate concrete moist-cured 7-days followed by drying and 50% relative humidity has a much lower modulus of rupture than continuously moist-cured beams. Sand and gravel show similar results; however, the percentage of strength reduction is considerably greater for lightweight concrete than for sand and gravel concrete. The loss of flexural strength is attributed to internal stresses caused by moisture gradients within the specimens. Shrinkage or rapid cooling will reduce tensile strength of an element because of the greater initial contraction of the surface. This leaves tensile stress at the surface, compressive stress in the interior.

Increasing the rate of application of stress results in an increase in the modulus of rupture. A straight-line relationship exists between the modulus of rupture and the log of the rate of increase of stress.

The flexure test yields a considerably higher value of tensile strength than either the direct or splitting tension test because the formulas to calculate modulus of rupture assume a linear tensile stress distribution which is known to be incorrect.

Flexural strength is substantially affected not only by the strength of the mortar, but, since failure is in tension, also by the tensile strength of the coarse aggregate particles and by the bond between the coarse aggregate particles and the mortar. Certain aggregates may produce high compressive strength but be incapable of producing high flexural strength. Similarly, certain aggregates are adequately strong in all respects, but may have surface characteristics that are not favorable to a good bond between the mortar and the coarse aggregate, and, therefore, have poor flexural strength producing ability. The relative level of tensile resistance of lightweight concrete is a definite characteristic of each particular aggregate when the concretes are compared on an equal compressive strength basis. The relationship between water–cement ratio and strength, which is helpful for compressive strength, is less definitive in its application to flexural strength.

Concrete exposed to temperatures well above normal room temperatures generally shows a deterioration in flexural strength, Fig. 6-11. The rate of strength deterioration is highest at temperatures above 400°F. Cycling of the concrete produces large reductions in tensile strength with cracks becoming noticeable on the concrete surfaces after a number of cycles. These cracks have their origin in the differential movements of paste and aggregate during the heating and cooling periods.

Flexural strength is not a constant proportional part of the compressive strength, but the proportionality ratio decreases from about 22% to 10% as compressive strength increases. Angular coarse aggregate improves the flexural strength much more than the compressive strength so that the tensile to compressive strength ratio is greater for concretes made with crushed coarse aggregate. Other factors influencing the ratio are age, mix proportions, properties of fine aggregate, and curing.

With given materials and procedures a relationship can be established between compressive and flexural strength, there being no general relationship between compressive and flexural strength that is valid for all conditions. Where

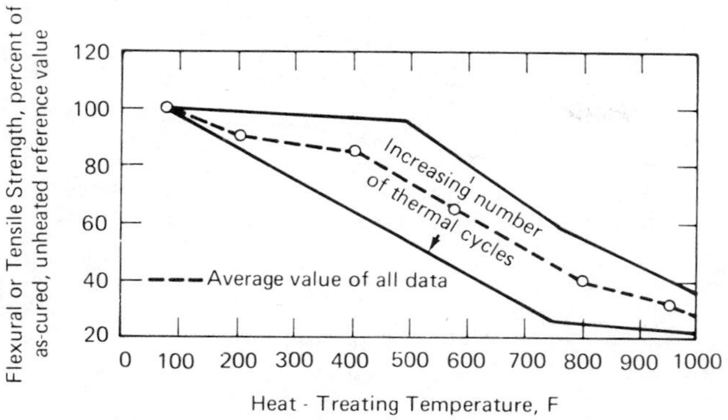

Fig. 6-11 Extremes of the influence of heat exposure on the flexural or tensile strength of concrete as determined by various investigators. (From: Lankard *et al.*, "Effects of Moisture Content on the Structural Properties of Portland Cement Concrete Exposed to Temperatures up to 500°F," ACI SP-25, *Temperature and Concrete*, 1971.)

flexural strength is of importance, the compressive strength corresponding to the required flexural strength might well be determined in the laboratory and compression tests used thereafter for field control and evaluation purposes. However, it is generally accepted that the modulus of rupture is approximately a function of the square root of compressive strength:

$$f_r = K\sqrt{f'_c}$$

where:

f_r = modulus of rupture, psi, third point loading
f'_c = compressive strength, psi
K = a constant usually between 8 and 10

For both lightweight and normal-weight high-strength concrete, the modulus of rupture falls in the range of 7.5 to $12\sqrt{f'_c}$. A value of 11.7 is recommended. The correlation between modulus of rupture and the compressive strength improves at later test ages, 28 and 95 days.

6.3.3 Splitting Tension

The splitting tensile strength of concrete cylinders (ASTM C496) is a convenient relative measure of tensile strength. The test is performed by application of diametrically opposite compressive loads to a concrete cylinder laid on its side in the testing machine. Fracture, or "splitting," occurs along a diametral plane that is subjected largely to a uniform tensile stress. The theory on which the splitting tensile strength is calculated assumes that the concrete is elastic and that a state of plain stress exists. Neither assumption is true, and the calculated splitting tension is slightly higher than the true axial tensile strength.

The cylinder splitting method is a more reliable measure of tensile strength (lower coefficient of variation) than the modulus of rupture beam test, particularly for concretes that are subjected to drying conditions. Perhaps the most important advantages of the split cylinder are the approximate uniformity of tensile stress over the diametral area of the cylinder and the simplicity of the test, which affords the opportunity to test economically a large number of specimens. The large number of samples may offset the variation of results that must always be expected in tensile investigations. The split cylinder test results will satisfactorily reflect such variables as compressive strength, age of concrete, and type of curing.

The tensile splitting strengths of continuously moist-cured lightweight and normal-weight concretes of equal compressive strength are equal. In all cases the tensile strength of steam-cured lightweight concrete is considerably less than that of normal weight concrete. The tensile strength of concretes that undergo drying is more relevant to behavior of concrete in structures. During drying of the concrete, moisture loss progresses at a slow rate into the interior of concrete members, resulting in the probable development of tensile stresses at the exterior faces and balancing compressive stresses in the still moist interior zones. Thus the tensile resistance of drying concrete will be reduced from that indicated by continuously moist-cured concrete. This effect is more pronounced for concrete made with lightweight aggregate than for concrete made with natural aggregates. The splitting tensile strengths of normal-weight concretes actually increase when subjected to air drying during the curing period, Fig. 6-12.

The splitting tensile strength of air-dried all-lightweight concrete varies from approximately 70% to 100% that of the normal weight concrete, when comparisons are made at equal compressive strength. The reduction is more pronounced for the higher strength concrete than for lower, and is also greater for all-lightweight concrete than for sand-lightweight. Replacement of lightweight fines by sand generally increases the splitting tensile strength of lightweight concrete subjected to drying. In some cases this increase is nonlinear with respect to the sand content, so that with some aggregates partial sand replacement is as beneficial as complete replacement. Also air-dried all-lightweight concretes with high splitting strengths show little improvement as the sand content is increased. The results of splitting tensile strength tests on air-dried lightweight concrete may be used as a reliable measure of the unit shear capacity of lightweight concrete beams and slabs.

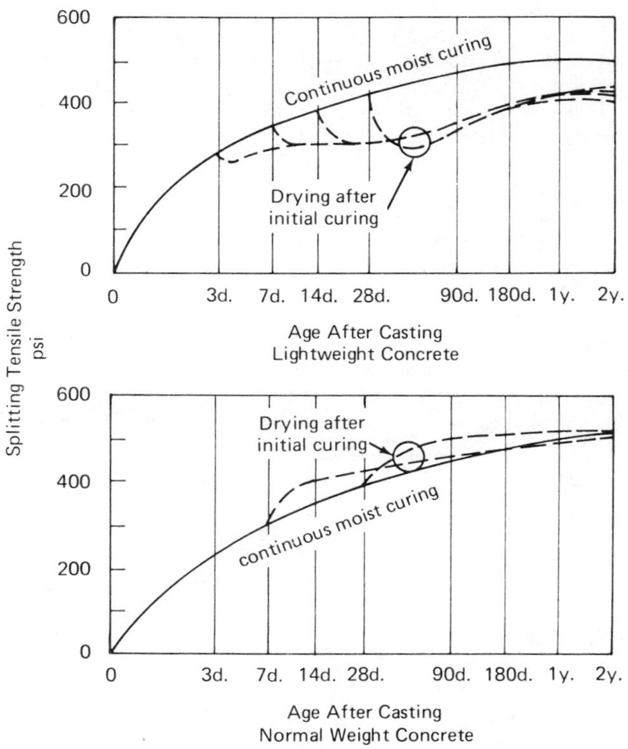

Fig. 6-12 Effect of moist curing and drying time on splitting tensile strength (Series I). (From: Hanson, J. A., "Effect of Curing and Drying Enivronments on Splitting Tensile Strength of Concrete," *ACI Journal Proceedings*, 65(7), July 1968.)

Tensile splitting tests are not required in ACI 318 if shear and torsion stresses, cracking moment, modulus of rupture, and bar development lengths are based on the reasonable assumption that, for a given compressive strength, the tensile strength of lightweight aggregate concrete (with or without sand replacement) is a fixed proportion of that for normal-weight concrete. The percentage of normal-weight concrete shear stress permitted is 75 if all-lightweight aggregate is used, or 85 if natural sand is combined with lightweight coarse aggregate to produce sand-lightweight concrete. Linear interpolation is used for partial sand replacement of fine aggregate. Alternatively these values may be upgraded if splitting tensile tests demonstrate that the tensile strength is higher than the assumed conservative percentages. In any case, the test for splitting tensile strength is used only for laboratory determination of its relationship to the compressive strength. It is not intended for control of, or acceptance of, strength properties of the concrete in the field. If use of the splitting tensile strength of lightweight aggregate concrete yields calculated permissible shear values greater, or bar development lengths less, than allowed for normal-weight concrete, the values for normal-weight concrete should be used.

Most structural lightweight aggregate producers have sufficient test data available to recommend splitting tensile strength value to the engineer for concrete in which the aggregate consists entirely of lightweight aggregate and for concrete containing lightweight aggregate and natural sand.

There is a general indication that the splitting test of a drilled core provides a better measure of the flexural strength of thick concrete pavements than can be obtained by testing beams sawed from the pavements. Since the cost of procuring and testing sawed beams as compared to drilled cores is in the order of 50 to 1, a much large number of cores can be tested to increase the reliability of the splitting strength.

Figure 6-13 shows the effect of elevated temperature on split-cylinder tensile strength of a siliceous aggregate concrete.

The results of splitting tension determinations at low temperatures are shown in Fig. 6-14(a) for various concrete mixtures in a moist condition. Although there is some scatter of points, the trends are similar for the different mixes. Tests on the 5.5-bag mix indicate very little difference in strengths between 75°F and 35°F, and it is felt that the same would be true for the other concretes. In general, strengths increase as the temperature is lowered from 75°F, reach maximum values in the range from -75°F to -150°F, then decrease as the temperature approaches -250°F. A rather abrupt increase in the splitting strengths is observed between 75°F and 0°F.

The increases in splitting strengths are related to the evaporable water contents, and Fig. 6-14(b) shows that the splitting strength of oven-dry concrete increases only slightly as the temperature is lowered from 75°F to -150°F, whereas

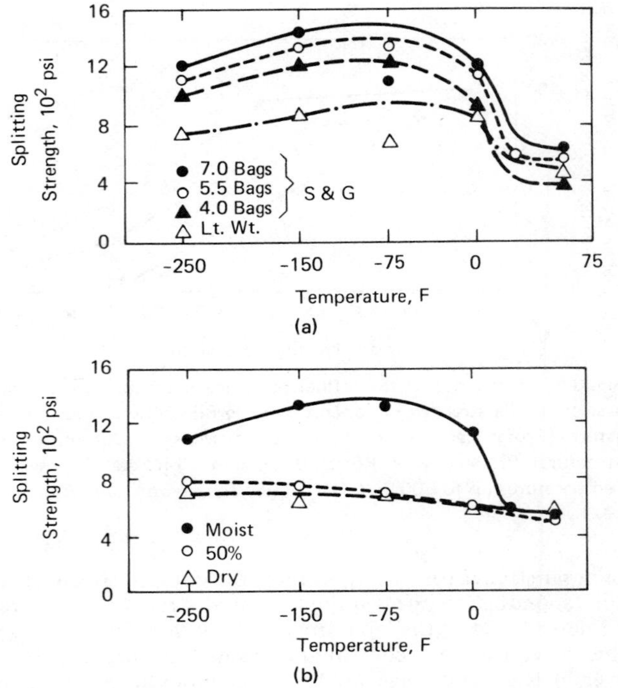

Fig. 6-14 (a) Effect of low temperatures on strength of moist concrete; and (b) effect of moisture condition on strengths at low temperatures. (From: Monfore and Lentz, "Physical Properties of Concrete at Very Low Temperatures," *PCA Research Bulletin 145*, 1962.)

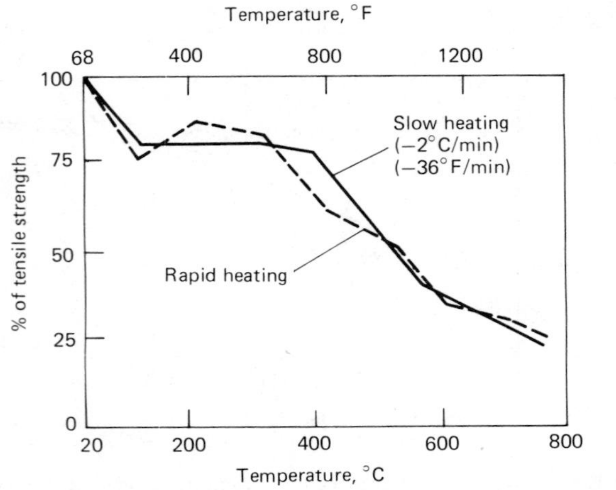

Fig. 6-13 Effect of temperature on split-cylinder tensile strength of a siliceous aggregate concrete. (From: "FIP/CEB Report on Methods of Assessment of Fire Resistance of Concrete Structural Members," Great Britain, 1978.)

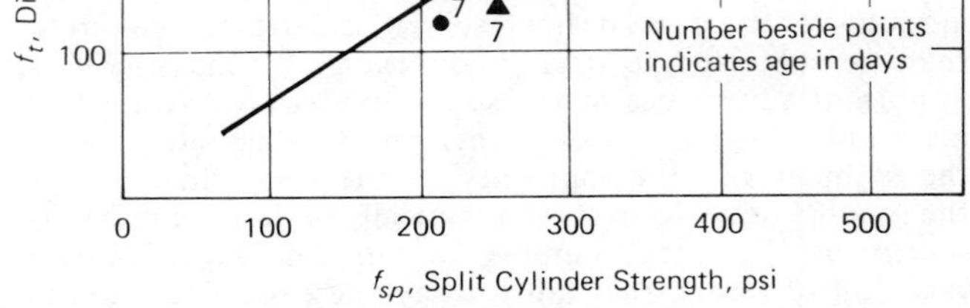

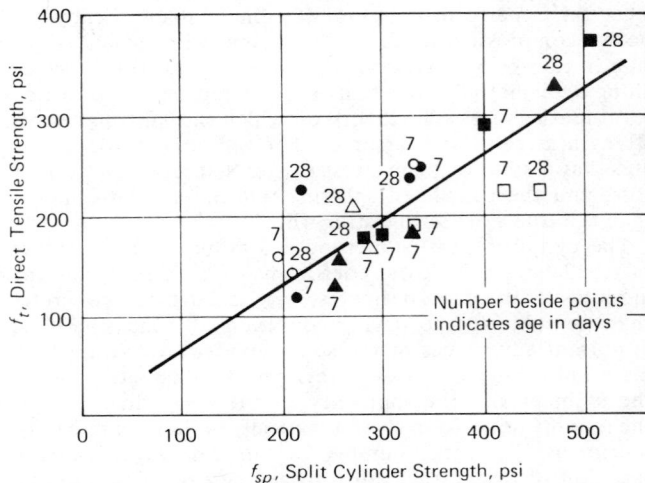

Fig. 6-15 Relationship between direct tensile and splitting strengths of lightweight concrete. (From: Ledbetter, W. B., and Thompson, J. N., "A Technique for Evaluation of Tensile and Volume Change Characteristics of Structural Lightweight Concrete," *ASTM Proceedings*, V. 65, 1965, pp. 712–726.)

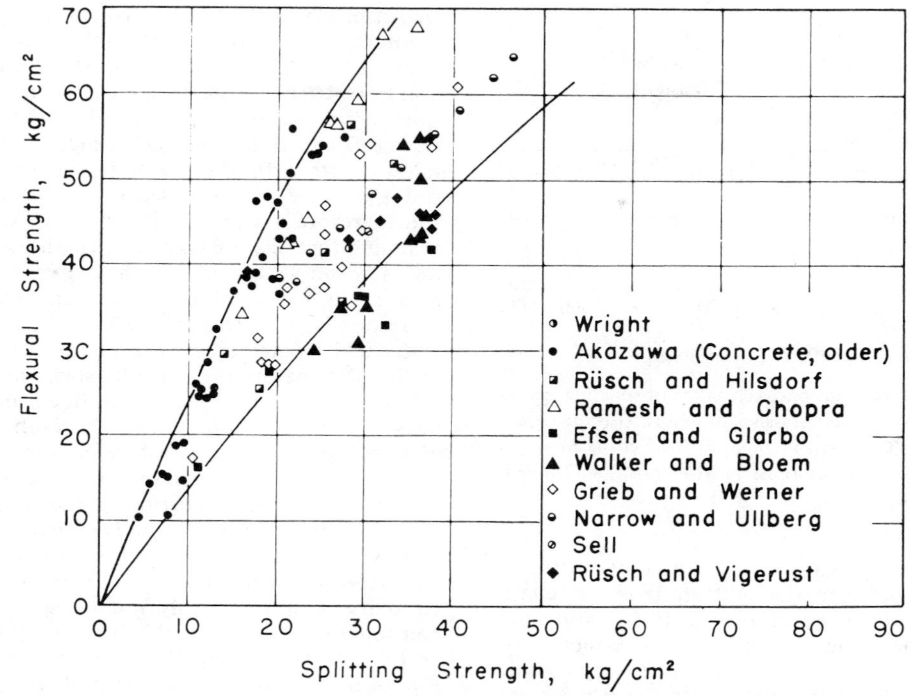

Fig. 6-16 Experimental results obtained by various investigators for the relationship between splitting strength and flexural strength of concrete (after Bonzel). Note: to convert to psi, multiply by 14.22 (From: Popovics, S., "Relations between Various Strengths of Concrete," *Highway Research Record* 210, 1967.)

the increase in comparable moist concrete is considerably greater. The increase for concrete equilibrated in an atmosphere of 50% relative humidity is only slightly greater than that for oven-dry concrete.

6.3.4 Relationships Between Tests

The ratio between the direct tensile strength and the splitting strength changes with variations in the composition of the concrete and depends to a considerable degree on the type and properties of the aggregates. The relationship for lightweight concrete is shown in Fig. 6-15.

The ratio of direct tensile strength to modulus of rupture increases with increasing specimen size and compressive strength. When the size of the modulus of rupture beam becomes very large, the strain gradient diminishes, and the modulus of rupture approaches the tensile strength.

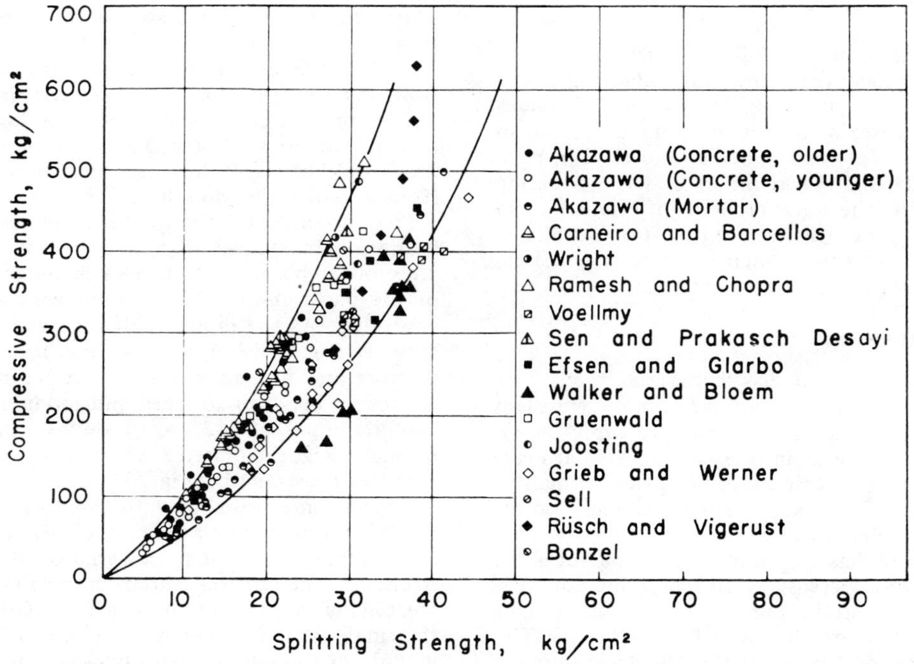

Fig. 6-17 Experimental results obtained by various investigators for the relationship between splitting strength and compressive strength of concrete (after Bonzel). Note: to convert to psi, multiply by 14.22. (From: Popovics, S., "Relations between Various Strengths of Concrete," *Highway Research Record* 210, 1967.)

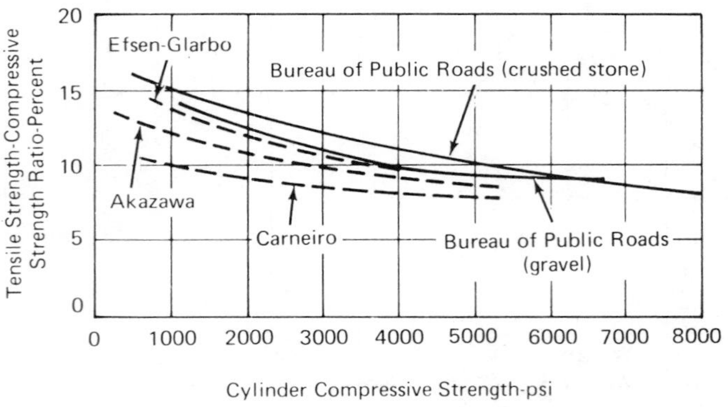

Fig. 6-18 Comparison of relation between ratios of splitting tensile to compressive strengths and compressive strengths. (From: Grieb, W. E., and Werner, G., "Comparison of Splitting Tensile Strength of Concrete with Flexural and Compressive Strengths," *ASTM Proceedings*, 62, pp. 972–990, 1962.)

For a given coarse aggregate and method of curing, a linear relation exists between the splitting tensile strength and the flexural strength of concrete, and this relationship is not affected by the cement content of the concrete or the age at test, Fig. 6-16. However, there is an appreciable reduction in the flexural–splitting strength ratio with increase in tensile splitting strength of the concrete. For a given tensile splitting strength, the flexural strength of concrete increases as the maximum size of aggregate decreases. Also for a given tensile splitting strength, the corresponding flexural strengths are higher for rounded aggregates than for angular aggregates, which in turn are higher than lightweight aggregate concrete.

The ratio of the splitting tensile strength to the compression strength is curvilinear, Fig. 6-17. The ratio decreases as the compressive strength increases (Fig. 6-18); the ratio is affected by both the cement content, and the age at test.

6.4 MODULUS OF ELASTICITY

The stress–strain relation and modulus of elasticity of concrete are important design properties. Static determinations of the modulus of elasticity provide one of the values needed for design purposes such as determining deformation and stress distribution between concrete and steel in reinforced or prestressed concrete members and the buckling effect in long columns. The static modulus of elasticity also is useful for calculating the stresses resulting from shrinkage, settlement, or other distortions such as floor slab deflections, and it affects camber control or prestress loss.

6.4.1 Stress-Strain Relation

Although concrete is not a truly elastic material, the theory of elasticity can be applied to it within limits of stress (up to 35 to 50% of ultimate stress) and time (creep effects). However, marked departure from linearity is noted at near-ultimate stresses. Concrete has no range of proportionality, no pronounced elastic limit, and no marked yield point, though it does creep under constant stress.

Unreinforced concrete has a certain amount of ductility. This ductility, however, decreases with increasing concrete strength. Typical stress–strain curves for normal weight and lightweight concrete are shown in Fig. 6-19(a–c). The shape of the ascending portion of the stress–strain curve is more linear and steep as the concrete strength increases, and the strain at the maximum stress is slightly higher for high-strength concrete. The slope of the descending branch becomes steeper for high-strength concrete. To obtain the descending branch of the stress–strain curve, it is generally necessary to avoid the specimen–testing system interaction (special precautions must be taken during testing); this is more difficult to do for high strength. It is not clear at what value of strength the material becomes classically brittle and the slope of the descending branch becomes vertical and therefore not practically usable in design. In design it is desirable that the concrete not crush or substantially enter the descending branch of the curve before the steel has reached its yield point. In steel, an elastic strain of 0.001 corresponds to a stress of 30 ksi. Therefore, reinforcement with yield strengths between 60 and 75 ksi will reach these yield points before the concrete starts weakening.

Stress–strain curves for unconfined and triaxial strength tests are shown in Fig. 6-20. The results indicate the effect of pressure at various moisture contents on the relation of stress to strain in concrete. Considering each mixture individually, the initial portion of the stress–strain curves for each confining pressure and moisture content approximately coincide. Since the modulus is usually computed from the initial portion of the curve, Young's modulus of elasticity would probably not be appreciably affected by confining pressure up to 1500 psi or by moisture condition at time of test if pore pressure effects were not present. Figure 6-21 shows that considerably higher strains can be attained as the lateral stress increases.

Figure 6-22 shows that with increasing temperature ultimate stress decreases and ultimate strain increases. Tests were conducted on specimens stored at 65% relative humidity and 20°C that were heated without any load.

6.4.2 Modulus of Elasticity of Normal-Weight Concrete

The modulus of elasticity of concrete, E_c, is defined as the ratio of normal stress to corresponding strain for tensile or compressive stresses. It is also known as elastic modulus, Young's modulus, or Young's modulus of elasticity. It depends on the modulus of the cement paste, the modulus of the aggregate, and the relative amounts of paste and aggregate. The modulus also varies according to its definition, whether initial tangent, tangent, chord, or secant modulus; rate of loading; number of cycles of load application; the time the load is sustained; shrinkage and creep; and to some extent all the other possible variable parameters of the concrete specimen. The secant modulus is commonly used, and because it decreases with an increase in stress, the stress at which the modulus is determined should be stated.

The modulus of the paste increases as the degree of hydration increases. Changes in modulus for a given concrete occur because of changes in the modulus of elasticity of the paste and increased bond with aggregate as curing continues. As the modulus of the paste increases, the concrete strength also increases, and for any given concrete mixture and curing condition there is a general empirical relationship between strength and modulus of elasticity. The formula, $E_c = w^{1.5}\ 33\ \sqrt{f'_c}$ defines this relationship for normal weight concrete, where w = unit weight and f'_c = compressive strength, (Fig. 6-23).

Concrete may comply with this formula only within ±15 to 20%. Modulus of elasticity for a given concrete exhibits a much higher coefficient of variation than the compressive strength. The greater variation results in part from the greater inaccuracies of the test procedures used to measure the small strains. Depending on how critically values for E_c will affect the design (as in buckling of long columns), the engineer should decide if the values determined by formula are sufficiently accurate, or if he should determine secant

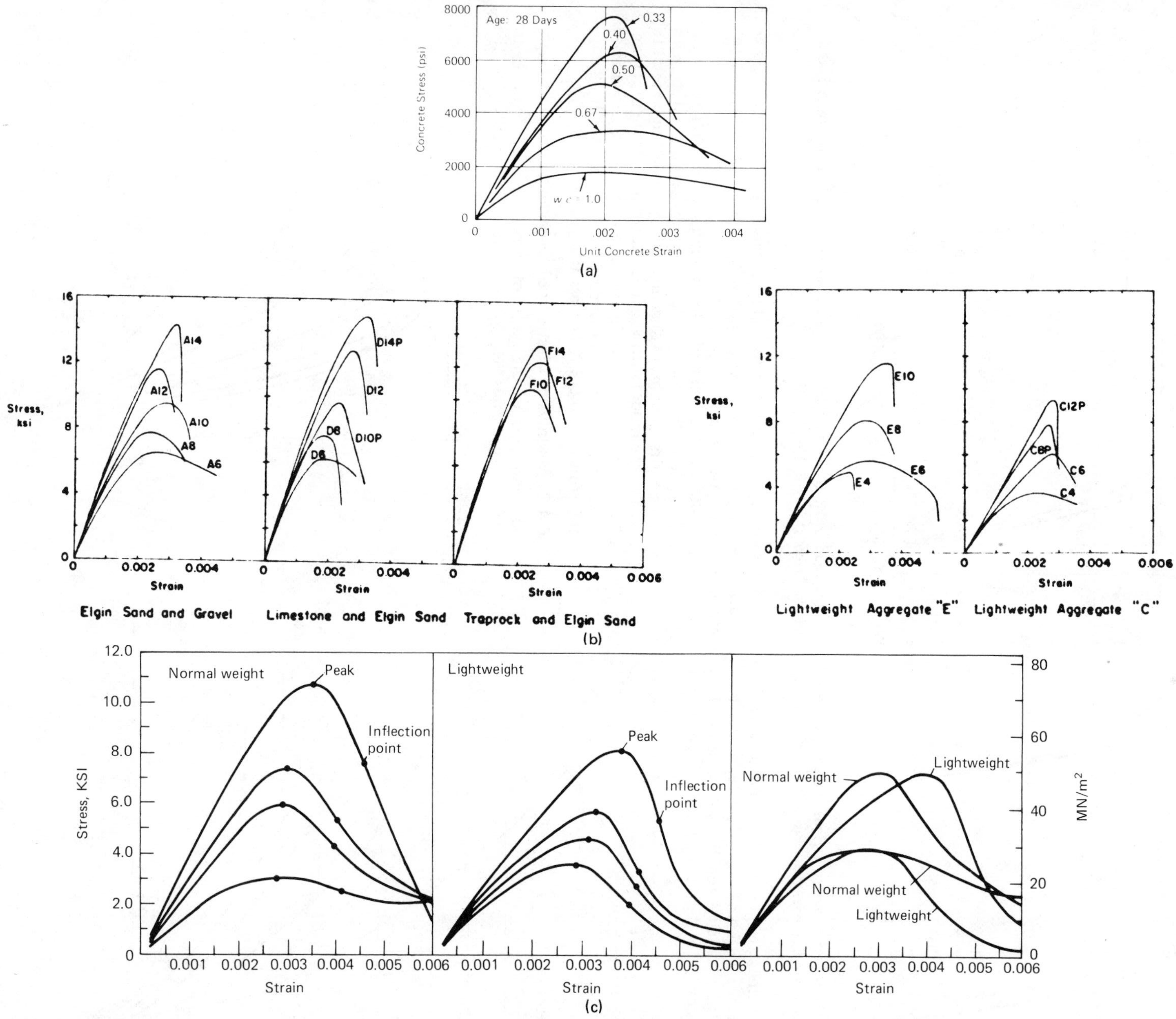

Fig. 6-19 Comparison of stress–strain curves for unconfined concrete of different strengths and aggregates. (a) Stress–strain curves for concrete from Hognestad, E., Hanson, N. W., and McHenry, D., "Concrete Stress Distribution in Ultimate Strength Design," *Journal of ACI 27 (4)*, 445–479, Dec. 1955; Proc. 52; PCA Research and Development Laboratory, Development Bulletin D6. (b) From Kaar, P. H., Hanson, N. W., and Capell, H. T., "Stress-Strain Characteristics of High-Strength Concrete," Research and Development Bulletin RD 051.01D, Portland Cement Association, 1977, 11 pp. (c) From Wang, P. T., Shah, S. P., and Naaman, A. E., "Stress-Strain Curves of Normal and Lightweight Concrete in Compression," *Journal of the American Concrete Institute*, Proc. V. 75, No. 10, Nov. 1978, pp. 603–611.

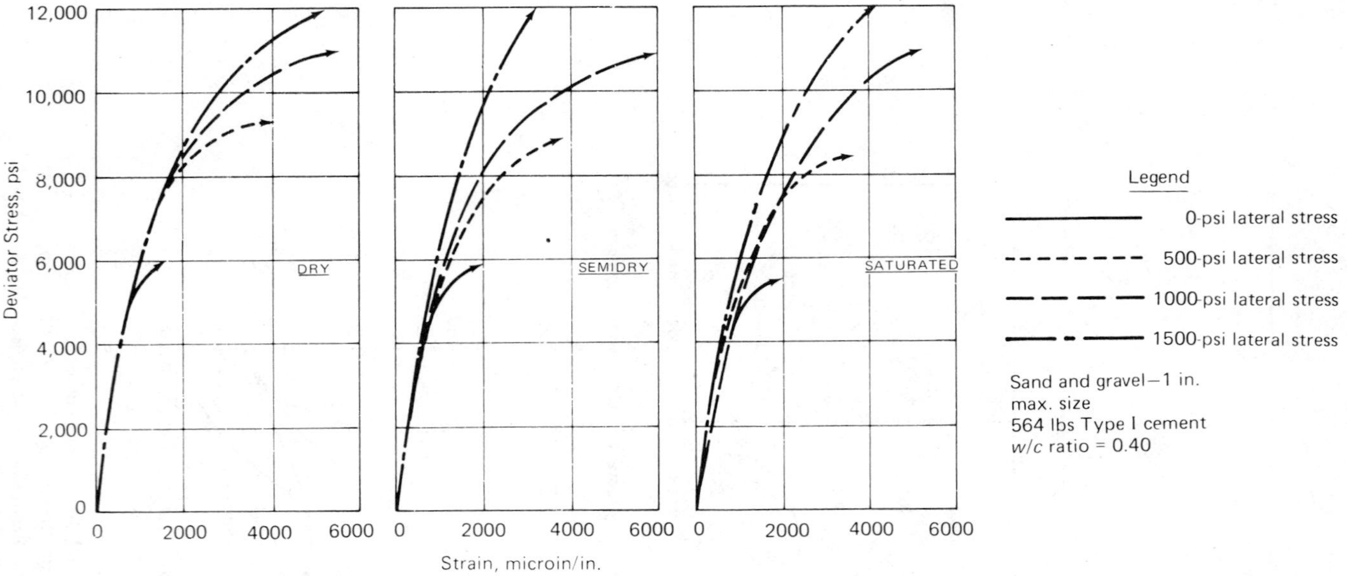

Fig. 6-20 Stress vs. strain at various moisture conditions. (From: Saucier, K. L., "Correlation of Hardened Concrete Test Methods and Results," U.S. Army Engineer Waterways Experiment Station Technical Report C69-2, March 1969.)

modulus values from tests made on the specified concrete in accordance with ASTM C469, "Method of Test for Static Modulus of Elasticity and Poisson's Ratio of Concrete in Compression." In general, normal weight concrete mixes would be expected to yield similar tensile and compressive secant moduli of elasticity values up to 40 to 50% of the ultimate strength.

The limited data on the modulus in the high-strength region have the same scatter shown by the normal strength test results. It appears that the same formula can be used to predict the modulus of elasticity in the high-strength region. The modulus of elasticity of mass concrete representative of various dams ranges from 3.5 to 5.5 $\times 10^6$ psi at 28 days and from 4.3 to 6.8 $\times 10^6$ psi at 1 year.

6.4.3 Modulus of Elasticity of Lightweight Concrete

The modulus of elasticity of concretes made with lightweight aggregates is not influenced materially by the

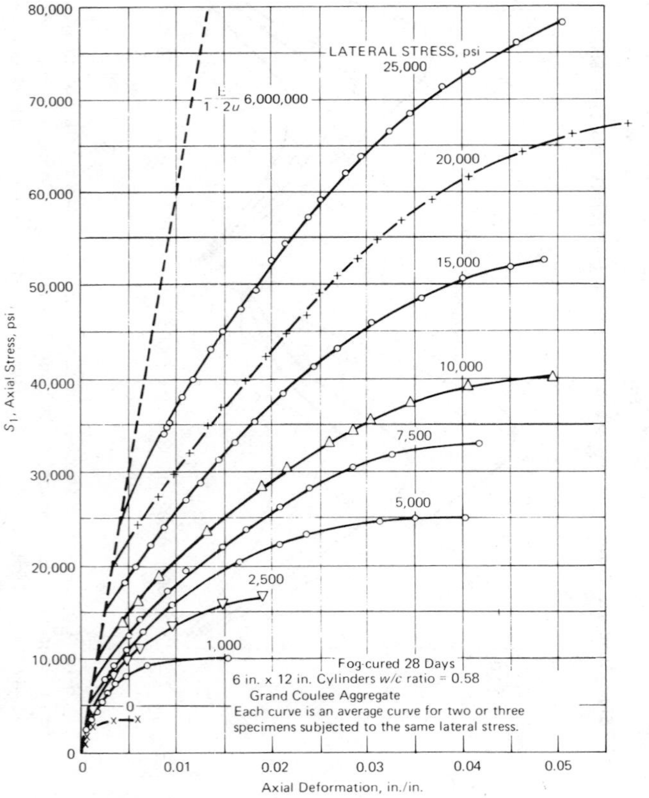

Fig. 6-21 Stress-strain curves for concrete under high triaxial stresses illustrating large strains. (From: Balmer, "Shearing Strength of Concrete under High Triaxial Stress—Computation of Mohr's Envelope as a Curve," *Structural Research Laboratory Report No. SP-23,* U. S. Department of the Interior, Bureau of Reclamation, Oct. 28, 1949.)

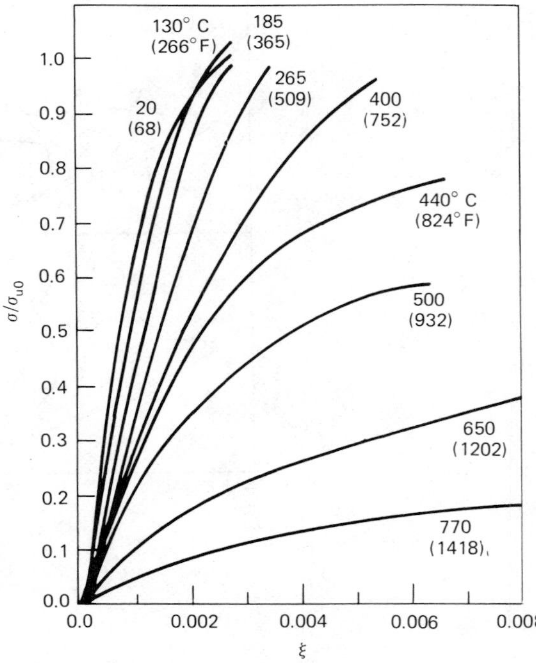

Fig. 6-22 Stress–strain relations for different temperatures.

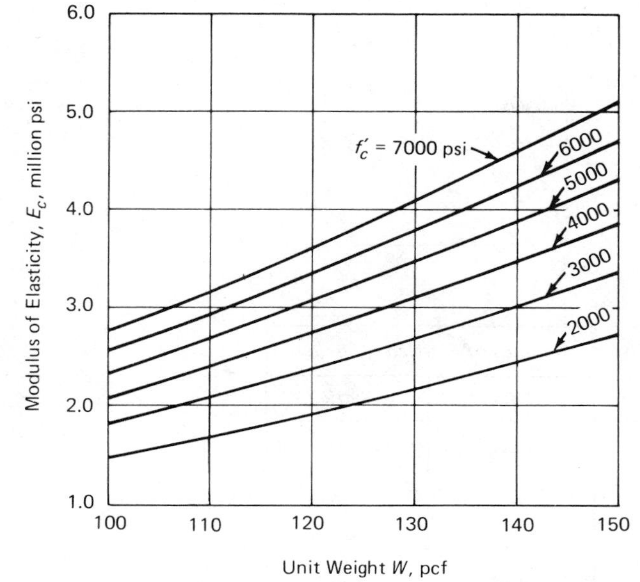

Fig. 6-23 Modulus of elasticity as a function of strength and air-dry unit weight of normal weight concrete.

volumetric concentration of aggregate. This is so because the modulus of the aggregate is generally about the same as that of the paste. Normal-weight sand often is used with lightweight aggregates to increase the modulus of elasticity. Depending on the type of lightweight aggregate and sand content, the modulus of elasticity of lightweight concrete is generally 20 to 50% lower than that for normal-weight concrete of equal strength, with the greater difference occurring in the low-strength range. A modulus of 3.5×10^6 psi has been obtained for sand-lightweight concrete of 7000 psi, 28-day strength, and 118 pcf wet concrete weight.

An approximate relationship that can be used to estimate the modulus of elasticity of structural lightweight concrete, E_c, in pounds per square inch, is:

$$E_c = Cw^{1.5}\sqrt{f_c'} \text{ psi}$$

in which w is the air-dry unit weight of the concrete in pounds per cubic foot, f_c' is the compressive strength in pounds per square inch as determined from 6×12 in. cylinders, and C is a factor dependent upon the value of f_c'.

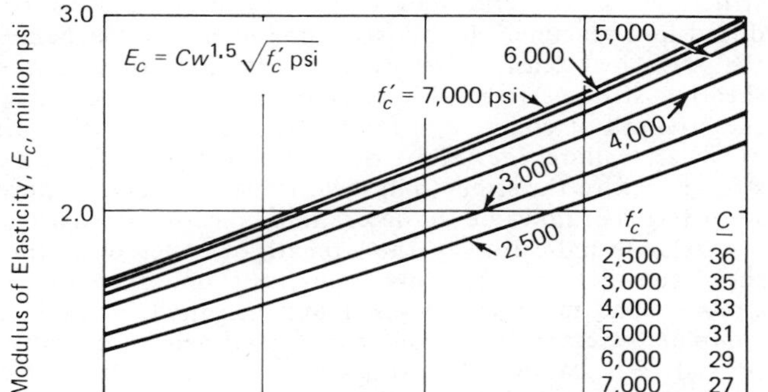

Fig. 6-24 Modulus of elasticity as a function of strength and air-dry unit weight of structural lightweight concrete. (From: Pfeifer, D. W., and Meinheit, D. F., "Structural Lightweight Aggregate Concrete–Modulus of Elasticity," Portland Cement Association, unpublished.)

Values of C corresponding to different values of f_c' are given in Fig. 6-24. This empirical formula is reasonably reliable for structural lightweight concretes with compressive strengths of 2500 to 7000 psi.

Static modulus and Poisson's ratio are of greater significance to the design engineer than the corresponding dynamic values (ASTM C215) because these elastic constants are stress-dependent. Static tests involve the application of stresses of the same order as those in practice, whereas the stresses induced in the dynamic tests are very small. The sustained modulus is approximately one-half that of the instantaneous modulus when load is applied at early ages and is a slightly higher percentage of the instantaneous modulus when the loading age is 90 days or greater. Structural lightweight concrete has a smaller difference than normal weight. Saturated concrete has a higher static modulus than dry or moist concrete, since pore water introduces a triaxial state of stress while the reverse is true for the dynamic modulus, so that as concrete dries, the difference between methods is reduced. Also the higher the compressive strength or modulus (around 6×10^6 psi), the closer the results.

For both normal weight and structural lightweight concretes, the modulus of elasticity increases faster than compressive strength at early ages, as shown by the fact that all curves are located above the line of equality in Fig. 6-25. The modulus of steam-cured concrete may vary from that of concrete moist-cured at normal temperatures.

6.4.4 Effect of High Temperature Exposure

Exposure to elevated temperatures causes a reduction in the elastic modulus of concrete, Fig. 6-26. At temperatures up to about 600°F the elastic modulus appears to be the structural property of concrete most sensitive to the effects of heat exposure. For a given concrete, after dehydration, the modulus tends to be independent of age or curing conditions. However, the lower the water–cement ratio, the higher the modulus of elasticity after dehydration. The effects of aggregate type are not clearly defined but appear to be small, provided the aggregate does not melt or decompose. If the concrete is subjected to cyclic temperatures, the decrease in modulus is even more severe, probably as a result of thermal stresses.

6.4.5 Effect of Low Temperature Exposure

Data on modulus of elasticity by the resonance frequency method (ASTM C215) on concretes cooled from room temperature to −250°F are listed in Table 6-2. The modulus of sand and gravel concretes in a moist condition increases about 50% as the temperature is decreased to −250°F. The 5.5-bag sand and gravel mix, in equilibrium with an atmosphere of 50% relative humidity, increases only 8% for this same temperature range; and an oven-dried specimen exhibits a constant modulus as the temperature decreases.

6.4.6 Repetitive Loading

The stress-strain relationship for concrete subjected to repeated compressive loads is cycle-dependent while the relationship for reinforcing steel is cycle-independent. During the first cycle the stress–strain curve for concrete is the same as that obtained in static tests, but after a few cycles the curve becomes almost linear. As the number of repetitions of load increases, the lines representing the stress-strain curve become concave upward. There also is a general trend toward a decrease in secant modulus (Fig. 6-27) and an increase in nonelastic strain with an increase in the number of loading cycles, presumably due to the development of microcracks in the concrete. Changes in modulus are also accompanied by a reduction in natural frequency.

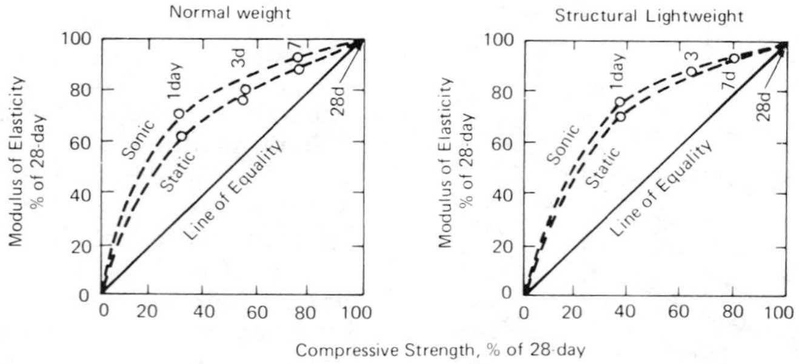

Concrete	Compressive Strength, 28 days	Modulus of Elasticity, 28 days	
		Sonic	*Static*
Normal Weight (145 pcf) Type I cement–517 lbs/c.y. sand and gravel–1 in. max. size w/c ratio–0.45 Slump–2 to 3 in.	6610 psi	6.09×10^6 psi	4.52×10^6 psi
Lightweight (99 pcf) Type I cement–611 lb/c.y. Expanded shale fine and coarse Air content–5.8% Slump–3 to 4 in.	5020	2.57	2.36

Fig. 6-25 Modulus of elasticity at early ages (From: PCA, unpublished data.)

6.5 POISSON'S RATIO

Poisson's ratio is the ratio of transverse (lateral) strain to the corresponding axial (longitudinal) strain resulting from uniformly distributed axial stress. Values of Poisson's ratio, μ, like modulus of elasticity, vary with the Poisson's ratio of the aggregate, the cement paste, and the relative proportions of the two. It also varies with moisture condition and age of the concrete. Values generally range between 0.11 and 0.28. For elastic strains under normal working stresses, Poisson's ratio is taken as 0.20. It is approximately the same for structural lightweight and normal weight concretes. With an increase in strength, age, or aggregate content, Poisson's ratio tends to decrease.

At room temperature, lower values of Poisson's ratio are obtained for higher-strength concrete. Results at higher temperatures are erratic, and no general trend for the effect of temperature on Poisson's ratio is indicated. Poisson's ratios, calculated from longitudinal and torsional resonant frequency tests, remain essentially constant as the temperature of the concrete is lowered from 75°F to –250°F.

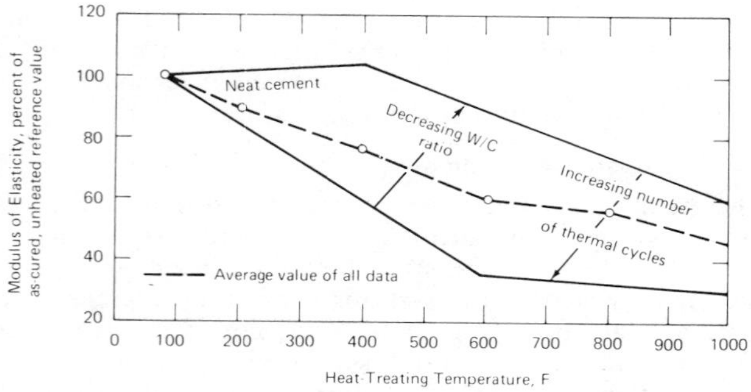

Fig. 6-26 Extremes of the influence of heat exposure on the modulus of elasticity of concrete as determined by various investigators. (From: Lankard et al., "Effects of Moisture Content on the Structural Properties of Portland Cement Concrete Exposed to Temperatures up to 500°F," ACI SP-25, *Temperature and Concrete*, 1971.)

Dynamic determinations of Poisson's ratio are consistently higher than the corresponding static values, especially at early ages. For saturated concretes, the dynamic values are 0.02 higher for lightweight and 0.03 higher for normal-weight concretes, with the difference becoming negligible for dry concretes.

6.6 SHEAR AND TORSION

The resistance of concrete to pure shearing stress has never been directly determined. In addition, the case of pure shear acting on a plane is seldom, if ever, encountered in actual structures. Whenever a state of pure shearing stress is produced in a specimen, it follows from the laws of mechanics that principal tensile stresses, equal in magnitude to the shearing stresses, must also exist on another plane. Since the strength of concrete in tension is less than its strength in shear, failure inevitably occurs as a result of tensile stresses before the strength in shear is reached. Thus, a pure shear test is of no value for determining shearing strength.

A reliable indication of the strength of concrete in pure shear can be obtained only from tests under combined stresses. The most widely used and accepted method of combined-stress testing consists of confined compressive triaxial tests. Mohr's theory has been successfully applied for analysis of triaxial results; however, some modification is required to explain the curvature of the failure envelope. The shearing strength determination obtained from the Mohr's concept appears valid, and, therefore, makes for a convenient and encompassing method of examining all strength parameters. It also more closely approximates the true shear strength of concrete than does that determined from direct tests.

In Mohr's theory, a combination of principal stresses at

TABLE 6-2 Young's Modulus of Elasticity at Various Temperatures*

Series No.	Aggregate	Cement Content, bags per cu yd	Resonance Modulus, 10^4 psi					
			75 F	35 F	0 F	-75 F	-150 F	-250 F
2	Sand and Gravel	7.0	6.6	6.6	7.0	8.0	9.0	9.0
5 (Moist)	*	5.5	5.5	5.8	6.6	7.6	8.4	8.5
8 (50%)	*	5.5	5.0	5.1	5.1	5.1	5.3	5.4
10 (Dry)	*	5.5	4.3	4.3	4.3	4.3	4.3	4.3
12	*	4.0	5.2	5.2	6.2	6.7	8.2	8.3
15	Expanded Shale	6.0	3.0	3.0	3.3	3.8	4.2	4.3

From: Monfore, G. E. and Lentz, A. E., "Physical Properties of Concrete at Very Low Temperatures," Portland Cement Association Research Bulletin *145, Portland Cement Association, Skokie, 1962.*

failure is transformed into a system of concurrent normal and shear stresses that define an envelope of failure conditions. A graphical solution for these stress combinations at failure may be obtained by plotting Mohr's circle for each of the test results on a common axis and then estimating the location of the common tangent of these circles.

The strength of concrete in pure shear—that is, when no normal stresses are present on the plane of failure—is the stress measured to the intersection of the rupture line or limiting curve with the vertical axis, Fig. 6-28. On the basis of the available data, the value of the strength of concrete in pure shear is approximately 20% of the compressive strength.

Shear failures can be abrupt in nature with little or no advance warning of their occurrence. Failure may occur with the formation of a critical diagonal crack, or if redistribution of internal forces is accomplished, failure may occur by shear-compression destruction of the compression zone at a higher load. Shear is resisted by the uncracked concrete or by concrete located above the inclined cracks, by aggregate interlock along the inclined crack, by dowel action of the reinforcement crossing the inclined cracks, and by contribution of any shear reinforcement. Uncracked concrete is very strong in direct shear; however, there is always the possibility that a crack will form in an unwanted location. The approach is to assume that a crack will form in an unfavorable location, and then to provide reinforcement that will prevent this crack from causing undesirable consequences. This discussion only considers the contribution of the concrete and not that of the reinforcement.

Shear stresses along a crack may be resisted by friction. The shear transfer strength is a function of the roughness of the fracture surface and is independent of the concrete strength. However, in some very high-strength concretes, fractures cross the aggregates leaving a relatively smooth fracture surface, and consequently a relatively low shear transfer at failure. When the crack is rough and irregular, the apparent coefficient of friction, μ, is 1.4 for concrete cast monolithically, 1.0 for concrete placed against hardened concrete, and 0.7 for concrete placed against as-rolled structural steel. To develop friction, however, a normal force must be present. This normal force may be obtained by placing reinforcing steel perpendicular to the assumed crack. As shear slip occurs along the crack, the irregularities of the crack will cause the opposing faces to separate, stressing the reinforcing steel in tension. A balancing compressive stress will then exist in the concrete, and friction will be developed along the confined crack.

The magnitude of shear at initial diagonal tension cracking is a function mainly of concrete strength, percentage of tension reinforcement, beam dimensions, and ratio of moment to shear. Diagonal tension is a combined stress problem in which horizontal tensile stresses due to bending as well as shearing stresses must be considered.

The resistance of concrete to diagonal tension stress is related to the tensile strength of concrete, which in turn may be approximated as a function of the square root of the compressive strength. This latter function is used for

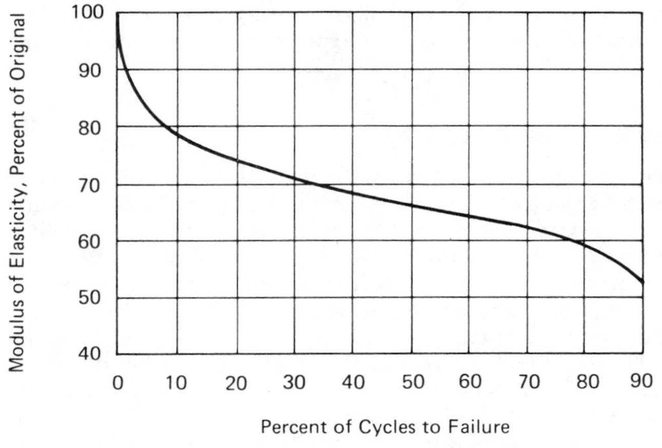

Fig. 6-27 Reduction in secant modulus of elasticity as the result of repeated loads. (From: Linger, D. A., and Gillespie, H. A., "A Study of the Mechanism of Concrete Fatigue and Fracture," *Highway Research News, No. 22,* 40–51, Highway Research Board, Feb. 1966.)

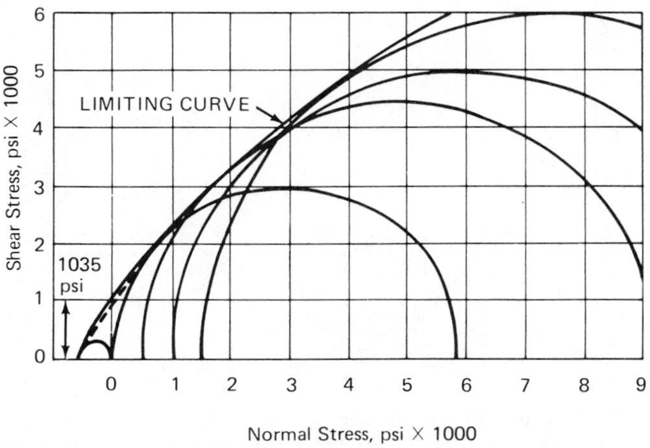

Fig. 6-28 Typical Mohr rupture diagram for normal weight concrete.

design, since only compressive strength is usually specified and controlled in construction of concrete structures.

The provision of ACI 318 for nominal shear stress and nominal torsion stress carried by the concrete applies to normal weight concrete. When structural lightweight aggregate concretes are used, provisions for shear and torsion must be modified.

For normal-weight concrete, the splitting tensile strength, f_{ct}, is approximately equal to $6.7\sqrt{f_c'}$. Therefore, when f_{ct} is specified and determined (ASTM C496) for a particular lightweight aggregate concrete, the value of $f_{ct}/6.7$ may be substituted for all values of $\sqrt{f_c'}$ affecting shear and torsion. If use of the splitting tensile strength of lightweight aggregate concrete yields calculated permissible shear values greater than allowed for normal-weight concrete, the values for normal-weight concrete must be used. Once this value is determined for a particular lightweight aggregate, it is representative of structural concrete made with that aggregate.

The modulus of elasticity in shear, G, which is the ratio of unit shearing stress to the corresponding unit shearing strain, may be calculated when the modulus of elasticity and Poisson's ratio are known: $G = E/2(1 - \nu)$. Typical values at ordinary temperatures, 75°F, (G_0) are shown in Fig. 6-29 for various aggregate types. The modulus at elevated temperatures, G_T, decreases markedly as temperature increases from 75°F to 1200°F. In the case of concrete members, G is difficult to measure, and it is not a constant because the stress–strain curve of concrete is not linear.

Behavior of concrete in pure torsion is analogous to behavior in pure shear. Members subjected to pure torsional loads develop diagonal tensile stresses due to the torsional shear stresses. The member fails by bending about an axis parallel to the wider face of the cross section and inclined at 45° to the longitudinal axis of the member. Failure is reached when the tensile stress induced by a 45° bending component of torque on the wider face reaches a reduced modulus of rupture. This tensile strength is related to some function of $\sqrt{f_c'}$.

The relationship between torsional strength and tensile strength of concrete is dependent on the size and shape of the torsion specimen and on the method used to calculate torsion strength. The torsion strength determined on specimens of circular section and calculated on the basis of rigid-plastic stress distribution (plastic theory) is 75% of the torsion strength calculated from a linear stress distribution (elastic theory). Therefore, the magnitude of the pure tensile strength of a concrete is expected to be less than its comparable torsion strength which was calculated from a linear stress distribution.

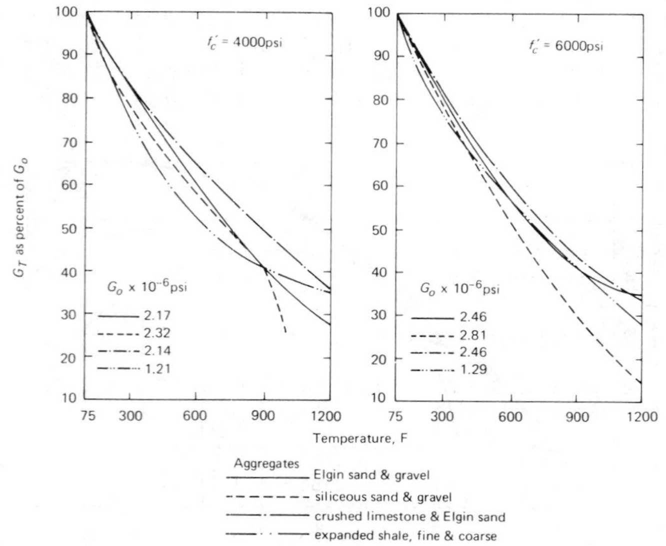

Fig. 6-29 Effect of temperature on the modulus of shear (G) of concrete. (From: Cruz, C. R., "Elastic Properties of Concrete at High Temperatures," *PCA Research Bulletin 191*, 1966.)

6.7 VOLUME CHANGES

Concrete changes in volume slightly during and after the hardening period. Understanding the nature of these changes is necessary for the designer. Frequently, stresses causing cracking can be prevented or minimized by controlling the variables that affect volume changes.

6.7.1 Factors Affecting Volume Changes

Normal volume changes of concrete are caused by variations in temperature and moisture (drying shrinkage), and by sustained stress (creep). Reliable information is available on the most important factors that affect the magnitude of volume changes. When several unfavorable factors act at the same time, the net result is the product, rather than the sum, of the individual effects.

The magnitude of volume changes in concrete is directly related to the properties of the constituent materials. By careful selection of materials and the proportions in which they are used, volume changes can be kept to a minimum. In areas where properties of economically available materials are such that concrete has inherently high volume change, the effects of additional adverse factors can be very critical.

If concrete were free to deform, uniform volume changes would be of little consequence, but since concrete is usually restrained by foundations, subgrades, steel reinforcement, or connecting members, significant stresses may develop. This is particularly true when tension is developed; thus restrained contractions causing tensile stresses in concrete are usually more important than restrained expansions which cause compressive stresses.

For convenience, the magnitude of volume changes is generally stated in linear rather than volumetric units. The volume changes that ordinarily occur are small, ranging in terms of change in length from a few up to about 1000 millionths. Changes in length are often expressed in "millionths" (1 millionth is simply 0.000001). For example, a change in length of 600 millionths may also be expressed as 0.000600. This can also be expressed as 0.06% or 0.72 in. per 100 ft.

6.7.2 Thermal Properties

Thermal properties of concrete are significant in connection with keeping differential volume or length changes at a minimum, extracting excess heat from the concrete, and dealing with similar operations involving heat transfer. The basic properties involved are coefficient of thermal expansion, or contraction, conductivity, diffusivity, and specific heat. The last three are largely interrelated.

The main factor affecting the thermal properties of a concrete is the mineralogic composition of the aggregate. Since the selection of the aggregate to be used is based on other considerations, little or no control can be exercised over the thermal properties of the concrete, and tests for thermal properties are conducted only for providing constants to be used in behavior studies. Specification requirements for cement, pozzolan, percent sand, and water content are modifying factors but with negligible effect. Entrained air is an insulator and reduces thermal conductivity, but other considerations that govern the use of the entrained air outweigh the significance of its effect on thermal properties.

1. Coefficient of Expansion or Contraction. The coefficient of thermal expansion or contraction, α, represents the change of concrete volume or, as usually measured on test specimens, the change in length with change in temperature.

An increase in temperature may cause concrete to expand or contract, the latter depending upon a change in moisture content. As with most materials, concrete tends to expand with increasing temperature; however, this normal thermal expansion may be overshadowed by shrinkage due to moisture loss.

Thermal expansion and contraction of concrete vary with factors such as type and amount of aggregate, richness of mix, water–cement ratio, temperature range, concrete age, and relative humidity (degree of saturation of concrete). Of these, aggregate type has the greatest influence.

Values of the coefficient of thermal expansion of cement paste may vary from 5.5 to 12.0×10^{-6} in./in. per °F, depending upon the differences in cement fineness, composition, water–cement ratio, age, and moisture content. The coefficient of the paste is generally higher than the coefficient of the aggregate. Pastes that are oven dry or saturated have similar coefficients of expansion. Pastes with intermediate moisture contents have higher coefficients of expansion with a maximum at relative humidities of 50 to 70% that may be twice as large. Since concrete is only partially paste, these effects, while evident, are moderated, and the variation in the coefficient of expansion of concrete is smaller.

The coefficient of thermal expansion of the most commonly used aggregates varies from about 2.0 to 7.0×10^{-6} in./in. per °F. Aggregates with a high quartz content, such as quartzite and sandstone, have the highest coefficients; aggregates containing little or no quartz, such as limestone, have the lowest coefficients; and aggregates with medium quartz content, such as igneous rocks, (granite, basalt, etc.) have intermediate values (see Table 6-3).

The coefficient of expansion of concrete varies approximately in proportion to the thermal coefficients and quantity of aggregate in the mixture. Since the coefficient of the paste is generally higher than that of the aggregate, the coefficient of the concrete will be somewhat higher than that of the aggregate, Table 6-3. The coefficients of expansion of concrete can be estimated from the coefficients of the paste and aggregates if due consideration is given to their relative solid volumes in the concrete.

Ranges for normal weight concretes are 5 to 7×10^{-6} for those made with siliceous aggregates and 3.5 to 5×10^{-6} for those made with limestone or calcareous aggregate. The values in each case depend on the mineralogy of specific aggregates. Approximate values for structural lightweight concretes are 3.6 to 6×10^{-6}, depending on the aggregate type and amount of natural sand used. When a precise value is not required, a coefficient of 5.5×10^{-6} is frequently used. If greater accuracy is needed, tests should be made on the specific concrete mix.

The thermal coefficient for steel is $(6.1 + 0.002\theta)\ 10^{-6}$ in./in./°F, where θ is the temperature in °F; which is comparable to that for concrete. The thermal coefficient for reinforced concrete can be assumed as six millionths per °F, the average for concrete and steel.

The effect of temperature on the thermal coefficient is difficult to separate from shrinkage effects. Only small changes take place in the coefficient as temperature is increased to 500°F or until the temperature at which nearly all the water is removed from the concrete and carbon dioxide is driven off from aggregates containing calcium carbonate. Above this temperature, the transition to a higher expansion coefficient occurs, Table 6-4. Concrete with lower coefficients of expansion have a great advantage in resisting stresses due to severe temperature changes or cycling over a short period of time. Thermal expansion is strongly reduced as the stress level is increased, as shown in Fig. 6-30.

TABLE 6-3 Average Coefficients of Linear Thermal Expansion of Rocks (Aggregates) and Concrete (Within Normal Temperature Ranges)

Type of Rock (Aggregate)	*Average Coefficient of Thermal Expansion* $\alpha \times 10^{-6}$ in./in per deg. F	
	Aggregate	*Concrete**
Quartzite, Cherts	6.1–7.0	6.6–7.1
Sandstones	5.6–6.7	5.6–6.5
Quartz Sands and Gravels	5.5–7.1	6.0–8.7
Granites and Gneisses	3.2–5.3	3.8–5.3
Syenites, Diorites, Andesite, Gabbros, Diabase, Basalt	3.0–4.5	4.4–5.3
Limestones	2.0–3.6	3.4–5.1
Marbles	2.2–3.9	2.3
Dolomites	3.9–5.5	–
Expanded Shale, Clay and Slate	–	3.6–4.5
Expanded Slag	–	3.9–6.2
Blast-Furnace Slag	–	5.1–5.9
Pumice	–	5.2–6.0
Perlite	–	4.2–6.5
Vermiculite	–	4.6–7.9
Barite	–	10.0
Limonite, Magnetite	–	4.6–6.0
None (Neat Cement)		10.3
Cellular Concrete		5.0–7.0
1:1 (Cement: Sand)**	–	7.5
1:3**		6.2
1:6**		5.6

**Coefficients for concretes made with aggregates from different sources vary from these values, especially those for gravels, granites, and limestones. Fine aggregates generally the same material as coarse aggregate.*
***Tests made on 2-yr old samples.*

Coefficients of thermal contraction for the range of temperature from 75°F to -250°F vary from 3.3×10^{-6} for lightweight aggregate concrete to 4.5×10^{-6} for 7.0-bag sand and gravel concrete, Table 6-5.

2. Conductivity. Conductivity, k, represents the uniform flow of heat through a unit thickness over a unit area of concrete subjected to a unit temperature difference between the two faces, Btu/ft hr sq ft °F.

Designers usually compute the heat transfer coefficient of an assembly on the basis of values for the thermal conductivity of the *dry* component materials. In service, some moisture is usually present. Moisture contents of materials in service are variable and unpredictable. It is difficult to measure heat transfer in moist materials, and the effect on heat transfer of a given amount of moisture is influenced by its distribution within the materials, which in turn is controlled by the properties of the materials, their arrangement in the construction, and variations of temperature imposed upon them by climatic exposure.

Generally, conductivity is essentially a function of the cement paste and aggregate, and, within these narrow limits, is a function of the unit weight of the concrete.

TABLE 6-4 Linear Coefficients of Thermal Expansion

Concrete Containing Elgin Sand and Gravel					
Water-Cement Ratio, by weight	Cement Content, sacks per cu yd	Linear Coefficient of Thermal Expansion at 28 days, per deg F		Linear Coefficient of Thermal Expansion at 90 days, per deg F	
		Below 500 F	Above 800 F	Below 500 F	Above 800 F
Moist Cured–average of three specimens per test					
0.4	7.82	4.2×10^{-6}	11.3×10^{-6}	3.6×10^{-6}	6.2×10^{-6}
0.6	5.52	7.1×10^{-6}	11.4×10^{-6}	4.7×10^{-6}	12.5×10^{-6}
0.8	4.43	6.1×10^{-6}	11.7×10^{-6}	9.3×10^{-6}	18.2×10^{-6}
Air Dried–one specimen per test					
0.4	7.82	4.3×10^{-6}	10.5×10^{-6}	6.8×10^{-6}	11.5×10^{-6}
0.6	5.52	4.3×10^{-6}	11.7×10^{-6}	4.9×10^{-6}	11.2×10^{-6}
0.8	4.43	5.3×10^{-6}	11.5×10^{-6}	6.5×10^{-6}	12.0×10^{-6}
Concrete Containing Expanded Shale Aggregate–one specimen per test					
Water-Cement Ratio	Cement Content, sacks per cu yd	Linear Coefficient of Thermal Expansion at 14 days, per deg F		Linear Coefficient of Thermal Expansion at 21 days, per deg F	
		Below 500 F	Above 800 F	Below 500 F	Above 800 F
Moist Cured					
0.68	6.36	3.4×10^{-6}	4.2		
Air Dried					
0.68	6.36	2.6×10^{-6}	5.4	2.8×10^{-6}	4.9

Note: Air-dried specimens were moist cured 3 days, then dried at 50 percent relative humidity.
From: Phillio, R., "Some Physical Properties of Concrete at High Temperatures," PCA Research Bulletin 97, *Portland Cement Association, Skokie, 1958.*

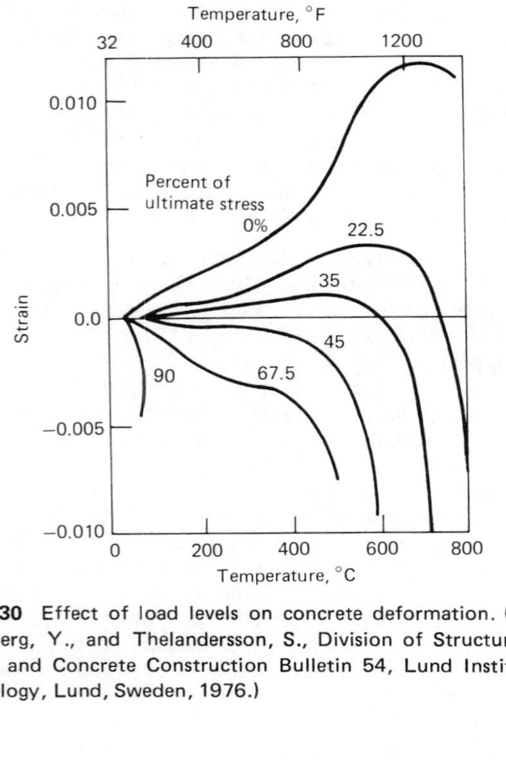

FIG. 6-30 Effect of load levels on concrete deformation. (From: Anderberg, Y., and Thelandersson, S., Division of Structural Mechanics and Concrete Construction Bulletin 54, Lund Institute of Technology, Lund, Sweden, 1976.)

Table 6-6 correlates conductivity with unit weight for oven-dry, normally dry, and saturated concretes. Conductivity of water at 75°F is quite high, about 56. Concretes, which vary in oven-dry unit weight from 20 to 150 pcf, will absorb from 43 to 7 pcf of water and produce saturated unit weights varying from 63 to 157 pcf. However, high unit weight of concrete does not necessarily indicate high conductivity. For instance, concrete made with barite aggregate has a k value of 0.79 with a unit weight of 182 pcf; while concrete made with hematite aggregate and with a unit weight of 177 pcf has a k of 1.52.

Mix proportions of cement, aggregate, and free water influence the conductivity of concrete. If the water content of concrete mixture is kept constant, the effect of richness of the mix will depend on the type of aggregate: in the case of lightweight aggregates, where k factor of aggregate is generally less than that of cement paste, the richer the mix the higher the conductivity; while in the case of conventional aggregate, where k factor of aggregate is greater than that of the paste, the richer the mix the lower the conductivity. Concrete mixes with high water-cement ratios will produce porous concrete with a low conductivity, especially when dry; however, strengths will be low.

Since moisture content decreases with increasing temperature and exposure time, the conductivity of concrete decreases with increasing temperature. At temperatures above that at which complete dehydration occurs, a gradual disintegration of cement paste occurs, resulting in a further decrease in thermal conductivity; see Fig. 6-31 and Table

TABLE 6-5 Thermal Contraction at Various Temperatures

Series No.	*Aggregate*	*Cement Content, bags per cu yd*	*Contraction, Millionths, in./in.*					
			75°F	*35°F*	*0°F*	*-75°F*	*-150°F*	*-250°F*
3	Sand and Gravel	7.0	0	210	410	680	1110	1470
6 (Moist)	*	5.5	0	250	400	740	1090	1420
8 (50%)	*	5.5	0	210	420	770	1020	1280
10 (Dry)	*	5.5	0	170	430	680	890	1240
13	*	4.0	0	170	250	350	720	1280
16	Expanded Shale	6.0	0	180	390	220	510	1060

From: Monfore, G. E., and Lentz, A. E., "Physical Properties of Concrete at Very Low Temperatures," PCA Research Bulletin 145, *Portland Cement Association, Skokie, 1962.*

6-7. Conductivity values obtained on cooled concrete test specimens, after being exposed to elevated temperatures, are higher by approximately 10 and 20%, respectively, than values obtained during heating.

Thermal conductivities of various concretes at temperatures ranging from -250°F to 75°F are listed in Table 6-8. The conductivity of lightweight aggregate concrete is nearly independent of temperature. An increase in moisture content causes an increase in conductivity of all the concretes and at all temperatures.

3. Specific Heat. Specific heat is the amount of heat required to raise the temperature of a unit mass of concrete by one degree (heat capacity of the concrete), and is identified by the symbol C.

The specific heat of concrete is equal to the summation of the weighted specific heats of the constituents. The common range of values for normal weight concrete is 0.20 to 0.28 Btu/lb-°F. The specific heat increases with a decrease in unit weight of concrete, but is affected very little by the mineralogical character of the aggregates or by variations in the aggregate or cement content of the mix. In general, the specific heat varies directly with variation in moisture content of the concrete because of the high specific heat of water. It also varies approximately linearly with an increase in temperature; over the range 68°F to 212°F, the increase in specific heat may be as much as 35%. Typical ranges for the "volumetric specific heat" (product of specific heat and density) for normal weight and lightweight concrete are shown in Fig. 6-32.

4. Diffusivity. Diffusivity is an index of the facility with which concrete will undergo temperature change, and it is identified by the diffusion constant a or h^2. It is expressed in m^2/hr or ft^2/hr and may be determined from the formula:

$$a = \frac{k}{C\rho}$$

where k = thermal conductivity; C = specific heat; and ρ = density of concrete in kg/m^3 or lb/ft^3.

TABLE 6-6 Correlation of Free Water, Unit Weight, and Conductivity of Concrete*

Concrete							
Oven Dry		*Normally Dry***			*Saturated*		
Unit Wt, ρ_o	Conductivity, k_o	Free Water, $(\Delta\rho)_n$	Unit Wt, ρ_n	Conductivity, k_n	Free Water, $(\Delta\rho)_s$	Unit Wt, ρ_s	Conductivity, k_s
20	0.60	3.0	23.0	0.77	43.0	63.0	4.2
30	0.81	2.8	32.8	1.02	34.0	64.0	4.3
40	1.05	2.6	42.6	1.28	27.0	67.0	4.5
50	1.30	2.4	52.4	1.58	21.7	71.7	4.9
60	1.60	2.2	62.2	1.91	17.3	77.3	5.5
70	1.94	2.0	72.0	2.30	15.0	85.0	6.5
80	2.34	1.8	81.8	2.80	13.2	93.2	7.7
90	2.82	1.7	91.7	3.40	12.0	102.0	9.5
100	3.47	1.6	101.6	4.18	11.0	111.0	12.0
110	4.30	1.5	111.5	5.40	10.1	120.1	15.0
120	5.62	1.4	121.4	7.10	9.3	129.3	19.0
130	7.30	1.3	131.3	9.30	8.6	138.6	24.0
140	9.40	1.2	141.2	12.05	7.8	147.8	29.0
150	12.00	1.1	151.1	15.40	7.0	157.0	34.0
160	15.00						

**From: Brewer, H. W., "General Relation of Heat Flow Factors to the Unit Weight of Concrete,"* PCA Development Bulletin 114, *Portland Cement Association, Skokie, 1967.*

***In equilibrium with normal ambient weather conditions, 50% R.H.*

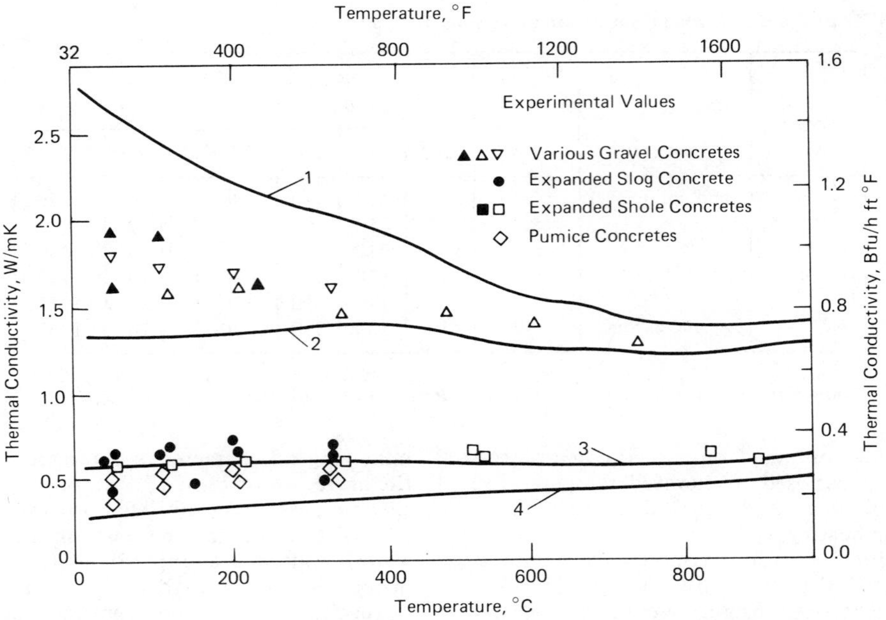

Fig. 6-31 Thermal conductivity of four "limiting" concretes (Nos. 1 and 2 are normal-weight concretes and 3 and 4 are lightweight concretes) and some experimental thermal conductivity data. (From: Harmathy, T. Z., *ASTM Journal of Materials*, V. 5, No. 1, 1970, 47.)

The value of diffusivity is largely affected by the aggregate type used in the concrete. Table 6-9 shows diffusivities for concretes made of a number of rock types. The range of typical values of diffusivity varies between 0.02 and 0.08 ft^2/hr.

The higher the value of diffusivity, the more readily heat will move through the concrete. If the rock type is not known, an average value of diffusivity can be taken as 0.04 ft^2 per hour, although as can be seen from the table the value of diffusivity varies from this average value. In general, the diffusivity increases with an increase in aggregate content or decrease in water–cement ratio; and it decreases with an increase in temperature of the concrete.

6.7.3 Adiabatic Temperature Rise

Heat of hydration is the heat generated when cement and water react. The amount of heat generated is dependent upon the chemical composition of the cement (heat of hydration of the cement used), water–cement ratio, and the cement content; the rate of heat generation is affected by fineness of cement and temperature of curing as well as cement chemical composition.

Adiabatic conditions are defined as occurring without loss or gain of heat; i.e., as isothermal. The extent to which hydration of the cement paste heats the concrete depends on the size of the structural element and its environment. Assumptions of adiabatic temperature rise become more significant as the size of the structural elements or dimensions of the placement become larger. Heat dissipation depends on the type of form, amount of exposed surface, and ambient temperatures at the various surfaces of concrete. The low thermal conductivity of concrete results in a slow rate of heat exchange between concrete and its surroundings. Therefore, at early ages, heat is generated in concrete at a higher rate than it can be transmitted to exposed surfaces. The result is a heat buildup approximating adiabatic conditions. Departure from adiabatic conditions is greatest near the cooling surfaces.

In certain structures such as those with considerable mass, the rate and amount of heat generated are important. If this heat is not rapidly dissipated, a significant rise in concrete temperature may occur accompanied by thermal expansion. Subsequent cooling of the hardened concrete to ambient temperature may create undesirable stresses. On the other hand, a rise in concrete temperature caused by

TABLE 6-7 Thermal Conductivity of Concretes Made with Type I Portland Cement and Various Aggregates, Btu/ft hr deg F[a*]

Expanded Slag (d = 98)[b]		*Gravel* (d = 144)		*Limestone* (d = 143)		*Sandstone* (d = 137)	
Mean Temp, deg F	*k*	*Mean Temp, deg F*	*k*	*Mean Temp, deg F*	*k*	*Mean Temp, deg F*	*k*
. . .	. . .	212	0.883	. . .	. . .	202	1.317
359	0.293	327	0.883	327	0.563	336	1.313
749	0.287	756	0.746	740	0.670	783	0.887

[a]*W/C ratio by weight, 0.65 ± 0.03. Aggregate-cement ratio by volume, 7.3 ± 0.07.*
[b]*d = density, lb/ft.*[3]
**From: Zoldners, N. G., "Effect of High Temperatures on Concrete Incorporating Different Aggregates,"* Proceedings, *American Society for Testing and Materials, Vol. 60, 1960.*

TABLE 6-8 Thermal Conductivity of Concrete at Various Temperatures*

Moisture Condition at Test	*Cement Content, bags/cu yd*	*Density lb/cu ft*	*Thermal Conductivity at Indicated Temperature, BTU/(hr) (sq ft) (F/in.)*				
			-250 F	*-150 F*	*-75 F*	*0 F*	*75 F*
Crushed Marble Concrete							
Moist	5.5	152	22	22	17	16	15
50% RH		148	21	18	16	16	15
Oven dry		143	13	14	13	13	12
Crushed Sandstone Concrete							
Moist	5.5	133	35	30	26	23	20
50% RH		124	17	16	16	14	15
Oven dry		120	11	11	11	10	10
Crushed Limestone Concrete							
Moist	5.5	141	24	20	18	16	15
50% RH		130	18	14	13	12	11
Oven dry		126	14	13	11	11	10
Elgin Sand and Gravel (3/4-inch max size)							
Moist	7.0	149	32	30	26	23	23
Moist	5.5	147	32	31	26	23	23
50% RH	5.5	143	25	26	21	20	19
Oven dry	5.5	141	18	18	17	15	16
Moist	4.0	143	34	31	26	24	21
Expanded Shale (3/4-inch max size)							
Moist	6.0	99	6.5	6.4	6.6	6.6	5.9
50% RH		96	5.1	5.6	5.6	5.3	5.5
Oven dry		89	3.4	3.6	3.7	3.9	4.3

**From: Lentz, A. E. and Monfore, G. E., "Thermal Conductivity of Concrete at Very Low Temperatures,"* PCA Research Bulletin 182, *1965 and "Thermal Conductivities of Portland Cement Paste, Aggregate and Concrete Down to Very Low Temperatures,"* PCA Research Bulletin 207, *Portland Cement Association, Skokie, 1966.*

heat of hydration is often beneficial in cold weather since it helps maintain favorable curing temperatures.

Typical heat of hydration curves for various portland cements are shown in Fig. 6-33. Most of the heat is generated within the first seven days, during which time the concrete also gains early strength. Both the rate and total adiabatic temperature rise in concrete differ among the various types of cement. In addition, low water–cement ratios reduce the rate of heat generation. An estimate of the temperature rise can be made using the equation:

$$T = \frac{CH}{S}$$

where:

T = Temperature rise of the concrete due to heat generation of cement under adiabatic conditions, °F
C = Proportion of cement in the concrete, by weight
H = Heat generation due to hydration of cement, Btu/lb (To convert calories per gram to Btu's per pound, multiply by 1.8; i.e., cal/g × 1.8 = Btu/lb)
S = Specific heat of the concrete, Btu/lb-°F

The temperature-rise equation provides the basis for computing maximum temperature rise and/or temperature rise at various time intervals corresponding to the rate of cement hydration time intervals. The maximum concrete temperature is determined by adding to the computed temperature the initial placement temperature of the concrete mix. Several methods of step-by-step integration have been devised for more detailed analysis of the temperature rise. For final temperature control studies, the actual heat generation should be obtained from laboratory tests. The laboratory study is made using the actual cement, concrete mix proportions, and specimen size that simulate adiabatic conditions for the element under consideration.

In high-strength concrete, the low water–cement ratio used counteracts the adverse effect of high cement content on the temperature rise. The temperature rise of high-strength concrete is approximately 11 to 15°F per 100 lb of cement per cu yd.

When a portion of the cement is replaced by a pozzolan, the temperature rise curves are greatly modified, particularly in the early ages. While the effects of pozzolans differ greatly, depending on the composition and fineness of the pozzolan and cement used in combination, a rule of thumb that has worked fairly well on preliminary computations has been to assume that pozzolan gives off about 50% as much heat as the cement that it replaces.

In general, the effects of water-reducing retarders in concrete are felt only during the first few hours after mixing and can be neglected in preliminary computations.

Placing concrete at lower temperatures (about 50°F) will

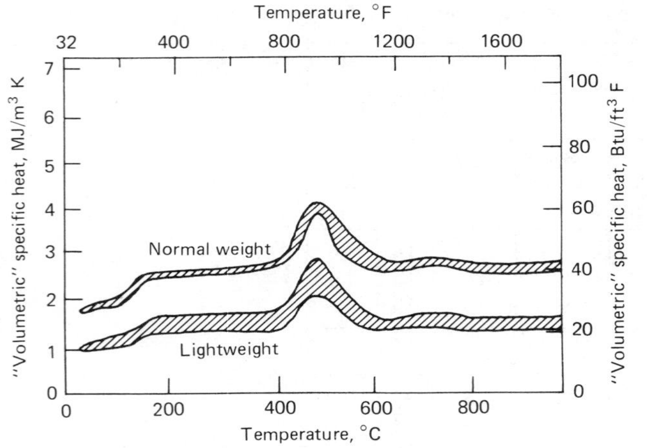

Fig. 6-32 Volumetric specific heats of normal weight and lightweight concretes. (From: Abrams, M. S., "Behavior of Inorganic Materials in Fire," ASTM STP 685, *Design of Buildings for Fire Safety*, American Society for Testing and Materials, 1979.)

Table 6-9 Typical Thermal Diffusivity Values

Type of aggregate in concrete	*Thermal diffusivity.*			
	m^2/h	*ft^2/h*	*ft^2/day[a]*	*m^2/day*
Quartz	0.0079	0.085	2.04	0.190
Quartzite	0.0061	0.065	1.56	0.146
Limestone	0.0055	0.059	1.42	0.132
Basalt	0.0025	0.027	0.65	0.060
Expanded Shale	0.0015	0.016	0.38	0.036

[a] *Convenient when computing heat flow in large structures.*

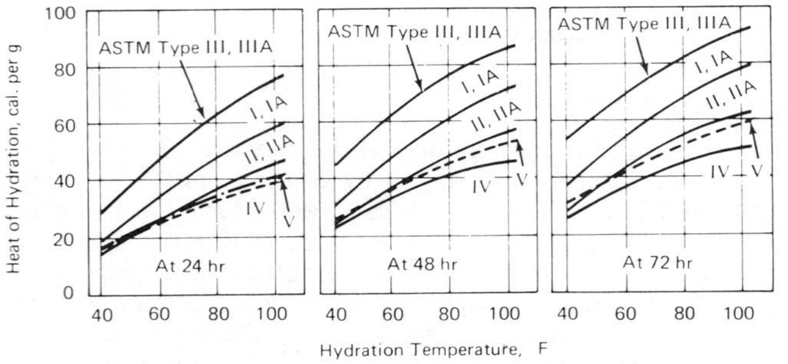

Fig. 6-33 Effect of temperature on the heat of hydration at early ages. Average heat of hydration of the different types of portland cement-conduction calorimeter. (From: Verbeck, G. J., and Foster, C. W., "Long-Time Study of Cement Performance in Concrete with Special Reference to Heats of Hydration," *PCA Research Bulletin 32*, 1949.)

lower the early rate of heat generation, but total heat generation will be equal in the long run. However, the lower initial temperature of the concrete results in a lower maximum temperature of concrete. The chief means for limiting temperature rise is controlling the type and amount of cementitious materials, although temperature rise may be controlled by embedded pipe cooling systems, placement in shallow lifts, or flooding the surface of the concrete for about a week to encourage maximum heat loss from the surface by evaporative cooling.

Factors, in addition to the above, subject to at least some degree of control by the designer include the overall temperature drop from the maximum concrete temperature to the final stable temperature, the rate of temperature drop, and the age of the concrete when the concrete is subjected to the temperature change.

6.7.4 Thermal Stresses and Cracking

Thermal stresses develop in two ways: (1) from temperature variations due to the heat of hydration, and (2) from temperature variations due to periodic cycles of ambient temperature.

The temperature in concrete and hence the thermal stresses depend upon the rate at which this heat is dissipated. The properties of concrete relevant to the determination of temperature distribution are: (1) thermal conductivity–for the determination of steady-state conditions; and (2) thermal diffusivity–for the determination of transient conditions. In general, thermal stresses are created due to restraint or in unrestrained member whenever the thermal gradient varies from linear within the concrete mass.

Heat can escape from a body inversely as the square of its least dimension. Consider a number of walls, made of average concrete and exposed to cool air on both faces. For a wall 6 in. thick, 95% of the heat in the concrete will be lost to the air in $1\frac{1}{2}$ hr. For a 5-ft-thick wall, this same amount of heat would be lost in a week. For a 50-ft-thick wall, which might represent the thickness of an arch dam, it would take two years to dissipate 95% of the heat stored, and for a 500-ft-thick dam, it would take 200 years to dissipate this amount of heat. Thus in most building structural members most of the heat generated by the hydrating cement is dissipated almost as fast as it is generated, and there is little temperature differential from the inside to the outside of the member.

Temperature drop may cause tensile stresses and eventual cracking if a member is restrained from shortening or if the cooling affects only part of a member (partially exposed walls or columns). Tensile stresses build up on the face as it is cooled. If the member is not restrained, uniform cooling causes only shortening but no stress. When cooling of an unrestrained member results in a linear gradient through the section, the result is shortening and curvature of the member, but no stresses are created.

Extreme differences between internal and outside temperatures may result in surface cracking. For example, temperature stresses occur as the temperature of the concrete rises because of heat of hydration and then drops essentially to the temperature of its surroundings. As the outer surface cools and tends to shrink, compressive stresses are set up in the center and tensile stresses in the cooler outer surfaces. When these tensile stresses become greater than the tensile strength of the concrete, cracking occurs. Concrete cracks at a strain of about 10^{-4} to 2×10^{-4}. The rate of cooling should be regulated to control the temperature drop to prevent extreme temperature differences. The rate may range from less than 1°F per day for periods of less than 25 days, with $\frac{1}{2}$ to $\frac{3}{4}$°F per day perferable for massive members, to 5°F in any 1 hr or 50°F in any 24-hr period for thin members.

The use of insulation on exposed surfaces of massive structural elements tends to decrease the temperature fluctuations within the mass; it also serves, however, to retain more heat so that a higher maximum temperature is usually attained.

The maximum permissible gradient for a nominally reinforced structure is generally given as less than 100°F. However, as percentage of reinforcement increases, maximum permissible temperature difference across the element

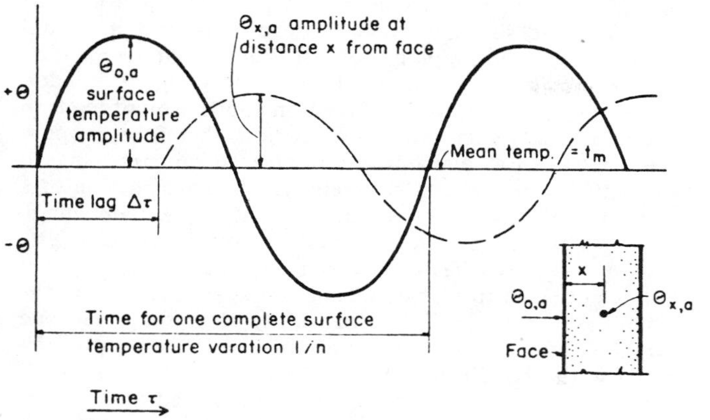

Fig. 6-34 Temperature amplitude at surface and at distance, x. (From: Fintel, M., and Khan, F. R., "Effects of Column Exposure in Tall Structures," PCA, EB018.01D, 1968.)

may be increased. This should be evaluated in terms of permissible deflections and restraint conditions.

Exposed structural members are affected by daily and annual cycles of temperature, as well as sustained elevated temperatures. Figure 6-34 gives a general idea of what happens in a concrete member subjected to an exterior sineform temperature oscillation with a frequency n. At a distance x from the face, the oscillation arrives with a time lag $\Delta\tau$. The surface amplitude, θ_{oa}, attenuates at the distance x to the amplitude θ_{xa}. The effects of time lag and attenuation can be combined into a single expression for the amplitude (range of temperature variation) at a distance x from the the surface at τ:

$$\theta_{xa} = \theta_{oa}\, e^{-x\sqrt{\pi n/h^2}} \sin\left(2\pi n\tau - x\sqrt{\frac{\pi n}{h^2}}\right)$$

where:

n = the frequency of temperature change
h^2 = thermal diffusivity

Since the interior reacts so much more slowly than the surface to cycles of temperature, it is as though the surface were restrained by the interior concrete.

In a radiation shield, attenuation of radiation results in a rise in temperature of the shielding concrete as the absorbed energy of radiation is converted to heat. The thermal distribution curve is assumed to be nonlinear with the maximum temperature occurring some distance in from the surface exposed to the radiation source, due to the dissipation of energy of the absorbed neutrons. In general, reinforced concrete should not be exposed to an incident energy flux greater than 2×10^{11} MeV/cm^2,* in order to ensure acceptable thermal stresses. This flux corresponds to approximately 100 Btu/ft^2/hr and would cause about 90°F temperature rise in the shield.

Tensile stresses across a section due to nonlinear temperature gradients can be calculated by the following equation and compared with the tensile strength of the concrete to determine if thermal cracking will occur:

$$\sigma = \frac{P + P_\theta}{A} + \frac{M + M_\theta}{I} y - E\alpha\Delta\theta(y)$$

*1 MeV (million-electron-volt) is the amount of energy that would be acquired by an electron in falling through a potential of 1,000,000 V. It is equal to 1.6×10^{-3} watt-sec.

where:

$$P_\theta = E\alpha \int_A \Delta\theta(y)\, dy$$

$$M_\theta = E\alpha \int_A \Delta\theta(y) y\, dy$$

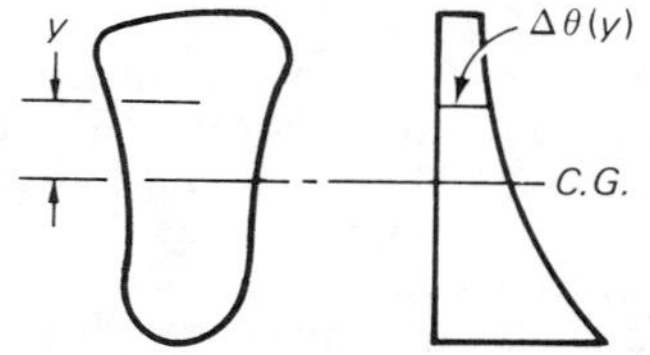

where:**

E = modulus of elasticity
α = thermal coefficient of expansion
$\Delta\theta(y)$ = distribution of rise in temperature
P, M = external axial load and moment
y = distance from center of gravity
A = total area of cross section
I = moment of inertia of cross section

Where the temperature variation, $\Delta\theta(y)$, cannot be expressed analytically, the indicated integrations can be performed numerically by the use of Simpson's rule.

Generally, creep modifies the thermal stresses considerably, but little quantitative information is available.

In the above discussion, the temperature was assumed to vary only vertically, that is, as a function of y. In reality, it can vary along the width of a beam as well as the depth. The equations could be extended to include the effects of two-dimensional distribution of temperature, but such a refinement is not considered warranted.

6.7.5 Drying Shrinkage and Creep

Concrete expands with a gain in moisture and contracts with a loss in moisture. If kept continuously in water, concrete slowly expands for several years, but the total amount of expansion is normally so small that it is unimportant (usually less than 150 millionths). Concrete that is not continuously wet is subject to water loss with resulting contraction or shrinkage. Since concrete exposed to the atmosphere loses some of its original water, it normally exists in a somewhat contracted state compared to its original dimensions.

The consideration of shrinkage in design may be highly critical for some structures and unimportant for others. Shrinkage can affect performance and appearance—extent of cracking, prestress loss, effective tensile strength, and warping. The amount of shrinkage that is tolerable depends on jointing and design details of the structure.

Unit length change due to drying shrinkage of small plain concrete (no reinforcement) specimens ranges from

**Equation reference is from: Gustaferro, A. H., Abrams, M. S., and Salse, E. A. B., *Fire Resistance of Prestressed Concrete Beams, Study C: Structural Behavior During Fire Tests* (RD009.01B), Portland Cement Association, 1971, and Townsend, C. L., "Control of Cracking in Mass Concrete Structures," U.S. Dept. of the Interior Bureau of Reclamation Engineering Monograph No. 34, 1965. See also "Addition to Commentary on Code Requirements for Nuclear Safety Related Concrete Structures (ACI 349-76)," *ACI Journal*, Aug. 1978, pp. 336–339.

about 400 to 800 millionths when exposed to air at 50% humidity. The extremes, however, may range from less than 200 millionths for low-slump lean mixes with good-quality aggregates to over 1000 millionths for rich mortars or some concretes containing poor-quality aggregates and an excessive amount of water. Concrete with a unit shrinkage of 600 millionths shortens about 0.72 in. per 100 ft while drying from a saturated condition to a state of moisture equilibrium in air at 50% relative humidity. This equals approximately the thermal contraction caused by a decrease in temperature of 100°F.

When concrete is loaded, the deformation caused by the load may be divided into two parts: a deformation that occurs immediately, and a time-dependent deformation that begins immediately but continues for years. This latter deformation is called creep.

Creep of concrete is the dimensional change or increase in strain with time due to a sustained stress. Magnitude of creep in tension or compression is the same. Creep of concrete may be either beneficial or detrimental, depending on the prevalent structural conditions. Concentrations of stress, either compressive or tensile, may be reduced by stress transfer through creep, or creep may lead to excessive long-time deflection, prestress loss, or growing camber. The effects of creep along with those of drying shrinkage should be considered (shrinkage may act with or against creep), and, if necessary, compensated for in structural designs. For example, in structures subjected to sustained thermal gradients, creep and shrinkage will reduce the initial thermal stresses but give rise to undesirable reversals of stress on cooling.

Creep consists of two components:

(a) Basic (or true) creep occuring under conditions of hygral equilibrium, which means that no moisture movement occurs to or from the ambient medium. In the laboratory basic creep can be reproduced by sealing the specimen in copper foil.
(b) Drying creep resulting from exchange of moisture between the stress member and its environment. Drying creep has its effect primarily during the initial period under load.

For structural engineering practice it is convenient to consider specific creep, ϵ_c', which is defined as the ultimate creep strain per unit of sustained stress. The ultimate magnitude of creep of plain concrete per unit stress (psi) can range from 0.2 to 2.0 millionths in terms of length, but is ordinarily about 1 millionth or less.

In newly placed concrete, the change in volume or length due to creep is largely unrecoverable. However, creep that occurs in old or dry concrete is largely recoverable, and creep recovery appears to be essentially independent of temperature and stress.

1. Important Parameters. Shrinkage and creep of concrete are related phenomena and are controlled by similar parameters. The needed parameters are those that are known or predictable by the designing engineer. These parameters affecting shrinkage and creep include: (1) water–cement ratio of the cement paste; (2) physical characteristics of the aggregate; (3) cement paste content and characteristics; (4) age of concrete when exposed to drying or when an external load is applied; (5) size and shape of the structural member; (6) amount of steel reinforcement; (7) environmental exposure conditions such as relative humidity, temperature, and carbon dioxide content of the air; and (8) curing conditions.

Most research data consider the basic creep and shrinkage characteristics of concrete as measured on small unreinforced prisms or cylinders in a controlled laboratory environment. However, the shrinkage and creep of full-scale structural elements are considerably reduced because of size effect, sequence of loading, and amount of reinforcement. In addition, most creep research is based on application of loads in one increment. Such creep information, therefore, is applicable to flexural elements of reinforced concrete and to elements of prestressed concrete. In the construction of a high-rise building, columns are loaded in as many increments as there are stories above the level under consideration, and research has shown that incremental loading over a long period of time makes a considerable difference in the magnitude of creep. Therefore, only after the creep and shrinkage strains have been determined separately and modified for the conditions of the designed structure can their combined effect on the structure be considered.

In comparing behavior under load of concretes made with different cements, the ratio of the applied stress to the strength at the time of loading should be considered. With respect to the cement, the magnitude and rate of creep strain is influenced by the strength attained by the cement paste at the time of loading. This is controlled to some extent by the chemical composition and fineness of the cement. For concretes loaded to a given stress, the creep will be less for a cement with the highest rate of strength gain. But for concretes loaded to a particular stress–strength ratio, the cement will be of less importance. Cements of different composition and fineness also have variable effects on drying shrinkage. Such differences have been moderated considerably in recent years in most cements by providing the optimum amount of gypsum in the cement. At optimum gypsum content, the type of cement per se does not significantly influence creep or drying shrinkage. The optimum amount of gypsum is unique for each cement and increases as the anticipated hydration temperature increases.

The role of the aggregate in concrete is to dilute the paste matrix and to restrain greatly the shrinkage and creep of the paste, thereby reducing the overall shrinkage and creep of the concrete. The effectiveness of aggregate in reducing shrinkage and creep increases as the volumetric fraction of coarse aggregate in the concrete increases. The mineralogical and physical properties of coarse aggregate are important in providing restraint to shrinkage and creep. Because of the great variation in aggregate within any mineralogical or petrological type, it is not possible to make a general statement about the magnitude of shrinkage or creep of concrete made with aggregates of different types. Hard aggregates with high density and high modulus of elasticity coupled with moderate porosity or absorption produce concrete with the lowest drying shrinkage and creep. Concrete for radiation shielding containing hydrous aggregates (limonite, bauxite, or serpentine) will shrink more than concrete containing aggregates such as granite or magnetite.

An excessive amount of clay or other fine filter material in an aggregate tends to increase shrinkage. Although an increase in water content results, the increase in drying shrinkage appears greater than can be accounted for on this basis alone. Clay particles or coatings on aggregates cause significant increases in creep, principally because of loss of bond, loss of restraint, and moisture movement.

Other aggregate variables such as grading, maximum size, and particles shape have their main influence on shrinkage or creep through the effect they have on the paste content required for adequate workability. For example, shrinkage can be minimized by using aggregate of the largest practicable size, since concrete mixes with large-size aggregates have a greater total aggregate content than mixes made with small-size aggregates.

Structural lightweight concretes made and cured at normal

temperatures have drying shrinkage ranging from about slightly less to 30% more and creep ranging from about the same to 50% more than that of some normal-weight concretes. High-strength lightweight concrete has about the same shrinkage and creep as comparable normal-weight concrete. Partial or full replacement of lightweight fines by a good grade of natural sand usually reduces shrinkage and creep for concretes made with most lightweight aggregates.

Little information is available on the effect of admixtures on drying shrinkage and creep, possibly because of the multiplicity of admixtures available and their frequent modification. At present, it is not possible to predict which combinations of admixtures and cements will influence shrinkage or creep. When drying shrinkage and creep are important, the admixtures should be evaluated using job materials, mixing, and curing. However, the use of accelerators such as calcium chloride and triethanolamine results in substantial increases in drying shrinkage and creep of concrete. Despite reductions in water content, some chemical admixtures of the water-reducing type also increase drying shrinkage and creep substantially (age of loading is important), particularly those that contain an accelerator to counteract the retarding effect of the admixture. Data on shrinkage and creep of concrete containing superplasticizers are rather limited. The available data indicate that the drying shrinkage and creep of concretes containing superplasticizers, in most cases, are equal to or slightly less than those of reference concretes prepared at equal slump (when these admixtures are used to lower the net water content of a concrete mixture). The materials commonly used to entrain air in concrete have little effect on shrinkage or creep.

Concrete containing pozzolanic admixtures that require more water may have increased shrinkage, but most low carbon fly ash will not appreciably affect shrinkage. However, pozzolans may increase creep because of the increase in paste content resulting from the increased volume of hydration products brought about by the reaction of cement hydration products and the pozzolans.

The influence of proportions is best understood by considering interrelated factors such as water–cement ratio, cement content, aggregate content, and total water content. Both the creep and the shrinkage of concrete increase with an increase in water content. Shrinkage is approximately proportional to the percentage of water by volume in the concrete, while creep varies with the water–cement ratio. Thus, shrinkage can be minimized by keeping the water content of the paste as low as possible and the total aggregate content of the concrete as high as possible. Use of low slumps and placing methods that minimize water requirements are thus major factors in controlling shrinkage. Any practice that increases the water requirements of the cement paste such as the use of high slumps, excessively high fresh concrete temperatures, or smaller coarse aggregate, increases shrinkage.

For a given stress, creep increases with an increase in water–cement ratio for a given paste content. However, for a constant water–cement ratio, creep increases with increasing paste content. Generally speaking for a particular stress, lean mixes creep more than rich mixes.

However, specimens loaded to the same stress–strength ratio and having the same paste content will have the same creep regardless of the water–cement ratio. Under this loading condition, a lean mix will exhibit less creep than a rich mix, although in a lean mix, a larger portion of the ultimate creep occurs at an early age.

The method of curing of concrete has a marked effect on the amount of shrinkage and creep. Atmospheric steam curing may reduce the shrinkage from 10 to 40% and creep about 25 to 40%. The effects of length of moist curing period on drying shrinkage are not as significant as their effects upon creep.

Shrinkage of concrete is caused by evaporation of moisture from the surface. The rate and amount of evaporation and consequently of shrinkage depend greatly upon relative humidity of the environment and ambient temperature. Since evaporation occurs only from the surface of a structural element, the volume-to-surface ratio (size and shape) of a member has a pronounced effect on the amount of its shrinkage.

The rate at which the internal relative humidity of concrete decreases as the member is subjected to a drying atmosphere is of considerable importance. Moisture migrates very slowly through concrete so that cement continues to hydrate beneath the outer surface for a long period of time after exposure to drying; hydration ceases when relative humidity is less than about 80%. Also, the relative humidity in concretes having low water–cement ratios decreases more slowly than concretes having high water–cement ratios, and moisture diffuses considerably slower in structural lightweight concrete than in normal-weight concrete.

The variation of internal relative humidity with time at various radial distances from the center of 6 × 12-in. solid cylinders is shown in Fig. 6-35. The internal humidity at any location dropped from the initial 100% relative humidity at a decreasing rate toward an asymptotic value of 50% relative humidity, the humidity of the environment. The initial rate of decrease of internal humidity is fastest near the surface, and later the rate of decrease is approximately the same for all locations. In a dry atmosphere, moderate-size members (24-in. diameter) may undergo 50–70% of their ultimate shrinkage within two to four months, while identical members kept in water may exhibit growth instead of shrinkage. In moderate-size members the inside relative humidity has been measured at 80% after four years of storage in a laboratory in 50% relative humidity.

Concrete, such as an interior column, subjected to a continuously dry atmosphere will exhibit greater drying shrinkage than concrete exposed to high humidities. Since the rate and amount of shrinkage greatly depend upon the relative humidity of the environment, the shrinkage specimen should be stored under conditions similar to those for

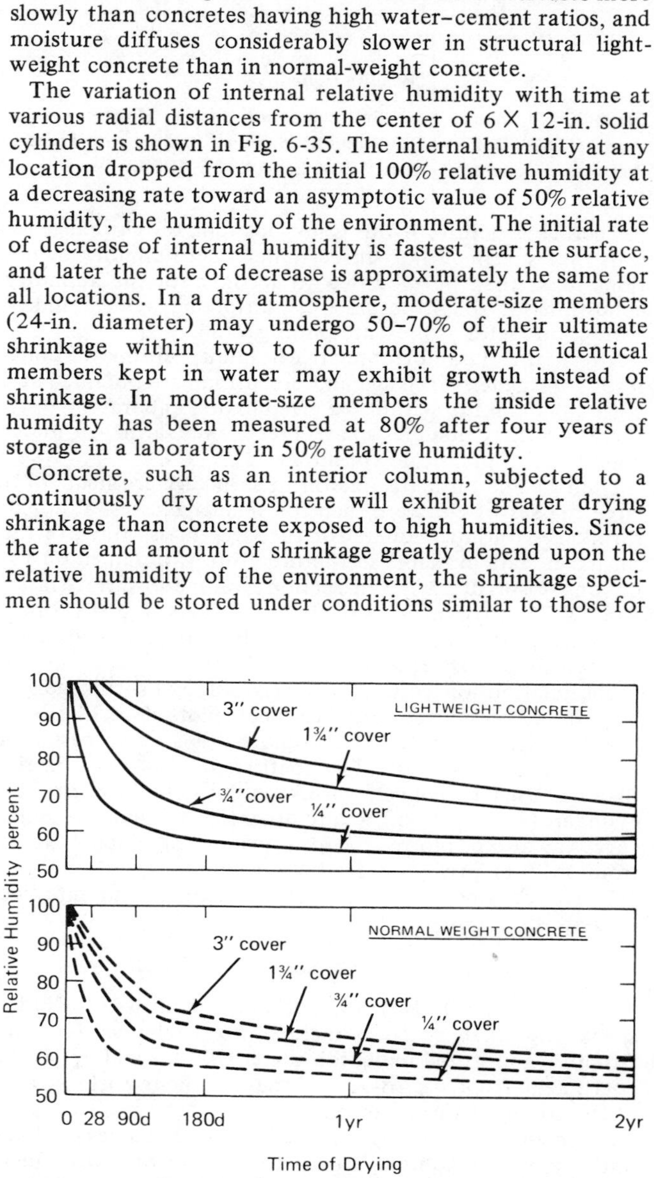

Fig. 6-35 Typical relative humidity distribution of 6 × 12 in. cylinders—moist-cured seven days, then dried at 73°F and 50% relative humidity. (From: Hanson, J. A., "Effects of Curing and Drying Environments on Splitting Tensile Strength of Concrete," *PCA Development Department Bulletin D141*, 1968.)

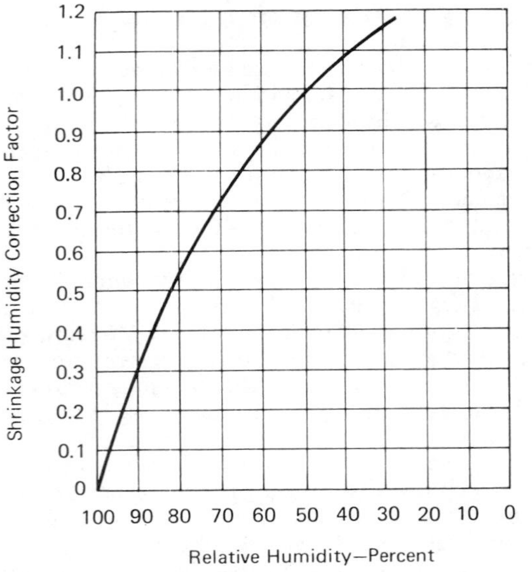

Fig. 6-36 Shrinkage humidity correction factor. (From: Freyermuth, C. L., "Design of Continuous Highway Bridges with Precast, Prestressed Concrete Girders," PCA EB014.01E, 1969.)

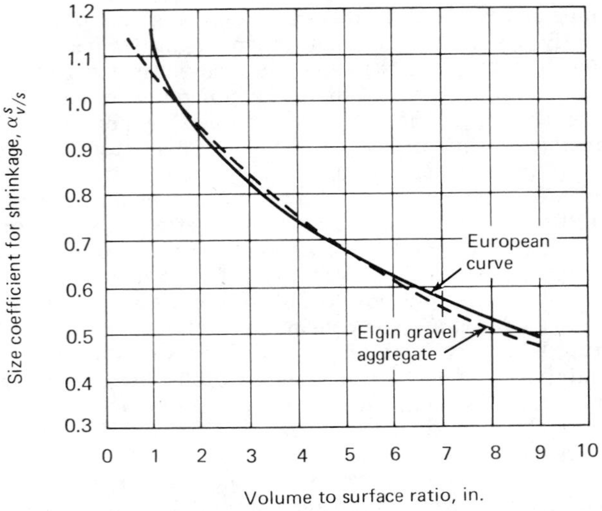

Fig. 6-37 Shrinkage vs volume to surface ratio. (From: Fintel, M., and Khan, F. R., "Effects of Column Creep and Shrinkage in Tall Structures—Prediction of Inelastic Column Shortenings," *ACI Journal Proceedings*, 66 (12), Dec. 1968.)

the actual structure. If this is not possible, the shrinkage results of a specimen not stored under field humidity conditions must then be modified to account for the humidity conditions of the structure by multiplying by a humidity correction factor as presented in Fig. 6-36.

Alternating the ambient relative humidity within two limits results in higher shrinkage and creep than are obtained at a constant humidity within the given limits. Laboratory tests, therefore, may underestimate creep and shrinkage under conditions of practical exposure.

The rate and ultimate amount of shrinkage are smaller for large masses of concrete than for smaller masses, although shrinkage continues longer for the large mass. Much of the shrinkage data available in the literature is obtained on 11-in.-long prisms of a 3 X 3-in. section (volume-to-surface ratio, V/S = 0.75 in.) or on 6-in.-diameter cylinders (V/S = 1.5 in.). Obviously, such data cannot be applied to usual-size members without considering the effect of size.

The relationship between the magnitude of shrinkage and the volume-to-surface ratio has been plotted in Fig. 6-37 based on laboratory data. Also plotted is a curve based on European investigations. The coefficient $\alpha^s_{v/s}$ shown in the figure is used to convert shrinkage data obtained on 6-in. cylinders (V/S = 1.5 in.) to any other size columns. A similar curve can also be plotted to convert laboratory data obtained from 3 X 3-in. prisms (V/S = 0.75 in.).

Thus, the amount of shrinkage, ϵ_s, of a nonreinforced column is:

$$\epsilon_s = \epsilon_{s,\,\text{test}}\,\alpha^s_{v/s}$$

where $\epsilon_{s,\,\text{test}}$ is the shrinkage obtained from 6-in. cylinder specimens made of the concrete mix to be used in the structure and stored under job-site conditions and $\alpha^s_{v/s}$ is the coefficient from Fig. 6-37 for the volume-to-surface ratio of the column being designed.

For many outdoor applications, concrete reaches its maximum moisture content during the season of low temperature. Thus, the volume changes due to moisture and temperature variations frequently tend to offset each other.

For constant relative humidity, changes in temperature have a negligible effect on shrinkage of concrete. Humidity of the air, however, tends to vary inversely with temperature. Hence shrinkage strains tend to be less for lower temperatures because of both the higher relative humidity and the lower rate of evaporation. In general, high temperatures tend to accelerate the rate of moisture removal from concrete. High temperature accompanied by drying may lead to initial shrinkage followed by expansion once the effect of thermal expansivity is greater than the shrinkage associated with moisture loss. There is a good correlation between internal relative humidity and drying shrinkage.

Creep is less sensitive to member size than shrinkage, since only the drying creep component of the total creep is affected by size and shape of members, whereas basic creep in independent of size and shape. It appears that drying creep has its effect only during the initial three months. Beyond 100 days, the rate of creep is equal to the basic creep. Creep of sealed specimens (basic creep) is about 20% less than similar unsealed specimens, although values up to about 65% have been obtained after three years of loading.

In Fig. 6-38 the relationship between creep and volume-to-surface ratio has been plotted. Also plotted is the curve based on European experience. The curves are almost identical. It is seen from the curves that in members with a volume-to-surface ratio of 10, drying creep is negligible, and only basic creep occurs. Only smaller members indicate any significant drying creep. In structural members of substantial dimensions, the rate of moisture diffusion is sufficiently slow so that their behavior approximates that of sealed laboratory specimens.

Creep increases as the ambient temperature increases. Sealed or water-stored specimens generally exhibit less creep than unsealed specimens, and creep decreases with increasing degree of hydration and increases with increasing moisture content of the specimen at loading. Creep at 120°F is approximately two to three times as great as creep at room temperatures. For temperatures of 120°F to about 212°F, some controversy exists about whether or not there is a further increase of total creep with increasing temperature. Some investigators have found a definite maximum of total creep in the range of 120°F to 180°F, but most have not, and have concluded that creep increases with temperature up to around 212°F, the creep at 212°F being on the order

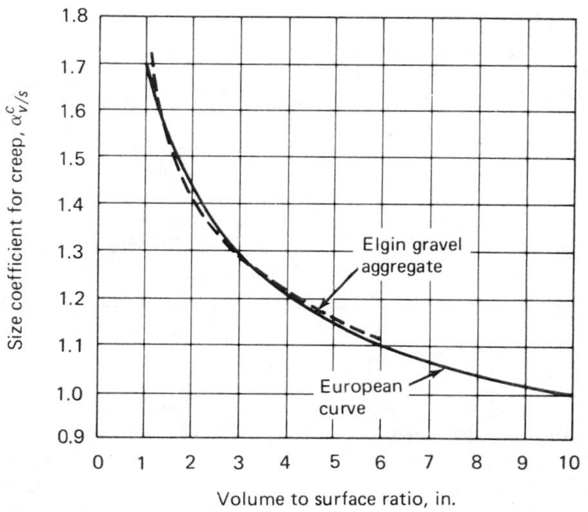

Fig. 6-38 Creep vs volume to surface ratio. (From: Fintel and Khan, "Effects of Column Creep and Shrinkage in Tall Structures—Prediction of Inelastic Column Shortenings," *ACI Journal Proceedings*, 66 (12), Dec. 1968.)

of four to six times as great as the creep at room temperature (at the end of a 60- to 100-day loading period). On the other hand, in an apparent contradiction, most investigators have also found a definite maximum for the creep rate between 120°F and 180°F, if the creep rate is computed for some period between 1 and 107 days under load. This seems to indicate that as the temperature increases, a larger portion of the (larger) total creep deformation occurs during the first few hours under load.

Limited data are available for creep at temperatures exceeding 212°F. Tests on unsealed specimens show no appreciable change in creep within the temperature range of 212°F to 280°F; the creep rate for a 1- to 100-day loading period appears to decline. Beyond 280°F, both creep rate and creep magnitude increase with temperature (unsealed specimens). For example, creep of concrete subjected to constant sustained compressive stress of 1800 psi for five hours while at high temperatures showed the following:

Creep of Concrete at High Temperatures Expressed as a Multiple of Creep at 75°F * (5-hr Test)

Temperature, °F	75	300	600	900	1200
Ratio	1.0	3.3	6.4	14.9	32.6

**From: Cruz, C. R., "Apparatus for Measuring Creep of Concrete at High Temperatures," PCA Research Bulletin 225, 1968.*

These changes in creep at high temperatures may be related to the inversions in strength and elasticity properties with increasing temperature. Creep at lower temperatures (−30°C) is about half of the creep at ambient temperatures (20°C).

Creep of concrete is a linear function of the stress–strength ratio for stress up to about 50% of the ultimate strength of the concrete. (Normal working stresses fall below 50%). For high-strength concrete, this linearity extends to a higher stress–strain ratio, possibly 70%. Beyond that level, creep becomes a nonlinear function of stress, increasing as a second or even third order power of the sustained stress. In the extremely high stress range (85 to 90% of ultimate strength), the creep rate increases rapidly and leads to failure at stresses well below the instantaneous ultimate strength. Under continuous load, creep continues for many years, but the rate decreases with time as the concrete increases in strength while under load. Drying shrinkage is not appreciably affected by the magnitude of the sustained stress.

Creep of concrete in tension is difficult to measure; thus, creep as measured in compression is assumed to apply to tension as well. Such an assumption can be considered as reasonable when the stress is below 50%.

During the initial period of loading the rate of creep is significant. The rate diminishes as time progresses until it eventually approaches zero. Figures 6-39(a) shows a typical creep versus time curve drawn on a standard scale. The same curve plotted on semilogarithmic graph paper is shown in Fig. 6-39(b), with time on the logarithmic abscissa.

For a given concrete mix the amount of creep depends not only upon the total stress but also to a great extent upon

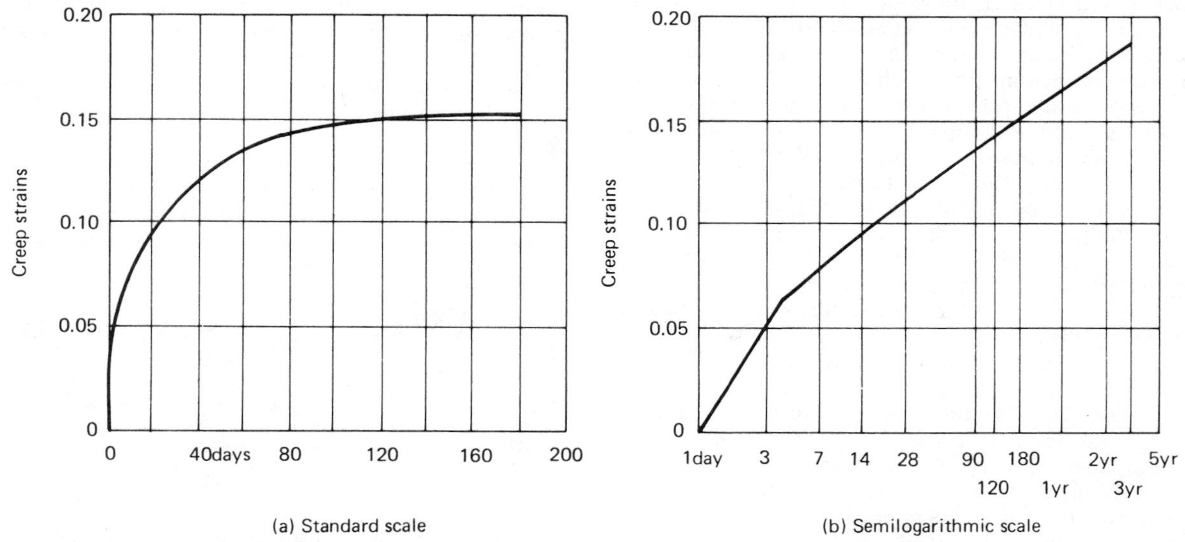

Fig. 6-39 Creep strains vs time under load. (From: Fintel and Khan, "Effects of Column Creep and Shrinkage in Tall Structures—Prediction of Inelastic Column Shortenings," *ACI Journal Proceedings*, 66 (12), Dec. 1968.)

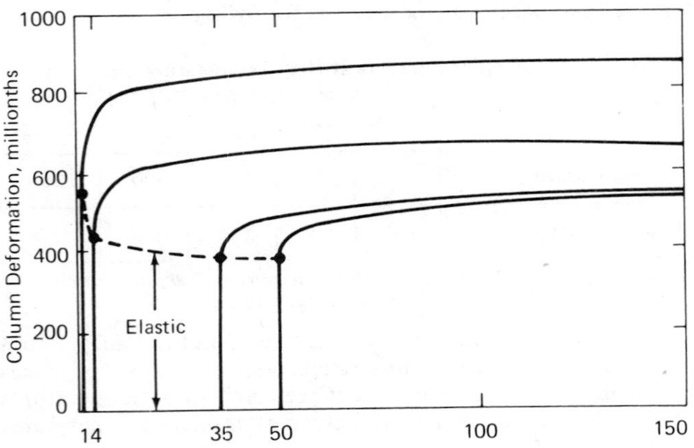

Fig. 6-40 Elastic and time-dependent creep deformation of instantaneously loaded reinforced columns. (From: Pfeifer, D. W., and Hognestad, E., "Incremental Loading of Reinforced Lightweight Concrete Columns," *Final Report of the Eighth Congress, International Association of Bridge and Structural Engineers*, New York, Sept. 1968, pp. 1055–1063.)

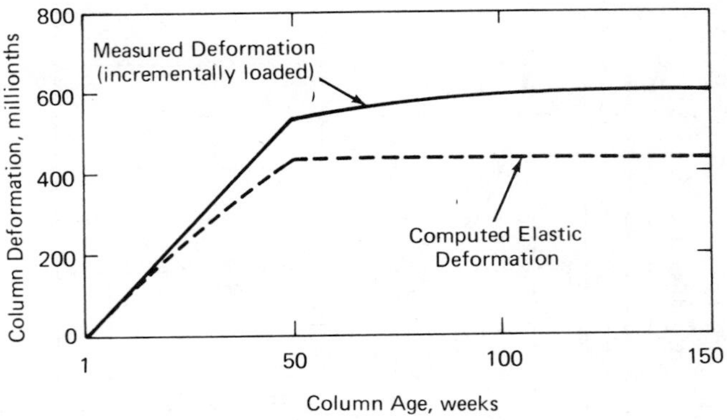

Fig. 6-41 Elastic and time-dependent creep deformation of incrementally loaded reinforced columns. (From: Pfeifer and Hognestad, "Incremental Loading of Reinforced Lightweight Concrete Columns," *Final Report of the Eighth Congress, International Association of Bridge and Structural Engineers*, New York, Sept. 1968, pp. 1055–1063.)

the loading history. A concrete element with its load applied at an early age exhibits a much larger specific creep than a specimen loaded later. This means that creep decreases with increase in strength-to-stress ratio at the time of loading. Also because of the gradual increase of the modulus of elasticity with age, the elastic shortening per unit stress of older concrete is smaller than that of concrete loaded at an earlier age. Therefore, concrete specimens of equal strength but of different age will have different creep characteristics. Figure 6-40 shows elastic and creep strains of columns loaded at various ages. The elastic response to load is in accord with elastic theory, and the significant increase in modulus of elasticity of concrete as a function of curing time is quite evident in the measured elastic response of the concrete. The measured time-dependent creep characteristics also reflect the influence of age of concrete at loading on the time-dependent behavior of concrete.

Loading history is particularly significant for columns of multistory buildings which are loaded in as many increments as there are stories above the level under consideration. Since creep decreases with age of the concrete at load application, each subsequent incremental loading contributes a smaller specific creep to the final average specific creep of the column, Fig. 6-41. The computed elastic shortening, taking into account the increased modulus of elasticity of the concrete, is also shown in the figure. It can be seen that the influence of creep is small when sealed reinforced columns are incrementally loaded during a long-time period.

The postulated and confirmed principle of superposition of creep states that: Strains produced in concrete at any time by a stress increment are independent of the effects of any stress applied either earlier or later. The stress increment may be either positive or negative, but stresses that approach the ultimate strength are excluded. Thus, each load increment causes a creep strain corresponding to the strength-to-stress ratio at time of its application, as if it were the only loading to which the column is subjected.

The specific creep values corresponding to the ages at which incremental loadings are applied in an intended multistory structure can be obtained by extrapolation from a number of laboratory samples prepared in advance from the actual mix to be used in the structure. It is obvious that sufficient time for such tests must be allowed prior to the start of construction, since reliability of the prediction improves with length of time over which creep is actually measured.

An alternate method to predict basic creep (without testing) from elastic modulus of elasticity has been proposed. Results of limited tests on normal-weight concrete indicate that creep can be predicted easily from the initial modulus at time of load application. Curves in Fig. 6-42 give the creep magnitude as related to the initial modulus of elasticity for different load durations. For design purposes the 20-year creep can be regarded as the ultimate creep. The dashed lines in Fig. 6-42 are extrapolations, while the solid lines are test results. Thus from the specified 28-day strength, the basic specific creep for loading at 28 days can be determined and then modified for construction time, member size, and percentage of reinforcement, as discussed later.

The creep–time relationship plotted on semilog graph paper in Fig. 6-39(b) consists of three straight-line segments. Such a curve is mathematically represented by an equation that is a sum of three exponential components. Only one of the three components is effective beyond about 10 days, the two others having an effect only during the very early days.

For the solution of creep due to incremental loading the primary interest is in the final creep value. The exponential expression for creep, represented graphically in Fig. 6-39(b), has a particular advantage for the structural engineer. It

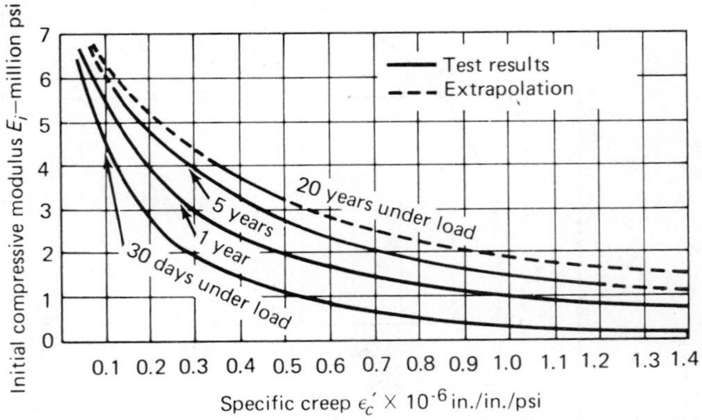

Fig. 6-42 Prediction of basic creep from elastic modulus. (From: Fintel and Khan, "Effects of Column Creep and Shrinkage in Tall Structures—Prediction of Inelastic Column Shortenings," *ACI Journal Proceedings*, 66 (12), Dec. 1968.)

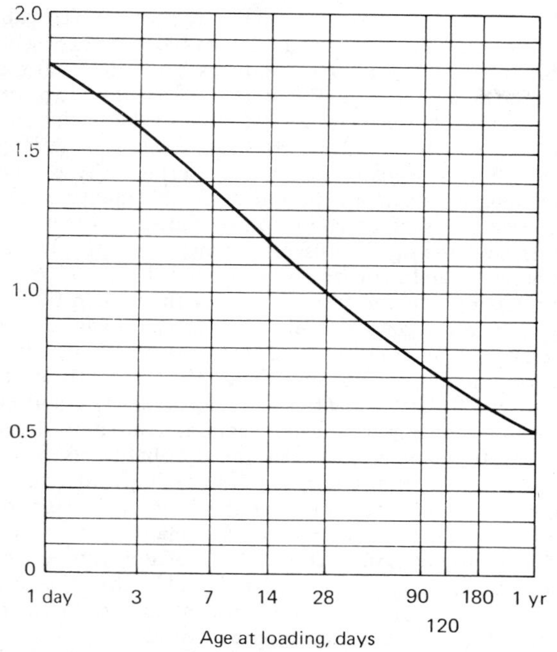

Fig. 6-43 Creep vs age at loading. (From: Fintel and Khan, "Effects of Column Creep and Shrinkage in Tall Structures—Prediction of Inelastic Column Shortenings," *ACI Journal Proceedings*, 66 (12), Dec. 1968.)

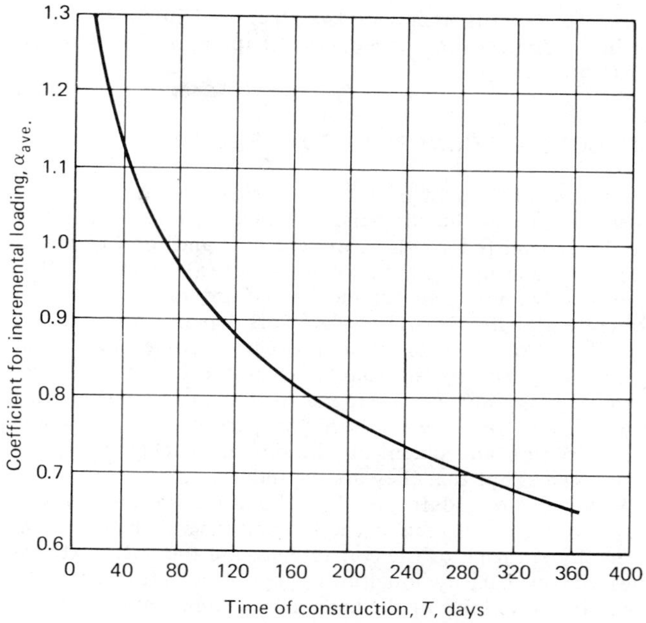

Fig. 6-44 Creep vs construction time. (From: Fintel and Khan, "Effects of Column Creep and Shrinkage in Tall Structures—Prediction of Inelastic Column Shortenings," *ACI Journal Proceedings*, 66 (12), Dec. 1968.)

allows interpolation and extrapolation with as few as two points, since beyond about 10 days it is represented as a straight line when time is plotted on a logarithmic scale.

The curve in Fig. 6-43 gives the relationship between creep and age at loading. The coefficient α_{age} relates the creep for any age at loading to the creep of a specimen loaded at the age of 28 days. The 28-day creep is used as a basis for comparison (with $\alpha_{age,\,28} = 1.0$). Thus, a loaded 6-in diameter standard cylinder wrapped in foil can supply the basic information which is subsequently modified to consider any age at loading. The wrapped specimen isolates the basic (true) creep excluding drying creep and shrinkage, thus simulating a very large column section. It is advantageous to use 6-in. standard cylinders for uniformity with other testing such as compression, split cylinder (tensile), and shrinkage, although any size specimens (foil wrapped) would produce the same specific creep for the same concrete mix and the same loading. Individual values for specific creep can be obtained from Fig. 6-42 or from the creep of a test specimen loaded at 28 days and then modified for the various ages at loading using the coefficient α_{age} from Fig. 6-43.

The coefficient, α_{ave}, plotted in Fig. 6-44 is used to convert the 28-day creep into the average specific creep for a column loaded with equal load increments at equal time intervals. The curve in this figure shows the relationship between creep and the total time of construction, T, during which N equal load increments have been applied. If the entire load were applied to the column at the age of seven days, it would have twice the creep ($\alpha_{ave} = 1.4$) than in incremental loading over a period of 262 days. It is evident from Fig. 6-44 that columns loaded in many increments during a long period of time have much smaller creep than columns loaded during a short period of time with only a few increments.

Both creep and shrinkage have a similarity regarding the rate of progress with respect to time. Figure 6-45 shows an average curve for the ratio of creep or shrinkage at any time to the final value at time t_∞.

It can be seen from Fig. 6-45 that at 28 days about 40% of the inelastic strains have taken place. After three and six months, 60 and 70%, respectively, of all the creep and shrinkage have taken place.

This curve can be used to extrapolate the ultimate creep and shrinkage values from laboratory testing of a certain duration of time. For example, if a specimen was measured to have a shrinkage strain of ϵ in./in. at 90 days, we can estimate the ultimate shrinkage to be:

$$\epsilon_s = \frac{\epsilon}{0.60}$$

Conversely, the curve can also be used to estimate the creep or shrinkage at any time from the given ultimate value.

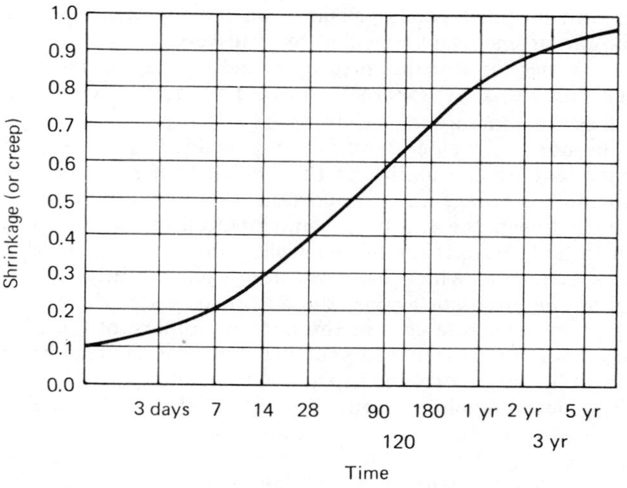

Fig. 6-45 Shrinkage or creep vs time. (From: Fintel and Khan, "Effects of Column Creep and Shrinkage in Tall Structures—Prediction of Inelastic Column Shortenings," *ACI Journal Proceedings*, 66 (12), Dec. 1968.)

For a further discussion of the structural implications of drying shrinkage and creep, see Chapter 10, "Multistory Structures."

6.8 BOND OF CONCRETE TO STEEL

Bond can be thought of as the shearing stress or force between a bar and the surrounding concrete. The force in the bar is transmitted to the concrete by bond, or vice versa. Bond is made up of three components: (a) chemical adhesion, (b) friction, and (c) mechanical interaction between concrete and steel. Bond of plain bars depends primarily on the first two elements, although there is some mechanical interlocking due to the roughness of the bar surface. Deformed bars, however, depend primarily on mechanical interlocking for superior bond properties. This does not mean friction and chemical adhesion are negligible for deformed bars, but that they are secondary.

With the use of deformed bars, bond failures could result from concrete crushing at the bearing face of the deformations; by shearing of the concrete around the outer extremities of the bar; by longitudinal splitting of the concrete cover in the vicinity of the bar; or by a combination of these three failure modes.

The bond of deformed bars is developed mainly by the bearing pressure of the bar ribs against the concrete. Bars having ribs with steep face angles (larger than about 40° with the bar axis) slip mainly by compressing the concrete in front of the bar rib. The concrete is crushed, and a concrete wedge forms in front of the bar rib. Bars with flat ribs (less than 30°), however, slip with the ribs sliding relative to the concrete.

With little confinement, large deformed bars may fail in bond by splitting of the concrete along the plane of the bar. This type of failure depends primarily on the load on the concrete and not much on the bar stress and the bar diameter or the bar perimeter. As the confinement around a bar improves, usually by the use of external concrete or transverse reinforcement, the ultimate load per unit length depends increasingly on the bar diameter. Small bars, top cast bars, or bars that are confined to the extent that bond failure generally occurs by shear failure of the concrete keys instead of splitting, will carry a maximum unit load proportional to the bar perimeter (hence to the bar diameter).

Based on the above reasoning, the maximum bond force per unit length depends primarily on the bar diameter, the amount of enclosure provided by transverse reinforcement (hoops, ties, or stirrups) placed around the bar, or confinement provided by external concrete (concrete cover), as well as the concrete tensile strength ($\sqrt{f_c'}$).

The bond of welded wire fabric embedded in concrete is dependent on the ability of the welded transverse wire to provide anchorage. The important variables are the size ratio between the transverse and longitudinal wire of the fabric and the quality of the weld connecting them. For deformed wire, which has been developed recently, crushing of the concrete against the deformations or the shearing of the concrete core at the outer periphery of the wire may be more critical than splitting. The bearing area also is important for controlling slip at a given load.

The development of bond in prestressed concrete is also attributed to adhesion, friction, and mechanical resistance. Friction between concrete and tensioned tendons (or grout and tendons) is the principal factor responsible for the transfer of the prestress force from steel to the concrete, and the other two factors are of less importance in bond development.

When the external prestressing force in the tendon is released, the tendon diameter tends to increase (so-called Hoyer effect), thus producing high radial pressure against the concrete, which in turn produces high frictional resistance in the transfer zone. Prestress transfer bond is present from the ends of a prestressed member to the beginning of a region in which the steel tension is constant. The length over which this transfer is made is termed the prestress transfer length, and is a function of the perimeter configuration, area and surface condition of the steel, the stress in the steel, and the method used to transfer the steel force to the concrete. Tendons with a slightly rusted surface can have an appreciably shorter transfer length than clean tendons. However, the danger of careless rusting allowing localized pitting should be considered. Gentle release of the tendon will permit a shorter transfer length than abruptly cutting the tendons. There is relative movement of steel and concrete, and accordingly adhesion cannot contribute to prestress transfer. Mechanical resistance probably contributes little to prestress transfer in the case of individual smooth wires, but it might be argued that strand will offer some mechanical resistance because of the helical grooves in the stranded configuration, but this type of deformation may not be depended upon for any appreciable contribution to the development of bond. A significant innovation in prestressing strand is the development of deformed strand, which is desirable for reducing the transfer length in some structural elements.

An increase in wire tension due to flexure reduces the diameter, relieves the radial pressure, and reduces the bond near the ends of a beam. Bond failure in prestress concrete also may occur due to too close spacing of the tendons.

Bond to concrete may be prevented for some pretensioned reinforcement in the end regions. Bond may be prevented by various means; one method is the use of plastic tubing which is often referred to as "blanketing."

Metal reinforcement should be free from loose, thick rust, mill scale, mud, oil, grease, paint, and loose dried mortar at the time concrete is placed, as these materials adversely affect or reduce bond.

A normal amount of rust increases bond. Normal rough handling of bars generally removes most of the loose rust and mill scale. However, in some instances it may be necessary to rub with a coarsely woven sack or to use a wire brush. Metal reinforcement, except prestressing steel, with rust, mill scale, or a combination of both, can be considered as satisfactory, provided the minimum dimensions, including height of deformations, and weight of a hand wire brushed test specimen are not less than applicable ASTM specification requirements. Prestressing steel should be clean and free of excessive rust, oil, dirt, scale and pitting. A light oxide coating that can be removed by a soft dry cloth is permissible.

Slip caused by relative movement (in addition to that caused by crushing) also occurs when the frictional properties of the rib face are reduced by grease. The extent to which slip properties are affected by grease depends on the face angle; ribs with flatter face angles are more affected by poor frictional properties.

Bond is increased by a tight adherent cement paste or mortar coating on the bars. In some cases, bond of hot dip galvanized reinforcement may be reduced due to the attack of fresh concrete on the zinc and evolution of hydrogen with the resultant formation of gas pockets next to the bars. Suspensions of bentonite and water are commonly used for construction convenience to support side walls of foundation trenches. Reinforcement in cages immersed in the suspension and then surrounded by tremie placed concrete may suffer reduced bond strength .

For epoxy-coated reinforcing bars (ASTM A775), a coating thickness of 12 mils is a practical upper limit at which there is no appreciable loss of bond of the coated bars in concrete. It is recommended that galvanized and epoxy-coated reinforcement be used with caution in conditions where they are subjected to cyclic loads or minimum development lengths or anchorage, as limited testing in concrete members is available. Based on the limited tests on No. 6 and 11 bars it appears that epoxy coated reinforcing has less slip resistance than normal mill scale reinforcing. Bond strength based on critical slip for the mill scale bars averages 32% greater than for epoxy coated bars. For development lengths greater than 12 in., mill scale bar critical slip strength is 15% greater than for epoxy-coated bars. Mill scale bars have 17% greater pullout strength than epoxy-coated reinforcing bars. In order to provide comparable performance with mill scale bars, a basic development length modification factor of 1.15 is proposed for epoxy-coated bars. Under bond fatigue loading in a working stress range, the slip behavior of the mill scale, epoxy-coated, and blast cleaned bars is essentially similar. The bond strength of epoxy-coated bars in comparison to mill scale bars is not adversely affected by up to 1.4 million cycles of loading in a working stress range. The relative difference based upon both flexural bond pullout strength and slip criteria is actually slightly less after the cyclic loading.

In freshly placed concrete, bleeding or water gain results in formation of a water (or water and entrapped air) space beneath the surfaces of solids, including the undersides of reinforcing bars. If reinforcing bars are rigidly positioned or restrained from settling with the concrete, an additional break of bond occurs at the underside of the reinforcement due to settlement shrinkage. This loss of bond may be expected to be about proportional to the depth of fresh concrete beneath the bars; the bond capacity of top bars is less than that of bottom bars. Settlement of concrete in the form results in better concrete consolidation on top of the lugs of a vertical bar than beneath the lugs. The early slippage and ultimate bond strength are thus more favorable when the bar is pulled against the direction in which the concrete settled.

During the earliest stages of hardening, after the concrete loses its plasticity, bond may be impaired if projecting reinforcement is subjected to impact or rough handling. Exposed portions of bars that are only partly embedded should not be struck or carelessly handled, and workmen should not be permitted to climb on bar extensions until the concrete is at least seven days old. Forms to which embedded parts are fastened or through which they protrude should not be stripped until the concrete has hardened sufficiently to avoid injury to bond.

Bond between concrete and deformed steel reinforcement is principally a function of the compressive strength of concrete or of its splitting tensile strength. Bond strength tests of some lightweight concretes yield bond strength values ranging from equal to 20% less than thos of normal-weight concrete of equal compressive strength. ACI 318, *Building Code Requirements for Reinforced Concrete*, provides that the basic development length of all-lightweight concrete be multiplied by 1.33 and when sand-lightweight concrete is used by 1.18. Results from a limited number of tests indicate that the bond strength of pretensioned strand in lightweight aggregate concrete is not different from that in normal-weight concrete.

The age of the concrete at the time of test affects the bond strength result. Concretes that reach the same compressive strength at different ages will indicate slightly different bond values with the same bar. Tests at early ages should give bond strength as a higher proportion of the concrete compressive strength. Concrete strength (up to 5500 psi) at transfer of prestress has little influence on the transfer length of clean seven wire strands up to and including 1/2 in. diameter. The average increase in transfer length over a period of one year following prestress transfer is quite low for all strand sizes, with the increase in transfer length with time independent of the concrete strength at the time of transfer.

The concrete mix and slump, rate of casting, and amount of vibration employed are each important factors affecting bond. Mix ingredients and proportions that result in low bleeding of concrete, such as the use of superplasticizers, are beneficial. Air entrainment improves bond of concretes that tend to bleed badly. For nominal amounts of entrained air (4 or 5% upper limit) no major effect on bond results, as compared with non-air-entrained concrete. With higher percentages of air, the bond strength of horizontal bars drops off rapidly as concrete strength decreases, and air bubbles may have the tendency to collect beneath the bars in exactly the same manner as does excess water. Low-slump concrete gives higher bond values than high-slump concrete even when the mixes are adjusted to give the same compressive strength. Proper vibration of concrete markedly increases bond. However, overvibration can increase water gain and reduce bond. Shrinkage of concrete may destroy the bond between the concrete and reinforcing steel and cause a high loss in bond strength at low values of slip. However, the ultimate bond strength is little affected when deformed steel is used, since with higher values of slip, the concrete is brought to bear against the protrusions of the reinforcing steel.

6.9 FATIGUE OF CONCRETE

Like other construction materials, concrete is subject to the effects of fatigue. Concrete, when subjected to repeated loads, may exhibit excessive cracking and may eventually fail after a sufficient number of load repetitions, even if the maximum stress is less than the static strength of a similar specimen. The fatigue strength of concrete is defined as a fraction of static ultimate strength that it can support repeatedly for a given number of cycles.

Fatigue is a process of progressive, permanent internal structural (microcracking) change in a material subjected to fluctuating stresses or strains. The internal changes may be damaging and result in progressive growth of cracks and complete fracture if stresses and fluctuations are sufficiently large. Fatigue fracture of concrete is characterized by a considerably larger microcracking and strains as compared to fracture of concrete under static loading.*

Although the design of flexural members of concrete is based almost entirely on data from static tests, many concrete structures, both plain and reinforced, are subjected to fatigue loadings. Included in this category are pavements, bridges, docks, crane girders, and other structures supporting oscillating machinery. However, the only known cases of fatigue failures have been in pavements and railway ties.

Fatigue data is usually shown by stress–fatigue life curves, known as S–N curves, Fig. 6-46. The repeated stress, S, expressed as a ratio or percentage of the static ultimate strength, is the ordinate, while the abscissa gives the life or number of cycles until failure on a logarithmic scale. The following may be observed from this figure: (1) the fatigue strength of concrete decreases with an increasing number of cycles; (2) a decrease of the range between maximum and minimum load results in increased fatigue strength for a

*Much of this discussion on fatigue is based on "Fatigue and Fracture of Concrete," Stanton Walker Lecture by C. E. Kesler, Nov. 18, 1970.

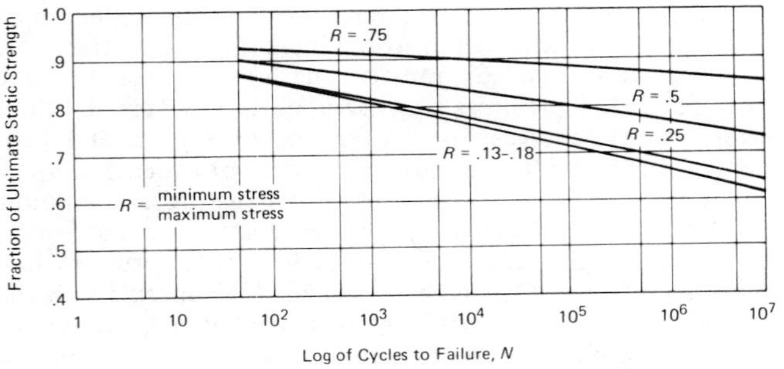

Fig. 6-46 *S–N* curves for plain concrete beams showing various ratios of minimum to maximum stress in the loading cycles. (From: Murdock, J. W., and Kesler, C. E., "Effect of Range of Stress on Fatigue Strength of Plain Concrete Beams," *ACI Journal*, 30 (2), Aug. 1958. Also, *Proceedings* 55, pp. 221–231.)

given number of cycles; and (3) plain concrete has no endurance limit up to 10 million cycles. (Endurance limit is the maximum fraction of ultimate static strength at which concrete exhibits essentially elastic behavior, thereby permitting unlimited stress repetitions without loss in fatigue resistance.)

Fatigue strengths of paste, mortar, and concrete are about the same when expressed as a fraction of their static ultimate strength. Many variables such as cement content, water–cement ratio, curing, entrained air, and aggregate that affect the static strength influence fatigue strength in a similar proportionate manner. Fatigue strength of concrete for a life of ten million cycles—for compression, tension, or flexure—is approximately 55% of static ultimate strength. This result is valid when the loads vary from near zero to some predetermined maximum and not when the minimum load is a significant percentage of the maximum load.

The results of fatigue tests usually exhibit substantially larger scatter than static tests. This inherent statistical nature of fatigue test results can best be accounted for by applying probabilistic procedures: for a given maximum load, minimum load, and number of cycles, various probabilities of failure can be calculated from the test results. By repeating this for several cycles, a relationship between probability of failure and number of cycles until failure at a given level of maximum load can be obtained. From such re-

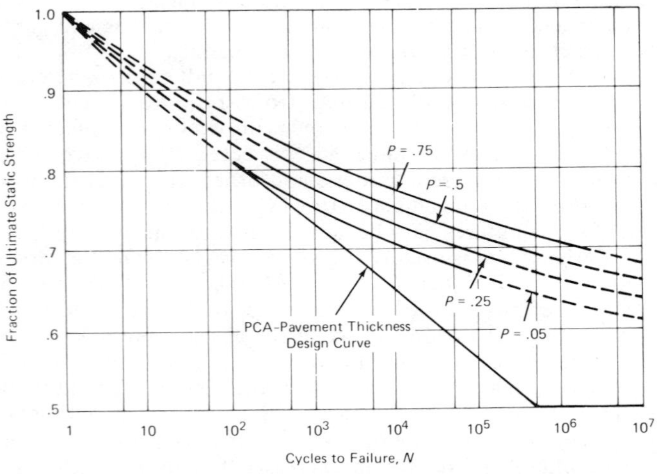

Fig. 6-47 *S-N* curve for constant probabilities of failure adjusted for use with Miner hypothesis. (Adapted from: Hilsdorf, H. K., and Kesler, C. E., "Fatigue Strength of Concrete Under Varying Flexural Stresses," *ACI Journal Proceedings*, 63, 1059–1076, 1966.)

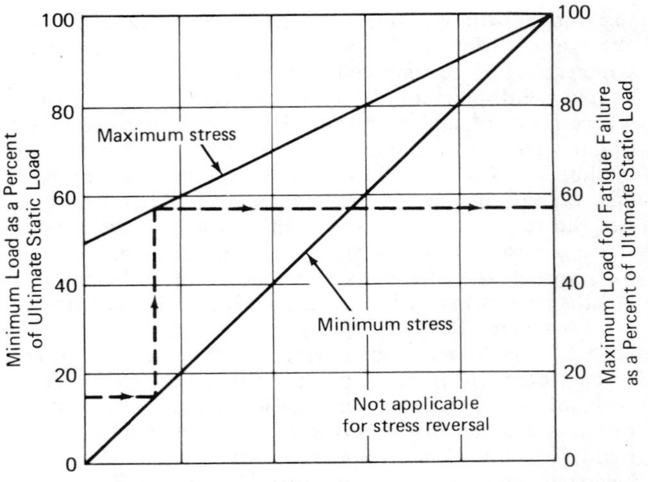

Fig. 6-48 Modified Goodman diagram showing fatigue strength of plain concrete in tension, compression, or flexure under 10 million cycles of repeated loading.

lationships, *S–N* curves for various probabilities of failure can be plotted, Fig. 6-47.

The usual fatigue curve is that shown for a probability of failure of 0.5. Design should probably be based on a lower probability of failure.

Design for fatigue is facilitated by use of a modified Goodman diagram as shown in Fig. 6-48. This diagram is based on the observation that the fatigue strength of plain concrete is essentially the same whether the mode of loading is tension, compression, or flexure, as long as no reversal of stress is involved. The diagram also incorporates the influence of range of loading. For a zero minimum stress level, the maximum stress level the concrete can support for ten million cycles without failure is taken conservatively as 50% of the static strength. As the minimum stress level is increased, the stress range that the concrete can support decreases.

From Fig. 6-48, one can determine the maximum stress in tension, compression, or flexure that concrete can withstand for 10 million repetitions and for a given minimum stress. For example, consider a structural element to be designed for 10 million repetitions. If the minimum stress is 15% of the static ultimate strength, then the maximum load that will cause fatigue failure is about 57% of static ultimate load. This value refers to 50% probability of failure.

Varying stress rates during fatigue tests change the fatigue response from that found when stress rates are constant. The fatigue strength and life of concrete subjected to repeated loads of varying magnitude are influenced by the sequence in which these loads are applied. Consider a test in which the maximum stress level is changed only once. Then the fatigue life of a specimen is larger if the higher stress level has been applied first compared to the fatigue life of a specimen in which the lower stress level was applied first. A relatively low number of cycles of high loads can in fact increase the fatigue strength of concrete under a lower load beyond the fatigue strength of concrete that has not been previously loaded. When the load is repeatedly varied, the fatigue strength decreases with increases in the ratio of the number of cycles at high stress level to those at low stress level. Data are not available showing the effect of randomly varying loads on fatigue behavior of concrete.

Accumulation of the effects of repeated loads of varying

stress can be made on the basis of the Miner hypothesis with sufficient conservatism to incorporate a very low probability of failure. The Miner hypothesis is that fatigue resistance not consumed by repetitions of one load is available for repetitions of other loads. Theoretically, the total fatigue used should not exceed 100%. Design curves, Fig. 6-47, incorporating the probability of failure have been developed so that when the Miner hypothesis is applied to them reasonable results are obtained. For example, the curve most applicable to pavement design has a constant probability of 0.05, which means that for 100 fatigue tests not more than five would show fatigue strengths below the curve. However, because the research to develop this curve was too limited to justify general use of the 0.05 curve for design, the PCA curve has been adopted for pavement thickness design.

Repetitions of loads with stress ratios below the fatigue strength (understress) increase concrete's ability to carry loads with stress ratios above the fatigue strength; that is, decrease in pavement deflection with increase in load repetitions. This understress may be a sustained load or rest period (up to five minutes) between repeated load cycles. No additional benefits are derived when rest periods extend beyond five minutes. Also, if the maximum stress at sustained load is above 80% of the static strength, then sustained loading may have detrimental effects on fatigue life.

The frequency of loading, between 70 and 900 cycles per minute, has no significant effect on fatigue strength, provided the maximum stress is less than 75% of the static strength. However, frequency as low as 10 cycles per minute may result in slightly lower fatigue lives because of increased significance of creep effects.

Stress gradient influences the fatigue strength of concrete in a manner similar to the influence of a gradient on static strength. Fatigue strength of eccentrically loaded specimens is 15 to 18% higher than concentrically loaded, otherwise similar specimens. For the purpose of design of flexural members limited by concrete fatigue in compression, it may be safe to assume that fatigue strength of concrete with stress gradient is the same as that of uniformly stressed specimens.

Few data are available on the fatigue strength of concrete for multiaxial stress states. It is known that saturated specimens subjected to biaxial stresses show ratios of fatigue to uniaxial strength higher than partially saturated specimens. Further, that difference increases as the ratio of the principal stresses approaches unity. There is no improvement in the fatigue capacity for lateral confining stress of 0.2 to 0.3 f'_c if the stress range approaches 0.7 f'_c. However, as the stress range decreases below that value, confining pressures become beneficial and at 0.6 f'_c approach values proportionate to those expected based on the increased axial strength caused by confinement.

While bond fatigue has generally not been a problem in reinforced concrete structures, it has not been thoroughly explored in research, and the limitations it could present are not known. Bond fatigue is the progressive deterioration of bond and the slip of tensile reinforcement under some form of repetitive loading. Such slip could possibly have the same effect as reinforcement with a reduced elastic modulus and could lead to premature concrete crushing and collapse. Based on the local conditions near the steel–concrete interface, it has been hypothesized that bond fatigue is related directly to the dissipation of energy in slippage.

Special fatigue tests are essential to evaluate the strength of structural elements, such as railroad crossties, where bond is critical (available transfer length of pretensioned tendons is short) and repeated loading is the main design consideration.

6.10 PERMEABILITY

Movement of water or air through concrete can be produced by various combinations of air or water pressure differentials, humidity differentials, and solutions of different concentrations (osmotic effects) or by temperature differentials. Various tests have been devised to determine permeability. Although these procedures may reveal the relative characteristics of the concretes involved, the quantitative value obtained may depend considerably upon details of the experimental conditions. This means that the conditions anticipated in service should be used for the test conditions.

6.10.1 Water Permeability

The water premeability of high-quality concrete depends mostly on the respective permeabilities of the cement paste and aggregate. Since the aggregate particles are surrounded by hardened cement paste, the permeability of concrete to water under hydrostatic pressure is principally a function of the permeability of the cement paste component of the concrete, provided the concrete is intact—not previously damaged by frost or rapid drying and not containing excessive under-aggregate fissures or honeycomb. All of the permeating water must pass through the paste component of the concrete (the continuous phase) regardless of the relative porosity or permeability of the aggregate. If the paste is of low permeability, the concrete will show similar characteristics.

The permeability to water of hardened cement paste is dependent primarily on the capillary porosity of the paste. The capillary porosity is a function of the original water–cement ratio and the length of the curing period (extent of hydration).

The water permeability of a well-cured paste is reduced approximately a thousandfold by reduction in water–cement ratios from 0.8 to 0.4 by weight. The change in permeability of a given paste with length of curing time also is enormous.

Introduction of aggregate particles into cement paste should tend to reduce the permeability by reducing the number of channels per unit cross section and by lengthening the path of flow per unit linear distance in the general direction of flow. However, during the plastic period the paste settles more than the aggregate, and thus fissures under the aggregate particles develop. In saturated concrete, these void spaces are paths of low resistance to hydraulic flow and thus increase the permeability of concrete. In general, with paste of a given composition, permeability of concrete is greater the larger the maximum size of the aggregate. In addition, the permeability of hardened concrete can never be as low as that of hardened cement paste because of the imperfect bond between aggregate and paste.

Significant quantities of air voids in the concrete should increase the permeability of saturated concrete roughly in proportion to their quantity, provided other factors remain constant. However, other factors seldom remain constant—it is commonly observed that air entrainment in most concretes will increase workability and reduce segregation and bleeding and permit reductions in the water–cement ratio—with the result that the concrete may actually be more impermeable despite the presence of the air voids. Therefore, all concrete that must be watertight should be air-entrained.

Figure 6-49 shows the effect of duration of moist curing on relative permeabilities of different concrete mixtures and something of the effects of cement content. Effects of

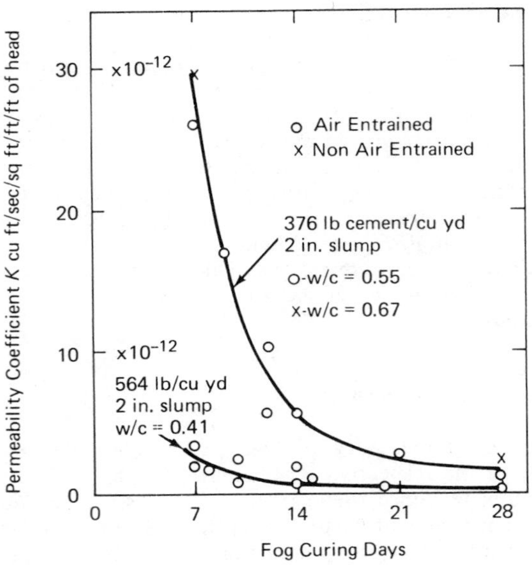

Fig. 6-49 Permeability vs curing time. (From: Tyler, I. L., and Erlin, B., "A Proposed Simple Test Method for Determining the Permeability of Concrete," *PCA Research Bulletin 133*, 1961.)

air entrainment do not seem to be large with cement contents held constant.

Test results obtained by subjecting non-air-entrained mortar discs to 20 psi water pressure are shown in Fig. 6-50. Mortar discs moist-cured for 7 days had no leakage when made with a water–cement ratio of 0.50. Leakage occurred in mortars made with higher water–cement ratios. In discs with a water–cement ratio of 0.80 the mortar still had leakage after curing for a month. For each water–cement

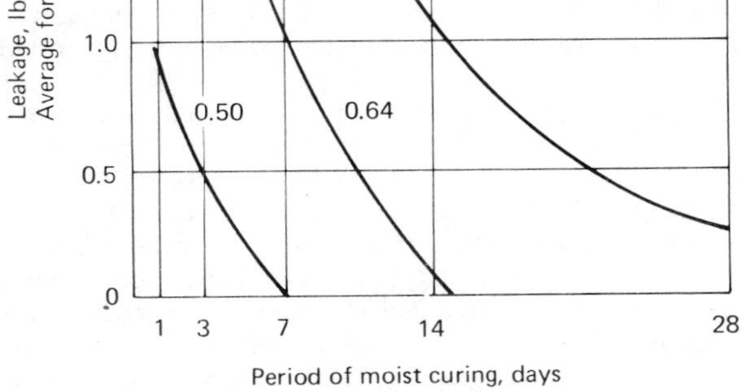

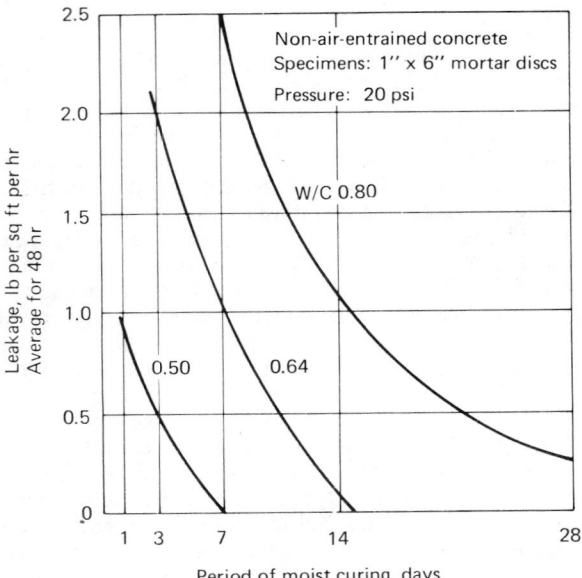

Fig. 6-50 Effect of water-cement ratio and curing on watertightness. Note that leakage is reduced as the water-cement ratio is decreased and the curing period increased. (From: *Design and Control of Concrete Mixtures*, 12th ed., Portland Cement Association, Skokie, Ill., 1979.)

ratio, leakage became less as the curing period was lengthened. The permeability of steam-cured concrete is generally higher than that of wet-cured concrete. Supplemental moist curing may be required to achieve an acceptably low permeability.

When made with normal-weight aggregate, concrete that is intended to be watertight should have a maximum water–cement ratio or water–cement plus pozzolan ratio of 0.48 for exposure to fresh water and 0.44 for exposure to seawater. In addition it should have a specified 28-day compressive strength of 3500 psi where concrete is not exposed to severe and frequent freezing and thawing, and 4000 psi where concrete is exposed to severe and frequent freezing and thawing, except where special structural or other considerations require concrete of greater strength. With lightweight aggregate, the specified compressive strength, f'_c, should be at least 3750 psi for exposure to fresh water and 4000 psi for exposure to seawater.

Concrete to be watertight needs adequate cement paste. Therefore it should have a minimum cement content as follows:

Coarse aggregate No.	*lb/yd^3*
467 (1 1/2 in. maximum)	517
57 (1 in. maximum) or 67 (3/4 in. maximum)	564

Admixtures known as permeability-reducing agents are usually either water-repellents or pozzolanic materials. They may reduce the permeability of concretes that have low cement contents and/or a deficiency of fines in the aggregate. However, their use in well-proportioned mixes with the above cement contents may increase the mixing water required, resulting in increased rather than reduced permeability. However, in lean mass concrete, age becomes a more important factor in permeability than cement content because of slow pozzolanic reaction.

Dampproofing admixtures, usually water-repellent materials, are sometimes used to reduce the capillary flow of moisture through concrete that is in contact with water or damp earth. Many so-called dampproofers are not effective, especially when used in concretes that are in contact with water under pressure.

Watertight concrete structures are not difficult to construct. Where there have been faulty structures, careful examination invariably has shown that leakage is confined to relatively small areas, and the rest of the structure is sound.

Principal causes of leaks are improper or careless construction methods rather than improper design or poor materials. Of the various imperfections caused by defective construction methods, segregation of ingredients stands out as the prime reason for nonuniformity and local porousness of concrete. Segregation can be minimized by proper handling, placing, and consolidation procedures. To be watertight concrete must also be free from honeycomb and cracks (or the crack width minimized). To aid in preventing honeycomb, concrete should have a minimum slump of 1 in.; with a 3 in. maximum for footings, caissons, substructure walls; and a 4 in. maximum for slabs, beams, reinforced walls, and columns.

In some cases, inadequate foundations have permitted unequal settlement with subsequent cracking and leakage.

Careful surveys of ground conditions will indicate proper procedures for preparing subgrades and foundation.

It is believed that the principal mode of transport of water through quality concrete is not ordinary laminar flow but a mode that might be called surface diffusion. An estimate of the amount of water passing through a concrete section can be obtained from Darcy's law for viscous flow (although viscosity cannot be treated as a constant but as a function of the mean size of the pores in the paste):

$$Q = KiA$$

where:

Q = total volume of fluid, percolating in unit time (cm^3)
K = permeability coefficient (Darcy or cm/sec)
i = hydraulic gradient or difference of energy heads in the fluid at any two points along a flow distance. It represents the energy lost through viscous friction as the fluid flows around the particles and through irregular void passages. Velocity head is neglected in this gradient. i can be regarded as the driving force, especially since phenomena like capillarity are not considered.
A = cross-sectional area of the element subject to percolation (cm^2)

or the equation:

$$Q = KA \frac{H}{L} T$$

where:

$\frac{H}{L}$ = hydraulic gradient
H = hydraulic head (cm)
L = thickness of concrete (cm)
T = time (sec)

TABLE 6-10 Permeability of Mass Concrete*

Dam Structure	*Permeability Kq***
Hoover	0.62×10^{-4}
Grand Coulee	–
Angostura	–
Kortes	–
Hungry Horse	1.85×10^{-4}
Canyon Ferry	1.93×10^{-4}
Monticello	8.20×10^{-4}
Anchor	45.2×10^{-4}
Glen Canyon	1.81×10^{-4}
Flaming Gorge	11.09×10^{-4}
Yellow Tail	1.97×10^{-4}†

**From: "Mass Concrete for Dams and Other Massive Structures" ACI Committee 207, ACI Journal, April 1970.*
***18 × 18-in. specimen, standard correction to age of 60 days. Kq is in cu ft/sq ft/yr/ft (head); it is a relative measure of the flow of water through concrete.*
†Preliminary mix investigations.

As can be expected for an empirical formula, the units present a confused picture and care must be taken in determining Q. Within the ranges of concretes normally used in practice, the coefficient of permeability, K, ranges from 10^{-12} to 10^{-10} cm/sec (the unit rate of discharge at unit hydraulic gradient with the dimensions of a velocity at a temperature of 70°F). Permeability values for mass concrete (dams) are shown in Table 6-10. Additional moist curing takes place as water penetrates, and this could conservatively reduce fluid outflow volume about ¼ or ⅕ in a year's time.

6.10.2 Water Vapor Permeability

Water vapor in air is a gas that exerts its own vapor pressure and that can diffuse through concrete, independently of the air with which it is mixed, until the partial pressures of the water vapor are equalized. However, when the air is moved suddenly or is heated or cooled, the water vapor present is similarly affected so that it is usually necessary to consider it as a part of an air–vapor mixture. Differences in total pressure of the air may result in a transfer of vapor with air, augmenting, and at times overriding, the effects of the flow produced by vapor-pressure gradients alone. This can be particularly important in the transfer of vapor through cracks and pinholes or through air-permeable building constructions. It will seldom be important in constructions without air spaces and having parged or plastered surfaces. It may, however, be an important means of vapor transfer through constructions lacking in air tightness.

Water vapor transmission through materials under the influence of a water vapor pressure gradient is called vapor diffusion. The design equation commonly used in calculating water-vapor transmission through materials is based on a form of Fick's law, and is as follows:

$$W = \bar{\mu} A \theta \frac{\Delta p}{l}$$

where:

W = total weight of vapor transmitted, grains
$\bar{\mu}$ = average permeability, grains per (hour) (square foot) (in. of mercury vapor pressure difference per in. of thickness)
A = area of cross section of the flow path, ft^2
θ = time during which the transmission occurred, hours
Δp = difference of vapor pressure between ends of the flow path, in. of mercury (at 73.4°F; 100–0 percent RH, $\Delta p = 0.82948$; for 100–20 percent RH, $\Delta p = 0.8(0.82948) = 0.66358$; for 100–50 percent RH, $\Delta p = 0.5(0.82948) = 0.41474$)
l = length of flow path (or thickness of specimen), in.

The actual transmission of vapor through a material is extremely complex, so that the coefficient, $\bar{\mu}$, is not a simple one but is actually a function of relative humidity and temperature, and may vary along the flow path through the concrete.

Whenever it is convenient to deal with a concrete of a thickness other than the unit thickness to which $\bar{\mu}$ refers, use may be made of the permeance coefficient M, where $M = \bar{\mu}/l$. The designation perm for the unit of permeance is now widely used and is a convenient substitute for the unit,

1 grain per (hour) (square foot) (in. of mercury vapor pressure difference). The corresponding unit of permeability is perm-inch, since it is the permeance of unit thickness.

The resistance of water vapor flow through concrete is the reciprocal of its permeance and is equal to the thickness divided by its permeability. The overall resistance of a section to water vapor flow is the sum of the resistances of the various materials in series comprising the section. The overall permeance of a section is equal to the reciprocal of the overall resistance.

The simplest method of finding the permeance of a specimen is to seal it over the top of a cup containing desiccant or water, place it in a controlled atmosphere, and weigh it periodically. The steady rate of weight gain or loss is normally the water vapor transfer. When the cup contains a desiccant, the procedure is called the dry-cup method, and when the cup contains water, the wet-cup method. Usually the surrounding atmosphere is held at 50% relative humidity, thus providing, in either method, substantially the same difference of vapor pressure, but the results obtained by the two methods for the same specimens are likely to be much different, the wet-cup method producing the higher values. Because concrete is sorptive, the rate of weight loss is high during the early period of the wet-cup test and decreases asymptotically as equilibrium conditions are obtained. The interval needed to establish diffusion equilibrium varies not only with the type of concrete but also with vapor pressure and specimen thickness. In the dry-cup test, virtually all transmission takes place by diffusion, and capillary flow is absent.

Typical permeability values for concrete are as follows:

*Concrete**	*Thickness in.*	*Permeance, M*	*Resistance 1/M*
Normal Weight Aggregate–cast as a plastic, workable mix:	Varies		
w/c = 0.40 by wt.	2 to 6	0.5	2.0
w/c = 0.50 "		0.8	1.25
w/c = 0.60 "		1.1	0.91
w/c = 0.70 "		1.3	0.77

**Assumes conventional curing and six months age. Concretes of lightweight aggregates having similar compressive strength may be regarded as having similar water vapor properties.*

Forces other than water vapor pressure, such as hydraulic pressure, absorption, adsorption, hygroscopicity, and capillarity, affect water vapor flow and cause moisture migration. These various forces are interrelated, but the actual relationship is very complex, and has not been expressed in a simple way for design use. Evidence to date indicates that in porous materials partially saturated with water there is also likely to be a migration of moisture to the cold side under the influence of the temperature gradient. This can occur by a process of evaporation, vapor flow, and condensation within the material.

The vapor flow calculations can be considered reasonably accurate when the primary mechanism for moisture migration is vapor diffusion and when coefficients are available that define the rate of vapor flow for the material under the conditions of use. When, however, the material is capable of holding substantial quantities of adsorbed water, as in concrete, the diffusion approach may be inadequate or even inappropriate, depending on the situation. Moisture is apparently transferred through partially dry concrete in the absorbed or condensed state by surface diffusion and does not move as vapor through concrete. Also, where good practice has been followed in the design and curing of concrete, it is believed that moisture does not move through the concrete by capillarity as it is usually understood. For example, the transfer process is considered to be the same where either liquid water or water vapor is present directly beneath a slab-on-ground. However, other conditions being similar, there is believed to be a much slower transfer to the absorbed state when vapor rather than liquid water is in contact with the slab. This results in a slower rate of moisture transfer through the slab where only vapor is present.

Owing to the hygroscopic nature of concrete, it is believed incorrect to consider the rate of moisture transfer proportional to the vapor pressure differential between a moist and a dry side of a slab. For water moving through concrete in a condensed state, the driving force is not computed from the difference between the two pressures but from the ratio between the two relative humidities.

It is recommended that the water vapor permeability of concrete not be expressed in terms of permeance. Computation of permeance is based on the assumption that moisture travels through partially dry concrete as vapor, whereas, under the conditions of concrete tests, it is transferred in the adsorbed state by surface diffusion. Motive force is proportional to the logarithm of the ratio between the vapor pressure on the moist side and that on the dry side; it is not equal to the difference between the two vapor pressures as assumed in the ASTM definition.

In a number of tests, vapor flow through concrete was measured as water vapor transmission (WVT), WVT being defined as the rate of migration of water through concrete, ($WVT = Wl/A\theta$) using a wet cup. The units of WVT are usually grains per in.^{2} per day or grains per ft^2 per hr.

Water vapor transmission of concrete decreases with increased age of the concrete and approaches an asymptotic value, Fig. 6-51. Also, shown in these figures is that WVT decreases with an increase in the strength (decrease in water-cement ratio) of the concrete. Admixtures have no appreciable effect on moisture migration when compared with plain concretes of the same w/c. This conclusion applies to concrete at all ages.

Moisture migration through good-quality concrete is less than 0.3 grains/hr/ft^2; but the flow through low-quality concrete often exceeds 2.0 grains. For slabs-on-ground, a gravel capillary break between the soil and the slab reduces WVT. No significant correlation exists between types of soil beds and moisture migration rates through slabs-on-ground. However, an increase in the slab thickness up to about 6 in. will decrease WVT.

Structural lightweight concretes show considerably less WVT than most sand-gravel concretes, possibly because of greater sorptive properties of the aggregate, Fig. 6-52. Slabs cast over soil show reduced inflow with increasing cement content. This relationship is not well defined for companion slabs cast over soil with a gravel capillary break, as all of the curves fall into a relatively narrow band. Comparison of the curves of Fig. 6-52 reveals the advantage of the gravel capillary break, particularly for slabs with lower cement content.

The outflow or rate of moisture loss from the surface of the same concretes is much higher at early ages—approximately four times the above values at 28 days, and ten times at seven days. This outflow at early ages is the flow that contributes to the problems encountered upon instal-

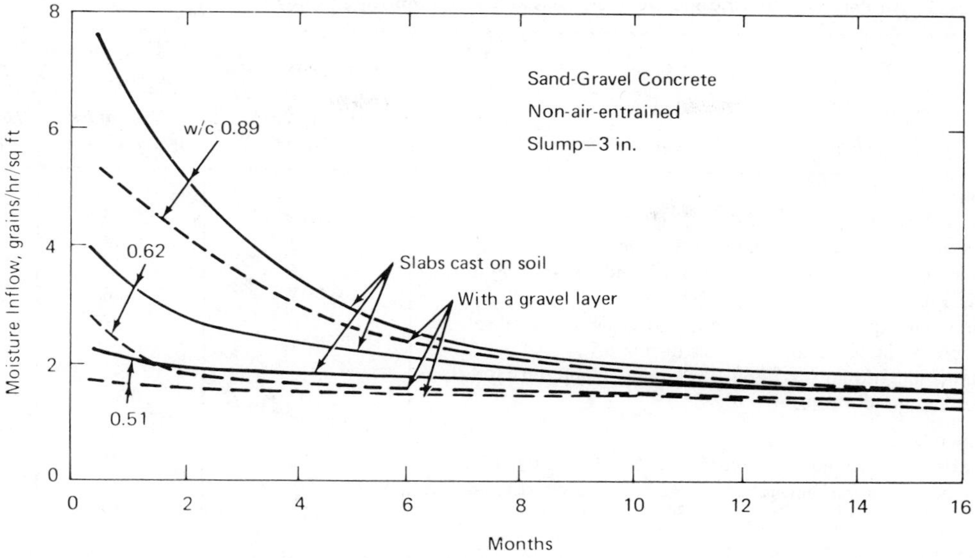

Fig. 6-51 Effect of water-cement ratio and gravel capillary layer. (From Brewer, H. W., "Moisture Migration—Concrete Slab-on-Ground Construction," *PCA Development Bulletin 89*, 1965.)

laton of floor coverings. For water-vapor exposure the loss from the surface is due almost entirely to loss of original mixing water, with very little moisture passing through the slab. After extended exposure periods, exceeding one year under the constant temperature and humidity conditions, inflow and outflow will become equal; at that time stabilized moisture migration through concrete is attained.

Concretes made with poor-quality aggregate show higher *WVT* rates than concretes made with high-quality aggregates, other factors being equal. *WVT* decreases with increases in maximum aggregate size.

Present knowledge and understanding regarding the transmission of water vapor through concrete is still rather incomplete. Although the values listed for concrete derive from standard test conditions, they may not adequately describe behavior under other conditions or test methods. For example, vapor permeability of concrete decreases as the mean relative humidity increases. The reason for this is that an increase in relative humidity decreases the air-filled pore space available for diffusion.

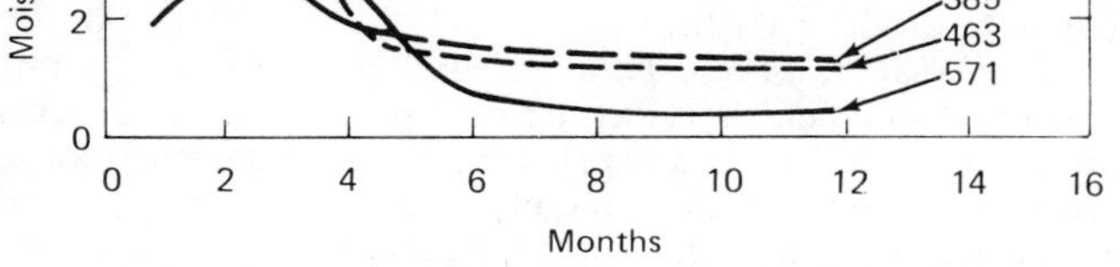

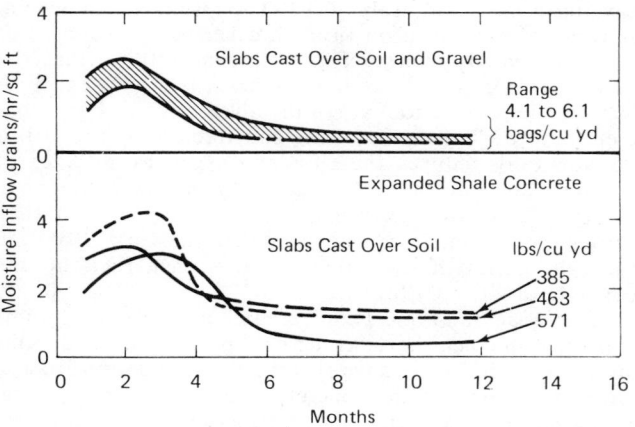

Fig. 6-52 Effect of cement content and gravel capillary layer for lightweight concrete. (From: Brewer, H. W., "Moisture Migration—Concrete Slab-on-Ground Construction," *PCA Development Bulletin 89*, 1965.)

6.10.3 Air Permeability

Air leakage is the uncontrolled movement of air through walls and roofs both into a structure (infiltration) and out of it (exfiltration) and the interchange of air within spaces in the building envelope. Air movement occurs as a result of air pressure differences produced by wind, chimney effects (when the temperature in a building differs from that outside, pressure differences occur between inside and outside as a result of the difference in density of the air), and the operation of mechanical ventilation systems.

The rate of flow of air through concrete depends on the thickness of the concrete, moisture content, and the pressure applied. There is no known relation between the air and water permeabilities of all concretes. The rate of flow of air through concrete having a low rate of flow may attain equilibrium within an hour, but, if the rate is high, it may continue to increase for many hours. The rate appears to be inversely proportional to the thickness of the slab and directly proportional to the pressure.

Concrete Thickness, in.*	4	7	8	9
Rate of flow, in.^{3}/psi-hr-ft^2	0.58	0.37	0.30	0.24

**Water-cement ratio = 0.62.*

Based on tests with pressures up to about 30 in. watergauge and for thickness from 4 in. to 9 in., the rate of seepage in cubic inches per square foot per hour is equal to 12½ times the ratio of the pressure (psi) to the thickness (in.).

A dense concrete having a minimum cement content of 500 lb per cu yd, a very good aggregate grading, a fly ash or

TABLE 6-11 Air Permeability of Concrete Specimens at Various Moisture Contents

Specimen Test Condition†	*Pressure Difference Across Specimen During Test (mm Hg)*	*Permeability Coefficient (K)*††	
		Air and Water Vapour	*Dried Air*
Specimen A			
Oven-dried beforehand, for 1 week at 105 C.	709	0.23	0.18
57% saturated beforehand, exposed to water vapour during test.	700	0.04	0.04
Vacuum end exposed for 4 hours beforehand, and during test, to water vapour at 1 psia.	701	0.06	0.06
Vacuum end exposed to water vapour, overnight at atmospheric pressure beforehand, and at 1 psia during test.	707	0.11	0.11
Vacuum end exposed for 7 hours beforehand, and during test, to water vapour at 1 psia.	707	0.11	–
Specimen B			
Oven-dried beforehand, for 1 week at 105 C, 48 hours at 70 C.	712	0.21	0.16
Vacuum end exposed beforehand for 2 hours to water vapour at 1 psia.	712	0.14	0.12
Specimen C			
Oven-dried beforehand, for 72 hours at 105 C, 48 hours at 70 C.	711	0.15	0.10
Vacuum end exposed beforehand for 65 hours to water vapour at atmospheric pressure.	711	0.09	0.07

(From: Loughborough, M. T., "Permeability of Concrete to Air," Ontario Hydro Research Quarterly, *First Quarter, 1966)*

†Concrete with 4000 psi compressive strength at 28 days made with 3/4 in. max size aggregate, 500 lb. cement per cu. yd and a water-cement ratio of 0.50.

††Cu in./min/sq ft area/in. of concrete/psi pressure differential.

pozzolan (where necessary to add fines), air entrainment, and a low water–cement ratio, can produce concrete impermeable to air under high pressures. Slump does not appear to affect permeability.

Prolonged curing reduces the air permeability but drying at any age significantly increases permeability; see Specimen A, Table 6-11. The numerous pores in concrete are normally lined with water—relatively immobile water—adsorbed on the pore walls. Presumably the pore area available for the relatively free flow of fluid is greatly reduced by this immobile lining, particularly in the small pores, where it may constitute most or all of the cross-sectional area of the pore. However, when concrete is air-dried, the coarser capillary spaces are completely emptied of water, and the finer capillary spaces and gel pores partially emptied owing to decreased water adsorption on the pore walls. This results in a drastically increased transmission area available for the movement of air or other gas. Concretes that have been dried are very permeable to air, thousands of times more permeable than they are to water.

TABLE 6-12 Calculated Diffusion Rates of Methane Through Concrete Slab at 30 psi Pressure

*Dry Slab**		
Thickness of slab	*Permeation Flow** std cu ft/ft²/day*	*Diffusion Rate** std cu ft/ft²/day*
5ft.	0.638	0.0193
1ft.	3.189	0.0964

**Methane would not flow through water saturated concrete as the 30 psi (206 kPa) is insufficient to displace the water.*

***std cu ft/(ft²) (day)* $= 0.304 \frac{m^2}{m^2}$ *day* $= 350 \times 10^{-4}\ m^2/m^2\,s$

Ref: Chen, L. C., and Katz, D. L., "Diffusion of Methane through Concrete," ACI Journal, V. 75, No. 12, Dec. 1978, pp. 673–679.

Little information is available on permeability of concrete to various gases, but even if a given concrete is relatively permeable to air, it may be substantially impermeable to some other gases.

The diffusion rates for methane gas through concrete have been measured, Table 6-12. The results are in general agreement with diffusion through other porous solids. For water saturated concrete Fig. 6-53 demonstrates that the thickness of concrete cover only has a relatively small effect on the penetration rate of dissolved oxygen. For a concrete of $w/c = 0.4$ an increasing concrete cover from 10 to 70 mm only reduces the flux of oxygen by a factor of approx. 2.6. For the same water–cement ratio, the availability of dissolved oxygen is lower through mortar than through concrete. It can be seen that the flux of oxygen through mortar of $w/c = 0.60$ is lower than that through concrete of $w/c = 0.40$.

Air leakage of a typical concrete structure occurs at joints and through cracks in the concrete. Specially detailed joints are required in low-leakage structures to minimize leakage, allow movement of the concrete to accommodate thermal expansion and to permit maintenance. Cracks in concrete will have a leakage rate according to their number, spacing, width, and penetration into the concrete. Leakage of air through cracks that do not completely penetrate the concrete can be computed by assuming that the leakage is only

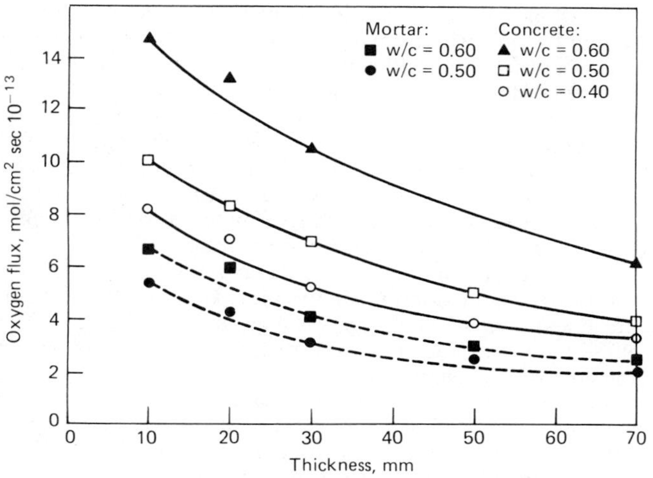

Fig. 6-53 Diffusion of dissolved oxygen through water saturated concrete. (From: Gjorv, O. E., Vannesland, Ø., and El Busaidy, A. H. S., "Diffusion of Dissolved Oxygen through Concrete," Paper 17, NACE Corrosion 76, March 1976, Houston, Texas, 13 pp.)

that which results from the thickness of the uncracked concrete.

When cracks penetrate completely through the concrete, the leakage rate through the cracks can be computed as follows:

$$q_T = 3.9 \times 10^4 \frac{b^3 LP}{x}$$

where:

q_T = volumetric leak rate (cfm)
b = crack width (in.)
L = crack length (in.)
x = crack depth (in.)
P = pressure differential (in. H_2O)

Mass flow of gases through joints or cracks is characterized by a flow proportional to the pressure differential or the square root of the pressure differential. Flow through a porous membrane such as concrete is more nearly represented by diffusion equations and is dependent on the partial pressure of the gas or vapor being considered on one side of the membrane compared to the other side. In such a case, it is possible for the flow of two different gases (air or water vapor) through a membrane to be in opposite directions if the partial pressure gradients are of opposite sign. Various combinations of diffusion and mass flow are possible.

6.11 DURABILITY

A durable concrete is one that will withstand the effects of service conditions to which it will be subjected, such as weathering, chemical action, and wear. Numerous laboratory tests have been devised for measuring durability of concrete, but it is extremely difficult to obtain a direct correlation between laboratory tests and field service.

6.11.1 Weathering Resistance

Disintegration of concrete by weathering is caused mainly by the disruptive action of freezing and thawing and by expansion and contraction, under restraint, resulting from temperature variations and alternate wetting and drying.

The most destructive factor of weather is freezing and thawing while the concrete is wet or moist. Deterioration may be caused by expansion of the water in the paste or in the aggregate particles, or by a combination of both. The resistance of concrete to the action of freezing and thawing is dependent upon the following factors: entrained air, cement content, total water content, and moisture condition of the aggregate. Entrained air improves the resistance of concrete to damage from frost action and should be specified for all concrete subject to cycles of freezing and thawing.

Melamine and naphthalene-based superplasticizers may cause some loss of entrained air, whereas lignosulfonate based materials sometimes result in an increase in air content. It is, therefore, important that correct quantities of air-entraining agents in superplasticized concrete be determined by trial mixes. Also, certain types of air-entraining agents may not be compatible with certain types of superplasticizers. The use of vinsol resin is generally recommended with superplasticizers.

Linear traverse measurements (ASTM C457) of air-entrained concrete mixtures containing superplasticizers indicate an increase in air-void spacing, a decrease in specific surface, and decreases in voids per inch and voids per cubic inch when compared to reference concretes with equal slump, cement content, and air content of the plastic concrete. However, in spite of these poor parameters, laboratory tests indicate satisfactory resistance to freeze-thaw and deicer scaling.

The effect of cement content and total water content on durability of normal weight or lightweight concrete is approximately the same in either case—increasing the cement content and decreasing the total water content improves durability.

Some research indicates that if high-strength concrete is to be used under saturated freezing conditions, air-entrained concrete should be considered despite the loss of strength due to air-entrainment. In contrast, other studies indicated excellent resistance to freezing and thawing of high-strength concretes, whether air-entrained or non-air-entrained. This was attributed to the greatly reduced freezable water contents and the increased tensile strength of high-strength concrete.

The effect of cement content and total water content on durability of normal weight or lightweight concrete is approximately the same in either case—increasing the cement content and decreasing the total water content improves durability.

Tests have indicated that the freeze-thaw resistance of structural lightweight concrete of compressive strengths less than 5000 psi may be increased by partial or full replacement of fine aggregates with normal-weight sand.

Moisture condition of lightweight aggregates at the time of mixing has a significant effect on freeze–thaw resistance of concrete. Non-air-entrained concrete made with air-dried aggregates usually is more resistant to freezing and thawing than concrete made with soaked aggregates. Vacuum-treated lightweight aggregates, when used in concrete subject to a freezing environment, must be allowed an extended air-drying period prior to freezing weather. The influence of moisture condition of aggregate is not as pronounced for air-entrained concretes.

To evaluate freeze–thaw resistance, laboratory freeze–thaw tests of concrete should be used, supplemented by field performance records.

6.11.2 Resistance to Deicers

Deicing chemicals* used for snow and ice removal can cause surface scaling. The damage is primarily a physical action and is not caused by chemical reactions or crystal pressures. Deicer scaling of inadequately air-entrained or non-air-entrained concrete is caused, as in normal freezing, by high hydraulic pressures. The presence of a deicer solution in concrete during freezing causes a buildup of hydraulic pressures in excess of the normal hydraulic pressures produced when water in concrete freezes. These pressures become critical and scaling will result when entrained air voids are not present to act as relief valves. The extent of scaling depends upon the amount of salts used and the frequency of application. Unfortunately, deicers can be applied indirectly in various ways, such as by drippings from the undersides of vehicles. Scaling is much more severe in poorly drained areas because the deicer solution is retained on the concrete surface during freezing and thawing.

Concrete requires a minimum curing period of seven days at or above 70°F to ensure adequate strength and a durable scale-resistant surface. When temperatures fall to near 40°F, a 14-day curing period may be necessary when using Type I cement. However, the required curing period can be reduced to 7 days by the use of Type III (High-Early-Strength) cement or Type I or II cement with an accelerator. In addition, air-entrained concrete's resistance to deicers is greatly increased by a period of air drying after the curing period. Concretes placed in the spring or summer have drying periods in the normal course of aging. Concretes placed in the fall season, however, often do not dry out enough before the use of deicing agents become necessary. This is especially true of fall paving cured by membrane-forming compounds. These membranes remain intact until worn off by traffic, and thus adequate drying may not occur before the onset of winter. Curing methods that allow drying at the completion of the curing period are preferable for fall paving on all projects where deicers will be used.

The required time for sufficient drying to take place cannot be pinpointed owing to variations in climate and weather conditions. A good rule of practice, however, is: Age for safe use of deicers—curing period plus 30 days.

If surface scaling (an indication of an inadequate air void system) should develop during the first frost season, a surface treatment may be used to protect the concrete against further damage. These treatments are usually made with linseed oil or neutral petroleum oil.

6.11.3 Chemical Resistance

Quality concrete must be assumed in any discussion of the effect of various substances on concrete and protective treatments. In general, achievement of adequate strength and sufficiently low permeability to withstand many exposures indefinitely requires proper proportioning, placing, and curing.

*Ammonium nitrate and ammonium sulfate have been sold as deicers. These materials in the presence of water react chemically with all forms of concrete and cause objectionable disintegration—even at room temperatures. Their use should be strictly prohibited.

1. Design Considerations Whenever concrete is to be coated for corrosion protection, the forms should be coated with materials that will remain on the forms when they are stripped. Hence, forms coated with form oils or waxes should not be used against surfaces to be coated. Curing membranes may be weakly bonded to the concrete and may in turn develop little or no bond to coatings applied over them. If form oils, waxes, or curing membranes are present, they should be removed by acid washing, sandblasting, scarifying, or other such processes.

Where spillage of corrosive substances is likely to occur, the floor should slope to drains approximately 1/8 to 1/4 in. per linear foot to facilitate washing down of the floor. The slope required depends on the distance between drains and the corrosive substance involved.

Many solutions that have no chemical effect on concrete, such as brines and salts, may crystallize upon drying. It is especially important that concrete subject to alternate wetting and drying of such solutions be impervious. When free water in concrete is saturated with salts, the salts crystallize in the concrete near the surface during the process of drying, and this crystallization may exert sufficient pressure to cause scaling. Structures exposed to brine solutions and having a free surface of evaporation should therefore be provided with a protective treatment on the side exposed to the solution.

In addition, movement of salts into the concrete may result in corrosion of reinforcing steel. It is important that sufficient concrete coverage be provided for reinforcement where the surface is to be exposed to corrosive substances. Metal chairs for support of reinforcement should not extend to the concrete surface. Deep recesses in the concrete should be provided for form ties, and they should be carefully filled and pointed with mortar.

Concrete in contact with soil or water containing moderate sulfate concentrations should be made with cement having less than 8% tricalcium aluminate (C_3A). A moderate sulfate condition exists when the water-soluble sulfate (as SO_4) in soil is from 0.10 to 0.20%, or the sulfate (as SO_4) in groundwater is from 150 to 1000 parts per million. Cement with tricalcium aluminate limited to 5% should be used when the water-soluble sulfate (as SO_4) in soil exceeds 0.20%, or sulfate (as SO_4) in groundwater exceeds 1000 parts per million. If this type is not available, a cement with a C_3A content between 5 and 8 may be used with a 10% reduction in water–cement ratio.

Sulfate resistance of concrete is also improved by use of entrained air. Concrete made with a low water–cement ratio, adequate cement content (660 lb/cu yd), entrained air, and cement having a low tricalcium aluminate content will be most resistant to attack from sulfate containing soil or waters, or seawater.

Deterioration of concrete in seawater exposures may occur through freezing and thawing action, corrosion of steel reinforcement, and chemical attack. Other causes for deterioration include wave action and erosion, and crystallization of salts brought within permeable concrete by capillary action and subsequent evaporation, particularly in and above tidal levels. Average seawater contains about 3% sodium chloride (NaCl) and 0.5% magnesium sulfate ($MgSO_4$). Solutions contianing as little as 0.5% $MgSO_4$ will attack concrete at warm temperatures. The presence of chlorides (as in seawater and some brines) minimizes or inhibits the expansion of concrete in sulfate solutions. Several plausible reasons have been offered to account for this, but they must be deemed conjectural.

Low C_3A cement (C_3A potential less than 8%) is indicated for concrete in seawater or exposed to sea spray.

Aggregates should be resistant to alkali–aggregate reaction. If this requirement cannot be met, then the cement should contain not more than 0.60% alkali (as Na_2O). If the concrete will be exposed to frost action, it should contain proper air entrainment. The use of seawater for mixing water should be prohibited unless the concrete is continually and completely submerged in seawater. Steel in reinforced concrete completely and permanently immersed in seawater will not corrode at any appreciable rate, as oxygen and carbon dioxide are virtually excluded. Steel reinforcement should be protected by at least 3 in. of concrete cover (AASHTO requires 4 in. cover for precast piles in seawater) except at corners, where 4 in. of cover should be allowed.

Portland-pozzolan and portland blast-furnace slag cements are not always resistant to sulfate solutions, and much depends on the C_3A content of the portland cement constituent. If the portland cement constituent of cements of either of these types contains sufficient C_3A as to make it, when used alone, vulnerable to chemical attack by sulfates or seawater, the replacement of some of such cement by a good pozzolan or good blast-furnace slag may be expected to improve the resistance of the composite cement. If, on the other hand, the portland cement constituent contains less than say 8% C_3A, then partial replacement by pozzolan or blast-furnace slag would not be expected to result in an increased resistance.

Acids attack concrete by dissolving both hydrated and unhydrated cement compounds as well as calcareous aggregate. In certain acid waters it may be impossible to apply an adequate protective treatment to the concrete, and the use of a "sacrificial" calcareous aggregate should be considered. Replacement of siliceous aggregate by limestone or dolomite having the equivalent of a calcium oxide concentration of at least 50% will aid in neutralizing the acid. The acid will attack the entire exposed surface more uniformly, reducing the rate of attack on the paste and preventing loss of aggregate particles at the surface. The use of calcareous aggregate will also retard expansion resulting from sulfate attack caused by some acid solutions.

The rate of attack on concrete may be directly related to the activity of the aggressive chemical. Solutions of high concentration are generally more corrosive than those of low concentration—but with some, the reverse is true. The rate of attack may sometimes be affected by the solubility of the reaction products of the particular concrete in the corrosive solution. Lowering of the hydrogen ion concentration, pH, generally causes more rapid attack in the concrete. Also, high temperatures usually accelerate any possible attack, and thus better protection is required than for normal temperatures.

2. *Surface Preparation.* Proper preparation of the concrete surface and good workmanship are essential for the successful application of any protective treatment. Concrete should normally be well cured (28 days to six months, depending on service conditions and coatings used) and dry before the protective coatings are applied. Moisture in the concrete may cause excessive internal vapor pressure that can result in the treatment's blistering and peeling.

Precautions should be taken to eliminate objectionable voids in the surface that may cause pinholes in the coating. Good vibration and placing techniques will reduce the number of these surface imperfections. The surface should be smoothed immediately after removal of forms by applying grout or mortar, or by grinding the surface and then working a grout into it.

It is important to have a firm base free of grease, oil, efflorescence, laitance, dirt, and loose particles. Removal of chemical contaminants must be accomplished before any other surface cleaning, such as acid-etching or sandblasting, takes place.

Concrete cast against forms is cometimes so smooth as to make adhesion of protective coatings very difficult to obtain. Such surfaces should be acid-etched, sandblasted lightly, or ground with silicon carbide stones to provide a slightly roughened surface.

3. *Choosing the Treatment.* Protective treatments for concrete are available for almost any degree of protection required. The coatings vary so widely in composition and performance that no one material will serve best for all conditions.

Every coating is formulated to render a certain performance under specified conditions. Its quality is not determined solely by the merits of any one raw material, since minor variations in formulation can make very substantial changes in performance. Coating performance also depends upon the surface preparation, method and quality of coating application, conditions during application, and film thickness. Any general discussion of chemical resistance and other properties of coatings must assume optimum formulation and proper use. The producers of the various coatings can provide valuable information on the merits of their products for a particular use and on the proper and safe procedure for application. Many coatings contain solvents that are fire, explosion, or toxic hazards.

The more common protective treatments are indicated in the following tables, which are taken from the publication "Effect of Various Substances on Concrete and Protective Treatments, Where Required" (IS001.05T), Portland Cement Association, 1981. For most substances, several treatments are suggested. They will provide sufficient protection in most cases. *The information in the tables is only a guide for determining when to consider various coatings for chemical resistance.*

The numbers and letters in the table refer to the following materials:

1- Magnesium fluosilicate or zinc fluosilicate
2- Sodium silicate (commonly called water glass)
3- Drying oils
4- Coumarone-indene
5- Styrene-butadiene
6- Chlorinated rubber
7- Chlorosulfonated polyethylene (Hypalon)
8- Vinyls
9- Bituminous paints, mastics, and enamels
10- Polyester
11- Urethane
12- Epoxy
13- Neoprene
14- Polysulfide
15- Coal tar-epoxy
16- Chemical-resistant masonry units and mortars:
  a. asphaltic and bituminous mortars
  b. epoxy resin mortars
  c. furan resin mortars
  d. hydraulic cement mortars
  e. phenolic resin mortars
  f. polyester resin mortars
  g. silicate mortars
  h. sulfur mortars
17- Sheet rubber
18- Resin Sheets
19- Lead sheet
20- Glass

ACIDS

Material	Effect on concrete	Protective treatments
Acetic		
<10%	Slow disintegration	1, 2, 9, 10, 12, 14, 16 (b, c, e, f, g, h)
30%	Slow disintegration	9, 10, 14, 16, (c, e, f, g)
100% (glacial)	Slow disintegration	9, 16, (e, g)
Acid waters (pH of 6.5 or less)	Slow disintegration.* Natural acid waters may erode surface mortar but then action usually stops	1, 2, 3, 6, 8, 9, 10, 11, 12, 13, 16, (b, c, e, f, g, h), 17
Arsenious	None	
Boric	Negligible effect	2, 6, 7, 8, 9, 10, 12, 13, 15, 16 (b, c, e, f, g, h), 17, 19
Butyric	Slow disintegration	3, 4, 8, 9, 10, 12, 16 (b, c, e, f)
Carbolic	Slow disintegration	1, 2, 16 (c, e, g), 17
Carbonic (soda water)	0.9 to 3 ppm of carbon dioxide dissolved in natural waters disintegrates concrete slowly	2, 3, 4, 8, 9, 10, 12, 13, 15, 16 (b, c, e, f, h), 17
Chromic:		
5%	None*	2, 6, 7, 8, 9, 10, 16 (f, g, h), 19
50%	None*	16 (g), 19
Formic:		
10%	Slow disintegration	2, 5, 6, 7, 12, 13, 16 (b, c, e, g), 17
90%	Slow disintegration	2, 7, 13, 16 (c, e, g), 17
Humic	Slow disintegration possible, depending on humus material	1, 2, 3, 9, 12, 15, 16 (b, c, e)
Hydrochloric:		
10%	Rapid disintegration, including steel	2, 5, 6, 7, 8, 9, 10, 12, 14, 16 (b, c, e, f, g, h), 17, 19, 20
37%	Rapid disintegration, including steel	5, 6, 8, 9, 10, 16 (c, e, f, g, h)
Hydrofluoric:		
10%	Rapid disintegration, including steel	5, 6, 7, 8, 9, 12, 16 (carbon and graphite brick; b, c, e, h), 17

In porous or cracked concrete, it attacks steel. Steel corrosion may cause concrete to spall.

Material	Effect on concrete	Protective treatments
75%	Rapid disintegration, including steel	16 (carbon and graphite brick; e, h), 17
Hypochlorous, 10%	Slow disintegration	5, 8, 9, 10, 16 (f, g)
Lactic, 5%	Slow disintegration	3, 4, 5, 7, 8, 9, 10 11, 12, 13, 15, 16 (b, c, e, f, g, h), 17
Nitric:		
2%	Rapid disintegration	6, 8, 9, 10, 13, 16 (f, g, h), 20
40%	Rapid disintegration	8, 16 (g)
Oleic, 100%	None	
Oxalic	No disintegration, It protects concrete against acetic acid, carbon dioxide, and salt water. POISONOUS, it must not be used on concrete in contact with food or drinking water.	
Perchloric, 10%	Disintegration	8, 10, 16 (e, f, g, h)
Phosphoric:		
10%	Slow disintegration	1, 2, 3, 5, 6, 7, 8, 9, 10, 11, 12, 13, 14, 15, 16 (b, c, e, f, g, h), 17, 19
85%	Slow disintegration	1, 2, 3, 5, 7, 8, 9, 10, 13, 14, 15, 16 (c, e, f, g, h), 17, 19
Stearic	Rapid disintegration	5, 6, 8, 9, 10, 11, 12, 13, 15, 16 (b, c, e, f, g, h), 17
Sulfuric:		
10%	Rapid disintegration	5, 6, 7, 8, 9, 10, 12, 13, 14, 15, 16 (b, c, e, f, g, h), 17, 19, 20
110% (oleum)	Disintegration	16 (g), 19
Sulfurous	Rapid disintegration	6, 7, 9, 11, 12, 13, 16 (b, c, e, h), 19, 20
Tannic	Slow disintegration	1, 2, 3, 6, 7, 8, 9, 11, 12, 13, 16 (b, c, e, g), 17
Tartaric, solution	None. See wine under "Miscellaneous."	

SALTS AND ALKALIES (SOLUTIONS)*

Material	Effect on concrete	Protective treatments
Bicarbonate:		
Ammonium, Sodium	None	
Bisulfate:		
Ammonium**, Sodium	Disintegration	5, 6, 7, 8, 9, 10, 11, 12, 13, 14, 15, 16 (b, c, e, f, h), 17
Bisulfite:		
Sodium	Disintegration	5, 6, 7, 8, 9, 10, 12 13, 16 (b, c, e, f, h), 17
Calcium sulfite solution)	Rapid disintegration	7, 8, 9, 10, 12, 13, 16 (b, c, e, f, h), 17

Dry materials generally have no effect.
**In porous or cracked concrete, it attacks steel. Steel corrosion may cause concrete to spall.*

Material	Effect on concrete	Protective treatments
Bromide, sodium	Slow disintegration	1, 2, 5, 6, 7, 8, 9, 10,
Carbonate:		
Ammonium, Potassium, Sodium	None	
Chlorate, sodium	Slow disintegration	1, 4, 6, 7, 8, 9, 10, 16 (f, g, h), 17, 19
Chloride:		
Calcium†, Potassium, Sodium†, Strontium	None, unless concrete is alternately wet and dry with the solution**	1, 3, 4, 5, 6, 7, 8, 9, 10, 11, 12, 13, 15, 16 (b, c, e, f, g, h), 17

†Frequently used as a de-icer for concrete pavements. If the concrete contains insufficient entrained air or has not been air-dried for at least 30 days after completion of curing, repeated application may cause surface scaling. See de-icers under "Miscellaneous."

SALTS AND ALKALIES (SOLUTIONS) *(Continued)*

Material	*Effect on concrete*	*Protective treatments*
Ammonium Copper Ferric (iron) Ferrous Magnesium Mercuric Mercurous Zinc	Slow disintegration**	1, 3, 4, 5, 6, 7, 8, 9, 10, 11, 12, 13, 15, 16 (b, c, e, f, g, h), 17
Aluminum	Rapid disintegration**	1, 3, 4, 5, 6, 7, 8, 9, 10, 11, 12, 13, 15, 16 (b, c, e, f, h), 17
Chromate, sodium	None	
Cyanide: Ammonium Potassium Sodium	Slow disintegration	7, 8, 9, 12, 13, 16 (b, c), 17
Dichromate: Sodium	Slow disintegration with dilute solutions	1, 2, 6, 7, 8, 9, 10, 11, 12, 13, 15, 16 (b, c, e, f, h), 17
Potassium	Disintegration	1, 2, 6, 7, 8, 9, 10, 11, 12, 13, 15, 16 (b, c, e, f, h), 17
Ferrocyanide, sodium	None	
Fluoride: Ammonium Sodium	Slow disintegration	3, 4, 8, 9, 13, 16 (a, c, e, h), 17
Fluosilicate, magnesium	None	
Hexametaphosphate, sodium	Slow disintegration	5, 6, 7, 8, 9, 12, 13, 15, 16 (b, c, e), 17
Hydroxide: Ammonium Barium Calcium Potassium, 15%‡ Sodium, 10%‡	None	
Potassium, 25% Sodium, 20%	Disintegration. Use of calcareous aggregate lessens attack.	5, 7, 8, 12, 13, 14, 15, 16 (carbon and graphite brick; b, c) 17
Nitrate: Calcium Ferric Zinc	None	
Lead Magnesium Potassium Sodium	Slow disintegration	2, 5, 6, 7, 8, 9, 10, 11, 12, 13, 16 (b, c, e, f, g, h), 17, 20
Ammonium	Disintegration**	2, 5, 6, 8, 9, 10, 11, 12, 13, 16 (b, c, e, f, g, h), 17, 20
Nitrite, sodium	Slow disintegration	1, 2, 5, 6, 7, 8, 9, 12, 13, 16 (b, c), 17
Orthophosphate, sodium (dibasic and tribasic)	None	
Oxalate, ammonium	None	
Perborate, sodium	Slow disintegration	1, 4, 7, 8, 9, 10, 13, 16 (d, f, g, h), 17
Perchlorate, sodium	Slow disintegration	6, 7, 8, 10, 16 (f, g, h), 17
Permanganate, potassium	None	
Persulfate, potassium	Disintegration of concrete with inadequate sulfate resistance	1, 2, 5, 7, 8, 9, 10, 12, 13, 16 (b, c, e, f, h), 17
Phosphate, sodium (monobasic)	Slow disintegration	5, 6, 7, 8, 9, 12, 15, 16 (b, c), 17
Pyrophosphate, sodium	None	
Stannate, sodium	None	
Sulfate: Ammonium	Disintegration**	5, 6, 7, 8, 9, 10, 11, 12, 13, 14, 15, 16 (b, c, e, f, g, h), 17
Aluminum Calcium Cobalt Copper Ferric Ferrous (iron vitriol) Magnesium (epsom salt) Manganese Nickel Potassium	Disintegration of concrete with inadequate sulfate resistance. Concrete products cured in high-pressure steam are highly resistant to sulfates.	1, 3, 4, 5, 6, 7, 8, 9, 10, 11, 12, 13, 15, 16 (b, c, e, f, g, h), 17
Potassium aluminum (alum) Sodium Zinc	Disintegration of concrete with inadequate sulfate resistance. Concrete products cured in high-pressure steam are highly resistant to sulfates.	1, 3, 4, 5, 6, 7, 8, 9, 10, 11, 12, 13, 15, 16 (b, c, e, f, g, h), 17
Sulfide: Copper Ferric Potassium	None unless sulfates are present	7, 8, 9, 10, 12, 13, 15, 16 (b, c, e, f, h), 17
Sodium	Slow disintegration	6, 7, 8, 9, 11, 12, 13, 15, 16 (b, c), 17
Ammonium	Disintegration	7, 8, 9, 12, 13, 15, 16 (a, b, c, e), 17
Sulfite: Sodium	None unless sulfates are present	1, 2, 5, 6, 7, 8, 9, 11, 12, 13, 15, 16 (b, c, e), 17
Ammonium	Disintegration	8, 9, 12, 15, 16 (b, c, e, h), 17
Superphosphate, ammonium	Disintegration**	8, 9, 12, 13, 15, 16 (b, c, e), 17, 19
Tetraborate, sodium (borax)	Slow disintegration	5, 6, 7, 8, 9, 10, 11, 12, 13, 15, 16 (b, c, e, f, g, h), 17
Thiosulfate Sodium	Slow disintegration of concrete with inadequate sulfate resistance	1, 2, 5, 6, 7, 8, 9, 10, 12, 13, 15, 16 (b, c, e), 17
Ammonium	Disintegration	8, 9, 12, 13, 15, 16 (b, c, e), 17

‡*If concrete is made with reactive aggregates, disruptive expansions may occur.*

PETROLEUM OILS

Material	*Effect on concrete*	*Protective treatments*
Heavy oil below 35° Baume* Paraffin	None	
Gasoline Kerosene Light oil above 35° Baume* Ligroin Lubricating oil* Machine oil* Mineral spirits	None. Impervious concrete is required to prevent loss from penetration, and surface treatments are generally used.	1, 2, 3, 8, 10, 11, 12, 14, 16 (b, c, e, f), 17, 19
Gasoline, high octane	None. Surface treatments are generally used to prevent contamination with alkalies in concrete.	11, 14, 17

**May contain some vegetable or fatty oils and the concrete should be protected from such oils.*

COAL TAR DISTILLATES

Material	*Effect on concrete*	*Protective treatments*
Alizarin Anthracene Carbazole Chrysen Pitch	None	
Benzol (benzene) Cumol (cumene) Phenanthrene Toluol (toluene) Xylol (xylene)	None. Impervious concrete is required to prevent loss from penetration, and surface treat treatments are generally used.	1, 2, 10, 11, 12, 16 (b, c, e, f, g), 19
Creosote Cresol Dinitrophenol Phenol, 5 to 25%	Slow disintegration	1, 2, 16 (c, e, g), 17, 19

SOLVENTS AND ALCOHOLS

Material	*Effect on concrete*	*Protective treatments*
Carbon tetrachloride	None*	1, 2, 10, 12, 16 (b, c, e, g)
Ethyl alcohol	None* (see de-icers under "Miscellaneous")	1, 2, 5, 7, 10, 12, 13, 14, 16, (b, c, e, f, g, h), 17, 19
Ethyl ether	None*	11, 12, 16 (c, e), 19
Methyl alcohol	None*	1, 2, 5, 7, 10, 12, 13, 14, 16 (b, c, e, f, g, h), 17, 19
Methyl ethyl ketone	None*	16 (c, e), 17, 19
Methyl isoamyl ketone	None*	16 (c, e), 17
Methyl isobutyl	None*	16 (c, e), 17
Perchloroethylene	None*	12, 16 (b, c, e)
t-Butyl alcohol	None*	1, 2, 5, 7, 10, 12, 13, 14, 16 (b, c, e, f, g, h), 17, 19
Trichloroethylene	None*	1, 2, 12, 16 (b, c, e, g)
Acetone	None.* However, acetone may contain acetic acid as impurity (see under "Acids").	1, 2, 10, 16 (c, e, g), 17, 19
Carbon disulfide	Slow disintegration possible	1, 2, 11, 16 (c, e, g)
Glycerin (glycerol)	Slow disintegration possible	1, 2, 3, 4, 7, 11, 12, 13, 16 (b, c, e, f, g), 17
Ethylene glycol**	Slow disintegration	1, 2, 7, 10, 12, 13, 14, 16 (b, c, e, f, g, h), 17

**Impervious concrete is required to prevent loss from penetration, and surface treatments are generally used.*
***Frequently used as de-icer for airplanes. Heavy spillage on concrete containing insufficient entrained air may cause surface scaling.*

VEGETABLE OILS

Material	*Effect on concrete*	*Protective treatments*
Rosin and rosin oil	None	
Turpentine	Mild attack and considerable penetration	1, 2, 11, 12, 14, 16 (b, c, e)
Almond China wood* Linseed* Olive Peanut Poppy seed Soybean* Tung* Walnut	Slow disintegration	1, 2, 8, 10, 11, 12, 14, 16 (b, c, e, f), 17. For expensive cooking oils, use 20.
Margarine	Slow disintegration–faster with melted margarine	1, 2, 8, 10, 11, 12, 13, 16 (b, c, e, f)
Castor Cocoa bean Cocoa butter Coconut Cottonseed Mustard Rapeseed	Disintegration, especially if exposed to air	1, 2, 8, 10, 11, 12, 14, 16 (b, c, e, f), 17

**Applied in thin coats, the material quickly oxidizes and has no effect. The effect indicated above is for constant exposure to the material in liquid form.*

FATS AND FATTY ACIDS (ANIMAL)

Material	*Effect on concrete*	*Protective treatments*
Fish liquor	Disintegration	3, 8, 10, 12, 13, 16 (b, c, e, f), 17
Fish oil	Slow disintegration with most fish oils	1, 2, 3, 8, 10, 12, 13 16 (b, c, e, f), 17
Whale oil	Slow disintegration	1, 2, 3, 8, 10, 12, 13, 16 (b, c, e, f), 17
Neat's-foot oil Tallow and tallow oil	Slow disintegration	1, 2, 3, 8, 10, 12, 13, 16 (b, c, e, f), 17
Beef fat Horse fat Lamb fat	Slow disintegration with solid fat—faster with melted	1, 2, 3, 8, 10, 12, 13, 16 (b, c, e, f), 17
Lard and lard oil	Slow disintegration—faster with oil	1, 2, 3, 8, 10, 12, 13, 16 (b, c, e, f) 17
Slaughterhouse wastes	Disintegration due to organic acids	8, 12, 13, 16 (b, c, e)

MISCELLANEOUS

Material	*Effect on concrete*	*Protective treatments*
Alum	See sulfate, potassium aluminum, under "Salts and Alkalies"	
Ammonia:		
Liquid	None, unless it contains harmful ammonium salts (see under "Salts and Alkalies")	
Vapors	Possible slow disintegration of moist concrete and steel attacked in porous or cracked moist concrete	8, 9, 10, 12, 13, 16 (a, b, c, f), 17
Ashes:		
Cold	Harmful if wet, when sulfides and sulfates leach out (see sulfate, sodium, under "Salts and Alkalies")	1, 2, 3, 8, 9, 12, 13, 16 (b, c, e)
Hot	Thermal expansion	16 (calcium, aluminate cement, fireclay, and refractory-silicate-clay mortars)
Automobile and diesel exhaust gases	Possible disintegration of moist concrete by action of carbonic, nitric, or sulfurous acid (see under "Acids")	1, 5, 8, 12, 16 (b, c, e)
Baking soda	None	
Beer	No progressive disintegration, but in beer storage and fermenting tanks a special coating is used to guard against beer contamination. Beer may contain, as fermentation products, acetic, carbonic, lactic, or tannic acids (see under "Acids").	8, 10, 11, 12, 16 (b, c, f), 17, coatings made and applied by Borsari Tank Corp. of America, 605 Third Ave., New York, N. Y. 10016
Bleaching solution	See the specific chemical, such as hypochlorous acid, sodium hypochlorite, sulfurous acid, etc.	
Borax (salt)	See Tetraborate, sodium, under "Salts and Alkalies"	
Brine	See chloride, sodium, or other salts under "Salts and Alkalies"	
Bromine	Disintegration if bromine gaseous—or if a liquid containing hydrobromic acid and moisture	10, 13, 16 (f, g)
Buttermilk	Slow disintegration due to lactic acid	2, 3, 4, 7, 8, 9, 10, 11, 12, 13, 16 (b, c, e, f), 17
Butyl stearate	Slow disintegration	8, 9, 10, 16 (b, c, e)
Carbon dioxide	Gas may cause permanent shrinkage. See carbonic acid under "Acids."	1, 2, 3, 6, 8, 9, 10, 11, 12, 13, 15, 16 (b, c, e, f, h), 17
Caustic soda	See hydroxide, sodium, under "Salts and Alkalies"	
Chile saltpeter	See nitrate, sodium, under "Salts and Alkalies"	
Chlorine gas	Slow disintegration of moist concrete	2, 8, 9, 10, 16 (f, g), 17
Chrome plating	Slow disintegration	7, 8, 9, 10, 16 (f, g), 20
Cider	Slow disintegration. See acid under "Acids."	1, 2, 9, 10, 12, 14, 16 (b, c, e, f, g), 17
Cinders cold and hot	See ashes above.	
Coal	None, unless coal is high in pyrites (sulfide of iron) and moisture. Sulfides leaching from damp coal may oxidize to sulfurous or sulfuric acid, or ferrous sulfate (see under "Acids" and "Salts and Alkalies"). Rate is greatly retarded by deposit of an insoluble film.	1, 2, 3, 6, 7, 8, 9, 12, 13, 16 (b, c, e, h), 17
Coke	Sulfides leaching from damp coke may oxidize to sulfurous or sulfuric acid (see under "Acids")	1, 2, 3, 6, 7, 8, 9, 12, 13, 16 (b, c, e, h)
Copper plating solutions	None	
Corn syrup (glucose)	Slow disintegration	1, 2, 3, 7, 8, 9, 12, 13, 16 (b, c, e), 17
De-icers	Chlorides (calcium and sodium), urea, and ethyl alcohol cause scaling of non-air-entrained concrete.	50% solution of boiled linseed oil in kerosene, soybean oil, modified castor oil, cottonseed oil, sand-filled epoxy, or coal-tar epoxy
Distiller's slop	Slow disintegration due to lactic acid	1, 8, 9, 10, 12, 13, 15, 16 (b, c, e, f, h), 17

MISCELLANEOUS *(Continued)*

Material	*Effect on concrete*	*Protective treatments*
Fermenting fruits, grains, vegetables, or extracts	Slow disintegration. Industrial fermentation processes produce lactic acid (see under "Acids").	1, 2, 3, 8, 9, 12, 16 (b, c, e), 17
Flue gases	Hot gases (400–1100° F) cause thermal stresses. Cooled, condensed sulfurous, hydrochloric acids disintegrate concrete slowly.	9 (high melting), 16 (g, fireclay mortar)
Formaldehyde, 37% (formalin)	Slow disintegration due to formic acid formed in solution	2, 5, 6, 8, 10, 11, 12, 13, 14, 16 (b, c, e, f, g, h), 17, 20
Fruit juices	Little if any effect for most fruit juices as tartaric and citric acids do not appreciably affect concrete. Sugar and hydrofluoric and other acids cause disintegration.	1, 2, 3, 6, 7, 8, 9, 11, 12, 16 (b, c, e), 17
Gas water	Ammonium salts seldom present in sufficient quantity to disintegrate concrete	9, 12, 16 (b, c)
Glyceryl tristearate	None	
Honey	None	
Hydrogen sulfide	Slow disintegration in moist oxidizing environments where hydrogen sulfide converts to sulfurous acid	1, 2, 5, 6, 7, 8, 9, 10, 11, 12, 13, 16, (b, c, e, f, g, h), 17, 19
Iodine	Slow disintegration	1, 2, 6, 12, 13, 16 (b, c, e, g), 17
Lead refining solution	Slow disintegration	1, 2, 6, 8, 9, 12, 16, (carbon and graphite brick; b, c, e, h), 17 20
Lignite oils	Slow disintegration if fatty oils present	1, 2, 6, 8, 10, 12, 16 (b, c, e, f)
Lye	See hydroxide, sodium and potassium, under "Salts and Alkalies,"	
Manure	Slow disintegration	1, 2, 8, 9, 12, 13, 16 (b, c, e)
Mash, fermenting	Slow disintegration due to acetic and lactic acids and sugar	1, 8, 9, 10, 12, 13, 16 (b, c)
Milk	None, unless milk is sour. Then lactic acid disintegrates concrete slowly.	3, 4, 8, 9, 10, 11, 12, 13, 16 (b, c, f), 17
Mine water, waste	Sulfides, sulfates, or acids present disintegrate concrete and attack steel in porous or cracked concrete	1, 2, 5, 8, 9, 10, 12, 13, 15, 16 (b, c, e, f, h), 17
Molasses	Slow disintegration at temperatures ⩾120° F	1, 2, 7, 8, 9, 12, 13, 16 (b, c, e), 17
Nickel plating solutions	Slow disintegration due to nickel ammonium sulfate	2, 5, 6, 7, 8, 9, 10, 13, 16 (c, e, f), 17
Niter	See nitrate, potassium, under "Salts and Alkalies"	
Ores	Sulfides leaching from damp ores may oxidize to sulfuric acid or ferrous sulfate (see under "Acids" and "Salts and Alkalies").	2, 9, 10, 12, 13, 15, 16 (b, c, e, f, g), 17
Pickling brine	Steel attacked in porous or cracked concrete. See salts, boric acid, or sugar.	1, 7, 8, 9, 12, 13, 16 (b, c, e, h), 17
Sal ammoniac	See chloride, ammonium, under "Salts and Alkalies."	
Sal soda	See carbonate, sodium, under "Salts and Alkalies."	
Saltpeter	See nitrate, potassium, under "Salts and Alkalies."	
Sauerkraut	Slow disintegration possible due to lactic acid. Flavor impaired by concrete.	1, 2, 8, 9, 10, 12, 13, 16 (b, c, e, f), 17
Seawater	Disintegration of concrete with inadequate sulfate resistance and steel attacked in porous or cracked concrete	1, 2, 5, 6, 7, 8, 9, 10, 11, 12, 13, 14, 15, 16 (b, c, e, f), 17
Sewage and sludge	Usually not harmful. See hydrogen sulfide above.	
Silage	Slow disintegration due to acetic, butyric, and lactic acids, and sometimes fermenting agents of hydrochloric or sulfuric acids	3, 4, 8, 9, 10, 12, 16 (b, c, e, f)
Sodium hypochlorite	Slow disintegration	7, 8, 9, 10, 13, 16 (d, f), 17
Sugar (sucrose)	None with dry sugar on thoroughly cured concrete. Sugar solutions may disintegrate concrete slowly.	1, 2, 3, 7, 8, 9, 10, 12, 13, 15, 16 (b, c, e, f), 17
Sulfite liquor	Disintegration	1, 2, 3, 5, 6, 8, 9, 10, 12, 13, 16 (b, c, e, f, h), 17, 19
Sulfur dioxide	None if dry. With moisture, sulfur dioxide forms sulfurous acid.	2, 5, 6, 8, 9, 10, 12, 13, 16 (b, c, e, f, g, h), 17, 19
Tanning bark	Slow disintegration possible if damp. See tanning liquor below.	1, 2, 3, 6, 8, 9, 11, 12, 13, 16 (b, c, e), 17
Tanning liquor	None with most liquors, including chromium. If liquor is acid, it disintegrates concrete.	1, 2, 3, 5, 6, 8, 9, 11, 12, 13, 16 (b, c, e), 17
Tobacco	Slow disintegration if organic acids present	1, 8, 9, 10, 12, 13, 16 (b, c, e, f), 17
Trisodium phosphate	None	
Urea	None (see de-icers)	
Urine	None, but steel attacked in porous or cracked concrete	7, 8, 12, 13, 16 (b, c, e)

MISCELLANEOUS (*Continued*)

Material	*Effect on concrete*	*Protective treatments*
Vinegar	Slow disintegration due to acetic acid	9, 12, 16 (b, c, e, h), 17
Washing soda	None	
Water, soft (<75 ppm of carbonate hardness)	Leaching of hydrated lime by flowing water in porous or cracked concrete	2, 3, 4, 8, 9, 10, 12, 13, 16 (b, c, e, f, h), 17
Whey	Slow disintegration due to lactic acid	3, 4, 5, 7, 8, 9, 10, 12, 13, 16 (b, c, e, f, h), 17
Wine	None–but taste of first batch may be affected unless concrete has been given tartaric acid treatment	For fine wines, 2 or 3 applications of tartaric acid solution (1 lb. tartaric acid in 3 pints water), 2, 8, 12, 16 (b), 20
Wood pulp	None	
Zinc refining solutions	Disintegration if hydrochloric or sulfuric acids present	8, 9, 10, 12, 13, 16 (b, c, e, f, h), 17
Zinc slag	Zinc sulfates (see under "Salts and Alkalies") may be formed by oxidation.	8, 9, 10, 12, 13, 16 (b, c, e, f, h), 17

6.11.4 Abrasion and Skid Resistance

Concrete surfaces must be capable of withstanding many abrasive forces with a minimum of wear. Use of quality concrete and proper attention to design, construction, and maintenance of the surfaces can do much to minimize or eliminate wear of the concrete.

1. Classification of Wear. Wear of concrete by abrasion can be classified as follows:

(a) Wear on concrete floors due to foot traffic, light trucking with small-wheeled carts, and the skidding or sliding of objects on the surface. This type of wear is essentially a rubbing action and is usually caused by the introduction of foreign particles, such as sand, metal scraps, or similar materials.
(b) Wear on concrete pavements due to heavy trucking and automobiles, with or without chains or metal studs. This type of wear is caused by a rubbing plus an impact-cutting action.
(c) Wear on underwater construction, due to the action of abrasive materials carried by flowing or turbulent waters. The abrasive materials are generally sand particles which often may be softer than the concrete, but the force exerted by rapidly moving water carrying the sand is such that a cutting action (erosion) is produced. Erosion varies with the type and amount of abrasive material, its velocity, the angle of contact, and changes in the direction of flow or the presence of eddies.
(d) Wear of concrete in hydraulic structures due to cavitation. Cavitation is very destructive, and even the best-quality concrete offers very little resistance to it. The magnitude of cavitation damage varies with the mean velocity of flow, raised, perhaps, to the sixth or seventh power. However, other factors, such as boundary roughness and the growth and form of boundary layers, together with stream turbulence, also have some influence on the intensity of cavitation damage.

2. Concrete Requirements. Quality concrete is a prerequisite to abrasion and erosion resistance and to the retention of pavement skid resistance. The compressive strength of concrete is the most important single factor related to wear. Wear resistance increases as compressive strength increases up to a certain point (usually about 6000 psi), Fig. 6-54, depending upon aggregate, mix composition, and type of test, but beyond this point increases in strength have very little effect on wear resistance. Concrete with large aggregate erodes less than mortar of equal strength. Soft aggregates are not desirable because after the film of paste has been worn away, the forces of erosion are resisted to a large extent by the aggregate. For best resistance to cavitation, the maximum size of aggregate near the surface should not exceed 3/4 in. because cavitation tends to remove large particles.

Comparisons of results of aggregate abrasion tests with those of abrasion resistance of concrete do not generally show a direct correlation. The wear resistance of concrete can be determined more accurately by abrasion tests of the concrete itself (ASTM C779).

Because of the porous structure of structural lightweight aggregates, the resistance of each thin wall or shell to load and/or impact may be low compared to the point load and impact resistance offered by concrete made with solid particles of similar composition. Therefore, the abrasion resistance of all-lightweight aggregate concretes may not be sufficient for steel-wheeled or exceptionally heavy industrial traffic. As the severity of wear lessens, the abrasion resistance should be as satisfactory as that of normal-weight concrete. The use of natural sand in lightweight concrete improves resistance to abrasion.

Operations and procedures such as proper finishing and adequate curing, removal of water by the vacuum process or absorptive form lining, low slump and low water–cement

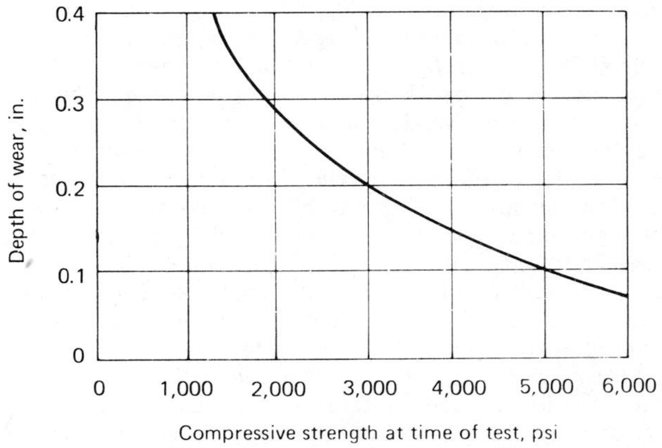

Fig. 6-54 Effect of compressive strength on the abrasion resistance of concrete. High-strength concrete is highly resistant to abrasion. (From: Sawyer, J. L., "Wear Tests on Concrete Using the German Standard Method of Test and Machine," ASTM Presentation at 60th Annual Meeting, June 1957.)

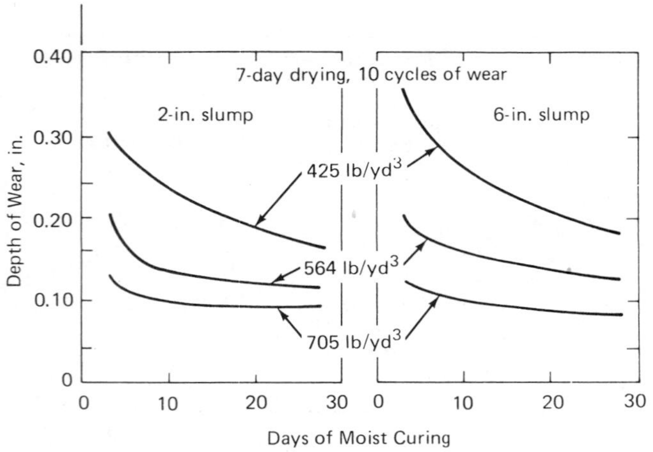

Fig. 6-55 Effect of cement content and cure on concrete wear. (From: Sawyer, J. L., "Wear Tests on Concrete Using the German Standard Method of Test and Machine," ASTM Presentation at 60th Annual Meeting, June 1957.)

ratio, all of which tend to increase the strength of the concrete at the surface or throughout the mass, also increase the abrasion resistance of the concrete.

Wear resistance of concrete increases as the cement content is increased, and when concrete surfaces are exposed to highly abrasive action or carry high traffic volumes, the cement content should be greater than 564 lb/cu yd. Also, significant reduction in wear occurs as the length of the moist-curing period is increased, Fig. 6-55. Wear resistance increases as the water-cement or voids (air plus water) to cement ratio is decreased, and the maximum water–cement ratio should not exceed 0.50. Air-entrained concrete is as resistant to wear as non-air-entrained concrete provided they are of equal compressive strength.

When conditions are such as to cause rapid loss of moisture from the freshly placed concrete surface, either a uniform fog spray of water, just enough to restore the surface sheen, or a monomolecular film to retard evaporation may be applied during finishing operations. The application of excessive amounts of water to the surface during finishing operations must be avoided. Excess water from any source reduces the possibility of obtaining wear-resistant surfaces.

Final finishing should be delayed as long as possible. The surface of a floor should be struck off slightly higher than the finished dimension, and when bleeding has stopped, the surface should be struck off to proper elevation. This will eliminate a possible weak top layer of mortar formed with excessive bleed water. Pressure applied to the surface by several hard trowelings after the concrete begins to set increases the surface density and abrasion resistance. Curing should be started as soon as possible without marring the surface and continued for at least seven days. Curing compounds, if used, should be applied after finishing operations are completed and between the end of the bleeding and the beginning of the drying of the concrete.

3. Wear on Floors. Floors that are to be subjected to severe abrasion should preferably be constructed with a heavy-duty concrete floor topping. Other satisfactory methods of densifying the surface are the use of ⅜-in. traprock cast on the surface, or the use of dry shakes consisting of iron filings, corundum, silicon carbide, or heat-treated aluminum oxides. The use of surface hardeners is more effective in increasing abrasion resistance on poorly cured or porous concrete surfaces than on high-density properly cured concrete.

Concrete that is subject to abrasion while in a moist condition is more liable to show wear than the same concrete maintained in a dry condition.

4. Cavitation. Quality concrete is not affected by a steady, tangential, high-velocity flow of water. On horizontal or sloping surfaces subjected to high-velocity flow (40 ft/sec in open channels or as low as 25 ft/sec in closed conduit) or on vertical surfaces past which water flows, an obstruction, or abrupt change in surface alignment, causes a zone of severe subatmospheric pressure to be formed against the surface immediately downstream from the obstruction or abrupt change. When the local absolute pressure approaches the ambient vapor pressure of the water, this zone is promptly filled with turbulent water interspersed with large, single voids, which later break up, or clouds of small fast-moving bubblelike cavities of water vapor. These cavities flow downstream with the water, and upon entering an area of higher pressure collapse with great impact. Water from the boundaries of the cavities rushes toward their centers at high speed when the collapse takes place, generating extremely high pressures (as high as 100,000 psi). Repeated collapse of the cavities near the surface of the concrete will pit it. Pitting resulting from cavitation can be easily distinguished by its honeycombed appearance as contrasted to the smoother surfaces resulting from damage by abrasion or erosion.

Although the advent of cavitation depends primarily on pressure changes (and consequently also on velocity changes), it is especially likely to occur in the presence of small quantities of undissolved air in the water. These bubbles of air behave as nuclei at which the change of phase from liquid to vapor can more readily occur. Dust particles have a similar effect. On the other hand, free air in large quantities (up to 2% by volume), while promoting cavitation, may cushion the collapse of the cavities and hence reduce the cavitation damage. Deliberate air entrainment in water may therefore be advantageous.

Cavitation can be reduced by constructing smooth and well-aligned surfaces free from irregularities such as depressions, projections, joints, and misalignments, and by the absence of abrupt changes in slope or curvature that tend to pull the flow away from the surface. For example, high-velocity flow passing through a gate opening causes the boundary layer of the flow to be disrupted; and a certain length of continuous surface contact, depending upon the velocity, is required for it to be reestablished. Surface irregularities must be completely eliminated for specific distances downstream from the ends of gate frames, by grinding to specified levels according to flow velocity, Table 6-13.

In extreme cases where low pressures cannot be avoided, critical areas are sometimes protected by facing with metal or resilient coatings that have good adhesion to concrete and an ability to be easily repaired.

5. Skid Resistance. Skid resistance of concrete pave-

TABLE 6-13 Offset and Grinding Tolerances for High-Velocity Flow*

Velocity Range, fps	*Distance of Treatment Downstream from Gate Frame, ft*	*Grinding Bevel, Ratio of Height to Length*
40 to 90	15	1 to 20
90 to 120	30	1 to 50
Over 120	50	1 to 100

*From: *Concrete Manual, U.S. Dept. of Interior, Bureau of Reclamation, 7th ed., 1963.*

ments, for simplicity, can be defined as the force, relative to the weight of the vehicle, that is available to permit stopping within a reasonable distance and to permit cornering without sideways displacement. It is influenced by both the tire and pavement surface. Regarding the tire, skid resistance is a function of its rubber characteristics and tread design. Tread design is not too significant on dry pavements because even smooth tires produce good skid resistance. However, when surfaces are wet, depth of tire tread gains in importance. The contribution of a concrete pavement to skid resistance is a function mainly of surface texture, concrete mix design, and aggregates needed to minimize wear. The durability of the surface texture is a function of the wear-resistant qualities of the concrete and the character and volume of the traffic.

The surface of a pavement should include both fine and coarse texture. The fine texture (grittiness) is formed by the sand in the cement-mortar layer and imparts the adhesion component in the tire–pavement interaction. The coarse texture is formed by the ridges of mortar left by the method of finishing and has the dual role of imparting the hysteresis* component and providing drainage under the tire.

Finishing of a concrete pavement has generally been accomplished using a burlap drag, a broom, or a belt. Texture depths obtained with a burlap drag vary considerably with the time of finishing. One way of obtaining a texture when finishing with a burlap drag would be to make three passes, each at a different time interval after casting. A pavement with a deep texture will retain skid resistance longer than a pavement with shallow texture; other factors being equal, however, the finishing method selected should be compatible with the environment, speed and density of traffic, topography and layout of the pavement, and economics of vehicle operations.

The skid resistance of a concrete pavement is controlled mainly by the fine aggregate in the mortar or wearing layer that is textured during finishing rather than the coarse aggregate which seldom functions as a portion of the surface. For example, a coarse aggregate with high polishing characteristics can be permitted in a mix if the mortar surface has adequate skid resistance. (Polish-resistant coarse aggregates include siliceous gravels, granites, diabase, quartzite, sandstones and expanded shale). To provide good skid resistance the proportion of fine aggregate in the concrete mix should be near the upper limit of the range that permits proper placing, finishing, and texturing. The siliceous particle content of the fine aggregate, as determined by an insoluble residue test, should be not less than 25%. The siliceous particle content is very important, and where economically feasible, a higher percentage should be required. Where suitable materials conforming to these requirements are not economically available, alternate methods of achieving a skid-resistant surface should be investigated. These might include blending of fine aggregates or applying wear-resistant particles to the surface. ASTM C33 cautions that certain manufactured sands produce slippery pavement surfaces and should be investigated for acceptance before use. Consideration should be given to replacing limestone fines with expanded shale fines.

To improve skid resistance at critical areas such as in toll plazas, near busy intersections, or where frequent braking, traction, or cornering occurs, ½ to 1 lb/sq yd of abrasives such as aluminum oxide or silicon carbide of 12 grit or smaller can be spread over the surface and floated into the plastic concrete.

*The hysteresis component is a function of the energy losses within the rubber as it is deformed by the textured pavement surface.

Methods of increasing the skid resistance of old or worn surfaces include acid-etching, mechanical abrading, sawing, and resurfacing. The choice of method depends on the economics and the desired result.

6.12 ACOUSTICAL PROPERTIES

Good acoustical design utilizes both absorptive and reflective surfaces, sound barriers, and vibration isolators. Some surfaces must reflect sound so that the loudness will be adequate in all areas where listeners are located. Other surfaces absorb sound to avoid echoes, sound distortion, and long reverberation times. Sound is isolated from rooms where it is not wanted by selected wall and floor–ceiling constructions. Vibrations generated by mechanical equipment must be isolated from the structural frame of the building.

6.12.1 Sound Absorption

Sound absorption control involves reduction of sound emanating from a source within a room. The extent of control depends on the efficiency of the room surfaces in absorbing rather than reflecting sound waves. The amount of sound absorbed by concrete depends upon its porosity. Sound waves reaching a porous surface cause air to flow into and from the surface. The friction of the air flowing through the pores converts a portion of the sound energy to heat.

The sound absorption coefficient is an indication of the sound-absorbing efficiency of a surface. A surface that could theoretically absorb 100% of impinging sound would have a sound absorption coefficient of 1. Similarly, a dense, nonporous concrete surface (cast-in-place) typically absorbs 1 to 2% of incident sound and has a coefficient of 0.015. (Most materials are tested for sound absorption at frequencies from 125 to 4,000 cps in octave steps.) Another designator, termed the noise reduction coefficient (NRC), is calculated by taking a mathematical average of the sound absorption coefficients obtained at frequencies of 250, 500, 1000, and 2000 Hertz (cps). It is expressed to the nearest integral multiple of 0.05. Cast-in-place concrete would have an NRC of 0.02, while precast units with lightweight aggregates may have an NRC of as high as 0.60. Where additional sound absorption is desired, cast-in-place concrete can be coated with spray-applied acoustical materials, acoustical tile can be applied with adhesive, or an acoustical ceiling can be hung below.

Many surfaces must be acoustically reflective, not absorptive, e.g., a band shell. Hard reflective surfaces on walls and ceiling near the stage in an auditorium serve as megaphones to direct sounds to the audience. Concrete is an excellent material for this use.

6.12.2 Sound Transmission Loss

Airborne sound reaching a wall, floor, or ceiling produces vibrations in the wall and is reradiated with reduced intensity on the other side. The ratio of sound pressure level on the source side to the sound pressure level on the reradiating side is known as the sound transmission loss (STL). The ratio is expressed in decibels, db.

Sound transmission loss measurements are made at 16 frequencies at one-third octave intervals covering the range from 125 to 4000 Hz (ASTM E90). Although test data are reported to a fraction of a decibel, a difference of three or less db is not especially significant because the human ear cannot detect a change in sounds less than three decibels. Single figure ratings are obtained by comparing plots of

STL vs. frequency data to standard contours (STC contour) (ASTM E413). The matching system permits performance to fall below the contour to a degree that is limited in amount at any single test frequency and for all 16 frequencies. The highest contour meeting the requirements is selected. The sound transmission class, or STC rating, is the STL of the STC contour at 500 Hz. The STC provides a convenient means for comparing various systems of construction, but for specific applications the data at specific frequencies may be of greater importance.

The STL of a wall or a floor–ceiling assembly is a function of the weight, stiffness, and vibration damping characteristics. Weight is concrete's greatest asset when it is used as a sound insulator. Stiffness increases the sound transmission loss at low frequencies. But sometimes the stiffness-controlled region is below the lowest test frequency. Resonant and coincident frequencies frequently may be present in the critical range. At these frequencies the sound transmission loss is reduced—there are dips in the sound transmission loss curve. If a wall or floor is highly damped, the dips are smaller. Small changes in weight or stiffness can shift the critical frequencies. Such shifts can either eliminate or introduce overlaps. Significant changes then result in sound transmission at those frequencies. This in turn can produce large changes in STCs.

The weight per square foot of a simple wall or floor construction can be used as a guide for determining sound transmission loss. Sound transmission loss should increase 6 db when either the weight of the panel or the frequency is doubled. This is known as the mass law, weight law, or limp wall law. Unfortunately the frequency range over which the mass law controls the sound transmission loss of most concrete walls and floors is not as great as the range that is important for sound control (100 to 4000 Hz.)

When the STCs of numerous constructions are plotted against weight per square foot, a practical mass law curve is obtained, Fig. 6-56. Concrete floor slabs cast on steel supporting members such as fluted or cellular steel decks or metal lath over bar joists perform acoustically very much like plain structural concrete floor slabs. However, dips result in STCs about three points lower than those of slabs not cast on steel supports of equal total weight (lower bound).

When a single concrete wall is used as a sound barrier, it is sometimes desirable to provide a resilient connection between the wall and the building frame to reduce the sound

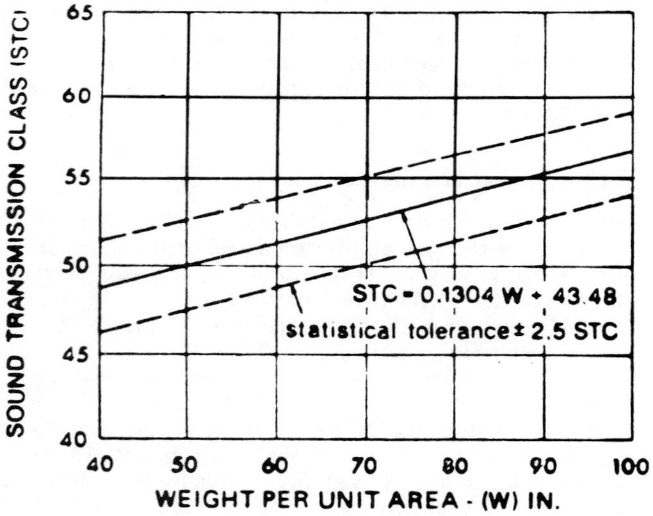

Fig. 6-56 Practical mass law curves.

TABLE 6-14 Typical Improvements for Wall, Floor, and Ceiling Treatments Used with Precast Concrete Elements

Treatment	*Increase in Ratings*	
	Air-borne (STC)	*Impact (IIC)*
Wall furring, ¾ in. insulation & ½ in gypsum board attached to concrete wall	3	0
Separate metal stud system, 1 ½ in. insulation in stud cavity & ½ in. gypsum board attached to concrete wall	5 to 10	0
2 in. concrete topping (24 psf)	3	0
Carpets and pads	0	43 to 56
Vinyl tile	0	3
½ in. wood block adhered to concrete	0	20
½ in. wood block and resilient fiber underlayment adhered to concrete	4	26
Floating concrete floor on fiberboard	7	15
Wood floor, sleepers on concrete	5	15
Wood floor on fiberboard	10	20
Acoustical ceiling resiliently mounted	5	27
if added to floor with carpet	5	10
Plaster or gypsum board ceiling resiliently mounted	10	8
with insulation in space above ceiling	13	13
Plaster direct to concrete	0	0

From: PCI Design Handbook—Precast and Prestressed Concrete, Prestressed Concrete Institute, Chicago, 1978.

transmission to adjacent walls, floors, and ceilings. A 5 to 7 db improvement can result. Usually a one-half inch fiber board will provide the desired isolation. When a resiliently mounted plaster or drywall surface finish is added to a concrete wall or ceiling, the STL is further improved. The effect of various floor and ceiling treatments can be predicted from results of previous tests. Table 6-14 indicates typical improvements. The improvements are usually additive, but in some cases the total effect may be slightly less than the sum. It should also be noted that carpets and acoustical ceilings can further reduce the sound level by absorbing sound in source or receiving room.

The performance of a building section with an otherwise adequate STL can be seriously reduced by a relatively small hole or any other path that allows sound to bypass the wall. Common flanking paths are openings around doors, pipes, electric boxes, and air ducts. Suspended ceilings in rooms where walls do not extend from the ceiling to roof or floor above allow sound to travel to adjacent rooms.

6.12.3 Impact Noise

Footsteps, dragged chairs, dropped objects, slammed doors, and plumbing generate impact noises. Even when airborne sounds are adequately controlled, there can be severe impact noise problems.

The test method used to evaluate systems for impact sound insulation is described in ASTM E492. For performance specification purposes the single number Impact Insulation Class (IIC) is used. The IIC of a floor–ceiling assembly is usually of the same magnitude as the STC.

IICs, like NRCs and STCs, should not be used as a complete substitute for performance data at specific frequencies. Sources of impact noise and acceptable noise spectra differ.

In general, thickness or weight per square foot of concrete does not greatly affect the transmission of impact sounds, as shown in the following table:

Thickness, in.	*Unit Weight of Concrete, pcf*	IIC
5	79	23
	114	24
	144	24
10	79	28
	114	30
	144	31

The most important floor construction or surface treatment items in the insulation of impact sounds are the resiliency, damping, and method of attachment to the frame. Structural concrete floors in combination with resilient materials effectively control impact sounds. One simple solution consists of good carpeting on resilient padding. The overall efficiency varies according to the characteristics of the carpeting and padding such as resilience, thickness, and weight. So-called resilient flooring materials such as linoleum, rubber, asphalt vinyl, vinyl asbestos, etc., are not entirely satisfactory directly on concrete, nor are parquet or strip wood floors when applied directly. Impact sounds also may be controlled by providing a discontinuity in the structure such as would be obtained by adding a resiliently mounted plaster or drywall suspended ceiling or a floating floor consisting of a second layer of concrete cast over resilient pads, insulation boards, or mastic. The thickness of floating slabs is usually controlled by structural requirements; however, 8 lb/sq ft is adequate in most instances.

The improvements to be made to a bare concrete floor having an IIC of 24 by adding various materials or constructions are shown in Table 6-14.

Solid connections between concrete floor slab, frame, and wall increase sound transmission at low frequencies. Impact sound can be reduced by soft joints between structural members or interior walls if structural stability is not impaired.

6.12.4 Vibration Isolation

The isolation of vibrations produced by equipment with unbalanced operating or starting forces can frequently be accomplished by mounting the equipment on a heavy concrete block placed on resilient supports. A slab of this type is called an inertia block. Inertia blocks can provide a desirable low center of gravity and compensate for thrusts such as those generated by large fans. For equipment with less unbalanced weight, a "housekeeping" slab is sometimes used below the resilient mounts to provide a rigid support for the mounts and to keep them above the floor where they remain cleaner and easier to inspect. This slab may also be mounted on pads of precompressed glass fiber.

The natural frequency of the total load on resilient mounts must be well below the frequency generated by the equipment. The required weight of an inertia block depends on the total weight of the machine and the unbalanced force. For a long-stroke compressor, five to seven times its weight might be needed. For high-pressure fans, one to five times the fan weight is frequently sufficient.

Simplified theory shows that for 90% vibration isolation a single resiliently supported mass should have a natural frequency about one-third the driving frequency. For 99% isolation the natural frequency should be one-tenth the driving frequency. Fortunately the natural frequency of a resiliently mounted mass is a function of the static deflection of the resilient supports; this makes determination of the natural frequency simple. A floor supporting resiliently mounted equipment must be stiff and heavy. If the static deflection of a floor supporting resilient mounts approaches the static deflection of the mounts, the floor then becomes a part of the vibrating system, and little vibration isolation is achieved. Therefore, floors supporting heavy rotating or reciprocating equipment often must be stiffer than floors designed to meet ordinary structural design criteria with floor deflection not more than about $1/6$ or $1/8$ of mount deflections. For example: a 20-ft span, with the usual $1/360$ of span deflection, will deflect 0.67 in. at maximum allowable load. This is the static deflection required of isolation mounts for 90% isolation of an 800 CPM machine. To obtain 90% isolation on such a floor the static deflection of the resilient mounts has to be increased by six to eight times. However, floors are seldom loaded to the design limits, and loading at center of span can often be avoided. With a moderate floor load good isolation is usually provided if the static deflection of the mounts is increased by an amount equal to $1/360$ of the floor span.

6.13 ELECTRICAL PROPERTIES*

Conduction of direct current or alternating current at audio frequencies through moist concrete is essentially conduction by an electrolyte. This may be visualized as the movement of ions in the evaporable water of the paste matrix, and in some cases in the pores of the aggregate particles.

The volume of evaporable water in paste found in the usual saturated concrete varies from about 60% at the time of mixing to about 40% when the portland cement is completely hydrated. This water contains ions, primarily Na^+, K^+, Ca^{++}, SO^{--}, and OH^-, whose concentrations vary with time. The concentrations of some ions increase, while those of others decrease. Any factor that affects the amount or properties of the liquid, or the kind or number of ions, will influence the resistance.

Moist concrete has a resistivity of the order of 10^4 ohm-cm, value in the range of semiconductors. Oven-dried concrete has a resistivity of the order of 10^{11} ohm-cm, a reasonably good insulator. Electrical resistivity of concrete increases with age and decreases with increasing salinity of mixing water. The resistivity of concrete is about five times that of paste. There are no definite values for the electrical resistivity of concrete, since these values depend on nature and amount of water in the concrete.

Various coatings have been tried on dry specimens to determine if the high resistance could be maintained. An epoxy coating is the most effective, but after 28 days of immersion in water, the resistance is only moderately greater than that of uncoated specimens. Some materials actually decreased the resistance of specimens.

*Most of the discussion on electrical properties is based on: Hammond, E., and Robson, T. D., "Comparison of Electrical Properties of Various Cements and Concretes," *The Engineer*, V. 199, Jan. 21, 1955, pp. 78–80; Jan. 28, 1955, pp. 114–115.

Natural phenomena, such as corrosion and current flows, generally involve dc. However, dc tends to polarize materials and electrodes, and thus to yield resistivity values that may or may not necessarily represent the true resistivity of the material. And, although ac does not polarize materials and thus yields their true resistivities, ac does not usually occur in nature.

Values discussed for electrical properties are intended to serve as a general guide. For any specific case, values may be affected by many variables, and should be determined by test.

6.13.1 Resistivity (dc)

This is the resistance in ohms, offered by a unit cube of concrete as measured between a pair of opposite faces under conditions such that surface leakage is negligible. Surface resistance of dried specimens is of a much higher order than volume resistance (see table).

dc Volume Resistivities (megohm-cm)

Specimen	*24h.*	*30d.*	*60d.*	*150d.*
Portland cement concrete	0.0023	0.05	0.1	0.3
Paste	0.0006	0.011	0.02	0.04

Specimens cured and dried at room temperature.

Volume resistivities of concretes *oven-dried* for several weeks and then cooled in a desiccator are shown in Fig. 6-57. Resistivities increase with voltage.

When oven-dried specimens are allowed to stand in air, resistivities decrease rather slowly, indicating that absorption of moisture into the interior is slow.

Surface resistivities of oven-dried and desiccator-cooled concretes decrease with increase in applied voltage (see table).

dc Surface Resistivities (megohms)

Concretes	*100 Volts*	*500 Volts*	*1000 Volts*
Portland cement			
Type I	5.0×10^3	1.9×10^3	1.6×10^3
Type III	5.0×10^3	2.5×10^3	2.3×10^3

Oven-dried specimens cooled to room temperature.

Storage of oven-dried specimens in the laboratory air greatly reduces surface resistivities (see table).

dc Surface Resistivities (megohms)

Concretes	*500 Volts*
Type I portland	92
Type III portland	80

Oven-dried specimens left in room air.

Surface resistivities of oven-dried specimens are much lower than the volume resistivities, in contrast to results for specimens dried in room air. This is logical, since specimens drying out in room air are more moist in the interior, whereas oven-dried specimens take up moisture from the air and become more moist on the surface. Since the surface resistance of the oven-dried specimens is (or becomes) very much less than the volume resistance, the former will, in general, be dominant in determining the effective resistance of most shapes.

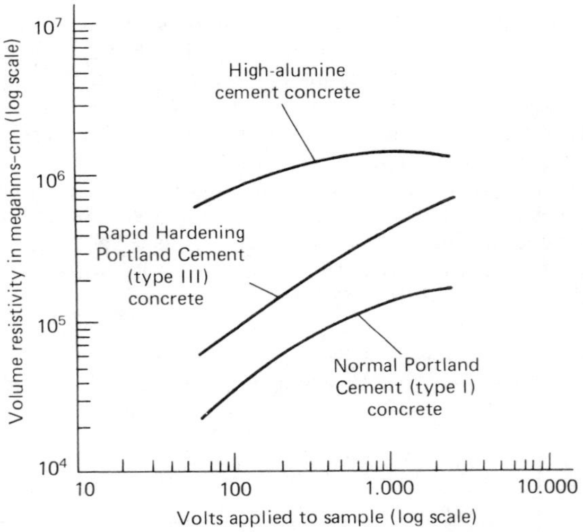

Fig. 6-57 Volume resistivities of 4:2:1 concrete after oven drying and cooling in a desiccator. (From: Hammond and Robson.)

The response of moist-cured concrete to the passage of direct current is shown in the left side of Fig. 6-58. The current due to an applied potential of 4 volts drops rapidly during the first minute and approaches equilibrium by 10 minutes. After 10 minutes, the potential increases to 6 volts and maintains this level for an additional 10 minutes. The current increases immediately with the increase in applied potential and rapidly reaches an equilibrium value. Twenty minutes after the first application of potential, the battery circuit was opened, and the potential of the electrodes in the concrete was measured during the following 10 minutes. This potential decreased rapidly at first, as shown by the right side of Fig. 6-58, but was still greater than 1 volt after 10 minutes.

The influences of normal curing in limewater at 73°F, steam curing, and autoclave curing are shown in Fig. 6-59. The initial resistance of the autoclaved specimens is quite high. Following the initial readings of the steam- and autoclave-cured specimens, they were stored in limewater at 73°F for additional measurements. At 90 days the resistances of the steam-cured specimens were highest. The steam- and autoclave-cured specimens contained fly ash, but the normal-cured specimens did not.

6.13.2 Resistance Temperature Coefficient

This is the change in resistance per degree of temperature change, and it may be based on any chosen temperature.

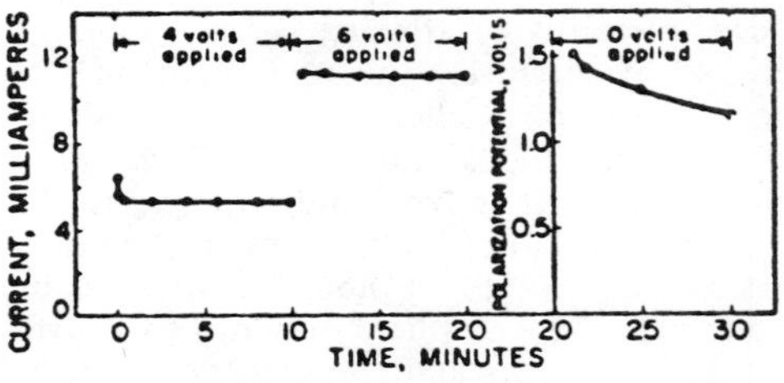

Fig. 6-58 Current-potential-time characteristics of concrete specimen. (From: Monfore, G. E., "The Electrical Resistivity of Concrete," *PCA Research Bulletin 224*, 1968.)

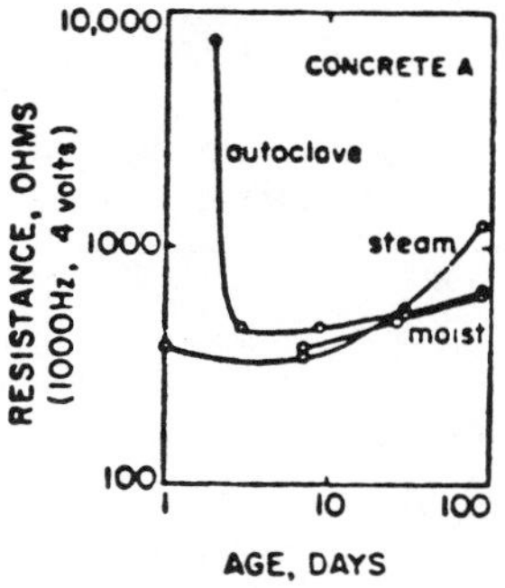

Fig. 6-59 Effect of type of cure. (From: Monfore.)

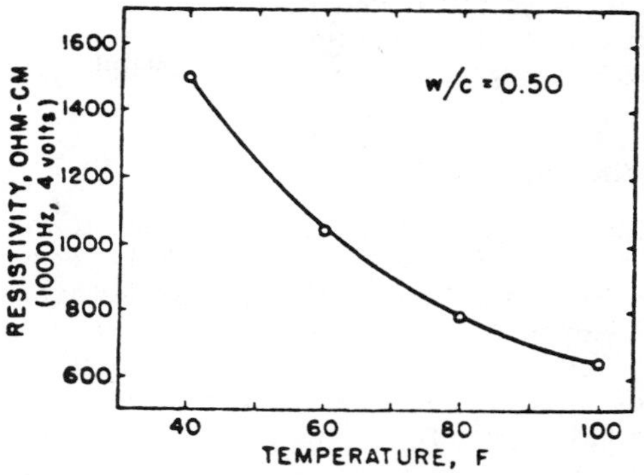

Fig. 6-60 Effect of temperature on resistivity of Paste A. (From: Monfore.)

Since the resistivity of electrolytes decreases with increasing temperature, the resistivity of moist paste, mortar, or concrete would also be expected to decrease with increasing temperature. Measurements of the resistivity of a paste over the temperature range from 40°F to 100°F are shown in Fig. 6-60. For this specific range, an average increase in temperature of 1°F caused an average decrease in resistivity of 1%.

6.13.3 ac Resistance

This is the true resistance responsible for loss of electrical energy, as distinguished from impedance, the apparent resistance, which is a function of resistance, capacitive reactance, and inductive reactance. The ac resistance of concrete is approximately equal to the impedance because the capacitive reactance is much greater than the resistance with which it is in parallel.

The agreement between the dc and ac determinations is quite satisfactory for the 50 c/s frequency. Agreement with the dc values is also good at all ac frequencies (up to 25,000 c/s, Table 6-15). On the other hand, aluminous cement specimens show a significant decrease in ac resistance with increase in frequency, and a smaller effect of this kind is indicated by some portland cement concretes. These effects

TABLE 6-15 Effect of Age and Current Frequency on the ac Characteristics of Concretes and Neat Cements (From: Hammond and Robson)

Type of Specimen	Type of Cement	Air-drying time after curing (days)	Resistance (megohms)				
			dc	50 c/s	500 c/s	5000 c/s	25,000 c/s
4 in. cubes 4:2:1 concrete External plate electrodes*	High alumina	5	—	0.0189	0.0173	0.0152	0.0139
		6	—	0.0216	0.0197	0.0171	0.0159
		18	—	0.0390	0.0351	0.0304	0.0275
		40	—	0.0652	0.0577	0.0494	0.0441
	Type I	7	0.00104	0.000938	0.000928	0.000925	0.000924
		42	—	0.00307	0.00306	0.00303	0.00302
		113	0.009	0.00820	0.00800	0.00761	0.00730
	Type III	39	—	0.00275	0.00273	0.00269	0.00269
6 in. cubes 4:2:1 concrete Embedded plate electrodes†	High alumina	138	—	0.0412	0.0360	0.0310	0.0280
		182	—	0.0526	0.0460	0.0394	0.0353
	Type I	126	—	0.00196	0.00194	0.00193	0.00194
	Type III	123	—	0.00158	0.00157	0.00155	0.00154
4 in. by 4 in. by 1 in. prisms. Neat cement. External plate electrodes.‡	High alumina	13	0.006	0.00549	0.00478	0.00408	0.00371
	Type I	9	0.00018	0.000151	0.000147	0.000143	0.000141
	Type III	9	0.000118	0.000118	0.000116	0.000114	0.000120

**Resistivity value-10 × measured resistances.*
†Resistivity values-30 × measured resistances (based on total cross section of specimen).
‡Resistivity values 40 × measured resistances.

can be attributed to the idiosyncrasies of imperfect dielectrics, but at least as far as the portland cement concretes are concerned, they seem to be of nearly negligible magnitude.

6.13.4 Capacitive Reactance

This is the opposition offered by capacitance to the flow of alternating current. Like resistance and impedance, it is measured in ohms. None of these quantities has any basic significance per se, but only in reference to a given set-up. Where the concern is simply with a given cross section and length of material like concrete at a given ac frequency, there would apparently be advantages in computing not only the resistance per unit cube (resistivity) but also the capacitive reactance per unit cube. At the given frequency, such a "specific" quantity would bear the same relation to the resistivity as the actual capacitive reactance of a given specimen bears to the resistance. Instead, however, the dielectric constant is computed, which is a ratio and hence dimensionless. It is simple to compute the ratio of the capacitive reactance to the resistance and to use this ratio along with the resistivity in estimating electrical behavior. This ratio enables one to determine, for example, when the impedance has practically the same value as the resistance and when it does not.

The values for capacitive reactance decrease markedly with an increase in the ac frequency. A large decrease of this nature is to be expected because of the reciprocal relationship shown by the formula, capacitive reactance $X = 1/2\ \pi f C$. However, the decrease in X is actually much less pronounced than the increase in f because the capacitance, C, decreases with increase in frequency. As indicated earlier, the ratio of X to R is of interest, and some values are given in the table.

Specimen	Cement	Air-drying, Days	Ratio of Capacitive Reactance to Resistance			
			50 c/s	*500 c/s*	*5000 c/s*	*25,000 c/s*
4-in. cubes of concrete	portland Type I	7	169	171	69	34
		42	208	149	35	21
		113	129	50	14	9
	portland Type III	39	290	146	39	24

It may be computed from the equation for impedance Z, namely $1/Z^2 = 1/X^2 + 1/R^2$, that when the ratio X/R is more than 10, the difference between R^2 and Z^2 is less than 1%.

6.13.5 Capacitance

In static electricity the capacitance measured in farads is the number of coulombs of electrical charge that a condenser will take per unit of voltage across its terminals. For alternating current electricity, a condenser of a given capacitance, C farads, produces capacitive reactance, X ohms, in accordance with the formula $X = 1/2\ \pi f C$, where f is the frequency of the alternating current in cycles per second. The capacitance decreases markedly with increase in frequency. Though the effect is not as great as the change in frequency, it is large. Capacitance per se has no fundamental significance for concrete, since it is naturally dependent on the dimensions of the specimen. The basic factor is the dielectric constant of the concrete. The change of the capacitance with frequency does have basic significance, but since it is proportional to the change of the dielectric constant with frequency, it will be examined in terms of the latter quantity.

6.13.6 Dielectric Constant

The dielectric constant of a substance is the factor by which the capacitance of a suitable condenser is multiplied when the substance is used between the plates or electrodes instead of empty space, i.e., vacuum. Since the concrete specimen responds to alternating current like a parallel arrangement of resistance and capacitance, the space between the electrodes can be regarded both as a resistance and as a parallel plate condenser. From the capacitance C, the area of a single plate, A, and the distance between plates, d, the dielectric constant can be calculated from the formula:

$$C = \frac{kA}{4.45d \times 10^6} \text{ microfarads}$$

when lengths are in inches. For accurate calculation, it is usually said that d should be small relative to the other dimensions. This confines the lines of force to the dielectric. However, as will be shown (see table), k turns out to be quite large for concrete (unless the ac frequency is quite high), and this would seem to obviate the need for a small d.

Specimen	Cement	Air-drying, Days	Dielectric Constant, k			
			50 c/s	*500 c/s*	*5000 c/s*	*25,000 c/s*
4-in. cubes of concrete	portland Type I	7	22,000	2,200	560	200
		42	5,600	780	330	110
		113	3,300	890	330	110
	portland Type III	39	4,400	890	330	110
6-in. cubes of concrete, embedded electrodes	portland, I	126	10,000	520	110	19
	portland, III	123	10,000	560	150	74
4 in. × 4 in. × 1 in. prisms of neat cement	portland, I	9	98,000	8,400	1,800	560
	portland, III	9	140,000	15,000	1,400	730

It is apparent that these calculated dielectric "constants" are not constant properties even for a given degree of drying but decrease very markedly with increase in frequency. Also, especially at the low frequencies, the values are extremely high relative to values of most dielectrics. The dielectric properties also are influenced by the salt content of the pore fluid, environmental conditions, and concrete thickness.

It would seem, in any case, that although the dielectric "constant" that is calculated is of unusual magnitude and unusual variability with frequency, it nevertheless has the usual significance from the practical point of view. A dielectric "constant" is, except for a constant factor, the capacitance of a unit cube of the material, and determines the capacitive reactance. That is the case here, and what is calculated apparently has the usual practical significance except that the dependence on frequency must be given due attention.

6.13.7 Dielectric Strength

This is the maximum voltage per unit thickness that a dielectric can stand before it "breaks down electrically," and a spark, or disruptive discharge, passes through it. Concretes made with different types of cement give much the same results when tested with dc and ac (50 c/s), and this is true irrespective of whether they are oven-dried or air-dried. Most values are between 10 and 15 kilovolts/cm.

6.13.8 Electrically Conductive Floors

Concrete floors in hospital operating rooms and other places, where sparking would be a hazard, may be made electrically conductive with an integral acetylene carbon-black admixture, or with specially prepared metallic aggregates or powders which are worked into the fresh concrete surface. For terrazzo work, 2% acetylene carbon-black by weight of cement has been used satisfactorily in a marble-chip topping over a sand bedding course that also contains 3% acetylene carbon-black by weight of cement. Detailed installation recommendations for conductive terrazzo are available from the National Terrazzo and Mosaic Association. Construction techniques are somewhat specialized for conductive floors.

The electrical resistance of the floors may be determined by reference to the U.S. Navy Bureau of Yards and Docks Specification 48Y, "Static Disseminating and Spark-Resistant Floors" or to *Code for the Use of Flammable Anesthetics* (Recommended Safe Practice, for Hospital Operating Rooms), Publication No. 56, National Fire Protection Association, Boston Mass. Electrically conductive floors often are also required to be spark resistant under abrasion or impact; a typical test for this property is given in the Bureau of Yards and Docks specification.

Concrete made for conductive flooring gives standard test resistance of about 0.03 to 0.3 megohms; regular concretes (perhaps approximately equilibrated at 50% humidity) give resistances of 8 to 80 megohms, as determined at one fixed voltage. High-alkali portland cements show lower resistance than low-alkali cements, and at 0.9% alkalies the high alkali cements gave 8 to 11 megohms. At 15% relative humidity the regular concretes gave values about five to eight times as high as at 50% humidity.

Cleaning procedures for conductive floors should not adversely affect the conductivity characteristics. It is preferable not to use any type of coating or wax on the floor unless it has been pretested to maintain conductivity. The floor should be prerinsed with plain water before each washing, and only a nonionic conductive floor cleaner should be used for washing. After each washing, the floor should be rinsed with plain water. Phenolic-based germicides should not be used, as they may form insoluble salts with the calcium in the floor, resulting in the buildup of an insulating layer.

6.14 REINFORCEMENT

6.14.1 Reinforcing Bars*

U.S. standard specifications for reinforcing bars are established by the American Society for Testing and Materials (ASTM). These standards govern strength grades, rib patterns, sizes, and markings of bars. All bars are furnished "deformed"—that is, with lugs, ribs or protrusions rolled into the bar, which increase bond performance with concrete.

1. Specifications and Sizes. Reinforcing bars are produced from three kinds of steel—new billet, axle, or rail—in four grades of yield strengths. The yield strength of a bar is its useful strength. The ASTM bar specifications are:

- Specifications for Deformed and Plain Billet-Steel Bars for Concrete Reinforcement (ASTM A615)
- Specifications for Rail-Steel Deformed Bars for Concrete Reinforcement (ASTM A616)
- Specifications for Axle-Steel Deformed and Plain Bars for Concrete Reinforcement (ASTM A617)
- Specifications for Low-Alloy Steel Deformed Bars for Concrete Reinforcement (ASTM A706)

Table 6-16 summarizes the math properties of these bars. If steel from a foreign source is to be used, its quality should be checked carefully, since it may fall below ASTM requirements for uniformity; it may also tend to be more brittle, thereby introducing serious problems in bending, when welding, or during handling at temperatures below 40°F.

There are 11 standard deformed reinforcing bar sizes. These are designated #3 through #11, #14, and #18. The size of reinforcing steel bars is designated by a number that refers to its nominal diameter (equivalent to the diameter of a plain bar having the same weight per foot as the deformed bar). For bar sizes #3 through #8 this designation represents the nominal diameter of the bar in eighths of an inch, Table 6-17.

Manufacturers mark the bar size, producing mill identification, and type of steel on the reinforcing bars. Only high-strength steel bars, Grades 60 and 75, are also identified by a grade marking.

2. Stress-Strain Curves. Ultimate strength design of reinforced concrete sections in accordance with the 1983 and previous *ACI Codes* is based on the assumption of an ideal elasto-plastic (flat-top) stress-strain relationship for the reinforcement. However, the actual stress-strain relationship for high-strength reinforcement may not be ideally elasto-plastic. Some reinforcing bars, particularly those meeting the ASTM specifications for Grade 75 steel, and some European and British steels do not always have a well-defined yield point and a flat plateau. Moreover, even those steels that do have a well-defined yield point, including most Grade 40 and some Grade 60 steels, exhibit strain-hardening; some of these have a very short flat plateau, and others a considerably long one.

Representative force-strain curves for five sizes in each grade of steel are shown in Fig. 6-61. While the curves for the Grade 60 bars are typically shown as having a flat yield plateau, some of the curves for these bars exhibit characteristics of the curves for the Grade 75 bars. The modulus of elasticity, E_s, for all reinforcing bars is practically the same and is taken as $E_s = 29 \times 10^6$ psi.

3. Bending. Some of the specified minimum bend radii permitted by the American Concrete Institute's *Building Code Requirements for Reinforced Concrete* (ACI 318) are more severe than those required by applicable ASTM specifications unless the Supplemental Requirements are specified. In such cases, occasional breakage can be expected unless the steel, by chance, is more ductile than required by the specifications. In any case, breakage may be expected when bending is attempted at low temperatures. Heating of bars to facilitate bending or straightening should be allowed only with the permission of designer.

Partially embedded bars can be successfully rebent (or

*Properties of welded wire fabric are not covered. For information refer to *Manual of Standard Practice—Welded Wire Fabric*, Wire Reinforcement Institute, 1972.

The ASTM specifications for reinforcing bars (A615, A616, A617, and A706) require identification marks to be rolled into the surface of one side of the bar to denote the producer's mill designation, bar size, and type of steel. Also required is a grade mark for Grade 60 and a mark indicating minimum yield strength for A706. Grade 40 and Grade 50 bars show only three marks (no grade or minimum yield strength mark) in the following order:

1st—Producing Mill (usually a letter)
2nd—Bar Size Number (#3 through #18)
3rd—Type Steel: N for New Billet
S for Supplemental Requirements A615
A for Axle
I for Rail
W for Low Alloy

The grade mark for Grade 60 and the minimum yield designation for A706 may be either the number 60 or one (1) single line.

A grade mark or minimum yield designation line is smaller and between the two main ribs which are on opposite sides of all U.S.-made bars. When a number is used for the grade mark or minimum yield designation, it is 4th in order.

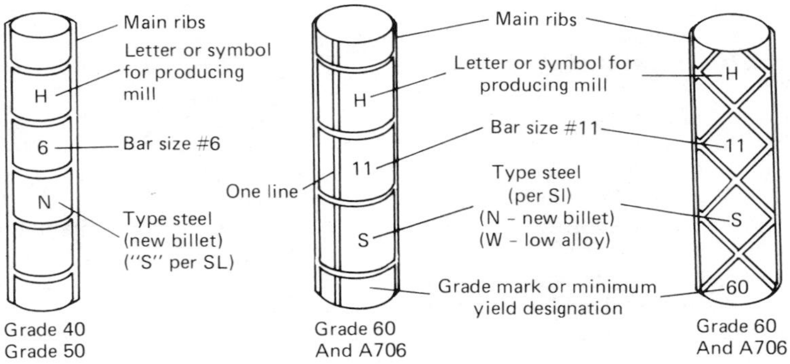

VARIATIONS: Bar identification marks may also be oriented to read horizontally (at 90° to those illustrated above).

Grade mark and minimum yield designation lines must be continued at least 5 deformation spaces.

Grade mark and minimum yield designation numbers may be placed within separate consecutive deformation spaces to read vertically or horizontally.

bent for the first time, which should be less critical) if they are first preheated to 1100–1200°F and then bent as gently and in as gradual an arc as possible. If there is no failure at the bend area, the reworked bars should perform as originally intended. Heating must be performed in a manner that will avoid damage to the concrete. If the bend area is within 6 in. or so of the concrete, some protective insulation may have to be applied. The heating operation should be done in such a manner that the temperature of the steel does not exceed 1200°F as measured by temperature-indicating crayons or other suitable means. The heated bars should not be artificially cooled (such as by water or forced air) until after cooling to at least 600°F. Rapid quenching of heated bars, such as might occur if the bars were dropped in snow or cold water, is detrimental to the steel.

There is the possibility of breakage or cracking for No. 11 bars when bending/straightening partially embedded bars. Certain deformation patterns perform better than others. The choice of successfully bending and/or straightening No. 11 bars increases if the bend area is preheated uniformly to 1400–1500°F, and extreme care is exercised in the bending operation.

4. Fatigue. Fatigue of steel reinforcing bars has not been a significant factor in their application as reinforcement in concrete structures. Fatigue strength of deformed bars has been investigated by testing samples of the bar, either by themselves in axial tension, or embedded in concrete beams in tension or flexural bending. The fatigue strength of a bar is a hypothetical value of stress for failure at exactly N cycles as determined from an S–N diagram; see Section 6.9. The hypothetical value of S referred to will be the stress range, S_r, at which 50% of the specimens of a given sample would be expected to survive N cycles. Disagreement exists as to whether a bar embedded in concrete has a lower or higher fatigue strength than one tested in air. However, there are indications that any difference should be small if the height and shape of the transverse lugs provide good bond between the steel and concrete.

While chemical composition, grain size, microstructure, and other metallurgical properties affect the fatigue strength of reinforcing bars, those effects are small compared with effects caused by the physical characteristics of the bar.

The main variables affecting fatigue strength are: (1) range of stress, (2) minimum stress, (3) bar size, (4) deformation geometry, (5) mechanical characteristics of the bar, and (6) fabrication procedures.

Reinforcing bars have a practical fatigue limit. The transition from the finite-life region to the long-life region of

TABLE 6-16 Physical Requirements for ASTM Deformed Reinforcing Bars

Type of Steel and ASTM Specification No.	*Size Nos. Inclusive*	*Grade*	*Tensile Strength Min., psi*	*Yield[1] Min., psi*	*Percent Elongation in 8" Minimum*	*Cold Bend Test Pin Diameter[3] (d = nominal diameter of specimen)*
Billet Steel A615	3–6	40	70,000	40,000	#3 ... 11 #4, #5, #6 ... 12	Under Size #6 ... 4d [2]per S1 ... $3\frac{1}{2}$d #6 ... 5d
	3-11 14, 18	60	90,000	60,000	#3, #4, #5, #6 ... 9 #7, #8 ... 8 #9, #10, #11, #14, #18 ... 7	Under Size #6 ... 4d [2]per S1 ... $3\frac{1}{2}$d #6 ... 5d #7, #8 ... 6d [2]per S1 ... 5d #9, #10, #11 ... 8d [2]per S1 ... 7d [2]#14, #18, (90°) ... 9d
Rail Steel A616	3-11	50	80,000	50,000	#3, #7 ... 6 #4, #5, #6 ... 7 #8, #9, #10, #11 ... 5	Under Size #9 ... 6d #9, #10 ... 8d #11 ... 8d (90°)
	3-11	60	90,000	60,000	#3, #4, #5, #6 ... 6 #7 ... 5 #8, #9, #10, #11 ... 4.5	Under Size #9 ... 6d #9, #10 ... 8d #11 ... 8d (90°)
Axle Steel A617	3-11	40	70,000	40,000	#3, #7 ... 11 #4, #5, #6 ... 12 #8 ... 10 #9 ... 9 #10 ... 8 #11 ... 7	Under Size #6 ... 4d #6 and Larger ... 5d
	3-11	60	90,000	60,000	#3, #4, #5, #6, #7 ... 8 #8, #9, #10, #11 ... 7	Under Size #6 ... 4d #6 ... 5d #7, #8 ... 6d #9, #10, #11 ... 8d
Low Alloy Steel[3] A706	3-11 14, 18	60[6]	80,000[4]	60,000[5]	#3, #4, #5, #6 ... 14 #7, #8, #9, #10, #11 ... 12 #14, #18 ... 10	Under Size #6 ... 3d #6, #7, #8 ... 4d #9, #10, #11 ... 6d #14, #18 ... 8d

[1]*Yield point or yield strength. See specifications.*
[2]*Under Supplemental Requirements of A615.*
[3]*Test bends 180 unless noted otherwise.*
[4]*Tensile strength shall not be less than 1.25 times the actual yield strength (A 706 only).*
[5]*Maximum yield strength 78,000 psi (A 706 only).*
[6]*Minimum yield designation.*

the stress range–fatigue life relationship occurs at about 1 million cycles; see Fig. 6-62. Stress range is the predominant variable affecting the fatigue strength of the test bars in the finite-life region, while stress range accounts for only a small part of the total variation in fatigue life of bars included in the long-life region. Therefore, that fatigue limit, for practical purposes, could be considered as a stress range at which an infinite number of loadings could be sustained.

The minimum stress level of a stress cycle has a sharply defined effect on fatigue strength in the finite-life region. When tensile, an increase in minimum stress causes a decrease in fatigue strength. When compressive, an increase in minimum stress causes an increase in fatigue strength. Changing the minimum stress of a stress cycle by 3 ksi was found to be equivalent to changing the stress range by about 1 ksi.

Effects of bar size and specimen geometry are interrelated because stresses are commonly considered in terms of midfiber depth. For large bars in shallow members, it is appropriate to consider the stress at the extreme fiber of the bar rather than the midfiber depth. However, even for axially stressed bars, the fatigue strength decreases as the bar size increases. The change is not marked with #5 bars showing strengths 10 to 20% greater and #11 bars 10 to 15% less than that of #8 bars.

While the deformations on a bar ensure good bond, they are also the principal factor causing a reduced fatigue strength. Most North American bars have lug base radii to lug height ratios of 0.1 to 0.2. If that ratio is increased to 1 to 2, the fatigue strength can be increased up to 20%. Further increases in that ratio decrease the bond strength and do not improve the fatigue strength. Other features of the

TABLE 6-17 ASTM Standard Reinforcing Bars

Bar Size Designation	Weight Pounds, per ft	Nominal Dimensions–Round Sections		
		Diameter, in.	Cross-Sectional Area, sq. in.	Perimeter, in.
#3	0.376	0.375	0.11	1.178
#4	0.668	0.500	0.20	1.571
#5	1.043	0.625	0.31	1.963
#6	1.502	0.750	0.44	2.356
#7	2.044	0.875	0.60	2.749
#8	2.670	1.000	0.79	3.142
#9	3.400	1.128	1.00	3.544
#10	4.303	1.270	1.27	3.990
#11	5.313	1.410	1.56	4.430
#14	7.65	1.693	2.25	5.32
#18	13.60	2.257	4.00	7.09

geometry of the bar, such as lug rise angle and lug width, cause variations in the stress range for failure of up to 5%.

The mechanical characteristics for the bar have little effect on the stress range for failure of deformed bars. Further maximum stresses exceeding the yield strength do not affect the fatigue strength, and cycling at stresses less than the endurance limit improves the fatigue strength slightly for subsequent stressing.

For plain bars, the presence of normal rust and mill scale does not affect fatigue resistance. However, in contrast to the situation for deformed bars, the stress range for failure increases as the bar grade increases.

A greatly reduced fatigue strength in both the finite- and long-life regions was observed in bars bent around a 6-in. mandrel. The sharper the bend, the greater was the reduction in fatigue strength. However, because of the type of

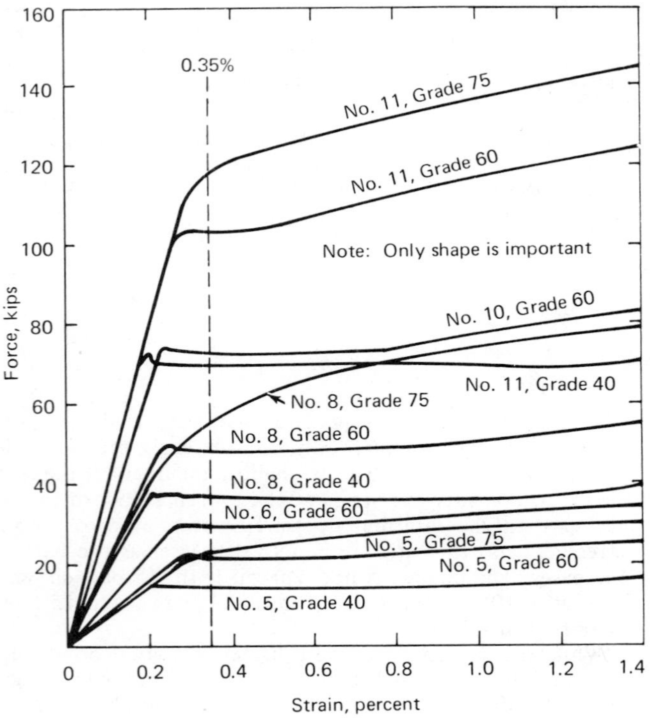

Fig. 6-61 Representative force-strain curves for test bars. (From: Hanson, J. M., et al., "Fatigue Strength of High-Yield Reinforcing Bars," *NCHRP Report Project No. 4-7*, PCA, Feb. 1970.)

specimens used, these tests cannot be considered representative of the conditions to which a bent-up bar in an ordinary reinforced concrete beam is subjected.

In fatigue tests on heavy girders containing bent-up bars, fatigue fracture of the bars occurred away from the bends, therefore, bending was not considered to have a detrimental effect on the fatigue strength of the bars.

A practical answer to the bending problem is simply to avoid bends where the stress range exceeds about 12 ksi. Alternatively, bends should be made hot or the bar stress-relieved subsequent to bending.

When proper welding procedures are followed, butt-welded reinforcing bars will have fatigue strengths as great as unwelded bars. Where welding procedures are careless or there is considerable weld spatter, fatigue strengths may be reduced by as much as 50% compared to strengths for properly welded bars.

The fatigue strength of bars with stirrups attached by tack welding is about one-third less than bars with stirrups attached by wire ties. All of the fatigue cracks are initiated at the weld locations when tack welding.

Even where proper welding procedures are followed and the fatigue strength of the bar is unaltered in the weld zone, the strength and ductility in that zone are likely to be reduced. Therefore, it is important only to bend welded bars remote from welds when they are used in applications where fatigue is a consideration.

In tests of welded joints in reinforcing bars and of joints in welded bar mats, the fatigue strength is considerably reduced from that of unwelded bars. However, the type of joint in welded bar splices is found to have a large effect. A 60° single-V butt joint appears to have the best fatigue characteristics.

Because North American reinforcing bars are specified only by minimum physical properties, some bars are unsuitable for galvanizing. Galvanizing can accelerate strain-age embrittlement effects so that bars become brittle virtually instantaneously. Galvanizing is unsuited to steels susceptible to strain embrittlement and therefore particularly to cold worked steels. Thus especially where fatigue is a consideration, galvanizing should be performed prior to bending. Unfortunately, such a restriction will often require double handling of a bar and retouching of the galvanizing in the bent region. Current good practice calls for galvanized bars to be chromated in order to passivate them and to eliminate any possible problems due to hydrogen absorption into the bar as a result of a reaction between the cement paste and the galvanized surface. The suitability of a bar lot for galvanizing can be rapidly checked by galvanizing a representative sample and testing for ductility in a reversed bend or hammer impact test.

In general, galvanizing slightly reduces the ultimate tensile strength, yield strength, and ductility of a bar. Increasing the pickling time beyond five minutes or use of abnormally high galvanizing temperatures further reduces strength and ductility. In general, galvanized bars passing ASTM A615 requirements for strength, ductility, and bend will have strengths averaging about 7% less than the strengths of similar untreated bars.

Epoxy-coated reinforcing bars have shown fatigue strengths about 10% higher than that of black uncoated bars from the same heat. The increase in strength is probably a result of the bars being sandblasted prior to the application of the epoxy.

For bars left exposed to the atmosphere in a marine environment, the fatigue strength of black bars decreases by about 10% in the first six months. In accelerated salt spray fog tests, of beams containing 0.01 to 0.02 in. wide cracks, corrosion was found to be confined primarily to the cracked

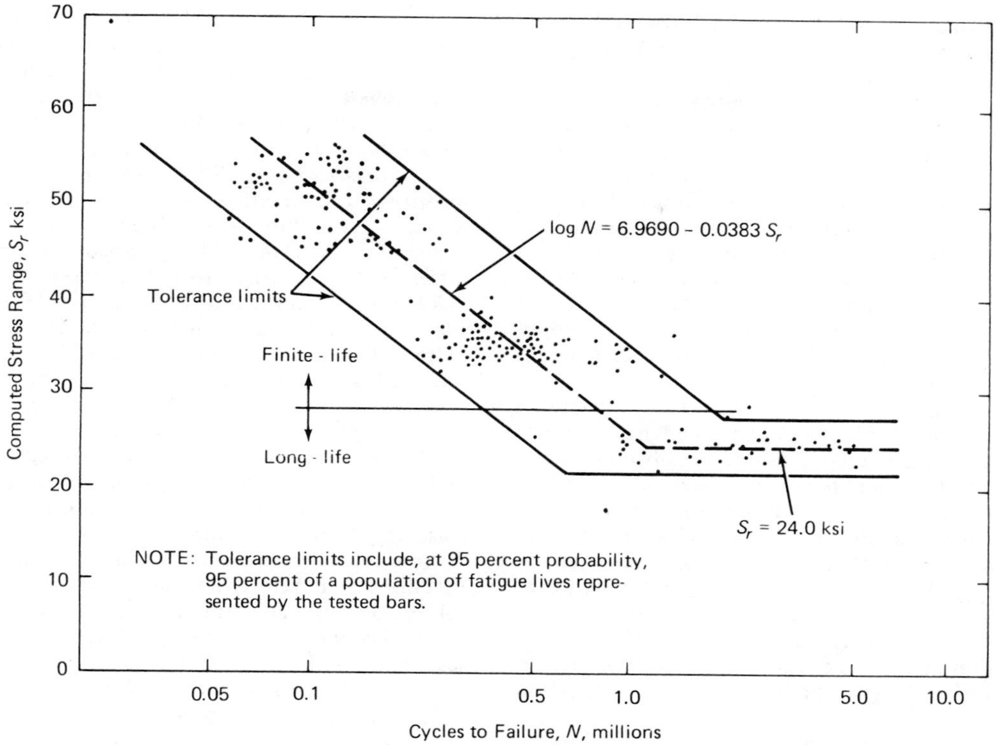

Fig. 6-62 Experimental test data on fatigue of deformed bars from a single U.S. manufacturer. Lower limit is also lower limit to previously published test data on North American-made bars. (From: Hanson et al., "Fatigue Strength of High-Yield Reinforcing Bars," *NCHRP Report Project No. 4-7*, PCA, Feb. 1970. See also: Helgason, T., Hanson, J. M. Somes, N. F. Corley, W. G., and Hognestad, E., *Fatigue Strength of High Yield Reinforcing Bars* (RD045.01E), Portland Cement Association, 1976.)

region, to progress rapidly for the first 30 days, and then to decrease in rate of development with time. For fatigue tests on those beams, the stress ranges for failure decreased by about 20% after 12 months' exposure. For similar beams with galvanized bars, the decrease in fatigue strength with time was less marked and about half that for black bars. However, the degree of improvement is less than that which can be achieved through selection of a bar with good lug geometry. Further, in any construction galvanized and black bars cannot be mixed. Black bars in the vicinity of the galvanized bars corrode owing to electrochemical actions.

The effective depth of a reinforced concrete beam had no direct influence on fatigue strength of the main reinforcement. Also, rates of loading up to several thousand cycles per minute and temperatures up to several hundred degrees C are normally not considered to have a significant effect on fatigue strength.

The stress range in straight deformed hot-rolled reinforcing bars, under repeated application of service loads, should not be permitted to exceed the value given by the following equation unless a higher value is substantiated by tests:

$$f_r = 21 - 0.33 f_{min} + 8(r/h)$$

where:

f_r = stress range, in ksi
f_{min} = corresponding minimum tensile stress (positive) or maximum compressive stress (negative), in ksi
$r:h$ = ratio of base radius to height of rolled-on deformation

When $r:h$ is not known, a value of 0.3 may be used.

For welded wire fabric and bar mats, the initial setting for the automatic welding heads is the principal factor dictating the fatigue characteristics. The fatigue characteristics of the welded intersection are dictated by stress concentration effects caused by the welding process. Under repeated loading, there is a rapid deterioration of the bond between the wire and the concrete, so that for wires stressed to the endurance limit all anchorage is provided by the cross wires after 10^4 cycles of loading.

A comparison of the *S–N* curves for wire fabric and bar mats with those for deformed bars indicates that an endurance limit may not be reached for the fabric and mats until about 5×10^6 cycles, whereas a limit is reached for the bars at about 1×10^6 cycles. However, the total amount of data in the long life range for fabric and mats is extremely limited and insufficient for reliable comparison. A realistic stress range for 2×10^6 cycles of loading for welded wire fabric is 15 to 17 ksi.

6.14.2 Prestressing Tendons

High-tensile steel is almost the universal material for producing prestress and supplying the compressive force to prestressed concrete. High-strength tendons are required to allow for stress losses due to inelastic deformation of the concrete and other causes. The tendon should behave elastically (stress–strain curve is straight line) up to at least the initial tensioning force, so that its behavior as a tendon is predictable and so that the measurement of elongation can be used to check the force in the tendon. A tendon should also have a large amount of plastic deformation before failure, to provide a steel that cannot fail abruptly without prior indication of distress.

High-tensile steel for prestressing (pretensioned and post-tensioned) takes one of three forms: wires, strands, or bars. The strength and bond of wire decreases with increasing diameter, leading to practical limits of about $^9/_{32}$ in. diameter in prestressing applications. For post-tensioning, wires are grouped in parallel into tendons utilizing as many as 186, $^1/_4$-in. wires. Strands are fabricated in the factory by twisting wires together. Strands also can be grouped into multistrand tendons. This decreases the number of units to be handled in tensioning operations and may reduce anchorage costs. By grouping strands it is also possible to obtain large steel cross sections. Strands have almost completely replaced wires in pretensioning operations. Heat-treated alloy steel bars are used for post-tensioning in many applications where the advantages of large diameter offset the reduced strength capability.

The stress–strain characteristics of wire in the as-drawn condition are nonelastic (Fig. 6-63), and the wire has a permanent curvature. To overcome these disadvantages the wire is mechanically straightened and then subjected to low-temperature heat treatment (about 600°F). This process, referred to as stress relieving, causes the wire to behave elastically up to a high proportion of the maximum stress. The as-drawn wire can also be subjected to combined low-temperature heat treatment and high tension. This process, known as stabilization, improves the maximum strength and the elongation at the maximum stress and typically reduces losses due to relaxation–produces low relaxation strand.

There is no yield point for prestressing steels as exhibited in the curve for structural carbon steel; at least, there is no marked yield point before the ultimate strength is reached. Instead, between the proportional limit and the ultimate strength, the elongation of the steel takes place in a regularly increasing manner, without any point where strain increases suddenly without a corresponding increase in stress. However, the idea of yield point strength is so well established in thinking of carbon steel members that an arbitrary means of fixing a yield strength, for comparing high-strength steels of the sort used for prestressing and for use in design equations, has been established. This is generally called the 0.2% offset yield strength. Specifications for prestressing wire and strand require that the yield strength be measured by the 1.0% extension under load method.

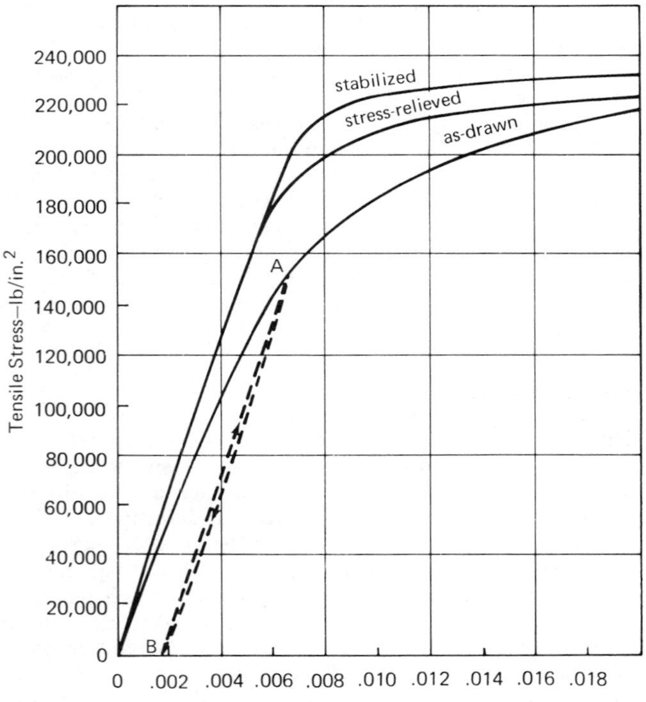

Fig. 6-63 Effect of postdrawing treatments on wire.

1. Specifications. Wire, strands, and bars for tendons in prestressed concrete conform to "Specifications for Uncoated Seven-Wire Stress-Relieved Strand for Prestressed Concrete" (ASTM A416), "Specifications for Uncoated Stress-Relieved Wire for Prestressed Concrete" (ASTM A421), or "Specifications for Uncoated High-Strength Steel Bar for Prestressing Concrete" (ASTM A722). Strands or wire not specifically itemized in ASTM A416 or A421 may be used, provided they conform to the minimum requirements of these specifications and have no properties that make them less satisfactory than those listed in ASTM A416 or A421. Wires used in making strands should be cold-drawn and either stress-relieved or low relaxation, in the case of uncoated strands, or hot-dip galvanized, in the case of galvanized strands.

Low relaxation strand (in some cases in the various specifications it is referred to as low relaxation strand and in others as stabilized strand) is now making its presence felt over the once commonly used stress-relieved material because it has significantly less loss of initial tension. This can result in improved and more predictable service performance, and, in many cases, will allow a higher load carrying capability.

Like stress-relieved strand, low relaxation strand goes through all the steps of normal strand processing. However, low relaxation strand is simultaneously tensioned while it is at the elevated temperature used in the stress relieving. The result is a permanent elongation of approximately 1% and an increase in yield strength of 5% over stress-relieved strand. The permanent elongation developed while the strand is at an elevated temperature gives it vastly superior resistance to relaxation. The internal flow of metallic structure, which takes place gradually in stress-relieved strand under tension, occurs almost simultaneously during the stabilization process for low relaxation strand. The result is that low relaxation strand has very little remaining capacity for creep or relaxation.

Specifications require that the low relaxation strand differ from ordinary stress-relieved strand in only two respects: first, it must meet certain relaxation loss requirements, as measured by ASTM E328 (ordinary stress-relieved strand has no such requirement); and second, the minimum yield strength, as measured by the 1% extension under load method, must be not less than 90% of the specified minimum breaking strength, as opposed to 85% for normal stress-relieved strand. All other requirements are the same.

2. Wire. The stress–strain characteristics in typical tensile tests of uncoated stress-relieved wires are shown in Fig. 6-64. Galvanized wire of 0.196 in. diameter with a minimum ultimate strength of 220,000 psi has also been used in prestressed concrete, although other sizes also can be manufactured. The modulus of elasticity of uncoated wire is approximately 29×10^6 psi. While this value is sufficiently accurate for some design calculations, it is recommended that the steel fabricator's load-elongation curve for the particular wire be used in computing the elongation required in jacking wires to a specified tension.

Table 6-18 shows a number of properties of prestressing wire used in the United States. Wires are manufactured according to the U.S. Steel Wire Gage, No. 2 of which has a diameter of 0.2625 in. and No. 6 a diameter of 0.1920 in. Neither of these is the exact equivalent of the millimeter counterparts. Hence, when the European types of anchor-

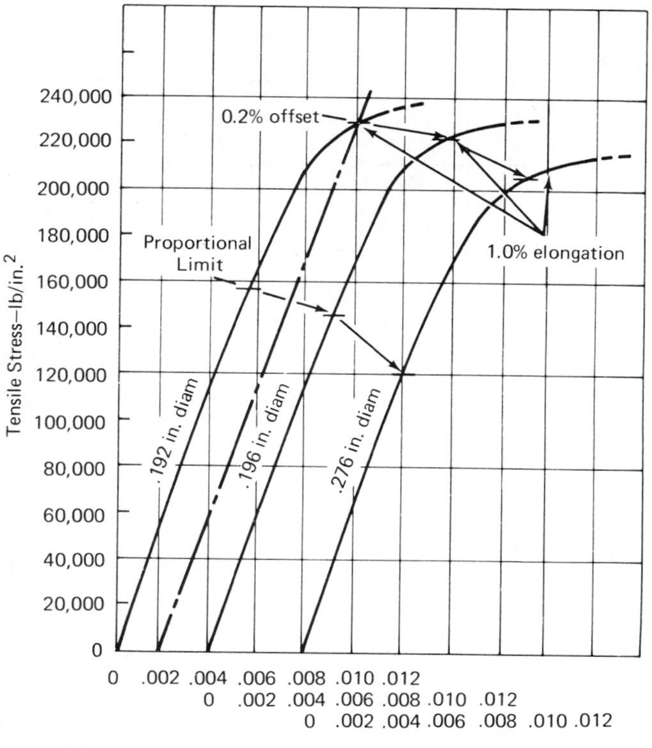

Fig. 6-64 Typical tensile tests in elastic region on uncoated stress-relieved wire for prestressed concrete.

ages are adopted, 0.276-in. and 0.196-in. wires are often specified. For post-tensioning systems developed in the United States, ¼-in. wires have been most commonly incorporated.

3. Strand. When strand is loaded, a considerable constructional stretch (a combination of direct tension, shear, torsion, and bending stresses) occurs due to the tendency of the helical wires to straighten and the resulting compaction of the strand. Since a strand with a long lay will have an appreciably higher E_s than a strand with a short lay, specifications require that the "strand have a uniform lay"; that is, the lay shall be constant for the full length of the strand. ("Lay" is the distance along the strand in which an outer wire makes one complete turn around the center wire.) For a given stress, a greater elongation is obtained than with a solid wire, and this gives a lower modulus of elasticity. For approximate calculations, a modulus of elasticity of 27×10^6 psi is often used for 250K grade and 28×10^6 psi for 270K grade.

Typical load–strain curves for 270 ksi seven-wire stress-relieved and stabilized strands are shown in Fig. 6-65, which are also typical for strands of all sizes. The properties of the various strands available in the United States are shown in Table 6-18.

For post-tensioning, multistrand tendons (with ½-in. or 0.6-in. strands) considerably larger than those listed in Table 6-18, $1^{11}/_{16}$ in. in diameter and above, are often employed. Beyond this size the problems of strand stiffness, heavy jacking equipment, etc., more than offset any saving due to handling fewer strands. In those few cases where larger strands must be used to get a large force into a small space, the strand fabricator should be consulted for advice on the best size to choose.

A galvanized prestressed concrete strand is composed of seven or more hot-dip galvanized wires. The temperature of the zinc bath and the speed of the wire passing through it are regulated so that the wire is stress-relieved as it is galvanized. Galvanizing can result in hydrogen embrittlement of prestressing steel, and therefore its use in structures where fatigue is a consideration is not recommended. The smallest seven-wire galvanized strand listed in Table 6-18 is 0.600 in. in diameter. Larger strands are made by adding extra layers of wire. A 1-in.-diameter strand is a 0.600-in.-diameter seven-wire strand with one more layer composed of 12 wires, for a total of 19 wires. Different sizes are made by varying either the number or size of the wires or both to produce the most economical strand. However, very little galvanized strand is being used today.

When the ultimate strengths given in Table 6-18 are divided by the areas, it is found that the stresses at ultimate load are only 202,000 to 214,000 psi instead of the much higher stresses given for uncoated strand. This is so partly because it is standard practice to list the gross area of the strand including the zinc, even though the zinc carries practically no load. Galvanizing may also increase the relaxation of the steel.

While seven-wire strands are stress-relieved after stranding to eliminate internal stresses due to stranding, it is not practical to stress-relieve the larger strands because the outer wires would reach the critical temperature and revert to a crystalline structure before the inside wires were stress-relieved. To achieve uniform elongation characteristics the large strands are prestretched. In this operation a full shop length (usually 3600 ft) is stretched in a tensioning rig to approximately 70% of its ultimate and held at that load a short time. During this period of high stress the fibers that are stressed beyond this yield point because of the applied load added to internal stresses will continue to elongate until they reach a point of stability. After the load is released, the strand has good elastic properties up to its prestretching load. The modulus of elasticity of these strands, when prestretched, ranges between 24,000,000 and 26,000,000 psi.

4. Bars. A typical stress–strain curve for bars is shown in Fig. 6-66, from which it can be noticed that a constant modulus of elasticity exists only for a limited range (up to about 80,000 psi stress) with a value between 25 and 28×10^6 psi. Some specifications determine the yield strength on the basis of 0.7% extension. Statistically the values so obtained are almost identical to those determined by the 0.2% offset method. Common sizes and properties of high-strength bars are listed in Table 6-18. Although only about 60% as strong per unit cross section, bars compete successfully with wire and strand for a range of applications.

5. Creep and Relaxation. In the design of prestressed concrete structures, a very important consideration is the loss of initial prestress force that may be expected. Losses are due to various factors that are part of the stressing operation as well as to the immediate and long-term strains in the structural materials.

The creep behavior of steel wires under long-term tensile stress is measured by one of two methods: constant stress (creep) or constant strain (relaxation). The creep phenomenon in steel varies with chemical composition, as well as with mechanical and thermal treatment applied during the manufacturing process.

Stress-relieved wire will show no appreciable creep when stressed up to 50% of its tensile strength. Beyond 50%, creep begins to become noticeable, Fig. 6-67, and gradually increases as the stress approaches tensile strength. The rule governing the rate of creep of stress-relieved high tensile wire is more complicated than a straight-line plot on a semilogarithmic scale. At low stresses, the curves are almost straight, becoming parabolic within the time span investi-

TABLE 6-18 Properties of Prestressing Tendons*

Prestressing Wire

Diameter	0.105	0.120	0.135	0.148	0.162	0.177	0.192	0.196	0.250	0.276
Area, sq in.	0.0087	0.0114	0.0143	0.0173	0.0206	0.0246	0.0289	0.0302	0.0491	0.0598
Weight, plf	0.030	0.039	0.049	0.059	0.070	0.083	0.098	0.10	0.17	0.20
Ult. strength, f_{pu}, ksi	279	273	268	263	259	255	250	250	240	235

Seven Wire Strand, f_{pu} = 270 ksi

Nominal Diameter, in.	3/8	7/16	1/2	9/16	0.600
Area, sq in.	0.085	0.115	0.153	0.192	0.215
Weight, plf	0.29	0.40	0.53	0.65	0.74

Seven Wire Strand, f_{pu} = 250 ksi

Nominal Diameter, in.	1/4	5/16	3/8	7/16	1/2	0.600
Area, sq in.	0.036	0.058	0.080	0.108	0.144	0.215
Weight, plf	0.12	0.20	0.27	0.37	0.49	0.74

Three and Four Wire Strand, f_{pu} = 250 ksi

Nominal Diameter, in.	1/4	5/16	3/8	7/16
No. of wire	3	3	3	4
Area, sq in.	0.036	0.058	0.075	0.106
Weight, plf	0.13	0.20	0.26	0.36

Galvanized Prestressing Strands

Diameter, in.	0.600	0.835	1.000	1 1/8	1 1/4	1 3/8	1 1/2	1 9/16	1 5/8	1 11/16
Area, sq in.	0.215	0.409	0.577	0.751	0.931	1.12	1.36	1.48	1.60	1.73
Weight, plf	0.737	1.412	2.00	2.61	3.22	3.89	4.70	5.11	5.52	5.98
Ult. strength, f_{pu}, ksi	46	86	122	156	192	232	276	300	324	352

Smooth Prestressing Bars, f_{pu} = 145 ksi

Nominal Diameter, in.	3/4	7/8	1.000	1 1/8	1 1/4	1 3/8
Area, sq in.	0.442	0.601	0.785	0.994	1.227	1.485
Weight, plf	1.50	2.04	2.67	3.38	4.17	5.05

Smooth Prestressing Bars, f_{pu} = 160 ksi

Nominal Diameter, in.	3/4	7/8	1.000	1 1/8	1 1/4	1 3/8
Area, sq in.	0.442	0.601	0.785	0.994	1.227	1.485
Weight, plf	1.50	2.04	2.67	3.38	4.17	5.05

Deformed Prestressing Bars

Nominal Diameter, in.	5/8	1	1	1 1/4	1 1/4	1 1/2
Area, sq. in.	0.28	0.852	0.852	1.295	1.295	1.630
Weight, plf	0.98	2.96	2.96	4.55	4.55	5.74
Ult. strength, f_{pu}, ksi	157	150	160	150	160	150

**From: Adapted from* PCI Design Handbook—Precast and Prestressed Concrete, *Prestressed Concrete Institute, Chicago. 1978.*

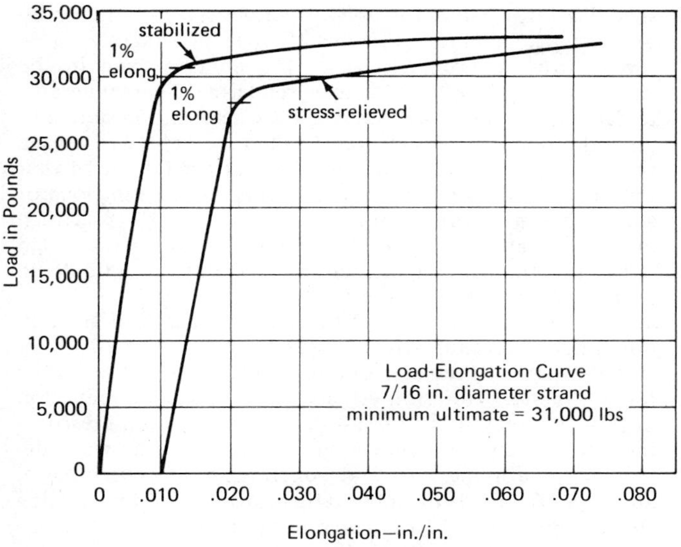

Fig. 6-65 Typical load-elongation curve for $^7/_{16}$-in. diameter stress-relieved and stabilized (low relaxation) strand.

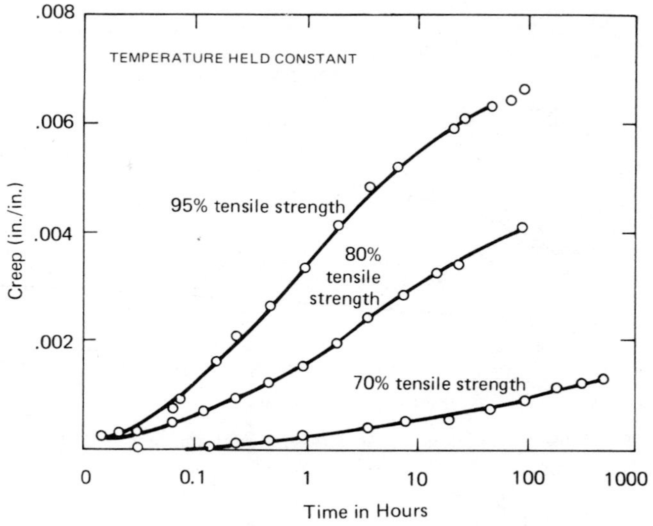

Fig. 6-67 Creep of stress-relieved wire loaded to different stress levels. (From: Podolny, W., Jr., and Melville, T., "Understanding the Relaxation in Prestressing," *PCI Journal*, 14 (4), 43–54, Aug. 1969.)

gated. At high stresses the curves are hyperbolic. The S-shaped curve indicates that the rate of creep reaches an inflection point and then tends to level off. Tests over longer periods of time indicate that the S-shaped curve (hyperbolic) also occurs at lower stress levels but over much longer periods of time.

Because of the logarithmic time nature of the creep phenomenon in wire and a doubt as to its validity when applied to conditions prevailing in prestressed concrete, a more valid approach to the creep behavior of prestressing steels for the designer is the measurement of stress relaxation when the wire is held at constant strain. As might be expected, relaxation, like creep, follows a substantially straight logarithmic law at normal stress and temperature.

For a given steel, the rate of relaxation is a function of initial stress, temperature, and duration of load application. At normal initial stress levels and temperatures, relaxation is predictable and is of relatively minor significance in terms of other losses imposed upon the structure or member. At elevated temperatures, the losses due to relaxation of the prestressing steel are of greater significance. The engineer must apply relaxation data to the actual conditions applying to the structure he is designing with rational reasoning and judgment. For all practical purposes, in prestressed concrete, 1000 hr seems to be a practical measure of an end point. If one considers that the shrinkage, creep, and elastic shortening of the concrete itself will reduce rather quickly the initial tension in the steel, and the approximate logarithmic nature of the steel relaxation, nothing very important can be expected to happen after 1000 hr.

In examining the 1000-hr value of stress relaxation for stress-relieved wires, Fig. 6-68, it is evident that relaxation

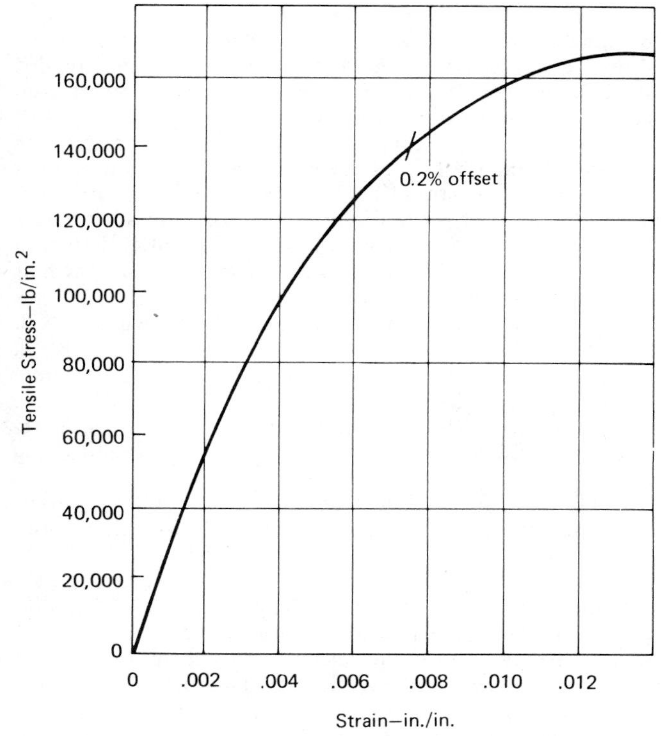

Fig. 6-66 Typical tensile test in elastic region for high-strength alloy bars.

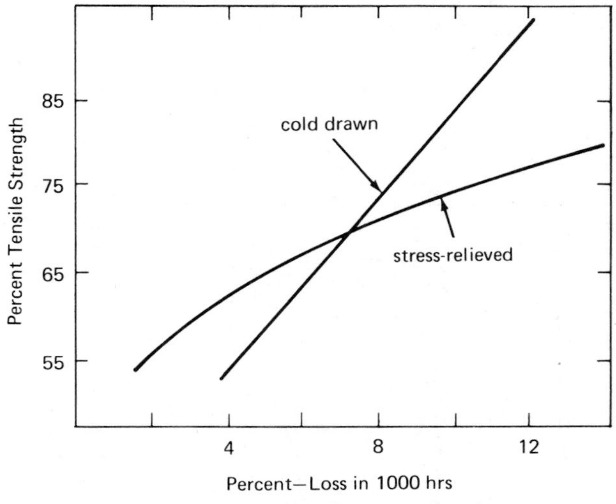

Fig. 6-68 Relaxation of high-strength, 0.196-in. diameter wires at various stress levels. (From: Podolny, W., "Understanding the Steel in Prestressing," *PCI Journal*, 12 (5), 54–56, Oct. 1967.)

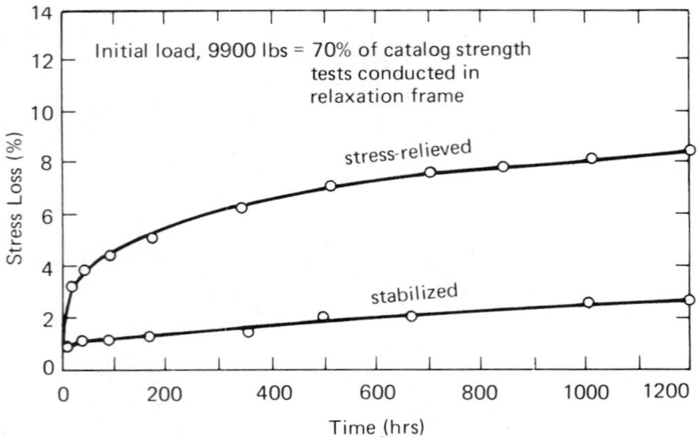

Fig. 6-69 Relaxation test in stress-relieved and stabilized (low relaxation), 0.276-in. diameter, prestressed-concrete wire.

is very small up to values of about 55% of ultimate strength. Beyond this point the rate of relaxation gradually increases, but reaches a peak at about 80% of tensile strength and then becomes less beyond that point. A stabilized wire or strand has considerably less relaxation than the stress-relieved tendon, as shown in Fig. 6-69.

Tendons composed of stress-relieved wires have relaxation losses of about the same magnitude as stress-relieved strand. Relaxation of strand is greater than in the straight constituent wire owing to the combined stress relaxation in the helical wires. In tendons at elevated temperatures, Fig. 6-70, or subjected to large lateral loads, relaxation loss is greater. Increasing temperatures has the same effect as increasing the initial value of stress.

In certain types of structures, where there may be considerable prestress losses due to causes other than relaxation, such as slip at anchorage, creep, shrinkage, elastic shortening of the concrete, and frictional loss due to intended or unintended curvature in the tendons, it may become economically feasible to re-stress tendons after some period of time. Relaxation losses may be considerably less after re-tensioning.

6. Fatigue. If the precompression in a prestressed concrete member is sufficient to ensure an uncracked section throughout the service life of the member, the fatigue characteristics of the prestressing steel and anchorages are not likely to be critical design factors. Consequently, fatigue considerations have not been a major factor in either the

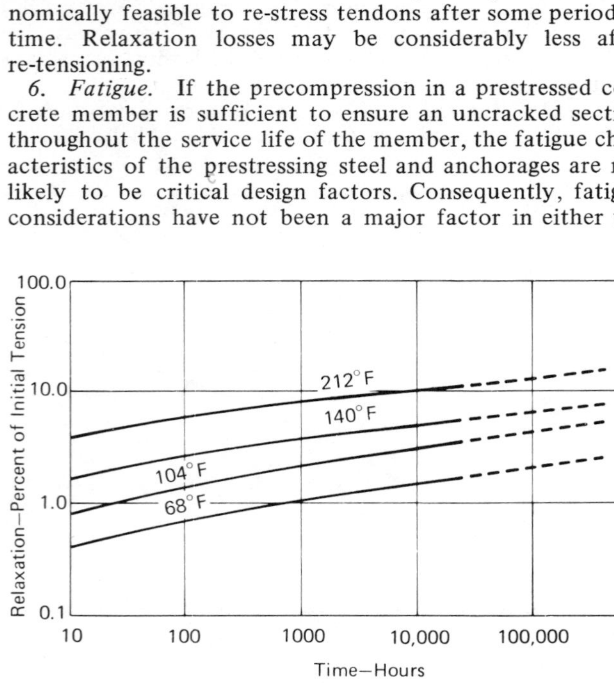

Fig. 6-70 Stress loss due to relaxation of stabilized strand at various temperatures. Initial tension = 70% of guaranteed ultimate strength.

specification of steel for prestressed concrete or the development of anchorage systems. Cracks presumably cause severe stress concentrations in the concrete and in the strands contiguous to the cracks, rendering them vulnerable to fatigue. It, therefore, appears necessary and desirable to limit the nominal tensile stress in the precompressed tensile zone to $6\sqrt{f_c'}$. Creep and shrinkage affect fatigue in that the prestress may be reduced, and thus the cracking load of a structure is also reduced—the stress range in the prestressing steel may be increased. Depressed strands (although stress in strands is increased by bending at hold-down points) do not cause a reduction in fatigue life.

In general, fatigue life decreases as the length of the tendon under constant stress increases. Loss of bond or increase in lateral pressures due to bends or curvatures in the tendon increase abrasion effects and also cause a reduction in fatigue strength. The bond developed between prestressing tendons and concrete is poorer than that between deformed reinforcing bars and concrete. As a consequence, abrasion effects result in the fatigue strength of prestressing steels, being about 15% lower in concrete where bond is partially destroyed than for the same steel in air.

For wires and strand, galvanizing reduces the ultimate and yield strength significantly and therefore also reduces the fatigue limit. For bars, galvanizing does not alter the static properties, but it does reduce the fatigue limit.

Tests have indicated that beams partially prestressed with pretensioned strand show no detrimental effects from repeated loads. However, more fatigue tests to evaluate bond fatigue are warranted, especially on partially prestressed beams and beams with strand anchorages away from zones of normal concrete compressive stress.

The characteristics of the anchorage in a post-tensioning system and not the prestressing system control the fatigue characteristics of the unbonded tendon. For unbonded construction, stress changes in the prestressing steel are transmitted directly to the anchorage. Although most anchorages can develop the static strength of the prestressing steel, they are unlikely to develop its fatigue strength. Further, bending at an anchorage can cause higher local stresses than those calculated from the tensile pull in the prestressing steel. Bending is likely where the prestressing steel is connected to the member at a few locations only throughout its length, or where there is angularity of the prestressing steel at the anchorage. Fatigue characteristics based on tests of single wire or strand anchorages are likely to overestimate the strength of multiwire or multistrand anchorages. Unless data to the contrary are available, the fatigue strength of anchorages should be taken as not greater than one-half the fatigue strength of the prestressing steel.

Limited tests on 0.25-in. (6.3-mm)-diameter wires of U.S. manufacture show a fatigue strength at 4×10^6 cycles in excess of 30 ksi. Fatigue characteristics appear to be independent of the wire diameter. Ribbing or crimping reduces the fatigue strength of a wire in a manner similar to the reduction in strength caused by deformations on a reinforcing bar.

The fatigue life of the high-strength $1/2$-in., 270-ksi, seven-wire prestressing strand compares favorably with that of the conventional $7/16$-in., 250-ksi strand, as shown in Fig. 6-71. At stresses in the working load range, the fatigue life of a specimen was in the vicinity of approximately 2,000,000 cycles of loading. There are indications of a decrease in fatigue strength with increasing wire size in the strand. Some prestressed concrete flexural members have shown that the prestressing strand failed in fatigue at a significantly lower number of cycles when cracks in concrete crossed the strand.

The overall scatter of fatigue data is of paramount impor-

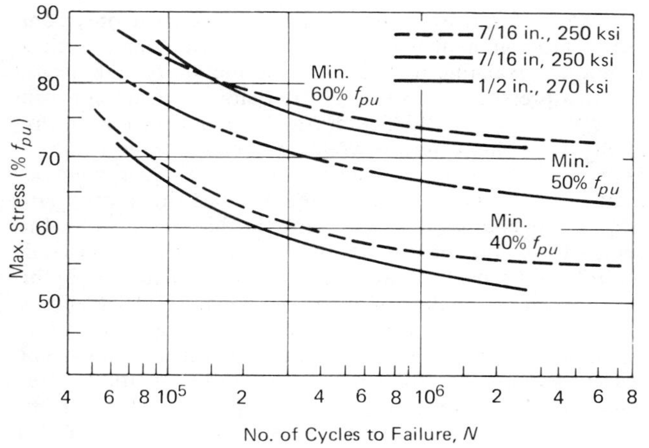

Fig. 6-71 Summary of fatigue tests on seven-wire, prestressing strand of U.S. manufacture. (From: Tide, R. H. R., and Van Horn, D. A., "A Statistical Study of the Static and Fatigue Properties of High Strength Prestressing Strand," Lehigh Univ., Fritz Eng. Lab. Report 309.2, June 1966.)

tance in defining the quality of the prestressing steel. For U.S. strand, a modified Goodman diagram has been developed for three discrete probability levels. These levels correspond to survival probabilities of 0.1, 0.5, and 0.9, and they were developed from data with minimum stress levels of 0.4, 0.5, and 0.6 times the static tensile strength. For the desired minimum stress and probability level, vertical intercepts define permissible stress ranges for failure for strands tested in the United States. (See Fig. 6-72.)

The fatigue life increases slightly as the temperature of the test specimen is reduced from 70°F to 0°F. This temperature effect is somewhat in agreement with the metallurgical concept that indicates an increase in fatigue life with lowered temperature, when the temperatures considered fall between 500°F and −250°F. The range of temperatures considered has not been great enough to produce a conclusive result.

Unless test data are available to justify higher values, the stress range in prestressed reinforcement should not exceed

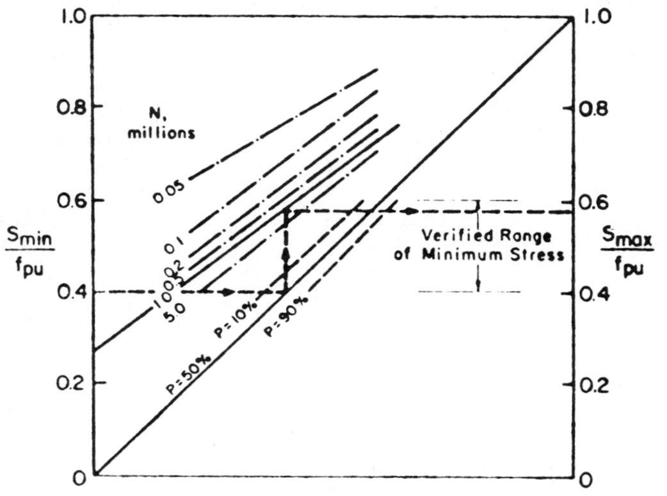

Fig. 6-72 Strength envelopes for strand tested in United States. (From: "Considerations for Design of Concrete Structures Subjected to Fatigue Loading," Reported by ACI Committee 215, *ACI Journal*, Mar. 1974, pp. 97–121.)

the following:

strand and bars	$0.10f_s'$
wires	$0.12f_s'$

The following specification is proposed for testing the capability of an unbonded tendon assembly (prestressing steel and anchorages) in resisting cyclic loading:

A dynamic test shall be performed on a representative specimen and the tendon shall withstand, without failure, 500,000 cycles of stressing varying from 60% to 66% of its minimum specified ultimate strength, and when the structure will be subject to earthquake loading another specimen shall withstand without failure 50 cycles of loading, corresponding to the following percentages of the minimum specified ultimate strength:

$$60 \pm \frac{2000}{L + 100}$$

where L is the length in feet of the tendon to be used in the structure. The period of each cycle involves the change from the lower stress level to the upper stress level and return to the lower.

Systems utilizing multiple strands, wires, or bars may be tested utilizing a test tendon of smaller capacity than the full-size tendon. The test tendon shall duplicate the behavior of the full-size tendon and generally shall not have less than 10% of the capacity of the full-size tendon. It is not necessary to require tests on samples taken from the material being used for the specific project; tests from lots made to the same specifications should be sufficient. With the recent advent of very large post-tensioning tendons, it was recognized that the availability of testing machines capable of performing a true fatigue test of full-capacity tendons was very uncertain. The main feature to be tested is the possible effect of the anchorage in reducing the fatigue strength of the prestressing steel itself. The design of the different anchorage components is a separate mechanical problem, which can be approached in an analytical manner. For these reasons, a reduced-capacity fatigue test is considered acceptable.

Dynamic tests are not required on bonded tendons, unless the anchorage is located or used in such manner that repeated load applications can be expected on the anchorage.

6.14.3 Splices

Splicing of reinforcing bars is only for the purpose of providing continuity. Splices are needed because of (1) limitations set up by the physical length of the bar, (2) transitions from a larger to a smaller bar, and (3) construction requirements, i.e., construction joints.*

The choice of methods available for splicing reinforcing bars has been widened in recent years. The traditional methods are lap splice by bond and welded lap or butt splices. Newer methods include couplers (for tension or compression) and end-bearing (for compression only). Semiautomatic butt-welded splices may be performed with commercially available equipment.

The entire responsibility for design, specifications, and performance of splices ultimately rests with the structural engineer. Only the person familiar with the structural analysis, probable construction conditions, and final conditions

*Most of the discussion on splices is based on "Reinforcement Anchorages and Splices," Concrete Reinforcing Steel Institute, Chicago, 1980, 40 pp, and "Mechanical Connections of Reinforcing Bars," reported by ACI Committee, *ACI Concrete International*, Jan. 1983, pp. 24–35.

of service can properly evaluate the variables to select the most efficient and economical splice method. The drawings, notes, and specifications should clearly show or describe all splice locations, types permitted or required, and, for lap splices, lengths of lap.

There is no such thing as a perfect splice or a perfect splicing system. Therefore, the designing engineer must carefully consider these points:

(a) The splice system needed to meet his design requirements.
(b) Availability for that particular project.
(c) Cost of material, supplies and tools.
(d) Cost of labor, insurance, taxes, etc.
(e) Equipment needed—welding machines, temporary electrical supply, etc.
(f) Time to make splices—for example, one floor.
(g) Inspection under job conditions.
(h) Cost of inspection.
(i) Union regulations.

1. Lap Splices. Lapped splices may be effected with bars either spaced or in contact. For bar-to-bar splices, contact splices are preferred for the practical reasons that, wired together, they are more easily secured against displacement during concreting. Caution must be exercised against designing lap-spliced bars spaced too widely apart, permitting a zigzag crack across an unreinforced section. Spacing of bars in spaced lap splices shall not exceed one-fifth the lap length or 6 in. Splices are usually staggered to prevent concentration in one plane, especially in seismic areas. Unless staggered, splices capable of carrying some tension are required to meet increasingly severe criteria. Lapped splices are generally used for bar sizes #11 and smaller, except that in columns #11 bars are frequently butt-spliced. If lapped splices are employed in critical tension, increased lap lengths or spirals must be provided at this location. When lapped splices are provided, the bars that extend up from below come inside of the bars above. Code requirements for lap splice length should be followed so that all spliced bars are fully utilized.

2. Welded Splices. Electric arc welding is the only commonly used manual welding process in the field. Various techniques available include shielded metal arc, submerged arc, and pressure gas welding. All welding should conform to AWS D1.4-79 "AWS Structural Welding Code—Reinforcing Steel" of the American Welding Society.

For welded splices it is necessary to know the chemical analysis of the bars being welded and use the appropriate welding procedures. The important consideration is that the specified welding procedure, including method, material, amount of preheat, if any, etc., be compatible with the chemical analysis. Bars with a carbon content exceeding about 0.35% cannot *easily* be welded in the field. Arc welding should be prohibited on bars with carbon equivalent greater than 0.55%. All bars except ASTM A706 are furnished to ASTM physical specifications only, and the chemical analysis can vary widely. This variation requires careful control of procedures at the job site. It may be desirable to restrict steel chemistry (usually carbon and manganese) to a given range to suit a specified welding procedure. Also, the ends of the bars to be butt-welded must be prepared with square and/or bevel cuts. Where arc welded splices are used, the following are required:

(a) Mill test analysis of the steel.
(b) Adjustment of welding techniques to suit analysis.
(c) Correct strength, grade, and size low-hydrogen electrodes kept oven-dry.
(d) Qualification tests to certify all welders before beginning a project and periodically during long projects.
(e) Continuous supervision of all welding operations.
(f) Magnetic particle, radiography, or other nondestructive inspection of welds.
(g) Occasional quality control tests of actual welds removed from structure when nondestructive test results are unsatisfactory.

Reinforcing bars conforming to ASTM A706, "Low-Alloy Steel Deformed Bars for Concrete Reinforcement," are recommended for use wherever important or extensive welding is required. In addition to restrictions on chemical composi-

TABLE 6-19(a) Compression-Only Splices

		Bolted Steel Sleeve			
		Solid Type Steel Sleeve	*Strap Type Steel Sleeve*	*Steel-filled Sleeve*	*Wedge-locking Sleeve*
Coupler	Bar size range	8–18	7–18	11–18	7–18
	Splices different bar sizes	Yes	Yes	Yes	Yes
Clear spacing required between bars (normal)	#18	$1\frac{1}{2}$ bar dia	$1\frac{1}{2}$ bar dia	5″	$1\frac{1}{2}$ bar dia
	#14	$1\frac{1}{2}$ bar dia	$1\frac{1}{2}$ bar dia	$4\frac{1}{2}$″	$1\frac{1}{2}$ bar dia
	#11	$1\frac{1}{2}$ bar dia	$1\frac{1}{2}$ bar dia	4″	$1\frac{1}{2}$ bar dia
	Minimum dowel length		6″	$4\frac{7}{8}$″	6″
#18 coupler installation requirements (normal)	Coupler length	12″*	12″	3″	12″
	Coupler maximum diameter/across corners	$2\frac{3}{4}$″	4″	$2\frac{5}{8}$″	$2\frac{3}{4}$″
	Coupler side wall thickness (nominal)	Nil	Nil	$\frac{9}{16}$″	Nil
	Cut square within $1\frac{1}{2}$ deg	Yes	Yes	No	Yes
Bar end preparation	Cleaning special cleaning	No	No	No	No
	Pre-drying/heating	No	No	Yes	No
	Threading	No	No	No	No
Installation tools	Hand held tools adequate	Yes	Yes	No	Yes
	Special tools required	No	No	Yes	No

**1 in. = 25.4 mm.*

TABLE 6-19(b) Tension–Compression Splices

		Cold Swaged Steel Sleeve	*Cold Swaged Sleeve with Threaded Ends*	*Forged Steel Sleeve*	*Mortar Filled Sleeve*	*Steel Filled Sleeve*	*Steel Sleeve Threaded for Special Bars*	*Taper-threaded Steel Coupler*	*Threaded Sleeve Rolled Threads on Bar*
Coupler	Bar size range	5–18	5–18	5–18	5–18	4–18	6–18	7–18	4–18‡
	Splices different bar sizes	Yes	Yes	Yes	Yes	Yes	Yes	Yes	Yes
Clear spacing required between bars (normal)	#18	4″§	2″	$1^1/_2$ bar dia	4.72″	(V) 5″	$1^1/_2$″ bar dia	$1^1/_2$″	2 bar dia
	#14	3″	$1^1/_2$″	$1^1/_2$ bar dia	3.90″	(V) $4^1/_2$″	$1^1/_2$″ bar dia	$1^1/_2$″	2 bar dia
	#11	$2^1/_2$″	1″	$1^1/_2$ bar dia	3.50″	(V) 4″	$1^1/_2$″ bar dia	$1^1/_2$″	2 bar dia
#18 coupler installation requirements (normal)	Minimum dowel length	12″	12″	(V)-18″ (H)-20″	18″	$4^7/_8$″	10″	$4^1/_2$″	4″
	Coupler length	12″	12″	9″	$36^1/_4$″	9″	$16^1/_2$″*	$8^1/_4$″ & $6^1/_2$	6″
	Coupler max. dia./ across corners	4″	4″	$3^1/_4$″	$4^3/_4$″	$3^3/_4$″	$3^1/_2$″	3″	3″
	Coupler side wall thickness (nominal)	$^5/_8$″	$^5/_8$″	$^1/_2$″	0.49″	$^9/_{16}$″	$^5/_8$″	NA	NA
Bar end preparation	Cut square within $1^1/_2$°	No	No	No	No	No	No	Cut, not sheared	No
	Cleaning-special cleaning	Remove loose part.	Remove loose part.	Remove loose part.	No	Yes	No	No	No
	Pre-drying heating	No	No	Yes Coupler	No	Yes	No	No	No
	Threading	No	No	No	No	No	No	Yes†	Yes
Installation tools	Hand held tools adequate	No	Yes	No	Yes	No	Yes: <= #11 No: >#11	Yes	Yes
	Special tools required	Yes	No	Yes	Yes Grout pump	Yes	Yes	No	No

*Requires alternate jam nut for load reversal conditions (tension-compression).
†Bar end threading normally done by bar fabricator.
‡ #6 – #14 meet ACI 318-83, Section 12.14.3.

tion including carbon, the carbon equivalent of each heat (50 to 200 tons) of ASTM A706 steel is limited to 0.55.

Connection of crossing bars or stirrups by small arc welds, known as tack welds, is not recommended. Unless these welds are made in conformance with all requirements of AWS D1.4-79, they tend to cause a metallurgical notch effect and may affect the ductility, fatigue, and strength of the bars. Tie wire will do the job without harm to the bars. Bars that have been joined by butt-welding also may have reduced fatigue strength. Good practice dictates that no welding should be done within four bar diameters of a cold bend to prevent loss in bar strength. In fact, it is preferable to weld only straight bars.

Thermite welding is a process in which the ends of the bars are fusion-welded. The bars are aligned with a gap between them. Refractory molds are assembled on the bars and sealed in place. Exothermic powders are filled into a separate cavity in the molds. The powders are ignited and burn with enough heat to form superheated molten steel. The steel flows through the gap between the bars, and some flows into a second cavity beyond the bars, preheating them. Subsequent flow completes the fusion. Finishing involves breaking off the gates.

Thermite welding is not as sensitive as arc welding to chemistry or end preparation of the bar. It has been used with success in making butt-welded joints in the large size bars, #14 and #18. This process has been successful in joining hard-to-weld steels because it welds the entire cross section at the same time and automatically provides preheat and slow cooling.

3. Couplers. These are mechanical devices joining two reinforcing bars to resist both tensile and compressive forces. The code requirement that such couplers develop 125% of the specified yield point usually controls design of these devices. Couplers are generally equivalent to butt welds although their resistance to fatigue, stress reversal, dynamic load, long-time creep, and other special conditions may vary. For specific performance information on the couplers considered, test data should be secured.

Three types of couplers are available and their requirements are shown in Table 6-19:

(a) Compression (only) couplers, also known as the "endbearing splice"
(b) 125% f_y tension couplers
(c) Minimum ultimate strength tension couplers

Prestressing tendon manufacturing and/or shipping conditions are often such that some tendons must be spliced. All splices have certain disadvantages, but they cannot be eliminated.

Couplings of tendons should be used only at locations specifically indicated and/or approved by the engineer. Couplings should not be used at points of sharp tendon curvature. All couplings should develop at least 95% of the minimum specified ultimate strength of the prestressing steel without exceeding anticipated set. The coupling of tendons should not reduce the elongation at rupture below the requirements of the tendon itself. Couplings and/or coupling components should be enclosed in housings long enough to permit the necessary movements, and fittings should be provided to allow complete grouting of all the coupling components in bonded tendons or should be completely protected with a coating material prior to final encasement in concrete in unbonded tendons.

Splices are not permitted in the wires of parallel-wire cables. There are two reasons for this. Since these wires are manufactured in lengths to 7000 ft, it is a simple matter to make a tendon without splices. When wires are spliced, they are spliced by welding, which damages the fibers and therefore lowers the strength of the wire. In a parallel-wire tendon each individual wire must carry its full share of the tension at all times. It cannot distribute part of its load to adjacent wires at a point where it is weak. As a result if a welded wire is used in a parallel-wire cable, it usually fails during the tensioning operation.

Seven-wire strands are produced in endless lengths and are most often shipped on reels containing 22,000 ft. In use the full length required is cut from the reel in one length, and no splice is needed. When the last length on the reel is not long enough to reach the full length of the casting bed, some operators find it desirable to salvage this length by splicing it to another short length. Splicing two short lengths of strand together in this manner is permissible if the coupling used will develop the full strength of the strand, will not cause failure of the strand under the fatigue loadings that will be applied to the structure being fabricated, and does not weaken the member by replacing too much of the concrete in the cross section. Strand tendon couplers are available and used occasionally for multistrand tendons in post-tensioning. The engineer must determine whether or not a particular splice has the necessary qualifications. Information on this subject is also available from some of the fabricators of seven-wire strands.

Splices are used in the individual wires of seven-wire strands during shop fabrication. ASTM Specification A416 says: "During fabrication of the strand, butt-welded joints may be made in the individual wires, provided there is not more than one such joint in any 150-ft section of the completed strand." Although the strength of the welded wire is reduced in the vicinity of the weld, the total strength of the strand is not seriously reduced. The welded wire is treated in the vicinity of the weld so that part of its load is transferred to the six other wires.

High-strength bars are spliced at the job site with couplers when necessary. Bars are fabricated in maximum lengths of up to 100 ft and are sometimes furnished in shorter lengths with splices because of shipping difficulties. Sleeve couplers are available to splice the bars to any desired length. The couplers are larger in diameter than the bar and are tapped to take a threaded bar in each end. They develop the full strength of the bar. The disadvantage in the use of couplers is the large hole that must be cored in the concrete at the location of the coupler. If concrete stresses are high, the net section should be checked with the area of the hole deducted.

6.14.4 Corrosion of Steel and Nonferrous Metals

Corrosion refers to the destruction of a metal or alloy by chemical change, electrochemical reaction, or physical dissolution due to its environment. Many years of experience have shown that reinforcing steel is protected from corrosion by concrete. The degree of protection afforded by the concrete depends, of course, upon the thickness of the concrete—the cover over the steel; and the quality of the concrete—its water–cement ratio, and hence, it permeability.

The exact mechanism by which concrete prevents corrosion of encased steel is not completely understood. When steel is encased in concrete, a protective iron oxide film forms at the steel–concrete interface as a result of the high alkalinity (pH) of concrete or mortar. The pH of concrete tends to be buffered at a value of about 12.5 corresponding to a saturated lime solution. Traces of sodium and potassium oxides may increase the pH to 13.2. As long as the alkalinity is maintained, this film is effective in preventing corrosion due to anions such as chlorides, sulfides, bromides, and iodides as long as bubbles of free oxygen do not occur at the steel surface.

Corrosion of metals in concrete is an electrochemical process and requires that moisture be present. The moisture conditions of concrete required to support active galvanic corrosion of susceptible metals are not known with sufficient accuracy. Moisture in concrete may be available from two sources: (1) When fresh concrete is placed, it contains mixing water which is only partly used up as the cement hydrates. The remaining water migrates to exposed surfaces and evaporates. The length of time it takes for concrete to dry out varies. It depends on such factors as relative humidity of the surrounding air, size and shape of the concrete mass, and concrete mix. (2) If hardened concrete is not dense and impervious, moisture may penetrate into the interior from the environment.

In dry concrete, oxygen gas can readily diffuse into the concrete and to the steel, and upon rewetting of the concrete, dissolve into solution and be immediately available for cathodic reactions. In wet concrete, oxygen gas must dissolve into solution at the surface of the concrete and then slowly migrate inward to the steel under the driving force of a concentration gradient (See Section 6.10.3) Oxygen solubility is greatly reduced if chloride ion is present in solution, and may in fact become so low as to result in early termination of cathodic reactions.

Galvanic corrosion is influenced by differences in: (1) the composition of the solution at the two electrodes; (2) the nature of the metals of the electrodes; and (3) environmental conditions, such as the presence of oxygen, chloride concentration, alkalinity, moisture, temperature, or large air- or liquid-filled spaces next to the metals. These conditions can be brought about by differences in external exposure, permeability, thickness, or uniformity of the concrete.

Anodic and cathodic electrochemical reactions must occur simultaneously at the surface of the steel to cause galvanic corrosion. Anodic reactions cause oxidation of metallic iron, while cathodic reactions result in reduction of dissolved oxygen to form hydroxyl ion. Sites of anodic reactions are visually recognizable by pitting of steel and by in situ deposits of dark green, black, or rust-colored corrosion products.

Oxidation of metallic iron at anodic sites is controlled by availability of oxygen for reduction at cathodic sites. Rate of corrosion is also partially determined by the size ratio of cathode area to anode area and by the cathode efficiency on the embedded steel. Other things being equal, larger ratios result in higher corrosion rates.

The two electrochemical reactions must be coupled electrically before they can form a viable corrosion cell. Metallic steel serves as one link between electrodes of the electric circuit. The other is provided by moist concrete of relatively low resistivity which is present in the vicinity of the steel. Corrosion of embedded steel is not a problem as long as the electrical resistivity exceeds a threshold level of $50-70 \times 10^3$ ohm cm. Even if the passivity of the steel is broken, corrosion of the steel will not occur if the concrete is less than 40% saturated, Fig. 6-73. For high moisture contents, the development of corrosion is controlled by the availability of oxygen and dissolved substances in the concrete, such as chloride ions, which reduce resistivity and more readily permit current flow between electrodes.

Localized galvanic corrosion may occur when one metal is placed in two different electrolytes, or in an electrolyte of varying concentration. For example, metal piping should not rest directly on a sand, earth, or insulating concrete base when concrete is to be placed over it. The piping in this situation makes contact with the base material and this contact can set up a flow of corrosion-causing current.

Corrosion reactions are strongly promoted by the presence of some of the halogen ions, particularly chlorides, which make moist concrete a strong electrical conductor electrolyte. Concentrations exceeding a certain value can, in the presence of oxygen and moisture, destroy the passivating oxide film on the surface of the steel, and as a result the steel rusts. The amount of calcium chloride that may be safely used depends upon the composition of the cement, particularly upon its tricalcium aluminate content. The following threshold values have been suggested as a general guide.

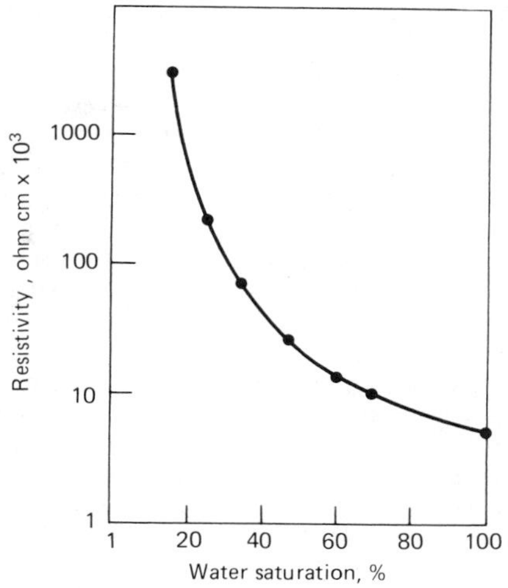

Fig. 6-73 Effect of water saturation on the electrical resistivity. (From: Gjørv, O. E., Vennesland, Ø., and El-Busaidy, A. H. S., "Electrical Resistivity of Concrete in the Oceans," Paper OTC 2803. Offshore Technology Conference 1977, 2-5 May 1977, Houston, Texas, pp. 581-588.)

*Chloride Ion Concentration, ppm**	*pH of Concrete*
71	11.6
710	12.4
8900	13.2

**Table reference from D. A. Hausmann, "Electrochemical Behavior of Steel in Concrete," Journal, ACI, Feb. 1964.*

Concentrations below these values probably do not cause corrosion of steel in the particular alkaline environment.

For corrosion protection, ACI 318 requires that the maximum water-soluble chloride ion concentration in hardened concrete at an age of 28 days contributed from the ingredients including water, aggregates, cementitious materials, and admixtures not exceed the following limits:

Type of Member	*Maximum Water Soluble Chloride Ion (Cl^-) in Concrete, percent by Weight of Cement*
Prestressed concrete	0.06
Reinforced concrete exposed to chloride in service	0.15
Reinforced concrete that will be dry or protected from moisture in service	1.00
Other reinforced concrete construction	0.30

At concentrations above these values, the steel may corrode if oxygen and moisture are available. When epoxy- or zinc-coated bars are used, the limits may be more restrictive than necessary.

Chlorides in concrete may originate from concrete admixtures, seawater, beach sand, or salt-containing aggregate. It is therefore recommended that any material for use in making concrete that is suspected of containing chloride be analyzed and rejected if chlorides present will exceed threshold value.

A chloride solution can rapidly migrate into dry concrete by absorption through emptied capillary (and gel) pores in the concrete. In wet concrete, dissolved chloride must slowly diffuse through solution under the driving force of a concentration gradient. Differential oxygen and chloride concentration cells may be established at the surface of the steel and increase the probability of corrosion. The presence of a crack in concrete would promote such nonuniform concentration.

The best protection against diffusion of chloride ions from the concrete exterior to the reinforcing steel is proper cover of good-quality concrete. For concrete exposed in marine environments or to deicer salts, corrosion after long-time exposure is always a possibility. Even the best concrete is not completely impermeable to chloride solutions, and given sufficient time, chloride ions will eventually reach the steel.

No data are available on the chloride transmission rates of concretes made with superplasticizers (high-range water reducers), but presumably, permeabilities to water and ions would be appreciably reduced at water–cement ratios of 0.30 or lower.

Porous concrete may permit the passage of carbon dioxide from the air into the concrete. This reacts with calcium hydroxide to form calcium carbonate, which lowers the pH. If this carbonation reaches the steel, the passivating oxide film can be impaired or destroyed. Normally, with good-quality concrete, the effect of carbonation does not penetrate more than $\frac{1}{2}$ in. even when exposed to the weather for a great many years. High water–cement ratios, low cement contents, thin cover over the reinforcing steel, and poor consolidation of the concrete should be avoided to preserve alkalinity. This also reduces the possibility of penetration to the steel of chlorides and acid-producing gases such as sulfides which are in the atmosphere in some industrial areas.

Where stress is associated with corrosion, the effect may be considerably more serious. Apart from the relatively simple case of stress-assisted corrosion, there could also be the phenomena of stress-corrosion and corrosion-fatigue. The distinctions between these phenomena are discussed below.

1. Stress-Assisted Corrosion. Steel wire that is not susceptible to stress corrosion may, under faulty conditions such as cracks in the concrete or a poorly made joint, undergo severe localized corrosion. The mechanism is relatively simple in that the penetration of the corrosive medium [usually nitrate (NO_3^-) and bisulfide (HS^-)] is assisted by the tendency of any initial corrosion pit to be enlarged or pulled open by the tensile stress. Provided the stressed steel is protected by well-made concrete, there is no danger of this type of corrosion. Where prestressing tendons are not bonded to concrete, such corrosion is quite possible if not guarded against.

2. Stress-Corrosion. The phenomenon of stress-corrosion is more complex. Not only do materials differ in their susceptibility, but also a limited number of corrosive media (i.e., chlorides, nitrates, and sulfides) are known to give rise to the phenomenon. Given the appropriate conditions of susceptible material (high-strength steels used in prestressing), stress, and corrosive medium, stress-corrosion is characterized by quite rapid development of deep cracks in the material. Quenched and oil-tempered, high-carbon steel wire is more susceptible to stress-corrosion cracking than cold-drawn wire. The mechanism of the attack is thought to be related to electrical potential differences between grains, or areas of the metal, which will assist intergranular attack. Under some circumstances, molecular hydrogen may be formed in the grain boundaries to exert pressures that weaken the material and encourage cracking under stress. For stress-corrosion to be progressive, it is necessary for the corrosive medium to be able to penetrate to the steel face. In general terms, if the corrosion product forms a hard but brittle film that will crack easily under stress, or if the corrosion products are removed in solution (e.g., acid attack), stress-corrosion may occur. If the conditions are such that the corrosion product forms a relatively soft blanket on the steel surface with consequent minimization of cracking or flaking of the film under stress, the possibility of stress-corrosion is minimized.

3. Corrosion-Fatigue. When a material is subjected to stress cycles in a corrosive environment, the effect of corrosion and fatigue acting together is synergistic and much more serious than that of the two effects acting independently. Not only may the corrosion produce pitting and give rise to stress concentration, but the alternating stresses tend to disperse the corrosion products and so enable the corrosive medium to be replenished at the metal surface. Once a fatigue crack has commenced, corrosion penetration combines to produce a rapid extension of the fissure. Under such circumstances there will be no true limiting fatigue-stress endurance limit, and low-amplitude stress cycles may eventually cause failure.

4. Corrosion Prevention. Various methods have been advanced for reducing the corrosion occurring on reinforcing steel. One method is the use of galvanized steel reinforcement, galvanized in accordance with ASTM A767. However, zinc is susceptible to attack by fresh concrete. This results in the evolution of hydrogen and the formation of calcium zincate, which occupies a greater volume than the original metal and may exert expansive pressures around the embedded element. The hydrogen may cause a stress-corrosion failure of prestressing steel.

The film of zinc covering reinforcement is so thin that the expansive pressures generally do not cause any damage to the surrounding concrete. Whether cracking of the concrete occurs subsequently or not depends on many factors such as strength of concrete, amount of concrete cover, size of galvanized member, exposure conditions, and chemical composition of cement. To what extent the steel will be corroded cannot be stated with certainty, but galvanizing does furnish sacrificial protection to the steel.

If it is desirable to avoid any reaction between zinc and alkaline fresh concrete, the metal should be protected by an organic coating or passivated by a chromate treatment.

The effect of addition of chromates to concrete on its performance is not yet fully known. It is therefore recommended, whenever a chromate treatment is required, to apply it to the galvanized surfaces, rather than add chromates to the concrete mixture. Chromate coating of galvanized surfaces can be readily done in most galvanizing plants. The chromate treatment is recommended for all galvanized elements that will be embedded in or will come in contact with concrete and mortar. Treated surface should be protected from water washing away coating before concrete is placed.

Admixtures containing any chloride, other than impurities from admixture ingredients, should not be used in prestressed concrete or in concrete with aluminum embedments. Concentrations of chloride ion may produce corrosion of embedded aluminum (e.g., conduit), especially if the aluminum is in contact with embedded steel and the concrete is in a humid environment. Serious corrosion of galvanized metal sheet and galvanized metal stay-in-place forms occurs, especially in humid environments or where drying is inhibited by the thickness of the concrete or coatings or impermeable coverings.

Nickel-plated steel will not corrode when embedded in chloride-free concrete and will provide protection to steel as long as no breaks or pinholes are present in the coating. The coating should be 3 to 5 mils thick to resist rough handling. Minor breaks in the coating may not be very detrimental in the case of embedment in chloride-free concrete; however, corrosion of the underlying steel would be strongly accelerated in the presence of chlorides owing to the large cathode/anode area, ratio that exists under such circumstances.

Cadmium coatings will satisfactorily protect steel embedded in concrete, even in the presence of moisture and normal chloride concentrations. Minor imperfections or breaks in the coating will generally not promote corrosion of the underlying steel.

Various nonmetallic coatings have been applied to reinforcing steel to reduce corrosion. Reinforcement epoxy-coated in accordance with ASTM A775 appears to significantly improve corrosion resistance. There is concern that macrocells might be developed with small anodes at damaged areas or pinholes on the epoxy coating connected to the larger cathodic area of the bottom uncoated reinforcing mat. At present, there are no constructions old enough to indicate whether or not this will be a problem.

Aside from the obvious problem of developing bond with some materials, the loss of the protection of the alkaline environment at the steel would seem to be a major shortcoming of all nonmetallic coatings except a dense cement mortar coating (similar to the approach of applying a cement slurry to stressed wires prior to grouting, or in the case of pretensioned members prior to casting of the concrete).

Where the penetration of chlorides into the interior of a reinforced or prestressed member is expected to be damaging, the use of low-water–cement-ratio overlays, waterproof membranes, sealants, or latex modified concrete to seal the concrete surface against penetration may be feasible. The same effect may be gained by using a thicker cover over the steel of up to 2 or $2\frac{1}{2}$ in.

Metals of dissimilar composition should not be embedded near or in direct contact with each other in moist or saturated concrete unless experience has shown that no detrimental galvanic action will occur. When it is not possible to separate the metals, impervious protective organic coatings such as bituminous coatings, phenolic varnish, chlorinated rubber, of coal tar–epoxies, should be used on the metal surfaces to prevent galvanic action. Chlorides, sulfides, nitrates, or other harmful chemicals must not be introduced by decomposition of the organic materials. In post-tensioning, it is important that no galvanic action occur between the duct material and tendon steel or the tendon anchorages. (Copper and aluminum should be avoided.)

The presence of stray currents flowing through the earth may lead to serious galvanic corrosion of long metallic elements encased in concrete in the ground. Corrosion by stray currents may be prevented by several methods such as conducting the stray currents away from the embedded metals back to their source or by interrupting the continuity of the embedded metallic elements by means of closely spaced insulating joints.

Cathodic protection, a process for reducing or eliminating corrosion on a metallic structure in contact with a corrosive electrolyte by introducing an electrical potential greater in strength and opposite in direction to the electrical potential that would result from electrolytic action, has been used successfully in a few instances to protect the steel in reinforced concrete structures where cover over the steel has been damaged to the extent that its protective value was lost. The application of such protection to prestressing cables does not appear to be justified economically or technologically. For reinforcing steel, a current requirement of 100 microamperes per square foot has been suggested but long-time polarization effects will reduce this requirement. (Polarization effects result from the chemical and physical changes at the surfaces of the anode or cathode, or both, and reduce the current flowing in the system. Reducing the current requirement reduces the long-time cost of the cathodic protection.)

Recently, an anodic corrosion inhibitor admixture incorporating 2 to 3% calcium nitrite in the concrete has been developed. The concrete can accommodate the inhibitor (with proper use of retarders) without significant loss of properties. However, there are several questions that are not yet answered concerning depletion of the inhibitor in the vicinity of the steel, the relative diffusion rates of nitrite and chloride ions to the steel surface, and the compatibility of the inhibitor with cathodic protection. There appears to be a critical chloride to nitrite ratio (less than 1.5) above which corrosion will occur.

Nonferrous metals embedded in concrete may corrode in two ways: (1) by direct oxidation in strong alkaline solutions normally occurring in fresh concrete and mortar; or (2) by galvanic currents that occur when two dissimilar metals are in contact in the presence of an electrolyte, or when an alloy or metal is not perfectly homogeneous, or when different parts of a metal have been subjected to different heat treatments or mechanical stresses.

Copper and copper alloys are practically immune to action from fresh concrete. No destructive action will occur on the embedded area even when the concrete is kept saturated with moisture. The presence of soluble chlorides, however, may lead to corrosion, and it is therefore advisable to avoid embedding copper in concrete containing chlorides, especially if the concrete and metal are to be exposed to moisture. Galvanic corrosion should be expected when copper and steel reinforcement are connected or in close proximity. Copper is strongly cathodic and will accelerate the corrosion of the steel if chlorides are present. When copper is used in conjunction with steel, it should be electrically insulated from the steel by means of an impervious organic coating or by use of short lengths of polyethylene tubing slit and slipped over the copper.

Copper coatings for the protection of a basic metal such as steel can protect only when they form a complete envelope. Breaks in the coating may create local galvanic cells and cause the exposed basic metal to corrode much more rapidly than it would if it were not coated.

Aluminum suffers attack when embedded in concrete. Initially, when aluminum is placed in fresh concrete, a reaction occurs resulting in the formation of aluminum oxide and the evolution of hydrogen. The greater volume occupied by these oxidation products causes expansive pressures around the embedded metal and may lead to damage to the surrounding concrete.

Corrosion will also occur if aluminum is galvanically connected to steel, with both metals embedded in concrete. If aluminum is to be embedded in reinforced concrete, it should be electrically insulated by a permanent coating. Bituminous paint, alkali-resistant lacquer such as methacrylate, or zinc chromate paint can be used.

If electrical insulation is not permanently maintained, the presence of chlorides will greatly accelerate corrosion of aluminum and lead to serious damage. Because of the difficulty of assuring the permanency of the coating at present, aluminum should not be embedded in or come in contact with concrete containing chlorides. Also, aluminum should not be used in concrete in or near seawater.

Where unpainted aluminum is not embedded but is in contact with concrete under conditions of condensation or dampness, corrosion may occur. This can be prevented by coating as above or by the use of a moisture-proof membrane such as plastic film, bitumen-impregnated paper, or felt.

Lead is attacked by fresh concrete and is converted to lead oxide or to a mixture of lead oxides. This corrosion tends to stop as the concrete cures and dries, but will continue in the presence of moisture and may cause total destruction of an embedded lead pipe in a few years. If the lead is coupled to reinforcing steeel in the concrete, galvanic cell action may be accelerated, and depending on the particular circumstances, either the lead or the steel will be attacked.

When lead is partially embedded in concrete and the remainder exposed to the air, a condition known as differential aeration occurs. The embedded lead has a different electrical potential than that exposed to the atmosphere and in the presence of water will form the anodic (positive) element of an electric cell. The portion in the air forms the cathodic (negative) element of the couple. The current flow will cause corrosion and gradual disintegration of the embedded lead.

Where it is necessary to embed lead in concrete, protection of the embedded portion with organic coatings is suggested. Where a lead strip connects two pieces of concrete such as for an expansion joint, the air space between the two sections should be as small as possible so as to reduce differential aeration.

Generally, no damage will be observed in concrete because of the softness of the lead which will absorb the expansive pressures caused by the formation of corrosion products.

6.14.5 Corrosion Protection of Bonded Tendons

The two main objectives when grouting the ducts of post-tensioned concrete members are: (1) to prevent corrosion of the prestressing steel, and (2) to provide efficient bond between the prestressing steel and the concrete member so as to control the spacing of cracks at heavy overload and thus increase the ultimate strength of a structural member. Both of these objectives require complete filling of the void space within the duct. Filling will be dependent on the production of a grout mix having the desired properties together with efficient equipment for its injection, and careful workmanship and supervision on the site.

Temporary coatings that are applied to the tendon steel to protect it from corrosion during transit, storage, and initial installation should allow the grout to bond uniformly to all parts of the tendon steel. Oil base coatings are not recommended for this application. The temporary corrosion prevention coating should be easily removed in the field with the use of nonchlorinated petroleum solvents for the installation of field-attached anchorages.

The essential properties of grout are sufficient consistency so it can be readily injected to completely fill the voids within the duct, low water content to ensure high strength and low shrinkage characteristics, low bleeding characteristics to prevent segregation and formation of water pockets which may later become air voids in contact with the steel and increase corrosion hazards, and expansion of grout during the first few hours after mixing.

For details and procedures for grouting refer to "Recommended Practice for Grouting of Post-Tensioned Prestressed Concrete," *Post-Tensioning Manual*, Post-Tensioning Institute, Phoenix, Arizona, 1981, 323 pp. The main cause of corrosion of unbonded tendons is the penetration of moisture to the tendon, either through the sheathing and coating or at the end anchorage.

Unbonded tendons should have the prestressing steel permanently protected against corrosion by a properly applied coating. While grease, wax, plastics, bitumen, and other materials have been used, it is recommended that the protective compound should take the form of a grease which also assists the free movement of the tendon during stressing.

Various types of greases are commercially available. However, most of them are not impervious to moisture, so that soluble harmful elements can migrate through the grease to reach the tendon unless the outer covering is completely waterproof.

Ordinary lubricating greases give sufficient protection in most circumstances, provided that the following basic requirements are met:

1. The sheathing must be completely waterproof and continuous for the full length of the tendon, up to and including the anchorages.

It is recommended that the plastic material used for sheathing be either high-density polyethylene or polypropylene. Both materials are tough, durable and nonreactive. High-density polyethylene is more flexible and less liable to embrittlement at extremely low temperatures, while polypropylene is more stable at high temperatures. Both materials have high resistance to abrasion and creep, although polypropylene is slightly superior in these respects.

2. The coating material should adhere to and be continuous over the entire tendon length to be protected and should completely fill the sheathing without air pockets.

3. The coating should remain ductile and free from cracks and should not become fluid over the anticipated range of temperatures during fabrication, transportation, storage, installation, concreting, and tensioning and while in service. In the absence of specific requirements, this range is usually taken as 0 to 160°F.

4. The coating material should not contain more than small traces of harmful impurities, such as chlorides, sulfides, or nitrates.

Special proprietary anticorrosive greases are available that have the following additional properties:

1. Provide a barrier to moisture and air.
2. Furnish a self-healing film and displace water.
3. Have reserve alkalinity for long-term acid neutralization.
4. Contain no solvents leaving trapped residue which may become a fire hazard or react with the sheathing material.

Special anticorrosive greases should always be used in structures subject to a corrosive environment.

The anchored end of a tendon is the most vulnerable location for corrosion attack, especially as the interstices between wires, or the wires forming a strand tendon, form capillaries that allow moisture to gain access to the most highly stressed parts of the tendon.

The anchorages and projecting strand ends usually are enclosed by fabric-reinforced plastic sheaths, which are secured around the female cones. The use of wax and grease may reduce the efficiency of wedge anchorages. Anchorage zones should be encased in drypack concrete, grout, or epoxy mortar, and the encasement should be free from any chlorides. Shrinkage cracks will permit moisture penetration; therefore, details of mix and application are important. When encasement cannot be used, anchorage and end fittings should be completely coated with a corrosion-resistant paint or grease equivalent to that applied to the tendons.

A suitable enclosure will then be necessary to prevent entrance of moisture or deterioration or removal of the coating.

APPENDIX

A6.1 SPECIFYING CONCRETE

"Specifications for Ready-Mixed Concrete," ASTM C94 provides three alternate bases for specifying concrete.

(1) Under Alternate 1, the so-called prescription basis for specifying quality is followed. In this case, the purchaser accepts responsibility for the design of the concrete mixture and specified cement content, aggregate sizes, maximum water–cement ratio, slump, and (in the case of air-entrained concrete) air content. In the case of governmental agencies with a staff of engineers and testing facilities available, this procedure may go so far as to specify batch weights of cement, of each of the sizes of the aggregates to be used, and of water. The philosophy followed under this procedure is that a properly designed prescription produced in specified equipment according to carefully laid-out procedures will give the desired results.

(2) Under Alternate 2, the ready mixed concrete producer is responsible for the design of the concrete mixture. The purchaser merely specifies the minimum compressive strength desired along with usual aggregate sizes, slump, and air content. This procedure comes as close as any to a performance-type specification. But, even here, the producer must furnish evidence in advance of delivery that the specified strength will be attained. The philosophy followed here is that, if the strength is adequate and requirements for aggregate size, slump, and air content are met, the desired properties of concrete will be obtained. Unfortunately, specified strengths can be obtained by means that may be prejudicial to other properties of concrete such as durability, shrinkage, resistance to abrasion, and long-time strength. Also, it is unfortunate that this procedure has been abused by producers who were overly optimistic in their estimates of what strengths can be obtained over a period of time from materials from given sources.

(3) Under Alternate 3, the ready mixed concrete producer is responsible for the design of the concrete mixture, but the purchaser specifies both a minimum cement content and a minimum compressive strength. It is suggested that the minimum cement content be at about the same level as would ordinarily be required for the specified strength, while at the same time the amount of cement should be sufficient to assure durability, watertightness, surface texture, and density for the expected service conditions. The philosophy followed under this procedure is to obtain the properties of concrete desired by the use of the right amount of cement while at the same time using strength tests to check on uniformity and quality.

A6.2 CONSTRUCTION PRACTICES*

Ready-mixed concrete should be batched, mixed, and transported in accordance with "Specifications for Ready-Mixed Concrete" (ASTM C94) or "Specification for Concrete Made by Volumetric Batching and Continuous Mixing" (ASTM C685). Plant equipment and facilities should conform to the "Check List for Certification of Ready Mixed Concrete Production Facilities" of the National Ready Mixed Concrete Association. Reference should also be made to ACI 614, "Recommended Practice for Measuring, Mixing, Transporting and Placing Concrete," *ACI Journal*, July 1972.

Cost of forming is a major item and must be kept to a minimum to achieve construction economy. This is accomplished by considering various types of forms for their flexibility, reusability, and ease of handling. Because proper design, construction, and removal of forms is an involved subject, refer to the work of ACI Committee 347 in "Recommended Practice for Concrete Formwork (ACI 347-78)" and *Formwork for Concrete*, ACI Special Publication No. 4, 1981.

Consolidation is very important in achieving the potential strength of concrete. Figure A6-1 shows the substantial effects that incomplete consolidation has on several properties of concrete.

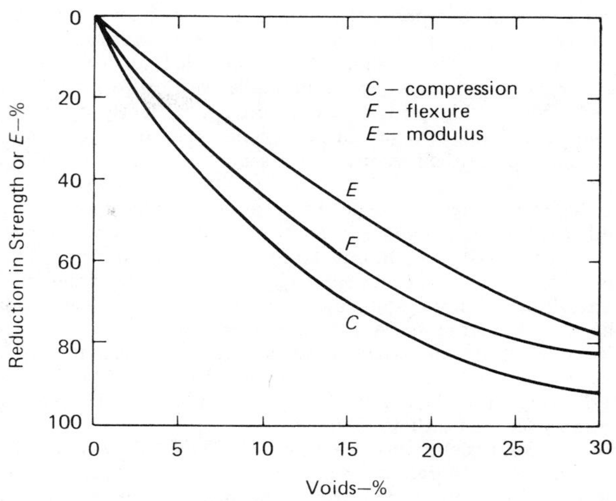

Fig. A6-1 Comparative effects of incomplete consolidation on concrete properties. (From: Kaplan, M.F., "Effects of Incomplete Consolidation on Compressive and Flexural Strength, Ultrasonic Pulse Velocity, and Dynamic Modulus of Elasticity of Concrete," *ACI Proceedings*, 56, 853–867, Mar. 1960.)

Jarring of concrete during hardening may occur from blasting, pile driving, railway train movements, or vehicular traffic; for example, a truck passing over a bridge while other sections of the deck are being paved.

The damage potential of intermittent or continuous jarring depends on the amplitude and frequency of the vibration and the age of the concrete. The particle velocity of earthborne vibrations, which relates amplitude and frequency, is a useful indicator of damage potential; damage potential is greater with increased velocity. Particle velocity decreases exponentially with depth below the surface of the earth. In addition, the particle velocity varies with the amount of energy transmitted to the soil, the physical properties of the soil, and the distance between the fresh concrete and the source of vibration. The safe level of particle velocity for continuous vibration is usually less than that of intermittent vibration; this is due to the possible magnification at resonance, which depends primarily on the inherent damping characteristics of the structure. However, no definite threshold value has been established for a damaging particle velocity.

Jarring of concrete or reinforcement generally will not be detrimental to compressive strength if such jarring does not continue beyond the time of initial set of the concrete. In fact jarring may actually be beneficial to strength due to improved concrete consolidation and reduced water–cement ratio from bleeding which is stimulated by the vibration. It may also increase the concrete-to-steel bond through the removal of entrapped air and water from underneath the reinforcing bars. However, violent impacts or vibration may cause fresh slipformed concrete to slump. Also, concrete is still very weak after initial set, and jarring can produce cracks that subsequently will not close and heal. Such cracking may be more of a problem with certain types of construction; for example, thin lightly reinforced slabs on ground would be more prone to damage than a heavy foundation.

A6.3 CURING

For most structural uses, the curing period for cast-in-place concrete is usually three days to two weeks, depending on such conditions as air temperature, cement type, mix proportions, etc. More extended curing periods are desirable for bridge decks and other slabs exposed to weather and chemical attack.

Forms should be removed in such manner as to ensure the complete safety of the structure. In determining the time for removal of forms, consideration should be given to the construction loads and to the possibilities of deflections. The construction loads are frequently at least as great as the design live loads. At early ages, a structure may

*The reader should refer to one of the many excellent tests on concrete technology such as "Design and Control of Concrete Mixtures," available from the Portland Cement Association.

be strong enough to support the applied load but may deflect sufficiently to cause permanent damage. Where the structure as a whole is adequately supported on shores, the removable floor forms, beam and girder sides, column forms, and similar vertical forms may be removed after 24 hr, provided the concrete is sufficiently strong not to be injured. Forms also should be designed and constructed with some thought as to their removal with a minimum of danger to the concrete.

For the development of sufficient strength to carry imposed loads the judgment of the design engineer may be guided by strength tests of job-cured cylinders or beams. Moisture retention measures may be discontinued when the average strength reaches 70% of the specified strength. Still more valid indications of the rate of strength gain can be obtained from strength-time curves which may have been developed for the set of materials used on the job. Pavements should not be opened to traffic until the concrete has attained a flexural strength of 550 psi when tested by the third-point loading method or a compressive strength of 3500 psi.

Without careful simultaneous reshoring, it is hazardous in freezing weather to remove shores even temporarily before suitable tests show conclusively that the specified strength has been attained. Ordinarily, for temporary removal of support from an entire panel during reshoring, attainment of 55 to 65% of the design strength is sufficient. The strength may be determined from field-cured test specimens or estimated for properly cured concrete by determining the length of the curing period and comparing it to the age of laboratory-cured specimens that reached the desired strength. The length of time the field concrete has been cured should be determined by the cumulative number of days or fractions thereof, not necessarily consecutive, during which the temperature of the air in contact with the concrete is above 50°F and the concrete has been damp or thoroughly sealed from evaporation and loss of moisture.

Reshores should be left in place as long as necessary to safeguard each member and, consequently, the entire structure. The number of tiers reshored below the tier being placed and the length of time reshores remain in place are dependent on the development of sufficient strength to carry dead loads and any construction loads with adequate factors of safety.

To ensure that concrete achieves the desired durability during cold weather, it should be maintained at the temperature shown in Table A6-1(A and B), for the period of time shown. The concrete should not be subjected to freezing in a saturated condition before reaching the design strength. The minimum curing period for adequate scale resistance to chemical deicers generally corresponds to the time required to develop the design strength of the concrete. An air-drying period of 30 days, which enhances resistance to scaling, should then elapse before application of deicing salts.

At the end of the necessary heating (protection) period, concrete should be cooled slowly. Rapid cooling can cause cracking (especially in mass concrete configurations such as bridge piers, abutments, dams, and large structural members). By simply shutting off the source of heat, concrete will cool gradually to the outside air temperature, but the depth of the different sections of concrete determines the cooling period. Concrete should be permitted to cool at a rate not greater than that shown in line 5, Table A6-1.

A6.4 QUALITY CONTROL

A comprehensive quality control program is required at both the concrete plant and the site to ensure that concrete quality and uniformity are acceptable and meet job requirements. Inspection of concreting operations from stockpiling of aggregate through completion of curing is important. Also, routine sampling of all mix materials may be necessary to control uniformity of the concrete. Usually, normal field control can be accomplished adequately by testing fresh concrete for slump, temperature, air content, and unit weight; the hardened concrete is tested for strength.

The architect or engineer should select the testing laboratory, and the owner should pay for testing by separate contract with the laboratory. This quality assurance to the owner in no way relieves the the contractor of his responsibility for quality materials and workmanship required to meet the specifications.

A6.4.1 Sampling

Samples must be taken on a strictly random basis if they are to measure properly the acceptability of the concrete. Samples should be obtained and handled in accordance with "Method of Sampling Fresh Concrete" (ASTM C172). The choice of times of sampling or the batches of concrete to be sampled must be made on the basis of chance alone within the period of placement in order to be representative. If batches to be sampled are selected on the basis of appearance, convenience, or other possibly biased criteria, statistical concepts lose their validity. Obviously, not more than one strength test (average of two cylinders made from a sample) should be taken from a single batch, and water may not be added after the sample is taken.

A predetermined sampling plan (chance approach) should be set up before the start of production or deliveries by establishing the intervals at which samples will be taken. The intervals may be set in terms of either time elapsed or yardage placed.

The number of samples will be based generally on the degree of control required. The degree of control that can be realized practically and the level of quality aimed at are directly linked to the economy of the situation. One can control to closer and closer tolerances, until a point is reached at which further control costs outweigh the benefits to be realized. Job conditions will determine the most practical number of samples required. The minimum number of samples is usually set forth in specifications. Requirements are usually a minimum of one sample for each 100 or 150 cu yd for each class of concrete, and at least one sample to be taken daily from each class of concrete. In addition, *Building Code Requirements for Reinforced Concrete, ACI 318* requires in the case of slabs or walls that samples be taken for each 5000 sq ft of surface area placed. In calculating surface area, only one side of the slab or wall should be considered. If the average wall or slab thickness is less than $9\frac{3}{4}$ in., more frequent sampling than once for each 150 cu yd placed will be required. For pavements, two beams should be made for each 2000 sq yd with not less than two beams per day. These values are a minimum and adequate for general construction. However, concrete used in vital structural sections such as columns, beams or slabs should be sampled more frequently. Additional specimens may be required when temperature or moisture content changes suddenly, or when materials or sources of materials are changed.

When the frequency of testing will provide less than five strength tests for a given class of concrete, tests should be made from at least five randomly selected batches or from each batch if fewer than five are used. When the total quantity of a given class of concrete is less than 50 cu yd, the strength tests may be waived if adequate evidence of satisfactory strength is provided, such as strength test results from the same type of concrete supplied on the same day by the same supplier and under comparable conditions in other work.

The slump test should be made at the start of operations each day, whenever the appearance of the concrete indicates a change in consistency, and whenever making strength and air content tests. Temperature of concrete should be recorded when strength tests are made, at frequent intervals in hot or cold weather conditions, and at the start of operations each day. The frequency of air content tests depends on the severity of exposure conditions. It may vary from one test for each truckload to one test for each five truckloads of concrete.

On work such as pavements and large bridge decks where very rapid rates of concrete placement occur, i.e., >50 cu yd per hour, it is impractical to continue a frequency of one test per truckload for very long. Under such conditions and once satisfactory control has been established, the frequency of testing can be reduced for as long as each test result fails within specification requirements. When a test result falls outside the specified limits, the testing frequency should revert to one test per load of concrete until satisfactory control is reestablished. A sudden loss in slump, the development of finishing difficulties, the appearance of bleeding, a change in temperature or aggregate grading, or a loss in yield calls for a check on the air content. Unit weight tests are made at the same time as the slump test.

A6.4.2 Consistency Tests

The slump test is the common method for controlling concrete consistency. It is also a good indicator of variations in water content from batch to batch. A change in slump of structural light-weight concrete may be indicative of a change in air content, moisture content of the aggregates, or a change in gradation or density. The slump test for consistency of concrete should be made in accordance with the "Method of Test for Slump of Portland Cement Concrete" (ASTM C143). The slump test gives satisfactory results when slumps are below 6 in., beyond which the use of the slump test can be questioned,

TABLE A6-1 Recommended Concrete Temperature for Cold-Weather Construction—Air-Entrained Concrete

Line	Condition		*Sections less than 12 in. (300 mm) thick*	*Sections 12 to 36 in. (300 mm to 0.9) thick*	*Sections 36 to 72 in. (0.9 to 1.8 m) thick*	*Sections over 72 in. (1.8 m) thick*
			°F	°F	°F	°F
1	Minimum temperature fresh concrete as *mixed* in weather indicated. °F	Above 30°F	60	55	50	45
2		0°F to 30°F	65	60	55	50
3		Below 0°F	70	65	60	55
4	Minimum temperature fresh concrete as *placed* and *maintained*		55	50	45	40
5	Maximum allowable *gradual* drop in temperature in first 24 hours after end of protection		50	40	30	20

*Placement temperatures listed are for normal-weight concrete. Lower temperatures can be used for lightweight concrete if justified by tests.

A. Recommended Duration of Recommended Concrete Temperature in Cold Weather—Air-Entrained Concrete

Permanent Service Category	*For Durability*		*For Safe Stripping Strength*	
	Conventional Concrete, Days*	*High-early-strength Concrete,** Days*	*Conventional Concrete,* Days*	*High-early-strength Concrete,** Days*
Lightly stressed, no exposure,† favorable moist-curing	2	1	2	1
Lightly stressed, exposed, but later has favorable moist-curing	3	2	3	2
Moderately stressed, exposed†	3	2	6	4
Fully stressed, exposed†	3	2	See B below	

B. Recommended Duration of Recommended Concrete Temperature for Fully Stressed, Exposed, Air-Entrained Concrete

Required Percentage of Design Strength f'_c	*Days at 50°F*			*Days at 70°F*		
	Type of Portland Cement			*Type of Portland Cement*		
	I	II	III	I	II	III
50	6	9	3	4	6	3
65	11	14	5	8	10	4
85	21	28	16	16	18	12
95	29	35	26	23	24	20

Adapted from Cold-Weather Concreting *(ACI 306-78). American Concrete Institute. Cold weather is defined as that in which average daily temperature is less than 40°F, except that if temperatures above 50°F occur during at least 12 hours in any day, the concrete should no longer be regarded as winter concrete and normal curing practice should apply.*

The values shown in B are approximations and will vary according to the thickness of concrete, mix proportions, etc. They are intended to represent the ages at which supporting forms can be removed.

NOTE: For concrete that is not air entrained, *ACI 306-78 states that protection for durability should be at least twice the number of days listed.*

**Made with ASTM Type I or II portland cement.*
***Made with ASTM Type III portland cement, or an accelerator, or an extra 20% of cement.*
†"Exposure" means subject to freezing and thawing.

for example, to characterize flowing concrete. The slump test should be made first before any other test when concrete arrives at the job, since it frequently determines at once whether a batch of concrete will be acceptable or should be rejected. A tolerance of up to 1 in. above the maximum specified slump when the average slump is 4 in. or less, or a tolerance of $1\frac{1}{2}$ in. when the slump is 4 to 6 in., may be allowed for individual batches provided the average for all batches or the most recent 10 batches, whichever is fewer, does not exceed the maximum slump. Concrete of lower than usual slump may be used provided it can be properly placed and consolidated.

Another method of measuring consistency is the ball penetration test, "Method of Test for Ball Penetration in Fresh Portland Cement" (ASTM C360). When calibrated for a particular set of materials, the results can be related directly to slump. However, the relationship must be checked at least once a day thereafter if the ball penetration test is used for control. This test has the advantages of being relatively simple and of not requiring a molded specimen. The test can be made on fresh concrete in any open container provided the minimum lateral dimension of the sample is about 18 in. and the depth at least 8 in. This test is not accurate in high-slump concrete (5 in. or more), concrete containing large coarse aggregate (2 in. or more), or thin slabs.

A6.4.3 Temperature Measurement

Because of the important influence of concrete temperature on the properties of fresh and hardened concrete, many specifications place limits on the temperature of fresh concrete. Temperature measurements are made primarily during hot or cold weather to make certain that the concrete arrives ready for placing at a temperature suitable for prevailing weather conditions. Although there is no standard method for measuring temperature of fresh concrete, certain simple precautions should be observed. Armored thermometers are available. The thermometer should be accurate to $\pm 2^{\circ}$F and should remain in a representative sample of concrete until the reading becomes stable.

A6.4.4 Tests for Air Content

Available methods for determining air entrainment in freshly mixed concrete measure only air volume and not the air void characteristics. However, it has been shown that the volume of entrained air in a concrete mixture is generally indicative of the adequacy of the air void system when using air-entraining materials meeting ASTM specifications.

A number of methods for measuring air content of fresh concrete are in use. Standards cover the pressure method (ASTM C231), the volumetric method (ASTM C173), and the gravimetric method (ASTM C138). Variations of the first two methods are also in use. Air content is best measured by the pressure method for normal-weight concrete and the volumetric method for structural lightweight concrete. Variations from the specified value of air content should not exceed ±1% for normal-weight concrete or ±1.5 percentage points for structural lightweight to avoid adverse effects on compressive strength, workability, or durability.

Rapid and frequent indications of the air content can be obtained with a Chace pocket air meter. The air content as measured by this method is an approximation, usually within one-half a percentage point of that measured by the pressure method. Therefore, the pocket air meter is used as a control to indicate changes in air content, but is should be supplemented by tests from another method, particularly when the air content approaches the specified limits. It may also be especially useful in checking air contents in small areas near the surface that may have suffered reductions in air content because of faulty finishing procedure.

A6.4.5 Unit Weight

The unit weight is a quick and useful measurement for controlling quality. A change in unit weight generally indicates a change in either air content or aggregate weight. For lightweight concrete a maximum allowable unit weight is usually specified. The unit weight of freshly mixed lightweight concrete is correlated with the 28-day (design) air-dry unit weight and used as a basis for placement and control and acceptance during construction. When unit weight measurements (ASTM C138 or C567) indicate a variation of more than a few pounds per cubic foot for normal-weight concrete or ±2% for structural lightweight concrete, the air content should be checked first to establish if the correct amount of air has been entrained. If air contents are correct, then a check should be made on the aggregates to make certain that the unit weight, gradation, and moisture content have not changed. Results of these checks generally will reveal the cause of the variations in unit weight of concrete and indicate what mix adjustments are in order.

A6.4.6 Tests for Strength

Strength specimens must be fabricated, cured, handled, and tested in strict conformance to standard procedures. Specimens should be made and cured in accordance with "Method of Making and Curing Concrete Test Specimens in the Field" (ASTM C31). Compression specimens should be cast in reusable steel molds or in disposable tin molds rather than paraffined cardboard molds (ASTM C470). Most cardboard molds appear to produce specimens with strengths 2 or 3% less than those molded in steel molds, and occasionally strength reductions of 10 to 15% may occur.

The strength of a specimen is greatly affected by disturbances, changes in temperature and exposure to drying, particularly within the first 24 hr after casting. Thus, cylinders should be cast in location where subsequent movement is unnecessary and where protection is possible. Cylinders should not be moved after they are 15 to 20 minutes old. Because of the danger of producing cracks and weakened planes, concrete with slumps less than about 1 in. should be protected from rough handling at all ages.

After cylinders are molded they must be stored carefully at 60°F to 80°F during the first 24 hr and moisture loss prevented by covering them with an oiled metal or glass plate and a double layer of wet burlap. Storage at lower temperatures but above freezing would tend to increase 28-day strengths, while storage at higher temperatures would produce lower 28-day strengths. In the wintertime, plywood curing boxes equipped with light bulbs or small thermostatically controlled electric heaters have been used with great success. Maximum-minimum thermometers should be used. Coverings of damp burlap are highly recommended in curing boxes. Curing boxes with a small cake of ice are a practical means of obtaining the desired temperature control in hot weather. When curing boxes are not used in the summertime, coverings of damp burlap will provide some cooling due to the evaporation of water in the shade. In the summertime, coverings of damp burlap with sheet polyethylene over the outside must be avoided, since the plastic sheeting prevents evaporation, and the burlap provides insulation for the retention of heat due to hydration during the first 24 hr.

Within 24 hr the specimens should be taken carefully to the laboratory and placed in standard moist curing. The failure to bring the specimens into the laboratory for standard curing within 20 ± 4 hr is probably the most frequent and perhaps the most serious of all potential violations of standard testing procedures. If tin can or cardboard molds are used, it is preferable to ship the cylinders in the molds for protection. Cylinders removed from the molds should be packed in wet sawdust, wet sand, or wet burlap to prevent drying. These materials will also act to cushion any jolts. Rough handling and lack of protection from too rapid drying will tend to produce erratic and low-strength results.

Cylinder ends should be capped or ground in accordance with the requirements of "Method of Capping Cylindrical Concrete Specimens" (ASTM C617).

The age of test specimens should be 28 days (unless using accelerated curing) but 7-day specimens may be used, provided that the relationship between the 7- and 28-day strength of the concrete is established by test for the materials and proportions used. Usually if three specimens are made, two are tested at 28 days for acceptance, and one is tested at 7 days for information.

Testing of specimens should be done in accordance with "Method of Test for Compressive Strength of Cylindrical Concrete Specimens" (ASTM C39) and "Method of Test for Flexural Strength of Concrete—Using Simple Beam with Third-Point Loading" (ASTM C78). Modulus of rupture tests by center-point (ASTM C293) or cantilever loading, or compressive strength tests, may be used for job control of pavements if relationships to third-point tests are determined before construction starts. Typical relationships are shown in Fig. A6-2.

The moisture content of the specimen has a considerable effect on the result; a saturated cylinder will show 20 to 30% lower strength

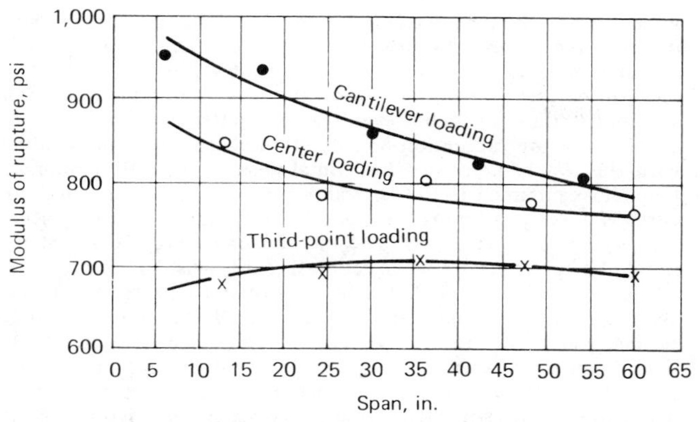

Fig. A6-2 Relationships between modulus of rupture test methods. (From: *Design and Control of Concrete Mixtures* 11th ed., Portland Cement Association, Skokie, Ill., 1968.)

than that of a companion specimen tested dry. This is important when comparing cores taken from concrete in service with molded specimens that are tested as they are taken from the moist room.

The flexural strength tests are very sensitive to deviations in specified standard moist curing conditions. Since the high stresses in a flexural specimen are at the surface, these specimens are highly sensitive to small amounts of drying which induce tensile stresses in the extreme fibers of the specimen, thereby greatly reducing the measured tensile stresses at failure. For this reason, when a moist specimen is removed from curing for testing, it must be maintained in a moist condition until testing is completed. As little as 30 minutes of drying may reduce measured flexural strength as much as 10%, and a period of drying prior to testing of three or four days reduces flexural strengths 40%.

Tests on concrete specimens when carefully conducted only show the potential strength of the concrete in the structure. When test specimens are to be cured entirely under field conditions in accordance with ASTM C31—as a check on the adequacy of curing and protection of the concrete in the structure—they should be molded at the same time and from the same samples as the laboratory-cured acceptance test specimens. Field-cured cylinders must be cured under conditions identical to those of the structure. If the structure is protected from the elements, the cylinder should be protected similarly. That is, cylinders related to members not directly exposed to the weather should be cured adjacent to those members and provided with the same degree of protection and type of curing. Obviously, the field cylinders should not be treated more favorably than the elements they represent.

Procedures for protecting and curing the concrete should be improved when the strength of field-cured cylinders at the test age designated for measuring f'_c is less than 85% of that of the companion standard laboratory moist-cured cylinders. This percentage has been set merely as a rational basis for judging the adequacy of field curing. The comparison is made between the actual measured strengths of companion job-cured and laboratory-cured cylinders, not between job-cured cylinders and the specified value of f'_c. However, results for the job-cured cylinders are considered satisfactory if they exceed the specified f'_c by more than 500 psi even though they fail to reach 85% of the strength of companion laboratory-cured specimens.

Field-cured specimens also are useful in determining when sufficient strength has developed to permit safe removal of forms or when a structure may be put into service.

Instead of using field-cured cylinders to approximate conditions of strength development in the structure, other procedures such as the following may be used:

(a) Tests of cast-in-place cylinders in accordance with "Tentative Methods of Test for Compressive Strength of Concrete Cylinders Cast-in-Place in Cylindrical Molds" (ASTM C873). (This method is limited to use in slabs where the depth of concrete is from 5 to 12 in.)
(b) Penetration resistance in accordance with "Standard Method of Test for Penetration Resistance of Hardened Concrete" (ASTM C803).
(c) Pullout strength in accordance with "Tentative Method of Test for Pullout Strength of Hardened Concrete" (ASTM C900).
(d) Maturity factor measurements and correlation in accordance with Section 7.3 of "Recommended Practice for Cold Weather Concreting" reported by ACI Committee 306.

Procedures (b), (c), and (d) require sufficient data using job materials to demonstrate correlation of measurements on the structure with compressive strength of molded cylinders or drilled cores.

Present-day speed of construction requires knowledge of concrete quality earlier than the usual 7, 14, or 28 days; otherwise erection will have proceeded far beyond the placement in question. Since strength develops with the hydration of cement, concrete specimens can be exposed to elevated temperatures and moisture conditions adequate to develop a significant portion of their ultimate strength within 24 to 48 hr.

ASTM in 1971 adopted "Method of Making and Accelerating Curing of Concrete Compression Strength Test Specimens" (ASTM C684), which includes three procedures. The procedure to use is chosen by the engineer on the basis of his experience, and local conditions. The three procedures are: (1) warm water method, (2) boiling water method and (3) autogenous curing method. All have some particular merits and all have certain disadvantages. Methods (1) and (2) have the merit of avoiding the need for a delay because, in boiling method, 24 hr of initial moist curing is used before the $3\frac{1}{2}$-hr boiling period; and in the warm water method, the water external in the specimen is kept at 95°F throughout the 24-hr curing period. However, both these methods require curing tanks and the compression testing machine to be on site or close by. Both provide results by next day, after about $28\frac{1}{2}$ hr for one and 24 hr for the other.

While it may be desirable to have strength results in about 24 hr, this is not essential on most jobs, and a compensating advantage of the autogenous method is that testing at 72 hr or 96 hr is possible. Appropriate correction factors can be developed that allow cylinders made on a Thursday or Friday simply to sit in their containers until Monday, thus avoiding all overtime weekend work. Valid relationships exist for nonstandard hydration conditions, and it is simply a matter of calibration. Starting temperatures over the normal range 50°F to 90°F have little effect. Though the initial container cost is high, the containers are designed for many reuses. At 100 reuses the extra cost is only 20 to 30 cents per test.

A6.4.7 Investigation for Structural Adequacy

Procedures used in the design of concrete structures allow for the normal variability of strength tests as well as for the favorable treatment of laboratory test specimens in comparison with the concrete in place by specifying strength levels greatly in excess of stresses to which the concrete will actually be subjected.

The concrete strength is considered to be satisfactory as long as averages of any three consecutive tests remain above the specified f'_c, and no individual test falls below the specified f'_c by more than 500 psi or the specified modulus of rupture by more than 75 psi. Strength tests failing to meet these criteria will occur occasionally (probably about once in 100 tests), even though strength level and uniformity are satisfactory. Allowance should be made for such statistically normal deviations in deciding whether or not the strength level being produced is adequate.

Ocassionally an isolated test result deviates so far from the average values as to be highly improbable. A test cylinder that is one of a set of three or more companion cylinders may be discarded if the compressive strength deviates more than 1500 psi from the average of the remaining cylinders. However, an attempt should be made to explain and correct the cause of the deviation. If the suspect cylinder is one of a set of two companion cylinders, it may be discarded only if the cylinder shows evidence of questionable practices in sampling or in molding, curing, capping, or testing.

The significance of low strength measurements should be investigated in the following sequence:

Step 1. Verifying testing accuracy—Sampling and testing of concrete in the field are often not in strict accord with the standard procedures prescribed for acceptance testing. Stern measures should be taken whenever deviations from standard procedures are discovered. Inaccuracies may lead to expensive delays or supplementary investigation. The testing laboratory should be held responsible for deficiencies in the testing procedures and services it is asked to provide. Laboratories should not be employed unless they meet the requirements of ASTM Recommended Practice E329 and unless they have been inspected within three years by the Cement and Concrete Reference Laboratory of the National Bureau of Standards and have corrected deficiencies disclosed in the inspection.

If the testing of the concrete is found to have deviated from the standard methods, it may be possible to terminate the investigation at this point. If the testing has been satisfactory, or if this phase of the findings is inconclusive, it may be necessary to continue with Step 2 or Step 3, or possible both simultaneously.

Step 2. Compare structural requirements with measured strength—Lower strength may, of course, be tolerated under many circumstances but this is a matter of judgment. The structural engineer must determine if the load-carrying capacity has been significantly reduced. If time and conditions permit, an effort can be made to improve the strength of the concrete in place by supplemental wet curing. Effectiveness of such a treatment must, of course, be verified by further strength evaluation using procedures to be discussed.

Step 3. Nondestructive tests—Several devices are available for obtaining estimates of concrete strength in place. Generally speaking, they do not provide values that can be translated directly into cylinder or core strengths. However, in the hands of a skilled operator, they can yield useful information on the concrete in place by comparing readings on portions of the structure represented by the low strength tests with other portions that are considered acceptable. Such readings can indicate quite convincingly whether or not the questioned concrete differs appreciably from concrete judged acceptable.

(*a*) *Rebound Hammer* (ASTM C805)

One such test involves the use of the Swiss (Schmidt) hammer—an impact testing device. The hammer is not intended to replace the standard cylinder nor is it intended to give an accurate measure of the compressive strength of concrete. Its limitations and proper use should be understood by all concerned prior to its acceptance as a testing tool. It is valuable for field use to determine relative levels of strength, uniformity of concrete in place, and where and when cores should be drilled. But it must be properly calibrated and it must be used competently.

The Swiss hammer, essentially a concrete hardness tester, measures the rebound of a spring loaded plunger after it has struck a smooth concrete surface. A reading then is taken from a scale on the side of the test hammer while holding the hammer firmly against the concrete. A button usually is provided to lock the pointer on the scale after impact if it is not convenient to take a reading while holding the hammer against the concrete. High rebound readings are associated with high concrete strength.

At least 10 individual readings are necessary to obtain a good representative mean of a given test. The significance of the mean is not improved much by taking more than 20 individual readings. The two or three highest and lowest readings should be rejected when evaluating the mean. The rebound number is considered reliable when the 10 readings do not deviate more than $\pm 2\frac{1}{2}$ to $3\frac{1}{2}$ on the scale. Whenever possible, the readings representing one test (between 10 or 20) should be confined to an area of about 1 sq ft. A rough grid should be marked out on a 2-in. spacing and the intersections used for test points. The test position should be 2 in. or more from the edge of the concrete, or low readings will result. A repeated test on the same spot or near it will result in lower readings, due presumably to local partial crushing of the concrete.

The rebound of the hammer is affected by the hardness of the surface; smoothness of the surface; moisture content of the near surface; age of the specimen; aggregate type and concentration; size, shape, and rigidity of the specimen; and the strength of the concrete.

The calibration curve provided with the instrument to indicate the compressive strength from the reading obtained from the test hammer scale is not always precise. To be most useful, the test hammer should be calibrated for the particular materials and mix designs on the job, or the hammer can be used only to indicate difference between questionable and acceptable portions of the structure.

One method of calibration is to obtain rebound readings on standard test cylinders or cores immediately prior to their being tested in compression. Fifteen readings should be taken on the middle two-thirds of the cylinder surface along three vertical lines, 120° apart, while the cylinder is anchored in the testing machine by a load of at least 300 psi. The average reading (from nine numbers) after discounting the three highest and three lowest readings then is assigned to the cylinder compressive strength and a curve developed. The test hammer can be used horizontally, vertically upward or downward, or at any intermediate angle; however, readings in these various directions require separate calibration or correction charts. Rebound readings with the test hammer in a horizontal position generally are higher than those obtained in the vertical position. Even after a calibration curve is developed, individual points will deviate from the relationship by ±500 to 1000 psi. The method of curing and age at test should be kept constant when calibrating the Swiss hammer.

The area to be tested should be smooth and uniform and preferably formed rather than finished. Areas exhibiting honeycombing, scaling, rough texture, or high porosity should be avoided. Rock, sand, or air pockets will result in very low readings, and the presence of air holes, large rock particles or reinforcement near the surface will cause erratic results. Concrete placed against wooden forms usually produces rebound numbers 10 to 20% lower than concrete cast against metal forms or finished with a metal trowel. This is especially evident with concrete of high compressive strength. However, the scatter of individual results on formed surfaces is lower. When necessary, a smooth surface may be produced by uniformly rubbing an area about 2 in. in diameter with a carborundum stone. This is a cumbersome procedure and can lead to large errors if not carried out uniformly.

Usually, old, dry concrete (age greater than 90 days) in which the surface is harder (carbonation effects) than the interior registers higher rebound numbers than normal. Conversely, new, moist concrete surfaces (age less than 7 days) give a low rebound value. It is recommended that the test hammer be used only on concrete from 7 to 90 days old. Within this general range of ages, a wet concrete surface will give about a 20% lower mean reading than a dry surface of the same concrete. The differences between wet and dry specimens are smaller at the earlier ages. It should be remembered that a standard moist-cured cylinder will show lower strength than that of a companion cylinder tested dry. However, the rebound number changes faster than strength with a change in moisture condition. Wetting concrete areas to be tested and covering them with damp burlap and polyethylene for several days can be used to minimize the effect when relatively young and relatively old concrete must be compared.

Different aggregates, in general, will give different calibration curves. A lightweight aggregate concrete will give a lower rebound reading then normal-weight concrete. Gravel coarse aggregate concrete often produces a higher reading than crushed stone concrete (particularly soft limestones) of the same compressive strength. In addition, the coefficient of variation of individual readings increases with the size and amount of coarse aggregate.

Low rebound numbers will result if small, long, or thin specimens (less than 4 in. thick) are not rigidly supported to prevent movement under impact.

The Swiss hammer is designed to measure compressive strength above 1000 psi and, therefore, is not applicable to concrete at very early ages. At early ages the increase in concrete surface hardness lags behind the gain in strength, but after three months outstrips it. Therefore, use of the hammer at early ages causes severe pocking and scarring of the concrete. The coefficient of variation of individual readings constituting a mean decreases with increase in the strength of the concrete. An understanding of all of these factors will aid in interpreting the results of a Swiss hammer test.

(*b*) *Penetration Probe* (ASTM C803)

The Windsor probe is a recent development that also appears to be basically a hardness tester. The technique of using the Windsor probe consists of forcing a hardened alloy probe into the concrete. The input energy is developed by the expanded gases of a precision-loaded industrial-type powder cartridge.

The test probe is used as follows: (1) Three probes, spaced in a 7-in. triangular pattern by means of a templet, are driven into the concrete. The driver is loaded separately for each probe. (2) The 9-in. equilateral triangular, lower gauge plate is placed over the three probes and can be secured by compression springs for overhead work. (3) An upper gauge plate is placed over the three probes. Both lower and upper plates provide a mechanical average of the concrete surface and the projection of the probe respectively. (4) The depth gauge is inserted through the appropriate hole in the upper gauge plate and the exposed height of the probes is recorded in inches. Finally, refer to a graph for compressive strength. However, reliance should be placed upon comparisons between accepted and questioned portions of the structure rather than upon numerical strength levels.

The variables affecting the test results have not been fully investigated. However, tests during World War II on projectiles shot into concrete showed that moisture content of concrete and aggregate type had significant effects on penetration. In addition, penetration decreased with an increase in volume, fineness modulus, maximum size, and crushing strength of the aggregate (Moh's hardness of aggregate). These same variables may be of significance with the Windsor probe.

Preliminary test results indicate that the probe should be calibrated for the particular materials used on the job and that:

1. The depth of penetration does indicate the development of strength.
2. The shape of the probe is important with a step (lightweight concrete) and nonstep probe (normal weight concrete) utilized.
3. Maximum size of aggregate does make a difference.
4. The probe will have a greater penetration into mortar than into concrete.
5. The probe will indicate a difference in wet and dry strength of concrete.
6. The probe can measure compressive strength in the range of 1500 to 7000 psi when the minimum thickness of the sample is around 6 in.

Since the Windsor probe is a hardness tester similar to the Swiss hammer, it probably will be affected by the same variables, and the advantages of one testing system over the other should be considered. The probe measures hardness to a much greater depth than the Swiss hammer, and test results may be influenced less by surface texture and carbonation effects. The initial cost of the Windsor probe test apparatus is more than the cost of the Swiss hammer. In addition, the probes and charges are a continual expense, and the Windsor system requires cleaning of the gun. The test may not be exactly nondestructive, since it may cause spalling. A spall may result when the probe is used closer than 5 or 6 in. from the edge of the concrete, or if the probe hits a flat or hard aggregate particle immediately beneath the surface and deflects. If the probes are withdrawn properly, little or no damage results. Surface damage usually consists of not over $1\frac{1}{8}$ in. disturbance on the surface and a $\frac{5}{16}$-in. hole in the center for the depth of the probe. The dimples left in the concrete by the test hammer may also be objectionable.

(c) *Pulse Velocity* (ASTM C597)

The condition of concrete in service may be obtained by measuring the velocity of pulses of sonic or ultrasonic vibrations transmitted through it. The pulse velocity method (ASTM C597) was developed to test large masses of concrete. This method also may be used to assess concrete uniformity but should not be considered as a means of measuring strength unless a velocity-strength calibration curve is developed from cylinders made with the concrete mix used on the job. The pulse velocity method is a specialized technique that requires experience in using the apparatus and in interpreting the results.

The pulse velocity method consists of measuring the time of travel of a pulse or train of waves through a measured path length in the concrete. The measurements may be made on any size or shape object. The receiving and transmitting transducers need not be positioned exactly opposite each other. Knowing the distance between transducers, transmission times can be measured across corners of structures or along one face. When using the amplitude of the received pulse as a criterion, the attenuation characteristics of the concrete can be determined. The higher frequency components of the pulse are more rapidly attenuated (dampened) than the lower frequencies, and the pulses's shape changes as it passes through the concrete, becoming more rounded as the distance increases. Therefore, choice of the transducers for a particular application is important. Care should be taken in coupling the transducers onto rough or uneven surfaces or it may be difficult to obtain reproducible, accurate measurements of pulse velocity and attenuation. Often it is recommended that a note be made of the signal strength for each observation for use in interpreting the pulse velocity data.

There are a number of factors affecting the pulse velocity. Some of these are: (1) mix proportions, (2) type of aggregate, (3) moisture content, (4) compaction, (5) reinforcement, and (6) age of concrete. There does not appear to be complete uniformity of opinion as to the degree of importance of these factors in the estimation of strength.

Concretes of different mix proportions may have the same compressive strength for different pulse velocities. If the water-cement ratio is held constant and the proportions of cement paste and aggregate varied, the strength will remain essentially constant, whereas the velocity will not. The more aggregate added, the more closely will the pulse velocity of the concrete approach that of the aggregate. An increase in aggregate-cement ratio results in an increase in pulse velocity. The fine aggregate does not affect the velocity as much as does the coarse aggregate. For a given change in compressive strength, the change in velocity is higher for concretes with a high aggregate-cement ratio. However, this increase is nominal when high-strength concrete is used.

Coarse aggregate has a velocity greater than that of the concrete in which it is contained. Pulse velocities through neat cement paste are appreciably lower than those through concrete or coarse aggregate. More cement increases the strength of concrete but tends to decrease the pulse velocity. For the same maximum size and type of aggregate, the pulse velocity decreases as the water-cement ratio increases.

For the same maximum size aggregate, water-cement ratio, and aggregate-cement ratio, the pulse velocity may vary considerably with aggregate type. This variation is due to the wide ranges in aggregate elasticity and density. The type of coarse aggregate also may have a larger effect on the damping coefficient than on the strength.

Pulse attenuation (damping) of concrete increases when the concrete dries out but is less sensitive to changes in moisture content than is pulse velocity. Pulse velocity may vary as much as 10%, depending on the amount of moisture in the concrete. Voids due to poor consolidation have much less effect on pulse velocity than on compressive strength owing to transmission through the coarse aggregate. The effect of steel on pulse velocity depends upon the proximity of the steel to the transmission path, the size and concentration of reinforcing, the distance between transducers, and the relative velocities in concrete and steel. Only a small amount of energy is coupled into reinforcement, but the pulse should not be propagated near the axis of thick reinforcing bars.

Pulse velocity increases with age at a decreasing rate similar to a strength age curve. At strengths less than 4000 psi, the relationship between velocity and strength appears to be less dependent on age and water-cement ratio than it does at higher strengths. Low pulse velocity at an early age forecasts low strength at later ages. It is generally agreed that very high velocities (more than 12,000 fps) are indicative of very good concrete, and that very low velocities (less than 10,000 fps) are indications of poor concrete. It is further agreed that periodic, systematic changes in velocity are indicative of similar changes in the quality of the concrete. When test cylinders are available for development of a calibration curve, the strength may be estimated to within ±10 to 25%. The pulse velocity test has particular application as an accurate and economical method of quality control in prestressing and precast operations because of the uniformity of materials.

In some cases, use of all of the nondestructive tests should be considered. A combination of several tests will improve the estimate of the concrete strength for which full information on composition and age is not available and will assist in locating areas of low, medium, and high strength prior to coring. The investigator of concrete quality in field structures must have accurate knowledge of the factors affecting the test results to adequately interpret the variations in results. However, it is dangerous to try to develop a general relationship that will hold for all concretes on the basis of any of these nondestructive tests.

Step 4. Core and beam tests—Compressive strength tests of cores drilled from the structure provide a measure of concrete strength in place, but not one that can be equated to or translated into an equivalent strength for molded cylinders. The testing of cores requires great care in the operation itself and in the interpretation of the results.

The procedures for obtaining, preparing, and testing cylindrical cores drilled from concrete are covered in ASTM C42, "Method of Obtaining and Testing Drilled Cores and Sawed Beams of Concrete." Controversy frequently develops over interpreting the results of core tests. Concrete in the area represented by the core tests should be considered structurally adequate if the average of three cores is equal to at least 85% of f_c', and if no single core is less than 75% of f_c'. To expect core tests to be equal to f_c' is not realistic, since differences in the size of specimens, conditions of obtaining samples, and procedures for curing do not permit equal values to be obtained.

The fact that core strengths may not equal the strength specified for molded cylinders should not be cause for concern. As mentioned earlier specified cylinder strengths allow a large margin for the unknowns of placement and curing conditions in the field as well as for normal variability. For cores actually taken from the structure, the unknowns have already exerted their effect, and the margin of measured strength above expected working stresses can logically be reduced. The structural engineer should examine cases where core strength values are below 85% to determine if, in his judgment, there is cause for concern. In many cases, values considerably below 85% might be obtained from acceptable concrete, due only to normal variation.

Three cores should be taken for each case of a cylinder test more than 500 psi below f_c'. The location of the core in the concrete member is important. The location of cores should be determined by the architect/engineer to least impair the strength of the structure. When possible, the cores should be cut so that the test load is applied in the same direction as the service load. Horizontally drilled cores may be up to 15% weaker than vertically drilled cores. Slabs generally show an increase in strength from top to bottom, possibly due to the migration of water toward the surface while the concrete is plastic. In columns, there is a sizable decrease (15 to 20%) in strength near the top 12 to 18 in. and an increase (5 to 7%) near the bottom. The strength in between is nearly constant. These differences are decreased when using a relatively low slump air-entrained concrete mix.

It has been found that sawing significantly reduces the flexural strength of beams below that obtained on the same size of molded specimens at all strength levels, Fig. A6-3. For similar curing conditions the modulus of rupture of the sawed specimens averages about 25–30% less than molded beams. Intermittent drying, such as might be experienced by specimens from a structure, is more damaging to sawed specimens than to molded specimens. The effect of sawing is not related to type of aggregate or size of beam but is related to the location of the sawed surface with respect to the tension side, the strength reduction being greatest when a sawed surface is in tension. Because of the importance of curing and moisture distribution, it is doubtful that sawed beam specimens will give reliable indications of the potential load-carrying capacity of a pavement structure.

The splitting tensile test (ASTM C496) may be used for determining the tensile strength of drilled cores. Certain aspects of this test appear to recommend it over the flexural test for determining tensile strength of concrete where cut specimens are involved. Because the zone of maximum stress in the splitting tensile test does not occur at the surface of the specimen, it is unlikely that there would be any effect on strength as a result of the cutting of aggregate particles during drilling. However, because the bearing edges of such a specimen would be along a cut surface, the effect of such an edge, or the need for special leveling treatment of the specimen in this regard, have to be determined.

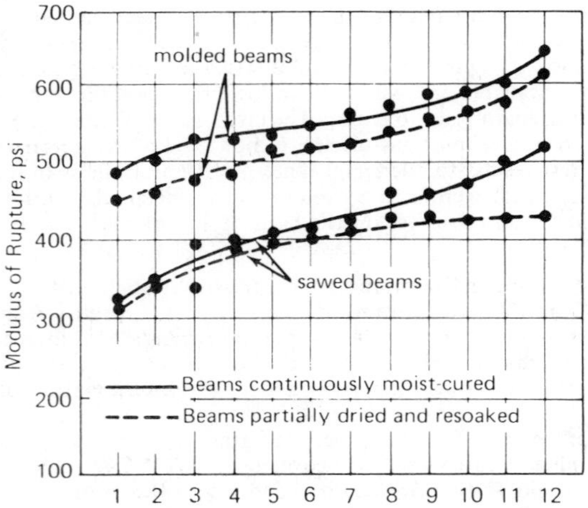

Fig. A6-3 Relationship between flexural strengths of sawed and molded beams of same concrete. (From: Walker, S., and Bloem, D. L., "Studies of Flexural Strength of Concrete," *ASTM Proceedings*, 57, 1957.)

Cores should be drilled with a diamond bit to avoid irregular cross section and damage from drilling. If possible, drill completely through the member to avoid having to break out the core. If the core must be broken out, use wooden wedges to minimize the likelihood of damage, and allow two extra inches of length at the broken end to permit sawing off ends to plane surfaces before capping.

The inclusion of reinforcing steel in the cores reduces the strength. The reductions tend to be larger for cylinders containing two bars rather than one. The cores, therefore, should be trimmed to eliminate the reinforcement whenever possible.

The length (L) of the core, when capped, should be twice its diameter (D). Seldom will it be possible to obtain this ratio of 2. ASTM C42 gives correction factors for converting the strength of a test core to that of a cylinder with an L/D ratio of 2. These correction factors apply to lightweight as well as to normal-weight concrete in the strength range of 2000 to 6000 psi. They are not necessarily applicable to concrete dry at the time of loading (greater corrections are necessary). Length to diameter ratios between 2.0 and 1.5 are to be preferred, and cores with an L/D ratio of less than 1.0 should be avoided.

The smallest core used should have a diameter of $2\frac{1}{2}$ in., while ASTM C42 states that the diameter of the core specimen should be at least three times the maximum nominal size of the coarse aggregate and must be at least twice the maximum coarse aggregate size. Some tests have shown that 2-, 4-, and 6-in.-diameter cores, all having an L/D ratio of 2, give essentially equal strength, while other tests showed that 6 × 12 in. cores have about 20% higher strength than 4 × 8 in. cores. The variability of compressive strength results of drilled cores increases with decreasing core diameter. For most structural concretes, $3\frac{1}{2}$- to 4-in.-diameter cores seem to be the most logical compromise between cost, reliability, and potential damage to the structural element.

There is a wide difference of opinion over the relative merits of soaking the cores in water for 40 hr to obtain a uniform moisture content and testing them wet, or approximating the moisture content existing in the structure. The moisture condition chosen will depend upon the purpose of the test, but it should be recognized that dry samples of concrete test 20 to 30% stronger than similar saturated samples. Usually, if the concrete in the structure will be dry under service conditions, the cores should be air-dried (temperature 60°F to 80°F, relative humidity less than 60%) for seven days before test and should be tested dry. If the concrete in the structure will be more than superficially wet under service conditions, the cores should be immersed in water for at least 40 hr and tested wet.

Curing is a critical factor in strength development of all concretes, but it affects thinner elements to a greater extent than massive sections. Poorly cured slabs may be 10 to 30% weaker than well-cured slabs, although both will be less than standard cylinders. Cores from porous lightweight aggregate concretes are less adversely affected by poor curing, possibly owing to absorbed moisture in the aggregate providing additional internal curing.

In addition to indicating compressive strength, an examination of a core may provide information on the water, cement, and air contents of the concrete. These data should help answer the question of why the strength was low initially.

If the results of properly made core tests are so low as to leave structural integrity in doubt, further action may be required as described in Steps 5 and 6.

Step 5. Load tests—As a last resort, a structural strength investigation by analytical methods or by means of load tests, or by a combination of these methods, may be required to check the capacity of structural members in which adequacy of strength is seriously in doubt. Generally, such tests are suited only to flexural members—slabs, beams, and the like—but they may in some cases be applied to other types.

In some cases the analytical procedure is preferable to load testing. In other cases, analytical evaluation may be the only practicable procedure. Certain members, such as columns and walls, may be difficult to load, and the interpretation of the load test results equally difficult unless severe damage or actual collapse occurs. In any event, the selection of the portion of the structure to be tested, the test procedure, and the interpretation of the results should be done under the direction of a qualified engineer experienced in structural investigations.

Load testing procedures and criteria for their interpretation are to be found in ACI 318.

Step 6. Corrective measures—In those rare cases where a structural element fails the load test or where structural analysis of untestable members indicates an inadequacy, appropriate corrective measures must be taken. The alternatives, depending upon individual circumstances, are:

(a) Reduce the load rating to a level consistent with the concrete strength actually obtained.
(b) Augment the construction to bring its load-carrying capacity up to original expectations. This might involve adding new structural members or increasing the size of existing members.
(c) Replace the unacceptable elements.

Step 7. Assignment of responsibility—Usually, violation of strength specifications is not of such a nature that repair or replacement of the concrete is required. However, with or without actual condemnation of the structural element, considerable expense may accrue. Assignment of financial responsibility for such expense can become very complicated because of overlapping responsibilities among the principals—the specifier, contractor, producer, and testing laboratory. The National Ready-Mixed Concrete Association has developed the following guidelines for determining responsibility:

A sensible method of adjudicating controversies over the consequences of low strength tests is described in the ASTM Specification for Ready-Mixed Concrete, Designation C94. In cases where direct negotiation does not provide a solution, C94 states that "A decision shall be made by a panel of three qualified engineers, one of whom shall be designated by the purchaser, one by the manufacturer (of the concrete), and the third chosen by these two members of the panel. The question of responsibility for the cost of such arbitration shall be determined by the panel. Its decision shall be binding, except as modified by a court decision." This system is intended for controversies between the concrete supplier and his customer, but the principle can logically be extended to include other parties to a dispute. If agreement to use such a procedure cannot be reached, it is recommended that controversies of this kind be submitted to the American Arbitration Association.

The following suggestions are offered as guidelines in the allocation of responsibility:

1. If the molding, curing, or testing of strength cylinders was demonstrably faulty and the concrete later found to be acceptable, whoever was responsible for the faulty procedures should absorb the costs of any supplemental testing required. The panel or arbitration referee should decide whether or not any related costs are also to be assigned to those responsible for the improper testing.
2. If the testing of the concrete as delivered is found to have been properly performed, the costs of additional testing required to verify the acceptability of the structure should be borne by the producer. Related costs, such as structural modifications if needed, should be assigned by the panel or arbitration referee to the producer or contractor, depending upon the extent to which low strength is attributable to improper manufacture of the concrete or to deficiencies in its handling, placing and curing.

Obviously, clear-cut establishment of responsibility is not always possible, and compromise may be the appropriate solution. The engineer must realize his responsibility to prepare lucid, noncontradictory specifications and to apply sound judgment in their application and interpretation. This factor should be taken into account in the assignment of responsibility for costs resulting from questioned strength.

7

Fire Resistance

ARMAND H. GUSTAFERRO*

7.1 INTRODUCTION

7.1.1 Definitions

The following definitions apply to certain terms used in this chapter:

Fire resistance—the property of a material or assembly to withstand fire or to give protection from it. As applied to elements of buildings, it is characterized by the ability to confine a fire or to continue to perform a given structural function, or both. (Defined in ASTM E 176)

Fire-resistive—having fire resistance. (ASTM E 176)

Fire endurance—a measure of the elapsed time during which a material or assembly continues to exhibit fire resistance under specified conditions of test and performance. As applied to elements of buildings, it is measured by the methods and to the criteria defined in the "Methods of Fire Tests of Building Construction and Materials" (ASTM Designation: E 119). (ASTM E 176)

Fire resistance rating (or "fire rating")—a legal term defined in building codes. Fire ratings are assigned by building codes for various types of construction and are usually given in half-hour increments. Fire ratings are usually based on fire endurances.

Fire rate—the amount of insurance premium per unit of valuation, usually expressed in cents per hundred dollars valuation. Fire rates are based on type of structure, occupancy, location, and to some extent, on the fire rating of component parts.

*Consulting Engineer, The Consulting Engineers Group, Inc., Glenview, Illinois.

7.1.2 Fire Rating Requirements in Building Codes

Table 7-1 summarizes some of the requirements for fire resistance ratings that appear in model building codes. For simplicity, only certain values are shown, and qualifying statements have been omitted. As a result, Table 7-1 should not be used to determine specific rating requirements. Instead, the table is included merely to indicate the range and trends in building code requirements.

The fire resistance ratings shown in Table 7-1 are given in hours, and are based on the criteria of ASTM E 119.

7.1.3 Standard Fire Tests of Building Construction and Materials (ASTM E 119)

The fire-resistive properties of building components are measured and specified according to this common standard. Performance is defined as the period of resistance to standard fire exposure elapsing before the first critical end point is reached.

The standard fire exposure is defined in terms of a time-temperature relationship of the fire shown in Fig. 7-1. The fire represents combustion of about 10 lb of wood (with a heat potential of 8000 BTU per lb) per sq ft of exposure area per hr of test. Actually, the fuel consumed during a fire test is dependent on the furnace design and on the heat capacity of the test assembly. For example, the amount of fuel consumed during a fire test of an exposed concrete floor specimen is likely to be 10 to 20% greater than that used for a test of a floor with an insulated ceiling, and considerably greater than that for a combustible assembly.

The standard, ASTM E 119, specifies minimum sizes of

TABLE 7-1 Typical Fire Resistive Rating Requirements in Model Building Codes

Structural Element	Type of Construction: Highest Fire Resistive	2nd Highest Fire Resistive	Highest Non-Combustible	Model Code*
Columns–supporting more than one floor	4 hrs	3 hrs	2 hrs	Basic
	4	3	2	Standard
	3	2	1	Uniform
Girders, Beams, and Trusses	4	3	2	Basic
	4	3	2	Standard
	3	2	1	Uniform
Floors	3	2	1.5	Basic
	3	2	1	Standard
	2	2	1	Uniform
Roofs	2	1.5	1	Basic
	1.5	1	1	Standard
	2	1	1	Uniform

**Basic* = Basic Building Code *promulgated by the Building Officials and Code Administrators International, Inc.*
Standard = Standard Building Code *promulgated by the Southern Building Code Congress.*
Uniform = Uniform Building Code *promulgated by the International Conference of Building Officials.*

specimens to be exposed in fire tests. For floors and roofs, at least 180 sq ft must be exposed to fire from beneath, and neither dimension can be less than 12 ft. For tests of walls, either loadbearing or nonloadbearing, the minimum specified area is 100 sq ft with neither dimension less than 9 ft. The minimum length for columns is specified to be 9 ft, while for beams it is 12 ft.

During fire tests of floors, roofs, beams, loadbearing walls, and columns, the maximum permissible superimposed load is applied. The standard permits alternate tests of large steel beams and columns in which a superimposed load is not required, but the test end point criteria are modified for such tests.

Floor and roof specimens are exposed to fire from beneath, beams from the bottom and sides, walls from one side, and columns from all sides.

End Point Criteria:

(a) Loadbearing specimens must sustain the applied loading—collapse is an obvious end point.
(b) Holes, cracks, or fissures through which pass flames or gases hot enough to ignite cotton waste must not form.
(c) The temperature of the unexposed surface of floors, roofs, or walls must not rise an average of 250°F or a maximum of 325°F at any one point.
(d) Walls must sustain a hose stream test (simulating in a specified manner a fire fighter's hose stream) and twice the superimposed load after the fire test.
(e) In alternate tests of large steel beams (not loaded during test) the end point occurs when the steel temperature reaches an average of 1000°F or a maximum of 1200°F at any one point.

In addition, ASTM E 119 lists criteria for floors, roofs, and beams fire tested in a "restrained" condition. Restrained in this case means that thermal expansion of the specimen is restricted during the fire test. Two classifications can be derived from fire tests of restrained specimens, unrestrained and restrained. Only unrestrained assembly classifications can be obtained from tests of unrestrained specimens. ASTM E 119 includes a guide for classifying constructions as restrained or unrestrained, Table 7-2. It can be noted that cast-in-place and most precast concrete constructions are considered to be restrained.

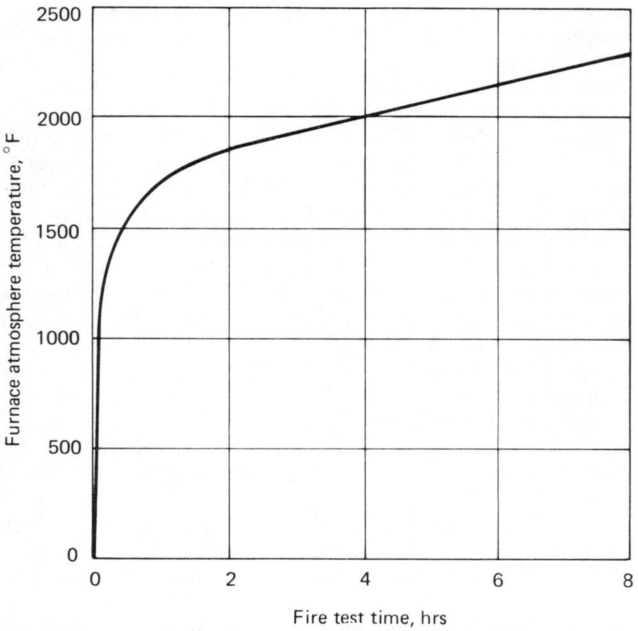

Fig. 7-1 Furnace atmosphere temperature specified in ASTM E119.[7-2] (Permission to reprint granted by the American Society for Testing and Materials, Copyright ASTM.)

7.2 PROPERTIES OF STEEL AND CONCRETE AT HIGH TEMPERATURES

The physical properties of steel and concrete are affected by temperature. At temperatures encountered in fires, strength and modulus of elasticity are reduced.

TABLE 7-2 Construction Classification Restrained and Unrestrained[7-2]

I. *Wall bearing:*	
Single span and simply supported end spans of multiple bays:[a]	
(1) Open-web steel joists or steel beams, supporting concrete slab, precast units, or metal decking	unrestrained
(2) Concrete slabs, precast units, or metal decking	unrestrained
Interior spans of multiple bays:	
(1) Open-web steel joists, steel beams or metal decking, supporting continuous concrete slab	restrained
(2) Open-web steel joists or steel beams, supporting precast units or metal decking	unrestrained
(3) Cast-in-place concrete slab systems	restrained
(4) Precast concrete where the potential thermal expansion is resisted by adjacent construction[b]	restrained
II. *Steel framing:*	
(1) Steel beams welded, riveted or bolted to the framing members	restrained
(2) All types of cast-in-place floor and roof systems (such as beam-and-slabs, flat slabs, pan joists and waffle slabs) where the floor or roof system is secured to the framing members	restrained
(3) All types of prefabricated floor or roof systems where the structural members are secured to the framing members and the potential thermal expansion of the floor or roof system is resisted by the framing system or the adjoining floor or roof construction[b]	restrained
III. *Concrete framing:*	
(1) Beams securely fastened to the framing members	restrained
(2) All types of cast-in-place floor or roof systems (such as beam-and-slabs, flat slabs, pan joists and waffle slabs) where the floor system is cast with the framing members	restrained
(3) Interior and exterior spans of precast systems with cast-in-place joints resulting in restraint equivalent to that which would exist in condition III (1)	restrained
(4) All types of prefabricated floor or roof systems where the structural members are secured to such systems flooring and the potential thermal expansion of the floor or roof systems is resisted by the framing system or the adjoining floor or roof construction.[b]	restrained
IV. *Wood construction:*	
All types	unrestrained

[a]Floor and roof systems can be considered restrained when they are tied into walls with or without tie beams, which walls are designed and detailed to resist thermal thrust from the floor or roof system.
[b]For example, resistance to potential thermal expansion is considered to be achieved when:
(1) Continuous structural concrete topping is used,
(2) The space between the ends of precast units or between the ends of units and the vertical face of supports is filled with concrete or mortar, or
(3) The space between the ends of precast units and the vertical faces of supports, or between the ends of solid or hollow core slab units does not exceed 0.25 per cent of the length for normal weight concrete members or 0.1 per cent of the length for structural lightweight concrete members.
[Permission to reprint granted by the American Society for Testing and Materials, Copyright ASTM.]

7.2.1 Steel

Figure 7-2 shows stress–strain diagrams for ASTM A-36 and A-421 steels at various temperatures.

Figure 7-3 shows the range of yield strengths at high temperatures for structural grade steels,[7-27] and Fig. 7-4 shows the relationship between temperature and ultimate strength of ASTM A-421 steel and high strength alloy steel bars.[7-9, 7-29] The information in these graphs can be used in calculations of the capacities of reinforced or prestressed concrete members at high temperatures. Note that for structural steels half the room temperature strength is retained at about 1100°F, while for cold-drawn prestressing steel the comparable temperature is 800°F.

Figure 7-5 shows the relationship between the temperature and modulus of elasticity of structural steels.[7-27]

Figure 7-6 shows the thermal expansion of steel at elevated temperature.[7-37]

7.2.2 Concrete

The properties of concrete at high temperatures are affected to some extent by the type of aggregate. Most structural concretes can be classified into three aggregate types: carbonate, siliceous, and lightweight. Carbonate aggregates include limestone and dolomite, and are grouped together because they undergo chemical changes at temperatures in the range of 1300°F to 1800°F. Siliceous aggregates, which include granite, quartzite, sandstones, schists, and other material consisting largely of silica, do not undergo chemical changes at temperatures commonly encountered in fire. Although an abrupt volume change accompanies the inversion of quartz from the α- to β-form at about 1060°F, siliceous aggregate concretes do not exhibit the same abrupt change in volume or other physical properties. Both the carbonate and siliceous aggregate concretes generally have unit weights in the 140 to 150 pcf range. Structural lightweight concretes have strenghts in excess of about 2500 psi and unit weights within the range of about 90 to 120 pcf. Lightweight aggregates are either manufactured by expanding shale, slate, clay, slag, or fly ash, or occur naturally. The expanded shales, clays, and slates are heated to about 1900°F to 2000°F during manufacture. At these temperatures the aggregates become molten. As a result, such lightweight aggregates near the surface of concrete subjected to standard fire tests begin to soften after about four hours of exposure. In most practical situations, this softening is not significant.

Figure 7-7 shows data on the strength of concrete at high temperatures.[7-28] The data show strengths of concrete specimens stressed to 40% of their compressive strength during the heating period. After the test temperature was reached, the load on the specimen was increased until failure occurred. Note that the strengths of carbonate and lightweight aggregate concretes are only slightly reduced at temperatures up to 1200°F. The strength of siliceous aggregate concrete is reduced to about 55% of 1200°F.

Figure 7-8 shows the modulus of elasticity of concretes at high temperatures,[7-13] and Fig. 7-9 shows data on thermal expansion of concretes. A comparison of Fig. 7-6 and Fig. 7-9 indicates that the thermal expansion of concrete and steel are about the same throughout the range encountered in fires.

At high temperatures, creep and relaxation of concrete increase significantly. Figures 7-10 and 7-11 show data on creep and relaxation of carbonate aggregate concretes.[7-23]

7.3 TEMPERATURES WITHIN CONCRETE MEMBERS DURING FIRES

Figure 7-12 shows data on temperatures within solid concrete slabs exposed to a standard fire (ASTM E 119) on one

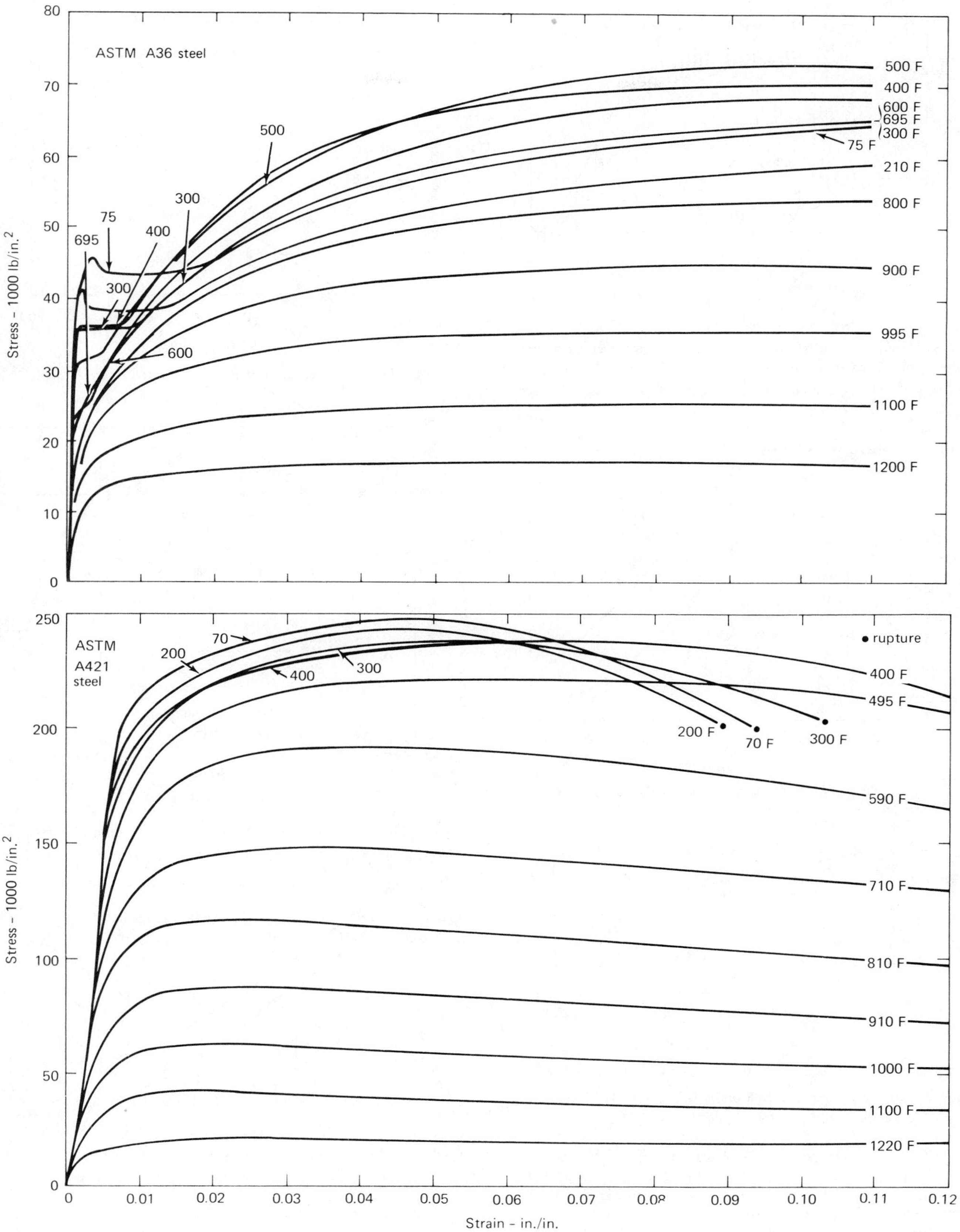

Fig. 7-2 Stress-strain curves for ASTM A36 and A421 steels at various elevated temperatures.[7-7] (Permission to reprint granted by the American Society for Testing and Materials, Copyright ASTM.)

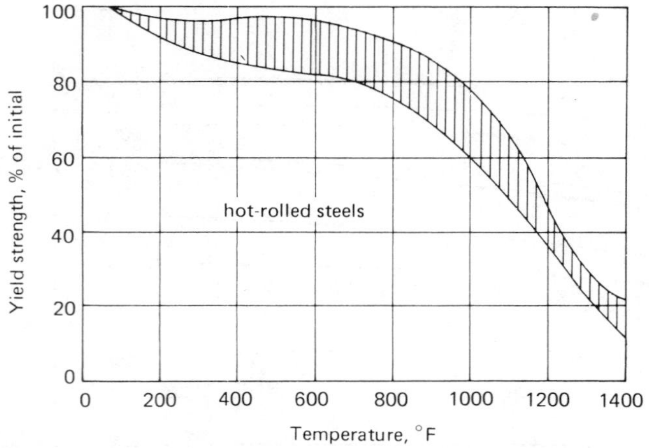

Fig. 7-3 Range of yield strengths of structural grade steels at high temperatures. Data may be used for reinforcing bars.[7-27] [Figures from USS Steel Design Manual (Copyright 1968).]

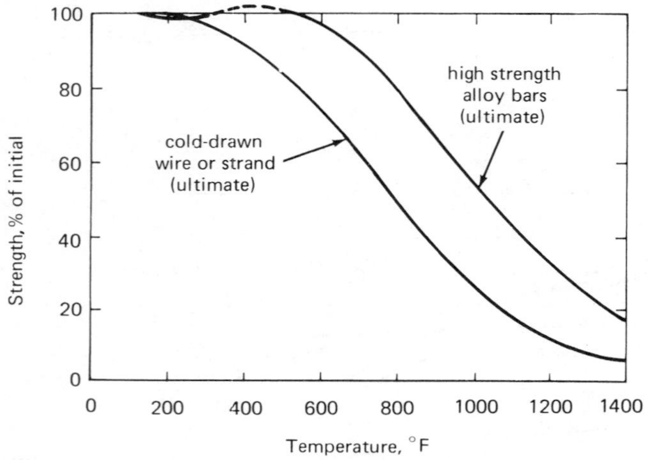

Fig. 7-4 Ultimate strength of prestressing steels at high temperatures.[7-9, 7-29]

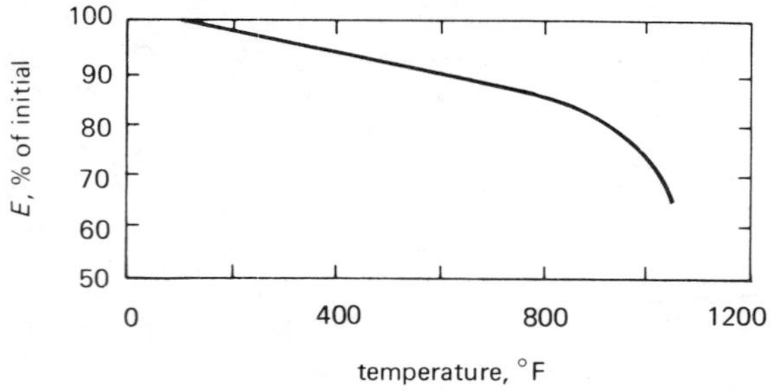

Fig. 7-5 Modulus of elasticity of steel at high temperatures.[7-27] [Figures from USS Steel Design Manual (Copyright 1968).]

side.[7-19] The curves in Fig 7-12 are applicable to slabs of any thickness provided that the slab thickness is at least about 1 in. thicker than the curve in question. If a steel bar is centered 1 in. above the underside of a carbonate aggregate concrete slab exposed to an ASTM E 119 fire from beneath, its temperature will reach 1100°F at about 2 hr 20 min. Thus, if the critical temperature of a bar is 1100°F, the fire endurance of the slab would be 2 hr 20 min.

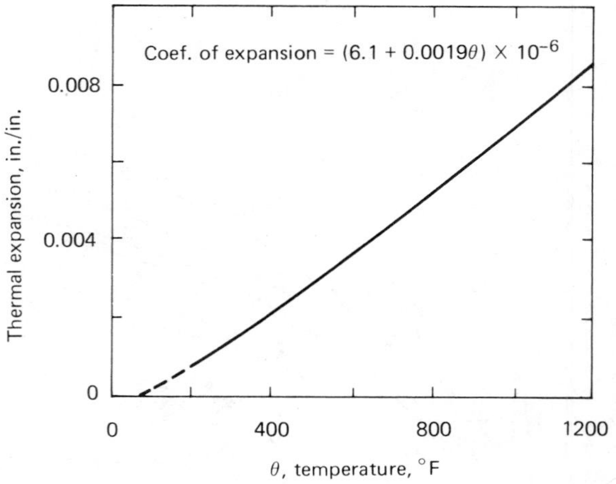

Fig. 7-6 Thermal expansion of steel at elevated temperatures.[7-37]

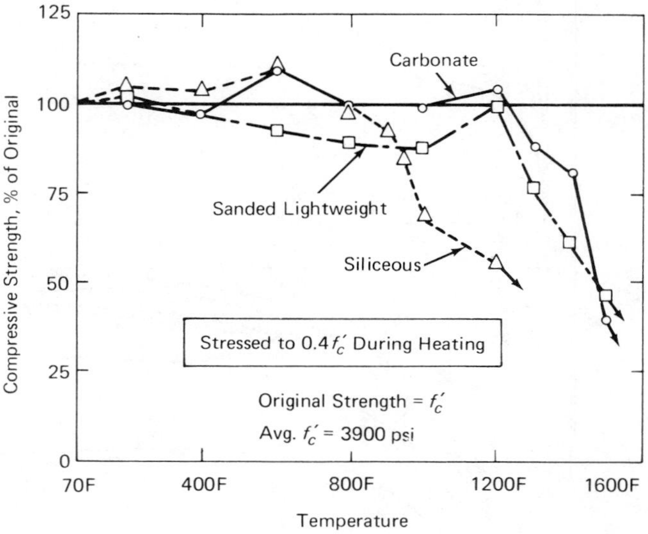

Fig. 7-7 Compressive strength of concrete at high temperatures.[7-28]

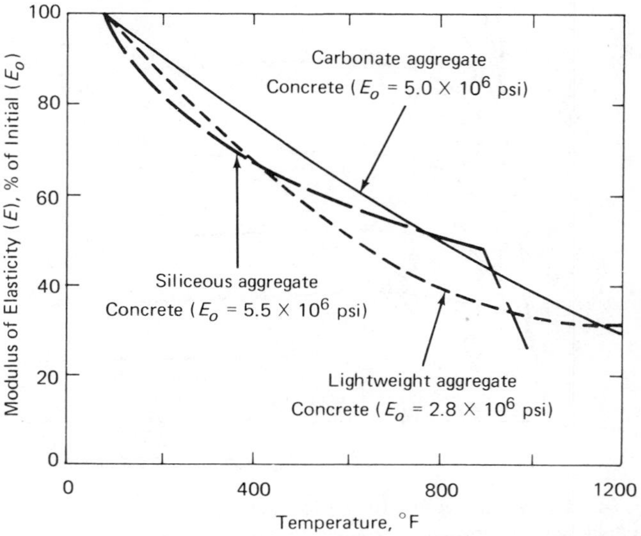

Fig. 7-8 Modulus of elasticity of concrete at high temperatures.[7-13]

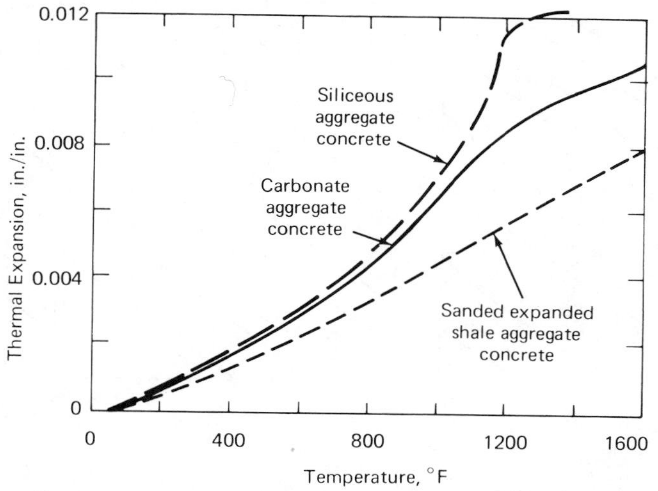

Fig. 7-9 Thermal expansion of concrete at high temperatures.

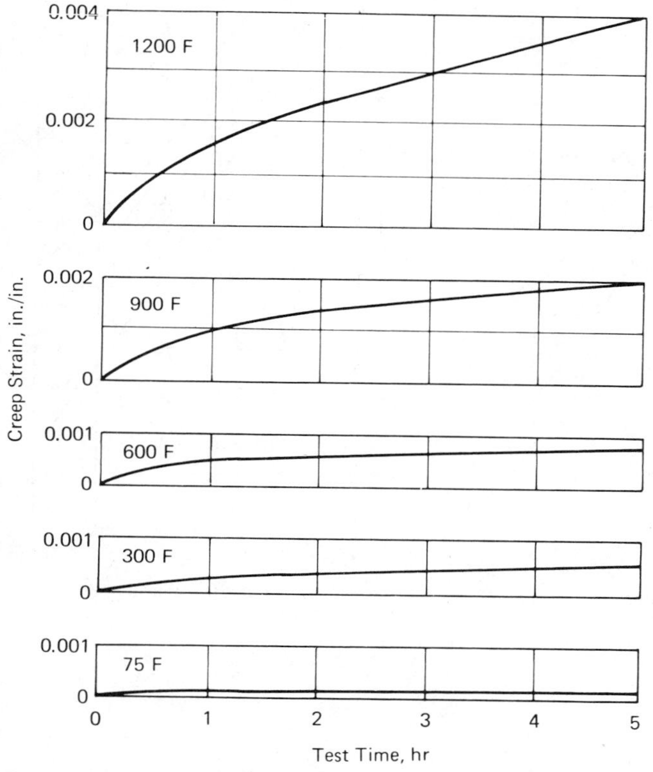

Fig. 7-10 Creep of a carbonate aggregate concrete at various temperatures (applied stress = 1800 psi, $f_c' = 4000$ psi.)[7-23]

Graphs of temperatures within concrete beams are not as simple as those for slabs because beams are heated from more than one face. In standard fire tests, beams are fired from beneath, so the sides as well as the bottom are generally exposed to fire.

Figures 7-13 and 7-14 show isothermal plots of temperatures within concrete beams after various periods of fire exposure. These are used to estimate temperatures within beams of any size during standard fire exposure by interpolation, if necessary.

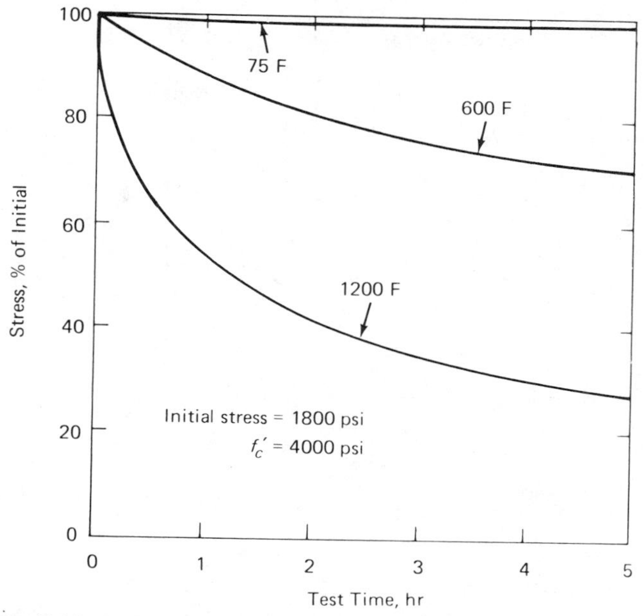

Fig. 7-11 Stress relaxation of a carbonate aggregate concrete at various temperatures.[7-23]

7.4 FIRE ENDURANCE DETERMINED BY 250° RISE OF UNEXPOSED SURFACE

The standard fire test method (ASTM E 119) limits the average temperature rise of the unexposed surface of floors, roofs, and walls to 250°F. Some building codes modify this requirement. For example, the *Wisconsin State Building Code* reduces the fire endurance requirement determined by heat transmission (i.e., 250°F rise) to half that required for structural integrity for many occupancies. For concrete assemblies, this temperature rise depends mainly on the thickness and unit weight of the concrete, but it is also influenced by aggregate type, concrete moisture condition, air content, maximum aggregate size, and aggregate moisture content at time of mixing. Figure 7-15 shows typical unexposed surface temperatures recorded during standard fire tests.[7-19] The data in Fig. 7-15 can be used to determine the fire endurance of concrete slabs as affected by the temperature rise of the unexposed surface. In addition, the data can be used for solving special problems. For example, in a storage vault for computer tapes it may be desirable to limit the unexposed surface temperature to, say 200°F for a 2-hr fire duration. If the ambient temperature is 70°F, the rise to 200°F is 130°F. To determine the floor slab thickness to meet these requirements, enter the charts in Fig. 7-15 with a time of 2 hr and a rise of 130°F. Slab thickness of about 6.75 in. for carbonate aggregate concrete, 7 in. for siliceous, or 5.5 in. for sanded lightweight concrete would satisfy the conditions in the example.

7.4.1 One-Course Slabs

Figure 7-16 shows the relationship between slab thickness and fire endurance for a wide range of structural concretes.[7-19] The values shown represent air-entrained concretes fire-tested when the concrete was at the standard moisture condition (75% R.H. at middepth), made with air-dried aggregates having a nominal maximum size of 0.75 in. It can be seen that, in general, fire endurance increases with a decrease in unit weight. Lightweight concretes generally weigh about 95 to 105 pcf, whereas sanded lightweight con-

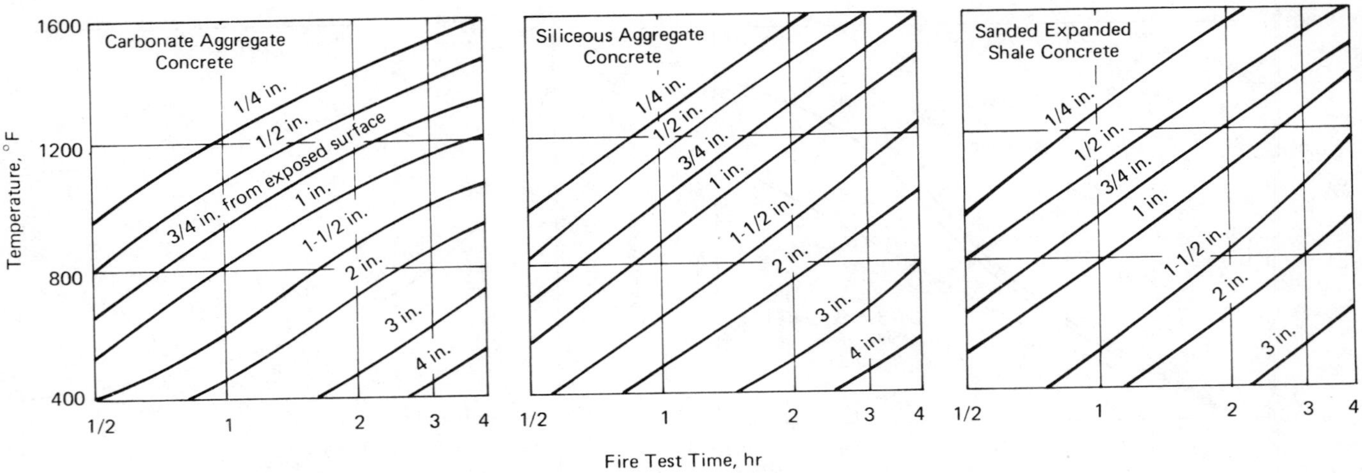

Fig. 7-12 Temperatures within concrete slabs during fire tests.[7-19]

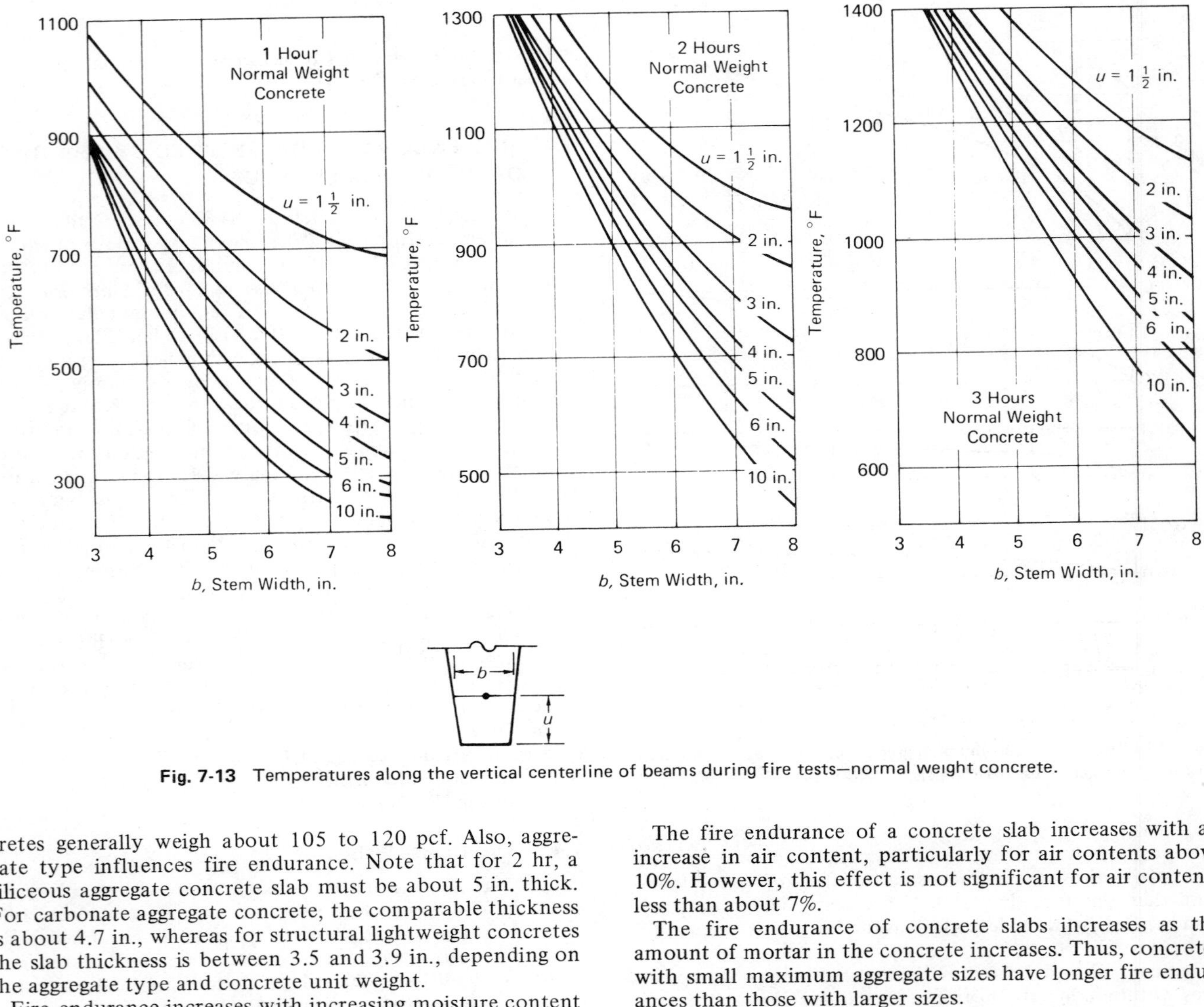

Fig. 7-13 Temperatures along the vertical centerline of beams during fire tests—normal weight concrete.

cretes generally weigh about 105 to 120 pcf. Also, aggregate type influences fire endurance. Note that for 2 hr, a siliceous aggregate concrete slab must be about 5 in. thick. For carbonate aggregate concrete, the comparable thickness is about 4.7 in., whereas for structural lightweight concretes the slab thickness is between 3.5 and 3.9 in., depending on the aggregate type and concrete unit weight.

Fire endurance increases with increasing moisture content of the concrete. An appendix to ASTM E 119 gives a method for calculating the effect of moisture on fire endurance.

The fire endurance of a concrete slab increases with an increase in air content, particularly for air contents above 10%. However, this effect is not significant for air contents less than about 7%.

The fire endurance of concrete slabs increases as the amount of mortar in the concrete increases. Thus, concretes with small maximum aggregate sizes have longer fire endurances than those with larger sizes.

Factors such as water–cement ratio, cement content, and slump have almost no influence on fire endurance within the normal ranges for structural concretes.

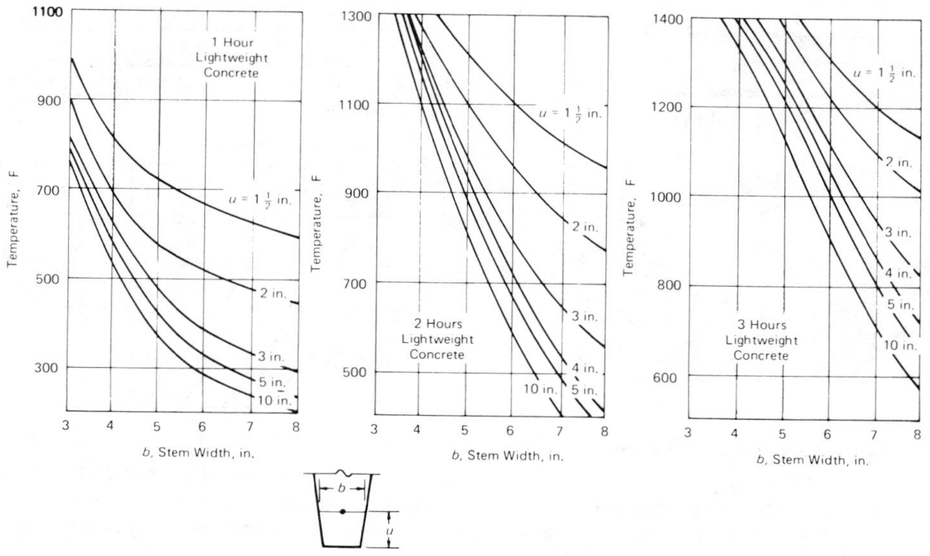

Fig. 7-14 Temperatures along the vertical centerline of beams during fire tests—lightweight concrete.

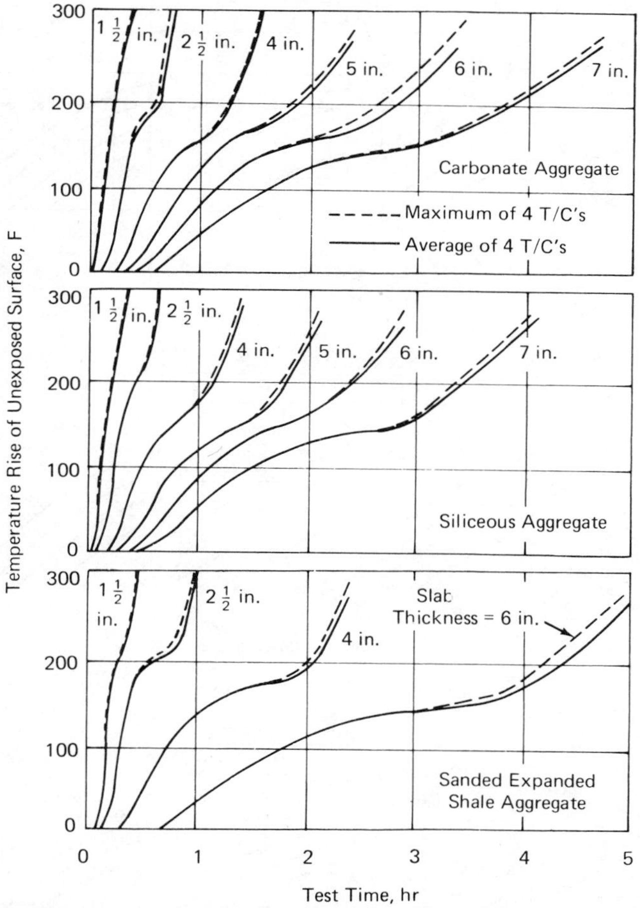

Fig. 7-15 Typical temperature rise of unexposed surface during fire tests of concrete slabs.[7-19]

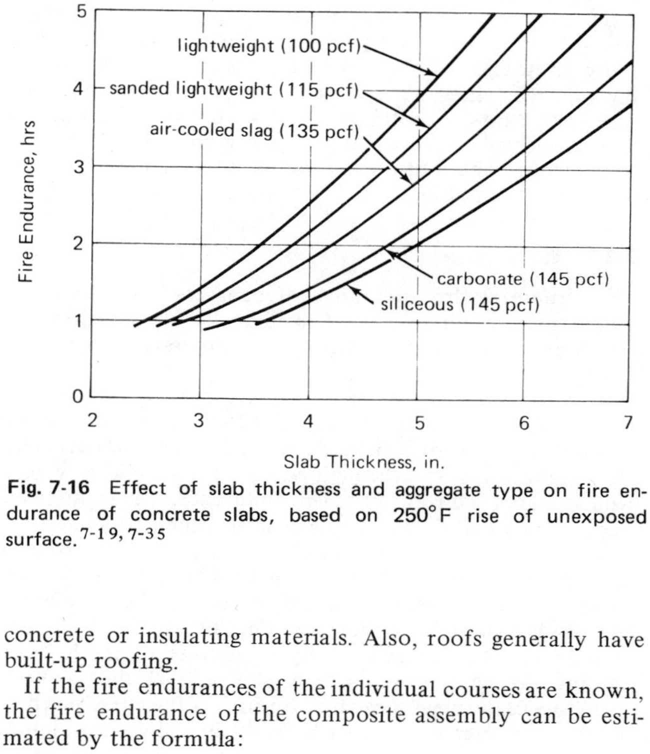

Fig. 7-16 Effect of slab thickness and aggregate type on fire endurance of concrete slabs, based on 250°F rise of unexposed surface.[7-19, 7-35]

7.4.2 Two-Course Floors and Roofs

Floors and roofs frequently consist of concrete base slabs with insulating undercoatings or overlays of other types of concrete or insulating materials. Also, roofs generally have built-up roofing.

If the fire endurances of the individual courses are known, the fire endurance of the composite assembly can be estimated by the formula:

$$R = (R_1^{0.59} + R_2^{0.59} + \cdots R_n^{0.59})^{1.7}$$

where R = the fire endurance of the composite assembly in minutes, and R_1, R_2, R_n = the fire endurances of the individual courses in minutes.[7-38]

This formula has shortcomings in that it does not account for the location of the course relative to the fired surface. The course nearest the fire has the greatest influence on the fire endurance. Thus, if a two-course floor consists of equal thicknesses of normal weight and lightweight concretes, the fire endurance would be longer if the lightweight concrete were exposed to fire, than if the courses were reversed. Nevertheless, the formula generally estimates the fire endurance within about 10% of a test value.[7-24]

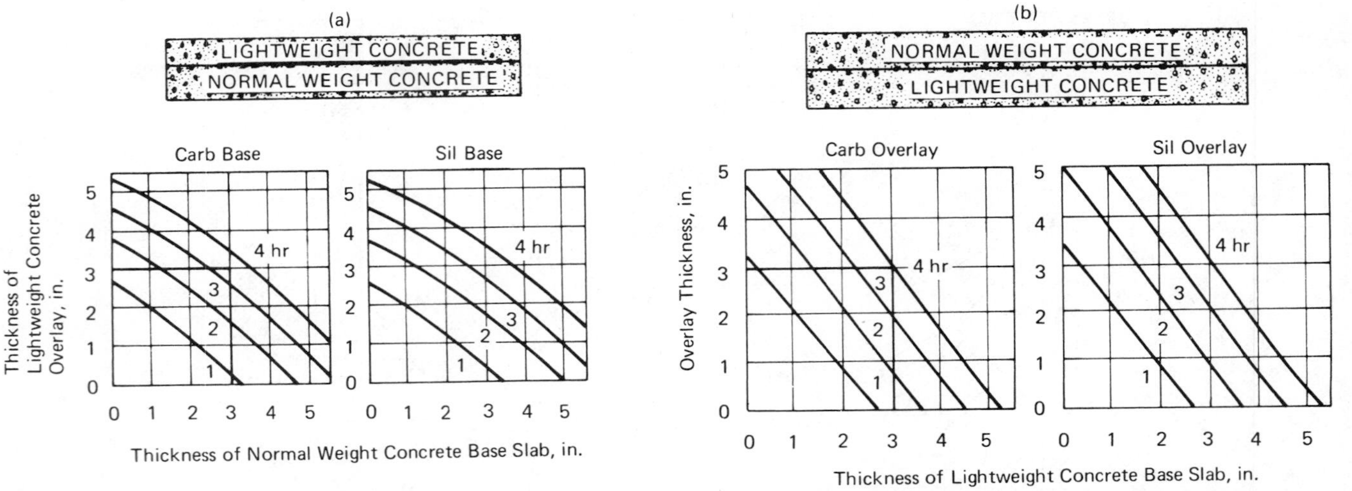

Fig. 7-17 Fire endurance of two-course floors—combinations of normal weight concrete and lightweight concrete. (a) Lightweight concrete overlays on normal weight concrete base slabs; and (b) normal weight concrete overlays on lightweight concrete base slabs.[7-24]

Figure 7-17 through 7-20 show fire endurances of various two-course floors.[7-24] From Fig. 7-17(a) it can be seen that a floor consisting of a 3-in. base slab of carbonate aggregate concrete and a 3-in. overlay of lightweight concrete will have a fire endurance of about 3.5 hr. With the courses reversed, Fig. 7-17(b), the fire endurance is about 4 hr. Standard three-ply built-up roofing adds about 10 to 20 minutes to most concrete roof assemblies, but with some rigid board insulations, the addition can be an hour or more. Asphalt floor tile does not affect the fire endurance significantly.

7.4.3 Walls

The structural fire endurance of concrete walls is seldom a governing factor because it is generally much longer than the fire endurance determined by temperature rise of the unexposed surface.

Walls made of cast-in-place or precast structural concrete have about the same fire endurance (250°F rise of unexposed surface) as floors or roofs made of the same materials. Thus the data in Figs. 7-16 through 7-19 are applicable for walls as well as floors and roofs. The fire endurance of masonry walls depends mainly on the equivalent thickness and aggregate type. Table 7-3 shows typical equivalent thicknesses for hourly fire endurances of masonry walls made of various aggregates. Equivalent thickness is determined by dividing the solid volume of a masonry unit by its face area. The volume can be determined by a number of methods including the immersion method, the sand or shot method, or by accurate measurements. Results of the three methods do not agree precisely, partly because the immersion method is influenced by the unit's surface texture and absorption, and the measurement method depends on accuracy of determining fillet radii and core taper. The shot method is reported to be accurate and reproducible to within about 1%.

EXAMPLE 7-1: What is the fire endurance of a wall made of 8 × 8 × 16-in. block if the actual dimensions of the block are 7.65 in. high by 15.60 in. long, and the volume is 471 cu in.? The aggregate is expanded shale.

SOLUTION:

$$\text{equivalent thickness} = \frac{471}{15.60(7.65)} = 3.95 \text{ in.}$$

From Table 7-3, the fire endurance can be estimated to be slightly greater than 2 hr.

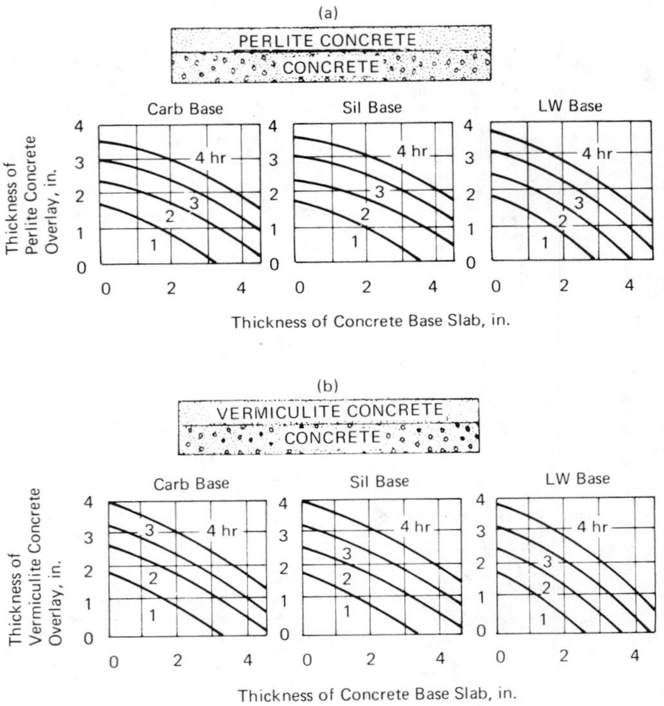

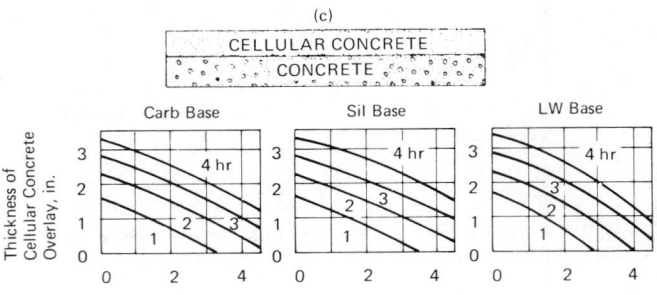

Fig. 7-18 Fire endurance of two-course roofs—concrete slabs with insulating concrete (30 pcf) overlays. (a) Perlite concrete overlays on concrete base slabs; (b) vermiculite concrete overlays on concrete base slabs; and (c) cellular concrete overlays on concrete base slabs.[7-24]

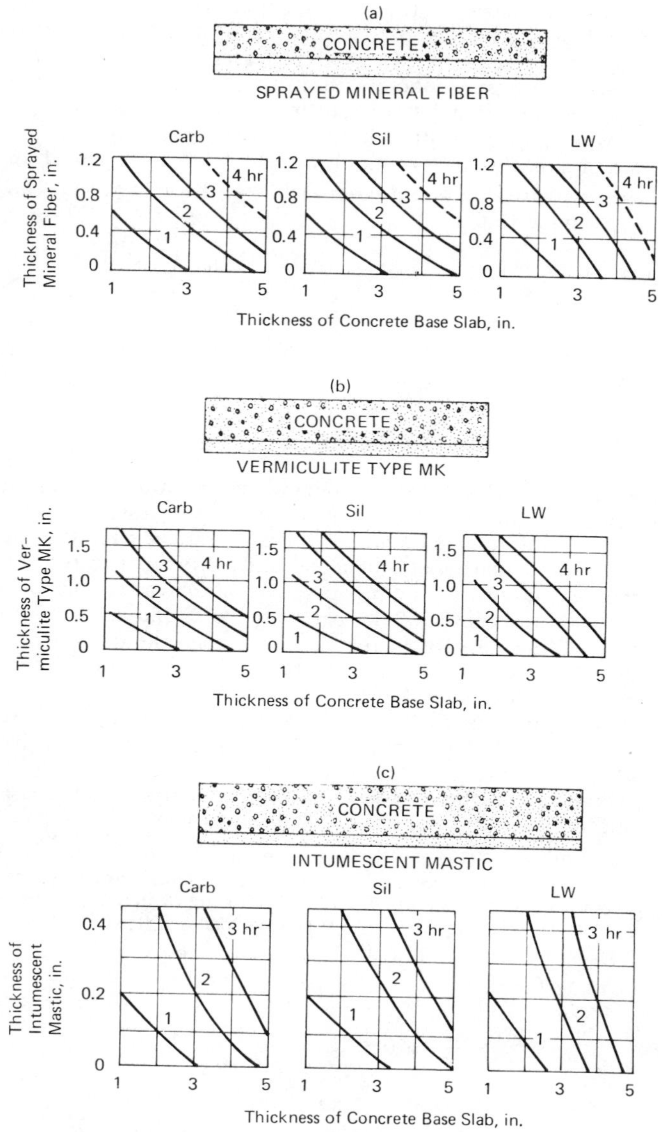

Fig. 7-19 Fire endurance of two-course floors—slabs with sprayed insulation. (a) Concrete floors undercoated with sprayed mineral fiber; (b) concrete floors undercoated with sprayed vermiculite type MK; and (c) concrete floors undercoated with sprayed intumescent mastic.[7-24]

7.5 FIRE ENDURANCE OF SIMPLY SUPPORTED (UNRESTRAINED) SLABS AND BEAMS

7.5.1 Structural Behavior

Figure 7-21 shows a simply supported reinforced concrete slab. The rocker and roller supports indicate that the ends of the slab are free to rotate and expansion can occur without resistance. The reinforcement consists of straight bars located near the bottom of the slab. If the underside of the slab is exposed to fire, the bottom of the slab will expand more than the top, resulting in a deflection of the slab. The strength of the concrete and steel near the bottom of the slab will decrease as the temperature increases. When the moment capacity is reduced to the applied moment, flexural collapse will occur.[7-16]

Figure 7-22 illustrates the behavior of a simply supported slab exposed to fire from beneath. If the reinforcement is

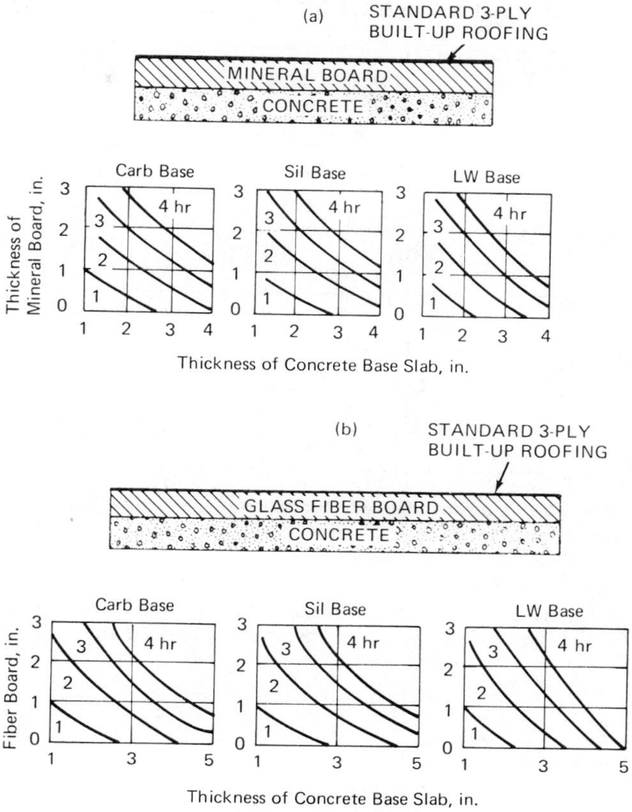

Fig. 7-20 Fire endurance of two-course roofs—rigid board insulation with built-up roofing. (a) Mineral board insulation on concrete roofs; and (b) glass fiber board insulation on concrete roofs.[7-24]

TABLE 7-3 Minimum Equivalent Thickness of Concrete Masonry Walls for Various Fire Endurances

Coarse Aggregate Type	*Minimum Equivalent Thickness, in., for Fire Endurance of* 1 Hr	2 Hr	3 Hr	4 Hr
Expanded Slag or Punice	2.1	3.2	4.0	4.7
Expanded Clay or Shale	2.6	3.8	4.8	5.7
Limestone or Air-Cooled Slag	2.7	4.0	5.0	5.9
Calcareous Gravel	2.8	4.2	5.3	6.2
Siliceous Gravel	3.0	4.5	5.7	6.7

(a) From Ref. 7-32.

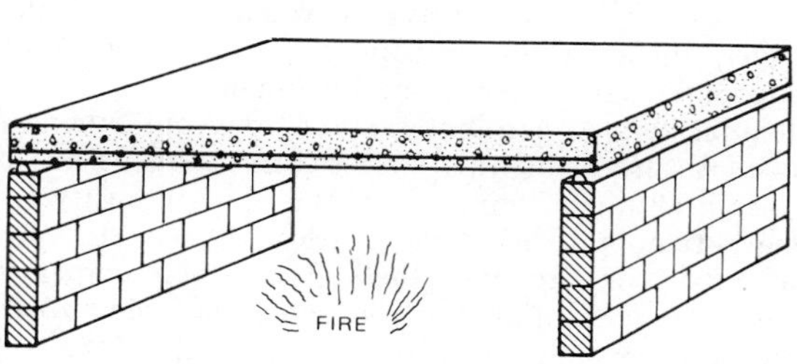

Fig. 7-21 Simply supported reinforced concrete slab subjected to fire from below.

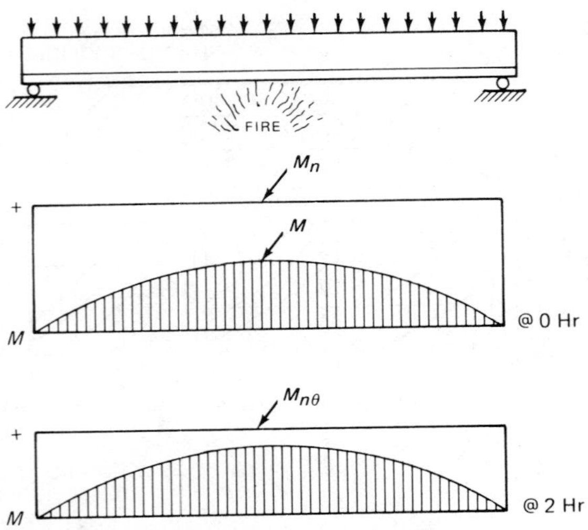

Fig. 7-22 Moment diagram for simply supported beam or slab before and during fire exposure.

straight and uniform throughout the length, the ultimate moment capacity will be constant throughout the length.

$$M_n = A_s f_y \left(d - \frac{a}{2}\right)$$

where:

A_s is the area of the reinforcing steel;
f_y is the yield strength of the reinforcing steel;
d is the distance from the centroid of the reinforcing steel to the extreme compressive fiber;
a is the depth of the equivalent rectangular compressive stress block at ultimate load, and is equal to $A_s f_y / 0.85 f'_c b$ where f'_c is the cylinder compressive strength of the concrete and b is the width of the slab.

If the slab is uniformly loaded, the moment diagram will be parabolic with a maximum value at midspan:

$$M = \frac{wl^2}{8}$$

where w = dead plus live load per unit of length, and l = span length.

It is generally assumed that during a fire the dead and live loads remain constant. However, the material strengths are reduced so that the retained moment capacity is

$$M_{n\theta} = A_s f_{y\theta} \left(d - \frac{a_\theta}{2}\right)$$

in which θ signifies the effects of elevated temperatures. Note that A_s and d are not affected, but $f_{y\theta}$ is reduced. Similarly a_θ is reduced, but the concrete strength at the top of the slab, f'_c, is generally not reduced significantly. If, however, the compressive zone of the concrete is heated, an appropriate reduction should be made.

Flexural failure can be assumed to occur when $M_{n\theta}$ is reduced to M. From this expression, it can be noted that the fire endurance depends on the load intensity and the strength–temperature characteristics of steel. In turn, the duration of the fire until the critical steel temperature is reached depends upon the protection afforded to the reinforcement. Usually the protection consists of the concrete cover, i.e., the thickness of concrete between the fire-exposed surface and the reinforcement. In some cases additional protective layers of insulation or membrane ceilings might be present.

For prestressed concrete the ultimate capacity formulas must be modified by substituting f_{ps} for f_y, and A_{ps} for A_s where f_{ps} is the stress in the prestressing steel at ultimate, and A_{ps} is the area of the prestressing steel. In lieu of an analysis based on strain compatability, the value of f_{ps} can be assumed to be:

$$f_{ps} = f_{pu} \left(1 - \frac{0.5 A_{ps} f_{pu}}{bd f'_c}\right)$$

where f_{pu} is the ultimate tensile strength of the prestressing steel.

7.5.2 Estimating Fire Endurance

Figure 7-23 shows the fire endurance of simply supported concrete slabs as affected by type of reinforcement (hot-rolled reinforcing bars and cold-drawn wire or strand), type of concrete (carbonate, siliceous, and lightweight aggregate), moment intensity, and u, where u is defined as the distance between the center of the bar, wire, or strand and the bottom of the slab (or in the case of beams, the nearest fire-exposed surface). The curves are applicable to hollow-core slabs as well as solid slabs.

The graphs in Fig. 7-23 can be used to estimate the fire endurance of simply supported beams by using "effective u" rather than u. Effective u accounts for beam width by assuming that the u values of corner bars or tendons are reduced by one half for use in calculating the effective u.

EXAMPLE 7-2: Given a simply supported one-way slab reinforced with #4 Grade 60 bars on 6-in. centers. Slab is made of lightweight aggregate concrete with f'_c = 4000 psi. Bars are centered 1 in. from bottom of 6-in. slab. Span is 15 ft and live load is 100 psf. Determine the fire endurance.

SOLUTION:

$$w_d = \frac{6}{12}(150) = 75 \text{ psf}$$

$$M = (0.075 + 0.100)(15)^2/8 = 4.92 \text{ ft-kips}$$

$$M_n = A_s f_y \left(d - \frac{a}{2}\right)$$

$$a = 0.4(60)/0.85\ (4)\ (12) = 0.59 \text{ in.}$$

$$M_n = 0.4(60)(5.00 - 0.30)/12 = 9.40 \text{ ft-kips}$$

$$\frac{M}{M_n} = \frac{4.92}{9.40} = 0.523$$

$$\omega = A_s f_y / bd f'_c = \frac{0.4}{12(5)} \frac{(60)}{4} = 0.10$$

From upper right graph of Fig. 7-23, with M/M_n = 0.523, u = 1 in., and ω = 0.10, the fire endurance is about 2 hr.

EXAMPLE 7-3: Determine the fire endurance for the prestressed concrete beam shown. Siliceous aggregate concrete with M/M_n = 0.50 and $\bar{\omega}_p = A_{ps} f_{pu} / bd f'_c = 0.306$

SOLUTION:

$$\text{Effective } u = \frac{6(2) + 2(1)}{8} = 1.75 \text{ in.}$$

From lower center graph of Fig. 7-23, fire endurance is about 2 hr.

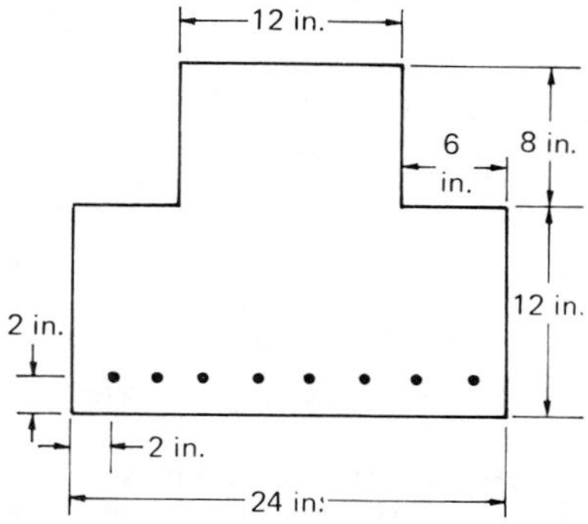

7.6 FIRE ENDURANCE OF CONTINUOUS BEAMS AND SLABS

7.6.1 Structural Behavior

Structures that are continuous or otherwise statically indeterminate, undergo changes in stresses when subjected to fire.[7-15, 7-17, 7-26, 7-43] Such changes in stress result from temperature gradients within structural members, or changes in strength of structural materials at high temperatures, or both.

Figure 7-24 shows a continuous beam whose underside is exposed to fire. The bottom of the beam becomes hotter than the top and tends to expand more than the top. This differential heating causes the ends of the beam to tend to lift from their supports, thus increasing the reaction at the interior support. This action results in a redistribution of moments, i.e., the negative moment at the interior support increases while the positive moments decrease.

During the course of a fire, the negative moment reinforcement (Fig. 7-24) remains cooler than the positive moment reinforcement because it is farther from the fire. Thus the increase in negative moment can be accommodated. Generally the redistribution that occurs is sufficient to cause yielding of the negative moment reinforcement. The resulting decrease in positive moment means that the positive moment reinforcement can be heated to higher temperature before failure will occur. Thus, it is apparent that the fire endurance of a continuous reinforced concrete beam is generally significantly longer than that of a simply supported beam having the same cover and loaded to the same moment intensity.

7.6.2 Detailing Precautions

It should be noted that the amount of redistribution that occurs is sufficient to cause yielding of the negative moment reinforcement. Since by increasing the amount of negative moment reinforcement, a greater negative moment will be attained, care must be exercised in designing the member to assure that a secondary type of failure will not occur. To avoid a compressive failure in the negative moment region, the amount of negative moment reinforcement should be small enough so that ω, i.e., $A_s f_y / bd f'_c$, is less than about 0.30 even after reductions due to temperature in f'_c, b, and d are taken into account. Furthermore, the negative moment reinforcing bars must be long enough to accommodate the complete redistributed moment and change in the location of inflection points. It is recommended that at least 20% of the maximum negative moment reinforcement in the span extend throughout the span.

7.6.3 Estimating Fire Endurance

The charts in Fig. 7-23 can be used to estimate the fire endurance of continuous beams and slabs. To use the charts,

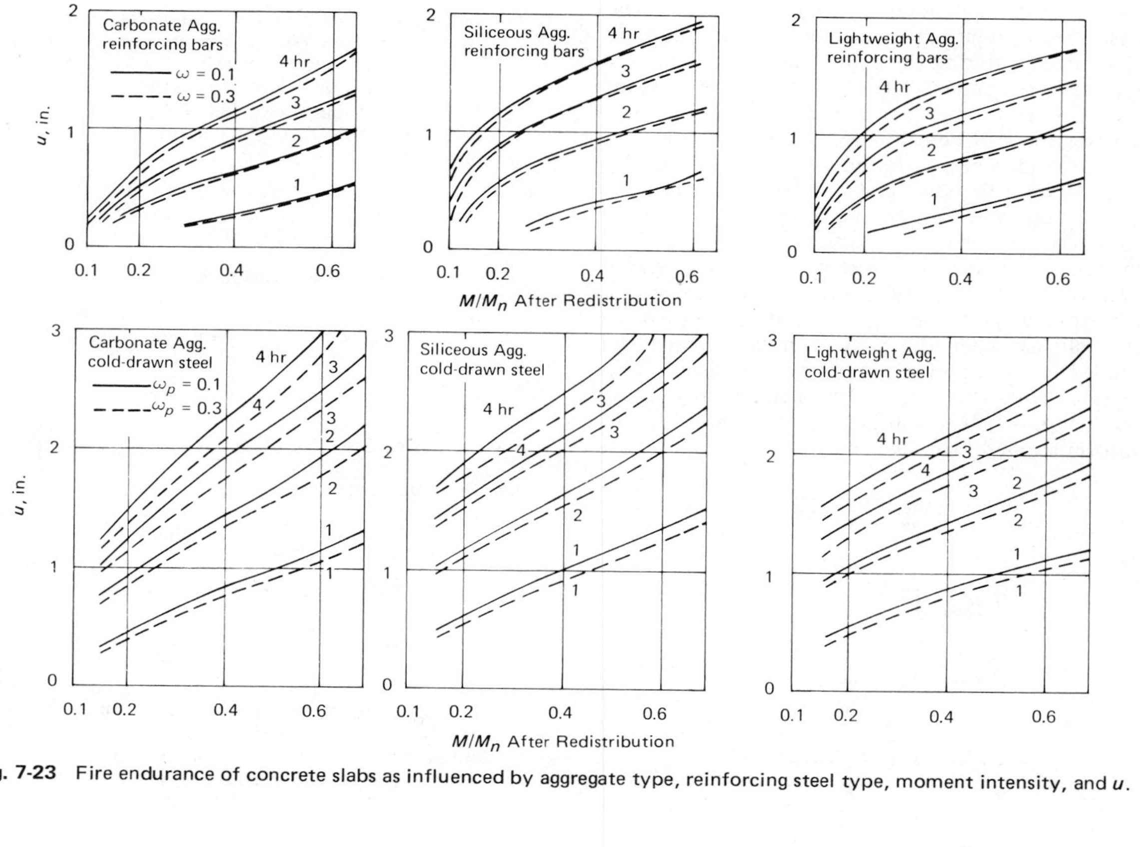

Fig. 7-23 Fire endurance of concrete slabs as influenced by aggregate type, reinforcing steel type, moment intensity, and u.

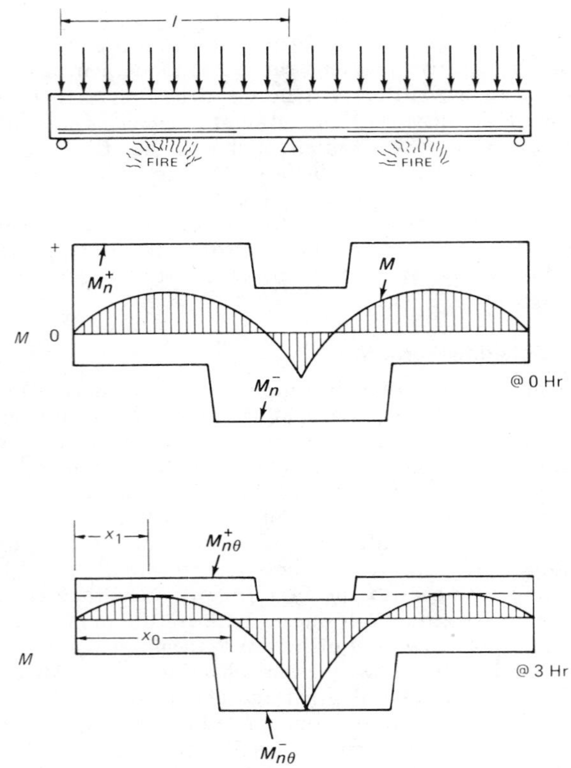

Fig. 7-24 Moment diagrams for continuous two-span beam before and during fire exposure.

first estimate the negative moment at the supports taking into account the temperatures of the negative moment reinforcement and of the concrete in compressive zone near the supports. Then estimate the maximum positive moment after redistribution. Entering the appropriate chart with the ratio of that positive moment to the initial positive moment capacity, the fire endurance for the appropriate u (positive moment region) can be estimated. If the resulting fire endurance is considerably different than that originally assumed in estimating the steel and concrete temperatures, a more accurate estimate can be made by trial and error. Usually such refinement is unnecessary.

It is also possible to design the reinforcement in a continuous beam or slab for a particular fire endurance period. From the bottom diagram of Fig. 7-24, the beam can be expected to collapse when the positive moment capacity, $M_{n\theta}^+$, is reduced to the value indicated by the dashed horizontal line, i.e., when the applied moment at a point x, from the outer support, $M_{x1}^+ = M_{n\theta}^+$.

For a uniform applied load, w:

$$M_{x_1} = \frac{wlx_1}{2} - \frac{wx_1^2}{2} + \frac{M_{n\theta}^- x_1}{l}$$

$$x_1 = \frac{M_{n\theta}^-}{wl} + \frac{l}{2}$$

and:

$$M_{n\theta}^+ = -\frac{wl^2}{2} + wl^2\sqrt{\frac{2M_{n\theta}^-}{wl^2}}$$

Also:

$$x_0 = 2x_1$$

For a symmetrical interior bay:

$$x_1 = l/2$$

$$M_{x_1}^+ = wl^2/8 + M_{n\theta}^-$$

or:

$$M_{n\theta}^- = M_{n\theta}^+ - wl^2/8$$

EXAMPLE 7-4: Given a two-span reinforced concrete slab 6 in. thick reinforced for positive moment with #4 Grade 40 bars on 6-in. centers with $u = 1$ in. Each span is 18 ft and superimposed load is 40 psf. Concrete is made with siliceous aggregate, and $f_c' = 4000$ psi. Determine amount of negative moment reinforcement required for a 2-hr fire endurance.

SOLUTION:

(a) Determining positive moment capacity at 2 hr. From Fig. 7-12, $\theta = 1170°$F. From Fig. 7-3, $f_{y\theta} = 0.38$, $f_y = 15.2$ ksi, $A_s = 0.20\ (12)/6 = 0.40$ in.^{2}/ft.

$$a_\theta = 0.40(15.2)/0.85\ (4)(12) = 0.15 \text{ in.}$$

$$M_{n\theta}^+ = 0.40(15.2)(5.00 - 0.08) = 29.9 \text{ in.-kip} = 2493 \text{ ft-lb}$$

(b) Determine required negative moment capacity at interior support.

$$M_{n\theta}^- = -\frac{wl^2}{2} \pm wl^2\sqrt{\frac{2M_{n\theta}^+}{wl^2}}$$

For a 1-ft strip, $w = 40 + 75 = 115$ lb/ft.

$$M_{n\theta}^- = -\frac{115(18)^2}{2} \pm 115(18)^2\sqrt{\frac{2(2943)}{115(18)^2}} = -3820 \text{ ft-lb}$$

$$= 45.8 \text{ in.-kips}$$

(c) Determine required amount of negative moment reinforcement. From Fig. 7-12, it can be estimated that the negative moment reinforcement will be about 200°F at 2 hr, so $f_y \approx$ 40 ksi. Neglect the compressive zone concrete with a temperature above 1300°F (see Fig. 7-7); $d = 4.25$ in., and assume that the average f_c' in the compressive zone is 3000 psi, and that $a_{\bar{\theta}} = 0.4$ in.

$$A_s = \frac{45.8}{40(4.25 - 0.20)} = 0.28 \text{ in.}^2/\text{ft}$$

use #4 Grade 40 on 8-in. centers. Check $a_{\bar{\theta}}$:

$$a_{\bar{\theta}} = \frac{0.30(40)}{0.85(3)(12)} = 0.39 \text{ in.} \quad \text{OK}$$

Check

$$\omega' = 0.3(40)/12\ (4.25)(3) = 0.08 < 0.30 \quad \text{OK}$$

(d) Determine lengths of top bars.

$$x_0 = 2x_1 = 2\left(\frac{M_{n\theta}^-}{wl} + \frac{l}{2}\right)$$

$$= 2\left(\frac{-3820}{115\ (18)} + \frac{18}{2}\right) = 14.3 \text{ ft}$$

Theoretically, bars could be cut off 18.0 – 14.3 = 3.7 ft (plus development length) on either side of the intermediate support, but it is suggested that 20% of the bars be continued to the support, 40% of the bars be cut off at 3.7 ft (plus development length), and the others at 1.85 ft (plus development length).

7.7 FIRE ENDURANCE OF FLOORS AND ROOFS IN WHICH RESTRAINT TO THERMAL EXPANSION OCCURS

7.7.1 Structural Behavior

If a fire occurs beneath a small interior portion of a large reinforced concrete slab, the heated portion will tend to expand and push against the surrounding part of the slab. In turn, the unheated part of the slab exerts compressive forces on the heated portion. The compressive force, or thrust, acts near the bottom of the slab when the fire first occurs, but as the fire progresses, the line of action of the thrust rises as the heated concrete deteriorates.[7-11] If the surrounding slab is thick and heavily reinforced, the thrust forces that occur can be quite large, but considerably less than those calculated by use of elastic properties of concrete and steel together with appropriate coefficients of expansion. At high temperatures, creep and stress relaxation play an important role. Nevertheless, the thrust is generally great enough to increase the fire endurance significantly. In most fire tests of restrained assemblies, the fire endurance is determined by temperature rise of the unexposed surface rather than by structural considerations, even though the steel temperatures are often in excess of 1500°F.[7-11, 7-14, 7-25, 7-36]

The effects of restraint to thermal expansion can be characterized as shown in Fig. 7-25. The thermal thrust acts in a manner similar to an external prestressing force, which, in effect, increases the positive moment capacity.

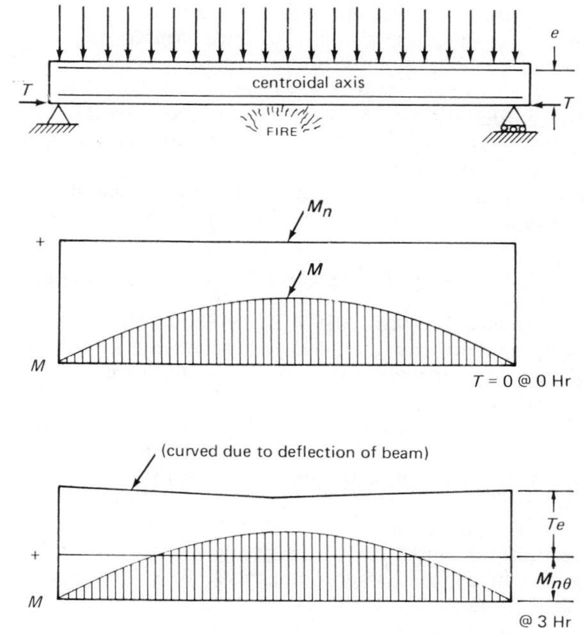

Fig. 7-25 Moment diagrams for axially restrained beam during fire exposure. Note that at three hours, $M_{n\theta}$ is less than M, and effects of axial restraint permit beam to continue to support load.[7-14]

7.7.2 Calculation Procedure

The increase in bending moment capacity is similar to the effect of "fictitious reinforcement" located along the line of action of the thrust.[7-39] It can be assumed that the fictitious reinforcement has a yield strength (force) equal to the thrust. By this approach, it is possible to determine the magnitude and location of the required thrust to provide a given fire endurance. The procedure for estimating thrust requirements is: (1) determine temperature distribution at the required fire test duration; (2) determine the retained moment capacity for that temperature distribution; (3) if the applied moment, M, is greater than the retained moment capacity, $M_{n\theta}$, estimate the midspan deflection at the given fire test time (if $M_{n\theta}$ is greater than M, no thrust is needed); (4) estimate the line of action of the thrust; (5) calculate the magnitude of the required thrust, T; (6) calculate the thrust parameter, T/AE, where A is the gross cross-sectional area of the section resisting the thrust and E is the concrete modulus of elasticity prior to fire exposure;[7-36] (7) calculate $Z = A/S$ in which S is the heated perimeter defined as that portion of the perimeter of the cross section resisting the thrust exposed to fire; (8) enter Fig. 7-26 with the appropriate thrust parameter and Z value and determine the strain parameter, L_a/L; (9) calculate L_a by multiplying the strain parameter by the heated length of the member; and (10) determine if the surrounding or supporting structure can support the thrust T with a displacement no greater than L_a.

The above explanation is greatly simplified because in reality restraint is quite complex, and can be likened to the

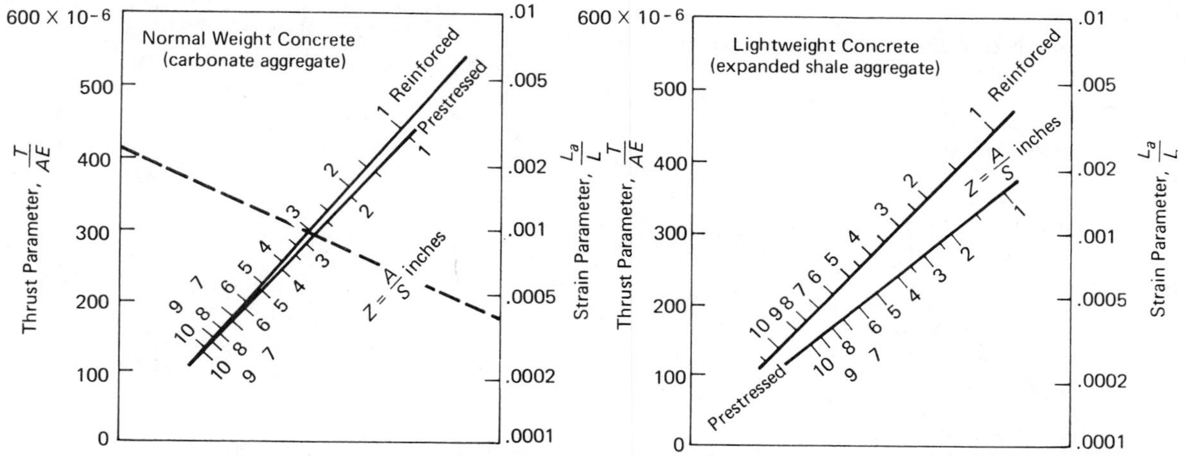

Fig. 7-26 Nomograms relating thrust, strain, and Z ratio.[7-36] (Permission to reprint granted by the American Society for Testing and Materials, Copyright ASTM.)

behavior of a flexural member subjected to an axial force. Interaction diagrams similar to those for columns can be constructed for a given cross section at a particular stage of a fire, e.g., 2 hr of a standard fire exposure.

The guidelines in ASTM E119-83 given for determining conditions of restraint are useful for preliminary design purposes. Basically, interior bays of multi-bay floors or roofs can be considered to be restrained, and the magnitude and location of the thrust are generally of academic interest only.

7.8 FIRE ENDURANCE OF CONCRETE COLUMNS

In a standard fire test, ASTM E119-83, a column is concentrically loaded and exposed to fire on all sides. The fire endurance is the duration of fire exposure prior to structural collapse.

Historically, reinforced concrete columns have had excellent performance records in fires. Because of structural requirements, building columns are seldom smaller than 12 in. in diameter or 12 in. square. Many building codes assign ratings of 4 hr for 12-in. columns made of carbonate or structural lightweight aggregate concrete, and ratings of 2 and 3 hr for similar columns made of siliceous aggregate concrete. Also, for siliceous aggregate concrete, the minimum dimension for a 4-hr rating is generally 16 in. in many building codes. Fire tests conducted in Europe indicate that smaller columns can be used for shorter fire endurances.

Even though the standard fire test for columns calls for fire exposure on all sides of a column, a more severe situation may exist when a column is exposed to one, two, or three sides. To evaluate the structural consequences of such a fire exposure, an estimate of the temperature distribution must be made. The isotherms in rectangular columns exposed to fire on three sides will be similar to those for beams of the same dimensions. For columns exposed to fire on one side only, the isotherms can be assumed to be similar to those in a slab. The temperature–strength relationships of steel and concrete can then be applied, and the capacity estimated.

During a standard fire test of a column, a constant concentric load is applied. In many situations the load on a column exposed to fire will increase due to thermal expansion of the column and the restraint of the members framing into the column. This situation is likely to occur in lower-story columns of multistory buildings. In general, such columns are large and can withstand fires of long duration while sustaining considerable overloads.

NOTATION

a	= depth of equivalent rectangular stress block at ultimate load
A	= gross cross-sectional area of a section
A_{ps}	= area of prestressing steel
A_s	= area of reinforcing steel
b	= width of the compressive zone of a flexural member
d	= distance between the centroid of the reinforcing steel and the extreme compressive fiber
e	= distance between the line of thrust and the centroidal axis of a member
E	= modulus of elasticity
E_0	= modulus of elasticity of concrete prior to heating
f_c'	= compressive strength of concrete
f_{ps}	= stress in prestressing steel at ultimate load
f_{pu}	= ultimate tensile strength of prestressing steel
f_y	= yield strength of reinforcing steel
l	= span length
L	= heated length
L_a	= thermal expansion of member
M	= applied dead plus live load moment
M_n	= ultimate moment capacity of a flexural member
M_n^+ or M_n^-	= positive or negative ultimate moment capacity of a flexural member
R	= fire endurance of composite assembly as determined by criteria for temperature rise of unexposed surface
R_1, R_2, or R_n	= fire endurance of individual courses of a composite assembly as determined by criteria for temperature rise of unexposed surface
S	= "heated perimeter," i.e., the heated portion of the perimeter of the cross section resisting the thrust
T	= thermal thrust
w	= applied load
w	= applied dead plus live load per unit of length
w_d	= applied dead load per unit of length
x_0 and x_1	= distances measured along a span from the exterior support (see Fig. 7-24) to points of zero moment and maximum positive moment, respectively
Z	$= A/S$
θ	= temperature (subscript θ signifies the effects of elevated temperatures)
ϕ	= capacity reduction factor = 0.9 for flexure
ω	$= A_s f_y / b d f_c'$
$\bar{\omega}_p$	$= A_{ps} f_{pu} / b d f_c'$

BIBLIOGRAPHY

7-1. "Fire Protection Handbook," 14th Ed., National Fire Protection Association, Boston, Mass.

7-2. ASTM Designation: E119-83, "Standard Methods of Fire Tests of Building Construction and Materials," Vol. 04.07, *1983 Annual Book of Standards*, American Society for Testing and Materials.

7-3. Kordina, K., and Bornemann, P., "Brandversuehe an Stahlbeton-platten," *Deutscher Ausschuss fur Stahlbeton*, Heft 181, 1966, William Ernst and Sohn, Berlin.

7-4. Harmathy, T. Z., and Stanzak, W. W., "Elevated-Temperature Tensile and Creep Properties of Some Structural and Prestressing Steels," *Fire Test Performance*, ASTM STP 464, 186–208, American Society for Testing and Materials, 1970.

7-5. Philleo, Robert, "Some Physical Properties of Concrete at High Temperatures," *Proceedings, American Concrete Institute*, **54**, April 1958, p. 857.

7-6. Thompson, John P., "Fire Resistance of Reinforced Concrete Floors," PCA Publication T140 (1963), 32 pp.

7-7. Commissie Voor Uitvoering Van Research (Neatherlands), "Fire Tests of Prestressed Concrete Beams," *CUR Report 13* (in Dutch), January 1958, 54 pp.

7-8. Ashton, L. A., and Bate, S. C. C., "The Fire Resistance of Prestressed Concrete Beams," *Journal of the American Concrete Institute*, **32**, May 1961, pp. 1417–1440.

7-9. Abrams, M. S., and Cruz, C. R., "The Behavior at High Temperature of Steel Strand for Prestressed Concrete," *Journal of the PCA Research and Development Laboratories*, **3** (3), Sept. 1961, pp. 8–19; *PCA Research Department Bulletin 134*.

7-10. Carlson, C. C. "Fire Resistance of Prestressed Concrete Beams, Study A—Influence of Thickness of Concrete Covering Over Prestressing Steel Strand," *PCA Research Department Bulletin 147*, July 1962.

7-11. Selvaggio, S. L., and Carlson, C. C., "Effect of Restraint on Fire Resistance of Prestressed Concrete," *Fire Test Methods, ASTM STP No. 344*, American Society for Testing and Materials, 1962.

7-12. Selvaggio, S. L., and Carlson, C. C., "Fire Resistance of Prestressed Concrete Beams. Study B—Influence of Aggregate and Load Intensity," *Journal of the PCA Research and Development Laboratories*, **6** (1), Jan. 1964, pp. 41–64, and **6** (2), May 1964. pp. 10–25; *PCA Research Department Bulletin 171*.

7-13. Cruz, C. R., "Elastic Properties of Concrete at High Temperatures," *Journal, PCA Research and Development Laboratories*, **8** (1), Jan. 1966, pp. 37–45; *PCA Research Department Bulletin 191*.

7-14. Carlson, C. C., et al., "A Review of Studies of the Effects of Restraint on the Fire Resistance of Prestressed Concrete," *Proceedings, Symposium on Fire Resistance of Prestressed Concrete*, Braunschweig, Germany, 1965, International Federation for Prestressing (F.I.P.). Bauverlag GmbH, Wiesbaden, Germany; *PCA Research Department Bulletin 206*.

7-15. Ehm, H., and vonPostel, R., "Tests of Continuous Reinforced Beams and Slabs Under Fire," *Proceedings, Symposium on Fire Resistance of Prestressed Concrete*; translation available at S.L.A. Translation Center, John Crerar Library, Chicago, Ill.

7-16. Gustaferro, A. H., and Selvaggio, S. L., "Fire Endurance of Simply Supported Prestressed Concrete Slabs," *Journal, Prestressed Concrete Institute*, **12**(1) Feb. 1967, pp. 37–52; *PCA Research Department Bulletin 212*.

7-17. Report No. B1-59-22, "Fire Test of a Simple, Statically Indeterminant Beam," TNO Institute for Structural Materials and Building Structures, Delft, Holland; translation available at S.L.A. Translation Center, John Crerar Library, Chicago, Ill.

7-18. Selvaggio, S. L., and Carlson, C. C., "Restraint in Fire Tests of Concrete Floors and Roofs," *ASTM STP 422*, American Society for Testing and Materials; *PCA Research Department Bulletin 220*.

7-19. Abrams, M. S., and Gustaferro, A. H., "Fire Endurance of Concrete Slabs as Influenced by Thickness, Aggregate Type, and Moisture," *Journal, PCA Research and Development Laboratories*, **10** (2), May 1968, pp. 9–24; *PCA Research Department Bulletin 223*.

7-20. Menzel, Carl A., "Tests of the Fire Resistance and Thermal Properties of Solid Concrete Slabs and Their Significance," *Proceedings, ASTM*, **43**, 1943, pp. 1099–1153.

7-21. Malhotra, H. L., "The Effect of Temperature on the Compressive Strength of Concrete," *Magazine of Concrete Research* (London), **8** (22), 1956, pp. 85–94.

7-22. Harmathy, T. Z., and Berndt, J. E., "Hydrated Portland Cement and Lightweight Concrete at Elevated Temperatures," *ACI Journal Proceedings* **63** (1), Jan. 1966, pp. 93–112.

7-23. Cruz, C. R., "Apparatus for Measuring Creep of Concrete at High Temperatures," *PCA Research Department Bulletin 225*.

7-24. Abrams, M. S., and Gustaferro, A. H., "Fire Endurance of Two-Course-Floors and Roofs," *Journal of the American Concrete Institute*, **66** (2), Feb. 1969, pp. 92–102.

7-25. Gustaferro, A. H., and Carlson, C. C., "An Interpretation of Results of Fire Tests of Prestressed Concrete Building Components," *Journal of the Prestressed Concrete Institute*, **7** (5), Oct. 1962, pp. 14–43.

7-26. Gustaferro, A. H., "Temperature Criteria at Failure," *Fire Test Performance*, ASTM STP 464, 68–84, American Society for Testing and Materials, 1970.

7-27. Brockenbrough, R. L., and Johnston, B. G., *Steel Design Manual*, U.S. Steel Corp., Pittsburgh, Pa., 1968, 246 pp.

7-28. Abrams, M. S., "Compressive Strength of Concrete at Temperatures to 1600 F," *Symposium on Effect of Temperature on Concrete*, American Concrete Institute Publication ST-25, Detroit, Mich., 1971.

7-29. Gustaferro, A. H., et al., "Fire Resistance of Prestressed Concrete Beams. Study C: Structural Behavior During Fire Tests," *PCA Research and Development Bulletin (RD009.01B)*, Portland Cement Association, 1971.

7-30. Allen, L. W., "Fire Endurance of Selected Non-Loadbearing Concrete Masonry Walls," *Fire Study No. 25 of the Division of Building Research*, National Research Council of Canada, Ottawa, Canada.

7-31. Abrams, M. S., and Gustaferro, A. H., "Fire Endurance of Prestressed Concrete Units Coated with Spray-Applied Insulation," *Journal of the Prestressed Concrete Institute*, Jan.–Feb. 1972.

7-32. "Fire Resistance Ratings," American Insurance Association, New York, N.Y.

7-33. "Design for Fire Resistance of Precast Prestressed Concrete," Prestressed Concrete Institute, Chicago, 1977, 81 pp.

7-34. "Fire Resistance Index," Underwriters' Laboratories, Inc., Northbrook, Ill., Jan. 1972.

7-35. Gustaferro, A. H., et al., "Fire Resistance of Lightweight Insulating Concretes," ACI Special Publication 29, *Lightweight Concrete*, American Concrete Institute, Detroit, Mich.

7-36. Issen, L. A., et al., "Fire Tests of Concrete Members: An Improved Method for Estimating Restraint Forces," *Fire Test Performance*, ASTM, STP 464, American Society for Testing and Materials, 1970, pp. 153–185.

7-37. *Manual of Steel Construction*, 7th Ed., American Institute of Steel Construction, New York, N.Y.

7-38. Report BMS 92, "Fire Resistance Classifications of Building Constructions," National Bureau of Standards, Washington, D.C., 1940, 70 pp.

7-39. Salse, E. A. B., and Gustaferro, A. H., "Structural Capacity of Concrete Beams During Fires as Affected by Restraint and Continuity," *Proceedings* 5th CIB Congress, Paris, France, 1971, available from Centre Scientifique et Technique du Batiment, Paris.

7-40. *Reinforced Concrete Fire Resistance*, Concrete Reinforcing Steel Institute, Schaumburg, IL, 1980, 256 pp.

7-41. Abrams, M. S., "Behavior of Inorganic Materials in Fire," *Deisgn of Buildings for Fire Safety,* ASTM STP 685, American Society for Testing and Materials, 1979. pp. 14–75.

7-42. "Guide for Determining the Fire Endurance of Concrete Elements," ACI Committee 216, *Concrete International*, Feb. 1981, pp. 13–47.

7-43. Lin, T. D., et al., "Fire Endurance of Continuous Reinforced Concrete Beams," RD 072.01B, Portland Cement Association.

8

Prestressed Concrete*

ANTOINE E. NAAMAN**

8.1 PRINCIPLE AND METHODS OF PRESTRESSING

8.1.1 Introduction

Prestressing is the deliberate creation of permanent internal stresses in a structure or system in order to improve its performance. Such stresses are designed to counteract those induced by external loadings. Prestressing generally involves at least two materials, the stressor and the stressee, which, when acting together, perform better than either one taken separately. Prestressing is a principle. The French mathematician Henri Poincaré once said: "A principle is neither true nor false, it is convenient." The principle of prestressing is indeed very convenient and has been widely applied. Its application to steel and concrete is relatively recent but has received widest application.

The application of prestressing to concrete is in a way a natural result. Concrete is strong in compression and weak in tension. For design purposes its tensile resistance is discounted. Prestressing the concrete produces compressive stresses, either uniform or nonuniform, which counteract tensile stresses induced by external loadings. Prestressing originally attempted to counteract tensile stresses entirely, thus producing a crack-free material during service. However, it has since evolved to counteract only in part externally induced tensile stresses, thus allowing tension and controlled cracking in a way similar to reinforced concrete. This has led to what is called partial prestressing.

Examples of prestressing are numerous among manufactured tools and products. Indeed, some are very old and illustrate the principle of prestressing. The hunter's bow is prestressed by the string to achieve a sharp recoil action during ejection of the arrow. The dried wooden staves forming a wooden barrel are prestressed by tightening metal bands around them. When the barrel is filled with liquid, the wooden staves expand, the prestress is increased, and leakage is prevented. To prevent the relative movement between the iron tire and the wooden rim of a cartwheel, the tire is fitted around the rim while in a heated state. Upon cooling, contraction of the tire produces a permanent prestress in the form of radial compression on the rim. The blade of a frame saw is prestressed (in tension) by twisting a rope at the opposite end of the wooden frame.

The first application of prestressing to concrete appears to be due to P. H. Jackson, an engineer from California. In 1886[8-1] he obtained a U.S. patent for tightening steel tie rods in artificial stones (concrete blocks) and concrete arches used for slabs and roofs. At about the same time, in 1888, C. E. W. Doehring from Germany also obtained a patent for prestressing concrete slabs with metal wires. However, the performance of the first prestressed concrete structural elements was hindered by the low steel strengths available at the time: because of the relatively low steel stresses used and the relatively high prestress losses due to creep and shrinkage of the concrete, the prestress would soon vanish. Retensioning was suggested by G. R. Steiner

*This chapter is adapted entirely from *Prestressed Concrete Analysis and Design-Fundamentals*, by Antoine E. Naaman, McGraw-Hill, New York, 1982, with permission from McGraw-Hill Book Co.

**Professor of Civil Engineering, Department of Civil Engineering, The University of Michigan, Ann Arbor, Michigan.

(United States, 1908) to overcome this problem, while other researchers such as J. Mandl and M. Koenen of Germany attempted to identify and quantify prestress losses.

However, it was the French engineer Eugene Freyssinet[8-2] who first grasped the importance of prestress losses and proposed ways to overcome them. Based on his experience in building arch bridges and prestressing them by external jacking at the crown to facilitate formwork removal, he suggested that very high strength steels and high elongations must be used in prestressed concrete.[8-2] High steel elongations would not be entirely counteracted by the shortening of the concrete due to creep and shrinkage. Later, in 1940, he introduced his first prestressing system, a wedge-anchored cable with 12 wires, which is still in use today.

Although prestressed concrete planks and fence posts were produced by R. E. Dill in the United States as early as 1925, it was only in 1949/1950 that the first prestressed concrete bridge, the 155-ft-span (47-m) Walnut Lane Bridge in Philadelphia, was built.

Simultaneously with and in continuation of the developments of E. Freyssient, many researchers contributed greatly to the full expansion of prestressed concrete. They include G. Magnel of Belgium,[8-3] Y. Guyon of France,[8-4] P. Abeles of England,[8-5] who developed the concept of partial prestressing, F. Leonhardt of Germany,[8-6] V. V. Mikhailov of Russia, and T. Y. Lin of the United States[8-7] to whom we owe the design method of load balancing which is so convenient for indeterminate structures.

Many prestressing systems and techniques were also developed, and today prestressed concrete is widely accepted and used. Numerous books and textbooks on the design and construction of prestressed concrete structures were written, and many associations and institutes contribute to the advancement of the state of the art of prestressed concrete.[8-8–8-21]

More than four decades of experience have given prestressed concrete a proven record of reliable performance. At present, applications of prestressed concrete essentially occur in every structural element or building system: bridges, building components such as beams, slabs, and columns, pipes and piles, pavements, ties, tanks, tunnels, stadia, nuclear power vessels, TV towers, floating storage, and offshore structures. Prestressed concrete bridges have reached span lengths previously considered exotic, and higher limits are expected. The Parrotts Ferry Bridge in California has a main span of 640 ft (195 m), while that of the Pasco-Kennewick cable stayed bridge is 981 ft (299 m). Bulk bridge applications in the United States were in the span range of 50 to 150 ft (15 to 46 m) with the precast prestressed I girder and box girder bridges extensively used in America's interstate highway system, as well as in most highways and secondary roads. Similar extensive usage of precast prestressed hollow-cored slabs, T and double T beams of spans of up to 100 ft (31 m) is observed in the U.S. building market, where numerous building systems are used, making prestressed concrete fully competitive in many sectors of the construction industry.

8.1.2 Prestressing Methods

Several methods and techniques of prestressing are available. However, except for chemical prestressing, most can be classified within two major groups: pretensioning and post-tensioning. Some methods are specifically identified with a particular application but nevertheless belong to one of the above groups.

1. Pretensioning

In pretensioning the prestressing tendons (wires, strands) are stretched to a predetermined tension and anchored to fixed bulkheads or molds. The concrete is cast around the tendons, and cured, and upon hardening the tendons are released. As the bond between the tendons and the concrete resists the shortening of the tendons, the concrete gets compressed. The prefix "pre" in pretensioning refers to the fact that the tendons are put in tension *prior* to hardening of the concrete.

In order to stretch the tendons, hydraulic jacks are generally used. Once the predetermined elongation is reached, the tendons are anchored to the bulkhead using anchors similar to those described for post-tensioning. Anchors for individual strands are also called chucks.

Depending on the pretensioned structural elements produced, the profile of the tendons is either straight (Fig. 8-1), as in hollow-cored slabs, or allows for one or two deflection points (also called draping or hold-down points), as in bridge girders (Fig. 8-2). Draping is generally achieved by pulling or pushing down part of the tendons to the desired position.

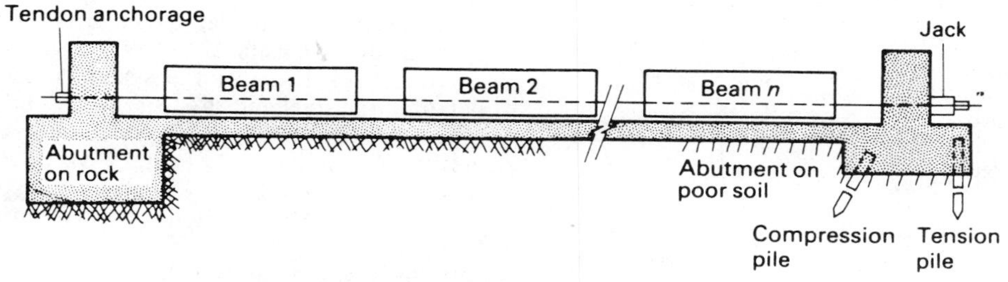

Fig. 8-1 Typical pretensioning bed and abutments showing beams with straight tendons.

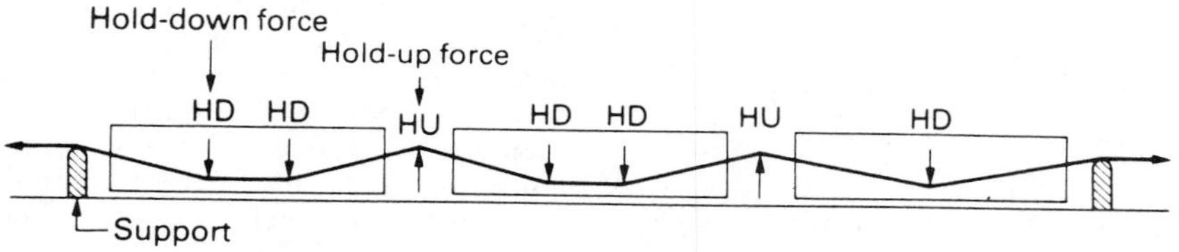

Fig. 8-2 Typical pretensioning tendons profile with one or two draping points.

Small-diameter tendons are generally used in pretensioning to allow for the bond between steel and concrete to develop over a short distance. Most popular sizes in the United States are the 3/8-in.- (9.5-mm-) and the 1/2-in.- (12.5-mm-) diameter strands.

Pretensioning is the method mostly used for the production of precast prestressed concrete elements in the United States because it offers great potential for mechanization. Efficient long-line production techniques with casting bed lengths of up to 600 ft (182 m) where individual elements are cast end to end are preferred, as they require a single tensioning operation. Elements of standardized cross sections are mass-produced yet customized by varying the length of each element and by placing inserts, holes, or blockouts for the mechanical or electrical distribution systems. Accelerated curing often permits early removal of the elements and daily reuse of the forms (24-hr production cycle). Excellent quality control and optimum use of labor and materials are achieved.

Typical elements and member cross sections aimed at particular applications were developed and standardized. The most common standard shapes in the United States are shown in Fig. 8-3. Spans of up to 150 ft (46 m) are not uncommon and are mostly limited by transportation and erection constraints. Load tables and charts were developed by the industry in which, for a given standard shape, external load, and span, the most appropriate section and tendon arrangement can be readily selected.[8-22]

2. Post-tensioning

In post-tensioning the tendons are stressed and anchored at the ends of the concrete member *after* the member has been cast and has attained sufficient strength.

Commonly, a mortar-tight metal tube or duct (also called sheath) is placed along the member before concrete casting. The tendons may have been preplaced loose inside the sheath prior to casting or could be placed after hardening of the concrete. After stressing and anchoring, the void between each tendon and its duct is filled with a mortar grout, which subsequently hardens. Grouting ensures bonding of the tendon to the surrounding concrete, improves the resistance of the member to cracking, and reduces the risks of corrosion for the steel tendons.

The above post-tensioning technique implies using what are commonly called "bonded tendons." If the duct is filled with grease instead of grout, the bond will be prevented throughout the length of the tendon, and the tendon force will apply to the concrete member only at the anchorages. This leads to "unbonded tendons." Unbonded tendons are generally coated with grease or bituminous material, wrapped with waterproof paper or placed inside a flexible plastic hose, and placed in the forms prior to concrete casting. When the concrete reaches sufficient strength, the tendons are stressed and anchored. They remain unbonded throughout their length and during service of the structure. This technique is being increasingly used in slab systems of residential, office, and parking structures with several bays because of its extreme efficiency and economy. The tendons are put in tension at the periphery of the slab and may span up to ten consecutive bays.

The tendons generally used in post-tensioning are made of wires, strands, or bars. Bars are tensioned one at a time; wires and strands can be tensioned singly or in groups. In one of the Freyssient systems, 12 wires or strands forming a tendon can be pulled simultaneously. Up to 170 wires with 0.25-in. (6.35-mm) diameter, can form a single tendon in the BBRV system, and up to 31 strands with 0.6-in. (15.2-mm) diameter can form a single tendon in the VSL system. These tendons carry very large forces. Tendons with a capacity of up to 1000 tons are commonly used in nuclear vessels. They often need specialized jacking and anchoring equipment. Hydraulic jacks are normally used and, with the tendons and anchorages, they all are often an integral part of the post-tensioning system selected.

Although post-tensioning can be used in precast prestressed operations, it is most useful in cast-in-place construction where large building and bridge girders cannot be transported, and for customized structures that need tensioning on the job site. Its application in unusual projects, such as nuclear power vessels, TV towers, and offshore structures, has become a must and will certainly continue to expand.

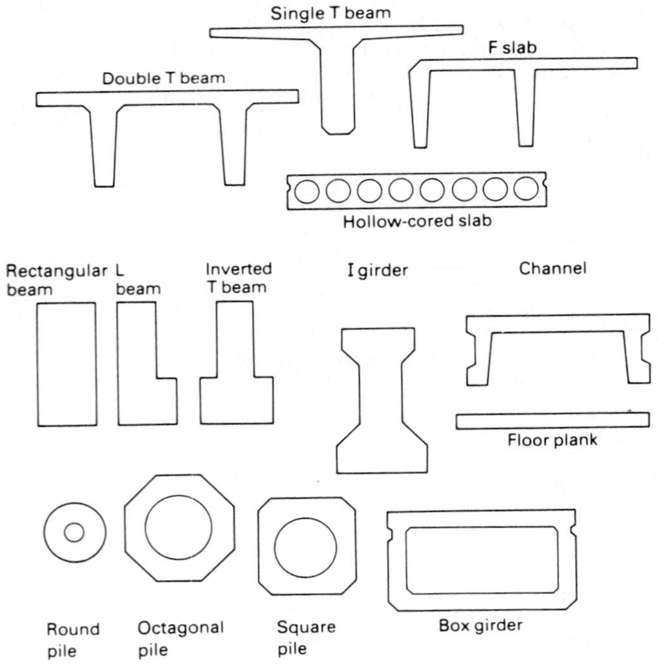

Fig. 8-3 Typical standard sections of precast prestressed concrete products in the United States.

8.1.3 Prestressing Systems

As mentioned above, tensioning the tendons may be achieved in several ways. The most common tensioning systems are mechanical. They are generally protected by patents. Knowledge of the system used is very helpful in detailing the steel and the location of end anchorages.

The basic principles used in these systems are few and essentially similar. The details vary. Patents have been taken on the method of applying the prestress, the type of jack used, the method or device used to anchor the tendons, the number and diameter of wires or strands forming a tendon, and the like.

Typical anchorage systems are shown in Fig. 8-4. Some are based on the principle of direct bearing. These include threaded bars anchored with nut and plate such as for the Dywidag system, or wires with preformed end buttons bearing on a plate through an anchor head such as the BBRV system. In the buttoned wires systems sufficient accuracy is needed in estimating the exact length of the tendons before and after tensioning. Such difficulty is overcome if an anchor system based on wedge action or wedge and grip action is used. The wedge may accom-

First Freyssinet wedge cone for 12 wires.

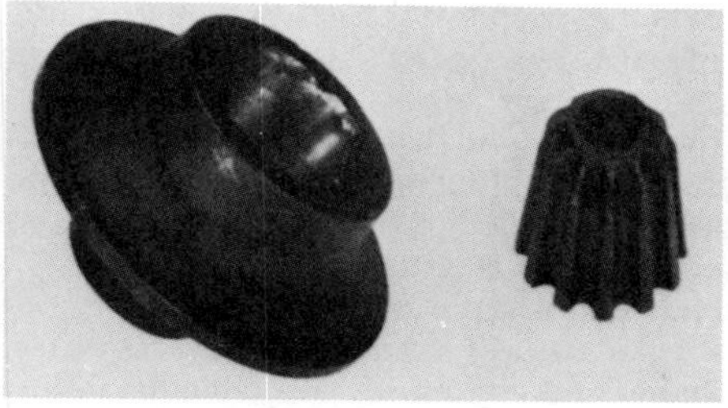
Freyssinet wedge cone for 12 strands.

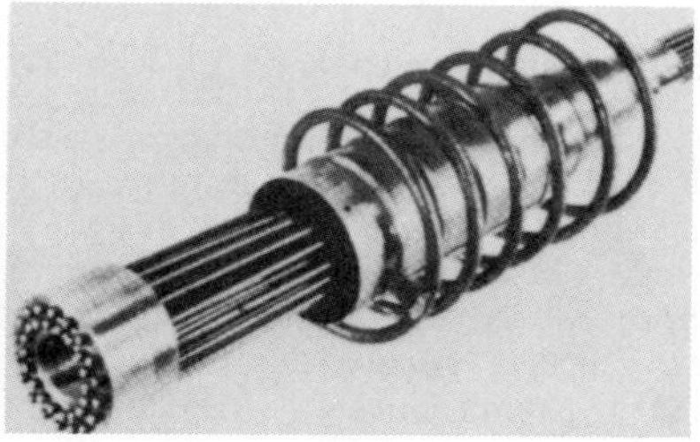
BBRV anchorage for buttonhead wires.

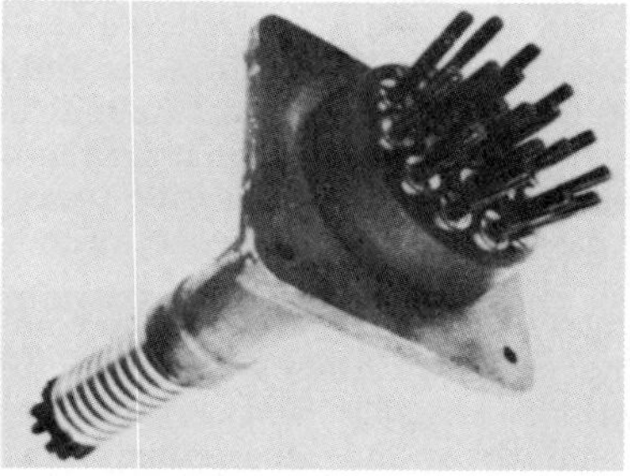
Freyssinet multistrand K range anchorage.

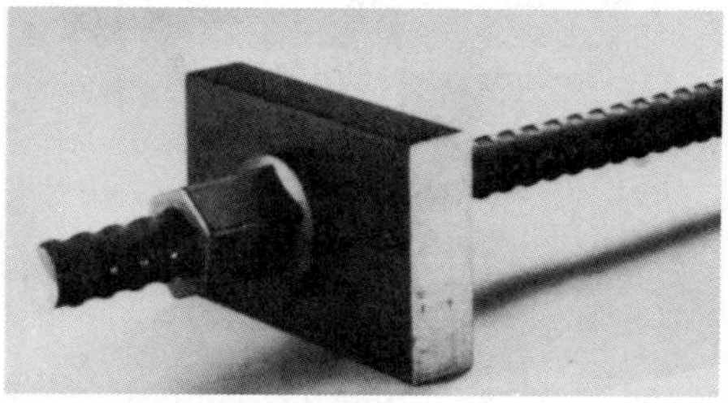
Dywidag threaded bar anchorage.

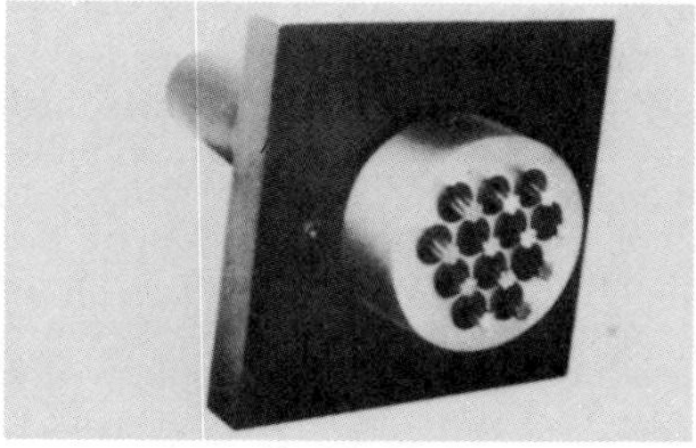
VSL multistrand type E anchorage.

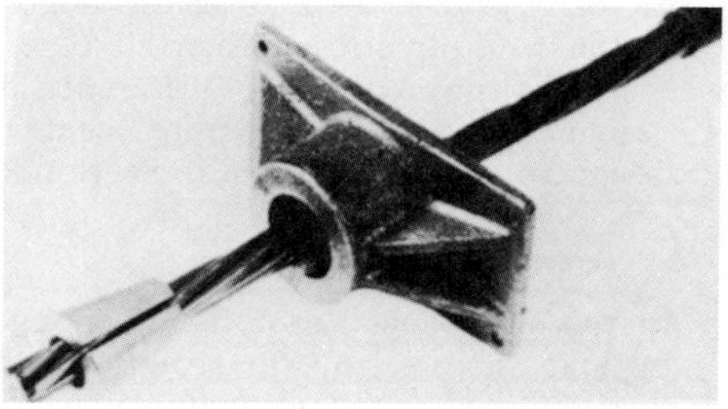
Inryco Cona monostrand anchorage

CCL systems multistrand anchorage.

Fig. 8-4 Typical anchorages used in various prestressing systems.

modate several wires or strands simultaneously on its outer periphery, as in the Freyssient system, or it may grip in a sandwich a single bar or strand, as in the Cona system. Several wedges holding one strand each can have the same anchor head, as in the VSL system. Dead anchors are also available. They are encased directly in the concrete and are generally used in short members when tensioning is needed at one end only. Additional information, technical data, and even design aids are frequently available from the various manufacturers of these systems.

8.2 PRESTRESSING MATERIALS

8.2.1 Prestressing Steels

The importance of using, in prestressed concrete, high-strength steels that permit high elongations was mentioned in Section 8.1.1. Otherwise, because of prestress losses, the steel stress will decrease substantially or vanish with time. As prestress losses in the steel (not including friction) can often approach 60 ksi (414 MPa), it is clear that minimum tensile strengths substantially higher than this value are needed to achieve effective prestressing. Other desirable properties are needed. Ideally a tendon material should not only have high strength but also (1) remain elastic up to relatively high stresses, (2) show sufficient ductility before failure, (3) have good bonding properties, low relaxations, and good resistance to fatigue and corrosion, and (4) be economical and easy to handle.

Three types of steel tendons are used in prestressed concrete: wires, strands (or cables) made with several wires, and bars. Typical shapes and commonly available diameters are given in Chapter 6 of this handbook. Prestressing tendons used in the United States are manufactured to meet minimum ASTM specifications described in the following standards: A421 for uncoated stress-relieved wires, A416 for uncoated seven-wire stress-relieved strands, and A722-75 for uncoated high-strength steel bars. Although these standards would guarantee minimum properties, actual properties depend greatly on the manufacturing process. Local manufacturers and suppliers should be consulted for particular details.

Most prestressing wires are produced by the cold working (drawing or rolling) process. Wires are manufactured with different cross-sectional shapes and surface conditions: round or oval, smooth or indented, ribbed, twisted, or crimped. When cut to size, round wires used in some post-tensioning systems can have button heads formed at their ends.

Prestressing strands are produced from several wires. In the seven-wire strand, six perhipheral wires are wound helically over a central wire that has a slightly higher diameter than the others. Because strands are made with relatively small-diameter wires, they are much easier to handle (more flexible) than a single tendon of the same nominal diameter, and they achieve superior properties due to better quality control. Most popular is the ½-in.-diameter strand with a minimum tensile strength of 270 ksi (1860 MPa).

Prestressing bars are manufactured with a smooth or ribbed surface. Smooth bars can be end-threaded mechanically to be used with anchoring systems based on nut and plate. The ribs in a ribbed bar can act as a thread, as in the Dywidag system. Thus the bar can be anchored anywhere along its length. Prestressing bars are generally made wit alloy steel heat-treated to achieve desirable properties.

8.2.2 Concrete

The compressive strength is the most important structural design property of concrete. Many other properties can be related to the compressive strength. For some applications only the desired compressive strength is specified in the design. In U.S. practice, the compressive strength is obtained from testing 6 × 12 in. cylinders at the age of 28 days. When high-early-strength cement (type III) is used, an age of 7 days is commonly considered. The compressive strength corresponds to the maximum stress at the peak point of the stress–strain curve. The strength obtained from a cylinder test falls within a range of 70 to 90% of the strength obtained from a cube test. The higher the strength, the higher the ratio. It is generally assumed in design that the actual compressive strength is equal to the design specified strength f_c'. In effect, the definition of f_c' is such that a random compression test on the material should lead with a probability of 90% to a strength higher than or equal to f_c'.

Most common values of f_c' for prestressed concrete structures in the United States are between 5 and 7 ksi (35 and 48 MPa) whether normal weight or lightweight concrete is used. Higher compressive strengths of 9 and 11 ksi (62 and 76 MPa) are used with normal weight concretes but to a much lesser extent. These strengths are on the average substantially higher than those specified for ordinary reinforced concrete construction.

Other useful mechanical properties of concrete can be related to its compressive strength. They include its tensile strength determined from a uniform tensile test, its tensile strength determined from the split cylinder test, its modulus of rupture or tensile strength determined from a flexural test, and its elastic modulus. Typical values are given in Chapter 6.

8.3 DESIGN CRITERIA

8.3.1 Design Approaches

The safety and serviceability of structures are generally achieved by satisfying a number of code requirements. In designing for safety, several design approaches can be followed. These approaches are generally based on theory, supported by experimental evidence. Currently encountered design approaches include: working stress design (WSD), strength design (SD), limit or plastic design, limit state design, nonlinear design, and probabilistic design. Although a single approach is generally sufficient, current practice in prestressed concrete involves the combination of working stress design (WDS) and strength design (SD).

1. Working Stress Design

In this approach, the stresses under working loads are limited to permissible values or allowable stresses, and the structure is analyzed assuming linear elastic behavior. Safety is ensured by selecting allowable stresses as relatively small fractions of the characteristic strengths of the component materials. Allowable stresses are specified in various codes and may vary from one code to another. For instance, the maximum permissible compressive stress on concrete flexural members may be taken as $0.45 f_c'$. This implies a safety factor of $(1/0.45) = 2.22$ against concrete failure. Note that in the working stress design all types of loads are treated the same, no matter how different their variability is. The design of prestressed concrete beams by the working stress design approach is covered in Section 8.5.

2. Strength Design

In this approach, the design working loads are multiplied by load factors, and the structure is designed to resist the factored loads at its ultimate capacity. The load factors associated with a type of loading are adjusted to reflect the degree of variability and uncertainty of that loading. This is more realistic than in the WSD approach where all loads are treated the same. The application of SD to prestressed beams in flexure and shear is covered in Sections 8.6 and 8.7.

8.3.2 Design Codes

Although basic engineering concepts and judgment can be used to design a structure, most of the guesswork can be reduced and better efficiency achieved if structural requirements set by design codes are satisfied. Design codes (such as Refs. 8-23 through 8-28) are written to protect the user and society as a whole. They provide information on methods of analysis and design, minimum design requirements, and minimum expected performance. They represent a summary of the collective opinion or agreed-upon state of knowledge of the profession.

In the United States most reinforced and prestressed concrete structures (except bridges) are designed in accordance with the *Building Code Requirements for Reinforced Concrete* (ACI 318-83) published by the American Concrete Institute. It is regarded as an authoritative statement of current good practice in the field of concrete structures. The *ACI Code* is incorporated entirely or in part in many municipal and regional codes in the United States and is used as a reference in many foreign countries.

Most prestressed concrete bridges for highways or railways in the United States are designed in accordance with two major codes: the AASHTO (American Association of State Highway and Transportation Officials) *Standard Specifications for Highway Bridges*[8-25] and the AREA (American Railway Engineering Association) *Manual for Railway Engineering*.[8-26] Except for some subtle differences, the sections of these codes related to prestressed concrete are essentially identical in scope with the corresponding sections of the *ACI Code*.

In this text, reference to these codes will often be made, and preferably their latest editions should be used. These codes contain specific information on analysis and design methods, as well as service loads, load factors, and allowable stresses.

8.4 PRESTRESS LOSSES

8.4.1 Sources of Loss of Prestress

The stress in the tendons (hence the prestressing force) of a prestressed concrete member continuously decreases with time. The total stress reduction during the lifespan of the member is called "total loss of prestress." It is essential to estimate the magnitude of the total loss of prestress, since it leads to the value of the effective prestressing force needed for design.

The total loss of prestress is generally attributed to the cumulative contribution of some or all of the following sources:

1. Elastic shortening. Because the concrete shortens when the prestressing force (in full or in part) is applied to it, the tendons already bonded to the concrete shorten, simultaneously losing part of their stress.
2. Relaxation (or creep) of the stressed tendons. Relaxation is a property of the steel and is described in Chapter 6.
3. Shrinkage of concrete. The gradual loss, with time, of free water from the concrete, called shrinkage, induces a shortening in the concrete that leads to a loss of stress in the attached tendons.
4. Creep of concrete. Creep is caused by the compressive stresses in the concrete. It induces a shortening strain in the concrete in excess of the elastic strain that increases with time and leads to a loss of stress in the attached tendons.
5. Friction. Friction loss occurs during tensioning of post-tensioned tendons. It represents the difference in stress between the jacking end of the tendon and a section along the member.
6. Anchorage set. Many post-tensioning anchorages of the wedge type require that the wedge "sets in" a certain distance in order to lock the tendon at end of jacking. This set (also called seating or slip) leads to a loss of stress in the tendon.
7. Other factors. These include restraining effects of adjoining elements and temperature effects, if any. As they depend on the type of structure, the corresponding stress loss cannot be covered here.

Each of the above sources leads to a separate prestress loss in the tendons. Prestress losses occur either instantaneously or with time. Instantaneous losses in pretensioned members are generally reduced to the effect of elastic shortening of the members at transfer (or release) of prestress. In post-tensioned members they include, in addition to the partial effect of elastic shortening, the effect of anchorage set and frictional losses between steel and concrete. Time-dependent losses include the effect of relaxation in the steel, as well as the effects of shrinkage and creep in the concrete. They affect all prestressed concrete elements.

The main sources of prestress losses, the stage at which they occur, and the corresponding designation of stress loss in the tendon are summarized in Table 8-1. Each loss is attributed a literal value with a subscript containing a capital letter representing the particular source. For instance, Δf_{pR} represents the total stress loss in the prestressing tendons due to relaxation. Two literal values are associated with time-dependent losses, one nondimensional and the other dimensional. The first value, such as Δf_{pR}, represents the total stress loss due to relaxation at the end of the life of the structure, while the other, such as $\Delta f_{pR}(t_i, t_j)$, represents the loss during a particular time interval (t_i, t_j).

It is generally assumed that the tendons of a member are tensioned to the same stress and have identical properties. The total loss of stress in the tendons Δf_{pT} is the sum of the total losses due to each source. The magnitude of the total loss of prestress in a tendon varies along the member. It is assumed here that prestress losses are being computed at a given section, most likely the critical section of interest.

TABLE 8-1 Sources of Prestress Losses

	Stage of Occurrence		*Tendon Stress Loss*	
Source of Prestress Loss	*Pretensioned Members*	*Post-tensioned Members*	*During Time Interval* (t_i, t_j)	*Total or During Life*
Elastic shortening of concrete (*ES*)	At transfer	At jacking	....	Δf_{pES}
Relaxation of prestressed tendons (*R*)	Before and after transfer	After transfer	$\Delta f_{pR}(t_i, t_j)$	Δf_{pR}
Shrinkage of concrete (*S*)	After transfer	After transfer	$\Delta f_{pS}(t_i, t_j)$	Δf_{pS}
Creep of concrete (*C*)	After transfer	After transfer	$\Delta f_{pC}(t_i, t_j)$	Δf_{pC}
Friction (*F*)	....	At jacking	....	Δf_{pF}
Anchorage set (*A*)	....	At transfer	....	Δf_{pA}
Total	Life	Life	$\Delta f_{pT}(t_i, t_j)$	Δf_{pT}

8.4.2 Methods for Estimating Prestress Losses

Many methods and design recommendations have been developed to predict prestress losses.[8-29, 8-30] For design purposes, they can be classified according to three levels of difficulty and related computational accuracy. The corresponding three categories are identified as follows:

1. Accurate determination of losses by the time-step method.
2. Lump sum estimate of the total loss of prestress Δf_{pT}
3. Lump sum estimates of the separate total loss due to each source such as creep and shrinkage, that is Δf_{pC}, Δf_{pS}, etc.

Each category is described in more detail in the following sections.

Note that the accurate determination of losses is more important for some structures than for others. The more accurate and detailed the technique of predicting losses, the more specific is the input information needed. While in the lump sum estimate of total loss (Δf_{pT}) one may only need to know if the structure is pretensioned or post-tensioned, in the time-step method data on the basic materials properties, the shape of the structure, the age at loading, environmental conditions, and the like are necessary.

1. Prestress Losses by the Time-Step Method

The time-step method is an accurate technique to predict losses and involves mostly time-dependent losses due to relaxation of the steel and shrinkage and creep of concrete. Instantaneous losses are computed as for the other methods.

In order to fully understand the time-step method, it is essential to realize that time-dependent losses are also interdependent. Figure 8-5 provides an overall diagram of how total losses are cumulated and illustrates the interfering causes and effects between the different sources of loss of prestress.

It is important to understand that the rate of stress loss

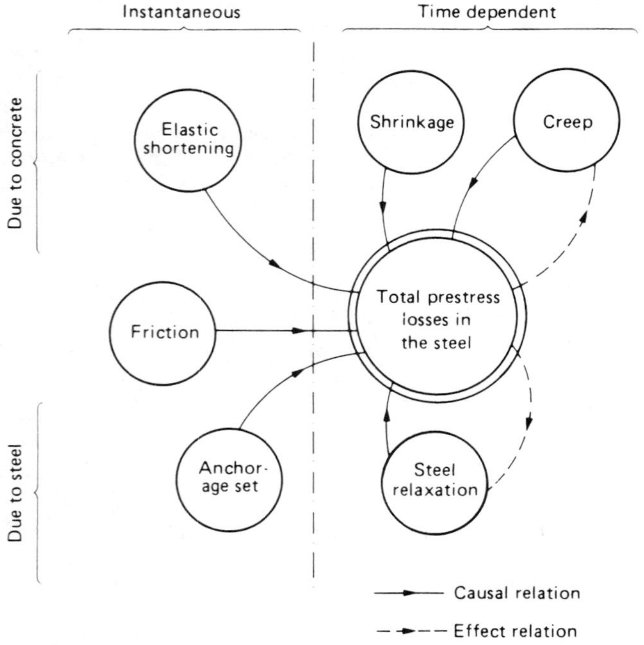

Fig. 8-5 Interrelationships of causes and effects between prestress losses.

TABLE 8-2 AASHTO Lump Sum Estimate of Prestress Loss for Routine Design[8-25]

	Total Loss	
Type of Prestressing Steel	$f'_c = 4{,}000$ *psi* (*27.58 MPa*)	$f'_c = 5{,}000$ *psi* (*34.47 MPa*)
Pretensioning		
Strand		45,000 psi (310.26 MPa)
Post-tensioning*		
Wire or Strand	32,000 psi (220.63 MPa)	33,000 psi (227.53 MPa)
Bars	22,000 psi (151.68 MPa)	23,000 psi (158.58 MPa)

**Losses due to friction are excluded. Friction losses should be computed according to Article 1.6.7(A).*

due to one effect such as relaxation is being continuously altered by changes in the steel stress due to the other effects such as creep and shrinkage. The rate of creep, in its turn, is simultaneously altered by the changes in the steel stress due to relaxation and shrinkage. To account for these interdependent effects with time, a repetitive computation procedure using successive time intervals is used. Generally, the lengths of these time intervals are not equal and vary with the age of the concrete. The stress in the steel at the beginning of any time interval is taken equal to that at the end of the preceding interval. It is assumed constant during a time interval and used to compute incremental losses during that interval. The steel stress at the end of a time interval is obtained by subtracting the incremental stress losses from the stress at the beginning of the interval. More accuracy can be obtained by increasing the number of intervals. A typical example of loss computations by the time-step method with a reasonable number of time intervals is covered in Ref. 8-15.

2. Lump Sum Estimate of Total Losses

Lump sum estimates of total prestress losses, excluding friction and anchorage set, are suggested in many references. Those given in the AASHTO specifications are reproduced in Table 8-2. They apply to bridges where average standard conditions prevail. Based on an extensive study of prestressed concrete building members, Zia et al.[8-30] recommended maximum limit values of total prestress loss applicable when the tendon stress immediately after anchoring does not exceed $0.83 f_{py}$ (where f_{py} is the yield stress of prestressing steel). These limits are summarized in Table 8-3.

It is this author's approach to use as a first approximation for normal weight concrete a lump sum total loss of 45 ksi

TABLE 8-3 Recommended Maximum Stress Loss Not Including Friction and Anchorage Set[8-30]

	Maximum Loss† ksi (MPa)	
Type of Strand	*Normal Weight Concrete*	*Lightweight Concrete*
Stress-relieved strand	50 (345)	55 (380)
Low-relaxation strand	40 (276)	45 (131)

†For $f_{pi} \leqslant 0.83 f_{py}$.

(311 MPa) for pretensioned members and 35 ksi (242 MPa) for post-tensioned members (excluding the effects of friction and anchorage set). An increment of 5 ksi (35 MPa) is used for lightweight concrete, and a reduction of 5 ksi (35 MPa) is used for low-relaxation strands. Total loss values are reassessed and revised, if necessary, in the final design.

3. Lump Sum Estimate of Separate Losses

Several methods can be found in the technical literature to estimate the separate contribution of each source of loss.[8-25, 8-29] The total loss of prestress is then obtained by summing up the separate contributions. Many of these methods have merit and can be used. However, because of the implied authority of the code, only the method described in the AASHTO specifications[8-25] and adopted by ACI Committee 343 on Bridge Structures[8-31] is recommended. It requires some basic information on materials properties and environmental conditions, but it is self-explanatory and simple to implement.

8.4.3 Prestress Losses In Design

Let us define by f_{pi} the initial stress in the prestressing steel at time of load transfer and f_{pe} the effective stress remaining after all losses have taken place. The corresponding prestressing forces are defined as F_i and F. F_i is the force that the concrete experiences under initial loading. Thus f_{pi} may mean the stress just after transfer for a pretensioned member and just before transfer for a post-tensioned member. The difference between f_{pi} and f_{pe} essentially represents the total prestress losses for a post-tensioned beam, while it represents time-dependent losses (excluding initial relaxation) for a pretensioned beam. In the preliminary design and dimensioning of the structure the above difference between f_{pi} and f_{pe} is not as useful as their ratio. Let us define:

$$\eta = \frac{f_{pe}}{f_{pi}} = \frac{F}{F_i} \tag{8-1}$$

where F_i is the initial and F the final or effective prestressing force.

Note that as the stress in the steel generally varies along the length of the member, the coefficient η will also vary. However, η is needed mostly at the critical section of a simply supported beam or at a limited number of critical sections of a continuous beam. For all practical purposes it could be assumed the same for all critical sections. As a first approximation in design the value of η can be taken between 0.75 and 0.85. Assuming a value of $f_{pi} = 0.70 f_{pu}$, the corresponding value of f_{pe}, the effective prestress, will be somewhere between $0.53 f_{pu}$ and $0.58 f_{pu}$. Such approximations will frequently be used in preliminary designs. However, for final designs a more exact assessment of prestress losses is recommended.

8.5 FLEXURE: WORKING STRESS ANALYSIS AND DESIGN

8.5.1 Notations for Flexure

In the most general case, the uncracked cross section of a prestressed concrete beam can be characterized for design purposes by a number of variables and geometric properties. Geometric properties can be determined directly from the dimensions of the cross section. Referring to Fig. 8-6, the following notations and definitions will be used:

A_c = area of concrete cross section (it may indicate the net or the gross area depending on the problem at hand and whether preliminary or final design are considered; practically, in pretensioned members it is taken as the gross area, while in post-tensioned members it is often the net or the transformed area)
A_{ps} = area of prestressing steel
d_p = distance from extreme compression fiber to the centroid of prestressing steel (or force)
d_c = concrete cover from the precompressed tensile fiber to the centroid of prestressing steel (d_c can generally be estimated in a preliminary design and revised for the final design)
$(d_c)_{min}$ = minimum feasible value of d_c
e_o = eccentricity of the prestressing force (or centroid of prestressing steel) with respect to the centroid of the concrete section [e_o varies along the member; thus $e_o(x)$ is used when needed]
F = final prestressing force or effective prestressing force after all losses
h = total height of concrete section ($h = y_t + y_b = d_p + d_c$)
I = moment of inertia of the section with respect to an axis passing by its centroid (it generally implies gross sectional inertia; for a final design it may indicate the transformed inertia)
y_t = distance from the centroid of the concrete section to the extreme top fiber
y_b = distance from the centroid of the concrete section to the extreme bottom fiber
ρ_p = A_{ps}/bd_p prestressed reinforcement ratio (b is the width of a rectangular section or flange width of T section)
σ = stress in concrete in general (it will be used unless a widely standard notation such as f_c' applies)
Z_t = I/y_t = section modulus with respect to the top fiber
Z_b = I/y_b = section modulus with respect to the bottom fiber

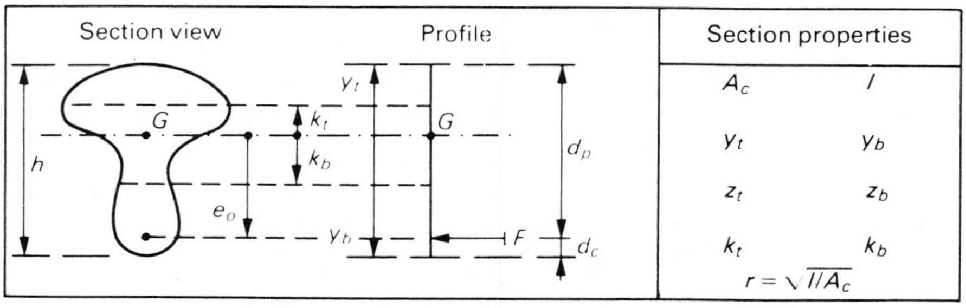

Fig. 8-6 Typical characterization of beam cross section.

$r = \sqrt{I/A_c}$ = radius of gyration of the section

$k_t = -I/A_c y_b = -Z_b/A_c = -r^2/y_b$ = distance from the centroid of the concrete section to the upper limit of the central kern

$k_b = I/A_c y_t = Z_t/A_c = r^2/y_t$ = distance from the centroid of the concrete section to the lower limit of the central kern

Note that $y_b = e_o + d_c$ and $d_p = e_o + y_t$. The central kern defines an area within the concrete section where a compressive force will not lead to tensile stresses on any part of the section. The upper limit of the central kern k_t has been given a negative sign to be consistent with the sign convention given for e_o (Section 8.5.2) and because k_t can be seen as a particular value of e_o. The reinforcement ratio of prestressed concrete members is generally less than 1%, and most values in beams and slabs practically fall between 0.1 and 0.5%.

8.5.2 Sign Convention

Although in prestressed concrete many effects such as compression or tension can be found by inspection, it is very important to provide a consistent sign convention in order to reduce error in any complex and systematic analysis; of course, this is essential in any computerized design. The following rules of sign convention will be generally followed in this chapter:

1. Plus (+) for compressive stresses in concrete
2. Minus (−) for tensile stresses in concrete

The two extreme fibers of a horizontal member (beam) will be referred to as top and bottom fiber. A vertical member is rendered horizontal by a clockwise rotation of $\pi/2 = 90°$. Thus the left and right fiber of a vertical member are identical from a stress analysis viewpoint respectively to the top and bottom fiber of a horizontal member. In order to evaluate the effect of a bending moment on stresses, the usual sign convention for moments as described in Fig. 8-7 will be used. Thus a moment that tends to bend a simply supported beam so that it will retain water is considered positive. For instance, the moment due to the self weight of a simply supported beam is positive, while the moment at the intermediate support of a two-span continuous beam is negative. Therefore:

3. Use plus (+) for the numerical value of the moment for positive moments and minus (−) for negative moments.

An extreme fiber stress due to a bending moment of magnitude M can be calculated from the expression M/Z_t or M/Z_b where Z_t and Z_b represent the section moduli with respect to the top and bottom fibers, respectively. In order

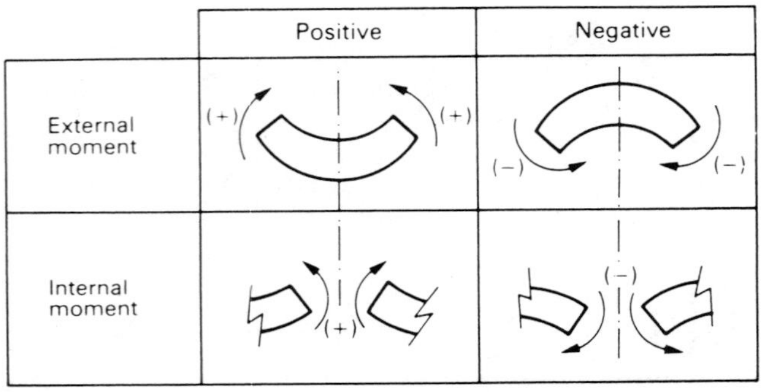

Fig. 8-7 Sign convention for moments.

for flexural stresses to show the correct sign corresponding to a tension or a compression the following rule is set:

4. Multiply the above expression by +1 if the stress calculated is on the top fiber, and by −1 if it is on the bottom fiber. Note that the same holds for stresses calculated at points "above" and "below" the neutral axis of bending of the section. For vertical members "left" replaces "top" or "above," and right replaces "bottom" or "below."

The prestressing force F acting on a concrete section with eccentricity e_o has the same effect as a concentric force F applied at the centroid of the section, thus inducing a uniform compressive stress (positive) and a moment of magnitude Fe_o. If e_o is positive downward as in Fig. 8-6, the moment is negative (convex), and its value is $(-Fe_o)$. The corresponding flexural stresses are given on the top fiber by:

$$+\left(\frac{-Fe_o}{Z_t}\right) = \frac{-Fe_o}{Z_t}$$

and on the bottom fiber by:

$$-\left(\frac{-Fe_o}{Z_b}\right) = \frac{Fe_o}{Z_b}$$

Another way to present these results is to follow the following rules:

5. Use the absolute value of the moment Fe_o (i.e., F and e_o are positive).
6. Multiply the stress due to Fe_o by +1 when it is computed on a fiber on the same side as e_o with respect to the neutral axis, and by −1 when it is on the opposite side.

Expressions for the stress on the top and bottom fibers due to a bending moment and to the prestressing force for a typical simply supported prestressed beam are summarized in Table 8-4. For each expression several equivalent forms can be used. However, some may be easier to handle than others in a particular problem.

The following sign convention will be used for the steel (prestressed or nonprestressed):

7. Plus (+) for tensile stresses
8. Minus (−) for compressive stresses

It may seem awkward to have different sign conventions for stresses in the steel and concrete. However, the above recommended sign convention is natural for prestressed concrete as it generally leads to positive stresses in the concrete (compression) and in the steel (tension). When confusion may arise in some equations where compression steel is present, the absolute value sign will be used to warn the reader.

8.5.3 Loading Stages

In the design of prestressed concrete members, loading refers not only to externally applied loads such as dead and live loads but often to a combination of these loads acting with the prestressing force on the concrete section. Several loading stages can be identified in the elastic range of behavior, among which the *initial* and the *final* loadings are generally most critical.

The *initial loading* refers primarily to the stage in which the prestressing force is transferred to the concrete and no external loads are present except the weight of the member. At this time the prestressing force is maximum as pre-

TABLE 8-4 Various Expressions of Stresses Due to a Moment or a Prestressing Force

Effect of	*Extreme Fiber*	*Various Expressions for Stresses*
Positive moment of magnitude M	Top	$\sigma_t = \frac{My_t}{I} = \frac{M}{Z_t} = \frac{M}{A_c k_b} = \frac{My_t}{A_c r^2}$
	Bottom	$\sigma_b = -\frac{My_b}{I} = -\frac{M}{Z_b} = +\frac{M}{A_c k_t} = -\frac{My_b}{A_c r^2}$
Prestressing force F at eccentricity e_o toward bottom fiber of beam	Top	$\sigma_t = \frac{F}{A_c} - \frac{Fe_o y_t}{I} = \frac{F}{A_c}\left(1 - \frac{e_o y_t}{r^2}\right) = \frac{F}{A_c}\left(1 - \frac{e_o A_c}{Z_t}\right) = \frac{F}{A_c}\left(1 - \frac{e_o}{k_b}\right) = \frac{F}{Z_t}(k_b - e_o)$
	Bottom	$\sigma_b = \frac{F}{A_c} + \frac{Fe_o y_b}{I} = \frac{F}{A_c}\left(1 + \frac{e_o y_b}{r^2}\right) = \frac{F}{A_c}\left(1 + \frac{e_o A_c}{Z_b}\right) = \frac{F}{A_c}\left(1 - \frac{e_o}{k_t}\right) = \frac{F}{Z_b}(e_o - k_t)$

stress losses have not yet taken place, and the concrete strength is minimum as the concrete is still young; consequently, the stresses in the concrete can be critical. In pretensioned members in order to speed production, the prestressing tendons are released simultaneously at a time when the strength of the concrete has reached 60 to 80% of its specified 28 days strength. By curing it at higher temperatures, these strengths can be achieved in less than 24 hr after the concrete is poured. In post-tensioned members, often the prestressing tendons are not tensioned all at the same time but rather in two or three steps to allow the concrete to reach its specified strength before the prestressing force is fully applied. In most cases the initial loading leads to critical stresses, and its effect must be carefully assessed.

The *final loading* stage refers here to the most severe loading under service conditions; it is then assumed that all prestress losses have occurred (i.e., the prestressing force has its final and smallest value), and that the most critical combination of external loadings is applied; such a combination includes the weight of the members, superimposed dead loads, live loads, impact, and the like. Load combinations are generally specified in various codes and specifications.

Although the initial and final loadings are often the two most critical loadings, some intermediate loadings may become critical in the design. For example, special conditions during handling, transportation, and erection of precast prestressed members may lead to stresses more critical than those induced by the initial and final loadings. Every particular case must be studied with care and if necessary integrated in the design.

Among all possible loadings applied to a prestressed section, two will bound the others in terms of flexural stresses, and will be identified as the *two extreme loading* conditions. In a majority of cases they are due to the effect of $(F_i + M_{min})$ and $(F + M_{max})$ where, under initial and final conditions, M_{min} and M_{max} are the minimum and maximum bending moments, respectively, and F_i and F are the initial and effective values of the prestressing force at the section considered. For example, M_{min} may be the same as the moment due to the weight of the member, while M_{max} may be the moment due to weight plus the superimposed dead load and live load. The relation between the initial and the final prestressing force is described by their ratio $\eta = F/F_i$; η can be estimated in a preliminary design and then more carefully assessed in the final design. Typical stress diagrams on the concrete section of a pretensioned member are shown in Fig. 8-8. The two extreme loading conditions are shown to induce flexural stress within allowable stress. Allowable stresses are represented by short vertical lines dashed on one side.

8.5.4 Allowable Stresses

Allowable stresses on the concrete section as well as in the steel are generally provided by the codes of practice or

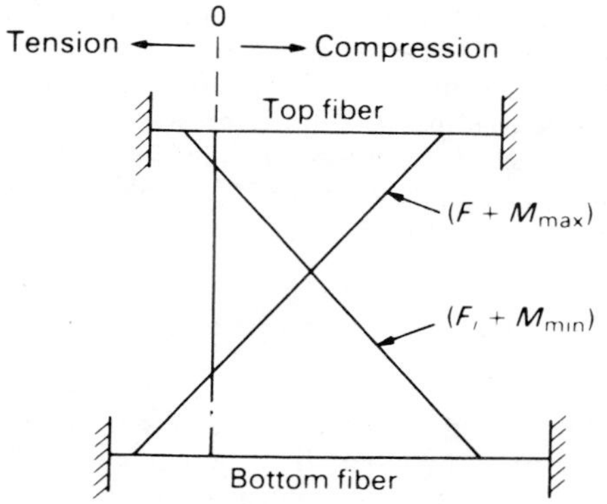

Fig. 8-8 Typical stress diagrams under extreme loadings.

specifications considered for a particular design. Typical values of allowable stresses are given in Refs. 8-23 and 8-25.

As generally two extreme loading conditions bound all others, at least four allowable stresses on the concrete section must be considered in the design: namely two (tension and compression) for the initial loading and two (tension and compression) for the most severe final loading. Without specifically referring to any particular numerical value of allowable stresses, the following notation will be used in describing them:

$\bar{\sigma}_{ti}$ = allowable temporary tensile stress in the concrete (initial most severe loading)
$\bar{\sigma}_{ci}$ = allowable temporary compressive stress in the concrete (initial most severe loading)
$\bar{\sigma}_{ts}$ = allowable service tensile stress in the concrete (final most severe loading)
$\bar{\sigma}_{cs}$ = allowable service compressive stress in the concrete (final most severe loading)

In a composite construction one additional allowable stress must be considered, namely:

$\bar{\sigma}_{c\,\text{slab}}$ = allowable service compressive stress in composite slab

Note that in order to identify allowable stresses for concrete a bar is placed on top of the literal value. The same notation is used without the bar for actual stresses under the same conditions. For the steel, allowable stresses are given directly as a fraction of ultimate strength; thus no literal values are needed.

8.5.5 Mathematical Basis for Flexural Analysis

For the analysis, it is assumed that materials behave elastically in the working range of stresses applied. The usual hypotheses of Hooke and Navier–Bresse are assumed valid, namely:

1. The materials (steel and concrete) are elastic and there is a proportional relationship between stresses and strains.
2. Plane sections remain plane after bending.
3. There is a perfect bond between steel and concrete.

This is equivalent to say that both the stress and strain diagrams along the section of concrete under bending are linear, and that the changes in strains in the steel and in the concrete at the level of the steel are identical. Also the load-deflection or moment-curvature curve is assumed linear for the loadings considered. Typical stress diagrams for the two extreme initial and final loadings have been described in Fig. 8-8. Note that the highest stresses in the section occur at the extreme top and bottom fibers.

As two extreme loadings are generally critical, and as for each two allowable stresses must be specified, at least four allowable stresses must be considered in the analysis. As under flexural loading maximum stresses occur on the two extreme fibers (top and bottom), eight inequality equations comparing actual stresses with allowable stresses can be derived. They are of the form:

$$(\text{Actual stress}) \begin{Bmatrix} \geqslant \\ \text{or} \\ \leqslant \end{Bmatrix} (\text{allowable stress})$$

Let us develop one of these equations for a pretensioned simply supported member. The actual stress on the top fiber under initial conditions must be more than or equal to the allowable initial tensile stress. Therefore:

$$\sigma_{ti} = \frac{F_i}{A_c} - \frac{F_i e_o}{Z_t} + \frac{M_{\min}}{Z_t} \geqslant \bar{\sigma}_{ti} \tag{8-3}$$

where $M_{\min}$ represents the dead load moment at the section considered. Equation (8-3) could also be rewritten in several different ways, one of which may be more suitable if a particular variable is to be emphasized, as, for example:

$$\begin{cases} F_i \leqslant (M_{\min} - \bar{\sigma}_{ti} Z_t)/(e_o - k_b) \\ e_o \leqslant k_b + (1/F_i)(M_{\min} - \bar{\sigma}_{ti} Z_t) \\ 1/F_i \geqslant (e_o - k_b)/(M_{\min} - \bar{\sigma}_{ti} Z_t) \end{cases} \tag{8-4}$$

As mentioned above, eight inequality equations (which will also be described as "stress conditions" or "stress constraints") can be derived and are similar in form to eq. (8-4). However, in actual design problems, out of the eight conditions four are generally nonbinding. For example, if for a simply supported member the tensile stress on the top fiber for the initial loading is of concern and is checked against allowable (as for eq. 8-3), there is certainly no need to check, against allowable, the compressive stress on the same fiber and for the same loading. If for the same loading we were checking the section at the intermediate support of a two-span continuous beam, we would check the stress on the top fiber against initial allowable compression, and that eliminates the need to check against initial allowable tension. Thus the number of inequality equations that must be considered in the analysis at a given section is essentially reduced to four; i.e., four of them are binding, while the four others are not. The four that are binding in a particular design depend on the sign of the applied moments.

The eight stress inequality equations written in various ways are shown four at a time in Tables 8-5 and 8-6. They have been numbered in roman notation I to IV and I′ to IV′. The coefficient η was defined in Section 8.4.3 and is the ratio of the final prestressing force after all losses to the initial prestressing force. For a cross section where all applied moments ($M_{\min}$ and $M_{\max}$) are positive, only stress inequalities I to IV need to be considered; similarly, if all applied moments are negative, stress inequalities I′ to IV′ become binding. When a particular section is subject to moments of different signs, it is possible, by inspection, to select out of the eight inequalities the four that would be binding; on the other hand, one can also check systematically the eight inequalities against allowable stresses and select the four that are binding.

Note that the stress inequalities I′ to IV′ given in Table 8-6 are described here as "complementary stress inequalities." This is so because often one does not need to use them. It can be shown that if all applied moments are negative, stress conditions I to IV can still be used, provided the concrete section is assumed in its inverted position (i.e., use properties of inverted section) and the sign of the moments is changed from negative to positive; the position of F within the cross section remains unchanged. It is because stress conditions I to IV can cover the majority of practical problems that they are often encountered alone in the technical literature and with no reference to the four others.

In Tables 8-5 and 8-6, a fifth condition, number V, has been included, which will be described as the "practicality condition." Essentially it states that the prestressing force must be inside the concrete section with an adequate cover $(d_c)_{\min}$. Thus the design eccentricity e_o must be less than or equal to a maximum practical value $(e_o)_{mp} = y_b - (d_c)_{\min}$. Although in an analysis or investigation problem condition V is obviously satisfied, in a design problem condition V can be binding and can be used with advantage in

TABLE 8-5 Useful Ways of Writing the Four Stress Inequality Conditions

Way	*Stress Condition*	*Inequality Equation*
1	I	$(F_i/A_c)[1 - (e_o/k_b)] + M_{min}/Z_t \geqslant \bar{\sigma}_{ti}$
	II	$(F_i/A_c)[1 - (e_o/k_t)] - M_{min}/Z_b \leqslant \bar{\sigma}_{ci}$
	III	$[(F \text{ or } \eta F_i)/A_c][1 - (e_o/k_b)] + M_{max}/Z_t \leqslant \bar{\sigma}_{cs}$
	IV	$[(F \text{ or } \eta F_i)/A_c][1 - (e_o/k_t)] - M_{max}/Z_b \geqslant \bar{\sigma}_{ts}$
2	I	$e_o \leqslant k_b + (1/F_i)(M_{min} - \bar{\sigma}_{ti}Z_t)$
	II	$e_o \leqslant k_t + (1/F_i)(M_{min} + \bar{\sigma}_{ci}Z_b)$
	III	$e_o \geqslant k_b + [1/(F \text{ or } \eta F_i)](M_{max} - \bar{\sigma}_{cs}Z_t)$
	IV	$e_o \geqslant k_t + [1/(F \text{ or } \eta F_i)](M_{max} + \bar{\sigma}_{ts}Z_b)$
3	I	$F_i \leqslant (M_{min} - \bar{\sigma}_{ti}Z_t)/(e_o - k_b)$
	II	$F_i \leqslant (M_{min} + \bar{\sigma}_{ci}Z_b)/(e_o - k_t)$
	III	$F = \eta F_i \geqslant (M_{max} - \bar{\sigma}_{cs}Z_t)/(e_o - k_b)$
	IV	$F = \eta F_i \geqslant (M_{max} + \bar{\sigma}_{ts}Z_b)/(e_o - k_t)$
4	I	$1/F_i \geqslant (e_o - k_b)/(M_{min} - \bar{\sigma}_{ti}Z_t)$
	II	$1/F_i \geqslant (e_o - k_t)/(M_{min} + \bar{\sigma}_{ci}Z_b)$
	III	$1/F = 1/\eta F_i \leqslant (e_o - k_b)/(M_{max} - \bar{\sigma}_{cs}Z_t)$
	IV	$1/F = 1/\eta F_i \leqslant (e_o - k_t)/(M_{max} + \bar{\sigma}_{ts}Z_b)$
All	V	$e_o \leqslant (e_o)_{mp} = y_b - (d_c)_{min}$ = maximum practical eccenticity

optimizing or simplifying a solution. This is why it has been included in the tables.

In an analysis or investigation problem, the above stress inequality equations can be directly checked as all quantities are known. In a design problem, however, these inequalities can be used to determine exactly or put bounds on some of the unknown variables such as prestressing force F, eccentricity e_o, and/or section properties. For example, if the concrete cross section is given, the stress conditions can be used to determine bounds on all the possible values of F and e_o that would be acceptable for the problem at hand. This is clarified in the next sections.

8.5.6 Geometric Interpretation of the Stress Inequality Conditions

The geometric interpretation of the stress conditions was first explored by Magnel.[8-3] As emphasized throughout this section, the geometric representation can be a very useful and powerful technique for the solution of many problems in which the working stress design approach is used.

Let us assume that the geometric properties of the concrete cross section are given; then (as η is estimated a priori) only two unknown variables remain in eqs. I to IV, namely e_o and F_i or F. One can plot on a two-dimensional scale the

TABLE 8-6 Useful Ways of Writing the Four Complementary Stress Inequality Conditions

Way	*Stress Condition*	*Inequality Equation*
1	I′	$(F_i/A_c)[1 - (e_o/k_b)] + M_{min}/Z_t \leqslant \bar{\sigma}_{ci}$
	II′	$(F_i/A_c)[1 - (e_o/k_t)] - M_{min}/Z_b \geqslant \bar{\sigma}_{ti}$
	III′	$[(F \text{ or } \eta F_i)/A_c][1 - (e_o/k_b)] + M_{max}/Z_t \geqslant \bar{\sigma}_{ts}$
	IV′	$[(F \text{ or } \eta F_i)/A_c][1 - (e_o/k_t)] - M_{max}/Z_b \leqslant \bar{\sigma}_{cs}$
2	I′	$e_o \geqslant k_b + (1/F_i)(M_{min} - \bar{\sigma}_{ci}Z_t)$
	II′	$e_o \geqslant k_t + (1/F_i)(M_{min} + \bar{\sigma}_{ti}Z_b)$
	III′	$e_o \leqslant k_b + [1/(F \text{ or } \eta F_i)](M_{max} - \bar{\sigma}_{ts}Z_t)$
	IV′	$e_o \leqslant k_t + [1/(F \text{ or } \eta F_i)](M_{max} + \bar{\sigma}_{cs}Z_b)$
3	I′	$F_i \geqslant (M_{min} - \bar{\sigma}_{ci}Z_t)/(e_o - k_b)$
	II′	$F_i \geqslant (M_{min} + \bar{\sigma}_{ti}Z_b)/(e_o - k_t)$
	III′	$F = \eta F_i \leqslant (M_{max} - \bar{\sigma}_{ts}Z_t)/(e_o - k_b)$
	IV′	$F = \eta F_i \leqslant (M_{max} + \bar{\sigma}_{cs}Z_b)/(e_o - k_t)$
4	I′	$1/F_i \leqslant (e_o - k_b)/(M_{min} - \bar{\sigma}_{ci}Z_t)$
	II′	$1/F_i \leqslant (e_o - k_t)/(M_{min} + \bar{\sigma}_{ti}Z_b)$
	III′	$1/(F \text{ or } \eta F_i) \geqslant (e_o - k_b)/(M_{max} - \bar{\sigma}_{ts}Z_t)$
	IV′	$1/(F \text{ or } \eta F_i) \geqslant (e_o - k_t)/(M_{max} + \bar{\sigma}_{cs}Z_b)$
All	V	$\lvert e_o \rvert \leqslant \lvert (e_o)_{mp} \rvert$ = maximum practical eccentricity

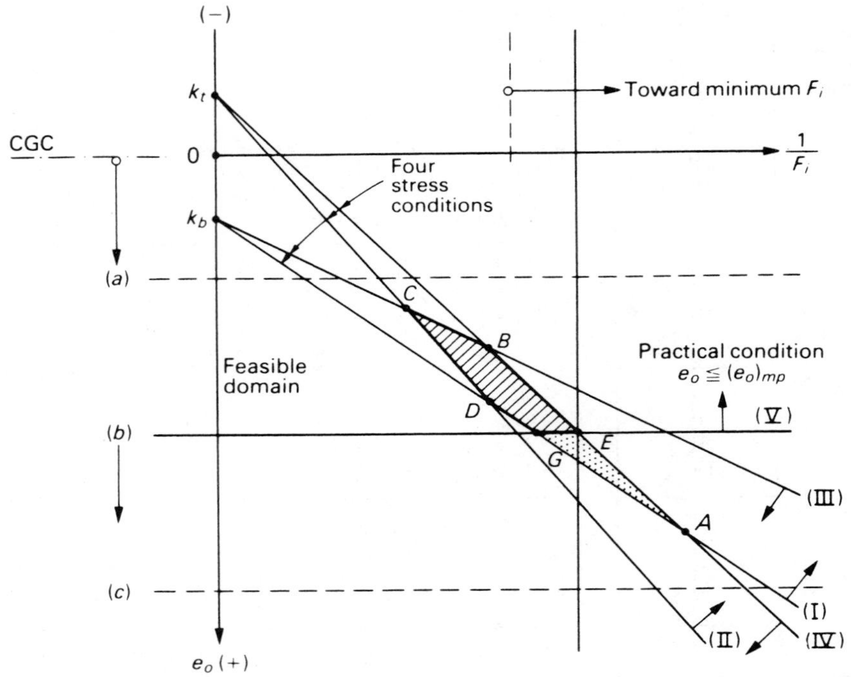

Fig. 8-9

curves corresponding to the four equations at equality; each curve will separate the plane into two parts, one where the inequality is satisfied and the other where it is not. If e_o is plotted versus F_i, the curves will be hyperbolas. However, if e_o is plotted versus $1/F_i$, then straight lines are obtained, and the geometric representation is much simplified. For this it is better to use the second way of writing the equations in Table 8-5, as they are written in the form $e_o = a(1/F_i) + b$ where b is the intercept and a the slope of the line. When plotted as shown in Fig. 8-9, the inequality equations delineate a domain of feasibility limited by a quadrilateral A, B, C, D. Essentially any point inside this domain has coordinates F_i and e_o that satisfy the four stress inequality conditions I to IV. The practicality condition V can also be represented at equality on the same graph by a horizontal line parallel to the $1/F_i$ axis. If this line intersects the quadragon A, B, C, D, such as case (b) of Fig. 8-9, then a new reduced feasibility domain is defined such as $EBCDG$. Any point inside this new domain would have satisfactory and practically feasible values of F_i and e_o. If the line representing condition V does not intersect the domain A, B, C, D, such as in cases (a) or (c) of Fig. 8-9, then either there is no practical solution (case a) or any point of the domain A, B, C, D represents a practically feasible solution (case c). In case (a) a new concrete cross section must be used leading to higher section moduli. In case (c), as any point of the domain A, B, C, D is feasible, one must select the one leading to the smallest prestressing force, i.e., point A, the intersection of lines representing conditions I and IV. The corresponding analytical solution is obtained by solving two equations, I and IV, to determine two unknowns, F_i and e_o. In case (b) the smallest value for the prestressing force is obtained at point E, the intersection of IV and V. The corresponding analytical solution is obtained by solving IV for F_i, after replacing e_o by $(e_o)_{mp} = y_b - (d_c)_{\min}$.

Note that the geometric interpretation of the stress conditions gives a very clear picture of what is or what should be done about a particular problem. For example, in a given analysis or design problem, one can plot the feasibility domain and check whether the proposed values of F and e_o are represented by a point inside the domain. If it is inside, there is no need to check the stresses; if it is not, one can spot right away the condition or conditions that are not satisfied and devise a corrective action. Other types of practical questions that can be best answered by using the above geometric representation are the following: (1) Given an eccentricity e_o, what are the maximum and minimum feasible values of prestressing force? (2) Given a prestressing force, what is the range of feasible eccentricities at a given section?

A typical example is treated next, in which the geometric representation is used in both an investigation problem and a design problem where the concrete section is given.

8.5.7 Example: Analysis and Design of a Prestressed Beam

This example is also continued in Sections 8.7.5 and 8.8.5.

Consider the pretensioned simply supported member shown in Fig. 8-10 with a span length of 70 feet. It is assumed that $f'_c = 5000$ psi, $f'_{ci} = 4000$ psi, $\bar{\sigma}_{ti} = -189$ psi, $\bar{\sigma}_{ci} = 2400$ psi, $\bar{\sigma}_{ts} = -424$ psi, $\bar{\sigma}_{cs} = 2250$ psi. Normal weight concrete is used; i.e., $\gamma_c = 150$ pcf, live load = 100 psf, and superimposed dead load = 10 psf. Assume that $\eta = F/F_i = 0.83$ and $(e_o)_{mp} = y_b - 4 = 23.1$ in.

(a) Investigate flexural stresses at midspan given: $F = 229.5$ kips (corresponding to ten ½-in.-diameter strands) and $e_o = 23.1$ in.
Note: $F_i = F/\eta = 276.5$ kips.

In order to calculate the stresses the geometric properties of the section (given in Fig. 8-10) and the applied moments are needed.

$$\text{Minimum moment } M_{\min} = M_G = 0.573(70^2/8)$$
$$= 350.962 \text{ kips-ft}$$

Additional moment due to superimposed dead load and live load $\Delta M = 0.44(70^2/8) = 269.5$ kips-ft.

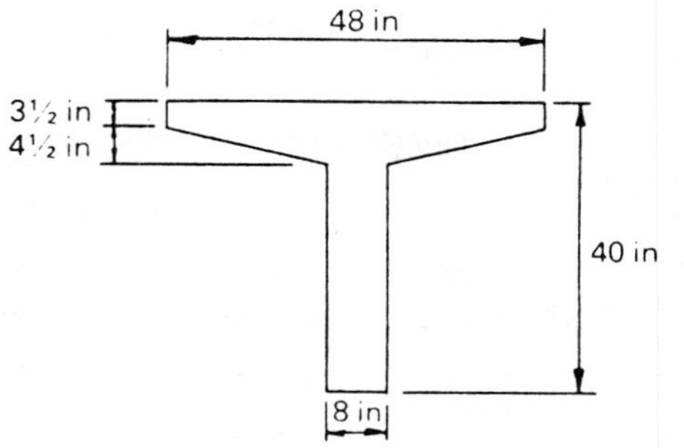

SECTION PROPERTIES

$A_c = 550$ in^2
$I = 82{,}065$ in^4
$y_t = 12.9$ in.; $y_b = 27.1$ in.
$Z_t = 6362$ in^3; $Z_b = 3028$ in^3
$k_t = -5.51$ in.; $k_b = 11.57$ in.
$w_G = 0.573$ *klf*

Fig. 8-10

Maximum moment $M_{max} = M_{min} + \Delta M = 620.462$ kips-ft

Referring to the four stress inequality equations given in Table 8-5 (way 1) and multiplying the values of moments by 12,000 in order to have units of pounds per square inch leads to:

Condition I: $$\sigma_{ti} = \frac{F_i}{A_c}\left(1 - \frac{e_o}{k_b}\right) + \frac{M_{min}}{Z_t}$$

$$\sigma_{ti} = \frac{276{,}500}{550}\left(1 - \frac{23.1}{11.57}\right) + \frac{350.962 \times 12{,}000}{6362}$$

$$\simeq 161 \text{ psi} > \bar{\sigma}_{ti} = -189 \text{ psi} \quad \text{OK}$$

The results for the other conditions are given as follows:

Condition II $\sigma_{ci} = 1219$ psi $< \bar{\sigma}_{ci} = 2400$ psi OK

Condition III $\sigma_{cs} = 754$ psi $< \bar{\sigma}_{cs} = 2250$ psi OK

Condition IV $\sigma_{ts} = -292$ psi $> \bar{\sigma}_{ts} = -424$ psi OK

Therefore the section is satisfactory with respect to flexural stresses.

(b) Plot the feasibility domain for the above problem and check geometrically if allowable stresses are satisfied.

The equations at equality given in Table 8-5 (way 2) are used to plot linear relationships of e_o versus $1/F_i$ on Fig. 8-11. They are reduced to the following form:

Condition I $$e_o = 11.57 + 5.410\left(\frac{10^6}{F_i}\right)$$

Condition II $$e_o = -5.51 + 11.4787\left(\frac{10^6}{F_i}\right)$$

Condition III $$e_o = 11.57 - 8.276\left(\frac{10^6}{F_i}\right)$$

Condition IV $$e_o = -5.51 + 7.424\left(\frac{10^6}{F_i}\right)$$

where e_o is in inches and F_i is in pounds. Also equation V showing $(e_o)_{mp} = 23.1$ in. is plotted in Fig. 8-11. The five lines delineate a feasibility domain $ABCD$.

Let us check whether the given values of F_i and e_o are represented by a point that belongs to the feasible region:

$$\frac{1}{F_i} = \frac{1}{276{,}500} \simeq 3.6 \times 10^{-6}$$

The representative point is shown in Fig. 8-11 as point A'. As it is on line AD, it belongs to the feasible region and therefore all allowable stresses are satisfied. Note that all stresses would still be satisfied if the eccentricity is reduced to approximately 21 in. for the same force. This is shown as point A'' on line AB and allows the designer to accept a reasonable tolerance on the value of e_o actually achieved during the construction phase.

(c) Assuming the prestressing force is not given, determine its value.

This is essentially a typical design problem where the concrete cross section is given. It can be solved directly analytically or from the graphical representation of the feasibility domain. Anyway, the graphical representation helps in the analytic solution. It dictates the choice of point A of Fig. 8-11 as the solution that minimizes the prestressing force. Point A corresponds to the intersection of the line representing $(e_o)_{mp}$ with that representing stress condition IV. The corresponding value of F is obtained by replacing e_o by $(e_o)_{mp}$ in eq. IV (way 3) of Table 8-5; that is:

$$F = \frac{M_{max} + \bar{\sigma}_{ts} Z_b}{(e_o)_{mp} - k_t} = \frac{620.462 \times 12{,}000 - 424 \times 3028}{23.1 + 5.51}$$

$$= 215{,}368 \text{ lb}$$

and:

$$F_i = \frac{F}{0.83} = 259{,}479 \text{ lb} \simeq 259.5 \text{ kips}$$

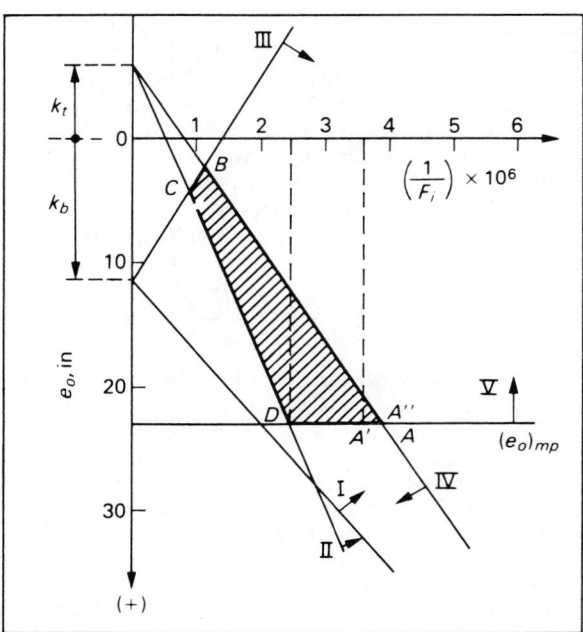

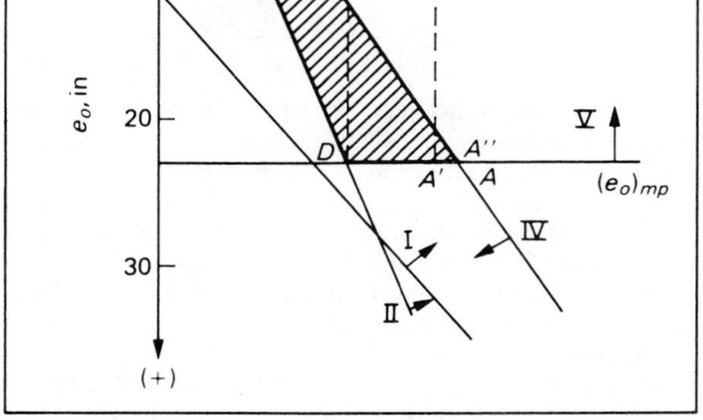

Fig. 8-11

Graphically the coordinates of A can be read on Fig. 8-11 as $e_o = 23.1$ in. and $1/F_i = 3.9 \times 10^{-6}$, which leads to $F_i \simeq 257{,}000$ lb = 257 kips. It can be seen that the graphical solution gives essentially the same answer as the analytical one. Note that the practical value of the prestressing force to use in the design should correspond to an integer number of tendons. In this case, exactly 9.38 strands, each with a final force of 22.95 kips, would be required. The number is rounded off to 10. The resulting higher prestressing force allows for an acceptable tolerance on the value of e_o.

(d) If the beam is to be used with different values of live loads, what is the maximum value of live load it can sustain?

Referring to the stress conditions, it can be observed that conditions I and II do not change, and therefore lines I and II on Fig. 8-11 are fixed. Increasing the value of the live load will increase the value of M_{max} and thus will change the slopes of lines III and IV so as to reduce the size of the feasible domain. Consequently, point A of the feasible domain will move in the direction of AD, and line BA tends to rotate toward CD. The maximum value of live load in this case is the one that will make lines II and IV have the same slopes. Therefore:

$$\frac{M_{max} + \bar{\sigma}_{ts} Z_b}{\eta} = 11.4787 \times 10^6$$

which leads to $M_{max} = 10{,}811{,}193$ lb-in. = 900.93 kips-ft. Subtracting from M_{max} the values of moments due to dead load and superimposed dead load, leads to the live load moment, from which the live load can be determined as 214.5 psf. The representative point on Fig. 8-11 is D, which shows the following coordinates: $e_o = 23.1$ in. and $10^6/F_i \simeq 2.5$, i.e., $F_i \simeq 400{,}000$ lb = 400 kips. The reader is encouraged to check numerically in this case that the two allowable stresses $\bar{\sigma}_{ci}$ and $\bar{\sigma}_{ts}$ are attained while the two others are satisfied, as the geometric representation indicates. Note that such a design may have to be revised if the assumed value of e_o cannot be practically achieved.

8.5.8 Some Preliminary Design Hints

If the concrete cross section is properly selected for the problem at hand (such as from a handbook of standard beams), it is very likely that the maximum practical eccentricity (Fig. 8-9, case b) will control the design. It becomes clear that the prestressing force can then be determined from point E, which represents the intersection of stress condition IV with the line corresponding to $(e_o)_{mp}$. Referring to Table 8-5, the prestressing force can be determined from eq. IV at equality in which e_o is replaced by $(e_o)_{mp}$, that is:

$$F = \eta F_i = \frac{M_{max} + \bar{\sigma}_{ts} Z_b}{(e_o)_{mp} - k_t} \tag{8-5}$$

The designer can then check whether the other three stress inequality conditions are satisfied for the values of F and e_o. A similar approach was used in the example Section 8.5.7c. If the other three equations are not satisfied, it is better then to build the feasibility domain as in Fig. 8-9 to see what needs to be done.

A very approximate method can often be used in cost estimate studies to estimate globally the value of the prestressing force. It is derived from observing that in order to have a feasible limit zone the upper eccentricity limit e_{ou} must be less than or equal to the lower eccentricity limit e_{ol}; thus:

$$k_t' + \frac{M_{max}}{F} \leqslant k_b' + \frac{M_{min}}{F_i} \tag{8-6}$$

Replacing F_i by F/η and solving for F leads to:

$$F \geqslant \frac{M_{max} - \eta M_{min}}{k_b' - k_t'} \tag{8-7}$$

which indicates that the minimum value of F is given by

$$(F)_{min} = \frac{M_{max} - \eta M_{min}}{k_b' - k_t'} = \frac{\Delta M + (1-\eta) M_{min}}{k_b' - k_t'} \tag{8-8}$$

Because (1) η is generally around 0.80, (2) the dead load moment is often of the same order as the live load moment, and (3) often no tension is allowed in the section, the following approximations can be made:

$$\begin{cases} M_{max} - \eta M_{min} \simeq 1.2\,\Delta M & (8\text{-}9) \\ k_b' - k_t' \simeq k_b - k_t = \gamma h & (8\text{-}10) \end{cases}$$

ΔM is independent of the beam cross section, and γ, the geometric efficiency of the section, can be estimated within a reasonable range depending on the section shape. Thus eq. (8-8) becomes:

$$(F)_{min} \simeq \frac{1.2\,\Delta M}{\gamma h} \tag{8-11}$$

Applied to example Section 8.5.7a, eq. 8-11 gives:

$$(F)_{min} \simeq \frac{1.2 \times 269.5 \times 12}{0.427 \times 40} = 227.2 \text{ kips}$$

which is very close to the answer obtained in the exact analysis. If we did not know the value of γ, we could estimate $\gamma \simeq 0.45$ for a T section and obtain $F \simeq 215.6$ kips.

8.5.9 Cracking Moment

Although this chapter deals essentially with uncracked prestressed concrete beams, the load or moment at which cracking occurs is needed in the design. The cracking moment is the moment for which the tensile stress on the extreme fiber of the concrete section reaches a value equal to the modulus of rupture of the concrete. For the bottom fiber of a prestressed concrete section, the cracking moment can be determined from satisfying the following equation:

$$\frac{F}{A_c}\left(1 - \frac{e_o}{k_t}\right) - \frac{M_{cr}}{Z_b} = f_r \tag{8-12}$$

where the first term of the left-hand side of the equation is the stress due to the prestressing force, and the second term is the stress due to the cracking moment just before cracking. Solving for M_{cr} gives:

$$M_{cr} = Z_b\left[\frac{F}{A_c}\left(1 - \frac{e_o}{k_t}\right) - f_r\right] = F\left(e_o + \frac{Z_b}{A_c}\right) - f_r Z_b$$
$$= F(e_o - k_t) - f_r Z_b \tag{8-13}$$

Note that the values of k_t and the modulus of rupture are negative.

Let us assume $f_r = -7.5\sqrt{f_c'}$; the cracking moment of the beam treated in Example 8.5.7a is given by:

$M_{cr} = 229{,}500\,(23.1 + 5.51) + 530 \times 3028 = 8{,}170{,}835$ lb-in.

or:

$M_{cr} = 680.9$ k-ft

8.6 FLEXURE: STRENGTH ANALYSIS AND DESIGN

8.6.1 Load-Deflection Response

Similarly to reinforced concrete, prestressed concrete offers a great versatility in its flexural behavior. Depending on the values of the design variables and parameters, a prestressed concrete member can be made either to exhibit a great ductility after cracking and before failure, or to fail altogether in a sudden manner. It can be designed to carry a relatively small or large load before failure; thus in order to achieve a good design, it is essential to understand the causal effects of important variables on the behavior at ultimate.

The overall behavior of a simply supported prestressed concrete beam subjected to a monotonically increasing load can be well described by its load-deflection curve. Such a typical curve is shown in Fig. 8-12 for an underreinforced beam with bonded tendons. Several points are marked on the curve and correspond to a particular state of behavior. Points 1 and 2 correspond to the theoretically predicted camber of the beam, assumed weightless, when the initial or the effective prestresses are applied respectively. However, when the prestress is applied, self-weight acts automatically. Point 3 represents the camber due to the combined effects of self-weight and the effective prestressing force assuming all prestress losses have taken place. Typical stress diagrams along the cross section of maximum moment corresponding to points 3 to 9 are also shown in Fig. 8-12. If additional load beyond self-weight is applied, several points of interest can be identified until failure: point 4 represents the point of zero deflection and corresponds to a uniform state of stress in the section (also called balanced state); point 5 represents decompression or zero stress at the bottom fiber; if cracking has already occurred due to prior loading, or if the tensile strength of the concrete is assumed nil, point 5 would represent the boundary between cracked and uncracked section behavior and thus would take the place of point 6; point 6 represents the onset of cracking in the concrete under first loading; beyond point 6 the prestressed concrete section behaves similarly to a cracked reinforced concrete section subjected to combined bending and compression; if the applied load on the beam keeps increasing, the stresses in the steel and the concrete extreme compressive fiber will continue to increase until either material reaches its nonelastic characteristics; this limit is represented by point 7 of Fig. 8-12. For increasing loads, the steel will first reach its yielding strength (bonded tendons) represented by point 8, and finally the maximum capacity of the beam is attained at point 9. Note that point 9 represents the point of maximum load, which is described as the ultimate load; generally beyond this point the beam still provides some resistance to increasing deflections but at values of loads less than the ultimate load.

8.6.2 Flexural Types of Failures and Reinforcement Limits

The typical load-deflection response of a prestressed concrete beam as described in Fig. 8-12 is a desirable type of

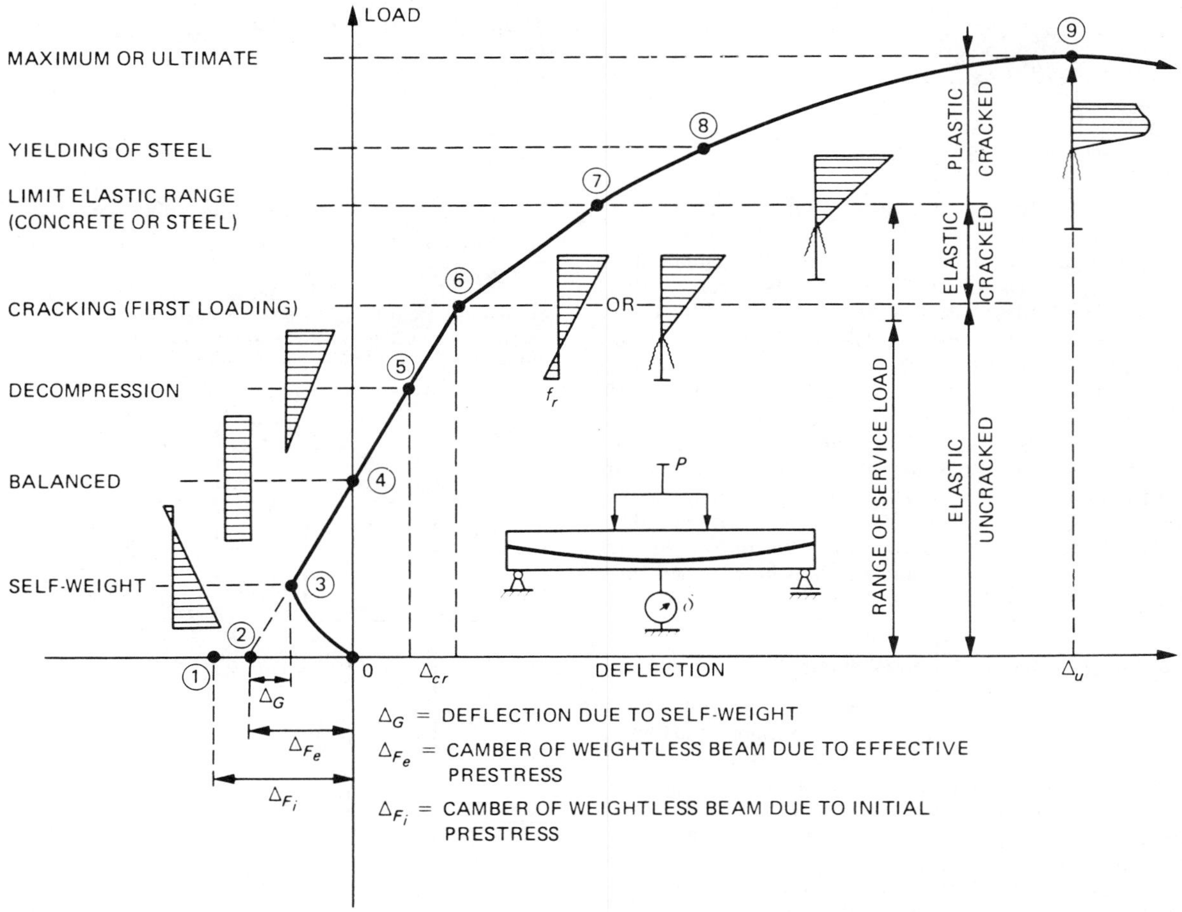

Fig. 8-12 Typical load-deflection curve of a prestressed beam (underreinforced; bonded tendons; first loading).

behavior; design limitations in various codes tend in general to ensure such behavior. Other types of behavior can, however, be observed. As the load is progressively increased on a simply supported prestressed concrete beam, the following types of flexural failures might occur, depending on the amount of steel reinforcement provided:

1. Fracture of the steel immediately after concrete cracking and thus sudden failure (insufficiently reinforced section)
2. Crushing of the concrete compressive zone, preceded by yielding and plastic extension of the steel (normally or underreinforced section)
3. Crushing of the concrete compressive zone before yielding of the steel (overreinforced section)

In order to avoid sudden failure immediately after cracking, the *ACI Code* recommends a minimum value of the reinforcement ratio for reinforced concrete members, whereas, for prestressed concrete, such a limit is indirectly set by requiring that the ultimate moment be 20% higher than the cracking moment.

In order to distinguish between underreinforced and overreinforced sections, the *ACI Code* uses a parameter called here the combined reinforced index q, or simply the reinforcing index, and defined as follows:

$$q = \omega_p + \omega - \omega' \tag{8-14}$$

where the partial reinforcing indexes ω_p, ω, and ω' are given by:

$$\omega_p = \frac{A_{ps} f_{ps}}{bdf'_c} = \rho_p \frac{f_{ps}}{f'_c} \tag{8-15}$$

$$\omega = \frac{A_s f_y}{bdf'_c} = \rho \frac{f_y}{f'_c} \tag{8-16}$$

$$\omega' = \frac{A'_s |f'_y|}{bdf'_c} = \rho' \frac{|f'_y|}{f'_c} \tag{8-17}$$

The absolute value of f'_y is used as it is a compressive stress in the steel and is negative. A_s and A'_s are the areas of the nonprestressed tensile and compressive reinforcement and d is the distance to the centroid of the tensile force. Note that for a purely prestressed section $d = d_p$, while for a purely reinforced section $d = d_s$. The 1977 *ACI Code* defines underreinforced rectangular sections as sections for which the value of q is less than or equal to $q_{max} = 0.30$. If q is larger than q_{max}, it is then assumed that the section is overreinforced and the ultimate moment capacity is limited by that of the concrete compressive block. For T section behavior the limiting value of the reinforcing index q should correspond to $q_w = 0.30$ where q_w is the reinforcing index of the web. A detailed evaluation of the limitations of the reinforcement for prestressed and partially prestressed sections is given in Refs. 8-15 and 8-32.

8.6.3 Analysis of the Section at Ultimate

The purpose of the analysis at ultimate or maximum load is generally to determine the nominal moment resistance (moment at ultimate behavior) of the section, assuming that cross-sectional dimensions, material properties, and amounts of reinforcement are given.

Whatever the assumptions and the particular approaches taken by different codes, two equations of statics pertaining to the equilibrium (force and moment) of the section at ultimate can be written. Assumptions used in the *ACI Code* for reinforced concrete are valid for prestressed concrete. Following is a derivation of the nominal moment resistance of fully prestressed concrete rectangular and T sections. The case where nonprestressed conventional reinforcement is used in addition to the prestressed reinforcement is covered in detail in Ref. 8-15.

1. Rectangular Section (or Rectangular Section Behavior)

Referring to Fig. 8-13, the force equilibrium equation can be set by writing that the compressive force in the concrete compression block is equal to the total tensile force in the steel, thus:

$$0.85 f'_c ba = A_{ps} f_{ps} \tag{8-18}$$

The moment equation states that the nominal moment resistance of the section is equal to the internal couple or moment of the above equal and opposite forces, thus:

$$M_n = 0.85 f'_c ba \left(d_p - \frac{a}{2}\right) \tag{8-19}$$

where d_p is the distance from the extreme compressive fiber of concrete to the centroid of the prestressing force.

In the above two equations, the following three variables are considered unknowns: a, f_{ps}, and M_n. In order to determine their values, a third equation is needed. Such an equation is given, for instance, in the *ACI Code*, which recommends an expression to estimate the value of f_{ps}, the stress in the prestressing steel at ultimate capacity of the section. Once f_{ps} is known, a can be computed from eq. (8-18), and M_n from eq. (8-19).

In lieu of a more accurate analysis based on strain compatibility, the 1977 *ACI Code* recommends the use of the following approximate values of f_{ps} if f_{pe}, the effective prestress, is not less than $0.5 f_{pu}$:

1. For members with bonded prestressing tendons:

$$f_{ps} = f_{pu}\left(1 - 0.5\rho_p \frac{f_{pu}}{f'_c}\right) \tag{8-20†}$$

where f_{pu} is the ultimate strength and ρ_p the reinforcement ratio of the prestressing steel given by A_{ps}/bd_p.

†Equations (8-20 and 8-21) have been modified in the 1983 ACI Code. The new equations would introduce only slight changes in the numerical results of examples 2 and 4.

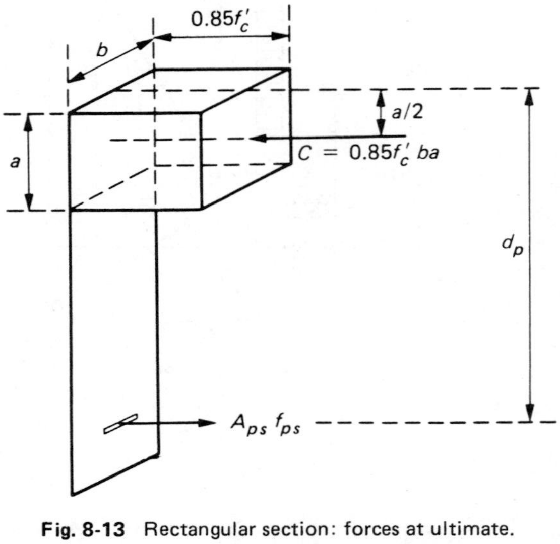

Fig. 8-13 Rectangular section: forces at ultimate.

2. For members with unbonded prestressing tendons:

$$\begin{cases} f_{ps} = f_{pe} + 10{,}000 + \dfrac{f'_c}{100\rho_p} \text{ in psi} \\ f_{ps} = f_{pe} + 69 + \dfrac{f'_c}{100\rho_p} \text{ in MPa} \end{cases} \qquad (8\text{-}21)^\dagger$$

but f_{ps} in eq. (8-21) shall not be taken greater than f_{py} or (f_{pe} + 60,000) psi or (f_{pe} + 414) MPa. Note that eq. (8-20) is valid in all systems of units. The reader is referred to the 1983 *ACI Code* where additional prediction equations for f_{ps} are given for slabs with unbonded tendons and for prestressed members containing a substantial amount of nonprestressed reinforcement.

Once the nominal moment resistance at ultimate M_n is determined, the design moment resistance acceptable by the *ACI Code* is obtained by multiplying M_n by the strength reduction factor ϕ, thus:

$$\text{Design moment resistance} = \phi M_n \qquad (8\text{-}22)$$

ϕM_n will be considered as the nominal moment capacity for the purpose of design and is often made equal to the strength design moment required for the given loading and termed M_u.

†Equations (8-20 and 8-21) have been modified in the 1983 ACI Code. The new equations would introduce only slight changes in the numerical results of examples 2 and 4.

It is worth noting that eq. (8-19), which leads to the value of the nominal moment resistance, is not unique in its form. It could be written in different ways; for instance, if the force in the steel is used instead of using the force in the concrete, eq. (8-19) becomes:

$$M_n = A_{ps} f_{ps} \left(d_p - \frac{a}{2}\right) \qquad (8\text{-}23)$$

Another way to write M_n is as a function of the reinforcing index q, that is:

$$M_n = f'_c bd^2 q(1 - 0.59q) \qquad (8\text{-}24)$$

This form is convenient because it shows only one unknown, q, instead of two, namely a and f_{ps}. However, it may be misleading because it suggests a direct relation between M_n and f'_c, whereas in reality M_n is generally only slightly sensitive to f'_c because f'_c is in the denominator of q.

To illustrate the use of the above equations, the following example of a rectangular section is developed.

Note also that a general flow chart illustrating the computations of nominal moments according to the steps described in this section and the *ACI Code* is given in Fig. 8-14. It applies to rectangular and T sections, reinforced,

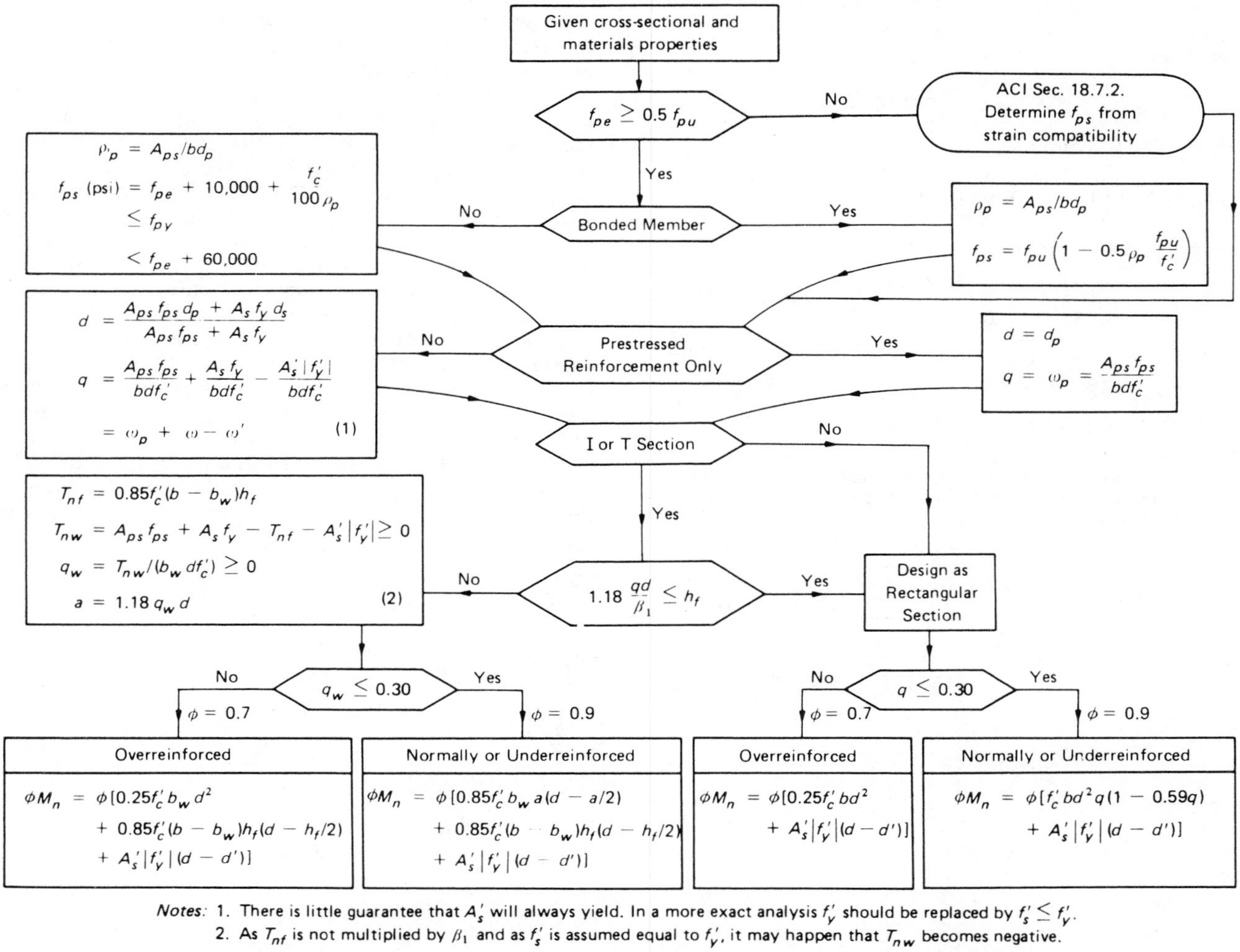

Fig. 8-14 Flow chart to determine the nominal moment resistance of prestressed and partially prestressed sections in accordance with the 1977 *ACI Code*.[8-15]

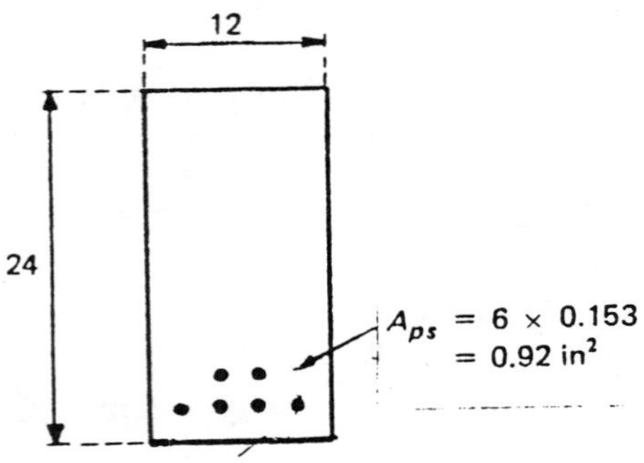

Fig. 8-15

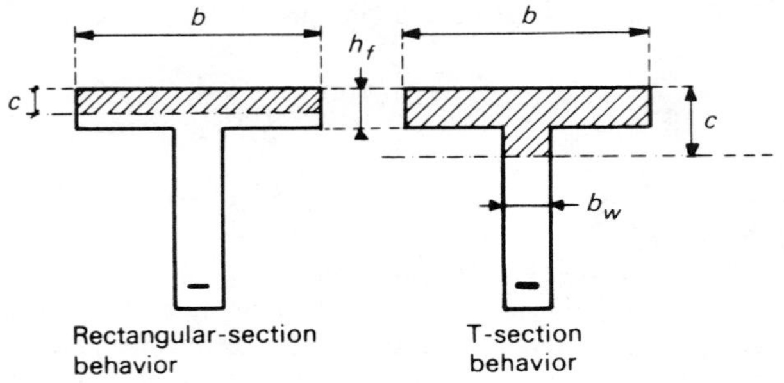

Fig. 8-16

prestressed, and partially prestressed with and without compressive steel, and allows for bonded or unbonded tendons.

2. Example: Analysis at Ultimate of a Rectangular Section†

(a) Determine the nominal moment resistance of a prestressed concrete rectangular section with dimensions given in Fig. 8-15 for which $f'_c = 5000$ psi, $f_{pu} = 270$ ksi, $f_{pe} = 0.55 f_{pu} = 148.5$ ksi, $f_y = 60$ ksi, $A_{ps} = 0.92$ in.^{2} or six $\frac{1}{2}$-in.-diameter prestressing strands, $d_p = 21.33$ in. The prestressing tendons are assumed bonded. The following steps are followed according to the procedure described above and in the flow chart (Fig. 8-14).

Compute:

$$\rho_p = \frac{A_{ps}}{bd_p} = 0.00359$$

From eq. (8-20) compute:

$$f_{ps} = 270\left(1 - 0.5 \times 0.00359 \times \frac{270}{5}\right) = 243.8 \text{ ksi}$$

From eq. (8-18) compute:

$$a = \frac{A_{ps} f_{ps}}{0.85 f'_c b} = \frac{0.92 \times 243.8}{0.85 \times 5 \times 12} = 4.4 \text{ in.}$$

From eq. (8-19) compute the nominal moment resistance:

$$M_n = 0.85 \times 5 \times 12 \times 4.4\,(21.33 - 4.4/2)$$
$$= 4293 \text{ kps-in.} \simeq 358 \text{ kips-ft}$$

†Equations (8-20 and 8-21) have been modified in the 1983 ACI Code. The new equations would introduce only slight changes in the numerical results of examples 2 and 4.

The moment resistance acceptable by the code is given by eq. (8-22):

$$\phi M_n = 0.9 M_n = 322 \text{ kips-ft}$$

(b) The reader may want to check that if for the same prestressed section, unbonded tendons were used, everything else being the same, the following results would have been obtained: $f_{ps} = 172.41$ ksi; $a = 3.11$ in.; $M_n = 261.4$ kips-ft; $\phi M_n = 235.2$ kips-ft.

3. T Section Behavior

According to the assumption that the tensile strength of concrete is neglected in the analysis, the portion of the concrete section that falls below the neutral axis is considered to offer no resistance and thus, in effect, is ignored in the computation of nominal moment. It could be of any shape, one of which is the shape corresponding to that of a rectangular section. In approaching the analysis particular to T sections, it is important to realize that if the neutral axis falls in the flange (Fig. 8-16), the section is treated exactly similar to a rectangular section of same width b, and the corresponding equations developed above apply without any modification. Thus, depending on the various variables and parameters, a T section does not necessarily imply T-section behavior at ultimate. In fact, as the depth of the neutral axis c at ultimate moment capacity of reinforced and prestressed concrete members is generally small in comparison to the depth of the member, T sections behave often as rectangular sections. This is particularly true for small values of the reinforcement ratio or the reinforcing index. If, however, the neutral axis falls outside the flange (i.e., in the web), a special analysis must be undertaken.

It is clear that the first step would be to determine the depth of the neutral axis c. For this, the T section is first

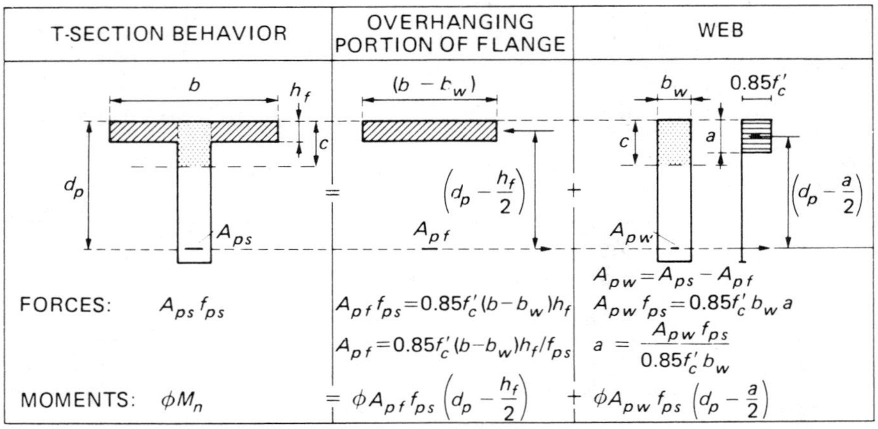

Fig. 8-17 Fully prestressed T section: *ACI Code* approach.

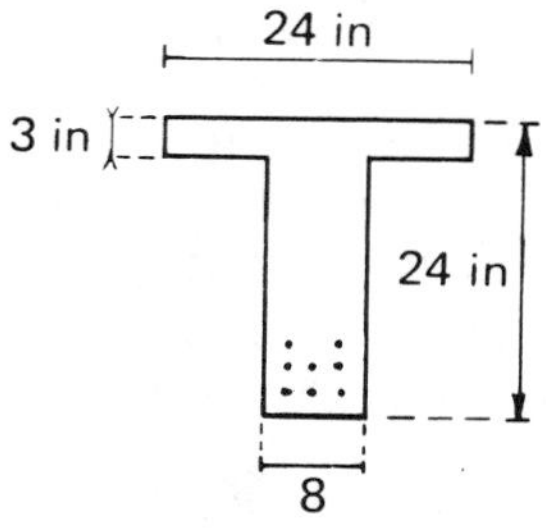

Fig. 8-18

assumed to behave as a rectangular section for which c at ultimate capacity is computed from $c = a/\beta_1$; c is then compared to the flange depth h_f of the T section, and if c is larger than h_f, the section is considered to behave as a T section, and the following approach is proposed according to the *ACI Code:* the T section is divided into two fictitious parts, one corresponding to the overhanging portion of the flange and the other to the web alone (Fig. 8-17). The analysis is then pursued wherein the two parts are treated separately as two rectangular sections and their nominal capacities are summed up to give the nominal capacity of the T section. The usual code approach is first to compute the areas of steel reinforcement needed to develop the ultimate strength of the overhanging portion of the flange and the web and then to determine corresponding forces and moments. This is illustrated in Fig. 8-17 for a fully prestressed section where most relevant equations are also given.

The following example illustrates the steps described above. An expanded computational chart for the calculation of nominal moments is also given in Fig. 8-14 and may be used as an alternative.

4. Example: Analysis at Ultimate of a T Section†

(a) Determine the nominal moment capacity of a prestressed concrete T section with dimensions described in Fig. 8-18 for which $A_{ps} = 1.377$ in.^{2} and $d_p = 20$ in. Material properties are exactly the same as for the first Example in Section 8.6.3. Assume bonded tendons are used.

Compute:

$$\rho_p = \frac{A_{ps}}{bd_p} = 0.00287$$

Estimate:

$$f_{ps} = 270\left(1 - 0.5 \times 0.00287 \times \frac{270}{5}\right) = 249.1 \text{ ksi}$$

First assume rectangular-section behavior and determine neutral axis c; using eq. (8-18):

$$a = \frac{1.377 \times 249.1}{0.85 \times 5 \times 24} = 3.36 \text{ in.}$$

and:

$$c = a/\beta_1 = \frac{3.36}{0.80} = 4.2 \text{ in.}$$

As c is larger than $h_f = 3$ in., the section should be treated as a T section.

Using the equations of Fig. 8-17:

$$A_{pf} = 0.85 \times 5 \times (24 - 8) \times 3/249.1 = 0.819 \text{ in.}^2$$

$$A_{pw} = 1.377 - 0.819 = 0.558 \text{ in.}^2$$

†Equations (8-20 and 8-21) have been modified in the 1983 ACI Code. The new equations would introduce only slight changes in the numerical results of examples 2 and 4.

$$a = \frac{0.558 \times 249.1}{0.85 \times 5 \times 8} = 4.09 \text{ in.}$$

$$\phi M_n = 0.9 \times 0.819 \times 249.1(20 - 3/2)$$
$$+ 0.9 \times 0.558 \times 249.1(20 - 4.09/2)$$

$$\phi M_n = 5643 \text{ kips-in.} = 470.25 \text{ kips-ft}$$

The reader may want to check that if unbonded tendons are used, everything else being the same, the following results would be obtained: $f_{ps} = 175.92$ ksi; $a = 2.36$ in.; $c = 2.96$ in.; rectangular-section behavior; $\phi M_n = 4103$ kips-in. = 341.92 kips-ft.

8.7 DESIGN FOR SHEAR

8.7.1 Prestressed Versus Reinforced Concrete in Shear

It is often said that, with respect to shear, prestressed concrete offers two major advantages over reinforced concrete, namely:

1. For the same external loading, everything else being equal, the shear force in prestressed concrete is smaller than that in reinforced concrete. This is illustrated in Fig. 8-19 where the sign convention for shear adopted in this test is also given. It is observed that at any section x, the difference in shear between a prestressed concrete beam and an otherwise identical reinforced concrete beam is essentially due to the vertical component of the prestressing force $V_F = F \sin \alpha$. V_F acts in a direction opposite to the external loads, thus re-

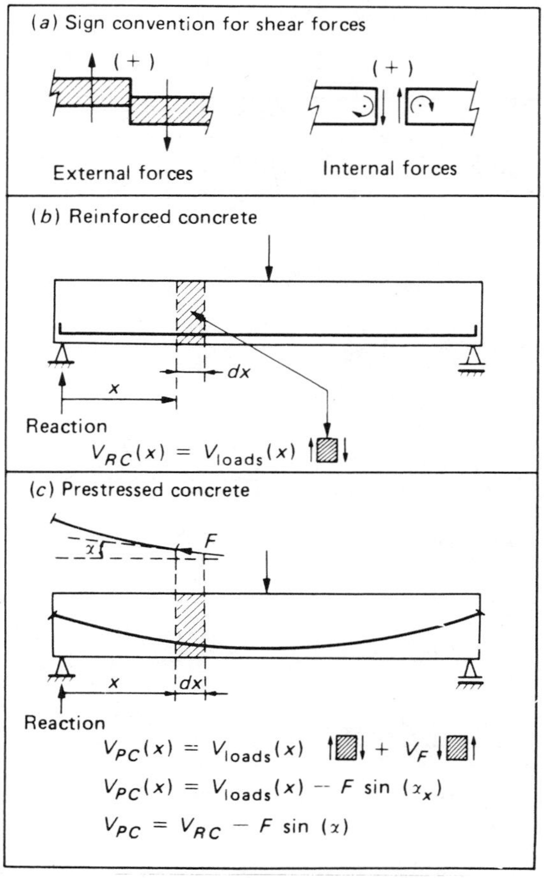

Fig. 8-19 Shear in reinforced and prestressed concrete.

ducing the shear force in prestressed concrete, which at any section x is given by:

$$V(x) = V_{\text{loads}}(x) - F \sin(\alpha_x) \qquad (8\text{-}25)$$

2. Owing to the compression induced by prestressing, the diagonal tension is smaller in prestressed concrete than in reinforced concrete. Furthermore, its angle of inclination with respect to the beam's axis is reduced. This implies that if cracking occurs and if, for safe design, the inclined crack is assumed to cross at least one stirrup, the stirrups' spacing in prestressed concrete would be larger and their required area smaller than in reinforced concrete. Thus, a more economical solution (for shear) is obtained.

8.7.2 Shear Cracking Behavior

Structural cracking occurs in concrete when its tensile strength is exceeded. As any external loading generally leads to combined effects such as flexure and shear, it is likely that more than one type of cracking will be critical, depending on the variables at hand.

An extensive number of investigations[8-33–8-35] have shown or confirmed that two types of shear-related cracks can develop in reinforced and prestressed concrete beams: flexure-shear cracks and web-shear cracks (Fig. 8-20a). The manner in which these cracks develop and grow strongly depends on the relative magnitude of shearing and flexural stresses.

Flexure-shear cracking is due to a combined effect of flexure and shear. The corresponding cracks start as flexural cracks (normal to the beam's axis). Then, owing to the increased effect of diagonal tension at the tip of the crack, they deviate and propagate at an inclined direction corresponding essentially to the inclination of the diagonal tension plane. Typical flexure-shear cracks are shown in Fig. 8-20(a). Flexure-shear cracking can lead to several types of failures, schematically illustrated in Fig. 8-20(b). Very slender beams generally fail in flexure either by their tensile reinforcement or by the concrete compressive zone.

However, in beams with smaller shear span-to-depth ratio, failure may occur by flexure-shear cracking before the flexural capacity is developed. In moderately slender beams one of the cracks may continue to propagate until it becomes unstable, reaching throughout the depth of the beam and leading to what is described as diagonal tension failure. In relatively deep beams a secondary crack triggered by a flexure-shear crack may propagate horizontally along the longitudinal reinforcement, leading to a loss of bond followed by a loss of anchorage near the support and subsequent failure described as shear tension failure. Alternatively, the concrete at the upper end of the crack may fail by crushing in what is described as shear compression failure.

Web-shear cracking (Fig. 8-20a) occurs when the magnitude of principal tension is relatively high in comparison to flexural stresses. It is characteristic of beams with narrow webs, such as I beams, where cracking due to diagonal tension develops before flexural cracking. Web-shear cracking may lead to the same types of failures described for flexure-shear cracking, namely, diagonal tension, shear tension, and shear compression failures. In addition (Fig. 8-20b), crushing under diagonal compression may occur within the web, leading to web-crushing failure.

It is preferable to avoid shear failure, as it is substantially more brittle than flexural failure. To supplement the shear resistance of concrete members and to ensure flexural failure prior to shear failure, shear reinforcement (also called web or transverse reinforcement) in the form of stirrups is generally provided.

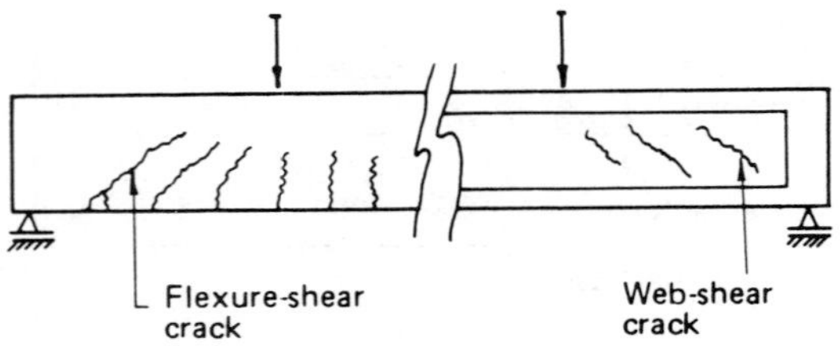

(*a*) Types of shear cracking

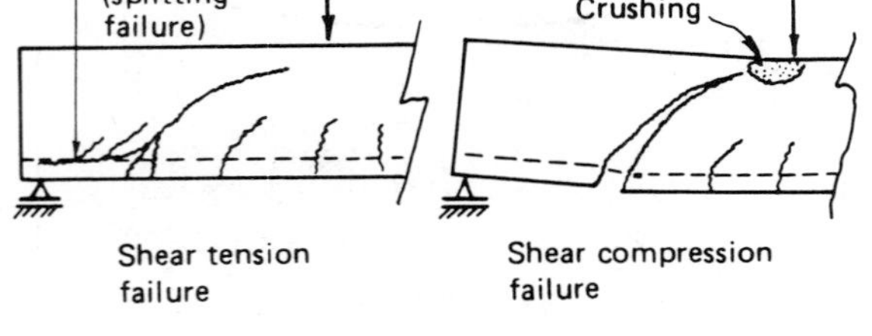

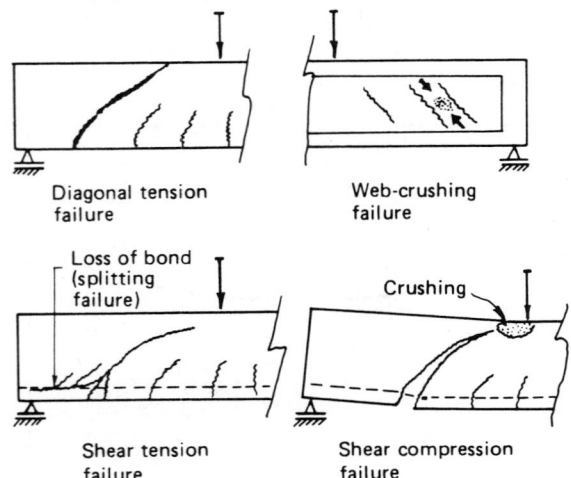

(*b*) Major types of shear failures

Fig. 8-20 Shear cracking and shear failures.

8.7.3 ACI Code Design Criteria for Shear

1. Basic Approach

The ACI design approach is based on ultimate strength requirements (factored loads). A design shear strength at each section of a member is required, and it is achieved by the combined contributions of concrete and shear reinforcement. The concrete is assumed to provide a prescribed shear resistance; the shear reinforcement (stirrups) provides the excess resistance, if any, needed to satisfy the required strength. In the 1971 edition of the code, shear-related equations were given in stress units. In the 1983 edition, they are given in force units where the force is obtained by multiplying the shear stress by $b_w d_p$. Although the approaches used and the results obtained are identical, the author prefers the use of stresses, as they are independent of the size of the member and thus can be easily correlated when different solutions are compared.

The required design shear strength at each section of the member should be less than or equal to its shear resistance, that is:

$$v_u \leqslant \phi v_n \qquad (8\text{-}26)$$

where:

v_u = design shear stress resulting from factored loads
v_n = nominal shear strength of the section
ϕ = strength reduction factor (equal to 0.85 for shear)

Equation (8-26) can be conveniently rewritten as:

$$\frac{v_u}{\phi} \leqslant v_n \qquad (8\text{-}27)$$

The design shear stress v_u is obtained from:

$$v_u = \frac{V_u}{b_w d_p} \tag{8-28}$$

where:

V_u = design shear force at section considered due to factored loads
b_w = width of rectangular section or web width of flanged section
d_p = d_p or $0.8h$, whichever is larger.

The provision to use $0.8h$ in case the actual value of d_p is smaller, is substantiated by test data. It is valid because the section is uncracked and the web area available to compute nominal shear stress is not a direct function of the location of the prestressing steel.

The nominal shear strength v_n is defined as the sum of the shear resistance of the concrete and the shear contribution of the web reinforcement at the section considered. That is:

$$v_n = v_c + v_s \tag{8-29}$$

where:

v_c = nominal shear strength provided by the concrete
v_s = nominal shear strength provided by the shear reinforcement

The values of v_c and v_s are defined in the next sections.

2. Shear Strength Provided by Concrete

According to the *ACI Code*, the value of v_c is to be taken as the smaller of v_{ci}, the resistance to flexure-shear cracking, and v_{cw}, the resistance to web-shear cracking. The equations given by the code for v_{ci} and v_{cw} are simplified approximations of equations derived either theoretically from analyzing the section (v_{cw}) or from semirational analysis supported by experimental evidence (v_{ci}).

The flexure-shear cracking stress v_{ci} is made of three components: (1) the shear stress needed to transform a flexural crack into an inclined crack, $0.6\sqrt{f'_c}$, (2) the shear stress due to the self-weight of the member, V_G/bd_p, and (3) the additional factored shear stress that initially will cause a flexural crack to occur, $\Delta M_{cr}/(\Delta M_u/\Delta V_u)/b_w d_p$. Hence, the shear resistance of the concrete, corresponding to flexure-shear cracking, can be taken as:

$$v_{ci} = 0.6\sqrt{f'_c} + \frac{V_G}{b_w d_p} + \left(\frac{\Delta V_u \times \Delta M_{cr}}{\Delta M_u}\right)\frac{1}{b_w d_p} \tag{8-30}$$

but not less than $1.7\sqrt{f'_c}$, where:

b_w = width of rectangular section or web width of flanged section
d_p = d_p or $0.80h$, whichever is larger at section considered
V_G = shear force due to self-weight of member at section considered
ΔV_u = factored shear force due to superimposed dead load plus live load at section considered under same loading as ΔM_u
ΔM_u = factored bending moment due to superimposed dead load plus live load at section considered
ΔM_{cr} = moment in excess of self-weight moment, causing flexural cracking in the precompressed tensile fiber at section considered ($=M_{cr} - M_G$)

The effect of self-weight shear V_G is considered separately for two reasons:

1. Self-weight is generally uniformly distributed, whereas live loads can have any distribution and can be concentrated loads. The ratio $\Delta M_u/\Delta V_u$, which appears in eq. (8-30), remains constant at a given section when external loads increase proportionately and are independent of load factors. Corresponding calculations are easier to handle.
2. By separating the effect of self-weight, eq. (8-30) remains applicable to composite members. For such members, V_G should include not only the weight of the precast unit but also that of the cast-in-place slab. However, note that the construction sequence in composite members (i.e., shored versus unshored) influences the value of ΔM_{cr}, as part of the load is resisted by the precast section alone and part by the composite section. Appropriate section properties should be used to determine ΔM_{cr}.

Note that the cracking moment increment ΔM_{cr} used in eq. (8-30) is to be computed assuming a modulus of rupture of $-6\sqrt{f'_c}$ instead of $-7.5\sqrt{f'_c}$. Using eq. (8-13) and assuming the beam is statically determinate, leads to:

$$\Delta M_{cr} = Z_b\left[6\sqrt{f'_c} + \frac{F}{A_c}\left(1 + \frac{e_o A_c}{Z_b}\right)\right] - M_G \tag{8-31}$$

The nominal resistance of the concrete, corresponding to web-shear cracking at the section considered, is to be taken as:

$$v_{cw} = 3.5\sqrt{f'_c} + 0.3\sigma_g + \frac{V_p}{b_w d_p} \tag{8-32}$$

where:

σ_g = compressive stress in the concrete at the centroid of the cross section or at the junction of web and flange if the centroid is in the flange, due to effective prestress
V_p = vertical component of prestressing force at section considered ($= F \sin \alpha$)
d_p = d_p or $0.8h$, whichever is larger

Alternatively, v_{cw} may be theoretically computed as the shear stress corresponding to a combination of dead load and live load that results in a principal tensile stress of $-4\sqrt{f'_c}$ at the centroid of the section or at the junction of web and flange when the centroid is in the flange.

The computation of the values of v_{ci} and v_{cw}, given by the above expressions, is time-consuming, especially as several sections along the span must be considered. An alternative, more conservative method for beams with uniform loads is suggested in the code and can be used when the effective prestress f_{pe} is not less than $0.4f_{pu}$. The shear resistance of the concrete can then be taken as:

$$v_c = 0.6\sqrt{f'_c} + 700\frac{V_u d_p}{M_u} \tag{8-33}$$

but not less than $2\sqrt{f'_c}$ and not more than $5\sqrt{f'_c}$. V_u and M_u are the factored shear and moment, and d_p is the depth to the centroid of the prestressing force at the section considered. The ratio $V_u d_p/M_u$ is not to be taken greater than 1 and the lower bound value of $0.8h$ on d_p, used elsewhere, does not apply. Once the value of v_c has been determined, the shear force resistance of the concrete can be obtained from:

$$V_c = v_c b_w d_p \tag{8-34}$$

3. Required Area of Shear Reinforcement

In order to determine the required area of shear reinforcement, the value of v_s is needed. The minimum design value of v_s is obtained from eqs. (8-26) and (8-27) at equality. That is:

$$\frac{v_u}{\phi} = v_n = v_c + v_s \tag{8-35}$$

from which:

$$v_s = \left(\frac{v_u}{\phi} - v_c\right) \tag{8-36}$$

where v_s is the required nominal shear resistance of the shear reinforcement. The area of shear reinforcement is obtained following an approach identical to that used for reinforced concrete. A shear crack extending at 45° to the beam's axis is assumed, and the force carried by the web reinforcement is considered to balance a principal tension across the crack equal in magnitude to the shear stress, v_s, balanced by the shear reinforcement. The following results are derived.

For vertical stirrups:

$$A_v = \frac{(v_u/\phi - v_c)\,b_w s}{f_y} = \frac{(V_u/\phi - V_c)s}{f_y d_p} \tag{8-37}$$

For stirrups inclined at an angle α to the beam's axis:

$$A_v = \frac{(v_u/\phi - v_c)\,b_w s}{f_y(\sin\alpha + \cos\alpha)} = \frac{(V_u/\phi - V_c)\,s}{f_y d_p\,(\sin\alpha + \cos\alpha)} \tag{8-38}$$

The above two equations can also be used to find the required spacing s for a given type of stirrup A_v:

For vertical stirrups:

$$s = \frac{A_v f_y}{(v_u/\phi - v_c)\,b_w} = \frac{A_v f_y d_p}{V_u/\phi - V_c} \tag{8-39}$$

For inclined stirrups:

$$s = \frac{A_v f_y\,(\sin\alpha + \cos\alpha)}{(v_u/\phi - v_c)\,b_w} = \frac{A_v f_y\,(\sin\alpha + \cos\alpha)\,d_p}{V_u/\phi - V_c} \tag{8-40}$$

4. Limitations and Special Cases

A number of design limitations relative to shear reinforcement are given in the *ACI Code:*

1. *Maximum spacing.* The spacing s of shear reinforcement measured parallel to the axis of the member shall not exceed $3h/4$ or 24 in. (61 cm). In common practice the spacing is taken not less than 3 in. (75 mm).
2. *Maximum shear.* The value of $(v_u/\phi - v_c)$ shall not exceed $8\sqrt{f'_c}$. If it does, the section is insufficient for shear resistance, and its dimensions should be changed (namely, b_w and/or d_p). If $(v_u/\phi - v_c)$ exceeds $4\sqrt{f'_c}$, the spacing obtained from eqs. (8-39) and (8-40) should be halved and should not exceed $3h/8$ or 12 in. (30 cm).
3. *Minimum shear reinforcement.* Shear reinforcement restrains the growth of inclined cracking and, hence, increases ductility and provides a warning of failure. Otherwise, in an unreinforced web, the sudden formation of inclined cracking might lead directly to sudden failure. Accordingly, the code requires a minimum area of shear reinforcement to be provided in all reinforced and prestressed concrete structural members when the factored shear stress v_u exceeds half the shear strength of concrete ϕv_c, except in (1) slabs and footings, (2) concrete joist construction for which the code allows a 10% increase in v_c, and (3) beams with total depth not greater than 10 in. (25 cm), $2^1/_2$ times the thickness of the flange, or one-half the web width, whichever is greater.
   When the effective prestress f_{pe} exceeds $0.40 f_{pu}$ (which is common in prestressed concrete), the minimum required area of shear reinforcement may be computed by:
   $$(A_v)_{\min} = \frac{A_{ps}}{80}\,\frac{f_{pu}}{f_y}\,\frac{s}{d_p}\sqrt{\frac{d_p}{b_w}} \tag{8-41}$$
   where d_p has a lower bound value of $0.80h$, and other notations are standard.
   A more general equation for the minimum required area of shear reinforcement is given in the code and applies to reinforced as well as prestressed concrete (with no limitation on the effective prestress). It is given by:
   $$(A_v)_{\min} = 50\,\frac{b_w s}{f_y} \tag{8-42}$$
   where b_w and s are in inches and f_y in pounds per square inch. Equation (8-42) will generally require greater minimum shear reinforcement in typical building members than will eq. (8-42).
4. *Lightweight concrete.* When prestressed lightweight concrete is used instead of normal weight concrete, the same procedure to determine the shear reinforcement is used with the following modifications prescribed by the code:
   (*a*) When the split cylinder strength f'_{tc} is specified for lightweight concrete and when $f'_{tc}/6.7$ is less than $\sqrt{f'_c}$, replace $\sqrt{f'_c}$ by $f'_{tc}/6.7$ in the equations giving v_c.
   (*b*) When f'_{tc} is not specified, the values of $\sqrt{f'_c}$ used in the expressions of v_c and ΔM_{cr} shall be multiplied by 0.75 for "all-lightweight" concrete and 0.85 for "sand-lightweight" concrete.

8.7.4 Design Expedients

Designing for shear can be very time-consuming, as one does not know a priori the locations of the critical sections that require more than the minimum amount of reinforcement. It is not uncommon to find (assuming uniform loading) that excess reinforcement is needed at two sections and is not needed at a section in between.

In most common design problems involving beams, it is very likely that the minimum amount of shear reinforcement $(A_v)_{\min}$ at the maximum allowable spacing of $0.75h$ or 24 in. ($\simeq$ 60 cm) will prevail throughout the span. If the alternative method to determine v_c is considered, its lower bound value of $2\sqrt{f'_c}$ can be used as a first approximation in design and checked against the value of v_u/ϕ at several sections. Only if significant differences exist will the more accurate analysis be pursued.

In order to ease the design procedure, a design flow chart for shear is proposed in Fig. 8-21. It applies to normal weight prestressed concrete beams and synthesizes most of the steps encountered if the *ACI Code* approach is followed. As in each complete iteration only one section is considered, it is appropriate to set the numerical results in a table, such as shown in the following example.

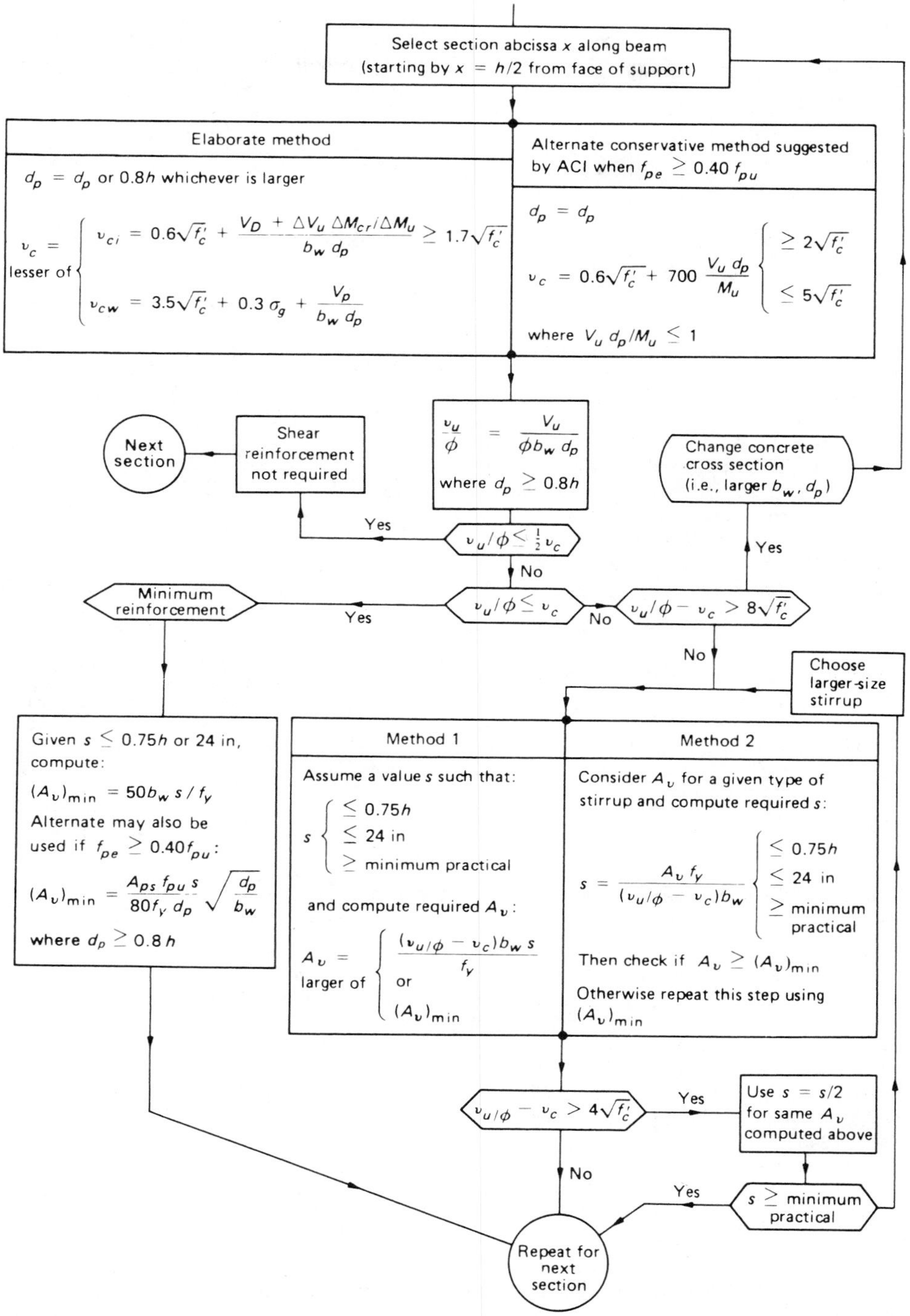

Fig. 8-21 Flow chart for the design of shear reinforcement in beams.

8.7.5 Example: Design of Shear Reinforcement

The T beam described in Section 8.5.7 is considered (Fig. 8-22). The design characteristics of interest to this example are: $h = 40$ in, $b_w = 8$ in., $A_c = 550$ in^2, $A_{ps} = 1.53$ in^2, $y_t = 12.9$ in., $Z_b = 3028$ in^3, $f'_c = 5000$ psi (normal weight concrete), $F = 229.5$ kips, $f_{pe} > 0.4 f_{pu}$, $l = 70$ ft, live load = 0.4 klf, superimposed dead load = 0.04 klf, self-weight $w_G = 0.573$ klf.

The final eccentricity of the prestressing force along the half span is given by:

$$\begin{cases} e_o = 21.7 \text{ in.} & \text{for } x \geqslant 28 \text{ ft} \\ e_o = 7.9 + 13.8 \dfrac{x}{28} & \text{for } 0 \leqslant x < 28 \text{ ft} \end{cases}$$

Let us define the following values of factored loads:

$$w_u = 1.4(0.573 + 0.04) + 1.7 \times 0.4 = 1.538 \text{ klf}$$

$$\Delta w_u = 1.4(0.04) + 1.7 \times 0.4 = 0.736 \text{ klf}$$

The first critical section to be analyzed is taken at $h/2 =$

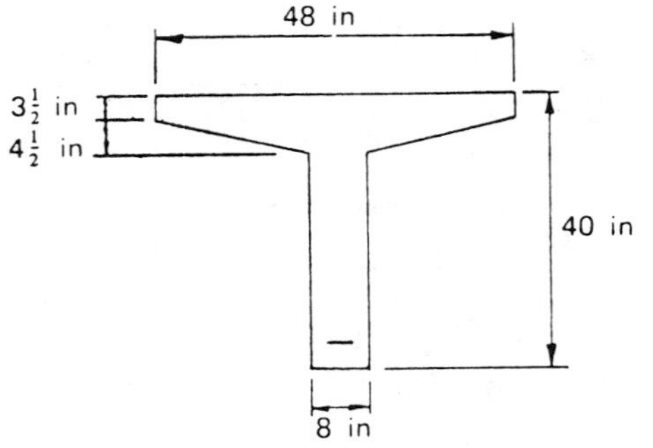

Fig. 8-22

20 in. from the face of the support. If a support pad 8 in. wide is used, the location of the section from the center of the support becomes $x = 20 + 8/2 = 24$ in. or 2 ft.

Two approaches will be followed, the more elaborate approach and the alternative conservative approach, according to the steps described in the flow chart, Fig. 8-21. The computations for the first critical section will be covered in detail. Results obtained for this section and four others are summarized in Tables 8-7 and 8-8.

1. Elaborate Approach

The eccentricity of the prestressing force at $x = 2$ ft is given by:

$$e_o = 7.9 + 13.8 \frac{2}{28} = 8.9 \text{ in.}$$

Thus:

$$d_p = e_o + y_t = 8.9 + 12.9 = 21.8 \text{ in.}$$

For the equations used in the elaborate approach, d_p is limited by $0.8h = 32$ in. Thus, use $d_p = 32$ in. Following the design flow chart, Fig. 8-21, let us first compute v_{ci}. For this we need the following values:

$$V_G = w_G \left(\frac{l}{2} - x\right) = 0.573 \times 33 = 18.91 \text{ kips}$$

$$\Delta V_u = \Delta w_u \left(\frac{l}{2} - x\right) = 0.736 \times 33 = 24.288 \text{ kips}$$

$$\Delta M_u = \Delta w_u \frac{x(l - x)}{2} = \frac{0.736 \times 2 \times 68}{2}$$

$$= 50.05 \text{ kips-ft} = 600.57 \text{ kips-in.}$$

$$M_D = w_G \frac{x(l - x)}{2} = \frac{0.573 \times 2 \times 68}{2}$$

$$= 38.964 \text{ kips-ft} = 467.57 \text{ kips-in.}$$

$$\Delta M_{cr} = Z_b \left[6\sqrt{f'_c} + \frac{F}{A_c}\left(1 + \frac{e_o A_c}{Z_b}\right)\right] - M_G$$

$$= 3028 \left[6\sqrt{5000} + \frac{229{,}500}{500} \cdot \left(1 + \frac{8.9 \times 550}{3028}\right)\right] - 467{,}570$$

$$= 4{,}123{,}153 \text{ lb-in.} = 4123 \text{ kips-in.}$$

TABLE 8-7 Summary of Shear Design Computations for Example 8.7.5, Case 1

Section, x	V_D, *kips*	ΔV_u, *kips*	ΔM_u, *kips-in.*	ΔM_{cr}, *kips-in.*	d_p *or* $0.8h$ *in.*	v_{ci}, *psi*	V_p, *kips*	v_{cw}, *psi*	v_c, *psi*	V_u, *kips*	v_u/ϕ, *psi*	$v_u/\phi - v_c$	A_v *at s or s for* A_v	$(A_c)_{min}$ *at* s_{max}
$0.5h$	18.91	24.288	600.57	4123	32	767.6	↑	↑	409.5	50.75	233.2			
$l/20$	18.05	23.184	1027.82	3958	32	461.3			409.5	48.45	222.6			(0.13 in^2
$2l/20$	16.044	20.608	1947.45	3638	32	255.3			255.3	43.06	197.9		use next	required)
$3l/20$	14.039	18.032	2758.9	3402	32	184	9.426	409.5	183	37.68	173.2	negative	column	(0.22 in^2
$4l/20$	12.033	15.456	3462.14	3250	32	146			146	32.30	148.4			provided)
⋮	⋮	⋮	⋮	⋮	⋮	⋮			⋮	⋮	⋮			
$l/2$							↓	↓						

TABLE 8-8 Summary of Shear Design Computations Using the Conservative Method; Example 8.7.5, Case 2

Section	*Distance, x ft*	d_p, *in.*	V_u, *kips*	M_u, *kips-in.*	$\frac{V_u d_p}{M_u} \leqslant 1$	v_c $\leqslant 5\sqrt{f'_c}$ $\geqslant 2\sqrt{f'_c}$ *psi*	$\frac{v_u}{\phi} = \frac{V_u}{\phi b_w d_p}$† *psi*	$\left(\frac{v_u}{\phi} - v_c\right)$ *psi*	A_v *for s or s for* A_v	$(A_v)_{min}$ *at* s_{max}
$0.5h$	2	21.8	50.75	1255.2	0.88	353.5	233.2	....		
$l/20$	3.5	22.52	48.45	2148.1	0.51	353.5	222.6	....		(0.13 in^2
$2l/20$	7	24.25	43.06	4070	0.26	224.4	197.9	....	use next	required)
$3l/20$	10.5	25.8	37.68	5765.9q	0.17	161.4	173.2	11.8	column	(0.22 in^2
$4l/20$	14	27.7	32.30	7235.7	0.12	141.4	148.4	7		provided)
⋮	⋮	⋮	⋮	⋮	⋮	⋮	⋮	⋮		
$l/2$										

†$d_p = d_p$ *or* $0.8h$ *whichever is larger.*

Thus:

$$v_{ci} = 0.6\sqrt{f_c'} + \frac{V_G + (\Delta V_u \times \Delta M_{cr})/\Delta M_u}{b_w d_p}$$

$$= 0.6\sqrt{5000} + 1000\left(\frac{18.91 + (24.288 \times 4123)/600.57}{8 \times 32}\right)$$

$$= 767.6 \text{ psi} > 1.7\sqrt{f_c'} = 120 \text{ psi} \qquad \text{OK}$$

In order to determine v_{cw}, we need the following:

$$\sigma_g = \frac{F}{A_c} = \frac{229{,}500}{550} = 417.3 \text{ psi}$$

(Note: The centroid is in the web.)

$$\sin\alpha \simeq \tan\alpha = \frac{(e_o)_{28} - (e_o)_{\text{support}}}{28 \times 12} = 0.04107$$

$$V_p = F\sin\alpha = 229{,}500 \times 0.04107 = 9426 \text{ lb}$$

Thus:

$$v_{cw} = 3.5\sqrt{f_c'} + 0.3\sigma_g + \frac{V_p}{b_w d_p}$$

$$= 3.5\sqrt{5000} + 0.3 \times 417.3 + \frac{9426}{8 \times 32} = 409.5 \text{ psi}$$

The shear resistance v_c is the smaller of v_{ci} and v_{cw}. Thus at $x = 2$ ft:

$$v_c = 409.5 \text{ psi}$$

In order to determine v_u, we need V_u:

$$V_u = w_u\left(\frac{l}{2} - x\right) = 1.538\,(35 - 2) = 50.75 \text{ kips}$$

Thus:

$$\frac{v_u}{\phi} = \frac{V_u}{\phi b_w d_p} = \frac{50{,}750}{0.85 \times 8 \times 32} = 233.2 \text{ psi}$$

The value of v_u/ϕ is to be compared to v_c. As $v_c = 409.5 > v_u/\phi$, only minimum reinforcement is needed at this section.

Let us select a stirrup with $f_y = 60$ ksi, at a maximum spacing $s = 24$ in. $< 0.75h$. As $f_{pe} > 0.4f_{pu}$, the minimum area of shear reinforcement is obtained from:

$$(A_v)_{\min} = \frac{A_{ps} f_{pu}\, s}{80 f_y\, d_p}\sqrt{\frac{d_p}{b_w}} = \frac{1.53 \times 270 \times 24}{80 \times 60 \times 32}$$

$$\cdot\sqrt{\frac{32}{8}} \simeq 0.13 \text{ in}^2$$

Note that a #3 U stirrup would provide 0.22 in² and thus would be adequate. The other sections covered in Table 8-7 lead to the same minimum reinforcement.

If, instead of selecting the spacing in the above equation, we had selected the area of a #3 U stirrup, the required spacing would have been $s = 40.9$ in. Of course, then the maximum allowable spacing of 24 in. would control, leading to the same result.

Providing a given area of reinforcement at a given spacing allows the determination of the maximum shear stress to be resisted by the reinforcement. In this case, we have:

$$\left(\frac{v_u}{\phi} - v_c\right) \leqslant \frac{A_v f_y}{b_w s} = \frac{0.22 \times 60{,}000}{8 \times 24} \simeq 69 \text{ psi}$$

2. Alternate Conservative Approach

In order to compute v_c we need:

$$V_u = 50.75 \text{ kips} \qquad \text{(calculated above)}$$

$$M_u = w_u\,\frac{x(l - x)}{2} = \frac{1.538 \times 2 \times 68}{2}$$

$$= 104.58 \text{ kips-ft} = 1255 \text{ kips-in.}$$

$$d_p = 21.8 \text{ in.}$$

$$\frac{V_u d_p}{M_u} = 0.88 < 1 \qquad \text{OK}$$

Thus:

$$v_c = 0.6\sqrt{f_c'} + 700\,\frac{V_u d_p}{M_u} = 658 \text{ psi}$$

v_c is larger than $2\sqrt{f_c'} = 141.4$ psi but should be taken less than $5\sqrt{f_c'} = 353.5$ psi. Thus, use $v_c = 353.5$ psi. Results for the other sections are summarized in Table 8-8.

If we compare v_u/ϕ and v_c we find out that, similarly to the previous case, a minimum reinforcement is sufficient.

3. Design for Increased Live Load

To illustrate a case where more than the minimum reinforcement is needed, let us assume that the specified live load for the previous example is increased from 0.4 klf to 1 klf. Let us also assume that the beam is designed as a partially prestressed beam, whereas the amount and profile of the prestressing steel are the same as described above, but, in order to balance the increase in the required resistance, nonprestressed reinforcement is provided. It amounts to four #9 bars with $A_s = 4$ in², $f_y = 60$ ksi, and $d_s = 36$ in. at midspan.

For this case, the values of V_G, d_p, M_{cr}, and V_p are the same as in case 1 and are given in Table 8-7. Because of the increase in live load, the following values of factored loads apply:

$$w_u = 1.4(0.573 + 0.04) + 1.7 \times 1 = 2.5582 \text{ klf}$$

$$\Delta w_u = 1.4 \times 0.04 + 1.7 \times 1 = 1756 \text{ klf}$$

The corresponding computations for ΔV_u, ΔM_u, v_{ci}, v_{cw}, v_c, and the required spacing (eq. 8-39) of #3 U stirrups are summarized in Table 8-9. It can be observed that, although the live load has been substantially increased, the required amount of shear reinforcement is still relatively small. If the alternative conservative approach had been used, the value of v_c would have been the same as shown in Table 8-8. This is so because the ratio V_u/M_u remains the same with an increase in uniform live load. The values of v_{ci}, v_{cw}, and v_c, as well as the required stirrups spacing, are plotted versus x in Fig. 8-23. The shaded area corresponds to the part of the beam where stirrups are theoretically needed. It can be seen that the greatest need for stirrups is not near the support but instead in a region where flexure-shear cracking may control. However, note that, althrough shear may not be critical near the support, additional stirrups are needed to contain tensile splitting cracks in the end zone.

8.8 DEFLECTION COMPUTATION AND CONTROL

8.8.1 Deflection: Types and Characteristics

Deflection is defined as the total movement induced at a point of a member from the position before application of

TABLE 8-9 Summary of Shear Design Computations for Example 8.7.5, Case 3

Section x	ΔV_u, *kips*	ΔM_u, *kips-in.*	v_{ci}, *psi*	v_{cw}, *psi*	v_c, *psi*	V_u, *kips*	v_u/ϕ, *psi*	$v_u/\phi - v_c$, *psi*	*Required* s *for* A_v†
$0.5h$	57.948	1,432.9	767.6	↑	409.5	84.42	388	....	24
$l/20$	55.314	2,452.3	418.9		409.5	80.58	370	....	24
$2l/20$	49.168	4,646.4	255		255	71.63	329	74	22.3
$3l/20$	43.022	6,582.4	183		183	62.68	288	105	15.7
$4l/20$	36.876	8,260.2	146	409.5	146	53.72	247	101	16.3
$5l/20$	30.73	9,680	121		121	44.77	206	85	19.4
$6l/20$	24.584	10.841.5	120		120	35.81	165	45	24
$7l/20$	18.438	11,745	120	↓	120	26.86	120	0	24
$8l/20$	12.292	12,390	120	373	120	17.91	76	....	24
$9l/20$	6.146	12,778	120	373	120	8.95	38	....	24
$l/2$	0	12,907	120	373	120	0	0	....	24

†For $A_v = 0.22\ \text{in}^2$ and $s \leqslant 24$ in.

the load to the position after application of the load. The maximum deflection, which, in uniformly loaded simply supported beams, occurs at midspan, is generally of main interest in design. A distinction is often made between "camber," which is the deflection caused by prestressing, and "deflection," which is that produced by external loads. They are identical in nature but generally opposite in sign. Typically, prestressing produces *upward* camber in a simply supported beam, while self-weight produces *downward* deflection. Their combination may produce an upward or a downward movement. In order to avoid confusion, the term "deflection" will be used in its most general form, unless the separate effect of camber is addressed.

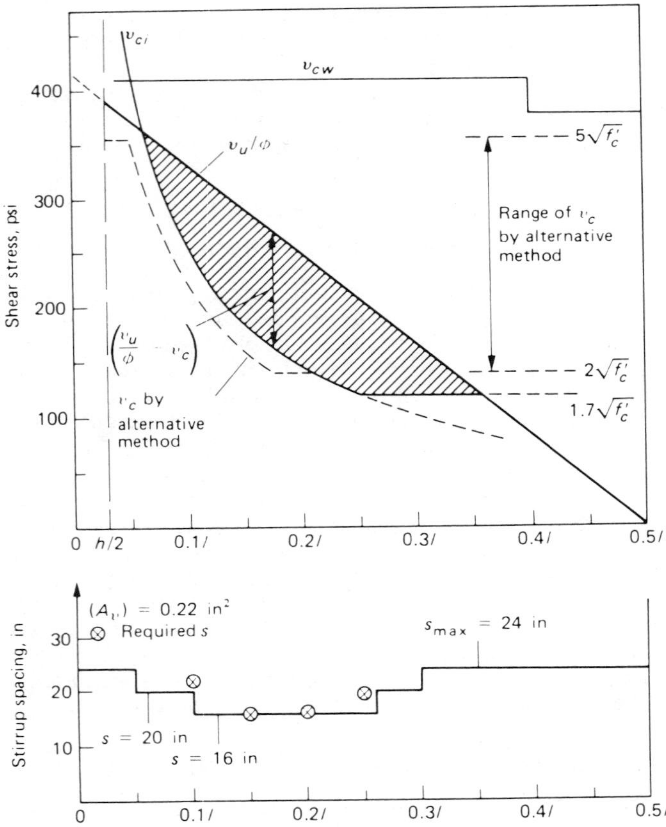

Fig. 8-23 Shear stress distribution and stirrups spacing for example beam.

The following sign convention will be followed: plus (+) for downward deflection and minus (−) for upward deflection.

In reinforced concrete beams, deflection is due to external loads and is always downward. In prestressed concrete, deflection depends on the combined effect of prestressing and external loading. It can easily be controlled by changing the magnitude and profile of the prestressing force. It is not uncommon, in partial prestressing, to achieve a zero deflection design. In both reinforced and prestressed concrete members, deflection under sustained loading continues to increase with time, mainly due to the effects of creep and shrinkage of concrete and relaxation of prestressing steel. Excessive deflections, especially those developing with time, are common causes of trouble and must be limited.

In computing deflection, one may differentiate between camber and deflection, short-term or immediate instantaneous deflection, and long-term or time-dependent deflection. The total deflection itself can be separated into two parts, an instantaneous part and an additional, time-dependent part. Furthermore, a different approach is followed, depending on whether the member is uncracked, as in fully prestressed members, or cracked, as in reinforced and partially prestressed members. Here only the case of prestressed uncracked sections is addressed.

8.8.2 Short-Term Deflections in Prestressed Members

Fully prestressed concrete members are uncracked under service loads and are assumed linear elastic. Their deflections can be determined using theoretical derivations such as the moment-area theorems.

Typical formulas for deflections are given in Fig. 8-24 for several profiles of the prestressing force and several types of loading. Simply supported beams with constant prestressing force and constant sectional properties are assumed. Because superposition is valid in computing deflections for uncracked members, many combinations are practically covered. Note that two expressions are given for each case, one in function of the prestressing force and the other in function of the curvatures at the midspan section and at the support. The deflection expressed in function of the curvatures has the following most general form:[8-36, 8-37]

$$\Delta = \Phi_1 \frac{l^2}{8} + (\Phi_2 - \Phi_1) \frac{a^2}{6} \tag{8-43}$$

vhere:

Φ_1 = curvature at midspan
Φ_2 = curvature at the support
a = length parameter which is a function of the tendon profile used

The values of a have been directly integrated in the expressions shown in Fig. 8-24. The curvature at any section an be computed from the strain distribution along the section as:

$$\Phi = \frac{\epsilon_{ct} - \epsilon_{cb}}{h} \tag{8-44}$$

vhere:

ϵ_{ct} = strain on top fiber of section
ϵ_{cb} = strain on bottom fiber of section
h = depth of section

quation (8-44) is illustrated in Fig. 8-25(a) for uncracked sections and in Fig. 8-25(b) for cracked sections where ϵ_{cb} vanishes and h is replaced by c.

Typical formulas for deflections in cantilever beams are given in Fig. 8-26.

In using all deflection expressions such as those given in Figs. 8-24 and 8-26, two important properties must be defined, namely, the modulus of elasticity of the concrete and the moment of inertia of the section (or of each section if variable depth is used).

The modulus of elasticity of the concrete material (secant modulus at $0.45f_c'$) can be estimated from the expression recommended in the *ACI Code* and given in Chapter 6.

However, the strength of concrete varies with age, and its modulus too. In computing initial camber or deflection, it is common to use the initial modulus E_{ci}, while E_c is considered for service load deflections.

The moment of inertia of the section depends on whether the section is cracked or uncracked. When the section is uncracked, it is customary to use the gross moment of inertia I_g for pretensioned members and the net moment

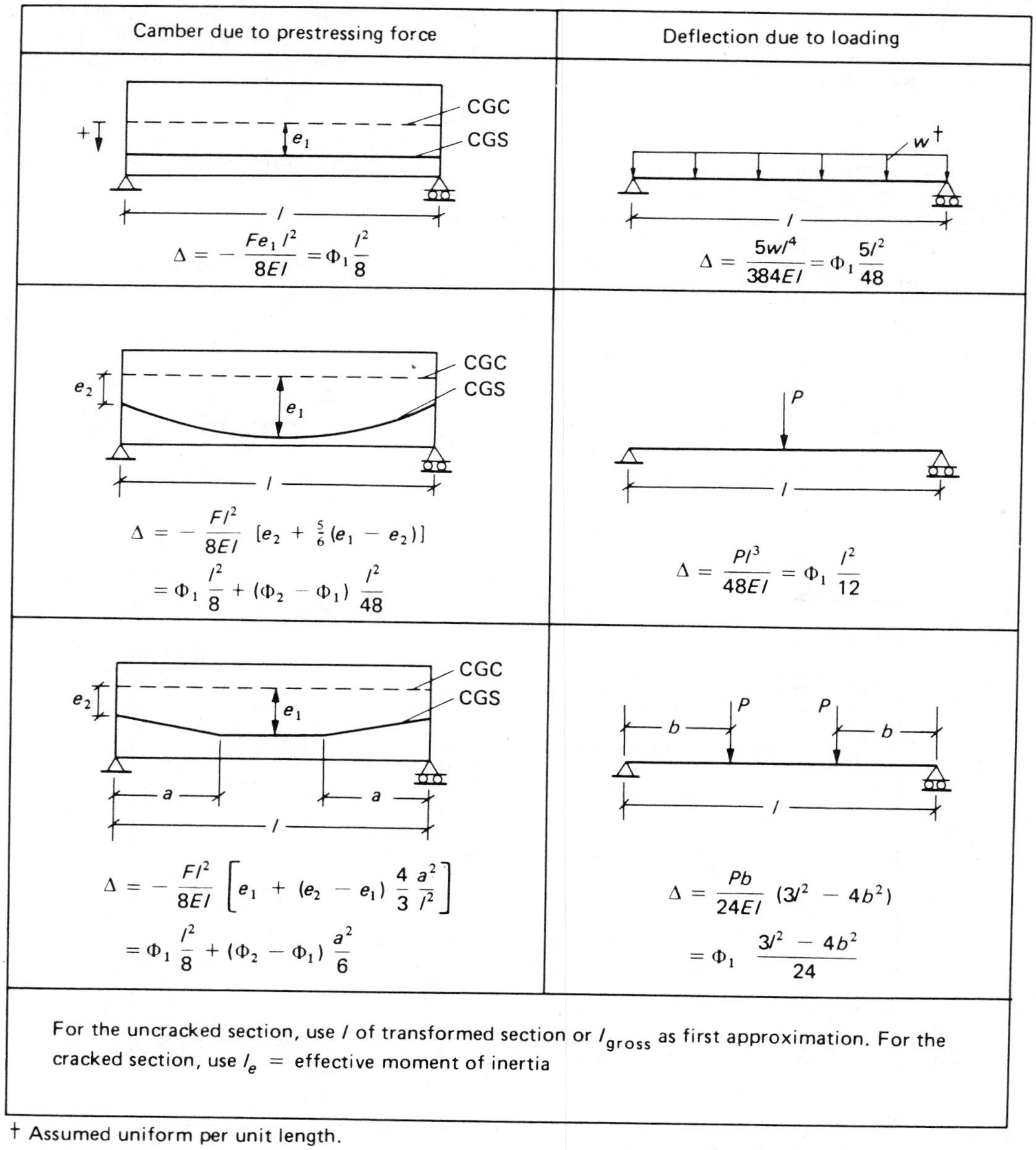

Camber due to prestressing force
Deflection due to loading
CGC
CGS
e_1
e_2
l
w†
P
b
a
$\Delta = -\frac{Fe_1 l^2}{8EI} = \Phi_1 \frac{l^2}{8}$
$\Delta = \frac{5wl^4}{384EI} = \Phi_1 \frac{5l^2}{48}$
$\Delta = -\frac{Fl^2}{8EI}\left[e_2 + \frac{5}{6}(e_1 - e_2)\right] = \Phi_1 \frac{l^2}{8} + (\Phi_2 - \Phi_1)\frac{l^2}{48}$
$\Delta = \frac{Pl^3}{48EI} = \Phi_1 \frac{l^2}{12}$
$\Delta = -\frac{Fl^2}{8EI}\left[e_1 + (e_2 - e_1)\frac{4}{3}\frac{a^2}{l^2}\right] = \Phi_1 \frac{l^2}{8} + (\Phi_2 - \Phi_1)\frac{a^2}{6}$
$\Delta = \frac{Pb}{24EI}(3l^2 - 4b^2) = \Phi_1 \frac{3l^2 - 4b^2}{24}$
For the uncracked section, use I of transformed section or I_{gross} as first approximation. For the cracked section, use I_e = effective moment of inertia

† Assumed uniform per unit length.

Fig. 8-24 Typical midspan deflections for simply supported beams.

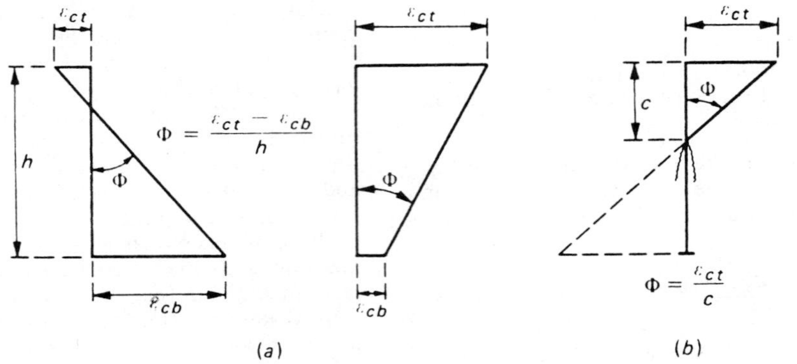

Fig. 8-25 Representation of curvature. (a) Uncracked section. (b) Cracked section.

of inertia I_n for post-tensioned members with unbonded tendons. In all cases with bonded tendons, the moment of inertia of the transformed section can be used. However, often the corresponding lengthier calculations do not result in a significant gain in accuracy.

8.8.3 Long-Term Deflection: Simplified Prediction Method

The deflections calculated from the expressions given in Figs. 8-24 and 8-26 are "instantaneous elastic" deflections. We assumed that they were associated with a short-term loading. If the load is sustained, as in the case of dead loads, the deflection increases with time, mainly because of the effects of creep and shrinkage of the concrete and relaxation of prestressing steel. In such a case, the deflection will be called the total deflection. The total deflection can be separated into two parts: an instantaneous elastic part and an additional long-term part. The first part is calculated as described in the previous sections and is assumed to remain constant for a given load. The additional part increases with time.

Continuous research efforts are being undertaken to predict time-dependent deflection in reinforced and prestressed concrete structures more reliably.[8-38–8-40] It has been customary in everyday design to estimate the additional long-

Camber due to prestressing	Deflection due to loading
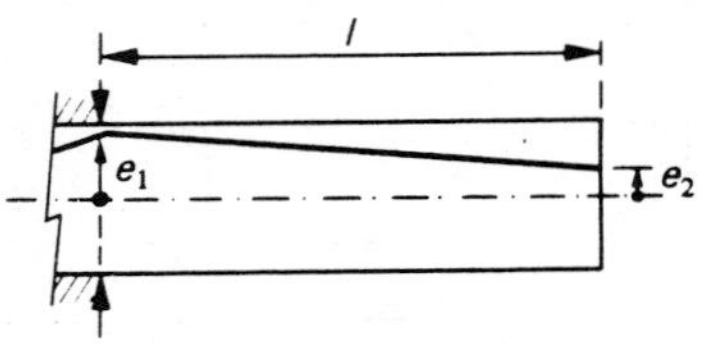 $-\dfrac{Fl^2}{2EI}\left(e_2 + \dfrac{2(e_1 - e_2)}{3}\right)$	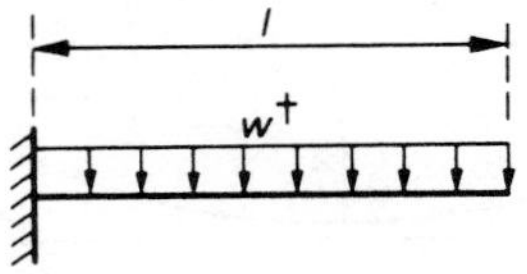 $\dfrac{wl^4}{8EI}$
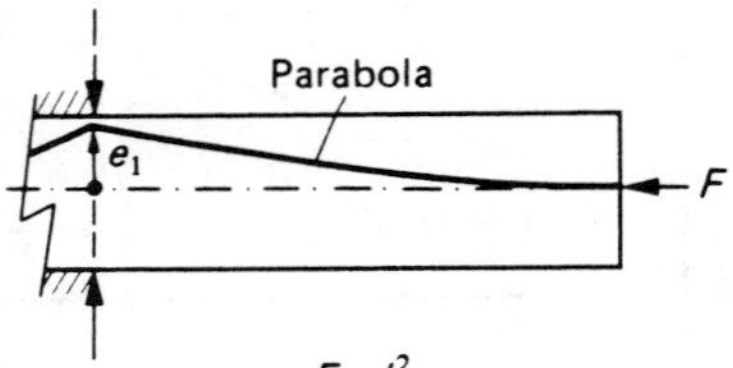 $-\dfrac{Fe_1l^2}{4EI}$	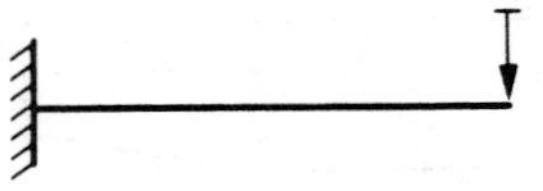 $\dfrac{Pl^3}{3EI}$
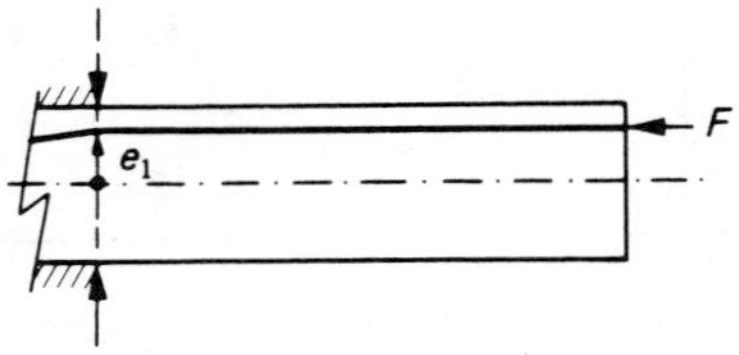 $-\dfrac{Fe_1l^2}{2EI}$	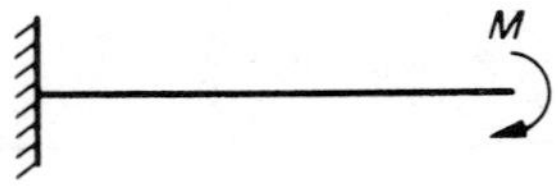 $\dfrac{Ml^2}{2EI}$

† Assumed uniform per unit length.

Fig. 8-26 Typical end deflections for fixed end cantilever beams.

term deflection by multiplying the instantaneous deflection Δ_i by an appropriate factor. The basic reason behind this approach comes from analyzing the effect of creep of concrete, which leads to the following relation:[8-15]

$$\Delta_U = \Delta_i + \Delta_i C_{CU} = \Delta_i(1 + C_{CU}) = \Delta_i + \Delta_{add} \tag{8-45}$$

in which:

$$\Delta_{add} = C_{CU}\,\Delta_i \tag{8-46}$$

and:

Δ_U = life total deflection for the sustained loading considered
Δ_{add} = additional long-term deflection
C_{CU} = ultimate creep coefficient
Δ_i = instantaneous elastic deflection

Thus, the additional long-term deflection can be estimated by multiplying the instantaneous deflection by a factor. According to eq. (8-46), the multiplier is equal to the ultimate creep coefficient. However, some other effects, such as those of shrinkage, relaxation, and the presence of nonprestressing steel, may be accounted for in the equation predicting Δ_{add}. One such prediction equation is discussed next.

Based on an extensive evaluation of parameters influencing Δ_{add} for typical precast prestressed members, Martin [8-40] suggested the following equation:

$$\Delta_{add} = \lambda\Delta_i = \eta\,\frac{E_{ci}}{E_c}\,k_r\,C_{CU}\,\Delta_i \tag{8-47}$$

where:

$\eta = F/F_i$ and F_i is the prestressing force immediately after transfer
$k_r = 1/(1 + A_s/A_{ps})$ when $A_s/A_{ps} \leqslant 2$
C_{CU} = ultimate creep coefficient
E_{ci} = elastic modulus of concrete at time of transfer

Different values of the multiplier λ were recommended for prestressed composite and noncomposite members, depending on the effects of self-weight, prestressing, and superimposed dead load, if any. They are given in Table 3.4.1 of the PCI handbook[8-22] and generally vary between 1.85 and 3.

Often a rule of thumb is used in evaluating the long-term deflection of standard building members. It has the following form:

$$\Delta_U = \lambda_1(\Delta_i)_{F_i} + \lambda_2(\Delta_i)_D \tag{8-48}$$

where λ_1 and λ_2 are factors derived from experience and are of the order of 2.4 and 2.8, respectively, for common design problems. In evaluating $(\Delta_i)_{F_i}$, the modulus of elasticity of concrete at the time of transfer is used.

8.8.4 Deflection Limitations and Control

Although structural members can be properly designed for strength, they may develop excessive cambers and deflections over time. Hence their behavior in service can be jeopardized. The *ACI Code* (Section 9.5) provides a number of deflection limitations for reinforced concrete members. In principle they apply to prestressed concrete members as well. Permissible deflections are also prescribed for steel bridges in the AASHTO specifications. Although they are more severe than those given by ACI, they can be used as guidelines for prestressed concrete bridges.

The deflection of prestressed and partially prestressed concrete members can be controlled to a great extent by proper selection of the magnitude and trajectory of the prestressing force. Appropriate actions can also be taken to reduce the time-dependent effects on camber or deflection. They include: (1) adding nonprestressed steel at appropriate locations, (2) precambering at casting, and (3) delaying or staging the application of prestress.

8.8.5 Example

Let us consider the prestressed pretensioned beam described in Example 8.5.7a. A prestressing steel profile having two draping points at 28 ft from the support is assumed with e_0 = 21.7 in. at midspan and e_0 = 7.9 in. at the supports.

Let us first compute the initial instantaneous elastic deflection due to the initial prestressing force and the self-weight of the beam. Using the initial elastic modulus of concrete and the equations of Fig. 8-24, we have:

$$(\Delta_i)_{F_i+G} = (\Delta_i)_{F_i} + (\Delta_i)_G$$

where e_1 and e_2 are the eccentricities of the prestressing force at midspan and at the supports, respectively. Thus:

$$(\Delta_i)_{F_i+G} = -\frac{276{,}500 \times (70 \times 12)^2}{8 \times 3.834 \times 10^6 \times 82{,}065} \cdot \left[21.7 + (7.9 - 21.7)\,\frac{4}{3}\left(\frac{28}{70}\right)^2\right] + \frac{5 \times (70 \times 12)^4 \times 573/12}{384 \times 3.834 \times 10^6 \times 82{,}065}$$

$$(\Delta_i)_{F_i+G} = -1.4537 + 0.9838 \simeq -0.47 \text{ in.}$$

Let us estimate the additional long-term deflection (here camber) from Eq. (8-47), assuming $\lambda = 2.3$:

$$\Delta_{add} = 2.3(-0.47) = -1.175 \text{ in.}$$

It satisfies the *ACI Code* limitation dealing with attached nonstructural elements likely to be damaged, that is:

$$(\Delta_{add}) = 1.175 \text{ in.} < \frac{l}{480} = 1.75 \text{ in.} \qquad \text{OK}$$

Note that in a final design the effect of the superimposed dead load should be considered. In the above computation such effect leads to a smaller additional camber.

8.9 ANALYSIS AND DESIGN OF COMPOSITE BEAMS

8.9.1 Types of Prestressed Concrete Composite Beams

Composite construction implies the use, in a single structure acting as a unit, of different structural elements made with similar or different structural materials. Common examples include steel–concrete beams or columns, orthotropic steel decks, and sandwich panel construction. The two structural materials are utilized in the best possible manner: in a steel–concrete composite beam the steel is designed to carry tension and the concrete to carry compression.

In a composite member where only concrete is used as a material, the concrete is placed in at least two separate stages generally leading to two different unit weights and properties. This is the case for composites made with a precast reinforced or prestressed concrete element combined with a concrete element cast in place at a different time. When the precast element is also prestressed, it es-

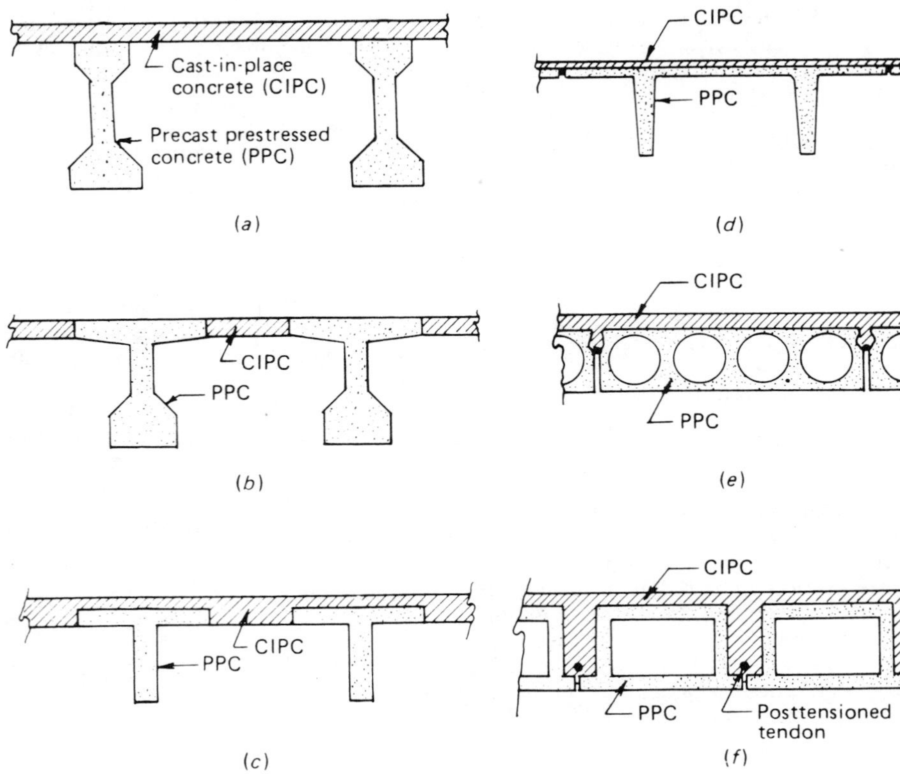

Fig. 8-27 Typical cross sections of composite beams.

sentially simulates the function of the steel in a steel–concrete system; it will carry the tension without cracking.

Typical cross sections of prestressed concrete composite beams are shown in Fig. 8-27. Several alternatives can be identified, whereas the cast-in-place (CIP) slab is cast either on top of the precast prestressed concrete (PPC) beams, or in between them, or both. The cast-in-place slab can act as a structural element allowing lateral transfer of loading as in Fig. 8-27(a–c) or as a topping having a continuous interface with the precast elements (Fig. 8-27(d, e). The former case is typical of bridge applications, whereas the latter is typical of building applications. An interesting application is also shown in Fig. 8-27(f), where the cast-in-place element is itself a beam that may be subsequently reinforced or post-tensioned.

8.9.2 Particular Design Aspects of Prestressed Composite Beams

The design of prestressed concrete composite beams can be essentially reduced to that of noncomposite beams, provided their differences are understood and accounted for. Particular design differences comprise:

1. The loading stages and their relation to whether the beam responds as a composite or noncomposite beam.
2. The transformed effective flange width and corresponding transformed section properties.
3. The horizontal shear at the interface between the precast beam and the cast-in-place slab.

Other design considerations such as flexure by allowable stresses or ultimate strength, shear, torsion, and deflection are to a great extent very similar to those of noncomposite monolithic beams.

Each of the above-mentioned differences and similarities is addressed respectively in the next sections.

8.9.3 Loading Stages, Shored Versus Unshored Beams

The various loadings affecting composite beams can be separated into two groups, one involving the precast section alone and the other involving the composite section (Table 8-10). Within each group two extreme loadings (a minimum and a maximum loading) can be identified for design. Loadings affecting the precast-prestressed beam are the prestressing force, the self-weight of the beam, and if the beam is *unshored*, the weight of the cast-in-place slab and other elements such as diaphragms cast at the same time as the slab.

Shoring implies the use, during slab casting and curing, of temporary supports or shores under the precast beam and/or the slab, to relieve the beam from supporting the weight of the slab by itself. After hardening, the shores are removed, the weight of the slab is released, and the beam acting now as a composite beam sustains the additional weight of the slab. Further loadings by superimposed dead loads and live loads are resisted by the composite beam. Differences in loading between the shored and unshored cases are identified in Table 8-10, where the symbols used are explained. As, depending on the loadings, either the precast or the composite section properties apply, the maximum external moment is given for each case. The maximum moment for the precast section is defined by M_p and the additional maximum moment on the composite section by M_c. Final stresses under combined loadings are obtained by superimposing the stresses induced by M_p on the precast section upon those induced by M_c on the composite section. Their determination is covered in Section 8.9.5.

TABLE 8-10 Loading Stages in Prestressed Composite Beams during Service

Resisting Section	*Extreme Loading Stage*	*Loading Combination*: *Unshored Construction*	*Loading Combination*: *Shored Construction*	*Incremental Maximum External Moment*
Precast section	1	$F_i + GP$	$F_i + GP$	. . . .
	2	$(F_i, F) + GP + S + A$	$(F_i, F) + GP$	M_p
Composite section	3	. . . .	$S + A$	. . . .
	4	$SD + L + I$	$S + A + SD + L + I$	M_c

Notation: (F_i, F) = initial prestressing force, effective prestressing force or any value in between; GP = precast girder self-weight; S = cast-in-place slab weight; A = additional weight acting with slab (diaphragms); SD = superimposed dead load; L = live load; I = impact load if any.

8.9.4 Effective and Transformed Flange Width and Section Properties

1. Effective Flange Width

Theoretically, when a monolithic beam with an infinitely wide flange is subjected to flexural loading, the compressive stress on the top fiber is not constant but varies transversely across the flange owing to what is known as "shear lag."[8-41] The stress is maximum at the level of the web and decreases away from it. It would be too complex to utilize the theoretical solution to calculate the compressive stresses in the flange. Instead, a simplified approach is adopted in design, whereas the flange is assumed to have an equivalent effective width b_e over which the flexural stress is assumed uniform transversely.

For practical purposes, the *ACI Code* and AASHTO specifications prescribe design values of the effective flange width b_e for reinforced and prestressed concrete noncomposite and composite flanged beams. They are summarized in Fig. 8-28 for beams with either a slab on one side only, such as end beams, or a slab on two sides such as interior beams. They are to be used in determining the part of the section that resists the loads. Values given by AASHTO apply to nonisolated beams only (which are part of a slab system); if the beam is isolated, AASHTO imposes more stringent values of b_e.

2. Transformed Flange Width

A prestressed composite beam is assumed to behave elastically under service load. The strain distribution along the

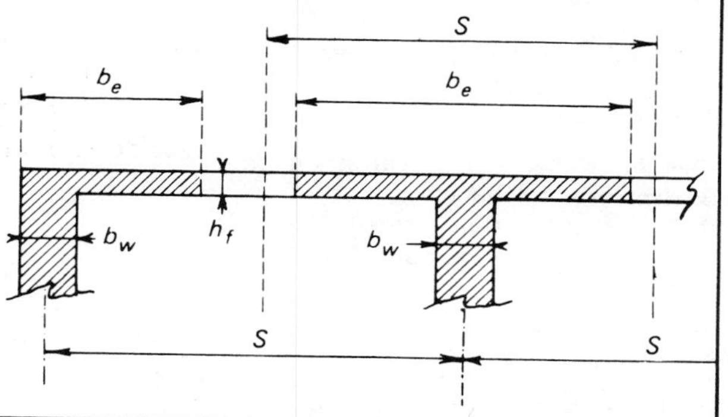

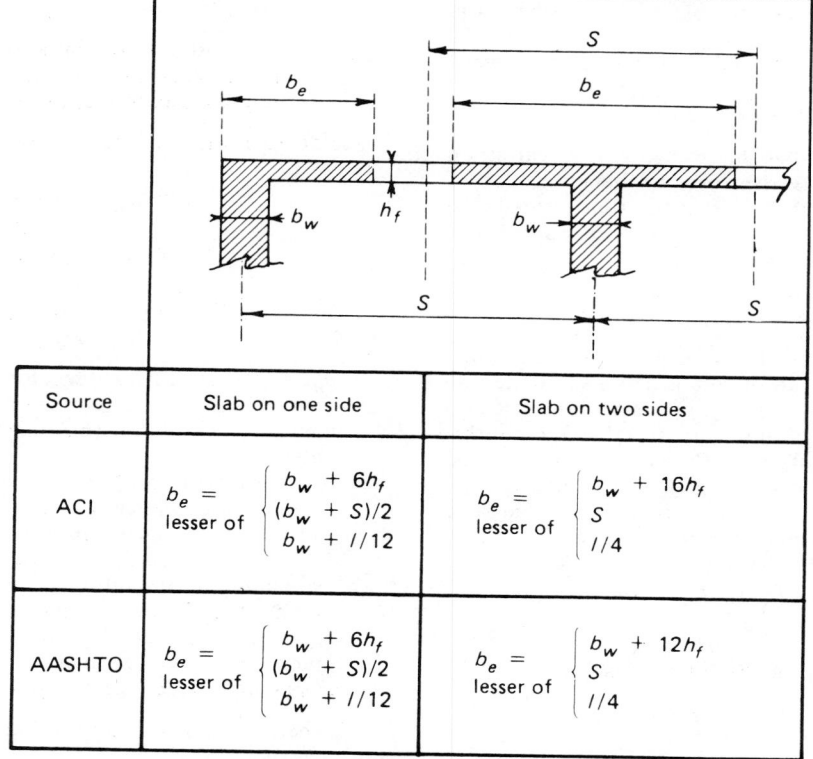

Source	Slab on one side	Slab on two sides
ACI	b_e = lesser of $\begin{cases} b_w + 6h_f \\ (b_w + S)/2 \\ b_w + l/12 \end{cases}$	b_e = lesser of $\begin{cases} b_w + 16h_f \\ S \\ l/4 \end{cases}$
AASHTO	b_e = lesser of $\begin{cases} b_w + 6h_f \\ (b_w + S)/2 \\ b_w + l/12 \end{cases}$	b_e = lesser of $\begin{cases} b_w + 12h_f \\ S \\ l/4 \end{cases}$

Fig. 8-28 Design values of effective flange width b_e.

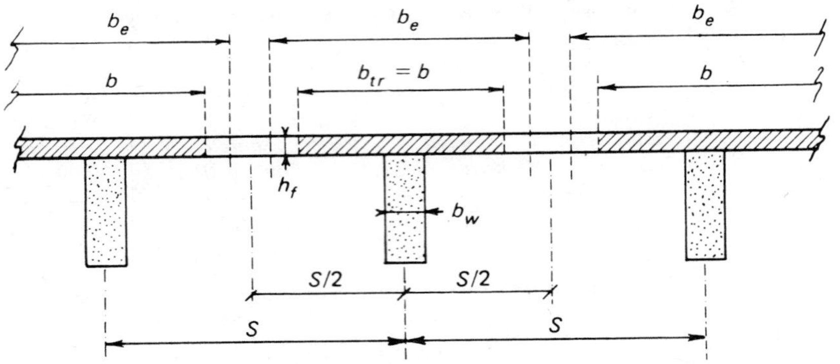

(c)

Fig. 8-29 Transformed versus effective flange widths.

entire section due to a bending moment is assumed linear, and the stresses are calculated from the strains using Hooke's law. As two different materials are involved, the precast beam and the cast-in-place slab, and as each has a different modulus of elasticity, different stresses are generated in each material for the same strain. To account for this difference in design, one of the two elements is transformed into a fictitious element having the same modulus of elasticity as the other one. This is done using the concept of transformed section generally applied to the cast-in-place slab. The slab section of depth h_f and width b_e (Fig. 8-29) is transformed into an equivalent section having same depth h_f and a transformed width b_{tr}. It can be easily shown that b_{tr} is given by:

$$b_{tr} = b_e \frac{(E_c)_{\text{CIPC}}}{(E_c)_{\text{PPC}}} = b_e n_c \qquad (8\text{-}49)$$

where:

$(E_c)_{\text{CIPC}}$ = modulus of elasticity of the cast-in-place concrete slab
$(E_c)_{\text{PPC}}$ = modulus of elasticity of the precast-prestressed concrete beam
n_c = ratio of the above moduli

Equation (8-49) ensures that, under bending (linear strain diagram), the total compressive force in the actual slab of width b_e is the same as the force in the transformed slab of width b_{tr} and having the same modulus of elasticity as the precast beam.

3. Cross-Section Properties

The use of a transformed width b_{tr} leads to a fictitious slab having the same strength and modulus as those of the precast beam. Consequently, the composite section can also be considered transformed into an equivalent monolithic (noncomposite) section having the same strength and modulus as the precast beam. Its geometric properties are determined and directly used in flexural design in the same manner as monolithic noncomposite sections (Fig. 8-30). In order to correlate with the notation used earlier to describe the flange width in general, let us call:

$$b = b_{tr} \qquad (8\text{-}50)$$

Thus the area of the composite section is given by:

$$A_{cc} = A_c + h_f b_{tr} = A_c + h_f b = A_c + h_f b_e n_c \qquad (8\text{-}51)$$

where A_c is the cross-sectional area of the precast concrete beam.

Other section properties listed in Fig. 8-30 are determined according to established procedures. The subscript c is used to differentiate them from those of the precast beam.

8.9.5 ACI Code Provisions for Horizontal Shear

Investigation of horizontal shear strength is required in all composite members and is treated similarly to other effects, such as vertical shear, bending, and the like.

Shear Transfer Resistance. As for vertical shear, torsion, and bending, the design for horizontal shear transfer is based on ultimate strength requirements. Shear forces are used by the *ACI Code*, but shear stresses are preferably used here.

The design is based on satisfying the following relation:

$$v_{uh} \leqslant \phi v_{nh} \qquad (8\text{-}52)$$

where:

v_{uh} = factored design horizontal shear stress
v_{nh} = nominal horizontal shear strength
ϕ = strength reduction factor = 0.85

The design horizontal shear stress is given by:

$$v_{uh} = \frac{V_u}{b_v d_{pc}} \qquad (8\text{-}53)$$

where V_u is the factored total shear force, and v_{uh} is the required design horizontal shear strength at the interface. The following limiting values of v_{nh} are recommended by the *ACI Code:*

1. v_{nh} = 80 psi (0.55 MPa) when contact surfaces are clean, free of laitance, intentionally roughened to a full amplitude of 1/4 in. (6.3 mm), and ties are not provided.
2. v_{nh} = 80 psi (0.55 MPa) when minimum ties are provided in accordance with eqs. (8-41) and (8-42), and the contact surfaces are not intentionally roughened but are clean and free of laitance. This means that the minimum shear reinforcement in the precast beam is extended to the cast-in-place slab.
3. v_{nh} = 350 psi (2.4 MPa) when minimum ties are provided, and the contact surfaces are intentionally roughened, clean, and free of laitance.

Tie spacing shall not exceed four times the least dimension of supported element (mostly slab thickness) or 24 in.

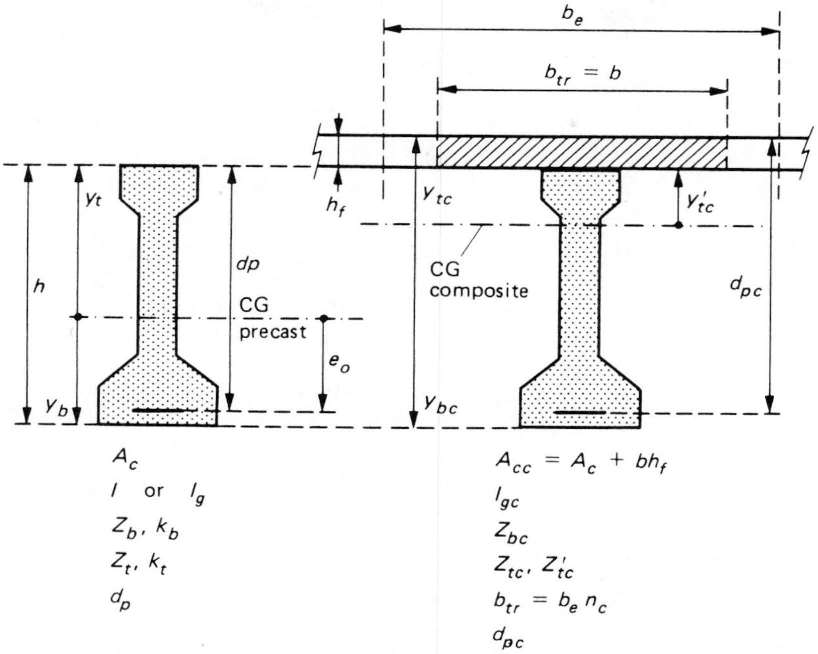

Fig. 8-30 Geometric properties of precast and composite sections.

(60 mm). Ties for horizontal shear may consist of simple bars or wires, multiple-leg stirrups, and vertical legs of smooth or deformed welded wire fabric. Ties must be fully anchored into all interconnected elements.

If v_{uh} exceeds 350 psi (2.4 MPa) at the section considered, the design shall proceed in accordance with section 11.7 of the *Code*, which is based on evaluating the shear friction at the interface. In any case, the value of v_{uh} shall not exceed $0.2f'_c$ or 800 psi (5.52 MPa), whichever is smaller. Otherwise, section dimensions have to be changed.

8.9.6 Flexure: Working Stress Analysis and Design

Most of the concepts developed in Section 8.5 for the analysis and design of noncomposite beams apply to composite beams. The section properties of both the precast beam and the transformed composite beam (Fig. 8-30) are used when needed. The adopted notation and terminology are to be correlated with those of Section 8.5.

1. Extreme Loadings

Whether the beam is shored or not, two extreme loading conditions for the composite system can be identified and essentially bound all others in terms of extreme fiber stresses on the precast beam (Fig. 8-31). The first extreme loading is the initial loading under initial prestressing force and self-weight of the precast beam, that is ($F_i + M_{GP}$ or M_{min}); the second extreme loading corresponds to the cumulative effects of the final prestressing force, self-weight of beam, weight of cast-in-place slab, additional dead load, and live load, that is ($F + M_{max}$). In between these two loadings the beam changes from noncomposite to composite, and the corresponding bending moments generate different types of stresses. In order to follow the variations of stresses, the bending moment M_{max} is broken down into two parts, M_p and M_c, representing the maximum bending moments on the precast and composite sections respectively (see also Table 8-10).

2. Stress Inequality Conditions

Five stress inequality conditions can be written for the composite beam and are of the form:

$$(\text{Actual stress}) \begin{Bmatrix} \geqslant \\ \text{or} \\ \leqslant \end{Bmatrix} (\text{allowable stress}) \qquad (8\text{-}54)$$

Four allowable stresses are binding for the precast beam and one for the cast-in-place slab. The two stress inequality conditions for the first extreme loading are identical to the

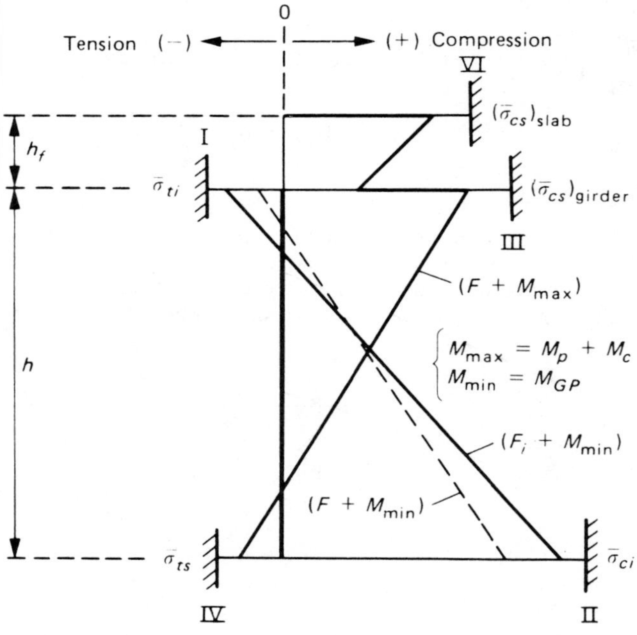

Fig. 8-31 Extreme loading's stresses in composite sections.

TABLE 8-11 Useful Ways of Writing the Stress Inequality Conditions for Composite Sections

Way	*Stress Condition*	*Inequality Equation*
1	I	$(F_i/A_c)[1-(e_o/k_b)] + M_{GP}/Z_t \geqslant \bar{\sigma}_{ti}$
	II	$(F_i/A_c)]1-(e_o/k_t)] - M_{GP}/Z_b \leqslant \bar{\sigma}_{ci}$
	III	$(\eta F_i/A_c)[1-(e_o/k_b)] + M_p/Z_t + M_c/Z'_{tc} \leqslant \bar{\sigma}_{cs}$
	IV	$(\eta F_i/A_c)[1-(e_o/k_t)] - M_p/Z_b - M_c/Z_{bc} \geqslant \bar{\sigma}_{ts}$
2	I	$e_o \leqslant k_b + (1/F_i)(M_{GP} - \bar{\sigma}_{ti} Z_t)$
	II	$e_o \leqslant k_t + (1/F_i)(M_{GP} + \bar{\sigma}_{ci} Z_b)$
	III	$e_o \geqslant k_b + (1/F_i)[(M_p + M_c Z_t/Z'_{tc} - \bar{\sigma}_{cs} Z_t)/\eta]$
	IV	$e_o \geqslant k_t + (1/F_i)[(M_p + M_c Z_b/Z_{bc} + \bar{\sigma}_{ts} Z_b)/\eta]$
3	I	$F_i \leqslant (M_{GP} - \bar{\sigma}_{ti} Z_t)/(e_o - k_b)$
	II	$F_i \leqslant (M_{GP} + \bar{\sigma}_{ci} Z_b)/(e_o - k_t)$
	III	$F = \eta F_i \geqslant (M_p + M_c Z_t/Z'_{tc} - \bar{\sigma}_{cs} Z_t)/(e_o - k_b)$
	IV	$F = \eta F_i \geqslant (M_p + M_c Z_b/Z_{bc} + \bar{\sigma}_{ts} Z_b)/(e_o - k_t)$
All	V	$e_o \leqslant (e_o)_{mp} = y_b - (d_c)_{min}$ = maximum practical eccentricity
	VI	$(M_c/Z_{tc}) \leqslant (\bar{\sigma}_{cs})_{slab}$

noncomposite beam case. They are given in Table 8-11 and are numbered I and II. Two other stress inequality conditions (numbered III and IV) can be written for the precast beam under the second extreme loading, and one condition can be written for the cast-in-place slab (numbered VI). In addition, the practicality condition which states that the prestressing steel must be placed inside the section also applies here and is given the same number (V) as in Tables 8-5 and 8-6.

Let us develop stress conditions III, IV, and VI for the composite beam assuming unshored construction first. Referring to the stress-versus-loading diagrams of Fig. 8-32 and summing up the stresses on the top fiber of the precast beam, we have:

$$\frac{\eta F_i}{A_c}\left(1-\frac{e_o}{k_b}\right) + (M_{GP}+M_S)/Z_t + (M_{SD}+M_L)\frac{y'_{tc}}{I_{gc}} \leqslant \bar{\sigma}_{cs} \quad (8\text{-}55)$$

where:

M_{GP} = bending moment due to self-weight of precast prestressed beam or girder
M_S = bending moment due to cast-in-place slab
M_{SD} = bending moment due to superimposed dead load
M_L = bending moment due to live load
y'_{tc} = distance from centroid of composite beam to top fiber of precast beam = $h - y_{bc}$

In the above equation we have used the final value of the prestressing force. However, the effect of time, when needed, should not be ignored; it is described as "time lapse" in Fig. 8-32, and the notation (F_i, F) is used to remind the reader that any value between F_i and F may also apply.

If the precast beam is shored during casting and curing of the cast-in-place slab, the slab weight will apply only when the section acts as a composite section. Referring to Fig. 8-33, the third stress inequality condition leads to:

$$\frac{\eta F_i}{A_c}\left(1-\frac{e_o}{k_b}\right) + M_{GP}/Z_t + (M_S+M_{SD}+M_L)\frac{y'_{tc}}{I_{gc}} \leqslant \bar{\sigma}_{cs} \quad (8\text{-}56)$$

Equations (8-14) and (8-15) can be rewritten in a single form emphasizing the separate effects of the moments acting on the precast beam and composite beam, respectively, that is:

$$\frac{\eta F_i}{A_c}\left(1-\frac{e_o}{k_b}\right) + \frac{M_P}{Z_t} + \frac{M_c}{Z'_{tc}} \leqslant \bar{\sigma}_{cs} \quad (8\text{-}57)$$

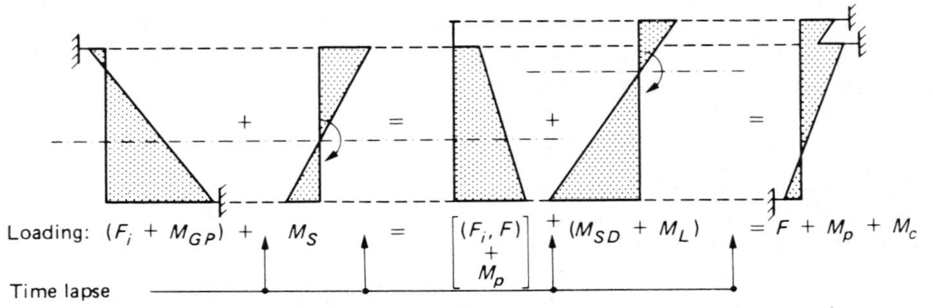

Fig. 8-32 Stress vs. loading diagrams: unshored composite beam.

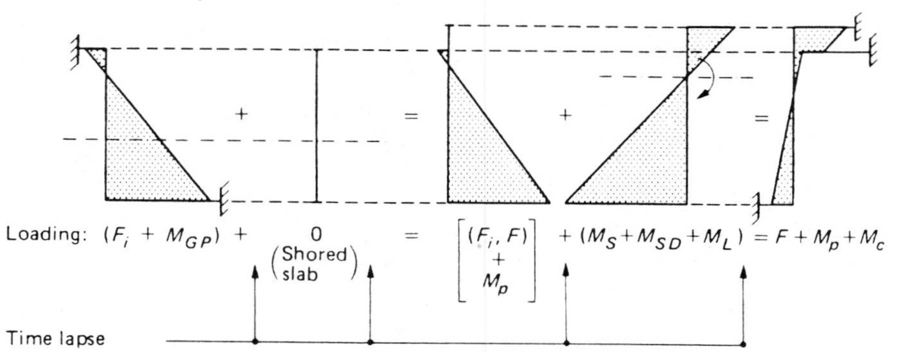

Fig. 8-33 Stress vs. loading diagrams: shored composite beam.

where:

M_p = sum of external bending moments acting on the precast beam

M_c = sum of external bending moments acting only on the composite beam

$Z'_{tc} = I_{gc}/y'_{tc}$

The values of M_p and M_c are given for the unshored case by:

$$M_p = M_{GP} + M_S \quad (8\text{-}58)$$

$$M_c = M_{SD} + M_L \quad (8\text{-}59)$$

and for the shored case by:

$$M_p = M_{GP} \quad (8\text{-}60)$$

$$M_c = M_S + M_{SD} + M_L \quad (8\text{-}61)$$

Any additional load such as weight of diaphragms or weight of attached elements not accounted for in eqs. (8-58) and (8-59) can be added to the values of M_p or M_c where appropriate (Table 8-10).

Following exactly the same approach, the fourth stress inequality condition related to the bottom fiber of the composite section at maximum load leads to:

$$\frac{\eta F_i}{A_c}\left(1 - \frac{e_o}{k_t}\right) - \frac{M_p}{Z_b} - \frac{M_c}{Z_{bc}} \geqslant \bar{\sigma}_{ts} \quad (8\text{-}62)$$

in which the notation is defined earlier.

The last stress inequaltiy condition states that the maximum compressive stress in the cast-in-place slab at full service load is less than or equal to the allowable compressive stress in the slab, thus:

$$\frac{M_c}{Z_{tc}} \leqslant (\bar{\sigma}_{cs})_{\text{slab}} \quad (8\text{-}63)$$

All the above stress inequality conditions are summarized in Table 8-11 and can be directly correlated with Tables 8-5 and 8-6 given for the noncomposite case. Three different ways of writing these equations are proposed, each emphasizing a particular variable, namely, the stress, the eccentricity e_o, and the prestressing force.

The sixth condition, corresponding to eq. 8-63, is largely satisfied in the great majority of cases. The five others can be treated in a manner exactly similar to their handling in Section 8.5. Thus the working stress analysis and design of composite beams are reduced to those of noncomposite beams, and the concepts developed in Section 8.5 apply with the equations given in Table 8-11.

8.9.7 Designing for Ultimate Strength, Shear, and Torsion

Failure in composite beams often occurs at the interface between the precast beam and the cast-in-place slab where excessive slip may develop under increased loading. If, however, an adequate connection is provided to assure shear transfer, prestressed concrete composite beams behave at ultimate in much the same way as noncomposite beams.

The simplest and fastest approach to analyze a composite beam at ultimate is to assume a monolithic section (with transformed slab width) similar to what was used in the working stress design approach of the previous section. In such a case the analysis and design at ultimate are identical to those of noncomposite beams, and the provisions developed in Section 8.6 as well as the design flow chart (Fig. 8-14) apply, provided d_{pc} is used instead of d_p.

The analysis and design for shear and torsion of composite beams are similar to those of monolithic noncomposite beams, provided adequate shear transfer is ensured at the interface. The design is based on strength resistance to factored loads and the provisions given in Section 8.7 for noncomposite beams as well as the design flow chart (Fig. 8-21) apply here.

8.10 CONTINUOUS BEAMS

8.10.1 Sign Convention and Special Notation

In dealing with indeterminate structures, it is very important to follow a consistent sign convention, as one does not know a priori the sign of the secondary moments (defined in Section 8.10.2). Their value and sign are derived from the analysis. The sign convention set in Section 8.5.2 remains valid here. In particular, the eccentricity of the prestressing force $e_o(x)$ at any section x is assumed positive when F is below the neutral axis and negative when it is above it (Fig. 8-34). For vertical members, positive is to the right and negative is to the left. The same sign convention holds for the C line.

The prestressing force F produces a compression in the concrete and is assumed positive. The primary moment generated by F at any section x and the corresponding eccentricity $e_o(x)$ have opposite signs.

The following notations (some of which are explained later) will be used:

e = eccentricity in general (used mostly in figures to reduce the burden of subscripts)

$e_o(x)$ = eccentricity of the centroid of the prestressing steel at section x

$e_c(x)$ = eccentricity of the C line at section x

$e_{oc}(x)$ = eccentricity of the Zero-Load-C line (ZLC line)

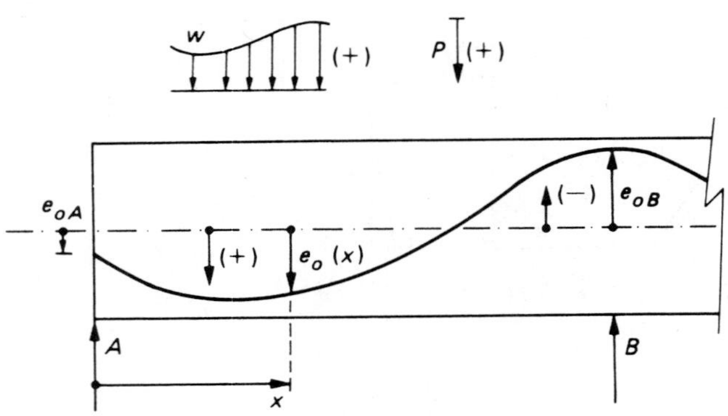

Fig. 8-34 Sign convention for tendon eccentricity.

at section x. The ZLC line is a statically indeterminate structure is due to the sole effect of prestressing (no external loads).

$M(x)$ = external moment, in general, at section x

$M_1(x)$ = primary moment due to the prestressing force at section x. $M_1(x) = -Fe_o(x)$

$M_2(x)$ = secondary moment at section x due to prestressing

$M_F(x)$ = total moment at section x due to prestressing. $M_F(x) = M_1(x) + M_2(x)$

The reference to section x is ignored in the above notation if it is obvious that no confusion will result in the mathematical treatment.

8.10.2 Secondary Moments and Zero-Load-C Line

The prestressing force F and an external moment M acting at any section of a simply supported beam can be resolved into a force $C = F$ acting at a distance $\delta = M/F$ from the line of action of F. The geometric lieu of the C force along the various sections of a member is defined as the C line or pressure line (also called thrust line). For a simply supported beam the eccentricity of the C line with respect to the centroid of the section is defined as:

$$e_c(x) = e_o(x) - \frac{M(x)}{F} \qquad (8\text{-}64)$$

where $e_o(x)$ is the eccentricity of the prestressing force and $M(x)$ the external moment at section x. Equation (8-64) suggests that if no external moment is applied, the eccentricity of the C force at any section is identical to that of the prestressing steel; hence, the C line coincides with the trajectory of the steel. Thus, in a simply supported beam (or statically determinate structure), the effect of prestressing is reduced to that of the tendons alone at each section. The supports do not provide any restraint to the deformation of the structure, and the prestressing moment is given by $M_F(x) = -Fe_o(x)$. This is not the case, however, for prestressed continuous beams where intermediate supports restrict the free deformation of the structure, hence leading to support reactions called secondary reactions. Secondary reactions act like concentrated loads on a simply supported beam. They generate at each section a moment called secondary moment (also called parasitic moment or hyperstatic moment). Therefore, under the sole effect of prestressing (i.e., in the presence of no external loads) two types of moments are generated at each section of a continuous beam: the primary moment defined as for a simply supported beam by $M_1(x) = -Fe_o(x)$ and the secondary moment $M_2(x)$ generated by the secondary reactions. The moment due to prestressing at each section x becomes:

$$M_F(x) = M_1(x) + M_2(x) = -Fe_o(x) + M_2(x) \qquad (8\text{-}65)$$

Since, in general, $M_F(x)$ is determined from the analysis of the structure, say by moment distribution, the secondary moment is derived from eq. (8-65) as:

$$M_2(x) = M_F(x) - M_1(x) = M_F(x) + Fe_o(x) \qquad (8\text{-}66)$$

Secondary moments are secondary in nature but not in magnitude. They can represent a significant portion of the prestressing moment and, hence, must be accounted for in design. Advantage can be taken of their presence and may lead to savings in the prestressing force.

Because of the existence of secondary moments, the C line under the sole effect of prestressing, defined here as the Zero-Load-C line (ZLC line), because no external load is applied,

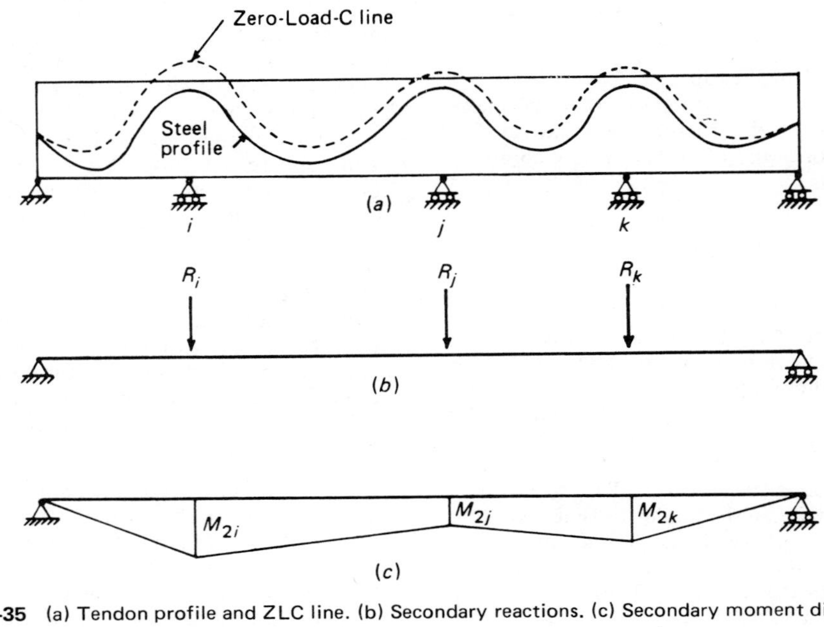

Fig. 8-35 (a) Tendon profile and ZLC line. (b) Secondary reactions. (c) Secondary moment diagram.

does not coincide with the prestressing steel. Its eccentricity at any section x is given by:

$$e_{oc}(x) = -\frac{M_F(x)}{F} = e_o(x) - \frac{M_2(x)}{F} \qquad (8\text{-}67)$$

Equation (8-67) suggests that the prestressing force acts as if it had an eccentricity $e_{oc}(x)$. Thus, $e_{oc}(x)$ can also be described as the effective eccentricity of the prestressing force. In the presence of an external moment $M(x)$, the eccentricity of the C line becomes:

$$e_c(x) = e_{oc}(x) - \frac{M(x)}{F} = e_o(x) - \frac{M(x) + M_2(x)}{F} \qquad (8\text{-}68)$$

which suggests that $M_2(x)$ acts like an additional external moment on the section. When the secondary moments are equal to zero, eq. (8-68) is reduced to eq. (8-64), used in simply supported beams. The tendon profile in the continuous beam is then said to be concordant.

If both the ZLC line and the steel profile are known, eq. (8-67) leads to:

$$\frac{M_2(x)}{F} = e_o(x) - e_{oc}(x) = \Delta e_{oc}(x) \qquad (8\text{-}69)$$

$\Delta e_{oc}(x)$ can be considered as a fictitious increase or decrease in the eccentricity of the prestressing steel at section x, due to the presence of secondary moments.

A typical tendon profile and corresponding ZLC line are plotted in Fig. 8-35(a). The secondary reactions and corresponding secondary moments are schematically shown in Fig. 8-35(b, c). As the secondary reactions are generated by prestressing, they form a system of forces with a null resultant. Since they act as concentrated forces at the supports, the variation of secondary moment between consecutive supports is linear.

8.10.3 Example: Secondary Moments and Concordancy Property

The existence of secondary moments can be illustrated by the simple example of a two-span continuous beam prestressed by a straight tendon with eccentricity e throughout (Fig. 8-36a). Assume that only the effect of prestressing is considered; that is, no external loads are acting. If the intermediate support B were nonexistent, the beam would be simply supported at A and C and would camber under the effect of prestressing. The presence of support B restrains the movement of the beam and hence generates a reaction R_B. The magnitude and direction of R_B are such that R_B should create a deflection at B equal and opposite to the camber created by prestressing. Referring to Fig. 8-36 and assuming a simply supported beam with span $2l$, the deflection due to R_B is given by:

$$\Delta_1 = \frac{R_B(2l)^3}{48EI}$$

The camber due to prestressing is given by:

$$\Delta_2 = -\frac{Fe(2l)^2}{8EI}$$

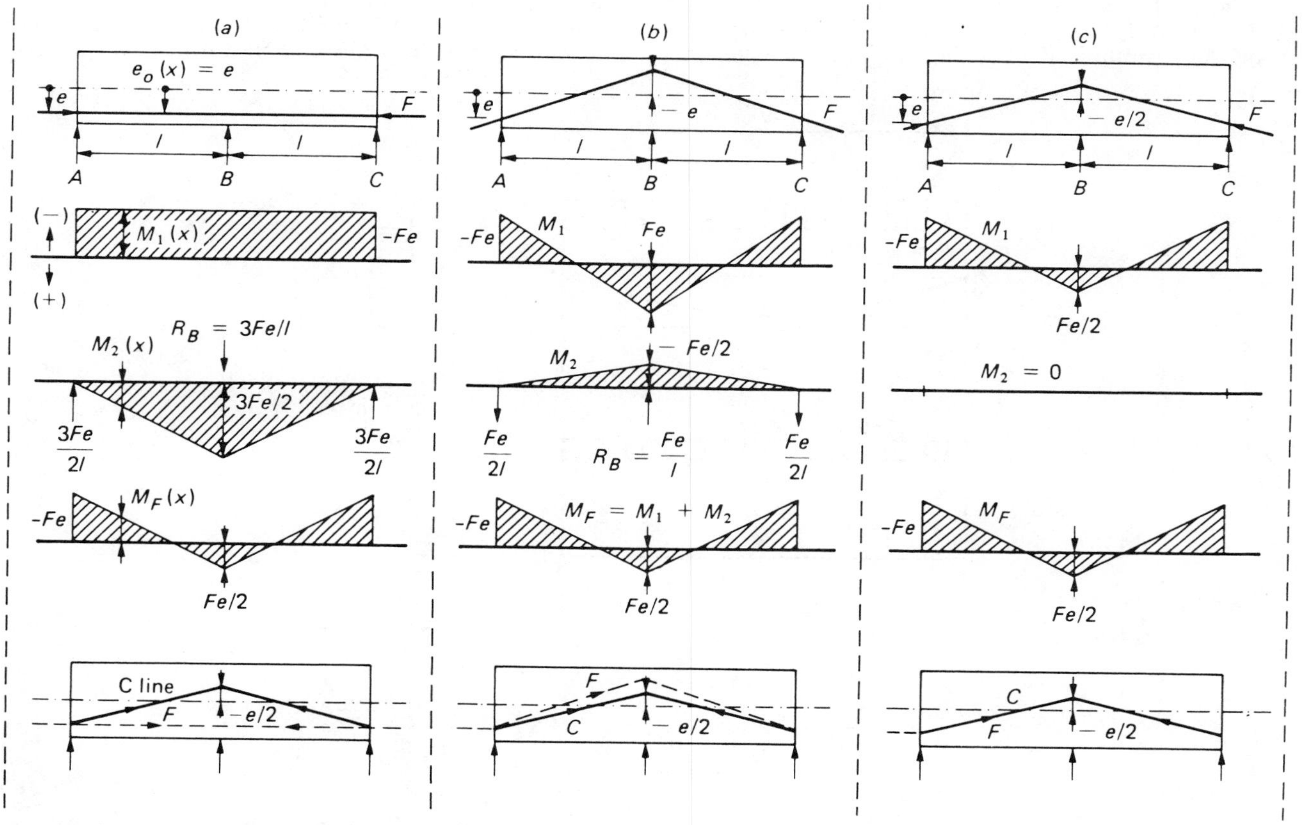

Fig. 8-36 Examples illustrating the variations of primary moments, secondary moments, prestressing moments, and ZLC line in continuous beams.

Setting the sum $(\Delta_1 + \Delta_2)$ equal to zero leads to:

$$R_B = \frac{3Fe}{l}$$

Prestressing the continuous beam ABC of Fig. 8-36(a) generates at each section x a primary moment $M_1(x) = -Fe$ and a secondary moment $M_2(x)$ induced by the secondary reaction R_B. $M_2(x)$ is obtained by treating the beam AC as simply supported with a concentrated force R_B at B. The primary and secondary moment diagrams for this case are plotted in Fig. 8-36(a), as well as their sum, the prestressing moment $M_F(x)$. Two observations can be made: (1) the secondary moment can be significant in magnitude, and (2) the prestressing moment in a continuous beam can be substantially different from the primary moment otherwise obtained if the beam was simply supported. Given the prestressing moment, the ZLC line can be determined from eq. (8-67) and is also shown in Fig. 8-36(a). Note that it deviates substantially from the steel profile. It has the same eccentricity e at the end supports and an eccentricity $-e/2$ at B.

Two other cases are covered in Fig. 8-36. In case b the eccentricity of the prestressing force at B is changed from $+e$ (case a) to $-e$, while the eccentricities at the supports are kept the same. Using the deflection equations given in Fig. 8-24, the reaction R_B is calculated in a manner similar to case a. The primary, secondary, and total moments are determined and plotted in Fig. 8-36(b). It can be observed that, although both the primary and secondary moment diagrams are different from case a, the resulting prestressing moment diagram is the same. Hence, the corresponding ZLC line is also the same.

Finally, in case c of Fig. 8-36 the steel profile is modified to show an eccentricity $-e/2$ at support B. This profile is the same as the ZLC line found in cases a and b. Following the same analytic steps, it is found that the secondary reaction R_B vanishes. Hence, the secondary moment is zero at any section, and the prestressing moment becomes equal to the primary moment (Fig. 8-36c). It is observed that the prestressing moment and the ZLC line are the same as for cases a and b. Moreover, in case c the ZLC line coincides with the trajectory of the steel. When this occurs, the tendon profile is said to be "concordant." Therefore, for case c we have a concordant steel profile, while for cases a and b we have nonconcordant steel profiles.

The above example suggests an additional important result: the three different steel profiles of cases a, b, and c in which only the eccentricity at the intermediate support B was varied, led to the same ZLC line. This result is due to a property of the "linear transformation" explained in Refs. 8-4 and 8-15.

8.10.4 Design by the Load-Balancing Method

The analysis of prestressed concrete statically indeterminate structures is hindered by the existence of secondary moments due to prestressing. To determine the secondary moments the effects of the prestressing tendons are replaced by those of external loads equivalent to the tendons, and the analysis of the structure is performed for such loads. The procedure is lengthy, especially during preliminary design. However, one method that overcomes this drawback is the load-balancing method proposed by T. Y. Lin.[8-42] It reduces the analysis of the prestressed structure to that of a nonprestressed structure in which consideration of secondary moments is essentially bypassed. It offers a particularly simple and elegant means to design continuous beams, slabs, shells, and frames.

Balancing the external load consists of selecting a prestressing force and steel profile that create a transverse load exactly equal and opposite to the external load. For instance, to balance a uniform load w_b in a simply supported beam, a parabolic steel profile with zero end eccentricities can be selected (Fig. 8-37). The prestressing force F needed

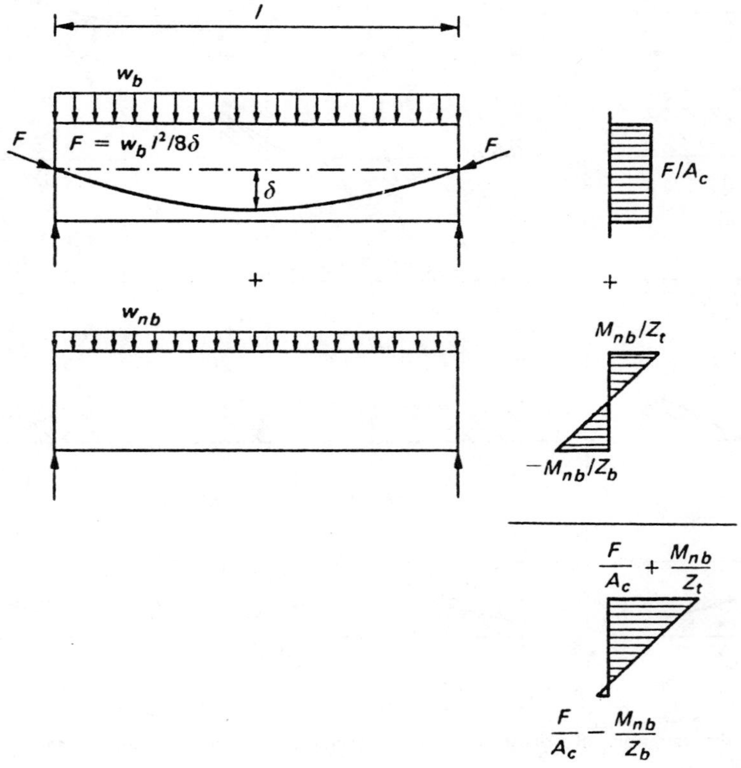

Fig. 8-37 Typical stresses under balanced and unbalanced loads.

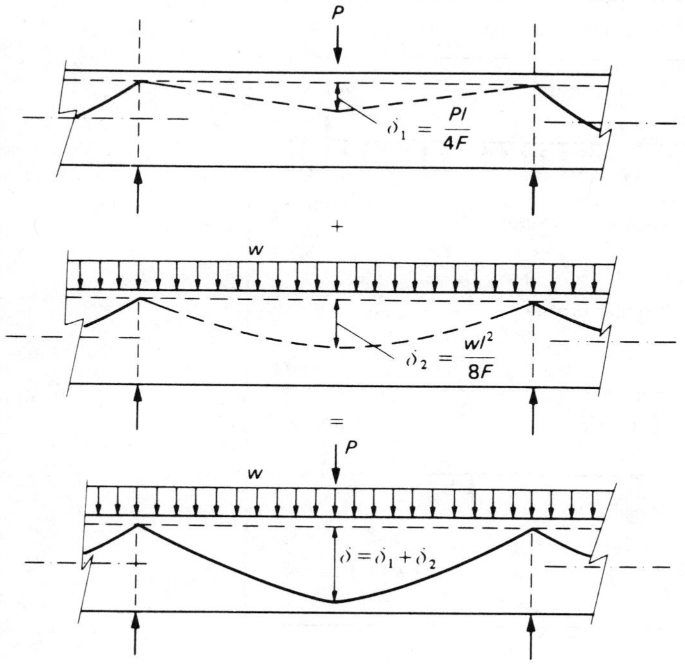

Fig. 8-38 Superposition in load balancing.

will be a function of the load to be balanced, w_b, and the acceptable sag δ of the tendon. As the transverse load created by the tendons balances exactly the external load, a uniform state of stress F/A_c develops throughout the beam. The beam remains essentially level, and no deflection or camber is observed. Note that, in order to balance the load, the end eccentricities were taken equal to zero; otherwise an end moment that disturbs the uniform state of stress is generated. The same approach for load balancing can be applied to a continuous beam, noting that eccentricities at intermediate supports are not necessarily equal to zero because in such a case the moments on each side of the support balance each other. If because of load balancing the continuous beam remains level, no secondary moments are generated (under F and the balanced load w_b), and the beam can be analyzed for the unbalanced load w_{nb} as if it were a continuous nonprestressed beam in which a uniform compression F/A_c were present. The moments $M_{nb}(x)$ induced by the unbalanced loads are then calculated by elastic analysis and the corresponding stresses at the extreme fibers, $M_{nb}(x)/Z_t$ and $-M_{nb}(x)/Z_b$, determined. Resulting stresses due to the uniform compression and the unbalanced moments are computed and compared with the allowable stresses. This is illustrated in Fig. 8-37 for the case of a simply supported beam.

The balanced load need not necessarily be uniform. It can be a concentrated load, a uniform load, or a combination thereof (Fig. 8-38). For a uniform load, a shallow parabolic tendon profile is generally selected, while a linear profile with a sharp directional change is used for a concentrated load. Assuming shallow members, the principle of superposition holds; hence, the combined profile of Fig. 8-38 would balance both the applied uniform and concentrated loads.

In an approach similar to that followed in Section 8.5, it will here also be assumed that the concrete cross section is given, and that the design is reduced to finding the prestressing force and its profile. Referring to a continuous beam (Fig. 8-39) where a typical end span, intermediate span, and cantilever span are shown, the following design steps are suggested, assuming uniform loadings and relatively shallow members:

1. Select the balanced load w_b. It is generally taken equal to the dead load plus the sustained part of the live load, if any. This will ensure a level structure even under long-term effects. Note that for each span a different value of the uniform balanced load may be used.
2. Select a steel profile made out of parabolas, having maximum practical eccentricities at the intermediate supports and maximum feasible sags in span. Zero eccentricities must be present at the end supports. At cantilever ends, if any, the slope of the tendon has also be zero.
3. Determine the prestressing force required in each span to balance the balanced load of that span. (Use the relations given in Fig. 8-40.) Select the highest value obtained, call it F, and assume it is adopted throughout the member. The sags are then adjusted (reduced) in those spans where the required force was less than F. Their final values are obtained from the relationship between F and the balanced load.

   Note that, in cantilever spans, because of the implied relationship between the sag and the eccentricity at the near-support section (Fig. 8-39), the near-support eccentricity may have to be adjusted to achieve the required sag, and this will influence the sag in the adjacent span.
4. Compute the unbalanced moment $M_{nb}(x)$ due to the unbalanced load w_{nb} by elastic analysis as if the beam were continuous nonprestressed.
5. Check whether the stresses at critical sections and other key sections are within allowable limits. Stresses are given by:

$$\frac{F}{A_c} + \frac{M_{nb}(x)}{Z_t} \tag{8-70}$$

   and:

$$\frac{F}{A_c} - \frac{M_{nb}(x)}{Z_b} \tag{8-71}$$

   If stresses are acceptable, the design can be pursued. If they are not, the design should be revised. Generally, either the prestressing force has to be increased, or the concrete cross section has to be modified.
6. Modify the theoretical tendon profile shown in Fig. 8-39 by providing smooth transitions over the supports, and check the effects of such modification. It is likely that secondary moment will be generated. These mo-

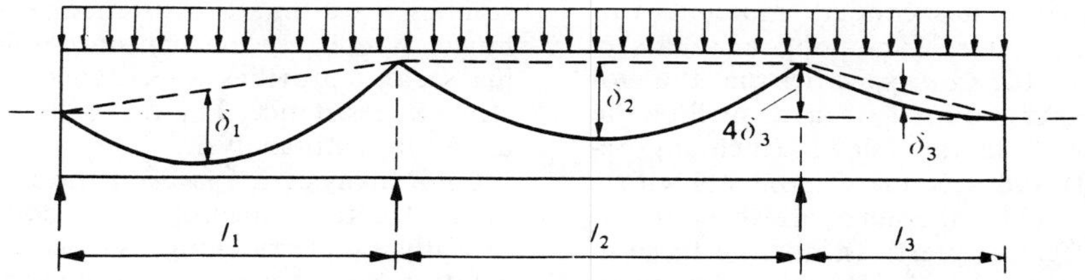

Fig. 8-39 Typical load balancing in a continuous beam.

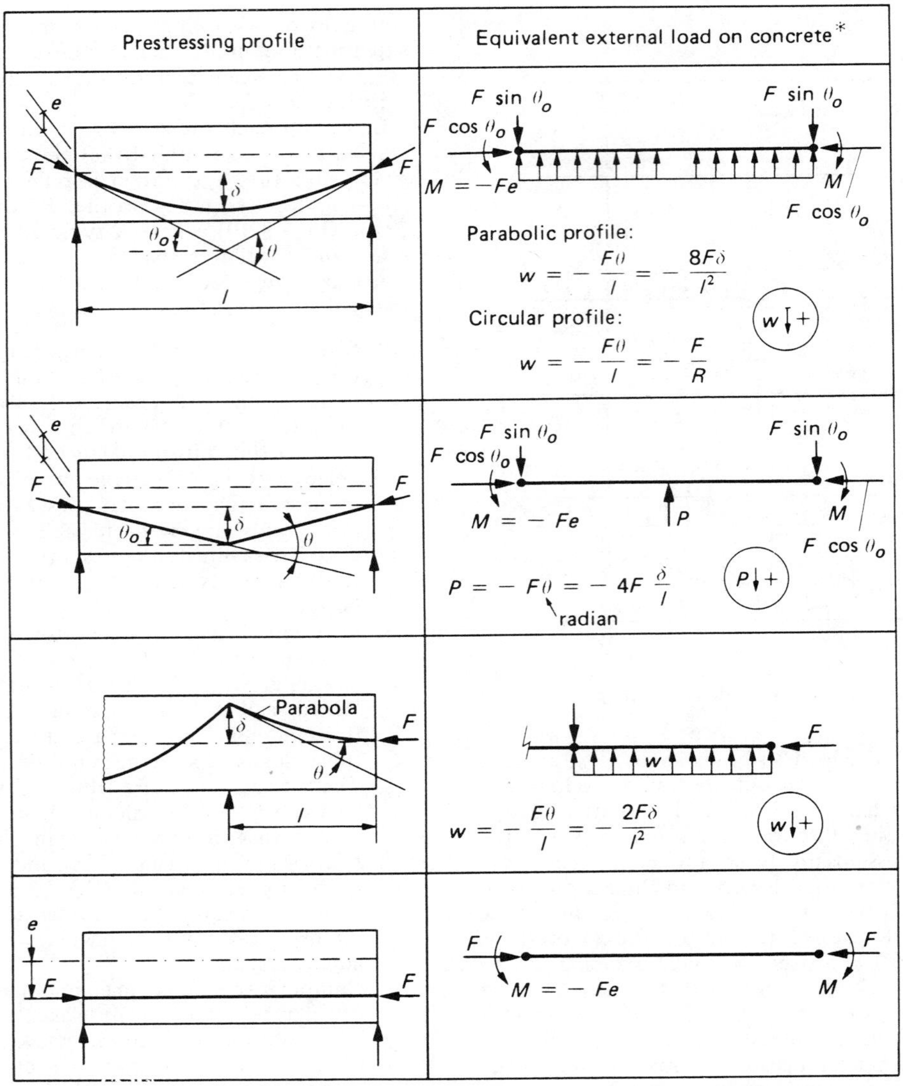

*Assuming shallow members for δ in function of θ.

Fig. 8-40 Equivalent load formulas for typical tendon profiles in beams.

ments are often neglected in slabs, but may be significant in beams.

8.10.5 Strength Analysis

The analysis of a statically indeterminate prestressed concrete structure differs from that of a statically determined structure by two main aspects: (1) secondary moments must be accounted for in the analysis; and (2) the attainment of ultimate moment resistance at one critical section does not necessarily lead to the collapse of the structure, because plastic hinges may form at several critical sections and redistribution of moments occurs.

It was pointed out earlier that secondary moments can be considered as additional external moments at each section. In section 18.10.3 the *ACI Code* specifies that the moments used in computing the required strength shall be the sum of the moments due to factored loads (including redistribution, if permitted) and secondary moments with a load factor of 1. The secondary moments are to be determined on the basis of effective prestress, not the force in the tendons at ultimate behavior.[8-43] Hence, referring to the required strength for one of the most common load combinations, we have:

$$M_u(x) = 1.4M_D(x) + 1.7M_L(x) + M_2(x) \qquad (8\text{-}72)$$

where $M_2(x)$ is the secondary moment at section x, obtained from elastic analysis. The ultimate moment resistance $\phi M_n(x)$ at a given section can be computed as described in Section 8.6.

If the most critical section in a continuous beam has sufficient rotational capacity at ultimate, it behaves as a plastic hinge. Failure of the member will not follow the formation of the first plastic hinge. Instead, redistribution of moments occurs, and, if the loading is increased, another section reaches its ultimate resistance, leading to another plastic hinge. A collapse mechanism develops in which each hinging section provides a resistance equal to its own ultimate strength resistance. Failure of the member can then be predicted by limit analysis.

Limit analysis is based on the formation of plastic hinges in a structure, leading to a collapse mechanism. Theoretically, a plastic hinge shows an elastoplastic moment-rotation response similar to the behavior of steel structures.

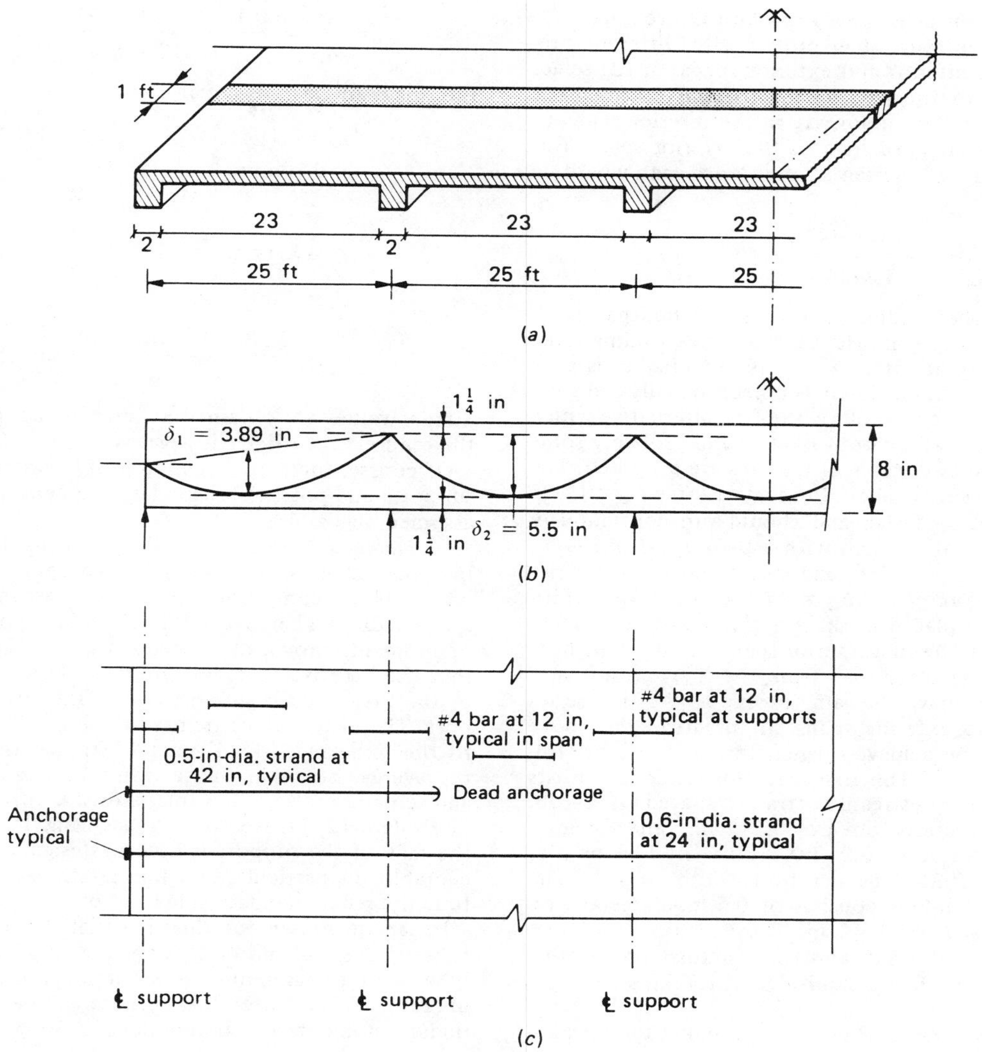

Fig. 8.41 Slab of example problem.

In reinforced and prestressed concrete the cracking of members at one critical section reduces substantially the corresponding stiffness, and a pseudoplastic response may develop. Such a response may simulate a plastic hinge if sufficient ductility exists at the critical section. Adequate ductility can be achieved if a low reinforcing index q is used, if nonpréstressed bonded reinforcement is added to the section, and if sufficient resistance to shear cracking is provided.

8.10.6 Example: One-Way Slab Design

A parking structure consists of a 60-ft-wide one-way slab system with five consecutive spans, 25 ft center to center. The slab is supported on columns through 2-ft-wide beams placed along column lines (Fig. 8-41a). Provide a preliminary design for the slab, assuming unbonded tendons are specified and given the following information: normal weight concrete, $\gamma_c = 150$ pcf, $f_c' = 5000$ psi, $f_{ci}' = 3500$ psi, $E_c = 4.287 \times 10^6$ psi, minimum concrete cover to the centroid of the prestressed reinforcement = $1^1/_4$ in., the tendons are strands with $f_{pu} = 270$ ksi, and assumed $f_{pe} \simeq 160$ ksi. A uniform live load of 100 psf is prescribed.

Let us select a slab depth h = 8 in. which corresponds to a span-to-depth ratio $l/h = 37.5$, and let us provide a design for a 1-ft-wide strip of slab, assuming it acts as a one-way continuous beam. The corresponding dead weight of the strip per linear foot is $w_G = 100$ plf.

Prestressing Steel. Load balancing will be used to determine the prestressing steel. A theoretical tendon profile made out of parabolas with maximum practical eccentricities or minimum cover at support sections and at midspan or near midspan sections is selected. The vertex of each parabola falls at midspan for a typical interior span and at $0.414l$ (near midspan) for the exterior span. The corresponding sags are shown in Fig. 8-41(b), and the prestressing force to balance a uniform load $w_b = w_G = 0.1$ klf is given by:

For the exterior span:

$$F_1 = \frac{w_b l_1^2}{8\delta_1} = \frac{0.1 \times \overline{25}^2}{8 \times 3.89/12} = 24.1 \text{ kips}$$

For a typical interior span:

$$F_2 = \frac{w_b l_2^2}{8\delta_2} = \frac{0.1 \times \overline{25}^2}{8 \times 5.5/12} = 17.05 \text{ kips}$$

One of two different alternatives can be followed, namely, either (1) keep the same steel profile with different prestressing forces at interior and exterior spans, or (2) select the highest value of the prestressing force F_1 throughout the slab and adjust the tendon sag in the interior spans to achieve the same balanced load as the exterior spans. For the second alternative, the sag at an interior span would be given by:

$$\delta_2 = \frac{w_b l_2^2}{8F_1} = \frac{0.1 \times 25^2}{8 \times 24.1} \times 12 = 3.89 \text{ in.}$$

This means that the tendon eccentricity at midspan is reduced from 2.75 to 1.14 in. Hence, the corresponding value of d_p is reduced by about 24%, and the nominal resistance of the cracked section at ultimate is at least equally reduced. Therefore, it is decided to follow the first alternative which leads to a lesser overall amount of steel and probably some cost savings. Thus, two series of tendons are proposed. The tendons of the first series should provide a force F_2 = 17.05 kips per foot width of slab and should run continuously through the length of the slab with a theoretical profile as shown in Fig. 8-41(b). The tendons of the second series should provide a force $F_1 - F_2$ = 7.05 kips per foot width of slab and will be placed in the exterior spans only. They will be anchored in the first interior span with dead anchors placed at about a quarter span from the first interior support, and they will have the same profile as the first series of tendons in the exterior spans. In practice, the above requirements can be achieved using, for instance, the following tendon layout. The first series of tendons consists of 0.6-in.-diameter prestressing strands spaced at 24 in. Each has a cross-sectional area of 0.217 in.^{2} and carries a final prestressing force of 0.217 × 160 = 34.72 kips that is equivalent to 17.36 kips per foot width of slab. The second series of tendons consists of 0.5-in.-diameter prestressing strands spaced at 42 in. They each have a cross-sectional area of 0.153 in^2 and carry a final prestressing force of 24.48 kips that is equivalent to 7 kips per foot width of slab.

Bonded Reinforcement. Because unbonded tendons are used, the *ACI Code* prescribes the addition of a minimum amount of bonded bar reinforcement at critical sections. The following required values per foot width of slab can easily be obtained assuming f_y = 60 ksi:

$$A_s = 0.004A = 0.004 \times 12 \times 4 = 0.192 \text{ in.}^2$$

This can be achieved using a #4 bar every 12 in. that is equivalent to A_s = 0.20 in^2. The length of these bars should be at least 7 ft 8 in. in both positive and negative moment regions, and they should have a minimum clear concrete cover of 3/4 in.

Reinforcement Layout. The final reinforcement layout in the main direction is schematically described in Fig. 8-41(c). It leads to the following summary of results per foot width of slab:

Exterior span: at $0.414l_1$ from exterior support:

$$A_{ps} = 0.1085 + 0.044 = 0.1525 \text{ in.}^2$$

$$A_s = 0.20 \text{ in.}^2$$

$$e_o = 2.75 \text{ in.}$$

$$d_p = 6.75 \text{ in.}$$

$$d_s = 7 \text{ in.}$$

Similar values are obtained at the first interior support except that the tendon eccentricity is negative. Note that the average prestress for the exterior span is given by:

$$\sigma_g = \frac{F}{A_c} = \frac{0.1525 \times 160 \times 10^3}{12 \times 8} \simeq 254 \text{ psi}$$

Interior span: at midspan:

$$A_{ps} = 0.1085 \text{ in.}^2$$

$$A_s = 0.20 \text{ in.}^2$$

$$e_o = 2.75 \text{ in.}$$

$$d_p = 6.75 \text{ in.}$$

$$d_s = 7 \text{ in.}$$

$$\sigma_g = \frac{0.1085 \times 160 \times 10^3}{12 \times 8} \simeq 181 \text{ psi}$$

Similar values are obtained at interior supports except that the tendon eccentricity is negative.

Of course, the actual tendon profile is smoothed over the supports and potential secondary moments so generated are assumed negligible.

Unbalanced Load. The unbalanced load is equal to the live load, that is, w_{nb} = 0.1 klf. Using available tables for moments in continuous beams and assuming alternative span loadings when needed, leads to the approximate values of moments shown in the second line of Table 8-12. Note that the negative moments are given both at the centerline of the first interior support and at the left face of the support. The latter values are used in design. The moments due to the unbalanced load generate stresses that are added to the average stress in the section under balanced condition and compared to allowable stresses. Stresses are shown in Table 8-12, lines 3 to 5. Considering that the moment at the face of the support is used in design, all stresses are acceptable. In particular, no tensile stress exceeds in magnitude the recommended code limit of $-6\sqrt{f_c'}$ or -424 psi.

It can be shown for this slab that the ultimate moment resistance, shear and deflections are all satisfactory. Values of strength design moments and nominal moments are given in Table 8-12. Note that because of load balancing, secondary moments are ignored in computing M_u.

8.11 PRESTRESSED CONCRETE SLABS

Prestressed concrete slabs are widely used as floors, roofs, and decks. They can be made to resist bending either primarily in one direction (one-way slabs) or in two directions (two-way slabs). Detailed design information can be found in Refs. 8-44–8-47.

8.11.1 One-Way Slabs

One-way prestressed concrete slabs do not necessarily imply a monolithically cast rectangular plate. They can be made of precast beams placed adjacent to each other between parallel lines of support, post-tensioned T-beams as for joist-type slabs, or a composite system made of precast beams and a cast-in-place slab. In these cases the slab is mostly called a "deck" or a "slab deck." The analysis and design of slab decks are essentially reduced to the analysis and design of beams. Monolithically cast one-way rectangular slabs, whether in single or continuous spans, are designed similarly to rectangular beams.

The design of these slabs starts generally by assuming a slab depth using the span as a guide. For common applications where a relatively light live load is used, the span-to-depth ratio ranges between 25 and 35 for single spans and between 30 and 45 for continuous spans. Values obtained

TABLE 8-12 Numerical Results for Example 8.10.6

		Exterior Span	*First Interior Support*		*Typical Interior Span*
Line	*Item*	*Near Midspan*	*Center of Support*	*Left Face of Support*	*Midspan*
1	Moment due to balanced load or dead load, $w_b = w_G = 0.1$ klf	$0.078 w_b l^2$	$-0.1071 w_b l^2$	$\frac{w_b(l-1)}{2}(1-0.21l)$ kips-ft	$0.364 w_b l^2$
2	Moment due to unbalanced load or live load $w_{nb} = w_L = 0.1$ klf	$0.10 w_{nb} l^2$	$-0.1205 w_{nb} l^2$	$\frac{w_{nb}(l-1)}{2}(1-0.24l)$ kips-ft	$0.0805 w_{nb} l^2$
3	Uniform stress due to balanced loading condition, psi	254	254	254	181
4	Maximum stresses due to unbalanced load, psi	± 586	∓ 706	∓ 563	± 472
5	Maximum stresses under service loads, top and bottom fiber, psi	840 -332	-452 960	-309 817	653 -291
6	Strength design moment, $\lvert M_u = 1.4M_D + 1.7M_L \rvert$, lb-in.	209,400	266,092	208,080	140,858
7	Design nominal moment resistance ϕM_n, lb-in.	225,262	225,262	225,262	182,539
8	A_{ps}, in^2 per foot width of slab	0.1525	0.1525	0.1525	0.1085
9	A_s, in^2 per foot width of slab	0.20	0.20	0.20	0.20

are rounded off to the next quarter- or half-inch. Other minimum requirements related to fire endurance should also be considered in selecting an appropriate depth. Once the depth is obtained, the designer proceeds by considering a unit-wide strip of slab as a beam and determining the prestressing force according to the procedures described in Section 8.10. To determine the prestressing force, the load-balancing approach is generally preferred for continuous slabs. The balanced load is mostly taken equal to the dead load, and a tendon profile made out of parabolic segments with zero eccentricities at the end supports and maximum practical eccentricities at intermediate supports and other critical sections is selected. The required prestressing force per unit width of slab is then determined and the steel profile adjusted accordingly. The force is then translated into tendons at a given spacing. For the final design, the tendon profile is smoothed over the intermediate supports. The distance over which the transition is made is taken equal to about $0.1l$ on each side of the support. The effect of smoothing the tendon on the possible generation of secondary moments is often neglected in practice for slabs.

When the main flexural reinforcement in one-way slabs consists of unbonded tendons, the *ACI Code* (section 18.9) recommends the addition of a minimum amount of ordinary bonded reinforcement. The purpose of such reinforcement is to provide an alternative load-carrying system in the event of a catastrophic failure or abnormal loading of one span (such as due to fire) of a continuous slab that leads to distress in the other spans. Additional information can be found in Ref. 8-44.

8.11.2 Two-Way Slabs

Two-way prestressed concrete slabs belong mostly to the following groups: edge-supported slabs, flat slabs, and flat plates. However, the flat plate slab construction is the most common form of prestressed concrete slab construction. Flat plates are extensively used in residential and commerical buildings of all types, including hotels, parking structures, and the like. They offer numerous advantages, including a minimal slab depth with an unobstructed bottom surface that can be formed and finished easily. Construction time is reduced, and substantial cost savings may be achieved. In the United States, most prestressed concrete flat plates are cast in place and post-tensioned using unbonded tendons. The tendons are generally placed from one edge to the opposite edge of the slab, spanning several panels. They are stressed from and anchored at the periphery of the slab. The connections with the supporting columns are generally cast monolithically with the slab. However, in the particular case of lift slabs, the flat plate is cast at ground level and then lifted to its final position along pre-erected steel or concrete columns. Connections using steel collars or corbels are then attached between the slabs and the columns.

Many analysis and design methods are applicable to two-way slabs, some of them simpler than others. For analysis, only one approximate method, the equivalent frame method (which is generally applicable to all types of two-way slabs), is considered acceptable by the *ACI Code* and recommended in Ref. 8-44. For design, the load-balancing approach is the fastest and most convenient method to treat prestressed concrete flat slabs. More than any other method, it allows a clear visualization of the relationship between the distribution of moments and the tendon arrangement to balance these moments. Load balancing is applied to each of the two principal directions of the slab in a manner similar to that used in one-way slabs. The design generally starts by assuming a slab depth using the span as a guide. The span-to-depth ratio of two-way slabs designed for live loads of the order of 50 psf (2.4 kN/m^2) commonly ranges from 40 to 55. Higher values are associated with the presence of edge beams or drop panels. The depth of two-way slabs may also be controlled by fire-rating requirements. For a given fire endurance, the thickness requirements of concrete slabs, whether plain, reinforced, or prestressed, are essentially the same. Recommended minimum thicknesses vary from about 3 to 7 in. (7.5 to 17.5 cm) for fire endurances of 1 to 4 hr and for various types of aggregates. For the same range of fire endurance, the recommended clear concrete cover over the reinforcement ranges from about 3/4 in. (19 mm) to 1 1/4 in. (32 mm) for restrained slabs and from about 3/4 to 2 in. (19 to 50 mm) for unrestrained slabs. Such fire safety requirements must be considered when the design would otherwise call for a very thin slab.

Similarly to one-way slabs, it is generally recommended to limit the total length of two-way slabs to about 150 ft (46 m) between construction joints in each direction. For longer lengths the effects of slab shortening on the attached columns and walls should be considered in design.[8-44]

8.12 PRESTRESSED CONCRETE TENSION MEMBERS

8.12.1 Types and Advantages

Prestressed concrete tension members are structural elements predominantly subjected to axial tension. They are mostly linear, circular, or parabolic (catenary) in shape. Linear tension elements, commonly called ties, include restraining ties for arch bridges, soil anchors for retaining structures, and truss members. Circular elements are part of any figure of revolution and include cylindrical tanks, silos, and pressure vessels. Parabolic prestressed concrete tension members, often described as stress ribbons, follow the same principles as catenary steel cables. They are used in inverted suspension bridges. Typical examples of tension members are shown in Fig. 8-42, where only the tensile member is emphasized in its position with respect to the rest of the structure, schematically described by a line configuration.

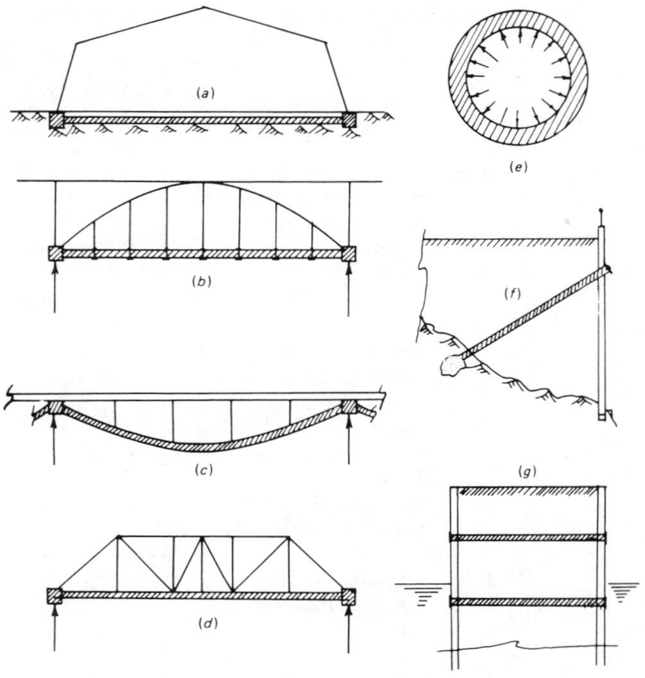

Fig. 8-42 Typical examples of tension members.

Prestressed linear tension members are similar to compressive members, except that they carry a much higher level of prestress. They are generally concentrically prestressed. However, when a combination of flexure (self-weight) and tension exists, the prestress may be slightly eccentric to balance the weight and keep a uniform state of stress in the section.

Because it can sustain tensile stress and remain crack-free, prestressed concrete is an ideal material for tanks, reservoirs, pipes, pressure vessels, and containers in general. Although early applications were mostly for water tanks, they have expanded to accommodate oil, gas, chemicals, slurries, liquids at cryogenic temperatures, and granular materials (silos). Nuclear containment vessels are among the latest and largest-scale applications.

The use of concrete in pure tensile members, also called ties, may seem paradoxical, as concrete is weak in tension. However, a prestressed concrete element can be treated as a single composite material that can sustain tension, with no reference to its components, steel and concrete. Prestressed concrete tension members can present a number of salient advantages over their counterparts made with either steel or reinforced concrete:

1. A prestressed concrete tie can be designed not to crack under normal service loads and acceptable levels of service overloads.
2. Because it is crack-free, it offers an excellent corrosion protection to the steel reinforcement.
3. The use of concrete offers an inherent fire resistance that, in some instances, may be an important design factor.
4. The most significant advantage of using a prestressed concrete tension member is that the total deformation necessary to develop the full resistance to the applied external load can be controlled practically to any desired degree. This is so because the member behaves essentially as a linear elastic crack-free material, and a

change in its cross section leads to a proportional change in its deformation under load. The total deformation (elongation or shortening) under load is a very important design variable. An excessive deformation may lead to distorsional distress and failure in the structure itself or in some of its elements. Everything else being equal, the deformation of a prestressed concrete tie can be more than an order of magnitude smaller than that of a steel tie.

8.12.2 Analysis and Design Considerations

The analysis of prestressed concrete tensile members presents little mathematical difficulty. It is generally assumed that the member behaves in a linear elastic manner under service loads, and that both the prestressing force and the external tensile load are concentric to the axis of the member, thus leading to uniform stresses in the section. However, unless the various design criteria are clearly understood and accounted for, the design of prestressed concrete tension members may often err on the unsafe side and may lead to inconsistent results. Most common design criteria include: maximum allowable compression, ultimate strength, safety against cracking, safety against decompression, minimum reinforcement, and maximum instantaneous and long-term deformations. Depending on the particular problem, other criteria may be added. For instance, in partially prestressed members, a maximum crack width criterion may be necessary. Each criterion leads to an analytic relationship to be satisfied by the design. Only two unknowns are needed in practice, the area of prestressing steel A_{ps} and the gross-sectional area of the concrete A_g. Other variables may be assumed at first and revised later, if necessary. Generally, it is not possible to know a priori which criterion or set of criteria will control a particular design.

A systematic procedure that guarantees a range of satisfactory designs and, if desired, an optimum design is developed in Refs. 8-15 and 8-48.

In a preliminary design the following approximation is suggested. Use a value of the final prestressing force F 20% higher than the maximum applied tensile load in service, and determine the gross-sectional area of concrete A_g based on a uniform compressive stress under F of $0.25\ f_c'$, that is, $A_g \simeq F/(0.25\ f_c')$. The area of prestressing steel A_{ps} is determined from F/f_{pe} where f_{pe} is the effective prestress. The member is then analyzed for the various design criteria, and the design is revised accordingly if needed.

8.13 PRESTRESSED CONCRETE COMPRESSION MEMBERS

8.13.1 Types and Advantages

Compression members are structural elements mostly of linear shape, such as columns, poles, and foundation piles. On a larger scale, TV towers and shafts of offshore structures are treated as compression members. The latter types are compressed not only in the longitudinal direction but also, because of hydrostatic pressure, in the circumferential direction.

It may seem irrational to prestress, that is, precompress, a compression member. However, compression members are seldom subjected to pure compression only. Columns, for instance, must be capable of resisting a variety of loads, including lateral loads from wind and earthquakes, from shearing forces transmitted by beams and slabs, and, if precast, from transportation and erection. Moreover, code provisions generally imply the existence of a minimum eccentricity, thus bending, even when pure compression is theoretically considered. In most cases the most critical loading combination of compression members involves substantial bending.

The use of prestressed versus nonprestressed steel in a column leads to a small reduction in its resistance to pure compression but increases significantly its resistance to first cracking. Consequently, its deflection in the uncracked state is substantially reduced, and its performance in service is improved. Prestressing allows the use of precasting and thus offers its related benefits, such as savings on forms and the use of high-strength concrete. As a column's capacity in compression is directly proportional to the concrete strength, this may be a very substantial advantage. Precast prestressed columns used in building structures are often designed to span several stories. They are connected in place by post-tensioning or other standard jointing techniques. The cost of connections is an important factor to consider in the early stages of design.

Because they can be made of a single element, prestressed concrete piles are very efficient structural members. They are widely used for marine structures and building foundations. Lengths of up to 120 ft (36 m) are common.[8-48] Prestressed concrete piles offer a number of important advantages that have made them competitive in all applications requiring piling. These advantages include durability, high load-moment resistance, ability to take uplift (tension), ability to penetrate hard strata, ease of handling and transportation, and economy.[8-11] Their use has been extended to fenders and sheet piling for waterfront bulkheads. Typical cross sections of piles and sheet piles are shown in Figs. 8-43 and 8-44.

Prestressed concrete poles are used for lighting, electric and telephone transmission lines, antenna masts, and the like. They are highly suited for urban installation.[8-48] Because they are often subjected to torsion in addition to compression and bending, their cross section is generally selected to achieve good torsional resistance. Typical cross sections of prestressed concrete poles are shown in Fig. 8-45.

8.13.2 Analysis and Design Considerations

The behavior of prestressed concrete columns subjected to monotonically increasing axial compression is very similar to that of reinforced concrete columns. It is important to realize that, although prestressed concrete columns are subjected to a compressive force F in addition to external loading, they are not more vulnerable to buckling than reinforced concrete columns. This is so because the tendons do not change position within the cross section, even when a lateral displacement is induced. Thus, contrary to Euler's case, a lateral displacement does not generate an additional moment due to F in the section.

For the purpose of analysis and design, columns are essentially classified into three categories, short, medium, and long.[8-49] The exact delineation of the three categories is clarified in section 10.11.4 of the *ACI Code*. In brief, a short column can be analyzed or designed from its cross section only; a medium column is essentially designed as a short column with due account given to slenderness effects; and the design of long columns is governed by instability criteria. The *ACI Code* does not specifically cover slenderness effects in prestressed concrete columns. Because at ultimate both reinforced and prestressed concrete columns show similar behavior, the provisions given in the *ACI Code* for reinforced concrete can be somewhat extended to prestressed concrete. This was essentially done in the PCI

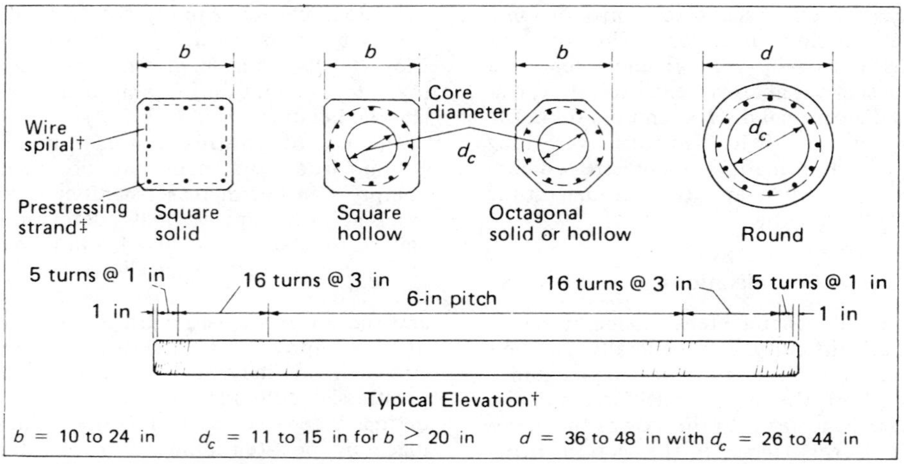

† Wire spiral varies with pile size.
‡ Strand pattern may be circular or square.

Fig. 8-43 Typical cross sections of prestressed piles.[8-22] (Courtesy of the Prestressed Concrete Institute.)

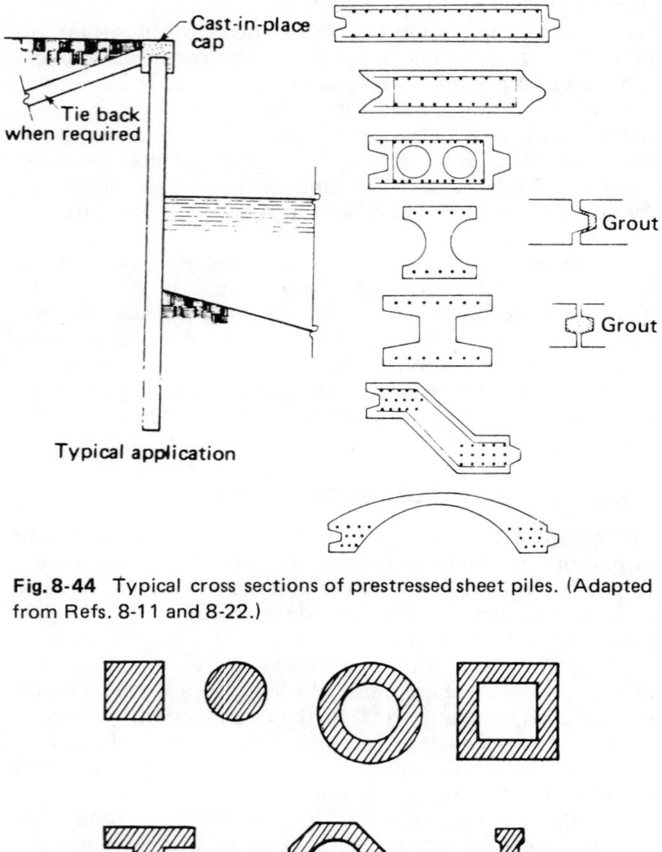

Fig. 8-44 Typical cross sections of prestressed sheet piles. (Adapted from Refs. 8-11 and 8-22.)

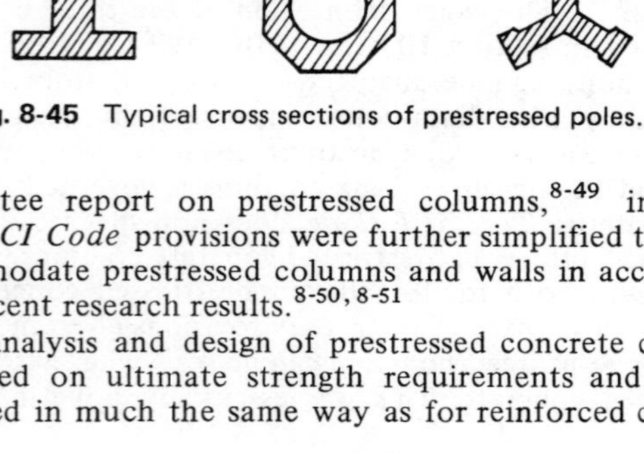

Fig. 8-45 Typical cross sections of prestressed poles.

committee report on prestressed columns,[8-49] in which some *ACI Code* provisions were further simplified to better accommodate prestressed columns and walls in accordance with recent research results.[8-50, 8-51]

The analysis and design of prestressed concrete columns are based on ultimate strength requirements and are approached in much the same way as for reinforced concrete columns. The analysis of short columns is generally reduced to determining the load-moment interaction diagram of the section and checking whether the diagram provides an envelop to factored loading combinations. A number of design requirements specific to prestressed columns are summarized below:

1. *Minimum longitudinal reinforcement.* Prestressed concrete compressive members (columns and bearing walls) should have an average effective prestress not less than 225 psi (1.55 MPa). This provision indirectly sets a minimum reinforcement ratio. Compressive members with lower prestress shall, like ordinary reinforced concrete, have a minimum nonprestressed reinforcement ratio of at least 1%.
2. *Lateral reinforcement.* Except for walls, for which section 10.15 of the *ACI Code* applies, members with average prestress equal to or greater than 225 psi (1.55 MPa) shall have all prestressing tendons enclosed by lateral ties or spirals in accordance with section 18.11 of the *ACI Code.* For walls with average prestress not less than 225 psi (1.55 MPa), minimum lateral reinforcement may be waived where analysis shows adequate strength and stability. Note that besides its practical role of holding the longitudinal steel together, lateral reinforcement provides a confinement that increases strength and significantly improves ductility. It increases the shear resistance of columns and limits the buckling of longitudinal bars.
3. *Minimum size of columns.* Contrary to previous editions of the *Code*, no minimum cross-sectional sizes are set for columns in the current *ACI Code.* However, slenderness effects, lateral deflections, and other practical considerations limit the size of prestressed concrete columns. In practice, a cross section of less than 8 × 8 in. (20 × 20 cm) is not desirable.

Additional design recommendations related to columns, bearing walls, and piles can be found in Refs. 8-22 and 8-49.

8.13.3 Design Aids

No closed-form solution has been developed for the design of columns. Generally a trial column is selected based on previous experience. Its design load-moment interaction diagram is then determined and compared to all points

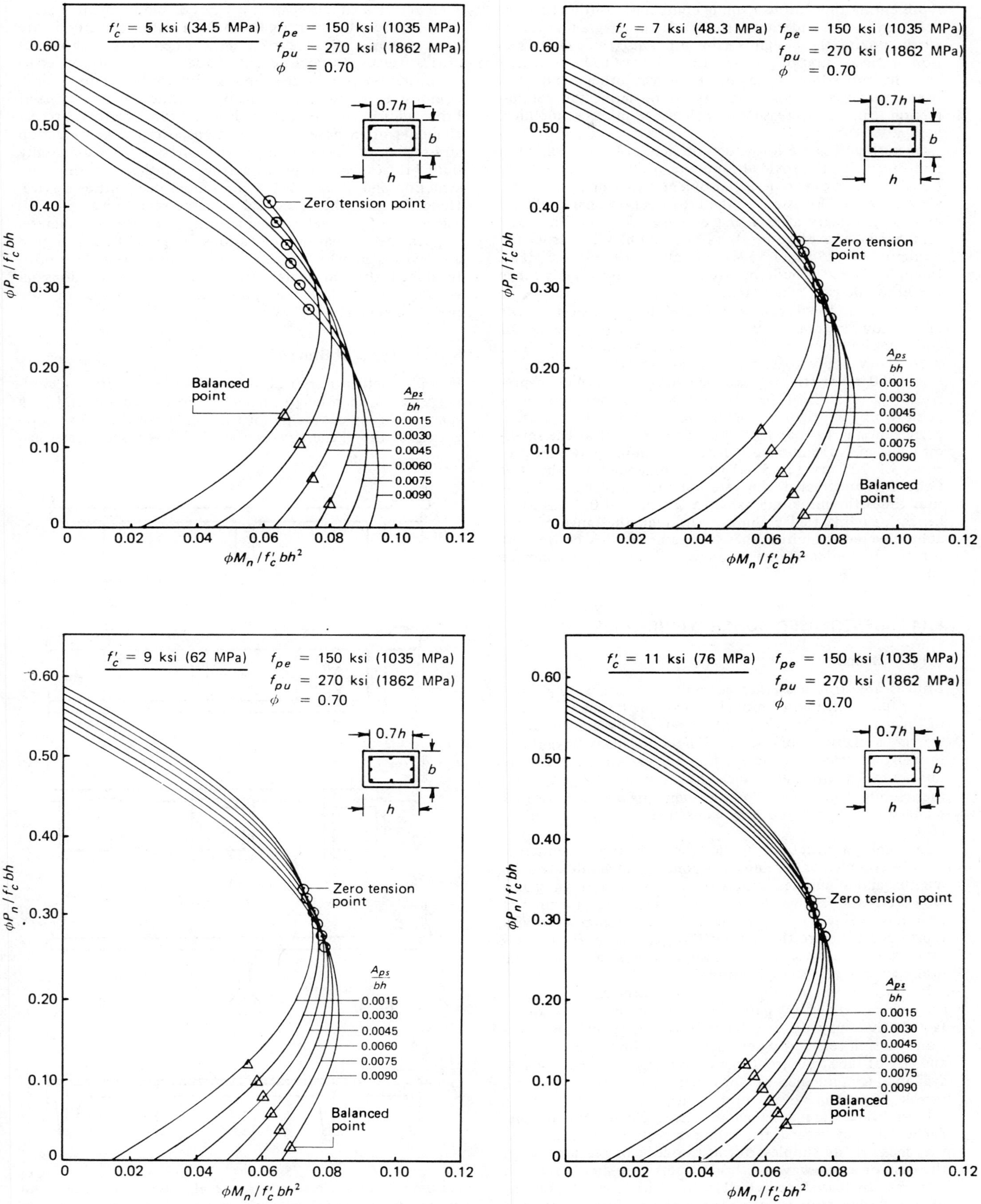

Fig. 8-46 Nondimensionalized load-moment interaction diagrams for prestressed concrete columns.

representing the various loading combinations (due account being given to slenderness effects). If the diagram envelops all the points, the design is safe. Otherwise the column section or the amount of prestress or both have to be increased and the procedure repeated. If, on the other hand, it is found that the margin of safety is too large, the column section can be decreased to achieve a more economical design.

Once the cross section is determined, at least a minimum reinforcement is provided, and the problem is essentially transformed from a design problem to an analysis or review problem. Revision of the cross-section characteristics may be necessary after a first design evaluation. The common range of average prestress in columns varies from the minimum of 225 psi (1.55 MPa) to about 900 psi (6.2 MPa). In applications for building structures, the most likely range is 300 to 400 psi (2 to 3 MPa).

To speed up the design, nondimensionalized load-moment interaction diagrams have been developed for prestressed concrete columns.[8-15, 8-22, 8-52] Some are shown in Fig. 8-46. They apply to square columns prestressed with 270 ksi (1860 MPa) strands, assumed having an effective prestress f_{pe} = 150 ksi (1035 MPa). The steel configuration is as shown in the figure. Four reinforcement ratios are used, corresponding to four levels of average prestress. The smallest value of 0.015 corresponds to the minimum average prestress of 225 psi (1.55 MPa) recommended by the *ACI Code* for prestressed concrete columns and the largest is four times the minimum value. A ϕ factor of 0.7 was used, hence assuming tied columns. Although the range of variables covered is limited, these diagrams are very convenient for the preliminary dimensioning of prestressed concrete columns.

8.14 PRESTRESSED CONCRETE BRIDGES

8.14.1 Scope

Bridges are structures that generally perform a single but major function: that of providing a simple means to "cross" or "reach" between two points separated by a deep valley, a river, a highway, or the like. There are many advantages in using prestressed concrete for bridges. Among them are minimum maintenance, increased durability, good aesthetics, and when factory precast elements are used, assured plant quality, fast and easy construction, and low initial cost.

In common structural applications, prestressed concrete usually complements reinforced concrete at moderate-span lengths and competes with structural steel at long spans. However, for bridge decks and where factory precast products are available, prestressed concrete competes with reinforced concrete throughout the span range. Hence, it is establishing, at least in the United States, a very strong dominance in bridge applications.

In spite of their unique common purpose, bridges are each characterized by particular site conditions and other factors that may dictate the type of design and construction solution selected. Such factors include the span length and size of the structure, types of loading, clearance, access, available technologies of construction or fabrication, site profile, importance of the bridge, and cost.

Most bridges are designed to carry vehicles and people, for which they offer a flat riding or walking surface called the deck. In its simplest form, the deck of a bridge, as in the case of a one-way slab bridge, acts as a simple flexural element. Increased analytical difficulties arise, depending on the design requirements, the type of construction, and the construction sequence. Examples include a continuous span versus a simple span or a series of simple spans, a statically indeterminate structure versus a statically determinate one, a skewed or a curved deck where torsion becomes critical versus a straight deck, a variable-depth bridge versus a constant-depth bridge, and a composite versus a noncomposite structure. The type of construction may by itself bring additional constraints. For instance, factory precast elements often have to sustain transportation and erection stresses more severe than service stresses. In segmentally built bridges, each segment acts as a cantilever during construction and as part of the continuous deck during service. Hence, it may be subjected to large stress reversals. The selection of a final solution, using prestressed concrete, implies the evaluations of various design and construction alternatives in which solutions involving cast-in-place post-tensioned structures are often compared to solutions involving factory precast pretensioned elements or site precast elements.

8.14.2 Types of Bridges

For all practical purposes, bridges can be classified according to their span length, namely, short, medium, and long spans. A clear cut does not exist between the above categories, as they generally overlap. Moreover, such classifica-

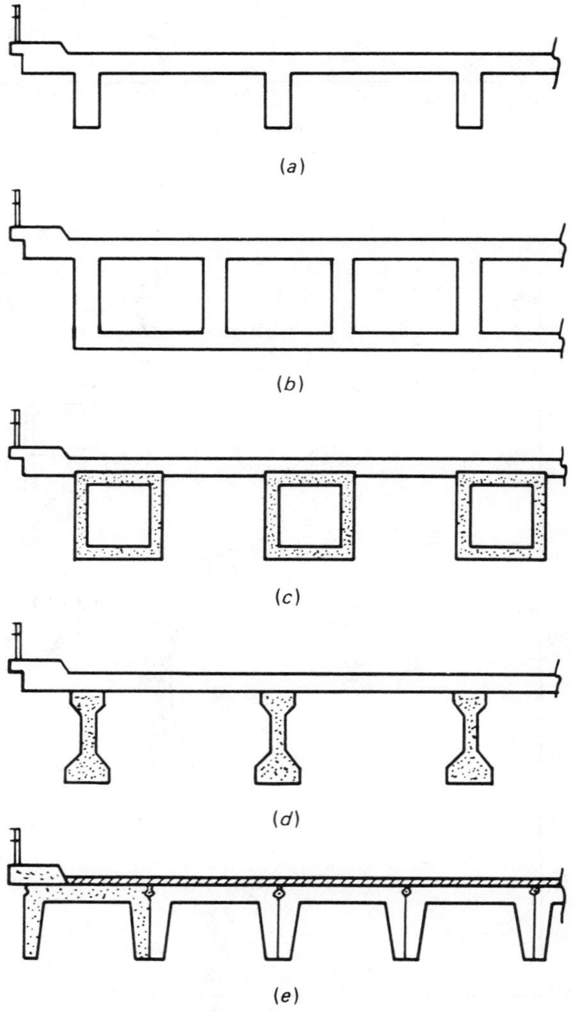

Fig. 8-47 Typical cross-sections of bridge decks. (a) T beams. (b) Box beams. (c) Spread box beams. (d) I beams. (e) Adjacent channel beams.

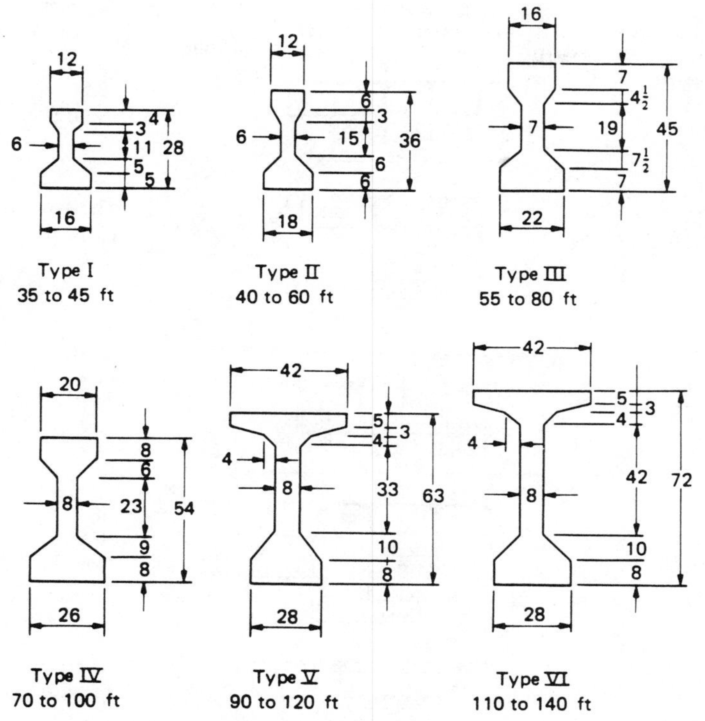

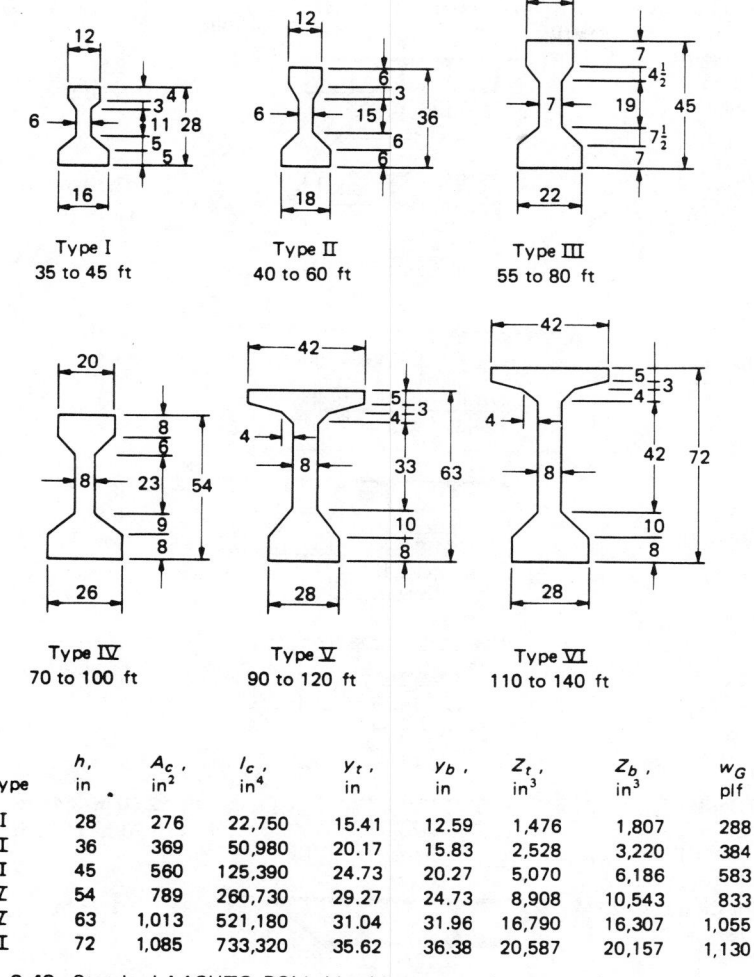

Type	h, in	A_c, in^2	I_c, in^4	y_t, in	y_b, in	Z_t, in^3	Z_b, in^3	w_G, plf
I	28	276	22,750	15.41	12.59	1,476	1,807	288
II	36	369	50,980	20.17	15.83	2,528	3,220	384
III	45	560	125,390	24.73	20.27	5,070	6,186	583
IV	54	789	260,730	29.27	24.73	8,908	10,543	833
V	63	1,013	521,180	31.04	31.96	16,790	16,307	1,055
VI	72	1,085	733,320	35.62	36.38	20,587	20,157	1,130

Fig. 8-48 Standard AASHTO–PCI bridge I beams: section properties and span range.

tion is subjective and relative. For instance, a span considered long for a factory precast element may be considered medium for a cast-in-place element. Similarly, the maximum length for a factory precast element using land transportation may be substantially smaller than an on-site precast solution using floating equipment.

Bridges spanning up to about 50 ft (15 m) are generally considered short, spans of between 50 and 100 ft (15 and 30 m) are considered moderate, whereas spans above 100 ft are considered long when factor precast elements are used. The length range for long-span bridges can again be divided, into long spans and very long spans that can exceed 1000 ft (305 m).

Typical bridge deck sections, either cast-in-place and post-tensioned or using precast pretensioned elements, are shown in Fig. 8-47.

1. Short-Span Bridges

In the United States short-span bridge decks are usually built with precast pretensioned beams that are transported to the site and erected. The beams, which have built-in shear keys, are generally placed adjacent to each other, as shown in Fig. 8-47(e). The shear keys are then filled with a mortar grout to provide for lateral resistance and load transfer. A topping of bituminous concrete or the equivalent, of about 2-in. (5-cm) thickness, is generally added to provide a wearing surface and for the purpose of leveling. Typical cross sections, design aids, and standard details can be found in Ref. 8-53.

2. Medium- and Long-Span Bridges Using Precast Beams

The decks of medium-span bridges using precast pretensioned beams are generally built either in a manner similar to short-span bridges where the beams are placed adjacent to each other (Fig. 8-47e), or as composite decks where the beams are transversely spaced and a cast-in-place slab is added to provide for lateral continuity (Fig. 8-47c, d). The most common composite bridge deck in the United States consists of precast prestressed I beams, such as those especially developed and standardized by AASHTO–PCI for highway bridges, with a structural cast-in-place concrete slab on top. Their dimensions, properties, and span ranges are shown in Fig. 8-48.

3. Long- and Very Long-Span Bridges

Long-span prestressed concrete bridges are generally cast-in-place or site precast and post-tensioned. Although length is relative, spans above 165 ft (50 m) are considered long. Very long spans are spans above about 500 ft (152 m). Long-span bridges are often built by a segmental construction technique. In segmental construction the deck is built by segments, one at a time. Segments can be precast or cast-in-place. Box beams are considered best suited for this type

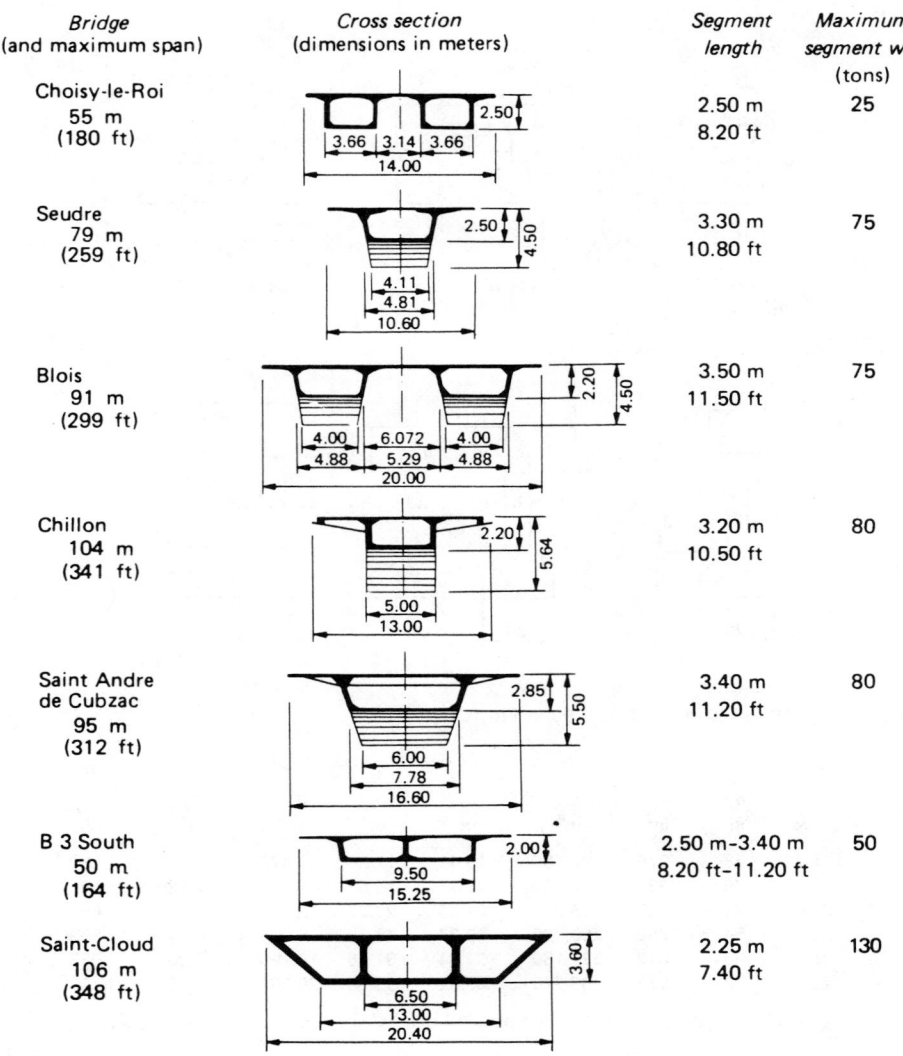

Bridge (and maximum span)	*Cross section* (dimensions in meters)	*Segment length*	*Maximum segment wt.* (tons)
Choisy-le-Roi 55 m (180 ft)	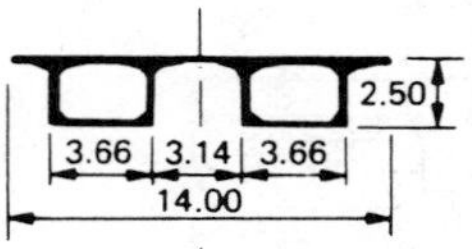	2.50 m 8.20 ft	25
Seudre 79 m (259 ft)	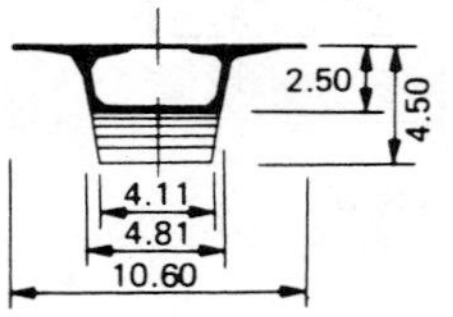	3.30 m 10.80 ft	75
Blois 91 m (299 ft)	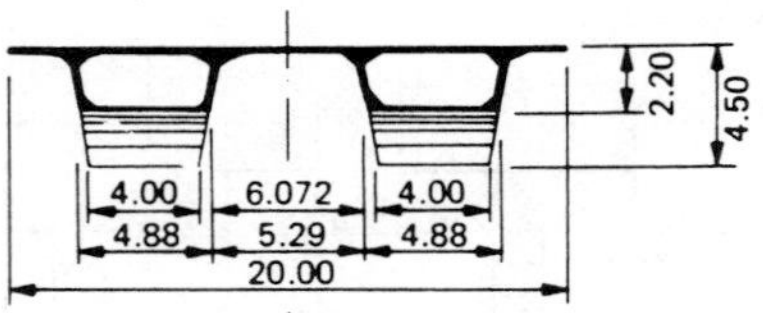	3.50 m 11.50 ft	75
Chillon 104 m (341 ft)	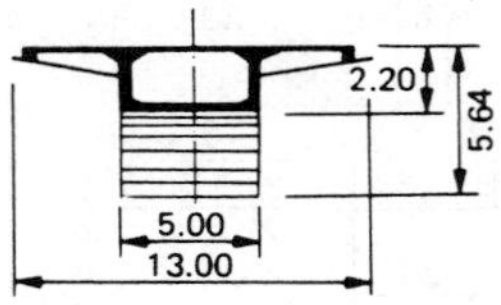	3.20 m 10.50 ft	80
Saint Andre de Cubzac 95 m (312 ft)	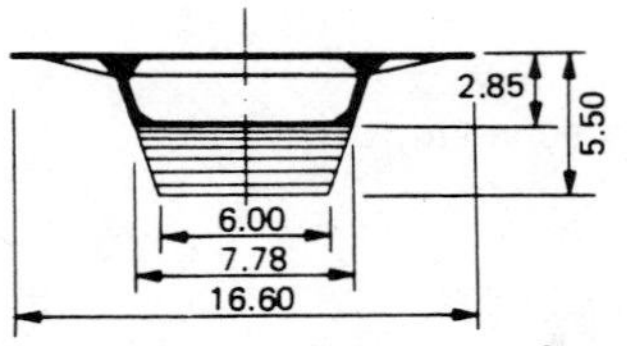	3.40 m 11.20 ft	80
B 3 South 50 m (164 ft)	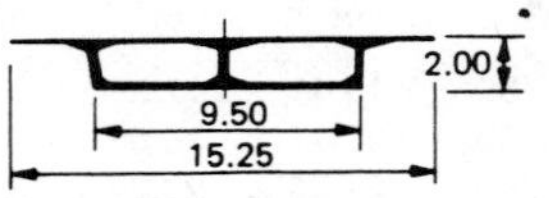	2.50 m-3.40 m 8.20 ft-11.20 ft	50
Saint-Cloud 106 m (348 ft)	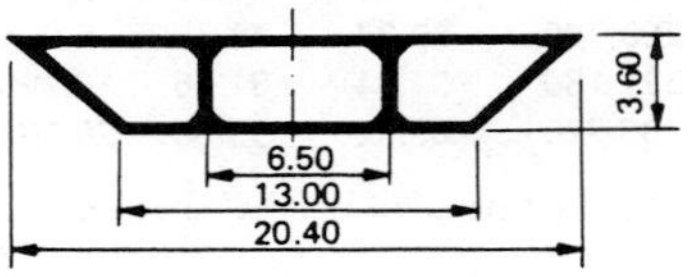	2.25 m 7.40 ft	130

Fig. 8-49 Evolution of typical sections for segmental bridges.[8-54] (Courtesy of the Prestressed Concrete Institute.)

of construction. They offer superior torsional rigidity and stability during construction and in service. Typical cross sections and their evolution are shown in Fig. 8-49, where actual dimensions are also given for several segmental bridges built in France.[8-54] It is observed that, for spans above about 200 ft (60 m), sections tend to have a variable depth.

Because the depth of the deck structure tends to increase substantially for very long spans, a cable-stayed solution can be considered as an alternative to segmental cantilever construction. In such a solution the prestressed concrete deck ribbon, generally of constant depth, is suspended from sloping stay cables emanating from a high tower. The sloping cables are stiff and dissipate all anchorage forces in the deck girder, so that a beneficial longitudinal compressive force occurs in the deck. The longest cable-stayed bridge in the United States is the Pasco-Kennewick Intercity Bridge, which was completed in 1978.[8-55] It has a main span of 981 ft (299 m) and a total length of 2503 ft (763 m).

8.14.3 Analysis and Design Considerations

In the United States highway bridges are generally designed according to the Standard Specifications for Highway Bridges of the American Association of State Highway and Transportation Officials (AASHTO).[8-25] Railway bridges are designed according to the provisions of the *Manual of Railway Engineering* of the American Railway Engineering Association (AREA).[8-26] Other local and regional codes must also be considered, in particular for secondary and special-purpose structures. The design provisions of the AASHTO and AREA specifications are very similar to those of the *ACI Code*. Although, in their basic design philosophy, they favor the working stress design over the ultimate strength design approach preferred by ACI, such difference does not much materialize when prestressed concrete is considered. The report of ACI Committee 343 on the analysis and design of reinforced concrete bridge structures provides a valuable document where both the AASHTO and ACI philosophies are accommodated at best.[8-31]

One of the main differences between the analysis of bridge beams (or decks) and that of building members is the assessment of the effects of live loads specified in various codes. Once live load effects have been determined, the analysis of bridge beams is pursued according to the basic methodology described in Sections 8.5–8.9. Examples of computation and design aids to evaluate live load effects can be found in Ref. 8-15.

Because a large proportion of bridges and highway overpasses are in a span range below about 150 ft (46 m), many attempts were made in the United States and elsewhere to standardize bridge deck sections for such bridges. In the United States this effort was particularly fostered by the precast prestressed concrete industry, the U.S. Department of Transportation, AASHTO, and many state departments of transportation. Cost savings and other benefits can be substantial. Computer programs were written for the analysis and design of bridges and were used to generate design charts for typical bridge deck configurations using standardized beams such as box beams, I beams, T beams, and the like.[8-53] Several studies have dealt with the use of composite decks made with precast prestressed concrete beams and a cast-in-place concrete slab. The use of the AASHTO I beams, described in Fig. 8-48, was particularly extensive. Some additional information on the analysis and design of bridges can be found in Refs. 8-56–8-58.

REFERENCES

8-1. Jackson, P. H., U. S. Patent No. 357999 (1886/88).

8-2. Freyssinet, E., and Seailles, J., French Patent No. 680 547 (1928).

8-3. Magnel, G., *Prestressed Concrete*, 3rd Ed., revised and enlarged, Concrete Publications Ltd., London, 1954.

8-4. Guyon, Y., *Prestressed Concrete*, V. 1 and 2, John Wiley & Sons, New York, 1953, 1960. Also *Limit State Design of Prestressed Concrete*, translated from French, Halstead and Wiley, New York, 1972.

8-5. Abeles, P., *An Introduction to Prestressed Concrete*, V. 1 and 2, Concrete Publications Ltd., London, 1964.

8-6. Leonhardt, F., *Prestressed Concrete Design and Construction*, English translation, Wilhelm Ernst and Sohn, Berlin, 1964 (1st Ed., 1955, 2nd Ed., 1962 in German).

8-7. Lin, T. Y., *Design of Prestressed Concrete Structures*, 2nd Ed., John Wiley & Sons, New York, 1963; also revised and enlarged 3rd Ed. by T. Y. Lin and N. Burns, 1981.

8-8. Preston, K., and Sollenberger, N., *Modern Prestressed Concrete*, McGraw-Hill Book Company, New York, 1967.

8-9. Mikhailov, V. V., *Prestressed Concrete Structures: Theory and Design*, translated from Russian by J. H. Dixon, Cement and Concrete Association, London, 1969.

8-10. Khachaturian, N., and Gurfinkel, G., *Prestressed Concrete*, McGraw-Hill Book Company, New York, 1969.

8-11. Gerwick, Ben C., Jr., *Construction of Prestressed Concrete Structures*, Wiley-Interscience, New York, 1971.

8-12. Ramaswamy, G. S., *Modern Prestressed Concrete*, Pitman Publishing, London, 1976.

8-13. Libby, J. R., *Modern Prestressed Concrete: Design Principles and Construction Methods*, 2nd Ed., Van Nostrand Reinhold, New York, 1977.

8-14. Nilson, A. H., *Design of Prestressed Concrete*, John Wiley & Sons, New York, 1978.

8-15. Naaman, A. E., *Prestressed Concrete Analysis and Design-Fundamentals*, McGraw-Hill Book Company, New York, 1982, 670 pp.

8-16. Federation Internationale de la Precontrainte (FIP), London, England.

8-17. Prestressed Concrete Institute, Chicago, Illinois 60606, U.S.A.

8-18. Post-Tensioning Institute, Phoenix, Arizona 85013, U.S.A.

8-19. American Concrete Institute, Detroit, Michigan 48219, U.S.A.

8-20. Portland Cement Association, Skokie, Illinois 60076, U.S.A.

8-21. Cement and Concrete Association, London, England.

8-22. *Design Handbook Precast Prestressed Concrete*, 2nd Ed., Prestressed Concrete Institute, Chicago, 1978.

8-23. *Building Code Requirements for Reinforced Concrete*, ACI 318-83, American Concrete Institute, Detroit, 1983.

8-24. *Commentary on Building Code Requirements for Reinforced Concrete*, ACI 318-83, American Concrete Institute, Detroit, 1983.

8-25. *Standard Specifications for Highway Bridges*, 12th Ed., American Association of State Highway and Transportation Officials (AASHTO), Washington, D.C., 1977, 496 pp.

8-26. *Manual of Railway Engineering*, American Railway Engineering Association (AREA), Washington, D.C., 1973.

8-27. *Building Code Requirements for Minimum Design Loads in Buildings and Other Structures*, American National Standards Institute (ANSI), A 58.1, 1972, New York, 1972.

8-28. *British Standard Code of Practice for the Use of Structural Concrete*, British Standards Institution, CP-110, London, 1972.

8-29. PCI Committee on Prestress Losses, "Recommendations for Estimating Prestress Losses," *PCI Journal*, V. 20, No. 4, July/Aug. 1975, pp. 43–75; also comments in V. 21, No. 2, Mar./Apr. 1976.

8-30. Zia, P., Preston, H. K., Scott, N. L., and Workman, E. B., "Estimating Prestress Losses," *Concrete International*, V. 1, No. 6, June 1979, pp. 32–38.

8-31. ACI Committee 343, "Analysis and Design of Reinforced Concrete Bridge Structures," Report 343-77 (also reprinted in ACI *Manual of Concrete Practice*), American Concrete Institute, Detroit, 1977, 118 pp.

8-32. Naaman, A. E., "A Proposal to Extend some Code Provisions on Reinforcement to Partial Prestressing," *PCI Journal*, V. 26, No. 2, Mar./Apr. 1981, pp. 74–91.

8-33. MacGregor, J. G., Sozen, M. A., and Siess, C. P., "Strength of Concrete Reams with Web Reinforcement," *ACI Journal*, V. 62, No. 12, Dec. 1965, pp. 1503–1519.

8-34. ACI Special Publication SP-10, *Commentary on Building Code Requirements for Reinforced Concrete* (ACI 318-63), by ACI Committee 318, American Concrete Institute, Detroit, 1965, 91 pp.

8-35. ASCE–ACI Joint Committee 426 Report, "The Shear Strength of Reinforced Concrete Members," *Journal of the Structural Division, ASCE,* V. 99, No. ST6, June 1973, Chaps. 1–4, pp. 1091–1197. (Also reproduced in ACI *Manual of Concrete Practice*.)

8-36. Tadros, M. K., "Designing for Deflection," reprint of a paper presented at the PCI Seminar on Advanced

Design Concepts in Precast Prestressed Concrete, PCI Convention, Dallas, Oct. 1979.

8-37. Branson, D. E., *Deformation of Concrete Structures*, McGraw-Hill Book Company, New York, 1977.

8-38. Branson, D. E., and Kripanarayanan, K. M., "Loss of Prestress, Camber, and Deflections of Non-Composite and Composite Prestressed Concrete Structures," *PCI Journal*, V. 16, No. 5, Sept./Oct. 1971, pp. 22-52.

8-39. Tadros, M. K., Ghali, A., and Dilger, W. H., "Time-Dependent Prestress Loss and Deflection in Prestressed Concrete Members," *PCI Journal*, V. 20, No. 3, May/June 1975, pp. 86-98.

8-40. Martin, L. D., "A Rational Method for Estimating Camber and Deflections," *PCI Journal*, V. 22, No. 1, Jan./Feb. 1977, pp. 100-108.

8-41. Sabnis, G. M., *Handbook of Composite Construction Engineering*, Van Nostrand Reinhold Co., New York, 1979.

8-42. Lin, T. Y., "Load Balancing Method for Design and Analysis of Prestressed Concrete Structures," *ACI Journal*, V. 60, No. 6, June 1963, pp. 719-742.

8-43. Lin, T. Y., and Thornton, K., "Secondary Moment and Moment Redistribution in Continuous Prestressed Concrete Beams," *PCI Journal*, V. 17, No. 1, Jan./Feb. 1972, pp. 1-20. See also discussion of above paper by A. H. Mattock and closure by the authors in *PCI Journal*, V. 17, No. 4, July/Aug. 1972, pp. 86-88.

8-44. ACI-ASCE Joint Committee 423, "Recommendations for Concrete Members Prestressed with Unbonded Tendons," *Concrete International*, V. 5, No. 7, July 1983, pp. 61-76.

8-45. Park, R., and Gamble, W. L., *Reinforced Concrete Slabs*, John Wiley, New York, 1980, 618 pp.

8-46. "Design of Post-Tensioned Slabs," Post-Tensioning Institute, Phoenix, Arizona, 1977, 52 pp.

8-47. Rice, P. F., and Hoffman, E. S., *Structural Design Guide to the ACI Building Code*, 2nd ed., Van Nostrand Reinhold, 1979.

8-48. Naaman, A. E., "Optimum Design of Prestressed Concrete Tension Members," *Journal of the Structural Division, ASCE*, V. 108, No. ST 8, Aug. 1982, pp. 1722-1738.

8-49. PCI Committee on Prestressed Concrete Columns, "Recommended Practice for the Design of Prestressed Concrete Columns and Bearing Walls," *PCI Journal*, V. 21, no. 6, Nov./Dec. 1976, pp. 16-45. Also, discussion by L. D. Martin and Committee Closure in *PCI Journal*, V. 23, No. 1, Jan./Feb. 1978.

8-50. Nathan, N. D., "Slenderness of Prestressed Concrete Columns," *PCI Journal*, V. 28, No. 2, Mar./Apr. 1983, pp. 50-77.

8-51. Alcock, W. J., and Nathan, N. D., "Moment Magnification Tests of Prestressed Concrete Columns," *PCI Journal*, V. 22, No. 4, July/Aug. 1977, pp. 50-61.

8-52. Salmons, J. R., and McLaughlin, D. G., "Design Charts for Proportioning Rectangular Prestressed Concrete Columns," *PCI Journal*, V. 27, No. 1, Jan./Feb. 1982, pp. 120-143.

8-53. *Precast Prestressed Concrete Short Span Bridges—Spans to 100 Feet*, 2nd Ed., Prestressed Concrete Institute, Chicago, Illinois, 1981.

8-54. Muller, J., "Ten years of Experience in Precast Segmental Construction," *PCI Journal*, V. 20, No. 1, Jan./Feb. 1975, pp. 28-61.

8-55. Grant, A., "The Pasco-Kennewick Intercity Bridge," *PCI Journal*, V. 24, No. 3, May/June 1979, pp. 90-109.

8-56. Libby, J. R., and Perkins, N. D., *Modern Prestressed Concrete Highway Bridge Superstructures: Design Principles and Construction Methods*, Grantville Publishing Co., San Diego, 1976, 254 pp.

8-57. *Precast Segmental Box Girder Bridge Manual*, joint publication by the Post-Tensioning Institute and the Prestressed Concrete Institute, Chicago, 1978, 116 pp.

8-58. Podolny, W., and Muller, J. M., *Construction and Design of Prestressed Concrete Segmental Bridges*, John Wiley, New York, 1982, 584 pp.

9

Post-Tensioned Slab Systems*

CLIFFORD L. FREYERMUTH**

9.1 INTRODUCTION

Post-tensioned concrete slabs have become a major factor in the construction of floor systems for commerical and residential buildings of all types. In their two major popular forms (one-way slabs, two-way flat plates) they have been found to be economical for structural applications in parking structures, apartment buildings, office buildings, hospitals, and industrial buildings of both the high-rise and low-rise type. In the past 15 years, surveys made by the Post-Tensioning Institute indicate that over 500 million sq ft of post-tensioned slabs were completed.

The first post-tensioned slabs in the U.S. were built in the mid-1950s. Most post-tensioned slabs constructed in the 1950s were associated with the lift-slab method of construction. In these applications, post-tensioning was introduced to solve weight, deflection, and cracking problems that arose with conventionally reinforced lift-slabs. Little was known at that time of the behavior of post-tensioned slabs. For this reason, design criteria tended to be conservative, and design techniques were awkward and time-consuming. Possibly the single most important reason for the growth of post-tensioned slab construction was the development and dissemination of techniques that greatly simplified the design and analysis of linearly indeterminate post-tensioned members. These "load-balancing" techniques, which were introduced by Professor T. Y. Lin,[9-1] involved visualizing the post-tensioning tendon as a system of loads acting on the concrete, similar to "familiar" ones such as dead and live loads. This concept eliminated much of the mystique and almost all of the mathematical drudgery previously associated with indeterminate analysis of post-tensioned members. The load-balancing concept is now by far the most widely used method for analysis and design of post-tensioned structures.

Other major factors that contributed to the growth of post-tensioned slab construction included:

1. Improvements and simplifications in post-tensioning hardware and field methods—making post-tensioned slab construction as easy for the contractor as conventionally reinforced slabs.
2. Improvements in forming systems that enhanced the overall economics of cast-in-place slab construction.
3. Testing programs on post-tensioned slabs that greatly expanded the understanding of their behavior, and led to improved code criteria and more economical, safe designs.
4. The recognition of the economy of post-tensioned slabs with short spans, i.e., 20 ft and less.

The body of this chapter presents the state-of-the-art in post-tensioned slab design.

*Material in this chapter has been extracted from *Design of Post-Tensioned Slabs* published by the Post-Tensioning Institute and is used by permission.

**Editor of material in this chapter, and Executive Director, Post-Tensioning Institute.

TABLE 9-1 Typical Span–Depth Ratios

	Continuous Spans		*Simple Spans*	
	Roof	*Floor*	*Roof*	*Floor*
One-way solid slabs	50	45	45	40
Two-way solid slabs (supported on columns only)	45–48	40–45		
Two-way waffle slabs (36" pans)	40	35	35	30
Beams	35	30	30	26
One-way joists	42	38	38	35

The above ratios may be increased if calculations verify that deflection, camber, and vibration frequency and amplitude are not objectionable.

9.2 PRELIMINARY SIZING OF MEMBERS

There are no set span–depth limits for post-tensioned members, but the values in Table 9-1 are provided as a guide to the preliminary sizing of members. Since the entire gross transformed section is effective in resisting service load deflections in a post-tensioned structure with design stresses below the modulus of rupture, the necessary structural depth is reduced by approximately one-third to provide stiffness comparable to a cracked flexural element with nonprestressed reinforcement. The resulting savings in concrete is often a significant factor in the relative economy of post-tensioned construction.

9.3 LOAD-BALANCING METHOD OF ANALYSIS

The basic concept of load balancing represents the influence of the tendons by equivalent loads. The concept is illustrated in Fig. 9-1(a), where the tendon is selected to directly counteract the imposed loading at the indicated eccentricity, e'. Since the moment induced by the tendon and the load offset each other, the net stress in the beam will be the axial compressive stress from the post-tensioning, P/A. If it is desired to design the beam for zero stress at the bottom fiber at center span (or any other value of stress less than the modulus of rupture), it is only necessary to reduce the amount of post-tensioning provided. The net stress on the section may be calculated from $P/A + M_n c/I$ where M_n is the net (unbalanced) bending moment on the section. For continuous designs, the tendon geometry would be assumed as shown in Fig. 9-1(b).

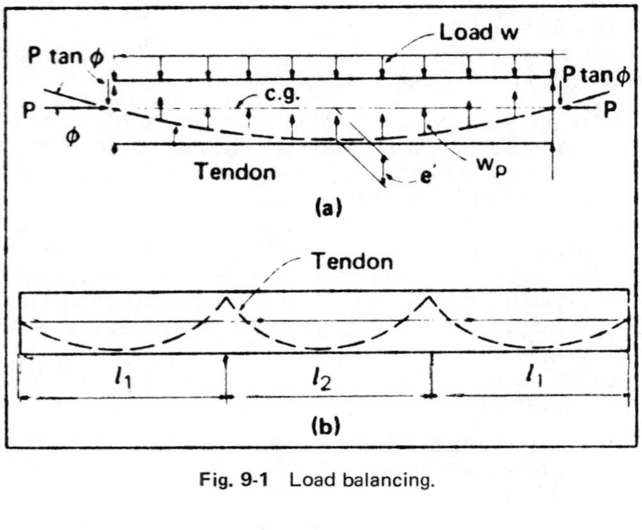

Fig. 9-1 Load balancing.

The above example serves to illustrate the salient features of load balancing. These are that the prestressing force is selected to balance or counteract some portion of the load, and that under the balanced loading conditions the structure will theoretically have no deflection and will be subjected only to an axial compressive stress P/A, from the tendon. The net moment in the structure at any point is that resulting from the load not "balanced" by the post-tensioning. This is an extremely powerful concept for visualizing the effect of the post-tensioning on the structure, and it greatly simplifies design calculations. Secondary post-tensioning moments can be readily obtained by subtracting the primary moment, Pe, from the moments caused by the balanced load at any point. Application of the load-balancing design procedure is illustrated by the flat plate design example presented in Section 9.12.

The load-balancing procedure is by far the most convenient design approach for most indeterminate post-tensioned structures, and is recommended as the basic design approach for most applications.

9.4 STRENGTH ANALYSIS

The required design capacity in moment and shear is determined by multiplying elastically determined service load moments and shears by the load factors specified in section 9.2 of ACI 318-83. The moments and shears so obtained may be modified for redistribution of moments due to inelastic behavior in accordance with the provisions of section 18.10.4 of ACI 318-83. Further, the dead and live load factored moments (design moments) are to be modified by the secondary moments due to reactions induced by post-tensioning (with a load factor of 1.0) as specified in section 18.10.3 of ACI 318-83.

9.5 FLEXURE AND SHEAR IN ONE-WAY SLABS

9.5.1 Flexure in One-Way Slabs

Post-tensioning tendons exert loads and moments on slabs, opposing gravity loads, moments, and their flexural stresses. Thus, the post-tensioned member must resist only a portion of the dead and live loads in flexure, with the remaining portion counteracted or balanced by the action of tendons. Through its anchors, the tendon also exerts an axial compressive stress that reduces the flexural tensile stress resulting from the load not balanced by the tendons. The amount of tendons is selected by satisfying flexural tensile stress limits prescribed by ACI 318-83. The maximum net tensile stress limit by ACI 318-83 is $12\sqrt{f_c'}$, but as this exceeds the modulus of rupture (conventionally assumed as $7.5\sqrt{f_c'}$), special considerations of deflection are necessary when tensile stresses exceed $6\sqrt{f_c'}$.

The behavior in flexure of one-way slabs at loads less than the cracking load is linear and elastic, and deflections can be predicted using gross section properties. Where one-way slabs are designed for service load tensile stresses greater than the modulus of rupture (between $7.5\sqrt{f_c'}$ and $12\sqrt{f_c'}$), behavior remains nearly linear, but at a stiffness less than predicted by gross section properties. The moment gradient in negative moment regions is steep, and cracking is confined to a very narrow zone, extending from the support face only a few slab thicknesses at most. Thus, the amount of bonded steel crossing the cracking zone can have little effect on the width of the cracking zone or on crack spacing. It is only necessary to provide a small amount of mild steel to improve flexural behavior of this narrow zone, as tension stresses drop off sharply away from the support face. Mid-span regions exhibit more gradual moment gradients than do support regions. Thus, a small amount of bonded mild steel can be more effective in distribution and controlling cracks in positive moment areas. Test results indicate that an area of mild steel equal to slightly more than 0.1% of the cross-sectional area is adequate to provide sufficient crack control and to ensure true beam-type behavior. By contrast, simple span beams post-tensioned with unbonded tendons and containing no bonded reinforcement behave after cracking as tied arches. With increasing load and rotation, the single initial crack increases in width and progresses toward the compression face until finally an abrupt compression failure occurs. This effect is eliminated in both slabs and beams by providing a small amount of conventional, bonded reinforcing steel.

Cracking due to high tensile stresses reduces stiffness, and prediction of deflections can no longer be based on elastic gross section properties. The bilinear approach to deflection prediction specified by ACI 318-83 section 18.4.2(c) does not adequately describe the behavior of continuous members. The bilinear approach assumes that the total deflection can be predicted as the sum of the deflection up to cracking calculated by elastic means, and an additional deflection based on the cracked section properties. This approach is approximately correct for simply supported single span members, since the properties of a single section dominate the effective stiffness, and a single load pattern dictates deflection behavior. However, for continuous members two or more sections and load patterns affect the behavior. For most continuous structures the maximum moments at a given section must be found by evaluating the effects of three load patterns; loads on all spans, loads on adjacent spans, and loads on alternate spans. Since these load patterns can occur in any order and at any time during the life of the structure, it is apparent that the slab must be considered to be cracked at support and midspan sections where tensile stresses exceed $7.5\sqrt{f_c'}$. This produces an effective stiffness at these sections less than that based on gross section properties. Tests, however, show that the effective stiffness is greater than that based on these very limited portions of the member. In addition, the calculation of cracked section properties is not straightforward. While the force in the unbonded tendons is not affected by local strains, the forces in bonded reinforcing are completely dependent on local strains. The behavior of a single given cross section can be modeled by considering it to be a column under an axial load equal to the effective prestress and subjected to moment. From equilibrium considerations, the crack depth and steel strain can be found by trial and error, and the transformed section properties calculated. Sections away from peak moments remain uncracked and elastic; so deflection behavior cannot be determined from the properties of a single section, and the method of section 18.4.2(c) of ACI 318-83 underestimates the stiffness of continuous members.

Results of testing give a better picture of the effective stiffness. Fig. 9-2 compares behavior at different load stages for a three-span slab designed for service tensile stresses of $9\sqrt{f_c'}$, but loaded to stress levels of $12\sqrt{f_c'}$. Degradation of stiffness was maximized by the increased cracking due to 33% service overloading. Had the slab been designed for the higher loads required to produce tensile stresses of $12\sqrt{f_c'}$, it would have contained additional bonded reinforcement sufficient to provide the corresponding higher

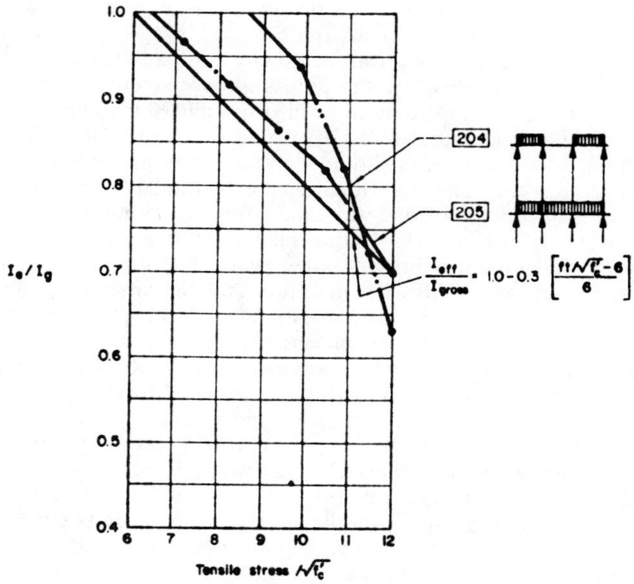

Fig. 9-2 Effective *I* for post-tensioned one-way slabs designed to tensile stress above $6\sqrt{f_c'}$.

required ultimate capacity. It therefore would have exhibited slightly higher stiffness due to the effects of higher transformed section properties. The straight line is a proposed approximation for effective stiffness of continuous rectangular members containing at least 0.1% of bonded reinforcement. As can be seen, Fig. 9-2 is conservative for the range of tensile stresses between $6\sqrt{f_c'}$ and $10\sqrt{f_c'}$, but somewhat overestimates the effective stiffness of $12\sqrt{f_c'}$. However, at $12\sqrt{f_c'}$ loads are higher than the design service load for the test specimens, and it is therefore reasonable to expect a loss of stiffness at that particular point for the test slab.

9.5.2 Shear in One-Way Slabs

Typically, one way slabs are subjected to low shear stresses, since high concentrated loading is uncommon. Compared to tee beams, the available shear area is large, and typical slab design shear stresses are well below the lower bound shear capacity of $2\sqrt{f_c'}$ specified by ACI 318-83. Where high shears are encountered, such as from high concentrated loading near supports, eq. (11-10) of ACI 318-83 may be used to more closely reflect available shear capacity. For conditions of high shear due to permanent concentrated loads, it is recommended that tendons be harped under the load to maximize the effective depth, *d*, and the shear carried by the tendon.

9.6 FLEXURE AND SHEAR IN FLAT PLATES

9.6.1 Flexure in Flat Plates

It is helpful to the understanding of post-tensioned flat plates to forget the arbitrary column strip, middle strip, and moment percentage tables that have long been familiar to the designer of reinforced concrete flat plates. Instead, let us first examine the mechanics of the action of the tendons.

The load-balancing approach is an even more powerful tool for examining the behavior of two-way systems than it is for one-way members. By the balanced-load approach, attention is focused on the loads exerted on the slab by the tendons, transverse to the plane of the slab. As for one-way slabs, this typically means a uniform load exerted upward along the major portion of the central length of a tendon span, and statically equivalent downward load exerted over a short length between the points of reverse curvature. In order to apply an essentially uniform upward load over the entire slab panel and to satisfy some rationality of statics, these downward loads should be reacted by another structural element. The additional element could be a beam or wall as in the case of one-way slabs, or columns in a two-way system. However, a look at a plan view of a flat plate reveals that columns provide an upward reaction for only a very small area. Thus, to maintain statical rationality, we must provide, perpendicular to the above ("primary") tendons, another set of tendons ("secondary") to provide an upward load under the downward load. Remembering that the downward load of the "primary" tendon system occurs over a relatively narrow width under the reverse curvatures and that the only available exterior reaction, the column, is also relatively narrow, it becomes obvious that the "secondary" tendon system should be in narrow strips or bands passing over the columns. There are two ways of accomplishing this two-part tendon system to obtain the nearly uniform upward load we desire for ease of analysis. Tendons of the "primary" system could be spaced uniformly in each of two directions and reacted by a "secondary" system along column grid lines in each direction. This method would fit within preconceived ideas of two-way slabs, as the "secondary" tendons are then analogous to two-way beams, but it would not recognize the full potential of the balanced-load approach, nor would it be practicable for structures whose columns are not aligned on rectangular grids. The second method does recognize the full potential and, by illustration, aids in the understanding of the contribution of the tendons.

The power of the second method becomes very clear if we examine a slab that has columns of irregular layout. For an example, with reference to Fig. 9-3, let us assume that the "odd"-numbered grid lines are offset one-half bay from the "even"-numbered grid lines, but columns on a given "letter" grid are aligned. If we reverted to the "column strip" approach illustrated in Fig. 9-4 as for conventionally reinforced slabs, we would find that each span that started in a "column strip" would end in a "middle strip," and tracing of load paths, a rational analysis, and proportioning reinforcement would become difficult if not impossible. So let us return to the basic idea of balancing loads with tendons. The "primary" system of tendons (parallel to "letter" grid lines) can be accomplished with little regard for column location. It is only necessary to place the high points of the tendon profile (where reverse curvature and downward load occur) at the intersection of the tendons with the "number" grid lines. This system is then reacted with the "secondary" tendons placed on the "number" grid lines as shown in Fig. 9-5. By this procedure, the portion of the gravity load balanced by the tendons is carried directly to columns, without any flexural action of the slab. Since this balanced load is typically a large portion of the permanent load on the slab, errors in analysis that are due to incorrect assumptions of load path are a function of relatively small loads, and thus are small. The possible consequences of such errors can be investigated by examining the behavior of the slab under overloads.

Tests[9-7, 9-8] and applications have demonstrated that a post-tensioned flat plate behaves as a flat plate regardless of tendon placement. The effects of the tendons are, of course, critical to the behavior, as they exert loads on the plate as well as provide reinforcement. Since the tendons exert transverse loads on the plate, these loads may be considered like any other dead or live loads. Since the tendon effect is opposite to the effect of gravity loads, the net loads causing bending are relatively small. An additional

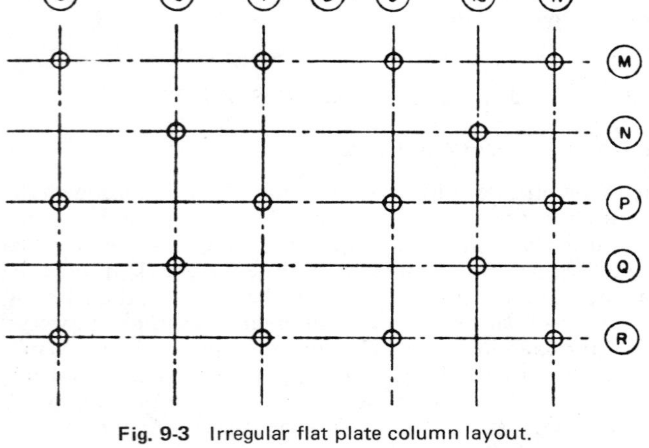

Fig. 9-3 Irregular flat plate column layout.

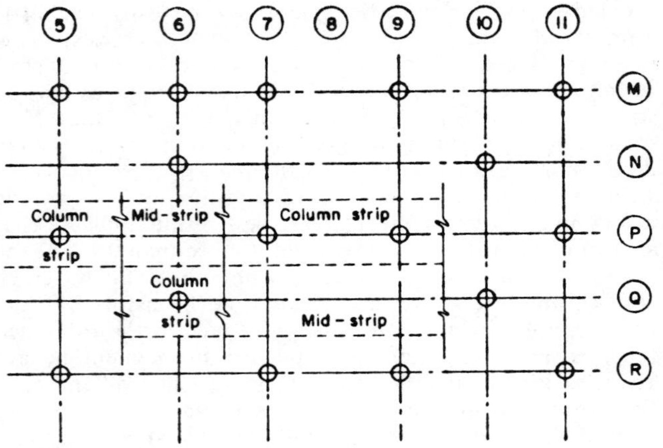

Fig. 9-4 Column strips and middle strips with irregular column layout.

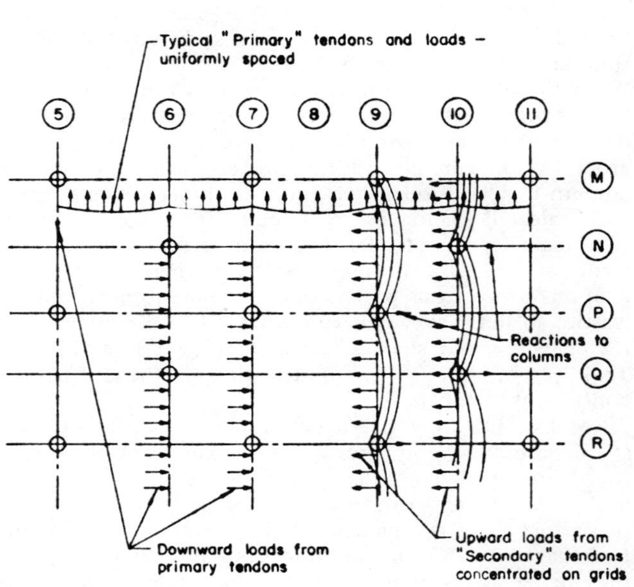

Fig. 9-5 Load balancing with bonded tendons for irregular column layout.

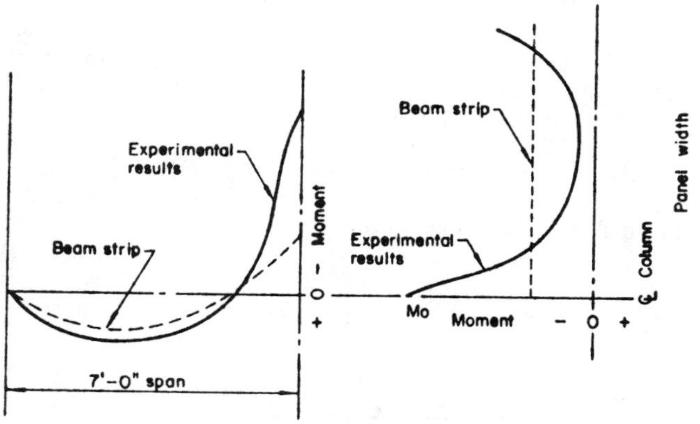

Moment of support (Adapted from Reference 9-3)

Fig. 9-6 Distribution of moments along column lines in a flat plate.

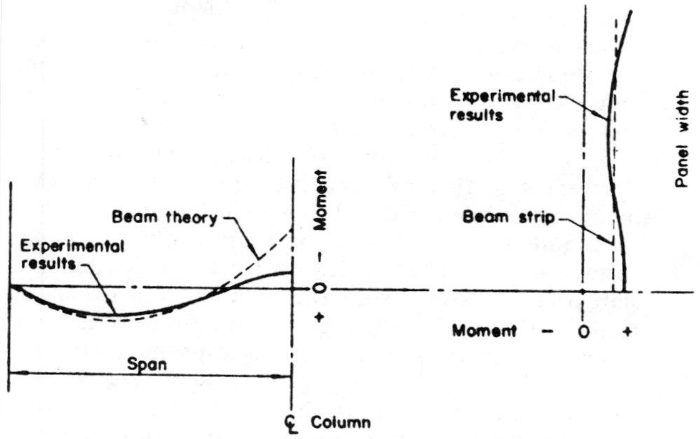

Moment of midspan (Adapted from Reference 9-3)

Fig. 9-7 Distribution of moments at midspan in flat plate.

effect of the tendons is the axial precompression which counteracts flexural tensile stresses. Therefore, at service dead load, the net downward loads which cause bending in the plate are normally very low, and the slab is essentially under uniform axial compression.

Examination of the distribution of moments in Fig. 9-6 reveals that negative moments are sharply peaked in the immediate vicinity of the column, and that the moment at the column face is several times the moment midway between columns. Since the largest moments are at the column faces, first cracking occurs there at rather low net loads. As the loading is further increased, these cracks extend out from the column faces, and positive moment cracks form at midspan. Finally, full yield line patterns are formed, signaling flexural failure.

Shear behavior will be more fully discussed later, but must be briefly discussed here as it is intimately linked to flexure. First cracking occurs at the column faces at an early load stage. For square panels and equal precompression in both directions, the first crack forms at the faces perpendicular to the reinforcing that has the smaller effective depth, regardless of the distribution of tendons. These cracks are quite small and have no significant effect on the stiffness of the slab. However, as cracking extends toward midpanel with increasing load, the initial cracks at the column face increase in width and depth. This zone is of utmost importance to the shear strength of the slab, and for this reason crack width and depth should be minimized. While compression due to prestress delays the formation of cracks, it is less efficient in controlling crack widths and spacings than nonprestressed bonded reinforcement, placed in the top of slabs immediately adjacent to and above the column. A small amount of bonded nonprestressed reinforcement controls and distributes cracks, and allows fuller utilization of the potential shear and flexural strength in the column zone. Bonded reinforcement uniformly distributed in the top of slabs within lines one and one-half times the slab thickness on either side of the column toughens the punching zone and is effective for maximizing the shear capacity. Bars placed outside that immediate column zone are of no value in controlling cracks in the shear zone. Thus, all mild steel required for support sections should be placed as close to the column vicinity as construction practice will allow. Although bonded tendons provide some control of cracking due to their strain compatibility, few slabs are constructed with bonded tendons, and very limited testing has been conducted on such slabs. Bonded reinforcement placed near midpanel does not become effective in crack control or contribute to the flexural capacity until the high load levels where yield lines start to develop. Even then, negative moment strains are much lower near midpanel than in the column region, so that the full yield force of such bars may never be realized.

In midspan regions, positive moments are almost uniformly distributed across the panel width, as shown in Fig. 9-7. First cracking, therefore, occurs much later at midspans than at supports. However, since such positive moment cracking is not confined to a small zone, there is a noticeable but not marked loss in stiffness caused by that cracking. The behavior of a member that is reinforced only by unbonded tendons has been described as similar to that of a shallow tied arch. This behavior has been observed in one-way statically determinant members, but it appears to be less significant for redundant, lightly prestressed structures such as continuous one-way slabs or two-way plates, slabs and waffles.[9-2–9-6] It has been shown in all two-way tests that bonded, mild reinforcing is required in positive moment regions only where ultimate flexural strength requirements exceed the flexural capacity provided by tendons alone, and that no minimum crack control bars are necessary for satisfactory behavior under service or ultimate loads.

9.6.2 Shear in Flat Plates

Until quite recently the ACI-318 *Building Code* did not contain specific provisions for punching shear in prestressed slabs. Likewise, until recently there was little research on moment transfer in prestressed slabs. The Committee 426 review of code provisions for shear for reinforced concrete slabs[9-12] has pointed out the need for careful consideration of the effects of transfer of moment as well as transfer of shear between a slab and a column. Thus, while typical interior slab/column connections may not be required by analysis to transfer unbalanced vertical load moments or lateral load moments, moment transfer may actually occur due to pattern live loading, temperature movements, or deformations under lateral loading caused by flexibility of the primary lateral resistance system. Further for conventional reinforced flat plates, and some prestressed plates as well, punching failure at a column can cause load and moment to be transferred to adjacent columns, and a horizontal progressive shear failure can then theoretically be propagated if the design shear capacity is based on vertical load only. With proper consideration of moment transfer and detailing, this possibility is avoided. The following steps are recommended for shear design of prestressed flat plates:

1. Use the equivalent frame procedure of ACI 318-83 section 13.7 (excluding sections 13.7.7.4 and 13.7.7.5)

to obtain moments for frame members including unbalanced moments at joints. Use of "prismatic" column stiffness rather than equivalent frame stiffness, while conservative for moment transfer at exterior columns, will lead to underestimation of moment transfer at interior columns, and will also lead to inadequate proportioning of flexural capacities at exterior spans. Moment capacity provided at an exterior column based on moments obtained using prismatic column stiffness cannot be mobilized owing to torsional flexibility of slab sections away from the column, and other sections of the spans will then be required to resist more moment than anticipated. An assumption of zero column stiffness will usually give conservative moment values in the slab, but is also unsatisfactory because there are no unbalanced moments that must be recognized for shear calculations. The equivalent column stiffness of section 13.7.4 of ACI 318-83 will give good approximation of frame member moments and unbalanced moments at joints, and should always be used.

2. Detail tendon and reinforcing bars for maximum design strength and reserve strength. To maximize design strength and reserve strength, control cracking, and toughen the punching zone, bonded steel reinforcement should be uniformly distributed within lines one and one-half times the slab thickness on either side of the column. To maximize reserve post-failure strength, a bundle of tendons should pass through the column within the column reinforcing cage in at least one direction, if not both. Since the unbonded tendon is not sensitive to local strains, this bundle will provide post-punching failure vertical load capacity in direct tension.[9-11] Thus it can allow large vertical deflections without rupture of the tendon, and prevent collapse of the slab.

An accepted method for evaluating total shear stress is that first described by DiStasio and Van Buren[9-13] and reflected in provisions of ACI 318-83. The critical section for shear is assumed to be at $d/2$ from the support face. The general expression for shear stress is:

$$\phi V_u = \frac{V_u}{b_0 d} + \frac{\alpha M_t c_3}{J_c}$$

where:

V_u = factored shear force at section
v_u = shear stress at design (factored) loads
b_0 = perimeter of shear section at $d/2$ from face of column as defined by ACI 318-83, Chapter 11
d = distance from centroid of tendon to compression face in direction of moment transfer, but need not be less than $0.8h$, where h is member thickness
ϕ = capacity reduction factor for shear, 0.85.
α = fraction of moment transferred by shear

$$\alpha = 1 - \frac{1}{1 + 2/3 \left(\dfrac{c_1 + d}{c_2 + d}\right)^{1/2}}$$

(ACI 318 differs from DiStasio and Van Buren here, since the latter had recommended that only 20% of the unbalanced moment be transferred by shear, regardless of the proportions of the critical section. For square columns, α will be 40%, twice the value of DiStasio and Van Buren.)

c_1 = support dimension in the direction of moment transfer
c_2 = support dimension perpendicular to c_1
c_3 = distance from centroid of critical shear section to extreme fiber in direction of moment transfer
M_t = net moment to be transferred to column
J_c = polar moment of inertia of critical section

For the simplest case, a square interior column, the shear stress is calculated as shown in Fig. 9-8 with additional notation and constants as follows:

$$c_1 = c_2; \quad c_3 = \frac{(c_1 + d)}{2}$$

$$b_0 d = 4d(c_1 + d)$$

$$J_c = \frac{d(c_1 + d)^3}{6} + \frac{(c_1 + d)d^3}{6} + \frac{d(c_2 + d)(c_1 + d)^2}{2}$$

For exterior columns, part of the unbalanced moment is transferred by the eccentricity of the centroid of the shear section, g, with respect to the column centroid. Thus, M_t, the moment to be transferred by shear and torsion, is the unbalanced moment minus $V_u g$, the moment transferred by shear eccentricity. The calculation procedure and equations for exterior columns are presented in Fig. 9-9. Similar calculation procedures may be used for corner columns.

Review of available vertical load punching tests indicate that eq. (11-13) of the ACI 318-83 conservatively predicts shear strength of prestressed two-way slabs. Figure 9-10 shows the results of single-column slab specimen punching shear tests, and results of multi-panel slabs tested in shear. Equation (11-13) (expressed in terms of shear rather than stress) is:

$$V_{cw} = b_0 d(3.5\sqrt{fc} + 0.3F/A) + V_p$$

where F/A = average precompression in the direction of moment transfer. Effective limit 500 psi, minimum 125 psi. For values of precompression less than 125, $v_c = 4\sqrt{f_c'}$ as for non-prestressed construction.

V_p = shear carried through critical section by tendons. For thin slabs, this term must be carefully evaluated, as field placing practices can have a great effect on the profile of the tendons through the critical section. Conservatively, this term may be taken as zero.

Recent tests at the University of Washington[9-11] and the University of Illinois[9-18] confirm the applicability of the

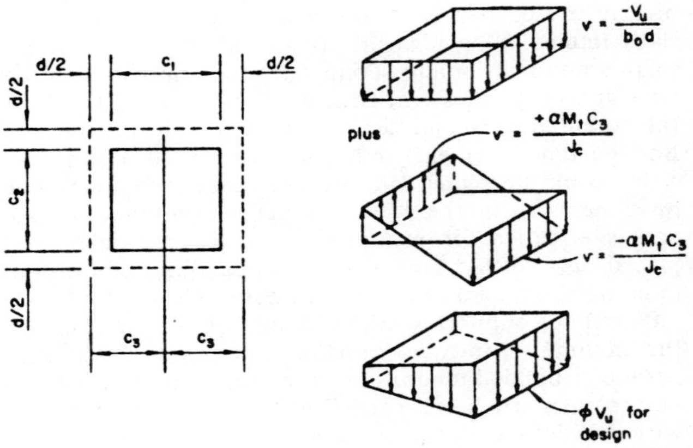

Fig. 9-8 Moment-shear interaction—relationships for interior columns.

above approach for calculations of the shear capacity of connections transferring moment as well as shear. The shear stress, v_{cw}, is compared to the maximum total vertical plus torsional shear stress, and F/A is the average precompression in the direction of moment transfer. Exterior column connections must be detailed to control cracking due to torsion and to prevent loss of stiffness. This requires that some tendon anchorages be placed within the column width, and that auxiliary closely spaced bonded reinforcement be placed through the column and as close as practicable to the column in each direction. The flexural capacity of the tendons and the bonded reinforcement within lines $3h/2$ on either side of the column must exceed the fraction of the moment not transferred by shear, $(1 - \alpha)M_t$, divided by the capacity reduction factor for flexure.

9.7 FIRE RESISTANCE

Complete discussion of the fire-resistive characteristics of post-tensioned slabs is presented in a publication available from the Post-Tensioning Institute.[9-14] For convenience, slab thickness and cover requirements for post-tensioned slabs are summarized in Tables 9-2 and 9-3, respectively. The cover to the prestressing steel at the anchor should be at least 1/4 in. greater than that required away from the anchor. Minimum cover to the bearing plate should be 3/4 in. in slabs.

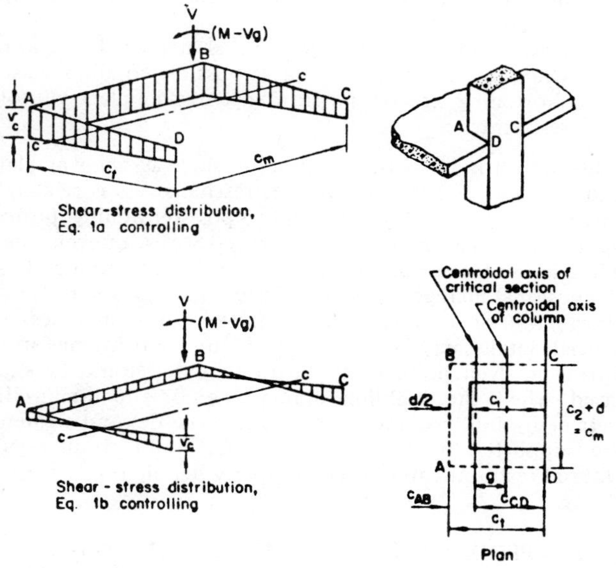

Shear-Stress Equations

$$v_c = \frac{V}{A_c} + \frac{\alpha(M - V_g)\,c_{AB}}{J_c}, \text{ (Eq.1a)}; \quad v_c = \frac{V}{A_c} - \frac{\alpha(M - V_g)\,c_{CD}}{J_c}, \text{ (Eq.1b)}$$

$$\alpha = 1 - \frac{1}{1 + 2/3\,(c_m/c_t)^{1/2}} \text{ (Eq.2)}$$

Critical-Section Properties:

$$A_c = d\,(c_m + 2c_t); \; c_{AB} = \frac{c_t^2 d}{A_c}, \; c_{CD} = c_t - c_{AB}; \; g = c_{CD} - \frac{c_1}{2}$$

$$J_c = \frac{dc_t^3}{6} + \frac{c_t d^3}{6} + c_m d c_{AB}^2 + 2\,c_t d(c_t/2 - c_{AB})^2$$

Fig. 9-9 Moment-shear interaction—relationships for edge column connections.

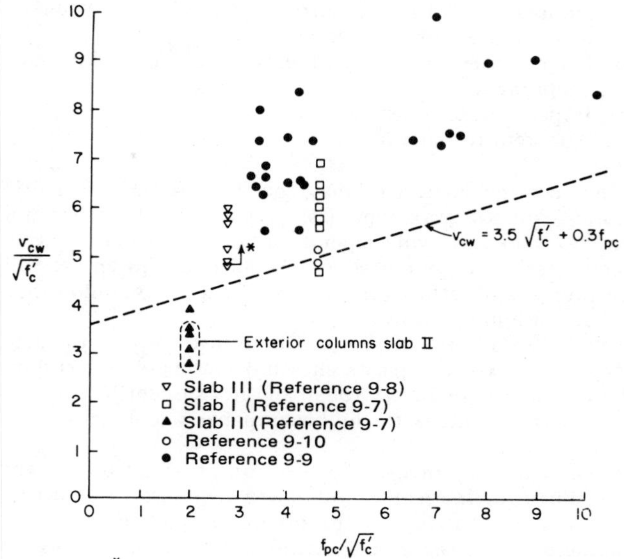

*Did not fail in punching shear; maximum stress attained

Fig. 9-10 Shear test data versus eq. (11-13) of ACI 318-83.

TABLE 9-2 Required Concrete Slab Thickness Requirements for Various Fire Endurances

Aggregate Type	*Slab Thickness, in., for Fire Endurance Indicated* 1 hr	$1\frac{1}{2}$ hr	2 hr	3 hr	4 hr
Carbonate	$3\frac{1}{4}$	$4\frac{1}{8}$	$4\frac{5}{8}$	$5\frac{3}{4}$	$6\frac{5}{8}$
Siliceous	$3\frac{1}{2}$	$4\frac{1}{4}$	5	$6\frac{1}{4}$	7
Lightweight	$2\frac{5}{8}$	$3\frac{1}{4}$	$3\frac{3}{4}$	$4\frac{5}{8}$	$5\frac{1}{4}$

TABLE 9-3 Required Concrete Cover Thickness for Slabs Prestressed with Post-Tensioned Reinforcement

Restrained or Unrestrained	Aggregate Type	*Cover Thickness, in., for Fire Endurance of* 1 hr	$1\frac{1}{2}$ hr	2 hr	3 hr	4 hr
Unrestrained	Carbonate	3/4	$1\frac{1}{16}$	$1\frac{3}{8}$	$1\frac{7}{8}$	—
Unrestrained	Siliceous	3/4	$1\frac{1}{4}$	$1\frac{1}{2}$	$2\frac{1}{8}$	—
Unrestrained	Lightweight	3/4	1	$1\frac{1}{4}$	$1\frac{5}{8}$	—
Restrained	Carbonate	3/4	3/4	3/4	1	$1\frac{1}{4}$
Restrained	Siliceous	3/4	3/4	3/4	1	$1\frac{1}{4}$
Restrained	Lightweight	3/4	3/4	3/4	3/4	1

9.8 CORROSION PROTECTION

Corrosion protection for unbonded tendons requires the following properties of the coating materials:

1. Free from cracks and not brittle or fluid over the entire anticipated range of temperatures. In the absence of specific requirements, this is usually taken as 20°C to 70°C.
2. Chemically stable for the life of the structure.

3. Nonreactive with the surrounding materials such as concrete, tendons, wrapping, or ducts.
4. Noncorrosive or corrosion-inhibiting for tendons and anchorages.
5. Impervious to moisture.
6. Adherent to tendons.

The coating materials may be bitumastics, asphaltic mastics, greases, wax, epoxies, or plastics. The minimum coating thickness will depend on the particular coating material selected, but it should be adequate to ensure full continuity and effectiveness with a sufficient allowance for variation in application.

Currently, additional protection against corrosion is provided by an extruded plastic sheath over the greased tendon. The plastic sheath is very tough and not susceptible to tearing as was the case with the heavy paper wrapping used in prior years.

Even with the corrosion protection provided by current procedures for fabricating unbonded tendons, care must be exercised in construction to obtain proper cover in the completed structure and to avoid exposure of tendons to corrosive environments while stored on the job site awaiting installation. Admixtures containing calcium chloride or aggregates containing chlorides should not be used in post-tensioned structures.

9.9 CONSTRUCTION JOINTS AND CLOSURE STRIPS[9-9]

The maximum length of a slab between construction joints should be limited to about 150 ft to minimize the effect of slab shortening, and to avoid excessive loss of prestress due to friction. When special consideration is given to reducing the effects of axial shortening on both the plate and the substructure elements, and where tendons with low friction losses are utilized, somewhat longer distances between construction joints may be used. Tendons should be stressed from both ends when their length exceeds 100 ft.

Open strips (closure strips) may temporarily separate adjacent slabs during construction. These strips are sometimes used to provide access for stressing of tendons, and they should preferably be left open for a sufficient length of time to help minimize the effect of slab shortening and reduce the restraint forces described in Section 9.10. Reinforcement, either prestressed or nonprestressed, should be provided to achieve continuity when the closure strip concrete is cast. The design of reinforcement should assure continuity, taking into consideration the amounts of deflection and camber that occur prior to casting the closure strip. Propping may be used to assure full continuity for both dead and live load.

9.10 RESTRAINT FORCES FROM SUPPORTING WALLS AND COLUMNS

When walls and columns are rigidly connected to the slabs, their stiffness should be taken into consideration in the equivalent frame (or other) analysis.

When columns and walls have significant stiffness in the direction of prestress, the effects of possible diversion of prestress into bending and/or translating substructure elements on the performance of the flat plate should be considered. The columns or walls may also restrain the slab from shortening, thus resulting in cracks. Likewise, the effects of the prestressing forces on stiff supporting elements may require investigation. However, design and construction options are available to reduce the effects of shortening on both the plate and the supporting elements,[9-14] and the moments or stresses that occur over a period of time due to creep and shrinkage shortening are themselves reduced approximately 50% by creep.[9-15]

The restraints due to dimensional changes can be accommodated or reduced in the following ways:

1. Design or locate supporting elements to minimize restraint. Relatively long flexible columns may reduce restraint forces to the point where they can be easily accommodated by column reinforcement. Lateral load resisting elements can often be located near the center of movements so that no restraint develops.
2. Segment the structure with pour strips or temporary joints to minimize the movement and restraint developed during post-tensioning and due to early volume changes.
3. Detail the connection between the flexural elements and columns to permit movement. Joint details to permit movement between a slab and a stairwell or elevator core are shown in Fig. 9-11. Figure 9-12 shows two examples of joint details for concrete slab construction that provide for rotation and displacement during post-tensioning and a lateral structural connection. The joint details shown in Fig. 9-12 may reduce the shear capacity somewhat. Such joints require careful detailing, and the possible reduction of shear strength should be considered during the design stage.
4. Shrinkage-compensating cement may be used to reduce the long-term dimensional change effects on the structure.
5. Delay connection of the flexural element and the restraining elements as long as possible to allow elastic shortening, creep, and shrinkage to take place without restraint.

In applications of post-tensioning such as one-way slabs and two-way flat plates, the prestressing stress is relatively small. For example, stresses due to post-tensioning in these applications may be on the order of 150 psi or even somewhat less. Stresses of this magnitude do not produce large dimensional changes in normal-sized buildings due to elastic shortening or due to concrete creep. In such cases, no special details may be required to minimize restraint forces. However, even in these applications, care should be exercised when the building dimensions, or the dimensions between joints become large, or when the flexural elements are supported by rigid elements that could produce substantial restraint forces if not properly detailed.

9.11 APPROXIMATE QUANTITIES FOR POST-TENSIONED ONE-WAY AND TWO-WAY SLABS

Quantities included: Concrete (c.y./s.f.), Post-tensioning (kips/linear foot), Rebar (lbs/square foot).

The quantity estimates presented in Table 9-4 have been prepared to assist architects, engineers, and contractors in evaluating the feasibility of post-tensioned slabs for specific projects.

For each span, two slab thicknesses are indicated. The smaller slab thickness is based on a span/thickness ratio of $l/45$ and utilizes the minimum cover requirements for slabs specified by the ACI-318 *Building Code* (3/4 in.). The larger slab thickness utilizes the same tendon profile as the thinner slab, but provides 1 in. additional cover. The use of thicker

TABLE 9-4 Approximate Quantities for Post-Tensioned One-Way and Two-Way Slabs

				Fe REQUIRED K/L.F.	*LBS/S.F. = REBAR*		
Span Ft.	*Location Int./Ext.*	*Slab "T" Inches*	*Concrete C.Y./S.F. $f'_c = 4000$*	*Reg. Wgt. Concrete*	*Two-Way Slab*	*One-Way Slab Span*	*Shrinkage & Temperature**
16	INT.	4.5	0.014	6.79	0.25	0.50	0.35
	EXT.	4.5	0.014	9.22	0.25	0.39	0.35
	INT.	5.5	0.017	8.39	0.25	0.61	0.43
	EXT.	5.5	0.017	11.26	0.25	0.48	0.43
18	INT.	5	0.015	8.10	0.25	0.52	0.37
	EXT.	5	0.015	10.80	0.25	0.43	0.37
	INT.	6	0.019	9.72	0.25	0.65	0.45
	EXT.	6	0.019	12.96	0.25	0.54	0.45
20	INT.	5.5	0.017	9.43	0.30	0.57	0.40
	EXT.	5.5	0.017	12.57	0.30	0.48	0.40
	INT.	6.5	0.020	11.14	0.30	0.68	0.47
	EXT.	6.5	0.020	14.86	0.30	0.57	0.47
22	INT.	6	0.019	10.89	0.35	0.69	0.43
	EXT.	6	0.019	14.52	0.35	0.54	0.43
	INT.	7	0.022	12.74	0.35	0.79	0.52
	EXT.	7	0.022	16.99	0.35	0.61	0.52
24	INT.	6.5	0.020	12.48	0.35	0.69	0.47
	EXT.	6.5	0.020	16.64	0.35	0.58	0.47
	INT.	7.5	0.023	14.40	0.35	0.77	0.56
	EXT.	7.5	0.023	19.20	0.35	0.67	0.56
25	INT.	7	0.022	13.13	0.35	0.72	0.51
	EXT.	7	0.022	17.50	0.35	0.67	0.51
	INT.	8	0.025	15.00	0.35	0.84	0.59
	EXT.	8	0.025	20.00	0.35	0.71	0.59
26	INT.	7	0.022	14.20	0.38	0.71	0.51
	EXT.	7	0.022	18.92	0.38	0.61	0.51
	INT.	8	0.025	16.22	0.38	0.83	0.59
	EXT.	8	0.025	21.63	0.38	0.71	0.59
27	INT.	7	0.022	15.31	0.40	0.70	0.52
	EXT.	7	0.022	20.41	0.40	0.60	0.52
	INT.	8	0.025	17.50	0.40	0.82	0.59
	EXT.	8	0.025	23.32	0.40	0.71	0.59
28	INT.	7.5	0.023	16.04	0.45	0.76	0.55
	EXT.	7.5	0.023	21.38	0.45	0.65	0.55
	INT.	8.5	0.025	18.17	0.45	0.82	0.62
	EXT.	8.5	0.025	24.23	0.45	0.70	0.62
30	INT.	8	0.025	18.00	0.50	0.82	0.58
	EXT.	8	0.025	24.00	0.50	0.70	0.58
	INT.	9	0.028	20.25	0.50	0.85	0.67
	EXT.	9	0.028	27.00	0.50	0.73	0.67
32	INT.	8.5	0.026	20.09	0.55	0.84	0.58
	EXT.	8.5	0.026	26.78	0.55	0.73	0.58
	INT.	9.5	0.029	22.45	0.55	0.95	0.69
	EXT.	9.5	0.029	29.93	0.55	0.82	0.69
34	INT.	9	0.028	22.29	0.60	0.85	0.67
	EXT.	9	0.028	29.73	0.60	0.75	0.67
	INT.	10	0.031	24.79	0.60	0.94	0.74
	EXT.	10	0.031	33.02	0.60	0.82	0.74
35	INT.	9.5	0.029	23.28	0.70	0.94	0.70
	EXT.	9.5	0.029	31.03	0.70	0.82	0.70
	INT.	10.5	0.032	25.73	0.70	1.00	0.77
	EXT.	10.5	0.032	34.30	0.70	0.87	0.77
36	INT.	9.5	0.029	24.62	0.75	0.93	0.70
	EXT.	9.5	0.029	33.83	0.75	0.82	0.70
	INT.	10.5	0.032	27.22	0.75	1.00	0.78
	EXT.	10.5	0.032	36.29	0.75	0.87	0.78

**Post-tensioning may be used in lieu of bonded shrinkage and temperature reinforcement by providing 100 psi compression (P/A), after all prestress losses, on the gross concrete section.*

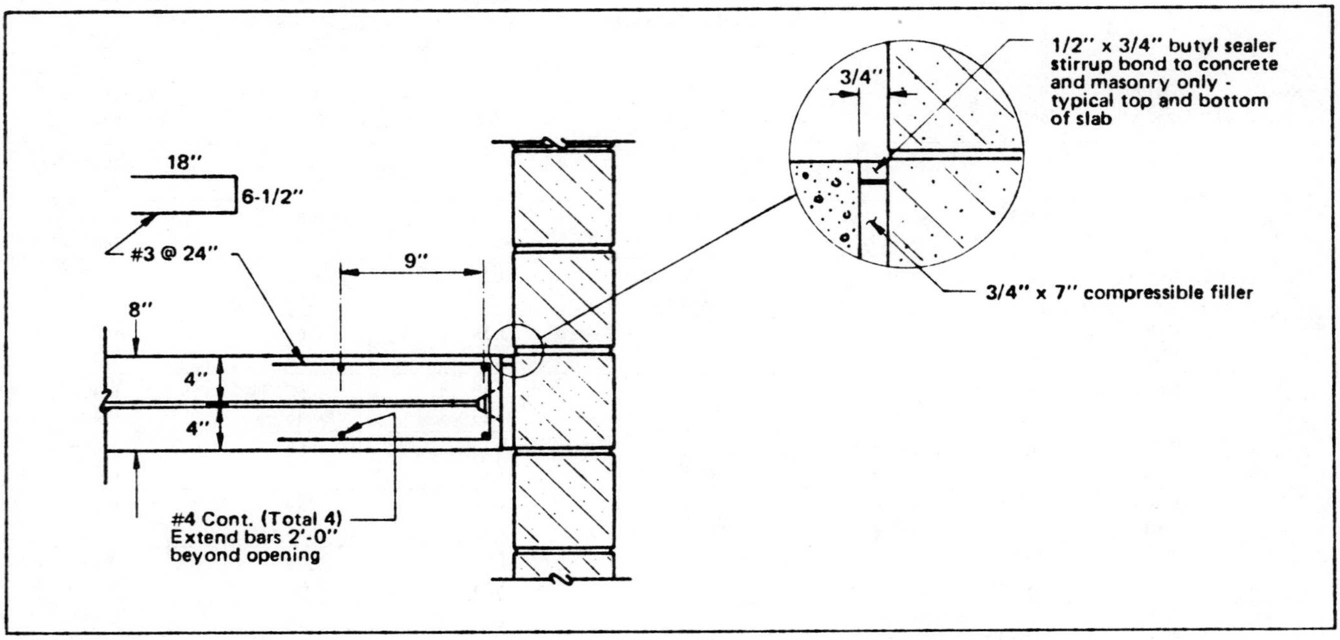

Fig. 9-11 Details for separation of slab from elevator core or stairwell.

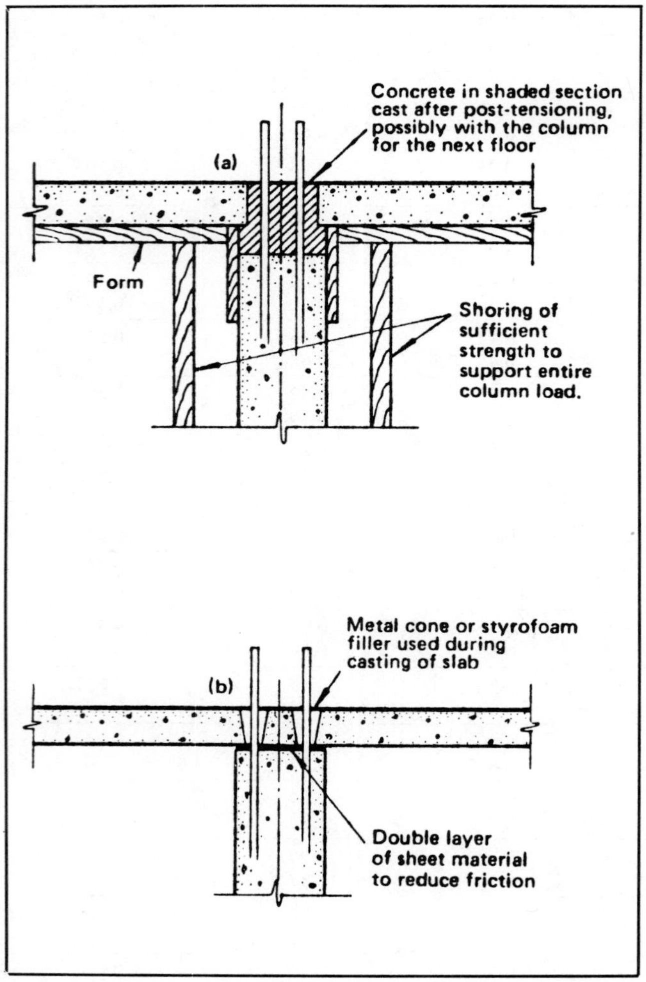

Fig. 9-12 Examples of joint details for concrete floor construction.

slabs with additional cover is recommended for applications exposed to corrosive environments such as parking structure slabs in areas where deicer chemicals are used.

To obtain approximate quantities, enter the table at the span selected, follow the directions listed below, obtain quantities, and apply unit prices representative of those in the area where the project is located.

Directions for use of tables:

1. The reinforcement quantities shown in these tables are for the spans in one direction only. For two-way systems, add the quantities shown for the spans in both directions. Rebar requirements for one-way slabs are the sum of the span rebar and shrinkage and temperature rebar.
2. Where it is desirable to use added post-tensioning for the end bays, use the Fe shown for Ext. Span (exterior span) in combination with the Fe shown for Int. Span (interior).
3. The rebar quantities shown are only approximate and should be modified for specific applications, particularly for design live loads in excess of those required for office building, apartment, or garage application.
4. Tendon forces were determined by load-balancing 80% of the slab dead load. In all cases, this provided a minimum P/A in excess of 125 psi. Stress-relieved 1/2" 270k strand tendons provide an effective force of 24.33k and weigh 0.52 pound per linear foot.
5. These tables are presented for estimating purposes only and should not be used as final design requirements without verification by a competent engineer.

9.12 FLAT PLATE APARTMENT DESIGN EXAMPLE

1. Design a typical transverse strip as described in the two drawings and the following calculations.

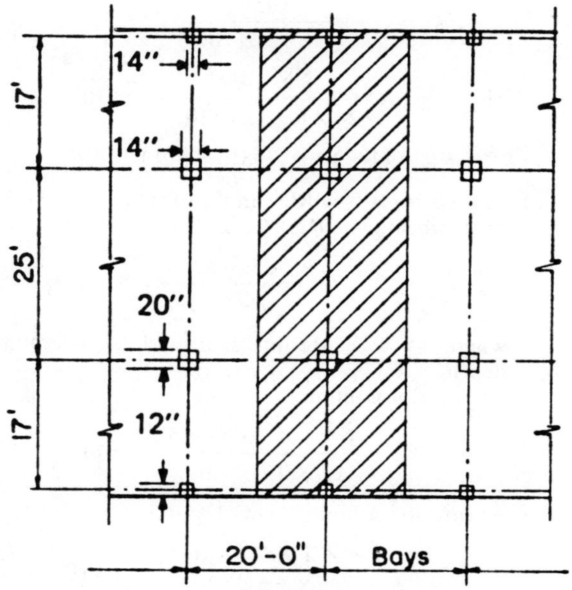

PART PLAN

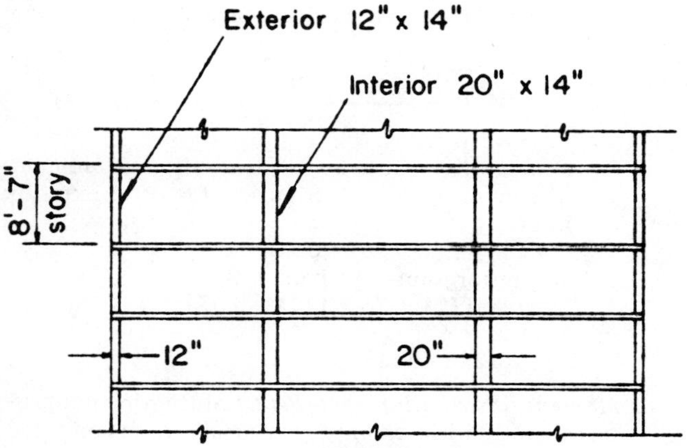

SECTION

Slab Thickness @ $L/45$:

$$\text{Longitudinal} = \frac{20 \times 12}{45} = 5.3''$$

$$\text{Transverse} = \frac{25 \times 12}{45} = 6.7''$$

Use $6\frac{1}{2}''$ slab.
Loads:

$6\frac{1}{2}''$ Slab =	81	
Partitions =	15	
Dead =	96 × 1.4 =	134
Live =	40 × 1.7 =	68
Total =	136 psf	= 202 psf, ultimate

Assume hardrock concrete $f_c' = 4000$ psi in slabs and columns

2. *Design procedure:* Assume a set of loads to be balanced by parabolic tendons. Analyze an equivalent frame subjected to the net downward loads, according to the principles of ACI 318-83 section 13.7. Check flexural stresses at critical sections, and revise load-balancing tendon forces as required to obtain net flexural tension stresses in accordance with ACI 318-83.

When final forces are determined, obtain frame moments for factored dead and live loads. Calculate secondary moments induced in the frame by post-tensioning forces, and combine with factored load moments to obtain design ultimate moments. Provide minimum mild steel reinforcement in accordance with ACI 318-83 section 18.9. Check ultimate flexural capacity and increase mild steel if required by strength criteria. Investigate shear strength, including shear due to vertical load and due to moment transfer by torsion; compare total to allowable value calculated from ACI 318-83 eq. 11.11.2.

3. *Load balancing:* Arbitrarily, a force corresponding to an average compressive stress of 175 psi, with maximum parabolic tendon profile, will be used for the initial estimate of balanced load.
Then:

$$F_e = 0.175 \times 6.5 \times 12 = 13.65 \text{ kips/ft}$$

Assuming $\frac{1}{2}''$ diameter, 270 ksi strand tendons, and 30 ksi long-term losses, effective force per tendon is $0.153 \times (0.7 \times 270 - 30) = 24.33^k$.

For a 20-ft bay, $20 \times 13.65/24.33 = 11.2$ tendons, say eleven.
Then:

$$F_e = 11 \times 24.33/20 = 13.38^k/\text{ft}$$

$$F/A = 13.38/78 = 0.172 \text{ ksi}$$

4. *Tendon profile:*

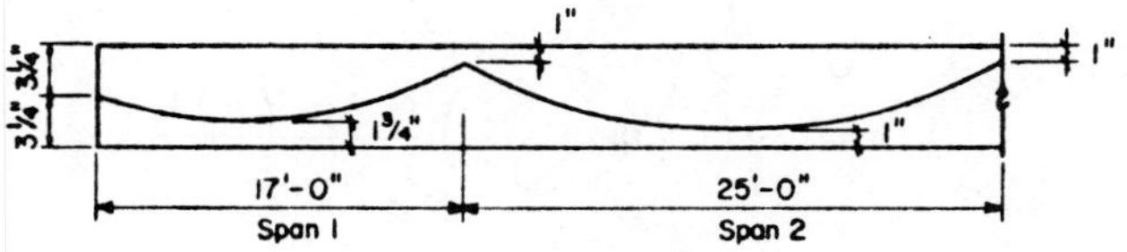

For spans 1 and 3:

$$a = (3\frac{1}{4} + 5\frac{1}{2})/2 - 1\frac{3}{4} = 2.625''$$

$$W_{\text{bal}} = 8 \times F \times a/(12 \times L^2)$$

$$= (13.38 \times 2.625)/(1.5 \times 289) = 0.081 \text{ ksf}$$

Net load causing bending = $W_{\text{net}} = 0.136 - 0.081$

$$= 0.055 \text{ ksf}$$

For span 2:

$$a = 6\frac{1}{2} - 1 - 1 = 4\frac{1}{2}''$$

$$W_{\text{bal}} = 0.064 \text{ ksf}$$

$$W_{\text{net}} = 0.072 \text{ ksf}$$

5. *Equivalent frame properties:* (See ACI 318-83, section 13.7)
   (a) *Equivalent columns:* The basic stiffness of columns, including the effects of "infinite" stiffness at joints, may be calculated by classical methods or by simplified methods that are in close agreement. The following formula for "approximate" stiffness is taken from *Equivalent Frames of Reinforced Concrete* by Cross and Morgan.[9-16]

$$K_c = \frac{EI}{L'}\left(1 + 3\frac{L}{L'}\right)$$

where I is taken at the column, L is center to center height, L' is clear height.

Carryover factors are approximated by $-\frac{1}{2} \times (1 + 3h)$ where h is the length of infinite I.

A simpler approximation is shown by Rice and Hoffman in *Structural Design Guide to the ACI Building Code:*[9-17]

$$K_c = \frac{4EI}{L - 2h}$$

The approximate formulas give results within 5% of the "exact" values, and considering the nature of assumptions necessary for design of the highly complex two-way flat plate, these formulae are completely adequate. Refer to Rice and Hoffman, ibid., for a comparison of approximate and classical methods.

Exterior column—14 X 12:

$$I = \frac{14 \times 12^3}{12} = 2016 \text{ in.}^4$$

$$\frac{E_{\text{col}}}{E_{\text{slab}}} = 1.0$$

$$K_c = \frac{4 \times 1.0 \times 2016}{103 - 2 \times 6.5} = 90$$

$$= 180 \text{ joint total } (90 \times 2)$$

Torsional stiffness of slab in column line, K_t, is calculated as follows:

$$C = \left(1 - 0.63\,\frac{x}{y}\right)\frac{x^3 y}{3}$$

$$= \left(1 - 0.63\,\frac{6.5}{12}\right)\frac{(6.5)^3 \times 12.0}{3}$$

$$= 724$$

$$K_t = \frac{\sum 9 \times C \times E}{L_2 \times (1 - c_2/L_2)^3}$$

$$= \frac{9 \times 724 \times 1}{20 \times 12(1.0 - 1.17/20)^3} + \frac{9 \times 724 \times 1}{20 \times 12(1.0 - 1.17/20)^3} = 65$$

Equivalent column stiffness is then obtained.

$$\frac{1}{K_{ec}} = \frac{1}{K_t} + \frac{1}{K_c}$$

$$K_{ec} = (1/65 + 1/180)^{-1} = 48$$

Interior column—14 X 20:

$$I = \frac{14 \times (20)^3}{12} = 9333 \text{ in.}^4$$

$$K_c = \frac{4 \times 1 \times 9333}{103 - 2 \times 6.5} = 415$$

$$= 830 \text{ joint total } (415 \times 2)$$

$$C = \left(1 - 0.63 \times \frac{6.5}{20}\right) \times \frac{(6.5)^3 \times 20}{3}$$

$$= 1456$$

$$K_t = \frac{9 \times 1456}{240(1 - 1.17/20)^3} + \frac{9 \times 1456}{240(1 - 1.17/20)^3} = 130$$

$$K_{ec} = (1/830 + 1/130)^{-1} = 112$$

(b) *Slab stiffness* (see Rice and Hoffman, ibid.), width of slab-beam = 20/2 + 20/2 = 20 ft:

$$K_s = \frac{4EI}{L_1 - c_1/2}$$

where L_1 is centerline span and C_1 is column depth.

At exterior column:

$$K_s = \frac{4 \times 1 \times 20 \times (6.5)^3}{12 \times 17 - 12/2} = 111$$

At interior column, spans 1 and 3:

$$K_s = \frac{4 \times 1 \times 20 \times (6.5)^3}{12 \times 17 - 20/2} = 110$$

Use single value of 111 for both ends of span since there is so little difference. At interior span 2:

$$K_s = \frac{4 \times 1 \times 20 \times (6.5)^3}{12 \times 25 - 20/2} = 76$$

(c) *Distribution factors for analysis by moment distribution.* Slab distribution factor at exterior joint:
= 111/(111 + 48) = 0.70
at interior joints for spans 1, 3:
= 111/(111 + 76 + 111) = 0.37
span 2:
= 76/298 = 0.25

6. *Moment distribution—net loads:* Since the nonprismatic section causes only very small effects on fixed end moments and carryover factors, fixed end moments will be calculated from $wl^2/12$ and carryover factors taken as 1/2. Span 1, 3 net load $FEM = 0.055 \times 289/12 = 1.32$. Span 2 $FEM = 0.072 \times 625/12 = 3.75$.

Moment Distribution:

0.70	0.37	0.25 symm.
−1.32	−1.32	−3.75 FEM
+0.92	−0.90	+0.61 dist
+0.45	−0.46	−0.30 C.O.
−0.32	+0.07	−0.05 dist
−0.27	−2.62	−3.49

7. *Check net tensile stresses:*
(a) At face of column:
Moment at column face is centerline moment $+ Vc/3$:

$$-M_{\max} = -3.49 + \frac{20}{3 \times 12}(12.5 \times 0.072)$$

$$= -3.49 + 0.50 = -2.99 \text{ ft-k}$$

$$S = bt^2/6 = 12 \times 6.5 \times 6.5/6 = 84.5 \text{ in.}^3$$

$$= 84.5 \text{ in.}^3 \frac{\text{ft}}{12 \text{ in.}} = 7.04 \text{ in.}^2 \text{ ft}$$

Then:

$$f_t = \frac{2.99}{7.04} - 0.172 = 0.425 - 0.172$$

$$= +0.253 \text{ ksi} \quad \text{OK, since } 6\sqrt{f_c'} = 380 \text{ psi}$$

(b) Check midspan tensile stress:

$$+M_{max} = 5.62 - 3.49 = +2.13 \text{ ft-kip}$$

$$f_t = \frac{2.13}{7.04} - 0.172$$

$$= 0.130 \text{ ksi} > 2\sqrt{f_c'}\text{, i.e., } 126 \text{ psi}$$

By requirements of ACI 318-83 section 18.9, when tensile stress of $2\sqrt{f_c'}$ is exceeded, the entire tensile force must be resisted by mild reinforcing at a stress of $f_y/2$.

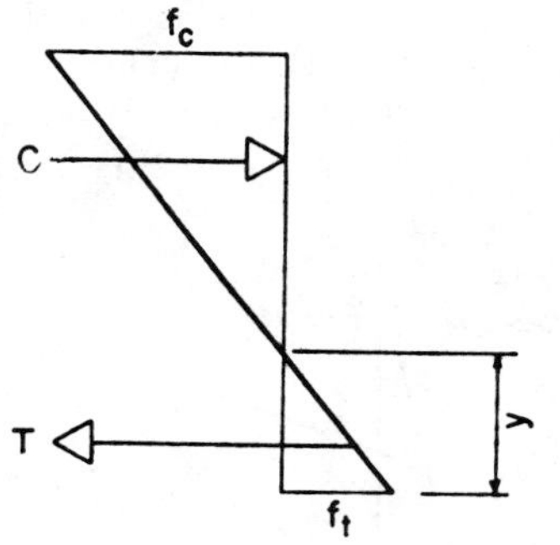

$$f_c = -0.302 - 0.172 = -0.474 < 0.45 \times (4000)$$

Then:

$$y = 6.5(0.130/(0.130 + 0.474)) = 1.40 \text{ in.}$$

$$T = 0.130 \times 1.40 \times \frac{12}{2} = 1.09^k/\text{ft}$$

$$A_s = \frac{1.09}{60/2} = 0.036 \text{ in.}^2/\text{ft}$$

Say 4 No. 4 bars at 60″ o.c. bottom of midspan 2, $A_s = 0.04$ in.^{2}/ft.

This completes the service load portion of the design, but the ultimate strength in flexure and shear must be verified to complete the design. Based on the service load flexural stresses, the design is conservative.

8. *Ultimate flexural capacity.*
   (a) *Calculation of design moments:* Design moments for statically indeterminate post-tensioned members are determined by combining frame moments due to factored dead and live loads with secondary moments induced into the frame by the tendons. The load balancing approach directly includes both primary and secondary effects, so that for service conditions only "net loads" need be considered.

   At ultimate, the balanced load moments are used to determine secondary moments by subtracting the primary moment, which is simply $F \times e$, at each support. For multistory buildings where typical vertical load design is combined with varying moments due to lateral loading, an efficient design approach would be to analyze the equivalent frame under each case of dead, live, balanced, and lateral loads, and combine these cases for each design condition with appropriate load factors. For this example the balanced load moments are determined by moment distribution as follows:

Span 1, 3 balanced load $FEM = 0.081 \times 289/12 =$ 1.95

Span 2 $FEM = 0.064 \times 625/12 = 3.33$

Moment Distribution Balanced Loads:

0.7	0.37	0.25 symm.
+1.95 −1.37	+1.95 +0.51	+3.33 FEM −0.35 dist
−0.25 +0.18	+0.68 −0.19	+0.17 C.O. +0.13 dist
+0.51	+2.95	+3.28

Since load balancing accounts for both primary and secondary moment directly, secondary moments can be found from the following relationship:

$$M_{bal} = M_1 + M_2, \quad \text{then} \quad M_2 = M_{bal} - M_1$$

The primary moment, M_1 is $F \times e$ at each support. Thus at exterior columns, the secondary moment, M_2, is:

$$M_2 = 0.51 - 13.38^k \times \frac{0 \text{ in.}}{12 \text{ in./ft}}$$

$$= 0.51 \text{ ft-k}$$

At interior column, span 1, 3

$$M_2 = 2.95 - 13.38 \times (3.25 - 1.0)/12$$

$$= 0.44 \text{ ft-k}$$

Span 2:

$$M_2 = 3.28 - 13.38 \times (2.25)/12$$

$$= 0.77 \text{ ft-k}$$

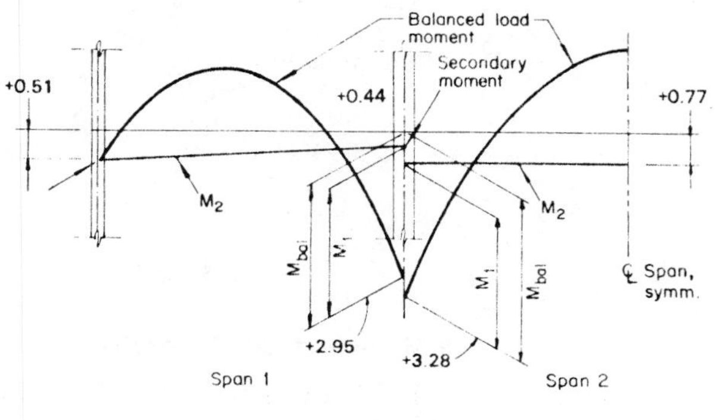

Factored load moments
Span 1, 3:

$$1.4 \text{ dead} + 1.7 \text{ live } FEM = 0.202 \times .289/12$$

$$= -4.86 \text{ ft-k}$$

Span 2:

$$FEM = 0.202 \times 625/12 = -10.52 \text{ ft-k}$$

Moment Distribution Factored Loads:

0.7	0.37	0.25 symm.
-4.86	-4.86	-10.52 FEM
+3.40	-2.09	+1.42 dist.
+1.05	-1.70	-0.71 C.O.
-0.74	+0.37	-0.25 dist.
-1.15	-8.28	-10.06

Combine factored load and secondary moments to obtain design moments:

	Span 1		Span 2
factored load moments	-1.15	-8.28	-10.06
secondary moments	+0.51	+0.44	+0.77
design moments at col	-0.64	-7.84	-9.29
moment reduction to face $Vc/3$	+0.42	+1.19	+1.40
design moments at critical section	-0.22	-6.65	-7.89

Calculate design midspan moments:
Span 1

$$V_{\text{exterior}} = \frac{.202 \times 17}{2} - \frac{(8.28 - 1.15)}{17}$$

$$= .172 - 0.42 = 1.30 \text{ kips/ft}$$

$$V_{\text{center}} = .172 + 0.42 = 2.14 \text{ kips/ft}$$

Point of zero shear and maximum moment:

$$x = \frac{1.30}{.202} = 6.45 \text{ ft from exterior column}$$

End span positive moment =

$$1.30 \times 6.45 = +.840$$
$$-.202\,(6.45)^2 = -4.20$$
$$\text{End moment} = -1.15$$
$$+3.05$$
$$M_2 = 0.43$$
$$+M_{\max} = +3.48 \text{ ft-kips/ft}$$

Span 2:

$$V = \frac{-202 \times 25}{2} = 2.52 \text{ k/ft.}$$

$$+ \text{Moment} = .202 \times 25^2/8$$

$$= 15.78 \text{ ft-kips/ft}$$

$$-\text{End moment} = -10.06$$
$$+5.72$$
$$M_2 = +.77$$
$$+M_{\max} = 6.49 \text{ ft-kips/ft}$$

(b) *Calculation of flexural strength:*
Check interior support section: Section 18.9 of ACI 318-83 requires a minimum amount of mild steel at the immediate column zone regardless of service load stress conditions, or strength, unless more than the minimum is required for ultimate flexural capacity. This minimum amount is to help ensure the integrity of the punching zone so that full shear capacity can be developed, and is determined by the following expression:

$$A_s = 0.00075 h \times L$$

The initial check of flexural strength will be made considering this steel.

$$A_s = 0.00075 \times 6.5 \times (17 + 25)/2 \times 12$$

$$= 1.22 \text{ in.}^2$$

Say 6 No. 4 Bars × 9 ft. Space at maximum 6″ o.c. so that bars are placed within a width of column plus 1½ slab thickness each side of column. Then for average 1-ft strip:

$$A_s = 6 \times 0.2/20 = 0.06 \text{ in.}^2/\text{ft}$$

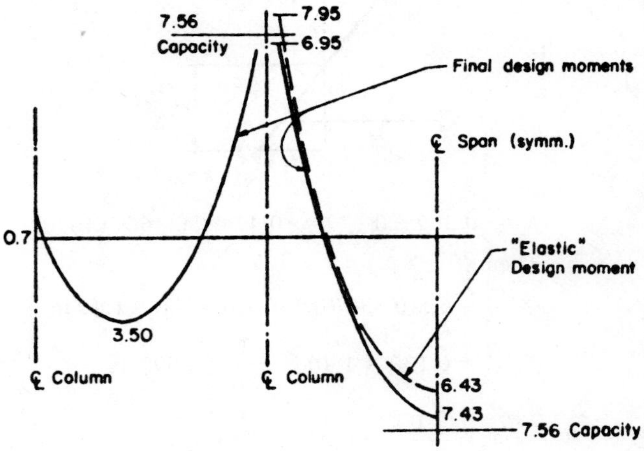

Calculate stress in tendon at ultimate: use ACI 318-83 eq. (18-5):

$$f_{ps} = f_{se} + \frac{f_c'}{300\rho_p} + 10 \text{ ksi}$$

Since we have eleven-½-in.-dia tendons in a 20-ft bay, each with area = 0.153 in.²:

$$\rho_p = \frac{A_{ps}}{bd} = \frac{11 \times 0.153}{20 \times 12 \times 5.5} = 0.00127$$

$$f_{se} = 0.7 \times 270 - 30 \text{ ksi losses} = 159 \text{ ksi}$$

$$f_{pe} = 159 + 10 + \frac{4.000}{0.00127 \times 300}$$

$$= 169 + 10.5 = 179.5$$

$$F_{su} = \frac{179.5 \times 0.153 \times 11}{20} = 15.10 \text{ k/ft.}$$

$F_u = 60 \times 0.06 =$ $\underline{3.60}$

$F =$ Total force = 18.7 k/ft

depth of compression block $a = \dfrac{F}{0.85\, b f_c'}$

$$a = \frac{18.7}{0.85 \times 12 \times 4} = 0.46 \text{ in.}$$

Since bars and tendons are in the same layer:

$$\left(d - \frac{a}{2}\right) = \left(5.5 - \frac{0.46}{2}\right)/12 = 0.44 \text{ ft}$$

moment capacity at column centerline =

$$M_u = 0.9 \times 0.44 \times 18.7 = 7.41 \text{ ft-kips/ft}$$

This value is less than the required capacity of 7.89 ft-kips/ft

Calculate available capacity at midspan and allowable inelastic moment redistribution at column. See ACI 318-83 section 18.10.

Allowable redistribution:

$$= 20\% \times \left(1 - \frac{\omega_p + \omega}{0.30}\right)$$

$$\sum \omega = \frac{18.7}{5.5 \times 12 \times 4} = 0.071 < 0.20$$

$$R = 20\left(1 - \frac{0.071}{0.30}\right) = 15.2\%$$

$$M_R = 0.152 \times 7.41 = 1.13 \text{ ft-k}$$

Since the midspan of span 2 requires 4 No. 4 bars from service load considerations, the flexural strength is:

$F_{su} =$ 15.10 k/ft

$F_u = 60 \times 0.04$ 2.40

$F =$ total force = 17.50

$$a = \frac{17.50}{0.85 \times 12 \times 4} = 0.43 \text{ in.}$$

$$\left(d - \frac{9}{2}\right) = \left(5.5 - \frac{0.03}{2}\right)/12 = 0.44 \text{ ft}$$

moment capacity at center of span:

$$M_u = 0.9 \times 0.44 \times 17.50 = 6.93 \text{ ft-kips/ft.}$$

The required moment capacity is 6.49 ft-k, which leaves 0.44 ft-k available to accommodate moment redistributed from the support section. If 0.44 ft-k is redistributed:

$-M = -7.89 + 0.44 = -7.45 > 7.41$ NG

$+M = +6.49 + 0.44 = 6.93 = 6.93$ OK

Thus minimum rebar and tendons are not adequate for strength at the column. Addition of 2 No. 4 bars at midspan will make mispan strength identical to strength at column (which also has 6 No. 4 bars) = 7.41 ft. kips/ft. In this case 7.41 ft. kips. -6.49 ft. kips/ft. or 0.92 ft. kips/ft. are available for redistribution. Redistributing 0.50 ft. kips =

$-M = -7.89 + 0.50 = 7.39 < 7.41$ OK

$-M = +6.49 + 0.50 = 6.99 < 7.41$ OK

Therefore, with addition of 2 No. 4 bars at midspan and redistribution of 0.50 ft. kips from the negative moment area to midspan, both sections are adequate for strength. Midspan sections of spans 1 and 3 have more than adequate capacity by comparison with span 2.

The flexural capacity at exterior columns is governed by moment transfer requirements. Since moment transfer also involves shear stresses, the two aspects will be treated under the heading of shear.

9. *Ultimate shear capacity:*
   (a) *Shear at exterior column:* Vertical shear at exterior column calculated above at 1.30 kip/ft. Total shear at exterior column = $20 \times 1.30 =$ 26.0 kips.

   Assume exterior skin is masonary and glass averaging 0.4 klf:

   Slab shear = 26.0

   $V_u = 1.4 \times 0.4 \times 20 = 11.2$

   Total shear = 37.2 kips

   (b) *Moment transfer:* At exterior columns, a portion of the total bay moment is transferred to the column by the eccentricity of the critical section relative to the column center. In order to evaluate these cases, the properties of the critical section must first be calculated.

   Shear section properties: The critical shear section is taken at $d/2$ from the face of the column as per Fig. 9-8. Referring to Fig. 9-8:

$$\text{Assume } d = 0.8 \times 6.5 = 5.2 \text{ in.}$$

$$c_1 = 12 \text{ in.}$$

$$c_2 = 14 \text{ in.}$$

$$c_m = 14 + 5.2 = 19.2 \text{ in.}$$

$$c_t = 12 + \frac{5.2}{2} = 14.6 \text{ in.}$$

$$A_c = 5.2(19.2 + 2 \times 14.6) = 252 \text{ in.}^2$$

$$c_{AB} = 14.6^2 \times 5.2/252 = 4.40 \text{ in.}$$

$$c_{CD} = 14.6 - 4.40 = 10.2 \text{ in.}$$

$$g = 10.2 - 12/2 = 4.2$$

$$\alpha = 1 - \frac{1}{1 + 2/3\left(\dfrac{c_t}{c_m}\right)^{1/2}}$$

$$\alpha = 1 - \frac{1}{1 + 2/3\left(\dfrac{14.6}{19.2}\right)^{1/2}}$$

$$\alpha = 1 - \frac{1}{1 + .58} = 1 - .63 = 0.37$$

$$J_c = \frac{dc_t^3}{6} + \frac{c_t d^3}{6} + c_m dc_{AB}^2$$

$$+ 2\,c_t d\left(\frac{c_t}{2} - c_{AB}\right)^2$$

$$J_c = \frac{5.2 \times 14.6^3}{6} + \frac{14.6 \times 5.2^3}{6}$$

$$+ 19.2 \times 5.2 \times 4.40^2$$

$$+ 2 \times 14.6 \times 5.2\left(\frac{14.6}{2} - 4.40\right)^2$$

$$J_c = 2697 + 342 + 1937 + 1277$$

$$= 6249 \text{ in.}^4$$

Total bay moment at column centerline:

$$M_u = 20 \times (-0.69) = -13.8 \text{ ft-kips}$$

Moment transferred by eccentricity of shear reaction:

$$Vg = 26.0 \times \frac{4.2}{12} = -9.10 \text{ ft-kips}$$

Net moment to be transferred $= M_t = -13.8 - (-9.10) = -4.70$ ft. kips

Amount to be transferred by shear $= \alpha M_t$

$$\alpha M_t = 0.37 \times -4.70 = -1.74 \text{ ft. kips}$$

$$v_c = \frac{37{,}200}{252 \times .85} + \frac{1{,}740 \times 12 \times 4.40}{6249 \times 0.85}$$

$$v_c = 173 + 17 = 190 \text{ psi}$$

Shear stress allowable according to equation (11-37) of ACI 318-83:

$$V_c = 4\sqrt{f_c'} = 4\sqrt{4000}$$

$$V_c = 253 \text{ psi}$$

Allowable shear stress exceeds calculated stress. Note that the research report Ref. 9-18, published following the writing of ACI 318-83 supports the use of equation (11-38) of ACI 318-83. In this case, Equation (11-38) would permit an allowable shear stress at the exterior column:

$$V_c = 3.5\ 4000 + 0.3 \times (172)$$

(ignoring V_p/A_c)

$$V_c = 221 + 52 = 273 \text{ psi}$$

Check flexural moment transfer: Although the flexural moment to be transferred is small, for illustrative purposes, calculate the capacity of the section of width equal to the width of the column plus 1½ slab thicknesses each side. Assume that of the ten tendons required for the 20-ft bay width, two are anchored within the column cage and are bundled together across the building. This minimum amount and location of post-tensioning force should be noted on the structural drawings. Besides providing flexural capacity, this prestress will act directly on the critical section for shear and enhance the strength. As previously shown, a minimum amount of mild steel is required at all columns. For this joint the area of rebar is:

$$A_s = 0.00075 \times 6.5 \times 12 \times 17 = 1.0 \text{ in.}^2$$

Say 5 No. 4 bars × 5 ft including standard hook

$$A_s = 1.0 \text{ in.}^2$$

Calculate stress in strand tendon:

$$b = 14 + 3 \times 6.5 = 33.5 \text{ in.}$$

$$f_{pe} = \frac{33.5 \times 3.25 \times 4}{300 \times 0.153 \times 2} + 169 = 172 \text{ ksi}$$

$$F_p = \frac{172}{159}\ 24.33 \times 3 = 79.0^k$$

$$F_y = 60 \times 5 \times 0.2 = 60.0^k$$

$$T_u = 139.0 \text{ kips}$$

$$a = \frac{139}{0.85 \times 4 \times 33.5} = 1.22 \text{ in.}$$

tendon $j_u d = (3.25 - 1.22/2)/12 = 0.22$ ft

rebar $j_u d = (5.5 - 1.22/2)/12 = 0.41$ ft

$\phi M_u = 0.9 \times (0.22 \times 79 + 0.41 \times 60) = 37.38$ ft-kips. Much greater than moment transfer requirements.

(c) *Shear at interior column:* Direct shear left and right of interior columns is calculated in section 8 above.

Total direct shear $= (2.14 + 2.52)20 = 93.20$ kips

Moment transfer $M_t = 20(9.29 - 7.84)$

$$= 29.00 \text{ ft-kips}$$

Shear section properties (see Fig. 9-7):

$$d = 6.5 - 1.0 = 5.5 \text{ in.}$$

$$d + c_1 = 5.5 + 20 = 25.5 \text{ in.}$$

at torsion faces, $d = 0.8 \times 6.5 = 5.2$ in.

$$d + c_2 = 5.2 + 14 = 19.2 \text{ in.}$$

$$b_0 d = 2 \times (25.5 \times 5.2 + 19.2 \times 5.5)$$

$$= 476 \text{ in.}^2$$

Polar moment of inertia:

$$J = 2\left(5.2 \times (25.5)^3/12 + 25.5 \times (5.2)^3/12 + 19.2 \times 5.5 \times \left(\frac{25.5}{2}\right)^2\right) = 49{,}300 \text{ in.}^4$$

$$\frac{J}{12 \times c} = \frac{49{,}300}{12 \times 25.5/2} = 322.2 \text{ in.}^2\text{-ft}$$

Portion of moment to be transferred by torsional shear:

$$\alpha = 1.0 - \frac{1.0}{1.0 + 2/3\sqrt{\frac{25.5}{19.2}}} = 43.3\%$$

$M_{vt} = 0.433 \times 29.00 = 12.56$ ft-kips

$M_{vf} = 29.00 - 12.56 = 16.44$ ft-kips

Shear stresses:

$$\text{direct shear} = v_u = \frac{93.20}{476} = 0.196 \text{ ksi}$$

$$\text{torsional shear} = v_t = \frac{12.56}{322.2} = 0.039$$

total shear = 0.235 ksi
divided by ϕ (= 0.85) = 0.276 ksi

Calculated allowable shear stress by ACI 318-83 equation (11-13):

$$v_{cw} = 3.5\sqrt{f'_c} + 0.3\,F/A + V_p/A_c$$

$$v_{cw} = 3.5\sqrt{4000} + 0.3 \times 0.172$$

$$= 0.273$$

0.273 < 0.276 but the V_p component of tendon force crossing the critical section will reduce or eliminate the slight deficiency.

Moment transfer by flexure:

$M_{vf} = 16.44$ ft-kips

$b = 14 + 3 \times 6.5 = 33.5$ in.

Say $F_p = 79.0$ kips as per exterior column.

$A_s = 0.00075 \times 6.5 \times 21.0 \times 12 = 1.22$ in.^{2}

Use 6 No. 4 bars, $A_s = 1.20$ in.^{2}

$F_y = 1.20 \times 60 = 72.0$ kips

$T_u = 79.0 + 72.0 = 151.0$ kips

depth of compression block,

$$a = \frac{151}{0.85 \times 4 \times 33.5} = 1.33 \text{ in.}$$

$$\left(d - \frac{a}{2}\right) = \left(5.5 - \frac{1.33}{2}\right)/12 = 0.40 \text{ ft}$$

$$\phi M_u = 0.9 \times 0.41 \times 151 = 54.4 \text{ kips}$$

Moment transfer capacity in flexure of 54.4 ft. kips is much greater than the required flexural moment transfer of 16.44 ft-kips.

It is strongly emphasized that tendons in each direction should pass directly through the column reinforcing cage. In addition to shear contribution at design loads, these tendons provide a residual shear capacity after complete failure of the punching section by acting in pure tension or "catenary" action. Shear reinforcement may be designed using ACI 318-83 section 11.11.

10. *Deflection:* Calculate live load deflection of a 1-ft strip in 25-ft span by the area moment procedure.

Live load = 0.040 ksf,

$I = 12 \times 6.5^3/12 = 275$ in.^{4}

$E = 3.8 \times 10^3$ ksi

$EI = 1045 \times 10^3$ k-in.^{2}

$$\text{Support moments} = \frac{0.040}{0.202} \times 10.06$$

$$= 1.99 \text{ ft-kips}$$

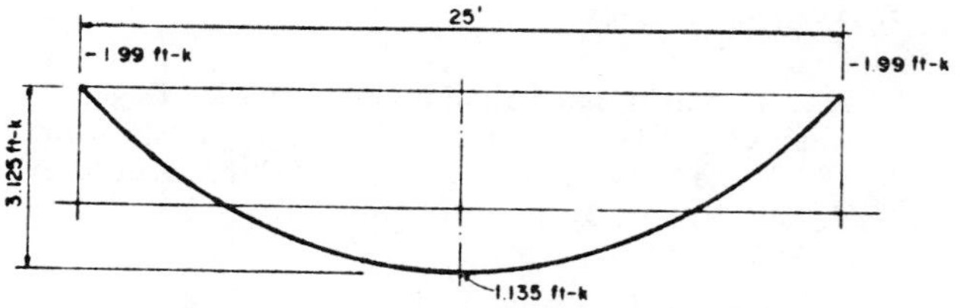

Slope at A and B is $\Sigma \dfrac{Mdsx}{EI}$.

$$-\text{Moment} = \frac{1.990 \times 12 \times 25 \times 12 \times .5}{1045 \times 10^3}$$

$$= -.00342$$

$$+\text{Moment} = \frac{3.125 \times 12 \times 25 \times 12 \times 2/3 \times .5}{1045 \times 10^3}$$

$$= +.00359$$

Slope at A and B = +.00017.

Positive slope means tangent slopes downward from left to right from Support A. This slope is increased by negative moment and decreased by positive moment.

Deflection at midspan:

$$+0.00017 \times 12 \times 12.5 + 1.99 \times 12^3 \times 12.5^2 \times 0.5/(1045 \times 10^3) = 0.283$$

$$-3.125 \times 12^3 \times 12.5^2 \times 2/3 \times 0.375/(1045 \times 10^3) = -0.202$$

total deflection = +0.081

approximate live load deflection in 20-ft spans interior span is similar to fixed end beam

$$= \frac{wL^4}{384EI} = \frac{0.04 \times 20^4 \times 12}{384 \times 1045 \times 10^3} = 0.027 \text{ in.}$$

total deflection at center of panel = 0.108 in.

Calculated live load to span deflection ratio = 25 × 12/0.108 = span/2778.

The net dead load in this span is 0.038 ksf; so by comparison with the live load deflection calculated above, the long-term dead load deflection, assuming a creep factor of 2, is approximately 0.24 in. and the total long-term dead plus instantaneous live load deflection at the center of the panel would be less than 0.36 in. Permissible long-term plus short-term deflection = span/480 = 0.63 in. The calculated deflections of this structure are satisfactory.

11. *Distribution of tendons:* In accordance with the discussion in Section 9.6.1 and the provisions of section 18.12 of ACI 318-83, the eleven tendons per 20-ft bay in this design will be distributed in a group of two tendons directly through the column with the remaining eight tendons spaced at 2 ft 3 in. centers (about four times the slab thickness). Tendons in the direction perpendicular to the tendons designed in this example are to be placed in a narrow band through and immediately adjacent to the columns.

REFERENCES

9-1. Lin, T. Y., "Load Balancing Method for Design and Analysis of Prestressed Concrete Structures," *Journal of the American Concrete Institute, Proceedings* V. 60, No. 6, June 1963, pp. 719–742 (ACI Title 60-36).

9-2. Scordelis, A. C., Pister, K. S., and Lin, T. Y., "Strength of a Concrete Slab Prestressed in Two Directions," *Journal of the American Concrete Institute*, V. 28 (*Proceedings*, V. 53), No. 3, Sept. 1956, pp. 241–256.

9-3. Scordelis, A. C., Lin, T. Y., and Itaya, R., "Behavior of a Continuous Slab Prestressed in Two Directions," *Journal of the American Concrete Institute, Proceedings* V. 56, No. 6, Dec. 1959, pp. 441–459.

9-4. Brotchie, J. F., and Beresford, F. D., "Experimental Study of a Prestressed Concrete Flat Plate Structure," *Civil Engineering Transactions*, Institution of Engineers, Sydney, Australia, Oct. 1967, pp. 276–282.

9-5. Odello, R. J., and Mehta, B. M., "Behavior of a Continuous Prestressed Concrete Slab with Drop Panels," Thesis, Division of Structural Engineering and Structural Mechanics, University of California, Berkeley, 1967.

9-6. Muspratt, M. A., "Behavior of a Prestressed Concrete Waffle Slab with Unbonded Tendons," *Journal of the American Concrete Institute*, Dec. 1969, pp. 1001–1004 (ACI Title 66-88).

9-7. Hemakom, R., "Strength and Behavior of Post-Tensioned Flat Plates with Unbonded Tendons," Ph.D. Dissertation, University of Texas at Austin, Austin, Texas, 1975.

9-8. Winter, Victor C. "Test of Four Panel Post-Tensioned Flat Plate with Unbonded Tendons," Research Paper, The University of Texas at Austin, Austin, Texas, 1977.

9-9. ACI-ASCE Committee 423, "Tentative Recommendations for Prestressed Concrete Flat Plates," *Journal of the American Concrete Institute*, Feb. 1974, pp. 61–71.

9-10. Burns, N. H., and Gerber, L. L., "Ultimate Strength Tests of Post-Tensioned Flat Plates," *Journal of the Prestressed Concrete Institute*, V. 16, No. 6, Nov.–Dec. 1971, pp. 40–58.

9-11. Hawkins, N. M., and Trongtham, N., "Moment Transfer between Unbonded Post-Tensioned Concrete Slabs and Columns," Progress Report to the Post-Tensioning Institute and Reinforced Concrete Research Council on Project #39, Structures and Mechanics Division, Department of Civil Engineering, University of Washington, Seattle, Nov. 1976.

9-12. ASCE–ACI Committee 426, "The Shear Strength of Reinforced Concrete Members-Slabs," *Journal of the Structural Division, American Society of Civil Engineers*, V. 100, No. ST8, Aug. 1974, pp. 1543–1591.

9-13. DiStasio, J., and Van Buren, M. P., "Transfer of Bending Moment Between Flat Plate Floor and Column," *Journal of the American Concrete Institute*, V. 32, No. 3, Sept. 1960, pp. 299–314 (ACI Title 57-14).

9-14. *Post-Tensioning Manual*, Post-Tensioning Institute, Glenview, Illinois, 1976.

9-15. Ghali, Amin, Dilger, Walter, and Neville, Adam M., "Time-Dependent Forces Induced by Settlement of Supports in Continuous Reinforced Concrete Beams," *Journal of the American Concrete Institute*, Nov. 1969, pp. 907–915.

9-16. Cross, Hardy, and Morgan, Newlin Dolbey, *Continuous Frames of Reinforced Concrete*, John Wiley & Sons, New York, 1954.

9-17. Rice, Paul F., and Hoffman, Edward S., *Structural Design Guide to the ACI Building Code*, Van Nostrand Reinhold, New York, 1979.

9-18. Sunidja, Harianto, Foutch, Douglas, A., and Gamble, William L., "Response of Prestressed Concrete Plate–Edge Column Connections," Structural Research Series No. 498, University of Illinois at Urbana-Champaign, Urbana, Illinois, March, 1982.

10

Multistory Structures

MARK FINTEL *

10.1 DEFINITION

The definition of a tall building depends first on such considerations as where the building is located. A 20-story building may be considered tall in Evanston, Illinois, but is not considered tall in New York City. It also depends on who is asking the question. For a transportation engineer it is tall when elevator and escalator problems are to be solved; for an architect, when tallness becomes a consideration in planning, aesthetics, and environment; for a fire prevention specialist when the building is taller than the fire-fighting equipment or when the building has to be divided into horizontal fire zones, etc.

For the structural engineer a tall building can be defined as one whose structural system must be modified to make it sufficiently economical to resist lateral forces due to wind or earthquakes within the prescribed criteria for strength, drift, and comfort of the occupants.

10.2 HISTORY OF DEVELOPMENT OF HIGH-RISE BUILDINGS

Throughout history, man has always tried to express his greatness by building monuments as high as he could practically achieve, with the Egyptian pyramids as the best example. The largest of these monumental tombs of the pharaohs is the pyramid of Cheops, which was constructed by thousands of slaves who piled huge masonry blocks one on top of the other to a peak of 481 ft—equivalent to a 40-story building. Only in 1907 was this record exceeded by the 47-story Singer skyscraper in New York, rising to a height of 612 ft.

Through the centuries there were two basic materials used in construction: wood and masonry. Neither of these two materials had qualities that were suited to reaching more than a few stories. Wood lacked the strength required for buildings more than 50 to 60 ft high, and any tall building of wood presented a fire hazard.

Masonry, on the other hand, had the strength and was fire-resistant—but had the drawback of weight. The mass of masonry is too great, thus putting too much weight on the lower supports of a tall structure. In the pyramids there is little usable space, since most of the area of the lower levels is used to support the parts above.

Masonry construction reached its peak in 1891 when the 17-story, 210-ft-high Monadnock Building was constructed in Chicago by the architects Holabird and Root, but the walls in the ground floor were up to 7 ft thick. The walls of the load bearing masonry structure needed the thickness to provide the strength required to support the building above and to resist the overturning due to wind, since the walls had little help for wind resistance from other parts of the building.

A wide utilization of multistory buildings for practical purposes had to wait until many of the technical problems

*Consulting Engineer, formerly Director, Advanced Engineering Services, Portland Cement Association, Skokie, Illinois.

blocking the commercial utilization and the construction of multistory buildings were solved. The most important of the technical developments were efficient and safe vertical elevators and construction materials with a higher strength.

The socio-economic problems that came with the industrialization of the nineteenth century and the insatiable need for space in the cities gave the big impetus to high-rise construction. By the latter part of the nineteenth century, land in the rapidly developing cities was already so scarce that the only way to go was up.

The first modern multistory building using a steel skeleton was constructed in 1883 in Chicago. It was the ten-story Home Insurance Building utilizing steel I-beams and masonry walls. The big advance in this building was that the weight of the outer shell was not carried on the masonry but was supported on the steel I-beams.

By the turn of the century the downtown business section of New York had established itself as the nation's financial center, and there was a great demand for office space around Wall Street. The first skyscraper at the turn of the century was the 20-story steel skeleton Battery Place Building, followed by the 21-story Trinity Building; and, further uptown, the 21-story triangular Flatiron Building was constructed. Other cities, too, had their skyscraper beginnings at that time; in Cincinnati in 1903 the first reinforced concrete skyscraper, the 16-story Ingalls Building rose to a height of 210 ft. After 80 years, the building is still in use.

In 1913 the 60-story, 792 ft-high Woolworth Building in lower Manhattan was completed. A product of the soaring imagination of architect Cass Gilbert and financing by Woolworth, this gothic cathedral-style building with a high-rise tower remained the queen of New York's skyline for two decades. After 60 years of service and some modernization (self-service elevators and air conditioning) the Woolworth Building is still as vigorous and useful as on the day it was completed. Concrete was used extensively for its foundations and in the floor slabs.

World War I slowed down the progress in high-rise construction, and only after the war's conclusion did a new and larger period of high-rise buildings start which continued through the 1920s and into the thirties. This period produced many excellent structures such as:

The 67-story 60 Wall Tower in New York
The 71-story Bank of Manhattan Building in New York
The 77-story, 1046-ft Chrysler Building in New York
The 33-story Tribune Tower in Chicago
The 32-story Saving Fund Society Building in Philadelphia

In this period the tall building became a symbol of American development in the eyes of the world, and at home every new skyscraper became a source of amazement and pride.

The crowning glory of this "golden age" of the skyscraper was, obviously, the Empire State Building in New York, which was announced on August 30, 1929. The 102 stories rise to a height of 1250 ft, and with its modern appendage, the TV antenna, it now has a height of 1472 ft. The building used 60,000 tons of steel and 62,000 cubic yards of concrete. Concrete was used for the floor slabs, staircases, and other places where strength, stiffness, and fire protection were needed.

Also, the Rockefeller Center, a new concept of a city within a city, started in this period—still during the Depression. The Center contains the towering RCA Building, with its 70 stories, a number of other high-rise office and commercial buildings, and the world's largest movie house.

The Depression put an end to the great era of skyscrapers, and only in the late 1940s in the wake of World War II did a new era of modern high-rise building begin.

Breaking with tradition, the Secretariat Building of the United Nations completely departed from the lofty towers of the early period, creating a slab-type tower with walls of glass.

As reinforced concrete penetrated the construction field at the turn of the century, its use in high-rise concrete structures also became more widespread. Although prior to World War I high-rise structures were mostly in the domain of structural steel, the foundations and, at times, floor slabs were concrete. After World War I, reinforced concrete multistory structures appeared sporadically, mostly for loft buildings using flat slabs with column capitals and in a very few instances for apartment buildings up to 12–14 stories in height. The structural systems of reinforced concrete buildings of those times were developed as an imitation of steel structures and consisted of columns, girders, beams, and slabs.

A quiet revolution was taking place in the field of concrete buildings. A group of reinforced concrete apartments, 12 to 14 stories high, constructed in the early forties in Brooklyn known as the Clinton Houses, used a beamless type of floor system: the flat plate. The Clinton Houses were a radical departure and set the pattern for the future development of flat plate construction. At the same time, shear walls were introduced as an economical efficient bracing system for multistory buildings. Both of these elements—the flat plate and shear walls—became the major structural systems in all residential buildings of any height.

Also, during the post-World War II period, and especially in the fifties, concrete structures developed a personality of their own, reflecting the properties inherent in the material, such as its strength, moldability, fire-safety, handsome appearance, and weather resistance. Contributing to this trend was the widespread use of flat-plate slabs, which utilize two-dimensional load transfer instead of the traditional one-way action. Also, many reinforced concrete buildings of this period abolished the conventional boxlike form and developed new concepts. An example of a structure with a strongly pronounced "concrete personality" is the Marina City complex in Chicago (Fig. 10-1), which, with its sculpture-like appearance, is a milestone in the development of concrete high-rise structures.

As a result of new technology of materials, new construction methods, and bigger cranes, by the early fifties reinforced concrete buildings reached 20–22 stories. In 1958 they went to 38 stories, and in 1962 the Americana Hotel in New York City rose to a 50-story height—while in the same year Marina City in Chicago rose to 60 stories.

In 1968 the 70-story Lake Point Tower apartment building in Chicago was completed as the tallest building utilizing the economical flat plate–shear wall combination.

The fifties and sixties witnessed an unprecedented surge in concrete high-rise buildings, which soared rapidly from record to record.

The development of framed tube-type buildings, applied first in the 43-story Chestnut De Witt apartment building in Chicago in 1965 and the 38-story CBS building in New York in 1965, was followed by the tube-in-tube concept used in the 38-story Brunswick Building in Chicago, and many more buildings, since, including the 52-story One Shell Plaza in Houston, Texas, completed in 1966.

A further milestone in concrete tubular structures was the introduction, in 1982, of bundled tubes for the 58-story One Magnificent Mile building in Chicago, with a height of 663 ft. And, the most recent development, also in 1982, in concrete high-rise structural systems was the introduction

Fig. 10-1 The 60-story Marina City Twin Towers built in Chicago in 1962.

of a diagonally braced framed tube for the slender, 50-story, 570-ft-high building at 780 Third Avenue in New York.

During the 1960s, a new structural type was developed by Fazlur R. Khan, the imaginative creator of most of the modern framing systems for high-rise buildings: the composite tubular structure. It combines a reinforced concrete peripheral framed tube, interior steel columns, and a steel slab system having a thin composite concrete slab. The first building utilizing this system was the 24-story Gateway Office Building in Chicago, constructed in 1968. The 51-story One Shell Square building in New Orleans, constructed in 1972, was the first very tall building (697 ft) utilizing the system. Since that time, a number of similar composite structures have been erected in many cities. The 55-story, 728-ft-tall Southeast Financial Center in Miami is presently (1983) under construction. Also under construction is the 75-story, about 1000-ft-tall Main Center in Dallas. And, an 84-story, 1100-ft-tall composite structure is now on the drawing boards.

The tallest reinforced concrete building at present is the 74-story Water Tower Place in Chicago, completed in 1976, with a height of 859 ft. The tower, on top of a 14-story base of commercial space, has a 20-story hotel, and 40 stories of condominiums. The wind resistance of the slender tower is supplied by the peripheral framed tube, which is stiffened by a shearwall connecting its two longitudinal faces.

In the field of steel structures, the late sixties brought the diagonally braced 100-story John Hancock Building in Chicago; the twin 110-story World Trade Center in New York City; and the crowning glory of the steel high-rise structures, the 120-story Sears Tower in Chicago, which utilizes bundled tubes as the structural system, soaring to a world record of 1450 ft.

Table 10-1 gives a partial list of concrete buildings taller than 400 ft throughout the world.

10.3 CONTRIBUTING FACTORS TO CONCRETE HIGH-RISE BUILDINGS

The tremendous blossoming of reinforced concrete multistory construction in the past several decades has been attributable in large part to four major factors, which are only briefly summarized below, since a complete description of these factors appears in other chapters throughout the *Handbook*: (1) development of high-strength materials; (2) development of new design concepts; (3) development of new structural systems; and (4) improved construction methods.

10.3.1 High-Strength Materials

Only 25 years ago, concretes of 2,000 to 3,000 psi cylinder strength at 28 days were commonplace. Today, in multistory construction for lower-story columns and shear walls, a strength of 5,000 psi is seen on most jobs, and occasionally up to 7,000 psi. Concrete with a compressive strength of 8,000 and 9,000 psi at 90 days has been used for lower-story columns and walls on many very tall buildings, primarily in Chicago since 1974, and also in New York, Minneapolis, Toronto, and Dallas. Strength of 11,000 psi was utilized in 1981 on two buildings in Chicago, and 14,000-psi concrete is currently (1983) undergoing a field test. The recently developed superplasticizers have played an important role in the implementation of high-strength concrete.

The development and acceptance of structural lightweight concrete has helped overcome some design difficulties of multistory structures. Dead load savings of up to 20% are important to critical column size, in earthquake design, and where poor soil conditions limit the weight of the structure. The use of lightweight concrete may prove more economical than normal weight concrete. The higher price for lightweight concrete must be offset by a saving in the slab and column reinforcement, and by savings in the cost of foundations.

Development of reinforcing bars is continuing in several ways. Steels with yield strengths of 60 and 75 ksi were developed and put on the market. Their use in multistory structures started with the introduction of ultimate strength design for the first time in the *ACI Building Code* of 1956. By the early 1970s the 60-ksi reinforcing steel had totally replaced the traditional 40-ksi steel. High steel stresses utilized in structures designed by this method result in con-

TABLE 10-1 World's Tallest Concrete Buildings (Unofficial), 400 Feet and Taller

No.	*Name and Location*	*Height*	*No. of Stories*	*Occupancy*	*Year*
1.	Water Tower Place, Chicago, IL	859	74	Commerical/ hotel/apts.	1976
2.*	MLC Center, Sydney, Australia	808	68	Office	1976
3.	Rennaissance I, Detroit, MI	739	73	Hotel	1977
4.	Peachtree Center Plaza, Atlanta, GA	723	71	Hotel	1975
5.*	Carlton Center, Johannesburg, S. Africa	722	50	Office	1973
6.	One Shell Plaza, Houston, TX	714	52	Office	1971
7.	Neiman Marcus, Chicago, IL	690	65	Commercial/ Residential	under constr.
8.	Southern Bell, Atlanta, GA	677	46	Office	1981
9.	Trump Tower, New York, NY	664	68	Office/res.	1982
10.	One Magnificent Mile, Chicago, IL	663	58	Office/res.	1982
11.*	Overseas Chinese Banking, Singapore	660	52	Office	1976
12.*	Parque Central Torre Oficinas, Caracas, Venezuela	656	56	Office	1979
13.	Michigan & Oak, Chicago, IL	650	60	Multiple	1981
14.	Lake Point Tower, Chicago, IL	645	70	Apartments	1965
15.	1000 Lake Shore Plaza, Chicago, IL	640	55	Apartments	1964
16.*	Belmont Centre, Kuala Lumpur	632	50	Multiple	1979
17.*	International Plaza, Singapore	624	50	Commercial/ Office	1976
18.**	Place Victoria, Montreal, Que.	624	47	Office	1964
19.	101 Park Avenue, New York, NY	618	46	Office	1982
20.*	Development Bank of Singapore, Singapore	612	50	Office	1975
21.*	Sun Hung Kau Centre, Hong Kong	600	50	Multiple	1980
22.*	National Westminster Bank, London, England	600	52	Office	1980
23.*	Australia Square, Sydney, Australia	600	51	Multiple	1968
24.*	Connaught Centre, Hong Kong	599	50	Multiple	1974
25.*	La Tour Fiat, Paris, France	590	45	Office	1974
26.	Museum of Modern Art, New York, NY	590	57	Museum/apts.	1982
27.	Marina City, Chicago, IL	588	60	Multiple	1962
28.	Mid-Continental Plaza, Chicago, IL	582	50	Office	1972
29.*	Coltejer, Medellin, Colombia	574	37	Office	1972
30.**	Hotel de Mexico, Mexico City, Mexico	573	37	Office	1972
31.*	Las Americas, Bogota, Colombia	572	47	Office	1974
32.	780 Third Avenue, New York, NY	570	50	Office	u.c.
33.**	Royal Bank Plaza, Toronto, Ontario	567	41	Office	1976
34.	Palace Hotel, New York, NY	563	51	Hotel	1980
35.*	La Nacional, Bogota, Colombia	561	47	Office	
36.	Newberry House, Chicago, IL	560	58	Apartments	1972
37.*	C P F, Singapore	560	45	Office	1978
38.*	Post Office Tower, London, England	560		Office	1965
39.	Harbor Point, Chicago, IL	558	54	Apartments	1975
40.*	Ministry of Foreign Affairs, Moscow, USSR	558	26	Office	1952
41.*	Palacio Zarzur Kogan, Sao Paulo, Brazil	557	50	Office	1960
42.	No. 3 Park Avenue, New York, NY	556	42	Multiple	1975
43.	Galleria, New York, NY	552	57	Multiple	1978
44.*	Nauru House, Melbourne, Australia	550	47	Office	1979
45.*	Nauru House, Sydney, Australia	550	47	Office	1973
46.	Xerox, Chicago, IL	550	45	Office	1979
47.**	Manufacturers Life Center, Toronto, Ontario	545	51	Apartments	1971
48.*	Rio Sul Center, Rio de Janiero, Brazil	538	50	Office	1980
49.	Frontier Towers, Chicago, IL	533	55	Apartments	1973
50.	Pillsbury, Minneapolis, MN	529	40	Multiple	1982
51.*	Avianca, Bogota, Colombia	528	41	Office	1969
52.*	Standard Bank Centre, Johannesburg, S. Africa	525	31	Office	1970
53.**	Manulife Centre, Toronto, Ontario	525	55	Multiple	1973
54.	River Plaza, Chicago, IL	524	56	Apartments	1978
55.	111 East Chestnut, Chicago, IL	520	57	Apartments	1973
56.*	Hong Leong, Singapore	520	45	Multiple	1976
57.	964 Third Avenue, New York, NY	516	39	Office	1969
58.	Grand Central Tower, New York, NY	516	43	Office	1982
59.	Brooks Tower, Denver, CO	504	42	Apartments	1965
60.*	Karl Marx Universitat, Leipzig, German Dem. Rep.	502	36	Office	1971

TABLE 10-1 *(Continued)*

No.	*Name and Location*	*Height*	*No. of Stories*	*Occupancy*	*Year*
61.*	Dubai International Grade Center, United Arab Emirate	500	38	Multiple	1978
62.*	Mandarin Hotel, Singapore	500	40	Hotel	1973
63.*	U.I.C., Singapore	500	40	Office	1974
64.	Americana Hotel, New York, NY	500	50	Hotel	1962
65.	1818 Market Street, Philadelphia, PA	500	40	Office	1974
66.	Mariott Hotel, Chicago, IL	500	45	Hotel	1978
67.	Walton Colonnade, Chicago, IL	500	44	Apartments	1974
68.	Edgewater Beach Apartments, Chicago, IL	499	39	Apartments	1972
69.	320 W. 57th Street, New York, NY	498	50	Apartments	1978
70.**	Place Desjardins la Tour Du Sud, Montreal, Quebec	498	40	Multiple	1976
71.**	Trizec, Winnipeg, Manitoba	494	32	Office	1980
72.*	U.A.F., Paris, France	492	41	Office	1973
73.*	Tour Cite Administrative, Brussels, Belgium	492	36	Office	1970
74.*	Banco do Estado, Sao Paulo, Brazil	492	35	Office	1947
75.	CBS, New York, NY	491	38	Office	1964
76.	Centre Square (2 Towers), Philadelphia, PA	490	38	Multiple	1973
77.	No. 5 Penncenter, Philadelphia, PA	488	36	Office	1970
78.**	Simpson Tower, Toronto, Ontario	486	33	Apartments	1968
79.	Excelsior House, New York, NY	483	47	Apartments	1964
80.	Industrial Valley Bank, Philadelphia, PA	482	32	Multiple	1969
81.**	Two Bloor Street West, Toronto, Ontario	480	34	Office/ Retail	1974
82.	1600 Belt Plaza, Seattle, WA	480	33	Office	1976
83.**	Harbour Centre, Vancouver, British Columbia	479	28	Office	1976
84.*	Bank & Foreign Trade Bldg., Warsaw, Poland	479	46	Office	1978
85.	Brunswick Building, Chicago, IL	478	36	Office	1964
86.	Park Tower, Chicago, IL	476	54	Apartments	1974
87.	Broadway & 64th Sts, New York, NY	471	42	Apartments	1969
88.*	Hochhaus Platz der Republik, Frankfurt, Fed. Rep. of Germany	469		Office	1973
89.*	Guys Hospital, London, England	468	35	Hospital	1971
90.*	Atlas, Buenos Aires, Argentina	462	42	Apartments	1950
91.	36 Central Park So. NYC	461	43	Apts/Hotel	1970
92.*	Communications Centre, Singapore	460	32	Office	1979
93.*	Colseguros, Bogota, Colombia	459	36	Office	1974
94.*	Trust Bank Centre, Johannesburg, S. Africa	459		Office	1970
95.*	Park Regis, Sydney, Australia	458	45	Multiple	1969
96.*	Tour Blanche, Paris, France	456	42	Apartments	1973
97.	347 W. 57th Street, New York, NY	456	45	Apartments	1982
98.	One Biscayne Corp., Miami, FL	456	40	Office	1979
99.*	African Eagle Life Centre, Johannesburg, S. Africa	456	30	Office	1976
100.*	Ed. Lineu de Paula Machado, Rio de Janiero, Brazil	453	34	Office	1980
101.*	1 Interaco Stawki 3 Street, Warsaw, Poland	453	38	Office	1975
102.*	Ardomore Park, Singapore	450	36	Apartments	1979
103.*	Concordia, Koln, Germany	450	49	Apartments	1972
104.	Mariott Hotel, New Orleans, LA	450	42	Hotel	1974
105.**	Holiday Inn Hotel, Montreal, Quebec	450	38	Hotel	1977
106.	Dag Hammarskjold Tower, New York, NY	446	47	Apartments	1982
107.*	Ed Banerj, Rio de Janiero, Brazil	446	34	Office	1965
108.	Xerox Tower, Rochester, NY	443	30	Office	1967
109.*	St. Johns Beacon, Liverpool, England	442		Restaurant	1965
110.**	A G T Tower, Edmonton, Alberta	441	34	Office	1972
111.*	Ministry of Transport, Moscow, USSR	440	20	Office	1953
112.	Harlem River NYC, NY	440	46	Apartment	1974
113.*	Shaw Tower, Singapore	440	34	Multiple	1974
114.*	Bank of Brazil, Sao Paulo, Brazil	440	24	Office	1955
115.**	Richardson, Winnipeg, Manitoba	439	34	Office	
116.*	Bangkok Bank, Bangkok, Thailand	439	35	Office	1979
117.	National Bank of Georgia, Atlanta, GA	439	32	Office	1961
118.	Bank of New Orleans, New Orleans, LA	438	31	Office	1970
119.*	Ed. Sao Sebastaio, Rio de Janiero, Brazil	438	45	Office	1980
120.	Parker Meridien Hotel, New York, NY	437	40 plus Penthouse	Hotel	1980

TABLE 10-1 *(Continued)*

No.	*Name and Location*	*Height*	*No. of Stories*	*Occupancy*	*Year*
121.*	Forschungshochhaus, Jena, German Dem. Rep.	435	30	Office	1976
122.*	City Hall, Kuala Lumpur	435	31	Office	1979
123.	168 N. LaSalle, Chicago, IL	435	35	Office	1975
124.	Gulf Life Tower, Jacksonville, FL	433	27	Office	1967
125.*	Stadt Berlin, Berlin, Germany	433	41	Hotel	1970
126.*	Volksas Bank Building, Praetoria, S. Africa	432	38	Office	1978
127.	150 S. Wacker, Chicago, IL	430	35	Office	1970
128.	Jerome Ave. NYC, NY	430	42	Apartment	1971
129.	Madison @ 89th- St. NYC, NY	430	38	Apartment	1971
130.**	Pl. Desjardin la Tour de l'Est, Montreal, Quebec	428	32	Multiple	1976
131.	East End Ave. NYC, NY	426	42	Apartment	1971
132.**	Four Seasons Sheraton Hotel, Toronto, Ontario	426	46	Hotel	1972
133.*	Cafetereis Band, Medellin, Colombia	426	36	Office	1975
134.**	La Port Royal, Montreal, Quebec	425	33	Apartments	1964
135.**	Le Cartier Apartments, Montreal, Quebec	425	32	Apartments	1964
136.	Citizens Plaza Bank, Louisville, KY	420	30	Office	1972
137.	Devonshire Building, Boston, MA	420	41	Office/Apts.	1982
138.*	Center Pierelli, Milan, Italy	420	35	Office	1958
139.**	Nonoalco Tlatelolco Tower, Mexico City, Mexico	417	25	Office	1962
140.**	Complexe G, Ottawa, Ontario	415	33	Office	1972
141.	Hyatt Regency, Kansas City, MO	413	42	Hotel	1980
142.*	Barbicon, London, England	412	40	Apartments	1970
143.	3rd & 92nd NYC, NY	411	42	Apartments	1974
144.**	Leaside Tower, Toronto, Ontario	410	43	Apartments	1970
145.*	Ponte Investment Apt., Johannesburg, S. Africa	410		Apartments	
146.*	Viamonte No. 147/153, Buenos Aires, Argentina	410	38	Apartments	1949
147.	1st Ave. @ 55th St. NYC, NY	409	38	Apartments	1970
148.*	B P Centre, Cape Town, South Africa	409	31	Office	1972
149.*	New Alexandria House, Hong Kong	408		Multiple	1978
150.*	Fustion Centre, London, England	407	34	Office	1970
151.	Harley Hotel, New York, NY	407	38	Hotel	1980
152.*	City Mutual, Melbourne, Australia	406	35	Office	1972
153.*	City Mutual, Sydney, Australia	406	35	Office	1972
154.	Peachtree Summit No. 1, Atlanta, GA	406	31	Office	1976
155.	World Trade Center, Baltimore, MD	406	28	Office	1977
156.	Broadway and 44th NYC, NY	405	33	Office	1972
157.	Coca Cola Headquarters, Atlanta, GA	403	28	Office	1979
158.*	Bogota Hilton, Bogota, Colombia	403	38	Hotel	1973
159.**	Harbour Square Apartment, Toronto, Ontario	403	34	Apartments	
160.*	Hotel Hilton, Kuala Lumpur	402	36	Hotel	1973
161.	Riverview Apartments, New York, NY	402	37 plus penthouse	Apartments	1982
162.*	Seguros Tequendama, Bogota, Colombia	400	38	Office	1970
163.	1100 S. Michigan, Chicago, IL	400	35	Apartments	1970
164.	Continental Plaza Hotel, Chicago, IL	400	27	Hotel	1974
165.	Outer Drive East Apartments, Chicago, IL	400	40	Apartments	1969
166.	Laclede Gas, St. Louis, MO	400	34	Office	1969

**Non-North American country.*
***North America excluding the United States.*

siderable savings in the amounts of reinforcing steel per square foot of floor area.

The use of prefabricated welded wire fabric and welded reinforcing mats has helped save time and labor on slabs, shells, walls, and other plate-type construction.

The use of bundled bars and the introduction of large-size bars—#14S and #18S with diameters of 1.70 and 2.26 in. respectively—help to satisfy spacing requirements, particularly in heavily reinforced members. A further help in the same direction is the use of column splices by welding and couplings. Several kinds of couplings are on the market. Some have only compression transfer capacity, whereas others are good also for the transfer of tensile forces.

10.3.2 New Design Concepts

Cast-in-place reinforced concrete structures, by virtue of their monolithic character, are designed with consideration of continuous frame action. This results in a more economical design than in structural systems in which most of the members are designed as simply supported.

Two recent developments are especially worthy of mention: ultimate strength design, and limit design. In USD, which has become the major design method of the 1971 *ACI Code* under the name of "strength design," the ultimate capacity of a section is computed to correspond to the service load increased by a selected factor of safety.

Limit design considers the behavior of a structural assembly, as it approaches its ultimate load capacity.

Although many developments of limit design methods have been advanced by researchers in recent years, the possibility does not appear that in the near future a formalized limit design method will be developed for high-rise reinforced concrete structures. The moment redistribution of up to 20% allowed in the *ACI Code* (to be made after an elastic analysis is carried out) is a partial limit design.

In both methods (strength design and limit design), the inelastic behavior of materials is considered rather than considering only elastic behavior according to the classical working stress design theory, the elastic analysis of structures. These new methods help to close the gap that exists between strength computations and actual test results.

The above design developments permit a better utilization of materials, through methods that more closely reflect the actual behavior of the materials and of the structure in which they are used. They allow a more uniform factor of safety throughout the structure than is possible with working stress design and elastic analysis.

10.3.3 New Structural Systems

Concrete started off as an imitation of structural steel systems, and the early buildings consisted of columns, beams, and slabs. It can be compared to the early stages in the automotive industry when the motor cars were patterned after the horse-drawn buggies.

The development and popularization of the flat plate has been an important contributor to the widespread use of multistory reinforced concrete apartment buildings. Its major technical advantages are a smooth, beamless ceiling that requires only a skim coat of plaster to provide a perfect finish; and reduced story height, which permits an additional story every 10 stories as compared with structural systems using beam and slab construction. The simplified formwork required for the flat plate results in an efficient job operation. No other structural system can offer all these technical and economic advantages to residential high-rise buildings. In recent years, casting two floors per week in tall structures has not been uncommon in metropolitan areas.

The introduction of shear walls as a bracing element in the late 1940s in housing projects in New York had an important impact on concrete high-rise buildings. Fulfilling simultaneously the functions of a load-carrying element, a wall within or between apartments, and an efficient bracing element, the shear wall became the most important element for all multistory buildings. Today there is rarely a building of more than 15–18 stories without shear walls as its major bracing element.

Utilization of shear panels—one- or several-story shear walls—scattered throughout the plan and shifted in location to offer architectural flexibility while supplying sufficient rigidity, is another way to reduce cost and adapt to the functional requirements.

The staggered wall–beam system developed in the late 1960s is another innovation suitable for residential buildings. Although only a few high-rise buildings have been built with this system, its advantages of large unobstructed areas in the typical floor, and column-free areas under the building for parking, may eventually lead to a broader application

Finally, the framed tube, a peripheral system of closely spaced exterior columns, interconnected with spandrel beams, creates a highly rigid structure for tall buildings. A further step is the tube-in-tube resulting from adding a core element to the framed tube. The tubular structure was still further modified by bundling several tubes together, and finally by adding diagonal bracing.

In the explosive development of high-rise structural systems, still only in its infancy, it appears safe to predict that many other new and revolutionary concepts will be developed by architects and engineers in the years ahead.

10.3.4 Improved Construction Methods

Improved construction techniques, a few of which are mentioned below, and improved equipment probably are the strongest contributors to the development of high-rise structures. Improvement of construction methods has provided better technical solutions to many problems and, in most cases, has resulted in reduced construction costs.

Lift-slab construction appeared on the scene in the early fifties. With this method, slabs are cast at the ground level, one on top of the other. Only side forms are necessary. Specially designed collars for column connections are cast in the slab. After the slabs attain the desired strength, they are lifted hydraulically upon the columns (Fig. 10-2) and fastened to the columns after reaching their final position. Buildings up to 15 floors in height have been constructed by this method. Stability of such structures usually is achieved by shear walls, or by reinforced concrete cores. Although the idea of lift slabs seems to eliminate expensive operations of shoring and forming, lift slabs have not attained the widespread utilization in the US warranted by their apparent advantages.

In some cases lift slabs have been designed as post-tensioned, prestressed structures.

Slipforming is a method of constructing walls, and occasionally columns, in which 3- to 4-ft-high forms are raised continuously, while concrete is placed between them. Progress of 10 to 12 in. per hour is regarded as average today. This method, originally developed for silos and bins, has been used recently on many high-rise structures in which the cores were slip-formed (Fig. 10-3). In the example shown in Fig. 10-4, a 25-story apartment building in Milwaukee, Wisconsin was built using slip-forming for the exterior and interior walls, at a speed of a story per day. The floor slabs followed three stories behind at the same speed, consisting of precast beams and slabs cast in place over nonrecoverable

Fig. 10-2 Lift-slab method of construction.

Fig. 10-3 Slipforming of a core for a 20-story building.

Fig. 10-4 Slipforming of exterior and interior walls of a 25-story building in Milwaukee, Wisconsin (Bay Shore Apartments).

plywood forms. Even the sculptured balconies were made by slip-form sliding past a stationary sculptured plastic form (Fig. 10-5).

Two advantages result from slipforming: saving of time and cost. However, to satisfy an economical slipforming operation, a certain minimum wall height is required, and this appears to be only 35 ft.

A combination of slipforming with lift slabs has a good potential for lowering the cost of multistory buildings. Figure 10-6 shows an apartment building in San Francisco with three slipformed cores and lift-slab floors.

Plastic-lined forms are another innovation that provides smooth concrete surfaces, and more reuses of formwork. Also, fiberglass-reinforced plastic forms provide an excellent concrete surface, and make possible the molding of the most complicated shapes. Such forms were still in perfect condition after 60 reuses each on the Marina City project (Fig. 10-7). An additional advantage is that no finish was required for column surfaces. Fiberglass forms are particularly advantageous where complicated shapes are to be executed, or where the same shape is to be repeated many times.

During the 1970s, flying forms came into wider use. Such forms, covering large areas (i.e., 20 X 60 ft), are assembled into one unit supported by lightweight trusses. The units are transferred from floor to floor in one piece by the crane. The multiple use of fiberglass flying forms permits the design of more elaborate structural sections for the slab system (such as complicated joist shapes, etc.), thus improving the structural efficiency of the slab system.

Also, pumping to deliver the wet concrete into the formwork of multistory buildings improved the economy of concrete construction. By 1981 concrete had been pumped to a height of over 1000 ft in Houston, Texas.

The use of prefabricated members, which has grown rapidly in the bridge, housing, and small structures field, has entered the field of multistory buildings. The tendency has been to bring production work into a plant where weather is immaterial, and working conditions are stable. However, precasting on or near the site has also found favor because of the greater quality control possible, and the savings possible through repetition in production and avoiding the

Fig. 10-5 Balconies in the Milwaukee building.

Fig. 10-6 Construction utilizing slipforming and lift slabs.

expensive formwork. Precast construction combines the advantages of reinforced concrete, such as weather, fire, and corrosion resistance, with the advantages of prefabrication, shorter construction time, and ease in obtaining accuracy in dimensions through an industrialized process.

Fig. 10-7 Reinforced fiberglass forms (Marina City).

Fig. 10-8 Palacio Del Rio, San Antonio, Texas.

In the 1960s the developments progressed to the point of precasting entire rooms or apartments, including almost all finishes, as in the example of the Palacio Del Rio in San Antonio, Texas (Fig. 10-8), thus leaving for the field only the fastening together of such units and connecting utilities. The majority of prefabricated multistory structures, however, consist of large panel components (slabs and walls) or systems in which prefabrication has included primarily the structural frame and exterior envelope, the finishes (some also prefabricated) being applied in the field. Figures 10-9 and 10-10 show structures consisting entirely of large precast elements except for cast-in-place cores and shear walls.

Fig. 10-9 Large panel structure.

Fig. 10-10 Large panel structure.

Between the two types of structural systems, fully prefabricated skeleton and the cast-in-place structure, there is a wide variety of composite structures, in which prefabricated elements (thin slabs or shells of columns) are used as formwork, sometimes containing the main reinforcement. The cast-in-place part, ranging from about 20 to 80%, ties the structure together into a monolithic unit.

Successful examples of such buildings are the many buildings in Hawaii (like the 27-story Ilikai Hotel, Fig. 10-11), the 27-story Gulf-Life building in Jacksonville, Florida (Fig. 10-12), the 40-story One Biscayne Tower in Miami, and others.

Another important factor contributing to the growing use of concrete for multistory structures is improvement in con-

Fig. 10-11 The 27-story Ilikai Hotel in Honolulu, Hawaii.

Fig. 10-12 Gulf-Life Building in Jacksonville, Florida.

struction equipment, such as hoists, cranes, self-rising cranes, and concrete mixing, conveying, and pumping equipment. This subject is discussed in detail in Chapter 24, "Construction Methods and Equipment."

10.4 TYPES OF OCCUPANCY

The function of a building has a considerable effect on the selection of the structural type. As a result of the particular function of the building, such considerations as the size of the open space required, mechanical services (in particular, the forced-air distribution system), and the need for future flexibility all enter the picture when the most adaptable structural framing system is chosen.

With regard to type of occupancy, the following categories can be distinguished as specifically affecting the structural systems of tall structures:

1. Residential buildings, including apartment buildings, hotels and dormitories
2. Office buildings and commercial space
3. Mixed occupancy—commercial and residential
4. Industrial buildings and parking garages

10.4.1 Residential Buildings

Residential buildings (apartment buildings, hotels, and dormitories) are characterized by the presence of partitions that are designed during the planning stage, are constructed with the progress of the structure, and usually remain in the same location for the life of the building. This is obviously the present method of construction. It is possible that in the future apartment buildings will be designed with movable partitions, which may then change our structural philosophy.

1. Structural Systems for Residential Buildings The presence of permanent partitions allows the column layout to correspond to the architectural plan. The present architectural layouts prevalent in the United States can be handled with spans up to 20–22 ft. This also includes the category of luxury apartments. It is worth noting that in Western Europe the average spans in residential occupancy are in the range of 12–15 ft.

Among mechanical systems affecting the structure, distribution of air for heating and cooling has the greatest influence on the selection of the structural system. In residential buildings owing to permanency of partition location, air is handled in vertical ducts from which a room or several rooms are fed using short horizontal stubs through the walls from the vertical ducts. Such air supply arrangement does not require ducts under the ceiling except sometimes in vestibule areas, thus the structural slab functions as the ceiling. As a result, the flat plate has become the most adaptable slab system for residential construction. The smooth ceiling, which needs only a skin coat of plaster, the possibility of having the columns in scattered locations to suit the architectural layout, and the absence of beams, which gives absolute freedom to the partitions layout, have made the flat plate the slab system that has allowed economical residential buildings up to 70 stories high.

Experience has shown that in tall buildings the best economy is achieved with the thinnest slab, as the slab weight also affects the load on columns and on foundations.

The typical modern apartment slab-type building with a double-loaded corridor is 50 to 60 feet wide, uses three transversal spans, and requires a 7- to $7\frac{1}{2}$-in. flat plate. Customarily, hotels also have the same transversal three-span layout, while in the longitudinal direction the columns are located 24 to 28 ft apart to accommodate two rooms. The 24 to 28 ft span may require a minimum slab thickness of 8 to 9 in.

When one-way joist systems are used for residential occupancy (usually with somewhat larger spans), a hung ceiling provides a smooth underside. Since the space between the webs is generally not needed for mechanical purposes in residential construction, the joist system is economical only if the cost of the hung ceiling is offset by economies of the joist slab against the flat plate.

Column size in residential buildings is usually not a critical design consideration (except for lobbies and public areas), since columns can be hidden in closets, kept in corners, blended into the architectural layout, and, in the extreme case, utilized as walls. Experience shows that in tall residential buildings columns with the lowest percentage of reinforcing result in the best overall structural economy. If larger columns are acceptable to the architect, they not only are economical but also add substantially to the lateral rigidity of the structure.

The lateral resistance of residential buildings is provided mostly by the frames in buildings up to 15–18 stories. Since flat plate slab systems are predominant, spandrel beams around the periphery carry a substantial part of the lateral forces. In recent years spandrel beams have been eliminated in the majority of apartment buildings to simplify formwork and to accommodate the use of the more economical flying forms. In such buildings, shear walls have been introduced to provide the required rigidity. Residential buildings above 20 stories, except for shear wall buildings, always

Fig. 10-13 The 70-story Lake Point Tower in Chicago.

have a shear wall–frame interactive system for resistance to lateral loads. The tallest building in this category is the 70-story Lake Point Tower (see Fig. 10-13), which has a shear wall–frame system in which an extremely large core resists the bulk of the wind forces, and thus the resistance provided by the frame amounts to an insignificant portion.

A major objective of the design engineer is to devise a structural system that is designed for vertical loads only while the resistance to lateral forces is provided by the 33% increase in allowable stresses. This can be achieved if the shear walls carry a sufficient amount of vertical loads, so that the overturning moments due to lateral loads do not result in substantial net tensile forces. Tensile resultants in shear walls cause serious difficulties at the foundation level, since they have to be either anchored into the ground, or the weight of adjacent columns has to be mobilized to overcome the tensile forces by designing very heavy girders to transfer the loads. The ideal case is created when all the gravity load of the building is used to counteract the overturning moment—which is possible usually only in shear wall buildings. The practical optimum case is to have sufficient shear walls to limit the resultant tension under gravity plus wind to an insignificant value. Such an approach should result in no premium for height in structures. Most of the well-engineered concrete buildings above 40 stories have been in this category.

10.4.2 Office Buildings

Office buildings are characterized by the absence of partitions during design and construction, since office space is designed rather than offices, with subsequent partitioning to accommodate the needs of a particular tenant. The partition layout usually changes when tenants change. This procedure requires flexibility in the supply of heating, cooling, power, and telephone connections to the entire area that can be used as office space. As a result, some services are usually carried vertically within the service core, and a distribution network of air ducts, telephone, and power supply lines is carried horizontally under the structural slab to distribute the services over the entire office space. Since a hung ceiling is required to cover up the ducts and lines, the structural slab does not require a smooth ceiling, and the slab system can be selected for structural efficiency and economy. The flat slab with drop panels is the prevalent system for spans up to 23–28 ft, owing to its low overall thickness and structural efficiency. For two-way spans in the range of 30–35 ft, the waffle slab provides good efficiency, either as a flat plate type with a solid portion around the columns or as a two-way slab with solid beams on the column lines in both directions of the building; such slabs have been used for spans up to 60 ft. The previous spacing of joists and waffle slabs of either 2 or 3 ft on centers, is slowly disappearing with 5 ft on centers introduced recently. The waffle slab provides a very attractive ceiling where a hung ceiling is not required.

In recent years many of the prestigious office buildings have been built without interior columns—the structure consisting of an interior core and an exterior peripheral tubular system of closely spaced columns interconnected with spandrel beams. The column-free span between core and periphery ranges between 35 and 40 ft. The Brunswick Building, shown in Fig. 10-14, is a typical example of this

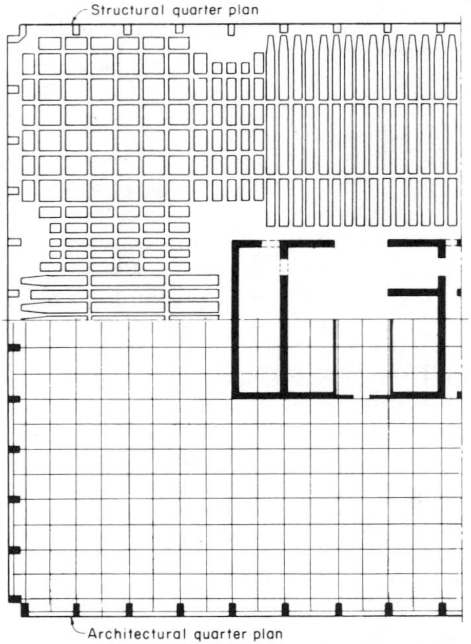

Fig. 10-14 Half of the typical floor of the Brunswick Office Building in Chicago (structural and architectural).

building type. The slab system consists of one-way joists connecting the core with the periphery, while the corners are two-way joists (waffle slabs). It is desirable to have a correlation between the module of the column spacing and the spacing of the joists.

A recent study[10-1] on the economics of long-span office buildings showed that slabs with spans in the range of 35 to 45 ft need only between 1.5 and 2.0 lb/ft^2 of additional reinforcing steel above that required for the traditional 20-ft spans. The study also showed that post-tensioned flat plates, flat slabs, and ribbed slabs are a very economical alternative to reinforced concrete and to structural steel for long-span office buildings.

1. Structural Systems for Office Buildings With the increasing height of buildings in the fifties and sixties, new concepts evolved to economically provide resistance to lateral forces due to wind and in many cases also due to earthquake. This evolution is shown in Fig. 10-15.

As can be seen in the figure, in buildings up to 15 stories, frame action usually suffices to provide lateral resistance, except for very slender or unusual buildings. The 33% increase in allowable stresses provided by codes is in most cases sufficient to accommodate the wind forces and moments.

When we approach the 20-story height, the rigidity of frame buildings is mostly insufficient, and sway due to wind may begin to control the design. Introduction of shear walls which interact with the frame (see Section 10.5.7) materially increases the total rigidity of the building beyond the sum of the two individual components. With proper spandrel beam exterior frames, such shear wall frame interactive systems can be used for office buildings of up to 50 stories or more if a large enough core is utilized.

The framed tube, first introduced in the sixties, consists of a closely spaced grid of exterior columns, connected with beams. It is an efficient system to provide lateral resistance without a premium for height in buildings without interior columns starting at 30–35 stories. The efficiency of this system is derived from the great number of rigid joints acting along the periphery, creating a large tube. The framed tube represents a logical evolution of the conventional frame structure, possessing the necessary lateral stiffness with excellent torsional qualities while retaining the planning flexibility of interior column free space.

To visualize the action of a framed tube, it is simple to start with a solid peripheral wall (a solid tube) which obviously will act as a cantilever with a moment deflection. When the wall is penetrated with small round openings (windows), it will still behave as a cantilever. When the openings become larger and rectangular, instead of round, part of the lateral forces are resisted by shear distortion of the columns and beams, and only the rest by moment (cantilever) deflection of the tube.

The ratio of moment deflection to shear deflection depends on the stiffness relationship between beams and columns. The two sides parallel to the wind act as frames, while the windward and leeward sides have a considerable drop off in participation. The rate of shear lag (drop-off from the corner column) depends on the beam-to-column stiffness ratio and the number of floors. A typical distribution of column axial forces in such a structure is shown in Fig. 10-16. The effect of shear lag will readily be noted. With increasing beam stiffness and increasing number of stories, a higher participation of the windward and leeward sides may bring the framed tube closer to a rigid tube. In addition, the framed tube has an unusually high torsional resistance owing to the location of the stiffness around the periphery.

When the sway or wind stresses begin controlling the design (it may be around 40 stories), the framed tube is supplemented by a core, to create the tube-in-tube system which is essentially a shear wall–frame interactive system with all its advantages. The tallest tube-in-tube is the 714-ft-tall, 52-story Shell Oil Building in Houston, Texas (see Fig. 10-17).

When we reach 70–80 stories, the tube-in-tube may no longer be a sufficiently rigid system, and the exterior tube must be made rigid to act in 100% flexure as a cantilever. Either a set of diagonal members within the peripheral tube creates an exterior cantilevered truss system, or interior connecting shear walls act as webs to tie the opposite faces of the tube into a single unit.

The use of diagonals in the exterior in a reinforced concrete building was investigated in a research project at the Illinois Institute of Technology in a 115-story building 1450 ft high. The thorough study[10-2] investigated all the architectural, structural, mechanical, construction, and cost aspects of the building with the conclusion that the building is technically feasible at an estimated cost lower than that of the 100-story John Hancock Building, its steel counter-

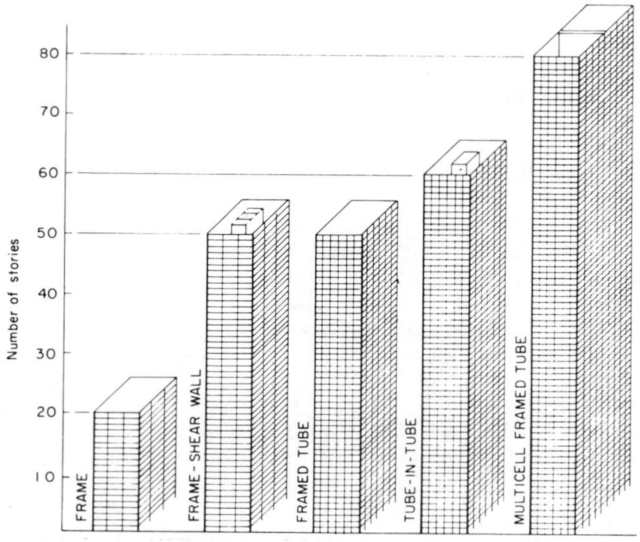

Fig. 10-15 Structural systems for various height office buildings.

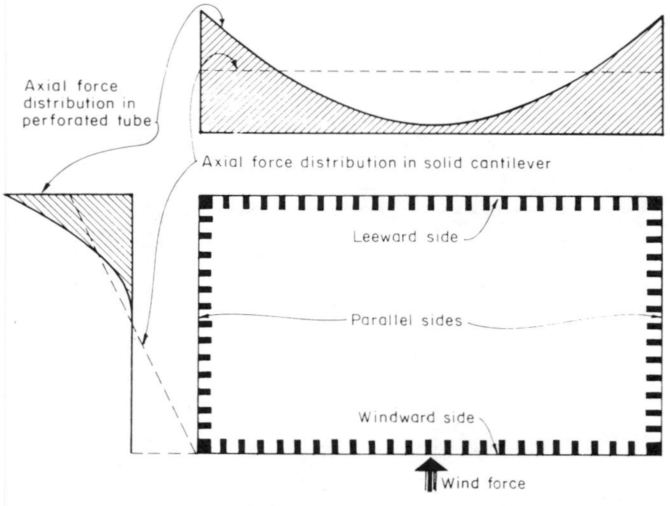

Fig. 10-16 Shear lag in the windward and leeward sides of a framed tube.

Fig. 10-17 The 714-ft-tall, 52-story Shell Oil Plaza Building in Houston, Texas.

part. A concrete diagonally braced framed tube for a slender 50-story (570-ft-high) structure was utilized in the 780 Third Avenue Building in New York. It is worth noting that for a given geometric shape, the diagonally braced tube provides the limit of rigidity. The alternate method of creating a rigid tube is a multicell arrangement as shown in Fig. 10-18 through the use of cross shear walls in one or both directions. The 77-story Water Tower Inn in Chicago utilizes a two-cell arrangement as shown in Fig. 10-18. Although the dividing shear wall is penetrated by door openings, it still provides close to 100% efficiency of the exterior tube.

10.4.3 Mixed Occupancy—Commercial, Residential

Although multistory parking garages under offices and apartment buildings have been used for many years, urban planning and financing considerations recently led to multistory buildings combining such occupancies as: offices and hotels, stores and apartments, theater and apartments, in addition to parking floors under all of these buildings with mixed occupancies. The usual architectural treatment is to place the residential occupancy (apartment or hotel) in the upper part of the building while using the lower part for office floors, department store floors, and parking underground.

Since commercial space requires larger spans than residential occupancy, the planner is faced with the dilemma of either choosing a span suitable for only one of the two occupancies, or seeking the difficult and costly technical solution of a transitional floor between the two occupancies. There are numerous examples of story-high transfer girders of reinforced or prestressed concrete picking up columns of multistory buildings up to 50 stories high. In most cases the entire frame above participates together with the transfer girder as a vierendeel girder carrying the load, regardless of whether the designer has considered this or not. The cost of transfer girders is very high and should be weighed against the long-term financial advantages offered by the improved function of a structure planned for anticipated occupancy with proper spans.

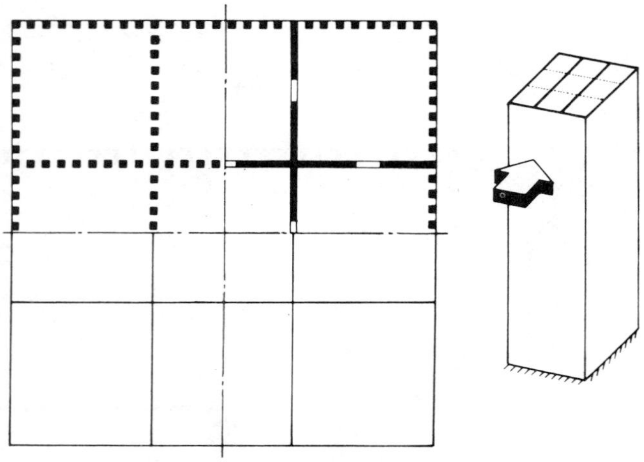

Fig. 10-18 Multicell framed tubes (plan).

10.4.4 Industrial Buildings and Garages

Industrial and warehouse multistory buildings are mostly characterized by heavy floor loads, sometimes up to 1000 lb/ft^2 of floor area, for which the flat slab with column capitals is very well suited. Also, two-way solid slabs supported on beams are efficient for this type of use. Only moderate spans are economically feasible for heavy floor loading.

Multistory parking garages have been using flat slabs with drop panels with or without column capitals extensively. In the 1950s and 1960s, spans of 29 to 30 ft were widely used, since they provide efficient right-angle parking with a 22-ft roadway. Also, one-way joists spanning 50 to 60 ft supported by shorter spanning beams and beams with one-way solid slabs have been used for parking garages. With the popularization of prestressing in recent years, many garages have been constructed using long-span prestressed, precast elements. Two-way post-tensioning of parking decks and use of shrinkage-compensating cements helps to reduce the problem of leakage on the cars on the floor below.

Lateral resistance of industrial buildings and garages is usually provided by frame action, since the number of stories is generally in the lower range. If cores are available in such buildings, they contribute substantially to the lateral resistance.

10.5 RESISTANCE TO WIND LOADS*

In the 1950s and 1960s buildings underwent a complete transformation in appearance and in the makeup of their components; as a result, their resistance to wind and earthquake forces was completely changed.

Traditionally, the primary concern of the structural engineer designing a building has been the provision of a structurally safe and adequate system to support vertical loads. This is understandable, since the vertical load-resisting capability of a building is the reason for its existence. Any calculations undertaken to check the adequacy of the design with regard to lateral loads were often cursory in nature

*Design for earthquake loads is handled in Chapter 12, "Earthquake Resistant Structures."

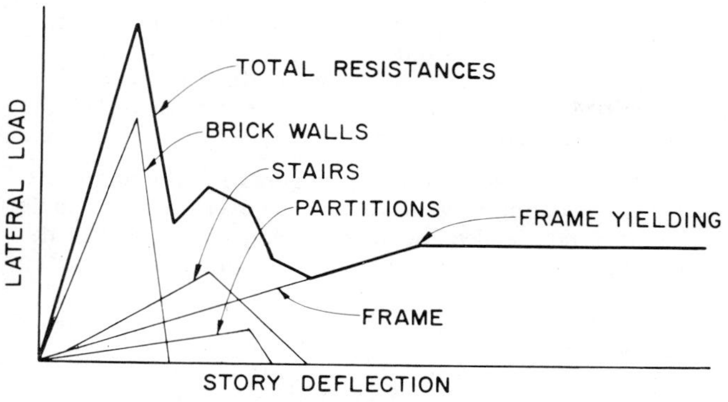

Fig. 10-19 Schematic plot of load versus story deflection in a conventional building.

and more as an afterthought than as an essential and integral part of the total design effort. This attitude did not appear to affect the resulting designs significantly, as long as the buildings involved were not too tall, were not in seismic zones, or were constructed with adequate built-in safety margins in the form of substantial nonstructural masonry walls and partitions.

With the increasing use of light curtain walls, dry-wall partitions, and high-strength concrete and steel reinforcement in tall buildings, the effect of wind loads have become more significant.

Figure 10-19 shows a schematic plot of the lateral resistance (as represented by the lateral load versus story deflection) in a conventional building. It can readily be seen that although the frame may have been designed to provide the lateral resistance, in reality the bulk of the lateral resistance was supplied by the heavy brick walls, stairs acting as diagonal braces, and partitions. It is evident that under these circumstances only a load many times the design wind load (such as may occur in an earthquake) could exhaust the strength of the so-called nonstructural elements, in which case the frame takes over the resistance of lateral loads and eventually yields after reaching the elastic limit at quite large story deflections.

Figure 10-20 shows a schematic plot of the lateral resistance in a modern structure in which the exterior enclosure is a light curtain wall, the partitions are movable, and the stairs are separated from the structure. As a result, the lateral resistance of the building is only slightly larger than that of the frame alone, and no reserve capacity against lateral loads is available.

If, in the past, we had neglected considering wind loads in our traditional buildings, they probably would still have behaved well even in an extremely intense wind, because

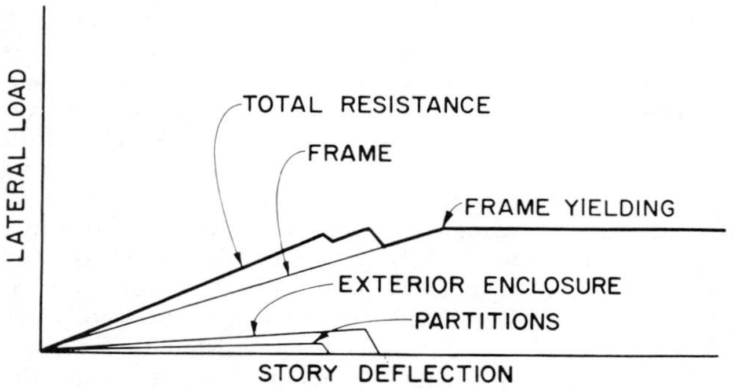

Fig. 10-20 Schematic plot of load versus story-deflection in a modern building.

of the enormous overstrength of the heavy cladding. In today's average structure, the sensitivity to wind forces has materially increased, and neglecting wind forces may cause distress. Despite the fact that we have stripped our buildings to the bare skeleton, we may still get away with neglect of wind forces in cases of wide structures, as a result of our still-generous factors of safety. However, in our multistory structures of tomorrow, the consideration of lateral resistance will be the prime consideration, since the gradual reduction of member stiffness will cause the secondary effects of today to become primary effects in our future structures.

10.5.1 Wind Forces

Wind is the general word for air naturally in motion which, by virtue of the mass and velocity, possesses kinetic energy. If an obstacle is placed in the path of the wind so that the moving air is stopped or deflected from its path, then all or part of the kinetic energy of the moving air is transformed into pressure. The intensity of pressure at any point on an obstacle depends on the shape of the obstacle, the angle of incidence of the wind, the velocity and density of the air, and the lateral stiffness of the engaged structure.

Under the action of a natural wind, a tall building will be continually buffeted by gusts and other aerodynamic forces. The structure will deflect about a mean position and will oscillate continuously.

If the wind energy that is absorbed by the structure is larger than the energy dissipated by structural damping, then the amplitude of oscillation will continue to increase and will finally lead to destruction; the structure will become aerodynamically unstable.

As discussed previously, the structural forms used today have greater flexibility combined with less mass and damping than those used for traditional structures of the past. These factors have increased the importance of wind as a design consideration. For estimations of the overall stability of a structure and of the local pressure distribution on the building, a knowledge of the maximum steady or time-averaged wind loads is usually sufficient.

The determination of wind design forces on a structure is basically a dynamic problem. However, for reasons of tradition and for simplicity, it has been usual practice to use a quasi-static approach and to treat wind as a statically applied pressure, neglecting its dynamic nature.

The designer's source of information on steady wind loads is usually a building code, but the data contained in such codes are of necessity presented in a generalized form and may not be adequate for a special case, especially if the structure departs from conventional building forms.

Some of the considerations that enter into the choice of a design wind pressure are:

(a) The anticipated lifetime of the structure and its relation to the return period of maximum wind velocity
(b) The duration of gusts
(c) The magnitude of gusts
(d) Variation of wind speed with height
(e) Angle of incidence of the wind
(f) Influence of the ground
(g) Influence of the architectural features
(h) Influence of internal pressures
(i) Lateral resistance of structure

ANSI Standard A58.1-1982[10-3] contains the most comprehensive and up-to-date provisions concerning wind loading on structures. Under ANSI A58.1-82, design pressures and forces can be determined from:

$$p = qGC, \text{ and } F = qGCA$$

where:

p = design pressure in psf; positive value means acting toward the surface, negative value means acting away from the surface.

F = design force in pounds.

q = velocity pressure in psf; q_z is the value determined at height z (ft) above ground, and q_h is the value determined at mean roof height h (ft).

G = gust response factor (dimensionless); G_z is the value determined at height z (ft) above ground, and G_h is the value determined at mean roof height h (ft).

A = area of structure or cladding and component (sq ft).

C = pressure coefficient (dimensionless); positive value means acting toward the surface, whereas negative value means acting away from the surface; C_p is external pressure coefficient, C_{pi} is internal pressure coefficient, and C_f is force coefficient.

Design pressures and forces are determined for each structure separately for main wind-force resisting systems and for components and cladding. There are two reasons for this: (1) it is recognized that the spatial extent of wind gust may engulf components, but not the entire structure; and (2) gust response characteristics of a component would be significantly different from that of the whole structure.

The velocity pressure, q, in psf is given by:

$$q_z = 0.00256 K_z (IV)^2$$

where V is the basic wind speed in mph, I is the importance factor, K_z is the velocity exposure coefficient, and 0.00256 is a constant reflecting air mass density.

Basic wind speed, V, is defined as fastest-mile wind speed at 33 ft (10 m) above ground of terrain exposure C (flat open country and grassland) and associated with an annual probability of occurrence of 0.02. V for any location in the country can be determined from a contour map included in ANSI A58.1-82.

The importance coefficient, I, modifies wind speed to 100-year or 25-year mean recurrence intervals. It has a value (away from the hurricane ocean line) of 1.0 for usual structures, 1.07 for essential facilities and buildings for public assembly, and 0.95 for buildings that represent a low hazard to human life in the event of failure (e.g., agricultural buildings, minor storage facilities, etc.).

The velocity pressure exposure coefficient, K_z, takes into account changes in wind speed with height above ground and with the nature of the surroundings (types of terrain). It is recognized that the wind speed varies with height because of ground friction, and that the amount of friction varies with the ground roughness.

$$K_z = 2.58 \left(\frac{z}{z_g} \right)^{2/d}$$

where:

z = elevation in feet.

z_g = gradient height in feet; at this height wind velocity becomes constant.

α = coefficient depending on exposure.

Four roughness categories or exposure conditions are considered:

1. Centers of large cities and very rough terrain, Exposure A (z_g = 1500 ft, $\alpha = 3$, $D_0 = 0.025$, where D_0 = surface drag coefficient, see below).
2. Suburban areas, towns, city outskirts, wooded areas, and rolling terrain, Exposure B (z_g = 1200 ft, $\alpha = 4.5$, $D_0 = 0.010$).
3. Flat open country and grassland, Exposure C (z_g = 900 ft, $\alpha = 7$, $D_0 = 0.005$).
4. Flat, unobstructed coastal areas directly exposed to wind blowing over bodies of water, Exposure D (z_g = 700 ft, $\alpha = 10$, $D_0 = 0.003$).

The gust response factor, G, accounts for the additional loading effects due to wind turbulence over the fastest-mile speed of wind. It also includes loading effects due to dynamic amplification of flexible buildings and structures.

$$G_z = 0.65 + 3.65\, T_z$$

where:

$$T_z = \frac{2.35 (D_0)^{1/2}}{(z/30)^{1/\alpha}}$$

The pressure and force coefficients, C, for buildings and structures and their components and cladding are given in several figures and tables in ANSI A58.1-82.

The quasi-static approach to wind load design has generally proved sufficient for most structures. However, a more detailed analysis, including wind tunnel studies, may be appropriate for special structures. The above approach to wind may not be satisfactory for ultra-high-rise buildings, especially with respect to comfort of the occupants (in very flexible structures) and the permissible horizontal movement, or drift, which might result in cracking of partitions and glass. These important factors are related to the frequency and amplitude of the vibrations, which depend on the natural frequencies of the building and gust fluctuations of the wind, rather than on steady wind pressure.

10.5.2 Serviceability Criteria

With respect to wind design, the following aspects have to be considered to ensure the satisfactory performance of a structure under service conditions:

(a) Lateral deflection of the structure, particularly as this affects its stability and the cracking of nonstructural elements and structural members.

(b) Motion of the structure, as it affects comfort of the occupants.

1. Lateral deflection or drift is the magnitude of displacement at the top of a building relative to its base. The ratio of the total lateral deflection to the building height, or the story deflection to the story height, is referred to as the "deflection index." The imposition of a maximum allowable lateral sway (drift) is based on the need to limit the possible adverse effects of lateral sway on the stability of individual columns as well as the structure as a whole, and the integrity of nonstructural partitions, glazing, and mechanical elements in the building. No systematic study has yet been published to determine the precise relationship between drift and the above factors. Cracking associated with lateral deflections of nonstructural elements such as partitions, windows, etc., may cause serious maintenance problems (loss of acoustical properties, leakage, etc.). Therefore, a drift limitation should be selected to minimize such cracking.

In the absence of code limitations in the past, buildings were designed for wind loads with arbitrary values of drift, ranging from about 1/300 to 1/600, depending on the judgment of the engineer. Deflections based on drift limitation of about 1/300 used several decades ago were computed assuming the wind forces to be resisted by the structural frame alone. In reality, as mentioned previously, the heavy masonry partitions and exterior cladding common to buildings of that period considerably increased the lateral stiffness of such structures. In constrast, in most buildings that have been constructed in recent years, the frame alone resists the lateral forces. The dry-wall interior partitions and the light curtain-wall exterior contribute little to the lateral resistance of modern buildings.

To date (1983) only the *Uniform Building Code*, BOCA, and the *National Building Code of Canada*, among North American model building codes, specify a maximum value of the deflection index of 1/500, corresponding to the design wind loading. Also, ACI Committee 435[10-4] recommends a drift limit of 1/500. In recent years many engineering offices, owing to competitive pressures, have somewhat relaxed the drift criterion by allowing an overall drift of slightly over $H/500$, with the maximum drift in any one story not to exceed $H/400$. Also, in cases where wind tunnel studies indicate wind forces in the building to be smaller than those specified in the code, designers take the liberty of applying the $H/500$ criterion to the smaller (wind tunnel) wind forces.

The performance of modern reinforced concrete buildings designed in recent years to meet this criterion appears to have been satisfactory with respect to the stability of the individual columns and the structure as a whole, the integrity of nonstructural elements, and the comfort of the occupants of such buildings.

Most of the modern tall reinforced concrete buildings containing shear walls have computed deflections ranging between $H/800$ and $H/1200$ due to the inherent rigidity of the shear wall–frame interaction.[10-5]

It is realized, of course, that the method of calculating the drift as well as the degree to which the assumptions used in such calculations corresponded to the actual structure, may have varied widely from one building to another. Since computing drift is a highly complex procedure, many simplifying assumptions are usually made, which cast doubt on the reliability of such computations. Even between two highly sophisticated computer programs (both incorporating axial shortening of columns), the magnitude of computed deflections may vary 30 to 40% as a result of one assuming center-to-center distances of its line network, and the other considering finite dimension of its joints. Neglecting the axial deformation in the columns of a frame leads to a stiffer structure. This results in computed values for the lateral deflection that are less than those that would be obtained if axial deformation were not suppressed. This effect, although relatively small (about 10–15%) for most multistory buildings, can be significant for tall slender buildings. The increasing availability of computer programs for lateral load analysis should help eliminate the differences in the manner of computing deflections and should lead to a more precise definition of an allowable drift.

To establish a realistic relationship between drift and the pertinent design parameters based on existing buildings, an elaborate field study would have to be undertaken to evaluate their behavior considering the actual wind forces, the method of calculation, and the assumptions made in computing lateral deflections.

For cases where excessive drift is expected, floating partitions with a capability to accept relative movement between skeleton and partition may be required. Floating partitions do not contribute to the lateral rigidity of buildings.

2. *Perception.* The sway motion of a tall building under turbulent wind, if perceptible, may produce psychological effects that render the building undesirable from the user's viewpoint. The reduction of such perceptible motion to acceptable levels may thus become an important criterion in the design of any tall building. It is apparent that the sensation of motion which can be disturbing to an occupant of a building can result either from the visual perception of relative displacement with respect to some reference object, or, if visual effects are excluded, from the acceleration of the floor on which the observer stands. A number of tests have confirmed the effect of acceleration in producing a sensation of motion. Also, if the acceleration is very small, but changes frequently from negative to positive the rate of change of acceleration (which is commonly known as jerk) can equally produce the sensation of motion.

Determination of minimum tolerable values of acceleration for the typical or normal person needs further studies. It is obvious that the acceptability of a design with respect to perception of sway motion can only be assessed through a dynamic analysis of the building under a set of a probable range of wind exposures. No perceptible motion has been reported in concrete buildings to date.

At present reliable procedures and criteria for limiting discomfort to occupants due to building motion are scarcely available. The supplement to the National Building Code of Canada, 1980 edition, contains expressions by which peak along-wind and across-wind accelerations of buildings can be calculated. According to the supplement, "it appears that when the amplitude of acceleration is in the range of 0.5 percent to 1.5 percent of the acceleration due to gravity, movement of the building becomes perceptible to most people." Based on this and other information, a tentative acceleration limitation of 1 to 3% of gravity once every 10 years is recommended for use in conjunction with the expressions for computation of accelerations. The lower value is thought suitable for apartment buildings, and the higher value for office buildings.

10.5.3 Analysis for Lateral Forces

The degree of sophistication to which a structural analysis is carried out obviously depends on the importance of a project. A wide range of approaches have been used for buildings of varying heights and importance, from simple approximate methods that can be carried out manually or with the aid of a desk calculator, to more refined techniques involving computer solutions, as well as model studies.

While the use of approximate methods for the lateral load analysis of frames may, be current standards of engineering practice, be adequate for regular frames with a few stories or for the preliminary analysis of tall frames, the increasing availability of computer programs argues against the use of such approximate methods when more reliable solutions are obtainable.

Only a few years ago most of the multistory buildings were analyzed by approximate methods such as the portal or the cantilever method. The recent advent of computers and the abundance of publicly available sophisticated programs has de-emphasized the approximate methods which at present are useful for preliminary estimating only.

Simplified methods, when used for the preliminary analysis of tall structures, should generally be followed by a computer analysis that includes the effects of axial deformations as well as bending.

For major high-rise structures, a computer analysis may be the most economical means of arriving at a reliable design. A logical procedure for lateral load analysis is:

(a) Carry out a simple preliminary analysis by longhand calculations with approximate methods and determine the required member sizes for the combination of gravity and lateral loads;
(b) If necessary, use a computer program to achieve a more accurate analysis.

In view of the increasing availability of computer programs for frame analysis, longhand methods that involve simplifying assumptions and lengthy arithmetical work must be considered as obsolete.

In investigating the use of a program, there are many features that can be included and may be important in a frame analysis. For example:

1. Plane frame or space frame
2. Flexible data input for irregular structures, but simplified input for regular structures
3. Axial deformation of vertical members
4. Shear deformation of members
5. Finite size of joints between members
6. Floors fully rigid in-plane
7. Nonlinear material behavior
8. Second-order geometry (i.e., P–Δ) effects

It is important to know the significance of any feature that causes a significant increase in data input, storage requirement, or execution time. For example, if column axial deformation is negligible, it is worthwhile to neglect it in the analysis by suppressing the vertical degrees of freedom of the structure.

At present, the scope of the analysis is normally governed by the features available in the program to be used. As more programs become available and further features can be included in the analysis, it will become increasingly important for the designer to be able to choose the program or program features that give an efficient solution to his problem.

The following assumptions are common to the analysis and design of all high-rise structural systems:

1. The floor slabs distribute load to the lateral load resisting units, mainly through forces in their own plane. The actual in-plane deformation of the floors will seldom have a significant effect on this distribution, and the assumption that the floors are fully rigid in-plane is widely used.

2. Out-of-plane bending of floor slabs can be important. A very small resistance to bending in a beam connecting two vertical units can have a significant effect on the behavior of the entire system. It is most important to ensure that a connecting member possesses the necessary flexural strength. Slab–column joints in flat-plate structures should be given careful attention.

3. The effect of torsion should be considered if the layout is unsymmetrical or if the stiff vertical units are located close to the center of the structure.

Any combination of frames, walls, or tubes can be idealized as a space frame. If a space frame computer program that considers all six degrees of freedom at each joint is used in the analysis, any torsional effects will automatically be accounted for. Although space frame analysis of multistory structures is still relatively costly, it is being used more frequently of late, particularly for nonrectangular structures. It should be noted that in the late 1970s and the early 1980s most of the prestigious office buildings constructed were no longer rectangular. The architectural trend has been to design complex plan shapes with many corners and protrusions. For such structures, the traditional two-dimensional analysis does not provide reliable results; a three-dimensional analysis is desirable, and sometimes necessary. The discussion in this chapter will be confined to analysis techniques that assume full in-plane rigidity of the floor slabs.

The effect of relative deflections due to temperature and creep can be important, as discussed further in this chapter. P–Δ moments may be significant in tall structures. A method of including these in analyses is described briefly below in the section on frames.

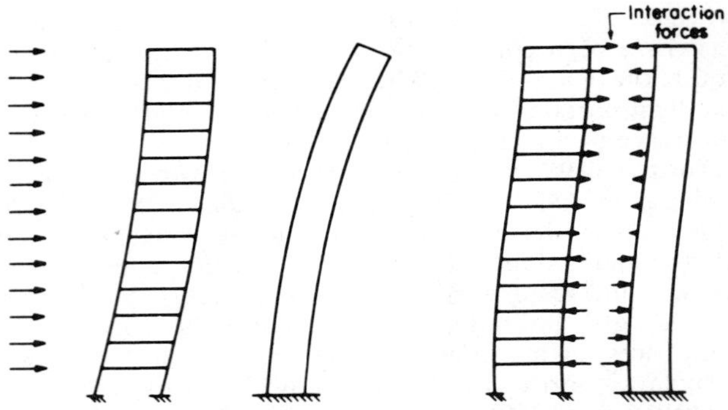

Fig. 10-21 Deformation modes.

With respect to lateral resistance, there are three basic types of structural systems: frames, shear walls, and tubes. The basis for the classification is the mode of deformation of the unit when subjected to lateral loading.

Frames deform in a predominantly shear mode where relative story deflections depend on the shear applied at the story level. Shear mode deformation is illustrated in Fig. 10-21(a).

Walls deform in an essentially bending mode as illustrated in Fig. 10-21(b).

Tubes, if unperforated, behave in the same way as walls. However, openings that are normally present in units of this type produce a behavior intermediate between that of a frame and that of a wall.

Any of the three types of units described above, singly or combined, form a *structural system* (for lateral loads). Generally, as the height of a building increases, a point is reached beyond which the lateral sway under wind loading, and hence consideration of stiffness and not strength, will govern the design of the structural system. The point, in terms of height, at which this condition is reached depends on the type of structural system. Ideally, that system should be chosen which does not require an increase in the sizes of the members beyond that required to support the design vertical loads, i.e., a "premium-free" building.

Table 10-2, from Ref. 10-6, is presented to guide the choice of a suitable structural system for a particular building. The ranges of applicability shown may vary somewhat depending upon the use of the building, the story heights, and the design live and wind loads.

TABLE 10-2 Guide to Selection of Structural Systems

	*Number of Stories**	
Structural System	*Office Buildings*	*Apartment Buildings, Hotels, etc.*
Frame	up to 15	up to 20
Shear Wall (egg crate)		up to 150
Staggered Wall Beam		up to 40
Shear Walls acting with Frames	up to 40	up to 70
Single Framed-Tube	up to 50	up to 60
Tube-in-tube and bundled tubes	up to 80	up to 100

**The values given here are based on present day practice as well as trends indicated by current thinking.*

The present methods of lateral load analysis for average buildings disregard the dynamic character of the wind–structure interaction. A major reason for this is that all of the presently used building codes specify loads that are to be considered as static loads upon a structure, and which the structure is assumed to resist within its elastic range. Only unusual structures may require a dynamic investigation (analytical and/or experimental) involving the periods of vibration.

The methods of analysis suitable for the individual types of structures will be discussed in the corresponding sections, together with other aspects of these structural systems.

10.5.4 Frame Structures

The term "frame" denotes a structure that derives its resistance to lateral loading from the rigidity of the connections between columns and beams or slabs.

1. Components of Drift. In a frame-type structure the drift may be thought of as consisting of two parts: one due to bending in the columns and beams, and the other due to axial deformation of the columns. As the height-to-width ratio of the structure increases, the effect of column axial deformation assumes greater significance.

The effect of secondary moments caused by the axial forces and the deflections (P–Δ) will also tend to increase the lateral deflection.

2. Hand Calculation Methods. Simplified methods of analysis for building frames are often used for purposes of preliminary designs. Such methods should be able to furnish rough numerical data with a minimum of effort.

Portal method: The basic assumption of the portal method is that points of contraflexure are located at mid-length of all columns and beams. In addition, an assumption concerning the distribution of shears in the columns of a story is made. These assumptions reduce a highly statically indeterminate problem to a statically determinate one. The method neglects the effect of axial deformation in the columns.

The assumptions associated with the portal method result in errors in the vicinity of the base and top of the frame and at setbacks or locations where significant changes in member stiffness occur. In these locations, large errors in the calculated member moments may be expected. This can be particularly serious in bottom stories, where the combination of large axial forces and moments in the columns can lead to instability problems. This type of error can be partially corrected by performing a more exact analysis for the localized discontinuity regions or by using tabulated information on the location of inflection points in the bottom and top stories.[10-7]

The errors resulting from disregarding column axial deformations increase with the increase in the number of bays and the number of stories in a frame, and are reflected most markedly in the moments in the exterior columns and girders of the upper stories of tall frames.

Drift calculated on the basis of moments obtained by the portal method is subject to errors as the assumed locations of the points of contraflexure lead to predictions of drift larger than that based on exact analysis. Additional errors arise from the neglect of axial column deformation.

3. Computer Programs for Frames. A number of computer programs have been developed in recent years. STRUDL, for example, is capable of carrying out elastic analysis of either plane frames or three-dimensional reticulated structures under static and dynamic loading conditions.

The more important frame program features that have relevance to lateral load analysis are as follows:

(a) *Axial deformation.* Axial deformation of columns may be important in tall slender frames or in frames with stiff connecting beams. While no definite rules can be given, the effect of column axial deformation will generally be important if the frame height-to-width ratio exceeds about three to four. Axial deformation of beams is always negligible.

(b) *Shear deformation.* This can usually be neglected.

(c) *Finite size of joints between members.* Manual methods and the early computer programs based structural analysis on the centerlines of members. This, however, results in erroneous moments and drifts, since the real spans are smaller than the center-to-center distances between joints. A number of possibilities exist (particularly in computer programming) to reduce these errors).

In most cases it is realistic to assign a greater stiffness to the area within the joint between beams and columns than that assigned to the connecting members.

For symmetrical members with equal end rotations, an equivalent EI value can be used:

$$(EI)_e = \frac{K\,(1 + C)\,L}{6}$$

where:

$(EI)_e$ = equivalent EI

K = end rotational stiffness

C = carry-over factor

L = center-to-center span of member

where K and C are evaluated for a member with infinite flexural stiffness within the joints.[10-8]

Where a computer program has the capability of considering members with variable moments of inertia, the joint rigidity can be simulated by increasing the moments of inertia of the column and beam sections within the joint area; they should be increased to about 10 or 20 times the normal column or beam moment of inertia to realistically simulate the joint rigidity. Excessive increase of stiffness in the joint area may lead to significant numerical error in the solution and should be avoided.

An efficient approach is to treat the ends of the members as being fully rigid, and then calculate the stiffness properties of the combined elements before the frame analysis is undertaken. Some frame programs incorporate a finite joint facility of this type.

(d) *Foundation movement.* Ability to include elastic spring supports is a useful program feature (especially for shear wall foundations).

(e) *Second order geometry. (P–Δ effects).* When a frame sways laterally by an amount, Δ, the columns are subjected to an eccentric moment equal to $P\Delta$, where P is the total vertical load at the level at which Δ is measured. This P–Δ effect can be significant, particularly in tall unbraced frames.

4. Effects of Masonry Infill Walls and Partitions. In many frame-type structures, walls and partitions are made of precast or masonry units. Although such elements are often considered to be nonstructural and perhaps—because of very light reinforcement—may not contribute significantly to the ultimate strength of the structure, they generally contribute substantially to the lateral stiffness of the structure under working load conditions. The behavior of multistory infilled frames, i.e., frames with the space between the members filled with a masonry or cast-in-place concrete wall, under lateral loading is essentially that of a vertical cantilever beam. The effect of walls and partitions can be particularly well observed on the response of structures subjected to earthquake motions. Walls filling the space between frame members not only tend to increase the stiffness, but may altogether alter the mode of response of the frame, changing it to a shear wall and, as a result, changing the entire structure and the resulting distribution of forces among the different frame components.

Studies[10-9] have indicated the possibility of taking full advantage of infill partitions by constructing them to ensure their participation in resisting frame distortions with particular reference to the satisfaction of the drift criterion.

The stiffening effect of the infill panel on the frame can be represented fairly well by a diagonal strut having the same thickness as the panel and an effective width that depends on a number of factors. Studies of frames in which the infilling was not bonded to the frame indicate that the effective width of the diagonal strut is influenced by the relative stiffness of the column and the infill; the height-to-length ratio of the infill panel; the stress–strain relationship of the infill material; and the diagonal load on the infill. The effective width of the strut increases with increasing column stiffness and panel height-to-length ratio, and decreases with increasing value of the load and modulus of elasticity of the infill material. Reference 10-9 gives curves for the approximate equivalent strut width in terms of the geometrical and physical properties of the frame and infill and the applied diagonal load. While it may be conservative to disregard the stiffening effect of infilling panels on the behavior of frames subjected to wind loading, neglecting their effect on the response of a structure subjected to earthquakes can be dangerous. Walls filling the space between frames not only increase the stiffness (and damping) and hence the frequency of vibration of the building, but also alter its mode of deformation under lateral load. A major effect of the change in behavior from a predominantly shearing type to a cantilever mode of deformation is a significant increase in the axial forces to which the exterior columns are subjected.

5. Effective Slab Width. Structures consisting of flat slabs or plates and columns are utilized as frames to resist lateral loads by considering the slabs as equivalent beams. The effective width of the slab determines the stiffness of the one-dimensional beam elements to be used in a plane frame analysis.

Analytical and limited experimental studies dealing with the problem of effective slab width have considered the c_1/l_1 ratio (c_1 = column dimension in the direction of analysis, l_1 = slab span in the direction of analysis) as the principal variable, and have indicated effective slab width ratios ranging from 0.20 to unity, the effective width increasing with increasing values of the c_1/l_1 ratio. The slab aspect ratio l_2/l_1 (l_2 = slab span transverse to the direction of analysis) is another important variable, the effective slab width increases with increasing values of l_2/l_1. Reduction in stiffness due to cracking of slab members is important. Other variables such as the c_2/c_1 ratio (c_2 = column dimension transverse to the direction of analysis), and the distribution of slab reinforcement, also have an influence on the effective slab width.

The problem of moment transfer between columns and slabs and of stiffness of an "equivalent beam" to properly model the action of a slab for use in a plane frame analysis has been studied, both experimentally and analytically, at the University of Washington in Seattle since the mid-1970's, and at other institutions. While definitive results have been derived for the strength aspects of moment transfer, the solution to the problem of equivalent stiffness is still not fully settled. Obviously, the equivalent stiffness depends upon the extent of cracking of the concrete and the yielding of the reinforcing bars around the column. Equivalence may also depend upon the purpose of the analysis. An analysis undertaken to provide values of slab design moments may require different equivalence criteria from an analysis intended to serve as a basis for column design or an analysis aimed at studying overall frame deformation.

It is presently customary among designers to consider one-quarter to one-half of the panel width as equivalent slab width for *stiffness* when carrying out an analysis for lateral loads. However, only one and a half times the slab thickness on either side of the column is considered effective for the transfer of the unbalanced moments between columns and slabs resulting from the analysis.

The joints between column and flat plate floors designed for vertical loads only have a limited capacity to resist lateral loads. This is one of the main reasons for adding shear walls in flat plate structures above 15 stories. To enhance the joint stiffness, drop panels or beams can be added. In some cases, the total floor thickness has been increased to satisfy lateral resistance requirements.

The slab–column joints of lift-slab structures are not normally assumed to participate in lateral load resistance.

6. Analysis. If all the plane frames parallel to the direction of loading are the same, the total lateral load in that direction can be distributed equally to each frame. The frames can be analyzed by the methods discussed previously in Section 10.5.3. Alternatively, a distribution of shear to each column based on relative column and beam stiffness[10-10] or column stiffness[10-11] can be used. Parallel frames (with no torsion) can be idealized in a manner similar to the shear wall–frame method, illustrated further in Fig. 10-32.

7. Range of Applicability—Number of Stories. Although buildings utilizing flat-plate-type slab systems of up to 75 stories have been built, stiffness considerations tend to limit the height of flat-plate structures without shear walls and without spandrel beams to about 14 to 16 stories. This limitation has become apparent with the increasing use of lightweight elements or glass for walls and lightweight partitions. For buildings of greater height, shear walls, or shear walls acting in conjunction with adjoining frames, generally provide a more efficient solution.

10.5.5 Shear Wall Structures

The term shear wall is actually a misnomer as far as high-rise buildings are concerned, since a slender shear wall when subjected to lateral forces has predominantly moment deflections and only very insignificant shear distortions.

Figure 10-22 shows various ways of utilizing shear walls in multistory buildings. The scheme of Fig. 10-22(a) has shear walls without openings across the building with access through a gallery running alongside the building. Each wall accepts a share of the lateral load proportional to its stiffness. The calculation of lateral stiffness is simple, and stresses in such shear walls without openings involve simple bending theory only.

Schemes (b) and (c) of Fig. 10-22 have interior corridors, and, therefore, the shear walls are interrupted by an opening on each floor, and they are interconnected either by slabs or beams. Although the major shear walls are usually in the transverse direction of the building, separating the individual apartments, stability in the longitudinal direction is normally provided by elevator shafts or some longitudinal shear walls. Such structural systems as shown in Fig. 10-22 are known as egg-crate or crosswall buildings; they are extremely rigid in the direction of the shear walls.

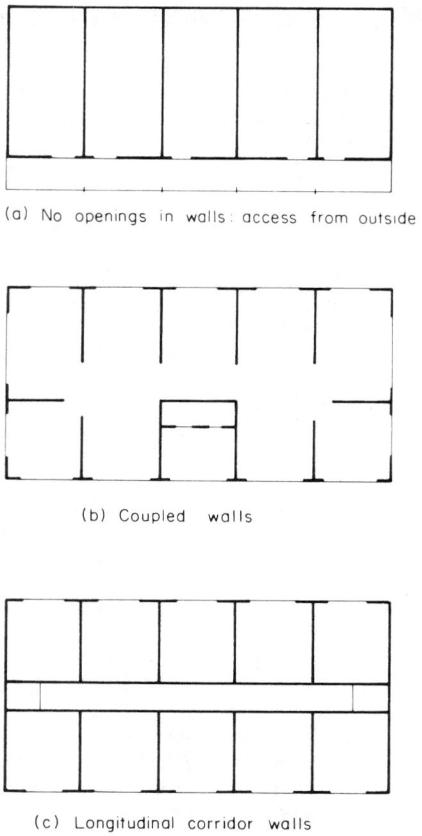

(a) No openings in walls: access from outside
(b) Coupled walls
(c) Longitudinal corridor walls

Fig. 10-22 Schematic layouts of shear wall structures.

10.5.6 Coupled Shear Wall Structures

Walls with openings present a complex problem to the analyst. Openings normally occur in vertical rows throughout the height of the wall and the connection between the wall segments is provided by either connecting beams or floor slabs, or a combination of both.

The terms "coupled shear walls," "pierced shear walls," and "shear wall with openings" are commonly used to describe such units as shown in Fig. 10-23.

If the openings are very small, their effect on the overall state of stress in a shear wall is minor. Larger openings have a more pronounced effect and, if large enough, result in a system in which typical frame action predominates. The degree of coupling between two walls separated by a row of openings has been conveniently expressed in terms of a geometrical parameter α (having a unit 1/length), which gives a measure of the relative stiffness of the connecting beams with respect to that of the walls. The parameter α appears in the basic differential equation of the so-called continuum approach.[10-12–10-15] The studies indicate that when the dimensionless parameter αH (H being the total height of the walls) exceeds 13, the walls may be analyzed as a single homogeneous cantilever. When $\alpha H < 0.8$, the walls may be treated as two separate cantilevers. For intermediate values of αH (i.e., $0.8 < \alpha H < 13$), the stiffness of the connecting beams should be considered.

The effectiveness of the coupling of shear walls can be clearly seen from Fig. 10-28 (Example 10-1), where the cantilever moment of the shear wall acting alone is compared with the moment reduced due to frame action caused by coupling.

Prior to the early 1960s no analytical techniques for coupled shear walls were available. During the 1950s and into the 1960s, coupled shear walls were designed as individual cantilevers, each carrying the amount of wind in proportion to its stiffness. Although the shear walls were greatly overdesigned, this approach resulted in underdesigned connecting links (beams or slabs). Since the 1960s a considerable amount of research information has become available, and a number of practical approaches have been suggested.[10-12–10-17]

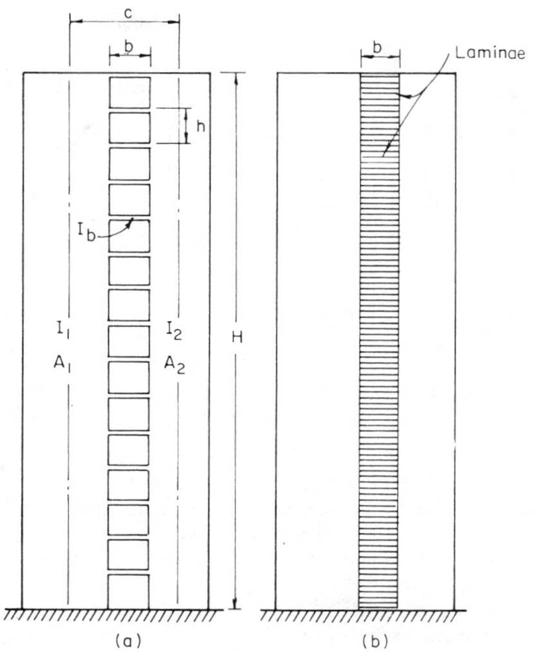

Fig. 10-23 Coupled shear walls.

1. Methods of Analysis. One of the more familiar approaches is the continuum approach. In its most basic form it assumes that elastic structural properties of the coupled wall system remain constant throughout, that both walls are founded in a common stiff footing, and that the points of contraflexure of all beams are at midspan.

In this method, the individual connecting beams of Fig. 10-23(a) are replaced by a continuous connection of laminae as in Fig. 10-23(b). Under horizontal loading, the walls deflect and induce shear forces in the lamina. A second-order differential equation is set up and solved to give shears, moments, and deformations throughout the wall. Several papers use this approach with differing choice of variables, all yielding essentially the same results.

Tests on plastic models have generally confirmed the accuracy of the continuum approach for walls conforming to the basic assumptions of the method. In practice, some of these assumptions do not hold. Slight inaccuracies in the method may stem from local wall deformations and reduction in stiffness of coupling beams due to their cracking.

Solutions are available for uniformly distributed loads, triangularly distributed loads, and concentrated point loads at the top of the building. Perhaps the most readily available and convenient are those developed by Coull and Choudhury.[10-14,10-15]

In some cases, elastic analysis will result in moments that cannot be developed by beam sections restricted in size for architectural reasons. The possibility of designing the beams for ultimate moments in proportion to the elastic moment distribution has been suggested[10-16] and a theory has also been developed for the elasto-plastic analysis of coupled shear walls[10-13] that would result in substantial simplifications and economies.

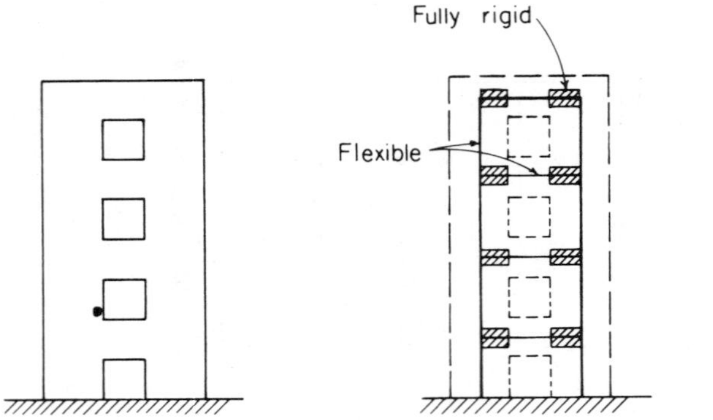

Fig. 10-24 Substitute frame.

Analytical methods employing computerized frame analysis are not only more accurate, but are considerably more flexible and can take into account many more variables than the continuum approach, which was originally developed for manual operation.

For analysis of multistory shear walls, a good computer program considers a frame with finite joints in which the coupled wall is analyzed as a frame except that the finite width of the columns in comparison with the beam is recognized. The analogy, in which the beams are assumed to be infinitely stiff from the centerline of the column (wall) to the edge of the actual opening, is illustrated in Fig. 10-24(b). The calculation can take into account changes in wall thickness, story height, and concrete strength at various locations within the height of the building.

Because of the popularity of shear walls coupled solely by slabs across the corridor in apartment buildings, researchers were attracted to investigate the interaction between the slab and walls, and, particularly, to try to device a simple approach to determine the equivalent slab width to be used for stiffness in frame analysis. Analytical studies[10-16] using elastic models produced a range of results varying from the full width of the bay to half the corridor opening (distance between walls). The consequences of selecting an effective width of slab should be understood by the designer, as it has implications for both the strength and the stiffness of the coupled system.

Recent studies using reinforced concrete prototype models of shear walls coupled solely by slabs were carried out at the University of Toronto.[10-17] The specimen represented present practice; monotonic loading was used simulating wind response. The results showed that using an equivalent stiffness of a cracked section having a width of one-half the corridor opening provides a good correlation with the observed behavior in laboratories. The shear capacity of the coupling slab can be predicted by using a U-shaped critical section around the toe of the wall and a shear stress of $4\sqrt{f_c'}$. The flexural strength of the coupling slab can be predicted by using a slab width equal to the corridor opening plus the wall thickness. The stiffness of the slab–wall connection depends to a large degree upon the extent of cracking. In the elastic range of behavior a considerable width of slab will participate in load transfer. However, with the increase in the lateral deformation, a loss of stiffness occurs. Recent studies[10-18] have shown that following a number of reversing cycles into the inelastic range, there is a severe stiffness degradation in the slab, making it questionable to use shear walls coupled solely by slabs as a primary system for energy dissipation in earthquakes. In contrast, shear walls coupled with beams are an extremely efficient system for dissipating the lateral forces of earthquakes.

2. Proportioning of Shear Walls. Although the analysis of shear wall systems has advanced considerably in the last decade, the proportioning (design) of shear walls has not kept pace with the analytical developments. A number of experimental studies[10-20, 10-21] ascertained the shear strength of such walls and the results were incorporated into sections 11.16 and A.8 of the ACI 318-71 code. As for flexural strength, it is obvious that the traditional method of treating the wall as a homogeneous section may be valid only for low eccentricities resulting in uncracked sections, while the slender walls of tall buildings which, by their geometry, are slender cantilevers, should be proportioned for moment similarly to any flexural element. Such a method of flexural treatment will lead to concentrations of flexural reinforcement at the extreme ends of the shear walls with minimum reinforcement both vertically and horizontally (for shear and shrinkage control) in the remainder of the section.

Tests[10-22] indicate that only very low and long shear walls with height-to-horizontal length ratios (H/l_w) less than one will fail in shear. Contrary to common preconception and the misleading name, the strength of taller shear walls, and particularly of those in multistory buildings, will invariably be controlled by flexure; they behave like slender cantilevers. As a matter of fact, it is quite difficult to provide sufficient flexural reinforcement to force a shear failure on a slender shear wall; only extremely heavily reinforced columns at the ends of thin shear walls (dumbbells) can force a shear failure under heavy lateral loads.

The tensile reinforcement, obviously, also acts in compression at moment reversals and should preferably be enclosed with ties (as in a column) to improve the strain capacity of the concrete in compression and at the same time improve the ductility of the shear wall.

All portions of a shear wall should be designed to resist the combined effects of axial load, bending, and shear, determined from the analysis of the structural system.

Flexural reinforcement should be provided in accordance with the requirements of the *ACI Building Code.* Walls proportioned such that a linear strain distribution does not apply should be designed as short cantilevers.

Minimum amount of reinforcement in the vertical direction should be as required by flexural calculations and as specified in the code for shear strength. A wall with near-minimum amount of vertical reinforcement is usually the most efficient. If much more than the minimum amount of vertical reinforcement is required to resist the design axial loads and moments, a change in shearwall proportions should be considered. Minimum amount of reinforcement in the horizontal direction should be as specified in the provisions for shear strength and shrinkage control. In addition to providing the necessary amounts of reinforcement, it is essential that reinforcement details in every shear wall receive careful attention to ensure optimum performance.

If there are tensile forces resulting from the most severe combination of vertical loads and overturning moments due to lateral loads, they must be anchored into the foundation medium, unless they can be overcome by gravity loads mobilized from neighboring elements.

3. Distribution of Lateral Loads. The proportion of the total lateral load that each wall resists depends on its stiffness relative to that of all walls or coupled wall systems in a building. The lateral stiffness of each wall or coupled wall can be based on the deflection at the top when subjected to a uniformly distributed unit lateral loading. Calculation of stiffnesses can be made quickly through the use of tables.[10-14, 10-15] The method of distributing the load is described in Ref. 10-23 for low structures. When the walls do not remain constant over the height of the structure, the use of a plane frame program with walls connected by link bars (as illustrated in Fig. 10-32) is advantageous for irregular configurations in multistory structures. Once the loading on each wall is determined, wall stresses can be calculated and thicknesses modified, if necessary.

4. Application. Shear wall buildings are used in apartment, hotel, and other residential buildings where walls are customarily spaced between 15 and 24 ft apart with floor slab thicknesses proportioned according to span. Spans up to 40 ft have been used with prestressed hollow-core concrete slabs.

The shear wall structure is used in buildings where permanent partitions and the lack of flexibility for future modifications can be tolerated. Its major advantages lie in the speed of construction, low reinforcing steel content and acoustical privacy.

In current North American practice, shear wall buildings are mainly cast-in-place, but trends to systems building are leading to an increase in the number of buildings being constructed using large panel, precast components for floors and/or walls.

Shear wall structures are well suited for construction in earthquake areas, and they have performed well during recent disasters.[10-24, 10-25] While costs vary from city to city, shear wall buildings usually become economical as soon as lateral forces affect the design and proportioning of flat plate or beam and column structures. Buildings of up to 70 stories have been built using shear walls. Feasibility studies for projects up to 200 stories utilizing shear walls have been made and found workable.

5. Coupled Shear Walls Supported on Exterior Columns Only. Parking areas under residential buildings require different spans from the apartments above. For this reason the shear walls must be stopped and supported on exterior columns, thus leaving the entire parking area column-free. The lower portions of such shear walls act as a deep beam spanning between the supporting columns.

The majority of shear wall buildings have coupled shear wall systems to accommodate corridors in the middle. A study carried out at the Portland Cement Association showed the feasibility of supporting a coupled shear wall on exterior columns as shown in Fig. 10-25. The computer study showed that the second floor beam (supporting the shear walls) acts like the tension member for the coupled shear wall above. The lintels over the doors for the next five stories act as compression struts; above that level the forces in the lintels are minimal, as can be seen in Fig. 10-25. The study also indicated that the shear wall must be supported during construction only until about the fifth floor is cast; after that the structure is self-supporting.

EXAMPLE 10-1: The 20-story apartment building designed with coupled shear walls is shown in its typical plan and elevation in Fig. 10-26. The typical wall section shows the pair of coupled shear walls

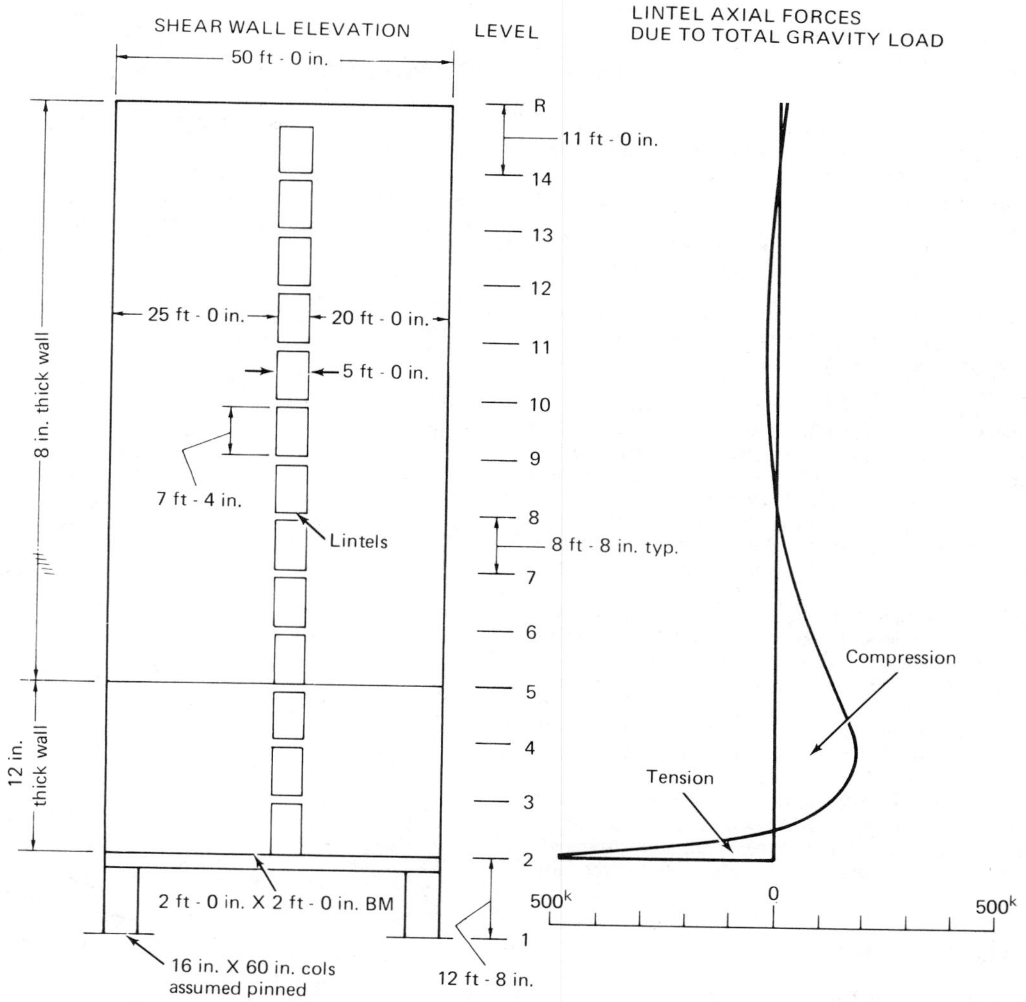

Fig. 10-25 Coupled shear walls supported on exterior columns.

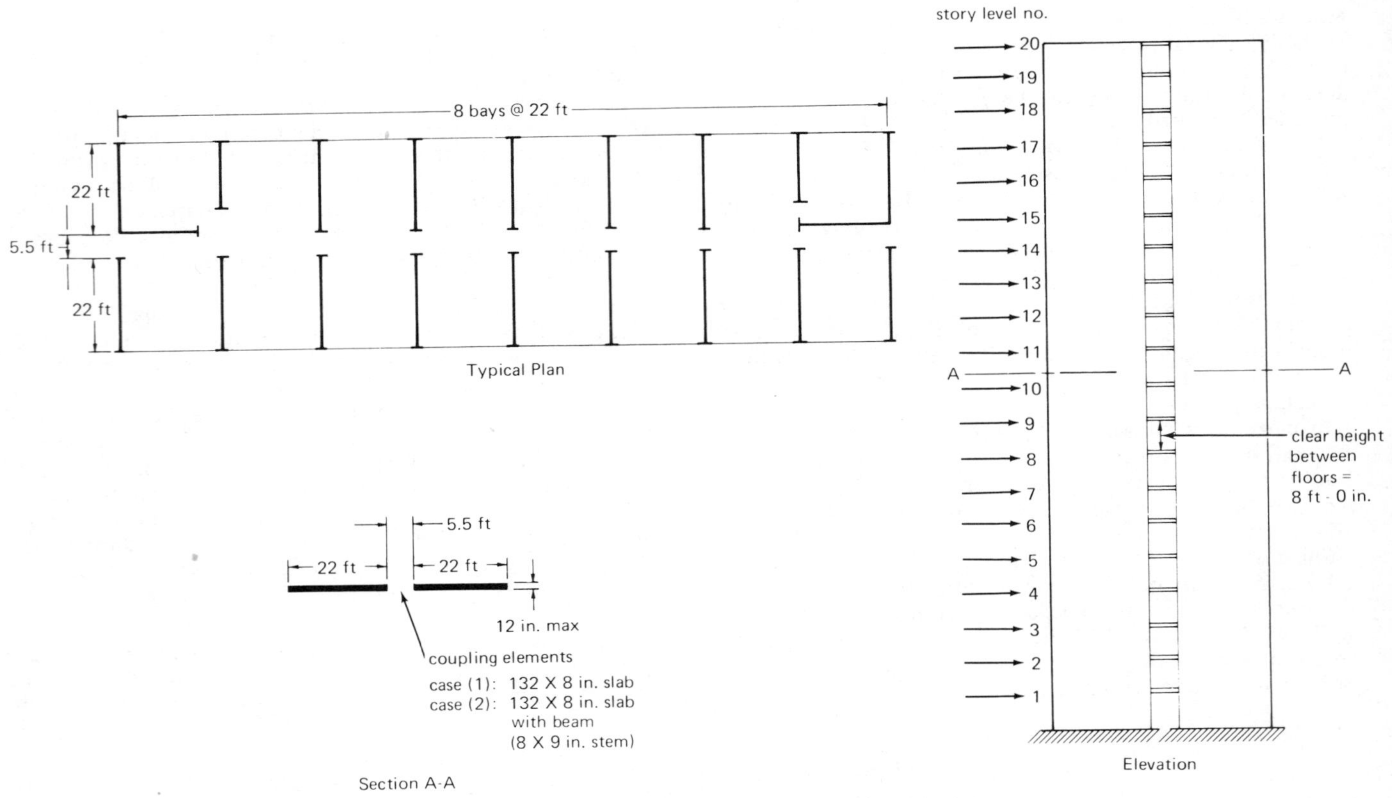

Fig. 10-26 Typical plan and elevation of a 20-story, coupled shear wall building.

each 22 ft long and having a thickness that varies from 8 in. for the top six stories to 10 in. for the intermediate seven stories, and 12 in. for the bottom seven stories.

The coupled shear wall linked by (1) 8-in. slab; (2) the same slab plus 8 × 9 in. stem.

The objective is to determine if there is a benefit in coupling the walls, and, to what degree the addition of a beam to the slab improves the coupling.

An analysis for a uniform wind load of 20 psf was carried out using a frame computer program,[10-19] which can accommodate the finite width of the shear walls. The results of the lateral load analysis are shown in Figs. 10-27 and 10-28.

Two cases are shown in Figs. 10-27 and 10-28. The first case has slabs for the connecting elements, while the second case has the same slabs but with an additional beam in the plane of the coupled walls, so that the coupling elements—having a T-section—have a stiffness twice that of the slab in the first case. The choice between the slab as coupling or beam coupling should be made on the basis of the capability of the linkage to transfer into the shear wall the coupling moments and shears, in addition to the moments and shears due to gravity.

Figure 10-27 shows, as expected, that increasing the stiffness of the coupling elements increases the moments, that these elements take. In this case, doubling the stiffness of the coupling elements increased the maximum moment in the coupling element by about 15%. Increasing the stiffness of the coupling elements in turn decreases the moment in the shear walls. This is shown in Fig. 10-28. Also shown in this figure are the overturning moments in a shear wall when acting as a free cantilever, i.e., without coupling. Note the very significant reduction (to about one-third) in the shear wall moments resulting from the coupling of the shear walls. The effect of the doubling of the coupling stiffness on the shear wall moments can be seen as only secondary within the total reduction of the free cantilever moments.

10.5.7 Shear Wall-Frame Buildings

Since the late 1940s, when the first shear walls were introduced, their use in high-rise buildings to resist lateral loads has been extensive, in particular to supplement frames that, if unaided, often could not be efficiently designed to satisfy lateral load requirements.

The great majority of multistory buildings today are, in fact, shear wall–frame structures, since elevator shafts, stairwells, and central core units of tall buildings are mostly

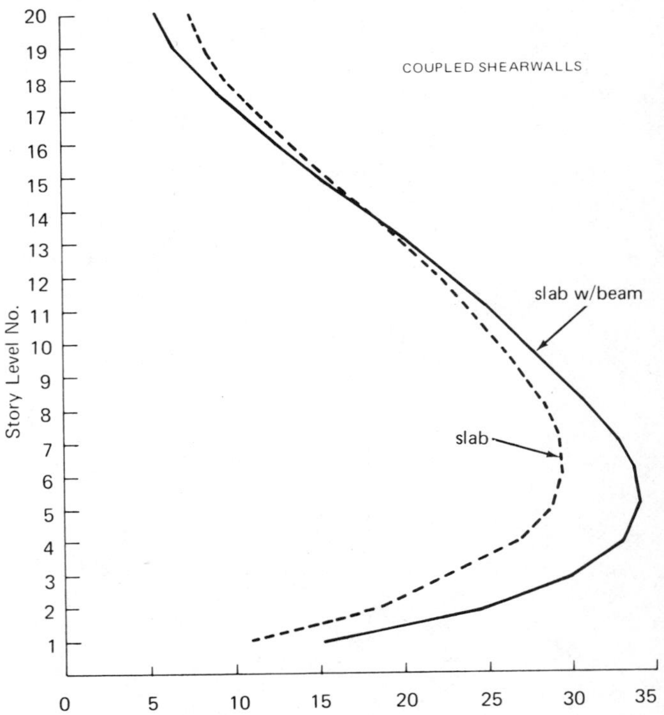

Fig. 10-27 Moments in the coupling elements.

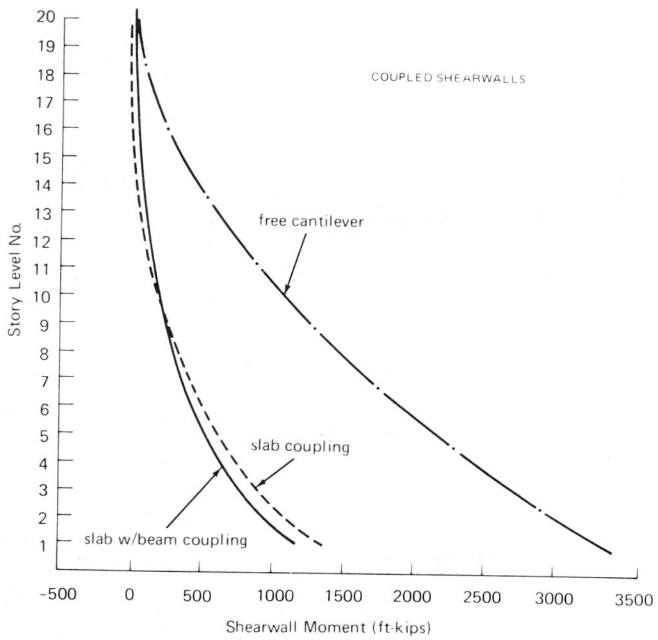

Fig. 10-28 Shear wall moments for the coupled system.

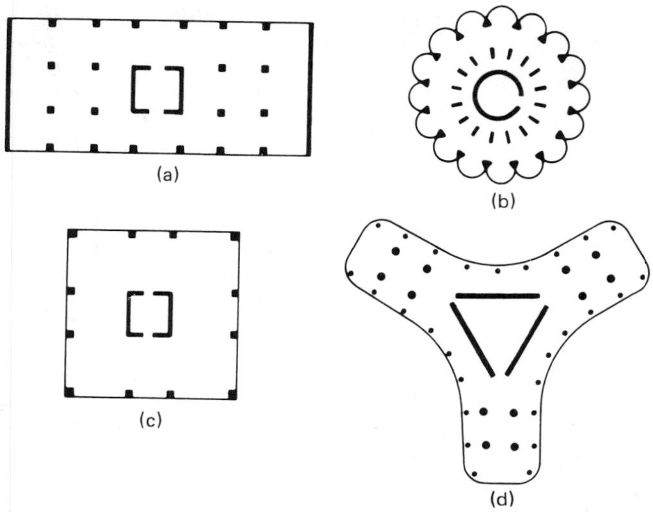

Fig. 10-29 Typical layouts of high-rise buildings with shear wall–frame interaction.

treated as shear walls, in addition to isolated concrete walls (if some are used). Frame structures depend primarily on the rigidity of member connections for their resistance to lateral forces, and they tend to be uneconomical beyond 15 to 20 stories. To improve the rigidity and economy, shear walls are introduced in buildings exceeding 15 to 20 stories in height.

The term shear wall-frame structure is used here to denote any combination of frames and shear walls. Included in this category would be the typical frame structure with shear walls appropriately located about the plan, as shown in Fig. 10-29 and the so-called hull-core or framed-tube-with-core structure shown in Fig. 10-29(c). The shear wall can have any plan shape and may be linear, angular, rectangular, or circular in plan.

The common assumption to neglect the frame and assume that all the lateral load is resisted by the shear walls may not always be conservative, since, owing to the interactive forces, the frame is usually subjected to forces higher than the exterior applied wind forces in the upper stories. Therefore, distributing the applied wind to the different resisting elements in proportion to their relative stiffness in the case of shear wall–frame interaction can lead to grossly erroneous results.

Consideration of shear wall–frame interaction leads to a more economical design. Since the shear wall moments are reduced, less reinforcing is needed as the frame takes over some of the lateral load moments. In most cases the frame can accept the additional moments due to lateral loads within the 33% increase in allowable stresses, except for the top stories of the frame, which often require additional reinforcing. Shear walls are efficiently utilized if they are distributed about the plan so that they carry their proportional share of the vertical load, rather than having them function mainly as lateral load resisting elements. This condition may, however, conflict with the desirability of locating the principal lateral load resisting elements along or near the periphery of a building.

The main function of a shear wall for the type of structure being considered here is to increase the rigidity for lateral load resistance. Shear walls also resist vertical load, and the difference between a column and a shear wall may not always be obvious. The distinguishing features are the much higher moment of inertia of the shear wall than a column and the width of a shear wall, which is not negligible in comparison with the span of adjacent beams. The moment of inertia of a shear wall would normally be at least 50 times greater than that of a column, and a shear wall would be at least 5 ft wide.

The introduction of deep vertical elements (shear walls) represents a structurally efficient solution to the problem of stiffening a frame system. The frame deflects predominantly in a shear mode shown in Fig. 10-21(a), while the shear wall deflects predominantly in a bending mode as shown in Fig. 10-21(b).

In a building, the in-plane rigidity of the floor slabs forces the deflection of the walls and the frames to be identical at each story level. To force the walls and the frame into the same deflected shape, internal forces are generated that equalize the deflected shape of each. Thus, the frame in the upper stories pulls back the wall, while in the lower stories the wall pushes back the frame. These internal interactive forces, shown in Fig. 10-21(c), greatly reduce the deflection of the overall combined system, creating a considerably higher overall stiffness than would be the sum of the individual components, each resisting a portion of the exterior loads. In that distinctive feature of increasing the stiffness through a set of internal forces lies the great advantage of shear wall–frame interactive systems.

1. Methods of Analysis. Methods of analysis for manual or desk calculator application of shear wall–frame interaction were first developed in the early 1960s.[10-11, 10-26] As distinct from many other methods, the Khan-Sbarounis[10-11] analytical treatment is based on a forced-convergence procedure that causes both systems to have the same deflected shape by the use of an iterative process. The authors present a family of curves for the distribution of the wind forces between the walls and the frames for a wide range of stiffness ratios between a shear wall and a one-bay equivalent frame.

To enter the charts, an equivalent shear wall–frame system must first be established by lumping together all the frames (columns and beams) into a one-bay equivalent 10-story frame and all shear walls into an equivalent shear wall. Then, the stiffness ratio of the shear wall to column is determined and a corresponding chart entered for either a uniformly or triangularly distributed lateral load. From the example chart (Fig. 10-30) the percentages of the base shear resisted by the frame at every level can be read and then translated into actual forces in the equivalent frame. Subsequently, these

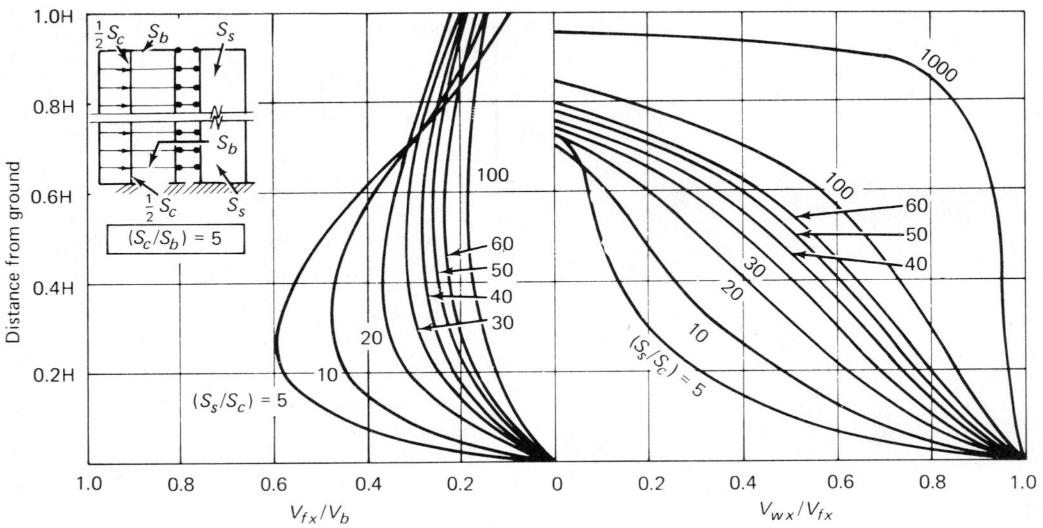

Fig. 10-30 Shear wall-frame lateral load distribution curves.

forces of the equivalent frame are distributed to the component elements from which the equivalent frame was composed. The left side of the chart in Fig. 10-30 gives the percentage of the lateral load at any particular level resisted by the shear wall. Thus, at a glance, the effectiveness of the shear wall within the system can be evaluated.

With the forces determined, the moments and the deflections of the frame and the shear wall are computed; the deflections of the two systems must obviously be the same if the procedure was carried out correctly.

An alternate method can be used with the information given in the same paper by reversing the order of the procedure. A family of deflected shapes for the interactive system is given for the various stiffness ratios between the shear wall and column and for the different loadings (see Fig. 10-31 for an example chart*). From the known deflected shape of the frame and the shear wall, the moments and forces can be computed that would cause these deflections.

*Both example charts are for a column-to-beam ratio $S_c/S_b = 5$ for the equivalent one-bay frame and for a uniform wind loading.

Although the procedure[10-11] may seem tedious by present-day standards, it offers the advantage of giving a clear physical understanding of the behavior when following through the computational techniques. If organized in a tabular form, it does not present insurmountable difficulties in an office routine. In the 1960s this method was the most popular with designers for manual handling of the shear wall–frame interaction problem.

Another method based on a number of simplifying assumptions was published by MacLeod in 1970[10-27] with the purpose of offering a less time-consuming procedure for office use. Its results are slightly less accurate; however, they are more than sufficient for designing the average building of average height.

In buildings with extremely large shear walls the frames take only a small proportion of the lateral load; a simplified analysis may be adequate, or the shear wall may be assumed to resist all the lateral load. When further analysis is required, a computer program should be used. The analysis of frame–shear wall structures under lateral load using one of a number of computer programs now available is a fairly straightforward procedure.

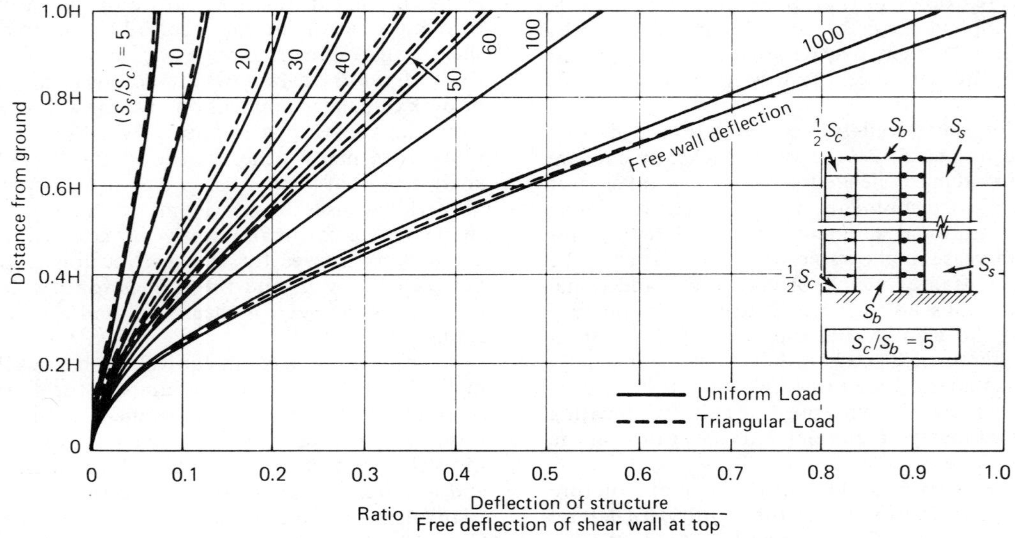

Fig. 10-31 Deflected shape of shear wall–frame interactive system.

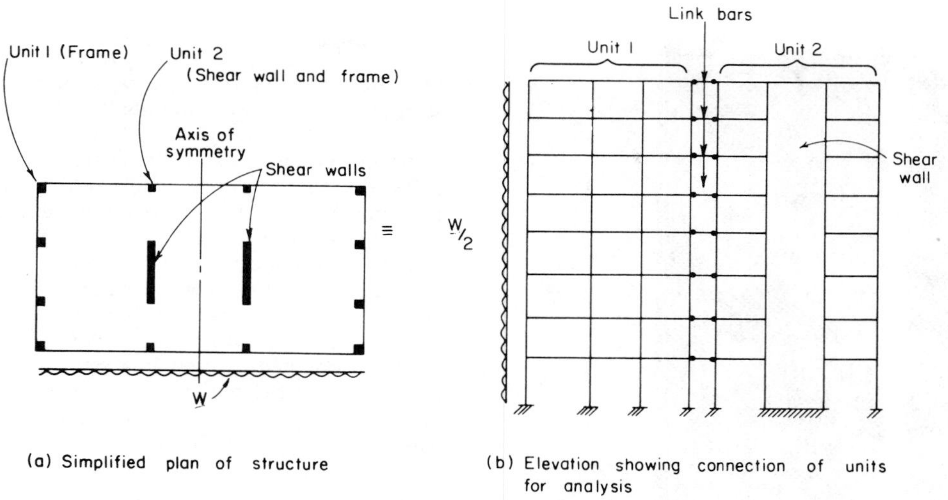

Fig. 10-32 Idealization for plane frame analysis.

Discontinuity of a shear wall (at both the top and botton) is usually accompanied by a shifting of large shear forces between supporting wall and columns through the slabs.

In frame–shear wall structures, neglecting the axial deformation in the columns will tend to throw a greater proportion of the lateral load to the frames, and underestimate the load carried by the shear wall.

Where torsion may be neglected, structures can be idealized as shown in Fig. 10-32 and plane-frame programs used, such as STRESS, STRUDL, PCA programs or the many other programs publicly available. Because torsion due to lateral loads can produce significant stresses, particularly in corner columns, a more-or-less symmetrical arrangement of the principal lateral load resisting elements with respect to the entire plan should be aimed for. In this connection, the torsional resistance of a structure would be much improved by locating the lateral load resisting elements where they do the most good, i.e., along the plan periphery of the building.

2. Shear Wall–Frame Interaction through Large Girders. A significant increase in lateral stiffness of tall buildings can be achieved by tying a centrally located shear wall with the peripheral columns through deep (usually story-high) flexural members at the top and possibly at other intermediate levels. Such linkage (outriggers), as in Fig. 10-33, mobilizes the longitudinal (axial) stiffness of the peripheral columns in resisting the lateral loads.

The first reinforced concrete building utilizing this concept was the 51-story Place Victoria Towers in Montreal (Fig. 10-34), constructed in 1964, in which the X-shaped core is linked at four levels (mechanical floors) by story-high girders to the heavy corner columns.

The U.S. Steel Building in Pittsburgh, completed in 1970,

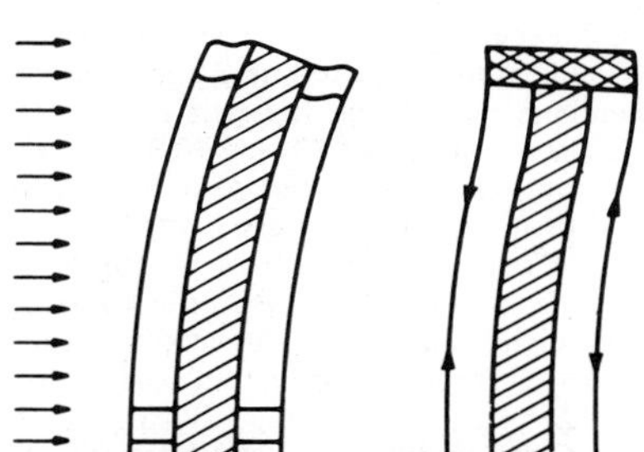

Fig. 10-33 Core connected with heavy girders at the top to exterior columns.

is another example with a heavy girder at the roof level connecting the core with the exterior columns.

The main objective is to cause the structure to act more as a vertical cantilever beam and so resist a larger proportion of the lateral loads by axial forces rather than by bending in the columns. Obviously, the axial load resistance represents the more efficient use of the structural material.

3. Application. The plan in Fig. 10-29(c) represents the typical modern hull-core type office building without interior columns and using the periphery either as a framed tube or with conventional column spacing. Nonetheless, many office buildings are still constructed with the traditional 25-ft span range in both directions. The concrete core is almost always present in office buildings as the backbone of the structural system. No other interior shear walls are used in office buildings, as they would interfere with the rentable space.

In apartment buildings, Fig. 10-29(a, b, d), the amount of interior shear walls depends on the architectural layout. Since shear walls fulfill the function of partitions or walls separating apartments, there is usually no conflict with the architectural layout of the typical floor, and the number of walls can be generous, thus providing sufficient rigidity.

Shear walls constructed with conventional forming techniques are more expensive than other types of walls, so there is usually a tendency to use the least amount of them. In addition, the walls may be an architectural obstruction, since even in apartment buildings where shear walls are part of the floor plan in the typical floor, they may create layout difficulties in the lower stories of the building in parking garages and lobbies. Also, in the case of future typical floor modifications, the presence of immovable shear walls may create difficulties. Experienced designers have a feel of how many shear walls are necessary for a particular building height and wind load intensity, based on their previous design experience and also observation of the performance of buildings. The final choice of the amount and location of shear walls usually represents a compromise between the architectural and structural requirements that meet the cost objective.

4. Amount and Size of Shear Walls. The least amount of shear walls structurally sufficient for a particular building is determined with the following considerations in mind.

(a) To provide adequate rigidity to meet the imposed deflection criteria.
(b) No tensile forces in the shear walls should result from

Fig. 10-34 The 47-story-high Place Victoria, Montreal, Canada.

the combination of gravity (dead) load and overturning wind moments. If there are resulting tensile forces, they must be resisted either by mobilizing load from adjacent columns using transfer girders (usually in the basement), or by anchoring the shear walls into the foundation medium. The economic consequences of the tensile forces should be investigated. If they are unavoidable, the size or number of shear walls needs to be increased.

Structures or elements of structures in which the ratio of wind stress to dead load stress is high are very sensitive to wind.

The most economical shear wall–frame structure (no premium for height) is achieved when there is a proper balance between the gravity load (dead load) and the overturning moment on each of the shear walls; ideally, if the overturning stresses can be accommodated within the 33% increase in the allowable gravity load stresses. Such a condition is achieved in shear wall structures where the shear walls carry the entire gravity load. At the other extreme, many buildings have only a central core as a shear wall that carries sometimes only a small portion of the dead load while resisting the majority of the overturning moments. If such buildings have a properly balanced shear wall-frame interaction, they may have sufficient rigidity to resist lateral forces. However, cases have been observed where a slipformed core had around it a one-bay frame in which the flat plate had neither sufficient connection to the exterior columns nor a moment connection to the core; the result was an intolerable flexibility of the total building in response to wind.

EXAMPLE 10-2: Shear wall-frame interaction. The objective of this example is to determine, through a step-by-step optimization, the minimum amount of shear walls required for a given apartment building. The typical half-floor plan, section and column sizes (36-story building) are shown in Fig. 10-35. The basic structural system denoted as structure "A" consists of a centrally located corewall extending throughout the entire height of the building and 10 three-bay open frames in the transverse direction. The assumed uniform wind load is 20 psf on the 60 × 220 ft building.

The structure was analyzed for wind using the computer program described in Ref. 10-19. Although structure "A" has more than sufficient stiffness (the computed drift is 1/850), the net tension in the extreme windward "fiber" at the base of the corewall is 700 psi.

To reduce this tension in the corewall, a 20-ft-wide, 24-story-high shear wall was introduced in an exterior bay of one of the open frames as shown in Fig. 10-36(a). This new structural layout was denoted structure "B". The analysis of this case (see Table 10-3) shows that although the tension at the base of the corewall is reduced significantly—to 395 psi from 700 psi in "A"—substantial tensile stresses occur at the base of the added shear wall (515 psi). This indicates that the added shear wall, due to its stiffness, attracted a high moment relative to its dead load—resulting in significant net tensile stresses.

In an effort to further reduce the tension at the base of the corewall a third structure, structure "C," with a pair of 20-ft-wide, 18-story-high shear walls along an exterior column line, Fig. 10-36(b), was next analyzed. As might be expected, this further stiffened the structure, bringing the drift (deflection index) down from 1/873 for structure "B" to 1/928. In addition, the stress at the base of the corewall was reduced to the point where a net compressive stress of 30 psi occurs in the extreme windward fiber. However, the tensile stresses at the base of the additional shear walls have increased to 615 psi—from 515 psi in structure "B."

When the shear walls in structure "C" are assumed to be located along an interior column line, such as line 2 in Fig 10-36(b), the net tensile stress at the base is reduced from 615 psi to 192 psi (as shown for the case "C-1" in Table 10-3). This is due to the added dead load on the shear wall. In structure "C" and "C-1" only the slab strips were considered as linking the additional pair of shear walls

Structure "D," a fourth structure considered, is essentially the same as structure "C-1" except that beams were introduced to link the additional shear walls along column-line 2 such that the stiffness of the coupling elements connecting the pair of shear walls is three times that of the slab strips in "C." Table 10-3 indicates that the increase in stiffness of the coupling between the pair of shear walls not only increased the compressive stress at the base of the corewall, but also slightly decreased the tensile stress under the additional (coupled) shear walls. The decrease of tensile stress in the coupled shear walls is only slight, since the reduced shear wall moments are accompanied by axial forces (tension and compression) resulting from coupling.

It is obvious that the tensile force resulting from the net tensile stresses has to be either anchored into the foundation material, or it must be shifted with the help of shear beams to the neighboring columns to be overcome by their gravity loads. If the tensile load

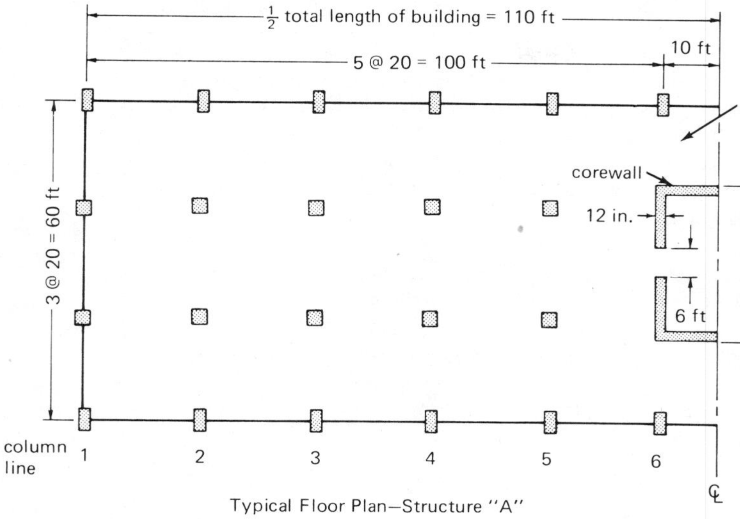

Column Sizes

Story Level	Interior Columns	Exterior Columns
1 – 12	22 in. square	16 in. X 20 ft
13 – 24	25 in. square	16 in. X 26 ft
25 – 36	30 in. square	16 in. X 36 ft

Shearwall is 12 in. thick throughout entire height of building.

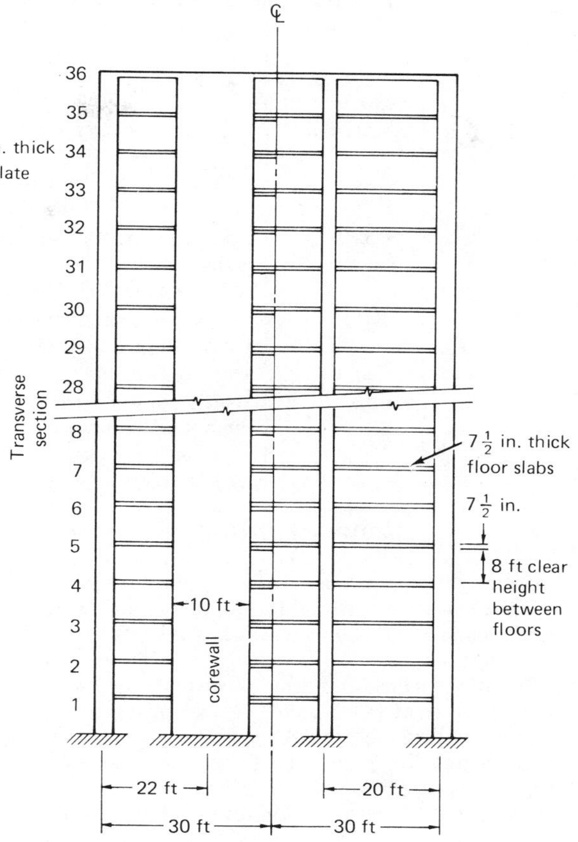

Fig. 10-35 Sturcture "A"—plan, section, and column sizes.

cannot be accommodated, the shear wall will rotate at the base, resulting in a different distribution of wind shears and moments throughout the structure.

Figure 10-37 shows the variation of the story shears in the corewall along the height of the building for the four cases considered, plotted in terms of the percentage of the total applied story shear. Note the relatively abrupt changes in the magnitude of the story shears at locations where changes in stiffness occur. Also shown in the figure is a tabulation of the overturning moments at the first floor resisted by the corewall in terms of percentages of the total overturning moment. It can be seen from the figure that in the upper stories the corewall has negative shears due to frame–shear wall interaction—which means that the internal interactive forces cause a considerable reduction of the overturning moments in the corewall. As can also be seen, the corewall in structure "A" carries 82% of the shear at the first floor, while carrying only 33% of the total overturning moment at this level; the remainder of the overturning moment is resisted by the overturning of the frame.

10.5.8 Shear Panel Buildings

Regular shear walls extending throughout the height of the building force a discipline on the architect, and it may be

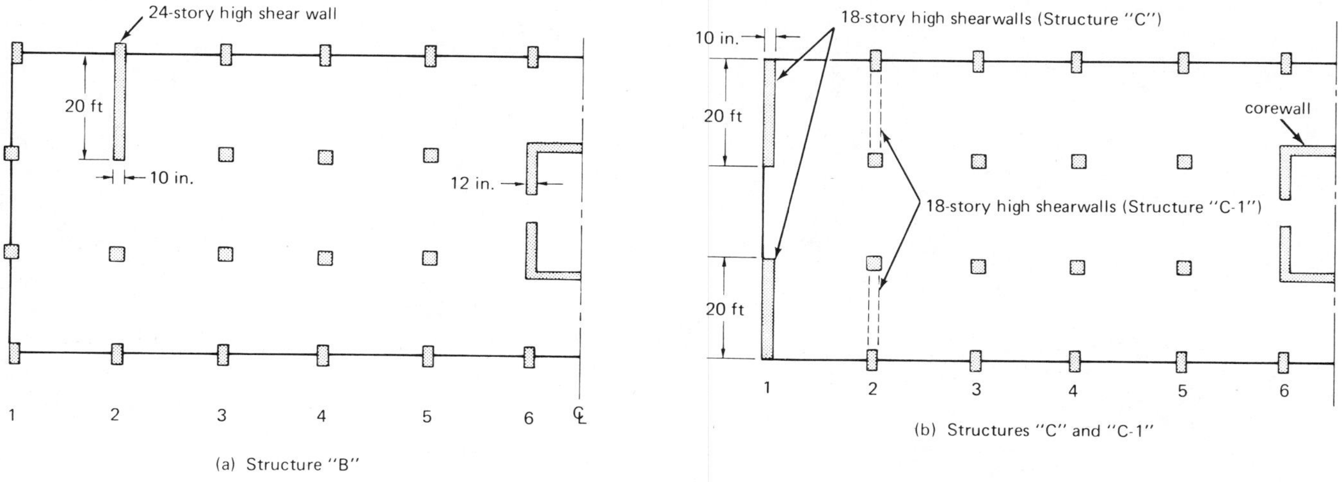

Fig. 10-36 Typical floor plans for structures "B" and "C."

TABLE 10-3 Summary of Results of Lateral Load Analysis of 36-Story Apartment Building

Structure designation	*Description*	*Drift*	*Stress in extreme fiber-windward side*	
			Corewall	*Additional shear wall*
A	Basic Structure 1 frame-shear wall 4 open frames	1/820	+700 psi (tension)	—
B	Basic structure "A" with one open frame replaced by frame shear wall (w/single 20′ wide, 24-story high shear wall)	1/873	+395 psi	+515 psi
C	"A" w/one open frame replaced by frame-shear wall (w/2-20′ wide 18 story high shear walls on column line 1)	1/928	–30 psi compression	+615 psi
C-1	Same shear walls as in "C" moved to interior column line 2	1/928	–30 psi compression	+192 psi
D	"C-1" with beams linking 18-story shear walls 3 times as stiff as in "C"	1/992	–130 psi	+180 psi

**Considering effect of dead load only.*

difficult, in some cases, to accommodate a suitable mix of apartment sizes required by the developer. A new system of shear panels has been introduced recently to provide more layout flexibility.

A shear panel building is defined as containing shear panels of reinforced concrete extending on one or several stories within the height of the building and scattered throughout the plan. The panels are used mainly as walls separating apartments. In lower buildings, say up to 15 stories, the same function of limiting the drift could be accomplished by clay or concrete masonry panels, if they fit tightly within the frame. It should be noted that in a shear panel building, resistance to lateral loads is provided primarily through their shear resistance. The panels can accept only limited overturning moments, since they have no vertical continuity throughout the height of the building. In this respect, shear panels are shear walls in the true meaning of the word, while continuous shear walls in tall buildings resist the lateral forces predominantly by moment resistance.

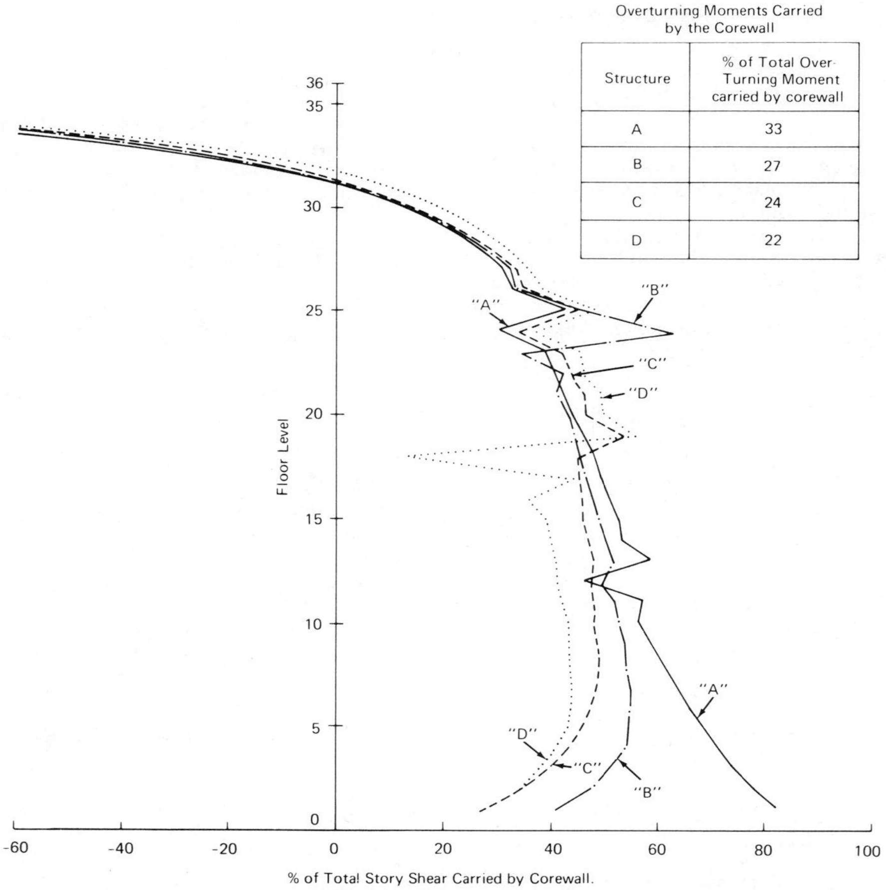

Fig. 10-37 Comparison of shears and moments in the corewall due to wind.

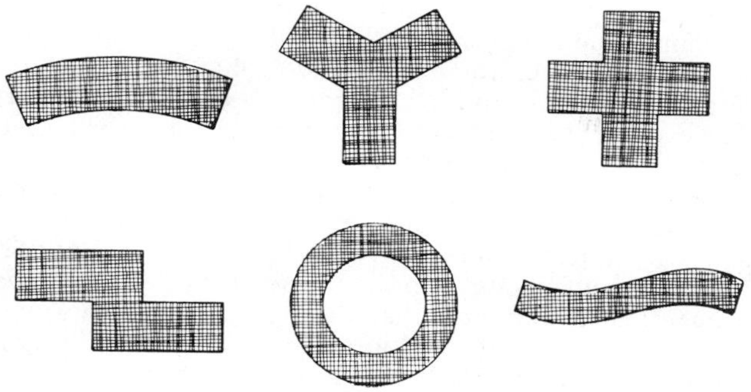

Fig. 10-38 Shapes of buildings affecting lateral resistance.

It is desirable to locate shear panels between columns so that their weight can be carried directly by the columns and not by the slabs in bending. If located between columns, they act as deep beams supporting their own weight and the slabs.

At tops and bottoms of shear panels, the slab acting as a membrane transfers shear from panel to panel. Thus, the slabs are subjected in their planes to large forces at every discontinuity of a shear panel. These forces may require reinforcement of the slab as a horizontal plate. If the panel layout is not symmetrical in plan, the torsion of the building should be considered.

Fig. 10-39 Toronto City Hall.

Since no shear wall–frame interaction can take place to any significant degree, the effectiveness of shear panels may be comparatively low.

Shear panels have been used on several flat plate apartment buildings in the 20-story range. Fulfilling the double function of a wall and a lateral resisting element for the building, they may be economically attractive, particularly if prefabricated in a precasting yard or on site and erected as the casting of the slabs progresses. Further studies are required on this potentially advantageous system before it can be widely utilized.

10.5.9 Lateral Resistance Contributed by the Shape of the Structure

It is well known that the shape of structures (other than rectangular blocks) has a substantial effect on their lateral resistance. Figure 10-38 shows a number of building shapes that, by the nature of their geometry, are assumed to increase lateral resistance compared to rectangular shapes. Until quite recently, this subject was still elusive to the structural designer, since in the majority of cases the analysis was based on two-dimensional plane frames, usually lumped together for computer-time economy. It is only with the recent availability of economical, three-dimensional computer programs that such structures can be analyzed efficiently, and it is possible to optimize the structural system by using auxiliary parametric studies of subassemblies. The fact that most of the prestigious buildings of the eighties are nonrectangular may be attributed in part to the ability of engineers to analyze them efficiently using three-dimensional analyses.

A structure in which the shape was the prime source of lateral resistance is the City Hall of Toronto (Fig. 10-39). The two towers were analyzed as vertical cylindrical shells stiffened by the slab system.

Fig. 10-40 Caromay Building after the 1967 earthquake in Caracas, Venezuela.

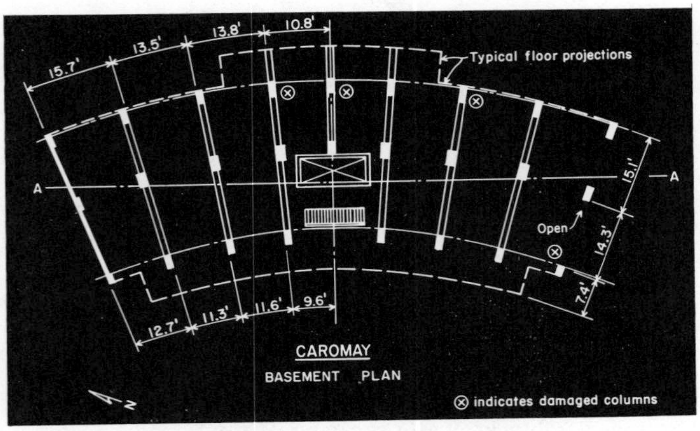

Fig. 10-41. Caromay framing plan.

The example of the curved 19-story Caromay building in Caracas, Venezuela (Fig. 10-40) is only one of many cases of structures that, during an earthquake, exhibited the influence of the shape on the lateral resistance. The structural system of the Caromay consists of radial frames as shown in Fig. 10-41, spanned between the frames by a light joist floor with hollow clay tile fillers. Although the building was designed with radial frames, it behaved as a vertical cylindrical shell as a result of its shape. The motion of the building was perpendicular to axis *A-A* (Fig. 10-41). As a result, the columns located farthest from the neutral axis of the building were subjected to high axial forces due to racking and failed in classical compression at their mid-heights. No evidence of frame moments could be seen at the tops and bottoms of the columns. These observations led to the conclusion that the lateral resistance of this structure was due to shell action, and no frame action (for which the building was designed and reinforced) took place.

10.6 DESIGN FOR VERTICAL LOADS

Although the lateral resistance is what constitutes the difference between a tall building and a low building, the design for vertical load is nevertheless the basic aspect of engineering, even for tall structures, in its effect on the economy of the project. Proper column layout that leads to economical spans while not impeding the functional requirements of the building is the key to an economical slab system to support the gravity loads. The slab system constitutes between 60 and 85% of the structural cost, depending upon the height and function of the building.

10.6.1 Analysis for Vertical Loads

Although computers have considerably advanced the sophistication of our designs for lateral loads, the soundest approach to design slab systems for gravity loads is the traditional method of considering each floor slab with its columns above and below assumed fixed at their remote ends. This procedure, which has been in use for a number of decades, proved itself both technically valid and easy to handle in a design office. Most of the available procedures for the design of the various slab systems are based on these assumptions.

Although many of the comprehensive computer programs can be used for gravity load analysis, there is an inherent susceptibility to receiving misleading results when such an analysis is carried out, since these programs are based on the input of the entire structural system with its loading, and then an analysis is performed for the entire structure as one unit. The dead load in a real building, however, is built up gradually and, for example, in a 40-story building, a dead load of the 10th floor cannot be resisted by a 40-story frame, since at the time the 10th floor load is applied, there is only a 10-story frame available to resist this load, and not a 40-story frame. Therefore, the procedure of simultaneous analysis of an entire structure is correct only for live loads and loads applied after the structure is completed. For dead loads a simultaneous analysis is correct only if all vertical elements of the structure have identical stress levels. Simulteneous loading of the structure can lead to large errors in cases where neighboring columns are designed to different stress levels that result in differential elastic shortenings. For example, if we consider a flat plate slab or a continuous beam in the 50th story, supported by a highly stressed column while the neighboring columns are lightly stressed, the differential elastic shortening over the height of the structure will result in the slab not having any negative moments over the highly stressed column; on the contrary, the analysis will show substantial positive moments over the column. In reality, when the 50th floor slab is cast, all the elastic shortening of its supports due to gravity from 49 slabs has already taken place, and no elastic settlement stress will affect the slab when it is cast. However, the slab will be subjected to elastic differential support shortening for all the dead loads that will be applied above the 50th floor.

Similarly, if a transfer girder is designed with the entire structure above considered as a vierendeel girder in a simultaneous analysis, the load on the transfer girder will be underestimated, since the lower story loads are carried mainly by the girder (no frame above is yet available at the time they are cast) and not by the entire vierendeel. (See Section 10.6.5)

Computer programs need to be developed that will treat construction time as an additional variable, which means the vertical load will be considered as applied gradually story by story while at the same time the framework is being progressively built up. Until the time when such programs are available, the traditional method of designing one story at a time produces much more reliable results.

10.6.2 Selection of Slab Systems

The choice of a slab system for vertical loads depends upon length of span, loading intensity, and the function of a building. As discussed in a previous section of this chapter, the need for a smooth ceiling in residential occupancies (apartments, hotels, dormitories) resulted in the flat plate being the most prevalent and economical slab type. Also, post-tensioned flat plates and flat slabs received wide usage in recent years, in particular for longer spans in apartments, commercial uses, and parking garages.

In office buildings, where a hung ceiling is used to cover mechanical and electrical ducts, a ribbed ceiling (either one-way joists or waffle slabs) may be economical.

Figures 10-42 and 10-43 show preliminary selection charts for flat plates, flat slabs with and without column capitals, and one-way joists. Figure 10-44 gives a preliminary selection chart for thicknesses of post-tensioned flat plates and flat slabs. While the previous selection charts are plots of thicknesses as functions of span versus load, Fig. 10-44 gives the thickness as a band versus span for the normal range of loadings. These selection charts are based on the long-time deflection limit of $L/360$ stipulated by the *Code*. The charts can be used for preliminary sizing only.

A survey of 36 office buildings[10-1] located throughout the United States and Canada featuring long spans (between 35 and 45 ft) showed that both reinforced and post-tensioned

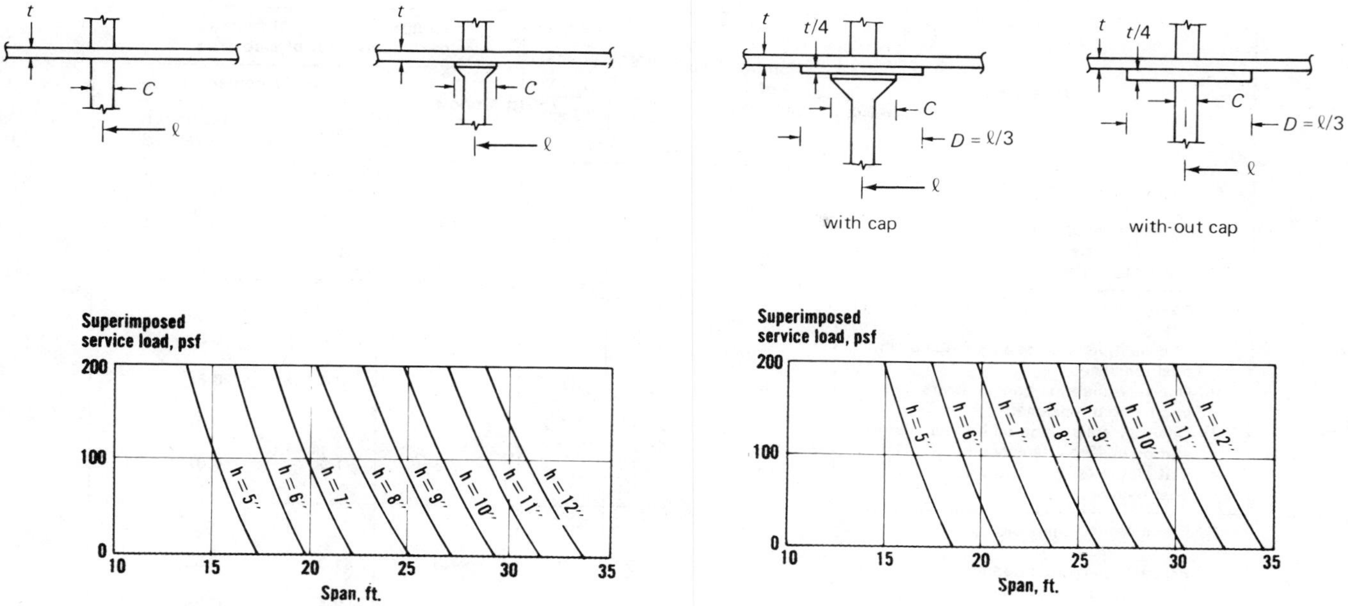

(a) Flat Plates—Square Bays

(b) Flat Slabs—Square Bays

Fig. 10-42 Preliminary selection charts for flat plates and flat slabs.

one-way joists, flat plates and flat slabs, and one-way slabs on beams are the most commonly used slab systems for cast-in-place concrete. The survey showed that the increase in structural material quantities for span increases from the traditional 20 ft to the 30–40 ft range requires adding only about 1 in. of concrete and 1.5 psf of steel. A summary averaging the material quantities for the slabs only (columns and walls not included to eliminate the effect of number of stories) is given in Fig. 10-45. The cast-in-place structures surveyed had slab systems of 4000 psi normal weight concrete. The reinforcing steel with a strength of 60 ksi was used, and post-tensioning steel had a strength of 270 ksi. These results are based on a limited sampling of several buildings per category, and should, therefore, be interpreted with caution, considering the large number of variables affecting specific designs (such as local codes, local construction practice, live load requirements, etc.).

10.6.3 Columns

Column loads are customarily determined on the basis of tributary areas, despite the fact that such effects as differential creep, differential elastic shortening, high differences in adjacent negative moments, and creep of the slab system may sometimes considerably upset the distribution by tributary areas.

The amount of column reinforcement has a very dominant effect on the economy of high-rise reinforced concrete buildings. Therefore, the lowest possible percentage of vertical reinforcement should be attempted. This can be achieved either by increasing the column size (if there is no conflict with architectural requirements), by increasing the concrete strength (up to 11,000 psi concrete has already been used), or by both. High strength reinforcement with a yield strength of 60 ksi is almost exclusively used for columns. Premiums sometimes have to be paid for the use of 75 ksi steel if the material is not available in stock.

Quite often in the upper stories of buildings the columns are much larger than required to carry the load. The oversized columns result either from architectural requirements or from economic considerations to have as few changes in the column section as possible. In such cases the *Code* stipulates that the percentage of reinforcement be less than the required minimum of 1% by considering only that part of the column section needed to carry the load.

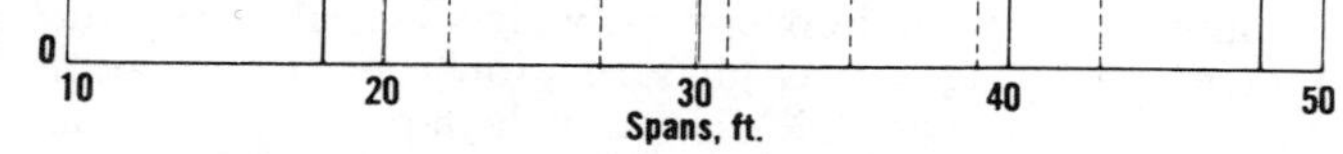

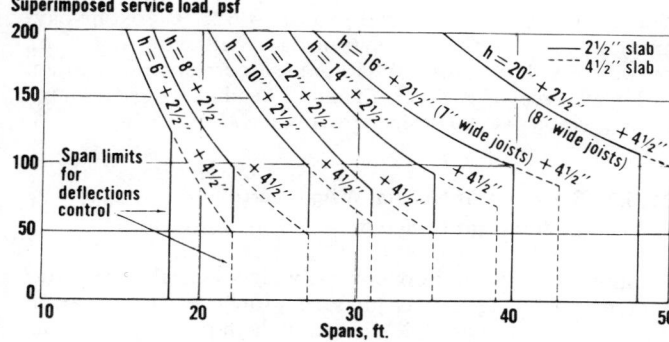

Fig. 10-43 Preliminary selection chart for one-way joist slabs.

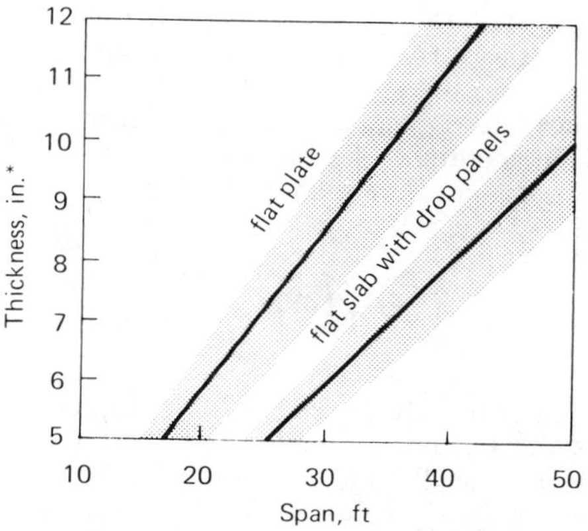

Fig. 10-44 Preliminary selection chart for post-tensioned flat plates and flat slabs.

Slab system	Number of buildings	Average spans, ft		Average quantities of materials per square foot of slab area		
				Concrete, in.	Reinforcement, lb	
		Slabs	Beams		Mild	Post-tensioned or prestressed
A. Cast-in-Place Nonstressed						
1. Conventional spans	2	19.3		8.2	2.7	
2. One-way joists on beams	8	37.8	31.6	8.5	4.3	
3. Flat plates (slabs)	2	27.5		9.6	3.8	
B. Cast-in-Place Post-Tensioned (p/t)						
1. One-way joists on beams, beams only post-tensioned	2	30.0	30.0	7.5	1.7	0.30
2. One-way joists on beams, joists and beams post-tensioned	2	40.7	23.1	6.9	2.0	0.82
3. One-way solid slabs on beams, slabs and beams post-tensioned	7	22.7 clear	38.9	8.4	2.1	0.87
4. Flat plates (slabs), two-way post-tensioned	3	30.7		8.3	1.0	0.87
C. Precast and Composite						
1. Precast, slabs and beams prestressed (ps)	4	34.3	32.4	7.5	1.2	0.63
2. Precast ps thin-slabs-composite	3	22.4 clear	28.3	7.8	2.4	0.38
3. Precast ps joists-composite	2	37.0	34.0	6.9	2.9	0.56

Fig. 10-45 Summary of average spans and structural quantities for long spans (from survey[10-1]).

10.6.4 Live Load Reduction for Columns, Beams, and Slabs

Most general building codes permit a reduction in live load for design of columns, beams, and slabs in structures to account for the probability that the total floor area "influencing" the load on a member may not be fully loaded simultaneously. Traditionally, the concept of reducing the amount of live load for which a member must be designed has been based on a tributary floor area supported by that member. With publication of the 1982 edition of the American National Standards Institute (ANSI) A58.1,[10-3] the concept of live load reduction is revised significantly to reflect an influence area rather than a tributary area for evaluating the amount of reduction permitted. For example, for an interior column the influence area is the total floor area of the four surrounding bays (four times the traditional tributary area). For an edge column, the two adjacent bays "influence" the load effects on the column, and a corner column has an influence area of one bay. For interior beams the influence area consists of the two adjacent panels, while for the peripheral beam, it is only one panel. For two-way slabs, the influence area is equal to the panel area.

The reduced live load, L, per square foot of floor area supported by columns, beams, and two-way slabs having an influence area of more than 400 ft^2 is:

$$L = L_0 \left(0.25 + \frac{15}{\sqrt{A_1}}\right)$$

where L_0 is the unreduced design live load per square foot, and A_1 is the influence area as described above. The reduced live load shall not be less than 50% for members supporting one floor, or less than 40% of the unit live load, L_0, otherwise. No reduction of live loads can be made for places of public assembly, for parking garages (except for garages limited to passenger cars only), and for roofs.

Using the above ANSI expression for reduced live load, the reduction multiplier (RM) as a function of influence area is plotted in Fig. 10-46. In multistory buildings, influence areas for columns supporting more than one floor are summed.

The live load reduction multiplier for beams and two-way slabs having an influence area of more than 400 ft^2 ranges from 1.0 to 0.5. For influence areas on these members exceeding 3600 ft^2, the reduction multiplier of 0.5 remains constant.

The live load reduction multiplier for columns in multistory buildings ranges from 1.0 to 0.4 for cumulative influence areas between 400 and 10,000 ft^2. For influence areas on columns exceeding 10,000 ft^2, the reduction multiplier of 0.4 remains constant.

For illustration, the live load reduction multiplier for columns of a 7-story building with 25-ft square bays has been tabulated in Table 10-4. It can be seen that for interior columns the maximum reduction has been reached at the third story. If the slab system consists of two-way slabs with beams between column supports, then the interior beams are designed with an $RM = 0.674$ (A_1 = 1250 ft^2 for two-bay areas); for the peripheral beams and the two-way slabs, an $RM = 0.850$ (A_1 = 625 ft^2 for one-bay area) is used. If the slab system is a flat plate, flat slab, or waffle slab, an $RM = 0.850$ is used corresponding to an influence area of 624 ft^2. Spandrel beams in this framing, if any, would also be designed with an RM of 0.850.

10.6.5 Selection of Normal Weight versus Lightweight Concrete

Structural lightweight concrete is widely available, and the present knowledge of its physical properties and behavior has made it a desirable factor in high-rise construction

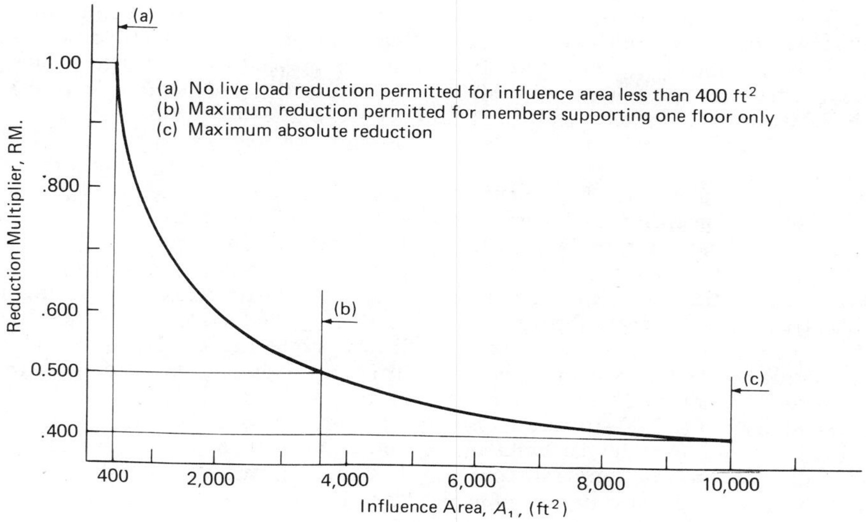

Fig. 10-46 Reduction multiplier (RM) for live loads $\left(0.25 + \frac{15}{\sqrt{A_1}}\right)$

under some circumstances. It is used in many and varied construction applications; however, for multistory construction its use may be particularly attractive. The use of normal weight versus lightweight concrete in multistory construction involves the study of several variables. In most areas of the country the lightweight concrete used in buildings has a weight of 110–115 lb per cubic yard, since the lightweight fines are replaced by natural sand.

The reduced dead load resulting from using lightweight concrete produces structural advantages such as: (a) a reduction of sizes of flexural members, columns, and foundations; (b) equivalent fire ratings obtained with thinner lightweight concrete sections; and (c) lower inertia forces in earthquake design.

The use of lightweight concrete for high-rise buildings was extremely popular in many cities of North America during the 1950s and 1960s. At that time the cost differential between normal and lightweight concrete was very low. In New York, for example, at times there was no cost premium for lightweight concrete, and in some metropolitan areas the cost differential was about $2.50 per cubic yard of concrete. As a result, many high-rise buildings were constructed with lightweight concrete.

With the drastic changes in energy costs following the oil embargo of 1973, prices for lightweight aggregates have substantially increased because most of them are manufactured by heat processes. As a consequence, the use of lightweight concrete in high-rise construction has declined.

Under the present circumstances, the weight reduction (when using lightweight concrete) must produce substantial savings in slab and column reinforcement to warrant the cost premium.

10.6.6 Transfer Girders

In recent years many buildings have been constructed with mixed occupancies, such as apartments or hotels in the upper stories while the lower stories contained commerical space, theaters, schools, or other nonresidential space. Additionally, in large metropolitan locations most of the new buildings contain parking garages in the lower stories.

TABLE 10-4 Live Load Reduction Multipliers (*RM*) for Columns of a 7-Story Building (25-ft square bays)

	Interior Columns		*Edge Columns*		*Corner Columns*	
Story	A_1	*RM*	A_1	*RM*	A_1	*RM*
7 (Roof)		*		*		*
6	2500	0.550	1250	0.674	625	0.850
5	5000	0.462	2500	0.550	1250	0.674
4	7500	0.423	3750	0.495	1875	0.596
3	10000	0.400**	5000	0.462	2500	0.550
2	12500	0.400**	6250	0.439	3125	0.518
1	15000	0.400**	7500	0.423	3750	0.495

**No reduction permitted for roof live loads.*
***Maximum reduction permitted.*

Often closely spaced peripheral columns of multistory buildings cannot continue within the ground story where wide open spaces are required in lobbies and entrances. In other cases buildings are constructed over highways or railroads, and the normally laid-out columns of a multistory building must be transferred to a system of columns compatible with the highway or railroad.

Functional considerations require different spans for the different parts of the building. On the other hand, cost per square foot rises with length of span. If there are many floors of shorter spans to be constructed over an area with larger spans, it is advisable to investigate the technical and economical feasibility of a transfer floor separating the two different structural systems.

There are numerous examples of buildings with transfer girders in reinforced concrete, prestressed concrete and structural steel. In one example an entire 60-story residential building has been placed over a commercial building with only a few of the upper columns continuing directly through the lower part, while most of the 60-story columns were supported by a story-high grid that, in turn, rested on columns laid out to the requirements of the commerical space.

In general, transfer of substantial loads is a costly operation and should be done only after careful consideration. The least costly transfer can be accomplished by using inclined columns within the transfer floor, as shown in Fig. 10-47. In this particular case, each of the quadrupeds "collected" four columns of the 10-story office building into one column below the ground floor. It is obvious that such transfer of forces is accompanied by tensile and compressive horizontal components in the slabs at both ends of the inclined columns. These forces, which can be very substantial, have to be accommodated within the slabs. Post-tensioning offers a very effective means, both technically and economically, to accommodate such tensile forces.

The usual way to transfer column loads is through transfer girders, sometimes one or several stories high, to give them sufficient rigidity not only to reduce the elastic deflections, but also to minimize the creep deflections that would cause distortions within the entire building above. Shear and creep considerations are important design aspects to keep in mind when designing transfer girders.

The transfer girders are usually designed to carry the entire dead and live load of the columns above. This condition of full load can actually occur only if the structure above consists of simply supported members. In reinforced concrete, where all members are usually continuous, the live load is carried not only by the transfer girder but by the entire structure above acting together with the transfer girder as a vierendeel structure. However, in supporting the dead load, participation of the structure above is only partial in sharing the load with the transfer girder, since the dead load is applied gradually, as the structure is being built up. Thus, for example, the load of the lower 10 stories of a 60-story frame resting upon a transfer girder is resisted by a structure consisting *only* of the transfer girder and 10 stories above (not 60 stories). Therefore, to use a frame program to simultaneously handle all the vertical loads for vierendeel action of the transfer girder with the entire structure

Fig. 10-47 Inclined columns as a means of changeover from regular to large spans (Australia Square Office Building in Sydney, Australia).

Fig. 10-48 Some of the ground-story columns have been eliminated and the entire elevation participates as a vierendeel girder (Lincoln Park Building).

will lead to an underestimated load on the transfer girder. Tables and curves to estimate the load carried by transfer girders carrying multistory columns with consideration of the above-mentioned loading history were prepared in Ref. 10-28.

Figure 10-48 shows an elevation of a 32-story building in which some of the ground floor columns have been eliminated and the entire elevation designed as a vierendeel girder. It is obvious that in such a case spandrel beams in a number of lower stories are subjected to extremely high shears to transfer the load gradually to the column below; consequently their shear reinforcing must be carefully designed, or, rather, overdesigned.

Figure 10-49 shows a "natural solution" for the problem of shear transfer by the spandrel beams of the lower stories, resulting in a tree-like appearance—the support columns branch out into the closely spaced columns of the structure above. The thickness changes of the spandrel beams followed the results of the analysis, while the shear stresses were kept about constant.

Owing to its size, the construction of a transfer girder may pose field problems. For example, the transfer girder of the Brunswick Building in Chicago (Fig. 10-50), which is about 7 ft thick by 26 ft high, had to be cast in three horizontal layers to avoid excessive generation of heat of hydration. The decision of such unusual horizontal layering of a girder was made only after a series of tests verified the validity of this procedure. The actually measured temperature of the concrete in the massive girder never exceeded 155°F.

10.7 STAGGERED WALL-BEAM STRUCTURES

This new structural system developed for both structural steel and reinforced concrete was introduced in the 1960s.[10-29, 10-30] The basic concept is illustrated in Fig. 10-51. The system employs story-high, pierced walls extending across the entire width of the building to lines of columns placed along the exterior faces. By staggering the locations of these wall beams on alternate floors, large clear areas are created on each floor, yet the floor slabs span only half the spacing between adjacent wall beams from the top of one to the bottom of the next. Within certain structural limitations, the wall beams can be pierced for corridor and door openings at locations required by architectural planning.

The staggered wall-beam system is best suited for rectangular plans, but it may be adapted almost equally to other plan shapes, e.g., Y-shape, cruciform, annular, serpentine, or broken rectangle. The system is suitable for most types of multistory construction having permanent interior partitions such as apartments, hotels, student residences, etc. Although the system was initially developed for cast-in-place construction, it can very well be built with precast elements for both the walls and the slabs, or either. Obviously very careful planning must take care of the fact that the main reinforcement for the beams is located in the slabs.

A great advantage of the wall-beam building is the ease with which a large open area can be created underneath for parking, for commercial use, or even to allow a highway to pass under the building. If such clear areas are desired, a nontypical slab will be created for the slab above the open space.

It is possible to maintain the basic module of say, 12 ft as the span of the structural slabs and exterior column spacing and still achieve different clear spaces from those in Fig. 10-51. Figure 10-52 shows some of the many possible variations in wall-beam layout to accommodate the architectural planning of various apartment units within a building. Although the wall-beam layouts in Figs. 10-52(a) and 10-52(b) are shown as regular both horizontally and vertically, the pattern can be changed along the length of the building to accommodate various combinations of apartments on any one floor. Examples of such variations are shown in Figs. 10-52(c) and 10-52(d). Thus, considerable

Fig. 10-49 The natural look of the transfer to a limited number of columns in the ground story.

Fig. 10-50 The Brunswick Building in Chicago.

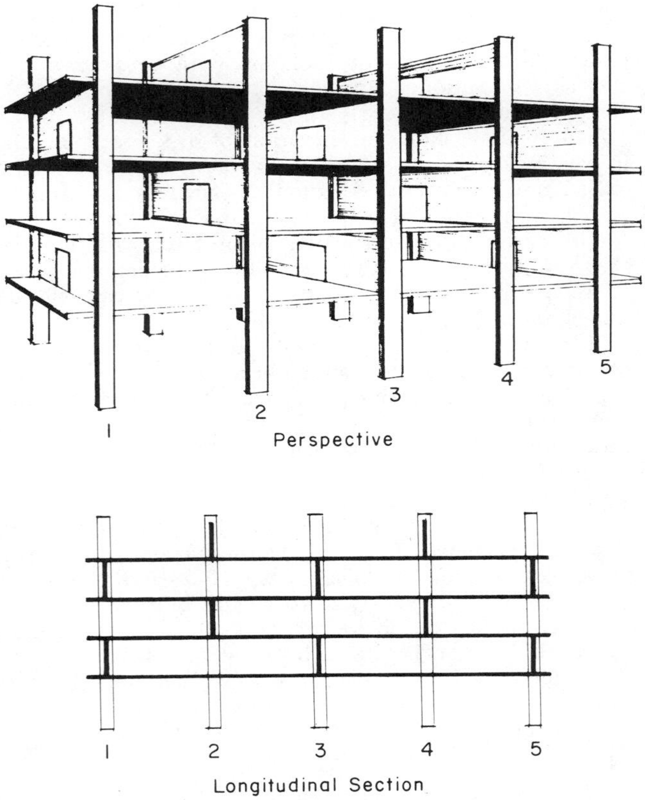

Fig. 10-51 Regularly staggered wall-beam building.

latitude in arrangement of units is possible within the basic structural requirements of the system, namely, wall beams on alternate floors at each column line (to keep floor slab spans on the basic module) and staggered with respect to those in the other—though not necessarily adjoining—column lines (to provide lateral rigidity).

10.7.1 Architectural Design

It can be seen from Figs. 10-53(a) and 10-53(b) that the wall-beam system will always have two typical floors, the layout on any one floor being repeated two floors above and two floors below. The existence of two separate typical floors makes it easy to create attractive building exteriors of either very simple or highly complex design. This

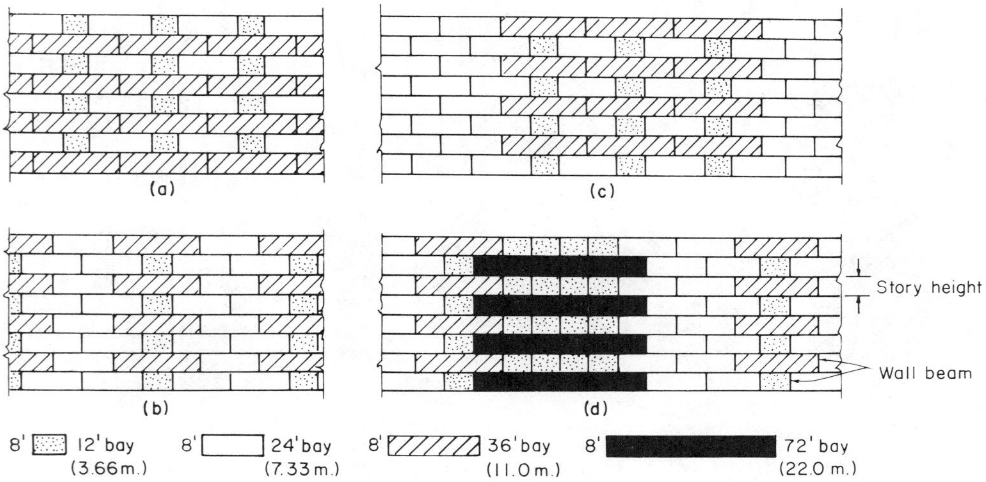

Fig. 10-52 Longitudinal section showing possible wall-beam layouts based on a 12-ft basic module.

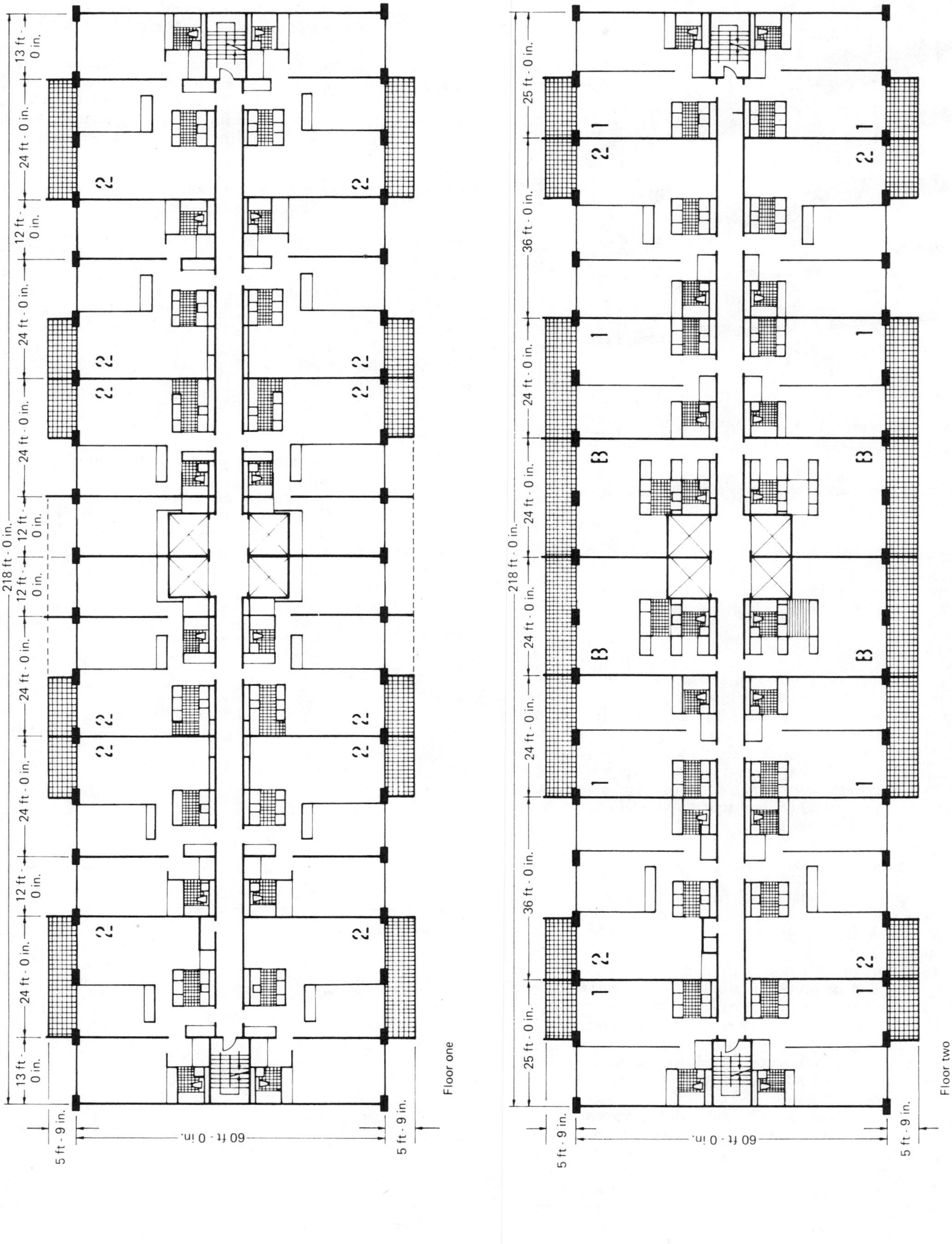

Fig. 10-53 Two typical floor layouts for staggered wall-beams.

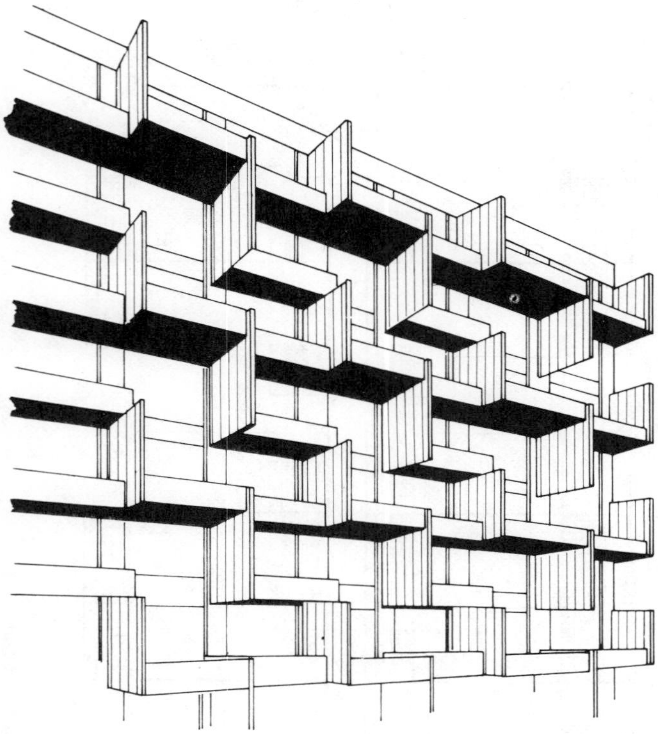

Fig. 10-54 A possible facade treatment for two-story units with balconies.

is true particularly with cantilevered balconies, as shown in Fig. 10-54.

The wall-beam is well suited for two-story apartment units to separate living and bedroom areas.

10.7.2 Mechanical Design

Since the apartment units on consecutive floors are staggered, more vertical plumbing lines are required, as kitchens and bathrooms are above one another on alternate floors and not consecutive floors. A detailed study of the plumbing, ventilating, heating and air conditioning requirements associated with an apartment building using the staggered wall-beam system shows that they can be satisfied at a cost only slightly higher than that required for a flat plate or a shear wall building.

10.7.3 Structural Analysis and Design

In the longitudinal direction, the floors act as continuous one-way slabs either resting on or hung from alternate wall beams. The slabs, in combination with the wall beams, form a concrete I-beam. The alternating or staggered system of beams results in each floor slab being alternately subjected to tension and compression forces in the transverse direction as it forms first the bottom and then the top flange of wall beams. As a result, shearing forces are created in the horizontal plane of the floor slabs that tend to reduce the external bending moment applied to the individual wall beam. These beneficial shearing forces are not considered in the analysis and design.

Virtually all wall beams will contain openings. The presence of an opening reduces the shear capacity of the beam at that section. Since the lower slab is relatively flexible, it can be assumed that the beam over the opening carries the entire vertical shear. The action of gravity loading on the wall beams creates a state of combined bending and axial compression in this beam.

In buildings of, say, less than 12 stories, owing to the relatively flexible columns, the wall-beams may be analyzed as simply supported beams.

Where the column stiffness is of a similar order of magnitude to that of the wall beam, it may be desirable to consider the restraining effect of the columns, particularly since this can result in a reduced area of reinforcement required at the center of the beam span. Expressions for the rotational stiffness of a wall beam with openings in its web and worked-out examples illustrating the analysis and design of a typical wall beam have been provided in Ref. 10-31.

10.7.4 Simply Supported Wall Beams with a Single Opening in Web

The analysis for the moments and forces in the sections of a simply supported wall beam with a web opening may be carried out by considering Fig. 10-55. The wall beam has an I-section consisting of a story-high wall acting as a web and portions of floor slabs forming the top and bottom flanges.

By assuming that the lower framing flange (LFF) acts mainly as a tension member, developing negligible moment and shear, the following relation holds at any vertical section through the opening:

$$M^{\circ} = M' + M''$$

where:

M° = total external moment at any section through the opening.

M' = Tf = primary moment, produced by the tension, T, in the LFF and the compressive force, C, in the UFF (upper framing flange), f being the effective lever arm of the C–T couple.

M'' = secondary bending moment in the UFF.

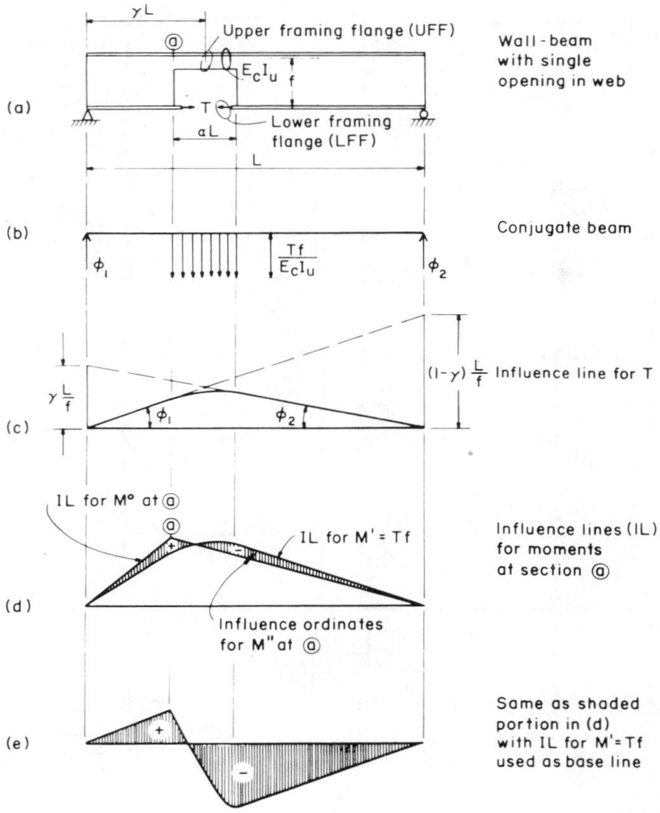

Fig. 10-55 Influence line for secondary moments at left end of upper framing flange.

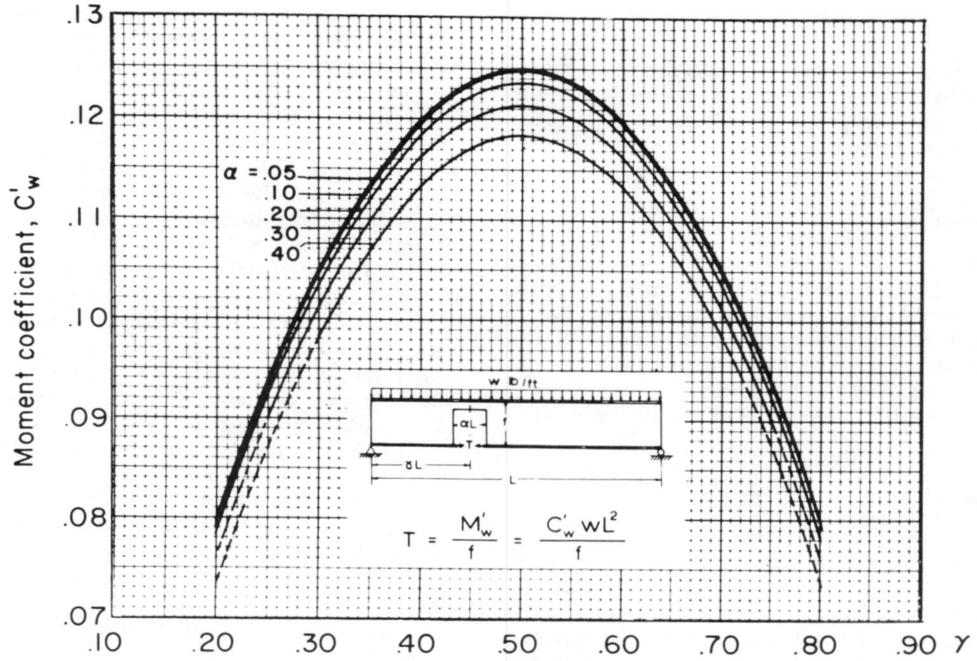

(a) Coefficient for the tension, T, in the lower framing flange due to a uniformly distributed load, w, over entire span.

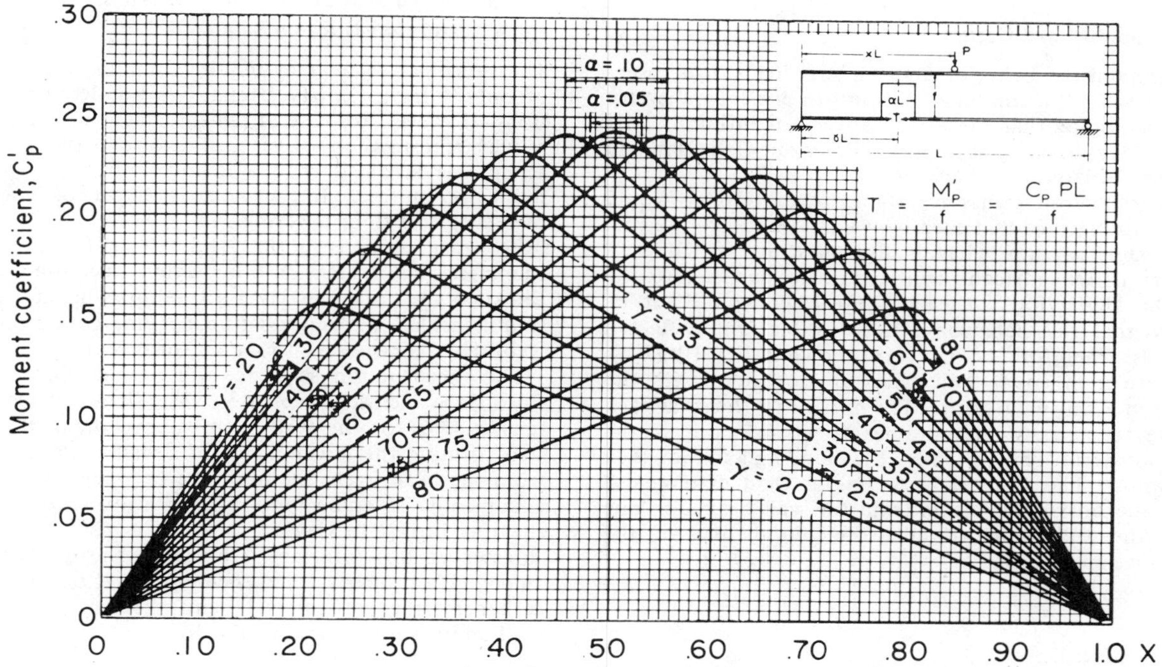

(b) Influence lines for the tension, T, in lower framing flange (connecting curves over openings for $\alpha \neq .05$ omitted).

Fig. 10-56 Coefficients for calculating the tension in lower framing flange.

It is convenient to carry out the analysis by determining the influence line for the uniform tension, T, in the lower framing flange. From the known value of T, the moment M''_x at any section in the upper framing flange may be obtained from the preceding relationship, i.e.:

$$M''_x = M^\circ_x - M' = M^\circ_x - Tf$$

where M°_x is the simple beam external bending moment at the section considered.

The influence line for the tension, T, in the lower framing flange reinforcing steel is obtained by determining the deflected shape of the beam axis due to a unit shortening of the lower flange. The deflected shape then represents the influence line for T (Muller-Breslau's principle).

Curves from Ref. 10-31 for calculating the tension T in the lower framing flange for a wall-beam with a single rectangular opening under a uniform load over the entire span are given in Fig. 10-56(a). In a wall-beam building, corridor walls will usually run longitudinally along each side of the corridor opening. The dead weight of these walls will represent concentrated loads on the wall beam. The value of the tension, T, corresponding to concentrated loads on the span may be obtained using the influence lines for T shown in Fig. 10-56(b).

Reference 10-31 also has curves for calculating moments in the upper framing flanges due to uniform loads over the span, for concentrated loads, for partial loads, and charts for stiffness of wall beams to be used in frame analysis.

10.7.5 Shear Stresses

The shear stresses in the beams above the web openings may be the critical design condition. Consequently, openings in the web should be so located as to avoid regions of high shear, which means preferably in the center of the span. Also wind action causes added shears in the UFF in particular in the lower part of the building. Shear strength considerations may require thickening of the UFF at times.

10.7.6 Deflections

The wall beam is an extremely rigid element and under usual loading conditions will present no deflection problem. The presence of openings in the beam causes a slight increase in the deflection due to vertical loads over what would be expected in a solid-web beam. In view of this, and to estimate possible camber requirements, a check of the deflections of the wall beam may be desirable.

10.7.7 Lateral Load Design

To understand the behavior under lateral loads, it is essential to consider the combined action of adjacent transverse frames. Assuming that the floor slabs act as infinitely stiff horizontal diaphragms, all points on any one floor slab will have equal horizontal deflections.

Considering each transverse frame separately, it would appear at first glance that each bent would undergo the stiff beam–flexible column behavior illustrated in Fig. 10-57(a). However, if the adjacent bents are also considered, the horizontal deflections at each floor level would not be equal, so that this behavior is not possible. The deflected shape must, therefore, be of a form shown in Fig. 10-57(b), which results in equal deflections at each floor with the columns in a single curvature. The horizontal shears at each floor are resisted by the wall-beam system acting in the manner illustrated in Fig. 10-57(b), inducing axial loads in the columns. The floor slabs, acting as rigid diaphragms, transfer the horizontal shears from the wall beams on one floor to the wall beams on adjacent column lines on the floors above and below.

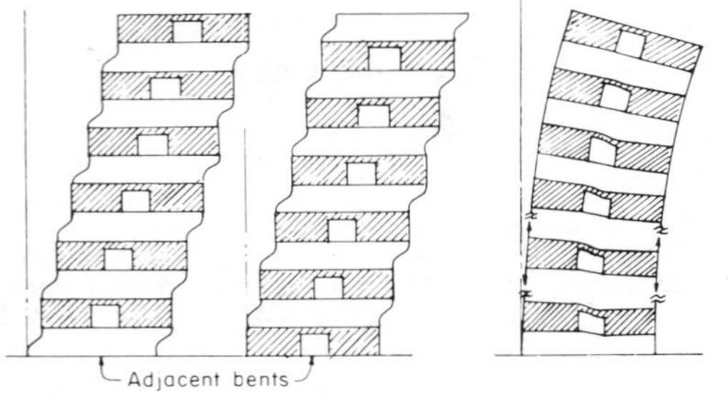

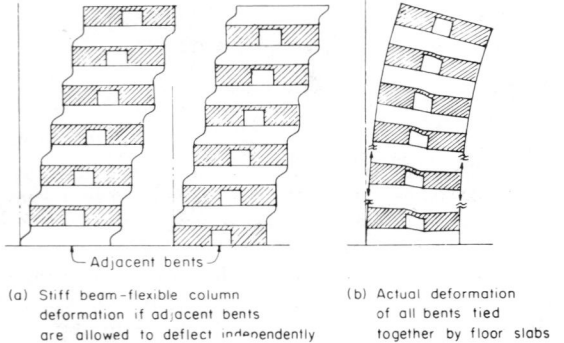

(a) Stiff beam-flexible column deformation if adjacent bents are allowed to deflect independently

(b) Actual deformation of all bents tied together by floor slabs

Fig. 10-57 Deformation of staggered wall-beam frames under transverse lateral loads.

The building may be analyzed by lumping together identical frames and considering each of these composite frames separately under lateral loads proportional to their respective tributary areas.

Because of the restraint provided by the adjacent wall beams, lateral (shear) deformation of the columns in each of the above composite frames is prevented. The frames thus act essentially as vertical cantilevers. Because the columns function principally as axially loaded members under lateral loads directed parallel to the wall beams, they can be oriented more effectively by having their long sides parallel to the longitudinal axis of the building. This alignment also will provide greater freedom in locating the interior partitions behind the columns.

Lateral loads applied in a longitudinal direction of the building can be handled in a number of ways. With the floor slabs, the outside columns constitute a frame that would be sufficient to carry the horizontal loading in a building of moderate height. For taller buildings, this frame could be stiffened by the introduction of shallow spandrel beams at the column lines. If these frames were insufficient to carry the horizontal loading, then the elevator shaft or stairwells could be made load-bearing to create interior shear walls. In this case, the analysis should take into account the interaction of the shear wall and the frame systems.[10-11,10-25,10-26]

10.7.8 Slab and Wall Minimum Thickness and Construction Sequence

The short slab spans may require a minimum slab thickness of only 4 in. to satisfy strength and deflection considerations. However, for ease of reinforcement placement and to reduce transmission of airborne noise between floors to acceptable levels, it is recommended that slabs be a minimum of 5 in. thick. Impact noise is almost solely a function of the type of floor covering.

For ease of construction and to satisfy the minimum thickness requirement, a 6-in. web is recommended. This thickness permits one layer of reinforcing and adequate sound insulation. Under certain loading conditions thicker webs may be necessary to satisfy shear requirements.

The construction sequence will usually involve casting floor slabs and webs separately. The slabs contain the main tensile steel for the wall beams above them. Because the slab forming the lower flange of a wall beam is suspended from the beam web, the interface between web and flange is subjected to some tension in addition to the horizontal shearing forces. In view of this, it is advisable to check the shresses at this interface. Generally, the connection provided by a properly roughened construction joint, tied together by the vertical web reinforcement, is adequate as verified in the tests described below.

10.7.9 Reinforcing Details

Figure 10-58 shows the reinforcement arrangement for a 60-ft wall beam with two openings in an apartment building as presented in the design example in Ref. 10-31. It is important to note the bars around the openings as in the flanges above the openings. The side opening had to be made lower to satisfy the critical shear condition in the flange above it. The flexural reinforcement of the wall beam is placed within the slab, so as not to interfere with the openings. Since the slab containing the flexural reinforcement of the wall is cast first, a cold joint is unavoidable. The joint must be carefully roughened and additional dowels provided through the interface.

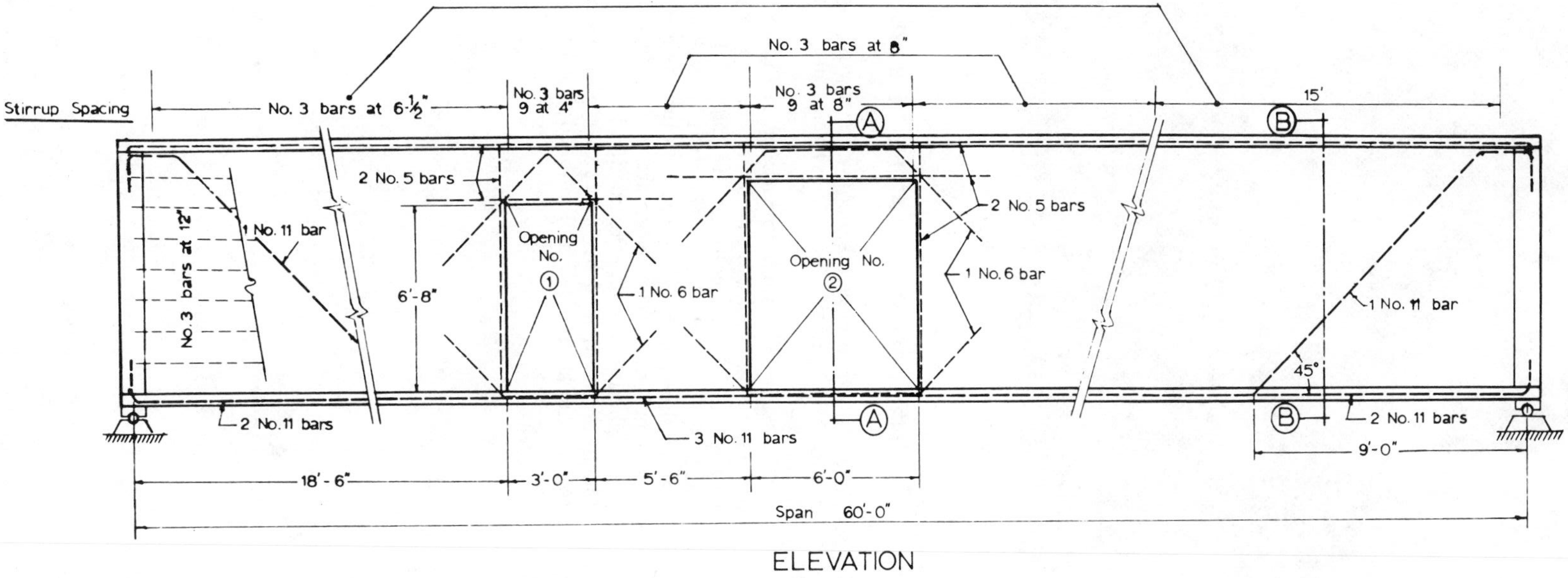

Materials:

1. Concrete: Lightweight, $f_c' = 3000$ psi
   v_c (assumed) $= 0.3\phi F_{sp}\sqrt{f_c'}$
2. Steel Reinforcement: A 432
   $f_y = 60{,}000$ psi

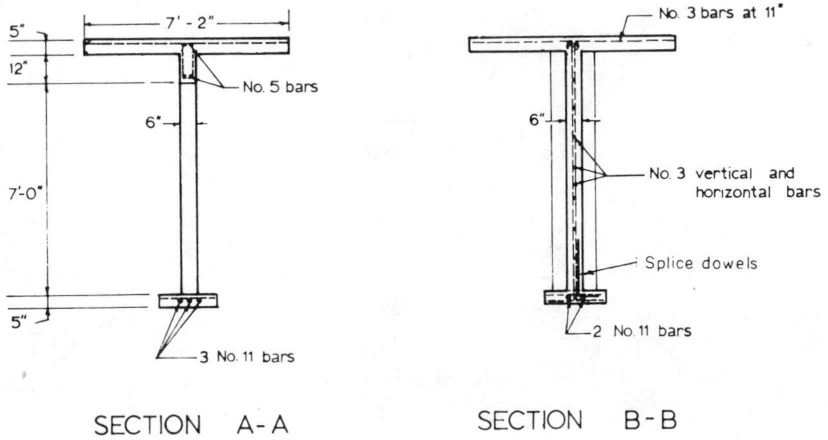

Fig. 10-58 Detail of wall beam with two openings in web.

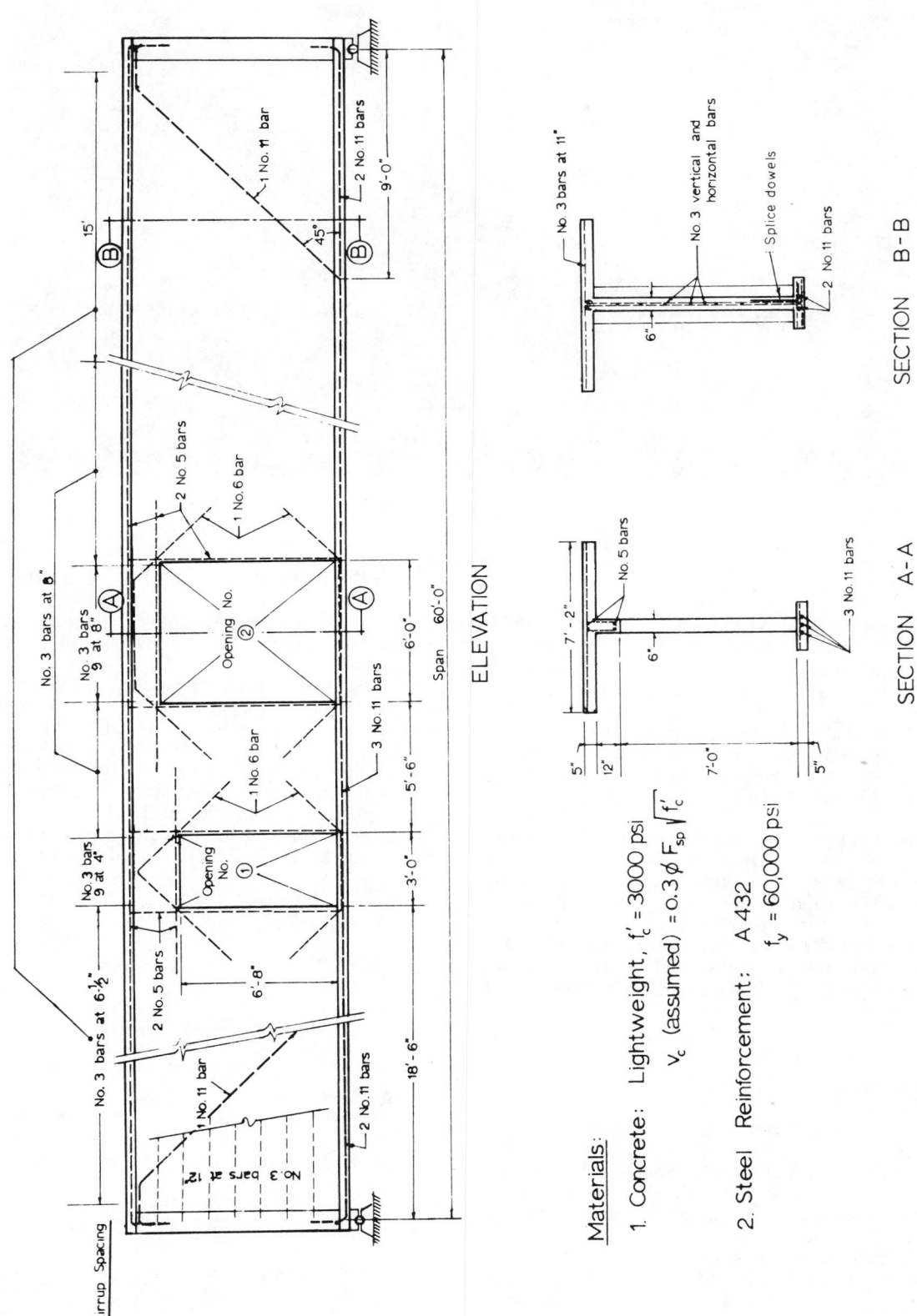

Fig. 10-59 Half-scale model of wall-beam during test to destruction.

10.7.10 Testing

The tests to failure that were carried out at the PCA laboratories[10-32] on a half-scale model (Fig. 10-59) showed the ultimate load capacity to be in excess of the flexural capacity predicted by the *Code* formulas. This apparent increase should be attributed to strain hardening of the reinforcing steel. Also the shear behavior above the doorway was by far better than predicted by theory. For the beam over the smaller doorway the maximum applied shear was nearly double the capacity implied by the *ACI Code* formulas.

A secondary but significant result of the tests was to indicate the desirability of increasing the web reinforcement in the solid web sections up to 100% above that required by the *Code* in order to reduce the width of diagonal tension cracks if and when they occur. The additional web reinforcement is not required for structural reasons but rather by aesthetic considerations of the crack width, if the wall has no other finishes than plaster or paint, since they are at eye level.

10.7.11 Economics

An economic study of apartment buildings ranging from 16 to 36 stories to compare the wall-beam system with the traditional flat plate and shear wall buildings showed that considering the reinforced concrete skeleton and walls between apartments, the wall-beam system uses by far the least amount of concrete and is, therefore, the lightest as far as foundations are concerned. The amount of reinforcement is about the same as in the flat plate system. The study also showed that since the wall beam system resists wind by cantilever action, the premium for height is smaller than in flat plate buildings. Figure 10-60 shows that each of the three systems is most economical for one quantity only: amount of formwork, volume of concrete, or amount of reinforcing steel.

10.8 PIPE COLUMNS—FLAT PLATE STRUCTURES

In search for improved economy for low-rise apartments in reinforced concrete a new system has recently been introduced.

The structural system, introduced in Chicago, consists of a 5 in. thick, reinforced concrete flat plate supported on pipe columns. The columns are capped by I-beam shear heads, located below the slab in the direction of the partition as shown in Fig. 10-61. The columns are 3-in.-diameter pipes, and the shear heads are 6 I standard (12.5 lb/ft) beams about 2 ft long. Connection between columns above the floor is provided by means of a pipe sleeve welded on top of the shear head and embedded within the concrete

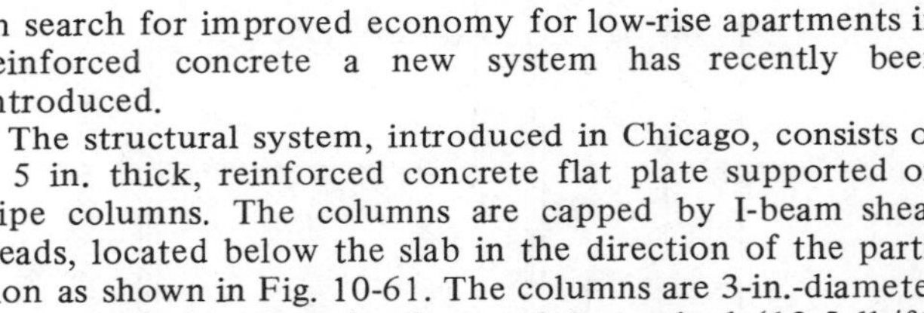

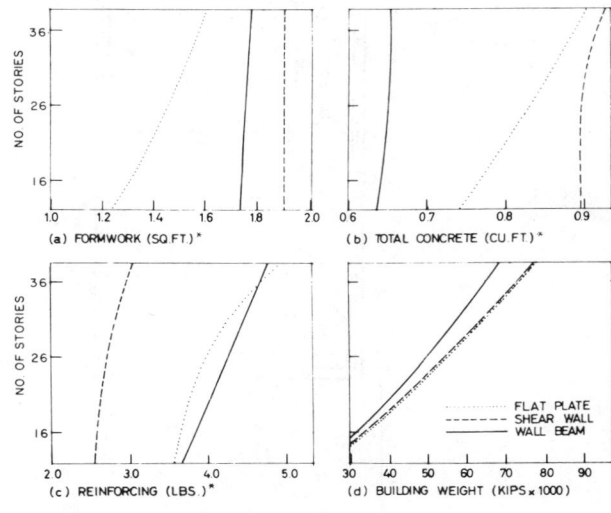

* Per square foot of floor area.

Fig. 10-60 Results of comparative study on quantities of structural material required for each of three building types.

Fig. 10-61 Pipe columns with I-beam shear heads.

floor thickness, serving as a receptable of the column above. This pipe sleeve is the only mechanical shear connector between the floor and the shear head. The shear strength, which is usually the governing design factor for thin flat plates supported on small columns, is entirely eliminated as a problem by the size of the shear heads.

In a subsequent modification of the system, a shear head (consisting of two channel-sections) has been located within the slab thickness, instead of under the slab. Thus, all the protrusions have been eliminated, resulting in a simpler operation.

The size of the pipe columns and the protruding shear heads (if any) are such that they can be incorporated within the thickness of the standard 4-in. partitions. Obviously, the orientation of the shear head must follow the direction of the partition containing the column. The 5-in. slab is theoretically sufficient for spans up to 15 ft. A number of buildings in which the system was utilized were designed with a column layout resulting in spans of about 13–14 ft without interference with the apartment layout.

The slabs were designed in accordance with the ACI requirements for flat plates. A simplification of the usual reinforcement arrangement was achieved by concentrating the entire negative reinforcement over the column heads, thus eliminating the middle strip negative reinforcement. To verify that no excessive cracking would occur in the areas where the negative reinforcement was removed, in violation of the accepted practice and current codes, a field test was carried out[10-33] which proved that neither deflection nor cracking was affected by the changed location of the negative reinforcement. The fact that strength of flat plates is not affected by steel distribution was verified some years ago in a series of tests in Australia.

The pipe column–flat plate system described above does not have any inherent lateral resistance. Therefore, in some of the buildings constructed, masonry walls in both directions proceeded simultaneously with the slabs, providing the necessary lateral strength, while in others a system of diagonal cables between the columns provided the stability during construction, supplemented later by the peripheral masonry.

The 3-in. columns are sufficient to carry about four to five floor slabs, depending upon the tributary areas of the slab. In 8-story buildings, double columns were used in some locations where the load was beyond the capacity of a single pipe column.

The forming and shoring of the slabs was accomplished with large-area prefabricated flying forms contributing considerably to the economy of the system.

10.9 EFFECTS OF VERTICAL MOVEMENTS DUE TO TEMPERATURE, CREEP, AND SHRINKAGE

In the 1950s considering the effects of temperature, creep, and shrinkage on the behavior of buildings was an academic subject. Although the researcher was familiar with some aspects of these subjects, the designer of buildings rarely considered them quantitatively in his designs.

Traditionally, the effect of temperature, creep, and shrinkage was considered in horizontal structures, such as long-span bridges. These effects were usually neglected in multistory concrete buildings, since, in the past, such structures seldom exceeded 20 stories. Some tall reinforced concrete buildings built in the 1960s without consideration of creep, shrinkage, and temperature effects in the columns and shear walls have developed partition distress, as well as structural overstress, and in some cases, cracking in horizontal elements. It has become necessary for the structural engineer to consider the various differential movements and to develop acceptable structural, as well as architectural, details for the satisfactory performance of the buildings.

The differential elastic and inelastic length changes are cumulative over the height of a structure. They start with zero at ground level and reach a maximum at the roof. Therefore, they become more critical with increasing height of a structure. For example, a realistic strain differential between an adjacent column and wall of 150×10^{-6} in./in. would produce the insignificant amount of about 0.016 in. per story of 9-ft height—which, however, for an 80-story building, would mean a maximum cumulative differential shortening of 1.3 in. at the roof level. Such distortions can obviously not be neglected, either structurally or in architectural details.

This sensitivity of ultra-high-rise buildings to the effects of volume changes brought about in the late 1960s, the development of a number of procedures to consider these effects in the design of buildings. Also, architectural and mechanical details have been developed to accommodate building distortions.

10.9.1 Vertical Movements Due to Temperature Changes of Exposed Columns

In the 1960s a large number of multistory apartments and office buildings were built with exterior columns partially, and in some cases fully, exposed to the weather. This was done initially for architectural expression of the structural frame and later for its significant economy, since about

15 to 20% of the elevation is replaced by structural elements. Also, removing protruding columns from the inside helped the trend of exposing columns to gain acceptance.

When subjected to seasonal temperature variations, exposed columns change their length relative to the interior columns which remain unchanged in a controlled environment. For low buildings this causes insignificant structural problems which can be ignored. However, in taller buildings with partially or fully exposed columns this creates distortions of the slabs in the exterior bays, and temperature stresses become significant and must be considered owing to the cumulative nature of column length changes.

In the mid-sixties, several 30-story buildings in the eastern United States developed serious partition cracking that focused attention on the problem of vertical movements due to column exposure.

The thermal movement of the exposed columns causes a racking of floor slabs and consequently a distortion of partitions. The column shortening within a story is only secondary in its effects on partitions. As can be seen in Fig. 10-62, the major effect on partitions is the rotation of the partition due to the cumulative change in length of the exposed column. The partition pivots around Point B when the exposed column shortens. This rotation causes separation between partition and exterior columns. The separation, δ, is the widest at Points A and C, decreasing gradually to zero.

The width of separation depends not only on the amount of movement of the exposed column, but also on the ability of the partition to distort elastically and inelastically. As could be observed from a number of buildings with partially exposed columns, there was generally insignificant distress in most of the common partition assemblies in the range below 20 stories.

It should be mentioned that all tall buildings with columns protruding beyond the glass line, clad or unclad, will experience temperature movements. Even some multistory apartment buildings with exterior columns completely within the building cladding developed partition distress in upper floors attributable to temperature movements.

Between the years 1965 and 1968 a comprehensive and relatively simple engineering procedure was established[10-34–10-36] for the solution of the structural and architectural considerations of column exposure. Also, at the same time, measurements of thermal movements were taken on a number of high-rise structures with exposed columns to verify the validity of the developed analytical procedure.

By the late sixties, the manufacturers of dry-wall partitions improved the resiliency and details so that partitions could take a sufficient amount of frame distortions (temperature, creep, wind) without visible signs of distress and without loss of their acoustical properties.

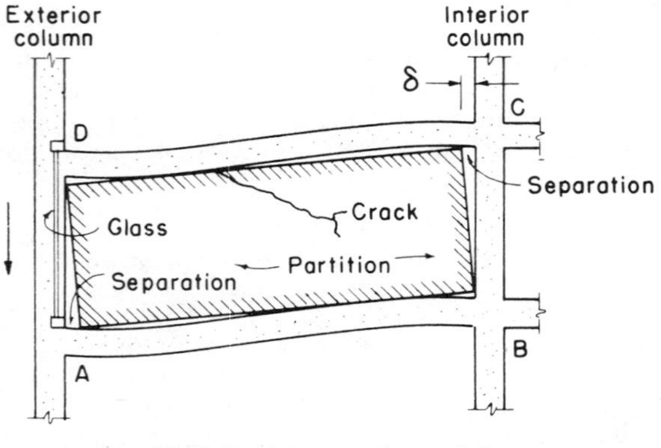

Fig. 10-62 Partition rotation and distress.

Two basic planning and design concepts have evolved for handling structures with exposed columns:

(a) To accept large thermal movements and to design the structure for the resulting stresses providing all the needed architectural details to accept distortions without distress.
(b) To limit thermal movement to an acceptable magnitude as related to the span by limiting the column exposure through changing the glass line or by applying insulation.

The implications and applications of these two concepts will be discussed in subsections 10.1 and 10.2 of this section (see below).

The engineering solution of the column exposure problem consists of:

(a) Determination of a realistic design temperature.
(b) Determination of isotherms, gradients, and average temperatures for exposed columns.
(c) Determination of length changes of exposed columns and bowing stresses as a result of thermal gradients.
(d) A frame analysis and design of the structure for moments and load transfer due to differential length changes of columns.
(e) Details for partitions, cladding, windows, and other structural elements subjected to distortions as a result of thermal movements.

A design example including determination of isotherms, gradients, and a complete analysis for moments, movements, load transfer, and stress is given in Ref. 10-35.

1. Design Temperature. The highest and lowest effective temperatures of all partially exposed columns or walls (having one face inside the building at a constant temperature and the other face outside subjected to ambient temperature variations) for the full seasonal cycle are to be established. Weather conditions rarely remain steady over a long period of time. Since a steady state of heat conduction through any section is only possible if the boundary temperatures remain constant for a sufficient time, an equivalent steady state has to be derived.

There are two factors that influence the equivalent steady state of temperature: (a) the time lag of penetration of exterior temperature fluctuation into the concrete, and (b) the attenuation of their intensity (damping) as the distance from the face increases.

The time lag and attenuation of amplitude depend upon the frequency of temperature change and the thermal properties of the material (concrete), such as conductivity and specific heat. Materials with a higher thermal conductivity respond more quickly to temperature change.

The attenuation of exterior temperature amplitudes at various distances from the face is such that rapid temperature changes penetrate only skin deep. Studies show that amplitude of the 24-hr cycle does not stay long enough to penetrate sufficiently into the concrete. For example, 6 in. away from the face the daily amplitude from the mean attenuates to only 28%; the 7-day amplitude to 62%; and the 90-day amplitude attenuates to 88%. Thus, only the amplitudes of slowly changing temperatures penetrate considerably into the depth of a member. Beyond a certain distance from the surface the temperature of the concrete is affected by fluctuations of long cycles only.

It follows that to use a peak temperature lasting for only a few hours as the exterior design temperature for reinforced concrete members is too severe, since the thermal inertia prevents the concrete from responding to quick changes of the surrounding temperatures.

Based on studies,[10-34] it was recommended to use the minimum mean daily temperature with a frequency of recurrence once in 40 years as the equivalent steady state exterior winter temperature for design purposes of reinforced concrete members of usual size range subjected to exterior temperature variations. Such climatic temperature data are readily available from the guide books for heating, ventilating, and air conditioning. Also, some local weather bureaus or building codes may be a good source for obtaining the mean temperature.

A study of the lowest temperatures with a 40-year recurrence shows that the lowest daily mean temperatures for the United States range from −40°F for parts of Montana, Wyoming, and North Dakota to +40°F for tips of California and Florida. The highest mean daily temperatures, however, show very little fluctuation over the entire United States; they range from 95°F to 105°F.

2. *Temperature Isotherms, Gradients, and Average Temperatures.* After an equivalent steady state temperature has been selected, isotherms and gradients through the partially exposed columns have to be determined. For sections other than the infinitely long wall, the temperature gradient under a steady state of conduction is not linear. The problem is to determine realistic isotherms and an average temperature of partially exposed columns as well as the lowest and highest surface temperatures.

One of the better-known methods of solving two- and three-dimensional steady-state heat flow problems is the relaxation method. The time involved in the analysis by the relaxation method may not always justify its use. A useful known graphical method of flow of water through soil was adapted for construction of isotherms and gradients. The application is based on the concept of flow net construction known from soil mechanics. The heat flow is represented by two sets of intersecting lines. One set of lines represents the heat flow lines; the other set of lines, perpendicular to the first, is the set of isothermal lines. An isothermal line is a contour of equal temperatures. The two sets of lines must intersect at right angles to form curvilinear squares with equal average sides in each direction. The difference in the final gradient between a very accurately drawn flow net and a less accurate net is insignificant. Usually, a few successive trial sketches are sufficient to construct a reasonably accurate flow net. The basic requirements are: (a) that the two sets of curves intersect at right angles, and (b) that each field be as nearly curvilinear square as possible. The construction of isotherms is shown in Fig. 10-63.

The average temperature is given theoretically in a three-dimensional representation by a horizontal plane that divides the total volume created by the isotherms into equal parts. In most cases of prismatic columns, this may be simplified into a two-dimensional problem. A mean gradient through the column is drawn, and the average temperature computed by a numerical summation as shown in Fig. 10-64.

$$t_{\text{average}} = \frac{\Sigma(\Delta d \times t)}{d} = \frac{\Sigma t}{n}$$

where Δd are vertical strips into which the column width, d, is subdivided. The graphically measured mean temperature of each strip is denoted t. If the column is subdivided into n equal strips, each d/n wide, the equation is simplified as indicated above.

To account for the thermal properties of the applied finishes and surface conductances, an equivalent column section instead of the actual section is used. The finishes and surface conductances are converted into equivalent

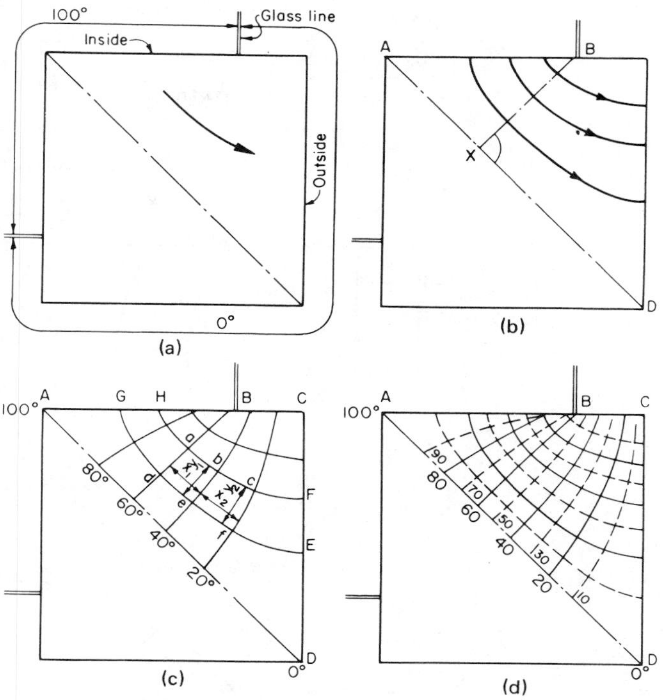

Fig. 10-63 Graphical construction of the heat flow net.

thicknesses of concrete (normal weight or lightweight) having the same thermal conductance. These equivalent thicknesses are added to the actual column section, to provide the equivalent column section used for construction of the isotherms. The equivalent thicknesses, t_e, of concrete are:

$$t_e = \frac{\text{conductivity of concrete, } k}{\text{conductance of finish or surface resistance, } C}$$

The following are some of the commonly assumed thermal properties of the concrete, finishes, and surface conductances:

Conductance of inside surface (still air)	$C = 1.46$
Conductance of outside surface (winter, 15 mph wind)	$C = 6.00$
Conductance of gypsum plaster $\frac{1}{2}$ in. thick	$C = 3.12$
Conductance of gypsum board $\frac{5}{8}$ in. thick	$C = 2.67$
Conductance of rigid polystyrene 1 in. thick	$C = 0.2 \div 0.25$
Conductance vertical spaces $\frac{3}{4}$ to 4 in. (winter)	$C = 1.03$
Conductivity lightweight concrete (110 psf)	$k = 5.0$
Conductivity normal weight concrete	$k = 13.0$

The units for thermal conductance, C, are B hr^{-1} sq ft^{-1} F^{-1}

The units for thermal conductivity, k, are B in. hr^{-1} sq ft^{-1} F^{-1}

The equivalent column section is used only to construct the isotherms. The temperatures at the face of the plaster are those along the line of equivalent concrete thickness of the plaster as shown in Fig. 10-64. Similarly, the outside concrete surface temperature is at the line of the concrete thickness. The inside finish in Fig. 10-64 consists of $\frac{1}{2}$ in. plaster. The temperatures used in the formula for the computations of the average temperature are at the actual column faces and not of the equivalent column. Comparison

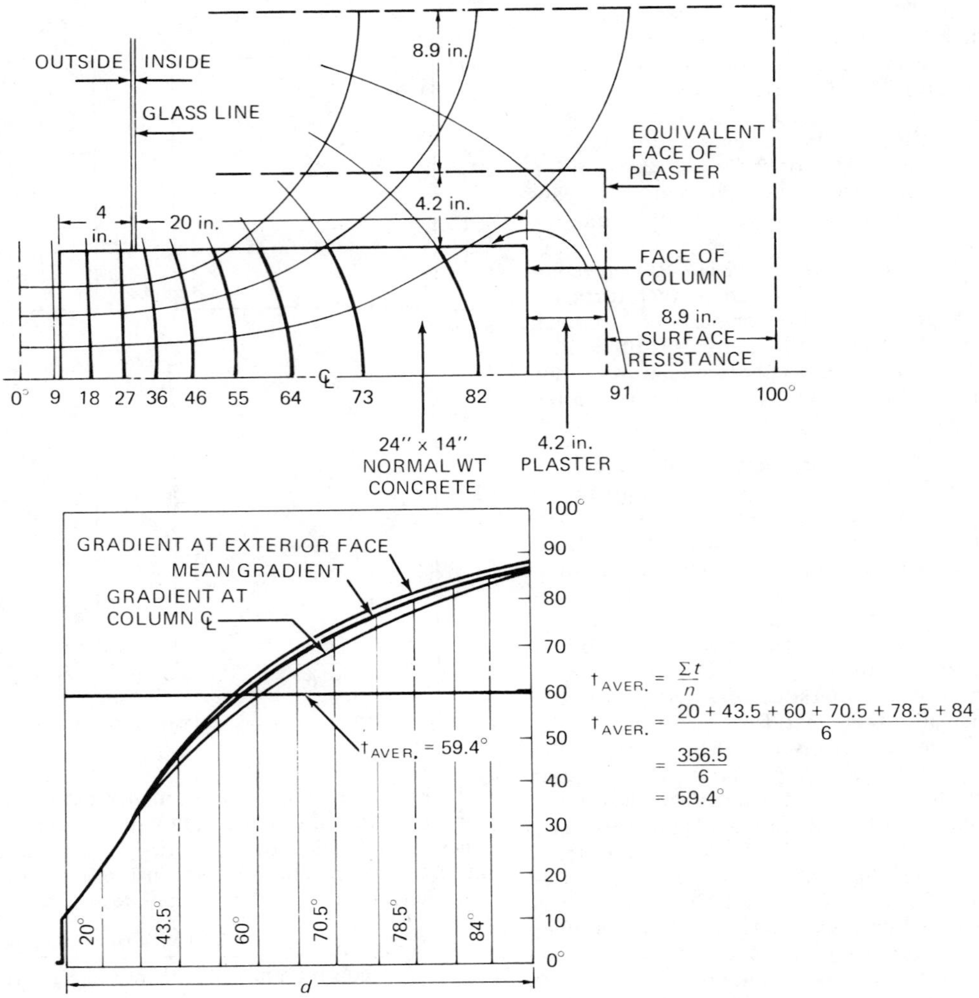

Fig. 10-64 Isotherms, gradients, and average temperature.

of lightweight and normal weight columns indicates that the surface temperature drop for lightweight columns is much smaller than for normal weight columns. Consequently, the average temperature of lightweight columns is higher. Because of lower thermal conductivity, lightweight columns have better insulation capacity.

For fully exposed columns no isotherms need be drawn, since all faces will have the same temperature. It is, however, necessary to choose a design temperature based on the section dimension and the time lag studies.[10-34] It is obvious that larger sections will have a mean temperature of a period longer than the mean daily temperature.

3. Condensation Control. The interior surface of an exterior, partially exposed column is much colder than the temperature inside the room. This is due to surface conductances which produce a temperature drop. However, this may not be objectionable, since adjacent windows are usually much colder, and within any space, condensation occurs on the surface with the lowest temperature. Insulation, air space, or plaster is used to raise the temperature of the inside surface to avoid condensation. This may not always be the best solution as far as the structural problems resulting from thermal movements are concerned. Columns with insulation on the inside will have a lower average temperature and larger thermal movements, and stresses will result. Studies show that no plaster or other finishes are required on columns in rooms with single glass windows; plaster on columns may be required in rooms with double-glazed windows; air spaces or rigid insulation may be necessary for windowless rooms.

4. Length Changes and Bowing. When a free-standing infinite wall is subjected to a temperature drop, a straight-line temperature gradient will result. A strip of wall, if not restrained, will bow as a result of different temperatures at opposing faces and will also shorten by Δl as shown in Fig. 10-65. The face with lower temperature shortens as compared with the warm face. If the wall is restrained, the bending moment caused by this restraint will be:

$$M = \frac{\Delta t \alpha EI}{d}$$

with:

Δt = temperature differential at opposing faces
α = coefficient of thermal expansion
d = wall thickness

When the wall element is forced back by an applied moment M of the same magnitude as restraint moment into its original shape, stresses will develop, and obviously there will be no resulting strains. If the wall is free-standing with no bending moments applied, strains will develop; however there will be no stresses.

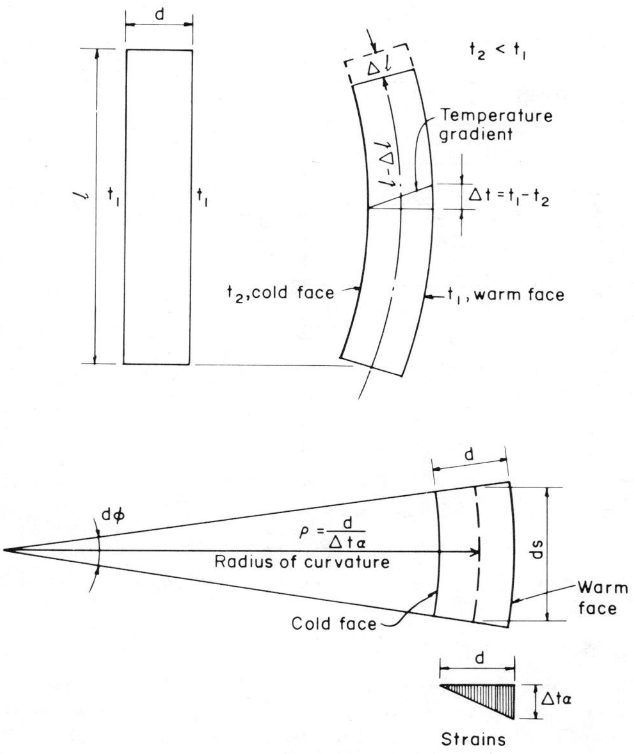

Fig. 10-65 Bowing of a wall strip, radius of curvature, and strains.

In an exposed column when the faces are subjected to different temperatures, a curvilinear gradient will result. If a restraint is applied as shown in Fig. 10-66(a), tensile forces will build up, since the column is prevented from shortening. When the longitudinal restraint is removed and replaced by rollers, Fig. 10-66(b), the column shortens, and compressive stresses build up on the opposite side until the compressive force equals the tension force; equilibrium of forces is established, and the shortening stops. The tensile and compressive forces P form a couple over the distance e' and establish the bowing moment $M_B = Pe'$, which is resisted by the rollers. In the building the entire frame prevents the bowing of the columns, and as a result the columns are subjected to tensile and compressive stresses on the cold and warm faces respectively, as shown in Fig. 10-66(b). The bowing moment depends upon the depth of the column and is independent of column length. Throughout the height of the structure bowing affects the frame only at the roof where there is no column above and where the exposed column changes its depth considerably.

If the rollers are removed, Fig. 10-66(c), the column is free to bow until a new equilibrium of forces and moments is established on the section.

The two effects caused by a temperature gradient through a column, the bowing and axial length change (shortening or elongation), should be considered separately and the resulting stresses superimposed.

5. Analysis and Design for Length Changes of Exposed Columns. The exposed columns respond to exterior temperature variations by changing their length to conform to the average temperature. This results in distortions of the exterior bay. The deflecting slabs, in turn, develop resistance shears that act on the exterior and interior columns, decreasing the length changes. This rebound of exterior and interior columns presents the actual resistance of the structure to the thermal distortions, and depends upon slab or beam stiffness and upon the axial stiffness of the columns.

In multistory frames with partially or fully exposed exterior columns, the effect of temperature movement of the exterior column is significant only in the elements of the exterior bay: the exterior column, the exterior bay slab or beam, and the first interior column.

An iterative analysis has been developed[10-35] to be performed in the following steps (see Fig. 10-67):

(a) Compute the free vertical movement, D_{fn}, of the exterior column. These are assumed as the initial relative movements between the exterior (1) and interior (2) columns at each floor.
(b) Compute fixed-end moments at horizontal floor members at each floor. Zero rotation of joints is assumed.
(c) Distribute moments using moment distribution or the slope-deflection method.

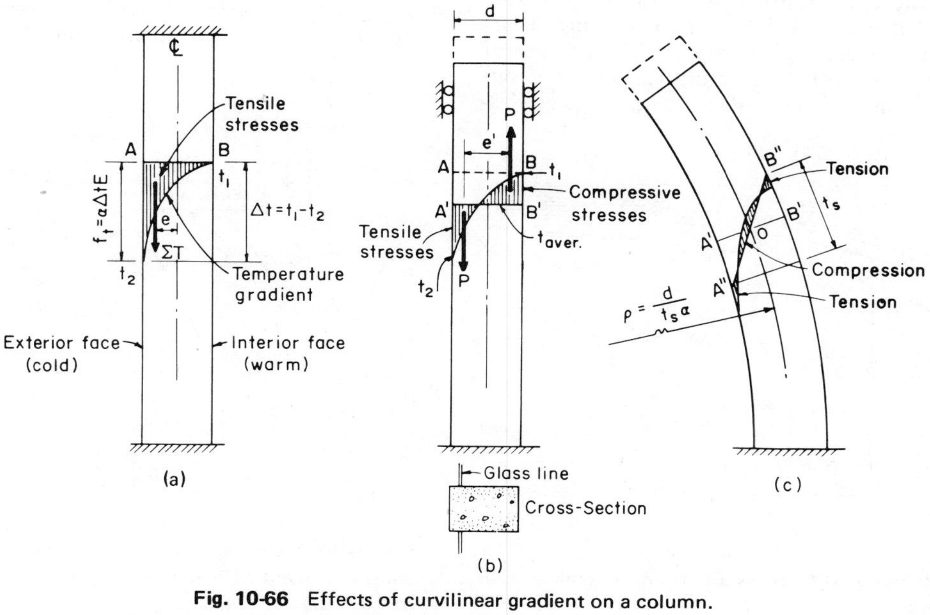

Fig. 10-66 Effects of curvilinear gradient on a column.

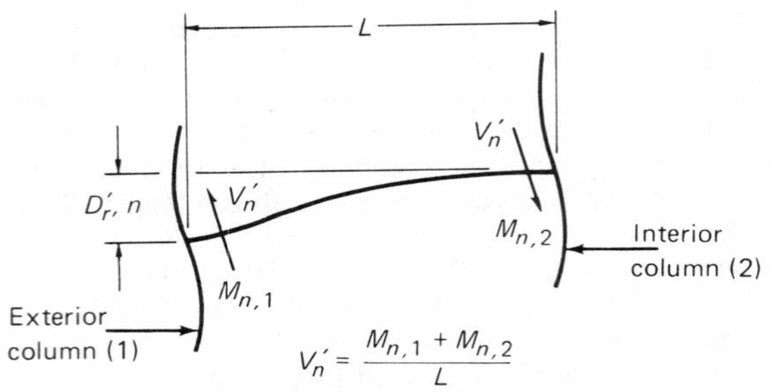

Fig. 10-67 Distributed moments and shears in beam.

(d) Compute resistance shears of the slabs applied to the columns, V_n'.
(e) Using shear forces as vertical loads at exterior and interior columns, compute the rebound of columns.
(f) Compute total story rebound, consisting of the sum of exterior and interior column rebounds.
(g) Apply convergence correction, and compute the corrected relative moments for the next cycle. The analysis is repeated from step (b) using the new corrected movements until convergence to within several percent is reached.

The presented iterative method can be used to obtain a fairly accurate solution within two or three cycles for buildings having relatively flexible girders. Even for very stiff girders, the convergence is quite rapid.

6. Simplified Design Method. A simplified design method has been developed[10-35] using the following three assumptions:

(a) Sizes of columns and horizontal members do not change drastically from floor to floor.
(b) Story heights are fairly constant.
(c) Points of contraflexure in all columns are at mid-height of each story.

In most buildings these assumptions are fulfilled fairly well. Using the above assumptions, the general iterative method is greatly simplified by introducing the equivalent shear stiffness, V_{eff}, which is the shear required to cause a unit relative moment of the two beam ends. Figure 10-68 shows curves for the equivalent shear stiffness for a wide range of practical stiffness ratios. The moments and shears at columns are then computed directly without the use of moment distribution. As a further simplification, the exterior and interior columns may be replaced by a single equivalent column as shown in Fig. 10-69, which has the same rebound as the total story rebound of the two columns. The simplified design method has been used in conjunction with the one-column, 10-story reference frame to develop curves for direct prediction of residual relative movements between the exterior and interior columns.

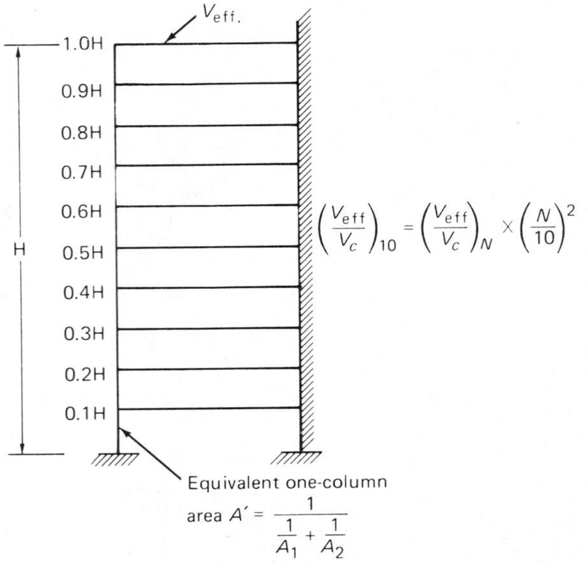

Fig. 10-69 One-column, 10-story reference frame for influence curves.

Since the exterior and interior columns are interconnected by the floor structure, the computed free moment of the exterior column will be reduced to a residual value, due to the stiffness of the floors. The resistance of the structure to length changes of the exterior columns depends not only upon the floor stiffness, but also upon the axial stiffness of the exterior and interior columns. Curves for prediction of residual relative movements between the exterior and interior columns are presented in Fig. 10-70 as a percent of the free movement. The residual movements as computed by the simplified design method differ only insignificantly from the more exact results of the iterative method. The differences increase toward the bottom of the structure where they are less significant, since the moments and stresses due to thermal movements are small.

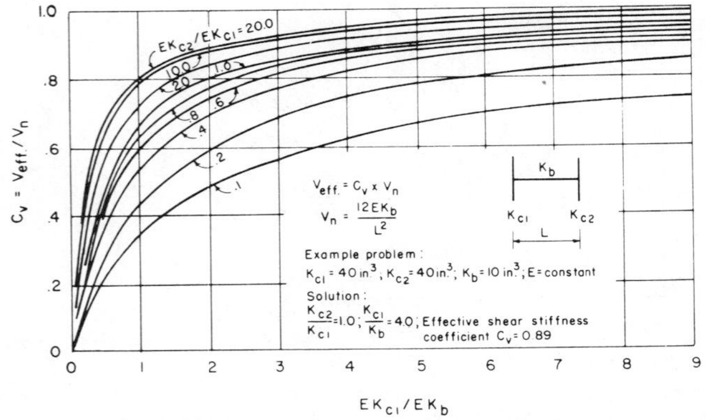

Fig. 10-68 Equivalent shear stiffness as a portion of nominal shear stiffness.

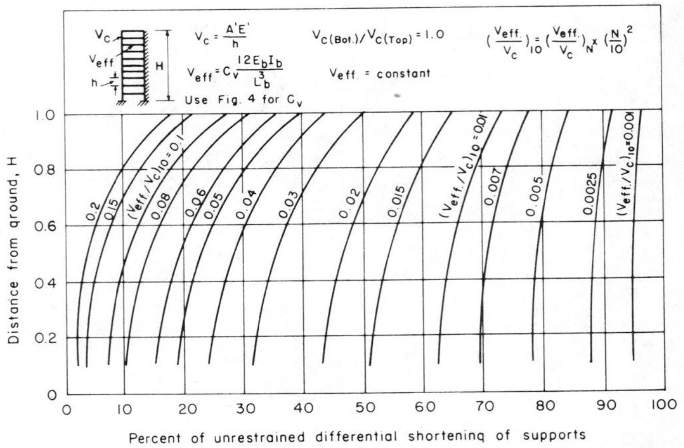

Fig. 10-70 Residual movements between exterior and interior columns as percent of free movement (10-story equivalent structure)—constant column stiffness.

The wider application of computerized methods in design offices during the late 1970s has made it possible to omit the cumbersome longhand analysis to determine the effects of thermal distortions. Thus, after the unrestrained vertical temperature movements of the columns are computed using the described methodology, a general frame computer program is used to determine the slab distortions, moments, shears, and load transfer.

7. Load Transfer between Exterior and Interior Columns. Vertical shears generated in the deflected slabs (see Fig. 10-67) cause transfer of loads from the shortened exposed column to the interior column. During the summer, when the exposed column elongates, the load transfer is reversed. The accumulation of shears starts at the roof level proceeding downward, the biggest accumulation of load occurring at the lowest level. The amount of load transfer can be extremely high for very rigid connecting beams. Flexible slab systems follow the column length changes more easily and, therefore, transfer only insignificant amounts of load from one column to the other; at the same time their restraining effect on the thermal movement of the exposed column is limited.

8. Stresses. Stresses in structural elements due to temperature length changes and bowing should be combined with effects of gravity loads and lateral loads due to wind or earthquakes. However, it would be unrealistic to design a structure for the full combined effect of extreme temperature, highest wind velocity, snow load, and gravity loads. To establish design criteria, consideration should be given to the following:

(a) The highest stresses due to temperature and wind do not usually occur in the same member within the structure. While temperature effects are highest in the upper parts of the frame, the wind stresses are usually highest in the lower parts.
(b) Although extreme temperature conditions may exist for a longer period, coincidence of full temperature and highest wind effects is extremely improbable during the lifetime of a structure.
(c) Some temperature effects, due to their occurrence over a longer period of time, are partially reduced by creep.

To assess the coincidence of high wind and extreme temperatures, local weather condition records should be studied. These conditions may be substantially different from one geographic location to another. Usually, severe temperatures are associated with milder winds; the stronger winds are associated with milder temperatures.

A study of the coincidence of high winds and low temperatures of the 40-year Chicago weather records[10-35] led to the conclusion that for a similar climate, in addition to the usual combination of vertical loads, the following combinations of wind and temperature (whichever is larger) should provide adequate safety:

(a) Full temperature effects plus 20% of wind effects.
(b) Full wind effects plus 25% of temperature effects.

9. Partition Behavior. The differential movement of exterior columns in high-rise buildings affects primarily partitions perpendicular to the exposed columns. Observations and studies of a number of existing structures with partially exposed columns[10-36] indicate only insignificant distress in buildings with less than 20 stories. The movements of the exposed column cause racking of floor slabs which may result in distress of partitions in upper floors (Fig. 10-62). Column shortening within a story is only secondary in its effects on partitions. However, the major effect on partitions is the rotation of the partitions due to the cumulative change in length of the exposed column. The partition pivots at the interior column when the exposed column shortens in winter, and separates from both columns. In winter the separation is widest at the lower corner of the exterior column and at the upper corner of the interior column; in summer, the separation is opposite to that in winter. The width of separation and possible magnitude of distress depend on the amount of movement of the exposed column, and on the ability of the partition to distort. Dry-wall partitions, composed of a series of individual vertical panels attached flexibly to studs, can absorb at each joint a certain amount of displacement that together with some distortion of each panel may provide sufficient adaptability to frame distortions. However, with large movements, hairline cracks may develop in the paint along the vertical joints of the panels. Rigid partitions constructed without consideration of vertical floor movements may suffer distress in the form of diagonal cracking.

There is a significant difference in behavior of partitions in apartment and office buildings. In apartment buildings, partitions are usually stacked over the entire height of the building. They are tightly fitted into the frame. Unless intentionally separated from the frame, the partitions will distort with the building and contribute to the rigidity of the structure in resisting any movements of the frame. In apartment buildings, solid partitions because of their rigidity (even if very small for some types) will reduce the residual movement of the exposed column. This additional resistance to thermal movements is usually not considered in the analysis.

In office buildings, on the other hand, partitions are mostly of the movable type and do not reach the underside of the slab. Consequently, they provide only insignificant resistance to distortions of the structure.

Some of the other possible causes of partition distress are:

(a) Volumetric changes of the partition material due to moisture changes.
(b) Wind, shrinkage, and creep distortions of the frame.
(c) Horizontal roof slab movements due to temperature changes of insufficiently insulated roof slabs.
(d) Deflection of slabs due to gravity loads.

Partition detailing should, therefore, have adequate provisions to accommodate the particular conditions not related to temperature movement of columns. Effects of thermal and other distortion of the frame on partitions should be anticipated by the designers and due consideration given in partition detailing.

Details of partitions and partition layout are of major influence on their behavior. For example, door openings carried to the ceiling may eliminate common location of cracks. The distortion of the structural frame induces shear and bending stresses in the partitions beyond their strength capacity, and, if not relieved by proper details, can cause unsightly as well as acoustically unacceptable cracking. In order to avoid such cracking of partitions, details around the edges of partitions should be provided to allow vertical as well as horizontal slippage. One of the simplest ways to achieve this is to provide a channel enclosure for partitions, where they meet columns and ceiling.

Manufacturers of dry-wall partitions have developed a number of details to let the building distort without straining the partitions. Details of the floating-type partitions are usually provided in upper floors only. Such details are described in the chapter "Joints in Buildings," and sketches given there in Fig. 4-2.

Although partition cracks seem to be the primary problem in some buildings, because they are visible, the structure itself can be considerably overstressed. Therefore, it is not sufficient to design and detail only the partitions for the temperature movements; it is also necessary to design and detail the slabs and columns for temperature deformations and stresses. It may also be desirable in some cases of very tall structures, or where the columns are fully exposed, to provide insulation at the outer faces of the columns if they are clad with stone or precast concrete panels.

10. Planning and Design Concepts. Buildings with exposed columns can be designed either (a) to accommodate large expected relative temperature movements, or (b) with controlled temperature movement.

10.1 Buildings designed for large movements. Large movements result usually when architectural considerations dictate a largely recessed glass line in tall buildings, or if the column exposure is complete even in moderately tall buildings.

Structurally there are two alternative solutions to solve problems in buildings designed with large exposure: (a) to provide strength for accommodation of relatively large movements of exterior columns, or (b) to provide details to relieve the stresses in floors that have excessive rotations.

To provide strength, heavy restraining girders connecting the exterior and interior columns may be provided at the roof, or at intermediate floors (mechanical floors). The restraining girders are generally very rigid and a full story in depth. These girders obviously also improve the resistance to wind loads, as discussed previously in Section 10.5.

The alternative is to provide details to relieve the stresses in the floor structure. Hinging of floors at the interior columns or shear walls for severe winter shortenings may assure proper functioning of the structural system.

The first building known to be designed for significant structural movements was the 38-story Brunswick Building in Chicago (Fig. 10-50). All typical exterior columns were exposed to about 70% of their area, resulting in an anticipated maximum computed movement of 1.25 in. at the top floor. To relieve high bending stresses, hinges between the slab and the shear wall were incorporated in the upper 12 floors, as shown in Fig. 10-71. The hinge details should include dowels to maintain lateral restraint between slabs and walls. By using elastomeric material around a portion of dowels, the slabs are free to rotate, but lateral restraint is provided. Observations indicate that the structure is performing as planned; no architectural or structural problems have been encountered.

10.2. Buildings designed for controlled movements. Although it is possible to accommodate structurally a considerable amount of movement of exposed columns, serviceability and economy may set an upper limit on the amount of thermal movement. Serviceability criteria are logically expressed as a ratio of the movement to the slab, or beam, length (angular distortion). Any movement limitation should relate to the type of structure and building material used. It is clear that apartment buildings will impose different movement limitations from warehouses, indistrial buildings and similar types of structures.

Structures with brickwork, masonry partitions, and plaster walls will require more stringent limitations than buildings with metal cladding. It should be noted that slab deflections due to column exposure are additive to deflections due to gravity loads. For buildings without masonry or plastered partitions, a thermal movement of $L/200$ (of exterior column relative to interior column) may be tolerable. For a thermal movement limitation of $L/200$, the center of the span will be subjected to a movement of $L/400$ only, which should be added to vertical load deflections and then compared to specified deflection limitations of the *Code*.

In some cases it may be desirable to develop more stringent criteria for maximum temperature movements. Reference 10-36 suggests that in office buildings a reasonable limit for temperature movements may be taken as 0.75 in. up or down from the horizontal position. Assuming clear span in office buildings of about 36 ft, a relative movement of 0.75 in. corresponding to $L/600$ between the exterior column and the shear core, will generally not cause excessive stresses requiring special structural details. These stresses normally will be less than the allowable overstress due to temperature as proposed earlier. Furthermore, a movement of $L/600$ can be reasonably taken care of by simple partition details without serious economic implications.

For the 52-story Shell Oil Building in Houston, the 0.75-in. limitation was accomplished by setting the glass line to achieve the desired average temperature of the column. Nominal modification of typical partition details was needed to avoid stresses in the partitions for such movements.

In apartment buildings where the bay sizes are considerably smaller, a relative movement between the exterior column and the interior column of 0.75 in., corresponding to $L/300$ for an 18-ft span, may require structural and partition modifications. To minimize the need for special structural and partition details in apartment buildings, it may be advisable to limit computed temperature movements to about 0.5 to 0.625 in. by placing the glass line as far out as possible. The actual movements in apartment buildings are usually smaller than computed, owing to the unaccounted contribution of partitions to the resistance of the frame.

10.9.2 Vertical Movements Due to Creep and Shrinkage

With the increasing height of buildings, the importance of time-dependent shortening of columns and shear walls becomes more critical owing to the cumulative nature of such shortening. It is known that columns with varying percentages of reinforcement and varying volume-to-surface ratios will have different creep and shirnkage strains. An increase in percentage of reinforcement and in volume-to-surface ratio reduces strains due to creep and shrinkage under similar stresses.

In a multistory building, adjacent columns may have different percentages of reinforcement due to different tributary areas or different wind loads. As a result, the differential elastic and inelastic shortening will produce moments in the connecting beams or slabs and will cause load transfer to the element that shortens less. As the number of stories increases, the cumulative differential shortening also increases, and the related effects become more severe. A common example is the case of a large, heavily reinforced column attracting additional loads from the adjacent shear wall which has higher creep and shrinkage due to a lower percentage of reinforcement and a lower volume-to-surface ratio. Significant differential shortening may also occur due to a time gap between a slip-formed core and the slabs. In this case the columns are subjected to the full amount of creep and shirnkage, while the core may have had the bulk of its inelastic shortening occurring prior to casting of the adjacent columns.

It is customary, at present, to neglect the effect on the frame of elastic and inelastic shortening of columns and walls. For low and intermediate-height structures this may be acceptable; however, neglecting the differential shorten-

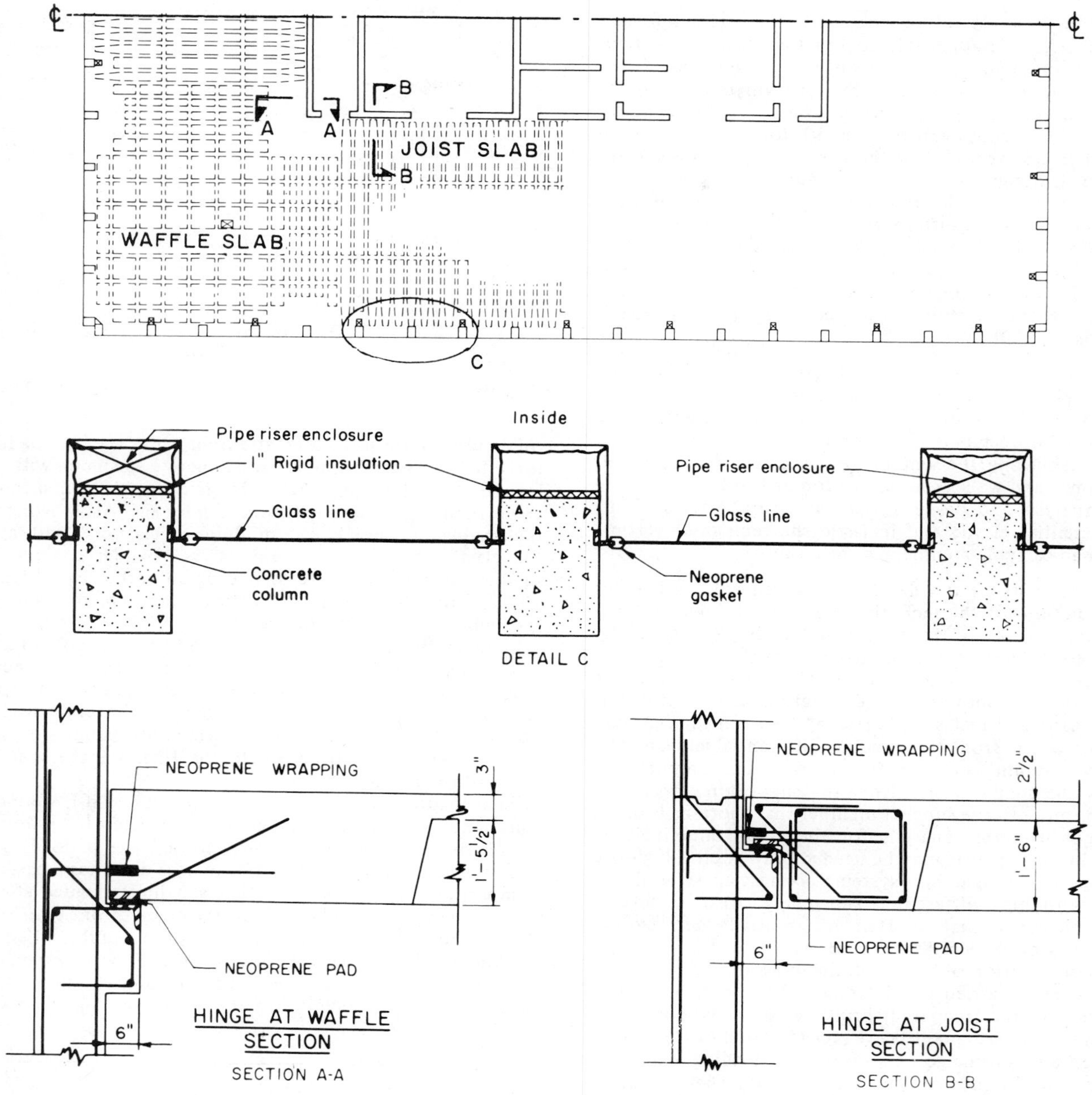

Fig. 10-71 Details of hinges and the glassline in the Brunswick Office Building.

ing in ultra-high-rise buildings may lead to distress in the structure and in nonstructural elements of the building.

In a number of tall buildings in the United States built in the early sixties, structural cracking and partition distress were observed as a result of differential creep between shear walls and highly reinforced columns in close proximity to each other. Another example of the reality of differential creep and shrinkage of vertical elements is a 50-story building in Australia in which the measured differential at the roof level between the concrete core and peripheral columns was 1.1 in. after about four and one-half years. Fortunately, no problems were experienced—the long spans of about 35 ft between the core and peripheral columns caused only small slab rotations. The elevator rails had to be adjusted twice over the years to accommodate the shortening of the elevator shafts.

Buildings up to 30 stories with flexible slab systems, such as flat plate slabs of average spans, or long-span joint systems, are usually not adversely affected structurally by differential shortening of supports. In these cases the knowledge of the total shortening is needed to make allowance in the architectural details to avoid future distress of partitions, windows, cladding, and other nonstructural elements. Differential shortening can be minimized by proportioning adjacent columns or walls to have similar stresses of the transformed section and similar percentages of reinforcement. The volume-to-surface ratio has a lesser effect on differential shortening.

Although a large amount of research information is available on shrinkage and creep strains, it is not directly applicable to columns of high-rise buildings. The available shrinkage data must be modified, since they are obtained on small standard prisms or cylinders stored in a controlled laboratory environment. The available creep research is based on application of loads in one increment. Such creep information, therefore, is applicable to flexural elements of reinforced concrete and to elements of prestressed concrete. In the construction of a high-rise building, however, columns are loaded in as many increments as there are stories above the level under consideration. If a 50-story building is constructed in 50 weeks, then the first-story columns receive 2% of their design load every week during the construction period. Incremental loading over a long period of time makes a considerable difference in the magnitude of creep.

An engineering procedure was established in the late 1960s[10-37, 10-38] for the solution of the structural considerations involved in the effects of differential column shortening due to elastic shortening, creep, and shrinkage in tall buildings, consisting of:

(a) Determining the amount of creep and shrinkage occurring in columns and walls with consideration of the loading history, size of the member, percentage of reinforcement, and environment.
(b) Establishing the amount of elastic shortening in columns and walls as necessary for analysis.
(c) Analysis and design for the structural effects of differential elastic and inelastic shortenings of vertical load-carrying members in the structure.

The first part of the developed procedure[10-37] handles the predictions of the inelastic (creep and shrinkage) shortening as a function of the incremental loading sequence, the volume-to-surface ratio, and the effect of the percentage of reinforcement. The second part[10-38] handles the analysis of multistory structures for the structural effects of differential elastic and inelastic shortening between adjacent columns or walls; from the residual differential movements, the corresponding moments in the frame are computed, as well as the load transfer from the support member that shortens more to the support member that shortens less.

A rigorous frame analysis for the elastic and inelastic strains in the supports may be needed only in ultra-high-rise buildings or in rigid slab systems connecting vertical elements with high differential shortening. Also, structures where elements that shorten differentially are closely spaced may need special investigation.

The computation of elastic strains does not present difficulties as it is carried out by established procedures. The prediction of the creep and shrinkage strains is more complex and requires a quantitative consideration of the effects on creep and shrinkage of section size, loading sequence, and amount of reinforcement. Example 10-3 (below) shows how to compute the creep and shrinkage strains for a reinforced concrete column protected from the weather.

1. Computation of Creep and Shrinkage Shortening. The nature of creep and shrinkage and their dependence upon the constituent materials are explained in Chapter 6 of this *Handbook.* Since creep depends upon sequence and intensity of loading while shrinkage proceeds independently of construction time, creep and shrinkage strains should be computed separately, and modified for the condition of the designed structure and then their combined effect on the structure considered.

For structural engineering practice, it is convenient to consider specific creep ϵ_c', which is defined as the ultimate creep strain per unit of sustained stress loaded at any specified age, say, 28 days.

Determination of specific creep can be either done in the laboratory on 6-in. cylinders or roughly predicted from the initial modulus of elasticity[10-39] from charts.

2. Effect of Construction Time. For a given mix of concrete, the amount of creep depends not only upon the total stress but also to a great extent upon the loading history. It is well established that a concrete specimen with its load applied at an early age exhibits a much larger specific creep than a specimen loaded at a later age. Also, owing to the gradual increase of the modulus of elasticity with age, the elastic shortening per unit stress of older concrete is smaller than of concrete loaded at an earlier age. Figure 10-72, from Ref. 10-40, shows elastic and creep strains of specimens loaded at various ages.

Loading history is particularly significant for columns of multistory buildings which are loaded in as many increments as there are stories above the level under consideration. Since creep decreases with age of the concrete at load application, each subsequent incremental loading contributes a smaller specific creep to the final average specific creep of the column.

Each load increment causes a creep strain as if it were the only loading to which the column is subjected.[10-41] In Fig. 10-73 the relationship of creep to the age of loading is shown using the 28-day specific creep as unity (as a basis of comparison). It can be seen that a cylinder loaded at the age of one year has only one-half the creep of one loaded at 28 days of age. If we load a column with a number of equal incremental loads over a period of time, T, we can then determine the average creep for the total load by a summation of the incremental load with the corresponding creep val-

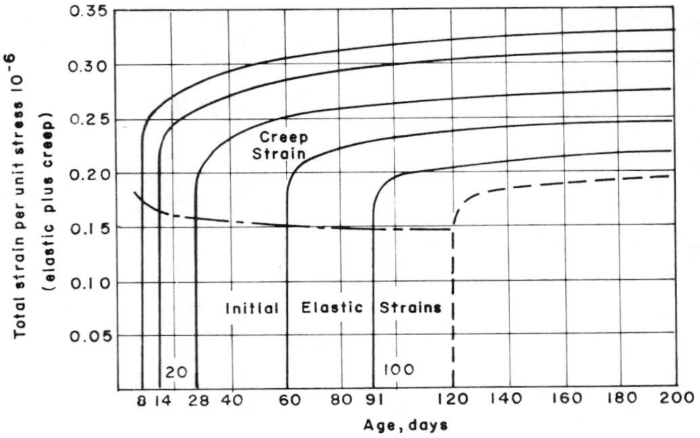

Fig. 10-72 Elastic and creep strains versus time, from Ref. 10-40.

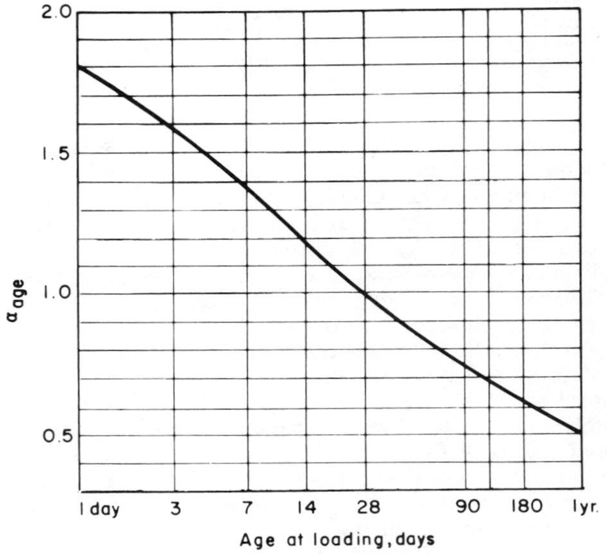

Fig. 10-73 Creep versus age at loading.

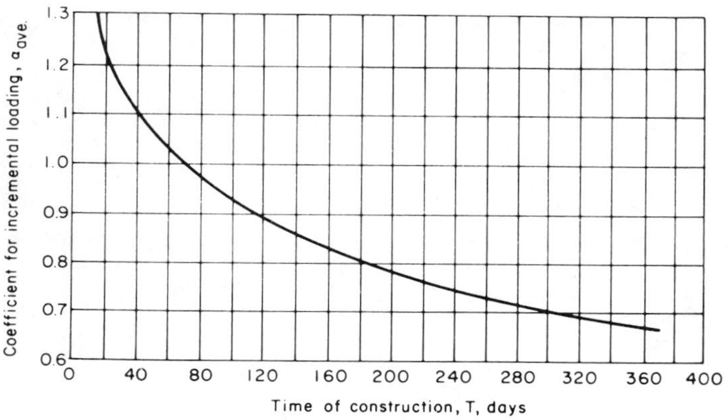

Fig. 10-74 Creep versus construction time.

ues. Figure 10-74 gives the coefficient, α_{ave}, to be used to convert the 28-day creep into the average creep for a specific length of construction time. We can see from the curve that if the entire load is applied at age of 7 days, it will have twice the creep it would have if applied in incremental loading over a period of 262 days.

3. Effect of Member Size. Both creep and shrinkage depend on the member size, not to the same degree. Creep is less sensitive to member size than shrinkage which is caused by evaporation of moisture from the surface. The rate and amount of evaporation and consequently of shrinkage depend greatly upon the relative humidity of the environment, the size of the member, and the mix proportions of the concrete. In moderate-size columns (30-in. diameter) the inside relative humidity has been measured at 80% after four years of storage in a laboratory in 50%-relative humidity. Since evaporation occurs only from the surface of members, the volume-to-surface ratio of a member has a pronounced effect on the amount of shrinkage.

In Fig. 10-75 the relationship between creep and the volume-to-surface ratio is plotted, based on Elgin gravel aggregate concrete. A plot of shrinkage versus the volume-to-surface ratio is given in Fig. 10-76. The European experience on the surface-to-volume effects on creep and shrinkage is also plotted for comparison, since the U.S. information is based on only one research investigation. The coefficient $\alpha^s_{v/s}$ is used to convert shrinkage data obtained on 6-in. cylinders (v/s = 1.5 in.) to any other size columns.

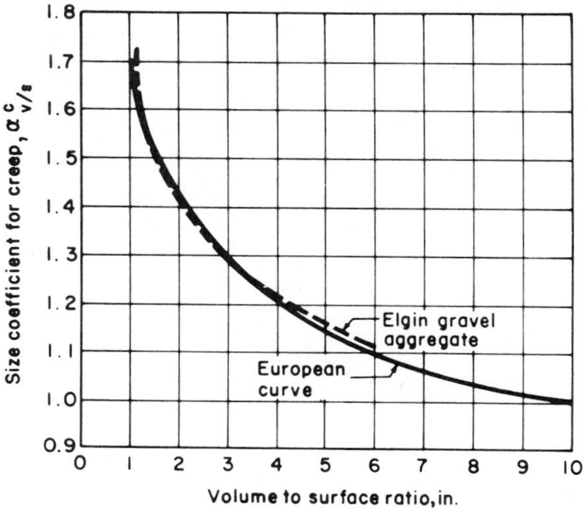

Fig. 10-75 Creep versus volume-to-surface ratio.

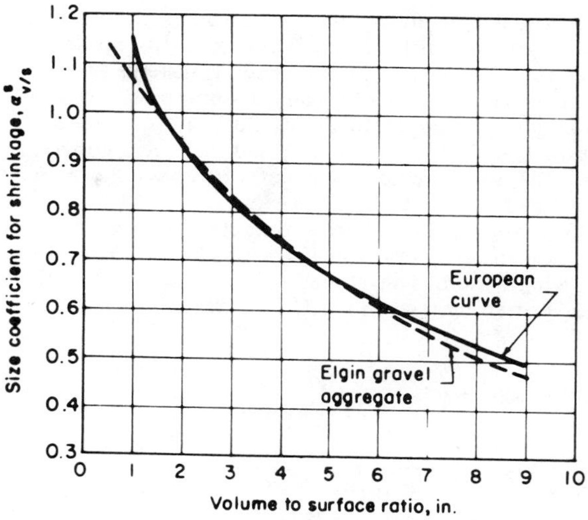

Fig. 10-76 Shrinkage versus volume-to-surface ratio.

4. Progress of Creep and Shrinkage with Time. Creep and shrinkage are similar regarding the rate of progress with respect to time. Figure 10-77 shows an average curve for the ratio of creep or shrinkage at any time to the final value at time t_∞. It can be seen from the curve that at 28 days about 40% of the inelastic strains have taken place. After three and six months, 60 and 70%, respectively, of all the creep and shrinkage have taken place. The curve can be used to extrapolate the ultimate creep and shrinkage values from laboratory testing of a certain duration of time. Conversely, the curve can also be used to estimate the creep or shrinkage at any time from the given ultimate value.

5. Effect of Reinforcement on Creep and Shrinkage. The longitudinal reinforcement has by far the most predominant restraining effect on creep and shrinkage. Although the magnitude of creep and shrinkage of plain concrete specimens may vary considerably depending on the concrete properties and climatic conditions, the final inelastic strains in reinforced concrete columns and walls have much less variation owing to the restraining effect of the reinforcement.

Tests have shown that when reinforced concrete columns are subjected to sustained loads, there is a tendency for additional stress to be gradually transferred to the steel with a simultaneous decrease in the concrete stress. Long-term tests by Troxell et al.[10-47] showed that in columns

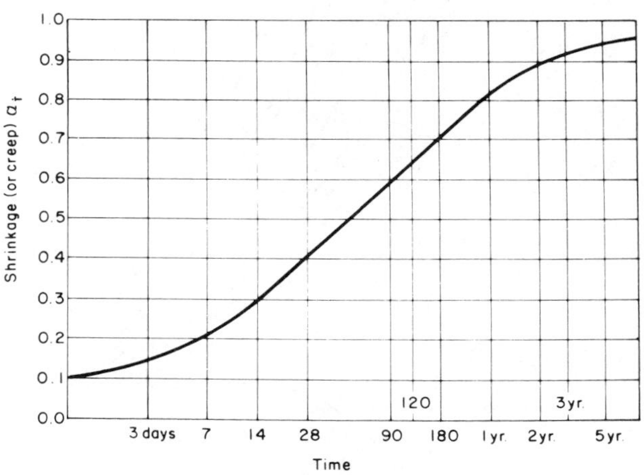

Fig. 10-77 Shrinkage or creep versus time.

with a low percentage of reinforcement the stress in the steel increased until yielding; while in highly reinforced columns after the entire load had been transferred to the steel, further shrinkage actually caused some tensile stresses and cracking of the concrete. It should be noted, however, that despite the redistribution of load between concrete and steel, the ultimate load capacity of the column remains unchanged.

The change in stress in the concrete, Δf_c, and in the steel, Δf_s, due to creep and shrinkage can be calculated with the following formulas developed by Dishinger[10-43] which have been verified by tests in the United States.[10-44]

$$\Delta f_c = \left(f_c + \frac{\epsilon_s}{\epsilon_c'}\right) 1 - e^{-\frac{pn}{1+pn}\epsilon_c' E_c}$$

$$\Delta f_s = \frac{\Delta f_c}{p} = \frac{f_c + \epsilon_s/\epsilon_c'}{p} \; 1 - e^{-\frac{pn}{1+pn}\epsilon_c' E_c}$$

$$= \frac{(f_c\epsilon_c' + \epsilon_s)}{p\epsilon_c'} \; 1 - e^{-\frac{pn}{1+pn}\epsilon_c' E_c}$$

in which:

f_c = initial elastic stress in the concrete
ϵ_s = total shrinkage strain of plain concrete adjusted for v/s ratio
ϵ_c' = ultimate specific creep of plain concrete (in./in./psi)
E_c = modulus of elasticity of concrete
p = reinforcement ratio of the section
n = modular ratio E_s/E_c

The ratio of residual creep and shrinkage strains of a reinforced column to the total creep and shrinkage strain of the identical column without reinforcement is presented in Fig. 10-78, based on the Dishinger formulas for various percentages of reinforcement, varying specific creep and modulus of elasticity of concrete. It is evident from the curves that the residual creep and shrinkage decreases with increased percentage of reinforcement.

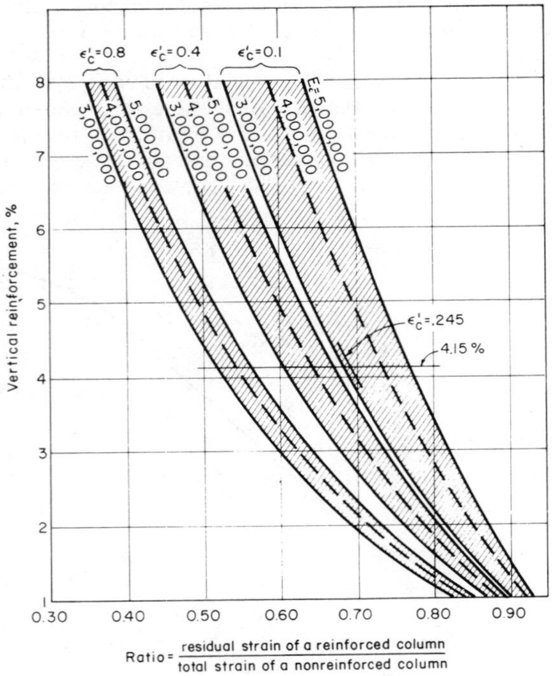

Fig. 10-78 Restraining effect of the reinforcement.

The function $1 - e^{-\frac{pn}{1+pn}\epsilon_c' E_c}$ has been plotted in Ref. 10-37 for convenience of numerical handling of the Dishinger equations given above for computing the changes in the concrete and steel stresses.

EXAMPLE 10-3: Assume an inside column 36 stories below the roof. Floor-to-floor height is 9.0 ft. The 20 × 49-in. column is reinforced with 26 #11 bars (4.15%) of A431; f_y = 75,000 psi.

Concrete:

$$f_c' = 5000 \text{ psi at 28 days, normal weight}$$

$$E_c = 33w^{3/2}\sqrt{f_c'} = 4.05 \times 10^6$$

$$n = \frac{E_s}{E_c} = 7.2$$

Transformed column area:

$$A_t = A_g + (n-1)A_s$$

$$= 20 \times 49 + (7.2 - 1) \times 40.6 = 1232 \text{ in.}^2$$

The planned construction progress is one floor in 8 calendar days—total time for 36 floors (load increments) is:

$$T = 36 \times 8 = 288 \text{ days}$$

The dead lood of the typical floor is 37 kips.

Twenty-year specific basic creep for loading at age of 28 days was estimated from previous experience to be $\epsilon_c' = 0.33 \times 10^{-6}$ in./in./psi for $E_c = 4.05 \times 10^6$ psi.

Shrinkage determined on the same mix of a previous job was 630×10^{-6} in./in. during the first 90 days. The 6-in. cylinders were moist-cured for seven days and then stored in the laboratory in 50% relative humidity and 70°F.

Required: To compute the ultimate residual creep and shrinkage strains of the reinforced concrete column and the additional stress in reinforcing steel.

The following steps will be carried out:

(a) Compute for the plain concrete column the total ultimate creep strains, considering effects of incremental loading of column size; and shrinkage considering volume-to-surface ratio.
(b) Compute for the reinforced concrete column the residual creep and shrinkage strains.
(c) Compute the additional stress in the vertical reinforcing steel due to creep and shrinkage.

SOLUTION:

(a) Creep strains for nonreinforced column:

Consersion of specific creep for loading at 28 days to consider incremental loading over a period T = 288 days using α_{ave} = 0.70 from Fig. 10-74:

$$\epsilon_{c,\,ave}' = \epsilon_{c,\,28}' \times \alpha_{ave}$$

$$= 0.33 \times 10^{-6} \times 0.70 = 0.231 \times 10^{-6} \text{ in./in./psi}$$

Modification of specific creep for size effect using Fig. 10-75:
Volume-to-surface ratio:

$$v/s = \frac{20 \times 49}{2(20+49)} = 7.1 \text{ in.} \rightarrow \alpha_{v/s}^c = 1.06$$

$$\epsilon_c' = 0.231 \times 10^{-6} \times 1.06 = 0.245 \times 10^{-6} \text{ in./in./psi}$$

Sustained stress on the concrete:

$$f_c = \frac{P}{A_t} = \frac{36 \text{ N } 37{,}000}{1232} = 1080 \text{ psi}$$

Total creep strain:

$$\epsilon_c = \epsilon_c' \times f_c = 0.245 \times 10^{-6} \times 1080 = 265 \times 10^{-6} \text{ in./in.}$$

Shrinkage strains for nonreinforced concrete column:

Conversion of 90-day measured shrinkage to ultimate shrinkage (coefficient from Fig. 10-77 representing the ratio of shrinkage at 90 days to ultimate shrinkage):

$$\epsilon_{s_\infty} = \frac{630 \times 10^{-6}}{0.60} = 1035 \times 10^{-6} \text{ in./in.}$$

Conversion of shrinkage measured on the 6-in. cylinder to account for size of real column using Fig. 10-76:

$$v/s = \frac{20 \times 49}{2(20 + 49)} = 7.1 \text{ in.} \rightarrow \alpha^s_{v/s} = 0.57$$

$$\epsilon_s = 1035 \times 10^{-6} \times 0.57 = 590 \times 10^{-6}$$

Total creep and shrinkage strains for nonreinforced column:

$$\epsilon = \epsilon_c + \epsilon_s = (265 + 590)\, 10^{-6} \text{ in./in.}$$

(b) Residual creep and shrinkage for columns from Fig. 10-78 for a reinforcing ratio of 4.15% for $E_c \approx 4 \times 10^6$, interpolating for $\epsilon_c' = 0.245$:

$$\text{Ratio} = \frac{\text{residual strain of a reinforced column}}{\text{total strain of a nonreinforced column}} = 0.69$$

(c) Additional stress in the reinforcing steel corresponds to the residual strain of the reinforced column; therefore Δf_s = residual strain $\times E_s = 590 \times 10^{-6} \times 29 \times 10^6 = 17{,}100$ psi. The same stress could be computed by solving the Dishinger formula for Δf_s. From the residual creep and shrinkage strain the shortening of the column over the entire height of the structure can be computed.

6. *Analysis for Elastic and Inelastic Differential Movement.* After the elastic and inelastic differential shortenings of the columns and walls are determined, a frame analysis can be carried out.

It should be pointed out that each slab is affected only by the differential shortenings of its support which will occur after the slab has become a part of the frame. All elastic and inelastic shortenings to which the supports were subjected prior to the casting of a slab are of no consequence to this particular slab. Thus, a fourth dimension enters the analysis; namely, construction time and sequence. The availability of powerful and sophisticated computer programs, since the 1970s, makes it possible now to use the computed unrestrained column shortenings as input into a frame analysis to determine the resulting moments, shears, and axial forces and the residual shortenings. Such analyses are advisable and warranted for complex structures. For structures having a simple layout, a simplified analysis for differential shortening is considered sufficient, in view of the existing uncertainties in the assumptions of magnitude of creep and shrinkage, and their still unexplored dependence upon changes in climate.

It is reasonable to suggest at present a simple frame analysis for differential shortening similar to that suggested in Sections 10.9.1.5 and 10.9.1.6 of this chapter for the effects of temperature changes in exposed columns, by using the charts in Figs. 10-68, 10-69, and 10-70.

7. *Effects of Slow Differential Settlements of Slabs.* The elastic and inelastic differential shortening of supports of a slab in a multistory building does not occur instantaneously. Elastic shortening occurs during the period it takes to construct the structure above the slab under consideration. Creep and shrinkage shortening continues for years at a progressively decreasing rate. During this time creep of the concrete of the slab will cause relaxation of the moments caused by differential shortening; in other words, the elastic moments in the slab (due to slab settlement) will continue "creeping out."

Experimental and analytical studies by Ghali et al.[10-45] have shown that for settlements applied over a period of more than 30 days the amount of relaxation is about 50%. Thus for practical design of buildings, it seems reasonable to assume the maximum value of the differential settlement moments at 50% of the elastic moments that would occur without creep relaxation of the slab. This reduced moment should then be used with appropriate load factors in combination with other loads. The 50% reduction accounts only for creep relaxation during the period of settlement. Beyond this time a further creeping-out of the settlement moments takes place.

8. *Load Transfer between Adjacent Differentially Shortening Elements.* The girders and slabs tend to equalize the column stresses; hence, unless the columns are specifically designed for equal axial stresses, columns adjacent to each other over short spans usually carry substantially different (greater or smaller) loads from what can be accounted for on the basis of tributary areas at each floor.

In a frame with a differential settlement of supports, the support that settles less will receive additional load from the support that settles more. The transferred load is:

$$V = \frac{M_1 + M_2}{L}$$

where M_1 and M_2 are the settlement moments (reduced due to creep relaxation) at the two ends of the horizontal element, and L is the span. The load transferred over the entire height of the structure is a summation of load transferred on all floors.

The load transfer is cumulative starting from the top of the structure and progressing down to its base.

9. *Stresses.* The stresses due to differential shortening should be treated as equivalent dead load stresses with appropriate load factors before combining them with other loading conditions. In choosing a load factor, it should be considered that the design shortening-moments in slabs or beams occur only for a short while during the life of the structure, and they continue to creep out.

10. *New Procedures.* The above discussed solutions for column exposure and for creep and shrinkage of columns were prepared during the late sixties. At that time the structures were in the 50-story range, and the engineering offices used manual methods, primarily, for analysis and design. In the late seventies and in the eighties, changes in the construction industry have required an extension of the previously developed methods for designing buildings approaching the 1000-ft range. Also, the wider application of computer usage in structural engineering offices has made it possible and desirable to use the computer to solve problems requiring a great deal of meticulous arithmetic and extensive "bookkeeping" of data. Computer utilization is particularly significant because consideration of shrinkage and creep requires extensive computations and summations, as every story-high column segment in a multistory building is loaded in as many increments as there are stories above, and for each loading increment each column segment has new time-dependent properties (modulus of elasticity, creep coefficients, and shrinkage coefficients), changing column sizes (volume-to-surface ratios), and varying reinforcement ratios. Because of this complexity, very few structures were analyzed for the effects of shrinkage and creep using previous manual methods. Although the column length changes due to volume changes were mostly within tolerable limits for buildings in the 30–40 story range, there were instances where neglect in considering these effects caused performance problems.

The developed computerized procedure,[10-46] applicable

to concrete, steel, and composite structures, considers separately the elastic and creep components due to gravity loads, and also the shrinkage shortening. Since structural effects result from differential distortions caused by column shortening after a slab has been installed, the procedure separates the shortenings of supports that occur before slab installation from those that occur after slab installation. Those taking place before slab installation are important mainly for detailing of steel columns. The column shortening after slab installation can cause tilt of slabs, and needs to be known beforehand so that compensation can be specified during construction. It should be noted that composite structures having both concrete and steel columns are particularly susceptible to differential column length changes. Although the total shortening of a steel column and of a concrete column may be in the same range, their respective elastic shortenings differ greatly from one another; also the inelastic shortening occurs only in the concrete column.

Figure 10-79, from a design example in Ref. 10-46, shows the "before" and "after" slab installation deformations of supports of an 80-story composite structure having an exterior concrete beam-column system and interior steel columns. Shortenings have been computed for potential low and high values of shrinkage and creep. In this structure, light erection columns are embedded in the peripheral concrete columns; their elastic shortening prior to embedment becomes part of the total differential shortening between the interior and exterior columns. The total shortening of the composite columns of more than 10 in. is of secondary significance, since it affects only the cladding, for which relief details are customarily provided at every story or every several stories. Only the differential shortenings between the interior and exterior columns are of importance, since they cause slabs to tilt with structural and serviceability consequences. To keep the tilt of the slab within acceptable limits, the differential shortenings should be compensated for.

In Fig. 10-79(c), the total differential between the interior and exterior columns is shown to be about 4.3 in. at the 60th story, for a high creep value. Length corrections of the interior columns can be made during fabrication at every 10th story, as shown on the figure. Consequently, after all the loads have been applied and after shrinkage and creep have taken place, the slabs will be horizontal.

Figure 10-80 shows computed shortenings of an exterior column and a shear wall segment of a 70-story reinforced concrete office building.[10-46] The deformations that occurred before slabs were cast are of no structural consequence. As the formwork for each slab is usually installed horizontally, the pre-installation slab differentials are automatically compensated for. Only the after-slab-installation deformations (right side of Fig. 10-80) may need to be compensated for, if the predicted amount is more than can be tolerated. In the case of this 70-story building, the maximum post-installation differential is computed to be 1.05 in. at the 56th floor. It may be desirable to compensate for such distortion by simply raising (cambering) the formwork at the columns relative to the shear walls; thus, after the anticipated elastic, creep, and shrinkage length changes have taken place, the slabs will be horizontal.

It is important to note that the elastic, shrinkage, and creep differential shortenings can be compensated for, since they occur only once in the life of the structure. On the

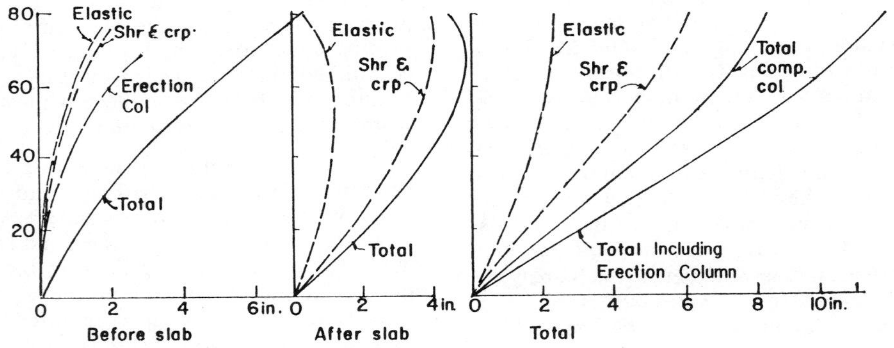

(a) Exterior composite column

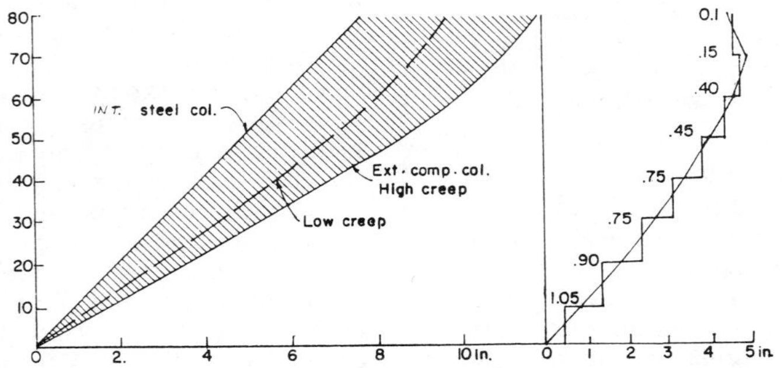

(b) Differential shortening between interior steel column and exterior composite column (c) Compensation

Fig. 10-79 Elastic, shrinkage, and creep shortening of vertical supports in an 80-story composite structure. Differential shortenings and compensation.

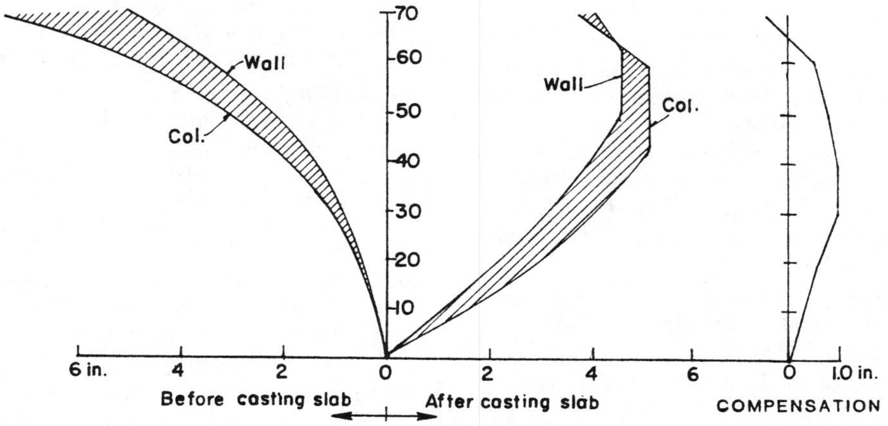

Fig. 10-80 Differential shortening between column and wall of a 70-story shearwall-frame structure.

contrary, length changes due to temperature and wind cannot be compensated for—they must be limited, since they occur repeatedly during the life of the structure.

REFERENCES

10-1 Fintel, M., and Ghosh, S. K. "Economics of Long-Span Concrete Slab Systems for Office Buildings—A Survey," SP 024.01B, Portland Cement Association, Skokie, IL., 1982.

10-2 Hodgkison, R. L. "An Ultra High-Rise Concrete Office Building," Master of Science thesis in School of Architecture, Illinois Institute of Technology, Chicago , Ill. June, 1968.

10-3 American National Standards, *Minimum Design Loads for Buildings and Other Structures*, ANSI A58.1-1982.

10-4 *ACI Committee 435*, "Allowable Deflections," *ACI Journal, Proceedings*, V. 65, No. 6, June 1968, pp. 433–444.

10-5 Fintel, Mark, "Deflections of High-Rise Concrete Buildings," *ACI Journal, Proceedings*, V. 72, No. 7, July 1975, pp. 324–328.

10-6 ACI Committee 442, "Response of Buildings to Lateral Forces," Mark Fintel, Chairman, *ACI Journal, Proceedings*, V. 68, No. 2, Feb. 1971, pp. 81–106.

10-7 "Frame Constants for Lateral Loads on Multistory Concrete Buildings," *Advanced Engineering Bulletin No. 5*, Portland Cement Association, 1962.

10-8 "Handbook of Frame Constants," T-32-2, Portland Cement Association, 1947, 32 pp.

10-9 Smith, C. S., and Carter, C., "A Method of Analysis for Infilled Frames," *Proceedings Inst. of Civ. Eng.* (London), V. 44, Sept. 1969.

10-10 "Continuity in Concrete Building Frames," 4th Ed., Portland Cement Association, 1959.

10-11 Khan, F. R., and Sbarounis, J. A., "Interaction of Shear Walls and Frames," *Proceedings, ASCE*, V. 90, ST 3, June 1964, pp. 285–335.

10-12 Beck, H., "Contribution to the Analysis of Coupled Shear Walls," *ACI Journal Proceedings*, V. 59, No. 8, Aug. 1962, pp. 1055–1070.

10-13 Pauley, T., "Coupling Beams of Reinforced Concrete Shearwalls," *Proceedings, ASCE*, V. 97, ST 3, Mar. 1971.

10-14 Coull, A., and Choudhury, J. R., "Stresses and Deflections in Coupled Shear Walls," *ACI Journal, Proceedings*, V. 64, No. 2, Feb. 1967, pp. 65–72.

10-15 Coull, A., and Choudhury, J. R., "Analysis of Coupled Shear Walls," *ACI Journal, Proceedings*, V. 64, No. 9, Sept. 1967, pp. 587–593.

10-16 Barnard, P. R. and Schwaighofer, Jr., "Interaction of Shear Walls Connected Solely through Slabs," *Tall Buildings*, Pergamon Press Limited, London, 1967, pp. 157–173.

10-17 Schweighofer, J., and Collius, M. P., "An Experimental Study of the Behavior of Reinforced Concrete Coupling Slabs," *ACI Journal, Proceedings*, V. 74, No. 3, Mar. 1977.

10-18 Pauley, T., and Taylor, R. G., "Slab Coupling of Earthquake-Resisting Shearwalls," *ACI Journal*, Pro*ceedings*, V. 78, Mar.-Apr. 1981.

10-19 Derecho, A. T., "Computer Program for the Analysis of Plane Multistory Frame-Shear Wall Structures Under Lateral and Gravity Loads," Portland Cement Association, 1971.

10-20 Cardenas, A. E., Hanson, J. M. Corley, W. G., and Hognestad, E., "Design Provisions for Shear Walls," *ACI Journal, Proceedings*, V. 70, No. 3, Mar. 1973.

10-21 Tsuboi, Y., Suenaga, Y., and Shigenobu, T., "Fundamental Study on Reinforced Concrete Shear Wall Structures—Experimental and Theoretical Study of Strength and Rigidity of Two-Directional Structural Walls Subjected to Combined Stresses M. N. Q.," *Transactions of the Architectural Institute of Japan*, No. 131, Jan. 1967; *PCA Foriegn Literature Study No. 536*, Nov. 1967.

10-22 Cardenas, A. E., and Magura, D. D., "Strength of High-Rise Shear Walls—Rectangular Cross Sections," *Response of Structures to Lateral Forces*, Publication SP-31, American Concrete Institute, Detroit, Mich., 1973.

10-23 Benjamin, J. R., *Statically Indeterminate Structures*, McGraw-Hill Book Company, New York, 1959, 347 pp.

10-24 Fintel, M., "The Behavior of Reinforced Concrete Structures in the Caracas Earthquake of July 29, 1967," XS6731, Portland Cement Association, 1968, 52 pp.

10-25 Fintel, Mark, "Ductile Shear Walls in Earthquake-

Resistant Multistory Buildings," *ACI Journal, Proceedings*, V. 71, No. 6, June 1974.

10-26 "Design of Combined Frames and Shear Walls," *Advanced Engineering Bulletin No. 14*, Portland Cement Association, 1965.

10-27 MacLeod, I. A., "Shear Wall-Frame Interaction–A Design Aid," Portland Cement Association, 1970.

10-28 Colaco, J. P., and Lombajian, Z. H., "Analysis of Transfer Girder System," *ACI Journal, Proceedings*, V. 68, No. 10, Oct. 1971.

10-29 "New Steel Framing System Promises Major Savings in High-Rise Apartments," *Architectural Record*, V. 139, June 1966, pp. 191–196.

10-30 Fintel, M., "Staggered Transverse Wall Beams for Multistory Concrete Buildings," *ACI Journal, Proceedings*, V. 65, No. 5, May 1968, pp. 366–378.

10-31 Fintel, M., Barnard, P. R., and Derecho, A. T., *Staggered Transverse Wall Beams for Multistory Concrete Buildings, A Detailed Study*, Portland Cement Association, Skokie, Ill., 1968.

10-32 Carpenter, J. E., and Hanson, N. W., "Tests of Reinforced Concrete Wall Beams with Large Web Openings," *ACI Journal, Proceedings*, V. 66, No. 9, Sept. 1969, pp. 756–766.

10-33 Cardenas, A. E., and Kaar, P. H. "Field Test of a Flat Plate Structure," *ACI Journal, Proceedings*, V. 68, No. 1, Jan. 1971.

10-34 Fintel, Mark, and Khan, F. R., "Effects of Column Exposure in Tall Structures, Temperature Variations and Their Effects," *ACI Journal, Proceedings*, V. 62, No. 12, Dec. 1965.

10-35 Khan, Fazlur R., and Fintel, Mark, "Effects of Column Exposure in Tall Structures, Analysis for Length Changes of Exposed Columns," *ACI Journal, Proceedings*, V. 63, No. 8, Aug. 1966.

10-36 Khan, Fazlur R., and Fintel, Mark, "Effects of Column Exposure in Tall Structures, Design Considerations and Field Observations of Buildings," *ACI Journal, Proceedings*, V. 65, No. 2, Feb. 1968.

10-37 Fintel, Mark, and Khan, F. R., "Effects of Column Creep and Shrinkage in Tall Structures–Prediction of Inelastic Column Shortening," *ACI Journal, Proceedings*, V. 66, No. 12, Dec. 1969.

10-38 Fintel, M., and Khan, F. R., "Effects of Column Creep and Shrinkage in Tall Structures–Analysis for Differential Shortening of Columns and Field Observations of Structures," *Designing for Effects of Creep, Shrinkage and Temperature in Concrete Structures*, Publication SP-27 American Concrete Institute, Detroit, Mich., 1971.

10-39 Hickey, K. B., *Creep of Concrete Predicted from Elastic Modulus Tests*, Report No. C-1242, U.S. Dept. of the Interior, Bureau of Reclamation, Denver, Colo., Jan. 1968.

10-40 Ross, A. D., "Creep of Concrete Under Variable Stress," *ACI Journal*, V. 54, No. 9, Mar. 1958, pp. 739–758.

10-41 McHenry, D., "A New Aspect of Creep in Concrete and its Application to Design," *Proceedings ASTM*, V. 43, 1943, p. 1069.

10-42 Troxell, G. E., Raphael, J. M., and Davis, R. E., "Long-Time Creep and Shrinkage Tests of Plain and Reinforced Concrete," *Proceedings ASTM*, V. 58, 1958, pp. 1101–1120.

10-43 Dishinger, F., *Der Baningenieur* (*Berlin*) V. 18 (39/40), Oct. 1937, pp. 595–621.

10-44 Pfeifer, D. W., "Reinforced Lightweight Concrete Columns," *Journal of the Structural Division, ASCE*, V. 95, ST 1, Jan. 1969, pp. 57–82; and Pfeifer, D. W. "Full-size Lightweight Concrete Columns," *Journal of the Structural Division, ASCE*, V. 97, ST 2, Feb. 1971, pp. 495-508.

10-45 Ghali, Amin, Dilger, W. and Neville, A. M.; Time-Dependent Forces Induced by Settlement of Supports in Continuous Reinforced Concrete Beams, *ACI Journal, Proceedings*, V. 66, No. 11, November 1969

10-46 Fintel, M., Iyengar, H., and Ghosh, S. K. "Column Shortening in Tall Structures–Predictions and Compensation," Portland Cement Association, 1984.

11

Tubular Structures for Tall Buildings

FAZLUR R. KHAN, Ph.D. *

11.1 INTRODUCTION

It is only in the last 30 years that reinforced concrete has found increasing use in the construction of tall buildings. In its initial development in the early 1900s reinforced concrete buildings were limited to only a few stories in height. The structural type used was the traditional beam-column frame system which made the construction of taller buildings relatively expensive and, therefore, economically unfeasible. In the early 1950s, the introduction of shear wall type of construction opened up the possibility of using concrete in apartment and office buildings as high as 30 stories. Taller buildings still remained economically unattractive, and technically inadequate, because the shear walls which were mostly used in the core of the building were relatively small in dimension compared to the height of such buildings, leading to insufficient stiffness to resist lateral loads. It was obvious that the overall dimensions of the interior cores were too small to economically provide the stability and stiffness for buildings over 30 or 40 stories.

The natural tendency then was to find new systems that would utilize the perimeter configurations of such buildings rather than to rely on the core configurations alone. The development of the framed tube system was, therefore, a logical outcome of this challenge. The framed tube system in its simplest form consists of closely spaced exterior columns tied at each floor level with relatively deep spandrel beams, thereby creating the effect of a hollow concrete tube perforated by openings for the windows. Since the system simulated a hollow tube using perimeter closely spaced frame elements, it is referred to as "framed tube" (Fig. 11-1). This system was apparently first applied on the design of the 43-story DeWitt-Chestnut apartment building in Chicago in 1963 (Fig. 11-2). Since then the system has received wide acceptance among designers all over the world, and many variations are being used in a number of buildings under construction.

From the point of view of construction economy, the framed tube compares favorably with the normal shear wall type of construction for medium-rise buildings, but provides a distinct economic advantage for taller buildings. Moreover, the closely spaced column system has the additional advantage of also being the window wall system, thus replacing the vertical mullions for the support of the glass windows. In some recent buildings the elimination of the traditional curtain wall with its metallic mullions was in itself the justification for choosing this structural system.

The center-to-center spacing of the exterior columns in the framed tube structural system is generally from 4 ft to a maximum of about 10 ft. Depending on the overall proportion and height of the building, the maximum center-to-

*Dr. Fazlur R. Khan, until his untimely death in March 1982 was Partner, Skidmore, Owings and Merrill, Chicago, Ill. See appendix at the end of this chapter by the Handbook Editor Mark Fintel.

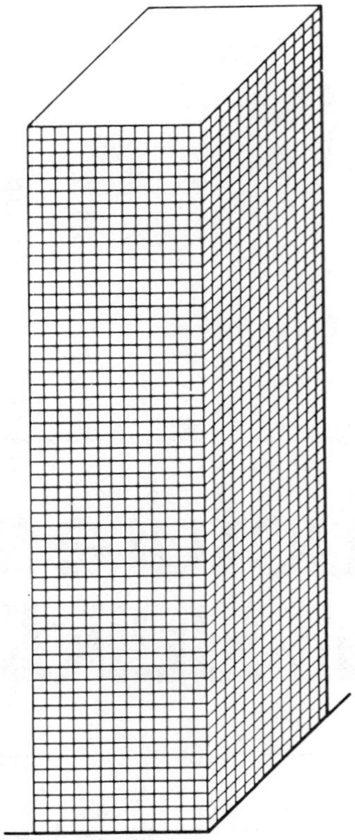

Fig. 11-1 Framed tube.

Fig. 11-2 The 43-story DeWitt-Chestnut apartment building in Chicago.

center spacing of the peripheral columns can probably be increased to 15 ft. The spandrel beams interconnecting the closely spaced columns generally vary from 2 ft in depth to about 4 ft in depth with widths from 10 in. up to 3 ft. In designing the framed tube structural system it is necessary to keep the proper balance of stiffness between the spandrels and the columns so that both of these elements are efficiently utilized to provide stiffness of the structure against lateral sway, and to assure the overall strength of the tube system to resist lateral forces. In most recent structures stiffness for limiting lateral sway controlled the proportions more often than strength requirements. It is obviously advisable to optimize the subsystem represented by two columns and two spandrels as shown in Fig. 11-3 to arrive at the optimum spacing of the columns and the proportion of the spandrels and the columns in terms of overall architectural program. In the DeWitt-Chestnut apartment building the columns were spaced on 5 ft 6 in. centers and the spandrels were 2 ft deep. The spacing of the columns in this case was also related to the module for interior planning of the apartment floors.

The framed tube structural system has expanded in its application and in its variation over the last five years. One of these variations is its application with an interior shear wall, commonly referred to as 'tube-in-tube' system, (discussed in section 11.8 below) as used in the 52-story One Shell Plaza Building in Houston, reaching a height of 714 ft. Another variation of the framed tube system has been used as the exterior envelope in conjunction with the traditional steel framing for the interior of the building. This system, known as the SOM Composite System, has been used in three major tall buildings in the country, namely: the 24-story CDC Building in Houston, the 35-story Union Station Building in Chicago, and the 50-story One Shell Square Building in New Orleans.

In view of the wide and varied application of the framed tube system, there is an obvious need for developing a preliminary analysis and design method for such a system. Of course, with availability of large computer programs, the final solution may be easily refined; however, the use of computers for preliminary design of even a moderate size framed tube may prove too expensive. Therefore, a method of preliminary analysis and design (using influence curves) for a wide range of proportions of the framed tube elements and geometry is presented.

This chapter is primarily intended for a clearer and better understanding of the behavior of the framed tube system.

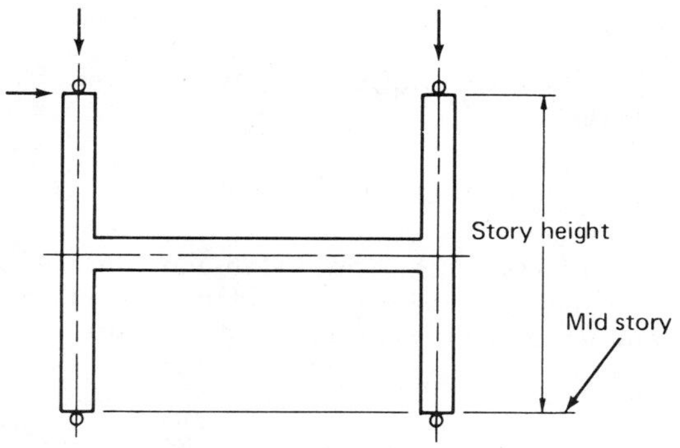

Fig. 11-3 Column-spandrel subsystem.

11.2 BEHAVIOR OF THE FRAMED TUBE SYSTEM

The framed tube system combines the behavior of a true cantilever, such as a shear wall, with that of a beam-column-frame. The overturning under lateral load is resisted by the tube form causing compression and tension in the columns, while the shear from the lateral load is resisted by bending in columns and beams primarily in the two sides of the building parallel to the direction of the lateral load (Fig. 11-4). Therefore, for all practical purposes the bending moments in these columns can be determined by judicious choice of the point of contraflexure in each story. While it is true that in the lower few stories, as well as in the upper few stories, the point of contraflexure does not remain in the middle of story height, the intermediate stories which constitute the major portion of the building generally have the point of contraflexure at mid-height of each story. It is, therefore, possible to compute the bending moments in these columns with reasonable accuracy for any known lateral shear at each story. One can, of course, make a simple iterative, Maney-Goldberg type, slope-deflection solution, or a modified moment distribution solution to determine more accurate moments in these columns. In fact, such an iterative solution will also give a good approximation of that portion of the total deflection which is caused by the frame action only. To this the additional overturning deflection caused by tension or compression in the column must be added to compute the total lateral deflection.

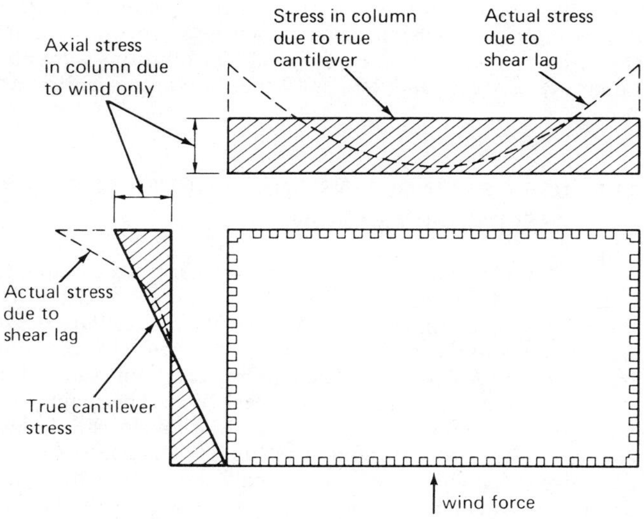

Fig. 11-4 Axial stress distribution in the columns due to wind—true cantilever vs actual stress due to shear lag.

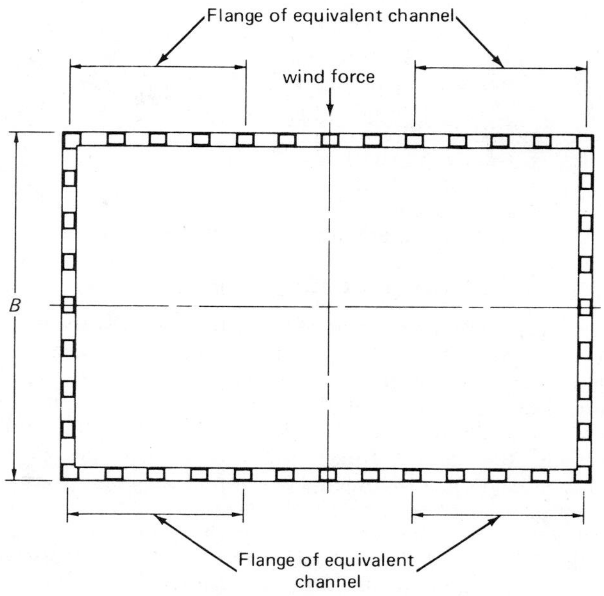

Fig. 11-5 Framed tube plan.

The cantilever tube type behavior becomes significant when the overturning of the entire building due to lateral load is considered. For analyzing the overturning of the entire framed tube, the exterior column system can be considered as part of a rigidly-diaphragmed hollow tube. However, in recognition of the fact that the webs of the hollow tube, that is, the two sides parallel to the direction of the lateral force, are not truly solid webs but are, in fact, grid frames, one must consider then the effect of loss of efficiency due to the flexibility of this web-frame causing what is known as shear lag, as shown in Fig. 11-4. For a very preliminary estimate of the overall resistance, as well as the deflection of the building, the effective configuration of the tube could be reduced to two equivalent channels resisting the total overturning moments (Fig. 11-5). Experience has indicated that for preliminary designs channel flanges normally should not be more than half the depth of the web (walls parallel to lateral load), or more than about 10% of the height of the building. These approximate rules have generally given conservative values of shear and moment as compared to the actual forces in the exterior columns obtained by the exact analysis performed subsequently by a generalized computer program such as STRESS, STRUDL, or others.

11.3 INITIAL PRELIMINARY DESIGN APPROACH

The overturning moment resisted by the two equivalent channels will produce axial forces in the closely spaced columns of these channels, as well as shear forces in the connecting spandrels. The preliminary estimate of the axial forces in the columns as well as the shears in the connecting spandrel can be based on the classical beam theory and can be expressed as follows:

$$P_w = \frac{M \times C \times A_c}{I_e}$$

and

$$V_s = \frac{V_w \times Q \times h}{I_e}$$

where

P_w = axial force due to wind
M = overturning moment
I_e = effective moment of inertia of the tube
V_s = spandrel shear
V_w = total wind shear
Q = sum of first moment of column areas about the neutral axis
C = distance of any column from the neutral axis
h = story height
A_c = cross sectional area of column

From the structural point of view, the framed tube differs from the solid tube by many large openings. Its behavior is, therefore, of a hybrid nature showing characteristics of the pure frame as well as of the pure tube. Even though one may expect the framed tube behavior to be more like a tube resisting the lateral forces through axial forces in the columns, significant moments develop in the columns in the two walls parallel to the wind direction.

In a framed tube-type structure the preliminary design method will generally indicate a reasonably uniform shear force in the spandrel beams along the two exterior walls parallel to the wind force. The preliminary moments in the spandrel beams are consequently derived from these shears. The preliminary design of the closely spaced columns should be based on the known dead and live loads added to the axial forces due to overturning; the moments caused by story shears should also be considered.

The study of a number of actual framed tube structures indicates that the lateral deflection, or sway, of these structures due to lateral load is primarily contributed by the frame action of the two walls parallel to the wind forces. For example, in the 43-story DeWitt-Chestnut Building, out of a total of 7 in. of lateral sway under the Chicago wind loading, approximately 5 in. (about 70%) was contributed by the frame action and only 2 in. was contributed by the overturning moment. This particular case shows that the efficiency of a framed tube diminishes as the building gets taller. As the height reaches a certain point, the lateral sway, and not the strength, of the building will control the design of the structural system. Therefore, the structural efficiency producing a premium-free building will gradually diminish with increasing number of stories.

11.4 PRELIMINARY DESIGN USING INFLUENCE CURVES

To achieve a more accurate design than the equivalent channel method proposed earlier for framed tube structures of any proportion and height within practical range, the author developed influence curves which can be directly used for a relatively accurate preliminary design. These curves have been developed on the basis of a number of computer runs on a 10-story equivalent framed tube with variable nondimensional parameters representing ratios of shear stiffness, S_b, of the spandrel beam to the axial stiffness, S_c, of the columns, and a linearly varying ratio of bending stiffnesses of columns to spandrels. A check of a number of framed tube buildings showed that this ratio ranged from 0.95 at the roof to 0.4 at the ground level. For this chapter a ratio of 0.75 at the roof to 0.5 at ground level is assumed. A separate computer run, with a constant ratio of bending stiffness of columns to spandrels of 1.0, did not change the qualitative results.

The significant structural properties affecting the tube action are

1. Bending stiffness:

$$K_c \text{ for column} = \frac{I_c}{H}$$

$$K_b \text{ for spandrel beam} = \frac{I_b}{L}$$

2. Shear stiffness of the spandrel beams (defined as the force required to displace one end of the spandrel a unit distance at right angles to the axis of the beam):

$$S_b = \frac{12\,EI_b}{L^3}$$

3. Axial stiffness of the column (defined as the axial force required to shorten the column a unit distance along the axis of the column):

$$S_c = \frac{A_cE}{H}$$

where

I_c = moment of inertia of the column
I_b = moment of inertia of the spandrel beam
A_c = cross-sectional area of the column
H = height of column
L = effective span of the spandrel beam
E = modulus of elasticity

The controlling parameters of framed tubes are.

$$\text{stiffness ratio} = \frac{K_c}{K_b}$$

$$\text{stiffness factor, } S_f = \frac{S_b}{S_c}$$

$$\text{aspect ratio, R} = \frac{\text{flange frame}}{\text{web frame}}$$

To plot influence curves, framed tubes have been analyzed for uniform lateral load and for the following ranges of variables: aspect ratio values of 0.5, 0.666, 1.0, 1.5 and 2.0; stiffness factor values of 0.1, 1.0, 10.0, and a linearly-varying stiffness ratio of 0.75 at roof to 0.5 at ground level for all values of aspect ratio and stiffness factor combinations.

Analysis of the equivalent framed tube from which the design curves have been developed was based on the configuration shown in Fig. 11-6, and direct solutions were obtained by converting the framed tubes into equivalent plane frames.

11.5 USE OF NONDIMENSIONAL CURVES FOR PRELIMINARY DESIGN

A total of nine nondimensional preliminary design curves, Figs. 11-7 to 11-15 are presented. These curves are primarily for computing column axial force coefficients for flange and web frame columns and shear force coefficients for the web frame beams. All these coefficients relate to unit values for the corresponding forces of the ideal tube. The main purposes of developing the curves was to provide the design engineer a tool to determine the tubular characteristics of any given framed tube and quickly compute the total deflection and bending moments and shears in the beams caused by the tubular nature of the entire structure. To make all the curves applicable to a wide range of realistic

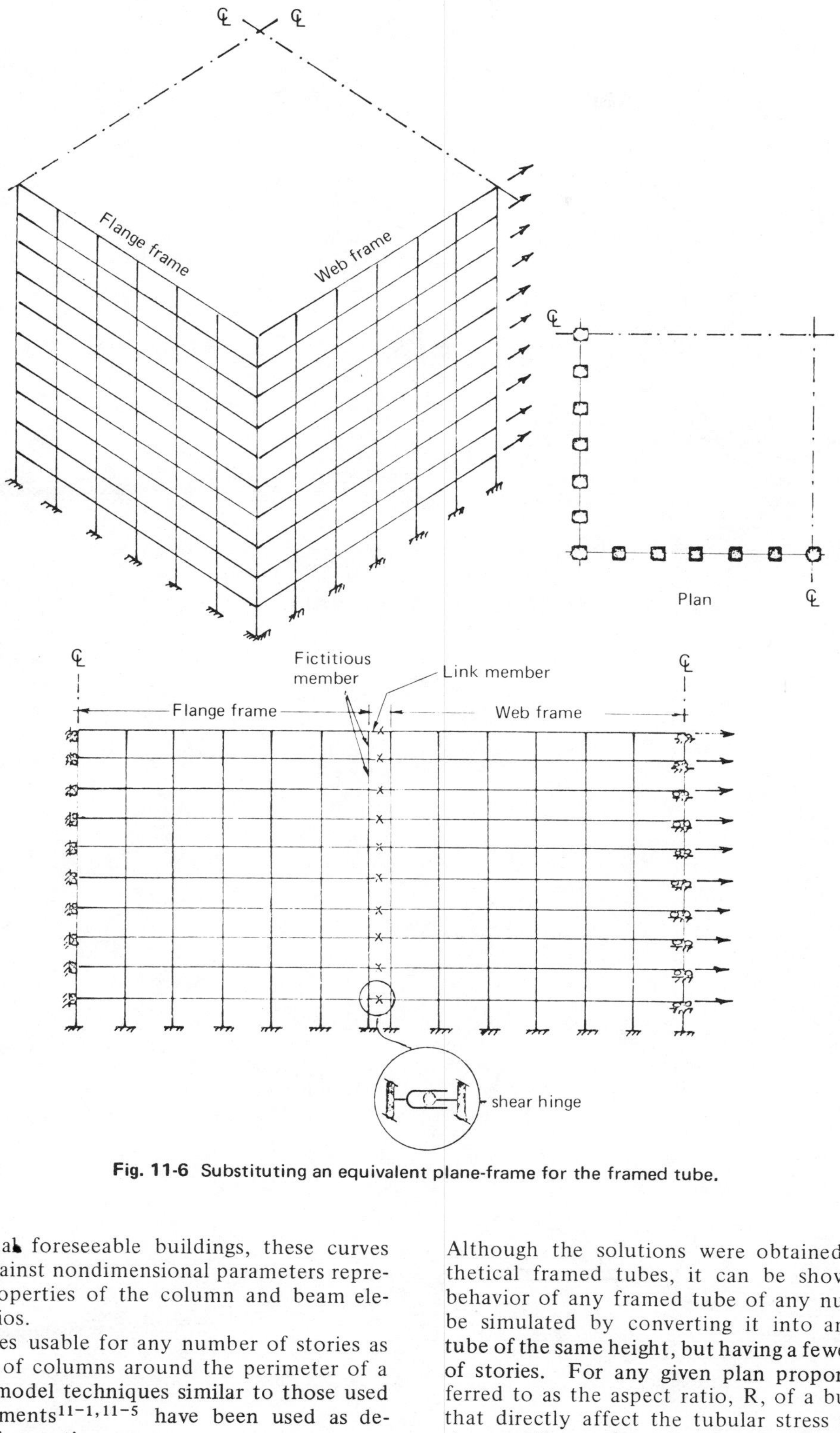

Fig. 11-6 Substituting an equivalent plane-frame for the framed tube.

proportions of actual foreseeable buildings, these curves have been plotted against nondimensional parameters representing the basic properties of the column and beam elements and aspect ratios.

To make the curves usable for any number of stories as well as any number of columns around the perimeter of a building, reduction model techniques similar to those used in previous developments[11-1,11-5] have been used as described in the following sections.

11.6 REDUCTION MODELING APPROACH

To achieve uniformity of reference to the design curves, computer solutions were obtained on a number of specific 10-story frame tubes, each representing stiffness factor, aspect ratios, and stiffness ratio indicated on the curves. Although the solutions were obtained on 10-story hypothetical framed tubes, it can be shown that the tubular behavior of any framed tube of any number of stories can be simulated by converting it into an equivalent framed tube of the same height, but having a fewer or greater number of stories. For any given plan proportion, commonly referred to as the aspect ratio, R, of a building, the variables that directly affect the tubular stress distribution are the shear stiffness, S_b, of the spandrel beams, and the axial stiffness, S_c, of the columns. For reasonably uniform spacing of perimeter columns, S_b and S_c in the graphs represent the sum total of all columns and beams around the perimeter. If, however, stiffness factor, S_f, varies on each face, an average value can be taken for a reasonable approximation.

It can be shown that any framed tube of N stories can be reduced to a 10-story equivalent tube of the same height by

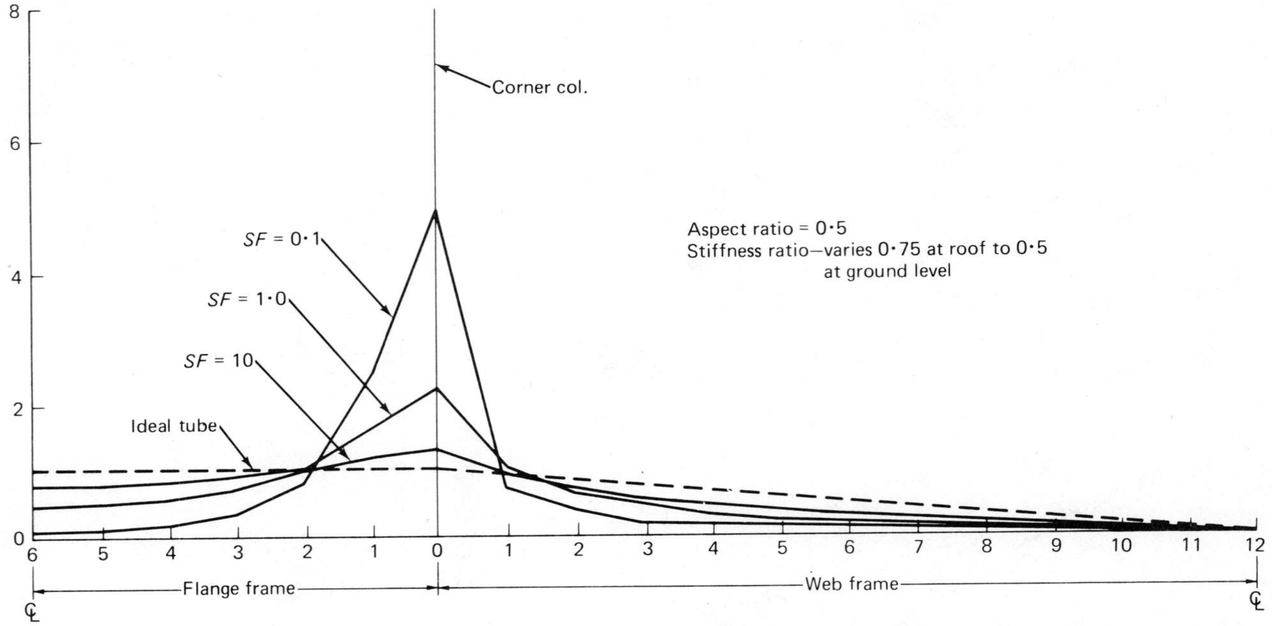

Fig. 11-7 Column axial force coefficients—level 1, aspect ratio = 0.5.

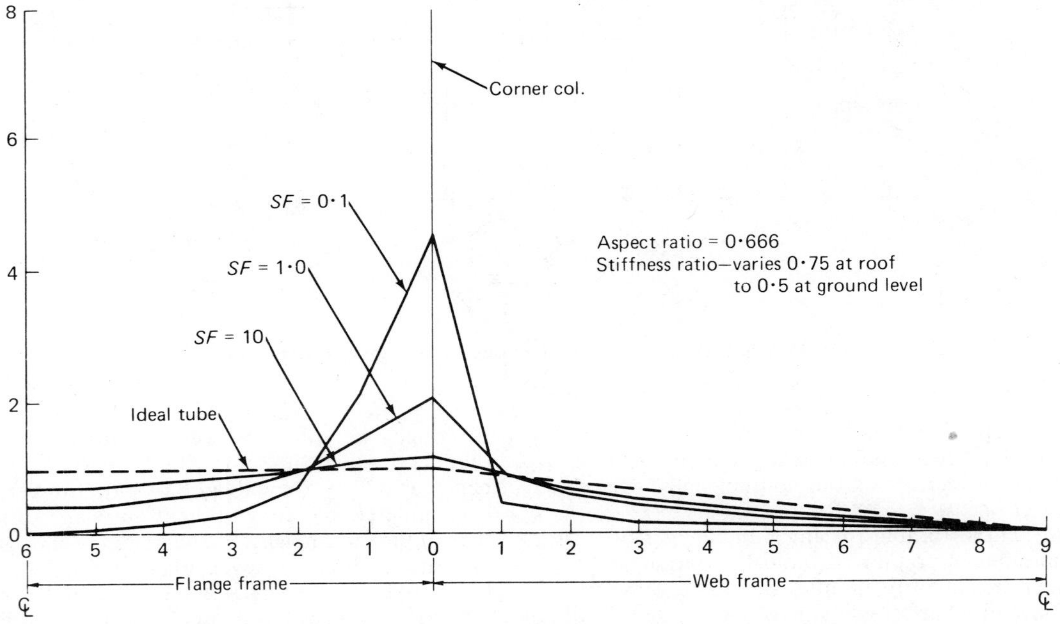

Fig. 11-8 Column axial force coefficients—level 1, aspect ratio = 0.666.

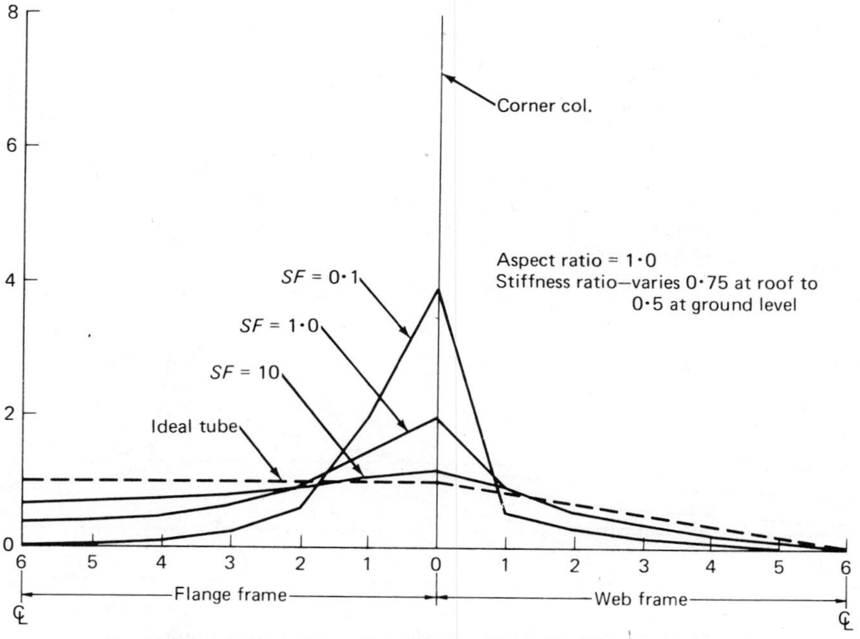

Fig. 11-9 Column axial force coefficients—level 1, aspect ratio = 1.0.

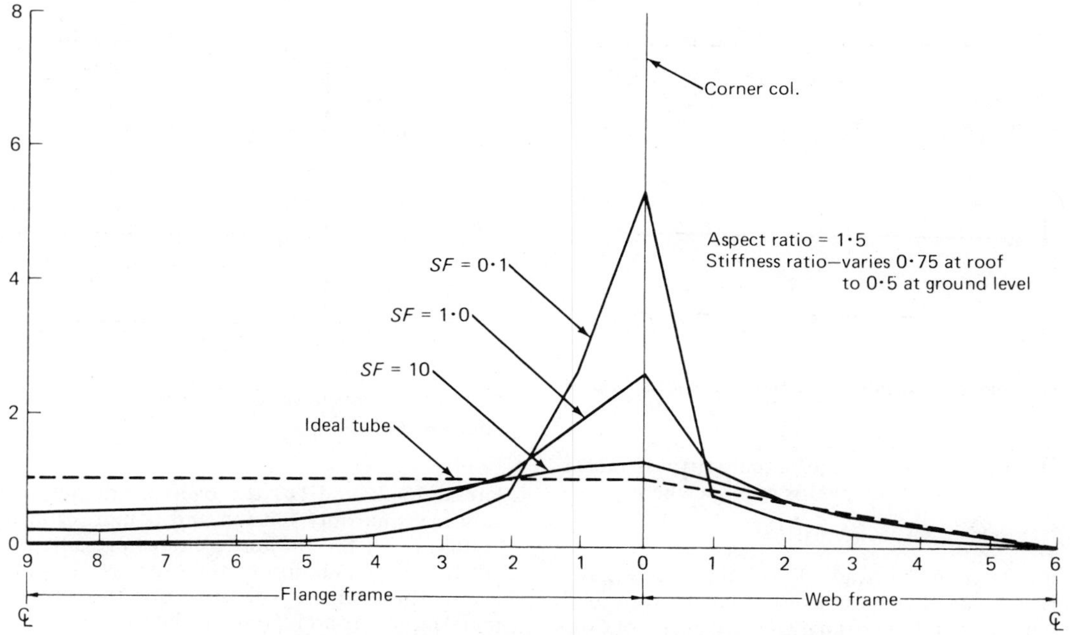

Fig. 11-10 Column axial force coefficients—level 1, aspect ratio = 1.5.

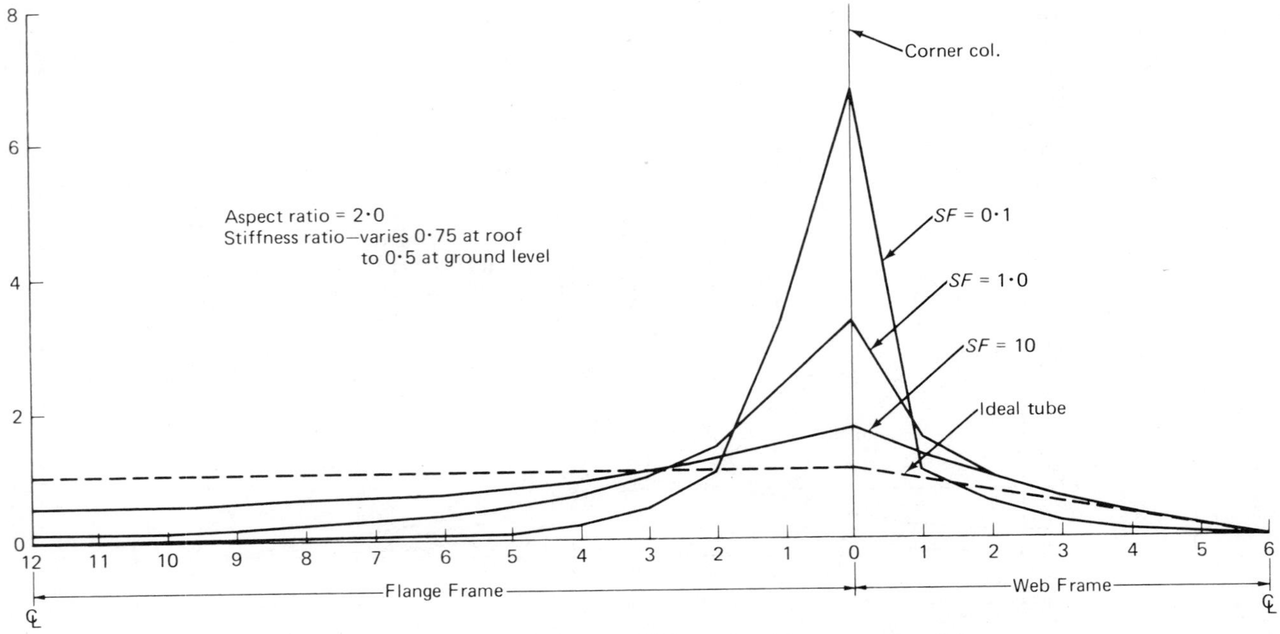

Fig. 11-11 Column axial force coefficients—level 1, aspect ratio = 2.0.

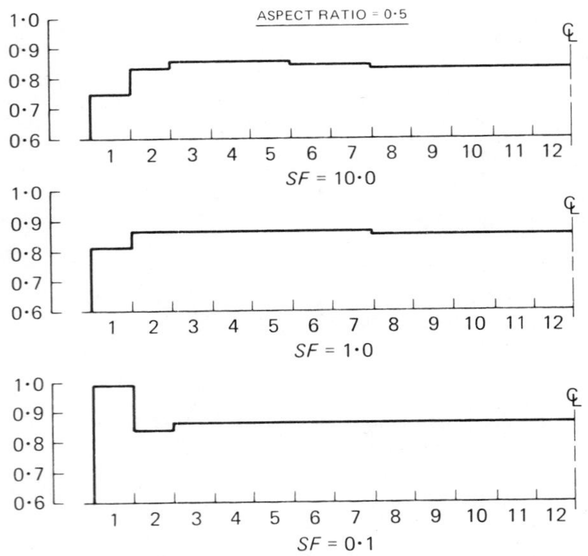

Fig. 11-12 Shear force coefficients in web-frame beams, aspect ratio = 0.5.

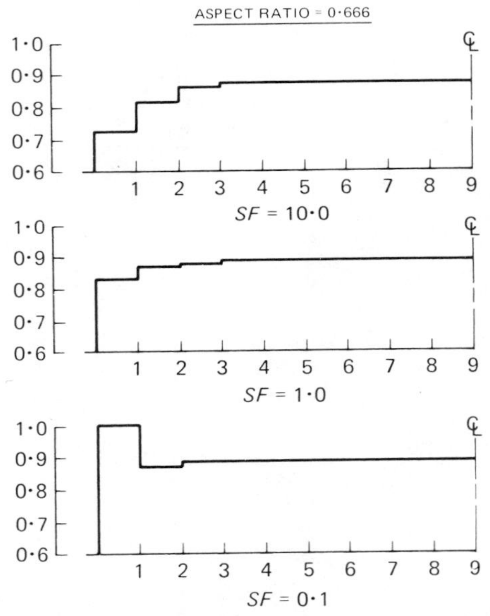

Fig. 11-13 Shear force coefficients in web-frame beams, aspect ratio = 0.666.

considering a transformation of the actual stiffness factor to a 10-story equivalent stiffness factor of S'_{f10} where

$$S'_{f10} = S_f \times (N/10)^2$$

Using this simple reduction model technique, any given actual framed tube can first be converted to an equivalent 10-story framed tube and then analyzed by using the influence curves.

11.7 TOTAL BEHAVIOR OF A FRAMED TUBE

As was pointed out earlier in this chapter, the framed tube system always has two components of its behavior: (a) the frame action of the two sides parallel to the direction of the lateral load, and (b) the overturning action of the entire tube causing only tension and compression in the exterior columns. All the influence curves presented in this chapter are for the evaluation of the tube action only, although the use of these curves will allow one to compute the approximate moments and shears in the spandrels which will define one boundary of the column moments. The other boundary of the column moments at any typical story should be obtained by assuming the point of contraflexure at mid-height of each story. For preliminary design the higher of the two values should be used.

To compute the total deflection, the additional deflection due to frame action must be calculated separately and added to the deflection caused by tube action.

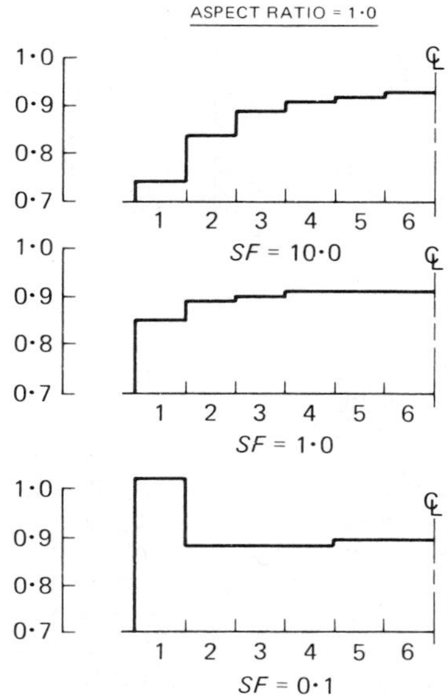

Fig. 11-14 Shear force coefficients in web-frame beams, aspect ratio = 1.0.

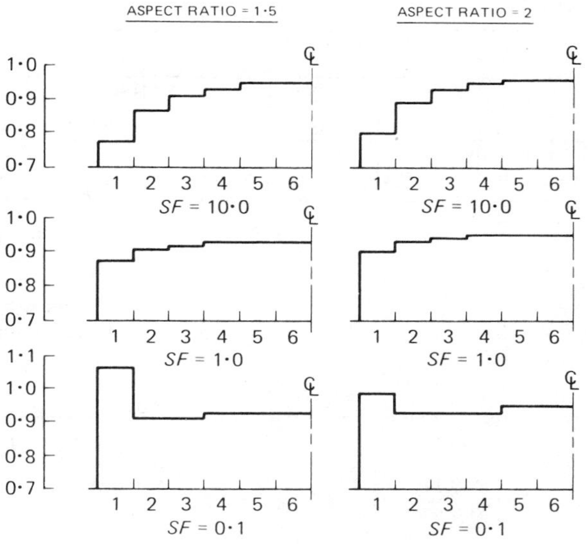

Fig. 11-15 Shear force coefficients in web-frame beams, aspect ratio = 1.5.

EXAMPLE 11-1: To illustrate the use of influence curves for preliminary designs, a hypothetical framed tube 50 stories high and 130 × 180 ft in plan with columns at 10 ft on center and a 13-ft floor to floor height is considered.

Before using the influence curves, the basic parameters of the structure, like aspect ratio, average stiffness factor, average stiffness ratio, and frame deflection, must be computed. Also, the deflection and forces in columns and beams for an ideal tube of the same Aspect Ratio must be known. Basic section properties and parameters are presented in Tables 11-1 and 11-2.

Frame deflection can be computed by any standard method and is approximately 3.79 in. From Table 11-2, by conjugate-beam method, cantilever deflection is computed as 7.5 in. Any other similar method could also give reliable results.

The actual forces in columns, shears in beams, and deflection of the system, can now be computed using the influence curves. For this example, Fig. 11-10, SF = 10 curve is used for computing the column axial forces in first story columns, and Fig. 11-15 (c), SF = 10 curve is used for computing the beam shears in the first story beams of a 10-story model. The above results are summarized in Table 11-3 and compared with the results obtained from a more exact computer solution.

The deflection of the total system is the sum of the frame deflection plus the ideal tube deflection times the magnification factor. This magnification factor is defined as the ratio of the sum of the ideal tube column forces to the actual column forces of the flange columns. For this example, the sum of the ideal tube column forces is 1280 × 10 = 12800 kips and the sum of the actual column forces is 10825 kips. Hence, the magnification factor is 1.17.

$$\Delta \text{ Total } = 3.79 + 7.5 \times 1.17 = 12.49 \text{ in.}$$

$$\Delta \text{ Actual} = 11.51 \text{ in.}$$

Comparing the values of the above example problem with an exact computer analysis shows that the use of influence lines gives fairly reliable results with an accuracy of 90 to 95%. The shear forces in the spandrel beams of the web frames (parallel to the wind direction) differ by 5% only from the exact analysis at the center of

TABLE 11-1

	Column Properties			*Spandrel Beam Properties*					
Floor	*Area in.*^{2}	*M. of I. in.*^{4}	*Axial Stiffness* $S_cK/in.$	*Area in.*^{2}	*M. of I. in.*^{4}	*Shear Stiffness* $S_bK/in.$	*Stiffness Factor* $S_f = S_b/S_c$	$S_{f10} = S_f(N/10)^2$	*Stiffness Ratio* = K_c/K_b
1-5	972	462000	24400	1455	452000	13000	0.53	13.2	0.795
6-10	810	447000	20400	1200	372000	10700	0.525	13.1	0.925
11-15	810	447000	20400	1200	372000	10700	0.525	13.1	0.925
16-20	810	447000	20400	1200	372000	10700	0.525	13.1	0.925
21-25	717	400000	18000	1058	330000	9520	0.525	13.1	0.935
26-30	615	363000	15400	940	282000	8140	0.52	13.0	1.0
31-35	615	363000	15400	940	282000	8140	0.52	13.0	1.0
36-40	510	307000	12800	754	234000	6750	0.52	13.0	1.01
41-45	510	307000	12800	754	234000	6750	0.52	13.0	1.01
46-50	510	307000	12800	754	234000	6750	0.52	13.0	1.01

Average Stiffness Factor for ten-story model = 13.05

Average Stiffness Ratio = 0.95

$$\text{Aspect Ratio} = \frac{180}{130} = 1.4$$

For this example use Stiffness Factor = 10.0

and Aspect Ratio = 1.5

TABLE 11-2

Floor	*1*	*5*	*10*	*15*	*20*	*25*	*30*	*35*	*40*	*45*
Overturning Moment, ft-kip	975000	789750	624000	477750	351000	243750	156000	87750	39000	9750
Tubular Moment of Inertia = Σ AR^2, ft^4	333444	277870	277870	277870	245966	210975	210975	174955	174955	174955

$$\text{Column axial force at Level 1} = \frac{M \times C}{I} \times A_c = 1280 \text{ kip}$$

$$\text{Spandrel beam shear at center line spandrel} = \frac{VQ}{I} \times H = 2850 \text{ kip}$$

TABLE 11-3

Flange Column Axial Forces			*Shear Force in Web Frame Beams*		
Col. Location	*Using Influence Curves kips*	*Actual, kips*	*Beam Location*	*Using Influence Curves, kips*	*Actual, kips*
0	1660	1583	1	2340	1954
1	1530	1426	2	2620	2257
2	1280	1237	3	2750	2450
3	1150	1100	4	2810	2565
4	1020	1001	5	2850	2640
5	900	929	6	2850	2690
6	830	877	7	2850	2700
7	830	841			
8	830	818			
9	795	806			

the frame; the difference, increasing progressively towards the corners. However, since in a typical framed tube structure the size of the spandrel beam is generally constant at any level, the inaccuracy of the shears in the corner spandrel beam during the preliminary design is of little consequence.

11.8 TUBE-IN-TUBE STRUCTURAL SYSTEM

Tall office buildings have generally a reasonably large service core and it is advantageous to use a shear wall enclosing the entire service core. Because of the demand for column-free office space, it is then a natural solution to eliminate all interior supports and provide closely spaced exterior columns, thus creating a framed tube. In most such recent office buildings this clear span generally varies from 35 to 40 ft. The resulting structural system consists of an inner tube created by the shear walls, and the outer tube consisting of the closely spaced column system, creating what may be called a tube-in-tube system.

The tube-in-tube system has the advantage of both the framed tube structure, as well as the shear wall type structure. In fact, the shear wall inner tube greatly enhances the structural characteristics of the exterior framed tube by considerably reducing the shear deflection of the columns of the framed tube. The tube-in-tube system is a refined and unique version of the shear wall-frame interaction type structures as described in section 10.5.7 of this handbook.

11.8.1 Approximate Analysis

Before making an approximate analysis for preliminary design of this type of structure, it is first necessary to define the factors contributing to its total lateral deflection. The tube-in-tube structure subjected to lateral loads will primarily act as a shear wall-frame interactive system, interaction being between the interior shear wall tube and the walls of the framed tube parallel to the direction of wind. An additional deflection will occur due to column shortening of the exterior tube caused by overturning moments.

For approximate analysis of the shear wall-frame interaction type behavior of this system, one of the most direct approaches would be to use the influence charts presented in Ref. 11-2. Results from these charts should be sufficient for a preliminary design, as the more sophisticated, exact analysis of actual structures have shown that these influence charts normally give results within about 5% accuracy. The column shortening of the framed tube can be taken into account with reasonable accuracy as suggested in Ref. 11-2. The method suggests compensation for the column shortening by adjusting the actual moment of inertia of the inner shear wall tube to an equivalent moment of inertia slightly larger than its actual value. This procedure may be summarized as follows:

$$I_e = I \times (1 + K')$$

where

I_e = equivalent increased moment of inertia of inner tube
I = actual moment of inertia of inner tube
K' = λ_1/λ_2
λ_1 = pure frame stiffness of framed tube
λ_2 = pure rigid tube stiffness of framed tube

After the shear wall-frame interaction charts are used to determine the lateral shear distribution between the shear wall and the exterior frame tube, the two separate systems, viz., the shear wall and the framed tube, can be designed

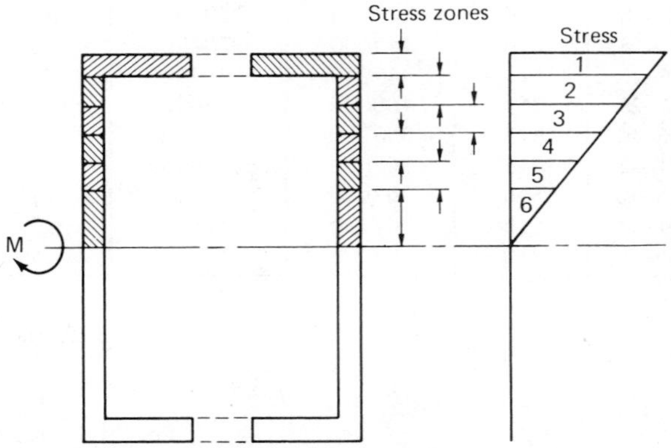

Fig. 11-16 Dividing a shear wall into stress zones.

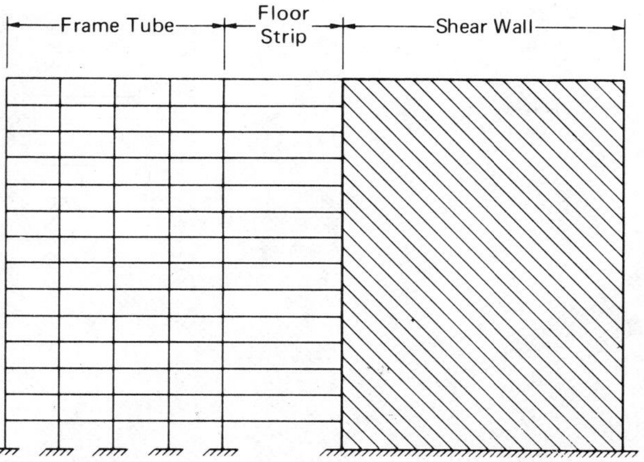

Fig. 11-17 Plane-frame model of tube-in-tube system.

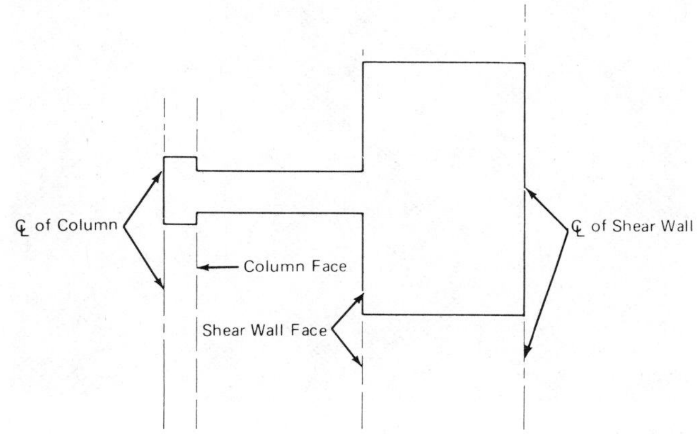

Fig. 11-18 Simulation of the link beam.

separately. The preliminary design of the exterior framed tube has already been discussed in detail earlier. The design of the shear wall may require distribution of the shear force in the shear wall to the different elements which make up the shear wall tube. In designing the shear wall, the overturning moment in the shear wall may cause large axial stresses which can be handled in the design by dividing the shear wall into small segments and considering each of these segments as a column unit as shown in Fig. 11-16. This rational design method will allow distribution of vertical reinforcement throughout the shear walls according to the expected stress intensity due to overturning moments. Another method is to treat the shear wall as a flexural reinforced concrete element and to concentrate the tensile reinforced the furthest distance away from the neutral axis as possible (see section 10.5.6.2).

11.8.2. Exact Method of Analysis

It is not practical at the present time to attempt a truly exact analysis in view of its complexity due to the large number of members and joints in the exterior framed tube, and due to an extremely large moment of inertia of the shear wall tube which is interconnected to the tube by a rather flexible floor structure. However, in recognition of the nature and philosophy of structural design, a somewhat idealized mathematical model may be used for the tube-in-tube structure. One possibility of analysis is to apply an iterative method using a forced-convergence technique[11-2, 11-4] to the idealized structure consisting of the framed tube interconnected with the inner shear wall tube.

While the presently available generalized type computer programs such as STRESS and STRUDL are easily adaptable to framed tube type structures as discussed earlier, their adaptation to tube-in-tube type structures becomes somewhat undesirable because of the problems involved in simulation of this system into a space-frame. It is possibly easier to simulate the tube-in-tube structure as a plane-frame composed of an equivalent frame representing the framed tube and an equivalent shear wall representing the inner tube; both connected by the floor structure represented by a link beam as shown in Fig. 11-17. The plane-frame modeling will generally give acceptable results except for the following two possible problems:

a. If the shear wall moment of inertia is extremely high compared to the columns of the framed tube, there may be a possibility of ill-conditioned matrix and round-off errors.
b. If the structural floor slab is sufficiently rigid, then the rotation of the shear walls will cause vertical displacements at the junction of the shear wall and the link beam which must be compensated. This can be done by simulation of the link beam as shown in Fig. 11-18.

REFERENCES

11-1 Khan, F. R., "Current Trends in Concrete High Rise Buildings," *Proceedings*, Symposium on Tall Buildings, University of Southhampton, England, April, 1966.

11-2 Khan, F. R., and Sbarounis, J. A., "Interaction of Shear Walls and Frames," *Proceedings, ASCE*, **90** (St 3), 285–335, June 1964.

11-3 Khan, F. R., "On Some Special Problems of Analysis and Design of Shear Wall Structures," *Proceedings*, Symposium on Tall Buildings, University of Southhampton, England, April, 1966.

11-4 Khan, Fazlur R., and Fintel, Mark, "Effects of Column Exposure in Tall Structures, Analysis for Length Changes of Exposed Columns," *ACI Journal, Proceedings*, **63** (8), August 1966.

11-5 Khan, F. R. and Amin, N. R., "Analysis and Design of Framed Tube Structures for Tall Concrete Buildings," SP36, ACI, Detroit, 1973.

APPENDIX

The untimely death of Dr. F. R. Khan, the author of this chapter, at the age of 52 was a severe loss to the engineering community. During the 1960s and 70s, Dr. Khan became the most prominent innovator in the area of highrise buildings, both in concrete and steel.

It was during the early sixties that he laid the groundwork for his later successes in the field of highrise buildings. This was a time when intense urbanization was bringing in its wake a new wave of highrise buildings. His 1964 ASCE paper[11-2], coauthored by J. A. Sbarounis, on shear wall-frame interaction was a milestone in the development of economical highrise buildings in both concrete and steel. With the methodology developed in this paper, the stiffness of frame buildings could be increased up to several times, without an increase in cost.

In the same period, Dr. Khan also initiated the tubular design concept, with its first application in the 43-story reinforced concrete Chestnut-Dewitt apartment building in Chicago in 1963 (see Fig. 11-2).

The next innovation, pushing still further economically feasible height of multistory buildings, was the application of shear wall-frame interaction principles to tubular structures, creating the tube-in-tube concept (a new phrase coined by Khan), applied first to the

Brunswick Building in Chicago (see Fig. 10-50). This concept was soon applied to many other structures, including the 52-story Shell Oil Plaza in Houston (Fig. 10-17) which was the tallest reinforced concrete building in the world at the time of its completion.

Also, in the 60s, came Khan's first steel version of the tubular structure: the diagonally braced 100-story John Hancock Building in Chicago. It became another milestone, particularly due to the strong expression of its dominant structural feature in the architectural facade of the building.

Then came the Sears Tower, using a further innovation–bundling nine tubes into a single structural system–with 110 stories and 1450 ft height, the world's tallest building. Like the John Hancock, it used about half the steel needed for a conventional design.

More innovations followed under Dr. Khan's direction, including composite buildings, combining the advantages of the rigidity of a concrete tubular structure and the speed of erection of steel slab systems and interior columns. Numerous ultra highrise composite structures up to 66 stories in height have been built since 1965.

A principal feature of Khan's work was to make highly efficient exterior tubular configurations carry the lateral loads imposed on multistory buildings, rather than assigning this role to less efficient interior frames which clutter the rentable space, as had been common to this time. The innovations introduced by Khan not only improved the rigidity of tall buildings, resulting in their superior

Fig. 11-19 Diagonally braced framed tube, 780 3rd Ave., New York (during construction, April 1983) (*courtesy R. Rosenwasser & Assoc., New York*).

Fig. 11-20 Bundled tubes. One magnificent mile, Chicago. (*Photo courtesy Shidmore, Owings & Merrill, Chicago*).

performance, but also resulted in substantial economies over the cost of buildings designed using traditional schemes. Most of the ultra highrise buildings today are built on principles introduced by Fazlur Khan. He became the master builder of tall structures of the 60s and 70s, and his influence will endure for decades to come. His buildings represent an economic answer to the needs of the day, utilizing advanced technology and the art of engineering. Khan believed that there is beauty and simplicity in the structural form of a building that is natural to it from an engineering point of view. Instead of going for a preconceived architectural expression, he let the natural structural form be the architectural expression of the building.

Dr. Khan's last accomplishments were the adaptation of the diagonally braced tube to concrete for the 50-story 780 3rd Avenue building in New York (shown in Fig. 11-19) and the bundled tube concept adapted to concrete for the 58-story One Magnificent Mile in Chicago, shown in Fig. 11-20. Construction on both of these projects has been completed since Dr. Khan's death in March of 1982.

Since the first edition of this Handbook was published in 1974, very few new technical developments have taken place in the area of tubular structures. The many efficient, large, sophisticated two- and three-dimensional computer programs have made the analysis of tubular structures much simpler, more economical, and accessible to design engineers.

MARK FINTEL
Editor

12

Earthquake-Resistant Structures

ARNALDO T. DERECHO, Ph.D.*, MARK FINTEL,** and S. K. GHOSH, Ph.D.***

12.1 INTRODUCTION

Although the incidence of earthquakes of destructive intensity has been confined to a relatively few areas of the world, the catastrophic consequences attending the few that have struck near centers of population have focused attention on the need to provide adequate safety against this most awesome of nature's quirks.

The satisfactory performance of a large number of reinforced concrete structures subjected to severe earthquakes in different areas of the world has demonstrated that it is possible to design such structures to successfully withstand earthquakes of major intensity.

Early attempts to provide for earthquake resistance in buildings were based on rather crude assumptions about structural behavior and were handicapped by a lack of proper analytical tools as well as reliable earthquake records. Observations of the behavior of reinforced concrete structures subjected to actual earthquakes, analytical studies, and laboratory experiments by a number of investigators over the last three decades or so have all contributed toward putting the subject of earthquake-resistant design on a firm rational basis.

The present state of knowledge regarding earthquake-resistant design and construction has benefited from efforts aimed at understanding the various aspects of the problem, the major ones being:

1. The earthquake phenomenon. Seismological and geophysical studies have not only yielded insight into the immediate cause of earthquakes but have also provided a basis for the prediction of earthquake occurrence. The continued accumulation of records on magnitude and geographical distribution of earthquakes and related crustal strains represents a progressively firmer basis for the statistical estimation of the likelihood of occurrence of earthquakes of specified ranges of magnitude in particular regions.
2. Behavior of structures and structural components under earthquake loading. Analytical studies of the dynamic response of idealized models of structures to earthquake excitation have provided much valuable information which has helped explain the observed behavior of actual structures subjected to earthquakes. Such dynamic response studies, using actual, modified, or artificially generated accelerograms as input data, represent one of the most useful tools for the design of earthquake-resistant structures. The sophistication

*Consultant, Wiss Janney, Elstner Associates, Inc., Northbrook, Illinois.
**Principal FINTEL GHOSH, Inc., Glenview, IL., formerly director, Advanced Engineering, Portland Cement Association, Skokie, Illinois.
***Principal FINTEL GHOSH, Inc., Glenview, IL., Associate Professor of Civil Engineering, University of Illinois at Chicago, Chicago, Illinois.

with which these analyses have been carried out has been made possible only by the availability of large electronic digital computers.

3. Experimental studies aimed at understanding the behavior of structural members subjected to earthquake-type loadings have helped provide the basis of the design specifications on earthquake-resistant construction.
4. On-site observations of structures subjected to earthquakes have provided valuable lessons on the effects of deficiencies in design and construction, and a basis for assessing the adequacy of earthquake design specifications.

Much still remains to be done, especially in the area of seismic regionalization. Various aspects of seismic structural response still await further clarification, such as damping, foundation–structure interaction, and three-dimensional nonlinear behavior; a large amount of information still must be accumulated in order to be able to treat earthquake analysis and design on a sound probabilistic basis. However, the experience of the last three decades has indicated that proper analysis and design, guided by a knowledge of the basic objectives to be achieved, and particular attention to design details, can provide safe and economical structures to resist earthquakes.

The material that follows will be concerned principally with earthquake engineering as it applies to the structural design of reinforced concrete buildings. It should be noted, however, that many of the observations made in relation to buildings also apply, to some extent, to other civil engineering structures.

12.2 EARTHQUAKES

12.2.1 Causes of Earthquakes

The exact chain of events that leads to earthquakes has been the subject of much theorizing and speculation. Seismologists agree on two distinct mechanisms which directly result in earthquakes. *Tectonic* earthquakes are disturbances (waves of distortion) resulting from a rupture or a sudden movement along an existing fault in the earth's crust,* while *volcanic* earthquakes are those associated with volcanic eruptions or subterranean movements of magma. The major portion of observed seismic activity has been tectonic in origin.

A widely held theory of the tectonic mechanism of earthquakes, considered well-substantiated for earthquakes accompanied by visible faulting, is the "elastic rebound theory" first proposed by H. F. Reid in 1906.[12-1] According to the theory, earthquakes are caused by the relatively sudden release of accumulated strains in the crustal rocks through fracture along areas of weakness. The release of energy produces waves of distortion that propagate in all directions, with accompanying reflections from the earth's surface as well as reflections and refractions as they traverse the earth's interior. The exact process by which forces are generated within the earth's interior that ultimately result in crustal upheavals (mountain building and earthquakes) is not yet well understood.

The point of fracture within the earth's interior where the initial slip or rupture occurs, triggering large-scale slippage sometimes extending for hundreds of miles, is called the earthquake *focus* or *hypocenter.* The point on the surface vertically above the focus is called the *epicenter* (Fig. 12-1).

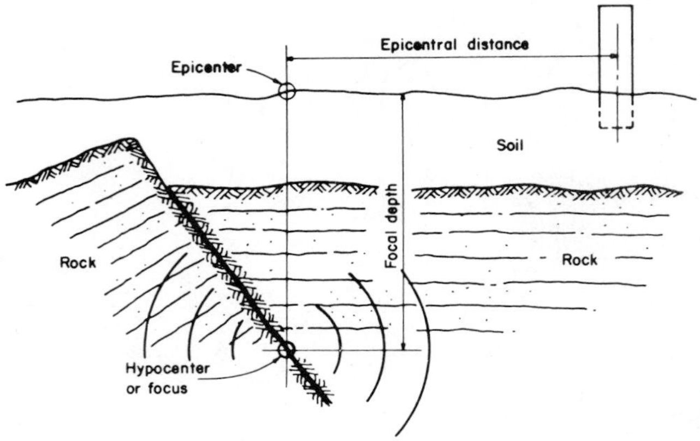

Fig. 12-1 Illustrating definition of common earthquake-related terms.

Earthquake foci may occur anywhere from near the surface to a depth of 700 kilometers or more. The term normal or shallow-focus earthquakes has been used to denote shocks with focal depths of about 60 km (0–44 miles) or less, while those with greater focal depths fall under the broad classification of deep-focus earthquakes. The latter include two subclasses: intermediate, with foci between 70 and 300 km (44–188 miles) and deep, for focal depths greater than 300 km. Most ordinary earthquakes have shallow foci; that is, they originate in the crustal layer, above the Mohorovicic discontinuity.

Shallow-focus earthquakes tend to produce only local effects. The ground motion in the localized area near the epicenter, however, can be expected to be more intense than that which would result if the focus were located deeper and the effects felt over a much wider area.

12.2.2 Geographic Distribution of Earthquakes

Almost all major earthquakes and approximately 95% of all earthquakes have occurred in two long and relatively narrow zones. The principal zone, which accounts for about 80% of all earthquakes, is the *circum-Pacific seismic zone*, which borders the Pacific ocean, running up the west coast of South, Central, and North America, through the Aleutians, and down the coast of Asia through Japan and the Phillippines, to New Zealand. Paralleling the circum-Pacific belt of major tectonic activity, but generally a hundred or so miles to one side of it, is the chain of active volcanoes making up what has often been referred to as the Pacific "circle of fire." The closeness of these two zones once led to the mistaken belief that earthquakes are due principally to volcanic activity.

The second major seismic zone, called the *Alpide zone*, runs in a generally east–west direction across Europe, from the Azores through Spain, Italy, Greece, Turkey, and northern India, turning southeastward through Burma and Sumatra to join the circum-Pacific belt in New Guinea. The Alpide belt accounts for about 15% of all earthquakes. A third zone, accounting for most of the remaining five or so percent of recorded earthquakes, is a fairly narrow zone along midoceanic rifts—along the center of the midoceanic ridges—particularly those in the Atlantic and Indian oceans.

A world map showing the epicenters of recent earthquakes

*A term usually applied to the upper layer of the earth, from the surface to the so-called Mohorovicic discontinuity, which separates the crust from the mantle. The thickness of the earth's crust varies from about 5 to 60 kilometers, being thicker in the continental areas and thinner under the sea. Another seismic discontinuity which has been detected at levels above the Mohorovicic discontinuity is the Conrad discontinuity.

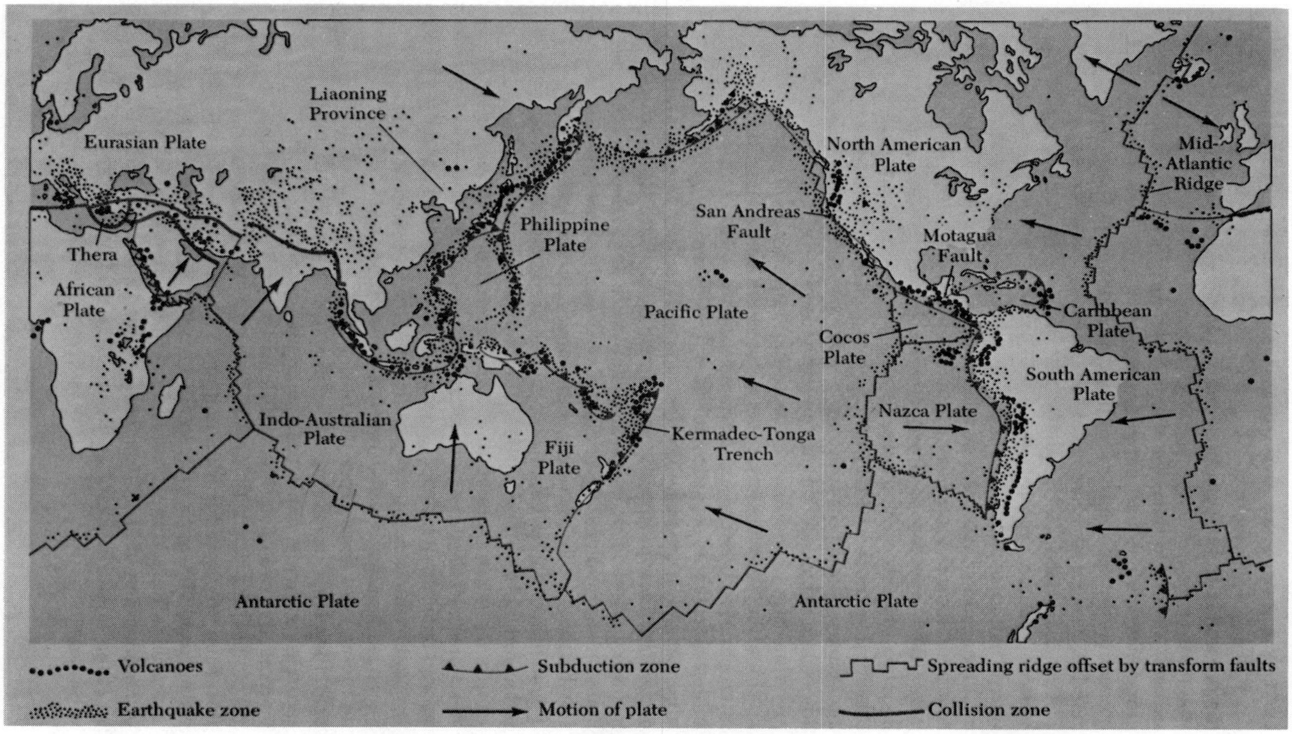

Fig. 12-2 World map showing epicenters of recent earthquakes (small dots).[12-2]

is presented in Fig. 12-2. The map is a good indicator of global seismicity.

12.2.3 Nature of Earthquake Motions—Seismic Waves

The effects of earthquakes are usually classified as primary, if due directly to the causative process, such as faulting or volcanic action; and secondary, when due to the ground motion resulting from the passage of seismic waves. Of the secondary effects, two types may be distinguished: those associated with landslides or soil consolidation or liquefaction triggered by earthquake motions, where inertial effects in structures are negligible, and others in which dynamic (inertial) effects are predominant. Except under unusual circumstances,* most of the damage associated with earthquakes has been due to the latter type of effects, and engineering efforts aimed at designing earthquake-resistant structures are concerned mainly with such effects.

The ground motion at a particular site is influenced by three factors: (a) source parameters, such as the earthquake magnitude (energy released), depth of focus, and geological conditions at and near the focus; (b) transmission path parameters, that is, epicentral distance and properties and geologic character of the intervening ground; and (c) local site parameters or the geologic configuration and properties of the ground at the site.

The complex and erratic motion of the ground characterizing an earthquake is the result of the passage of waves of distortion through the affected area. Such motions recorded by seismographs are juxtapositions of a number of different wave types,** among the more significant of which are longitudinal (compressional-dilatational) or P-waves, transverse (shear) or S-waves, Rayleigh or R-waves, and Love or Q-waves.

Both P- and S-waves are body waves, traversing the interior of the earth and making up the high frequency components of strong ground motion near the epicenter. P-waves travel at velocities ranging from 5 to 13 km/sec, compared to a range of from 3 to 8 km/sec for S-waves, the velocity in each case increasing with depth.*** In a P-wave, the particles vibrate in the direction of propagation, while in an S-wave the particles oscillate in a plane perpendicular to the direction of propagation.

The R- and Q-waves are surface waves that propagate along the surface of the earth at velocities only slightly less than that of S-waves. Rayleigh waves are generated whenever waves propagating through a continuous medium (such as the body waves, P and S) reach a free surface. The energy input to structures on or near the ground surface comes predominantly from Rayleigh waves. These tend to attenuate rapidly with distance from the surface. Love waves result from the presence of a distinct discontinuity at some depth from the free surface of the medium through which body waves propagate. Whenever several layers having different material properties occur, other types of waves are generated, for example, the so-called guided waves in layered media. In the R-wave the vibration is in the direction of propagation, while in the Q-wave the particle dis-

*Such as occurred in the city of Niigata, Japan, during the earthquake of June 16, 1964, when a number of buildings were rendered useless—though some were practically undamaged—by excessive tilting due to the liquefaction of the sand on which many of the buildings were founded (see Fig. 12-3).

**The dominance of one type of wave in the combination, often producing a distinct characteristic or "phase" in the recorded signal.

***The velocity increase occurs from the surface to the boundary of the mantle and the core, at which depth an abrupt decrease in wave velocity occurs. From this point down to the inner core, there is again a slight increase in the P-wave velocity. No shear wave rays are known to have traversed the earth's outer core, which is a basis for the belief that the material in this region is liquid in character.

(a)

(b)

Fig. 12-3 (a) Apartment buildings at Kawagishi-cho, Niigata, Japan, after the Niigata earthquake of June 16, 1964.[12-3] (b) Close-up of building shown in (a), Niigata, Japan.[12-4]

placement is transverse to the direction of travel, with no vertical or longitudinal component. Surface waves form a significant portion of the strong ground motion in epicentral regions of shallow-focus earthquakes and also make up the dominant long-period phases of shallow-focus teleseisms, i.e., earthquakes recorded at distances over 1000 km (621 miles) from the epicenter.

Generally, the amplitude of recorded ground motion diminishes with increasing distance from the epicenter. Under certain geologic conditions, particularly in areas where a deep layer of saturated alluvium overlies bedrock, a substantial amplification of wave motion transmitted from the bedrock may occur. Also, the short-period (or high frequency), high-energy components of seismic waves tend to be damped out much faster than the longer-period waves, so that at some distance from the epicenter, the ground motion consists predominantly of long-period oscillations.

The ground motion at a particular point is completely determined by three translational and three rotational components along three perpendicular directions. The rotational components of earthquake motions are, however, usually not significant. Most seismographic stations are equipped with instruments designed to record the horizontal components along two perpendicular directions and the vertical component of the ground motion.

12.2.4 Strong-Motion Earthquake Records

Instrumental records of earthquake ground motion close to the epicenter are of particular interest to the structural engineer. These usually consist of acceleration traces of the components of motion along two perpendicular horizontal directions and in the vertical direction. Such records are obtained using strong-motion accelerographs (SMACs).

Strong-motion seismographs or accelerographs are basically simple, damped oscillators, the recorded response (i.e., the relative displacement between the movable mass and a fixed base) being proportional to the input acceleration. Unlike the more sensitive standard seismographs used by seismologists to record motions resulting from distant earthquakes, SMACs are less sensitive instruments designed to record motions in the epicentral regions of earthquakes. In such locations, the high magnification of a conventional seismograph usually results in a jumbled trace, if not a broken instrument. Also, while standard seismographic stations usually have their instruments set up on massive concrete blocks founded on solid rock, SMACs, by the very nature of their purpose, are set up at widely different locations, from open fields to basements and upper floors of buildings. Another difference is that while conventional seismographic records are continuous (the recording paper being set at a comparatively low speed), with particular emphasis placed on the time at which signals are received, SMACs set up for engineering purposes do not start recording until triggered by an initial motion, with the time scale used mainly for the proper reduction of the data. The first modern strong-motion seismograph was introduced in 1931, and the first strong-motion records were obtained during the March 1933 Long Beach (California) earthquake.

Although ground motions recorded at a given location can hardly be expected to be repeated in the future, strong-motion records, if available over a long period, can provide valuable insight into the general character of the ground motion and the effect of geologic conditions at a particular site. Where a number of accelerograms are available for a limited region, a set obtained by judicious sampling from the available data can serve as the basis of dynamic response studies of proposed structures, when such studies are deemed desirable. In an attempt to remedy the present lack of strong-motion records for structural response studies, several methods have recently been proposed for generating accelerograms artificially, using nonstationary stochastic models, that is, filtered "shot noise."[12-5,12-6] The methods consist essentially of assuming certain properties of actual earthquake accelerograms as basic characteristics and using these parameters as constraints in the generation of random numbers representing ground acceleration amplitudes.

A set of acceleration traces that has gained widespread popularity through repeated use in dynamic response studies is that of the Imperial Valley (California) earthquake of May 18, 1940. This was recorded at the El Centro instrument site which rests on some 5000 feet of alluvium and is located some 4 miles from the causative fault break. The 1940 El Centro record represents the strongest ground motion thus far recorded. Although several other earthquakes had magnitudes and maximum recorded ground accelerations greater than the 1940 El Centro (see next section and Table 12-2), none so far* has exhibited the combination of high-frequency, large-amplitude pulses lasting over as long a duration as the El Centro record.

A plot of the variation of the north–south component of horizontal ground acceleration during the first 30 seconds of the 1940 El Centro earthquake is shown in Fig. 12-4. Also shown are plots of the ground velocity and displacement as obtained from the acceleration trace by successive integration. The maximum value of the recorded ground acceleration was about $0.33g$, ($1.0g$ being equal to the acceleration due to gravity), the maximum calculated ground velocity about 13.7 in./sec, and the corresponding recorded vertical acceleration about $0.28g$.

The reason for the use of acceleration traces instead of velocity or displacement traces in strong-motion records is the greater accuracy with which the latter two can be obtained from the former by a process of integration rather than vice versa by differentiation. This will be obvious from an examination of Fig. 12-4.

Since actual strong-motion accelerograph records do not show a zero-acceleration line, such a base line has to be assumed when reducing the data for digitization and integration purposes. Several methods, satisfying different criteria, have been used to adjust the accelerogram base line. The method used in such a base line adjustment can affect the calculated maximum displacement significantly although the maximum velocity and response spectra (see Section 12.3.3.3) derived from the resulting accelerogram are relatively insensitive to the method of base line adjustment. Significant errors in the calculated net ground displacement can also be expected as a result of the uncertainty concerning the initial part of the shock, which is not recorded due to accelerograph triggering and start-up delays.

12.2.5 Earthquake Magnitude and Intensity

Magnitude and intensity scales have been developed in attempts to describe earthquakes in quantitative terms.

The magnitude of an earthquake is a measure of its size in terms of the energy released and radiated in the form of seismic waves. It is determined on the basis of instrumental (i.e., microseismic) data, particularly, the amplitude of ground motion at a specified distance from the epicenter.

*Except perhaps the record taken at the abutment of Pacoima Dam during the San Fernando earthquake of February 9, 1971. There is, however, some doubt as to the cause of the extremely high accelerations recorded at this particular site, where the rock on which the instrument rested exhibited fractures resulting from the earthquake.

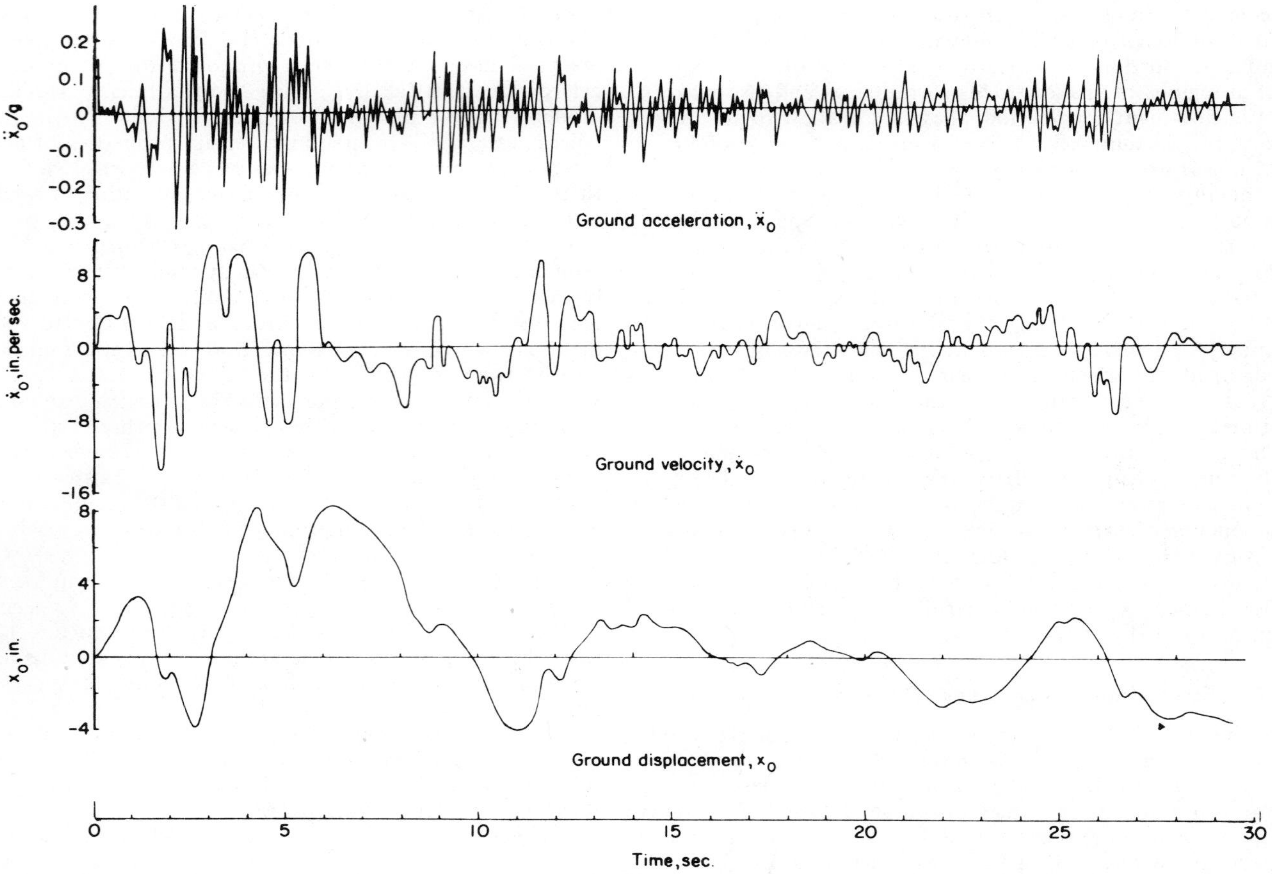

Fig. 12-4 May 18, 1940 El Centro (Calif.) Earthquake accelerogram, N-S component—and corresponding derived velocity and displacement plots.[12-7]

Of several magnitude scales that have been proposed, the best known is the Richter scale. On this scale, magnitude is defined in terms of the amplitude of the trace recorded by a standard seismograph located 100 km from the epicenter. Estimates of the radiated energy based on the recorded amplitudes and periods of seismic waves during a number of earthquakes have led to correlations between the arbitrarily defined magnitude and the energy released during an earthquake. The accepted relationship has changed over the years with the accumulation of more data; the current one for shallow-focus California earthquakes given by:

$$\log E = 11.4 + 1.5M$$

where E is the energy released in ergs and M is the magnitude on the Richter scale. The above equation indicates a 32-fold increase in released energy for every unit increase in magnitude.

The largest earthquakes recorded* have had a Richter magnitude of 8.9. This is believed to be the largest possible magnitude on the Richter scale, the limit being set by the maximum amount of strain energy that can be stored by crustal rocks without rupture.

Although earthquakes of magnitude less than about 5 are believed to be incapable of producing damaging effects, the magnitude of an earthquake, per se, does not provide an indication of its destructiveness at any given location. The destructiveness of an earthquake, although dependent partly on its magnitude, is a function of other equally important factors such as focal depth, epicentral distance, local geology, and structural characteristics including resistance to damaging oscillations.

The word "intensity" when used to describe earthquakes is generally intended to denote the potential destructiveness of an earthquake at a particular location. If it were possible to predict intensity with respect to specific structural types, the structural engineer would have a good basis for assessing his design earthquake loads. Since the destructiveness of earthquake motions is determined not only by the character of the ground motion at a site but also by the type of structure and geology of the site, it is readily seen that intensity as a measure of potential destructiveness is not easy to define, much less predict, quantitatively.

If one considers only the ground motion at a given site, the principal parameters characterizing it (particularly as it affects the dynamic response of structures) are the amplitude of the motion, its frequency characteristics and the duration of the large-amplitude pulses. Early attempts at quantifying earthquake intensity have sought to relate the maximum ground acceleration with the observed effects on structures and other natural features at specific locations.

The most commonly quoted intensity scale, at least on the American continent, is the Modified Mercalli (MM) scale, which is given in Table 12-1. The scale, which has 12 grades of intensity, was adopted, with modifications to suit conditions in America, by H. O. Wood and F. Neumann in 1931 from a proposal originally devised by G. Mercalli for use in Italy. Also shown in the table are approximate

*One on the Colombia–Ecuador border in 1906 and another in Japan in 1933.

TABLE 12-1 The Modified Mercalli Intensity Scale and corresponding approximate ground accelerations.[12-8]

Intensity	Ground Acceleration a, $\frac{cm}{sec}$	Ground Acceleration a, $\frac{a}{g}$
I - Not felt except by a very few under especially favorable circumstances.		
II - Felt only by a few persons at rest, especially on upper floors of buildings. Delicately suspended objects may swing.	2	
III - Felt quite noticeably indoors, especially on upper floors of buildings, but many people do not recognize it as an earthquake. Standing motor cars may rock slightly. Vibration like passing truck. Duration estimated.	3, 4, 5, 6	.005g
IV - During the day felt indoors by many, outdoors by few. At night some awakened. Dishes, windows, doors disturbed; walls make creaking sound. Sensation like heavy truck striking building. Standing motor cars rocked noticeably.	7, 8, 9, 10	.01g
V - Felt by nearly everyone; many awakened. Some dishes, windows, etc., broken; a few instances of cracked plaster; unstable objects overturned. Disturbances of trees, poles, and other tall objects sometimes noticed. Pendulum clocks may stop.	20	
VI - Felt by all; many frightened and run outdoors. Some heavy furniture moved; a few instances of fallen plaster or damaged chimneys. Damage slight.	30, 40, 50, 60	.05g
VII - Everybody runs outdoors. Damage negligible in buildings of good design and construction; slight to moderate in well-built ordinary structures; considerable in poorly built or badly designed structures; some chimneys broken. Noticed by persons driving motor cars.	70, 80, 90, 100	0.1g
VIII - Damage slight in specially designed structures; considerable in ordinary substantial buildings, with partial collapse; great in poorly built structures. Panel walls thrown out of frame structures. Fall of chimneys, factory stacks, columns, monuments, walls. Heavy furniture overturned. Sand and mud ejected in small amounts. Changes in well water. Disturbs persons driving motor cars.	200	
IX - Damage considerable in specially designed structures; well-designed frame structures thrown out of plumb; great in substantial buildings, with partial collapse. Buildings shited off foundations. Ground cracked conspicuously. Underground pipes broken.	300, 400, 500, 600	0.5g
X - Some well-built, wooden structures destroyed; most masonry and frame structures destroyed with foundations; ground badly cracked. Rails bent. Landslides considerable from river banks and steep slopes. Shifted sand and mud. Water splashed over banks.	700, 800, 900, 1000	1g
XI - Few, if any, (masonry) structures remain standing. Bridges destroyed. Broad fissures in ground. Underground pipelines completely out of service. Earth slumps and land slips in soft ground. Rails bent greatly.	2000	
XII - Damage total. Waves seen on ground surfaces. Lines of sight and level distorted. Objects thrown upward into the air.	3000, 4000, 5000, 6000	5g

TABLE 12.2 Statistics Of Some Recent Earthquakes

Name of Earthquake	*Date of Occurrence*	*Richter Magnitude*	*Maximum MM Intensity*	*Max. Recorded Hor. Grnd. Acceleration*
San Francisco, California	April 18, 1906	8.3	XI	
Imperial Valley, (El Centro) California	May 18, 1940	6.7–7.1	X	0.33 g
Olympia, Washington	April 13, 1949	7.1	VIII	0.31 g
Kern County (Taft), California	July 21, 1952	7.7	XI	0.18 g
Chile (Concepcion & Southern Chile)	May 22, 1960	7.5–8.5		
Agadir, Morocco	February 29, 1960	5.6		
Alameda Park, Mexico	May 11, 1962			0.098 g
Skopje, Yugoslavia	July 26, 1963	5.4–6.0	VIII	
Prince William Sound, Alaska	March 27, 1964	8.4		
Niigata, Japan	June 16, 1964	6.2–7.5	VII	0.19 g
Parkfield, California	June 27, 1966	5.5	VIII	0.50 g
Lima, Peru	October 7, 1966			0.40 g
Caracas, Venezuela	July 29, 1967	5.7–6.5		
Koyna, India	December 11, 1967	6.25–7.5	VIII	0.63 g
San Fernando, California	February 9, 1971	6.6	VIII	1.20 g (Pacoima Dam)

values of the ground acceleration, a, corresponding to the different intensity ranges, as obtained from the relation $I = 3 \log a + 1.5$ proposed by Gutenberg and Richter.[12-8]

The estimated magnitudes and reported MM intensity ratings of some recent earthquakes are shown in Table 12-2.

The Modified Mercalli (MM) intensity of an earthquake is usually assessed by distributing questionnaires to or interviewing persons in affected areas, and through observation of earthquake effects (i.e., macroseismic data) by experienced personnel. It is readily apparent that the determination of the intensity at a particular site involves considerable subjective judgment and is influenced greatly by the type and quality of structures as well as the geology of the area. Thus, comparisons of intensity ratings made by different persons in different countries or during different periods of time can be misleading. However, intensity ratings, when assessed uniformly over a sufficiently large area affected by a single earthquake, can yield valuable information on the geology of the site. Irregularities in the shape of lines connecting plotted points of equal intensity (i.e., isoseismals) may indicate unusual geologic conditions.

From the point of view of the structural engineer, reported Modified Mercalli intensities can be considered only as a very crude quantitative measure of the destructiveness of an earthquake, since they do not provide, in terms of relevant structural parameters, specific information on damage related specifically to structures of interest.

The recognition of the shortcomings of the MM intensity scale, particularly in relation to structural engineering applications, has prompted a number of investigators to propose other measures of earthquake intensity.

12.2.6 Seismic Regionalization

In designing for earthquakes, the expected character (i.e., intensity, duration, and frequency content) of the ground motion at a particular location, as well as its anticipated frequency of occurrence, are of primary interest. The estimation of seismic risk on the basis of historical records is sometimes referred to as the statistical method of predicting earthquakes. Most seismic risk studies are based on characterizing earthquake destructiveness in terms of a single parameter (e.g., peak ground acceleration). Seismic risk is usually expressed in terms of the probability of a specified earthquake intensity or peak ground acceleration occurring or being exceeded in a given time interval, called the recurrence interval or return period. Because of meager historical data and the need to exercise judgment in the interpretation of such data, some inconsistencies are bound to arise. For instance, seismic regionalization maps for the United States and Canada, prepared by different groups, show contiguous regions with widely different seismic risk. With improved record keeping and a more comprehensive data base, more reliable seismic regionalization maps should result.

The subdivision of an area into regions each associated with a known or assigned seismic risk has served as a useful basis for the implementation of code provisions on earthquake-resistant design. The present seismic risk maps used in the U.S. and Canadian model building codes show each area divided into zones of approximately equal seismic risk. Associated with each zone is a factor that enters into the expression for determining the so-called equivalent lateral load for use in design. State-of-the-art ground shaking regionalization maps for the United States have been included in "Tentative Provisions for the Development of Seismic Regulations for Buildings" drawn up by the Applied Technology Council (document ATC3-06).[12-9] The maps seek to take into account the distance from anticipated earthquake sources. This consideration necessitated the use of two separate ground motion parameters and the preparation of two maps. Also, the maps are based on the premise that the probability of exceeding the design ground shaking should, as a goal, be roughly the same in all parts of the country. This contrast with the zoning maps until recently in use in the United States which have been based upon estimates of the maximum ground shaking experienced during the recorded historical period, without consideration of how frequently such motions might occur. The ATC has chosen to represent the intensity of ground shaking by two parameters: effective peak acceleration (EPA) for near earthquakes and effective peak velocity (EPV) for distant earthquakes. The EPA and EPV are related to peak ground acceleration and peak ground velocity, but are not the same as or even proportional to peak acceleration and velocity. The EPA and EPV for a motion may be either greater or smaller than the peak acceleration and velocity, although generally the EPA will be smaller than the peak acceleration, whereas EPV will be larger than the peak velocity. EPA is expressed in terms of a dimensionless coefficient, A_a, which is equal to EPA expressed as a fraction of g (e.g., if EPA = $0.2g$, $A_a = 0.2$). EPV is expressed in terms of a dimensionless parameter, A_v, which is a velocity-related acceleration coefficient. The ATC seismicity maps divide the United States into seven map areas. The A_a and A_v coefficients for each map area are given. The probability is estimated at 90% that the recommended EPA and EPV at a given location will not be exceeded during a 50-year period. This probability is equivalent to a mean recurrence interval of 475 years, or an average annual risk of 0.002 event per year.

The Seismic Risk Map for the contiguous 48 United States shown in Fig. 12-5 is a slightly modified and smoothed version of a map developed by Algermissen and Perkins.[12-10]

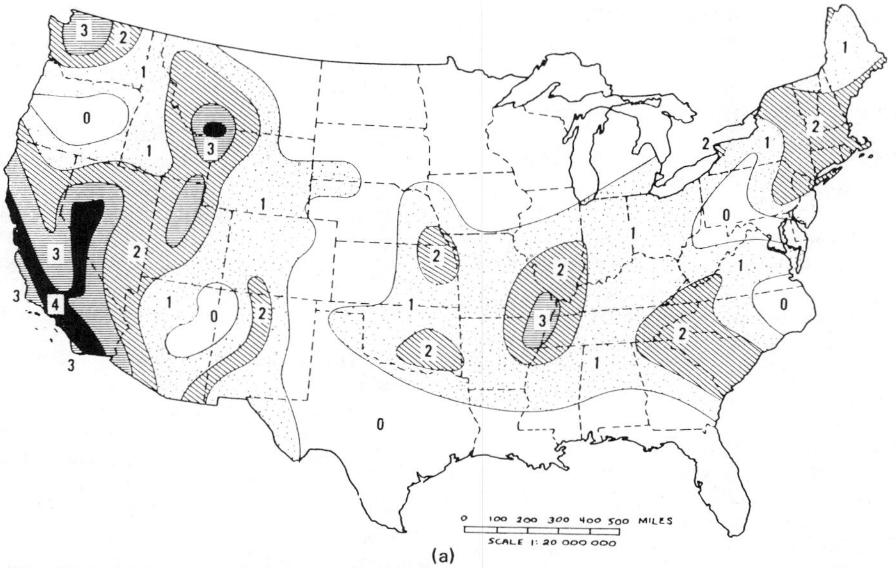

(a)

Fig. 12-5 (a) Seismic risk map of the contiguous 48 United States—ANSI A58.1-1982[12-12]

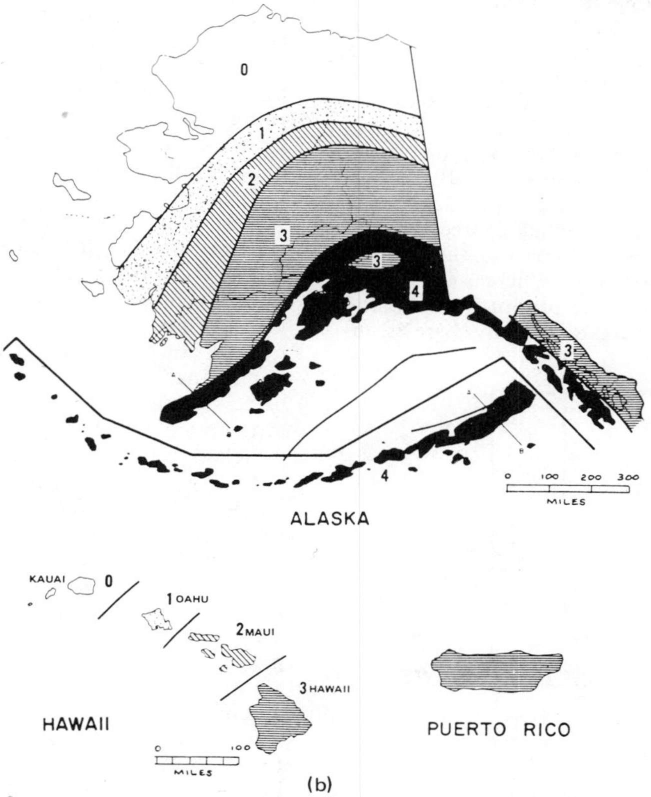

(b)

Fig. 12-5 (b) Seismic risk map of Hawaii, Alaska, and Puerto Rico—ANSI A58.1-1982[12-12]

The Algermissen–Perkins 1976 map formed the basis of the EPA map in ATC3-06. Maps of horizontal accelerations and velocities caused by earthquakes for exposure times of 10, 50, and 250 years at the 90% probability level of non-exceedence for the contiguous United States have recently become available from the U.S. Geological Survey.[12-11] The map of Fig. 12-5 was adopted by the American National Standards Institute (ANSI) in the 1982 edition of its "Minimum Design Loads for Buildings and Other Structures (ANSI A58.1-1982)."[12-12] A similar map, based on the earlier work of Algermissen,[12-13] appears in the 1982 edition of the *Uniform Building Code* (UBC-82).[12-14] The earlier work of Algermissen showed the estimated maximum intensity of ground shaking without regard to frequency of occurrence.

12.3 RESPONSE OF STRUCTURES TO EARTHQUAKE MOTIONS

12.3.1 General

The forces that a structure subjected to earthquake motions is called upon to resist result directly from the distortions induced by the motion of the ground on which it rests. The response (i.e., the magnitude and distribution of forces and displacements) of a structure resulting from such a base motion is influenced by the properties of both the structure and the foundation, as well as the character of the exciting motion.

A simplified picture of the behavior of a building during an earthquake may be obtained by considering Fig. 12-6. As the ground on which the building rests is displaced, the base of the building moves with it. However, the inertia of the building mass resists this motion and causes the building to suffer a distortion (greatly exaggerated in the figure). This distortion wave travels along the height of the structure in much the same manner as a stress wave in a bar with a free end. The continued shaking of the base causes the building to undergo a complex series of oscillations.

It is important to draw a distinction between forces due to wind and those produced by earthquakes. Occasionally, even engineers tend to think of these forces as belonging to the same category, just because codes specify design wind as well as earthquake loads in terms of equivalent static forces. Although both wind and earthquake forces are dynamic in character, a basic difference exists in the manner in which they are induced in a structure. Whereas wind loads are external loads applied, and hence proportional, to the exposed surface of a structure, earthquake forces are essentially inertial forces. The latter result from the distortion produced by both the earthquake motion and the inertial resistance of the structure. Their magnitude is then a function of the mass of the structure rather than its exposed surface. Also, in contrast to structural response to essentially static gravity loading or even to wind loads, which can often be validly treated as static loads, the dynamic character of the response to earthquake excitation can seldom be ignored. Thus, whereas in designing for static loads one would feel greater assurance about the safety of a structure made up of members of heavy section, in the case of earthquake loading the stiffer and heavier structure does not necessarily represent the safer design.

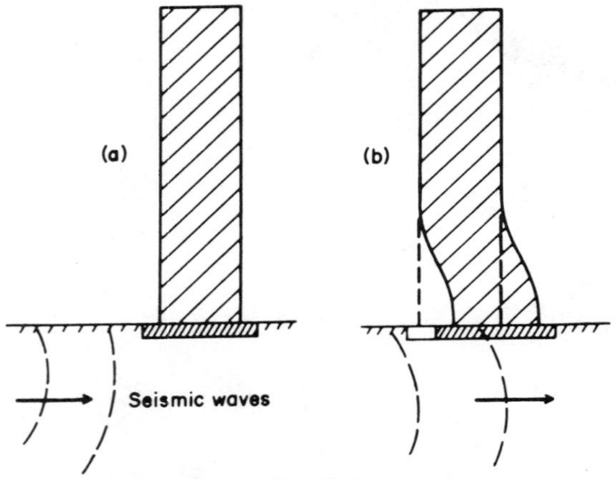

Fig. 12-6

Uncertainties in the determination of the proper earthquake design loads to be used for a proposed structure arise from a number of factors, among the more important of which are:

1. The difficulty of predicting the character of the critical earthquake motions (i.e., amplitude, frequency characteristics, and duration) to which a proposed structure may be subjected during its lifetime.
2. The difficulty of ascertaining the values of the structural parameters affecting the dynamic response (e.g., stiffness and damping), as well as the dynamic properties of the soil or supporting medium.

Insofar as the earthquake motion is concerned, the quantity most commonly used in analysis is the timewise variation of the ground acceleration in the immediate vicinity of a structure. For earthquakes of tectonic origin, the character of the variation of the surface acceleration is primarily dependent on the magnitude and focal depth of the earthquake, the epicentral distance of the site considered, and the properties of the intervening as well as the surrounding ground. The most significant parameters characterizing the ground motion at a site that influence structural response are intensity, frequency characteristics, and duration of the large-amplitude pulses. Earthquake motions are random in character, and there is no way to ascertain the exact nature of any future earthquake at a particular site. However, continuing studies of the seismic history and geology of a region should yield valuable estimates of the expected range of the significant ground motion parameters. The Algermissen–Perkins acceleration and velocity contour maps[12-11] corresponding to different return periods represent a good start toward the quantification of regional seismic risk. With information from such maps as a basis and some knowledge of the geologic conditions at a proposed site, a set of accelerograms[12-5, 12-6] representative of probable future earthquakes can be artificially generated to serve as input in dynamic analyses. The range of variation to be considered for each significant parameter in such a set of input accelerograms obviously depends on the degree of uncertainty attached to that parameter.

At any point, the ground acceleration may be described by horizontal components along two perpendicular directions and a vertical component. In addition, rocking and twisting (rotational) components may be present. As mentioned earlier, the rotational components of earthquake ground motion are usually negligible. Rocking and torsional effects in structures due to the horizontal components of ground motion do occur, however, as a result of ground compliance and the noncoincidence of the centers of mass and of rigidity.

Because buildings and most structures are more sensitive to horizontal or lateral distortions, it has been the practice in most instances to consider structural response to the horizontal components of ground motion only. The effects of the vertical component of ground motion* have generally not been considered significant enough to merit special attention. Here, the principal motivation has been simplification and the reduction of the analytical work involved, which can be considerable, even when only the horizontal components of ground motion are considered. For most structures, experience seems to have justified this practice. In most instances, a further simplification of the actual three-dimensional response of structures is made by assuming the design horizontal acceleration components to act nonconcurrently in the direction of each principal plan-axis of a building. It is tacitly assumed that a building designed by this approach will have adequate resistance against the resultant acceleration acting in any direction.

Insofar as a structure is concerned, the principal properties that affect its dynamic response are its mass, stiffness, and damping characteristics. Under certain conditions, the effect of the foundation or supporting medium may have to be considered.

When yielding is expected to occur under moderate to strong earthquake excitation, the yield level or, alternatively, the ratio of yield-to-design moments becomes a significant factor. Under this condition, the capacity of a structure to deform well beyond the initial yield displacement without significant loss of strength (i.e., ductility) assumes utmost importance.

The determination of the mass and initial stiffness properties of a particular structure generally does not present any difficulty. However, the effective stiffness and damping of a structure, and hence its characteristic period of vibration, can change during vibration as a result of cracks occurring in members, even before the onset of yielding. Yielding further increases the period of vibration of a structure.

The evaluation of the nature, magnitude, and distribution of a structure's damping has, up to the present, been a rather vague process and is a problem that is only beginning to receive the systematic study that it deserves. Convenience in mathematical modeling and solution of the dynamic response problem has required the assumption of a viscous-type damping in place of the actual mechanism. Vibration tests of actual structures have yielded estimates of the equivalent viscous damping ranging from 5% to 20% of critical.**

Most dynamic analyses of multistory structures have assumed the free-field ground motion to be applied, unaltered, to the building foundation, which is assumed to be rigid. While this approximation may be valid for most structures founded on soil that is relatively stiff with respect to the structure, cases arise where soil–structure interaction effects produce significantly different results from those based on a rigid foundation assumption. The effect of soil–structure interaction on earthquake response can vary depending upon the properties of the soil and the structure as well as the character of the input motion. The effect is largest on the fundamental mode of vibration and diminishes rapidly in the higher modes. (See Sections 12.3.2 and 12.3.5 for definitions of modes and mode shapes.)

12.3.2 Dynamic Analysis of Single-Degree-of-Freedom (SDF) Systems

An often convenient way of studying the dynamic response of structures is by considering the total response in terms of

*In several instances, the vertical component of ground acceleration has been of the same order of magnitude as the horizontal components.

**The value of damping that would just cause an initial displacement to decay to zero without any oscillation.

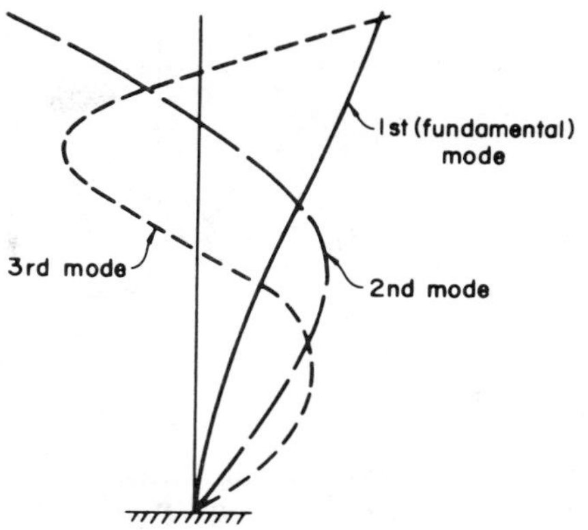

Fig. 12-7 Typical shapes of the first three natural modes of vibration of a multistory structure.

component modal responses. The response in the elastic range, and during sufficiently short intervals in the inelastic range, may be thought of as the superposition of the responses of the modes of vibration characterizing the system. In general, a system will have as many modes of vibration as it has significant degrees of freedom.*** In each such mode of vibration, all the masses vibrate in phase; that is, they maintain the same positions relative to each other. Associated with each mode is a characteristic frequency or period of vibration. Each mode of vibration may thus be considered as a single-degree-of-freedom system having a particular period or frequency of vibration. Figure 12-7 shows typical shapes for the first three modes of a multistory building where the significant degrees of freedom are related to the lateral displacements of the floors. The first or fundamental mode corresponds to the longest or fundamental period (or lowest frequency) of vibration.

The response of most multistory buildings is determined largely by the first few modes, with the higher modes contributing relatively little to the total response, except at the top of relatively flexible buildings. Studies of the elastic response of multistory frame buildings indicate that for most buildings the fundamental mode contributes about 80%, and the second and the third modes contribute about 15%, of the total response. Because of the dominant influence of a single (i.e., the fundamental) mode of vibration on the response of multistory structures and the comparative ease of analyzing the response of single-degree-of-freedom (SDF) systems, much can be learned of basic response characteristics of actual structures from a consideration of the dynamic response of simple systems. This is the main justification for studying the response of SDF systems here. Of course, as mentioned earlier, each of the higher modes of vibration can be considered as an SDF system also.

12.3.3 Linear SDF Systems

The force–deformation curve for a linearly elastic system is shown in Fig. 12-8(a). Also shown, in Fig. 12-8(b), is the corresponding curve for an inelastic bilinear system. The solid-line curve is often used as a convenient approximation of the actual force–deformation curve shown in dashed lines on the same figure. The unloading branch of the idealized inelastic curve is usually assumed to have the same slope as the initial loading branch.

1. Formulation of Equation of Motion. A schematic

***The number of displacement components required to specify the positions of all significant mass particles in a structure. In the lumped-mass idealization, the mass of the structure is assumed concentrated at a number of discrete locations.

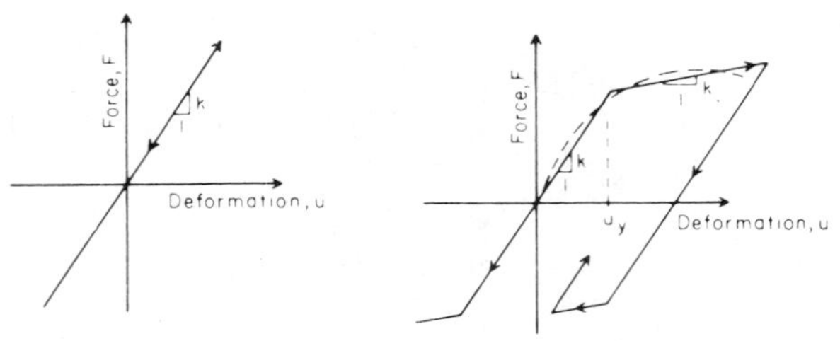

Fig. 12-8 (a) Linear and (b) bilinear restoring-element force–deformation relationships.

representation of an SDF system with a fixed base is shown in Fig. 12-9(a). The system consists of a rigid mass m, a weightless resisting element with spring constant k, and a dashpot with damping coefficient c. The dashpot, which represents the energy-absorbing mechanism in the system, is assumed here to be of the viscous type (in which the damping force is proportional to the velocity of only one direction, that is, along the direction of the time-varying applied force $P(t)$. The system may be thought of as representing the single-story frame shown in Fig. 12-9(b) with the mass of the frame assumed concentrated in the infinitely stiff girder, where the only motion considered significant is the lateral displacement of the mass.

Although the fixed-base structure shown in Fig. 12-9 does not correspond to the earthquake situation where the base of the structure undergoes motion, it is considered here in order to point out the similarity between the externally applied force in a fixed-base system and the equivalent effective force corresponding to the moving base system.

The differential equation of motion obtained by considering the equilibrium of the free body shown in Fig. 12-9(c) is given by:

$$m\ddot{x} + c\dot{x} + kx = p(t) \tag{12-1}$$

where the dot ($\dot{}$) indicates differentiation with respect to time.

Consider next the moving-base SDF system shown in Fig. 12-10 with the displacement of the base described by

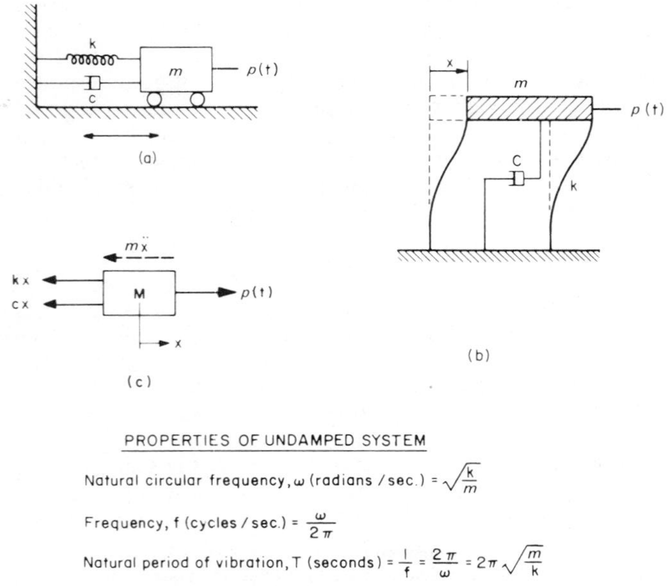

Fig. 12-9 Fixed-base, single-degree-of-freedom (SDF) system subjected to a time-varying force.

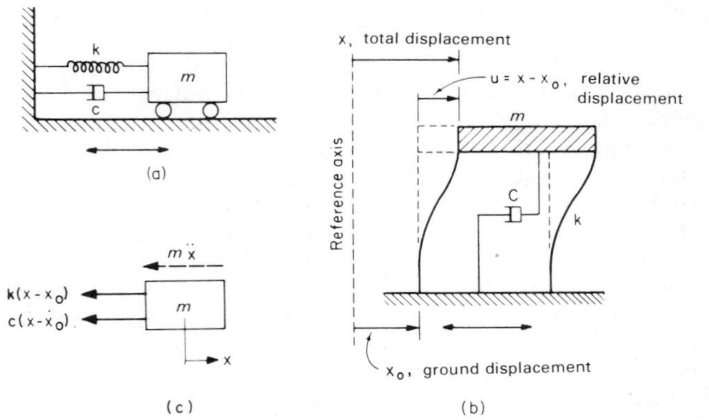

Fig. 12-10 Movable-base SDF system subjected to a time-varying base acceleration.

x_o. Note that there is no applied force acting on the mass; this is similar to the situation in a structure subjected to earthquake motion. Equilibrium of the free body in Fig. 12-10 now gives

$$m\ddot{x} + c(\dot{x} - \dot{x}_o) + k(x - x_o) = 0 \tag{12-2}$$

Expressing eq. (12-2) in terms of the relative displacement of the mass with respect to the base, $u = x - x_o$, yields

$$m\ddot{u} + c\dot{u} + ku = -m\ddot{x}_o(t) \tag{12-3}$$

A comparison of eqs. (12-2) and (12-3) indicates that the relative displacement, u, of the mass in a system subjected to a time-varying base acceleration, $\ddot{x}_o(t)$, is equal to the absolute displacement, x, of the mass in a fixed-base system in which the mass is subjected to a time-varying force equal to $-m\ddot{x}_o(t)$. The quantity $-m\ddot{x}_o(t)$ may be called the effective load resulting from the base motion.

2. *Closed Form Solution of Equation of Motion.* For the purpose of solving eq. (12-3) the following notation will be introduced for convenience:

$$\omega = \sqrt{\frac{k}{m}} \quad \text{and} \quad \beta = \frac{c}{2\omega m} \tag{12-4}$$

where ω represents the circular frequency of undamped vibration (in radians/sec) and β the damping ratio relative to the critical value of the damping coefficient. ω is related to the natural frequency, f, in cycles per second, and period of vibration, T, by the relations:

$$f = \frac{\omega}{2\pi} \quad \text{and} \quad T = \frac{1}{f} = \frac{2\pi}{\omega} \tag{12-5}$$

With the above notation, eq. (12-3) may be rewritten as:

$$\ddot{u} + 2\beta\omega\dot{u} + \omega^2 u = -\ddot{x}_o(t) \tag{12-6}$$

(a) Homogeneous case. The solution for the homogeneous (free vibration) case of eq. (12-6), when the right-hand side representing the applied force is equal to zero, is given by:

$$u(t)_h = e^{-\beta\omega t}(A \cos \omega_d t + B \sin \omega_d t) \tag{12-7}$$

where $\omega_d = \omega\sqrt{1 - \beta^2}$ may be referred to as the damped circular frequency of vibration. The solution given by eq. (12-7) corresponds to the case when $\beta < 1$. When $\beta = 1$, the system is said to be critically damped (so that $c_{cr} = 2\omega m$); when $\beta > 1$, it is overdamped. In either case, the motion is nonoscillatory. Since for most civil engineering structures β ranges from about 0.02 to 0.20, the preceding two cases need not be considered here.

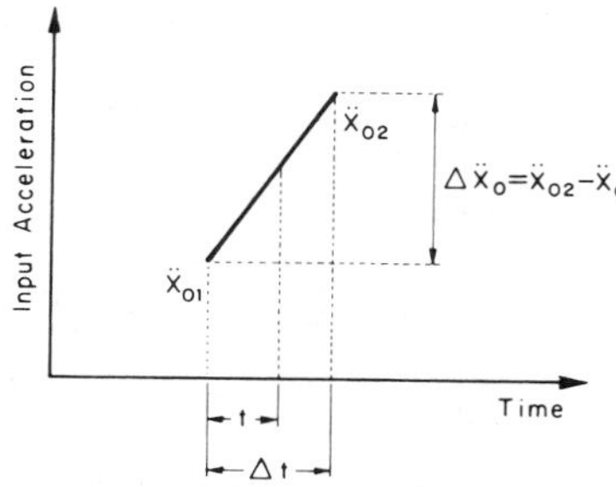

Fig. 12-11 A linearly-varying acceleration pulse representing a typical segment of an earthquake acceleration input.

(b) Particular solution for a linearly varying base acceleration. The input motions used in earthquake response analyses usually consist of digitized data based on actual strong-motion earthquake accelerograph records or artificially generated accelerograms. The data consist of time-acceleration coordinates defining the accelerogram, with the variation of the acceleration assumed to be linear between adjacent points of definition. A general case of an earthquake input acceleration segment may thus be described by the linearly varying acceleration pulse of duration Δt shown in Fig. 12-11 and described by the relation:

$$\ddot{x}_o(t) = \ddot{x}_{o1} + \frac{t}{\Delta t}(\ddot{x}_{o2} - \ddot{x}_{o1}) \qquad (12\text{-}8)$$

where t is reckoned from the beginning of the pulse. Solution of eq. (12-6) with the expression for $\ddot{x}_o(t)$ as defined by eq. (12-8) substituted on the right-hand side, yields:

$$u(t)_p = -\frac{1}{\omega^2}\left[\ddot{x}_{o1} + \left(\frac{t}{\Delta t} - \frac{2\beta}{\omega \Delta t}\right)\Delta\ddot{x}_o\right] \qquad (12\text{-}9)$$

where:

$$\Delta\ddot{x}_o = \ddot{x}_{o2} - \ddot{x}_{o1}$$

The complete solution of the differential equation of motion for a SDF system with damping ratio β, when subjected to a piecewise linear base acceleration, is given by the sum of the right-hand sides of eqs. (12-7) and (12-9); that is:

$$u(t) = u(t)_h + u(t)_p$$
$$= e^{-\beta\omega t}(A\cos\omega_d t + B\sin\omega_d t)$$
$$-\frac{1}{\omega^2}\left[\ddot{x}_{o1} + \left(t - \frac{2\beta}{\omega}\right)\frac{\Delta\ddot{x}_o}{\Delta t}\right] \qquad (12\text{-}10)$$

Note that if the relative displacement at the end of the interval is desired, as is usually the case, $t = \Delta t$.

Using the initial conditions $u = u_1$ and $\dot{u} = \dot{u}_1$ at $t = 0$, the constants of integration, A and B, are determined, as follows:

$$A = u_1 + \frac{1}{\omega^2}\left(\ddot{x}_{o1} - \frac{2\beta}{\omega\Delta t}\Delta\ddot{x}_o\right) \qquad \text{(a)}$$

and: (12-11)

$$B = \frac{1}{\omega_d}\left(\dot{u}_1 + \beta\omega A + \frac{\Delta\ddot{x}_o}{\omega^2\Delta t}\right) \qquad \text{(b)}$$

where the A appearing in eq. (12-11b) is given by eq. (12-11a).

Equation (12-10)—with A and B substituted from eq. (12-11)—gives the relative displacement of the mass at any time during the interval Δt. Successive differentiations of eq. (12-10) yield the corresponding expressions for the relative velocity, $\dot{u}(t)$, and relative acceleration, $\ddot{u}(t)$. The dynamic response analysis of a linearly elastic SDF system subjected to earthquake motion may be undertaken by repeated application of eq. (12-10) and the corresponding expressions for $\dot{u}(t)$ and $\ddot{u}(t)$ for each segment of the base acceleration record, the response values obtained for the end of one time interval being used as the initial values of the succeeding interval. The calculations are most conveniently carried out with the aid of a digital computer. For completeness, the expressions for the relative velocity and acceleration are given below.

$$\dot{u}(t) = -\,[(A\omega_d + B\beta\omega)\sin\omega_d t$$
$$+ (A\beta\omega - B\omega_d)\cos\omega_d t]\,e^{-\beta\omega t} - \frac{\Delta\ddot{x}_o}{\omega^2\Delta t} \qquad (12\text{-}12a)$$

$$\ddot{u}(t) = (\beta^2\omega_d^2)(A\cos\omega_d t + B\sin\omega_d t)\,e^{-\beta\omega t}$$
$$+ 2\beta\omega\omega_d(A\sin\omega_d t - B\cos\omega_d t)\,e^{-\beta\omega t} \qquad (12\text{-}12b)$$

where A and B are given by eq. (12-11).

3. *Response Spectra, Pseudo-velocity, and Pseudo-acceleration.* A response spectrum is a plot showing the variation of the maximum value of a particular response parameter (e.g., displacement, velocity, acceleration, stress, etc.) with the frequency (or period) of a linear SDF system subjected to a specific forcing function. In earthquake response studies, the forcing function is generally the earthquake accelerogram record. Although the term response spectrum was originally used, and is still commonly used only with reference to linear SDF systems, its use has been extended in later work to linear as well as nonlinear MDF systems. A point on a response spectrum curve is obtained by analyzing a particular SDF system—as defined by its frequency and damping ratio—using a specific base acceleration record. The maximum value of the response parameter of interest, which generally occurs within the duration of the exciting motion, is then noted. For a fixed value of the damping ratio, β, the spectral value of the parameter and the associated frequency (or period) of the system together define a point on the response spectrum curve. Other points are obtained by using different values of the frequency. A response spectrum thus shows the maximum response of different SDF systems, as defined by their frequencies and damping ratios. In the following discussion, the term response spectrum will be assumed to refer to linear SDF systems, unless otherwise noted.

An examination of eqs. (12-10) and (12-12) shows that the response of a SDF system to base excitation is dependent on the properties of the system as given by its natural circular frequency, ω, and its damping ratio, β, as well as the character of the exciting motion $\ddot{x}_o(t)$. It is thus seen that for the same value of β and the same range of frequencies, the response spectra for different earthquake motions may be expected to differ from one another. As such, they provide a means of characterizing earthquake motions.

Furthermore, because the dynamic behavior of more complex systems, such as multistory buildings, is strongly influenced by the fundamental mode response, spectral plots provide the designer with a convenient means of assessing the response of a structure of known period to a specified ground motion.

Figure 12-12 shows acceleration spectra for different values of the damping ratio, β, corresponding to the 1940 El Centro, California, earthquake (N-S component).

The sharp peaks in certain ranges of frequency shown in Fig. 12-12, especially for the case of $\beta = 0$, reflect the resonant behavior of the system in these frequency ranges. In this respect, the lightly damped SDF system acts as a nar-

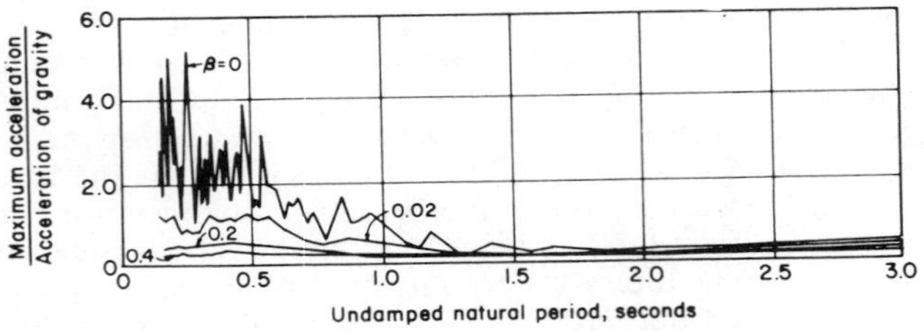

Fig. 12-12 Acceleration response spectra for 1940 El Centro, Calif., earthquake (N-S component).[12-15]

row band filter that amplifies the input motion in ranges where the dominant frequency approaches that of the system. A response spectrum plot may thus be used to infer the character of the input motion, with particular regard to frequency content. It is significant to note, however, that even a moderate amount of damping tends not only to reduce the response but also to smooth out the jagged character of the spectral plot. Since most structures possess some amount of damping, the sharp peaks do not have much importance in practice. Also, where response spectra are used for design purposes, it is customary to allow for earthquake motions of varying frequency characteristics by using smoothed or averaged spectra based on a number of earthquake records—all reduced or normalized to a reference intensity level.

The response parameters most often considered in earthquake response studies and in terms of which response spectra are plotted are the maximum relative displacement, the maximum relative velocity, and the maximum absolute acceleration of the mass. The maximum relative displacement assumes importance in design, since the strains in a structure are proportional to the relative displacement. The maximum relative velocity provides a measure of the elastic energy stored in a system, while the absolute acceleration can be shown to be directly related to the seismic or lateral force coefficient appearing in most design codes. Also, where experimental measurements are to be taken as a check on calculations, the absolute acceleration serves as a good basis for comparison, since it is the structural response most easily measured.

In the study of the earthquake response of simple systems, it has been found useful and convenient to present response spectra in terms of velocity and acceleration quantities that are not quite equal to the actual quantities they are meant to represent. These have, however, been shown to provide fairly good measures of the quantities they represent and bear such a simple relationship to each other and to the spectral relative displacement as to allow convenient representation.

Thus, if the maximum (spectral) value of the relative displacement is denoted by S_d, the spectral pseudo-velocity is defined as:

$$S_v = \omega S_d \qquad (12\text{-}13)$$

and the spectral pseudo-acceleration as:

$$S_a = \omega^2 S_d \qquad (12\text{-}14)$$

For zero damping, the difference between the maximum relative velocity and the spectral pseudo-velocity, S_v, is generally negligible. For exciting motions of very short duration containing relatively short-period peaks, however, the difference between these two quantities can be significant over some frequency ranges. As damping increases, a deviation of the order of 20% may be expected. Generally, the spectral velocity, S_v, is nearly equal to the maximum relative velocity in the higher frequency range, with the difference tending to be greater for the low-frequency (long-period) systems.

The spectral velocity can be shown to be directly related to the maximum energy per unit mass in a system. Thus, denoting the strain energy in the spring by U, we have:

$$\frac{U_{max}}{m} = \frac{\frac{1}{2} k (x - x_o)^2_{max}}{m} = \frac{1}{2} \omega^2 S_d^2 = \frac{1}{2} S_v^2$$

The spectral acceleration is actually the maximum absolute acceleration of the mass in an undamped SDF system. Thus, for such a system, using Newton's second law of motion:

$$F_{max} = k(x - x_o)_{max} = kS_d = m\ddot{x}_{max}$$

$$\ddot{x}_{max} = \frac{k}{m} S_d = \omega^2 S_d = S_a$$

The spectral pseudo-acceleration, however, does not differ greatly from the maximum absolute acceleration for systems with moderate damping, for the entire frequency range. The relation between the spectral accelerations, S_a, and the maximum force acting on the spring, which in codes is usually expressed as a seismic or lateral force coefficient, C, multiplied by the weight of the structure, W, may be shown as follows:

$$F_{max} = k(x - x_o)_{max} = kS_d = CW$$

$$C = \frac{k}{W/g} \cdot \frac{S_d}{g} = \frac{\omega^2 S_d}{g} = \frac{S_a}{g}$$

Figure 12-13 shows averaged pseudo-velocity response spectra for several ground motions, all adjusted to have the same spectral intensity* as the N-S component of the 1940 El Centro, California, earthquake. It will be noted that with the relationships given by eqs. (12-13) and (12-14), all three response parameters can be obtained when one of them is known. Thus, both spectral displacement and spectral pseudo-acceleration can be obtained from the spectral pseudo-velocity plot, the first by dividing and the second by multiplying the velocity spectrum ordinate by the corresponding circular frequency ω.

Because of the linear relationship between the logarithms of both the spectral displacement and the spectral acceleration with the logarithms of the spectral velocity and circular frequency ω, as indicated by eqs. (12-13) and (12-14), it is possible to have a single plot showing the variation of all three response quantities with the frequency (or period). Figure 12-14 shows such a plot, with log scales on all axes, the spectral displacement and the spectral acceleration being read along diagonal scales.

In Fig. 12-14, the scales have been nondimensionalized to indicate the ratios of the spectral values of the response parameters to the corresponding maximum ground motion parameters. The plot shown is typical of earthquake response spectra and provides analytical support to some aspects of the dynamic response of simple systems that would appear intuitively obvious. Thus, with low-frequency (long-period) systems—corresponding to a heavy mass supported by a light spring—the mass remains practically motionless when the base is subjected to an earthquake-type

*First defined by Housner[12-16] as the area under the velocity spectrum curve between ordinates representing periods of 0.1 to 2.5 seconds. The spectrum intensity provides a measure of the maximum stress induced in linearly elastic structures by the particular ground motion.

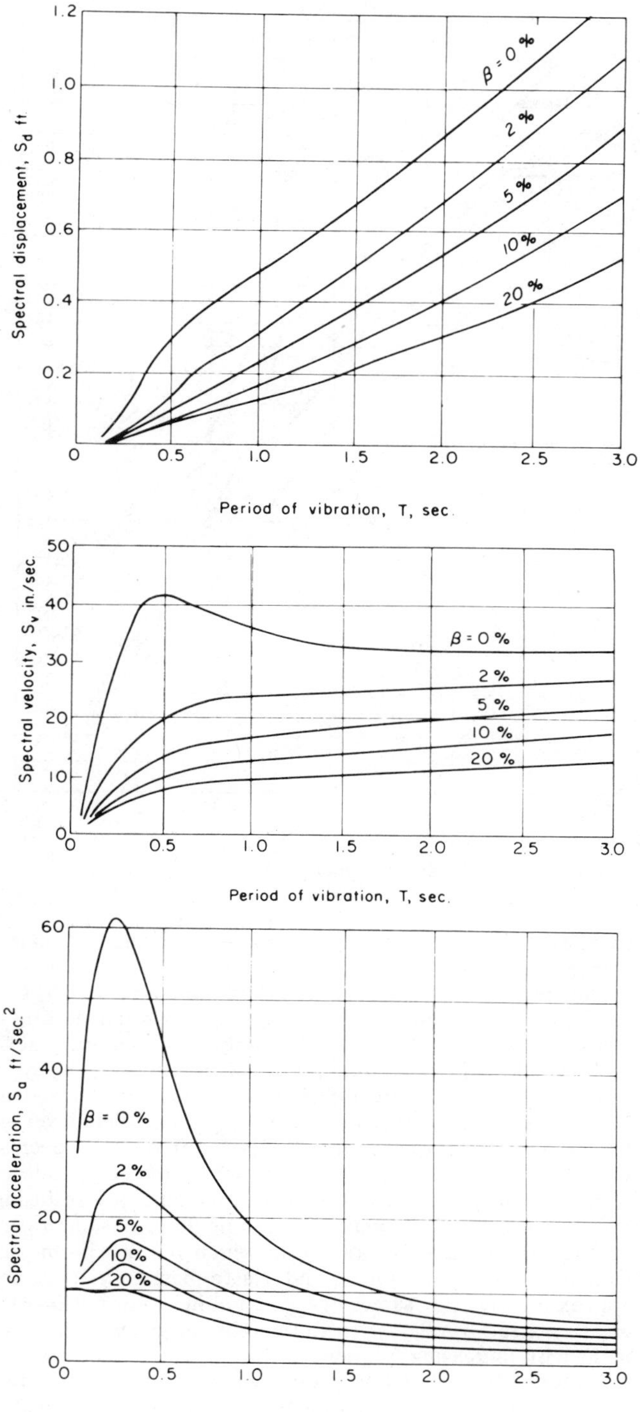

Fig. 12-13 Averaged response spectra.[12-16]

motion. For this case, the relative displacement of the mass or the deformation in the spring is essentially equal to the base displacement. This is indicated by the spectral lines that approach the diagonal line $S_d/(x_o)_{max} = 1.0$ on the left side of the plot. On the right side of the plot, that is, for high-frequency systems, the spectral lines are shown converging toward the diagonal line $S_a/(\ddot{x}_o)_{max} = 1.0$, indicating that for systems exemplified by a light mass supported by a very stiff spring, the mass simply moves with the base, with the spring hardly deforming. Hence the mass acquires essentially the same absolute acceleration as the base.

In the intermediate frequency range, some modification occurs of the response parameters characterizing the motion of the mass relative to that of the base. Generally, the amplification factor for displacement is less than that for velocity, which in turn is less than that for acceleration. Figure 12-14 indicates that for linear systems with damping ratios of 5 to 10%, subjected to the N–S component of the 1940 El Centro motion, the maximum amplification factors for displacement, velocity, and acceleration are about 1.5, 2.0, and 2.5, respectively.

A typical response spectrum curve in Fig. 12-14 may be represented by three line segments marking the approximate envelope of response for systems having a particular damping ratio and subjected to a specfic ground motion. The three line segments, shown as the dashed broken line *a-b-c-d* in the figure for systems with 5 to 10% damping and subjected to the 1940 El Centro (N–S) motion, effectively divides the frequency axis into three regions: a low-frequency region from *a* to *b*, a medium-frequency region from *b* to *c*, and a high-frequency region from *c* to *d*. The width of each frequency region will generally vary with the nature of the ground motion.

12.3.4 Bilinear Hysteretic SDF Systems

When the force–deformation characteristics of the restoring element (spring) in a SDF system is other than linear (i.e., when k is not constant throughout the response), the governing differential equation of motion of the system, corresponding to eq. (12-1), becomes nonlinear. However, during sufficiently small intervals of time, the system may be assumed to be linear, with the value of the spring stiffness depending upon the magnitude and direction of the deformation during the interval considered. A step-by-step solution using such an approach allows the practical evaluation of the response of nonlinear systems for arbitrary force–deformation characteristics of the systems and for any base motion.

1. Some Results of Analyses of Bilinear, Inelastic (Hysteretic) SDF Systems. A system having a force-deformation curve similar to that shown in Fig. 12-9(b) differs from a linearly elastic system in two respects: (1) the hysteretic action which occurs after yielding allows it to dissipate energy (as measured by the area under the *F–u* curve), an effect that tends to decrease the maximum deformation of the system; and (2) it represents a softer system, with a lower effective frequency of vibration than a linear system having the same initial stiffness or small-amplitude frequency. The reduction in effective frequency that occurs after yielding may either decrease or increase the maximum deformation, depending upon the frequency and the yield level, (the force required to produce first yield) of the system.

The determination of the response of nonlinear systems involves a significantly greater amount of time than that required for linear systems. Because of this and the availability of extensive data on linear systems, efforts have been directed toward formulating approximate design rules relating the behavior of nonlinear systems to that of linear systems.

A special case of the bilinear hysteretic system is one in which the slope of the second, post-yield, branch of the force–deformation curve, k_2, is equal to zero. This is commonly referred to as an elasto-plastic system.

Figure 12-15, from Ref. 12-18, shows the variation of the maximum spring deformation, u_{max}, with the undamped

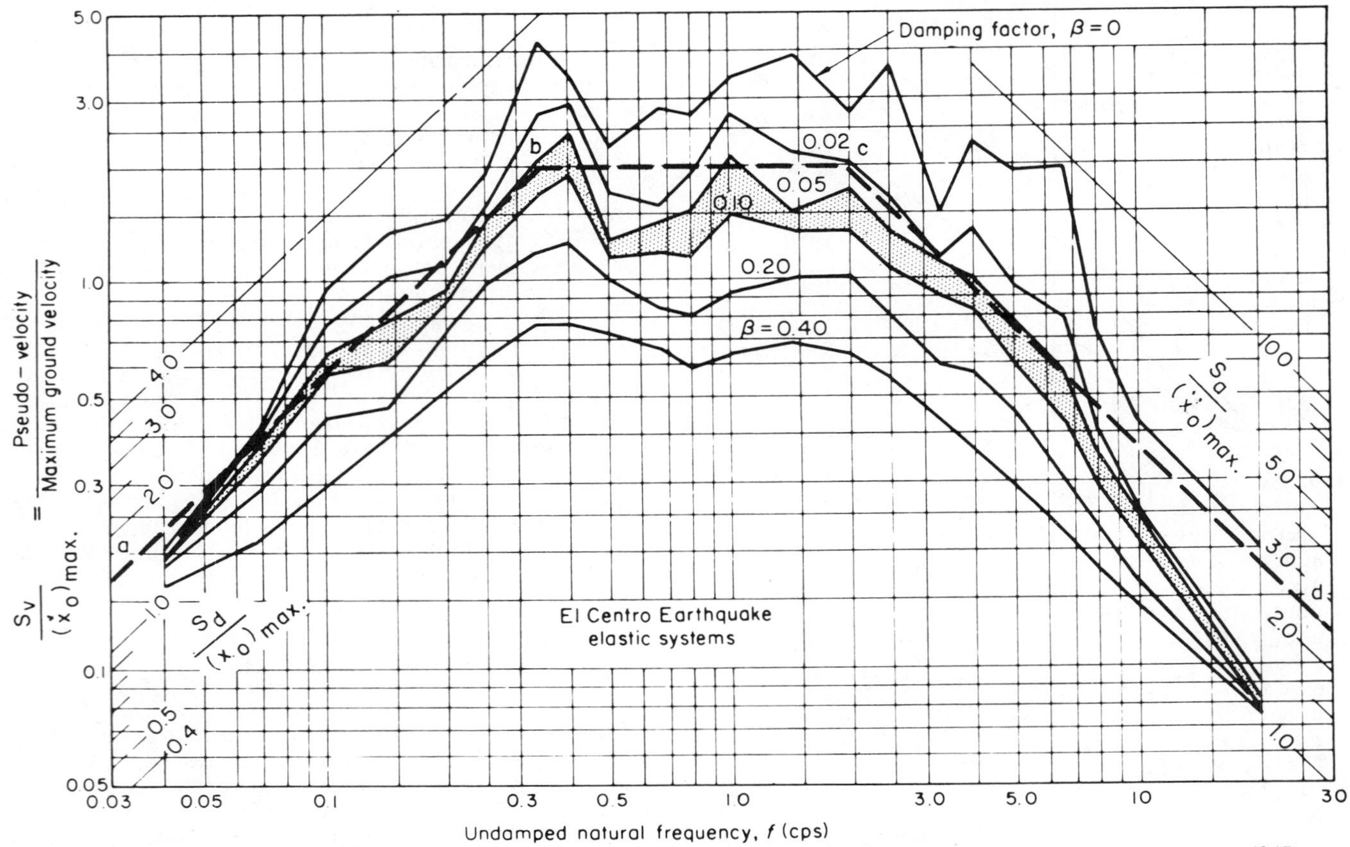

Fig. 12-14 Deformation spectra for linearly elastic SDF systems subjected to 1940 El Centro earthquake (N–S component).[12-17]

small-amplitude frequency of elasto-plastic systems subjected to the 1940 El Centro (N–S) earthquake, for different values of the yield level. A constant damping ratio, $\beta = 0.02$, has been used throughout. The maximum spring deformation is given as the ratio $u_{max}/(x_o)_{max}$, while the yield level is expressed as the ratio, $F_y/(F_e)_{max}$, of the yield force in the elasto-plastic system to the maximum

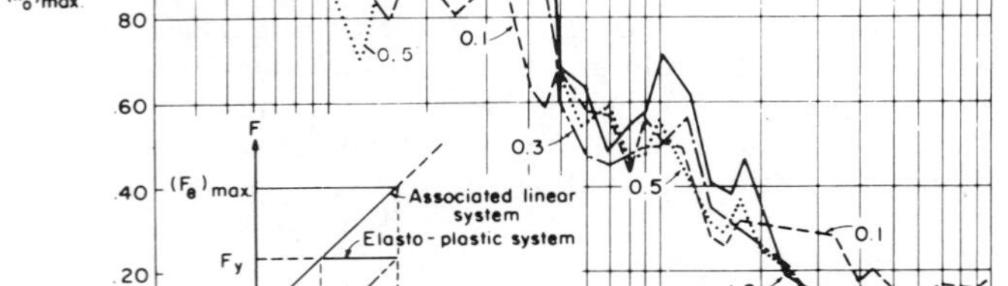

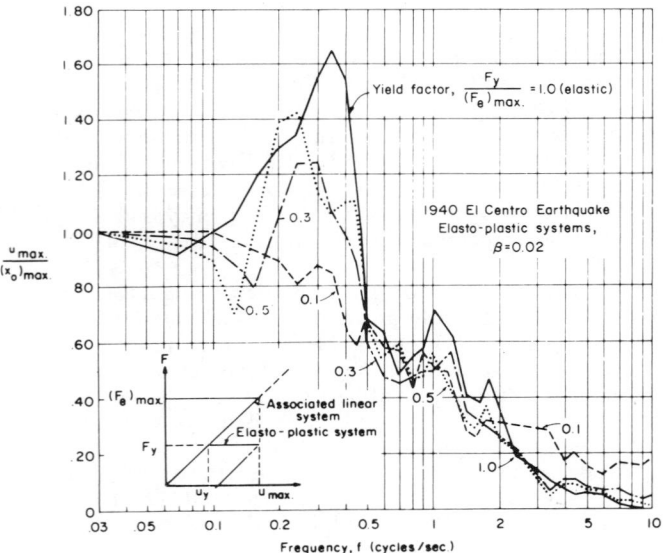

Fig. 12-15 Displacement spectra for elasto-plastic SDF systems having different yield levels—1940 El Centro (N–S) earthquake.[12-18]

force in the associated linear system. Note that the curve marked $F_y/(F_e)_{max} = 1.0$ corresponds to a linear system.

The following points may be noted from Fig. 12-15. For extremely low-frequency systems, the maximum deformations of linear as well as elasto-plastic systems are practically equal to the maximum ground displacement, $(x_o)_{max}$. Thus, in this frequency range, the softness of the elasto-plastic system overshadows any effect of the yield level on the response. As one moves to the right on the spectral plot, there is a region in which the effect of the yield level is most pronounced, a lower yield level generally resulting in lower values of the maximum spring deformation, u_{max}. Farther to the right is a region in which the maximum deformations of both linear and elasto-plastic systems are approximately the same, the maximum deformation decreasing steadily relative to the maximum ground displacement with increasing frequency.

It should be noted that while the effect of a decreasing yield level is to decrease the maximum deformation in the moderately low- and medium-frequency regions, the same decrease in yield level produces a slight increase in the maximum deformation of high-frequency systems. Thus, for very stiff systems, a lowering of the yield level below that required for elastic response may produce significant inelastic components of deformation. This observation points to a qualitative distinction between the effect of inelastic action and that of viscous damping on a linear system, since the latter tends to decrease the maximum deformation of a linear system over the entire frequency range.

The bilinear hysteretic system in which the slope of the second, post-yield, branch of the F–u curve is nonzero and positive represents an intermediate case between the elasto-plastic and the linear cases just discussed. As might be expected, the spectral curves for bilinear systems with low values of the ratio k_2/k_1 do not differ appreciably from those for elasto-plastic systems. As the value of the ratio k_2/k_1 increases, that is, as the bilinear F–u curve approaches the linear case, the response of the bilinear system tends to approach that of the associated linear system. The response of bilinear systems appears to be the most sensitive to changes in the k_2/k_1 ratio in the high-frequency region, where a slight increase in the value of the ratio above zero can produce a significant decrease in the maximum deformation.

As the yield level in a bilinear hysteretic system approaches zero, the maximum deformation of the system tends to the value corresponding to a linear system having a stiffness equal to k_2.

12.3.5 Linear Multi-Degree-of-Freedom (MDF) Systems

1. Some Terms Related to the Dynamic Analysis of MDF Systems. Space does not allow a detailed discussion of the dynamic analysis of structures with multiple degrees of freedom (MDF). For this, the reader can refer to any of a number of textbooks on structural dynamics.[12-19–12-21] Reference 12-22 is the first exhaustive study of the dynamic analysis of frame structures.

In order to provide the reader with some background on the basic ideas involved in the dynamic analysis and response of MDF systems, a few of the terms commonly encountered in the literature on the subject are briefly discussed below.

The number of degrees of freedom of a system is defined as the least number of independent displacement coordinates necessary to completely determine the configuration of the system at any instant. For frame structures, the significant degrees of freedom are generally associated with the displacement components of the joints defined by the intersection of member axes. Thus, for a plane frame, the degrees of freedom are usually identified with the rotation and the vertical and horizontal displacement components of the joints. Often, owing to the relatively high axial stiffness of the floor system, only one horizontal degree of freedom is assumed for all joints of a particular floor.

The use of lumped-mass models to represent the actual distributed mass structure is a convenient device for reducing the infinite number of degrees of freedom of the structure to a manageable few. This makes possible the formulation of the force equilibrium of the system in terms of a set of ordinary differential equations instead of the partial differential equations that would be required for the continuous system. With proper care in the lumping procedure, model response values corresponding closely to those of the real structure should be obtainable. In response studies of multistory building, for instance, it is generally sufficient to assume the masses as concentrated at the floor levels and to formulate the problem in terms of the displacements of these masses.

The natural or principal modes of vibration of a linear structure are free, undamped, periodic vibrations, a unique linear combination of which makes up the actual motion of the structure at any instant. Each mode of vibration is characterized by the masses, or points defining the geometry of the structure, all vibrating in phase, that is, with the displacements of the masses, as measured from the initial position, always having the same relationship with one another. Thus, all points on a structure vibrating in one of its natural modes pass the equilibrium position at the same time and reach their extreme positions at the same instant. A structure can be made to vibrate freely in one of its natural modes alone if carefully started in that particular mode.

A structure has as many natural modes of vibration as it has degrees of freedom—each mode being associated with a characteristic frequency or period of vibration. The mode having the lowest frequency (or longest period) is called the first or fundamental mode. Generally, each mode of vibration and the associated frequency are distinct from all the others, although cases do arise in which two or more modal frequencies have values very close to each other.

Figure 12-7 shows typical shapes for the first three modes of vibration of a multistory building in terms of the lateral displacements of the floors. Note that the curves intersect the vertical axis at as many points or nodes (counting the one at the base) as the number of the mode. The displacements defining a particular mode shape are only relative and do not indicate a specific magnitude; are always in the same proportion, regardless of the cause of the vibration.

A damped MDF system possesses natural modes of vibration only if the character of the damping satisfies a certain condition.* Among systems satisfying this condition are those in which the damping matrix is proportional to either the stiffness or the mass matrix. For such systems, the same (normal mode) coordinates that uncouple the equations of motion of the undamped system also uncouple the equations corresponding to the damped system.

2. Two Connotations of Dynamic Analysis as Used in Earthquake Response Studies. The term dynamic analysis as applied to MDF systems, particularly in earthquake engineering, has been associated with two slightly different analytical approaches. In the first one, referred to as mode superposition using response spectra, approximate values of the maximum response parameters are determined from the maximum responses corresponding to the different modes of vibration behaves as an independent SDF system with a characteristic frequency, the maximum modal response values may be obtained from appropriate response spectra for SDF systems (see Section 12.3.3.3).

Several methods of combining the modal contributions have been proposed to allow for the fact that the individual modal maxima generally do not occur simultaneously. A commonly used method takes the square root of the sum of the squares of the modal maxima. This method tends to yield low values of response quantities when the frequencies corresponding to two or more significant modes of vibration are close to each other. Taking the sum of the absolute values of the modal contributions provides an upper bound on the response. Often, only the contributions of the first few modes are considered, since these usually make up the major part of the response.

The above method has been employed by many for design purposes using averaged or smoothed response spectra based on a number of earthquake records, all reduced to a standard or reference intensity. The use of average spectra provides, in one simple step, for the probable variation in response to different earthquake input motions. It should be noted that mode superposition is applicable only to linearly elastic systems, and, if such systems are damped, only when the damping satisfies a certain condition so that independent uncoupled modes exist.

The second approach to dynamic analysis involves the determination of the time history of response of the mathematical model of a structure to a particular earthquake. The analysis may employ modal superposition or direct numerical integration[12-22] of the equations of motion. In both cases, the calculations are carried out for the total response at the end of a short interval of time, the analysis proceeding in a step-by-step fashion using the conditions prevailing

*See eq. (12-15).

at the end of one time interval as the initial conditions for the succeeding interval.

In modal analysis, the displacement, $u_i(t)$, of the ith mass of an n-degrees-of-freedom system is expressed as a linear combination of the characteristic modal displacements and a time-varying function, $q_j(t)$; that is:

$$u_i(t) = \sum_{j=1}^{n} \phi_{ij} q_j(t)$$

where j is the index denoting mode number. In matrix notation, the vector of displacements

$$\{u(t)\} = [\Phi]\{q(t)\}$$

where $[\Phi]$ is the $(n \times n)$ modal matrix, each column, $\{\phi_j\}$ of which represents the characteristic displacements corresponding to a natural mode of vibration. The time-varying function $q_j(t)$ corresponding to the jth mode of vibration essentially gives the response of a SDF system having the same frequency as the jth mode when subjected to an effective load (see eq. 12-3) given, in matrix notation, by:

$$\frac{\{\phi_j\}^T [m] \{1\}}{\{\phi_j\}^T [m] \{\phi_j\}} \ddot{x}_0(t)$$

In the above expression, $[m]$ is the $(n \times n)$ mass matrix and $\ddot{x}_0(t)$ is the actual input base acceleration. The superscript T in $\{\phi_j\}^T$ indicates a transpose of the vector (or matrix).

As with mode superposition using response spectra, modal analysis is applicable only to linearly elastic systems in which damping, if present, satisfies certain conditions. Also, where external forces are applied to a structure, these must all have the same time variation in order for the uncoupling, which characterizes modal analysis, to be possible.

For a viscous-damped system, the decomposition of the response of a structure into its component modal responses is possible when the damping matrix satisfies the following relation:

$$([m]^{-1}[c])([m]^{-1}[k]) = ([m]^{-1}[k])([m]^{-1}[c]) \tag{12-15}$$

where $[m]$, $[c]$ and $[k]$ represent the mass, damping, and stiffness matrices of the system, respectively, and the (-1) superscript indicates the inverse of a matrix. Equation (12-15) essentially states that the product of $[m]^{-1}$ with $[k]$ and $[c]$ commutes. Among the forms of the damping matrix that satisfy the above condition are those in which it is proportional to either the stiffness or the mass matrix of the system, or is a linear combination of these, that is:

$$[c] = \alpha_1 [k] + \alpha_2 [m]$$

where α_1 and α_2 are scalar proportionality constants.

For a linearly elastic system with viscous damping which allows the uncoupling of the equations of motion, modal analysis permits a significant saving in computing time, since the determination of the modes of vibration and the associated frequencies is carried out only once. The rest of the calculations involve the determination of the responses of single-degree-of-freedom systems.

In the direct (response history) approach,[12-22] the set of simultaneous differential equations in incremental form is transformed into a set of algebraic equations in terms of one of the response parameters, using the assumption of a linearly varying response acceleration. The resulting set of algebraic equations, which can be cast in tri-diagonal or banded form, is then solved using a recursion relation. The direct integration method, which does not require the uncoupling of the equations of motion, is necessary in nonlinear dynamic analysis.

The time history response analysis of a typical multistory structure, which will generally require the use of a large computer, can be time-consuming and expensive, particularly when inelastic response is considered. Its use may be justifiable only for a few important projects. It does, however, provide more reliable values than mode superposition using response spectra, and is the only means of determining the deformability requirements of members corresponding to a particular design and earthquake input motion.

For design purposes, the use of a single earthquake record as input may not provide sufficient assurance of the behavior of a proposed structure under future earthquakes. For this purpose, the use of a set of earthquake records, possibly including artificially generated accelerograms, would be advisable.

3. Shear and Flexure Beam Lumped-Mass Models for Multistory Buildings. A shear beam model is one in which the shear force acting on any mass depends only on its displacement relative to adjacent masses or supports. This model, shown schematically in Fig. 12-16(a), is typified by the idealized flexible column–rigid girder frame shown in Fig. 12-16(b). The assumption of rigid girders, which implies no rotation at the joints and the neglect of axial deformation in the columns, restricts the coupling between stories and results in a relatively simple model. Because of its simplicity, the shear beam model has often been used to approximate the behavior of low, open-frame structures.

The effect of assuming no joint rotation can result in errors on the order of 100% in both the mode shapes and frequencies, particularly of low-frequency structures, while the neglect of axial deformation in the columns can lead to errors of about 10% in the calculated frequencies. Both of the above assumptions result in a stiffer structure and hence tend to yield frequencies of vibration that are greater than the actual.

A flexure beam model corresponds to a typical cantilever beam with masses concentrated at discrete points along its length, and represents an example of a far-coupled system. As indicated schematically in Fig. 12-16(c), the force acting on any mass is dependent on its displacement relative to all the other masses (and supports) of the system. An analytical model of a multistory building that considers joint rotations as well as axial deformation in the columns would correspond to a flexure beam model.

12.3.6 Nonlinear MDF Systems

For analyses considering nonlinear response, the procedure outlined for SDF systems may be used. This approach considers the response of a nonlinear system as being essentially linear over small intervals of time. The value of the stiffness during each short interval is taken as the slope of the local tangent to the force–displacement curve. Thus as yielding occurs in members and the stiffness of a struc-

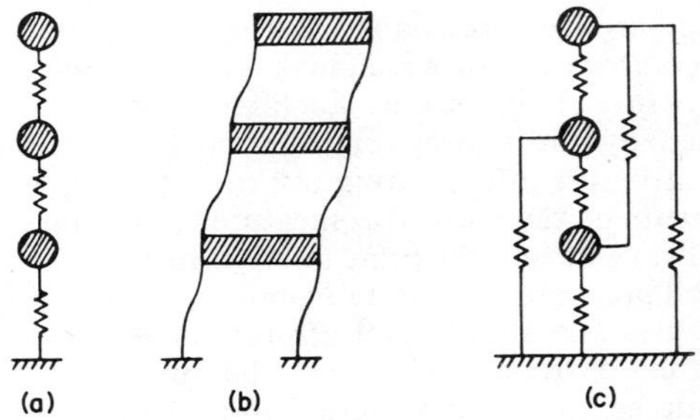

Fig. 12-16 (a) and (b) Shear beam model; and (c) flexure beam model.

ture changes, the response of the nonlinear system is obtained as the response of successively different linear systems. Since each change in member stiffness, which may occur when a member yields or unloads from a post-yield condition, theoretically changes the stiffness of the structure, it is readily seen that the nonlinear dynamic analysis of even a plane multistory, multibay structure can be quite time-consuming.

Computer programs developed to undertake the nonlinear dyanmic analysis of multistory frames can be made to generate a variety of useful information such as maximum deformations and corresponding forces at all significant locations, as well as deformability requirements and time history records or plots of deformations at particular points in the structure. In spite of the limitations of the method, associated mainly with uncertainties in the values of stiffness and damping to be used in the model, nonlinear dynamic analysis represents a powerful tool in the study of the earthquake response of structures. Its use has provided insight into the basic mechanism of earthquake resistance of framed structures as well as a measure of the magnitude of the structural response to particular earthquakes.

The relatively large amount of numerical work involved and considerable computer storage capacity required in a typical inelastic dynamic analysis have made it necessary, in most instances, either to assume fairly simple force–deformation characteristics for the members of multistory, multibay frames or to consider only single-bay frames a few stories high. Allowance is usually made only for yield hinges forming at the ends of members, the hinging being governed by the magnitude of the bending moment only. The neglect of axial load–moment interaction effects on the assumed force–deformation curves of members would be reasonable in strong column–weak beam designs where little or no yielding is expected in the columns. However, where the axial loads on the columns are relatively high, the effect of axial load on the yield moment could have appreciable effects on the response.

12.4 SOME RESULTS OF ANALYTICAL STUDIES ON THE NONLINEAR BEHAVIOR OF MULTISTORY STRUCTURES SUBJECTED TO EARTHQUAKE EXCITATION

Some of the more significant results of analytical studies of the earthquake response of multistory frame, wall, coupled-wall, and frame–shear wall structures are summarized below. In view of the limited scope of these studies, particularly with respect to the range of parameters considered, the observations drawn from their results can serve only as indications of what might be expected under closely similar conditions. The trends exhibited by the results presented below, however, should provide useful guides in the preliminary planning of multistory structures.

The principal basis of comparison used in the following discussion is the ductility requirement (defined in the course of the discussion), associated with each parameter examined.

12.4.1 Open Frame Structures

The configuration and relative member stiffnesses of the basic 20-story frame structure considered in a study by Clough and Benuska,[12-22] from which most of the results presented below are drawn, are shown in Fig. 12-17. The typical tapering of the member stiffnesses from the base to the top will be noted. The frames are assumed spaced at

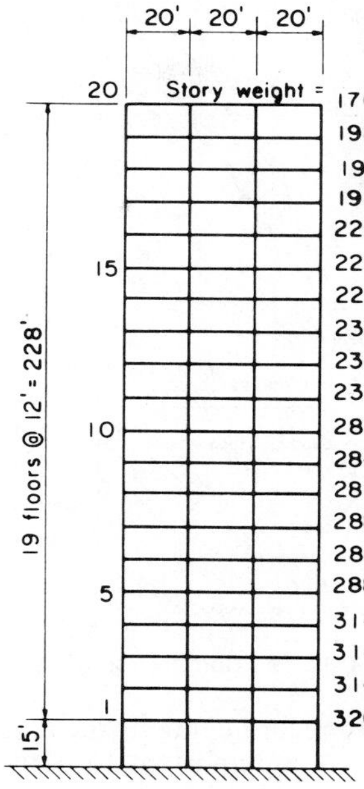

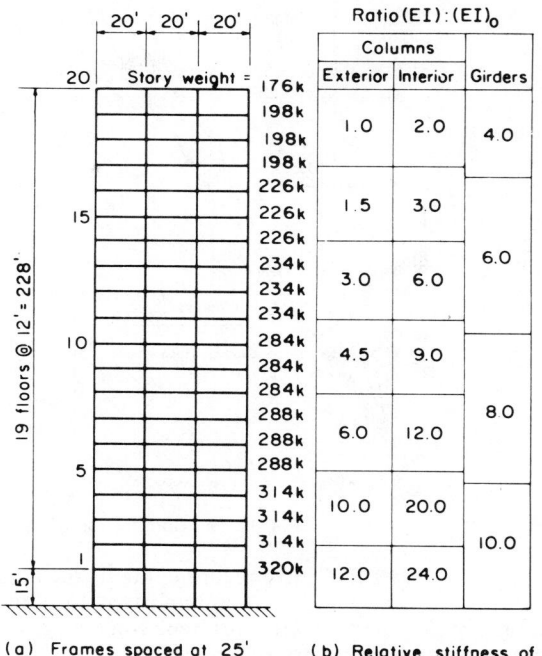

(a) Frames spaced at 25'

Ratio (EI):(EI)$_0$

Columns		
Exterior	Interior	Girders
1.0	2.0	4.0
1.5	3.0	6.0
3.0	6.0	
4.5	9.0	8.0
6.0	12.0	
10.0	20.0	10.0
12.0	24.0	

(b) Relative stiffness of columns and girders

Fundamental period, 2.2 sec.

$(EI)_0 = 133{,}500$ k-ft^2

Fig. 12-17 Properties of standard open-frame structure studied in Ref. 12-22.

25-ft intervals, all the lowest columns being assumed fixed at the base. The frames were designed for vertical loads plus the lateral forces prescribed by the 1964 edition of the *Uniform Building Code*, using simple approximate procedures. Then, using the resulting relative stiffness properties indicated in the figure, a computer analysis for these same statically applied forces was carried out. The computed member forces were designated as design forces. The yield moments were then taken as twice the design values for the girders and six times the corresponding design values for the columns. The effect of axial load on the yield capacity of the columns was not considered in the study.

In the nonlinear dynamic response analysis, the moment–rotation characteristics of the members were assumed to be of the bilinear type with the second, post-yield branch having a slope equal to 5% that of the first (elastic) branch. The term ductility factor as used in the following discussion is defined as the ratio of the maximum total (elastic + plastic) rotation at the end of a yielded member to the yield rotation angle. The yield rotation angle is defined as that corresponding to a moment acting at the end of a simply-supported member having the same section but with a span equal to one-half that of the actual member. The use of a half-span is based on the antisymmetrical mode of deformation of members in a frame due to lateral displacement.

Where damping was considered, an equivalent damping ratio, β_1, equal to 10% of the critical for the first mode of vibration was assumed, with smaller damping in the higher modes (proportional to the period of vibration).

Except where the duration of the earthquake was the parameter considered, the analytical results were obtained by subjecting the structures to the first four seconds of the 1940 El Centro earthquake (N–S component) applied directly to the base. As can be seen in Fig. 12-4, the first four

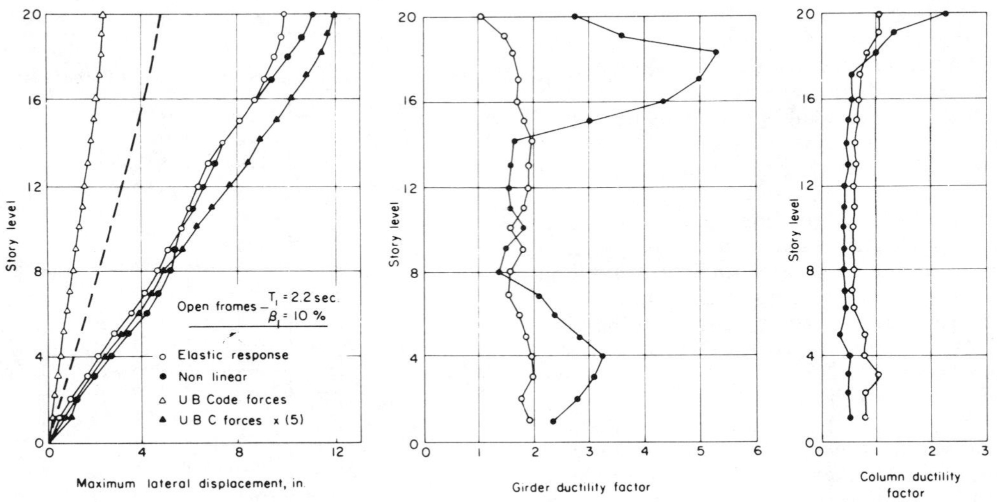

Fig. 12-18 Comparison of elastic and nonlinear seismic response of frames.[12-22]

seconds of the 1940 El Centro record contains the most intense portion of the motion. It is important to note that other earthquake records with different frequency characteristics may produce results which vary significantly from those presented.

1. Comparison of Linear (Elastic) and Nonlinear Response. As with SDF systems (see Fig. 12-15), the maximum lateral displacements for both linearly elastic and nonlinear frames having the same period* and damping are approximately equal over a broad range of fundamental period values. Figure 12-18(a) shows the maximum lateral displacements for elastic and nonlinear frames having a fundamental period, T_1 = 2.2 seconds and a damping ratio β_1 = .10. This similarity in the maximum displacement envelopes should not, however, be interpreted as implying that similar maximum deformations will be developed in corresponding members of the two frames. Figure 12-18(b) shows the girder ductility requirement for the nonlinear case varying from 2 at midheight to 5 at the top, compared to a maximum-to-yield moment ratio of about 2 for the elastic case. Clearly, deducing member ductility requirements on the basis of elastic response values can be grossly unconservative. An analysis assuming completely elastic response will generally overestimate the inelastic deformations in the columns and underestimate these in the girders of a typical strong column–weak beam code-designed frame (Fig. 12-18c).

Also shown in Fig. 12-18(a) is the computed lateral deflection of the elastic frame when subjected to the static lateral forces specified by the *Uniform Building Code*, 1964 Edition.** For this particular frame, the deflection under the code-specified lateral forces is only about one-fifth of the maximum dynamic displacement produced by the N–S component of the 1940 El Centro earthquake. The dashed curve in Fig. 12-18(a) shows the approximate deflected shape corresponding to first yield in the statically loaded, UBC-designed frame. The curve was obtained by multiplying the displacement values from a static analysis under UBC-forces (the -Δ- curve shown in the same figure) by the ratio of yield to design stress of the weakest element—in this case, 2.0. If one were to use the average ratio of maximum displacements for either elastic or nonlinear dynamic response to the above "static first-yield" displacements as an indication of ductility requirements, one would arrive at a value of about 2.5 for this case. Figure 12-18(b) clearly indicates that it is important to draw a distinction between the ductility factors associated with the lateral displacement of a frame and the member ductility factors. Since the former is achieved through inelastic deformations at the critically stressed portions of a relatively few members (mostly beams), the corresponding member ductility factors will generally be greater than those associated with the lateral displacement.

In Fig. 12-18(c), as well as in the following figures, a ductility factor less than unity indicates the ratio of the maximum moment to the yield moment in a member.

2. Effect of Period of Vibration. Two 20-story frames having fundamental periods of 1.6 and 2.8 seconds were considered in addition to the standard 2.2-second frame. All three frames were the same except that the basic stiffness parameter, EI_o, was varied to obtain the different periods. The results, shown in Fig. 12-19, indicate that the maximum lateral displacement, in the period range considered, increases almost in proportion to the building period (though not the flexibility, which varies as the square of the period). The general tendency of the maximum displacement to increase with the period of a structure (or with a decrease in its fundamental frequency) in the period range considered is also apparent from the response spectra for both elastic and elasto-plastic SDF systems shown in Fig. 12-15.

There is a slight decrease in girder ductility requirements for the more flexible (long-period) structures. However, this particular trend can be reversed in the case of earthquake records with significantly different frequency characteristics, as indicated later in Section 12.4.1.8.

A probabilistic study by Ruiz and Penzien[12-23] of the response of eight-story, shear-beam models subjected to a number of artificially generated accelerograms showed that in structures with a fundamental period of about 0.5 second, the ductility requirements tend to decrease toward the top of the structure. This contrasts with the variation typical of the more flexible frames shown in Fig. 12-19(b), in which the influence of the higher modes of vibration causes a significant increase in the ductility requirements in the

*The (fundamental) period of the nonlinear frame, as used in this discussion, refers to the period corresponding to the initial stiffness, before yielding occurs.

**In the 1964 *Code*, the total base shear was distributed over the height of a building in a triangular manner, with a maximum intensity at the top and zero intensity at the base. Subsequent editions of this *Code* incorporated one of the principal recommendations resulting from the Clough–Benuska study, and provided for the application of a portion of the specified base shear as a concentrated load at the top of slender structures, with the balance to be distributed over the height of the building in a triangular fashion.

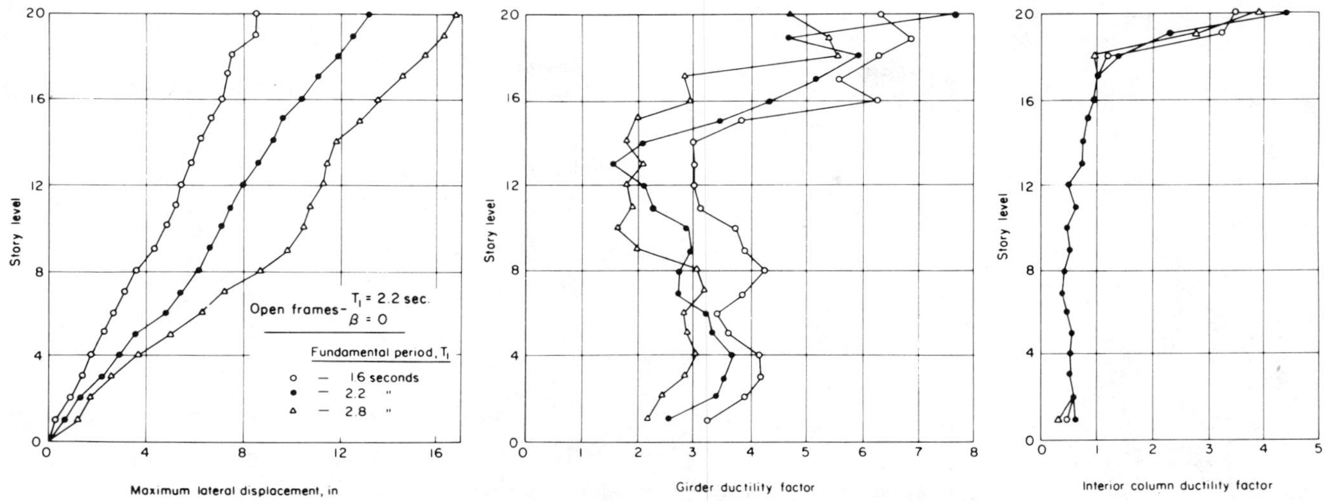

Fig. 12-19 Effect of period of vibration on seismic response of frames.[12-22]

top stories. The same study by Ruiz and Penzien showed that the ductility requirements at the base of a stiff structure can be significantly greater than those for a flexible structure subjected to the same excitation.

An interesting result of the Ruiz–Penzien study was that the whiplash effect in flexible structures with fundamental periods of approximately 2.0 seconds, due to the influence of the higher modes of vibration, still occurred even though the model frame was designed with 10% of the UBC-specified base shear applied at the top of the frame.

3. Effect of Number of Stories. Figure 12-20 shows the results obtained from analyzing a 10-story and a 30-story

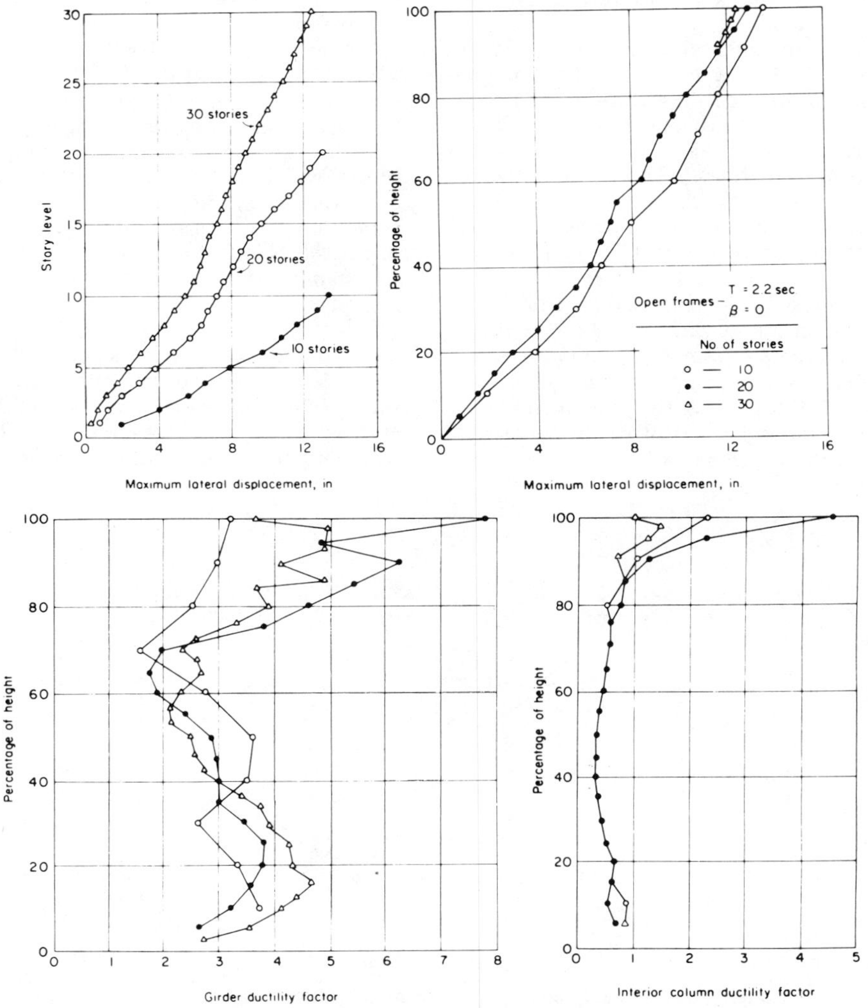

Fig. 12-20 Effect of building height on seismic response of frames.[12-22]

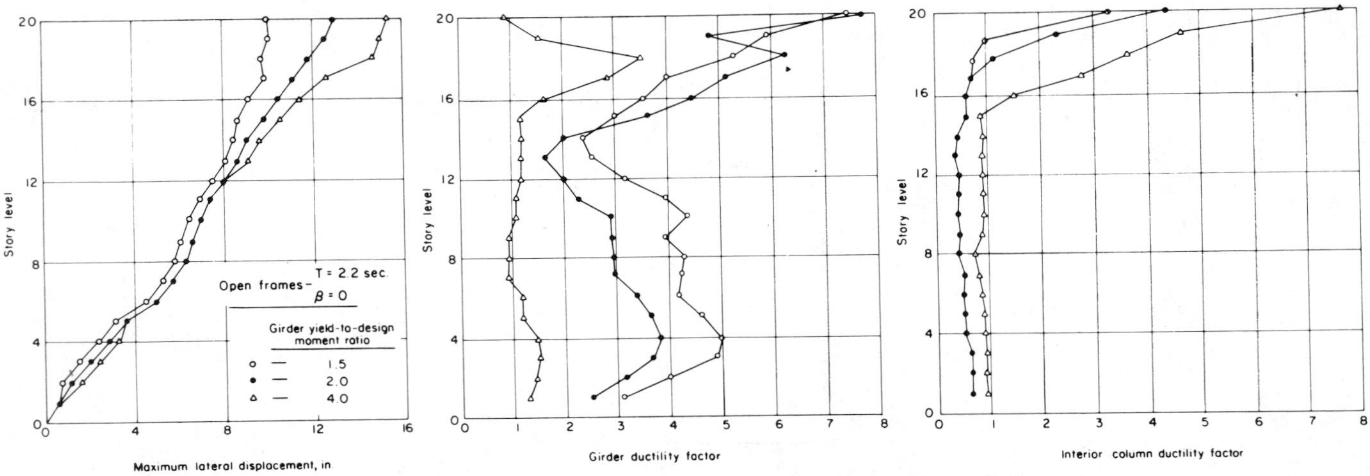

Fig. 12-21 Effect of girder strength on seismic response of frames.[12-22]

frame in addition to the standard 20-story frame—all having fundamental periods of 2.2 seconds. All three frames have about the same maximum deflection at the top, indicating a decreasing relative story displacement with increasing height of frame. When the maximum deflection and ductility requirements are plotted as a function of relative story height, however, the curves for the three frames are very similar. The more pronounced whiplach effect on the member ductility requirements of the taller structures is evident in the figure.

In actual practice, the fundamental period of vibration generally increases with the height of a building. On the basis of results presented in the preceding section, the maximum lateral displacement can be expected to increase with the height of a building.

4. Effect of Strength of Girders. Three frames were considered: the standard one with a girder yield-to-design moment ratio of 2.0 and two other frames, identical to the first, except that the yield moments were 1.5 and 4.0 times the design moments.

Figure 12-21 shows the results of analyses. Figure 12-21(a) indicates that the maximum lateral deflections of the three frames are practically the same except in the upper stories where an increase in deflection accompanies an increase in girder strength. Figure 12-21(c) shows that this increase in deflection in the upper stories is a result of the increased inelastic deformation in the columns of these stories, indicating that the maximum deflection is primarily controlled by the columns rather than the girders.

As might be expected, the girder ductility requirements decrease with increasing girder strengths. This is shown in Fig. 12-21(b). More significant, however, is the fact that the increase in girder strength forces more of the inelastic deformation to occur in the columns, as indicated in Fig. 12-21(c). In general, decreasing the yield strength of one member type (i.e., columns or girders) with respect to another tends to attract inelastic deformations toward the weaker members, resulting in reduced yielding in the stronger member type.

5. Effect of Column Strength. This variable was studied by considering two frames, having column yield-to-design moment ratios of 2.0 and 10.0, in addition to the standard frame, which had a moment ratio of 6.0. Figure 12-22 indicates that increasing the column strength beyond that corresponding to a ratio of 6.0 does not materially affect the response. This follows from the fact that the columns in the standard building remain essentially elastic during the response.

Figure 12-22(c), however, shows that a reduction in column strength can have a significant effect on the distribution of ductility requirements. The results indicate that if the columns do not have a sufficient margin of strength above the design strength, most of the inelastic deformation will tend to occur in the columns. Because of the danger

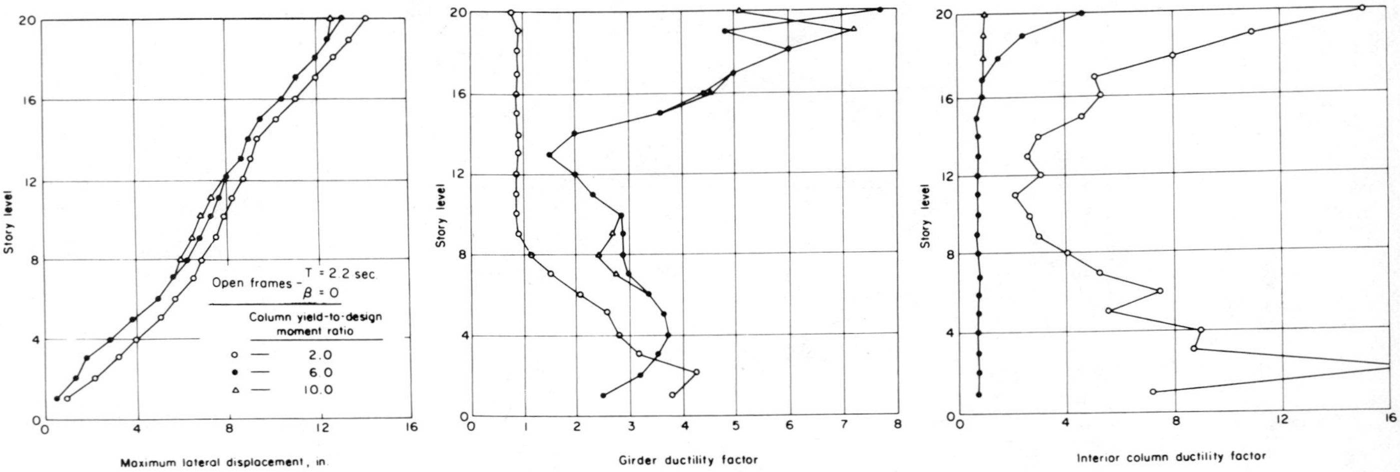

Fig. 12-22 Effect of column strength on seismic response of frames.[12-22]

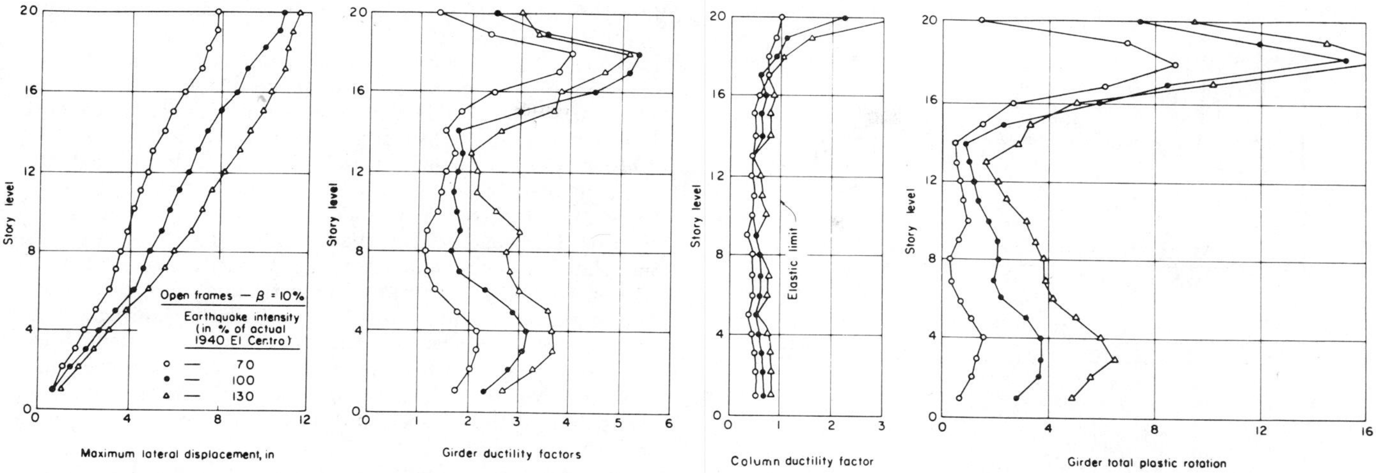

Fig. 12-23 Effect of earthquake intensity on seismic response of frames.[12-22]

of instability associated with excessive yielding in the columns, such a condition should be avoided.

6. Effect of Intensity of Earthquake. The results shown in Fig. 12-23 were obtained by subjecting the standard open frame building to the first four seconds of the 1940 El Centro (N–S) earthquake, the intensity—here, used to denote the amplitude of the ground acceleration—being varied by multiplying the actual ground acceleration by 0.7 in one case and by 1.3 in another. As might be expected, there is a general increase in response with increasing intensity. Except in the top stories, the columns remain elastic even under an earthquake with intensity equal to 1.3 times that of the 1940 El Centro (N–S).

7. Effect of Duration of Earthquake. To study this effect, the standard building was subjected to a synthetic 12-second base motion consisting of the first 6 seconds of the El Centro earthquake (N–S component) followed by the same 6-second motion and also to the first 8 seconds of the El Centro N–S record.

Figure 12-24 shows the results including those corresponding to the first 4 seconds of the El Centro N–S record for comparison. Figure 12-24(a and b) indicates that the effect of the second 4 seconds of the El Centro motion, which have relatively smaller-amplitude pulses, is not significant as far as maximum displacement and ductility requirements are concerned. However, the 12-second synthetic earthquake of large-amplitude pulses produced a significant increase in both maximum lateral displacements and girder ductility requirements. The effect of duration is also apparent on the total plastic rotation or accumulated ductility requirements of the girders, as shown in Fig. 12-24(c). The total plastic rotation provides an indication of the number of times yielding occurs and/or the extent of such yielding, and is significant in view of the deterioration in strength that may accompany repeated loading into the plastic range, or the instability that may result from extensive yielding. The damage potential of an earthquake thus increases with the duration of large-amplitude motion. This observation is also borne out by the results of a statistical study by Husid[12-24] of the earthquake response of simple (SDF) bilinear systems subjected to gravity loads, which indicated that the intensity of earthquakes of short duration required to produce collapse would have to be greater than that of longer-lasting earthquakes.

For the structure and base motions considered, the column ductility requirements are not significantly affected by the duration of ground motion. This follows from the fact that inelastic deformation occurs mostly in the girders, with columns remaining essentially elastic except at the top stories.

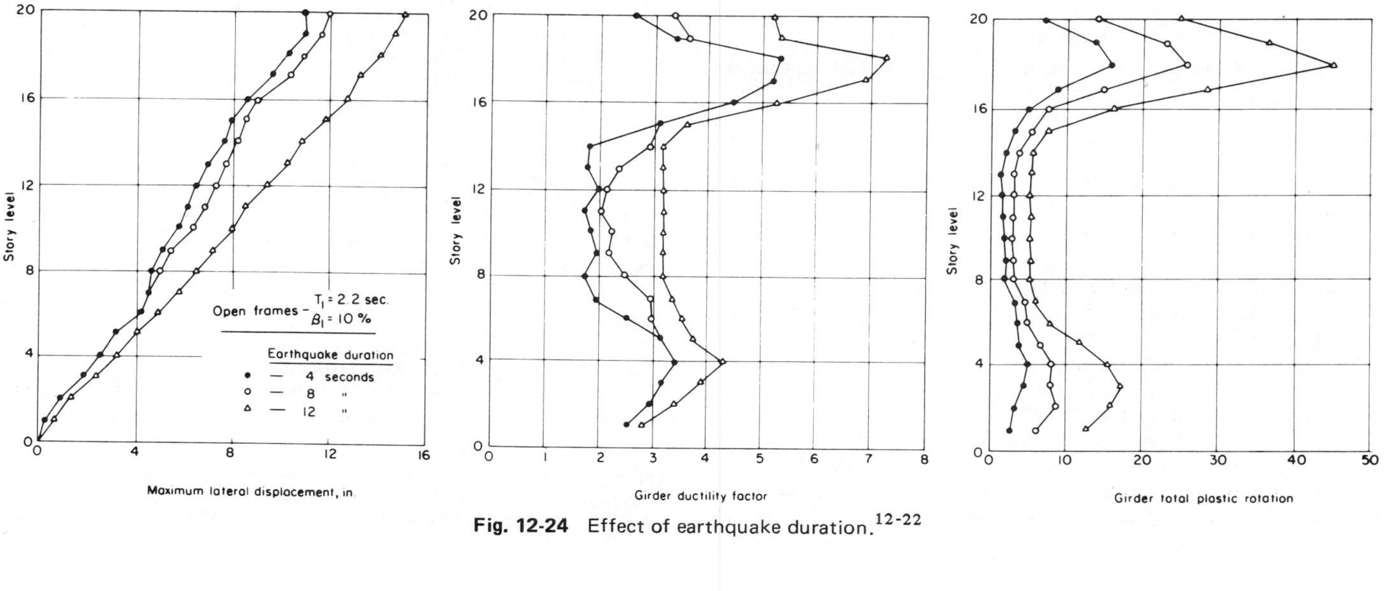

Fig. 12-24 Effect of earthquake duration.[12-22]

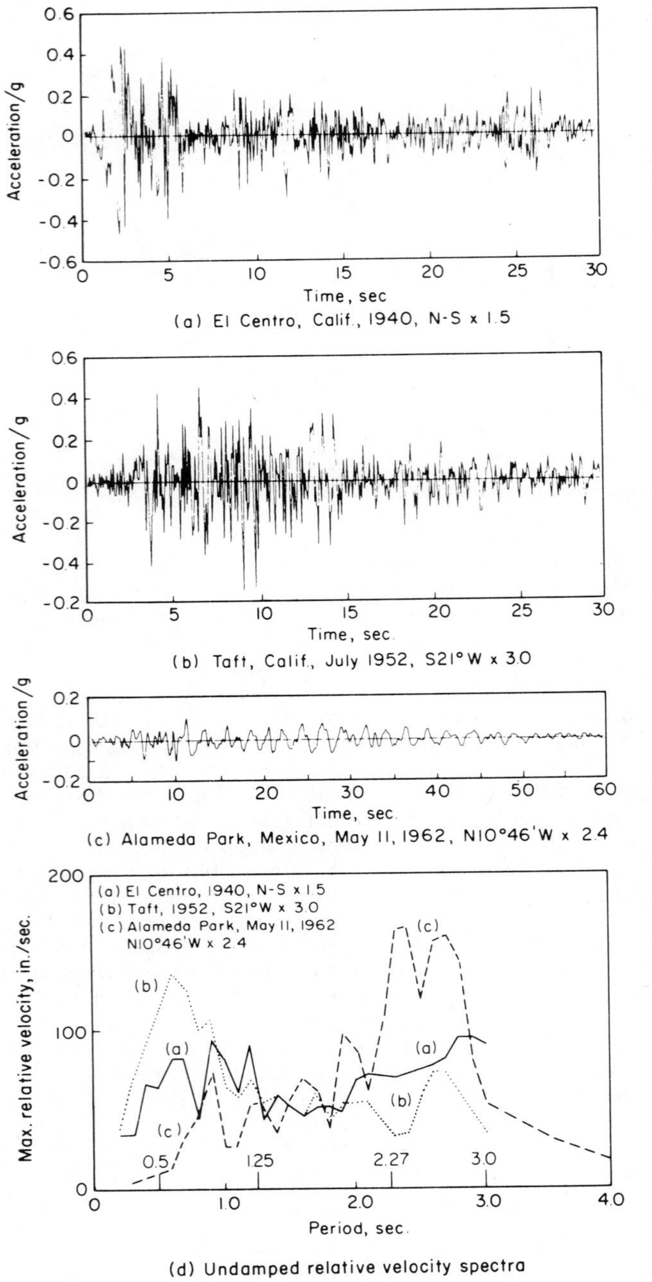

Fig. 12-25 Frequency characteristics of three sample earthquake accelerograms.

8. *Effect of Frequency Characteristics of Earthquake Motion.*—The Clough–Benuska study[12-22] considered the effect of two of the three major characteristics of the earthquake input motion, namely, intensity (i.e., acceleration amplitude) and duration of large-amplitude pulses. The frequency characteristics of the motion, which generally vary with each earthquake, may have a significant effect on the response of a structure, depending upon the significant periods of vibration of the latter.

Figure 12-25(a–c) from Ref. 12-25, shows the acceleration traces of three earthquakes having distinct frequency characteristics. The amplitudes of the 1952 Taft (S 21°W) and the 1962 Alameda Park, Mexico (N 10°46′W) accelerograms shown have been multiplied by factors of 3.0 and 2.4, respectively, to give them about the same spectral intensity* as the 1940 El Centro (N–S) accelerogram magnified by a factor of 1.5.

Figure 12-25(d) shows that the 1952 Taft (S 21°W) accelerogram has its dominant frequencies in the short-period range (0.5–1.0 second) while the 1962 Alameda Park record has its peaks in the longer-period range (2–3 seconds). The 1940 El Centro record, on the other hand, has a relatively uniform distribution of peaks over the period range of interest (i.e., 0.5 to 3.0 seconds) compared to the other two records. It is readily seen that a strong ground motion having the frequency characteristics exhibited by the 1940 El Centro N-S record more likely to excite a greater number of the vibration modes that contribute significantly to the response of a structure than an earthquake in which the dominant frequencies are restricted to a narrow range. Note that the Alameda Park record has a 60-second duration compared to the 30-second duration of the other two records shown in Fig. 12-25(a–c).

As mentioned in Section 12.4.1.2, the trend of decreasing column and girder ductility requirements with increasing period of vibration observed for the 1940 El Centro (N–S) earthquake is reversed in the case of the Alameda Park earthquake. This is shown in Fig. 12-26, from the study by Goel and Berg[12-25] of the response of 10-story, single-bay, symmetrical shear frames to the three earthquake motions shown in Fig. 12-25. The columns in this study were assumed to behave elastically throughout the response, with inelastic action allowed only in the girders. The whiplash effect, which is apparent in the column maximum-to-yield moment ratio curves for the El Centro and Taft earthquakes, is practically absent in the Alameda Park response. Since the whiplash effect reflects the contribution of the higher modes of vibration and none of the structures considered have higher modes with periods in the 2–3-second range where the Alameda Park velocity spectrum has its dominant peaks, the results shown in Fig. 12-26 would be expected. The relatively smaller ductility requirements associated with the Alameda Park earthquake, even for the 2.27-second structure, indicate that the higher modes of vibration also play a significant role in determining the magnitude of member ductility requirements.

It is significant to note that in spite of the smaller amplitude of the acceleration pulses in the Alameda Park earthquake, the maximum lateral displacement of the 2-27-second structure is greater for this earthquake than for either the El Centro or the Taft earthquake. This clearly indicates the major influence of the fundamental mode on the maximum lateral displacement—since for this structure the period of the fundamental mode falls within the range of the dominant frequency components of the Alameda Park earthquake. The same trend of increasing maximum displacements with increasing period will be noted for all three earthquakes.

For design purposes, the variability in frequency characteristics of earthquake motions may be provided for by using either average response spectra or, where a time-history response is desirable, a set of input accelerograms with frequency characteristics covering the range of expected variation.

12.4.2 Walls

The results presented are from the report of an investigation conducted at the Portland Cement Association (PCA).[12-26] The basic structure considered in this study is a

*Spectral intensity is defined as the area under the undamped velocity response spectrum between ordinates including period values of practical interest. In this study[12-25], the period range used was from 0.5 to 3.0 sec.

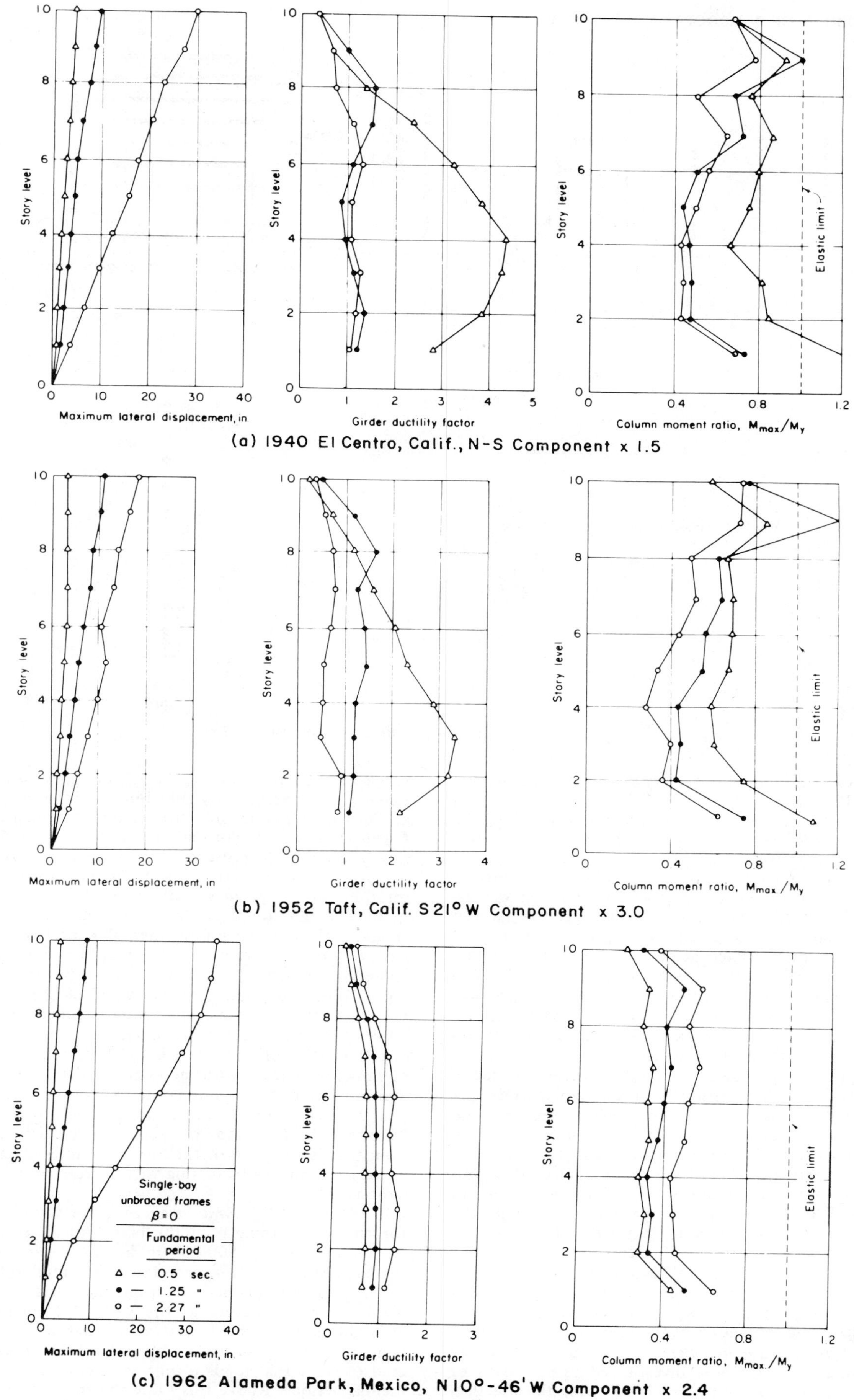

Fig. 12-26 Effect of frequency characteristics of earthquake motion.[12-25]

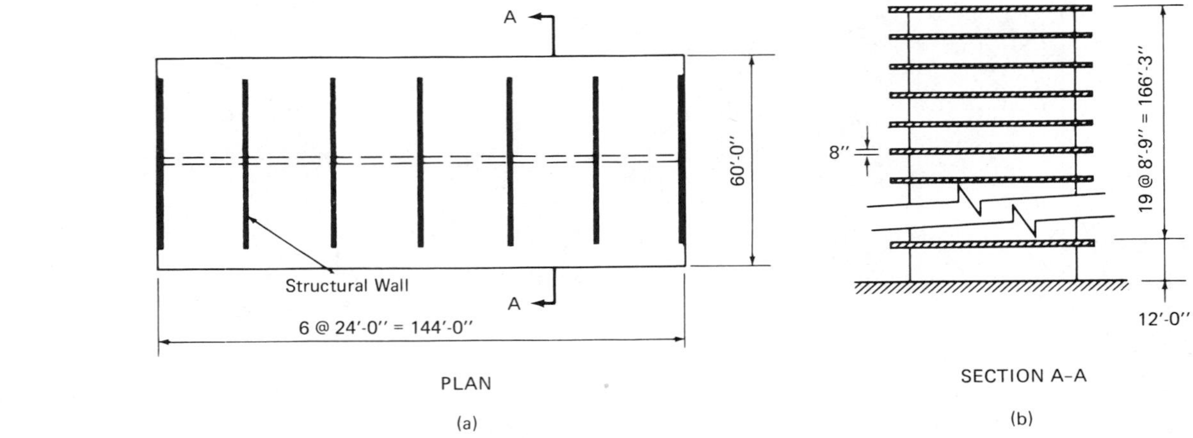

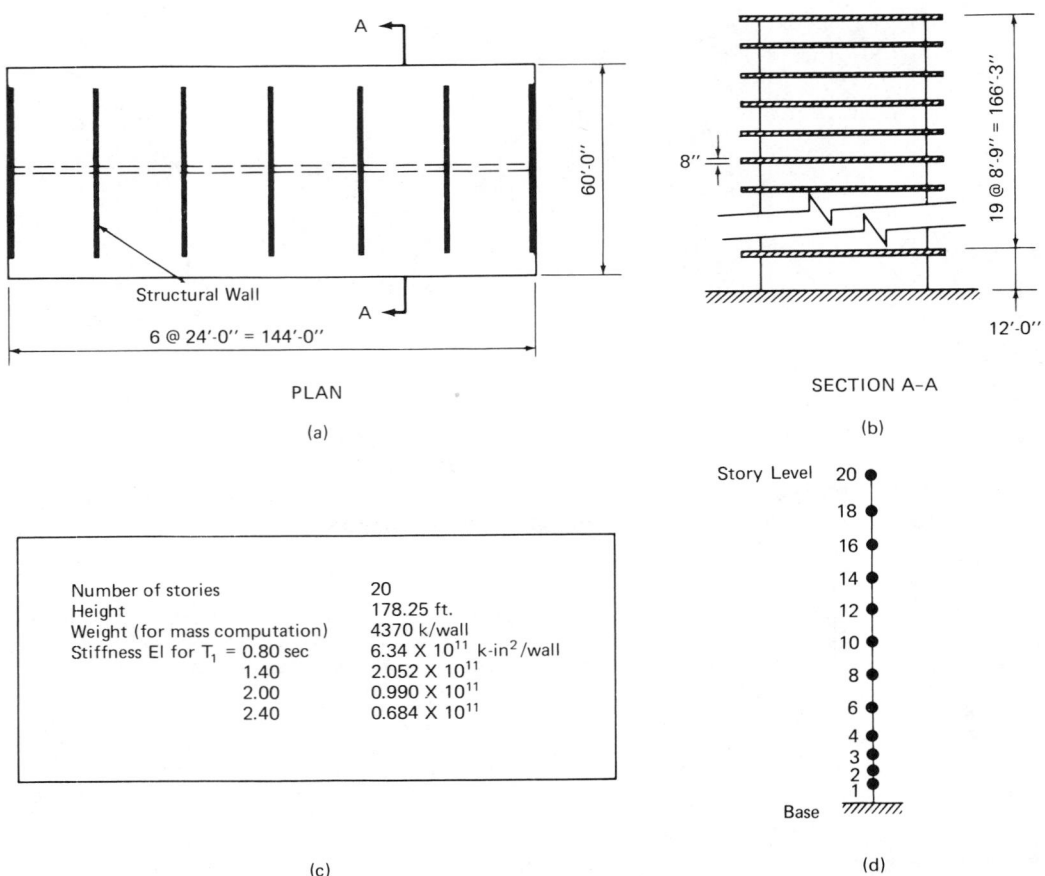

Number of stories	20
Height	178.25 ft.
Weight (for mass computation)	4370 k/wall
Stiffness EI for T_1 = 0.80 sec	6.34 X 10^{11} k-in^2/wall
1.40	2.052 X 10^{11}
2.00	0.990 X 10^{11}
2.40	0.684 X 10^{11}

(c)

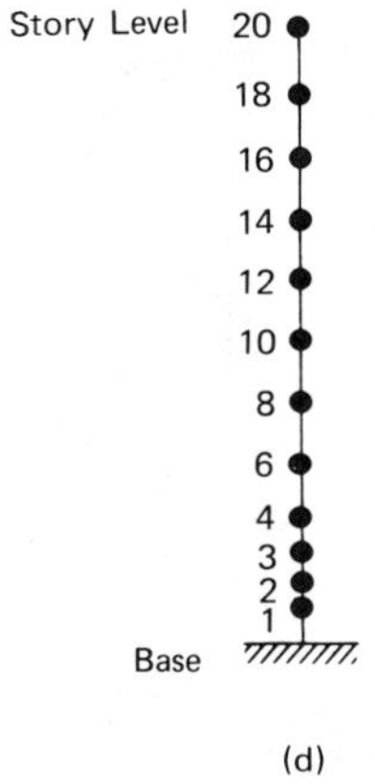

Fig. 12-27 Properties of standard shear wall structure considered in Ref. 12-26.

20-story building consisting mainly of a series of parallel walls (Fig. 12-27a–c).

The stiffness of the walls was assumed uniform along the height. A constant wall cross section through the height was also assumed. However, a reduction in the yield strength of sections above the base was included to reflect the effects of axial loads on moment capacity. The building was assumed to be fully fixed at the base. Inelasticity was allowed in dynamic analyses by means of concentrated flexural "point hinges" that formed at the ends of elements when the yield moment was exceeded at these points. The hysteretic moment–rotation relationship for these hinges was an extended version of Takeda's model,[12-27] which accounts for the observed decrease in reloading stiffness in reinforced concrete members subjected to reversed inelastic loading. A 12-mass model of the 20-story walls was used in analyses (Fig. 12-27d), with the masses concentrated at each floor level in the first four floors where most inelastic action usually took place.

The ground motion used in analyses had the same frequency characteristics as the E–W component of the 1940 El Centro record. The duration of the motion was set at 10 seconds. The intensity was normalized to 1.5 times the spectrum intensity corresponding to the first 10 seconds of the N–S component of the 1940 El Centro record.

Ductility was defined on the basis of nodal rotations as being equal to θ_{max}/θ_y, where θ_{max} was the maximum computed rotation at the node, and θ_y was the nodal rotation corresponding to yielding at the base.

1. Effect of Fundamental Period. The effect of the initial fundamental period was investigated using values of 0.80, 1.40, 2.00, and 2.40 seconds to cover the practical range for 20-story buildings. Each period was investigated under varying values of yield strength of the critical section at the base (M_y), in order to examine the relationship between these two major variables and the response quantities. Figure 12-28 presents maximum horizontal displacements, ductility requirements, and cumulative plastic hinge rotations along the height of the walls for M_y = 500,000 in.-k. The horizontal displacements show a consistent increase with increasing fundamental period (or decreasing stiffness) of the structure. The ductility requirements become greater with decreasing fundamental period (increasing stiffness). Beyond a certain value of the fundamental period, however, the ductility requirements do not decrease significantly with an increase in period.

It is worth noting that although rotational ductility requirements increase with increasing structural stiffness, the absolute value of maximum rotation for a stiff structure is less than that for a more flexible structure (Fig. 12-29). Thus, Fig. 12-28(c) shows the cumulative plastic hinge rotation, that is, the sum of the absolute values of the inelastic hinge rotations, as increasing with increasing period of the structure.

2. Effect of Flexural Strength. The values considered for the yield strength of the base critical section ranged from 500,000 to 1,500,000 in.-k. The results for the particular case of T_1 = 1.4 seconds are presented in Fig. 12-30. For the same fundamental period, the horizontal displacements decrease sharply as the yield level increases from 500,000 in.-k to the largest value that would still cause yielding at the base under the earthquake motion considered (a value that tends to decrease with increasing fundamental period). Above this value the trend is reversed, and

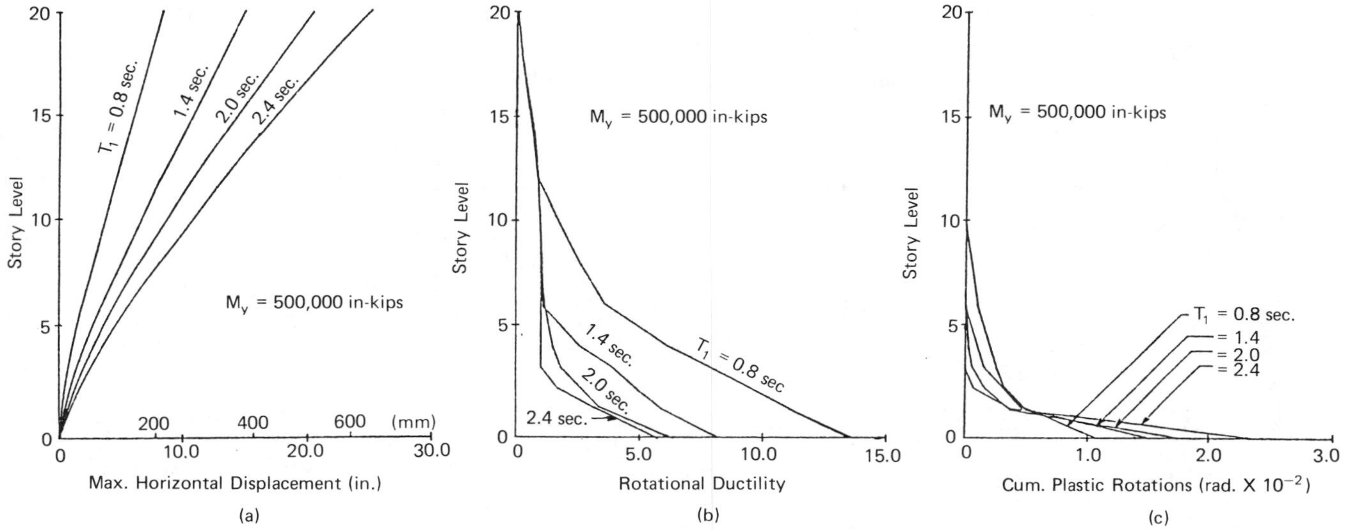

Fig. 12-28 Effect of fundamental period of vibration on the seismic response of structural walls.[12-26]

an increase in yield level is accompanied by an increase in horizontal displacements. It can be seen that the ductility requirements increase significantly as the yield level decreases. The cumulative plastic hinge rotations also increase as the yield level decreases.

3. Effect of Intensity of Earthquake. Three sets of analyses were carried out corresponding to different combinations of the fundamental period, T_1, and the yield level, M_y. In all cases, the input motion used was the first 10 seconds of the E–W component of the 1940 El Centro record, normalized to 0.75, 1.0, as well as 1.5 times the spectrum intensity corresponding to the first 10 seconds of the N–S component of the same earthquake. Figure 12-31 shows the results for T_1 = 1.4 seconds and M_y = 500,000 in.-k. A consistent increase in response with increasing intensity is evident. The displacements and ductility requirements increase almost proportionately with increasing intensity.

4. Effect of Duration of Earthquake. In studying the effect of earthquake duration, the response of the reference structure, with T_1 = 1.4 seconds and M_y = 500,000 in.-k, to the first 10 seconds of the E–W component of the 1940 El Centro record was compared with its response to an accelerogram with a 20-second duration.

The 20-second accelerogram consisted of the first 12.48 seconds of the E–W component of the 1940 El Centro record followed by the portion of the same record between 0.98 seconds and 8.5 seconds. Because the 1940 El Centro E–W component has its peak acceleration at about 11.5 seconds, it was decided to include this peak in the composite record and add a segment from the more intense portion of the first 10 seconds to make a 20-second record. This altered the velocity response spectrum so that a normalizing factor (1.54) smaller than that (1.88) used for the 10-second input motion was indicated to obtain a spectrum intensity equal to 1.5 times that of the 1940 El Centro N–S component. The amplitude of the pulses in the 10-second input was thus larger than in the first 10 seconds of the 20-second composite accelerogram by a factor of 1.88/1.54, or 1.22.

In the response envelopes of Fig. 12-32, the curves corresponding to the 20-second composite accelerogram described above are marked "20 sec.-(a)." This figure shows that the displacements and ductility requirements are greater for the 10-second record than for the 20-second composite record having the same spectrum intensity. The cumulative plastic rotations in the hinging region, however, are greater for the 20-second-long record, as shown in Fig. 12-32(c). This follows from the fact that the structure goes through a greater number of inelastic oscillations when subjected to longer excitation.

The curves in Fig. 12-32 marked "20 sec.-(b)" represent the response of the structure to the 20-second composite accelerogram when scaled by the same factor (1.88) used in normalizing the 10-second record.

Figure 12-32 indicates that the major effect of increasing the duration of the large-amplitude pulses in the input accelerogram is to increase the number of cycles of large-

M_y = 500,000 in. kips

	Yield Rotation θ_y (rad.)	Max. Rotation θ_{max} (rad.)
1	.00014	.00135
2	.00070	.00448

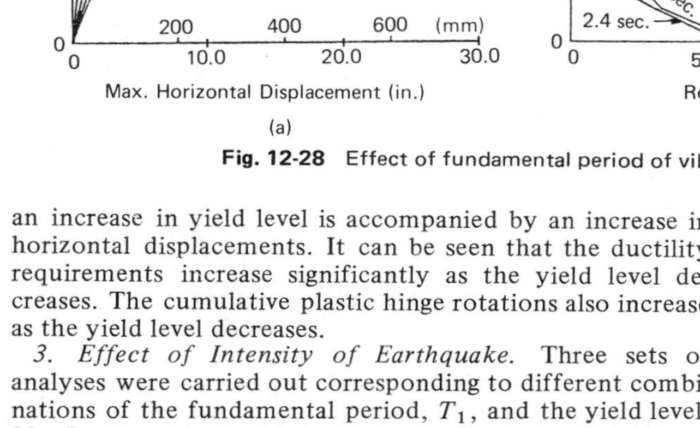

Nodal Rotation 1st Floor Level

MEASURES OF DEFORMATION

Ductility Ratio vs. Absolute Rotations

$$\mu_1 = \frac{\theta_{max.}^1}{\theta_y^1} = 9.6 \qquad \text{but note that:}$$

$$\mu_2 = \frac{\theta_{max.}^2}{\theta_y^2} = 6.4 \qquad \theta_{max}^2 = 3.3\,\theta_{max}^1$$

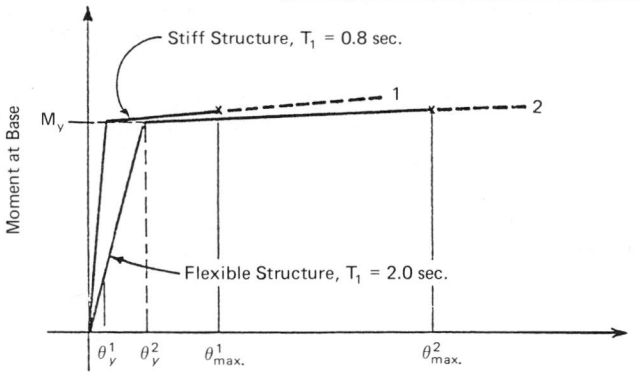

Fig. 12-29 Rotational ductility ratio vs. maximum absolute rotation as measures of deformation.

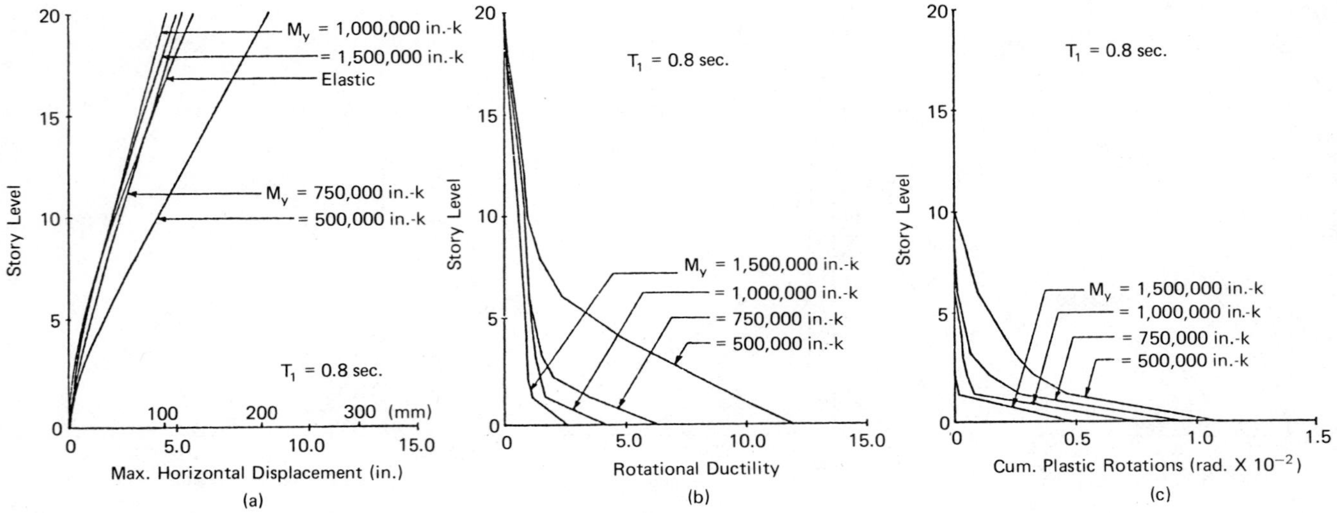

Fig. 12-30 Effect of yield level, M_y on seismic response of structural walls.[12-26]

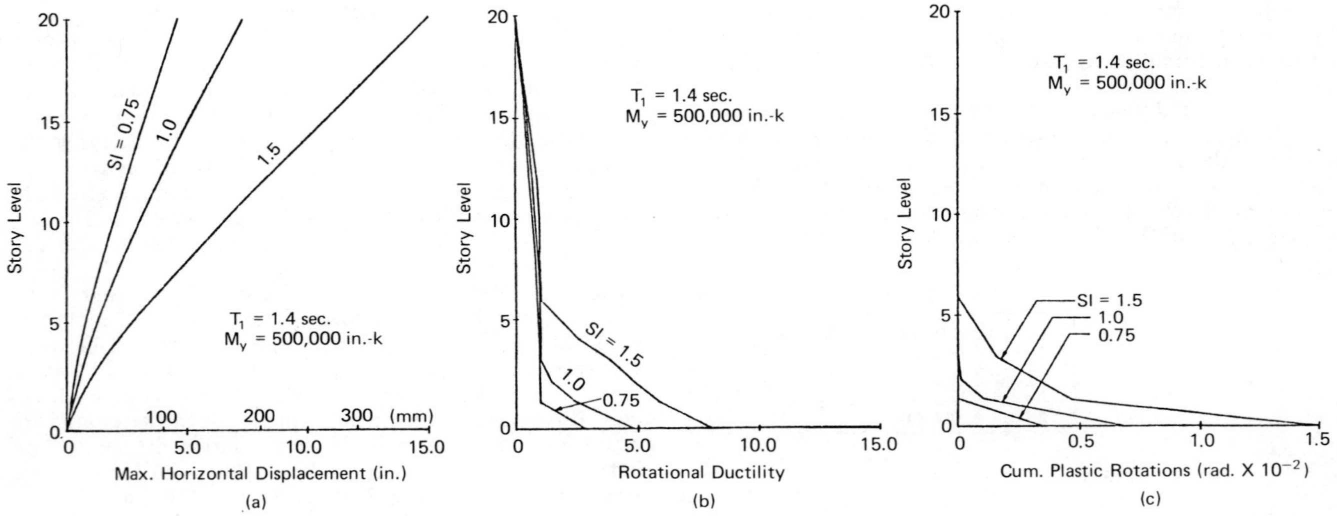

Fig. 12-31 Effect of intensity of ground motion on seismic response of structural walls.[12-26]

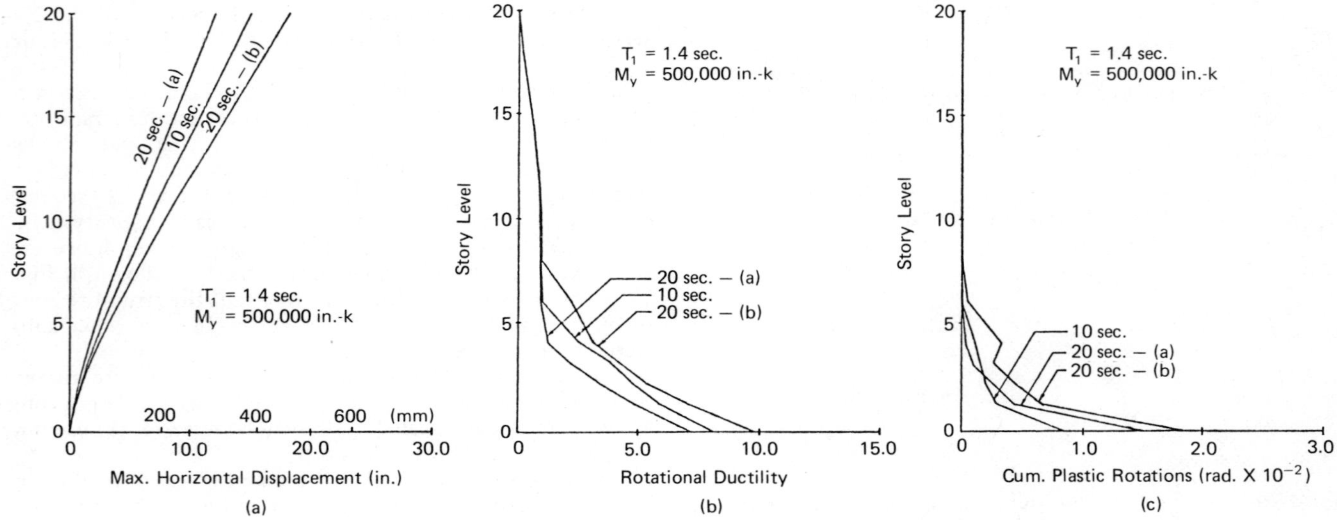

Fig. 12-32 Effect of duration of ground motion on seismic response of structural walls.[12-26]

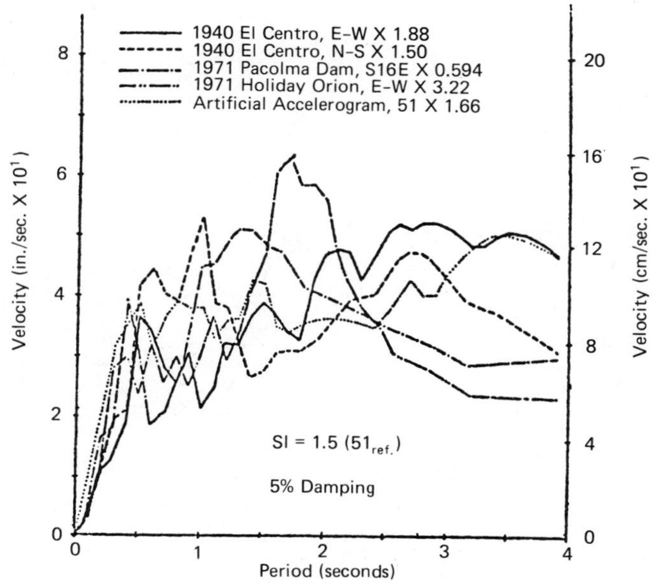

Fig. 12-33 Relative velocity responses spectra for first 10 seconds of normalized input motions.[12-26]

amplitude deformations that a structure will undergo. This conclusion assumes that the intensity and frequency characteristics of the additional motion do not differ significantly from those of the shorter-duration input.

5. Effect of Frequency Characteristics of Earthquake Motion. To determine the effect of frequency content of earthquake input motion, three sets of analyses were carried out. A total of five accelerograms were used. The corresponding normalized 5%-damped velocity spectra are shown in Fig. 12-33. The accelerograms used in Sets (a) and (b) were normalized with respect to intensity to 1.5 times the 5%-damped spectrum intensity of the N-S component of the 1940 El Centro record, which is referred to as SI_{ref}. Those used in Set (c) were normalized to 0.75 times SI_{ref}.

Set (a). Envelopes of response values for a structure with a period of 1.4 seconds and a yield level, M_y = 500,000 in.-k are shown in Fig. 12-34. Figure 12-34(a and c) indicates that the E-W component of the 1940 El Centro record, classified as "broad-band ascending" with respect to its response spectrum (which in turn is indicative of frequency characteristics), produces larger maximum displacements and ductility requirements than the other three input motions considered. However, the same record produces the least value of the maximum horizontal shear. The artificial accelerogram S1 produces the largest shear. Because all the structures yielded and the slope of the second, post-yield branch of the assumed moment-rotation curve was relatively flat, the moment envelopes for this case did not show any significant differences among the four input motions used.

As yielding progresses and the effective period increases, it is the "broad-band ascending" type of accelerogram (in this case, the El Centro E-W component) that excites the structure most severely. Response to the other types of accelerograms—and particularly the "peaking" accelerograms—tends to be less severe.

Set (b). Figure 12-35 shows response envelopes for a structure with fundamental period, T_1 = 0.8 second and a yield level, M_y = 1,500,000 in.-k. As can be seen, the peaking accelerogram (N-S component of the 1940 El Centro record) consistently produces a greater response in the structure than a broad-band record. A comparison of Fig. 12-35 and Fig. 12-34 indicates that the ductility requirements are significantly less for this structure with a high yield level. In addition, yielding does not progress as high up the wall as in the case of the structure with period T_1 = 1.4 seconds and a low yield level.

The greater response of the structure under the peaking N-S component of the 1940 El Centro record follows from the fact that the dominant frequency components for this motion occur in the vicinity of the period of the structure. In this region the E-W motion has relatively low-power components. Also, because of the high yield level of the structure, yielding was not extensive, particularly under the E-W component. Apparently, the effective period of the yielded structure did not shift into the range where the higher-powered components of the E-W motion occur.

Set (c). In the third set, the same structure considered in Set (a) was used, but with the intensity of ground motion reduced by one-half. The two motions used were the 1971 Pacoima Dam, S16E record (peaking) and the 1940 El Centro E-W (broad-band).

Calculated envelopes of response are shown in Fig. 12-36. These support the observation that when yielding in a structure is not extensive enough to cause a significant increase in the effective period, the peaking-type accelerogram is likely to produce the more critical response. Figure 12-36 shows that yielding in the structure does not extend far above the base when compared to Set (a) where the input motion was twice as intense. It will be noted that in Set (a), the 1940 El Centro E-W motion (a broad-band accelero-

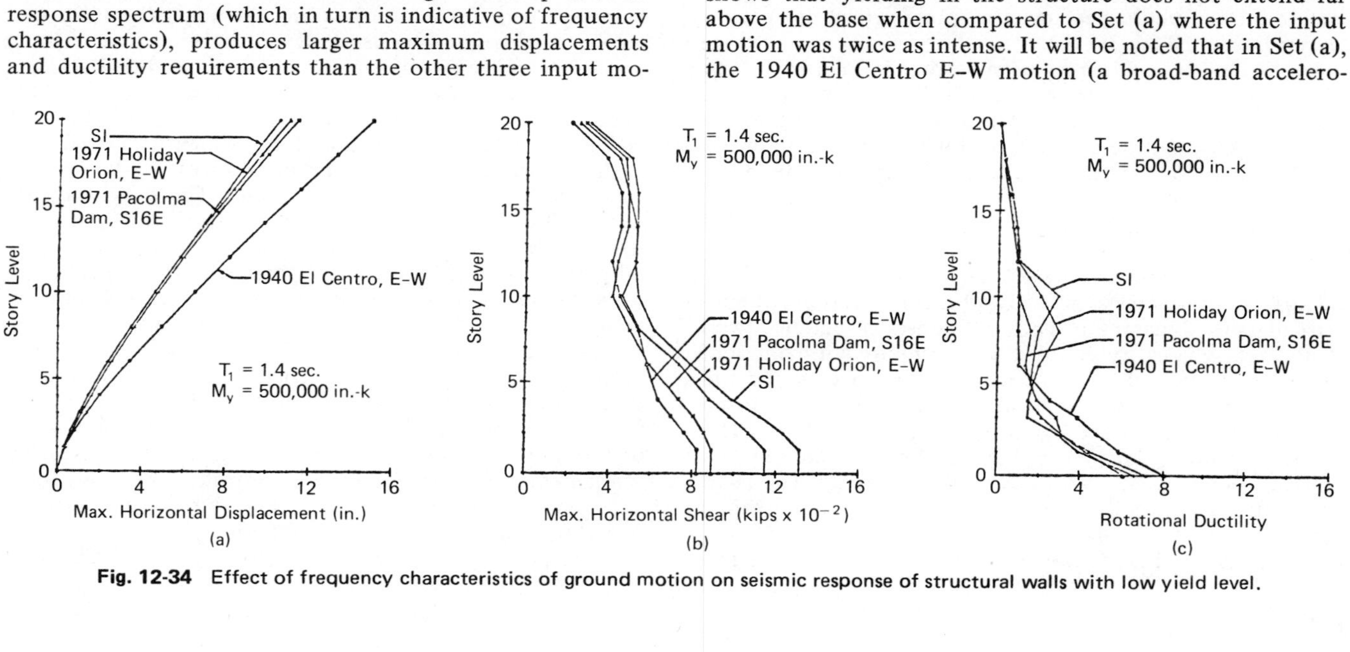

Fig. 12-34 Effect of frequency characteristics of ground motion on seismic response of structural walls with low yield level.

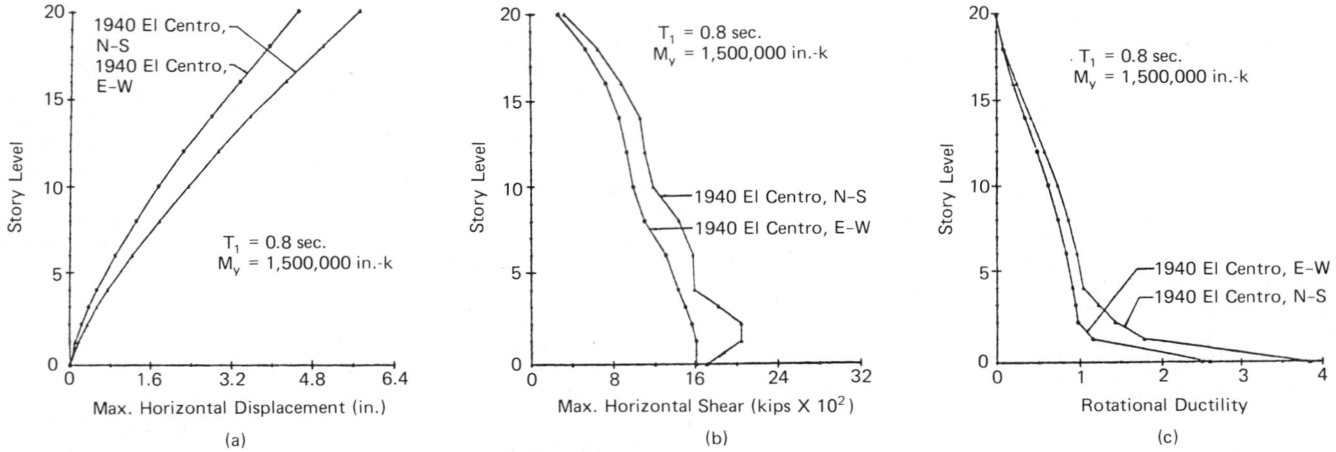

Fig. 12-35 Effect of frequency characteristics of ground motion on seismic response of structural walls with high yield level.

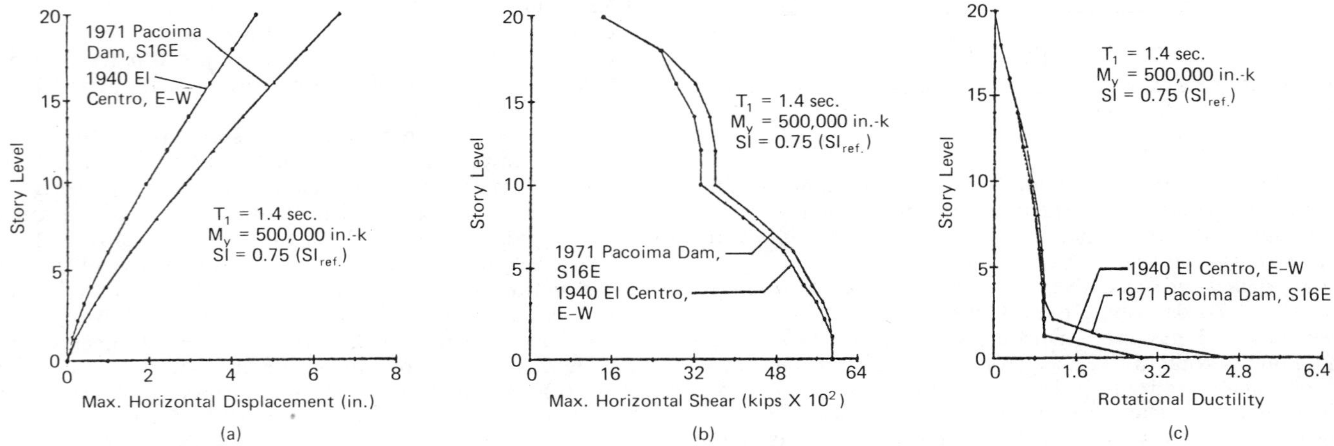

Fig. 12-36 Interacting effects of frequency characteristics, intensity, and yield level.

gram) with intensity equal to $1.5(SI_{ref.})$ represented the critical motion, while the Pacoima Dam record (a peaking motion) produced a relatively smaller response. By reducing the intensity of the motions by one-half in Set (a), yielding in the structure is significantly reduced. Consequently, the Pacoima Dam record becomes the more critical motion, as shown in Fig. 12-36.

The extent of yielding in a structure is influenced by the yield level, M_y, as well as the intensity of the input motion. For this reason, both parameters should be taken into account in selecting the appropriate type of motion to use as input with particular reference to frequency characteristics.

12.4.3 Coupled Wall Systems

Saatcioglu[12-28] carried out a parametric study to investigate the effects of several structural and ground motion parameters on the dynamic inelastic response of coupled wall structures. A 20-story coupled wall structure with an initial fundamental period of 1.8 seconds was selected for the parametric investigation. The floor plan was chosen to be symmetric in two directions as shown in Fig. 12-37(a). The 20-story prototype structure (Fig. 12-37b) was reduced to a 10-story model (Fig. 12-37c) by vertical lumping. The 1940 El Centro E-W record was selected for use in dynamic analyses.

The modified Takeda model as used in the PCA investigation of isolated walls (Section 12.4.2 and Ref. 12-27) applies to members under constant axial force. Coupled walls generally undergo substantial changes in the level of axial force during their response to earthquake motions. Because of this continuous change in axial force and the interaction between axial force and bending moment, the yield moment changes continuously; the effective stiffness in the post-yield range is also affected. The modified Takeda model was further modified for this investigation to include axial force–flexure interaction effects. The definition of rotational ductility for wall and beam segments that was used in this investigation is given in Fig. 12-37(e).

1. *Effect of Fundamental Period.* The previously selected 20-story structure was analyzed with three different stiffness values corresponding to three different fundamental periods of 1.2, 1.8, and 2.4 seconds. Stiffness properties were found such that the beam-to-wall stiffness ratio remained constant for each structure. Results indicated a consistent increase in maximum horizontal displacement with increasing fundamental period (Fig. 12-38), because of a decrease in yield rotation caused by an increase in stiffness while yield moment was kept unchanged. Absolute values of maximum rotation were higher for structures with longer fundamental periods (more flexible structures). Shear force and bending moment envelopes were not significantly affected by changes in fundamental period.

2. *Effect of Beam-to-Wall Stiffness Ratio.* Changing

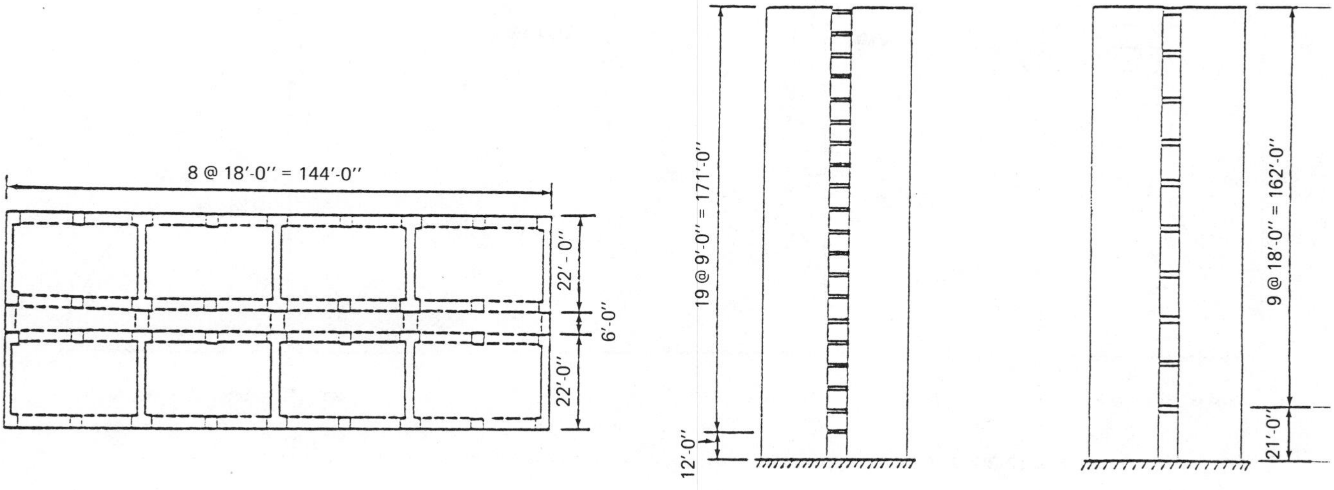

Plan

(a)

Elevation

20-Story Prototype

(b)

10-Story Model

(c)

Fundamental period	1.8 sec
Number of stories	20
Height	183 ft
Wall width	22 ft
Clear beam span	6 ft
Wall stiffness parameters:	
EI (million k-in.^{2})	38,600
GA (million kips)	1.13
EA (million kips)	6.70
Stiffness taper[1]	1.00 EI at base
(Step variation)	0.80 EI at 6th floor
	0.65 EI at 12th floor
Beam stiffness parameters[2]:	
EI (million k-in.^{2})	26.0
GA (million kips)	0.091
EA (million kips)	0.85
Wall yield moment, M_y	400,000 k-in.
Strength taper[3]	1.00 M_y at base
(Step variation)	0.75 M_y at 6th floor
	0.50 M_y at 12th floor
Beam yield moment[2]	4,000 k-in.
Post-yield stiffness on	5% of elastic for walls
Primary curve	7% of elastic for beams
Weight	2533 k/wall
Weight for inertia forces	4565 k/wall
Damping	5% of critical
Base fixity condition	fully fixed
Ground motion	El Centro 1940, E-W
Intensity of ground motion[4]	1.5 El Centro 1940, N-S

Notes:

[1] *The same taper also applies to "GA" and "EA."*

[2] *Stiffness parameters and beam yield moment must be multiplied by 2.0 to obtain values for 10-story model.*

[3] *Field moments are also adjusted at every floor on the basis of weight.*

[4] *Based on spectrum intensity.*

(d)

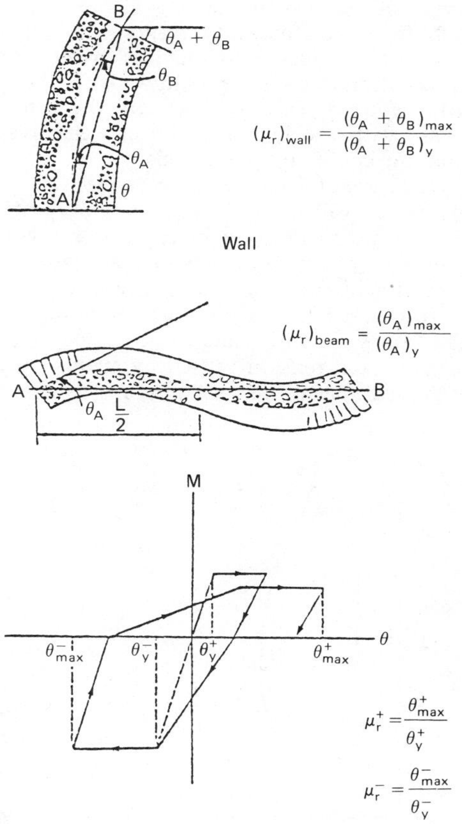

(e)

Fig. 12-37 Properties of coupled wall structure studied in Ref. 12-28.

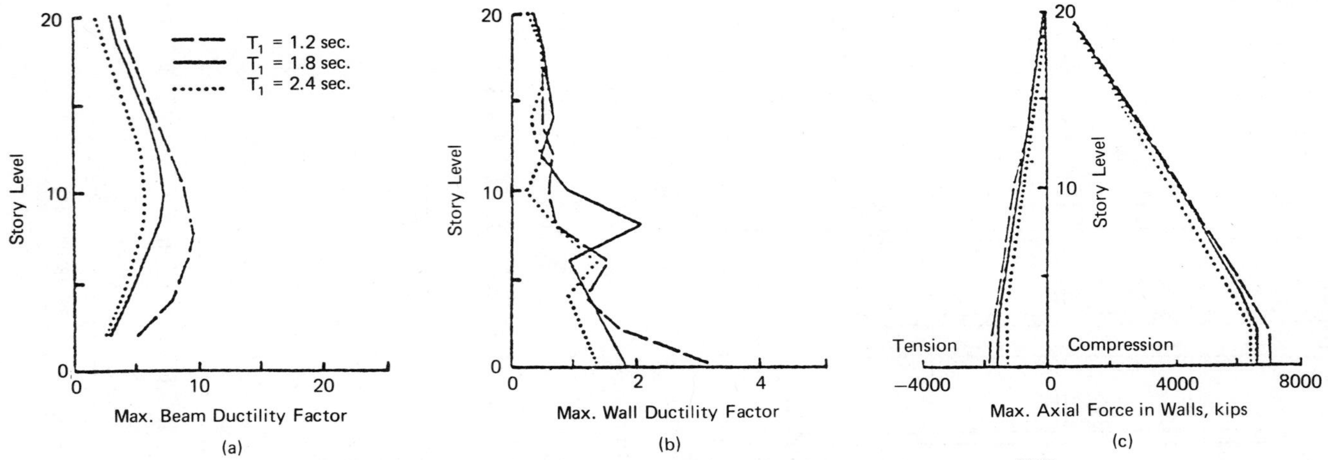

Fig. 12-38 Effects of fundamental period on seismic response of coupled walls.[12-28]

beam stiffness while leaving wall stiffness unchanged, to achieve different beam-to-wall stiffness ratios, changes fundamental period. Therefore, both the wall and the beam stiffness were changed to arrive at different stiffness ratios while maintaining the same fundamental period. It was found that below a certain value of wall stiffness, increase in beam stiffness does not affect the initial fundamental period of a coupled wall structure.

The structures considered in Fig. 12-39 were in the range of high wall stiffness where any increase in beam stiffness had to be compensated for by a significant reudction in wall stiffness to keep the fundamental period unchanged. Structures with high beam-to-wall stiffness ratios showed a high force and ductility response of the beams. Wall response was affected by changes in wall stiffness as well as by variation of axial forces in walls. Wall ductilities decrease with reduced wall stiffness. On the other hand, they increase with the high axial tension that results from strong coupling caused by stiff beams.

3. Effect of Beam-to-Wall Strength Ratio. Beam strength relative to wall strength plays an important role in achieving the desired yielding sequence within a coupled wall structure. Except for variations in beam strength, the structures considered in Fig. 12-40 were identical.

Structures with strong coupling beams exhibited undesirable behavior, with yielding occurring first in the walls, due to high axial tension caused by strong coupling. Structures with weak beams also showed undesirable behavior. In this case, although all beams yielded prior to yielding in walls, beam ductility demands were unrealistically high.

Wall ductility demand increased at high and low extremes of beam strength. When coupling beams are strong, increased tension and accompanying reduction in flexural capacity of walls results in high wall ductility demands. As beam strength is reduced, more yielding takes place in beams, and wall ductility demands are reduced. However, as beam strength is further reduced, inelastic action in beams becomes excessive, rendering them ineffective as coupling members. Beyond this point, walls start acting as isolated walls; ductility demands keep increasing with continued exposure to ground shaking.

4. Effect of Wall Strength. A series of analyses was carried out on the coupled wall structure of Fig. 12-37(c) with different levels of wall yield moment. In each analysis, the beam yield moment was adjusted so that the beam-to-wall strength ratio remained unchanged. Results of analyses showed (Fig. 12-41), as expected, that reduced strength with consequent early yielding in walls and beams increased ductility requirements.

5. Effect of Intensity of Earthquake. The 20-story coupled wall structure selected for parametric investigation was analyzed under three different earthquake intensity levels. The E–W component of the 1940 El Centro record was used in all cases. Figure 12-42 illustrates the corresponding response envelopes. Structural response was found to increase almost linearly with increasing earthquake intensity.

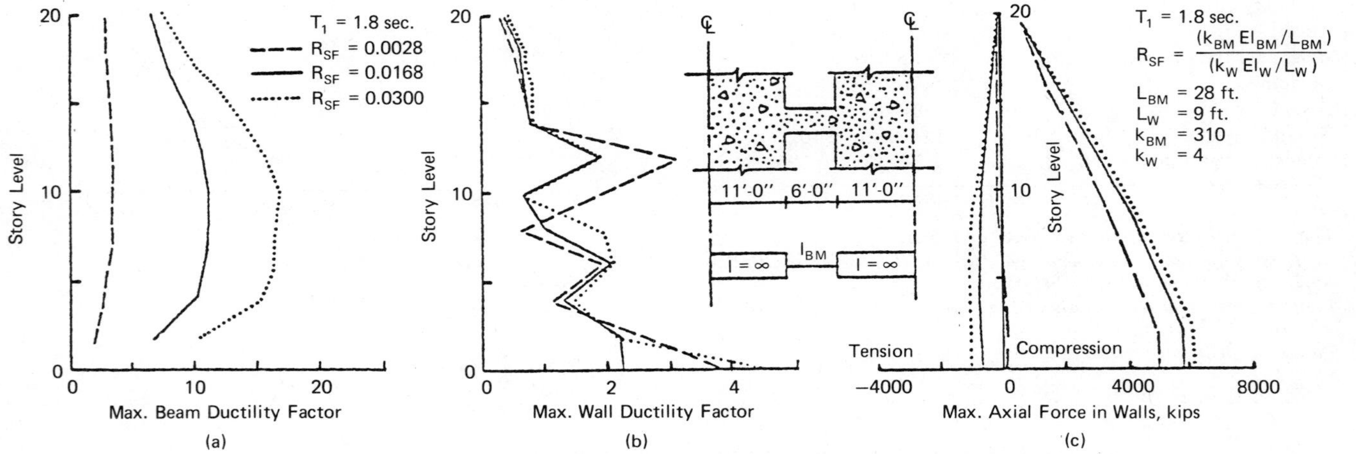

Fig. 12-39 Effects of beam-to-wall stiffness ratio on seismic response of coupled walls.[12-28]

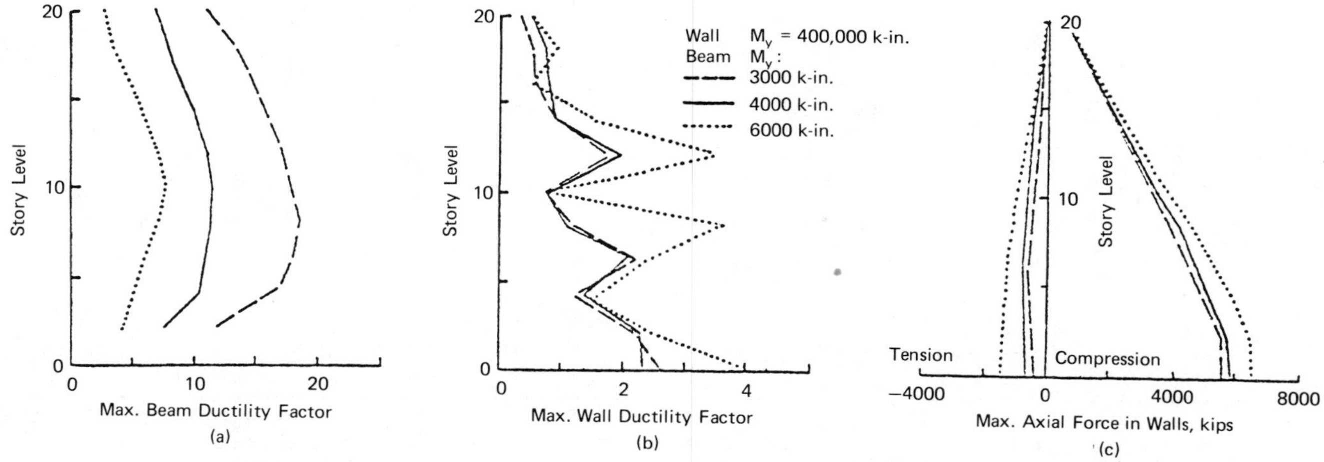

Fig. 12-40 Effects of beam-to-wall strength ratio on seismic response of coupled walls.[12-28]

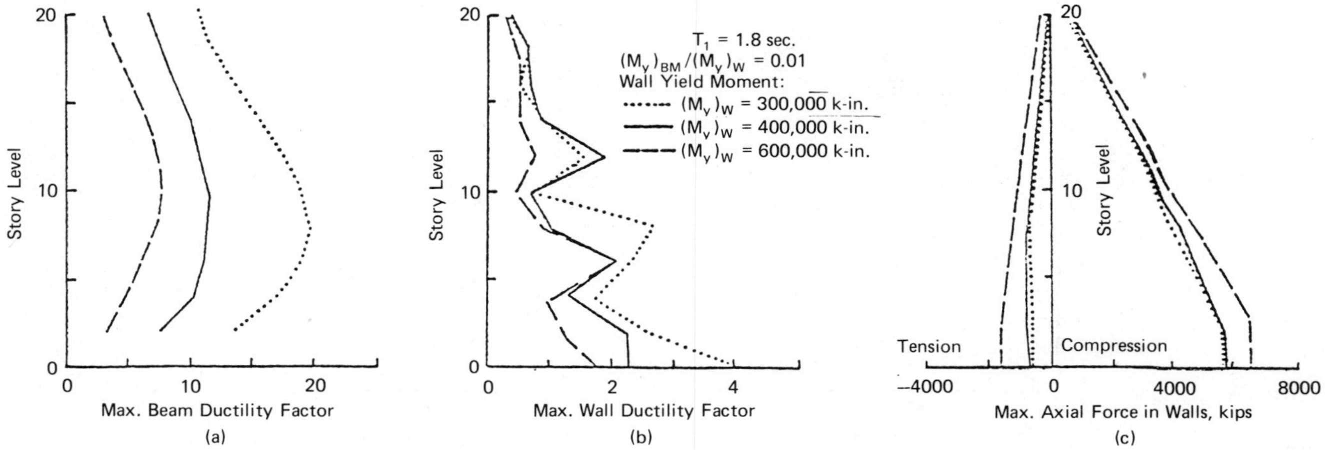

Fig. 12-41 Effects of wall strength on seismic response of coupled walls.[12-28]

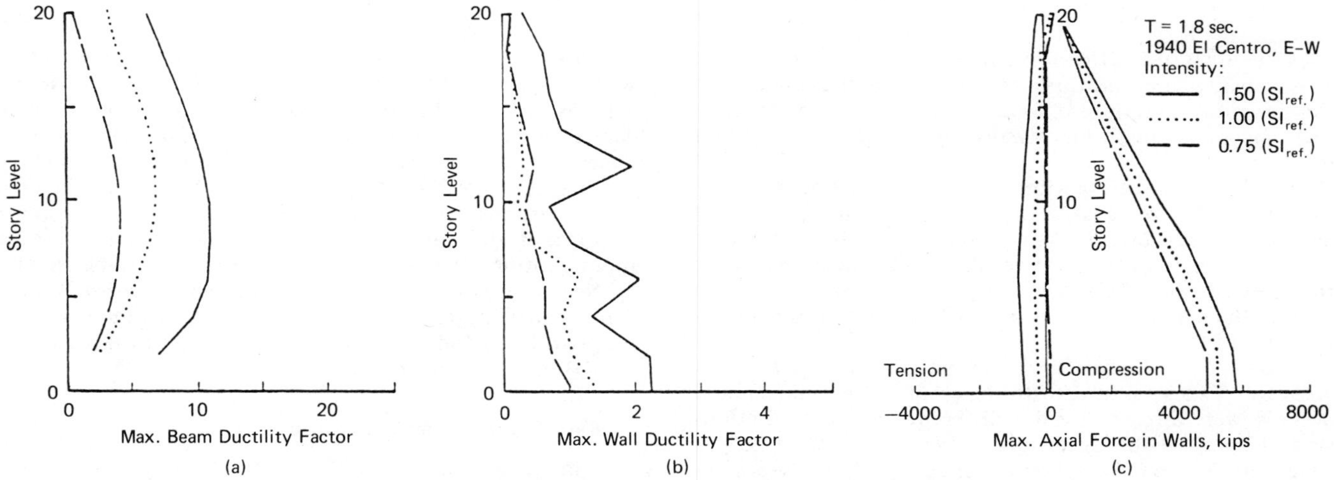

Fig. 12-42 Effects of intensity of earthquake on seismic response of coupled walls.[12-28]

6. Effect of Duration of Earthquake. Two analyses were carried out to investigate this aspect. The E–W component of the 1940 El Centro record was the input motion. The structure analyzed had an initial fundamental period of 1.8 seconds. The response spectrum of the selected input motion indicated that it would remain critical for a period lengthened from 1.8 to 4.0 seconds. The two analyses were carried out with a 10-second and a 20-second duration of earthquake motion. The 20-second duration used in the second analysis was obtained by repeating the 10-second accelerogram used in the first analysis. Response envelopes from the two analyses (Fig. 12-43) show a considerable increase in wall ductility at the base with increasing duration. In general, a longer duration caused an increase in all other

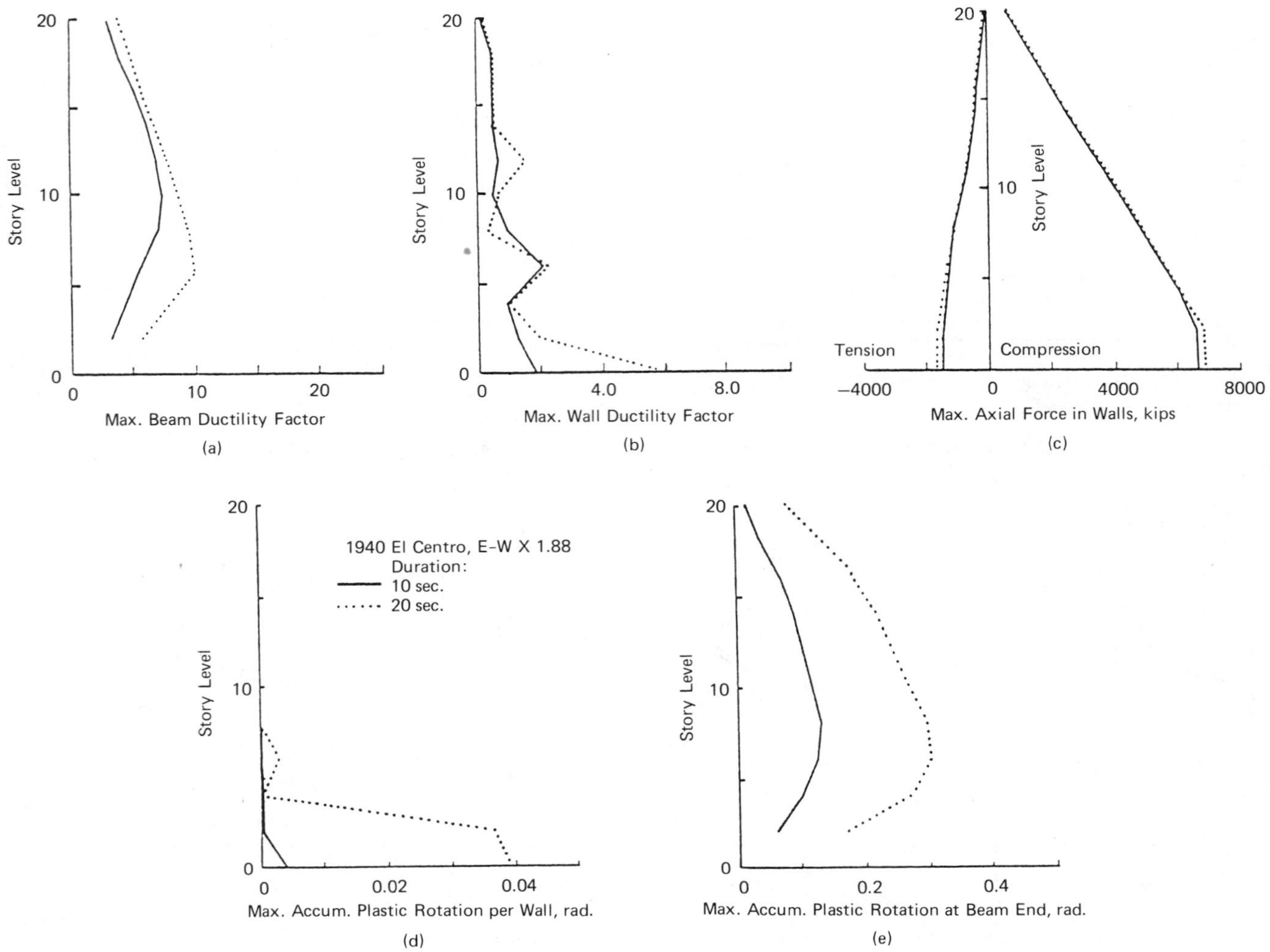

Fig. 12-43 Effects of earthquake duration on seismic response of coupled walls.[12-28]

response quantities. However, in view of the fact that the artificial 20-second motion consisted of intense oscillations for the entire duration, it would be misleading to conclude that increased duration always causes larger structural response.

Cumulative plastic rotations along the height of the structure from the two analyses are compared in Fig. 12-43(d and e). The expected trend is observed. As the earthquake duration increases, cumulative plastic rotations and hence the amount of dissipated energy always increase, even though the maximum values of response quantities may not be affected.

7. Effect of Frequency Characteristics of Earthquake Motion. Two structures with different initial fundamental periods were analyzed under different ground excitations (Fig. 12-44a). Each accelerogram was normalized to 1.5 times the 5% damped spectrum intensity of the N–S component of the 1940 El Centro record.

The first set of analyses was made for a structure with an initial fundamental period, T_1 = 1.0 second. Results indicated that the S16E component of the 1971 Pacoima Dam record created the most critical force, displacement, and ductility response. The 5%-damped velocity response spectrum for this accelerogram shows a pronounced peak at or close to the fundamental period of the structure considered (Fig. 12-44a).

The second set of analyses was conducted on a structure with fundamental period, T_1 = 1.8 seconds. Response envelopes shown in Fig. 12-44 indicate that none of the three accelerograms considered for the particular structure produces critical response in all the response quantities along the height of the structure. While one input motion created strong response at the base, the others produced stronger response at different levels along the height. However, those input motions that did not produce critical response in a particular response quantity produced near-critical values. Thus, single-degree-of-freedom velocity response spectra do provide a good basis for the choice of critical input motion(s) for a particular structure.

12.4.4 Frame-Shear Wall Structures

The buildings considered[12-22] consist of a parallel arrangement of essentially identical frames and shear walls such that in the direction of motion, and neglecting torsional effects, four frames may be assumed to act together with a shear wall. In Fig. 12-45, the four frames have been lumped into the single frame shown coupled with the shear wall. Also shown in Fig. 12-45 are the relative stiffnesses of the members of the standard structure in terms of a reference $(EI)_0$. The value of $(EI)_0$ has been adjusted to give the standard structure a fundamental period of 2.2 seconds.

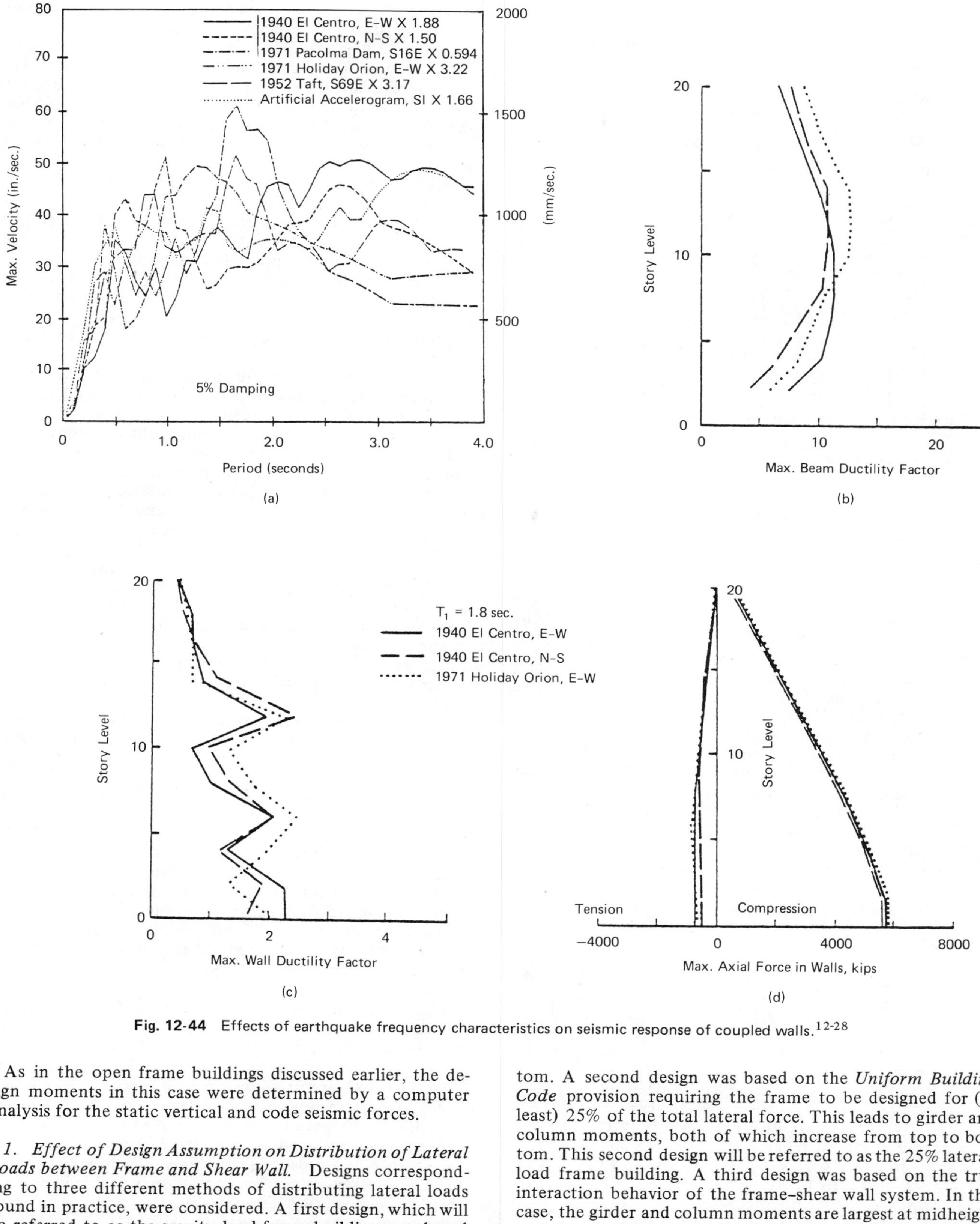

Fig. 12-44 Effects of earthquake frequency characteristics on seismic response of coupled walls.[12-28]

As in the open frame buildings discussed earlier, the design moments in this case were determined by a computer analysis for the static vertical and code seismic forces.

1. Effect of Design Assumption on Distribution of Lateral Loads between Frame and Shear Wall. Designs corresponding to three different methods of distributing lateral loads found in practice, were considered. A first design, which will be referred to as the gravity-load frame building, was based on the assumption that the entire lateral load is carried by the shear wall. The frame for this building is thus designed only for vertical loads, with the girder moments being uniform and the column moments increasing from top to bottom. A second design was based on the *Uniform Building Code* provision requiring the frame to be designed for (at least) 25% of the total lateral force. This leads to girder and column moments, both of which increase from top to bottom. This second design will be referred to as the 25% lateral-load frame building. A third design was based on the true interaction behavior of the frame–shear wall system. In this case, the girder and column moments are largest at midheight and decrease both upward and downward. This will be called the interaction-frame building.

Because of the different deformation patterns exhibited by frames and shear walls when acting separately under lateral loads, interacting forces which vary in magnitude and di-

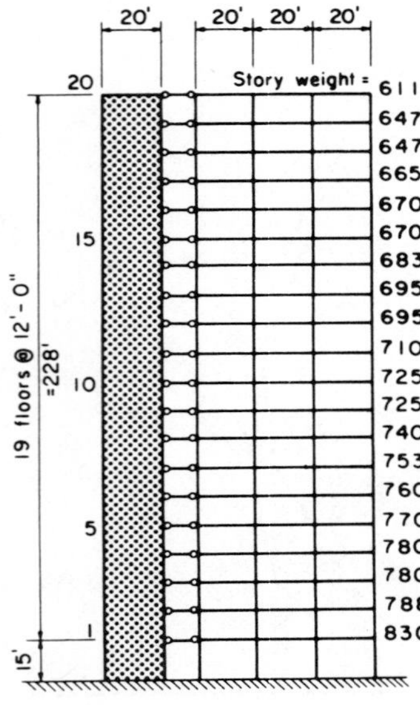

(a) 4 frames spaced at 25' per shear wall

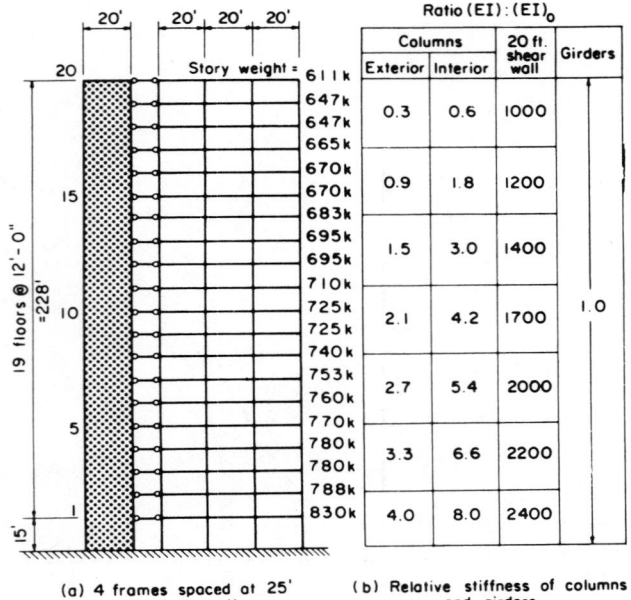

Ratio (EI) : (EI)$_0$

Columns: Exterior	Columns: Interior	20 ft. shear wall	Girders
0.3	0.6	1000	1.0
0.9	1.8	1200	
1.5	3.0	1400	
2.1	4.2	1700	
2.7	5.4	2000	
3.3	6.6	2200	
4.0	8.0	2400	

(b) Relative stiffness of columns and girders

Fundamental period, 2.2 sec.

$(EI)_0 = 390{,}000$ k-ft^2

Fig. 12-45 Properties of standard frame-shear wall structure considered in Ref. 12-22.

rection along the height of the structure are developed when both elements are tied together by floor slabs. The floor slabs are usually assumed to be stiff enough in their planes to produce equal lateral displacements at the floor levels, in the absence of torsion. This is illustrated in Fig. 12-46. It will be noted that the shear wall behaves essentially as a vertical cantilever beam, while the frame exhibits the deformation typical of a shear beam under transverse loads. The interaction between the two elements is such that the frame tends to reduce the lateral deflection of the shear wall at the top, while the wall helps support the frame near the base.

Because of the different bases of design used in proportioning the members of each structure, a different set of member yield moments corresponds to each building. In each case the ratio of yield-to-design moments was set equal to 2 for the girders and 6 for the columns and shear

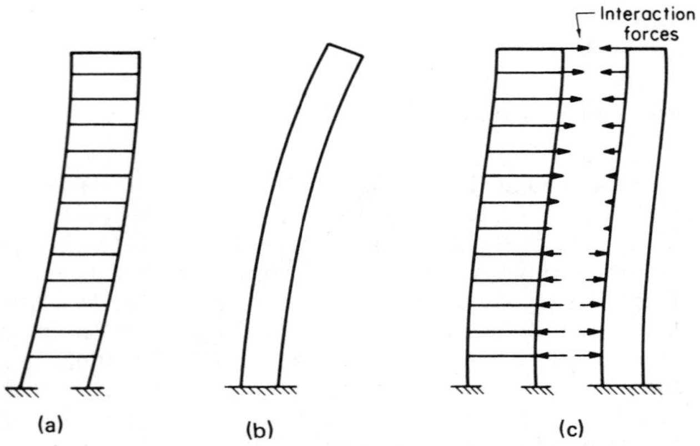

Fig. 12-46 (a) Rigid frame: shear mode of deformation; (b) shear wall: bending mode of deformation; and (c) interconnected frame and shear wall (equal deflection at each story level).

walls. In all cases, the reference stiffnesses were adjusted to yield a fundamental period of 2.2 seconds.

Figure 12-47 shows the maximum lateral displacements corresponding to the three cases considered. All three types of buildings have essentially the same maximum displacement at the top. The flexure beam mode is apparent in both the gravity-load frame building and the 25% lateral-load frame building—indicating the predominant influence of the shear wall relative to the frame. The curve for the interaction-frame building, on the other hand, indicates an almost linear variation with height, representing an intermediate case between the flexural mode of the shear wall and the shear-beam mode of the frame.

The girder and column ductility requirements corresponding to the three buildings considered are shown in Fig. 12-47 (b and c). The very favorable distribution of strength in the interaction-frame building, resulting in significantly lower ductility requirements for girders over the entire height and for columns at the top, is evident. The relatively low design strength of the frame in the gravity-load frame building is reflected in the high girder ductility requirements. It is worth noting that designing for frame–shear wall interaction tends to eliminate yielding of the columns at the top stories.

Figure 12-47(d) shows that the maximum overturning moments in the shear wall for the three buildings considered are roughly of the same order of magnitude. It is significant to note that, for the yield-to-design moment ratio assumed, none of the shear walls was stressed beyond the elastic range.

An idea of the relative strengths of the different frames corresponding to the three design assumptions used is given in Fig. 12-47(e), which shows the maximum column axial forces in the various frames. Since the magnitude of the column axial forces developed is a direct reflection of the strength of the girders, it is readily seen that the interaction frame is almost twice as strong as the 25% lateral-load frame and about three times as strong as the gravity-load frame.

The above comparisons clearly demonstrate the desirability of considering the true interaction behavior in designing frame–shear wall structures.

2. *Effect of Period of Vibration and Frame-to-Shear Wall Stiffness Ratio.* Two structures with fundamental periods of 1.6 and 2.6 seconds, were considered in addition to the standard 25% lateral-load frame building, having a fundamental period of 2.2 seconds. The 1.6- and 2.6-second-period buildings were obtained by varying the width of the shear wall in the standard building. Thus, the 2.6-second building had a 10-ft-wide shear wall with a stiffness ratio relative to the shear wall in the standard building of 0.2, while the 1.6-second building had a 38-ft-wide shear wall and a stiffness ratio of 5.0.

The maximum lateral displacements for the three buildings considered are shown in Fig. 12-48(a). The increase in displacement with increasing flexibility (longer period) observed earlier in the case of open frames is also evident here. Figure 12-48(b and c) shows the girder and column ductility requirements, respectively. The latter does not show any significant difference in behavior among the three buildings. A slight decrease in girder ductility requirements occurs in the stiffer (shorter-period) structures. This trend is contrary to that observed in open frame structures. It should be noted that the periods of the three structures considered differ not because of a change in stiffness of the frames, as was the case in the open frame study discussed earlier, but because of a change in the width and hence the stiffness of the shear walls. In all three structures, the frame portions were identical. The difference in girder ductility requirements observed

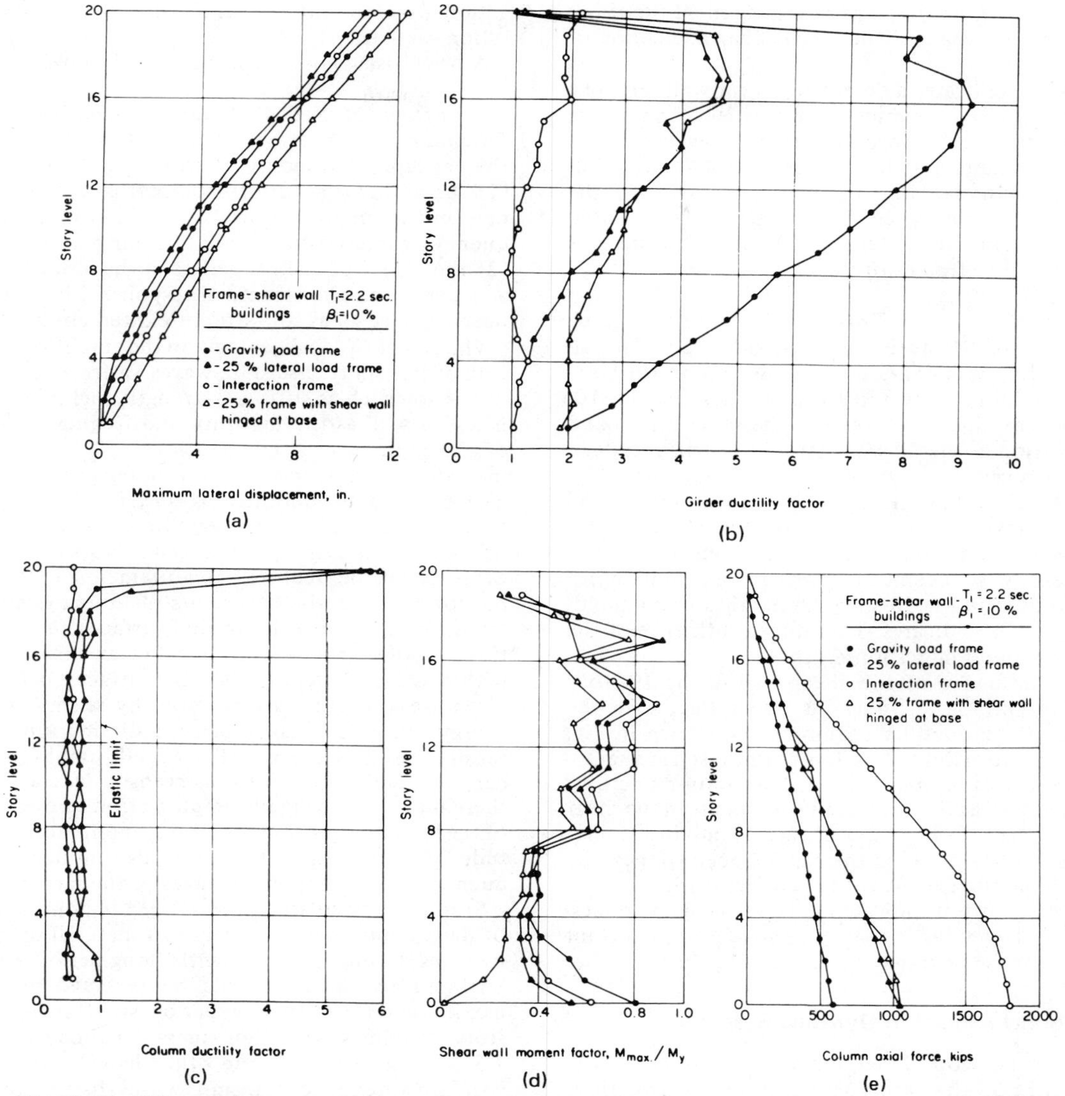

Fig. 12-47 Effect of design assumption on distribution of lateral forces between frame and shear wall.[12-22]

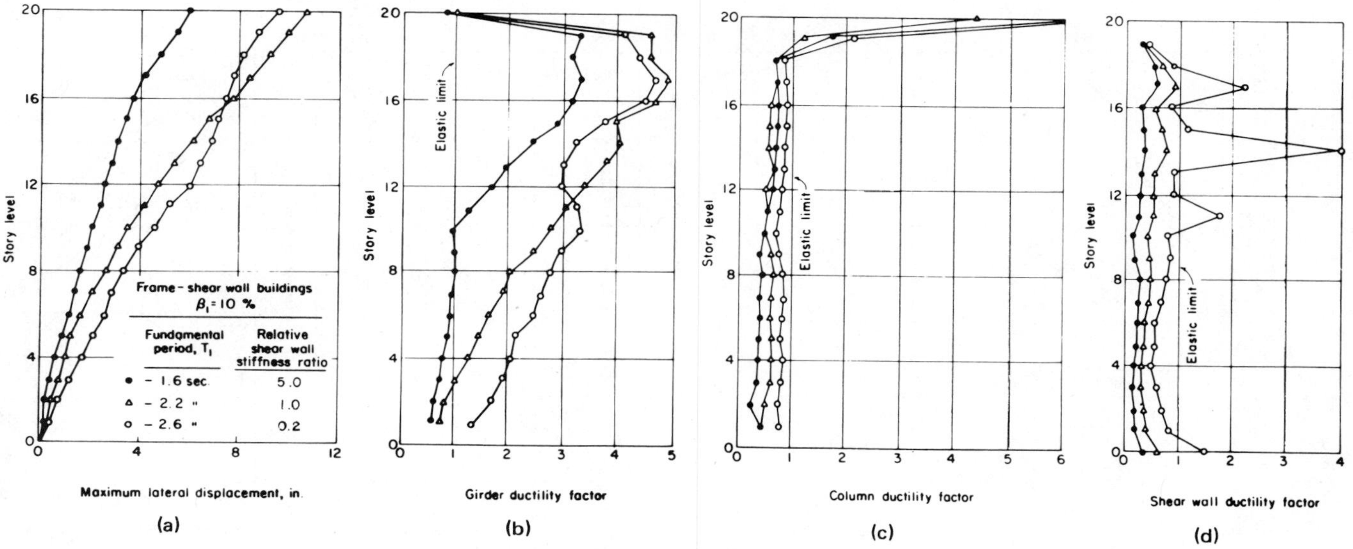

Fig. 12-48 Effect of period of vibration on seismic response of frame-wall structures.[12-22]

above could thus be interpreted as reflecting more the effect of the shear wall-to-frame stiffness ratio than the period of vibration.

Figure 12-48(d) indicates a decreasing shear wall ductility requirement (M_{max}/M_y for ductility ratios less than unity) for the stiffer structures. More important, however, is the relatively large ductility requirement indicated for the 2.6-second structure compared to the elastic behavior of the other two structures considered. This points to the potential danger of rupture in the stiffening elements of structures with low shear wall-to-frame stiffness ratios.

3. Effect of Number of Stories. A 10-story building consisting of the top 10 stories of the standard 25% lateral-load frame building was considered in addition to the standard 20-story building. The reference stiffness of the 10-story building was adjusted to give it the same 2.2-second fundamental period as the 20-story structure. This results in a structure that is very flexible for a 10-story building.

The results are shown in Fig. 12-49. The maximum lateral displacements, shown plotted against relative story height, are similar for both structures. The nearly linear variation of displacement with height in the 10-story building, a behavior also noted in the 20-story interaction-frame building (see Fig. 12-47), indicates the reduced influence of the shear wall on the building response.

The reduced influence of the shear wall in the 10-story building evident in Fig. 12-49(a) is also reflected in the corresponding girder ductility requirements, shown in Fig. 12-49(b), which are significantly less than those for the 20-story structure with the same period. The column ductility requirements, as well as the overturning moments in the shear wall, shown in Fig. 12-49(c and d), are essentially similar for both cases. As was noted in the case of open frames, the columns remain elastic except at the top few stories.

The above comparison indicates that, as in open frames, the number of stories by itself is a relatively unimportant factor in the response of frame–shear wall structures.

12.4.5 Additional Aspects of Dynamic Response

Among variables not considered above that may have a significant influence on the earthquake response of structures are: (a) effect of soil conditions including soil–structure interaction; (b) effect of gravity loads; (c) three-dimensional action—torsion; and (d) effect of irregular building configuration—setbacks.

A brief discussion of these factors follows.

1. Effect of Soil Conditions, Including Soil–Structure Interaction. The effect of ground conditions at a site on the earthquake response of structures involves two aspects: (1) the amplifying (or attenuating) effect of the local geology on the intensity as well as its filtering effect on the frequency characteristics of the transmitted seismic waves, and (2) the effect of soil properties in the immediate vicinity of a structure on the structural response. The latter aspect has been referred to as soil–structure interaction.

The effect of local geology on the amplification or diminution of transmitted seismic waves varies with the properties of the soil such as stiffness, strength, and layering characteristics, as well as the intensity and frequency characteristics of the input waves. Observations[12-29, 12-30] have generally indicated a tendency of the ground displacement and acceleration to be amplified as seismic waves pass from bedrock to softer surface layers. This observed amplification has been particularly pronounced in cases where deep alluvial deposits overlie bedrock. Measurements[12-31] of earthquake motion made in the basements of buildings situated at different depths below the ground surface and at different locations in the Tokyo area indicated an increase in intensity with decrease in depth below the surface. On the other hand, when weak soils are overstressed by seismic waves, so that energy absorption takes place, a diminution of motion intensity may be expected. Thus, while amplification may occur for input motions of low intensity in relatively soft soils, there may be a negligible amplification or even a diminution of the incoming motion for intense input motions. The amplitude of the resulting motion is also influenced by the frequency of the component waves of the incoming motion.

Studies by Seed and Idriss[12-32, 12-33] point to the tendency of deep deposits of relatively soft material to transmit wave motions having predominantly long-period characteristics. Such surface motion would be expected to produce their maximum effect on long-period structures such as multistory buildings with fundamental periods in the range of 1.5 to 3 seconds. On the other hand, shallow deposits of stiff soils tend to transmit waves having predominantly short-period characteristics which would have their largest effect on short-period structures.

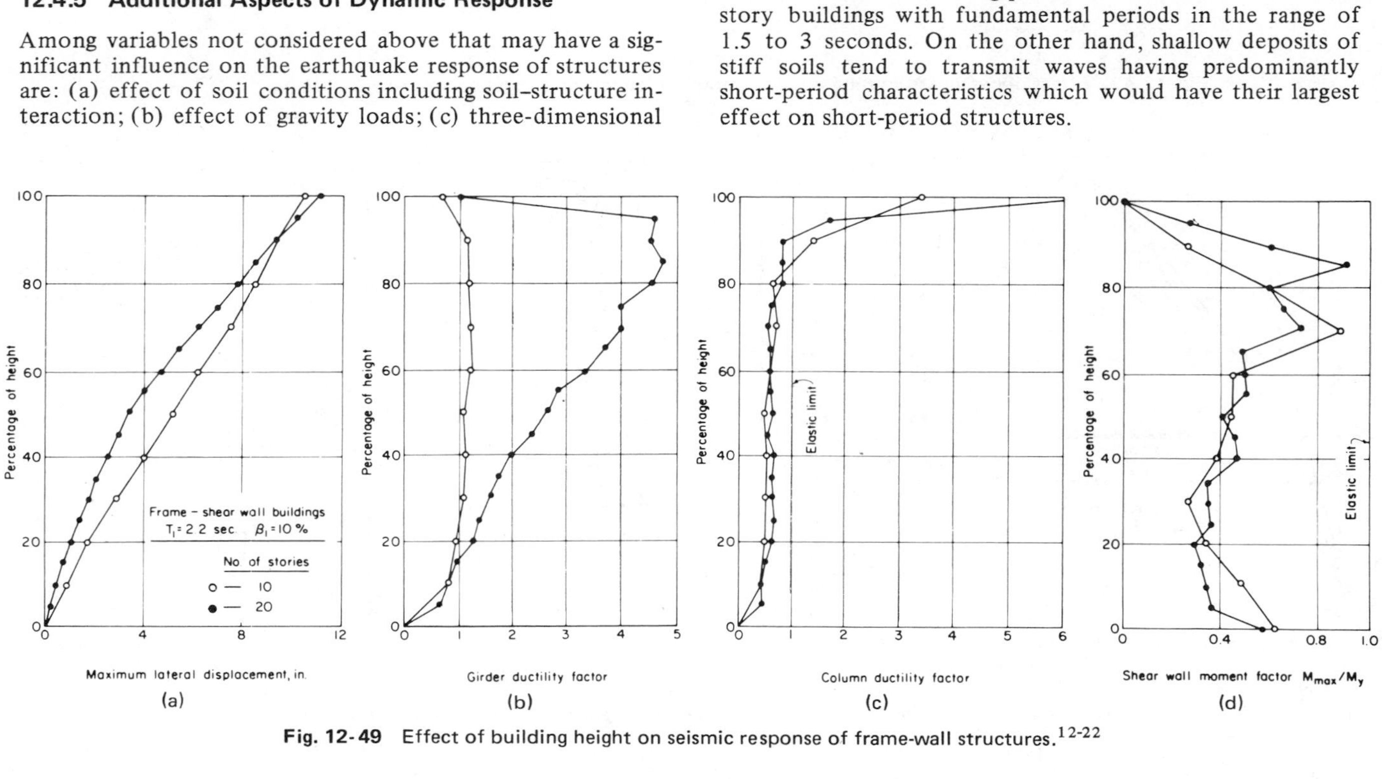

Fig. 12-49 Effect of building height on seismic response of frame-wall structures.[12-22]

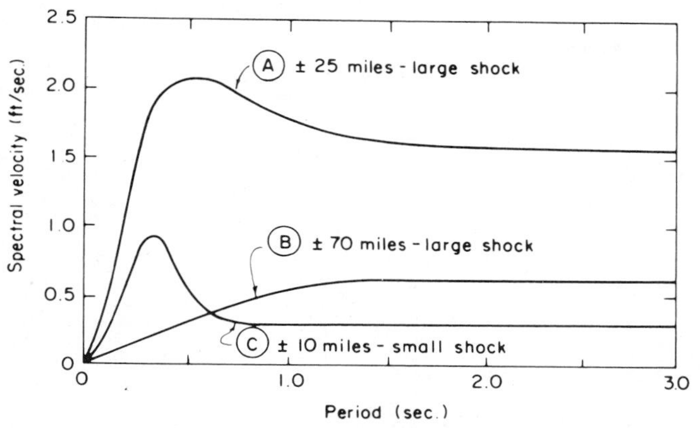

Fig. 12-50 Effect of epicentral distance on frequency characteristics of earthquake motion.[12-16]

Closely related to local geology is the probable epicentral distance. Records of earthquake motions at different sites indicate that short-period components tend to be filtered out faster as seismic waves spread out from the energy source, with the result that longer-period waves tend to predominate at greater distances from the epicenter. The effect of epicentral distance on the period or frequency characteristics of seismic waves is illustrated in Fig. 12-50, from Ref. 12-16. This figure shows smoothed undamped velocity response spectrum curves corresponding to the July 21, 1952 Kern County (Tehachapi) earthquake with a Richter magnitude of 7.7 (curves A, B) and the March 22, 1957 San Francisco earthquake with a magnitude of 5.5 (curve C). Curve A, based on a record taken at Taft, Calif., about 25 miles from the epicenter, shows relatively large response in the short-period range (indicating the dominance of component waves in this period range). Curve B, based on a record obtained at the Hollywood Storage Co. Building, some 70 miles from the epicenter, shows a flattening of the spectrum in the short-period range, and clearly demonstrates the attenuation of the high-frequency components of ground motion with increasing distances from the epicenter. Curve C is based on a record obtained about 10 miles from the epicenter of the relatively small 1957 San Francisco earthquake.

The above discussion suggests that for shocks originating from a particular region, the ground motion at a given site may be expected to exhibit dominant periods within a relatively narrow range. For instance, in a large part of Mexico City which is built on soft volcanic sediments to a depth of about 1000 ft, recorded motions originating from the Acapulco region some 260 km away have indicated dominant periods in the 1.8- to 3.0-second range (see Fig. 12-25(d)). Most of the damage during the July 28, 1957 Mexico earthquake (magnitude = 7.5), for instance, occurred in tall, long-period buildings.[12-34] This and similar observations[12-35] indicate the desirability of avoiding resonance effects, whenever possible, by designing buildings in which none of the significant modes of vibration (usually the first few) have periods close to that of the underlying soil deposits.

The above observations should be borne in mind when assessing the probable effects of local geology on the expected character of the ground motion at a particular site. Whenever possible, use should be made of local measurements and observations of the behavior of existing structures under previous earthquakes to gain an indication of the effect of local geology on the expected earthquake motion.

The effect of the foundation medium in modifying the response of a structure, in relation to its behavior when founded on an essentially rigid base, has been the subject of a number of analytical studies.[12-36–12-38] A good review of the subject is available in the commentary to Chapter 6 of Ref. 12-9. For buildings founded on firm or moderately firm ground, the effect of soil–structure interaction is relatively small. This is particularly true for multistory buildings which are relatively flexible compared to the supporting medium. For such cases, the assumption of a structure fixed to a rigid base to which the input motion is applied is justifiable.

For stiff structures, particularly those resting on relatively soft ground, the effects of soil–structure interaction can be significant. No definite trends, however, are shown by the available analytical results. Thus, soil–structure interaction can produce an increased response in one case and a decreased response in another, depending upon the properties of the structure, the supporting soil, as well as the earthquake excitation. It is known that the coupling (or the lack of it) between a structure and its supporting soil generally results in a system that has a longer fundamental period than the same structure fixed to a rigid base. Also, the damping in the soil–structure system, which consists of internal damping within both the structure and the soil immediately adjacent to it and radiation damping resulting from the dissipation of energy through dispersion to the surrounding ground, will generally be greater than that for the structure alone. On the basis of the above observations it has been suggested by Whitman[12-39] that from the design standpoint, the effect of soil–structure interaction on the stresses in a structure is always beneficial. This statement can best be understood by considering Fig. 12-51, which shows a sketch of response spectra corresponding to a single earthquake for a structure founded on a rigid base and also for a soil–structure system. Note that because of the greater damping in the soil–structure system, the corresponding curve everywhere lies below the one associated with a rigid-based structure. If the structure has a period corresponding to a low point in the spectrum, such as point *a* in Fig. 12-51, the result of considering soil–structure interaction may be a slight increase in the response. On the other hand, if the point on the spectrum corresponding to the period of the system lies near a peak, such as point *b* in the figure, soil–structure interaction will tend to decrease the response. Since, in designing for earthquakes, it is necessary to allow for slightly different earthquake motions, the effect of which may be approximated by smoothed or averaged spectra, the shift in period indicated in Fig. 12-51 will have little effect on the maximum response. With such smoothed spectra, consideration of soil–structure interaction will always tend to produce a decrease in response.

It should be pointed out that because of the predominant influence of the rocking component of motion of the structure, soil–structure interaction may cause the absolute motion of the top of a structure to increase slightly. This will have to be considered in relation to connecting or nearby structures.

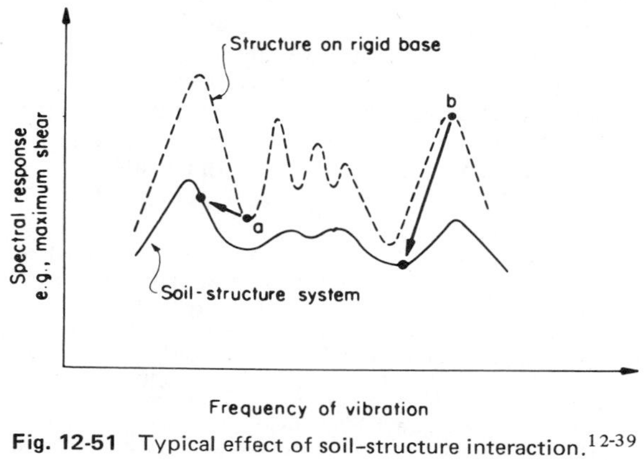

Fig. 12-51 Typical effect of soil–structure interaction.[12-39]

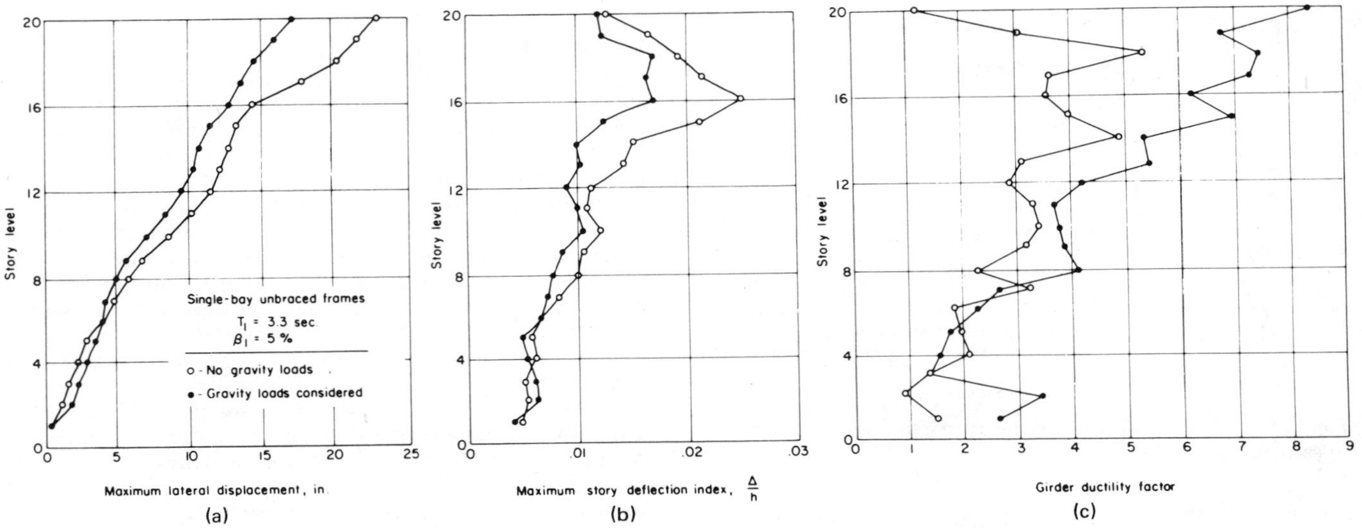

Fig. 12-52 Effect of gravity loads on seismic response of frames.[12-40]

2. *Effect of Gravity Loads.* A study by Anderson and Bertero[12-40] on the response of 10- and 20-story, single-bay, unbraced frames showed that the inclusion of gravity loads, as well as the effects of the associated *P*-Δ moments* and flexure–axial load interaction in the columns, causes yielding to occur much earlier in the response, so that the stiffness of the frame is reduced earlier and more energy is dissipated through inelastic deformation. For the base excitation considered in the study, an average of about 80% of the input energy was dissipated through inelastic action and damping, the former making up more than half of the dissipated energy. The increased inelastic deformation is reflected in a significant increase in the girder ductility requirements, compared to a case when gravity loads are absent. A substantial increase in the column ductility requirements, particularly in the lower stories, can also occur as a result of the action of gravity loads and the reduction in effective yield moment due to flexure-axial load interaction.** The presence of gravity loads does, however, have some beneficial effects in that it tends to reduce the maximum lateral displacements as well as the whiplash effect in the top stories. A reduction in the maximum lateral displacement due to the *P*-Δ moments was also observed by Hanson and Fan[12-41] in a study of the earthquake response of braced multistory single-bay frames.

Figure 12-52, from Ref. 12-40, shows the dynamic response of 20-story, single-bay frame designed according to the *Uniform Building Code* (1968 Edition) and subjected to the 8-second portion of the 1940 El Centro record (N–S component) between 0.72 and 8.72 second. An equivalent viscous damping equal to 5% of the critical for the fundamental mode, with lower values for the higher modes, was assumed. Figure 12-52(a) shows the maximum lateral displacements for idential frames with and without gravity loads. The reduction in the whiplash effect is apparent from Fig. 12-52(b), which shows the variation of the story deflection index, Δ/h, that is, the ratio of the relative story displacement to the story height, along the height of the frame.

*The secondary moment produced by a vertical column load, *P*, acting through a relative lateral displacement, Δ, between the column ends.

**The flexure-axial load interaction curve assumed in Ref. 12-40 corresponds to steel members whose flexural capacities are always reduced by the presence of axial load. The shape of the flexure-axial load interaction curve for a typical reinforced concrete member is different, for values of axial load below the balanced point, an increase in the axial load is accompanied by an increase in the ultimate moment capacity of a "short-column" section.

The ductility factor shown plotted in Fig. 12-52(c) is expressed in terms of curvature rather than rotation as in the Clough–Benuska study.[12-22]

In frames in which the columns are weaker than the girders, the maximum lateral displacements may increase owing to the presence of gravity loads.

The frames originally considered in Ref. 12-40 were designed to satisfy strength requirements only; hence the large deflections under the code-specified forces in Fig. 12-52(a). For these frames, the effect of the *P*-Δ moments varied from 4% to 8% of the overturning forces. When the frames were stiffened to satisfy the drift criterion (i.e., $\Delta/h \approx 0.0025$ under working loads), the *P*-Δ effect became negligible. The same conclusion was arrived at by Goel[12-42] in a study of the response of 10- and 25-story, single-bay symmetrical frames to the 1940 El Centro and the 1952 Taft, Calif., earthquakes (N–S and S21°W components, respectively), with the acceleration ordinates magnified by 1.5 and 3.0, respectively. This study showed that although the maximum lateral displacements were increased by about 10% by the *P*-Δ effect when the response was elastic, the effect became a negligible 1% when the girders were assumed to respond inelastically, with the columns remaining elastic throughout the response.

In a study of the effect of gravity loads on the collapse of simple (SDF) bilinear systems, Husid[12-24] noted that gravity loads significantly increased the permanent set that a structure suffered when subjected to strong ground motion.

3. *Effect of Three-Dimensional Action: Torsion.* The analysis of a structure by considering separate sets of mutually orthogonal planar frames and subjected to horizontal components of base motion parallel to their respective planes assumes no interaction between the forces acting on members common to two perpendicular frames, and neglects torsional effects. In addition, a two-dimensional analysis can allow only for an approximate consideration of the stiffness contribution of elements lying normal to the plane considered. Such an approach is consistent with present design practice and should yield reasonable results for most cases, particularly if a predominantly linear response can be assumed and no appreciable torsional effects are present. The UBC requires that a structure be designed to withstand the specified lateral forces considered as acting nonconcurrently in the direction of each of the main axes of the structure.

In considering inelastic response, however, the interaction

of forces resulting from components of motion parallel to each of the principal planes in a structure may cause early yielding in some members and modify the response of the structure significantly. The determination of the interaction effects in such a case will require a three-dimensional analysis of the response of the entire structure. A study by Nigam and Housner[12-43] of the earthquake response of a simple, single-story, three-dimensional frame model showed that the interaction of moments along two mutually perpendicular directions resulted in early yielding in the columns with a consequent reduction in the input energy and the response velocity. The interaction also tended to produce greater permanent lateral displacement.

The study of dynamic torsional effects in buildings, particularly in multistory structures where this effect is more pronounced, has been possible only with the recent development of programs for the dynamic analysis of three-dimensional frame structures.[12-44–12-49] Torsion occurs when the center of mass does not coincide with the center of rigidity in a story level. This can be the result of a lack of symmetry in the building plan or a random disposition of live loads in an otherwise symmetrical structure. Torsion can also be induced in symmetrical structures by the rotational component of ground motion.

The UBC requires the consideration of a minimum torsional moment equal to the story shear acting at any level with an eccentricity of 5% of the maximum building dimension at that level in buildings that depend on the diaphragm action of the floor slabs to distribute the horizontal shears to the various vertical lateral-load-resisting elements. A study by Newmark[12-50] of the elastic torsion in symmetrical buildings, based on an approximate analysis of torsional ground displacements corresponding to different building plan dimensions and shear wave velocities,* indicated that buildings with short periods of torsional vibration should be designed for larger values of "accidental" eccentricity than buildings with longer periods. Newmark further suggested that an accidental design eccentricity of 10% of the longer plan dimension would appear reasonable for frame buildings with periods shorter than about 0.6 second or shear-wall buildings with periods shorter than about 1.0 second.

The magnitude of the torsional shear induced in the exterior elements of a structure will obviously depend on the disposition of the various resisting elements in plan. Clearly the most efficient arrangement for torsion is the fully symmetrical building with as many of the lateral-load-resisting elements located along the periphery as is practicable. This, of course, assumes that the floor slabs provide an effective means of transmitting the horizontal forces to the major resisting elements and make these act as a unit. A study by Koh et al.[12-51] of the elastic dynamic response of four actual buildings in Japan ranging from 15 to 40 stories indicated that torsional effects arising from an eccentricity of from 5% to 10% of the longer plan dimension of a building may produce increases in the shears acting on the exterior vertical elements of up to 20% over that which would occur if torsion were absent.

In buildings with L-, Y-, U-, H-, or T-shaped plans which are built integrally as units, large forces may develop at the junction of the "arms" as a result of vibrational components directed normal to the axes of the arms. In addition, there are horizontal torsional effects on each arm arising from the differential lateral displacements of the two ends of each arm.

Yielding in a corner column or end shear wall in a building due to torsional stresses tends to destroy the (stiffness) symmetry in an originally symmetrical building or increase the eccentricity in an unsymmetrical building, as the center of resistance moves away from the yielded member. The increase in eccentricity causes yielding to develop further. This tendency toward magnification of torsional effects by yielding in corner or end elements suggests that such elements should be designed more conservatively than other members where torsional vibrations can be significant.

*Velocities of propagation in the supporting medium.

4. Effects of Setbacks in Buildings. Setbacks or appreciable changes in the overall plan size or dimensions in the elevation of buildings represent major discontinuities in geometry, mass, and stiffness and, as such, give rise to force concentrations at and in the immediate vicinity of the discontinuity. The increase in the maximum dynamic shears at the base of a setback portion of structure can be considerable, particularly when the fundamental period of the setback is close to the period of the first lateral vibration mode of the supporting base structure. This resonance effect occurs because the motion transmitted to the base of the setback structure is an essentially forced harmonic vibration which is influenced to a large extent by the dynamic properties of the base structure. In many instances, the magnitude of the acceleration in multistory structures at levels above the ground is significantly greater than that at the base.

A study by Blume and Jhaveri[12-52] of the elastic response of shear beam models of 15-story buildings, with symmetrically disposed setbacks having different heights and masses relative to the base structure, showed that the ratio of the maximum shear at the base of the setback to the maximum shear at the same height in a uniform building (without a setback) tends to increase with decreasing height and mass of the setback.

In addition to the increase in shear at the base of the setback structure, the same study showed that an increase in the maximum shear at the base of the building over that which would occur in a uniform building of the same total height can result from the presence of a setback.

It is clearly desirable to avoid major setbacks in multistory buildings in seismic areas, particularly unsymmetrical setbacks where torsional forces can further aggravate an unfavorable situation.

12.5 SOME RESULTS OF EXPERIMENTAL INVESTIGATIONS ON THE SEISMIC BEHAVIOR OF REINFORCED CONCRETE MEMBERS AND CONNECTIONS

The essence of rational, earthquake-resistant design of structures is the correlation of demand with capacity. The demand, expressed in terms of forces and deformations in critical regions, is best obtained through dynamic analysis. Strength and deformation capacities are derived from laboratory tests. Experimental studies of the behavior of reinforced concrete structural elements under simulated seismic loading have expanded our knowledge rapidly in recent times. The work that has been done cannot be reviewed adequately in the brief space available. The principal purpose of this review is to point out those aspects of element behavior that are the most relevant to the design of earthquake-resistant reinforced concrete buildings.

12.5.1 Behavior of Basic Structural Elements

It may appear ideal to test real buildings under the actual loading conditions to which they may be subjected during their service lives. Such tests usually are not economically feasible. There are also technical drawbacks to testing real buildings.[12-53, 12-54] First, the seismic response of a bare structural system is altered by the participation of nonstructural elements. This complicates extrapolation of test results to different structural systems, or even to the same structural system with different arrangements of nonstructural elements. It thus becomes desirable to study the behavior of basic structural elements under loads or deformations similar to those expected from severe seismic ground motion, and the effects of nonstructural components on such behavior. Second, the information gathered by testing a complete building or structure is difficult to evaluate and is usually insufficient to forecast the structure's behavior. Prediction necessitates the development of a theoretical model, a logical starting point for which is an understanding of, and work toward a realistic idealization of, the behavior of simple elements.

There is another important reason to study the behavior of basic structural elements.[12-55] Under monotonic loading, the stress–strain properties of steel and concrete, and steel to concrete bond characteristics, can be combined to predict section response in the form of moment–curvature, or member response in the form of moment–rotation or load–deflection relationships. These can then be combined to predict the response of frames or frame subassemblages. It is possible to experimentally check the validity of the model used to predict response at the material, section, member, or structure level. A particular aspect of response at any level may not be predictable using lower-level response characteristics, and empirical relationships may have to be relied on, as is the case with the shear strength of reinforced concrete members.

When a structure is subjected to reversed or cyclic loads producing large inelastic deformations, the response at any level becomes dependent on the loading path. Many attempts, none successful, have been made to use material properties to predict section, member, and finally structure response under cyclic loading. Much more success has been achieved by modeling force–deformation relationships for members under cyclic loads without attempting an analytical transition from lower-level response characteristics.

12.5.2 Loading History

Laboratory tests of reinforced concrete structural elements (and assemblages and subassemblages) have been of three basic types.

A *static monotonic load test* is performed by applying load or deflection in increments until a specimen is destroyed. Such a test is simple and relatively inexpensive, and also represents a "control" or "reference" point for other types of tests. However, monotonic test data are not sufficient to describe the response of structures and structural elements to earthquake-type loads.

A *static reversing load test* is performed using load or deflection control; a specimen is subjected to a predefined series of load cycles such that the sequence from elastic behavior through inelastic behavior to destruction of the specimen can be monitored. A complete control of the loading sequence is possible. It is also possible to closely observe overall behavior, crack patterns, and the physical condition of the specimen at any time during a test. Static reversing loads permit detailed evaluation of the hysteretic behavior of a test specimen. However, because loads are applied slowly, characteristics sensitive to the rate of loading are neglected. Also, the inertia forces on a structure are not properly simulated.

A *dynamic earthquake simulator test* requires an earthquake simulator consisting of a servo-equipped hydraulic system, a power supply, a test platform, an electronic control, and monitoring equipment. Model structures attached to the test platform are subjected to base motions simulating particular earthquake records. This type of test is more suitable for model structures than for structural elements. Both the initial and the operating costs of the test facility are high. Limits on the size of the test structure also constitute an important disadvantage.

By far the largest and most useful volume of information on the seismic response of reinforced concrete structural elements has been obtained from static reversing load tests. In the past a variety of loading sequences have been used, making a comparison of test results difficult. A few example loading histories are illustrated in Fig. 12-53. The differences are readily apparent. In some cases, the load is cycled between prescribed deflection limits until failure or severe distress is observed (Fig. 12-53a, Loading Type A). In other cases, the deformation limit increases after the application of a number of load cycles at a given limit, simulating a steadily intensifying earthquake (Fig. 12-53a, Loading Type B). Derecho et al.[12-56] have recommended a loading program for use in quasi-static tests of isolated shear walls to simulate response to earthquakes. This was derived from results of inelastic dynamic analysis of isolated walls.

Figure 12-54 illustrates the effect of different loading histories on the behavior of reinforced concrete columns. It is obviously important that users of experimental results be cognizant of the influence of loading history on response in evaluating experimental data.

12.5.3 Seismic Behavior of Ordinary R.C. Beams in Moment-Resisting Frames

Studies on the seismic behavior of ordinary reinforced concrete (R.C.) beams (cast in situ using normal weight concrete) in moment-resisting frames have been reviewed by Bertero.[12-53, 12-54, 12-57] A distinction was made between the behavior of beams with critical regions subject predominantly to flexure, and that of beams with critical regions subject to high shear in addition to flexure. The following is a digest of Bertero's review.

1. Beams with Critical Regions under Low Shear. The effect of shear on the hysteretic behavior of beam critical regions designed and detailed according to present seismic codes can be neglected even for the most severe seismic ground motion as long as the nominal unit shear stress in such regions is less than $3\sqrt{f_c'}$. Several aspects of the seismic behavior of flexure-controlled critical regions are discussed below.

Flexural strength. In general, flexural strength can be predicted accurately if the mechanical characteristics of the reinforcement under uniaxial tension and compression, and of the concrete (confined and unconfined) under uniaxial compression, are known. Two areas of uncertainty in such evaluation are: (1) the effect of the strain rate, and (2) the effective width of the floor slabs participating in the development of flexural capacity.

Deformation and energy dissipation capacities. Deformation and energy dissipation capacities are controlled by the buckling of the main reinforcing bars, which is difficult to predict. To assure sufficient deformation and energy dissipation capacities, most seismic codes contain stringent provisions governing the proportioning and detailing of flexural critical regions. Such provisions set limits on the amount of longitudinal reinforcement, positive movement capacity at column connections, and transverse reinforcement. The latter provides confinement of concrete, lateral restraint to main longitudinal reinforcing bars, and shear resistance. The current U.S. seismic code provisions in this regard are reviewed in Section 12.6.

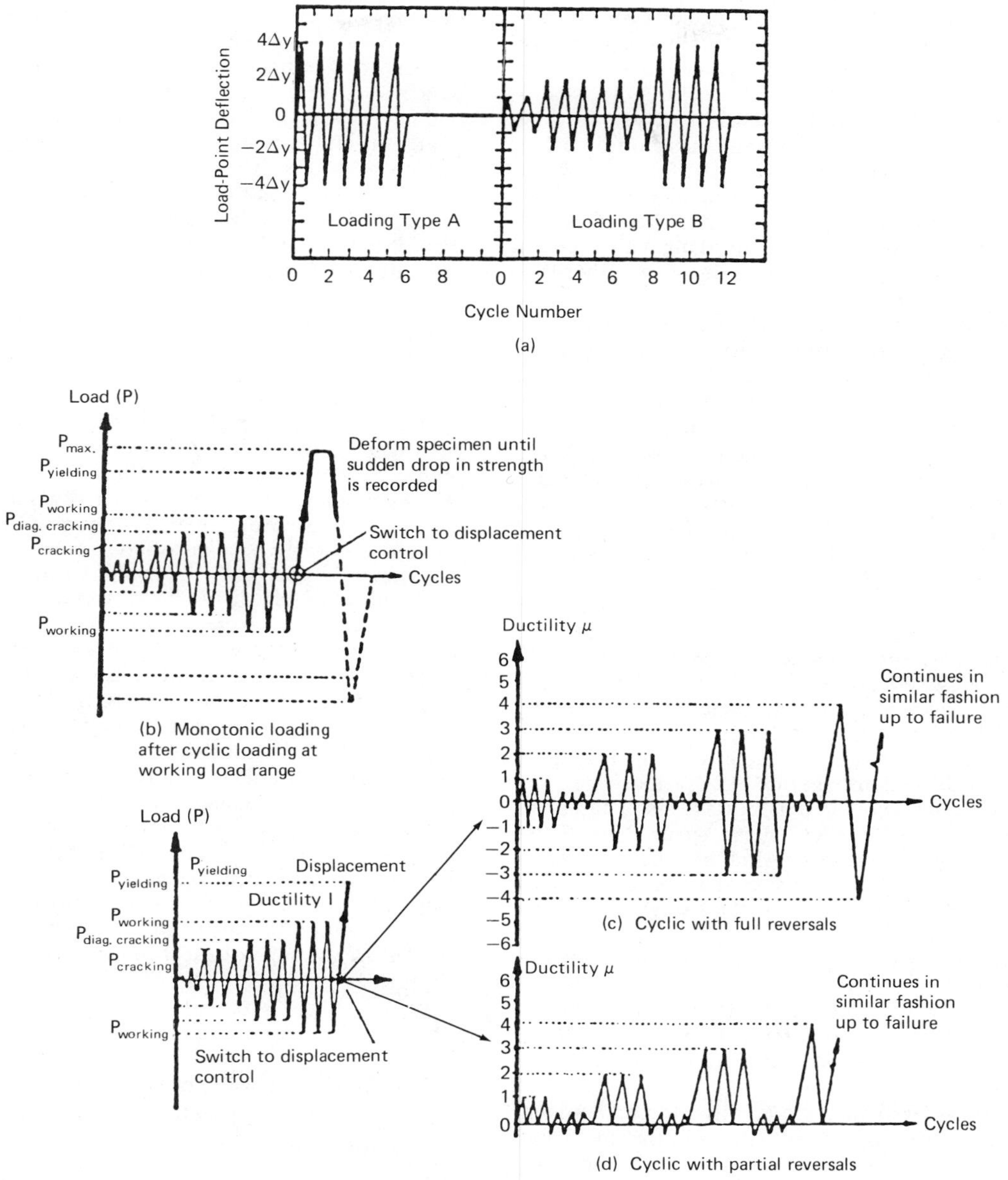

Fig. 12-53 Various types of loading history.[12-55, 12-57]

Stiffness. To estimate the strength, deformation, and energy dissipation capacities needed in a structure to withstand an earthquake, it is necessary to predict the dynamic characteristics of the structure at the time the earthquake hits. The stiffness of a ductile reinforced concrete moment-resistant frame is usually controlled by the stiffness of the beams rather than that of the columns. Thus the initial stiffness of the beams needs to be estimated as accurately

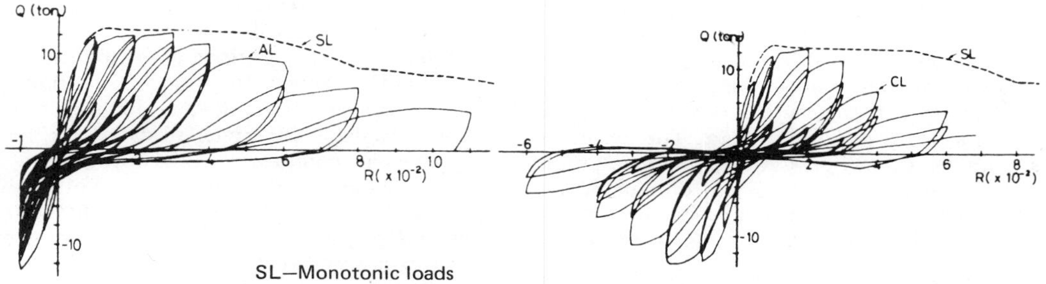

SL—Monotonic loads
AL—Alternating loads (deflection in one direction limited)
CL—Cyclic loads (equal deformation levels in both directions)

Fig. 12-54 Influence of load history.[12-55]

as possible. Because the time of occurrence of future earthquakes and the history of excitation to which a structure has been subjected are uncertain, it is best to consider a range of probable stiffness values in analysis. It is also possible that certain beam regions in a structure may be critically stressed owing to service conditions alone, leaving these beams in a state of reduced stiffness. Unfortunately, no satisfactory practical rules have yet been offered for prediction of initial stiffness values. Neither the effect of cracking nor the effective width of the flanges to account for floor slab contribution is well defined in either the code or the available literature.

2. *Beams with Critical Regions Under High Shear.* As the nominal shear stress on a beam critical region, designed and detailed according to the present U.S. seismic codes, exceeds $3\sqrt{f_c'}$, the critical region can still develop maximum flexural strength and flexural deformation capacities under monotonically increasing loads. However, these critical regions, under repeated moment reversals and particularly under full rotation reversals, undergo a degradation in stiffness and energy absorption and dissipation capacities considerably larger than the degradation of a similar critical region with low shear stresses. Degradation in strength also starts to occur as the number of similar loading cycles inducing reversal of rotation increases. Although such regions are capable of developing flexural yield strength, an early shear failure mechanism (sliding shear) starts to develop after one cycle of full bending reversals beyond the yield strength level.

Ma et al.[12-58] have investigated the effects of high shear on flexural critical regions. Figure 12-55 illustrates the results obtained from two of their beam tests. The beams were identical except for their shear spans. The figure clearly shows the pinching effect induced by high shear on the load–displacement relationship. This pinching effect resulted in a two-thirds reduction in the energy dissipation capacity.

Studies on the effects of shear on flexural critical regions indicate that:

1. When maximum nominal shear stress induced during inelastic reversals is high in both directions (around $5.3\sqrt{f_c'}$), the degree of shear stiffness degradation becomes very significant.

2. Shear in cracked critical regions of reinforced concrete members is resisted through shear stresses in the uncracked part of the section, aggregate interlock and friction along crack faces, web reinforcement across inclined cracks, and dowel action of the longitudinal reinforcement. As a beam is subjected to several loading reversals, flexural and/or flexure-shear cracks may extend through the beam section, so that shear may have to be resisted without the help of any uncracked concrete. Dowel action, and aggregate interlock and friction, become less effective as crack widths increase, and as concrete crushes in the compression zone. As a result, large shear distortions could occur. It should be pointed out that the above degradation occurs because of the opening of cracks induced by yielding of the main reinforcement, and is therefore a combined flexure–shear type of degradation mechanism, with contribution from bond slippage of the main reinforcing bars. The deformation pattern in the critical region during reloading, at advanced stages of inelastic deformations, is dominated by shear deformation at cracks that remain open throughout the beam section. This behavior has been named "shear sliding."

3. Recorded shear force–shear distortion diagrams indicate that after flexural yielding occurs in both loading directions, the degradation of shear resistance and the amount of shear distortion increase with the magnitude of applied load and/or deformation as well as with each repeated cycle of reversal. The possible shear degradation mechanisms include: (a) the opening of cracks due to yielding and/or slippage of the main reinforcement; (b) the spalling of the concrete cover around the periphery of the flexural critical region; (c) the loss of stirrup–tie anchorage due to large strain variations where a tie is crossed by inclined cracks, and/or by the splitting and spalling of the concrete cover; (d) the crushing and grinding of concrete at crack surfaces leading to less effective aggregate interlock; and (e) the local disruption of bond between the longitudinal steel and concrete due to dowel action along the open cracks. The most important factors affecting shear stiffness degradation appear to be reduced aggregate interlock along the large cracks, and loss of bond due to dowel action of the longitudinal steel.

Experimental results[12-59] have shown that the behavior of flexural critical regions under high shear stress can be significantly improved by the addition of diagonal reinforcement. This is illustrated in Fig. 12-56, from Ref. 12-59, which compares hysteresis loops obtained with different types of web reinforcement. The use of diagonal reinforcement is an effective means of controlling sliding shear.

Scribner and Wight[12-60] have presented significant results concerning the effect of shear in flexural critical regions of beams. Specimens were designed with a variety of longitudinal beam reinforcement and tested using four different shear spans, such that maximum shear stresses varied from $2\sqrt{f_c'}$ to $6\sqrt{f_c'}$. From the results, the following conclusions were drawn: (a) the repeatability of member hysteretic behavior is related to maximum beam shear stress; (b) intermediate longitudinal shear reinforcement provides significant increases in energy dissipation and repeatability of hysteretic response for beams with shear stresses between $3\sqrt{f_c'}$ and $6\sqrt{f_c'}$; (c) beams with shear stresses below the above range perform satisfactorily without intermediate longitudinal shear reinforcements, whereas beams with shear stresses higher than $6\sqrt{f_c'}$ do not perform satisfactorily regardless of the type of shear reinforcement used.

12.5.4 Seismic Behavior of Ordinarily Reinforced Columns, Subjected to Significant Axial Forces, in R.C. Moment-Resisting Frames

The behavior under lateral load of columns subjected to very low axial forces is similar to that of beams, except that the columns may be subjected to higher shear when they are shorter. This section summarizes Bertero's review[12-53, 12-54, 12-57] of the behavior of ordinarily reinforced concrete columns in moment-resisting space frames whose critical regions are subjected to significant axial forces. In seismic analysis and design, a frame structure can be modeled as a planar frame and subjected independently to the horizontal component of ground motion acting in the plane of the frame. Discussion here is limited to the behavior of columns loaded along one principal plan axis.

Most available experimental data have been obtained under constant compressive axial forces, with shear and bending applied along one principal plane of a column.

Columns are the elements most susceptible to failure in destructive earthquakes. This is not surprising because, as Bertero has pointed out,[12-54, 12-57] the real nominal shear stress in a column can be four or more times the value obtained using the current U.S. seismic code procedures, the value for which the shear reinforcement in a column is designed.

Ohmori[12-61] has reported on tests of columns with effec-

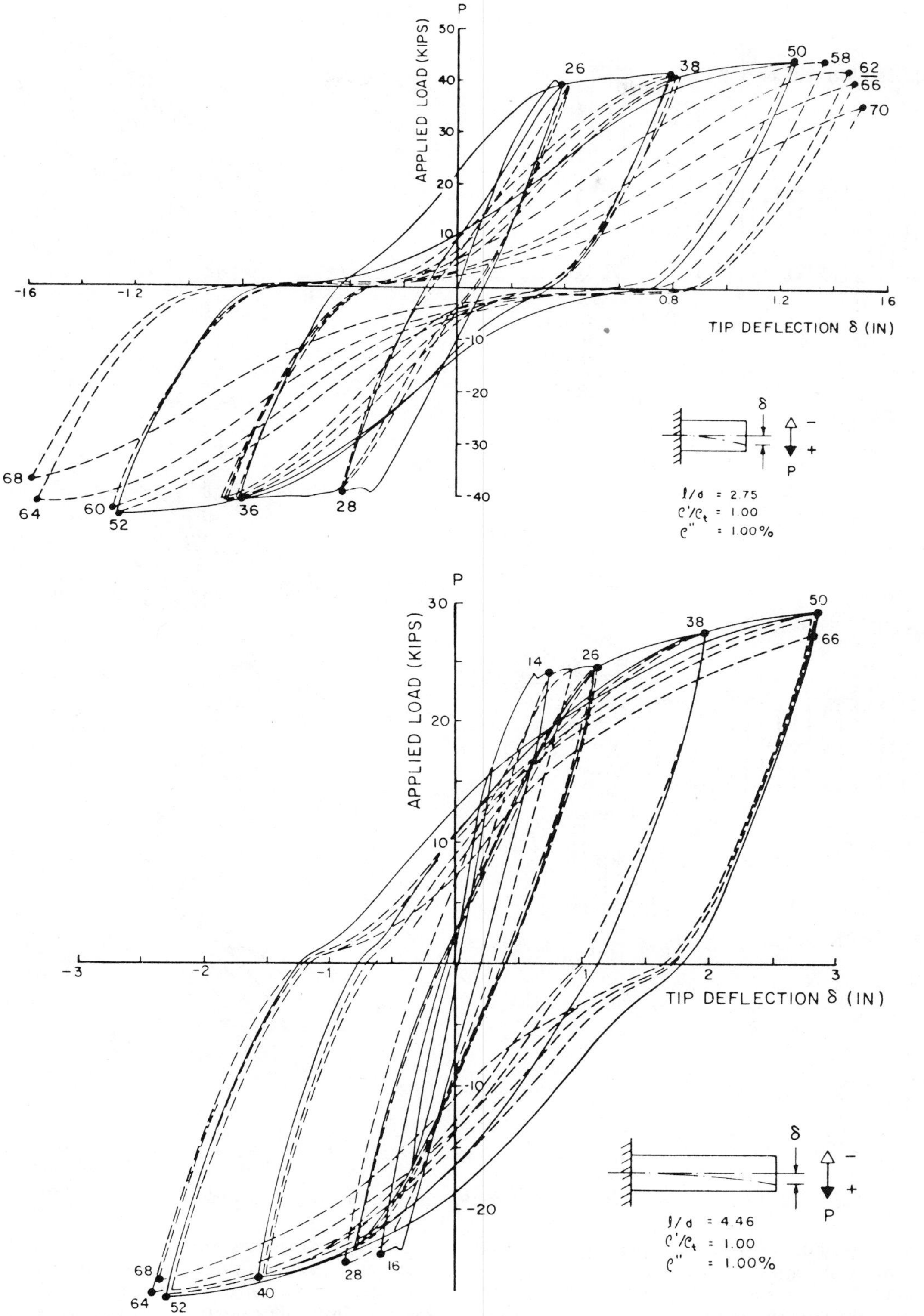

Fig. 12-55 Comparison of hysteretic behavior of flexural members as affected by different amounts of shear forces.[12-58]

tive transverse reinforcement arrangements that for the first time enabled the design and construction of high-rise concrete buildings (18 stories) in Japan. A combination of spiral and hoop reinforcements was developed, and was shown to result in very stable and ductile hysteretic behavior. The other two types of transverse reinforcement that showed good behavior for the same reinforcement ratio were the spirals and the hoops with cross-ties.

From an analysis of results of the experimental investigations carried out to date, it can be concluded that:

1. Short (nonslender) reinforced concrete columns, if designed and detailed according to code recommendations for ductile moment-resisting frames, can develop moderate inelastic deformations prior to shear failure or significant shear degradation, when subjected to significant constant axial loads and to cyclic shear reversals. The inelastic defor-

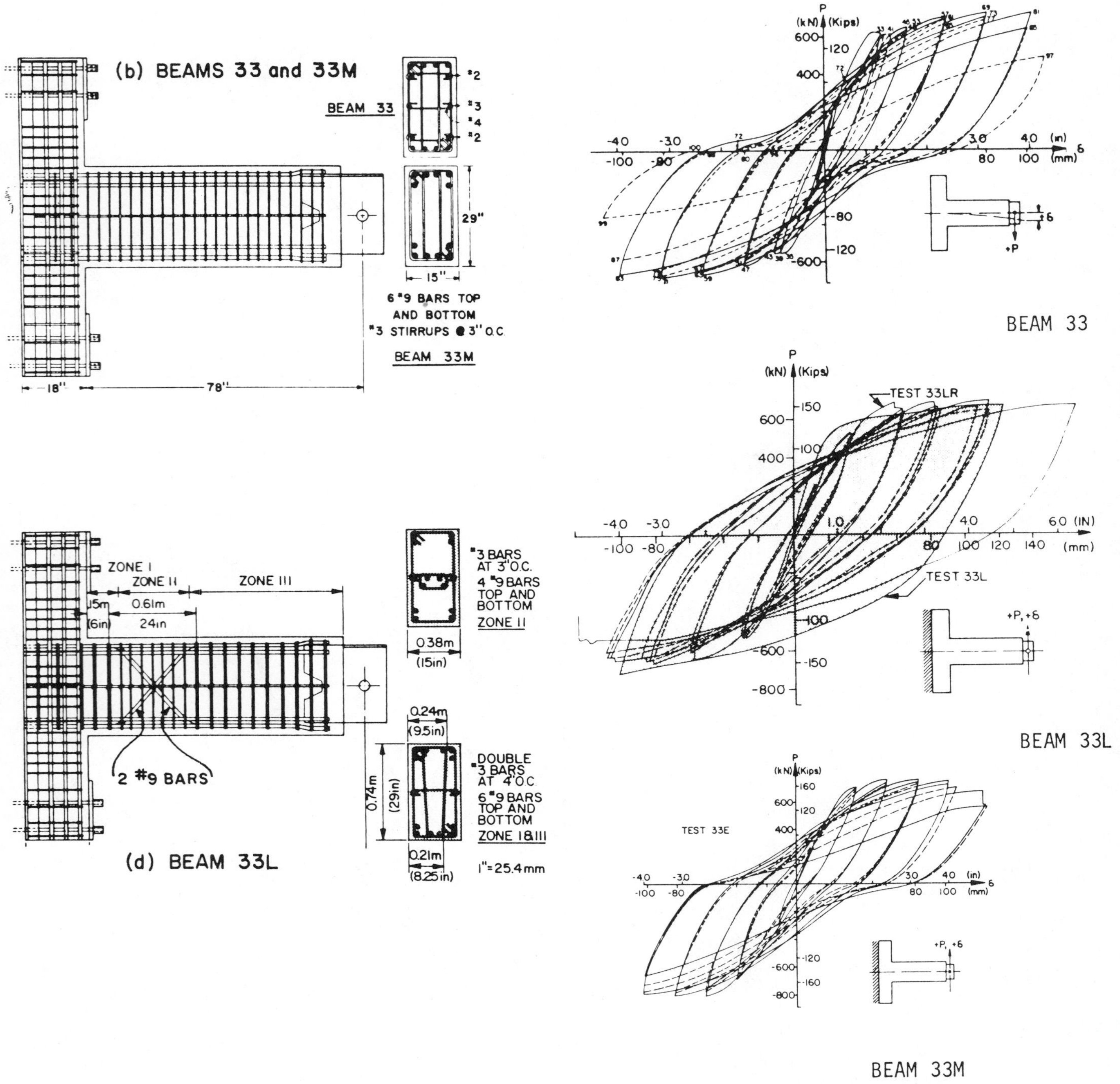

Fig. 12-56 Effect of different types of web reinforcement on hysteretic behavior of R.C. beams.[12-59]

mation capacities would usually be adequate with respect to the magnitude and nature of inelastic deformation demands expected in columns of weak girder–strong column frames. However, these deformation capacities may be insufficient in columns forming parts of soft stories in certain frames. Also, the above observations are valid for nearly constant column axial forces. The change from a relatively ductile shear–compression failure mode in columns with significant compressive axial forces to a brittle diagonal tension failure mode in similar columns with tensile or low compressive axial forces indicates the need to investigate the inelastic behavior of short columns in which the axial forces vary.

2. Shear capacities computed according to code provisions, using actual rather than assumed material properties, are likely to be available even with several reversals of *moderate* inelastic deformations. However, degradation associated with *large* inelastic cyclic deformation results in an entirely different stress state from that on which the code recommendations are based. To develop such large deformations and still maintain shear strength, the contribution of concrete to shear carrying capacity should be ignored unless the core concrete can be kept effectively confined, even under very large deformations.

3. Reversal of deformations reduces the maximum inelastic deformation a member can develop in a given direction.

This phenomenon is important where design is controlled by inelastic deformation demands. It may be necessary to specify not only the expected deformation level, but also the number and type of reversals (partial, full) expected. Moderate ductile behavior and high shear stresses are compatible. However, it is necessary to provide sufficient and properly detailed transverse reinforcement.

4. A comparison of the behavior of columns with different types of transverse reinforcement indicates that the circular spiral is the most effective in maintaining a member's shear strength. Its continuity and relatively close spacing provide excellent confinement to the core concrete and restrain the width of inclined shear cracks. However, the close spacing of the spiral causes significant spalling through the height of the column, reducing the area of concrete in contact with the longitudinal steel and thus contributing to bond deterioration along this reinforcement.

12.5.5 Seismic Behavior of Reinforced Concrete Beam–Column Joints

Surveys of earthquake damage usually show little or no evidence of joint failure, except in cases of very poor detailing and construction. However, because of numerous failures in beams, and particularly in columns, recent seismic codes have much more stringent design and detailing requirements for these two elements. Therefore, the joint may now become the weakest link in the beam–joint–column subassemblage. This presumption has been corroborated by recent experimental results in laboratories and in the field. In many cases, although there is no visible sign of distress in the joint, it has failed internally with a loss of the required anchorage to the main reinforcing bars of the beams and/or columns.[12-54]

The most important concerns in the design and detailing of reinforced concrete beam–column joints are: [12-62]

1. To preserve the integrity of the joint so that the ultimate strength and deformation capacities of the connecting beams and columns can be developed.
2. To prevent excessive degradation of the joint stiffness under seismic loading by minimizing cracking of the joint concrete and the loss of bond between joint concrete and the loss of bond between joint concrete and longitudinal beam and column reinforcement.
3. To prevent brittle shear failure of the joint.

The first of these concerns requires adequate confinement of the joint concrete. The second concern requires adequate transverse joint reinforcement that does not exhibit a flat yield plateau, and adequate anchorage or development of the longitudinal beam and column reinforcement. The third concern dictates that the joint should have adequate shear strength to resist the forces imposed on it by the connecting members.

The above concerns can also be viewed in terms of the following failure modes[12-63] that are likely in a connection:

1. *Shear failure*: Fig. 12-57 shows typical load-deflection curves for specimens exhibiting joint shear failure. After the first loading cycle, the stiffness near the origin is reduced. A shear failure may preclude the development of yielding in the beams in some cases (Fig. 12-57a); in others the beam may reach yield, but with cycling the shear capacity decreases, and shear failure occurs (Fig. 12-57b).
2. *Anchorage failure*: The load-deflection curve for a specimen with anchorage problems may be similar to that of a specimen exhibiting shear problems. The crack pattern may provide an indication of the failure mode. Figure 12-58(a) shows an exterior beam-column joint in which the anchorage of beam bars within the joint has failed. A more difficult problem is that of bond deterioration of longitudinal beam or column bars passing through a joint. Because the bars are subjected to a very large stress gradient across the joint, bond along the bars degrades rapidly with load reversals, and anchorage must be provided in the members outside the joint. The stiffness is reduced markedly near the origin. The strength may remain nearly constant; however, large deformations must be imposed before the peak strength is reached. Figure 12-58(b) shows the results of a test where anchorage through the joint has deteriorated with load reversal.
3. *Beam or column hinging*: If the joint shear capacity is sufficient, inelastic deformation is developed by hinging in the beams or columns. Figure 12-59 illustrates load-deflection behavior for member hinging. Continued application of inelastic load reversals may produce shear degradation which eventually reduces the capacity below that at flexural hinging of the members.
4. *Compression failure through the joint*: Although not reproduced in laboratory tests, the problem is of concern and is more likely to occur in an exterior or corner joint where confinement is not provided by members on all four sides of the joint.

It should be pointed out that the modes of behavior pre-

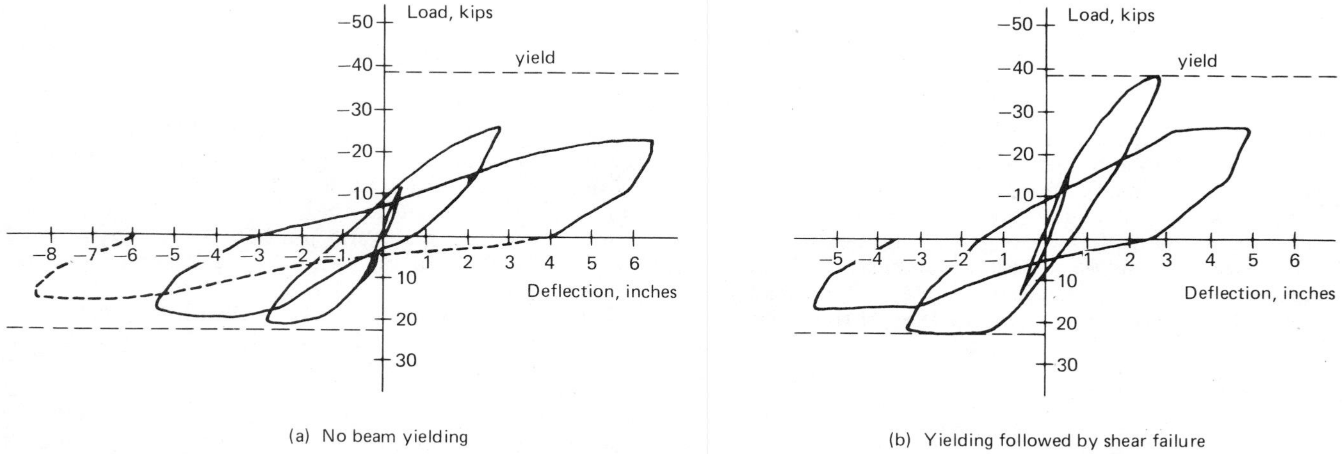

(a) No beam yielding

(b) Yielding followed by shear failure

Fig. 12-57 Load-deflection curves—shear failure.[12-55]

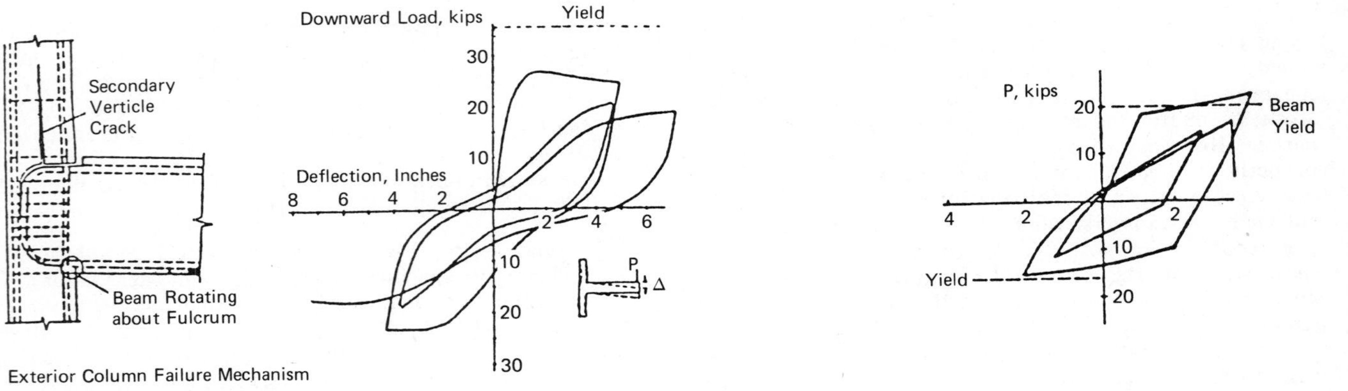

(a) Beam anchorage (b) Column anchorage

Fig. 12-58 Load-deflection curves—anchorage failure.[12-55]

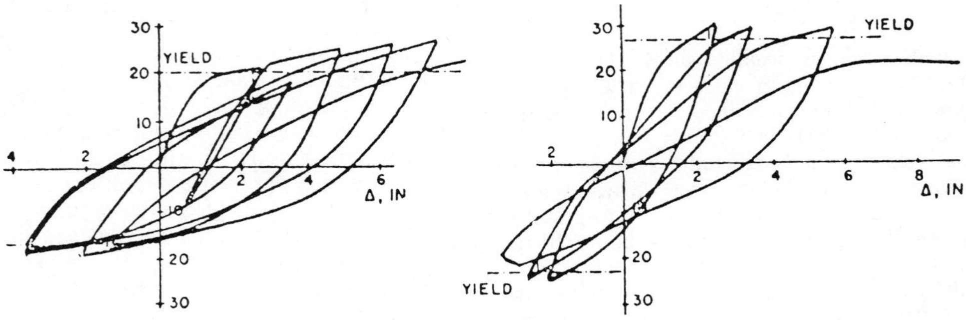

(a) Beam hinging (b) Column hinging

Fig. 12-59 Load-deflection curves—member hinging.[12-55]

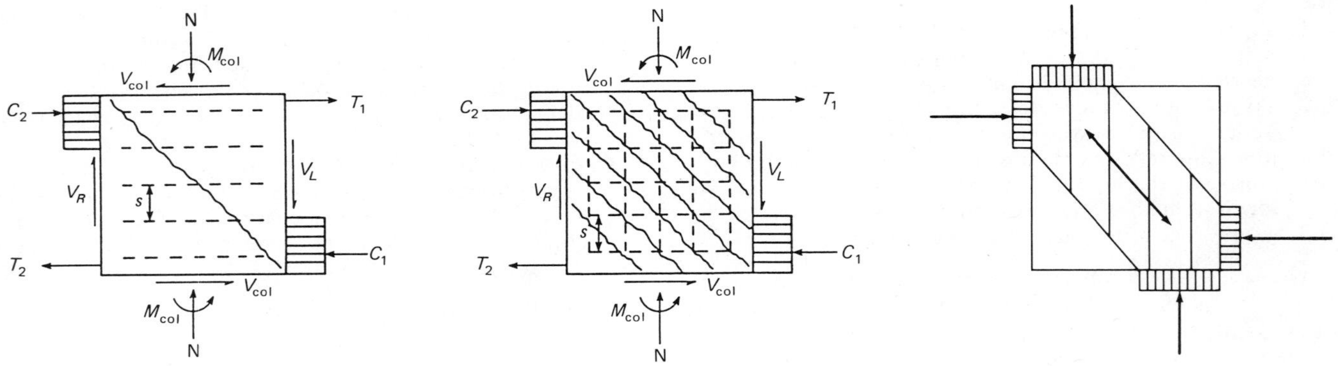

(a) Beam Shear Mechanism (b) Joint Truss Mechanism (c) Compression Strut Mechanism

Fig. 12-60 Mechanisms of joint shear resistance.[12-55]

sented in the foregoing figures were observed in the laboratory under different load histories.

1. Available Joint Strength. The determination of strength (in shear) of reinforced concrete beam-column joints requires the definition of mechanisms by which a joint resists the various possible combinations of forces imposed on it by the connecting members. Joint resistance is currently explained in terms of three different mechanisms:[12-62–12-64]

1. Beam shear mechanism
2. Joint truss mechanism
3. Compression strut mechanism

Current design procedure in the United States follows the 1976 ACI-ASCE Committee 352 Report,[12-65] which is based on the beam shear mechanism (Fig. 12-60a). A diagonal crack is assumed to extend through the joint. The joint shear strength is computed as the sum of the contributions of concrete and the transverse joint steel crossing the crack. The contribution of the transverse steel is computed from truss analogy.

The 1982 *New Zealand Code*[12-66] utilizes compression strut (Fig. 12-60c) and joint truss (Fig. 12-60b) mechanisms to determine joint shear capacity. A diagonal concrete compression strut is assumed to resist a portion of the joint shear; this constitutes the contribution of concrete to joint shear capacity. The contribution of the horizontal ties and the vertical joint reinforcement (longitudinal column bars) to shear resistance are computed using the joint truss mechanism. The horizontal ties are assumed to resist the horizontal component of forces imposed on the joint, while the vertical forces are resisted by compression forces in the column including the vertical reinforcement.

In the ATC 3-06[12-9] recommendations, the contribution of transverse reinforcement to joint shear capacity is considered to be marginal, while ultimate capacity is mainly determined by the confined compressive strength of concrete in accordance with the compressive strut mechanism.

Maximum allowable shear stress. The maximum allowable nominal shear stress in joints is of extreme concern to designers, as this value may determine the joint and hence the column size. Unfortunately, widely varying values are prescribed by various recommendations. For example, ATC 3-06[12-9] specifies a maximum average shear stress of $12\sqrt{f_c'}$ for joints not confined by transverse beams and $16\sqrt{f_c'}$ for joints confined on all four sides. ACI-ASCE 352 (1976)[12-65] allows a maximum average shear stress of $20\sqrt{f_c'}$, while the 1983 draft version of the same report specifies values varying between $12\sqrt{f_c'}$ and $20\sqrt{f_c'}$. The capacity reduction factors to go with the above allowable stresses are also different in the ATC and the other recommendations. The joint area to be used in shear stress computations also varies from recommendation to recommendation.

Another significant issue is the effect of column axial load on joint shear capacity. In Ref. 12-65, the contribution of concrete to joint shear capacity is a function of the minimum compressive axial load on the joint. However, on seismic joints, the axial load is taken equal to zero, because it is influenced by overturning forces and vertical accelerations. While vertical acceleration affects all columns, overturning primarily affects corner columns and exterior columns of transverse or longitudinal bays. It would thus seem appropriate to differentiate between exterior and interior columns rather than to totally neglect axial load effects for all columns. The 1983 ACI-ASCE Committee 352 report assigns a higher allowable nominal shear stress to interior beam–column joints than to exterior beam–column joints.

The *New Zealand Code*[12-66] attaches a much higher importance to column axial load; for cases where the column axial load is less than 10% of $f_c' A_g$, the *New Zealand Code* ignores the contribution of concrete to joint shear strength completely.

Transverse joint reinforcement. Among the three sets of recommendations discussed earlier, the *New Zealand Code* requires the maximum amount of joint reinforcement, the ATC 3-06 recommendations generally provide the minimum reinforcement, and the 1976 ACI-ASCE recommendations require an intermediate amount of reinforcement.

An analysis of test data by Zhang and Jirsa[12-64] indicates that increasing transverse reinforcement beyond a certain volumetric ratio does not necessarily increase the joint shear strength. Therefore, considering congestion problems, an excessive amount of transverse reinforcement is not always justified as a means of increasing shear capacity. However, transverse reinforcement should provide adequate confinement for the concrete. A minimum ratio of 0.01 is suggested, with a more reasonable value being around 0.03 or 0.04.[12-62] The transverse reinforcement is recommended to consist of closed hoops, well anchored in the confined joint area, and the steel should not have a flat yield plateau. The ATC 3-06 recommendations on seismic beam–column joints are in accord with the above discussion. So are the recommendations in the tentative 1983 ACI-ASCE Committee 352 report. As long as the design horizontal nominal shear stress on the joint core does not exceed an allowable value, the amount of transverse reinforcement is determined by confinement considerations. New Zealand researchers strongly dispute the assumption that as long as a joint is properly confined, no additional transverse shear reinforcement in the joint is needed.

2. Required Deformability. One of the reasons for the difference between the U.S. and New Zealand approaches to the estimation of shear capacity available in a joint is a corresponding difference of opinion as to the deformability required in concrete frame structures subjected to earthquakes. The New Zealand approach is to provide a design that would enable a building as a whole to deflect laterally through at least four compete load cycles, with the total horizontal deflection at the top reaching at least four times that at first yield, without the horizontal load carrying capacity being reduced by more than 20%. The ACI-ASCE Committee 352 report and ATC 3-06 use less rigorous requirements and do not specify performance criteria in terms of load reversals at certain ductility levels.

A precise determination of deformability demands in terms of a number of excursions at a certain ductility level is not simple. For example, on the West Coast of the United States, a surface wave magnitude 7.5 earthquake would probably produce strong motion for at least 20 seconds. Thus, a building with a fundamental period of 2 seconds would go through ten cycles of large deformations plus residual swaying (depending on damping); however, the number of cycles beyond the elastic limit is not known. Some studies indicate that even though buildings may go through many deformation cycles within the elastic range, the number of damaging deformation reversals at high ductility levels may be limited to three cycles or less.

3. Stiffness Characteristics. The degradation of joint stiffness due to concrete cracking (diagonal cracking), yielding of transverse reinforcement, and loss of bond between concrete and the longitudinal beam and column reinforcement is of concern. Reference 12-65, recommends in favor of full development of the longitudinal beam and column reinforcement and warns of added rotation in the members due to loss of bond and bar pull-out under cyclic loading. However, no specific provisions are made for the consideration of these effects.

Loss of bond between longitudinal beam reinforcement and joint concrete under cyclic load reversals and the resulting bar pull-out or pull-through have been thoroughly researched. In exterior joints, the crushing of concrete inside the hook and bond deterioration along the lead reinforcement before the hook can cause beam bars to pull out at the column face. This phenomenon results in a reduction of beam-end fixity and increased beam rotations at column faces. At interior joints, the longitudinal beam reinforcement—pulled on one side and pushed on the other side of the joint—loses bond, with resulting beam end rotations and, sometimes, a complete pull-through of the beam bars. This phenomenon occurs under repeated load reversals at or beyond yielding of the reinforcement. The major consequence of this is a significant increase in frame deformations, or drift, which may lead to frame instability. Also, the distribution of moments along structural members due to combined dead, live, and seismic loads changes as beam ends lose their fixity at columns. Several conceptual remedies, with few quantitative guidelines, have been recommended. These include:

1. The use of small-diameter bars for longitudinal beam reinforcement.
2. The use of approximately equal amounts of top and bottom reinforcement at beam ends.
3. Forcing the location of beam plastic hinges away from the column face. This can be accomplished by designing haunched beams, cutting off or hooking some of the

longitudinal beam reinforcement some distance away from the column face, or bending and diagonally crossing some of the longitudinal beam reinforcement at a distance from the column face. Diagonal crossing of the flexural reinforcement provides for an increased beam shear capacity at the plastic hinge, but is not easy to place in the field.
4. The use of fiber reinforcing or special anchorage devices within the joint.

The present embedment requirements for reinforcing bars in or through joints have been developed by ACI Committee 408, and consider confinement and cyclic loading. The differences between requirements for exterior and interior joints are based on whether hooked or straight bars are developed. The latest Committee 352 report requires 20 bar diameters as a minimum column dimension when straight beam bars are to be developed, while this increases to 35 diameters under the *New Zealand Code*. Further, for interior columns, the location of the plastic hinge should be moved away from the face of the column. To achieve this, the continuous top and bottom beam reinforcement through interior columns should be at least two-thirds of that required at the column face. The remaining one-third should be hooked after extension through the column, and at least one-sixth of the clear span (but not less than the beam depth) beyond the face of the column.

For all columns of lateral load resisting frames in high seismic zones, confinement should be provided by the required transverse ties and/or perpendicular beams that have continuous top and bottom longitudinal bars through the joint.

Column details. Energy dissipation in reinforced concrete frames should be primarily through plastic hinges developed in beams. The columns should not be subjected to moments beyond their yield capacity as beams reach their maximum inelastic capacities. In computing maximum beam capacities, strain hardening characteristics of the beam steel should be accounted for; an increase of 25% beyond specified yield strength of the reinforcement is recommended. In addition, in computing the column flexural capacity, only the confined core section of the column, that is, the section within the transverse confinement reinforcement, should be used. Also, for computing bar anchorage and development requirements, only the portion of bars within the confined core should be considered. Another recommendation, based on recent research, is to equally distribute vertical column reinforcement around the perimeter with a minimum of eight bars. This seems to help increase the joint confinement, improve the anchorage of beam bar extensions, and resist the vertical forces required by the truss analogy model.

4. Bidirectional Joint Loading. Most of the current design recommendations and recent research considers only unidirectional loading, that is, only forces and moments in the plane of the frame. In reality, joints are subjected to simultaneous action of forces resulting from gravity loads and seismic loads along all three orthogonal axes. This topic requires further research, and designers must treat corner joints or other high-eccentricity situations as special cases.

The following conclusions were drawn in Ref. 12-62 concerning the design of joints:

5. Joints in Regions of High Seismicity. The concrete strength must be at least 4 ksi for exterior joints and may need to be increased to 5 ksi, while interior joints require 5 ksi strength.

The exterior column dimension is determined by the bar size and embedment into the joint, rather than by shear through the joint.

In the event that confinement by beams is not provided, column dimensions may need to increase by about 15%.

Generally, more transverse ties are required through the joint when shear, rather than confinement, is the criterion for design.

6. Joints in Regions of Low Seismicity. Only the exterior column joints appear to be critical. Design on the basis of shear rather than confinement results in more transverse ties.

The exterior column dimensions are determined by embedment length of hooked bars. Top bars of the beam reinforcement should be kept to smaller sizes to result in smaller columns.

At interior joints, 3-ksi concrete would suffice. Although 3-ksi concrete could be used for exterior joints also, then the embedment of beam longitudinal bars might require larger column sizes.

12.5.6 Behavior of Reinforced Concrete Shear Walls

1. Observed Behavior and Modes of Failure. The behavior as well as the eventual mode of failure of a wall subject to load reversals is a function of the magnitude of the applied shear stresses. The nominal shear stress, v, is usually computed as V/hd, where V is the base shear, h is the wall thickness, and d is the effective depth between the extreme compression fiber and the centroid of the rebars in tension. Code requirements allow use of the value $0.8l_w$ for d, where l_w is the total length of the wall.

The behavior of a wall subjected to $v_{max} \leqslant 3\sqrt{f'_c}$ is characterized by the formation of a predominantly horizontal crack pattern in the lower hinging region of the wall after a few inelastic deformation reversals. After yielding of the vertical steel, forces are transferred predominantly by interface shear across horizontal cracks and by dowel action. The capacity of this shear transfer mechanism is adequate to develop a flexural failure mode. Final failure is by longitudinal bar fracture, crushing of concrete in the compression zone, or lateral instability of the compression zone.[12-67, 12-68]

Vertical reinforcement in walls must be adequately anchored to develop reinforcement stresses relied on in design. Bar fracture can be the result of severe inelastic load reversals that cause alternate tensile yielding and compressive buckling. Special transverse bars around vertical reinforcement are effective in delaying buckling and limiting damage.

Well-proportioned walls are generally under reinforced, so that crushing in compression zones is not expected. However, in slender rectangular walls that contain a high percentage of vertical reinforcement and carry significant axial compressive stresses, crushing can occur. This mode can also be important for an unsymmetrical section such as a T-shape, the web of which can be heavily stressed under load reversals. In design, vertical reinforcement ratios in boundary elements can be limited to avoid concrete crushing. Transverse reinforcement in the boundary element can be utilized to increase the strain capacity of the concrete. Additionally, vertical load stresses can be limited as they are for columns.

The capacity of a rectangular wall can also be limited by out-of-plane instability of the compression zone. This is likely to occur only after several reversals of large inelastic deformations. Limits on minimum wall thickness are used to avoid out-of-plane instability.

The behavior of a wall subjected to high nominal shear stresses is characterized by the development of inclined cracks crisscrossing the web to form relatively symmetrical compression strut systems for the two reversible directions of loading. Shear transfer is largely by truss action, which provides a stiffer mechanism than that for walls exhibiting flexural behavior. Capacities of walls subjected to high nominal shear stresses can be limited by diagonal tension, sliding shear, or web crushing.[12-67, 12-68]

Diagonal tension failure is a phenomenon associated with monotonic loading. As lateral loads on a wall are increased, a major diagonal tension crack develops, and the yield stress of transverse reinforcement crossing this crack is exceeded. Eventually, shear reinforcement fractures before flexural capacity of the wall can be reached. Current *ACI Code* provisions for shear design of walls were developed to prevent this type of failure.

Sliding shear failure occurs under increasing numbers of inelastic load reversals as interface shear transfer along cracks deteriorates. Eventually shear can no longer be transferred, and the upper portion of a wall slides over the lower portion. The possible occurrence of such sliding depends on the attainment of flexural yielding, the number of load reversals, the level of axial loads, and, most important, the crack pattern that develops. If cracks under load reversals intersect to form basically horizontal planes, it is possible that shear transfer capacity across such planes will be lost after a large number of repeated inelastic cycles.

Generally, in walls subjected to low nominal shear stresses ($\leqslant 3\sqrt{f_c'}$), there is adequate shear transfer capacity even though the cracks may intersect to form basically horizontal planes. In walls capable of resisting high nominal shear stresses ($\geqslant 7\sqrt{f_c'}$), inclined cracks form a diagonal strut system that precludes development of sliding shear. Therefore, failure by sliding shear is a possibility only in the range of nominal shear stresses between $3\sqrt{f_c'}$ and $7\sqrt{f_c'}$. Vertical wall boundary elements help in resisting sliding shear. They function as large dowels that transfer much of the horizontal shear. The use of diagonal web reinforcement has also been suggested as a means of countering sliding shear failure.

Web crushing failure (prevalent in walls subjected to $v_{max} \geqslant 7\sqrt{f_c'}$) is preceded by the development of inclined cracks that, under load reversals, form a diagonal compressive strut system in each direction of loading. As load reversals progress and higher shears are applied, crushing of the diagonal struts limits wall capacity. Crushing is initiated by the formation of a relatively small zone of high compressive stresses in the web of the wall. Web crushing capacity has been found to depend on concrete compressive strength and on maximum shear distortions applied to a wall.

2. *Strength and Deformability.* The effects of several important factors on wall strength and deformation characteristics are discussed below.[12-67–12-69]

(a) *Load history.* Walls behaving predominantly in flexure develop much larger deformation capacities under monotonic loading than under a large number of inelastic load reversals. This should be evident from Fig. 12-61, which shows a comparison between the load–displacement relationships of two nominally identical walls—one subjected to monotonic and another to cyclic loading. Considerably more energy is dissipated under load reversals.

For walls subjected to repeated inelastic load reversals that impose low nominal shear stresses ($v_{max} \leqslant 3\sqrt{f_c'}$), flexural capacities can be up to 15% less than monotonic flexural capacities.

In walls with shear-dominated behavior, reversing loads appear to have a less significant effect on strength and deformation capacity then for walls in flexure. The diagonal strut systems that resist high nominal shear stresses under reversible loads do not permit such deterioration in shear stiffness as is observed in walls governed by flexure. Also, walls subjected to high nominal shear stresses generally do not undergo as many inelastic cycles as walls subjected to low nominal shear stresses.

Oesterle et al.[12-67] have pointed out that shear wall behavior under load reversals is dependent not so much on the entire previous load history as on the maximum level of deformations previously reached. One reversal at a rotational ductility of five has been observed to result in a wall equivalent to one that has sustained three complete reversals each at ductilities of 1, 2, 3, 4, and 5.

(b) *Nominal shear stress level.* The effects of nominal shear stress level on wall behavior and failure mode have been discussed previously. Probably as a result of those effects, nominal shear stress level has a strong influence on wall deformation capacity. Deformation capacity data for wall specimens tested at the Portland Cement Association Laboratories are illustrated in Fig. 12-62. Deformation capacity is expressed in terms of rotational ductility defined in the figure. Maximum rotation was taken from the last cycle in which at least 80% of the previous maximum observed load was sustained throughout the cycle. Yield level was defined as corresponding to the yielding of all the main tensile reinforcement in one of the boundary elements. All specimens tested possessed substantial inelastic rotational capacity under reversing loads. Ductility decreases as the maximum shear stress level increased. The presence of axial loads, indicated by open symbols in Fig. 12-62, increased ductility in specimens subjected to high shear stresses.

(c) *Section shape.* There is a limit to the amount of reinforcement that can be physically placed in the end regions of a rectangular wall. Therefore, the maximum attainable flexural capacity is low. For equivalent moment-to-shear ratios, the level of shear stresses also is lower than in the webs of walls with end columns or flanges. A thin rectangular wall is susceptible to lateral instability of the compression zone under severe load reversals.

In a wall with end columns, the column boundary elements provide relatively large in- and out-of-plane stiffness. These elements also resist sliding shear by acting as large dowels. Since the boundary elements provide space for reinforcement, relatively high flexural capacities can be developed. Therefore, relatively high nominal shear stresses can be generated in the webs. Shear capacity is usually limited by web crushing.

A flanged wall can develop high shear stresses in the web. Residual capacity, after web crushing, depends on the design and detailing of the boundary elements.

(d) *Moment-to-shear ratio.* The nominal shear capacity of a wall increases as the moment-to-shear ratio decreases. This is primarily because at low moment-to-shear ratios, flexural yielding is unlikely to precede web crushing. Wall tests have shown that web crushing shear capacity is a function of shear distortions as well as concrete strength. Shear distortions increase significantly as flexural yielding takes place.

(e) *Flexural reinforcement.* The amount of vertical reinforcement controls the moment capacity of a wall section and, thus, the maximum shear stress that can develop in that section. In seismic design it is important to recognize that shear stresses in walls are related to the actual, rather than the design, flexural capacity. The actual flexural capacity can be estimated considering that the actual yield stress of reinforcement normally exceeds the specified minimum, and that reinforcement strain-hardens. Distributed vertical web reinforcement should also be considered in calculating actual flexural strength.

For the same total amount of vertical reinforcement, walls having bars concentrated near their ends develop higher moment capacity and, an important consideration, higher deformability than walls with uniformly distributed reinforcement.[12-70]

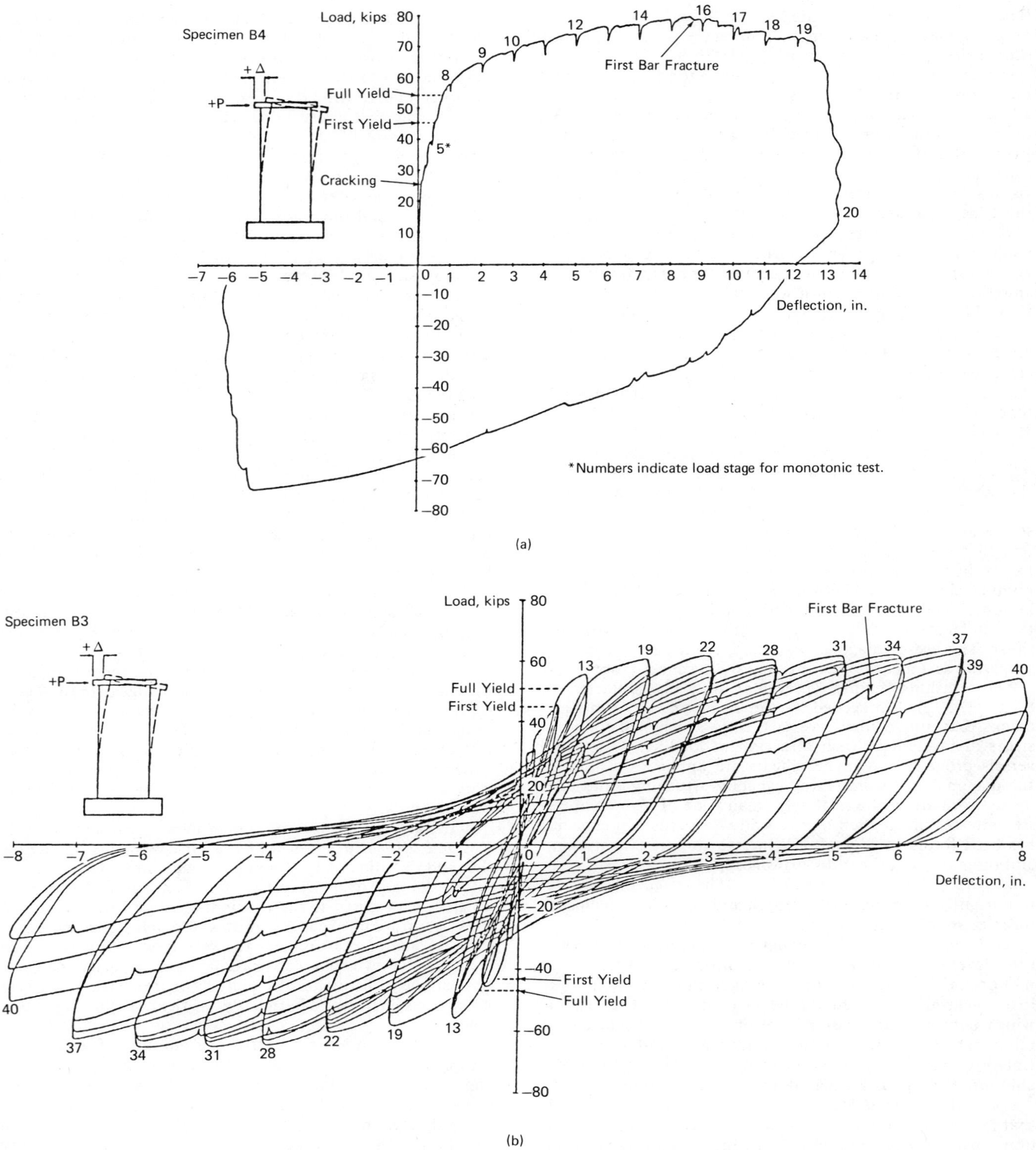

Fig. 12-61 Load-deflection relationship for wall subjected to (a) monotonic load, (b) load reversals.[12-68]

(f) *Shear reinforcement.* Horizontal reinforcement serves primarily to prevent diagonal tension failure. Horizontal bars are ineffective in resisting sliding shear. The onset of web crushing depends on the degree of shearing distortions. Although additional horizontal steel improves inelastic performance by reducing shear deformations in specimens subjected to high shears, the improvement is not sufficient to significantly influence overall behavior or web crushing strength.

Wall tests under cyclic loads have shown that measured strengths exceed design strengths as determined by present code provisions which seek to prevent diagonal tension failures and assume that shear is partly resisted by concrete and partly by horizontal shear reinforcement.[12-68] It has been suggested that the contribution of concrete to shear resistance be ignored in the seismic design of walls, because of concern over potential degradation of the concrete contribution when reversed inelastic cyclic loads occur. The

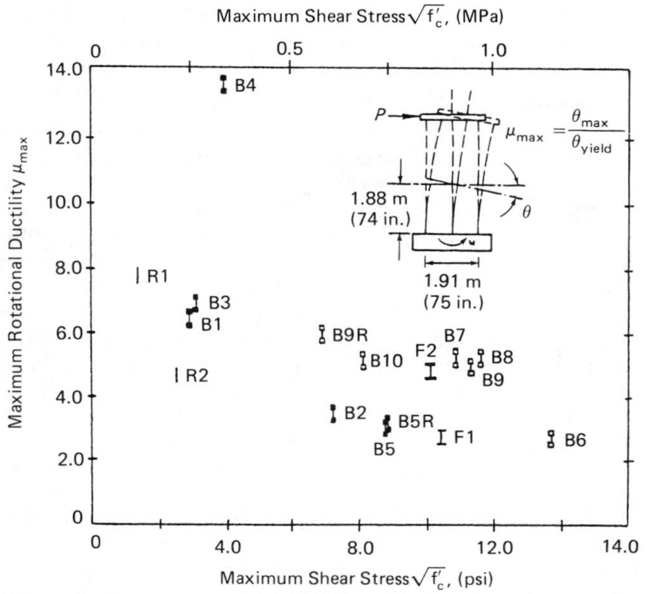

Fig. 12-62 Observed rotational ductilities versus maximum nominal shear stress for shear walls.[12-67]

suggestion obviously would have the effect of adding horizontal reinforcement. The results of recent PCA wall tests[12-67] do not support the need for additional horizontal reinforcement.

(g) *Diagonal (web) reinforcement.* The use of diagonal reinforcement in webs of shear walls has been proposed to reduce shear distortions and to resist sliding shear. Improved hysteretic response with the use of diagonal reinforcement has been observed in tests.[12-69] Since diagonal reinforcement is more difficult to place than conventional reinforcement, a diagonal arrangement of bars should be considered only when anticipated energy dissipation demands are severe.

(h) *Special transverse reinforcement.* Current building codes require earthquake-resistant shear walls to be detailed with special transverse reinforcement in boundary elements over the entire wall height. Design criteria are based on providing confinement to increase concrete strain capacity. In reality, transverse reinforcement serves at least three other functions: (1) it supports vertical reinforcement against inelastic buckling; (2) together with vertical bars it contains fractured concrete within the core; and (3) it improves shear capacity and boundary element stiffness.

The confining function of transverse reinforcement is important in walls with relatively low concrete strength, high percentages of vertical reinforcement, and significant axial compression, that is, in walls with relatively large neutral axis depths (exceeding about 15% of the horizontal lengths of walls). The other functions of transverse reinforcement are not important until interstory drifts become rather large (about 2%). However, in view of the uncertainty associated with the prediction of both earthquake input motion and deformation of concrete structures, it is advisable to furnish all shear walls in regions of high seismicity with special transverse reinforcement. Such reinforcement will provide reserve deformation capacity, and thus additional safeguards, against severe damage during an unanticipated catastrophic event. Bertero has found spiral and square transverse reinforcement to be more effective than rectangular hoops.[12-69] In regions of low to moderate seismicity, use of transverse reinforcement for functions other than confinement is not justified. Ordinary column ties may provide adequate reserve deformation capacity.

In all cases, special transverse reinforcement is needed only in wall segments where concentrated inelastic rotations are expected. In an actual structure, potential hinging and damage are most likely to occur at the wall bases and at discontinuities resulting from abrupt changes in strength or stiffness.

(i) *Concrete strength.* Wall performance is influenced in several ways by concrete strength, which affects the onset of crushing at the extreme compression fiber, web crushing capacity, and abrasion resistance along crack interfaces. The first effect is accounted for in conventional design for flexure and axial loads.

Web crushing capacity is commonly considered to be a function of concrete strength alone. Recent PCA tests[12-67] have shown that web crushing is related to deformation levels as well. For example, comparison specimens, one of 3-ksi and the other of 7-ksi concrete, sustained maximum shear loads within approximately 15% of each other. Failure of both specimens was by web crushing. The primary difference in test results was that the wall with the lower concrete strength had a significantly lower deformation capacity.

(j) *Compressive axial loads.* Axial loads significantly affect the stiffness, strength, deformability, and energy dissipation capacity of shear walls, the presence of moderate compression being highly beneficial. This is so largely because compressive axial stresses increase shear stiffness (reduce shear distortions for equivalent deflections).

(k) *Construction joints.* Construction joints perform satisfactorily when constructed following standard practice. Such practice requires roughening and cleaning the surface of previously cast concrete to remove laitance and loose particles.

12.5.7 Seismic Behavior of R.C. Coupling Beams in Shear Wall Structural Systems

Coupled shear walls, when suitably located in multistory reinforced concrete buildings, can offer particularly advantageous resistance against earthquake attack. Their stiffness minimizes damage during moderate disturbances. To meet the demands of very large earthquake excitations, suitably detailed coupled shear walls can be expected to be ductile enough to assure survival without collapse. The major source of energy dissipation should be the coupling beams because they will give some protection to the walls, which are more difficult to repair than the beams and in which large ductilities may be difficult to develop without generating large compression strains in the confined concrete.

Paulay[12-71] has made a detailed examination of the observed performance of coupling beams in coupled shear wall structural systems. He has pointed out that these beams are often deep relative to their span. For this reason, large shear forces are generated that dominate the inelastic behavior of these beams. Furthermore, the deformation capacity (ductility), the number of yield excursions, and the number of plastic rotation reversals that are demanded of these coupling beams are very large when compared with those encountered in beams of ductile-moment-resisting frames.

Paulay[12-71] has analyzed and tested the behavior of coupling beams that have been designed and reinforced according to conventional procedures and those that have been designed and reinforced on the premise that the shearing force resolves itself into diagonal compression and tension forces, intersecting each other at midspan where no moment is to be resisted. This last procedure results in diagonally reinforced beams. A summary of Paulay's findings may be found in Ref. 12-57.

12.5.8 Seismic Response of R.C. Slab Systems

The *Uniform Building Code* requires all buildings designed with a horizontal force factor, K, of 0.67 or 0.8 to have ductile-moment-resistant space frames, and such frames are required around the perimeter of all frame buildings in seismic zones 2, 3, and 4. If the building exceeds 160 ft in height, those frames must be capable of resisting not less than 25% of the required seismic forces for the structure as a whole. All framing elements not required by design to be part of the lateral force-resisting system must be adequate for vertical load-carrying capacity and induced moment at a distortion $3/K$ times that resulting from the *Code*-required lateral forces. The UBC[12-14] provisions are intended to exclude the use of flat plate–column frames as ductile-moment-resistant space frames. Such framing is, however, widely used in the eastern and central parts of the United States, which include many Zone 2 and 3 regions.

In Ref. 12-72, Hawkins has presented an excellent discussion of the following topics: (1) the use of flat plate construction as a vertical or lateral load-carrying system in seismic zones; (2) the effectiveness of slabs as coupling members between shear walls; and (3) the diaphragm action of the slab as it transmits seismic forces to the vertical elements resisting lateral loads. In view of space limitations and the fact that the use of flat plate–column frames as *primary* lateral load resisting systems is not to be recommended at least in seismic zones 3 and 4, the seismic response of R.C. slab systems is not discussed here. The reader should consult Ref. 12-72 and the literature cited therein for information on the subject.

12.6 CURRENT PRACTICE IN DESIGN FOR EARTHQUAKE MOTIONS

12.6.1 General

Economical earthquake-resistant design should aim at providing appropriate dynamic characteristics to structures so that acceptable response levels would result under the design earthquake(s). The structural properties over which the designer exercises some degree of control and which he can modify to achieve the desired results are the magnitude and distribution of stiffness and mass, relative strengths of members, and their ductilities. Perhaps the most convenient measure of response, particularly insofar as it affects the design of a structure, is the maximum lateral displacement of the structure as well as the maximum deformations of its component elements. The range of deformation, relative to the force–deformation characteristic of a particular structure or element, that may be considered acceptable will vary with the type of structure and/or its function.

In some structures, such as slender free-standing towers or smoke stacks which depend for their stability on the stiffness of the single element making up the structure, or in nuclear containment buildings, where a more-than-usual conservatism in design is required, yielding of the principal elements in the structure cannot be tolerated. In such cases, the design needs to be based on an essentially elastic response to moderate-to-strong earthquakes, with the critical stresses limited to the range below yield.

In most buildings, particularly those consisting of frames and other multiply-redundant systems, however, economy is achieved by allowing yielding to take place in some members under moderate-to-strong earthquake motion.

The performance criteria implicit in most earthquake code provisions require that a structure be able to:[12-73]

1. Resist earthquakes of minor intensity without damage; a structure would be expected to resist such frequent but minor shocks within its elastic range of stresses.
2. Resist moderate earthquakes with minor structural and some nonstructural damage; with proper design and construction, it is expected that structural damage due to the majority of earthquakes will be limited to repairable damage.
3. Resist major catastrophic earthquakes without collapse.

The above performance criteria allow only for the effects of a typical ground shaking. The effects of slides, subsidence, or active faulting in the immediate vicinity of the structure, which may accompany an earthquake, are not considered. Of course, proper design should aim at minimizing the effects of these other factors, when investigation reveals a reasonable chance of such actions occurring.

While no clear quantitative definition of the above earthquake intensity ranges has been given, their use implies the consideration not only of the actual intensity level but also of their associated probability of occurrence with reference to the expected life of a structure.

The prinicipal concern in earthquake-resistant design is the provision of adequate strength and ductility to assure life safety, that is, to prevent collapse under the most intense earthquake that may reasonably be expected at a site during the life of a structure. Observations of building behavior in recent earthquakes, however, have made engineers increasingly aware of the need to ensure that buildings housing facilities essential to post-earthquake operations, such as hospitals, power plants, fire stations, and communication centers, not only survive without collapse but remain operational after an earthquake. This means that such buildings should suffer a minimum of damage. Thus, damage control has been added to life safety as a second design criterion.

Often, damage control becomes desirable from a purely economic point of view. The extra cost of preventing severe damage to the nonstructural components of a building, such as partitions, glazing, ceilings, elevators, and other mechanical systems, may be justified by the savings realized in replacement costs and from continued use of a building after a strong earthquake.

12.6.2 Design Philosophy Behind Earthquake Code Provisions for Buildings

The availability of dynamic analysis programs[12-22,12-44–12-49] designed for use with large electronic computers has made possible the analytical determination of earthquake-induced forces in reasonably realistic models of most structures. However, except perhaps for the relatively simple analysis by modal superposition using response spectra, such dynamic analyses, which can range from an elastic time-history analysis for a single earthquake record to nonlinear analyses using a representative ensemble of accelerograms, can be costly and may be economically justifiable as a design tool only for a few large and important structures. For most practical applications, reliance will usually have to be placed on the simplified prescriptions found in most codes.[12-73] Although necessarily approximate in character—a feature imposed by the need for simplicity and ease of application—the provisions of such codes and the philosophy behind them gain in reliability as design guides with continued application and modification to reflect the lessons derived from observations of actual structural behavior under earthquakes. Code provisions must, however, be viewed in the proper perspective, that is, as minimum requirements covering a broad class of structures of more or less conventional configuration. Unusual structures must still be designed with special care and may call for procedures beyond those normally required by codes.

The basic form and features of modern code provisions on earthquake-resistant design have evolved from rather simplified concepts of the dynamic behavior of structures and have been influenced greatly by observations of the performance of structures subjected to actual earthquakes. It has been noted, for instance, that many structures built in the 1930s and designed on the basis of more or less arbitrarily chosen lateral forces have successfully withstood severe earthquakes. The satisfactory performance of such structures has been attributed to one or more of the following:[12-75,12-76] (1) yielding in critical sections of members—yielding not only increased the period of vibration of such structures but allowed them to absorb greater amounts of the input energy from an earthquake; (2) the greater actual strength of such structures resulting from the so-called nonstructural elements which were generally ignored in analysis, and the significant energy-dissipation capacity that cracking in such elements represented; and (3) the reduced response of the structure due to yielding of the foundation.

The code-specified design lateral forces have the same general distribution as the typical envelope of maximum horizontal shears indicated by an elastic dynamic analysis. However, the code forces, which are assumed to be resisted by a structure within its elastic (working) range of stresses, are substantially smaller than those that would be developed in a structure subjected to an earthquake of intensity equal to that of the 1940 El Centro, if the structure were to respond elastically to such ground excitation. Thus, buildings designed under the present codes would be expected to undergo fairly large deformations (four to six times the lateral displacements resulting from the code-specified, statically applied shears—see Fig. 12-18) when subjected to an earthquake with the intensity of the 1940 El Centro. These large deformations will be accompanied by yielding in many members of the structure, and, in fact, such is the intent of the codes. The acceptance of the fact that it is economically unwarranted to design buildings to resist major earthquakes elastically and the recognition of the capacity of structures possessing adequate strength and ductility to withstand major earthquakes by responding inelastically to them, lies behind the relatively low forces specified by the codes. These reduced force levels must be and are coupled with additional requirements for the design and detailing of members and their connections in order to ensure sufficient deformation capacity in the inelastic range.

The capacity of a structure to deform in a ductile manner, that is, to deform beyond the yield limit without significant loss of strength, allows such a structure to absorb a major portion of the energy from an earthquake without serious damage. Laboratory tests [12-77–12-80] have demonstrated that reinforced concrete members and connections designed and detailed by the present codes do possess the necessary ductility to allow a structure to respond inelastically to earthquakes of major intensity without significant loss of strength.

Because of the relatively large inelastic deformations a building designed by the present codes may undergo during a strong earthquake, proper provision must be made to ensure that the structure does not become unstable under the vertical loads. The codes thus prescribe the so-called strong column–weak beam design with the intent of confining yielding to the beams while the columns remain elastic throughout their seismic response. It is required that the sum of the moment strengths of the columns meeting at a joint, under the design axial loads, be greater than the sum of the moment strengths of the beams framing into the joint in the same plane. A comparative study by Anderson and Bertero[12-40] of the dynamic response of frames designed according to this philosophy with that of frames designed using allowable stress methods with yielding allowed in the columns and also of frames designed to achieve minimum weight (hence minimum stiffness) showed that the present code philosophy in this regard results in better structural performance under earthquake loading than the other design methods.

The design provisions contained in the main body of the *ACI Building Code*[12-81] as well as the regular provisions in most other codes do in fact provide some ductility which should be sufficient for structures subjected only to minor earthquakes that may occur frequently, such as those associated with UBC Zone 1 area. For structures that may be subjected to earthquakes of moderate intensity (Zone 2) some additional confinement, anchorage, and shear reinforcement details may be required. For structures that may be subjected to strong-intensity earthquakes (Zone 3 and 4), appreciable inelastic deformations can be expected, so that substantial ductility is required. The design provisions contained in the *SEAOC Recommendations*[12-73] and the UBC Zone 3 and 4 requirements,[12-14] as well as Appendix A of ACI 318-83,[12-81] are intended to provide this additional ductility.

The discussion that follows will be concerned mainly with the design of structures that have to satisfy special ductility requirements in order to ensure satisfactory performance under moderate-to-strong earthquake intensities.

12.6.3 Isolation Concepts for Earthquake Resistance

Apart from problems associated with the current approach to earthquake-resistant design, there is the important question of whether we have chosen the best means of adapting our structures to the earthquake phenomenon. The concern over the soundness of our traditional approach to earthquake-resistant design and the results of observation of structures subjected to recent earthquakes have brought about the examination of earthquake-adaptive systems and the revival of the shock-isolation concept as applied to earthquake-resistant structures. This concept represents a radical departure from current aseismic design practice. Its successful implementation on a broad scale promises significant simplifications in the design of tall reinforced concrete structures.

In contrast to the present philosophy of designing an entire structure to withstand the distortions resulting from earthquake motions, an adaptive system is designed to isolate the upper portions of a structure from destructive vibrations by confining the severe distortions to a specially designed portion at its base. A number of isolation devices or mechanisms have been proposed including a soft story with hinging columns,[12-82] a combination ball bearing and rod system,[12-83] steel balls on ellipsoidal cavities,[12-84] etc.[12-85] References 12-86 and 12-87 contain up-to-date information on the subject.

An isolation system for a multistory structure should be a resisting element that exhibits essentially linear elastic behavior under the maximum wind loading, but yields when subjected to earthquake forces slightly greater than that corresponding to the maximum wind loading. By allowing the isolating mechanism at the base of a structure to yield at a predetermined lateral load, the structure above it is effectively isolated or shielded from forces that would otherwise cause inelastic deformations. The isolating mechanism thus sets an upper limit to the forces that can be transmitted to the structure from the foundation. The structure supported on an isolating mechanism then need only be designed for vertical and wind loads, with special attention for earthquake resistance focused only on the isolating mechanism at its base. This clearly offers economic and technical advantages when compared to the present method of analyzing a complex structure under earthquake motion and providing ductile members and connections throughout the entire structure.

The isolating effect that a yielding base can provide to a superstructure during earthquakes has been borne out by observations of buildings in earthquake-damaged areas. In a number of very illuminating cases, the upper portions of multistory buildings in which a lower story apparently suffered extensive inelastic deformations early in the shaking remained virtually unscathed.

An effective isolation system not only allows the structure above to remain elastic during a strong earthquake but spares the nonstructural elements from extensive distress. Since the nonstructural components (e.g., partitions, glazing, mechanical equipment, etc.) in a typical multistory structure account for about 80% of the building's cost, a positive means of preventing distress in such elements during strong earthquakes would represent significant savings in the repair and replacement costs that would otherwise be required.

12.6.4 Preliminary Design—Some Guides to Planning and Detailing

A good preliminary design of an earthquake-resistant building is a structure proportioned to satisfy the requirements of gravity and wind loads. The planning and layout of the structure, however, must be undertaken with proper consideration of the dynamic character of earthquake response. Thus, modifications in both configurations and proportions to anticipate earthquake requirements should be incorporated at the outset into the design for gravity and wind. For the purposes of such modification, the results of the analytical studies presented in the preceding sections will serve as useful guides. Also essential to the finished design is a particular attention to detail that can often mean the difference between a severely damaged structure and one with but minor repairable damage.

To help in the preparation of a preliminary design for earthquake resistance, the items below have been collected for ready reference. The list reflects the results of studies discussed in the preceding sections as well as some of the more important lessons gained from observations of the behavior of structures subjected to actual earthquakes. The items listed should be considered where deemed applicable, in addition to the minimum requirements implied in earthquake code provisions. The list is by no means exhaustive, nor is it generally applicable to all types of buildings.

(1) A building that is simple in both plan and elevation, with a minimum of setbacks or changes in section, is generally preferable to an irregularly shaped structure. This is so because the force concentrations that occur at major discontinuities in either geometry or stiffness even under static loads tend to be aggravated under dynamic conditions. The required deformability at such regions of discontinuity is usually substantially greater than in other portions of a structure.

Although it may not be practicable to plan a fully symmetrical building, any effort to reduce the eccentricity of the effective inertial force due to the noncoincidence of the centers of mass and rigidity will pay off in reduced torsional stresses, which can be critical in corner columns and end walls. Particular attention should be given to the effects of torsional response of projecting wings of buildings in relation to the motion of the building as a whole.

Horizontal torsional effects are best accommodated by having the major lateral stiffening elements—whether shear walls or grids of closely spaced frame elements—located near or along the plan periphery of a building. The so-called

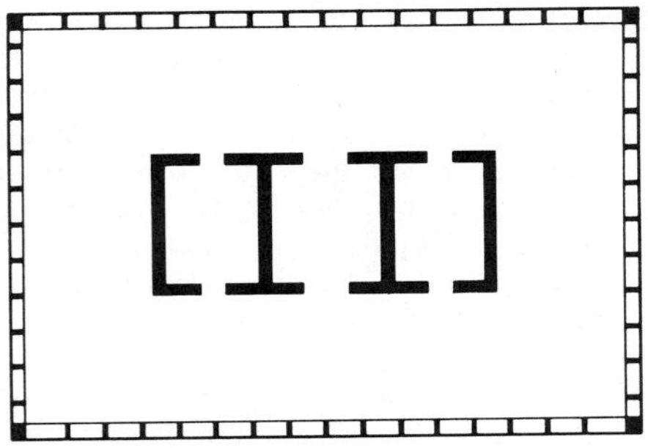

(a)

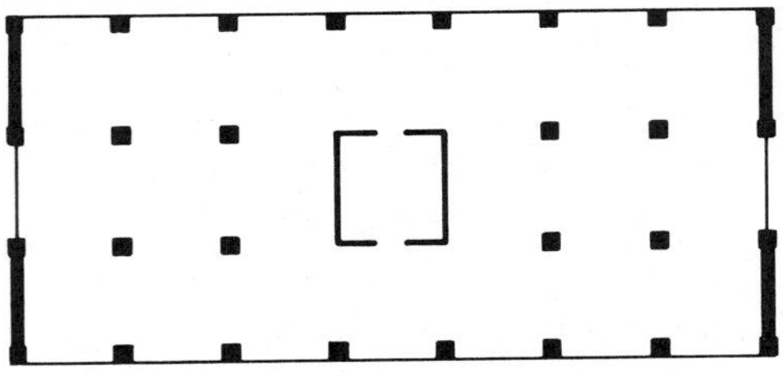

(b)

Fig. 12-63 (a) Framed tube with core and (b) frame–shear wall structure.

hull-core or framed-tube-with-core structure laid out on a nearly square plan, as shown schematically in Fig. 12-63(a), is well suited for high-rise buildings. For buildings of moderate height, a frame–shear wall arrangement such as is shown in Fig. 12-63(b) may be considered. For relatively low buildings, a structure consisting exclusively of conventional frame elements may be used. The introduction of shear walls will generally be dictated by the need to limit the drift or lateral deflection under wind as well as earthquake loading to safe and tolerable values.

(2) The useful range of deformation of the principal resisting elements in a structure should be of about the same order of magnitude, when practicable. This makes for an efficient design wherein all the resisting elements participate in carrying the induced forces in proportion to their relative stiffnesses over the entire range of deformation. This is illustrated in Fig. 12-64(a), for a structure with two resisting elements ① and ②. In Fig. 12-64(b), the resisting elements ① and ② not only possess different initial stiffnesses (as given by the slopes of the curves) but, more importantly, exhibit different ductilities or deformation capacities. In such a case, the following points have to be considered.

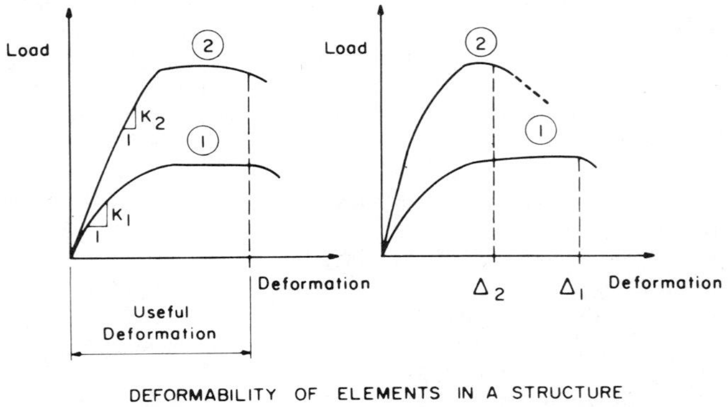

Fig. 12-64 Deformability of elements in a structure.

1. If the structure cannot survive with only element ① supporting it, care must be taken to ensure that the maximum probable deformation or lateral displacement under dynamic conditions does not exceed the deformation capacity of element ②, Δ_2.
2. If the expected maximum deformation could exceed Δ_2, then the resisting element ① must be so designed that it can support the additional load that may come upon it when element ② fails.

It should be borne in mind that the failure or partial failure of a major element in a structure will alter its dynamic characteristics and hence its response to a particular ground motion. A partial failure of an element that does not lead to general instability will usually result in a significant increase in the period of vibration (due mainly to a decrease in stiffness) as well as damping.

The case illustrated in Fig. 12-64(b) is typical of a combination frame–shear wall structure. In this connection, the use of slitted walls[12-88, 12-89] instead of the solid-panel shear wall may be considered. The "slitted wall" consists essentially of columns placed side by side with small gaps in between. The closeness of the columns and hence the considerable rotational restraint provided by the horizontal connecting members obviously makes the columns laterally stiffer than they would be if connected by fairly flexible beams. Figure 12-65 shows the load–deflection curve for an experimental slitted wall panel reported in Ref. 12-89, as compared to a solid-wall panel. Note the significantly greater deformation capacity, but reduced stiffness and strength, of the slitted-wall panel.

(3) The need to adequately tie together all the structural elements making up a building or a portion of it that is intended to act as a unit cannot be overemphasized. This applies to superstructure as well as foundation elements, particularly in buildings founded on relatively soft soil. Here, attention should be focused on the design of the segments of elements at and near the joints, since these are generally the regions that are most critically stressed.

Adequate connections should be provided across construction joints* that may be required between portions of a building or between the main portion of a structure and an appendage (e.g., stairway enclosure, carport, etc.).

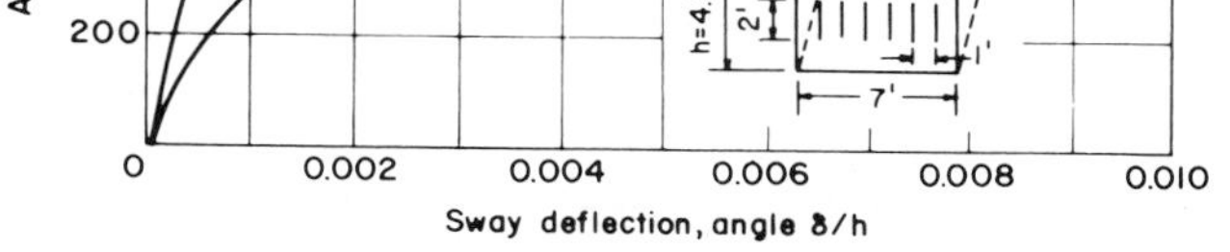

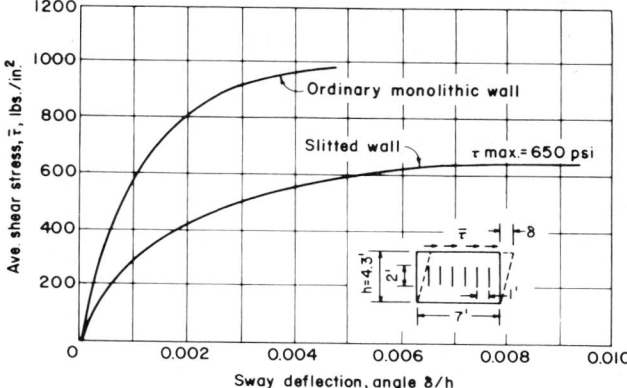

Fig. 12-65 Experimental load-deflection curves for solid and slitted walls.[12-89]

*A distinction should be made between a construction joint and an expansion joint. The former marks the termination of one placement of concrete and the beginning of the next placement, the stoppage usually being required by the impracticality of doing a large concreting job in one continuous operation. An expansion joint, on the other hand, is a plane of separation specifically provided to allow free movement between parts of a structure.

(4) The different portions of a building should either be tied together adequately or separated from each other by a sufficient distance to prevent their hammering against each other.

Expansion or similar joints used to separate parts of a building that differ considerably in height, plan size, shape, or orientation should be of sufficient size to allow for their swaying independently of or out of phase with each other without impact. Any required passageway, corridor or bridge linking structurally separated parts of a building should be so detailed as to allow their free, unhindered movement during an earthquake (Fig. 12-66). Multidirectional rollers or pads of low-friction material (e.g., Teflon) under one end of such connecting elements may be used.

In order to avoid hammering between adjoining buildings or separate portions of a building when vibrating out of phase with each other, a gap (perhaps filled with readily crushable materal) equal to from four to six times the sum of the calculated lateral deflections of the two structures under the design (code) seismic forces or the sum of the maximum deflections of the two structures as indicated by a dynamic analysis, would be desirable.

(5) The use of very stiff walls to fill the spaces between relatively more flexible frame elements should be considered carefully during the preliminary design stage.

If the infilling material is intended to act in combination with the enclosing frame, then it should be designed and constructed to ensure this composite action. Proper reinforcement and connection to the enclosing frame are essential. The analysis should likewise consider the increased stiffness of the infilled frames.

If the infill is made of fairly brittle material, such as glass or lightly reinforced brick masonry, and cannot be expected to contribute significantly to the lateral resistance of the frame, then it should be effectively isolated from the surrounding frame by gaps or readily crushable or yielding material (e.g., deep rubber gaskets for glass panels) to allow sufficient relative movement between the frame and such elements. Because the nonstructural elements in a building usually represent a major percentage of the total cost, huge losses can result from the failure of such elements. In addition, the failure of many of these brittle elements, when subjected to compressive forces, can sometimes be explosive in nature and hence potentially injurious to occupants and passersby.

The isolation of a masonry wall may be accomplished by providing a gap between the wall and the surrounding structural members. A channel attached to the latter may be used to provide lateral support to the wall while allowing free movement within the concealed gap (see Fig. 12-67). Alternatively, masonry walls may be run along a line slightly offset from the plane of the columns, with a gap provided between the top of the wall and the floor above. The *SEAOC Recommendations*,[12-73] for example, call for connections and joints between exterior precase non-load-bearing wall panels or similar elements and the supporting frame members to allow for relative movement between stories of not less than two times the drift caused by wind or seismic forces, or ¼ in., whichever is greater.

The effect of introducing low walls between columns, as shown in Fig. 12-68, should be noted. The reduction in height of the columns increases their stiffness with respect to bending in the plane of the wall. This will cause the columns to be subjected to greater horizontal shears than they would be expected to develop if the walls were absent. This is in addition to the effect which a decrease in the period of vibration of the structure—due to the increase in the stiffness of the columns—would bring. The reduction in height also reduces the lateral deformation capacity of the

(a)

(b)

Fig. 12-66 (a) The Olive View Medical Center in Los Angeles after the February 9, 1971 San Fernando earthquake; and (b) close-up of one of three toppled stair towers, Olive View Medical Center, LA.

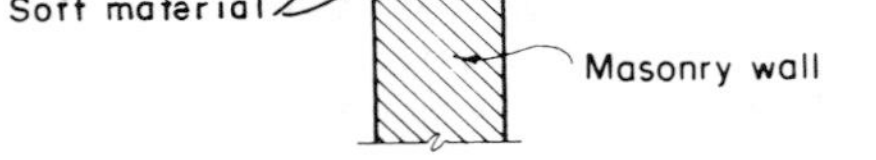

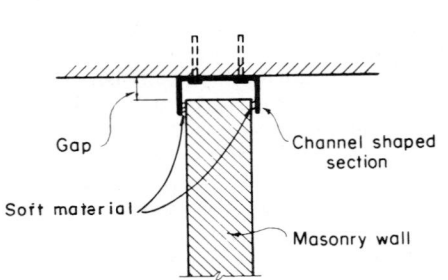

Fig. 12-67 Detail of gap between masonry wall and adjoining structure.

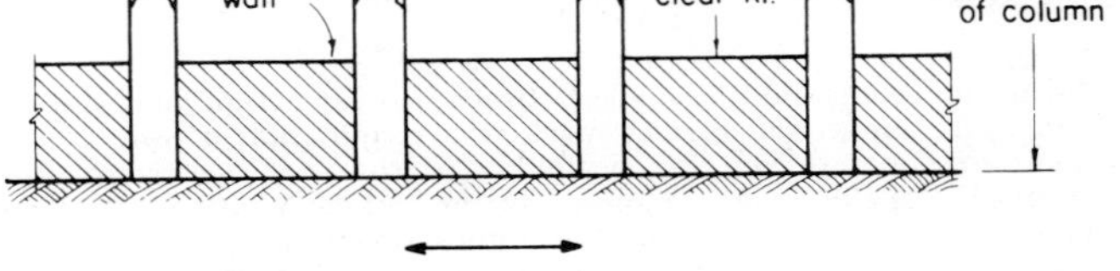

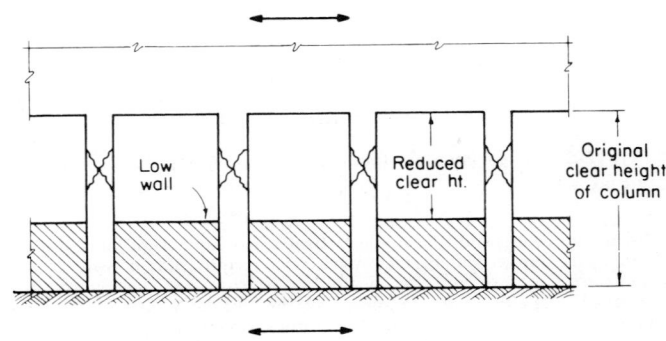

Fig. 12-68 Effect of introducing low walls between columns originally designed for longer clearance height.

columns in the plane of the wall. The use of such walls without allowing for their effects on the columns has been known to cause severe distress in portions of the columns above the wall.

(6) In order to avoid excessive internal forces in the structure due to differential vertical displacements resulting from foundation yielding or liquefaction, buildings, particularly multistory buildings, should be founded on a firm stratum.

12.6.5 Providing Against Non-ductile Failure

In buildings subjected to earthquake motion, both beams and columns are loaded in transverse shear in addition to bending, axial load (significant only in columns), and possibly torsion. Because of the abrupt nature of a failure due to shear (i.e., diagonal tension) or loss of anchorage of reinforcement, members should be proportioned so that their strength and ductility are controlled by flexure. Code provisions are intended to provide members with an inclined cracking load greater than their flexural capacity. Flexural failure is ensured to be ductile by limiting the maximum amount of longitudinal tensile reinforcement, so that failure is initiated by the yielding of such reinforcement. Codes also specify a minimum amount of flexural reinforcement, to prevent a sudden brittle-type flexural failure in lightly loaded members when tension cracking occurs.

Effective interaction between compression-resisting concrete and tension-carrying steel in reinforced concrete members requires the provision of adequate anchorage to develop the strength of the reinforcement. The abrupt character of bond or anchorage failure requires careful attention in design to avoid such failure before the flexural capacity of a member is reached.

The inclined cracking load of a beam without shear reinforcement is determined by the tensile strength of the concrete, the amount of longitudinal reinforcement, and the ratio of the applied shear to moment acting at the critical section. The presence of a compressive load increases the apparent tensile strength of the concrete and hence increases the inclined cracking load. Once inclined cracking starts, the load on a beam is transferred across a crack by a combination of shear on the compression zone, aggregate interlock along the crack, and dowel action of the longitudinal reinforcement. At this stage, the integrity of a beam and its capacity to carry more loads must be maintained through the use of shear reinforcement.

12.6.6 Confinement Reinforcement

Apart from the transverse reinforcement required by shear (diagonal tension), sufficient transverse confinement reinforcement needs to be provided at critical regions of a structure where inelastic action might occur (see Section 12.5). This is particularly true for structures that may experience moderate-to-strong earthquakes; the critical regions may be subjected to repeated, reversed cycles of deformation well into the inelastic range.

The use of sufficient confinement reinforcement in critical regions ensures their ductile performance by increasing the strength and strain capacity of the confined concrete in compression and by reducing the disruptive effect of large amplitudes of reversed loading on the concrete core. In beam–column connections, such confinement assures the maintenance of the vertical load-carrying capacity of the joint as well as adequate anchorage of any longitudinal beam steel embedded within the core. The increase in compressive strength of the confined concrete also helps to compensate for strength loss due to the spalled outer shell.

Transverse reinforcement that confines also acts as shear reinforcement, and vice versa. Shear reinforcement lying in planes transverse to the longitudinal axis of a member is the most effective, since it resists shear in either direction and can also serve as hoops to confine the concrete and as ties to restrain the longitudinal reinforcement. Such reinforcement may be in the form of circular or rectangular spirals or individual stirrups. Tests of reinforced concrete beams under cyclic loading have clearly demonstrated the considerable energy-absorbing capacity of members well provided with transverse reinforcement.

12.6.7 Effect of Rate of Loading

The relatively greater rate of loading associated with earthquakes, as compared to static loading, results in a slight increase in the strength of reinforced concrete members, due primarily to the increase in the yield strength of the reinforcement. The calculation of both strength and deformation capacity (ductility) of reinforced concrete members for earthquake loads on the basis of material properties obtained by static tests is thus reasonable and conservative.

12.6.8 Principal Earthquake Design Provisions of ANSI A58.1-1982 and Appendix A of ACI 318-83

1. Principal Design Steps. The principal steps involved in the design of cast-in-place reinforced concrete buildings, with particular reference to the application of the earthquake-resistant design provisions of nationally accepted model codes or standards, will be discussed below. The minimum design loads specified in ANSI A58.1-1982, "Minimum Design Loads for Buildings and Other Structures,"[12-12] and the design and detailing provisions contained in Appendix A of ACI 318-83, "Building Code Requirements for Reinforced Concrete,"[12-81] will be used as bases for the discussion. Except for minor differences, the earthquake design provisions of the "National Building Code of Canada-1980" (NBCC-80)[12-90, 12-91] are essentially the same as those found in ANSI A58.1.

In 1978, a comprehensive document on earthquake-resistant design, entitled "Tentative Provisions for the Development of Seismic Regulations for Buildings," (ATC 3-06),[12-9] was published by the Applied Technology Council. The effort of the Council was sponsered by the National Science Foundation and the National Bureau of Standards. ATC-3, as the document is commonly referred to, represents the collective effort of a large body of contributors including design professionals and researchers from universities, building industries, and government agencies, as well as building code officials. The document is intended to serve as the basis of a nationally recognized model code for earthquake-resistant design. Since its publication, ATC-3 has been under review. It is expected that in a few years the provisions of ATC-3, perhaps with some modifications, will find their way into the codes. The discussion that follows, although confined to the provisions of ANSI A58.1 and ACI Appendix A, makes occasional references to parallel provisions of ATC-3.

The load requirements in ANSI have, for the most part, been adopted from UBC-79 (the requirements are essentially the same as those appearing in UBC-82[12-14]), with a few elements taken from ATC-3. Many of the design provisions in UBC are based on the SEAOC Recommendations.[12-73]

The lateral forces specified in ANSI are intended as equivalent static loads. As its tittle indicates, ANSI A58.1 is primarily a load standard, defining minimum design loads for structures but otherwise leaving out material and member detailing requirements. ACI Appendix A, on the other hand, does not specify the manner in which earthquake loads are to be determined, but sets down the requirements by which to proportion and detail reinforced concrete members in structures expected to undergo inelastic deformations during earthquakes.

Design of a reinforced concrete building in compliance with the current U.S. seismic code provisions consists of the following principal steps:

(a) Determination of design "earthquake" forces:
  (i) Calculation of base shear corresponding to the computed or estimated fundamental period of vibration of a structure (a preliminary design of the structure is assumed here);
  (ii) Distribution of the base shear over the height of the building.
(b) Analysis of the structure under the (static) lateral forces calculated in step (a), as well as under gravity and wind loads, to obtain member design forces. The lateral load analysis can be carried out most conveniently by using a computer program for frame analysis.
(c) Designing members and joints for the most unfavorable combination of gravity and lateral loads. The emphais here is on the design and detailing of members and their connections to ensure their ductile behavior.

Changes in section dimensions of some members may be indicated during the design phase under step (c) above. However, unless the required changes in dimensions are such as to materially affect the overall distribution of forces in the structure, a reanalysis of the structure using the new member dimensions need not be undertaken. Uncertainties in the actual magnitude and distribution of the seismic forces as well as the effects of yielding in redistributing forces in the structure would make such refinement unwarranted. It is, however, most important to design the members and their connections to ensure their ductile behavior and thus allow the structure to sustain without collapse the severe distortions that may occur during a major earthquake. The code provisions intended to ensure adequate ductility in structural elements represent the major difference between the design requirements for conventional, non-earthquake-resistant structures and those located in regions of high earthquake risk.

The major details involved in the above steps are discussed below.

1.1 *Design base shear:* This respresents the total horizontal seismic force (service load level) that may be assumed acting parallel to the axis of the structure considered. The force corresponding to the other horizontal axis of the structure is assumed to act nonconcurrently, and may thus be considered separately. The total lateral force or base shear, V, is given by:

$$V = ZIKCSW \qquad (12\text{-}16)$$

where:

Z = A numerical coefficient whose value depends on the seismic zone in which the structure is located, as determined from the seismic risk maps of Fig. 12-5 (for the United States). ANSI-82 assigns a value of $1/8$ to Z for areas in Zone 0, $3/16$ for areas in Zone 1, $3/8$ for those in Zone 2, $3/4$ for locations in Zone 3, and 1.0 for those in Zone 4.

I = Occupancy importance factor, reflecting the greater conservatism in design desirable for buildings housing areas of public assembly or essential facilities that must remain operational after a major earthquake, such as hospitals, fire and police stations, public communications stations, power stations, and structures having critical national defense capabilities. ANSI assigns a value of 1.5 to I for essential facilities, 1.25 for buildings where the primary occupancy is for assembly of more than 300 persons, and 1.0 for all other buildings.

K = A factor depending on the structural system used. This factor is intended to account for differences in the available ductilities or energy-dissipation capacities of various structural systems. The values of K assigned to the different structural systems have been influenced to a large extent by observations of the performance of these systems in actual earthquakes. Values of K corresponding to different types of structures are listed in Table 12-3, and range from 0.67 for a ductile moment-resisting space frame designed to resist the entire lateral load to 2.5 for elevated tanks. Note that ANSI makes a distinction between two types of frame–shear wall systems. Where the space frame is designed primarily to carry the vertical loads while the shear wall is designed to resist all the lateral (seismic) loads, a value of 1.0 is assigned to K. In so-called dual systems, with $K = 0.8$, the space frame must be capable of resisting at least 25% of the prescribed seismic forces.

C = A coefficient related to the flexibility of a structure:

$$C = \frac{1}{15\sqrt{T}}, \text{ but not greater than } 0.12 \qquad (12\text{-}17a)$$

where T is the elastic undamped fundamental period of vibration of the building (in seconds) in the direction of motion considered. The fundamental period of vibration, T, to be used in eq. (12-17a) is to be established using the structural properties and deformation characteristics of the resisting elements in a properly substantiated analysis. A suggested expression for the fundamental period is:

$$T = 2\pi\sqrt{\left(\sum_{i=1}^{n} w_i\delta_i^2\right)\Big/\left(g\sum_{i=1}^{n} f_i\delta_i\right)} \qquad (12\text{-}17b)$$

where:

w_i = weight of portion of structure located at level i
f_i = lateral force at level i—distributed along the height of the structure according to eqs. (12-18) and (12-19)
δ_i = elastic lateral displacement at level i due to the forces f_i
g = acceleration due to gravity
n = uppermost level in main portion of structure

In the absence of a precise determination of T, such as by use of eq. (12-17b), the fundamental period of a building may be estimated using the following expressions:

(1) For shear walls or exterior concrete frames utilizing deep beams or wide piers or both:

$$T\ (\text{sec}) = \frac{0.05\, h_n}{\sqrt{D}} \qquad (12\text{-}17c)$$

where:

h_n = height (ft) of structure above the base
D = dimension (ft) of building in direction parallel to the applied forces

TABLE 12-3 Horizontal Force Factor, *K*, for Buildings and Other Structures* (From Ref. 12-12)

Arrangement of Lateral Force-Resisting Elements	*K*
Bearing wall system: A structural system with bearing walls providing support for all, or major portions of, the vertical loads. Seismic force resistance is provided utilizing:	
Unreinforced masonry walls	–†
Reinforced concrete or reinforced masonry walls or braced frames	1.33
One-, two-, or three-story light wood or metal frame–wall systems	1.00
Building frame system: A structural system with an essentially complete space frame providing support for vertical loads. Seismic force resistance is provided by shear walls or braced frames.[(a)]	1.00
Moment-resisting frame system: A structural system with an essentially complete space frame providing support for vertical loads. Seismic force resistance is provided by a moment-resisting frame system in conformance with:	
Requirements for ordinary concrete frames	–†
AISC specifications for ordinary steel frames	1.00
Requirements of Appendix A to ACI 318-83[12-81] or Part II of the AISC specifications.[12-92] for special frames	0.67
Dual system: A structural system with an essentially complete space frame providing support for vertical loads. Seismic force resistance is provided by a combination of a special moment-resisting frame system capable of resisting at least 25% of the prescribed seismic forces and shear walls or braced frames satisfying the requirements of Appendix A to ACI 318-83[12-81] or Part II or the AISC specification.[12-92]	0.80
Elevated tanks: Tanks plus full contents, where tanks are supported on four or more cross-braced legs and are not supported on a building.	2.50‡
Structures other than buildings.	2.00

**Where specified wind loads produce higher stresses, such wind load shall be used instead of the loads resulting from earthquake forces.*
†*This system is not considered appropriate for buildings in seismic Zones 2, 3, and 4. Buildings in Zone 0 and those in Zone 1 with I less than 1.5 may make use of this system under certain conditions.*
Zone 1 with I less than 1.5 may make use of this system under certain conditions.
‡*The minimum value of KC shall be 0.12 and the maximum value of KCS need not exceed 0.29 or 0.23 for Soil Profile 3 in Seismic Zones 3 and 4, respectively. The tower shall be designed for accidental torsion as specified in Section 9.5.5g Ref. 12-12. Elevated tanks that are supported by buildings or do not conform to the type or arrangement of supporting elements as described above shall be design using $C_p = 0.3$ (C_p is horizontal force factor for elements of structures and nonstructural components.*
(a)*Designed in accordance with provisions in the main body of ACI 318-83[12-81] (for concrete) or the AISC Specifications (for steel).[12-92]*

(2) For isolated shear walls not interconnected by frames or braced frames:

$$T = \frac{0.05\, h_n}{\sqrt{D_s}} \qquad (12\text{-}17d)$$

where D_s = longest dimension (ft) of a shear wall or braced frame in direction parallel to the applied forces

(3) In buildings in which the lateral-force-resisting system consists of moment-resisting space frames capable of resisting 100% of the required lateral forces and the frames are not enclosed or adjoined by more rigid elements tending to prevent them from resisting the lateral forces:

$$T = C_T h_n^{3/4} \qquad (12\text{-}17e)$$

where $C_T = 0.025$ for concrete frames. (C_T = 0.035 for steel frames.)

Equations (12-17c) through (12-17e) are intended to provide realistic conservative estimates of the fundamental period of a building. They represent the lower bound of periods recorded for a number of buildings for which data were obtained during the 1971 San Fernando, California earthquake and are close to the ambient periods recorded for these same buildings. Where the more exact eq. (12-17b) is used, the calculated period may not exceed the value obtained by the applicable expression from eqs. (12-17c) through (12-17e) by more than 20%.

The variations of coefficient C with $h_n/\sqrt{D}$ in eq. (12-17c) and with h_n in eq. (12-17e) are shown in Fig. 12-69.

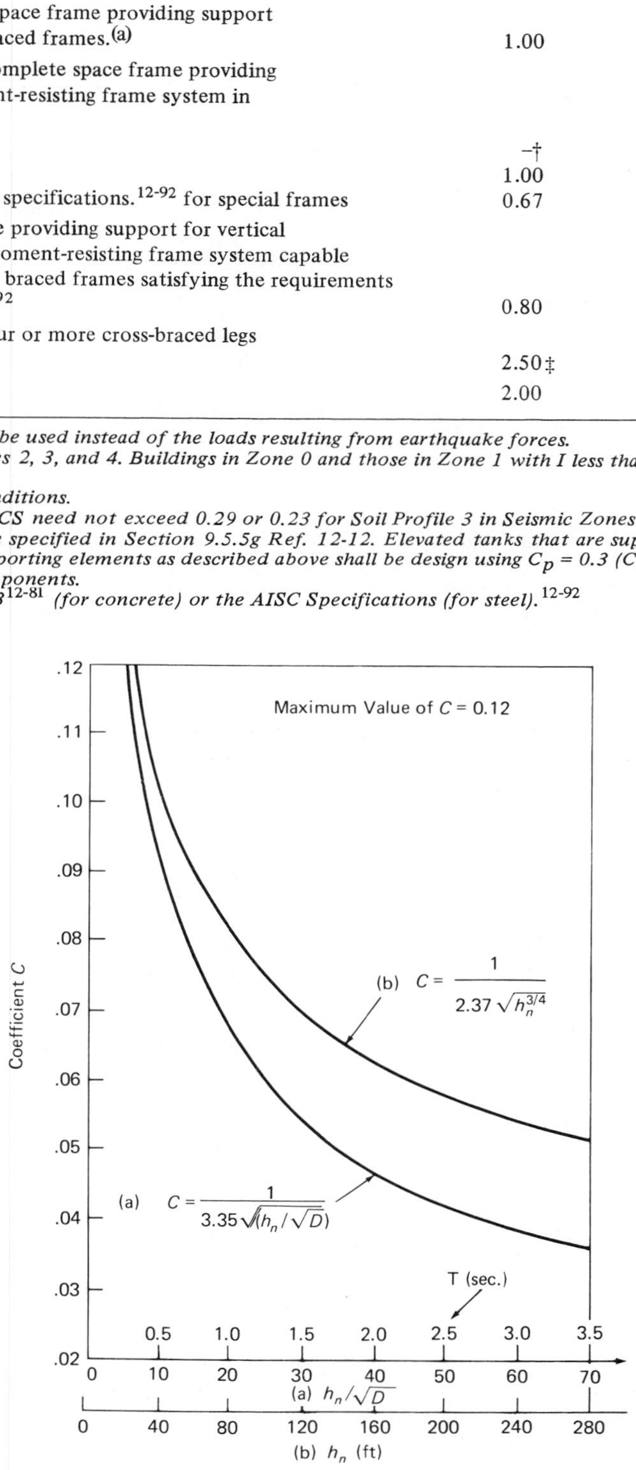

Fig. 12-69 Variation of seismic coefficient C with h_n and $h_n/\sqrt{D}$.

S = Soil factor, designed to account for possible magnification of dynamic structural response due to adverse soil conditions, such as may occur when flexible buildings are founded on a deep layer of relatively soft material. The manner in which the effect of soil conditions on the base shear is determined in ANSI is based on the ATC-3 recommendations, rather than the UBC. It is based mainly on studies of the shapes of response spectra for motions actually recorded at the surface of different types of soil.

ANSI recognizes three different types of soil profile: S_1, S_2, and S_3. The value of the soil factor or soil profile coefficient, S, is 1.0 for soil profile type S_1, 1.2 for soil profile S_2, and 1.5 for soil profile type S_3. The three types of soil profile are defined as follows:

(1) Soil Profile Type S_1 is a profile that includes either:
   (a) Rock of any characteristic, either shalelike or crystalline in nature. Such material may be characterized by a shear wave velocity greater than 2500 ft/sec.
   (b) Stiff soil conditions where the soil depth is less than 200 ft and the soil types overlying rock are stable deposits of sands, gravels, or stiff clays.

(2) Soil Profile Type S_2 is a profile with deep cohensionless deposits or stiff clay conditions, including sites where the soil depth exceeds 200 ft and the soil types overlying rock are stable deposits of sands, gravels, or stiff clays.

(3) Soil Profile Type S_3 is a profile with soft to medium-stiff clays and sands, characterized by 30 ft or more of soft to medium-stiff clays without intervening layers of sand or other cohesionless soils.

In Zones 3 and 4, the base shear tends to be less for stiff buildings founded on Soil Profile 3 as compared to the same buildings on Soil Profiles 1 and 2. This is due to the reduction in peak acceleration that occurs in heavily shaken soft soil deposits where damping associated with large strains can be significant. Thus for some buildings the product CS will be largest for Soil Profile 3, while for other buildings CS will be largest for Soil Profile 2. Where there is doubt concerning the proper choice of soil profile, the larger value of CS should be used.

W = The total dead load, except that in storage and warehouse occupancies, W shall include the total dead load plus 25% of the floor live load. Also, where the ground snow load is greater than 30 lb/sq ft, the full snow load shall be included in W, unless a reduction (maximum, 25%) is warranted due to short snow load duration.

1.2 *Distribution of base shear:* for structures having regular shapes or framing systems, the total lateral force or base shear, V, is to be divided into two parts: a concentrated load, F_t, applied at the top of the structure (to simulate the whiplash effect due to the higher modes of vibration) and the balance, $(V - F_t)$, to be distributed over the entire height of the building, generally as concentrated loads at the floor levels. The latter is to be distributed in a triangular manner (a distribution corresponding essentially to the fundamental mode of vibration), increasing from zero at the base to a maximum at the top.

The top load, F_t, is given by:

$$F_t = \begin{cases} 0.07\,TV \leqslant 0.25\,V & \text{when } T > 0.7 \text{ sec} \\ 0 & \text{when } T \leqslant 0.7 \text{ sec} \end{cases} \tag{12-18}$$

where T is the fundamental period, and V is the base shear.

The magnitude of the distributed forces, F_x, making up the balance of the total lateral force, that is, $(V - F_t)$, is given by:

$$F_x = \frac{(V - F_t)\, w_x h_x}{\sum_{i=1}^{n} w_i h_i} \tag{12-19}$$

where:

F_x = lateral force at level x above the base

w_x, w_i = that portion of the total weight, W, located at or assigned to level x or i, respectively

h_x, h_i = height above the base to level x or i respectively

Level n = the uppermost level in the main portion of the structure

At each level x, the force F_x is to be applied over the area of the building in accordance with the mass distribution at that level. At the top floor, the total horizontal load will consist of the sum of F_t and the distributed load F_n. Figure 12-70 shows a typical distribution of forces for a multistory structure having a uniformly distributed dead mass over its height.

As an alternative to eqs. (12-16) and (12-17), ANSI-82 allows the use of design lateral forces established on the basis of properly substantiated data, such as those obtained from dynamic analysis considering an appropriate model of the structure. The analysis, which would have to be approved by the local authority, may take the form of modal superposition using a smoothed response spectrum or response history analysis using an ensemble of ground acceleration time histories as input. Several restrictions relating to the calculated base shear and drift apply, one of these being that the computed base shear shall not be less than 90% of that obtained using eqs. (12-16) and (12-17).

1.3 *Setbacks:* When the plan dimension of a setback structure in each direction is at least 75% of the corresponding plan dimension of the supporting structure, the codes allow the determination of the design seismic forces by considering the building as uniform.

Setbacks that have dimensions less than 75% of the corresponding dimensions of the base structure are to be

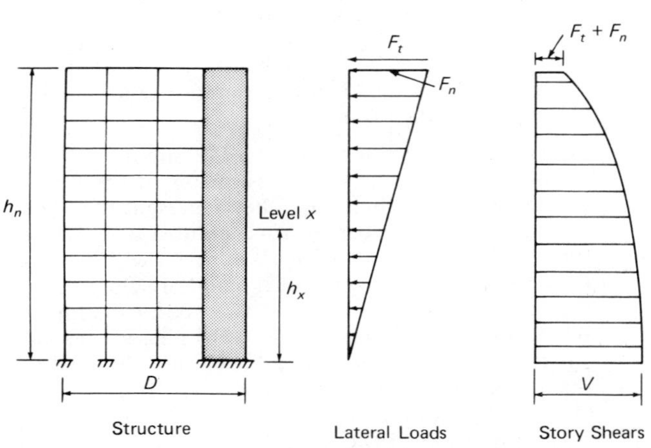

Fig. 12-70 Typical distribution of code-specified static lateral forces and story shears in a building with uniform mass-distribution.

designed as separate structures, using for the base shear the larger value obtained by considering the tower or setback either as a separate building over its own height or as part of the overall structure.

Appendix C to the *SEAOC Recommendations*[12-73] gives more details concerning the treatment of buildings with setbacks.

1.4 *Requirements relating to certain structural systems:* ANSI-82[12-12] does not include any limitation on the height of buildings using various structural systems. UBC-82,[12-14] however, requires that all buildings exceeding 160 ft in height and located in seismic Zones 3 and 4* have ductile moment-resisting space frames capable of resisting not less than 25% of the specified lateral forces on the entire structure. Ductile moment-resisting space frames are space frames satisfying the special ductility requirements for earthquake resistance given in the code or in ACI Appendix A. Box-type or shear wall buildings, consisting of assemblages of interconnected walls without the 25%-lateral-load frames, may not have a height exceeding 160 ft in Zones 3 and 4. No height limit is imposed on frame buildings or on frame–shear wall buildings that have a moment-resisting space frame satisfying the above 25%-lateral-load-resisting-capacity requirement. Buildings in seismic Zones 1 and 2 may have concrete shear walls or braced frames designed for a K-value of 1.0 or 1.33 instead of the 25%-lateral-load ductile moment-resisting space frame.

ANSI-82 allows the use of the basic provisions for conventional concrete construction found in the main body of the *ACI Code* for proportioning building frames designed for a factor $K = 1.0$. However, such frames are to be provided with shear walls or vertical bracing trusses to resist the lateral earthquake forces.

Buildings designed using a K-value of 0.67 must have special (i.e., ductile) moment-resisting space frames conforming to the provisions of ACI Appendix A. Dual systems, consisting of a vertical-load-carrying space frame and a lateral-force-resisting system composed of a ductile moment-resisting frame and shear wall or braced frame, shall have a moment-resisting frame capable of resisting at least 25% of the prescribed lateral forces.

Members of braced frames (i.e., truss systems or their equivalent which resist lateral forces primarily by axial forces in their members) located in seismic Zones 3 and 4 are to be designed for 1.25 times the lateral forces determined from eq. (12-16). In addition, the lateral reinforcement in such members shall conform to the requirements of ACI Appendix A. Similar requirements apply to braced frames in seismic Zone 2 having an importance factor, I, greater than 1.0.

Special ductility requirements also apply to structural elements below the bases of buildings designed for $K = 0.67$ and $K = 0.80$ which trasmit forces resulting from the prescribed lateral loads to the foundation.

1.5 *Drift:* In multistory buildings, the deflection index, that is, the ratio of the drift or lateral displacement at the top of a building to its total height or, alternatively, the maximum ratio of the relative interstory displacement to the story height (the latter will generally be greater than the former) has been used as a convenient measure for determining the appropriate lateral stiffness of a structure. In design for wind loading, the considerations involved in choosing the maximum deflection index relate to (a) the stability of the individual columns as well as the structure as a whole, (b) the integrity of nonstructural partitions and glazing, and (c) the comfort of occupants of the building. Under strong earthquake motions, the major consideration concerning drift has been the stability of the structure under the action of gravity loads, when undergoing significant lateral displacements.

In recent years, an awareness of the need to keep essential facilities operational after a major earthquake has pointed to the desirability of including damage control as a primary design criterion, in addition to life safety. The need to minimize damage to the structure as well as the nonstructural components of a building during a major earthquake dictates a minimum lateral stiffness to limit interstory distortion. The maximum permissible interstory displacement will depend on the maximum tolerable deformation of the most critical component in the structure. This may be governed by considerations of damage affecting movement of people within a building or mechanical equipment malfunction.

In multistory buildings, satisfying drift limitations may require the provision of additional lateral stiffness, beyond that required for gravity loads and wind loading. A frame may be stiffened laterally by increasing the frame member sizes (particularly the beam sizes in the lower stories) or more efficiently by introducing structural walls or shear walls judiciously positioned in plan.

ANSI-82, like UBC-82, sets a maximum limit of 1/200 on the ratio of story drift to story height, unless it can be demonstrated that a greater value can be tolerated. The drift to be used in calculating the ratio is that obtained by application of the specified lateral forces, multiplied by $1.0/K$, K being the factor used in the expression for base shear and listed in Table 12-3. The ratio $1.0/K$ may not be less than unity.

2. *Analysis for Member Forces.* With the lateral loads at the floor levels determined, the next step is an analysis of the structure for the corresponding member forces. Before undertaking an analysis of the structure under the design seismic loads, however, it is advisable to determine the corresponding design wind loads and compare them with the former. In many instances, it becomes readily apparent which of the two loading cases is the more critical, so that the lateral load analysis of the structure need be carried out only for the critical loading. Where the critical loading condition is not immediately apparent because of the different distribution of loads that are of about the same order of magnitude, analyses of the structure considering both loading cases may be desirable.

For relatively low buildings, an approximate analysis, such as by the portal method, may be carried out. When the portal method is used, however, allowance must be made for the fact that in regions of discontinuity, such as the base and the top of a frame, the points of contraflexure in the columns (ordinarily assumed at midheight) generally do not occur at midheight. The results of the portal method for these regions may be improved by adjustments in the assumed locations of the points of contraflexure in the columns. For this purpose, the data found in Ref. 12-93 will be found useful.

In structures consisting of frames and shear walls or of linked shear walls with significantly different variations in stiffness along the height, the distribution of the applied lateral load in proportion to the relative stiffnesses of the resisting elements at each story level can lead to erroneous results. For these cases, the interaction between the connected elements having different deformation patterns under lateral loads must be considered.

Figure 12-71 shows the results of a static elastic analysis of a typical 20-story frame–shear wall structure under lateral loads. The curves indicate the distribution of the hori-

*The UBC-82 Seismic Risk Map indicating earthquake zones differs slightly from the one used in ANSI-82 (Fig. 12-5).

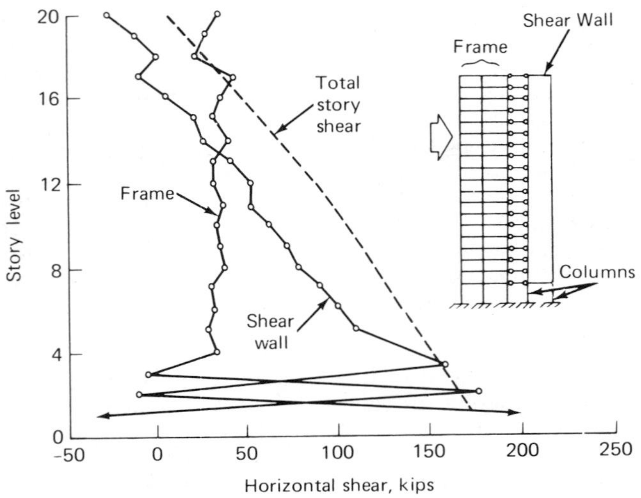

Fig. 12-71 Distribution of horizontal story shears in a typical frame–shear wall structure.[12-94]

zontal story shears between the frame and the shear wall along the height of the structure. It is worth noting that the total shear carried by the frame in the top stories can exceed the applied story shear at these levels. It is clear that distributing the applied shear at each level in proportion to the relative stiffnesses of the resisting elements can lead to erroneous and unconservative results in this case. Assuming that the shear wall carries all the lateral load can also result in under-designed frame members in the upper stories.

Figure 12-71 also shows the sizable increase in horizontal shears that occurs in the lower stories of a system in which the shear wall is supported at the base by columns. The shears shown for the lowest story of the shear wall are those corresponding to the two supporting columns. The character of the shear distribution curves shown for the lower stories is typical of the force concentrations that occur in regions of major geometric or stiffness discontinuities in any structural system. While yielding of the columns will tend to relieve the high stresses accompanying such large shears, the fact that these forces can exceed the total base shear is worth noting. In structures subjected to earthquakes, discontinued shear walls can lead to early yielding in the columns supporting the shear walls, with correspondingly large ductility requirements. ACI Appendix A requires that columns supporting discontinued stiff members or walls be provided with transverse reinforcement spaced not more than 4 in. apart throughout their entire height if the factored axial compressive force due to earthquake effects exceeds $(A_g f'_c/10)$.

For tall structures, an analysis that takes account not only of the axial deformations in columns but also of the finite width of members would be desirable. For this purpose, the use of a computer program for frame analysis would be most convenient.

2.1 *Overtruning moment:* The lateral loads determined from a distribution of the design base shear are intended to represent, to some measure, the envelope of maximum forces at the different story levels. Earlier versions of the UBC allowed a reduction in the overturning moments calculated on the basis of thse forces, because these maximum forces generally do not occur simultaneously. Such a reduction would reflect the effect of the higher modes of vibration on the response of a structure, since in the higher modes, the lateral displacements, and hence the associated forces, change direction at some nodes.

Since 1971, the *SEAOC Recommendations*[12-73] and the *Uniform Building Code*[12-14] no longer allow such a reduction of the overturning moment calculated on the basis of the distributed base shear. Thus, the response of a structure is essentially assumed to be described entirely by the fundamental mode response.

ANSI-82 allows some reduction in the overturning moment calculated from the distributed base shear for tall buildings. For these structures, the overturing moment calculated from the distributed base shear tends to overestimate the actual moments. Thus, the overturning moment at story x, given by:

$$M_x = F_t(h_n - h_x) + \sum_{i=x}^{n} F_i(h_i - h_x) \qquad (12\text{-}20)$$

where the terms are as defined earlier, may be reduced by a factor k, having a value depending on the location of the story, as follows:

$$k = \begin{cases} 1.0 \text{ for the top ten stories} \\ \text{a value linearly varying from 1.0 for the 10th story} \\ \text{from the top to 0.8 for the 20th story below the top} \\ 0.8 \text{ for the 20th story from the top and those below it} \end{cases}$$

Any uplift caused by the overturning moment at the base of a structure should be combined with the minimum expected dead load to check the adequacy of both the vertical-load-carrying elements and the foundation. It is pointed out that the shear capacity of a reinforced concrete member is adversely affected by the presence of axial tension.

The preceding discussion has considered the determination of design forces on the elements forming a structure under the action of loads directed parallel to one of the principal plan axes of the structure. The analysis of the structure under loads acting in the perpendicular direction can be carried out similarly. For this purpose, the structure may be assumed to consist of two sets of mutually perpendicular plane frames, each set of frames being analyzed separately for the corresponding lateral loads. The codes require that a structure be designed for lateral loads, acting nonconcurrently, in the direction of each of the main axes of the structure. Members of a structure common to both directions (i.e., columns) may be proportioned for the lateral loads acting along one axis (the more critical direction) and then checked for adequacy with respect to the lateral loads acting in the perpendicular direction. Note, however, that columns, and particularly corner columns, will generally be subjected to biaxial moments.

2.2 *Horizontal torsional moments:* When the vertical lateral-load-resisting elements are connected by a floor system that is very stiff in its own plane, such as a monolithically cast reinforced concrete slab, the elements located near the plan periphery can be subjected to significant shear due to torsional response associated with noncoincidence of the center of mass and the center of rigidity. This is in addition to the direct shears parallel to the principal plan axes.

Where known eccentricities exist in some stories of a multistory building, the entire structure will tend to oscillate in torsion, producing torsional moments even in stories that have nominally zero eccentricity between the centers of mass and rigidity. To account for this effect, ATC-3 recommends that the design torsional moment for any story be not less than either (a) one-half the product of the story shear and the maximum computed eccentricity in the stories below the story considered, or (b) one-half the maximum computed torsional moment in the stories above.[12-14]

In addition to torsion in a story due to any known eccentricity of the center of mass with respect to the center of rigidity, and even in nominally symmetrical buildings, codes require that a torsional moment corresponding to a minimum eccentricity be considered in the design of vertical lateral-load-resisting elements. This "accidental eccentricity" is intended to account for such factors as the rotational component of ground motion, unaccounted for effects due to unfavorable distribution of dead and live load masses, and differences between computed and actual values of stiffnesses and yield strengths. When the floor system or diaphragm is very flexible in its plane relative to the attached vertical elements, horizontal torsional moments will be insignificant.

The UBC requires diaphragm-connected shear-resisting elements to be designed for an accidental horizontal torsion corresponding to an eccentricity of 5% of the maximum building dimension between the centers of mass and rigidity at any level. ANSI requires only that the 5% eccentricity be based on the dimension of the building perpendicular to the direction of the applied forces.

The total torsional moment in a story will thus consist of any explicitly calculated torsion due to a known eccentricity and that due to accidental torsion.

The additional shears due to horizontal torsional moments may be calculated approximately by assuming the vertical elements at each story to be fixed at their ends to parallel rigid plates. The torsional shear force acting on each element may then be taken as proportional to its lateral stiffness and its distance from the center of rigidity of the story considered.

The location of the center of rigidity or center of rotation, c_r, under the above assumption, is determined by a method similar to that used to obtain the center of rotation of a rivet group, except that one uses the lateral stiffness* of the vertical elements in place of the cross-sectional area of the rivets. This recognizes the flexural action of the vertical elements as the principal mode of resistance to horizontal torsion. Thus, referring to Fig. 12-72, if k_x and k_y are the lateral stiffnesses of a particular element along the x- and y-axis, respectively, then the coordinates of the center of rotation, x_r and y_r, with respect to an arbitrary origin, such as point O in the figure, is given by:

$$x_r = \frac{\Sigma k_y x}{\Sigma k_y} \text{ and } y_r = \frac{\Sigma k_x y}{\Sigma k_x} \qquad (12\text{-}21)$$

the summation being taken over all the vertical elements in the story.

The shear due to torsion along each coordinate axis resisted by a particular element will then be proportional to the lateral stiffness of the element, relative to the total rotational stiffness of the story, and its distance from the center of rigidity. The total rotational stiffness of the story, J_r, about the center of rigidity, is given by:

$$J_r = \Sigma\ (k_x \bar{y}^2 + k_y \bar{x}^2) \qquad (12\text{-}22)$$

where $\bar{x}$ and $\bar{y}$ are the coordinates of the centroid of a particular element (in plan) referred to the center of rigidity. Note the similarity between J_r and the polar moment of inertia of a rivet group about its centroid. The shears along the two axes due to a horizontal torsional moment $M_r = He$ (H being the seismic story shear at the level considered), for a particular element with stiffnesses k_x and k_y are then obtained from:

$$V_x = \frac{M_r \bar{y}}{J_r} k_x \text{ and } V_y = \frac{M_r \bar{x}}{J_r} k_y \qquad (12\text{-}23)$$

2.3 *Lateral forces on elements, nonstructural components, and diaphragms:* Floor and roof slabs that are relatively stiff in their own planes serve to distribute horizontal forces among the various vertical lateral-load-resisting elements in a building. ANSI requires that such horizontal diaphragms, which have to be adequately tied to the connected elements, be designed to resist forces, F_{px}, given by:

$$F_{px} = \frac{\sum_{i=x}^{n} F_i}{\sum_{i=x}^{n} w_i} w_{px} \leqslant 0.30\, ZI\, w_{px} \qquad (12\text{-}24)$$

where:

F_i = applied lateral force at level i
w_i = portion of the total dead load, W, assigned to level i
w_{px} = weight of floor or roof diaphram at level x
Z and I are the zoning factor and occupancy importance factor, respectively, as defined earlier in connection with the calculation of base shear.

Sometimes a diaphragm is required to transfer substantial lateral forces through in-plane action from connected vertical elements. This can occur when the vertical lateral-load-resisting elements above a diaphragm are offset with respect to those below it, or when significant changes occur in the stiffness distribution of the vertical elements above the diaphragm relative to that of the elements below it. In these cases, ANSI requires that the diaphragm be designed for the appropriate force associated with the load transfer, which shall not be less than $0.14ZI\ w_{px}$, in addition to the force F_{px}, defined above.

Although the failure of ornamentation, veneers, parapets, and similar appendages seldom affects the structural integrity of a building, it represents a serious menace to the safety of the occupants as well as passersby. Numerous cases of failure of building elements such as ceilings and parapets have clearly indicated the need to have such elements tied securely to the rest of the structure.

To ensure continued operation of essential services, mechanical and electrical equipment installed in buildings should be designed to withstand earthquake-induced forces and displacements. For example, elevator cars and their counterweights have been known to jump off their guide rails as a result of earthquake motions. Likewise, the mountings and supports of motors and other machinery should be capable of resisting the forces transmitted through these components.

ANSI specifies design forces for parts or portions of structures, nonstructural components, and their anchorage to

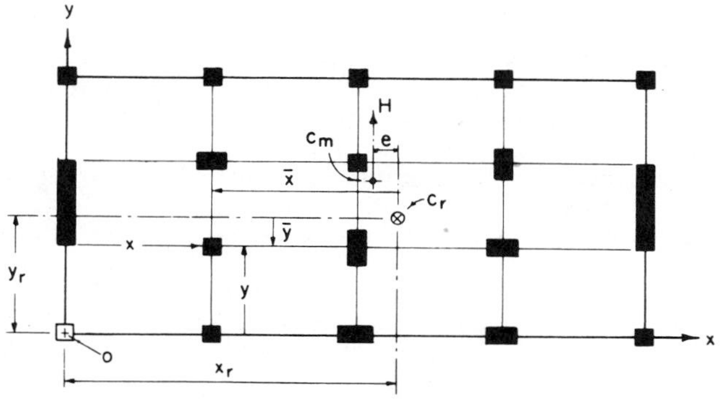

Fig. 12-72 Calculating shears due to horizontal torsional moments.

*That is, the force required to displace one end of a vertical element a unit horizontal distance relative to the other end.

TABLE 12-4 Horizontal Force Factor, C_p, for Elements of Structures and Nonstructural Components (from Ref. 12-12)

Part or Portion of Building	*Direction of Horizontal Force*	C_p
Exterior bearing and nonbearing walls; interior bearing walls and partitions; interior nonbearing walls and partitions; masonry or concrete fences over 6 ft high	Normal to flat surface	0.3*
Cantilever elements:		
Parapets	Normal to flat surface	0.8
Chimneys or stacks	Any direction	0.8
Exterior and interior ornamentations and appendages	Any direction	0.8
When connected to part of, or housed within, a building:		
Penthouses, anchorage and supports for chimneys, and stacks and tanks, including contents	Any direction	0.3†, ‡
Storage racks with upper storage level at more than 8 ft in height, plus contents	Any direction	0.3†, ‡
All equipment or machinery	Any direction	0.3†, ‡
Suspended ceiling framing systems (applies to seismic Zones 2, 3, and 4 only)	Any direction	0.3 §
Connections for prefabricated structural elements other than walls, with force applied at center of gravity of assembly	Any direction	0.3**

**C_p* for elements laterally self-supported only at the ground level may be two-thirds of value shown.*
†W_p for storage racks shall be the weight of the racks plus contents. The value of C_p for racks over two storage support levels in height shall be 0.24 for the levels below the top two levels.
Where a number of storage rack units are interconnected so that there are a minimum of four vertical elements in each direction on each column line designed to resist horizontal forces, the design coefficients may be as for a building with K-values from Table 12-3; CS = 0.2 for use in the formula V = ZIKCSW; and W equal to the total dead load plus 50% of the rack-rated capacity.
‡For flexible and flexibly mounted equipment and machinery, the appropriate values of C_p shall be determined with consideration given to both the dynamic properties of the equipment and machinery and to the building or structure in which it is placed but shall be not less than the listed values. The design of the anchorage of the equipment and machinery is an integral part of the design and specification of such equipment and machinery.
For essential facilities and life saftey systems, the design and detailing of equipment that must remain in place and be functional following a major earthquake shall consider drifts in accordance with Section 9.13 of ANSI A58.1-1982.[12-12]
§Ceiling weight shall include all light fixtures and other equipment that is laterally supported by the ceiling. For purposes of determing the lateral force, a ceiling weight of not less than 4 lb/ft² shall be used.
***The force shall be resisted by positive anchorage and not by friction.*
NOTE: Seismic restraints may be omitted from the following installations:
(a) Gas piping less than 1-in. inside diameter
(b) Piping in boiler and mechanical rooms less than 1 1/4-in. inside diameter
(c) All other piping less than 2 1/2-in. inside diameter
(d) All electrical conduit less than 2 1/2-in. inside diameter
(e) All rectangular air-handling ducts less than 6 ft² in cross-sectional area
(f) All round air-handling ducts less than 28 in. in diameter
(g) All piping suspended by individual hangers 12 in. or less in length from the top of the pipe to the bottom of the support for the hanger
(h) All ducts suspended by hangers 12 in. or less in length from the top of the duct to the bottom of the support for the hanger

the main structural system in terms of a coefficient, C_p, and the weight of the part considered, W_p, as follows:

$$F_p = ZIC_pW_p \qquad (12\text{-}25)$$

In the above equation, F_p is the design earthquake force, and Z and I are as defined earlier. An exception to the values of I specified earlier is made for anchorage of machinery and equipment required for life safety systems, for which I is to be taken equal to 1.5 for all buildings.

Values of the coefficient C_p are given in Table 12-4. Note that the design force on walls and cantilevered parapets are to be applied normal to the plane of such members, while for other parts or appendages the specified force is to be applied in any direction. In all cases, the intent is to design the part and its anchorage for the most unfavorable loading condition.

3. Load Factors and Loading Combinations Used as Bases for Design; Strength Reduction Factors. Codes generally require that the strength or load-resisting capacity of a structure and its component elements be at least equal to or greater than the forces due to any of a number of loading combinations that may reasonably be expected to act on it during its life. Concrete structures are commonly designed using the ultimate* strength method. In this approach, structures are proportioned so that their (ultimate) capacity is equal to or greater than the required (ultimate) strength. The required strength is determined as caused by a combination of factored loads, that is, specified service loads mutliplied by appropriate "load factors." The capacity of a structure, on the other hand, is obtained by applying a "strength reduction factor," ϕ, to the nominal resistance of the element determined from basic mechanics.

Load factors are intended to account for the variability/uncertainty in the magnitude of the specified service loads. Different load factors are used for different types of loads (or the forces that they produce) before combining these loads to obtain a design ultimate load. To allow for the lesser likelihood of certain types of loads occurring simultaneously, reduced load factors are specified for some loads when considered in combination with other loads.

ANSI-82[12-12] requires that structures, their components, and foundations be designed so that their strength exceeds

*Since ACI 318-71, the term "ultimate" has been dropped so that what used to be referred to as "ultimate strength design" is now simply called "strength design."

the effects of the relevant combination of factored loads from the list below:

$$U = \begin{Bmatrix} 1.4D \\ 1.2D + 1.6L + 0.5(L_r \text{ or } S \text{ or } R) \\ 1.2D + 1.6(L_r \text{ or } S \text{ or } R) + (0.5L \text{ or } 0.8W) \\ 1.2D + 1.3W + 0.5L + 0.5(L_r \text{ or } S \text{ or } R) \\ 1.2D + 1.5E + (0.5L \text{ or } 0.25S) \\ 0.9D - (1.3W \text{ or } 1.5E) \end{Bmatrix} \quad (12\text{-}26)$$

For garages, places of public assembly, and all areas where the live load is greater than 100 psf, the load factor on L in the third, fourth, and fifth combinations in eq. (12-26) is to be taken equal to 1.0. Also, where the effects of F, H, P, or T, as defined below, are significant, these are to be considered in design as the following factored loads: $1.3F$, $1.6H$, $1.2P$, and $1.2T$.

In the above equations:

U = required strength to resist the factored loads
D = dead loads
L = live loads, including impact effects, where appropriate
L_r = roof live loads
S = snow load
R = rain load
W = wind load
E = earthquake load
F = load due to fluids with well-defined pressures and maximum heights
H = load due to soil pressure
P = load due to ponding
T = load due to effects of temperature, shrinkage, moisture changes, creep, differential settlement, or combinations thereof

ACI 318-83[12-81] specifies slightly different load factors for some load combinations, as follows:

$$U = \begin{Bmatrix} 1.4D + 1.7L \\ 0.75[1.4D + 1.7L \pm (1.7W \text{ or } 1.87E)] \\ 0.9D + (1.3W \text{ or } 1.43E) \\ 1.4D + 1.7L + (1.7H \text{ or } 1.4F) \\ 0.9D - (1.7H \text{ or } 1.4F) \\ 0.75(1.4D + 1.7L + 1.4T) \geqslant 1.4(D + T) \end{Bmatrix} \quad (12\text{-}27)$$

where the notation are as defined earlier.

As mentioned, the capacity of a structural element is calculated by applying a strength reduction factor, ϕ, to the nominal strength of the element. The factor ϕ is intended to account for variations in material strength, the nature of the expected failure mode, as well as the importance of a member to the safety of the structure as a whole. For conventional reinforced concrete structures, ACI 318-83 specifies the following values of the strength reduction factor ϕ:

0.90 for flexure, with or without axial tension
0.90 for axial tension
0.75 for spirally reinforced members subjected to axial compression, with or without flexure
0.70 for other reinforced members (tied columns) subjected to axial compression, with or without flexure (with an increase in the ϕ-value for members subjected to combined axial load and flexure, as the loading condition approaches the case of pure flexure)
0.85 for shear and torsion
0.70 for bearing on concrete

Appendix A to ACI 318-83 specifies the following exceptions to the above values of the strength reduction factor: (a) For structural members other than joints, a value of $\phi = 0.60$ is to be used for shear when the nominal shear strength of a member is less than the shear corresponding to the development of the nominal flexural strength of the member. For shear in joints, $\phi = 0.85$. (b) For frame members subjected to axial compression and flexure, $\phi = 0.5$ is to be used when the axial compressive force exceeds $(A_g f'_c/10)$— where A_g is the gross cross-sectional area—and the transverse reinforcement provided does not conform to the more stringent requirements for lateral reinforcement for earthquake-resistant structures (see Section 12.6.8.4.3). Exception (a) above applies primarily to low-rise walls or portions of walls between openings. Exception (b) is intended to discourage the use of ties as lateral reinforcement in columns designed for earthquake resistance.

4. *Code Provisions Designed to Ensure Ductility in Reinforced Concrete Members.* The principal code provisions designed to ensure the necessary ductility in reinforced concrete structures that may be subjected to strong earthquake motions as contained in Appendix A of ACI 318-83[12-81] will be discussed below.

Special provisions concerning the design of earthquake-resistant structures first appeared in the 1971 edition of the *ACI Code*. The provisions of Appendix A supplement or supersede those in the main body of the *Code* and deal with the design of ductile moment-resisting space frames and shear walls of cast-in-place reinforced concrete.

Neither the main body of the *Code* nor Appendix A specifies the magnitude of the earthquake forces to be used in design. The Commentary to the Code states that its provisions are intended to enable structures to sustain a series of oscillations in the inelastic range.

The last section of Appendix A is intended for frames located in areas of moderate seismic risk. In structures located in regions of low seismic risk and designed for the estimated earthquake forces, very little inelastic deformation may be expected. In such cases, the ductility provided by designing to the provisions contained in the main body of the *Code* will generally be sufficient.

A major objective of the design provisions in Appendix A as well as in the main body of the *Code* is to have the strength of a structure governed by a ductile type of flexural failure mechanism. The main difference lies in the relatively greater range of deformation, with yeilding actually expected, and hence the greater ductility required in designs for resistance to major earthquakes. The need for greater ductility results from the design philosophy that allows the use of reduced forces in proportioning members and provides for the inelastic deformations that are expected under severe earthquakes by special ductility requirements. This philosophy is based on the recognition that it is generally uneconomical to proportion structures, and buildings in particular, to resist major earthquakes within their elastic range of stresses. The strongest justification of this philosophy has been the satisfactory performance of many structures designed by it that have been subjected to actual earthquakes.

The provisions of ductility given in Appendix A are aimed at preventing the brittle or abrupt types of failure associated with inadequately reinforced and overreinforced members failing in flexure, as well as by shear (i.e., diagonal tension) and anchorage failures. In addition, because the critical regions in frame structures generally occur at and near joints, Appendix A requires the use of adequate confinement reinforcement to ensure the proper performance and integrity of these regions when subjected to large amplitudes of reversed loading.

A provision unique to earthquake-resistant design of frames is the requirement that the sum of the moment strengths of the columns in a beam–column connection be greater than 1.2 times that of the beams framing into the joint. This is intended to ensure that yielding in such frames would occur in the beams rather than in the columns, and would thus preclude any instability effects that might result from plastic hinges forming in the columns. This requirement often results in column sizes that are larger than would otherwise be required, particularly in the upper floors of multistory buildings with appreciable beam spans.

The major provisions of Appendix A of ACI 318-83 will be discussed below.

4.1 Limitations on material strengths: Appendix A requires a minimum specified concrete strength, f'_c, of 3,000 psi and a maximum specified yield strength of reinforcement, f_y, of 60,000 psi. These limits are imposed as reasonable bounds on the variation of material properties, particularly with respect to their unfavorable effects on the sectional ductility of members in which they are used. A decrease in the concrete strength and an increase in the yield strength of the tensile reinforcement tend to decrease the ultimate curvature and hence the sectional ductility of a member subjected to flexure. Also, an increase in the yield strength of reinforcement is generally accompanied by a decrease in the ductility—as measured by the maximum deformation—of the material itself.

Appendix A requires that reinforcement for resisting flexure and axial forces in frame members and wall boundary elements be ASTM 706 Grade 60 low alloy steel intended for application where welding or bending, or both, are important. However, ASTM 615 billet steel bars of Grade 40 or 60 may be used in these members if the following two conditions are satisfied:

$$\text{actual } f_y \leqslant \text{specified } f_y + 18{,}000 \text{ psi}$$

$$\frac{\text{actual ultimate tensile stress}}{\text{actual } f_y} \geqslant 1.25$$

The first requirement helps to limit the magnitude of the actual shears that can develop in a flexural member above that computed on the basis of the specified yield value when plastic hinges form at the ends of a beam. The second requirement is intended to ensure steel with a sufficiently long yield plateau.

In the "strong column–weak beam" frame intended by the *Code*, the relationship between the moment capacities of columns and beams may be upset if the beams turn out to have much greater moment capacity than intended by the designer. Thus, the substitution of 60-ksi steel of the same area for specified 40-ksi steel in beams can be detrimental. The shear strength of beams and columns, which is generally based on the condition of plastic hinges forming at the ends of the members, may become inadequate if the moment capacity of member ends, M_y, should be greater than intended as a result of the steel having a substantially greater yield strength than specified.

4.2 Flexural members (beams): These include members, having a clear span greater than four times the effective depth, that are subjected to a factored axial compressive force not exceeding $(A_g f'_c/10)$, where A_g is the gross cross-sectional area. The significant provisions relating to flexural members are:

(a) Limitations on section dimensions:

$$\text{width-to-depth ratio} \geqslant 0.3$$

$$\text{width} \begin{cases} \leqslant 10 \text{ in.} \\ \leqslant [\text{width of supporting column} \\ \quad + 1.5 \text{ (depth of beam)}] \end{cases}$$

(b) Limitations on flexural reinforcement ratio (see also Fig. 12-73):

$$\rho_{\min} = \begin{cases} 200/f_y \\ \text{two continuous bars at both top} \\ \quad \text{and bottom of member} \end{cases}$$

$$\rho_{\max} = 0.025$$

(c) Moment capacity requirements:
At beam ends:

$$M_y^+ \geqslant 0.50 M_y^-$$

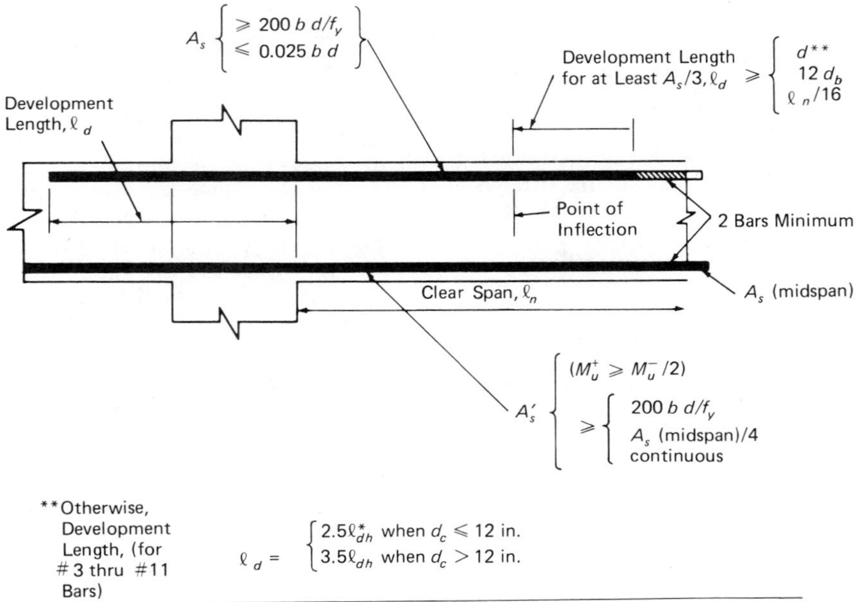

* ℓ_{dh} as defined for bars with a standard 90 hook in normal weight concrete (see Fig. 12-74)
# d_c = Depth of concrete cast in one lift beneath bar

Fig. 12-73 Longitudinal reinforcement requirements for flexural members.

At any point in beam span:

$$M_y^+ \text{ or } M_y^- \geqslant 0.25 M_y^{max} \text{ at beam ends}$$

(d) Restrictions on lap splices: Lap splices shall not be used:
  (1) Within joints.
  (2) Within $2h$ from face of support, where h is depth of beam.
  (3) At locations of potential plastic hinging.

  Lap splices are to be confined by hoops or spiral reinforcement with maximum spacing or pitch of $d/4$ or 4 in.

(e) Restrictions on welding of longitudinal reinforcement: Welded splices and mechanical connectors may be used provided:
  (1) They are used only on alternate bars in each layer at any section.
  (2) The distance between splices of adjacent bars $\geqslant$ 24 in.

(f) Development length requirements for longitudinal steel in tension:
  (1) For bar sizes #3 through #11 with a standard 90-degree hook (as shown in Fig. 12-74) in normal weight concrete:

$$\text{development length, } l_{dh} \geqslant \begin{cases} f_y d_b / 65\sqrt{f_c'} \\ 8d_b \ (d_b \text{ is bar diameter}) \\ 6 \text{ in.} \end{cases}$$

  (2) When bars of the same size are embedded in lightweight-aggregate concrete, l_{dh} is to be at least 1.25 times the above value.
  (3) The 90-degree hook shall be located within the confined core of a column or other boundary element.
  (4) For straight bars of sizes #3 through #11:

l_d = 2.5 (l_{dh} specified for bars with 90-degree hooks) when the depth of concrete cast in one lift beneath the bar $\leqslant$ 12 in.

l_d = 3.5 (l_{dh} specified for bars with 90-degree hooks) when the depth of concrete cast in one lift beneath the bar $>$ 12 in.

  (5) If a bar is not anchored by means of a 90-degree hook within the confined column core but is extended into a boundary element, the portion of the required straight development length not located within the confined core shall be increased by a factor of 1.6.

(g) Transverse reinforcement requirements for confinement and shear: Transverse reinforcement in beams must satisfy requirements associated with their dual function as confinement reinforcement and shear reinforcement (see Fig. 12-75).
  (1) Confinement reinforcement in the form of hoops is required:
    (i) Over a distance $2d$ from faces of support (where d is the effective depth of the member)
    (ii) Over distances $2d$ on both sides of sections where flexural yielding may occur due to earthquake loading.
  (2) Hoop spacing:
    (i) First hoop at 2 in. from face of support.

$$\text{(ii) Maximum spacing} \leqslant \begin{cases} d/4 \\ 8 \times \text{(diameter of smallest longitudinal bar)} \\ 24 \times \text{(diameter of hoop bar)} \\ 12 \text{ in.} \end{cases}$$

  (3) Where hoops are not required, tie spacing $\leqslant d/2$.
  (4) Shear reinforcement is to be provided so as to preclude shear failure prior to development of plastic hinges at beam ends. Design shears for determining shear reinforcement are to be based on a condition where plastic hinges occur at beam ends due to the combined effects of lateral displacements and factored gravity loads (see Fig. 12-76). The "probable flexural strengths," M_{pi}, associated with plastic hinges are to be computed using a strength reduction factor, $\phi = 1.0$, and assuming a stress in the tensile reinforcement, $f_s = 1.25 f_y$.
  (5) In determining the required shear reinforcement, the contribution of the concrete, V_c, is to be neglected if the shear associated with the "probable flexural strengths" at the beam ends is greater than one-half of the total design shear,

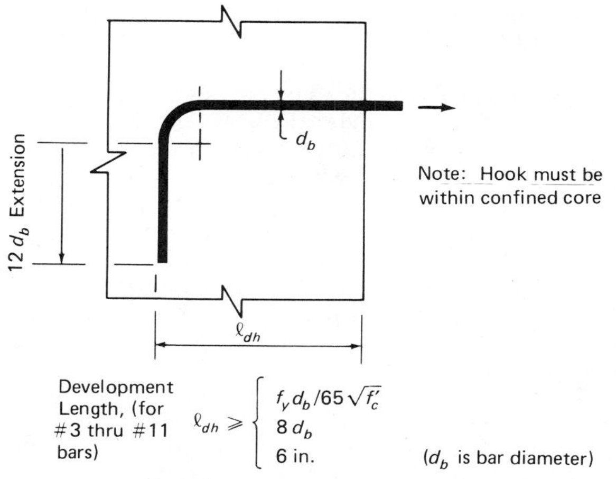

Fig. 12-74 Standard 90-degree hook.

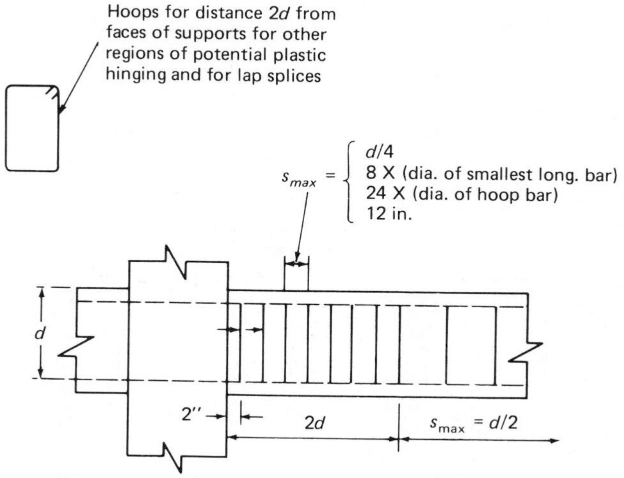

Fig. 12-75 Transverse reinforcement limitations for flexural members. Minimum bar size—#3.

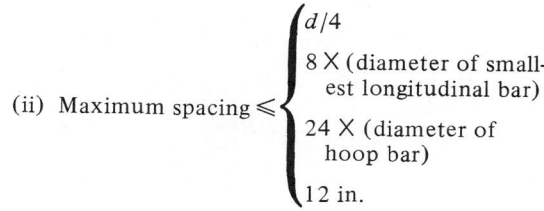

$$V_a = \frac{M_{p1} + M_{p2}}{\ell} + 0.75\left(\frac{1.4D + 1.7L}{2}\right)$$

$$V_b = \frac{M_{p1} + M_{p2}}{\ell} - 0.75\left(\frac{1.4D + 1.7L}{2}\right)$$

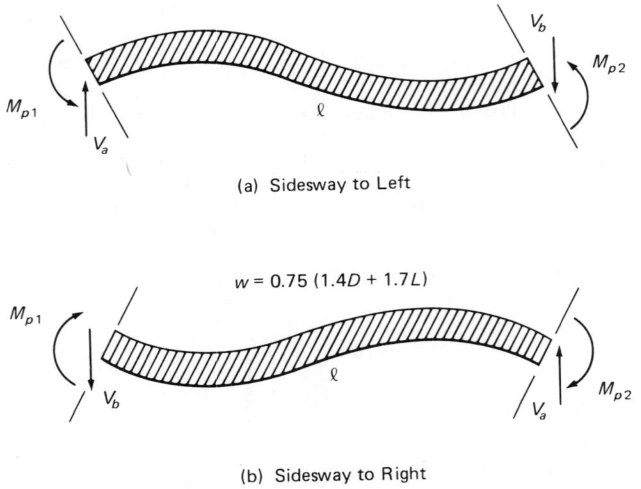

Fig. 12-76 Loading cases for design of shear reinforcement in beams—uniform gravity loads.

and the factored axial compressive force including earthquake effects is less than $(A_g f_c'/20)$.

(6) Shear reinforcement must be in the form of hoops in regions where confinement is also required, as discussed above under confinement reinforcement (see Fig. 12-77). Otherwise, stirrups or ties may be used.

(7) The transverse reinforcement provided must satisfy the requirement for confinement or shear, whichever is larger.

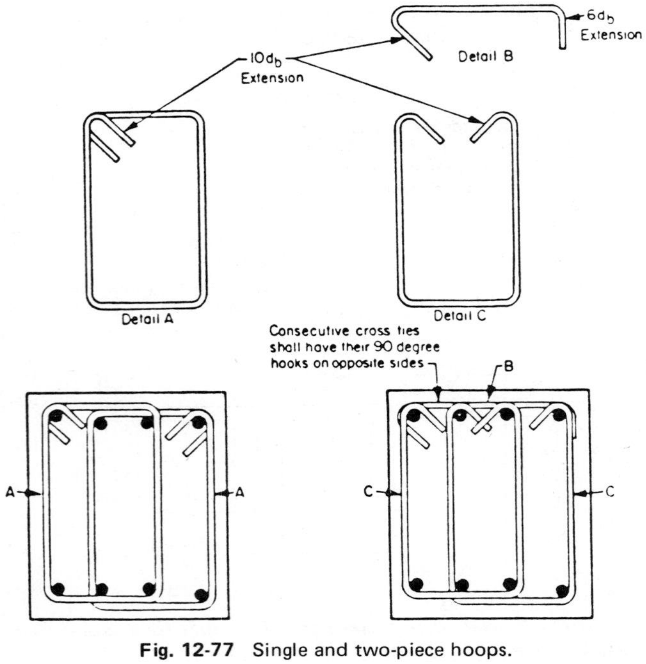

Fig. 12-77 Single and two-piece hoops.

Discussion:

(a) Limitations on section dimensions: These limitations have been guided by experience with test specimens subjected to cyclic inelastic loading.

(b) Flexural reinforcement limitations: Because the ductility of a flexural member decreases with increasing values of the reinforcement ratio, Appendix A limits that maximum reinforcement ratio to 0.025. The use of a limiting value based on the "balanced condition" as given in the main body of the *Code*, while applicable to members subjected to monotonically increasing loading, fails to describe conditions in a flexural member subjected to reversals of inelastic deformation. The limiting ratio of 0.025 is based mainly on considerations of steel congestion and also on limiting shear stresses in beams of typical proportions. From a practical standpoint, low steel ratios should be used whenever possible. The requirement of at least two bars, top and bottom, is dictated by construction rather than behavioral requirements.

The selection of the size, number, and arrangement of flexural reinforcement should be made with full consideration of construction requirements. This is particularly important in relation to beam-column connections, where construction difficulties can arise as a result of reinforcement congestion. The preparation of large-scale drawings of the connections, showing all beam, column, and joint reinforcement will help eliminate unanticipated problems in the field. Such large-scale drawings will pay dividends in terms of lower bid prices and a smooth-running construction job.

(c) Positive moment capacity at beam ends: To allow for the possibility of the positive moment at the end of a beam due to earthquake-induced lateral displacements exceeding the negative moment due to gravity loads, the *Code* requires a minimum positive moment capacity at beam ends equal to 50% of the corresponding negative moment capacity.

(d) Lap splices: Lap splices of flexural reinforcement are not allowed in regions of potential plastic hinging, since such splices are not considered to be reliable under reversed inelastic cycles of deformation. Hoops are mandatory for confinement of lap splices at any location because of the likelihood of loss of shell concrete.

(e) Welded splices and mechanical connectors: Welded splices and mechanical connectors shall conform to the requirements given in the main body of the *Code*. A major requirement is that the splice should develop at least 125% of the specified yield strength of the bar.

(f) Development length: The expression for l_{dh} given above already includes the coefficients 0.7 (for concrete cover) and 0.80 (for ties) that are normally applied to the required basic development length, l_{hb}. This is so because Appendix A requires that hooks be embedded in the confined core of the column or other boundary element. The expression for l_{dh} also includes a factor of about 1.4, representing an increase over the development length required for conventional structures, to provide for the effect of load reversals.

Except in very large columns, it is usually not possible to develop the yield strength of a reinforcing bar from a framing beam within the width of a column. Where beam reinforcement can extend through a column, its capacity is developed by embedment in the column and within the compression zone of the beam on the far side of the connection (see Fig. 12-73). Where no beam is present on the opposite side of a column, such as in exterior columns, the flexural reinforcement in a framing beam has to be developed within the confined region of the column. This is usually done by means of a standard 90-degree hook plus whatever extension is necessary to develop the bar, the development length being measured from the near face of the column, as shown in Fig. 12-74.

Appendix A makes no provision for the use of #14 and #18 bars because of a lack of sufficient information on the behavior at anchorages of such bars when subjected to load reversals simulating earthquake effects.

(g) Transverse reinforcement: Because the ductile behavior of earthquake-resistant frames designed to current codes is premised on the ability of the beams to develop plastic hinges with adequate rotational capacity, it is essential to ensure that shear failure does not occur before the flexural capacity of the beams has been developed. Transverse reinforcement is required for two related functions: (a) to provide sufficient shear strength so that the full flexural capacity of a member can be developed, and (b) to help ensure adequate rotation capacity at plastic hinging regions by confining the concrete in the compression zone and by providing lateral support to the compression steel. To be equally effective with respect to both functions under load reversals, the traverse reinforcement should be placed perpendicular to the longitudinal reinforcement.

Shear reinforcement in the form of stirrups or stirrup-ties is designed for the shear due to the factored gravity loads and the shear corresponding to plastic hinges forming at both ends. Plastic end moments associated with lateral displacement in either direction should be considered (Fig. 12-76). It is important to note that the required shear strength in beams (as in columns) is determined by the flexural strength of the frame member (as well as the factored loads acting on the member) rather than by the factored shear force calculated from a lateral load analysis.

Because of the direct dependence of the required web reinforcement on the yield strength of the flexural reinforcement, any unintended substantial overstrength in the latter could result in a nonductile shear failure preceding the development of the full flexural capacity of a member. The limitations on the actual strength of steel reinforcement mentioned earlier, as well as the use of $\phi = 1.0$ and $f_s = 1.25 f_y$ in calculating the probable strength, M_{pi}, of a beam end, are all designed to reduce the chances of a shear failure preceding flexural yielding. The use of $f_s = 1.25 f_y$ reflects the strong likelihood of the deformation in the tensile reinforcement entering the strain-hardening range.

To allow for load combinations unaccounted for in design, a minimum amount of web reinforcement is required throughout the length of all flexural members. Within regions of potential hinging, stirrup-ties or hoops are required. A hoop may be made of two pieces of reinforcement: a stirrup having 135-degree hooks with ten-diameter extensions anchored in the confined core and a crosstie to close the hoop (see Fig. 12-77). Consecutive crossties shall have their 90-degree hooks on opposite sides of the flexural member.

4.3 Frame members subjected to axial load and bending: Appendix A[12-81] makes the distinction between columns or beam–columns and flexural members on the basis of the magnitude of the factored axial load imposed on a member. Thus, when the factored axial load does not exceed $(A_g f'_c/10)$, the member falls under the category of flexural members, as discussed in the preceding section. When the factored axial force on a member resisting earthquake-induced forces exceeds $(A_g f'_c/10)$, the member is considered a beam–column. Design of such members is governed by the requirements given below.

(a) Limitation on section dimensions:

shortest cross-sectional dimension $\geqslant$ 12 in.
(measured on line passing through geometric centroid)

$$\frac{\text{shortest dimension}}{\text{perpendicular dimension}} \geqslant 0.4$$

(b) Limitations on longitudinal reinforcement:

$$\rho_{\min} = 0.01, \qquad \rho_{\max} = 0.06$$

(c) Flexural strength of columns relative to beams framing into a joint ("strong column–weak beam" provision):

$$\sum M_e \geqslant \tfrac{6}{5} \sum M_g \tag{12-28}$$

where:

ΣM_e = sum of design flexural strengths of columns framing into joint. Column flexural strength shall be calculated for the factored axial force, consistent with the direction of lateral loading considered, that results in the lowest flexural strength.

ΣM_g = sum of design flexural strengths of beams framing into joint.

(d) Restriction of lap splices: Lap splices are to be used only within the center half of the column length and shall be designed as tension splices.

(e) Welded splices or mechanical connectors in longitudinal reinforcement: Welded splices or mechanical connectors may be used at any section of the column, provided that:
  (1) They are used only on alternate longitudinal bars at a section.
  (2) The distance between splices along the longitudinal axis of reinforcement $\geqslant$ 24 in.

(f) Transverse reinforcement for confinement and shear: As in beams, transverse reinforcement in columns must provide confinement of the concrete core as well as shear resistance. In columns, however, the transverse reinforcement must all be in the form of closed hoops or continuous spiral reinforcement. Sufficient reinforcement should be provided to satisfy the requirement for confinement or shear, whichever is larger.
  (1) Confinement requirements (see Fig. 12-78) are:
    (a) Volumetric ratio of spiral or circular hoop reinforcement:

$$\rho_s \geqslant \begin{cases} 0.12 \dfrac{f'_c}{f_{yh}} \\ 0.45 \left(\dfrac{A_g}{A_{ch}} - 1 \right) \dfrac{f'_c}{f_{yh}} \end{cases} \tag{12-29}$$

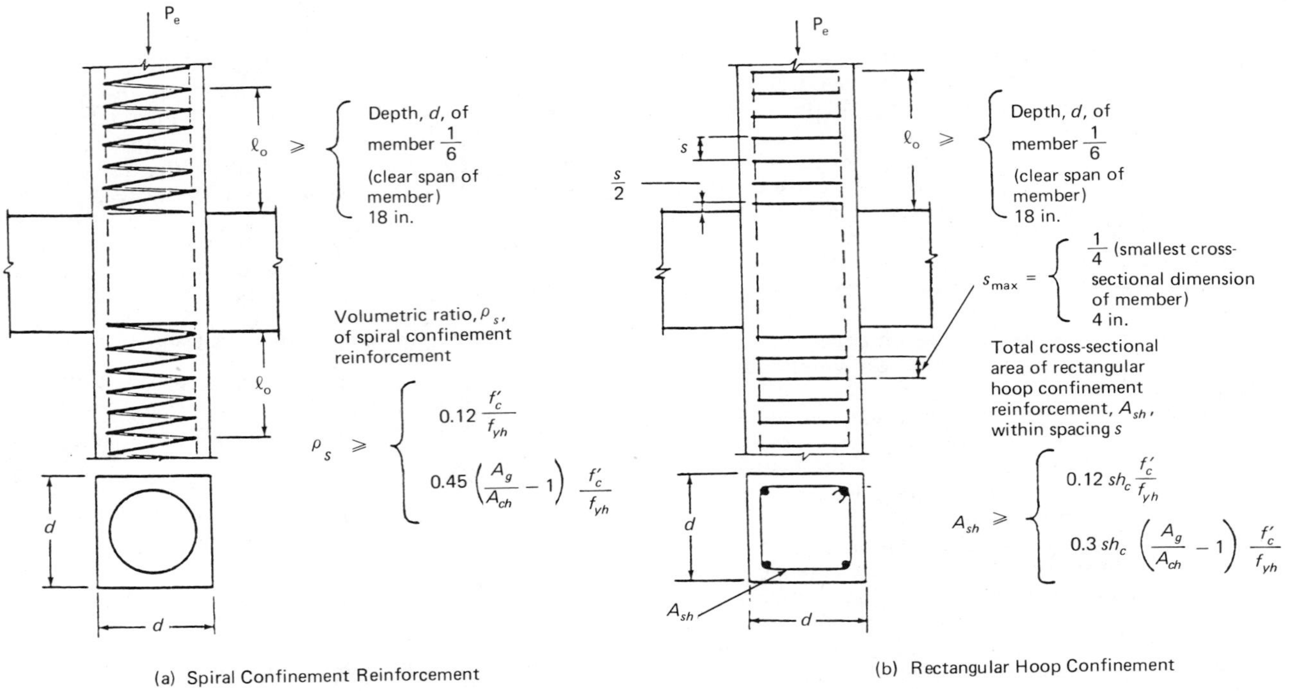

Fig. 12-78 Confinement requirements at column ends.

where:

f_{yh} = specified yield strength of transverse reinforcement, in psi

A_{ch} = core area of column section, measured to the outside of transverse reinforcement, in sq in.

(b) Rectangular hoop reinforcement, total cross-sectional area, within spacing s:

$$A_{sh} \geqslant \begin{cases} 0.12\, sh_c \dfrac{f'_c}{f_{yh}} \\ 0.3\, sh_c \left(\dfrac{A_g}{A_{ch}} - 1\right) \dfrac{f'_c}{f_{yh}} \end{cases} \tag{12-30}$$

where:

h_c = cross-sectional dimension of column core, measured center-to-center of confining reinforcement

s = spacing of transverse reinforcement measured along longitudinal axis of member, in in.

$$s_{max} = \begin{cases} \frac{1}{4} \text{ (smallest cross-sectional dimension of member)} \\ 4 \text{ in.} \end{cases}$$

Maximum spacing in plane of cross-section between legs of overlapping hoops or cross ties (see Fig. 12-78) = 14 in.

(2) Confinement reinforcement is to be provided over a length l_0 from each joint face or over distances l_0 on both sides of any section where flexural yielding may occur in connection with inelastic lateral displacements of the frame, where:

$$l_0 \geqslant \begin{cases} \text{depth, } d\text{, of member} \\ \frac{1}{6} \text{ (clear span of member)} \\ 18 \text{ in.} \end{cases}$$

(3) Transverse reinforcement for shear in columns is to be determined for the shear associated with the largest nominal moment strengths of the column (using $f_s = f_y$ and $\phi = 1.0$) ends, calculated for the factored axial compressive force resulting in the largest moment strengths.

Discussion:

(b) Reinforcement ratio limitation: Appendix A specifies a reduced upper limit for the reinforcement ratio in columns from the 8% of Chapter 10 of the *Code* to 6%. However, construction considerations will in most cases place the practical upper limit on the reinforcement ratio, ρ, near 4%. Convenience in detailing and placing reinforcement in beam column connections make it desirable to keep the column reinforcement low.

The minimum reinforcement ratio is intended to provide for the effects of time-dependent deformations in concrete under axial loads as well as maintain a sizable difference between cracking and yield moments.

(c) Relative column-to-beam strength requirements: To ensure the stability of a frame and maintain its vertical load-carrying capacity while it undergoes large lateral displacements, the *Code* requires that inelastic deformations be generally restricted to the beams. This is accomplished by requiring that the sum of the flexural strengths of the columns meeting at a joint, under the design axial loads, be equal to or greater than $^6/_5$ times the sum of the moment strengths of the framing beams in the same plane. As indicated in Fig. 12-79, the signs of the flexural moments of col-

$$(M^p_{ct} + M^p_{cb}) \geqslant \frac{6}{5}(M^p_b + M^p_{br})$$

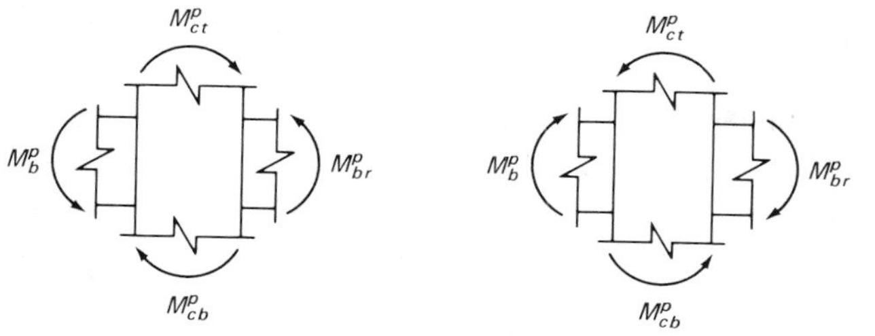

Fig. 12-79 "Strong column–weak beam" frame requirements.

umns and beams are to be such that the column moments oppose the beam moments. Also, the "strong column–weak beam" relationship has to be satisfied for beam moments acting in both directions.

If eq. (12-28) is not satisfied at a joint, columns supporting reactions from that joint are to be provided with transverse reinforcement over their full height. Columns not satisfying eq. (12-28) are to be ignored in calculating the strength and stiffness of the structure. However, since such columns contribute to the stiffness of the structure before they suffer severe loss of strength due to plastic hinging, they should not be ignored if neglecting them results in unconservative estimates of design forces. This may occur in determining the design base shear or in calculating the effects of torsion in a structure. Columns not satisfying eq. (12-28) should satisfy the minimum requirements for "members not proportioned to resist earthquake-induced forces," discussed under Section 12.6.8.4.6.

(f) Transverse reinforcement for confinement and shear: Sufficient transverse reinforcement in the form of rectangular hoops or spirals should be provided to satisfy the larger requirement for either confinement or shear.

Circular ties represent the most efficient form of confinement reinforcement. The extension of such spirals into the beam–column joint, however, may cause some construction difficulties.

Rectangular hoops, when used in place of spirals, are considered less effective with respect to confinement of the concrete core. Their effectiveness is increased, however, by the use of supplementary crossties, each end of which has to engage a peripheral longitudinal bar. Crossties of the same bar size and spacing as the hoops may be used. Consecutive crossties are to be alternated end for end along the longitudinal reinforcement (Fig. 12-80). As indicated in Fig. 12-80, crossties or legs of overlapping hoops are to be spaced no farther than 14 in. apart in a direction perpendicular to the longitudinal axis of the member. The requirement of having the crossties engage a longitudinal bar at each end would almost preclude placing them before the longitudinal bars are threaded through.

In addition to satisfying confinement requirements, the transverse reinforcement in columns must resist the maximum shear associated with the formation of plastic hinges in the frame. Although the "strong column–weak beam" provision governing the relative moment strengths of beams and columns is intended to have most of the inelastic deformation occur in the beams of a frame, the *Code* recognizes that hinging can occur in the columns. Thus, the shear reinforcement in columns is to be based on the condition that the nominal moment strength (i.e., plastic or yield moment, with strength reduction factor, $\phi = 1.0$) is developed at the ends of a column. The value of these plastic moments—obtained from the *P–M* interaction diagram for the column section—is to be maximum consistent with the possible factored axial compressive forces on the column. Moments associated with lateral displacements of the structure in both directions, as indicated in Fig. 12-81, are to be considered. The axial load corresponding to the maximum moment capacity should then be used in computing the permissible shear stress in concrete, v_c.

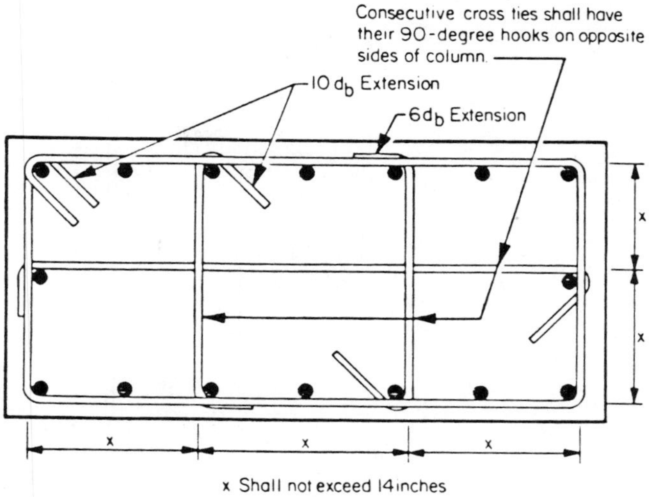

Fig. 12-80 Transverse reinforcement in columns.

(g) Columns supporting discontinued walls: Columns supporting discontinued shear walls or stiff partitions tend to be subjected to large shear (see Fig. 12-71) and compressive forces, and can be expected to suffer significant inelastic deformations during strong earthquakes. In recognition of this, the *Code* requires confinement reinforcement throughout the height of such columns (see Fig. 12-82) whenever the axial compressive force due to earthquake effects exceeds $(A_g f'_c/10)$.

4.4 Beam–column connections: In conventional reinforced concrete buildings, the beam–column connections normally are not designed by the structural engineer. Detailing of bars within the joints is usually relegated to a draftsman or detailer. In earthquake-resistant frames, however, the design of beam–column connections requires as much attention as the design of the members themselves, since the integrity of the structure may well depend on the proper functioning of such connections. As mentioned earlier, beam–column joints represent regions of geometric and stiffness discontinuities in a frame and as such tend to be subjected to relatively high force concentrations. A substantial portion of the damage in frame structures subjected to strong earthquakes has been observed to occur at these connections. This has been particularly evident where attention to their proper design apparently has been inadequate.

Because of the congestion of reinforcement that may occur as a result of too many bars converging within the limited space of the joint, the proportioning of the frame columns and beams should be undertaken with due regard

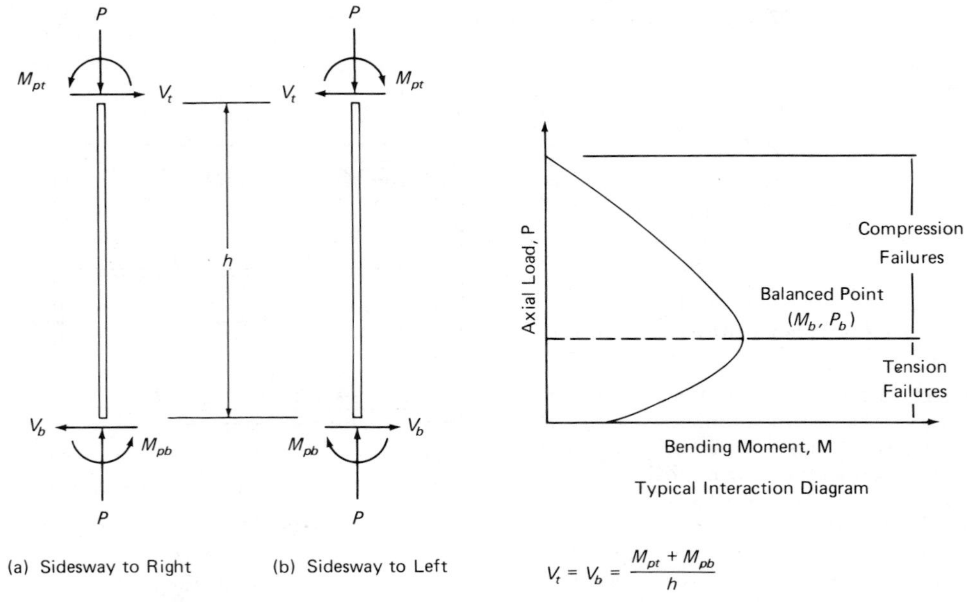

Fig. 12-81 Loading cases for design of shear reinforcement in columns.

to the design of the beam–column connection. Usually little difficulty is encountered if the amount of longitudinal reinforcement used in the frame members is kept low. Also, the preparation of large-scale detailed drawings showing bar arrangements within the joints will be of much help in avoiding unexpected difficulties in the field.

The provisions of Appendix A[12-81] relating to beam–column connections have to do mainly with:

(a) Transverse reinforcement for confinement: Minimum confinement reinforcement of the same amount required for potential hinging regions in columns, as defined by eqs. (12-29) and (12-30), must be provided through beam–column joints around the column reinforcement. For "confined" joints, a 50% reduction in the required amount of confinement reinforcement is allowed, the required amount to be placed within the depth of the shallowest framing member. A confined joint is defined as one with beams framing into all four sides and where each beam has a width equal to at least three-fourths of the width of the column face into which it frames.

(b) Design for shear: Shear force in a joint is to be calculated by assuming stress in tensile reinforcement of framing beams equal to $1.25 f_y$. Shear strength of the connection is to be computed as:

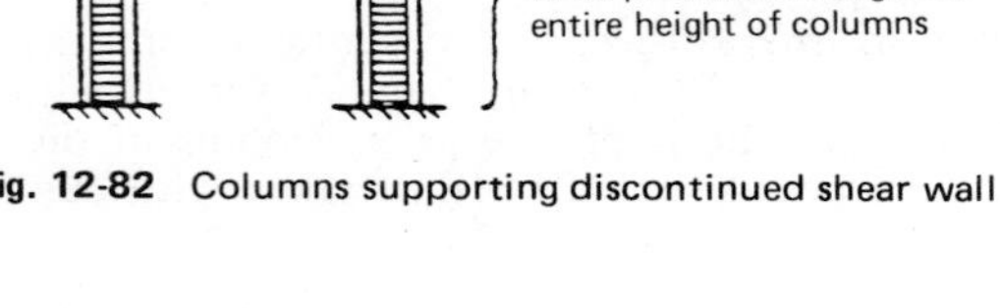

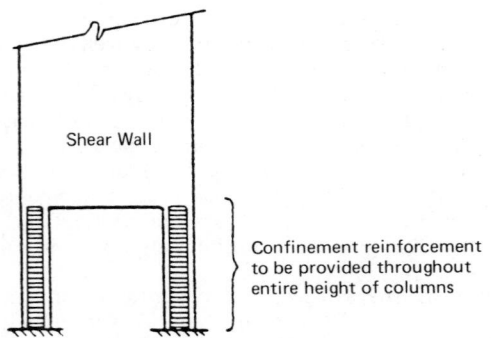

Fig. 12-82 Columns supporting discontinued shear wall.

$$\underset{\text{(for normal-weight concrete)}}{\varphi V_c} = \begin{cases} \varphi\, 20\sqrt{f_c'}\, A_j & \text{for confined joints} \\ \varphi\, 15\sqrt{f_c'}\, A_j & \text{for unconfined joints} \end{cases}$$

where:

φ = 0.085 for shear

A_j = minimum cross-sectional area of joint in a plane parallel to the axis of the reinforcement generating the shear force

For lightweight concrete, V_c shall be taken as three-fourths the value given above for normal-weight concrete.

(c) Anchorage of longitudinal beam reinforcement terminated in a column must be within the confined column core.

Discussion:

(a) Transverse reinforcement for confinement: The transverse reinforcement in a beam–column connection is intended to provide adequate confinement of the concrete to ensure its ductile behavior and allow it to maintain its vertical load-carrying even after spalling of the outer shell. It also helps resist the shear forces transmitted by the framing members and improves the bond between steel and concrete within the connection.

The minimum amount of confinement reinforcement, as given by eqs. (12-29) and (12-30), must be provided through the joint regardless of the magnitude of the calculated shear force in the joint. The 50% reduction in the amount of confinement reinforcement allowed for joints having horizontal members framing into all four sides recognizes the beneficial effect provided by these members in resisting the bursting pressures generated within the joint.

(b) As indicated in Fig. 12-83, the design shear is based on the most critical combination of horizontal shears transmitted by the framing beams and columns. Tests have indicated that plastic hinging at the ends of beams, for deformations associated with response to

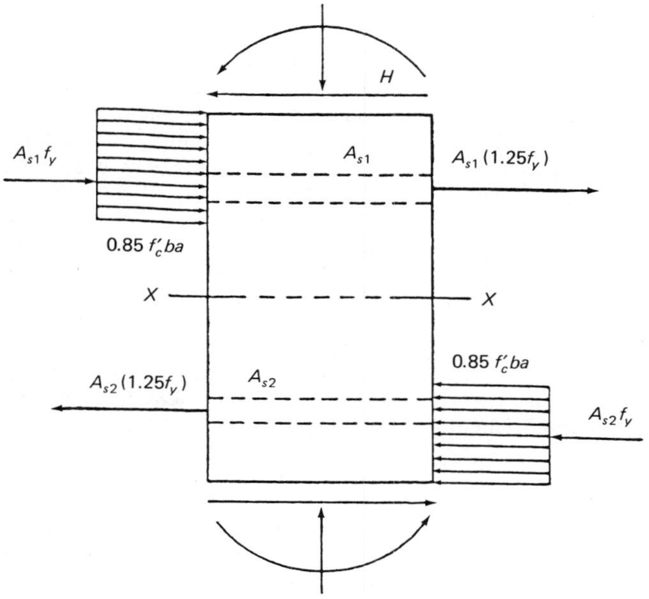

Fig. 12-83 Horizontal shear in beam–column connection.

strong earthquakes, impose strains in the flexural reinforcement well in excess of the yield strain. Because of the likelihood of strains in the tensile reinforcement going into the strain-hardening range, and to allow for the actual yield strength of the steel exceeding the specified values, the *Code* requires that the horizontal shear in the joint be determined by assuming the stress in the flexural tensile steel to be equal to $1.25f_y$.

Results of tests reported in Ref. 12-95 indicate that shear strength of joints is not too sensitive to the amount of transverse (shear) reinforcement. Based on these results, the 1983 edition of ACI Appendix A makes the shear strength of beam–column connections a function only of the cross-sectional area of the joint, A_j, and f'_c (see Section 12.5.5).

When the design shear in the joint exceeds the shear strength of the concrete, the designer may either increase the column size or increase the depth of the beams. The former will increase the shear capacity of the joint section, while the latter will tend to reduce the required amount of flexural reinforcement in the beams, with accompanying decrease in the shear transmitted to the joint.

(c) The anchorage or development length requirements for beam reinforcement in tension have been discussed earlier under flexural members. Splicing of main reinforcement within a joint should be avoided whenever possible.

4.5 Shear walls: When properly proportioned so that they possess adequate lateral stiffness to reduce interstory distortions due to earthquake-induced motions, shear walls or structural walls reduce the likelihood of damage to the nonstructural elements of a building. When used with rigid frames, walls form a system that combines the gravity-load-carrying efficiency of the rigid frame with the lateral-load-resisting efficiency of the structural wall.

Observations of the comparative performance of rigid-frame buildings and buildings stiffened by structural walls during recent earthquakes[12-96] have pointed to the consistently better performance of the latter. The performance of buildings stiffened by properly designed structural walls has been better with respect to both safety and damage control. The need to ensure that critical facilities remain operational after a major tremor and the need to reduce economic losses from structural and nonstructural damage, in addition to the primary requirement of life safety (i.e., no collapse), has focused attention on the desirability of introducing greater lateral stiffness into earthquake-resistant multistory structures. Structural walls, which have long been used in designing for wind resistance, offer a logical and efficient solution to the problem of lateral stiffening of multistory buildings.

Shear walls are normally much stiffer than regular frame elements and are therefore subjected to correspondingly greater lateral forces during response to earthquake motions. Because of their relatively greater depth, the lateral deformation capacity of walls is limited, so that, for a given amount of lateral displacement, shear walls will tend to exhibit greater apparent distress than frame members. However, over a broad period range, a shear wall structure, which is substantially stiffer and hence has a shorter period than a frame structure, will suffer less lateral displacement than the frame, when subjected to the same ground motion intensity. Shear walls with a height-to-depth ratio of about 3 behave essentially as vertical cantilever beams and should therefore be designed as flexural members, with their strength governed by flexure rather than by shear.

Primarily because of the greater stiffness of shear wall structures, but also because of earlier concerns about the deformation capacity of shear walls, codes have specified larger *K*-factors for determining the design base shears for such structures.

Isolated shear walls or individual walls connected to frames will tend to yield first at the base where the moment is the greatest. Coupled walls (i.e., two or more walls linked by short, rigidly connected beams at the floor levels), on the other hand, have the desirable feature that significant energy dissipation through inelastic action in the coupling beams can be made to precede hinging at the bases of the walls.

The principal provisions of ACI Appendix A relating to structural walls (and diaphragms) are as follows (see also Fig. 12-84):

(a) Walls (and diaphragms) are to be provided with shear reinforcement in two orthogonal directions in the plane of the wall. Minimum reinforcement ratio for both longitudinal and transverse directions:

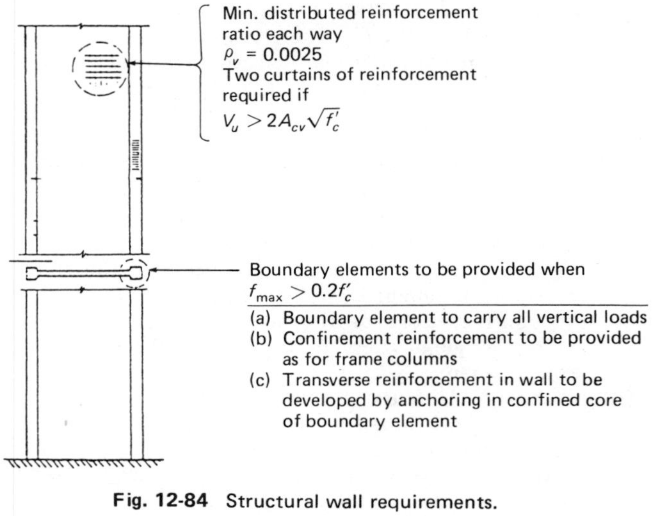

Fig. 12-84 Structural wall requirements.

$$\rho_v = \frac{A_{sv}}{A_{cv}} = \rho_n \geqslant 0.0025 \quad \left\{ \begin{array}{l} \text{reinforcement to be} \\ \text{continuous and} \\ \text{distributed uniformly} \\ \text{across shear area} \end{array} \right.$$

where:

A_{cv} = net area of concrete section, i.e., product of wall thickness and length of section in direction of shear considered

A_{sv} = projection on A_{cv} of area of shear reinforcement crossing the plane of A_{cv}

ρ_n = ratio of distributed shear reinforcement on a plane perpendicular to plane of A_{cv}

Maximum spacing of reinforcement = 18 in.

At least two curtains of reinforcement—each having bars running in the longitudinal and transverse directions—are to be provided if the in-plane factored shear force assigned to the wall exceeds $2A_{cv}\sqrt{f'_c}$.

(b) Shear strength of walls (and diaphragms): For walls with height-to-horizontal length ratio, $h_w/l_w \geqslant 2.0$, the shear strength is to be determined from the expression:

$$\phi V_n = \phi A_{cv}(2\sqrt{f'_c} + \rho_n f_y)$$

where:

ϕ = 0.60, unless nominal shear strength provided exceeds the shear, corresponding to development of nominal flexural capacity of wall

A_{cv} = as defined earlier

h_w = height of entire wall or of segment of wall considered

l_w = length of entire wall or of segment of wall considered in direction of shear force

For walls with $h_w/l_w < 2.0$, the shear strength may be determined from:

$$\phi V_n = \phi A_{cv}(\alpha_c\sqrt{f'_c} + \rho_n f_y)$$

where the coefficient α_c varies linearly from a value of 3.0 for $(h_w/l_w) = 1.5$ to 2.0 for $(h_w/l_w) = 2.0$.

Where a wall is divided into several segments by openings, the value of the ratio (h_w/l_w) to be used in calculating V_n for any segment shall not be less than the corresponding ratio for the entire wall.

The nominal shear strength, V_n, of all wall segments or piers resisting a common lateral force shall not exceed $8A_{cv}\sqrt{f'_c}$, where A_{cv} is the total cross-sectional area of the walls. The nominal shear strength of any individual vertical or horizontal segment of wall shall not exceed $10A_{cp}\sqrt{f'_c}$, where A_{cp} is the cross-sectional area of the wall segment.

(c) Development length and splices: All continuous reinforcement is to be anchored or spliced in accordance with the provisions for reinforcement in tension, as discussed for flexural members.

Where boundary elements are present, the transverse reinforcement in walls is to be anchored within the confined core of the boundary element to develop the yield stress in tension of the transverse reinforcement.

(d) Boundary elements: Boundary elements are to be provided, both along vertical boundaries of the wall and around the edges of openings, if any, when the maximum extreme-fiber stress in the wall due to factored forces including earthquake forces exceeds $0.2f'_c$. The boundary members may be discontinued when the calculated compressive stress is less than $0.15f'_c$.

Boundary elements need not be provided if the *entire* wall is reinforced in accordance with the provisions governing transverse reinforcement for members subjected to axial load and bending, as given by eqs. (12-29) and (12-30) and the related spacing requirements.

Boundary members of structural walls are to be designed to carry all the factored vertical loads on the wall, including self-weight and gravity loads tributary to the wall, as well as the vertical force required to resist the overturning moment due to factored earthquake loads. Such boundary elements are to be provided with confinement reinforcement in accordance with eqs. (12-29) and (12-30) and the related spacing requirements.

Discussion:

(a) The use of two curtains of reinforcement in walls subjected to significant shears (i.e., $> 2A_{cv}\sqrt{f'_c}$) serves to reduce fragmentation and premature deterioration of the concrete under load reversals into the inelastic range. Distributing the reinforcement uniformly across the height and horizontal length of the wall helps control the width of inclined cracks.

(b) Appendix A allows calculation of the shear strength of any structural wall using a coefficient $\alpha_c = 2.0$. However, advantage can be taken of the greater observed shear strength of walls with low height-to-horizontal length (h_w/l_w) ratios by using an α_c-value of up to 3.0 for $h_w/l_w = 1.5$ or less.

To effectively restrain the growth of inclined cracks, the required shear reinforcement should be appropriately distributed along the height and horizontal length of the wall. It should be noted that the vertical reinforcement in the boundary elements (or reinforcement concentrated near the edges of the wall when no boundary elements are used) for resisting flexure in the wall is not to be included in determining satisfaction of the requirements for ρ_v or ρ_n.

Appendix A limits the average nominal unit shear strength of structural walls to $8\sqrt{f'_c}$, with allowance for exceeding this average in any individual wall in a group of walls or wall segments, provided that the unit shear in the individual wall does not exceed $10\sqrt{f'_c}$. This upper bound on strength that may be developed in any individual segment is intended to limit the degree of shear redistribution among several connected wall segments. A wall segment refers to a part of wall bounded by openings or by an opening and an edge.

It is important to note that Section A.2.3.1 of ACI 318-83 requires the use of a strength reduction factor (ϕ) for shear of 0.6 for all members (except joints) where the nominal shear strength is less than the shear corresponding to the development of the nominal flexural strength of the member. In the case of beams, the design shears are obtained by assuming plastic end moments corresponding to a tensile steel stress of $1.25f_y$. Similarly, for a column, the design shears are determined not by applying load factors to shears obtained from a lateral load analysis, but from the consideration of maximum developable moments, consistent with the axial force on the column, occurring at the column ends. This approach to shear design is intended to ensure that even when flexural hinging occurs at member ends due to earthquake-induced deformations, no shear failure would develop. Under the above conditions, the *Code* allows the use of the normal strength reduction factor for shear of 0.85. When design shears are not based on the condition of flexural strength being developed at member ends, the *Code* requires the use of a lower shear strength reduction factor to achieve the same result, that is, prevention of premature shear failure.

In the case of structural walls, a condition similar to that used for the shear design of beams and columns is not so readily established. This is so primarily because the magnitude of the shear at the base of a wall (or at any level above) is influenced significantly by the forces and deformations beyond the particular level considered. Unlike the flexural behavior of beams and columns in a frame, which can be considered as close-coupled systems (i.e., with the forces and deformations in the members determined primarily by the displacements in the end joints), the state of flexural deformation at any section of a structural wall (a far-coupled system) is influenced significantly by the displacements of points far removed from the section considered. Results of dynamic inelastic analyses of isolated structural walls under earthquake loading[12-26] also indicate that the base shear in such walls is strongly influenced by the higher modes of response.

A distribution of static lateral forces along the height of a wall, corresponding to the fundamental mode, such as is assumed by the codes, may produce flexural yielding at the base, if the section at the base of the wall is designed for such yielding. However, other distributions of lateral forces, having a resultant closer to the base, can produce yielding at the base only if the magnitude of the resultant horizontal force, and hence the base shear, is increased. Results of the study of isolated walls referred to above,[12-26] which would also apply to frame–shear wall systems in which the frame is flexible relative to the wall, in fact indicate that for a wide range of wall properties and input motions, the resultant of the dynamic horizontal forces producing yielding at the base of the wall generally occurs well below the two-thirds-of-total height level associated with fundamental mode response. This would imply significantly larger base shears than those due to lateral forces distributed according to the fundamental mode response. The study of isolated walls mentioned above shows ratios of maximum dynamic shears to "fundamental mode shears" (i.e., shears associated with horizontal forces distributed according to the *Code*) ranging from 1.3 to 4.0, the value of the ratio increasing with fundamental period.

(c) Actual forces in longitudinal bars of stiff members may exceed calculated forces. Because of this likelihood, and the importance of maintaining the flexural capacity of a wall, the *Code* requires that all continuous reinforcement be developed fully.

Similarly, the horizontal reinforcement in walls requiring boundary elements is called upon to function as web reinforcement. Because of this, the *Code* requires that such bars be fully anchored in the boundary elements (which act as flanges of vertical cantilever beams). Standard 90-degree hooks should be used whenever possible. Such hooks minimize the loss of bond that may otherwise result owing to the occurrence of large transverse cracks in the boundary elements when subjected to large inelastic deformations.

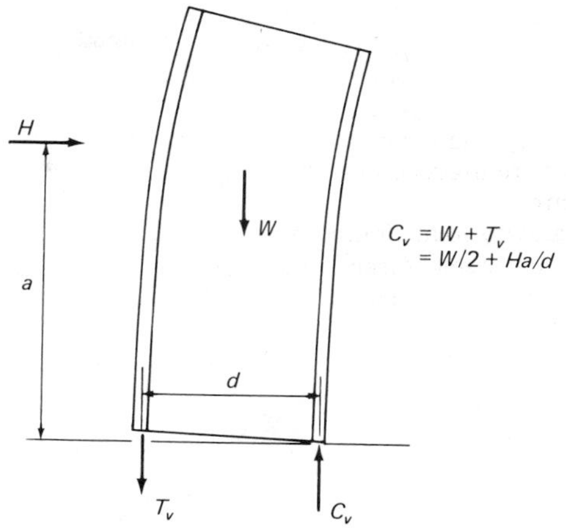

Fig. 12-85 Loading condition assumed for design of structural wall boundary elements.

(d) The *Code* uses a concrete stress of $0.2f_c'$, calculated using a linearly elastic model based on gross sections of structural members and factored forces, as indicative of significant compression. Structural walls subjected to compressive stresses exceeding this value are generally required to have boundary elements.

The condition assumed in requiring that boundary members be designed for all gravity loads as well as the vertical forces associated with overturning of the wall due to earthquake forces is illustrated in Fig. 12-85. This requirement assumes that the boundary member alone may have to carry all the vertical (compressive) forces at the critical wall section when the maximum horizontal earthquake force acts on the wall. Under load reversals, this loading condition imposes severe demands on the concrete in the boundary element. Hence the requirement for confinement reinforcement similar to those for members subjected to axial load and bending.

The design of the boundary element is carried out by considering it as an axially loaded short column subjected to the factored compressive axial force at the critical section.

Diaphragms of reinforced concrete, such as floor slabs, that are designed to transmit horizontal forces through bending and shear in their plane, are treated in much the same manner as structural walls.

Truss elements of reinforced concrete are also covered, although very briefly, in the 1983 edition of Appendix A. A major requirement for truss elements relates to the provision of special transverse reinforcement when the compressive stress exceeds $0.2f_c'$.

4.6 Frame members not forming part of lateral-force-resisting system: Frame members that are not relied on to provide lateral resistance to earthquake-induced forces need not satisfy the stringent requirements governing lateral-load-resisting elements. This refers mainly to the requirements for transverse reinforcement for confinement and shear. Except for the requirements noted below, non-lateral-load-resisting elements, whose primary function is the transmission of vertical loads to the foundation, need comply only with the minimum reinforcement requirements of Appendix A, in addition to those found in the main body of the *Code*.

A special requirement for non-lateral-load-resisting elements is that they be checked for adequacy with respect to a lateral displacement twice that calculated for the factored lateral forces. Under this requirement, the gravity-load system should be capable of maintaining its vertical-load-carrying capacity, without reduction, under the specified lateral displacement. Although elements of the gravity-load system need not be designed for moments related to the lateral forces, they may have to be provided with adequate confinement reinforcement in regions where plastic hinging can occur.

For gravity-load frame members subjected to factored axial compressive forces exceeding $(A_g f'_c/10)$, the following requirements relating to transverse reinforcement have to be satisfied:

$$\text{Maximum tie spacing, } s_0 \text{ (over length } l_0 \text{ from face of joint)} \leqslant \begin{cases} 8 \text{ (diameter of smallest longitudinal bar)} \\ 24 \text{ tie diameters} \\ \frac{1}{2} \text{ least cross-sectional dimension of column} \end{cases}$$

where:

$$l_0 \geqslant \begin{cases} \frac{1}{6} \text{ clear height of column} \\ \text{maximum cross-sectional dimension of column} \\ 18 \text{ in.} \end{cases}$$

The first tie is to be located within a distance of $s_0/2$ from the face of the joint. Maximum tie spacing in any part of the column = $2s_0$.

4.7 Frames in regions of moderate seismic risk: Although ACI Appendix A does not define "moderate seismic risk" in terms of a commonly accepted quantitative measure, it assumes that the probable ground motion intensity in such regions would be a fraction of that expected in a high seismic risk zone, to which the bulk of Appendix A is addressed. By the above description, an area of moderate seismic risk would correspond to Zone 2, as defined in ANSI-82.[12-12]

For regions of moderate seismic risk, the provisions for the design of structural walls given in the main body of the *ACI Code* are considered sufficient to provide the necessary toughness. The requirements of Appendix A for structures in moderate-risk areas relate mainly to frames.

The same axial compressive force of $A_g f'_c/10$, used to distinguish flexural members from columns in high seismic risk zones, also applies in regions of moderate seismicity.

For shear design of beams, columns or two-way slabs resisting earthquake effects, the magnitude of the design shear should not be less than either of the following:

(a) The sum of the shear associated with the development of nominal moment strength at each restrained end and that due to factored gravity loads. This is similar to the corresponding requirements for high-risk zones and illustrated in Fig. 12-76, except that the stress in the flexural reinforcement is taken as f_y rather than $1.25 f_y$.

(b) The maximum factored shear corresponding to the application of design gravity and earthquake forces, but with the earthquake effect taken twice the calculated value. Thus, if the critical load combination consists of dead load (D) + live load (L) + earthquake effects (E), then the design shear is to be computed from:

$$U = 0.75[1.4D + 1.7L + 2(1.87E)]$$

Detailing requirements for beams: The positive moment strength at the face of a joint shall not be less than one-third the negative moment capacity at the same section. (This compares with one-half for beams in areas of high-seismic risk.) The moment strength—positive or negative—at any section is to be no less than one-fifth the maximum moment strength at either end of the beam.

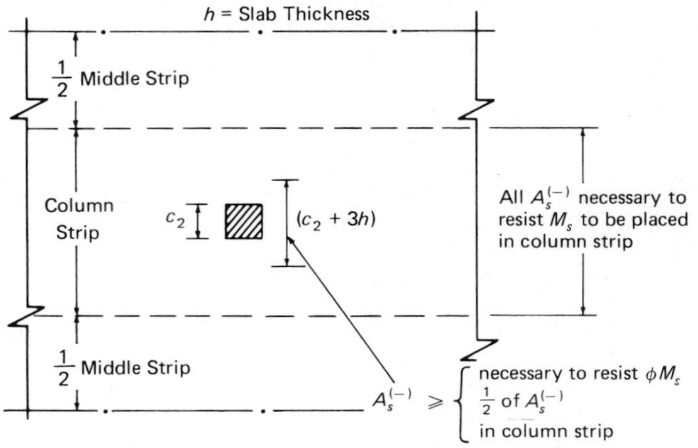

Fig. 12-86 Requirements relating to location of reinforcement in slabs without beams—frames in regions of moderate seismic risk.

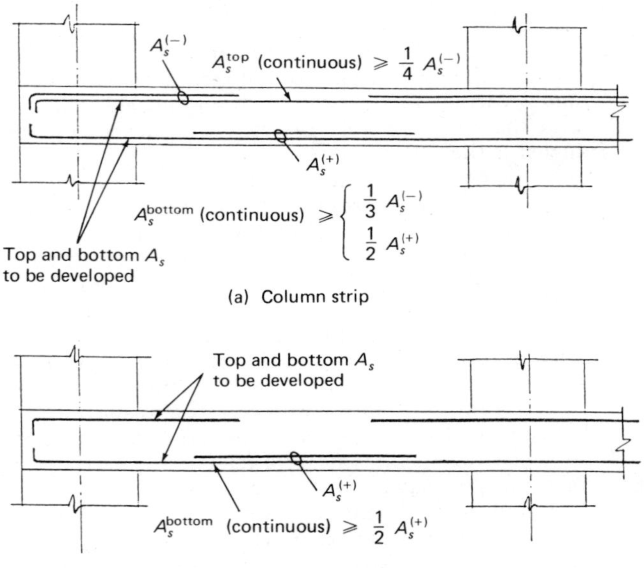

(a) Column strip

(b) Middle strip

Fig. 12-87 Requirements relating to arrangement of reinforcement in slabs without beams—frames in regions of moderate seismic risk.

Stirrup spacing requirements are identical to those for beams in regions of high seismic risk (Fig. 12-75).

Detailing requirements for columns: The tie spacing requirements for columns are identical to those for gravity load-carrying members, as given in Section 12.6.8.4.6.

Detailing requirements for two-way slabs without beams: It is worth noting that requirements for two-way slabs without beams are covered in Appendix A for frames in regions of moderate seismic risk only. This suggests that the *Code* considers the use of properly designed two-way slabs without beams as acceptable components of the lateral-load-resisting system in regions of moderate risk only.

The requirements for slabs without beams are illustrated in Figs. 12-86 and 12-87. The moment M_s in Fig. 12-86 is the portion of the factored slab moment balanced by the support moment. The factor γ_f represents the fraction of the unbalanced moment at a joint transferred by flexure, as defined in Chapter 13 of the *Code*; that is:

$$\gamma_f = \frac{1}{1 + \frac{2}{3}\sqrt{\frac{c_1 + d}{c_2 + d}}}$$

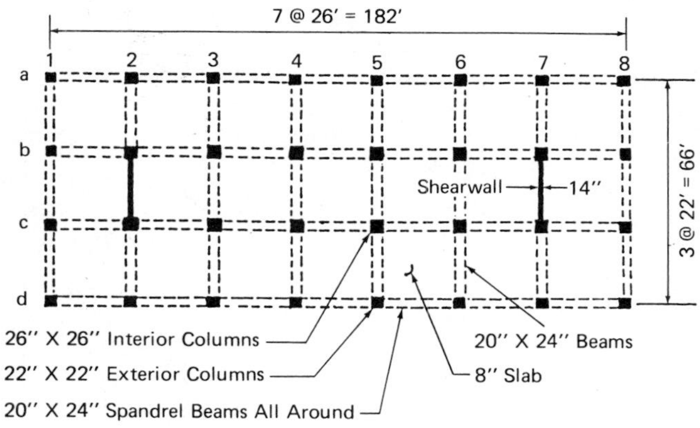

Fig. 12-88(a) Typical plan of building considered.

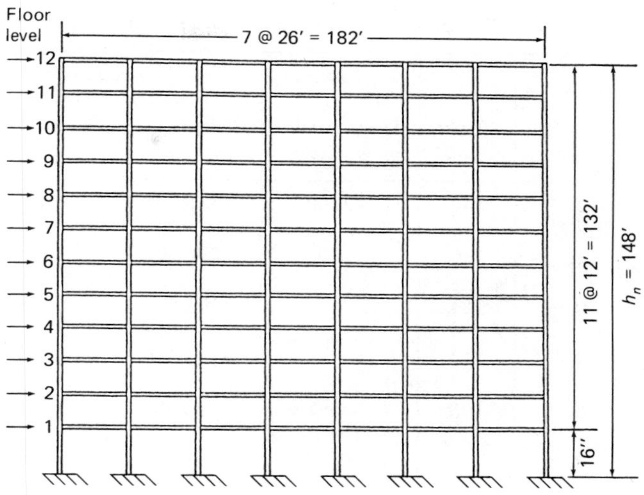

Fig. 12-88(b) Longitudinal section.

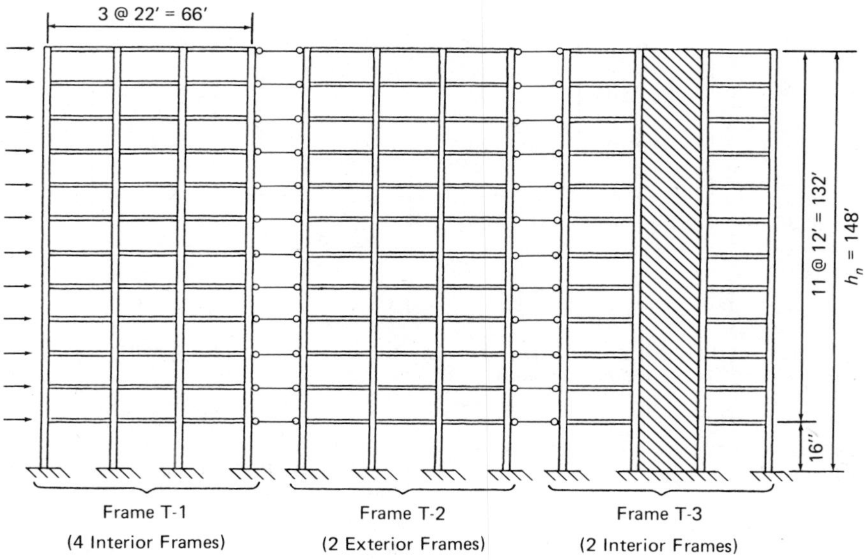

(c) Analytical model of building for lateral load analysis in transverse direction.

Fig. 12-88(c) Analytical model of building for lateral load analysis in transverse direction.

where:

c_1 = dimension of column cross section in the direction of the span for which moments are determined

c_2 = dimension of column cross section measured transverse to direction of span

d = effective depth of slab

12.7 DESIGN EXAMPLE—SAMPLE DESIGNS FOR ELEMENTS OF A 12-STORY FRAME-SHEAR WALL BUILDING

The application of the earthquake-resistant design provisions of ANSI-82[12-12] with respect to design loads and of Appendix A of ACI 318-83[12-81] to the design of typical members of a 12-story frame–shear wall building located in seismic Zone 3 will be illustrated below. Except for minor variations in local areas, the division of the United States into different seismic zones (i.e., 0, 1, 2, 3, and 4), as adopted by ANSI-82, is very similar to that used in UBC-82. In UBC seismic Zone 3, major damage associated with a Modified Mercalli intensity of VII or greater is expected. Included in seismic Zone 4 are those areas within Zone 3 that are close to certain major fault systems.

The typical plan and elevation of the structure considered are shown in Fig. 12-88(a and b). The columns and structural walls have constant cross sections throughout the height of the building,* the bases of the lowest story segments being assumed fixed. The beams and slabs also have the same dimensions at all floor levels. Although the element dimensions in this example are within the practical range, the structure itself is a hypothetical one and has been

*The uniformity in member dimensions used in this example has been adopted mainly for simplicity. The beneficial effect of such uniform stiffness along height on the dynamic response of a structure is discussed in Section 12.6.4.

TABLE 12-5 Design Lateral Forces in Transverse (Short) Direction (corresponding to entire structure)

			Seismic Forces			Wind Forces		
Floor level (from base)	*Height h_x ft*	*Story weight. w_x kips*	*$w_x h_x$ ft-kips*	*Lateral force. F_x kips*	*Story shear. ΣF_x kips*	*Wind pressure (average) psf*	*Lateral force. H_x kips*	*Story shear. ΣH_x kips*
12 (Roof)	148	2100	311,000	300*	300	23.7	25.9	25.9
11	136	2200	299,000	196	496	23.1	50.5	76.4
10	124	↓	273,000	179	675	22.6	49.4	125.8
9	112		246,000	161	836	22.0	48.1	173.9
8	100		220,000	144	980	21.4	46.7	220.6
7	88		193,000	126	1106	20.6	45.0	265.6
6	76		167,000	109	1215	19.8	43.2	308.8
5	64		141,000	92	1307	19.0	41.5	350.3
4	52		114,000	75	1382	18.2	39.8	390.1
3	40		88,000	58	1440	17.1	37.3	427.4
2	28		61,500	40	1480	15.9	34.7	462.1
1	16	2200	30,800	20	1500	14.4	36.7	498.8
		26,300	2,144,300	1500			498.8	

*Representing the sum $(F_t + F_{12})$.

Base shear, $V = ZIKCSW$, *where* $C = \dfrac{1}{15\sqrt{T}}$ *and* $T = \dfrac{0.05\,h_n}{\sqrt{D}}$.

In transverse direction, $h_n = 148'$ *and* $D = 66' \Rightarrow T = 0.91$ *sec and* $C = 0.07$. *Thus,* $V = (1.0)(1)(0.8)(0.07)(1)W = 0.56W = 0.056(26{,}300) = 1473$, *say* <u>1500 kips.</u> $F_t = 0.07TV = (.07)(.91)(1500) = 96$ *kips.*

chosen mainly for illustrative purposes. Other pertinent design data are as follows:

Service loads—vertical:

Live load:
- basic = 50 psf
- additional average load to allow for heavier basic load on corridors = 25 psf
- thus, total average live load = 75 psf

Superimposed dead load:
- average for partitions = 20 psf
- ceiling and mechanical = 10 psf
- thus, total average superimposed dead load = 30 psf

Material properties:

Concrete: $f'_c = 4\,000$ psi; $w_c = 145$ pcf

Reinforcement: $f_y = 60$ ksi

12.7.1 Determination of Design Lateral Forces

On the basis of the given data and the dimensions shown in Fig. 12-88, the weights of a typical floor* and the roof were estimated and are listed in Tables 12-5 and 12-6. The calculation of the base shear, V, using eq. (12-16), for the transverse and longitudinal directions are shown at the bottom of Tables 12-5 and 12-6, respectively. For this example, the importance factor, I, and the soil factor, S, have been assigned values of unity. The period of the structure in the transverse direction is computed using eq. (12-17c), while that in the longitudinal direction is obtained using eq. (12-17e). Note that a value of $K = 0.8$ has been used in the transverse direction where one has a frame-shear wall structure,** while a value of $K = 0.67$ has been used in the longitudinal direction, where the structure consists of moment-resisting frames.

Calculation*** of the undamped natural periods of vibration of the structure in the transverse direction, using the story weights listed in Table 12-5 and member stiffnesses based on gross concrete sections, gave a value for the fundamental period of 1.34 seconds (compared to the T-value of 0.91 second given by the approximate formula in the *Code*). The calculated mode shapes as well as the corresponding periods of the first five modes of vibration of the structure in the transverse direction are shown in Fig. 12-89. Note

*The weight of a typical floor includes that of all elements located between two imaginary parallel planes passing through midheight of the columns above and below the floor considered.

**The requirements relating to the use of $K = 0.8$ for dual systems are that the moment-resisting space frame be capable of resisting at least 25% of the prescribed seismic forces and that the individual elements satisfy the requirements of ACI Appendix A for special ductile moment-resisting frames and structural walls.

***Using the computer program described in Ref. 12-97.

TABLE 12-6 Design Lateral Forces in Longitudinal Direction (corresponding to entire structure)

		Seismic Forces				*Wind Forces*		
Floor level (from base)	*Height h_x ft*	*Story weight. w_x kips*	*$w_x h_x$ ft-kips*	*Lateral force. F_x kips*	*Story shear. ΣF_x kips*	*Wind pressure (average psf)*	*Lateral force. H_x kips*	*Story shear. ΣH_x kips*
12 (Roof)	148	2100	Same as for transverse direction (arrow from 12 to 1)	239*	239	19.7	7.8	7.8
11	136	2200		149	388	19.3	15.3	23.1
10	124	↓ (arrow from 10 to 2)		136	524	18.8	14.9	38.0
9	112			122	646	18.0	14.3	52.3
8	100			109	755	17.2	13.6	65.9
7	88			96	851	16.6	13.2	79.1
6	76			83	934	16.0	12.7	91.8
5	64			70	1104	15.0	11.9	103.7
4	52			57	1061	14.0	11.1	114.8
3	40			44	1105	13.0	10.3	125.1
2	28			30	1135	11.8	9.4	134.5
1	16	2200		15	1150	10.1	9.3	143.8
		26,300	2,144,300	1150			143.8	

$^*(F_t + F_{12})$.

In longitudinal direction, $T = C_T h_n^{3/4}$, *where* C_T *(concrete frames)* $= 0.025$

$$= 0.025\,(148)^{3/4} = 1.06 \text{ sec.}$$

$$C = \frac{1}{15\sqrt{T}} = \frac{1}{15\sqrt{106}} = 0.65.$$

Base shear, $V = ZIKCSW = (1.0)(1)(0.67)(0.065)(1)\,W = 0.0436\,W$
$= 0.0436(26{,}300) = 1146$, *say* $\underline{1150 \text{ kips}}$.

$F_t = 0.07TV = (.07)(1.06)(1150) = 85$ *kips.*

that the mode shapes indicate only the relative displacements of the story masses (assumed concentrated at the floor levels). The maximum displacement for each mode has been set equal to unity.

The lateral seismic design forces resulting from the distribution of the base shear in accordance with eq. (12-19) are listed in Tables 12-5 and 12-6. As an example, the seismic lateral force F_x at the 10th floor level in the transverse direction is given by:

$$F_{10} = \frac{(V - F_t)\,w_x h_x}{\sum_{i=1}^{n} w_i h_i} = \frac{(1500 - 96)(124)(2200)}{2{,}144{,}300}$$

$$= 179 \text{ kips}$$

Also shown in the tables are the story shears corresponding to the distributed seismic forces.

For comparison, the wind forces and story shears corresponding to a basic wind speed of 75 mph and Exposure B (urban and suburban areas), computed as prescribed in ANSI-82, are shown for each direction in Tables 12-5 and 12-6.

Analyses of the structure in both directions under the respective seismic and wind loads, and assuming no torsional effects, were carried out using a plane frame computer program.[12-98] For the purpose of analyzing the structure in the transverse direction, the model shown in Fig. 12-88(c) was used. This model consists of three different frames linked by hinged rigid bars at the floor levels to impose equal horizontal displacements at these levels. (This device is used to model the effect of floor slabs which may be assumed as rigid in their own planes.) Frame T-1 represents the four identical interior frames along lines 3, 4, 5, and 6 which have been lumped together in this single frame, while Frame T-2 represents the two exterior frames along lines 1 and 8. The third frame, T-3, represents the two identical frame–shear wall systems along lines 2 and 7.

In the longitudinal direction, two linked frames, each similar to the frame shown in Fig. 12-88(b), were used to represent the two identical exterior frames L-1 along lines a and d and the two identical interior frames L-2 along lines b and c (see Fig. 12-88a).

The lateral displacements due to both seismic and wind forces listed in Tables 12-5 and 12-6 are plotted in Fig.

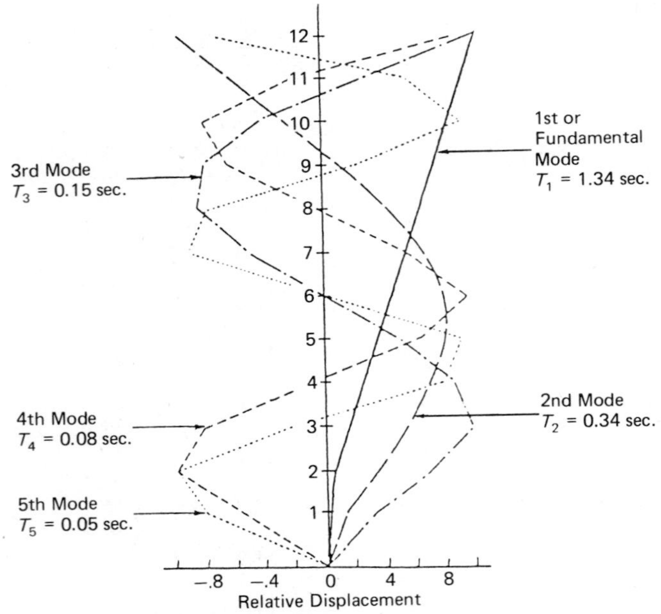

Fig. 12-89 Undamped natural modes and periods of vibration of structure in transverse direction.

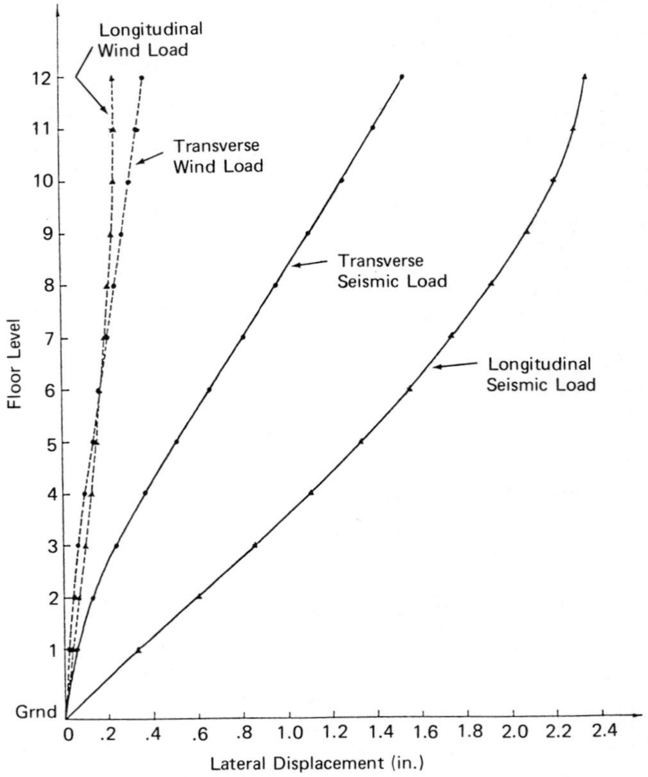

Fig. 12-90 Lateral displacements under seismic and wind loads.

12-90. Although the seismic forces used to obtain the curves of Fig. 12-90 are approximate, the results shown still serve to draw the distinction between wind and seismic forces, that is, the fact that the former are external forces the magnitudes of which are proportional to the exposed surface, while the latter are inertial forces depending primarily on the mass and stiffness properties of the structure. Thus, while the ratio of the total wind force in the transverse direction to that in the longitudinal direction (see Tables 12-5 and 12-6) is about 3.5, the corresponding ratio for seismic forces is only 1.3. As a result of this and the smaller stiffness of the structure in the longitudinal direction, the displacement due to seismic forces in the longitudinal direction is significantly greater than that in the transverse direction, although the displacements due to wind in both directions are about the same. The typical deflected shapes associated with predominantly cantilever or flexure type structures (as in the transverse direction) and shear (open frame) type buildings (as in the longitudinal direction) are evident in Fig. 12-90. The average deflection indices, that is, the ratio of the lateral displacement at the top to the total height of the structure, are 1/4790 for wind and 1/1170 for seismic loads in the transverse direction. The corresponding values in the longitudinal direction are 1/7510 for wind and 1/760 for seismic loads.

An idea of the distribution of lateral loads among the different frames making up the structure in the transverse direction may be obtained from Table 12-7, which lists

TABLE 12-7 Distribution of Horizontal Seismic Story Shears among the Three Transverse Frames Shown in Fig. 12-88(c)

	Frame T-1 (4 interior frames)		*Frame T-2 (2 exterior frames)*		*Frame T-3 (2 interior frames with shear walls)*		
Story Level	*Story Shear*	*% of Total*	*Story Shear*	*% of Total*	*Story Shear*	*% of Total*	*Total Story Shear kips*
12	364	121	153	51	−217	−72	300
11	255	51	115	23	126	26	496
10	297	44	129	19	249	37	675
9	304	36	133	16	399	48	836
8	316	32	138	14	526	54	980
7	321	29	140	13	645	58	1106
6	319	26	140	12	756	62	1215
5	307	24	134	10	866	66	1307
4	285	21	124	9	973	70	1382
3	242	17	105	7	1093	76	1440
2	205	14	88	6	1187	80	1480
1	48	3	20	1	1432	96	1500

the portion of the total story shear at each level resisted by each of the three lumped frames. Note that at the top (12th story level), the lumped frame T-1 takes 121% of the total story shear. This reflects the fact that in frame–shear wall systems of average proportions, interaction between frame and wall under lateral loads results in the frame "supporting" the wall at the top, while at the base most of the horizontal shear is taken by the wall. Table 12-7 indicates that for the structure considered, the two frames with walls take 96% of the shear at the base in the transverse direction.

To illustrate the design of two typical beams on the sixth floor of an interior frame, the results of analysis of the structure in the transverse direction under seismic loads have been combined for the beams, using eq. (12-27), with results obtained from a gravity load analysis of a single-story bent including those beams (using the computer program described in Ref. 12-98). The results are listed in Table 12-8. Similar values for typical exterior and interior columns on the second floor of the same interior frame are shown in Table 12-9. Corresponding design forces for the structural wall section at the first floor of Frame T-3 of Fig. 12-88(c) are listed in Table 12-10. The last column in Table 12-10 lists the axial load on the boundary elements (the 26″ X 26″ columns forming the flanges of the structural walls) calculated in accordance with the ACI requirement that these be designed to carry all factored vertical loads on the walls, including self-weight, gravity loads, and vertical forces due to earthquake-induced overturning moments. The loading condition associated with this requirement is illustrated in Fig. 12-85. In both Tables 12-9 and 12-10, the additional forces due to the effects of horizontal torsional moments corresponding to the minimum ANSI-prescribed eccentricity of 5% of the building dimension perpendicular to the direction of the applied forces have been included. These additional forces were calculated using the procedure described in Section 12.6.8.2.2.

It is pointed out that for buildings located in seismic Zones 3 and 4, (i.e., high seismic risk areas) the detailing requirements for ductility prescribed in ACI Appendix A have to be met even when the design of a member is governed by wind loading rather than by seismic loads.

TABLE 12-8 Summary of Design Moments for Typical Beams on Sixth Floor of Interior Transverse Frames along Lines 3 through 6 (Fig. 12-88a)

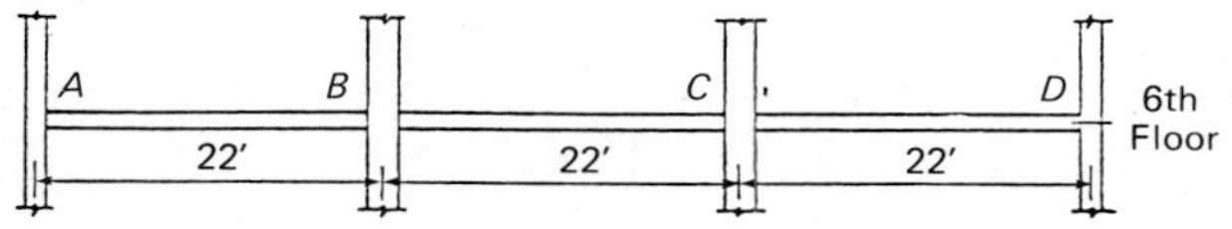

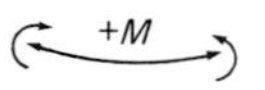

$$U = \begin{cases} 1.4D + 1.7L & (12\text{-}27a) \\ 0.75\,(1.4D + 1.7L \pm 1.87E) & (12\text{-}27b) \\ 0.9D \pm 1.43E & (12\text{-}27c) \end{cases}$$

		Design Moment in ft-kips		
Beam AB		*A*	*Near Midspan of AB*	*B*
Eq. (12-27a)		−86	(+130)	−179
Eq. (12-27b)	sidesway to right	+127	+126	(−324)
	sidesway to left	(−257)	+70	+56
Eq. (12-27c)	sidesway to right	(+162)	+79	−265
	sidesway to left	−230	+22	(+125)

Beam BC		*B*	*Midspan*	*C*
Eq. (12-27a)		−179	(+93)	−179
Eq. (12-27b)	sidesway to right	+89	+80	(−357)
	sidesway to left	(−357)	+60	+89
Eq. (12-27c)	sidesway to right	(+157)	+47	−297
	sidesway to left	−297	+26	(+157)

12.7.2 Design of Flexural Member AB

The aim is to determine the flexural and shear reinforcement for the beam AB on the sixth floor of a typical interior transverse frame. The beam carries a dead load of 3.7 kips/ft of span and a live load of 1.95 kips/ft. The design (factored) moments are as indicated in the sketch accompanying Table 12-8. The beam has dimensions $b = 20$ in. and $d = 21.5$ in. The slab is 8 in. thick. Use $f'_c = 4{,}000$ psi and $f_y = 60{,}000$ psi.

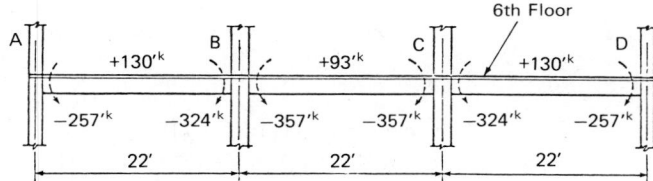

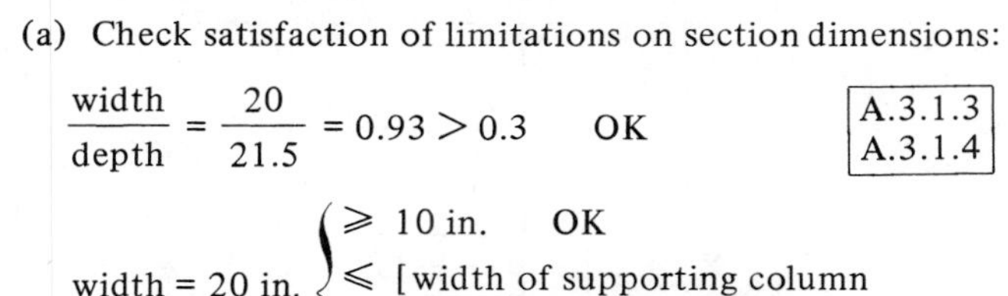

In the following solution, the boxed-in section numbers at the right-hand margin correspond to ACI 318-83.

(a) Check satisfaction of limitations on section dimensions:

$$\frac{\text{width}}{\text{depth}} = \frac{20}{21.5} = 0.93 > 0.3 \qquad \text{OK}$$ A.3.1.3 A.3.1.4

$$\text{width} = 20 \text{ in.} \begin{cases} \geqslant 10 \text{ in.} \qquad \text{OK} \\ \leqslant [\text{width of supporting column} + 1.5 \times \text{depth of beam}] \\ = 26 + 1.5(21.5) = 58.25 \text{ in.} \qquad \text{OK} \end{cases}$$

(b) Determine required flexural reinforcement:

(1) Negative moment reinforcement at support B: Since the negative flexural reinforcement for both beams AB and BC at joint B will be provided by the same continuous bars, the larger negative moment at joint B will be used. In the following calculations, the effect of any compressive reinforcement will be neglected.

From $C = 0.85 f'_c ba = T = A_s f_y$:

$$a = \frac{A_s f_y}{.85 f'_c b} = \frac{60 A_s}{(.85)(4)(20)} = 0.882 A_s$$

TABLE 12-9 Summary of Design Moments and Axial Loads for Typical Columns on Second Floor of Interior Transverse Frames along Lines 3 through 6 (Fig. 12-88a)

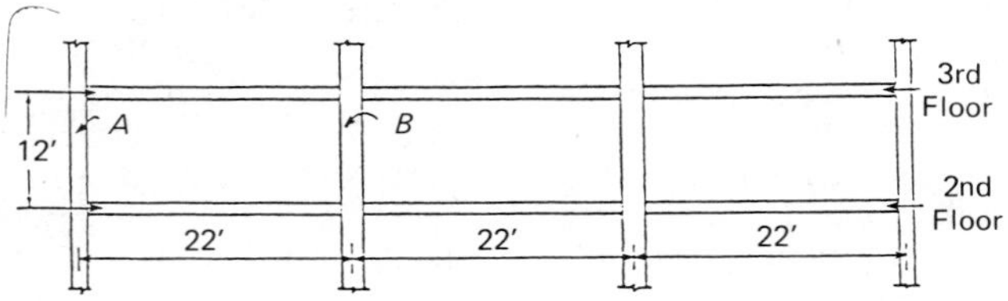

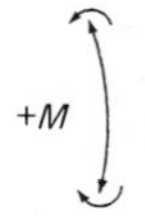

$$U = \begin{cases} 1.4D + 1.7L & (12\text{-}27a) \\ 0.75\,(1.4D + 1.7L \pm 1.87E) & (12\text{-}27b) \\ 0.9D \pm 1.43E & (12\text{-}27c) \end{cases}$$

		Exterior Column A			*Interior Column B*		
		Axial Load, kips	*Moment, ft-k**		*Axial Load, kips*	*Moment, ft-k**	
			Top	*Bottom*		*Top*	*Bottom*
Eq. (12-27a)		(-930)	-95	+95	(-1678)	+53	-53
Eq. (12-27b)	sidesway to right	-526	-15	-9	-1225	(+173	-207)
	sidesway to left	-870	(-133	+151)	-1293	-93	+127
Eq. (12-27c)	sidesway to right	-328	+25	-44	-857	+146	-181
	sidesway to left	-680	-101	+120	-925	-126	+161

**Including moments corresponding to horizontal torsional shears due to minimum story shear eccentricity.*

$$M_u \leqslant \phi M_n = \phi A_s f_y [d - a/2]$$

$$(357)(12) = (.90)(60)A_s[21.5 - (.5)(.882A_s)]$$

$$A_s^2 - 48.75A_s + 179.9 = 0$$

or:

$$A_s = 4.02 \text{ in.}^2$$

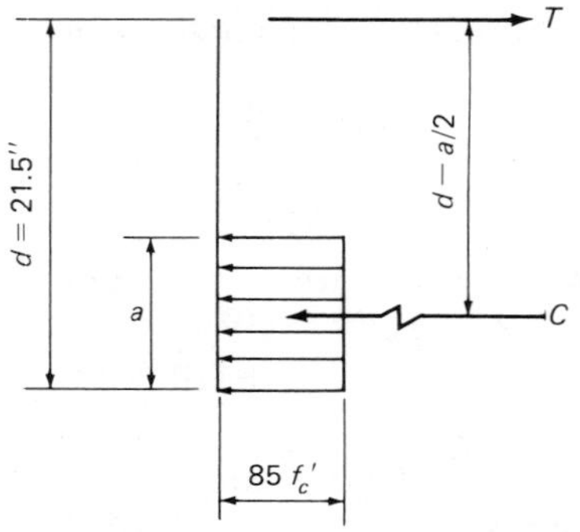

Alternatively, convenient use may be made of design charts for singly reinforced flexural members with rectangular cross section given in standard references.[12-99, 12-100] These charts generally present curves relating the coefficient of resistance, $R_u = M_u/bd^2$ and the reinforcement ratio, $\rho = A_s/bd$.

Use four No. 9 bars, $A_s = 4.00$ in.^{2} This gives a negative moment capacity at support B of $\phi M_n = 355$ ft-kips.

Check satisfaction of limitations on reinforcement ratio:

$$\rho = \frac{A_s}{bd} = \frac{4.0}{(20)(21.5)}$$

$$= 0.0093 > \rho_{min} = \frac{200}{f_y} = 0.0033$$

$$\text{and} \quad < \rho_{max} = 0.025 \quad \text{OK}$$

A.3.2.1

(2) Negative moment reinforcement at support A:

$$M_u = 257 \text{ ft-kips}$$

As at support B, $a = 0.882A_s$. Substitution into:

$$M_u = \phi A_s f_y [d - a/2]$$

yields $A_s = 2.82$ in.^{2}

Use three No. 9 bars, $A_s = 3.0$ in.^{2} This gives a negative moment capacity at support *A* of $\phi M_n = 272$ ft-kips.

(3) Positive moment reinforcement at supports: A positive moment capacity at the supports equal to at least 50% of the corresponding negative moment capacity is required; i.e., A.3.2.2

TABLE 12-10 Summary of Design Loads on Structural Wall Section at First Floor Level of Transverse Frame along Line 2 (or 7) (Fig. 12-88a)

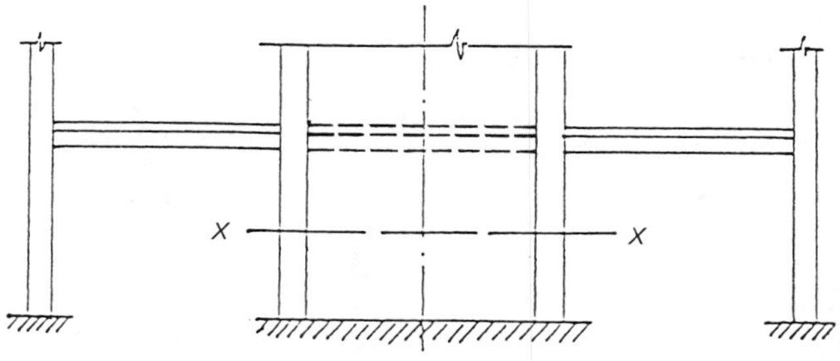

$$U = \begin{cases} 1.4D + 1.7L & (12\text{-}27a) \\ 0.75\,(1.4D + 1.7L \pm 1.87E) & (12\text{-}27b) \\ 0.9D \pm 1.43E & (12\text{-}27c) \end{cases}$$

	Design Forces Acting on Entire Structural Wall			
	Axial Load, kips	*Bending* (overturning) Moment, ft-kips*	*Horizontal* Shear, kips*	*Axial Load** on Boundary Element, kips*
Eq. (12-27a)	4598	nominal	nominal	2326
Eq. (12-27b)	3449	53,280	1129	4166*
Eq. (12-27c)	2506	54,464	1153	3744*

**Includes effect of horizontal torsion due to accidental eccentricity of story shear (ecc. = 0.05 × 182 = 9.1 ft.)*

***Based on loading condition illustrated in Fig. 12-85.*

$$\text{min } M_u^+ \text{ (at support A)} = \frac{272}{2} = 136 \text{ ft-kips}$$

which is less than M_{max}^+ at A (see Table 12-8) = 162 ft-kips.

min M_u^+ (at support B for both spans AB and BC)

$$= \frac{355}{2} = 178 \text{ ft-kips}$$

Note that the above required capacities are greater than the design positive moments near the mid-spans of both beams AB and BC.

(4) Positive moment reinforcement at midspan—to be made continuous to supports:

$$a = \frac{A_s f_y}{.85 f_c' \, b} = \frac{60A_s}{(.80)(4)(60)} = 0.294A_s$$

Substituting into:

$$M_u = (162)(12) = \phi A_s f_y [d - a/2]$$

yields A_s required at A = 1.7 in.^{2} Similarly, corresponding to M_u^+ (required capacity at support B) = 178 ft-kips, A_s (required) = 1.85 in.^{2}

Use three No. 7 bars continuous through both spans. A_s = 1.80 in.^{2} This provides a positive moment capacity of 172 ft-kips—about 3% less than that required at support B.

Check:

$$\rho = \frac{1.8}{(20)(21.5)} = 0.0042 > \rho_{min} = \frac{200}{f_y}$$

$$= 0.0033 \quad \text{OK} \quad \boxed{10.5.1}$$

(c) Calculate required length of anchorage of flexural reinforcement in exterior column:

$$\text{Development length, } l_{dh} \geqslant \begin{cases} f_y \, d_b / 65\sqrt{f_c'} \\ 8\, d_b \\ 6 \text{ in.} \end{cases}$$

(plus standard 90° hook located in confined region of column) $\boxed{\text{A.6.4.1}}$

For the No. 9 (top) bars (bend radius ⩾ 5 d_b):

$$l_{dh} \geqslant \begin{cases} (60{,}000)(1.128)/65\sqrt{4000} = \underline{17 \text{ in.}} \\ (8)(1.128) = 9.0 \text{ in.} \\ 6 \text{ in.} \end{cases}$$

For the No. 7 (bottom) bars (bend radius ⩾ 4 d_b):

$$l_{dh} \geqslant \begin{cases} (60{,}000)(.875)/65\sqrt{4000} = \underline{13 \text{ in.}} \\ (8)(.875) = 7 \text{ in.} \\ 6 \text{ in.} \end{cases}$$

See the sketch for detail of flexural reinforcement anchorage in the exterior column. Note that the development length l_{dh} is measured from the near face of the column to the far edge of the vertical 12-bar-diameter extension (see Fig. 12-74).

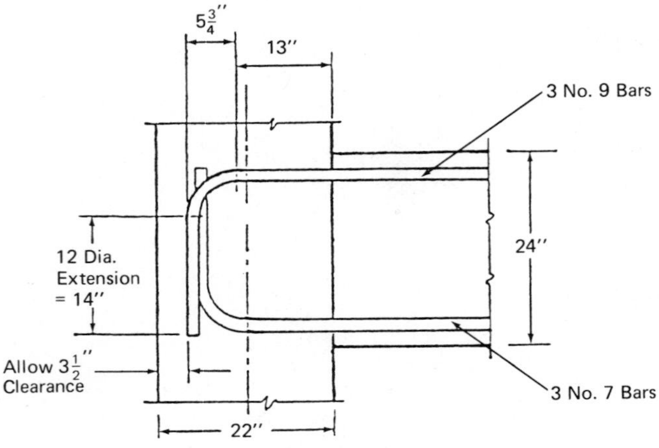

(d) Determine shear reinforcement requirements: Design for shears corresponding to end moments obtained by assuming the stress in the tensile flexural reinforcement equal to $1.25f_y$ and a strength reduction factor, $\phi = 1.0$, plus factored gravity loads (see Fig. 12-76). A.7.1.1

Table 12-11 shows values of design end shears corresponding to the two loading cases to be considered. In the table:

$$w_u = 0.75(w_D + w_L) = 0.75[1.4(3.7) + 1.7(1.95)]$$
$$= 6.37 \text{ kips/ft}$$

Appendix A requires that the contribution of concrete to shear resistance, V_c, be neglected if the earthquake-induced shear force (corresponding to the "probable flexural strengths" at beam ends calculated using $f_s = 1.25\ f_y$ and $\phi = 1.0$) is greater than one-half of the total design shear, and the axial compressive force including earthquake effects is less than $(A_g f_c'/20)$. A.7.2.1

For sidesway to the right, the shear at end B due to the plastic end moments in the beam (see Table 12-11) is:

$$V_B = \frac{238 + 482}{20} = 36 \text{ kips}$$

which is less than 50% of the total design shear, V_u = 99.7 kips. Therefore, the contribution of concrete to shear resistance can be considered in determining shear reinforcement requirements.

At right end B, V_u = 99.7 kips. Using $V_c = 2\sqrt{f_c'}\ b_w d = 2\sqrt{4000}\ (20)(21.5)/1000$ = 54.4 kips: 11.3.1.1

$$\phi V_s = V_u - \phi V_c = 99.7 - 0.85 \times 54.4$$
$$= 53.5 \text{ kips}$$
$$V_s = 62.9 \text{ kips}$$

TABLE 12-11 Determination of Design Shear Forces for Beam Spans

	$V_u = \dfrac{M_{AB}^{\pm} + M_{BA}^{\mp}}{l} + 0.75\left(\dfrac{w_u l}{2}\right)$	
Loading	*A*	*B*
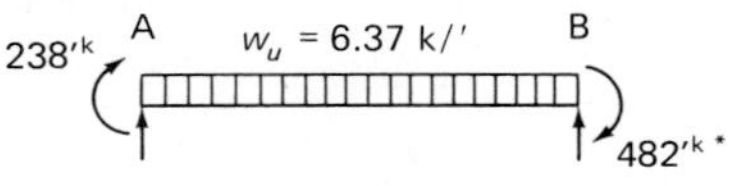 Sidesway to right	27.7 kips	(99.7 kips)
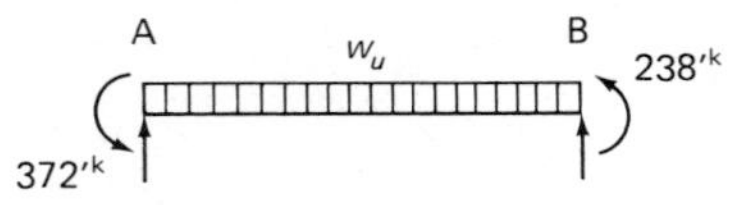 Sidesway to left	(94.2 kips)	33.2 kips

**When compression reinforcement is considered, a value about 0.5% greater than the above value is obtained.*

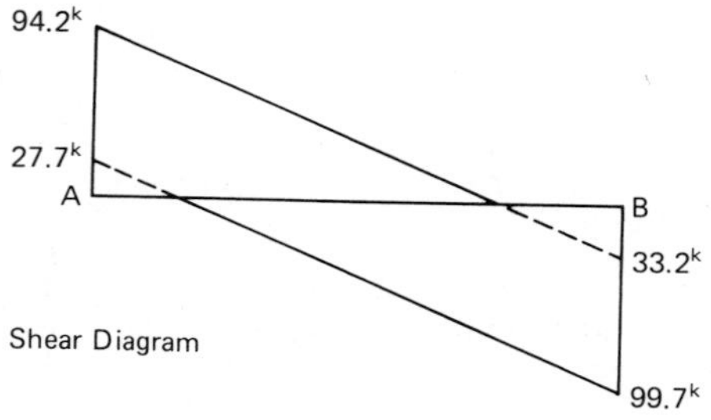

Shear Diagram

Required spacing of # 3 closed stirrups (hoops)—A_v (2 legs) = 0.22 in.^{2}:

$$s = \frac{A_v f_y d}{V_s} = \frac{(.22)(60)(21.5)}{62.9}$$

$$= 4.5 \text{ in.}$$ 11.5.6.2

Maximum allowable hoop spacing within distance $2d$ = 2(21.5) = 43 in. from faces of supports:

$$s_{max} = \begin{cases} d/4 = 21.5/4 = 5.4 \text{ in.} \\ 8 \times \text{(dia. of smallest long. bar)} = 8(.875) = 7 \text{ in.} \\ 24 \times \text{(dia. of hoop bars)} = 24(.375) = 9 \text{ in.} \\ 12 \text{ in.} \end{cases}$$

A.3.3.2

Beyond distance $2d$ from the supports, maximum spacing of stirrups:

$$s_{max} = d/2 = 10.5 \text{ in.}$$ A.3.3.4

Use #3 stirrups spaced as sketched.

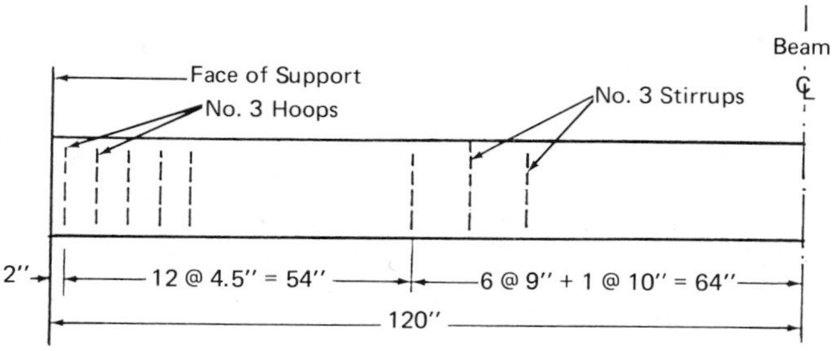

Where the loading is such that inelastic deformation may occur at intermediate points within the span (e.g., due to concentrated loads near midspan), the spacing of hoops will have to be determined in a manner similar to that used above for segments near supports. In the present example, the maximum positive moment near midspan (see Table 12-8) is well below the positive moment capacity provided by the three No. 7 continuous bars. A.3.3.1

Note that lap splices in longitudinal reinforcement should not be used within joints or within a distance of $2d$ from the faces of supports. Where used outside of these regions, such splices are to be confined over the length of the lap by hoops or spirals with a maximum spacing or pitch of $d/4$ or 4 in. A.3.2.3

(e) Negative reinforcement cutoff points: For the purpose of determining cutoff points for the negative reinforcement, a moment diagram corresponding to plastic end moments and 0.9 times the dead load will be used. The cutoff point for two of the four No. 9 bars at the top, near support B of beam AB, will be determined.

With the negative moment capacity of a section with two No. 9 top bars = 186 ft-kips (calculated using $f_s = f_y$ = 60 ksi and ϕ = 0.9), the distance from the face of of the right support B to where the moment under the loading considered equals 186 ft-kips is readily obtained by summing moments about section a–a in the figure below and equating these to −186 ft-kips. Thus:

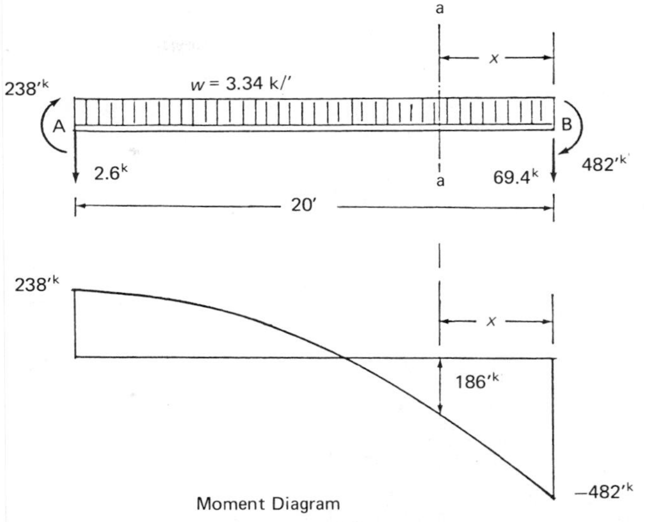

$$69.4x - 482 - 3.34\frac{x^2}{2} = -186$$

Solution of the above equation gives x = 4.82 ft. Hence, two of the four No. 9 top bars near support B may be cut off at (noting that for a No. 9 bar, l_{dh} = 17 in.):

$$x + 3.5\,l_{dh} = 4.8 + (3.5)(17)/12$$

$$= 9.8 \text{ ft from face of right support B}$$

A.6.4.2

Since this indicated cutoff point is practically at midspan, examine the alternative of cutting only one of the four No. 9 top bars near support B and making the three No. 9 bars continuous along the top of the beam.

By a procedure similar to that given above, the cutoff point for one of the four No. 9 top bars is obtained as 8.25 ft. This scheme will be adopted, *using a cutoff point for one No. 9 top bar 8.5 ft from the face of support B and making three No. 9 bars continuous along the top.*

(f) Flexural reinforcement splices: Lap splices of flexural reinforcement should not be placed within a joint, within a distance $2d$ from faces of supports, or at locations of potential plastic hinging. A.3.2.3

Note that all lap splices have to be confined by hoops or spirals with a maximum spacing or pitch of $d/4$ or 4 in. over the length of the lap. A.3.2.3

(1) Bottom bars, No. 7: The bottom bars along most of the length of the beam may be subjected to maximum stress. Use Class C splice. Required length of splice = $1.7 l_d \geqslant 12$ in., where: 12.16.1

$$l_d = 0.04 A_b f_y/\sqrt{f'_c} \geqslant 0.0004\, d_b f_y$$

$$= (.04)(.60)(60{,}000)/\sqrt{4000}$$

$$\geqslant (.0004)(.875)(60{,}000)$$

$$= 22.8 \text{ in (governs)} > 21 \text{ in.}$$ 12.2.2

Class C splice length = (1.7)(22.8) = <u>39 in.</u>

(2) Top bars, No. 9: Since the midspan

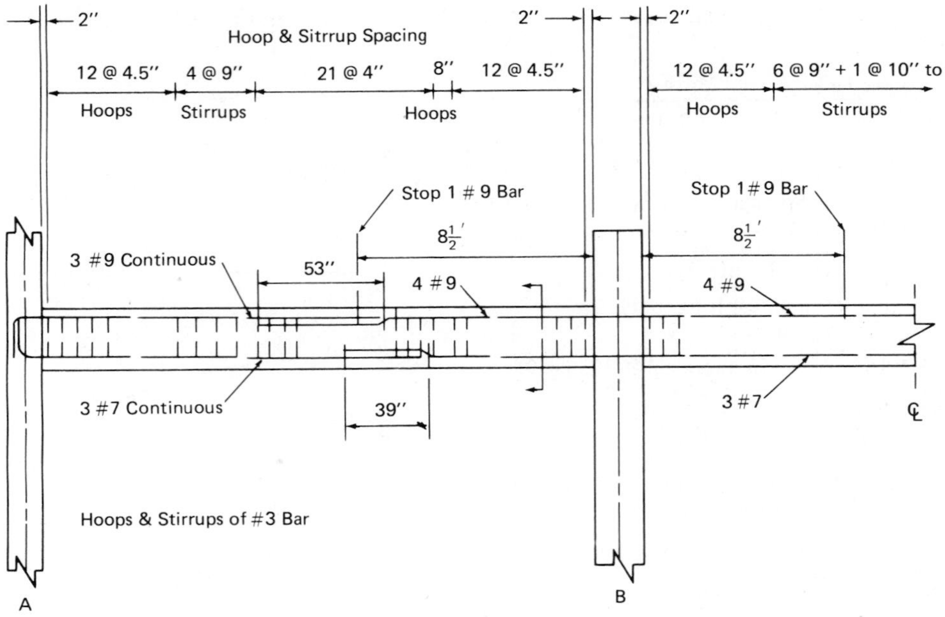

Fig. 12-91(a) Flexural member reinforcement.

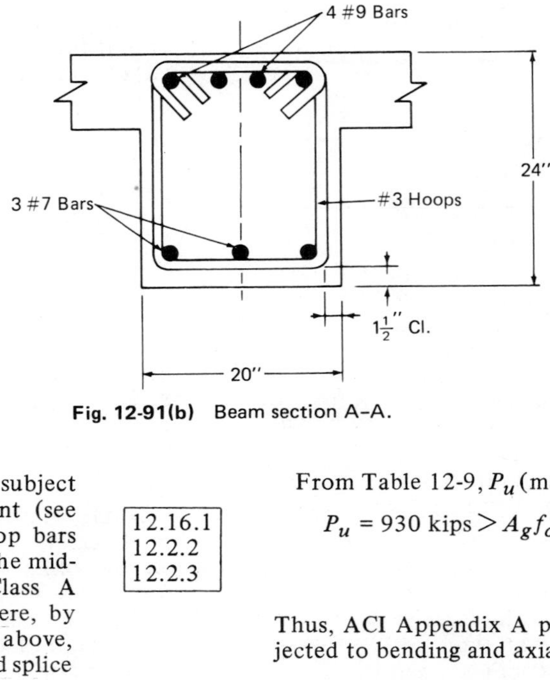

Fig. 12-91(b) Beam section A–A.

portion of the span is always subject to a positive bending moment (see Table 12-8), splices in the top bars should be located at or near the mid-span. Required length of Class A splice = $1.4\, l_d \geqslant 12$ in., where, by using the same expression given above, one obtains $l_d = 38$ in. Required splice length = 1.4(38) = 53 in.

12.16.1
12.2.2
12.2.3

(g) Detail of beam. See Fig. 12-91.

12.7.3 Design of Frame Column A

The aim is to design the transverse reinforcement for the exterior tied column on the second floor of a typical transverse interior frame, that is, one of the frames in Frame T-1 of Fig. 12-88(c). The column dimension has been established as 22 in. square and, on the basis of the different combinations of axial load and bending moment corresponding to the three loading conditions listed in Table 12-9, *eight No. 7 bars arranged in a symmetrical pattern* have been found adequate.[12-101, 12-102] Assume the same beam section framing into the column as considered in Section 12.7.2 Use $f_c' = 4{,}000$ psi and $f_y = 60{,}000$ psi.

From Table 12-9, P_u (max) = 930 kips:

$$P_u = 930 \text{ kips} > A_g f_c'/10 = (22)^2(4)/10 = 194 \text{ kips}$$

A.4.1

Thus, ACI Appendix A provisions governing members subjected to bending and axial load apply.

(a) Check satisfaction of vertical reinforcement limitations and moment capacity requirements:

(1) Reinforcement ratio:

$$0.01 \leqslant \rho \leqslant 0.06$$

$$\rho = \frac{A_{st}}{A_g} = \frac{8(.60)}{(22)(22)} = 0.01 \quad \text{OK}$$

A.4.3.1

(2) Moment strength of columns relative to that of framing beam in transverse direction:

$$\sum M_e \text{ (columns)} \geqslant \tfrac{6}{5} \sum M_g \text{ (beams)}$$

A.4.2.2

From Section 12.7.2, ϕM_n^- of beam at A = 272 ft-kips, corresponding to sidesway to left.

From Table 12-9, maximum axial load on Column

A at the second floor level for sidesway to left, $P_u = 870$ kips. Using the P–M interaction charts given in ACI SP-17A,[12-101] the moment capacity of the column section, corresponding to $P_u = \phi P_n = 870$ kips, $f'_c = 4$ ksi, $f_y = 60$ ksi, $\gamma^* = 0.75$, and $\rho = 0.01$, is obtained as $\phi M_n = M_e = 284$ ft-kips. With the same-size column above and below the beam, total moment capacity of columns = 2(284) = 568 ft-kips. Thus:

$$\sum M_e = 568 > \tfrac{6}{5} M_g = (6)(272)/5 = 326 \text{ ft-k} \qquad \text{OK}$$

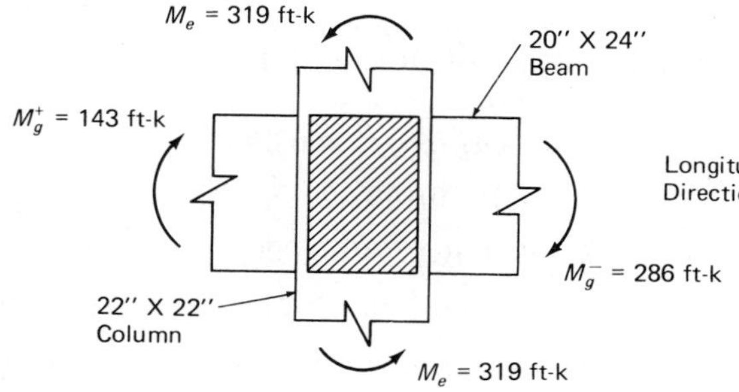

(3) Moment strength of columns relative to that of framing beams in longitudinal direction: Since the columns considered here are located in the center portion of the exterior longitudinal frames, the axial forces due to seismic loads in the longitudinal direction are negligible. (Analysis of the longitudinal frames under seismic loads indicated practically zero axial forces in the exterior columns of the four transverse frames represented by Frame T-1 in Fig. 12-88.) Under an axial load of $0.75\,[1.4D + 1.7L + 1.87\,(E = 0)] = (.75)(930) = 698$ kips (see Table 12-9), the moment capacity of the column section with eight No. 7 bars is obtained as $\phi M_n = M_e = 319$ ft-kips.

If we assume a ratio for the negative moment reinforcement of approximately 0.0075 in the beams of the exterior longitudinal frames ($b_w = 20$ in., $d = 21.5$ in.):

$$A_s = \rho b_w d = (\text{approx.})\,(.0075)\,(20)\,(21.5)$$

$$= 3.23 \text{ in.}^2$$

Assume four No. 8 bars, $A_s = 3.16$ in.^{2}

Negative moment capacity of beam:

$$a = \frac{A_s f_y}{0.85 f'_c b_w} = \frac{(3.16)(60)}{(.85)(4)(20)} = 2.79 \text{ in.}$$

$$\phi M_n^- = M_g^- = \phi A_s f_y (d - a/2)$$

$$= (.90)(3.16)(60)[(21.5 - 1.39)/12]$$

$$= 286 \text{ ft-kips}$$

Assume a positive moment capacity of the beam on the opposite side of the column equal to one-half the negative moment capacity calculated above, or 143 ft-kips.

*γ = ratio of distance between centroids of outer rows of bars and dimension of cross-section, in the direction of bending.

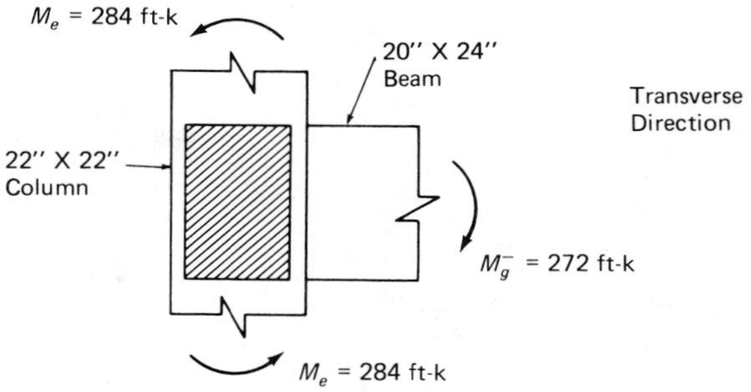

Total moment capacity of beams framing into joint in longitudinal direction, for sidesway in either direction:

$$\sum M_g = 284 + 143 = 429 \text{ ft-kips}$$

$$\sum M_e = 2(319) = 638 \text{ ft-kips} >$$

$$\tfrac{6}{5} \sum M_g = \tfrac{6}{5}(429) = 515 \text{ ft-kips} \qquad \text{OK}$$

A.4.2.2

(b) Determine transverse reinforcement requirements:

(1) Confinement reinforcement (see Fig. 12-78): Transverse reinforcement for confinement is required over distance l_0 from column ends, where:

$$l_0 \geqslant \begin{cases} \text{depth of member} = \underline{22 \text{ in.}} \text{ (governs)} \\ \tfrac{1}{6} \text{ clear height} = (10 \times 12)/6 = 20 \text{ in.} \\ 18 \text{ in.} \end{cases}$$

A.4.4.4

Maximum allowable spacing of rectangular hoops:

$$s_{\max} = \begin{cases} \tfrac{1}{4} \text{ smallest dimension of column} = 22/4 \\ \qquad = 5.5 \text{ in.} \\ \underline{4 \text{ in.}} \text{ (governs)} \end{cases}$$

A.4.4.2

Required cross-sectional area of confinement reinforcement in the form of hoops:

$$A_{sh} \geqslant \begin{cases} 0.12\, s h_c \dfrac{f'_c}{f_{yh}} \\ 0.3\, s h_c \left(\dfrac{A_g}{A_{ch}} - 1\right) \dfrac{f'_c}{f_{yh}} \end{cases}$$

A.4.4.1

where:

s = spacing of transverse reinforcement (in.)
h_c = cross-sectional dimension of column core, measured c–c of confining reinforcement (in.)
A_{ch} = core area of column section, measured outside to outside of transverse reinforcement (in.^{2})
f_{yh} = specified yield strength of transverse reinforcement (psi)

For a hoop spacing of 4 in., $f_{yh} = 60{,}000$ psi, and tentatively assuming No. 4 bar hoops (for the purpose of estimating h_c and A_{ch}), required cross-sectional area (see sketch) is:

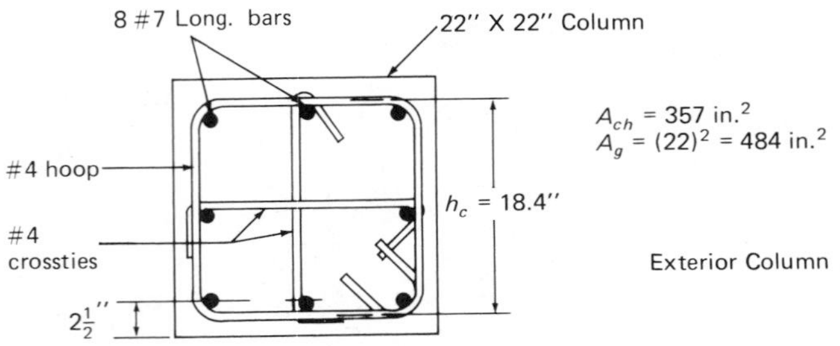

$$A_{sh} \geqslant \begin{cases} (.12)(4)(18.4)(4000)/60{,}000 \\ \quad = \underline{0.59 \text{ in.}^2} \text{ (governs)} \\ (.3)(4)(18.4)\left(\dfrac{484}{357} - 1\right)\dfrac{4000}{60{,}000} = 0.52 \text{ in.}^2 \end{cases}$$

A.4.4.3

No. 4 hoops with one crosstie, as shown in the sketch, provide $A_{sh} = 3(.20) = 0.60$ in.^{2}

(2) Transverse reinforcement for shear: As in the design of shear reinforcement for beams, the design shear in columns is based not on the factored shear forces obtained from a lateral load analysis but rather on the nominal flexural strength provided in the columns. ACI Appendix A requires that the shear be determined from the largest nominal moment strength consistent with the estimated axial forces on the column. A.7.1.2

Assume that an axial force close to $\phi P_b = 484$ kips (corresponding to the "balanced point" on the interaction diagram for the column section considered—which would yield close to if not the largest moment strength) can occur (see Table 12-9). On this basis:

$$M_u \text{ (at column ends)} = \phi M_b = 355 \text{ ft-kips}$$

from which (see Fig. 12-81):

$$V_u = \frac{2M_u}{l} = \frac{2(355)}{10} = 71 \text{ kips}$$

Assume:

$$V_c = 2\sqrt{f_c'}\, bd^*$$
$$= 2\sqrt{4000}\,(22)(19.5)/1000 = 54 \text{ kips}$$

Required spacing of #4 hoops with $A_v = 2(.20) = 0.40$ in.^{2} (neglecting crosstie) and $V_s = (V_u - \phi V_c)/\phi = 29.5$ kips:

$$s = \frac{A_v f_y d}{V_s} = \frac{(2)(.20)(60)(19.5)}{29.5} = 15.9 \text{ in.}$$

11.5.6.2

Thus, the transverse reinforcement spacing over the distance $l_0 = 22$ in. near the column ends is governed by the requirement for confinement rather than shear.

*This lower bound value for V_c is assumed here primarily for convenience. Advantage can be taken of the increased shear resistance of concrete due to axial compression, when needed, by using the appropriate expression given in ACI 318-83.

Maximum allowable spacing of shear reinforcement = $d/2$ or 9.7 in. 11.5.4.1

Use No. 4 hoops and crossties spaced at 4 in. within a distance of 24 in. from the column ends and No. 4 hoops spaced at 9 in. or less over the remainder of the column.

(c) Minimum length of lap splices of column vertical bars: ACI Appendix A limits the location of lap splices of column bars within the center half of the member length, the splices to be designed as tension splices. Since generally all of the column bars will be spliced at the same location, a Class C splice will be required. Required length of splice = 1.7 l_d, where: A.4.3.2 12.16.2 12.16.1

$$l_d = 0.04A_b f_y/\sqrt{f_c'} \geqslant 0.0004\, d_b f_y$$
$$= (.04)(.60)(60{,}000)/\sqrt{4000}$$
$$\geqslant (.0004)(.875)(60{,}000)$$
$$= 23 \text{ in. (governs)} > 21 \text{ ins.}$$

12.2.2

Thus, required splice length = 1.7(23) = $\underline{39 \text{ in.}}$ $\underline{\text{Use 40-in. lap splices.}}$

(d) Detail of column. See Fig. 12-92.

12.7.4 Design of Exterior Beam-Column Connection.

The aim is to determine the transverse reinforcement and shear strength requirements for the exterior beam-column

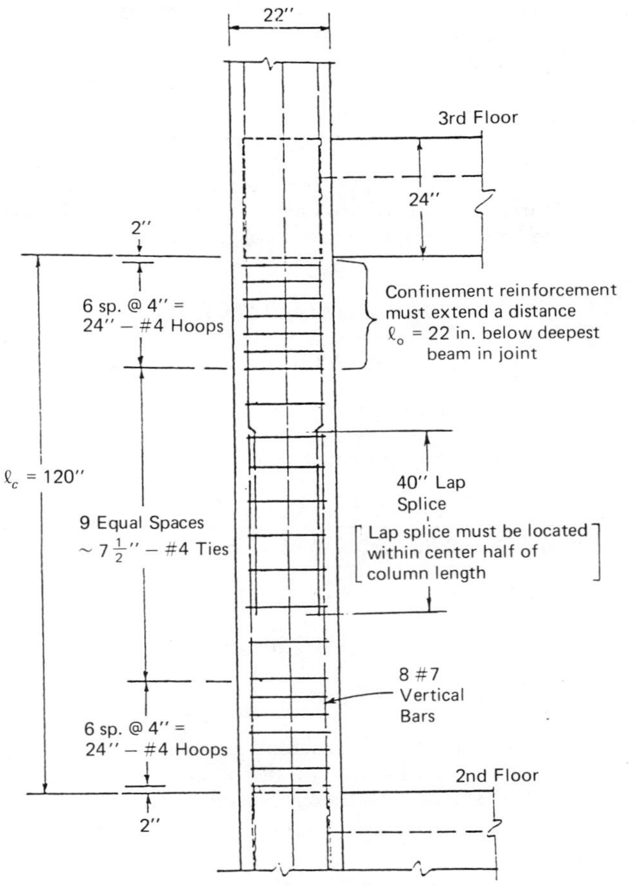

Fig. 12-92 Column reinforcing details.

connection between the beam considered in Section 12.7.2 and the column of Section 12.7.3. Assume the joint to be located at the sixth floor level.

(a) Transverse reinforcement for confinement: Appendix A requires the same amount of confinement reinforcement within the joint as for the length l_0 at column ends, unless the joint is confined by beams framing into all vertical faces of the column. In the latter case, only one-half of the reinforcement required for unconfined joints need be provided. [A.6.2.1 / A.6.2.2]

In the case of the beam–column joint considered here, beams frame into only three sides of the column so that the joint is considered unconfined.

In Section 12.7.3, confinement requirements at column ends were satisfied by No. 4 hoops with crossties, spaced at 4 in.

(b) Check shear strength of joint: The shear across section x–x (see sketch) of the joint is obtained as the difference between the tensile force from the top flexural reinforcement of the framing beam (stressed to 1.25 f_y) and the horizontal shear from the column above.

$$\text{Tensile force from beam } [A_s \ (3 \ \#9 \text{ bars}) = 3.0 \text{ in.}^2] = (3)(1.25)(60) = 225 \text{ kips}$$

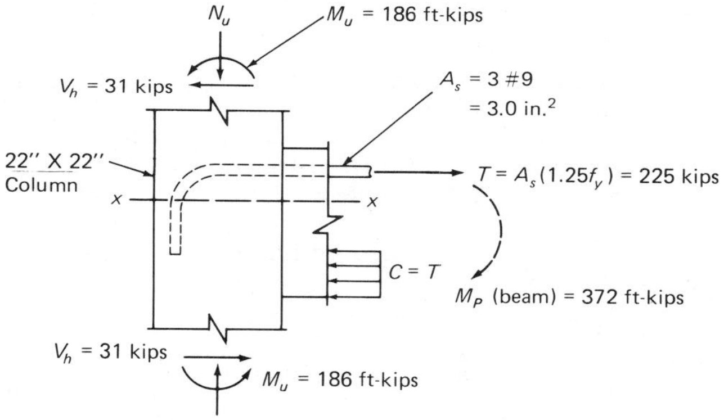

An estimate of the horizontal shear from the column, V_h, can be obtained by assuming that the beams in the adjoining floors are also deformed so that plastic hinges form at their junctions with the column, with M_p(beam) = 372 ft-kips (see Table 12-11, for sidesway to left). By further assuming that the plastic moments in the beams are resisted equally by the columns above and below the joint, one obtains for the horizontal shear at the column ends:

$$V_h = \frac{M_p\text{(beam)}}{\text{story height}} = \frac{372}{12} = 31 \text{ kips}$$

Thus, net shear at section x–x of joint = 225 − 31 = 194 kips. ACI Appendix A makes the nominal shear strength of a joint a function only of the area of the joint cross section, A_j, and the degree of confinement by framing beams. For the unconfined joint considered here:

$$\phi V_c = \phi \, 15\sqrt{f_c'}\,A_j$$

$$= (.85)(15)(\sqrt{4000})(22)^2/1000$$

$$= 390 \text{ kips} > V_u = 194 \text{ kips} \quad \text{OK}$$

[A.2.3.1 / A.6.3.1]

Note that if the shear strength of the concrete in the joint as calculated above were inadequate, any adjustment would have to take the form of (since transverse reinforcement is considered not to have a significant effect on shear strength) either an increase in the column cross section (and hence A_j) or an increase in the beam depth (to reduce the amount of flexural reinforcement required and hence the tensile force T).

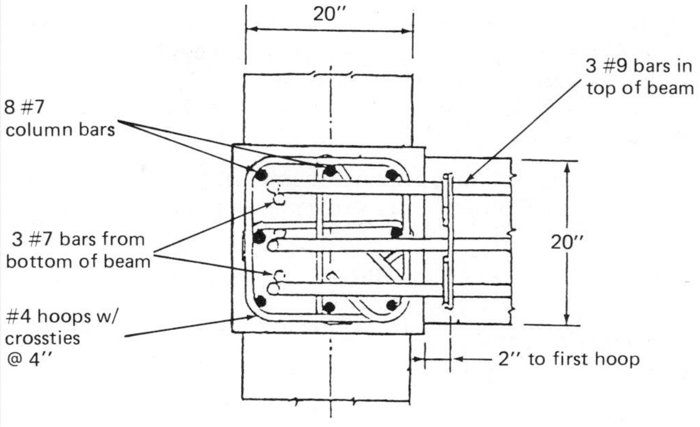

(a) Plan

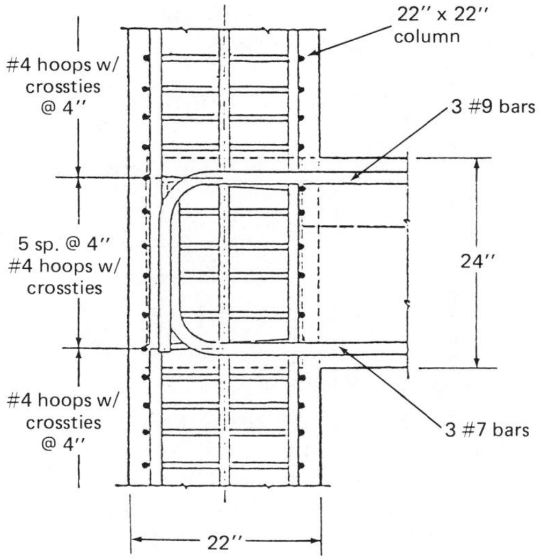

(b) Section

Fig. 12-93 Detail of exterior beam–column connection.

(c) Detail of joint. See Fig. 12-93. (The detail should be checked for adequacy in the longitudinal direction.) *Note:* The use of crossties within the joint may cause some placement difficulties. To relieve the congestion, No. 6 hoops spaced at 4 in. but without crossties may be considered as an alternative. Although the cross-sectional area of confinement reinforcement provided by No. 6 hoops at 4 in. (A_{sh} = 0.88 in.^{2}) exceeds the required amount (0.59 in.^{2}), the requirement of Section A.4.4.3 relating to a maximum spacing of 14 in. between crossties or legs of overlapping hoops (see Fig. 12-80) will not be satisfied. However, it is believed that this should not be a serious shortcoming in this case, since the joint is restrained by beams on three sides.

12.7.5 Design of Interior Beam–Column Connection.

The objective is to determine the transverse reinforcement and shear strength requirements for the interior beam–

column connection at the sixth floor of the interior transverse frame considered in previous examples. The column is 26 in. square and is reinforced with eight No. 11 bars. The beams have dimensions b = 20 in. and d = 21.5 in. and are reinforced as noted in Section 12.7.2 (see Fig. 12-91).

(a) Transverse reinforcement requirements (for confinement)—Maximum allowable spacing of rectangular hoops:

$$s_{max} = \begin{cases} \frac{1}{4} \text{ smallest dimension of column} \\ = 26/4 = 6.5 \text{ in.} \\ \underline{4 \text{ in.}} \text{ (governs)} \end{cases}$$

A.4.4.2

For the column cross section considered and assuming No. 4 hoops, h_c = 21.9 in., A_{ch} = $(22.4)^2$ = 502 in.^{2}, and $A_g = (26)^2$ = 676 in.^{2} With a hoop spacing of 4 in., the required cross sectional area of confinement reinforcement in the form of hoops is:

A.4.4.1

$$A_{sh} \geqslant \begin{cases} 0.12\, s h_c \dfrac{f_c'}{f_{yh}} = (.12)(4)(21.9)(4000)/(60{,}000) \\ = \underline{0.70 \text{ in.}^2} \text{ (governs)} \\ 0.3\, s h_c \left(\dfrac{A_g}{A_{ch}} - 1\right) \dfrac{f_c'}{f_{yh}} \\ = (.3)(4)(21.9)\left(\dfrac{676}{502} - 1\right) \dfrac{4000}{60{,}000} \\ = 0.61 \text{ in.}^2 \end{cases}$$

Since the joint is framed by beams [having widths (20 in.) ⩾3/4 width of column] on all four sides, it is considered confined, and a 50% reduction in the amount of confinement reinforcement indicated above is allowed. Thus, A_{sh} (required) ⩾ 0.35 in.^{2}

A.6.6.2

No. 4 hoops with crossties spaced at 4 in. o.c. provide A_{sh} = 0.60 in.^{2} (See Note at end of Section 12.7.4.)

(b) Check shear strength of joint: Following the same procedure used in Section 12.7.4, the forces affecting the horizontal shear across a section near middepth of the joint shown in the sketch are obtained.

$$\phi V_c = \phi\, 20\sqrt{f_c'}\, A_j$$
$$= (.85)(20)(\sqrt{4000})(26)^2/1000$$
$$= 726 \text{ kips}$$
$$> V_u = 375 \text{ kips} \quad \text{OK}$$

A.6.3.1

12.7.6 Design of Structural Wall

The aim is to design the structural wall (shear wall) section at the first floor of one of the identical frame–shear wall systems in Frame T-3 (see Fig. 12-88c). The preliminary design, as shown in Fig. 12-88, is based on a 14-in.-thick wall with 26-in.-square vertical boundary elements, each of the latter being reinforced with eight No. 11 bars.

Preliminary calculations indicated that the cross section of the structural wall at the lower floor levels needed to be increased. In the following, a 20-in.-thick wall section with 32 in. × 50 in. boundary elements reinforced with 24 No. 11 bars is investigated, and other reinforcement requirements are determined.

The design forces on the structural wall at the first floor level are listed in Table 12-10. Note that because the axis of the shear wall coincides with the centerline of the transverse frame of which it is a part, lateral loads do not induce any vertical (axial) forces on the wall.

The calculation of the maximum axial force on the boundary element corresponding to eq. (12-27b), P_u = 4166 kips, shown in Table 12-10, involved the following steps:

At base of wall:
Dead load, D = 2823 kips
Live load, L = 412 kips
Moment at base of wall due to seismic load (from lateral load analysis of transverse frames)—including a moment of 1806 ft-kips due to accidental eccentricity, M_b = 38,056 ft-kips

Referring to Fig. 12-85, and noting the load factors used in eq. (12-27b):

$$W = 0.75\,(1.4D + 1.7L)$$
$$= (.75)\,[(1.4)(2823) + (1.7)(412)] = 3489 \text{ kips}$$
$$Ha = 1.4M_b = (1.4)(38{,}056) = 53{,}278 \text{ ft-kips}$$
$$C_v = W/2 + Ha/d$$
$$= 3489/2 + 53{,}278/22 = 4166 \text{ kips}$$

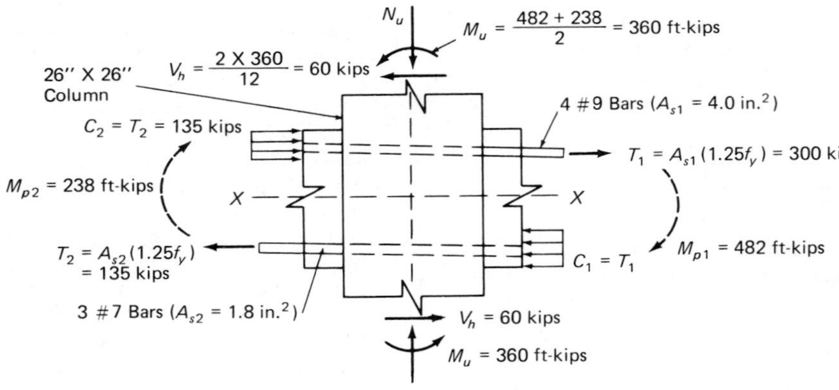

$$\text{Net shear across section } x\text{-}x = T_1 + C_2 - V_h$$
$$= 300 + 135 - 60$$
$$= 375 \text{ kips} = V_u$$

Shear strength of joint, noting that joint is confined:

(a) Check whether boundary elements are required: Appendix A requires boundary elements to be provided if the maximum compressive extreme-fiber stress under factored forces exceeds $0.2f_c'$, unless the entire wall is reinforced to satisfy Sections

A.5.3.1

A.4.4.1 through A.4.4.3 (relating to confinement reinforcement).

It will be assumed that the wall will not be provided with confinement reinforcement over its entire section. For a homogeneous rectangular wall 26.17 ft long (horizontally) and 20 in. (1.67 ft) thick:

$$I_{\text{n.a.}} = \frac{(1.67)(26.17)^3}{12} = 2494 \text{ ft}^4$$

$$A_g = (1.67)(26.17) = 43.7 \text{ ft}^2$$

Extreme-fiber compressive stress under M_u = 53,280 ft-kips and P_u = 3449 kips (see Table 12-10):

$$f_c = \frac{P_u}{A_g} + \frac{M_u h_w/2}{I_{\text{n.a.}}} = \frac{3449}{43.7} + \frac{(53{,}280)(26.17)/2}{2494}$$

$$= 358.4 \text{ ksf} \quad \text{or} \quad 2.49 \text{ ksi} > 0.2f_c' = (.2)(4) = 0.8 \text{ ksi}$$

Therefore, *boundary elements are required*, subject to the confinement and special loading requirements specified in Appendix A.

(b) Determine minimum longitudinal and transverse reinforcement requirements for wall:

(1) Check whether two curtains of reinforcement are required: Appendix A requires that two curtains of reinforcement be provided in a wall if the in-plane factored shear force assigned to the wall exceeds $2A_{cv}\sqrt{f_c'}$, where A_{cv} is the cross-sectional area bounded by the web thickness and the length of section in the direction of the shear force considered. From Table 12-10, the maximum factored shear force on the wall at the first floor level is V_u = 1153 kips. [A.5.2.2]

$$2A_{cv}\sqrt{f_c'} = (2)(20)(26.17 \times 12)(\sqrt{4000})/1000$$

$$= 839 \text{ kips}$$

$$< V_u = 1153 \text{ kips}$$

Therefore, two curtains of reinforcement are required.

(2) Required longitudinal and transverse reinforcement in wall: Minimum required reinforcement ratio:

$$\rho_v = \frac{A_{sv}}{A_{cv}} = \rho_n \geqslant 0.0025 \text{ (max. spacing = 18 in.)}$$

[A.5.2.1]

With A_{cv} (per foot of wall) = (20)(12) = 240 in.^{2}, required area of reinforcement in each direction per foot of wall = (.0025)(240) = 0.60 in.^{2}/ft.

Required spacing of No. 5 bars [in two curtains, A_s = 2(.31) = 0.62 in.^{2}]:

$$s(\text{required}) = \frac{2(.31)}{.60}(12) = 12.4 \text{ in.} < 18 \text{ in.}$$

(c) Determine reinforcement requirements for shear: [Refer to discussion of shear strength design for structural walls under Section 12.6.8.4.5(b).] To allow for increased shear reinforcement requirements above the minimum indicated above, assume two curtains of No. 6 bars spaced at 14 in. o.c. both ways.

Shear strength of wall (h_w/l_w = 148/26.17 = 5.66 > 2):

$$\phi V_n = \phi A_{cv}(2\sqrt{f_c'} + \rho_n f_y) \qquad \text{[A.7.3.2]}$$

where:

$$\phi = 0.60$$

$$A_{cv} = (20)(26.17 \times 12) = 6281 \text{ in.}^2 \qquad \text{[A.2.3.1]}$$

$$\rho_n = \frac{2(.44)}{(20)(12)}\left(\frac{12}{14}\right) = 0.00314$$

Thus:

$$\phi V_n = (.60)(6281)\,[2\sqrt{4000} + (.00314)(60000)]/1000$$

$$= 3768.6\,[126.4 + 188.4]/1000$$

$$= 1186 \text{ kips} > V_u = 1153 \text{ kips} \quad \text{OK}$$

Therefore, *use two curtains of No. 6 bars spaced at 14 in. o.c. in both horizontal and vertical directions.* [A.7.3.5]

(d) Check adequacy of boundary element acting as a short column under factored vertical forces due to gravity and lateral loads (see Fig. 12-85): From Table 12-10, maximum compressive axial load on boundary element, P_u = 4166 kips. [A.5.3.3]

With boundary elements having dimensions 32 in. × 50 in. and reinforced with 24 No. 11 bars:

$$A_g = (32)(50) = 1600 \text{ in.}^2$$

$$A_{st} = (24)(1.56) = 37.4 \text{ in.}^2$$

$$\rho_{st} = 37.4/1600 = 0.0234$$

Axial load capacity of boundary element acting as a short column:

$$\phi P_{n(\max)} = 0.80\phi[0.85f_c'(A_g - A_{st}) + f_y A_{st}]$$

$$= (.80)(.70)\,[(.85)(4)(1600 - 37.4) + (60)(37.4)] \qquad \text{[10.3.5.2]}$$

$$= (.56)\,[5313 + 2246] = 4233 \text{ kips} > P_u = 4166 \text{ kips} \quad \text{OK}$$

(e) Check adequacy of structural wall section at base under combined axial load and bending in the plane of the wall: From Table 12-10, the following combinations of factored axial load and bending moment at the base of the wall are listed, corresponding to eq. (12-27a, b, and c):

Eq. (12-27a): P_u = 4598 kips, M_u = small

Eq. (12-27b): P_u = 3449 kips, M_u = 53,280 ft-kips

Eq. (12-27c): P_u = 2506 kips, M_u = 54,464 ft-kips

Figure 12-94 shows the ϕP_n–ϕM_n interaction diagram (obtained using a computer program for generating *P–M* diagrams) for a structural wall section having a 20-in.-thick web reinforced with two curtains of reinforcement each having No. 6 horizontal and vertical bars spaced at 14 in. o.c. and 32 in. × 50 in. boundary elements reinforced with 24 No. 11 vertical bars, with f_c' = 4,000 psi, f_y = 60,000 psi (see Fig. 12-95). The design load combinations listed above are shown plotted in the figure. The point marked *a* represents the *P–M*

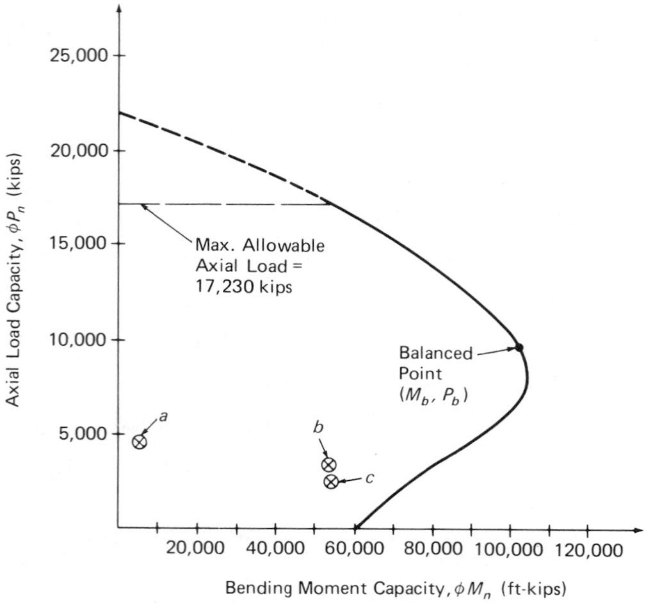

Fig. 12-94 Axial load–moment interaction diagram for structural wall section.

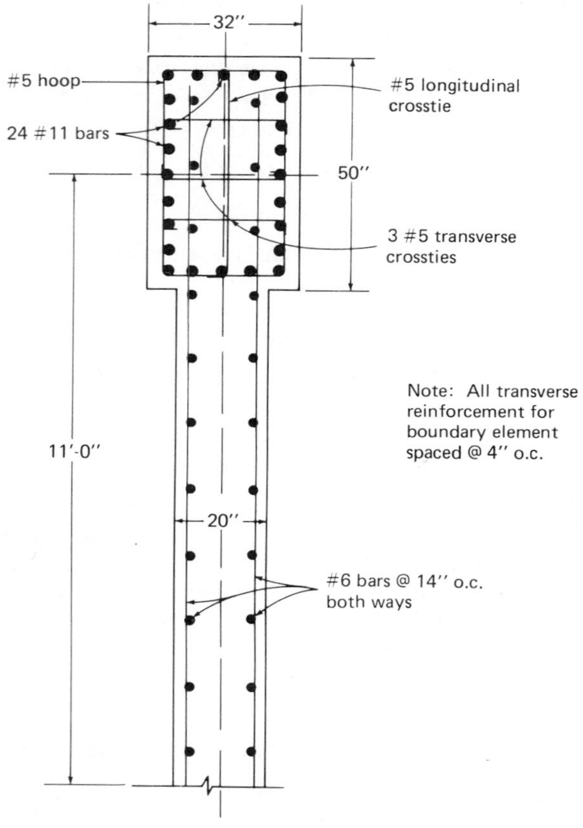

Fig. 12-95 Half-section of structural wall at base.

combination corresponding to eq. (12-27a), with similar notation for the other two combinations.

It is seen in Fig. 12-94 that the three design loadings represent points inside the interaction diagram for the structural wall section considered. *Therefore, the section is adequate with respect to combined bending and axial load.*

Incidentally, the "balanced point" in Fig. 12-94 corresponds to a condition where the compressive strain in the extreme concrete fiber is equal to $\epsilon_{cu} = 0.003$ and the tensile strain in the row of vertical bars in the boundary element farthest from the neutral axis (see Fig. 12-95) is equal to the initial yield strain, $\epsilon_y = 0.00207$.

(f) Determine lateral (confinement) reinforcement requirements for boundary elements (see Fig. 12-95):

$$\text{max. spacing, } s_{max} = \begin{cases} \frac{1}{4} \text{ smallest dimension of boundary element} = \frac{32}{4} = 8 \text{ in.} \\ \underline{4 \text{ in.}} \text{ (governs)} \end{cases}$$

A.5.3.2
A.4.4.2

(1) Required cross-sectional area of confinement reinforcement in short direction:

$$A_{sh} \geqslant \begin{cases} 0.12 s h_c \dfrac{f_c'}{f_{yh}} \\ 0.3 s h_c \left(\dfrac{A_g}{A_{ch}} - 1\right) \dfrac{f_c'}{f_{yh}} \end{cases}$$

A.4.4.1

Assuming No. 5 hoops and crossties spaced at 4 in. o.c. and a distance from centerline of No. 11 vertical bars to face of column of 3 in., we have:

h_c (for short direction)

$$= 44 + 1.41 + 0.625 = 46.04 \text{ in.}$$

$$A_{ch} = (46.04 + 0.625)(26 + 1.41 + 1.25)$$
$$= 1337 \text{ in.}^2$$

A_{sh} (required in short direction)

$$\geqslant \begin{cases} (.12)(4)(46.04)(4/60) \\ \quad = \underline{1.47 \text{ in.}^2} \text{ (governs)} \\ (.3)(4)(46.04)\left(\dfrac{(32)(50)}{1337} - 1\right)\left(\dfrac{4}{60}\right) \\ \quad = 0.72 \text{ in.}^2 \end{cases}$$

With 3 crossties (i.e., five legs, including outside hoop):

$$A_{sh}(\text{provided}) = 5(.31) = 1.55 \text{ in.}^2 \qquad \text{OK}$$

(2) Required cross-sectional area of confinement reinforcement in long direction:

h_c (for long direction)

$$= 26 + 1.41 + 0.625 = 28.04 \text{ in.}$$

$$A_{ch} = 1337 \text{ in.}^2$$

A_{sh} (required in long direction)

$$\geqslant \begin{cases} (.12)(4)(28.04)(4/60) \\ \quad = \underline{0.90 \text{ in.}^2} \text{ (governs)} \\ (.3)(4)(28.04)(1.196 - 1)(4/60) \\ \quad = 0.44 \text{ in.}^2 \end{cases}$$

With one crosstie (i.e., three legs, including outside hoop):

$$A_{sh}(\text{provided}) = 0.93 \text{ in.}^2 \qquad \text{OK}$$

(g) Determine required development and splice lengths: ACI Appendix A requires that all continuous reinforcement in structural walls be anchored or spliced in accordance with the provisions for reinforcement in tension as given in the Appendix. [A.5.2.4]

(1) Lap splice for No. 11 vertical bars in boundary elements (the use of mechanical connectors may be considered as an alternative to lap splices for these large bars):

Assuming that 50% or less of the vertical bars are spliced at any one location, a Class B splice may be used. [12.16.2]

Required length of splice = $1.3 l_d$, where: [12.16.1]

$$l_d \geqslant 2.5 \times \begin{cases} f_y d_b / 65\sqrt{f_c'} \\ \quad = (60{,}000)(1.41)/(65)(\sqrt{4000}) \\ \quad = \underline{21 \text{ in.}} \text{ (governs)} \\ 8d_b = (8)(1.41) = 12 \text{ in.} \\ 6 \text{ in.} \end{cases}$$ [A.6.4.2]

Thus required splice length = (1.3)(2.5)(21) = <u>68 in.</u>

(2) Lap splice for No. 6 vertical bars in wall "web": Again assuming no more than 50% of bars spliced at any one level so that a Class B splice may be used, and using the same expression for l_d as above, l_d = 11 in. Hence, required length of splice = (1.3)(2.5)(11) = <u>36 in.</u>

(3) Development length for No. 6 horizontal bars in wall, assuming no hooks are used within boundary element: Since it is reasonable to assume that the depth of concrete cast in one lift beneath a horizontal bar will be greater than 12 in., the required factor of 3.5 to be applied to the development length, l_{dh}, required for a 90°-hooked bar will be used (Section 12.6.8.4.2f). [A.6.4.2]

$$l_d = 3.5 l_{dh} \geqslant 3.5 \times \begin{cases} f_y d_b / 65\sqrt{f_c'} = \dfrac{(60{,}000)(7.5)}{(65)(\sqrt{4000})} \\ \quad = \underline{11 \text{ in.}} \text{ (governs)} \\ 8d_b = (8)(.75) = 6 \text{ in.} \\ 6 \text{ in.} \end{cases}$$

Thus, required development length, $l_d = 3.5(11)$ = <u>39 in.</u> This length can be accommodated within the confined core of the boundary element so that no hooks are needed, as assumed. Required lap splice length for these No. 6 horizontal bars, assuming 50% or less of bars spliced at any one location, = $1.3 l_d$ = (1.3)(39) = <u>51 in.</u>

(h) Detail of structural wall: See Fig. 12-95. It will be noted in Fig. 12-95 that the No. 6 vertical wall "web" reinforcement, required for shear resistance, has been carried into the boundary element. The Commentary to ACI Appendix A specifically states that the concentrated reinforcement provided at wall edges for bending shall not be included in determining shear reinforcement requirements. The area of vertical shear reinforcement located within the boundary element could, if desired, be considered as contributing to axial load and bending capacity.

12.8 ALTERNATIVES TO ASEISMIC DESIGN BY BUILDING CODES

12.8.1 Inelastic Analysis

A major earthquake can cause significant inelasticity in any standard multistory building designed by normal code procedures. The true dynamic behavior of such a structure can be determined only if allowance is made for nonlinearity in the analysis procedure.

For analysis considering nonlinear response, the equations of motion developed in Section 12.3 need to be used directly. The response of a nonlinear system may be considered as being essentially linear over small intervals of time. As yielding occurs in members and the stiffness of a structure changes, the response of the nonlinear system is obtained as the response of successively different linear systems.

Computer programs developed to undertake nonlinear dynamic analysis of multistory frames[12-40, 12-103, 12-104] can generate a variety of useful information, such as maximum deformations and forces at all significant locations, as well as time history records of deformations and forces at particular points in the structure. However, there are uncertainties in the values of stiffness and damping to be used in conjunction with the structural model.

The volume of numerical work and the computer storage capacity required for inelastic dynamic analysis make it necessary to assume fairly simple force–deformation characteristics for structural members. Allowance is usually made only for yield hinges forming at the ends of members, the hinging being governed by the magnitude of the bending moment.[12-27, 12-105, 12-106] More elaborate force–deformation relationships for structural members incorporating axial load–moment interaction effects and inelastic shear behavior have recently become available, and have been used to a limited extent.[12-28, 12-107]

12.8.2 Design Considering Inelastic Response

Three approaches to design that accounts for inelastic structural response are currently available, and are discussed below.

12.8.2.1 Design Based on Inelastic Response Spectra Newmark[12-108, 12-109] has demonstrated use of the design spectrum to approximate inelastic behavior. The UBC-82[12-14], ATC 3-06,[12-9] and other code approaches to seismic design are in principle based on the concept of the inelastic design spectrum. In ATC 3-06, the response modification factor, R, is used, in effect, to derive inelastic design spectra from the corresponding elastic spectra. The earthquake force derived from the inelastic design spectrum is the design base shear of the codes.

Shibata and Sozen[12-110] have proposed a "substitute structure" method, where, instead of deriving an inelastic response spectrum from an elastic spectrum, a substitute structure is derived from the actual structure with properties adjusted to account for inelastic behavior, and the elastic design spectrum is directly used.

12.8.2.2 Shortcomings of Design Approach Based on Inelastic Design Spectrum (Including Code-Recommended Design) These shortcomings are:

1. Internal forces determined from elastic analysis under static loads specified in codes or derived from inelastic

design spectra differ from forces caused by actual inelastic seismic structural response.

2. The distribution and magnitude of inelastic deformations in structural members cannot be determined through elastic analysis under static forces given in codes or derived from inelastic spectra. Thus, special ductility details must be provided throughout the structure, although inelasticity may actually occur only in certain locations. Also, there is no certainty that the ductility provided through conformance with code-prescribed detailing requirements will always suffice.
3. Elastic story drifts under static forces given in codes or derived from inelastic response spectra, even when multiplied with a ductility factor or a response modification factor, may differ from actual inelastic story drifts. Thus, the intended damage control and safety against instability may not be achieved.
4. Manipulating the relationship between column and beam strengths at a joint, on the basis of results of gravity load and static lateral load analyses, may also be ineffective.[12-111] Actual earthquakes can cause axial forces in columns and other internal forces that may substantially alter the strength relationship intended by the designer.

12.8.2.3 Inelastic Design Based on Energy Considerations Housner[12-112] proposed a simplified design method based on equating the energy input from an earthquake with the energy absorbed in a structure. Blume et al.[12-7] proposed a more elaborate and formalized "reserve energy technique" and applied it to at least one 24-story reinforced concrete frame building. More recently, Larson[12-113] has discussed the need for considering the earthquake energy input, and its absorption, in design. Work toward a simple energy method for seismic design of structures is in progress at the University of British Columbia.[12-114] A great deal of judgment and intuition is needed to successfully apply the available energy methods.

12.8.2.4 Design Utilizing Inelastic Response History Analysis Information on the amount and distribution of internal forces and deformations in yielding structures can be obtained only through inelastic response history analyses of structures subjected to earthquake input motions. Such analysis has often been used in research studies of seismic structural response.[12-22,12-26,12-28]

The Muto Institute has developed a Dynamic Design Procedure (DDP)[12-115,12-116] that utilizes dynamic analysis. The linear and nonlinear properties of equivalent "stick" models are determined from tests on segments of the actual frames. In applications of DDP, peak ground accelerations of $0.1g$, $0.3g$, and $0.5g$ have been taken as corresponding to "minor or moderate," "severe," and "worst" earthquakes, respectively. Dynamic (elastic or inelastic, as appropriate) analyses of structures under actual recorded ground motions (normalized to the above peak ground accelerations) have been carried out, and compliance with several postulated design criteria has been checked.

Fintel and Ghosh[12-117,12-118] have proposed an Explicit Inelastic Dynamic Design Procedure (EIDDP) that uses earthquake accelerograms as loading, dynamic inelastic response history analysis to determine member forces and deformations, and resistances from tests for proportioning the members.

1. Features of EIDDP.

1. The systematic selection of input motions that are likely to excite critical response in a particular structure forms an integral part of EIDDP.
2. Nonlinear models of member hysteretic response, generalized from test results, are used in response history analysis.
3. The designer exerts control over where in the structure the bulk of seismic energy dissipation for the selected input motion(s) takes place.
4. Two reference stages are usually considered: a "design earthquake" and a "hypothetical maximum credible earthquake," which are comparable with the serviceability and ultimate strength limit states. Just one or more than two reference stages can also be considered. Performance criteria for each reference stage may be chosen to suit the designer's particular requirements.

EIDPP has become feasible because of the development of computer programs with which two-dimensional inelastic response history analyses can be carried out relatively inexpensively. To date, EIDDP has been applied to the design of several multistory reinforced concrete buildings.[12-119–12-123]

2. Principal Elements of EIDDP. The suggested design procedure entails:

1. Preliminary layout and design of the structural system for gravity loads and for code-specified wind and earthquake forces.
2. Modeling of the structure for dynamic analysis in two orthogonal directions using multibay–multistory models, and determination of the natural undamped periods of the structure in each direction.
3. Selection of design earthquake accelerograms with frequencies that may critically excite the structure. By subjecting the model to these records (normalized for intensity according to local seismicity), the one or two producing the maximum response can be chosen. Quite often, the maximum displacements and the largest bending moments or shear forces are caused by different accelerograms. In such cases, more than one design earthquake accelerogram must be used in further computation.
4. Specifying performance criteria under the design and the hypothetical maximum credible earthquakes.
5. Determination of internal forces and deformations caused by the design earthquake using inelastic response history analysis. Repeat of such analysis with altered strength levels for the structural members until all performance criteria are satisfied and an optimum relationship between strength and deformability demands is obtained.
6. Proportioning of members for strength and deformability.
7. Checking that the structure has the needed margin of ductility to meet the chosen safety criteria corresponding to the hypothetical maximum credible earthquake. Checking for stability may become necessary if there is yielding in the columns or walls.

3. Input Motion. For response history analysis, input motion accelerograms, rather than design response spectra, are needed. A procedure for the selection of input motions to critically excite a given structure has been developed at the Portland Cement Association.[12-124] The procedure attempts to match the frequency content of input motions, as indicated by their velocity response spectra, with the frequency characteristics of a structure. The procedure is virtually site-independent and may, therefore, be somewhat conservative. The hypothetical maximum credible earthquake for a site is considered to differ from the design earthquake only in intensity.

4. Dynamic Analysis. The computer program DRAIN-2D[12-103] was selected for inelastic response history analysis,

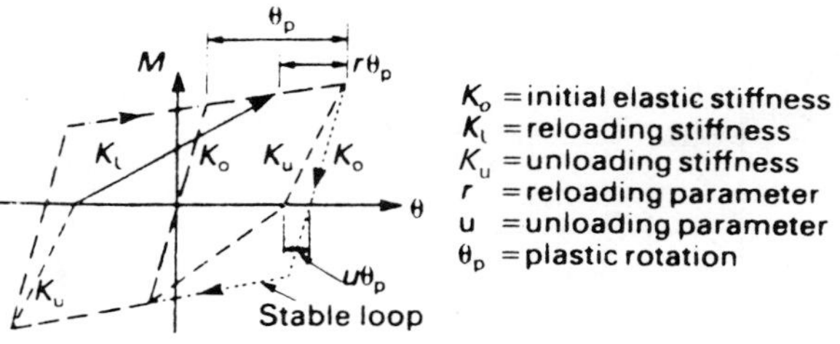

Fig. 12-96 Modified Takeda model of the moment-rotation relationship of a reinforced concrete member segment.

based on a comparative evaluation of several available programs. DRAIN-2D determines the seismic response of plane inelastic structures, with masses lumped at nodal points, through step-by-step integration of the coupled equations of motion, based on the assumption of a constant response acceleration during each time step. Viscous damping having both mass-proportional and stiffness-proportional components is used.

Program DRAIN-2D accounts for inelastic effects by allowing the formation of "point hinges" at the ends of elements where the moments exceed the specified yield moments. The moment versus end rotation characteristics of elements are defined in terms of basic bilinear relationships. These develop into hysteretic loops following the modified Takeda model,[12-27] which incorporates post-yield decreases in unloading and reloading stiffnesses (Fig. 12-96). P-delta effects can be considered in DRAIN-2D analysis, but are seldom significant in concrete structures containing shear walls.

5. Control of Energy Dissipation. In a building structure, the beams are not crucial to structural stability and are easy to repair. A structure can be designed such that, under severe earthquake-induced excitation, all inelastic deformations will be confined to designated beams, while the more vital elements such as columns and walls will remain elastic throughout their seismic response. This is best achieved by intentionally making the designated members weaker in relation to others. Intentionally weakened beams act as structural fuses that (Fig. 12-97): (a) preset the commencement of yielding during an earthquake, thus ensuring controlled energy dissipation in selected members; (b) reduce the moments, M_y, transferred from beams into columns, thus protecting the columns from hinging; (c) reduce the seismic axial forces $(M_y^- + M_y^+)/l$ (tension and compression) in the columns or walls; and (d) decrease the congestion of reinforcement, as well as shear stresses, in the joints.

Beam strength reductions, as discussed above, can be made only if the corresponding deformations and internal forces do not violate the chosen performance criteria. If the initially chosen strength levels result in unacceptable deformations, shears, or axial forces, then adjustments must be made and the structure reanalyzed, until all performance

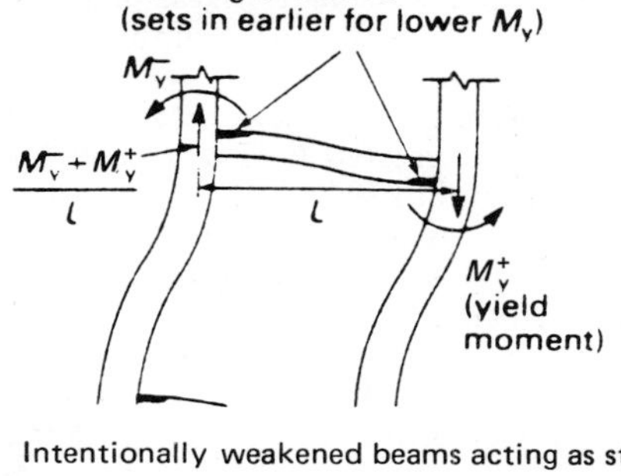

Fig. 12-97 Intentionally weakened beams acting as structural fuses.

criteria are satisfied. This iteration process can usually be accomplished with three to five analyses.

6. Definition of Ductility. Rotational ductility for wall and beam segments is defined in EIDPP as shown in Fig. 12-37(e).

7. Performance Criteria. For the design earthquake, serviceability criteria need to be formulated; for the maximum credible earthquake, safety against collapse must be assured.

Under the design earthquake, inelasticity may be disallowed, or may be confined only to the beams, with limits put on the inelastic deformations. The interstory drifts may be limited to assure damage control. Shear stresses in structural members, axial forces in vertical members, and so on, may also be limited.

Under the maximum credible earthquake, some inelasticity may be allowed in columns and walls, although this should be kept to a minimum because of stability considerations. The maximum deformations in all members must be below the available deformation capacities. The total lateral deflection should be limited to assure stability.

Suitable limits on deformations, shear stresses, and so on, constituting performance criteria, can be established only through correlation with observed deformations, cracking, and damage in laboratory tests of structural members.

8. Proportioning of Members–Combination of Load Effects. Member proportioning has been discussed in Ref. 12-118.

9. Advantages of EIDDP.

1. The magnitude and location of inelastic deformations are determined so that ductility details can be provided only where required and usable.
2. A desired balance between strength and ductility can be designed into the various structural members.
3. A predetermined sequence of plastification can be designed into a structure, thus regulating the onset and progress of energy dissipation.
4. Progressively more relaxed performance criteria may be specified for increasing intensities of seismic excitation.
5. The designer can creatively devise various structural configurations and distribute the energy-dissipating elements where they will be the most effective without endangering structural stability.

10. Shortcomings of EIDPP.

1. Two-dimensional rather than three-dimensional analysis is used so that torsion has to be considered separately. The approach is limited to reasonably symmetrical structures for which two-dimensional modeling is realistic. This shortcoming is common to all current design approaches.
2. Shear forces obtained from dynamic analysis fluctuate a lot owing to higher mode effects. When these are compared with shear capacities based on tests carried out under essentially static (monotonic or cyclic) loading, an unintended conservatism results.
3. The use of an inelastic approach, although relatively inexpensive in terms of computer cost, requires special knowledge on the part of the designer. It is not intended at this time for routine designs.

12.8.3 Design Example Using Explicit Inelastic Dynamic Design Procedure

12.8.3.1 Building Studied and Its Structural System A typical floor layout of the 16-story apartment building chosen for this design example is shown in Fig. 12-98(a). The structure is a composite precast and cast-in-place sys-

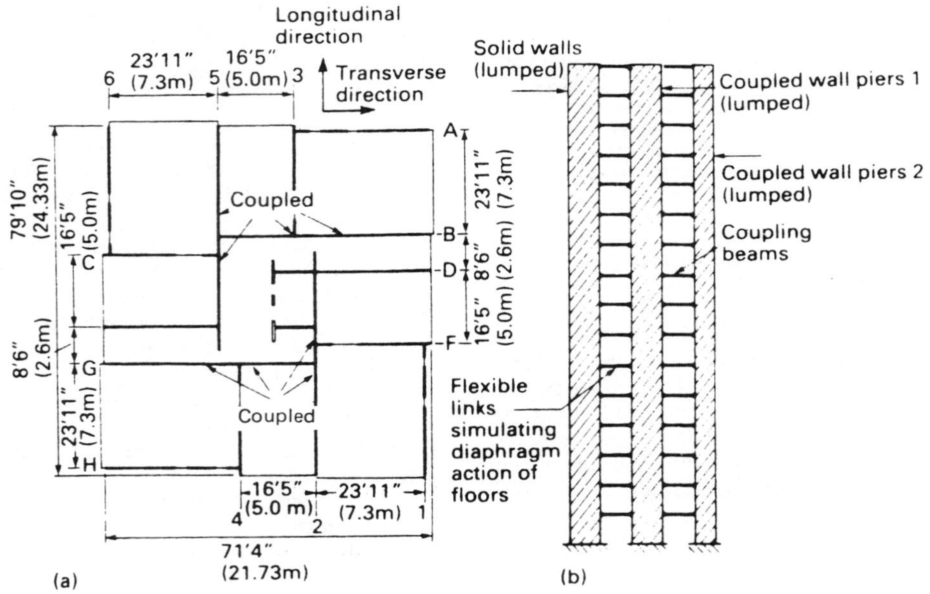

Fig. 12-98 (a) Floor plan; (b) schematic section of building studied.

tem. The vertical hollow cores of the 8-in.-thick precast walls carry vertical continuity reinforcement, and are filled with concrete. Most of the slabs consist of 4-in.-thick prestressed precast units plus a $3\frac{1}{2}$-in.-thick cast-in-place topping that is composite with the precast units. The cast-in-place connections in the structural system provide for good continuity between walls and slabs, between adjacent slabs, and between successive lifts of walls. However, the precast walls have no vertical connections at their junctions. Two pairs of walls, in each orthogonal direction, separated by doorways, are coupled by the door lintels having a span-to-depth ratio of 1.43.

12.8.3.2 Partial Design Problem Considered The coupling beams or door lintels were chosen to act as structural fuses. It was decided that under the design earthquake, only the coupling beams should yield and dissipate energy, while the walls should remain fully elastic. The coupling beams themselves were required to remain elastic up to at least $0.75 \times 1.7 = 1.3$ times the design wind loads.

Of major concern in the design process was the determination of appropriate strengths to be built into the walls and the coupling beams, in order to limit the deformation demands in critical regions to reasonable levels, while keeping the maximum moments, shear forces, and axial forces in these regions acceptable. The determination of design strength levels for the coupling beams in the longitudinal direction only is discussed below.

12.8.3.3 Modeling of the Structure For purposes of analysis, the isolated walls in each orthogonal direction were lumped into a single solid wall, the area and moment of inertia of the lumped wall being equal to the sum of those of the individual walls. The coupled walls in each direction were similarly lumped into a single coupled wall. The lumped solid wall and the lumped coupled wall were connected through flexible links as shown in Fig. 12-98. These links simulated the function of slabs that transmit little bending.

12.8.3.4 Input Motion For the analyses reported in this chapter, the input motion selected was the first 10 seconds of the E–W component of the 1940 El Centro record. The acceleration ordinates were multiplied by a factor to give the resulting accelerogram a spectrum intensity equal to 0.67 times that of the N–S component of the 1940 El Centro record for the design earthquake. The 1940 El Centro N–S record has a peak ground acceleration of 0.33*g*.

12.8.3.5 Results of Dynamic Analysis Seven inelastic dynamic analyses of the structure in the longitudinal direction under the design earthquake were carried out for different yield levels of the coupling beams. The lowest level of yield moments was about equal to the maximum elastic moments caused in these beams by twice the UBC[12-14] 25 psf zone wind forces. At the other extreme, the coupling beams were elastic. In all seven analyses, the walls were kept elastic throughout their seismic response. Five percent of critical damping was assumed.

The results obtained from the seven analyses are presented in Fig. 12-99. Figure 12-99(a) shows that the ductility requirements in the coupling beams increase drastically as the coupling beam strength levels decrease. The elastic coupling beams obviously require no ductility. Figure 12-99(b) shows that there is a trade-off between the shear capacity and the ductility requirements. The shear capacity requirements increase as the coupling beam strength levels increase, the shear capacity demands being the highest for the elastic coupling beams. Figure 12-99(c) shows the axial forces due to coupling in the coupled wall piers. These forces also increase with increasing coupling beam strength and may cause unduly high tensile and compressive forces in the piers of strongly coupled walls.

12.8.3.6 Choice of Coupling Beam Strength Figure 12-99 clearly shows that it is advantageous in design to choose the lowest level of coupling beam strength, provided the corresponding required ductilities can be supplied. In this study, in view of the high ductility and shear capacity requirements in the coupling beams, a decision was made to reinforce them diagonally, as suggested by Paulay and Binney.[12-125] Any shear force such beams may be subjected to is carried in truss action directly by the diagonal reinforcement so that relatively high shear force levels cease to be a problem. Also, the limit of available ductility in diagonally reinforced coupling beams is relatively high, and may be set at 15, based on available test results[12-125]

Based on a consideration of the actual diagonal reinforce-

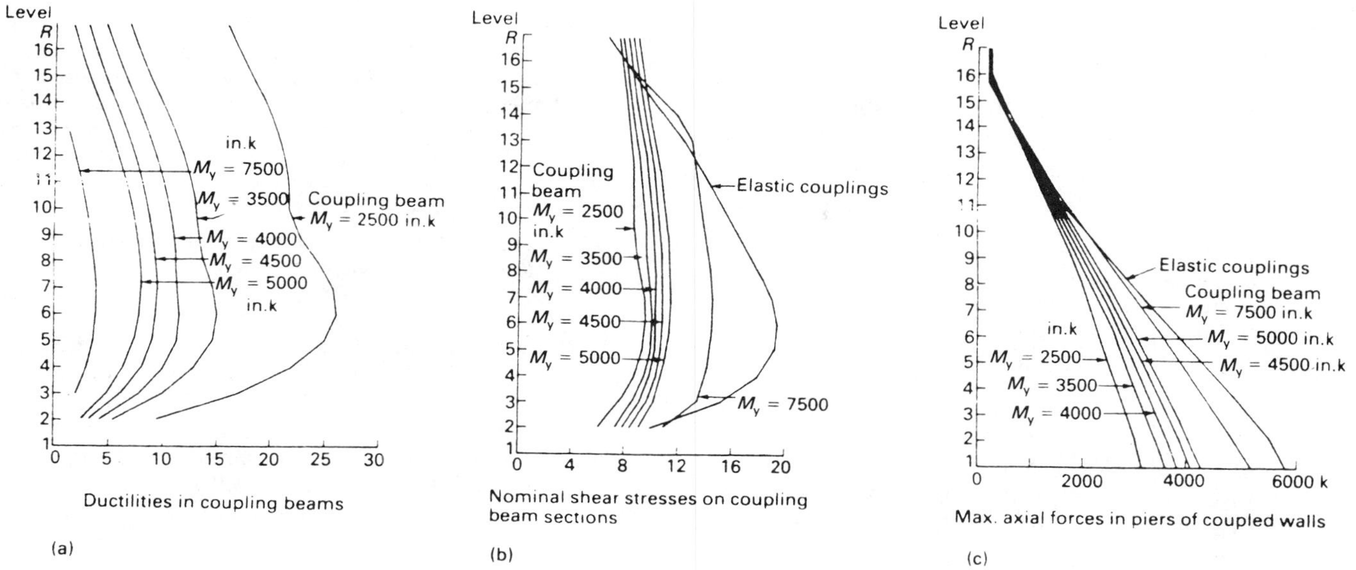

Fig. 12-99 Choice of coupling beam strength: results of dynamic analysis.

ment that can conveniently be protruded from the precast walls into the coupling beams, a yield level of 4000 in. k was selected for the lumped coupling beams. This appears to be a good choice in view of the results presented in Fig. 12-99. The corresponding ductilities required in the coupling beams are reasonable (the highest value is 11.6). The corresponding axial forces in the coupled wall piers are also at a reasonable level. Further increases in coupling-beam strength would substantially increase these forces, causing higher tension and compression.

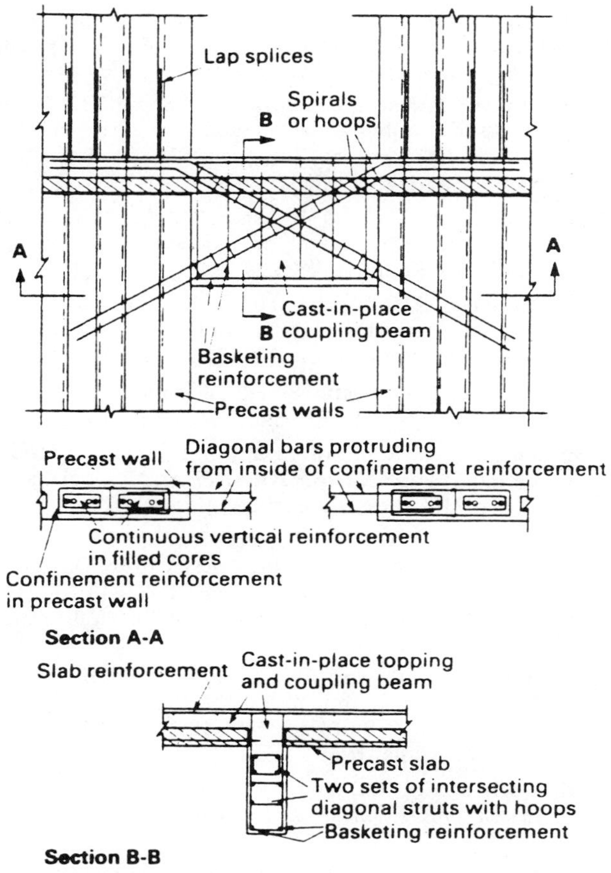

Fig. 12-100 Coupling beam–precast wall joint details.

12.8.3.7 Proportioning of Coupling Beams The diagonally reinforced coupling beams were proportioned as suggested in Ref. 12-125. The 4000 in. k yield level of the lumped coupling beams (2000 in. k for an individual beam) required the provision of two No. 6 Grade 60 bars on each face of each beam in each direction (totaling four bars in each direction or eight to a beam, Fig. 12-100). The four bars in each direction form a diagonal strut. It is important to confine the concrete within each diagonal strut by ties or spirals embracing all bars of the strut. Such ties should be spaced no more than 4 in. apart.

REFERENCES

The following abbreviations will be used to denote commonly recurring reference sources:

BSSA *Bulletin of Seismological Society of America*
JEMD *Journal of Engineering Mechanics Division, American Society of Civil Engineers (ASCE)*
JSTR *Journal of the Structural Division, ASCE*
WCEE World Conference on Earthquake Engineering

12-1 Richter, C. F., *Elementary Seismology*, W. H. Freeman & Co., San Francisco, 1958.

12-2 Bolt, B. A., *Earthquakes–A Primer,* W. H. Freeman & Co., San Francisco, 1978.

12-3 Japan National Committee on Earthquake Engineering, "Niigata Earthquake of 1964," V. III, *Proc.* 3rd WCEE, New Zealand, 1965.

12-4 Kawasumi, H. (Editor), *General Report on the Niigata Earthquake of 1964*, Tokyo Electrical Engineering College Press, Mar. 1968.

12-5 Jennings, P. C., Housner, G. W., and Tsai, N. C., "Simulated Earthquake Motions for Design Purposes," V. 1, *Proc.* 4th WCEE, Santiago, Chile, 1969.

12-6 Ruiz, P., and Penzien, J., "Probabilistic Study of the Behavior of Structures During Earthquakes," Report No. EERC 69-3, Earthquake Eng. Res. Center, University of California, Berkeley, Mar. 1969.

12-7 Blume, J., Newmark, N. M., and Corning, L., "Design of Multistory Reinforced Concrete Buildings for Earthquake Motions," Portland Cement Association, Skokie, Ill., 1970.

12-8 Gutenberg, B., and Richter, C. F., "Earthquake Magnitude, Intensity, Energy and Acceleration," *BSSA*, 32(3), July 1942.

12-9 Applied Technology Council, *Tentative Provisions for the Development of Seismic Regulations for Buildings*, ATC Publication ATC 3-06, US Government Printing Office, Washington, D.C., 1978, 505 pp.

12-10 Algermissen, S. T., and Perkins, D. M., "A Probabilistic Estimate of Maximum Acceleration in Rock in the Contiguous United States," Open-File Report 76-416, U.S. Department of the Interior–Geological Survey, 1976.

12-11 Algermissen, S. T., Perkins, D. M., Thenhaus, P. C., Hanson, S. L., and Bender, B. L., "Probabilistic Estimates of Maximum Acceleration and Velocity in Rock in the Contiguous United States," Open-File Report 82-1033, U.S. Department of the Interior–Geological Survey, 1982, 99 pp.

12-12 American National Standards Institute, *Minimum Design Loads for Buildings and Other Structures*, ANSI A58.1-1982, New York.

12-13 Algermissen, S. T., "Seismic Risk Studies in the United States," V. I, *Proc.* 4th WCEE, Santiago, Chile, 1969.

12-14 International Conference of Building Officials, *Uniform Building Code*, Whittier, Calif., 1982, 780 pp.

12-15 Housner, G. W., Martel, R. R., and Alford, J. L., "Spectrum Analysis of Strong-Motion Earthquakes," *BSSA*, 43(2), Apr. 1953.

12-16 Housner, G. W., "Behavior of Structures during Earthquakes," *JEMD*, Paper No. 2220, Oct. 1959.

12-17 Newmark, N. M., "Current Trends in the Seismic Analysis and Design of High-Rise Structures," Chap. 16 of *Earthquake Engineering*, R. Wiegel, Editor, Prentice-Hall, Englewood Cliffs, N.J., 1970.

12-18 Veletsos, A. S., "Maximum Deformations of Certain Nonlinear Systems," V. II, *Proc.* 4th WCEE, Santiago, Chile, 1969.

12-19 Biggs, J. M., *Structural Dynamics*, McGraw-Hill Book Co., New York, 1964.

12-20 Hurty, W. C., and Rubinstein, M. F., *Dynamics of Structures*, Prentice-Hall, Englewood Cliffs, N.J., 1965.

12-21 Clough, R. W., and Penzien, J., *Dynamics of Structures*, McGraw Hill Book Co., New York, 1975.

12-22 Clough, R. W., and Benuska, K. L., "FHA Study of Seismic Design Criteria for High-Rise Buildings," Report HUD TS-3, Federal Housing Administration, Washington, D.C., Aug. 1966.

12-23 Ruiz, P., and Penzien, J., "Stochastic Seismic Response of Structures," *JEMD*, Paper No. 8050, Apr. 1971.

12-24 Husid, R., "The Effect of Gravity on the Collapse of Yielding Structures with Earthquake Excitation," V. II, *Proc.* 4th WCEE, Santiago, Chile, 1969.

12-25 Goel, S. C., and Berg, G. V., "Inelastic Earthquake Response of Tall Steel Frames," *JSTR*, Paper No. 6061, Aug. 1968.

12-26 Derecho, A. T., Ghosh, S. K., Iqbal, M., Freskakis, G. N., and Fintel, M., *Structural Walls in Earthquake-Resistant Buildings–Analytical Investigation, Dynamic Analysis of Isolated Structural Walls–Parametric Study*, Report on research sponsored by National Science Foundation (RANN), Grant no. ENV74-14766, Portland Cement Association, Skokie, Ill., 1978, 233 pp.

12-27 Takeda, T., Sozen, M. A., and Nielsen, N. N., "Reinforced Concrete Response to Simulated Earthquake," *JSTR*, 96(ST12), Dec. 1970, pp. 2557–2573.

12-28 Saatcioglu, M., "Inelastic Behavior and Design of Earthquake-Resistant Coupled Walls," Ph.D. thesis, Northwestern University, Evanston, Ill., 1981, 272 pp.

12-29 Gutenberg, B., "Effects of Ground on Earthquake Motion," *BSSA*, **47**(3), July 1957.

12-30 Alcock, E. D., "The Influence of Geologic Environment on Seismic Response," *BSSA*, **59**(1), Feb. 1969.

12-31 Ohsaki, Y., and Hagiwara, T., "On Effects of Soils and Foundations upon Earthquake Inputs to Buildings," Bldg. Res. Inst. (Japan) *Research Paper No. 41*, June 1970.

12-32 Seed, H. B., and Idriss, I. M., "Influence of Soil Conditions on Ground Motions during Earthquakes," *Journal of the Soil Mech. and Foundation Div., ASCE*, Paper No. 6347, Jan. 1969.

12-33 Seed, H. B., and Idriss, I. M., "Influence of Soil Conditions on Building Damage Potential during Earthquakes," *JSTR*, Paper No. 7909, Feb. 1971.

12-34 Duke, C. M., and Leeds, D. J., "Soil Conditions and Damage in the Mexico Earthquake of July 28, 1957," *BSSA*, **49**(2), Apr. 1959.

12-35 Minami, J. K., and Sakurai, J., "Some Effects of Substructure and Adjacent Soil Interaction on the Seismic Response of Buildings," V. III, *Proc.* 4th WCEE, Santiago, Chile, 1969.

12-36 Merritt, R. G., and Housner, G. W., "Effect of Foundation Compliance on Earthquake Stresses in Multistory Buildings," *BSSA*, **44**(4), Oct. 1954.

12-37 Parmelee, R. A., Perelman, D. S., and Lee, S. L., "Seismic Response of Multiple-Story Structures on Flexible Foundations," *BSSA*, **59**(3), June 1969.

12-38 Khanna, J., "Elastic Soil–Structure Interaction," V. III, *Proc.* 4th WCEE, Santiago, Chile, 1969.

12-39 Whitman, R. V., "Soil-Structure Interaction," *Seismic Design for Nuclear Power Plants*, R. J. Hansen, Editor, MIT Press, Cambridge, Mass., 1970.

12-40 Anderson, J. C., and Bertero, V. V., "Seismic Behavior of Multistory Frames Designed by Different Philosophies," Report No. EERC 69-11, Earthquake Eng. Res. Center, University of California, Berkeley, Oct. 1969.

12-41 Hanson, R. D., and Fan, W. R. S., "The Effect of Minimum Cross Bracing on the Inelastic Response of Multi-story Buildings," V. II, *Proc.* 4th WCEE, Santiago, Chile, 1969.

12-42 Goel, S. C., "P-Δ and Axial Column Deformation in Aseismic Frames," *JSTR*, Paper No. 6738, Aug. 1969.

12-43 Nigam, N. C., and Housner, G. W., "Elastic and Inelastic Response of Frame Structures During Earthquakes," V. II, *Proc.* 4th WCEE, Santiago, Chile, 1969.

12-44 Weaver, W., Jr., Nelson, M. F., and Manning, T. A., "Dynamics of Tier Buildings," *JEMD*, Paper No. 6293, Dec. 1968.

12-45 Wen, R. K., and Farhoomand, F., "Dynamic Analysis of Inelastic Space Frames," *JEMD*, Paper No. 7621, Oct. 1970.

12-46 Utku, S., "ELAS—Program for the Analysis of Structures," Dept. of Civil Engineering, Duke University, 1968.

12-47 Wilson, E. L., "SAP—A General Structural Analysis Program," Structural Eng. Lab., University of Calif., Berkeley, Sept. 1970.

12-48 NASTRAN (NASA Structural Analysis Computer System), Computer Software Management and Information Center (COSMIC), University of Georgia, Athens, Georgia.

12-49 Wilson, E. L., and Dovey, H. H., "TABS—Static and Earthquake Analysis of Three-Dimensional Frame and Shearwall Buildings," Report No. EERC 72-1, Earthquake Eng. Res. Center, University of California, Berkeley, May 1972.

12-50 Newmark, N. M., "Torsion in Symmetrical Buildings," V. II, *Proc.* 4th WCEE, Santiago, Chile, 1969.

12-51 Koh, T., Takase, H., and Tsugawa, T., "Torsional Problems in Design of High-Rise Buildings," V. II, *Proc.* 4th WCEE, Santiago, Chile, 1969.

12-52 Blume, J. A., and Jhaveri, D., "Time-History Response of Buildings with Unusual Configurations," V. II, *Proc.* 4th WCEE, Santiago, Chile, 1969.

12-53 Bertero, V. V., "Experimental Studies Concerning Reinforced, Prestressed and Partially Prestressed Concrete Structures and Their Elements," Introductory Report of the Symposium on Ultimate Deformability of Structures Acted on by Well Defined Repeated Loads, International Association for Bridge and Structural Engineering, Lisbon, 1973.

12-54 Bertero, V. V., "Seismic Behavior of Structural Concrete Linear Elements (Beams and Columns) and Their Connections," *Proceedings* of the A.I.C.A.P.–C.E.B. Symposium on Structural Concrete Under Seismic Actions, Rome, 1979, V. I, pp. 123–212.

12-55 Jirsa, J. O., "Behavior of Elements and Subassemblages—R. C. Frames," *Proceedings* of a Workshop on Earthquake-Resistant Reinforced Concrete Building Construction, Berkeley, July 1977, V. III, pp. 1196–1214.

12-56 Derecho, A. T., Iqbal, M., Ghosh, S. K., Fintel, M., and Corley, W. G., "Structural Walls in Earthquake-Resistant Buildings, Dynamic Analysis of Isolated Structural Walls—Representative Loading History," Report to the National Science Foundation, Portland Cement Association, Skokie, Ill., Aug. 1978.

12-57 Bertero, V. V., "Seismic Behavior of Linear Elements (Beams and Columns)," *State-of-the-Art in Earthquake Engineering 1981*, Turkish National Committee on Earthquake Engineering, Ankara, Oct. 1981, pp. 323–364.

12-58 Ma, S. Y., Bertero, V. V., and Popov, E. P., "Experimental and Analytical Studies on the Hysteretic Behavior of Reinforced Concrete Rectangular and T-Beams," Report No. EERC 76-2, Earthquake Eng. Res. Center, University of California, Berkeley, 1976.

12-59 Bertero, V. V., and Popov, E. P., "Hysteretic Behavior of R.C. Flexural Members with Special Web Reinforcement," *Proceedings*, The U.S. National Conference on Earthquake Engineering, Ann Arbor, Mich., 1975, pp. 316–326.

12-60 Scribner, C. F. and Wight, J. K., "Delaying Shear Strength Decay in Reinforced Concrete Flexural Members Under Large Load Reversals," Report UMEE 78R2, Department of Civil Engineering, The University of Michigan, Ann Arbor, Mich., 1978.

12-61 Ohmori, N., "The 18-Storied Shiinamachi Building, Cast-in-Field Reinforced Concrete Systems," *Proceedings*, Workshop on Earthquake-Resistant Reinforced Concrete Building Construction, University of California, Berkeley, July 1977, V. II, pp. 756–769.

12-62 "Cyclic Loading of Reinforced Concrete Frame Joints: Recent Research Developments and Applications to Design," Draft Report, ATC-11 Project, Applied Technology Council, Berkeley, Calif., Feb. 1983.

12-63 Jirsa, J. O., "Seismic Behavior of R.C. Connections (Beam–Column Joints)," *State-of-the-Art in Earthquake Engineering 1981*, Turkish National Committee on Earthquake Engineering, Ankara, Oct. 1981, pp. 365–374.

12-64 Zhang, L. and Jirsa, J. O., "A Study of Shear Behavior of Reinforced Concrete Beam–Column Joints," PMFSEL Report No. 82-1, The University of Texas at Austin, Feb. 1982.

12-65 ACI-ASCE Committee 352, "Recommendations for Design of Beam-Column Joints in Monolithic Reinforced Concrete Structures," *ACI Journal, Proc.*, 73(7), July 1976, pp. 375–393.

12-66 Standards Association of New Zealand, *Code of Practice for the Design of Concrete Structures*, NZS 3101:1982, 1982.

12-67 Oesterle, R. G., Aristizabal-Ochoa, J. D., Fiorato, A. E., Russell, H. G., and Corley, W. G., "Earthquake Resistant Structural Walls—Tests of Isolated Walls—Phase II," Report to the National Science Foundation, Portland Cement Association, Skokie, Ill., Oct. 1979.

12-68 "Cyclic Loading of Reinforced Concrete Shear Walls: Recent Research and Design Implications," Draft Report, ATC-11 Project, Applied Technology Council, Berkeley, Calif., Mar. 1983.

12-69 Bertero, V. V., "Seismic Behavior of R.C. Wall Structural Systems," *State-of-Art in Earthquake Engineering 1981*, Turkish National Committee on Earthquake Engineering, Ankara, Oct. 1981, pp. 375–382.

12-70 Cardenas, A. E., Hanson, J. M., Corley, W. G., and Hognestad, E., "Design Provisions for Shear Walls," *ACI Journal, Proc.*, 70(3), Mar. 1973, pp. 221–230.

12-71 Paulay, T., "Coupling Beams of Reinforced Concrete Shear Walls," *Proceedings*, Workshop on Earthquake-Resistant Reinforced Concrete Building Construction, University of California, Berkeley, July 1977, V. III, pp. 1452–1460.

12-72 Hawkins, N. M., "Seismic Response Constraints for Slab Systems," *Proceedings*, Workshop on Earthquake-Resistant Reinforced Concrete Building

Construction, Berkeley, July 1977, V. III, pp. 1253–1267.

12-73 Seismology Committee, Structural Engineers Association of California, "Recommended Lateral Force Requirements and Commentary," San Francisco, 1980.

12-74 Clough, R. W., "Dynamic Effects of Earthquakes," *Transactions, ASCE*, V. **126**, Part II, Paper No. 3252, 1961.

12-75 Blume, J. A., "Structural Dynamics in Earthquake-Resistant Design," *Transactions, ASCE*, V. **125**, Part I, Paper No. 3054, 1960.

12-76 Berg, G. V., "Response of Multistory Structures to Earthquake," *JEMD*, Paper No. 2790, Apr. 1961.

12-77 Hanson, N. W., and Conner, H. W., "Seismic Resistance of Reinforced Concrete Beam–Column Joints," *JSTR*, Paper No. 5537, October 1967.

12-78 Bertero, V. V., and Bresler, B., "Seismic Behavior of Reinforced Concrete Framed Structures," V. I., *Proc.* 4th WCEE, Santiago, Chile, 1969.

12-79 Agrawal, G. L. Tulin, L. G., and Gerstle, K. H., "Response of Doubly Reinforced Concrete Beams to Cyclic Loading," *ACI Journal, Proc.*, **62**(7), 823–835, July 1965.

12-80 Burns, N. H., and Siess, C. P., "Repeated and Reversed Loading in Reinforced Concrete," *JSTR*, **92**(ST5), 65–78, Paper No. 4932, Oct. 1966.

12-81 ACI Committee 318, *Building Code Requirements for Reinforced Concrete* (ACI 318-83), American Concrete Institute, Detroit, Mich. 48219.

12-82 Fintel, M., and Khan, F. R., "Shock-Absorbing Soft-Story Concept for Multistory Earthquake Structures," *ACI Journal, Proc.*, **66**(5), 381–390, May 1969.

12-83 Caspe, M. S., "Earthquake Isolation of Multistory Concrete Structures," *ACI Journal, Proc.*, **67**(11), Nov. 1970.

12-84 Anonymous, "Ball Bearing Seismic Resistance" (on Matsushita system), *Engineering News Report*, 73, Mar. 16, 1967.

12-85 Newmark, N. M. and Rosenblueth, E., *Fundamentals of Earthquake Engineering*, Prentice-Hall, Englewood Cliffs, N.J., 1971.

12-86 Kelly, J. M., "Aseismic Base Isolation: Its History and Prospects," *Joint Sealing and Bearing Systems for Concrete Structures*, V. 1, American Concrete Institute, SP-70, 1981, pp. 549–586.

12-87 Kelly, J. M., Beucke, K. E., and Skinner, M. S., "Experimental Testing of a friction damped base isolation system with fail-safe characteristics," Report EERC 80-18, Earthquake Eng. Res. Center, University of California, Berkeley, July 1980.

12-88 Muto, Kiyoshi, "Earthquake Resistant Design of 36-Storied Kasumigaseki Building," Special Report, 4th WCEE, Santiago, Chile, 1969.

12-89 Muto, Kiyoshi, "Earthquake Proof Design Gives Rise to First Japanese Skyscrapers," *Civ. Eng.—ASCE*, Mar. 1971, pp. 49–52.

12-90 Associate Committee on the National Building Code, *National Building Code of Canada*, NRCC No. 17303, National Research Council of Canada, Ottawa, 1980.

12-91 Associate Committee on the National Building Code, *National Building Code of Canada*, NRCC No. 17724, National Research Council of Canada, Ottawa, 1980.

12-92 American Institute of Steel Construction, *Specification for the Design, Fabrication and Erection of Structural Steel for Buildings*, Chicago, 1978.

12-93 "Frame Constants for Lateral Loads on Multistory Concrete Buildings," *Advanced Engineering Bulletin No. 5*, Portland Cement Association, 1962.

12-94 Derecho, A. T., "Frames and Frame–Shearwall Structures," paper presented at Tall Building Symposium, ACI Fall Convention, Buffalo, N.Y., Nov., 1971.

12-95 Meinheit, D. F., and Jirsa, J. O., "Shear Strength of R/C Beam–Column Connections," *JSTR*, 107(ST11), Nov. 1981.

12-96 Fintel, M., "Ductile Shear Walls in Earthquake Resistant Multistory Buildings," *J. Am. Concr. Inst., Proc.* 71(6), June 1974, pp. 296–305.

12-97 Fugelso, L. E., and Derecho, A. T., "Program DYFRQ—for the Determination of the Natural Frequencies and Mode Shapes of Plane Multistory Structures," Portland Cement Association, Skokie, Ill., 1975 (unpublished).

12-98 Derecho, A. T., "Analysis of Plane Multistory Frame–Shear Wall Structures Under Lateral and Gravity Loads," Publication XL097D, Portland Cement Association, Skokie, Ill., 1971.

12-99 *Design Handbook in Accordance with the Strength Design Method of ACI 318-77: Vol. 1—Beams, Slabs, Brackets, Footings, and Pile Caps*, Publication SP-17 (81), American Concrete Institute, Detroit, 1981.

12-100 Winter G., and Nilson, A. H., *Design of Concrete Structures*, 9th Ed., McGraw-Hill Book Co., New York, 1979.

12-101 *Design Handbook in Accordance with the Strength Design Method of ACI 318-77: Vol. 2—Columns*, Publication SP-17A (78), American Concrete Institute, Detroit, 1978.

12-102 *CRSI Handbook*, Concrete Reinforcing Steel Institute, Schaumburg, Ill., 1982.

12-103 Kanaan, A. E., and Powell, G. H., "DRAIN-2D: A General Purpose Computer Program for Dynamic Analysis of Inelastic Plane Structures," Reports No. EERC 73-6 and EERC 73-22., Earthquake Eng. Res., Center, University of California, Berkeley, Apr. 1973 (revised Sept. 1973) and Aug. 1975, 273 pp.

12-104 Otani, S., "SAKE—A Computer Program for Inelastic Response of R/C Frames to Earthquakes," *Civ. Eng. Stud., Univ. Ill. Struct. Res. Ser.* No. 413, Nov. 1975.

12-105 Clough, R. W., "Effect of Stiffness Degradation on Earthquake Ductility Requirements," Report No. EERC 66-16, Earthquake Eng. Res. Center, University of California, Berkeley, 1966, 196 pp.

12-106 Riddell, R., and Newmark, N. M., "Statistical Analysis of the Response of Nonlinear Systems Subjected to Earthquakes," *Civ. Eng. Stud., Univ. Ill. Struct. Res. Ser.* No. 468, Aug. 1979.

12-107 Takayanagi, T., and Schnobrich, W. C., "Non-linear Analysis of Coupled Wall Systems," *Earthq. Eng. & Struct. Dyn.*, 7(1), Jan.–Feb. 1979, pp. 1–22.

12-108 Newmark, N. M., and Hall, W. J., "A Rational Approach to Seismic Design Standards for Structures," *Proc.* 5th WCEE, Rome, 1974, V. 2, pp. 2266–2275.

12-109 Newmark, N. M., Design of Structures to Resist Seismic Motions," *Proc.* Earthq. Eng. Conf., University of South Carolina, Jan. 1975, pp. 235–275.

12-110 Shibata, A., and Sozen, M. A., "Substitute-Structure Method for Seismic Design in R/C," *JSTR*, 102 (ST1), Jan. 1976, pp. 1–18.

12-111 Biggs, J. M., Lau, W. K., and Persinko, D., *Aseismic Design Procedures for Reinforced Concrete Frames*, Publication No. R-79-21, Department of Civil Engineering, Massachusetts Institute of Technology, Cambridge, Mass., 1979, 79 pp.

12-112 Housner, G. W., Limit design of structures to resist earthquakes, *Proc.* 1st WCEE, Berkeley, Calif., 1956, pp. 5.1–5.13.

12-113 Larson, M. A., "Needed Improvements in Aseismic Design," *Struct. Moments*, No. 3, Structural Engineers Association of Northern California, San Francisco, Feb. 1980, 4 pp.

12-114 McKevitt, W. E., Anderson, D. L., Nathan, N. D., and Cherry, S., "Towards a Simple Energy Method for Seismic Design of Structures," *Proc.* 2nd U.S. Nat. Conf. Earthq. Eng., Stanford, Calif., Aug. 1979, pp. 383–392.

12-115 Muto Institute of Structural Mechanics, "Structural Design of Shinjuku Mitsui Building (SMB)," Tokyo, 1972, 5 pp.

12-116 Muto Institute of Structural Mechanics, "Structural Design of International Telecommunications Center Building," Tokyo, 1973, 7 pp.

12-117 Fintel, M., and Ghosh, S. K., "The Structural Fuse: An Inelastic Approach to Seismic Design of Buildings," *Civ. Eng.–ASCE*, 48–51, Jan. 1981.

12-118 Fintel, M., and Ghosh, S. K., "Explicit Inelastic Dynamic Design Procedure for Aseismic Structures," *J. Am. Concr. Inst., Proc.* 79(2), Mar.–Apr. 1982, pp. 110–118.

12-119 Fintel, M., and Ghosh, S. K., *Seismic Resistance of a 16-Story Coupled Wall Structure: A Case Study Using Inelastic Dynamic Analysis*, Publication EB82.01D, Portland Cement Association, Skokie, Ill., 1980, 31 pp.

12-120 Fintel, M., and Ghosh, S. K., "Case Study of Aseismic Design of a 16-Story Coupled Wall Structure Using Inelastic Dynamic Analysis," *J. Am. Concr. Inst., Proc.* 79(3), May–June 1982, pp. 171–179.

12-121 Fintel, M., and Ghosh, S. K., Seismic Resistance of a 31-Story Shear Wall–Frame Building Using Dynamic Inelastic Response History Analysis, *Proc.* 7th WCEE, Istanbul, 1980, V. 4, pp. 379–386.

12-122 Fintel, M., and Ghosh, S. K., "Application of Inelastic Response History Analysis in the Aseismic Design of a 31-Story Frame–Wall Building," *Earthq. Eng. & Struct. Dyn.*, 9(6), Nov.–Dec. 1981, pp. 543–556.

12-123 "Inelastic Seismic Design Idea Simplifies Concrete High-Rise: Novel Technique Relieves Rebar Congestion," *Eng. News Rec.*, 28 Aug. 1980, pp. 58–59.

12-124 Derecho, A. T., Fugelso, L. E., and Fintel, M., *Structural Walls in Earthquake-Resistant Buildings —Dynamic Analysis of Isolated Structural Walls—Input Motions*, Report on research sponsored by National Science Foundation (RANN), Grant No. ENV74-14766, Portland Cement Association, Skokie, Ill., 1977, 54 pp.

12-125 Paulay, T., and Binney, J. R., "Diagonally Reinforced Coupling Beams of Shear Walls," *Shear in Reinforced Concrete*, Publication SP-42, 1974, American Concrete Institute, pp. 579–598.

13

Large Panel Structures

IAIN A. MacLEOD, Ph.D.*

13.1 INTRODUCTION

For the purpose of this chapter a large panel building is one that derives its main support from story-height precast concrete wall units. Figure 13-1 is a photograph of a 13-story large panel building and Fig. 13-2 shows a large panel being placed.

This structural form was developed in Europe in the 1960s based on the philosophy that by precasting, good quality control of fabrication of the units can be achieved, and site labor is minimized. It has been extensively used since then, particularly in Eastern Europe. In the United States and Canada, large panel structures are used occasionally for both high-rise and low-rise structures. In the United Kingdom large panel buildings are seldom now used for high-rise construction and the housing market in general, but are still popular for hotel and office blocks up to about 9 stories. There are two main reasons why high-rise large panel blocks are not being built in the United Kingdom. First, some of the early buildings of this type were badly designed and have suffered from either severe serviceability problems or, in one case, collapse due to a gas explosion. Second, they were mainly used to solve local authority housing problems, and the high-rise format was not popular with many tenants, especially those with young families.

The latter problem is not difficult to solve if the principle that only those who want to live in a high-rise block should do so is adopted. Also, provided lessons learned from previous failures are properly interpreted and catered for in new designs, large panel structures can be built that will perform well in service and have an adequate factor of safety against collapse.

It is inappropriate here to make any general statement about the cost of large panel structures. The cheapest method of construction at a given location is dependent upon local cost and availability of materials, and, above all, on labor sources. Efficient large panel construction needs specially trained personnel in both manufacture and erection, since quality control at all stages of manufacture and erection is of prime importance. Thus, if it is to be adopted in an area, continued use of this type of construction would be most advantageous.

13.1.1 Layout

It has been common in European practice to use load bearing walls as partitions between rooms within an apartment. American practice has tended to have load bearing walls only between apartments, thus giving longer floor spans. In Europe floors have been precast or cast in situ, one-way or two-way, whereas in America floors constructed using precast prestressed planks are the norm.

Three main types of layout can be identified:

1. Cross wall—Fig. 13-3(a).
2. Spine wall—Fig. 13-3(b).
3. Mixed system—Fig. 13-3(c), a "European" layout.

The adoption of a large panel format does place restrictions on the architectural layout, since modular repetition is necessary to keep costs down. However, there is scope for

*Professor of Structural Engineering, University of Strathclyde, Glasgow, Scotland.

Fig. 13-1 Mass. Pike Towers in Boston, Mass. Courtesy Sepp Firnkass, Eng'g.

inventiveness in layout, provided principles of good design are not ignored.

In earthquake zones, fairly rigid layout criteria must be imposed. In such areas, good design would preserve a uniformity of structure with height, special care being taken with any variation at the ground floor level. The structure should be symmetrical in plan with the stiffer elements of lateral resistance being placed near the periphery rather than toward the center of the building. Such criteria are useful to follow even with no potential earthquake loading; but architectural requirements often cause a move away from the best structure, and many buildings throughout the world have performed satisfactorily with layouts that would not be the first choice from the structural point of view. Thus in areas of potentially serious earthquake risk, the structural design has to impose strict limitations on the architecture; but where earthquake loading is not a problem, the architect can be given greater scope in layout. However, the value of both the architect and engineer getting together to resolve their interdependent problems as early as possible cannot be overemphasised.

Fig. 13-2 Erection of external panel.

Large panel buildings normally have high lateral stiffness.

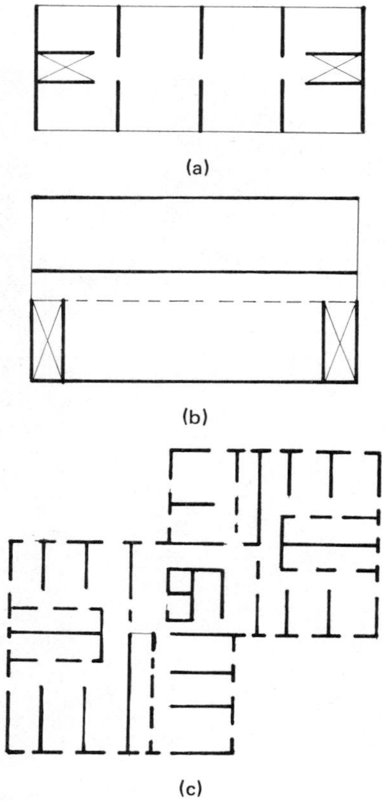

Fig. 13-3 Typical layouts (a) Cross wall system. (b) Spine wall system. (c) "European layout."

It is good practice wherever possible to organize the layout so that lateral load-resistant units can be linked together. For example, if two walls lie in a plane, efforts should be made to connect them by beams having sufficient strength and stiffness to promote composite action between them. Also walls at right angles to a plane of loading can profitably be designed to act as "flanges" to walls in the plane in order to improve strength and stiffness. This is further discussed in Section 13.3.

13.2 LOADING

In this section only those aspects of loading that have special relevance to large panel structures are discussed. A thorough discussion on all aspects of loading is given in Ref. 13-1. "Loading" in this context denotes all sources of normal and abnormal stress.

Progressive Collapse. In 1967 a gas explosion occurred at the 18th story of the 22-story large panel Ronan Point building in London, England. This caused the side panels of the apartment at the 18th story to be blown away, resulting in collapse of the wall and floor panels above the explosion. The debris loading from above then carried away most of the wall and floor panels in the corner of the building down to ground level (Fig. 13-4). This failure is of great significance in the development of structural design. Prior to this event, progressive collapse was given very little attention in the literature or in the codes of practice. Thus the Ronan Point collapse was taken to be due, in part at least, to lack of code provision for this type of structure. There is an important lesson here. Code provisions should be viewed only as minimum requirements, and structural designers, especially if they are being in any way innovative, should look beyond the code provisions to seek out potentially critical situations. This does not mean that they should design for excess safety. It means that they should have a proper grasp of the behavior of the structure so as to avoid future difficulties. The Ronan Point failure well illustrates this. In the 1960s in the United Kingdom, many designers recognized the potential problems with connections in precast structures and were designing large panel buildings that would now be considered adequate with respect to structural integrity. Unfortunately others were less thoughtful about potential hazards and felt that if their designs complied with code provisions, they would be satisfactory. One should not blame the code because they were wrong. The code is only an aid to good design and not a guarantee.

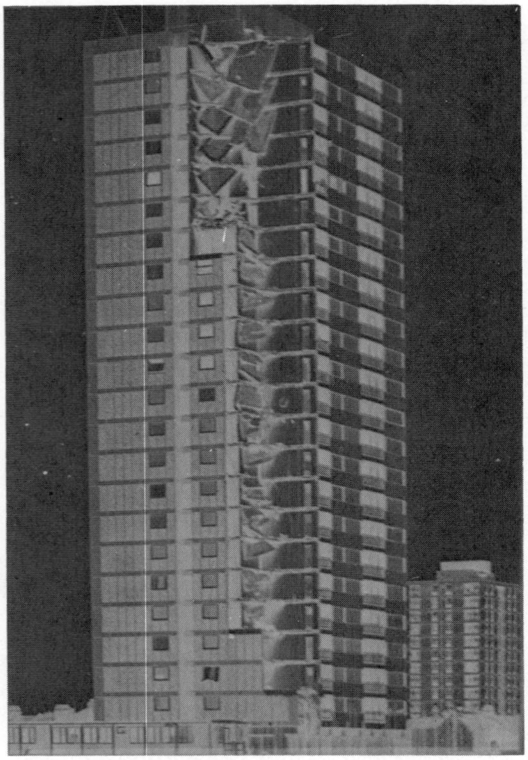

Fig. 13-4 Ronan Point Building after collapse.

It is thought that the explosion at Ronan Point occurred at the worst possible position in the building. If it had occurred at a higher story, the debris loading might have been insufficient to take away the lower panels. At a lower level the vertical load in the panels might have prevented the walls from blowing out. Nevertheless, after lengthy litigation the designers were held to be liable. Although other problems were identified, the main fault lay in the design of the connections between the panels and lack of tie steel to hold the units together.[13-11] These problems, although not exclusive to large panel structures, are clearly identified areas in which special care is needed in design of this type of structure.

However, there has probably been an overreaction to the progressive collapse problem, and large panel buildings are only susceptible to it if they are badly detailed.

It is normally too expensive to design a structure so that all the supporting members remain intact after a severe gas explosion. It is, therefore, arranged that if part of the support structure (e.g., one wall panel) becomes "ineffective," secondary support systems will be mobolized so that further collapse will not occur. It is, of course, not possible to provide an absolute guarantee against progressive collapse. Any building can be demolished with sufficient explosive power. The aim in design is to give reasonable security that local loss of support due to an explosion or similar accident will not trigger a major collapse.

Lateral Load. Lateral load is mainly due to wind or the horizontal component of an earthquake motion. These factors are assumed to act nonconcurrently and are important in the design of large panel buildings. The analysis of large panel buildings under lateral load is described in Section 13.3.

Earthquake Loading. An earthquake imposes a special kind of abnormal loading on a structure. Although it would be unlikely to cause a single panel to become ineffective, the interconnection of the components of the structure assumes a very special importance. Cast-in-situ shear wall structures are known to behave well in earthquakes;[13-10] therefore, provided the connection detailing is good, large panel structures should behave similarly. In modern philosophy for earthquake design, ductility is the prime criterion, and there is some difference of opinion as to whether the interpanel connections can provide adequate ductility. Vertical joints in a large panel building do have the potential to be used as energy dissipators, but can they continue to transmit sufficient load after a series of high-intensity load reversals? It is believed that "wet" joints with shear keys and loop reinforcement within the height of the panel (Fig. 13-17) will give the most satisfactory performance under severe earthquake conditions.[13-14] Other types of joint may be less satisfactory in this respect. The structural requirements to resist progressive collapse are favorable for earthquake design but will not necessarily provide sufficient earthquake resistance.

Reference 13-15 gives useful information on earthquake design for large panel buildings.

Vertical Load. In a building whose vertical support is provided by interconnected walls, a significant amount of redistribution of the vertical load may occur. In fact with a layout of the type shown in Fig. 13-3(c), stress in the walls at the base of the building will tend to be uniform irrespective of the direction of floor span. This may be important in both the design of the walls and the foundations.

Differential Settlement of Foundations. Large panel structures normally have high vertical stiffness, so that a small amount of differential movement would cause appreciable stress and cracking. However, provided the connections between the walls have sufficient strength, such structures should be highly resistant to differential foundation support conditions, and we know of no records of cracking due to settlement of large panel structures. The relationship between the stiffness of the foundations, stiffness of the structure, and stiffness of the soil has an important effect on the distribution of vertical load as well as on the differential settlement. For large panel buildings the structure would normally have a controlling influence, and this should be taken into account in designing the foundations.

Fire. The Ronan Point report[13-11] suggested that fire could cause serious movement at horizontal joints in the external walls due to "arching" of the slab. CEB[13-12] suggests that thermal gradients in walls whose ends are restrained from rotation could significantly reduce their buckling strength.

One of the most important factors governing the fire resistance of a panel is the amount of in-plane restraint provided by the adjacent structure.[13-13] Interior panels therefore have a much greater fire resistance than panels adjacent to the exterior of the building. The provision of tie reinforcement within the joints (Section 13.9) will have some effect in restraining in-plane movement of the slabs and walls, and hence will enhance their fire resistance. Fire is another source of local failure that could result in progressive collapse. In general, however, the fire resistance of large panel structures should be most favorable.

13.3 ANALYSIS

The structure of a large panel building is a system of interconnected walls and floor slabs. In some cases the layout is fairly simple, and division into identifiable shear walls to resist lateral load poses little problem. In other cases such a division may be rather arbitrary. For example, in the layout shown in Fig. 13-3(c) the degree of interconnection between the walls is high, and it is difficult to define an analytical model of it that would be both practical and an accurate representation of the real behavior.

However, the fact that the analytical modeling is difficult is often an indication of good strucutral layout; in that the greater the degree of interconnection between the walls, the better the structural form and the more diffiuclt the modeling.

Two main classes of analytical models can be identified. In the more common type, shear walls are identified, analyzed separately, and then combined to simulate the behavior of the complete structure. In the second approach, a complete three-dimensional model of the structure is formulated.

13.3.1 Division of Structure into Shear Walls

For lateral loading in a given direction (parallel to a major axis of the structure) a system of shear walls needs to be defined that are assumed to resist the load. Of the layouts shown in Fig. 13-3(a) and 13-3(b) there is little difficulty involved in identifying shear walls for the cross wall and spine wall types, but the layout of Fig. 13-3(c) is not so straightforward.

The main aim should be to recognize as much interconnection between the walls as practicable. A connecting beam between two walls that lie in a plane may be formed from a lintel beam or a floor slab, or a combination of both. Such connecting beams can significantly improve the strength and stiffness of adjacent walls, provided they can develop the required strength and stiffness. Figure 13-5 shows the effect of the connection stiffness on the shear in the connection between two walls. Beyond a certain value of stiffness the behavior of the assembly is not sensitive to the stiffness of the connection, and errors in estimates of this parameter may not be important. However, for lower values of stiffness small variations significantly affect the overall behavior, so that extra care is needed.

Factors that affect the connecting beam stiffness are:

1. *The effective width of the floor slab.* If the slab can be assumed to participate in transmitting vertical shear, then it is normal to treat it as an equivalent beam (where there are no lintel beams) or as a flange to the lintel beam. In either case an equivalent width is assumed. Several papers give charts to aid choice in this. It is advisable to underestimate the effective width, since measurements of real stiffness of reinforced concrete members tend to produce lower values than generally found using conventional theories. An effective width not greater than one quarter of the opening width or one quarter of the distance to the next parallel wall on either side should give a reasonable estimate.
2. *The connection between the floor slab and the wall.* In precast structures a full moment connection between the slab and the wall may not be provided, and some support flexibility may need to be assumed. For example, one end of the connection may not take any moment and may have to be modeled as a pin.
3. *Cracking, shrinkage, and creep.* Apart from other considerations, it is notoriously difficult to accurately estimate the stiffness of a reinforced concrete member because of cracking, shrinkage, and creep. Therefore, while one should make as good an estimate as possible, one should guard against using a sophisticated procedure to model a particular aspect of behavior when its effect may be overpowered by another aspect that is only crudely modeled. Such a philosophy in modeling is, of course, not confined to this particular problem.

13.3.2 Flange Effect of Walls at Right Angles

Walls at right angles to the plane of the loading form flanges to the shear walls, provided the connections between them

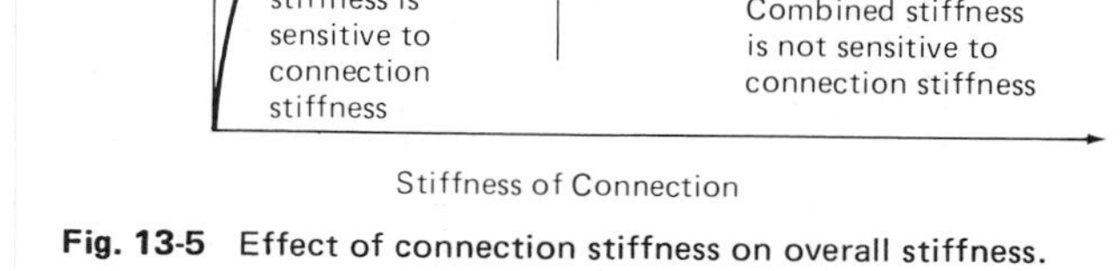

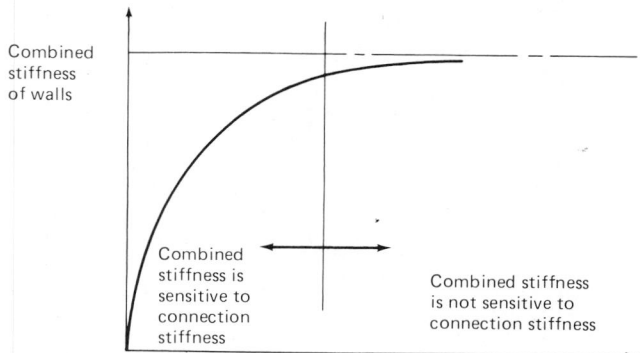

Fig. 13-5 Effect of connection stiffness on overall stiffness.

possess adequate strength and stiffness. If the "flange" walls have no openings, then the effective width for high walls would be taken as one-half of the distance to the next wall on either side. With walls having a low height-to-width ratio, this may overestimate the effect due to shear lag in the flange wall.

If the flange wall has openings, the effective width may be taken up to the edge of the nearest opening on either side, or the shear transfer across the openings can be treated as described in Section 13.3.5.

13.3.3 Analysis of Separate Shear Walls

Walls that do not have openings can be modeled as vertical cantilevers to which bending theory is applicable.

For walls with openings a frame model is the best approach. For example, to treat the wall of Fig. 13-6(a) as a frame, the effective widths of columns of the frame must be taken into account (Fig. 13-6c), which is normally done by specifying rigid ends to the beams. This is best accomplished using a special element,[13-16] and although use of high but finite stiffness can be successful, the possibility of incurring numerical instability in the solution by so doing is not negligible. Axial deformation of the columns must be included, but it is advantageous to neglect axial deformation of the beams. Shear deformation of all members may be included, and springs can be used to model foundation movement. Another useful feature is a rotational spring connection between the flexible and rigid parts of the beams to model the flexibility of the beam-to-column connection.

For walls with uniform properties, as a further assumption to the frame model a row of openings can be treated as a continuous shear connection (Fig. 13-6b). A second-order differential equation is formed and solved for different load cases.

Solutions are relatively simple for walls with a single row of openings (or two symmetrical rows) and constant properties with height. The charts of Ref. 13-17 are useful. Solutions have been devised for walls with several rows of openings and other variations.[13-18–13-20]

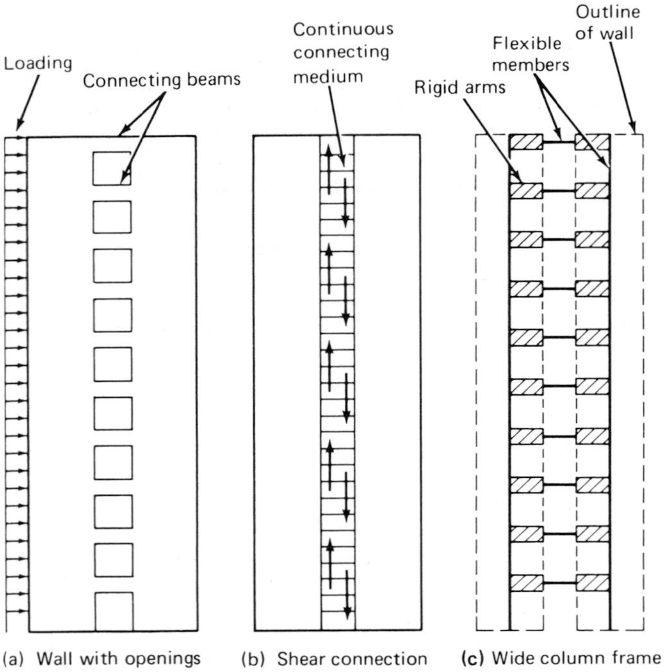

Fig. 13-6 Analysis techniques for shear walls with openings.

In certain cases where the layout of the openings is irregular, a plane stress finite element mesh may be the only practical model. Use of plane stress finite elements should be kept to a minimum, since such analysis is much more costly than frame analysis. If part of a wall can be realistically modeled as a frame element, then it will be needlessly expensive to treat it as a system of finite elements. When parts of a wall need to be modeled using finite elements, the main part can be treated as a frame, and the two types of element combined in one analysis using a rigid transition between the two types of element.[13-21] Some practical considerations in modeling shear walls are discussed in Ref. 13-22.

13.3.4 Distribution of Load to Units

In the treatment that is now described the floors are assumed to be fully rigid in their own planes so that movements of all walls at each floor level are related by a rigid body relationship; for example, with no torsion all movements are the same in a given direction. This is a reasonable assumption for tall buildings but may be less accurate for shorter buildings, near the base of tall buildings, and close to major changes in the plan layout.[13-23]

To illustrate the techniques used, the 20-story structure shown in Fig. 13-7 is used. This is not a realistic structure but is used here to simplify the presentation.

If none of the three shear walls has any openings, the distribution with height of the lateral load will have the same form as that of the applied load. For example, for a uniformly distributed applied load W lb/ft, the loading will be distributed in the proportion:

$$\frac{I_1}{\Sigma I} W : \frac{I_2}{\Sigma I} W : \frac{I_3}{\Sigma I} W$$

to each wall. I_i is the moment of inertia of wall i, and $\Sigma I = I_1 + I_2 + I_3$ (shear deformation is neglected here).

Now suppose the center wall has a single row of openings, and the uniformly distributed load is applied again. The distribution of load to each wall is no longer uniform and will take the form illustrated in Fig. 13-7(d) (computer analysis). This is so because the walls behave differently but are constrained by the floor slabs to take up the same deflection at each story level. Therefore, for best accuracy an analysis should account for this fact.[13-24–13-27] However, it is normally found to be sufficiently accurate to connect the walls analytically at only a few stories and, in fact, preferably (as far as the arithmetic is concerned) only at one location, namely, the top. In other words the stiffness, K_w, of each shear wall is defined as the load to cause unit deflection at the top; the lateral load is distributed in proportion to these stiffnesses.

Another way to approach the same solution is to calculate an equivalent EI for the walls with openings:

$$(EI)_e = \frac{H^3 K_w}{8} \tag{13-1}$$

where H is the height of the wall, and $(EI)_e$ is the equivalent EI. The load is then distributed in proportion to these EI values.

EXAMPLE 13-1: The idealized 20-story structure of Fig. 13-7(a) is composed of two plain walls (Fig. 13-7c) and one wall with openings (Fig. 13-7b). Denoting the wall with openings as Wall A and the two plain walls together as Wall B, we want to calculate the distribution of lateral load to these walls and hence calculate the stresses. Consider a uniformly distributed load of 10 lb/ft height on the structure, and assume that this does not induce any torsion. Units used are feet and pounds, $E = 5 \times 10^8$ lb/ft^2, and shear deformation is neglected.

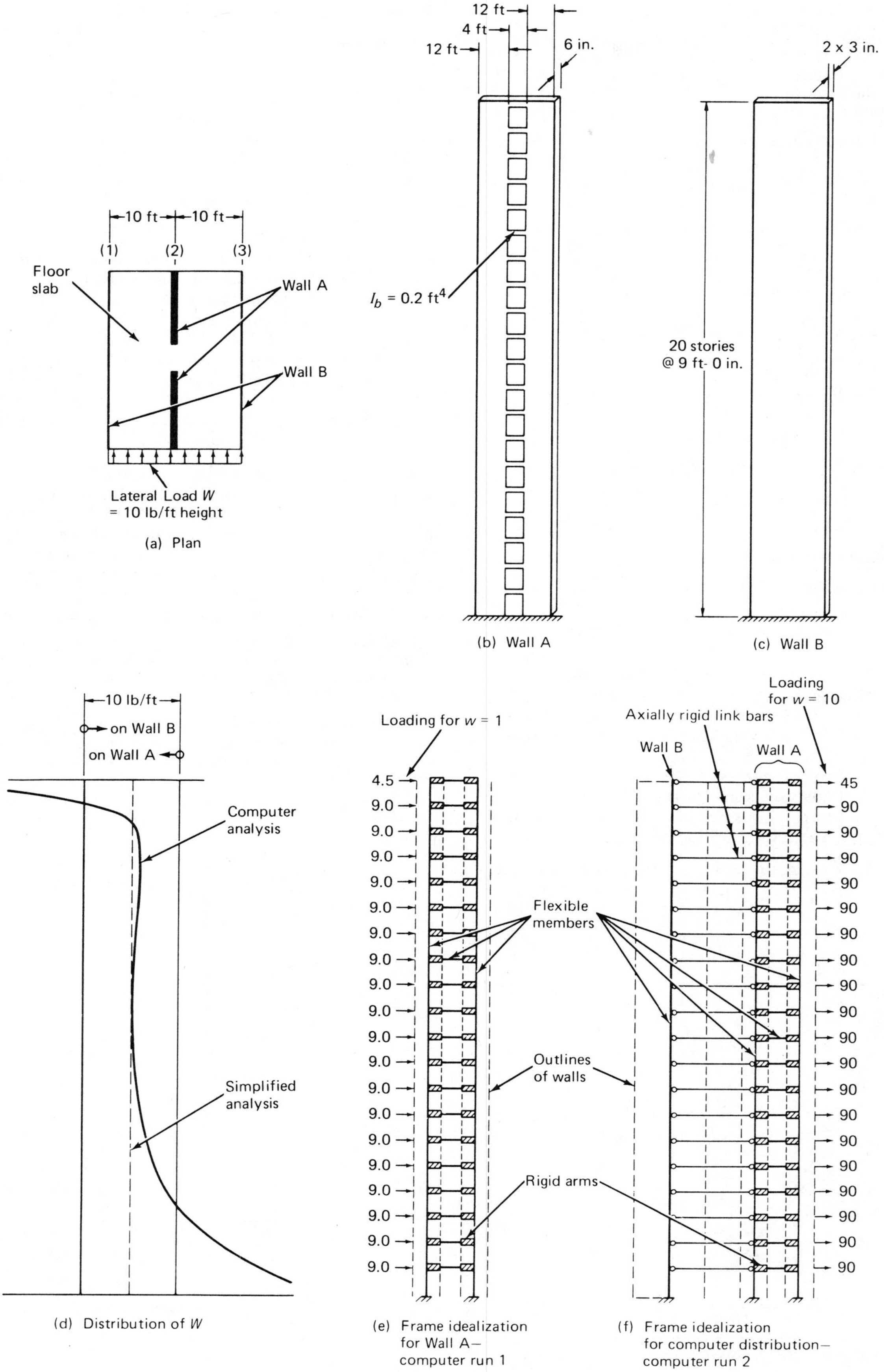

Fig. 13-7 Interconnected shear wall problem.

SOLUTION: Simplified Method.

Step 1: Analyze Walls A and B separately. Wall B analysis is straightforward, but Wall A is more complex. The charts of Ref. 13-17 could be used, but since a frame program was readily available, an analysis of the frame idealization of Fig. 13-7(e) was carried out. This is called Run 1.

Step 2: Calculate wall stiffness:

$$K_{WA} = \frac{10^4}{3.082} \text{ (from Run 1)} = 3245$$

$$K_{WB} = \frac{8EI}{H^4} = \frac{8 \times 5 \times 10^8 \times 0.5 \times 28^3}{180^4 \times 12} = 3486 \qquad \Sigma K = 6731$$

Step 3: Distribute load to walls:

$$\text{to Wall A} = \frac{K_{WA}}{\Sigma K} W = \frac{3245}{6731} \times 10 = 4.82 \text{ lb/ft}$$

$$\text{to Wall B} = \frac{K_{WB}}{\Sigma K} W = \frac{3486}{6731} \times 10 = 5.18 \text{ lb/ft}$$

This distribution is shown in Fig. 13-7(d).

Step 4: Calculate the top deflection of structure:

$$\Delta = \frac{\text{Load on Wall A}}{K_{WA}} = \frac{4.82}{3245} = 1.482 \times 10^{-3} \text{ ft}$$

Step 5: Calculate stresses. For example:

Max. bending stress in Wall B

$$= \frac{5.18 \times H^2}{2} \times \frac{6}{0.5 \times 28^2} = 1284 \text{ lb/ft}^2$$

1. Computer Distribution. Figure 13-7(f) shows how a plane frame program can be used to carry out a no-torsion analysis of a shear wall structure taking account of the interaction at each story level. The distribution of load to each wall is illustrated in Fig. 13-7(d).

2. Comparison of Accuracy. The distribution of load by the simplified analysis is significantly different from the computer distribution. This is not important from the point of view of design. The important factors are as follows:

(a) Vertical stress in the walls: Fig. 13-8 shows the vertical edge stress in Wall A. Accuracy for Wall B is slightly better than this (not shown).

(b) Maximum beam shear: Table 13-1 shows that the accuracy is adequate.

(c) Deflection: Table 13-1 shows that the accuracy for estimation of top deflection is very good. The closeness of the results here is to some extent fortuitous, but estimates of deflection tend always to be more accurate than the moments and forces derived from them.

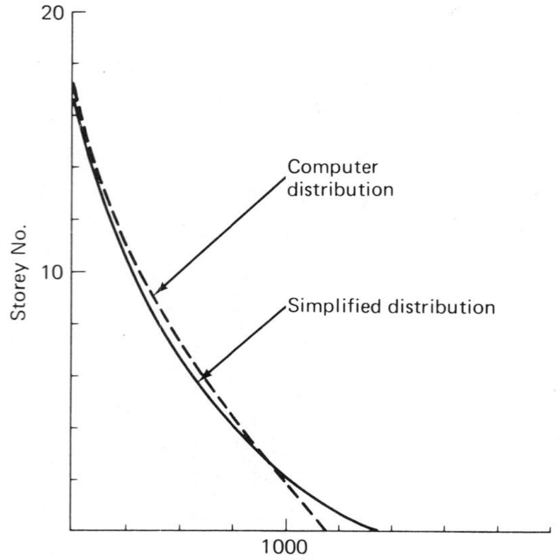

Fig. 13-8 Vertical edge stress in Wall A.

TABLE 13-1 Accuracy of Simplified Analysis

	(a) *Simplified Analysis*	*(b)* *Computer Run 2*	*% difference* $\frac{(a)-(b)}{(b)} \times 100$
Maximum Beam Shear (Wall A)	313.0	269.2	+16%
Top Deflection	1.479×10^{-3}	1.482×10^{-3}	-0.2%

The properties of the structure used for this example are favorable to accuracy for the simplified method. The results shown in Fig. 13-8 and Table 13-1 for the simplified method would be suitable for design. However, they probably represent an upper limit to the accuracy for the method when walls with and without openings are connected. Techniques for determining when the simplified method is inadequate have not yet been established.

EXAMPLE 13-2: *Torsion.* An analysis with torsion can be made in a similar fashion to the distribution previously described. For example, assume the load acts off-center on the structure of Fig. 13-7(a), i.e., as in Fig. 13-9(a).

SOLUTION: Consider the top of the structure as a rigid beam supported by springs and loaded as shown in Fig. 13-9(b). Movement of this beam can be defined by a deflection at 1 (Δ) and a rotation (Θ).

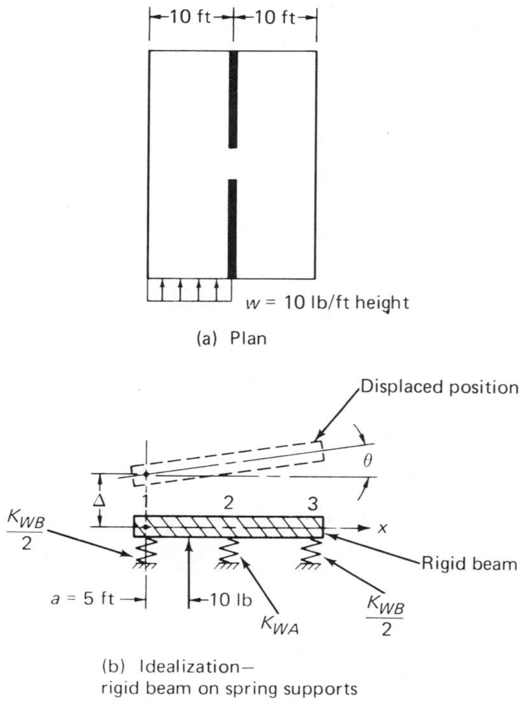

Fig. 13-9 Example problem with torsion.

Equilibrium gives the simultaneous equations:

$$\Sigma K_i(\Delta + x_i\Theta) = W \quad (13\text{-}2)$$

$$\Sigma K_i x_i(\Delta + x_i\Theta) = W_a \quad (13\text{-}3)$$

These can be solved for Δ and Θ. The load on each wall is then:

$$W_i = WK_i(\Delta + x_i\Theta) \quad (13\text{-}4)$$

The calculations for the system of Fig. 13-9(b) are set out in Table 13-2. The equilibrium equations, (13-2) and (13-3), are then:

$$6731\ \Delta + 67310\ \Theta = 10$$

$$67310\ \Delta + 102170\ \Theta = 10 \times 5$$

Solving this gives:

$$\Delta = 61.0 \times 10^{-5}$$

$$\Theta = 8.76 \times 10^{-5}$$

Hence the loading on the separate walls, using eq. (13-4), will be:

$$W_1 = 10 \times 1734 \times 10^{-5}\ (61.0 + 0) \qquad = 1.06\ \text{lb/ft}$$

$$W_2 = 10 \times 3245 \times 10^{-5}\ (61.0 + 10 \times 8.76) = 4.84\ \text{lb/ft}$$

$$W_3 = 10 \times 1734 \times 10^{-5}\ (61.0 + 20 \times 8.76) = 4.10\ \text{lb/ft}$$

These are the approximate loads on the walls assumed to be uniformly distributed.

The procedure gives satisfactory answers when the free deflection of the walls is close to that of a cantilever. When openings induce frame behavior in some of the walls, comparison of results between this simplified analysis and one where all the floors are considered may show significant differences.

13.3.5 Complete Models

An obvious possibility for modeling a system of interconnected walls is to treat the complete system as a space structure using frame elements, and/or plane stress finite elements.

This may be unjustifiably expensive, and simplications to reduce the size of the model should be sought.

For example, out-of-plane bending of the walls can be eliminated, and the floors can be treated as rigid planes. The latter assumption may not reduce the solution time owing to increased bandwidth of the structural stiffness matrix, depending on the solution technique used.

A useful technique is to treat the walls as plane frames and to provide only a vertical connection at nodes of walls where they connect at right angles.[13-28] This gives good overall results but can produce anomalies when relatively flexible columns are thus connected.

A good practical way of providing a space model of a large panel structure is to treat the shear in all vertical connecting lines as continuous functions.[13-29,13-30] Provided a stable solution routine is found for the resulting differential equations, low solution times will be obtained as compared with the use of a discrete element model.

TABLE 13-2 Analysis with Torsion

Wall	K_i	X_i	X_i^2	K_iX_i	$K_iX_i^2$
1	1743	0	0	0	0
2	3245	10	100	32450	324500
3	1743	20	400	34860	697200
Σ =	6731			67310	102170

13.4 STRUCTURAL AND FUNCTIONAL DESIGN—GENERAL

In the following sections, the main problems associated with the design of large panel structures are discussed. In practice the design of the wall and floor panels cannot be considered separately from the design of the vertical and horizontal connections. They are discussed separately here in order to emphasize the importance of the connection design. The word "connection" is used here to describe the region where panels are connected and the "joint" is specifically the area between the connected parts.

The discussion covers mainly structural problems, which cannot, however, be considered in isolation from functional factors such as thermal, weather, and sound insulation.

13.5 DESIGN OF WALL PANELS

13.5.1 General

In order to limit differential vertical movements, it is important that all the load bearing walls be constructed from the same type of concrete within each story height. For the same reason, the vertical stress due to dead load should ideally be constant in all the walls. This will not normally be possible, but at least precast and cast-in-situ load bearing walls should not be present together at the same story level.

The thickness of a wall will be dependent on other factors as well as strength requirements. Sound insulation criteria normally result in a minimum thickness of 6 in. Reference 13-6 recommends that the thickness should be not less than height/25.

Standard code provisions are in general applicable to large panel walls, and Refs. 13-31 and 13-32 give detailed recommendations.

It would be uncommon in a large panel building not to be able to assume that all walls are effectively braced. However, structures of this type have been built for which stability in the longitudinal direction did rely on special moment connections between the panels and the floor slabs.[13-33]

13.5.2 Eccentricity

The Polish *Code*[13-34] recommends that the average of the resultant top and bottom eccentricities on a wall panel be used for design. This is also recommended in Refs. 13-6 and 13-35. To calculate the eccentricity of a simply supported floor slab on a wall, it is normally assumed that the resultant of the floor reaction acts at one-third of the depth of the bearing area from the loaded face. For the design of a given load panel only the eccentricities at floor levels above and below are considered.

13.5.3 External Walls

It has been common for construction in cold climates to use a sandwich panel for external walls (Fig. 13-10). Such panels tend to be the most expensive part of the structural system, and cheaper alternatives may be possible. In the sandwich, the structural part is kept to the inside, thus partially isolating it from temperature changes. Some designers now feel that in order to provide more effective insulation a solid external wall with insulation on the inside face is preferable.

In a sandwich panel, ties (which might be more accurately described as shear connectors) must be provided to support the external leaf. These must be strong enough in shear to

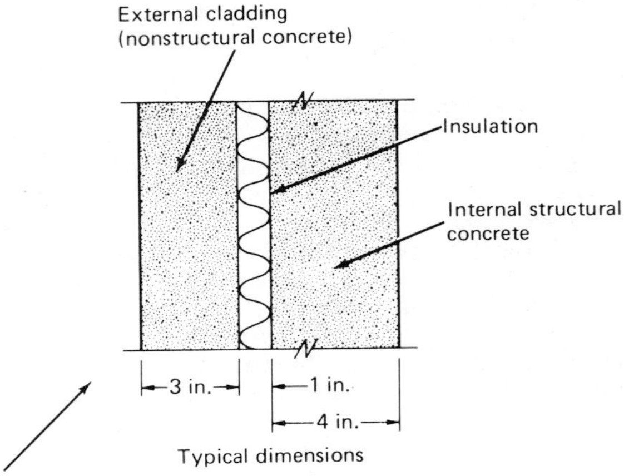

Fig. 13-10 Section through external sandwich wall panel.

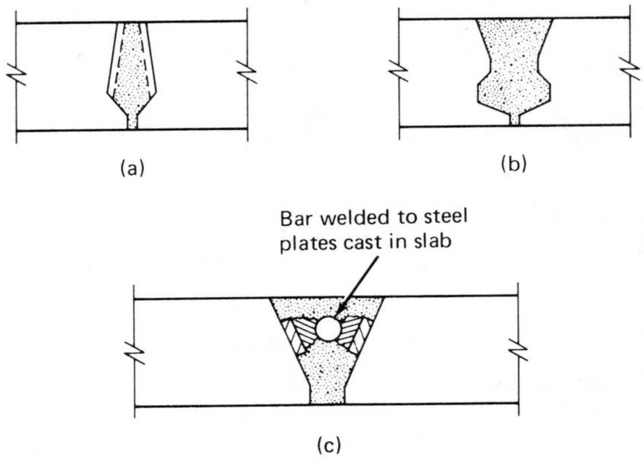

Fig. 13-11 Joints between floor units—parallel to span.

support the external leaf and will decrease the slenderness. If a new system is being devised, the efficiency of the tie arrangement should be proved by suitable tests.

13.5.4 Reinforcement

Reference 13-59 recommends that a reinforced wall be defined as one having greater than 0.1% reinforcement both vertically and horizontally. Reference 13-31 gives this figure as 0.4% vertically. In a wall panel, an unreinforced wall will have some horizontal steel in the top and bottom (e.g., one No. 5 bar each 4 in. of wall thickness[13-6]), and there will be vertical ties that will act as reinforcement. Openings should be provided with reinforcement—diagonal reinforcement across the reentrant corners is particularly useful in shear walls.[13-36] For heavily loaded walls where the layout of the openings is complex, a plane stress finite element analysis might be considered and reinforcement specified, based on the rules given in Ref. 13-37.

13.6 DESIGN OF FLOOR SLABS

Unlike wall design, the design of the floors for large panel buildings involves consideration of several alternative possibilities. The designer must weigh these interacting factors and decide on a system that will satisfy design criteria of strength, serviceability, sound insulation, and economy.

The possible alternatives for floor construction include the following:

1. *Cast-in-situ floors.* Systems do exist that have cast-in-situ floors with precast wall panels. The main advantage of casting the floors in situ is that they can easily be made continuous over the wall supports—hence they can be made thinner, or less reinforcement can be used. However, the slab thickness is often governed by sound insulation requirements and may exceed the optimum thickness based on purely structural considerations.

   Some of the points made in the discussion that follows are pertinent to cast-in-situ construction, but this chapter is concerned specifically with precast work.
2. *One-way or two-way span.* A two-way span is obviously more efficient structurally than a one-way system but can only be used (a) if the layout permits it, and (b) if room-size floor units are used.

   With less than room-size floor units, the unsupported joint between adjacent units should include a shear key. Tests by Lewicki[13-38] show that this is essential for spreading load and preventing undesirable cracking in the line of the connection. The actual shape of this key does not appear to be of great importance. The main thing is that some shear transfer is provided. Figure 13-11 shows some typical keys.
3. *Continuous or simply supported.* When a connection is designed to be simply supported, cracking at the top of the joint between the floor concrete and the in-situ floor slab adjacent to the wall may prove troublesome. Some continuity is desirable to prevent this.

   The provision of moment continuity at the wall–floor junction will obviously enhance its resistance to progressive collapse. It may, however, induce undesirable eccentricities in the wall panels.

   The platform type of connection shown in Fig. 13-13(a) will result in some moment continuity at the support, and it is common with this type of detailing to provide top steel in the slabs to prevent cracking due to live load moment.
4. *Normal weight or lightweight concrete.* Lightweight concrete might be used in preference to normal weight for two reasons:
   (a) To reduce vertical stress. This is seldom necessary in large panel structures, since the walls are not normally fully stressed. For high blocks where wall stresses are significant, there could be some advantage is using lightweight concrete, but sometimes *extra* dead load is an advantage in limiting any tendency for tension to be developed in the walls.
   (b) To allow larger precast units to be erected for a given crane capacity. This is discussed in Section 13.12.

   Any advantage the lightweight concrete might have must, of course, be set against its normal disadvantages of increased cost and decreased rigidity.
5. *Hollow or solid.* In order to reduce weight, one-way span slabs are normally cored. The hollow cores can be used as service ducts and for anchoring tie reinforcement as discussed in Section 13.9.
6. *Reinforced or prestressed.* The need for prestressing is, of course, dependent on the span. It is advantageous to keep the floor slab thickness at each story level the same. After it is decided what this is to be, the longer spans can be prestressed to limit deflection. In the early eighties use of long-span prestressed units has become the most common solution for precast floors.

In the United States and Canada most of the prestressed hollow core slabs are placed upon continuous bearing pads of a plastic material as shown in Fig. 13-14. This allows some slab shortening due to shrinkage and creep. Also slab rotation due to gravity deflection does not shift the reaction toward the edge of the wall. Prior to this practice, when concrete was bearing on concrete, cracking of the slabs along the walls was often observed.

13.7 DESIGN OF HORIZONTAL CONNECTIONS

Floors and walls are connected by a horizontal joint as shown diagrammatically in Fig. 13-12. This connection must transmit the following force actions:

(a) Vertical load (including vertical resultants of lateral load)
(b) Horizontal shear due to lateral load
(c) Shear in the plane of the floors
(d) Bending stress in the plane of the floors
(e) Tie forces
(f) Transverse bending due to floor loadings
(g) Shear due to floor loadings

If all these stress actions act concurrently (as they can do), then the state of stress within the joint will be most complex. Thus, it is not rational to treat the stresses resulting from the different loadings separately. However, very little published information is available on wall–floor connections, and there is no design procedure available that will account for the interaction of even two of these force actions. Horizontal connections are therefore designed mainly on an empirical basis.

13.7.1 Vertical Load (a)

This is probably the most important of the loadings mentioned above. The presence of the connection may cause the wall panel to fail at a lower load than it would otherwise take. Laboratory testing of the connection under vertical load is normally recommended. If experimental results are not available, Ref. 13-12 (Clause 47.21) recommends (somewhat tentatively) values for a reduction factor on the strength of the concrete in the wall panels to estimate the strength of the joints. It is implied that the separate strengths of the wall and joint concretes should be similar.

The bearing of the panel on the joint should be as uniform as possible. To achieve this (1) in the direction of the thickness of the panel the joint should not be stepped or consist of different materials (see discussion of floor support later in this section), and (2) the erection technique is important (Section 13.12).

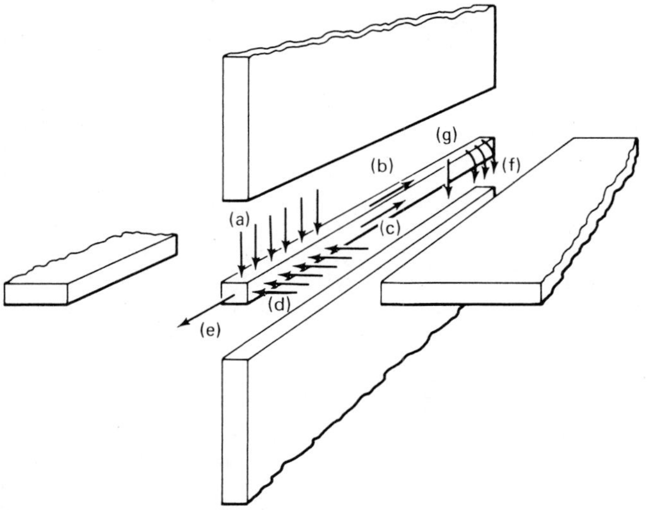

Fig. 13-12 Exploded view of horizontal connection showing force actions.

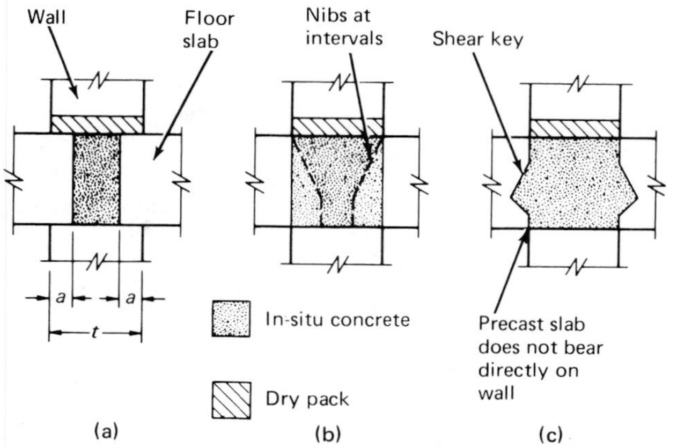

Fig. 13-13 Vertical section through horizontal connections (reinforcement not shown). (a) Platform support, (b) nibs, and (c) shear transfer via in-situ concrete.

Two important factors affecting the vertical strength of the connection are the ratio of strength of the in-situ concrete to that of concrete in the wall, and the method of supporting the floor slabs. These two factors are related in that the floor slab support detail will affect the confinement and, hence, the strength of the in-situ material.

Three types of connection are shown in Fig. 13-13. Of these, the platform detail of Fig. 13-13(a) is the only one for which design rules have been formulated.

Tests by the Portland Cement Association[13-8] for this type of connection have resulted in design rules that appear to give a better prediction of the strength of such connections under vertical load than previously used methods. Their proposed expression for strength of a grouted connection (Fig. 13-14) is:

$$P_u = \frac{tL}{k} f_u C R_e \tag{13-5}$$

where:

P_u = strength of connection
t = width of grout column
L = horizontal length of connection
f_u = maximum usable grout strength defined as follows: compressive strength of grout, f_g, or wall concrete, $f'_{c(\text{wall})}$, whichever is less, but not more than 0.8 $f'_{c(\text{wall})}$, unless the walls are reinforced against splitting and slab cores are filled with grout, and not more than 1.25 $f'_{c(\text{wall})}$ in any case
f_g = compressive strength of grout in the connection measured on 6 X 12-in. (152 X 305 mm) cylinders
$f'_{c(\text{wall})}$ = compressive strength of the wall concrete
k = factor representing the fraction of load transferred through the grout column

$$= 0.65 + \left[\frac{f_g - 2500}{50{,}000}\right]$$

for f_g is psi, or for f_g in MPa

$$= 0.65 + \left[\frac{f_g/0.0069 - 2500}{50{,}000}\right]$$

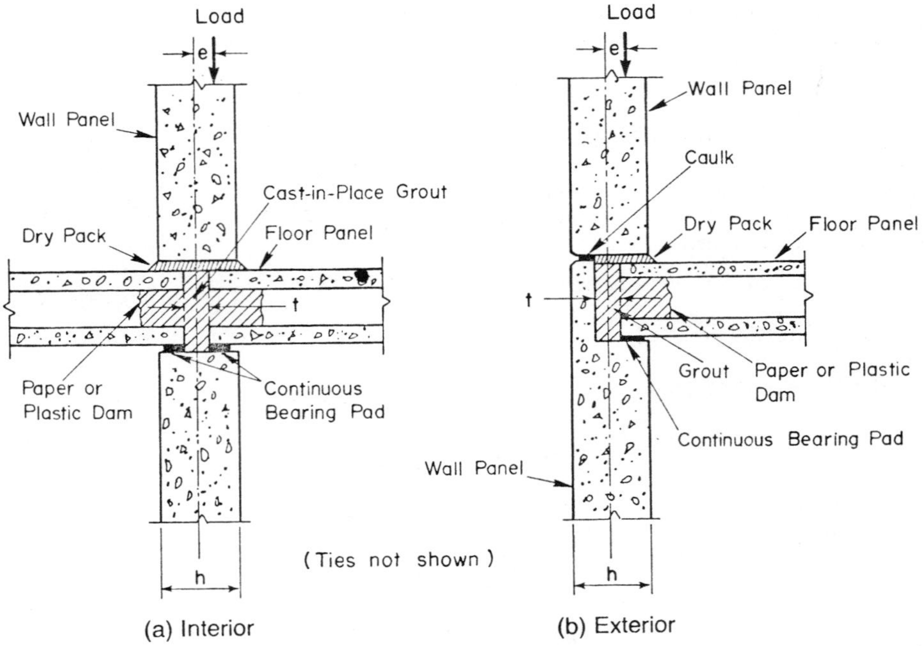

Fig. 13-14 Grouted horizontal connections.

R_e = reduction factor for eccentrically loaded walls = $1 - 2e/h$

e = eccentricity of vertical load measured from point of load application to center of grout column

h = wall thickness

C = filled core factor determined as follows:

For filled slab cores:

$$C = r\sqrt{\frac{2500}{f_g}} \quad \text{for } f_g \text{ in psi, or}$$

$$= r\sqrt{\frac{2500}{f_g/0.0069}} \quad \text{for } f_g \text{ in MPa}$$

but not less than 1.0 in either case

For unfilled slab cores:

$C = 1.0$

$r = 1.4$ for interior connections

$= 1.2$ for exterior connections

Equation (13-5) is based on a limited series of tests, and the following constraints on the design are advised if using it:

1. Grout strength, f_g should be in the range 3000 to 6800 psi.
2. If f_g exceeds 0.8 $f'_{c(\text{wall})}$, transverse reinforcement should be provided in the wall, and 0.06 in^2/ft length of wall adequately anchored within 1 in. of the top and bottom surfaces of the wall should be a suitable amount of such reinforcement.
3. The compressive strength of the dry-pack mortar should be greater than f_g and greater than $f'_{c(\text{wall})}$.
4. The following order of dimensions should be maintained:

Thickness of dry-pack mortar	1 in.
Slab bearing panels:	
Thickness	1/8 in.
Width	2 in.
Width of grout column	3 in.
Wall thickness	8 in.
Extension of slab into connection	2 1/2 in.

5. When slab cores are grouted, they should be filled to at least 1 in. beyond the face of the wall.
6. The elastic modulus of bearing panels should be less than 50 ksi.

Reference 13-39 suggests that the vertical strength of an ungrouted internal vertical connection can be based on:

$$P_u = 0.85\, A_e f'_{c(\text{slab})} R_e \tag{13-6}$$

where:

P_u = strength of connection

A_e = effective bearing area of floor slabs = $b_b t_r$

b_b = total length of end-bearing of floor planks

t_r = sum of the web thicknesses of hollow-core planks on either side of connection

$f'_{c(\text{slab})}$ = compressive strength of slab concrete

R_e = reduction factor for eccentrically loaded walls as determined in eq. 13-5

13.7.2 Transmission of Shear Due to Lateral Load (b)

Horizontal shear due to lateral load must be transmitted along the crack (preformed by the method of construction) between the bottom of the wall and the joint. This shear must be transmitted by friction along the bottom of the wall panels and by dowel action of the steel in the vertical joints and tie steel in the walls. Reference 13-12, Clause 54.51, allows the use of a higher coefficient of friction for joints formed entirely in concrete (0.35) than for joints with a layer of mortar (0.2).

The mechanism of shear transfer in the wall is not as simple as this, of course. The tie steel in the horizontal joints will act as shear reinforcement in a similar fashion to stirrups in a conventional reinforced concrete beam. Care must be taken if there is a likelihood of vertical tension in the wall, but normally shear failure of shear walls is most unlikely.

The transmission of vertical shear at the junction with a vertical joint is discussed in Section 13.8.

13.7.3 Force Actions in the Plane of the Floors due to Lateral Load (c), (d), and (e)

Lateral load is distributed between the vertical units by in-plane actions in the floor slabs. These actions are seldom if ever calculated on a mathematical basis, and the in-plane reinforcement is specified on an empirical basis. (See below, subsection 13.7.5.)

13.7.4 Transverse Bending and Shear due to Floor Loadings (f) and (g)

The provision of moment continuity between floor slabs and walls is discussed in Section 13.6.

Figure 13-13 shows three common details for shear transfer between floor slabs and walls.

The detail of Fig. 13-13(a) where the floor slab rests continuously on top of the wall is commonly used. It has the advantage of providing an even bearing stress along the length of the connection and is easy to cast and erect. However, it does not leave much space for in-situ joint concrete and reinforcement.

The intermittent nib support, Fig. 13-13(b), does provide space for more in-situ joint material and can allow adequate connection strength. It does not provide uniform bearing stress on the wall, and the casting molds are more complicated than in the former case. A positive advantage is that it provides keys that will resist shear forces in the plane of the slabs.

A detail of the type shown in Fig. 13-13(c) has also been used. The floor panel either rests initially on special removable leveling devices or is supported by the projecting bars which bear on the top of the wall panels. The in-situ concrete is then cast, and when set will allow vertical shear to be transmitted via the shear key onto the wall. The floor slab does not rest directly on the wall. Thus a uniform bearing stress will be achieved, together with adequate space for joint concrete and reinforcement.

13.7.5 Reinforcement of Horizontal Connections

This dealt with in Section 13.9.

Note that in Ref. 13-14 the authors warn that special precautions should be taken when reinforcement is welded. They recommend that carbon content should be less than 0.5% and low hydrogen electrodes should be used. Welding should not be closer than 8 in. from cold bends, and tack welding should be avoided.

13.8 DESIGN OF VERTICAL CONNECTIONS

Vertical shear will be transmitted between adjacent walls, and it is worthwhile to develop as much shear transfer as is practicable in order to induce the greatest amount of monolithic action in the wall system. The vertical shear will be transmitted at the wall–floor junction and along the height of the vertical joint (Fig. 13-15). The effect of the wall–floor junction may be significant, but it is not easy to calculate, so principal attention is given to the latter component.

It is common in European practice to provide a "wet" joint using infilled mortar, whereas American practice has tended toward the use of "dry" joints using welded or bolted connections. The dry joints are preferred for speed of erection.

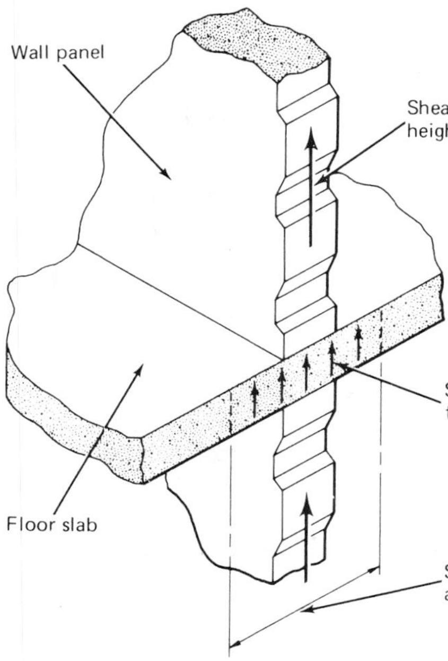

Fig. 13-15 Transmission of vertical shear.

13.8.1 Dry Joints

There are many different types of dry joint, and one cannot make general comments on behavior. Design should be based on physical tests. The shear in the joint due to lateral load is normally based on bending theory assuming full interaction across the joint. The shear stiffness of the joint will normally be less than the monolithic stiffness, but this reduction is normally neglected.

13.8.2 Wet Joints

Figure 13-16 shows the structural action of a vertical joint. The vertical shear, V, causes a horizontal tie action. If there are no shear keys, this tension is thought of as being due to shear friction,[13-40] using the expression:

$$V = H\mu$$

where V is the applied vertical shear within the panel height,

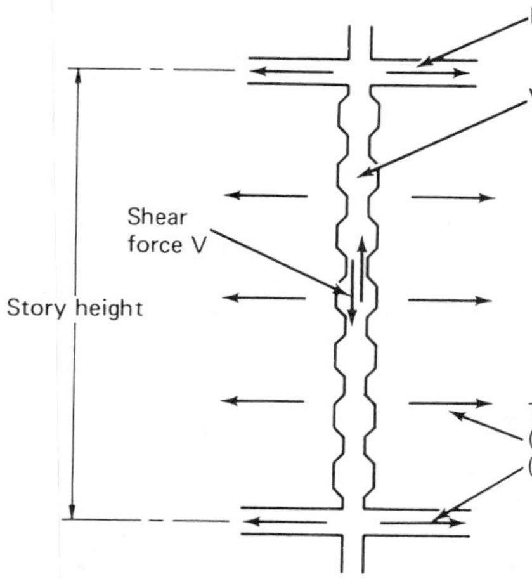

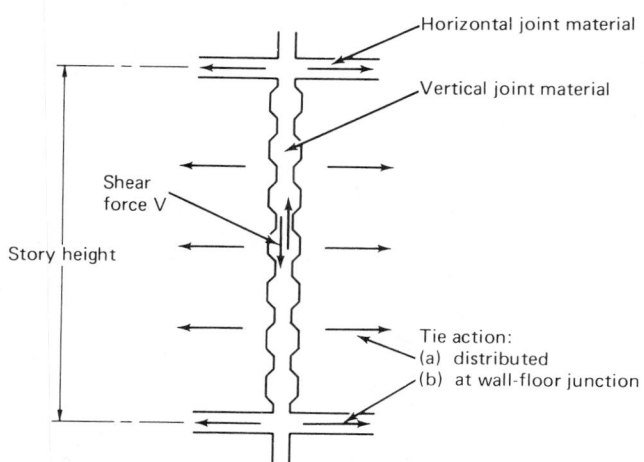

Fig. 13-16 Force actions for transmission of shear at vertical connections.

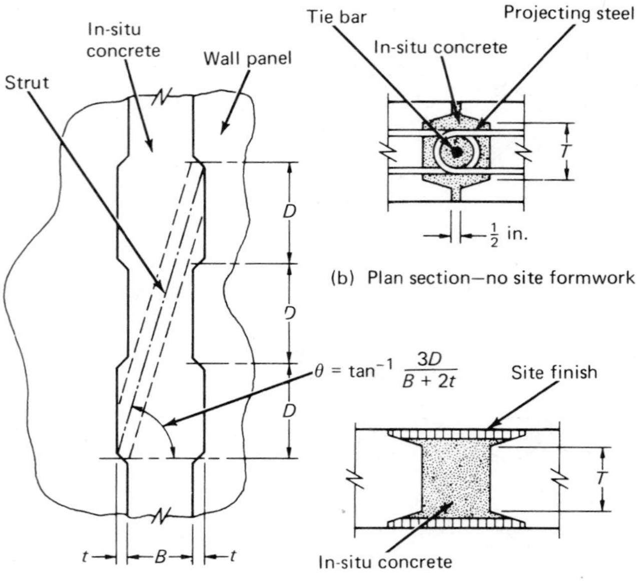

(a) Elevation (c) Plan section—site formwork required (reinforcement not shown)

Fig. 13-17 Wet joints with shear keys.

H is the horizontal tensile force, and μ is a shear-friction coefficient, normally in the range 1.4–1.7 for this purpose.

With shear keys a strut hypothesis[13-12,13-41] gives a better physical interpretation although this is not basically different from the shear-friction model. Inclined struts are assumed to form within the joint as shown in Fig. 13-17(a). The inclined strut reaction may be resolved into horizontal and vertical components. The horizontal thrust, T, must be resisted by tie steel distributed as discussed above. The vertical component must be resisted by shear keys formed in the edges of the wall panels. Thus, three important factors are (1) the angle of inclination of the struts, (2) the tie steel, and (3) the shear keys.

1. Angle of Inclination of the Struts. Consider the two rigid planes AB and CD (Fig. 13-18) connected by n struts with angle of inclination θ and vertical spacing d. If a total shear of V is applied as shown, then the shear per strut, Q, is given by:

$$Q = \frac{V}{n} \tag{13-7}$$

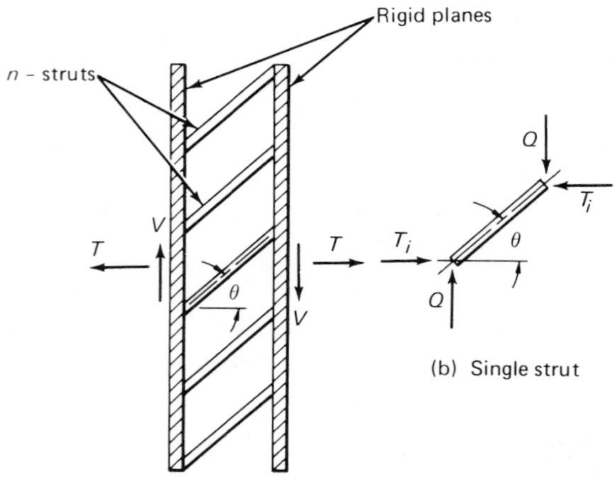

(a) Struts between rigid planes

Fig. 13-18 Transmission of shear by strut action.

This causes:

$$\text{Axial load in the strut} = Q/\sin\theta \tag{13-8}$$

with the resulting horizontal component:

$$H_i = Q/\tan\theta \tag{13-9}$$

Therefore the total horizontal component is:

$$H = nQ/\tan\theta = V/\tan\theta \tag{13-10}$$

This is the horizontal tensile force, which must be resisted by tie reinforcement. The smaller the assumed θ, the larger will be the predicted tie and strut forces. CEB[13-12] recommends $\theta = 45°$. Test results[13-42] indicate that $\theta = \tan^{-1}(3D/(B + 2t))$ may be acceptable for castellated joints (see Fig. 13-17a).

2. Tie Steel. If V is the total vertical shear over a story height as calculated using bending theory, then eq. (13-10) gives the total tie force over the story height. Steel to resist this can be placed in the horizontal joint and should be continuous across the line of the vertical joint. In addition, loops within the vertical joint will also resist the tie forces (see below, subsection 13.8.3).

3. Shear Keys. The most common type of shear key is the castellated type shown in Fig. 13-17. A considerable amount of testing work has been carried out on such joints.[13-42–13-45] It appears that within normal limits the dimensions of the keys are not critical to the strength of the connection. The width of the joint B should be kept as small as possible consistent with adequate anchorage for loop reinforcement and satisfactory placing of concrete. A depth of castellation, D, of the order 3 to 12 in. is normal. The distance, t, should not be less than 0.5 in., and the sloping edges to the key allow better filling of the joint without significantly affecting its strength.

The thickness of the shear key, T, depends on the method of forming the joint. The detail of Fig. 13-17(b) has been common in the past. This requires a minimum of site formwork but has not proved altogether satisfactory. A detail of the type shown in Fig. 13-17(c) is becoming popular. This allows almost the full thickness of wall to act as a shear key. Extra site finishing is required, but this is not felt to be a disadvantage.[13-46]

The strength of such a joint may be expressed as:[13-42]

$$Q = \beta D T f_{ct} \tag{13-11}$$

where f_{ct} is the tensile splitting strength of the concrete and β is an empirical constant. For cracking shear (Q_c), $\beta = 0.5$, and for ultimate shear (Q_u), $\beta = 1.0$ should give conservative estimates.

Formerly the shear strength of keys was assessed on the basis of:

$$Q_{\text{allowable}} = DTf_s \tag{13-12}$$

where f_s is the allowable shear stress in the concrete. This procedure did not recognize the basic behavior of the shear transmission but was conservative. The use of eq. (13-11) with the above values of β will normally be equivalent to using a slightly higher allowable shear stress.

13.8.3 Reinforcement of Vertical Connections

1. Loop Bars (Fig. 13-17b). These will contribute to the tie resistance and help to limit long-term cracking as discussed below. Some designers allow a contribution to the shear strength of the joint from loop bars acting as dowels. Tests[13-42] indicate that the presence of the loops may not significantly affect the cracking strength but may increase the ultimate strength and make the behavior more ductile.

2. Vertical Tie Bars (Fig. 13-17b). These bars in combination with the loops will form a mechanical connection between adjacent units that will help to prevent progressive collapse. They are therefore worth having.

3. Long-Term Movements. Trouble has been experienced in some large panel buildings with long-term cracking, particularly at the vertical joints. The mortar infill in a vertical joint is normally placed with a wet consistency to ensure that the joint is properly filled. This will tend to cause shrinkage. Other sources may be creep, differential settlement, wind stresses, or temperature movements.

The provision of loop reinforcement within the vertical joint will help to reduce such effects. Practical experience tends to support this conclusion.[13-46] Also there is some evidence to suggest that cracking is more prominent in vertical joints that do not have castellations than in those that do (they certainly have a much lower cracking strength).[13-42]

13.9 TIE STEEL

Tie steel is needed in the planes of the floors and walls to:

1. Tie the panels together to resist collapse due to accidental loading.
2. Contribute to the mechanisms for distribution of lateral load, vertical load, and thermal movement.

The normal means of specifying the amount of tie steel, ranging from pure guesswork to use of simplified mechanisms that are unlikely to occur in isolation, do not fully account for real behavior. Thus one should not treat any design rule as an absolute requirement. In a given situation it may be sensible from a structural point of view to use more or less tie steel than rules or calculations indicate. Legal considerations may of course govern in some situations where code provisions are applicable.

13.9.1 Tie Steel in the Planes of the Floor Slabs

Three types of ties are normally specified in the planes of the floors (Fig. 13-19):

1. Peripheral ties at or close to the external walls.
2. Longitudinal ties which are parallel to the floor spans.
3. Transverse ties which are at right angles to the floor spans.

References 13-6, 13-31, and 13-35 all give different recommendations for the required areas of these ties. Table 13-3 shows the variation in minimum values. The recommendations of Ref. 13-6 are probably based on more sound engineering principles than those of Refs. 13-31 and 13-35, but some fairly sweeping assumptions were nevertheless necessary to derive them.

Longitudinal Ties over an Interior Support. In Ref. 13-6 it is assumed that if an interior support wall becomes ineffective, the floor above this wall will not be able to hang from the wall above; so to prevent it from falling down, a secondary mechanism must develop. The assumed mechanism is shown in Fig. 13-20. There will be a contribution from shear transfer between the floor units as shown in Fig. 13-20(a), but the main action is assumed to be the suspension mechanism due to extension of the longitudinal tie shown in Fig. 13-20(b). To achieve a stable situation of this type, large deformations are required, and bond slip in the tie must occur. A suitable combination of tie elongation and tie strength must be chosen. To allow the elongation,

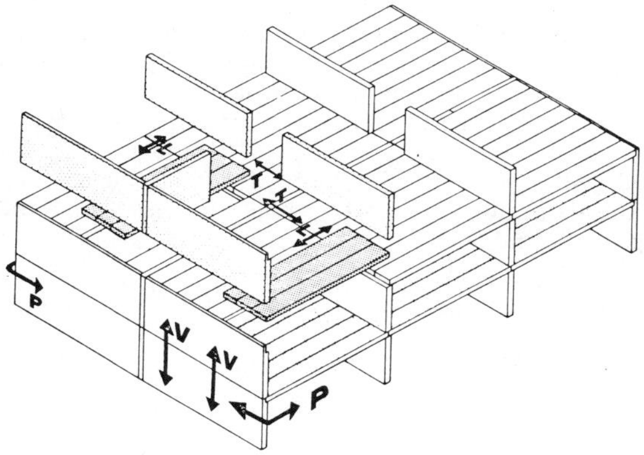

(a)

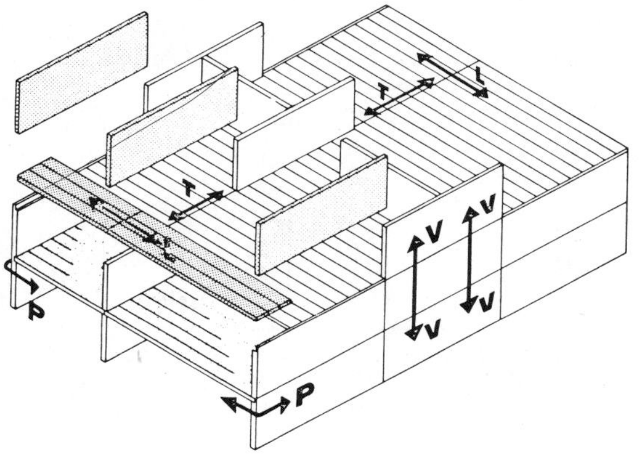

(b)

Fig. 13-19 Tie steel directions. (a) Cross wall structure, T: transversal, L: longitudinal, V: vertical, P: peripheral, (b) Spine wall structure.

TABLE 13-3 Minimum Horizontal Tie Steel for a 10-story Building

	Peripheral	*Transverse*	*Longitudinal*
Ref. 13-31 (CP110)	13 kips	4 kips/ft	4 kips/ft
Ref. 13-35 (PCI)	16 kips	1.5 kips/ft	1.5 kips/ft
Ref. 13-6 (PCA)	Lower limit not specified	18 kips/tie	Lower limit not specified

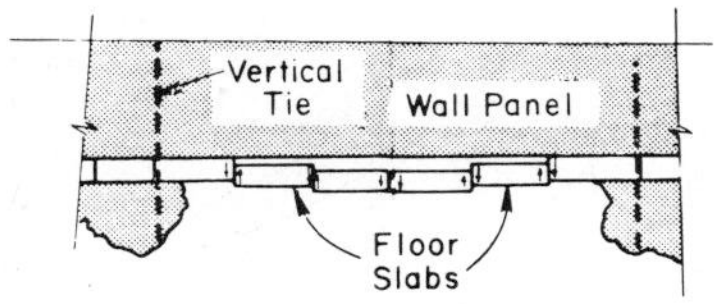

(a)

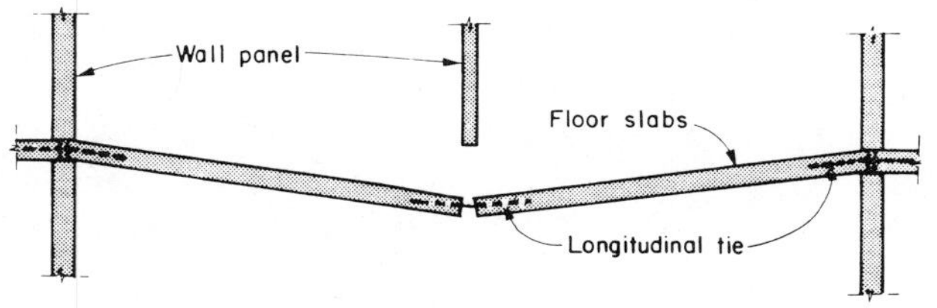

(b)

Fig. 13-20 Support mechanism for floor above ineffective wall, (a) support through membrane action in shear, (b) suspension mechanism by means of longitudinal ties.

it is believed that 7-wire unstressed prestressing strand with specified embedments can provide the necessary bond slip while still maintaining a satisfactory ultimate load capacity. The ties are to be located in the longitudinal floor joints between adjacent slabs, and are not continuous, contrary to the recommendations of Ref. 13-31, which states that they should be effectively continuous for the length of the building. However, the tie action will be transferred from the tie bars to the floor slabs, which will then act as ties to transmit horizontal forces.

Figure 13-21 can be used to select a suitable tie size.

The following definitions are used:

w = factored vertical load of floor slab = $D + 0.5L$
l_b = larger of spans that must be supported by suspension mechanism
b_p = spacing of longitudinal ties, usually equal to width of precast floor plank
$\lambda = w_b b_p$
F_l = actual tie force
F_Q = ultimate strength of tie
δ_l = total extension of one side of suspension system.
Δ_Q = maximum attainable pullout while maintaining the ultimate tie force, F_Q

The procedure is:

1. Select a size and embedment length of tie bar.
2. From the table in Fig. 13-21 find the embedment length, usable capacity (F_Q), and maximum extension, Δ_Q.
3. Calculate λ.
4. Enter the graph in Fig. 13-21 at the level of F_Q and extend horizontally to the appropriate value of λ. Draw down from this point to the appropriate value of l_b and then horizontally find a value of $\delta_l/2$. If this is less than Δ_Q, the chosen size of tie is adequate.

The longitudinal tie should be placed below the transverse tie (Fig. 13-22).

Longitudinal Tie at External Support. If an external load bearing wall is lost, it is assumed that the transverse tie will act in catenary to support the slab (Fig. 13-23). The longitudinal tie must then have a mechanical connection to the transverse tie. It should, therefore, loop over the transverse tie as shown in Fig. 13-23(c). To function thus it is best to specify a mild steel bar. This tie must also be able to resist the tensile force of a floor suspension system if the adjacent interior wall becomes ineffective. Reference 13-6 recommends that the strength of this tie be not less than 0.75 times the strength of the longitudinal tie in the adjacent interior support, or 0.71 wb_pl_b. These ties must, of course, be adequately anchored.

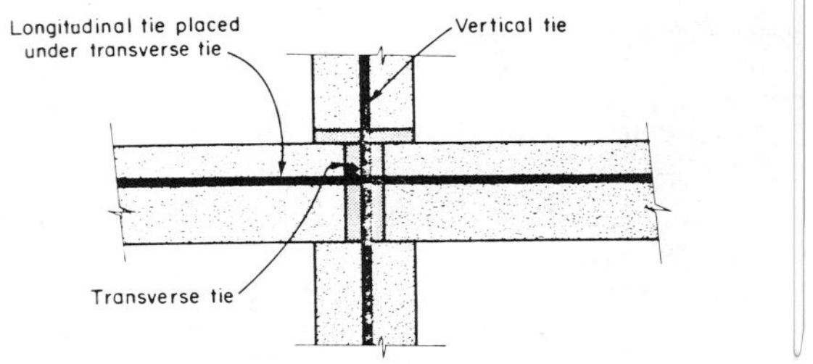

Fig. 13-22 Relative positions of longitudinal and transverse ties.

While the above recommendations are sensible, one might feel that complete loss of support for a slab above an ineffective wall would not be unacceptable, provided the debris loading did not trigger further collapse. Some large panel systems use looped bars grouted into the slab cores as longitudinal ties. These bars are looped around a tie in the transverse joint, and tests show that the in-situ mortar can be adequately vibrated into the core (a stop end in the core is not needed), so that the bar will fail in tension and not bond.

Transverse Ties. To estimate the requirement for transverse ties, a cantilever mechanism due to a wall at right angles to the span becoming ineffective is postulated in Ref. 13-7. Figure 13-24 shows the mechanism. It is assumed that there is a very short compression block at the base of the cantilever, and that the variation of strain in the transverse ties is linear with height. The load tributary to the cantilevered wall is calculated, and a relatively simple calculation gives the maximum force in the transverse ties. The required strength is found from:

$$F_x = \alpha_x \frac{W_s l_d^2}{2h_s} \qquad (13\text{-}13)$$

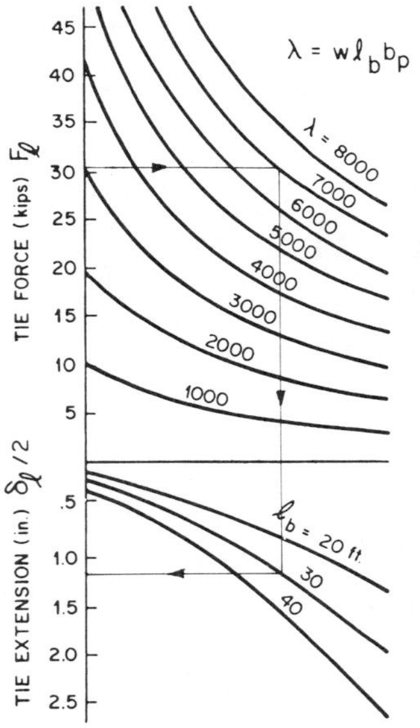

Strand (Dia.-f_{pu})	Embedment Length (inches)	Useable Capacity F_Q (kips)	Maximum Extension Δ_Q (inches)
1/4-250	18	6.8	0.75
5/16-250	24	11.0	0.94
3/8-250	30	15.0	1.13
3/8-270	30	17.3	1.13
7/16-270	36	23.3	1.31
1/2-270	42	31.0	1.50
0.6-270	48	44.0	1.80

For a particular set of design conditions, i.e., w, l_b and b_p, the tie must be chosen such that the following relationships are satisfied:

$$F_Q \geqslant F_l$$

$$\Delta_Q \geqslant \delta_l/2$$

Fig. 13-21 Longitudinal tie design guide.

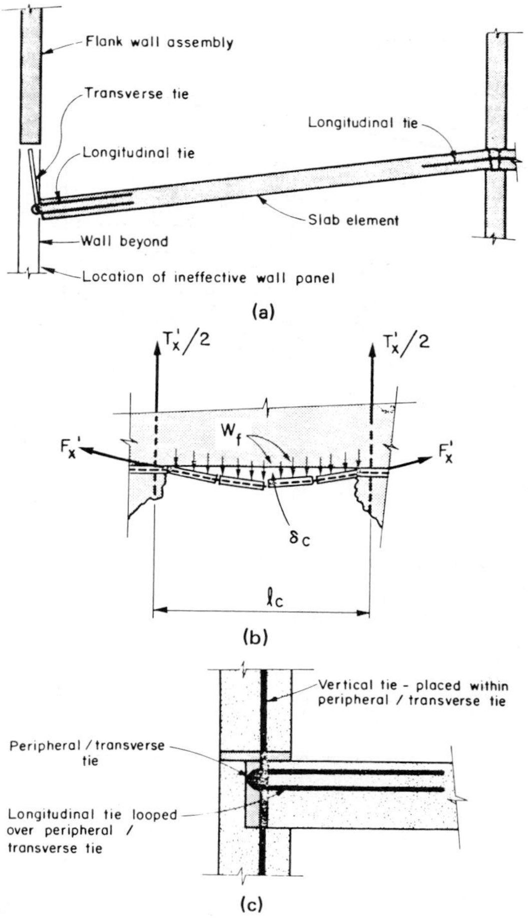

Fig. 13-23 Support mechanism for floor above ineffective external wall. (a) Longitudinal section, (b) Transverse tie in catenery. (c) Detail of connection .

where:

α_x = a coefficient found from Fig. 13-25
W_s = the average total story load per foot of wall $(D + 0.5L)$
l_d = the unsupported length of the wall panel.
h_s = the average story height

Tests reported in Ref. 13-7 showed good agreement between observed values of the tie force and values calculated using the above equation. The tests were carried out on a model of the mechanism, and it is probable that the cantilever mechanism would be inhibited by other support conditions in a real structure. One would therefore expect that transverse tie requirements based on this mechanism would be quite conservative. In order to develop the compression zone, if the vertical joint is not grouted for other reasons, the lowest quarter of the joint at each story level should be filled with dry pack* (according to the recommendations of Ref. 13-6). The minimum tie force of 18 kips for transverse ties quoted in Table 13-3 is simply the strength of a ⅜-in. prestressing tendon taken to be the minimum size likely to be used for such a purpose. The transverse ties are located in the joint between the walls and the floors. Use of unstressed prestressing tendons is recommended, and the ties are continuous across the building.

*Dry-pack mortar has a high cement content and a dry consistency (water/cement ratio of the order of 0.35).

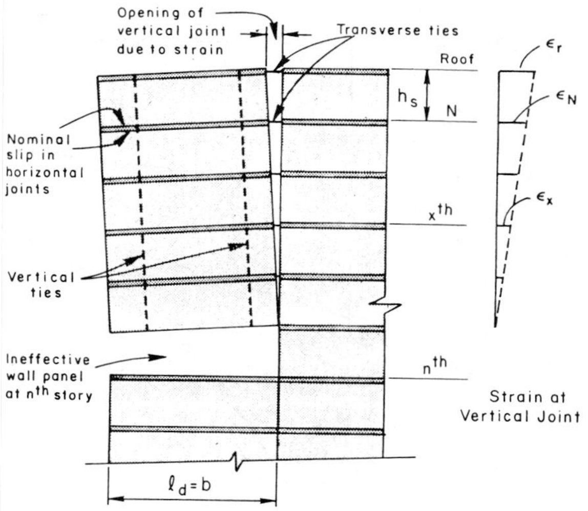

Fig. 13-24 Cantilever mechanism.

If the calculations give tie steel areas that seem rather high, one should consider whether for the structure being designed this cantilever mechanism is realistic.

Peripheral Ties. References 13-31 and 13-35 both give specific recommendations for extra tie bars around the periphery of the building at each floor level. Reference 13-6 gives guidance for the design of the exterior longitudinal tie (as described in this section) which is a peripheral tie, but no extra steel is recommended at the transverse periphery, provided that there is a transverse tie within 18 in. of the outermost edge of the structure.

Special care should be taken to anchor the peripheral ties at the corners of the building.

13.9.2 Vertical Ties

For the cantilever mechanism of Fig. 13-24 to function, a shear force must be transmitted across the horizontal joints. The maximum value of this shear is found from:

$$V_{hx} = \frac{\beta_x W_s l_d^2}{2h_s} \qquad (13\text{-}14)$$

where V_{hx} is the horizontal shear at story x due to the cantilever mechanism, β_x is a coefficient obtained from Fig. 13-26, and l_d and h_s are as defined for eq. (13-13).

If the cantilever mechanism is adopted for the design, then vertical ties in the walls should be provided to withstand this value of shear. Although the tensile forces in the vertical ties measured in the tests described in Ref. 13-7 were very small, it is reasonable to ensure that they can support in tension the total wall and the floor load from the story above an ineffective panel ($= W_s \times l_d$). Some designers in the United Kingdom feel that the requirements for vertical tie steel in Ref. 13-31 are excessive. Two vertical ties per panel are normally used. Various methods have been devised for making the vertical ties continuous through the horizontal joint either by using connected steel rods or by lapping the coupling with reinforcement in the wall panel. The coupling usually incorporates a leveling device for the walls (Fig. 13-27; see Section 13-12).

Post-tensioning of vertical ties may improve the performance under earthquake loading.[13-15]

13.9.3 Alternate Load Paths

If a panel becomes ineffective, then the load previously tributary to it must find some other means of support.

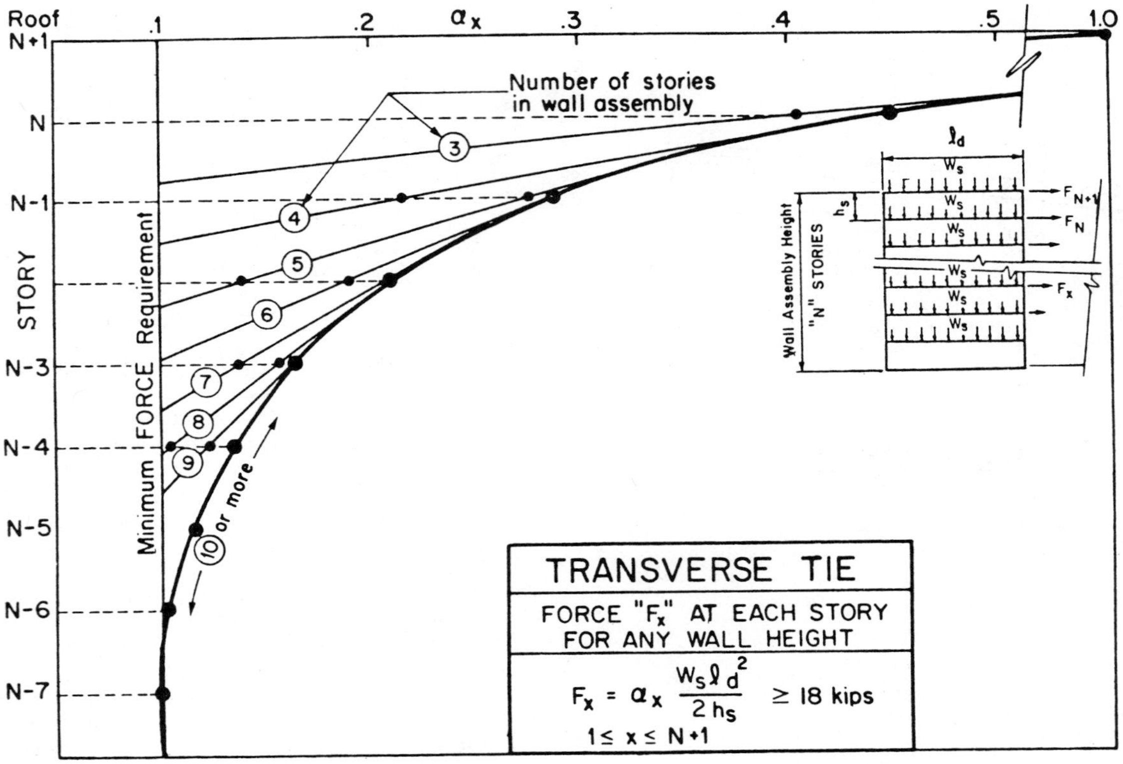

Fig. 13-25 α_X values for eq. (13-13).

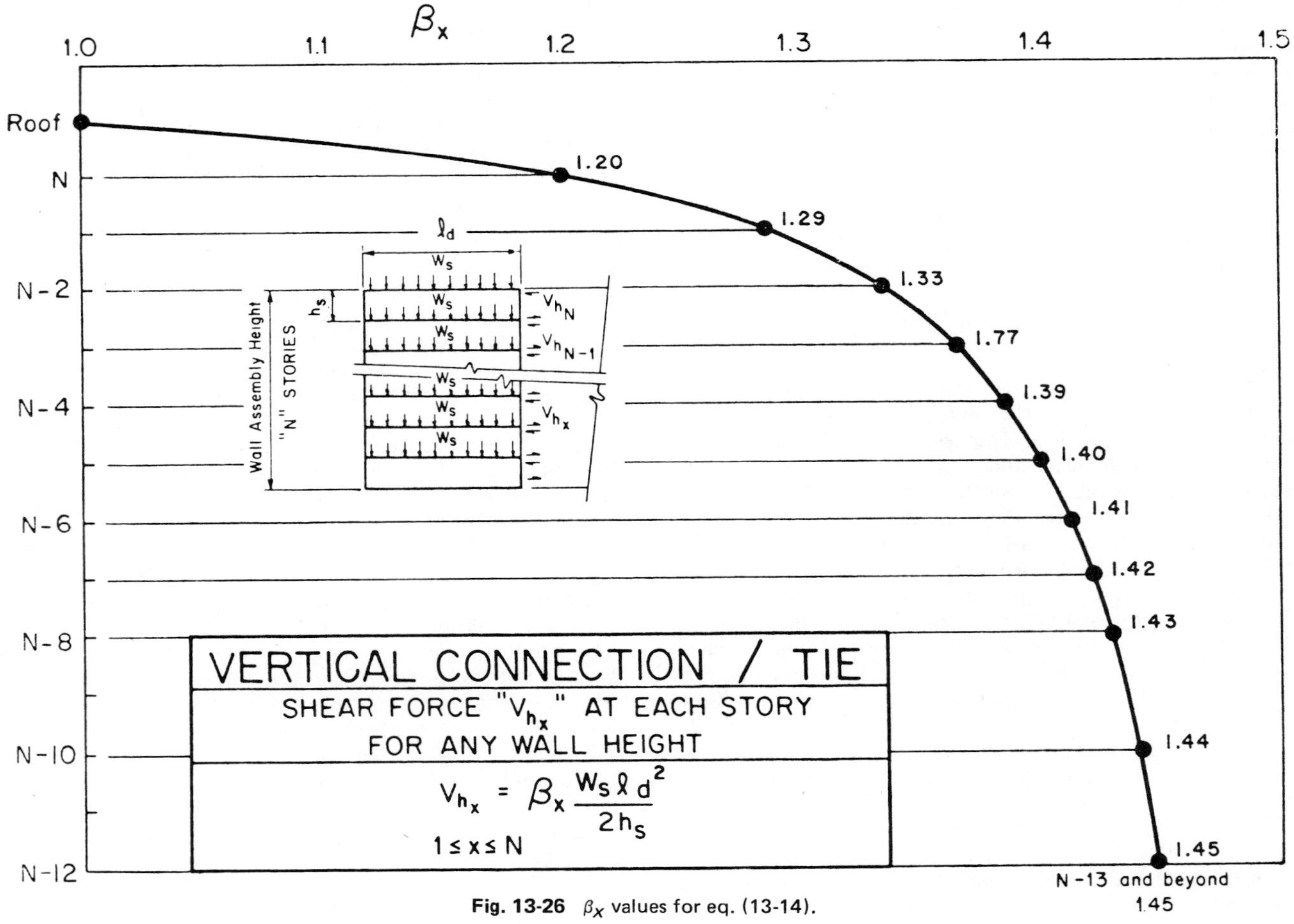

Fig. 13-26 β_X values for eq. (13-14).

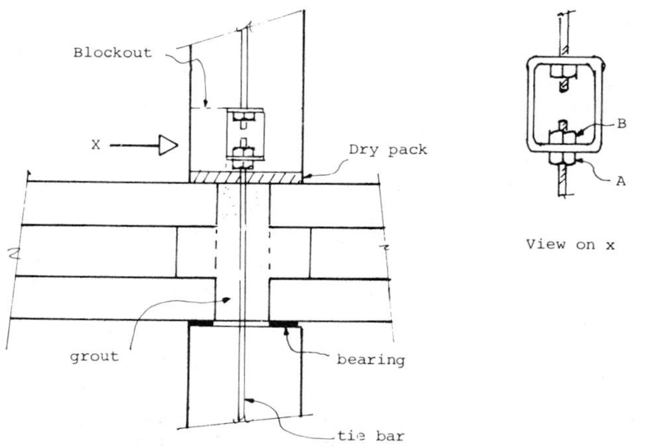

Fig. 13-27 Typical vertical tie connection.

If the cantilever mechanism described in Section 13.9.2 is postulated, then two basic situations can arise. The cantilever may be supported by an adjacent wall, in which case the vertical load is transmitted in shear through the vertical joint (Fig. 13-28a). If, however, the root of the cantilever is adjacent to an opening (Fig. 13-28b), the vertical load is not so easily provided for. In the latter situation, Ref. 13-6 suggests the use of return walls or "integral columns" that are designed to the redistributed vertical load.

The integral columns are built into the precast units. It is recommended that the return walls or integral columns should have a width of not less than one-third of the story height or 10% of the unsupported horizontal length of the wall. They must, of course, be designed as "strong" elements; i.e., they must not fail in an explosion. This is a difficult criterion to satisfy, since the real behavior of the load wall system in an explosion is not easy to predict (a maximum design pressure of 720 lb/ft^2 is recommended in Ref. 13-6).

It should be remembered that the cantilever mechanism shown in Fig. 13-24 does not stand alone. Other walls including non-load-bearing walls and slab suspension mechanisms in the floors above may also provide alternative load paths. A bridging mechanism is, of course, a better situation if it can be realized.

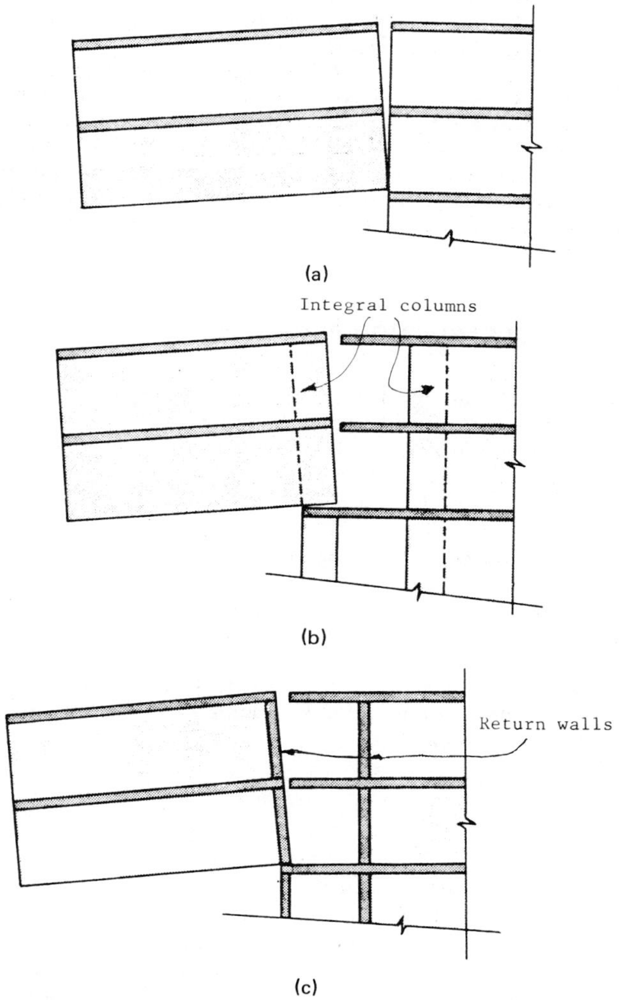

Fig. 13-28 Types of cantilever mechanism. (a) Adjacent walls. (b) Adjacent opening with integral columns. (c) Adjacent openings with return walls.

13.10 DESIGN OF TRANSFER BEAMS

A common feature of all types of tall buildings is to have a system of support at the first-story level that is different from that of the upper floors. This is done for architectural considerations, often to facilitate public spaces, lobbies, and parking. In the case of large panel structures, the shear walls are often discontinued at second-floor level, and a transfer beam distributes the loading to a column structure. Figure 13-29 shows a wall support of this type.

13.10.1 Structural Action of Transfer Beams

It is not realistic to consider the transfer beams simply as bending members carrying the weight of the wall. For the system of Fig. 13-29 there must be a flow of stress from the uniformly distributed value to more concentrated values above the supports. This involves an "arching" action, and hence, tensile stresses in the transfer beam. The tensile action tends to predominate over any bending action that may develop.

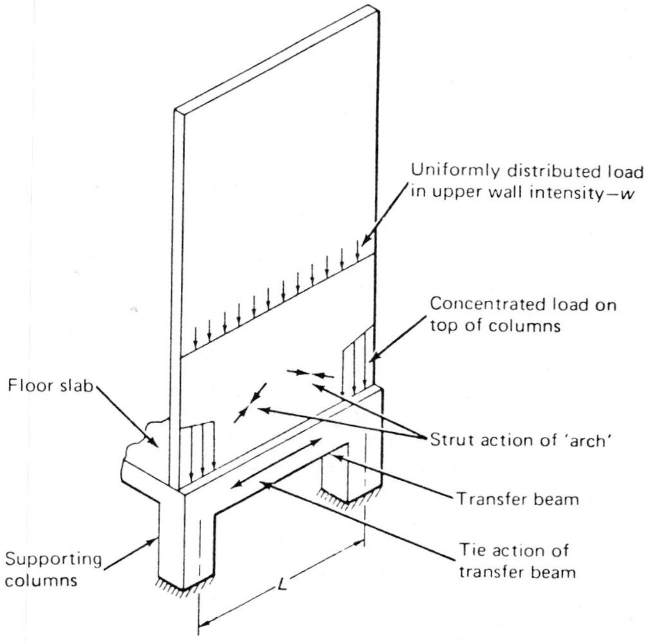

Fig. 13-29 Transfer beam at base of wall.

Therefore, as far as design is concerned, it may be necessary to reinforce the wall above the columns to resist the concentrated loads there. It will also be necessary to reinforce the transfer beam to take the tension.

As a very rough guide to this tie force, the formula[13-47]

$$T = \frac{wL}{4} \tag{13-15}$$

will normally give a conservative estimate. T is the tie force, L is the distance between the column centers, and w is the uniformly distributed vertical load. The reinforcement to take this load should extend right across the beam and should be suitably anchored at the ends. Ref. 13-48 gives more general formulas for this purpose together with formulas for estimating beam bending moment and wall stress concentrations. These formulas are also quoted in Ref. 13-35.

Where the transfer beam is continuous over several supports, differential settlement may have an important effect, and a finite element analysis may be necessary.[13-48] Such an analysis may also be desirable to assess the effect of lateral loading in the region of the transfer beam.

13.11 SERVICEABILITY

As mentioned in Section 13.1, some large panel structures have suffered from serviceability problems in the United Kingdom. Such problems result from either bad design or poor workmanship, and are avoidable in all cases. Some serviceability problems are peculiar to certain geographical locations and to the life style of the occupants.

Moisture Problems. A major problem has been excessive moisture within large panel buildings. This is due either to rain penetration through the external walls or roof or to condensation within the building. The severity of the rain penetration problem is highly dependent on geographical location. In the past an open drained joint was thought to be the best solution for providing a watertight seal in vertical joints. This type of joint, however, had the main sealing element at the inside face of the outer leaf of the external sandwich panel. In this location it is inaccessible, and if the original workmanship is not good, problems of rectifying faulty sealing become severe. Having the main seal at the outer face of the panel offers the distinct advantage that inspection and remedial work are at least possible.

Another difficulty has been with the insulation layer in external sandwich panels becoming wet and giving reduced thermal resistance. Use of a nonporous insulation will overcome this; otherwise the insulation must be protected by suitable sealing.

The possibility of casting window frames into the external panels is attractive in reducing site labor, but this was found to cause water penetration through gaps formed by drying shrinkage of timber window frames. This difficulty has been overcome by fitting the windows (from the inside) after erection of the panels using flexible sealants. Water penetration may only cause problems at sites subject to driving rain, but it should be considered as a design parameter, in detailing the connections.

Condensation. Another significant problem in large panel buildings in the United Kingdom, condensation seems to cause difficulty only where houses are not kept warm throughout the day when the external temperature is low. The pattern of heating in the United Kingdom variable. When the heating level is controlled from within the apartment and paid for directly by the occupant, it is not uncommon to allow the inside air temperature within an apartment to cool down significantly during the day. When the occupant returns home after work and everything is switched on—electric kettle, stove, washing machine—the cold internal face of the external wall becomes a dehumidifier, removing the excess water vapor from the internal air. This is a most undesirable situation, since these walls can become very wet. The problem can be overcome either by ensuring that the minimum internal temperature is maintained as a suitable level, or by providing insulation at the inner face of the external walls without any cold bridges. The cold bridges may occur at the vertical and horizontal joints where it is difficult to provide a continuous insulation barrier.

Another source of condensation has been lack of insulation in internal walls that separate warm and cold areas. For example, in some layouts internal access passages are not protected from the external air, and walls adjoining them must be insulated.

Thus condensation may or may not be a potential problem, but again it must be put on the design checklist.

Repaired Units. Before a large panel is placed in its final position, it is susceptible to damage. Repair jobs on panels must be carefully done if future difficulities with broken panels are to be avoided. A good mechanical key to the repaired area should be provided by cutting back to existing reinforcement or by doweling. Also the use of calcium chloride to speed the setting time of the repair is to be discouraged. If there is doubt about the efficiency of a repair, the panel should be condemned.

Corrosion of Ties in Sandwich Panels. The ties that connect the two leaves of a sandwich panel have given trouble in some cases due to corrosion. The severity of this problem is a function of the amount of moisture that will penetrate to the tie and the chemical content of the local atmosphere. Bronze ties have been observed to suffer from stress corrosion due to chemical attack in wet conditions, and galvanized ties have also given trouble.

13.12 CONSTRUCTION PROBLEMS

As with all buildings, factors affecting the organization of the construction of large panel buildings are important for the economic success of the structures. In this section we briefly discuss some of these factors.

13.12.1 Location of Casting Plant

Some contractors have movable casting plants that can be set up on the job site. This minimizes transport cost but requires a fairly large work area at the job site. Transport costs do not appear to be a critical factor, and the use of a fixed casting plant is the most common solution. Haulage distances of 100 miles are not uncommon.

13.12.2 Size of Panels

The amount of site connection should be minimized in order to keep site work down and to maintain even wall and floor surfaces. Hence, the larger the precast unit, the better. The capacity of the erection crane used on the site and the physical size of the units to be transported are factors that limit the size of the units. Also, if a large unit is damaged when being moved, it will be more expensive to replace than a smaller one. Wall panels 25 ft long are not uncommon.

13.12.3 Erection Details

The stability of the structure under erection it most important. The wall panels are normally temporarily propped by stays that are fixed to the slab lifting-hooks and to special brackets on the walls. These stays should be carefully designed to withstand wind loading, since the consequences of a heavy panel coming adrift can be serious.

13.12.4 Leveling of the Units

The most common approach is to level the walls carefully and then set the floors on them directly. With the type of floor support detail shown in Fig. 13-13(c), it becomes possible to level the floors using special temporary support brackets after the walls have been positioned.

The leveling and alignment of the external wall units is most important, since the joints between the exposed faces are normally featured and must line up. This is not only governed by aesthetic considerations; the joints may form a drainage path that must not be blocked.

Walls are normally leveled using the vertical tie connection. Fig. 13-27 shows a typical detail.

The construction sequence here would be:

1. The lower panel is positioned.
2. The floor slabs are placed.
3. The in-situ concrete is placed.
4. Nut A is set to the correct level.
5. The upper wall panel is placed, and nut B is tightened.
6. The dry-pack mortar is placed.
7. When the dry-pack has set, nut A is slackened off.

The slackening of nut A is most important so that the wall panel has an even bearing on the dry-pack and does not transmit its load only through the vertical tie.

If wedges or shims are used for leveling the walls, it is especially important that they be removed after the dry-pack has set. There is a case on record where this was not done, so that there was extensive cracking in the walls due to stress concentrations.

13.12.5 Method of Casting

1. Vertical. A common method of casting wall panels is in vertical battery molds. A series of vertical, double-faced molds are laid side by side so that the back of one mold forms the casting face for the next unit. This is especially useful when the casting is done on site where space is normally restricted. Disadvantages are that it is difficult to seal the molds from leakage of wet concrete at projecting bars, and the placing of the concrete is difficult to control. Also, there may be a marked variation in the concrete strength within the height of the panel if cast vertically.

2. Horizontal. This is the best casting method in principle, but requires a larger working area.

Detailed consideration of factors associated with casting procedures is beyond the scope of this chapter.

13.12.6 Handling of Units

The handling stresses in large panels is an important factor in the design of the reinforcement, and there is scope for ingenuity in devising methods that keep such stresses to a minimum. For wall panels it is obviously advantageous to minimize out-of-plane bending by keeping the units vertical at all times.

Floor panels must be suspended horizontally for erection, and the position of the lifting hooks is a most important factor. Ideally extra reinforcement to cover handling stresses should not be required.

13.12.7 Crainage for Erection

Optimization of the erection procedure is dependent on the choice of crane system used. Factors to be considered are:

1. Number of cranes: normally only one is used but for large buildings two cranes might prove advantegeous.
2. The weight of the precast units in combination with the required reach.
3. The crane base, which can be fixed or can move on rails.
4. Lateral support for the crane against the building, which it may be desirable to provide.

13.13 FURTHER READING

The comprehensive series of reports from the Portland Cement Association[13-1–13-9] provides a detailed guide to recent North American experimental and analytical studies. References 13-38 and 13-66 do the same for East European practice. Further sources are given in the bibliography. Reference 13-6 is written in the form of a code, and although the recommendations given are sound, they should not be accepted as the last word on the matter. Anyone who proposes to adopt a new system should make an extensive study of the literature and try to obtain information directly from others who have experience on this area.

BIBLIOGRAPHY

13-1 Schultz, D. M., and Fintel, M., "Design and Construction of Large Panel Structures," Report No. 1, "Loading Conditions," Office of Policy Development and Research, Department of Housing and Urban Development, Washington, D.C., Apr. 1975, PCA Report No. EB091.01D.

13-2 Fintel, M., Schultz, D. M., and Iqbal M., "Design and Construction of Large Panel Structures," Report No. 2, "Philosophy of Structural Response to Normal and Abnormal Loads," Office of Policy Development and Research, Department of Housing and Urban Development, Washington, D.C., Aug. 1976, PCA Report No. 092.01D.

13-3 Kripanarayanan, K. M., and Fintel, M., "Design and Construction of Large Panel Structures," Report No. 3, "Wall Panels: Analysis and Design Criteria," Office of Policy Development and Research, Department of Housing and Urban Development, Washington, D.C., PCA Report No. EB093.01D.

13-4 Schultz, D. M., Burnett, E. F. P., and Fintel, M., "Design and Construction of Large Panel Structures," Report No. 4, "A Design Approach to General Structural Integrity," Office of Policy Development, Washington, D.C., PCA Report No. EB094.01D.

13-5 Schultz, D. M., Johal, L. S., Derecho, A. T., and Iqbal, M., "Design and Construction of Large Panel Structures," Report No. 5, "Special Topics," Office of Policy Development and Research, Department of Housing and Urban Development, Washington, D.C., PCA Report No. EB095.01D.

13-6 Schultz, D. M., "Design and Construction of Large Panel Structures," Report No. 6, "Design Methodology," Office of Policy Development and Research, Department of Housing and Urban Development, Washington, D.C., PCA Report No. EB096.01D.

13-7 Hanson, N. M., and Burnett, E. F. P., "Design and Construction of Large Panel Structures," Supplemental Report A, "Wall Cantilever and Slab Suspension Tests," Office of Policy Development and Research, Department of Housing and Urban Development, Washington, D.C., PCA Report EB097.01D.

13-8 Johal, L. S., and Hanson, N. M., "Design and Construction of Large Panel Structures," Supplemental Report B, "Horizontal Joint Tests," Office of Policy Development and Research, Department of Housing and Urban Development, Washington, D.C., PCA Report EB098.01D.

13-9 Hanson, N. W., "Design and Construction of Large Panel Structures," Supplemental Report C, "Seismic Tests of Horizontal Joints," Office of Policy Development and Research, Department of Housing and Urban Development, Washington, D.C., PCA Report EB099.01D.

13-10 "Report on the Behaviour of Reinforced Concrete structures in the Caracas, Venezuala Earthquake of July 29, 1967," Portland Cement Association, Skokie, Ill.

13-11 Griffiths, H., Pugsley, A., and Saunders, O., "Report of the Inquiry into the Collapse of Flats at Canning Town," H.M.S.O., London, U.K., 1968.

13-12 "International Recommendations for the Design and Construction of Large Panel Structures," Comite European du Beton (CEB), 1967. English Translation No. 137, Cement and Concrete Association, 52 Grosvener Gardens, London, SW1, U.K.

13-13 Selvaggio, S. L., and Carlson, C. C., "Restraint in Fire Tests of Concrete Floors and Roofs," *Research Department Bulletin No. 220,* Portland Cement Association, Skokie, Ill.

13-14 Lewicki, B., and Pauw, A., "Joints, Precast Panel Buildings," *Planning and Design of Tall Buildings,* V. 3, ASCE 1972, pp. 171–189.

13-15 Proceedings ATC-8 Seminar, January 1982. "Design of Prefabricated Concrete Buildings for Earthquake Load," Applied Technology Council, Berkeley, Calif., Suite 806, 2150 Shattuek Ave., Berkeley, Calif. 97404.

13-16 MacLeod, I. A., "Lateral Stiffness of Shear Walls with Openings," *Tall Buildings* (Coull and Stafford Smith, Eds.), Pergamon Press, Elmsford, N.Y., 1967, pp. 223–244.

13-17 (a) Coull, A., and Choudhury, J. R., "Analysis of Coupled Shear Walls," *ACI Journal, Proceedings,* V. 64, Sept. 1967, pp. 587–593.
(b) Coull, A., and Choudhury, J. R., "Stresses and Deflections in Coupled Shear Walls," *ACI Journal, Proceedings*, V. 64, Feb. 1967, pp. 65–72.

13-18 Rosman, Riko, "Aproximate Analysis of Shear Walls Subject of Lateral Loads," *ACI Journal, Proceedings,* V. 61, June 1964, pp. 717–732.

13-19 Rosman, R., "Tables for the Internal Forces of Pierced Shear Walls Subject to Lateral Load," *Bauingenieur Praxis,* Heft 66, W. Ernst und Sohn, Berlin, 1966 (in German and English).

13-20 Rosman, R., *Statik und Dynamik der Scheibensysteme des Hochbaues*, Springer Verlag, Berlin, 1968 (in German).

13-21 MacLeod, I. A., Green, D. R., Wilson, D. R., and Bhatt, P., "Two Dimensional Treatment of Complex Structures," Technical Note 78, *Proceedings Inst. of Civ. Engrs.*, Part 2, V. 53, De. 1972, pp. 589–596.

13-22 Poland, C. D., "Practical Application of Computer Analysis to the Design of Reinforced Concrete Structures for Earthquake Forces," *Reinforced Concrete Structures Subject to Wind and Earthquake Forces*, ACI publication SP-63, 1980.

13-23 Unemori, A. L., Roesset, J. M., and Becker, J. M., "Effect of Inplane Floor Slab Flexibility on the Response of Crosswall Building Systems," *Reinforced Concrete Structures Subject to Wind and Earthquake Forces*, ACI publication SP-63, 1980.

13-24 Winokour, A., and Gluck, J., "Lateral Loads In Asymmetric Multistory Structures," *Proceedings ASCE*, V. 94, ST3, Mar. 1968, 645–656.

13-25 Webster, J., "The Static and Dynamic Analysis of Orthogonal Structures Composed of Shear Walls and Frames," *Tall Buildings* (Coull and Stafford Smith, Eds.), Pergamon Press, Elmsford, N.Y., 1967, pp. 377–395.

13-26 Coull, A., and Irwin, A. W., "Analysis of Load Distribution in Multistory Shear Wall Structures," *The Structural Engineer*, V. 8, No. 8, Aug. 1970, pp. 301–306.

13-27 Clough, R. W., King, I. P., and Wilson, E. L., "Structural Analysis of Multistory Buildings," *Journal of the Structural Division, ASCE*, V. 90, ST3, 1964, pp. 19–34.

13-28 MacLeod, I. A., "Analysis of Shear Wall Buildings by the Frame Method," *Proceedings Inst. of Civ. Engrs.*, V. 55, Sept. 1976, pp. 593–603.

13-29 Rosman, R., "Analysis of Spatial Concrete Wall Systems," *Proceedings Inst. of Civ. Engrs.*, Suppl., 1970, pp. 131–152.

13-30 Petersson, H., "Analysis of Loadbearing Walls in Multistory Buildings," Chalmers Institute of Technology, Gothenberg, Sweden, 1974.

13-31 *Code of Practice for "The Structural Use of Concrete*," CP110, British Standards Institution, London, 1972.

13-32 *Building Code Requirements for Reinforced Concrete*—ACI 318-77. American Concrete Institute, Detroit, Mich., 1977.

13-33 Harley-Haddow, T., "Building Structures," Chap. 2 of *Structural Engineering in Scotland* (MacLeod, Green, and Wilson, Eds.), London Pentech Press, 1981.

13-34 *The Structures of Large Panel Buildings*, Polish Standard PN68 B-03253, Warsaw, Poland, 1969.

13-35 PCI Committee on Precast Bearing Buildings, "Considerations for the Design of Precast Bearing Wall Buildings to Withstand Abnormal Loads," *Journal PCI*, V. 21, No. 2, Mar./Apr. 1976.

13-36 Paulay, T., "Coupling Beams of Reinforced Concrete Shear Walls," *Journal of the Structural Division, ASCE*, V. 97, ST9, Sept. 1971, pp. 2407–2419.

13-37 Clarke, L. A., "The Provision of Tension and Compression Reinforcement to Resist In-plane Forces," *Magazine of Concrete Research*, V. 28, No. 94, Mar. 1976, pp. 3–12.

13-38 Lewicki, B., *Building with Large Prefabricates*, Elsevier Publishing C9., Amsterdam, The Netherlands, 1966.

13-39 Johal, L. S., and Hanson, N. M., "Design for Vertical Load on Horizontal Connections in Large Panel Structures," *PCI Journal*, V. 27, No. 1, Jan. 1982, pp. 63-79.

13-40 Birkeland, P. W., and Birkeland, H. W., "Connections in Precast Concrete Construction," *ACI Journal*, Mar. 1966, p. 345.

13-41 Despeyroux, J., Comments on paper "Shear Wall Construction in System Building," *Tall Buildings*, pp. 129–130.

13-42 Bhatt, P., and Nelson, H. M. "Strength and Deformation of Castellated Vertical Joints in Shear Walls," *Proceedings, Conference on Joints in Structures*, University of Sheffield, U.K., July 1970.

13-43 Hansen, K., and Olesen, S. O., "Failure Load and Failure Mechanism of Keyed Shear Joints," Report No. 69/22, Damarks Ingeniorakademi, Civil Engineering Department, Structural Laboratory, 10 Olster Volgade, Copenhagen, Denmark.

13-44 von Haslasz, R., and Tantow, G., "Grosstafelbauten—Konstruction und Berechnung," *Bauingenieur Praxis*, Heft 55, W. Ernst, Berlin, 1966 (general treatment on design of large panel buildings, in German).

13-45 Pommeret, M., "Les Joints Verticaux Organises Entre Grands Panneaux Coplanaires" (Vertical Joints between Coplanar Large Panels), *Annales de L'Institute Technique de Batiments et des Travaux Publics*, V. 22, No. 258, June 1969, pp. 997-999.

13-46 Peacock, J. D., "Large Precast Wall Panels. A Manufactureres Review of Jointing Problems—with some Solutions," *Proceedings*, International Symposium on Load Bearing Walls held in Warsaw, June 1969.

13-47 Green, D. R., MacLeod, I. A., and Girardan, R. S., "Force Actions in Shearwall Supports Systems," *Response of Multistory Concrete Structures to Lateral Load*, ACI Publication SP-36, 1971.

13-48 Green, D., "The Interaction of Solid Shear Walls and their Supporting Structure," *Building Science*, V. 7, 1972, pp. 239–248.

13-49 Diamant, R. M. E., *Industrialized Buildings*, 3 V., Iliffe Books, London, 1964.

13-50 *Proceedings*, International Conference on Load Bearing Walls, Warsaw, 1969.

13-51 Shapiro, G. A., and Sokolov, M. E., "The Strength and Deformation of Horizontal Joints in Large Panel Buildings," Library Communication No. 1217, Aug. 1964, Building Research Station, Watford, U.K., (translated from Russian).

13-52 Pommeret, M., "Contreventement des Batiments par Grands Panneaux" (Wind Bracing of Large Panel Buildings), *Annales de L'Institut Technique du Batiments et des Travaux Publics*, V. 20, No. 234, June 1967, pp. 795–796, and V. 21, No. 246, June 1968, pp. 940–941.

13-53 Cattaneo, L. E., "Structural Tests of Mechanical Connections for Concrete Panels," Report COM-73-10854, Material Technical Information Service, U.S. Dept. of Commerce, 5285 Pat Royal Rd., Springfield, VA., Nov. 1972.

13-54 Armer, G. T. S., and Kumar, S., "Tests on Assemblies of Large Precast Concrete Panels," *Precast Concrete*, V. 3, No. 8, Sept. 1972, pp. 541–546.

13-55 Latta, J. K., "Precast Concrete Walls—Problems with Conventional Design," Canadian Building Digest No. CBD 93, Natural Research Council, Ottawa, Canada, 1967.

13-56 Latta, J. K., "Precast Concrete Walls—A New Basis for Design," Canadian Building Digest No. CBD 94, National Research Council, Ottawa, Canada, Oct. 1967.

13-57 Pollner, E., Tso, W. K., and Heidebrecht, A. C., "Analysis of Shear Walls in Large Panel Construction," *Canadian Journal of Civil Engineering*, V. 2, No. 3, 1975, pp. 357–367.

13-58 Petersson, H., "Stresses and Deformations in Damaged Large Panel Structures," Dept. of Building Construction, Univ. of Technology, Gothenberg, Sweden, Oct. 1973.

13-59 Speyer, I. W., "Considerations for the Design of Precast Concrete Bearing Walls to Withstand Abnormal Loads," *PCI Journal*, Mar. 1976, pp. 19–51.

13-60 Zeck, V. I., and Becker, J. M., "Joints in Large Panel Concrete Structures," Publication No. R76-16, Dept. of Civil Engineering, Massachusetts Institute of Technology, Cambridge, Mass., Jan. 1976.

13-61 Breen, J. E., "Developing Structural Integrity in Bearing Wall Buildings," *PCI Journal*, Jan.-Feb. 1980, pp. 43–73.

13-62 Harris, H. G., and Iyengar, S., "Full Scale Tests on Horizontal Joints of Large Panel Structures," *PCI Journal*, Mar.-Apr. 1980, pp. 72–92.

13-63 Cholewicki, A., "Loadbearing Capacity and Deformability of Vertical Joints," *Building Science*, V. 6, 1971.

13-64 Hansen, K., Kavyrchine, M. Mechorn, G., Olsen, S., Pume, D., and Schwine, H., "Design of Vertical Keyed Shear Joints in Large Panel Buildings," *Building Research and Practice*, July/Aug. 1974.

13-65 Watson, V., and Hirst, M. J. S., "Experiments for the Design of a System Using Large Precast Concrete Panel Components," *The Structural Engineer*, V. 50, No. 9, 1972.

13-66 Sebestyen, G., *Large Panel Buildings*, Publishing House of the Hungarian Academy of Sciences, Budapest, 1965.

13-67 Berndt, K., *Die Montagebauarten des Wohnungsbaues in Beton*, Bauverlag GMBH, Weisbaden, 1969 (details of German large panel systems, in German).

14

Thin Shell Structures

W. C. SCHNOBRICH, Ph.D.*

14.1 INTRODUCTION

A precise classification of a structure as being a thin shell or a nonshell is not possible, as many structures could be classified either way. Within this chapter a thin shell is considered to be a surface structure, one of whose dimensions, its thickness, is much smaller than the other dimensions. This structure is constructed as a curved, faceted, or folded surface so that geometry activates axial forces to become a significant load-carrying stress-system. Thin shells can be classified according to several different systems, two of which are their shape and their method of formation.

14.1.1 Shell Geometry

Shells may have a curvature of the surface in one or two directions. The singly curved surfaces are developable. Cylinders and cones are examples. The doubly curved surfaces are nondevelopable. They are of positive curvature if for any point on the surface the origin of both principal radii of curvature of that point are on the same side of the surface, that is, curvatures are in the same direction. If the radii are on opposite sides of the surface, the shell has a negative curvature. Hyperbolic paraboloids and conoids are examples of this latter type of shell. Positive curvature shells react to pressures normal to the surface with direct forces of the same sign in any two orthogonal directions, whereas negative curvature shells have stresses of opposite sign. The sign of the curvature is therefore indicative of the structural behavior.

Negative curvature shells have two sets of asymptotic lines, singly curved and zero curvature shells have one such set of lines, whereas the positive curvature shells have no real asymptotic lines. If an edge of the shell is one of the asymptotic lines of the surface, there is a tendency for any disturbance or force applied to that edge to propagate deep into the shell.

Shells may also be classified according to the manner of generating them. Rotational shells are formed by revolving a plane curve about an axis in its plane. Translational shells are generated by moving one curve along another. Classification on the basis of the method of generation, however, tells little about the structural behavior and is not very meaningful. All translational shells do not behave in a similar fashion; elliptical and hyperbolic paraboloids, although formed in the same manner, have markedly different load-carrying characteristics.

14.1.2 Shell Usage

Concrete thin shells are used for many structures. Roof, foundation, and containment structures have all been built using thin shell systems. Their use as roof structures is motivated primarily by span and aesthetic considerations. Employment as foundation and containment structures is done primarily for economic reasons.

The cost factors for roof shells vary considerably with shape, support conditions, etc. An extensive discussion of cost factors for various shells is presented in a paper by Gensert, Kirsis, and Peller.[14-18] The high cost of formwork or scaffolding can severely restrict the size of spans considered economic. Form reuse can considerably lower the cost for the shell.

As the span length increases, buckling may become the

*Professor of Civil Engineering, University of Illinois at Urbana.

controlling factor. Use of boxlike or hollow cross sections then becomes necessary to further extend the possible span sizes.[14-14]

14.1.3 Shell Behavior

The ideal behavior of the shell is to carry its load by only in-plane or membrane forces and have these forces nearly constant over the shell. For arches, this requires that the structure be funicular with the load. For shells, this is not completely necessary. They will carry load predominately by membrane forces if the support conditions are appropriate. Sharp variations in either loading or shell stiffness will result in bending moments developing to either carry the load or restore compatibility. How widespread this region of bending is depends upon the shell geometry.

Positive curvature, domelike shells transmit loads to the supports primarily by compressive arching forces, provided some support exists along each edge. Disturbances applied to the edges tend to damp out quickly. Negative curvature shells utilize in-plane shear as a prime mechanism; singly curved shells behave as curved beams when longitudinal edges are not supported, or as arches if those edges are supported. These latter shells tend to propagate edge disturbances in the form of moments much farther into the shell than do the positive curvature shells.

14.2 SHELLS OF REVOLUTION

The most frequent use of the shell of revolution is for a containment or reservoir structure, but it has also been employed for roof structures when a circular plan is being used.

14.2.1 Surface Equation

A surface of revolution is obtained by rotating a plane curve about an axis lying in that plane. The curve is called the meridian; its plane the meridian plane. Planes normal to the axis of revolution intersect the surface on parallel circles. For these shells of revolution the lines of principal curvature are the meridians and the parallels; therefore, positions of points on these lines are convenient coordinates to use.

The radii of principal curvature are those of the meridian denoted by R_1 and the distance along the normal from the axis of revolution to the surface denoted by R_2 (Fig. 14-1).

Table 14-1 presents expressions for the radii of curvature for several common shell types.

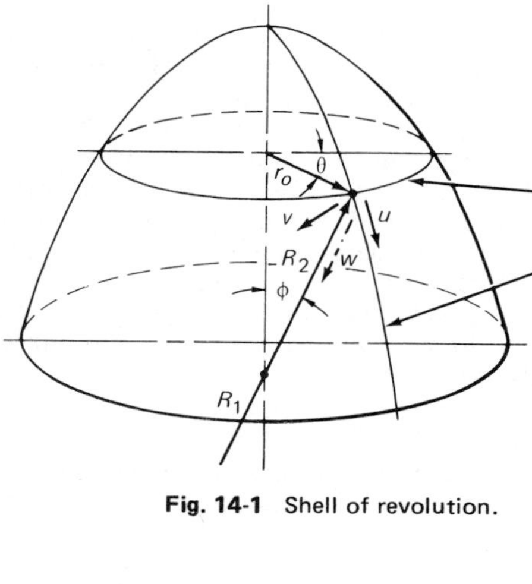

Fig. 14-1 Shell of revolution.

14.2.2 Method of Analysis

The method of analysis that can be used depends upon the geometry of the shell. If the shell is shallow, opening angle $\phi < 30$, or if the rise, H, is less than one-eighth the base diameter, then special shell equations must be used.[14-16, 14-24] If the shell is steep, it can be analyzed in the following manner:

1. Divide the shell into segments, one segment for each change in geometry.
2. For each of the segments compute membrane forces and displacements.
3. The values of these quantities along junctions between segments will in general violate continuity conditions or support conditions along those edges.
4. Compute the influences of applying unit values of edge reaction and of edge moment to common junctions and edges.
5. Write the compatibility equations for the edges or supports, and solve for the necessary corrective forces.
6. Superimpose these corrective force solutions upon the original membrane solutions.

In view of the complexity of the computations involved even in the simplified approach described above, a number of computer programs have been written to solve this problem.[14-5, 14-26] Most recent general-purpose finite element programs include shell of revolution elements for use on symmetric problems.

14.2.3 Membrane Theory

The stress resultants or internal forces acting in the shell are shown in Fig. 14-2. Since there are only three resultants, the membrane equilibrium equations are determinate. For the axisymmetric case, axial equilibrium requires:

$$N_\phi = -\frac{P}{2\pi r_o \sin \phi_o} \tag{14-1}$$

where P is the integrated axial force resultant of all loads applied to the shell above the parallel section ϕ_o. Radial equilibrium of the element shown in Fig. 14-2 gives:

$$\frac{N_\phi}{R_1} + \frac{N_\theta}{R_2} = -p_z \tag{14-2}$$

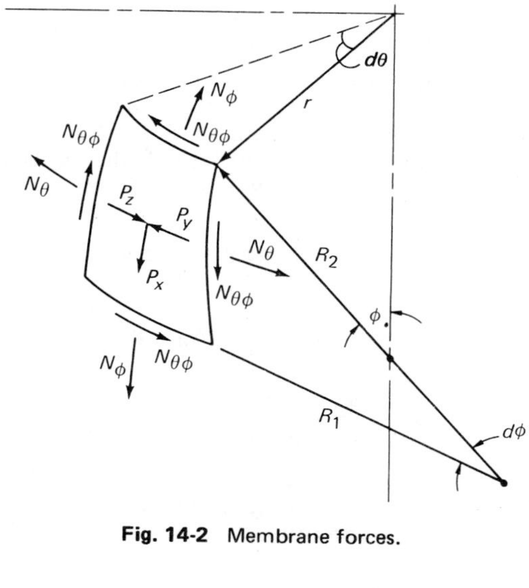

Fig. 14-2 Membrane forces.

TABLE 14-1 Geometry of Shells of Revolution

Doubly Curved Shells of Revolution

Increment of Arc

$$ds^2 = \alpha_1^2 d\phi^2 + \alpha_2^2 d\theta^2$$

where:

$\alpha_1 = R_1$
$\alpha_2 = R_2 \sin \phi = R$
a_1 = radius at base of ellipsoid
b_1 = rise of ellipsoid
a_2 = throat radius of hyperboloid
b_2/a_2 = slope of asymptotic line
R_o = radius of curvature at $\phi = 0$
γ = shell shape parameter

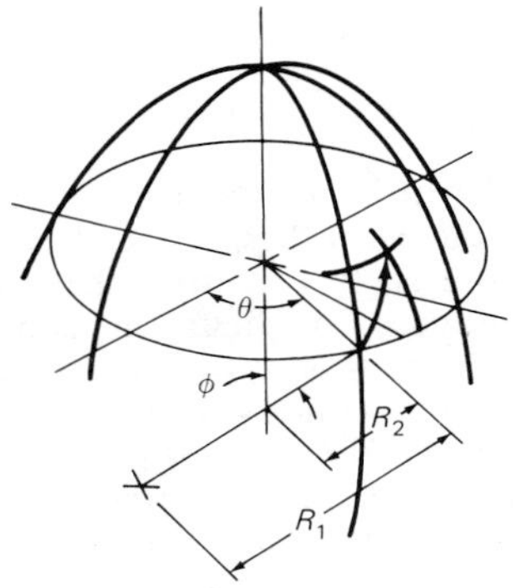

Shell Shape	*Shape Parameter*	R_1	R_2
sphere	$\gamma = 0$	R_o	R_o
paraboloid	$\gamma = 1$	$\dfrac{R_o}{(1+\gamma \sin^2 \phi)^{3/2}}$	$\dfrac{R_o}{(1+\gamma \sin^2 \phi)^{1/2}}$
ellipsoid	$\gamma > -1$	$\dfrac{a_1^2 b_1^2}{(a_1^2 \sin^2 \phi + b_1^2 \cos^2 \phi)^{3/2}}$	$\dfrac{a_1^2}{(a_1^2 \sin^2 \phi + b_1^2 \cos^2 \phi)^{1/2}}$
hyperboloid	$\gamma < -1$	$-\dfrac{a_2^2 b_2^2}{(a_2^2 \sin^2 \phi - b_2^2 \cos^2 \phi)^{3/2}}$	$\dfrac{a_2^2}{(a_2^2 \sin^2 \phi - b_2^2 \cos^2 \phi)^{1/2}}$

Developable Shells of Revolution

Increment of Arc

$$ds^2 = \alpha_1^2 dx^2 + \alpha_2^2 d\theta^2$$

where

$\alpha_1 = 1$
$\alpha_2 = R_2$

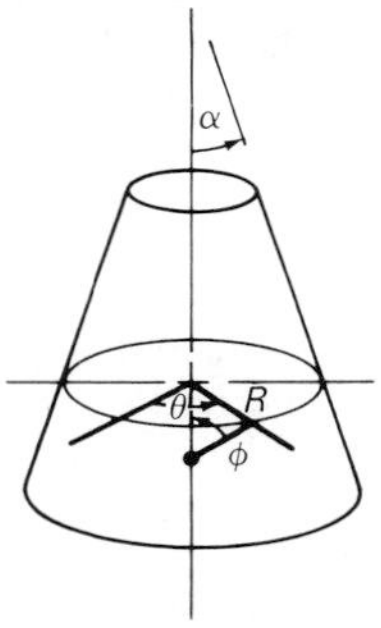

	R_1	R_2
cylinder	∞	R
cone	∞	$\dfrac{R}{\sin(90° - \alpha)}$

From these equations the membrane solutions for a variety of shell types and loadings can be determined (Table 14-2). Further tables can be found in Ref. 14-1.

With the membrane forces established, the displacements can be calculated as shown in Ref. 14-16 from:

$$u = \left[\int \frac{f(N)}{\sin \phi} d\phi + C\right] \sin \phi \qquad (14\text{-}3a)$$

$$w = u \cot \phi - \frac{R_2}{Et}(N_\theta - \nu N_\phi) \qquad (14\text{-}3b)$$

$$\Delta_\phi = \frac{u}{R_1} + \frac{dw}{R_1 d\phi} \qquad (14\text{-}3c)$$

where $f(N) = (1/Et)\ [N_\phi(R_1 + \nu R_2) - N_\theta(R_2 + \nu R_1)]$, ν is Poisson's ratio, and C is a rigid body motion along the axis to adjust to support conditions.

14.2.4 Edge Corrections

If the forces and/or displacements, as computed from the above equations, do not agree with the support conditions of the structure, a correction must be applied. This correction involves solving the governing equations of a bending theory. Several approximate solutions[14-3, 14-23] to such theories have been reported in the literature. If the opening angle exceeds 30 degrees,[14-24] the Geckeler or equivalent cylinder solution is adequate. Table 14-3 gives the important internal forces and displacements when unit edge forces are applied. Reference 14-1 contains a more extensive set of tables drawn from many sources. The functions F_1 to F_4 are given in Table 14-4.

14.2.5 Stability Considerations

Buckling has long been an area of active experimental and analytical research.[14-21, 14-22] Since the buckling of many shells is very sensitive to initial imperfections and the type of support conditions, sizable differences can exist between computed and actual buckling values. For reinforced concrete, shrinkage and tensile cracking, as well as creep, can further reduce the buckling load. Therefore generous factors of safety are recommended.

The buckling pressure for a spherical shell can be expressed as:

$$P_{cr} = CE\left(\frac{t}{R}\right)^2 \tag{14-4}$$

where E is the tangent modulus for the material and C is a coefficient to account for boundary conditions, imperfections, etc. The classical value of C for an ideal clamped shell is $2/\sqrt{3(1-\nu^2)}$. This value is reduced to a recommended value of 0.35 for metal shells. For concrete shells a further reduction of no less than one-half is recommended. The tangent modulus value can be that suggested by Griggs.[14-21]

$$E = \frac{2\sigma_o}{\epsilon_o}\sqrt{\left(1 - \frac{\sigma}{\sigma_o}\right)} \tag{14-5}$$

where the subscript o indicates the maximum quantity.

A comprehensive summary of the details of the determination of the buckling loads for shells of revolution can be found in Ref. 14-1. Modifications necessary to adapt these techniques to reinforced concrete shells can be found in the ACI special publication SP-67.[14-11]

EXAMPLE 14-1: Analyze a spherical shell of constant thickness.

SOLUTION: For a spherical shell loaded by its own weight, membrane theory (Table 14-2) predicts stresses of:

$$N_\phi = -\frac{p_E r}{1 + \cos\phi}$$

$$N_\theta = p_E r\left(\frac{1}{1 + \cos\phi} - \cos\phi\right) \tag{14-6}$$

where p_E is the weight of the shell in force per unit area of shell surface. If the opening angle of the shell is less than 52° then the entire shell would be in compression, provided the support conditions are as required by membrane theory. Figure 14-3 shows the variation of these membrane forces along the meridian of the shell.

If the spherical shell is loaded by a load uniformly distributed over a plane perpendicular to the axis of the shell, the membrane stresses are:

$$N_\phi = -\frac{p_s r}{2}$$

$$N_\theta = -\frac{p_s r}{2}\cos 2\phi \tag{14-7}$$

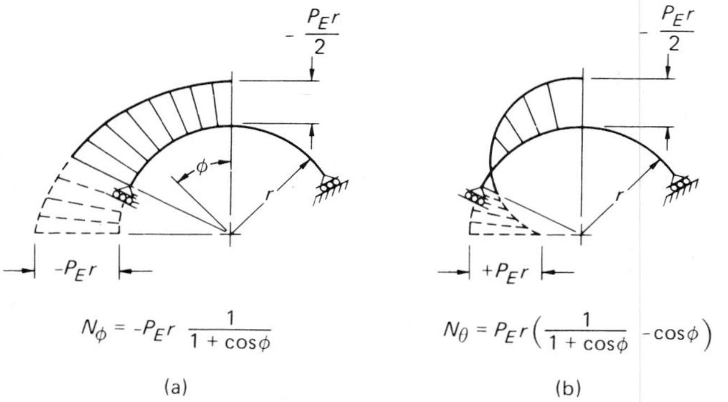

Fig. 14-3 Membrane stresses due to dead weight. (a) Meridianal stress; and (b) hoop stress.

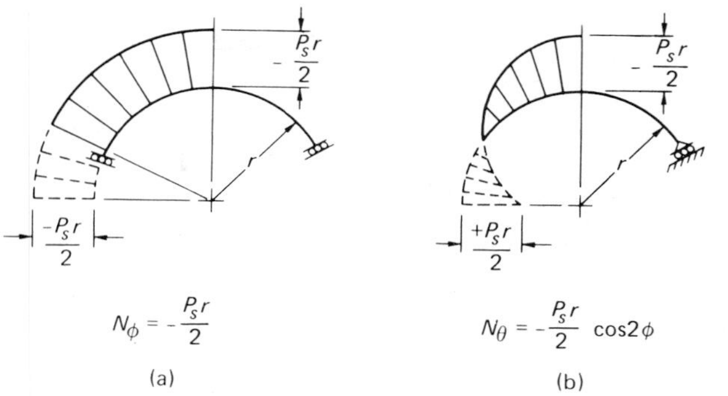

Fig. 14-4 Membrane stresses due to snow load. (a) Meridianal stress; and (b) hoop stress.

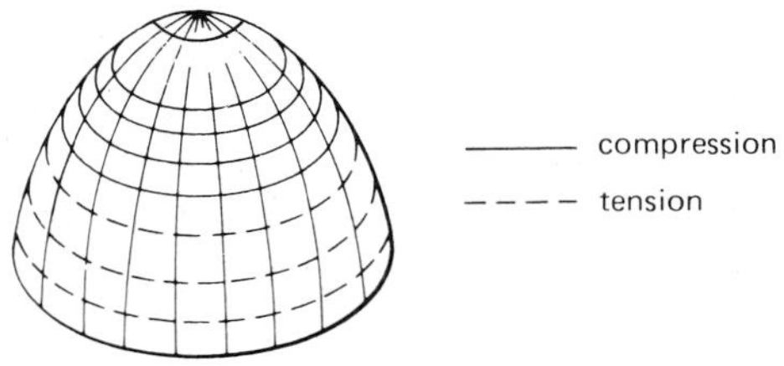

Fig. 14-5 Stress trajectories.

If the opening angle is less than 45°, then, again, the entire shell is in compression. The variation of the stresses along the meridian is shown in Fig. 14-4. The stress trajectories of such a system are as shown in Fig. 14-5.

The displacements of the spherical shell under dead weight, found from eqs. (14-3a, b, c) are:

$$u = \frac{p_E r^2(1+\nu)}{Et}\left[\ln\frac{1+\cos\phi}{1+\cos\phi_c} + \frac{1}{1+\cos\phi_c} - \frac{1}{1+\cos\phi}\right]\sin\phi$$

$$w = \frac{p_E r^2(1+\nu)}{Et}\left[\ln\frac{1+\cos\phi}{1+\cos\phi_c} - \frac{1+\cos\phi_c - \cos\phi}{\cos\phi(1+\cos\phi_c)}\right]\cdot\cos\phi$$

$$+\frac{p_E r^2}{Et}\cos\phi \tag{14-8}$$

so that the displacements of the edge become:

$$\Delta_H = -\frac{p_E r^2}{Et}\left(\frac{1+\nu}{1+\cos\phi_c} - \cos\phi_c\right)\sin\phi_c$$

$$\Delta_\phi = -\frac{p_E r}{Et}(2+\nu)\sin\phi_c \tag{14-9}$$

A similar set of displacemens can be found for the projected load.

EXAMPLE 14-2: *Ring-beam-supported shell.* The shell shown in Fig. 14-6 is to be analyzed for the dead load stresses. The dimen-

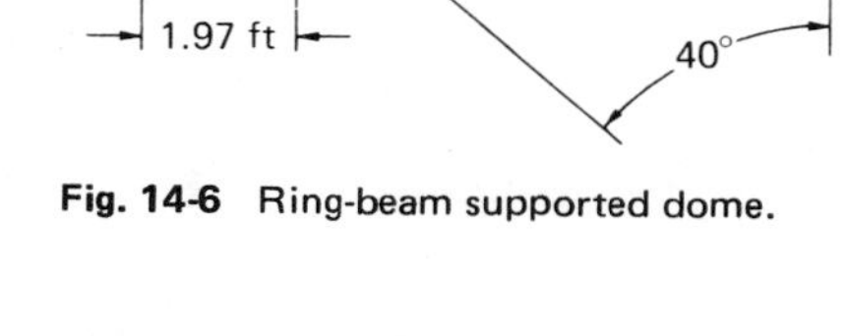

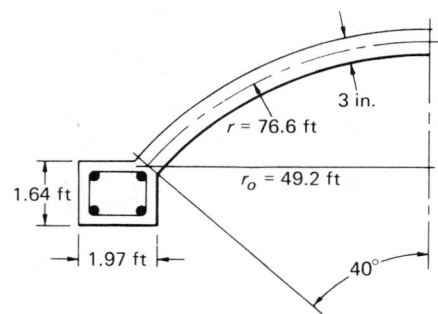

Fig. 14-6 Ring-beam supported dome.

TABLE 14-2 Membrane Stress Resultants—Spherical Shell

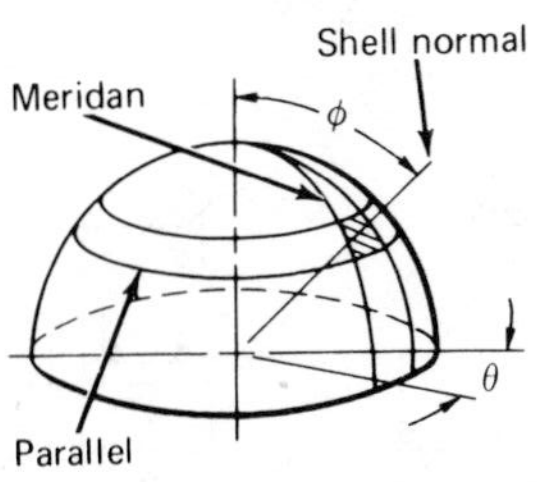

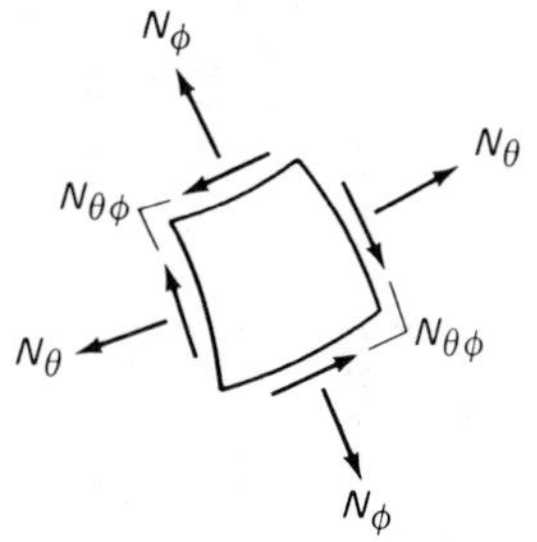

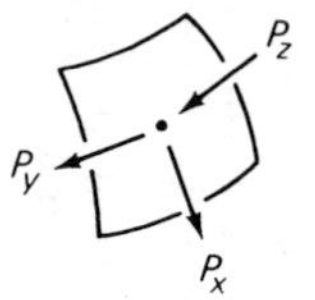

N_θ, N_ϕ Unit normal forces
$N_{\theta\phi}$ Unit central shear
p_x, p_y, p_z Surface load components

System	Loading	N_ϕ	N_θ	$N_{\theta\phi}$
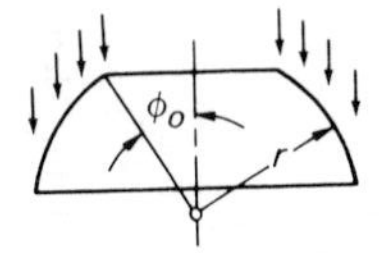	$p_x = p_E \sin\phi$ $p_z = p_E \cos\phi$	$-p_E r \dfrac{\cos\phi_o - \cos\phi}{\sin^2\phi}$	$p_E r \left(\dfrac{\cos\phi_o - \cos\phi}{\sin^2\phi} - \cos\phi\right)$	0
		For $\phi_o = 0$ (no vertex opening)		
		$-p_E r \dfrac{1}{1+\cos\phi}$	$p_E r \left(\dfrac{1}{1+\cos\phi} - \cos\phi\right)$	0
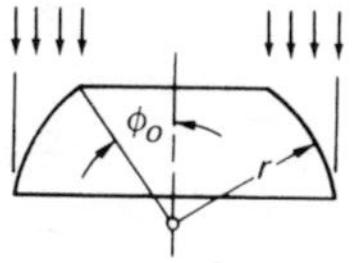	$p_x = p_s \sin\phi \cos\phi$ $p_z = p_s \cos^2\phi$	$-p_s \dfrac{r}{2}\left(1 - \dfrac{\sin^2\phi_o}{\sin^2\phi}\right)$	$p_s \dfrac{r}{2}\left(1 - \dfrac{\sin^2\phi_o}{\sin^2\phi} - 2\cos^2\phi\right)$	0
		For $\phi_o = 0$ (no vertex opening)		
		$-p_s \dfrac{r}{2}$	$-p_s \dfrac{r}{2} \cos 2\phi$	0
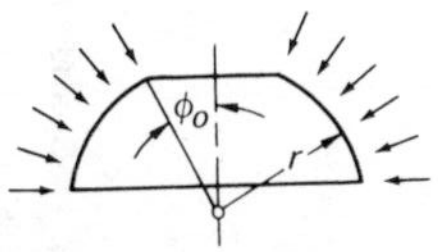	$p_z = p$	$-p \dfrac{r}{2}\left(1 - \dfrac{\sin^2\phi_o}{\sin^2\phi}\right)$	$-p \dfrac{r}{2}\left(1 + \dfrac{\sin^2\phi_o}{\sin^2\phi}\right)$	0
		For $\phi_o = 0$ (no vertex opening)		
		$-p \dfrac{r}{2}$	$-p \dfrac{r}{2}$	0

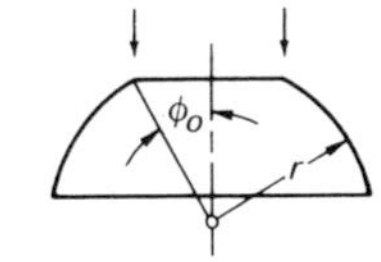	Edge Load p_L	$-p_L \dfrac{\sin\phi_o}{\sin^2\phi}$	$p_L \dfrac{\sin\phi_o}{\sin^2\phi}$	0
		For $\phi_o = 0$ (no vertex opening)		
	Vertex Load P_L	$-P_L \dfrac{1}{2\pi r \sin^2\phi}$	$P_L \dfrac{1}{2\pi r \sin^2\phi}$	0
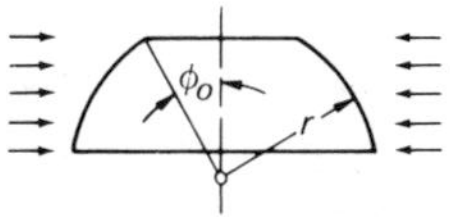	$p_z = p_w \sin\phi \cos\theta$	$-p_w \dfrac{r}{3} \dfrac{\cos\phi\cos\theta}{\sin^3\phi} \times [3(\cos\phi_o - \cos\phi) - (\cos^3\phi_o - \cos^3\phi)]$	$p_w \dfrac{r}{3} \dfrac{\cos\theta}{\sin^3\phi} \times [\cos\phi(3\cos\phi_o - \cos^3\phi_o) - 3\sin^2\phi - 2\cos^4\phi]$	$-p_w \dfrac{r}{3} \dfrac{\sin\theta}{\sin^3\phi} [3(\cos\phi_o - \cos\phi) - \cos^3\phi_o + \cos^3\phi]$
		For $\phi_o = 0$ (no vertex opening)		
		$-p_w \dfrac{r}{3} \dfrac{\cos\phi\cos\theta}{\sin^3\phi} \times (2 - 3\cos\phi + \cos^3\phi)$	$p_w \dfrac{r}{3} \dfrac{\cos\theta}{\sin^3\phi} (2\cos\phi - 3\sin^2\phi - 2\cos^4\phi)$	$-p_w \dfrac{r}{3} \dfrac{\sin\theta}{\sin^3\phi} (2 - 3\cos\phi + \cos^3\phi)$

Membrane Stress Resultants—Other shells of revolution with curved meridian

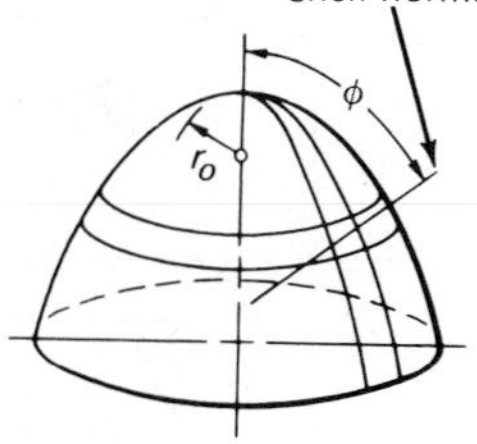

r_o Radius of curvature at vertex
p_x, p_z Surface load components
N_θ, N_ϕ Unit normal forces

	System	Loading	N_ϕ	N_Θ	$N_{\Theta\phi}$
Parabola	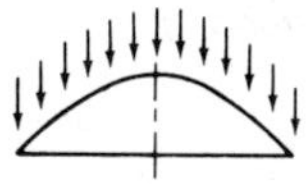	$p_x = p_E \sin\phi$ $p_z = p_E \cos\phi$	$-p_E \dfrac{r_o}{3} \dfrac{1 - \cos^3\phi}{\sin^2\phi \cos^2\phi}$	$-p_E \dfrac{r_o}{3} \dfrac{2 - 3\cos^2\phi + \cos^3\phi}{\sin^2\phi}$	0
	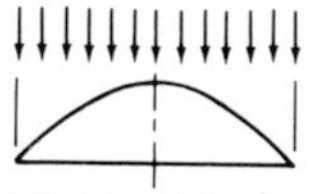	$p_x = p_s \sin\phi\cos\phi$ $p_z = p_s \cos^2\phi$	$-p_s \dfrac{r_o}{2} \dfrac{1}{\cos\phi}$	$-p_s \dfrac{r_o}{2} \cos\phi$	0
	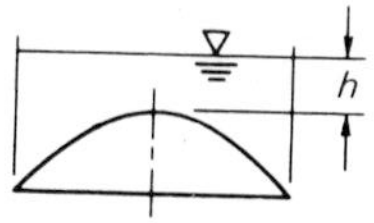	$p_s = \gamma\left(h + \dfrac{r_o}{2}\tan^2\phi\right)$	$-\gamma \dfrac{r_o}{2}\left(h + \dfrac{r_o}{4}\tan^2\phi\right)\dfrac{1}{\cos\phi}$	$-\gamma \dfrac{r_o}{2}\left[h(2\tan^2\phi + 1) + r_o \tan^2\phi\left(\tan^2\phi + \dfrac{3}{4}\right)\right]\cos\phi$	0

TABLE 14-2 *Continued*

System	*Loading*	N_ϕ	N_θ	$N_{\theta\phi}$
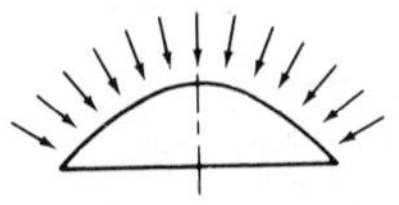	$p_z = p$	$-p \dfrac{r_o}{2} \dfrac{1}{\cos\phi}$	$-p \dfrac{r_o}{2} \dfrac{1 + \sin^2\phi}{\cos\phi}$	0
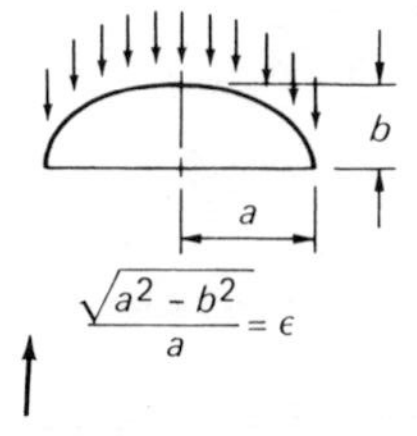 $\dfrac{\sqrt{a^2 - b^2}}{a} = \epsilon$	$p_x = p_E \sin\phi$ $p_z = p_E \cos\phi$	$-\dfrac{p_E}{2} \dfrac{\sqrt{a^2 \tan^2\phi + b^2}}{a^2 \sin\phi \tan\phi} \Big[a^2 -$ $-\dfrac{a^2 b^2 \sqrt{1 + \tan^2\phi}}{b^2 + a^2 \tan^2\phi} +$ $+ \dfrac{b^2}{\epsilon} \ln \dfrac{(1+\epsilon)\sqrt{b^2 + a^2 \tan^2\phi}}{b(\epsilon + \sqrt{1 + \tan^2\phi})} \Big]$	$p_E \Big[\dfrac{(b^2 + a^2 \tan^2\phi)^{3/2}}{2 \tan^2\phi \sqrt{1 + \tan^2\phi}} \times$ $\times \left(\dfrac{1}{\epsilon a^2} \ln \dfrac{(1+\epsilon)\sqrt{b^2 + a^2 \tan^2\phi}}{b(\epsilon + \sqrt{1 + \tan^2\phi})} + \right.$ $\left. + \dfrac{1}{b^2} - \dfrac{\sqrt{1 + \tan^2\phi}}{b^2 + a^2 \tan^2\phi} \right) - \dfrac{a^2}{\sqrt{b^2 + a^2 \tan^2\phi}} \Big]$	0
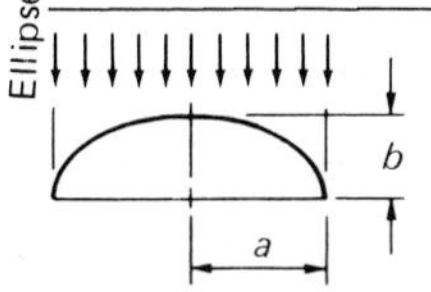	$p_x = p_s \sin\phi \cos\phi$ $p_z = p_s \cos^2\phi$	$-\dfrac{p_s}{2} \dfrac{a^2 \sqrt{1 + \tan^2\phi}}{\sqrt{b^2 + a^2 \tan^2\phi}}$	$-\dfrac{p_s}{2} \dfrac{a^2}{b^2} \dfrac{b^2 - a^2 \tan^2\phi}{\sqrt{b^2 + a^2 \tan^2\phi} \sqrt{1 + \tan^2\phi}}$	0

sions of the shell are as shown in the figure. The thickness of 3 in. is used as a reasonable minimum. The dome is constrained by a ring beam that is in turn supported on closely spaced slender columns. The solution is achieved first from the membrane state, then corrected for the boundary conditions. The load is assumed to be 50 psf to include roofing, etc. in the dead weight. Also Poisson's ratio ν is assumed zero.

SOLUTION:
Membrane forces and displacements:

$$N_\phi = -\frac{p_E r}{1 + \cos\phi} = -2168 \text{ lbs/ft (edge)}$$

$$= -1915 \text{ lbs/ft (crown)}$$

$$N_\theta = p_E r\left(\frac{1}{1 + \cos\phi} - \cos\phi\right) = -766 \text{ lbs/ft (edge)}$$

$$= -1915 \text{ lbs/ft (crown)}$$

$$\Delta_H = (r \sin\phi_c)\,\epsilon_\theta = \frac{rN_\theta}{Et}\sin\phi_c$$

(For membrane theory outward positive.)

$$E\Delta_H^S = \frac{(49.2)\,12}{3}(-766) = -150{,}749 \text{ lb/ft (inward)}$$

$$\Delta_\phi = \frac{1}{R}\left(\frac{dw}{d\phi} + u\right) = -\frac{2p_E r}{Et}\sin\phi_c$$

$$E\Delta_\phi^S = -\frac{2(50)(76.6)(12)}{3}(0.643)$$

$$= -19{,}700 \text{ lb/ft}^2 \text{ clockwise or flattening}$$

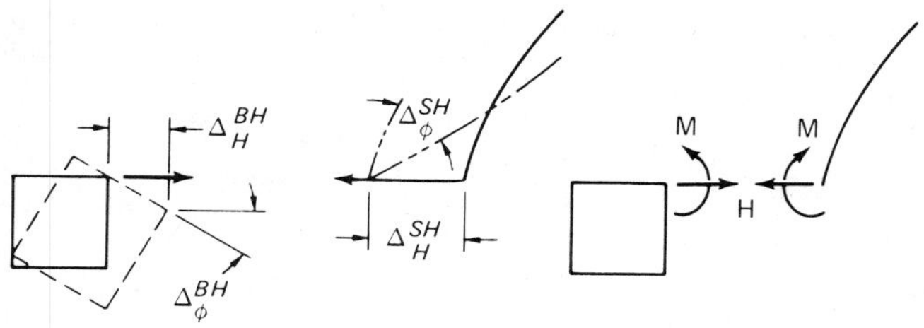

Fig. 14-7 Compatibility analysis at edge. (a) Displacements due to $H = 1$; and (b) forces to restore compatibility.

Horizontal thrust at edge: The edge or ring beam is placed so that the thrust line for N_ϕ passes through the centroid of the beam. The horizontal component.

$$\bar{H} = N_\phi \cos\phi_c = (2168)(\cos 40°) = 1661 \text{ lb/ft (outward)}$$

Horizontal displacement of ring beam from $\bar{H}$ outward thrust:

$$E\Delta_H^B = \frac{\bar{H}r_c^2}{bd} = \frac{(1661)(49.2)^2}{(1.97)(1.64)} = 1{,}244{,}300 \text{ lb/ft (outward)}$$

$E\Delta_\phi^B = 0$ since thrust goes through centroid of ring beam

Compatibility: The shell and the ring beam no longer have compatible displacements at junction. Additional edge moments and horizontal forces are applied to both the shell and the beam to enforce compatibility (Fig. 14-7):

$$E\Delta_H^S + E\Delta_H^{SH} + E\Delta_H^{SM} = E\Delta_H^B + E\Delta_H^{BH} + E\Delta_H^{BM}$$
$$E\Delta_\phi^S + E\Delta_\phi^{SH} + E\Delta_\phi^{SM} = E\Delta_\phi^B + E\Delta_\phi^{BH} + E\Delta_\phi^{BM}$$

TABLE 14-3 Influence of Edge Forces on Forces and Displacements of Shell of Revolution

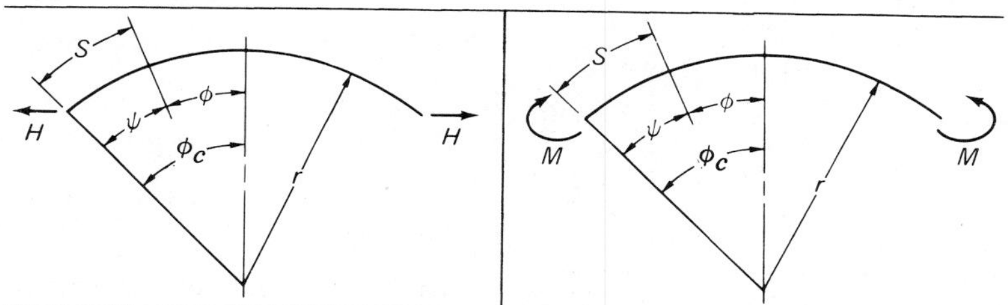

N_ϕ	$F_2 H \sin\phi_c \cot(\phi_c - \psi)$	$\frac{2\sqrt{3}}{\beta t} F_4 M \cot(\phi_c - \psi)$
N_θ	$2\beta F_3 H \sin\phi_c$	$\frac{2\sqrt{3}}{t} MF_2$
Q_ϕ	$HF_2 \sin\phi_c$	$-\frac{2\sqrt{3}}{\beta t} MF_4$
M_ϕ	$\frac{\beta t H}{\sqrt{3}} F_4 \sin\phi_c$	MF_1
Δ_ϕ	$-\frac{2\beta^2 H}{Et}\sin\phi_c$	$-\frac{4\sqrt{3}\,\beta}{Et^2} M$
Δ_H	$-\frac{2\beta r H}{Et}\sin^2\phi_c$	$-\frac{2\beta^2 M \sin\phi_c}{Et}$

where:

$$\psi = \phi_c - \phi \qquad \beta^4 = 3(1 - \nu^2)\left(\frac{r}{t}\right)^2$$

and the functions F_1 to F_4 are given in Table 14-4. Δ_H inward positive.

The displacements and rotations of the ring beam as a result of applying horizontal force and moment along the top connection with the shell are

$$E\Delta_H^{BH} = \frac{4Hr^2}{bd} = \frac{4(49.2)^2}{(1.97)(1.64)} H = 2997H \text{ (inward)}$$

$$E\Delta_H^{BM} = -\frac{6Mr^2}{bd^2} = -\frac{6(49.2)^2}{(1.97)(1.64)^2} M = -2741M \text{ (outward)}$$

$$E\Delta_\phi^{BH} = -\frac{6Hr^2}{bd^2} = -2741H \text{ clockwise}$$

$$E\Delta_\phi^{BM} = \frac{12Mr^2}{bd^3} = 3342.8M \text{ counterclockwise or raises}$$

TABLE 14-4 Table of Functions F_1 to F_4

$\beta\phi$	F_1	F_2	F_3	F_4
0	1.0000	1.0000	1.0000	0
0.1	0.9907	0.8100	0.9003	0.0903
0.2	0.9651	0.6398	0.8024	0.1627
0.3	0.9267	0.4888	0.7077	0.2189
0.4	0.8784	0.3564	0.6174	0.2610
0.5	0.8231	0.2415	0.5323	0.2908
0.6	0.7628	0.1431	0.4530	0.3099
0.7	0.6997	0.0599	0.3798	0.3199
0.8	0.6354	-0.0093	0.3131	0.3223
0.9	0.5712	-0.0657	0.2527	0.3185
1.0	0.5083	-0.1108	0.1988	0.3096
1.1	0.4476	-0.1457	0.1510	0.2967
1.2	0.3899	-0.1716	0.1091	0.2807
1.3	0.3355	-0.1897	0.0729	0.2626
1.4	0.2849	-0.2011	0.0419	0.2430
1.5	0.2384	-0.2068	0.0158	0.2226
1.6	0.1959	-0.2077	-0.0059	0.2018
1.7	0.1576	-0.2047	-0.0235	0.1812
1.8	0.1234	-0.1985	-0.0376	0.1610
1.9	0.0932	-0.1899	-0.0484	0.1415
2.0	0.0667	-0.1794	-0.0563	0.1230
2.1	0.0439	-0.1675	-0.0618	0.1057
2.2	0.0244	-0.1548	-0.0652	0.0895
2.3	0.0080	-0.1416	-0.0668	0.0748
2.4	-0.0056	-0.1282	-0.0669	0.0613
2.5	-0.0166	-0.1149	-0.0658	0.0492
2.6	-0.0254	-0.1019	-0.0636	0.0383
2.7	-0.0320	-0.0895	-0.0608	0.0287
2.8	-0.0369	-0.0777	-0.0573	0.0204
2.9	-0.0403	-0.0666	-0.0534	0.0132
3.0	-0.0423	-0.0563	-0.0493	0.0071
3.1	-0.0431	-0.0469	-0.0450	0.0019
3.2	-0.0431	-0.0383	-0.0407	-0.0024
3.3	-0.0422	-0.0306	-0.0364	-0.0058
3.4	-0.0408	-0.0237	-0.0323	-0.0085
3.5	-0.0389	-0.0177	-0.0283	-0.0106
3.6	-0.0366	-0.0124	-0.0245	-0.0121
3.7	-0.0341	-0.0079	-0.0210	-0.0131
3.8	-0.0314	-0.0040	-0.0177	-0.0137
3.9	-0.0286	-0.0008	-0.0147	-0.0140

$F_1 = e^{-\beta\varphi}(\cos\beta\varphi + \sin\beta\varphi)$
$F_2 = e^{-\beta\varphi}(\cos\beta\varphi - \sin\beta\varphi)$
$F_3 = e^{-\beta\varphi}(\cos\beta\varphi)$
$F_4 = e^{-\beta\varphi}(\sin\beta\varphi)$

The displacements and rotations at the edge of the shell due to the horizontal force and moment are (from Table 14-3):

$$E\Delta_H^{SH} = -\frac{2\beta r}{t} H \sin^2\phi_c = -\frac{2(23.0)(76.6)}{0.25}(0.643)^2 H$$
$$= -5827H \text{ (outward)}$$

$$E\Delta_H^{SM} = -\frac{2\beta^2 M}{t}\sin\phi_c = -\frac{2(23)^2}{0.25} M(0.643)$$
$$= -2721.2M \text{ (outward)}$$

$$E\Delta_\phi^{SH} = -\frac{2\beta^2 H}{t}\sin\phi_c = -2721.2H \text{ clockwise or flattening}$$

$$E\Delta_\phi^{SM} = -\frac{4\sqrt{3}\,\beta}{t^2} M = -\frac{4\sqrt{3}\,(23)}{(0.25)^2} M$$
$$= -2550M \text{ clockwise or flattening}$$

Writing the compatibility equations for the shell beam junction line:

$$-150{,}749 + 5827H + 2721M = 1{,}244{,}300 - 2997H + 2741M$$
$$-19{,}700 - 2721H - 2550M = 0 - 2741H + 3343M$$

Thus:

$$H = 158.5 \text{ lb/ft}$$
$$M = -2.8 \text{ lb-ft/ft}$$

The stress resultants anywhere in the shell can now be determined using Tables 14-3 and 14-4 plus the membrane solution.

Hoop force (N_θ) at the edge is:

$$N_\theta = 2\beta H \sin\phi_c + \frac{2\sqrt{3}}{t} M - 766 \text{ lb/ft}$$
$$= 4688 - 39 - 766 = +3883 \text{ lb/ft}$$

This hoop tension decays to approximately zero when

$$2\beta F_3 H \sin\phi_c = 766 \quad \therefore \quad F_3 = 766/4688 = 0.1633$$
$$\text{or } \beta\phi \cong 1.0 \quad \therefore \quad \phi = 0.0435 \text{ or } 2°30'$$

This means that the tension stresses do not get beyond $3^1/_3$ ft from the edge.

This large tension force can be suppressed by prestressing the ring beam. The inclusion of a prestressed ring beam is readily accomplished by considering the prestressing force of F as a ring pressure of F/r.

14.3 CYLINDRICAL SHELLS

A special form of the shell of revolution is that of the cylindrical shell. It is used for a variety of structures ranging from pressure vessels and containment, or storage units, to roof structures. There are several reasons for singling out the cylindrical shell for special discussion. The prime reason is the extensive use made of this shell configuration. From a structural point of view the more important aspect is the fact that the shell is developable and, as stated in Section 14.1.1, this can have a marked effect on the manner in which the shell responds to load. Bending stresses are especially important for this shell geometry.

In order for the bending effects of an edge disturbance to rapidly damp out, the structural behavior of strips parallel to the edge must act as a stiff, elastic restraint on the normal displacement of sections, or strips, running perpendicular to the edge. If the edge is a straight or asymptotic line of the surface the elastic restraining or foundation effect of strips parallel to the edge is small, and bending penetrates

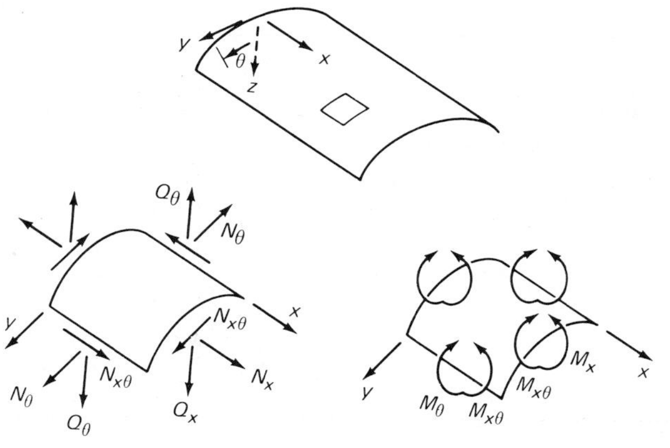

Fig. 14-8 Positive stress resultants.

deeply. If the cylinder is closed, the restraining sections are rings, and disturbances applied to the curved edges damp out rapidly, as described in Section 14.2.

Although the cross section of the cylinder can have any configuration, the most common, by far, is the circular cylinder.

The analysis of the cylindrical shell can be achieved by solution of the governing differential equations or by approximate methods based in part upon structural intuition. In order to be computationally feasible, these procedures frequently require extensive systems of tabulated numerical data if the analysis is to be performed with a simple calculator. An excellent source of tabulated data for cylindrical shells is Ref. 14-35. However, with the wealth of computer programs available capable of solving the cylindrical shell problem, i.e., NASTRAN, SAP, FINITE among others, their use is recommended. This is particularly true when investigating problems involving the superposition of a large number of load components such as occurs with the wind load on a complete cylinder. See Ref. 14-44 for wind load description.

The positive definitions of stress resultants and displacements to be used in the analysis of the cylindrical shell are shown in Fig. 14-8. The nomenclature used is as shown in that figure.

14.3.1 Membrane Theory

The importance of membrane theory varies with the configuration. For complete, circular cylindrical shells under symmetric or asymmetric load, membrane stresses predominate. For cylindrical panels, stresses computed by membrane theory may represent only a small part of the total stress picture. The membrane stresses for a number of loading conditions are given in Table 14-5. For the liquid container, for example, only hoop stress in the amount:

$$N_\theta = p_z r = \gamma r x \tag{14-10}$$

is developed. The resulting displacements are:

$$w = p_z r^2/Eh = \gamma r^2 x/Et$$
$$\Delta\phi = \frac{dw}{dx} = \gamma r^2/Et \tag{14-11}$$

The structure is thus assumed to function as a system of rings sitting atop each other and carrying the fluid pressure by hoop tension. At the base some bending correction will be necessary in order to return the shell edge to its proper position.

For the antisymmetric load, such as shown as case 3 in Table 14-5(b) and considered as a quite gross approximation to the wind load, the membrane solution yields:

$$N_x = \frac{p_w}{2r}(L - x)^2 \cos\theta$$
$$N_\theta = -p_w r \cos\theta \tag{14-12}$$
$$N_{xy} = -p_w(L - x)\sin\theta$$

with x measured from the base of the shell. The corresponding displacements are

$$Eu = \frac{p_w x}{6rt}\,[3L^2 - 3Lx + x^2]\cos\theta$$
$$Ev = \frac{p_w}{24r^2t}[12r^2(4L_x - 2x^2) + x^2(6L^2 - 4Lx + x^2)]\sin\theta \tag{14-13}$$
$$Ew = -\frac{p_w}{24r^2t}[12r^2(4Lx - 2x^2) + x^2(6L^2 - 4Lx + x^2)]\cos\theta$$

where L is the length of the shell. This solution is equivalent to a cantilever beam of circular cross section. N_x performs the function of the axial fiber stresses, and N_{xy} plays the role of the beam shear.

14.3.2 Bending Theory

For the complete cylindrical shell, bending theory is used primarily to adjust the normal displacement, w, and the rotation Δ_θ at the edge of the segment to make those displacements compatible with the adjacent segments or with known displacement conditions. These bending solutions are of the beam-on-elastic-foundation type. They are listed in Table 14-6 for the case where a disturbance is applied to one end, and this disturbance has a negligible effect at the far end. Although these equations were developed for the symmetric case, they can also serve as a first approximation to the antisymmetric case. The edge effect solution for shells of revolution as listed in Table 14-3 can be looked on as the solution for an equivalent cylinder. Therefore Table 14-6 can be derived from Table 14-3 with an appropriate changes in definition of β. For more waves of loading in the circumferential direction, use of a computer program is recommended. Application of the bending analysis to symmetric cylindrical shells is demonstrated in Chapter 18 (Fig. 18-31).

The application of bending theory to cylindrical panels or roof structures, frequently referred to as barrel shells., is much more complex. A number of books [14-12, 14-19, 14-25] have been written on the development and solution of the equations. Even texts covering the broad class of shell structures[14-2, 14-16] place a good bit of emphasis on the cylindrical roof structure. *ASCE Manual 31* was published to provide a tabulated form of solution of the complex differential equation system in order to make the analysis feasible within a design office. The tables in the manual were found to be limiting in their range, so they were supplemented by additional tables by PCA.[14-9]

With current computer capabilities the use of such elaborate tables is now considered unnecessary. Present analysis methods fall into two categories—beam methods and direct stiffness methods. Included in this latter grouping is the finite element method.

For the complete circular cylinder, axisymmetric solid or

TABLE 14-5 Membrane Stresses in Developable Shells

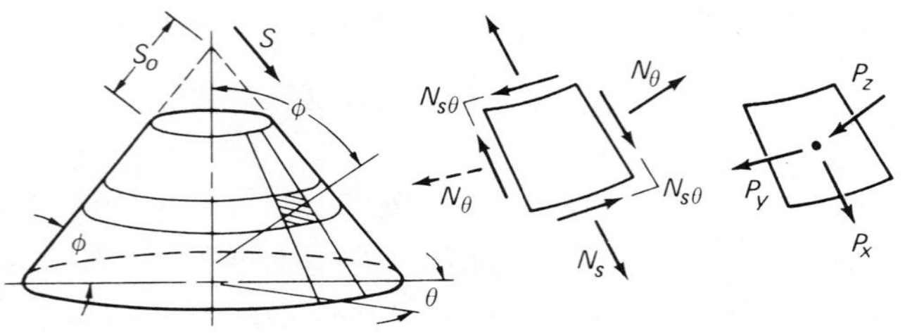

N_s, N_θ	Unit normal forces
$N_{s\theta}$	Unit central shear
p_x, p_y, p_z	Surface load components

(a) Conical shell

System	*Loading*	N_s	N_θ	$N_{s\theta}$
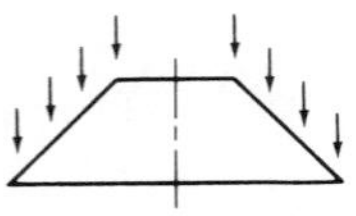	$p_x = p_E \sin\phi$ $p_z = p_E \cos\phi$	$-p_E \dfrac{s^2 - s_o^2}{2s} \dfrac{1}{\sin\phi}$	$-p_E\, s \dfrac{\cos^2\phi}{\sin\phi}$	0
		For $s_o = 0$ (complete cone)		
		$-p_E \dfrac{s}{2} \dfrac{1}{\sin\phi}$	$-p_E\, s \dfrac{\cos^2\phi}{\sin\phi}$	0
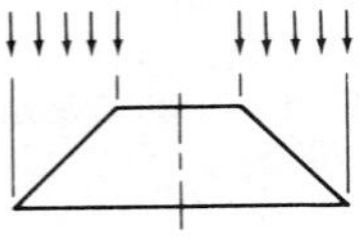	$p_x = p_s \sin\varphi \cos\phi$ $p_z = p_s \cos^2\phi$	$-p_s \dfrac{s^2 - s_o^2}{2s} \cot\phi$	$-p_s\, s \dfrac{\cos^3\phi}{\sin\phi}$	0
		For $s_o = 0$ (complete cone)		
		$-p_s \dfrac{s}{2} \cot\phi$	$-p_s\, s \dfrac{\cos^3\phi}{\sin\phi}$	0

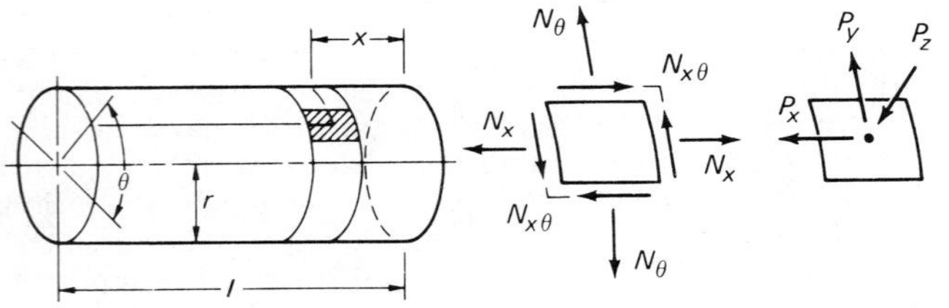

N_x, N_θ	Unit normal forces
$N_{x\theta}$	Unit central shear
p_x, p_y, p_z	Surface load components

(b) Circular cylindrical shell

System	*Loading*	N_x	N_θ	$N_{x\theta}$
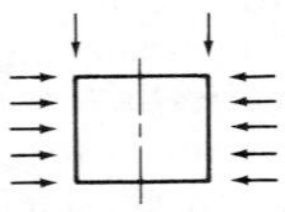	$p_x = p_E$ $p_z = p$	$-p_E x$	$-pr$	0
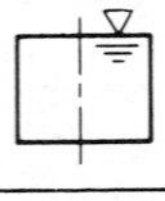	$p_z = -\gamma x$	0	$\gamma r x$	0
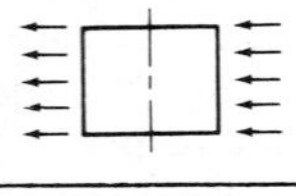	$p_z = p_w \cos\theta$	$p_w \dfrac{x^2}{2r} \cos\theta$	$-p_w\, r \cos\theta$	$-p_w\, x \sin\theta$
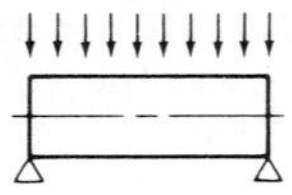	$p_y = -p_E \cos\theta$ $p_z = p_E \sin\theta$	$-p_E \dfrac{x}{r} (l - x) \sin\theta$	$-p_E\, r \sin\theta$	$-p_E (l - 2x) \cos\theta$
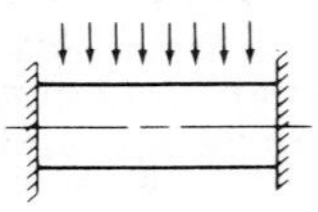	$p_y = -p_E \cos\theta$ $p_z = p_E \sin\theta$	$p_E \left[\dfrac{l^2}{6r} - \nu r - \dfrac{x}{r}(l - x)\right] \sin\theta$ ν = Poisson's Ratio	$-p_E\, r \sin\theta$	$-p_E (l - 2x) \cos\theta$

TABLE 14-6 Edge Corrections for a Closed Cylindrical Shell

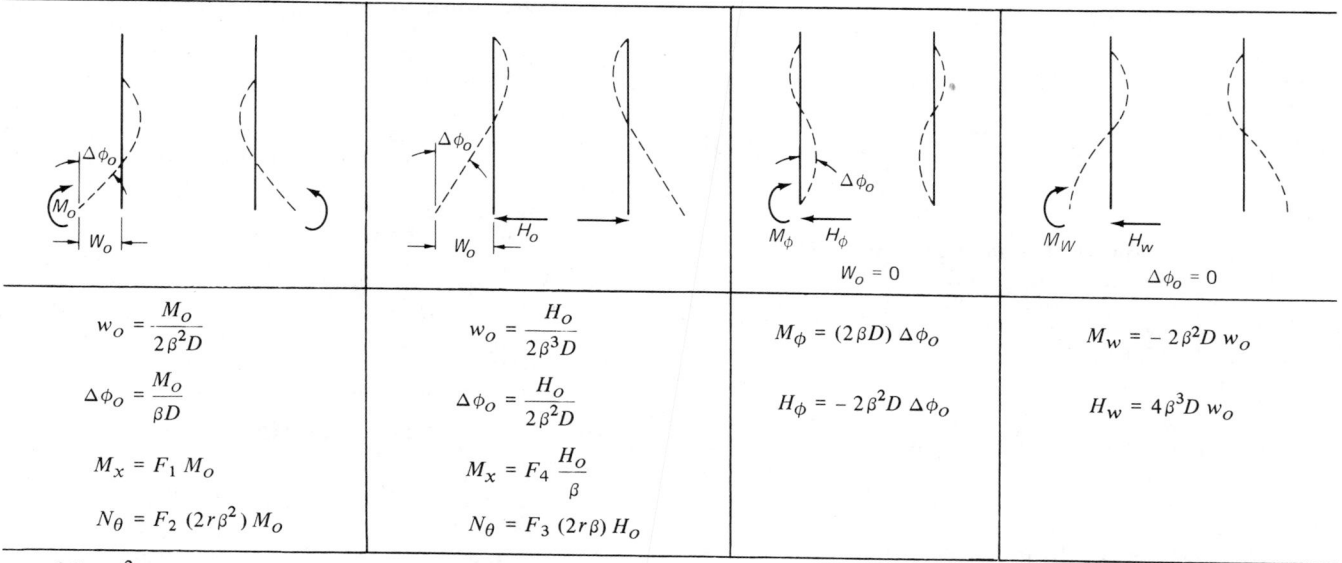

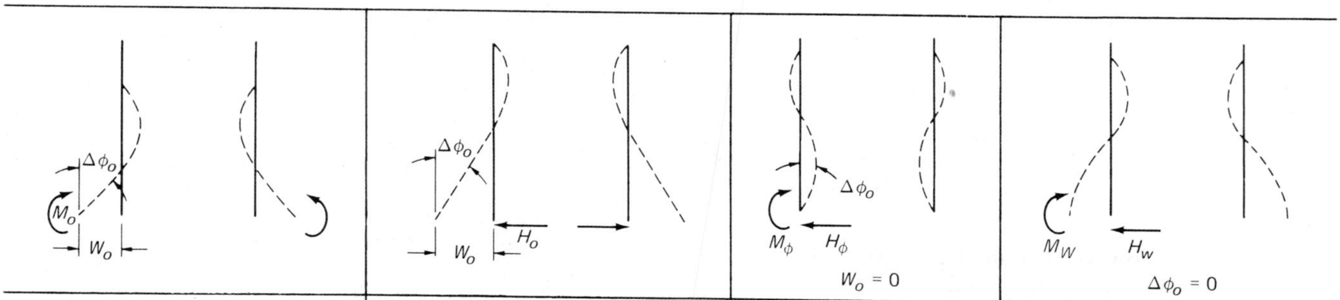

$w_o = \dfrac{M_o}{2\beta^2 D}$	$w_o = \dfrac{H_o}{2\beta^3 D}$	$M_\phi = (2\beta D)\,\Delta\phi_o$	$M_w = -2\beta^2 D\, w_o$
$\Delta\phi_o = \dfrac{M_o}{\beta D}$	$\Delta\phi_o = \dfrac{H_o}{2\beta^2 D}$	$H_\phi = -2\beta^2 D\,\Delta\phi_o$	$H_w = 4\beta^3 D\, w_o$
$M_x = F_1 M_o$	$M_x = F_4 \dfrac{H_o}{\beta}$		
$N_\theta = F_2\,(2r\beta^2)\,M_o$	$N_\theta = F_3\,(2r\beta)\,H_o$		

$\beta^4 = \dfrac{3(1-\nu^2)}{r^2 t^2}$ and F_1 to F_4 as defined in Table 14-4 with ϕ replaced by x.

shell elements can be applied. Because of the rapid damping out of edge effects for complete cylinders, the use of a flexibility or simple edge correction approach is normally more efficient than a stiffness method as done in the finite element approach.

14.3.3 Barrel Shells

Barrel shells are known to carry the loads applied to them by two types of actions. Because of the stiffness of an arch, the first tendency of the barrel is to try to transfer the load from the crown toward the edges by transverse arch action. Ultimately the load must be transferred longitudinally, and this is done by beam action. If the arch direction is short relative to the longitudinal span, the so-called long barrel, the transverse arch very quickly feels the absence of support along the longitudinal edge. For this case arching plays a minor role and ordinary, or shallow, beam theory is a good approximation (Fig. 14-9a). As the transverse arch increases relative to the longitudinal span, a short barrel results. Here transverse arching has equal importance to longitudinal beam action. The stress patterns that develop are similar to those of a deep beam (Fig. 14-9b).

The important stress resultants in the design of a barrel shell are: (1) the longitudinal stress, N_x, which is equivalent to the fiber stresses in a beam; (2) the membrane shear, $N_{x\theta}$, which is equivalent to the usual shear forces in a beam; (3) the hoop force, N_θ, significant primarily for short shells; and (4) the hoop moment, M_θ.

1. Beam Method. For the long barrel the assumption that plane sections remain plane is reasonable. Therefore, the linear variation of longitudinal stress, N_x, with vertical position in the cross section is assumed to exist. The application of ordinary beam theory for the determination of stresses in the cylinder was first presented by Lundgren. Chinn[14-7] prepared the general formulas, while Parme and Connor[14-30] adapted correction factors and tables to make the design of long-barrel shells a relatively easy task. The beam method can be used for symmetric shells under uniform load for the following cases:

(a) Single shells without edge beams, if the span to radius ratio, L/r, is greater than 5.
(b) Single shells with ordinary edge beams, if $L/r > 3$.
(c) Typical interior shell of a multiple shell system with butterfly edges, if $L/r > 2$.
(d) Typical interior shell of a multiple shell system with edge beams, if $L/r > 3$.

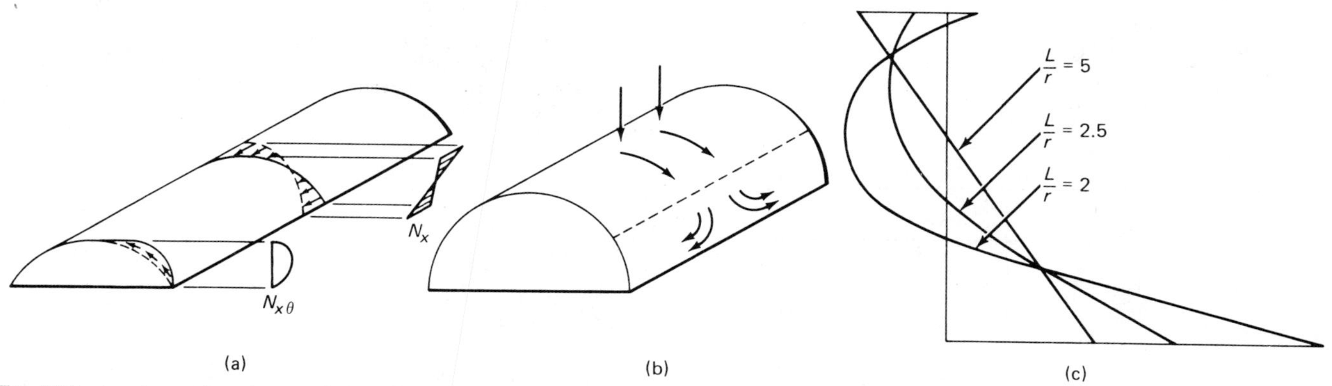

Fig. 14-9 Load-carrying characteristics of barrel shell. (a) Long barrel, ordinary beam theory; (b) short barrel, deep beam theory; and (c) longitudinal stress, N_x.

Computation of the longitudinal and shear stress is by standard beam equations:

$$N_x = \frac{Myt}{I} \tag{14-14}$$

$$N_{x\theta} = \frac{VQ}{Ib}\,t \tag{14-15}$$

where:

I = cross section moment of inertia (Table 14-7)
M = beam bending moment, $\frac{1}{8}\,WL^2$ or appropriate value
V = total shear on the cross section
b = thickness widths of shell at a horizontal section through the shell
y = distance from neutral axis to location of stress point

To compute the transverse hoop force and moments, the equations presented by Chinn or a tabular form presented in the example can be used. These tend to be laborious, so Parme and Connor developed Tables 14-8 for single and for typical interior shells for L/r ratios quoted above.

As the shell gets longer there is a tendency for the shell to flatten under uniform load. Table 14-8(b and c) allows this flattening to be corrected out of the solution.

For longitudinally continuous shells the above limits should be raised slightly. They are reasonable approximations for the intermediate regions between inflection points; however, the regions near the interior supports function like deep beams, since the spacing between inflection points there is small relative to the clear span.

In order for the $N_{x\theta}$ forces to exist as predicted by beam theory, a stiff diaphragm or transverse arch must be placed between the column supports. If these transverse members are not provided and what is commonly called a ribless shell is used, then the shell must provide its own transverse arch. Analysis methods to handle such cases are presented in Refs. 14-30 and 14-43.

In reinforcing long-barrel shells, the philosophy of the steel placement also follows that of the beam. Longitudinal reinforcement is provided to carry the total tensile, N_x, force and is placed as low near the longitudinal edge of the shell, or, if an edge beam is used, as deep in the beam as possible. Diagonal steel at approximately 45° is placed near the support to function as stirrups or bent-up bars as in a beam. Hoop reinforcement is provided for the transverse shell moments, M_θ.

If the barrel shell is too shallow, that is, if the span to depth ratio is outside the region considered good beam proportions, edge beams are added to form T-beam action. These edge beams can be prestressed if architectural requirements force a still-too-shallow section. The prestressing also serves to control deflections and cracking. If architectural considerations demand that no edge beams be used, so that feather or butterfly edges are employed, use of prestressing cables placed in the shell surface is highly desirable to control cracking and minimize the possibility of leakage.

2. *Direct Stiffness Method.* For the short and intermediate-length barrel shells, the analysis must utilize the solution of a complex differential equation system such as described in Ref. 14-2. Restricting consideration to only shells simply supported at the transverse ends, the theory of Ref. 14-19 was developed into a computer program described in Ref. 14-40. This program, entitled MULEL,[14-41] is available with instructions from ACI headquarters.

TABLE 14-7 Moment of Inertia

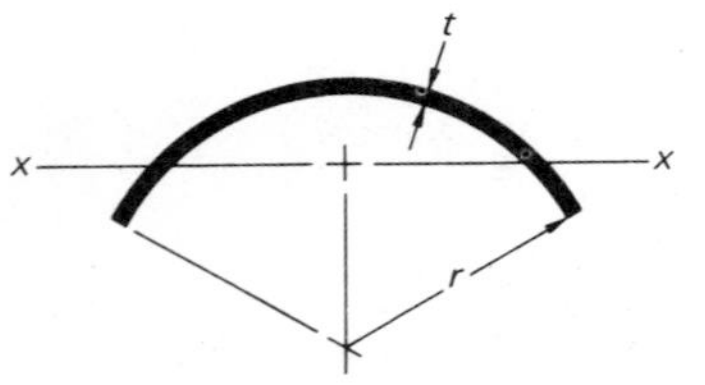

ϕ_k, deg	(11)	ϕ_k, deg	(11)
22.5	0.00041	37.5	0.00502
25.0	0.00068	40.0	0.00687
27.5	0.00110	45.0	0.01216
30.0	0.00168	50.0	0.02017
32.5	0.00249	55.0	0.03174
35.0	0.00358	60.0	0.04782

From Ref. 14-7.

EXAMPLE 14-3: A multiple barrel shell is to be used to span 50 ft. The loads for this shell, shown in Fig. 14-10, are:

Shell (P_d) (dead load plus roofing)	= 40 psf
Snow (P_u)	= 30 psf
Total	= 70 psf

The dead load is distributed uniformly on the surface, while the snow load is taken uniform on the horizontal projection. For this shell Table 14-8 applies.

SOLUTION: The ΔH correction is negligible. Therefore:

$$N_x = \frac{L}{r}(L)\,[p_u \text{ col } (1) + P_d \text{ col } (5)] = 100\,[30 \text{ col } (1) + 40 \text{ col } (5)]$$

$$N_{x\theta} = -L\,[p_u \text{ col } (3) + P_d \text{ col } (7)] = -50\,[30 \text{ col } (3) + 40 \text{ col } (7)]$$

$$M_\theta = r^2\,[p_u \text{ col } (4) + P_d \text{ col } (8)] = 625\,[30 \text{ col } (4) + 40 \text{ col } (8)]$$

The results are tabulated in Table 14-9.

The shear force, $N_{x\theta}$, is handled by steel placed on a 45° diagonal from the transverse supports down toward the valley of the shell. This means #5 @ 8 in. in the region of maximum shear.

The total tensile steel can be approximated by assuming linear variation from the valley to the neutral axis. The neutral axis is located at:

$$\bar{y} = \frac{r \sin \theta_k}{\theta_k} - r \cos \theta_k = 2.2225 \text{ ft}$$

from the valley edge. This represents a tension zone of 5.56 ft of arc from the valley. A conservative estimate of total tensile force is therefore:

$$T_s = \tfrac{1}{2}\,(5.56)(47.5) = 132 \text{ kips}$$

The total area of steel necessary is 6.6 sq in. (assuming an allowable stress of 20 ksi for the steel) placed as low in the shell as cover and spacing limits allow. The reinforcing is shown in Fig. 14-11.

EXAMPLE 14-4: The typical single shell with edge beams to be analyzed is shown in Fig. 14-12. The shell is to be analyzed for a loading of live load plus dead load of 50 psf of shell surface and the beam weight of 600 plf.

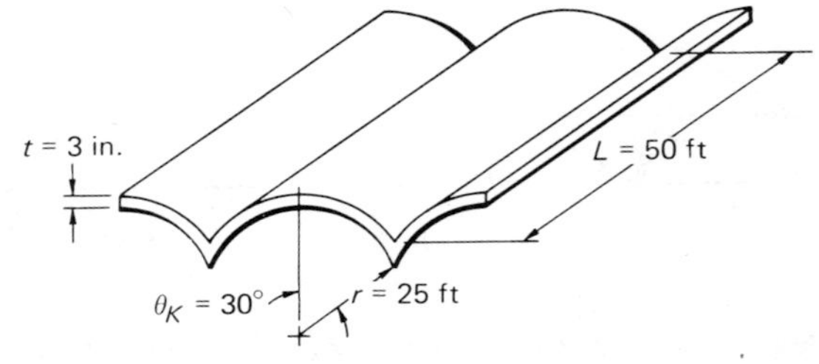

Fig. 14-10 Typical interior shell.

TABLE 14-8a Stresses in Barrel Shells from Vertical Loads*

$$N_x = \frac{L^2}{r}\,[p_u \text{ col (1)} + p_d \text{ col (5)}]$$

$$N_\theta = r\,[p_u \text{ col (2)} + p_d \text{ col (6)}] + \frac{p_u t^2}{6r}\,[\text{col (9)} \times \Delta H/(p_u r^2/Et)]$$

$$N_{x\theta} = -L\,[p_u \text{ col (3)} + p_d \text{ col (7)}]$$

$$M_\theta = r^2 [p_u \text{ col (4)} + p_d \text{ col (8)}] + \frac{p_u t^2}{6}\,[\text{col (10)} \times \Delta H/(p_u r^2/Et)]$$

Uniform transverse load

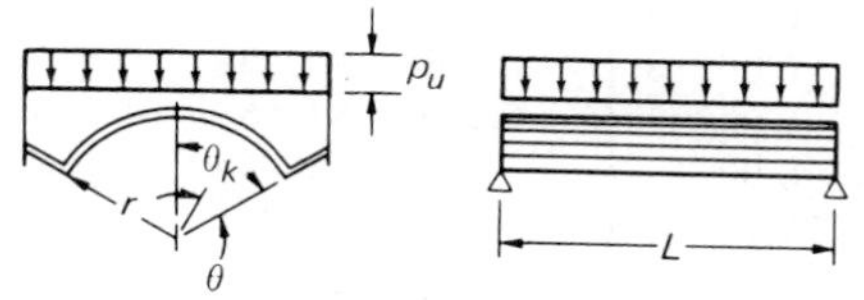

Dead weight load

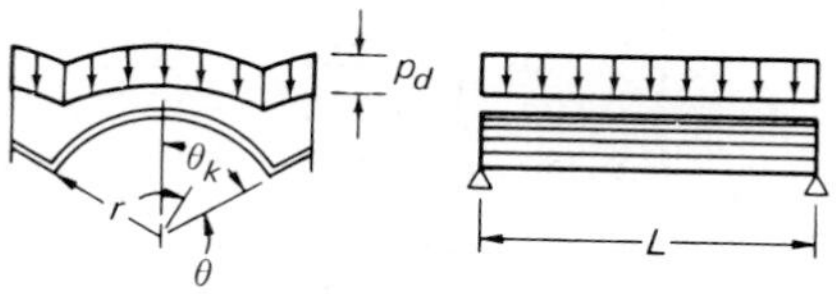

θ deg.	θ	N_x (1)	N_θ (2)	$N_{x\theta}$ (3)	M_θ (4)	N_x (5)	N_θ (6)	$N_{x\theta}$ (7)	M_θ (8)
	θ_k	-6.010	-1.411	0.000	-0.00292	- 6.167	-1.433	0.000	-0.00309
	$0.75\theta_k$	-4.875	-1.189	2.211	-0.00112	- 5.003	-1.205	2.269	-0.00118
22.5	$0.50\theta_k$	-1.482	-0.614	3.533	0.00232	- 1.521	-0.615	3.626	0.00245
	$0.25\theta_k$	4.137	0.049	3.084	0.00235	4.245	0.065	3.165	0.00249
	0	11.927	0.361	0.000	-0.00662	12.239	0.384	0.000	-0.00702
	θ_k	-4.855	-1.402	0.000	-0.00353	- 5.012	-1.430	0.000	-0.00378
	$0.75\theta_k$	-3.937	-1.182	1.985	-0.00135	- 4.064	-1.202	2.049	-0.00145
25.0	$0.50\theta_k$	-1.193	-0.612	3.170	0.00280	- 1.232	-0.613	3.273	0.00300
	$0.25\theta_k$	3.342	0.044	2.765	0.00282	3.451	0.064	2.855	0.00304
	0	9.617	0.347	0.000	-0.00797	9.929	0.374	0.000	-0.00857
	θ_k	-4.000	-1.393	0.000	-0.00417	- 4.158	-1.426	0.000	-0.00453
	$0.75\theta_k$	-3.242	-1.175	1.799	-0.00159	- 3.370	-1.199	1.869	-0.00173
27.5	$0.50\theta_k$	-0.980	-0.609	2.871	0.00331	- 1.018	-0.610	2.985	0.00360
	$0.25\theta_k$	2.755	0.038	2.503	0.00332	2.863	0.063	2.602	0.00363
	0	7.908	0.331	0.000	-0.00938	8.220	0.363	0.000	-0.01025
	θ_k	-3.350	-1.383	0.000	-0.00482	- 3.508	-1.422	0.000	-0.00533
	$0.75\theta_k$	-2.714	-1.166	1.643	-0.00183	- 2.842	-1.195	1.720	-0.00203
30.0	$0.50\theta_k$	-0.817	-0.606	2.622	0.00384	- 0.856	-0.607	2.746	0.00424
	$0.25\theta_k$	2.308	0.032	2.284	0.00383	2.417	0.061	2.392	0.00426
	0	6.608	0.314	0.000	-0.01082	6.920	0.352	0.000	-0.01204
	θ_k	-2.844	-1.372	0.000	-0.00548	- 3.002	-1.418	0.000	-0.00618
	$0.75\theta_k$	-2.303	-1.158	1.511	-0.00207	- 2.431	-1.191	1.595	-0.00235
32.5	$0.50\theta_k$	-0.691	-0.603	2.410	0.00438	- 0.729	-0.603	2.544	0.00492
	$0.25\theta_k$	1.960	0.026	2.098	0.00434	2.069	0.660	2.215	0.00492
	0	5.596	0.297	0.000	-0.01227	5.908	0.339	0.000	-0.01393
	θ_k	-2.442	-1.361	0.000	-0.00615	- 2.601	-1.414	0.000	-0.00707
	$0.75\theta_k$	-1.977	-1.148	1.397	-0.00232	- 2.105	-1.186	1.488	-0.00268
35.0	$0.50\theta_k$	-0.591	-0.599	2.227	0.00491	- 0.629	-0.600	2.372	0.00565
	$0.25\theta_k$	1.684	0.019	1.938	0.00484	1.793	0.058	2.064	0.00561
	0	4.794	0.278	0.000	-0.01370	5.105	0.326	0.000	-0.01591
	θ_k	-2.118	-1.349	0.000	-0.00679	- 2.278	-1.409	0.000	-0.00800
	$0.75\theta_k$	-1.714	-1.138	1.298	-0.00255	- 1.842	-1.181	1.396	-0.00302
37.5	$0.50\theta_k$	-0.510	-0.596	2.069	0.00544	- 0.548	-0.596	2.224	0.00640
	$0.25\theta_k$	1.461	0.012	1.798	0.00532	1.571	0.057	1.933	0.00632
	0	4.146	0.260	0.000	-0.01509	4.458	0.312	0.000	-0.01796
	θ_k	-1.853	-1.335	0.000	-0.00742	- 2.013	-1.404	0.000	-0.00897
	$0.75\theta_k$	-1.498	-1.127	1.211	-0.00277	- 1.627	-1.176	1.315	-0.00337
40.0	$0.50\theta_k$	-0.443	-0.592	1.929	0.00595	- 0.482	-0.592	2.095	0.00719
	$0.25\theta_k$	1.279	0.005	1.675	0.00578	1.389	0.055	1.819	0.00705
	0	3.616	0.241	0.000	-0.01641	3.928	0.297	0.000	-0.02006
	θ_k	-1.449	-1.307	0.000	-0.00853	- 1.610	-1.393	0.000	-0.01096
	$0.75\theta_k$	-1.170	-1.104	1.065	-0.00316	- 1.299	-1.165	1.183	-0.00408

TABLE 14-8a *Continued*

θ deg.	θ	N_x (1)	N_θ (2)	$N_{x\theta}$ (3)	M_θ (4)	N_x (5)	N_θ (6)	$N_{x\theta}$ (7)	M_θ (8)
45.0	$0.50\theta_k$	-0.343	-0.585	1.694	0.00688	- 0.381	-0.583	1.882	0.00883
	$0.25\theta_k$	1.001	-0.011	1.468	0.00657	1.112	0.052	1.630	0.00854
	0	2.809	0.202	0.000	-0.01872	3.120	0.266	0.000	-0.02437
50.0	θ_k	-1.160	-1.276	0.000	-0.00939	- 1.322	-1.380	0.000	-0.01301
	$0.75\theta_k$	-0.935	-1.079	0.947	-0.60344	- 1.065	-1.152	1.079	-0.00480
	$0.50\theta_k$	-0.271	-0.578	1.504	0.00762	- 0.308	-0.574	1.713	0.01054
	$0.25\theta_k$	0.802	-0.029	1.300	0.00714	0.914	0.049	1.481	0.01002
	0	2.232	0.164	0.000	-0.02042	2.543	0.234	0.000	-0.02871
55.0	θ_k	-0.946	-1.242	0.000	-0.00989	- 1.109	-1.367	0.006	-0.01506
	$0.75\theta_k$	-0.761	-1.053	0.849	-0.00358	- 0.892	-1.139	0.995	-0.00549
	$0.50\theta_k$	-0.217	-0.572	1.347	0.00807	- 0.255	-0.563	1.578	0.01277
	$0.25\theta_k$	0.655	-0.048	1.161	0.00742	0.767	0.045	1.360	0.01144
	0	1.805	0.128	0.000	-0.02130	2.115	0.201	0.000	-0.03293
60.0	θ_k	-0.783	-1.205	0.000	-0.00992	- 0.947	-1.352	0.000	-0.01705
	$0.75\theta_k$	-0.629	-1.025	0.766	-0.00355	- 0.761	-1.124	0.927	0.00613
	$0.50\theta_k$	-0.177	-0.566	1.213	0.00815	- 0.214	-0.552	1.467	0.01398
	$0.25\theta_k$	0.543	-0.068	1.043	0.00734	0.656	0.042	1.261	0.01275
	0	1.481	0.095	0.000	-0.02118	1.790	0.167	0.000	-0.03688

From Ref. (14-7).

TABLE 14-8b Effect of a Unit Horizontal Displacement

θ_k deg	θ	T_θ (9)	M_θ (10)
22.5	θ_k	2463.0	62.82
	$0.75\theta_k$	2451.1	50.96
	$0.50\theta_k$	2415.6	15.49
	$0.25\theta_k$	2356.9	- 43.34
	0	2275.5	-124.66
25.0	θ_k	1461.9	45.95
	$0.75\theta_k$	1453.2	37.26
	$0.50\theta_k$	1427.2	11.30
	$0.25\theta_k$	1384.3	- 31.63
	0	1324.9	- 91.02
27.5	θ_k	912.9	34.65
	$0.75\theta_k$	906.4	28.09
	$0.50\theta_k$	886.8	8.49
	$0.25\theta_k$	854.4	- 23.86
	0	809.8	- 68.50
30.0	θ_k	594.6	26.80
	$0.75\theta_k$	589.5	21.71
	$0.50\theta_k$	574.3	6.54
	$0.25\theta_k$	549.3	- 18.46
	0	514.9	- 52.86
32.5	θ_k	401.2	21.17
	$0.75\theta_k$	397.2	17.14
	$0.50\theta_k$	385.2	5.14
	$0.25\theta_k$	365.4	- 14.59
	0	338.4	- 41.66
35.0	θ_k	279.0	17.03
	$0.75\theta_k$	275.8	13.79
	$0.50\theta_k$	266.1	4.12
	$0.25\theta_k$	250.3	- 11.74
	0	228.6	- 33.43

θ_k deg	θ	T_θ (9)	M_θ (10)
37.5	θ_k	199.2	13.92
	$0.75\theta_k$	196.5	11.26
	$0.50\theta_k$	188.6	3.35
	$0.25\theta_k$	175.7	- 9.60
	0	158.0	-27.24
40.0	θ_k	145.5	11.53
	$0.75\theta_k$	143.3	9.32
	$0.50\theta_k$	136.7	2.76
	$0.25\theta_k$	126.0	- 7.96
	0	111.4	-22.50
45.0	θ_k	82.2	8.20
	$0.75\theta_k$	80.7	6.62
	$0.50\theta_k$	76.0	1.94
	$0.25\theta_k$	68.4	- 5.66
	0	58.2	-15.89
50.0	θ_k	49.6	6.06
	$0.75\theta_k$	48.4	4.88
	$0.50\theta_k$	44.9	1.41
	$0.25\theta_k$	39.3	- 4.19
	0	31.9	-11.65
55.0	θ_k	31.5	4.62
	$0.75\theta_k$	30.6	3.72
	$0.50\theta_k$	27.9	1.06
	$0.25\theta_k$	23.7	- 3.20
	0	18.1	- 8.81
60.0	θ_k	20.9	3.62
	$0.75\theta_k$	20.2	2.91
	$0.50\theta_k$	18.1	0.82
	$0.25\theta_k$	14.8	- 2.51
	0	10.5	- 6.84

TABLE 14-8c Membrane Horizontal Displacement for Uniform Loads

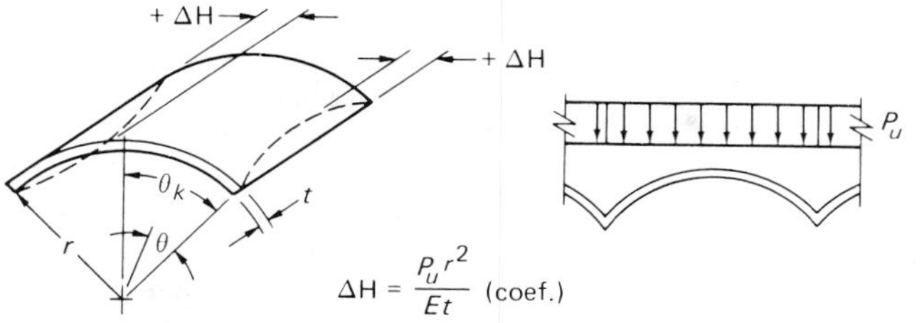

$$\Delta H = \frac{P_u r^2}{Et} \text{ (coef.)}$$

	r/L						
	0.100	0.125	0.150	0.175	0.200	0.225	0.250
θ_k deg	$\Delta H/(P_u r^2/Et)$						
22.5	− 91.83	− 38.37	− 18.88	− 10.37	− 6.16	− 3.87	− 2.52
25.0	− 123.80	− 51.79	− 25.54	− 14.09	− 8.42	− 5.33	− 3.52
27.5	− 161.58	− 67.67	− 33.43	− 18.49	− 11.09	− 7.07	− 4.71
30.0	− 205.26	− 86.02	− 42.55	− 23.59	−14.20	− 9.08	− 6.09
32.5	− 254.82	−106.85	− 52.91	− 29.38	−17.72	−11.38	− 7.66
35.0	− 310.10	−130.69	− 64.47	− 35.84	−21.66	−13.94	− 9.42
37.5	− 370.84	−155.63	− 77.17	− 42.94	−25.99	−16.77	−11.37
40.0	− 436.65	−183.31	− 90.94	− 50.65	−30.70	−19.83	−13.47
45.0	− 581.48	−244.21	−121.26	− 67.61	−41.05	−26.59	−18.13
50.0	− 739.50	−310.68	−154.34	− 86.14	−52.37	−33.98	−23.22
55.0	− 904.36	−380.03	−188.87	−105.47	−64.18	−41.70	−28.54
60.0	−1068.79	−449.20	−223.31	−124.77	−75.97	−49.41	−33.85

From Ref. 14-7.

SOLUTION: The analysis is performed using a beam method. The properties of the cross section are as listed in Table 14-10. The shell is divided into 12 segments, each of length 2.182 (Fig. 14-13). Properties of cross section determined for unit thickness then corrected. Values listed in the table are to the midpoint of each segment.

$$\delta \text{ Area} = \text{arc length (thickness)} = (2.182)(0.25) = 0.5455 \text{ sq ft}$$

$$\text{Area cross section} = 6\ (0.5455) + 4 = 7.273 \text{ sq ft}$$

$$\text{Centroid, } \bar{\eta}\text{: Area} \times \bar{\eta} = (\delta \text{ Area} \times \eta - 8) = 0.5455\ (13.3805) - 8$$

$$\text{Area} \times \bar{\eta} = -0.701 \qquad \bar{\eta} = -0.0964$$

It is desirable for the neutral axis to fall within the beam to minimize chances of flexural cracking causing leakage. The beam moment of inertia is:

$$I = \sum (y^2)\ \delta \text{ Area} + I_{0\text{Beam}} + A_{\text{Beam}} y^2 = 40.83 \text{ ft}^4$$

Load:

$$\text{Shell} + LL = 50\ (6)(2.182) = 654.6 \text{ plf}$$

$$\text{Beam} = 600 \text{ plf}$$

$$\text{Total} = 1254.6 \text{ plf}$$

Moment:

$$\tfrac{1}{8} WL^2 = \tfrac{1}{8}(1.255)(100)^2 = 1568 \text{ ft-kip}$$

$$N_{x\text{Crown}} = \frac{My}{I} t = \frac{1568}{40.83}(0.25)[25(1 - 0.866) + 0.096]$$

$$= 33.09 \text{ kip/ft}$$

$$N_{x\text{ Beam}} = \frac{1568}{40.83}(1)(3.9036) = 149.91 \text{ kip/ft}$$

The shear is computed at the center of each segment by using the unbalanced horizontal force. (Table 14-11)

Shear:

$$\frac{1}{2}\frac{L}{2} S = \sum \frac{M}{I} y\ \delta \text{ Area} \quad \therefore S = \frac{4}{L} \sum \frac{M}{I} y\ \delta \text{ Area}$$

$$S_{\text{seg }3} = \frac{4}{L}\left[292.416 - 13.007 - 32.316 - \frac{1}{2} 47.944\right] = 8.924$$

$$\Delta V' = \frac{dS}{dx} (\delta \text{ arc}) \sin \theta$$

TABLE 14-9 Stresses in Example 14-3 Shell

	@ $x = L/2$ N_x	@ $x = 0, L$ $N_{x\theta}$	@ $x = L/2$ M_θ
θ_k	−24,082	0	−223.6
$\frac{3}{4}\theta_k$	−19,510	−5904	−85.1
$\frac{1}{2}\theta_k$	−5875	−9425	+178.0
$\frac{1}{4}\theta_k$	16,592	−8210	+178.3
0	47,504	0	−503.9

TABLE 14-10 Section Properties of Example 14-4 Shell

Seg.	θ	Sin θ	Cos θ	η	y	y^2
Beam	30	0.5000	0.8660	0	−1.9036	3.6237
1	27.5°	0.4618	0.8870	0.5245	0.6209	0.3855
2	22.5°	0.3827	0.9239	1.4462	1.5426	2.3796
3	17.5°	0.3007	0.9537	2.1922	2.2886	5.2377
4	12.5°	0.2164	0.9763	2.7568	2.8532	8.1408
5	7.5°	0.1305	0.9914	3.1353	3.2317	10.4439
6	2.5°	0.0436	0.9991	3.3255	3.4219	11.7094
Totals				13.3805		38.2969

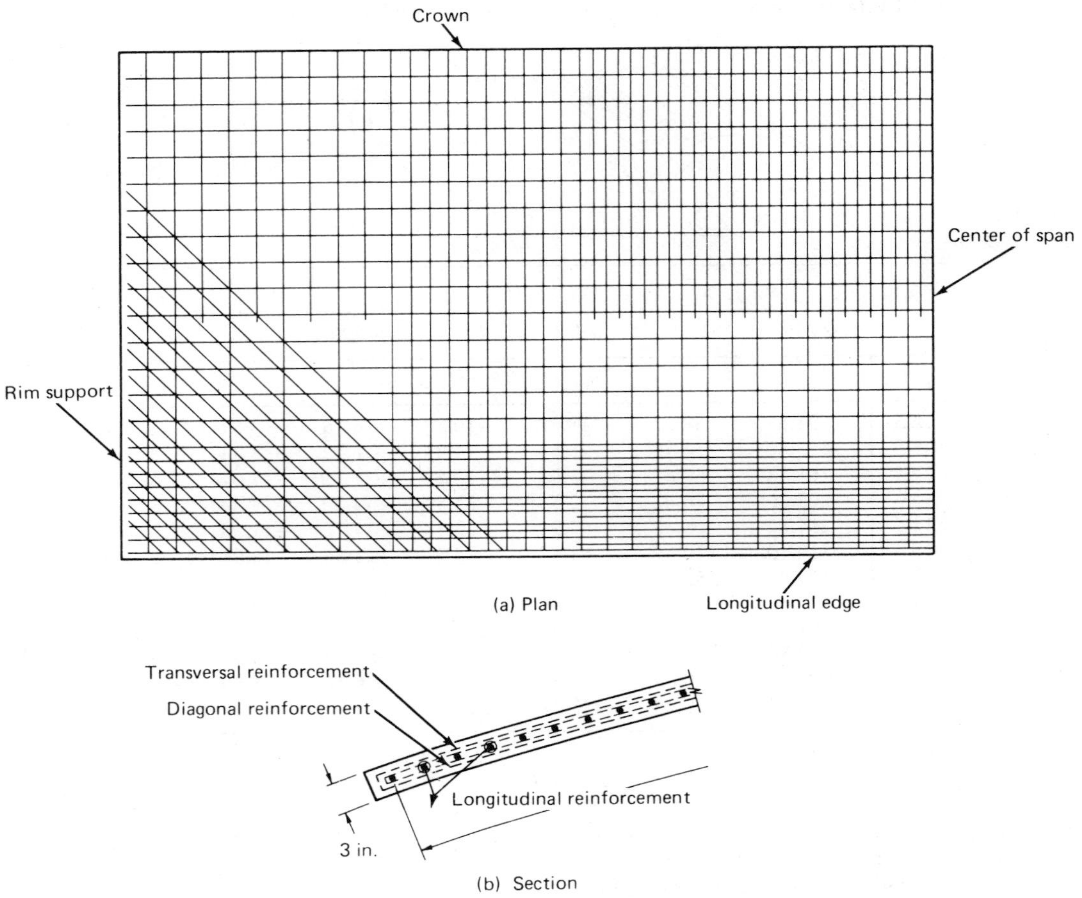

Fig. 14-11 Reinforcing.

Since $\Delta V'$ does not sum up to total vertical load of 1.2546 klf, a new shear ΔV is calculated by: (Table 14-11)

$$\Delta V = \frac{1.2546}{1.2457} \Delta V'$$

The magnitude of the loading within one segment is:

$$(50)(2.182) = 0.109 \text{ kip}$$

This is added to ΔV to find V.

With the net horizontal and vertical forces per unit length established, it is possible to compute shell moment M_θ. In order to compute the transverse moment M_θ, it is necessary to assume moment conditions at the junction of the shell and the beam. If the torsional moment of the beam is negligible and its resistance to horizontal forces is negligible, the moment starts at zero, and the system is determinate. Multiplying the V- and the H-forces of Table 14-11 by the moment arms Δx and Δy of Table 14-12, and then summing yields the values shown in the tabulation of M_θ.

If the beams along the edge were very stiff against rotation and/or horizontal movement, or, if the shell was a typical interior shell, the above method would still be applicable. Using the above-computed M_θ, compute M_θ/EI, then the horizontal displacement and rotation of the edge with M_θ/EI as load. Apply corrective forces as in any typical flexibility approach. The moment at the junction of the shell and the beam is like the fixed-end moment in a beam. It is not necessary for statics but occurs from compatibility. The magnitude depends upon how much reinforcement is provided. Examples in the literature range from the equivalent of full-fixity to pin-supported.

Three networks of reinforcing are necessary for this shell (see Fig. 14-14):

(a) Longitudinal steel to carry the usual beam tension force. Total area of steel:

$$\tfrac{1}{2}(149.91)(3.9036)/20 = 14.63 \text{ sq in.}$$

Use 19 #8 bars places as low in beam as possible, i.e., (4, 4, 4, 4, 3) (allowable steel stress assumed @ 20 ksi).

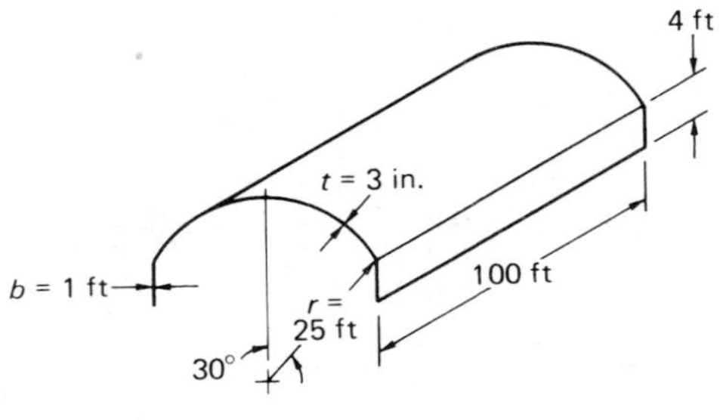

Fig. 14-12 Single shell.

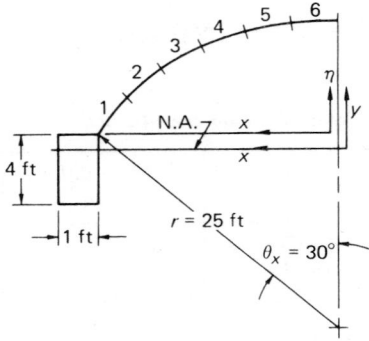

Fig. 14-13 Cross section.

TABLE 14-11 Increment Forces

Seg.	$\frac{M}{I} y\ \delta\ Area$	*Shear, S*	$\frac{dS}{dx}\ \delta\ Arc$	$\Delta V'$	ΔV	*Total V* ΔV + *Load*	*H*
Beam	292.416	8.920 11.696		0.6317	0.6362	0.0362	
1	-13.007	11.436	0.4991	0.2305	0.2321	0.1231	0.4427
2	-32.316	10.529	0.4595	0.1758	0.1771	0.0681	0.4245
3	-47.944	8.924	0.3894	0.1171	0.1179	0.0089	0.3714
4	-59.772	6.770	0.2954	0.0639	0.0644	-0.0446	0.2884
5	-67.701	4.220	0.1842	0.0240	0.0242	-0.0848	0.1826
6	-71.685	1.433	0.0625	0.0027	0.0027	-0.1063	0.0624
Total	+0.009			1.2457	1.2546	~0	

(b) Shear steel placed in shell running diagonally from the transverse support down to the beam:

Steel @ seg.

1	2	3	4
#5 @ 6	#5 @ 6	#5 @ 8	#5 @ 8

Use #4 @ 12.

(c) Steel placed in the hoop direction to reinforce against the hoop moment M_θ:

This steel should be placed as the top layer. It should be placed over the full length of the shell.

14.4 TRANSLATIONAL SHELLS

One form of shell used almost exclusively for roof systems is the translational shell. There have been cases where these shells have been used as foundation structures. This category of shell classification contains probably the largest variety of different shell surfaces. Principal shells of the translational type are the elliptical parabola, hyperbolic parabola, and the conoid. One feature that explains the popularity of these shells is the range and variety of appearances that can be achieved with the same basic shell configurations.

TABLE 14-12 Increment Distances

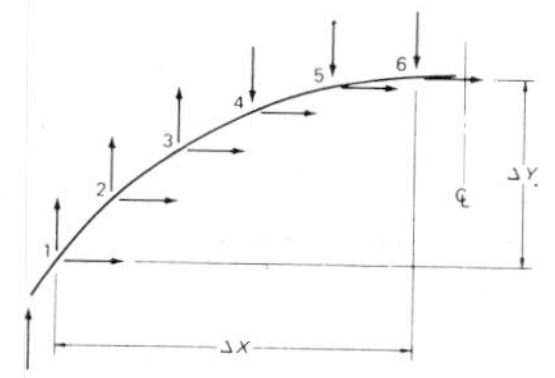

a) Horizontal Distance, Δx

To / *From*	6	5	4	3	2	1	*x*
Edge	11.4095	9.2368	7.0890	4.9822	2.9330	0.9562	12.5
1	10.4533	8.2806	6.1328	4.0260	1.9768		11.5438
2	8.4765	6.3038	4.1560	2.0492			9.5670
3	6.4273	4.2546	2.1068				7.5178
4	4.3205	2.1478					5.4110
5	2.1727						3.2632
6							1.0905

b) Vertical Distance, Δy

To / *From*	6	5	4	3	2	1	*x*
Edge	5.3255	5.1353	4.7568	4.1922	3.4462	2.5245	-1.9036
1	2.8010	2.6108	2.2323	1.6677	0.9217		0.6209
2	1.8793	1.6891	1.3106	0.7460			1.5426
3	1.1333	0.9431	0.5646				2.2886
4	0.5687	0.3785					2.8532
5	0.1902						3.2317
6							3.4219

M_θ at center of segment

Seg	1	2	3	4	5	6	Units
M_θ	0.0346	-0.0585	-0.2394	-0.4409	-0.6072	-0.7001	ft-kips/ft

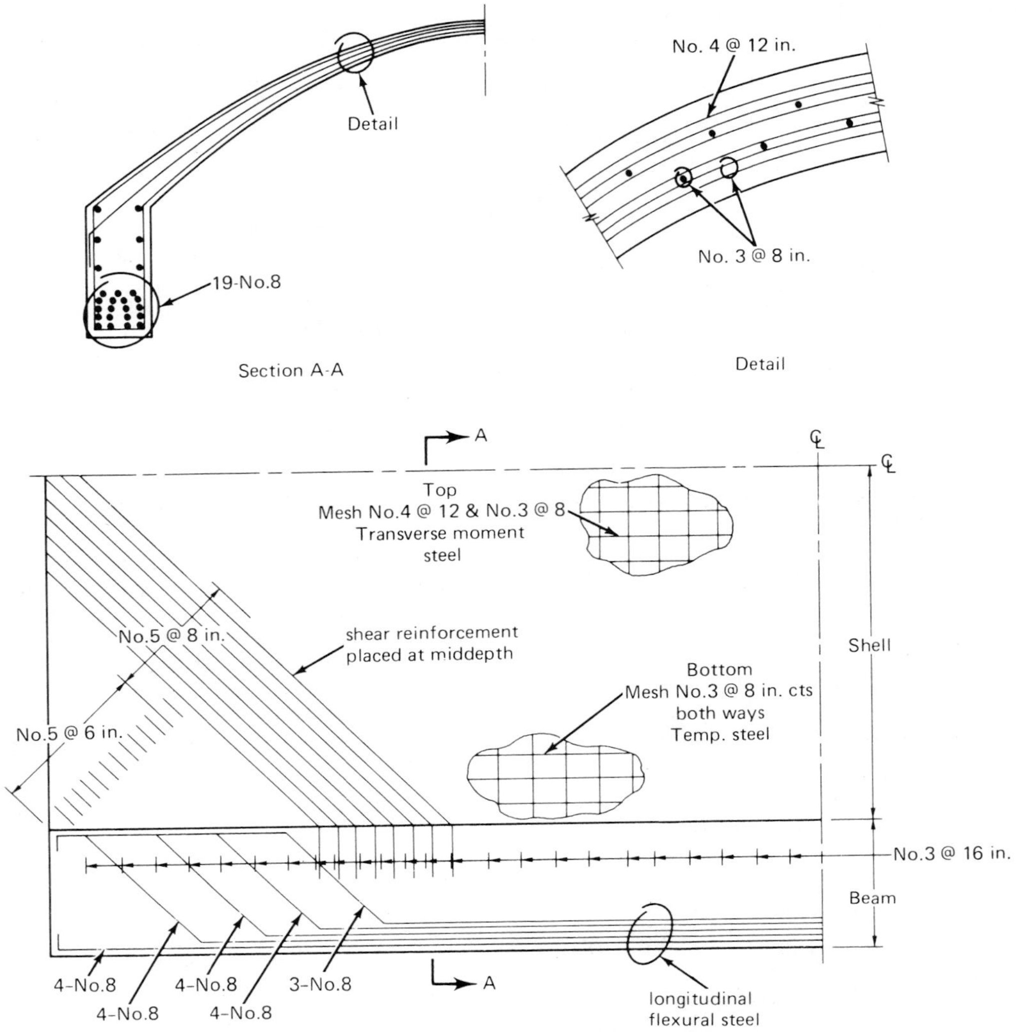

Fig. 14-14 Sketch of main reinforcing of cylindrical shell of Example 14-4.

14.4.1 Introduction

The basic construction of the shells in this category is to take a curve fixed in space and then translate a second curve along the first. This sweeps out the desired shell surface. Depending upon the selection of curves, the surface can have positive, zero, or negative curvature. The surface thus can have zero, one, or two systems of real asymptotic lines.

14.4.2 Elliptical Parabola

1. Surface Equation. This shell surface is formed by taking two identical parabolic arches, setting them in parallel planes a given distance apart, and placing a third parabolic arch to span the distance between the origin arches. This third parabola is then moved along the origin parabolas, sweeping out the continuous surface shown in Fig. 14-15. The equation of the surface is the sum of the equations of the parabolas:

$$z = h_x\left(\frac{x}{a}\right)^2 + h_y\left(\frac{y}{b}\right)^2 \tag{14-16}$$

where h_x and h_y are the rise of the arches in the x- and y-directions, respectively, and $2a$ and $2b$ are the spans of the parabolas in the x- and y-directions. The surface thus formed has a domelike appearance.

2. Membrane Theory. Structurally the elliptical parabola is a very efficient shell. This shell carries load primarily by membrane stresses as long as there is some form of support along its edges. The shell functions as a system of interconnecting arches that transfer the load toward the edges

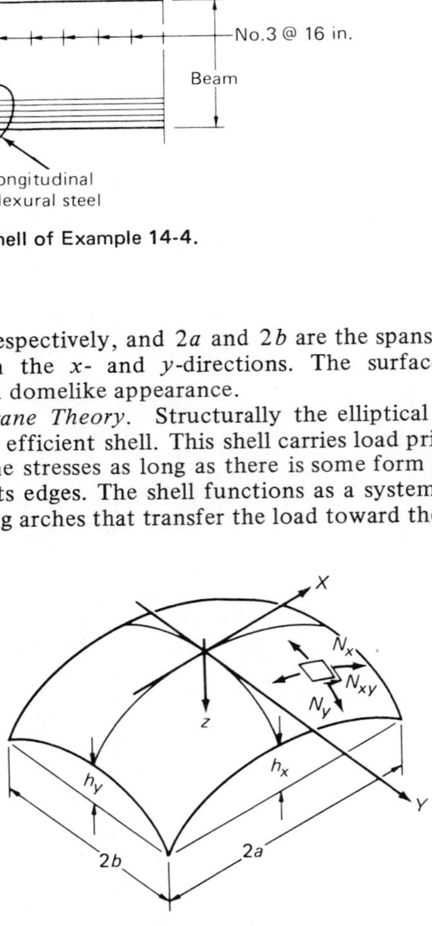

Fig. 14-15 Elliptical parabola.

of the shell. The membrane equation specialized to the elliptical parabola loaded by a vertical load is:

$$\frac{2h_y}{b^2}\frac{\partial^2 F}{\partial x^2} + \frac{2h_x}{a^2}\frac{\partial^2 F}{\partial y^2} = p \qquad (14\text{-}17)$$

where F is a stress function set by the internal forces. One solution of this equation was achieved by Parme[14-29] in the form of a series of trigonometric and hyperbolic functions. The series solution for F is then integrated to determine the stress resultants, which are expressible as:

$$N_x = -\frac{pa^2}{h_x}k\left[\text{function I of } \frac{h_x}{h_y}, \frac{x}{a}, \frac{y}{b}\right]$$

$$N_y = -\frac{pb^2}{kh_y}\left[\text{function II of } \frac{h_x}{h_y}, \frac{x}{a}, \frac{y}{b}\right] \qquad (14\text{-}18)$$

$$N_{xy} = -\frac{pab}{\sqrt{h_x h_y}}\left[\text{function III of } \frac{h_x}{h_y}, \frac{x}{a}, \frac{y}{b}\right]$$

where:

$$k = \sqrt{\frac{1 + 4h_x^2\left(\dfrac{x}{a^2}\right)^2}{1 + 4h_y^2\left(\dfrac{y}{b^2}\right)^2}}$$

and:

p = vertical load uniformly distributed over the horizontal projection

Tables 14-13 and 14-14 give the numerical values for the functions in the brackets evaluated at several points on the shell for various h_x/h_y ratios.

The membrane theory presumes edge-supporting members in the plane of the edge while flexible normal to that plane. Such a supporting edge is obtainable, for example, by a truss. Frequently, for architectural reasons, the edge beams are tied arches, thus departing some from membrane boundaries.

The membrane solution results in just shear forces existing around the boundary of the shell surface. These shear forces accumulate as direct forces in the edge-supporting members.

3. Edge Corrections. Because the edge of the membrane shell and its supporting members want to deflect in the vertical plane in different amounts, some flexure develops in the region of the edge. It has been shown that this flexure region closely resembles that of a beam on a stiff, elastic foundation. Thus the application of the Geckeler approximation, discussed earlier for shells of revolution and closed cylinders, is a reasonable means to assess the bending magnitude and penetration into the shell.

The application of edge shear and moment to the shell along the edge $x = \pm a$ results in a deflection and rotation of:

$$w = -2\beta^2\frac{\Gamma_y^2}{Et}M_{xo} - 2\beta\frac{\Gamma_y^2}{Et}Q_{xo} + \frac{p\Gamma_y^2}{Et}$$

$$\frac{dw}{dx} = 2\beta^2\frac{\Gamma_y^2}{Et}Q_{xo} + 4\beta^3\frac{\Gamma_y^2}{Et}M_{xo} \qquad (14\text{-}19)$$

where Γ_y is the radius of curvature of the shell along the edge and the remaining terms are as defined in Section 14.2.4. Compatibility demands that these displacements match the corresponding quantities for the supporting member. Since there is a significant difference in stiffness of the support members and the shell, conservative assumptions are to consider the edge clamped in the region near the column support while simply supported over the region of midspan of the edge. Near the column support, therefore, $w = dw/dx = 0$, and we find:

$$Q_{xo} = \frac{p}{\beta} \approx pC$$

$$M_{xo} = -\frac{p}{2\beta^2} \approx -\frac{1}{2}pC^2 \qquad (14\text{-}20)$$

where $C = \dfrac{1}{\beta} = 0.76\sqrt{tr_y}$. The penetration of the moment is $M_x = M_{xo}F_2$. In the midspan region:

$$Q_{xo} = \frac{P}{2\beta} \approx \frac{1}{2}pC$$

$$M_{xo} = 0 \qquad (14\text{-}21)$$

The penetration of the bending into the shell is obtained by using these values of Q_{xo} and M_{xo} in the equations of Table 14-6 with $Q_{xo} = H_o$. Thus the positive moment is:

$$M_x = F_4\frac{p}{2\beta^2} = \frac{1}{2}pC^2F_4 \qquad (14\text{-}22)$$

where F_4 is found from Table 14-4 by replacing $\beta\phi$ by βy, as was done for the cylinder.

It should be remembered these stress resultants are necessary for compatibility but not equilibrium.

4. Buckling Considerations. Since this shell form carries the load primarily through compressive arching forces, buckling of the shell surface must be considered (see Section 14.2.5). A conservative estimate of the buckling pressure[14-36] is:

$$p_{cr} = 0.06\frac{Et^2}{\Gamma_x\Gamma_y} \qquad (14\text{-}23)$$

A generous factor of safety should be applied to this pressure to assure that imperfections, creep displacements, and other such effects do not result in premature shell buckling.

If the boundary conditions represent a significant departure from the diaphragm supports of the membrane theory, a nonlinear bending solution using the finite element procedure is recommended.

5. Bending Solutions. Sharp departures from the membrane theory result if edge supports do not receive adequate horizontal restraint at the column heads. The shell acts as a very flexible cross-sectioned curved beam. Reliable stress evaluations can only be obtained from solutions to the equations of shallow shell theory. Numerical procedures such as finite element present the most direct solution procedure. This is most easily achieved using one of the general-purpose programs such as NASTRAN, STRUDL, SAP, FINITE, or other such programs.

EXAMPLE 14-5: Given an elliptical parabola spanning a 100 × 80 ft rectangular plan. The rise is 20 ft with $h_x = h_y = 10$ ft. Load is 70 psf uniform on the horizontal projection. The maximum variation of k is from 0.89 to 1.08. Assume $k = 1$.

SOLUTION:

$$N_x = -\frac{70(50)^2}{10}\text{(Table Coeff.)} = -17{,}500\text{ (Coeff.)}$$

$$N_y = -\frac{70(40)^2}{10}\text{(Table Coeff.)} = -11{,}200\text{ (Coeff.)} \qquad (14\text{-}24)$$

$$N_{xy} = \frac{70(50)(40)}{10}\text{(Table Coeff.)} = -14{,}000\text{ (Coeff.)}$$

Computed maximum normal stress resultants are

$$N_x = -8750\text{ lb/ft @ } x = 0, \quad y = 40\text{ ft}$$

$$N_y = -5600\text{ lb/ft @ } x = 50, \quad y = 0$$

TABLE 14-13 Coefficients for Computing Force Components of Elliptical Paraboloid Shell*

x/a	Force component	*Value of y/b*									
		(a) $h_x/h_y = 1.0$					(d) $h_x/h_y = 0.8$				
		0	0.25	0.50	0.75	1.0	0	0.25	0.50	0.75	1.0
0.00	N_y	0.250	0.233	0.182	0.101	0	0.289	0.270	0.213	0.119	0
	N_x	0.250	0.267	0.318	0.399	0.500	0.211	0.230	0.287	0.381	0.500
	N_{xy}	0	0	0	0	0	0	0	0	0	0
0.25	N_y	0.267	0.250	0.199	0.111	0	0.304	0.285	0.228	0.130	0
	N_x	0.233	0.250	0.301	0.389	0.500	0.196	0.215	0.272	0.370	0.500
	N_{xy}	0	0.029	0.068	0.096	0.108	0	0.034	0.069	0.100	0.114
0.50	N_y	0.318	0.301	0.250	0.150	0	0.347	0.331	0.277	0.169	0
	N_x	0.182	0.199	0.250	0.350	0.500	0.153	0.169	0.223	0.331	0.500
	N_{xy}	0	0.068	0.140	0.210	0.244	0	0.065	0.139	0.215	0.255
0.75	N_y	0.399	0.389	0.350	0.250	0	0.416	0.406	0.369	0.270	0
	N_x	0.101	0.111	0.150	0.250	0.500	0.084	0.094	0.131	0.230	0.500
	N_{xy}	0	0.096	0.210	0.356	0.465	0	0.091	0.201	0.353	0.480
1.0	N_y	0.500	0.500	0.500	0.500	0	0.500	0.500	0.500	0.500	0
	N_x	0	0	0	0	0	0	0	0	0	0
	N_{xy}	0	0.108	0.243	0.465	∞		0.101	0.229	0.443	∞
		(b) $h_x/h_y = 0.6$					(c) = $h_x/h_y = 0.4$				
0.00	N_y	0.336	0.316	0.252	0.143	0	0.395	0.374	0.307	0.180	0
	N_x	0.164	0.184	0.248	0.357	0.500	0.105	0.126	0.103	0.320	0.500
	N_{xy}	0	0	0	0	0	0	0	0	0	0
0.25	N_y	0.348	0.329	0.267	0.555	0	0.403	0.383	0.319	0.192	0
	N_x	0.152	0.171	0.233	0.345	0.500	0.097	0.117	0.181	0.308	0.500
	N_{xy}	0	0.031	0.067	0.103	0.120	0	0.026	0.060	0.101	0.125
0.50	N_y	0.383	0.367	0.312	0.197	0	0.425	0.410	0.357	0.235	0
	N_x	0.117	0.133	0.188	0.304	0.500	0.075	0.090	0.143	0.265	0.500
	N_{xy}	0	0.060	0.132	0.216	0.265	0	0.049	0.115	0.208	0.274
0.75	N_y	0.436	0.426	0.392	0.296	0	0.459	0.451	0.419	0.331	0
	N_x	0.064	0.074	0.108	0.204	0.500	0.041	0.049	0.081	0.169	0.500
	N_{xy}	0	0.081	0.185	0.342	0.494	0	0.065	0.156	0.316	0.506
1.00	N_y	0.500	0.500	0.500	0.500	0	0.500	0.500	0.500	0.500	0
	N_x	0	0	0	0	0	0	0	0	0	0
	N_{xy}	0	0.089	0.208	0.413	∞	0	0.070	0.173	0.363	∞
		(c) $h_x/h_y = 0.2$									
0.00	N_y	0.462	0.446	0.388	0.248	0					
	N_x	0.038	0.054	0.112	0.252	0.500					
	N_{xy}	0	0	0	0	0					
0.25	N_y	0.465	0.451	0.396	0.261	0					
	N_x	0.035	0.049	0.104	0.239	0.500					
	N_{xy}	0	0.014	0.040	0.088	0.128					
0.50	N_y	0.473	0.462	0.414	0.303	0					
	N_x	0.027	0.038	0.086	0.197	0.500					
	N_{xy}	0	0.027	0.074	0.174	0.280					
0.75	N_y	0.485	0.480	0.456	0.383	0					
	N_x	0.015	0.020	0.044	0.117	0.500					
	N_{xy}	0	0.034	0.098	0.246	0.510					
1.00	N_y	0.500	0.500	0.500	0.500	0					
	N_x	0	0	0	0	0					
	N_{xy}	0	0.038	0.108	0.262	∞					

From Ref. 14-25.

TABLE 14-14 Shear Along the Edges of Elliptical Paraboloid Shell*

y/b	h_x/h_y				
	1.0	0.8	0.6	0.4	0.2
			At $x = \pm a$		
0.0	0.0000	0.0000	0.0000	0.0000	0.0000
0.1	0.0419	0.0389	0.0342	0.0307	0.0137
0.2	0.0854	0.0793	0.0701	0.0550	0.0286
0.3	0.1319	0.1231	0.1096	0.0872	0.0481
0.4	0.1836	0.1721	0.1546	0.1254	0.0731
0.5	0.2432	0.2294	0.2081	0.1728	0.1075
0.6	0.3204	0.3066	0.2859	0.2493	0.1818
0.7	0.4071	0.3897	0.3627	0.3173	0.2296
0.8	0.5363	0.5178	0.4887	0.4400	0.3443
0.85	0.6279	0.6090	0.5791	0.5292	0.4306
0.9	0.7570	0.7378	0.7074	0.6667	0.5659
0.95	0.9777	0.9582	0.9276	0.8763	0.7741
1.0	∞	∞	∞	∞	∞
x/a			At $y = \pm b$		
0.0	0.0000	0.0000	0.0000	0.0000	0.0000
0.1	0.0419	0.0444	0.0468	0.0488	0.0500
0.2	0.0854	0.0903	0.0950	0.0990	0.1014
0.3	0.1319	0.1391	0.1460	0.1519	0.1553
0.4	0.1836	0.1930	0.2019	0.2095	0.2140
0.5	0.2432	0.2545	0.2652	0.2743	0.2798
0.6	0.3204	0.3317	0.3425	0.3516	0.3571
0.7	0.4071	0.4213	0.4348	0.4463	0.4532
0.8	0.5363	0.5515	0.5659	0.5782	0.5855
0.85	0.6279	0.6434	0.6582	0.6707	0.6782
0.9	0.7570	0.7728	0.7878	0.8005	0.8081
0.95	0.9777	0.9935	1.0087	1.0215	1.0290
1.0	∞	∞	∞	∞	∞

**From Ref.* (14-29).

The membrane shear for points along the edge $x = a$, and the principal tension stress:

$$N = \frac{N_y}{2} + \sqrt{\left(\frac{N_y}{2}\right)^2 + (N_{xy})^2} \qquad (14\text{-}25)$$

at the same points are given in Table 14-15.

The approximate area of steel to be placed on a diagonal line between adjacent edges is also shown in the table. A minimum reinforcement of at least 0.0018 bt should be provided (ACI 318, Section 7.12.2.1).

The edge bending effects are found from eqs. (14-20) and (14-21). The radius of curvature is:

$$R_x = \frac{\left[1 + \left(\dfrac{\partial z}{\partial x}\right)^2\right]^{3/2}}{\partial^2 z/\partial x^2} \qquad (14\text{-}26)$$

For the edge $y = b$, the curvature at the crown ($x = 0$) is 125 ft, while at the support the curvature is 156 ft ($x = a$). Therefore, the maximum bending moment near the crown is:

$$M_x = \tfrac{1}{2} p(0.76)^2 \, tR_xF_2 = 0.16(70)(0.76)^2 \, (0.25)(125)$$
$$= 202 \text{ ft-lb/ft}$$

The maximum negative edge moment is:

$$M_x = -\tfrac{1}{2} p(0.76)^2 \, tR_x = -0.5(70)(0.76)^2 \, (0.25)(156)$$
$$= -788 \text{ ft-lb/ft}$$

Steel to sustain these moments should be provided. The tie force that must be sustained at the corners can be found using Simpson's integration on the normal stress resultants at the crown (using the N_x values computed from Table 14-13, $x/a = 0$):

$$P_{tie} = \frac{1}{3}\frac{40}{4}\,[4375 + 8750 + 4(4672 + 6982) + 2(5565)]$$
$$= 236{,}200 \text{ lb}$$

14.4.3 Hyperbolic Parabola

1. *Surface Equation.* The equation defining the hyperbolic paraboloid surface is:

$$z = h_y\left(\frac{y}{b}\right)^2 - h_x\left(\frac{x}{a}\right)^2 \qquad (14\text{-}26)$$

when the coordinate lines are aligned with the principal curvatures. This surface is generated by translating a parabola spanning in the y-direction along a parabola spanning in the x-direction. The y-parabola is concave downward, while the x-parabola is concave upward.

If the coordinate lines are selected to match the asymptotic or straight characteristic lines of the surface, the defining equation is:

$$z = k\,xy \qquad (14\text{-}27)$$

where $k = f/ab$. This equation applies regardless of the angle of intersection ω_1 of the asymptotic lines. Figure 14-16 shows the shell formed in this manner.

To be effective the shell should have the product kt (warping times the thickness) not less than 0.003. Some shells have been built with this ratio as low as 0.002, but then care must be exercised that the shell is not subjected to adverse loading.

The hyperbolic parabola is one of the most versatile of the shell forms, being adaptable into the shapes shown in Fig. 14-17. Additional, more complex arrangements are shown in Fig. 14-18.

2. *Membrane Theory.* With the surface defined by eq. (14-27), one membrane equilibrium equation relates the shear force directly as a function of the loading:[14-6, 14-42]

$$N_{xy} = \left(\frac{y}{2}\,p_x + \frac{x}{2}\,p_y - p_z\,\frac{ab}{2f}\right)\sin\omega \qquad (14\text{-}28)$$

where p_x, p_y, and p_z are the load intensities in the x-, y-, and z-directions. The shear force is determined by the load intensity at the point and by the warping geometry of the shell surface. The only way to change the magnitude of the shear force is to change the geometry of the shell.

TABLE 14-15 Membrane Shear and Principal Tension Stress Along Edge $x = a$

y/b	0	0.1	0.2	0.3	0.4	0.5	0.6	0.7	0.8	0.85	0.9	0.95
N_{xy}	0	−587	−1196	−1847	−2570	−3405	−4486	−5699	−7508	−8791	−10598	−13688
N	0			554	1001	1609	2488	3550	5210	6430	8510	11160
A_s								#4 @ 12	#4 @ 9	#4 @ 7	#4 @ 5	#4 @ 4

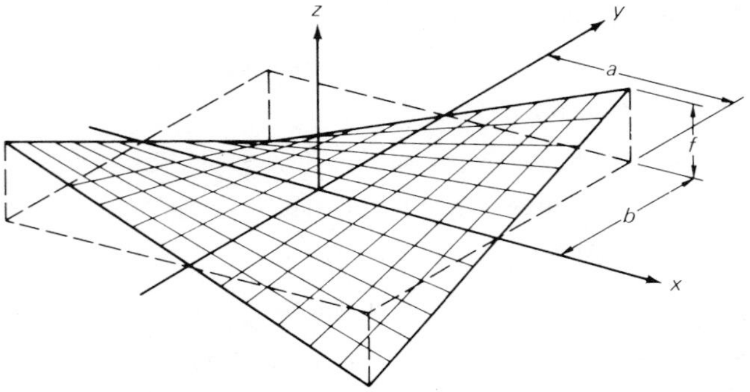

Fig. 14-16 Geometry of hyperbolic paraboloid, $z = kxy$ ($k = f/ab$).

The normal forces are determined from:

$$N_x = -\int \left(\frac{\partial N_{xy}}{\partial y} + p_x \sin \omega \right) dx + f_1(y) \quad (14\text{-}29)$$

$$N_y = -\int \left(\frac{\partial N_{xy}}{\partial x} + p_y \sin \omega \right) dy + f_2(x) \quad (14\text{-}30)$$

These direct forces are the true normal forces only for an ω of 90°. If the axes are nonorthogonal, these are forces parallel to the axes directions.

For a shallow shell under a constant vertical load, N_{xy} is essentially constant; $N_{xy} = p_z ab/2f$. The direct forces, N_x

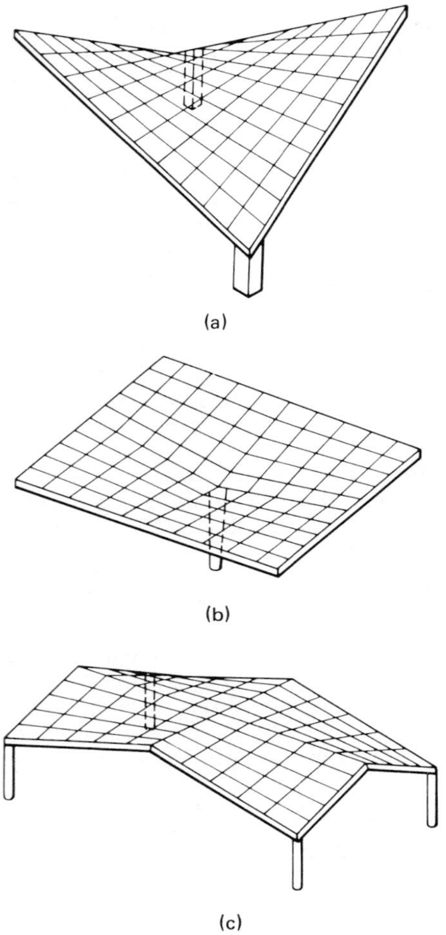

Fig. 14-17 Some fundamental configurations of hyperbolic paraboloids. (a) Saddle shell; (b) inverted umbrella; (c) hipped roof.

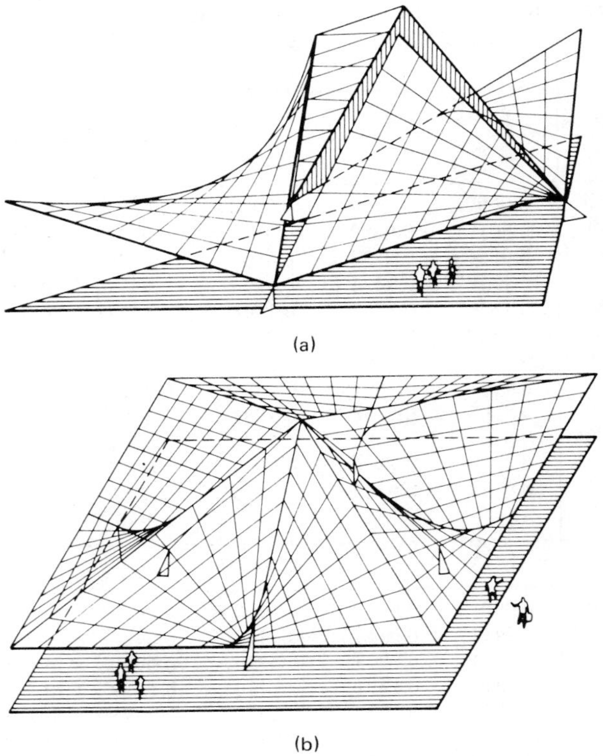

Fig. 14-18 More complex surface formed from simple panels. (a) Three-panel shell; (b) eight-panel shell.

and N_y, depend upon the boundary restraint and are normally negligible. The principal stresses for this case are:

$$N_1 = -N_2 = p_z \frac{ab}{2f} \quad (14\text{-}31)$$

if ω is 90°. In this case they represent a system of intersecting arches: a parabolic compression arch under uniform stress and a draping, or inverted, parabolic arch in uniform tension.

The membrane theory predicts the existence of shear forces around the boundary of the shell surface and assumes edge members capable of accepting these forces as a reaction (Fig. 14-19). These shear forces accumulate in the edge members as axial forces

$$N_A = \int N_{xy} dx + N_{corner} \quad (14\text{-}32)$$

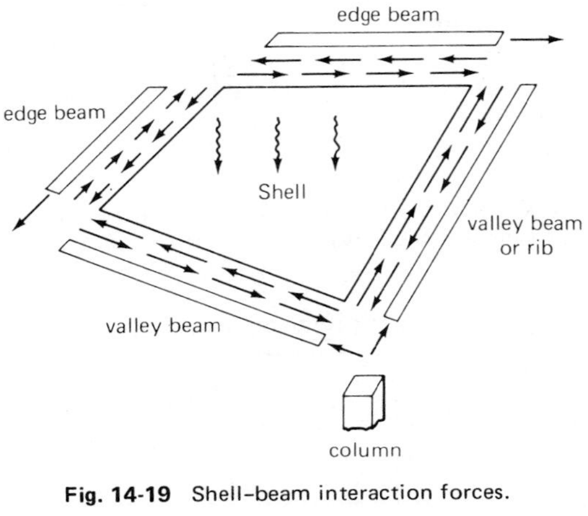

Fig. 14-19 Shell–beam interaction forces.

where N_{corner} is the initial axial force existing in the edge member at its beginning. This initial axial force is normally negligible.

If the loads on the shell vary significantly over the shell surface, or if the shell has a very small k-value, then a bending analysis is necessary.

The membrane stresses for various load conditions are given in Table 14-16.

3. Unsymmetrical Load. An unsymmetrical load on a hyperbolic shell cannot be resisted solely by membrane forces, since such a theory predicts a sharp change in forces in the supporting members. For example, the inverted umbrella of Fig. 14-17(b) would have an axial force of magnitude N in the edge member of the loaded panel. Just beyond that point, in the unloaded panel the continuation of the same edge member has no load. To account for such unbalanced loads, Ref. 14-46 suggests that ribs between the loaded panels be designed as cantilever beams for a load at least one half the unbalanced load.

4. Edge Corrections. An edge correction similar to that discussed for elliptical parabolas can also be applied to the hyperbolic parabola (H.P.), but this correction must be re-

TABLE 14-16 Membrane Stresses for Hyperbolic Paraboloid

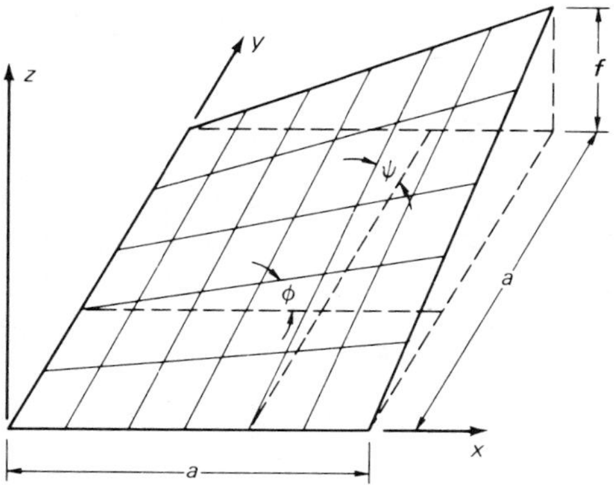

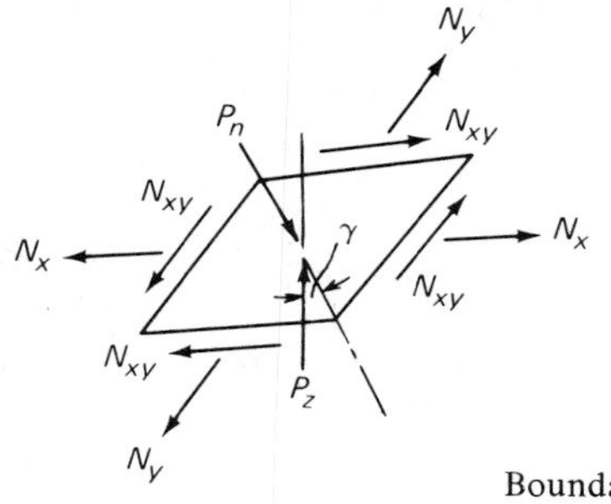

Equation of the surface

$$z = xy/n; \; n = a^2/f$$

N_x, N_y Unit normal forces
N_{xy} Unit central shear
$P_n P_z$ Surface load components along the normal and the z axis

Boundary Conditions: $x = 0, N_x = 0, y = 0, N_y = 0$

System	*Loading*	N_x	N_y	N_{xy}
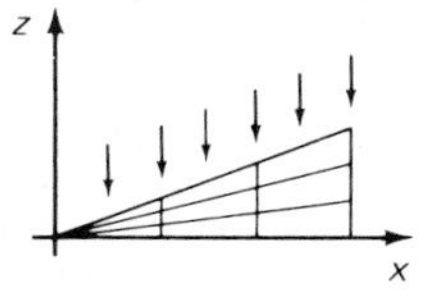	$p_z = -p_E$	$-p_E \frac{y}{2} \ln \frac{x+\sqrt{x^2+y^2+n^2}}{\sqrt{y^2+n^2}} \cdot \frac{\cos\psi}{\cos\phi}$	$-p_E \frac{x}{2} \ln \frac{y+\sqrt{x^2+y^2+n^2}}{\sqrt{x^2+n^2}} \cdot \frac{\cos\phi}{\cos\psi}$	$\frac{p_E}{2}\sqrt{x^2+y^2+n^2}$
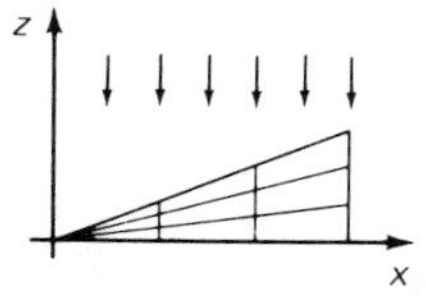	$p_z = p_s \cos\gamma$	0	0	$p_s \frac{a^2}{2f}$
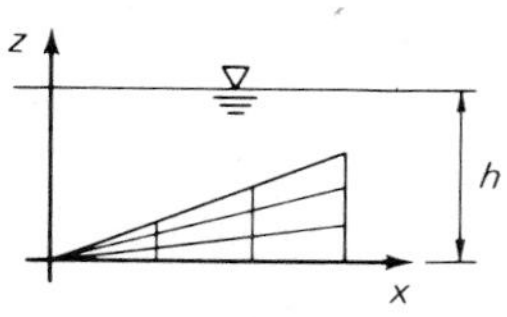	$p_n = \gamma\left(h - \frac{xy}{n}\right)$	$\frac{\gamma}{2n^2}\left[\frac{x^4}{4}+\frac{x^2}{2}(5y^2+n^2) - 4nhxy\right]\frac{\cos\psi}{\cos\phi}$	$\frac{\gamma}{2n^2}\left[\frac{y^4}{4}+\frac{y^2}{5}(5x^2+n^2) - 4nhxy\right]\frac{\cos\phi}{\cos\psi}$	$\frac{\gamma}{2n}\left(h-\frac{xy}{n}\right)\cdot(x^2+y^2+n^2)$
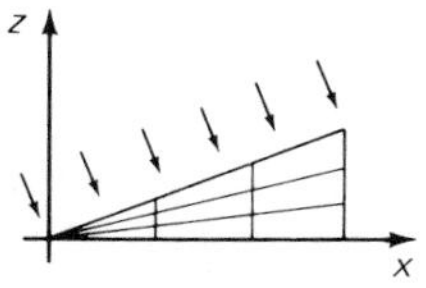	$p_n = p$	$-p\frac{2xy}{n}\frac{\cos\psi}{\cos\phi}$	$-p\frac{2xy}{n}\frac{\cos\phi}{\cos\psi}$	$p\frac{x^2+y^2+n^2}{2n}$
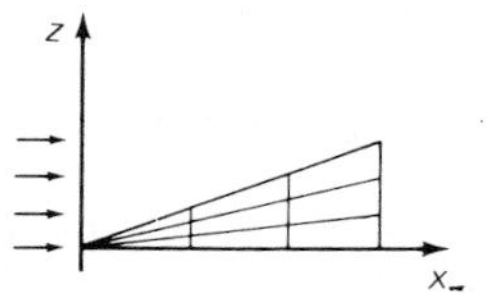	$p_n = p_w \frac{y^2}{x^2+y^2+n^2}$	$-p_w\frac{y}{n}\left(x+\frac{y^2}{y^2+n^2}\cdot\arctan\frac{x}{\sqrt{y^2+n^2}}\right)\frac{\cos\psi}{\cos\phi}$	$-p_w\frac{x}{y}\left(y-\sqrt{x^2+n^2}\cdot\arctan\frac{y}{\sqrt{x^2+n^2}}\right)\frac{\cos\phi}{\cos\psi}$	$p_w\frac{y^2}{2n}$

stricted to moderate and small shells. This procedure is particularly useful in determining the bending moments in the shell in the region near a rib separating loaded panels.

A bending analysis is particularly important to establish the stresses in the supporting beams. The supporting member should initially be sized to sustain the forces predicted by membrane theory, but then a bending analysis must be performed (see Section 14.4.3.6). For most H.P. shells the maximum axial force in the supporting beams is usually close to that predicted by membrane theory. This is not true of the ridge beam of a hipped roof H.P., however. In the hipped-roof configuration of Fig. 14-17(c), the crown beams do not receive the axial force but only one-half to one-third that predicted by membrane theory.[14-37] Furthermore bending from its dead weight is important.

5. Buckling Considerations. The buckling load of a hyperbolic parabola has been investigated by Reissner,[14-34] who predicted the critical load for a simply supported H.P. as:

$$p_{cr} = 2E(kt)^2/\sqrt{3(1-\nu^2)} \qquad (14\text{-}33)$$

This equation is adequate to predict the buckling load of H.P. shells, provided account is taken of imperfections, and creep deformations on the geometry. The supporting members can be designed against buckling by ignoring any stabilizing influence of the shell. The membrane shear load causing buckling of the edge member is:

$$N_{xy,\,\text{crit}} = 18.95\,\frac{EI}{a^3} \qquad (14\text{-}34)$$

For shells covering larger spans, even as little as 100 × 100 ft, more complete studies should be performed to properly establish the possibility of buckling.

6. Bending Analysis. Although membrane theory has been found to be adequate for small- to moderate-span H.P. shells, bending increases in importance as the spans of the shell increase. This is particularly true for the supporting or edge members. The shell and the edge members should no longer be considered as separate structural members. The weight of the supporting members can have a significant effect upon stresses.[14-38] The critical aspect in the design of the structure becomes the design of the support members. Design of the edge member to carry its own weight is an impossible task. The shell assists in carrying some of this weight. To evaluate the bending stresses in the supporting members requires a complex computer program utilizing finite element,[14-32] finite difference,[14-27] or other similar numerical method. Reference 14-8 provides a discussion of the available computer methods.

EXAMPLE 14-6: A hyperbolic paraboloid in the form of an inverted umbrella has a plan of 40 × 50 ft with a rise of 5 ft. Shell thickness is 3.5 in. Average thickness including edge beams is approximately 4.5 in.; $f'_c = 3000$ psi; $f_s = 20{,}000$ psi; loading D.L. + L.L. + roofing is 80 psf.

SOLUTION: Stresses in the umbrella are:

$$N_{xy} = \frac{pab}{2f} = \frac{(80)(20)(25)}{2(5)} = 4000 \text{ lb/ft}$$

Required tensile reinforcement in the directions parallel to the edges:

$$A_s = \frac{N_1}{f_s} = \frac{1}{2} \times \frac{4000}{20000} \times 1.414 = 0.14 \text{ in.}^2 \text{ both ways}$$

Shell reinforcement: use #3 @ 9 cts. both ways (0.146 in.^{2})

Outer edge beams:

$$T = 4000 \times 25 = 100{,}000 \text{ lb}$$

which requires:

$$A_s = \frac{100000}{20000} = 5 \text{ in.}^2 \qquad (4\text{-}\#10 \text{ or } 5\text{-}\#9)$$

For a larger shell it might be desirable to prestress the edge beams to eliminate the tensile stresses.

Valley beam designed as tied column:

Max comp.:

$$c = 2(4000)(25)\,\frac{\sqrt{25^2 + 5^2}}{25} = 204{,}000 \text{ lb}$$

assuming $p_g = 0.01$:

$$A_g = \frac{c}{0.8(.225f'_c + p_g f_s)} = \frac{204{,}000}{0.8(.225 \times 3000 + 0.01 \times 20{,}000)}$$

$$= 291 \text{ in.}^2$$

Required reinforcement:

$$0.01(291) = 2.91 \text{ in.}^2 \qquad \text{Use 5 \#7}$$

with #3 ties @ 12 in. cts. with 16 × 18 section at column head. For an unsymmetrical load on one-half the shell (one-half the design live load of 20 psf assumed):

$$\text{Moment} = \tfrac{1}{2}\,p_L(2ab)\left(\frac{b}{2}\right) = \frac{1}{2}\,p_L ab^2$$

$$= \tfrac{1}{2}(20)(20)(25)^2 = 125{,}000 \text{ ft-lb or } 1{,}500{,}000 \text{ in-lb}$$

Required:

$$wd^2 = M/K = 1{,}500{,}000/236 = 6356 \text{ in.}^3$$

Use $w = 16$ in.; then:

$$d^2 = 6356/w = 397$$

Increase d to:

$$d = 20 \text{ in.}$$

$$A_s = M_s/f_s jd = 1{,}500{,}000/(20{,}000 \times 0.866 \times 20) = 4.33 \text{ in.}^2$$

requires another 4 #8 and 2 #7. Use 4 #8; and 2 #7 at the top with 5 #7 in the bottom.

Check secondary stresses per Ref. 14-11.

Cantilever moment in edge beams:

$$M = WL/3 \quad \text{with } L = 80 \times .707 = 56.56 \text{ in.}$$

$$M = 80 \text{ psf} \times \frac{1}{2}\,\frac{80 \times 80}{144} \times \frac{56.56}{3} = 33{,}600 \text{ in.-lb}$$

Check 12 × 12 perimeter beam with 4 #6 bars:

$$M_c = Kwd^2 = 2 \times 236 \times 12 \times 10^2 = 566{,}000 \text{ in.-lb}$$

$$M_s = A_s f_s jd = 2 \times 0.88 \times 20{,}000 \times 0.866 \times 10$$

$$= 304{,}000 \text{ in.-lb} \qquad \text{adequate}$$

Check punching shear around column head.

7. Groined Vault. One particularly interesting form of the hyperbolic paraboloid is the groined vault configuration shown in Fig. 14-20. This arrangement allows the designer to have a stress-free edge. Therefore a thin trim edge can be exposed to the observer, and the dramatic thinness of this form of construction can be displayed. The roof acts as a system of cross arches with the valleys acting as the primary structure to get the loads to the supports. Tables for stress evaluation are given in Ref. 14-13. Also, Ref. 14-6 presents an analysis procedure capable of handling the membrane stresses even for deep groins.

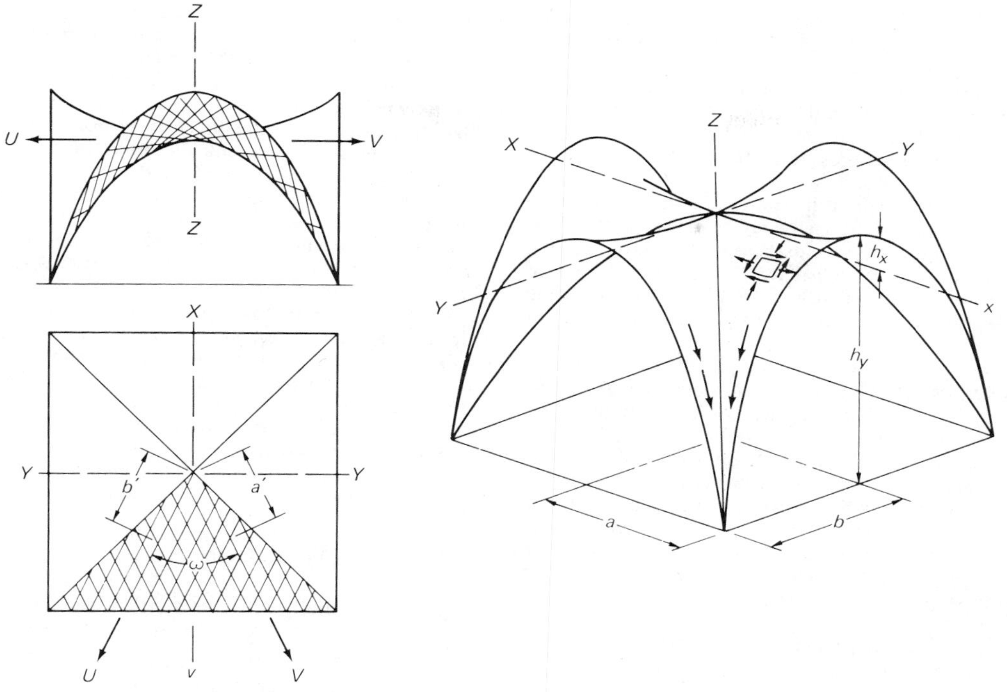

Fig. 14-20 Groined vault.

14.5 COMPUTER METHODS

Currently a number of computer programs are available for solving shell problems as elastic structure problems. An elastic solution is the commonly accepted basis for determining stresses, displacements, and stability of thin reinforced concrete shells. The majority of the available programs utilize the finite-element method. A large variety of element choices exist. This is partly the result of a lack of consensus as to what constitutes an adequate shell element. Most current-generation finite-element programs offer several shell element alternatives. Flat elements normally must be triangular, and absolutely must be if used for general shells. Rectangular elements are possible only for cylinders. Flat elements are developed by combining plane stress elements with plate bending elements. If no shell elements exist in the program, the analyst can achieve the same effect by a simple overlaying placement of the two types of elements over the grid pattern selected for the structure. Although, for shells, curved elements generally are preferable to flat elements, the fact that the element is curved does not mean that the element model is a geometric reproduction of the real shell. Up to a point, the higher the number of degrees of freedom per element, usually the more efficiently that element performs the stress analysis. However, elements with more than three nodes in any direction are not to be recommended because of the increased computational effort to solve the equations. The quadratic (midside node) isoparametric element is a good curved shell element to use. Flat triangular elements such as in SAP function adequately. The Semiloof and Lagrangian shell elements[14-17] are excellent elements but are not readily accessible because they are not in most major element programs. To be useful to the shell designer, the program must contain the ability to include beam members placed eccentric to the shell. The analysis of the supporting beam members is normally the critical aspect of the structure design.

Because of the reluctance of some finite elements to deform into the deflection shapes that some shells bend into, the accuracy of any finite-element solution should be checked with static checks on gross equilibrium. The suitability of the elements to the type of shell problem should be verified. Many elements are based upon shallow-shell theory and should not be applied to problems well outside that class. Extreme care should be exercised in interpreting results in regions of rapid changes in curvature, or at abrupt changes in direction of the surfaces' tangent planes, such as groins, etc.

If two opposite edges of the shell can be considered simply supported, the most efficient solution is by the direct stiffness method described in Refs. 14-40 and 14-41. MULEL[14-40] is such a program written specially for cylindrical shells.

14.6 PROPORTIONING AND REINFORCEMENT

Shell thickness is frequently not based on strength considerations but rather on construction, stability, code, or reinforcing steel cover requirements. Three inches is about the minimum. Four inches may be required if the enclosed environment has high humidity as, for example, over a swimming pool.

Stress concentrations resulting from abrupt changes in thickness should be avoided. Gradual thickening should be used.

Guidelines on procedures for reinforcing shells were reported by the ACI committee in their "Practice and Commentary."[14-10] Because of the prominent role that cracking plays in the serviceability of shells, the determination of the reinforcement requirements is still done on a working stress basis. The adequacy of this approach is based on experience. In rare instances, the use of factored loads with a limit state determination of carrying capacity as described by Gioncu[14-20] is possible. These guidelines are:

403. Shell reinforcement

(a) The stress in the reinforcement may be assumed at the allowable value independently of the strain in the concrete.

(b) Where the tensile stresses vary greatly in magnitude over the shell, as in the case of cylindrical shells, the reinforcement capable of resisting the total tension may be concentrated in the region of maximum tensile stress. Where this is done, the percentage of crack control reinforcing in any 12 in. width of shell shall be not less than 0.35 percent throughout the tensile zone.

(c) The principal tensile stresses shall be resisted entirely by reinforcement.

(d) Reinforcement to resist the principal tensile stresses, assumed to act at the middle surface of the shell, may be placed either in the general direction of the lines of principal tensile stress (also referred to as parallel to the lines of principal tensile stress), or in two or three directions. In the regions of high tension it is advisable, based on experience, to place the reinforcing in the general direction of the principal stress.

(e) The reinforcement may be considered parallel to the lines of principal stress when its direction does not deviate from the direction of the principal stress more than 15 deg. Variations in the direction of the principal stress over the cross section of a shell due to moments need not be considered for the determination of the maximum deviation. In areas where the stress in the reinforcing is less than the allowable stress a deviation greater than 15 deg can still be considered parallel placing; a stress decrease of 5 percent shall be considered to compensate for each additional degree of deviation above 15 deg. Wherever possible, such reinforcing may run along lines considered most practical for construction, such as straight lines.

(f) Where placed in more than one direction, the reinforcement shall resist the components of the principal tension force in each direction.

When the reinforcement is placed in two directions at right angles to each other, the principal stress resultant N_p is

$$N_p = f_{sy}(A_{sy} \cos^2 \delta + A_{sx} \sin^2 \delta \tan \delta) \quad (14\text{-}35)$$

where δ is the angle measured in a counterclockwise direction from the face on which N_x acts (angle from the x-axis to the principal stress plane),

$$\tan 2\delta = \frac{2N_{xy}}{N_x - N_y} \quad (14\text{-}36)$$

A_{sy} is the area of steel in the y-direction, A_{sx} that in the x-direction, f_{sy} is the steel stress in the y-direction. The steel stress in the x-direction is:

$$f_{sx} = f_{sy} \tan \delta \quad (14\text{-}37)$$

(g) In those areas where the computed principal tensile stress is the concrete exceeds 300 psi, placement of at least one layer of the reinforcing shall be parallel to the principal tensile stress, unless it can be proven that a deviation of the reinforcing from the direction parallel to the lines of principal tensile stress is permissible because of the geometrical characteristics of a particular shell and because for reasons of geometry only insignificant and local cracking could develop.

(h) Where the computed principal tensile stress (psi) in the concrete exceeds the value $2\sqrt{f_c'}$ (where f_c' is also in psi), the spacing of reinforcement shall not be greater than three times the thickness of the thin shell. Otherwise the reinforcement shall be spaced at not more than five times the thickness of the thin shell, nor more than 18 in.

(i) Minimum reinforcement shall be provided as required in the Building Code (ACI 318) even where not required by analysis.

(j) The percentage of reinforcement in any 12 in. width of shell shall not exceed $30 f_c'/f_s$. However, the maximum percentage shall not exceed 6 percent if $f_s = 20{,}000$ psi, 5 percent if $f_s = 25{,}000$ psi, or 4 percent if $f_s = 30{,}000$ psi when the latter values are acceptable. If the deviation of the reinforcing from the lines of principal stress is greater than 10 deg, the maximum percentage shall be one-half of the above values.

(k) Splices in principal tensile reinforcement shall be kept to a practical minimum. Where necessary they shall be staggered with not more than one-third of the bars spliced at any one cross section. Bars shall be lapped only within the same layer. The minimum lap for shell reinforcing bars, where draped, shall be 30 diameters with a 1 ft 6 in. minimum unless more is required by the Building Code (ACI 318), except that the minimum may be 12 in. for reinforcement not required by analysis. The minimum lap for welded wire fabric shall be 8 in. or one mesh, whichever is greater, except that Building Code requirements shall govern where the wire fabric at the splice must carry the full allowable stress.

(l) The computed stress of the shell reinforcing at the junction of shell and supporting member or edge member shall be developed by anchorage within or beyond the width of the member.

(m) Reinforcement to resist bending moments shall be proportioned and provided in the conventional manner with proper allowance for the direct forces.

(n) The concrete cover over reinforcement at surfaces protected from weather and not in contact with the ground shall be at least 1/2 in. for bars (3/8 in. when precast), 3/8 in. for welded wire fabric, and 1 in. for prestressed tendons. In no case shall the cover be less than the diameter of bar, prestressed tendon, or duct.

(o) If greater concrete cover is required for fire protection, such cover requirements shall apply only to the principal tensile and moment reinforcement whose yielding would cause failure.

14.7 CONSTRUCTION

Since thin shells are frequently used because of their high structural efficiency, due care must be exercised in their construction so that conditions not anticipated in design do not occur. In order to accommodate the steel requirements, it is rare that a shell less than 3 in. can be cast economically. The design engineer should investigate the effect of variations in shell surface as a result of possible variations in the formwork. Tolerances should be set to define acceptable variations.

Of particular importance is the sequence of decentering. This should be specified by the engineer in order to avoid the occurrence of an unanticipated support condition resulting in a highly concentrated force system. Generally decentering should begin at points of maximum deflections and proceed toward points of minimum deflection. The decentering of edge members should proceed along with that of the adjoining shell. If the edge members are to be reshored, the spacing and sequence should be specified by the design engineer. Support members should be decentered after decentering the shell.

If the shell surface has maximum slopes of less than 45°, use of formwork on both faces is unnecessary. In those regions where the slope exceeds 45° the use of a top form is desirable. In addition, slump of the concrete used, the method of placement, the amount of steel, and the steel pattern have a considerable influence on the behavior of concrete placed on slopes. For shells with very steep slopes, shotcreting or plastering can be used to eliminate the need for the top form.

Relative cost figures for shell construction can be found in Ref. 14-18.

If the shell cannot be cast in one day, it should be sectioned into two or more regions. Concrete placed in one region should be allowed two or three days' curing before the next region is cast. This procedure is necessary to reduce shrinkage stresses and cracking.

Construction joints should be shown on the design drawings and should be located if possible in regions of compressive stress.

The construction aspects suggested by the ACI committee are:

501. Aggregate size

(a) The maximum size aggregate shall not exceed one-half the shell thickness, nor the clear distance between reinforcement bars, nor 1 1/2 times the cover. Where formwork is required for two faces, the maximum size of aggregate shall not exceed one-quarter the minimum clear distance between the forms nor the cover over the reinforcement.

502. Forms

(a) Removal of thin shell concrete forms shall be considered a matter of design and the form removal sequence shall be specified or approved by the engineer.

(b) The minimum strength of concrete f'_c, based on field-cured cylinders, at the time of decentering and of reshoring, when required, shall be designated by the engineer.

(c) Where, in the opinion of the designing engineer, stability of short- or long-time deflections are important factors, the modulus of elasticity at the time of decentering, based on field-cured beams, shall be specified by the engineer. The proportions and loading of these specimens shall insure action which is primarily flexural.

(d) The batter on vertical elements, or other elements, if desired or required for stripping, and the construction tolerances shall be designated by the engineer. When movable forms are used, a batter of 1/8 in. per ft minimum is recommended for vertical surfaces to permit easy removal.

NOTATION

a, b = one half the span of elliptical and hyperbolic paraboloids in the x- and y-directions, respectively
E = Young's modulus
F = Airy stress function used in the analysis of shallow shells
F_1 to F_4 = functions expressing the decay of boundary disturbances with distance along a coordinate line of a shell of revolution
H = horizontal component of edge force
h_x, h_y = rise of elliptical parabola in the x- and y-directions respectively
I = moment of inertia of cross section of beam or cylindrical shell analyzed as a beam
k = conversion factor to transform projected force components back into shell surface for elliptical parabola
also twist of surface for hyperbolic parabola
L, l = span of shell
M_ϕ = bending moment in the shell
M_θ = bending moment in the shell
N_ϕ = normal face in meridian direction
N_θ = normal force in hoop or circumferential direction
$N_{\theta\phi}$ = in plane shear force
R_1, R_2 = principal radii of curvature
Q = transverse shear force
r = radius of curvature
s = coordinate along generator of conical shell
t = shell thickness
u, v = displacement components in tangent plane
w = displacement component normal to shell surface
x = coordinate along generator of cylindrical shell
β = constant whose value is function of the radius to thickness ratio
γ = specific weight
Δ_H = horizontal displacement of edge of shell
Δ_ϕ = rotation of normal at edge of shell
θ = angular coordinate in circumferential direction for shell of revolution and in hoop direction for cylinder
ν = Poisson's ratio
ϕ = angular coordinate measuring the inclination of the normal to the shell surface from the axis of revolution
ω = angle between referenced axes for hyperbolic parabola

BIBLIOGRAPHY

14-1 Baker, et al., *Structural Analysis of Shells*, McGraw-Hill, New York, 1972.

14-2 Billington, D. P., *Thin Shell Concrete Structures*, 2nd. Ed., McGraw-Hill, New York, 1982.

14-3 Bleich, H. H., and Salvadori, M. G., "Bending Moments on Shell Boundaries," *Journal of the Structural Division, ASCE*, V. 85, ST8, Oct. 1959.

14-4 Bouma, A. L., "On Approximate Methods of Shell Analysis, A. General Survey," World Conference on Shell Structures, San Francisco, 1962.

14-5 Budiansky, B., and Radkowski, P. P., "Numerical Analysis of Unsymmetrical Bending of Shells of Revolution," *AIAA Journal*, V. 1, No. 8, Aug. 1963.

14-6 Candela, F., "General Formulas for Membrane Stresses in Hyperbolic Paraboloidical Shells," *Journal American Concrete Institute*, V. 32, No. 4, Oct. 1960.

14-7 Chinn, J., "Cylindrical Shell Analysis Simplified by Beam Method," *Journal American Concrete Institute*, V. 30, No. 11, May 1959. Also discussion by Parme and Conner, V. 31, No. 6, Dec. 1959.

14-8 Clough, R. W., and Johnson, C. P., "Finite Element Analysis of Arbitrary Thin Shells," *American Concrete Institute Special Publication SP-28*, Symposium on Concrete Thin Shells.

14-9 "Coefficients for Design of Cylindrical Shell Roofs," Portland Cement Association, Skokie, Ill., 1959.

14-10 "Concrete Shell Structures, Practice and Commentary," *Journal American Concrete Institute, Proceedings*, V. 61, No. 9, Sept. 1964.

14-11 "Concrete Shell Buckling," *Special Publication SP-67*, American Concrete Institute, Detroit, Mich., 1981.

14-12 "Design of Cylindrical Concrete Shell Roofs," *ASCE Manual of Engineering Practice No. 31*, ASCE, New York, 1952.

14-13 "Elementary Analysis of Hyperbolic Paraboloid Shells," Portland Cement Association, Chicago, 1960.

14-14 Esquillan, N., "The Shell Vault of the Exposition Palace Paris," *Journal of the Structural Division, ASCE*, V. 86, ST1, Jan. 1960.

14-15 Fischer, L., *Theory and Practice of Shell Structures*, Wilhelm Ernst & Sohn, Munich, 1968.

14-16 Flugge, W., *Stresses in Shells*, Springer Verlag, Berlin, 1960.

14-17 Gallagher, R. H., "Problems and Progress in Thin Shell Finite Element Analysis," *Finite Elements for Thin Shells and Curved Members* (Ashwell and Gallagher, Editors), John Wiley, New York, 1976.

14-18 Gensert, R. M., Kirsis, U., and Peller, M., "Economic Proportioning of Cast-in-Place Concrete Thin Shells," *ACI Special Publication SP-28*, Symposium on Concrete Thin Shells.

14-19 Gibson, J. E., *The Design of Cylindrical Shell Roofs*, E. & F. N. Spon, Ltd., London, 1961.

14-20 Gioncu, V., *Thin Reinforced Concrete Shells*, John Wiley, New York, 1979.

14-21 Griggs, P. H., "Buckling of Reinforced Concrete Shells," *J. of EMD, ASCE*, V. 97, EM3, June 1971.

14-22 Haas, A. M., and Van Koten, H., "The Stability of Doubly Curved Shells Having a Positive Curvature Index," *Herron*, V. 17, No. 4, 1970–71.

14-23 Hetenyi, M., "Spherical Shells Subjected to Axial Symmetrical Bending, *IABSE Publications*, V. 4, 1937–38.

14-24 Hoff, N. J., "The Effect of Meridian Curvature on the Influence Coefficients of Thin Spherical Shells," *Problems of Continuum Mechanics*, Soc. Ind. and Applied Math., 1961.

14-25 Holland, I., *Design of Circular Cylindrical Shells*, Oslo University Press, Oslo, 1957.

14-26 Jones, R. E., and Strome, D. R., "Direct Stiffness Method Analysis of Shells of Revolution Utilizing Curved Elements," *AIAA Journal*, V. 4, No. 9, Sept. 1966.

14-27 Mohraz, B., "A Lumped Parameter Element for the Analysis of Hyperbolic Paraboloid Shells," *Int. J. for Numerical Methods in Engineering*, V. 4, 1972.

14-28 NASA, "Collected Papers on Instability of Shell Structures-1962," *NASA Technical Note D-1510*.

14-29 Parme, A. L., "Shell of Double Curvature," *Transactions ASCE*, V. 123, 1958.

14-30 Parme, A. L., and Conner, H. W., "Design Constants for Ribless Concrete Cylindrical Shells," *IASS Bulletin No. 18,* June 1964.

14-31 Parme, A. L., and Conner, H. W., "Design Constants for Interior Cylindrical Concrete Shells," *Journal American Concrete Institute*, July 1961.

14-32 Pecknold, D. A., and Schnobrich, W. C., "Finite Element Analysis of Skewed Shallow Shells," *Journal of the Structural Division of ASCE*, V. 95, ST4, Apr. 1969.

14-33 Ramaswany, G. S., *Design and Construction of Concrete Shell Roofs*, McGraw-Hill, New York, 1968.

14-34 Reissner, E., "On Some Aspects of the Theory of Thin Shells," *Journal of Boston Society of Civil Engineers*, V. 42, No. 2, Apr. 1955.

13-35 Rudiger, D., and Urban, J., *Circular Cylindrical Shells*, B. G. Teubner Verlag, Leipzig, 1955.

13-36 Schmit, H., "Ergebnisse von Beulversuchen mit Doppelt Gerkrümmten Schalenmodellen aus Aluminum," *Proceedings* of the Symposium on Shell Research, North Holland Publishing Co., Amsterdam, 1961.

13-37 Schnobrich, W. C., "Analysis of Hipped Roof Hyperbolic Paraboloid Structures," *Journal of the Structural Division of ASCE*, V. 97, ST7, July 1972.

14-38 Schnobrich, W. C., "Analysis of Hyperbolic Paraboloid Shells," *American Concrete Institute Special Publication SP-28*, Symposium on Concrete Thin Shells.

14-39 Schnobrich, W. C., "Prediction of Nonlinear Behavior of Reinforced Concrete Shells," *Nonlinear Behavior of Reinforced Concrete Spatial Structures*, V. 3 (Mehlhorn, Ruhle, and Zerna, Editors), Werner Verlag, Dusseldorf, 1981.

14-40 Scordelis, A. C., and Lo, K. S., "Computer Analysis of Cylindrical Shells," *Journal American Concrete Institute of Proceedings*, V. 61, No. 5, May 1969.

14-41 Scordelis, A. C., "Analysis of Cylindrical Shells and Folded Plates," *Concrete Thin Shells*, American Concrete Institute Publication SP 28.

14-42 Scordelis, A. C.; Ramirez, H. D.; and Ngo, D., "Membrane Stresses in Hyperbolic Paraboloid Shells Having an Aribtrary Quadrilateral Shape in Plan," *Journal American Concrete Institute*, *Proceedings*, V. 67, No. 1, Jan. 1970.

14-43 Tedesko, A., "Multiple Ribless Shells," *Journal of Structural Division ASCE*, V. 87, ST7, Oct. 1961.

14-44 Toebes, G. H., Discussion of "Wind Stresses in Domes," *Trans. ASCE*, Part I, 1962, pp. 854–861.

14-45 Tester, K. G., "Bertrag zur Berechnung der Hyperbolishen Paraboloidschale," *Ingenieur Archiv*, V. 16, 1947.

14-46 Yu, C. W., and Kriz, L. B., "Tests of a Hyperbolic Paraboloid Reinforced Concrete Shell," *Proceedings* of World Conference on Shell Structures, San Francisco, 1962.

15

Reinforced Concrete Chimneys

WADI S. RUMMAN, Ph.D.*

15.1 INTRODUCTION

The construction of tall, reinforced concrete chimneys has been on the increase in the last two decades, owing primarily to the increasing demand for air pollution control. Chimneys in the range of 1250 ft in height have been built (see Figs. 15-1 and 15-2), and there is every reason to believe that this trend toward the construction of taller chimneys will continue.

This increasing demand for tall chimneys together with the use of slipform construction has emphasized the advantages of reinforced concrete construction. One primary advantage in a reinforced concrete chimney is that its resistance to wind vibrations is much greater than that of a corresponding steel chimney. In fact, very few self-standing steel chimneys exceed 350 ft in height.

Because of the recent changes in the proportions of chimneys, many structural problems, such as the response to earthquake and wind forces, become more critical. However, the availability of the digital computer makes it possible to analyze chimneys for such forces and to compute stresses for many loading conditions in a quick and economical manner.

Some, but not all, of the advances in chimney design in the last ten years are incorporated in the recent ACI Standard 307-79, *Specification for the Design and Construction of Reinforced Concrete Chimneys.*[15-1] This specification contains *the basic information necessary* to compute stresses due to dead weight, static wind, earthquake forces, and temperature. Reference to this specification is made throughout the chapter. However, many problems, which are not now incorporated in this specification, will be presented.

*Professor of Civil Engineering, University of Michigan, Ann Arbor.

15.2 PROPORTIONING

The height of the chimney as well as the diameter at the top are normally chosen so that exit velocity and dispersion of gases are within the specified limits. The bottom diameter, however, is more frequently controlled by the structural requirements of both the concrete shell and the foundation. For example, a small-base diameter can create design problems in the overhang part of the foundation and can also cause high stresses in the concrete stack. Experience has shown that, in general, a height-to-outside-base-diameter ratio in the range of 12–13 to 1 will provide good proportions for the design of both a tapered stack and its foundation. Once the top and bottom diameters are established, the outside diameters can vary linearly along the chimney to provide easier construction without significant loss in economy. The above remarks on proportioning may not apply for chimneys with multiple flue openings having large diameters with little or no taper.

Fig. 15-1 The 1250-ft-tall chimney of the International Nickel Co. of Canada, in Sudbury, Ontario. The chimney was designed and constructed by the M. W. Kellogg Co.

The thicknesses of the concrete are usually determined by satisfying the concrete stresses due to static wind and dead weight, and by satisfying the minimum thickness as specified by the ACI chimney specification, which states: "The minimum shell wall thickness for any chimney with an internal diameter of 28 ft or less shall be 8 in. When the internal diameter exceeds 28 ft, the minimum thickness shall be increased 1/4 in. for each 2 ft increase in internal diameter." Although considerable adjustments in the reinforcing steel may be required for earthquake and dynamic wind loading, the thicknesses of the concrete, as controlled by the static wind loading and the minimum requirements, will remain practically unchanged.

Table 15-1 lists typical existing chimneys, giving the height, outside top diameter, outside bottom diameter, and total weight of the concrete.

15.3 LINERS

The liner, whose primary function is to protect the concrete from the corrosive elements of the flue gases and to provide an adequate exit velocity of the gas, is a very important part of a reinforced concrete chimney. Most of the original concrete chimneys were constructed as shown in Fig. 15-3(a), using the corbel-supported brick lining and fiberglass or fused silica insulation between the lining and the concrete shell. However, many chimneys of this type have shown signs of acid attack on the concrete shell which is the result of hot gases working their way through the brick lining and their condensation on the relatively cold concrete. The condensate is acidic because of the sulfur content of the fuel. A good solution to the acid problem is the use of independent steel liners, as shown in Fig. 15-3(b). Many chimneys have been built with steel liners in the last 20 years or so, and their performance insofar as furnishing protection against acid attack has been excellent.

Most of the steel liners have been supported at, or near, the bottom, although a few are supported (hung) near the top. One of the advantages of top support is that the liner is in tension rather than compression; but a disadvantage is that the expansion due to temperature will have to be taken up by a joint near the flue opening. Bottom-supported liners, on the other hand, are susceptible to buckling above the flue opening, not only due to the loading, but also due to temperature differentials between one side of the liner and another.

Other types of liners are the fiber reinforced plastic (FRP) liner, the independent concrete liner, and the independent brick liner. The fiber reinforced plastic liner and the independent brick liner are specially useful in installations where scrubbers are used. The presence of scrubbers would cause excessive condensation due to the low temperature of the gases. Such condensation could cause corrosion in steel liners.

15.4 WIND LOADING

The wind loading, in both alongwind and acrosswind directions, is an important factor in chimney design. The present ACI chimney standard (ACI307-79) provides for the design of chimneys in the alongwind direction by specifying wind pressures for different map areas. This will be referred to as the ACI 307-79 static wind loading.

15.4.1 Static Wind

The design static wind pressures for circular chimneys are specified in the ACI chimney standard, ACI307-79, and are given in Table 15-2.

The standard specifies that "in special locations where records or experience indicate that these pressures are inadequate, higher pressures may be used at the discretion of the owner's engineers," and also adds that "the job specifications should indicate the wind pressure zone to be used for design of the chimney, to avoid misinterpretations of borderline map locations."

15.4.2 Radial Pressure

In addition to static wind pressures which are applied on the projected area of the chimney, the designer is also confronted with the effects of the radial pressure distribution around the circumference of the chimney. Experiments conducted on circular cylinders and stacks[15-2, 15-3] have shown that the distribution of the wind pressures around the circumference is as indicated in Fig. 15-4.

The radial forces, which will have a resultant force (F) acting to the right, are resisted by shearing forces around the circumference, as shown in the sketch. These shearing forces, which are assumed to vary sinusoidally around the circumference, should have a resultant equal to F and acting opposite to it.

Setting:

$$S_\theta = A \sin \theta \qquad (15\text{-}1)$$

Fig. 15-2 The chimney in Fig. 15-1 during slipforming.

we obtain from the equilibrium condition:

$$F = 2\int_0^{\pi} (A \sin\theta)\, r\, d\theta \qquad (15\text{-}2)$$

from which:

$$A = \frac{F}{\pi r} \qquad (15\text{-}3)$$

The critical bending moments in the ring as obtained by different sources are tabulated in Table 15-3. Erdei and Ghosh[15-4] used a radial distribution of pressure based on the experiments of Roshko.[15-3] Diver[15-5] used the radial pressure distribution recommended by the French *Code*,[15-6] whereas the writer has used radial pressure distribution based on the experiments of Dryden and Hill.[15-2] The ASCE Task Committee on Wind Forces[15-7] recommends similar pressure distribution.

15.4.3 Alongwind Vibration

The approach developed in a recent thesis at The University of Michigan[15-8] and summarized in a paper[15-9] presented at the 2nd Chimney Design Symposium and Exhibition, Edinburgh, March 1976, will be used to generate simple and practical procedures for alongwind vibrations.

The wind speed $V(Z, t)$, at any height Z, is taken equal

TABLE 15-1 Basic Data for Typical Chimneys

Chimney No.	*Height (ft)*	*Top Outside Diameter (ft)*	*Bottom Outside Diameter (ft)*	*Total Weight of Concrete Stack (kips)*	*Ratio of Height to Outside Base Diameter*
1	352	23.58	30.90	4530	11.4
2	450	16.33	35.79	6740	12.6
3	534	18.67	35.03	8370	15.2
4	622	23.33	47.26	12530	13.2
5	700	31.50	59.75	15390	11.7
6	800	36.56	65.00	25030	12.3
7	900	34.25	72.00	33800	12.5
8	1000	34.25	77.13	33890	13.0
9	1103	42.29	83.87	51260	13.2
10	1200	37.00	95.29	64550	13.1

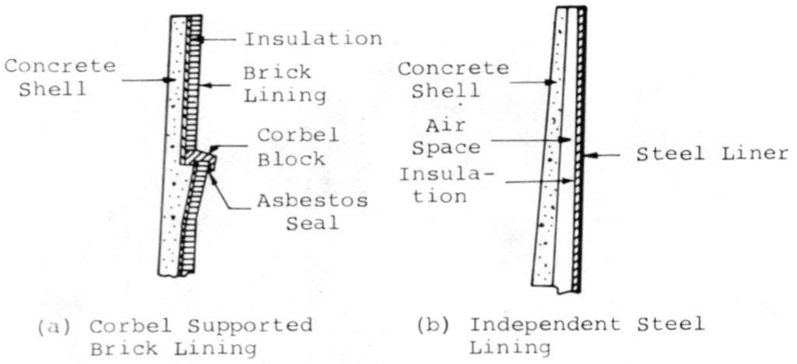

Fig. 15-3 Concrete chimneys with two types of lining.

to the sum of two components, the mean part $V(Z)$ and the fluctuating component $v(Z, t)$; that is:

$$V(Z, t) = \bar{V}(Z) + v(Z, t) \qquad (15\text{-}4)$$

$\bar{V}(Z)$ is assumed to follow the power low profile using an open terrain exposure factor:

$$\bar{V}(Z) = \left(\frac{Z}{Z_{\text{ref}}}\right)^{1/7} \times \bar{V}_{\text{ref}} \qquad (15\text{-}5)$$

The fluctuating component $v(Z, t)$ is expressed in terms of the standard deviation σ, a gust factor G, and a random number γ; that is:

$$v(Z, t) = G(Z) \cdot \sigma(T_1, T_2) \cdot \gamma(Z, t) \qquad (15\text{-}6)$$

The standard deviation $\sigma(T_1, T_2)$, would vary with the mean speed, the duration of the simulated wind record T_1, and the time interval between two data points T_2. The gust factor, $G(Z)$, is modeled to vary with height. The random number, $\gamma(Z, t)$, varies with height and time and is modeled to vary between 0 and ± 1 with normal distribution.

The force $F(Z, t)$, at any level Z, and at any time t, is given by:

$$F(Z, t) = \tfrac{1}{2} \rho C_D V^2(Z, t) D(Z) \qquad (15\text{-}7)$$

in which the mass density of air ρ would be taken as .00000238 k-sec^2/ft^4 and the drag coefficient C_D would be taken as 0.7. $D(Z)$ is the outside diameter at any height Z, in feet, and $V(Z, t)$ is the total wind speed in ft/sec.

The above forcing function was used to obtain the response of many chimneys using many runs because of the use of the random number γ. The damping coefficient that was used in the response was taken as 0.02 of critical.

Based on the results of the above approach, the following procedure would give a good estimate of alongwind vibration.

The load on the chimney, at any height Z, due to the mean wind profile, is given by eq. (15-8) and is obtained

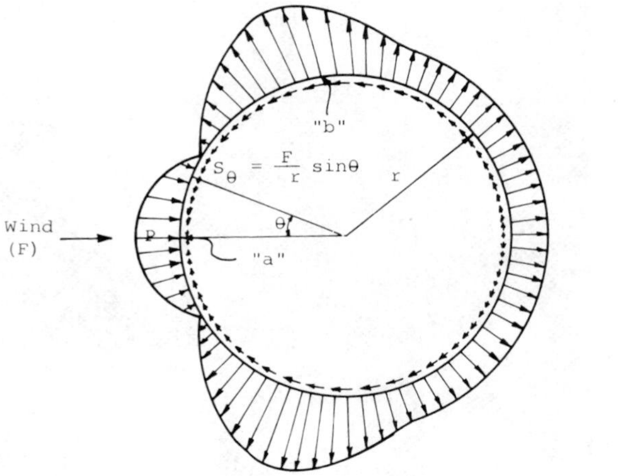

Fig. 15-4 Distribution of wind pressure on ring.

from eq. (15-7) by using $\bar{V}(Z)$ of eq. (15-5) for $V(Z, t)$:

$$w(Z) = 6.8 \times 10^{-7}\, (Z)^{2/7}\, (\bar{V}_{\text{ref}})^2\, D(Z) \qquad (15\text{-}8)$$

where $\bar{V}_{\text{ref}}$ is the mean speed in MPH at the reference height of 30 ft. Both Z and $D(Z)$ should be in units of feet. The load $w(Z)$ would be in kips/ft.

The mean loading $w(Z)$ as given by eq. (15-8) would then be applied as a static loading on the chimney to obtain shears and moments. The fluctuating part of the wind as well as the dynamic effects of the wind are accounted for by multiplying the shears and moments obtained from the static mean profile by the following multipliers:

Moment multipliers, J_m:

$$J_{m\,(\text{base})} = 1.75 \qquad \text{for } T_1 \geqq 3.0 \text{ sec/cycle}$$
$$= 1.75 + 0.35(3 - T_1) \qquad \text{for } T_1 \leqq 3.0 \text{ sec/cycle} \qquad (15\text{-}9)$$

$$J_{m\,(\text{top})} = 3.1 \qquad \text{for } T_1 \geqq 3.0 \text{ sec/cycle}$$
$$= 3.1 + 0.30(3 - T_1) \qquad \text{for } T_1 \leqq 3.0 \text{ sec/cycle} \qquad (15\text{-}10)$$

$$J_m(Z) = J_{m\,(\text{base})} + \frac{J_{m\,(\text{top})} - J_{m\,(\text{base})}}{H} \cdot Z \qquad (15\text{-}11)$$

in which H = height of chimney, Z = distance at any level from the base, and T_1 = the period of the first mode in seconds per cycle.

Shear multipliers, J_v:

$$J_{v\,(\text{base})} = 1.4 \qquad \text{for } T_1 \geqq 3.0 \text{ sec/cycle}$$
$$= 1.4 + 0.30(3 - T_1) \qquad \text{for } T_1 \leqq 3.0 \text{ sec/cycle} \qquad (15\text{-}12)$$

$$J_{v\,(\text{top})} = 3.0 \qquad \text{for } T_1 \geqq 3.0 \text{ sec/cycle}$$
$$= 3.0 + 0.40(3 - T_1) \qquad \text{for } T_1 \leqq 3.0 \text{ sec/cycle} \qquad (15\text{-}13)$$

TABLE 15-2 Design Pressures for Chimneys with Circular Cross Section

	Wind Pressure (psf)–Map Area per UBC				
Height Zone (ft)	*30 or less*	*35*	*40*	*45*	*50*
0–100	23	23	26	29	32
100–500	31	33	36	42	45
500–1200	34	36	42	48	54
1200–over	36	42	48	54	60

TABLE 15-3 Maximum Ring Moments

	Section a (Fig. 15-4) Tension on Inside	*Section b (Fig. 15-4) Tension on Outside*
Erdei & Ghosh	$0.354\, pr^2$	$0.311\, pr^2$
Diver	$0.284\, pr^2$	$0.256\, pr^2$
Rumman	$0.314\, pr^2$	$0.272\, pr^2$

$$J_v(Z) = J_{v\,(\text{base})} + \frac{J_{v\,(\text{top})} - J_{v\,(\text{base})}}{H} \cdot Z \qquad (15\text{-}14)$$

Both the first and second modes of vibration were considered in establishing the multipliers for shears and moments.

The first mode of the chimney, T_1, which was based on a previous paper,[15-10] can be fairly accurately obtained by using the following formula:

$$T_1 = 5\,\frac{H^2}{d_0}\sqrt{\rho/E}\left(\frac{t_0}{t_n}\right)^{.22}\left(\frac{d_0}{d_n}\right)^{1.1} \qquad (15\text{-}15)$$

when d_0 = mean diameter at top; d_n = mean diameter at bottom; t_0 = thickness at top; t_n = thickness at bottom; ρ = mass density of concrete; E = modulus of elasticity; H = height of chimney.

15.4.4 Acrosswind Vibration and Ovaling

The lateral or acrosswind vibration and ovaling of the circular cross section due to vortex shedding have occurred in steel stacks, especially in welded stacks without gunite. Although failures have occurred in steel stacks due to resonant conditions, no such failures have been reported in concrete chimneys. The variable diameter, the larger mass, and the higher damping are factors contributing to the resistance of concrete chimneys to the lateral swaying and ovaling.

1. Acrosswind Vibration. In recent years some observations of large lateral deflections have been reported, and although the chimney itself was not impaired, damage was observed in the liner due to excessive deflections. In addition, interference effects could cause a downwind chimney in a line of chimneys to double or even triple its movement in the isolated condition. Many articles[15-7, 15-11–15-13] have been written on acrosswind vibration, and many chimneys have been successfully designed[15-14, 15-15] based on these articles. The equations that are presented here to estimate the shears and bending moments due to lateral vibrations have been obtained by using the procedures outlined in Refs. 15-8 and 15-9. The forcing function on the chimney representing the lateral forces due to vortex shedding is modeled as:

$$F(Z,t) = \tfrac{1}{2}\,\rho C_L\,\bar{V}^2(Z)\,D(Z)\,\sin\left[2\pi f_s(Z) + \phi(Z)\right] \qquad (15\text{-}16)$$

where C_L = lift coefficient; $\phi(Z)$ is a random angle uniformly distributed between 0 and 2π whose use will produce lateral sinusoidal forces at the different levels that will be randomly out of phase; and $f_s(Z)$ is the shedding frequency (cycles/sec) at any level Z, determined from the strouhal number relationship, given as follows:

$$f_s(Z) = \frac{\bar{V}(Z)}{D(Z)}\,S \qquad (15\text{-}17)$$

Many chimneys were analyzed using the above procedure. For each chimney seven runs were made because of the use of the random angle $\phi(Z)$. A typical curve for chimney No. 5 of Table 15-1 is given in Fig. 15-5. Both the first and second mode responses are given in the figure. Figure 15-5 shows the average response as well as the average plus one standard deviation of the seven computer runs used. It is of interest to note that the peak of the first mode would take place when the wind profile is such that this shedding frequency (as given by eq. 15-17) at about 2/3 of the height from the base coincides with the first mode frequency of the chimney. The first peak of the second mode curve would take place when the wind profile is such that the shedding frequency near the top coincides with the second mode frequency of the chimney, whereas the second peak would occur when the wind profile creates a shedding frequency at about midheight that coincides with the second mode frequency of the chimney.

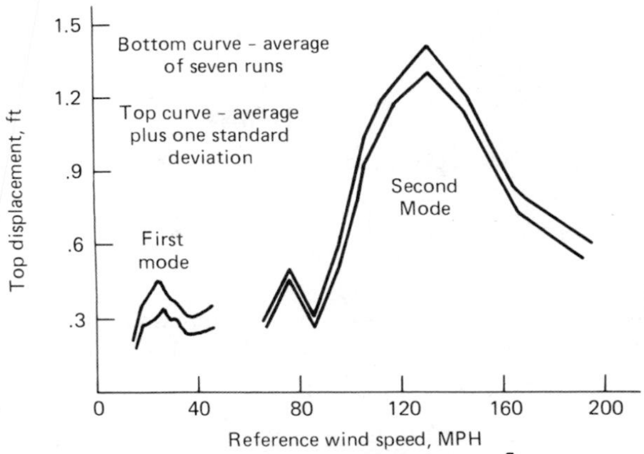

Fig. 15-5 Acrosswind response for 1000-ft chimney (first and second modes, β = 0.02).

As a result of establishing similar curves for many chimneys using a strouhal number S = 0.2, the following procedure can be used to estimate the forces due to first mode resonance.

The modal multiplier, q_1, for the average plus one standard deviation response can be obtained from:

$$q_1 = \frac{\rho}{16\pi^2 S^2}\,\frac{C_L}{\beta^{2/3}}\,2.20\,\frac{D_c^3\,\phi_c(Z_2 - Z_1)}{\int_0^H m\phi_1^2\,dZ} \qquad (15\text{-}18)$$

where ρ = mass density of air, C_L = lift coefficient, S = strouhal number (use 0.2), m = mass per unit length, ϕ_1 = first mode shape, and β = fraction of critical damping. The value:

$$(Z_2 - Z_1) = \frac{0.15 D_c}{\text{taper}} \qquad \text{(taper = change of diameter per unit length)}$$

but is not to exceed $3D_c$ or 175 ft. (See Fig. 15-6 for Z_1, Z_2, Z_c, D_c, etc.) The maximum value of the multiplier q_1 would occur when $D_c^4\,\phi_c$ is maximum. In case Z_2 exceeds H, locate D_c so that Z_2 is equal to H. The mean wind speed

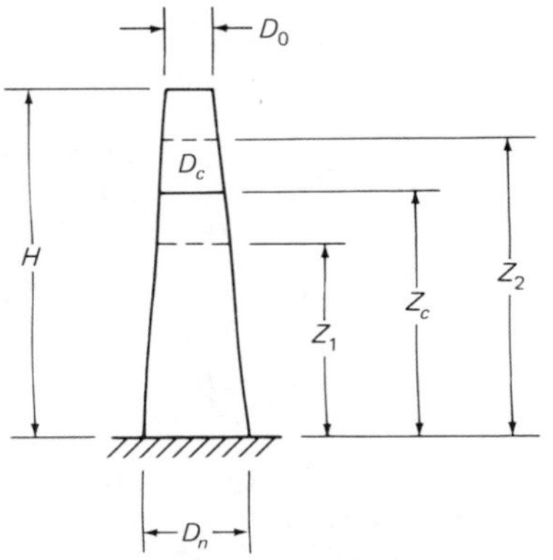

Fig. 15-6 Symbols for acrosswind analysis.

at Z_c from the base would be:

$$\bar{V}_c = \frac{1}{T_1} \frac{D_c}{S} \tag{15-19}$$

and the reference mean wind speed would be:

$$\bar{V}_{ref} = \left(\frac{Z_{ref}}{Z_c}\right)^{1/\alpha} \bar{V}_c \tag{15-20}$$

Use $\alpha = 7$ (open terrain) and $Z_{ref} = 30$ ft.

The modal multiplier, q_2, for the acrosswind response in the second mode can be approximated by the following formula:

$$q_2 = \frac{\rho}{16\pi^2 S^2} \frac{C_L}{\beta^{2/3}} 3.08 \frac{D_c^3 \phi_c(Z_2 - Z_1)}{\int_0^H m\phi_2^2 \, dZ} \tag{15-21}$$

The definition of terms and the limitations given for the first mode response would also apply for this case. The first peak of the second mode acrosswind response (see Fig. 15-5) would probably occur with $Z_2 = H$. In this case:

$$(Z_2 - Z_1) = \frac{0.15 D_0}{0.925 \times \text{taper}}$$

Again $(Z_2 - Z_1)$ must not exceed $3D_c$ or 175 ft.

A formula similar to that of eq. (15-15) has been developed for the period of the second mode. This formula is given in the following equation:

$$T_2 = 0.82 \frac{H^2}{d_0} \sqrt{\rho/E} \left(\frac{t_0}{t_n}\right)^{.09} \left(\frac{d_0}{d_n}\right)^{.78} \tag{15-22}$$

The lift coefficient, C_L, to be used for acrosswind response has been found to vary with the aspect ratio (height/diameter). Based on tests by Vickery,[15-16] the following formulas can be used for the lift coefficient:

$$C_L = 0.67 \qquad \text{for } \frac{H}{D'} \geqq 20$$

$$C_L = 0.67 - 0.005\left(20 - \frac{H}{D'}\right)^{1.45} \qquad \text{for } \frac{H}{D'} \leqq 20 \tag{15-23}$$

where D' = average outside diameter over the top third of the chimney.

In addition to the above approaches for predicting the alongwind and the acrosswind responses, some probabilistic methods have been proposed. These approaches are given in Refs. 15-17 through 15-20.

2. *Ovaling of Circular Cross Sections.* If the period of a circular ring coincides with half of the period of the shedding of vortices, then ovaling of the cross section could occur. It has been shown[15-15] that the velocity that could cause ovaling vibrations in the fundamental mode is given by the following equation:

$$V = 366\left(\frac{t}{r}\right) \tag{15-24}$$

in which V = the wind velocity in mi/hr; r = the radius of the chimney cross section in feet; and t = the thickness of the cross section in inches.

15.5 EARTHQUAKE LOADING

With the availability of computers, chimneys can be analyzed for accelerograms obtained from actual earthquakes.[15-21, 15-22] Although such a method or a method that utilizes response spectra for earthquakes is recommended for final design, the designer now has at his disposal certain simplified procedures that are at least useful for preliminary design. One such procedure is contained in ACI Standard 307-79; and another procedure, which is based on studies made on the response of many actual chimneys to the response spectrum summarized in Table 15-4, will be presented in this chapter. This response spectrum is for a 50% probability level and 5% damping as generated by Newmark, Blume, and Kapur.[15-23] It is the spectrum that has been adopted for use in the design of steel chimney liners for earthquakes.[15-24]

The main steps in this procedure are to determine the period of the fundamental mode of vibration and the base shear, and then to distribute the base shear as lateral forces along the chimney, from which shears and bending moments are computed at different sections. These shears and moments are then multiplied by factors to obtain values that are comparable to those obtained from the response spectrum of Table 15-4.

TABLE 15-4 Spectral Values for Zone 3

Frequency f, Cycles/sec	Displacement Spectrum, in.	Velocity Spectrum, in./sec	Acceleration Spectrum, g's
Horizontal Direction			
$f \leqslant 0.25$	16.90	$106.1858f$	$1.7285f^2$
$0.25 \leqslant f \leqslant 2.5$	$\frac{3.462}{f^{1.1436}}$	$\frac{21.7524}{f^{0.1436}}$	$0.35409f^{0.8564}$
$2.5 \leqslant f \leqslant 9.0$	$\frac{8.439}{f^{2.1158}}$	$\frac{53.0238}{f^{1.1158}}$	$\frac{0.86314}{f^{0.1158}}$
$9.0 \leqslant f \leqslant 33.0$	$\frac{21.29}{f^{2.5369}}$	$\frac{133.7690}{f^{1.5369}}$	$\frac{2.1775}{f^{0.5369}}$
$f > 3.330$	$\frac{3.256}{f^2}$	$\frac{20.4581}{f}$	0.333
Vertical Direction			
$f \leqslant 0.25$	11.26	$70.7487f$	$1.1517f^2$
$0.25 \leqslant f \leqslant 3.6$	$\frac{2.322}{f^{1.13933}}$	$\frac{14.5896}{f^{0.13933}}$	$0.23749f^{0.86067}$
$3.6 \leqslant f \leqslant 9$	$\frac{7.674}{f^{2.07257}}$	$\frac{48.2172}{f^{1.07257}}$	$\frac{0.78490}{f^{0.07257}}$
$9 \leqslant f \leqslant 33$	$\frac{21.29}{f^{2.5369}}$	$\frac{133.7690}{f^{1.5369}}$	$\frac{2.1775}{f^{0.5369}}$
$33 \leqslant f \leqslant 50$	$\frac{98.75}{f^{2.9758}}$	$\frac{620.4645}{f^{1.9758}}$	$\frac{10.100}{f^{0.9758}}$
$f > 50$	$\frac{2.1705}{f^2}$	$\frac{13.6377}{f}$	0.222

15.5.1 Period

The period of the fundamental mode can be easily computed using the Stodola process. If the chimney is assumed to be linearly tapered in both diameter and thickness, then the curves obtained by the writer[15-10] or those obtained by Housner and Keightley[15-25] can be used. Equation (15-15) is based on the curves of Ref. 15-10 and gives values that are in excellent agreement with those computed by more exact methods. The example at the end of the chapter will utilize the Stodola process in computing the fundamental period and will compare the answer to that obtained by the formula.

15.5.2 Base Shear

From studying the response of many chimneys to the spectrum of Table 15-4, a fairly accurate relationship was established between the ratio of base shear to total weight and the period of the fundamental mode. This relationship can be expressed by the following equation:

$$\frac{V_b}{W} = Z\frac{0.35}{\sqrt{T}} \tag{15-25}$$

in which V_b = base shear; W = total weight; T = first mode period in sec/cycle; and Z = zone coefficient as defined by the 1982 *Uniform Building Code.*[15-26] For location in Zone 1, $Z = 3/16$. For location in Zone 2, $Z = 3/8$. For location in Zone 3, $Z = 3/4$. For location in Zone 4, $Z = 1.0$.

15.5.3 Base Shear Distribution

The base shear is distributed along the chimney so that the force intensity at any level is proportional to the weight intensity at that level and the square of the distance of that level from the base. Referring to Fig. 15-7, the force dF applied at any level x would be:

$$dF = \frac{[w(x)\,dx]\,x^2}{\int_0^H [w(x)\,dx]\,x^2} V_b \tag{15-26}$$

where $w(x)$ is the intensity of weight per unit length.

15.5.4 Shears and Bending Moments

Using the above force distributions, shears and bending moments at any height h are then computed as follows:

$$V_h = J_v \int_h^H dF \tag{15-27}$$

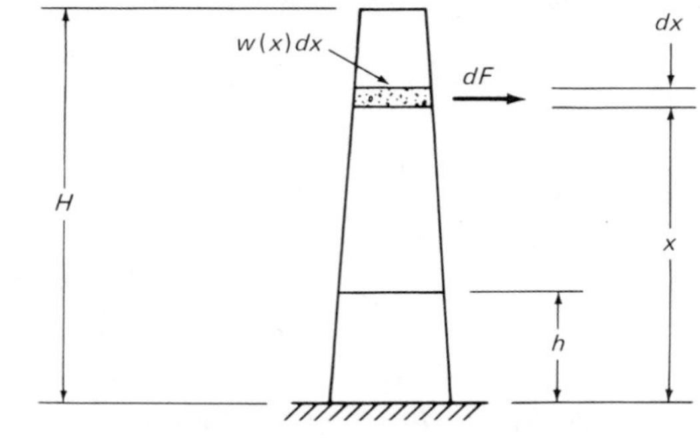

Fig. 15-7 Symbols for earthquake analysis.

$$M_h = J_m \int_h^H (dF)(x - h) \tag{15-28}$$

The correction factors J_v and J_m are selected to make the shear and moment curves comparable to those obtained from the response spectrum of Table 15-4. The following equations seem to give a satisfactory representation of these factors for chimneys with small or no taper:

$$J_v = 8[J_{v\,(\text{top})} - J_v(0.5H)]\left[\frac{h}{H} - 0.5\right]^3 + J_v(0.5H)$$

$$\text{for } 0.5 \leqslant \frac{h}{H} \leqslant 1.0 \tag{15-29}$$

$$J_v = 1 - 2[1 - J_v(0.5H)]\,\frac{h}{H} \tag{15-30}$$

$$\text{for } 0 \leqslant \frac{h}{H} \leqslant 0.5$$

$$J_v(0.5H) = 0.43 + 0.5/T^{3/2} \tag{15-31}$$

$$J_{v\,(\text{top})} = 0.96 + 0.23T \tag{15-32}$$

$$J_m = 2.78[J_m(0.9H) - J_m(0.3H)]\left[\frac{h}{H} - 0.3\right]^2 + J_m(0.30H)$$

$$\text{for } 0.3 \leqslant \frac{h}{H} \leqslant 1.0 \tag{15-33}$$

$$J_m = 11.1[J_{m\,(\text{base})} - J_m(0.3H)]\left[0.3 - \frac{h}{H}\right]^2 + J_m(0.3H)$$

$$\text{for } 0 \leqslant \frac{h}{H} \leqslant 0.3 \tag{15-34}$$

$$J_m(0.9H) = 1.15 + 0.025T^2 \tag{15-35}$$

$$J_m(0.3H) = 0.30 + 0.004(6 - T)^3 \tag{15-36}$$

$$\text{for } T > 6 \text{ use } 0.30$$

$$J_{m\,(\text{base})} = 0.40 + \frac{(6 - T)^3}{300} \tag{15-37}$$

$$\text{for } T > 6 \text{ use } 0.40$$

For chimneys with taper it is suggested that the above J_v and J_m values be multiplied by the factors shown in Tables 15-5 and 15-6.

15.6 STRESSES

The following formulas are incorporated in the ACI standard for chimneys and are repeated here for the convenience of the reader. For the complete coverage the reader should refer to ACI Standard 307-79.

15.6.1 Stresses Due to Dead Weight and Moment

For a horizontal section subjected to a bending moment about axis 1-1 in Fig. 15-8 and also to a normal force (W), the maximum compressive stress in the concrete and the maximum tensile stress in the steel are computed by the following formulas:

$$f_{c1} = \frac{W(\cos\beta - \cos\alpha)}{2rt[(1-p)(\sin\alpha - \alpha\cos\alpha) - (1-p+np)(\sin\beta - \beta\cos\alpha) - np\pi\cos\alpha]} \qquad (15\text{-}38)$$

$$f_c = f_{c1}\left[1 + \frac{t}{2r\cos\beta(\cos\beta - \cos\alpha)}\right] \qquad (15\text{-}39)$$

$$f_s = nf_{c1}\left[\frac{1+\cos\alpha}{\cos\beta - \cos\alpha}\right] \qquad (15\text{-}40)$$

where:

W = total normal load acting on the section under consideration
r = mean radius of chimney shell at section under consideration
t = thickness of chimney shell at section under consideration
p = ratio of total area of vertical reinforcement to total area of concrete of chimney shell at section under consideration
n = ratio of modulus of elasticity of the reinforcement to the modulus of elasticity of the concrete
α = one-half of the central angle subtended by the neutral axis as a chord on the circle of radius r
β = one-half the central angle subtended by the opening as a chord on the circle of radius r
$e = M/W$

The angle α is determined from the following equation:

$$\frac{e}{r} = \frac{(1-p)(\alpha - \sin\alpha\cos\alpha) - (1-p+np)(\beta + \sin\beta\cos\beta - 2\sin\beta\cos\alpha) + np\pi}{2[(1-p)(\sin\alpha - \alpha\cos\alpha) - (1-p+np)(\sin\beta - \beta\cos\alpha) - np\pi\cos\alpha]} \qquad (15\text{-}41)$$

The formulas listed above can also be used when a second opening diametrically opposite to the one in Fig. 15-8 exists. In case of no openings, the value of β is equal to zero, and the above equations are considerably simplified.

15.6.2 Stresses Due to Temperature

The vertical stress in the concrete (f_{CTV}) and the vertical stress in the steel (f_{STV}) due to a temperature differential, T_x, can be computed by the following formulas:

$$f_{CTV} = Lk\,T_x\,E_c \qquad (15\text{-}42)$$

and:

$$f_{STV} = L(z-k)T_x\,E_s \qquad (15\text{-}43)$$

where:

L = thermal coefficient of expansion of the concrete and of the reinforcing steel, to be taken as 0.0000065 per °F
E_c = modulus of elasticity of the concrete
E_s = modulus of elasticity of the steel
$k = -pn + \sqrt{pn\,(pn + 2z)}$
p = ratio of total area of vertical temperature reinforcement to total area of concrete of chimney shell at section under consideration
z = ratio of distance between inner surface of chimney shell and vertical temperature reinforcement to total shell thickness, t

T_x, which is the magnitude of the temperature differential across the concrete, depends on such factors as the type of liner, the insulation, the thickness of the concrete and the space between the liner and the concrete shell. The ACI chimney specification gives formulas for different conditions. For example, the temperature differential for the lined chimney with unventilated space between the lining and the shell, may be computed from the formula:

$$T_x = \frac{tD_{bi}}{C_c D_c}\left[\frac{T_g - T_o}{\frac{1}{K_1} + \frac{t_b D_{bi}}{C_b D_b} + \frac{D_{bi}}{K_r D_s} + \frac{tD_{bi}}{C_c D_c} + \frac{D_{bi}}{K_2 D_{co}}}\right] \qquad (15\text{-}44)$$

where:

T_g = maximum temperature of gas inside chimney, °F
T_o = minimum temperature of outside air surrounding chimney, °F
D_{bi} = inside diameter of lining, ft
D_b = mean diameter of lining, ft
D_s = mean diameter of space between lining and shell, ft
D_c = mean diameter of concrete chimney, ft
D_{co} = outside diameter of concrete chimney, ft
C_c = coefficient of thermal conductivity of the concrete chimney shell, Btu per sq ft per in. of thickness per hr per °F difference in temperature (use 12 for concrete)
K_1 = coefficient of heat transmission from gas to inner surface of chimney lining, Btu per sq ft per hr per °F difference in temperature
C_b = coefficient of thermal conductivity of chimney lining, Btu per sq ft per in. of thickness per hr per °F difference in temperature
t = thickness of concrete shell, in.
t_b = thickness of lining, in.
K_r = coefficient of heat transfer by radiation between outside surface of lining and inside surface of concrete chimney shell, Btu per sq ft per hr per °F difference in temperature ($K_r = T_g/120$)
K_2 = coefficient of heat transmission from outside surface of chimney shell to surrounding air, Btu per sq ft per hr per °F difference in temperature (K_2 = 12)

The horizontal stresses in the concrete and the steel due to the temperature are computed by similar formulas to those used for vertical stress computations.

15.6.3 Stresses Due to Combined Temperature and Load

The maximum vertical stresses in the concrete due to the bending moment and the axial load combined with temperature are computed by the following formulas:

$$f_{c,comb} = \frac{f_{CTV}\,k_{comb}}{k}; \quad k_{comb} \leqslant 1 \qquad (15\text{-}45)$$

$$f_{c,comb} = f_{c1} + \frac{f_{CTV}}{k}\left[\frac{2\,pnz + 1}{2(1+pn)}\right]; \quad k_{comb} \geqslant 1 \qquad (15\text{-}46)$$

where:

$$k_{comb} = -pn + \sqrt{pn(pn+2z) + 2k(1+pn)\frac{f_{c1}}{f_{CTV}}} \qquad (15\text{-}47)$$

TABLE 15-5 Multipliers for J_V

h/H	0.05	0.15	0.25	0.35	0.45	0.55	0.65	0.75	0.85	0.95
Multipliers	1.03	0.99	0.89	0.83	0.88	0.97	0.92	0.86	0.94	1.01

TABLE 15-6 Multipliers for J_m

h/H	0	0.1	0.2	0.3	0.4	0.5	0.6	0.7	0.8	0.9
Multipliers	0.94	0.91	0.89	0.92	0.90	0.82	0.76	0.78	0.87	0.97

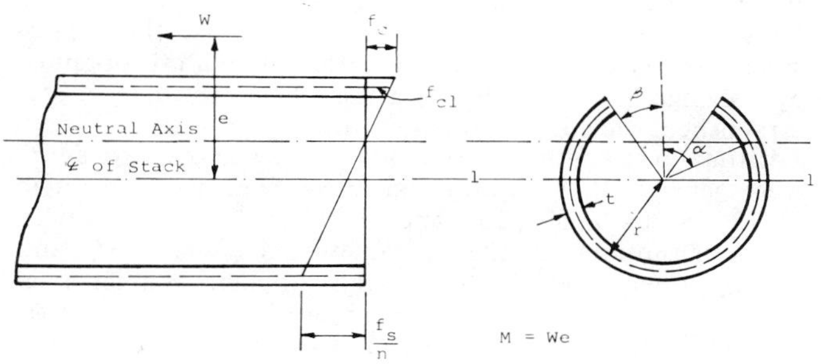

Fig. 15-8 Cross section subjected to bending and axial forces.

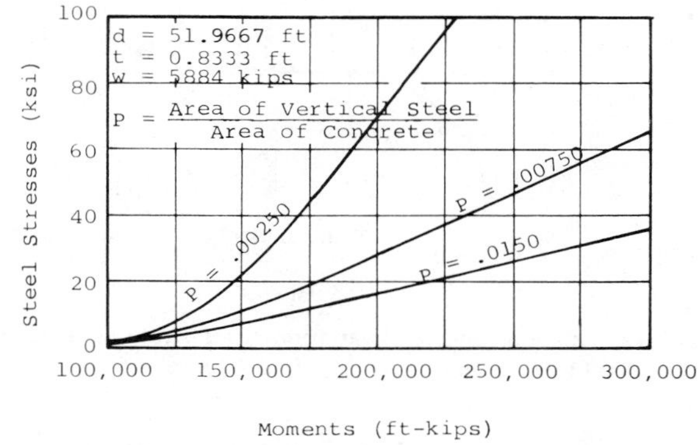

Fig. 15-9 Steel stresses vs moments of steel ratio, *P*.

The maximum combined vertical stress in the steel is calculated from:

$$f_{s,comb} = \frac{f_{STV}}{z - k} \times \left[z + pn - \sqrt{pn(pn + 2z) - 2pn(z - k)\frac{f_s}{f_{STV}}} \right] \quad (15\text{-}48)$$

except that when $f_s/f_{STV} \geqslant z/(z - k)$, the stress is the same as that given by eq. (15-40).

The formulas of Sections 15.6.2 and 15.6.3 do not account for any inside steel. The effects of the inside steel are incorporated in the formulas of ACI 307-79.

15.7 STRENGTH OR ULTIMATE DESIGN

The moments and allowable stresses as specified in the chimney standard ACI307-79 as well as the moments that were obtained in the previous sections are used in conjunction with a working stress design.

This section will concentrate on ultimate design. The ultimate design is more desirable than a working stress design because of the nonlinear variation of the stresses (especially in the steel) with respect to the change in the bending moment. This nonlinearity of the stresses is primarily due to a change in the bending moment without a corresponding change in the axial force. Figure 15-9 illustrates how the stress in the steel can increase at a much faster rate than the increase in the bending moment.

For the strength design it is recommended that load factors of 0.9 and 1.7 be used for the axial load and the bending moment, respectively.

The ultimate bending moment capacity of the hollow circular cross section as presented here[15-27, 15-28] is based on the stress–strain curve for concrete suggested by Hognestad[15-29] and shown in Fig. 15-10. The stress–strain relationship for the reinforcement is assumed to be of the bilinear type, as shown in Fig. 15-11.

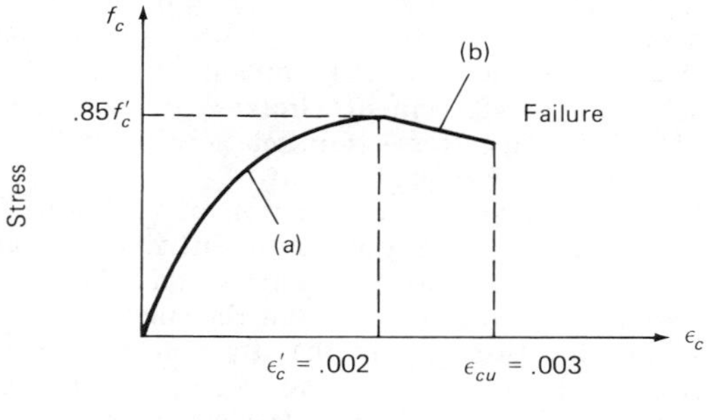

$$(a): f_c = 0.85 f_c' \left[2\frac{\epsilon_c}{\epsilon_c'} - \left(\frac{\epsilon_c}{\epsilon_c'}\right)^2 \right]$$

$$(b): f_c = 0.85 f_c' \left[1 - 0.15 \frac{\epsilon_c - \epsilon_c'}{\epsilon_{cu} - \epsilon_c'} \right]$$

Fig. 15-10 Parabolic and declining variation of stress–strain curve for concrete (as suggested by Hognestad).

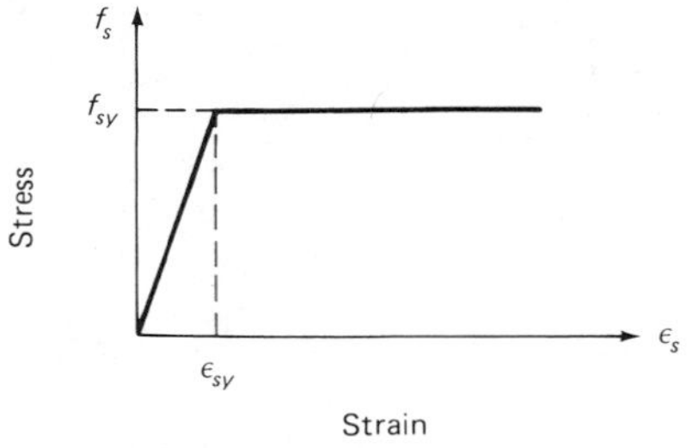

Fig. 15-11 Idealized stress–strain curve for reinforcing steel.

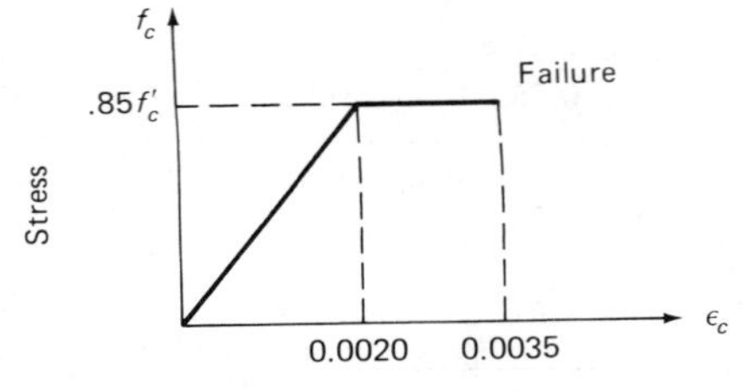

Fig. 15-12 Bilinear stress–strain curve for concrete (as used by Brettle).

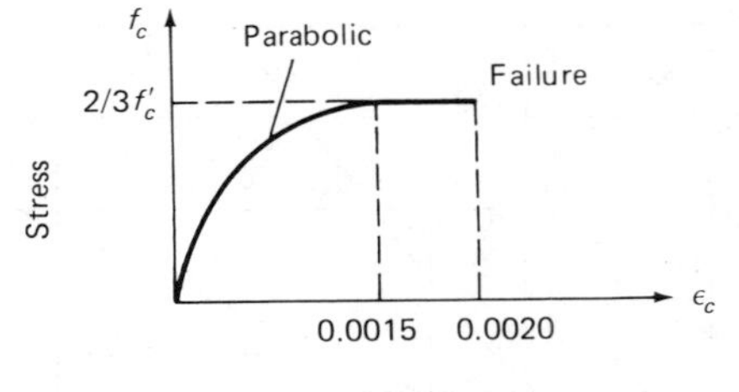

Fig. 15-13 Parabolic and level variation stress–strain curve for concrete (as used by Greiner)

Other researchers used different assumptions for the stress–strain relationship in concrete. Brettle[15-30] assumed a bilinear curve as shown in Fig. 15-12 and considered the criteria for failure to be at a maximum strain of 0.0035 in the concrete. He also incorporated the compressive reinforcement in the calculations. Cannon[15-31] utilized the compression block as specified by ACI Standard 318 and ignored the compression reinforcement. Greiner[15-32] used the curve of Fig. 15-13 for the stress–strain relationship and incorporated the compressive reinforcement.

Figure 15-14 shows a cross section of a chimney with a flue opening and gives the strain distribution and the stresses at ultimate in both the concrete and the steel. The angle α locates the neutral axis, and the angle β defines the opening. The locations where the strain in the concrete is 0.002, the strain in the compression steel is equal to the yield strain, and the strain in the tensile steel is equal to the yield strain, are designated by the angles θ_1 θ_2, and θ_3, respectively.

From the two equilibrium conditions, $\Sigma F = 0$ and $\Sigma M = 0$, the following equations are obtained:

$$W + T - C_c - C_s = 0 \qquad (15\text{-}49)$$

and:

$$Wr \cos \alpha - M_n + C'_c + C'_s + T = 0 \qquad (15\text{-}50)$$

where C_c C_s, and T are the forces in the concrete, the compression steel, and the tensile steel, respectively. C'_c, C'_s, and T' are the moments of these forces about the neutral axis.

For any cross section and for any value of the axial force W the ultimate nominal bending moment M_n is obtained by solving eqs. (15-49) and (15-50). Because the expressions involved are rather complex, the value of α is obtained by solving eq. (15-49) by iteration. Once the value of α is established, then M_n is obtained from eq. (15-50).

Ultimate strength curves based on the above procedure were generated[15-27] and are given in Figs. 15-15 through 15-18, for $\beta = 0$, 10°, 20°, and 30°. The curves are given in terms of the dimensionless parameters, M_n/Wr, W/rtf'_c, and $p(f_{sy}/f'_c)$, where M_n is the ultimate moment, W is the axial force on the section, r is the mean radius, t is the thickness, p is the steel ratio, f_{sy} is the yield strength of the steel, and f'_c is the strength of the concrete.

The values of the curves for ultimate design can be approximately represented by the following equation:

For $\beta = 0$ (no openings)

$$\frac{M_n}{Wr} = 1 + \left[6 - 2\left(\frac{W}{rtf'_c}\right)^2\right] \frac{p(f_{sy}/f'_c)}{\dfrac{W}{rtf'_c}} \qquad (15\text{-}51)$$

For a cross section with one opening multiply the right-hand side of the above equation by the factor:

$$\left[1 - \frac{\beta^{1.15}}{200}\right]$$

where β is in degrees (see Fig. 15-14).

The criteria for ultimate design can be expressed by the following relationships:

$$M_u \leqq \phi M_n \qquad (15\text{-}52)$$

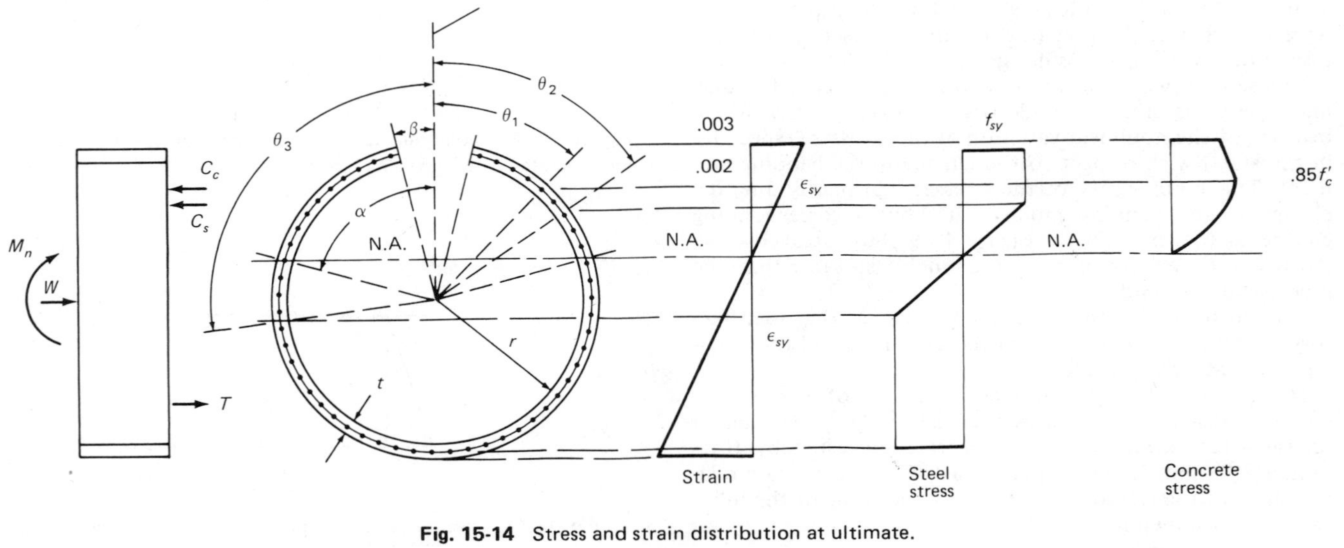

Fig. 15-14 Stress and strain distribution at ultimate.

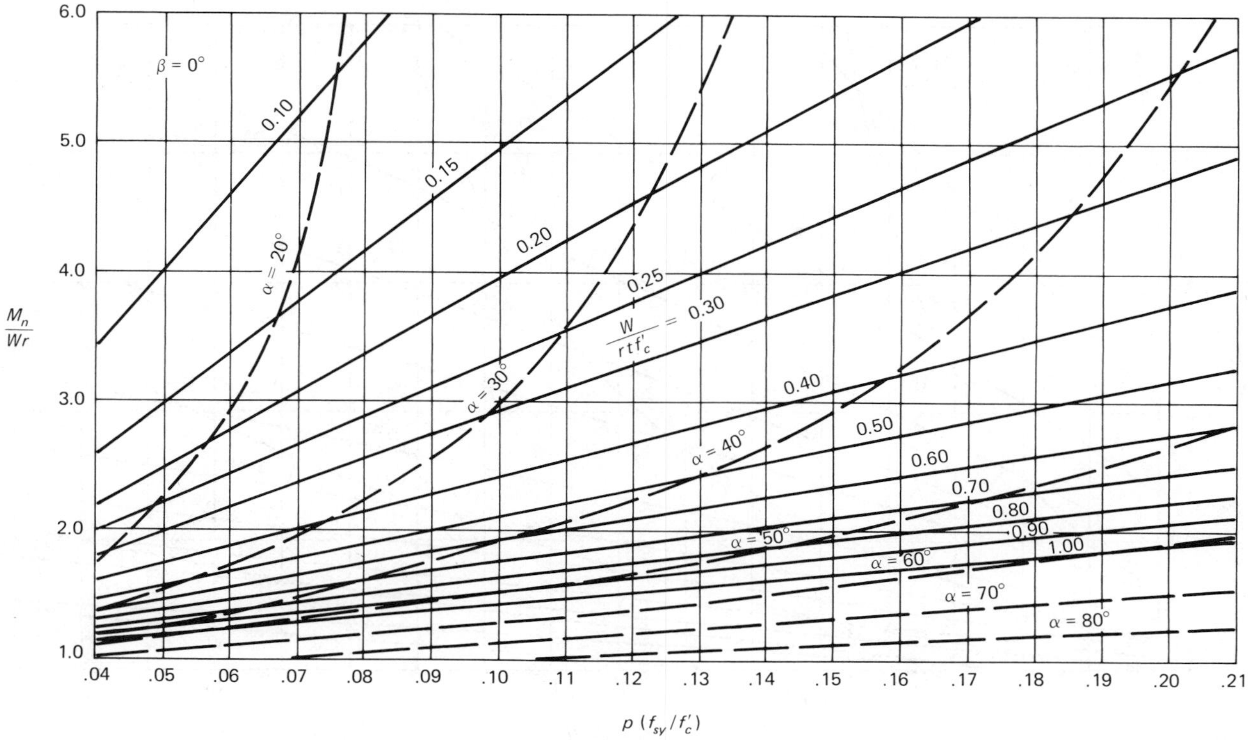

Fig. 15-15 Ultimate strength design chart with no opening ($\beta = 0$).

where:

$$M_u = 1.7M \qquad (15\text{-}53)$$

where M is the bending moment due to wind or earthquakes. M_n is obtained by using the curves or equations and by using 0.9 of axial load for W.

The value for the capacity reduction factor ϕ would fall between 0.7 and 0.9 as stipulated by ACI318-83. It is the writer's opinion that a value of ϕ close to 0.9 can be justified, since test results[15-33] have shown moment capacities larger than the computed values by about 15 to 20%.

EXAMPLE 15-1: To illustrate the use of the equations and curves

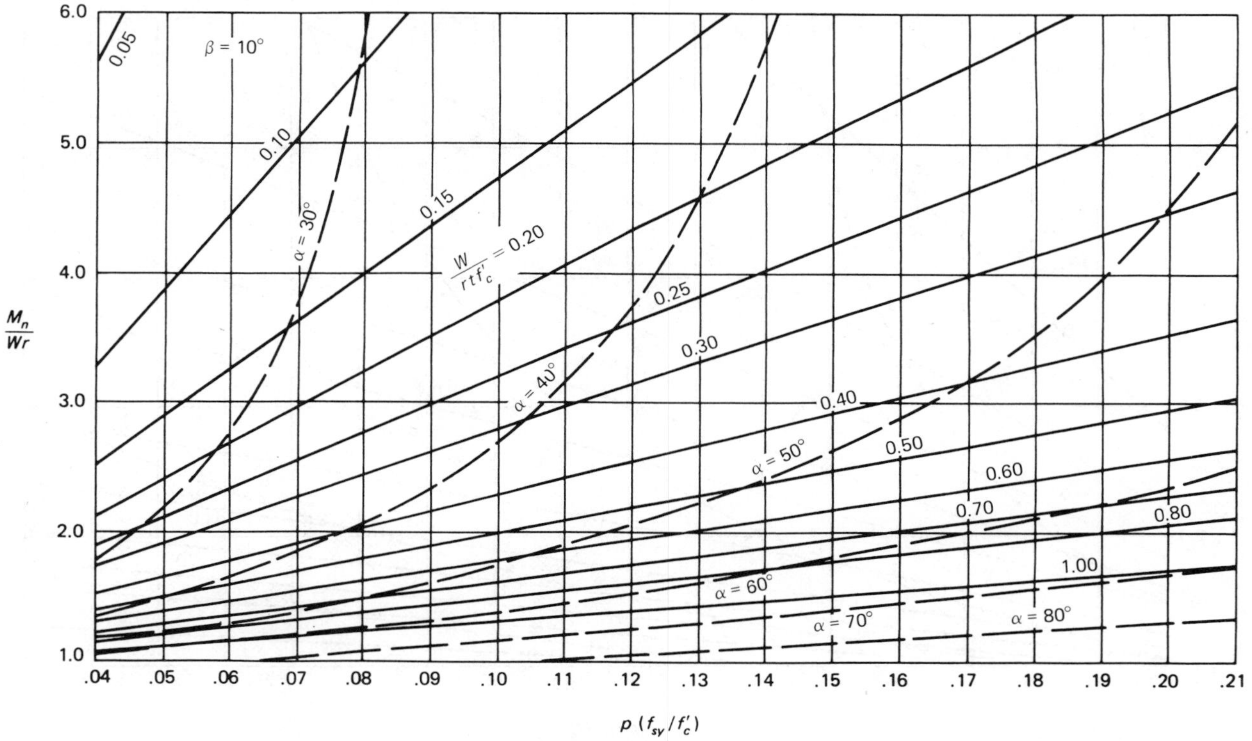

Fig. 15-16 Ultimate strength design chart with an opening ($\beta = 10°$).

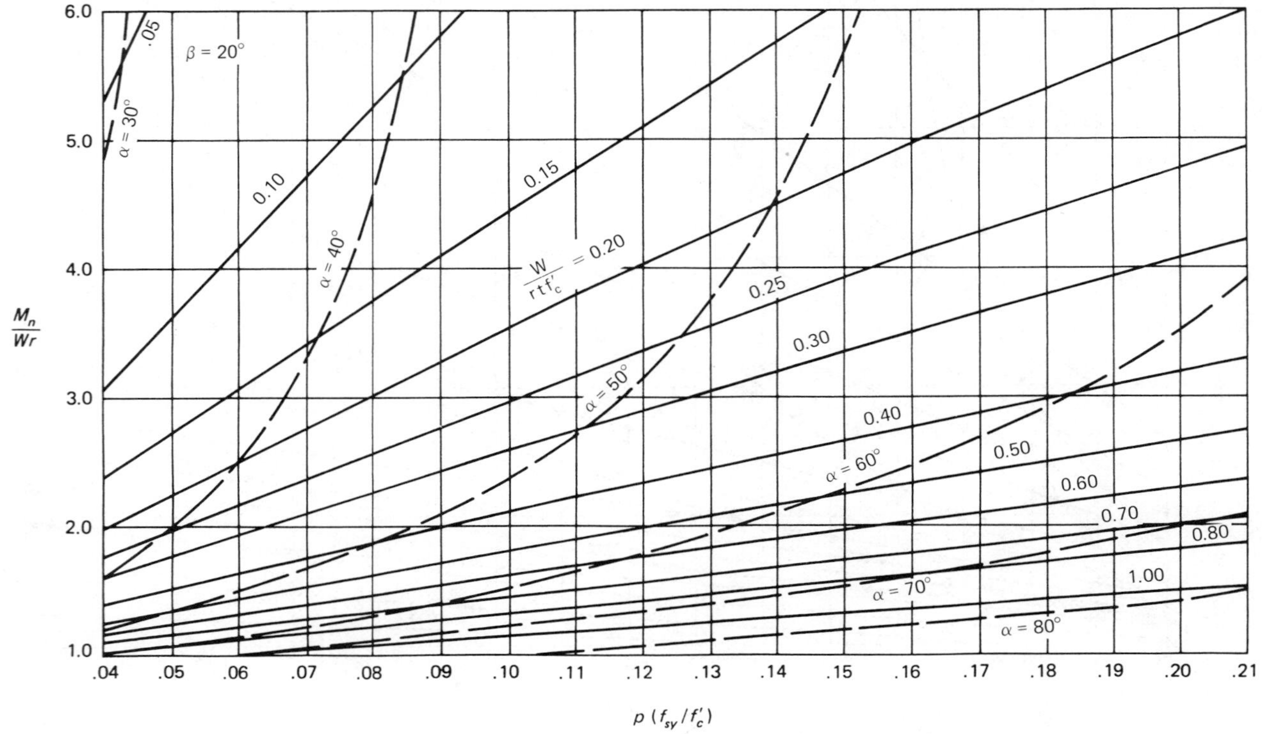

Fig. 15-17 Ultimate strength design chart with an opening ($\beta = 20°$).

presented in this chapter, the following example of chimney computations is given. The example chimney is 1000 ft high and has a constant taper with outside diameters varying from 34.25 ft at the top to 77.13 ft at the bottom. The outside diameters, thicknesses, and moments of inertia are tabulated at 11 equally spaced locations in Table 15-7.

SOLUTION:

1. First mode; mode shape, shears, and moments: The computations for the first mode of vibration are given in Fig. 15-19. The computations are primarily based on Newmark's "Numerical Procedure,"[15-34] which is very well adapted to this type of problem. The elements of this method will be summarized here for convenience.

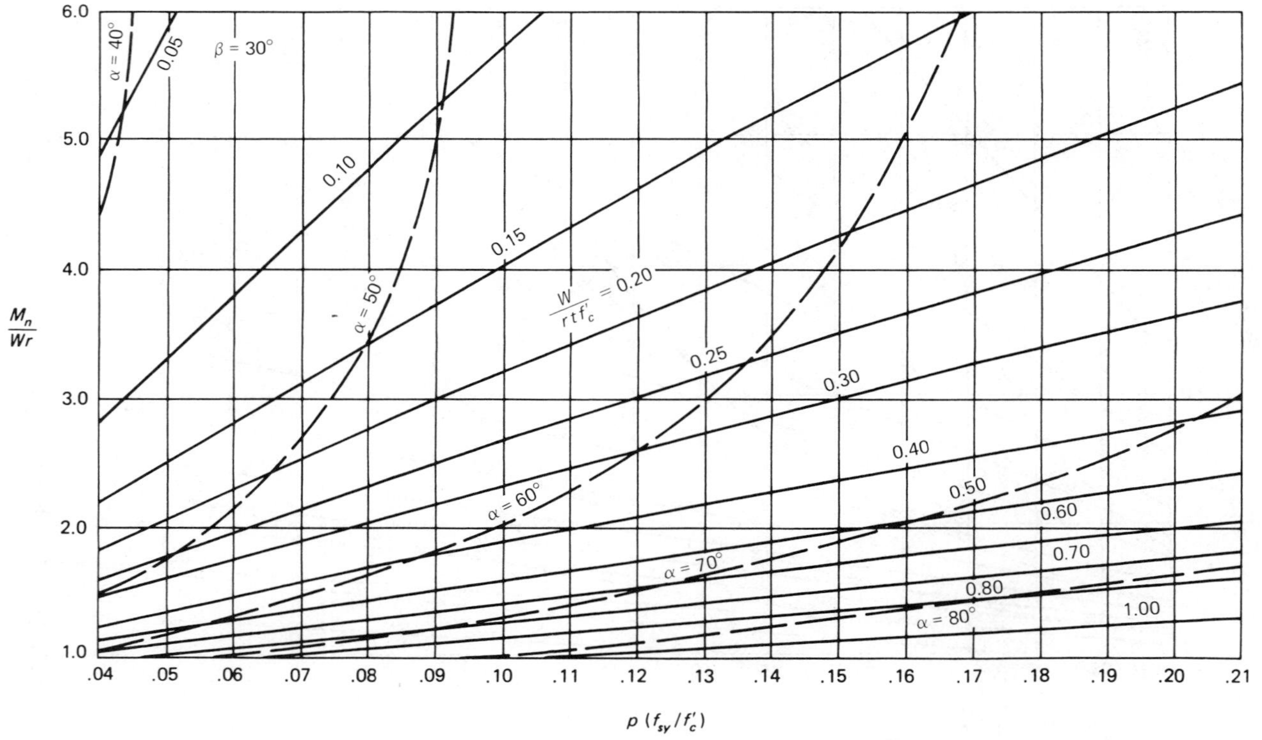

Fig. 15-18 Ultimate strength design chart with an opening ($\beta = 30°$).

TABLE 15-7 Properties of Example Chimney

Distance from Top (ft)	*Outside Diameter (ft)*	*Thickness (in.)*	*Moment of Inertia (ft^4)*
0	34.25	9.00	11080
100	38.54	10.59	18520
200	42.83	12.25	29320
300	47.11	12.25	39270
400	51.40	12.25	51280
500	55.69	12.25	65520
600	59.98	12.79	85620
700	64.27	15.96	130280
800	68.55	19.36	190120
900	72.84	22.43	262570
1000	77.13	31.00	420770

If a beam, as, for example, a cantilever beam, is subjected to a distributed loading as shown in Fig. 15-20(a), then the shears and bending moments in the beam can be obtained by dealing with the equivalent concentrated loads as shown in Fig. 15-20(b). Although the shapes of the shear and bending moment diagrams are not the same in the two systems, it should be apparent that the shears and moments at locations 1, 2, 3, and 4 in the actual system would be the same as those in the equivalent system. Once the values of the equivalent concentrated loads are determined, the computations for the shears and bending moments become rather simple. Another simplification is used in which the equivalent concentrated load at a location is given in whole rather than in parts. In this case, although the shear values cannot be determined at the locations, their values at sections midway between the locations would be approximately equal to the constant shear value in the equivalent system. The equivalent concentrated loads for any load distribution can be obtained by assuming a parabolic curve for the distribution and by using equal elements. The formulas for these concentrated loads are given in Fig. 15-21.

In Fig. 15-19, the 1000-ft chimney is divided into ten equal segments. The chimney is subjected to the weight of the concrete given in line 1.

The process involves assuming a mode shape (line 2) and computing inertia loads (line 3). The inertia load is equal to the load times the deflection times the square of the frequency, ω (radians per second), and divided by the acceleration of gravity, g. The equivalent concentrated loads given in line 4 are used to calculate the deflections given in line 11. This deflection curve is then normalized (line 12) by making the top displacement equal to unity. This normalized shape is then compared to the assumed shape of line 2. If line 12 agrees reasonably well with line 2, then the process is terminated, and the frequency, ω is obtained by equating the values of line 2 to those of line 11. If the shape of line 12 does not agree with the shape of line 2, then the process is repeated, starting with line 3 with an assumed shape equivalent to line 12.

In most cases one cycle is enough, especially if a reasonable shape is assumed, and if ω is computed as the average of values obtained from different locations along the chimney.

The period of the chimney as computed in Fig. 15-19 will be compared to the value obtained using eq. (15-15). The equation gives a value of:

$$5\,\frac{(1000)^2}{33.5}\sqrt{\frac{4.66}{5.472\times 10^8}}\left(\frac{9}{31}\right)^{.22}\left(\frac{33.5}{74.55}\right)^{1.1} = 4.35 \text{ sec/cycle}$$

The value of T_1 computed in Fig. 15-19 is 4.48 sec/cycle.

2. *Alongwind moments:* The alongwind moments will be generated for the example chimney using a reference mean wind speed of 70 MPH. The computations are given in Fig. 15-22. The load intensity, w due to the mean wind profile is computed from eq. (15-8) and is given in line 3 of Fig. 15-22. To account for the fluctuating part of the wind and for the dynamic effect, the shear multiplier, J_v (line 7) and the moment multiplier, J_m (line 8), have been calculated from eqs. (15-9) through (15-14). Using these multipliers and using the shears and moments due to the mean wind profile (lines 5 and 6, respectively) the alongwind shears and moments are calculated for the total wind and are given in lines 9 and 10. The deflections given in line 16 are based on the moments of line 10.

3. *Acrosswind moments:* The acrosswind moments will be generated by computing the multipliers for first and second modes expressed by formulas (15-18) and (15-21), respectively.

First mode computations: The critical section is at the location where $D^4\phi$ is maximum. Using the information in Fig. 15-19 (line 12) and the diameters as given in Table 15-7, the critical location for acrosswind vibration in the first mode would be at about 600 ft from the base where the critical diameter, D_c is 51.40 ft. Using $D_c = 51.40$ ft, strouhal number $S = 0.2$, and the first mode frequency $f = .223$ cycles/sec, the mean wind speed at 600 ft from the base that would give a shedding frequency f_s equal to .223 would be (from eq. 15-17 or 15-19) equal to $0.223 \times 51.4/0.2 = 57.3$ fps. The mean wind speed at the reference height of 30 ft would be (from eq. 15-20) equal to:

$$\left(\frac{30}{600}\right)^{1/7} * 57.3 = 37.35 \text{ fps} = 25.5 \text{ MPH}$$

The critical resonant zone, $(Z_2 - Z_1) = 0.15D_c/\text{taper}$, would be equal to $0.15 \times 51.4/0.0429 = 179.7$ ft. This distance should not exceed $3D_c = 3 \times 51.4 = 154.2$ ft or 175 ft. The integral $\int m\phi_1^2 dZ$ can be easily obtained by using Simpson's rule. The values $m\phi^2$ at the 11 equally spaced locations starting from the top would be:

$$\frac{1}{32.2}\,[11.84, 10.74, 8.76, 5.563, 3.124, 1.490, 0.6051, 0.2279,$$
$$0.0489, 0.00306, 0] \text{ k-sec}^2$$

$$\therefore \int m\phi_1^2 dZ = \frac{100}{3\times 32.2}(11.84 + 0) + 4(10.74 + 5.565 + 1.490$$
$$+ 0.2279 + 0.00306) + 2(8.76 + 3.124$$
$$+ 0.6051 + 0.0489)$$
$$= 112.9 \text{ k-ft-sec}^2$$

The lift coefficient C_L would be 0.67, since $H/D' = 1000/41.4 = 24$ is greater than 20.

Using formula (15-18) with $(Z_2 - Z_1) = 154.2$ ft, $D_c = 51.4$ ft $C_L = 0.67$, $\int m\phi_1^2 dZ = 112.9$ k-ft-sec^2, $\rho = .00000238$ k-sec^2/ft^4, and using 0.02 for β and 0.2 for S, one obtains:

$$q_1 = \frac{0.00000238 * 0.67}{16\pi^2 * (0.2)^2 * (0.02)^{2/3}} * 2.20\,\frac{(51.4)^3 * 0.359 * (154.2)}{112.9}$$
$$= 0.503$$

The modal displacements, shears, and moments given in Fig. 15-19 are multiplied by 0.503 to give the first mode acrosswind values for the chimney. The computed top displacement of 0.503 ft and the reference critical mean speed of 25.5 MPH compare well with the values shown in Fig. 15-5.

Second Mode Computations: Approximate modal characteristics for the second mode are obtained by computing the period from formula (15-22) and by using a reasonable mode shape. These modal characteristics are generated in Fig. 15-23. The shape shown in line 2 of the figure is believed to be a good approximation for the second mode shape of chimneys and should give good results. The inertia load intensities of line 3 are obtained by multiplying ω_2^2/g by the values of the product of lines 1 and 2.

As indicated previously, the first peak of the second mode resonance would occur when the wind profile is such that the shedding frequency near the top coincides with that of the second mode frequency of the chimney. For this reason it is reasonable to use $Z_2 = H$. The value of $(Z_2 - Z_1)$ would therefore be equal to:

$$\frac{0.15D_0}{0.925 * \text{taper}} = \frac{0.15 * 34.25}{0.925 * 0.0429} = 129.5 \text{ ft}$$

For $(Z_2 - Z_1)$ not to exceed $3D_c$, one obtains the following rela-

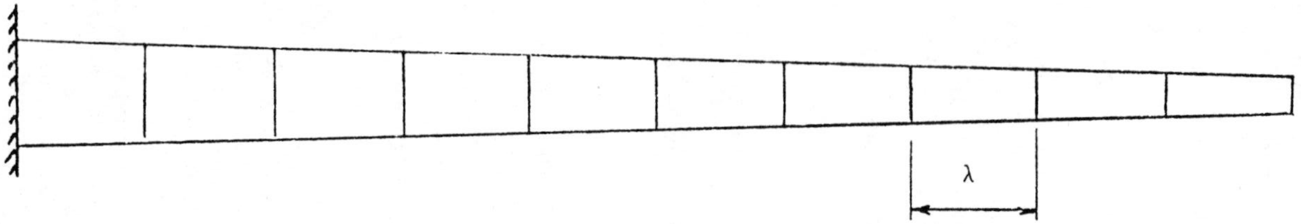

												Multiplier	
Weight Intensity	90.75	62.51	50.89	39.45	29.59	26.30	24.24	22.17	20.11	15.66	11.84	k/ft	1
Assumed defl. curve	0	.01	.03	.08	.14	.24	.36	.50	.66	.83	1.00	Y_t	2
Inertia load intensity	0	.63	1.53	3.16	4.14	6.31	8.73	11.09	13.27	13.00	11.84	$\frac{\omega^2}{g} Y_t \frac{k}{ft}$	3
Equiv. conc. load		7.83	19.09	37.27	50.87	75.97	104.70	132.90	156.79	155.11	73.81	$\frac{\lambda}{12} \frac{\omega^2}{g} Y_t \frac{k}{ft}$	4
V	814.34	806.51	787.42	750.15	699.28	623.31	518.61	385.71	228.92	73.81		$\frac{\lambda}{12} \frac{\omega^2}{g} Y_t \frac{k}{ft}$	5
M	5688	4874	4067	3280	2530	1830	1207	688	303	74	0	$\frac{\lambda^2}{12} \frac{\omega^2}{g} Y_t \frac{k}{ft}$	6
I	42.08	26.26	19.01	13.03	8.56	6.55	5.13	3.93	2.93	1.85	1.11	I_0	7
M/EI	135.2	185.6	213.9	251.7	295.6	279.4	235.3	175.1	103.4	40.0	0	$\frac{\lambda^2}{12EI_0} \frac{\omega^2}{g} Y_t \frac{k}{ft}$	8
Angle change	923	2205	2576	3027	3487	3325	2808	2090	1249	503		$\frac{\lambda^3}{144EI_0} \frac{\omega^2}{g} Y_t \frac{k}{ft}$	9
Chord slope	923	3128	5704	8731	12218	15543	18351	20441	21690	22193		$\frac{\lambda^3}{144EI_0} \frac{\omega^2}{g} Y_t \frac{k}{ft}$	10
Deflection	0	923	4051	9755	18486	30704	46247	64598	85039	106729	128922	$\frac{\lambda^4}{144EI_0} \frac{\omega^2}{g} Y_t \frac{k}{ft}$	11
Defl. shape	0	.007	.031	.076	.143	.238	.359	.501	.660	.828	1.00	Y_t	12
M	289708	248248	207145	167061	128861	93208	61476	35042	15433	3769	0	$Y_t \cdot k$	13
V	415	411	401	382	356	317	264	196	117	38		Yt k/ft	14

$\lambda = 100$ ft $\quad E = 3.8 \times 10^6$ psi $\quad I_0 = 10000$ ft^4

$$128922 \frac{\lambda^4}{144EI_0} \frac{\omega^2}{g} Y_t = Y_t \qquad \omega^2 = 1.97 \text{ (radians/sec)}^2,\ \omega = 1.40 \text{ radians/sec}$$

Fig. 15-19 First mode computations.

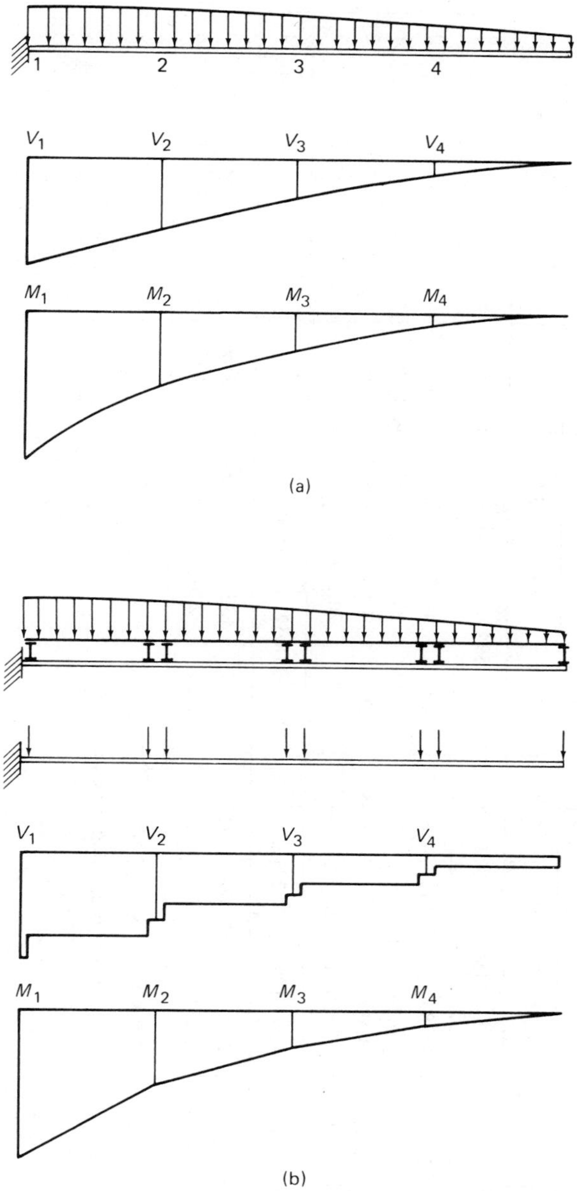

Fig. 15-20 Actual and equivalent loading systems.

tionship:

$$D_c = 34.25 + 0.0429 * \frac{3D_c}{2}$$

which gives $D_C = 36.6$ ft and $(Z_2 - Z_1) = 109.8$ ft.

The mass-shape factor, $\int m\phi_2^2\, dZ$, is computed from the information in Fig. 15-23 to be equal to 97.4 k-ft-sec^2.

Using formula (15-21), the second mode multiplier would be:

$$q_2 = \frac{0.00000238}{16\pi^2(0.2)^2}\ \frac{0.67}{(0.02)^{2/3}} * 3.08\ \frac{(36.6)^3 * 0.724 * 109.8}{97.4}$$
$$= 0.423$$

Since the frequency of the chimney (from formula 15-22) is 0.923 cycles/sec, the critical mean wind speed at a height of 1000 − 109.8/2) = 945 ft would be 0.923 * 36.6/0.2 = 168.9 fps, which gives a reference mean wind speed of:

$$\left(\frac{30}{945}\right)^{1/7} * 168.9 = 103.2 \text{ fps} = 70.0 \text{ MPH}$$

4. *Earthquake moments:* The earthquake moments for the 1000-ft chimney will be obtained for Zone 3 by using the procedure described in Section 15.5.

Period: From formula (15-15), $T_1 = 4.35$ sec/cycle.

Base shear: The base shear, which is a function of the period, the zone coefficient, and the total weight of the chimney, will be computed from formula (15-25). The weight of the chimney can be obtained by integrating the weight intensities at the 11 stations using Simpson's rule. This would give:

$$W = \frac{100}{3}[(90.75 + 11.84) + 4(62.51 + 39.45 + 26.30$$
$$+ 22.17 + 15.66) + 2(50.89 + 29.59$$
$$+ 24.24 + 20.11)]$$
$$= 33887 \text{ kips}$$

$$\therefore\ V_b = 33887 * \frac{3}{4} * \frac{0.35}{\sqrt{4.35}} = 4265 \text{ kips}$$

The base shear is distributed along the height of the chimney according to the rules expressed by formula (15-26). The shears and bending moments are then obtained by using eqs. (15-27) through (15-37). The computations are shown in Fig. 15-24. The shear values in row 12 are obtained by modifying the values of row 10 by the multipliers of Table 15-5. Similarly, the moment values in row 13 are calculated by multiplying the values of row 11 by the factors of Table 15-6.

A more rational method to generate shears and bending moments in a chimney is to use the response spectra method. For chimneys, three or four modes would be required, which would make it necessary to use a computer to obtain the modal characteristics. In this example, the shears and moments will be calculated by using the response spectra of Table 15-4 and by incorporating only the first and second modes of vibration. The modal characteristics of the first mode are given in Fig. 15-19, while approximate values for the shears and bending moments of the second mode are given in Fig. 15-23. The use of the spectrum techniques would require establishing the value:

$$\alpha_j = \frac{\int_0^H m\phi_j\, dx}{\int_0^H m\phi_j^2\, dx}$$

for the first and second modes of vibration ($j = 1, 2$).

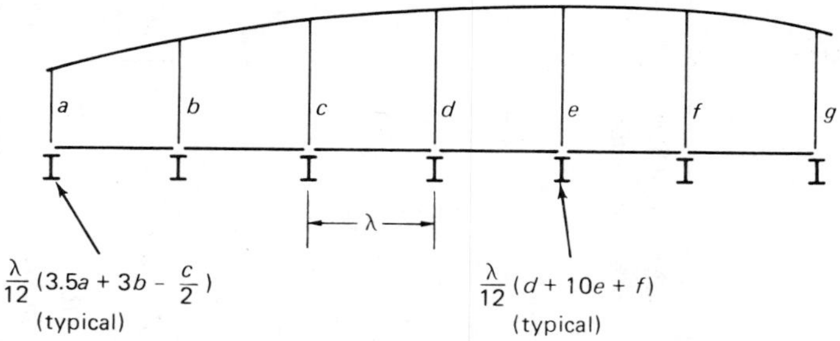

Fig. 15-21 Formulae for equivalent concentrated loads.

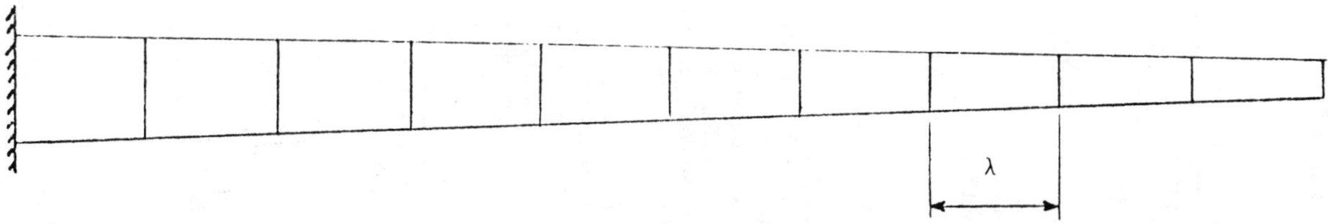

												Multiplier	
Z (height above base)	0	100	200	300	400	500	600	700	800	900	1000	ft	1
D (outside diameter)	77.13	·72.84	68.55	64.27	59.98	55.69	51.40	47.11	42.83	38.54	34.25	ft	2
w	0	0.905	1.038	1.093	1.107	1.096	1.065	1.020	0.964	0.897	0.821	k/ft	3
Equiv. conc. load		10.088	12.378	13.075	13.259	13.132	12.766	12.229	11.557	10.755	5.083	λ/12 k/ft	4
Shear $\bar{V}$	114.322	104.234	91.856	78.781	65.522	52.390	39.624	27.395	15.838	5.083		λ/12 k/ft	5
Moment $\bar{M}$	595.05	480.72	376.49	284.63	205.85	140.33	87.94	48.32	20.92	5.08	0	$λ^2$/12 k/ft	6
J_v	1.48	1.64	1.80	1.96	2.12	2.28	2.44	2.60	2.76	2.92			7
J_m	1.75	1.89	2.02	2.16	2.29	2.43	2.56	2.70	2.83	2.97	3.10		8
$V = J_v\bar{V}$	1410	1425	1378	1287	1158	995	806	594	364	124		k	9
$M = J_m\bar{M}$	867781	757134	633758	512334	392830	284168	187605	108720	49336	12573	0	k-ft	10
I	420770	262570	190120	130280	85620	65520	51280	39270	29320	18520	11080	ft^4	11
M/EI	2.062	2.884	3.333	3.933	4.588	4.337	3.658	2.769	1.683	0.679	0	$1/E$ k/ft^3	12
Angle change	14.203	34.235	40.147	47.251	54.150	51.616	43.686	33.031	20.278	8.473		$λ/12E$ k/ft^3	13
Chord slope	14.20	48.44	88.59	135.84	189.99	241.60	285.29	318.32	338.60	347.07		$λ/12E$ k/ft^3	14
Deflection	0	14.20	62.64	151.23	287.07	477.06	718.66	1003.95	1322.27	1660.87	2007.94	$λ^2/12E$ k/ft^3	15
Deflection	0	0.022	0.095	0.230	0.437	0.727	1.094	1.529	2.014	2.529	3.058	ft	16

λ = 100 ft $\quad$ $E = 3.8 \times 10^6$ psi

Fig. 15-22 Alongwind response.

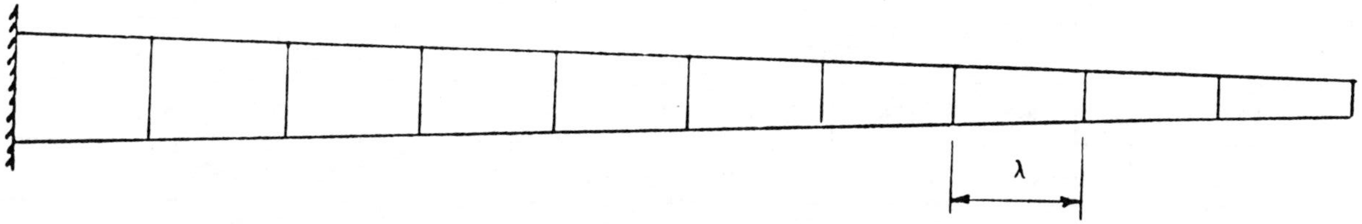

												Multiplier	
Weight intensity	90.75	62.51	50.89	39.45	29.59	26.30	24.24	22.17	20.11	15.66	11.84	$\frac{k}{ft}$	1
Mode shape	0	-0.03	-0.13	-0.27	-0.41	-0.49	-0.46	-0.29	0.05	0.50	1.00	ft	2
Inertia load intensity	0	-1.959	-6.912	-11.128	-12.674	-13.463	-11.649	-6.717	1.050	8.180	12.369	$\frac{k}{ft}$	3
Equiv. conc. load		-26.50	-82.21	-130.87	-151.33	-158.95	-136.67	-77.77	11.96	95.22	67.31	$\frac{\lambda}{12} \cdot \frac{k}{ft}$	4
V	-589.81	-563.31	-481.10	-350.23	-198.90	-39.95	96.72	174.49	162.53	67.31		$\frac{\lambda}{12} \cdot \frac{k}{ft}$	5
M	-1722.25	-1132.44	-569.13	-88.03	262.20	461.10	501.05	404.33	229.84	67.31	0	$\frac{\lambda^2}{12} \cdot \frac{k}{ft}$	6
V	-4915	-4694	-4009	-2919	-1658	-333	806	1454	1354	561		k	7
M	-1435208	-943700	-474275	-73358	218500	384250	417542	336942	191533	56092	0	k-ft	8

$\int m\phi_2^2 dz = 97.4$ k-ft-sec^2 $\lambda = 100$ ft

From formula (15-22), $T_2 = 1.083$ sec/cycle ($\omega_2 = 5.80$ radians/sec).

Fig. 15-23 Second mode characteristics.

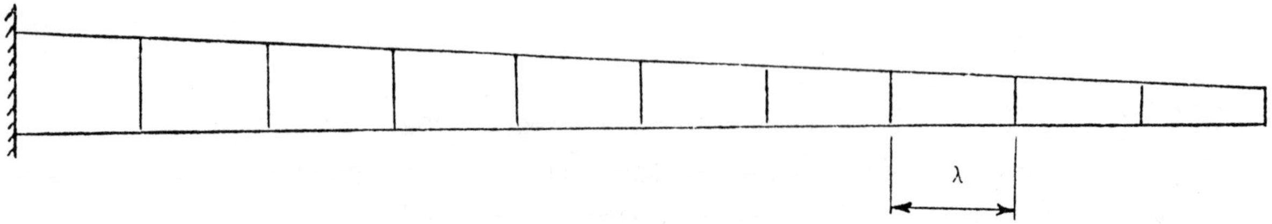

												Multiplier	
Weight intensity, w	90.75	62.51	50.89	39.45	29.59	26.30	24.24	22.17	20.11	15.66	11.84	$\frac{k}{ft}$	1
Distance from base, x	0	1	2	3	4	5	6	7	8	9	10	λ	2
wx^2	0	62.51	203.56	355.56	473.44	657.50	872.64	1086.3	1287.0	1268.5	1184.0	$\lambda^2 \frac{k}{ft}$	3
Equiv. conc. load of wx^2, F	85.75	828.66	2453.2	4227.5	5747.0	7921.1	10470	13023	15225	15156	7306	$\frac{\lambda^3}{12} \frac{k}{ft}$	4
$\frac{V_b}{\sum F} = .05173 \times 12 \frac{ft}{\lambda^3}$													5
V_h^1	4260	4218	4091	3872	3575	3165	2623	1950	1162	378		k	6
M_h^1	29294	25034	20816	16725	12853	9278	6113	3490	1540	378	0	$\lambda \cdot k$	7
J_v	0.949	0.846	0.743	0.640	0.537	0.486	0.525	0.670	0.991	1.561			8
J_m	0.415	0.361	0.329	0.318	0.354	0.463	0.645	0.898	1.225	1.623	2.096		9
$V_h = J_v V_h^1$	4043	3568	3040	2478	1920	1538	1377	1307	1152	590		k	10
$M_h = J_m M_h^1$	12157	9037	6848	5319	4550	4296	3943	3134	1887	613	0	$\lambda \cdot k$	11
Corrected V_h	4164	3532	2706	2057	1690	1492	1267	1124	1083	596		k	12
Corrected M_h	11428	8224	6095	4893	4095	3523	2997	2445	1642	595	0	$\lambda \cdot k$	13

$V_b = 4265$ k $\lambda = 100$ ft

Fig. 15-24 Earthquake shears and moments for example chimney (Zone 3).

TABLE 15-8 Evaluation of α_1 and α_2

$\frac{h}{H}$	w (k/ft)	ϕ_1	ϕ_2	$w\phi_1$	$w\phi_1^2$	$w\phi_2$	$w\phi_2^2$
1.0	11.84	1.00	1.00	11.84	11.84	11.84	11.84
0.9	15.66	0.828	0.50	12.97	10.74	7.83	3.92
0.8	20.11	0.660	0.05	13.27	8.76	1.01	0.05
0.7	22.17	0.501	−0.29	11.11	5.57	−6.43	1.86
0.6	24.24	0.359	−0.46	8.70	3.12	−11.15	5.13
0.5	26.30	0.238	−0.49	6.26	1.49	−12.89	6.32
0.4	29.59	0.143	−0.41	4.23	0.61	−12.13	4.97
0.3	39.45	0.076	−0.27	3.00	0.23	−10.65	2.88
0.2	50.89	0.031	−0.13	1.58	0.05	−6.62	0.86
0.1	62.51	0.007	−0.03	0.44	0	−1.88	0.06
0	90.75	0	0	0	0	0	0

The quotients α_1 and α_2 are evaluated by Simpson's rule using the tabulated values in Table 15-8:

$$\alpha_1 = \frac{(11.84+0)+4(12.97+11.11+6.26+3.00+0.44)+2(13.27+8.70+4.23+1.58)}{(11.84+0)+4(10.74+5.57+1.49+0.23+0)+2(8.76+3.12+0.61+0.05)}$$

$$= 1.857$$

$$\alpha_2 = \frac{(11.84+0)+4(7.83-6.43-12.89-10.65-1.88)+2(1.01-11.15-12.13-6.62)}{(11.84+0)+4(3.92+1.86+6.32+2.88+0.06)+2(0.05+5.13+4.97+0.86)}$$

$$= -1.511$$

The first mode period (from Fig. 15-19) is 4.49 sec/cycle, and the second mode period as computed from formula (15-22) is 1.08 sec/cycle. Using the values for T_1 and T_2, the displacement spectra for the first and second modes (Table 15-4) are 16.90 in. (1.408 ft) and $3.462/.926^{1.1436} = 3.78$ in. (0.315 ft), respectively. The modal multipliers for the first and second modes would be $1.857 * 1.408 = 2.615$ and $1.511 * 0.315 = 0.476$, respectively.

Table 15-9 gives the calculations for the earthquakes moments due to first and second mode responses. The combined response is obtained by using the square root of the sum of the squares of the modal responses.

Table 15-10 gives the earthquake moments in the chimney as obtained by three different procedures: (a) equivalent "static" approach using the formulas presented in this chapter, (b) spectrum techniques using two modes with the second mode approximated as suggested in this chapter, and (c) spectrum techniques using a computer program and utilizing five modes of vibration. The spectrum values given in Table 15-4 were used for procedures (b) and (c). It is apparent from Table 15-10 that more than two modes of vibration are required for earthquake response, especially in the upper portions of the chimney. This would be especially true for the taller and more flexible chimneys.

5. *Selection of vertical reinforcement:* The vertical reinforcement in the 1000-ft chimney will be obtained using ultimate strength design. The bending moments due to the following loading conditions will be obtained: (a) alongwind response due to a reference mean wind speed of 70 MPH, (b) acrosswind first mode response, (c) acrosswind second mode response, and (d) earthquake response that would be obtained by using the equivalent "static" procedure and by using the average response of Zones 2 and 3. The bending moments of the above four cases are given in Table 15-11.

The selection of the reinforcement can be obtained by using the charts given in Figs. 15-15 through 15-18, or by using formula (15-51). In this example the formula will be used.

The ultimate strength design will be based on using a load factor of 1.7 for the moment and 0.9 for the axial load W. The capacity reduction factor will be taken as 0.8. The yield stress of the steel will be taken as 60 ksi (8640 ksf), and the strength of the concrete will be taken as 4444 psi (640 ksf). The value for f_c' was selected to be compatible with the modulus of elasticity of the concrete, which was considered to be 3800 ksi in previous calculations.

Based on the above constants, formula (15-51) will reduce to the form (using the units of kips and feet):

$$\frac{(1.7\,M/0.8)}{0.9\,Wr} = \left\{1 + \left[6 - 2\left(\frac{0.9\,W}{640\,rt}\right)^2\right] \frac{13.5\,P}{(0.9\,W/640\,rt)}\right\} \left\{1 - \frac{\beta^{1.15}}{200}\right\}$$

Table 15-12 gives, at different sections along the chimney, the values of the axial load (W), the mean radius (r), the thickness (t), the criti-

TABLE 15-9 Earthquake Spectrum Moments (k-ft)

$\frac{h}{H}$	First Mode Moments, M_1'	Second Mode Moments, M_2'	First Mode Response, M_1	Second Mode Response, M_2	$\sqrt{M_1^2 + M_2^2}$
1.0	0	0	0	0	0
0.9	3769	56092	9856	26700	28461
0.8	15433	191533	40357	91170	99703
0.7	35042	336942	91635	160384	184716
0.6	61476	417542	160760	198750	255627
0.5	93208	384250	243739	182903	304733
0.4	128861	218500	336972	104006	352657
0.3	167061	73358	436865	34918	438258
0.2	207145	474275	541684	225755	586845
0.1	248248	943700	649169	449201	789431
0	289708	1435208	757586	683159	1020119

First mode multiplier = 2.615
Second mode multiplier = .476

TABLE 15-10 Earthquake Moments (k-ft) for Zone 3 as Obtained by Different Procedures

$\frac{h}{H}$	*(a) Equivalent "Static" Approach*	*(b) Spectrum Techniques Using Two Modes With Second Mode Approximated*	*(c) Spectrum Techniques Using Five Modes and Using a Computer Program*
1.0	0	0	0
0.9	59500	28461	57428
0.8	164200	99703	150138
0.7	244500	184716	217694
0.6	299700	255627	267226
0.5	352300	304733	321766
0.4	409500	352657	389173
0.3	489300	438258	470300
0.2	609500	586845	590019
0.1	822400	789431	804535
0	1142800	1020119	1119538

TABLE 15-11 Wind and Earthquake Bending Moments (k-ft)

$\frac{h}{H}$	*(a) Alongwind*	*(b) 1st Mode Acrosswind*	*(c) 2nd Mode Acrosswind*	*(d) Earthquake Moments for Average of Zones 2 and 3*
1.0	0	0	0	0
0.9	12573	1896	23727	44625
0.8	49336	7763	81018	123150
0.7	108720	17626	142526	183375
0.6	187605	30922	176620	224775
0.5	284168	46884	162538	264225
0.4	392830	64817	92426	307125
0.3	512334	84032	31030	366975
0.2	633758	104194	200618	457125
0.1	757134	124869	399185	616800
0	867781	145723	607093	857100

cal bending moment (M), and the angle of half the opening (β). Column seven of the table lists the steel ratios (P) as computed by the above formula. It is of interest to note that a change of the "reduction capacity factor," ϕ, from 0.8 to 0.9 would change the steel ratios to the following values (starting from the section at 0.9 H from the base): 0.00378, 0.00759, 0.00871, 0.00796, 0.00797, 0.00940, 0.00323, 0.00674, 0.007980, and 0.00358. The last column of Table 15-12 gives the minimum steel ratios that are required to ensure that the nominal moment, M_n, is well above the cracking moment, M_{cr}. The values are twice those required to make M_n equal to M_{cr}. The chimney standard ACI 307-79 stipulates a minimum steel ratio of .0025. It is obvious that this ratio would be inadequate for the upper portions of the chimney. For evaluating the nominal moment at the top of the chimney where W is equal to zero, formula (15-51) could be written in the following form:

$$M_n = \left\{Wr + \left[6 - 2\left(\frac{W}{rtf_c'}\right)^2\right] f_{sy} r^2 tP\right\} \left\{1 - \frac{\beta^{1.15}}{200}\right\}$$

For $W = 0$ and $\beta = 0$:

$$M_n = 6 f_{sy} r^2 tP$$

The cracking moment, M_{cr} is obtained from the following formula:

$$\frac{M_{cr}}{\pi r^2 t} - \frac{W}{2\pi rt} = f_r$$

where f_r is the modulus of rupture.

In this particular example, the critical reference mean wind speed for second mode resonance was 70 MPH, which is exactly the design speed. However, in stiffer chimneys the critical wind speed would probably be well above the design value. In this case it would be reasonable and on the safe side to multiply the value, q_2, obtained from eq. (15-21) by the square of the ratio of ($V_{\text{design}}/V_{\text{critical}}$).

6. *Selection of circumferential reinforcement:* The selection of the circumferential reinforcement is primarily controlled by the circumferential bending moments caused by the wind pressure distribution shown in Fig. 15-4. At any location along the chimney the maximum bending moment causing tension on the inside is equal to $0.314\ pr^2$, while the maximum bending moment causing tension on the outside is $0.272\ pr^2$. The pressure, p, will be based on 1.28 times the mean wind profile, having a reference wind speed of 70 MPH. The factor of 1.28 is used to account for the gusting effects. The value of p at any height, Z, would be equal to:

$$p = \frac{1}{2}\rho \left[\left(\frac{Z}{30}\right)^{1/7} \times 1.28 \times 70 \times \frac{5280}{3600}\right]^2$$

$$= \frac{1}{2} \times 0.00000238 \times 6535\ Z^{2/7}$$

$$= 0.007777\ Z^{2/7}\ \text{ksf} \quad (Z \text{ is in feet})$$

Table 15-13 gives the pressures along the chimney and tabulates the maximum circumferential moments per foot of height. The tabulated inside and outside steel ratios are based on using ultimate design according to ACI 318-83 with a moment load factor of 1.7

TABLE 15-12 Vertical Steel Ratios

$\frac{h}{H}$	*w (kips)*	*r (ft)*	*t (ft)*	*M (k-ft)*	β	*Steel Ratio P*	*Min. Steel Ratio*
1.0	0	16.75	.750	0	0	–	0.00872
0.9	1370	18.83	.883	44625	0	0.00443	0.00718
0.8	3178	20.90	1.021	123150	0	0.00886	0.00596
0.7	5293	23.04	1.021	183375	0	0.01031	0.00458
0.6	7613	25.19	1.021	224775	0	0.00964	0.00324
0.5	10129	27.33	1.021	284168	0	0.00983	0.00194
0.4	12869	29.46	1.066	392830	0	0.01157	0.00096
0.3	16308	31.47	1.330	512334	0	0.01020	0.00138
0.2	20823	33.47	1.613	633758	0	0.00852	0.00146
0.1	26355	35.49	1.869	757134	21°	0.00994	0.00496*
0	33887	37.27	2.583	867781	9°	0.00486	0.00331*
Total vertical steel in sq in./circumferential ft = 12t × steel ratio, where t is in inches							

M_{cr} is based on solid section.

TABLE 15-13 Circumferential Steel

Height above base (ft)	*r (ft)*	*t (in.)*	*Pressure, p (ksf)*	*Moment Causing Tension on Inside (k-ft)/ft of Height*	*Moment Causing Tension on Outside (k-ft)/ft of Height*	*Inside Steel Ratio*	*Outside Steel Ratio*
1000	16.75	9.00	0.0560	4.933	4.274	0.00351	0.00303
900	18.83	10.59	0.0543	6.045	5.237	0.00280	0.00242
800	20.90	12.25	0.0525	7.201	6.238	0.00231	0.00200
700	23.04	12.25	0.0505	8.418	7.292	0.00271	0.00234
600	25.19	12.25	0.0484	9.643	8.354	0.00311	0.00269
500	27.33	12.25	0.0459	10.765	9.325	0.00349	0.00301
400	29.46	12.79	0.0431	11.746	10.174	0.00342	0.00295
300	31.47	15.96	0.0397	12.346	10.694	0.00210*	0.00182*
200	33.47	19.36	0.0353	12.417	10.756	0.00135*	0.00117*
100	35.49	22.43	0.0290	11.469	9.935	0.00089*	0.00077*
0	37.27	31.00	–	–	–	–	–
Note that area of steel in sq in. per ft of height = steel ratio × 12 × *d*							

**Use min. value of 0.00222.*

and with a capacity reduction factor of 0.9. In computing the steel ratios, f_c' was taken equal to 4444 psi, f_{sy} equal to 60,000 psi, and d equal to $(t - 2.25)$ inches. The steel ratios given in the last two columns of Table 15-13 should not be less than a specified minimum value. The minimum can be established on the basis that the nominal moment M_n must be greater than the cracking moment M_{cr}. For beams, the minimum steel ratio of ACI 318-83 was obtained by doubling the steel required to make M_n equal to M_{cr}. However, the chimney shell under circumferential bending can be regarded as a slab-type structure where distribution of loading could occur. For this reason, it is recommended that the minimum circumferential steel ratio on either face be $1\frac{1}{3}$ that required to make M_n equal to M_{cr}. This stipulation would give a minimum ratio of approximately $400/3\ f_{sy}$, in which f_{sy} is in lb./sq in.

For vertical steel and for sections with no openings, the steel ratio which is twice that required to make $M_n = M_{cr}$ can be approximated by the following formula:

$$P_{\text{min.}} = \frac{\pi f_r}{3 f_{sy}} - 0.012 \frac{W}{rtf_c'}$$

REFERENCES

15-1 *Specification for the Design and Construction of R/C Chimneys (ACI 307-79)*, American Concrete Institute, Detroit, Mich., 1979.

15-2 Dryden, H. L. and Hill, G. C., "Wind Pressure on Circular Cylinders and Chimneys," Research Paper No. 221, *Bureau of Standards Journal of Research*, V. 5, Sept. 1930.

15-3 Roshko, A., "Experiments on the Flow Past a Circular Cylinder at Very High Reynolds Numbers," *Journal of Fluid Mechanics*, Cambridge University Press, May 1961.

15-4 Erdei, C. and Ghosh, J., "The Effects of Wind on Large-Diameter Chimneys and Shafts," *Concrete*, Sept. 1967.

15-5 Diver, M., "Étude des Cheminées en Béton Arme," *Annales de l'Institute du Bâtiment et des Travaux Publics*, Paris, May 1966.

15-6 *Règles Définissant les Effects de la Neige et du Vent Sur les Constructions*, Société de Diffusion des Techniques du Bâtiment, Paris, 1965.

15-7 ASCE Task Committee on Wind Forces, "Wind Forces on Structures," *Transactions, ASCE*, V. 126, Part II, 1961, pp. 1124–1198.

15-8 Vanasup, J., "Response of Tall Concrete Chimneys Subjected to Fluctuating Wind," Ph.D. thesis, Department of Civil Engineering, The University of Michigan, June 1975.

15-9 Rumman, Wadi S. and Vanasup, J., "Alongwind and Acrosswind Response of Concrete Chimneys," 2nd Chimney Design Symposium and Exhibition, University of Edinburgh, Mar. 1976.

15-10 Rumman, W. S., "Vibrations of Steel-lined Concrete Chimneys," *Journal of the Structural Division, ASCE*, V. 89, ST5, Proc. Paper 3661, Oct. 1963, pp. 35–63.

15-11 Dockstader, E. A., Swiger, W. F., and Ireland, E., "Resonant Vibration of Steel Stacks," *Transactions, ASCE*, V. 121, 1956, pp. 1088–1111.

15-12 "Wind Effects on Building and Structures," *Proceedings*, Conference held at the National Physical Laboratory, Teddington, Middlesex, England, II, June 1963.

15-13 Walter, L. D. and Woodruff, G. B., "The Vibrations of Steel Stacks," *Transactions, ASCE*, V. 121, 1956, pp. 1054–1087.

15-14 Maugh, L. C. and Rumman, W. S., "Dynamic Design of Reinforced Concrete Chimneys," *ACI Journal, Proceedings*, V. 64, No. 9, Sept. 1967, pp. 558–567.

15-15 Rumman, Wadi S., "Basic Structural Design of Concrete Chimneys," *Journal of the Power Division, ASCE*, V. 96, PO3, Proc. Paper 7334, June 1970, pp. 309–318.

15-16 Vickery, B. J., "Load Fluctuations in Turbulent Flow," *Journal of the Engineering Mechanics Division, ASCE*, V. 94, EM1, Proc. Paper 5782, Feb. 1968, pp. 31–46.

15-17 Vickery, B. J. and Kao, K. H., "Drag or Alongwind Response of Slender Structures," *Structural Division, Proceedings, ASCE*, V. 98, ST1, Jan. 1972, pp. 21–36.

15-18 Vickery, B. J. and Clark, A. W., "Lift or Acrosswind Response of Tapered Stacks," *Structural Division, Proceedings, ASCE*, V. 98, ST1, Jan. 1972, pp. 1–20.

15-19 Vickery, B. J., "A Model for the Prediction of the Response of Chimneys to Vortex Shedding," *Proceedings* 3rd Int. Chimney Design Symposium, Munich, 1978.

15-20 (a) Vickery, B. J. and Basu, R. I., "Across-Wind Vibration of Structures of Circular Cross-Section, Part I: Development of a Mathematical Model for Two-Dimensional Conditions," *Journal of Wind Engineering and Industrial Aerodynamics*, 12 (1983), pp. 49–73.

15-20 (b) Basu, R. I. and Vickery, B. J., "Across-Wind Vibration of Structures of Circular Cross-Section, Part II: Development of a Mathematical Model for Full-Scale Application," *Journal of Wind Engineering and Industrial Aerodynamics*, 12 (1983), pp. 75–79.

15-21 Rumman, Wadi S., "Earthquake Forces in Reinforced Concrete Chimneys," *Journal of the Structural Division, ASCE*, V. 93, ST6, Proc. Paper 5650, Dec. 1967, pp. 55–70.

15-22 Rumman, W. S. and Maugh, L. C., "Earthquake Forces Acting on Tall Concrete Chimneys," *International Association for Bridge and Structural Engineering*, Sept. 1968.

15-23 Newmark, Nathan M., Blume, J. A., and Kapur, K., "Seismic Design Spectra for Nuclear Power Plants," *Journal of the Power Division, ASCE*, V. 99, PO2, Proc. Paper 10142, Nov. 1973, pp. 287-303.

15-24 Task Committee on Steel Chimney Liners, Fossil Power Committee, Power Division, "Design and Construction of Steel Chimney Liners," published by ASCE, 1975.

15-25 Housner, G. W. and Keightley, W. O., "Vibrations of Linearly Tapered Cantilever Beams," *Journal of the Engineering Mechanics Division, ASCE*, V. 88, EM2, Proc. Paper 3101, Apr. 1962, pp. 95–123.

15-26 *Uniform Building Code*, International Conference of Building Officials, Whittier, Calif., 1982.

15-27 Sun, Ru-Tsung, "Inelastic Behavior of Reinforced Concrete Chimneys," Ph.D. thesis, Department of Civil Engineering, University of Michigan, Ann Arbor, Jan. 1974.

15-28 Rumman, Wadi S. and Sun, Ru-Tsung, "Ultimate Strength Design of Reinforced Concrete Chimneys," *ACI Journal, Proceedings*, V. 74, No. 4, Apr. 1977.

15-29 Hognestad, Eivind, "A Study of Combined Bending and Axial Load in Reinforced Concrete Members," *Engineering Experimental Station Bulletin No. 399*, University of Illinois, Urbana, 1951, 128 pp.

15-30 Brettle, H. J., "Theory and Ultimate Strength Design for Reinforced Concrete Chimneys," *Publications, International Association for Bridge and Structural Engineering (Zurich)*, V. 32-I, 1972, pp. 1–36.

15-31 Cannon, R. W., "Ultimate Strength Design Charts for Design of Reinforced Concrete Chimneys," Civil Engineering Design Research Report, Tennessee Valley Authority, Knoxville, 1971, 14 pp.

15-32 Greiner, G., "Ein Beitrag zur Berechnung der Bruchsicherheit von Kreisringquerschnitten für Beigung mit Axialcruck," *Der Bauingenieur (Berlin)*, V. 37, No. 11, Nov. 1962, pp. 413–418.

15-33 Mokrin, Z. A., "Reinforced Concrete Members of Hollow Circular Section Under Monotonic and Cyclic Bending," Ph.D. thesis, Department of Civil Engineering, University of Michigan, Ann Arbor, Dec. 1978.

15-34 Newmark, N. M., "Numerical Procedure for Computing Deflections, Moments, and Buckling Loads," *Transactions, ASCE*, V. 108, 1943.

16

Silos and Bunkers

SARGIS S. SAFARIAN and ERNEST C. HARRIS*

16.1 INTRODUCTION

Bins for storing granular materials are of two main types—silos (also called deep bins) and bunkers (or shallow bins). The important difference between the two is in the behavior of the stored material. This behavior difference is influenced by both bin geometry and characteristics of the stored material. Material pressure against the walls and bottom are usually determined by one method for silos and by another for bunkers. (The present ACI 313 Silo Standard,[16-6] however, uses the same method for both silos and bunkers.)

Silos and bunkers are made from many different structural materials. Of these, concrete is probably the most frequently used. Concrete can offer the necessary protection to the stored materials, requires little maintenance, is aesthetically pleasing, and is relatively free of certain structural hazards (such as buckling or denting[16-1]) that may be present in silos or bunkers of thinner materials.

Silos and bunkers may be of various plan shapes and may occur singly or connected in groups. (See Fig. 16-1.) Figure 16-2 shows typical group arrangements. These arrangements may be used for either silos or bunkers.

Many spectacular silo collapses have occurred.[16-2] The many causes of structural distress may be classified as follows:

1. Design errors, including the use of incorrect loads, failure to consider critical combinations of loaded and unloaded cells, and specifying improper details.
2. Construction errors, such as misplacing or omitting reinforcing bars, poor concrete quality control, and improper slipform techniques.
3. Operational errors, including "improvements" in rate of discharge, changing the type of stored material, or attaching items that deliver unexpected lateral loads to the silo.
4. Combinations of the above.

16.2 DESIGN CONSIDERATIONS

The design process for silos and bunkers is of two types—functional and structural. Functional design must provide for adequate volume, proper protection of the stored materials, and satisfactory methods of filling and discharge. Structural considerations are stability, strength, and control (minimizing) of crack width and deflection. Loads to be considered include the following:

1. Dead load of the structure itself and items supported by the structure
2. Live loads, as follows:
   (a) Forces from stored material
   (b) Changes in above due to filling and emptying
   (c) Wind
   (d) Snow
   (e) Seismic forces on structure and stored material
3. Thermal stress due to stored material (especially important in long silo groups)

16.3 DESIGN METHOD—USD OR WSD

Either design method: strength (USD) or working stress (WSD)—may be used for silos and bunkers. In the materials

*Sargis S. Safarian, P. E., President, S M H Engineering, Inc., Lakewood, Colorado, and Ernest C. Harris, Ph.D., P.E., Professor of Civil Engineering, University of Colorado at Denver, Colorado.

Fig. 16-1 Grain elevator- group of hexagonal and circular silos, Denver, Colorado. (Courtesy of Cargill, Inc.)

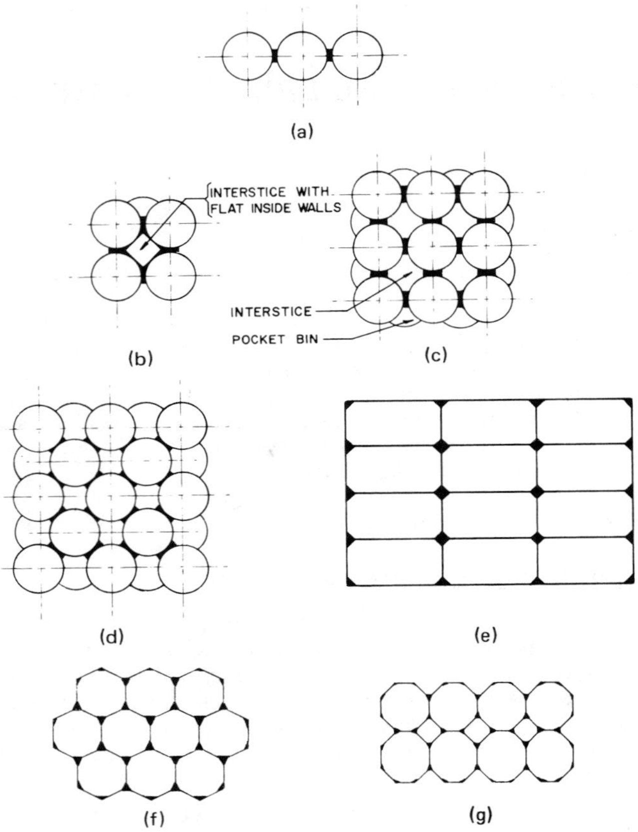

Fig. 16-2 Typical silo or bunker groups.

that follow, most formulas for forces or moments give ultimate values of force or moment. However, by merely setting K_l and K_g equal to 1.0, these same formulas may be used to solve for WSD forces and moments.

The *ACI Code* values of load factor, K, for dead and live loads should be used for silo or bunker design by USD. The weight of stored material should be considered as live load. Strength reduction factors ϕ should be as recommended by ACI 318, except that in the case of slipformed concrete, which is made without continuous inspection, the factor should be reduced to 0.95 times the ϕ-value shown by ACI 318.

The authors personally prefer USD, and in all numerical examples herein, USD is used.

16.4 PROPERTIES OF STORED MATERIALS

The properties of the material to be stored affect the intensity of pressure loadings. In addition, they influence material flow and must be considered in selecting the outlet shape and size and the type of unloading system.[16-3–16-5]

Table 16-1 shows the properties of commonly stored ma-

TABLE 16-1 Typical Design Properties of Granular Materials (Use only in the absence of the actual values.)*

Material	*Weight lb/cu ft (kg/m³)*	*Angle of Repose*	*Coefficient of Friction*	
			Against Concrete	*Against Steel*
Cement, Portland	100 (1600)	25°	0.466	0.30
Cement, Clinker	88 (1410)	33°	0.60	0.30
Peas	50 (800)	25°	0.296	0.263
Wheat	50 (800)	25°	0.444	0.414
Beans	46 (740)	31.5°	0.442	0.366
Barley	39 (620)	37°	0.452	0.376
Corn	44 (700)	27.5°	0.423	0.374
Oats	28 (450)	28°	0.466	0.412
Sugar granular	63 (1000)	35°	0.431	
Sand dry	100 (1600)	35°	0.70	0.50
Sand moist	113 (1810)	40°	0.65	0.40
Sand saturated	125 (2000)	25°	0.45	0.35
Flour	38 (610)	40°	0.30	0.30
Lime (burned pebbles)	56 (900)	35°	0.50	0.30
Lime, powder	44 (700)	35°	0.50	0.30
Coal, bituminous	50–65 (800–1040)	32–42°	0.50	0.30
Coal, anthracite	60–70 (960–1120)	24–35°	0.50	0.30
Coke	38 (600)	40°	0.80	0.50
Gravel dry	113 (1810)	35°	0.45	
Gravel wet	125 (2000)	25°	0.40	0.75
Manganese ore	125 (2000)	40°		
Iron ore	165 (2640)	40°	0.50	0.364
Clay dry	106 (1700)	40°	0.50	0.70
Clay damp	113 (1810)	25°	0.30	0.40
Clay wet	138 (2200)	15°	0.20	0.30
Lime burned, fine	57 (910)	35°	0.50	0.30
Lime burned, coarse	75 (1200)	35°	0.50	0.30
Gypsum in lumps, limestone	100 (1600)	40°	0.50	0.30

**The properties listed here are illustrative of values that might be determined from physical testing. Ranges of values show the variability of some materials. Design parameters should preferably be determined by test and the values above used with caution.*

terials. These values should be used only in the absence of test data for the actual material to be stored. Caution and good judgment must be used in selecting the properties to be used, as pressures are quite sensitive to variations of those properties. The designer must be alert to the possibility of large variations from the tabulated values—for example, the possibility of a material ranging from dry, to damp, to saturated. If the expected condition is not certain, the designer should assume the worst condition for that material.

The designer should be alert to possible varying usage—the storage of different materials in the same silo or bunker. Which material should be used in design? Perhaps more than one, since one material may exert the greater lateral pressure while another may cause the more severe vertical pressure. For each type of pressure sought, the "worst" material should be considered.

16.5 SILO OR BUNKER?

Unless he elects to use *silo* static pressure equations for each, the designer must classify the structure as either a *silo* (deep bin) or *bunker* (shallow bin).

Empirical approximations are preferred by many engineers. Two such approximations are:

a. By Dishinger,[16-6] $H > 1.5\sqrt{A}$
b. By the *Soviet Code*,[16-7]
   $H > 1.5D$ for circular silos
   $H > 1.5a$ for rectangular silos

If the bin in question satisfies either of the above, it is considered as a silo. If it satisfies neither rule, it is considered to be a bunker.

A second method is based on the position of the plane of rupture of the stored material. Figure 16-3 shows bins of two different depths. The plane of rupture is determined by the Coulomb theory. If friction against the wall is neglected for the case of a vertical wall and horizontal top surface, the Coulomb plane of rupture is midway between the angle of repose (ρ) and the vertical wall. If the rupture plane intersects the top surface of the stored material, the bin is a bunker, otherwise it is a silo.

However, engineers do not agree on the location of the plane of rupture. Some would start the plane at the bottom of the hopper, point C of Fig. 16-3(b), while others would pass it through point D. Thus, by one interpretation the bin would be a silo—by the other, a bunker. Fortunately, for such borderline cases, exact classification is not critical.

16.6 SILOS

16.6.1 Silo Loadings from Stored Material

Material stored in a silo applies lateral forces to the side walls, vertical force (through friction) to the side walls, vertical forces to horizontal bottoms, and both normal and frictional forces to inclined surfaces. The static values of these forces, resulting from materials at rest, are all modified during withdrawal of the material. In general, all forces will increase, so that loads during withdrawal tend to control the design.

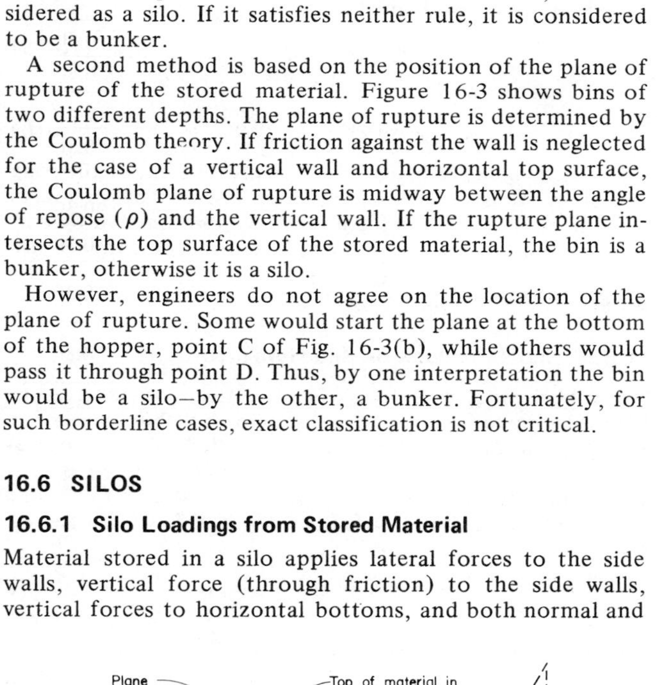

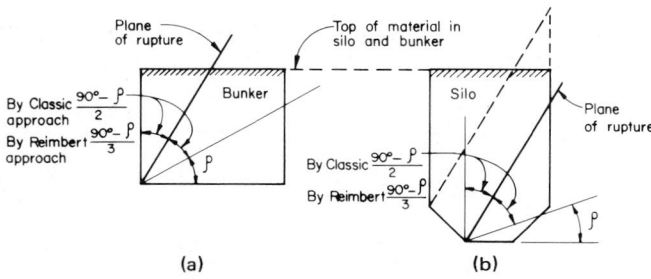

Fig. 16-3 Classification of bins using plane of rupture.

Forces applied by stored materials may also be affected by moisture changes, by compaction, and by settling which may accompany alternate expansion and contraction of the walls during daily or seasonal temperature change.

A rigorous approach to calculation of silo loads would involve the conditions of material flow during emptying. Until such an approach is perfected, equations derived for static forces may be combined with experimental data to approximate the pressure increases occurring during material withdrawal. To compute design loads, the method recommended in the American Concrete Institute Standard ACI 313-77[16-6] for design and construction of concrete bins, silos, and bunkers is used herein. The procedure involves determining the static pressures (or forces) and then multiplying these by "overpressure" factors, C_d, to obtain design pressures (or forces).

16.6.2 Computation of Static Pressures—Lateral and Vertical

Two methods for determining static pressures are Janssen's classic method[16-8] and Reimbert's method.[16-9] Figure 16-4 shows the silo dimensions used for each. Janssen's method is the more popular in the United States. However, experiments[16-9] show Janssen's method to be unconservative in some cases, whereas Reimbert's method is reported to give pressures agreeing closely with test results.

Janssen's Method: Vertical static unit pressure at depth Y below the surface is:

$$q = \frac{\gamma R}{\mu' k}\left(1 - e^{-\mu' kY/R}\right) \qquad (16\text{-}1)$$

Lateral static unit pressure at depth Y is:

$$p = \frac{\gamma R}{\mu'}\left(1 - e^{-\mu' kY/R}\right) = qk \qquad (16\text{-}2)$$

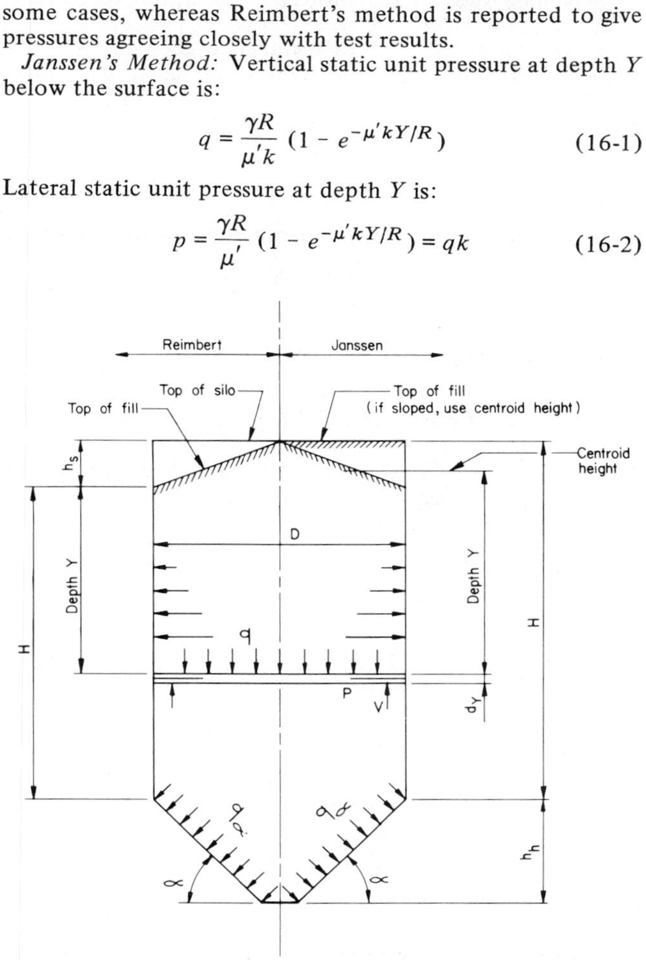

Fig. 16-4 Silo dimensions for use in Reimbert's and Janssen's equations.

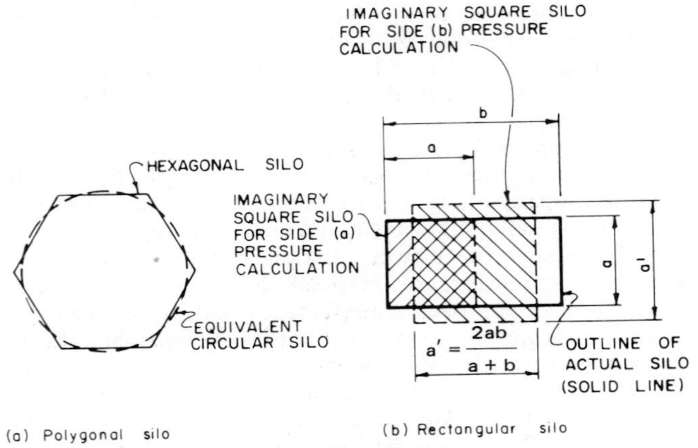

Fig. 16-5 Equivalent silo shapes.

where k is assumed to be:

$$k = \frac{p}{q} = \frac{1 - \sin \rho}{1 + \sin \rho} \tag{16-3}$$

In the above, R is the hydraulic radius (area/perimeter) of the horizontal cross section of the inside of the silo.

For circular silos, $R = D/4$

For polygonal silos, $R = D/4$ for a circular shape of equivalent area may be used (Fig. 16-5).

For square silos or the shorter wall, "a," of rectangular silos, use $R = a/4$.

For the long wall, "b," of rectangular silos, use $R = a'/4$ where a', the length of side of an imaginary square silo, is:

$$a' = \frac{2ab}{a + b} \tag{16-4}$$

Reimbert's Method:
Vertical static unit pressure at depth Y below the surface is:

$$q = \gamma \left[Y \left(\frac{Y}{C} + 1 \right)^{-1} + \frac{h_s}{3} \right] \tag{16-5}$$

Lateral static unit pressure at depth Y is:

$$p = p_{max} \left[1 - \left(\frac{Y}{C} + 1 \right)^{-2} \right] = z p_{max} \tag{16-6}$$

Values of p_{max} (maximum lateral unit pressure) and C ("characteristic abscissa") for use in eqs. (16-5) and (16-6) vary by silo shape, as follows:

For circular silos:

$$p_{max} = \frac{\gamma D}{4\mu'} \tag{16-7}$$

$$C = \frac{D}{4\mu' k} - \frac{h_s}{3} \tag{16-8}$$

For polygonal silos of more than four sides:

$$p_{max} = \frac{\gamma R}{\mu'} \tag{16-9}$$

$$C = \frac{L}{\pi} \left[\frac{1}{4\mu' k} \right] - \frac{h_s}{3} \tag{16-10}$$

(Use R as defined above for Janssen's method.)
For rectangular silos—on shorter wall, "a":

$$(p_{max})_a = \frac{\gamma a}{4\mu'} \tag{16-11}$$

$$C_a = \frac{a}{\pi \mu' k} - \frac{h_s}{3} \tag{16-12}$$

For rectangular silos—on longer wall, "b":

$$(p_{max})_b = \frac{\gamma a'}{4\mu'} \tag{16-13}$$

$$C_b = \frac{a'}{\pi \mu' k} - \frac{h_s}{3}, \tag{16-14}$$

where a' is as defined for the Janssen method.

For design purposes, the granular material is usually assumed level at the top of the silo ($h_s = 0$).

16.6.3 Static Pressure on Flat Bottoms

Pressure on flat bottoms is given by eqs. (16-1) and (16-5). For rectangular silos, however, these equations give different bottom pressures for areas next to the short and long sides. An approximation frequently used is to assume the pressure, q_a—computed using R (Janssen) or C (Reimbert) for the short side, "a"—to act on the area A_a, as shown in Fig. 16-6. Similarly, pressure q_b, computed using R or C for side "b," is assumed to act on area A_b.

16.6.4 Static Forces—Vertical Friction

For round, square, and regular polygonal silos, the total static frictional force per foot-wide vertical strip of wall above depth Y is approximately:

By Reimbert's method:

$$V = (\gamma Y - q) R \tag{16-15}$$

By Janssen's method:

$$V = (\gamma Y - 0.8q) R \tag{16-16}$$

For rectangular silos, lateral pressures differ for the long and short walls; hence, side friction and vertical pressures also differ. The friction loads may be approximated by eqs. (16-15) and (16-16) when terms q and R, respectively, are substituted by q_a and R_a for side "a," and by q_b and R_b for side "b." (See Fig. 16-6.)

16.6.5 Static Pressures on Silo Hoppers

Static horizontal pressures, p, and vertical pressures, q, on inclined hopper walls are calculated by the Janssen or Reimbert formulas. The hydraulic radius, R, may be reduced within the hopper depth, but usually is assumed constant and equal to that of the silo. The static unit pressure normal to the inclined surface at depth Y from the top of the fill is:

$$q_\alpha = p \sin^2 \alpha + q \cos^2 \alpha \tag{16-17}$$

16.6.6 Overpressure and Overpressure Factors, C_d

Equations (16-1) through (16-17), above, are for static pressures only, due to stored material at rest, before with-

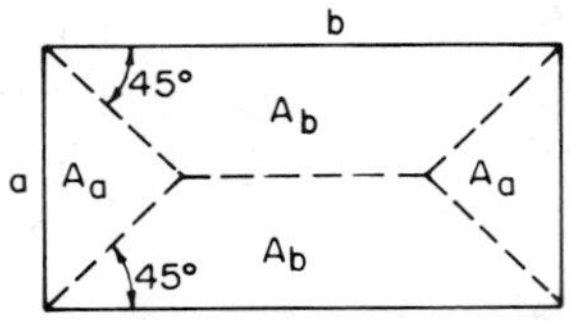

Fig. 16-6 Assumed distribution of vertical pressure on horizontal plane of rectangular silo.

drawal is begun. During withdrawal, these pressures may increase.[16-9–16-11, 16-15, 16-16] The increases are sometimes called dynamic effects, but the term "overpressure" is preferred, since the increases include both static and dynamic effects. Among these are pressure increase due to material withdrawal, arching of the material (causing higher wall pressure and vertical friction), and collapse of the arched material (causing increased vertical pressure due to impact).

Locally, the total pressure may be several times larger than[16-12, 16-13] the static pressure computed by the Janssen or Reimbert equations. Silos have been designed ignoring overpressure. Many silos so designed have failed, and many others not yet showing signs of distress may have dangerously low margins of safety.

TABLE 16-2 Values of Overpressure Factor, C_d

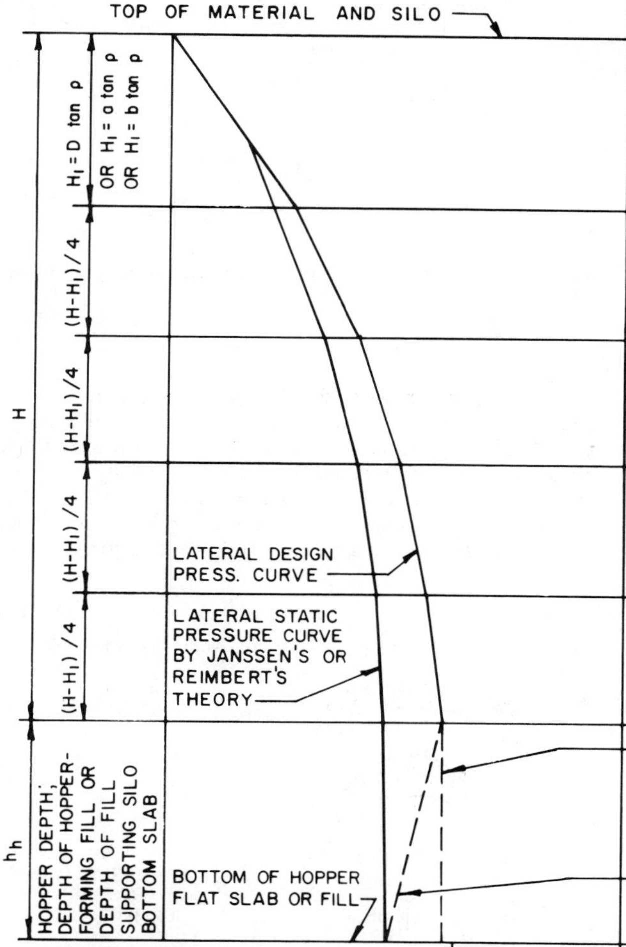

Height zone	OVERPRESSURE FACTOR C_d: $\frac{H}{D}$ (or $\frac{H}{d}$) ≤ 2, JANSSEN	REIMBERT	$\frac{H}{D}$ (or $\frac{H}{d}$) = 3, JANSSEN	REIMBERT	$\frac{H}{D}$ (or $\frac{H}{d}$) ≥ 4, JANSSEN	REIMBERT	FOR POWDERY—COHESIVE LIKE CEMENT OR FLOUR, WHEN EMPTYING IS DONE PNEUMATICALLY: $\frac{H}{D}$ (or $\frac{H}{d}$) = 4, JANSSEN	REIMBERT	$\frac{H}{D}$ (or $\frac{H}{d}$) ≥ 5, JANSSEN	REIMBERT
$H_1 = D \tan \rho$ OR $H_1 = a \tan \rho$ OR $H_1 = b \tan \rho$	1.35	1.10	1.45	1.20	1.50	1.25	1.60	1.30	1.65	1.35
$(H-H_1)/4$	1.45	1.20	1.55	1.30	1.60	1.35	1.70	1.40	1.75	1.50
$(H-H_1)/4$	1.55	1.45	1.65	1.55	1.75	1.60	1.80	1.70	1.90	1.75
$(H-H_1)/4$	1.65	1.65	1.75	1.75	1.85	1.85	1.90	1.90	2.00	2.00
$(H-H_1)/4$	1.65	1.65	1.75	1.75	1.85	1.85	1.90	1.90	2.00	2.00
h_h (hopper depth)	USE SAME PRESSURE WITHIN HOPPER HEIGHT OR, IF DESIRED, REDUCE PRESSURES IN ACCORDANCE WITH HYDRAULIC RADIUS CHANGE. IF DESIRED, PRESSURES MAY BE REDUCED FROM TOP OF FILL TO TOP OF FLAT SLAB AS SHOWN									
OVERPRESSURE FACTOR C_d FOR USE IN CALCULATING DESIGN BOTTOM PRESSURES IN SILOS (SEE NOTE 6) — CONCRETE BOTTOM	1.35	1.50	1.35	1.50	1.35	1.50	1.35	1.50	1.35	1.50
STEEL BOTTOM	1.50	1.75	1.50	1.75	1.50	1.75	1.50	1.75	1.50	1.75

*C_d *values given in this table are inadequate for the higher loads associated with mass flow.*

NOTES:

1. *C_d factor for lateral pressures is given for the bottom of each height zone shown.*
2. *In the region of a flow-correcting insert (e.g., Buhler-Nase) lateral pressures may be many times greater than static, and the C_d values above are not sufficient.*
3. *Silo bottom pressures need not be considered larger than the pressure caused by 100% of weight of silo contents.*
4. *If $H_1 < H \leqslant 2H_1$, use the second C_d value from the top for the entire silo height H.*
5. *Values of factor C_d for H/D between those given in Table C.1 should be determined by linear interpolation.*
6. *Values of C_d factor given in Table C.1 for calculating design bottom pressures shall be multiplied by 0.75 for noncohesive material except for homogenizing silos in which pneumatic withdrawal is used.*
7. *The C_d factors shown in Table C.1 are minimum recommended values. However, lower C_d factors may be used, but only for particular cases for which the designer can demonstrate that such lower C_d factors are satisfactory.*

Overpressure is not yet well enough understood to consider by rational methods. However, its effect can be approximated using overpressure factors, C_d, to convert from computed static pressures to design pressures. In general:

$$\text{design pressure} = C_d \times \text{static pressure} \qquad (16\text{-}18)$$

Table 16-2 shows a lateral pressure design curve and gives values of overpressure factors, C_d. The factors for use with Janssen's and Reimbert's methods are from the ACI 313-77 Standard (Table C.1).[16-6]

To promote better flow of material, designers sometimes use a flow-improvement device such as the Bühler Nose in the silo or bunker directly above the hopper. Tests show that this device causes large additional local overpressures, beyond the normal overpressures without the device.[16-28, 16-34] Local overpressures caused by such devices should be determined. A method for computing such overpressures may be found in the author's book on silos and bunkers[16-28] or in other published sources.[16-34–16-36] When the device is used and its effect is not determined, then it is suggested that respective overpressure factors, C_d, for walls at a height equal to three times the device depth (centered about the device) should be at least 50% higher than shown by Table 16-2.

16.6.7 Design Pressures

For silos with centrally located discharge openings, design pressures due to stored material are:

$$q_{des} = C_d q \qquad (16\text{-}19)$$

$$p_{des} = C_d P \qquad (16\text{-}20)$$

$$q_{\alpha,\,des} = p_{des} \sin^2 \alpha + q_{des} \cos^2 \alpha \qquad (16\text{-}21)$$

The maximum wall friction force per unit length of wall is simultaneous with the minimum, or static, q at depth Y. Therefore, for concrete silos:

$$V_{des} = V \qquad (16\text{-}22)$$

16.6.8 Effects of Eccentric Discharge

Eccentric discharge may be considered by adding a correction, p_{ecc}, to the lateral design pressure, p_{des}, computed at depth Y by either the Janssen or Reimbert formula. Assume this design-pressure increase p_{ecc} to be constant from the top of the hopper (or hopper-forming fill) to a height equal to D (or a or b for rectangular silos) and to reduce linearly from that point to the top of the silo. The value of p_{ecc} should remain at maximum for the full depth of the hopper. The design pressure at any depth Y below the surface of the stored material would be the sum of p_{des} without eccentricity plus the correction for eccentricity, or:

$$p_{des} = C_d p + \frac{Y}{H - D}(p_{ecc}) \qquad (16\text{-}23)$$

The amount of p_{ecc} in the lower range of height may be taken as an arbitrary percentage of static pressure (the *Commentary* to ACI 313 suggests 25%), or it may be estimated from Reimbert's experimental data, or by the following method involving "imaginary silos." In the latter method, the full correction at depth H (top of hopper) is:

$$p_{ecc} = p_i - p_{st} \ (\text{at } Y = H) \qquad (16\text{-}24)$$

where p_i is the lateral static pressure at depth H in an imaginary silo, as shown by Fig. 16-7 or Fig. 16-8.

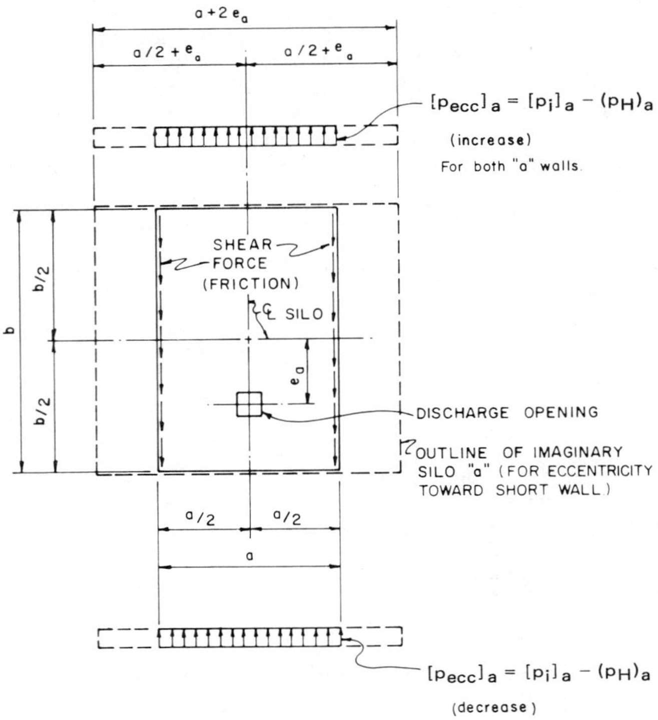

Fig. 16-7 Pressure change due to eccentric discharge in rectangular silo.

For rectangular silos, the imaginary silo is determined as shown by Fig. 16-7. When the opening is displaced toward side "a," correction p_{ecc} for sides "a" is computed using an imaginary silo measuring $(a + 2e_a) \times b$. (If e_a is larger than a, the imaginary silo should measure $3a \times b$.) Similarly, if the opening is eccentric toward side "b," the imaginary silo will measure $(b + 2e_b) \times a$. If both eccentricities occur,

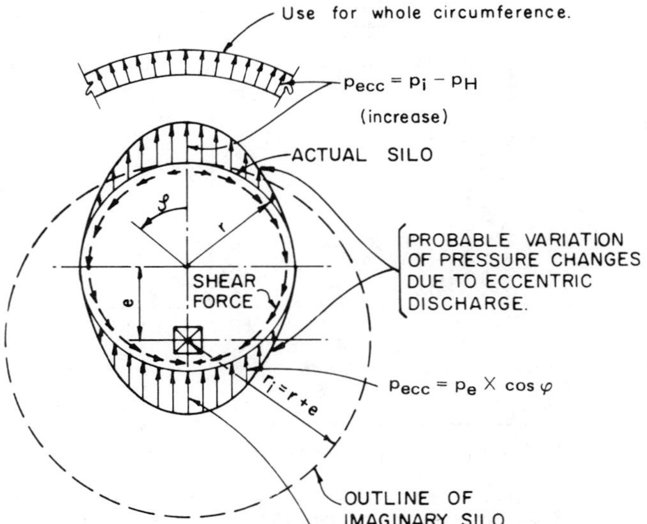

Fig. 16-8 Pressure change due to eccentric discharge in circular silo.

each correction is computed separately, using the first described imaginary silo to determine p_{ecc} for sides "a," and the second for sides "b."

The imaginary circular silo of Fig. 16-8 is centered on the discharge opening and has a radius equal to that of the actual silo plus the eccentricity.

Where multiple discharge openings occur, even though the group is centrally located, eccentric discharge is always possible and should be considered.

16.6.9 Lateral Pressure Design Curve

Table 16-2 shows different C_d values for lateral pressure in various depth zones. The lower portion of the table provides C_d values for bottom pressures, one for concrete, the other for steel bottoms. If bunker design pressures are determined from the Rankine method, then the C_d factors of Table 16-2 are not applicable. The authors suggest, instead to increase the lateral pressure values obtained by the Rankine method by a minimum of 25%. Factor C_i (for impact) shown in Table 16-12 may control bunkers into which material is dumped suddenly. See Sect. 16.14.2.

16.6.10 Wind and Earthquake Loads

Wind may affect the stability of empty silos, particularly tall, narrow silos or silo groups. Foundation pressures and column stresses, however, may be worse with wind acting on the full silo. Wind load reduction for cylindrical shape may be applied to single circular silos, but not to silo groups.

Earthquake loads may affect stability and strength. Columns and walls supporting silos may be particularly vulnerable and foundations may also be affected. In the absence of better codes covering the seismic design of silos, the authors suggest the earthquake requirements of the ACI 313-77[16-6] be used in conjunction with the latest *Uniform Building Code.*[16-19] In this approach, seismic forces are assumed to act in any horizontal direction, but vertical acceleration forces are usually neglected.

There is test evidence[16-20] that only a fraction of the stored material weight need be considered when computing earthquake forces. The authors and ACI 313 use not less 80% of the weight of the stored material as an effective live load, W_{eff}, from which to determine seismic forces. Lower percentages have been suggested, but recent research suggests that nothing lower than 80% be used.[16-36]

The centroid of this effective live load is assumed to coincide with the centroid of the actual volume of stored material.

When silo bottoms are on supports independent from the silo walls (bunker bottoms are usually not independently supported), these two independent structures will share the lateral force from seismic action on the effective weight of the stored material. It is extremely difficult if not impossible to predict and determine the actual distribution of the lateral seismic force between silo walls and the bottom structure. As an approximate solution, the authors compute the above distribution as follows:

1. To the independent bottom structure, the *larger* of A and B, plus C:
   A. A portion of the lateral force due to W_{eff}, divided according to the relative stiffness of the two structures. This condition usually will not control if the silo walls are extended down and the bottom system is independently supported on columns, because in that case the relative rigidity of the columns is usually negligible compared to that of the walls. Bottom systems supported on walls or silo walls supported on columns will result in a more significant distribution of the lateral forces from stored material.
   B. One-third of the product of coefficient of internal friction times the effective weight, W_{eff}, of stored material above the hopper.
   C. Lateral force due to the actual (not effective) weight of material within the hopper.
2. To the silo structure, the *larger* of the following:
   D. The lateral force due to total W_{eff} minus the lesser items A and B, above.
   E. The lateral force due to 50% of W_{eff}.

The authors present the above load distribution only as an approximation. If a more accurate solution is available, it should be used in lieu of the above.

In computing stability against overturning due to earthquake, the entire weight of the stored material may be used. However, if the silo has an independently supported bottom, the material weight must be divided between silo walls (friction) and the bottom structure.

Based on the above assumptions and for simplification of the computation of complex seismic forces, the authors recommended that the minimum total lateral seismic force or base shear be computed by the UBC method (also adopted by ACI 313-877):

$$H_e = ZC_p(W_g + W_{eff}) \qquad (16\text{-}25)$$

in which W_g is the weight of the structure; Z is an earthquake zone factor equal to $3/16$, $3/8$, $3/4$, and 1 for zones 1, 2, 3, and 4, respectively (see the 1982 edition of the *Uniform Building Code*, United States map showing earthquake zones 1, 2, 3, and 4); and C_p is the product of a group of factors, assumed to vary from 0.2 to 0.1, categorized as follows: All silos with material stored on silo bottoms above the ground and subject to seismic loads should be designed for $C_p = 0.2$ unless an earthquake dynamic analysis shows adequate strength and stability of the structure under a lesser value of C_p but not less than 0.1. Silos in which the walls extend to the ground and the stored material rests directly on the ground should be treated as tanks resting on the ground, using $C_p = 0.10$. For cases intermediate between these two extremes, C_p may be obtained by linear interpolation. As an example of an intermediate case, see Fig. 16-19(b). More on seismic design of silos may be found in Ref. 16-28.

16.7 CIRCULAR SILOS

16.7.1 Wall Design Procedure

Ordinarily the following sequence is applicable:

1. Determine hoop reinforcing steel required for design pressure for stored material.
2. Determine the wall thickness required.
3. Check crack width and modify the design as necessary.
4. Modify the hoop steel for temperature effect and for horizontal bendings, as necessary.
5. Determine the vertical steel.

16.7.2 Horizontal (Hoop) Reinforcing

If the radial pressure from stored material is known to be uniform along the inside circumference at any given depth, the circular silo is treated as though subject to tension only.

The ultimate hoop tensile force per unit height is

$$F_u = K_l p_{des}(D/2) \qquad (16\text{-}26)$$

The required steel area per unit height is

$$A_s = F_u/(\phi f_y) \qquad (16\text{-}27)$$

Hoop steel can be in one layer or two. The required area will vary with height, being proportional to the design pressure. Hoop steel area, as required at the bottom of the pressure zone, but excluding temperature steel, should be continued at least 3 ft below that zone (i.e., below the top of the hopper, ring beam, or slab) or to the bottom of the slab or ring beam, whichever is farther. (Where hopper-forming fill is used, the full required hoop steel area should be continued to at least 3 ft below the top of the fill, and one-half of that steel area per unit width to the bottom of the fill.)

Horizontal bending moment due to eccentric filling or discharge, or to connection to other walls or structures must be considered in combination with the computed hoop tension when determining the required wall reinforcement.

16.7.3 Wall Thickness

An isolated circular silo under uniform radial load gets all its strength from the horizontal steel wherever the concrete is cracked. Were the loading truly uniform, wall thickness would matter little, provided it was sufficient for vertical load and for lap splicing of the steel. However, since transient nonuniform load may occur, but the moments are not computable, some other approach is needed. One approach is the PCA formula,[16-21] eq. (16-28), below:

$$h_{min} = \left(\frac{mE_s + f_s - nf_{c,ten}}{f_s f_{c,ten}}\right) PD/2 \qquad (16\text{-}28)$$

In this equation, m is a concrete shrinkage coefficient, n is the ratio E_s/E_c, and f_s is the allowable (WSD) steel stress. PCA suggests using $0.1 f_c'$ for allowable stress $f_{c,ten}$. Consistent units must be used.

Unless steps are taken to reduce friction between the forms and concrete, the authors recommend that slip-formed silo walls be not less than 6 in. thick. Walls 9 in. or thicker should be doubly reinforced.

Vertical compressive strength should also be checked. Suggested limits for circular silos are:

$$F_{vert} = 0.225 f_c' bh \text{ for WSD methods} \qquad (16\text{-}29)$$

or

$$F_{vert} = 0.385 f_c' bh \text{ for USD method} \qquad (16\text{-}30)$$

Having a tentative thickness, the next step is to check for crack width. A limit of 0.008 in. is suggested for grain or cement storage silos and other silos exposed to the weather.[16-2] One method of checking crack width due to axial tension is presented by Lipnitski and Abramovitsch.[16-22] In this method, the walls are assumed subject to pure tension and to have centrally located reinforcement. The total outside width w_{cr} of a vertical crack caused by simultaneous action of short- and long-term loadings is:

$$w_{cr} = w_1 - w_2 + w_3 \qquad (16\text{-}31)$$

where:

w_1 = crack width due to short-term application of total loading, F_{tot} (including overpressure).

w_2 = that portion of w_1 due to static loading F_{st} alone, during the time that F_{tot} acts. (w_2 is included in w_3; thus $w_1 - w_2$ is the width increase due to short-term overpressure.)

w_3 = crack width due to long-term loading only.

Values of w_1, w_2, and w_3 are each calculated using the formula:

$$w_n = \Psi_1 s_{cr} f_s / E_s \qquad (16\text{-}32)$$

In the above, s_{cr} (crack spacing) is:

$$s_{cr} = A\beta/\Sigma o \qquad (16\text{-}33)$$

where A is the area of wall per unit height (sq in. per ft) and $\beta = 0.7$ and 1.0 for deformed and plain bars, respectively. Factor Ψ_1 of eq. (16-32) is determined as follows:

$$\Psi_1 \text{ (for short term)} = 1 - 0.7\left[\frac{0.8Af_t'}{F_{st} \text{ or } F_{tot}}\right] \qquad (16\text{-}34)$$

but not less than 0.3,

$$\Psi_1 \text{ (for long term)} = 1 - 0.35\left[\frac{0.8Af_t'}{F_{st}}\right] \qquad (16\text{-}35)$$

but not less than 0.65, where F_{tot} includes overpressure and is computed using p_{des}; and F_{st} is the static tension only, computed using the static pressure p. Both F_{st} and F_{tot} are unfactored, i.e., not ultimate values.

Note: When the term in brackets of eq. (16-34) or (16-35) exceeds 1.0, it should be set equal to 1.0. The authors use $f_t' = 4.5\sqrt{f_c'}$.

Crack width must be checked at various depths Y.

16.7.4 Temperature Reinforcement for Circular Silos

Temperature and shrinkage steel requirements of ACI 318 apply to silos. In addition, hot stored materials may cause thermal stresses too high to be ignored.

The approximate method illustrated below was developed specifically for cement storage silos. Its principles, however, should apply also to silos for storage of other hot granular materials. In this method:

1. Tensile strength of the concrete is neglected.
2. Wall temperatures are assumed to vary only radially.

In buildings, the usual practice is to ignore a certain amount of inside–outside temperature difference. For silos and bunkers, the authors usually neglect not more than the first 80°F of difference. The design inside temperature of the stored material is then:

$$T_{i,des} = T_i - 80°\text{F} \qquad (16\text{-}36)$$

The temperature of hot granular material in silos may drop appreciably near the inside face of the wall.[16-23] With hot cement, for example, the authors suggest that an 8-in. thickness of cement be considered as insulating material across which the temperature varies linearly. Figure 16-9 shows the temperature variation in the cement and the silo wall. The temperature difference ΔT between inside and outside of the wall is then:

$$\Delta T = K_t(T_{i,des} - T_0) \qquad (16\text{-}37)$$

K_t is determined by heat transfer principles; for stored cement and various wall thicknesses its value is given by Fig. 16-10.

The horizontal ultimate bending moment due to ΔT using plane-strain analysis[16-25] with Poisson's rate = 0.2 is:

$$M_{xt,u} = 1.25 E_c h^2 \alpha_c \Delta T K_g \text{ (in.-lb/ft)} \qquad (16\text{-}38)$$

Unit will be in.-lb/ft if E_c is in psi.

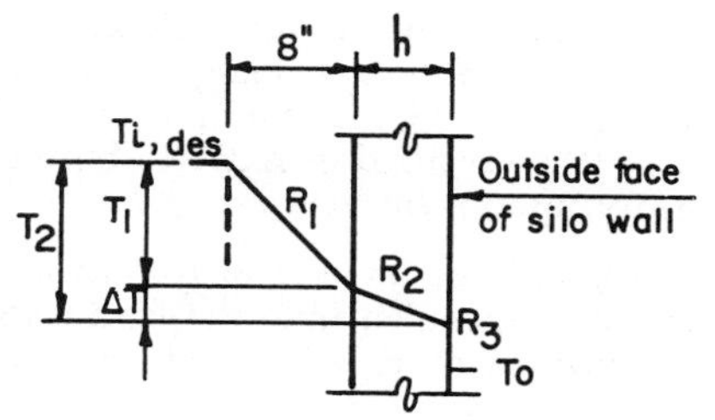

$\Delta T = [T_{i,\,des} - T_o] \times K_t$

T_i = Temperature of stored cement (°F)
$T_{i\,des}$ = (T_i – 80°F) Design temperature. of stored cement.
T_o = Design winter dry-bulb temperature (F)
h = Silo wall thickness

Fig. 16-9 Computation of ΔT for wall of cement storage silo.

Equation (16-38) is for walls in which the radius is considerably larger than the thickness.[16-26] The approach of section 9.2.7 of ACI 318-83[16-27] is not recommended, since if hot material is stored, the worst combination of dead, pressure, and temperature loadings is not improbable or short-term, but an *assured* combination.

The bending moment due to temperature gradient is derived for an uncracked section. The added reinforcing, however, is determined for a cracked section. The required additional horizontal steel area, $A_{s,t}$ to resist moment $M_{xt,u}$ should be added to that required for material pressure alone.

The added steel, $A_{s,t}$, should be located near the colder face. In singly reinforced walls, it should be added to the main hoop steel, ordinarily near the outer face. In doubly reinforced walls, the entire amount, $A_{s,t}$, should be added to the outside layer.

16.7.5 Vertical Reinforcement for Circular Silos

Both exterior and interior silo walls should have vertical reinforcement. Vertical bars serve several purposes: cantilever bending of the silo under wind or seismic load; distribution of lateral load concentrations to ensure participation of hoop steel above and below; supports for hoop steel and embedded items; and resistance to vertical bending due to radial temperature gradient.

The authors suggest that vertical bars in the storage zone be #4 or larger; that the vertical steel area be not less than the wall area (bh) times 0.0020 for exterior walls, or 0.0015 for interior walls; and that the spacing of vertical bars in each layer not exceed 16 in. for exterior, or 18 in. for interior walls. Embedded jackrods should not be considered as vertical reinforcement. The area of vertical steel in silo walls below the pressure zone should be neither less than used above nor less than 0.0020 bh.

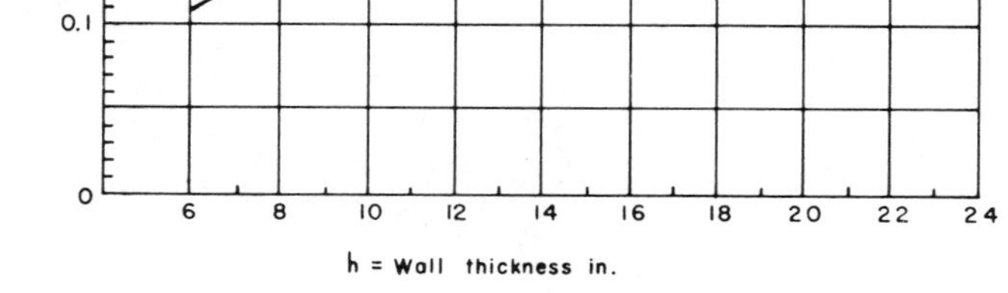

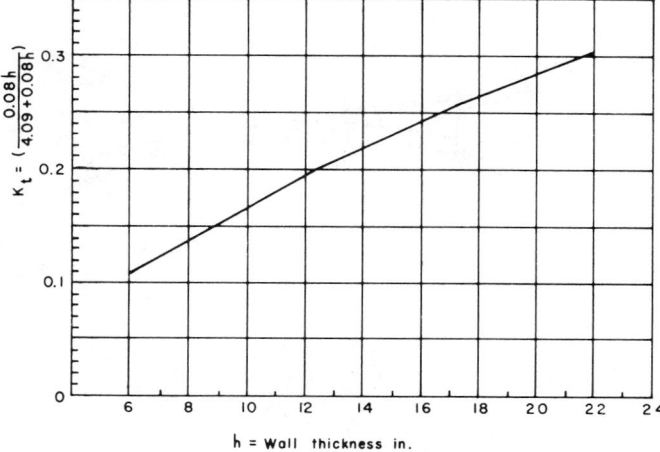

Above curve is based on the following assumptions.[16-24]

1. Resistance of 8" (203 mm) cement (considered to act as insulating material) = 3.92.
2. Resistance of 1" (25 mm) thick concrete = 0.08.
3. Resistance of outer surface film = 0.17.

Fig. 16-10 Determination of K_t for use in computing ΔT for wall of cement storage silo.

16.7.6 Horizontal Reinforcement for Circular Silos

The authors suggest that minimum horizontal reinforcement (excluding temperature steel) in pressure or nonpressure zones should not be less than 0.0030 times the gross concrete area per unit height of wall. This is slightly over that required by ACI 313-77.

16.8 GROUPED CIRCULAR SILOS

Figures 16-1 and 16-2 show circular silos in connected groups. Individual circular cells of such groups are first analyzed and designed as though isolated, with interstices empty and interaction ignored. Portions having significant horizontal bending are then modified so that the horizontal reinforcement can resist the combined hoop tension and bending.

Interstice walls may be assumed fixed at the point where adjacent circular silos are joined.

While it would be better to analyze the entire cross section of the silo group as a rigid frame, complexity of the analysis where all members are curved encourages the use of approximate methods.

16.8.1 Interstice Cells—(Fig. 16-2)

The portion of wall common to both the circular cell and the interstice must be designed for the following conditions:

1. Circular cell full and interstice empty. This causes ring tension if the wall is circular; tension and bending if it is flattened.
2. Circular cells empty and interstice full. If the four intersection points are treated as points of fixed position, each curved interstice wall will behave as a fixed-end arch and will be subject to both axial compression and bending.
   If the walls are flattened, they will have bending moment and axial tension. Their edges could be considered fixed at the junction to the circular walls or as members of a rectangular rigid frame.

Design pressures in interstice cells are determined by the same procedure as for silos. The interstice cross section is replaced by an imaginary circle or rectangle of area equal to that of the interstice.

16.8.2 Pocket Bins

Pocket bins (Fig. 16-2) complicate structural analysis of the horizontal frame and cause appreciable bending moment and axial load in the walls of the silos to which they attach. The main silo walls will almost certainly require more horizontal steel in some locations, and perhaps greater thickness than would be required if no pocket bin were present.

16.8.3 Circular Silos With Internal Crosswalls

Both the cylindrical walls and crosswalls of such silos are subject to combined horizontal tension and bending. In estimating static pressures due to stored material, each cell may be assumed replaced by a circle or rectangle of equal area. The static pressures are then multiplied by factors, C_d (as given for interstices), to obtain design pressures, q_{des} and p_{des}.

The walls of the cross section can be analyzed as rigid frames[16-28] to determine horizontal bending moments, $M_{x,u}$, and the horizontal axial tensions, F_u. Various load patterns (certain cells full, others empty) must be used. Crosswalls will have vertical dead load and friction load. If not supported at their bottoms, they must act as deep beams to transfer those loads to the main walls.

16.9 DETAILS—CIRCULAR SILOS AND SILO GROUPS

16.9.1 Horizontal Reinforcement

Horizontal reinforcement in circular silo walls should provide for the direct tension, bending moment (if any), and temperature gradient (if any). When only one layer is used, it should be placed at the center and staggered or nearer the outside face of the wall. Walls 9 in. or thicker should have two layers. Hoop reinforcing usually is continued below the pressure zone as noted in Section 16.7.2.

16.9.2 Horizontal Reinforcement at Wall Intersections

Reinforcement at wall intersections must be properly anchored to prevent separation of the walls. Bar forces are usually maximum at wall intersections; thus, bars must extend *each way* from the intersection a sufficient distance for full anchorage by bond. It is generally a good practice to determine reinforcement requirements at the intersection by analyzing the frame formed by the group of cells. However, the complexity of analysis frequently dictates the use of approximations. In such cases, extra horizontal steel should be provided at areas such as wall intersections where bending moment is expected in combination with the computed tension.

16.9.3 Vertical Reinforcement

Vertical reinforcement should be provided as described in Section 16.7.5. In addition, the thickened portion of walls at intersections below the hopper support or silo bottom slab should have vertical steel equal to not less than 0.0025 times the column cross section area. These "intersection columns" usually have horizontal ties spaced vertically at no more than 24-in. centers.

16.9.4 Reinforcing Bar Splices

Where fixed forms (nonmoving) are used, bar splices should conform to the *ACI Code*. The ACI 313-77 Standard states the following minimum requirements for the length of lap in horizontal reinforcement in the pressure zone of silo walls to be built by slipforming:

For noncircular silos with unstaggered splices, the length specified by ACI 318 for Class C splices. For circular silos (or any cell with circular reinforcing) with staggered splices and with continuous inspection, the length specified by ACI 318 for Class B splices, plus 6 in. (150 mm).

In determining the lengths of lap, bars in slipformed structures shall be assumed as bottom bars. (However, horizontal bars shall be treated as top bars in jumpformed structures.) Concrete thickness covering the reinforcement at lap splices shall be at least that specified in ACI 318 but not less than 1 in. (25 mm) for interior surfaces.

Lapped splices should not be used in zones where the concrete is in tension perpendicular to the lap unless adequate reinforcing is provided to prevent cracking parallel to the lap.

The authors concur with the above lap splice requirements, but prefer to use additional lengths of 9 in. instead of the 6 in. recommended by ACI 313.

Bar splices, both horizontal and vertical, should be staggered in circular silos. Hoop-reinforcing splices in the pressure zone should be offset by not less than one lap length or 3 ft and should not coincide in vertical array any more frequently than every third bar.

16.9.5 Dowels

Engineers do not agree as to the value of dowels joining silo walls to the foundation. The authors' preferences are as follows:

1. Always provide dowels for separate columns and for "intersection columns." The dowels should match the

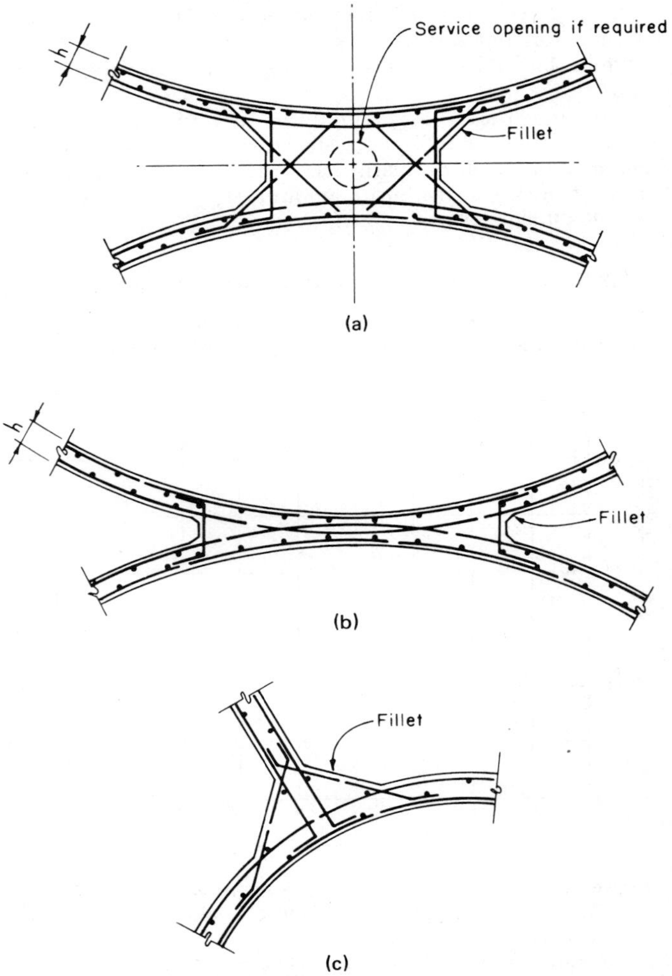

Fig. 16-11 Typical reinforcement pattern at intersecting walls of circular silos.

column steel and have sufficient embedment of complete development in both column and foundation.
2. Include dowels at other points along the bottom of the wall as needed to prevent uplift of the wall from earthquake or wind loading.

16.9.6 Fillets

Fillets serve several purposes. They eliminate sharp corners, permitting smoother flow of the stored material and reducing the volume of dead storage. They make slipforming easier. They reduce stress concentrations and bending moments, helping to prevent cracks along wall intersection lines.

At intersecting walls, fillets should be provided, as shown by Fig. 16-11. Fillets 9 X 9 in. or larger, especially in a tensile zone, should be properly reinforced. Within the pressure zone, fillet steel should be of area per unit width at least matching that of the near face of the adjoining walls and should be extended into the far face of each wall at least the length required for anchorage.

16.9.7 Reinforcement at Wall Openings

Wall openings in the pressure zone should be kept small. Unless areas of stress concentrations of openings are analyzed and reinforcement provided accordingly, wall reinforcement must be added as follows:

1. Horizontal (hoop) reinforcing terminated by the opening must be replaced by adding at least 1.2 times the terminated hoop steel area, one-half above and one-half below the opening.
2. Added vertical reinforcement adjacent to each side of the opening should be determined in the manner given for openings in the nonpressure zone below.

Openings below the pressure zone range from very small ones to those large enough to pass trucks or railroad cars. Around openings below the pressure zone, walls should have extra reinforcement, as follows (see Fig. 16-12):

1. Horizontal reinforcement should be added above the opening to enable the wall above, acting as a deep beam (curved as straight), to span the opening and carry all vertical loads applied to the wall above. The amount of added horizontal reinforcement above and below the opening, however, should be not less than the normal horizontal steel area for a height of wall equal to one-half the opening width.
2. Additional vertical reinforcement must be added to the wall on each side of the opening. The added steel should be computed by assuming a narrow strip of wall (not over $4h$ in width) on each side of the opening to act as a column, unsupported within the opening height and carrying its own share of the vertical load plus one-half of the wall loads occurring over the wall opening within a height equal to the opening width. The added steel area for each column strip should be at least two bars of the size in the wall and should have an area equal to at least one-half of the vertical steel area eliminated by the opening. Added

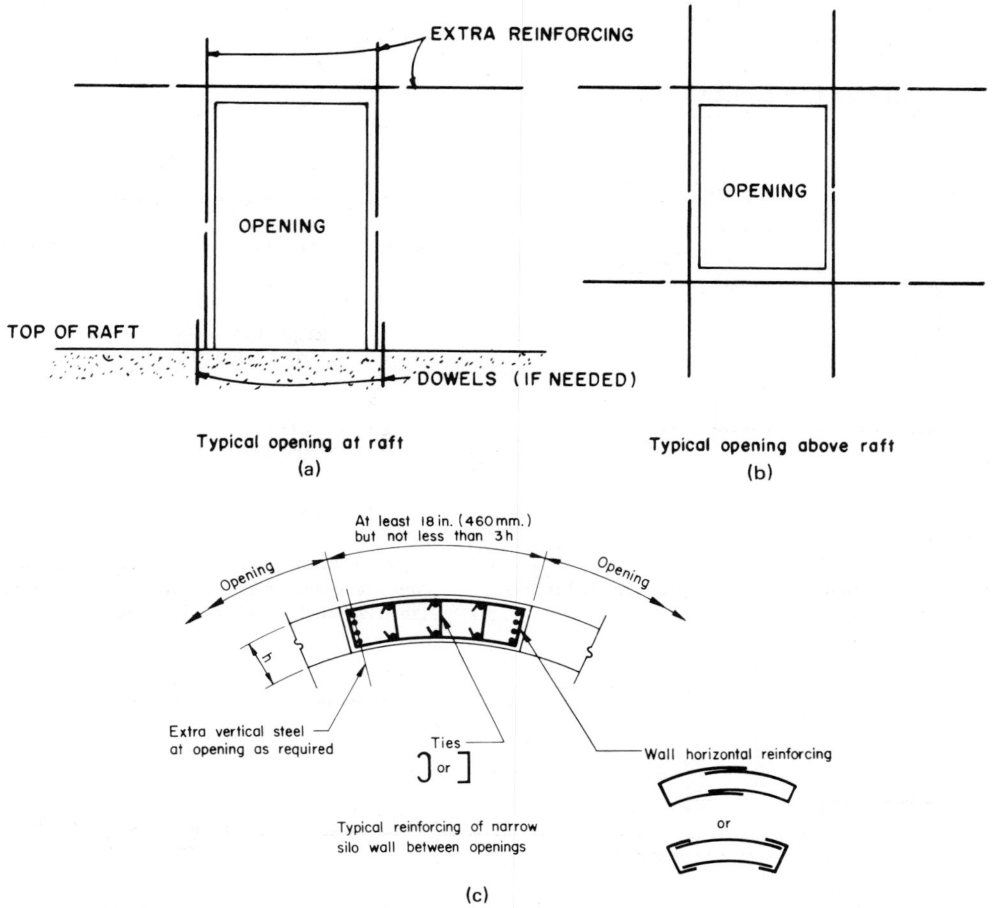

Fig. 16-12 Extra steel at openings.

horizontal and vertical reinforcing to replace load-carrying steel that is interrupted by an opening should extend in each direction beyond the opening. The extension each way should be sufficient to develop the full yield strength through bond, but neither less than one-half of the opening dimension in a direction perpendicular to the reinforcing bars in question nor less than 24 in. (610 mm).

3. Walls between closely spaced openings, as shown in Fig. 16-12(c), should be designed as tied columns.

16.10 SQUARE AND RECTANGULAR SILOS

16.10.1 Horizontal Forces Due to Stored Material

The cross section of a rectangular silo, Fig. 16-13(a), is a rigid frame subject to outward pressure, p_{des}, varying with depth. The resulting moment diagram will be as shown by Fig. 16-13(b), each wall having axial tension, bending moment, and shear.

The horizontal ultimate tensile forces per unit height at depth Y are:

$$F_{a,u} = K_l p_{b,\text{des}} (b/2) \quad \text{(for wall "}a\text{")} \quad (16\text{-}39)$$

$$F_{b,u} = K_l p_{a,\text{des}} (a/2) \quad \text{(for wall "}b\text{")} \quad (16\text{-}40)$$

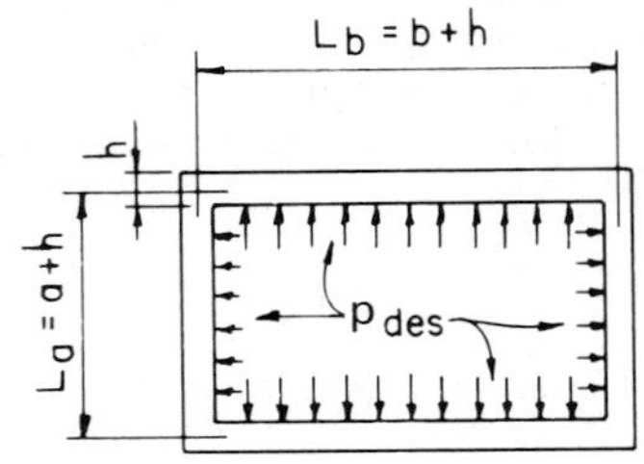

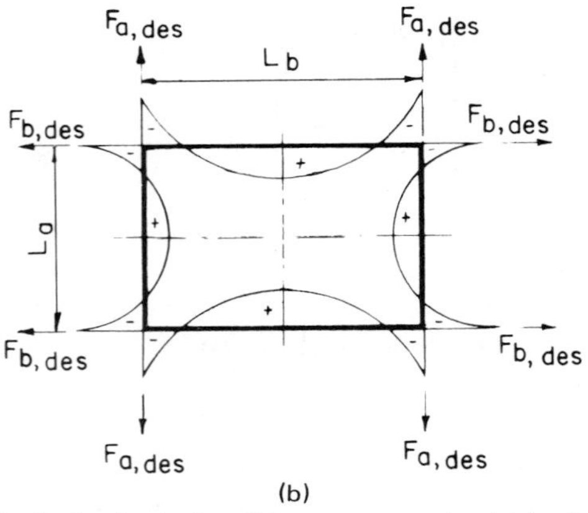

Fig. 16-13 Design loads, bending moment, and axial loads for rectangular silo.

The ultimate bending moments, $M_{x,u}$, may be reduced from the value computed at the frame centerline to the value at the inner face of the wall or toe of the wall fillet.

16.10.2 USD Approach for Combined Tension and Bending

The silo wall must be reinforced to resist combined tension and bending. The combination may be replaced by an eccentric tension, with $e = M_u/F_u$. The eccentric force, F_u, may lie in the space between the layers of steel (Fig. 16-14) or outside that space (Fig. 16-15). A different design approach is needed for each case.

CASE I: Small eccentricity; $e \leqslant \frac{h}{2} - d''$ (Fig. 16-14):

By taking moments about either layer of steel and ignoring the concrete strength:

$$\text{Reqd } A_s = \frac{F_u e'}{\phi f_y (d - d')} \quad (16\text{-}41)$$

$$\text{Reqd } A_s' = \frac{F_u e''}{\phi f_y (d - d')} \quad (16\text{-}42)$$

CASE II: Large eccentricity, $e > \frac{h}{2} - d''$ (Fig. 16-15):

Determine *Code* limit for ratio y/d from Table 16-3. Compute y_L, limiting depth of compression block.

Assuming compressive steel to be needed, compute f_s' (effective compressive steel stress) in ksi:

$$f_s' \text{ effective} = 87 \left(\frac{y_L - \beta_1 d'}{y_L} \right) - 0.85 f_c' \quad (16\text{-}43)$$

but not over:

$$f_y - 0.85 f_c'$$

If f_s' is negative, compression steel will be ineffective. If a singly reinforced member would not be adequate, either depth d must be increased, or A_s' moved to a location where it will be effective.

$$\text{Reqd } A_s' = \frac{F_u(e''/\phi) - 0.85 f_c' b y_L \left(d - \frac{y_L}{2} \right)}{(f_s')_{\text{eff}} (d - d')} \quad (16\text{-}44)$$

If eq. (16-44) gives a positive A_s', compressive steel is required. In that case, the required tensile steel is:

$$\text{Reqd } A_s = \frac{F_u/\phi + 0.85 f_c' b y_L + A_s' (f_s')_{\text{eff}}}{f_y} \quad (16\text{-}45)$$

If eq. (16-44) gives a negative A_s', no compressive steel is needed. Then, whether or not negative steel is provided, the wall is designed as singly reinforced. In this case, the required tensile steel

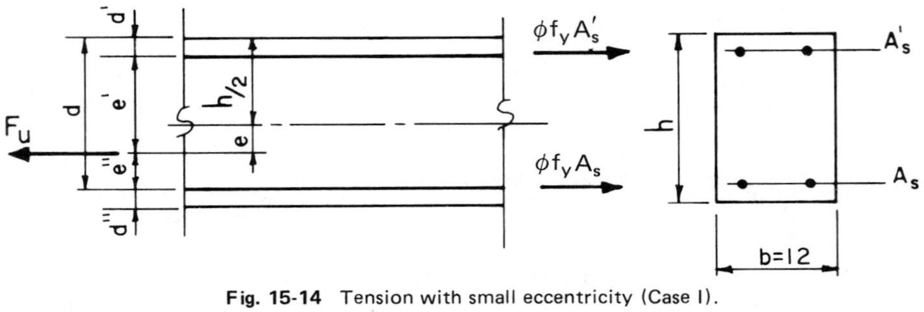

Fig. 15-14 Tension with small eccentricity (Case I).

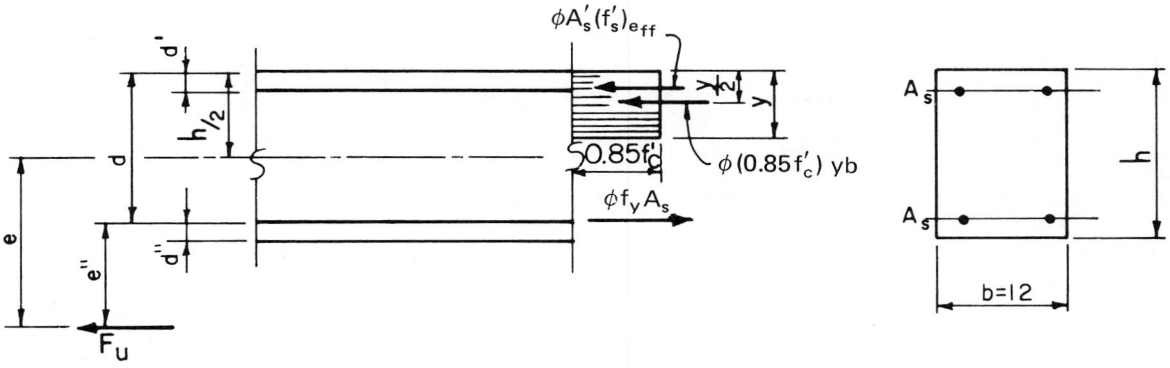

Fig. 16-15 Tension with large eccentricity (Case II).

TABLE 16-3 Code-Limit Values of *y/d*

Concrete Strength f'_c psi	Steel Yield Strength, psi		
	40,000	50,000	60,000
Up to 4,000	0.436	0.405	0.378
5,000	0.411	0.381	0.355
6,000	0.386	0.357	0.333

area is:

$$A_s \text{ reqd} = \frac{F_u/\phi + 0.85 f'_c by}{f_y} \qquad (16\text{-}46)$$

where y, actual depth of a compression block, is:

$$y \simeq d - \sqrt{d^2 - \frac{2 F_u e''}{0.85 \phi f'_c b}} \qquad (16\text{-}47)$$

16.10.3 Wall Thickness—Rectangular Silos

Walls of rectangular silos must have sufficient thickness to satisfy the following requirements:

1. For vertical compression, the USD requirements of Section 14.2 of ACI 318-83.[16-27]
2. Flexural strength under combined tension and bending.
3. Shear. The required nominal shear strength, V_n, at the face of the support for each wall should not exceed:

$$V_c = 2(1 - 0.002 F_u/A_g) \sqrt{f'_c}\, b_w d, \qquad (16\text{-}48)$$

4. In addition, the thickness should not be less than 6 in. for slipform construction unless special precautions are taken to reduce friction and avoid lifting the concrete.

16.10.4 Crack Width—Rectangular Silos

Crack width limits should be the same as for circular silos. Where the wall tensile force is between the two layers of steel (Case I, Fig. 16-14), the method shown for circular silos may be used to estimate the actual crack width.

With larger eccentricity (Case II, Fig. 16-15), the cracks are larger on one side but are closed where the concrete is in compression. Thus, the stored material should receive adequate protection at sections of large bending moment.

16.10.5 Temperature Effect—Rectangular Silos

If hot material is to be stored, extra steel may be needed. Computation of the thermal moments is complex. However, for the simplest case, a single square silo cross section, edge moments tend to keep the walls flat, and the same moment will exist at all points between the edges. The horizontal ultimate moment, with Poisson's ratio = 0.2, will be:

$$M_{xt,u} = M_{yt,u} = 1.25\, K_g E_c h^2 \alpha_t \Delta T \text{ (in.-lb/ft)} \qquad (16\text{-}49)$$

16.10.6 Other Vertical Forces and Moments

Vertical bending moments will occur near the bottom of walls that are continuous with a silo bottom slab or with hopper walls. The wall thickness and vertical steel must provide for such bending combined with the vertical forces from dead load, roof load, and friction from the stored material. Bottom slabs or hoppers, suspended from the walls, will transfer vertical tensile forces to the lower portions of the silo walls. In such cases, the walls must also act as deep beams, transferring their vertical loads to corner columns by shear and in-plane bending.

16.10.7 Groups of Rectangular Silos

Figure 16-16 shows a typical group of rectangular silos with pyramidal hoppers. To compute the wall tensions, shears, and bending moments for such groups, various combinations of loaded and unloaded cells must be considered.

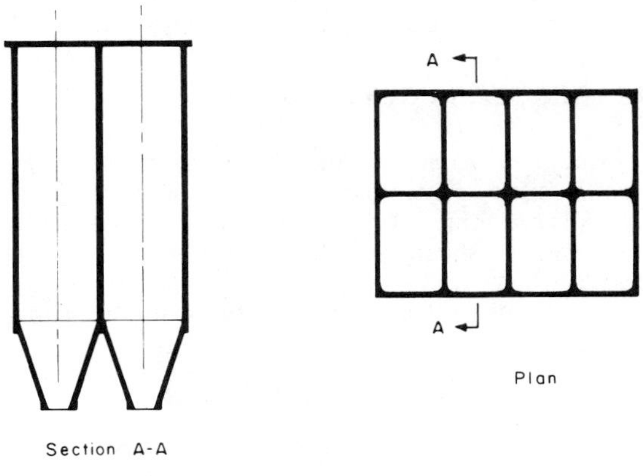

Fig. 16-16 Typical group of rectangular silos.

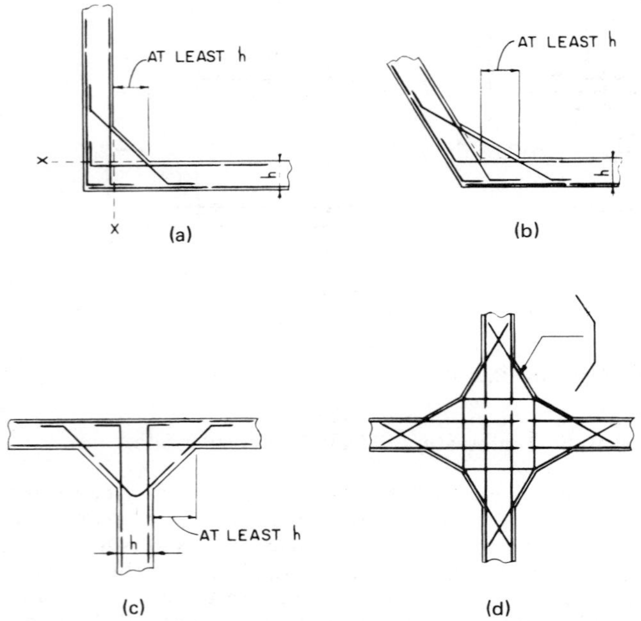

Fig. 16-17 Corner and intersection details for rectangular and polygonal silos.

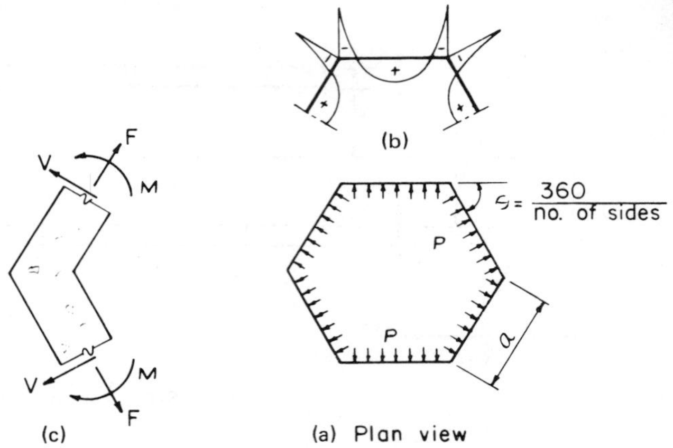

Fig. 16-18 Polygonal silo wall forces.

16.11 DETAILS—RECTANGULAR SILOS AND SILO GROUPS

All detail requirements given earlier for circular silos apply equally to rectangular silos.

16.11.1 Horizontal Steel

Ordinarily the horizontal steel (designed for combined tension and bending) will be in two layers. For all walls, the ACI requirements for continuing a portion of the positive steel over the full span and for bar extension and anchorage at supports and points of contraflexure must be observed.

16.11.2 Corner Details

Fillets and corner reinforcing patterns are shown by Fig. 16-17. The fillet size should be at least equal to the wall thickness (but not less than 9 in.) and should have reinforcement equal to that computed for combined bending and axial load at the face of the wall or toe of the fillet. The fillet steel should extend past its intersection with the inner wall steel a sufficient distance for anchorage and should terminate on the opposite side of the wall.

16.12 REGULAR POLYGONAL SILOS

The principles of design and detail of single or grouped polygonal silos are similar to those for rectangular silos. Their walls are subject to similar loading and to similar horizontal tension, shear, and bending moment, and vertical forces.

The horizontal ultimate tensile force in the wall of any single regular polygonal silo (Fig. 16-18) is:

$$F_u = K_l \, p_{des} \left[\frac{a}{2}\right]\left[\frac{\sin\theta}{1 - \cos\theta}\right] \qquad (16\text{-}50)$$

This equation reduces to:

$$F_u = 0.866 \, K_l a \, p_{des} \qquad \text{(for hexagonal silos)} \qquad (16\text{-}51)$$

$$F_u = 1.207 \, K_l a \, p_{des} \qquad \text{(for octagonal silos)} \qquad (16\text{-}52)$$

Bending moments $M_{x,u}$ and $M_{y,u}$ are determined as for rectangular silos. Reinforcement and details are as for rectangular silos, except that for polygons with eight or more sides, wall intersection fillets may be omitted.

16.13 SILO BOTTOMS

Silo bottoms are supported by the silo walls extended down to the foundation or independently supported by columns resting on either the same foundations as the walls or independent foundations. (See Fig. 16-19.)

16.3.1 Loads

Silo bottom structures should be designed for load imposed by the stored material, their own dead load, the weight of any equipment, platforms, etc., hung from the silo bottom, and occasionally earthquake and temperature effects.

16.13.2 Effect of Eccentric Discharge

The authors suggest that eccentric discharge be considered in the design of hoppers or of bottoms with hopper-forming fill, since they are subjected to lateral pressures. It is recommended to assume the maximum increase for both the side near and the side opposite the eccentric discharge for square or rectangular silos and all around the perimeter for other shapes. The influence of eccentric discharge on vertical pressures is unknown, and current practice is to ignore eccentric discharge in the design of flat bottoms.

The maximum value of p_{ecc} (at depth H) is computed based on silo dimensions at the top of the hopper and is assumed constant for the depth of the hopper or hopper-forming fill.

Eccentric discharge should be considered in any case with multiple openings.

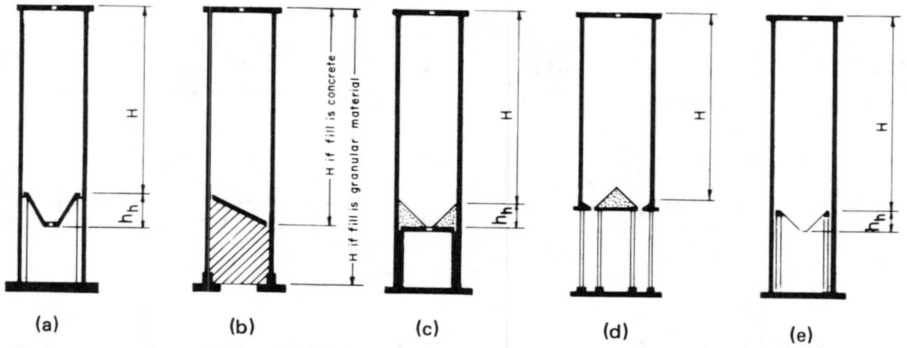

Fig. 16-19 Typical vertical cross sections of silos. (a) Silo on raft foundation, independent hopper resting on pliasters attached to wall; (b) silo with wall footings and independent bottom slab supported on fill; (c) silo with hopper-forming fill and bottom slab supported by thickened lower walls; (d) silo with multiple discharge openings and hopper-forming fill resting on bottom slab, all supported by columns—raft foundation has stiffening ribs on top surface; and (e) silo on raft foundation, with hopper independently supported by a ring beam and column system.

16.13.3 Earthquake Forces on Silo Bottoms

Where independent bottom supports are provided, lateral earthquake forces should be divided between the walls and the bottom support structure as suggested earlier. In addition, since wall friction may be destroyed by quake action, the entire vertical effective weight, W_{eff}, of the stored material should be considered applied to the bottom structure.

16.13.4 Flat Bottoms

The simplest flat bottom is a slab of uniform thickness. The flat bottom may also be a ribbed slab or a beam-and-slab system. To assure good flow of material, hopper-forming fill is usually constructed around the discharge opening. Concentric slab openings should not exceed one-third of the silo diameter.

For a slab without hopper-forming fill, the design loads are dead load and pressure, q_{des}, computed at the top of the slab. For USD methods, the vertical load per unit area would be:

$$w_u = K_g g + K_l q_{\text{des}} \tag{16-53}$$

With earthquake, vertical friction at the wall is assumed to be zero, so that the ultimate vertical pressure on the bottom is:

$$w_u = 0.75\,(K_g g + K_l \gamma H) \tag{16-54}$$

For a slab with hopper-forming fill, the weight of the fill itself and of the material stored in the hopper are added as dead load, and q_{des}, for use in eq. (16-53), is computed at the top of the fill.

Slab shear stresses should be checked carefully.

16.13.5 Moments and Deflections for Circular Slab Bottoms

Radial and tangential bending moments for circular slabs with (or without) a central hole can be computed by the methods of Timoshenko and Woinowsky-Krieger[16-29] or from tables.[16-28] When a concentric opening is not over one-tenth of the silo diameter (or, for rectangular silos, the largest side of the opening does not exceed one-tenth of the short side of the silo), the opening is frequently ignored in computing slab bending moments. In this case, extra bottom reinforcement to replace that eliminated by the hole and added top reinforcement should be provided around the opening, or the methods of Examples 16-3 and 16-4 can be used.

16.13.6 Rectangular Flat Bottoms

Economy usually limits the wall length in rectangular silos to a maximum of about 20 ft. Thus, the bottom slab span is small, and such slabs are usually supported along their boundaries only.

Moments and deflections of simple and fixed-edge rectangular slabs may be calculated using the methods presented by the *ACI Building Code* for analysis of two-way slabs. The effect of circular or rectangular discharge openings in the bottom slabs should be considered in the design. The extra reinforcing added to compensate for that intercepted by the opening should be continuous from support to support.

16.13.7 Additional Load at Openings

The edges of bottom slab openings usually receive added line loads comprising the weight of the hopper and equipment supported by it, the weight of stored material in the hopper and the design vertical pressures of material above the top of the hopper. Equations by Griffel[16-30] and tables by the authors[16-28] and others[16-22,16-31] may be used.

16.13.8 Conical Hoppers

The design pressure, $q_{\alpha,\text{des}}$, may be computed from eq. (16-21). In computing $q_{\alpha,\text{des}}$, engineers usually use the dimensions at the top of the hopper to obtain R or C for the Janssen or Reimbert formulas, ignoring the reductions of cross section within the hopper.

The conical hopper shell is subject to two tensile membrane forces. The meridional force, F_m, is parallel to the generator line of the cone. The tangential force, F_t, is in the plane of the shell and horizontal. The meridional force per unit width at depth Y is computed from equilibrium of the loads on the cone below that depth. These loads, shown in Fig. 16-20(a), are the resultant of vertical pressures, q_{des} (at depth Y) and W, the combined weights of the hopper itself and material stored below depth Y plus any equipment supported by the hopper. (See Notation.)

For design by USD:

$$F_{mu} = K_l \left[\frac{q_{\text{des}} D}{4 \sin \alpha} + \frac{W_l}{\pi D \sin \alpha}\right] + K_g \left[\frac{W_g}{\pi D \sin \alpha}\right] \tag{16-55}$$

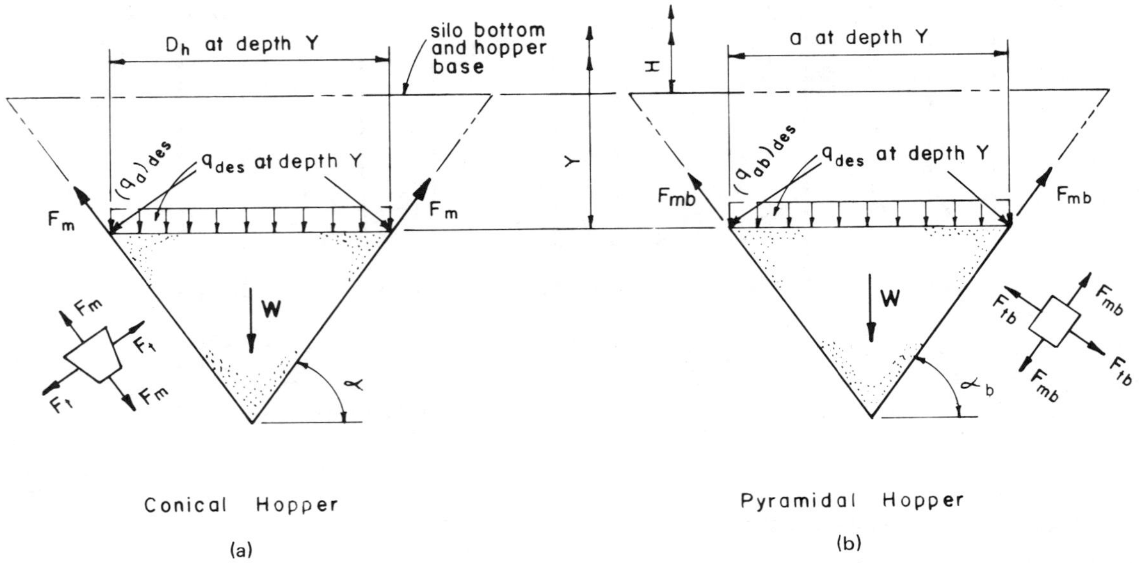

Fig. 16-20 Forces in conical and pyramidal hoppers.

$$F_{tu} = K_l \left[\frac{q_{\alpha,\,des} D}{2 \sin \alpha} \right] \tag{16-56}$$

Both forces are maximum at the upper edge of the hopper, and approach zero at the lower edge.

The minimum acceptable thickness for the cone should be determined considering the acceptable crack width.[16-2, 16-26] The authors prefer, however, that the shell thickness never be less than 5 in.

The required reinforcement area per unit width of shell is:

$$A_s \text{ reqd} = F_{mu}/(\phi f_y) \quad \text{(meridional direction)} \tag{16-57}$$

$$A_s \text{ reqd} = F_{tu}/(\phi f_y) \quad \text{(horizontal)} \tag{16-58}$$

A conical hopper is usually supported at its upper end by a ring beam. This is merely a thickened portion of the conical shell with a cross section to satisfy the supporting condition and the loading. The ring-beam depth should not be less than one-tenth of the hopper diameter.

If simply supported all around by the silo walls, ring-beams are usually designed for the horizontal component of F_{mu} only. If the hopper wall is eccentric to the centroid of the ring-beam, the beam will also receive uniform bending moment. The monolithically cast ring-beam and conical shell are very stiff, however, and this moment is usually neglected in design of the ring-beam with all-around support.

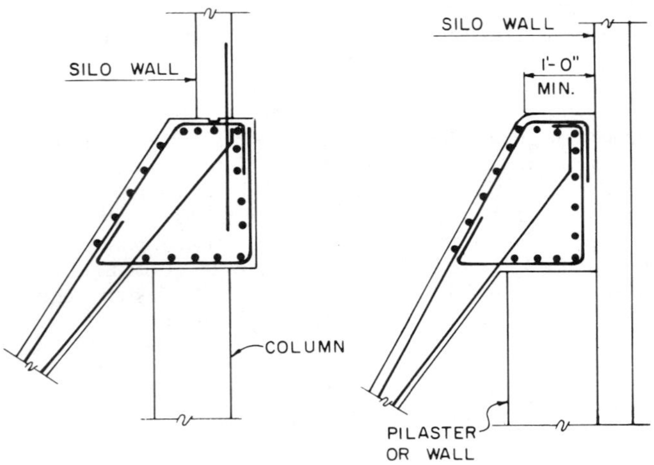

Fig. 16-21 Typical details of hopper-supporting beam.

The area of longitudinal steel in such ring-beams is arbitrary, but should be not less than 0.5% of the cross-sectional area of the ring-beam. (See Fig. 16-21.)

If the upper edge of the hopper is keyed all around to the silo walls or monolithic with the walls, adequate negative reinforcement should be provided in the hopper wall at the intersection of hopper and silo walls. The negative steel should be extended into the silo walls for complete anchorage by bond.

The ring-beam and upper edge of a conical hopper supported at isolated points along its boundary by columns, pilasters, or wall pockets may be designed in the manner shown later herein for a concrete ring-beam supporting a steel hopper. If desired, the stiffness and strength added by the concrete hopper shell may be approximated by considering a width of hopper shell (the authors suggest four times the thickness) to act as a part of the ring-beam.

16.13.9 Pyramidal Hoppers

Loads for pyramidal hoppers are the same as for conical hoppers. (See Fig. 16-20b.) Pyramidal hopper walls, however, are subject to bending as well as tensile membrane forces. The bending always includes two-way, plate-type bending and may include significant in-plane bending.

All of the following is for symmetrical (or nearly symmetrical) hoppers only. The angles of slope for walls "a" and "b" are α_a and α_b, respectively.

The membrane tensile forces in the inclined walls are assumed to consist of horizontal forces, F_t, and forces, F_m, in the plane of the wall and normal to F_t. Forces F_m per unit width vary in intensity along the width of the wall. For simplicity, however, they are usually assumed to be uniform along any one wall. Material pressures for each wall are assumed as those computed using the properties of the silo wall to which it is attached, plus a share of the dead load. In the hoppers of Fig. 16-22, for example, material pressures applied to area A_a are assumed to affect the value of F_m for wall "a," and those applied to area A_b to affect F_m for wall "b." If the area of the discharge opening is ignored, and if some arbitrary factors c_a and c_b are selected to define the division of the load to area A_a and A_b (such that, for a symmetrical case, $2c_a + 2c_b = 1.0$), then the ultimate meridional forces per unit width of wall are:

$$F_{mau} = \frac{K_l(c_a W_l + A_a q_{a,\,des}) + K_g c_a W_g}{a \sin \alpha_a} \tag{16-59}$$

$$F_{mbu} = \frac{K_l(c_b W_l + A_b q_{b,\,des}) + K_g c_b W_g}{b \sin \alpha_b} \qquad (16\text{-}60)$$

(In the above, $q_{a,\,des}$ is computed using the properties for the short wall and $q_{b,\,des}$ using the imaginary square silo. For simplicity, some engineers ue the average of $q_{a,\,des}$ and $q_{b,\,des}$ for all walls of the hopper.)

The ultimate horizontal forces, F_{tau} and F_{tbu}, per unit width of wall at depth Y may be approximated by:

$$F_{tau} = K_l \left(\frac{b}{2}\right) q_{\alpha b,\,des} \sin \alpha_a \qquad (16\text{-}61)$$

$$F_{tbu} = K_l \left(\frac{a}{2}\right) q_{\alpha a,\,des} \sin \alpha_b \qquad (16\text{-}62)$$

Moments from two-way, plate-type bending (sometimes called local bending) may be computed using Tables 16-4 and 16-5 for inclined triangular walls with fixed sides and subject to either uniform or triangular loading, and (by superposition) for trapezoidal loading. Tables 16-4 and 16-5 may also be used for trapezoidal walls in which the length of the bottom edge is not over one-fourth of the top edge, by substituting for the trapezoid an imaginary triangle of slant height:[16-22]

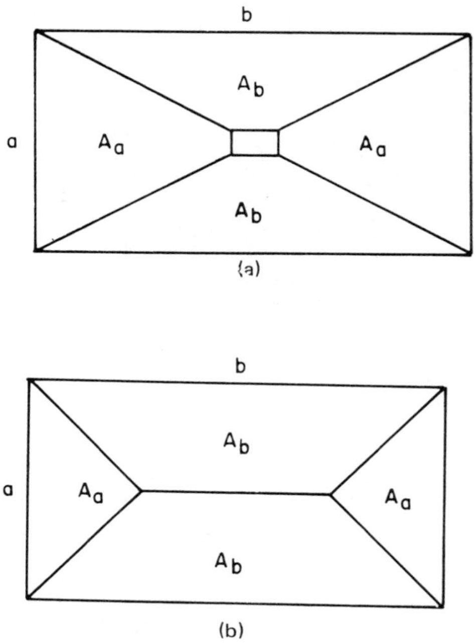

Fig. 16-22 Plan view of pyramidal hoppers.

TABLE 16-4 Isosceles Triangular Slab, Fixed Edges, Uniform Load

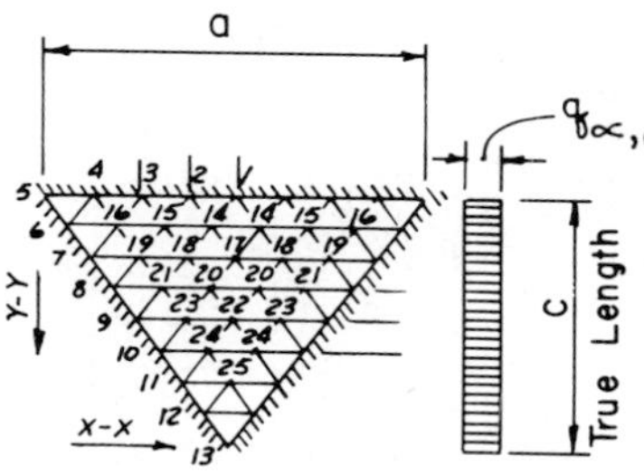

Bending Moments: $M_x = n_x \, q_{\alpha, des} \cdot a^2/64$ (per unit width)

$M_y = n_y \, q_{\alpha, des} \cdot a^2/64$ (per unit width)

Reaction per unit length of support: $R = \gamma \, q_{\alpha, des} \cdot a/8$

* Deflection: $\Delta = \dfrac{\lambda \, q_{\alpha, des} \, a^4 (1 - \nu^2)}{4096 \, EI}$

* Consistent units must be used in above equations. To obtain Δ in inches when a = feet, q_α = lb/sq ft, E = psi and I = in.^{4}/ft, multiply the above expression by 1,728.

Ratio a/c = 0.75					Ratio a/c = 1.00					Ratio a/c = 1.50				
Point	λ	n_x	n_y	γ	Point	λ	n_x	n_y	γ	Point	λ	n_x	n_y	γ
1	0	−0,2330	−1,3977	2,444	1	0	−0,2091	−1,2547	2,219	1	0	−0,1536	−0,9215	1,823
2	0	−0,2008	−1,2044	2,184	2	0	−0,1787	−1,0721	1,968	2	0	−0,1288	−0,7729	1,603
3	0	−0,1185	−0,7111	1,428	3	0	−0,1039	−0,6232	1,294	3	0	−0,0715	−0,4291	1,061
4	0	−0,0343	−0,2055	0,442	4	0	−0,0297	−0,1784	0,481	4	0	−0,0195	−0,1169	0,478
5	0	0	0	0	5	0	0	0	0	5	0	0	0	0
6	0	−0,5758	−0,1710	1,148	6	0	−0,2241	−0,0958	0,577	6	0	−0,0526	−0,0122	0,318
7	0	−1,3155	−0,3559	2,175	7	0	−0,6748	−0,2598	1,515	7	0	−0,1901	−0,1073	0,780
8	0	−1,6061	−0,4476	2,531	8	0	−1,0164	−0,4016	2,118	8	0	−0,3791	−0,2513	1,313
9	0	−1,3521	−0,3939	2,240	9	0	−1,1018	−0,4276	2,204	9	0	−0,5271	−0,3761	1,708
10	0	−0,8259	−0,2561	1,633	10	0	−0,8148	−0,3305	1,796	10	0	−0,5128	−0,3998	1,774
11	0	−0,3147	−0,1180	1,016	11	0	−0,4169	−0,1754	1,116	11	0	−0,3970	−0,2961	1,393
12	0	−0,0600	−0,0330	0,493	12	0	−0,0984	−0,0302	0,481	12	0	−0,1717	−0,1108	0,613
13	0	0	0	0	13	0	0	0	0	13	0	−0,0858	0,0375	0
14	1,24235	0,3990	0,3901	—	14	0,62733	0,2066	0,1750	—	14	0,20177	0,0622	−0,0120	—
15	0,89879	0,2337	0,3524	—	15	0,44479	0,1141	0,1954	—	15	0,13276	0,0284	0,0193	—
16	0,37532	−0,1088	0,2113	—	16	0,17844	−0,0600	0,1530	—	16	0,05195	−0,0201	0,0811	—
17	1,99139	0,8284	0,6706	—	17	1,17677	0,5347	0,5865	—	17	0,45050	0,2466	0,3756	—
18	1,62283	0,5979	0,5655	—	18	0,95205	0,3819	0,5125	—	18	0,35592	0,1670	0,3412	—
19	0,73625	−0,1062	0,2396	—	19	0,42236	−0,0605	0,2531	—	19	0,14802	−0,0242	0,2013	—
20	1,58247	0,8117	0,3348	—	20	1,10230	0,6009	0,4303	—	20	0,49613	0,3154	0,4033	—
21	0,80196	0,0384	0,0702	—	21	0,53797	0,0119	0,1719	—	21	0,24086	0,0210	0,2103	—
22	0,98360	0,7271	0,0487	—	22	0,81779	0,6222	0,1576	—	22	0,45203	0,3831	0,2662	—
23	0,61334	0,2260	−0,0681	—	23	0,51131	0,2013	0,0131	—	23	0,27768	0,1209	0,1227	—
24	0,33246	0,3069	−0,0981	—	24	0,32460	0,2989	−0,1003	—	24	0,22829	0,2211	−0,0051	—
25	0,11205	0,2037	−0,0551	—	25	0,11584	0,2065	−0,1122	—	25	0,10763	0,1902	−0,1145	—

NOTE: For trapezoidal loading, use results from Table 16-4 ± results from Table 16-5. Table reconstructed after Lipnitski, Ref. 16-22.

TABLE 16-5 Isosceles Triangular Slab, Fixed Edges, Uniformly Varying Load

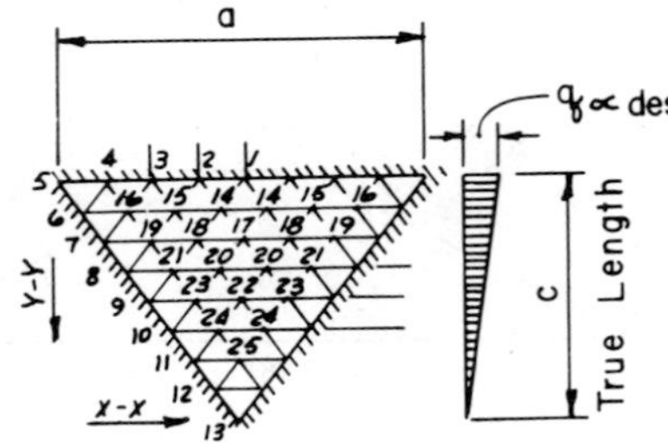

Bending Moments: $M_x = n_x q_{\alpha,\text{des}} \cdot a^2/64$ (per unit width)

$M_y = n_y q_{\alpha,\text{des}} \cdot a^2/64$ (per unit width)

Reaction per unit length of support: $R = Y q_{\alpha,\text{des}} \cdot a/8$

$$\text{*Deflection: } \Delta = \frac{\lambda q_{\alpha,\text{des}} \cdot a^4}{4096 EI}$$

Ratio a/c = 0.75					*Ratio a/c = 1.00*					*Ratio a/c = 1.50*				
Point	λ	n_x	n_y	γ	Point	λ	n_x	n_y	γ	Point	λ	n_x	n_y	γ
1	0	−0.1850	−1.1097	2.106	1	0	−0.1595	−0.9571	1.849	1	0	−0.1104	−0.6625	1.452
2	0	−0.1604	−0.9622	1.908	2	0	−0.1377	−0.8261	1.672	2	0	−0.0942	−0.5663	1.310
3	0	−0.0973	−0.5835	1.331	3	0	−0.0825	−0.4948	1.174	3	0	−0.0550	−0.3300	0.936
4	0	−0.0294	−0.761	0.511	4	0	0.0246	−0.1473	0.510	4	0	−0.0160	−0.0960	0.470
5	0	0	0	0	5	0	0	0	0	5	0	0	0	0
6	0	−0.5627	−0.1342	1.009	6	0	−0.1844	−0.0755	0.605	6	0	−0.0432	−0.0100	0.296
7	0	−1.0046	−0.2677	1.682	7	0	−0.5120	−0.1950	1.185	7	0	−0.1419	−0.0848	0.629
8	0	−1.1097	−0.3094	1.706	8	0	−0.7243	−0.2774	1.446	8	0	−0.2601	−0.1799	0.914
9	0	−0.8132	−0.2421	1.246	9	0	−0.6843	−0.2670	1.282	9	0	−0.3142	−0.2401	1.027
10	0	−0.3781	−0.1327	0.668	10	0	−0.4447	−0.1803	0.818	10	0	−0.3090	−0.2287	0.895
11	0	−0.1195	−0.0473	0.251	11	0	−0.1765	−0.0783	0.316	11	0	−0.2018	−0.1465	0.525
12	0	−0.0154	−0.0085	0.051	12	0	−0.0311	−0.0158	0.031	12	0	−0.0817	−0.0420	0.077
13	0	0	0	0	13	0	0	0	0	13	0	−0.0356	−0.0156	0
14	0.98637	0.3159	0.3659	–	14	0.47856	0.1581	0.1844	–	14	0.14721	0.0467	0.0283	–
15	0.72419	0.3245	0.3245	–	15	0.34753	0.0990	0.1905	–	15	0.10402	0.0285	0.0649	–
16	0.31303	0.0631	0.1938	–	16	0.14728	−0.0277	0.1425	–	16	0.04267	−0.0051	0.0785	–
17	1.46605	0.5976	0.5139	–	17	0.83876	0.3741	0.4481	–	17	0.30398	0.1627	0.2819	–
18	1.20279	0.4470	0.4328	–	18	0.68480	0.2768	0.3896	–	18	0.24447	0.1179	0.2556	–
19	0.55394	−0.1769	0.1769	–	19	0.31266	−0.0246	0.1969	–	19	0.10752	−0.0031	0.1530	–
20	1.05655	0.5338	0.1953	–	20	0.72519	−0.3878	0.2707	–	20	0.31453	0.1954	0.2574	–
21	0.54094	0.0274	0.0168	–	21	0.37265	0.0354	0.0948	–	21	0.15771	0.0224	0.1294	–
22	0.56125	0.4060	−0.0323	–	22	0.47715	0.3519	0.0442	–	22	0.26222	0.2136	0.1294	–
23	0.34967	0.1188	−0.0926	–	23	0.29998	0.1133	−0.0369	–	23	0.16348	0.0707	0.0464	–
24	0.14304	0.1225	−0.0997	–	24	0.15539	0.1329	−0.1090	–	24	0.11764	0.1066	−0.0464	–
25	0.02884	0.0483	−0.0465	–	25	0.03653	0.0563	−0.0884	–	25	0.04468	0.0704	−0.0991	–

NOTE: For trapezoidal loading, use results from Table 16-4 ± results from Table 16-5. Table reconstructed after Lipnitski, Ref. 16-22.
**See note on Table 16-4.*

$$c_i = \frac{ca}{a - a_1} \tag{16-63}$$

In the above, c is the actual slant height of the trapezoid, and a_1 and a are the bottom and top edge lengths, respectively.

Approximate solutions for bending moments in trapezoidal walls with bottom edges exceeding $a/4$ (either symmetrical or slightly unsymmetrical) may be obtained by solving for equivalent rectangular plates of width, a_{eq}, and height, c_i, as follows:[16-32]

$$a_{eq} = \frac{2(a^2 + 2a + a_1)}{3(a + a_1)} \tag{16-64}$$

$$c_i = c - \frac{a(a - a_1)}{6(a + a_1)} \tag{16-65}$$

16.13.10 Thickness and Reinforcement of Pyramidal Hopper Walls

Wall thickness and reinforcement must be selected to keep crack width within acceptable limits. However, in pyramidal hoppers, combined bending and axial tension in both the meridional and horizontal directions must be considered. The authors prefer that the minimum wall thickness be neither less than required for bending alone nor less than 6 in. Walls of pyramidal hoppers should be checked for shear.[16-27] See eq. (16-48).

16.13.11 Pyramidal Hopper Details

Pyramidal hoppers should have fillets at the inside of hopper wall intersections. The fillet size usually matches that of the silo walls at the top of the hopper and tapers to zero at the discharge opening. (See Fig. 16-23.)

Hopper wall steel is placed either (1) at both faces of the wall, or (2) continuously at the outer face with inside-face steel near the corners and upper and lower edges only, where required for negative moment. The first approach seems more practical in the United States though the second is common in Europe.

The steel areas (each way) should be not less than 0.003 times the wall area. Concrete cover for reinforcing bars near the inside face should be at least 1 in., but may be reduced

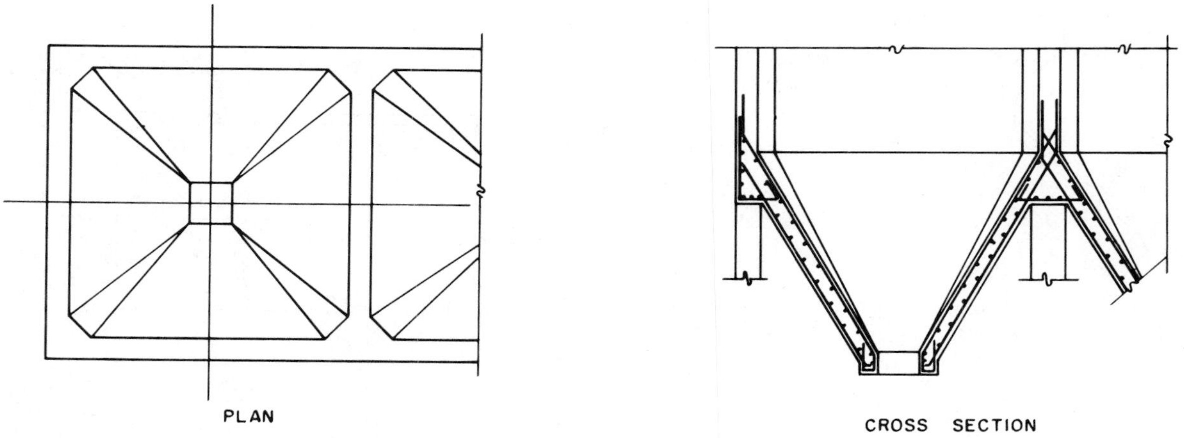

Fig. 16-23 Typical pyramidal hopper.

to 0.75 in. if a special inside lining is provided. Cover for other bars should conform to ACI 318.

The discharge end should be properly reinforced and detailed. Concrete flanges are usually provided around the opening with embedded anchoring devices to fasten discharge gates, chutes, etc.

16.13.12 Circular Concrete Ring-Beam-and-Column System Supporting a Steel Hopper

A steel hopper can do little to stiffen a concrete ring-beam; therefore, all of the loads it applies to the beam—torsion included—must be considered. The tables and equations that follow cover the design of such ring-beams and columns having an even number (4 to 12) of equally spaced columns, fixed at their bases, rigidly connected to the ring-beam and eccentric with the centroid of the ring-beam. (See Fig. 16-24.)

Equations (16-66) and (16-67) are for working design loads (WSD) rather than ultimate. In this type of structure live load is much larger than dead load. Consequently, it is simpler and sufficiently accurate to apply a common load factor, K_l, to the final answers in order to determine the USD values.

The external design loads acting on the ring-beam (Figs. 16-25 and 16-26) are approximately:

$$F_x = \frac{F_{mu} \cos \alpha}{1.7}, \text{ using } D = \text{silo inside diameter in eq. (16-55) to compute } F_{mu} \qquad (16\text{-}66)$$

$$F_y = g_r + \frac{F_{mu} \sin \alpha}{1.7} \qquad (16\text{-}67)$$

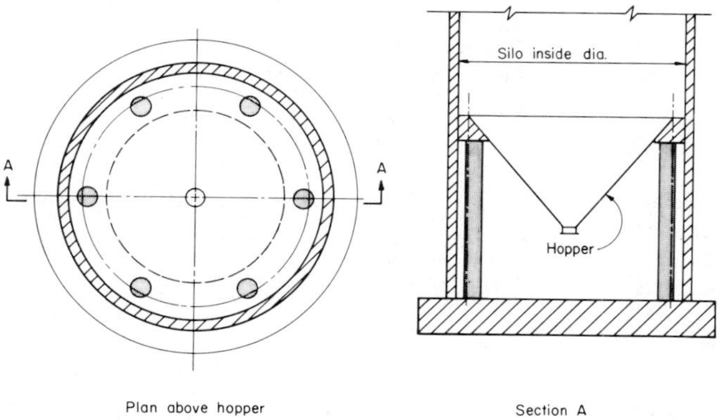

Fig. 16-24 Silo bottom—steel hopper supported on concrete ring-beam-and-column system.

where g_r is the weight of the ring-beam per unit length. If the vertical pressure (by stored material) acting on the face of the ring-beam is significant, it should be added to F_y (See Example 16-2.)

The WSD uniform torsional moment is:

$$M_t = F_m e \text{ (ft-lb/linear ft)} \qquad (16\text{-}68)$$

The cross-sectional area of the ring-beam (Fig. 16-26) is:

$$A_r = a_1 b_1 - \frac{b_2 a_2}{2} \qquad (16\text{-}69)$$

Coordinates of the centroid measured from origin O are:

$$\bar{x} = \frac{a_1 b_1^2/2 - (a_2 b_2/2)(b_1 - b_2/3)}{A_r} \qquad (16\text{-}70)$$

$$\bar{y} = \frac{a_1^2 b_1/2 - (a_2 b_2/2)(a_1 - a_2/3)}{A_r} \qquad (16\text{-}71)$$

An equivalent rectangle (dotted on Fig. 16-26) of height "a" and width "b" is substituted for the pentagon:

$$a = 2\bar{y} \qquad (16\text{-}72)$$

$$b = A_r/a \qquad (16\text{-}73)$$

The column shear, H_A, and upper end moment, M_A, are found by solving simultaneously eqs. (16-74) and (16-75):

$$\frac{F_x r^2}{A_r E_r} + \frac{H_A r^3 K_2}{2E_r I_{ry}} + \frac{H_A L^3}{3E_{col} I_c} + \frac{M_A L^2}{2E_{col} I_c} = 0 \qquad (16\text{-}74)$$

$$\frac{12 M_t r}{E_r a^3 \ln (r_2/r_1)} + \frac{C_2 r (M_A - R e_c)}{E_r I_{rx}} + \frac{C_3 F_y r^3}{E_r I_{rx}} + \frac{H_A L^2}{2E_{col} I_c} + \frac{M_A L}{E_{col} I_c} = 0 \qquad (16\text{-}75)$$

where K_2, C_2 and C_3 are numerical coefficients given by Tables 16-6, 16-8 and 16-9, respectively. Table 16-7 provides torsional properties of rectangular cross sections.

For design, values of thrust, shear, torque and bending moment in ring-beam are required at various locations. Tables 16-10 and 16-10A show equations for vertical bending moment and torque at various angular positions from the columns. Table 16-11 gives values at two locations for axial compression, vertical and horizontal shear and horizontal bending moment.

The WSD moment at the column base is then:

$$M_B = M_A + H_A L \qquad (16\text{-}76)$$

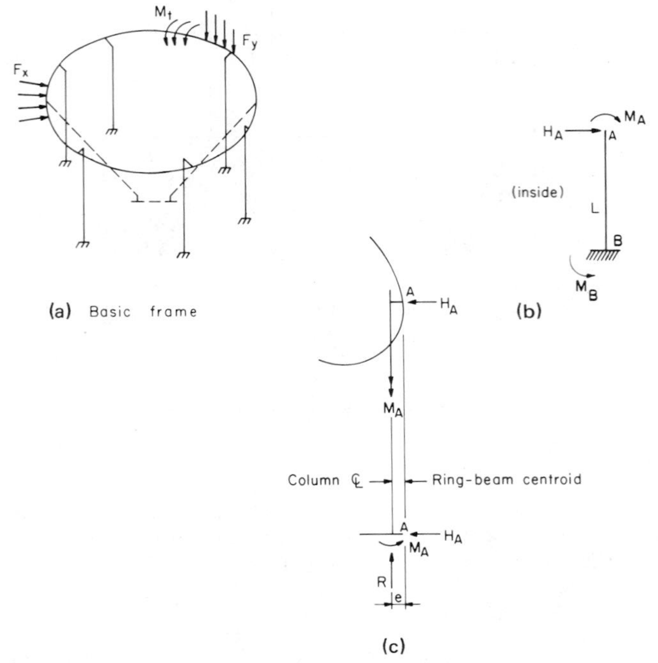

Fig. 16-25

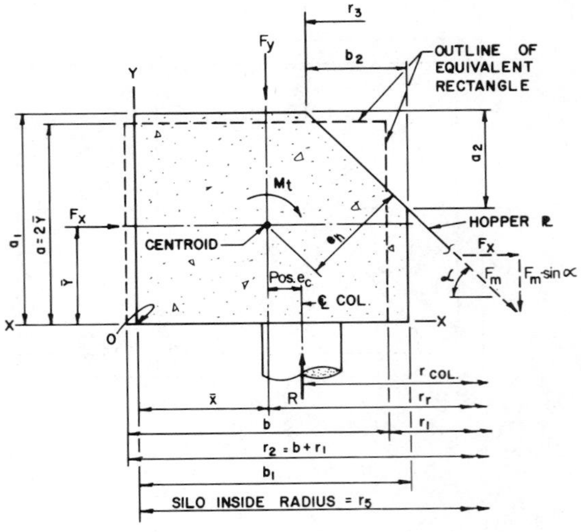

Fig. 16-26 Ring-beam cross section.

TABLE 16-6 Numerical Coefficients K_2

Number of Supports	K_2
4	.012159
6	.003364
8	.001387
10	.000701
12	.000404

For USD, the final forces and moments given by the preceding equations or Tables should be multiplied by the load factor, K_l. In all of the above equations, consistent units must be used. For example: F_x = lb/ft; M_t = ft-lb/ft; r, h_c, a, and b = ft; I = ft^4; E = lb/sq ft; A = sq ft; and M_T = ft-lb.

16.13.13 Column Design

Columns supporting silos or silo bottoms have large ratios of live to dead load. When unloaded after long periods of full loading, such columns may crack horizontally. To minimize such cracking, Lipnitski and Abramovitsch[16-22] suggest 1.5% for the upper limit of vertical reinforcing. The authors use a 2% limit unless a higher limit can be justified by analysis of the concrete behavior under long-term load.

TABLE 16-7 Torsional Properties of Rectangular Cross Sections

Section	*Torsional Constant in^4, (cm)^4*	*Torsional Section Modulus in^3, (cm)^3*	*Points of Max. Shear Stresses lb/in^2 (kg/cm^2)*	*Values of Coefficients σ, β, δ*			
				$m = \frac{h}{a}$	σ	β	δ
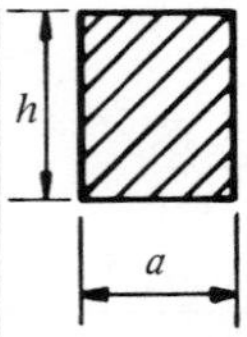	$K = \sigma a^4$	$Z_t = \beta a^3$	Middle of Long Sides $\tau_{max} = \frac{T}{Z_t}$ Middle of Short Sides $\tau = \delta \times \tau_{max}$. At Corner $\tau = 0$	1.0	0.140	0.208	1.00
				1.5	0.294	0.346	0.859
				2.0	0.457	0.493	0.795
				3.0	0.790	0.801	0.753
				4.0	1.123	1.150	0.745
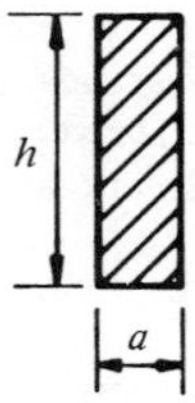	$K = \frac{(m - 0.63)a^4}{3}$	$Z_t = \frac{(m - 0.63)a^3}{3}$	All Points on Long Side except Corners $\tau_{max.} = \frac{T}{Z_t}$ Middle of Short Sides $\tau = 0.74 \times \tau_{max.}$	6.0	1.789	1.789	0.743
				8.0	2.456	2.456	0.742
				10.0	3.123	3.123	0.742

TABLE 16-8 Values of Factor C_2

λ	*Number of Equally-Spaced Columns* 4	6	8	10	12
1.0	0.785	1.047	1.341	1.645	1.954
1.1	0.800	1.056	1.347	1.650	1.959
1.2	0.814	1.065	1.354	1.656	1.963
1.3	0.828	1.074	1.361	1.661	1.967
1.4	0.842	1.091	1.367	1.666	1.972
1.5	0.857	1.092	1.374	1.671	1.976
1.6	0.871	1.102	1.381	1.677	1.981
1.7	0.885	1.111	1.388	1.628	1.985

$\lambda = E_r I_{rx}/GK$, *which* K = *torsion constant from Table 16-7*
$\theta_2 = C_2 rM'/E_r I_{rx}$

TABLE 16-9 Values of Factor C_3

λ	*Number of Equally-Spaced Columns* 4	6	8	10	12
1.0	.0191	.00353	.00117	.000496	.0002417
1.1	.0200	.00371	.00122	.000510	.0002509
1.2	.0210	.00389	.00128	.000524	.0002601
1.3	.0219	.00407	.00134	.000538	.0002693
1.4	.0229	.00425	.00139	.000553	.0002785
1.5	.0238	.00442	.00145	.000567	.0002877
1.6	.0248	.00460	.00151	.000581	.0002969
1.7	.0257	.00478	.00156	.000595	.0003061

$\lambda = E_r I_{rx}/G K$, $\theta_3 = C_3 F_y r^3/E_r I_{rx}$

16-14 DESIGN AND DETAILING OF REINFORCED CONCRETE BUNKERS

Bunkers may be square, rectangular, circular, or polygonal —single or in groups (Fig. 16-2). Bunkers may be symmetrical or nonsymmetrical about either or both of the principal axes, according to the location of the discharge opening. Circular bunkers usually have conical bottoms and single round discharge openings. Rectangular ones usually have pyramidal bottoms with one or more square, circular, or elongated discharge openings. Flat bottoms are seldom used, since the undesirable dead storage reduces the efficiency of material flow. This can be overcome, however, by hopper-forming concrete fill.

A bunker may consist of the pyramidal or conical hopper only, without vertical walls. In this case a continuous horizontal supporting beam is located along the upper edge and resists both the vertical and horizontal forces and torsion applied by the hopper shell.

Bunkers generally require a protective surface to guard against abrasion by falling sliding material. Usually, lining is necessary only on the sloping bottoms, whose surfaces receive the greatest abrasion and impact.

16.14.1 Loads and Forces

Bunkers are designed for loads imposed by stored material, their own weight, and the weight of equipment and platforms carried by the structure. Impact, roof loads, wind, and earthquake may also need to be considered.

Material pressures on bunker walls and bottoms may be determined by the ACI 313-77 method assuming bunkers as silos (neglecting height-to-width ratio limitation), pre-

TABLE 16-10 Equations for Vertical Bending Moment at Angle θ from Support. (+ = Compression Top)

No. of Supports	*Equation*
4	$M = M'(\frac{1}{2}\sin\theta + \frac{1}{2}\cos\theta) - F_y r^2(1\text{-}0.7854\sin\theta - 0.7854\cos\theta) + M_t r$
6	$M = M'(\frac{1}{2}\sin\theta + 0.866\cos\theta) - F_y r^2(1\text{-}0.5236\sin\theta - 0.9069\cos\theta) + M_t r$
8	$M = M'(\frac{1}{2}\sin\theta + 1.207\cos\theta) - F_y r^2(1\text{-}0.3927\sin\theta - 0.9481\cos\theta) + M_t r$
10	$M = M'(\frac{1}{2}\sin\theta + 1.539\cos\theta) - F_y r^2(1\text{-}0.3142\sin\theta - 0.9669\cos\theta) + M_t r$
12	$M = M'(\frac{1}{2}\sin\theta + 1.866\cos\theta) - F_y r^2(1\text{-}0.2618\sin\theta - 0.9770\cos\theta) + M_t r$

$M' = M_A - Re_c$

TABLE 16-10A Equations for Torque at Angle θ from Support

No. of Supports	*Equation*
4	$T = M'(\frac{1}{2}\cos\theta - \frac{1}{2}\sin\theta) + F_y r^2(\theta - 0.7854\sin\theta + 0.7854\cos\theta - 0.7854)$
6	$T = M'(\frac{1}{2}\cos\theta - 0.866\sin\theta) + F_y r^2(\theta - 0.9069\sin\theta + 0.5236\cos\theta - 0.5236)$
8	$T = M'(\frac{1}{2}\cos\theta - 1.207\sin\theta) + F_y r^2(\theta - 0.9481\sin\theta + 0.3927\cos\theta - 0.3927)$
10	$T = M'(\frac{1}{2}\cos\theta - 1.539\sin\theta) + F_y r^2(\theta - 0.9669\sin\theta + 0.3142\cos\theta - 0.3142)$
12	$T = M'(\frac{1}{2}\cos\theta - 1.866\sin\theta) + F_y r^2(\theta - 0.9770\sin\theta + 0.2618\cos\theta - 0.2618)$

$M' = M_A - Re_c$

TABLE 16-11 Summary of Axial Force (Thrust), Shear and Horizontal Bending Moment in Ring-Beam

Number of Supports	*Location*	*Compressive Thrust*	*Shear*		*Horizontal Bending M.*
			Vertical	*Horizontal*	
4	Support	$F_x r$	$0.7854\ F_y r$	$H_A/2$	$-0.1366\ H_A r$
	Midspan	$F_x r$	0	0	$0.0705\ H_A r$
6	Support	$F_x r$	$0.5236\ F_y r$	$H_A/2$	$-0.455\ H_A r$
	Midspan	$F_x r$	0	0	$0.0451\ H_A r$
8	Support	$F_x r$	$0.3927\ F_y r$	$H_A/2$	$-0.773\ H_A r$
	Midspan	$F_x r$	0	0	$0.0333\ H_A r$
10	Support	$F_x r$	$0.3142\ F_y r$	$H_A/2$	$-1.092\ H_A r$
	Midspan	$F_x r$	0	0	$0.0265\ H_A r$
12	Support	$F_x r$	$0.2618\ F_y r$	$H_A/2$	$-1.410\ H_A r$
	Midspan	$F_x r$	0	0	$0.0220\ H_A r$

sented earlier in this chapter, or by the Rankine Method, which follows. This theory is not entirely accurate, since it is based on assumptions not fully met and ignores boundary conditions. Friction forces on the walls are neglected, and pressures are assumed to be normal to the surface against which they act.

If the top surface of the stored material is horizontal:

Vertical static unit pressure at depth Y below the surface is:

$$q = \gamma Y \tag{16-77}$$

Horizontal static unit pressure at depth Y is:

$$p = k\gamma Y \tag{16-78}$$

where $k = (1 - \sin \rho)/(1 + \sin \rho)$.

Unit static pressure normal to an inclined surface at depth Y is:

$$q_\alpha = \gamma Y (\cos^2 \alpha + k \sin^2 \alpha) \tag{16-79}$$

If the surface of the stored material slopes at the angle of repose, approximately equal to ρ (see Fig. 16-27):

Vertical static unit pressure at depth Y is:

$$q = \gamma (Y + a_0 \tan \rho) \tag{16-80}$$

Horizontal static unit pressure at depth Y is:

$$p = \gamma Y \cos^2 \rho \tag{16-81}$$

Unit static pressure normal to the inclined surface of the hopper wall at depth Y is:

$$q_\alpha = \gamma \cos^2 \alpha (Y + a_0 \tan \rho) + \gamma Y \cos^2 \rho \sin^2 \alpha \tag{16-82}$$

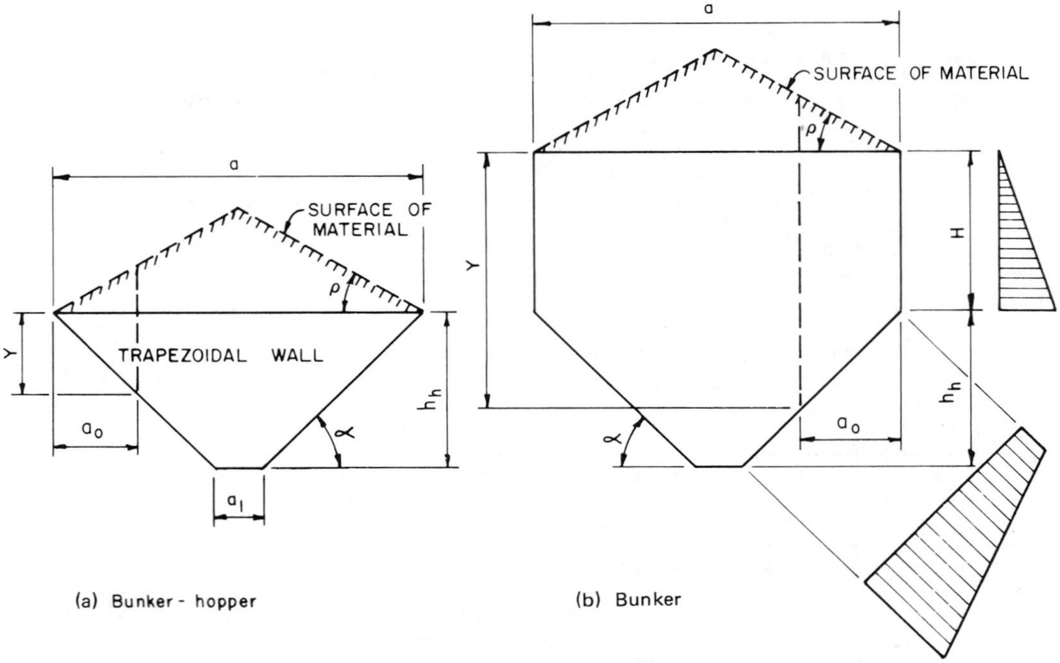

Fig. 16-27 Bunker and hopper dimensions.

(Note: Many engineers prefer to compute p, q, and q_α by the methods given for design values for silos.)

For small trapezoidal walls, it is sometimes sufficiently accurate to use average pressures. The average pressure on an inclined trapezoid of upper edge a and lower edge a_1 is:

$$q_{\alpha,\,\text{avg}} = \frac{a(5q_{\alpha,\,\text{top}} + 4q_{\alpha,\,\text{bot}}) + a_1(q_{\alpha,\,\text{top}} + 2q_{\alpha,\,\text{bot}})}{6(a + a_1)} \quad (16\text{-}83)$$

Pressure p_{avg} for a vertical trapezoid is computed by the same equation, but with $q_{\alpha,\,top}$ and $q_{\alpha,\,bot}$ substituted by p-values computed from eq. (16-78) or (16-81).

16.14.2 Design Pressures

Design pressures are:

$$p_{\text{des}} = C_d p \quad (16\text{-}84)$$

$$q_{\text{des}} = C_d q \quad (16\text{-}85)$$

$$q_{\alpha,\,\text{des}} = C_d q_\alpha \quad (16\text{-}86)$$

Sudden filling (as in a dump bunker) may cause considerable impact on the bunker bottom or sloping walls. Withdrawal, on the other hand, causes overpressure, above and beyond the computed static pressure. For all bunkers, overpressure should be considered. But for those in which impact due to fast filling is significant, the designer should use whichever is more severe: static pressure times C_d factor, or static pressure times an impact factor, C_i (see Table 16-12).

16.14.3 Wall Forces—Circular Bunkers with Symmetrical Conical Hoppers

Vertical walls of circular bunkers are subject to horizontal (hoop) tensile forces due to lateral pressure from the stored material, and frequently also to in-plane bending, vertical shear, and vertical membrane forces, depending upon the manner of support for the bunker.

The required ultimate hoop tensile force per unit width of vertical wall is:

$$F_u = K_l\, p_{\text{des}}(D/2) \quad (16\text{-}87)$$

in which p_{des} is computed from eq. (16-84) and either eq. (16-78) or (16-81).

For the conical hopper, the required ultimate tensile forces per unit width are:

Meridional:

$$F_{mu} = K_l \left[\frac{q_{\text{des}} D}{4 \sin \alpha} + \frac{W_l}{\pi D \sin \alpha}\right] + K_g \left[\frac{W_g}{\pi D \sin \alpha}\right] \quad (16\text{-}88)$$

Horizontal:

$$F_{tu} = K_l \left[\frac{q_{\alpha,\,\text{des}} D}{2 \sin \alpha}\right] \quad (16\text{-}89)$$

In the above, q_{des} and $q_{\alpha,\,des}$ are computed using eqs. (16-77) to (16-86) of this section, as applicable.

TABLE 16-12 Values of Impact Coefficient C_i for Bunkers (by S.P. Abramovitsch, Ref. 16-22).

Ratio of Volume Dumped in One Load to Total Silo Capacity		*1:2*	*1:3*	*1:4*	*1:5*	*1:6 and Less*
Impact factor C_i	Concrete bottom	1.4	1.3	1.2	1.1	1.0
	Steel bottom	1.75	1.60	1.5	1.3	1.25

16.14.4 Wall Forces—Rectangular Bunkers and Symmetrical Pyramidal Hoppers

Vertical walls of rectangular bunkers are subject to horizontal membrane tensile forces and two-way bending, and frequently also to in-plane bending, vertical shear, and vertical membrane forces, depending on the manner of support used.

The ultimate horizontal tensile forces per unit width of vertical wall are:

$$F_{au} = K_l\, p_{\text{des}}(b/2) \quad (16\text{-}90)$$

$$F_{bu} = K_l\, p_{\text{des}}(a/2) \quad (16\text{-}91)$$

where p_{des} is from eq. (16-84) and either eq. (16-78) or (16-81).

WSD bending moments for rectangular panels may be computed using tables published by the Portland Cement Association,[16-21] Fischer,[16-31] or Lipnitski and Abramovitsch,[16-22] and correcting to USD by multiplying by the appropriate load factors, K.

When bunkers are not supported continuously, the vertical walls will have vertical tensile forces and also in-plane bending due to their action as horizontal beams in transferring vertical loads to the supports. Where the sloping walls are monolithic with the vertical walls, both work together in providing this beam action. In-plane bending should also be considered in the trapezoidal walls of bunkers having no vertical walls.

In addition to in-plane bending or the participation with the vertical walls in resisting vertical bending, inclined walls of pyramidal hoppers are subject to horizontal and inclined membrane tensile forces, F_t and F_m, and to two-way bending moment. For symmetrical hoppers, the ultimate forces are:

Horizontal:

$$F_{tau} = K_l \left(\frac{b}{2}\right) [q_{\alpha,\,b}]_{\text{des}} \sin \alpha_a \quad (16\text{-}92)$$

$$F_{tbu} = K_l \left(\frac{a}{2}\right) [q_{\alpha,\,a}]_{\text{des}} \sin \alpha_b \quad (16\text{-}93)$$

Meridional:

$$F_{mau} = \frac{K_l(c_a W_l + A_a q_{\text{des}}) + K_g c_a W_g}{a \sin \alpha_a} \quad (16\text{-}94)$$

$$F_{mbu} = \frac{K_l(c_b W_l + A_b q_{\text{des}}) + K_g c_b W_g}{b \sin \alpha_b} \quad (16\text{-}95)$$

WSD values of bending moment may be computed using

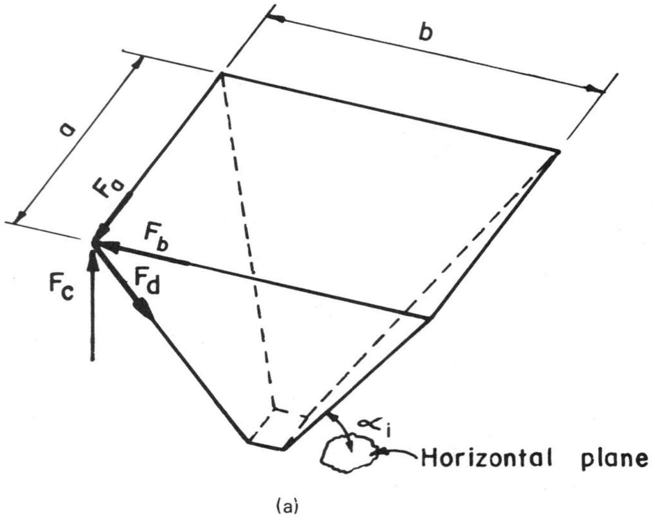

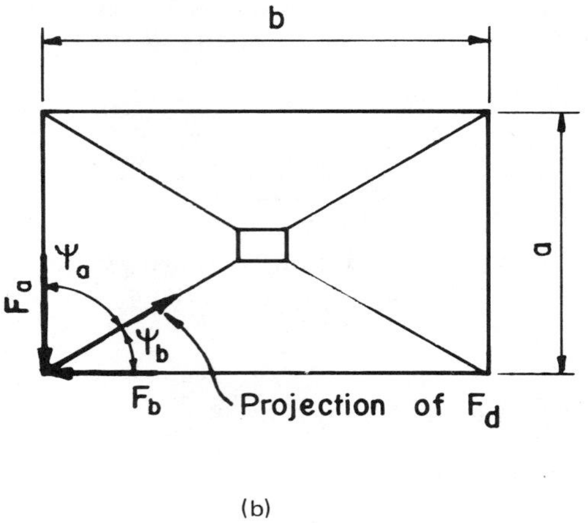

Fig. 16-28 Corner forces in pyramidal bunker-hopper.

Tables 16-4 and 16-5, and eqs. (16-63) and (16-83). These moments may be converted to ultimate moments by multiplying by K_l. For unsymmetrical hoppers, see Refs. 16-12, 16-31, and 16-32.

16.14.5 In-Plane Bending and Combined Action of Vertical and Inclined Walls

Engineers often disregard in-plane bending and combined action of vertical and inclined walls. The authors feel, however, that they should be disregarded *only* when their effects are shown to be negligible. Unfortunately, their analysis is complex, and many simple approximations are substituted for rational methods.

In the case of a simple bunker-hopper, having no vertical walls, the trapezoidal wall may be assumed (conservatively) to act as a simple beam, spanning from support to support, and carrying the in-plane forces, F_{mu}, computed from eq. (16-94) or (16-95). The USD bending moment at the centerline, for wall "*a*" would then be:

$$M_u = F_{mau}\, a^2/8 \qquad (16\text{-}96)$$

The tensile force (per unit width of wall) computed for the in-plane bending must be added to the membrane force F_{tau}, and the reinforcement must be chosen for the combined force acting together with bending moment M_{xu}.

The upper edge of the trapezoidal wall is subject to compression due to bending moment M, but this compressive force is probably more accurately estimated by considering the equilibrium of forces at the corner. Referring to Fig. 16-28, column force F_c is balanced by the resultant of a tensile force F_d along the intersection of walls "*a*" and "*b*" and compressive forces F_a and F_b at the upper edges of the walls:

$$F_d = F_c/\sin \alpha_i \qquad (16\text{-}97)$$

where α_i is the angle between the horizontal plane and the wall intersection line.

$$F_a = F_c \cot \alpha_i \cos \Psi_a \qquad (16\text{-}98)$$

$$F_b = F_c \cot \alpha_i \cos \Psi_b \qquad (16\text{-}99)$$

where Ψ_a and Ψ_b are the angles between the horizontal projection of F_d and edges "*a*" and "*b*" respectively.

Multiplying by appropriate load factors, K, these forces may be converted to ultimate values, F_{au} and F_{bu}. These ultimate compressive forces may be assumed applied to a narrow strip (of width not exceeding one-tenth of the edge length) near the upper edge, or to an edge beam if one is provided (preferable). If applied to a narrow strip, the compressive force is combined with the tensile force F_{tu}. That upper edge strip must also carry the bending moments M_{xu} due to normal pressure loading. If a top edge beam is used, it will carry not only a compressive force (F_{au} or F_{bu} as shown by Fig. 16-28) but also a share of the tensile forces F_{tau} or F_{tbu} determined for the hopper shell. In addition, the edge beam will carry vertical shear and bending, horizontal shear and bending, and (if the inclined wall is eccentric to the beam) torsion.

Extra steel to carry force F_{du} should be located parallel to the intersection of the inclined walls.

When the depth of the vertical wall exceeds one-half its span (from column to column), participation of the inclined wall in resisting vertical-plane bending is usually neglected. In this case, the vertical walls are assumed to transfer all vertical loads (dead, live, and impact) from the bunker and hopper to the columns.

16.14.6 Suggested Procedure for Design of Cast-in-Place Bunkers

The following procedure is recommended for designing cast-in-place bunkers:

1. Determine the properties of the material to be stored.
2. Establish basic bunker size and supporting system. (For preliminary purposes the wall thickness may be assumed equal to 1/25 of the span but not less than 6 in. for hoppers for vertical walls.)
3. Determine necessary geometric data—angle of inclined walls, wall sizes, center of gravity of material, etc.
4. Determine design pressures of granular material on walls and bottom, considering dynamic effect if any. Sketch pressure diagrams.
5. Determine tensile forces in all vertical and inclined walls.
6. Determine bending moments in the walls due to local bending.
7. Determine moments and shears due to in-plane bending.
8. Design bunker components:
   (a) Determine required thickness and reinforcement.
   (b) Check crack width and modify thickness, bar size, or bar spacing as required to ensure that the crack width not exceed a tolerable limit.

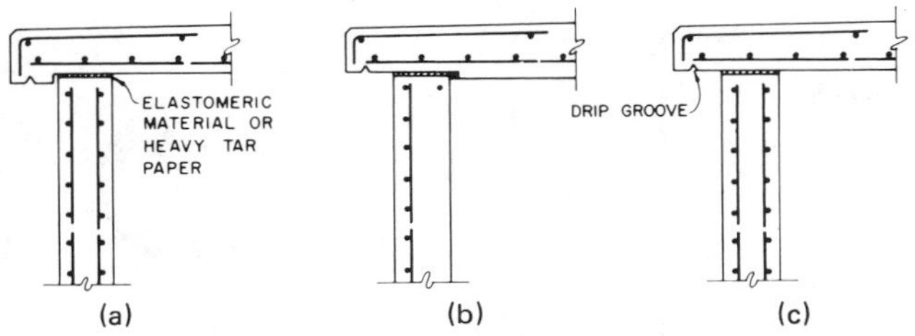

Fig. 16-29 Typical silo roof-to-wall details.

16.15 ROOFS FOR SILOS AND BUNKERS

The silo concrete roof slab is either doweled to the walls (providing full or partial continuity of walls and roof) or is supported in a manner permitting free expansion and contraction and slight movement during earthquake or temperature change or due to other lateral forces. For a free-floating roof, means must be provided to limit the movement to a tolerable amount. Possible means include fixing the roof at one interior location (such as at an elevator tower), using shear pins between roof and walls, an inverted curb with clearance to the outer wall surface, or a thickened portion of slab with clearance to selected inside surfaces. The latter two methods are illustrated by Fig. 16-29(a) and (b).

Single silo roofs more often are fixed to walls. Fixing the entire roof of monolithically cast large silo groups, however, may cause serious expansion and contraction problems. Therefore, in such cases, the roof is usually divided into independent sections, providing expansion joints between sections or simply allowing controlled expansion of the roof as described above.

Steel beams, if used, are usually incorporated in the slipform platform serving also as the bottom form for for the roof.

When a bunker is cast with stationary forms and has a concrete slab roof, the slab is usually keyed or doweled to the wall, so that is provides lateral support to the top of the wall. In this case, the slab is subjected to combined bending and tension. If the tension is significant, it must be considered in design of the slab.

16.16 FOUNDATIONS FOR BUNKERS AND SILOS

Chapter 5 of this *Handbook* covers the selection of foundation type and design methods. This section gives only those requirements peculiar to bunker and silo foundations.

Loads to be considered are:

1. Dead load—weight of structure, equipment, and supported backfill (if any).
2. Live load—full weight of stored material and snow. Overpressure does not effect silo foundations.
3. Wind.
4. Earthquake.

Fig. 16-30 Around-the-clock slipforming of cement storage silos, Portland, Colorado. (Courtesy of Ideal Cement Co.)

Load combinations to be considered in design of the foundation for strength, stability, and reactive pressure include:

1. Dead + live (all cells full).
2. Dead + live (all cells full) + earthquake.
3. Dead + wind, all cells empty.
4. Unsymmetrical loading of silo or bunker groups. Assume half of the cells full and the other half empty.
   (a) Dead + live.
   (b) Dead + live + wind or earthquake (whichever controls).
5. Dead + live in all cells but one; one cell empty. (May control for bending for the foundation slab.)

16.17 SLIPFORM CONSTRUCTION OF SILOS

Slipforming is well adapted to construction of silos. (See Fig. 16-30.) However, it is essential that specifications for slipform construction recognize the special problems of silo slipforming. Specifications for conventional reinforced concrete construction are usually inadequate for slipform work.

Within the pressure zone and for approximately 3 ft below, the horizontal bars should be secured to the vertical reinforcing during construction. Laying hoop bars directly on freshly poured concrete results in inaccuracy of bar spacing and bar lapping, especially when vibrators are used. It also makes bar placing completely dependent on the crew depositing concrete into forms.

Continuous, around-the-clock inspection is recommended during slipforming. The inspection should be by the engineer or a competent person appointed by the engineer. Continuous inspection is desirable, since the work to be inspected will be covered very quickly, making errors hard to correct. Inspection to assure proper lap length and vertical location of horizontal reinforcing is particularly important.

16.18 EXAMPLES

(Equation and table numbers in parentheses at right refer to equations and tables in text.)

EXAMPLE 16-1: Figure 16-31 shows the basic dimensions of a circular silo for clean coal storage. Its conical hopper is supported be eight pilasters, monolithic with the lower walls. The total load from the roof (dead, live and equipment) is 2000 lb/ft. Design the walls and hopper, using $f_c' = 4{,}000$ psi and $f_y = 60{,}000$ psi (40,000 psi for ties and stirrups) and assuming that the silo walls will be slipformed. The maximum tolerable crack width is 0.010 in.

SOLUTION:
For clean coal:

$$\gamma = 50 \text{ pcf};\ \rho = 35°;\ \mu' = 0.50 \qquad \text{(Table 16-1)}$$

For use in Janssen's equation:

$$k = (1 - \sin 35°)/(1 + \sin 35°) = 0.271 \qquad (16\text{-}3)$$

$$R = D/4 = 40/4 = 10.0 \text{ ft}$$

For lateral design pressure curve:

$$H_1 = D \tan \rho = 28 \text{ ft}$$

$$(H - H_1)/4 = (120 - 28)/4 = 23.0 \text{ ft}$$

At $Y = 120$ ft, by Janssen's equation (16-2):

$$p = \frac{\gamma R}{\mu'}\,[1 - e^{-\mu' kY/R}]$$

$$= \frac{50 \times 10}{0.50}\,[1 - e^{-(0.5 \times 0.271 \times 120/10)}]$$

$$= 803 \text{ lb/sq ft (psf)}$$

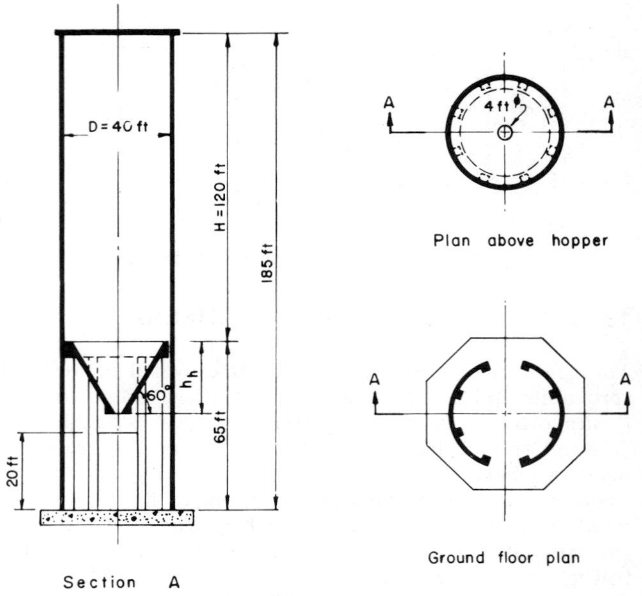

Fig. 16-31 Silo for Example 16-1.

From Table 16-2, C_d at the bottom = 1.75.

$$p_{des} = C_d p = 1.75 \times 803 = 1405 \text{ psf} \qquad (16\text{-}18)$$

Vertical pressure at $Y = 120$ ft is:

$$q = p/k = 803/0.271 = 2960 \text{ psf}$$

In the conical hopper at depth $Y = 130$ ft, $D = 27.5$ ft, $p = 828$ psf, $q = 3050$ psf, and (Table 16-2) $C_d = 1.35$.

$$p_{des} = 1.35 \times 828 = 1120 \text{ psf}$$

$$q_{des} = 1.35 \times 3050 = 4120 \text{ psf}$$

$$q_{\alpha,des} = 1.120 \sin^2 60° + 4120 \cos^2 60° = 1870 \text{ psf} \qquad (16\text{-}17)$$

Similarly, design pressures are calculated at various depths and are tabulated in Table 16-13.

Silo Wall Design:
Wall forces and hoop steel areas at $Y = 120$ ft:

$$F_u = 1.7 \times 1406 \times 20 = 47{,}800 \text{ lb/ft} \qquad (16\text{-}26)$$

$$\text{Reqd } A_s = 47{,}800/(0.90 \times 60{,}000) = 0.89 \text{ sq in./ft} \qquad (16\text{-}27)$$

Try #6 bars @ $5\frac{1}{2}$ in. c/c.

Wall thickness:

$$h_{min} = \left[\frac{0.0003 \times 29 \times 10^6 + 24{,}000 - 8 \times 400}{12 \times 24{,}000 \times 400}\right] \frac{803 \times 40}{2}$$

$$= 4.12 \text{ in.} \qquad (16\text{-}28)$$

Assume $h = 8$ in. Check vertical compressive stress at foundation.

Vertical loads:

wt of wall = 185 × 150 × 8/12 = 18,500 lb/ft

friction, V_{des} (at $Y = 120$) = (50 × 120 − 0.8 × 2,960) (10.0)

= 36,300 lb/ft (16-16, 16-22)

Roof (given) = 2000 lb/ft (treat as live load)

$$F_{u,vert} = 1.4(18{,}500 + 1.7(38{,}300) = 90{,}900 \text{ lb/ft}$$

$$f_{c,vert} = 90{,}900/(12 \times 8) = 947 \text{ psi}$$

permissible $F_{u,vert} = 0.385 \times 4000 \times 8 \times 12$

= 147,800 lb/ft O.K. (16-30)

(The above assumes that all hopper vertical load is carred by the pilasters.)

TABLE 16-13 Silo Wall Lateral Pressures and Hoop Reinforcing (Example 16-1)

Depth Y (ft)	*Static lateral press. p (lb/sq ft)*	*Overpress. factor C_d (FROM TABLE) 16-2*	*Design lateral press. $p_{des.}$ (lb/sq ft)*	$F_u = 1.7 r p_{des}$ (k/ft)	*Required hoop Steel A_s (in. sq/ft)*
$H_1 = 28$	316	1.45	458	15.6	0.29
51	499	1.55	773	26.3	0.49
74	633	1.65	1,044	35.5	0.66
97	731	1.75	1,280	43.5	0.81
120	803	1.75	1,406	47.8	0.89

Check crack width at $Y = 120$ ft:

$$F_{tot} = p_{des}D/2 = 28{,}120 \text{ lb/ft}; \; F_{st} = pD/2 = 16{,}060 \text{ lb/ft}$$

$$s_{cr} = (8 \times 12)(0.7)/5.14 = 13.1 \text{ in.} \qquad (16\text{-}33)$$

For short-term total load:

$$\psi_1 = 1 - 0.7 \left[\frac{0.8(8 \times 12)(285)}{28{,}120}\right] = 0.455 \qquad (16\text{-}34)$$

$$f_s = F_{tot}/A_s = 28{,}120/0.96 = 29{,}290 \text{ psi}$$

$$w_1 = (0.455 \times 13.1 \times 29{,}290)/(29 \times 10^6) = 0.00602 \qquad (16\text{-}22)$$

Similarly, for short-term static load:

$$\psi_1 = 0.046; \; f_s = 16{,}730 \text{ psi}; \; w_2 = 0.0035 \text{ in.} \qquad (16\text{-}34, 16\text{-}32)$$

and for long-term static load:

$$\psi_1 = 0.523; \; w_3 = 0.00395 \text{ in.} \qquad (16\text{-}35, 16\text{-}32)$$

Total crack width:

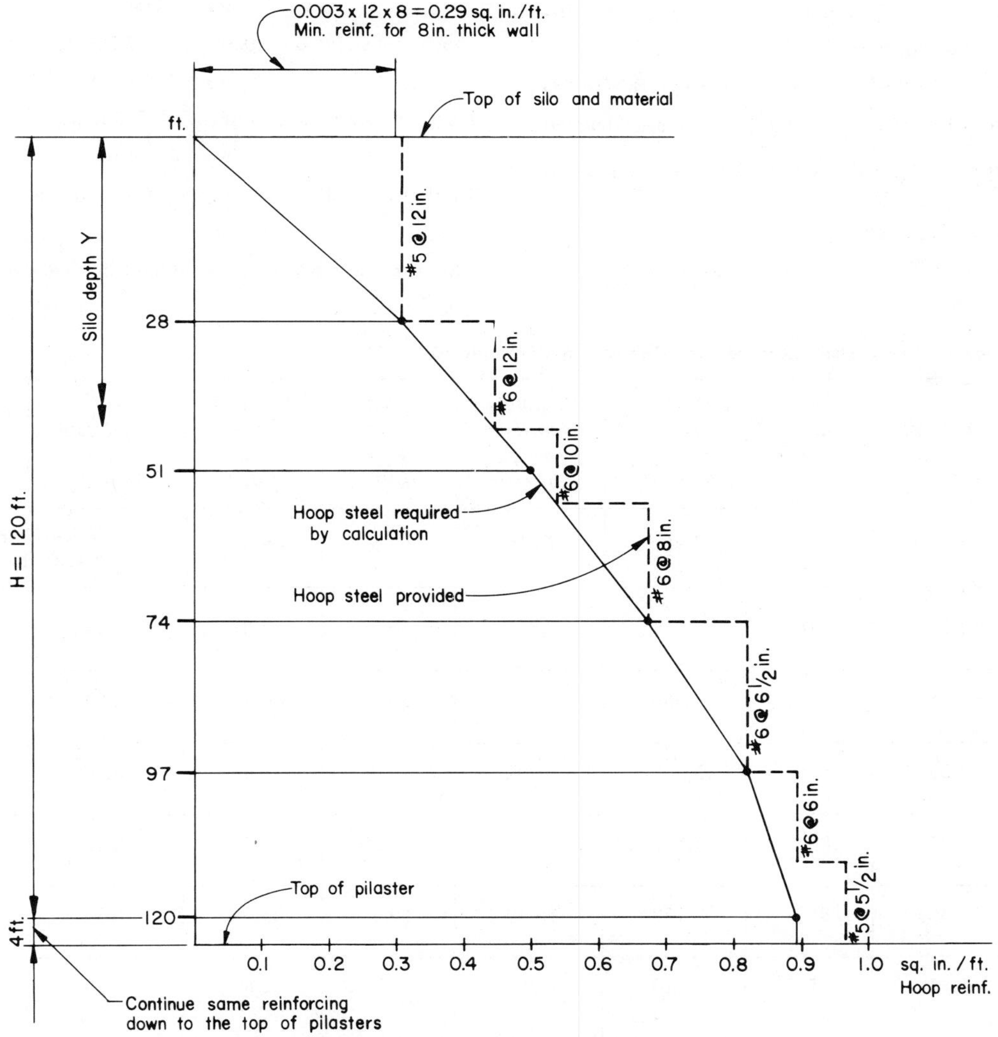

Fig. 16-32 Required and furnished hoop steel for silo walls (Example 16-1).

$w_{cr} = w_1 - w_2 + w_3 = 0.0096 < 0.010$ in. O.K. (16-31)

Crack width was checked at several levels and found OK. Use 8-in. wall. (See Table 16-13 and Fig. 16-32 for hoop steel.) Vertical steel area required in the pressure zone is

$$A_s = 0.002 \times 8 \times 12 = 0.192 \text{ sq in./ft.}$$

Use #4 at 12 in. centers. Place vertical bars on wall centerline. Place hoop bars alternately inside and outside of the vertical, tying them to the verticals.

In this example, uniform lateral pressure around the circumference was assumed. If non-uniform pressure can occur, use a thicker wall and two-layer reinforcement.

Conical Hopper Design:

Wall forces at depth $Y = 130$ ft (20 ft above outlet, 24 ft above inside vertex of cone). Assume an 8-in. average wall thickness.

$$\text{Inside } D = 27.5 \text{ ft}; \quad D_{avg} = 28.2$$

$$W_l = (\pi/3)13.75^2 \times 24 \times 0.05 = 237.6 \text{ k}$$

$$W_g = \pi \times 14.1 \times 28.2 \times 0.1 = 127.6 \text{ k}$$

Meridional force and required reinforcing:

$$F_{mu} = 1.7\left[\frac{4.12 \times 28.2}{4 \sin 60^\circ} + \frac{237.6}{28.2\pi \sin 60^\circ}\right] + 1.4\left[\frac{127.6}{28.2\pi \sin 60^\circ}\right] = 64.6 \text{ k/ft} \quad (16\text{-}55)$$

$$\text{Reqd } A_s = 64.6/(0.9 \times 60) = 1.20 \text{ sq in./ft}$$

Tangential force and required tangential reinforcing:

$$F_{tu} = 1.7(1870)(27.5)/2 \sin 60 = 50{,}500 \text{ lb/ft} \quad (16\text{-}56)$$

NOTE: Above uses inside diameter.

$$\text{Reqd } A_s = 50{,}500/(0.9 \times 60{,}000) = 0.94 \text{ sq in./ft} \quad (16\text{-}58)$$

The results of similar calculations at several elevations are tabulated in Table 16-14.

Check hopper wall for crack width at $Y = 130$ ft. Both directions must be considered.

Meridional force (horizontal cracks):

$$A_s(64 \text{ \#8 and } 64 \text{ \#9}) = 114.6/27.8\pi = 1.31 \text{ sq in./ft}$$

$$\Sigma o = 4.84 \text{ in./ft}$$

$$F_{m,tot} = 38{,}200 \text{ lb/ft}$$

$$F_{m,st} = 33{,}540/C_d + 4760 = 29{,}600 \text{ lb/ft}$$

$$s_{cr} = (8 \times 12)(0.7)/4.84 = 13.9 \text{ in.} \quad (16\text{-}33)$$

For short-term total load, $\psi_1 = 0.60$, $f_s = 29{,}200$ psi:

$$w_1 = 0.0084 \text{ in.} \quad (16\text{-}34, 16\text{-}32)$$

For short-term static, $\psi_1 = 0.34$, $f_s = 23{,}280$ psi:

$$w_2 = 0.0038 \text{ in.} \quad (16\text{-}34, 16\text{-}32)$$

For long-term static, $\psi_1 = 0.67$, $f_s = 23{,}280$ psi:

$$w_3 = 0.0075 \text{ in.} \quad (16\text{-}35, 16\text{-}32)$$

Total crack width, $w_{cr} = w_1 - w_2 + w_3 = 0.0121$ in. (16-31)

The width exceeds 0.010 in. and seems excessive. If the remaining outside bars are extended by about another 2 ft, the steel at this height will be 128 #8 and 64 #9. Recomputation of crack width then gives w_{cr} much less than 0.010 in. OK

Tangential force (vertical cracks):

$$F_{t,tot} = 50{,}500/1.7 = 29{,}700; \quad F_{t,st} = 29{,}700/C_d = 22{,}000 \text{ lb/ft}$$

$$A_s = 1.23 \text{ sq in./ft}; \quad \Sigma_o = 6.6 \text{ in./ft}$$

$$s_{cr} = 96(0.7)/6.6 = 10.2 \text{ in.} \quad (16\text{-}33)$$

Short-term total, $\psi_1 = 0.48$, $f_s = 24{,}150$ psi:

$$w_1 = 0.0041 \text{ in.} \quad (16\text{-}34, 16\text{-}32)$$

Short-term static, $\psi_1 = 0.30$, $f_s = 17{,}200$ psi:

$$w_2 = 0.0019 \text{ in.} \quad (16\text{-}34, 16\text{-}32)$$

Long-term static, $\psi_1 = 0.65$, $f_s = 17{,}900$ psi:

$$w_3 = 0.0040 \text{ in.} \quad (16\text{-}35, 16\text{-}32)$$

Total crack width, $w_{cr} = w_1 - w_2 + w_3 = 0.0062$ in. < 0.010 in.

OK (16-31)

Similar crack width checks should be made at several locations

TABLE 16-14 Conical Hopper Pressures, Forces, and Reinforcing (Example 16-1)

Depth Y (ft) / *Ins. dia D (ft)*	*Overpress. factor* C_d	*Static lateral press. p (lb/sq ft)*	*Static vertical press.* q_{des} *(lb/sq ft)*	*Des. press. on inclined face* q_α *(lb/sq ft)*	*Ult. tang. force* F_{tu} *(k/ft)*	*Ult. merid. force* F_{mu} *(k/ft)*	*Req'd. tang. steel (sq in./ft)*	*Req'd. merid. steel (sq in./ft)*
120 / 38.0	1.35	803	2,960	1,810	67.5	89.3	SEE RING BEAM CALCULATIONS	
125 / 32.5	1.35	816	3,010	1,840	60.5	77.8	1.12	1.44
130 / 26.6	1.35	828	3,060	1,870	50.5	64.6	0.94	1.20
135 / 21.0	1.35	839	3,100	1,900	41.0	51.6	0.76	0.96
140 / 15.5	1.35	850	3,140	1,920	30.2	37.6	0.56	0.70
145 / 10.0	1.35	860	3,170	1,940	19.0	23.0	0.35	0.43
150 / 4.0	1.35	870	3,210	1,960	9.6	12.6	0.18	0.23

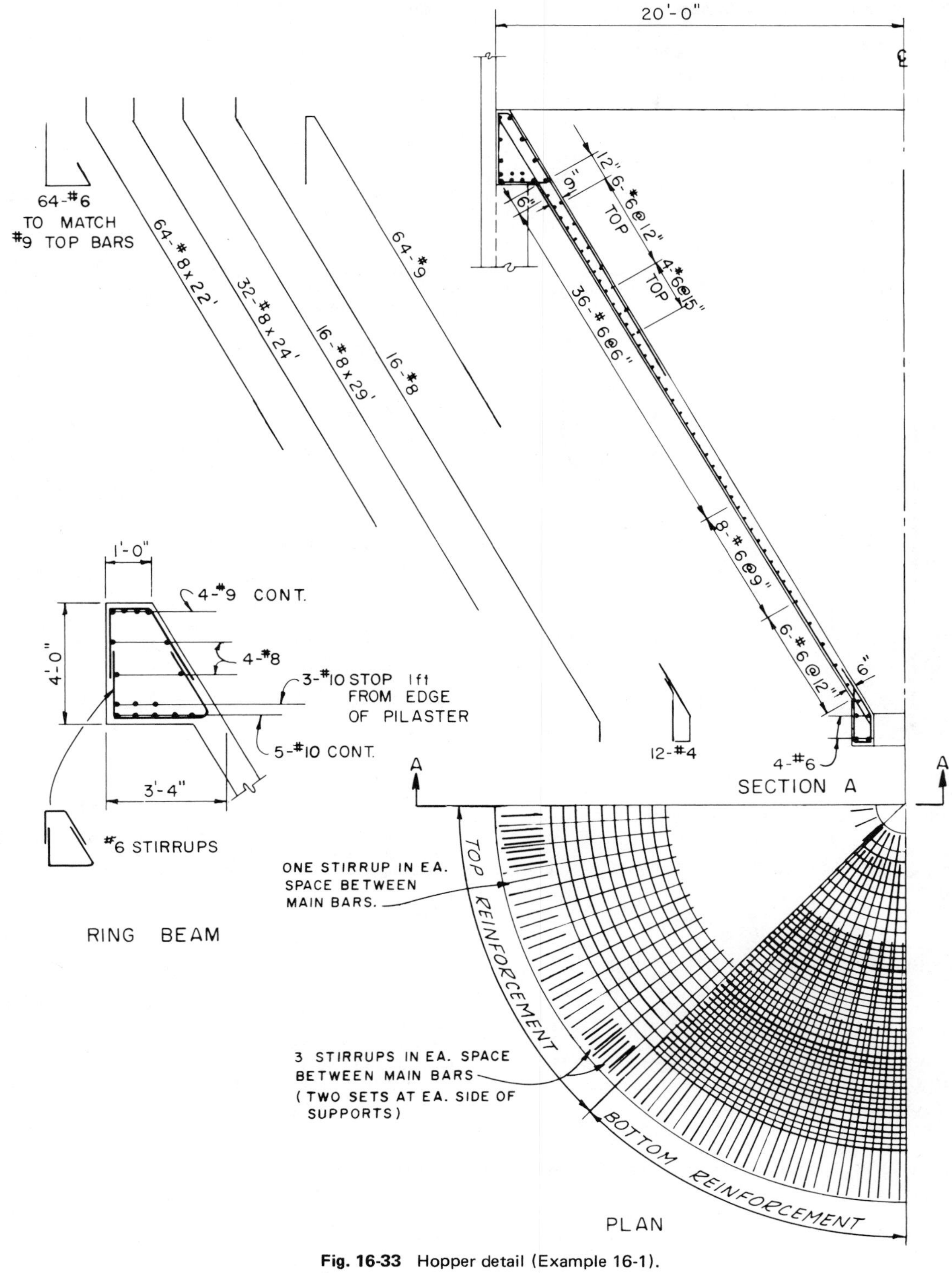

Fig. 16-33 Hopper detail (Example 16-1).

within the hopper, especially at the bottom of the ring-beam and points where meridional bars are terminated. See Fig. 16-33 for hopper detail. The ring-beam is analyzed by the procedure in Section 16.13.12. The ring-beam forces are summarized in Table 16-15.

EXAMPLE 16-2: The cement storage silo of Fig. 16-34 has a conical steel hopper, supported by a concrete ring-beam on six independent columns. The walls extend down to the raft foundation and have two opposite openings. Cement is stored at 200°F, while the outdoor design temperature is −5°F.

Design the silo and hopper supports using: $f_c' = 5000$ psi for the slipformed walls and hopper supports; $f_y = 60{,}000$ psi (40,000 psi for stirrups and ties). Assume the total service load from the roof to be 1000 lb/ft of wall.

SOLUTION:
For cement:

$$\gamma = 100 \text{ pcf}; \; \rho = 25°; \; \mu' = 0.466 \qquad \text{(Table 16-1)}$$

Use Reimbert's method. Assume material surface flat ($h_s = 0$).

TABLE 16-15 Summary of Forces in Ring Beam (Example 16-1)

Location	Vertical Shear (kips)	Compressive Force Due to F_x (kips)	Bending Moment (ft-k)		Torsional Moment Due to F_y (ft-k)
			Due to M_t	Due to F_y	
Support	345	495	675	−858	0
Midspan	0	495	675	431	0
9°33′ from support	198	495	675	0	66

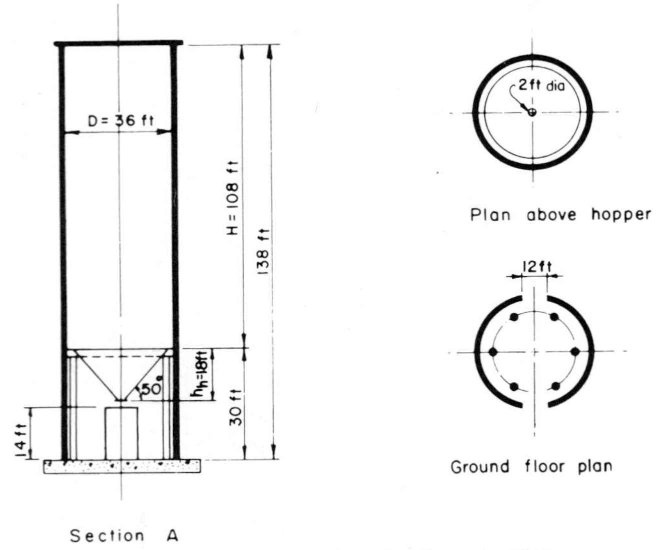

Fig. 16-34 Dimensions of silo for Example 16-2.

$$p_{max} = 100(36)/(4 \tan 25°) = 1930 \text{ lb/sq ft} \qquad (16\text{-}7)$$

$$C = \frac{36}{4 \times 0.466 \times 0.406} = 47.5 \text{ ft} \qquad (16\text{-}8)$$

At depth $Y = 108$ ft, $Y/C = 108/47.5 = 2.27$; $z = 0.906$.

$$q = 100(108)/3.27 = 3300 \text{ psf} \qquad (16\text{-}5)$$

$$p = zp_{max} = 1750 \text{ psf} \qquad (16\text{-}6)$$

Similarly, p and q are computed at various depths Y. Within the hopper, $C = 47.5$ ft is used throughout, even though the diameter changes with Y. (See Tables 16-16 and 16-17.)

Depth $H_1 = 36 \tan 25° = 16.8$ ft, say 18.0 ft.

$$\frac{H - H_1}{4} = \frac{108 - 18}{4} = 22.5 \text{ ft}$$

Overpressure factors, C_d, are taken from Table 16-2, using:

$$\frac{H}{D} = \frac{108}{36} = 3$$

Design pressures, computed by eqs. (16-19), (16-20), and (16-21), are tabulated in Tables 16-16 and 16-17.

Silo Wall Design:

Wall forces and tentative hoop steel areas (computed as in Example 16-1) are shown by Table 16-16. Two layers of hoop steel are needed to permit practical spacing. Therefore, assume a 9-in. wall thickness.

Check vertical compressive stress at bottom ($Y = 138$ ft):

$$\text{Wall wt} = 138 \times 150 \times 9/12 = 15{,}500 \text{ lb/ft}$$

Friction load:

$$V_{des} \text{ (at } Y = 108 \text{ ft)} = 67{,}500 \text{ lb/ft} \qquad (16\text{-}15)$$

From roof (treat as live) 1,000 lb/ft

Total live load = 68,500 lb/ft

$$\text{Ultimate load} = 1.7(68{,}500) + 1.4(15{,}500) = 138{,}200 \text{ lb/ft}$$

Permissible $F_{u,\text{vert}} = 0.385 \times 5000 \times 9 \times 12 = 207{,}900$ lb/ft OK (16-30)

Check crack width (using the procedure shown in Example 16-1):

At $Y = 108$ ft,

$$w_{cr} = 0.0042 \text{ in.} < 0.0008 \quad \text{OK}$$

(Other elevations must be checked also.)

Design temperature steel:

$$T_{i,\text{des}} = 200 - 80 = 120°\text{F} \qquad (16\text{-}36)$$

$$K_t = 0.15;\ \Delta T = [120 - (-5)]\ (0.15) = 18.8°\text{F}$$

(Figs. 16-9 and 16-10 and eq. 16-37)

$$M_{xt,u} = 1.25\ (4.03 \times 10^6)\ (9^2)\ (6 \times 10^{-6})\ (18.8)\ (1.4)$$
$$= 64{,}400 \text{ in.-lb/ft} \qquad (16\text{-}38)$$

$$\text{Reqd } (A_s)_t = 0.161 \text{ sq in./ft}$$

This amount is added to the hoop steel in each face in the bottom two-thirds of the silo height. One-half as much is added to each face for the upper one-third. The required total hoop steel areas are plotted in Fig. 16-35.

For vertical reinforcing, the procedure shown by Example 16-1 is used.

TABLE 16-16 Silo Wall Lateral Pressures and Hoop Reinforcing (Example 16-2)

Depth Y (ft)	Static Lateral Press. p (lb/sq ft)	Overpress. Factor C_d (From Table 16-2)	Design Lateral Press. p_{des} (lb/sq ft)	$F_u = 1.7rp_{des}$ (k/ft)	Required Hoop Steel A_s (sq in./ft)
$H_1 = 18$	913	1.2	1096	33.5	0.62
40.5	1366	1.3	1776	54.3	1.00
63	1572	1.55	2437	74.6	1.38
85.5	1683	1.75	2945	90.1	1.67
108	1750	1.75	3063	93.7	1.73

TABLE 16-17 Ring-Beam and Hopper Pressures and Forces (Example 16-2)

Depth Y (ft) / *Inside Rad. R (ft)*	*Overpress. Factor C_d for Bottom*	*Static Lateral Press. p (lb/sq ft)*	*Static Vertical Press. q (lb/sq ft)*	*Design Lateral Press. p_{des} (lb/sq ft)*	*Design Vertical Press. q_{des} (lb/sq ft)*	*Design Pressure on Inclined Face q_{des} (lb/sq ft)*	*Tangential Force F_t (k/ft)*	*Meridional Force F_m (k/ft)*	
108 / 16.27	1.75	1750	3300	3060	5780	4180	88.8	67.9	For Ring-Beam
111 / 13.5	1.75	1755	3330	3070	5830	4210	74.2	56.1	For Hopper Design
116 / 19.4	1.75	1760	3360	3080	5880	4240	52.0	38.3	
121 / 5.25	1.75	1775	3420	3105	5990	4290	29.4	21.2	
126 / 1.00	1.75	1785	3450	3125	6040	4330	5.7	3.9	

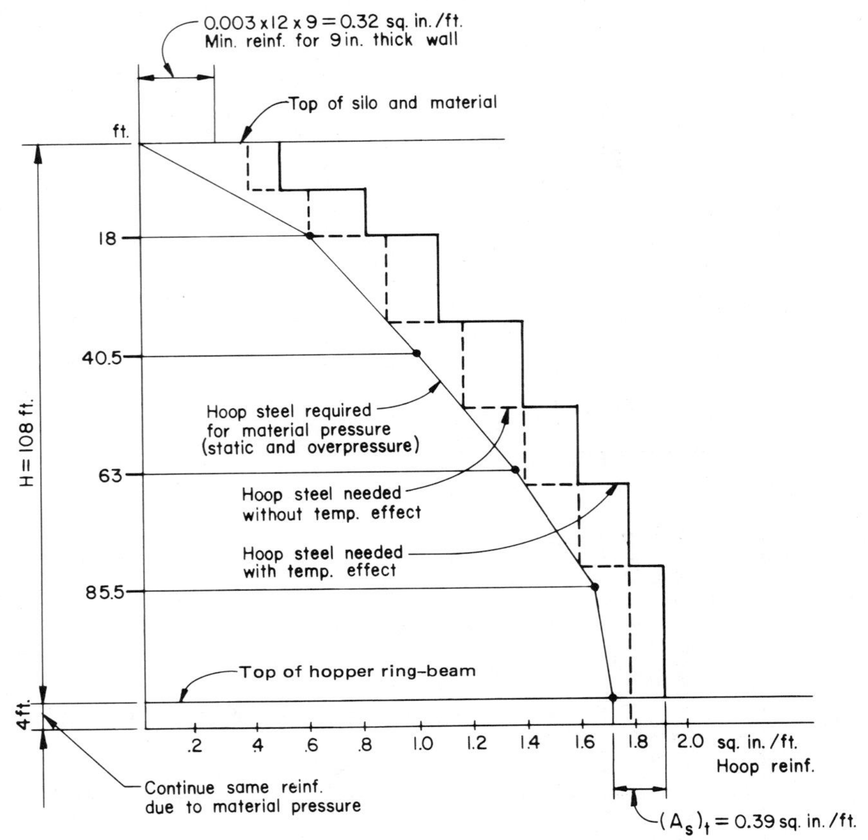

Fig. 16-35 Hoop reinforcement (Example 16-2).

Design of Hopper Supports (Ring Beam and Columns):
Assume ring beam and column sizes as in Fig. 16-36. Section properties are:

$$A_r = 13.62 \text{ sq ft};\ I_{rx} = 16.92 \text{ ft}^4;\ I_c = 3.16 \text{ ft}^4 \text{ (for 36-in. dia)}$$

Torsional constant: $K = 26.0 \quad \lambda = 1.56$

$$\lambda = 0.159 \qquad \text{(Table 16-7)}$$

$$K_2 = 0.003364;\ C_2 = 1.098;\ C_3 = 0.00 \qquad \text{(Tables 16-6; 16-8; 16-9)}$$

Ring-beam loads, neglecting weight of hopper (all WSD):

$$F_m = \frac{5780\,(2 \times 16.23)}{4 \sin 50°} + \frac{(100)\,(16^2)\,(19\pi)/3}{\pi(2 \times 16.27) \sin 50°} = 67{,}900 \text{ lb/ft}$$

$$F_x = 67{,}900 \cos 50° = 43{,}700 \text{ lb/ft} \qquad (16\text{-}66)$$

$$F_y = (150 \times 13.62) + 67{,}900 \sin 50° + 5780\,(18^2 - 16.23^2)/(2 \times 16.23) = 64{,}800 \text{ lb/ft} \qquad (16\text{-}67)$$

$$M_t = 67{,}900 \times 1.6 = 108{,}600 \text{ ft-lb/ft} \qquad (16\text{-}68)$$

Substitution in eqs. (16-74) and (16-75), using $h_c = 27.79$ ft, gives:

$$H_A = 32.54 \text{ kips, and } M_A = -609.9 \text{ ft-kips}$$

These values are used with Tables 16-10, 16-10A and 16-11 to compute the shears, bending moments, ring compressive forces, and torsion values shown in Table 16-18. All values shown are WSD values; they must be multiplied by K_l to obtain values for the USD method.

Column design:
Moment at top:

$$M_A = -609.9 \text{ ft-kip}$$

Bottom moment:

$$M_B = M_A - H_A L = -609.9 + 32.54 \times 27.79 = 294.4 \text{ ft-kip}$$

M_A is negative, meaning that it is opposite to the direction shown by Fig. 16-25. The column is bent in double curvature.

Column axial load:

$$F = 2 \times \text{ring-beam shear at support} + \tfrac{1}{6} \text{ of ring-beam weight}$$
$$= (2 \times 550) + 17 = 1117 \text{ k}$$

Column reinforcing:

$$M_u = K_l M_A = 1037 \text{ ft-kip};\quad F_u = K_l F = 1899 \text{ kip}$$

Reinforcing required for 34-in. column = 22 sq in.
For this area, $p = 0.0205 \simeq 0.02$. ∴ O.K.

EXAMPLE 16-3: Figure 16-37 shows dimensions of a circular silo for storing wheat. The silo bottom is a circular slab with central opening, simply supported on the lower walls and carrying hopper-forming concrete fill. Roof loads, total, are 1000 lb/ft of wall.
Design the slipform walls using $f_c' = 3000$ psi and $f_y = 40{,}000$ psi, and the slab using $f_c' = 4000$ psi and $f_y = 60{,}000$ psi.

SOLUTION:
For wheat:

$$\gamma = 50 \text{ pcf};\ \rho = 25°;\ \mu' = 0.444 \qquad \text{(Table 16-1)}$$

Silo Wall Design:
Table 16-19 shows design pressures, computed by Janssen's formulas. A 6-in. wall thickness was assumed and steel selected as shown. A check showed crack width to be 0.005 in., which is acceptable. The procedures for wall design are all as illustrated by Example 16-1. Final hoop steel areas are shown in Fig. 16-38. The minimum vertical reinforcement area is:

$$A_s = 0.002(6)(12) = 0.144 \text{ sq in./ft}$$

Bottom Slab Design:
Vertical pressure at top of fill ($Y = 120$ ft):

$$k = (1 - \sin \rho)/(1 + \sin \phi) = 0.41 \qquad (16\text{-}3)$$

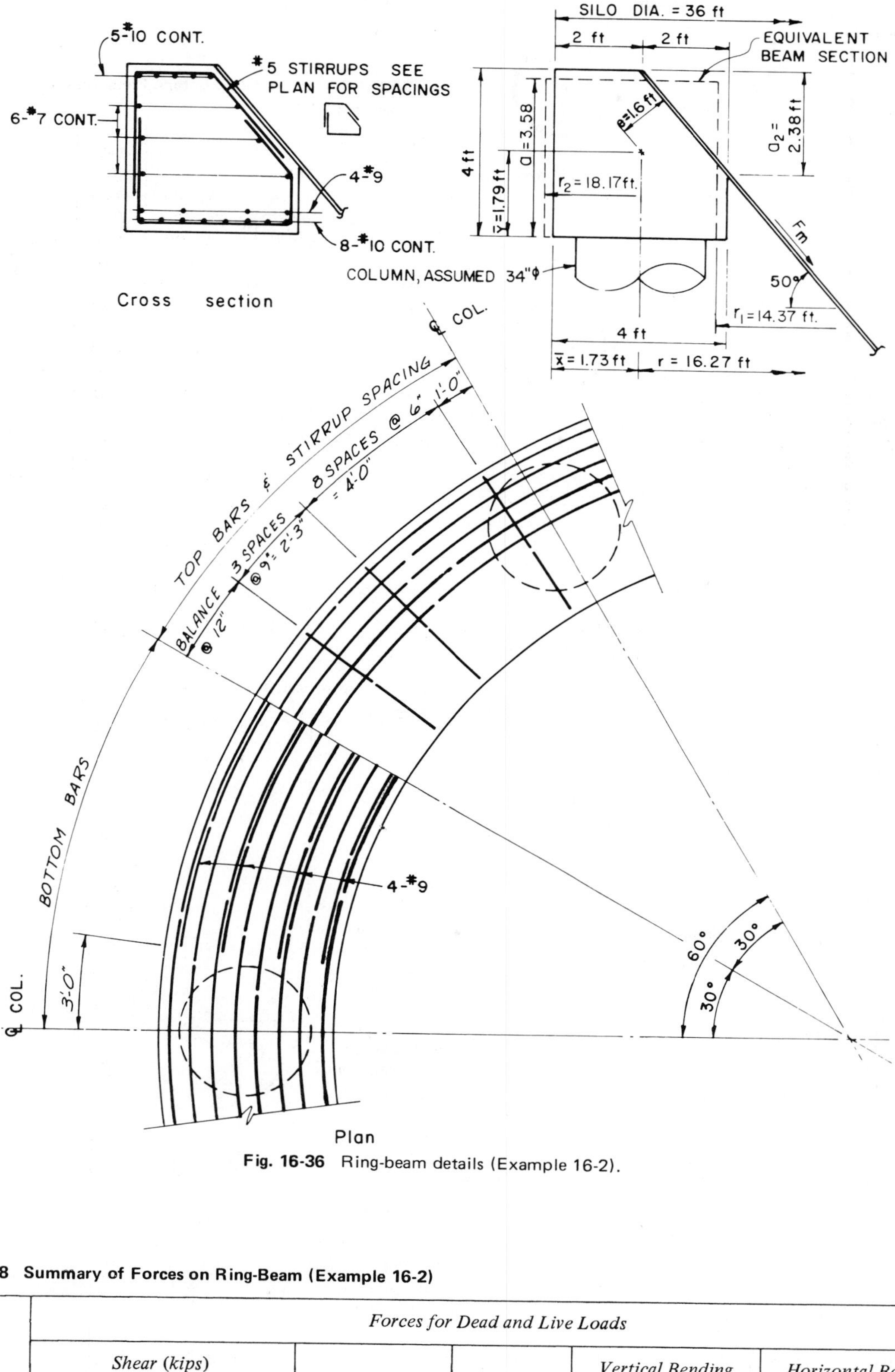

Fig. 16-36 Ring-beam details (Example 16-2).

TABLE 16-18 Summary of Forces on Ring-Beam (Example 16-2)

	Forces for Dead and Live Loads					
	Shear (kips)					
Location	*Vertical*	*Horizontal*	*Compressive Force (kips)*	*Torque (ft-k)*	*Vertical Bending Moment (ft-k)*	*Horizontal Bending Moment (ft-k)*
Support	551.0	16.0	708.0	305.0	−350.0	−240.0
Mid-span	0	0	708.0	0	1,956.0	24.0

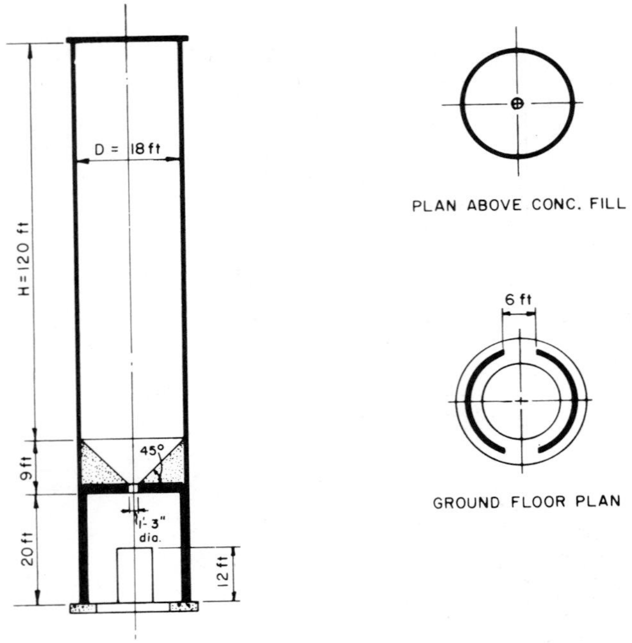

TABLE 16-19 Silo Wall Lateral Pressures and Hoop Reinforcing (Example 16-3)

Depth *Y* (*ft*)	Static Lateral Press. *p* (*lb/sq ft*)	Overpress. Factor C_d (From Table 16-2)	Design Lateral Press. p_{des} (*lb/sq ft*)	$F_u = 1.7rp_{des}$ (*k/ft*)	Required Hoop Steel A_s (*sq in./ft*)
$H_1 = 10$	167	1.50	250	3.9	0.10
37.5	394	1.60	630	9.7	0.37
65	470	1.75	821	12.6	0.35
92.5	495	1.85	916	14.0	0.39
120	503	1.85	931	14.2	0.40

Fig. 16-37 Dimensions for silo of Example 16-3.

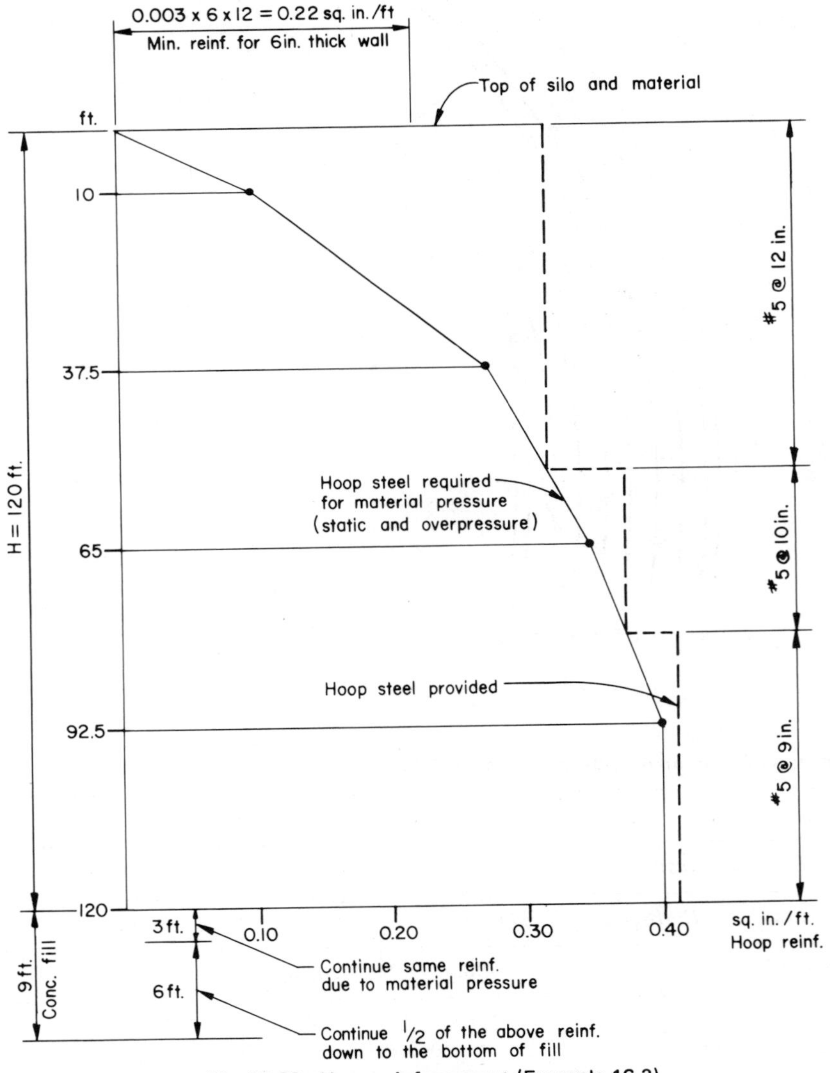

Fig. 16-38 Hoop reinforcement (Example 16-3).

$$q = p_{120}/k = 505/0.41 = 1230 \text{ psf} \qquad (16\text{-}1)$$

$$q_{des} = C_d q = 1.35 \times 1230 = 1660 \text{ psf} \qquad (\text{Table 16-2, eq. 16-19})$$

Load from wheat in hopper (assumed uniform)

$$= \frac{\pi(9^2)(9)(50)}{3\pi(9^2)} = 150 \text{ lb/sq ft}$$

Total design live load = 1660 + 150 = 1810 lb/sq ft

Load from weight of hopper-forming fill (assumed uniform)

$$= \frac{2\pi(9^2)(9)(150)}{3\pi(9^2)} = 900 \text{ lb/sq ft}$$

Slab weight–Assume 18-in. slab, wt. = 225 lb/sq ft

Total dead load = 900 + 225 = 1125 lb/sq ft

Ultimate load = w_u = 1.7(1810) + 1.4(1125) = 4650 lb/sq ft

Slabs with holes may be designed in either of two ways

1. By computing bending moments for a slab with no hole, and reinforcing with a steel member of adequate strength and of stiffness equal to that of the removed slab.
2. By considering the hole, and reinforcing for bending moments obtained using tables of radial and tangential bending moments[16-28] or Timoshenko's equations.[16-29]

When the hole is small, the first method may be the better. That method is shown here.

Ultimate bending moments per unit width, computed by eqs. (69) and (70) of Ref. 16-29, neglecting the hole, are tabulated in Table 16-20.

TABLE 16-20 Silo Bottom Slab Moment and Reinforcing Steel (Example 16-3)

r/r_o	$(M_r)_u$ (*k/ft*)	*Required Steel* (*sq in./ft*)	$(M_t)_u$ (*k/ft*)	*Required Steel* (*sq in./ft*)
0.0	70.8	1.12	70.8	1.12
0.1	70.2	1.1	70.5	1.11
0.2	68.1	1.06	69.1	1.08
0.3	64.7	1.00	67.4	1.1
0.4	59.5	0.88	64.3	1.00
0.5	53.3	0.83	60.9	0.95
0.6	45.4	0.70	56.4	0.92
0.7	36.1	0.54	50.9	0.79
0.8	25.4	0.38	44.7	0.68
0.9	13.4	0.23	37.8	0.58
1.0	0	0	30.1	0.47

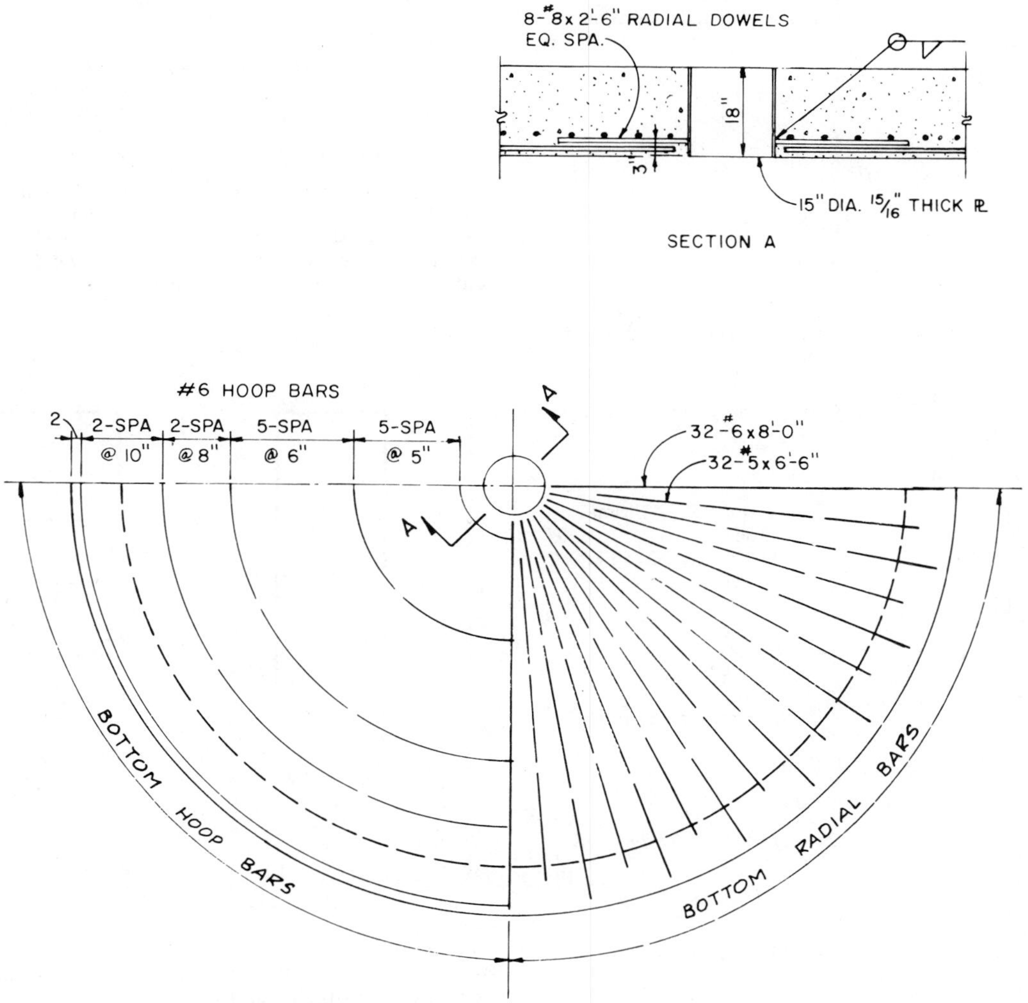

Fig. 16-39 Reinforcing of silo bottom slab (Example 16-3).

The slab stiffness and strength at the hole are replaced by the steel plate ring shown by Fig. 16-39. To provide plate stiffness equaling that of the missing concrete, the plate thickness is chosen using the transformed-area principle:

$$n = E_s/E_c = 8$$

$$\text{total plate thickness} = \text{opening width}/n$$

$$= 15/8 \text{ in. or } 15/16 \text{ in. per side}$$

Use 15/16 plate.

Assuming a radius of 8.75 ft to the center of the slab support, and using eq. (69) of Ref. 16-29, the radial ultimate bending moment at the edge of the hole is 70.8 ft-k/ft. The ring cross section (two rectangular sections) must have strength to resist the total unfactored moment on the 15-in. opening width.

$$M \simeq 1.25 \times (70.8/1.7) = 52.1 \text{ ft-kip}$$

$$S = 2 \times \tfrac{1}{6} \times \tfrac{15}{16} \times 18^2 = 101 \text{ in.}^3$$

$$f_b = 12M/S = 6.19 \text{ ksi} << F_b \qquad \text{(Plate thickness OK)}$$

Radial reinforcing shown by Table 16-20 must transmit its inner end load to the center ring plate. This transfer is through lap splicing to radial dowels which are welded to the ring plate. (See Fig. 16-39.) The total A_s required at the connection to the ring is approximately $1.11 \times 15\,\pi/12$, or 4.36 sq in. For #7 dowels, the number needed is then 4.36/0.60, or 8. Use eight dowels (or more) as shown.

Elastic deflection of the center of the plate is:

$$\Delta = \frac{0.058(4650)(8.60^4)(1728)}{(29/8)(10^6)(18^3)(12/12)} = 0.12 \text{ in.}$$

The long-term deflection is approximately 3.0 × 0.12, or 0.36 in.

Shear per unit width for critical section at distance d from the inside edge of the wall is:

$$V_u = \frac{4650\,(9 - 1.25)^2\,\pi}{2\pi(9 - 1.25)} = 18.0 \text{ k/ft}$$

$$V_n = \frac{V_u}{\phi} = 21.2 \text{ k/ft}$$

$$V_c = 2\sqrt{f'_c}\;bd = 2\sqrt{4000} \times 12 \times 15$$

$$= 22.8 \text{ k/ft} > 21.2 \text{ k/ft} \qquad \text{OK}$$

Lower Wall Design:

Assume a 12-in. thickness. Bearing width of slab on wall = 6 in. Compression stress at bearing of slab wall =

$$f_{c,\text{vert}} = \frac{4650\,(9^2)\,\pi}{144\,\pi(9^2 - 8.5^2)} = 290 \text{ psi} < 0.385\,f'_c \qquad \text{OK}$$

Total reaction at bottom of wall:

	Weight	K	Ultimate
from roof	56		
material above hopper	1525	1.7	2750 kips
material in hopper	38		
hopper-forming fill	229		
bottom slab	57	1.4	1397
upper wall	563		
lower wall, 0.15 (20)(18π) − 0.15 (6 × 12 × 2)	149		
			Total = 4147 kips

$$\text{Net wall width} = 18\pi - 12 = 44.5 \text{ ft}$$

$$\text{Avg compression} = 4147/44.5 = 93.2 \text{ k/ft}$$

$$\text{Permissible } F_u = 0.385\,(5)(12 \times 12) = 277 \text{ k/ft} > 93.2 \qquad \text{OK} \qquad (16\text{-}30)$$

Vertical steel, lower wall:

$$A_s = 12 \times 12 \times 0.002 = 0.29 \text{ sq in./ft}$$

Use #4 @ 16 in. c/c each face.

Horizontal steel, lower wall:

$$A_s = 12 \times 12 \times 0.0030 = 0.43 \text{ sq in./ft}$$

Use #4 @ 12 in. c/c each face.

At lower wall openings:
Vertical–Add (to each side) $A_s = 3 \times 0.29 = 0.87$ sq in.
Add 2 #7 each side of opening.
Horizontal–Add 2 #7 above opening.

EXAMPLE 16-4: Design a single rectangular concrete silo for storing peas. Dimensions are shown in Fig. 16-40. The bottom is a symmetrical pyramidal hopper. The silo walls rest on the hopper base, which is supported by four columns. The roof applies a total load of 3000 lb/ft to the top of the walls. The walls are to be slipformed. Use $f'_c = 4000$ psi and $f_y = 60{,}000$ psi. Slipforming will be done without continuous inspection.

SOLUTION:

Silo Wall Design:

For peas:

$$\gamma = 50 \text{ lb/cu ft}; \quad \rho = 25^\circ; \quad \text{and } \mu' = 0.296 \qquad \text{(Table 16-1)}$$

Using Janssen's equations:

$$k = (1 - \sin 25^\circ)/(1 + \sin 25^\circ) = 0.406 \qquad (16\text{-}3)$$

For the shorter wall:

$$R_a = a/4 = 3.0 \text{ ft}$$

For the longer wall:

$$a' = 2 \times 12 \times 20/(12 + 20) = 15.0 \text{ ft}$$

$$R_b = a'/4 = 3.75 \text{ ft}$$

$$H/a = 80/12 = 6.7$$

$$H_1 = a \tan \rho = 12 \tan 25^\circ \simeq 6.0 \text{ ft}$$

$$(H - H_1)/4 = (80 - 6)/4 = 18.5 \text{ ft}$$

Using these R-values and the above material properties, static lateral pressures are computed at various depths, Y, using eq. (16-2). The pressure values are tabulated in Table 16-21. Next, overpressure factors, C_d, are selected from Table 16-2, and design pressures computed by eq. (16-18) are added to Table 16-21.

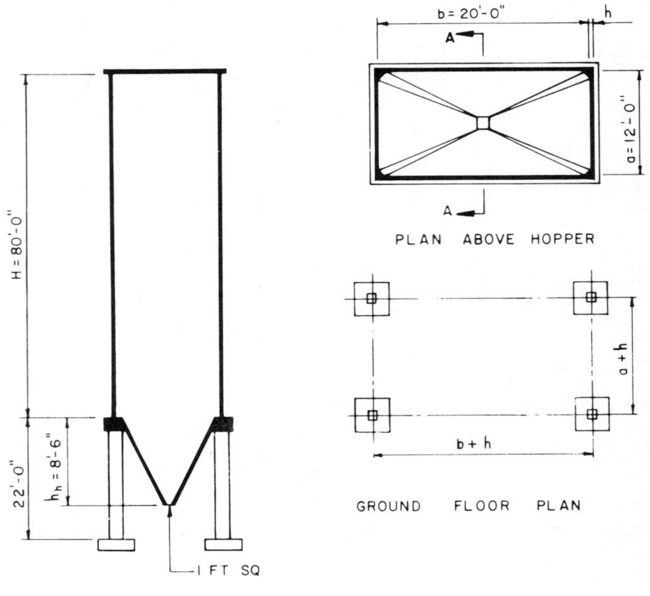

Fig. 16-40 Single rectangular silo with pyramidal hopper, supported on four columns (Example 16-4).

TABLE 16-21 Pressure on Silo Walls (Example 16-4)

Depth Y (ft)	Overpress. Factor C_d (Table 16-2)	Wall "a" Static Lateral Press. p (lb/sq ft)	Wall "a" Design Lateral Press. p_{des} (lb/sq ft)	Wall "b" Static Lateral Press. p (lb/sq ft)	Wall "b" Design Lateral Press. p_{des} lb/sq ft
6	1.50	108	162	111	167
24.5	1.60	317	507	345	551
43	1.75	416	728	474	830
61.5	1.85	464	858	545	1010
80	1.85	486	899	585	1080

Vertical wall loads due to friction:

$$A_a = 12 \times \tfrac{6}{2} = 36; \quad A_b = \tfrac{1}{2}(240 - 2A_a) = 84$$

At the bottom, $Y = 80$ ft, so that for wall a:

$$V_{a,\,des} = (50 \times 80 - 0.8 \times 486/0.406) \times 3 = 9130 \text{ lb/ft} \qquad (16\text{-}17, 16\text{-}23)$$

and for wall b:

$$V_{b,\,des} = (50 \times 80 - 0.8 \times 636/0.406) \times 84/20 = 11{,}550 \text{ lb/ft} \qquad (16\text{-}16, 16\text{-}22)$$

Wall tension and bending moment:

For the short wall at depth $Y = 80$ ft:

$$F_{a,u} = 1.7 \times 1080 \times 20/2 = 18{,}400 \text{ lb/ft} \qquad (16\text{-}39)$$

Similarly,

$$F_{b,u} = 9170 \text{ lb/ft} \qquad (16\text{-}40)$$

Assume wall thickness $h = 10$ in. with $d = 8.5$ in. Bending moments are computed by moment distribution for an idealized rectangular frame measuring 20′10″ by 12′10″. Outward pressures on the frame are the design pressures of Table 16-21. Corner moments so solved are $M_u = 49.1$ ft-kip/ft. Reduced to the edge of the fillet, the moments are:

Wall a:

$$M_{au} = -38.0 \text{ ft-kip/ft}$$

Wall b:

$$M_{bu} = -26.6 \text{ ft-kip/ft}$$

TABLE 16-22 Ultimate Forces and Moments and Horizontal Steel for Vertical Walls (Example 16-4)

Depth Y (ft)	Wall "a" Tension F_{au} (k/ft)	Wall "a" At Support M_u (ft-k/ft)	Wall "a" At Support A_s (sq in./ft)	Wall "a" At Midspan M_u (ft-k/ft)	Wall "a" At Midspan A_s (sq in./ft)	Wall "b" Tension F_{au} (k/ft)	Wall "b" At Support M_u (ft-k/ft)	Wall "b" At Support A_s (sq in./ft)	Wall "b" At Midspan M_u (ft-k/ft)	Wall "b" At Midspan A_s (sq in./ft)
6	2.84	−5.80	—	−2.13	—	1.65	−4.32	—	7.61	0.23
			0.20		0.09			0.14		—
24.5	9.37	−19.2	—	−7.73	—	5.17	−14.0	—	25.3	0.80
			0.66		0.32			0.46		—
43	14.1	−29.1	—	−12.6	—	7.43	−20.8	—	38.5	1.26
			1.02		0.52			0.69		—
61.5	17.2	−35.5	—	−16.0	—	8.75	−25.1	—	47.1	1.57
			1.25		0.65			0.83		—
80	18.4	−38.0	—	−17.6	—	9.17	−26.6	—	50.5	17.1
			1.35		0.78			0.88		—

Positive moment indicates tension on outside surface; negative, tension inside.
Upper half of block for steel area shows required steel area near outer face; lower half shows area for inner face.

At midspan:

wall a, M_{au} = −17.6 ft-kip/ft; wall b, M_{bu} = +50.5 ft-kip/ft

These values, and similar values computed for other depths, are tabulated in Table 16-22.

Horizontal reinforcement for combined tension and bending:
For wall "a," at the toe of the fillet, at depth Y = 80 ft:

$$e = M_{au}/F_{au} = 38.0 \times 12/18.4 = 24.8 \text{ in.}$$

$$e > h/2 - d'', \therefore \text{Case II} \qquad \text{(Fig. 16-15)}$$

$$\text{Limiting } y/d \text{ value} = 0.378 \qquad \text{(Table 16-3)}$$

$$y_L = 0.378\,(8.5) = 3.22 \text{ in.}$$

$$\phi = 0.95 \times 0.90 = 0.85$$

$$f_s' \text{ effective} = 87{,}000 \left[\frac{3.22 - 0.85\,(1.5)}{3.22}\right] - 0.85\,(4000) = 49{,}100 \text{ psi} \qquad (16\text{-}43)$$

$$< (f_y - 0.85 f_c')$$

Compressive steel will be effective.

$$e'' = 24.8 - 3.5 = 21.3 \text{ in.};\; b = 12 \text{ in.}$$

$$\text{Reqd } A_s' = \frac{18.4\,(21.3/0.85) - 0.85\,(4)(12)(3.22)(8.5 - 3.22/2)}{49.1\,(8.5 - 1.5)} = \text{negative} \qquad (16\text{-}44)$$

Even though negative steel would be effective, none is required, as shown by the negative answer.

Tensile reinforcement (inside face):

$$\text{Actual } y = 8.5 - \sqrt{(8.5)^2 - \frac{2 \times 18.4 \times 21.3}{0.85 \times 0.85 \times 4 \times 12}} = 1.45 \text{ in.} \qquad (16\text{-}47)$$

$$\text{Reqd } A_s = \frac{18.4/0.85 + 0.85\,(4)(12)(1.45)}{60} = 1.35 \text{ sq in./ft} \qquad (16\text{-}46)$$

For midspan of wall "a," F_{au} = 18.4 kip/ft and M_{au} = −17.6 ft-kip/ft.

By computations like those above, compressive steel is not needed, and the required tensile steel area is A_s = 0.78 sq in./ft. (There is no point of contraflexure in wall "a"; thus, this steel, too, is near the inside face.)

Similar computations for wall "b" show no compressive steel to be required either at the fillet or at midspan. The required tensile steel areas at Y = 80 ft are:

$$A_s \text{ (at corner, inside)} = 0.88 \text{ sq in./ft}$$

$$A_s \text{ (midspan, outside)} = 1.71 \text{ sq in./ft}$$

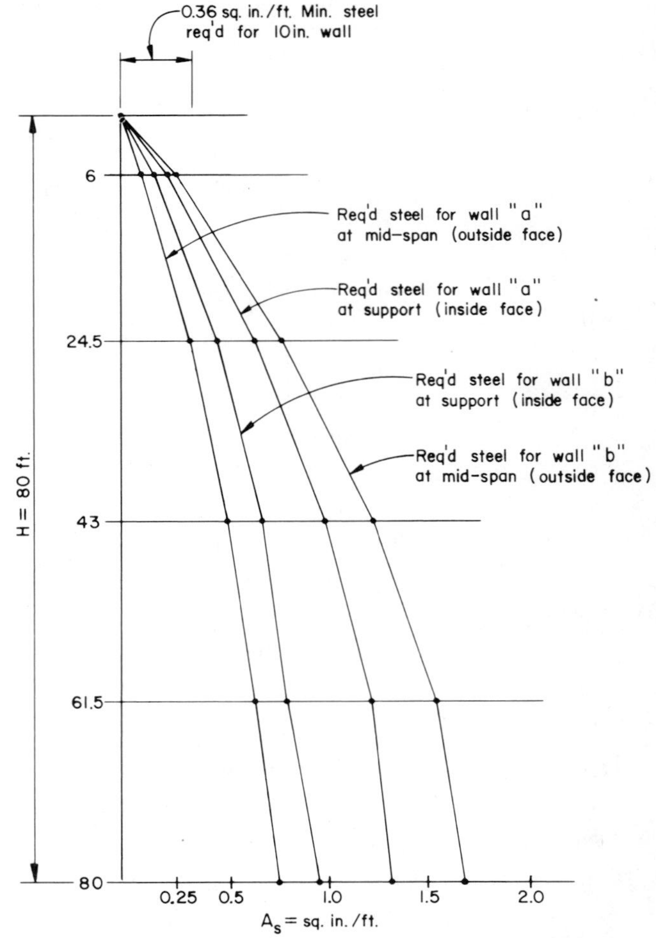

Fig. 16-41 Horizontal wall steel for rectangular silo (Example 16-4).

In this case there *is* a point of contraflexure, so the latter steel area must be provided near the outside face.

The results of similar computations at various depths Y are tabulated in Table 16-22. The minimum recommended horizontal reinforcing is 0.003 × 10 × 12 = 0.36 sq in./ft (Fig. 16-41).

Vertical reinforcement:

$$A_s \text{ reqd} = 0.02 \times 12 \times 10 = 0.24 \text{ sq in./ft}$$

Use #5 @ 16 in. c/c.

TABLE 16-23 Membrane Forces in Wall of Pyramidal Hopper (Example 16-4)

Depth	*Wall "a"*					*Wall "b"*				
	Design Pressures (psf)			*Ult. Forces (k/ft)*		*Design Pressures (psf)*			*Ult. Forces (k/ft)*	
Y (ft)	*Lateral* $p_{a,\,des}$	*Vertical* $q_{a,\,des}$	*Normal* $q_{\alpha,\,a,\,des}$	*Tang.* F_{tau}	*Merid.* F_{mau}	*Lateral* $p_{b,\,des}$	*Vertical* $q_{b,\,des}$	*Normal* $q_{\alpha,\,b,\,des}$	*Tang.* F_{tbu}	*Merid.* F_{mbu}
80	655	1620	1190	12.8	22.9	790	1950	1130	10.2	12.5
82	↓	↓	↓	9.9	17.8	↓	↓	↓	8.0	9.8
84				7.1	12.6				5.8	7.1
86				4.2	7.5				3.6	4.9
88				1.4	2.4				1.4	1.9

Hopper Design:
For all depths, use R equal to the hydraulic radius at the top of the hopper. In this case, the pressure changes very little with depth, so use the pressures for Y = 80 ft throughout.

Vertical design pressures at Y = 80 ft are:

$$q_{a,\,\mathrm{des}} = 1.35 \times 486/0.406 = 1620 \text{ lb/sq ft} \qquad (16\text{-}1, 16\text{-}3, 16\text{-}19)$$

$$q_{b,\,\mathrm{des}} = 1.35 \times 585/0.406 = 1950 \text{ lb/sq ft} \qquad (16\text{-}1, 16\text{-}3, 16\text{-}19)$$

Angles of slope are:

$$\alpha_a = 41.8^\circ \text{ and } \alpha_b = 57.1^\circ$$

Design pressures normal to the inclined walls are:

$$q_{\alpha a,\,\mathrm{des}} = 660 \sin^2 41.8^\circ + 1620 \cos^2 41.8^\circ = 1190 \text{ lb/sq ft} \qquad (16\text{-}17)$$

$$q_{\alpha b,\,\mathrm{des}} = 790 \sin^2 57.1^\circ + 1950 \cos^2 57.1^\circ = 1130 \text{ lb/sq ft} \qquad (16\text{-}17)$$

Horizontal ultimate membrane forces (tensile) are:

$$F_{tau} = 1.7(20/2)(1130) \sin 41.8^\circ = 12{,}800 \text{ lb/ft} \qquad (16\text{-}61)$$

$$F_{tbu} = 1.7(12/2)(1190) \sin 57.1^\circ = 10{,}190 \text{ lb/ft} \qquad (16\text{-}62)$$

The weight of the hopper and its contents are estimated to be 41,200 lb and 34,000 lb, respectively. A_a and A_b are each one-fourth of the hopper area, or (20 × 12)/4 = 60 sq ft.

Meridional ultimate forces (tensile) are computed assuming c_a and c_b to each be 1/4.

$$F_{mau} = \frac{1.7(0.25 \times 34{,}000 + 60 \times 1620) + 1.4(0.25 \times 41{,}200)}{12 \sin 45^\circ}$$

$$= 22{,}880 \text{ lb/ft} \qquad (16\text{-}59)$$

$$F_{mbu} = \frac{1.7(0.25 \times 34{,}000 + 60 \times 1950) + 1.4(0.25 \times 41{,}200)}{20 \sin 60^\circ}$$

$$= 13{,}150 \text{ lb/ft} \qquad (16\text{-}60)$$

Table 16-23 shows the membrane forces at various depths Y.

Hopper wall bending moments are computed using Table 16-4. For hopper wall "a," a = 13 ft (measured to center of walls), and c = 13 ft. Use a/c = 1.0.

At the center of the upper edge (point 1):

$$n_x = -0.209; \quad n_y = -1.255 \qquad \text{(Table 16-4)}$$

$$M_{xau} = 1.7[0.209(1190)(13)^2/64] = 1120 \text{ ft-lb/ft}$$

$$M_{yau} = 1.7\,[1.255(1190)(13)^2/64] = 6700 \text{ ft-lb/ft}$$

Similar moment computations were made for several points of the triangular wall. From these and the corresponding membrane forces (Table 16-23), the reinforcement was computed following the procedure shown for combined tension and bending. The final pattern of reinforcing steel selected is shown by Fig. 16-42.

Hopper Edge Beam and Bottom Reinforcement of Vertical Walls:
Dowels are provided to transfer the vertical forces from the hop-

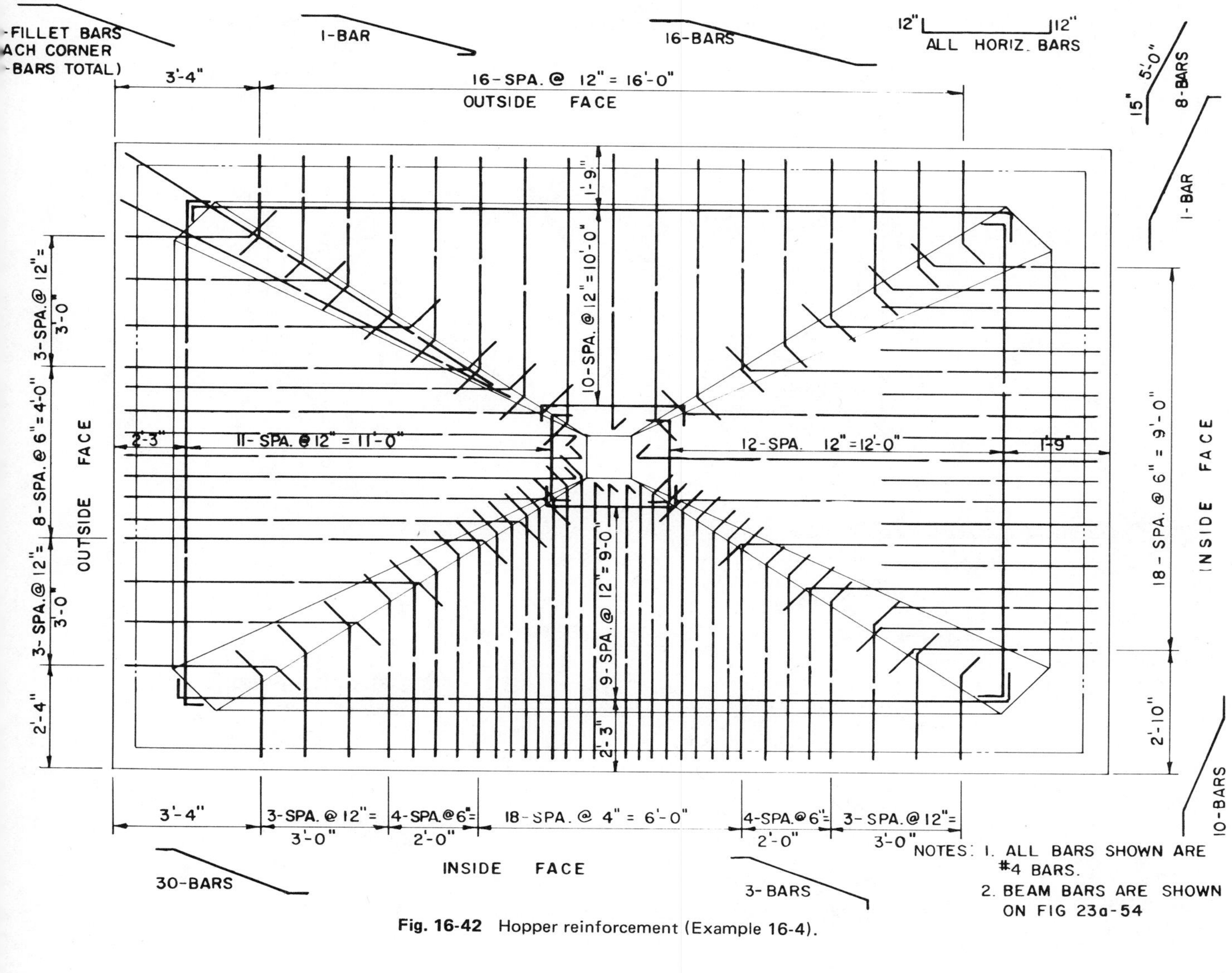

Fig. 16-42 Hopper reinforcement (Example 16-4).

per beam into the vertical walls. For wall "a," for example, the vertical load to be transferred is:

$$P_u = F_{mau} \sin \alpha_a = 15.3 \text{ kip/ft}$$

$$A_s \text{ reqd} = 15.3/(0.95 \times 0.9 \times 60) = 0.30 \text{ sq in./ft}$$

By scaling from a detail sketch of the walls and edge beam, it was determined that force F_{mau} is about 6 in. from the centroid of the edge beam. The resulting twisting moment can be resisted by the vertical wall, provided extra dowels are used. For this purpose, dowels having $A_s = 0.34$ sq in./ft are needed.

The total dowel steel required near the inner face is:

$$A_s = 0.30/2 + 0.34 = 0.49 \text{ sq in./ft}$$

and near the outer face:

$$A_s = 0.30/2 = 0.15 \text{ sq in./ft.}$$

The dowel requirement is met by using #7 "hairpin bars" @ 14 in. c/c as shown in Fig. 16-43.

Because of dowels joining it to the vertical wall, the edge beam has negligible vertical shear and torsion. The upper wall shear and horizontal components of the hopper wall meridional force and shear are assumed to be in equilibrium; thus, no horizontal transverse load is carried by the edge beam. Its only purpose is to simplify construction. Minimum longitudinal steel and stirrups are provided (Fig. 16-43).

The vertical walls are analyzed as deep girders to carry vertical load as follows:

For wall "a":

from dowels to hopper,	15.3 kip/ft
friction, 1.7(9.13)	= 15.5
wall weight,	
1.4(0.83)(80)(0.15)	= 13.9
Total	= 44.7 kip/ft

Similarly, for wall "b," the ultimate vertical load is 42.6 k/ft.

In this example, the vertical walls were analyzed as deep beams. That analysis showed bottom steel to be needed, as follows:

Bottom of wall "a," 2 #9

Bottom of wall "b," 3 #9

Column Design:

Load per column:

	Dead	Live
material stored in hopper		8,500 lb
mateiral stored above hopper		240,000
walls	168,750 lb	
roof (principally dead load)	17,000	
hopper	10,300	
Totals	196,050 lb	248,500 lb

$$\text{Ultimate load} = P_u = 1.4\ (196.1) + 1.7\ (248.5) = 698 \text{ kips}$$

A 20 in × 20 in. tied column with 8 #7 bars is satisfactory.

(Fig. 16-43)

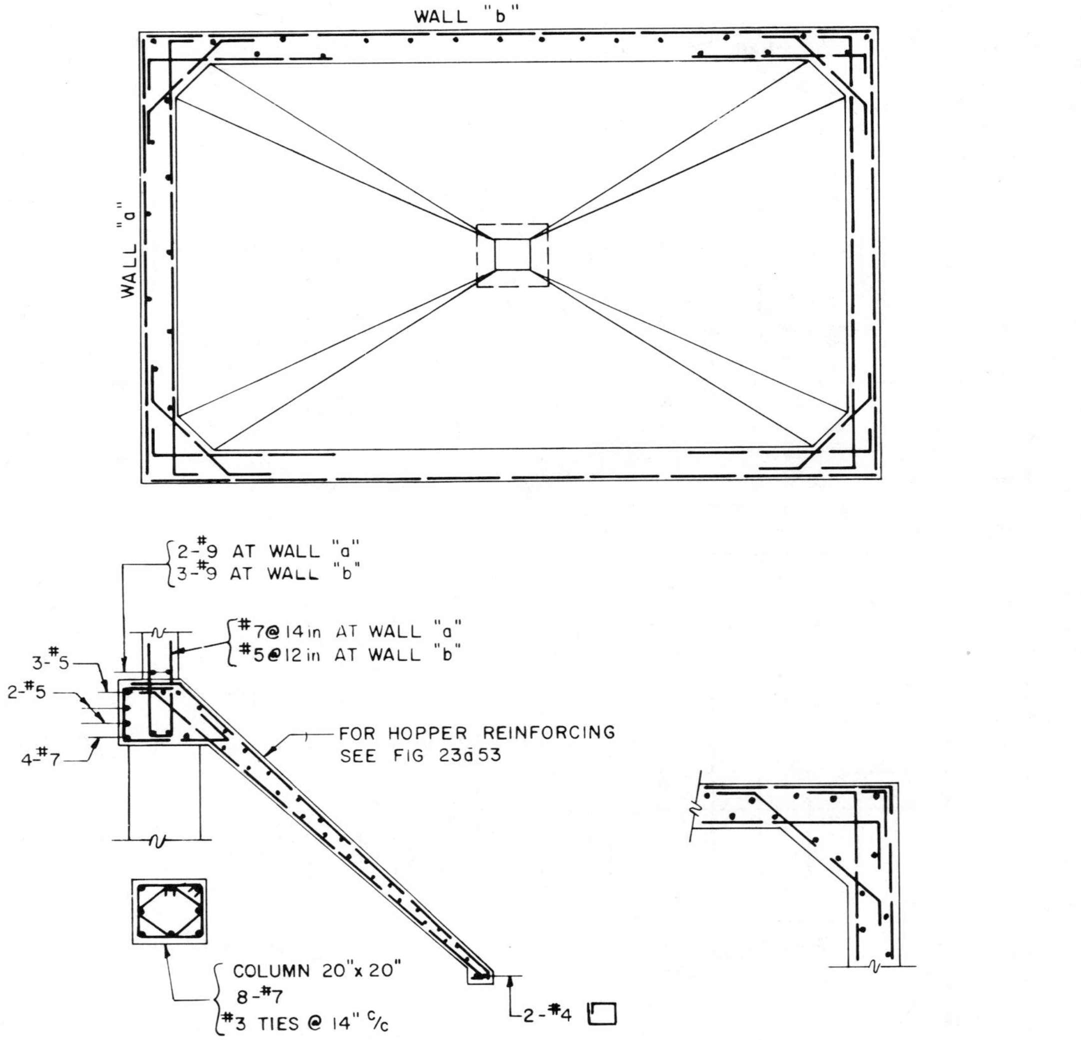

Fig. 16-43 Hopper reinforcement details (Example 16-4).

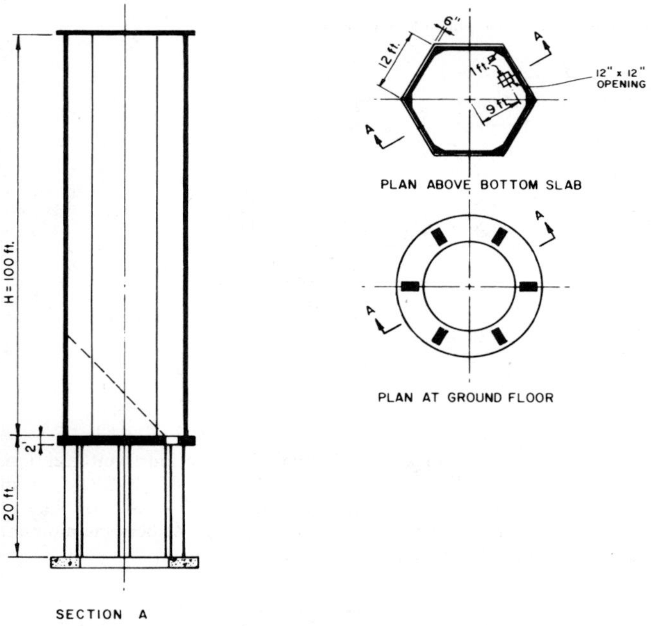

Fig. 16-44 Hexagonal silo dimensions (Example 16-5).

Earthquake zone 0 was assumed; therefore seismic forces were not considered. If seismic forces were to be considered, further analysis would be necessary. A stronger column, preferably spirally reinforced, would be required.

EXAMPLE 16-5: Figure 16-44 shows the basic scheme and dimensions of a hexagonal silo for storing dry sand. The bottom is a horizontal slab with one eccentric discharge opening. The structure is supported on six columns. Roof loads total 1200 lb per foot of wall. Slipforming is to be used. Design the walls, bottom, and columns using $f_c' = 4000$ psi and $f_y = 60{,}000$ psi.

SOLUTION:
For dry sand:

$$\gamma = 100\ pcf;\ \rho = 35^\circ;\ \mu' = 0.70 \qquad \text{(Table 16-1)}$$

Using Reimbert's method and assuming a flat top surface:

$$k = (1 - \sin 35^\circ)/(1 + \sin 35^\circ) = 0.271$$

The total area is $6 \times 0.5 \times 12 \times 12 \sin 60^\circ = 374.1$ sq ft, and the diameter of an equivalent circle is $2\sqrt{374.1/\pi} = 21.8$ ft.

$$R = 21.8/4 = 5.45 \text{ ft}$$

$$p_{max} = 100 \times 5.45/0.7 = 779 \text{ lb/sq ft} \qquad (16\text{-}9)$$

The term "L" in eq. (16-10) is the circumference of the equivalent circle, or 21.8π.

TABLE 16-24 Lateral Pressures on Silo Walls (Example 16-5)

Depth Y (ft)	C_d	*Static Lateral Press. p (lb/sq ft)*	*Lateral Press. Due to Eccent. (lb/sq ft)*	*Design Lateral Press.* p_{des} *(lb/sq ft)*
16	1.25	459	104	680
37	1.35	630	241	1090
58	1.60	694	378	1490
79	1.85	724	510	1850
100	1.85	740	510	1880

$$C = \frac{21.8\pi}{\pi}\left[\frac{1}{4 \times 0.7 \times 0.271}\right] - 0 = 28.7 \qquad (16\text{-}10)$$

$$H_1 = 21.8 \tan 35^\circ = 15.3 \text{ ft, say } 16 \text{ ft}$$

$$(H - H_1)/4 = (100 - 16)/4 = 21 \text{ ft}$$

$$H/D = 100/21.8 = 4.6$$

C_d factors are read from Table 16-2.

Table 16-24 shows overpressure factors and lateral pressures at various depths. The computations are illustrated below for depth $Y = 100$ ft.

$$Y/C = 100/28.7 = 3.48$$

$$p = 779\ [1 - (3.48 + 1)^{-2}] = 740 \text{ lb/sq ft} \qquad (16\text{-}6)$$

$$q = 100\ [100\,(3.48 + 1)^{-1} + 0] = 2230 \text{ lb/sq ft} \qquad (16\text{-}5)$$

$$p_{des} = 740 \times 1.85 = 1370 \text{ lb/sq ft} \qquad (16\text{-}20)$$

$$q_{des} = 1.50 \times 2230 = 3340 \text{ lb/sq ft} \qquad (16\text{-}19)$$

$$V_{des} = (100 \times 100 - 2230)\,5.45 = 42{,}300 \text{ lb/ft} \qquad (16\text{-}15)$$

All of the above are for a silo with central discharge.

The lateral design pressure must now be corrected for eccentricity of discharge. At $Y = 100$ ft:

$$Y = H;\ \therefore \text{ use full correction, } P_e$$

An imaginary silo will be selected by substituting a circle for the hexagon and proceeding as shown by Fig. 16-8. A circle of area equal to that of the hexagon has a radius of 10.9 ft. The radius of the imaginary silo is then:

$$r_i = 10.9 + 9 = 19.9 \text{ ft} \qquad \text{(Fig. 16-8)}$$

The corresponding hexagonal imaginary silo has all linear dimensions equal to (19.9/10.9) times those of the actual hexagon. Thus:

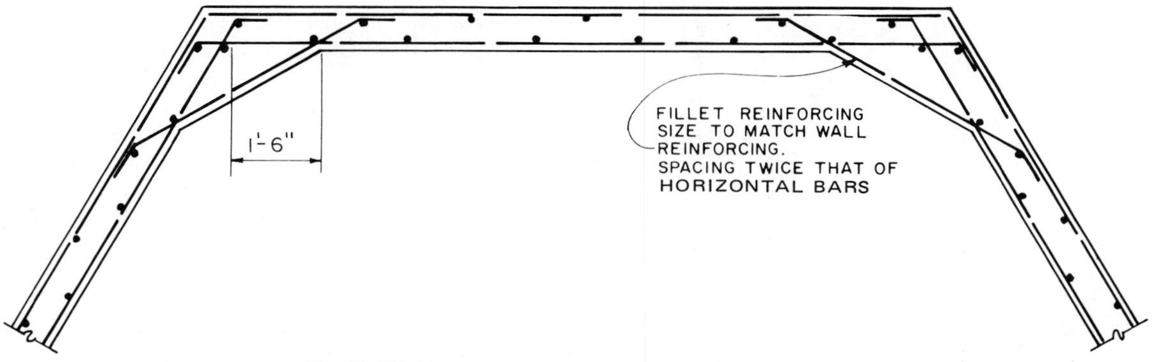

Fig. 16-45 Hexagonal silo wall reinforcing (Example 16-5).

TABLE 16-25 Horizontal Forces, Bending Moments, and Reinforcing Steel Areas (Example 16-5)

Depth Y (ft)	*Ultimate Tensile Force F_u (k/ft)*	*Ult. Moment at Edge of Fillet (ft-k/ft)*	*Required Steel Area Inside Face (sq in./ft)*	*Ultimate Moment at Mid-span (ft-k/ft)*	*Required Steel Area Outside Face (sq in./ft)*
16	12.0	4.3	0.26	7.4	0.35
37	19.3	6.9	0.42	11.9	0.57
58	26.3	9.4	0.57	16.3	0.78
79	32.7	11.7	0.71	20.2	0.97
100	33.2	11.9	0.72	20.6	0.99

$$R_i = 5.45\ (19.9/10.9) = 9.95 \text{ ft}$$

$$C = 28.7\ (19.9/10.9) = 52.4 \text{ ft}$$

$$Y/C = 100/52.4 = 1.91$$

$$p_{max} = 100\ (9.95)/0.7 = 1420 \text{ lb/sq ft} \qquad (16\text{-}9)$$

$$p_i = 1250 \text{ lb/sq ft} \qquad (16\text{-}6)$$

$$p_{ecc} = 1250 - 740 = 510 \text{ lb/sq ft} \qquad (16\text{-}25)$$

The corrected p_{des} at $Y = 100$ ft is then:

$$p_{des} = 1370 + (510) = 1880 \text{ lb/sq ft} \qquad (16\text{-}24)$$

At depths Y of less than (100 − 21.8) ft, the correction to p_{des} at that depth is:

$$p_{ecc} = 510\ (Y/78.2)$$

Corrections and corrected design pressures for various depths are tabulated in Table 16-24.

Wall force and bending moment at $Y = 100$ ft:

$$F_u = 0.866\ (1.7)\ (12)\ (1880) = 33{,}200 \text{ lb/ft} \qquad (16\text{-}51)$$

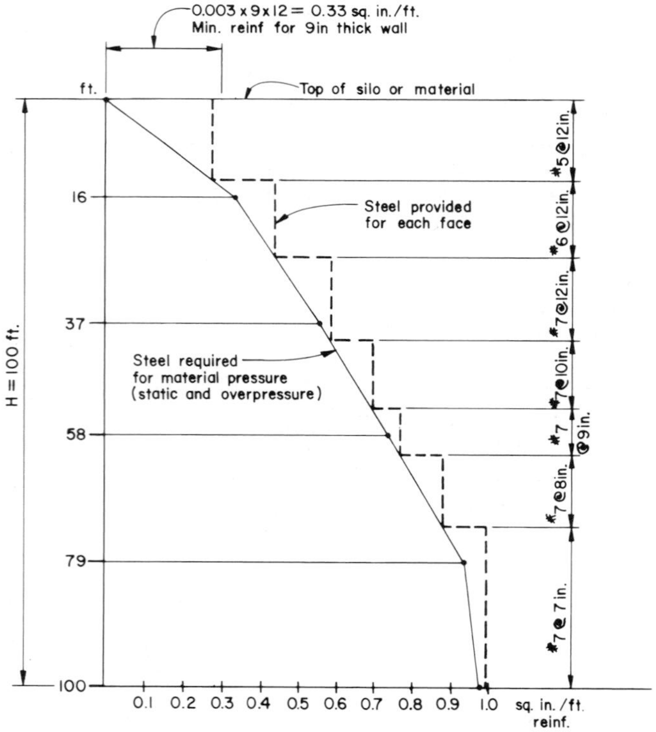

Fig. 16-46 Horizontal reinforcing for silo wall (Example 16-5).

Assume a 9-in. wall thickness. The span, intersection-to-intersection, is then 12.43 ft. The edge of the fillet (Fig. 16-45) is 1.71 ft from the intersection of walls. Fillet toes are 12.43 − 3.42, or 9.01 ft apart. Assuming a parabolic moment diagram, the bending moment at any depth, reduced to the edge of the fillet, is:

$$M_u = \frac{1.7\ p_{des}}{8}\ (9.01^2 - 12.43^2/3) = 6.31\ p_{des}$$

or at $Y = 100$ ft:

$$M_u = 6.31\ (1880) = 11{,}860 \text{ ft-lb/ft}$$

At midspan:

$$M_u = 1.7 \times (12.43)^2 p_{des}/24$$

Ultimate wall forces and bending moments for various depths, as well as the required areas of positive and negative steel (selected for combined tension and bending), are shown by Table 16-25. The reinforcing chosen is shown in Fig. 16-46.

Bottom Slab Design:

By hanging the bottom slab from the vertical walls and taking

TABLE 16-26 Ultimate Bending Moments and Required Steel Areas for Bottom Slab (Example 16-5)

$\frac{r}{r_o}$	$(M_r)_u$ ft.-K	REQ'D STEEL in/ft	$(M_t)_u$ ft.-K	REQ'D STEEL in/ft
0.0	121.5	1.35	121.5	1.35
0.1	120.5	1.34	121.0	1.34
0.2	116.5	1.29	118.5	1.32
0.3	111.5	1.23	115.5	1.29
0.4	102.0	1.12	110.5	1.22
0.5	91.5	1.00	104.5	1.15
0.6	78.0	0.85	97.0	1.07
0.7	62.0	.68	87.3	0.96
0.8	43.6	.47	76.6	0.84
0.9	23.0	.44	65.0	0.71
1.0	0.0	.44	52.0	0.57

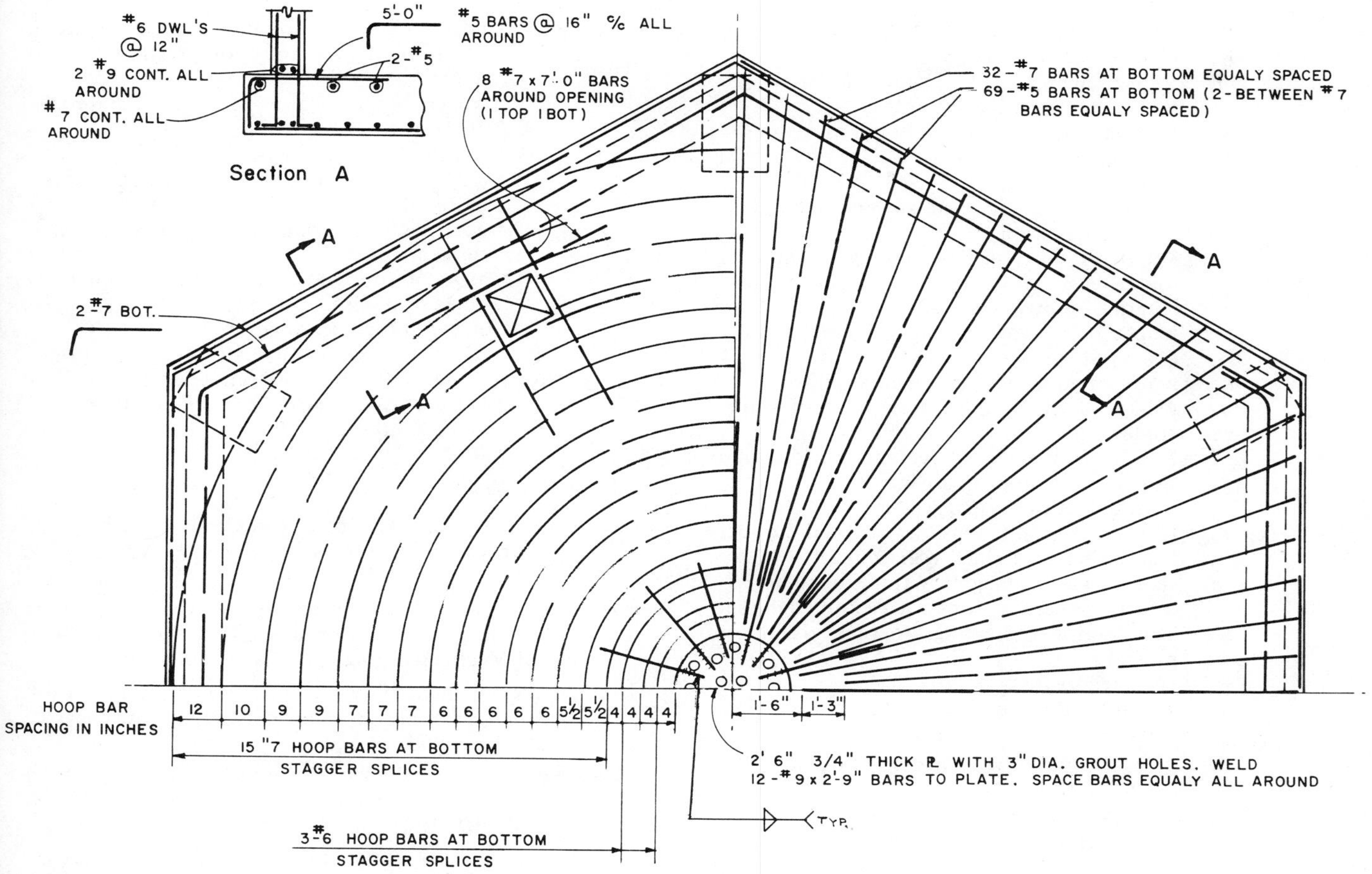

Fig. 16-47 Bottom slab reinforcing (Example 16-5).

advantage of their deep-beam action to transfer vertical loads to the six columns, the bottom slab can be designed as though supported all along the wall line. Since the discharge opening is small, tables for circular slabs may be used to obtain approximate radial and tangential bending moments. To do this, a circle of 10.9-ft radius is substituted for the actual hexagonal shape.

Bottom slab design loads, assuming a 24-in. slab, are q_{des} + 300, or 3740 lb/sq ft. Table 16-26 shows the bending moments and steel areas required (computed assuming simple support). Fig. 16-47 shows the slab reinforcement. Note that a circular/radial pattern is actually used. Around the discharge opening, added reinforcement must be provided to (at least) substitute for that which is intercepted by the hole. Dowels to suspend the slab from the wall must carry an ultimate load of:

$$F_u = 1.7\,(3.74)(374)/72 = 33.0 \text{ kips per foot of wall}$$

The required dowel area is

$$A_s = 33.0/(60 \times 0.90) = 0.61 \text{ sq in. per ft of wall}$$

The dowels must have anchorage for full development by bond, both in the slab and in the wall.

Finally, the walls are checked for bending as deep beams, and the columns are designed.

NOTATION

A = area
A_r = area of cross section of ring-beam
A_s = reinforcing steel area (usually per unit width of wall)
C, C_1, C_2, C_3 = constants, explained where used
C_d = overpressure factor, for converting static pressures to design pressures
D = diameter (inside, unless noted)
E_s, E_c = modulus of elasticity for steel and concrete, respectively
F, F_x, F_y = force, x and y components of force
F_b = allowable bending stress for structural steel
F_m = meridional membrane force per unit width of hopper wall
F_t = horizontal membrane force per unit width of hopper wall
H = height of silo (see Fig. 16-4); horizontal force
H_a = shear in column supporting ring-beam
I = moment of inertia
K = load factor (*ACI Code*); torsion factor
K_t = factor for computing temperature difference (Fig. 16-10)
L = perimeter of horizontal (inside) cross section; column height
M = bending moment, usually per unit width of wall
R = hydraulic radius of horizontal cross section of storage space
T_i = temperature inside mass of stored material
T_0 = temperature of air outside of silo wall
T_u = torque, ultimate
V = in eqs. (16-15) and (16-16), total vertical frictional force on a unit width of wall above the section in question: shear force
V_c = permissible ultimate shear by concrete alone

V_n = nominal shear strength of member
V_u = ultimate shear force
W = total weight of stored material, or weight of a hopper and its contents
W_g = that portion of W due to hopper weight
W_l = that portion of W due to weight of material stored in the hopper
Y = depth from surface of stored material to point in question
a = length of short side of rectangle; length of upper edge of inclined wall; inside wall length of polygonal silo
a' = length of side of imaginary square silo, eq. (16-4) or ring-beam
a_{eq} = width of equivalent rectangular plate
a_1 = length of bottom (shorter) edge of trapezoidal wall
b = length of long side of rectangle; width of a flexural section or ring-beam
c = slant height of hopper wall
c_a, c_b = factors defining division of total hopper load between inclined walls a and b
c_i = slant height of imaginary trapezoidal area
d = effective depth of concrete flexural member
des = a subscript indicating "design" force or pressure
e = eccentricity
f_b = actual bending stress
$f_{c,\text{vert}}$ = vertical compressive stress
f'_c = ultimate compressive strength of concrete
f_s = steel stress, tension
f'_s = effective compressive steel stress
f'_t = tensile strength of concrete
f_y = yield strength of steel
g = dead load per unit area (or a subscript indicating gravity load)
g_r = weight of ring-beam, per unit length
h = wall thickness
h_h = height of hopper
h_s = height of sloping top surface of stored material
k = ratio of p to q
l = a subscript meaning "live load"
m = shrinkage coefficient, generally 0.0003
n = modular ratio, E_s/E_c
p = horizontal static pressure due to stored material
p_{ecc} = a correction to p_{des} for effect of eccentricity of discharge
p_i = value of p in imaginary silo
q = static vertical pressure due to stored material
q_α = unit static pressure normal to surface inclined at angle α to horizontal
s_{cr} = crack spacing
st = a subscript meaning "static"
t = a subscript indicating "temperature" or "thermal"
u = a subscript meaning "ultimate"
w_1, w_2, w_3 = crack width increments, defined in Section 16.7.3
w_{cr} = crack width, total
w_u = ultimate load per unit area, normal to the wall surface in question
y = depth of compression block in USD
α = angle of inclination of hopper wall
α_i = angle between hopper wall intersection line and a horizontal plane
α_t = thermal coefficient of expansion of concrete (in./in./°F)
β_1 = ratio, depth of rectangular compression block to depth from compression face to neutral axis
γ = weight per unit volume for stored material
θ = angle of rotation, radians
$\lambda = EI_{rx}/GK$ for ring beam
μ = angle of friction between walls and stored material
μ' = coefficient of friction between stored material and wall
ρ = angle of internal friction, or (approximately) angle of repose
ϕ = strength reduction factor
ψ_1 = factor for use in crack width computation
ψ_a, ψ_b = angle between horizontal projections of hopper wall intersection line and edges "a" and "b," respectively

BIBLIOGRAPHY

16-1 Jenike, A. W., "Denting of Circular Bins with Eccentric Drawpoints," *Proceedings, ASCE*, V. 93, ST1, Feb. 1967, pp. 27–35.

16-2 Theimer, O. F., "Failures of Reinforced Concrete Grain Silos," *Publication 68-MH-36*, American Society of Mechanical Engineers, New York, 1968.

16-3 Jenike, A. W., "Flow of Bulk Solids," *Bulletin 64*, Utah Engineering Experiment Station, University of Utah, Mar. 1954.

16-4 Theimer, O. F., "Ablauf fördernde Trichterkonstructionen von Silozellen," *Aufberechmungs–Technik*, No. 10, Oct. 1969, pp. 547–556.

16-5 Kvapil, R., and Taubmann, H. J., "Flow and Extraction of Solids from Bins," *Publication 68-MH-32*, ASME, New York, 1968.

16-6 ACI Committee 313, *Recommended Practice for Design and Construction of Concrete Bins, Silos and Bunkers for Storing Granular Materials* (ACI 313-77) and *Commentary*, American Concrete Institute, Detroit, 1977.

16-7 *Soviet Concrete and Reinforced Concrete Code* (*Stroitelnie Normi i Pravila Tshast II Razdel B Glava 1 Betonnie i Zelezobetonnie Konstruktsi SNKP*), II-B.1-62, Publication of the Committee of Soviet Ministers USSR of Construction Works, Moscow, 1962.

16-8 Janssen, H. A., "Versuche Über Getreidedruck in Silozellen," *VDI Zeitschrift* (Düsseldorf), V. 39, Aug. 31, 1895, pp. 1045–1049.

16-9 Reimbert, M. and Reimbert, A., *Silos–Theory and Practice*, Trans Teach Publications, Clausthal, Germany, 1976.

16-10 Petrov, B. A., "Experimental Determination of Cement Pressure on Reinforced Concrete Silo Walls (Experimentalnoe Opredelenie Davlenia Cementa na Stenki Zhelezobetonnich)," *Cement*, No. 2, 1958, pp. 21–25.

16-11 Pieper, K., and Wenzel, F., *Druckverhältnisse in Silozellen*, Verlag von Wilhelm Ernst und Sohn, Berlin, Munich, 1968.

16-12 Deutsch, G. P., and Schmidt, L. C., "Pressures on Silo Walls," *Publication 68-MH-24,* ASME, New New York, 1968.

16-13 Takahashi, K., "Silo Design," *Concrete Journal*, V. 5, No. 9, Aug. 1969, Japanese National Council on Concrete (in Japanese).

16-14 "Instructions for Design of Silos for Granular Materials (Ukazania Po Proectirovaniu Silosov Dlia Siputschich Materialov," *Soviet Code CH-302-65*, Gosstroy, USSR, Moscow, 1965.

16-15 Pieper, K., "Investigation of Silo Loads in Measuring Models," *Publication 68-MH-30*, ASME, New York, 1968.

16-16 Pieper, K., and Wagner, K., "Der Einfluss Verschiedsher Auslaufarten aur die Saitendrucke in Silozellen," *Aufbereitungs–Technik*, No. 10, Oct. 1969, pp. 542–546.

16-17 Platonov, P. N., and Kovtun, A. P., "The Pressure of Grain on Silo Walls (Davlenie Zerna na Stenki Silosov Elevatorov)," *Mukomolno Elevatornaia Promyschlennost* (*Moscow*), V. 25, No. 12, Dec. 1959, pp. 22–24.

16-18 Leonhardt, Dr. F., "Zur Frage der Sicheren Bemessung von Zement-Silos," *Beton und Stahlbetonbau* (Berlin), V. 55, No. 3, Mar. 1960, pp. 48–58.

16-19 *Uniform Building Code*, Vol. I, International Conference of Building Officials, Whittier, Calif. 1982.

16-20 Chandrasekaran, A. R., and Jain, P. C., "Effective Live Load of Storage Materials Under Dynamic Conditions," *Indian Concrete Journal*, Sept. 1968, pp. 369–385.

16-21 "Circular Concrete Tanks without Prestressing," *Publication ST-57*, Structural Bureau, Portland Cement Association, Chicago.

16-22 Lipnitski, M. E., and Abramovitsch, S. P., *Reinforced Concrete Bunkers and Silos* (*Zhelezobetonnie Bunkera i Silosi*), Izdatelstvo Literaturi Po Stroitelstvu, Leningrad, 1967.

16-23 Bohm, F., "The Calculation of Circular Silos for the Storage of Cement," *Beton und Stahlbetonbau* (*Berlin*), V. 51, No. 2, Feb. 1956, pp. 29–36.

16-24 *ASHRAE Guide and Data Book*, American Society of Heating, Refrigerating and Air Conditioning Engineers, New York, 1963.

16-25 Boley, B. A., and Weiner, J. H., *Theory of Thermal Stresses*, John Wiley and Sons, Inc., New York, 1960.

16-26 Safarian, S. S., and Harris, E. C., "Determination of Minimum Wall Thickness and Temperature Steel in Conventionally Reinforced Circular Concrete Silos," *ACI Journal, Proceedings*, V. 67, No. 7, July 1970, pp. 539–547.

16-27 ACI Committee 318, *Building Code Requirements for Reinforced Concrete* (*ACI-318-83*), American Concrete Institute, Detroit, 1983.

16-28 Safarian, S. S., and Harris, E. C., *Design and Construction of Silos and Bunkers*, Van Nostrand Reinhold, New York, 1985.

16-29 Timoshenko, S., and Woinowsky-Krieger, S., *Theory of Plates and Shells*, 2nd Ed., McGraw-Hill, New York, 1959.

16-30 Griffel, W., *Plate Formulas*, Frederick Ungar Publishing Company, New York, 1968.

16-31 Fischer, W., *Silos und Bunker in Stahibeton*, Veb Verlag fur Bauwesen, Berlin, 1966.

16-32 Ciesielski, R., et al., *Behalter, Bunker, Silos, Schornstein, Fernsehturme und Freileitungsmaste*, Verlag von Weilhelm Ernst und Sohn, Berlin, 1970.

16-33 Ichikawa, K., Isobata, O, Mitani, K., and Sugita, M., "Measurements of Lateral Pressures in Flour Silo," The Research Laboratory of Shimuzu Construction Co., Ltd., Japan, June 1973.

16-34 Paduart, André, "Pressures at the Level of Constructions in Silos," *International Symposium on Silos*, No. 14, Wroclaw, June 1973.

16-35 Theimer, O. F., "Zur Berechung Von Mehlsilozellen mit Nasenauslauf," *Die Muehle* (magazine), Muenchen, Germany, June 1975.

16-36 von Hod, J. D., *Experimental Determination of Effective Mass of Stored Granular Material in Silos Subject to Horizontal Oscillations*, doctoral thesis, Univ. of Colorado at Denver, 1984.

17

Reinforced Concrete Masonry Construction

WALTER L. DICKEY*

17.1 INTRODUCTION

17.1.1 General

This section is to assist the engineer in the general design of modern masonry with emphasis on reinforced concrete masonry. Although masonry is historically one of man's oldest permanent construction materials, the introduction of newer engineering methods has been relatively slow, overshadowed somewhat by arbitrary and customary uses, or traditional design.

The design of reinforced masonry was developed early for the *Uniform Building Code*[17-1] because of the disastrous frequency of earthquakes. Hence there has been a relatively long service record. ACI Committee 531[17-2] has since developed an excellent similar code for design of concrete masonry, which has been adopted in many areas. The two are very similar, and design under either will be similar, with a few minor differences as in the treatment of the h/t limits. However, the provisions of this chapter will be based on UBC detail provisions because of their longer experience record. Also the first complete textbook, *Reinforced Masonry Design,*[17-2] has been published based on the UBC.

The design is currently based on linear elastic assumptions although there is considerable research to develop information for rational ultimate design that may be used as alternates.

One of the first uses of ultimate design for UBC was developed by a test program and analysis as an alternate method of design of tall walls. The program was conducted jointly by the ACI local chapter and SEAOSC. It has been accepted by many building code jurisdictions and is pending acceptance for UBC. It provides a rational ultimate capacity design, and rational deflection limits, without arbitrary h/t limits. It also demonstrated very dramatically the great capacity of masonry and concrete walls at heights far beyond the conventional h/t limitations.[17-7]

17.1.2 Reinforced versus Unreinforced

The historically poor seismic performance of traditionally constructed masonry, as well as the growing sophistication in earthquake- or tornado-resistant design, has introduced reinforcing, which forms a new and distinct type of masonry with respect to performance and resistance to loads; some reinforcing methods are clarified in the following. The historically poor performance of traditional materials has served to give masonry a negative image where earthquake resistance is concerned. Reinforced masonry, however, is a material that will perform well in earthquakes and is not subject to the same failures as previous unbraced masonry, or any unbraced material for that matter. It has performed

*Consulting Structural Engineer (CEAC, ACEC), Los Angeles, California; SEAONC, Past President; SEAOSC; ACI, Comm. 531; ASCE, Life Member; ACI/ASCE, 530; ASTM, C15, E5, E6; TMS, Past President, Code Committee; ICBO.

so well that there are indications that it has considerable damping and ductility that should be evaluated.

17.1.3 Masonry Materials

Among the materials being used today we find the following:

1. Stone masonry and clay brick masonry, which have been the materials used the longest in construction and demonstrate great permanence.
2. Concrete brick, which is used in ways similar to the older uses of (clay) brick, with certain improvements in shape that are possible in casting or fabrication.
3. Concrete block, which is a newer development and is especially well adapted for economy and efficiency of use, in both reinforced and unreinforced masonry.
4. Hollow (clay) brick, which is even newer development and is similar to concrete block.
5. Tile, another traditional masonry for which there are many reinforced techniques possible that may make it a very different and effective material. GSU (glazed structural units) provide not only a beautiful, durable finish, but excellent structural capacity.

However, masonry has been modified by the addition of a strong bonding portland cement type of mortar for the "new masonry," as well as a bonding portland cement grout. The earlier mortars were not intended to bond the masonry together but merely to keep the units apart the right distance and take up the spaces due to inequalities of size and dimension.

17.1.4 Design Considerations

Proper masonry design must include consideration of the following:

1. Detailing, a very important aspect in the proper use of masonry elements. One of the advantages of masonry unit construction is flexibility of construction shape. However, because hand work is involved, thought must be given to proper placement and modular dimension. Proper detailing requires that structural consideration and requirements be translated into proper placement of elements for field construction, that is, a combination of theory and practical installation.
2. Waterproofing, an important aspect of masonry whose details are included as an important facet of proper design.
3. Fire resistance and endurance, two very desirable features of masonry construction.
4. Seismic design considerations, which in this section are based on UBC provisions. They are included because structural engineers are increasingly aware of seismic factors. Also, it is recognized that all areas in the world are liable to some seismic activity. In the eastern part of the United States there has generally been a feeling that seismic factors were not important in structural design. However, records show that one of the largest earthquakes in the North American continent was in the central part of the United States. Therefore, seismic resistance should be considered in any area, even though traditionally and economically it has not been considered feasible to include it in the so-called nonseismic areas. When one considers the factors for Zone 1 and Zone 2, proper consideration of adequate seismic resistance will add relatively little to the cost of a structure, since about the only difference will be in the provisions of connections and in details. Seismic-resistant design is becoming more common; therefore, this factor is expanded in the design example of masonry buildings.

17.2 MASONRY ASSEMBLAGES

17.2.1 General

Masonry construction has been defined as a type of construction in which masonry units are placed by hand with mortar between them. The classification of types used in the UBC is as follows: unburned clay masonry, gypsum masonry, glass masonry, stone masonry, cavity wall masonry, hollow unit masonry, solid masonry, grouted masonry, reinforced grouted masonry, and reinforced hollow unit masonry. These have developed in different ways and are used in various manners due to their varying and specific characteristics.

The types of masonry construction are explained in more detail in the UBC.[17-1] The reader is referred to other building codes such as ACI 531[17-2] for similar applicable descriptions.

17.2.2 Unburned Clay Masonry

This is the descendant of early masonry such as that used in the early California Missions. It is an older type of masonry that has been used in many parts of the world.

17.2.3 Gypsum Masonry

Gypsum masonry is made with gypsum block or tile and gypsum mortar. Gypsum masonry is not to be used in bearing walls, or where exposed directly to the weather.

17.2.4 Glass Masonry

Glass block masonry consists of glass blocks manufactured for this purpose, which are used in nonload-bearing exterior or interior walls and in openings that otherwise might be filled with windows.

17.2.5 Stone Masonry

Stone masonry is that form of construction made with natural or cast stones laid and set in mortar.

17.2.6 Cavity Walls

Cavity wall masonry is that type of construction made with brick, structural clay tile, or concrete masonry units, or any combination of such units in which facing and backing are completely separated, except for the ties, which serve as bonding.

In these walls the facing and backing walls are generally from 3 in. to 4 in. thick, but may be thicker. There are arbitrary requirements and limits for spacing of ties, type of ties, thicknesses, and heights as limited by the UBC or ACI,[17-2] which should be referred to for detail.

These are generally designed as traditionally specified unreinforced elements, but they can be designed as reinforced masonry. They provide excellent weather-resistant walls.

17.2.7 Hollow Unit Masonry

Hollow unit masonry is that type of construction made with hollow masonry units in which units are laid in mortar. Generally, this is for unreinforced masonry construction. However, reinforcement may be embedded in the mortar joints. See "Reinforced Hollow Unit Masonry."[17-1]

17.2.8 Solid Masonry

Solid masonry consists of clay or concrete brick or solid load-bearing concrete masonry units laid with full mortar joints. There are arbitrary requirements outlined in the ACI and UBC for the thicknesses, bonds, headers, ties, and so forth. This also is a conventional type of construction that has performed satisfactorily and is subject to traditional limitations, generally as unreinforced masonry, or partially reinforced.

17.2.9 Grouted Masonry

Grouted masonry is constructed of clay brick or concrete brick units in which interior joints or areas are filled by pouring grout between the outer and inner wythes.

In the low-lift grouting method, the grout fill is placed as the wythes are installed.

In the high-lift grouting, the outer wythes are built to their full height, and grout is placed later. The end result is very nearly the same, the difference being merely that of a method of installation. Grouted construction is an effective method of incorporating reinforcing into masonry, either high-lift, low-lift, solid, hollow, or what not, and is especially applicable to earthquake areas or tornado areas.

1. The Low-Lift Grouting Method. Developed initially in California after the Long Beach earthquake, this method consists of laying up two outer solid tiers as separate wythes. One tier may be carried up 18 in. before grouting, but the other tier shall be laid and grouted in lifts not to exceed six times the width of the grouting space or a maximum height of 8 in. The rate at which the wythe should be raised is governed by the size of the units and the absorption of the units.

When work is stopped for over an hour, the horizontal joints are formed by stopping the tiers all at the same elevation and keeping the grout approximately 1 in. below the top, so that there is a keyed joint.

2. High-Lift Grouting. One of the newer methods of masonry construction, this procedure is particularly applicable to reinforced masonry. It is outlined in more specific detail in the UBC for two wythe grouted walls as follows:

(a) Tiers are laid up full height.
(b) The two tiers are tied with ties of not less than #9 wire in the form of a rectangle 4 in. wide and 2 in. less than the overall thickness of wall. One wythe or tier shall be built up not more than 18 in. higher than the other. Ties shall be spaced not more than 24 in. on center horizontally and 18 in. on center vertically for running bond, and not more than 12 in. on center vertically for stack bond.
(c) Cleanouts shall be provided by leaving out every other unit in the bottom tier of the section being poured, or by holes in the foundation. The inside shall be cleaned by a high pressure stream of air or water.
(d) The longitudinal vertical grout space shall be not less than 3 in. in width if it contains steel, not less than 2 in. if it contains no horizontal steel, and not less than the thickness required by proper clearances of reinforcing within the grout space.
(e) Vertical grout barriers or dams shall be built of solid masonry across the grout space for the entire height of the wall to control the flow of grout horizontally at not more than 30 ft apart.
(f) Grout shall be a plastic mix suitable for placing without segregation.
(g) Grouting shall be done in a continuous pour in lifts not exceeding 6 ft. It shall be consolidated at the time of placing and reconsolidated after excess moisture has been absorbed but before plasticity is lost.

Admixture to improve fluidity and to decrease water volume loss is highly desirable, especially for highly absorbent units.

17.2.10 Reinforced Grouted Masonry

This is grouted masonry with reinforcement added. The space between masonry units and reinforcement shall be not less than 1/4 in. except that 1/4-in. bars may be laid in horizontal mortar joints at least 1/2 in. thick. Steel wire reinforcement may be laid in horizontal mortar joints at least twice the thickness of the wire diameter.

17.2.11 Reinforced Hollow Unit Masonry

This type is made of hollow masonry units in which certain cells are continuously filled with grout and in which reinforcement is embedded. Only Type S and Type M mortar shall be used. It is generally assumed to be composed of concrete masonry units but may be of clay units. The hollow shapes used most, for a considerable period of time, have been the hollow two-cell units, which provide for easy placement of grout and reinforcement.

Hollow unit masonry is built to provide unobstructed vertical continuity of the cells that are to be filled with grout. The head joints shall be solidly filled with mortar for a distance in from the face of the wall or unit not less than the thickness of the longitudinal face shells. This is done to provide greater water resistance by breaking capillary action, and because the stress in the face shells is transmitted better this way. Bond shall be provided by lapping units in successive courses or by providing equivalent mechanical anchorage.

Cells to be filled shall be not less than 2 × 3 in.

Vertical reinforcement shall be held in position at the top and bottom and at intervals not exceeding 192 diameters of the reinforcement.

All cells containing reinforcement shall be solidly filled with grout. Units shall be laid up 4 ft maximum in height and filled with grout, except for high-lift grouting.

In so-called high lift grouting the wall is laid up full height with cells for grout kept clean of mortar by use of plugs, such as sponges, or by providing cleanouts. The grout may then be poured full height in walls not exceeding 8 ft (as for residences) or in lifts not exceeding 6 ft. These are puddled or vibrated upon pouring and then reconsolidated after water volume loss.

17.2.12 Veneer

Veneer is generally classified as a nonstructural material because it must not carry load other than itself; however, it is a frequently used masonry material, and the structural engineer may be required to consider it in design, particularly since the UBC requires that veneers be designed to certain criteria. In the revised *Uniform Building Code*, "Veneer" chapter, there are two types of veneer defined, "adhered veneer" and "anchored veneer." The requirements for the design are that the adhered veneer develop 50 psi in shear and that the anchored veneer attachments develop twice the weight of the veneer being attached.

The UBC also provides that "in lieu of the design required

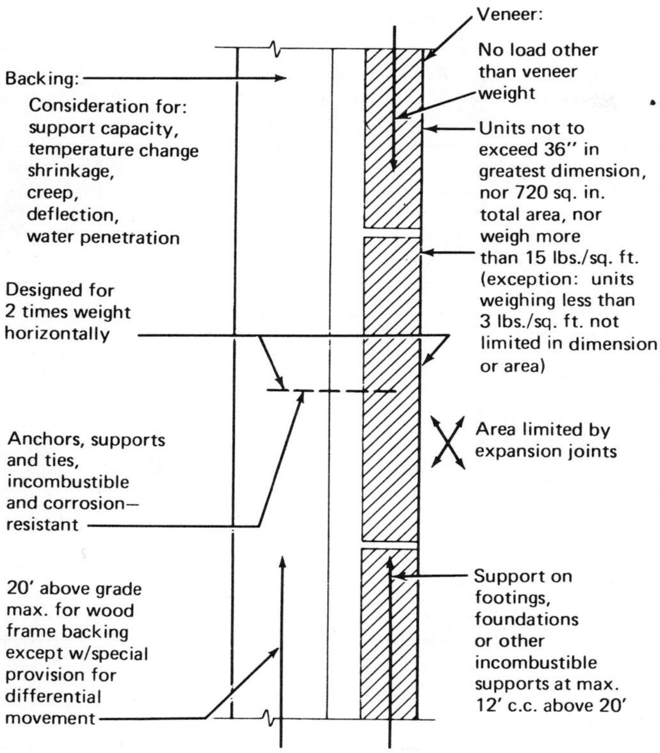

Fig. 17-1 Veneer units anchored to backing—general requirements. Reference: *Uniform Building Code*, Chap. 30, "Veneer."

by the section 3004 veneer may be applied by one of the methods specified in the *UBC Standard*." In that *Standard* there are listed many of the types of veneer application that have proved satisfactory for many years and hence would not require design. They are spelled out in adequate detail in the *Standard*. In the 1982 edition these methods have been moved from the *Standard's* volume into the Code, chapter 30. The anchored types are (see Figs. 17-1, 17-2):

1. Masonry or stone veneer (5 in. max.) on concrete or masonry
2. Masonry or stone veneer (5 in. max.) on stud construction
3. Slab type (2 in. max.) on concrete or masonry
4. Slab type (2 in. max.) on stud construction
5. Terra cotta on concrete or masonry

Other detail methods may be used if proven adequate, as by local service records.

17.3 MASONRY MATERIALS

The list of masonry units available to the engineer for masonry construction is repeated here, with certain comments pertinent to the design, to call attention to certain items that might otherwise be overlooked. Also, the appropriate ASTM number is mentioned, as well as certain physical properties of each type.

17.3.1 Brick from Clay or Shale

These are fired from clay or shale and are to be in compliance with ASTM C62 or C216. A summary of physical properties is listed in Table 17-1. There are three grades, SW, MW, and NW, that is, for severe-weathering, moderate-weathering, and negligible-weathering conditions. The physical properties and requirement classifications are based on these basic areas of use. It is to be noted that certain of the requirements of durability are waived for SW brick in areas where severe weathering may not occur, that is, where the weathering index is less than 100. In particular, the CB ratio is waived. The ratio of the cold water absorption to the boiling water absorption is a measure of the durability, and, if this is not required for freezing and thawing resistance, it is waived.

It is also noted that these units are considered solid even when they have core holes up to a volume of 25% of the total. When the volume is more than 25%, up to 40% coring, the units are known as hollow-brick and are to comply with ASTM C652, or with standards developed by the Western States Clay Products Association. If the void is greater than 40% of the units, they may be designed by one of the tile specifications, that is, C34, C126, C212.

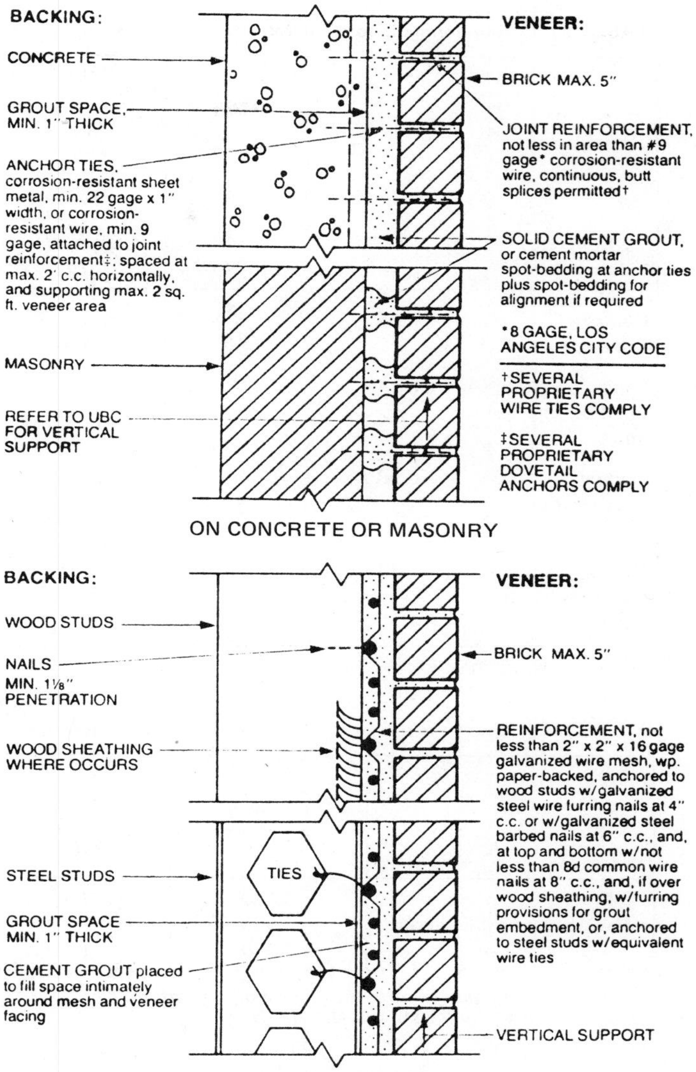

Fig. 17-2 Masonry veneer units anchored to concrete, masonry, or plaster backing. Reference: *Uniform Building Code*.

17.3.2 Concrete Block

These are the conventional hollow units or concrete block. These are as specified in ASTM "C90—Hollow Load-Bearing

TABLE 17-1 Requirements for Concrete Masonry Units

Type of Concrete Masonry Unit	ASTM Designation	Grade of Unit	Minimum Compressive Strength in psi on Average Gross Area		Maximum Water Absorption in lb/ft^3 for Units of Different Weight Classifications Based upon Oven Dry Unit Weight			
					Lightweight		Medium Weight	Normal Weight
			Average of Three Units	Individual Unit	Less than 85 lb/ft^3	Less than 105 lb/ft^3	105 to 125 lb/ft^3	125 lb/ft^3 or more
Concrete Brick	C 55	N	3500*	3000*		15	13	10
		S	2500*	2000*		18	15	13
Solid Load-bearing units	C 145	N	1800	1500	20	18	15	13
		S	1200	1000				
Hollow Load-bearing units	C 90	N	1000	800	20	18	15	13
		S	700	600				
Hollow Non-load-bearing units	C 129		350	300	NO ABSORPTION LIMIT			

Concrete Brick Tested Flatwise

Physical Requirements for Building Brick**

Grade	Compressive Strength, flat, min, psi		Water Absorption, 5-hr Boil, Max, %		Saturation* Coefficient, Max, %	
	Avg of 5	Individual	Avg of 5	Individual	Avg of 5	Individual
SW, severe weathering	3,000	2,500	17.0	20.0	0.78	0.80
MW, moderate weathering	2,500	2,200	22.0	25.0	0.88	0.90
NW, no exposure	1,500	1,250	No limit	No limit	No limit	No limit

**Ratio of 24-hr cold absorption to 5-hr boil absorption.*
***Facing Brick, Grades SW and MW similar.*

Physical Requirements for Structural Clay Tile

Type and Grade	Absorption, % (1 hr Boiling)		Compressive Strength, psi (Based on Gross Area)			
			End-Construction Tile		Side-Construction Tile	
	Avg of 5 Tests	Individual	Min Avg of 5 Tests	Individual	Min Avg of 5 Tests	Individual
Load-bearing (ASTM C34):						
LBX	16	19	1,400	1,000	700	500
LB	25	28	1,000	700	700	500
Non-load-bearing (ASTM C56),						
NB	–	28				
Floor tile (ASTM C57):						
FT1	–	25	3,200	2,250	1,600	1,100
FT2	–	25	2,000	1,400	1,200	850
Facing tile (ASTM C212):						
FTX	9	11				
FTS	16	19				
Standard	–	–	1,400	1,000	700	500
Special-duty	–	–	2,500	2,000	1,200	1,000
Glazed units (ASTM C126)	–	–	3,000	2,500	2,000	1,500

Concrete Masonry Units." For conditions of lesser stress requirements they are under "C129–Hollow Non-Load-Bearing Concrete Masonry Units." It is to be noted that the compressive stress required in C90 is 1000 or 700 psi on the gross area. However, the designer's concern is with the net, or actual strength of masonry. This testing and approval at 1000 psi on the gross area permits a manufacturer to supply a shape of minimum area at high strength or a maximum area of lower-strength material. This appears to add some confusion, but in practice it does not cause too much difficulty.

The following is an example to clarify the varying

strength relationships. A conventional 8-in. concrete block develops 1000 psi on the gross area. This is 2200 psi on net area, since they are less than one half solid. These then develop 1350 psi in masonry assemblages, even though the mortar and grout are also stronger than 2000 psi. The engineer must carefully distinguish between the strength of block units, the strength of grout, and the strength of masonry required for his design.

17.3.3 Concrete Brick

These are solid units of portland cement and appropriate aggregate and would be governed by ASTM "C55—Concrete Building Brick" or "C145—Solid Load-Bearing Concrete Masonry Units." These units are classified for use as Type I (moisture-controlled units), and Type II (non-moisture-controlled units). They come in grade N, for use as architectural veneer and facing units in exterior walls, for use where high strength and resistance to moisture penetration and severe frost are desired; and grade S, for general use where moderate strength and resistance to frost action and moisture penetration are required.

17.3.4 Hollow Clay Units

These are sometimes called clay block and generally would be specified under ASTM C652 "Hollow Brick." These requirements are similar to face brick but with 40% void permitted rather than 25% coring.

There are also hollow units of Grade LB, similar to ASTM C34 except that face shells are $1^1/_4$ in. minimum thickness. These are included in the UBC and considered in design the same as hollow concrete block units.

17.3.5 Partition Tile

Although they are generally nonbearing, they must be considered by the structural designer as to safety, stability, deflection, and anchorage.

17.3.6 Mortar

Mortar consists of portland cement and/or masonry cement, lime, and sand in proper proportions and generally specified under ASTM C270. However, the specification strength of the mortar is not as important to the designer as the strength of the masonry assemblage that will be constructed with that mortar afterward. Mortar is covered further in Section 17.4, below.

17.3.7 Grout

This consists of portland cement and aggregate to which is added enough water to make a fluid mixture. Care must be exercised to avoid excess water which might cause a segregation of the particles during pouring. This is specified according to ASTM C476 and is covered further in Section 17.4.

17.3.8 Reinforcing Bars

The reinforcing bars are as specified for reinforced concrete, that is, in ASTM A615, and so on.

17.3.9 Wire Reinforcing

This is used for joint reinforcing and is specified according to ASTM A82, that is, for the higher design stress of 30,000 psi.

17.3.10 Precast Concrete

Precast concrete panels are frequently classed and installed as masonry. These are discussed in detail elsewhere.

17.4 TESTING AND CONTROL

17.4.1 General

After the design, the next most important step is the control of the construction in the field to see that it accomplishes the goals of the design. Control or inspection is especially important in reinforced masonry because, in general, the designer has utilized the materials more efficiently than in some of the unreinforced traditional uses. If proper testing and control are not effective, the efforts of the design may be invalidated.

17.4.2 Mortar

This is one of the items that causes considerable question and confusion in reinforcement and interpretation.

ASTM C270 and many codes have certain requirements for mortar, namely, (1) proportions required, as shown by Table 17-2; and (2) development of compressive strength. Either proportion or performance may govern.

The performance, or strength, as shown in ASTM, is not applicable to the field control. The psi noted in the table

TABLE 17-2 Mortar Proportions (Parts by Volume) (UBC Table 24-A—Chapter 24)

Mortar Type	*Portland Cement*	*Hydrated Lime or Lime Putty*[1]		*Masonry Cement*	*Shovel Count*[2] *(at 7–8/ft³)*		*Parts*[2] *(ft³)*		*Damp Loose Aggregate*
		Min.	*Max.*		*$2\frac{1}{4}$ to 3*		*$2\frac{1}{4}$ to 3*		
M	1	—	$\frac{1}{4}$	—	21	28	2.81	3.75	Not less than $2\frac{1}{4}$ and not more than 3 times the sum of the volumes of the cement and lime used
	1	—	—	1	34	45	4.5	6	
S	1	$\frac{1}{4}$	$\frac{1}{2}$	—	21–25	28–34	2.8–3.4	3.7–4.5	
	$\frac{1}{2}$	—	—	1	25	34	3.4	4.5	
N	1	$\frac{1}{2}$	$1\frac{1}{4}$	—	25–38	34–51	3.4–5.1	4.5–6.7	
	—	—	—	1	17	22	2.25	3	
O	1	$1\frac{1}{4}$	$2\frac{1}{2}$	—	38–59	51–79	5.1–7.9	6.7–10.5	

[1] *When plastic or waterproof cement is used as specified in Section 2403(p), hydrated lime or putty may be added but not in excess of one-tenth of volume of cement.*

[2] *Not a part of Table 24-A, but added to aid field use.*

is for a cube (not a cylinder) of the material at a consistency that could not be used in the field. Consequently, there developed a compromise method, one recognizing that mortar should be wet to provide a good, intimate bond with the masonry units. Then while the mortar is in place, the masonry units will absorb the excess water and reduce the water/cement ratio; the end result will be much stronger.

It is emphasized that the values in the ASTM table are not the strengths of mortar that one might attain from field tests; there is no field testing provision in ASTM for mortar or for grout. Those ASTM strengths in the table are the strengths that would be developed by a laboratory mix of the proposed material, mixed to a specified flow that, incidentally, would not be used in the field. It would be too stiff for proper field use and for proper bond. Also, the values are those that will be developed by 2-in. cubes of the laboratory mix, which would be quite different from 2 in. × 4 in. cylinders that are taken as mortar samples in the field.

This change in water/cement ratio, influenced by various types of masonry units, and its influence on the strength of a wall, caused the California State Division of Architecture (now the Office of Architecture and Construction) to develop a field sampling method that is now incorporated in the *UBC Standards.*[17-4] The ASTM does not have such a method.

17.4.3 Grout

It is to be noted also that the ASTM standards for grout do not contain any requirements for determining the field strength of that grout. The premise is simply that the proposed material mixed in those proportions will provide adequately strong grout.

The then State Division of Architecture now called Office of Architecture and Construction (OAC) also developed a field grout sampling method to assure that the grout going into a project would be satisfactory rather than merely that it had the right proportions. This is also included as a *UBC standard*[17-4] and is essentially a field sampling method for obtaining grout samples that will approach a simulation of the grout as placed in a wall. The samples are cast in molds of the same material as the masonry in the wall (Fig. 17-3). The grout so sampled and tested is required by OAC to meet a compressive strength of 2000 psi. It will do more than that in the wall because of the longer period of absorption there.

17.4.4 Prism Testing

The f'_m, or ultimate compressive strengths of masonry at 28 days, upon which the design stresses are based as certain

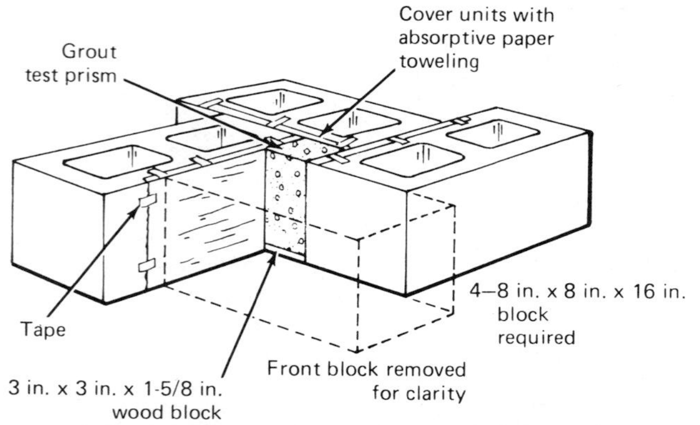

Fig. 17-3 Principal details of absorptive mold for making test prisms of grout.

fractions thereof, are normally achieved without difficulty by the conventional use of materials for f'_m as shown in Tables 17-3 and 17-4. For masonry of 2600 psi and higher strength it is necessary that prism tests be made. A prism test is the making and load-testing of an actual assemblage of mortar, grout, brick, block, and workmen that will be intended for the walls of the structure. This method has been specified in the *Uniform Building Code* for some time in Section 2404. Although the detail method of making, storing, and testing the samples is well described, the specifier or designer must make sure that the specimens are handled and transported in a proper manner; otherwise the procedure may give erroneous results.[17-1]

The sample of wall to be tested was originally specified as 16 in. long by 16 in. high; however, it was recognized that in some areas there may not be test facilities available to test a prism that large, and it now permits the length of the prism to be as short as 4 in. It is recommended, however that the prism not be made shorter than its width or thickness because that would change the orientation of the shorter dimension of the prism and might change the type of failure. This would give an erroneous, probably lower (or more conservative), value than the true one.

It must be emphasized that f'_m is the masonry assemblage strength. Certain fractions of this are used for various allowable stresses. This is not the grout strength, the brick strength, the mortar strength, or the block strength; it is generally less than the strength of those component portions. An example of this has been shown in the older version of UBC Table No. 24-H. For example, brick of 3000 psi, mortar of 2000 psi, and grout of 2000 psi were assumed to have, and generally did have, an f'_m of slightly over 1800 psi. Hollow concrete block masonry strength, or f'_m, was to be 1350 psi. This was provided by blocks of from 2,000 to 2,200 psi on the net area (the gross area strength was 1000 psi minimum), and the mortar and grout were assumed at 2000 psi. The prisms would develop two-thirds to three-fourths of the unit strength, that is, 1350 psi.

17.4.5 Detailing

If the detailing of masonry is, perhaps, more important than that of other materials, that is so partly because the units are hand-placed. Not only must the structural and aesthetic needs be fulfilled, but the practicality of placement is important.

One of the early developments was the modular size, for example, the original 2, 4, 8 in. size of the common brick. Modern units, as concrete block, are generally units for 4 in. modular dimensions; that is, a unit plus a mortar joint is a multiple of 4 in.

Examples of workable details are shown to aid detailing. (See Figs. 17-4 to 17-9.)

Another important part of detailing is the assurance of weather resistance, flashing or waterproofing, and the following recommendations are made to help complete the masonry design for its intended use.

17.5 RECOMMENDED WATERPROOFING

17.5.1 Parapets and Masonry Projections above the Roof

(a) Provide a metal cap at the top of the wall or a dense capping unit, sloped.

(b) Where cast-in-place concrete caps are specified, they shall be a minimum of 2 in. thick and shall be reinforced.

(c) The inside face of the parapet and all masonry above

TABLE 17-3 Allowable Working Stresses in Reinforced Masonry†

MAXIMUM WORKING STRESSES (pounds per square inch)
Reinforced Solid & Hollow[1] Unit Masonry

Type of Stress \ Type of Unit		Hollow Clay Units[1] Grade LB or Hollow Concrete Units Grade N		Grouted Solid, Hollow Units: Concrete, Grade N Clay, Grade LB or Solid Units 2500 psi on Gross		Solid Units 3000 psi on Gross		Special Testing[5] f'_m Established by Prism Tests								
Ultimate compressive strength	f'_m	675	1350	750	1500	900	1800	2000	2700	3000	3500	4000	4500	5000	5300	6000
Special inspection required		No	Yes	No	Yes	No	Yes	Yes	Yes	Yes	Yes	Yes	Yes	Yes	Yes	Yes
*Compression–Axial walls	$0.2\ f'_m$	135	270	150	300	180	360	400	540	600	700	800	900	1000	1060	1200
Compression–Axial, columns	$.18\ f'_m$	122	244	135	270	162	324	360	486	540	630	720	810	900	954	1080
Compression–Flexural	$.33 f'_m$	225	450	250	500	300	600	667	900*	–	–	–	–	–	–	–
Shear																
No shear reinforcement[2]	$.02\ f'_m$	15	27	15	30	15	36	40	50*	–	–	–	–	–	–	–
Reinforcement taking entire shear;																
Flexural members	$.05\ f'_m$	50	67	50	75	50	90	100	120	–	–	–	–	–	–	–
Shear walls	$.04\ f'_m$	30	54	30	60	30	72	75	–	–	–	–	–	–	–	–
Shear as revised for 1973 edition of UBC[6]																
No shear reinforcement,																
Flexural[2]	$1.1\sqrt{f'_m}$	25	40	25	43	25	47	49	50*	–	–	–	–	–	–	–
Shear walls																
$M/Vd \geqslant 1$	$.9\sqrt{f'_m}$	17	33	17	34	17	34*	–	–	–	–	–	–	–	–	–
$M/Vd = 0$	$2.0\sqrt{f'_m}$	25	50	25	50	25	50*	–	–	–	–	–	–	–	–	–
Reinforcing taking all shear,																
Flexural	$3.0\sqrt{f'_m}$	75	110	75	116	75	127	134	150*	–	–	–	–	–	–	–
Shear walls																
$M/Vd \geqslant 1$	$1.5\sqrt{f'_m}$	35	55	35	58	35	64	67	75*	–	–	–	–	–	–	–
$M/Vd = 0$	$2.0\sqrt{f'_m}$	52	73	55	77	60	85	89	104	110	118	120	–	–	–	–
Modulus of elasticity[3]	$1{,}000\ f'_m$	$.675 \times 10^6$	1.35×10^6	$.75 \times 10^6$	1.5×10^6	$.9 \times 10^6$	1.8×10^6	2.0×10^6	2.7×10^6	3.0×10^6	–	–	–	–	–	–
Modular ratio–$n = Es/Em$	$30{,}000/f'_m$	44	22	40	20	33	17	15	11	10*	–	–	–	–	–	–
Modulus of rigidity[3]	$400\ f'_m$	$.27 \times 10^6$	$.54 \times 10^6$	$.3 \times 10^6$	6×10^6	$.36 \times 10^6$	$.72 \times 10^6$	$.8 \times 10^6$	1.08×10^6	1.2×10^6	–	–	–	–	–	–
Bearing on full area[4]	$.25\ f'_m$	170	340	187	375	225	450	500	675	750	875	1000	1125	1250	1325	1500
Bearing on ⅓ or less of area[4]	$.30\ f'_m$	200	400	225	450	270	540	600	810	900	1050	1200	1350	1500	1590	1800
Bond–Plain bars		30	60	30	60	30	60	60	–	–	–	–	–	–	–	–
Bond–Deformed		100	140	100	140	100	140	140	–	–	–	–	–	–	–	–

†Table No. 24H, 1982 UBC, tabulates the allowable design stress factors, and they have been extended herein.

[1] Stresses for hollow unit masonry are based on net section.

[2] Web reinforcement shall be provided to carry the entire shear in excess of 20 psi whenever there is required negative reinforcement and for a distance of one-sixteenth the clear span beyond the point of inflection.

[3] Where determinations involve rigidity considerations in combination with other materials or where deflections are involved, the moduli of elasticity and rigidity under columns entitled "yes" for special inspection shall be used.

[4] This increase shall be permitted only when the least distance between the edges of the loaded and unloaded areas is a minimum of one-fourth of the parallel side dimension of the loaded area. The allowable bearing stress on a reasonably concentric area greater than one-third, but less than the full area, shall be interpolated between the values given.

[5] Special testing shall include preliminary tests conducted as specified in Section 2404 (c) to establish "f'_m" and at least one field test during construction of walls per each 5000 square feet of wall but not less than three such field tests for any building.

[6] For seismic stresses in shear walls, these stresses shall be halved, then increased ⅓ i.e., multiplied by ⅔.

**Maximum value permitted by UBC.*

TABLE 17-4 Allowable Working Stresses in Unreinforced Unit Masonry

	Type M	Type S	Type M or Type S Mortar				Type N		
Material	*Compression*[1]	*Compression*[1]	*Shear or Tension in Flexure*[2 3]		*Tension in Flexure*[4]		*Compression*[1]	*Shear or Tension in Flexure*[2 3]	
Special inspection required	No	No	Yes	No	Yes	No	No	Yes	No
Solid brick masonry									
4500 plus psi	250	225	20	10	40	20	200	15	7.5
2500–4500 psi	175	160	20	10	40	20	140	15	7.5
1500–2500 psi	125	115	20	10	40	20	100	15	7.5
Solid concrete unit masonry									
Grade A (N)	175	160	12	6	24	12	140	12	6
Grade B (S)	125	115	12	6	24	12	100	12	6
Grouted masonry									
4500 plus psi	350	275	25	12.5	50	25			
2500–4500 psi	275	215	25	12.5	50	25			
1500–2500 psi	225	175	25	12.5	50	25			
Hollow unit masonry[5]	170	150	12	6	24	12	140	10	5
Cavity wall masonry solid units[5]									
Grade N or 2500 psi	140	130	12	6	30	15	110	10	5
Grade S or 1500–2500 psi	100	90	12	6	30	15	80	10	5
Hollow units[5]	70	60	12	6	30	15	50	10	5
Stone masonry									
Cast stone	400	360	8	4	–	–	320	8	4
Natural stone	140	120	8	4	–	–	100	8	4
Gypsum masonry	20	20	–	–	–	–	20		
Unburned clay masonry	30	30	8	4	–	–			

[1] *Allowable axial or flexural compressive stresses in pounds per square inch gross cross-sectional area (except as noted). The allowable working stresses in bearing directly on concentrated loads may* ***be 50% greater than these values.***

[2] *This value of tension is based on tension across a bed joint, i.e., vertically in the normal masonry work.*

[3] *No tension allowed in stack bond across head joints.*

[4] *The values shown here are for tension in masonry in the direction of running bond, i.e., horizontally between supports.*

[5] *Net area in contact with mortar or net cross-sectional area.*

The above is Table No. 24-B of the 1973 edition of the UBC.

the roof shall be waterproofed by hot mopping with roofing asphalt or by covering with roofing paper or other approved waterproofing.

(d) No through-wall flashing shall be allowed (in seismic areas).

(e) Flashing and counterflashing shall be set one-half inch into the mortar joint and the portion of wall above the flashing shall be hot-mopped with roofing asphalt or other approved waterproofing.

(f) The exterior face of masonry above the roof shall be waterproofed in the same manner as the exterior wall.

(g) All joints shall be checked for tightness, and where cracks are visible, mortar shall be chipped out, tuck-pointed, and tooled.

(h) All flashing, heads, jambs, sills, inserts, and similar points shall be thoroughly caulked.

(i) Waterproofing shall be compatible with the masonry units and shall be guaranteed by the applicator and waterproofing manufacturer with full knowledge of the units used and the condition of the wall at the time of the application of the waterproofing.

(j) Waterproofing shall consist of a minimum of two coats with at least 24 hr between applications.

(k) Walls shall be waterproofed as frequently as recommended in the guarantee by the waterproofing manufacturer.

(l) The first coat of waterproofing shall be a filler coat unless otherwise specified.

17.5.2 Walls Below Grade

(a) All walls below grade shall be waterproofed on the exterior surface extending from the foundation pad to above grade.

(b) Waterproofing shall be either asphalt conforming to ASTM D449, Type A, or coal-tar pitch conforming to ASTM D450, Type B, or as otherwise approved.

(c) Surfaces to be waterproofed shall be clean, and dry, and shall be given either a priming coat of creosote oil conforming to ASTM D43 and two mop coats of hot coal-tar pitch, or a priming coat of asphalt primer conforming to ASTM D41 and two mop coats of hot asphalt. Mop coats shall be applied uniformly using not less than 20 lb of tar or asphalt per 100 sq ft per coat and shall provide a continuous impervious coating free from pinholes or other voids.

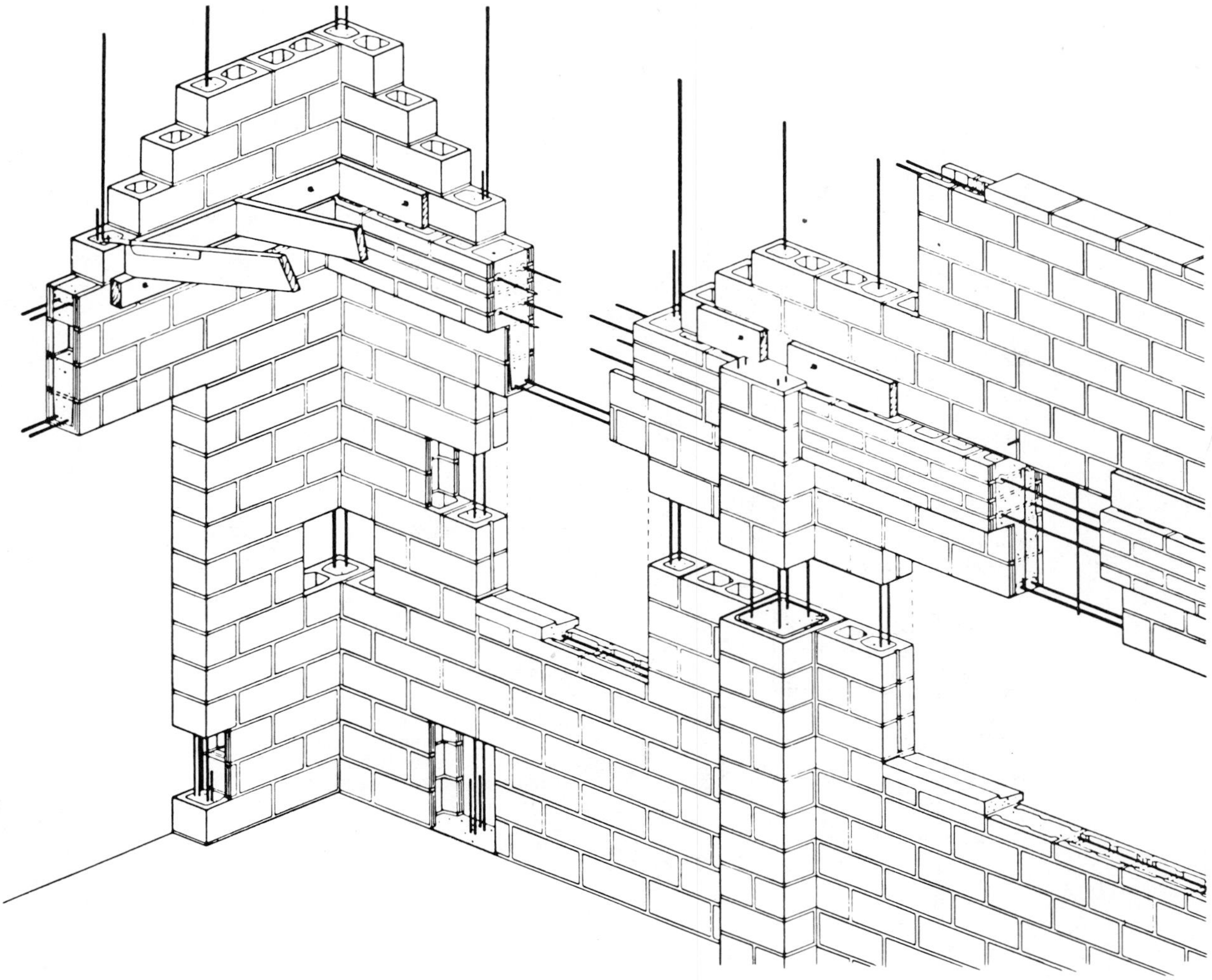

Fig. 17-4 Typical commercial construction.

(d) Where known water is present, membrane waterproofing shall be used as specifically noted on the plans.

17.5.3 Exposed Walls Above Grade

(a) All exterior masonry walls shall be waterproofed, except as approved otherwise.
(b) At the time of waterproofing the wall shall have been completed at least one month and shall be in a dry condition.

17.6 STRUCTURAL DESIGN

17.6.1 General

There are two rather different areas of masonry design, reinforced and unreinforced. Reinforced is based on principles similar to the elastic method of reinforced concrete, masonry functioning in compression and the steel in tension. Reinforced is used in areas subject to earthquakes because the steel provides for the excellent ductility and utilization of damping and energy absorption. The reinforcing changes the type of failure from a brittle type to a ductile type, desirable for earthquake resistance. There are certain arbitrary requirements for steel percentage in order to assure such performance.

Partially reinforced masonry as defined in UBC is similar, but requires only that the steel be put in the places and amounts required for performance.

These two types are also recommended for areas subject to tornadoes.

Unreinforced masonry may be considered in two manners. One is as traditional masonry, following certain arbitrary rules and limits; another is as the so-called engineered masonry in which loads and stresses are calculated based on fundamental principles, modified somewhat by correlation with test data.

These areas are discussed in the following, with primary emphasis on reinforced masonry as it gives promise of strength assurance with construction economies. A number of design aids are available.[17-3]

The most common or simple design procedure is to estimate, analyze to verify the adequacy, revise if necessary, and analyze again. This is frequently simple with reinforced masonry, since one must have a certain amount of steel because of arbitrary requirements, and often this minimum percentage is adequate.

17.6.2 Reinforced Masonry Principles

The design principles have been based on the same assumptions as the earlier development of reinforced concrete design, on the basis of elasticity assumptions.[17-3]

There has been a reluctance to accept ultimate design

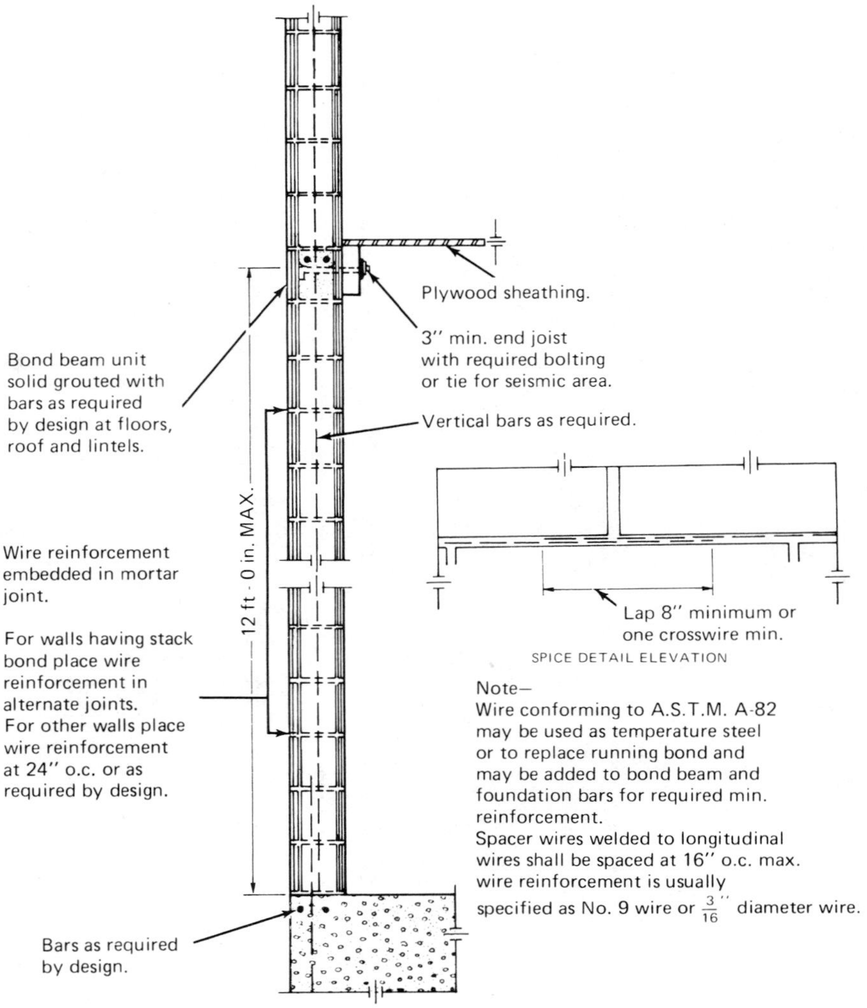

Plywood sheathing.
3" min. end joist with required bolting or tie for seismic area.
Bond beam unit solid grouted with bars as required by design at floors, roof and lintels.
Vertical bars as required.
12 ft - 0 in. MAX.
Wire reinforcement embedded in mortar joint.
For walls having stack bond place wire reinforcement in alternate joints. For other walls place wire reinforcement at 24" o.c. or as required by design.
Lap 8" minimum or one crosswire min.
SPICE DETAIL ELEVATION
Note– Wire conforming to A.S.T.M. A-82 may be used as temperature steel or to replace running bond and may be added to bond beam and foundation bars for required min. reinforcement. Spacer wires welded to longitudinal wires shall be spaced at 16" o.c. max. wire reinforcement is usually specified as No. 9 wire or 3/16" diameter wire.
Bars as required by design.

Fig. 17-5 Typical wall with wire reinforcement in mortar joints.

principles in masonry because of the lack of adequate data for proof and refinement of the methods and also because of the scatter of results that frequently occurs in masonry, partly due to individual workmanship differences and partly due to variations in the ingredient materials.[17-7]

The assumptions of elastic design used here are:

1. Plane sections before bending remain plane, and stress is directly proportioned to strain.
2. The moduli of elasticity of the masonry, mortar, and grout are constant within the member in the range of working stresses.
3. Stress in reinforcing is uniform over its area.
4. The member is straight and of uniform cross section.
5. External forces are in equilibrium.
6. In reinforced masonry the masonry carries no tensile stress.
7. For bending, the span of the member is large compared to the depth.

These elastic assumptions are adequately conservative in view of the low working stress limits imposed on masonry. Based on the above, the following equations are developed from straight-line theory for reinforced flexural members:

$$n = E_s/E_m = 30{,}000{,}000/1000 f'_m = 30{,}000/f'_m$$

$$v = \frac{V}{bjd}; \quad u = \frac{V}{\Sigma ojd}$$

$$A_v = \frac{V_s}{f_v jd}; \quad A_v = \frac{V}{f_v \sin \alpha}$$

$$p = \frac{A_s}{bd}; \quad k = \sqrt{2np - (np)^2} - np$$

$$k = \frac{1}{1 + \dfrac{f_s}{nf_m}}; \quad j = 1 - \frac{k}{3}; \quad A_s = \frac{M}{f_s jd}$$

$$f_m = \frac{M}{bd^2}\,\frac{2}{jk}; \quad M = K\,bd^2; \quad K = \frac{1}{2} f_m jk$$

17.6.2 Reinforced Bearing Walls

Bearing walls must be considered for combined stresses due to axial load and bending. Eccentricity of load application as well as stress due to lateral forces of wind and earthquake must be considered.

The empirical values of minimum wall thicknesses for various masonry materials are given in Table 17-7.

The code limitation on h/t in the so-called buckling formula is obviously an empirical type of equation. For one thing, it applies to horizontal spacing of lateral support as well as vertical spacing of lateral support. Also it makes no clear consideration of end conditions, although it is recognized in strength of materials that the end conditions in-

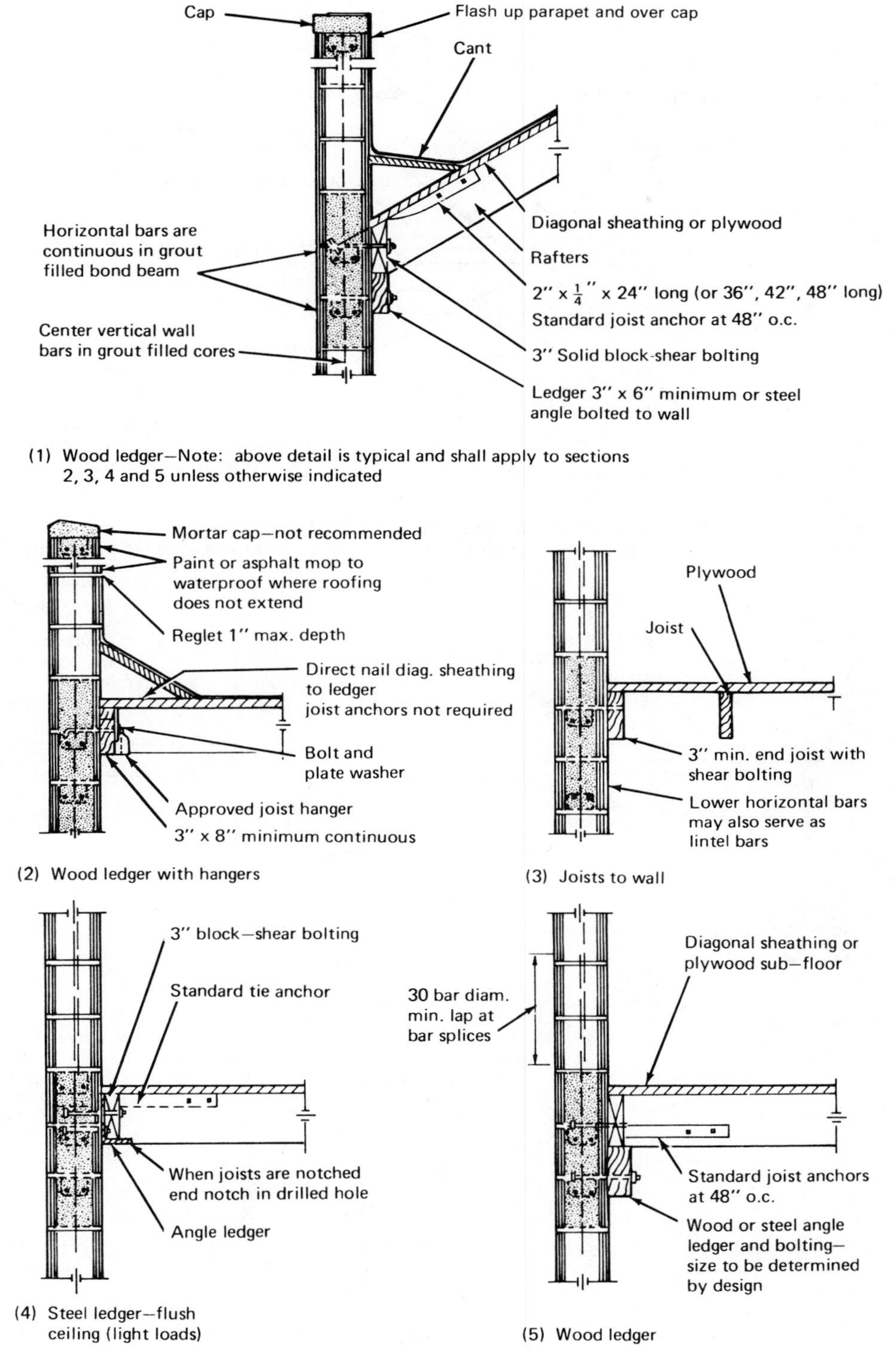

Fig. 17-6 Roof and floor details—commercial.

fluence very greatly the capacity of any wall element to resist buckling forces and lateral bending loads.

Simple practice in design as shown in Example 17-2 is to guess and then check the guess, that is, by the following conventional steps of analysis. Tables[17-3] and charts facilitate these steps.

1. $p = A_s/bd$
2. $n = E_s/E_m = E_s/1000 f'_m = 30{,}000/f'_m$
3. $k = \sqrt{2np + (np)^2} - np$
4. $j = 1 - k/3$
5. $M_s = A_s f_s jd$
6. $M_m = fm \times bxd^2/(2/kj)$

These steps are clarified and related as in Fig. 17-10, showing stresses and strains at a section subjected to simple bending.

The *Uniform Building Code* contains an exception to the empirical limits which states:

> *EXCEPTION:* The height or length to thickness ratio may be increased and the minimum thickness may be decreased when data is submitted which justi-

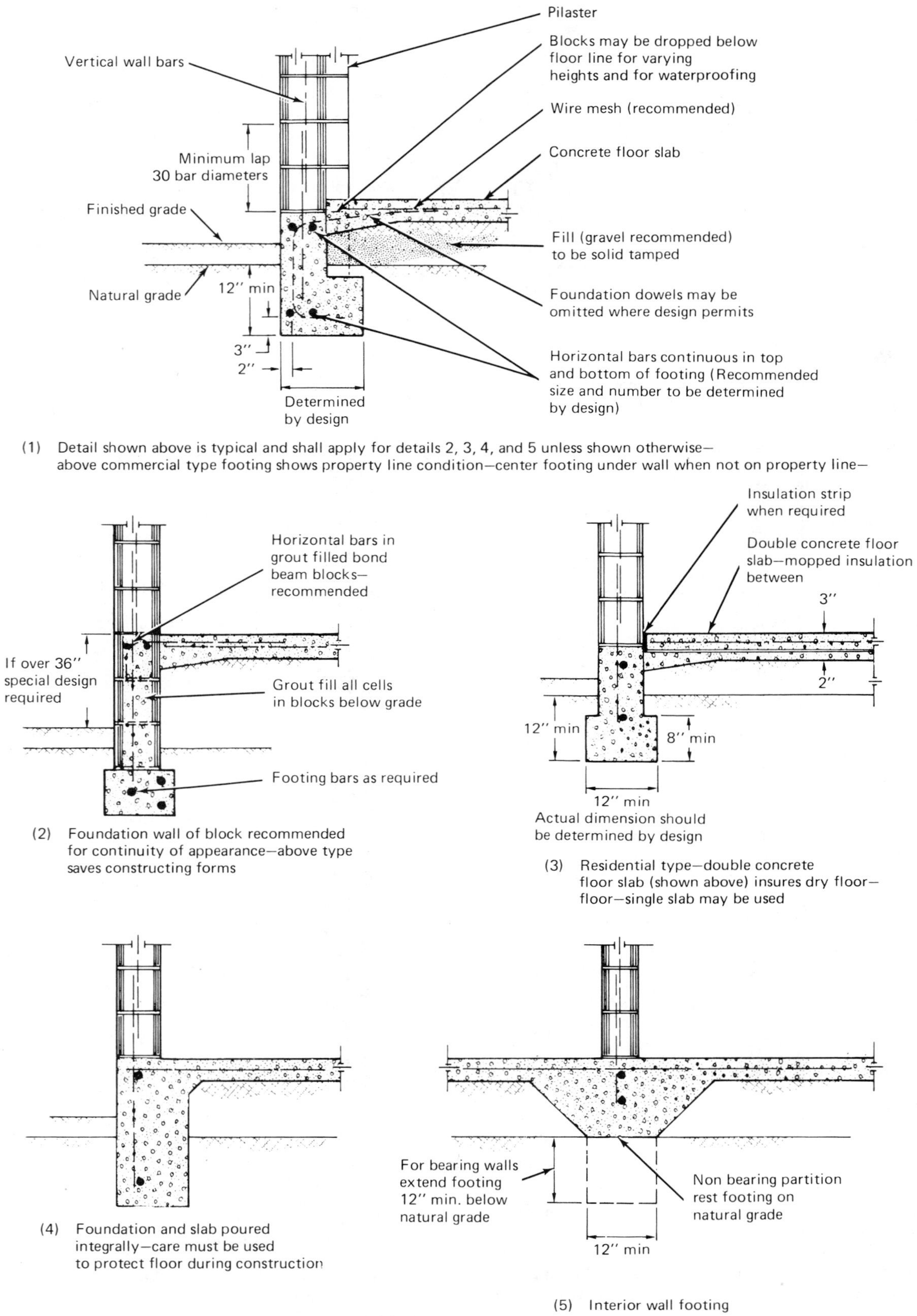
Pilaster
Blocks may be dropped below floor line for varying heights and for waterproofing
Vertical wall bars
Wire mesh (recommended)
Minimum lap 30 bar diameters
Concrete floor slab
Finished grade
Fill (gravel recommended) to be solid tamped
Natural grade
12″ min
Foundation dowels may be omitted where design permits
3″
2″
Horizontal bars continuous in top and bottom of footing (Recommended size and number to be determined by design)
Determined by design
(1) Detail shown above is typical and shall apply for details 2, 3, 4, and 5 unless shown otherwise—above commercial type footing shows property line condition—center footing under wall when not on property line—
Horizontal bars in grout filled bond beam blocks—recommended
If over 36″ special design required
Grout fill all cells in blocks below grade
Footing bars as required
(2) Foundation wall of block recommended for continuity of appearance—above type saves constructing forms
Insulation strip when required
Double concrete floor slab—mopped insulation between
3″
2″
12″ min
8″ min
12″ min
Actual dimension should be determined by design
(3) Residential type—double concrete floor slab (shown above) insures dry floor—floor—single slab may be used
(4) Foundation and slab poured integrally—care must be used to protect floor during construction
For bearing walls extend footing 12″ min. below natural grade
Non bearing partition rest footing on natural grade
12″ min
(5) Interior wall footing

Fig. 17-7 Foundation details—concrete floor.

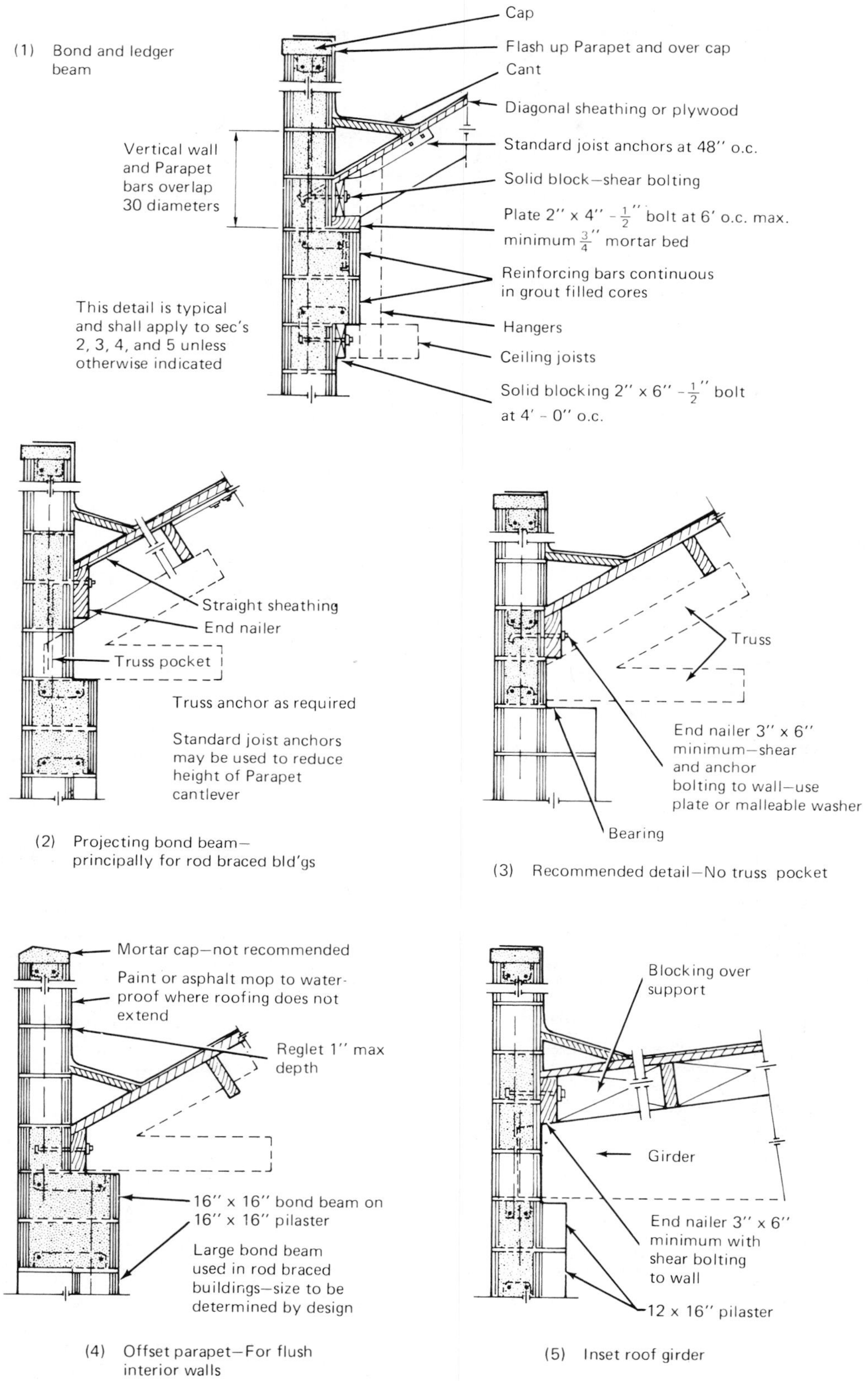
(1) Bond and ledger beam
Cap
Flash up Parapet and over cap
Cant
Diagonal sheathing or plywood
Standard joist anchors at 48″ o.c.
Solid block—shear bolting
Vertical wall and Parapet bars overlap 30 diameters
Plate 2″ x 4″ - 1/2″ bolt at 6′ o.c. max. minimum 3/4″ mortar bed
Reinforcing bars continuous in grout filled cores
This detail is typical and shall apply to sec's 2, 3, 4, and 5 unless otherwise indicated
Hangers
Ceiling joists
Solid blocking 2″ x 6″ - 1/2″ bolt at 4′ - 0″ o.c.
Straight sheathing
End nailer
Truss pocket
Truss anchor as required
Standard joist anchors may be used to reduce height of Parapet cantlever
(2) Projecting bond beam—principally for rod braced bld'gs
Truss
End nailer 3″ x 6″ minimum—shear and anchor bolting to wall—use plate or malleable washer
Bearing
(3) Recommended detail—No truss pocket
Mortar cap—not recommended
Paint or asphalt mop to waterproof where roofing does not extend
Reglet 1″ max depth
16″ x 16″ bond beam on 16″ x 16″ pilaster
Large bond beam used in rod braced buildings—size to be determined by design
(4) Offset parapet—For flush interior walls
Blocking over support
Girder
End nailer 3″ x 6″ minimum with shear bolting to wall
12 x 16″ pilaster
(5) Inset roof girder

Fig. 17-8 Roof details—commercial.

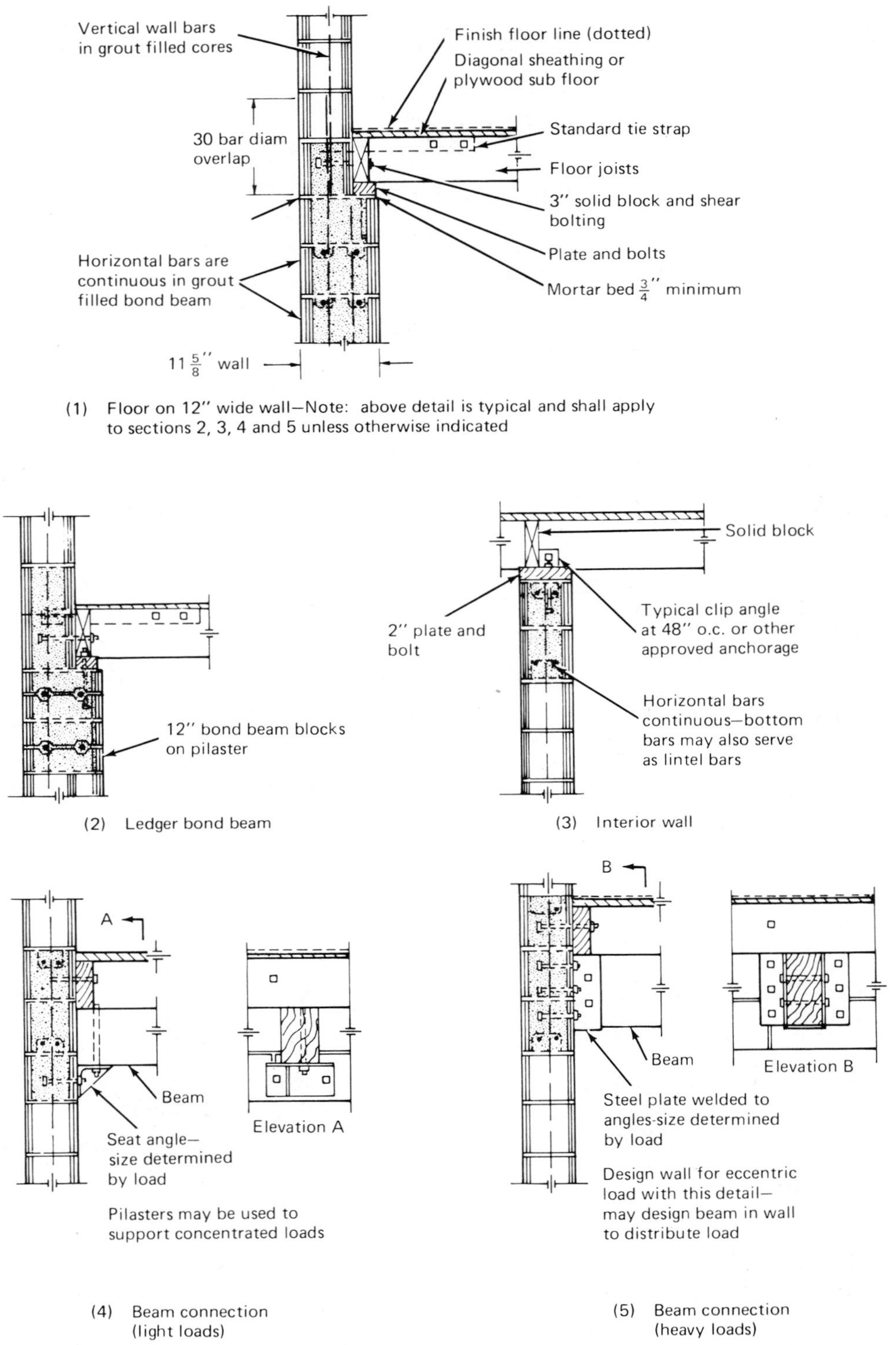

Fig. 17-9 Floor and beam details.

fies a reduction in the requirements specified in this Section.

The most common use of this exception has been the proper consideration of end restraint of walls (Fig. 17-11) as shown in theory of strength of materials and consideration of Euler's principle, just as has been done in steel and in concrete design. This consideration of end conditions requires assurance of such end restraint or that calculation of actual end rotation be made.

Tests have demonstrated that arbitrary h/t limits for engineered masonry are not valid, and codes will undoubtedly be revised. Meanwhile, careful consideration of end restraint and plate action is necessary for more effective utilization of masonry as an engineered building method.

Also, other items may govern rather than buckling and

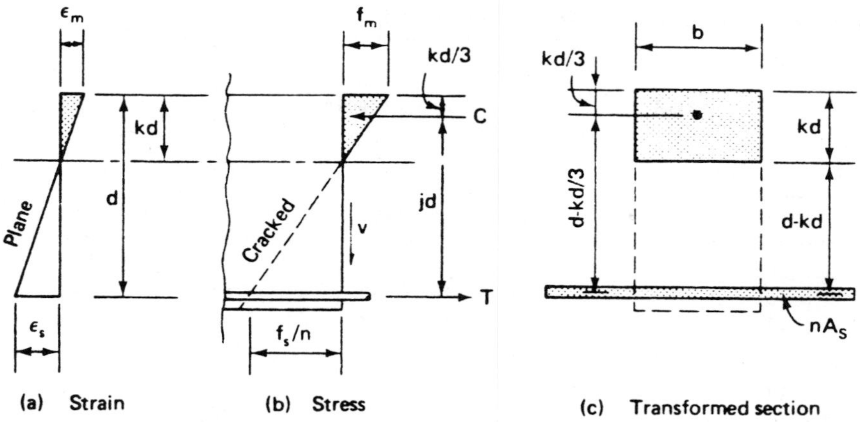

Fig. 17-10 Stress/strain patterns—cracked system.

must also be calculated; for example, wind, quakes, eccentricity, and so on, may cause excessive stresses.

Wall vertical capacity is based on the masonry material allowable capacity reduced by a reduction factor; that is:

$$f_m = 0.20 f'_m \left[1 - \left(\frac{h}{40t}\right)^3\right]$$

Both columns and walls must be designed with consideration of the lateral or bending loads which may occur simultaneously. The philosophy might be considered as providing that the fraction of strength the member has consumed in vertical load plus the proportion it has consumed in bending must not be more than 1, or fully consumed:

$$\frac{f_a}{F_a} + \frac{f_b}{F_b} \quad \text{shall not exceed 1}$$

where:

f_a = computed axial unit stress due to total axial load on effective area

F_a = permitted axial unit stress, for axial load only, including stress reduction, or increase due to combination with seismic or wind load

f_b = computed flexural unit stress

F_b = bending unit stress permitted if member were carrying bending only, including increase due to combined stresses of wind or seismic loads

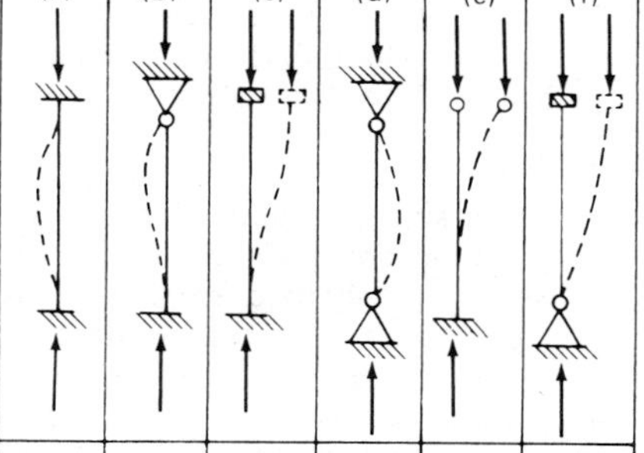

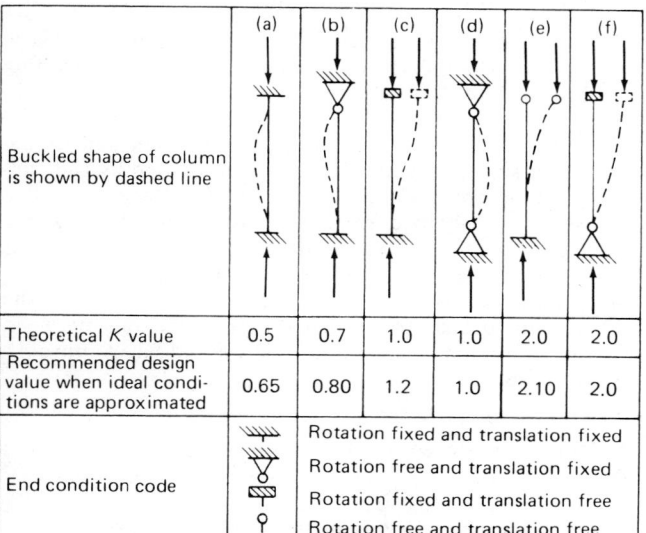

	(a)	(b)	(c)	(d)	(e)	(f)
Buckled shape of column is shown by dashed line						
Theoretical K value	0.5	0.7	1.0	1.0	2.0	2.0
Recommended design value when ideal conditions are approximated	0.65	0.80	1.2	1.0	2.10	2.0
End condition code	Rotation fixed and translation fixed; Rotation free and translation fixed; Rotation fixed and translation free; Rotation free and translation free					

Fig. 17-11 End restraint effects.

Stresses may be increased one-third for combination of vertical load with seismic or wind stresses, except that the capacity so indicated must not exceed the capacity when not so combined.

The design is, in practice, simplified by this factor. In general, if the wall is adequate for bending alone, the increase of vertical capacity due the one-third increase will keep the member satisfactory. Or, if the vertical capacity is adequate for the static loads, it will generally still be adequate for addition of the lateral loads.

Another saving factor for combined stresses is that when the compressive load on the wall is great enough that the compressive unit stress is greater than the tension produced by the bending, the section will be entirely in compression; there will be no tension. The full actual section of the wall will act and:

$$\frac{P}{A} \pm \frac{Mc}{I} = f$$

The steel at the center will not act under flexure and hence is not a design problem.

For more precise design one is referred to charts or tables to simplify the work.[17-3] These investigate the actual stress condition and provide a more correct, and generally greater, allowable capacity.

17.6.3 Reinforced Concrete Masonry Columns

The following provisions from the UBC guide the design of reinforced columns:

Every portion of a bearing wall whose length is less than three times its thickness must be designed as a column.

The least dimension of every masonry column shall be not less than twelve inches, unless designed for one-half the allowable stresses, in which case the minimum least dimensions shall be eight inches. No masonry column shall have an unsupported effective length greater than 20 times its least dimension.

The reinforcing ratio (p_g) shall be not less than 0.5% nor more than 4%. The number of bars shall be not less than four, nor the diameter less than three-eighths inch (3/8").

Where lapped splices are used, the amount of lap shall be sufficient to transfer the working stress by bond, but in no case shall the length of splice be less than 30 bar diameters, and welded splices shall be full butt welded.

Lateral ties shall be at least one-fourth inch (1/4") in di-

TABLE 17-5 Concrete Masonry Column Loads: P, in kips = $0.18f'_m + 0.65f_sA_s$; for columns 1-ft high

Column Size			Load on Steel Min = $0.005\, f_sA_g$ Max = $0.04\, f_sA_g$				Load on Masonry = $0.18\, f'_mA_g$												
			f_s = 16 ksi		f_s = 24 ksi														
nominal		A_g	Min	Max	Min	Max	675	750	900	1350	1500	1800	2000	2700	3000	35000	4000	5000	6000
	8	58	3	24	56	36	7	7.8	9.4	14.1	15.7	18.8	20.9	28.2	31.4	36.6	41.8	52.2	62.6
8	16	119	6	49	9	74	14	16	19	29	32	39	43	58	64	75	86	107	128
	24	180	9	75	14	112	22	24	29	44	48	58	65	87	97	113	130	162	194
	12	135	7	56	11	84	16	18	22	33	36	44	48	65	74	85	97	121	145
12	16	182	9	76	14	113	22	24	29	44	49	59	65	88	98	115	131	164	196
	24	275	14	114	22	171	33	37	44	67	74	89	99	134	148	173	198	247	297
	32	368	19	153	29	230	44	49	59	89	99	119	132	179	198	231	265	330	397
	16	244	12	101	19	152	29	33	39	59	66	79	88	118	131	153	175	220	263
16	24	368	19	153	29	230	44	49	59	89	99	119	132	179	198	231	265	330	397
	32	494	25	206	39	310	60	67	80	120	133	160	178	241	267	312	356	446	533
	40	620	32	264	49	388	75	83	100	151	167	201	223	301	335	390	447	558	670
	24	558	29	232	44	349	67	75	90	135	151	181	201	271	302	353	402	503	604
24	32	747	39	311	59	467	90	101	121	181	202	242	269	363	404	471	538	673	807
	40	936	48	390	74	585	113	126	151	227	253	303	337	442	505	590	673	841	1010
	48	1136	59	472	90	710	137	153	184	276	307	368	409	553	613	715	818	1022	1226

ameter and shall be spaced apart not over 16 bar diameters, 48 tie diameters, or the least dimension of the column. Lateral ties shall be placed not less than one and one-half inches (1½") and not more than five inches (5") from the surface of the column, and may be against the vertical bars, or placed in the horizontal bed joints. For seismic considerations and anchorages, additional ties must be provided.

The allowable axial load on columns shall not exceed:

$$P = A_g(0.18 f'_m + 0.65\, p_g f_s)\left[1 - \left(\frac{h}{40t}\right)^3\right]$$

where:

P = maximum concentric column axial load
A_g = the gross area of the column
f'_m = ultimate compressive masonry strength; the value of f'_m shall not exceed 6000 psi
p_g = ratio of the effective cross-sectional area of vertical reinforcement to A_g
f_s = allowable stress in reinforcement
t = least thickness of column, in.
h = clear height, in. (or effective height if calculated)

ACI provisions are slightly different.

It must be recognized that this equation for P is not compatible with the basic linear elastic assumptions; so great precision of calculation is not valid. The following example illustrates this:

Assume:

$$f'_m = 1350; \quad f_s = 24{,}000$$

Then:

$$F_m = 0.18 \times 1350 = 244 \text{ psi}$$

$$n = 30{,}000/1350 = 22$$

$$.65 f_s = 16{,}000$$

When the masonry stress is 244 psi, the steel embedded therein and adjacent will have deflected the same amount, and consequently the steel stress should then be $n \times 244$ or 5368 psi, a physically impossible phenomenon. The actual column strength stems from other factors such as ties, etc., and the column strengths are quite conservative. However, bending calculations based on elastic assumption combined with stress limits of arbitrary compressive equations must be carefully considered.

The above equation for P is easily solved by adding the basic capacity of the masonry to the capacity of the steel, that is $(0.18 f'_m A_g) + (0.65 A_s f_s)$, and applying the appropriate reduction factor depending on height, end restraint, and thickness. The thickness for use in h/t consideration is the nominal thickness, whereas the thickness for area or for stress determination is the actual or net thickness.

For solution of the above equation including exponential reduction factors, the rather simple design procedures shown in Table 17-5 may be used as follows:

1. Adjust the concentric load for the column to a basic or equivalent load for slenderness ratio 0 by dividing the load by the reduction factor, $[1 - (h/40t)^3]$, obtained from Fig. 17-12.
2. Enter Table 17-5 for the size of column and design prism strength, and select the load carried by the concrete masonry.
3. Subtract the load carried by the concrete masonry from the total load to determine the amount that must be carried by the steel reinforcement.
4. On the left side of Table 17-5 check to see that the amount of load to be carried by the steel falls between the minimum and maximum values, and select appropriate bars.

See design example following.

EXAMPLE 17-1 – Reinforced Masonry Column:
Given:

load = 150,000 lb
column = 16 × 16 in. (nominal)
height = 20 ft (= 240 in.)
material, or prism, strength, f'_m = 1500 psi
reinforcing, f'_s = 24,000 psi

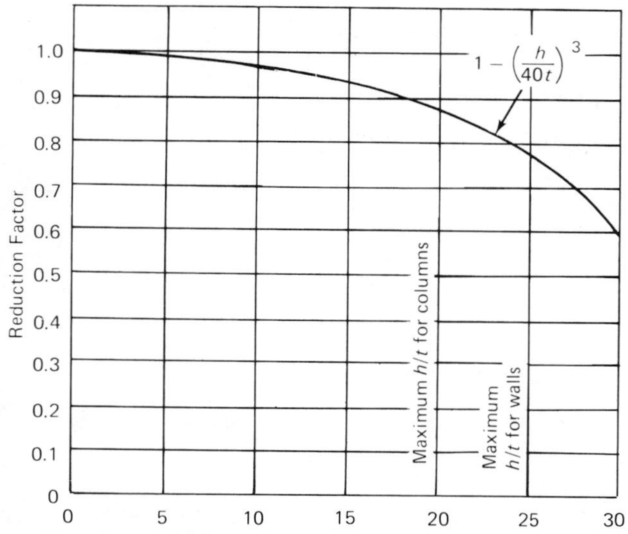

Fig. 17-12 Axial-load reduction factor curve.

SOLUTION:

$h/t = 240/16 = 15$
reduction factor = 0.94 (from curve 17-12)
equivalent or basic load = 150/(0.94) = 160 kips
masonry capacity = 66 kip (from table)

Steel must carry 160 − 66 = 94 kips. This is between the limits of maximum and minimum of 152 and 19 (from table). Select 8 #8 bars @ 99 kip (from table) to carry the 94 kips.

17.6.4 Combined Loads

Combined bending and direct stresses in reinforced masonry columns is considered most easily and conservatively by the simple unity equation as in walls:

$$\frac{f_a}{F_a} + \frac{f_b}{F_b} \text{ shall not exceed 1}$$

For conventional design the procedure is to assume the size of column and reinforcing and check compliance, revising as indicated by the analysis. For greater ease of such tedious calculations of walls and columns, large books of tables are available, such as Refs. 17-5 and 17-6.

17.6.5 Fire Endurance

Fire endurance is one of the excellent properties of masonry and will have an influence on design. For example, certain fire ratings for various occupancies require specific minimum thicknesses of material; these will then be considered in the stress analysis. Also, one must check whether the structure as designed for stress provides the thickness required by the functional design for fire resistance, sound resistance, thermal insulation, and so on.

The ratings for hollow block of various aggregate types for various thicknesses are shown in Table 17-6. The thickness is the "equivalent thickness" of hollow units, as listed in Table 43B of the UBC.[17-1]

The UBC properly allows inclusion of cement plaster, lath, and wall board and 1.5 times the thickness of gypsum as part of the required thickness (this last because the heat penetration resistance of gypsum is about twice as good as masonry of the same thickness). Grout is considered as contributing to equivalent thickness only when all cells are filled solid, not when filled intermittently. This is generally a conservative assumption.

Footnote 2 in Table 17-6 may be used to determine a veneer fire rating, by substituting the masonry thickness of 1.5 times the thickness of gypsum used in an approved installation of gypsum board or plaster.

EXAMPLE 17.2. Given the following:

8-in. block wall, nonbearing, 18 ft high, exterior,
reinforcement #5 @ 4 ft in direction of vertical span
Wind, $p = C_e C_q Q_s I = .7 \times 1.2 \times 17 \times 1 = 14.28$ psf
stresses for hollow concrete masonry construction, not continuously inspected

with h/t limits of:

25, for bearing walls
30, for exterior nonbearing
48, for interior nonbearing

Check if wall is satisfactory.

SOLUTION:

Calculation:	*Notes to designer:*
$h/t = \dfrac{18 \times 12}{8 \text{ in. (nominal)}} = 27$, OK	Less than 30
$w = 46$ psf	Assume grout @ 4ft o.c. in reinforced cells
$0.3w = 13.8$ psf	Seismic factor for Zone 4
∴ wind @ 15 psf governs	Use 15 psf, as round no.; wind = $p = C_e C_q Q_s I$
$f'_m = 675$	From Table 17-3, for no continuous inspection,
$0.2 f'_m = 135$	Bearing on wall
$0.33 f'_m = 225 \times \frac{4}{3} = 300$	Compression in flexure, increased one-third for wind stress
$n = 44$	$n = E_s/E_m = 30{,}000{,}000/1000 f'_m$
$f_s = 24{,}000 \times \frac{4}{3} = 32{,}000$	For steel with $f_y = 60{,}000$, increase of $\frac{1}{3}$ for wind stress
$A_s = \#5$ @ 4 ft = 0.077 sq. in/ft.	A_s reqd = 12 × 7.65 in. × 0.007 = 0.064 OK

$$M = \frac{15 \times 18 \times 18 \times 12}{8} = 4000 \text{ in. lb}$$

Vertical moment for wind of 15 psf, since wind governs

$$p = 0.077/12 \times 3.8 = 0.00168$$

$$np = 44 \times 0.0017 = 0.074$$

$$j = 0.894$$

Referring to np Table 17-8 (below)

$$2/kj = 7.04$$

$$f_s = \frac{4000}{0.077 \times 0.894 \times 3.8} = 15{,}700 \text{ psi} \qquad f_s = \frac{M}{A_s jd}$$

$$f_m = \frac{4000 \times 7.04}{12 \times 3.8^2} = 165 \text{ psi} \qquad f_m = \frac{M(2/kj)}{bd^2}$$

Therefore, the wall is ok for wind load imposed, since stresses are less than the maximum permitted.

EXAMPLE 17-3. Assume the above wall has vertical load of 5 kips/ft imposed from framing above. It is now a load bearing wall; thus:

Calculation: / *Notes to designer:*

$$h/t = \frac{18 \times 12}{8} = 27$$

This is greater than 25 permitted by code for simply supported bearing walls. Therefore wall thickness must be increased or the end conditions revised for more support; i.e., assume fixed at foundations or effective $h = 0.7$ clear height.

$$\frac{h'}{t} = \frac{0.7 \times 18 \times 12}{8} = 18.9$$

$$\therefore F_a = 0.89 \times 135 = 120$$

From Fig. 17-12, curve of reduction factor

$$f_a = \frac{P}{A} = \frac{5000}{12 \times 7.6} = 54.5$$

$$\frac{54.5}{120} + \frac{165}{300} = 0.45 + 55 = 1 \qquad \frac{f_a}{F_b} + \frac{f_b}{F_b} = 1$$

Wall is OK, especially since some code wording on combined stresses indicates that all allowable stress is increased one-third when combined with wind or seismic loads.

$$\frac{54.5}{135} + \frac{165}{300} = 0.40 + 55 = 0.95$$

It is also to be noted that fixing of the base will reduce the M at the midheight for combination with a reduced F_a value. The maximum M occurring as it does with P/A at the bottom will be with an F_a not reduced by h/t factor.

The above stress-check for bending could also have been determined by the method of Fig. 17-13; i.e.:

1. Enter chart @ $d = 3.8$ in.
2. Go horizontally to between $pn = 0.06$ and $pn = 0.08$.
3. Go down to M ft-lb (= 4065 ÷ 12) = 340.
4. Go left to $f_m = 165$ psi.

17.7 MULTISTORY LOAD-BEARING WALLS

17.7.1 High Rise

The high-rise load-bearing concept has led to a widely used method. Masonry bearing wall structures have been used since the time of early mankind. They are the earliest permanent construction material with many tried and true benefits and advantages, and this new scheme is merely the adaptation of the newer techniques of design to this old method of construction.

The design principles utilized here might impose certain design disciplines on the design team; however, by recognizing these limitations and taking advantage of them, a more efficient functional solution can be provided for the client. The scheme uses the high compressive bearing values of masonry for high-rise construction to aid the architect in his goal of providing the client with the most value and construction facility for the least cost. This scheme takes advantage of the sound insulation of masonry and its fireproofing qualities. Thermal inertia utilizes the heat absorption of the masonry due to its mass as an important factor in leveling off the load requirements, hence requiring less mechanical equipment and cost.

17.7.2 Structural Design

The vertical load requirements are simply those of ordinary statics as in nonseismic areas.

There are two basic facets of earthquake-resistant design. One is that of the so-called general considerations, that is, providing for symmetry of location of resisting elements, consideration of relative deflections, discontinuities, and similar factors. The other facet is the calculation of the magnitude of the assumed design forces to be imposed on the structure.

The general aspects are to provide rigidities in both directions and to avoid those arrangements that may cause trouble during an earthquake; that is, to visualize what might occur if the structure is shaken by ground motion in various directions. For example, we might consider the influence of the intersection of wings of buildings of varying

TABLE No. 17-6 Rated Fire-Resistive Periods for Various Walls and Partitions*

Material	*Construction*[1]	*Minimum Finished Thickness Face-to-Face (in.)*			
		4 hr	*3 hr*	*2 hr*	*1 hr*
Concrete masonry units[2]	expanded slag or pumice	4.7	4.0	3.2	2.1
	expanded clay or shale	5.7	4.8	3.8	2.6
	limestone, cinders or air cooled slag	5.9	5.0	4.0	2.7
	calcareous or siliceous gravel	6.2	5.3	4.2	2.8

**From UBC.*
[1] Thicknesses shown for concrete masonry units are "equivalent thicknesses" as defined in U.B.C. Standard No. 24-4. Thickness includes plaster, lath and gypsum wallboard where mentioned and grout when all cells are solidly grouted.
[2] See also Footnote No. 1. The equivalent thickness may include the thickness of portland cement plaster or 1.5 times the thickness of gypsum plaster applied in accordance with the requirements of Chapter 47 of the Code.

periods of vibration or the introduction of discontinuities and variation in pier sizes or shear walls, the stiffening effect of towers, stairs, and so forth. If potential elements of distress can be eliminated before design, the solution is much simpler.

The calculations herein are concerned primarily with the manipulation of numbers and equations in accordance with the *Uniform Building Code* requirements. There are several types of structural stystems which respond differently to earthquake motion, and accordingly, there are different code coefficients. The load-bearing type of building is a "box" system, that is, bearing and shear wall systems for the vertical loads, with diaphragms to carry the horizontal loads to those vertical shear walls. The assumptions of design are that loads are applied laterally to the structure at their centers of gravity and carried down through the system to the foundation.

The factors determining the magnitude of those forces are largely set on the basis of probability, as in Table 23H of the UBC.[17-1]

We may accept the summaries in the governing codes of earthquake design as the best judgment to follow at present. These, however, are to be considered as economic minimums, and the engineer is advised to use his judgment in designing to those numerical minimums. In some cases, he may find it advisable to increase them. In other instances, he may wish to revise a placement or location of elements to minimize the possibility of damage.

TABLE 17-7. Minimum Thickness of Masonry Walls

Type of Masonry	*Maximum Ratio—unsupported height or length to thickness*	*Nominal Minimum Thickness (in.)*
Bearing Walls:		
unburned clay masonry	10	16
stone masonry	14	16
cavity wall masonry	18	8
hollow unit masonry	18	8
solid masonry	20	8
grouted masonry	20	6
reinforced grouted masonry	25	6
reinforced hollow unit masonry	25	6
Nonbearing Walls:		
exterior unreinforced walls	20	2
exterior reinforced walls	30	2
interior partitions unreinforced	36	2
interior partitions reinforced	48	2

Most of the local code provisions, such as in major cities like Los Angeles, and the *Uniform Building Code*, are based on the Lateral Force Recommendations of the Structural Engineers Association of California, based on many years of study, exercise of judgment, and compromise.

17.7.3 Wall Design

Walls and parapets are designed to resist the lateral loads that would be due to their own inertia (seismic loads at right angles to the plane of the wall), while subjected to simultaneous vertical loads. This involves the problem of combining bending and direct stress. Superficially, this may seem to be a complicated problem; in practice, we recognize that the walls are designed with considerably more steel than the minimum, and they are subjected to considerable compressive stress that will minimize the possibility of tension. Also, they are relatively short in vertical spans for bending, say, a story height of 8 to 10 ft; therefore, a bending stress that might cause tension on one face and a cracked masonry section that would complicate the design will not generally occur due to this type of loading, especially in the lower stories.

A check should be made of the upper story walls, say, those carrying roof load only, or for an exceptionally high story, to confirm that the tension effect caused by wind or seismic load normal to the face would not overcome the compressive value contributed by the P/A, or direct stress.

Essentially, the designer will not be concerned with the forces perpendicular to the wall. His major concern will be those walls that serve as cantilever beams acting in the direction of the wall plane. This involves combined stresses of bending and compression. The *Code* stipulates that we may use the equation:

$$\frac{f_a}{F_a} + \frac{f_b}{F_b} \text{ shall not exceed 1}$$

Although it is recognized that this equation may not be correct, particularly when considering cracked sections, it is adequately on the safe side, within the range of permitted design stresses, and is safe and relatively simple to use. A more correct and precise method is, generally, not justified in view of the inaccuracy of load assumptions. For special conditions warranting closer, more precise calcula-

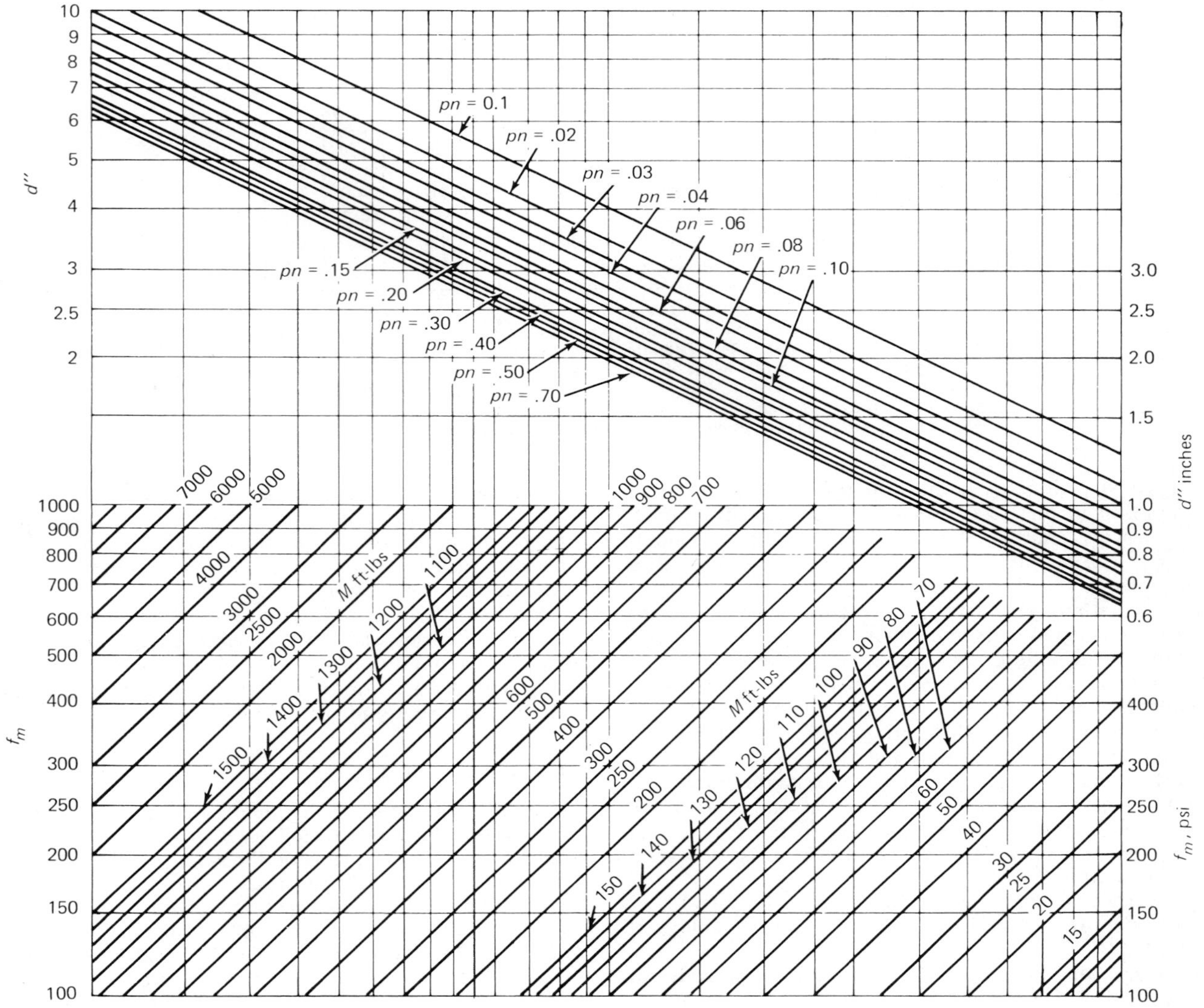

Fig. 17-13 Beam chart.
Starting from d'', go horizontally to *np* line, go vertically down to M, go horizontally to f_m #/sq in.
For d'' larger than 10" use $d''/10$ and M'#/100, then read f_m#/sq in. direct.
Check shear, axial, bond, and steel stresses separately from this chart.

tion, the designer is referred to books with charts such as Ref. 17-3.

17.7.4 Seismic Design Synopsis

The following design synopsis is an outline that one might follow through the earthquake portions of the *Uniform Building Code*. It does not include the vertical load design detail consideration, since this is similar to conventional vertical load consideration. Also, it does not cover the wind load design procedure, since that is simply a static lateral load assumption similar to the vertical load design. This synopsis is in the form of a CSI outline specification, with the "Notes to Designer" enclosed in parentheses.

SEISMIC DESIGN SYNOPSIS

Outline

Occupancy

(The type of occupancy, with the area required, will govern the type of construction and fire rating that must be used throughout the structure.)

Area

Type of construction

Required fire rating

(These will be based on the above as defined in the UBC.)

Walls

Floors

Floor load
live load
dead load

(These would be as determined by probable loads or those listed in load tables of the *Code*.)

Partition load

(Partitions are generally fixed in this type construction so the specific load as applied may be used in lieu of an average floor load which is sometimes imposed.)

Equipment load

(Based on actual loads.)

Roof load

live load

(This would be reduced according to the appropriate formula.)

dead load

(Actual loads.)

Vertical wall load

Wind load

(As determined by the site, height of building, shape, and so forth, based either on required loads or upon loads calculated by velocity determination.)

Shape factor

(This will generally be appropriate only in specific instances for special structures.)

Z or Zone factor

(The Zone factors in UBC and SEAOC are Zone 4 = 1 Zone 3 = 3/4, Zone 2 = 3/8, Zone 1 = 3/16. California is generally Zone 4, so that Z would be 1.)

K (Table 23H in the UBC)

shearwall $K = 1.33$

25% frame $K = .80$

100% frame $K = .67$

(This factor K is determined according to the scheme of framing. In the box system, considered most appropriate for masonry, the K-factor is 1.33, on the assumption that a stiff building will be subject to higher seismic forces than some other structural systems.)

$$T = \frac{0.05 h_n}{\sqrt{D}}$$

(This is the empirical determination of T, or period, based on dimension. Other type of calculation to determine the building period could be used in lieu of this simplification, and may be more correct, particularly in special structures.)

$$C = \frac{1}{15\sqrt{T}}$$

(Having determined T, coefficient C is determined, influencing the magnitude of force. Table below is provided to simplify this calculation. The C-factor for exterior bearing and nonbearing walls themselves is 0.30, or 30%g.)

$$V = ZIKCSW$$

(The above factors can now be substituted in this basic equation for the total base shear, V. This is to be distributed up the building according to the forces determined in the equations below.)

$$F_t = 0.07TV$$

(This represents the assumed portion of the shear which is to be concentrated at the top, because of the so-called whiplash effect.)

$$F_x = \frac{(V - F_t)\, w_x h_x}{\sum_{i=1}^{n} w_i h_i}$$

(This represents the assumed force to be applied to any level x and may be simply calculated by tabular form or by summation.)

$$M = F_t h_n + \sum_{i=1}^{n} F_i h_i$$

(This represents the overturning moment, at the base.)

$$M_x = \left[F_t(h_n - h_x) + \sum_{i=x}^{n} F_i(h_i - h_x)\right]$$

(Given a load distribution and magnitude, the design is simply that of structural design of the elements such as the cantilever piers carrying the lateral load.

It is to be noted that the above semidynamic technique of seismic design is changing rapidly with improvements in appropriate dynamic analysis.

One method of simplifying the problem and probably obtaining a more stable structure is to stagger the openings in one floor above the other so that the long wall would become a perforated plate rather than a series of coupled cantilevers, and, in this way, the stresses will generally be much less throughout the wall.

TABLE 17-8 Average Weight of Completed Wall

		Lightweight Aggregate			*Sand-gravel Aggregate*	
Wall Thickness	6"	8"	12"	6"	8"	12"
Solid grouted wall	56	77	118	68	92	140
Vertical cores grouted at 16" o.c.	46	60	90	58	75	111
24" o.c.	42	53	79	53	68	99
32" o.c.	40	50	73	51	65	93
40" o.c.	38	47	70	50	62	89
48" o.c.	37	46	68	49	61	87
No grout in wall	31	35	50	43	50	69

Note: The above table gives the average weights of completed walls of various thickness in pounds per square foot of wall face area. An average amount has been added into these values to include the weight of bond beams and reinforcing steel. Grout and mortar are assumed to use sand–gravel aggregates.

TABLE 7-9 Table of Flexure Coefficients

$$p = \frac{A_s}{bd} \quad n = \frac{E_s}{E_m} \quad k = \sqrt{2np + (np)^2} - np$$

$$j = 1 - \frac{k}{3} \quad f_m = \frac{M}{bd^2}\left(\frac{2}{kj}\right)$$

np	*k*	*j*	*2/kj*	*np*	*k*	*j*	*2/kj*
0.050	0.2702	0.910	8.14	0.180	0.4464	0.851	5.26
0.060	0.2914	0.903	7.60	0.185	0.4507	0.850	5.22
0.070	0.3106	0.897	7.19	0.190	0.4550	0.848	5.18
0.080	0.3279	0.891	6.85	0.195	0.4592	0.847	5.14
0.090	0.3437	0.885	6.58	0.200	0.4633	0.846	5.11
0.100	0.3583	0.881	6.34	0.220	0.4789	0.480	4.96
0.105	0.3651	0.878	6.24	0.240	0.4932	0.836	4.85
0.110	0.3718	0.876	6.14	0.260	0.5066	0.831	4.75
0.115	0.3782	0.874	6.04	0.280	0.5190	0.827	4.66
0.120	0.3844	0.872	5.97	0.300	0.5307	0.823	4.58
0.125	0.3904	0.870	5.89	0.350	0.5569	0.814	4.41
0.130	0.3962	0.868	5.82	0.400	0.5799	0.807	4.27
0.135	0.4019	0.866	5.75	0.450	0.6000	0.800	4.17
0.140	0.4074	0.864	5.68	0.500	0.6181	0.794	4.07
0.145	0.4127	0.862	5.62	0.550	0.6342	0.789	4.00
0.150	0.4179	0.860	5.56	0.600	0.6490	0.784	3.94
0.155	0.4229	0.859	5.51	0.650	0.6624	0.779	3.88
0.160	0.4278	0.857	5.46	0.700	0.6748	0.775	3.82
0.165	0.4326	0.856	5.41	0.750	0.687	0.770	3.78
0.170	0.4374	0.854	5.36	0.800	0.700	0.767	3.74
0.175	0.4419	0.853	5.31				

TABLE 17-10 Steel Areas

Factor UBC @ .002 TITLE 21 @ .003	Area of Steel Required (by UBC or Title 21) — Wall Thickness ½″	1″	2″	3″	4″	5″	6″	7″	8″	9″	10″	11″	12″
⅓ × .002	.004	.008	.016	.024	.032	.040	.048	.056	.064	.072	.080	.088	.096
⅓ × .003	.006	.012	.024	.036	.048	.060	.072	.084	.096	.108	.120	.132	.144
⅔ × .002	.008	.016	.032	.048	.064	.080	.096	.112	.128	.144	.160	.176	.192
1.0 × .002													
⅔ × .003	.012	.024	.048	.072	.096	.120	.144	.168	.192	.216	.240	.264	.288
1.0 × .003	.018	.036	.072	.108	.144	.180	.216	.252	.288	.324	.360	.396	.432

Bar			Area of Steel per Foot — Bar Spacing																
			8″	12″	16″	20″	24″	28″	32″	36″	40″	44″	48″	56″	64″	72″	80″	88″	96″
Size	*Diam.*	*Area*	0′-8″	1′-0″	1′-4″	1′-8″	2′-0″	2′-4″	2′-8″	3′-0″	3′-4″	3′-8″	4′-0″	4′-8″	5′-4″	6′-0″	6′-8″	7′-4″	8′-0″
2 - #9	.148	.0345	.052	.034	.026	.021	.017	.015	.013	.012	.010	.009	.0086	–	–	–	–	–	–
2 - #8	.162	.0412	.062	.041	.031	.025	.021	.018	.015	.014	.012	.011	.010	–	–	–	–	–	–
2 3/16	.1875	.0552	.083	.055	.041	.033	.028	.024	.021	.018	.017	.015	.014	–	–	–	–	–	–
2 1/4	.250	.098	.147	.098	.073	.059	.049	.042	.037	.033	.029	.027	.024	–	–	–	–	–	–
2 5/16	.312	.152	.229	.15	.114	.092	.076	.065	.057	.051	.046	.042	.038	–	–	–	–	–	–
#2	¼	.049	.073	.05	.036	.029	.024	.021	.018	.016	.015	.013	.012	–	–	–	–	–	–
#3	⅜	.110	.165	.11	.083	.066	.055	.047	.041	.037	.033	.030	.027	.024	.021	.018	.016	.015	.014
#4	½	.196	.293	.20	.147	.118	.098	.084	.073	.065	.059	.054	.049	.042	.037	.033	.029	.027	.024
#5	⅝	.307	.460	.31	.230	.184	.154	.132	.115	.102	.092	.084	.077	.066	.057	.051	.046	.042	.038
#6	¾	.442	.663	.44	.332	.265	.221	.189	.166	.147	.133	.120	.110	.095	.083	.074	.066	.060	.055
#7	⅞	.601	.900	.60	.450	.361	.300	.258	.226	.200	.180	.164	.150	.129	.112	.100	.090	.082	.075
#8	1.0	.786	1.180	.79	.590	.471	.392	.337	.295	.261	.236	.214	.196	.168	.147	.131	.118	.107	.098
#9	1.128	1.000	1.50	1.00	.750	.600	.500	.428	.375	.333	.300	.273	.250	.214	.187	.167	.150	.136	.125

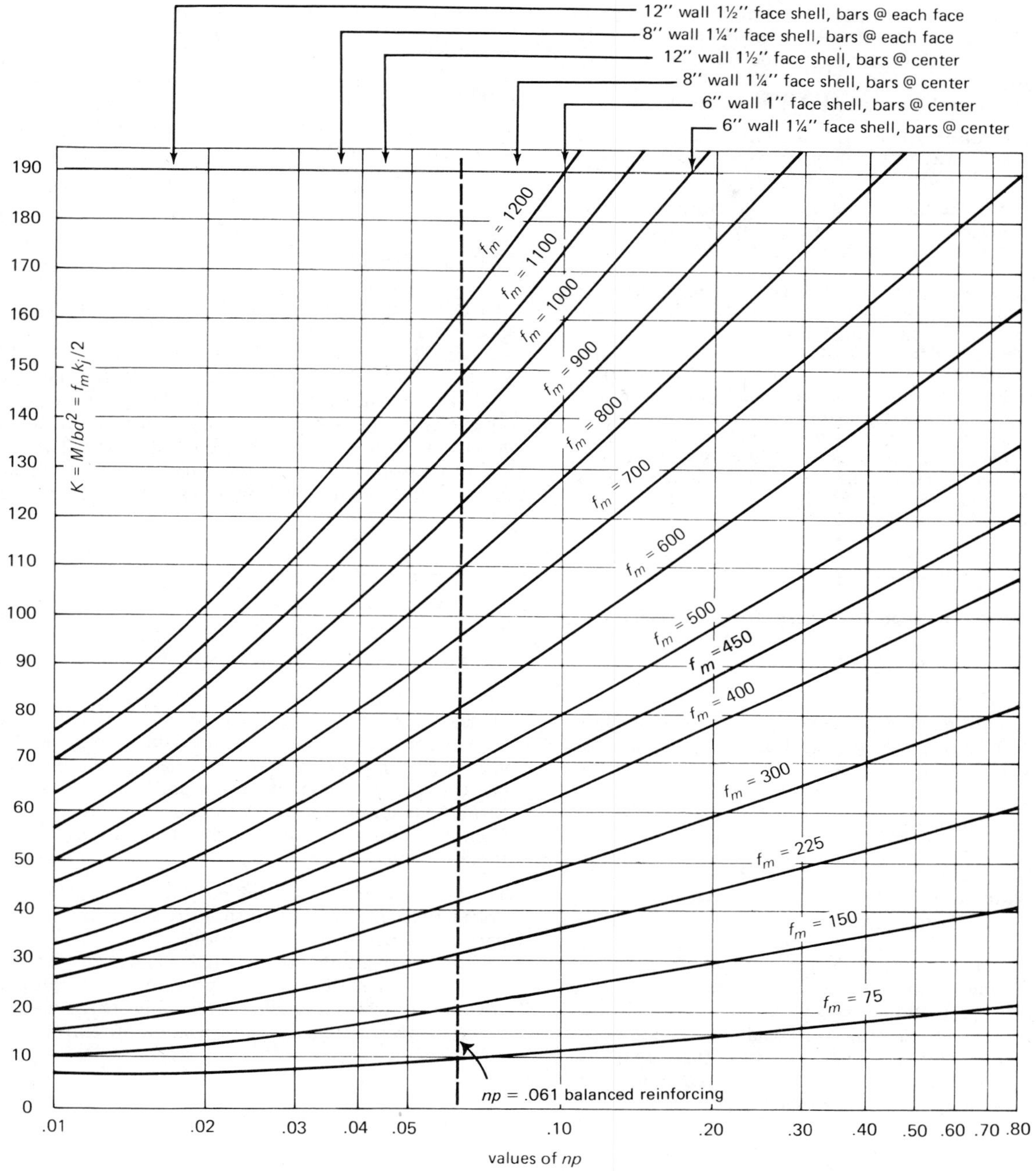

Fig. 17-14 Flexural coefficient "K." This is developed for rectangular beam sections. The limits of hollow or partially grouted walls, where the face shells would function as T-beams, are shown at the top. The area to the right of the arrow requires consideration of T-beam action. For closer design in these cases, T-beam design should be used, though the difference will not be great and is not frequently justified.

17.8 TABLES AND CHARTS

Some examples of design aids are shown here to be of help in design, but also to serve as suggestions for charts the designer may make for his own special needs. Table 17-5 for column design is used in Example 17-1, in "Structural Design" (Section 17.6). Wall weights are given in Table 17-8.

Two techniques of flexural aids are shown below. One is the "*np*" table and one is the *K* chart. This recognizes that some engineers prefer tables and some prefer curves, a personal preference. (See Table 17-9 and Fig. 17-14.)

Table 17-10 is to aid in determining the areas of steel required for masonry for various jurisdictions. The *Uniform Building Code* requires the area of steel in masonry to be not less than 0.002 times the wall area, with *not less* than *one-third* of the steel in either direction.

Title 21 of the *California Administrative Code* requires 0.003 times the wall area for total steel.

The values in the upper table are for the conditions of $^1/_3$, $^2/_3$, and *total area* as required above.

The desirable area of steel for one direction having been determined by use of the lower spacing table, this amount can be subtracted from the total required, giving the amount of steel required in the other direction. The most desirable spacing and size can then be selected.

The double wires listed are for the "ladder bar" type of reinforcement, and the area given is the area of both wires, that is, the contribution of the double wire to the total steel in that direction.

The wall thickness shown is actual, not nominal.

EXAMPLE 17-4. Given a nominal 8-in. block wall. Select ladder bar reinforcing and vertical bar reinforcing to be suitable for walls under the *Uniform Building Code.* Use Table 17-10.

SOLUTION: Enter the table of "Area of Steel Required" under 8 in. thickness and on the line for factor of $\frac{1}{3} \times 0.002$ find the value of 0.064 (or @ 7 in. = 0.056 + ½ in. = 0.004 which equals 0.060).

Provide 2 $\frac{3}{16}$ in. round wires at a spacing of 16 in., which, from the "Area of Steel per Foot Chart" is 0.041. This leaves 0.064 − 0.041 or 0.023 sq in. of steel required per foot. Provide also #5 bars horizontally at 8 ft for A_s = 0.038 in./ft, which is more than the 0.023 required and therefore is OK. (0.038 + 0.041 = 0.079)

Enter the table of "Area of Steel Required" for total; namely 1.0×0.002. For the 8-in. wall find 0.192 as total steel required (or @ 7 in. = 0.168 + ½ in. = 0.012 = 0.180). Subtracting the 0.079 provided horizontally from the total of 0.192 gives 0.113 required for the vertical steel. Entering the area table for a desirable 4-ft spacing, we find that #6 bars at 4 ft will give 0.110, which is satisfactory.

This could have been done on the basis of steel required for 7 in. plus ½ in. with a more economical end result, as shown in the parentheses.

Figure 17-14 is a convenient aid for a simple block design. Given a moment M, b, d, A_s, and type of material, one may determine M/bd^2 and f_m for the np values, or, determine the np and A_s for the allowable f_m.

17.9 SPECIFICATIONS

The following are guide specifications to aid and guide the specifier in development of a specification for a specific project. This section is written in the form of a CSI Guide Spec, i.e., with the spec statement on the left-hand side and "Notes" on the right. Care in providing the proper spec is especially important in masonry where there are certain deviations from past or traditional methods. The following is for reinforced grouted masonry; unreinforced masonry specs would be implied by omissions of items (see Ref. 17-1).

Division 4—Guide Specification, Reinforced Grouted Masonry

Guide Specifications	*Notes to Specifier*
Division 4—Masonry The General Requirements of the General Conditions, Supplementary Conditions and Division I of the specifications apply to this section.	Division 1, General Requirements is that portion of the CSI format for building specifications which contains all provisions pertaining to the job as a whole, including the Special Conditions pertinent to that particular job. In the CSI format, General Conditions and Supplementary General Conditions are not placed within the divisions of the specifications.
1. *Work Included* as part of the reinforced masonry contract. a. The work of this section shall include all labor, materials, equipment, and appliances required to complete the masonry work, including the setting and incorporating of the steel reinforcing thereof, as indicated on the drawings and as herein specified or necessary, so that other materials and work may be installed and/or performed and the whole work completed in accordance with the contract document. b. The setting and incorporating into the masonry of all bolts, anchors, metal attachments, nailing blocks, inserts, etc., as indicated on the drawings, as furnished by, and as located by, the General Contractor or others. c. The building in of all door and window frames, vents, conduits, pipes, and so forth, as furnished and set by the General Contractor or others.	1. *Work Included* and *Work not Included* are listed here to define these facets, though they are sometimes not made part of the Division Specifications. If these paragraphs are included, the designer and specifier must check that there are not portions overlapping, or not covered elsewhere and whether local jurisdiction or custom dictates specific items, e.g., furnishing of reinforcing bars.
d. (Specify here if any plain or unreinforced masonry is included, such as paving, and if furnishing and/or setting of stone, veneer, terra cotta, or similar materials is included.) e. The removal of and the repair of such sections of the masonry for inspection as specified. f. The cleaning of the masonry as necessary. g. The removal of all surplus masonry materials and waste upon completion of the masonry work.	d. These items are masonry work but are not incorporated into all projects.
2. *Work not Included* as part of the reinforced masonry contract. a. Carpenter work or concrete work of any description, except setting of precast concrete elements.	

Guide Specifications

b. The furnishing or placing of any anchorage set in concrete.

c. The furnishing or placing of reinforcing or dowel steel in concrete or in other material.

d. The furnishing or locating of any bolts or anchors for door frames, window frames, partitions, trusses, beams, girders, or other attachments.

e. The furnishing or placing of any shoring, templates, or similar carpentry work.

f. The furnishing or placing of any structural steel.

g. Water and electricity used during construction.

h. Continuous inspection or testing.

3. *Materials*

a. Cement—The cement shall conform to the specification for portland cement ASTM Designation C150 Type I or Type II, low alkali; or shall conform to the specification for masonry cement ASTM Designation C91.

b. White Cement—This shall be white portland cement.

c. Aggregate

(1) Sand for mortar

The aggregate used in mortar shall conform to the "Standard Specifications for Aggregate for Masonry Mortar, ASTM Designation: C144."

(2) Sand for GSU

The aggregate for pointing mortar of Glazed Structural Units shall be 80 mesh ground silica sand.

(3) Sand and Gravel for Masonry Grout

The aggregate for masonry grout shall conform to the "Standard Specifications for aggregate for masonry grout ASTM C404."

d. Water—Water used in mortar and grout shall be taken from a supply distributed for domestic purposes, and at the time of mixing shall be clean and free of acids, alkalies, or organic materials.

e. Quick Lime—The quick lime shall conform to the "Standard Specifications for Quick lime for Structural Purposes." ASTM Designation: C5.

f. Hydrated Lime—Hydrated lime shall conform to the "Standard Specifications for Hydrated Lime for Masonry Purposes, ASTM C207," Type S.

g. Mortar Coloring—The color shall be a standard commercial brand of chemically inert coloring material accurately measured in a definite manner for each batch of mortar to produce a consistently even (specify color) ________ color of mortar.

h. Admixture—No admixture shall be used in mortar or grout except by specific consent of the architect or engineer and the local building department.

i. Building Brick—Building brick shall comply with ASTM Designation C62.

(1) Grade: Brick shall be Grade (NW, MW, or SW).

(2) Type: Brick shall be type (FBS, FBX or FBA). (Specify type required.)

(3) Color: Shall be approved in job samples which include the range or variation in color.

j. Face Brick—Face brick shall comply with ASTM Designation C216.

(1) Grade: Brick shall be grade (MW or SW). (Specify only one grade.)

(2) Type: Brick shall be type (FBS, FBX, or FBA). (Specify type required.)

Notes to Specifier

b. The masonary contractor may choose to furnish veneer tie inserts so that he may be assured the ties and inserts are compatible.

a. The low-alkali requirement is included in order to reduce tendency for efflorescence. It has less soluble content to dissolve and contribute to the material that will be transported to the surface for efflorescence deposit.

Air entraining portland cement complying with ASTM C175 may be used satisfactorily but is not listed here because it is not customarily used. Only portland cement is permitted in UBC for reinforced masonry.

This is generally modified to use the maximum limit of the fines to give a more workable mortar.

g. The order of placing into the batch may affect the shade of some colors, as will climatic conditions and water content.

The best are mineral oxides, either natural or synthetic.

(1) The specifier shall check which of the three is to be used.

(2) The specifier should check the local market to determine the availability and proper name and size, actual or nominal, so he may specify correctly.

(3) The color will be normally terra cotta red unless otherwise specified and approved.

(2) FBS, or Face Brick Standard, has a wider color range and greater variation of size, and may be more economical. FBX, Face Brick Excellent, has a narrow color

Guide Specifications | *Notes to Specifier*

range, minimum variation of size, and high degree of mechanical perfection for high-quality face brick work. FBA, Face Brick Architectural, has a nonconformity of size, color, and texture for architectural blends, clinkers, and other specialties.

When a type is not specified, FBS will be supplied.

(3) Size: Face brick shall be ________________ (specify either nominal or actual size as supplied by the manufacturer) and as indicated on the drawings.

(3) The specifier must check the local market to determines sizes commercially available and specify accordingly.

(4) Colors and Textures: (Must be specified as by one of the options below.)

Same as approved wall sample constructed at job site.

Equal to that made by a certain manufacturer ________

Similar to that used on ____________ buildings as made by ______________

(4) The construction of a wall sample is a recommended procedure. It can serve as a sample for joints, bonding, workmanship, tolerance of units.

The specific brick desired should be named.

The reference to a building or part of a building is similar to the use of the sample panel.

k. Glazed Structural Units—Where the term "Glazed Structural Units" is used in these specifications or indicated on the drawings, it shall mean a GSU made of burned fireclay body with a ceramic glazed surface. Quality and physical requirements shall be in accordance with ASTM Designation C126.

k. The subject Glazed Structural Units was made subject of a Limited Scope Study by CSI.

(1) Grade: GSU shall be Grade S.

(1) Only in rare instances will Grade G, or ground joints, be necessary, since the manufacture of these units is such that Grade S is generally satisfactory for even the highest-quality work.

(2) Type: GSU shall be Type I.

(3) Size: GSU shall be nominal ____ X ____ X ____ inches.

(4) Glazed: GSU shall be ceramic glazed color as selected from Standard Palette.

l. Load bearing Tile—Tile shall be in accordance with ASTM C34. Type ________ and Grade ________.

l. These are not used customarily to a great extent in seismic areas.

m. Hollow brick shall have face shells of fired clay or shale in accordance with (MW, SW) grade brick as described in C652.

Color and texture shall be according to approved sample in the architect's office (or other designated location).

m. Samples shall consist of five specimens showing the range of color, size, and texture.

n. Hollow Concrete Units—shall comply with (ASTM C90, if load bearing) Grade (N-I) (N-II) (S-I) (S-II)

(ASTM C129, if non-load-bearing) (Type I if moisture-controlled) (Type II, if non-moisture-controlled).

n. The specifier shall select the appropriate grade desired, i.e., N for unprotected exposure, S for protected, and whether I, Moisture-controlled, or II, non-moisture-controlled.

o. Concrete Brick—shall comply with (ASTM C145) (ASTM C55) (Grade N-I) (N-II) (S-I) (S-II)

o. C55 is generally used because a stronger unit is required for compliance. The specifier shall check which is appropriate.

p. All concrete masonry units shall have a maximum linear shrinkage of not less than (   ) when tested in accordance with (   ).

p. This should be selected according to the custom of the area, e.g., 0.06 of 1% from saturated to oven-dry condition when tested in accordance with the quality control standards of the California Concrete Masonry Association.

q. Reinforcing Steel

(1) All reinforcing steel bars conform to ASTM A615.

(2) Wire shall conform to ASTM A82.

(3) Ladder bar prefabricated reinforcing shall be manufactured of wire conforming to ASTM A82.

(4) Detailing, fabrication and furnishing shall be in accordance with "Recommended Practice for Supplying Masonry Steel" of the Western Reinforcing Steel Institute.

(4) The industry has outlined methods and details in a "Recommended Practice" which should be included for reference.

r. Storage of Materials

(1) All mortar materials shall be stored under cover in a dry place.

(2) Units shall be piled off the ground in a dry location. During freezing weather all masonry units shall be protected with tarpaulins or other suitable materials.

(3) No damaged or contaminated material shall be used in this work.

Guide Specifications

4. *Mortar*

a. Proportions—Mortar shall be type M or S. The ingredients shall be measured accurately.

When partial batches are mixed, care shall be used in proportioning all the ingredients to maintain the same ratios.

b. Mixing—Mortar shall be mixed in a mechanically operated mortar mixer for at least three minutes after all ingredients are in the drum and at least long enough to make a thorough, complete intimate mix of the materials.

c. Tempering—The consistency of mortar shall be adjusted to the satisfaction of the mason, and water shall be added as is necessary or convenient in using the mortar. This should be done by forming a basin in the mortar, adding water and mixing it in—not by splashing water over the surface.

Mortar in which a final set has begun so that it has become harsh shall not be used.

5. *Grout (Low-Lift)*

a. Proportions

(1) Grout for wall spaces 2 in. or less shall be a mixture of one part cement, three parts sand, with water added to produce a consistency for pouring without separation of the materials (not more than one-tenth part of lime may be added).

(2) Grout for wall spaces exceeding 2 in. may be a mixture of one part cement and three parts masonry sand and two parts $^3/_8$ in. maximum pea gravel (not more than 5% of the pea gravel shall pass a number 8 sieve). As much water as possible shall be added to produce a consistency for pouring without segregation of ingredients.

(3) Grout for a wall space exceeding 4 in. shall be a mixture of one part cement and four to four and a half parts combined aggregate conforming to ASTM C404.

b. Mixing—All ingredients shall be measured according to the specified proportions for the batch and mixed in a mechanically operated batch mixer. The grout shall be mixed for a period of at least three minutes after all ingredients for the batch are in the drum. The drum must be completely emptied before the succeeding batch of materials is placed therein.

c. Tempering—The consistency of grout shall be adjusted so it will flow immediately into place without segregation of ingredients. Water may be added to compensate for loss, but grout that has begun final set and become harsh shall not be used.

6. *Grout (High-Lift)*

a. Proportions—The mix shall be one part portland cement, three parts sand, and not more than two parts pea gravel, with approved admix; or may be one part portland cement, two and one half parts sand with not more than one and one half parts pea gravel, measured by volume, damp loose.

b. Mixing—Grout shall be mixed for at least three minutes after all ingredients for the batch are in the drum. In any event, it shall be mixed long enough for complete

Notes to Specifier

If the project is for public works (or some other agency requiring it) the measuring box must be specified. For most work, however, that extra time (and cost) is to be avoided by use of shovel count. The Inspector may calibrate a shovel count initially by using a box.

b. Some specifications limit the time that mortar may be used after mixing. This is not valid because it makes no consideration of the temperature or dryness of the day. On a cold day mortar can be used for a long time with actually an increase in strength on remixing, whereas on an extremely hot, dry day the time limit set by some specifications may not be short enough. Actually, tests have shown that mortar and grout will increase in strength with continued mixing.

c. This factor is influenced greatly by temperature, humidity, etc., as mentioned above.

This factor is the governing one for length of time.

(1) Lime is not recommended, but may be permitted.

(2) This will generally result in a slump of from 8 to 10 in.

c. The time of set is dependent upon temperature and humidity. Also, if grout is kept continually agitated, as in the mixer, it may be used for relatively long periods, even increasing in ultimate strength.

a. An approved admix, such as Suconem Grout Aid, must be used for brick with a high initial rate of absorption in order to overcome the factors of excessive grout loss. Such admix will generally improve the quality of the grouted spaces in hollow unit masonry also even of lower absorption, but must be the proper type.

Guide Specifications

intimate mixing. The drum must be completely empty before the succeeding batch of material is placed therein.

Transit-mixed grout shall be used whenever possible. Transit-mix shall be continually rotated from the time the water is added until the grout is placed.

c. Admix—When the leaner mix is used, and when the brick are high absorption, an admix must be used to reduce water requirement and reduce volume loss.

7. *Workmanship*

a. Foundations—The foundation on which the wall is to be built shall have a clean, level surface prior to start of work by the masonry contractor. Sandblasting may be necessary if the surface contains laitance or other foreign material lodged in the pores or on the surface.

The foundation elevation shall be such that the bed joint is not to vary more than from 1/4″ to 3/4″. The foundation edge shall be true to line so that the masonry does not project over more than 1/4″.

Projecting dowels shall be clean of loose scale, dirt, concrete or other material that will inhibit bond. Dowels shall be in proper location prior to start of work by masonry contractor.

b. Wetting of Brick—All clay brick shall be wetted until they have an initial absorption rate not exceeding 0.25 ounces per square inch per minute determined in accordance with ASTM C67. When being laid, the brick shall have suction sufficient to hold the mortar and to delete the excess water from the mortar and grout. The brick shall be sufficiently damp so that the mortar will remain plastic enough to permit the brick to be leveled and plumbed immediately after being laid, without destroying bond.

c. Concrete brick and block shall not be wet, except in extremely hot dry weather, and the only by light spray.

d. Laying

(1) All masonry shall be plumb, level and true to line, and all corners and angles shall be square unless otherwise indicated on the drawings.

(2) Line blocks shall be used whenever possible. When it is absolutely necessary to use a line pin, the hole in the joint shall be filled with mortar immediately when the pin is withdrawn.

(3) All units shall be clean and free of dust, dirt, or other foreign materials before laying.

(4) All pattern work, bonds, or special details indicated on the drawings shall be accurately executed. Bonding headers shall not be used in grouted masonry.

(5) Mortar for all bed joints shall be beveled. Beveled bed joints shall be sloped toward the center of the wall in such a manner that the bed joint will be filled when the unit is finally brought to line. Furrowing of bed joints will not be permitted. Fins of bed joint mortar that protrude into the grout space are to be avoided. If they occur, they shall be left in palce, if not projecting more than the bed joint thickness. In no case shall they be cut off and dropped onto the grout below.

(6) All head joints of brick work, regardless of thickness, shall be completely filled with mortar or grout.

Head joints of hollow units shall be full for the thickness of the face shells.

(7) All units in stretcher courses shall be shoved into place.

(8) Both outer wythes of a brick wall shall be laid to a line, Units that are moved or shifted shall be relaid in fresh mortar.

Notes to Specifier

c. Suconem Grout Aid is one admix approved for this use as it reduces the excess water and provides for expansion to offset the volume loss.

a. The foundation may be brought to proper elevation by chipping, by pouring additional concrete, or, possibly, the unit may be cut to fit.

Dowels may be sloped not more than 1:6 where bending is necessary. Dowels improperly located shall be removed, and replaced with approved inserts.

b. This rate may be checked approximately, by drawing a circle with a wax pencil around a quarter, dropping 20 drops of water therein and noting the time for the water to disappear. This should take approximately 90 sec. If the time is much less the brick should be wetted to reduce its initial rate of absorption.

c. Wetting will cause expansion with subsequent shrinkage.

(5) This is to prevent void spaces in the bed. The beveled bed may leave some space at the back unfilled with mortar, but the grout will fill that. This is better than having excess fins protruding into the grout space.

(6) The grouting can fill the back portion of the head joint with grout, as well as or better than the mortaring.

(7) This is done to improve the bond of the head and bed joints.

Guide Specifications

(9) Except at the finishing course, all grout shall be stopped approximately 1 in. below the top of the last course. At the finished course, the last grout pour shall be brought flush with the top of the brick.

(10) Whenever possible, grouting shall be done from the inside face of the masonry. Extreme care shall be used to prevent any grout or mortar from straining the face of masonry to be left exposed or unpainted. If any grout or mortar does contact the fact of such masonry, it shall be removed immediately. All sills, ledges, offsets, etc., shall be protected from droppings of mortar. Door jambs and corners shall be protected from damage during the masonry work.

8. *Low-Lift Grouting*

a. One tier of two wythe walls may be carried up 18 in. before grouting, but the other exterior tier shall be laid up and grouted in lifts not to exceed 6 times the width of the grout space, with a maximum of 8 in.

b. Hollow units shall be laid up not higher than 6 ft before grouting.

9. *High-Lift Grouting*

a. Masonry shall be laid full story prior to grouting.

b. Cleanout holes shall be provided at the bottom of all cores containing vertical reinforcement in hollow unit masonry and in two-wythe masonry shall be provided by omitting alternate units on the first course of one wythe.

c. Mortar projections and mortar droppings shall be washed out of the grout space and off the reinforcing steel with a jet stream of water as required to clean the space, or by a stick and air.

d. All grout shall be consolidated at time of pouring by puddling or vibrating and then reconsolidated by later puddling before the plasticity is lost.

e. The minimum dimension of the grout space shall be 2 in.

f. Two-wythe masonry shall cure at least three days, and hollow unit masonry shall cure at least 24 hours before grouting.

g. Grout shall be poured to not more than 4-ft depths, then wait approximately one hour and pour another 4-ft depth. The full height in each section of the wall shall be poured in one day.

h. Vertical grout barriers or dams shall be built across the grout space of two-wythe masonry the entire height of the wall to control the flow of the grout horizontally. These barriers shall be less than 30 ft on center.

i. All reinforcing steel shall be inspected in place before grouting, and there shall be continuous inspection during the grouting operation.

j. The two-wythe masonry, wire ties, consisting of #9 wire rectangles, shall connect the wythes and shall be spaced not more than 12 in. o.c. vertically for stacked bond, not more than 24 in. o.c. vertically for running bond, and not more than 24 in. o.c. horizontally.

k. Grout spaces shall not be wetted down prior to pouring of the grout.

1. Unless otherwise indicated on the working drawings, the grout shall have 8 to 10 in. of slump.

m. Approved admix shall be used for high-lift grouting in highly absorbent masonry units

10. *Joinery*

a. General—At the time of laying, all masonry that is not to be plastered shall have the mortar joints finished as

Notes to Specifier

8. a. Minimum grout space is 3/4" or the thickness of steel plus clearances.

9. b. The cleanout holes may be provided in the foundation, and they may be closed by a form board or by masonry units. Sand or polyethelene film has been used as bond prevention to aid the cleaning of the bottom.

d. If the timing is proper, the consolidation of one course can serve as the reconsolidation of the pour placed earlier.

f. This is to assure that the wall wythes will not be too weak to resist the pressure, since blowouts are to be avoided.

g. The actual rate of pouring depends upon the absorption of the masonry and on the climate of the day.

k. Wetting will reduce the suction of the brick and reduce the strength of the grout.

l. This depends upon the fluidity of the grout and its flow into place.

m. This may be determined in the sample panel. The admix should reduce water, retain water, and provide for expansion to reduce volume loss, as Suconom Grout Aid.

10. a. The specifier shall select the appropriate paragraph below for a specific project.

Guide Specifications

specified. The mason shall cut out and repoint defective joints. Those having holes such as made by line pins shall be pointed and tooled properly. Regardless of other jointing herein specified, all jointing in all masonry surfaces that are exposed to the weather shall be tooled, making solid, smooth, watertight compacted joints.

Tooling shall be done when the mortar is partially set but still sufficiently plastic to bond. All tooling shall be done with a tool that compacts the mortar, pressing the excess mortar out of the joint rather than dragging it out.

b. Flush Joints—All joints (or specify which ones) _____ shall be "Flush Joints" made by cutting off the mortar flush with the face of the work with a trowel.

c. Tooled Joints—All joints (or specify which ones) ____________ shall be "Tooled Joints" made by striking them with a metal jointing tool to produce (specify "Concave" or "Flush" or "V" or other) ____________ finished joints.

d. Raked Joints—All joints (or specify which ones) ____________ shall be recessed by raking out the mortar a distance of 3/8" from the face and then tooling with a square jointer or the blade of a tuck pointer.

e. Struck Joints

f. (Specify other types of joints) ____________

11. *Pointing and Cleaning*

a. At the completion of the work, all holes or defective mortar joints in exposed masonry shall be pointed. Where necessary, defective joints shall be cut out and repointed.

b. Exposed masonry shall be protected against staining from wall grouting or other sources, and excess mortar shall be cleaned off the surfaces as the work progresses.

c. At the completion of the work, all exposed masonry shall be clean.

d. Sandblasting—In the event ordinary cleaning is not adequate, special methods such as sandblasting, chipping, and so forth must be used to clean the surface.

12. *Cleanup*

All waste and surplus masonry materials shall be removed from the job, and all stains or dirt from this operation affecting other materials adjacent shall be cleaned satisfactorily.

Notes to Specifier

c. Tooled joints provide best strength and weather resistance.

d. The raked joint is the one most likely to cause trouble from leakage. If the deep shadow of the raked joint is desired, a deep V, 70° instead of 90° apex, may be used as an alternate to give the depth while compacting the surface and pressing it tightly against the brick.

f. This is to provide for special or unusual joints that may be desired in design, such as beaded or weeping.

REFERENCES

17-1 *Uniform Building Code*. International Conference of Building Officials, 5260 South Workman Mill Road, Whittier, Calif. 90601, 1982.

17-2 *Building Code Requirements for Concrete Masonry Structures* (ACI 531) and *Commentary ACI 531*, American Concrete Institute, Box 19150 Redford Station, Detroit, Mich. 48219, 1980.

17-3 Schneider, R., and Dickey, W. L., *Reinforced Masonry Design*, Prentice-Hall, Englewood Cliffs, N.J., 1979.

17-4 *Uniform Building Code Standards*, International Conference of Building Officials, 5260 South Workman Mill Road, Whittier, Calif. 90601, 1982 Ed.

17-5 *TEK Notes*, National Concrete Masonry Association, P.O. Box 781, Herndon, VA 22070.

17-6 *Publication Catalogue*, National Concrete Masonry Association, P.O. Box 781, Herndon, Va. 22070.

17-7 *Test Report On Slender Walls, Committee Report, SEAOSC-ACI, September 1982*, Structural Engineers Association of Southern California, 2550 Beverly Boulevard, Los Angeles, California 90057.

18

Sanitary Structures—Tanks and Reservoirs

JOHN F. SEIDENSTICKER* and EDWARD S. HOFFMAN**

18.1 GENERAL DESIGN CONSIDERATIONS

Sanitary structures include water treatment and wastewater treatment structures. Since the purpose of these structures is to contain water, the watertightness of the basins is of paramount consideration. For water treatment plants, contamination of the water supply by groundwater must be prevented. For wastewater treatment plants, contamination of the groundwater is of concern, in addition to environmental effects, freezing and thawing, and so forth, on the exposed containment structure. Thus, watertight containments of maximum serviceability are required, leading to the choice of high-quality concrete as structural material.

The impermeability of the containment structure is also affected by construction sequence and joint details. Drying shrinkage is normal behavior for fresh concrete, and since it cannot usually be eliminated, the construction sequence should be devised to minimize this effect. Joints should be detailed to accommodate and control shrinkage, and thus to control any resultant cracking. However, the most effective way to control shrinkage is to design quality concrete mixes to minimize this effect, and to establish control procedures in the field to assure that quality is maintained as the concrete is placed.

*Principal Structural Engineer, Greeley and Hansen, Engineers.
**Vice President, Klein and Hoffman, Consulting Engineers.

Water and wastewater treatment structures are designed to perform similar unit processes (sedimentation, filtration, storage, etc.) and are somewhat similar hydraulically. Thus, the structural design is based on loading conditions that are also similar. In general, the raw water or wastewater flows through the plant under the force of gravity, with a resultant hydraulic grade line (HGL). Pumping is used where necessary to raise the HGL to a suitable elevation. For most treatment processes, where velocity is low, the HGL represents the top of the water level in the plant, and drops slowly due to energy losses as water moves through the various stages (except at filters where there is a greater loss of head under normal operating conditions).

A careful consideration of the various loads on the separate structures should include at least the following:

1. *Normal operating conditions.* Water loads can be determined by the design HGL throughout the plant. See Fig. 18-1 for a schematic representation of the HGL in a water treatment plant.
2. *Normal maintenance procedures.* For example, a single tank in a group of settling tanks will be emptied periodically for cleaning, and common walls will become pressure walls. Where high groundwater conditions exist, uplift pressures must be accommodated during such maintenance.
3. *Unusual or unplanned occurrences.* A shutdown of

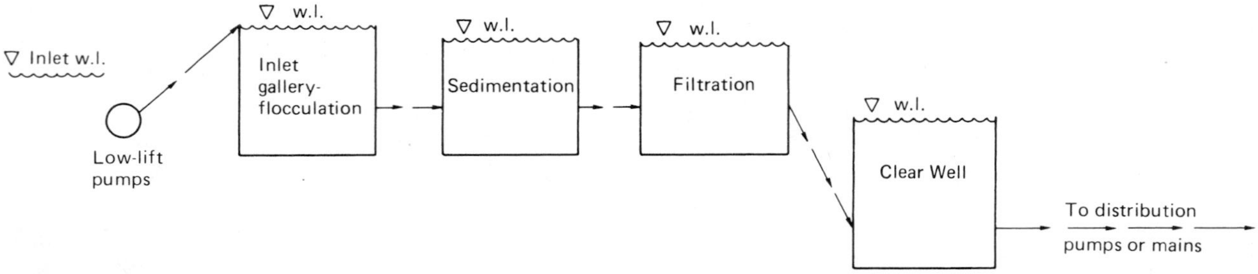

Fig. 18-1 Flow diagram—water treatment plant.

some plant element, such as final pumps, may cause the HGL to rise above the normal level, creating exceptional loads on structural elements. Such loads should be considered together with the likelihood that normal maintenance is also being performed.

The major design parameters are based on water containment structures where the water is under atmospheric pressure and dynamic conditions of flow are secondary. Certain elements, particularly where water is transferred between processes, such as in piping and flumes, will involve pressures above atmospheric, dynamic loadings, and energy–velocity relationships. In addition, supporting functions, such as pumping and chemical treatment, will have their own type of loadings. Careful consideration need be given to the inclusion of all loading elements, and with careful attention to principles of statics, errors can be avoided.

18.1.1 Water Treatment Plant—General Description

The function of a water treatment plant is to provide a potable and aesthetically pleasing water supply by removal of turbidity, hardness, or other undesirable minerals and the elimination of pathogenic microorganisms. Because of present-day requirements for efficiency and water quality, virtually all current designs are based on rapid sand filtration, and water flow through the plant is by gravity. After the raw water is raised by pumps to a suitable elevation for flow through the plant, it passes through the processes of coagulation and flocculation, sedimentation, filtration, and disinfection, and is stored in a clear well. Some chemical pretreatment is generally given to aid in flocculation. A commonly used flow diagram is illustrated in Fig. 18-1.

A typical cross section through a water treatment plant is shown in Fig. 18-2. From the HGL, the water pressure on the containing walls of the structure will be as shown in Fig. 18-3. Note that although normal operation of filters involves head loss, the structural design condition is based on a full hydrostatic head which occurs during backwashing of the filters.

Water treatment plants are generally enclosed to protect the quality of the water, particularly in the clear well. The filters will be enclosed with a one-story structure to provide for inspection and control. A waterproof membrane can be used to prevent groundwater from entering the tanks, and this is protected with 2 or 3 in. of concrete with light mesh reinforcement before backfilling.

Other elements of the plant include water conduits, flumes, etc., subject to hydraulic loads which may include dynamic forces, and pipe galleries, laboratories, chemical storage, etc. The loads to be considered in the design of these elements are included below in the respective sections on those elements.

18.1.2 Wastewater Treatment Plant—General Description

The function of a wastewater treatment plant is to improve the sanitary quality of wastewater by the reduction of suspended solids, biochemical oxygen demand (BOD), pathogenic microogranisms, and other matter such that the effluent quality is improved sufficiently for discharge into the waterway. The sludge created by the treatment processes must also be safely disposed of. The efficiency of the plant is commonly referred to in terms of percent removal of suspended solids and reduction of BOD. This is accomplished by treatment processes that can be described as preliminary, primary, secondary, and advanced (or tertiary) treatment.

Primary treatment involves a settling basin with a sludge collection trough (see Fig. 18-4). Settled sludge is removed by a collector or scraping device that moves it into the collection trough. This basin is usually not covered except for aesthetic reasons.

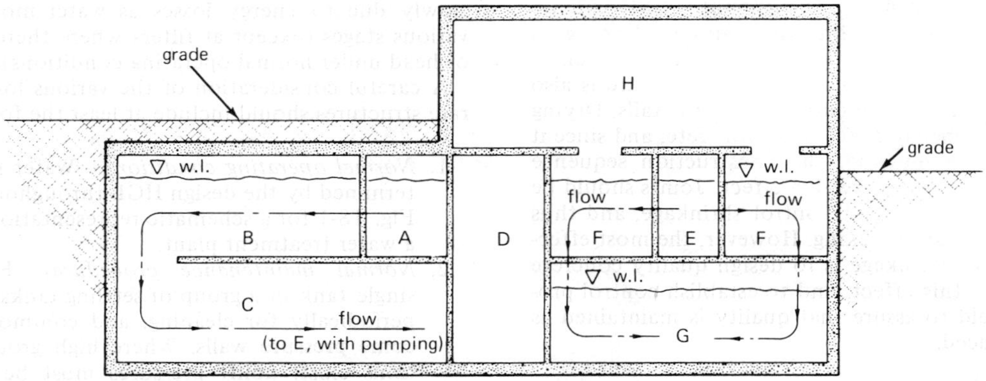

Fig. 18-2 Cross section—water treatment plant. (A) Intake gallery and flocculation; (B) upper story of sedimentation basin; (C) lower story of sedimentation basin; (D) pipe gallery; (E) filter inlet gallery and wash-water trough; (F) filters; (G) clear water well; and (H) filter gallery.

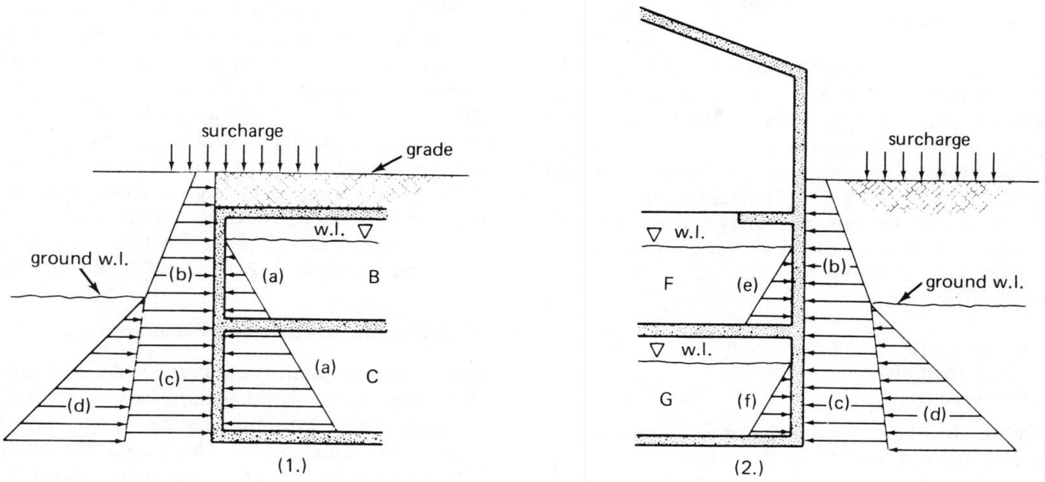

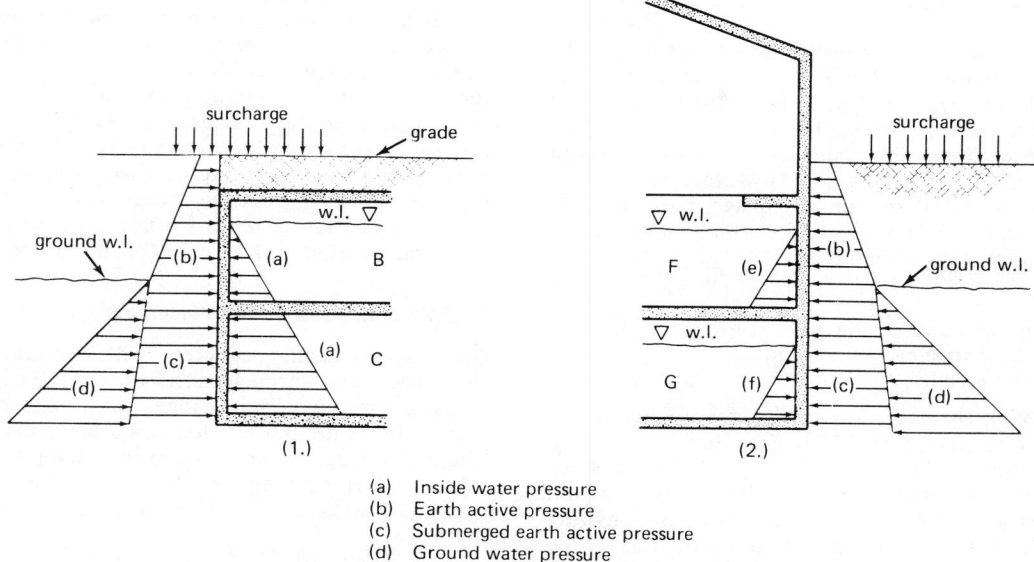

(a) Inside water pressure
(b) Earth active pressure
(c) Submerged earth active pressure
(d) Ground water pressure
(e) Filter box water & submerged filter material
(f) Clear well water pressure

Fig. 18-3 Pressures on walls—water treatment plant.

Secondary treatment may involve trickling filters, rotating biological contactors (RBC's), lagoons, or some form of activated sludge process, and, in most cases, is aerobic. Trickling filters are generally round. Aeration tanks used in the activated sludge process may be circular or rectangular structures. Where a pure oxygen process is used, the tanks must be enclosed with a roof sufficiently airtight to maintain the process pressure and prevent loss of oxygen.

Advanced wastewater treatment involves either biological, chemical, or physical processes to achieve treatment beyond that available by conventional secondary processes. These processes may include nitrification–denitrification, carbon absorption, sand-filtration, etc. Wastewater tanks for each process are similar in nature to previous descriptions.

The sludge collected from the treatment processes described above must undergo further treatment for stabilization of solids, reduction in pathogen levels, reduction of volume, and increased concentration of solids. Anaerobic digesters are covered, having either fixed or floating roofs for gas collection. The effluent from the various treatment processes is usually disinfected by chlorine before discharge into the waterway.

Some general characteristics of wastewater treatment plants are as follows:

1. Tanks will generally have sloping or hopper bottoms leading to collection troughs to facilitate the collection of sludge. Settling or sedimentation tanks will have power-driven equipment to move settled sludge to such troughs.
2. Where aerobic processes are involved, tanks may be open-top. Anaerobic digestion tanks will be closed-top for control of the process.
3. Wastewater treatment plants are usually at relatively low elevations with respect to the water table. They will therefore be subject to water pressures outside of tank walls as well as inside, so that flotation of the entire plant may be a problem. Where dead load of the tank is not sufficient to overcome buoyancy, relief valves must be provided in the basin floor or walls, or a tie-down system used.

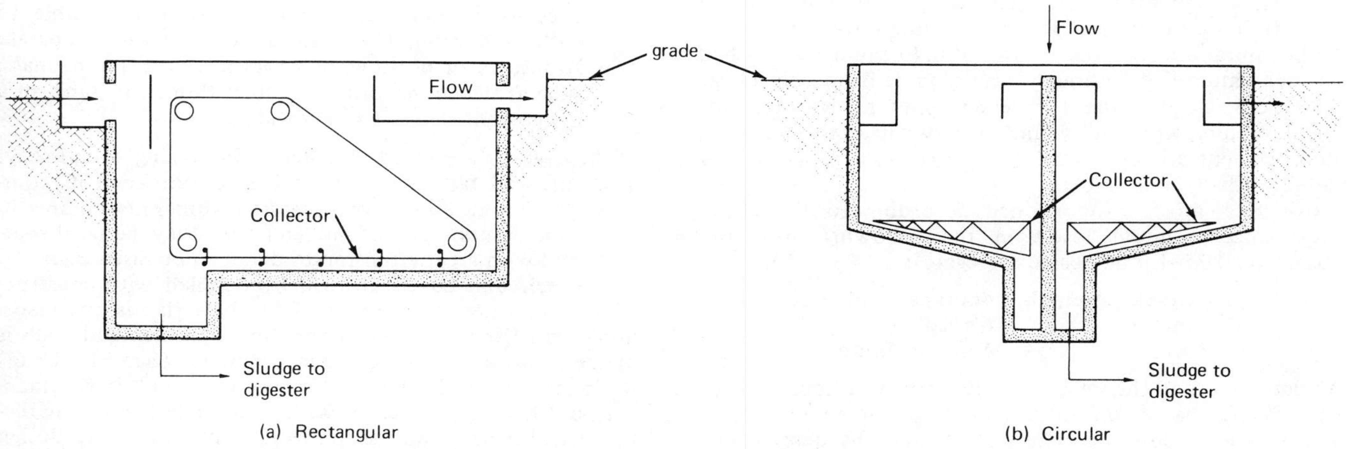

Fig. 18-4 Settling or sedimentation tanks.

18.2 WATERTIGHTNESS

Good-quality concrete is practically impervious to liquids. Leaks in basins will occur either at cracks or at joints in the concrete. Cracks are generated by stresses due to restrained drying shrinkage and/or temperature variations. A close spacing of contraction joints, such as 25 ft apart, would presumably eliminate cracking.[18-5] However, since every joint is itself a potential source of leakage, their number should be minimized, and the solution is thus a compromise. The elements that the designer can control to minimize cracking are as follows:

1. Concrete quality and curing
2. Joint type and spacing
3. Reinforcing percentage

The total drying shrinkage of unreinforced concrete has been shown to vary from 4×10^{-4} to 8×10^{-4} in./in. when exposed to air at 50% relative humidity.[18-6] Approximately 34% of the total shrinkage is realized in the first month, and 90% in the first year. Given reasonable limits for construction time, it is apparent that most of the potential drying shrinkage will take place before the basin is filled. After filling, at least one face of basin walls will be continuously wet, and swelling will occur, although not usually to equal the extent of the previous shrinkage. It has been reported that the swelling that occurs during continuous wetting over a period of years is about one-third of the shrinkage of air-dried concrete for the same period.

The most important factor affecting drying shrinkage is the amount of water per unit volume of concrete. Water can be reduced by use of both air-entraining and plasticizing admixtures, and by using the minimum amount of cement consistent with the desired structural quality. The concrete mix should have the largest practical coarse aggregate, as this will increase the total aggregate content, and reduce the required amount of cement. Harsh graded aggregates should be avoided.

Volume changes from temperature variations will also induce shrinkage cracking. Concrete placement during excessively warm weather or with excessively high plastic concrete temperatures increases shrinkage. Too rapid filling of a basin with water cooler than the basin concrete will induce shrinkage cracks.

18.2.1 Concrete Quality and Curing[18-7]

It has been recommended that concrete for tanks should have a water–cement ratio no higher than 0.53 (equivalent to 6 gal/bag); for thin sections this is reduced to 0.44 (equivalent to 5 gal/bag).[18-8] The use of water-reducing and air-entraining admixtures will enable the water–cement ratio to be minimized such that a ratio of 0.45 can be established as a maximum.[18-9] Although it is desirable from a shrinkage standpoint to minimize the cement content, requirements for durability, workability, and density will establish a minimum cement content of 5.5 to 6 bags (518 to 564 lb) per cubic yard.

Concrete mixes proportioned according to the above requirements should result in the following properties (based on 0.75–1.0 in. coarse aggregate):

compressive strength:	4000 psi minimum
air content:	6 ± 1%
slump:	4 in. maximum

Variations from the recommended concrete mix may be considered, based on local conditions or requirements, but should produce concrete of equivalent density and durability.

During the curing period, strength is gained and porosity is reduced very rapidly during the first few days. Continuous wet-curing will minimize shrinkage during the time of strength build-up, and should be considered preferable to the use of curing compounds during this initial period. Wall forms may be loosened and left in place so that a continuous flow of water may pass over the fresh concrete surface. For horizontal surfaces, burlap or other water-retaining coverings may be used, continuously moistened. After the initial curing period (seven days for normal cements), the use of membrane curing may be used for the subsequent curing.

18.2.2 Joint Details and Placement

Given good–quality concrete with a minimum of drying shrinkage, some shrinkage stresses will still exist. To control cracking from volume change, the designer can use steel reinforcing and joints. These two elements must be considered together, as an increase in spacing between control joints will require an increase in steel percentage. Three types of joints are defined below. (See Fig. 18-5.)[18-10]

1. *Construction joint.* Placed to define the end of a placement, this joint is basically a cold joint in the concrete. Reinforcing is continuous, and either a water stop is used, or the new concrete is bonded to the old. Properly bonded horizontal construction joints can be made watertight. Bonding of vertical construction joints is difficult, and waterstops are used instead. Although reinforcing is continuous across a vertical construction joint, the absence of bonding presumably creates a weak link, and defines the location of shrinkage cracks rather than random cracking. This theory has met with limited success. A recess can be placed at the joint to create a neat line.
2. *Contraction joint.* Reinforcing is omitted, and curing compound or other bond breaker is used to prevent restraining forces. Water bars are used across the joint. Unbonded dowels or keys may be used to transfer shear forces. Properly spaced contraction joints will presumably interrupt restraining forces such that other random cracking is eliminated. The choice of dumbell or labyrinth waterbar depends on the amount of movement anticipated at the joint. The labyrinth is easier to install, and less easily displaced by concrete placing operations. However, large joint movements can cause failure in concrete surrounding the labyrinth waterbar.
3. *Expansion joint.* This joint eliminates both tension and compression forces. Its use in walls of tanks is limited, since swelling would not normally exceed the drying shrinkage. Unless compressive stresses become excessive, good-quality concrete should be able to withstand them. This joint should be used to separate structures or portions of structures of different mass, such as the connection of an inlet flume to a tank, or a lightweight slab at a heavy wall.

These joints are listed in order of increasing complexity. The potential failure of a joint is in the order of its complexity. The authors have witnessed numerous failures in expansion joints, and recommend that they be used sparingly and with great attention to detail and construction.

Large circular basins are usually detailed with construction joints only. The center of the base slab is one placement, and the remainder of the tank base slab and walls is divided into a number of equal segments. (See Fig. 18-6.) Basins up to 150 ft in diameter have been built in this manner, and have proved to be watertight. The practice in rectangular basins is not as uniform. Three separate design approaches can be described as follows:

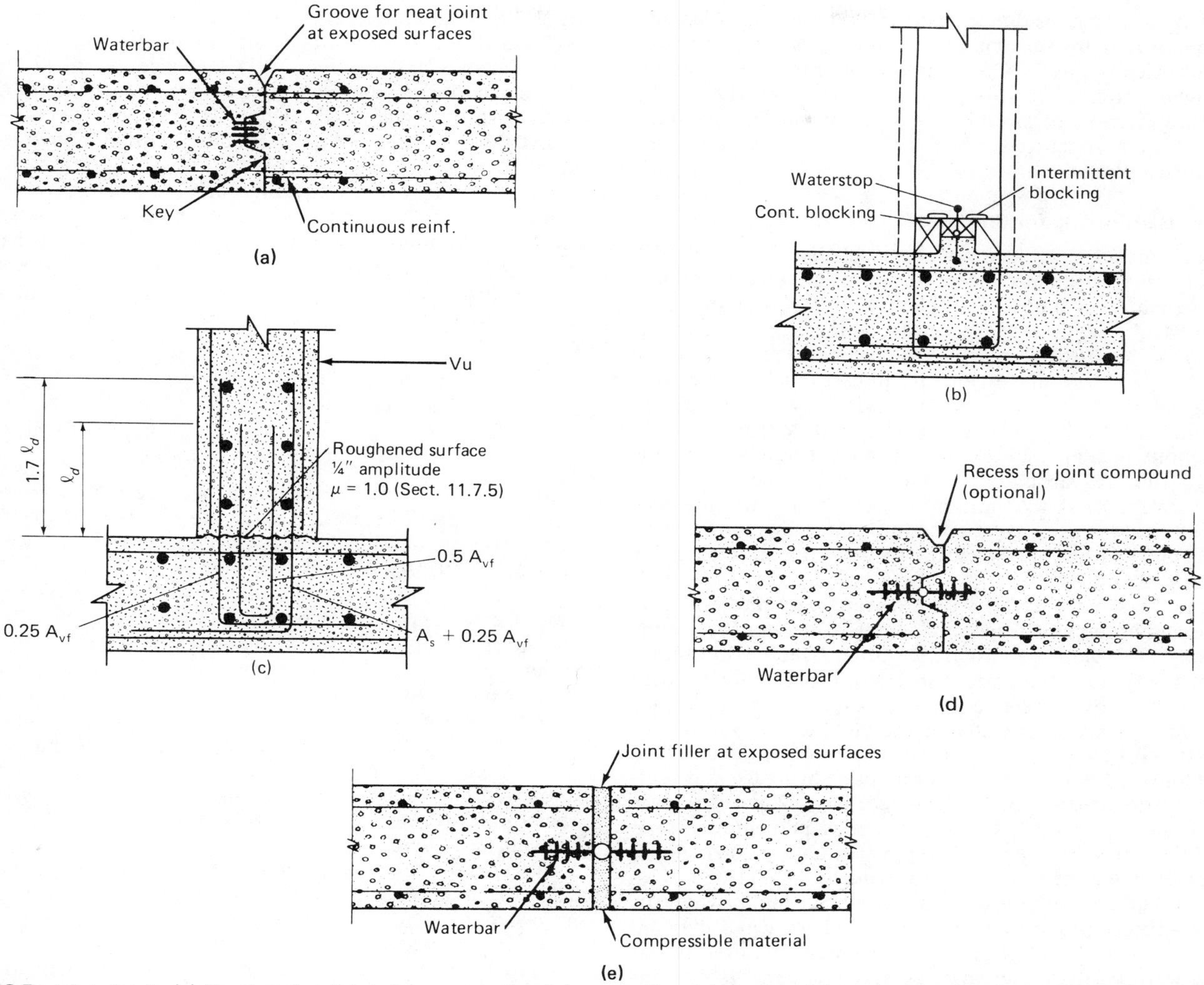

Fig. 18-5 Joint details (a) Construction joint; (b) construction joint at wall and base slab; (c) construction joint using shear friction; (d) contraction or control joint; and (e) expansion joint.

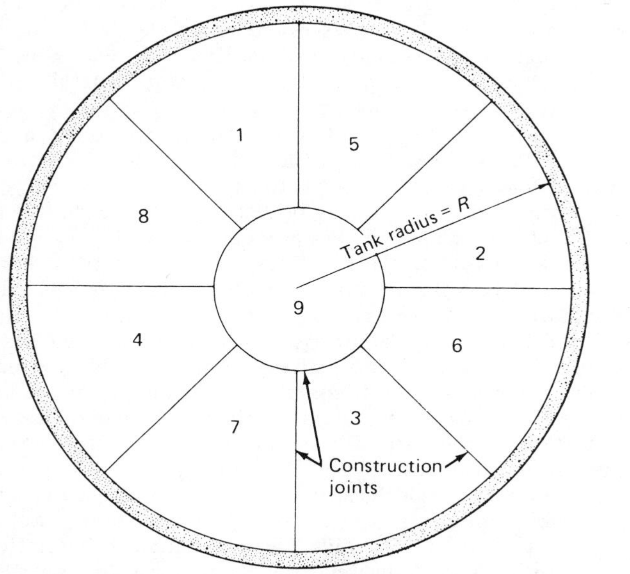

Fig. 18-6 Construction joint layout—circular basin. All areas are approximately equal. Sequence of concrete placement indicated by numbers.

1. Expansion joints placed 100 to 150 ft apart, in both directions. Between these joints, contraction joints are placed at about 25-ft intervals.
2. Expansion or contraction joints placed 50 to 60 ft apart, with no other joints.
3. "Continuous designs" with construction joints placed 40 to 60 ft apart, with additional shrinkage reinforcing based on the placement length. Expansion joints used sparingly, such as at changes in mass or where basin length exceeds 300 ft.

The use of method (3) recognizes that random shrinkage cracking will occur. This cracking is controlled by using higher steel reinforcing percentages than in the other two methods. The job-specific conditions of basin size, environment, and function will determine the design method used. Where practical, the authors prefer the use of method (3) for the following reasons:

1. It is the most consistent with the concept of concrete as a structural material.
2. It balances minor leakage through random cracking versus failed contraction or expansion joints. Controlled random cracks will seal themselves by a process known as autogenous healing, whereas leaky joints are difficult to repair, if the source of leakage can be found.

3. Lateral pressures due to fluid in a rectangular basin can be carried by tension reinforcing between end walls. Severing the basin with contraction or expansion joints means lateral pressures must be carried by the soil in base friction or lateral wall pressure. Spectacular failures have occurred when this support was removed during plant expansion.

18.2.3 Reinforcing for Shrinkage

The amount of reinforcing steel necessary to completely control cracking when contraction joints are eliminated has been estimated at 0.65%.[18-11] The use of 0.25% as specified by the *ACI 318 Code*[18-12] is not adequate in most cases. Observation by the authors on structures and conduits built with 0.25% or less reinforcement indicates shrinkage cracks will form at about 25-ft intervals with a width of about 0.06 in. This crack width is excessive; it will not permit autogenous healing, and if exposed to freezing may severely damage the area around the crack. Experience with reinforcing ratios of 0.40% minimum for Grade 60 reinforcement indicate this is a good compromise on cost versus serviceability. For a 12-in. wall, #5 @ 12 in. in both faces yields a ratio of 0.43%.

Figure 18-7 from Ref. 18-18 is a graphical representation of the minimum shrinkage and temperature reinforcement that we have found by experience to be effective in obtaining relatively watertight structures. Note that Grade 60 bars seem to be more effective for shrinkage and temperature steel, but not proportionally to the yield ratio, 60/40. For a maximum construction joint spacing of 60 ft (18 m), and with Grade 60 reinforcement, minimum shrinkage and temperature steel is 0.4% of the concrete area. When the spacing is reduced to 40 ft (12 m), the steel percentage drops to 0.33%, and when reduced to approximately 20 ft (6 m), the steel percentage of 0.18% conforms to ACI 318-83.

The designer should also consider cost in his decision on the construction method. For cases where leaks may be very damaging, the additional reinforcement required to ensure a watertight system may be more expensive than installation of additional contraction joints. For very large basins, the elimination of expansion joints is not practical.

18.3 DESIGN PARAMETERS

18.3.1 Loads

Live loads for use in structural design of the various elements can usually be determined by rational consideration of the type of material. For instance, water and raw sewage will weigh 62.4 lb/cu ft. Sediment in a water treatment basin (area B and C, Fig. 18-2) may vary from 40 to 70 lb/ cu ft. Sludge may vary from 65 to 75 lb/cu ft. A detailed list of weights of chemicals is given in the report of ACI Committee 350, "Concrete Sanitary Engineering Structures."

Live loads in offices and laboratories will often be governed by local building codes, but these should be recognized as minimum. The designer should make a load analysis based on use and proposed equipment to determine if heavier loads are justified. Some commonly used minimums are as follows (from the *American Standard Building Code*):[18-13]

offices	80 psf
public spaces	100 psf
railings	50 plf (applied horizontally at top of railing)

Heavy equipment loads should be determined from the manufacturer. Where such equipment rests on the floor, consideration should be given to the potential two-way structural action of the slab system,[18-14] and to the necessary space provided around each machine. For instance, a machine with a 4 X 5-ft base weighing 20,000 lb will load the floor at 1000 psf over its base area only. However, if no other machine is within 6 ft, and the slab can be shown to be capable of spreading the load, then the floor area can be taken as (4 + 6) by (5 + 6) or 110 sq ft, and the design load is more nearly 180 psf. It can also be shown that an HS 20-44 highway loading on a 20-ft span slab is equivalent to about 250 psf over the entire span, thus again illustrating the ability of properly designed concrete slabs to spread concentrated loads.

In general, when equivalent uniform live loads exceed 100 psf, a careful analysis of slab behavior and actual loads is in order. Arbitrarily excessive live loads will result in a structure of excessive cost.

Dead loads can be determined from known densities of structural materials. For other materials, such as concrete block walls, tile floors, roofing, etc., reference can be made to manufacturers catalogs, or to other lists of weights of building materials.

Knowledge of groundwater elevations and of floods of adjacent bodies of water is essential for the design of exterior walls and for analysis of uplift on the entire basin. Where tne basins rest on competent soils, lateral earth pressures are commonly taken as the active soil pressure plus maximum water level. Where basins are founded on rock or where the walls are otherwise restrained, earth pressures will be greater than the active pressure.

Buoyancy of the basin may be resisted by the dead weight of the structure and soil alone or combined with tie downs such as rockbolts, tension piles, or other soil anchors. The effective soil mass may be increased by extending the foundation mat beyond the exterior walls. Where this method is not practical, provisions for flooding of the basin should be made by relief valves in the basin floor or side walls. When buoyancy is controlled by flooding of the entire basin, a back-up system involving planned failure of a portion of the base slab should be provided to prevent loss of the entire basin.

When buoyancy of the structure is resisted by dead weight of structure and soil alone or combined with soil anchors, a safety factor of 1.25 should be provided against flotation based on high groundwater level or recurring floods. When checking the design for floods exceeding a 100 yr period, this safety factor may be reduced to 1.10. Similarly, in designing walls for exterior lateral pressures based on greater than 100 yr floods, the load factor may be reduced.

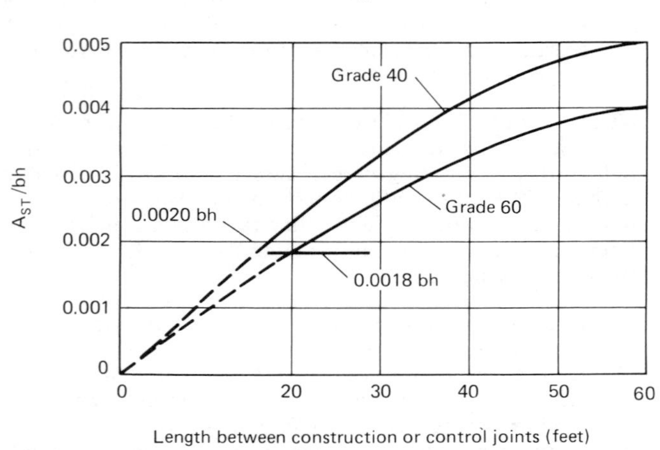

Fig. 18-7 Minimum shrinkage and temperature reinforcement graph.

18.3.2 Foundations

In the selection of foundation type, it is important to consider the minimizing of differential settlement, as this could lead to cracking in basin walls and loss of watertightness. Uniform settlement, if not excessive, may not be harmful. Under certain conditions, if the structure rests on a stiff- or hard-clay stratum of sufficient uniform thickness underlain by a soft stratum, some small uniform settlement may be anticipated and may be acceptable. However, connecting piping and conduits must be detailed to accommodate such settlement.

In general, it would appear unwise to combine foundation types, except under extreme subsoil conditions and carefully controlled investigation. Combining rigid foundations such as piles, caissons, or rock bearing with foundations on sandy or clayey soils can be expected to yield some differential settlement.

A good design approach is to include in the design team a professional who specializes in soils investigation to determine foundation requirements and to influence the choice of foundation type. Borings should always be taken under the guidance of this person as to number, depth and spacing.

Basin foundations on competent soils require special attention to the resultant pressure distribution under the base slab and to the variations in soil loadings caused by alternate or partial use of adjoining basins.[18-15,18-16]

18.3.3 Flexural Crack Control

Beginning about 1975 and culminating with the publication of the current ACI 350 report (November-December 1983), Committee 350 developed new recommendations utilizing higher working stresses with Grade 60 bars.[18-9] This breakthrough was made possible by the rather obscure ACI Building Code section on crack control (section 10.6 "Distribution of Flexural Reinforcement in Beams and One-Way Slabs"), which says that flexural reinforcement shall be well distributed within maximum flexural tension zones to limit the computed width of cracks. ACI 318 established a reinforcement distribution factor, Z (see Fig. 18-18):

$$Z = f_s \sqrt[3]{d_c A}$$

or $S = 0.5\,(Z/f_s)^3/d_c^2$, where S is bar spacing.

ACI 318 limits the crack widths for "interior" and "exterior" building exposures by limiting Z to 175 ksi (31,300 kgf/cm^2) and 145 ksi (25,900 kgf/cm^2), respectively, for these conditions. Determination of a Z value representing the limiting computed average surface crack width acceptable for watertight structures exposed to corrosive (sewage)

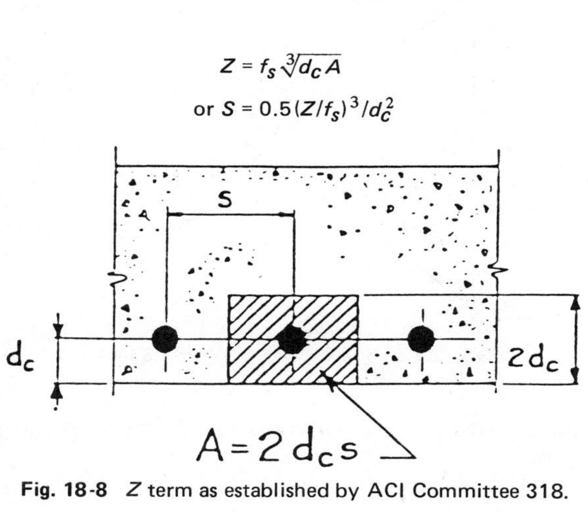

Fig. 18-8 Z term as established by ACI Committee 318.

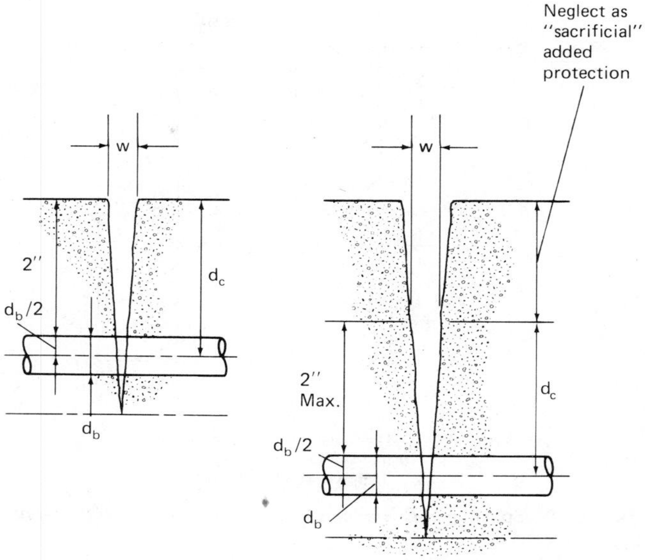

Fig. 18-9 A maximum of 2 in. (50 mm) is established for clear cover.

liquid was easily accomplished by inserting the maximum service load stress, maximum bar size (#11) and spacing (12 in.), with the cover employed in past designs showing good performance.

A maximum of 2 in. (50 mm) for clear cover (Fig. 18-9) should be used in the *ACI 318 Code* equation for Z, whenever specified cover is greater. The increase in computed values for Z with added cover simply reflects a trigonometric projection of the surface width, with no particular significance upon durability. Added cover in excess of 2 in. (50 mm), however, has proved its value for long-time protection, and it may be helpful to consider it simply as "sacrificial" protection against erosion of the concrete to the depth of the steel. At first, Committee 350 concluded that a Z-factor of 115 ksi (20,600 kgf/cm^2) was appropriate for "normal" sanitary exposure. They also recognized "*very severe*" sanitary exposures for which they concluded that a conservative Z-factor equal to 95 ksi (17,000 kgf/cm^2) was appropriate. The reinforcement distribution factor Z is related to exposure condition and crack width as follows:

Exposure condition	*Z Reinforcement distribution factor ksi*	*Crack width*
Exterior	145 (25,900 kgf/cm)	0.013 in. (0.33 mm)
Normal sanitary	115 (20,600 kgf/cm)	0.010 in. (0.25 mm)
Severe sanitary	95 (17,000 kgf/cm)	0.008 in. (0.20 mm)

The ACI Committee 350 now believes, that "very severe" exposures should be resisted by impervious liners or protective coatings, thus reducing the conditon to a "normal sanitary exposure." Very severe corrosive conditions can destroy the cover itself and cannot be controlled merely by lower service loads stresses or smaller bar spacings.

It is the Committee's opinion that "normal sanitary exposure" can be defined as concrete exposed to water with a pH greater than 5, concrete exposed to water with a sulfate concentration less than 1500 ppm, and air-entrained concrete exposed to wet/dry and freeze/thaw cycles.

The report of Committee 350 provides for the use of Grade 60 bars and permits higher service load stresses with smaller bar spacings than have been used in the past with Grade 40 bars. These stresses and bar spacings have been

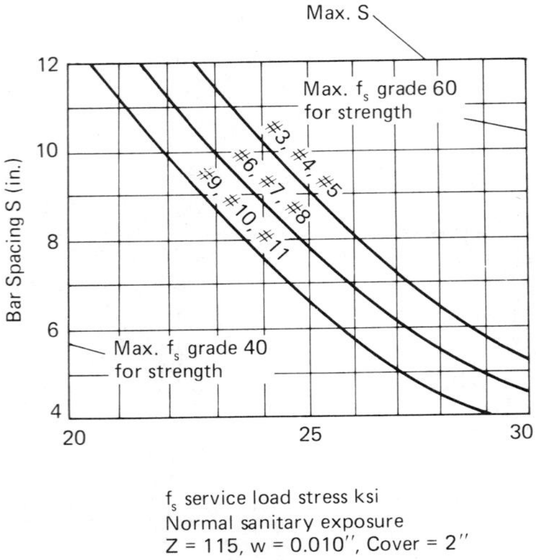

Fig. 18-10 Spacing for Grade 60 reinforcing varies with service load stress.

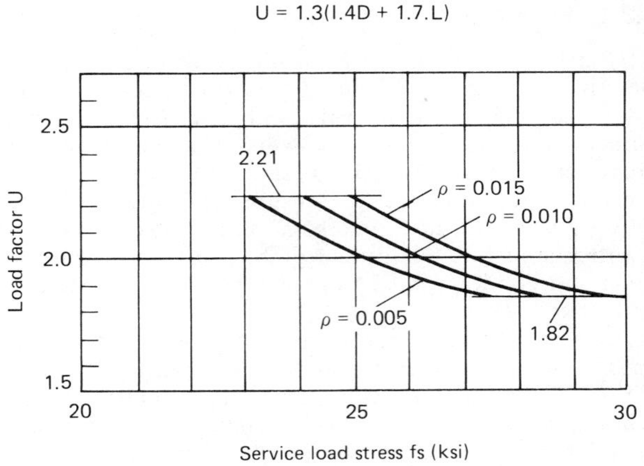

Fig. 18-11 Service load stress.

established to control crack widths to a maximum of 0.010 in. (0.25 mm) for normal sanitary exposure.

Fig. 18-10 shows how spacing for Grade 60 bars varies with service load stress for limiting crack widths to 0.010 in. (0.25 mm). Note that bar spacings are limited to a maximum of 12 in. (300 mm). Maximum service load stresses recommended are 20 ksi (1400 kgf/cm²) for Grade 40 bars and 30 ksi (2100 kgf/cm²) for Grade 60 bars.

18.3.4 Strength Design

The Committee 350 report recommends that structural members for sanitary structures be proportioned on the basis of strength design. This is done by utilizing an additional "sanitary exposure" coefficient, which results in increased load factors. The basic reason, aside from convenience, for conversion to strength design is that this method gives a more realistic evaluation of structural safety and actual stress distribution than working stress design. As we know, concrete is not the perfectly elastic material assumed in working stress design based upon elastic or "straight-line" theory. High, long continued elastic stresses in concrete diminish with creep, transferring load to steel, if available, or to adjacent lightly loaded concrete fibers, changing the assumed elastic "straight-line distribution." These sanitary exposure coefficients, together with an increase in the conventional load factor for fluids from 1.4 to 1.7, *increase* all load factors from ACI 318 a total of 30% for flexural reinforcement, 65% for direct tension reinforcement (such as ring tension), and 30% for stirrup or diagonal tension requirements.

1. Flexural Reinf. — Req'd. Strength $\geqslant 1.3\ U$
   $\phi M_n \geqslant 1.3(1.4M_D + 1.7M_L + 1.7M_F)$
2. Direct Tension Reinf. — Req'd. Strength $\geqslant 1.65\ U$
   $\geqslant 1.65(1.4\ T_D + 1.7\ T_L + 1.7\ T_F)$
3. Stirrup Reinf. — $\phi V_s \geqslant 1.3(V_c - \phi V_c)$
4. Concrete Shear & Compr. — Req'd. Strength $\geqslant 1.0U$

No increase is required in load factors for concrete shear, bond, or compression strength, so that proportioning member depths or thickness will be unchanged (Fig. 18-11). For flexure, the proposed increase in load factors results in a maximum load factor of 1.3 times 1.7 = 2.21 for normal live and water and earth load and a minimum load factor of 1.3 times 1.4 = 1.82 for all dead load. In conjunction with ϕ-factors prescribed in ACI 318, these new load factors result in flexural service load stresses in the reinforcement between 24 and 29 ksi, consistent with allowable stresses for working stress design in the current report by ACI Committee 350. The same limits on bar spacing developed for the 1983 report will, of course, apply equally well with use of strength design.

Flexural design equations are not prescribed directly by the 1983 *ACI Code*, which only lists the necessary assumptions for design. The exact solution for direct design is extremely complicated for use without the aid of a computer or programable calculator.

$$A_s = 1.7f_c'bd/2f_y - \frac{1}{2}\sqrt{\frac{2.89(f_c')^2b^2d^2}{f_y^2} - \frac{6.8f_c'bM_\mu}{\phi f_y^2}}$$

Figure 18-12 or Table 18-1 can be used to solve this equa-

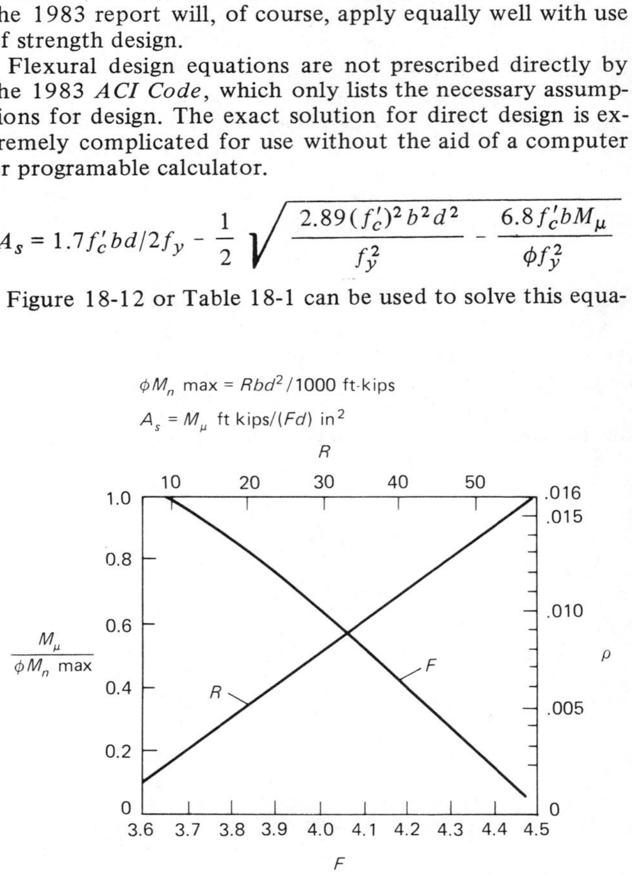

Fig. 18-12 Flexural design constants for $f_c' = 3000$ psi and $f_y = 60{,}000$ psi.

tion for the flexural constants familar to designers. These constants provide quick direct solutions for steel area A_s or size of section bd required for use without compression reinforcement.

18.3.5 Concrete Protection

Concrete protection for reinforcement should also be conservatively specified. Section 7.7 of ACI 318-83 specifies clearances under various conditions of service. For tanks with open tops, both faces of walls and walkway slabs should be considered as exposed to the weather. For ease in design and construction, the required protection is often set at 2 in. for all reinforcing, rather than specifying 1.5 in. for #5 and smaller bars. The reinforcing is placed within tolerances, and the consequences of a bar being placed too close to a water-containing surface may result in corrosion of the bar, and spalling of the surface. In addition, greater protection is of value for these surfaces where flexural cracking may occur, providing a path for water to reach and corrode reinforcing.

18.3.6 Economy

It has been the experience of the authors that a reinforcement ratio of about 1% produces economical structures. The figure below from Ref. 18-18 demonstrates this point and shows how the square foot cost of a 20 ft high wall with a fixed base and a hinged top varies with different wall thicknesses and steel percentages. For a service load moment of 60 ft kips the cost of a wall with 2% steel is 15% more than with 1% steel.

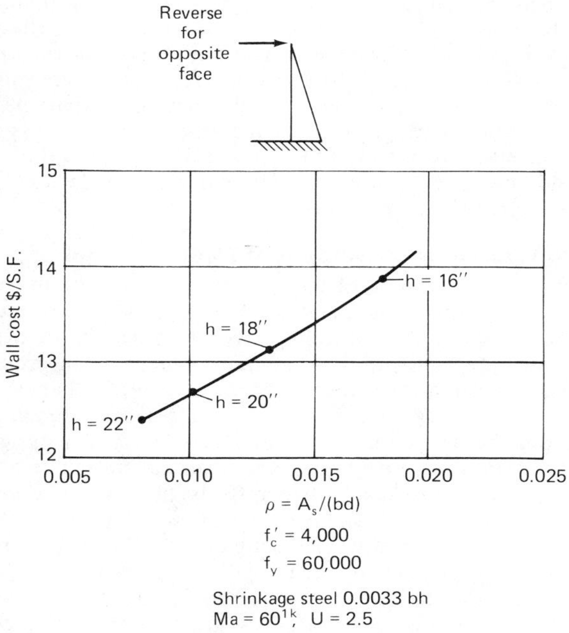

Costs vary with different wall thicknesses and steel percentage.

18.3.7 Vibrations

The mechanical equipment associated with water and wastewater treatment plants does not generally involve impact loads under normal operating conditions. The impact loads involved in hoists and cranes for removal or maintenance of mechanical equipment will be discussed later (Section 18.6.3 on pump houses). Much mechanical equipment, such as scrapers and clarifiers, operates at slow speeds such that dynamic loadings to the structure are negligible. Other equipment, such as centrifugal pumps, centrifuges, and compressors, operates at high rates of speed, and the dynamic loading to the structure is important.[18-19]

The determination of the dynamic load–structure interaction is a very complicated process primarily because of the difficulty in accurately predicting the response of the structure. The major thrust of the problem is to avoid conditions where the machinery frequency approaches that of the supporting structure or foundation. Design methods have been developed to estimate the natural frequency of the structure, but these should be used with caution because of the necessary simplifications and assumptions in the methods.

Because of these difficulties and the potential severity of problems, some rules have received wide acceptance. Where possible, the foundation is insulated or otherwise separated from the structure. Where the machine rests on a structure rather than an independent foundation, attempts should be made to determine natural frequencies of the structure in order to avoid resonance. In addition, the machine base may be supported on suitable vibration isolation equipment where practical.

For existing structures where problems exist, or for testing of newly built structures, it is possible to determine a structure transfer function with the use of excitation devices. Together with the knowledge of the machine dynamic input, a computer simulation of the interaction of machine and structure may be made. Variations in machine base and vibration isolation equipment may than be set into the computer, leading to a problem solution.

TABLE 18-1 Flexural Design Constants for f'_c = 3000 psi and f_y = 60,000 psi

$\dfrac{M_u}{\phi M_n \text{ max}}$	ρ	F	R
0.1	0.00132	4.46	5.85
0.2	0.00268	4.36	11.71
0.3	0.00410	4.28	17.56
0.4	0.00556	4.20	23.42
0.5	0.00710	4.12	29.27
0.6	0.00869	4.04	35.12
0.7	0.01037	3.95	40.98
0.8	0.01214	3.86	46.83
0.9	0.0140	3.76	52.69
1.0	0.0160	3.65	58.54

18.4 DESIGN OF RECTANGULAR TANKS

The shape of the tank is determined by its function. Circular tanks are structurally more efficient than rectangular tanks, both in terms of wall area per unit volume and economy of materials. However, for tanks within buildings or for batteries of tanks, the rectangular shape is preferred. Sidewalls of rectangular tanks and reservoirs can be designed as cantilever, propped cantilever, or two-way slabs, depending on the tank configuration. A typical project will often involve combinations of these structural systems.

18.4.1 Loadings and Analysis of Walls

Basin walls are designed for hydrostatic pressures on the inside, and earth pressures on the outside, plus hydrostatic pressure where the water table is high (Figs. 18-2, 18-3).

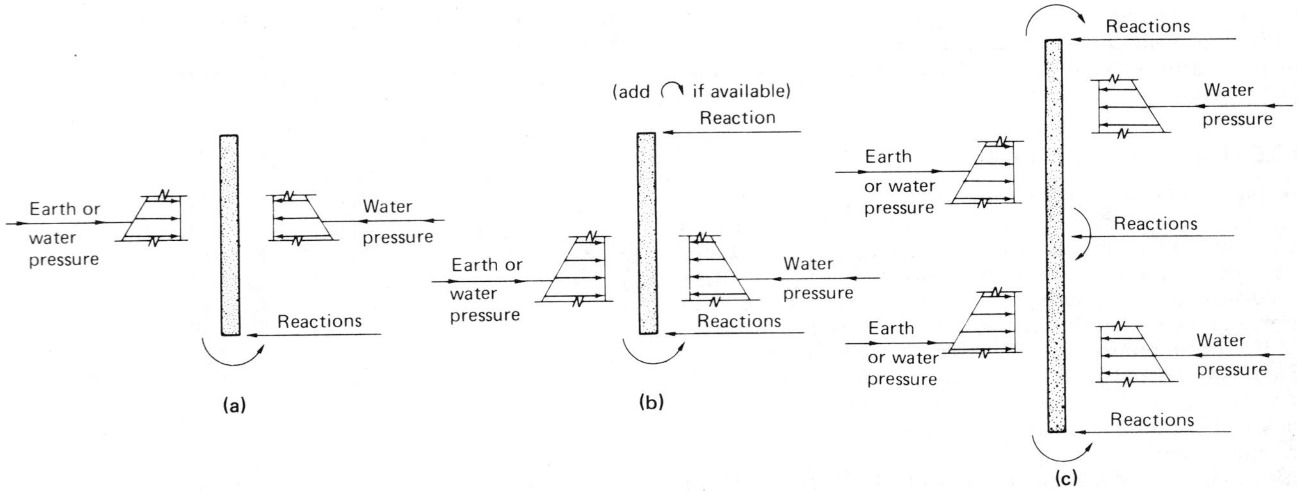

Fig. 18-13 Loading on basin walls. (a) Cantilever; (b) propped cantilever; and (c) two-span slab.

The inside water pressure has a triangular distribution, but the designer is cautioned to watch for possible overpressure, which may arise from vapor pressure in a closed tank, or from surges in a conduit (such as intake piping) producing a liquid level above the top of the tank.

For rectangular tanks, design is usually based on full continuity between the various elements. For small tanks, walls may be supported on three or four sides, depending on the existence of a top slab. Procedure for the design of such tanks including coefficients and design examples is given in "Rectangular Concrete Tanks,"[18-20] by the Portland Cement Association and in "Moments and Reactions for Rectangular Plates," Engineering Monograph 37, by the United States Department of the Interior Bureau of Reclamation. The walls are designed as two-way elements with varying edge supports, depending on construction details and the designer's choice. Coefficients in Ref. 18-20 are tabulated for wall length to height ratios up to 3, but it is recommended that their use be limited to ratios $\leqslant 2$. For ratios greater than 2, design should be based on one-way action. For single-story open basins, the wall will cantilever from the base except for the vicinity of corners, or may be a propped cantilever with lateral support at the top provided by a walkway slab-beam. For single-story closed basins, the wall will act as a vertical slab with continuity generally at both top and bottom. For two-story basins, the wall will become a two-span vertical slab (see Fig. 18-13). Moment distribution can be readily used to obtain effects of continuity with base and top slabs. Reinforcing details at corners should be considered to handle the local effects of continuity around the corner, where the walls revert to two-way action.[18-21]

Interior walls of a basin may or may not be designed for water pressure. Depending on the design or operating criteria: if the interior walls are to be baffles only, it is not necessary for the wall to support full hydrostatic head on either side alone. In wastewater treatment plants, for flexibility in operation and the requirement of servicing equipment in the basin, interior walls should be designed for such pressures. In water treatment plants, some part of the plant must always be in use, and this consideration will determine design criteria for interior walls.

The exterior walls are subjected to earth and groundwater pressures on the outside, and water pressure inside (see Fig. 18-13). These walls should be designed for either loading condition, as it is reasonable to anticipate that the basin may be empty when backfilled, or that backfill may not be in place (e.g., expansion of facility, etc.). Often testing of the structure is desirable to check for watertightness, and such testing should be done before backfilling so that leakage could be observed. Construction specifications for the project should indicate that backfilling operations take place after top slabs are in place, so that walls are not required to act as cantilevers, unless, of course, they are designed for this condition.

It is important for the designer to recognize the fixity that actually exists at the base of the wall. For a cantilever wall, which is statically determinate, full fixity at the base is required to provide stability. However, this does not mean that the joint itself does not rotate. Any rotation of the joint causes added deflection of the cantilever. Thus, for the propped cantilever, which is once redundant, continuity at the base will induce a moment generally less than the full fixed-end moment. The amount of the rotation at the joint, due to deformation in the bottom slab, can be determined from slope-deflection equations directly, or, using the results of a moment-distribution analysis, but is of little practical interest. However, this structural behavior should be considered in the design of basin end walls, in determining the moment between wall and slab.

18.4.2 Loadings and Analysis of Floors and Roof

From the standpoint of functional structure, the most efficient form for the design of these elements is usually some form of flat plate–flat slab. It is advantageous to design intermediate and roof slabs as thin as practical in order to limit excavation, and to avoid creation of air pockets which may become stagnant. For both these reasons, the flat slab is often the best choice for structure. The roof is designed to support itself, membrane, topping, and fill where applicable, and a live load that reflects probable maximum loading condition. This live load should be no less than 100 lb/sq ft for plants and people, unless it can be positively established that such loads will never exist. Consideration should be given to the possibility that trucks or other heavy equipment may cross the area, requiring additional live load capacity.

The intermediate slab of a two-story settling basin will be subjected to a live load of settled material, plus incidental people load. The sludge load can be determined from hydraulic design of the basin, at 62.4 lb/ cu ft when the basin is emptied. This slab is not subject to a water load, except as a tie between exterior walls, as the water pressure above and below the slab is roughly the same. Where the intermediate slab forms the base of filter boxes, the loading is vastly different (discussed later in the section on filter-boxes).

18.4.3 Foundations

Where the foundation is piles, drilled piers, or other discrete elements, the base slab will be designed as a structural element carrying all basin loads (i.e., concrete, water, superstructure, etc.) to the foundation. The magnitude of these loads may require incorporating large beams in the slab.

Where the base slab rests directly on sand or clay, it can be designed as a mat foundation carrying all loads except for the water, as the water load will be carried directly by the soil through the slab. Where possible, the bottom of this slab should be reasonably level for ease in excavation, and the top surface must drain, leading to the flat slab design. Drop panels can be incorporated in the top surface by sloping up or raising the slab at columns. The actual distribution of soil pressure will depend on relative soil–slab stiffness, and the conservative approximation is to assume uniform soil pressure. However, a beam on elastic foundation analysis should be performed to avoid underdesign of mid-height moment in walls supported or restrained at the top. This error does not occur with cantilever walls.

The design of foundations for open basins is handled in a manner similar to that for closed basins. If the foundation is piles or drilled piers, all loads must be supported on these elements, and the base slab becomes the structure to deliver the load to the piles. Where the soil has adequate bearing capacity, the only vertical loads to be supported are from the walls and any supported mechanical equipment on top walkways, and associated live loads. These loads can be distributed into soil pressure similarly to foundation loads for closed basins. The effect of soil stiffness relative to foundation stiffness should again be considered, and a sloping haunch may be provided in the top or bottom of the slab. Effects of continuity between walls and base slab will require careful attention to corner details. On such structures as aeration tanks, where interior walls are designed for loading on one side only, alternate loading conditions are considered in the determination of maximum base slab design moments.

EXAMPLE 18-1: The example given in this section is a multicell rectangular tank, such as may be part of a battery of aeration tanks. The system consists of four tanks 25 ft wide and 140 ft long with a design groundwater depth of 16.5 ft. The tank is designed for a fluid weighing 62.4 lb/cu ft. General tank layout and cross section are given in Figs. 18-14 and 18-15. For this design example, inlet and outlet structures, wiers, and gates are omitted in order to simplify the problem. The steps in the analysis are as follows:

1. Determine thickness of a vertical wall.
2. Determine thickness of the base slab.
3. Analyze tank—determine design moments for walls and base.
4. Determine reinforcing for walls and base.
5. Analyze and design walkways on long walls.
6. Analyze and design lateral walkways and tie beams.

Thicknesses of the walls and slabs in this example have been determined to be 14 in. and 18 in., respectively. The resulting dead load of the tank is found to be 6950 kips. Including a small average live load on the walkways and 16.5 ft of water, the average soil pressure under the base slab is 1400 psf. In order to use a mat foundation, the soil must then have a safe minimum bearing capacity of 1400 psf.

The dead load available to resist uplift is equivalent to 6950/(105.83 × 142.33 × 0.0624 × 1.25) = 5.9 ft of water displaced by

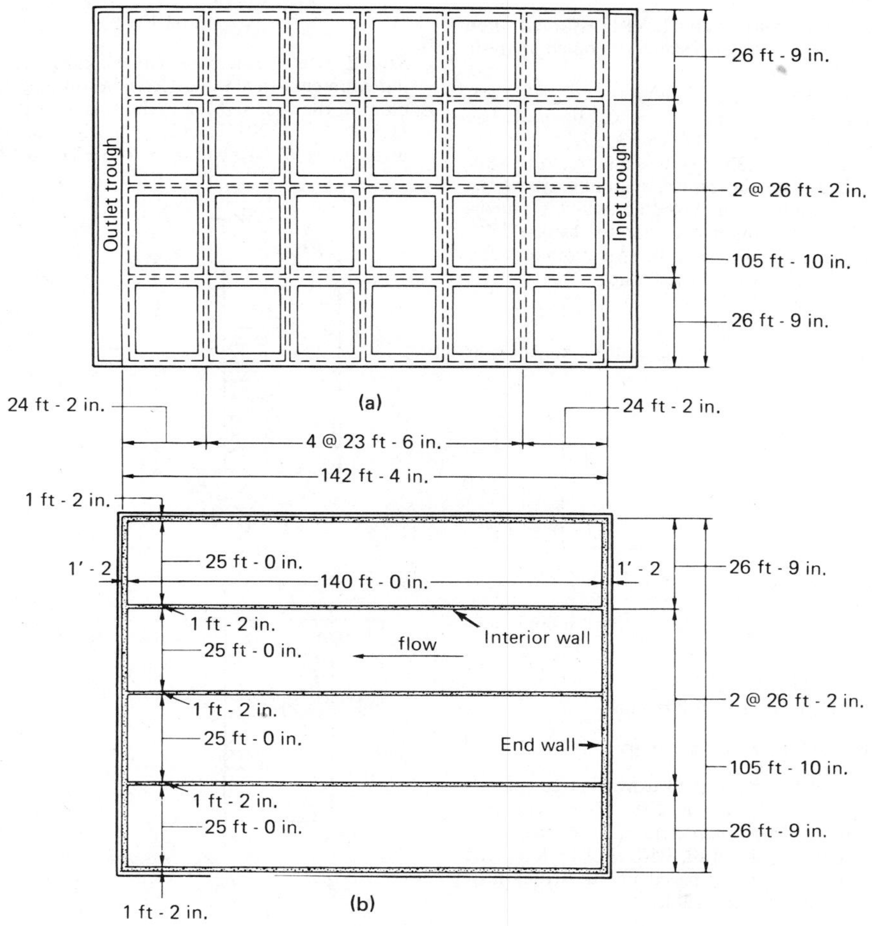

Fig. 18-14 Tank plans. (a) Top plan and (b) foundation plan.

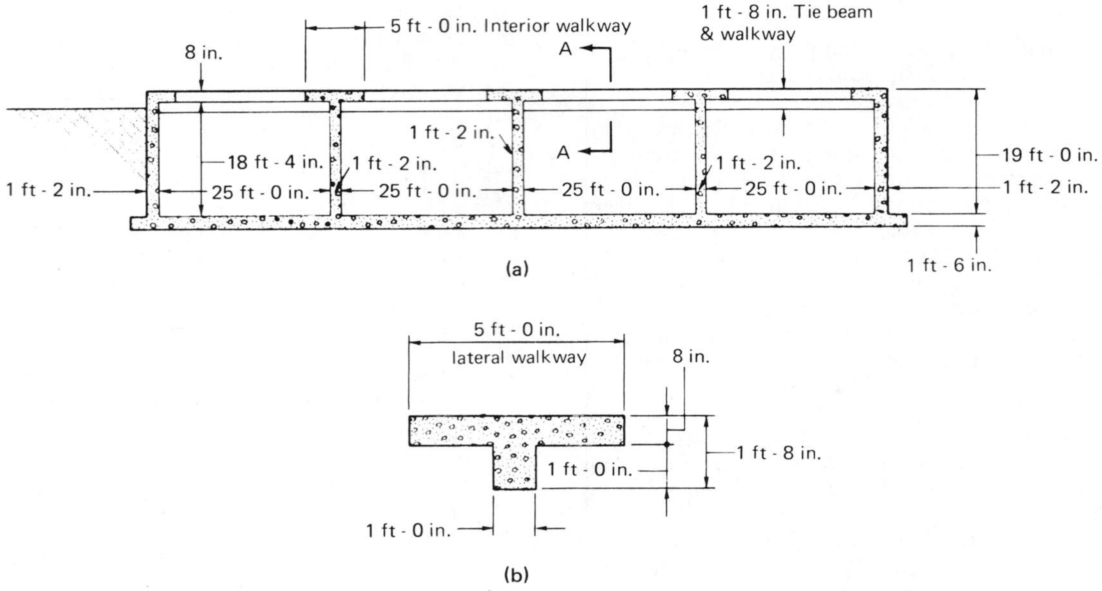

Fig. 18-15 Tank details. (a) Tank cross section and (b) section A-A.

the tank, where 1.25 is a load factor. If the groundwater level is higher than this, additional hold-down capacity may be obtained by extending the base slab to engage surrounding soil. For instance, extending the base slab to 6 ft from outside walls increases the tank dead load to 7650 kips, and 18 ft of soil above the slab yields an additional 6700 kips. Allowable ground water level is now 14350/ (117.83 × 154.33 × 0.0624 × 1.25) = 10.1 ft above bottom of base slab.

For this example problem the groundwater level is assumed to be below 5.9 ft above the bottom of the base slab. A detailed analysis for interior and end walls only is given.

Design parameters for this example will be concrete: $f_c' = 3000$ psi, and steel: $f_y = 60$ ksi. Strength design procedures will be used. The long wall panels have an aspect ratio of 140/18.33 = 7.65, and for the end wall panels, the ratio is 25/18.33 = 1.36. Thus the long walls and base slab will be analyzed as plane frames, spanning across the tank (Fig. 18-15a). The end walls be analyzed as two-way panels, using the PCA publication "Rectangular Concrete Tanks."[18-20] Clearance for all wall reinforcing is set at 2 in. All moments and shears are in terms of a 1-ft width of wall or base slab.

SOLUTION:

Step 1. Determine vertical wall thickness. The wall loadings are as follows:

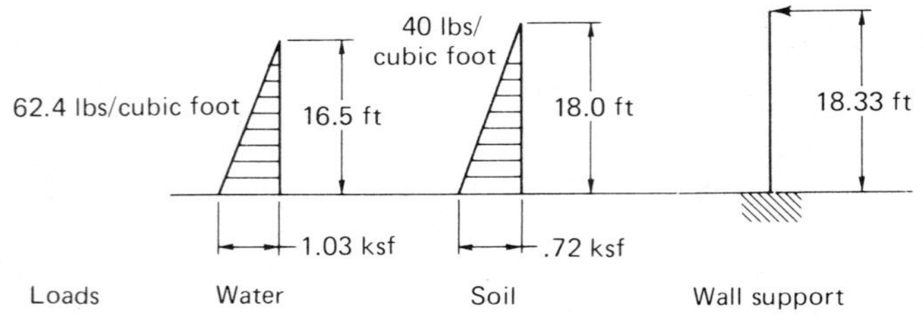

Fixed end moments, M_u:

$$\text{Water } (20.9)(1.3)(1.7) = 46.2 \text{ ft-kips}$$

$$\text{Soil } (23.8)(1.3)(1.7) = 52.6 \text{ ft-kips}$$

Formulas for fixed end moment for propped cantilever for partial uniform and triangular loadings are given in Fig. 18-16. The wall span is 18.33 ft, and for a thickness-to-span ratio of 1/16, the thickness is 18.33/16 = 1.14 ft. For a 14-in. wall, with d of 11.5 in., the maximum nominal moment strength M_n based on the strength of the concrete is, according to Fig. 18-11 or Table 18-1:

$$\phi M_n \max = Rbd^2 = (58.54)(12)(11.5)^2/1000 = 92.90 \text{ ft-kips}$$

Since the final end moment for the walls will be less than 46.20 ft-kips (the maximum fixed end moment), the 14-in. wall thickness will provide a more than adequate concrete section.

Calculate the shear strength of the wall.

$$\phi V_n = \phi V_c = \phi 2\sqrt{f_c'}\, b_w d$$

$$= (0.90)(2)\sqrt{3000}\,(12)(11.5)/1000$$

$$= 13.61 \text{ kips}$$

Step 2. Determine base slab thickness. The total factored load causing moments in the mat foundation equals tank dead load (*not* including the base slab), and the small live load on walkways, or (3560) (1.3)(1.4) + (120)(1.3)(1.7) = 6744 kips. The assumed uniform base pressure is 6744/(106.83 × 143.33) = 0.440 ksf. An approximate

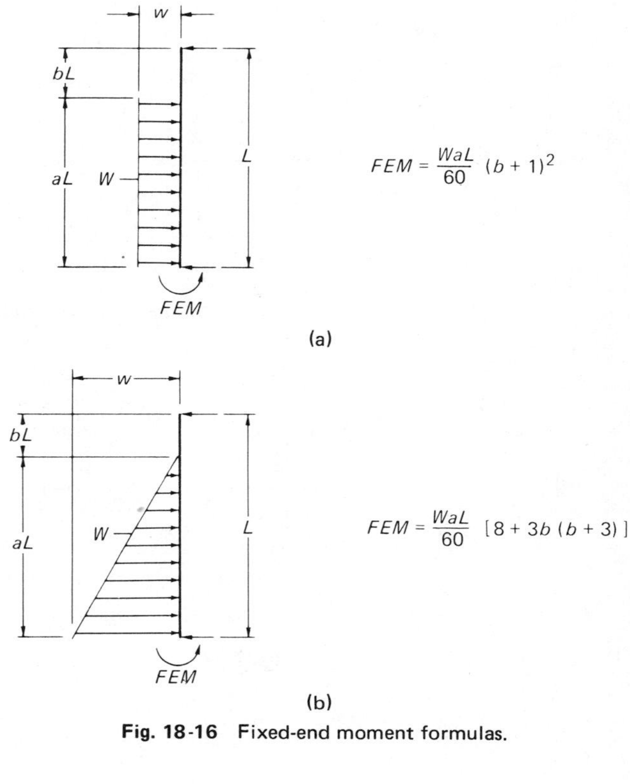

Fig. 18-16 Fixed-end moment formulas.

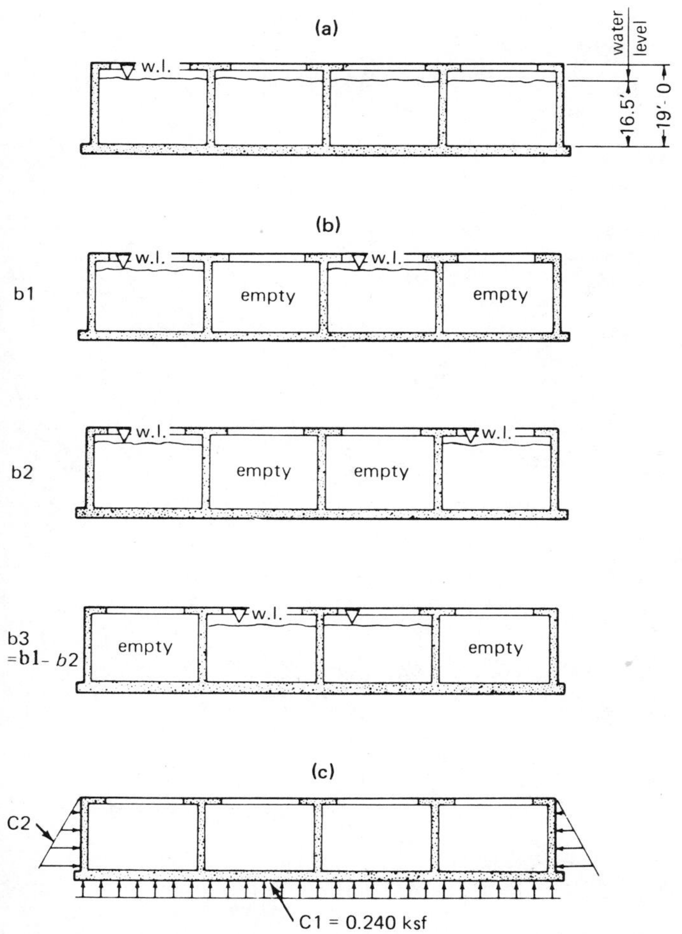

Fig. 18-17 Loading conditions. (a) Full load; (b) partial load; and (c) foundation load.

value for the maximum base moment may be determined by adding together:

1. Moment due to soil pressure using an assumed maximum moment coefficient of 1/11.
2. Moment due to support of interior walls, assuming three-fourths of fixed end moment to remain after distribution.

Thus:

$$M_\mu \approx \frac{1}{11} \times 0.440 \times 25^2 + \frac{3}{4} \times 46.2 = 59.65 \text{ ft-kips}$$

A depth-to-span ratio of 1/16 yields an 18.8-in. slab. For an 18-in. slab, the maximum concrete strength ϕM_n is $58.54 \times 12 \times 14.5^2/1000 = 147.70$ ft-kips, or approximately 147% more than required. It is determined that the tank analysis will be based on 14-in. walls and 18-in. base slab, but the designer should be ready to change these thicknesses if the following analysis suggests this to improve the design.

Step 3. Analyze tank. Normal tank operation will be with all cells filled (loading condition a, Fig. 18-17). To provide for the possibility of partial tank usage, three partial loadings are analyzed (loading conditions b1, b2, b3). Note that only two analyses need be made, since b3 = a – b2. The foundation cases are handled separately; the soil pressure becomes case c1 and lateral pressure on walls case c2 (assumed same as inside water pressure). The results of analysis of these loading conditions are given in Fig. 18-18, in terms of ft-kips per ft of width of walls and slab.

Long wall design. From Fig. 18-18, the maximum moment in ft.-kips at the base of the long walls are as follows:

Loading	*Exterior Wall*	*Interior Wall*	*Center Wall*
a + c1	14.1 c	3.1 c	0
a + c1 + c2	3.1 cc	3.7 c	0
b1 + c1	10.0 c	27.1 cc	32.1 c
b1 + c1 + c2	13.2 cc	29.5 cc	32.1 c
b2 + c1	11.1 c	30.4 cc	0
b2 + c1 + c2	12.1 cc	32.8 cc	0
b3 + c1	12.9 cc	39.5 c	0
b3 + c1 + c2	36.1 cc	37.1 c	0
c1	15.9 cc	3.0 c	0
c1 + c2	39.1 cc	0.6 c	0

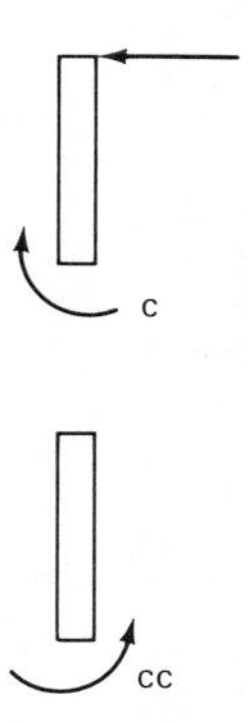

The interior walls are engineered for one design to simplify construction. Maximum positive moment in these three walls is established for an end moment of 27.1 ft-kips. The moment and shear curves are plotted for comparison in Fig. 18-19.

A wall design based on a fixed bottom may be unsafe. The reinforcing at the bottom would be more than adequate, but the midspan region could be underdesigned, and in particular, the lateral reaction at the top would be much smaller than the correct value, leading to a serious underdesign of the long walkways. These errors may be avoided by designing for the maximum of both simple and fixed end conditions, but resulting in an overdesign and unnecessary extra cost. The correct approach is to analyze the separate loading conditions as plane frames, as illustrated in this example, to determine accurately the maximum design moments and reactions.

Step 4. Determine reinforcing for walls and base. Reinforcing for the interior wall is determined as follows:

Maximum bottom moment = 39.5 ft-kips.

$$M_\mu/\phi M_n \text{ max} = 39.5/92.9 = 0.42$$

From Table 18-1, $F = 4.18$.

$$A_s = M_\mu/(Fd) = 39.5/(4.18 \times 11.5)$$
$$= 0.82 \text{ sq in./ft; use \#6 @ 6 dowels}$$

Maximum span moment = 29.1 ft-kips.

$$M_\mu/\phi M_n \text{ max} = 29.1/92.9 = 0.31;\ F = 4.27$$
$$A_s = 29.1/(4.27 \times 11.5)$$
$$= 0.59 \text{ sq in./ft; use \#5 @ 6}$$

Dowel splice length must meet the following ACI 318-83 *Code* requirements:

1. Min. of Class "C" splice length for #6 bars other than top bars spaced 6 in. or more on centers:

$$1.7 l_d = (1.7)(0.8)(0.04 A_b f_y/\sqrt{f_c'})$$
$$= (1.7)(0.8)(0.04)(0.44)(60{,}000)/\sqrt{3{,}000})$$
$$= 26 \text{ in.}$$

2. Dowels must extend to where the shear resulting from $U = 1.7F$ is two-thirds of ϕV_c. This condition occurs 1'8" above the top of the base slab. (See Fig. 18-19.)
3. Dowels must extend a minimum distance of $12 d_b$ or design depth d beyond where they are no longer required for moment. [See Fig. 18-19; (0.72)(12) + 11.5 = 20 in.]

Longitudinal reinforcement in these walls can be determined for two conditions:

a. The set of tanks considered a continuous design (see Section 18.2.2), with recommended minimum shrinkage reinforcement ratio of 0.004. To this should be added additional reinforcement area to resist tension from water load on end walls. Assuming deflection of a vertical section of end wall to be similar to that of the long walls, the maximum tension load will occur near midheight or at about 8-ft water depths (see Fig. 18-15).

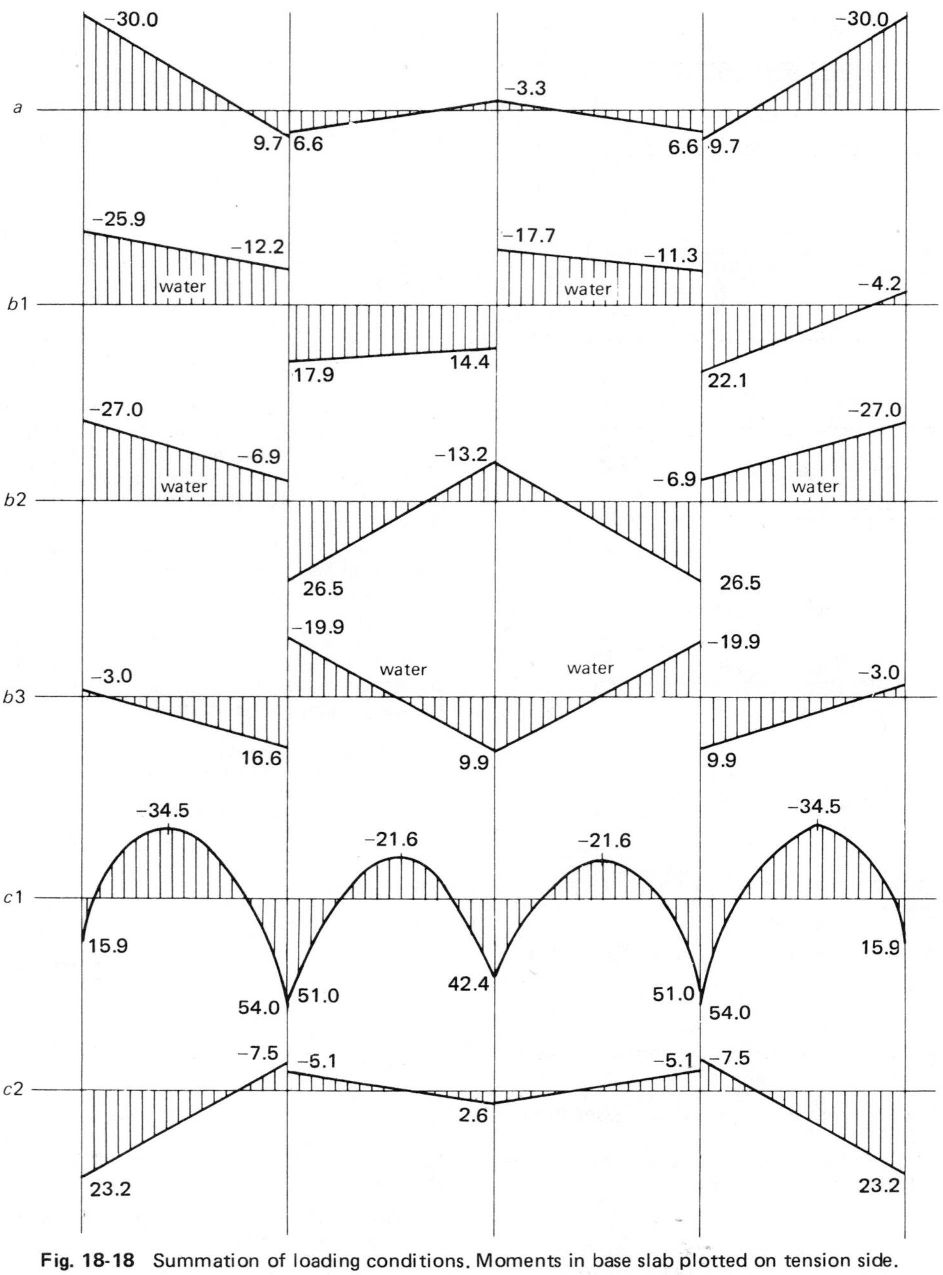

Fig. 18-18 Summation of loading conditions. Moments in base slab plotted on tension side.

The added load in the lower portion of end wall will be delivered to the base slab from continuity with the base slab. Using coefficients from Fig. 13 of "Moments and Reactions for Rectangular Plates," the assumed maximum value of this tension load is $0.2616 \times 0.624 \times 17.11^2 \times 1.65 \times 1.7 = 13.40$ kips/ft for one side, or $13.40 \times 2 = 26.8$ kips total.

$$A_s = 26.80/(0.9 \times 60) = 0.49 \text{ sq in./ft}$$

To this is added shirinkage reinforcement, as discussed in Section 18.2.3, of $0.0040 \times 12 \times 14 = 0.67$ sq in. per ft. Total horizontal reinforcing for both faces is 1.16 sq in/ft. Use #6 @ 9 each face horizontally.

b. Moment at the intersection with end walls, where two-way action will occur. The coefficient in Fig. 13 of "Moments and Reactions for Rectangular Plates" for walls with long aspect ratio $a/b = 0.75$ is 0. + 0.02503, and can be considered reasonably accurate as the wall approaches the corner. The value of a must be adjusted somewhat to enter the table owing to freeboard of wall above water level: a reasonable approximation is to use $16.5 + \frac{1}{3}(18.33 - 16.5) = 17.11$. Then $a/b = 12.5/17.11 = 0.73$. The longitudinal moment is $.0302 \times 0.0624 \times 17.11^3 \times 1.3 \times 1.7 = 20.86$ ft-kips, and A_s req'd = 0.46 sq in./ft. The solution earlier was to provide #6 @ 9 horizontally in both faces of the long walls. At the corners add additional #6 @ 9 dowel bars to carry the locally high bending moment from two-way structural action. The placement of reinforcing for interior walls is shown in Fig. 18-20.

The design of the exterior walls will follow a similar procedure. The maximum bottom moment is 3.91 ft-kips on the outside face and 36.6 ft-kips on the inside face. The maximum span moment is 38.4 ft-kips on the outside face and 36.6 ft-kips on the inside face. The value of the horizontal reaction at the top varies for interior and exterior walls. For the interior walls it is a maximum of 4.15 kips, and for exterior walls it is 5.09 kips (based on interior water load). (See Fig. 18-19.) These reactions will be used in the design of the walkways.

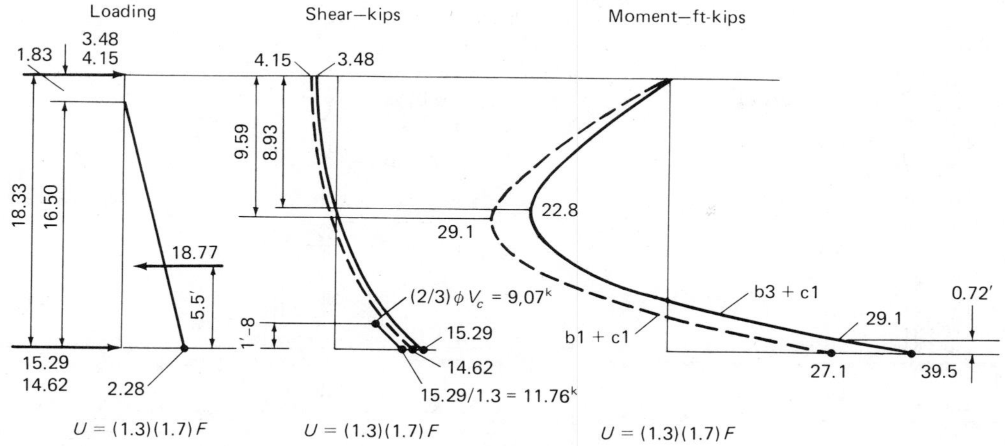

Fig. 18-19 Long interior wall.

Reinforcing for the base is determined as follows:

From Fig. 18-18, the maximum moment conditions are:
Tension top of slab (occurs between walls):

$$M_\mu = 53.6 \text{ ft-kips (Case b1 + c1)}$$

$$\phi M_n \max = Rbd^2 = (58.54)(12)(15.5)^2/1000 = 168.8 \text{ ft-kips}$$

$$M_\mu/\phi M_n \max = 53.6/168.8 = 0.32$$

$$F = 4.26 \text{ from Table 18-1}$$

$$A_s = 53.6/(4.26 \times 15.5) = 0.81 \text{ sq in/ft}$$

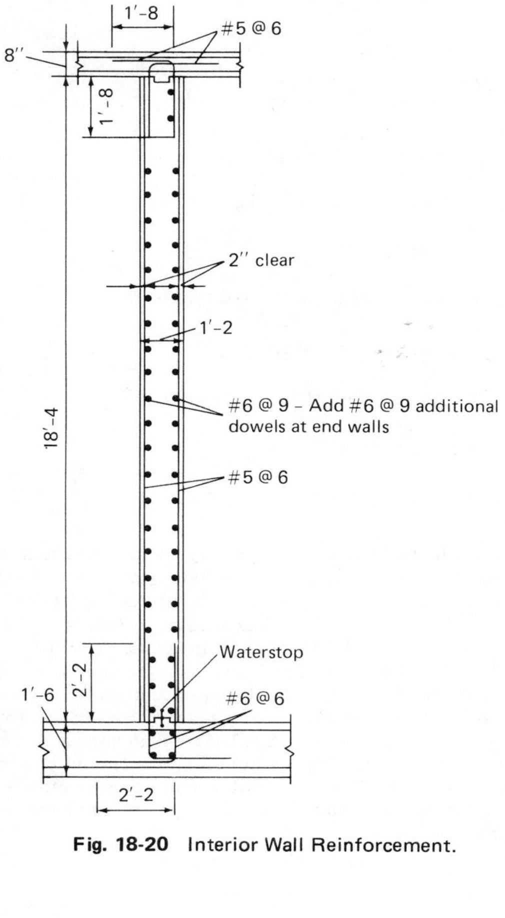

Fig. 18-20 Interior Wall Reinforcement.

Tension bottom of slab (occurs at interior wall):

$$M_\mu = 77.5 \text{ ft-kips (Case b2 + c1)}$$

$$A_s = 77.5/(4.16 \times 14.5) = 1.28 \text{ sq in./ft}$$

Tension top of slab at exterior wall:

$$M_\mu = 14.1 \text{ ft-kips (Case a + c1)}$$

$$A_s = 14.1/(4.44 \times 15.5) = 0.20 \text{ sq in./ft}$$

$$A_s(\min) = 200/f_y bh = (200/60{,}000)(12)(18)(1/2) = 0.36 \text{ sq. in./ft.}$$

Tension bottom of slab at exterior wall:

$$M_\mu = 39.1 \text{ ft-kips (Case c1 + c2)}$$

$$A_s = 39.1/(4.34 \times 14.5) = 0.62 \text{ sq in./ft}$$

To these reinforcement areas must be added area for tension steel to carry bottom thrust, equal to maximum shear at bottom of wall:

$$V_\mu \max = (15.29)(1.65/1.3) = 19.41 \text{ kips}$$

$$A_s\text{req'd} = 19.41/(2 \times 0.90 \times 60) = 0.18 \text{ sq in./ft}$$

in top and bottom layers. The required reinforcement areas are then:

Top:

$$A_s\text{req'd} = 0.81 + 0.18 = 0.99 \text{ sq in./ft; use \#7 @ 7}$$

Bottom:

$$A_s\text{req'd} = 1.28 + 0.18 = 1.46 \text{ sq in./ft; use \#7 @ 5}$$

Reinforcement arranged as shown in Fig. 18-21 will provide #7 @ 10 in bottom of slab in center of span and at exterior edge, and these areas must be checked. For the center of span, the moment capacity is (0.73 − 0.18) × 4.38 × 14.5 = 34.9 ft-kips and should be checked against case b1 + c1 and case b2 + c1 to determine bar length. For the exterior wall, the A_s required is 0.62 + 0.18 = 0.80 sq in. in the bottom of the slab and 0.20 + 0.18 = 0.38 sq in. in the top of the slab.

Consistent with wall design (see Section 18.2.2), bottom slab longitudinal reinforcement should be 0.0040-ratio shrinkage reinforcement plus tension for end walls.

Shrinkage:

$$0.0040 \times 12 \times 18 = 0.86 \text{ sq in.}$$

Tension:

$$(15.29)(1.65/1.3)/(0.9 \times 60) = 0.36 \text{ sq in.}$$

Total required reinforcement = 1.22 sq in./ft; use #6 @ 8½ top and bottom. See Fig. 18-21 for details of base slab reinforcement.

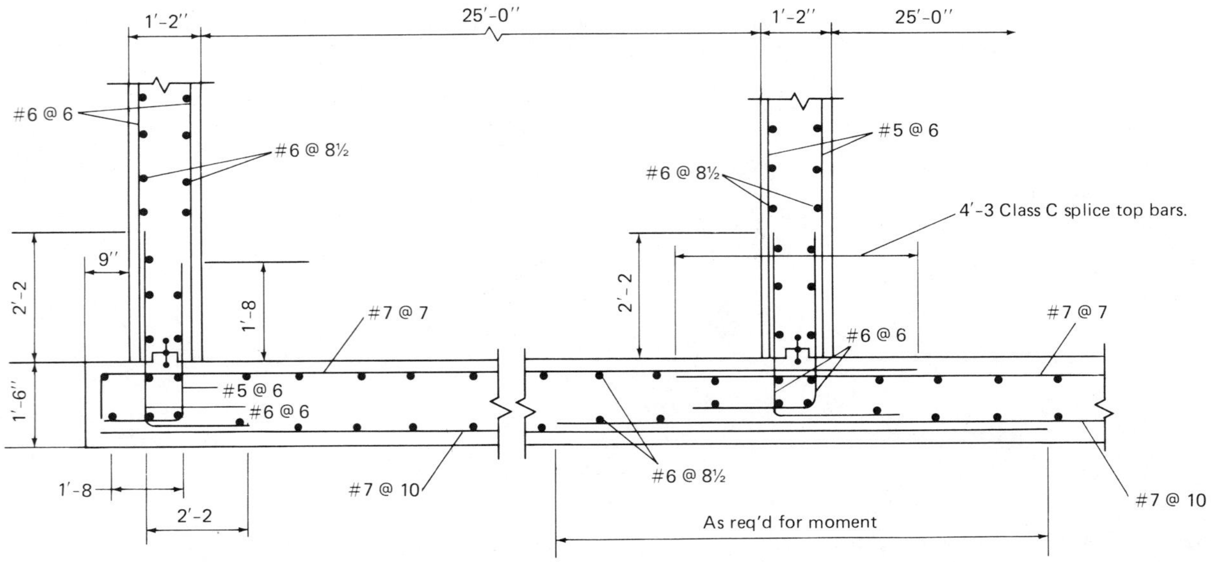

Fig. 18-21 Slab reinforcing details.

End wall design: Reinforcing for the end walls and wall intersections can be determined with the aid of coefficients in Fig. 13 in "Moments and Reactions for Rectangular Plates."

As developed earlier, the effective depth used was taken as 17.11 ft; then $a/b = 12.5/17.11 = 0.73$ and the coefficients are as follows:

y/b	Wall Intersection $x = 0$				Midspan $x = a$		
	R_x	R_y	M_x	M_y	R_y	M_x	M_y
1.0	−.0671	−.0671	0	0	.1061	0	0
0.8	.1275	—	.0181	.0036	—	−.0074	−0.113
0.6	.2274	—	.0301	.0060	—	−.0119	−.0196
0.4	.2616	—	.0302	.0060	—	−.0109	−.0198
0.2	.1479	—	.0155	.0031	—	−.0032	−.0021
0	−.0155	−.0155	0	0	.3828	.0101	.0505

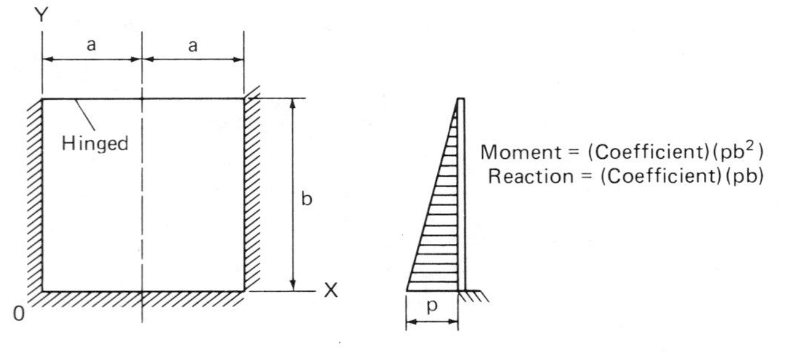

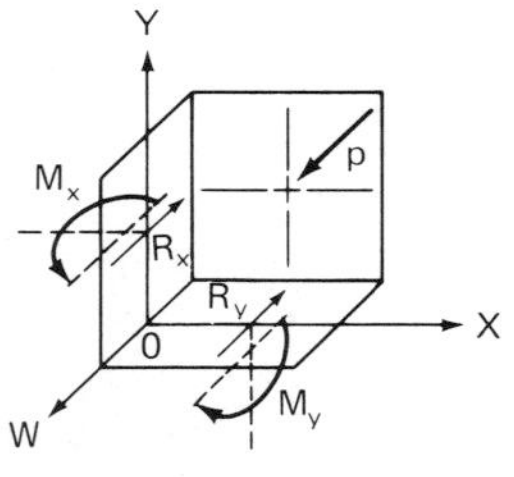

Positive sign convention

The resulting wall moments and reinforcing are as follows:

Vertical span at $x = a$ and $y = 0$:

$$M_u = .0101 \times (.0624 \times 17.11) \times 17.11^2 \times 1.3 \times 1.7 = 6.91 \text{ ft-kips}$$

$$A_s = 6.91/(4.46 \times 10.5) = 0.15 \text{ sq in./ft; use \#5 @ 10}$$

Vertical span at $x = a$ and $y = 0.66$:

$$M_u = -.0119 \times (.0624 \times 17.11) \times 17.11^2 \times 1.3 \times 1.7$$

$$= 8.22 \text{ ft-kips}$$

$$A_s = 8.22/(4.40 \times 10.5) = 0.18 \text{ sq in./ft; use \#5 @ 10}$$

Horizontal span at $x = 0$ and $y = 0.4b$

$$M_u = .0302 \times (.0624 \times 17.11) \times 17.11^2 \times 1.3 \times 1.7$$

$$= 20.86 \text{ ft-kips}$$

$$A_s = 20.86/(4.23 \times 11.5) = 0.43 \text{ sq in./ft}$$

Horizontal span at $x = a$ and $y = 0.4b$

$$M_u = 0.0198 \times (.0624 \times 17.11) \times 17.11^2 \times 1.3 \times 1.7$$

$$= 13.68 \text{ ft-kips}$$

$$A_s = 13.68/(4.34 \times 11.5) = 0.27 \text{ sq in./ft}$$

For the horizontal span, it is necessary to add some tension area to account for lateral reaction from long walls. From Fig. 13 of "Moments and Reactions for Rectangular Plates," the value for this reaction at $y = 0.4b$ and $x = 0$ is $0.2616 \times (.0624 \times 17.11) \times 17.11 \times 1.65 \times 1.7 = 13.40$ kips and added area $13.40/(0.9 \times 60) = 0.25$ sq in. each wall face. Total area required for the continuous reinforcing is then $0.43 + 0.25 = 0.68$ sq in. and #6 @ 7½ are used.

The details of this reinforcing are shown in Fig. 18-22, for inside pressure only. It is assumed in this layout that outside design pressure was the same, and that horizontal reinforcing in both faces is then the same. Note that horizontal reinforcing is moved to the outer layer in the end walls, since these moments are larger than the vertical moments.

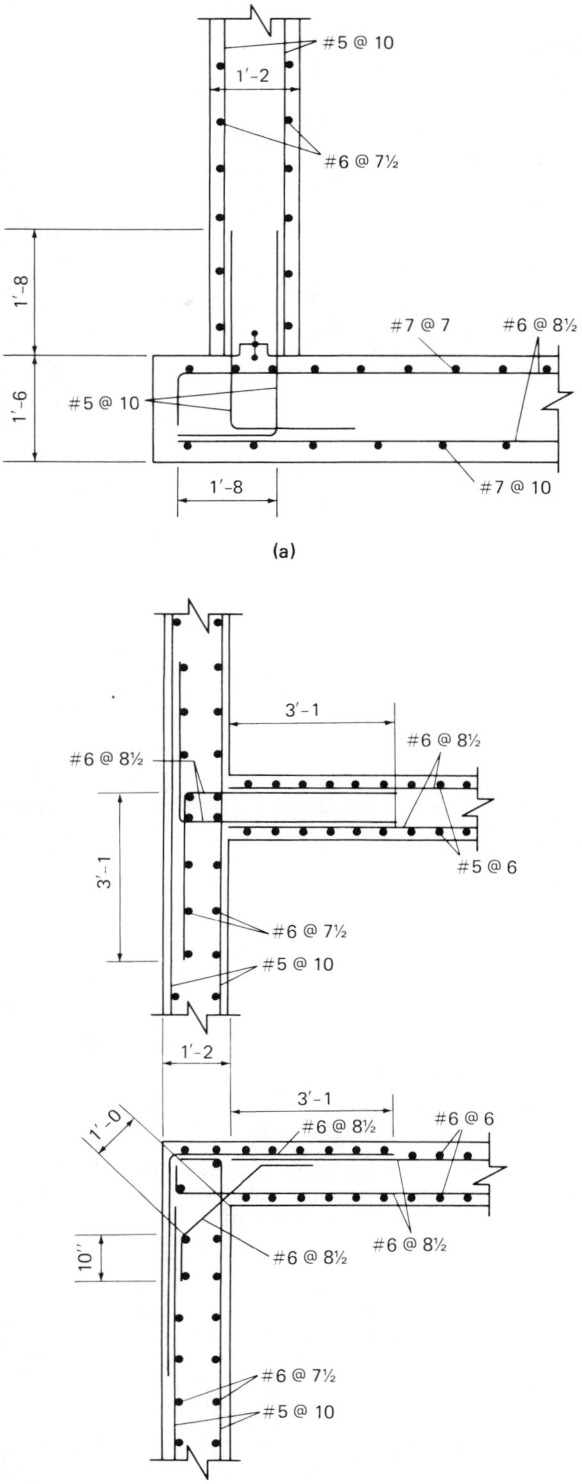

(b)

Fig. 18-22 End wall reinforcing details. (a) Vertical section; (b) plan at corners.

Design of longitudinal walkways: The design of interior walkways is presented; exterior walkways follow a similar pattern. These walkways provide lateral support for the top of the long walls, and span between the lateral walkways and tie beams. The maximum horizontal reaction for the interior walkways is 4.15 kips per foot of length. This is a uniform constant load, thus *ACI Code* coefficients are adequate for analysis:

$$\text{Maximum moment} = 1/10 \times 4.15 \times 22.5^2 = 210.1 \text{ ft-kips}$$

$$\text{Maximum shear} = 1.15 \times 4.15 \times 22.5/2 = 53.7 \text{ kips}$$

The minimum practical thickness for these elements is 8 in. to provide room for two bars vertically in each edge in addition to ties. The reinforcing clearance is set at 1.5 in. to ties. Then, assuming an 8 × 60 section, the design strengths are:

Moment:

$$\phi M_n \max = 58.54 \times 12 \times 57.5^2/1000 = 2322 \text{ ft-kips}$$

$$A_s = 210.1/(4.43 \times 57.5) = 0.82 \text{ sq in.; use 1 \#8}$$

Shear:

$$\phi V_n = \phi V_c = \phi 2\sqrt{f_c}\, b_w d = 0.85 \times 2 \times \sqrt{3000} \times 12 \times 57.5/1000$$

$$= 64.2 \text{ kips}$$

The moment and shear capacities are more than adequate. Minimum reinforcing is 0.0033 × 8 × 57.5 = 1.52 sq in. Since this minimum exceeds the 1.14 sq in. from analysis, an alternate minimum requirement of one-third more than analysis may be supplied, or 1.33 × 0.82 = 1.09 sq in. Note that partial loads may reverse bending moments; thus the same reinforcing is provided in both edges, and laps should fully develop the reinforcing and preferably be made at the quarter-point of the span. Additional longitudinal reinforcing is added in the 60-in. width to provide a total reinforcing ratio of 0.004, consistent with the minimum used for the basin walls and slab.

The overhanging slab is designed for a uniform load of 100 lb/sq ft or a concentrated load at the edge of the slab, and it is determined to use #4 @ 12 ties. A reinforcing detail is shown in Fig. 18-23.

Design of lateral walkways and tie beams: These elements provide the support for longitudinal walkways, in addition to their own weight and a reasonable live load. Since they are continuous across the tank cells, the maximum condition will occur in the exterior bays, and the lateral reaction from the exterior wall will govern. Further, the critical condition may occur when the exterior cell is full, and the soil has been removed on the outside. For this condition the beam is required to act as a tension member with moment. The wall reaction at the top is 5.09 kips/ft, and the tension is then 127.2 kips.

The significant live load on this beam will be caused by temporary loading from equipment removal or installation, etc. A minimum of 100 lb/sq ft should be considered acceptable, but potential concentrated loads should be checked. Allowing for a 12 × 20 beam section, the maximum support shear and moment are as follows:

slab	0.100 × 5 × 1.3 × 1.4	= 0.91
live	0.100 × 5 × 1.3 × 1.7	= 1.11
stem	0.150 × 1.3 × 1.4	= 0.27
		2.29 kips/ft

$$V_\mu = 2.29 \times 23.5/2 \times 1.15 = 30.9 \text{ kips}$$

$$M_\mu = 2.29 \times 23.5^2/10 = 126.5 \text{ ft-kips}$$

$$\phi M_n \max = 58.54 \times 12 \times 17.5^2 = 215.1 \text{ ft-kips}$$

$$A_s = 126.5/(4.05 \times 17.5) = 1.78 \text{ sq in.}$$

Check the tie beam for a combination of a tension force T_μ = 127.2 kips and a moment M_μ = 126.5 ft-kips. As shown in the load-moment interaction diagram in Fig. 18-23, 4 #11 bars are satisfactory.

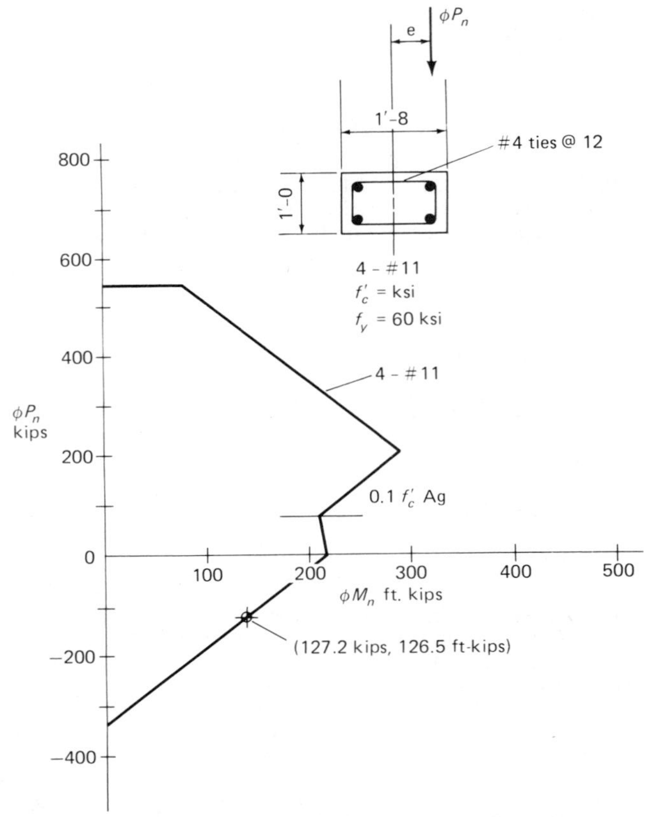

Fig. 18-23 Column load-moment interaction diagram.

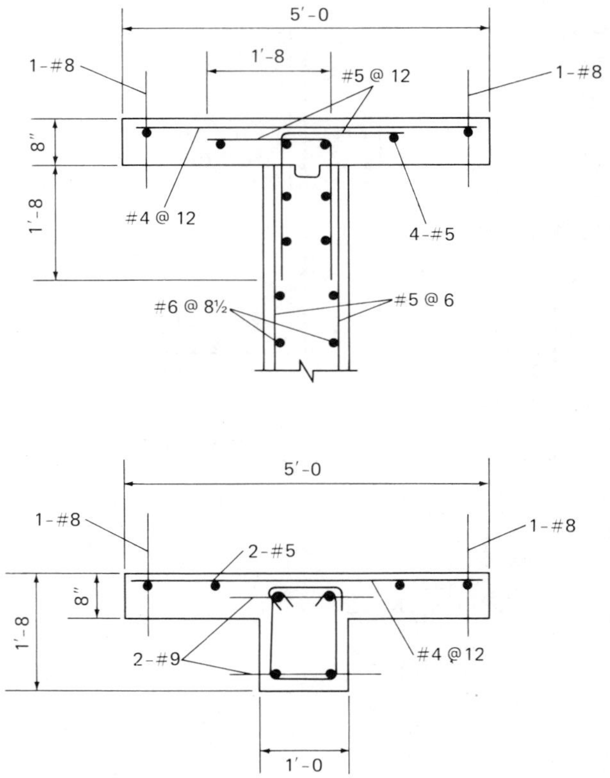

Fig. 18-24 Walkway reinforcing details. (a) Longitudinal walkway; (b) lateral walkway.

Beam shear requirements will require #3 stirrups at a spacing not to exceed $d/2$. For the 12 X 20 section:

$$\phi V_c = 0.85 \times 2\sqrt{3000} \times 12 \times 17.25/1000 = 19.3 \text{ kips}$$

The shear at d out is 30.9 − 3.7 = 27.2 kips and ϕV_s = 27.2 − 19.3 = 7.9 kips. For a maximum spacing, #3 @ 8 yields:

$$\phi V_s = A_v f_y d/s = 0.22 \times 60 \times 17.25/8 = 28.5 \text{ kips}$$

which is greater than 7.9 kips and OK. The reinforcing detail for this element is shown in Fig. 18-24.

EXAMPLE 18-2: This example concerns the design of a two-story settling basin, part of a water treatment plant (areas B and C, Fig. 18-2). The major point of this example is to illustrate the loads on the structure and the general approach to the wall and slab design. The detailed analysis of the slab design is omitted, but follows *ACI Code* (318-83) procedures, and differs little from an analysis of any other flat slab structure according to Chap. 13 of this *Code*.

The layout of the settling basin is shown in Fig. 18-25. The slabs should be designed for the following loads:

Top slab:

12-in concrete slab	150 psf
water proof membrane	10 psf
3-in concrete protection	38 psf
2-ft dirt cover	240 psf
Total DL	438 psf
LL	100 psf (minimum)

Intermediate slab:

DL–9-in. concrete slab	113 psf
LL–sludge, misc.	75 psf

Base slab (design pressure):

DL of supported slabs	551 psf
add for walls, columns	140 psf
Total DL	691 psf
LL	175 psf

SOLUTION: To determine total base soil pressure, add:

18-in. base slab	225 psf
17.75 ft water	1109 psf
Total pressure	2200 psf

For an underground basin such as this one, where the final ground level is about the same as the original ground surface, the weight of the removed soil = 22.5 X 120 = 2700 psf, and exceeds the final total soil pressure. Thus, settlement problems would not usually occur, but a soil investigation should nevertheless be carefully performed to verify uniform and adequate soil support.

The basin is analyzed as a continuous structure. Plane frames are considered in both directions, with the width of each frame 25 ft, to conform with the flat slab analysis procedures. It is recommended that the basin be filled before backfilling, and an analysis should be performed for this loading. During normal operation, the basin will be periodically emptied for cleaning, establishing a second major loading condition. Careful consideration should be given to the proper lateral soil pressure on the walls, to act together with the soil load on the top slab. If this lateral pressure may be removed at a future date, another loading condition is established. Possible flotation of the empty basin should be checked as described in Example 18-1. The dead load of this basin alone is equivalent to a ground water level 9 ft above the base slab, or 13 ft below grade.

The base slab may be established at uniform thickness, or drop panels may be incorporated in the top of the slab to provide additional shear capacity at the columns. In this example, drop panels are used, sloping up from the edges to the column in order to provide for easier cleaning of the basin. A collection trough leading to a sump pit should be provided to facilitate the removal of settled material.

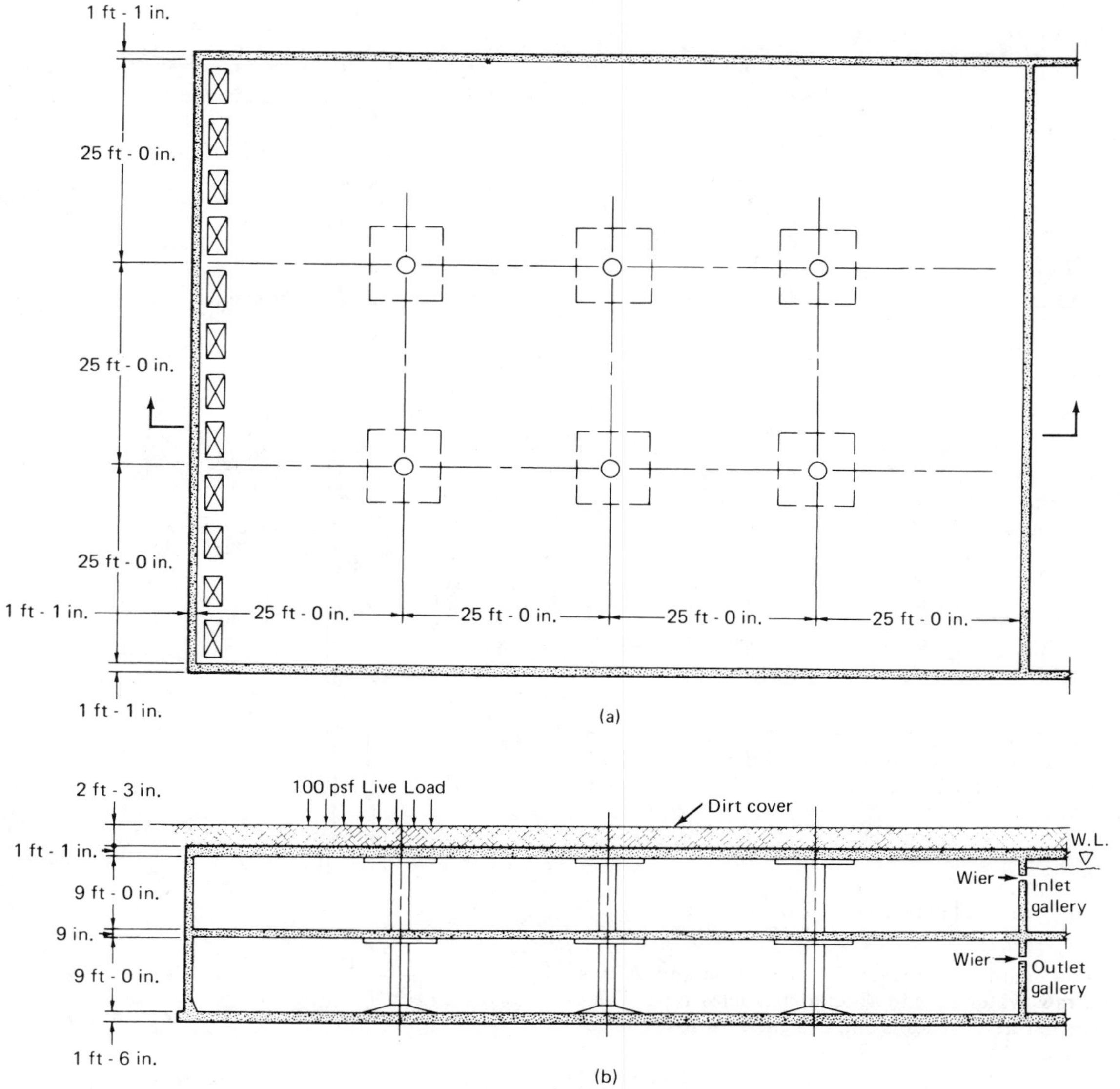

Fig. 18-25 Two-story settling basin. (a) Plan—intermediate level and (b) section.

The wall loadings and design moments are shown in Fig. 18-26. Lateral soil pressure is assumed to be greater than active due to the increased wall stiffness from support at top and intermediate levels. Wall reinforcing for a section in the middle of the basin is given in Fig. 18-27. Tension resulting from inside water pressure is carried in the slabs. In the vicinity of the corners, two-way action will deliver tension reaction to the wall, but it can be shown that the wall acting as a vertical plate element will easily deliver this load to the slabs. Thus lateral reinforcement in the wall is for shrinkage crack control only, at the ratio of 0.0040 established previously for continuous structures (see Section 18.2.3).

The effect of two-way action at the wall corners requires special consideration, as described in Example 18-1. Additional horizontal wall reinforcing is required to handle the corner moments, similar to that shown in Fig. 18.22.

18.4.4 Filter Boxes

Under normal operating conditions, water or waste water passing through filters will be subject to loss of hydraulic head (see Fig. 18-3). However, for backwashing or flushing cycles, the walls will be subjected to normal hydraulic pressure plus the submerged lateral pressure from filter material. Thus walls will be designed in similar manner to other basins except for somewhat higher lateral pressures. For instance, with sand as filter medium with unit dry weight of 120 lb/sq ft and ϕ of 30°, the maximum unit lateral pressure will be $0.25(120 - 62.4) + 62.4$, or about 80 lbs/sq ft/ft.

Filter boxes in water treatment plants are placed in banks of square or rectangular tanks. The sediment collected on the filter medium is removed by backwashing into the supply gullets, and then removed from the plant. These boxes are often supported above clear wells, into which the filtered water drains by gravity. Thus weight of filter medium and water is of primary concern. A common rapid sand filter will be 10 to 11 ft deep, with 4 ft of filter media (sand and gravel) underlain by underdrains. The total weight of media and water will then be 1035 lb/sq ft, as follows:

6 ft water @ 62.4	=	375 psf
4 ft sand @ 120	=	480 (70% solid)
4 ft water @ 0.30 × 62.4	=	80
underdrain system and gravel; approx.	=	100
		1035 psf

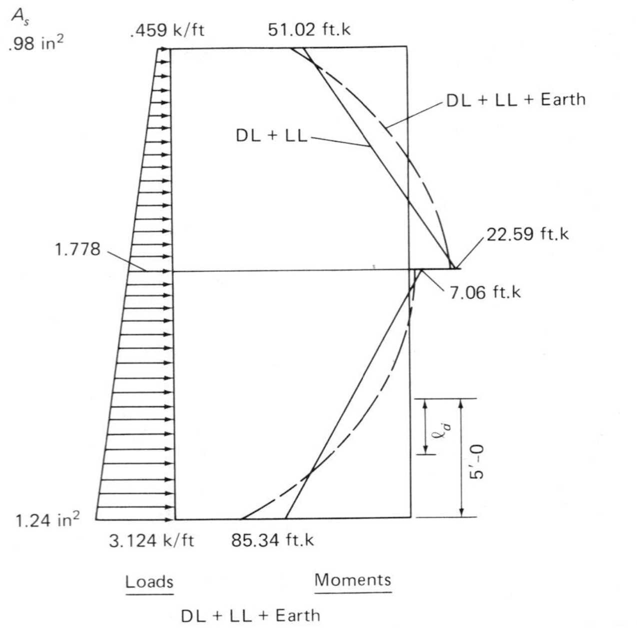

Fig. 18-26 (a) Wall loads and (b) design moments (ft-kips/ft of width).

Some newer media and proprietary media will have different specific gravities, which may increase the vertical load. For instance, in a mixed media filter, the following materials may be used:

	spec. grav.
garnet	4.0
sand and gravel	2.65
anthrafilt	1.60

These boxes require walkways for inspection and gullets for moving raw water in and filtered particles out. Walkways are cantilevered off walls and designed for loads as described previously in Example 18-1.

Wastewater trickling filters are generally round tanks with sloping bottoms. In order to accomplish even dispersal of the wastewater over the filter surface, rotating distributors are used. These tanks should be designed as outlined in the section on circular basins, but the designer is cautioned to be aware that they will be flooded periodically for fly control, at which time maximum lateral pressures will be achieved.

18.5 DESIGN OF CIRCULAR TANKS

Circular tanks are, in general, more economical than rectangular tanks. The circle has the least perimeter, in terms of area enclosed, and a major portion of the structural action is in hoop tension. However, whereas the circular tank may be preferred for economy, the tank shape used will largely be based on function and plant layout.

18.5.1 Edge Conditions

The detail and performance of the joint between the side wall and base slab have a significant effect on the structural behavior of the wall. Three types of joints are defined, as illustrated in Fig. 18-28. The sliding joint enables all lateral pressure to be resisted by hoop tension, whereas the hinged

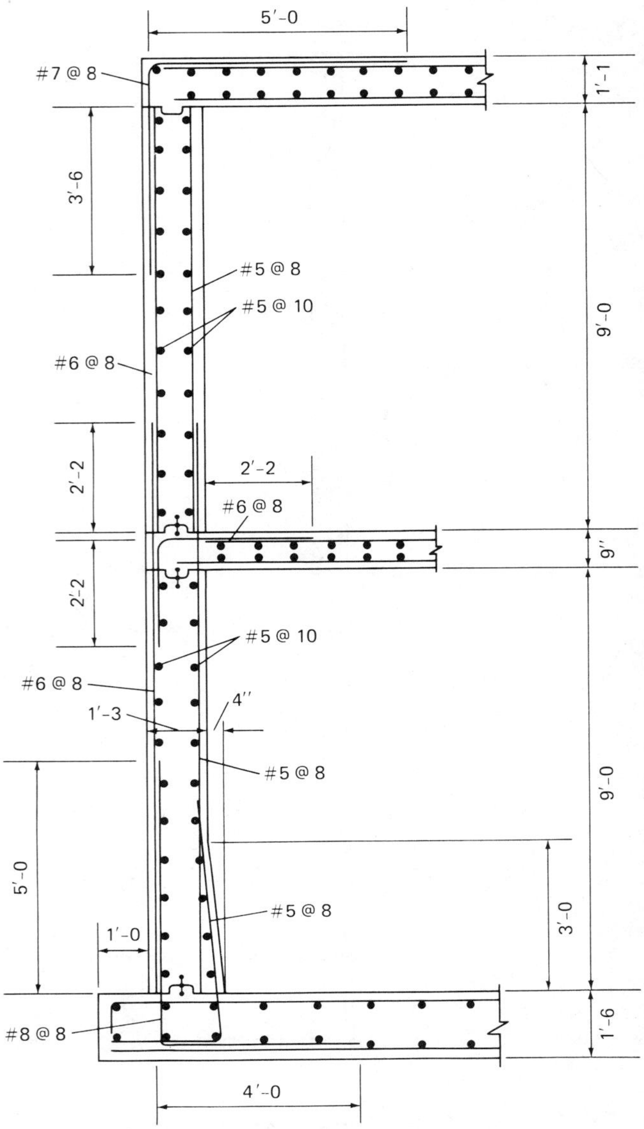

Fig. 18-27 Wall reinforcing detail.

or the continuous joint creates two-way action in the wall. The governing equation is of the same form as that for beams on elastic foundation. Tables of coefficients are available[18-22] to aid in determination of hoop tension and wall moments.

When the design utilizes a hinged or a continuous base joint, the ability of the foundation or base slab to deliver the required restraint must be carefully checked. Since axial tension is a relatively stiff structural system, a small radial displacement in the foundation or base slab will relieve the base shear, and in effect the base joint becomes more nearly a sliding joint. As an example, the radial displacement in Example 18-3, to follow, may be studied (see Fig. 18-29). The maximum hoop tension for a sliding joint would be $62.4 \times 33 \times 55 = 113{,}200$ lb/ft, and the reinforcement area is $113{,}200/20{,}000 = 6.70$ sq in./ft. The transformed section is $12 \times 22 + 9.2 \times 6.70 = 325.6$ sq in., and the unit stress is $113{,}200/325.6 = 406$ psi. The corresponding unit strain equals $f_c/E_c = 406/3{,}160{,}000 = 0.00013$, and the radius has received an elongation of $55 \times 12 \times 0.00013 = 0.086$ in., or a little over 1/16 in. Thus, if the wall is designed for a sliding base joint, the joint must accommodate this move-

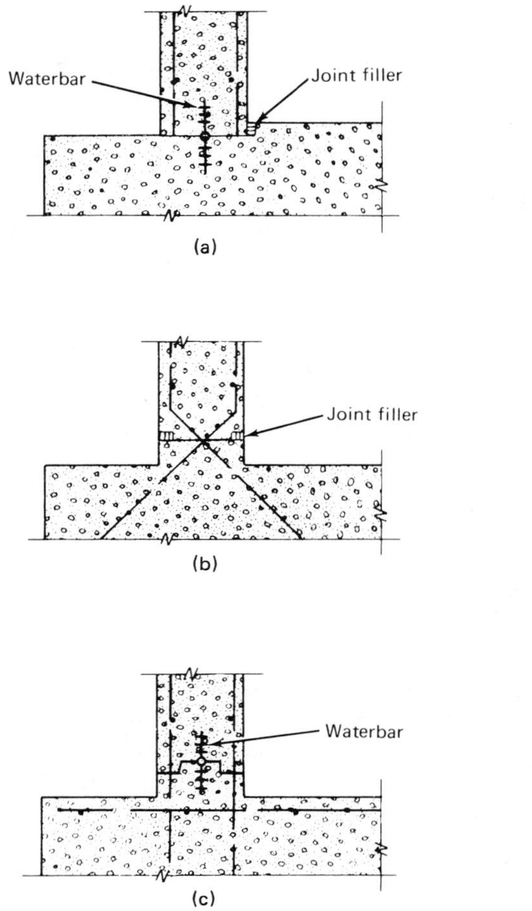

Fig. 18-28 Base joint details. (a) Sliding base; (b) hinged base; and (c) continuous base.

ment. Conversely, if it is designed for a hinged or continuous base joint, the foundation or base slab must be capable of restraining this movement.

18.5.2 Side Walls—Shrinkage and Tension

The design of circular tank walls is handled in two steps. First, the horizontal steel is provided for all the ring tension at an allowable stress, f_s, as though one were designing for a cracked section. Then the concrete stress is determined for the transformed section due to combined shrinkage and tension.

The formula for combined shrinkage and tension stress in a reinforced concrete ring can be developed as follows.[18-11] At a point near the center of an uncracked section of the wall, the shortening per unit length of steel and concrete must be equal, and strain capatibility yields:

$$\frac{f_{ss}}{E_s} = C - \frac{f_{cs}}{E_c}$$

where f_{ss} and f_{cs} are steel and concrete stresses due to shrinkage, and C is the coefficient of shrinkage of concrete. From stress consistency:

$$A_c f_{cs} = A_s f_{ss}$$

Combining these two expressions yields the following concrete tension due to shrinkage:

$$f_{cs} = CE_s \frac{A_c}{A_c + nA_s}$$

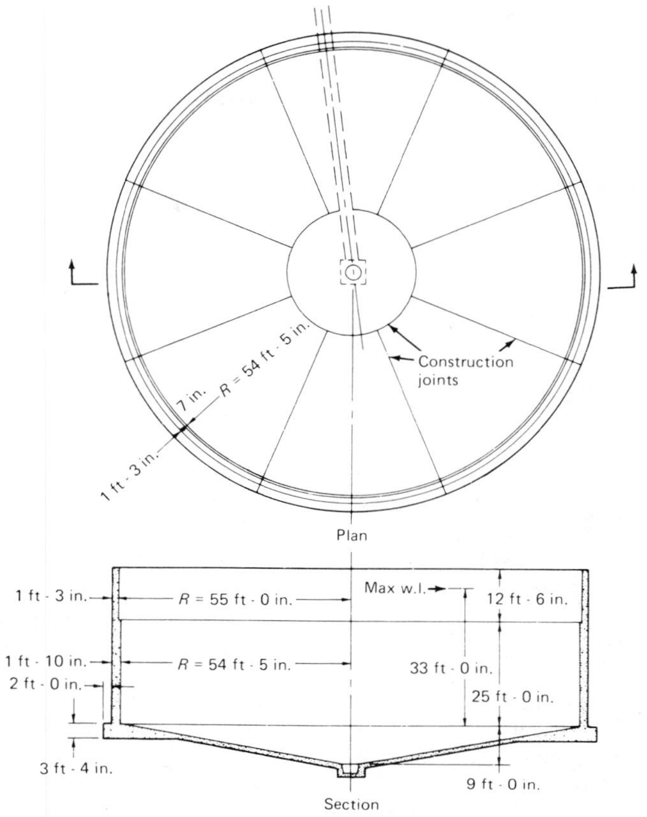

Fig. 18-29 Tank in Example 18-3.

The concrete stress in the transformed section due to tension, T, is very nearly equal to $T/(A_c + nA_s)$, and the combined concrete stress becomes:

$$f_c = \frac{CE_s A_s + T}{A_c + nA_s}$$

The limitation on concrete stress in the transformed section is established at 350 psi, in consideration of the quality concrete mix recommended in Section 18.2.1. The coefficient of shrinkage of concrete is generally assumed to be 0.0003. The allowable steel stress in ring tension has been subject to very conservative limitations in the past. Unless bond and crack width are carefully examined, simply reducing steel stresses may not prevent leakage.

For pure tension, the structural durability coefficient of 1.65 results in a service load stress for liquids of approximately 20 ksi:

$$f_s = \frac{\phi f_y}{(1.65)(1.7)} = \frac{(0.9)(60)}{2.80} = 20 \text{ ksi } (1400 \text{ kgf/cm}^2)$$

The Committee 350 report at the present time recommends a maximum service load stress of 14,000 ksi (1000 kgf/cm^2). The authors do not agree with this recommendation. It is the authors opinion that the proposed change to 20,000 psi (2000 kgf/cm^2) is proper. Increasing the service load hoop tension stress tends to make the concrete crack less often because the lower the steel stress, the greater the area of steel provided, with more restraint against shrinkage, resulting in higher tensile stresses in the concrete. This is demonstrated in Fig. 18-30 for a 100-ft (30-m) diameter sludge digestion tank using the theory set forth by the PCA for hoop tension in concrete and steel.

Limiting the service load hoop tension stress to a maxi-

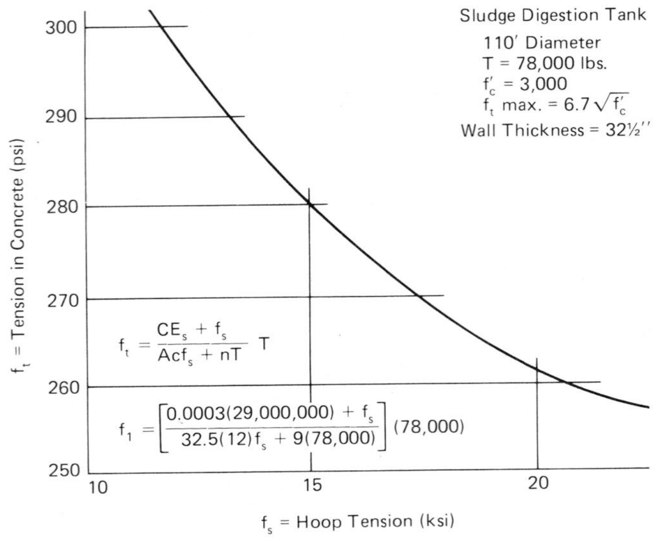

Fig. 18-30 Service load hoop tension stress.

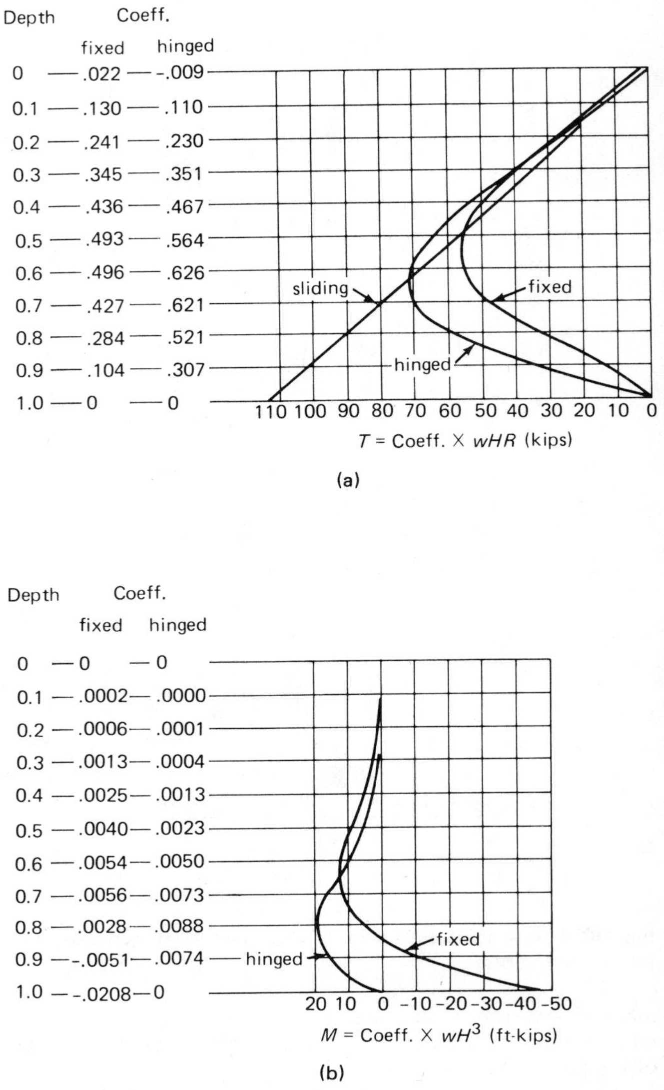

Fig. 18-31 (a) Hoop tension and (b) moment versus coefficients for determining ring reinforcing in Example 18-3.

mum value of 20 ksi (2000 kgf/cm^2) corresponds to a tension stress in the concrete of approximately 265 psi (18 kgf/cm^2). This general practice has been followed by an increasing number of designers, beginning as early as 1950, when modern "high bond" reinforcing bar deformations were standardized.

EXAMPLE 18-3: This example concerns the design of a medium-sized digestion tank. The tank plan and cross section are shown in Fig. 18-29. The lower wall thickness has been determined to be 22 in. Corbels are provided for support of clarifier equipment, and above this point the wall thickness is reduced to 15 in. The wall projects 4.5 ft above the maximum water level, and it is assumed that this projection corrects for the loss of circumferential stiffness in the thinner top section. Thus, for design purposes, the top of the tank is assumed to be at the maximum water level, with a uniform wall thickness of 22 in. Design parameters for this example will be concrete: $f_c' = 3000$ psi, concrete tension on transformed section: 350 psi maximum, $U = 1.65 \times 1.7$, and clearance to reinforcing = 2 in. except at bottom of bottom slab, 3 in.

The design of the tank wall and slab is accomplished with the aid of coefficients from the Portland Cement Association publication "Circular Concrete Tanks Without Prestressing."[18-22] When coefficients are used, the applicable table is listed. The steps in the analysis are as follows:

1. Establish preliminary wall thickness.
2. Determine ring reinforcing, check concrete stress, and verify wall thickness.
3. Determine vertical wall reinforcing.
4. Determine base slab thickness and reinforcing.

The design is given for interior water pressure only. The design for loading from exterior soil and groundwater pressure is not shown. However, it can be handled easily, as the tank ring will be in compression, and vertical wall moments only need be determined. For a tank of this size, design to resist flotation with tank dead load is not practical. Relief valves should be provided, together with a back-up system as described in Section 18.3.1.

Step 1. Establish wall thickness. The formula for concrete stress due to shrinkage and tension may be used to establish preliminary wall thickness. Substituting $A_s = T/f_s$ and $A_c = 12\,t$ yields:

$$t = \frac{CE_s + f_s - nf_c}{12\,f_c f_s}\,T$$

It is determined that this tank design shall utilize a continuous joint at the wall and base slab. Then the maximum hoop tension will occur at about 0.65 × water depth, and is 0.65 × 62.4 × 33 × 55 = 74,700 lb/ft. The estimated wall thickness is:

$$t = \frac{0.0003 \times 29{,}000 + 20 - 9.2 \times .350}{12 \times .350 \times 20} \times 74.7$$

$$= 22.3 \text{ in.; use 22 in.}$$

To enter the tables in Ref. 18-22, compute the quantity $H^2/D\,t = 33^2/(110 \times 1.83) = 5.4$.

Step 2. Determine ring reinforcing. With coefficients from Tables I and II, the hoop tension is determined for both a fixed and a hinged base, and is shown together with the tension for a sliding base in Fig. 18-31(a). Because of the difficulty in maintaining perfectly fixed conditions at the base, as discussed previously, the conservative design approach is to design for the maximum condition. Thus, $T_{max} = 0.626 \times 62.4 \times 33 \times 55 \times 1.65 \times 1.7 = 198{,}900$ lb/ft. Reinforcing at this level is $A_s = 198.9/(0.9 \times 60) = 3.68$ sq in./ft; use #7 @ 4 in two curtains ($A_s = 3.60$ sq in.). The concrete tensile stress based on the transformed section based on service load stress of 198,900/(1.65 × 1.7) = 71,000, is:

$$f_c = \frac{0.0003 \times 29 \times 10^6 \times 3.60 + 71{,}000}{22 \times 12 + 9.2 \times 3.60} = 345 \text{ psi}$$

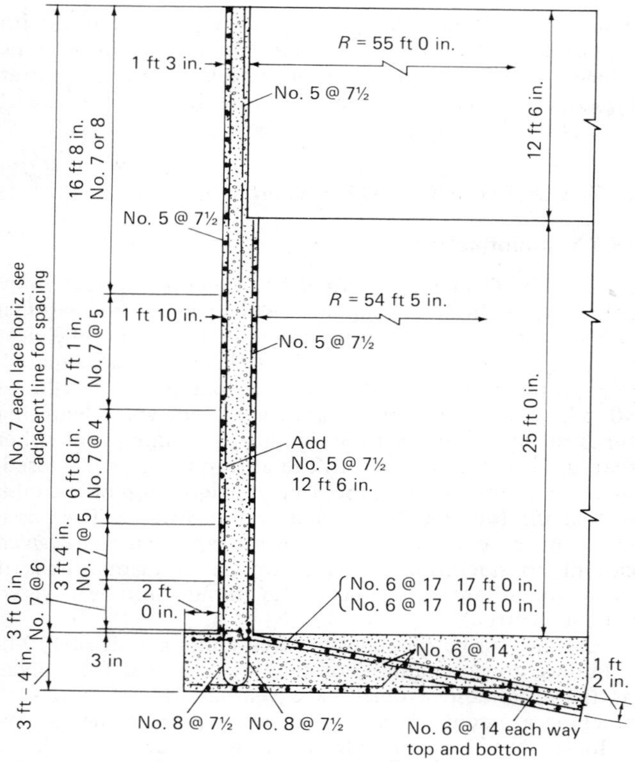

Fig. 18-32 Reinforcing details.

Since 350 psi is considered allowable, the 22-in. wall thickness is acceptable.

Hoop reinforcing is reduced above and below the point of maximum tension in accordance with requirements in Fig. 18-31(a). The details of this reinforcing are shown in Fig. 18-32.

Step 3. Determine vertical wall reinforcing. With coefficients from Tables VII and VIII, vertical wall moments are determined for both a fixed and a hinged base, and are plotted in Fig. 18-31(b). The designer is again required to choose proper design values. Because of the continuous joint design, water pressure on the base can be expected to provide rotational restrant at the joint. However, some rotation must occur, together with some strain in the base slab from the horizontal reaction at the wall base. The conservative design approach is again to provide reinforcing for maximum conditions. At the inside of the wall at the base, the moment per foot is $M_\mu = .0208 \times 0.0624 \times 1.3 \times 1.7 \times 33^3 = 103.1$ ft-kips. Reinforcing, $A_s = 103.1/(4.22 \times 19.5) = 1.25$ sq in. Use #8 @ 7½. This exterior vertical reinforcing is extended to the base, and should be checked for moment due to exterior soil and ground water load. The reinforcing is reduced to #5 @ 7½ where applicable. The details of this reinforcing are shown in Fig. 18-32.

Step 4. Determine base slab thickness and reinforcing. The thickness of the base slab at the edge must satisfy the following requirements:

1. Provide anchorage for #8 vertical dowels.
2. Provide stiffness and moment capacity for continuity with the wall.
3. Act as footing for the wall.

It is determined that a 3 ft 4 in. thickness will be adequate. The wall shear delivered to the base, from Table XVI for fixed base, is $0.207 \times 0.0624 \times 33^2 \times 1.65 \times 1.7 = 39.5$ kips/ft. Reinforcing is determined as for hoop tension, $A_s = 39.5/(0.9 \times 60) = 0.73$ sq in./ft; use #6 @ 14 in two layers. The minimum base thickness, near the center of the tank, is set at 14 in., consistent with tank size and minimum reinforcement ratio.

The top reinforcing in the thickened edge of base slab is determined for wall base moment = 103.1 ft-kips: $A_s = 103.1/(4.43 \times 37.5) = 0.62$ sq in./ft; use #6 @ 8½. The length of this reinforcement is set by cantilever slab span required using base water pressure = 33 × 62.4 × 1.3 × 1.7 = 4550 psf, with safety factor of 2: $L = 2 \times \sqrt{2 \times 103.1/4.55} = 13.5$ ft. Alternate #6 @ 17 bars are extended to R = 40 ft and lapped with #6 @ 14 bars from center portion. Bottom reinforcement in the thickened edge is provided by the wall dowels.

18.6 OTHER STRUCTURES

18.6.1 Superstructures

Superstructures for offices, laboratories, pump houses, etc., are designed for normal live loads, in addition to any special loads such as crane bridges, in similar manner to other buildings and structures. Roof design for covering of tanks, filter beds, etc., should take into consideration the exposure to high humidity and possible corrosive atmosphere. Concrete members are usually preferred to steel for reasons of permanence and maintenance in such environments. Reinforced concrete members should be designed in accordance with the strength design provisions of *Building Code Requirements for Reinforced Concrete,* ACI 318-83. If steel framing is used, purlins are preferable to bar joists for ease in painting and to limit inaccessible pockets. Similarly, precast planks are preferred to formed metal decks.

When precast, prestressed members are used, adequate covering should be provided for prestressing strands, and particular caution should be used in areas exposed to chlorine vapor. Steel connection and bearing plates should be adequately protected against corrosion. When cored slabs are used, the ends should be sealed to prevent the accumulation of moisture.

18.6.2 Control Structures

Sluice gates are used to control flow in large channels or to shut off or divert flow through various parts of the treatment plant. The operation of these gates is generally slow; thus, the final static head in closed position is the maximum lateral design load. As the gate is lowered into position, the final movement will be to wedge the gate tight within the guides at the sides and bottom. The gate stand is supported on a bracket or similar device, and this bracket is designed to support the full vertical load of the gate and controls. An additional important design criterion, which is sometimes forgotten, is the possibility of upward force on the bracket as the gate is wedged into position. This force can be computed from the mechanical advantage of the screw, the diameter of the control wheel, and the driving force. Motor-driven gates should be equipped with an automatic shut-off device to prevent this upward force from becoming excessive.

Bar racks and screens are designed for lateral loads equivalent to a completely clogged condition, although with either mechanical or manual cleaning the operating head loss should not exceed a couple of feet. Since the bar rack precedes the screen, and its purpose is to stop heavy items, the screen can be designed for full hydrostatic head. Consideration should be given, however, to the design of the bar rack to absorb energy required to stop heavy items. This requirement can be partially avoided in water treatment plants by careful planning of the intake. In rivers the bar rack and intake can be located perpendicular to the flow, and thus any blows to the rack will not be direct; that is, the rack will not be required to completely stop any item. Then, from knowledge of stream velocity and assuming a large log (at least equal to a railroad tie), it is possible to compute the energy required to redirect such a log through a reasonable angle, such as 45°

Since waste treatment plants must accept the entire flow of the incoming sewer, bar screen location cannot be employed to reduce impact loadings. Velocities of flow in sewers generally range from 2 to 6 ft/sec, depending on sewer diameter, with the larger velocities designed for larger sewers. Again, assumptions must be made regarding size and weight of the impact item.

Weirs as control devices will not have any particular loads as major design parameter; rather, the connection detail for surface weirs or the quality control for all such elements subject to higher water velocities will be the major consideration. For submerged weirs used to create uniform flow in a basin with a minimum of head loss, consideration must be given to construction details so that a dense uniform surface will be obtained throughout the length of the weir.

18.6.3 Pumping Stations

The structural design of the wet well is governed, under operating conditions, by the same sort of hydraulic loads described previously. However, given a considerable length of intake piping, there will be momentum of the intake water such that at shut-off of intake pumps a surge will develop. The height of this surge is one of the major hydraulic design elements of pumping stations. It can be moderated by overflow provisions, but may still result in considerable increased head applied to the wet well. For instance, in a plant designed with 3600 ft intake in 60-in. diameter pipe, with water velocity set at 5 ft/sec, the surge without overflow reached 19 ft above the static water level. A 36-in. overflow was provided with invert 5 ft above static water level, and the surge was reduced to 10 ft.

Although not strictly a structural design criterion, the layout and arrangement of baffles around the pump intakes is of prime importance for long and efficient plant life. Cavitation, eddy formation, and excessive local velocities can erode the highest-quality concrete. In large plants, such as Chicago's Central District Filtration Plant (capacity 150 mgd), model studies have been employed to indicate areas of redesign or the effect of improved contours.

Ventilation of the wet well is a necessary element, primarily because of the potential surge in the intake line. In water treatment plants, this can be accomplished through connection to the dry well and following air vents, and the volume of air to be moved must be considered as part of the surge calculation. On plants with direct lake intakes, surge due to wind and wave action has been known to cause sufficient air pressure in the dry well to break windows. Specifications for waste treatment plants[18-4] prescribe separation of the wet well and the dry well. In this case, any surge in the intake line and resultant air pressure must be dissipated through air vents in the wet well. However, normal design of waste treatment plants is such that pumping is a result of high flows and surges are not of the same order of magnitude as in long line water treatment plant intakes.

Dry wells and pump houses contain pumps, controls, and associated elements, and the design for structural support of these elements follows standard practice of AISC[18-23] and similar applicable codes. Other than providing structural support for the entire pump housing, it is necessary to provide superstructure with clearance and capability for removal of pumps, etc., for maintenance. Crane rail forces will depend on the weight of equipment to be moved, and the designer should use his judgment in assessing impact forces depending on the type and capacity of power-driven elements. In general, hand-operated hoists and travelers need little or no impact, whereas motor-driven equipment should include impact percentages in accordance with applicable codes. Standard practice in the design of crane bridges is to use double flanged wheels to provide better alignment of the bridge: in this case half of the lateral force is carried on each rail.

18.7 CORROSION PROTECTION

18.7.1 Chlorination

In virtually all plants, the final treatment is sterilization by chlorination. In water supplies, where the desired residual is low, chlorination is accomplished by diffusers in the clear well, and vapor problems in the well are not anticipated. For waste treatment, where the desired residual is 20 to 40 ppm, chlorine contact tanks are used, with detention times ranging from 15 to 30 minutes. It is important either that these tanks be open to the atmosphere, or, if closed, that the hydraulic line should be above the top of the tank so that the top concrete surface contacts water. The possibility of accumulation of chlorine vapor must be given careful consideration in waste treatment plants. Thus if the contact tanks are open, they should not be covered with superstructure.

Chlorine can be obtained in three container sizes: 150-lb cylinders, ton containers, or tank cars. Rooms for storage of these containers must have ventilation capabilities to preclude accidents from accumulation of chlorine gas. Interlocks should be provided with the doors and lighting such that positive ventilation is provided whenever the room is occupied. Emergency alarms are also recommended. Other than these safety precautions, no special consideration need be given to the structural design of the room, except for ease in handling and weighing the containers.

The design of rooms for handling of other chemicals is planned in a similar manner. Consideration must be given to ease of handling, and to safety where required. The conduit or piping for the chemicals is specially designed for that purpose, such as PVC, rubber, or stainless steel, and if properly installed should serve to separate the chemicals from the structure.

18.7.2 Coatings

Concrete in water treatment and domestic sewage plants seldom requires special protection, except for the handling of chemicals such as chlorine and liquid alum. A detailed list of chemicals used in treatment plants, including their effect on concrete, is given in the report of ACI Committee 350,[18-9] and in Portland Cement Association publications.[18-24]

Industrial waste treatment often involves high temperature and low pH, such that special protection is required. Coatings commonly used are special brick or tile, epoxies, and plastic or rubber sheets. Care should be taken to create a clean, sound surface for bonding of the protective coating. Detailed recommendations of the choice and application of proper coating may be obtained from the report of ACI Committee 515,[18-25] or from the manufacturer.

BIBLIOGRAPHY

18-1 "Water Treatment Plant Design," *Manual of Engineering Practice,* No. 19, American Soc. of Civil Eng., New York, 1940.

18-2 *Recommended Standards for Water Works,* Great Lakes–Upper Mississippi River Board of State Sanitary Engineers, 1982.

18-3 "Sewage Treatment Plant Design," *Manual of Engineering Practice,* No. 36, American Soc. of Civil Eng., New York, 1977.

18-4 *Recommended Standards for Sewage Works,* Great Lakes–Upper Mississippi River Board of State Sanitary Engineers, 1978.

18-5 "Causes, Mechanism, and Control of Cracking in Concrete," *ACI Special Publication SP-20,* American Concrete Inst., Detroit, 1968.

18-6 "Volume Changes of Concrete," *Information Sheet IS 018.02T,* Portland Cement Assn., Skokie, 1967.

18-7 *Manual of Concrete Practice,* Parts I, II and III, 1983, American Concrete Inst., Detroit.
NOTE: This *Manual* contains many papers and reports on concrete mix design, admixtures, aggregates, curing, placement, etc., not listed here separately.

18-8 "Watertight Concrete," *Information Sheet IS002. 02 T,* Portland Cement Assn., Skokie, 1969.

18-9 ACI Committee 350, "Concrete Sanitary Engineering structures," *ACI Journal, Proc.,* V. 80, November-December 1983.

18-10 Wallace, G. B., "Joints and Cracks in Concrete Water-holding Structures," *Proceedings* ACI Specialty Conf. on Concrete Construction in Aqueous Environment, Washington, D.C., 1962, p. 21.

18-11 Vetter, C. P., "Stresses in Reinforced Concrete Due to Volume Changes," *Transactions, ASCE,* V. 98, 1933, p. 1039.

18-12 *Building Code Requirements for Reinforced Concrete,* ACI 318-83, American Concrete Inst., Detroit, 1983.

18-13 *American National Standard Minimum Design Loads in Buildings and Other Structures*; ANSI A 58.1-1982.

18-14 Morris, C. T., and Carpenter, S. T., *Structural Frameworks,* John Wiley & Sons, New York, 1943.

18-15 Davies, J. D., "Influence of Support Conditions on the Behavior of Long Rectangular Tanks," *ACI Journal, Proc.,* V. 59, No. 4, Apr. 1962, p. 601.

18-16 Davies, J. D., "Analysis of Long Rectangular Tanks Resting on Flat Rigid Supports," *ACI Journal, Proc.,* V. 60, No. 4, Apr. 1963, p. 487.

18-17 Ghali, A., *Circular Storage Tanks and Silos,* E. & F. N. Spon Ltd., London, 1979.

18-18 Klein, F., Hoffman, E. S., and Rice, P. F., "Application of Strength Design Methods to Sanitary Structures," *Concrete International,* V. 3, No. 4, April 1981, p. 35.

18-19 Rothbart, H. A., Editor, *Mechanical Design and Systems Handbook,* McGraw-Hill, New York, 1964.

18-20 "Rectangular Concrete Tanks," *Information Sheet IS003 D,* Portland Cement Assn., Skokie, 1981.

18-21 Davies, J. D., and Cheung, Y. K., "Bending Moments in Long Walled Tanks," *ACI Journal, Proc.,* V. 64, No. 10, Oct. 1967, p. 685.

18-22 "Circular Concrete Tanks Without Prestressing," *Information Sheet IS 072.01D,* Portland Cement Assn., Skokie, 1942.

18-23 "Specification for the Design, Fabrication and Erection of Structural Steel for Buildings," American Institute of Steel Construction, New York, 1978.

18-24 "Effect of Various Substances on Concrete and Protective Treatments, Where Required," *Information Sheet IS 001 T,* Portlant Cement Assn., Skokie, 1980.

18-25 ACI Committee 515, "Guide for the Protection of Concrete Against Chemical Attack by Means of Coatings and Other Corrosion-Resistant Materials," *ACI Journal, Proc.,* V. 63, No. 12, Dec. 1966, p. 1305.

18-26 "Underground Concrete Tanks," *Information Sheet IS 071 D,* Portland Cement Assn., Skokie, 1973.

18-27 "Concrete for Hydraulic Structures," *Information Sheet IS 012.03,* Portland Cement Assn., Skokie, 1969.

18-28 "Concrete for Waste Water Treatment Works," *Information Sheet PA 063 W,* Portland Cement Assn., Skokie, 1976.

18-29 "Concrete for Water Treatment Works," *Information Sheet PA 069.01W,* Portland Cement Assn., Skokie, 1963.

18-30 ACI Committee 311, "Manual of Concrete Inspection," *ACI Special Publications SP-2,* 7th Ed., American Concrete Inst., Detroit, 1981.

18-31 "Shotcreting," *ACI Special Publication SP-14,* American Concrete Inst., Detroit, 1968.

18-32 Hurd, M. K., "Formwork for Concrete," *ACI Special Publications SP-4,* 4th Ed., American Concrete Inst., Detroit, 1981.

18-33 ACI Committee 344, "Design and Construction of Circular Prestressed Structures," *ACI Journal, Proc.,* V. 67, No. 9, Sept. 1970, p. 657.

18-34 ACI Committee 504, "Guide to Joint Sealants for Concrete Structures," *ACI Journal, Proc.,* V. 67, No. 7, July 1970, p. 489.

18-35 "Gravity Sanitary Sewer Design and Construction," *Manual of Engineering Practice,* No. 60, American Soc. of Civil Eng., New York, 1982.

18-36 "Concrete Manual," U.S. Bureau of Reclamation, Denver, 1963.

18-37 Allen, E. A., and Higginson, E. C., "Waterstops in Articulated Concrete Construction," *ACI Journal, Proc.,* V. 52, No. 1, Sept. 1955, p. 83. Also, Discussion, p. 1149.

18-38 Betz, J. M., "Repair of Corroded Concrete in a Waste-Water Treatment Plant," *Journal, Water Poll. Cont. Fed.,* Mar. 1964, p. 332.

18-39 Biczok, I., *Concrete Corrosion and Concrete Protection,* Chemical Publishing Co., New York, 1967.

18-40 Davies, J. D., and Long, J. E., "Behavior of Square Tanks on Elastic Foundations," *Journal, Engineering Mechanics Div.,* ASCE, V. 94, EM 3, June 1968, p. 753.

18-41 Hetenyi, M., *Beams on Elastic Foundation,* University of Michigan Press, Ann Arbor, 1946.

18-42 Norris, C. H., et al., *Structural Design for Dynamic Loads,* McGraw-Hill, New York, 1959.

18-43 Richart, F. E., Jr., Woods, R. D., and Hall, J. R., Jr., *Vibrations of Soils and Foundations,* Prentice-Hall, Englewood Cliffs, N.J., 1970.

18-44 Timoshenko, S., and Woinowski-Krieger, S., *Plates and Shells,* McGraw-Hill, New York, 1969.

19

Concrete Pipe

JOHN G. HENDRICKSON, JR.*

19.1 INTRODUCTION

19.1.1

Present concrete pipe is made in sizes ranging from 4 in. inside diameter to over 16 ft. Pipes are used to carry fluids in gravity flow such as in highway culverts, storm drains, and sanitary sewers. They are used for pressure lines such as sewer force mains, irrigation lines, and water supply lines.

Precast pipe is produced in elliptical and arch shapes for special uses. Precast boxes are also produced in a wide range of sizes and strengths.

Gravity flow pipes of smaller sizes are either plain or reinforced, but all larger sizes (greater than 24 in. i.d.) are reinforced against crushing. Some pipes are made with telescoping ends designed for mortar or mastic as a jointing material. However, to satisfy demands for leakproof but flexible joints, close-tolerance pipe ends designed for use with rubber gaskets are commonly produced.

Nonreinforced pressure pipe is made largely for carrying irrigation water at low pressures. Concrete pressure pipe for water lines is either reinforced or prestressed. Some designs incorporate a light-gauge steel cylinder in the wall of the pipe as a waterproof diaphragm. For high internal pressure, pipe ends are formed with steel end rings. When these are telescoped together, a round rubber gasket is compressed in a groove to form a flexible, watertight seal.

*Specifications Coordinator, Greeley and Hansen Engineers, Chicago, Illinois.

19.1.2 Specifications

Most concrete pipe are made to *ASTM Specifications* except pressure water pipe which is usually made in accordance with specifications of the American Water Works Association. Many agencies such as state highway departments and others have their own specifications which are closely patterned after *ASTM* or *AWWA Specifications.*

Commonly used sizes of concrete pipe are made on ingenious machines that mechanically compact concrete of a low water–cement ratio. Large sizes and more elaborate designs are made by a casting process utilizing external vibration. Curing with moist steam is more commonly used, but water curing and curing compounds are used, sometimes in conjunction with steam curing.

19.1.3 Acceptance Tests

The commonly used acceptance tests for concrete pipe designed to withstand backfill loading are the absorption test and the three-edge bearing test, as set forth in *ASTM Specification C497*, "Standard Methods of Test for Determining Physical Properties to Concrete Pipe or Tile." The absorption test is intended to check the density and imperviousness of the concrete, whereas the three-edge-bearing test verifies the structural strength of the pipe.

In the three-edge-bearing test the test pipe is placed horizontally on two closely spaced wooden strips. The test load is applied to the top of the pipe through a rigid longitudinal bearing beam.

Test load requirements and absorption limits are listed in the particular pipe specification. For nonreinforced pipe the minimum strength is specified that the pipe must withstand before failure. For reinforced pipe the minimum load is specified at which a longitudinal crack 0.01 in. wide and continuous over a 1-ft length will occur; also ultimate loads are specified. Some specifications require only tests to the 0.01-in. crack. As an alternate to the three-edge-bearing test, core tests are used to check the design of large-diameter pipe.

Pipe that carries both internal pressure and the weight of the backfill may be required to pass a three-edge-bearing test and a hydrostatic test.

19.2 NONREINFORCED CONCRETE PIPE

19.2.1 Scope

Concrete pipe for use in culverts, sanitary and storm sewers is made in conformance with *ASTM Specification C14*, "Standard Specifications for Concrete Sewer, Storm Drain and Culvert Pipe." Standard sizes of the inside diameter range from 4 in. through 24 in. Lengths are usually 5 or 6 ft. However, the largest sizes may be supplied in longer lengths by some manufacturers.

Three strength classes are available under the current specification. The pipe may be supplied for use with mortar or mastic jointing material. If pipe is used with rubber gaskets, the pipe ends must also conform to the tolerances of *ASTM Specification C443.*

19.2.2 Irrigation and Drainage Pipe

Concrete pipe for irrigation lines where mortar joints are to be used is made in conformance with *ASTM Specification C118*, "Standard Specifications for Concrete Pipe for Irrigation and Drainage." The same pipe installed with open joints is used for farm or highway subsurface drainage.

Standard concrete irrigation pipe under the specification is produced in sizes from 4 in. through 24 in. internal diameter and is used to carry irrigation water under pressure. Minimum wall thicknesses, three-edge-bearing test strengths, and hydrostatic test pressures for each size are set forth in the specification.

Standard and heavy duty drainage pipe produced under the specification is also made in sizes from 4 in. through 24 in. Wall thickness and three-edge-bearing test strengths are set forth in the specification for each class. This pipe is laid with open joints to provide underground drainage. The heavy duty pipe is stronger and can be laid at greater depths.

19.2.3 Nonreinforced Rubber Gasket Irrigation Pipe

Nonreinforced concrete pipe to be used with rubber gasket joints to carry irrigation water at low pressures is made in conformance with *ASTM Specification C505*, "Standard Specification for Nonreinforced Concrete Irrigation Pipe with Rubber Gasket Joints." Inside diameters are from 6 in. through 24 in. The specified working pressure for this pipe is 30 ft of head, but the required hydrostatic test pressure is 40 psi. Three-edge-bearing test loads are also specified.

19.2.4 Concrete Pipe for Subsurface Drainage

Three types of concrete pipe are made for substrate drainage: concrete drain tile, perforated concrete pipe, and porous concrete pipe.

1. *Concrete drain tile* is made in conformance with *ASTM Specification C412*, "Standard Specifications for Concrete Drain Tile." This pipe is widely used for agricultural drainage and also for foundation drains and highway underdrains. The pipe is available in three classes: Standard Quality, Extra Quality, and Special Quality. Sizes range from 4 in. through 24 in. Butt end joints are commonly used.

2. *Perforated concrete pipe* is made in conformance with *ASTM Specification C444*, "Standard Specification for Perforated Concrete Pipe." Sizes are from 4 in. through 27 in. or larger inside diameter. Type I has circular perforations, whereas Type II has slotted perforations. The size and spacing of the perforations is set forth in the specification for each size of pipe. Perforated pipe is installed with perforations at the bottom of the pipe.

3. *Porous concrete pipe* is made in conformance with *ASTM Specification C654*, "Standard Specification for Porous Concrete Pipe." The pipe is usually made with tongue-and-groove ends. The barrel section is made with a special concrete mix with the aggregate so graded as to produce a porous or honeycombed texture through which groundwater can pass into the pipe interior. The user is warned that the pipe may be subject to detrimental leaching where soft or acid water occurs.

19.3 REINFORCED CONCRETE PIPE—NONPRESSURE

19.3.1 Reinforced Circular Pipe

Reinforced concrete pipe for the construction of culverts, storm drains, and sanitary sewers is made in conformance with *ASTM Specification C76* "Standard Specification for Reinforced Concrete Culvert, Storm Drain, and Sewer Pipe." The size range covered by the specification is 12 in. through 144 in. However, special designs for larger sizes are accepted under this specification and pipe 204 in. in diameter have been made.

Five strength classes, Class I through V, are provided for. The 0.01-in. and ultimate strength requirements are set forth by D-load specified for each class. D-load is the load per foot of pipe diameter per linear foot for a pipe tested in three-edge bearing. The specified D-loads are:

	0.01-in.	ULT
Class I	800 D	1200 D
II	1000 D	1500 D
III	1350 D	2000 D
IV	2000 D	3000 D
V	3000 D	3750 D

Three wall thicknesses are available under the specification: Wall A is the thin wall design; Wall B, the most widely used; and Wall C, the heavy wall design. Standard designs are not available in all sizes, strengths, and wall thicknesses. Wall A designs for Class V strengths are not shown in the tables of standard designs, since a special design is required for thin wall pipe in the high strength range. All pipe larger than 108-in. require special or modified designs and may with the permission of the purchaser utilize the provisions of ASTM C655.

Small-diameter pipe up to about 30 to 33 in. is reinforced with a single circular cage or with a cage made slightly elliptical so as to better reinforce for the tensile stresses developed under load. Larger sizes are reinforced with an inner and outer circular cage or a single elliptical cage. Pipe 102 and 108 in. in diameter, if reinforced with an elliptical cage, also have an inside circular cage. The circular cage facilitates handling pipe of this size and type.

Reinforced concrete pipe is also supplied in accordance with *ASTM Specification C655*, "Standard Specification for Reinforced Concrete D-Load Culvert, Storm Drain, and Sewer Pipe." Under this specification the pipe manufacturer supplies the design of the pipe for the acceptance of the purchaser. Acceptance is based on tests or empirical evaluations.

19.3.2 Reinforced Arch Pipe

Reinforced concrete arch pipe is made for use in culverts, sanitary sewers, and storm sewers where vertical head room is limited. This pipe is made in conformance with *ASTM Specification C506.*

Standard sizes range from the hydraulic equivalent of a 15-in. circular pipe to that of a 108-in. circular pipe. The required dimensions and proportions of the arch pipe sections are set forth in the specification.

Three strength classes are specified. As for C76 pipe, these are specified by D-load. For arch pipe the D-load is the load per foot of nominal inside span per linear foot in three-edge bearing. The three standard strength classes for this type of pipe are:

	0.01-in.	ULT
Class II	1000 D	1500 D
III	1350 D	2000 D
IV	2000 D	3000 D

19.3.3 Reinforced Elliptical Pipe

Reinforced concrete elliptical pipe is also made for use in sewers and culverts. When placed with the long axis horizontal, it is called horizontal elliptical pipe or HE-pipe. Like arch pipe it is used where vertical headroom is limited.

The pipe may also be placed with the long axis vertical and is correspondingly called vertical elliptical or VE-pipe. Its use is in locations with limited lateral space or where very high backfill loading occurs.

1. Horizontal Elliptical Pipe. The pipe is made in conformance with ASTM Specification C507. Standard sizes range from the hydraulic equivalent of an 18-in. circular pipe through 108-in. equivalent circular.

Minimum wall thicknesses and steel areas are specified for five strength classes as follows:

	0.01 in.	ULT
Class HE-A	600 D	900 D
HE-I	800 D	1200 D
HE-II	1000 D	1500 D
HE-III	1350 D	2000 D
HE-IV	2000 D	3000 D

As with arch pipe, the D-load applies to the nominal inside horizontal diameter in feet.

2. Vertical Elliptical Pipe. The pipe is also made in conformance with *ASTM Specification C507.* The wall thickness and shape of vertical elliptical pipe are the same as those for horizontal elliptical pipe. However, since load requirements are quite different, the steel areas and concrete strengths are set forth in different tables.

Standard sizes for this pipe are from 36-in. equivalent round to 108-in. equivalent round.

Strength requirements are given for five classes as follows:

	0.01 in.	ULT
Class VE-II	1000	1500
VE-III	1350	2000
VE-IV	2000	3000
VE-V	3000	3750
VE-VI	4000	5000

19.3.4 Noncircular Pipe Joints

Joints for arch pipe and elliptical pipe are the tongue-and-groove type. Jointing materials commonly used are cement mortar or mastic. Rubber gaskets compressed between the telescoping joint surfaces are not used with these precast noncircular concrete pipe. To seal joints against infiltration and provide flexibility, external sealing bands are sometimes used. External sealing bands are rubber bands, 7 to 13 in. wide, depending on the pipe size, which are tightly wrapped around the joint exterior. The bands are held in place by the backfill and with an adhesive applied to the band. External sealing bands are supplied in conformance with *ASTM Specification C877*, "External Sealing Bands for Noncircular Concrete Sewer, Storm Drain and Culvert Pipe."

Where external sealing bands are used, the interior surface of the joint may be filled with mortar and troweled smooth for better hydraulic properties. This is usually done after backfilling to be fully effective.

19.4 PRECAST CONCRETE BOXES

There are two ASTM specifications for precast concrete boxes. Single-cell boxes for fill heights varying from 2 ft to a maximum of about 18 ft and for a wide range of sizes are covered by *ASTM Specification C789*, "Precast Reinforced Concrete Box Selections for Culverts, Storm Drains and Sewers." Single-cell boxes for less than 2 ft of cover and subjected to highway loadings are covered by *ASTM Specification C850*, "Precast Reinforced Concrete Box Sections for Culverts, Storm Drains, and Sewers with Less Than 2 Feet of Cover Subjected to Highway Loadings."

Sizes covered by these specifications range from a minimum of 3 ft by 2 ft to a maximum of 12 ft by 12 ft.

19.4.1 Precast Boxes for Cover in Excess of 2 Ft

There are three types of reinforced concrete boxes covered by *ASTM C789*, Table 1 gives design requirements for earth loads and HS20 live loads. Table 2 gives design requirements for earth loads and interstate live loads. Table 3 gives design requirements for earth loads only. Design requirements include concrete strength, wall thickness, and steel areas at critical sections for various heights of fill and horizontal spans.

Design criteria used to develop Tables 1, 2, and 3 are given in the appendixes of *ASTM C789.* Special or modified designs for sizes and loads not covered by the specification may be furnished if approved by the purchaser.

Acceptance of the manufactured sections is determined by concrete compressive strength tests, material tests, and inspection of the finished product for conformance with the specifications. Load tests are not required, as precast boxes are designed for the installed condition.

19.4.2 Precast Boxes for Less than 2 Ft of Cover

Two types of reinforced concrete boxes are covered by *ASTM C850.* Table 1 gives design requirements for precast boxes with less than 2 ft of cover and subject to HS20 highway loading. Table 3 gives design requirements for less than 2 ft of cover and subject to interstate live loading. Design criteria are outlined in the appendix. Impact factors have been included.

Acceptance criteria are the same as for *ASTM C789* precast boxes.

19.4.3 Joints for Precast Boxes

Precast boxes are made with tongue-and-groove type joints. As with precast arch and elliptical sections, the jointing materials commonly used are cement mortar, mastic, or external sealing bands. Joints on the interior may be mortared and troweled smooth.

19.5 REINFORCED CONCRETE PIPE FOR LOW INTERNAL PRESSURES

19.5.1 Reinforced Concrete Low-Head Pressure Pipe

This type of reinforced concrete pipe is intended for use with low internal hydrostatic pressure and is made in conformance with *ASTM Specification C361*, "Reinforced Concrete Low-Head Pressure Pipe." The maximum hydrostatic pressure usually does not exceed 125 ft of head. The use of this type of pipe is largely for irrigation and sewer force mains.

The specification lists standard designs for hydrostatic heads of 25, 50, 75, 100, and 125 ft in combination with earth fill loading over the top of the pipe for 5, 10, 15, and 20 ft. Specific installation conditions for which these designs apply are given in the appendix of the specification.

Sizes covered in this specification are 12 in. through 72 in. internal diameter.

Joints for this type of pipe utilize a solid rubber gasket to seal the joint against leakage and still provide some flexibility. The ends of the pipe may be all concrete, or steel end rings may be used. A characteristic of the pressure type of joint is that the gasket is confined in a groove in the spigot end of the pipe when the joint is closed (Fig. 19-1).

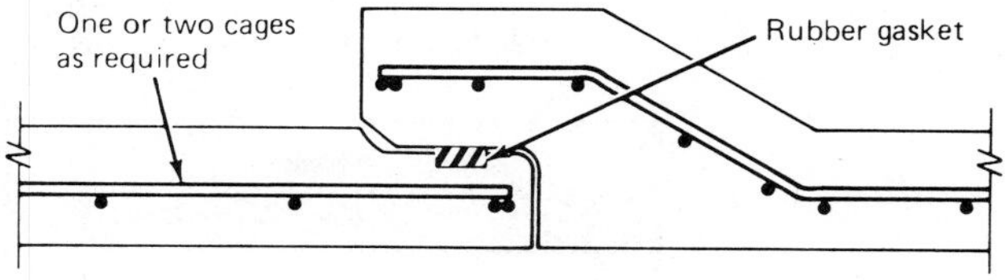

Fig. 19-1

19.5.2 Reinforced Concrete Noncylinder Water Pipe

Reinforced concrete water pipe for use with low internal pressures is made in conformance with *AWWA Specification C302*, "American Water Works Association Standard for Reinforced Concrete Water Pipe—Noncylinder Type, Not Prestressed." This type of pipe is largely used for low-head transmission lines, reservoir connections, sewer force mains, and lines not subject to possible higher pressures or shock. Maximum design pressure for this type of pipe is 45 psi.

The sizes covered by this specification are 12 in. through 96 in. internal diameter. Minimum lengths are usually 8, 12, or 16 ft. Maximum lengths are as follows:

Size, in.	*Max Length, ft*
12–15 (inclusive	$8^1/_2$
16–21 (inclusive)	12
24 and larger	16

Joints for this type of pipe commonly employ a steel bell ring and a steel spigot ring with a groove in which a rubber gasket is confined as shown in Fig. 19-2. However, all-concrete joints are also used.

Pipe designs are based on allowable stresses set forth in the specification. However, minimum designs are specified.

19-6 REINFORCED CONCRETE PIPE FOR MODERATE AND HIGH INTERNAL PRESSURES

19.6.1 Reinforced Concrete Cylinder Water Pipe

Concrete water pipe for design pressures varying from a minimum of 40 psi to 260 psi is manufactured according to *AWWA Specification C300*, "Reinforced Concrete Water Pipe—Steel Cylinder Type, not Prestressed." Sizes covered by the standard are 24 through 96 in. inclusive. Larger sizes have been made using special designs.

This pipe consists of a thin steel cylinder with steel joint rings welded to its ends. The cylinder is surrounded with one or more cages of reinforcing made from bars, wire, or

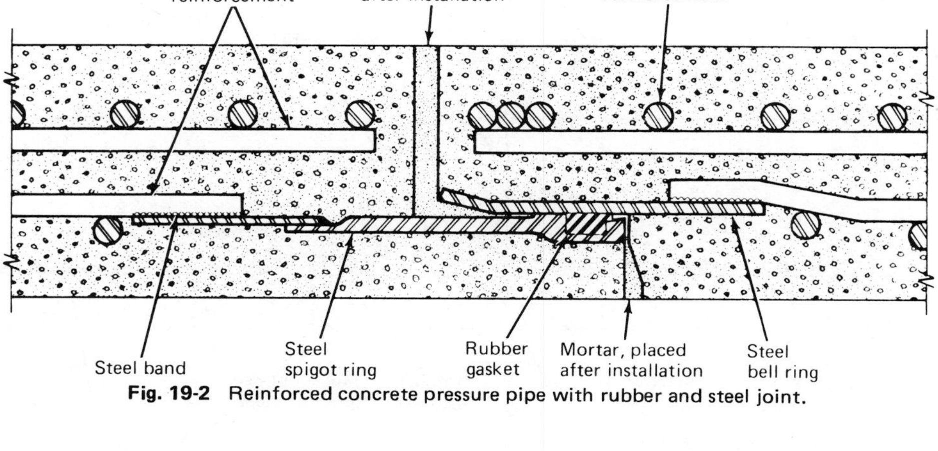

Fig. 19-2 Reinforced concrete pressure pipe with rubber and steel joint.

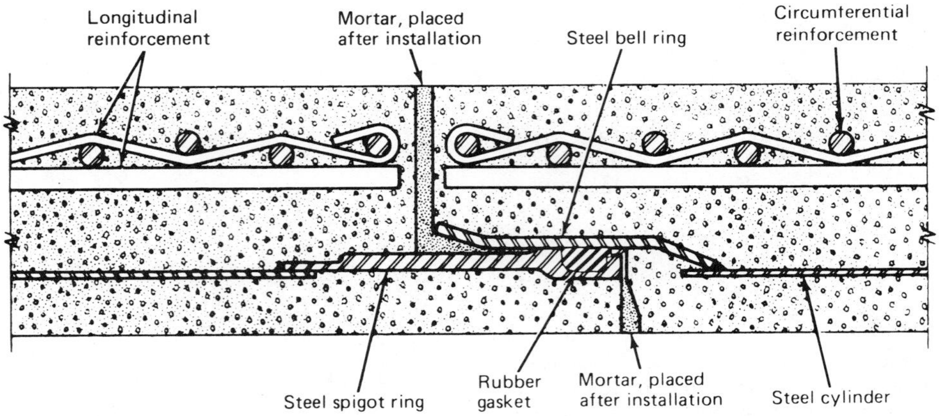

Fig. 19-3 Reinforced concrete cylinder pipe with rubber and steel joint.

welded wire fabric. This assembly is encased in a wall of dense concrete covering the cylinder both inside and out. The steel joint rings form a self-centering joint sealed with a round rubber gasket as shown in Fig. 19-3.

Manufacturers of this type of pipe supply design data on the allowable backfill loading combined with internal pressure. Such information has been developed largely by tests.

19.6.2 Prestressed Concrete Water Pipe

Prestressed concrete water pipe is manufactured according to *AWWA Specification C301*, "Prestressed Concrete Pressure Pipe, Steel Cylinder Type for Water and Other Liquids." Standard sizes covered by this specification are from 16 in. through 144 in., but prestressed pipe as large as 201 in. internal diameter has been built. Design pressures depend on the design of the pipe but are in the range 200 to 275 psi. Standard lengths are 16 or 20 ft.

Essential features of this pipe include a light-gauge steel cylinder with steel joint rings welded to each end. This unit is either lined with concrete or embedded in concrete to form a concrete core. High-tensile-strength wire is wound around the outside of the core at a predetermined stress. A dense coating of cement mortar or concrete is placed on the exterior of the core to protect the prestress wires.

Cores formed by placing the concrete entirely within the steel cylinder are used in sizes up to and including 20 in. and may be used up to and including 48 in. Cores with the steel cylinder embedded in concrete may be used in pipe sizes 24 in. through 48 in. and must be used in larger pipe sizes.

The steel joint rings form a self-centering joint sealed with a rubber gasket as shown in Figs. 19-4 and 19-5. This is the general type of joint used with all concrete pressure pipe.

19.6.3 Pretensioned Concrete Cylinder Pipe

This type of pipe is made according to *AWWA Specification C303*, "Reinforced Concrete Water Pipe–Steel Cylinder Type, Pretensioned." Standard sizes are from 10 in. to 42

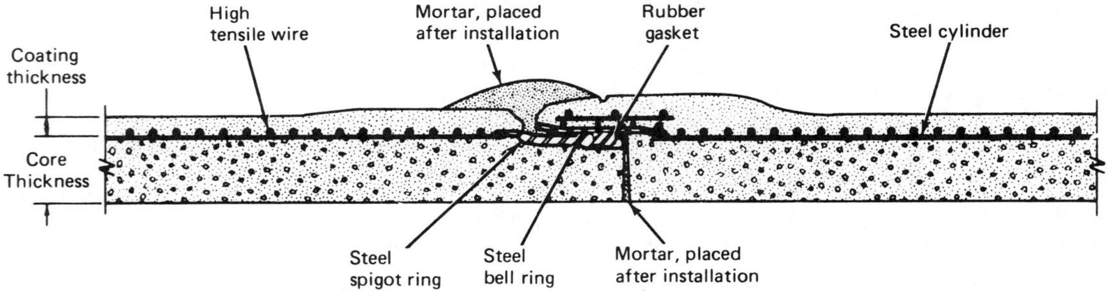

Fig. 19-4 Prestressed concrete cylinder pipe with rubber and steel joint.

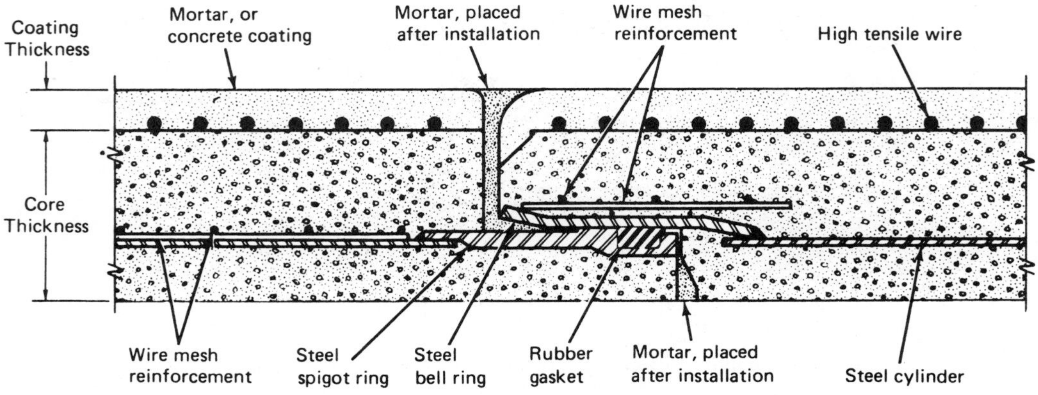

Fig. 19-5 Prestressed concrete embedded cylinder pipe with rubber and steel joint.

in., inclusive. Larger sizes can be furnished by agreement with the purchaser. Maximum design pressures are 400 psi. Lengths range from 24 ft to 40 ft. However, 21-in. and smaller pipe cannot exceed 32 ft in length.

The pipe is made with a welded thin-steel cylinder with joint rings welded to each end. This assembly is lined with a layer of cement mortar centrifugally spun into place. Reinforcement consisting of steel rod is wound on the outside of the cylinder at a predetermined stress of about 11,000 psi. The rod is welded to the steel joint ring at each end of the cylinder. A dense mortar coating is placed over the cylinder and rods.

The joint is formed by steel joint rings similar to that for other types of concrete pressure pipe, as shown in Fig. 19-4.

19.7 LOADS ON PIPE IN TRENCHES

Most sewers, waterlines, irrigation lines, and underdrains are constructed by excavating a trench in undisturbed soil. The pipe is placed on the bottom of the trench and covered with earth backfill to the original ground surface (See Fig. 19-6).

19.7.1 Trench Load Theory

In calculating backfill loads on rigid pipe it is assumed that the load develops as the backfill settles, since it is not compacted to the density of the original ground.

The load on the pipe equals the weight of the backfill material above the top of the pipe minus the shearing or friction forces developed along the sides of the trench as the backfill settles. These shearing forces are calculated in accordance with Rankine's theory. Cohesion between the backfill and the sides of the trench is assumed negligible.

The backfill along the sides of the pipe is assumed to carry none of the vertical load.

The load is calculated according to the Marston formula:

$$W_d = C_d w B_d^2$$

where:

W_d = backfill load in lb per linear ft of conduit
C_d = load coefficient
w = unit weight of the backfill in lb per cu ft
B_d = width of the trench in ft at the elevation of the top of the pipe.

The coefficient C_d depends on the coefficient of internal friction of the fill material, μ; the coefficient of sliding friction along the sides of the trench, μ': and the ratio of active lateral pressure to vertical pressure, K; as calculated from Rankine's formula. Values of C_d for various values of $K\mu$ or $K\mu'$ can be determined from Fig. 19-7.

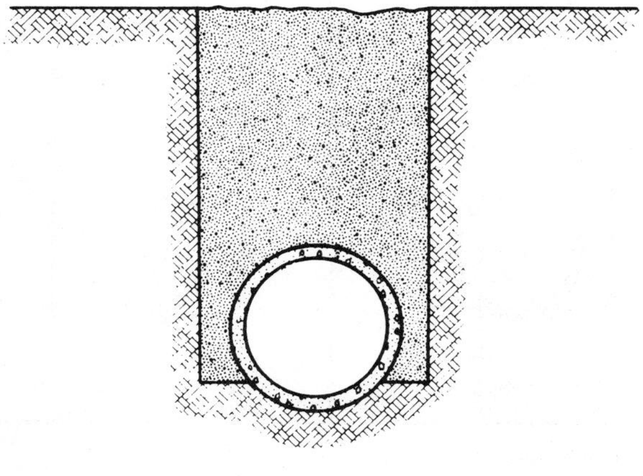

Fig. 19-6

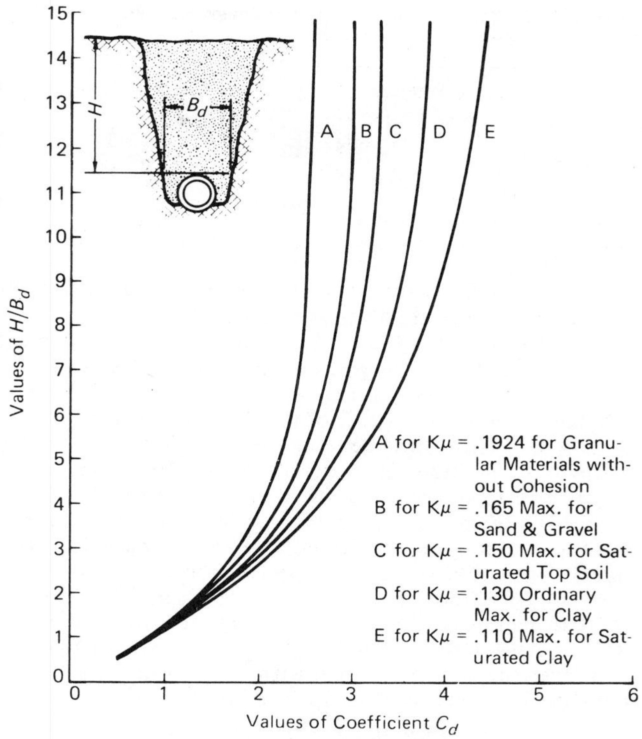

Fig. 19-7 Trench installations.

EXAMPLE 19-1: Consider a 60-in. pipe in a trench 8 ft wide, with a depth of 30 ft from the top of the pipe to the top of the natural ground. Backfill material is sand and gravel with a density of 120 lb/cu ft.

SOLUTION: Using the diagram of Fig. 19-7:

$$\frac{H}{B_d} = \frac{30}{8} = 3.75$$

$$C_d = 2.15$$

$$W_d = C_d w B_d^2$$

$$= 2.15 \times 120 \times 8^2$$

$$= 16512 \text{ lb/ft}$$

19.7.2 Concrete Pipe Bedding

The load carrying capacity of a rigid pipe depends on the distribution of the bedding on the underside of the pipe and the development of active lateral pressure at the sides of the pipe. For rigid pipe in trenches no active lateral pressure is assumed.

The effectiveness of the bedding is given by a bedding factor. This is applied to the three-edge-bearing strength of the pipe to calculate its load supporting strength. For reinforced pipe, the three-edge-bearing strength of the pipe at the 0.01-in. crack is used to calculate its supporting strength. For nonreinforced pipe the ultimate three-edge-bearing strength with a factor of safety of from 1.25 to 1.5 is used.

The four commonly used types of bedding are shown in Table 19-1 with the bedding factors determined experimentally at Iowa State College.*

*"The Supporting Strength of Rigid Pipe Culverts" by M. G. Spangler, Iowa State College *Engineering Experiment Station Bulletin 112*, 1933.

TABLE 19-1 Bedding Classes and Experimentally Determined Bedding Factors

Class of Bedding	*Bedding Factor*
Class A–Concrete Cradle	2.25–3.4
Class B–First Class	1.9
Class C–Ordinary	1.5
Class D–Impermissible	1.1

Bedding methods for pipes in trenches as determined at Iowa State College are illustrated in Fig. 19-8.

Class A–Concrete Cradle Bedding–requires embedment of the lower part of the pipe in plain or reinforced concrete of suitable thickness and extending up the sides of the pipe a distance not less than 25% of the depth of the pipe.

Class B–First Class Bedding–requires the pipe to be placed on fine granular materials on an earth foundation shaped to fit the underside of the pipe for at least 60% of its outside diameter. The trench is then filled to 1 ft over the top of the pipe with granular materials placed in 6-in. layers and tamped to fill all the space around the pipe.

Class C–Ordinary Bedding–requires placing the pipe on an earth bedding shaped to fit the exterior of the pipe for 50% of its external diameter. The trench is then filled to 6 in. over the top of the pipe with granular material placed and shovel-tamped to fill all space around the pipe.

Class D–Impermissible Bedding–allows the pipe to be placed on the bottom of the trench with no effort to shape the trench to fit the pipe. Fill is placed around the pipe.

Alternate bedding procedures are illustrated in Fig. 19-9. In addition, excellent bedding has been obtained by filling under and around the pipe with soil cement grout. On other installations lean concrete has been used to ensure uniform and solid support.

The American Concrete Pipe Association recommends two types of bedding for precast box culverts in trenches.

Class B Bedding is obtained by installing the box section on a minimum thickness of fine granular material at least 2 in. thick. Compacted granular material is placed on each side of the box section and up to at least 1 ft over the top of the box.

Class C Bedding is obtained by placing the box section on a flat trench bottom. Lightly compacted fill is placed on each side of the box section and up to 6 in. over the top of the box.

The objective in either case is to provide the load distribution assumed in design.

EXAMPLE 19-2: From the previous example the load on a 60-in. pipe was 16,512 lb/ft. Installing the pipe with Class B bedding, the required 0.01-in. strength would be:

$$\frac{16512}{1.9} = 8690 \text{ lb/ft}$$

$$\text{D-load} = \frac{8690}{5} = 1738 \text{ lb}$$

A Class IV pipe is required.

19.7.3 Effect of Trench Widths

In accordance with the preceding analysis the backfill load on a pipe will increase as the trench width increases. However, there is a limiting width beyond which there is no further increase in load. This width depends on the pipe diameter, the depth of the trench, and possible settlement of the pipe into the bedding at the bottom of the trench as discussed for embankment conduits in Section 19.8.1.

Figure 19-10 is used to determine the limiting trench width.

EXAMPLE 19-3: In the preceding example, H/B_c for the 60-in. pipe was 3.75. Assuming Class B bedding in granular material, a reasonable value of r_{sdP} is 0.50. From Fig. 19-10, these values give a B_d/B_c of about 2.0. Since the ratio in the example of B_d/B_c is 8.0/6.0 = 1.33 and less than 2.0, the transition-width ratio, the installation is correctly considered a trench-type installation.

A ratio of B_d/B_c greater than the transition-width ratio would indicate that the load calculation should have followed the procedure for embankment installations.

19.7.4 Combined Loading

Concrete pressure pipe is designed to carry simultaneously the external backfill load and the internal pressure. Since pipelines carrying liquid under pressure do not have to maintain a uniform grade line, it is customary to install them in relatively shallow trenches where the backfill is in the range of 5 to 6 ft. In such cases the internal pressure will be the major factor in the design. However, some installations may carry 15 to 20 ft of cover and must be designed accordingly. Backfill loads are calculated by the trench load theory.

1. Reinforced concrete pressure pipe made in conformance with *ASTM Specification C361* is designed for relatively low pressure (about 50 psi or less) and backfill cover varying from 5 ft to 20 ft.

Tables in the specification give the required steel areas for pipe sizes varying from 12 in. through 108 in., with backfill in 5-ft intervals from 5 ft to 20 ft, and for pressure heads varying in 25-ft intervals from 25 ft to 125 ft. These standard designs are provided for both circular and elliptical reinforcing cages and assume about a Class B bedding.

Two bedding procedures which are satisfactory for these designs are described in the appendix of *ASTM C361*.

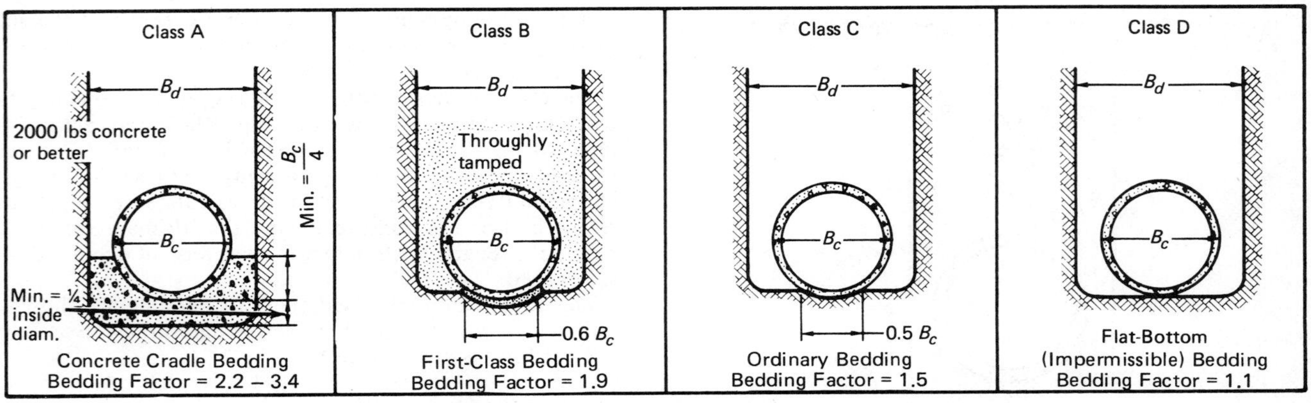

Fig. 19-8 Bedding method for trench conduits.

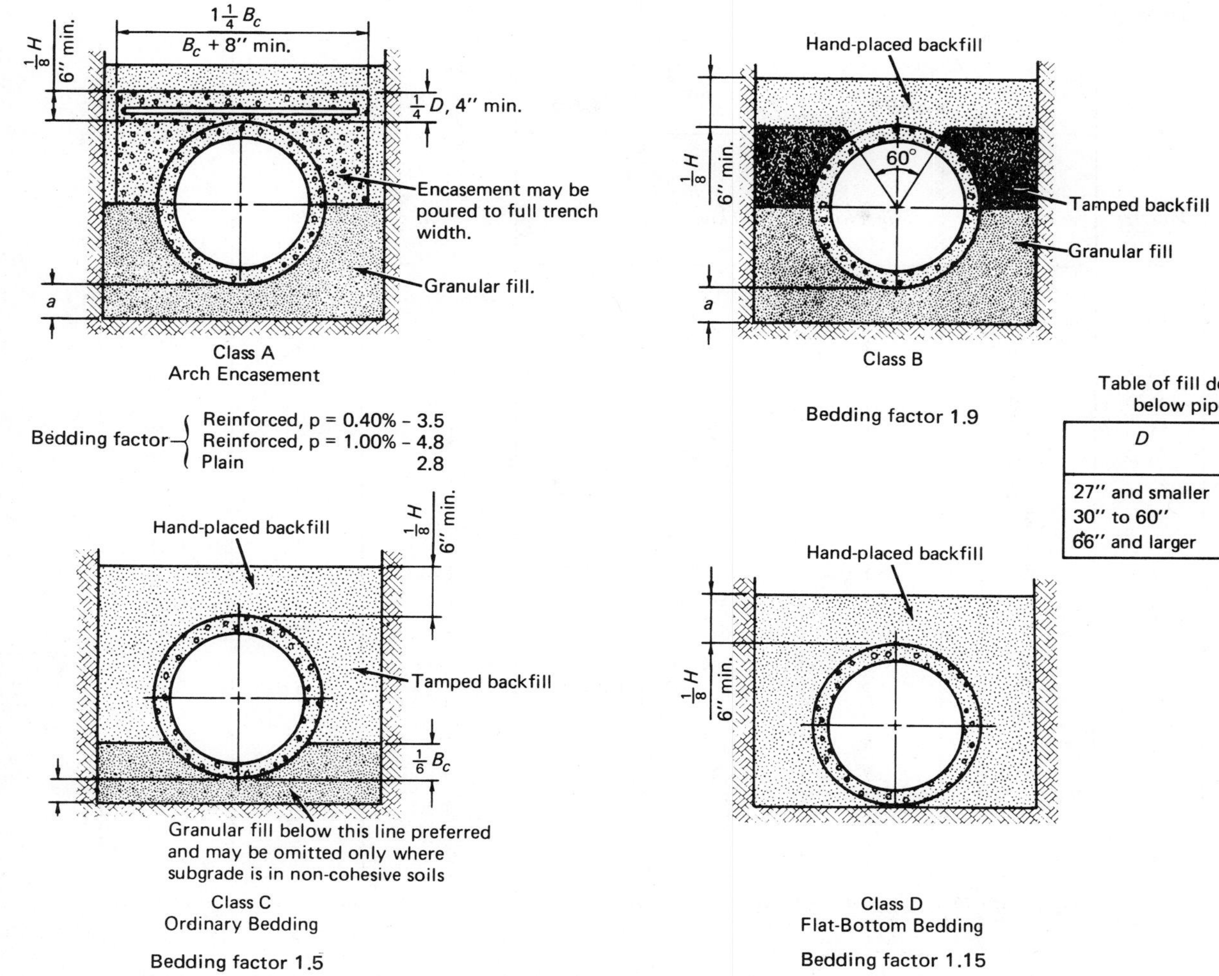

Table of fill depths below pipe

D	a min.
27" and smaller	3"
30" to 60"	4"
66" and larger	6"

Fig. 19-9 Alternate trench bedding.

2. Reinforced concrete water pipe made in conformance with *AWWA C302* must satisfy minimum requirements set forth in the following table taken from the specification.

The minimum longitudinal steel for each pipe shall be six bars and shall provide an area equal to six 0.25-in. round bars for pipe 8 ft long, six 0.375-in. round bars for pipe 12 ft long, and six 0.5-in. bars for pipe 16 ft long. The minimum longitudinal area shall also not be less than the equivalent of 0.5-in. round bars spaced 42 in. on centers. In pipe with two reinforcing cages, the longitudinals are to be divided equally between the two cages.

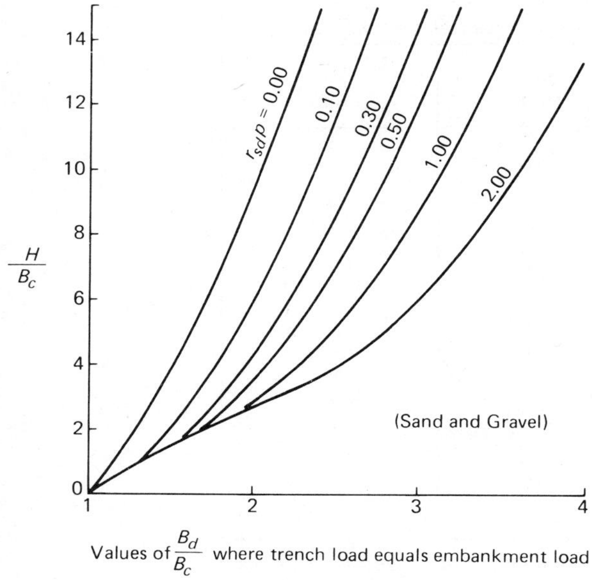

Fig. 19-10 Transition-width ratio curves.

The area of circumferential steel in the reinforcement cage or cages is based on a stress of 12,500 psi when the pipe is subjected to the design pressure specified by the purchaser with no allowance for tension in the concrete.

When external load is the determining factor in the design of the pipe, the design is based on the combined internal pressure and external load designated by the purchaser. Tensile stress in the tension steel is not to be greater than 22,000 psi and compressive stresses in the concrete are not to be greater than 45% of the 28-day compressive strength. The minimum circumferential steel cannot be less than given in Table 19-2 taken from *AWWA C302*.

3. Reinforced concrete pressure pipe made in conformance with *AWWA Specification C300* combines a light-gage steel cylinder with one or more reinforcing cages. Certain minimum requirements are given in Table 19-3.

For pipe designs based on internal pressure, the steel stresses are not to exceed 12,500 psi for the design pressure specified by the purchaser. The area of reinforcing bars is not to be less than 40% of the combined area of the cylinder and reinforcing cages.

When the design of the pipe must be based on combined

TABLE 19-2 Requirements for Pipe of Various Sizes*

Pipe ID (in.)	*6,000-psi Centrifugal Concrete* Min. Pipe Wall Thickness (in.)	*6,000-psi Centrifugal Concrete* Min. No. of Cages	*4,500-psi Poured Concrete* Nominal Pipe Wall Thickness (in.)	*4,500-psi Poured Concrete* Min. No. of Cages	*Circumferential Reinforcement Spacing* Min. (in.)	*Circumferential Reinforcement Spacing* Max. (in.)	Min. Total Steel Area per Lineal Foot (sq in.)
12	2	1			1¼	4	0.08
15	2	1			1¼	4	0.11
16	2⅛	1			1¼	4	0.12
18	2¼	1			1¼	4	0.14
20	2⅜	1			1¼	4	0.16
21	2⅜	1			1¼	4	0.17
24	2½	1	3	1	1¼	4	0.20
27	2⅝	1	3¼	1	1¼	4	0.23
30	2¾	1	3½	1	1¼	4	0.25
33	2⅞	1	3¾	2	1¼	4	0.28
36	3	1	4	2	1¼	4	0.30
42	3½	2	4½	2	1¾	5	0.35
48	4	2	5	2	1¾	5	0.40
54	4½	2	5½	2	1¾	5	0.45
60	5	2	6	2	1¾	5	0.50
66	5½	2	6½	2	2¼	6	0.61
72	6	2	7	2	2¼	6	0.71
78	6½	2	7½	2	2¼	6	0.81
84	7	2	8	2	2¼	6	0.90
90	7½	2	8	2	2¼	6	1.00
96	8	2	8½	2	2¼	6	1.09

**For pipe larger than 96 in. in diameter, dimensions and details of design shall be subject to approval by the purchaser.*

internal pressure and external load, the design is developed by the manufacturer subject to approval by the purchaser.

4. Prestressed concrete cylinder pipe can be designed for internal pressures as high as 200 psi, or higher, in combination with backfill loads. Standard design procedures provide for water hammer equal to at least 40% of the design pressure and for live load, including impact, equal to AASHTO H-20 loading.

According to *AWWA C301*, the wrapping stress in the high tensile wire is not to exceed 75% of the ultimate strength of

TABLE 19-3 Requirements for Pipe of Various Sizes*

Pipe ID (in.)	*Min. Thickness* Pipe Wall (in.)	*Min. Thickness* Concrete Lining (in.)	*Circumferential Reinforcement Spacing* Min. (in.)	*Circumferential Reinforcement Spacing* Max (in.)
20	3¼	1	1¼	4
24	3½	1	1¼	4
30	3½	1	1¼	4
36	4	1	1¼	4
42	4½	1	1¾	5
48	5	1¼	1¾	5
54	5½	1¼	1¾	5
60	6	1¼	1¾	5
66	6½	1½	2	6
72	7	1½	2	6
78	7½	1½	2	6
84	8	1½	2	6
90	8	1¾	2¼	6
96	8½	1¾	2¼	6

**For pipe larger than 96 in. in diameter, dimensions and details of design shall be subject to approval by the purchaser.*

the wire. The initial compression in the concrete at the time of wrapping is not to exceed 55% of the compressive strength of the concrete at the time of wrapping.

The design procedure for prestressed concrete cylinder pipe is described in detail in the paper "Standardization of Design Procedures for Prestressed Concrete Pipe," by R. E. Bald, *AWWA Journal*, Nov. 1960.

Tests have shown that for prestressed concrete cylinder pipe a cubic parabola best shows the relationship for internal pressure combined with external load. The design curve is defined by the equation:

$$w = \frac{W_o}{\sqrt[3]{P_o}} [\sqrt[3]{P_o - p}]$$

where:

w = the three-edge-bearing load equivalent of the design backfill loading (W_d divided by bedding factor)
p = maximum design pressure occuring in combination with w but not greater than 0.8 P_o
W_o = nine-tenths of the three-edge-bearing load producing incipient cracking in the core with no internal pressure (incipient cracking is 0.001-in. crack)
P_o = the internal pressure required to overcome all compression in the core concrete with no external load.

The wire area, tensile stress, wire spacing, and concrete stress in the core can be varied to produce a desired P_o. W_o must be determined from the manufacturer's test data. A transient load capacity beyond the design curve is available to carry water hammer and live load impact. These details are discussed in Appendix A of *AWWA C301*.

Appendix B of *AWWA C301* also provides a stress analysis method for combined load design.

5. The design procedure for pretensioned, concrete, cylinder pipe is discussed in detail in Appendix A of *AWWA C303*. For combined loads the design of the pipe must consider its structural characteristics under combined loading together with the bedding and compaction of material around the pipe to prevent injurious deflection. The maximum deflection under ground after installation should not exceed $D^2/4000$ in. Based on the loads on the pipe and estimates of the effectiveness of the bedding and backfilling, the required wall stiffness can be calculated. This can be compared with the calculated wall stiffness for the wall section chosen.

19.8 LOADS ON PIPE UNDER EMBANKMENTS

19.8.1 Embankment Installations

The typical case of a load on a pipe under an embankment is a highway culvert where the pipe is placed on the natural ground and the fill built up over it. Pipe culverts may be constructed by any of four different installation methods (see Fig. 19-11). The load will vary with each method. Therefore, the method of installation must be selected and the pipe designed accordingly.

1. Positive projecting conduits. By this method of installation the pipe culvert is installed on the surface of the natural ground with the top of the pipe above the adjacent surfaces of the ground. The embankment is then built up over the culvert to the predetermined elevation.

2. Induced trench conduit. In this type of installation the pipe is installed and the fill built up over the pipe as discussed for positive projecting conduits. However, the fill is first built to an elevation of one-half, one, or two outside pipe diameters plus 1 ft above the top of the pipe.

A trench equal in width to the outside pipe diameter is then dug with vertical sides to within 1 ft of the top of the pipe. Immediately after excavation the trench is refilled with loose compressible material. The remainder of the embankment is constructed in the normal method.

This method of installation reduces the backfill loading on the pipe by inducing arching of the embankment over the pipe. It is economically feasible only with high fills that would require special pipe designs with most other methods of installation.

3. Zero projecting conduit. As the name implies, the pipe is installed in a narrow trench in the natural ground of such depth that the top of the pipe is approximately level with the adjacent natural ground surface. After the trench is backfilled, the embankment is constructed in the normal manner.

4. Negative projecting conduit. The pipe is installed in a narrow trench in the natural ground of such depth that the top of the pipe is below the adjacent natural ground surface. The trench is backfilled with material that is thoroughly compacted to the elevation of the top of the pipe. Between the top of the pipe and the surface of the natural ground, loose highly compressible fill is used. The balance of the embankment is built up in the normal manner.

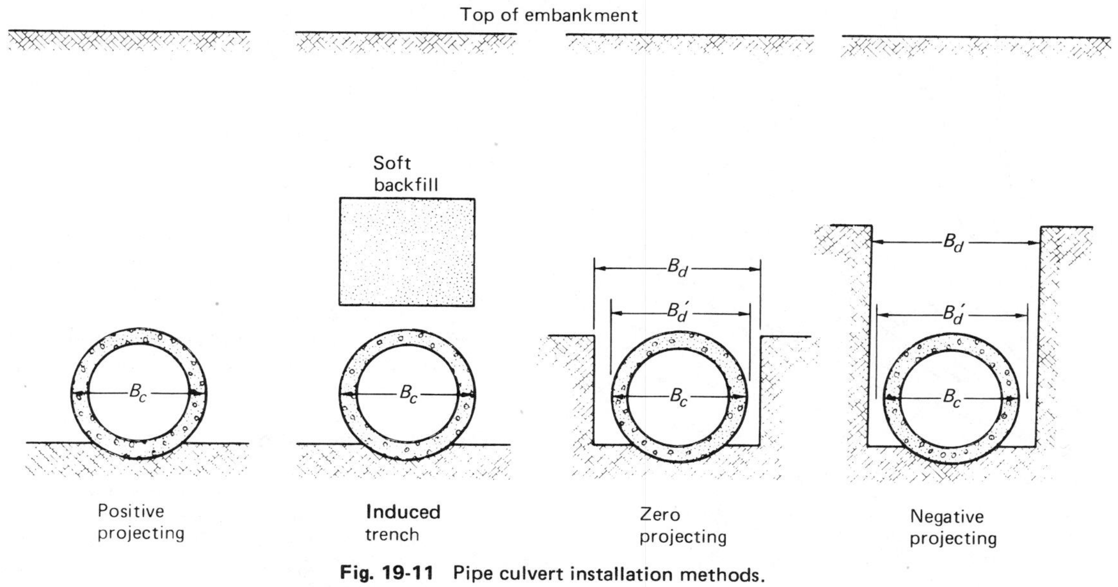

Fig. 19-11 Pipe culvert installation methods.

It will be noted that this method is somewhat similar to the induced trench method, and similarly the load on the pipe is reduced. However, the reduction is not as great as for the induced trench method.

The illustrations and figures shown are for circular pipe. However, the same theories apply to other shapes such as arch pipe or precast boxes.

19.8.2 Embankment Load Theory

The formulas used to calculate the load on embankment conduits have been developed at Iowa State College by research and field tests. For each type of installation the formulas are as follows:

Positive projecting:

$$W_c = C_c w B_c^2$$

Induced trench:

$$W_c = C_n w B_c^2$$

Zero and negative projecting:

$$W_c = C_c w B_d B_d'$$

where:

W_c = load per unit length of pipe
w = unit weight of fill material
B_d = width of trench at top of pipe
B_c = outside diameter of pipe
B_d' = average of B_c and B_d
C_c and C_n = load coefficient

19.8.3

The values of C_c and C_n can be determined from the curves of Figs. 19-12, 19-13, 19-14, and 19-15 for the various installation methods. The values depend on the ratio of height of fill over the conduit to the horizontal width of conduit or trench, projection ratio, settlement ratio, and the internal friction coefficient of the soil.

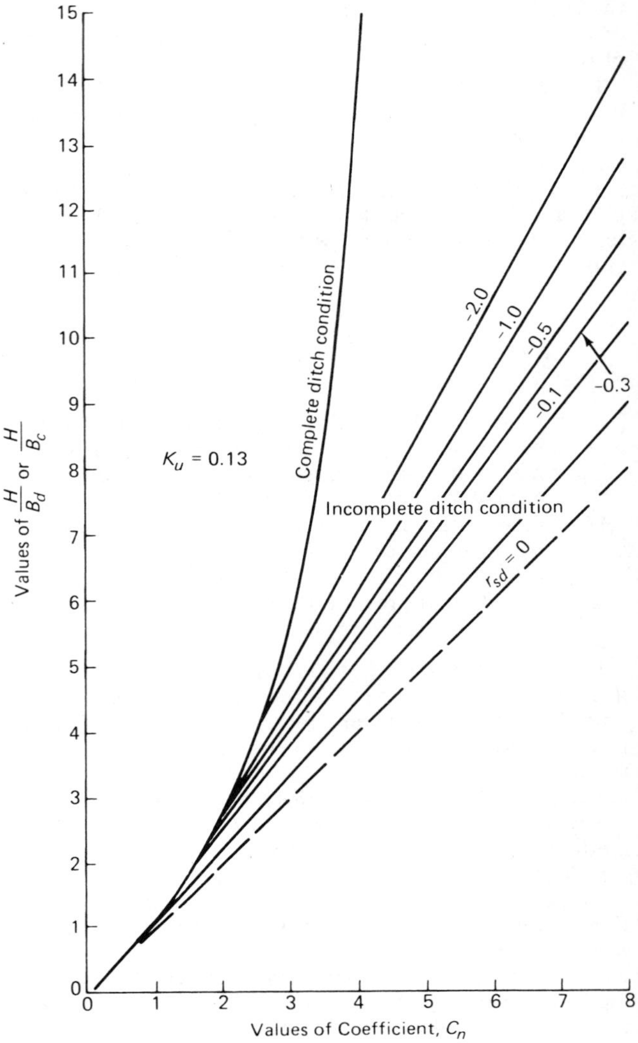

Fig. 19-13 Diagram for coefficient C_n for negative projecting conduits and induced trench conduits when p' = 0.5.

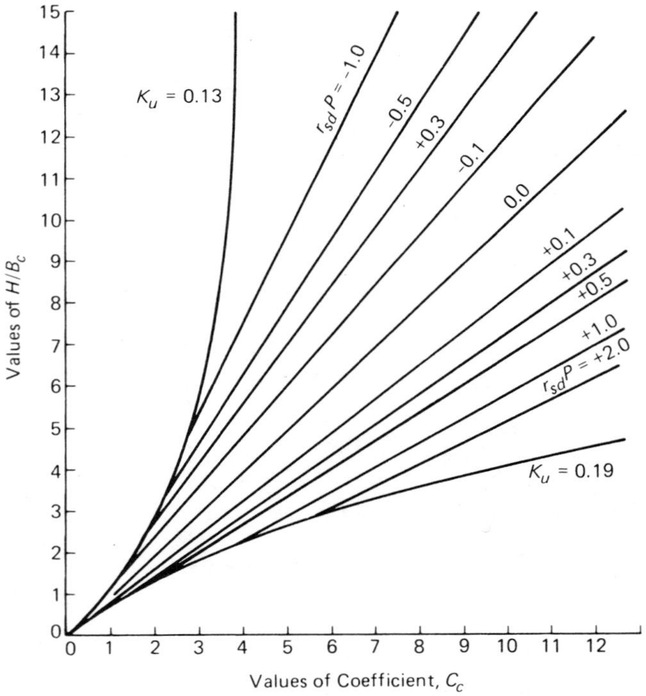

Fig. 19-12 Positive and zero projecting installations.

19.8.4

The projection ratio, p, for positive projecting conduits is illustrated in Fig. 19-16(a). It may vary from 1 to 0, but for most installations with Class B bedding it is assumed to be 0.7. For imperfect trench conduits and negative projecting conduits the projection ratio, p', can have any convenient value as shown in Fig. 19-16(b, c). The curves of Figs. 19-13, 19-14, and 19-15 are for values of p' of 0.5, 1.0, and 2.0.

19.8.5 Settlement Ratio

The settlement ratio, r_{sd}, is an abstract quantity depending on the settlement and deflection of the pipe itself and the compression of the soil adjacent to the pipe. The ratio determines whether friction forces acting on the column of soil over the pipe will add to or reduce the load on the pipe.

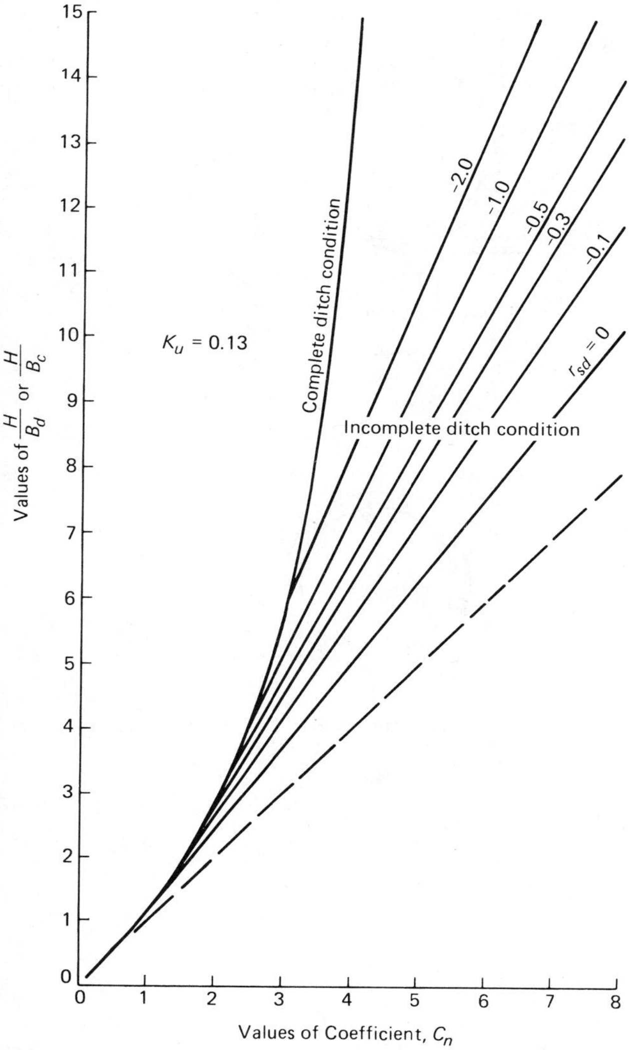

Fig. 19-14 Diagram for coefficient C_n for negative projecting conduits and induced trench conduits when $p' = 1.0$.

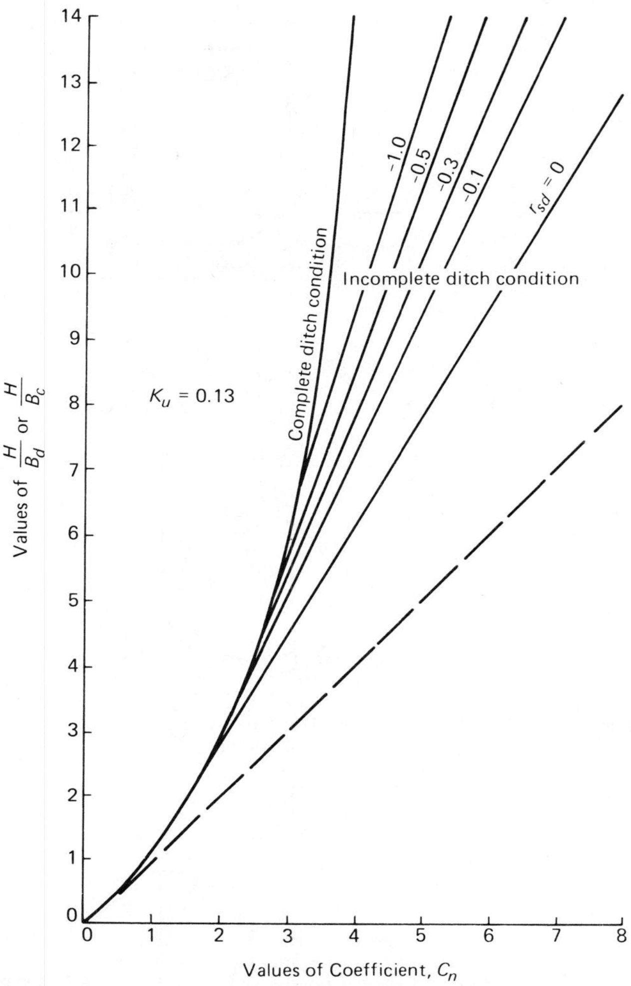

Fig. 19-15 Diagram for coefficient C_n for negative projecting conduits and induced trench conduits when $p' = 2.0$.

Based on the limited test data, the following values are suggested for concrete pipe culverts:

Pipe on rock or unyielding soil	$r_{sd} = +1.0$
Pipe on ordinary soil foundation	$r_{sd} = +0.5$ to $+0.8$
Pipe on yielding foundations as compared to the adjacent natural ground	$r_{sd} = 0$ to $+0.5$
For negative or imperfect trench conduits	$r_{sd} = 0$ to -1.0

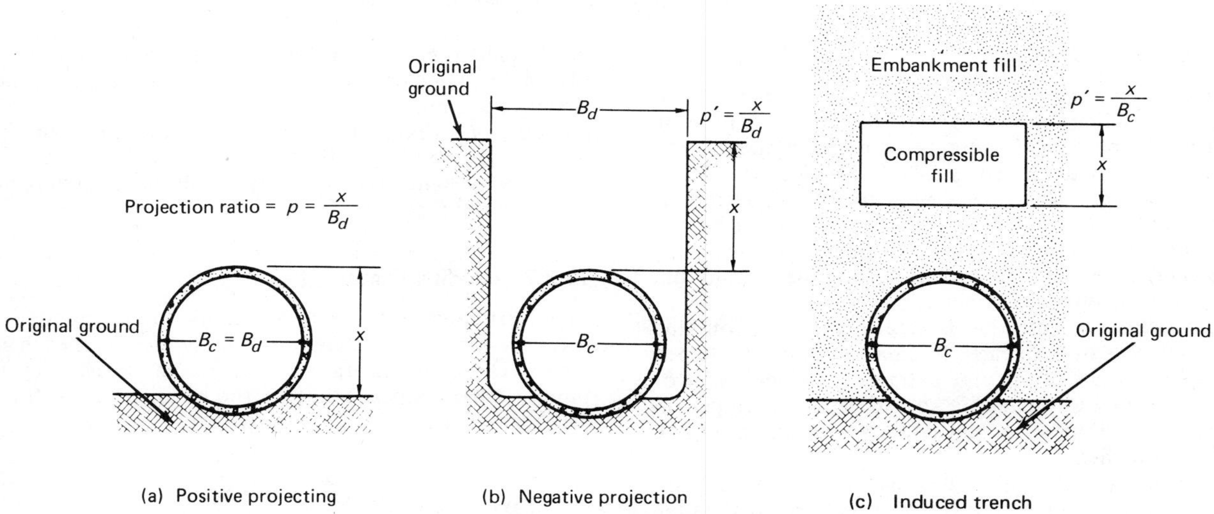

Fig. 19-16 Projection ratio for embankment conduits.

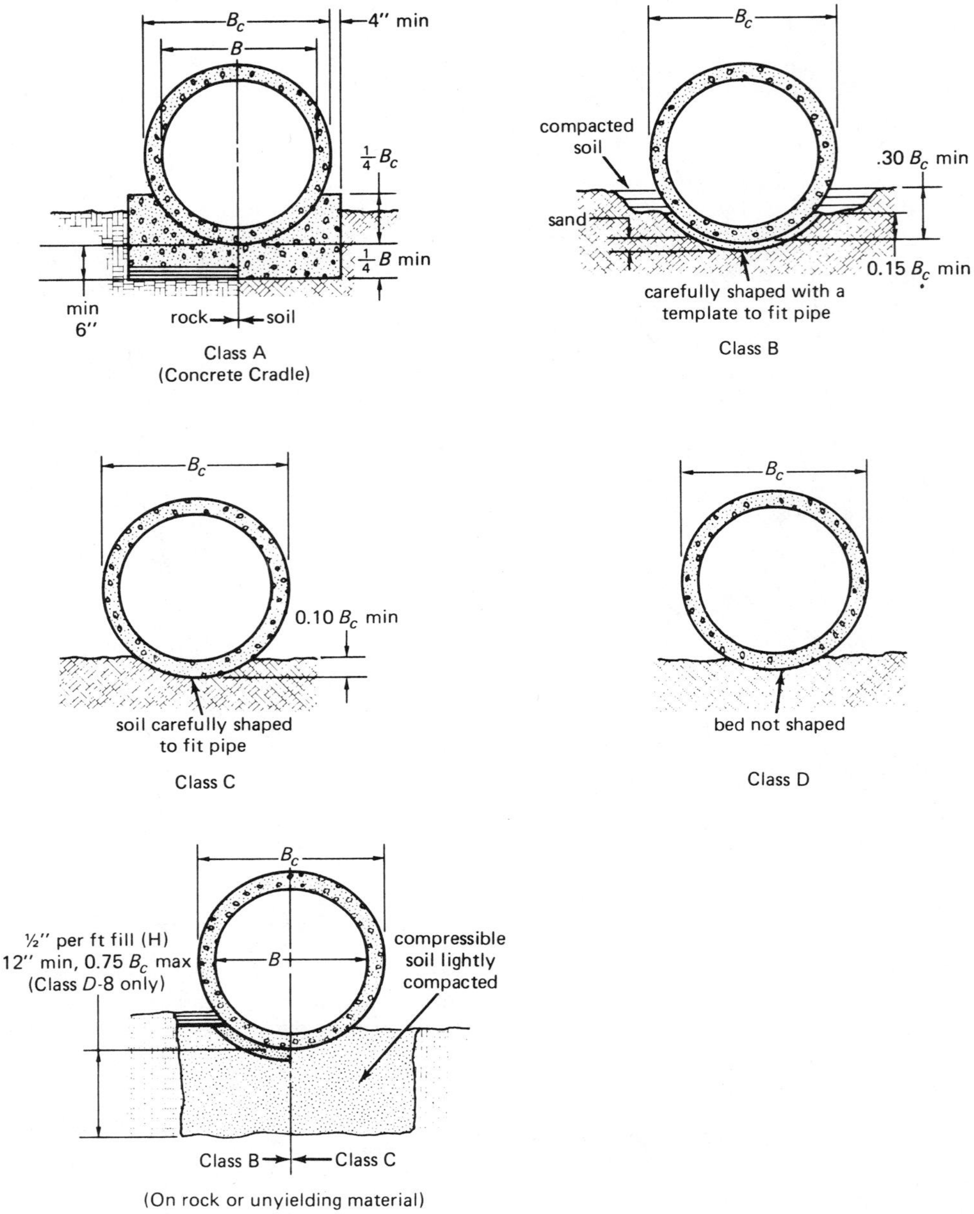

Fig. 19-17 Classes of bedding for embankment conduits.

19.8.6 Supporting Strength

The supporting strength of concrete pipe culverts is dependent on the bedding under and around the pipe and on the strength of the pipe itself. Four types of bedding are commonly used and are described as follows (see Fig. 19-17):

Class A bedding: The pipe is embedded in a continuous concrete cradle constructed of 2000-lb concrete or better; with a thickness under the pipe of one-fourth the inside diameter and extending up alongside the pipe a minimum of one-fourth the outside diameter.

Class B bedding: The pipe is carefully bedded on fine granular materials over an earth foundation carefully shaped to fit the lower part of the pipe exterior for at least 15% of its overall height. Compactible soil material is tamped in layers not more than 6 in. thick for the remainder of the lower 30% of its height.

Class C bedding: The pipe is bedded with "ordinary" care in a soil foundation shaped to fit the lower part of the pipe with reasonable closeness for at least 10% of its overall height. The remainder of the pipe is surrounded by material placed by hand to completely fill all spaces under and adjacent to the pipe.

Class D bedding: The pipe is placed without any effort to shape the bedding to fit the pipe. Material is placed around the pipe by hand to completely fill all spaces under and adjacent to the pipe.

19.8.7 Bedding Factor

In culvert construction practice, unlike trench construction, it is necessary to calculate the bedding factor for each installation. The active lateral pressure acting on the sides of the pipe is taken into account in the bedding factor formula. The formula is:

$$L_f = \frac{1.431}{N - xq}$$

where:

N = a function of the distibution of vertical load and vertical reaction

x = a function of the area of the vertical projection of the pipe on which the active lateral pressure of the fill material acts

q = the ratio of the total lateral pressure to the vertical load

$$q = \frac{MK}{C_c}\left(\frac{H}{B_c} + \frac{M}{2}\right)$$

Here M is the fractional part of B_c over which the lateral pressure acts. It may or may not be equal to the projection ratio, p.

When Class A (concrete cradle) bedding is used, X' rather than X is used in the bedding factor formula.

Values of N for Various Classes of Bedding

Bedding Class	*N*
Class A	0.505
Class B	0.707
Class C	0.840
Class D	1.310

Values of X and X′ for Various M Ratios

M	*X*	*X′*
0.0	0.000	0.150
0.3	0.217	0.743
0.5	0.434	0.856
0.7	0.594	0.817
0.9	0.655	0.678
0.1	0.638	0.638

EXAMPLE 19-4: Using data of previous examples, calculate loads and supporting strengths.

SOLUTION: Load calculation for 30 ft of fill over a projecting 60-in. culvert with $p = 0.7$:

$$r_{sd} = 0.7;\quad w = 120 \text{ lb/cu ft}$$
$$H = 30;\quad B_c = 60 + 2(6) + 72 \text{ in.} = 6 \text{ ft}$$
$$\frac{H}{B_c} = \frac{30}{6} = 5.0$$
$$pr_{sd} = 0.7 \times 0.7 = 0.49$$

From Fig. 19-12, $C_c = 7.5$.

$$W_c = C_c w B_c^2$$
$$W_c = 7.5 \times 120 \times 6^2$$
$$= 32400 \text{ lb/linear ft of pipe}$$

Supporting strength calculation:

$$L_f = \frac{1.431}{N - xq}, \quad \text{where } q = \frac{MK}{C_c}\left(\frac{H}{B_c} + \frac{M}{2}\right)$$

From preceding problem with $K = 0.33$:

$$q = \frac{0.7 \times 0.33}{7.5}\left(5.0 + \frac{0.7}{2}\right)$$
$$= 0.165$$

For Class B bedding, N = 0.707, x = 0.594:

$$L_f = \frac{1.431}{0.707 - 0.549 \times 0.165}$$
$$= 2.35$$

Assuming a factor of safety of 1.33, the required ultimate three-edge-bearing strength for the installation is

$$\text{three-edge} = \frac{W_c \times FS}{L_f} = \frac{32{,}400 \times 1.33}{2.35} = 18{,}350 \text{ lb/ft}$$

$$\text{three-edge D-load} = \frac{18{,}350}{5.0} = 3670 - D$$

Class V pipe is required.

19.8.8 Charts of Allowable Fill Heights

Figures 19-18(a) and (b) are taken from Charts I(a) and (b) of "Reinforced Concrete Pipe Culverts—Criteria for Structural Design and Installation" by the Bridge Division, Office of Engineering and Operations, U.S. Bureau of Public Roads, Aug. 1963.

These charts can be used to rapidly determine the required D-load strength for concrete pipe culverts under various fill heights and for various installation procedures.

19.8.9 Supporting Strength of Precast Boxes

ASTM Specifications C789 and *C850* provide complete designs for various standard sizes and for various fill heights. These designs are based on the installed condition. Therefore it is unnecessary to calculate loads, three-edge-bearing strengths, or bedding factors when using these specifications. It is important that the installation provide uniform support under the bottom slab and uniform side support. The American Concrete Pipe Association, in their "Concrete Pipe Handbook," recommends two types of bedding for embankment installations of precast boxes.

Class B bedding requires that a minimum of 2 in. of fine granular material be placed under the box, and that compacted soil be placed against the sides for at least 0.3 the outside width of the box. The remainder of the embankment is constructed in the normal manner.

Class C bedding requires that the natural ground be shaped flat to fit the underside of the box with about 0.1 the outside width of the box placed below the natural ground surface. The embankment is then constructed in the normal manner but must be brought up evenly on each side of the box.

19.9 LIVE LOADS ON BURIED PIPE

In addition to the weight of the backfill material, surface wheel loads from trucks, aircraft, railroad locomotives, etc. produce pressure on the top of buried conduits. This must be added to the backfill load to determine the total load on the pipe. The magnitude of the live load decreases rapidly as the cover over the pipe increases. For fills greater than 8 ft over highway culverts, the live load effect can be neglected. Rigid pavements also greatly reduce the effect of the live load.

Live load pressures on buried pipes can be calculated using the tables from the Portland Cement Association publication, "Vertical Pressures on Culverts under Wheel Loads of Concrete Pavement Slabs." No distinction is made between trench conduits and embankment conduits in calculating live load pressures.

Supporting strengths of pipe for live loads can be conservatively calculated using the load factors for trench-type installations.

19.10 HYDRAULIC CAPACITY OF CONCRETE PIPE SEWERS

Flow by gravity through a sewer depends on the slope of the sewer, the cross section of the sewer, depth of flow, character of the interior surface, whether flow is steady or intermittent, and whether there are any obstructions, bends, or connections.

Storm sewers are normally designed to flow just full but not under pressure for periods of maximum flow. Sanitary sewers are normally designed to flow half full. It is generally assumed that the hydraulic grade line is parallel to the pipe invert in calculating flow through sewers. It is also assumed that uniform flow will occur.

When the velocity of flow is known, the capacity of the sewer can be calculated from the formula:

$$Q = VA$$

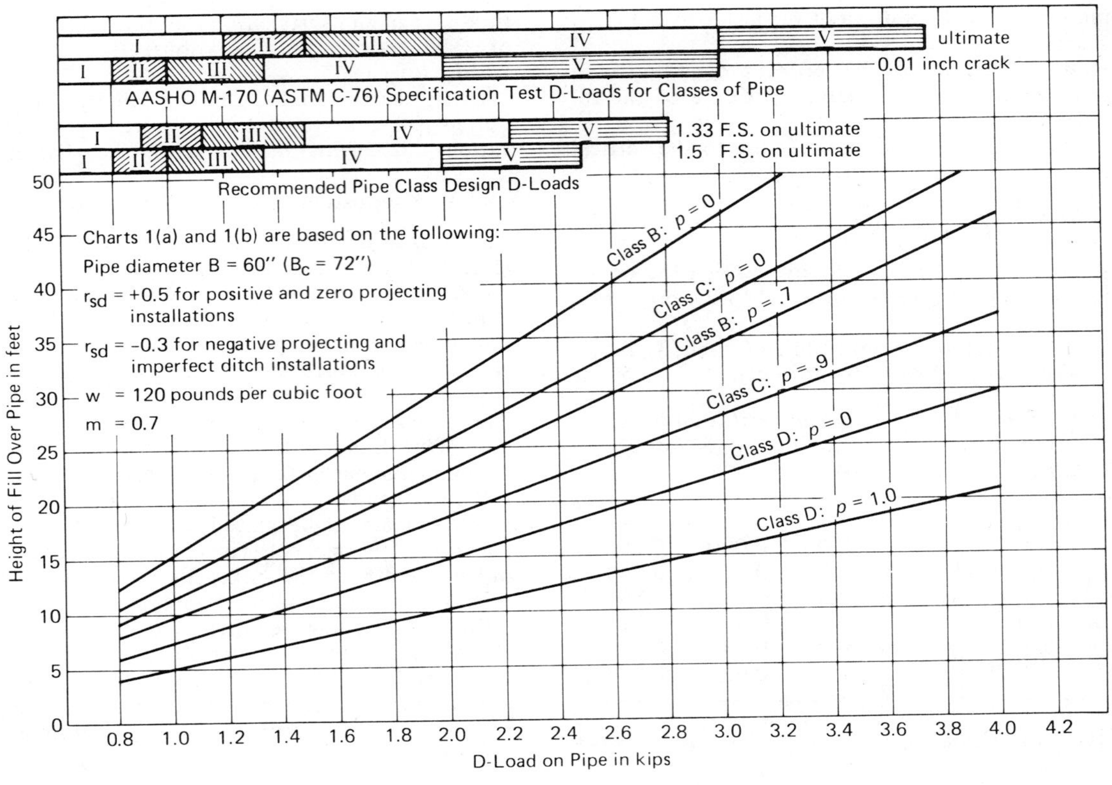

(a)

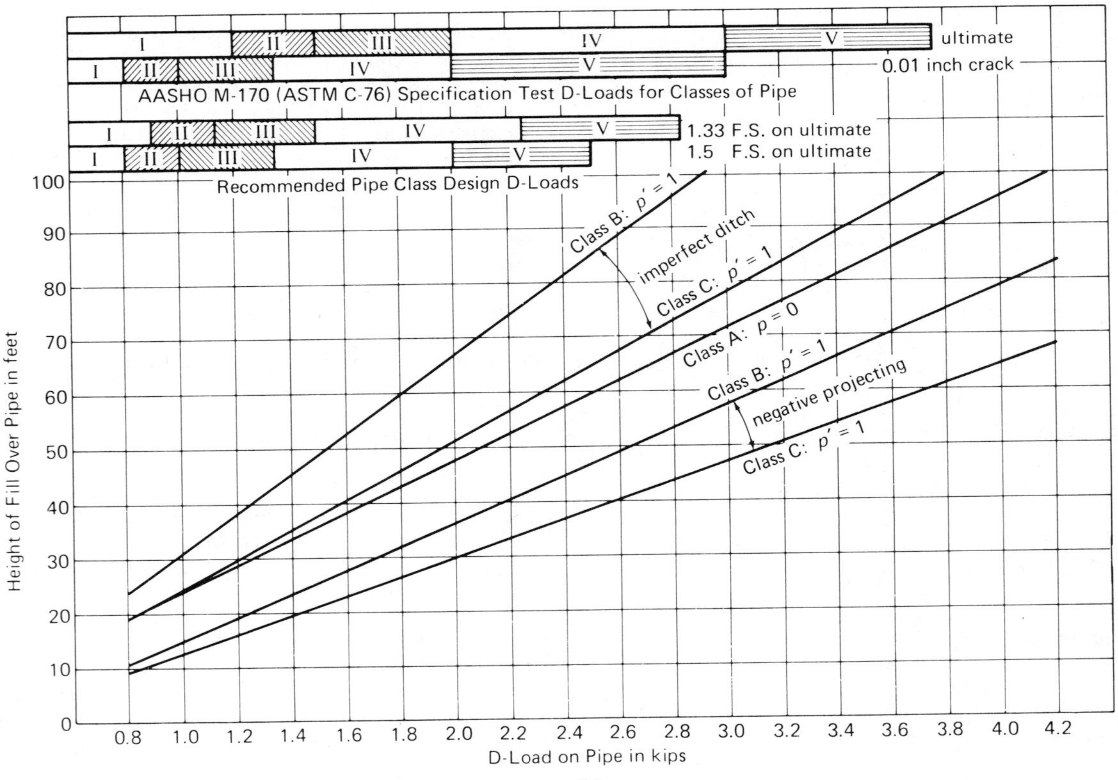

(b)

Fig. 19-18 (a) Positive and zero projecting installations and (b) Class A, induced trench, and negative projecting installations.

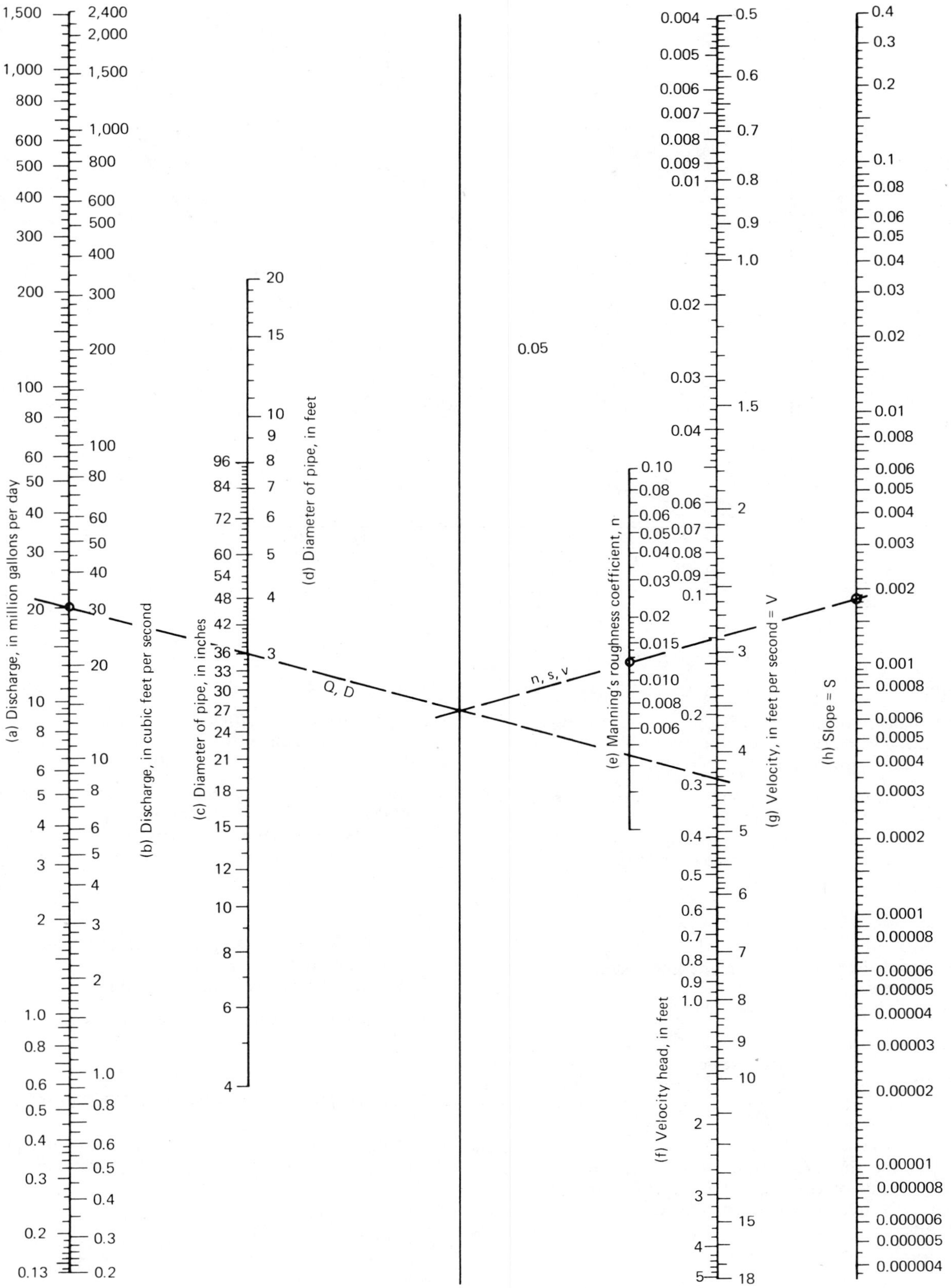

Fig. 19-19 Nomograph for flow in round pipe—Manning's formula.

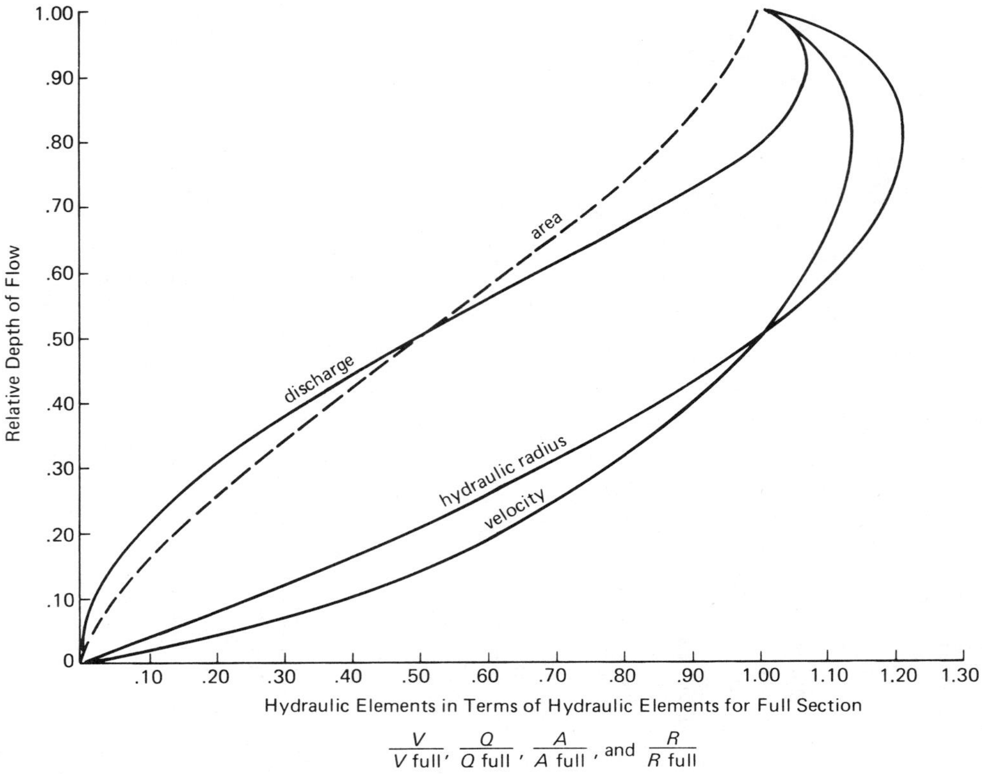

Fig. 19-20

where:

Q = quantity of flow in cu ft/sec
V = mean velocity of flow in ft/sec
A = cross-sectional area of the conduit in sq ft

The velocity of flow in gravity-flow sewers is usually calculated by the Manning formula:

$$V = \frac{1.486}{n} R^{2/3} S^{1/2}$$

where:

R = hydraulic radius, ft
S = slope of the hydraulic grade line (usually the slope of the sewer)
n = Manning roughness coefficient

The value of n commonly used for concrete pipe storm sewer is 0.012. For sanitary sewers a value of 0.013 is used.

The Manning formula is used for sewers for any cross section flowing either full or partly full. A nomograph shown in Fig. 19-19 is used to solve the Manning formula for circular sewers flowing full. Figure 19-20 can be used to determine the hydraulic properties of circular sections for any depth of flow. Using Figs. 19-19 and 19-20, the velocity in circular sewers for any depth of flow can be determined.

19.11 HYDRAULIC CAPACITY OF CONCRETE PRESSURE PIPE

Flow in irrigation lines operating under pressure is calculated by the Scobey formula:

$$V = C_5 H_f^{0.5} d_i^{0.625}$$

where:

C_s = roughness coefficient
d_i = inside diameter of the pipe, in.
H_f = loss of head in ft per 1000 ft of pipe

Values of C_s for concrete irrigation pipe are 0.345 for average pipe and 0.370 for smooth pipe.

The Hazen-Williams formula is widely used to compute the velocity of flow in concrete pressure water pipe:

$$V = 1.318\, C_i R^{0.63} S^{0.54}$$

where:

C_i = the Hazen-Williams roughness coefficient
R = mean hydraulic radius
S = slope of the hydraulic grade line

For concrete pressure pipes the value of C_i commonly used is 140.

19.12 HYDRAULIC CAPACITY OF CULVERTS

The hydraulic analysis of a culvert is complicated in that different types of flow can occur, depending on such factors as:

Type of inlet
Length and slope of culvert
Size of culvert opening
Barrel roughness
Depth of headwater
Depth of tailwater

19.12.1 Inlet Control

If a concrete pipe culvert is on a sufficiently steep slope (0.5% ± depending on size and length), it operates with

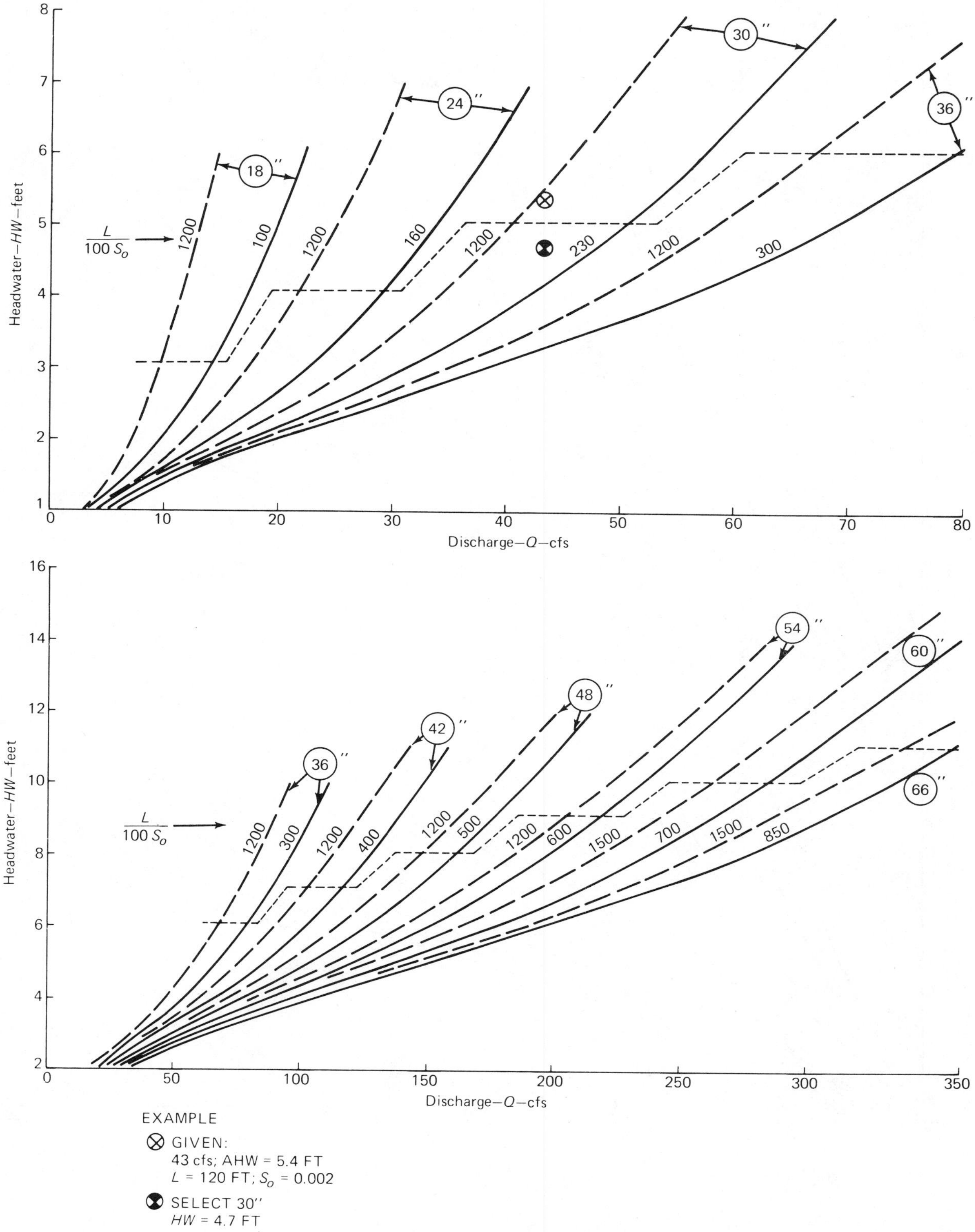

Fig. 19-21 Culvert capacity for circular concrete pipe, groove-edged entrance, 18 to 60 in.

"inlet control." With inlet control due to the slope of the culvert, the water is carried away from the inlet faster than it enters the culvert. Thus the inlet acts as an orifice. Flow through the culvert is governed by hydraulic characteristics of the inlet, culvert size, and the headwater depth. The barrel of the culvert always flows partly full under inlet control even with the inlet submerged.

19.12.2 Outlet Control

For flatter slopes, the flow through the culvert barrel will more nearly equal the rate of flow through the inlet. If the slope is sufficiently flat, the flow through the barrel will be less than that which can theoretically enter the culvert. When this occurs, the pipe begins to flow full if the inlet is

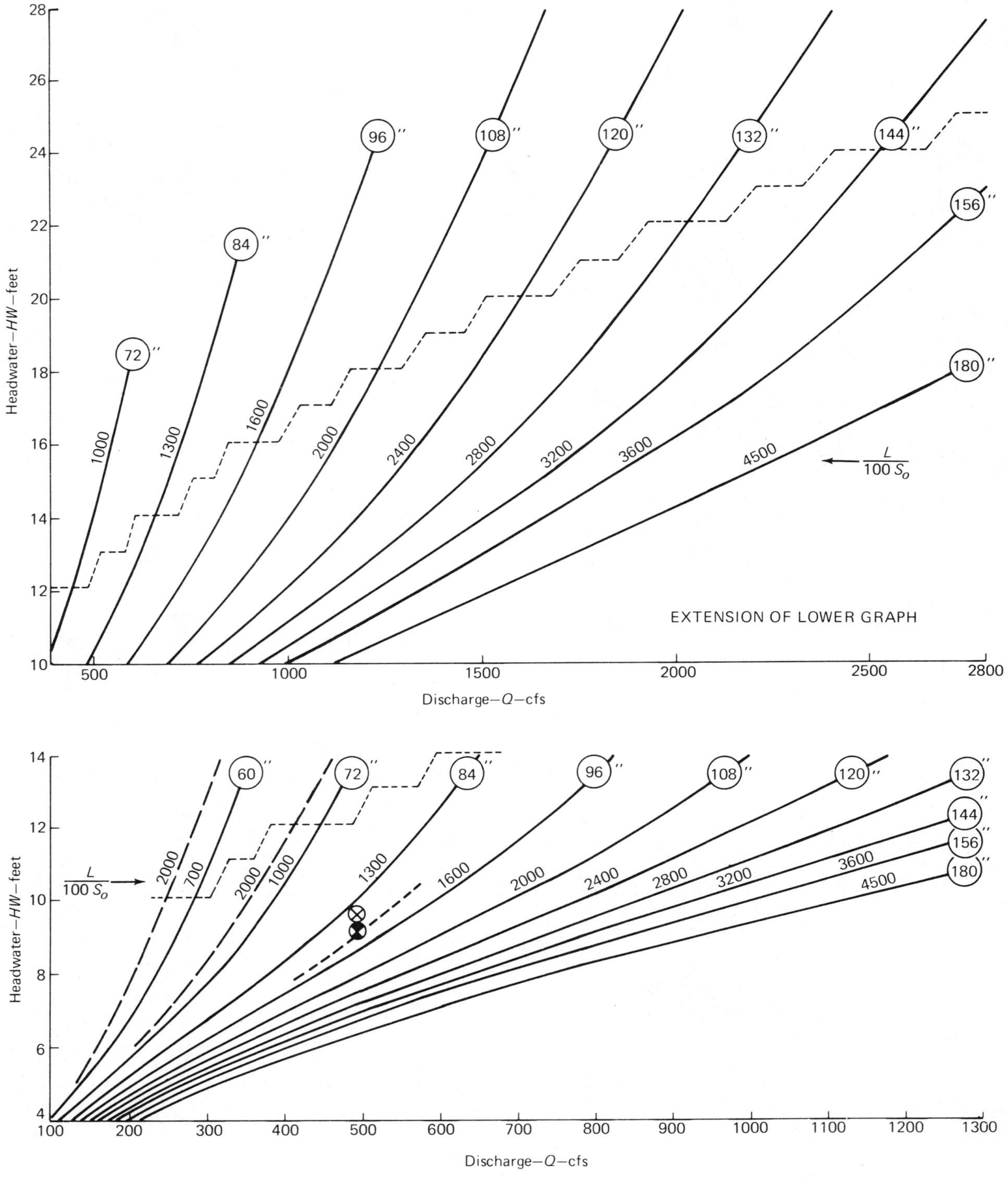

EXAMPLE

⊗ GIVEN:
490 cfs; AHW = 9.6 FT
L = 60 FT; S_o = 0.000

SELECT 90'' $\left(\frac{L}{D} = 8\right)$
HW = 9.2 FT

Fig. 19-22 Culvert capacity for circular concrete pipe, groove-edged entrance, 60 to 180 in.

submerged. The discharge through the culvert then depends on all the variables such as type of inlet, culvert size, headwater depth, length, slope, and barrel roughness. The culvert is then said to operate with "outlet control."

19.12.3 Unsteady Flow

Between inlet control and outlet control, the unsteady flow condition exists in which the culvert rapidly vascillates from one to the other. The average discharge lies between the two and can be obtained from charts by interpolation.

19.12.4 Hydraulic Analysis Charts

To simplify the hydraulic analysis of culverts, charts and nomographs have been developed and presented in various publications.

Figures 19-21 and 19-22 are Charts 19-2 and 19-3, taken from "Capacity Charts for the Hydraulic Design of Highway Culverts," Bureau of Public Roads, *Hydraulic Engineering Circular No. 10*, U.S. Government Printing Office, Washington D.C., 1965, and can be used for most concrete culvert pipe installations.

Each chart has a vertical scale giving the depth of headwater above the bottom of the pipe in feet. The horizontal scale gives the discharge in cubic feet per second. On Fig. 19-21 each pipe size is represented by two curves, a solid-line curve and a dashed-line curve. The solid-line curve gives the headwater-discharge relationship for that size culvert operating under inlet control as previously described. The dashed-line curve represents the headwater-discharge relationship for that size culvert operating under outlet control.

An index number is given on each curve. The index number equals the length (L) divided by the slope in percent ($100S_o$). If the index number calculated for the installation is equal to or less than that on the solid-line curve, the culvert operates with inlet control and the headwater-discharge relationship is given by that curve. If the index number is between that on the solid line curve and that on the dashed line curve, the headwater-discharge relationship can be determined by interpolation.

If the index number is greater than that on the dashed-line curve, the culvert operates with outlet control. The headwater-discharge relationship can be approximately determined by an extrapolation, but a more rigorous analysis is desirable, particularly if the difference in index numbers is large. Procedure is given in *Hydraulic Engineering Circular No. 10* previously referred to in Section 19.12.4.

On Fig. 19-22, each pipe size is represented by a single solid-line curve. For the large diameters shown on this chart and for practical headwater depth and culvert lengths, the curves representing inlet control and outlet control are very nearly alike. Consequently, a single curve is given, as most culverts in these sizes will operate with inlet control.

For most culverts, the length, slope, type of pipe, and allowable headwater depth are determined by the topography at the installation site. The problem is to determine the required size. For example, for a given location we might have $L = 400$ ft, $Q = 180$ cfs, and $S = 0.004$. Then, allowable $HW = 7$ ft.

Using Fig. 19-21, these quantities indicate a 54-in. pipe as a trial size. The index number is:

$$\frac{400}{100 \times .004} = 1000$$

This value falls between the solid-line curve and the dashed-line curve. Interpolation, however, shows that an allowable headwater of 7.0 ft will have a corresponding discharge of only 165 cfs. A 60-in. pipe will carry the required discharge with a headwater depth of about 6.4 ft and therefore is a satisfactory solution.

20

Marine Structures

BEN C. GERWICK, JR.*

20.1 GENERAL CONSIDERATIONS

20.1.1 Adaptability of Concrete for Marine Structures

Concrete is an almost ideal material for marine structures of many types—harbor, coastal, and ocean; and for all marine environments—submerged, partially and completely exposed, temperate, tropical, and arctic. It is economical, durable, strong. As will be shown in later sections, the design and placement techniques for marine applications assume great importance for optimum results.

The basic limitations on the use of concrete for marine structures have been essentially related to structures subjected to high cyclic bending forces, such as ocean storm waves, and impacts, such as those from collision or ice. Even in the deep ocean, however, concrete structures are now gaining increased acceptance through utilization of different structural concepts (e.g., caissons) in lieu of the more typical jacket frames. Concrete, therefore, appears to have a major role to play in the development and exploitation of the oceans.

Reinforced and prestressed concrete in typical marine applications possesses such inherent properties as high resistance to compressive loading, ability to conform to sustained imposed deformations, and rigidity. The interior surface of submerged concrete structures is generally free from weeping and condensation. It has excellent fatigue properties under cyclic loading such as waves. Under overload due to impact, grounding, or explosion, well-reinforced concrete does not tear and rip, but cracks and crushes locally, leading to a ductile mode of failure. At extreme low temperatures, it maintains a nonbrittle, ductile behavior.

Repairs to damaged sections are easily effected; in many cases, repairs can be accomplished underwater without the need for dewatering or drydocking. Concrete is readily molded into any desired form or shape, including doubly curved surfaces. It has high resistance to abrasion from moving gravel, sand, and ice.

Finally, concrete is inherently economical. It can be locally produced in most parts of the world, using predominantly indigenous materials and labor. Thus, it has been inevitable that concrete would come to play a significant role in marine structures. Like all materials, it requires the employment of proper design and construction techniques if the full advantages of this material are to be realized in service.

20.1.2 Environmental Considerations

Concrete for marine structures is subjected to structural criteria and performance requirements that are common to all concrete structures. However, the environment in which it must perform differs significantly from other environments; the marine environment exerts extreme demands on the concrete from both durability and loading aspects.

1. Submergence and Pressure. Concrete structures that are submerged in seawater must consider the effects of the continuous presence of water under hydrostatic pressure. This not only exerts a structural loading—it also means that the concrete face, at least, will be continuously wet, and that water will penetrate microcracks existing between the aggregate particles and the cement paste matrix, moving toward the inner face. If the inner face is at atmospheric

*Professor of Civil Engineering, University of California at Berkeley.

pressure, which occurs in many structures, then there is a continuous slow penetration.

The concrete will thus absorb water and increase its density. This absorption of water can have a dramatic effect on structures that are designed for flotation, with the dead weight being approximately offset by the buoyancy. In such a case, an increase in density of 2% due to absorption can increase the net (buoyant) weight by as much as 40 to 50%.

The uniform temperature of seawater in a particular locality tends to eliminate the short-term expansion and contraction of submerged concrete.

The continuously saturated environment of submerged concrete eliminates drying shrinkage, and permits long-term chemical crystallization processes to be completed, thus generally leading to higher strengths. Autogenous healing of microcracks and even hairline cracks may also take place due to secondary crystallization. Concrete placed under pressure (i.e., poured in a submerged state) benefits greatly in strength and density. The concrete is given additional compaction by the hydrostatic loading.

Recent data indicates that concrete continuously submerged in seawater becomes increasingly impermeable, apparently because of secondary chemical compound formation.

2. Seawater Chemistry. The chloride ions in seawater are of particular concern for concrete structures. The cement in concrete initially forms a protective coating on the embedded reinforcing steel, which prevents corrosion, in effect, passivating it. Chloride ions, penetrating to this coating, form a soluble compound, thus de-passivating the steel and rendering it subject to oxidation.

When seawater moves through permeable concrete and evaporates from the other face, salt cells are formed in the concrete pores. These may set up a battery action with the reinforcement, leading to electrolytic corrosion.

A piece of steel suspended bare in the sea–air splash zone will corrode less than the same steel encased in highly permeable concrete, hung in the same environment. This emphasizes the importance of making the concrete as impermeable as possible.

Seawater can also attack the concrete itself. Magnesium ions from the seawater can replace calcium ions present in the cement. This leads to a general softening and loss of strength. Sulfate ions can similarly cause disruption of the concrete, reacting with the tricalcium aluminate in the cement (C_3A) and with unsound aggregates. These effects can be minimized by limiting the C3A content. The aggregates must be selected for durability and soundness under sulfate attack. The dissolved gases in seawater, particularly oxygen, contribute to corrosion. Since the oxygen content decreases with depth, it follows that corrosion of the steel is usually also reduced with depth.

In some seas, H_2S (hydrogen sulfide) is present, dissolved in the seawater. H_2S may also be formed from a bacterial reaction with anaerobic bacteria. Its attack on the concrete can be lessened by making the concrete dense and impermeable and by using a rich cement factor.

Where acid, sewage, or other aggressive chemicals are discharged directly through or onto concrete structures, additional coatings such as epoxies or bitumastic may be required.

3. Movement of Seawater. This gives rise to many of the design loadings on the concrete structures, affecting their overall stability.

Currents in the ocean generally run from 0.5 knot to a usual maximum of perhaps 2 knots, although higher velocities occur in restricted channels and inlets. Contrary to popular belief, it is frequently found in specific locations that the bottom current may be almost as high as the surface current. In estuaries where fresh and salt water meet, it is not unusual to find the salt water flooding in at the bottom while the surface is still ebbing.

At some locations, the direction of the currents rotates during the tidal cycle. Prolonged winds in one direction may produce substantial currents even in the open sea.

The total current load may be computed by the formula:

$$P = \frac{Av^2 wK}{2g}$$

where:

P = total current load, lb
A = area exposed normal to the current, sq ft
v = velocity of current, fps
g = 32.2 ft/sec^2
w = unit weight of water, 62.4 or 64 lb/cu ft
K = shape constant (1.33 for rectangular, 0.75 for circular)

In the English system of units, this reduces to $P = Av^2K$.

The current forces can alternatively be computed by a naval architecture approach in which the components of surface friction (P_f) and eddy action (P_e) are added:

$$P = P_f + P_e = v^2 \sin\theta \,(sc + 3.2A)$$

in which:

θ = angle that direction of flow makes with the face
c = a constant, taken as 0.5
s = immersed surface, sq ft
v = velocity of current in knots

and the remainder of symbols are as stated above.

It is noted that this second formula gives values about 50% greater than the first approach. Observations from actual structures indicate that the true values are best represented by this second formula.

Current exerts pressure on the structure. It also may cause an eddy action by flow over the top, tending to lift a structure that is on the bottom. Conversely, when a structure is suspended just above the bottom, the current tends to increase its effective weight, pulling it down. The increased velocity of current underneath and around the structure may also cause erosive scour of the bottom sediments. Scour may be aggravated by the cyclic pore pressure differentials created in the soil by waves, which place sediments in suspension where they are readily moved by the waves.

Scour may also take place at the corners of a structure set on the bottom. Erosion may be more serious at the downstream corners than at the upstream nose.

Waves exert a dynamic and cyclic loading on a submerged structure. At the surface, the waves may actually break against the surface with enormous force. Local impact pressures as high as 3 tons/sq ft have been measured. Below the surface, large pressure differentials may occur. For structures completely submerged, dynamic uplift forces occur during the wave cycle which fortunately are usually not in phase with the maximum lateral forces.

The pressure differentials on the two sides of a bottom-supported structure may lead to tunneling underneath and consequent serious erosion of the bottom sediments.

If the pressure differential can penetrate into a composite structure such as a breakwater, back pressure may lift out the individual components with almost explosive force due to the hydraulic ram effect. This same phenomenon, in which wave force is changed into kinetic energy, is being used intentionally to absorb and dissipate wave energy. Holes have been provided in the vertical walls of caissons for quay walls and breakwaters, as well as offshore storage vessels in the North Sea, in order to absorb a portion of the wave energy.

Nonbreaking waves reflecting from a structure may form standing waves (clapotis) directly in front of the structure leading to erosion of the soils below the crest of the standing wave. Splashing waves deposit salt above the waterline and lead to alternate wetting and drying (see Section 20.1.2.7). The waves also throw salt into the atmosphere where it is carried onto the exposed superstructure by spray and fog.

Surf produces cavitation in the surface of concrete structures that can only be minimized by extreme care to achieve a hard, dense surface, free from fins and projections. The use of special treatments to densify the surface may be required in extreme cases.

Moving water also moves sand and gravel, causing abrasion. Even silt-laden waters moving at high speeds can cause abrasion of the concrete surfaces. Moving water, whether due to waves or currents, seriously affects the underwater placement of fresh concrete. It tends to pull the cement paste out of the mix by the Venturi effect. Small gaps in caissons or sheet pile cofferdams frequently have led to serious leaching of cement.

4. Marine Organisms. A great deal of publicity has been given to the boring clam, the sea urchin, and the rock-boring mollusc. However, these organisms are normally able to attack only weak, porous concrete. The barnacle exerts high local pressure on the concrete and may destroy bitumastic coatings and roughen the concrete surface.

In a few extreme cases (e.g., the Arabian Gulf), the acid from the mollusc has been of sufficient strength to dissolve the cement. In such cases, acid-resistant epoxy coatings may be necessary.

Fouling (growth of marine bio-organisms) of the hulls of concrete barges and ships is apparently less serious than with steel hulls, but, nevertheless, it does occur. It is reportedly easier to remove such fouling from concrete surfaces than from other hull types.

Fouling of concrete intake and discharge structures has received considerable attention. Periodic heating of the discharge may kill young mussels. Chemical impregnation of the concrete is under study, but is complicated by the obvious problem posed by progressive leaching out of the chemicals. Coatings such as dense polyurethane have been successful in preventing or minimizing the attachment of marine growth.

Barnacles and mussels may increase the effective diameter of concrete piles and do produce an adverse shape factor, thus increasing resistance to the waves and the consequent load on the piles.

5. Ice and Debris. Moving ice and debris may exert direct impact loads on concrete structures. Such material may pile up, increasing the effective area and thus the load. If the ice exists in solid sheet form, then it must fail by crushing, unless suitable ice breakers are installed to cause the ice to ride up and fail in tension.

Ice cakes and debris frequently become wedged between adjoining piles. The resulting torsion may break the piles. In one case in Alaska, a large ice mass that had formed on the piles slid down at low tide during the spring thaw and broke the adjoining batter piles. Ice cakes may also adhere to concrete breakwater units, wedging them apart and lifting them at high tide.

While the loading of ice and debris is much the same on concrete structures as on those of steel and timber, the resultant effects differ substantially, depending on the structural and physical characteristics in the particular case. Concrete piles may be more vulnerable to impact and torsion, whereas reinforced concrete caissons or sea walls may be more resistant because of their mass, local rigidity, and nonbrittle behavior at the low temperatures involved.

When water freezes inside a concrete structure, it exerts a typical bursting pressure. The design of a hollow-core cylinder pile or caisson must incorporate means of preventing freezing, such as a wood or styrofoam log floated inside and/or vents beneath the surface connecting to the unfrozen water. Otherwise, serious bursting may take place.

6. Seismic Effects, Tsunamis, and Explosions. A marine structure, whether of concrete or other material, is subjected to seismic accelerations like any other structure, plus the acceleration of the added mass of water. Because concrete is frequently chosen for very large structures (offshore oil storage, caissons, bridge piers, etc.), this dynamic effect must be given special consideration.

Tsunamis, or tidal waves, produce great uplift forces. Because of its dead weight, concrete is an ideal materal for the decks of structures in areas subject to tsunamis.

Explosive forces cause serious overpressures in water that tend to collapse a submerged structure. Concrete, by virtue of its inherent rigidity, is an ideal material to resist such overpressure. However, the designer of such structures should take care to eliminate, insofar as possible, reentrant angles and holes, or cavities, which might concentrate the blast effect.

Recent studies, still in a very preliminary phase, indicate that earthquakes may produce overpressure effects up to 100% of the hydrostatic loads, with the maximum effect occurring at about 300 ft of water depth.

All submerged structures are subjected to an additional added mass effect under dynamic loading due to the acceleration of the surrounding water by the structure. While each case must be studied individually, this added mass generally varies from a factor of 1 (100% additional) for small structures (e.g., a 5-ft-diameter cylinder) down to 0.4 for very large and rigid concrete structures (e.g., 400-ft-diameter structures).

7. Sea-Air Interface: The Splash Zone. This zone is subjected to both the salt-laden waters and the oxygen and CO_2 of the atmosphere. With alternate wetting and drying, severe corrosion of the reinforcement may occur. The reinforcement then expands owing to the corrosion products, spalling off more concrete and allowing corrosion to proceed further. Obviously, concrete designed for use in this zone must be made as impermeable as possible.

This zone is also subject to severe freeze–thaw attack. The combination of salt water splash and freeze–thaw is particularly disastrous. Air entrainment is essential for this exposure. Prestressing is also beneficial in eliminating hairline cracking and microcracking due to shrinkage.

8. Ship Impact. Until recently, little attention was given in concrete structure design to damage from ship impact, other than to provide fendering systems as required. The growth in size of vessels and the importance of critical marine structures has now led to greater consideration of the problem and to ways to absorb energy and minimize damage. Heavy mass helps, as does the provision of diaphragm action, prestressing, and well-distributed reinforcement.

9. Thermal Effects. Because concrete possesses low heat conductivity, and structural concrete members are relatively thick as compared with steel elements, significant thermal gradients occur. These may be due to the sun's radiant heat on above-water surfaces, contrasting with the cooler submerged zones, or to arctic sub-zero air temperatures contrasting with relatively warm submerged zones. Gradients occur when hot oil is stored in concrete vessels or flows through well casings or piping in the interior of concrete structures.

These gradients tend to be nonlinear through the concrete

cross section, dropping rapidly as they near the cold face, thus leading to cracking. The cracking relieves the strains by reducing the stiffness; hence thermal stresses tend to be self-limiting.

Severe, if temporary, thermal strains can also be set up due to heat of hydration, followed by differential cooling of the relatively thick cross sections typical of many concrete marine structures. The most effective way of minimizing these strains is to insulate the forms and to blanket the surface immediately after stripping the forms.

20.2 DURABILITY AND CORROSION PROTECTION

20.2.1 Corrosion of Reinforcement

This is the most common source of distress in concrete marine structures. The cement paste in concrete acts to protect reinforcement by passivating it, forming a protective oxide coating on the steel and giving a pH of 13 or 14 adjacent to the steel. The pH of seawater is about 8. Corrosion of the reinforcement occurs below a pH of 11.

The pH is lowered by the presence of chloride ions in the concrete. These may be introduced through seawater penetration or through salt in the mixing water or the curing water. Chlorides may have been present on the surface of the aggregates. Beach sands and desert sands subjected to salt fogs (as in some parts of Arabia) are particularly bad in this regard. The corrosion process also requires the presence of oxygen, permeating through microcracks in the concrete cover to the steel and thus setting up an electrochemical reaction. This process is aggravated by the existence of voids such as entrapped air pockets adjacent to the steel.

Salt-cell electrolytic corrosion is increased by salt water migrating into and through permeable concrete, then evaporating from the surface, leaving tiny deposits of salt that set up a potential with the reinforcement.

The corrosion process is much more rapid in hot climates and warm water than in cold. Proper protection against corrosion of the reinforcement by factors described above can be provided by the following means:

1. Dense impermeable concrete. This is achieved by a low water/cement ratio (below 0.40), a dense but workable mix design, good consolidation and curing, and a smooth dense surface finish. Low water/cement ratios diminish the bleed water that may be trapped at the cement-paste/aggregate interface and thus initiate microcracking. The addition of pozzolans such as fly ash, natural pozzolans or condensed silica fumes will increase the impermeability by several orders of magnitude.
2. Adequate cover. For marine exposures, a cover of 2 to 2.5 in. is generally recommended by most authorities. However, a cover as low as 1.5 in. has been used successfully on many prestressed concrete piles, etc., and has apparently been satisfactory owing to the dense concrete and rich mix employed. Even smaller covers, down to 0.375 in., have been used on concrete ships and boats, where especial care was taken to achieve a dense cement paste cover. A more rational approach to cover requirements relates the cover to the size of the coarse aggregate: since microcracks occur along the interfaces with the aggregate, the cover should be 1½ to 2 times the nominal maximum size of coarse aggregate. The surface "skin" of cement paste has also been found to be disproportionally beneficial.
3. Rich cement factor. Six to eight sacks per cubic yard is recommended.
4. Salt-free aggregates. Beach and desert sands may require washing with fresh water.
5. Minimum chloride content in mixing and curing water and in admixtures. The use of calcium chloride in reinforced marine structures is not recommended, and it is prohibited for prestressed constructions.
6. Selecting the proper type of cement. While low C_3A content may enhance the durability of the concrete, it causes a reduced resistance to chloride permeability and thus may not give as good a corrosion protection for the reinforcing steel. Thus, a moderate C_3A content is the best solution for marine exposure of reinforced or prestressed concrete: ASTM Type II cement is recommended. With rich cement mixes, a C_3A content of 6 to 10% is optimum.
7. Consolidation of cement around the steel. Vibration techniques should be adopted that eliminate voids near the steel while producing as impermeable a surface as possible. In grouting posttensioned ducts of marine structures, particular pains must be taken to prevent the formation of voids (see Section 20.3.4).
8. Adequate curing to ensure complete hydration. The forms and surface are insulated until the concrete has attained a relatively uniform temperature across the section.
9. In special cases, it may be desirable to seal the surface, as with bitumastic paint, epoxy paint, or other protective treatment.[20-5] Epoxy-coated reinforcing steel and ducts for prestressing tendons should be used in zones of extreme exposure.

Embedded metals other than steel, such as copper or aluminum, may set up electrolytic corrosion, especially where an electrolyte such as seawater is present. However, galvanized and cadmium-coated inserts have been successfully employed on a great many marine structures with satisfactory results.

Cracks in the concrete are undesirable in marine structures. In the atmospheric and splash zones, they permit the penetration of CO_2 (which lowers the pH), chlorides, and oxygen. However, small hairline cracks at the surface [0.12 mm (0.005 in.) width] apparently seal themselves with secondary crystallization in many cases, and have not proved nearly as serious as general permeability. Cracks of moderate width, up to 0.5 mm (.02 in.) or so, located well below the waterline, are usually not serious because of the low oxygen content of the seawater. Such cracks become the anode; however, the electrochemical process also requires an active cathode, namely, penetration of oxygen to the reinforcing steel in the uncracked zones. Since this steel area is large, theoretically the cathode could be very active. However, in practice the dense and impermeable protection of good-quality concrete and the low oxygen content of the seawater limit the cathodic potential and usually starve the reaction.

When corrosion of reinforcement has been a serious problem in marine structures, it has usually shown up within a very short time. One of the most notorious cases was the original San Mateo-Hayward Bridge across San Francisco Bay. The concrete was very permeable, leading to continuous electrochemical corrosion. Serious problems developed within one year, even before the bridge was completed. During subsequent years, extensive repairs were caried out, amounting in cost to more than the original value of the bridge. However, the salt-cells were spread throughout the concrete, so that repairs merely served to change cathodic zones to anodic, and vice versa.

However, another project in San Francisco Bay, a wharf, was successfully repaired by chipping off the undersurface, guniting it, and coating it with heavy coal tar paint. The upper surface had always been sealed by asphalt. In this case, apparently, the moisture electrolyte was effectively sealed off and passivated.

20.2.2 Chemical Attack on Concrete

Protection of concrete against chemical attack in a marine environment may utilize both internal and external means.

Internally, protection consists of selecting components that will remain inert and resist the seawater attack. These steps include:

1. Use of low alkali cement (less than 0.65% Na_2O and K_2O), such as ASTM Type II.
2. Use of nonreactive aggregates (ASTM tests C289 and C277).
3. Use of cement containing moderate C_3A content, 6 to 10%, such as ASTM Type II.
4. Aggregates to be sound—not subject to sulfate attack (see ASTM C88).
5. Adequate curing. Optimum results appear to be obtained with normal moist curing followed by a period of drying before immersion in salt water.
6. Good consolidation of concrete, and good surface finish.
7. Rich cement factor, low water/cement ratio.
8. Use of air entrainment, 4 to 6%, which appears to be beneficial although it is not always felt to be essential. In marine environments subject to freezing, air entrainment is essential.
9. Addition of pozzolans to the mix.

External means of protection include:

1. Epoxy coatings. There are numerous formulations available, including some that may be applied underwater. Some of them provide protection against acidic substances.
2. Bitumastic coatings. A hot-applied coal tar has been found especially effective. However, for practical reasons, most bitumastic coatings are sprayed or painted on cold. Some of these compounds are very toxic, requiring adequate protection for personnel engaged in applying the compound.
3. Dense polyurethane coatings, which have been found effective for protection against aggressive substances such as sewage, etc.
4. Other coatings and protective treatments, as described in Ref. 20-5. Epoxy coating of reinforcing steel and ducts.

Acids, even highly diluted, as from discharges of refineries and smelters, will attack concrete surfaces. Under such exposure, an external coating giving complete encasement is the only suitable answer.

20.2.3 Freeze-Thaw Salt-Water Attack

This is an extremely severe environment condition requiring special consideration. It occurs in the tidal and splash zones.

Air entrainment, 6 to 8%, is essential for durability. Water/cement ratios should be low (0.45 or less) and cement factors high (eight sacks per cubic yard). Also, particular care should be taken to achieve complete consolidation and a smooth dense surface free from fins, rock pockets, etc. Any patches or coatings should be permeable to water vapor. Otherwise the water vapor in the concrete, migrating to the cold face, will become trapped and may freeze, spalling the concrete just under the coating.

The exposed face of the concrete should be well reinforced with closely spaced bars in both directions or mesh.

Aggregates should have low water absorption, less than 4 to 5%.

The quality of the air entrainment should be verified by microscopic examination of the hardened concrete to ensure close spacing of air voids, e.g. less than 0.01 in. (0.25 mm). See ASTM C 457-80.

Jackets of treated timber, and wrought iron, have been extensively used in the past, as has granite facing. When properly done, these give excellent protection, but their cost is high. It is believed that careful attention to the internal steps listed in the preceeding section will give satisfactory results.

20.2.4 Fatigue Endurance

Reinforced and prestressed concrete have high resistance to fatigue. Since marine structures are often subject to cyclic loading under such lateral forces as waves, ship berthing, ice crushing, etc., fatigue endurance has been extensively studied in recent years.[20-22] Some general statements may be made about this complex subject:

1. Prestressing, by prevention of cracking, significantly improves endurance.
2. Reinforcing steel, concrete, and prestressing steel have good endurance when working at normal stress levels.
3. Reinforced concrete, which is submerged at depth, shows a reduction in endurance when cycled into cracking, due to both internal hydraulic ram effects and washing action.
4. Membrane shear capacity is reduced by cyclic loads that extend into the cracking range.
5. The corrosive atmosphere of the sea reduces the endurance of reinforcing steel.

Thus while fatigue endurance needs to be considered for concrete structures in the sea, especially offshore structures and floating vessels, proper design, including prestressing, may lead to superior performance under cyclic dynamic loading.

20.3 CONSTRUCTION METHODS AND PROCEDURES

20.3.1 Concrete Mix Design

1. Normal Weight Concrete. Normal weight concrete for marine use must be proportioned so as to achieve maximum durability for both the concrete and its embedded reinforcement.

Aggregates should be selected in accordance with the normal test requirements (i.e., ASTM C33), and should be nonreactive, sound- and abrasion-resistant. They should be free from salt and dust deposits. The total amount of soluble chlorides in sand and aggregates should not exceed 0.04% by weight.

Satisfactory concrete has been produced from beach sands by washing with fresh water. Properly selected coral aggregates can be utilized but generally produce concrete of reduced strength. Proportioning should be such as to ensure a dense compacted mix of minimum permeability.

Water must be free of excessive chlorides (not more than 600 ppm) and sulfates (not more than 1000 ppm).

Seawater can be used for unreinforced concrete in a marine exposure such as concrete breakwater units and for reinforced concrete that is fully and permanently submerged. The cement content must be especially rich. A retarding admixture should be used to prevent flash set. However, seawater has generally proved unsatisfactory for reinforced concrete construction, the chlorides having lowered the pH near the steel below the corrosion point. Prior coating of reinforcement with a cement slurry made with fresh water has proved helpful. In India, a phosphate wash on the steel has proved similarly effective. Epoxy coated reinforcement should be relatively free from corrosion. While these steps help in prolonging the life of reinforced concrete exposed to the action of both air and sea (or salt fog), seawater is definitely a choice of last resort for all reinforced marine concrete.

Cement should have a moderate C_3A content and low alkali content (less than 0.65% Na_2O and K_2O). ASTM Type II is generally to be preferred. Cement content should be high; six to eight sacks per cubic yard is recommended.

The *water/cement* ratio must be kept as low as possible to minimize permeability (w/c ratio of 0.45 or less by weight). (See Section 20.3.2.3 for special requirements for concrete placed underwater.)

Air entrainment of 4 to 8% is beneficial, and is essential for concrete that will be exposed to freezing and thawing.

Admixtures may often be beneficial in reducing the water/cement ratio, in preventing segregation, and in reducing the heat of hydration in mass pours. Admixtures should not contain more than a trace of chlorides.

Reinforcing steel should be well distributed.

Cover of 1 ½ to 2 in. should be provided over the reinforcement. Excessive cover (over 3 in.) is detrimental. A more rational approach to cover is to size it to be 1 ½ to 2 times the nominal maximum size of coarse aggregate and similarly 1 ½ to 2 times the diameter of the reinforcing bar. The surface finish should be as smooth and dense as practicable, since many attacks initiate at relatively minor defects. For concrete cast in the dry or precast, thorough curing is essential. Steam curing, if used, should be followed by water curing or two coats membrane curing compound. One should be applied immediately after removal from the steam, the other 8 hr later. Water curing alone should be 7 to 10 days. A period of drying-out afterward is desirable.

2. Heavyweight Concrete. Heavyweight concrete may be employed in certain marine structures to give maximum net negative buoyancy. Typical examples are pipeline encasement, to keep the cover from getting excessively thick and heavy in air; and bottom-founded structures where the heavy weight is used to offset uplift and tension due to wave forces. Thus, it is possible to produce concretes having a submerged effective weight of 130 to 160 lb/cu ft as compared with the 90 lb/cu ft obtained with normal concrete. (See ASTM C637.)

High density should be of particular advantage in breakwater units (tetrapods, etc.) where the high density will greatly improve resistance to displacement by waves.

The heavyweight mix can be obtained by using magnetite ore as aggregate. Other materials are ilmenite, steel shot, steel punchings, barite, and limonite. The mix must be carefully proportioned, and checked by trial mixes to ensure that excessive segregation will not occur. Good grading, rich cement content, low water/cement ratio, and air entrainment are useful in preventing segregation.

3. Lightweight Concrete. Lightweight concrete is of especial interest for floating structures. The net submerged weight may be one-half or less, as compared with normal concrete. Since minimum dimensions frequently control design (e.g., wall thickness for impermeability, rigidity, ability to place, space and cover for reinforcement, etc.), lightweight concrete may show substantial benefits.

Lightweight concrete made with expanded shales, slates, or clays, has a successful history of durability and performance at sea. Its impermeability and durability may in fact be made equal to that of normal concrete with good engineering control.

Certain specific steps are required with lightweight structural aggregate concrete. Air entrainment, 4 to 6%, is definitely desirable to prevent segregation. The maximum size of coarse aggregate particles should be one-half the cover over the reinforcing steel. The use of approximately 30 to 100% natural sand will improve consistency and performance in diagonal tension in the dry and semidry state. When it is fully submerged, the diagonal tension of this lightweight concrete is approximately equal to that of normal concrete. For typical marine structural elements, the cement factor should be at least seven sacks per cubic yard, and the water/cement ratio should be below 0.40.

4. Cyclopean Concrete. Cyclopean concrete employs a combination of large natural rock with underwater concrete binder, in order to form seawalls, quay walls, mat foundations, and caisson fill.

Clean rocks, all greater than 12 in. in diameter, are placed in layers approximately 3 ft thick. The voids, usually running 30%, are then filled with tremie concrete placed by tremie pipe or bottom-dump bucket. Unless great care is taken with the mix and placement procedure, segregation and laitance formation may lead to voids and weakened zone. Use of a thixotropic admixture helps to prevent segregation, as does a cohesive mix, obtained by a sand-to-total-aggregate ratio approaching 50% and a high cement factor. If the concrete is placed by the tremie method, the pipe must be emptied and resealed after each move. If it is placed by the bucket method, the bucket must have its top covered, and the bucket must not be opened until it is well seated on the cyclopean rock. The cover must be opened to the water at time of discharge: otherwise water will wash up through the concrete to fill the void on top.

5. Sacked Concrete. Underwater walls and slope protection have been successfully built using bagged or sacked concrete. In the author's opinion, the best results are obtained with burlap sacks, two-thirds filled with concrete of a low water/cement ratio. Once placed, these tend to knit together.

Dry mix in burlap sacks has also been employed; however, hydration is not always complete, owing to the impermeability of the dry mix. Sacks and bags may be placed in stretcher and header courses so as to interlock. Fabric forms of nylon have been developed that are spread over the surface to be protected and then pumped full of grout. The fabric itself may be porous to act as a combined filter and protective layer.

20.3.2 Underwater Concrete

1. General. High-quality concrete can be constructed underwater, provided the required engineering and construction techniques are exactly followed. With care, the resultant concrete structure can have strengths and durabilities equal to those obtained in the best practice in the dry. The problem is that carelessness and inexpertise can lead to much greater disasters underwater than in the dry. If the mix and placement techniques are violated, then the results will be very bad. The cost of repair or replacement will be excessive.

All techniques for underwater concrete placement are designed to prevent the water from mixing with the concrete and washing out the cement paste. Since the concrete placed underwater normally must be moved downward through a considerable height, steps must be taken to prevent segregation.

The first requirement may be met by conveying the concrete in a closed bucket or tube and depositing it immediately on top of or *under* the already poured concrete, so that the new concrete has minimum contact with the water.

The second requirement is additionally aided by selection of a mix that flows easily within itself, yet has great cohesiveness and richness.

Concrete, containing normal portions of coarse aggregate, has been successfully placed underwater to depths approaching 300 ft. Grout-intruded concrete and grout itself have been successfully placed to even greater depths, 500 to 1000 ft. Major structural constructions such as bridge piers and drydocks, with embedded reinforcement and structural steel, have been accomplished on a regular and consistent basis. Precast concrete elements have been joined underwater in full composite action by means of underwater-placed concrete.

2. Forms for Underwater Concrete. Side forms for underwater structures may usually best be formed from precast concrete. These are rigid enough to withstand excessive deflection from the liquid head of the concrete as it is poured. When the joining face between precast and underwater concrete is steeply inclined or vertical, the two concretes will obtain a high degree of bond and may even chemically knit.

Steel side forms have also been extensively used, especially on bridge piers. They are most efficient for cylindrical configurations where the steel can be used in tension. Steel sheet piles are commonly and extensively employed as side forms for bridge piers, intake structures, etc.

Stone-filled or steel-weighted timber cribs have been much used in the past. Even dikes of well-graded rock can be used to restrain a mass pour. Sand bags or concrete-filled burlap sacks or bags of cement may be placed to form dikes.

Bottom forms must be adequate to support the weight of the fresh concrete, and prevent leakage out or water erosion in. The most common bottom form is a blanket of graded gravel. Precast concrete bottom forms are frequently employed where the natural bottom is soft mud or exposed to the water.

Top forms are seldom required. When needed (e.g., to form a special seat for a gate), they should be of precast concrete or a combination of structural steel and precast concrete. The under surface must be convex downward, so as to promote a flow of the concrete up and around. Otherwise, laitance, bleed water, and weak material may be trapped underneath.

Frequently, all that is necessary to achieve a suitable top surface is to over-flow the top and then screed it off, working to rigid screeds. However, in most bridge piers and cofferdams the top surface will be later exposed in the dry. Then it can be chipped to proper grade and a topping lift placed in the dry.

Underwater forms must be tight, especially if the water is moving. Currents and waves will leach cement out of even small cracks, producing a weak zone thay may extend to a considerable depth. Sealing between side forms and the bottom is often accomplished by diver-placed sand bags, or by weighted canvas. Canvas and rubber hose rings have been used to seal horizontal joints. Pressures at such joints are usually very small; the full fluid head of the fresh concrete generally is dissipated by the friction loss through the joint.

The minimum cover provided over reinforcement for underwater-placed concrete should be of the order of 4 to 6 in., taking into account the nature of any permanent forms, the size of aggregate, and the accuracy of construction. Reinforcing bars should be well-spaced to permit flow between them. If necessary, bars may be bundled to permit increased gaps.

3. Underwater Concrete Mixes

3.1 Tremie Concrete. The recommended mix is as follows:

Coarse aggregate: for very large pours—1.5 in. maximum; for normal pours—0.75 in. maximum; for very restricted pours—0.375 in. maximum.

Always use rounded aggregates (gravels of spheroidal shape). Avoid elongated or sharp-edged aggregates. Maximum size of coarse aggregate should also be related to space between reinforcement to permit ready flow without vibration.

Fine aggregate: 42 to 45% fine sand; cement—as a rich mix is required, 7 to 8 sacks/cu yd, ASTM Type II cement; water—to produce 6 to 8 in. slump. The W/C ratio should be 0.45 or less. Pozzolans may replace 15% of the cement.

Admixtures: Air entrainment of 4% is usually desirable. Water-reducing and plasticising admixtures are employed with excellent results. (See Figs. 20-1 and 20-2.) However, not all those commercially available and marketed for this purpose are satisfactory. Further, specific admixtures with a satisfactory history of performance with the cements in question are to be preferred. When inadequate prior experience is available, a field test should be made to verify performance.

3.2 Grout-Intruded Aggregate. Coarse aggregate is selected to ensure approximately 20 to 30% voids. Fines below the minimum size (e.g., 0.625 in.) *must* be positively eliminated. This may require rescreening just prior to placement. On one major project where the aggregate was transported by barge, fines accumulated from load to load and worked their way to the deck of the barge. Every so often, an over-zealous operator cleaned off the deck with his clamshell bucket, placing it in the structure and thereby forming a layer of relatively impervious fines that later did not get properly cemented.

The grout may be a sand-cement or neat-cement grout. Sand-cement grouts usually are richer than a one-to-one mix, sometimes two parts cement to one part sand. Various

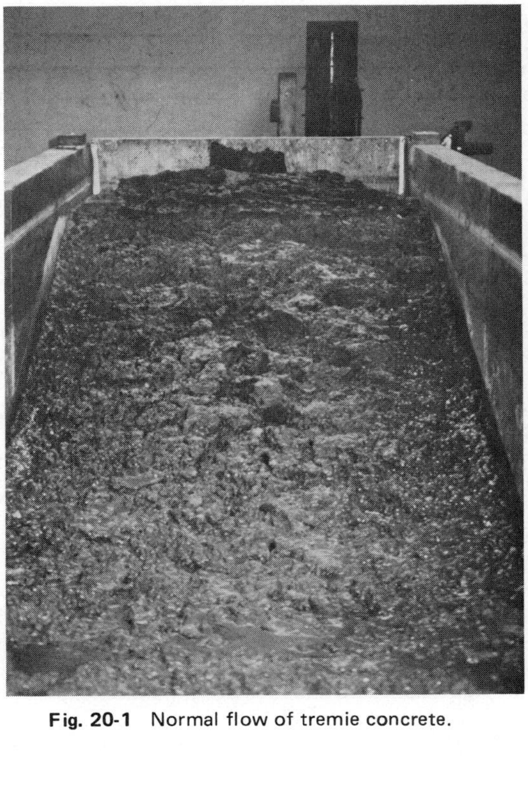

Fig. 20-1 Normal flow of tremie concrete.

Fig. 20-2 Superior flow of tremie concrete.

admixtures are employed to prevent segregation. One process uses a colloidal mixing technique to prevent segregation and improve flowability.

Various wetting agents may be employed to promote penetration of the grout. These admixtures and agents are proprietary, and it is essential that the manufacturer's recommendations be followed in detail. Techniques and admixtures should either have had extensive satisfactory performances on past structures or be thoroughly proved in field tests prior to use on the structure itself.

3.3 A Typical Tremie Grout. The mix may be as follows:

Aggregate: well-graded sand with a moisture content of 7% or less, from 500 to 1500 lb/cu yd, with sufficient fines to "carry" the coarse particles and prevent segregation; cement–Type II, from 10 to 16 sacks/cy yd; water–to give the consistency of thick soup.

Admixtures: air-entraining, water-reducing, and plasticising.

Neat-cement grout slurries (15 or 16 lb cement/gallon of water) have been successfully placed at great depths underwater, but may develop high heat.

4. Underwater Placement[20-9]

4.1 Tremie Process. The concrete is placed through a tube (see Fig. 20-3) whose lower end is always kept immersed in the fresh concrete. The tube may run from 2.5 in. in diameter for tremie grout up to 12 in. in diameter for large tremie concrete pours. Pipes are generally of steel, with flanged joints. The joints must be well gasketed and sealed; otherwise the water may leach cement from the concrete. Ocassionally, rubber hoses are employed, especially with tremie grout. Tremie pipes are spaced so as to give flow distances that will intersect with reasonable overlap, and to prevent excessive segregation and laitance formation (see Fig. 20-4). Design flow distances have varied up to 70 ft, the latter being with very deep pours. In general, a spacing of three times the depth of pour seems appropriate. The surface profile of good-quality, unsegregated, plastic tremie concrete will run from 1 on 6 to 1 on 10.

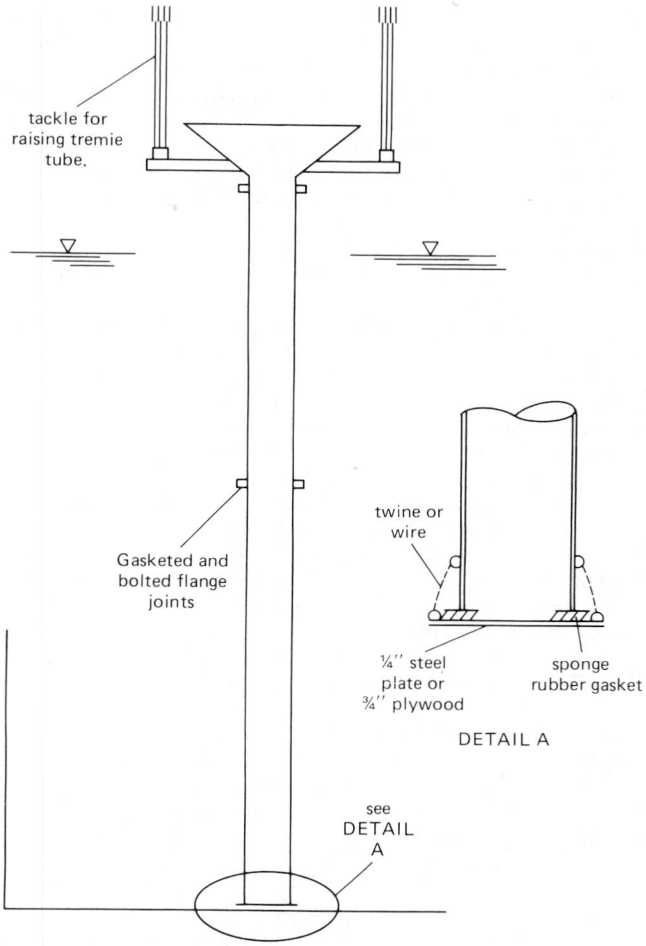

Fig. 20-3 Recommended method for starting tremie concrete placement.

The bottom end of the tremie tube must be kept below the surface a minimum practical distance, say 2 ft, yet must not become fixed in set concrete. Thus, it must be carefully raised. The raising must be done smoothly and gradually.

On one major bridge pier, because of errors of sounding, the pipe was raised several feet above the surface of the previously poured concrete. This resulted in a layer of sand and gravel, the cement being completely washed out. Then

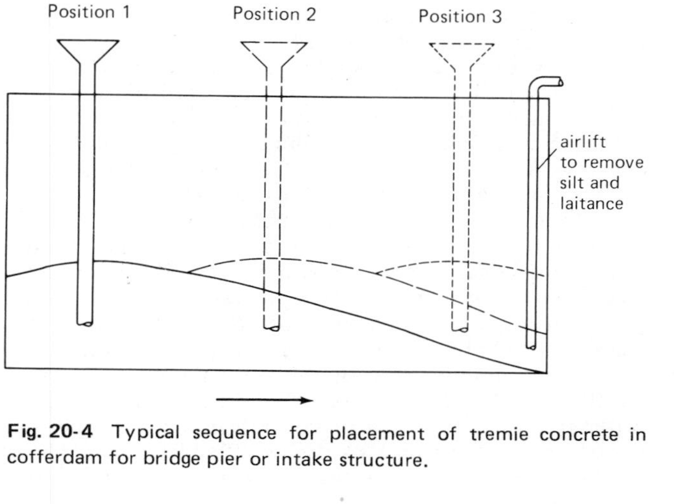

Fig. 20-4 Typical sequence for placement of tremie concrete in cofferdam for bridge pier or intake structure.

when the gravel once more mounded around the end of the tube, good underwater concrete was obtained, thus covering the loose sand and gravel with a structural shield. Upon dewatering of the cofferdam, the hydrostatic head caused a blow in the bottom. The eventual removal and replacement cost several million dollars.

On another project, the tremie tube was left too low too long, and the fresh concrete was injected below an already set plug. This acted as a hydraulic ram, raising the previously set concrete and cracking it.

The above examples indicate the importance of controlling the tube's embedment.

To start the pour initially, a seal (e.g., gasketed plywood, as shown in Fig. 20-3, Detail A) is affixed to the bottom, and the tube is lowered to the bottom and filled with concrete (in the dry); then the tube is raised a few inches, the weight of concrete breaks the bottom seal, and the concrete flows out. If the concrete surface drops too far in the tube, the tube is lowered back. Flow into and through the tube should be as continuous and regular as possible. Thus, the feed to the hopper on the tremie pipe should be by conveyor, or by a controlled bucket (hydraulic or air-controlled), or by pump—not by sudden opening of a bucket. For a very deep pour, exceeding 50 ft or so, the initial filling in the tremie tube should be graduated: first 2 to 3 ft of cement-sand mortar, followed by the regular tremie mix. While the latter may tend to segregate in free fall, it will remix in the mortar at the bottom.

When the water is deep, the tremie tube may float while empty. The tube may be weighted. In the past is has been common practice to set an open-ended pipe, extending to the bottom of the excavation, then insert an inflated ball in the top, and start pouring the concrete. The weight of the concrete forces the ball down the tube, pushing the water out ahead.

Although this method can be used successfully on moderate-depth pours, it has led to difficulties and can prove disastrous.

First the water being forced out the bottom can scour a hole in the soil, raising a mound of sand elsewhere. If the method is used to restart a pour in progress, the column of water will wash the cement out of the previously placed concrete.

Typical balls are inflated to about 11 psi, which is the equivalent of 25 ft of water head. At deeper depths the ball collapses and becomes ineffective as a seal. Later it may pop back to the surface and regain its shape, thus obscuring what actually took place.

If a "go-devil" is to be used, the old-fashioned bale of hay, or, for modern times, a pipeline pig, is far better for the initial start. Neither is satisfactory, however, for the restarting of a pour into unset previously placed concrete.

A newly developed means is to place the main tremie pipe and seat it firmly on the bottom, still full of water; then a smaller tremie pipe, with plugged end, is inserted in the main tremie pipe, and the small pipe is filled with concrete. It is then slowly withdrawn, filling the larger pipe, which is then subsequently used in the regular manner.[20-2]

In placing tremie concrete, the seal may be lost if the pipe is raised too high, or if the control of the pipe and the feed of the concrete are erratic and jerky. Then it is necessary to go through the starting process once again, taking particular pains not to erode or disturb the fresh previously placed concrete by a rush of water or agitation of the pipe.

A recent development in the Netherlands is a rubber tremie tube. The tube is collapsed flat and held this way by water pressure. As a slug of concrete is fed into the top, it opens the tube ahead of itself as it slides down. Above the slug, the tube is closed again by water pressure. For this case, the concrete may have a somewhat lower slump, say, 4 in. Progressive slugs are fed in at the top. This system has apparently worked satisfactorily on shallow (30 ft), thin placements, but appears to have limitations for major placements.

Valves have often been proposed and tried on the bottom of tremie pipes. They are not successful because the concrete is a mix rather than fluid. They should not be used. Many attempts have been made to pump concrete all the way, that is, down to the point of final placement. Most attempts have ended up unsatisfactorily or disastrously. In addition to the general physical problems of different mix requirements and water absorption, hence slump loss, there are the problems of segregation on the downward leg and excessive disturbance to the previously placed concrete because of high-velocity discharge.

Pumps may be used successfully to transport the concrete to the hopper, whence gravity feed (open to the atmosphere) may be used for vertical transport.

4.2. Tremie Grout. This should be well mixed in a machine mixer and placed through a pipe at relatively low pressure, sufficient only to accomplish uniform flow. For placement at great depths, the gravity head may cause the concrete to run out too fast, creating a partial vacuum and sucking air or water through the joints. Therefore, all joints should be sealed tight, and the pipe should be properly sized to pass the grout while developing a high enough friction head to keep the pipe full. Tremie grout has been successfully placed by pumping directly, although care must be taken to avoid excessive head as noted.

4.3. Underwater Buckets. Special bottom-dump buckets are available that have covers over the top to prevent contact with seawater. They are lowered to the bottom and then opened so that the concrete flows out over the surface. The bucket cover must provide for water flow into the space above the concrete as the bucket is emptied. This method produces much more laitance than does concrete that is properly placed through tremie tubes; thus, its use is primarily adapted to mass pours, cyclopean walls, etc. The bucket method is generally not suitable where there are piles extending up from the bottom, as these will force the concrete to fall through the water. The method is particularly useful for placing concrete in moving water or the surf zone, etc., as for filling a pipe trench. It can be placed with much lower slump and larger aggregate, then covered with weighted canvas, etc., to prevent erosive scour.

4.4. Grout-Intruded Aggregate. Grout pipes and recording pipes are preplaced in the forms or tied to the reinforcing steel; then the coarse aggregate, having no fines, is placed. Grout is then pumped in. The recording tube signals (by an electrical resistance gauge) that the grout has reached the prescribed level.

For best results, a blanket of aggregate should be kept 2 to 3 ft above the maximum grout level for that stage. This is particularly important where the surface is exposed to moving water. For deep pours, pipes, aggregate, and grout are placed in several stages.

It is essential to protect the preplaced aggregate from contamination by silt or sewage, etc., between the time of placement and grouting. In any event, the time interval should be kept as short as possible.

Grout-intruded aggregate is particularly well adapted to concreting structural elements containing many inserts and embedded items, as it causes minimum disturbance.

The grout pipes are usually placed on 5 to 7$\frac{1}{2}$ ft centers, and the usual maximum raise is 10 to 15 ft per stage. Recent studies in Japan show that the grout flow tends to be up along the pipe, then down into the aggregate. It is this flow pattern that limits the effective height in any one stage.

Grout-intruded aggregate placements have been used on the Mackinac Straits Bridge in Michigan and the massive piers of the Honshu-Shikoku Bridges in Japan. For the latter, a specially built mortar barge has been constructed, which provides large volumes of high-quality cement/sand mortar.

As noted earlier, the success of the method depends to a major extent on the maintenance of adequate voids between the aggregate particles—thus the need to ensure against inadvertent placement of fines and blocking by silting or marine growth.

4.5 Underbase Grouts. For filling the voids under the bases of large gravity structures such as those in the North Sea, special underbase grouts have been used. These are selected to flow well horizontally in restricted areas, without segregation, and to develop minimum heat of hydration, since such pours are usually massive and confined. For some cases, thixotropy is desirable to prevent escape into the sea, and hence thixotropic admixtures are employed.

Often a low strength is all that is required, and a low modulus is desirable to conform readily to local zones of high bearing.

Mixes that have been successfully used include: bentonite, cement, sand, and water; rock flour, cement, and water; and foamed concrete, made by salt water, cement, sodium silicate, and stabilizing admixtures.

4.6 Underwater-Setting Cement. A number of proprietary cements have been developed that set rapidly underwater. These are mixed in small batches, and then lowered in a covered can to a diver, who uses them for patches and repairs. They may gain their set in a matter of minutes, thus requiring no external protection against the water.

20.3.3 Precast Concrete

This technique is particularly well adapted for marine structures of all types: piling, sheet piles, breakwaters, overwater and submerged structures, etc.

The conceptual design should consider the available size of equipment for transport and erection or installation. Precast elements must be designed for adequate structural strength during these phases as well as for service in the permanent structure.

Connections of precast elements require extremely careful detailing. Not only must the connections perform adequately as in conventionally reinforced concrete, but they must accommodate the greater tolerances generally imposed by the marine environment (waves, current, survey difficulties, etc.) and must be durable in themselves against corrosion.

1. Cast-in-place connections, with overlapping or welded reinforcing steel, have been widely employed in the past. The surfaces should be sand-blasted or wire-brushed to remove laitance prior to setting. After setting, these surfaces and the reinforcing steel should be washed with fresh water to remove any salt. The concrete should have a very low water/cement ratio to minimize shrinkage, and should be well vibrated. Particular care should be taken with curing in order to prevent shrinkage cracking.
2. Thin concrete joints are much used in English practice. These are generally 3 in. thick, poured with a fine concrete with 0.375 in. maximum coarse aggregate, well-vibrated.
3. Matching joints (dry joints) in which the faces are cast against each other in the casting yard produce excellent results. The surfaces may be coated with a thin coat of epoxy just before joining in the field.
4. Grout pads—cement mortar spread thinly over a joint—are useful only in horizontal joints where the weight aids in achieving uniform compaction. Even here, however, some degree of nonuniform bearing may result.

20.3.4 Prestressed Concrete

Prestressed concrete assures the highest quality of performance in the marine environment. The resultant structure will be essentially crack-free, and have high resistance to cyclic and impact loadings.

With pretensioned concrete, thorough vibration will cause the cement paste to fill the interstices between the wires of strands. Exposed ends of pretensioning strands should be daubed with epoxy, although corrosion normally will proceed only about 0.125 to 0.5 in. deep.

With posttensioned concrete, the ducts should be washed clean with fresh water, then blown out with air before placement of the tendons. The tendons should be clean, dry, and free from salt. The author recommends the dusting of a vapor phase inhibitor powder (VPI) as the tendon is pulled into the duct, if the tendon is to be left ungrouted more than a few hours. The ends should then be sealed until the time of stressing.

A more recent method is to coat the tendons with water-soluble oil as they are inserted into the duct, then seal the ends. This water-soluble oil is removed during the grouting phase by pumping continuously until the discharge is free from discoloration by the oil. It has been found that this effectively removes the oil from the tendons.

In any case, the grouting should be carried out as soon as possible after stressing. However, grout cannot be allowed to freeze, so grouting should not be carried out until the structural concrete is at a temperature above freezing—which usually is interpreted as requiring that the ambient temperature be above freezing.

The end anchorages should be flush-type (recessed) anchorages. These recessed pockets should be filled flush with cement mortar, bonded to the pocket by a vapor-permeable bonding compound such as latex. The patch should be tied to the structural concrete by ties (wire or small reinforcing steel bars). Care must be taken to ensure that there is no shrinkage or bleed gap between the patch and the structure; this can be prevented by the "window-box" method of forming and placing on a vertical surface, for example. The joints between the fill and the structural concrete should then be sealed by epoxy painted over the joint.

Posttensioning may be very usefully employed to achieve continuity across joints and to make the structure act as a whole. The ducts must be made continuous across the joints by means of snugly fitting sleeves. These should be taped with a waterproof edge.

20.3.5 Ferro-Cement and Fiber-Reinforced Concrete

Ferro-cement is extensively utilized in concrete boat building, and more recently has been applied to small oceangoing barges.[20-6] Essentially, it consists of many layers of wire mesh, each coated with a thin layer of cement mortar placed by hand or by shotcrete. The total amount of steel is proportionally large, 800 lb/cu yd or so.

Ferro-cement is essentially crack-free (the tiny hair cracks are arrested by the mesh). It is durable, although minor rust staining may occur. It can readily be formed to complex double-curved surfaces. Thin ferro-cement panels are very flexible.

A recent related development is steel wire fibers, in which finely divided wires are mixed with the concrete.[20-7] Nylon fibers have been similarly mixed; these are not as effective in resisting cracks but do give excellent toughness under impact.

20.3.6 Polymer-Impregnated Concrete and Silica Fumes

Polymer-impregnated concrete holds forth excellent promise for marine structures, particularly vessels, deep-sea habitats, and arctic structures subjected to ice abrasion. Precast panels are impregnated with a monomer, preferably under vacuum, and then subjected to either irradiation or thermal treatment to change the monomer to a polymer.

The resultant material is extremely strong (15,000 psi and higher), impermeable, very durable against freeze–thaw and salt-water attack, and highly abrasion-resistant.

Silica fumes are another recent development that produces very high-strength, impermeable concrete. Their use offers great promise for marine structures, for which the added cost may be small in relation to the benefits.

The addition of 100 lbs. per cu. yd. or so of condensed silica fumes, combined with use of a superplasticizing admixture, has recently been applied to several major marine concrete structures. Compressive strengths of 12 000 psi have been obtained with hard rock aggregate concrete, and 7000 to 8000 psi with high quality lightweight aggregate. The resultant concrete is highly impermeable, and possess high abrasion resistance.

20.4 SPECIAL DESIGN CONSIDERATIONS FOR MARINE STRUCTURES

Structural design should be carried out in accordance with the standard provisions of ACI-318, ACI-357, and other relevant ACI publications. Specific consideration should be given to the matters discussed below.

Piling must be designed for column action, with consideration of the degree of fixity at the head and in the soil; accidental eccentricity; lateral loading, as from waves or soils; and dead weight bending in the case of batter piles. In certain framing schemes, the piles may be subjected to combined moment and axial compression or tension.

Because of the corrosive nature of the marine environment and because of the frequent need to prevent leakage, cracking must be strictly limited. This may dictate prestressing or a lower-than-normal stress level in reinforcing steel. The frequency of cracking loads must also be considered in evaluating stress levels for design.

Face reinforcement must be provided in both orthogonal directions; including the provision of adequate spiral in piling. The amount of reinforcement should be proportioned to ensure that if the surface zone is cracked for any reason—structural, thermal, or impact—there will be sufficient steel area to remain below yield to close the crack. The zone is usually taken as equal in depth to the cover plus eight times the rebar diameter, and the steel area times steel yield is set equal to or greater than the area of the surface zone times its direct tensile strength. This usually results in a requirement of about 1% steel area to surface zone area in both directions, with a percentage of 0.6 to 0.8% usually sufficient in practice.

The amount of cover and special requirements for materials, mix, placement, and protective coatings must be specified, as outlined in Section 20-2. All connections and fittings must be properly protected from corrosion.

Hollow structures, partially or wholly submerged, must be designed for hydrostatic loading, taking into account not only the basic shape of the structure but also the effect of tolerances in out-of-roundness, etc., due to form tolerances and plastic flow.

Deep-submerged structures must consider implosion and in-plane lamination (cracking) as failure mechanisms.[20-17]

Because loads due to water follow the deformations and because of the possible adverse effects of water under pressure in cracks, shears must be carefully considered, with no reduction for the compression zone adjacent to the support.

In the case of significant repeated cyclic shear, all shear should be taken by the steel. Membrane shears frequently dominate the design of some elements: the effect of crack width in degrading stiffness, etc., must be considered, especially when principal tensile stresses are at a significant angle with the reinforcing steel grid.

20.5 SPECIFIC APPLICATIONS

The application of concrete to marine structures has been so widespread that it is impracticable to attempt to discuss all past applications in detail. However, a brief description of main categories may be useful in indicating the prime considerations, precautions, and results in specific areas, as well as furnishing a background for the future applications to the entire field of harbor, coastal, and ocean structures.

More detailed and extensive information for specific areas is contained in the references listed at the end of this chapter.

20.5.1 Wharves, Jetties, Moles, and Trestles

These structures represent one of the largest uses of marine concrete.

1. Piling. Prestressed precast concrete piling, ranging in size from 12 to 36 in. square and from 24 to 72 in. round have been extensively utilized for bearing piles, batter piles, and moment-resisting, vertical structural columns. The larger sizes (generally those above 24 in.) are usually hollow-cored with wall thicknesses of 4.5 to 6 in.

Prestressing is usually designed to give a minimum effective prestress of at least 750 psi, in order to permit handling and driving without damage. For very long piles, batter piles, or moment-resisting piles, higher prestress values are desirable, usually 1000 to 1200 psi.

Hollow-cored piles should have sufficient spiral (0.6% steel area or more) to prevent vertical cracking, which may arise from a wide variety of causes. If the hollow-core piles are driven open-ended, then a vent or vents should be provided just below the pile head so as to relieve the hydraulic pressures that can build up internally, especially in driving through soft soils. These top vents should be plugged after installation, especially in areas subject to freezing. In freezing areas, a vent is normally also provided below waterline to ensure that the pile will drain properly, and not trap water which may subsequently freeze. In cases of extremely severe exposure, a log of styrofoam or even wood may be placed inside to absorb the expansive forces.[20-15]

Prestressed, concrete cylinder piles, 54 in. in diameter and up to 160 ft in length, were employed at Baton Rouge. (See Fig. 20-5.)

2. Deck Structures. The concrete caps must accommodate driving tolerance in the piling, yet assure proper structural behavior (e.g., fixed-end or tension connections to the piles) in accordance with the design. When precast concrete caps are used, it is good practice to make them over-width and over-length to enable them to be used under the most adverse combination of tolerances in pile location.

Concrete deck slabs are usually designed for continuity.

Fig. 20-5 Prestressed cylinder piles, Port of Baton Rouge, La.

They often incorporate the pile cap within the deck, thus minimizing formwork. One of the most efficient schemes of wharf construction, for example, utilizes a flat soffit and constant-thickness deck, the cap being obtained by reinforcement within the confines of the slab.

Precast deck slabs are extensively employed. These may be full-depth slabs, or simple-span construction, or half-depth slabs, designed to work in composite action with a poured-in-place top slab, thus permitting the attainment of continuity over the cap.

3. Concrete Edge Beams, Curbs, Pipe Beams, Fire Walls, and Catwalks. These comprise most of the remaining components of a modern wharf structure.

4. Fender Piles. A relatively new but well proven application of prestressed concrete, fender piles have been installed on major wharves in Kuwait, Singapore, and Los Angeles (Fig. 20-6). They should be designed for energy absorption and deflection.[20-15]

The over-water extensions of La Guardia Airport in New York represent one of the largest and most successful applications of precast, pretensioned, and posttentioned concrete, designed and constructed to meet extremely severe design criteria in an adverse marine environment.

20.5.2 Offshore Platforms

Concrete caisson structures are now extensively employed for the support of drilling, production, and storage of oil and gas in some of the world's most severe environments. They represent an extension of the technology originally developed for lighthouses in the Baltic and Irish seas and Eastern Canada. These lighthouse structures were constructed on and near shore, in stages—the first stage base

Fig. 20-6 Prestressed concrete fender piles.

structure being built in a basin, then launched for the second stage construction afloat, and finally towed to the site and seated on the sea floor by ballasting.

The Ekofisk caisson (Figs. 20-7 and 20-8) was the first of the new generation of offshore platforms, whose spectacular development has taken place primarily in the North Sea.[20-20] These structures have now reached dimensions of 165m (520 ft) in diameter and have been installed in water depths up to 150 m (500 ft). As of 1983, fourteen such structures were in place in the North Sea and a fifteenth was under construction. Two have been installed off the Brazilian shore; others are in the planning stage for various areas around the world.

The typical construction procedure follows these stages:

1. Construction of the base raft in a basin, kept dry by a dike.
2. Flooding of the basin, breaching of the dike, and tow to a deep-water site.
3. Construction of a large caisson, incorporating storage cells, while the structure is afloat. (See Figs. 20-9, 20-10 and 20-11).

Fig. 20-7 Ekofisk oil storage caisson under construction afloat in Stavanger Harbor, Norway; 86,000 cu m of prestressed, precast, and cast-in-place concrete destined for service in the North Sea.

Fig. 20-8 Ekofisk oil storage caisson during storm.

Fig. 20-9 Statfjord C platform: main storage cells.

Fig. 20-10 Ninian central platform under construction.

Fig. 20-11 Shell Cormorant platform: base caisson.

Fig. 20-12 Beryl A platform: slipforming of shafts.

4. Construction of shafts upward from the caisson proper, to support the deck. (See Figs. 20-12 and 20-13.)
5. Submergence of the caisson and shafts to enable the deck to be floated in over the top.
6. Deballasting, to raise the deck.
7. Tow to the site.
8. Ballasting down to the sea floor.

Such structures must resist very high external pressures during these construction stages, up to 160 tons/m^2. In service they must resist up to 2×10^8 cycles of dynamic loading due to waves.

High-strength concrete of low permeability is required. Reinforcing steel requirements are unusually high, averaging 350 lb/cu yd of concrete and locally reaching twice that amount. Obviously bundling of bars is required, and the concrete mix and placement procedures need special care in order to encase the reinforcement thoroughly.

Prestressing is used extensively, especially to resist cyclic loading and to increase shear resistance. Assistance can be gained by maintenance of internal pressures lower than those of the sea outside, thus precompressing the structure as a whole.

The mechanical systems for such large and complex structures are very extensive and must provide a high degree of reliability. Instrumentation also is extensive, in order to monitor both structural and mechanical performance.

The design considerations for such structures have been developed to include not only performance under operating loads and environmental loads but also a wide range of accidental loadings of low probability but potentially severe consequences. These include ship collision, internal explosion (deflagration) and fire, and dropped objects such as drill collars and casing and equipment.

The dynamic response of these structures is heavily influenced by the responses of the foundation; hence soil-structure interaction plays a dominant role.

Durability and resistance to fatigue are given special attention. Thermal gradients due to hot stored oil must be considered. High local forces due to breaking waves on the shaft or "hard points" of soil reaction under the base must be provided for.

Construction operations require consideration of the hydrostatic and dead loads at each stage of construction. Buoyancy (draft and freeboard), stability, dynamic forces

Fig. 20-13 Statfjord A platform: shafts.

Fig. 20-14 Concrete ciassons for Arctic Ocean platforms.

due to roll, and damage control in event of a mishap involve naval architectural considerations. Construction management plays a very large role in integrating the many suppliers, the equipment, and the large number of workers needed for effective completion within what is usually a very short time schedule.

Concrete offshore platforms appear to have an important continuing role in the development of remote and hostile environments such as the Bering Sea and northern North Sea because of their ability to carry the deck structure and all its equipment, as well as the short time required for installation.

20.5.3 Arctic Caissons

Concrete caissons appear to have a major role to play in the development of the Arctic Ocean because of their inherent ability to resist the intense local and global pressures due to sea ice. (See Fig. 20-14.)

The Tarsuit caissonn–retained island was constructed using four large caissons of structural lightweight concrete, selected in order to permit the delivery of the caissons on board a large submersible barge. They have successfully withstood both the ice and the freeze–thaw attack on the saturated concrete at the waterline. (See Figs. 20-15, 20-16, and 20-17.)

Many of the concrete caissons now being planned for the Arctic are being designed in lightweight concrete to reduce draft for the passage around Point Barrow and to enable installation in relatively shallow waters.

On the eastern coast of Canada, concrete platforms are being designed to resist iceberg impact, utilizing their mass and large sea floor bearing area to resist the impact of icebergs up to several million tons. Once deemed impossible, the development of new methodologies for analyzing soil-structure response under impact and the dissipation of energy by the crushing of ice, now show the concept to be practicable, at least for moderate water depths.

The intense local pressures imposed by ice, both Arctic sea ice and icebergs, must be resisted by the peripheral walls of the caisson in both flexure and shear. Use of high quantities of circumferential steel and through-wall stirrups is indicated. Nonlinear analyses, which reflect the change in properties due to cracking, show that thick-walled concrete sections have much better resistance to these high concentrated loads than are given by conventional formulas.[20-24]

Among the favorable properties of reinforced and prestressed concrete structures for Arctic environments are durability, abrasion resistance, ductile behavior at low temperatures, and local rigidity.[20-23]

20.5.4 Floating Structures

The first recorded use of reinforced concrete was the concrete boat that Lambot built about 1858 in France.

Reinforced concrete ships were constructed in substantial numbers in World Wars I and II, and a considerable number of concrete floating drydocks and moored floating docks are in service throughout the world. Concrete caissons for bridge piers have been launched and towed to position, to be later sunk into the bottom. Some, such as the Martinez-Benecia Bridge in California and the 1st Avenue Bridge in Seattle, Washington, utilize permanent buoyancy to carry much of the dead load.

The more recent advent of prestressing makes possible more efficient structural designs and has given new impetus to this development, since prestressing offers superior performance along with substantial economy.

With prestressing, the entire wall thickness is able to work to resist local bending stresses. Cracking, with its implications for durability, fatigue, and watertightness, is essentially prevented. The ability to resist impact from collision or other accident is generally improved. Concrete placement and consolidation are facilitated. Most modern floating structures are therefore prestressed to a greater or lesser degree.

1. Barges and Ships. Almost 20 large pretensioned concrete barges, designed by Honolulu-based engineer Al Yee, have been constructed in the Philippines and have seen practically continuous ocean service since 1964. First costs were reported competitive with steel, actual maintenance has been substantially less, and experience in collision, etc., has been definitely superior. These barges are generally

Fig. 20-15 Prestressed lightweight concrete caissons constructed in Vancouver, B.C., for floating to the Arctic Ocean.

of 2000-ton capacity and carry both dry cargo and petroleum products.

Recently a number of barges and dredge hulls have been built in New Zealand of prestressed concrete for service in the South Pacific.

Prestressed concrete appears especially well suited for the storage and transport of cryogenic substances such as LPG and LNG, because of its favorable properties at extremely low temperatures. Recent tests have shown the ability of prestressed lightweight concrete to resist high cyclic load-

Fig. 20-16 Prestressed lightweight concrete caissons in the Arctic Ocean.

Fig. 20-17 Summer storm waves attack Tarsuit caisson.

ings even after being subjected to extreme low temperatures (-160°C, -260°F) on one side.

The Arco Ardjuna floating terminal, a 62,000-ton displacement vessel, was constructed of prestressed concrete in Tacoma, Washington and towed across the Pacific to the Java Sea, where it has given highly satisfactory service in the refrigeration, storage, and transfer of LPG. (See Figs. 20-18 and 20-19.)

Reinforced and partially prestressed concrete compressor and production barges have been built in substantial numbers in Louisiana, towed as far as Mexico and Nigeria, and sunk in shallow water as permanent stations (Fig. 20-20).

The Rogamex I, a large prestressed concrete floating production plant designed to beneficiate phosphates, was built in Singapore and towed to Mexico.

Economic studies show that the greatest percentage sav-

Fig. 20-18. Construction of prestressed concrete floating terminal. Displacement of 6200 tons.

Fig. 20-19. ARCO SEKI ARDJUNA prestressed concrete floating LPG terminal under tow from Tacoma, Washington to the Java Sea.

Fig. 20-20 Concrete barge for service as production and processing platform.

ings can be achieved in large hulls, where the ratio of skin surface to volume offsets to a cosiderable degree the adverse effect of concrete's higher dead load. Thus, a number of designers are currently investigating very large barges.

Structural lightweight concrete was utilized with excellent results and durability in some of the ships from World Wars I and II; it appears that prestressed lightweight concrete may be an ideal material for concrete vessels.

Guidelines for the design of such floating structures have been developed by the classification societies such as Det Norske Veritas and American Bureau of Shipping, and are under development by the American Concrete Institute. They recognize the complex loading conditions and design aspects required by naval architectural considerations. One item of particular importance to the expanded use of concrete floating vessels is modification of the rules requiring drydocking for inspection at frequent intervals.[20-21]

2. Offshort Airports. The La Guardia Airport extension was constructed over water on a piled platform, using prestressed concrete for its 40 acres of runways (Fig. 20-

Fig. 20-21 Over-water extension of runways at La Guardia Airport, New York.

Fig. 20-22 Hood Canal replacement bridge pontoons.

21). Many proposals for floating airports have since been prepared, along with feasibility studies. These generally propose the use of concrete pontoons, prestressed together to act as a huge flexible mat, thus distributing concentrated loads from aircraft over a large area.

The pontoons may well be prestressed within themselves to care for local stresses and hydrostatic loading. They are then posttensioned together using tendons in ducts in the concrete or in separate tubes inside the pontoons. In the latter case, the tubes of course must be protected from corrosion.

3. Floating Bridges and Docks. A number of concrete floating bridges have been built, taking advantage of the durability, fatigue resistance, strength, and economy of reinforced and prestressed concrete. The earliest, the first Lake Washington Bridge, has seen more than 40 years of successful service. A second Lake Washington bridge has been built, and a third is under construction. Another such bridge crosses Lake Okanogan, in British Columbia.

Two of the floating bridges have failed in service, primarily, in this author's opinion, because of failure to properly assess the dynamic nature of the response of floating structures to storm waves.

The Derwent River floating bridge in Hobart, Tasmania was designed as a horizontal arch to resist the river current, but suffered repeated damage under storm waves impinging from seaward.

The Hood Canal Bridge, a long reinforced concrete structure, failed catastrophically during a 100-year storm because of a combination of flooding through openings, dragging of anchors, and dynamic response in the form of undulations. The replacement bridge has been designed to a more adequate criterion, and is being constructed of prestressed concrete (Fig. 20-22).

A prestressed concrete floating dock for container transfer was constructed in two 350-ft-long sections and towed from Puget Sound, Washington to Valdez, Alaska, where it was joined by posttensioning.

4. Ocean Thermal Energy Platforms (OTEC). Based on the utilization of the approximately 40°F (22°C) temperature differential between surface water and deep water in the tropical regions of the world's oceans, extensive design and planning efforts have been directed toward the development of the basic platform and its 3000-ft-long suspended cold water pipe. The platform itself is a large floating structure not unlike the offshore oil production platforms of the North Sea, and would follow a similar pattern for design and construction.

The cold water pipe, 100 ft or so in diameter and 3000 ft long, has been envisioned as constructed of special lightweight concrete, thus having a submerged unit weight of only 20 to 30 lb/cu ft.

5. "Compliant" Structures. These are conceived either as floating structures, moored on a spread or tension-leg mooring, or as articulated towers, held in place by a universal joint attached to a base. As of this writing, one articulated loading platform has been installed in the North Sea, for the transfer of oil to tankers, and others have been similarly designed.

While these structures follow closely the same design principles as offshore fixed structures and floating vessels, they are subject to more dynamic action, so that fatigue endurance achieves a dominant role, not only in the structural concrete but especially in the attachment to joints, moorings, etc. Such attachments are prestressed to the concrete, and special quality control of the tendons and anchorages is exercised.

By appropriate configuration, the response of such structures to storm waves can be planned (e.g., to minimize roll or heave to acceptable limits). In two versions planned for the arctic, the dynamic response to the moving ice will be amplified to aid in breaking the ice.

Compliant structures appear well suited to deep-water production of oil, to areas of weak soils, and to areas of high seismicity.

20.5.5 Marine Bridge Piers

Piers for major over-water bridges have long been constructed of concrete. A few of the types most widely employed at present are discussed in the following subsections.

1. Cofferdams. The conventional construction method is to drive sheet piles or install cribs, and then seal the bottom with a course of underwater concrete placed by either the tremie or the grout-intruded method. Then the cofferdam is dewatered, the seal cleaned, and the structural footing block poured in the dry. Some of the tremie concrete seals are very large—those of the Delaware Memorial Bridges were up to 30 000 cu. yds.

A newer method, which makes substantial economies possible, combines the seal course and footing block. After the sheet piles are installed and the interior excavated to grade, reinforcing steel cages, often combined with structural steel, are lowered into position, and high-quality underwater concrete is placed. Such a procedure reduces the required penetration of the sheet piles, reduces the excavation, reduces the bracing, eliminates the seal, and may reduce the area to be dewatered to just that of the pier shafts.

2. Caissons for Bridge Piers. These are generally constructed in stages. The lower lift or cutting edge is constructed on a ways or in a shallow basin, then launched and floated out to an outfitting dock. Here the walls are raised using panel forms, slip-forms, or precast panels. Then the caisson is towed to the site where it is progressively sunk by raising the top walls and excavating from within. When it reaches its design depth, the bottom is usually plugged with a course of underwater concrete.

Large concrete caissons have been used for such major bridges as the Lillebaelt Bridge in Denmark and the Carquinez Bridge in San Francisco.

The Honshu-Shikoku bridges in Japan utilize large steel caissons filled with grout-intruded preplaced aggregate concrete. (See Fig. 20-23.)

3. Cylinder Piles. Very large cylinder piles of prestressed concrete are increasingly being employed to support major bridges. These are then capped above water to form the bridge pier. The prestressed cylinder piles employed for the San Diego-Coronado Bridge were 54 in. in diameter. Fourteen-foot-diameter prestressed concrete cylinder piles weighing 660 tons were utilized for the Oosterschelde Bridge in The Netherlands. These were sunk by internal excavation and external weighting, and then plugged with a course of grout-intruded aggregate.

Recently 2000-ton caisson-piles were placed by a huge shearlegs for the piers for the bridge at Hiroshima, Japan (Fig. 20-24).

4. Bell Piers. This is a form of box-caisson construction utilizing precast concrete or steel shells to form the outer portion of the pier. These are lowered into place, fixed, and then filled with structural tremie concrete.

The pier itself is never dewatered.[20-4] This scheme has been used on a number of very large projects: the Narragansett Bay Bridge at Newport, Rhode Island; the Richmond-San Rafael and San Mateo-Hayward bridges in California; and three bridges across the Columbia River (Fig. 20-25).

Extensive cores from the San Mateo-Hayward Bridge and the I-205 Columbia River Bridge have proved that with proper materials, mixes, and placement techniques and control, high-quality structural concrete can be consistently obtained underwater.

Fig. 20-23 Grouting the preplaced aggregate for Honshu-Shikoku Bridge.

Fig. 20-24 A 2000-ton concrete caisson for the Hiroshima Bridge, Japan.

Fig. 20-25 Concrete bell-pier units for the Columbia River Bridge, Astoria, Oregon.

5. Storm Surges and Tidal Barriers. Two major storm surge barriers have been constructed, one across the lower Thames River in the United Kingdom and the other across the Oosterschelde Estuary in The Netherlands. The former utilized cofferdam construction for the gate piers and large precast concrete sill units. The latter was comprised of 65 mammoth pier caissons, each weighing about 24,000 tons. These were constructed in a basin, then carried one-by-one by a huge floating gantry crane and seated on a prepared base. Construction was carried out to extremely exacting tolerances. The concrete caissons were prestressed triaxially and designed for maximum durability.

Multiple cylindrical caissons were used to construct a tidal barrier in France, in order to produce hydroelectric power. In the Soviet Union a large concrete caisson combined in one structure the barrier and the power plant; it was constructed in a warm water port and towed to an Arctic passage where it was seated on a screeded base.

20.5.6 Subaqueous Tunnels

These include some of the largest and most challenging structural achievements of modern engineering: the John F. Kennedy tunnel at Antwerp (Fig. 20-26), the Chesapeake

Fig. 20-26 Prestressed concrete segments for the John F. Kennedy Subaqueous Tunnel, Antwerp, Belgium.

Bay tunnels, the Oakland-Alameda and BARTD tubes in San Francisco Bay, the LaFontaine tunnel at Montreal, and the East River (New York), Mobile (Alabama), and Baltimore tunnels in the United States.

The general pattern has been to construct these tunnels in segments, either in a dewatered basin or afloat, then tow them to position and accurately lower them to grade by controlled flooding. Individual segments may consist of 30,000 tons or more of concrete. Once in place on the bottom, they are joined, usually by a procedure that includes a tremie concrete seal. The huge size of these units and the great effect that minor deviations in wall thickness, water absorption, etc., may have on the net buoyancy just before sinking, make it desirable to provide means for making positive adjustment (e.g., weight boxes to be filled with concrete to a computed depth) prior to final positioning and sinking.

20.5.7 Other Subaqueous Structures

Similar in principle to the above are the many underwater structures currently being actively engineered and proposed: underwater oil storage vessels, sub-sea oil well completion chambers, habitats, mining chambers, etc.

Because of the hydrostatic pressures involved, great care must be exercised to achieve tolerance and quality control, high strength, and impermeability. External or internal liners of steel or epoxy mastic may be required. Manufacturing methods must be chosen that will minimize the effects of creep and shrinkage during construction, when the support conditions and effective weight are basically different from those in the final location. Such structures may fail by implosion and lamination rather than by conventional compression, so this phenomenon requires careful consideration.

Subaqueous floating tunnels have been planned for the crossing of deep fjords in Norway and a deep channel in Greece, and for the Straits of Messina in Italy. Such structures would be positioned well below ship keels and sea ice, reacting upward against tensioned anchors drilled into the deep sea floor. These will require extremely careful tolerance control in order to achieve the proper buoyancy. They are subject to consideration of dynamic forces, damage control, and accidental loadings similar to those of the compliant structures and floating bridges previously described.

20.5.8 Coastal Structures and Terminals

Concrete has been extensively utilized for seawalls and to a lesser extent for breakwaters. It is abrasion-resistant, heavy in unit weight, and durable. Concrete caissons have been employed in coastal waters for offshore terminals and breakwaters (Figs. 20-27 and 20-28).

1. Seawalls. These structures frequently employ concrete sheet piles, which in modern construction are usually prestressed, to provide greater durability and crack resistance, as well as inherent economy. Prestressing may be somewhat eccentric or may be supplemented by mild steel to take care of the moment conditions, particularly negative moment at the wale or tie-back level.

Tie-backs to the anchor are usually steel rods, but more recently prestressed concrete has been employed because of its reduced elongation under load and its durability. These concrete ties act as combination deadmen as well, gaining their pull-out resistance from embedment in the backfill,

Fig. 20-27 Offshore terminal caissons in Australia.

Fig. 20-28 Concrete caisson for coal shipping terminal, Australia.

without special deadmen. Such a principle was employed on the anchorage for the Lillebaelt suspension bridge in Denmark.

2. Breakwaters. Where extreme exposure conditions exist or where natural quarried rock is unavailable, precast concrete breakwater units are employed. Special shapes have been developed to give proper keying action, wave energy dissipation, resistance to erosion, and prevention of back pressure. These include the tetrapod, tribar, stabit, hollow-block, dolos, etc.[20-1, 20-11]

As breakwaters are extended into deeper water, and thus subject to more intense wave forces, unreinforced concrete armor units of refined shape, such as the dolos, have been subjected to considerable breakage as they have been dislodged and rolled by the waves. As a result, one trend has been to more compact shapes, such as 100-ton rectangular blocks or the tetrapod, even though their hydraulic efficiency is less. Another approach is to provide reinforcing, to hold the unit together after initial cracking. The author recommends consideration of higher-density aggregates in order to achieve a greater underwater unit weight. For example, some trap rock in the eastern United States will produce a concrete having a density of 180 lb/cu ft (116 lb/cu ft underwater) as compared to the more usual 148 lb/cu ft in air, 84 lb/cu ft underwater. Even greater densities are achievable with the use of iron ore as aggregate.

Massive concrete elements such as these generate great heat during hydration. Unless special steps are taken as noted in Section 20.1.2.9, cracking may be initiated even during manufacture, to be propagated during installation. In particular, the forms used in the casting must not restrain expansion and contraction.

Where unreinforced concrete is used, consideration may be given to the use of seawater as mixing water. A retarding admixture should be added in order to prevent flash set.

3. Seawall Caissons. Concrete caissons have been used as cribs, breakwaters, and seawall units. At Baie Comeau in Quebec, the walls of the caisson were perforated to absorb wave energy, the same concept being employed on the Ekofisk oil storage caisson (Section 20.5.2). The Japanese have used concrete caissons to form the core of their deepwater breakwater at Tomokomai, on Hokkaido, and a similar arrangement has been proposed for the breakwater protecting the offshore nuclear power plant off the coast of New Jersey.

20.5.9 Underwater Pipelines

1. Concrete pipe is extensively employed for outfalls for disposal of waste and for intakes for salt water for the cooling of power plants or industrial uses. Such pipes are generally precast, the large sizes being cast vertically, then set underwater, snugged tight, and joined.

To reduce the weight of these units for ease in handling, lightweight concrete has been employed. The bell and spigot ends were made of standard hard rock concrete, blended into the lightweight, in order to minimize spalling.

Often, pipe segments have been prejoined to facilitate setting. For the Hyperion Outfall in Los Angeles, the units were prejoined to 96-ft lengths.

Structural box sections, prestressed longitudinally, were used for an underwater interceptor near Seattle. The long units were set on underwater pile-supported piers.

In New Zealand, concrete pipe sections have been assembled into continuous pipelines by posttensioning, and then floated or pulled into position just like welded steel pipe.

A special arrangement has been adopted for the Ocean Outfall in San Francisco, which extends across the San

Andreas Fault. To accommodate displacements normal to the outfall, special joints are employed, which permit rotation and extension, but which prevent separation, thus forcing other joints farther from the fault zone also to participate in the articulation.

2. Concrete encased steel pipe is extensively employed for submarine oil, water, and gas pipelines. The concrete coating serves as weight coating, and protection to the inner corrosion coating of the steel.

Both conventional concrete and heavyweight concrete coating are used. The latter is useful in keeping down the total outside diameter of pipelines, and thus facilitating shipment, installation, and burial, as well as reducing current drag forces, "and handling weight in the air".

20.5.10 Repairs

With the increasing use of concrete in important marine structures, it has been necessary to develop appropriate methods of repair that can be carried out underwater. To repair cracks, sealing them and restoring compressive, tensile, and shear strength across them, epoxy injection methods have been extended to underwater locations. A hydrophobic component in the epoxy permits the epoxy to bond to wet surfaces; thus as the epoxy is injected, it displaces the water ahead of it. Since the leading edge of the epoxy may be contaminated by contact with the water, it is important to inject enough to ensure that this contaminated zone is expelled from a vent.

When the cracks are not crossed by reinforcement, stitch bolts or other confinement should be installed; otherwise the crack may be propagated by the hydraulic fracturing effect.

Grout-intruded aggregate and tremie concrete have been extensively employed for underwater repair of larger cavities. Small pockets may be filled with special cements designed to set quickly underwater and to resist leaching of the cement.

20.5.11 Conclusion

The strong surge of interest in the development of the resources of the continental shelf has in turn intensified interest in the utilization of concrete for major sea structures. On the one hand, the facilities desired and the environmental criteria are demanding ever larger and more sophisticated structures. Fortunately new technological developments are making available ever better materials and techniques. It is clear that concrete is destined to play an ever increasing role in the sea.

BIBLIOGRAPHY

20-1 Myers, J. J., Holm, C. H., and McAllister, R. F., *Handbook of Ocean and Underwater Engineering,* McGraw-Hill, New York, 1969.

20-2 Gjørv, Odd E., "Durability of Reinforced Concrete Wharves in Norwegian Harbours," Ingeniørforlaget A/S Oslo, 1968.

20-3 Gerwick, Ben C., Jr., "Placement of Tremie Concrete," Symposium, Concrete Construction in Aqueous Environments, *American Concrete Institute Publications SP-8,* 1964.

20-4 Havers, J. A., and Stubbs, F. W., Jr., Section 28, "Cofferdams and Caissons," *Handbook of Heavy Construction,* 2nd Ed., McGraw-Hill, New York, 1971.

20-5 "Effect of Various Substances on Concrete and Protective Treatments, Where Required," *Concrete Information Bulletin 1 S 001.03T,* Portland Cement Association, Skokie, Ill., 1968.

20-6 "Ferro-Cement Boats," Concrete Report (CR010.-01G), Portland Cement Association, 1969.

20-7 Monfore, G. E., "A Review of Fiber Reinforcement of Portland Cement Paste, Mortar, and Concrete," *Research Department Bulletin 226,* Portland Cement Association, 1968.

20-8 Gerwick, Ben C., Jr., "Controlled Sinking of Large Concrete Ocean Structures," *American Society of Mechanical Engineers Publication 71-UnT-6,* 1971.

20-9 Gerwick, B. C., Jr., "Underwater Concrete Construction," *American Society of Mechanical Engineers Publication 71-WA/UnT-8,* 1971.

20-10 Gerwick, B. C., Jr., "Construction of Large Concrete Shells for Ocean Structures, Techniques and Methods," *Proceedings,* Symposium on Hydromechanically Loaded Shells, International Association of Shell Structures, 1971.

20-11 *Proceedings,* Conference on Port and Ocean Engineering, Norwegian Technical University, Trondheim, Norway, 1971.

20-12 Gerwick, B. C., Jr., "Ocean Structures of Prestressed Concrete," *Journal of the Prestressed Concrete Institution,* Mar.–Apr. 1971.

20-13 Singer, R. H., "Prestressed Concrete Used for Offshore Barges, Storage," *Offshore,* Nov. 1970.

20-14 "Study of Construction Methods for Large Undersea Concrete Structures," CR72.002, Naval Civil Engineering Laboratory, Port Hueneme, Calif., 1971.

20-15 Gerwick, B. C., Jr., *Construction of Prestressed Concrete Structures,* Wiley-Interscience, New York, 1971.

20-16 "Concrete Polymer Material," *Fifth Topical Report, BNL 50275(T-602),* Brookhaven Natural Laboratory and Bureau of Reclamation.

20-17 *Proceedings,* Symposium on Concrete Sea Structures, Federation Internationale de la Precontrainte, London, Apr. 1973. Available through Cement and Concrete Association, Wexham Springs, Slough SL3 6PL, England.

20-18 Jackson, G. W., and Sutherland, W. M., *Concrete Boatbuilding,* John de Graff, Inc., New York, 1969.

20-19 Gerwick, B. C., Jr., "Design and Construction of Prestressed Concrete Vessels," *Preprints,* 1973 Offshore Technology Conference, Houston, May 1973.

20-20 Gerwick, B. C., Jr., and Hognestad, E., "Concrete Oil Storage Tank Placed on North Sea Floor," *Civil Engineering,* Aug. 1973.

20-21 Gerwick, G. C., Jr., Mansour, A., Price, E., and Thayamballi, A., "Feasibility and Comparative

Studies for the Use of Prestressed Concrete in Large Storage/Processing Vessels," *Journal, The Society of Naval Architects and Marine Engineers,* Nov. 1978.

20-22 Gerwick, B. C., Jr., and Venuti, W. J., "High and Low Cycle Fatigue Behavior of Prestressed Concrete in Offshore Structures," *Transactions, ASME,* Mar. 1980.

20-23 Gerwick, B. C., Jr., and Jahns, H., "Conceptual Design of Floating Drilling, Production, and Storage Caisson for Arctic Waters," *Engineering Structures,* Jan. 1981.

20-24 Gerwick, B. C., Jr., Litton, R. W., and Reimer, R. B., "Resistance of Concrete Walls to High Concentrated Ice Loads," *OTC Reprint-4111,* Offshore Technology Conference, May 1981.

21

Parking Structures

H. CARL WALKER*

21.1 INTRODUCTION

The concrete parking structure is actually a lightly loaded bridge built in the configuration of a building.

The uniqueness of a parking structure is that it is usually constructed on a building site, regulated by building codes, and designed by engineers with building design experience, but it must encounter, like a highway bridge, dynamic loadings, severe temperature changes, weathering, and deicing salts. Therefore, the engineer, when designing a parking structure, must keep in mind the uniqueness of the structure.

Because of the desire to provide maximum flexibility for parking space layouts, and to provide for the opportunity for changing the parking space layout without column interference, most parking structures are designed as clear span structures with spans ranging from 50 to 60 ft depending on the design car length and the parking space angle (Fig. 21-1). Shorter span structures are often used where the parking is below another-use structure such as an office or apartment building.

Since most modern freestanding parking structures are clear span design, they are designed for economy and durability using prestressed concrete—either precast (Fig. 21-2) or cast-in-place posttensioned (Fig. 21-3). When using prestressing, the designer must take into account the volume change factors of elastic shortening, shrinkage, and creep, in addition to temperature effects.

The real challenge in parking structure design is to provide for the contradictory requirements of designing a rigid structure to resist external lateral loads of wind and earthquake, while designing a limber structure to relieve the internal loads due to volume changes. In short, a parking structure must be a limber-rigid structure!

*Partner, Walker & Cagley, Consulting Engineers-Parking Consultants, Kalamazoo, Michigan. Founder, CWA Walker, Inc., Kalamazoo, Michigan.

21.2 STRUCTURAL SYSTEMS

Short span parking structures are generally of two-way slab design, usually flat plate, either posttensioned or mild steel reinforced. Sometimes, drop panels or column capitals are used to increase the shear capacity of the slab at the column.

Posttensioned long span beams supporting a one-way posttensioned slab are the common cast-in-place clear span structural system (Fig. 21-4). Long span pretensioned double tees, either topped or untopped, are the common precast structure (Fig. 21-5). The topped tee uses a composite cast-in-place concrete wear surface, usually 3 in. thick. The untopped system uses a double tee with the flange thickened to 4 in. or more with no additional wear surface added.

Other systems using long span joists, either precast or cast-in-place, and waffle-type slabs may be economical in certain geographical areas.

21.3 DESIGN LOADS—GENERAL

In addition to vehicle live loads and lateral loads due to wind or earthquake, the structure must be designed for bumper wall loads due to impacting of car bumpers against barrier walls, and for the horizontal loads due to volume changes, including temperature change effects.

While the vertical vehicle wheel loads are dynamic in nature, their magnitude is very small compared to the mass of the structure. Therefore the effect of dynamic vehicle wheel loading is neglected in parking structure design.

Fig. 21-1 Clear span parking structure interior eliminates column interference and provides maximum parking space layout flexibility. Renaissance Center Parking Structure, Detroit, Mich. Owner—Center Parking Associates, Detroit, Mich. Engineers-parking consultants—CWA Walker, Inc., Kalamazoo, Mich. General Contractor—Etkin, Johnson, Korb, Inc., Oak Park, Mich.

21.3.1 Vertical Loads

The building code live load requirement for passenger cars is usually 50 psf. The actual load of the vehicle is seldom greater than 30 psf when distributed over a parking space. Therefore the 50 psf design load includes a 20 psf impact load allowance. If the local building code provides for live load reduction, it is recommended that live loads be reduced to not less than 30 psf for the beams, girders, and columns. On the roof structure, snow loads should be added to the *reduced* live load. The roof slab should be designed for the reduced live load plus snow load.

21.4 LATERAL LOADS

Seismic loads on an open parking structure are resisted in the structural frame by frame action or shear walls. Special attention must be given to the additional building code requirements at connections and foundations for seismic load

Fig. 21-2 Precast prestressed concrete can be erected quickly—2800 spaces in less than seven months. Joe Lewis Arena Parking Structure, Detroit, Mich. Owner—City of Detroit, Mich. Engineers-parking consultants—CWA Walker, Inc., Kalamazoo, Mich. Construction managers—Barton-Marlow, Inc., Detroit, Mich.

Fig. 21-3 Posttensioning enhances the durability of cast-in-place concrete. City of Billings Park Two, Billings, Mont. Owner—City of Billings, Mont. Engineers-parking consultants—CWA Walker, Inc., Minneapolis, Minn. Consulting architects—Johnson-Graham Associates, Billings, Mont. General contractors—Sletten Construction Company, Great Falls, Mont.

resistance. Wind loads are applied to the gross area of the exterior elevation without reduction due to openings in the building facade unless a more rigorous analysis is made.

21.4.1 Lateral Bracing Systems

Some typical parking structure lateral bracing systems are shown in Fig. 21-6. Most cast-in-place reinforced concrete and posttensioned concrete facilities are designed as rigid frames to resist lateral loading (Fig. 21-6a). With sloped floor continuous ramps (Fig. 21-6b), the engineer can use the sloping floors to create a vertical truss in the structure. The truss action will carry the lateral loads to grade. Another practice is to allow the sloped floor slabs to create a truss-stiffened box supported by one-story columns (Fig. 21-6c). The truss-stiffened box on one-story columns (Fig. 21-6c) tends to eliminate the torsional action created by the eccentric resistance of the offset floor slab in the two bay structure in Fig. 21-6b.

Where a double interlocking helix is used (Fig. 21-6d), a space frame is developed that will provide lateral resistance using continuous columns and pinned beam connections throughout the structure. A computer analysis of the space frame concept may show large lateral deflections. However, because the structure will usually remain as a skeleton, these

Fig. 21-4 Posttensioning creates economical long spans in cast-in-place concrete. City of Kalamazoo, Parking Structure #1, Kalamazoo, Mich. Owner—City of Kalamazoo, Mich. Engineers-parking consultants—CWA Walker, Inc., Kalamazoo, Mich. Consulting architect—G. E. Diekema Associates, Kalamazoo, Mich. General contractor—Miller-Davis Company, Kalamazoo, Mich.

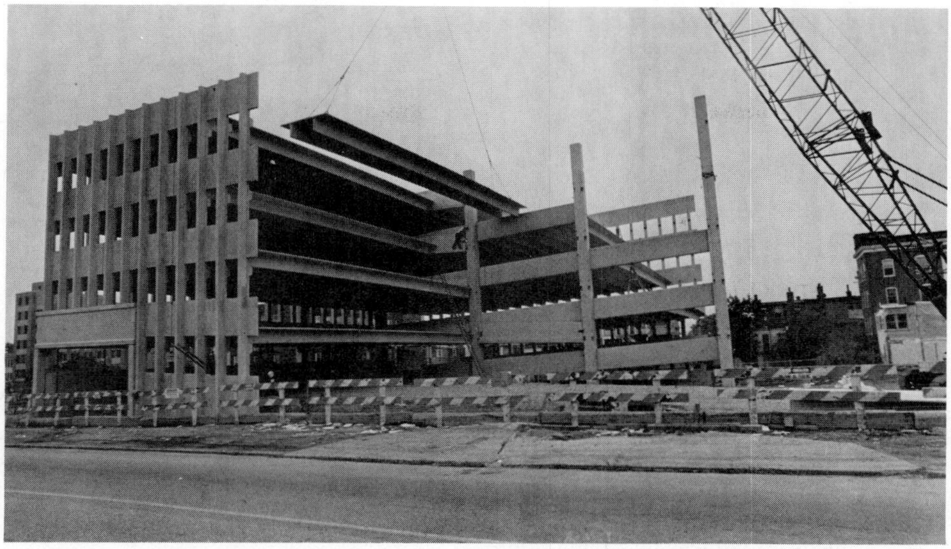

Fig. 21-5 Precast bearing walls combine structural and architectural requirements in a single unit. Wilmington Medical Center Parking Structure, Wilmington, Del. Owner—Wilmington Medical Center. Engineers-parking consultants—CWA Walker, Inc., Kalamazoo, Mich. General contractor—Etkin/Healy Joint Venture, Wilmington, Del.

large deflections may be acceptable even though they may be larger than would be acceptable if the building frame contained nonstructural brittle elements.

On low-rise facilities, sometimes it is more economical to simply cantilever the continuous columns off the foundations and provide pin connections between the beams and floors above Fig. 21-6e).

If shear walls are used to resist lateral loading (Fig. 21-6f), they are generally located at the center of mass of the facility to minimize the possibility of volume change restraints. Shear walls should be pierced with holes to aid ventilation flow and provide through visibility for security.

21.5 VOLUME CHANGE LOADS

Volume change loadings of elastic shortening, creep, shrinkage, and temperature change are particularly critical in the

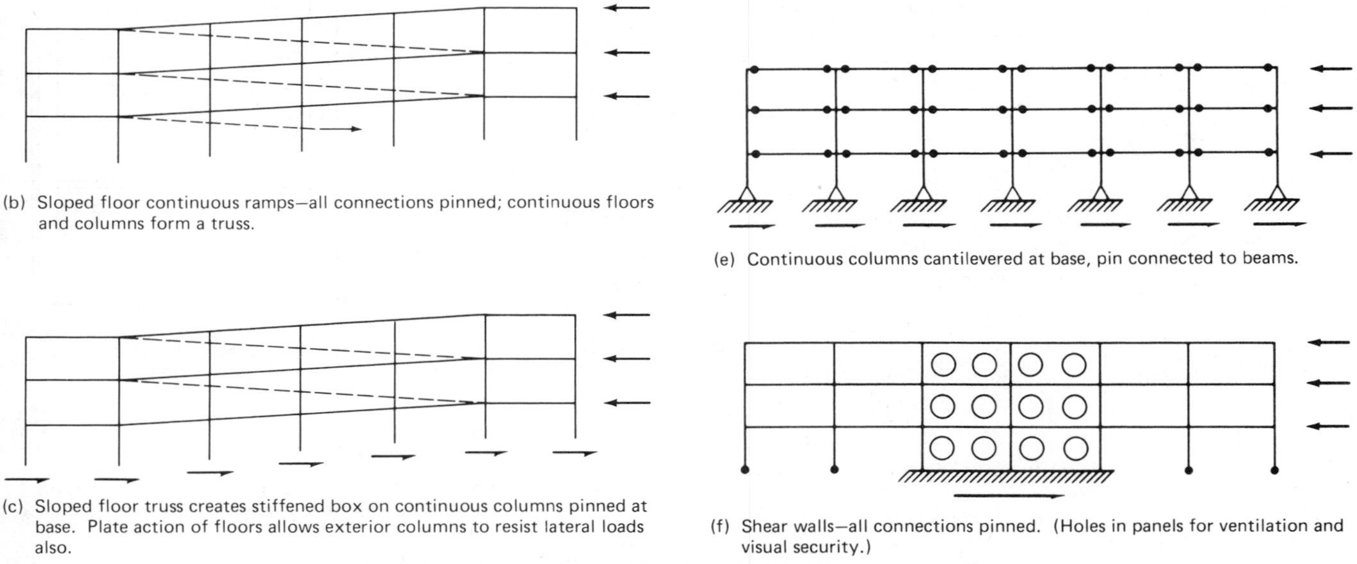

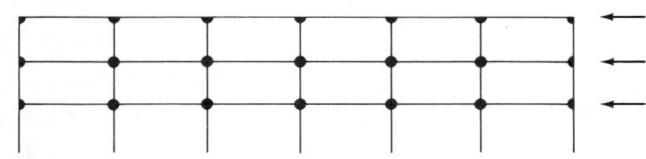

(a) Rigid frame—pinned at base.

(b) Sloped floor continuous ramps—all connections pinned; continuous floors and columns form a truss.

(c) Sloped floor truss creates stiffened box on continuous columns pinned at base. Plate action of floors allows exterior columns to resist lateral loads also.

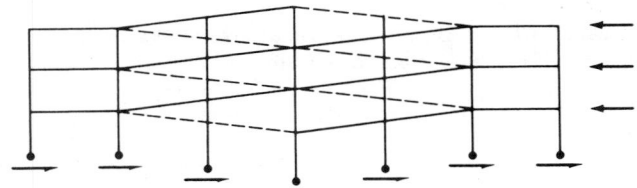

(d) Double interlocking helix space frame, continuous columns and floor slabs. Frame is stable in all directions with pinned connections.

(e) Continuous columns cantilevered at base, pin connected to beams.

(f) Shear walls—all connections pinned. (Holes in panels for ventilation and visual security.)

Fig. 21-6 Lateral load resisting structures (two bay structures shown).

design of columns and connections. Failure to include these loadings in the design can be very detrimental to the performance and safety of the structure.

21.5.1 Design for Volume Changes

Time-dependent volume changes are the most severe in cast-in-place posttensioned structures because all the volume change must be absorbed by the structure. In precast concrete, the elastic shortening and much of the creep and shrinkage take place prior to the element's being erected at the job site, and therefore, does not affect the structure as much.

Because the effects of elastic shortening, shrinkage, and creep cause a sustained shortening load in the structure, the concrete columns will creep to relieve themselves of the load created by the shortening. The computed volume change can be reduced (divided) by a factor (K_l) of 2 for crack-free cast-in-place posttensioned structures and by a factor (K_l) of 4 to 5 for precast and cast-in-place reinforced concrete structures, to obtain an equivalent volume change[21-2]. Posttensioned structures have a smaller volume change reduction factor than other concrete structures because the posttensioning compression in the concrete maintains a crack-free integrity, while other types of structures "breathe" as microcracks and joints open and close, thus relieving volume change forces.

Volume changes due to seasonal and daily temperature changes vary with geographical location. In certain parts of the world, annual mean temperature changes of 130°F can occur. Volume changes due to temperature cause instantaneous loads to columns and are not relieved to the same extent as the time-dependent volume change loads.

The qualitative effects of volume changes on the three basic types of concrete parking structures are shown in Table 21-1.

TABLE 21-1 The Qualitative Effect of Volume Changes on Various Types of Concrete Parking Structure Frames

Type of Volume Change	*Type of Parking Structure* / K_l *Factor* ③ — *Reinforced Concrete*	*Precast Prestressed Concrete*	*Cast-In-Place Posttensioned Concrete*
Elastic Shortening	None	None	All
K_l	–	–	2.0
Shrinkage	Very Little	Some	All
K_l	5.0	4.0	2.0
Creep	None	Some	All
K_l	–	4.0	2.0
Temperature Change	Some ①	Some ②	All
K_l	2.0	1.5	1.0

① *Shrinkage cracks act as mini-expansion joints and absorb some of the temperature volume change.*
② *Joints between precast units and at connections absorb some of the temperature volume change.*
③ *Equivalent volume change = calculated volume change ÷ K. See Ref. 21-2.*

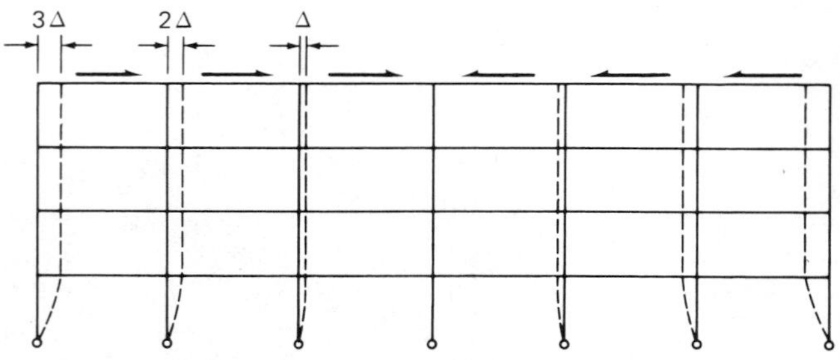

Fig. 21-7 Volume change effects on the frame, longitudinal direction.

21.5.2 Volume Change Effects on Frames

Volume change effects tend to change the length of a parking structure as a unit. The effects to the columns are the most severe between the foundation and the first supported slab (Fig. 21-7). Grade slabs and retaining walls must be kept free of the structure to prevent detrimental restraints.

Stair and elevator towers should also be kept free of the structural frame. On tall structures, lateral support of tall slender stair/elevator towers can be accomplished by attaching the tower to the frame at the deck immediately below the roof deck. Attachments are not made to the roof deck because of sun heating that causes independent roof expansion.

Because of the shortening between the foundation and first supported slab, positive moments can occur near the supports in rigid frames (Fig. 21-8), and tensile reinforcement should be put in the bottom of beams in what would normally be considered a negative moment region.

Heating of the roof slab by the sun while the beam remains cool can cause what is called "sun cambering." This

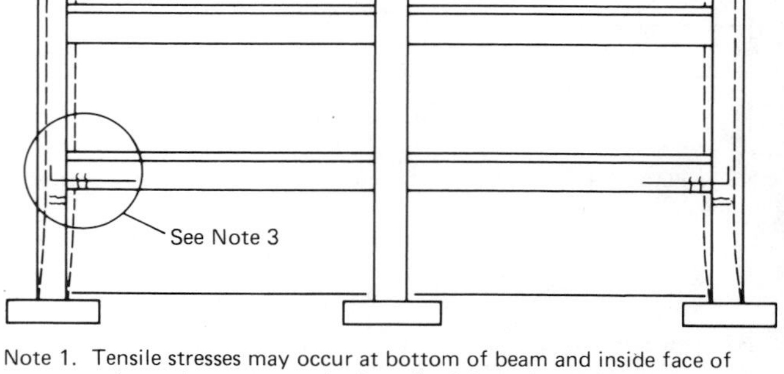

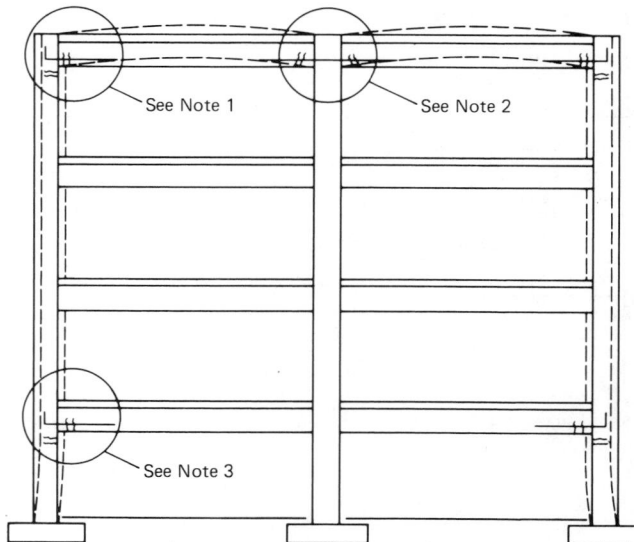

Note 1. Tensile stresses may occur at bottom of beam and inside face of column due to sun cambering. Hook beam bottom bars into column.

Note 2. Tensile stresses may occur at bottom of beam due to sun cambering. Run beam bottom bars continuous through column.

Note 3. Tensile stresses may occur at bottom of beam and inside face of column due to frame horizontal shortening. Hook beam bottom bars into column.

Fig. 21-8 Rigid frame sun cambering and volume change shortening effects.

can also cause bending moment reversals at the supports and should be resisted by bottom tensile reinforcement (Fig. 21-8).

21.6 POUR STRIPS AND EXPANSION JOINTS

In order to reduce the effect of volume changes on the structure, pour strips or expansion joints can be introduced.

Since most of the volume change in prestressed structures is shortening, expansion joints are really contraction joints. In cast-in-place posttensioned structures these joints are often constructed tight, since the shortening due to elastic deformation, creep, and shrinkage is greater than expansion due to temperature change.

DESIGN EXAMPLE 21-1: Horizontal deflection in columns caused by volume changes in a cast-in-place posttensioned structure. Reference: *PCI Design Handbook*. 2nd Ed., Part 4.

Two-tier structure located in Birmingham, Alabama, U.S.A:

150 pcf hardrock concrete
45°F temperature change
70% average relative humidity

$L = 300$ ft
$f'_{ci} = 2500$ psi
$f'_c = 4000$ psi
$f_{pe} = 200$ psi—average prestress
5″ slab ($v/s = 2.5$)—volume/surface ratio
$E_c = 3 \times 10^6$ psi

Trial #1—Full length structure, no pour strips or expansion joints:

(at infinite time)

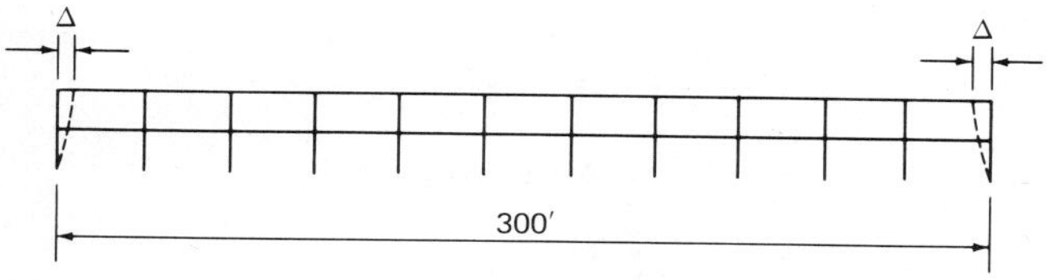

e_e elastic strain = .000067 in./in. $\left(\dfrac{f_{pe}}{E_c} = \dfrac{200}{3 \times 10^6}\right)$

e_c creep strain = .000111 $(524 \times 10^{-6})(.24)(.88)$

e_s shrinkage strain = .000459 $(510 \times 10^{-6})(.9)$

.000637 in./in.

e_t temperature strain = .000202 in./in.

(Refer to Tables 4.4.1, 4.4.2, and 4.4.4 in Ref. 21-2.)

$$\Delta = \frac{L}{2}\left(\frac{e_e + e_c + e_s}{K_L} + \frac{e_t}{K_t}\right)$$

Equivalent volume change factors

$$= \frac{(300)(12)}{2}\left(\frac{.000637}{2} + \frac{.000202}{1}\right)$$

$\Delta = 0.937$ in. equivalent horizontal deflection in column

Trial #2—Full length structure, pour strip at midlength:

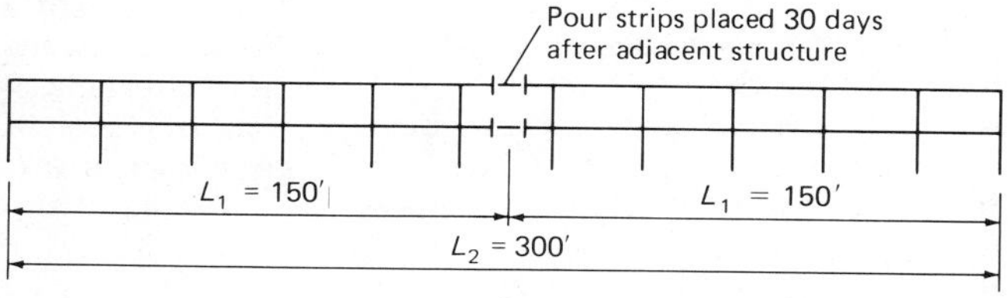

0–30 days:

e_e elastic strain = .000067 in./in.

e_c creep strain = .000039 $(229 \times 10^{-6})(.24)(.71)$

e_s shrinkage strain = .000169 $(235 \times 10^{-6})(.72)$

.000275 in./in.

(Refer to Tables 4.4.1, 4.4.2, and 4.4.4 in Ref. 21-2.)

$$\Delta_{30} = \frac{L_1}{2}\left(\frac{e_e + e_c + e_s}{2}\right)$$

$$= \frac{(150)(12)}{2}\left(\frac{.000275}{2}\right)$$

$$= 0.124 \text{ in.}$$

Movement after closure of pour strip:

30 days to ∞ (Trial #1 strains less 30 day strains)

e_c creep strain = .000072 in./in. $(111 - 39)(10^{-6})$

e_s shrinkage strain = .000290 $(459 - 169)(10^{-6})$

.000362 in./in.

e_t temperature strain = .000202

$$\Delta_{30-\infty} = \frac{L_2}{2}\left(\frac{e_c + e_s}{K_L} + \frac{e_t}{K_t}\right)$$

$$= \frac{(300)(12)}{2}\left(\frac{.000362}{2} + \frac{.000202}{1}\right)$$

$$= 0.689 \text{ in.}$$

$$\Delta = \Delta_{0-30} + \Delta_{30-\infty} = 0.108 + 0.689 = 0.813 \text{ in.}$$

Trial #3—Expansion joint at midlength:

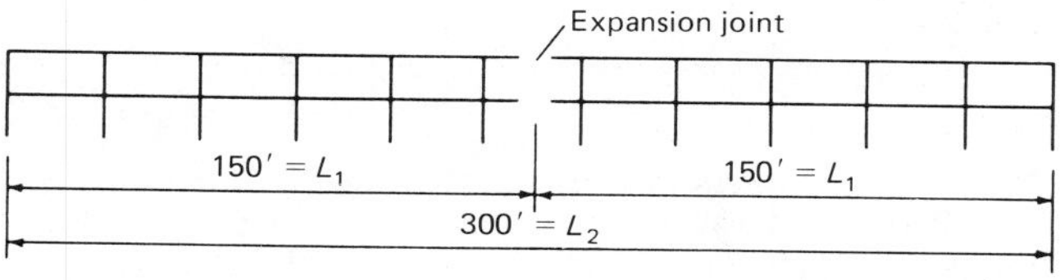

See Trial #1 for strains:

$$\Delta = \frac{L}{2}\left(\frac{e_e + e_c + e_s}{K_l} + \frac{e_t}{K_t}\right)$$

$$= \frac{150\ (12)}{2}\left(\frac{.000637}{2} + \frac{.000202}{1}\right)$$

$$= 0.469 \text{ in.}$$

Comparison:

End Column Horizontal Deflection

Trial #1 *No joints*	*Trial #2* *Pour strip*	*Trial #3* *Expansion joint*
0.937 in.	0.813 in.	0.469 in.

Design Example 21-1 compares the volume change effects of a 300-ft-long monolithic structure with a structure having a pour strip at midlength and with a structure having an expansion joint at midlength. With a pour strip (Fig. 21-9a), the structure is built in segments. Each segment absorbs the elastic shortening and part of the creep and shrinkage as a smaller unit, thus reducing volume change loads in the columns at the extremities of the structure. Before the facility is put into service (at least 28 days after the

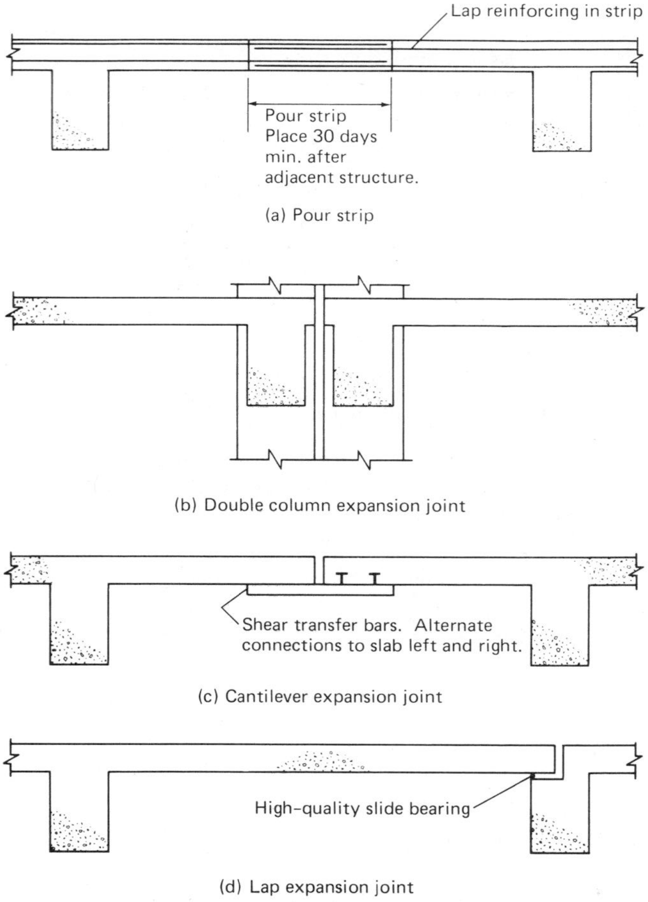

Fig. 21-9 Volume change control details.

adjacent sections have been posttensioned), the pour strips are closed, and the remaining creep and shrinkage plus all temperature movement must be absorbed by the completed structure.

Expansion joints are the surest way to relieve volume change forces. However, expansion joints normally require significant maintenance over the life of the structure.

The most reliable performing expansion joint is the double column expansion joint (Fig. 21-9b). The double column joint is more expensive than the other types of expansion joints; however, it truly allows the building to be separated into independent sections without reliance upon the adjacent structure for support.

The cantilever expansion joint (Fig. 21-9c) provides for architectural continuity. However, careful attention must be given to the placement of the negative moment reinforcing. The slabs may be overstressed and deflect if the negative moment reinforcing steel is placed improperly. It is recommended that sleeved shear transfer bars be installed on a cantilever-type joint to prevent detrimental impacting and prevent adverse wear at the joint seal as the wheel loads move from one side of the joint to the other.

The lap-type expansion joint (Fig. 21-9d) eliminates the possibility of deflection that may occur with a cantilever joint; however, the lap requires a high-quality slide bearing mechanism to ensure freedom of movement in the joint. The lap-type joint is more difficult to construct than the cantilever type, and great care must be given to ensure that adequate bearing is provided after all time-dependent volume changes have taken place. Also the lap joint must be kept sealed to prevent the accumulation of dirt in the joint.

21.7 BUMPER WALL DESIGN

Bumper walls must be designed for the impact loading of a vehicle hitting the wall. In lieu of a dynamic loading criterion, Ref. 21-1 recommends an equivalent static loading design criterion of a 10,000-lb point ultimate load at 18 in. above the floor to create a safe design for bumper loadings. This horizontal load is to be applied over a 1-ft-square area with the load distributed through the bumper wall system into the main structural elements in a manner that is logical and appropriate for the bumper wall system. (See below, Design Example 21-2.)

DESIGN EXAMPLE 21-2: *Bumper wall design example.*

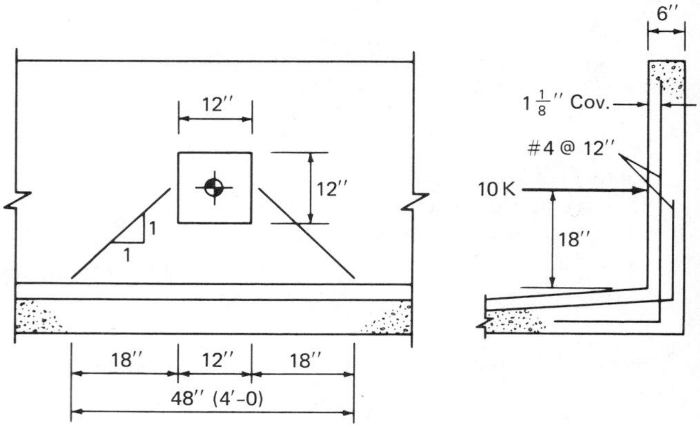

$$M_R = \frac{(10)(18)}{4} = 45.0 \text{ in.-k/ft}, \qquad f'_c = 4 \text{ ksi}$$

Check #4 @ 12" ($f_y = 60$ ksi), $A_s = 0.20$ in²/ft.

$$d = 6 - 1\tfrac{1}{8}^{*} - \tfrac{1}{2}(\tfrac{1}{2}) = 4\tfrac{5}{8}'' = 4.625''$$

$$a = \frac{A_s f_y}{0.85\, b\, f'_c} = \frac{(0.20)(60)}{0.85\,(12)(4)} = 0.29 \text{ in.}$$

$$M_u = \phi A_s f_y \left(d - \frac{a}{2}\right) = (0.9)(.20)(60)\left(4.625 - \frac{.29}{2}\right)$$

$$= 48.38 \text{ in.-k} > 45.0 \qquad \text{OK}$$

21.8 DURABILITY

The structural engineer can enhance the durability of a parking structure by developing a crack-free design. In a posttensioned structure, this can be accomplished by maintaining an average prestress of 175 psi in the structural direction and 100 psi in the nonstructural (temperature) direction. While ACI 318-83[21-3] allows the concrete to be placed in flexural tension at full live load, many engineers do not allow flexural tension in the slabs or beams at a reduced live load of 30 psf.

When the snow load is added to the reduced live load on the roof, the designer can provide additional mild steel reinforcing to increase the ultimate moment capacity, provided the tensile stresses do not exceed $12\sqrt{f'_c}$ under service load conditions.

Precast concrete structures should be designed so that a 2000-lb wheel (working) load placed on a 1-ft square on the edge of a flange (assuming no support by adjacent flanges) will not cause stresses in plain concrete exceeding 50% of the modulus of rupture ($7.5\sqrt{f'_c}$). This provides a crack-free environment which is desirable to prevent rapid deteri-

*Cover = 150% × 3/4" for weather resistance.

oration of the structure. Of course, reinforcing, usually in the form of welded wire fabric, is used to provide the flexural reinforcing in these slabs.

Deicing salts are very detrimental to parking structures. Where salt is used, the top concrete cover over reinforcing steel should be increased as recommended by ACI 318-83. Because flexure cracking of the top surface in the negative moment region allows access of the salts to the reinforcing, additional protection of epoxy coating of the top reinforcing is suggested. As a final protection to prevent penetration of deicing salts in the concrete, a high-quality surface sealer should be applied to the floor slabs.

REFERENCES

21-1 *Recommended Building Code Provisions for Open Parking Structures*, National Parking Association Parking Consultants Council, July 1980.

21-2 *PCI Design Handbook*, 2nd Ed., Section 4.4.2, "Equivalent Volume Change," Prestressed Concrete Institute, 1978.

21-3 *Building Code Requirements for Reinforced Concrete* (ACI 318-83), American Concrete Institute, Detroit, Mich., 1983.

22

Concrete Pavement Design

ROBERT G. PACKARD*

22.1 CONCRETE PAVEMENT DESIGN

Since several aspects of concrete pavement design are interrelated, it is essential to review their functions in minimizing the causes of uncontrolled cracking of slabs, loss of uniform support in the foundation, and loss of surface integrity. Control of these factors involves four aspects of design:

1. Concrete quality—selection of suitable materials and proportions of materials for a concrete pavement that will have adequate durability and strength.
2. Subgrade–subbase design—Proper subgrade preparation techniques are required to ensure reasonably uniform support for the slab and control subgrade volume changes caused by expansive soils or frost action. Where traffic volume is heavy, the use of a subbase layer is required to prevent mud-pumping; this layer must be a nonconsolidating granular or stabilized material to prevent joint faulting.
3. Thickness design—Slab thickness is determined so that flexural stresses due to traffic loads are kept within safe limits.
4. Joint design—The jointing arrangement determines where cracks will form due to restrained shrinkage and temperature stresses. These stresses are minimized by the use of short joint spacings in plain pavements; for reinforced pavements with longer joint spacings, intermediate cracks will form between the joints, but these are not detrimental because they are held tightly together by the reinforcing steel.

*Manager, Paving and Transportation Department, Portland Cement Association, Skokie, Illinois.

Details of these aspects of design for highway, street, and airport pavements are presented in the following sections.

22.2 CONCRETE QUALITY

Fundamentals of mix design for quality concrete are given in Chapter 6, "Properties of Materials for Reinforced Concrete." In addition to those fundamentals, the following recommendations are made specifically for paving mixtures to obtain (1) the durability needed to resist the effects of climate and traffic, and (2) the flexural strength required to carry the expected weights and numbers of traffic loads.

In frost-affected areas, concrete pavements are subjected to many cycles of freezing and thawing and to the application of deicing salts. To protect concrete pavements against the action caused by these agents, it is essential to have a mix with a low water–cement ratio, an adequate cement factor, and sufficient quantity of entrained air. The amounts of entrained air needed to produce weather-resistant concrete vary with the maximum size aggregate.* Recommended percentages of entrained air are:

Maximum size aggregate, in.	*Entrained air, percent*
$1\frac{1}{2}$, 2, $2\frac{1}{2}$	5 ± 1
$\frac{3}{4}$, 1	6 ± 1
$\frac{3}{8}$, $\frac{1}{2}$	$7\frac{1}{2}$ + 1

*Generally, maximum size aggregate should not exceed one-fourth the pavement thickness.

The amount of mixing water also has a critical influence on the durability and weather resistance of hardened concrete. The least amount of mixing water with a given cement content that will produce a plastic, workable mix will result in the greatest durability in the hardened concrete. Laboratory and field experience with air-entrained concrete shows that for satisfactory pavement durability the water–cement ratio should not exceed 0.54. The cement factor should be not less than 517 lb/cu yd. In areas where severe frost and deicing agents are common, the water–cement ratio should not exceed 0.49 with a minimum cement factor of 564 lb/cu yd.

22.3 SUBGRADES AND SUBBASES

For satisfactory performance a concrete pavement must be provided with a uniform foundation. The basic objective is to construct a foundation that is, and will remain, reasonably uniform under the climatic and traffic conditions that will prevail. This objective is achieved economically by proper soil grading operations and subgrade treatments during construction and, where needed, by the use of a subbase layer of high-quality material placed directly beneath the pavement. Building up strong support in the foundation is not essential and is usually not economically justified.

The principle of uniform support, rather than strong support, is explained by consideration of the properties of the concrete slab. The rigidity of concrete enables it to distribute loads over large areas of the subgrade; deflections are small, and pressures on the subgrade are low.[22-1–22-4]

To design a subgrade and subbase that provide reasonably uniform support for the slab, the three major causes of nonuniform support that need to be controlled are:

1. Expansive soils
2. Frost action
3. Mud-pumping

Effective control of high-volume-change soils and frost action is most economically achieved through appropriate subgrade preparation techniques, while prevention of mud-pumping requires a thin subbase layer. Although a subbase also provides some control of high-volume-change soils and frost action, the use of thick subbase layers for substantial control of these factors is no more effective than subgrade work and usually costs more.

22.3.1 Subgrades

Where the subgrade conditions are not reasonably uniform, correction is most economically achieved by proper subgrade preparation techniques such as selective grading, crosshauling, mixing at abrupt transitions, and moisture–density control of subgrade compaction. Particular attention is needed for the control of expansive soils and differential frost heave.

1. Expansive soils. Identification of the types of soils that are expansive and the mechanism of soil volume change has been gained through research and experience. A summary of the technology of expansive soils, including tests for identification, is given in Ref. 22-5. The simpler tests provide indices, such as plasticity index, shrinkage limit, and bar shrinkage, that serve as useful guides to identify approximate volume change potential of soils. For example, the following table shows approximate expansion plasticity relationships:

Degree of Expansion	*Percentage of Swell (ASTM D1883)*	*Approximate Plasticity Indices (ASTM D424)*
nonexpansive	2 or less	0 to 10
moderately expansive	2 to 4	10 to 20
highly expansive	more than 4	more than 20

Most soils sufficiently expansive to cause distortion of concrete pavements are in the AASHTO A-6 or A-7 groups. By the Unified Soil Classification system, soils classified as CH, MH, or OH are considered highly expansive.

Expansive soils have been found to be controlled effectively and economically by the following subgrade treatments:

a. *Subgrade grading operations.* Excessive swell can be controlled by placing the more expansive soils in the lower parts of the embankments and crosshauling less expansive soils to form the upper part of the subgrade in both embankments and excavations. Selective grading also makes it possible to have reasonably uniform soil conditions in the upper part of subgrade, with gradual transitions between soils with varying volume change properties.

In deep cuts into highly expansive soils, considerable expansion may occur due to the removal of the natural surcharge load and the consequent absorption of additional moisture. Since this expansion takes place slowly, it is essential to excavate these deep cuts well in advance of other grading work.

b. *Compaction and moisture control.* Volume changes are further reduced by adequate moisture and density controls during subgrade compaction. It is critically important to compact highly expansive soils at 1 to 3% above optimum moisture (*AASHTO T99*). Where embankments are of considerable height, compaction moisture contents can be increased from slightly below optimum in the lower part of the embankment to above optimum in the top 1 to 3 ft. Expansive soils compacted slightly wet of optimum expand less, have higher strengths after wetting, and absorb less water.

After pavements are placed in service, most subgrades reach a moisture content approaching their plastic limit. That is, the natural moisture content reached is close to and slightly above the standard optimum (*AASHTO T99*). When this moisture content is obtained in construction, the subsequent changes in moisture will be much less, and the subgrade will retain the reasonably uniform stability needed for good pavement performance.

c. *Nonexpansive cover.* In areas with prolonged periods of dry weather, highly expansive subgrades may require a cover layer of low-volume-change soil placed full width over the subgrade. This will minimize changes in the moisture content of the underlying expansive soil and will also have some surcharge effect. A low-volume-change layer with a low to moderate permeability is not only more effective but usually is less costly than a permeable, granular soil. Highly permeable open-graded subbase materials are not recommended as cover for expansive soils, since they permit greater changes in subgrade moisture content.

Local experience with expansive soils is the best guide for adequate depth of cover.

2. Frost Action. Uniquely, field experience with concrete pavements has shown that frost action damage due to inadequate design is a result of frost heave—in the form of abrupt, differential heave. Subgrade softening on thaw is not a design consideration for concrete pavements, since a strong subgrade support is not required. Design for concrete is concerned with reducing the nonuniformity of subgrade soil and moisture conditions that lead to objectionable differential heaving—especially where subgrade soils vary abruptly from non-frost-susceptible sands to the highly frost-susceptible silts at cut–fill transitions, where the ground-water is close to the surface, or where water-bearing strata are encountered.

2.1. Frost Heave. For frost heave to occur, all of three conditions must be present: (1) a frost-susceptible soil; (2) freezing temperatures penetrating the subgrade; (3) a supply of water.

Heaving is caused by the growth of ice lenses in the soil. When freezing temperatures penetrate a subgrade, water from the unfrozen portion of the subgrade is attracted to the frozen zone. If the soil is susceptible to high capillary action, the water moves to ice crystals initially formed, freezes on contact, and expands. If a supply of water is available, the ice crystals will continue to grow, forming ice lenses of appreciable thickness that lift or heave the overlying pavement.

2.2. Frost-Susceptible Soils. Criteria and soil-classifications used for identifying frost-susceptible soils usually reflect susceptibility to softening on thaw as well as to heaving. The major concern is to reduce heaving, especially differential heaving. Control of spring softening is not a consideration for concrete pavements. Thus, specific criteria or classifications identifying frost-susceptible soils should be reviewed with differentiation in classification between siols susceptible to heave and those susceptible to thaw softening.

The worst heaving usually occurs in fine-grained soils subject to capillary action. Low-plasticity soils with a high percentage of silt-size particles (0.05 to 0.005 mm) are particularly susceptible to frost heave. These soils have pore sizes small enough to develop capillary potential (suction) but large enough for rapid travel of water to the frozen zone. Coarser soils have higher rates of flow, but do not have the potential to lift enough moisture for heaving. More cohesive soils, although developing high capillarity, have low permeability, and moisture moves too slowly for growth of thick ice lenses.

2.3. Spring Subgrade Softening. In the spring, frozen subgrade thaws both from the surface downward and from the bottom upward. As a result, thawing is usually more rapid than freezing. When thawing starts, the moisture content of the subgrade may be high due to the previous moisture increase during freezing and due to surface water infiltration. This water in the upper thawed layer cannot drain downward because of the frozen zone below. In addition, ice lensing or simple expansion may have caused a loss in density. Under these conditions, there is a sharp reduction in subgrade support during the thaw period.

The periods of reduced subgrade support that accompany thawing have very little effect on concrete pavements. This is so because concrete reduces pressures on soft subgrades to safe limits by distributing loads over large areas and because concrete pavements are designed for fatigue stresses due to load repetitions.

Concrete pavements designed on the basis of normal weather subgrade strengths have ample reserve capacity for the periods of reduced support during spring thaws. Because of their reserve load-carrying capacity, concrete pavements are exempted from load restrictions during periods of spring thaw.

Further evidence that concrete pavements designed with uniform support are not influenced by spring thaw is shown by the results of the AASHO Road Test.[22-6] The pavement performance and the equations written to relate the design variables to traffic loads show that concrete pavements, with or without a subbase, were not affected by the spring thaw periods.

2.4 Control of Frost Heave. As in the case of expansive soils, a large degree of control of frost heaving is accomplished most economically by appropriate grading operations and by controlling subgrade compaction and moisture. These include:

a. *Grade elevation.* Grade lines are set high and side ditches are constructed deep so that highly frost-susceptible soils are farther from the capillary range of groundwater tables. Where groundwater is near the surface, the grade may be kept 4 or 5 ft above ditch bottom in cuts and natural ground in fills.

b. *Selective grading.* Highly frost-susceptible soils are placed in the lower portions of embankments, and less susceptible soils are crosshauled to form the upper portion of the subgrade. Crosshauling and mixing are also used at cut–fill transitions to correct abrupt changes in soil type.

c. *Mixing.* Where soils vary widely or frequently in texture, and where nonuniform conditions are not clearly defined, mixing of the soils is effective in preventing differential frost heave. With modern construction equipment, the mixing of nonuniform soils to form a uniform subgrade is often more economical than importing select materials from borrow pits.

d. *Removal of silt pockets.* Where highly frost-susceptible soils are pocketed in less susceptible soils, they are excavated and backfilled with soils like those surrounding the pocket. Moisture and density conditions for the replacement soil should be as similar as possible to those of the adjacent soils. At the edges of the pocket, the replacement soil should be mixed with the surrounding soil to form a tapered transition zone, just as in cut–fill transitions.

e. *Compaction and moisture control.* After reasonable uniformity has been achieved through the grading operations, additional uniformity is obtained by proper subgrade compaction at controlled moisture contents. The permeability of most fine-grained soils is substantially reduced when they are compacted slightly wet of *AASHTO T99* optimum moisture. Reducing soil permeability retards the rate of moisture flow to the frozen zone and consequently reduces frost heaving. Research[22-7, 22-8] confirms that less frost heave occurs at the wetter condition.

f. *Drainage.* Where high grades are impractical, subgrade drains are used to lower groundwater tables. Where wet spots are encountered in the grade due to seepage through a permeable stratum underlain by an impervious material, intercepting drains are used. Pipe backfill should meet filter criteria[22-9] so that neither soil infiltration nor clogging of pipe openings will occur.

g. *Non-frost-susceptible cover.* Layers of clean gravel or sand will reduce frost heave but are not required solely for this purpose when the less costly grading operations are properly employed. The benefit of the use of thick subbase layers is somewhat diminished, since coarse soils permit slightly deeper frost penetration than do fine-grained soils at their higher moisture content.[22-10, 22-11] When a subbase layer is required to prevent mud-pumping, it also provides some protection against frost action. However, these layers are more effective in preventing loss in subgrade support on thaw which is not a design consideration for concrete pavements.

22.3.2 Subbases

A subbase is not required for the sole purpose of obtaining strong or uniform support for the pavement. Also, a subbase is not necessarily required for purposes of control of expansive subgrades or for control of frost action. For these purposes, although a subbase layer is beneficial, proper subgrade preparation techniques should also be utilized as the most economical means of obtaining good pavement performance. The essential function of a subbase is to prevent mud-pumping of fine-grained soils.

1. Prevention of Mud-Pumping. Mud-pumping is the forceful displacement of a mixture of soil and water that occurs under slab joints, cracks, and pavement edges. It is caused by the frequent deflection of slab edges by heavy wheel loads when fine-grained plastic subgrade soils are saturated. Continued uncontrolled mud-pumping eventually leads to the displacement of enough soil that uniformity of support is destroyed and slab edges are left unsupported.

Subbase studies showed that three factors are necessary for mud-pumping to occur:

1. A subgrade soil that will go into suspension
2. Free water between the pavement and subgrade, or subgrade saturation
3. Frequent passage of heavy loads

These studies also showed that mud-pumping did not occur on natural subgrades or granular subbases with less than about 45% passing the No. 200 sieve, and a plasticity index of 6, or less.

The performance of test sections with no subbase at the AASHO Road Test shows that adequately designed pavements without subbases are suitable for many city streets, county roads, light-traffic highways, and light-duty airports. Subbase surveys conducted by highway departments show that:

1. Pavements designed to carry not more than 100 to 200 heavily loaded trucks per day do not require subbases to prevent pumping damage.
2. Soils with less than 45% passing a No. 200 sieve and with a *PI* of 6, or less, are suitable for moderate volumes of heavy truck traffic.
3. Subbases meeting *AASHTO Specification M155* effectively prevent mud-pumping in pavements carrying the greatest volumes of traffic.

AASHTO Specification M155 to prevent pumping states:

Granular material for use as subbase under concrete pavement may be composed of sand, sand-gravel, crushed stone, crushed or granulated slag, or combinations of these materials. The material shall meet the following requirements:

Maximum size:	not more than 1/3 the thickness of the subbase
Passing No. 200 sieve:	15% maximum
Plasticity index:	6 maximum
Liquid limit:	25 maximum

NOTE: Materials with a higher percentage passing No. 200 sieve or with a higher plasticity index than 6 or a higher liquid limit than 25 may be used, provided that a stabilization method found to be locally suitable is used. The material shall be graded suitably to permit compaction to such a density that a minimum increase in densification will occur after the pavement is in service.

2. Subbase Thickness. Since the primary purpose of a subbase is to prevent pumping, it is neither necessary nor economical to use thick subbases. Experimental projects have shown that a 3-in. depth of subbase will prevent mud-pumping under very heavy traffic. When a subbase is required, a depth of 4 to 6 in. is commonly specified for highway pavements and a depth of 6 to 9 in. for heavy-duty airport pavements.

TABLE 22-1 Grading Requirements for Subbase Materials

Sieve designation	*Percentage by weight passing square mesh sieves*					
	Grading A	*Grading B*	*Grading C*	*Grading D*	*Grading E*	*Grading F*
2-inch	100	100	–	–	–	–
1-inch	–	75–95	100	100	100	100
3/8-inch	30–65	40–75	50–85	60–100	–	–
No. 4	25–55	30–60	35–65	50–85	55–100	70–100
No. 10	15–40	20–45	25–50	40–70	40–100	55–100
No. 40	8–20	15–30	15–30	25–45	20–50	30–70
No. 200	2–8	5–20	5–15	5–20	6–20	8–25

3. Subbase Gradation Control. While a wide variety of locally available materials have performed well as subbases for concrete pavements, the subbase for an individual project should have a reasonably constant gradation. This makes it possible for compaction equipment to produce the uniform support that is essential for good pavement performance. *AASHTO Specification M155* is suitable for establishing minimum subbase requirements, but it does not afford acceptable gradation control for an individual project.

One method for overcoming this problem is to adopt a specification such as *AASHTO M147.* This specification is shown in Table 22-1. The subbase for a particular project is kept within the limits of a single gradation.

4. Subbase Compaction. Granular materials are subject to consolidation from the action of heavy truck traffic after pavements are placed in service. To prevent a detrimental amount of consolidation, subbases muct be compacted to high densities.

Performance experience and research furnish convincing support for the following recommendations:

1. Untreated (not stabilized) subbases should have a minimum of 100% standard density (*AASHTO T99*). On projects that will carry large volumes of traffic, the specified density should be not less than 105% of standard or 98 to 100% of modified density (*AASHTO T180*).
2. When subbase depths are increased beyond the minimum needed to prevent pumping, there is an increasing risk of subbase consolidation from heavy traffic.

5. Cement-Treated Subbases. The use of cement-treated subbases* for concrete pavement has become a standard practice in many parts of the United States and Canada. Several factors are responsible for their increasing use. An important one is the growing scarcity of good granular subbase materials at a time when tremendous quantities of these materials are needed for building new pavements.

The principal benefits to be derived from use of cement-treated subbases are:

1. A uniform and strong, nonconsolidating support is provided for the pavement.
2. Firm support for the slip-form paver or side forms contributes to the construction of smoother pavements.
3. A stable working base expedites construction operations.

6. Lean Concrete (Econocrete) Subbases. In the last few years there has been an increase in the use of lean concrete subbases.[22-14] These are constructed in the same manner as

*Design and construction practices for cement-treated subbases are given in Refs. 22-12 and 22-13. Data on the properties of cement-treated subbases are given in Refs. 22-2, 22-15, 22-16, and 22-26.

normal concrete, usually with low cement contents. The aggregates may range from normal concrete aggregates to substandard local materials or crushed, recycled concrete.

22.4 THICKNESS DESIGN FOR HIGHWAY AND STREET PAVEMENTS

The thickness design procedure of the Portland Cement Association[22-17] is presented as appropriate for a wide range of conditions with proper assessment of the design variables.

22.4.1 Basis for Design

The thickness design method is based on knowledge from the following sources:

1. Theoretical studies of pavement slab behavior by H. M. Westergaard,[22-19–22-21] G. Pickett and G. K. Ray,[22-23, 22-24] and others.
2. Model and full-scale tests such as the Arlington Test conducted by the Bureau of Public Roads[22-25] and tests made at the PCA laboratories.[22-1–22-4, 22-26]
3. Experimental pavements subjected to controlled test traffic, such as the Bates Test Road,[22-27] the Pittsburg, Calif. Test Highway,[22-28] the Maryland Road Test,[22-29] and the AASHO Road Test.[22-6, 22-31]
4. The performance of normally constructed pavements subject to normal mixed traffic.

All these sources of knowledge are useful. However, the knowledge gained from performance of normally constructed pavements is the most important.

22.4.2 Subgrade–Subbase Strength

The degree of subgrade or subgrade-subbase support is in terms of k, Westergaard's modulus of subgrade reaction. The k-value is determined by in situ plate bearing tests with a 30-in.-diameter plate. Procedures are described in *ASTM Method D1196* and the Department of the Army (U.S.), *Technical Manual TM-5-824-3* and are discussed in Ref. 22-32.

If time and equipment are not available for plate-load testing, k can be estimated by correlation to soil classification or laboratory strength test as shown in Fig. 22-1. Tables 22-2 and 22-3 show the effect of subbase thickness on the design k-value.

22.4.3 Load Stresses

Bending of a concrete pavement under wheel loads produces both compressive and tensile (or flexural) stresses. Although compressive stresses are not critical, the ratio of flexural stresses to flexural strength is high, often exceeding values of 0.5. As a result, slab thickness design is controlled by keeping flexural stresses due to traffic loads within safe limits.

For street and highway pavements with the usual 12-ft pavement lanes, the critical stress location is considered to be the transverse joint edge rather than at the outside pavement corner or edge. This conclusion is based on studies of truck placements[22-33] across pavement lanes and an analysis of the effect of load position.[22-34, 22-35] Accordingly, the charts shown in Figs. 22-2 and 22-3 were developed from influence charts[22-24] to be used for determining stresses at this load position.

While load stresses are critical at the transverse joint edge, shrinkage and temperature stresses are low at this location and usually are not additive to load stresses. For this reason, load stresses only are considered. Conservatively, no beneficial allowance is made for load transfer across the joint.

22.4.4 Flexural Strength of Concrete

Flexural strength is determined by modulus of rupture (M-R) tests, usually made on 6 × 6 × 30-in. beams. The modulus of rupture is determined by third-point loading tests (*ASTM C78*). The 28-day M-R values are currently used for thickness design of streets and highways. However, this practice is conservative, since concrete continues to gain strength for many years (see Fig. 22-14).

The following approximate relationship between flexural and compressive strength is sometimes useful in preliminary design stages; however, the final design should be based on modulus of rupture test data:

$$MR = K\sqrt{f_c'}$$

where:

MR = flexural strength (modulus of rupture), psi
K = a constant between 8 and 10
f_c' = compressive strength, psi

22.4.5 Fatigue

Like other construction materials, concrete is subject to the effects of fatigue. Since the critical stresses in concrete are flexural, fatigue due to flexural stress is used for thickness design. Stress ratio is defined as the ratio of flexural stress to modulus of rupture. For example, if an axle load causes a flexural stress of 500 psi and the modulus of rupture is 700 psi, then:

$$\text{stress ratio} = \frac{500}{700} = 0.71$$

Flexural fatigue research on concrete has shown that, as stress ratios decrease, the number of repetitions to failure increases. Allowable load repetitions for stress ratios between 0.50 and 0.85 are shown in Table 22-4. The values are conservatively based on fatigue research on concrete.[22-36, 22-37]

The design example in this section shows how Table 22-4 fatigue data are used for thickness design. This usage is based on the Miner hypothesis[22-38] that fatigue resistance not consumed by repetitions of one load is available for repetitions of other loads.

Theoretically, the total fatigue used should not exceed 100%. For designs based on the 28-day modulus of rupture, fatigue consumption can be increased to about 125%. This increase takes account of the strength gain after 28 days.

22.4.6 Load Safety Factors

Research[22-6, 22-29] shows that moving loads cause less stress than static loads. Therefore, the formerly used concept of assigning an impact factor to increase the effect of loads is more accurately classified as a load safety factor. In the design procedure axle loads are multiplied by load safety factors of 1.0, 1.1, or 1.2, depending on pavement class as follows:

1.0 = for highways, residential streets, and other streets that will carry small volumes of truck traffic
1.1 = for highways and arterial streets where there will be moderate volumes of truck traffic

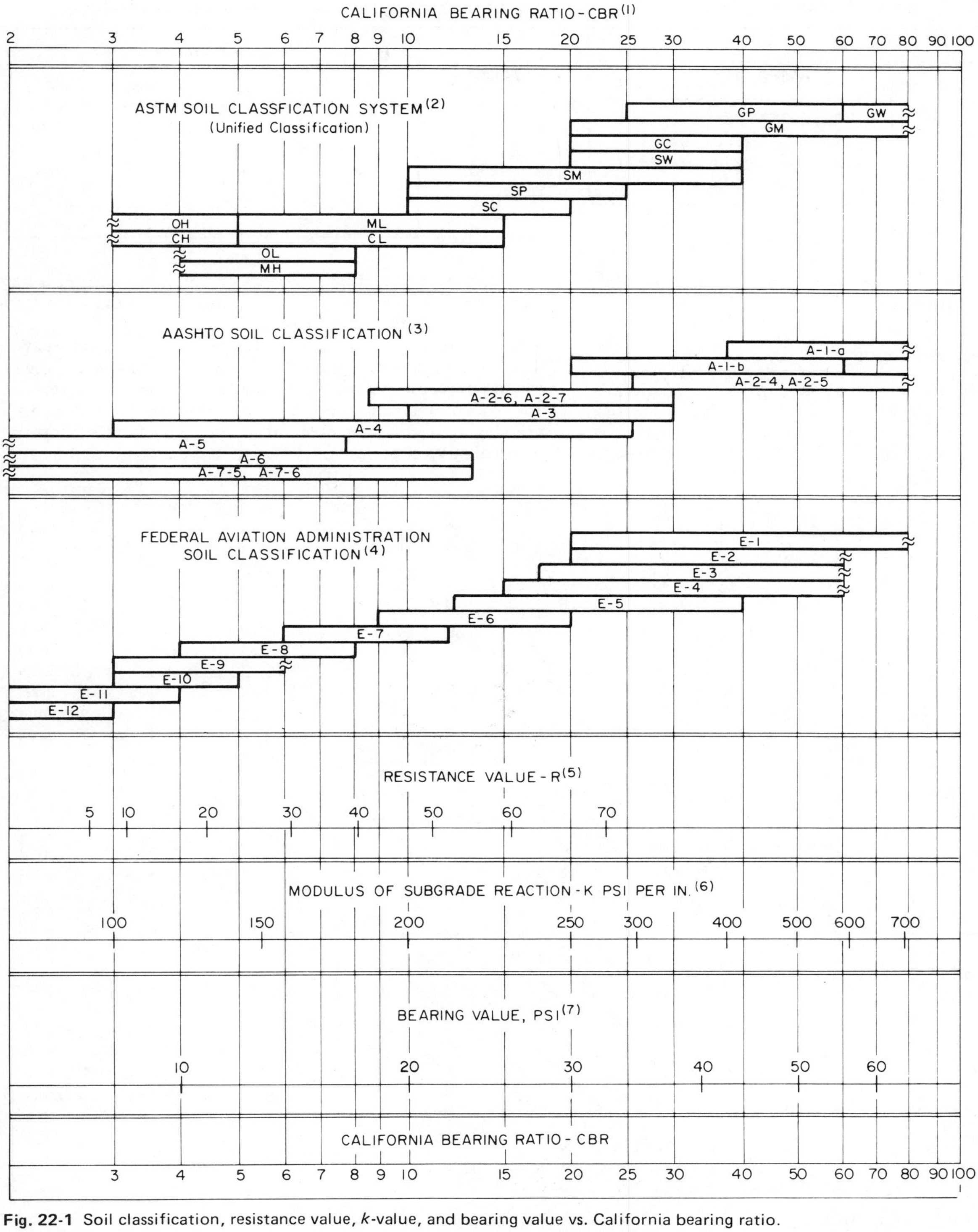

Fig. 22-1 Soil classification, resistance value, *k*-value, and bearing value vs. California bearing ratio.

(1) For the basic idea, see O. J. Porter, "Foundations for Flexible Pavements," Highway Research Board *Proceedings of the Twenty-second Annual Meeting*, 1942, Vol. 22, pages 100-136.
(2) ASTM Designation D2487.
(3) "Classification of Highway Subgrade Materials," Highway Research Board *Proceedings of the Twenty-fifth Annual Meeting*, 1945, Vol. 25, pages 376-392.
(4) *Airport Paving*, U.S. Department of Commerce, Federal Aviation Agency, May 1948, pages 11-16. Estimated using values given in FAA *Design Manual for Airport Pavements*.(Formerly used FAA Classification; Unified Classification now used.)
(5) C. E.Warnes, "Correlation Between *R* Value and *k* Value," unpublished report, Portland Cement Association, Rocky Mountain-Northwest Region, October 1971 (best-fit correlation with correction for saturation).
(6) See T. A. Middlebrooks and G. E. Bertram, "Soil Tests for Design of Runway Pavements," Highway Research Board *Proceedings of the Twenty-second Annual Meeting*, 1942, Vol. 22, page 152.
(7) See item (6), page 184.

TABLE 22-2 Effect of Untreated Subbase on *k* Values

Subgrade k value	*Subbase k value, pci*			
	4 in.	*6 in.*	*9 in.*	*12 in.*
50	65	75	85	110
100	130	140	160	190
200	220	230	270	320
300	320	330	370	430

TABLE 22-3 Design *k* Values for Cement-Treated Subbases

Subgrade k value	*Subbase k value, pci*			
	4 in.	*6 in.*	*8 in.*	*10 in.*
50	170	230	310	390
100	280	400	520	640
200	470	640	830	–

1.2 = for interstate and other multilane projects where there will be uninterrupted traffic flow and high volumes of truck traffic.

22.4.7 Traffic Analysis Period

The term "pavement life" is not subject to precise definition. Some engineers and highway agencies consider the life of a concrete pavement ended when the first overlay is placed. Based on this concept, the life of concrete pavements may vary from less than 20 years on some projects with design, material, or construction defects, to more than 50 years on other projects where these defects are absent.

A more reasonable basis for engineering analysis and pavement design is to recognize that the useful life of a concrete pavement does not end when the first—or second—overlays are placed. Instead, the concrete continues to serve as the primary load-carrying element in the pavement structure. Accordingly, a period of 40 years is often assumed as the

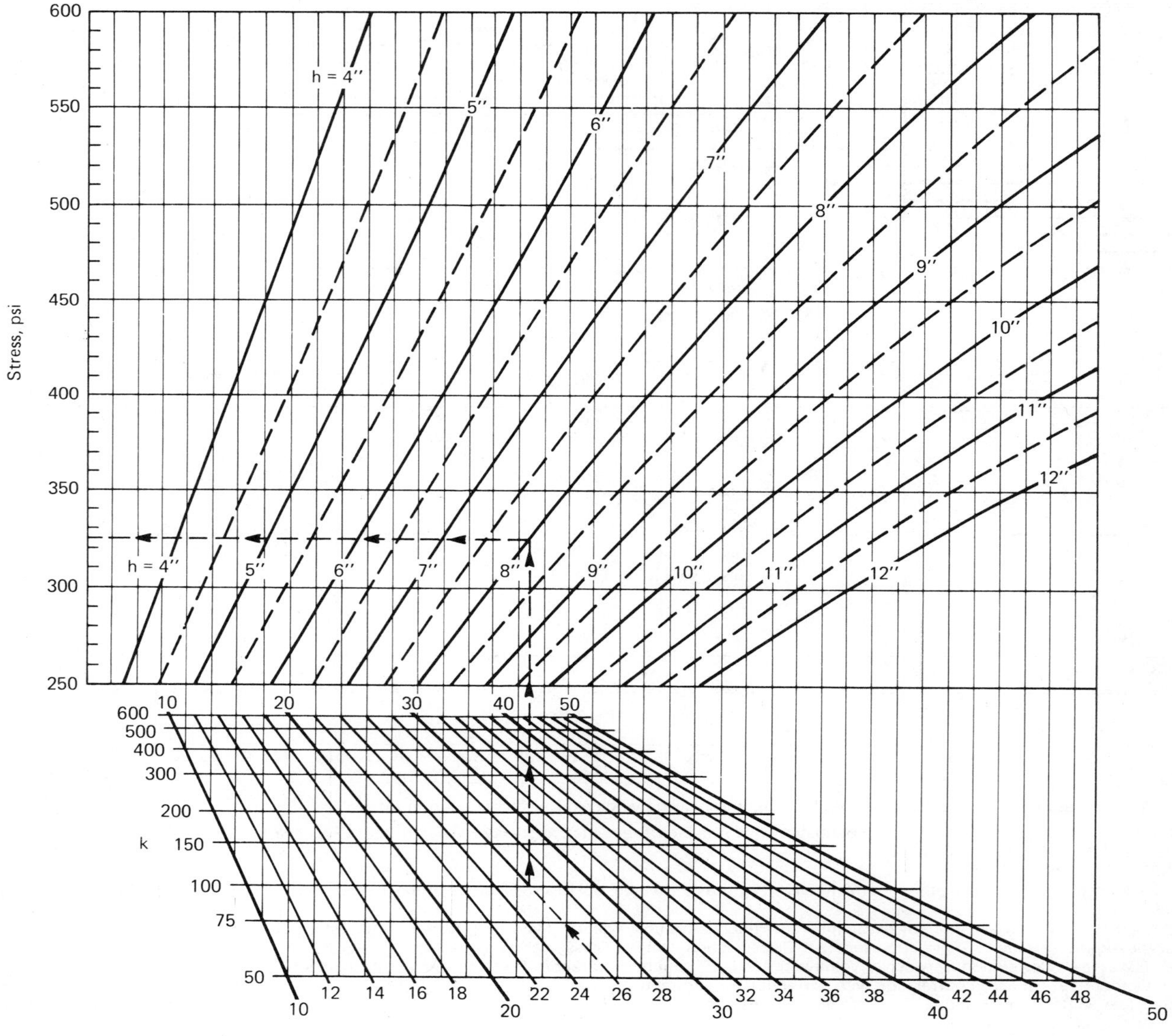

Fig. 22-2 Stress chart for single-axle loads.

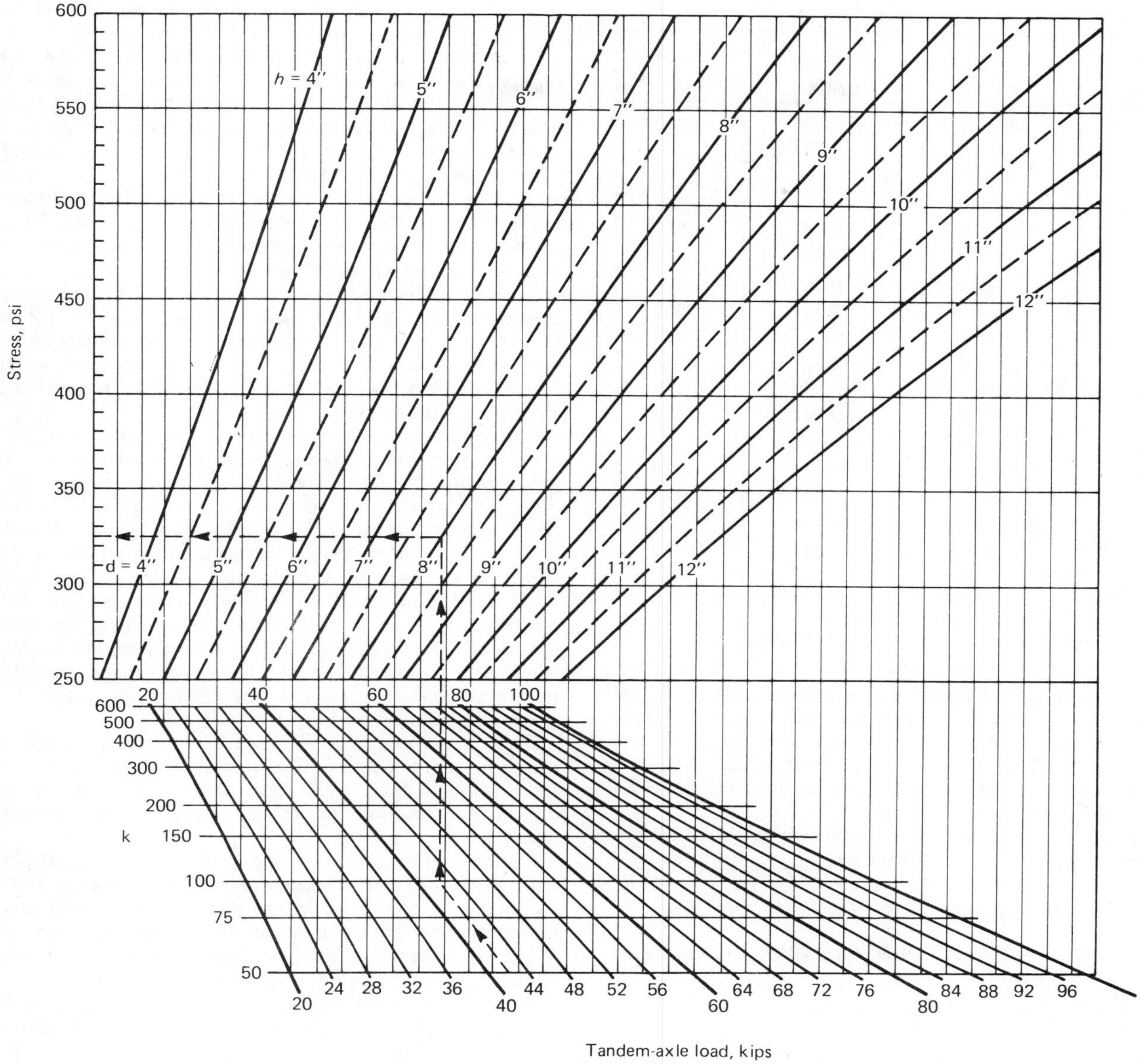

Fig. 22-3 Stress chart for tandem-axle loads.

traffic analysis period for design purposes. Periods of 20 years or 30 years are also used.

22.4.8 Traffic Analysis

The numbers and weights of heavy axle loads are major factors in the design procedure. The total number of trucks anticipated during the traffic analysis period is estimated. These are distributed into axle weight groupings by the use of special traffic loadometer studies made for the project, or by using loadometer surveys made by the highway department. Methods for detailed traffic analysis are given in Ref. 22-17. The end result of the traffic analysis is to estimate the numbers of single and tandem axle loads in separate weight categories that will use the pavement during the design period.

22.4.9 Design Procedure

An example of the design procedure and calculations is shown in Fig. 22-4 for an urban, interstate pavement. The axle loads and expected number of load repetitions (columns 1 and 6) were first estimated for a period of 40 years.

From the axle load data, stress and fatigue calculations are made for the trial thickness. Load stresses are determined from Figs. 22-2 and 22-3, and allowable load repetitions are listed in Table 22-4. The design thickness is selected to prevent flexural fatigue failure of the slab. In this case, at a trial thickness of 8.5 in., the total fatigue resistance used is 83%. This would be an adequate design. Normally, a slab thickness is selected so that fatigue consumption does not exceed 125% for the 28-day modulus of rupture.

22.5 JOINT DESIGN FOR HIGHWAY AND STREET PAVEMENTS

Joints are placed in concrete pavements to control cracking due to (1) tensile stresses resulting from restrained shrinkage; (2) compressive stresses that may occur when the slab expands; and (3) the combined effects of loads and re-

TABLE 22-4 Stress Ratios and Allowable Load Repetitions

Stress ratio*	*Allowable repetition*	*Stress ratio*	*Allowable repetition*
0.51**	400,000	0.69	2,500
0.52	300,000	0.70	2,000
0.53	240,000	0.71	1,500
0.54	180,000	0.72	1,100
0.55	130,000	0.73	850
0.56	100,000	0.74	650
0.57	75,000	0.75	490
0.58	57,000	0.76	360
0.59	42,000	0.77	270
0.60	32,000	0.78	210
0.61	24,000	0.79	160
0.62	18,000	0.80	120
0.63	14,000	0.81	90
0.64	11,000	0.82	70
0.65	8,000	0.83	50
0.66	6,000	0.84	40
0.67	4,500	0.85	30
0.68	3,500		

**Load stress divided by modulus of rupture*
***Unlimited repetition for stress ratios of 0.50 or less.*

strained warping. Restrained warping may occur when the slab attempts to warp upward or downward because of temperature and moisture differences between the upper and lower part of the slab. Stress occurs when the slab is restrained from warping by its weight or by frictional resistance of the subgrade or subbase.

Joints are also used to divide the pavement into suitable increments for construction purposes.

22.5.1 Basis for Design

Joint design is based on knowledge from the following three major sources:

1. Theoretical studies of concrete behavior[22-20, 22-25, 22-39–22-46]
2. Experimental pavements and model and full-scale laboratory tests[22-6, 22-25, 22-27, 22-29, 22-30]
3. The performance of highway and street pavements subject to normal mixed traffic.

Slab dimensions and jointing details for individual projects should reflect careful study of performance of pavements similar to the project being designed.

A detailed discussion of joint design is given in Refs. 22-47 and 22-48.

22.5.2 Transverse Joints

State highway departments use different types of transverse joint designs;[22-49] each state prefers one, or perhaps two, of the following types:

Undoweled, unreinforced slabs with short joint spacing of 15 ft or less
Doweled, unreinforced slabs with short joint spacings of 15 or 20 ft
Doweled, reinforced slabs with joint spacings from 30 to 50 ft
Continuously reinforced slabs with no transverse joints

Low-truck-traffic facilities such as parking lots and residential streets usually are constructed as unreinforced, undoweled pavements with short joints spacings.

In the range of slab thicknesses used for highway pavements, transverse joint spacing for plain pavements (unreinforced) should be about 15 to 20 ft to control cracking due to restrained contraction and restrained warping. For street pavements, 5 or 6 in. thick, joint spacings should be in the range of 10 to 15 ft. Surveys of pavements in service show that the correct spacing will vary in different localities and climates, and with different types of aggregate. Because of these variations, local experience is the best guide as to whether greater spacings can be used to control cracking effectively. Where experience is not available, the spacings shown in Table 22-5 are used. These spacings are for average soil and climatic conditions; the maximum values are used only where environmental conditions are favorable. For thin pavements, the spacings should be somewhat less than those tabulated.

For reinforced pavements (mesh-dowel pavements) longer joint spacings are used, and cracking is controlled by placing distributed steel between the joints. The purpose of the distributed steel is not to prevent cracking but to hold slab edges firmly together after cracks have formed. Based on both the economics and performance of mesh-dowel pavements, it is desirable to limit joint spacing to about 30 or 40 ft. Cost studies[22-50] for highway pavements indicate an optimum spacing in this range. In addition, the performance of all pavements depends strongly on the effectiveness of the joint seals. At joint spacings greater than 40 ft, the joint seals are usually not effective because of the greater fluctuations in joint width.

There are three general types of transverse joints: contraction joints, construction joints, and expansion joints. In addition to their special functions, expansion joints and untied construction joints also function as contraction joints.

1. Contraction Joints. The purpose of a contraction joint is to control cracking caused by shrinkage and the combined effect of loads and warping. Since contraction joints must be free to open, distributed steel reinforcement, if any, is interrupted at these joints. For the same reason, contraction joints are not tied by deformed tiebars.

Contraction joints also provide some relief from expansive forces, since the initial shrinkage of the concrete opens the joint slightly and thereby provides for subsequent expansion.

Recommended details for contraction joints are given in the following paragraphs and are illustrated in Fig. 22-5.

In these joints a plane of weakness is created by means of a groove formed while the concrete is plastic, or by insertion of a strip in the plastic concrete, or by sawing a groove after the concrete has hardened. The concrete subsequently cracks, and the irregular slab faces below the groove provide load transfer by aggregate interlock.

The depth of the groove should not be less than one-fourth of the slab thickness. Also, the depth of sawed grooves should not be less than the diameter of the largest size coarse aggregate. If the groove is formed, the corners should be edged to a radius of $1/8$ in.

The width of the groove should be such that horizontal slab movements will not increase or decrease the width excessively. (See Section 22.5.5.) This is necessary to facilitate sealing of the joint and to ensure that the joint sealant is not unduly strained. In some mild climate areas, the sealing of closely spaced joints may not be required.

Mechanical load-transfer devices are usually recommended where the joint spacing exceeds 20 ft, and even at a lesser spacing where service conditions are severe. The requirements for mechanical load transfer devices are given in

Project Design Two-B

Type Urban Interstate-Level Terrain No. of Lanes 4

Subgrade k 100 pci., Subbase 4-in. Granular Untreated

Combined k 130 pci., Load Safety Factor 1.2 (L.S.F.)

Procedure

1. Fill in Col. 1, 2, and 6, listing axle loads in decreasing order.
2. Assume 1st trial depth. Use ½-in. increments.
3. Analyze 1st trial depth by completing columns 3, 4, 5, and 7.
4. Analyze other trial depths, varying M.R., slab depth and subbase type.

Trial depth 8.5 in. M.R.* 700 psi k 130 pci

1	2	3	4	5	6	7
Axle Loads kips	Axle Loads × 1.2 L.S.F. kips	Stress psi	Stress Ratios	Allowable Repetitions No.	Expected Repetitions No.	Fatigue Resistance Used** percent
Single Axles						
30	36.0	367	.52	300,000	3,700	1
28	33.6	353	.51	400,000	3,700	1
26	31.2	328	<.50	Unlimited	7,400	0
24	28.8		"	"	195,000	0
22	26.4		"	"	764,000	0
		[From Fig. 22-2]		[From Table 22-4]		[Col. 6 ÷Col. 5 × 100]
Tandem Axles						
54	64.8	413	.59	42,000	3,700	9
52	62.4	398	.57	75,000	3,700	5
50	60.0	387	.55	130,000	36,270	28
48	57.6	375	.54	180,000	36,270	20
46	55.2	361	.52	300,000	57,530	19
44	52.8	346	<.50	Unlimited	179,790	0
42	50.4		"	"	"	0
40	48.0		"	"	"	0
		[From Fig. 22-3]		[From Table 22-4]		[Col. 6 ÷Col. 5 × 100]

Total = 83

*M.R. = Modulus of Rupture for 3rd pt. loading.
**Total fatigue resistance used should not exceed about 125%.

Fig. 22-4 Design procedure and calculations.

Section 22.8.2. Dowels are most commonly used as the load-transfer device. The sliding end of the dowel is coated or lubricated to permit free movement. Dowels are installed at middepth of the slab.

Experience has shown that dowels, or other load-transfer devices, are not needed under certain favorable conditions. The need for dowels depends on the amount of heavy truck traffic. For low-truck-traffic facilities such as secondary roads, residential streets, and parking lots, dowels are not needed, since load transfer by aggregate interlock is sufficient. Several state highway departments use undoweled joints in combination with stabilized or lean concrete subbases and short joint spacings of 15 to 20 ft or less. These subbases provide joint support[22-16] because their rigidity offers considerable resistance to deflection, bending and shear. This adds to the aggregate-interlock load-transfer capacity.

2. *Transverse Construction Joints.* Transverse construction joints are designed for planned interruptions to paving operations such as at the end of each day's paving or for emergency interruptions caused by inclement weather or equipment breakdown.

Planned construction joints are installed at a normal joint interval and consist of a butt-type joint as shown at the top of Fig. 22-6. In an emergency, such as equipment breakdown or non-delivery of materials, it may be necessary to install a joint that is not located at the regular joint spacing. There may be two types of these emergency joints:

1. For lane-at-a-time construction, a keyed (tongue-and-

TABLE 22-5 Spacing of Contraction Joints for Unreinforced Pavements

Type of coarse aggregate	
Crushed granite	25–30 ft
Crushed limestone	20–30 ft
Crushed flinty limestone	20–25 ft
Calcareous gravel	20–25 ft
Siliceous gravel	15–20 ft
Gravel less than ³/₄-in. size	15–20 ft
Slag	15–20 ft

groove) joint is required as shown at the bottom of Fig. 22-6. This joint is tied with deformed steel tie-bars to prevent joint movement that would otherwise cause sympathetic cracking in the adjacent lane.

2. For full-width paving, either a doweled butt joint or a tied keyed joint may be used.

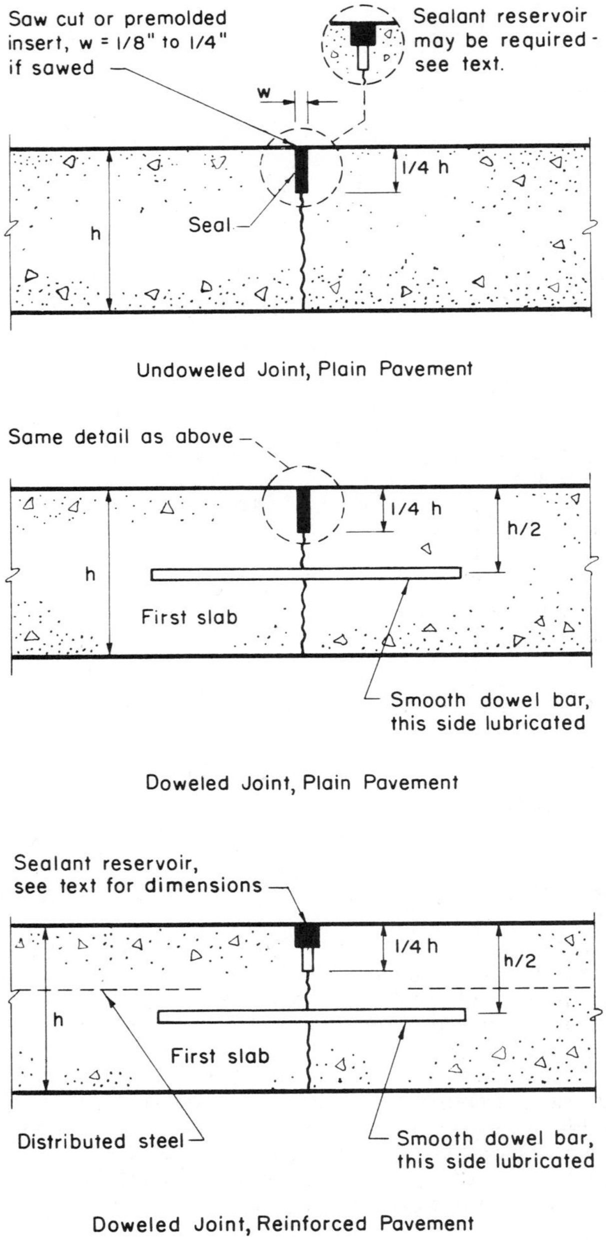

Fig. 22-5 Contraction joints.

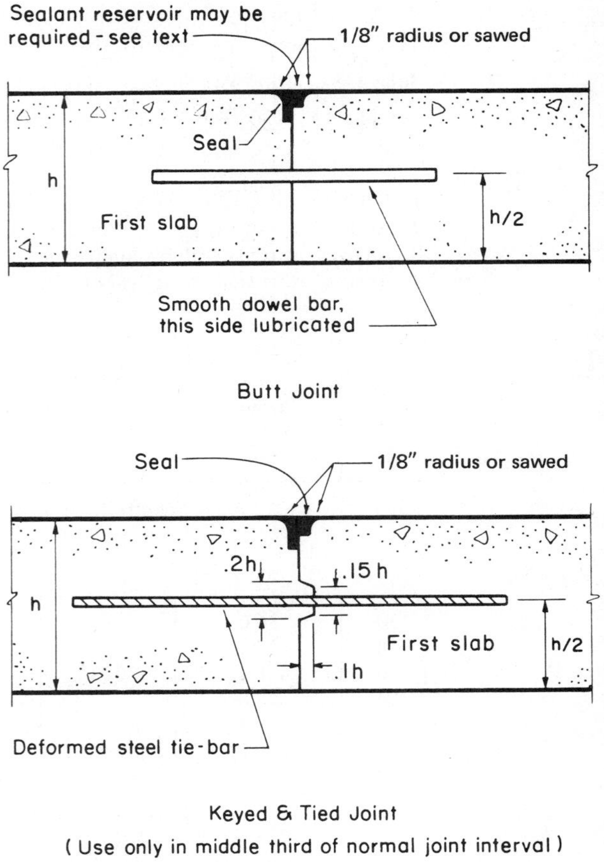

Fig. 22-6 Construction joints.

Whenever construction joints are tied, it is essential to edge or saw the joint so that spalling will not develop.

3. Expansion Joints. Design details for expansion joints are shown in Fig. 22-7. Studies[22-51–22-53] of pavements in service have shown that, except at structures and unsymmetrical intersections, expansion joints are not needed if:

1. The pavement is built with materials that have normal expansion characteristics.
2. The pavement is constructed during periods when temperatures are well above freezing.
3. Contraction joints are closely spaced so that intermediate cracks do not form.
4. Contraction joints are properly maintained to prevent infiltration of incompressible materials.

22.5.3 Longitudinal Joints

For control of longitudinal cracking, longitudinal joints are usually spaced at 12 ft to coincide with traffic lane markings. Recommended details for longitudinal joints are illustrated in Fig. 22-8.

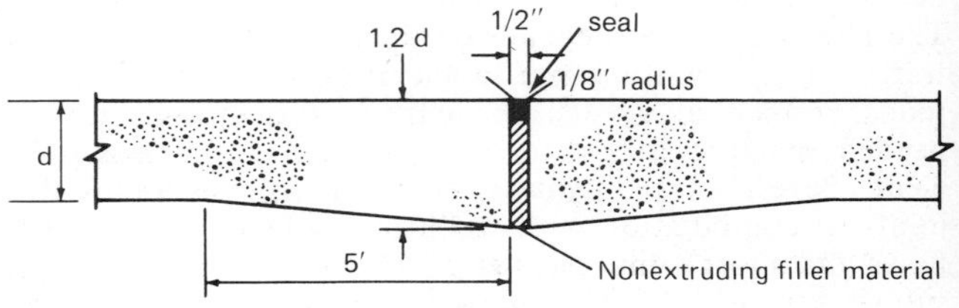

Fig. 22-7 Expansion joint.

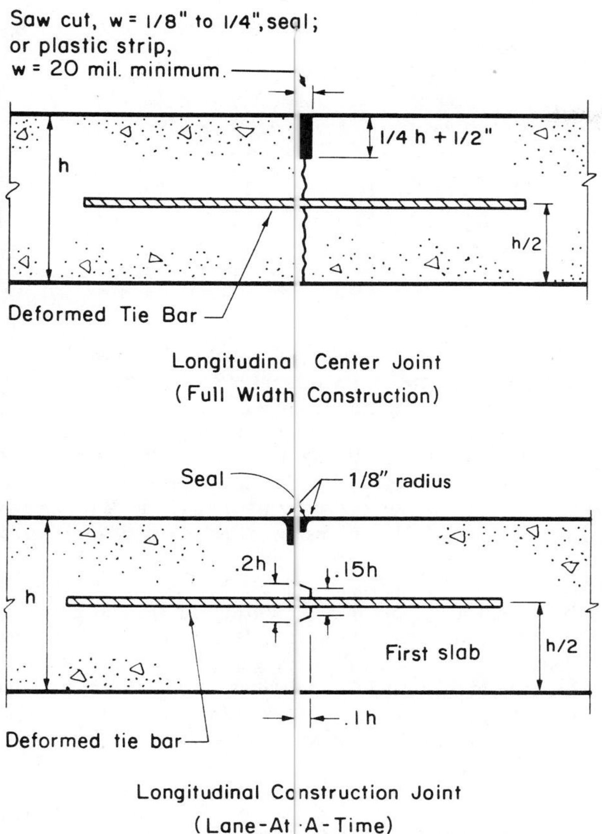

Fig. 22-8 Longitudinal joints.

Where two or more pavement lanes are constructed at one time, a weakened plane joint as shown at the top of Fig. 22-8 is specified. This is normally used for full-width construction and is called a longitudinal center joint. In these joints the groove is made by inserting a strip of plastic, metal, or other material in the fresh concrete, or by sawing a groove after the concrete has hardened. To ensure that the pavement cracks at the joint, it is essential that the depth of the groove be at least one-fourth of the slab thickness plus 0.5 in. Some agencies specify one-third of the slab thickness.

A keyed joint is specified for lane-at-a-time paving and where the total pavement width is not placed in one operation. This is called a longitudinal construction joint and is shown at the bottom of Fig. 22-8.

Both of these are hinged joints tied together with deformed tiebars that hold abutting slabs together, thus assuring load transfer at the joint by aggregate interlock, in the first case, and by the keyway, in the latter case. Recommended tiebar sizes and spacings for these joints are given in Section 22.8.3.

Tiebars may be omitted from longitudinal construction joints where there is enough transverse confinement, such as curbs and gutters, or paved shoulders at pavement edges, to prevent slab separation. Tiebars are also omitted from the interior joints of parking lots and other wide pavement areas, since these are confined by surrounding slabs. The usual practice is to tie only the longitudinal joints within 37.5 ft of the outside pavement edge.

22.5.4 Jointing Layout for Pavement Intersections

A typical jointing arrangement for pavement intersections is shown in Fig. 22-9. This figure illustrates the location of various joint types and the application of the principles presented previously in this section.

22.5.5 Joint Shapes Required for Effective Joint Sealing

As a pavement expands and contracts due to temperature and moisture changes, the pavement joints open and close. As a result, sealants placed in the joints must be capable of withstanding this repeated extension and compression[22-54–22-57] while maintaining their function of preventing the intrusion of incompressible solids. In order to maintain an effective seal, the joint width must be large enough so that later changes in joint width will not put undue strain on the sealant. To obtain the required joint width, the top of the joint may need to be sawed wider, forming a reservoir for the sealant material as shown in Fig. 22-10.

For poured joint sealants, the shape factor (depth-to-width ratio) of the sealant reservoir has a critical effect on the sealant's capacity to withstand extension and compression. Within certain practical limitations, the lower the

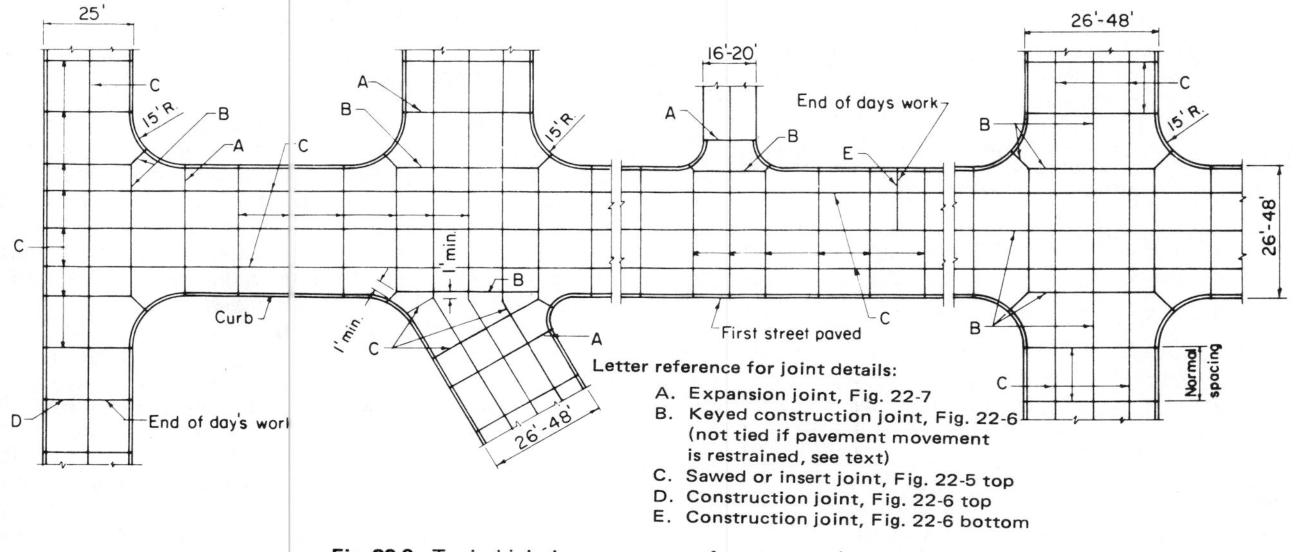

Fig. 22-9 Typical jointing arrangement for pavement intersections.

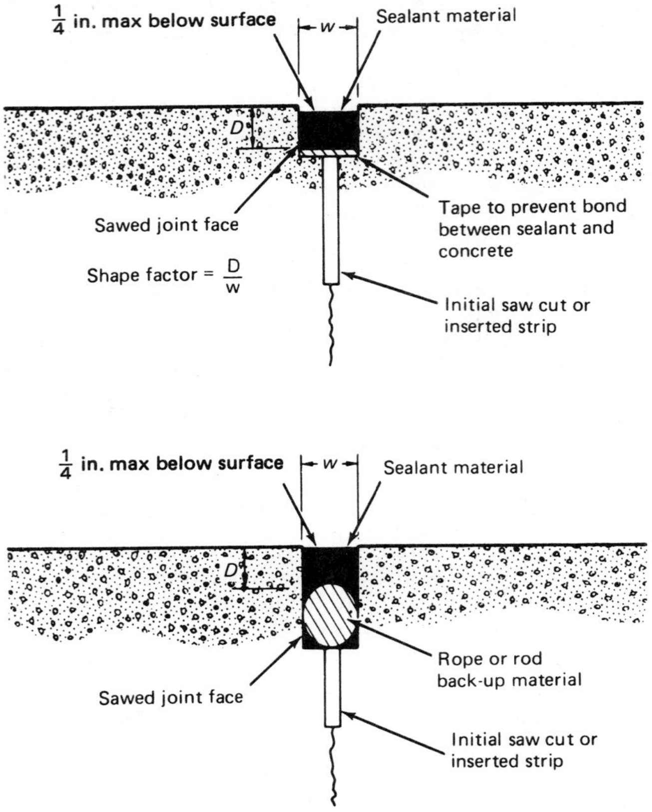

See text for dimensions of sealant reservoir

Fig. 22-10 Joint sealant reservoir and shape factor.

depth-to-width ratio, or shape factor value, the lower the strain on the sealant under a given joint movement. The required shape factor will depend on the properties of the sealant and the amount of joint movement; the latter in turn is related to the panel length or joint spacing and to the maximum expected seasonal temperature change in the slab.* Table 22-6 lists recommended depths and widths of the sealant reservoir for poured sealants frequently used such as those meeting current federal and ASTM specifications. A stiff, self-adhering plastic strip is applied to the bottom of the reservoir to break the bond between the sealant and the bottom concrete surface.

Frequently, a butyl or polypropylene rope is placed in the bottom of the sealant space to break bond and prevent loss of sealant into the crack below the joint filler. In this

*Actual joint movements are reported[22-55, 22-56] as approximately the same as theoretical values when a value of 5×10^{-6} in./in./°F is used for the thermal coefficient of concrete.

TABLE 22-6 Joint Width and Depth for Poured Sealants

Joint Spacing, ft	Sealant Reservoir Shape	
	Width, in.	*Depth of Sealant, in.*
15 or less	1/4	3/4
20	3/8	3/4
30	1/2	3/4
40	5/8	1

TABLE 22-7 Joint and Seal Width for Preformed Seals

Joint Spacing, ft	*Joint Width, in.*	*Sealant Width, in.*
20 or less	1/4	7/16
30	3/8	5/8
40	7/16	3/4
50	1/2	7/8

case, it is necessary to saw the reservoir deeper an extra amount equal to the rope diameter so that the shape factors of the sealant listed in Table 22-6 are maintained.

For preformed, compression-type seals, recommended joint width and width of seals are listed in Table 22-7. The depth of saw cut is such that the compression seal is installed about 0.25 in. below the pavement surface. With special equipment, the seals are installed, without stretching, to a compressed width of about half of their uncompressed width.

Tables 22-6 and 22-7 are presented for general guidance. Sealant manufacturers publish aids for selecting reservoir dimension to suit their products. For residential city streets and parking lots, with joint spacings of 15 ft or less, movement at the joints is small, resulting in small sealant strains, so that the joint widths shown above may be lessened without serious consequence.

22.6 THICKNESS DESIGN FOR AIRPORT PAVEMENTS

In determining the thickness of concrete pavements for airports, several major factors are considered: the types and weights of aircraft, and their effect on flexural stresses in the slab; the flexural strength of the concrete; the supporting strength of the subgrade (and subbase if one is used); the type of facility being designed (taxiway, runway, apron, hangar floor, etc.); and the frequency of aircraft operations expected on the pavement.

Complete descriptions of the current design procedures of several agencies are given in Refs. 22-58 through 22-60 and 22-67; the methods and results are quite similar as pointed out in Ref. 22-61. The procedure of the Portland Cement Association is described in this section.

22.6.1 Load Stresses

Flexural stresses caused by aircraft loads are computed on the assumption that the load is placed near the middle of a large expanse of pavement. This can be assumed when adequate load transfer is provided at all joints. Aircraft wheel loads do not travel near the free outside edges of pavements; if free edge butt joints are used in the interior of a pavement expanse, the edge thickness is increased to compensate for the lack of load transfer.

The flexural stress caused by a given aircraft load is determined by the use of influence charts[22-24] or by computer program.[22-62] Aircraft manufacturers publish the required data on loads, wheel spacings, and tire contact areas that are needed in the analysis. Design charts are then prepared for the specific aircraft that will use the pavement facility. Charts for several aircraft are shown in Figs. 22-11 through 22-13.

Design charts for most civil and military aircraft are published by the Portland Cement Association.

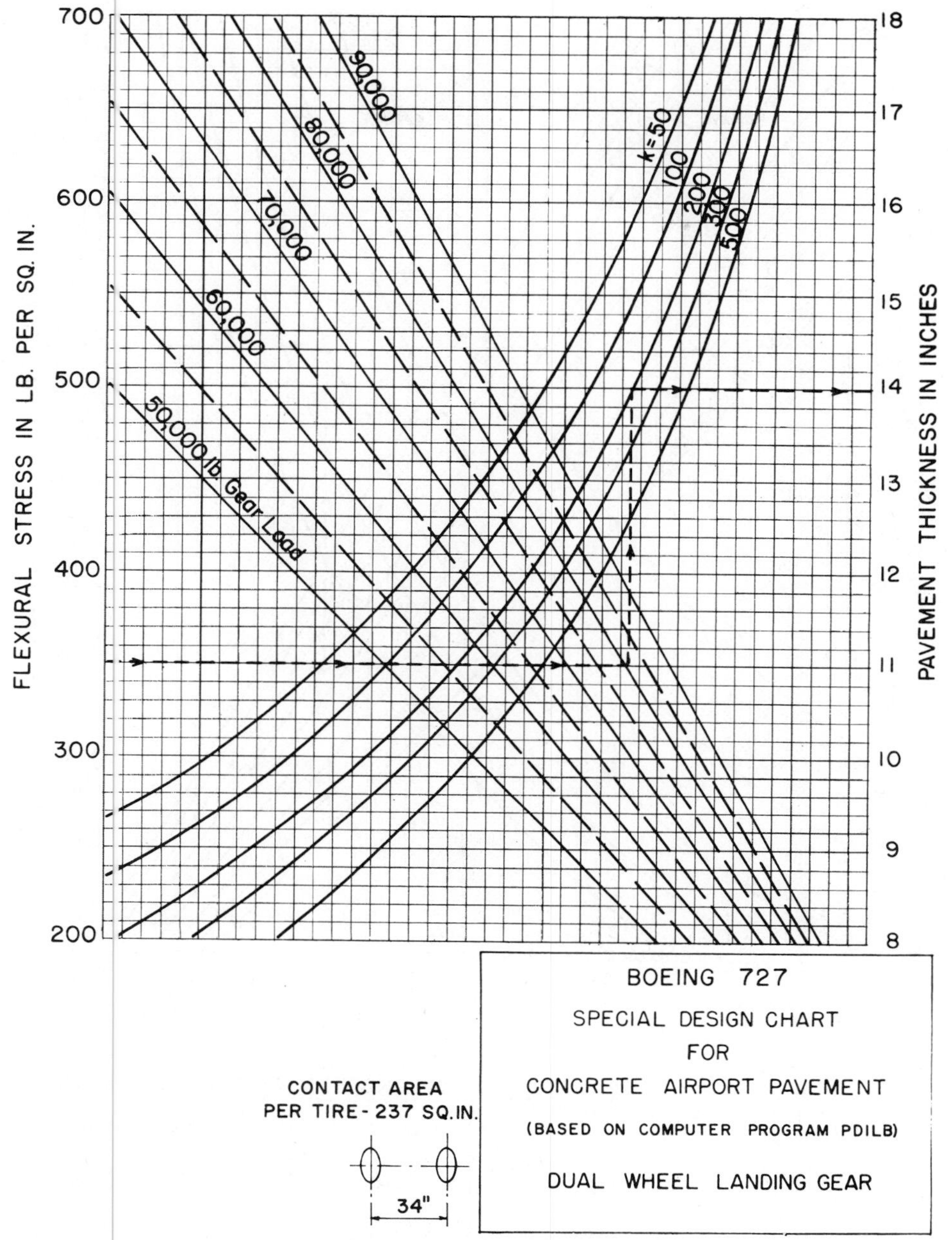

Fig. 22-11 Concrete airport pavement design chart for Boeing 727.

The design of pavements for lightweight, general aviation aircraft is discussed in Ref. 22-63.

22.6.2 Flexural Strength

Concrete strength is determined by third-point flexural strength tests (ASTM C78) performed at 90 days for airport pavement design, rather than the 28-day strength test used for highway design. Even in 90 days, an airport pavement is subjected to much fewer operations with less channelization of traffic compared to highway pavements for a 28-day period.

Because of the continued strength increase during the early years of pavement life, the modulus of rupture soon exceeds the 90-day value. Figure 22-14 shows a conservative relationship between flexural strength and age of concrete.

22.6.3 Subgrade–Subbase Strength

The supporting strength of the subgrade and subbase, Westergaard's modulus of reaction (k), is determined by plate loading tests or by correlation to soil type or laboratory strength tests as described in Section 22.4.2. Where subbase layers* will be used, the k-value on top of the subbase layer may be estimated from Tables 22-2 and 22-3. On large projects, these values should be verified by plate-loading tests.

22.6.4 Safety Factor

The safety factors (ratios of working stress to design modulus of rupture) used for airport pavement design represent

*For pavements serving heavy, multiple-wheel gear aircraft, the effect of stabilized subbases is discussed in Appendix B of Ref. 22-58.

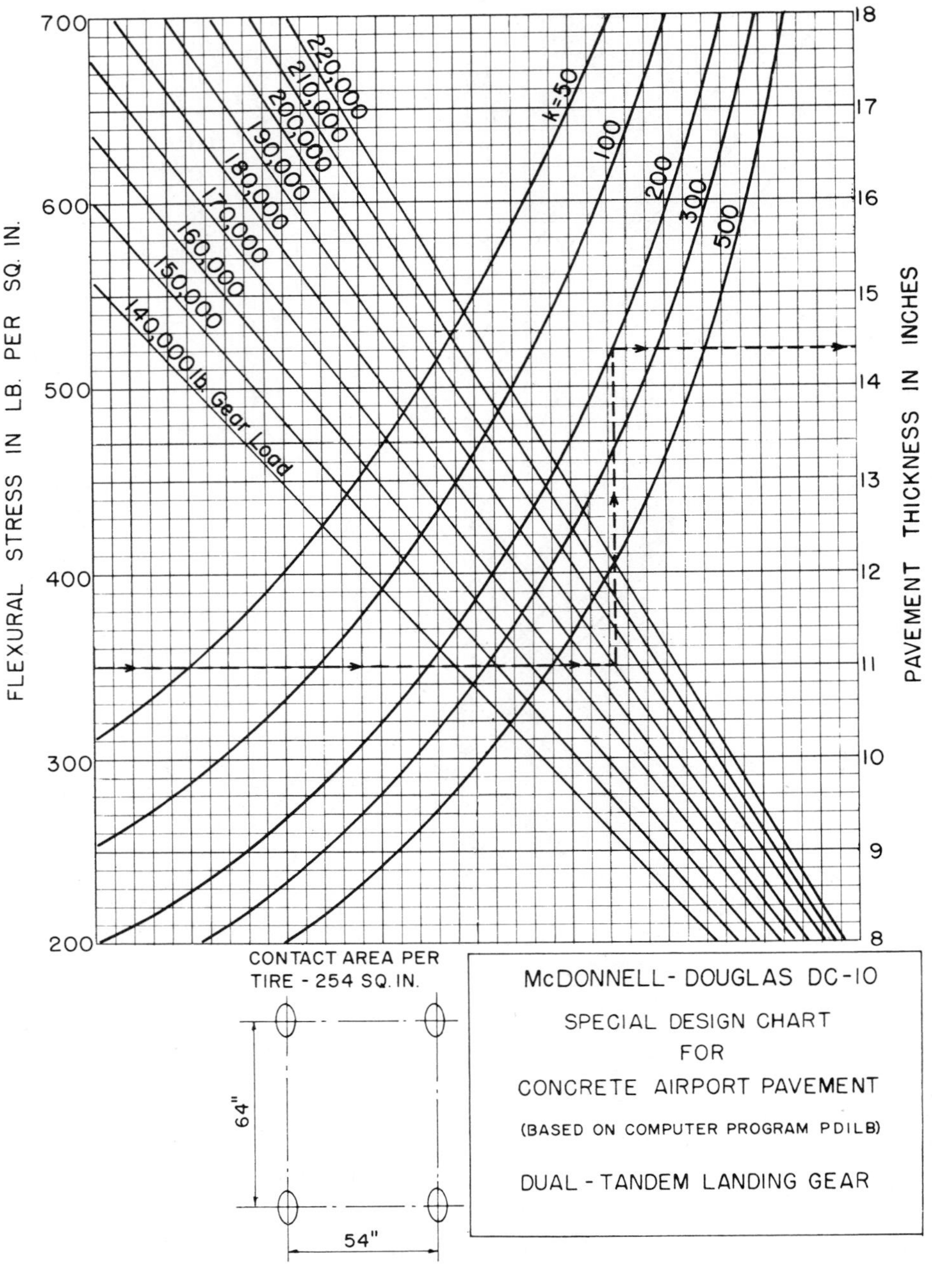

Fig. 22-12 Concrete airport pavement design chart for DC-10.

an assessment of the expected frequency of traffic operations and the channelization of these on runways, taxiways, and aprons. Undoubtedly, one of the most important factors in airport pavement design is the estimation of the loads of future, heavier aircraft and increased numbers of operations that the pavements will serve. Good design practice requires the gathering of data on expected future operating and load conditions from several sources, including commercial airline forecasts, information from airport operating officials, and loading data for future aircraft projected by aircraft manufacturers.*

Based on this information, a conservative safety factor is selected and used to determine the allowable working stress in the design charts. The following ranges of safety factors are recommended:

Installation	Safety Factor
Aprons, taxiways, hard standings, runway ends (for distance of 1000 ft), hangar floors	1.7–2.0
Runways (central portion)	1.4–1.7

The lower safety factors for runways are permissible because most runway traffic consists of fast-moving loads that are partly airborne. In addition, the aircraft wheel loads are distributed transversely over a wide pavement area in such a manner that the number of stress repetitions in any one spot is quite small. Also, the number of stress repetitions at one location on a runway will be much lower than on taxiways, even on a one-runway airport.

*When a specific forecast is made of the mixed aircraft that will operate during the design life, fatigue methods discussed in Appendix A of Ref. 22-58 may be used for the design or evaluation of pavements.

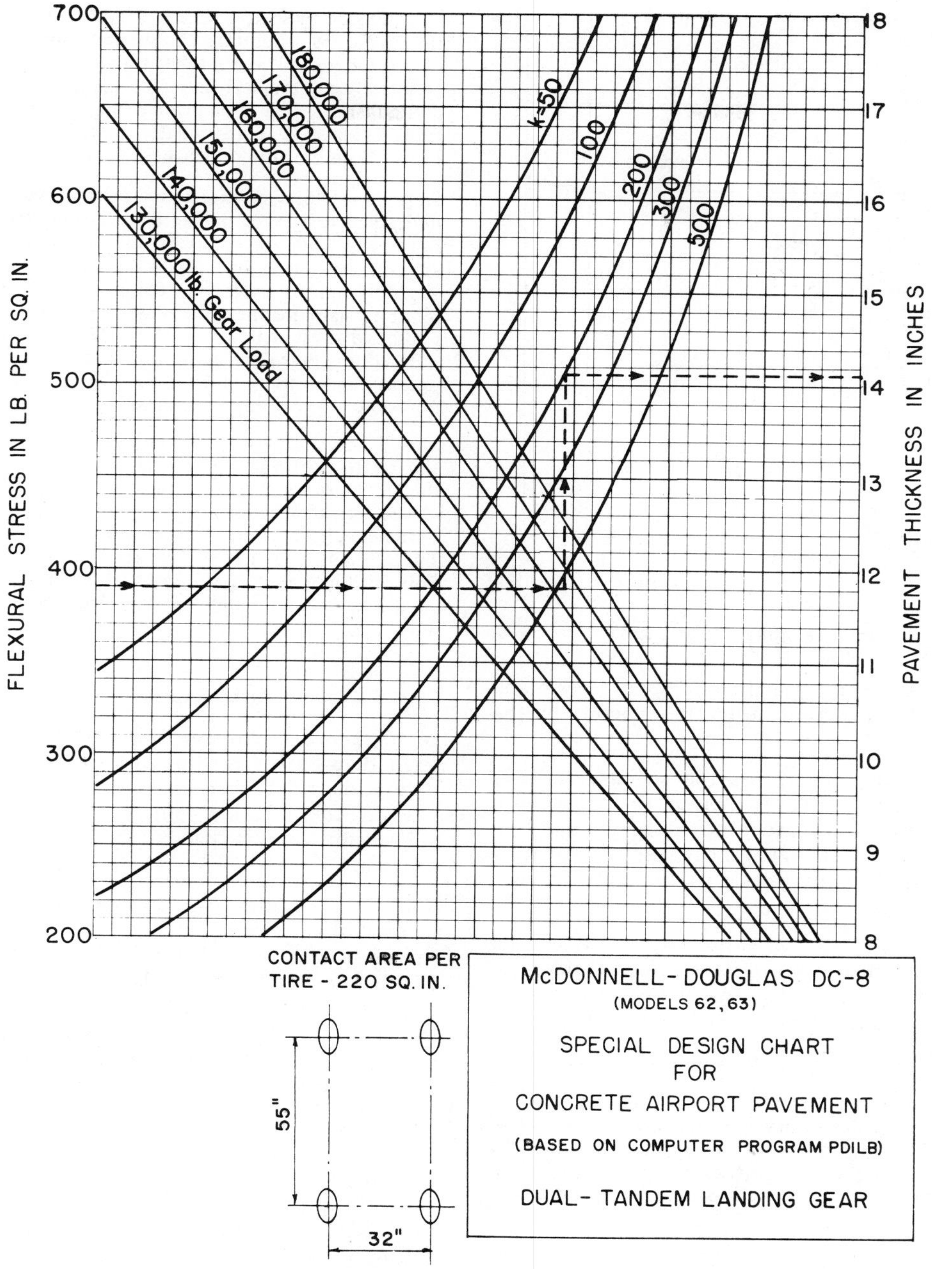

Fig. 22-13 Concrete airport pavement design chart for DC-8.

Where taxiways intersect runways, the runway for a short distance each way should be of the same thickness as the taxiway. Any portions of runways which will serve as taxiways must also be designed as taxiways. Because of traffic channelization and spot parking on aprons, the frequency of stress repetitions in one spot may approach that in the taxiways.

At airports with a large number of operations by planes with critical loads, safety factors near the top of the suggested range should be used. On fields with only occasional operations by planes with critical loads, safety factors near the bottom of the range should be used. Those fields with a few daily operations by these loads should use some intermediate value. Even though there may be a large number of operations by lighter aircraft, the fatigue limit of the concrete will not be used up. The safety factor of 2.0 results in pavement adequate for an unlimited number of operations.

For heavy-duty runways serving large volumes of traffic, designers sometimes select a "keel-section" design where the center section of the runway is thicker than the outside pavement edges.

22.6.5 Design Procedure

Determination of slab thickness is done in the following steps:

Step 1. The k-value is determined by plate loading tests or by correlation to subgrade soil test data (see Section 22.4.2).

Step 2. A careful estimate of future, as well as present, operating and local conditions is made, and an appropriate, conservative safety factor is selected as discussed in Section 22.6.4.

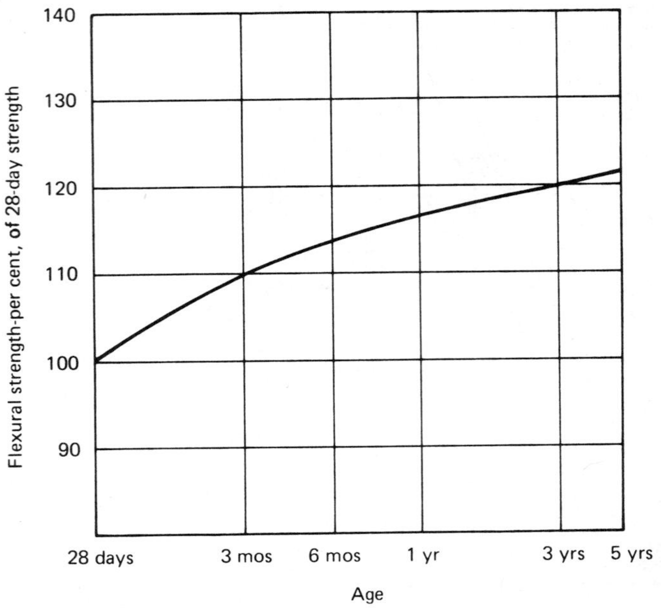

Fig. 22-14 Flexural strength-age relationship.

Step 3. Working stress for a specific aircraft is determined by dividing the modulus of rupture of the concrete by the safety factor chosen.

Step 4. From the design chart for the specific aircraft, determine the pavement thickness. Enter the chart with the allowable working stress and proceed to the right to intersect the gear load line; then proceed up or down to the k-value; from this point project to the right to read slab thickness.

Step 5. Repeat the process for other aircraft of critical loads, again selecting new, appropriate safety factors for the level of operations expected for these aircraft, and select a design thickness for the must critical condition.

The following example illustrates the use of the design procedure.

EXAMPLE 22-1: Assume that a new runway and taxiway are to be designed to serve frequent operations of aircraft of which the B-727 and DC-10 produce the most critical loading conditions. In addition, the runway is expected to carry occasional operations of DC-8-63's.

SOLUTION:

Step 1. The existing subgrade is a pumpable sandy clay soil on which several plate loading tests have indicated a k-value of 170 pci. As discussed in Section 22.3.2, a subbase will be required to prevent the traffic from pumping the subgrade soil. Since it is not necessary or economical to use a thick subbase layer, 6 in. is selected as an effective and practical subbase thickness. Based on these data, a preliminary design k-value of approximately 200 pci is estimated from Table 22-2. (This is later verified by plate tests on subbase test sections constructed during the design stages of the project.)

Steps 2-4. Data on concrete strengths made with local aggregates indicate that it is reasonable to specify a 90-day design modulus of rupture of 700 psi. Based on the previous discussion, appropriate safety factors are selected and shown in columns 4 and 7 of Table 22-8. Working stresses are computed (columns 5 and 8), and the required slab thicknesses (columns 6 and 9) are determined from published stress charts (Fig. 22-11–22-13 in this example).

Based on these data, a slab thickness of 14.5 in. is required for the taxiway and runway ends, while 13.0 in. is required for the central portion of the runway (thickness increased to next half-inch).

Step 5. Pavement stresses induced by other, less critical aircraft in the traffic forecast are next determined by the use of published design charts for these specific aircraft. (For this purpose, use of the charts is in the reverse order to that indicated previously; start with pavement thicknesses on the right-hand scale, and find slab stress on the left-hand scale.) If stresses for these other aircraft are less than 350 psi, these aircraft will not fatigue the pavements (safety factor of 2.0 or greater), and the slab thicknesses established above are adequate.

22.6.6 Pavement Evaluation

In addition to designing new pavements, it is sometimes desired to evaluate existing airport pavement installations to determine what gear loads can be carried without overstressing the slab.

The steps required in making such an evaluation are as follows:

Step 1. Determine the k-value, or values, of the subgrade beneath the present pavement. This may be done by loading the existing pavement as explained in Refs. 22-64 through 22-66, removing slab panels, and making plate-bearing tests directly on the subgrade or subbase course (see discussion of subgrade-subbase strength, Section 22.4.2). The results should be based on several tests, the number depending on the engineer's judgment concerning the requirements of the site.

Step 2. Determine the modulus of rupture of the concrete by cutting beams from the pavement and making flexural tests. If this is impractical, strengths may be estimated by making suitable age–strength increase on the 28-day strengths taken from construction records. Figure 22-14 may be used to estimate the increase in flexural strength with age. If there is any evidence of deterioration of the

TABLE 22-8 Example Calculations for Thickness Design

Design k-value - 200 pci

Design MR = 700 psi

Aircraft	Gear Load, Lbs.	Operations	Pavement Facility: Taxiway & Runway Ends: Safety Factor	Working stress, psi (MR ÷ Col. 4)	Slab Thickness, in.	Runway, central portion: Safety Factor	Working stress, psi (MR ÷ Col. 7)	Slab Thickness, in.
(1)	(2)	(3)	(4)	(5)	(6)	(7)	(8)	(9)
B-727	80,000	Frequent	2.0	350	14.0	1.7	412	12.5
DC-10	190,000	Frequent	2.0	350	14.4	1.7	412	12.7
DC-8-63	165,000	Occasional	1.8	389	14.1	1.5	467	12.4

concrete, specimens cut from the pavement should be tested.

Step 3. Find the thicknesses of the different installations (runways, taxiways, etc.) from construction plans and records or from concrete cores taken with a core drill.

In determining the strengths of the concrete and subbase–subgrade, use of correlative materials tests is increasing. These include compressive strength tests and tensile splitting tests on concrete specimens; dynamic, nondestructive tests on the surface of the pavement; and field and laboratory soil strength tests on the subgrade and subbase materials. These data supplement and reduce the number of conventional tests required.

With the thickness and flexural strength of the concrete and the *k*-value for the subgrade known, the load-carrying capacity of the pavement may be determined from the design charts. Either an allowable load for a given safety factor, or the safety factor that will be in effect under a given load, may be determined.

22.6.7 Future Pavement Design

Designers of airport pavements are faced with the need for evaluating existing design procedures and experience to apply to designs for future aircraft loadings and operational conditions.

Current design procedures are based on the prevention of flexural failure of the concrete slab. The safe limits or working stresses have been established by considerable performance experience on both civil and military airports for a wide range of slab thicknesses and aircraft loadings—extending to slab thickness of from 18 to 26 in. on military airfields designed for the more than 500,000-lb gross weight B-52 aircraft.

This experience indicates that heavy loads, per se, and large numbers of load repetitions are conditions that have been successfully handled with present design methods. This experience suggests that the flexural stress is a valid and critical design criterion—that when these stresses are kept within defined limits, other possible modes of distress such as slab deflections, subgrade pressures, and shearing stresses at joints are also kept within safe, but undefined, limits.

However, this experience is for single, dual, and dual-tandem gear aircraft—where an effective load area represented by four wheels, or fewer, affects the pavement and subgrade. For future aircraft and airport pavements, there are three factors that greatly increase the effective size of loaded area:

1. Increase in number of wheels so that as many as 8 or 12 wheels may be acting together to affect stresses and deflections.
2. Increase in pavement thickness; stiffer pavement systems will spread out the zone of interaction to encompass additional wheel loads.
3. Increase in size of tire contact areas.

As the gross weights of aircraft have increased, increases in the number of wheels and size of tires have been made to satisfy the allowable flexural stress in the concrete slab. With further increases in the number of wheels interacting, the question arises whether deflections and subgrade strains should be considered in addition to flexural stresses. It thus seems appropriate, for heavy, multiwheel aircraft, to also consider deflections and subgrade pressures and, through experience and research, establish safe working values.

As an aid to designers, deflection charts may be published by PCA for heavy, multiwheel aircraft that will cause critical deflection. The magnitudes of subgrade pressures and comparative load–transfer requirements for different aircraft can be closely related to relative deflection values. For this reason, deflections may be selected as the supplemental design factor.

22.7 JOINT DESIGN FOR AIRPORT PAVEMENTS

A discussion of the function of joints and the general basis for joint design is given in Sections 22.1 and 22.5 and cited references. Jointing practices for airport pavements are given in the design manuals of the Portland Cement Association,[22-58] the U.S. Army Corps of Engineers,[22-59] the Federal Aviation Administration,[22-60] and the U.S. Navy.[22-67]

The purpose of joints in concrete pavements is to control transverse and longitudinal cracking due to restrained shrinkage, and the combined effects of restrained warping and aircraft loads. Joints are also used to divide the pavement into suitable increments for construction, and to accommodate slab movements at intersections with other pavements and at structures.

Adequate transfer of loads across the joint must be provided to satisfy basic thickness design principles. Depending on the type of joint, load transfer is obtained by dowels, keyways, or the aggregate interlock of slabs with short joint spacings. Substantial joint support is also provided by the use of a stiff, cement-treated or lean concrete subbase under the slab. Where no load transfer mechanism is provided, the joint edges are thickened to compensate so that stresses and deflections at these free edges are controlled within the design limits.

Joint widths must be made large enough that the joints may be effectively sealed, as discussed in Section 22.5.5.

22.7.1 Transverse Joints

In airport pavements, especially aprons, intersections, and turnarounds, it is sometimes difficult to distinguish between longitudinal and transverse joints. Transverse joints are considered to be those joints perpendicular to the lanes of construction; their direction within the pavement will depend upon the plan of construction.

1. Contraction Joints, Unreinforced Pavements. The spacing of transverse contraction joints, as well as longitudinal joints, depends on the shrinkage properties of the concrete, subgrade soil conditions, climatic conditions, and slab thickness.

Table 22-5, for unreinforced (plain) pavements, may be used as a guide in the selection of joint spacings to reflect the shrinkage properties of concretes made with different aggregate types. These spacings are for average soil and climatic conditions and for slab thicknesses of 12 in. and less.

Pavement performance shows that the allowable joint spacing for crack control increases directly with slab thickness. Unless environmental and shrinkage conditions are especially severe, a joint spacing (in ft) not greatly exceeding twice the slab thickness (in in.) will control cracking. For example: a 15-ft spacing would probably be used for an 8-in slab under average conditions, and 25 ft for a 12-in. slab. This rule of thumb may be modified according to local experience.

Panels 25 by 25 ft have given very satisfactory service for thick pavements under extremely heavy traffic conditions. Generally, it is desirable to have panels with approximately equal transverse and longitudinal joint spacings. Experience[22-59] indicates that adequatly designed slabs are not

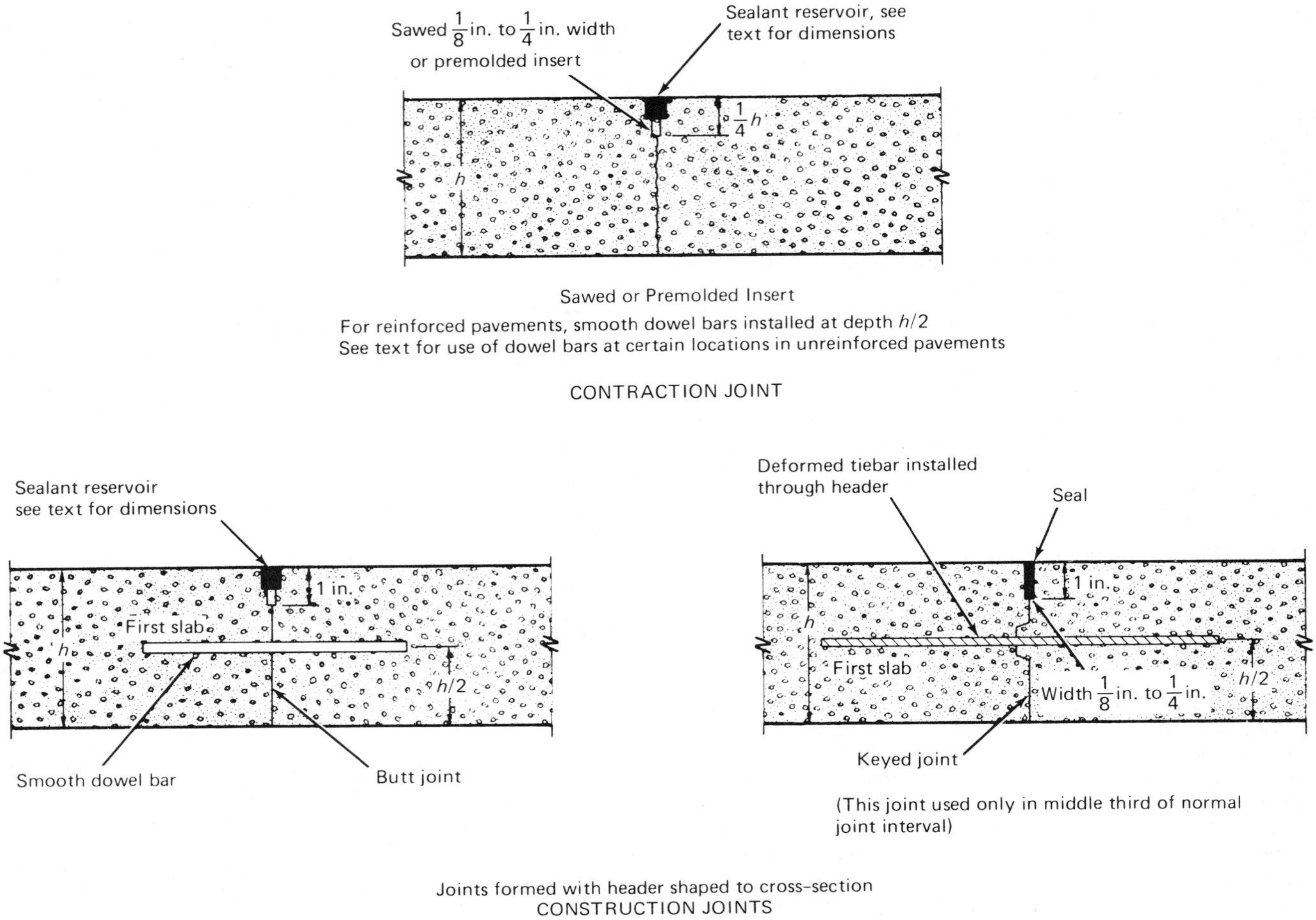

Fig. 22-15 Transverse joints, airport pavements.

likely to develop an intermediate crack if the length-to-width ratio does not exceed 1.25.

Transverse contraction joint details for unreinforced pavements are shown at the top of Fig. 22-15. Load transfer at contraction joints will be provided by the aggregate interlock between the fractured faces below the groove. This will be true only if the joints are kept quite tight by omission of expansion joints within the pavement and by using a contraction-joint spacing not exceeding those recommended above.

Airport performance studies have shown that dowels are not required at transverse joints with short joint spacing. However, dowels must be placed across contraction joints near the free ends of the pavement and near expansion joints if any are used. Observations have shown that for a distance of about 100 ft back from each free end and 60 ft back from each expansion joint, the joints will gradually open to a point where the aggregate interlock may not be effective. Dowels are required in these few joints to ensure load transfer. The size and spacings of dowel bars are given in Section 22.8.2.

2. Contraction Joints, Reinforced Pavements. Reinforced pavements (mesh-dowel pavements) have been used with joint spacings ranging from 30 to 75 ft. Design of these joints is the same as contraction joints for unreinforced pavements, except that dowels are required because joints open wider, making load transfer by aggregate interlock less effective. (Design for the amount of reinforcement and for dowels is given in Section 22.8.)

Based on the performance of mesh-dowel pavements, it is desirable to limit joint spacing to about 30 to 40 ft for airport pavements less than 12 in. thick and to about 50 ft for thicker pavements. Importantly, the performance of all pavements depends strongly on the effectiveness of the joint seals. For longer joint spacings, joint seals are usually not as effective because of the greater fluctuations in joint width.

3. Construction Joints. Transverse construction joints are necessary at the end of each day's run or where paving operations are suspended for 30 minutes or more. If the construction joint occurs at or near the location of a transverse contraction joint, the butt-type joint with dowels is recommended. If the joint occurs in the middle third of the normal joint interval, a keyed joint with tiebars should be used. These construction joints should be constructed in accordance with one of the details shown at the bottom of Fig. 22-15.

22.7.2 Longitudinal Joints

The spacing of longitudinal joints depends on the construction equipment used, the overall width of the pavement, and pavement thickness.

In the past, equipment has been best suited to paving widths of 20 to 25 ft, and these joint spacings have been commonly used for pavement thicknesses of 12 in. or more. Intermediate longitudinal joints are necessary in thinner pavements to prevent the formation or irregular longitudinal cracks. Pavement performance experience indicates the following as a guide to longitudinal spacings:

1. For all pavements thinner than 12 in., and for pavements 12 to 15 in. thick carrying channelized traffic,

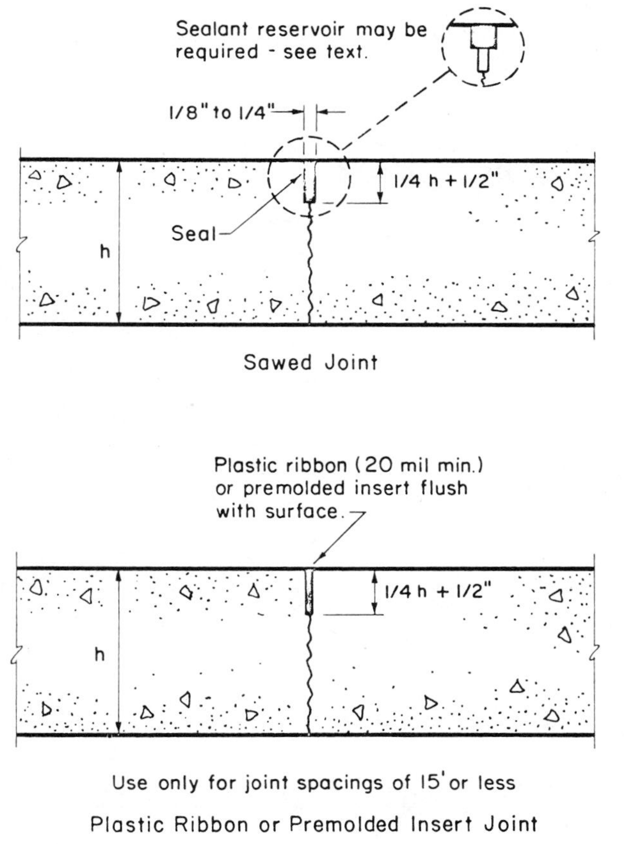

Note: Deformed tiebars at depth h/2 should be used across these joints where called for in text.

Fig. 22-16 Intermediate longitudinal joints, airport pavements.

longitudinal joints should not be more than 12.5 ft. apart.

2. For pavements thicker than 15 in., and for pavements 12 to 15 in. thick not carrying channelized traffic, intermediate joints at 12.5 ft are not required. In this case, convenient joint spacings are selected not to exceed contraction joint spacings for unreinforced pavements.

Recent developments in equipment permit paving widths up to 50 ft. This provides the opportunity to select joint spacings to satisfy specific design situations. For example, 37.5- or 50-ft-wide construction lanes may be used with intermediate joints at 12.5, 18.75, or 25 ft, depending on pavement thickness. These intermediate joints are often called center joints and may be the surface-groove type such as a saw kerf or premolded insert.

The longitudinal joints within 37.5 ft of pavement edges have generally been hinged (held together by deformed steel tiebars). It is necessary to tie these outside joints to prevent progressive joint opening. Figure 22-16 illustrates the common types of intermediate longitudinal joints.

Longitudinal construction joints at the edge of each construction lane may be of the keyed (tongue-and-groove) type to provide load transfer at this location. Details of these joints are shown in Fig. 22-17. The keyed joint may be extruded with a slip-form paver or formed with a shaped metal strip attached to the forms to produce a groove along the edge of the slab. When the adjacent slabs are placed, the new concrete will form the key portion of the joint. Performance experience and research[22-68] have shown that it is important to use the keyway dimensions shown in Fig. 22-17, upper right. Larger keys reduce the strength of the joint and may result in keyway failures. The key must also be located in the middepth of the slab.

A keyway may be constructed along the outside edge of all pavement areas to provide for load transfer for future pavement expansions. A thickened outside edge may be used for the same purpose. One-half of a tiebolt may be installed to permit tieing the added lanes at a later date.

In narrow taxiways (75 ft or less) all longitudinal joints should be tied (provided with deformed tiebars or tiebolts) to prevent excessive openings and loss of load transfer. In wider pavement areas it is necessary to tie only longitudinal joints within 37.5 ft of the free edge. Recommended sizes and spacings of tiebars are given in Section 22.8.3.

22.7.3 Expansion Joints

When contraction joints are spaced as outlined previously, expansion joints are not required transversely or longitudinally within airport pavements except under special conditions. The practice of omitting expansion joints tends to hold the interior of the pavement areas in restraint; prevents cracks and joints from opening, and adds to the strength of the joint.[22-51]

Expansion joints (0.75 to 1.5 in.) must be provided between concrete pavements and all buildings or other airfield structures. Expansion joints may sometimes be required at intersections of runways, taxiways, and aprons (see Section 22.7.4). Expansion joints used within the pavement may be doweled or provided with thickened edges. Figure 22-18 shows types of expansion joints to be used in the pavement and between pavements and structures.

Requirements of filler materials for expansion joints are given in current specifications of ASTM, AASHTO, and federal agencies.

22.7.4 Jointing Arrangements for Runways and Intersections

Figure 22-19 shows typical jointing arrangements for concrete airport pavement. This figure illustrates the location of various joint types and shows where tiebars and dowels are required.

The layout of joints at pavement intersections in airports presents many unusual problems. Because of the large, irregularly shaped areas of pavement involved and the possibility of any angle of intersection between two or more facilities, it is impossible to establish any universal joint pattern. Figure 22-20 illustrates some typical intersections and shows suggested jointing layouts for each.

The main body of pavement (runway or apron) will not contain expansion joints but will be free to move at each end. Intersecting pavements (taxiways, runways, etc.) must be provided with expansion joints to allow for longitudinal expansion without damage to either pavement. Since expansion joints are omitted in the main body of pavement (runways or aprons), longitudinal movement may occur along both edges. Both the expansion joint requirements and the free longitudinal movement may be provided by the expansion joints arrangements shown in Fig. 22-20.

22.8 USE OF STEEL IN JOINTED PAVEMENTS

Steel in jointed concrete pavements may be used in the form of distributed steel (i.e., welded wire fabric or bar mats distributed throughout the concrete); and in the form of deformed tiebars and smooth dowels across certain

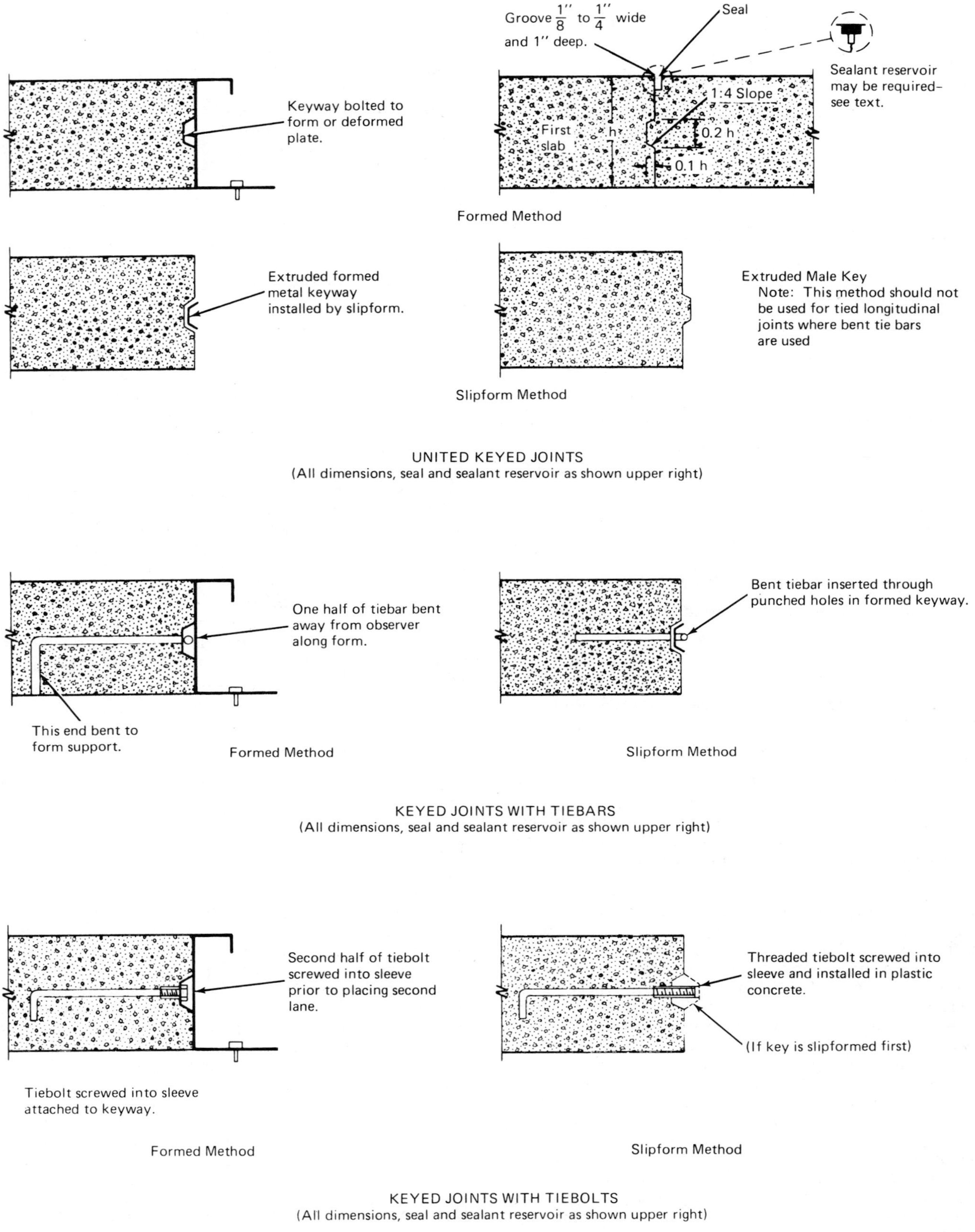

Notes: Tiebars or tiebolts are used only at certain locations - see text
Keyway and tiebars at depth h/2

Fig. 22-17 Longitudinal construction joints, airport pavements.

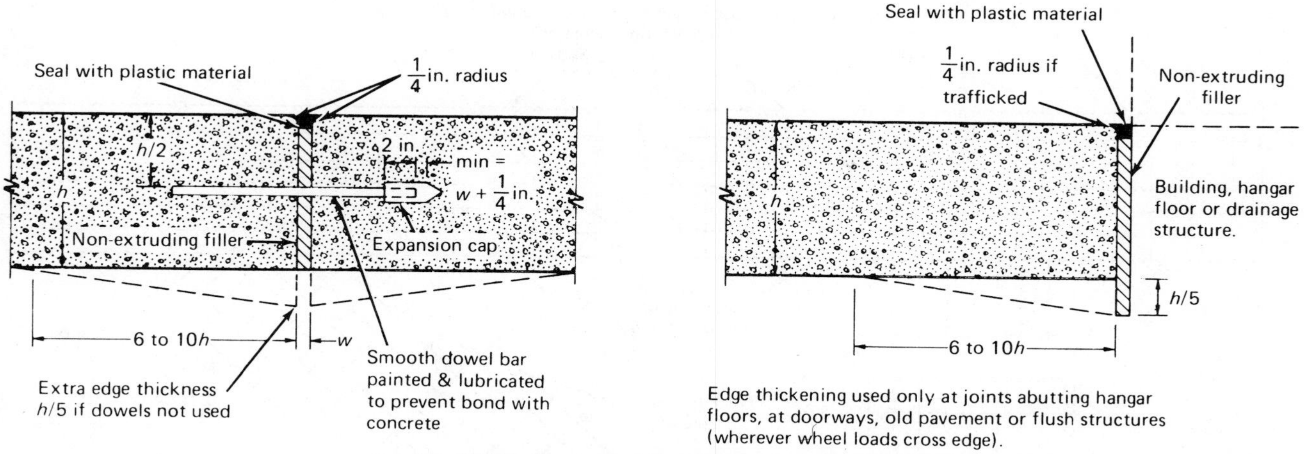
Seal with plastic material
$\frac{1}{4}$ in. radius
h/2
2 in.
min = $w + \frac{1}{4}$ in.
h
Non-extruding filler
Expansion cap
6 to 10h
w
Extra edge thickness h/5 if dowels not used
Smooth dowel bar painted & lubricated to prevent bond with concrete
Seal with plastic material
$\frac{1}{4}$ in. radius if trafficked
Non-extruding filler
Building, hangar floor or drainage structure.
h/5
6 to 10h
Edge thickening used only at joints abutting hangar floors, at doorways, old pavement or flush structures (wherever wheel loads cross edge).

Fig. 22-18 Expansion joints, airport pavements.

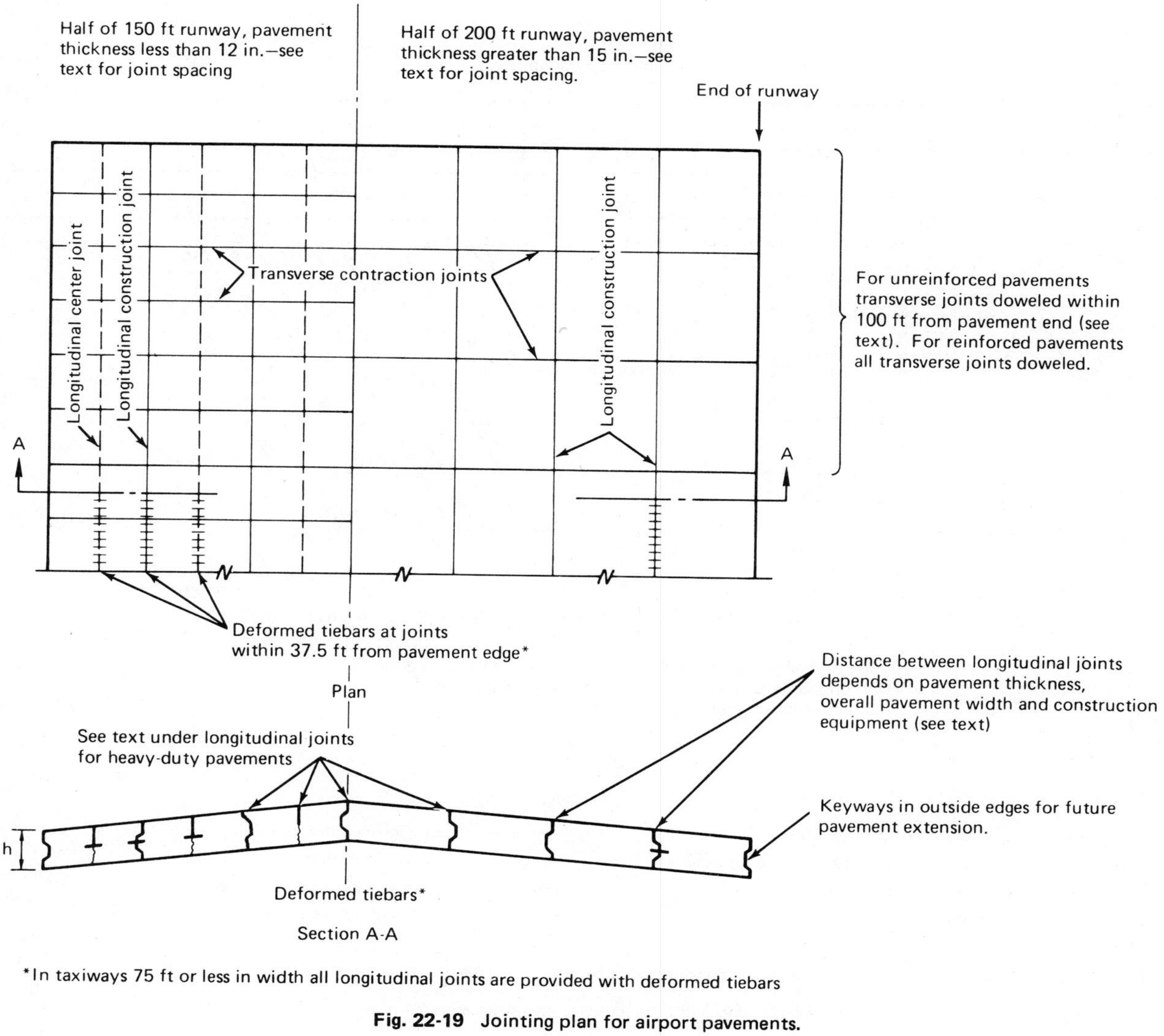
Half of 150 ft runway, pavement thickness less than 12 in.—see text for joint spacing
Half of 200 ft runway, pavement thickness greater than 15 in.—see text for joint spacing.
End of runway
Longitudinal center joint
Longitudinal construction joint
Transverse contraction joints
Longitudinal construction joint
For unreinforced pavements transverse joints doweled within 100 ft from pavement end (see text). For reinforced pavements all transverse joints doweled.
A
A
Deformed tiebars at joints within 37.5 ft from pavement edge*
Plan
See text under longitudinal joints for heavy-duty pavements
Distance between longitudinal joints depends on pavement thickness, overall pavement width and construction equipment (see text)
Keyways in outside edges for future pavement extension.
h
Deformed tiebars*
Section A-A
*In taxiways 75 ft or less in width all longitudinal joints are provided with deformed tiebars

Fig. 22-19 Jointing plan for airport pavements.

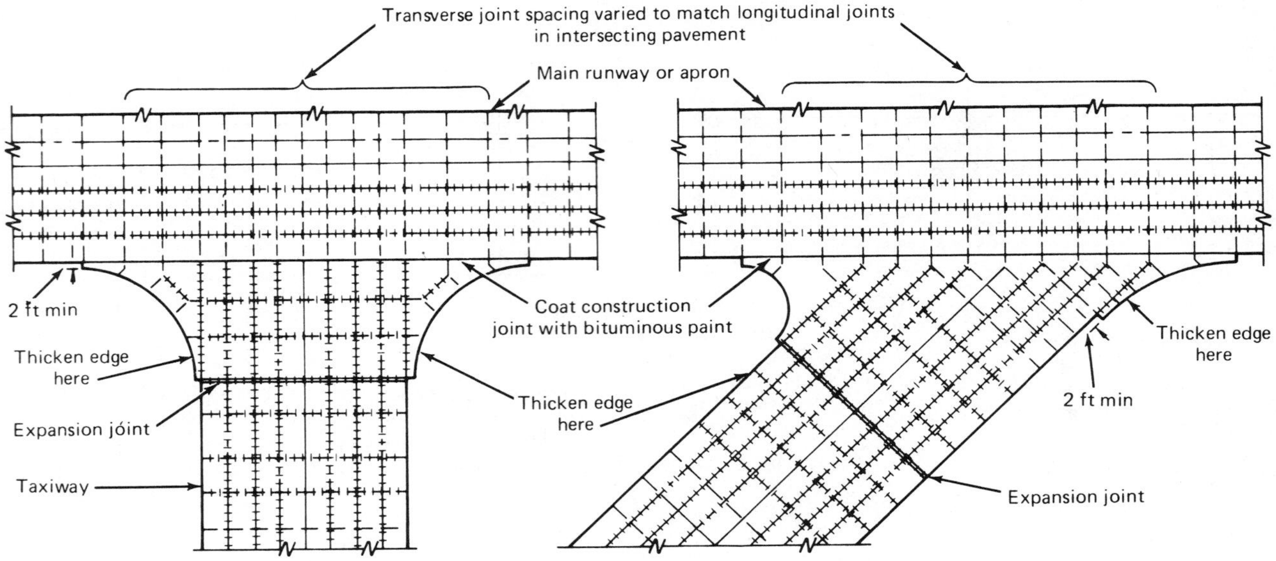

LAYOUTS EMPLOYING UNTIED KEYED CONSTRUCTION JOINT AT INTERSECTION

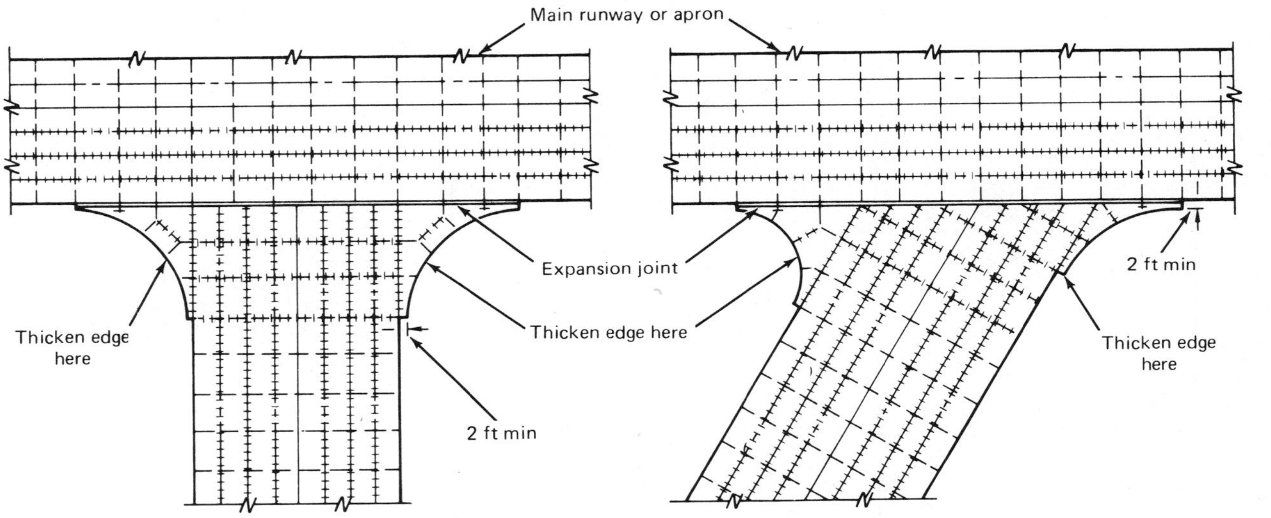

LAYOUTS EMPLOYING UNDOWELED THICKENED–EDGE EXPANSION JOINT AT INTERSECTION

Longitudinal joints tied within 37.5 ft of free pavement edges
Unreinforced pavements-transverse joints doweled on each side of expansion joint (reinforced pavements-all transverse joints doweled).

LEGEND

Keyed longit construction joint	Transverse contraction joint
Keyed longit construction joint with tiebars	Transverse contraction joint with dowels
Lontitudinal center joint	Transverse expansion joint with dowels
Longitudinal center joint with tiebars	Thickened-edge expansion joint at intersection

Note: For conditions requiring dowels, tiebars, expansion joints and thickened edges–see text.

Fig. 22-20 Typical intersections and suggested jointing layouts.

joints.* The basis for design of distributed steel, tiebars and dowels is given in Refs. 22-41, 22-45, and 22-69 through 22-72. The use of steel in continuously reinforced pavements is discussed in Section 22.9.

Where the pavement is jointed to form short panel lengths that will control intermediate cracking, distributed steel is not necessary; however, where joints are placed to form longer panels in which some intermediate cracking may be expected, distributed steel is used. In this case, dowels are used at all transverse joints to ensure adequate load transfer, since larger joint openings will result.

22.8.1 Distributed Steel

The function of distributed steel in jointed pavements is to hold together the fractured faces of slabs if cracks should form. The quantity of steel used may vary from 0.06 to 0.20% of cross-sectional area of the pavements, depending on joint spacing, slab thickness, and other factors.

Structural capacity across the cracks is achieved by the interlocking action of the rough faces of the slabs, and the infiltration of foreign material into the cracks is minimized. Distributed steel does not significantly increase flexural strength when it is used in quantities that are considered to be within the range of practical economy. When joint spacings are in excess of those that will control cracking, distributed steel is used to control the opening of intermediate cracks.

Since the steel is intended to keep cracks tightly closed, it must have sufficient strength to hold the two slabs together during contraction of the concrete. The maximum tension in the steel members across a crack is computed as equal to the force required to overcome friction between pavement and subgrade from the crack to the nearest free joint or edge. This force will be greatest when the crack occurs at the middle of a slab. For practical reasons, the steel is usually made the same weight throughout the length of the slabs.

The factors that must be considered in the design of distributed steel include the weight of the concrete that must be moved, the coefficient of subgrade resistance, and the tensile strength of the steel to be used. The amount of steel required per 1-ft width of slab is given in Fig. 22-21 as computed by the following formula:

$$A = \frac{LC_f wh}{24 f_s} \tag{22-1}$$

where:

A = area of steel required per foot of width of slab, sq in.
L = distance between free (untied) joints, ft
C_f = coefficient of subgrade (or subbase) resistance to slab movement
w = weight of concrete, lb/cu ft (150 lb/cu ft for normal weight concrete)
h = slab thickness, in.
f_s = allowable working stress in the steel, psi

When this formula is used to calculate longitudinal steel, L will be the distance between free transverse joints. A value of C_f of 1.5 is most commonly used for design.

The allowable working stress in the steel, f_s, will depend on the type of steel used and should provide a small factor of safety. However, safety factors need not be as high as those used for building and bridge design. Figure 22-21 shows steel areas for working stresses of 45,000 (welded wire fabric and 60,000 psi yield strength bar mats) and 35,000 psi (50,000 psi yield strength bar mats).

*Joint locations requiring tiebars or dowels are discussed in Sections 22.5 and 22.7.

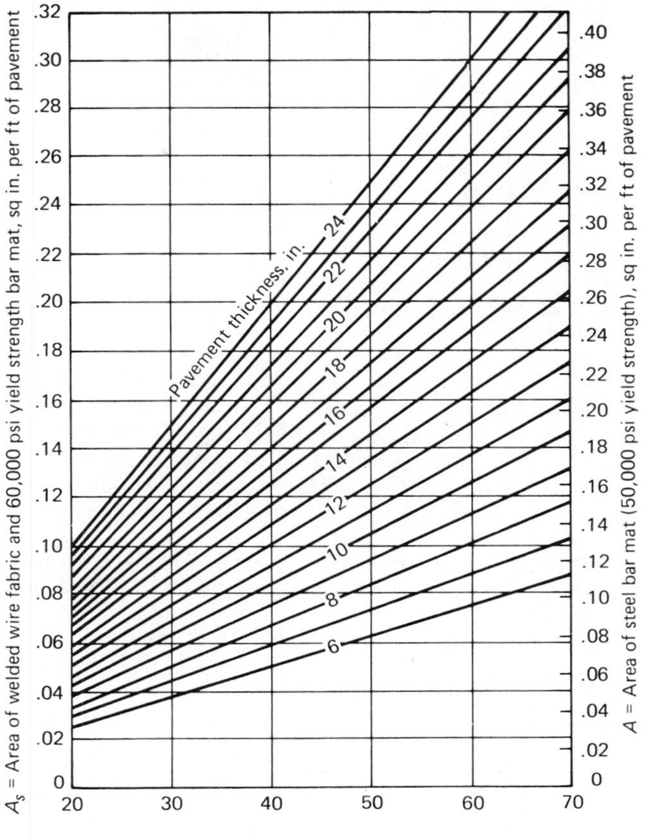

Fig. 22-21 Design chart for distributed steel.

Because longitudinal joint spacings are sufficiently close to control intermediate cracking (see Sections 22.5.3 and 22.7.2), transverse steel does not have to be as heavy as that required by eq. (22-1). Transverse steel needs to be only a sufficient amount to serve as spacers for the longitudinal steel.

Since distributed steel is not intended to act in flexure, its position within the slab is not critical, except that it should be adequately protected from corrosion, with a minimum cover of 2 in. of concrete. The steel may be placed at middepth of the slab (the neutral axis) or higher—up to one-third the slab thickness below the top surface.

22.8.2 Design of Dowels

Dowels are installed across joints in a concrete pavement to act as load-transfer devices that permit the joint to open and close. Their function is to distribute part of the load to the adjacent slab, thus reducing the deflection and stress in the loaded slab.

The locations at which dowels are used are discussed in Sections 22.5 and 22.7 on jointing. Correct alignment and lubrication of dowels are essential for proper joint function.

Several methods of theoretical analysis have been proposed for the design of dowels. Most of these design methods will result in dowel sizes and spacings that give satisfactory service. Condition surveys of existing pavements and extensive tests on full-scale slabs have shown no clear

TABLE 22-9 Recommended Size and Spacing of Dowel Bars

Slab Depth, in.	*Dowel Diameter, in.*	*Total Dowel Length*, in.*	*Dowel Spacing, in. c. to c.*
5–6	¾	16	12
7–8	1	18	12
9–11	1¼	18	12
12–16	1½	20	15
17–20	1¾	22	18
21–25	2	24	18

**Allowance made for joint openings and minor errors in positioning of dowels.*

cases of dowel failures where the pavement slab itself is adequate for the loads carried.

For an economical system of load transfer, the dowel size should be in correct proportion to the load for which the pavement is designed. Since the pavement thickness is also in proportion to the load, dowel design may be related to pavement thickness.

Table 22-9 lists suggested dowel sizes and spacings. It is based on studies of highway pavements and airport pavement experience.

Dowels are installed at the middepth of the slab.

22.8.3 Design of Tiebars

Tiebars or tiebolts are deformed steel bars. They are used across the joints of concrete pavement where it is necessary to hold the faces of the slab in firm contact. Tiebars by themselves are not designed to act as load-transfer devices. The load transfer across a joint in which they are used is provided by aggregate interlock or a keyway.

Tiebars are designed to overcome the resistance of the subgrade (or subbase) to horizontal slab movement when the pavement is contracting. The required cross-sectional area of tiebar per foot length of joint is given by the following formula:

$$A = \frac{b C_f w h}{12 f_s} \tag{22-2}$$

where:

- A = cross-sectional area of steel required per foot length of joint, sq. in.
- b = distance between joint in question and the nearest untied joint or free edge, ft
- C_f = coefficient of resistance between pavement and subgrade, or subbase (usually taken at 1.5)
- w = weight of concrete, lb/cu ft (150 lb/cu ft for normal weight concrete)
- h = slab thickness, in.
- f_s = allowable working stress in the steel, psi (usually taken as 2/3 of yield strength)

Tiebars should be long enough that anchorage on each side of the joint will develop the allowable working strength of the tiebar. In addition, an allowance of about 3 in. should be made for inaccurate centering of the tiebar. Expressed as a formula, this becomes:

$$L_t = \frac{1}{2} \frac{f_s \times d_b}{350^*} + 3 \tag{22-3}$$

*The maximum working stress for the bond in deformed bars is generally taken as 0.10 of the compressive strength of the concrete, up to a maximum of 350 psi. It is permissible to use this maximum value in the design of tiebars because paving concrete should have a compressive strength in excess of 3,500 psi.

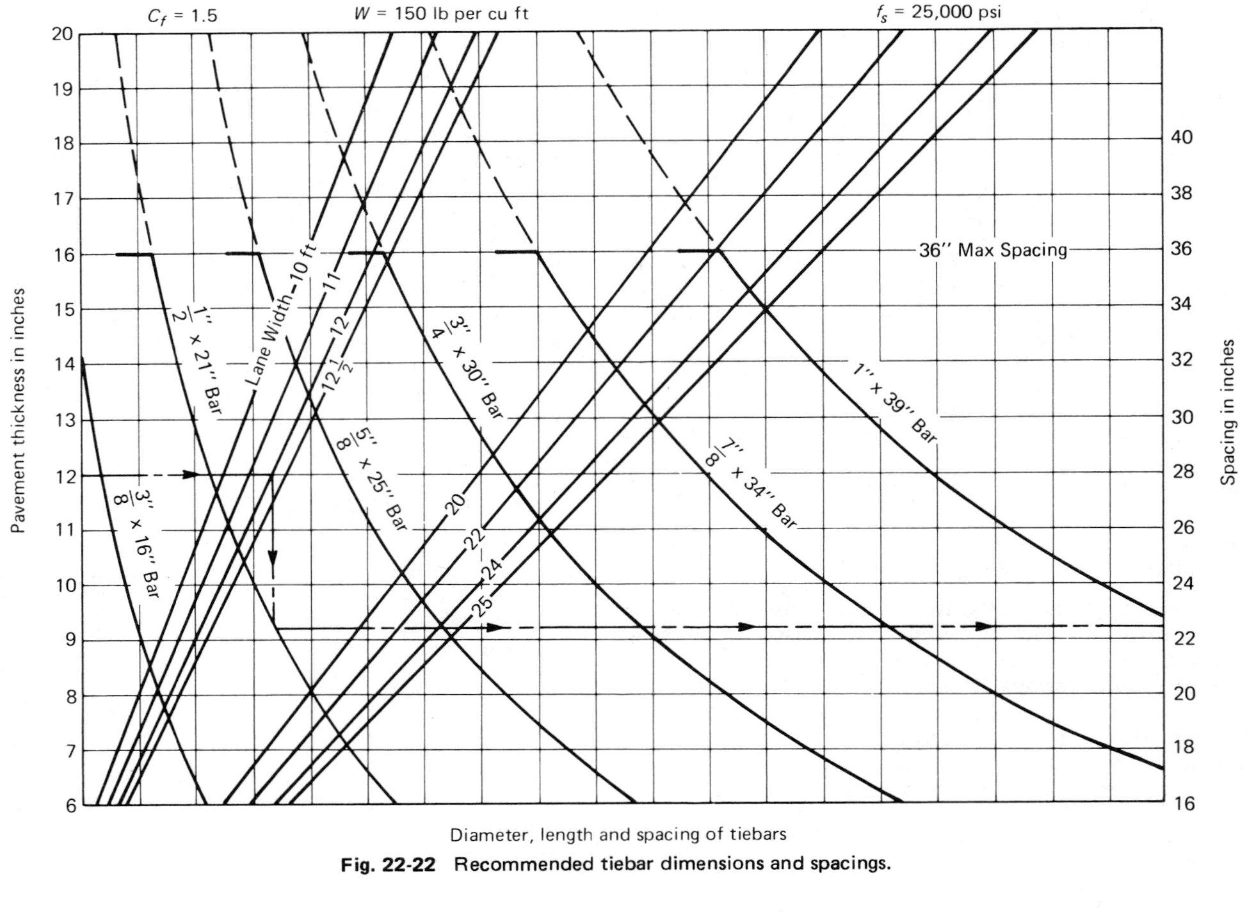

Fig. 22-22 Recommended tiebar dimensions and spacings.

where:

L_t = length of tiebar, in.
f_s = allowable working stress in steel, psi (same as that used in eq. 22-2)
d_b = diameter of tiebar, in.

Recommended tiebar dimensions and spacings are given in Fig. 22-22. Since a major use for tiebars is across longitudinal joints, it is worthwhile to standardize the length and spacing of tiebars in order to simplify construction procedures and reduce overall pavement costs.

The tiebar dimensions given in Fig. 22-22 satisfy eqs. (22-2) and (22-3) when the following factors are used:

$$C_f = 1.5; \quad w = 150 \text{ lb/cu ft}; \quad f_s = 25{,}000 \text{ psi}^*$$

22.9 CONTINUOUSLY REINFORCED PAVEMENTS

A continuously reinforced concrete pavement contains a relatively large amount of continuous steel, usually 0.6 to 0.7%, in the longitudinal direction. It is built without transverse joints throughout its length except at the locations where it intersects or abuts existing pavements or structures. Thus, there are no transverse joints to seal and maintain. In this type of pavement, transverse cracks develop at close intervals (averaging between 3 and 7 ft) and are held tightly together by the continuous, longitudinal steel.

The fundamental principles of design for this type of pavement are: (1) to provide adequate pavement thickness for traffic loads; and (2) to provide enough longitudinal reinforcing steel that the transverse cracks are kept tightly closed. Sources of information on the design of continuously reinforced pavements are given in Refs. 22-73 through 22-76 and Ref. 22-40.

22.9.1 Pavement Thickness

Thickness design procedures for continuously reinforced pavements are the same as those for jointed pavements described in Sections 22.4 and 22.6.

22.9.2 Longitudinal Steel

1. Amount. The amount of reinforcing steel required to control volume changes is dependent primarily on the thickness of the concrete, tensile strength of the concrete, and yield strength of the reinforcement. Other factors that influence the amount of steel are contraction due to temperature drop, shrinkage due to drying, and modulus of elasticity of concrete and steel.

The controlling factor is the crack width. If insufficient steel is used, the crack widths will be too wide, permitting intrusion of solids and water. Crack-width criteria have not been firmly established, but good performance has resulted when crack spacings average between 3 and 7 ft. Since crack spacing is directly related to crack width and is more readily observed than crack width, the design has indirectly become a matter of determining the amount of steel required to obtain desirable crack spacings. Formulas relating the amount of longitudinal steel and theoretical crack spacing have been proposed by several authors.[22-75–22-77]

In general, however, the amount of steel is based on data obtained from experimental pavements and pavements in service. It is the usual practice to specify steel having a minimum yield strength of 60,000 psi, and to require 0.6% of the gross cross-sectional area of the pavement. In severe climates, where freezing and thawing occur or where unusually heavy traffic prevails, it may be desirable to give consideration to the use of a somewhat higher percentage.

The following formula may be used to check the minimum amount of steel, 0.6% under normal conditions, or to compute a minimum amount based on special concretes or steels that might be selected:

$$\rho = \left(\frac{f_t'}{f_s - nf_t'}\right) 100 \qquad (22\text{-}4)$$

where:

ρ = steel area ratio, percent (total cross-sectional area of steel divided by gross cross-sectional area of concrete times 100)
f_t' = tensile strength of concrete, psi (assumed to equal 0.4 times the modulus of rupture)
f_s = allowable working stress of steel, psi (0.75 times the yield strength)
$n = E_s/E_c$ (ratio of elastic moduli, steel-to-concrete)

Having established the required percentage of longitudinal steel, the steel area may be computed by the formula.

$$A_s = \frac{bh\rho}{100} \qquad (22\text{-}5)$$

where:

A_s = total cross-sectional area of longitudinal steel, sq. in.
b = width of slab, in.
h = slab thickness, in.
ρ = steel area ratio, percent

2. Size and Spacing. Size and spacing of longitudinal steel members are interrelated and dependent on a number of factors. The minimum size should be such that the spacing between the bars will be large enough to permit easy placement of concrete. The minimum size should be that which will provide a clear space between bars of at least twice the size of the aggregate being used, but in no case less than 4 in.

The maximum size is governed by the percentage of steel, the maximum spacing permitted, bond strength, and load transfer considerations. For good load transfer and bond strength, it is believed that the spacing should not exceed 9 in.

Size and spacing are related by:

$$S_w = \frac{A_b}{h} \cdot \frac{100}{\rho} \qquad (22\text{-}6)$$

where:

S_w = spacing, center to center, in.
A_b = cross-sectional area of one steel bar or wire, sq in.
h = slab thickness, in.
ρ = steel area ratio, percent

At the locations where longitudinal steel is spliced, it is important that the length of splice be adequate to resist the tensile forces caused by shrinkage of the concrete at early ages. Recommendations for the length of lap splices are given in Refs. 22-73 through 22-75.

3. Position of Longitudinal Steel. Since the primary function of reinforcement in continuously reinforced pavements is to hold transverse cracks tightly closed, its design position vertically in the slab is not extremely critical. Practice has varied somewhat in this respect; pavements have been built with the center of the longitudinal steel ranging

*A working stress of 25,000 psi is used for steels with yield strengths of 33,000 and 45,000 psi, which have been normally specified if tiebars are to be bent and later straightened.

from 2.5 in. below the surface to middepth of the slab. Placement at middepth results in less steel stress at cracks due to wheel loads and temperature drops than for other placement positions. Another approach used is that the steel should be placed above middepth to reduce the surface width of cracks. The steel, of course, must have sufficient cover to preclude the development of cracks over the longitudinal reinforcement and prevent corrosion.

Although the middepth of slab has generally been chosen as the maximum depth of the longitudinal reinforcement, this has not been firmly established by experience; the suggested minimum design depth is one-third of slab depth and should provide a 2.5 in. cover over the longitudinal steel.

22.9.3 Transverse Steel

A relatively small amount of transverse steel is commonly, but not always,* used in continuously reinforced pavements to maintain the spacing of longitudinal bars to which it is tied or welded. In the case of pre-set reinforcement, the transverse steel aids in supporting the longitudinal steel above the subbase. Transverse steel may also serve the purpose of tie bars across longitudinal joints.

The subgrade drag theory, eq. (22-1), is used to compute the required amount of transverse steel; it is based on providing a sufficient amount of steel to hold chance longitudinal cracks tightly closed.

22.9.4 Transverse Joints for Continuously Reinforced Pavements

Some aspects of the design of transverse joints for continuously reinforced pavements are discussed in this section, since they apply in particular to this pavement type.

The two types of transverse joints in continuously reinforced pavements are construction joints, which are placed at the end of a day's work or when paving operations are temporarily stopped, and expansion joints, which are located at intersections with other pavements and at fixed objects. Jointing practices and design for longitudinal joints are the same as for jointed pavements and are discussed in Sections 22.5 and 22.7.

Construction joints, because of their smooth faces, do not have as high load-transfer capacity as natural cracks in which aggregate interlock supplements the shear strength of the longitudinal steel. Therefore, it is necessary to strengthen them. This is done by installing additional deformed bars of the same size as the longitudinal reinforcement. It is recommended that these extra bars be installed at a reasonably uniform spacing across the pavement, and in sufficient number to increase the area of steel across the joint at least one-third. These bars should be at least 3 ft long.

Expansion joints for continuously reinforced pavements are installed only at bridges, structures, and where the pavement adjoins a section of jointed pavement. These joints must accommodate large seasonal movements. Two types of expansion joints that have been successfully used are the wide-flange beam and the "bridge-finger" type. The joints should be designed to accommodate seasonal movements of 2 to 3 in. depending on climatic conditions. If the pavement ends are anchored by lugs or piles built into the subgrade, experience shows that about 1 in. of end movement needs to be accommodated by conventional expansion joints.

*Transverse steel may not be required where the longitudinal reinforcement is placed in fresh concrete by tube-feed or a method that will ensure its proper spacing and depth.

A less expensive terminal provision is to install several conventional expansion joints at pavement ends. However, this treatment should be used only where experience has indicated that sufficient movement is accommodated to prevent pavement growth that will damage adjoining structures or pavement.

Additional details on the design of terminal joints and end anchorages are given in Ref. 22-73 through 22-75.

22.10 CONCRETE OVERLAYS

A pavement overlay is required when the condition of the existing pavement is no longer serviceable, or when a pavement is being strengthened to carry greater loads than it was designed for. In such cases it is both practical and economical to strengthen by the use of a concrete overlay.

This section discusses the design of the three basic types of overlays (partially bonded, unbonded, and bonded) on existing concrete pavements. Concrete overlays for flexible pavement are also discussed. Basic information on the design of the overlays is given in Refs. 22-81 through 22-87.

The use of concrete overlays has been receiving increasing attention in the United States because of emphasis placed on the 3-R program (resurfacing, restoration, and rehabilitation) in recent years. As a result of this increasing experience, the design of overlays is in an emerging state-of-the-art that undoubtedly will develop to supersede the existing methods discussed herein. The type and thickness of overlay depend largely on the condition of the existing pavement, and this is one area in which methods for pavement evaluation and correction are now being developed. Reference 22-89 is a guide to the current state-of-the-art.

22.10.1 Flexural Strength

As explained in Sections 22.4.4 and 22.6.2, the design for a new pavement is based on a 28-day flexural strength for highway pavement and a 90-day strength for airport pavement. For overlay design, strengths at these ages are also used for determining the single-slab thickness, h, in eqs. (22-7), (22-8), and (22-9).

This practice is conservative, since the design of a concrete overlay should involve consideration of the relative flexural strength of the overlay and base slabs. Since the base slab may be several years old, the strength gain during this period should result in flexural strengths well above the 28- or 90-day design value (see Fig. 22-14). If desired, the strength values of the two layers may be considered in design by a method proposed by Mellinger.[22-87]

22.10.2 Condition of Existing Pavement

The condition of the existing pavement is an important factor in the selection of the type of concrete overlay. In subsequent overlay design charts and formulas, a coefficient C is used to express the structural condition* of the pavement as follows:

$C = 1.0$, when the existing pavement is in good overall structural condition
$C = 0.75$, when the existing pavement has initial joint and corner cracks, but no progressive structural distress or recent cracking
$C = 0.35$, when the existing pavement is generally badly cracked or structurally shattered

*C-values apply to structural condition only, and should not be influenced by surface defects.

Careful consideration must be given to the assignment of the *C*-value in terms of which cracks or other structural defects will influence the performance of the restructured pavement. Cracking may or may not represent a failed condition. For example, cracking due to warping stresses is not progressive and not structurally detrimental, since load transfer is provided by aggregate interlock.

Cracking due to nonuniform foundation support (subgrade pumping, subbase consolidation, subgrade settlement) may not be detrimental if the condition has reached equilibrium, and, through cracking, the slabs have settled so that uniform support is again provided.

Progressive structural defects, cracks, or joints where load transfer has been lost, rocking slabs, or progressive foundation settlement are conditions that will seriously affect performance of the overlay and should be carefully evaluated.

Thus, the selection of type and design of overlay, and preliminary repair work, should be based on a thorough knowledge of the pavement condition and the causes of structural defects.

22.10.3 Partially Bonded Overlays

Experience both with actual pavements in service and full-scale test pavements has shown that the use of a separation course* between the existing slab and the overlay slab leads to greater deflections and more breaking in the overlay slab than where such separation courses are not used. As a result, pavements with an overlay slab placed directly over the existing slab are stronger when no separation course is used. This direct overlay is termed a "partially-bonded" overlay.

*"Separation course," as used here, refers to any material between the two slabs that will break the bond (bituminous coating, plastic sheet, granular layer, or asphaltic concrete layer.)

Based on studies of overlay pavements, the U.S. Army Corps of Engineers[22-82] uses this formula for design of overlays placed directly on the existing pavement:

$$h_r = \sqrt[1.4]{h^{1.4} - Ch_e^{1.4}} \qquad (22\text{-}7)$$

where:

h_r = thickness of overlay, in.
h = thickness of required single slab, in. (as determined in Sections 22.4 and 22.6)
h_e = thickness of existing slab, in.
C = a coefficient indicating the structural condition of the existing pavement (Section 22.10.2)

The equation recognizes that friction between the two slabs or the development of some degree of bond provides somewhat greater capacity than when a separation course is used. As a result, thinner overlay sections are obtained than for unbonded overlays.

The required thicknesses for concrete overlay pavements may be taken directly from the curves of Fig. 22-23. Where thicknesses are required for other values of C, they may be found by interpolation between the three sets of curves.

Partially bonded overlays are not used when the existing pavement is in extremely poor condition unless the struc-

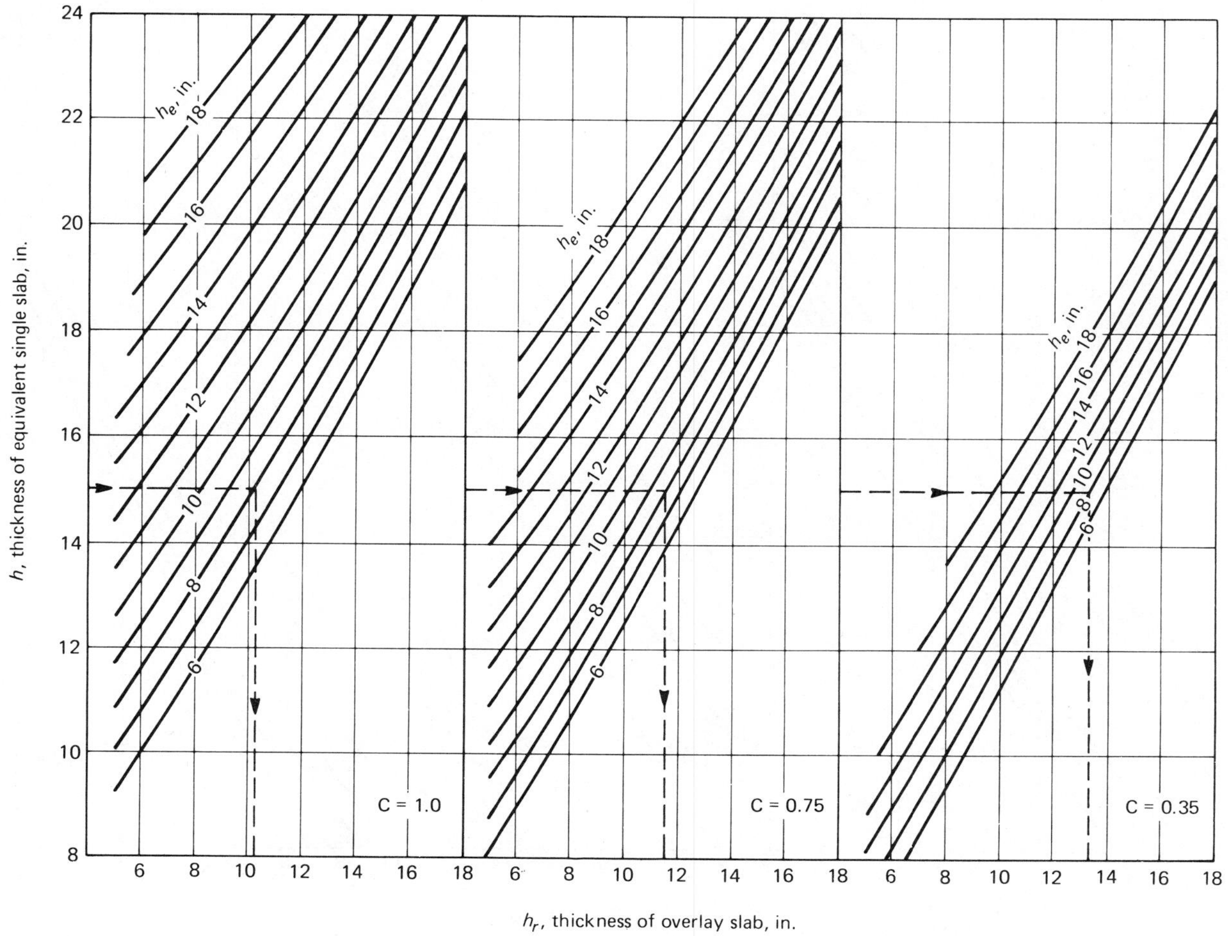

Fig. 22-23 Design chart for partially bonded overlays, $h_r = \sqrt[1.4]{h^{1.4} - Ch_e^{1.4}}$.

tural defects can be repaired so that the C-value is significantly better than 0.35.

22.10.4 Unbonded Overlays

A separation course between the slabs may sometimes be needed if the existing slab has an irregular surface or is in poor condition, or when the grade line is to be raised appreciably. This is called an unbonded overlay, and the thickness is determined from the following equation (or Fig. 22-24):

$$h_r = \sqrt{h^2 - Ch_e^2} \qquad (22\text{-}8)$$

where the symbols have the same meaning as in eq. (22-7). Equation (22-8) recognizes that there is not as much interaction between the slabs in the form of friction or bond and gives greater thickness than for partially bonded overlays.

In some cases where the grade line is to be raised and a thick layer of material is necessary between slabs, it may be more economical to determine a new k on top of the layer by plate-bearing tests and to design a new full-depth slab as described in Sections 22.4 and 22.6. This may result in a thinner slab than indicated by eq. (22-8), particularly if the separating layer consists of well-graded and well-compacted granular material.

22.10.5 Bonded Overlays

A bonded overlay is a resurfacing placed on a carefully cleaned and prepared pavement surface with a bonding agent consisting of a sand-cement grout or epoxy mixture.[22-83, 22-84]

Bonded overlays have been used on large areas of pavements both to correct surface defects and to increase the structural capacity of pavements. When used on a structurally sound slab to increase its load-carrying capacity, the overlay and base slab should have a combined thickness equal to a single slab of adequate design for the planned loading. In this case:

$$h_r = h - h_e \qquad (22\text{-}9)$$

Of the different overlay types, the thinnest section will be obtained for bonded overlays, since the resurfacing and the existing pavement act together as a monolithic slab. The economy of the thinner section, however, is somewhat offset by the extra cost of surface preparation and grouting required for the bonded overlay.

For bonded overlays, design thickness may be based on flexural strength of the existing pavement because of the monolithic slab action. Bonded overlays are recommended for use only where the existing slabs are in good structural condition, or structural defects have been repaired. Reports on the performance of bonded overlays are given in Refs. 22-85 and 22-86.

22.10.6 Slab Replacement or Repair

For a pavement with a few localized areas of structural defects, the C-value may sometimes be raised appreciably by a limited program of repair, patching, or slab replace-

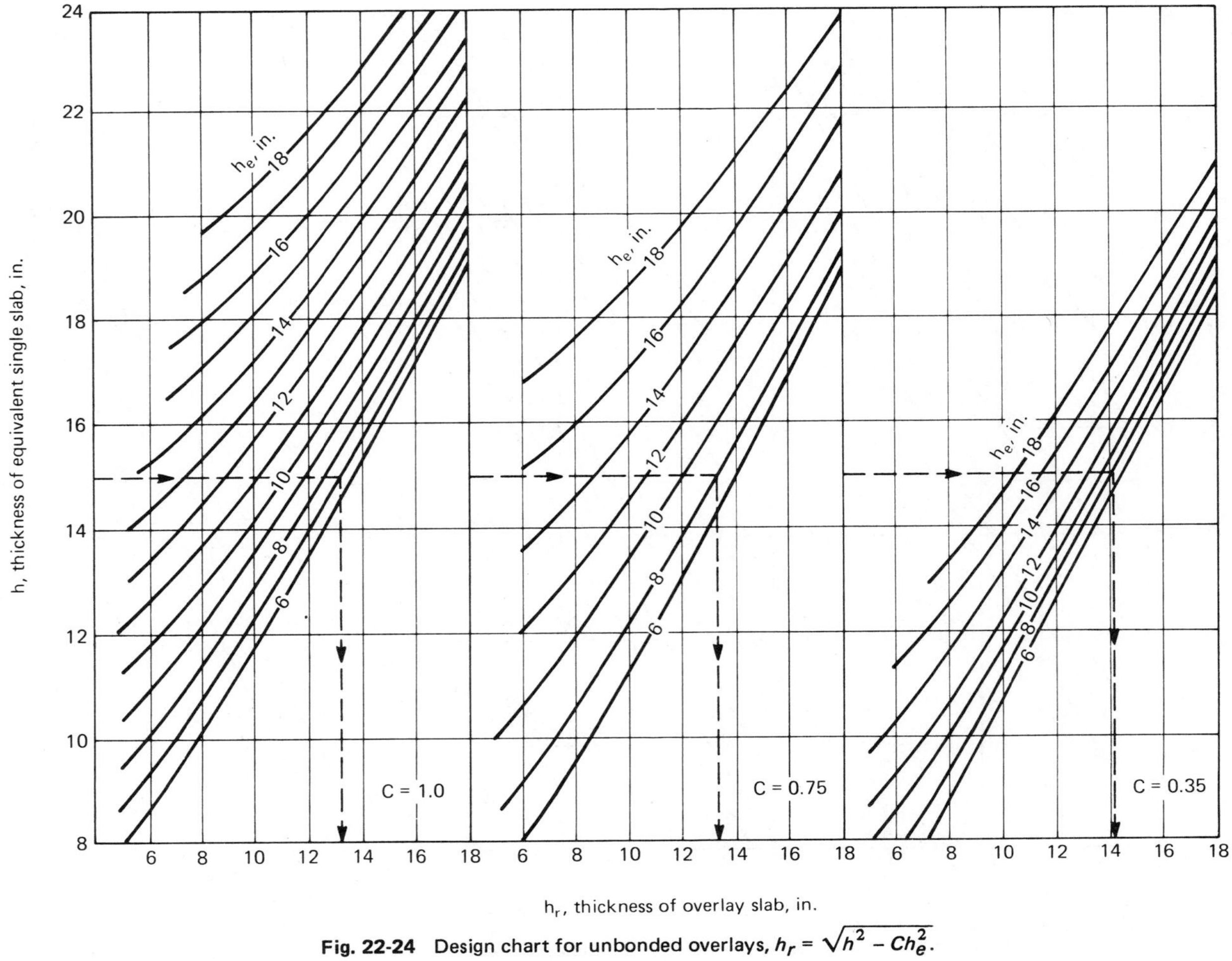

Fig. 22-24 Design chart for unbonded overlays, $h_r = \sqrt{h^2 - Ch_e^2}$.

ment. The increased C-value results in a thinner overlay for the entire pavement which may more than offset the costs of localized repairs.

22.10.7 Joint Location

Joints and random cracks in the base pavement will be reflected in partially bonded and bonded overlays unless some preventive measures are taken. The practice used to prevent joint reflection is to match the joint locations in the overlay with those in the existing slab.

Many old concrete pavements were built with expansion joints at relatively short intervals. Expansion joints can usually be omitted in the overlay with contraction joints placed over the expansion joint location. This applies only to partially bonded overlays. When a bonded or monolithic overlay is used, it is imperative that joints in the overlay match those in the base pavement precisely in location, and that the same type and width of joint be used (expansion or contraction).

When a separated or unbonded overlay is used in which the separation course is of a substantial thickness, or when a continuously reinforced overlay is used, it is not necessary to match the transverse joints in the overlay concrete to those in the existing concrete in either the location or type. This is one of the advantages of the separated overlay, and this type is required frequently when the joint pattern of the existing pavement is irregular and it is not desirable to repeat it in the overlay pavement.

In unbonded overlays with thin separation courses and partially bonded overlays that are not continuously reinforced, contraction joints can be placed directly over existing expansion joints, contraction joints, or construction joints. If this does not result in slab lengths short enough to control cracking, additional intermediate contraction joints should be placed to form equal slab lengths that are short enough to control cracking, as described in Sections 22.5 and 22.7. Load transfer at transverse joints is provided by aggregate interlock, except near the ends of pavements where dowels should be used. A plain (unreinforced) overlay with short joint spacings may be placed on a reinforced slab with long joint spacings, provided that the intermediate cracks in the existing pavement are tightly closed and in good condition. In this case, dowels are used in the overlay pavement only at locations matching the existing doweled joints.

For all overlay types, longitudinal joints in the resurfacing should also match the joints in the base. The longitudinal construction joints should be provided with a key* for load transfer to the adjacent slab. Tiebar requirements at longitudinal joints in concrete resurfacing are the same as explained in Sections 22.5 and 22.7 for full-depth pavement.

22.10.8 Tiebars, Dowels, and Distributed Steel

The need and use of tiebars, dowels, and reinforcing steel are the same as discussed in Sections 22.5 and 22.7. Where used, tiebars, dowels, and reinforcing steel are designed as outlined in Section 22.8. If they exist in the base slab at the proper location and are functioning adequately, tiebars, dowels, and distributed steel in the overlay are designed based on the overlay slab thickness. Otherwise, tiebar and distributed steel design, in bonded and partially bonded overlays, is based on the total thickness of old and new slab; dowel design is based on the thickness of an equivalent slab, h_e.

*For thin, bonded overlays, keyways or other load-transfer devices are usually omitted.

Reinforcement serves the same purpose in concrete overlays as it does in original construction. It is not required when short joint spacings are used but is needed for pavements with longer joint spacing to restrain cracks that occur so that they do not open sufficiently to present a maintenance problem.

If the old pavement is extensively cracked, the use of distributed steel or continuous reinforcement may be the most dependable method of minimizing uncontrolled cracking in unbonded or partially bonded overlays.

22.10.9 Continuously Reinforced Overlays

Because they are less susceptible to reflective cracking, continuously reinforced overlays appear to offer an advantage over other overlay types. For these overlays, a separation course is normally used over the existing pavement, and the overlay thickness is determined as described in Section 22.10.4. The amount of reinforcing steel is based on the overlay pavement thickness requirements; other design details are described in Section 22.9.

A few partially bonded (no separation course), continuously reinforced overlays have been constructed. Partially bonded overlays should be used only if the existing pavement is in fairly good condition, and joint spacings are short so that joint movements will not greatly affect the overlay. In this case, a thinner overlay could be used on the basis of the partially bonded overlay. However, additional steel may be required to prevent excessive crack opening in the overlay.

22.10.10 Separation Courses

Some success has been experienced in preventing reflective cracking by the use of separation courses between the base slab and overlay. There are not sufficient data available, however, to indicate the minimum thickness of separation course that will be completely effective. There are indications that any type of bond breaker will reduce the amount of reflective cracking. As discussed previously, the use of a bond breaker or separation course creates the requirement for a thicker overlay.

22.10.11 Concrete Overlays for Flexible Pavement

Concrete overlays on flexible pavements have been used for several years. These overlays have performed well and demonstrated the feasibility of this type of construction.[22-88]

Concrete overlays for flexible pavement should be designed in the same manner as a concrete pavement on grade. The modulus of subgrade reaction, k, should be determined by plate-bearing tests made on the surface of the flexible pavement. Several agencies apply the limitation that no k-value greater than 500 lb/cu in. will be used in designing rigid overlays for flexible pavements. This appears, however, to be a rather arbitrary limitation, and more development work should be done on this point to realize fully the advantages of this composite design.

BIBLIOGRAPHY

22-1 Childs, L. D., and Kapernick, J. W., "Tests of Concrete Pavements on Gravel Subbases," Paper 1800, *Proceedings of the American Society of Civil Engineers*, V. 84, HW3, Oct. 1958.

22-2 Childs, L. D., and Nussbaum, P. J., "Pressures at Foundation Soil Interfaces under Loaded Concrete and Soil-Cement Highway Slabs," *Proc., ASTM*, V. 62, 1962, pp. 1243–1263.

22-3 Childs, L. D., and Kapernick, J. W., "Tests of Concrete Pavements on Crushed Stone Subbases," *Proc., ASCE*, Paper 3497, V. 89, HW1, Apr. 1963, pp. 57–80.

22-4 Childs, L. D., Colley, B. E., and Kapernick, J. W., "Tests to Evaluate Concrete Pavement Subbases," *Research and Development Laboratory Bulletin D11*, Portland Cement Association. Reprinted from Paper 1297, *Proc., ASCE*, V. 83, HW3, July 1957.

22-5 "A Review Paper on Expansive Clay Soils," Woodward-Clyde & Associates, Los Angeles, Calif., 1967.

22-6 "The AASHO Road Test, Report 5, Pavement Research," *Highway Research Board Special Report 61E*, Washington, D.C., 1962.

22-7 Springenschmid, R., "The Influence of Water Content on the Frost Behavior of Consolidated and Stabilized Soils" (in German), *Rock Mechanics and Engineering Geology*, V. 3, No. 3/4, 1965, 113–121, Abstract in *Highway Research Abstracts*, V. 36, No. 9, Sept. 1965, p. 5.

22-8 Beskow, G., "Soil Freezing and Frost Heaving with Special Attention to Roads and Railroads," The Swedish Geological Society, Series C, No. 375, 26th Yearbook, No. 3, 1935. Translated by J. O. Osterberg, Northwestern University. Published by Technological Institute, Northwestern University, 1947.

22-9 "Investigation of Filter Requirements for Underdrains," U.S. Army Corps. of Engineers, *Technical Memorandum 183-1*, U.S. Waterways Experiment Station, Vicksburg, Miss., 1941.

22-10 "Frost Action in Roads and Airfields," *A Review of Literature 1765–1951, Highway Research Board Special Report 1*, 1952.

22-11 Miyakawa, I., and Koyama, M., "On the Residual Frost Subgrade Underneath the Select Fill in the Alleviation Practice for Frost Damage," Civil Engineering Institute, Hokkaido Development Bureau, Hiragishi, Sapporo, Japan. Reprinted from *Soil and Foundation*, V. 3, No. 1, Sept. 1962.

22-12 "Subgrades and Subbases for Concrete Pavements," Portland Cement Association, Skokie, Ill., 1971.

22-13 Kawala, E. L., "Cement-Treated Subbase Practice in U.S. and Canada," ASCE Proceedings Paper 4947, *Journal of the Highway Division*, V. 92, HW2, Oct. 1966, pp. 75–98.

22-14 Yrjanson, W. A., and Packard, R. G., "Econocrete Pavements—Current Practices," *Performance of Pavements Designed with Low Cost Materials*, Transportation Research Record 741, Transportation Research Board, Washington, D.C., 1980.

22-15 Colley, B. E., and Nowlen, W. J., "Performance of Concrete Pavements Under Repetitive Loading," *Research and Development Laboratory Bulletin D13*, Portland Cement Association, Skokie, Ill., 1958.

22-16 Colley, B. E., and Humphrey, H. A., "Aggregate Interlock at Joints in Concrete Pavements," *Highway Research Record, No. 189*, 1967, pp 1–18.

22-17 "Thickness Design for Concrete Pavements," Portland Cement Association, Skokie, Ill., 1974.

22-18 "Design of Concrete Pavement for City Streets," Portland Cement Association, Skokie, Ill., 1974.

22-19 Westergaard, H. M., "Computation of Stresses in Concrete Roads," Highway Research Board *Proceedings*, Fifth Annual Meeting, Part 1, Washington, D.C., 1925, pp. 90–112.

22-20 Westergaard, H. M., "Stresses in Concrete Pavements Computed by Theoretical Analysis," *Public Roads*, V. 7, No. 2, Apr. 1926, pp. 25–35.

22-21 Westergaard, H. M., "Theory of Concrete Pavement Design," Highway Research Board *Proceedings*, Seventh Annual Meeting, Part 1, Washington, D.C., 1927, pp. 175–181.

22-22 Westergaard, H. M., "Analytical Tools for Judging Results of Structural Tests of Concrete Pavements," *Public Roads*, V. 14, No. 10, Dec. 1933, pp. 185–188.

22-23 Pickett, G., Raville, M. E., James W. C., and MacCormick, F. J., "Deflections, Moments, and Reactive Pressures for Concrete Pavements," *Kansas State College Bulletin No. 65*, Kansas State College, Manhattan, Kans., Oct. 1951.

22-24 Pickett, G., and Ray, G. K., "Influence Charts for Concrete Pavements," Paper No. 2425, *Transactions, ASCE*, V. 116, 1951, pp. 49–73.

22-25 Teller, L. W., and Sutherland, E. C., "The Structural Design of Concrete Pavements," *Public Roads*, V. 16, Nos. 8–10, 1935; V. 17, Nos. 7, 8, 1936; and V. 23, No. 8, 1943.

22-26 Childs, L. D., "Tests of Concrete Pavement Slabs on Cement-treated Subbases," *Highway Research Record, No. 60*, 1964, pp. 39–58.

22-27 Older, C., "Highway Research in Illinois," *Proc., ASCE*, Feb. 1924, pp. 175–217.

22-28 Aldrich, L., and Leonard, I. B., "Report of Highway Research at Pittsburg, California, 1921–1922," California State Printing Office, Sacramento, Calif., 1923.

22-29 "Road Test One—MD," *Highway Research Board Special Report 4*, Transportation Research Board, 1952.

22-30 "Tests of Joints for Concrete Pavements," Portland Cement Association, Skokie, Ill., 1975.

22-31 Fordyce, P., and Teske, W. E., "Some Relationships of the AAHSO Road Test to Concrete Pavement Design," *Highway Research Record No. 44*, 1963, pp. 35–70.

22-32 Fordyce, P., and Yrjanson, W. A., "Modern Design of Concrete Pavements," Proceedings Paper 6726, *Transportation Engineering Journal, ASCE*, V. 95, TE3, Aug. 1969, pp. 407–438.

22-33 Taragin, A., "Lateral Placement of Trucks on Two-Lane and Four-Lane Divided Highways," *Public Roads*, V. 30, No. 3, Aug. 1958, pp. 71–75.

22-34 Fordyce, P., and Packard, R. G., "Concrete Pavement Design," paper presented at the Forty-ninth Annual Meeting, American Association of State Highway Officals Committee on Design, Portland, Ore., Oct. 1963.

22-35 "Load Stresses at Pavement Edge," Portland Cement Association, Skokie, Ill., 1969.

22-36 Kesler, C. F., *Fatigue and Fracture of Concrete*, Stanton Walker Lecture Series of the Materials Sciences, National Sand and Gravel Association and National Ready-Mixed Concrete Association, Washington, D.C., 1970.

22-37 Ballinger, C. A., "The Cumulative Fatigue Damage Characteristics of Plain Concrete," *Highway Research Record 370*, Transportation Research Board, Washington, D.C., 1971, pp. 48–60.

22-38 Miner, M. A., "Cumulative Damage in Fatigue," *Transaction, ASME*, V. 67, 1945, pp. A159–A164.

22-39 Westergaard, H. M., "Analysis of Stresses in Concrete Roads Caused by Variations in Temperature," *Public Roads*, V. 8, No. 3, May 1927, pp. 201–215.

22-40 Friberg, B. F., "Frictional Resistance Under Concrete Pavements and Restraint Stresses in Long, Reinforced Slabs," *Proc., Highway Research Board*, Washington, D.C., 1954.

22-41 Kelley, E. F., "Applications of Results of Research to the Structural Design of Concrete Pavements," *Public Roads*, V. 20, No. 5, July 1939.

22-42 Harr, M. E., and Leonards, G. A., "Warping Stresses and Deflections in Concrete Pavements," *Proceedings* of the Thirty-eighth Annual Meeting of the Highway Research Board, Washington, D.C., 1959.

22-43 Wiseman, J. F., Harr, M. E., and Leonards, G. A., "Warping Stresses and Deflections in Concrete Pavements, Part 2," *Proceedings* of the Thirty-ninth Annual Meeting of the Highway Research Board, Washington, D.C., 1960.

22-44 ACI Committee 325, Subcommittee II, "Considerations in the Selections of Slab Dimensions," *ACI Journal*, V. 28, No. 5, Nov. 1956.

22-45 Bradbury, R. H., *Reinforced Concrete Pavements*, The Wire Reinforcement Institute, Washington, D.C., 1938.

22-46 Moore, J. H., with discussions by E. C. Sutherland and W. Harwood, "Thickness of Concrete Pavements," Paper 2834, *Transactions, ASCE*, V. 121, 1956.

22-47 "Joint Design for Concrete Pavements," Portland Cement Association, Skokie, Ill., 1980.

22-48 "Concrete Streets: Typical Pavement Sections and Jointing Details," Portland Cement Assocation, Skokie, Ill., 1980.

22-49 "A Charted Summary of Concrete Highway Pavement Practices–1982," Portland Cement Association, Skokie, Ill., 1982.

22-50 Halm, H. J., "An Analysis of Factors Influencing Concrete Pavement Cost," *Highway Research Bulletin 340*, Transportation Research Board, 1962.

22-51 Anderson, A. A., "Expansion Joint Practice in Highway Construction," Paper 2384, *Transactions, ASCE*, V. 114, 1949.

22-52 Finney, E. A., and Oehler, L. T., "Final Report on Design Project, Michigan Test Road," *Proceedings* of the Thirty-eighth Annual Meeting of the Highway Research Board, 1959, pp. 241–285.

22-53 "Joint Spacing in Concrete Pavements," Transportation Research Board, *Research Report 17-B*, 1956.

22-54 Schutz, R. J., "Shape Factors in Joint Design," *Civil Engineering*, V. 32, No. 10, Oct. 1962.

22-55 Tons, E., "Factor in Joint Seal Design," *Highway Research Record No. 80*, Transportation Research Board, 1965.

22-56 Tons, E., "Materials and Geometry in Joint Seals," *Adhesive Ages*, V. 8, No. 9, Sept. 1965, pp. 22–28.

22-57 Cook, J. P., and Lewis, R. M., "Evaluation of Pavement Joint and Crack Sealing Materials and Practices," *National Cooperative Highway Research Program Report No. 38*, 1967.

22-58 Packard, R. G., *Design of Concrete Airport Pavement*, Portland Cement Association, Skokie, Ill., 1973.

22-59 "Rigid Pavements for Airfields Other than Army," *TM 5-824-3*, *AFM 88-6* Chap. 3, Departments of the Army and Air Force, Jan. 1970.

22-60 "Airport Paving," AC 150/5320-6C, Department of Transportation, Federal Aviation Administration, Dec. 7, 1978.

22-61 Ray, G. K., Cawley, M. L., and Packard, R. G., "Concrete Airport Pavement Design–Where Are We?" presented at ASCE/AOCI Airports Specialty Conference, Atlanta, Ga., April 16, 1971. Published by ASCE in *Airports, Key to the Air Transportation System*, ASCE, New York, 1972.

22-62 Packard, R. G., "Computer Program for Concrete Airport Pavement Design," Portland Cement Association, Skokie, Ill., 1967.

22-63 "Concrete Pavements for General-Aviation Airports," Portland Cement Association, Skokie, Ill., 1977.

22-64 Palmer, L. A., "Field Loading Tests for the Evaluation of the Wheel-Load Capacities of Airport Pavements," *Symposium on Load Tests of Bearing Capacity of Soils*, ASTM Special Technical Publication No. 79, 5-40, June 1947.

22-65 "The AASHO Road Test, Report 6, Special Studies," *Highway Research Board Special Report 61F*, Transportation Research Board, 1962.

22-66 "Results of Modulus of Subgrade Reaction Determination at The Road Test Site by Means of Pavement Volumetric Displacement Test," U.S. Army Engineer Division, Ohio River, Corps of Engineers, Ohio River Division Laboratories, Cincinnati, Ohio, Apr. 1962.

22-67 "Design Manual–Airfield Pavements," NAVFAC DM-21, Department of the Navy, Naval Facilities Engineering Command, Alexandria, Va., June 1973.

22-68 "Model Tests to Determine Optimum Key Dimensions for Keyed Construction Joints," U.S. Army Corps of Engineers, Ohio River Division Laboratory, Cincinnati, Ohio, 1954.

22-69 ACI Committee 325, Subcommittee III, "Structural Design Considerations for Pavement Joints," *ACI Journal*, V. 53, No. 1, July 1956.

22-70 Friberg, B. F., "Design of Dowels in Transverse Joints of Concrete Pavements," *Transactions, ASCE*, V. 66, No. 5, May 1940.

22-71 Bradbury, R. D., "Design of Joints in Concrete Pavements," *Proc., Highway Research Board*, 1932.

22-72 Timoshenko, S., and Lessels, J. M., *Applied Elasticity*, Westinghouse Technical Night School Press, Pittsburgh, Pa., 1925.

22-73 "Design and Construction, Continuously-Reinforced Concrete Pavement," Continuously-Reinforced Pavement Group, sponsored by Concrete-Reinforcing Steel Institute and Committee of Steel Bar Producers of the American Iron and Steel Institute, 1968.

22-74 ACI Committee 325, Subcommittee VII, "A Design Procedure for Continuously-Reinforced Concrete Pavements for Highways," Title 69-32, *ACI Journal*, V. 69, No. 6, June 1972, pp. 309–319.

22-75 McCullough, B. F., *Design Manual for Continuously-Reinforced Concrete Pavement*, United States Steel Corporation, Pittsburgh, Pa., 1970.

22-76 Vetter, C. P., "Stresses in Reinforced Concrete Due

to Volume Changes," *Transactions, ASCE*, V. 98, 1933.

22-77 Zuk, W., "Analysis of Special Problems in Continuously-Reinforced Concrete Pavements," *Highway Research Board Bulletin 214*, Transportation Research Board, 1959.

22-78 AASHO Committee on Design, "AASHO Interim Guide for the Design of Rigid Pavement Structures," American Association of State Highway and Transportation Officials, Washington, D.C., 1972, Chap. III, Rev. 1981.

22-79 McCullough, B. F., and Ledbetter, W. B., "LTS Design of Continuously-Reinforced Concrete," Paper 3357, *Transactions, ASCE*, V. 127, Part IV, 1962.

22-80 "Test Investigations of Lap Splices of Reinforcing Steel in Continuously-Reinforced Concrete Pavement," CRSI Committee on Continuously-Reinforced Pavement, *Bulletin No. 3*, Chicago, Ill., May 1963.

22-81 ACI Committee 325, Subcommittee VIII, "Design of Concrete Overlays for Pavements," Title No. 64-40, *ACI Journal*, V. 64, Aug. 1967, pp. 470–474.

22-82 "Rigid Airfield Pavements," *Engineering and Design Manual*, EM1110-45-303, U.S. Department of the Army, Office of the Chief of Engineers, Feb. 1958.

22-83 "Bonded Concrete Resurfacing," Portland Cement Association, Skokie, Ill., 1960.

22-84 ACI Committee 503, "Guide for the Use of Epoxy Compounds with Concrete," *ACI Journal, Proceedings*, V. 59, No. 9, Sept. 1962, pp. 1121–1142.

22-85 Gillette, R. W., "A Ten-Year Report on the Performance of Bonded Concrete Resurfacings," *Highway Research Record No. 94*, Transportation Research Board, 1965, pp. 61–76.

22-86 Gillette, R. W., "Performance of Bonded Concrete Overlays," *ACI Journal, Proceedings*, V. 60, No. 1, Jan. 1963, pp. 39–50.

22-87 Mellinger, F. M., "Structural Design of Concrete Overlays," *ACI Journal, Proceedings*, V. 60, No. 2, Feb. 1963, pp. 225–236.

22-88 Westall, W. G., "Concrete Overlays on Asphalt Pavements," *Highway Research News* No. 22, Transportation Research Board, Feb. 1966, pp. 52–57.

22-89 "Guide to Concrete Resurfacing Designs and Selection Criteria," Portland Cement Association, Skokie, Ill., 1981.

23

Preparation of Structural Drawings as Related to Detailing of Reinforced Concrete*

PAUL F. RICE** and WILLIAM C. BLACK***

23.1 GENERAL CONSIDERATIONS

All contract documents include a complete set of plans and specifications. Contract plans or drawings consist of architectural, structural, heating/ventilating, and electrical drawings. These drawings, along with the specifications, plus bid proposal, form the documents on which the general contractor makes his bid.

For all major building requiring reinforced concrete, separate structural drawings are required to show the reinforced concrete design. Generally accepted practice in building construction requires the structural drawings to show number or spacing, size, and position of all reinforcement, with sufficient dimensions so that length and bend points can be computed for each bar. In bridge design, usual practice is for the designer also to list for each type of bar required, the mark numbers, bending dimensions and lengths, and number required. In all reinforced concrete design, structural drawings should also indicate location of all splices, and construction, expansion, and contraction joints. Many of these requirements can be transmitted for typical conditions by means of general notes on the drawing combined with specification provisions. Grade of bars is usually designated in specifications—perhaps only in specifications where only one grade is to be used, although best practice is to repeat this requirement in general notes on the structural drawings. Where more than one grade of bars is employed, structural drawings must also indicate location of each.

This chapter will concern itself with the structural drawings and specifications on the drawings. The civil and/or structural engineer is generally responsible for the structural drawings and the part of the specifications that is pertinent. Basically, structural drawings are prepared for the direction of construction by the general contractor. The first use will be for the preparation of a cost estimate for bidding. The

*Illustrations in this chapter are courtesy of Concrete Reinforcing Steel Institute.

**Vice President, Engineering, Concrete Reinforcing Steel Institute, Shaumburg, Illinois.
***Consulting Structural Engineer, Bethlehem, Penna.

precision of cost estimates depends upon the precision of drawings. Maximum benefits from competitive bidding can be achieved only by providing complete information required for the estimates. The information necessary for an estimate includes dimensions of the concrete for computation of form areas and concrete quantities. Strength and type of concrete required must be prescribed and if more than one, elements utilizing each must be identified so that quantities may be computed.

All concrete outlines must be clearly defined and coordinated with the architectural drawings. Errors or omissions in drawings will encourage faulty estimates and can result in costly delays of subsequent construction. One common error in drawings is contradictory or incompatible structural and architectural details. Columns must be properly oriented—schedules must show relative location. Wall and column footing elevations must be shown; steps in wall footings must be completely defined. Brick ledges properly need to be shown in wall elevations, to define clearly the exact elevations and steps at changes in elevation.

All openings, whether in walls or floor systems, must be completely defined; that is, sizes of opening and relative location must be dimensioned. Usually, this information is provided on the architectural drawings, but if not, it should be included on the structural drawings.

The designer should indicate locations at which construction joints will be acceptable to him, overall limits on length cast at one time, if any, and any other limitations imposed by the design on extent of work that may be scheduled for one casting operation. Each general contractor's bid will be predicated on a schedule best suited to his operation, where the choice is open.

23.2 ALTERNATE DESIGNS

Where the contract documents call for a base bid plus alternates, it is most important to delineate the extent of the base bid and then define each alternate, whether it is an addition or a deduction from the base bid. Frequently, confusion arises because the alternate not only adds or deducts, but also modifies some element already covered under the base bid. The estimator must then, in effect, deduct the original element from the base bid and add the modified element effect before adding or deducting the other elements that merely add or deduct. An illustration of this is shown in Fig. 23-1.

The area covered under the base bid is shown as Area 1. Alternate #1 added Area A plus modified Area B from the original, and the contract requirements were that the Alternate #1 price was to include only the additional monies bid over and above the base bid. The estimator would have to first estimate Area 1 for the Base Bid. Then he would estimate Area A for addition under Alternate #1. Then he would separate costs/quantities for Area B as originally designed and deduct this amount under Alternate #1. Finally, he would estimate Area B as modified under Alternate #1, for addition to same. The material requirements for Alternate #1 would then be computed as follows:

Alternate #1 = Area A + modified Area B − original Area B

The price for Alternate #1 would be based upon these material requirements.

Reinforcing bars and welded wire fabric need to be completely defined for the reinforced concrete estimate. This requirement can be presented in plan, in section, in separate schedules, or in any combination of the three—the important point is that inclusion of complete data is mandatory.

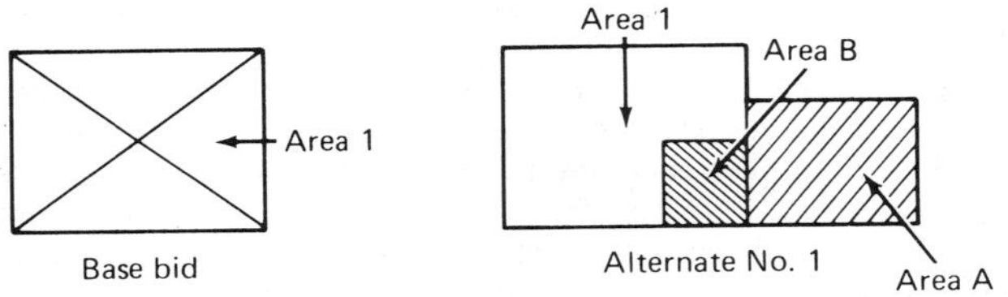

Fig. 23-1 Alternate bid areas.

23.3 REINFORCING BARS, SPIRALS, AND WELDED-WIRE FABRIC

All reinforcing bars and related material should be shown on the structural drawings, including the grade of steel. Even though the architectural drawings and specifications are part of the contract documents, complete reinforcing information should be shown on the structural drawings, as the reinforcing steel estimator is working primarily from structural drawings, and, under pressure of bid date deadlines, could easily overlook critical information located elsewhere.

Reinforcing materials are almost always furnished, and sometimes placed, under subcontracts. Different types of reinforcing steel, such as prestressing steel and fittings, conventional reinforcing bars, welded wire fabric, structural steel shearheads, etc., are often supplied and/or placed under separate subcontracts. Each subcontract supplier must estimate his specialty, and the chance of error is heightened by the sheer diversity. Omissions in estimates can be minimized if structural drawings are complete and clear, even if repetition of some specification provisions is necessary. Erroneous estimates can be costly to all parties involved. Ambiguous or incomplete drawings result in varying "allowances" by the estimators, and usually result in a wide scatter of subcontract bid prices.

23.4 ESTIMATING

The single item of reinforcing material is usually the largest subcontracted. Separate estimates to furnish reinforcement are ordinarily prepared by supplier subcontractors. The information required for a complete reinforcing steel estimate is needed quickly. It must be shown completely and without ambiguity if close estimates are to be secured. Time usually is not available to secure any information lacking on the drawings or clarifications of details on drawings. For the estimator, the drawings must clearly show:

1. Grade and size of all bars. Where more than one grade of bars is specified, the extent of each must be clearly shown.
2. Quantity as in columns and beams, or spacing as in slabs and walls. It is important to indicate clearly limits of reinforcing where spacing is given, especially where the reinforcing may change size and/or spacing (extent of each welded wire fabric style must also be shown).
3. Length of bars that do not run continuously, such as negative moment reinforcing in a beam.
4. Any unusual tolerance requirements in fabrication or placing. See Figs. 23-2 through 23-5 for standard tolerances and certain special tolerances for bar fabrication and placing.
5. Required bending. The estimator must separately compute quantities of rebars, straight in each size, light and heavy bending, and special fabrication for bending

STANDARD FABRICATING TOLERANCES

For bar sizes #3 through #11; see Fig. 23-3 for tolerance symbols.

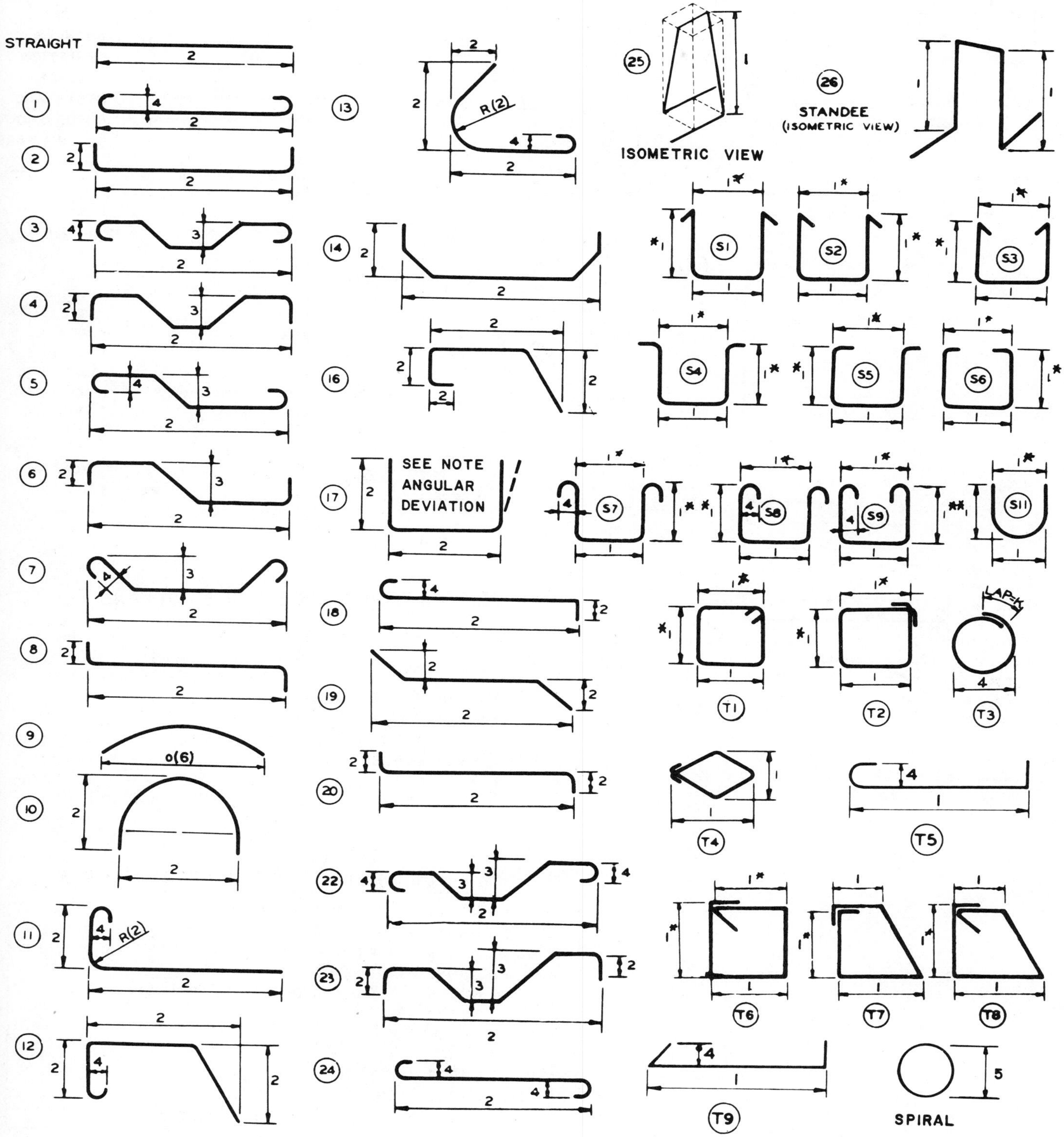

imensions on this line are to be within tolerance shown but are not to differ from the opposite parallel dimension more than ½".
ll tolerances single plane and as shown.
ngular Deviation — maximum ±2½° or ±½"/ft., but not less than ½", on all 90° hooks and bends.
olerances for Types S1-S[illegible], S11, T1-T9 apply to bar sizes #3-#8 inclusive only.

Fig. 23-2 Standard bending tolerances. Bar sizes #3 through #11.

STANDARD FABRICATING TOLERANCES

For bar sizes #14 and 18.

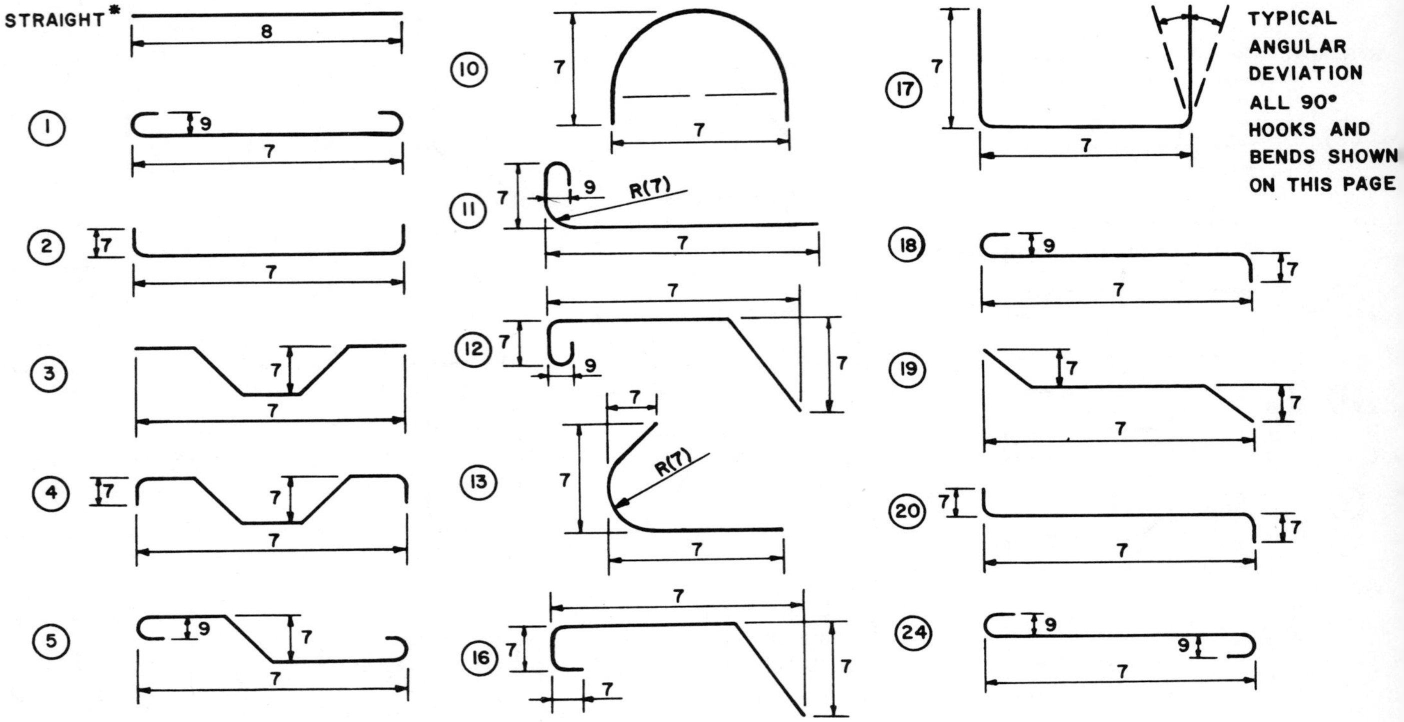

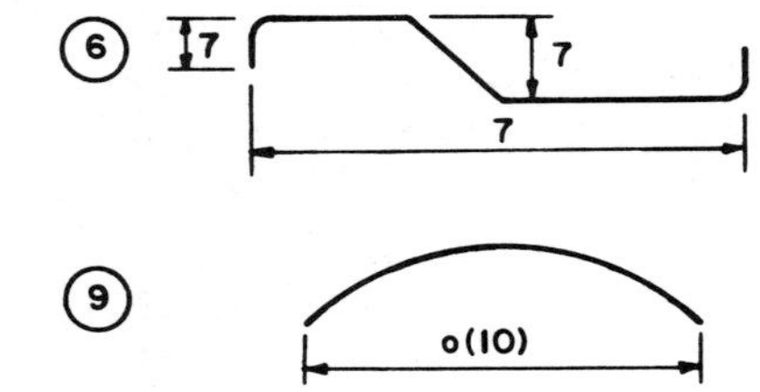

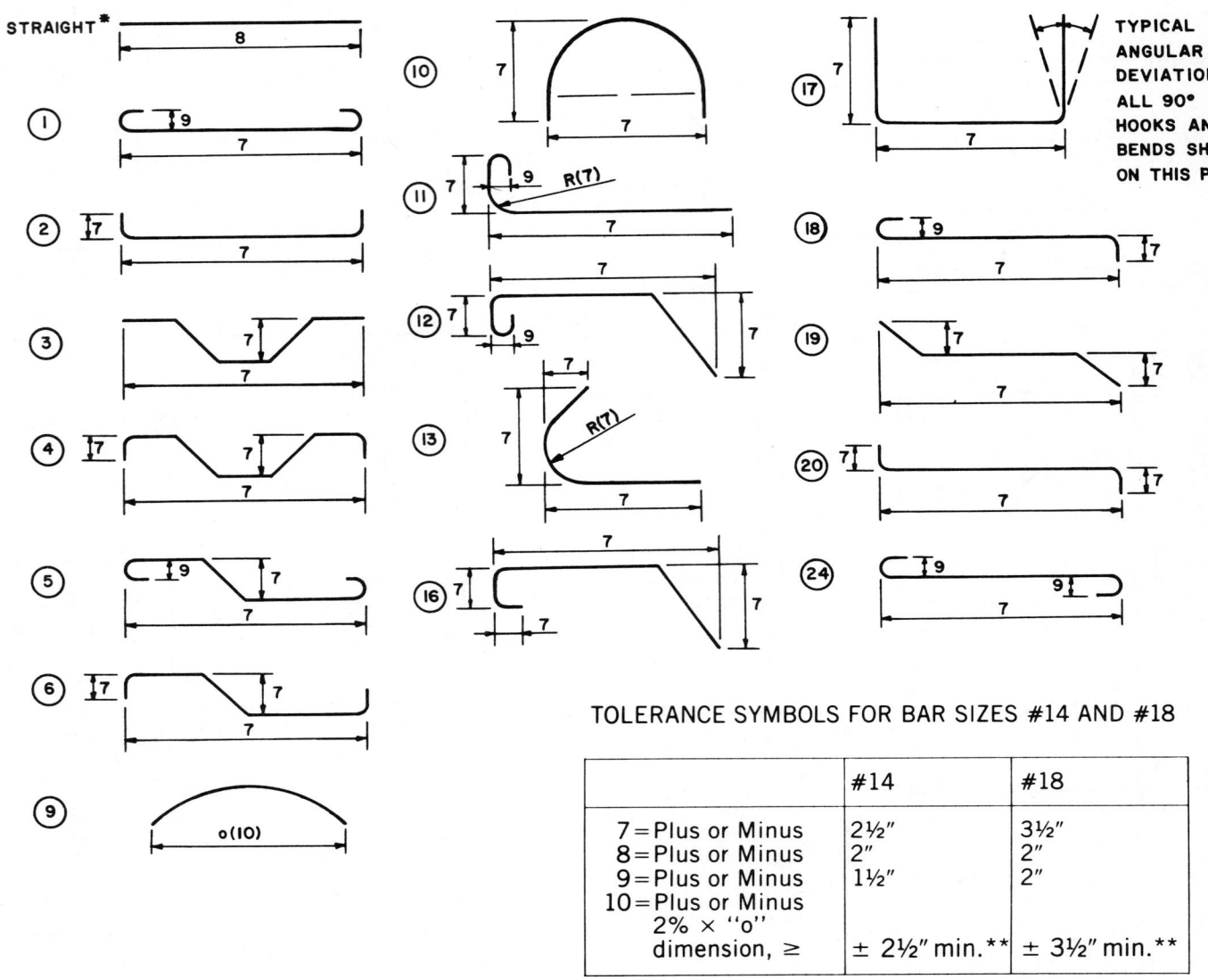

TOLERANCE SYMBOLS FOR BAR SIZES #14 AND #18

	#14	#18
7 = Plus or Minus	2½″	3½″
8 = Plus or Minus	2″	2″
9 = Plus or Minus	1½″	2″
10 = Plus or Minus 2% × "o" dimension, ≥	± 2½″ min.**	± 3½″ min.**

TOLERANCE SYMBOLS FOR BAR SIZES #3 THROUGH #11

1 = Plus or Minus ½″ for bar sizes #3, #4, and #5 (gross length < 12′-0″)
1 = Plus or Minus 1″ for bar sizes #3, #4, and #5 (gross length ≥ 12′-0″)
1 = Plus or Minus 1″ for bar sizes #6, #7 and #8
2 = Plus or Minus 1″
3 = Plus 0″, Minus ½″
4 = Plus or Minus ½″
5 = Plus or Minus ½″ for diameter ≤ 30″
5 = Plus or Minus 1″ for diameter > 30″
6 = Plus or Minus 1.5% × "o" dimension, ≥ ± 2″ minimum**

All tolerances single plane and as shown.
Angular Deviation — maximum ±2½° or ±½″/ft on all 90° hooks and bends.
*Saw-cut both ends — overall length ±½″.
**If application of positive tolerance to Type 9 results in a chord length ≥ the arc or bar length, the bar may be shipped straight.

Fig. 23-3 Standard bending tolerances. Bar sizes #14 and #18.

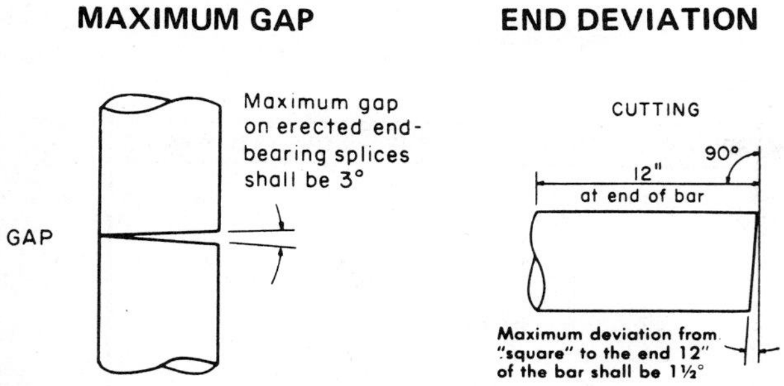

Fig. 23-4 Nominally square saw-cut ends. Maximum gap tolerance for spliced bars which transmit compressive stresses through direct end-bearing: for adequate structural performance, the total angular deviation of the gap shall not exceed 3°, as shown.

In order to achieve a proper fit in the field, the ends of the bars must be saw-cut, or otherwise cut in such a manner as to provide a reasonably flat surface. It is recommended that the end deviation of an individual bar from "square" not exceed 1.5° when measured from a right angle to the end 12 in. of the bar as shown. Relative rotation or other field adjustment of the bars may be necessary during erection to secure a fit that falls within the recommended gap limits.

to nonstandard radii. See Fig. 23-6 for standard end hooks and bend radii.

Clear and specific details are mandatory on splicing of reinforcing bars and welded wire fabric. Under ACI 318-83, it is necessary for the designer to show first, the type of splice required (i.e., butt or lapped); whether in tension or compression; and, if butt spliced, whether mechanical and/or arc welded splices are permitted. In the case of lapped splices, he must also indicate length or class of lap for all conditions and specifically identify where each applies. For long runs of continuous bars, such as temperature reinforcement, the designer must indicate where splices are permitted, if restricted, and arrangements for stagger, if required; or conversely, state that splices may occur at the contractor's option.

Welded (smooth and deformed wire) fabric splices must be dimensioned. Locations where they will be allowed must be clearly stated. End anchorages of all fabric should be shown in details.

When prestressing is involved, the structural drawings must indicate quantity, type, and special fittings, conduits, or sheaths needed, as well as general requirements such as initial force, assumed losses, final force, grouting, etc., that might restrict use of various proprietary systems.

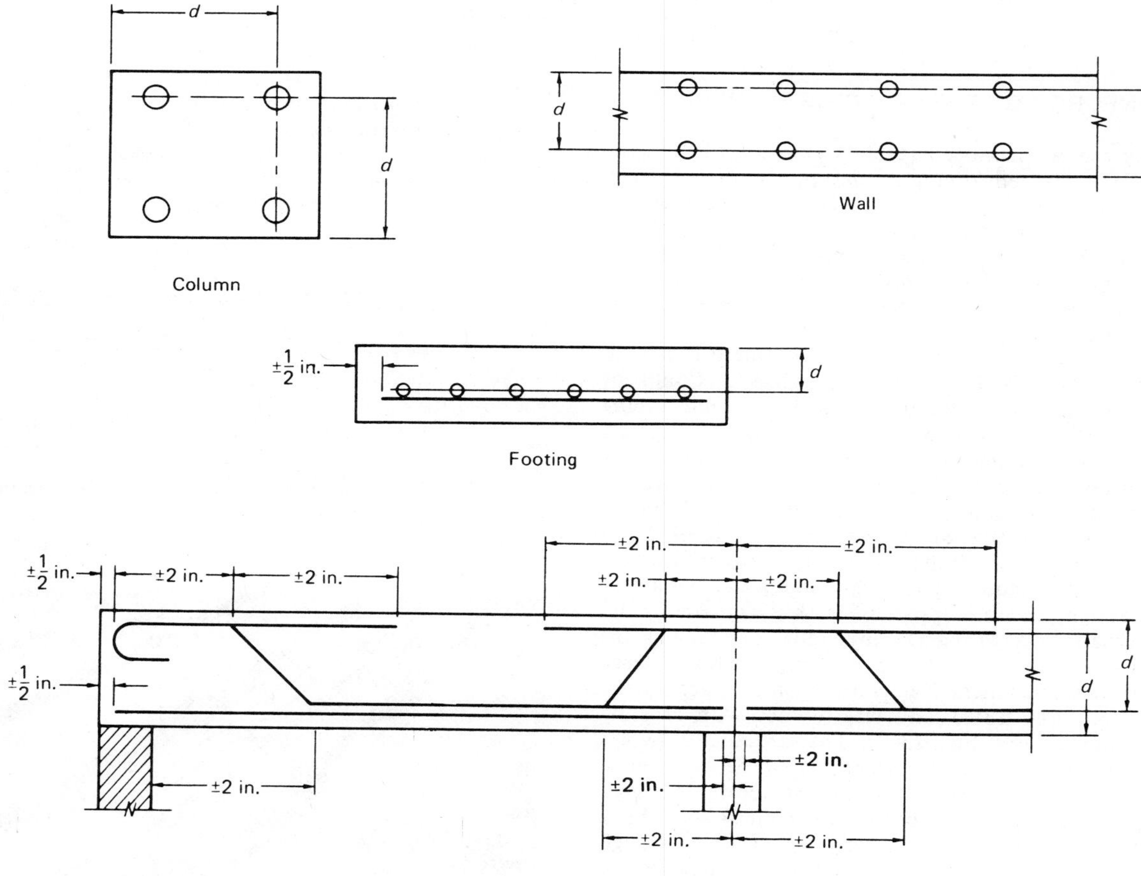

Beam, slab, joist, etc.

Tolerance on dimension "*d*" and clear cover where:

$d \leqslant 8$ in. $\pm \frac{3}{8}$ in. on d; $-\frac{3}{8}$ in. on min. cover

$d > 8$ in. $\pm \frac{1}{2}$ in. on d; $-\frac{1}{2}$ in. on min. cover

Fig. 23-5 Standard placing tolerances (ACI 318-83).

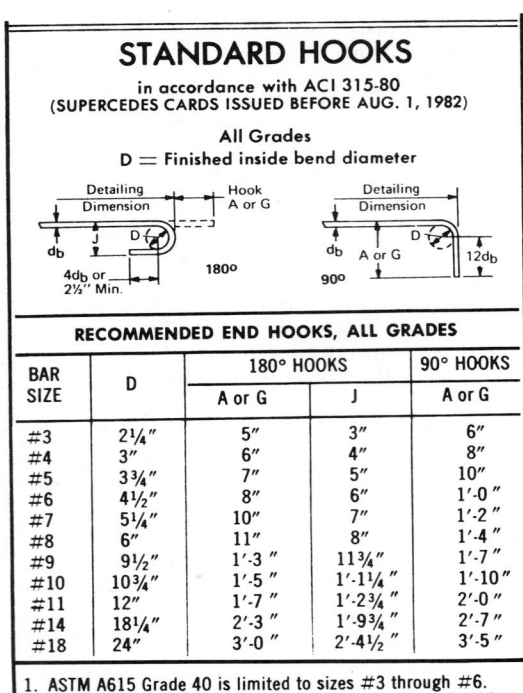

STANDARD HOOKS

in accordance with ACI 315-80
(SUPERCEDES CARDS ISSUED BEFORE AUG. 1, 1982)

All Grades

D = Finished inside bend diameter

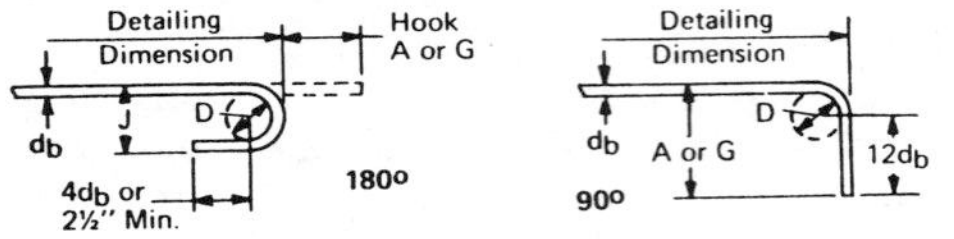

RECOMMENDED END HOOKS, ALL GRADES

BAR SIZE	D	180° HOOKS		90° HOOKS
		A or G	J	A or G
#3	2¼"	5"	3"	6"
#4	3"	6"	4"	8"
#5	3¾"	7"	5"	10"
#6	4½"	8"	6"	1'-0"
#7	5¼"	10"	7"	1'-2"
#8	6"	11"	8"	1'-4"
#9	9½"	1'-3"	11¾"	1'-7"
#10	10¾"	1'-5"	1'-1¼"	1'-10"
#11	12"	1'-7"	1'-2¾"	2'-0"
#14	18¼"	2'-3"	1'-9¾"	2'-7"
#18	24"	3'-0"	2'-4½"	3'-5"

1. ASTM A615 Grade 40 is limited to sizes #3 through #6.
2. Check availability with local suppliers for #14 and #18.

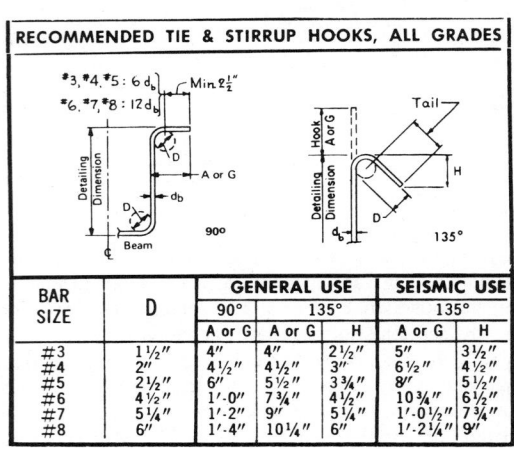

RECOMMENDED TIE & STIRRUP HOOKS, ALL GRADES

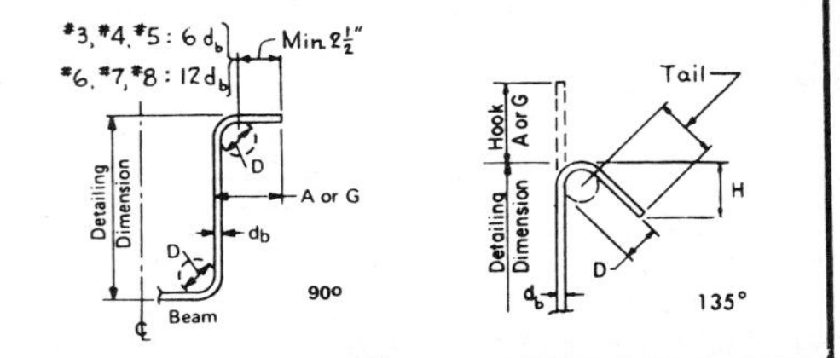

BAR SIZE	D	GENERAL USE			SEISMIC USE	
		90°	135°		135°	
		A or G	A or G	H	A or G	H
#3	1½"	4"	4"	2½"	5"	3½"
#4	2"	4½"	4½"	3"	6½"	4½"
#5	2½"	6"	5½"	3¾"	8"	5½"
#6	4½"	1'-0"	7¾"	4½"	10¾"	6½"
#7	5¼"	1'-2"	9"	5¼"	1'-0½"	7¾"
#8	6"	1'-4"	10¼"	6"	1'-2¼"	9"

Fig. 23-6 Standard end hooks.

23.5 REINFORCING STEEL PLACING DRAWINGS

Once an agreement has been reached between the general contractor and the reinforcing steel supplier, the contract documents are turned over to the supplier's engineering department for processing. The exact sequence of building the structure must now be determined to establish the sequence in which the placing drawings are prepared for the rebar or other reinforcing materials and the sequence for fabrication and shipment.

The reinforcing steel detailer needs complete detail information to prepare the placing drawings and bills of material. It is his task to define each and every reinforcing bar, spiral, piece of welded wire fabric, or prestressing unit required, as the case may be, for the entire contract. This precision is only possible when the contract drawings are clear, complete, and concise. Typical details should really be "typical"! Details that are similar to the typical detail drawn should all have dissimilarities noted as exceptions.

Many difficulties requiring redesign, due to inadequate or improper design, become apparent at this stage as the detailer is actually calculating the individual bar dimensions and lengths, and must check for shipping limitations (see Fig. 23-7), standard unless special arrangements are made), clearances (see Fig. 23-8), and interference with crossing bars and conduits (see Fig. 23-9). The structural drawings should always be checked for these points before release to bidders. Typical situations that the structural engineer preparing the drawings or responsible for same should check are: (1) width clearance for grider bars passing through columns; (2) depth clearance and cover of crossing slab and

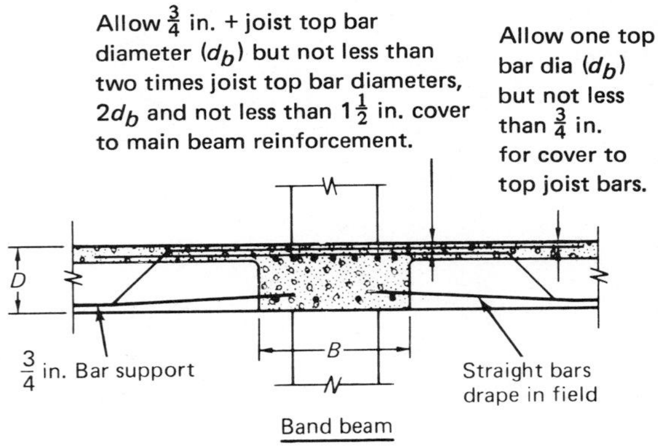

Fig. 23-8 Clearances for joist–band beam connections.

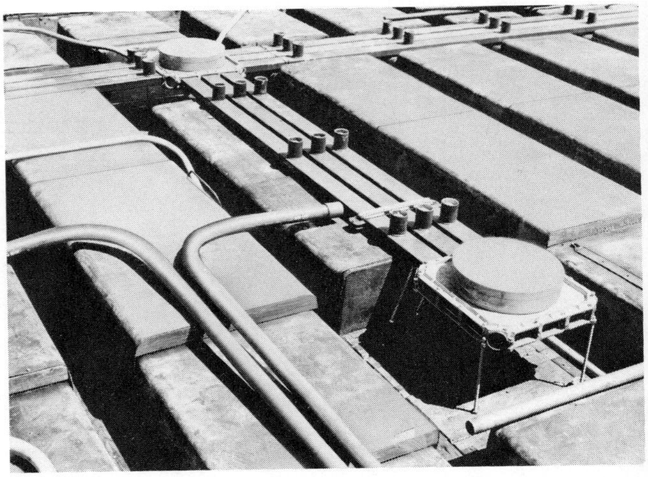

Fig. 23-9 Conduit in place, ready for reinforcement.

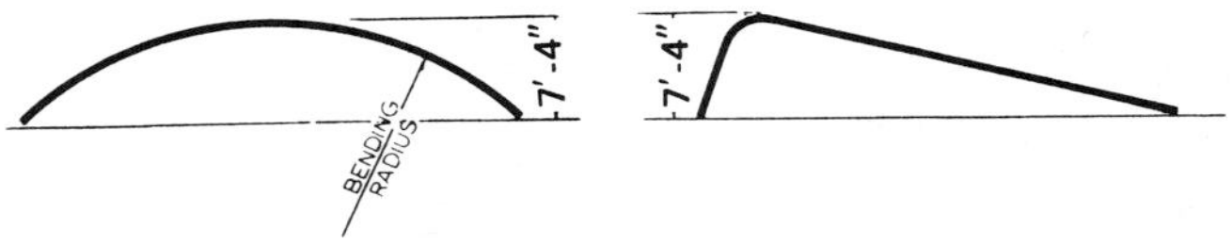

Fig. 23-7 Maximum dimension of bent bars for truck delivery.

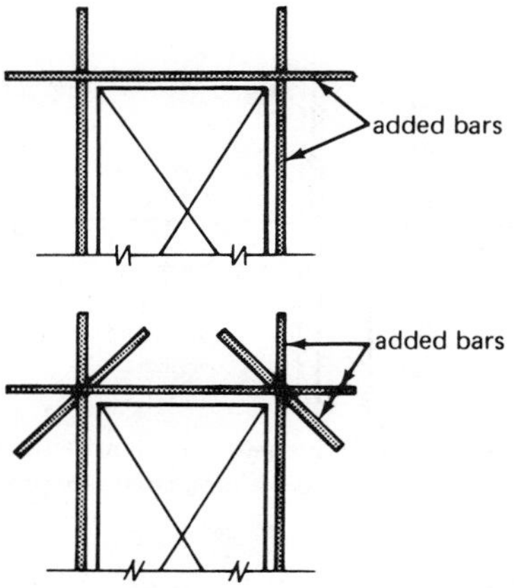

Fig. 23-10 Two common details of special reinforcement at openings in walls or slabs.

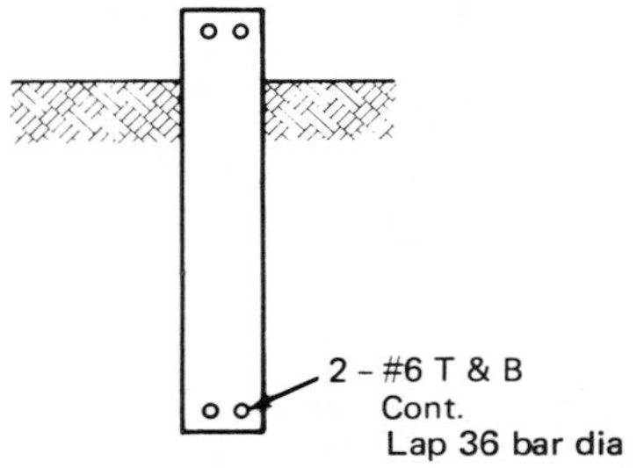

Fig. 23-11 Grade beam. NOTE: Whenever word "continuous" appears in bar description, additional explanation is required in General Notes, or in description of bar.

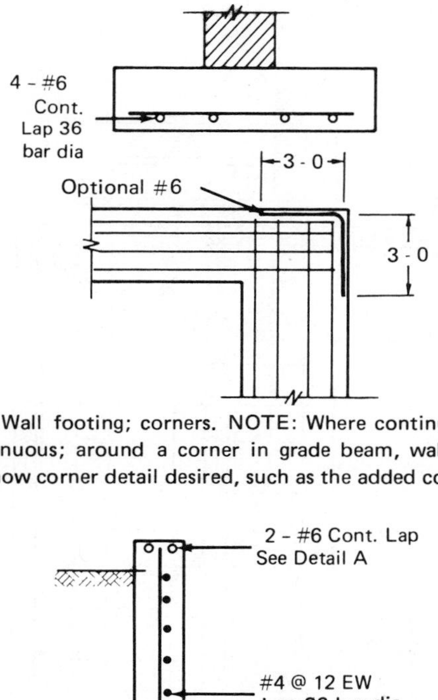

Fig. 23-12 Wall footing; corners. NOTE: Where continuous bar is to be continuous; around a corner in grade beam, wall footings, wall, etc., show corner detail desired, such as the added corner bar.

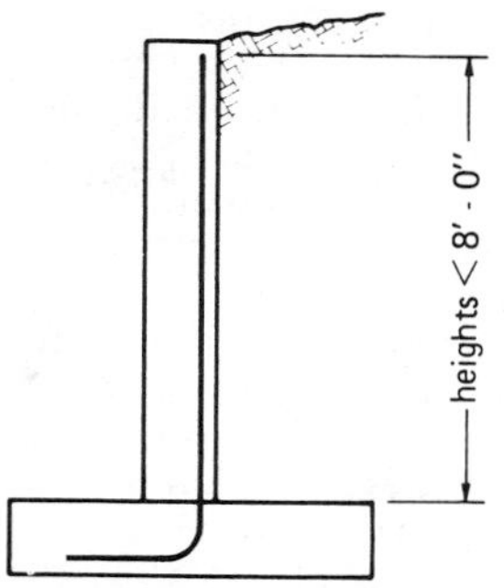

Fig. 23-14 Low retaining wall. Avoid splicing main steel. (Other bars not shown).

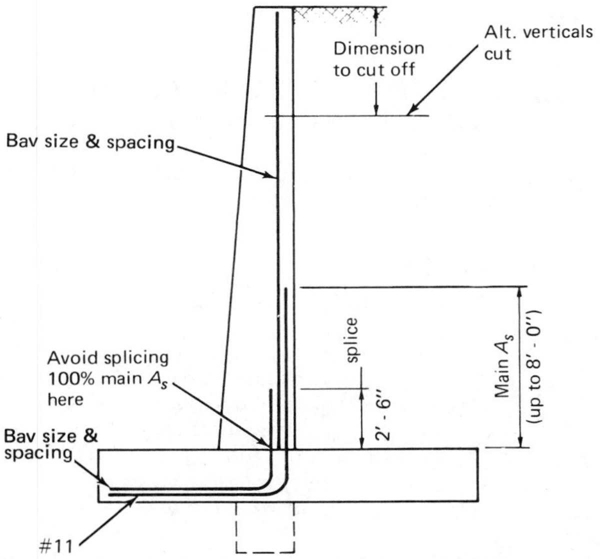

Fig. 23-15 Retaining wall requiring splices in main steel. Simple pattern to stagger splices.

beam, joist and beam, or beam and grider bars; (3) cover and depth for end hooks on truss bars in shallow slabs or joists; (4) interference with conduit by other trades; and (5) bar detail for holes in wall or slab.

23.6 APPROVALS

It is standard practice to submit placing drawings for approval prior to fabrication.

Placing drawings are submitted by the reinforcing steel

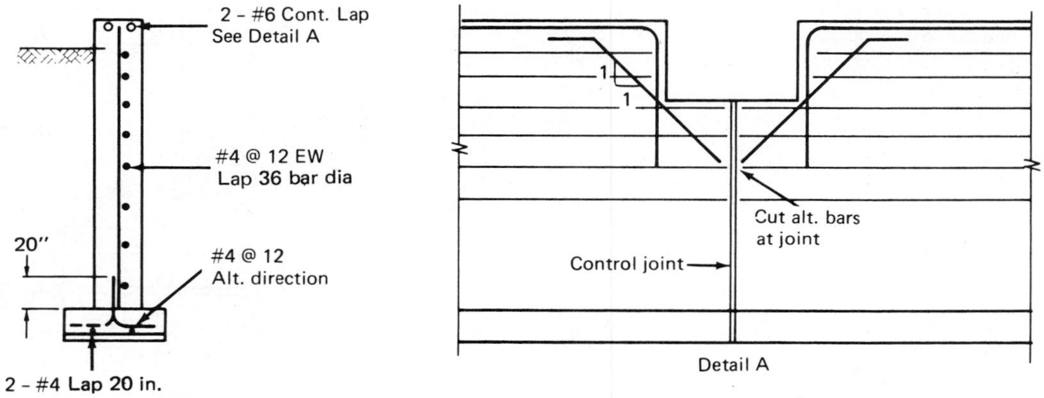

Fig. 23-13 Continuous bars require explanatory detail at interruptions.

fabricator (supplier) to the general contractor for approval by the structural engineer. At this time, the designer reviews the interpretation of his design by the detailer and either approves or notes any corrections, changes, or additions, Two important considerations at this time are: (1) changes necessary to revise, complete, or refine the design; and (2) delays in approval.

If changes in the design require more reinforcing material or additional time for reestimating or redetailing, the question of extra payment is involved. Undue delay in return of approved placing drawings may result in extended construction schedules which also increase costs. Changes during construction are always costly. Cost of changes to correct errors in material, field practice, ambiguous details, etc., are usually negotiated. The obvious best policy is to minimize problems by preparing complete, clear, concise contract drawings.

23.7 PRACTICAL CONSIDERATIONS FOR MOST EFFICIENT PRESENTATION OF DESIGNS

At this point various reinforced concrete elements will be discussed in some detail:

23.7.1 Foundations

Wall and column footing elevations must be shown and clearly marked as to whether they are top or bottom of footing. If intended to be cast separately, construction joints should be shown, reinforcement designed accordingly, and casting sequence indicated if essential to design. The designer must show (1) whether continuous bars in wall footings run through the column footing or (2) extend into the column footing some specific distance, or (3) whether separate dowels are to be provided out of the column footing. If corner bars are required in continuous footings or walls, they should be shown. Continuous large-size bars are frequently shown at the top and bottom in foundation walls. Where pockets occur at top of wall, the designer should indicate how the continuous reinforcement is to be detailed at these points. Brick ledges in foundation walls

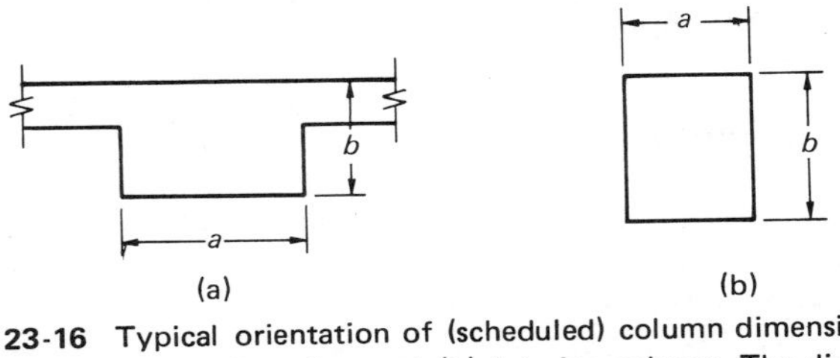

Fig. 23-16 Typical orientation of (scheduled) column dimensions. (A) Wall column orientation and (b) interior column. The dimensions are a = first dimension scheduled, and b = second dimension scheduled. Identify dimensions by a note, such as, (a) is N–S dimension and (b) is E–W dimension.

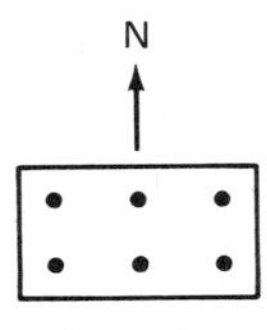

Fig. 23-17 Rebar locations in columns. NOTE: If nontypical, rebar plan sketch must be shown for each column at each change in rebars in schedule. If typical, General Note may be used, such as, "All column verticals in N–S faces of columns."

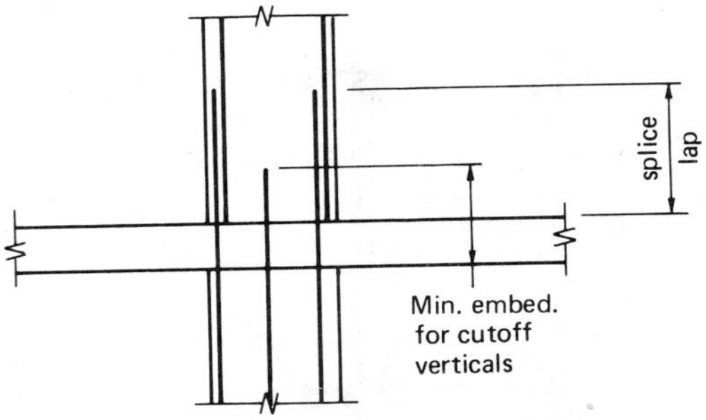

Fig. 23-18 Show minimum embedment required where standard cutoff 3 in. below finished floor will not satisfy design.

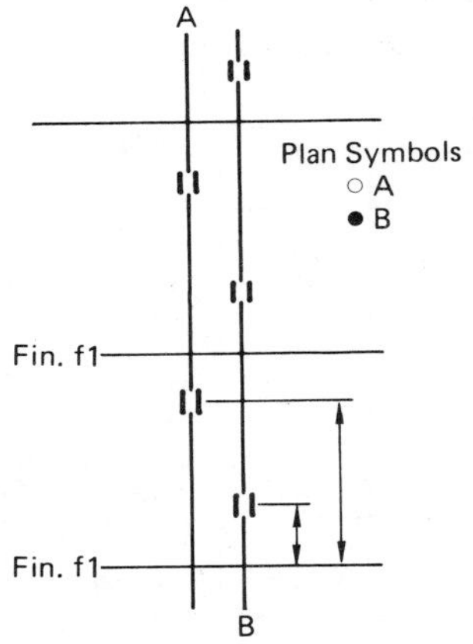

Fig. 23-19 Column elevation to show staggered splices. (Show dimensions to locate from finished floor or top of footing in schedule.)

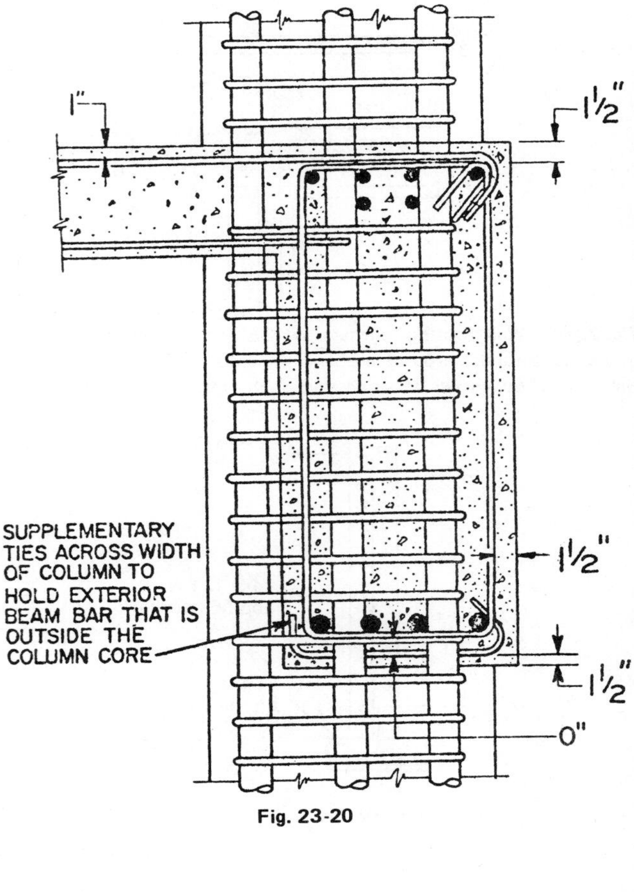

Fig. 23-20

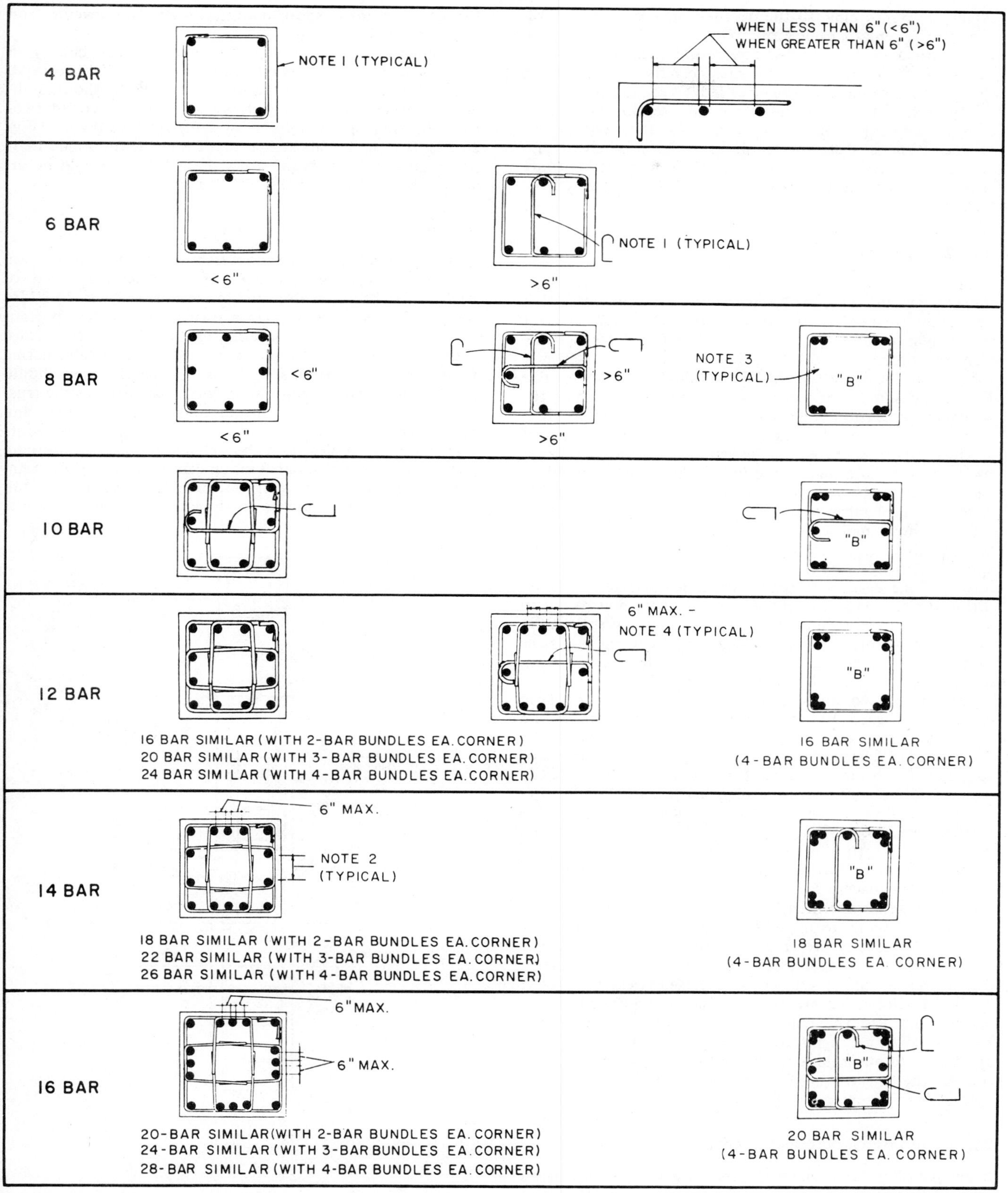

***Applicable for either preassembled cages or erection in place on free-standing butt spliced vertical bars.**

Fig. 23-21 Universal standard column ties.*

affect the reinforcing; so the detailer must know the elevation and extent of each. Columns and piers in foundation walls with brick ledges must be carefully located by the designer. The contract drawings must clearly show whether the column size includes the brick ledge width or not. Notes on typical sections through walls and section arrows on plan or elevation should be used to show the exact extent of each typical section. Cantilever walls must clearly indicate where splices are permitted and arrangement of any alternating vertical bars. (See Figs. 23-10 through 23-15, inclusive.)

23.7.2 Columns

Columns are usually shown on contract drawings in schedule format with cross sections, typical or otherwise. It is important to identify clearly column faces, especially for rectangular columns. One scheme that facilitates designation of direction for long side–short side in a separate schedule is shown in Fig. 23-16. The small key drawing like Fig. 23-16 should be placed on each sheet containing the schedule or portions of a schedule to which it refers. It is also important for the designer to indicate the number of vertical bars in each face. Where more than one bar size is used, location in a plan detail is essential. Such plan details should usually be in the schedule at each story where bar arrangement changes. (See Fig. 23-17.)

Where more than one size of rebar is required in a column, an elevation detail is necessary to locate the bars and splices. Usually a typical joint splice detail will suffice; however, the designer should provide for nonstandard conditions (Fig. 23-18) and show exactly where dowels are to be cut off for embedment or continuation by splices. He must specify splice and anchorage lengths for all bar sizes involved, and indicate when and how far to anchor terminated bars from a column below into one above.

Butt-spliced columns frequently require a schematic elevation of each column bar (Fig. 23-19) to show the relationship of staggered splices. As the column schedule is filled in, the designer should check spaces and number of spaces at the beam–column intersection to ensure that the bars will physically pass by each other. Figure 23-20 shows a careful scale drawing to illustrate such a check at a difficult intersection. Where such a study is required by the designer simply to select bar sizes, cover, etc., it should be added to his structural drawing for information to the detailer and the placer.

Column tie arrangements must be specifically shown by the designer, as there are several permissible (ACI) arrangements. It is usually preferable to use the ACI universal tie arrangement for ease of erection. (See Fig. 23-21.) Ninety-degree hooks are preferred for closed ties. Normally, column verticals that are not terminated at a floor level are offset bent into the column above. If the offset occurs below confining concrete on all sides, recommended design practice is to place three sets of ties at 3 in. (see Fig. 23-22) right at the offset bend. If it is desired, the designer should show this arrangement.

23.7.3 Beams, Girders, and One-Way Joists

Beams, girders, and one-way joists are usually shown on schedules, but sometimes are drawn in elevation and/or section. Extent of reinforcing as a percentage of the span is satisfactory as long as the detail really applies to *all* beams. The designer must check to determine if multiple layers of principal bars are required, owing to spacing limitations, and so indicate on his drawings. When stirrups are required, the designer must show quantity, size, and spacing, and ACI-bend types (see Fig. 23-23).

It is acceptable and often convenient, particularly in design schedules to designate bent bars by ACI bend type. When closed stirrups are required, the designer should indicate whether he will accept the ACI recommended two-piece units, or preferably, show same on his drawing (Fig. 23-24). If support bars at the top of the stirrups are desired, (in addition to bars required by design) they should be indicated by the designer (Fig. 23-25).

23.7.4 Slabs

Slab reinforcing may be shown right on the plan, in section, or in schedule format for one-way slabs. See Fig. 23-26 for design of slabs in schedule and in section, and Fig. 23-27 for design of same slabs on plan and in section. It is important to clearly indicate the alternating bars in one-way slabs. Figure 23-28 shows the wrong way to show alternating truss bars. The bottom straight bars in the plan view should not be indicated as continuous with truss bars, as the truss bars should be staggered in adjacent bays to provide uniformly spaced negative moment reinforcing. The schedule does not make it clear that the slab design requires alternate straight and truss bars at 9 in. on center. Temperature reinforcement should be clearly located—if in the bottom full width of slab, it should be so shown. If the designer wants

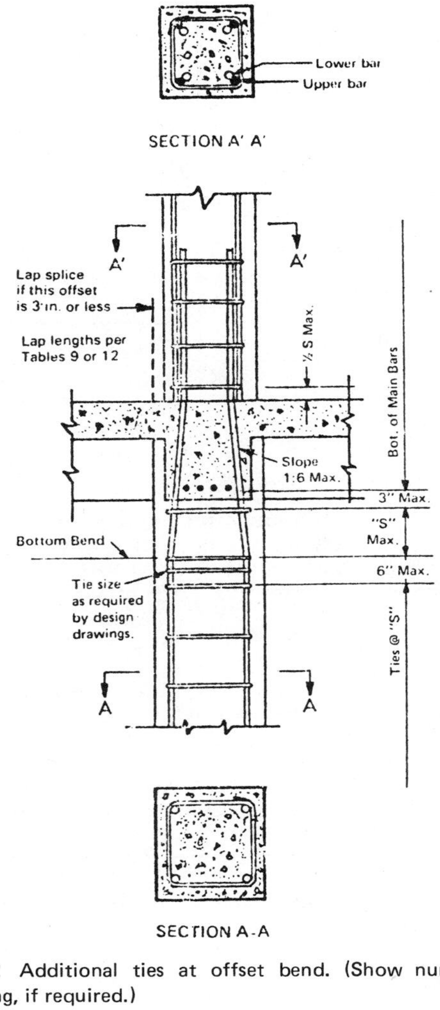

Fig. 23-22 Additional ties at offset bend. (Show number, size, and spacing, if required.)

TYPICAL BAR BENDS

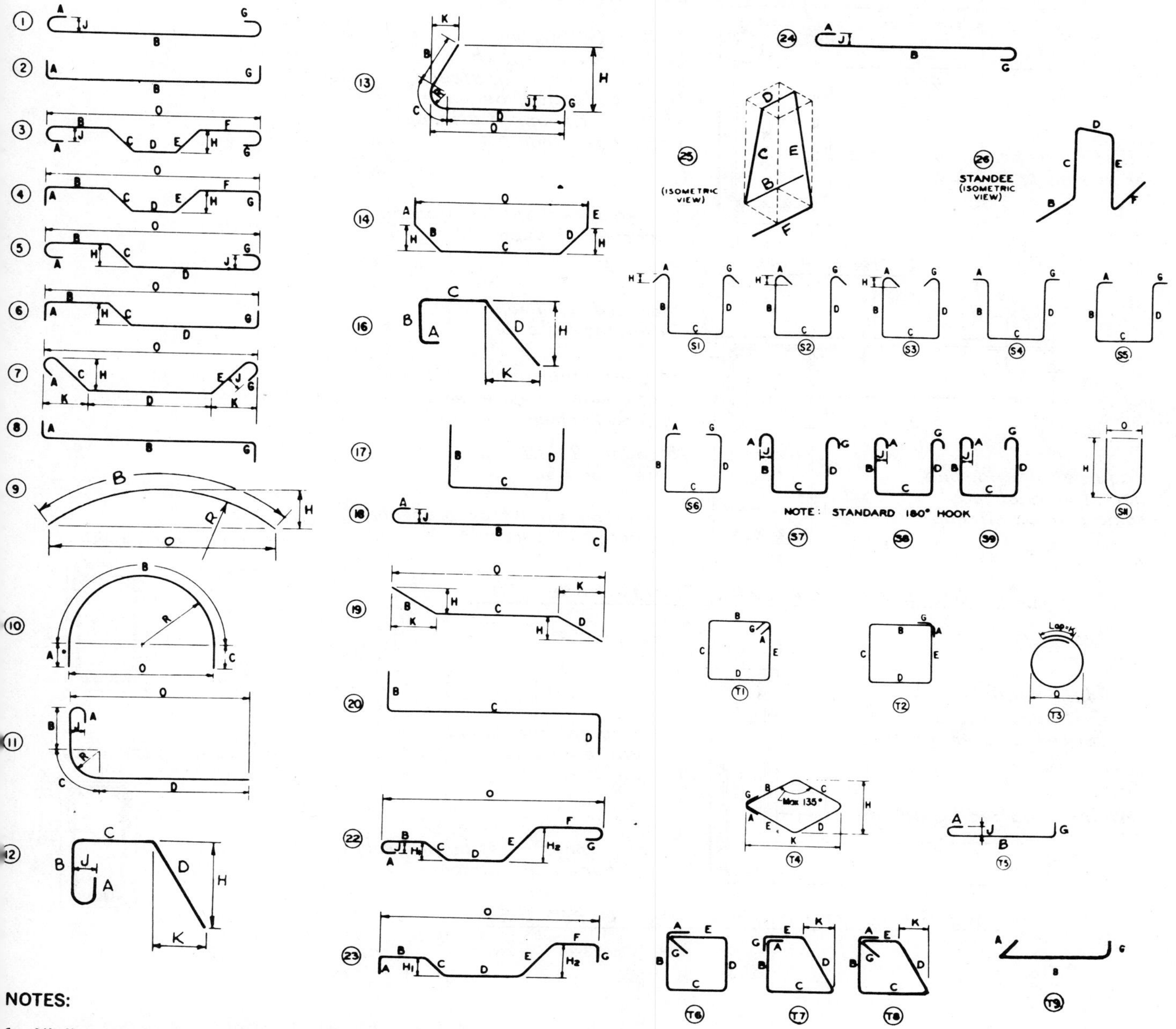

NOTES:

1. All dimensions are out-to-out of bar except "A" and "G" on standard 180° and 135° hooks.
2. "J" dimension on 180° hooks to be shown only where necessary to restrict hook size, otherwise standard hooks are to be used.
3. Where "J" is not shown, "J" will be kept equal to or less than "H" on truss bars. Where "J" can exceed "H", it should be shown.

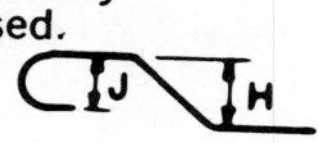

4. "H" dimension stirrups to be shown where necessary to fit within concrete.
5. Where bars are to be bent more accurately than standard bending tolerances, bending dimensions which require closer fabrication should have limits indicated.
6. Figures in circles show types.
7. For recommended diameter "D", of bends, hooks, etc., see tables, page 6-4.
8. Type S1-S9, S11, T1-T9 apply to bar sizes #3 through #8.

Unless otherwise noted diameter D is the same for all bends and hooks on a bar.

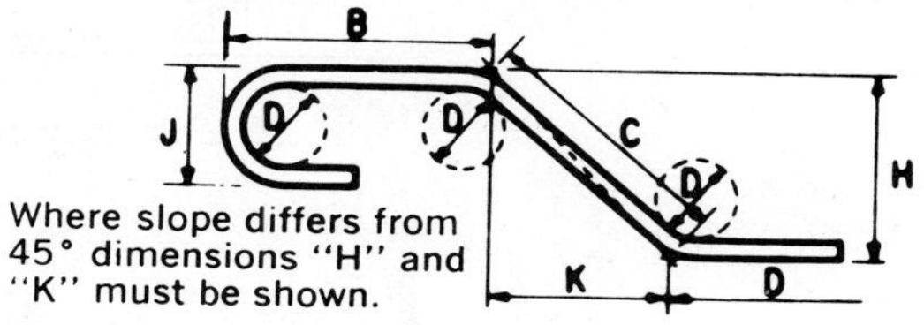

Where slope differs from 45° dimensions "H" and "K" must be shown.

ENLARGED VIEW SHOWING BAR BENDING DETAILS

Fig. 23-23 Typical bar bends

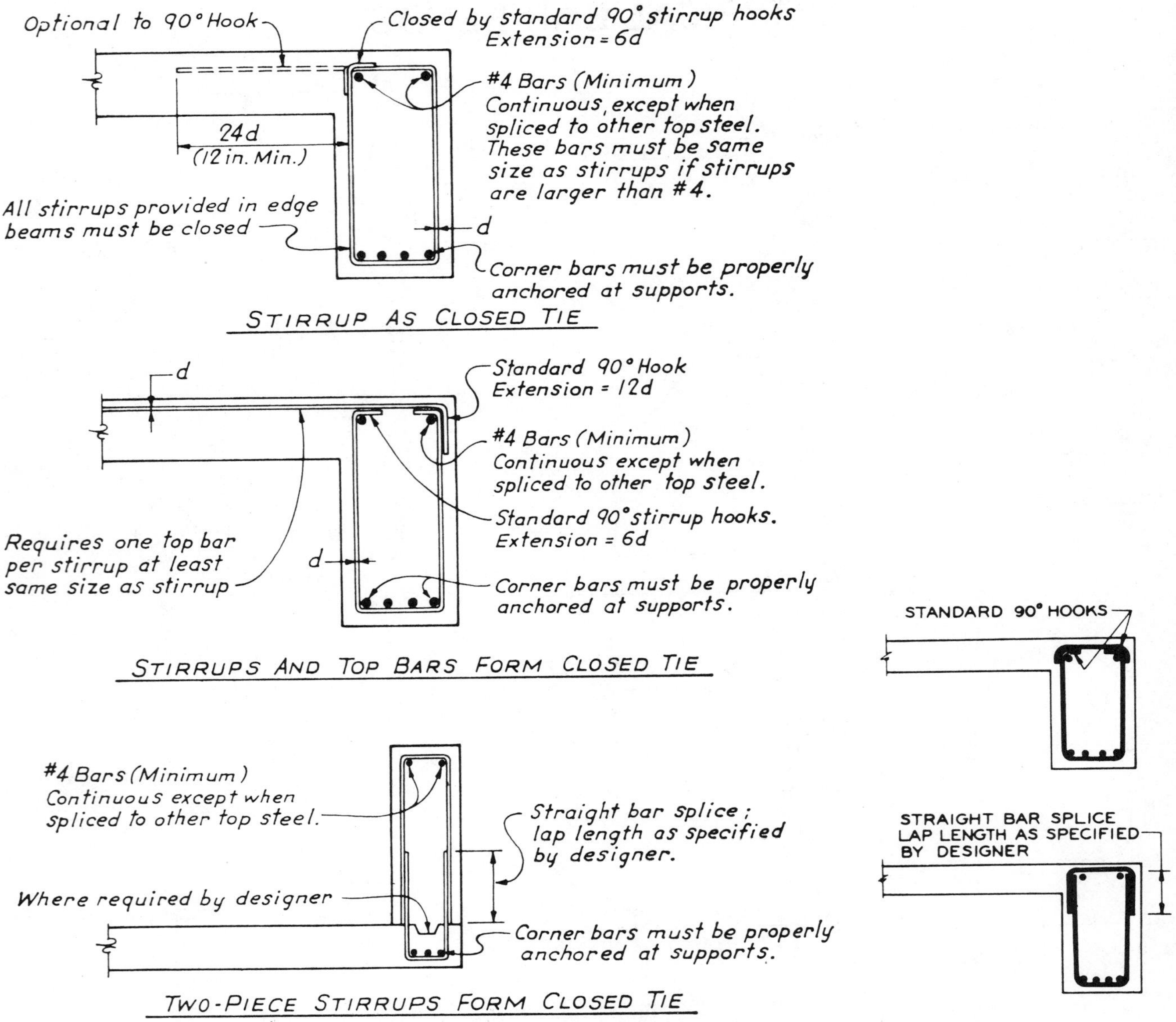

Fig. 23-24

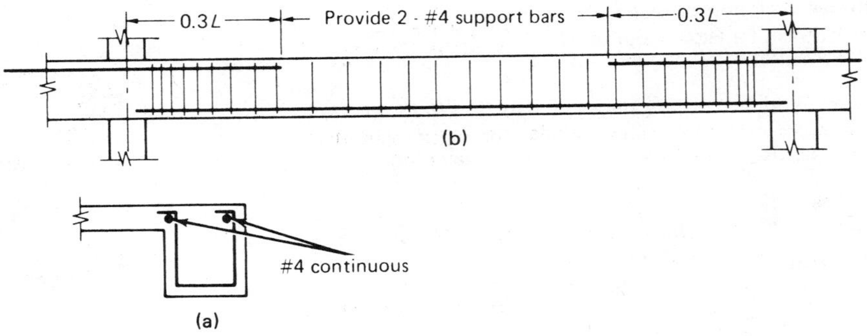

Fig. 23-25 If designer wishes to provide positive support for ties between cutoff points for main top bars, two alternative provisions are shown: (a) smaller corner bars, continuous as part of main steel; and (b) note for top "support bars" which will not be lap spliced for continuity.

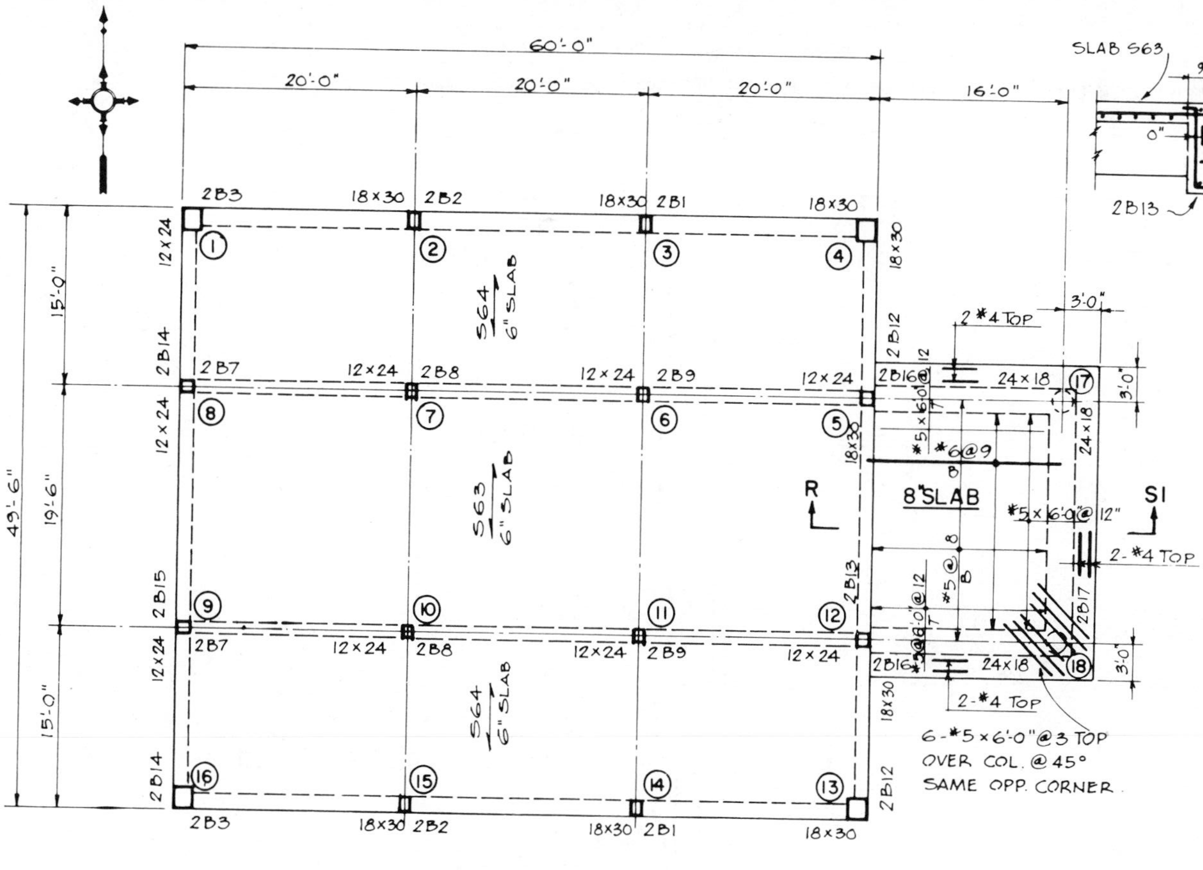

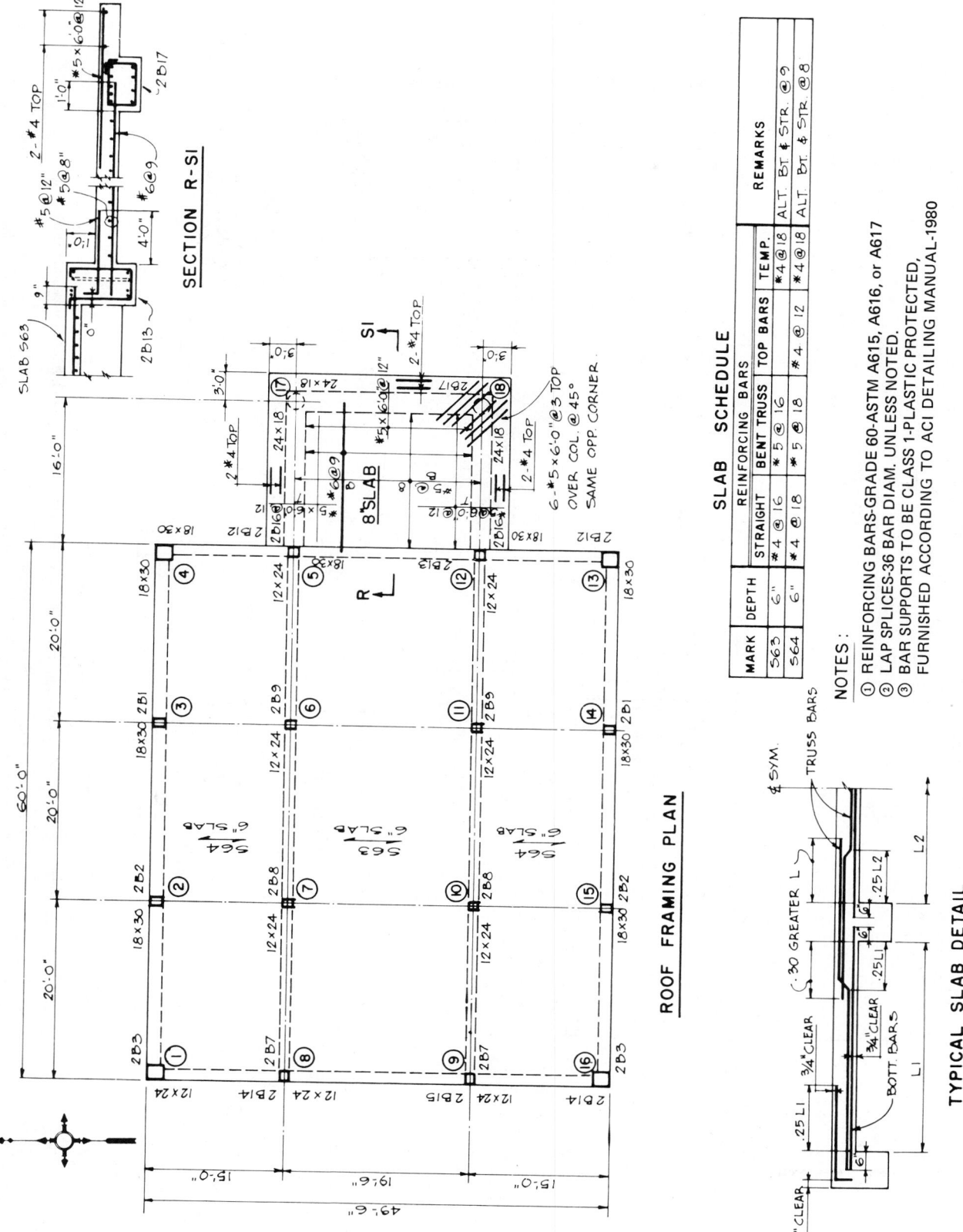

ROOF FRAMING PLAN

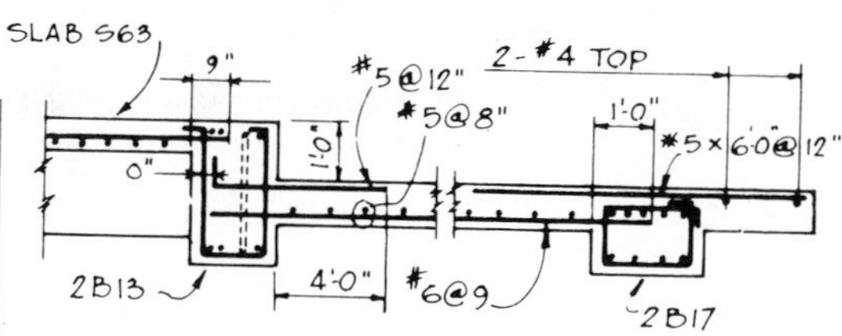

SECTION R-S1

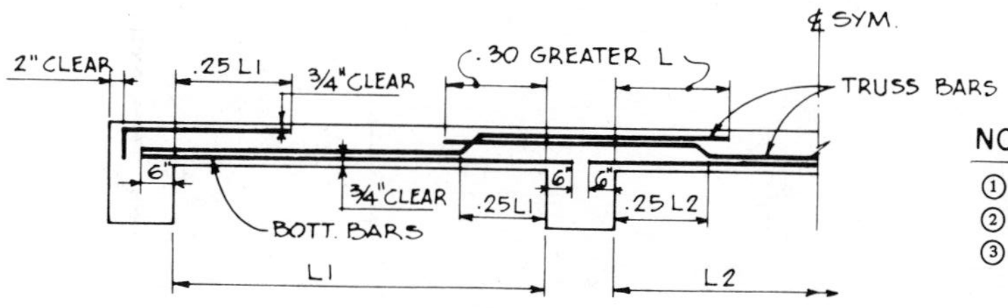

TYPICAL SLAB DETAIL

SLAB SCHEDULE

MARK	DEPTH	REINFORCING BARS				REMARKS
		STRAIGHT	BENT TRUSS	TOP BARS	TEMP.	
563	6"	#4 @ 16	#5 @ 16		#4 @ 18	ALT. BT. & STR. @ 9
564	6"	#4 @ 18	#5 @ 18	#4 @ 12	#4 @ 18	ALT. BT. & STR. @ 8

NOTES:

① REINFORCING BARS-GRADE 60-ASTM A615, A616, or A617
② LAP SPLICES-36 BAR DIAM. UNLESS NOTED.
③ BAR SUPPORTS TO BE CLASS 1-PLASTIC PROTECTED, FURNISHED ACCORDING TO ACI DETAILING MANUAL-1980

Fig. 23-26 Engineering drawing—Main slab reinforcement in schedule and special slab (RS-1) on detail section.

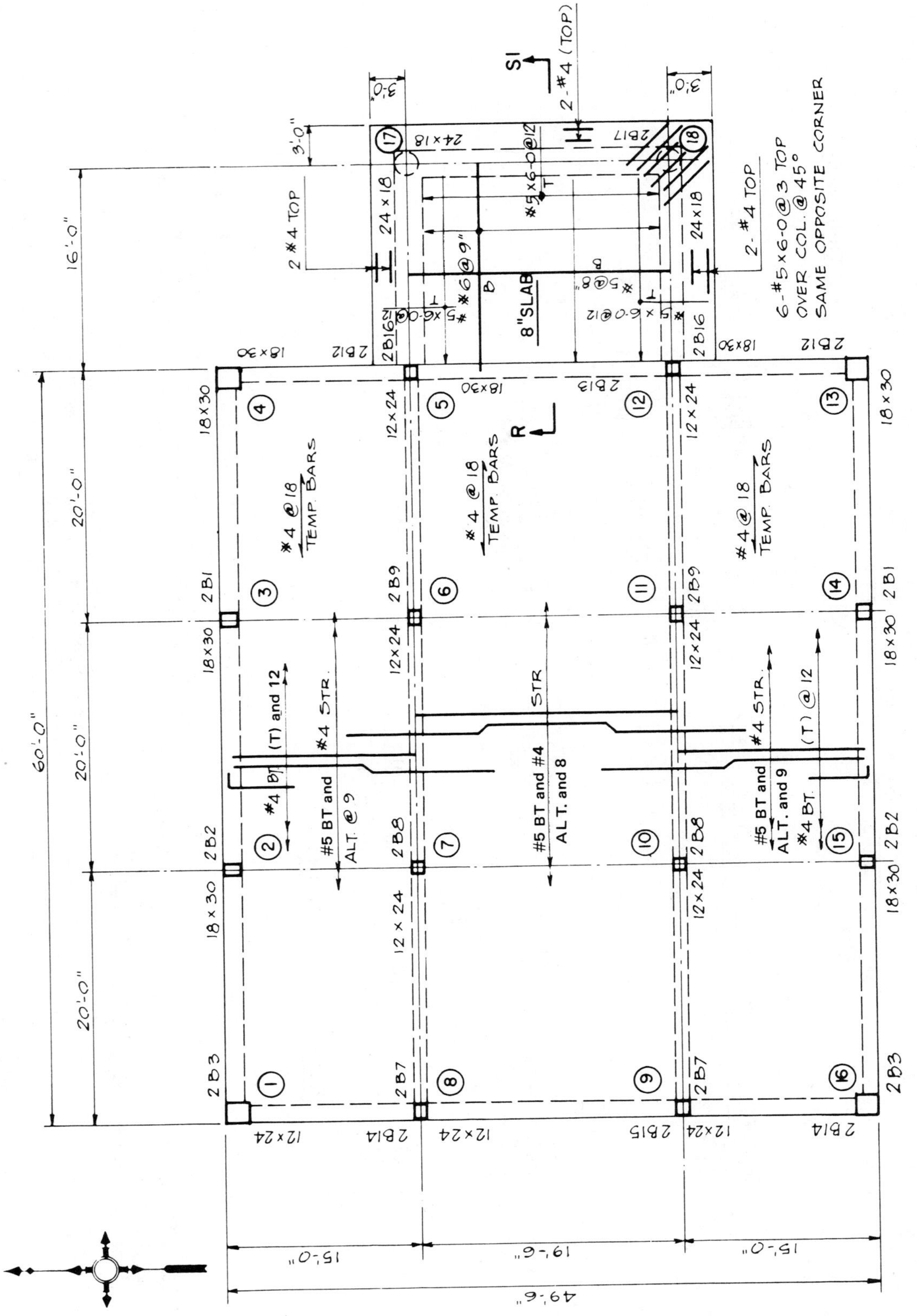

Fig. 23-27 Engineering drawing—Main slab reinforcement shown on plan. NOTE: See Fig. 23-26 for Section R-S1.

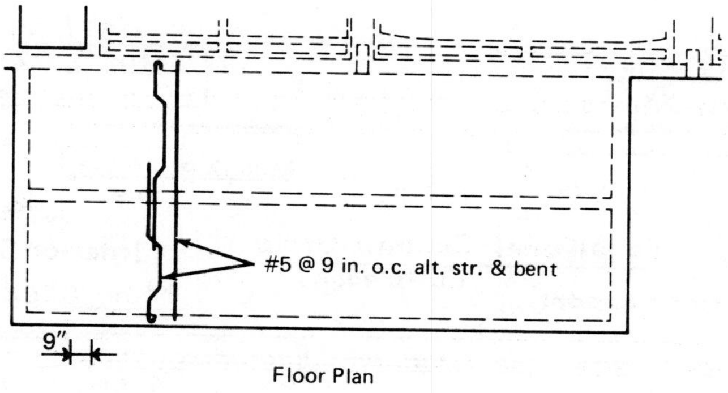

Slab Schedule			
Mark	Main	Temp	Remarks
S7	#5 @ 9 in.	#3 @ 11 in.	truss bars

Fig. 23-28 Wrong way to show alternate bars.

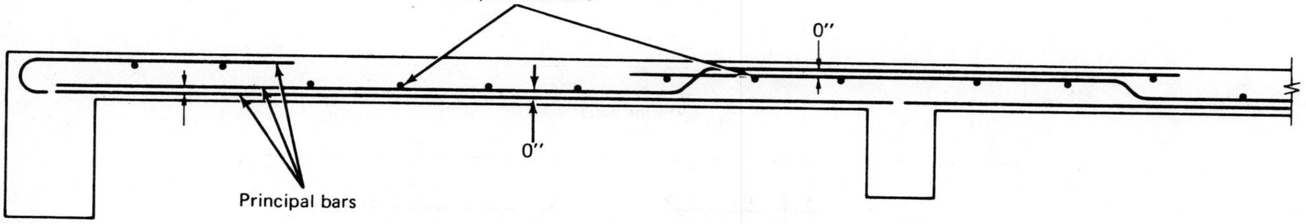

Fig. 23-29 Typical one-way solid slab.

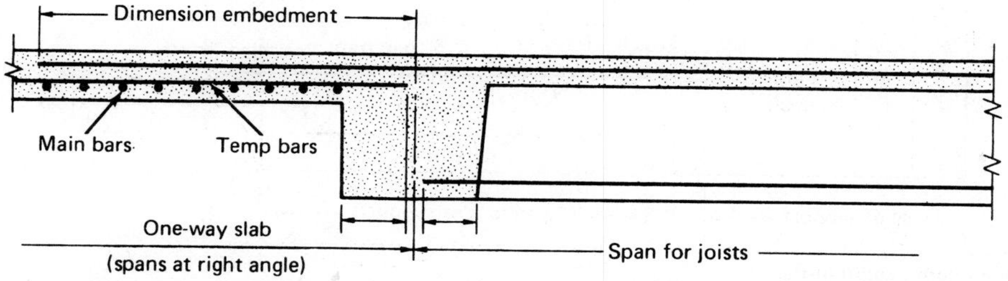

Fig. 23-30 Changing direction of span and/or floor system used. NOTE: Indicate all extensions, embedments, etc. required for design.

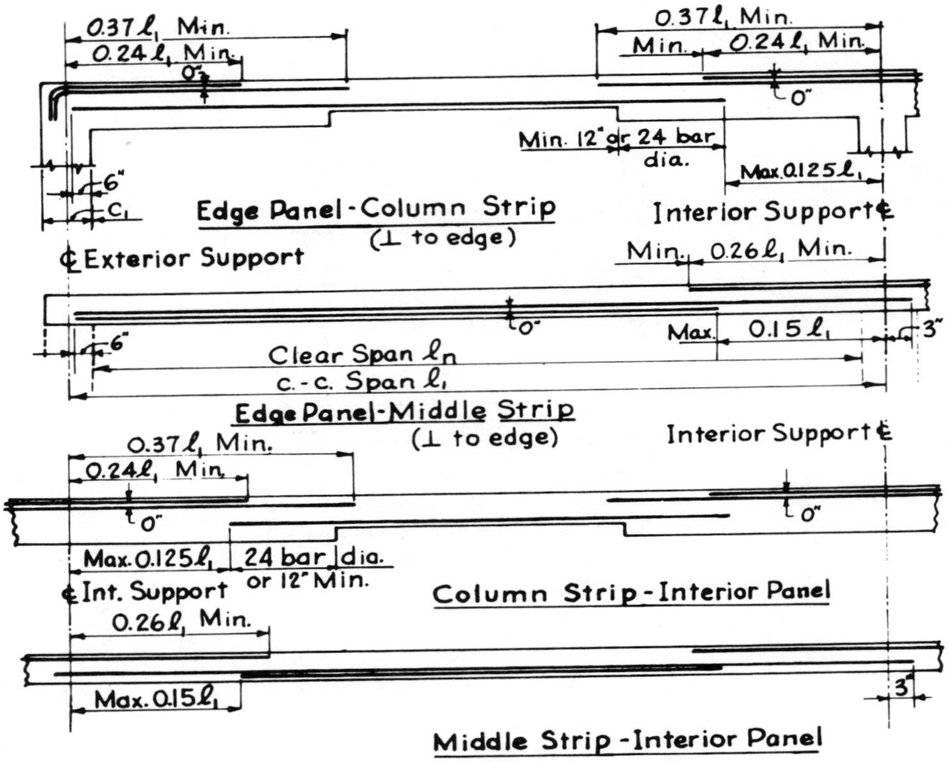

Fig. 23-31 Recommended bar details for solid flat slabs with drop panels.

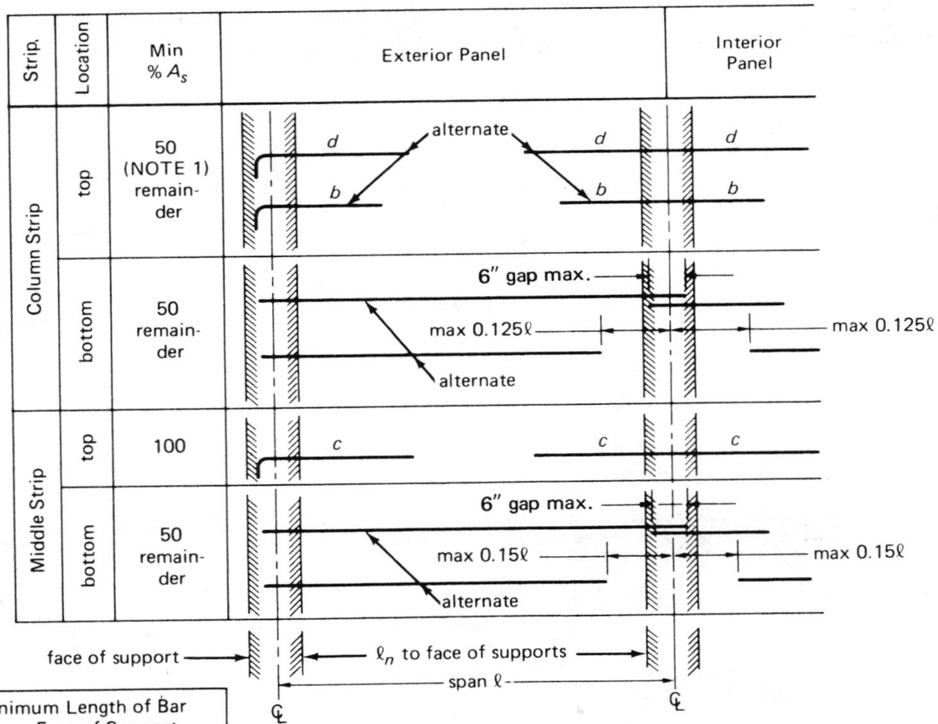

Minimum Length of Bar from Face of Support			
Mark	b	c	d
Length	$.20\ell_n$	$.22\ell_n$	$.30\ell_n$

Fig. 23-32 Code requirements for length of straight bars.

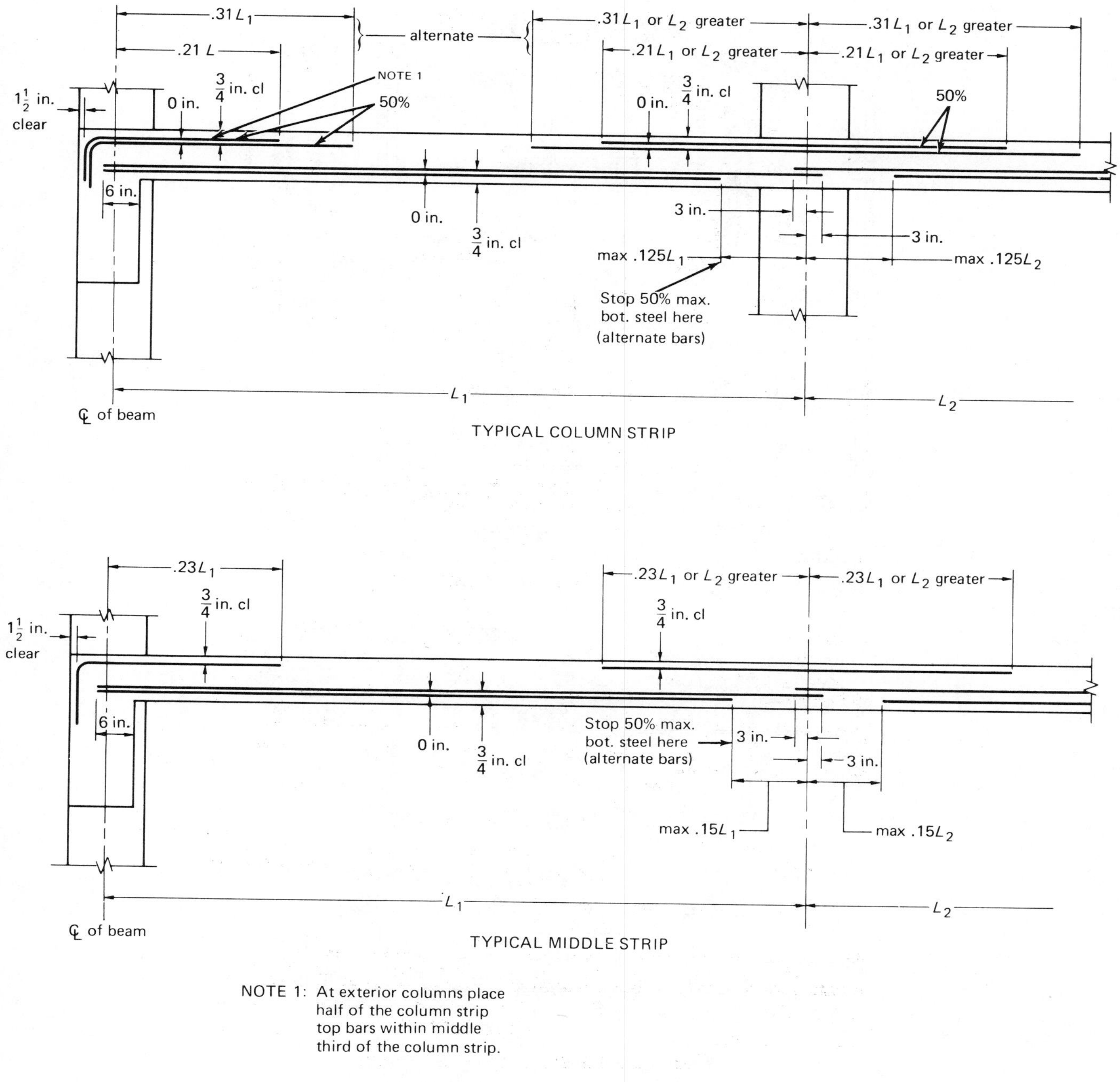

Fig. 23-33 Bar length details for flat plates.

the temperature bars to follow the tension face of the concrete, he must specifically show it thus (Fig. 23-29).

The designer's requirements where a slab changes direction should be clearly shown (see Fig. 23-30). The detailer must know whether to terminate the top bar and hook into the support or extend a specific anchorage distance into the adjacent span.

Two-way flat slabs have only minor differences from flat plates. Flat slabs with drop panels generally follow the ACI/CRSI standard rebar arrangement (see Fig. 23-31). It is usually satisfactory for the designer to show typical column and middle strips and use the foregoing arrangement.

Flat plates may or may not follow the standard rebar arrangement. Frequently, these involve erratic column layouts, and the designer then must show clearly the exact extent of the column and middle strips and the reinforcing included therein. He may show this right on the plan or in schedule format. Straight bars only are customary for flat plates. Figure 23-32 shows the *ACI Building Code* requirements for "straight bar" design. All the dimensions shown in tabular form (as Kl_n) are to the clear span. When column sizes do not vary significantly, it is practical to convert these dimensions to column centerline dimensions as follows:

$$K'l = Kl_n + .5 \times \text{col. dim.}$$

where K' is the modified constant, l the column centerline span dimension, K the tabulated constant (Fig. 23-32), and l_n the clear span. Assuming a 20-ft typical span and average

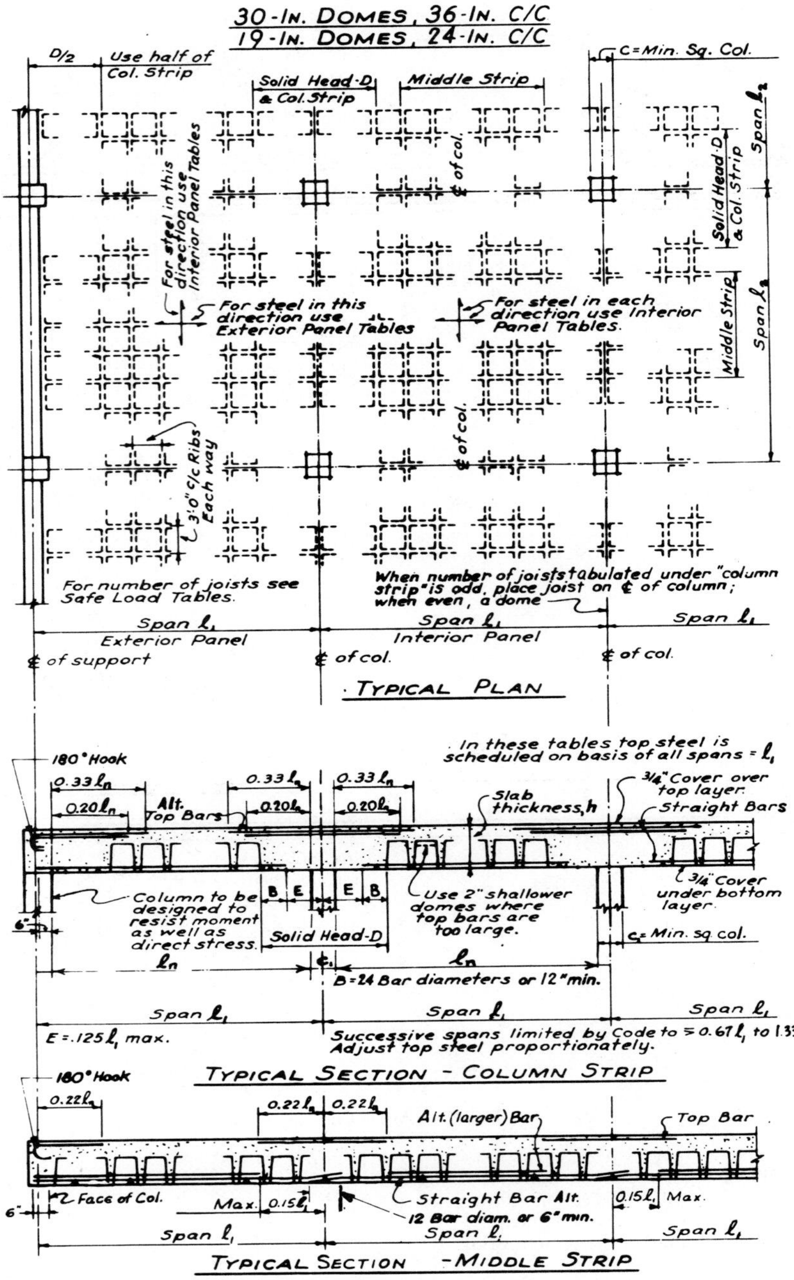

Fig. 23-34 Two-way waffle flat-slab.

12-in.-square columns, the typical detail on the structural drawing would appear as in Fig. 23-33.

In some cases it may be more convenient not to introduce column and middle strips formally on the plans. In such cases a uniform reinforcing mat over the entire slab is used for the bottom reinforcement, and loose bars are added on column lines only where necessary; the negative reinforcement is handled in the customary way with column and middle strips laid out on the plan view.

Waffle slabs, although designed as flat slabs, are really made up of joists in two directions and are usually scheduled. (See Fig. 23–34.)

24

Construction Methods and Equipment

J. F. CAMELLERIE*

24.1 CONSTRUCTION ORIENTATION OF DESIGN

The responsibility of the engineer, as well as the architect, is to provide a building or installation to meet the client's need in the required time and without undue cost. To design beams, walls, and footings sufficient to take the anticipated loads is only the beginning and most elementary part of the engineer's responsibility; unintelligent computers can perform much of this function. The science of engineering also includes combining the structural elements with each other, with the environment, and with the attitudes, capabilities, and tools available to the constructors. Structures must be built by men who cannot be controlled like robots, and they must be built in a natural environment that is not yet subservient to man.

Historically, man learned to build before he learned to draw or to compute. In our highly specialized society we have developed a breed of engineers who can compute and draw, but have only hazy notions of how to build. Structures designed by such engineers are "successfully" completed only after much anguish and modification in the field and at unnecessary cost in time and money to the owner and society.

The engineer must orient his design to the men and machines who will build the structure and to the materials, times, and environment in which it will be built. He must not complete the design on the assumption that the contractor will have the construction knowledge that he himself lacks, or that changes such as going from precast to cast-in-place concrete, or vice versa, are minor field modifications. Nor is the engineer's responsibility finished once the building department or other regulatory agency has approved the drawings for construction. The construction of the structure is, in fact, the implementation of his design. He must participate in this implementation, not merely as an inspector assuring compliance to the letter of the contract, but as a cooperative, guiding intelligence taking advantage of, or overcoming, unexpected developments as the work progresses.

Some of the specific aspects involved in the construction orientation of design will be discussed briefly in the pages that follow, but the concept and philosophy of the engineer and architect as "master builders" underlie all of the specific facets that follow.

24.1.1 Simplicity and Repetition

Simplicity and repetition are much to be desired in design. There is a natural tendency to make each beam and column the smallest size that meets stress requirements. This procedure will result in maximum economy of material but will certainly not result in minimum cost of the finished product. Using a great number of sizes increases the time and

*Manager of Construction, Advanced Technology and Industrial Projects, Ebasco Services, New York.

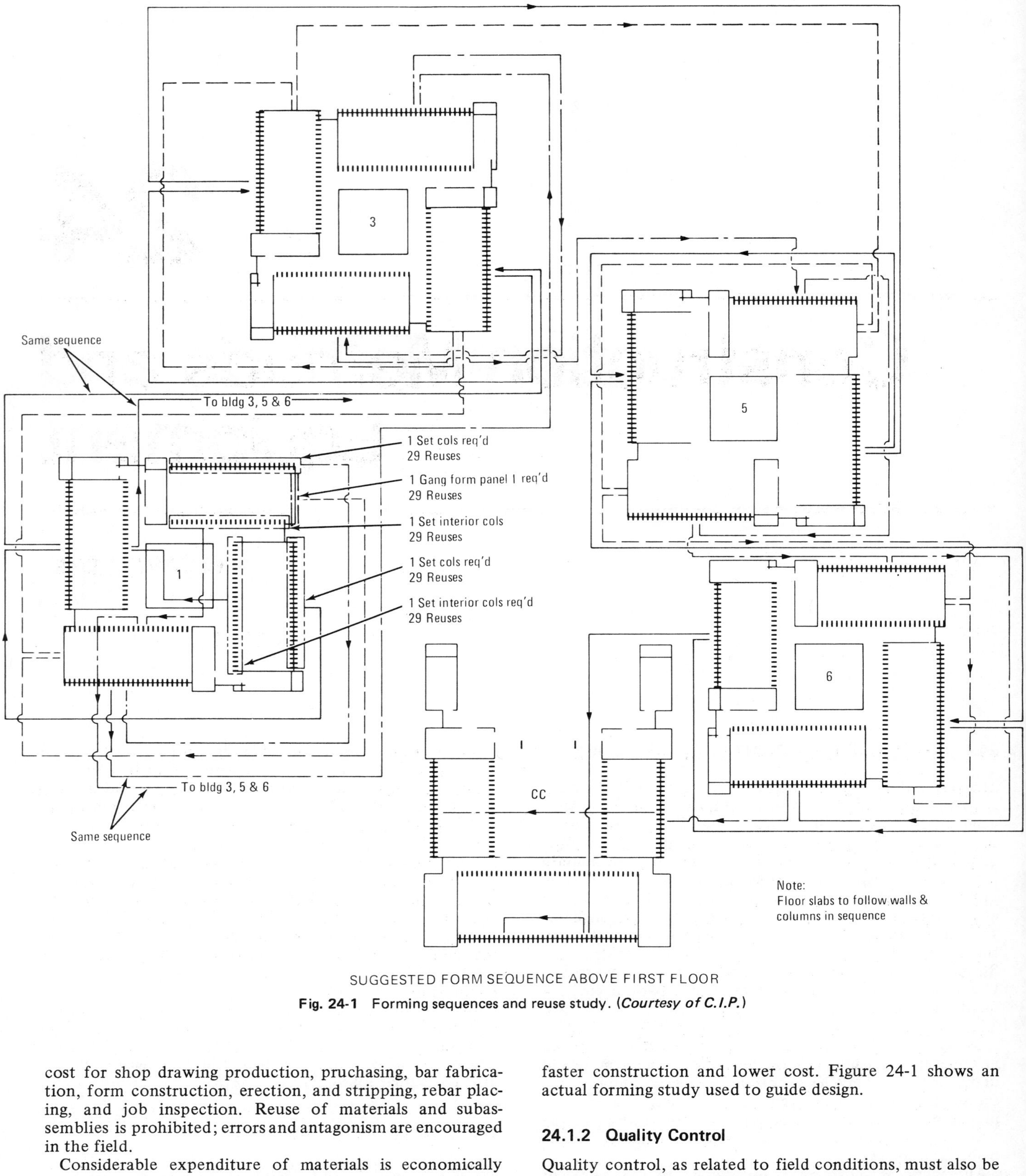

Fig. 24-1 Forming sequences and reuse study. (*Courtesy of C.I.P.*)

cost for shop drawing production, pruchasing, bar fabrication, form construction, erection, and stripping, rebar placing, and job inspection. Reuse of materials and subassemblies is prohibited; errors and antagonism are encouraged in the field.

Considerable expenditure of materials is economically justifiable in obtaining simplicity and repetition of design; this standardization will allow the use of assembly-line methods, prefabrication, subassemblies, specialized equipment, and reuse of more expensive materials, resulting in faster construction and lower cost. Figure 24-1 shows an actual forming study used to guide design.

24.1.2 Quality Control

Quality control, as related to field conditions, must also be integrated into the design. There is a great tendency to design walls to the least possible thickness and to obtain the required strength by using high-strength concrete and very high percentages of steel. In the first place, placing concrete in a 12-ft lift of 6-in. wall with 4% steel is like passing the concrete through a sieve. The mix will be completely separated by the time it reaches the bottom. To help matters along, the steel will physically prevent the proper use of vibrators. Structural draftsmen should be required to show reinforcing steel diameters to scale and with proper bend radii rather than the sharp pencil lines and right angle bends so neatly appearing on the drawings.

In addition, high-strength concrete is fast in setting; on a hot day it is very difficult to place this concrete before it takes initial set. It is not an ideal concrete to run through the steel "sieve." The physical problems associated with strength of concrete, percentage of steel, height of lift, and ambient temperatures must all be considered in the structural design, or the strength requirements will be satisfied only in theory but not in actuality.

The characteristics of concrete specified must be considered in relation to the type of structure being cast, climatic condition, placing techniques, method of mixing, and distance of transport from mixer to form. An increase of cement content will increase the strength of concrete in laboratory-cured cylinders. The effect may be opposite for the same concrete in the field, caught in downtown traffic, retempered with water, dumped into hoppers, raised up on a material tower, dropped into a buggy, wheeled a few hundred feet, and finally "placed" in a form. The designer's mind must be diverted from that 28-day moist-cured cylinder in the lab to the concrete that will be physically in his structure. A preoccupation with low slumps can also be very dangerous if not considered in the light of field placing.

24.1.3 Tolerances

Tolerances are a critical segment of design. It is absolutely essential that the design be such as to allow for size variation of members as fabricated. Many engineers have never taken the trouble to add up the various tolerances allowed by ASTM for steel beam fabrication. They include camber, sweep, rolling, out-of-square, web off-center, and length tolerances. A 40-ft beam set perfectly to plan and elevation at its ends can be 0.75 in. out of plan and elevation at the center, be 0.625 in. too long, and still be within acceptable ASTM tolerances. In is also interesting to note AISC tolerance for plumb in a steel building. This tolerance is 500 to 1 or 1 in. in 42 ft, with a maximum of 3 in. away from and 2 in. toward the building line for buildings 36 stories high and over. This tolerance is an erection tolerance and is additive to the fabrication tolerance.

The ACI *Standard Recommended Practice for Concrete Formwork* (ACI 347-78) recommends a tolerance in plan of 1/4 in. in any 10 ft and 1 in. for the entire structure (one-half this tolerance for exposed corner columns). These tolerances are quite optimistic for tall buildings when compared to AISC tolerances and to tolerances actually obtained in concrete frame construction. The same ACI standard suggests a tolerance of 1 in. per 50 ft of height above the base with a maximum of 3 in. deviation for slipformed concrete. In the opinion of the author, this 3-in. standard is more practical than the 1-in. or 1/2-in. requirement applied to building cores, slipformed or not. A practical vertical tolerance of less than 2 in. should not be seriously expected.

To illustrate the effect of tolerance on design, we may imagine a 40-story building with steel-column framing on the outside and a concrete core at the center. Structural steel beams span from columns to concrete core walls, a span of 40 ft. The tolerances at the top of the building may add up as follows:

℄ steel colums	3 in. toward building ℄
Fabrication tolerances for columns	0.75 in. toward building ℄
Beam length tolerances	0.625 in. too long
Concrete core tolerance	2 in. away from building ℄
Span may be actually	6.375 in. shorter than theoretical

The span can also be 5.375 in. longer than theoretical and still be within acceptable tolerances.

Actually this is a very pessimistic and somewhat exaggerated analysis. The author does not suggest designing for a tolerance of 6 in. It is obvious, however, that provision for span differentials of at least 2 in. should be considered.

Structural steel tolerances, plate glass tolerances, and all other fabrication and erection tolerances must be considered in concrete design as the concrete elements must be coordinated with and must accommodate the other structural and functional elements comprising the total structure. In addition to this, let the engineer not forget the dramatic movements that occur in the structure under construction due to sun or wind on one side, or seasonal thermal expansions and contractions.

It is essential that all concrete elements and joints be desgined to provide the minimum amount of tolerance and flexibility required to meet the expected field variations. Whenever possible, additional tolerance should be provided to meet unexpected problems and to allow for human error. An engineer must first determine that the tolerances he specifies are actually attainable in the field. No amount of construction care and no type of legal coercion can result in a tolerance beyond the capabilities of man and his tools working in a "cruel" world. Beyond this, fabrication and erection costs are directly related to the tolerance refinement required. It is therefore economically desirable and practically safer to design with maximum flexibility and to keep tolerance requirements as liberal as possible.

24.1.4 Design to Suit Available Equipment

An important construction aspect that must be considered in design is the increased capabilities offered and the limitations imposed by available construction equipment.

Availability of equipment should not only mean that the equipment is available in time for use on the job, but also that site conditions are such as to make the equipment usable at all required locations; that there are no government ordinances prohibiting or limiting the use of the equipment; that there is acceptance by local unions; and, finally, that maintenance and spare parts are available. The arsenal of construction equipment is increasing daily in variety, capacity, and reliability. There is also great capability for designing and fabricating equipment as required for special operations. This potential should certainly be considered in the design concept, since it is now possible to produce structures that were economically impossible to produce a few years ago.

Great as this equipment arsenal is, the limitations are very real and must be carefully investigated, as many an engineer discovered when his structure went out for bids. Confined urban construction sites on main traffic arteries require the most careful methods and equipment studies.

24.1.5 Labor Considerations

In a democratic country with well-developed trade unions, the strength, attitude, political influence, and business affiliations of the unions in any particular area must be care-

TABLE 24-1 Contractor Reaction Summary by Region

Contractor	*Interest by Region*									*Remarks*
	I	*II*	*III*	*IV*	*V*	*VI*	*VII*	*VIII*	*IX*	
Contractor "A" New York, N.Y.	X	X	X	X	X	X	X	X	X	
Contractor "B" Pittsburgh, Pa.	X	X	X	X	X	X	X	X	X	
Contractor "C" Kansas City, Mo.	X	X	X	X	X	X	X	X	X	
Contractor "D" Buffalo, N.Y.		X								Both slip form and precast
Contractor "E" Chicago, Illinois	X	X	X	X	X	X	X	X	X	
Contractor "F" Minneapolis, Minn.	X	X	X	X	X	X	X	X	X	
Contractor "G" Philadelphia, Pa.		X								
Contractor "H" Hutchinson, Kansas	X	X	X	X	X	X	X	X	X	
Contractor "J" Dallas, Texas				X		X	X			
Contractor "K" Philadelphia, Pa.	X	X	X	X	X	X	X			
Contractor "L" Spokane, Washington								X	X	

fully considered in the design of a project. In countries of strong government control, appraisals of official assistance and/or hindrance must be made.

In all areas, the labor market must be studied as to availability of men in the various trades, skill level of the labor, prevailing wages, fringe benefits, shifting and overtime practices, and competition for labor with other projects. The possibilities and economics of importing skilled and unskilled labor may also be considered. Such studies of the labor market may very well effect design decisions as to materials, type of structure, construction techniques, sequence of construction, and starting and completion dates.

The contractor is normally the person or organization that brings the equipment and labor together to actually perform the work. A check is advisable to determine the attitude of local contractors to the work in general and to the techniques in particular. The willingness and capability of specialist contractors to work in the specific area must also be looked into. There are many ways of doing this, and none of them is guaranteed as to accuracy. Nevertheless, this type of study is very valuable. Table 24-1 shows a typical study made to determine contractor reactions to a specific project and to various techniques. Table 24-2 shows a typical study to determine availability to specialist contractors by region.

24.1.6 Construction Schedules

Construction schedules are often academic, reflecting the wishful thinking of the designers and completely ignored by the contractors. Schedules should be carefully arrived at, considering construction methods, operation sequences, and climatic handicaps. There is no point in setting a schedule if the schedule is not implemented by meaningful

TABLE 24-2 Contractor Reaction Summary by Speciality

	Interest					Preference				Increase in Contingency			
Contractor	*General*	*Conventional Forming*	*Precast*	*Slip Form*	*Prestress*	*Conv. Form*	*Precast*	*Slip Form*	*Prestress*	*Pre-Cast %*	*Slip Form %*	*Pre-stress %*	*Remarks*
Contractor "A"	Great	Yes	Yes	Yes	Yes	–	–	√	–	–	–	–	
Contractor "B"	Some	Yes	Yes	Yes	Yes	–	–	–	–	5	10	5	
Contractor "C"	Great	Yes	Yes	Yes	Yes	√	–	–	–	2	5	5	
Contractor "D"	Some	Yes	Yes	No	No	√	√	–	–	3	5	5	
Contractor "E"	Some	Yes	Yes	Yes	Yes	√	–	–	–	2	6	3	
Contractor "F"	None	–	–	–	–	–	–	–	–	–	–	–	
Contractor "G"	Some	Yes	Yes	Yes	Yes	–	–	–	–	–	–	–	
Contractor "H"	Some	Yes	Yes	No	Yes	√	–	–	–	5	10	5	
Contractor "I"	Some	Yes	Yes	Yes	Yes	√	–	–	–	10	10	10	
Contractor "J"	Great	Yes	Yes	Yes	Yes	√	–	–	–	5	5	8	
Contractor "K"	None	–	–	–	–	–	–	–	–	–	–	–	not active

contractual penalties and rewards. Completion behind schedule may actually cost the owner considerable sums in rental, force him to miss a seasonal market opportunity or an educational semester, or get him into serious problems in the financing of the project. On the other hand, early completion may increase the cost of the project because of limited reuse of forms and equipment and possible payment of premium wages to labor. The cost value of early and late completion should certainly be considered in the early design stages.

24.2 SCHEDULING AND CONTROL OF CONSTRUCTION

24.2.1 Subdivision of Construction into Phases

Small projects automatically fall into phases by virtue of the necessary sequences of operation and the trades involved. Large, complicated projects have more problems, more possible choices of phasing, and more engineering decisions to make. The major considerations effecting the phasing of the job follow:

(a) Availability of funds
(b) Methods of financing
(c) Capability of contractors
(d) Impact on labor market
(e) Fabrication time
(f) Critical prerequisite sequences
(g) Interference between concurrent phases
(h) Use of equipment
(i) Weather considerations
(j) Economic and use alternates to determine completion date
(k) Design time

On very large projects to be constructed for government agencies or large private institutions, fiscal policies often impose very rigid limitations on how much of the total commitment will be made available in each fiscal year, and usually require that the money be spent during the year that it is budgeted. Failure to spend the money in time could mean loss of the money.

Jobs financed by private money, such as bank or insurance company investments, often dovetail the construction investment with other investments. In addition, payments of funds for construction are often tied into a series of completions. Obviously the construction schedule must be tied into the financial aspects of funding.

Some projects are of such a size that few, if any, contractors have the financial and personnel capacity to undertake them. Those contractors physically capable of doing the job may feel it is too much of a risk to take on one project alone. They would rather divide that amount of risk among three or four different projects. Too large a project may also drain the labor market and cause "labor piracy," inefficiency, and premium cost. In order to obtain enthusiastic bids in sufficient numbers, large construction jobs are often divided into several smaller sections and separate bids solicited for each phase.

In solving the above problems, construction management problems are created that must be skillfully solved. Those operations that are prerequisite to others must be completed before the next operations are started. Parallel and supporting operations must be phased in so as to be completed without holding up the critical work or adding time to the intermediate completion dates. Certain items that require long fabrication and/or delivery dates may have to be ordered very early and contract documents must allow and ensure such early purchases.

The most popular tool for controlling and dovetailling these operations is the critical-path chart. By showing all planning, procurement, fabrication, construction, and testing operations as vectors to a time dimension, these operations are so arranged that all operations follow immediately after prerequisite operations and as concurrently as possible. When the chart is properly done, one can tell the minimum construction time required, critical schedules that must be held, slack time in operations not on the critical path, effect of speeding certain operations, etc. This chart is no better than the input it receives, the contract documents governing the work, or the construction managers implementing it.

Design is prerequisite to all operations. Usually the time required for completion of the entire design will set back the construction schedule excessively. It is therefore necessary to carry out the design concurrent with construction with only a limited lead time. This necessitates dovetailling the design into the critical-path chart and following design sequences catering to construction schedules rather than those sequences more natural to design methods.

24.2.2 Dovetailing of Operations

Once a job is divided among several contractors, each having his own interests to promote, it becomes necessary to schedule and to cover the work by document and field management in such a way that interferences between operations are eliminated or minimized. There are, also, the interests of the client, such as an airport authority, which must maintain operation during construction; transportation and municipal agencies responsible for the streets and highways that the contractors must use; and, of course, the neighbors.

If the interaction of these problems is left to be solved "in the field," the result will be litigations, delays, and excessive extra costs. The architect or engineer must plan the construction procedures in considerable detail, with the help of construction consultants, if necessary. He must indicate in the contract documents:

(a) What site areas will be available, and when.
(b) What access roads are available, and when.
(c) What limitations exist on hours of work.
(d) What limitations exist on noise and other by-products.
(e) What equipment may be used, where, and when.
(h) What equipment may not be used.
(i) What cooperation is required to others.
(j) What access is required to others.
(k) Firm completion dates with penalty.
(l) Safety requirements for public and work force.
(m) Protection for existing installation and installation by others.
(n) Predictable delays or discontinuation of the work.

24.2.3 Common Use of Site and Equipment by Contractors

It is almost always necessary, especially in vertical construction, that the contractors use the same site area of the job. Hopefully this can be managed in a chronological order; often it is concurrent or at least with a time overlap. This division by time and area must be clearly defined, taking into consideration the realistic needs of the contractors involved. Very often equipment such as cranes, hoists, personnel elevators, utility lines, and permanent installations temporarily used for construction are required by more than one contractor in sequence, or concurrently.

If each contractor can have his own equipment fully under his own control, this is the ideal condition. Often space requirements prohibit this. There is obviously an unnecessary cost that always accrues to the owner if each contractor erects and dismantles the same tower crane on the same spot as he performs his particular phase of the work. Economy and site efficiency can best be served by specifying equipment capable of performing all phases and operations of the work, scheduling the hours of use to the contractors involved, and arranging for an equitable sharing of cost through a rental or allowance arrangement.

24.2.4 Good Administration Procedures and Payment Policies

In Section 24.2.1 above, we touched on the financial capabilities of contractors. Obviously, the financing required to do work for a client who pays promptly as the work is completed is not as great as the financing required to do work for a client who withholds 10% of all payments until total completion, delays paying large vouchers on the basis of small disagreements, and generally has administration procedures that take months to process the simplest unquestioned voucher. Such clients eliminate many competent construction firms from bidding, encourage brokers whose staff is composed of lawyers but few, if any, construction men, and definitely incur the cost of financing which the contractor passes on. Very often, such organizations also fail to make decisions on problems holding up work in the field. Architects and engineers should make every effort to promote contract documents that ensure reasonably quick payment for work performed, rapid approval of shop drawings, and helpful inspection procedures. Good administrative procedures attract better contractors, improve their cooperation, lower costs, and help meet schedules.

24.2.5 Inspection

Inspection is more difficult and more critical than ever. The industry is using new materials, new techniques, higher allowable stresses, and lower safety factors (ultimate strength design). There is more room for error and less allowance for it. All too often, the only qualification for an inspector is that he is willing to work for a very low salary. This is a very dangerous procedure for modern sophisticated construction. Inspectors must have some training in construction technology, must be given special instruction by the architect and engineer in charge, and must have open lines of communication to their superiors and to the contractor.

The design architect and engineer must visit the job often to check on and to instruct the resident staff. In some instances, when specialized knowledge is not available in-house, it is necessary to obtain the field assitance of construction consultants, manufacturers' technicians or engineering specialists.

The serious inspector will spend any free time he gets on the job studying the plans and specifications and reading relevant technical books and papers. A little homework would not hurt either. But remember, if this type of intelligent conscientious person is required on the job, the remuneration must be commensurate. Very often inspection degenerates into enforcing the letter of the law without regard to the way in which the instructions of the contract documents are affecting the work. The inspector is, and must be, a policeman. To be really effective, however, he must know the limits within which he may make his own decisions and must know when to request help from the design engineer. In his turn, the design engineer must clearly instruct his inspector in the field as to what is inflexible, what is flexible, and to what extent, and when to call for help.

One item that receives very little attention (except in cold weather) is concrete temperature. In the author's opinion, concrete thermometers should be commonplace on the job, and records of temperatures and slump at the truck discharge and in the forms should be kept all year round. Also keep in mind that slump control at the truck discharge does not necessarily result in proper slump at the placement area.

A few representative hints for new inspectors:

(a) Keep a daily diary.
(b) Check grade marking on rebars, bolts, wood, plywood, and accessories.
(c) Check concrete delivery tickets carefully to ensure that the truck is at the right job.
(d) Prevent unauthorized persons from requesting and getting water added at truck mixers.
(e) Check that concrete is not kept in truck mixers too long (check delivery ticket).

(f) Check that concrete is not moved laterally by vibrators or dropped more than 5 ft vertically without drop chutes or elephant trunks.
(g) If vibrators are allowed to sink into concrete under their own weight without additional force, the vibrator penetration will be about right.
(h) Check curing procedures to ensure that there are no misses and no discontinuity.
(i) Check that no debris (wood shavings, oil, coffee containers, bottles, etc.) falls into concrete or is left in forms before filling.
(j) Check forms for loose ties, improperly wedged shores and/or undue movement during placing.
(k) Spot-check layouts and elevations.
(l) Check that screeds, rebars, and prestress conduit are properly secured against movement during placing.
(m) Be especially careful about concrete cover on reinforcement.
(n) Check bent rebars, especially smaller sizes with high yield points, for cracks.
(o) Check rebar laps and connections.
(p) Check for broken conduit in posttension work.
(q) Beware of contractors buying coffee at critical times.
(r) In fact, be careful of diversionary tactics always.

24.2.6 Climatic Considerations

Certain operations, such as steel erection, can be carried out in any season without undue cost or delay. Other operations, such as concreting, are vulnerable to both extreme heat and cold. Excavation and foundation work is vulnerable to heavy rain. As obvious as this seems, many construction schedules ignore and seasonal effects and start "as soon as possible" and then allow no additional time for the resulting handicaps. In some instances, starting the job sooner, delaying the start, or even closing down the job for a period may result in considerable savings in cost—possibly without changing the completion date. Different systems, materials, and/or construction techniques should be considered in order to "beat the season," "close in the job," or otherwise adapt to the adverse conditions. Often, the cost of overtime is justified in order to avoid the cost of inefficiency due to inclement weather.

24.3 CONCRETE CONSTRUCTION TECHNIQUES

As long as men have been on earth, they have been devising ways of moving things, raising them, connecting them together, and generally building things where they wanted them to be. In less organized societies each community devised and passed on from man to man "tricks of the trade." In our time we have great capabilities for the exchange of knowledge, equipment, and materials, bringing these together from near and far, and erecting structures in the most difficult locations. Construction technology is the science of construction involving the use of the most advantageous materials and equipment available, including the necessary planning, preparation, and execution. Some of the more sophisticated techniques commonly used in the United States are discussed in the following paragraphs.

24.3.1 Precast Concrete

Concrete may be cast in place in its final shape and location, or it may be precast (i.e., cast somewhere other than where it will be used and utilimately assembled in the final location) (see Fig. 24-2). There are several advantages in using precast sections, the most significant being the application of factory methods to an otherwise custom-type industry.

Fig. 24-2 Placing precast concrete.

Precasting concrete members and panels at central plants permits the use and resue of highly productive automated equipment, weather protection during casting, stabilized experienced labor, concurrent operations removed from the site and from each other, excellent quality control, simplification, and standardization. The last item, standardization, is the key to ultimate success in the use of precasting. The industry is in the process of developing off-the-shelf items such as single-tees, double-tees, highway girder sections, slabs, and panels. As the various manufacturers coordinate their operations more and more closely, a catalog similar to the AISC catalog of available shapes will develop.

Precasting can be accomplished at plants as described above and shown in Fig. 24-3, or at the job site. On large

Fig. 24-3 A precasting yard.

jobs, when economic considerations so indicate, a temporary plant can be set up at the job site. This casting yard may be equipped with very sophisticated equipment or may utilize very simple forming and casting procedures.

Because precasting often enables vertical surfaces to be cast in a horizontal position, wall panels of very excellent aesthetic qualities can be produced in unlimited variations. These panels may have any type of troweled or scarified finish, molded designs, exposed aggregate finish, or open-space filligree effects. They will be free from tie holes and air bubbles. It is more difficult to obtain the same quality with vertical in situ casting. In planning precast construction, the following aspects must be carefully studied:

Proximity of precast suppliers
Transportation costs
Access to site
Equipment required to erect sections
Possibility of on-site precasting
Use of standard available sections
Simplicity of sections for manufacture
Largest pieces manageable
Simple but effective jointry
Lifting devices and strength during placing
Adequate tolerances
Weatherproofing of building

The four most common pitfalls are:

1. Difficulty or impossibility of placing due to size, weight, site obstructions, or height.
2. Difficulty of placing due to insufficient tolerance and/or inflexible jointry.
3. Intricate jointry requiring extensive welding or difficult grouting.
4. Failure to specify lift procedures and to design for loads that occur during lifting.

Summarizing the advantages of precast concrete, they may include lower costs, quicker construction, early structural strength, less congested construction sites, high quality, and relatively easy rejection of imperfect pieces. The disadvantages may include difficult and expensive jointry of the component parts, heavy transportation costs, weight limitations, and labor jurisdictional problems.

24.3.2 Prestressed Concrete

Prestressed concrete is widely used in this country in both cast-in-place construction (posttensioning) and precast construction (pretensioning). It allows longer spans without excessive depths, more crack-free structures, lighter members, and better joining of concrete subsections of buildings, whether cast-in-place or precast. In some cases, prestressing will allow the construction of structures that otherwise could not have been built. In all cases, it takes advantage of the higher-strength steels and concrete available, thereby effecting reductions in concrete and weight and the costs related to handling the plastic or precast concrete.

Prestressing is a technique that is well mated to precasting and helps to solve many of the weight and jointry problems that militate against precast concrete. By the same token, the casting yard operation used for precasting brings out the best economics in the use of prestressed concrete. The precasting yard lends itself to the economical use of sophisticated tensioning equipment and anchors, high-strength concrete, and accelerated curing.

One difficulty that is associated with the use of prestressed concrete is deflection control during stressing and during the later life of a member. It is difficult to estimate accurately the amount of deflection to be expected at various ages of a prestressed member, since the deflection depends on a varying modulus of elasticity, shrinkage and creep of the concrete, relaxing of the steel tendons, etc. This may result in connection difficulties and in unsightly malclosures or arch effects. The use of shear connections is often advisable.

One major problem in posttensioning work is the construction time lost when the tendons are tensioned in place. Posttensioning cannot be effected until the concrete has reached sufficient strength. Unless the construction sequences embody enough time for this concrete strength gain, the successive delays will lead to serious lengthening of the construction time and high project costs.

Another problem, particularly in high-rise building, is the effect of posttensioning on other components of the structure. For instance, the posttensioning of floor systems may lead to serious flexural stress in the columns. This possibility must always be carefully studied.

24.3.3 Slip-Form Construction

Vertical elements such as walls, columns, and piers lend themselves to construction by the slip-form method. In this method a form 3.5 to 6 ft high is erected and coupled to a jacking device capable of lifting the form together with working decks and scaffolds. The advantage of this system lies in the fullyautomated continuous reuse of the same form without stripping or reerecting. There is also a considerable economy in the reduction of scaffolding, automatic templating of rebars, and frequent elimination of hoist towers.

Probably the most important advantage is the speed of construction, which is a floor a day on daily slides, and up to 24 ft per day for round-the-clock slides. The low-cost-per-square-foot contact area of the forms is offset by the premium time associated with slipforming. Even on daily slips, a skeleton crew on overtime is required to clear the forms.

Like all other techniques discussed here, slipforming requires that the design be oriented to the construction method. Most often slip-form operations are performed on structures not specifically designed for this method with the contingent loss in speed and economy. Those items that contribute most significantly to the economy of this method are:

(a) Maximum concrete yardage per foot of height (5 to 20 yd/ft and more is in competitive range)
(b) Maximum height of structure
(c) Minimum number of inserts and openings
(d) Maintenance of uniform wall dimensions throughout height
(e) Willingness of trades to accept shifting for round the clock operation

24.3.4 Tilt-up Construction

Wall elements of limited heights may be economically and quickly constructed by the tilt-up method with many aesthetic finishes obtainable at affordable cost. Actually, panels up to four stories high (40 to 60 ft) have been cast and tilted up as one unit. This method of construction is a variation of site precasting. Side forms are erected on the ground, or more probably on the floor slab of a structure, the bottom edge of the wall preferably close to the final position. The panel is then cast in a horizontal position, and once the concrete has acquired sufficient strength, it is tilted up into the vertical position, usually by rotating it upward about its bottom edge. In some instances these panels have been cast in considerable lengths using road paving techniques and equipment, and later saw-cut into manageable lengths for crane handling. The panels are finally connected to steel or concrete columns, to each other, or to cast-in-place columns cast after the panels are in place, to form a continuous structure. The panels may be load-bearing with integral columns, plain curtain walls, or insulated sandwich panels.

The obvious advantages of tilt-up construction are: simplicity, minimal amount of formwork and stripping, absence of form ties, and ease in placing and finishing concrete, placing openings, inserts rebars, etc., and, finally, the relatively simple equipment necessary to raise it into position. Some of the economic advantage of this system is lost in the "teaspoon-type" concrete placing of cast-in-place columns; precast columns are somewhat expensive because of the fussy joinery. During the erection stage, the panels must be designed for dead weight, impact, and cohesion to the casting surface. These handling stresses may be up to four times the in-place design stresses. In addition, the panels must be designed as uncracked sections with maximum stress below allowable concrete stresses in tension in order to prevent cracks. Layout of tilt-up inserts must be carefully worked out in conjunction with strong backs and spreader beam rigging to minimize erection stresses and prevent stress concentrations.

The following factors must be carefully considered in deciding to use tilt-up walls:

1. Suitability of the floor slab as a casting bed
2. Adequate room to allow crane to be properly positioned for each panel tilt-up
3. Sufficient number of panels to justify cost and capacity of crane, including erection crew
4. Access for ready-mix concrete trucks
5. Adequate bracing for erected panels

Maximum economy has been obtained in the construction of one-story buildings having extensive wall areas with clear working space all along the walls. Tilt-up construction can also be used, when justified, in multistory structures by casting and tilting-up the wall panels on the floor slab. A very interesting modification of this method was used in the Westyard Distribution Center in New York City (Fig. 24-4). A very large number of panels were required. With three sides on heavily trafficked streets and one side over the entrance to the Lincoln Tunnel, the parking of trucks and the stationing of cranes around the building was a serious problem. A movable tilt-up casting table was designed that ran on rollers, cast the wall panels at hip level, and then raised them into position using a built-in hydraulic system.

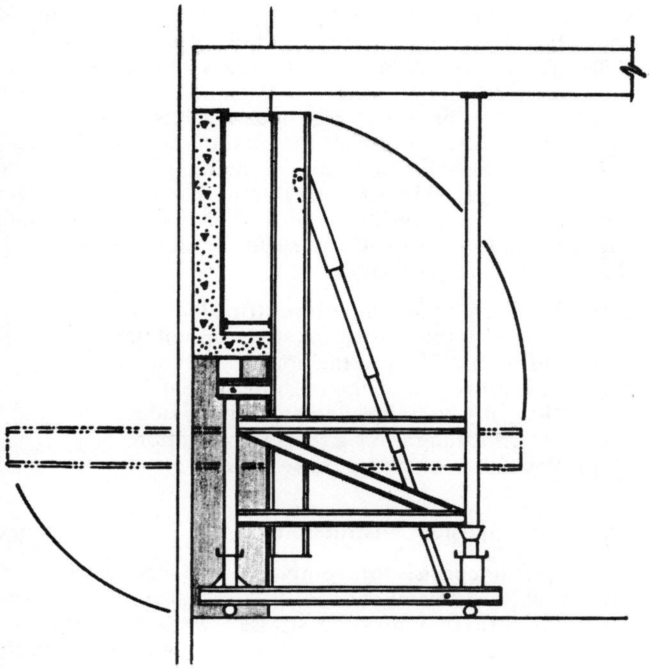

Casting Operations

1. Adjustable tilt-up form is rolled into position.
2. The form is tilted to the vertical position for fit to the concrete frame.
3. It is then rotated down to the horizontal position for rebar and concrete placing.
4. The concrete is steam cured overnight and the form with the panel is tilted up into position and the panel bolted to the frame.
5. The form is then moved to the next bay.

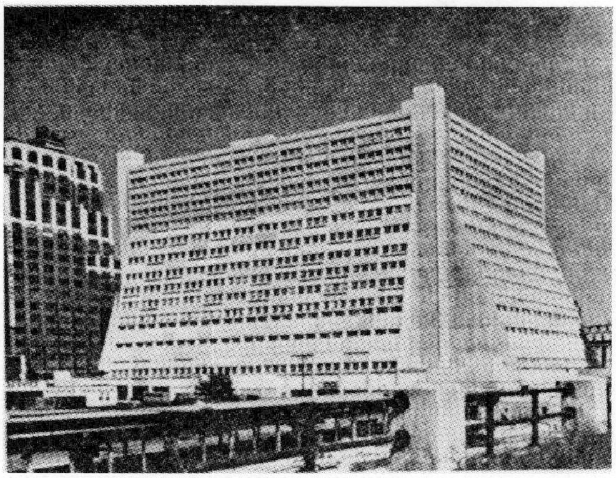

Westyard Distribution Center

Placing Concrete in Forms

Fig. 24-4 Precast, tilt-up construction. (*Courtesy of C.I.P.*)

Fig. 24-5 A lift-slab operation.

24.3.5 Lift-Slab Construction

This technique is used for casting several stories of floor slabs at ground level and then jacking them into position at proper floor height (Fig. 24-5). This system normally utilizes steel columns that are temporarily braced in place. The lift-slabs are cast one upon another on top of the basement or first-floor slab. A bond-breaker is used between slabs; only side forms are required. Structural steel shear heads are cast into the slabs around the columns to provide a lift point for raising into position and to act as the final device for stress transfer from the slabs into the columns. When the concrete in the slabs have reached sufficient strength, the slabs are jacked into position at the proper floor levels following a carefully preplanned sequence.

This technique offers considerable advantages costwise and timewise when the floor system lends itself to sectioning off into independent wings that can be raised straight up into position. An elevator and stair core is often used to furnish lateral stability, cast-in-place floor sections being used for connection. This system is somewhat limited as to height; it lends itself very well to medium-rise structures to 10 stories, although buildings up to 20 stories have been built. Lift-slabs also lend themselves very well to posttensioning techniques. All the slabs may be stressed in one operation while they are still at ground level, which results in a very economical posttensioning operation.

A recent innovation to this system is showing great promise to further reduce cost and construction time and to increase stability of the structure. Concrete bearing walls are alternately interleaved between floors and hinged to the floor slabs so that when a particular slab is jacked into position, hinged supporting walls automatically rotate into a vertical position under the slab to support the slab. Since the lifted walls are bearing walls, the cast-in-place, precast, or steel columns normally required are eliminated, saving also permanent lifting collars and eliminating shear transfer problems at column supports. However, this system is limited to, at the most, half the number of stories raised using the basic system. In addition, the floor plan must allow space for casting sufficient bearing walls.

Another innovation to the basic lift-slab system having similar advantages is the use of precast angular or channel-shaped wall columns, as well as prefabricated bathrooms and kitchens, which are slipped between the lifted slabs and become the final walls for the building. The slabs are raised slightly higher than the final elevation and then lowered down on to the precast wall columns. Improvements in available bond-breakers and application techniques have greatly benefited all lift-slab systems.

24.3.6 Drop-Slab Construction

This method of casting slabs is similar and opposite in technique to the lift-slab method. A core is constructed the full height of the building, and some form of transfer girders placed at the top, cantilevering out over the core. Rods are hung from the girders, and the slab forms are attached to the rods. Once the slab is placed and has acquired sufficient strength, the form is dropped downward into place for the next floor. This method can be used for high-rise structures, as there is no height limitation. Transfer girders may be provided at several levels in very high buildings. One such system is shown in Fig. 24-6.

The form reuse without the necessity of disassembly and reassembly offers considerable economic advantage. In conventional forming the forms must be disassembled into small units, carried out past the building line, and brought up over the cast slab and into the building for reassembly. This is a slow, expensive operation. There are, however, some expenses involved in the drop-slab system that militate against the time- and cost-saving features of this sytem. These disadvantages include:

(a) Necessity for floor closure strip all around
(b) Necessity for waiting for strength build-up
(c) Difficulties in breaking bond
(d) Possibility of a tricky re-shore system
(e) Necessity to bring concrete in from side
(f) Difficulty in raising girders into position
(g) Possibility of an expensive core

24.3.7 Composite Construction

This system consists of combining a concrete slab with structural steel beams, bar joists, or steel forms. The concrete takes the compression stresses, while the steel takes the tension stresses. This system offers good opportunities for economy by using each material in the most efficient manner. Care must be taken during construction, remembering that the steel elements must take deadload plus construction loads without the help of the concrete until such time as the concrete has gained sufficient strength. In addition, the steel sections used tend to be somewhat unstable laterally without the concrete. As a result, additional shoring and bracing is required and adds to the cost along with other special requirements of the system such as placing shear studs. These tend to reduce the savings that accure from the decrease in concrete and steel quantities. Engineers must be careful to show complete instructions for shoring and bracing on the structural drawings.

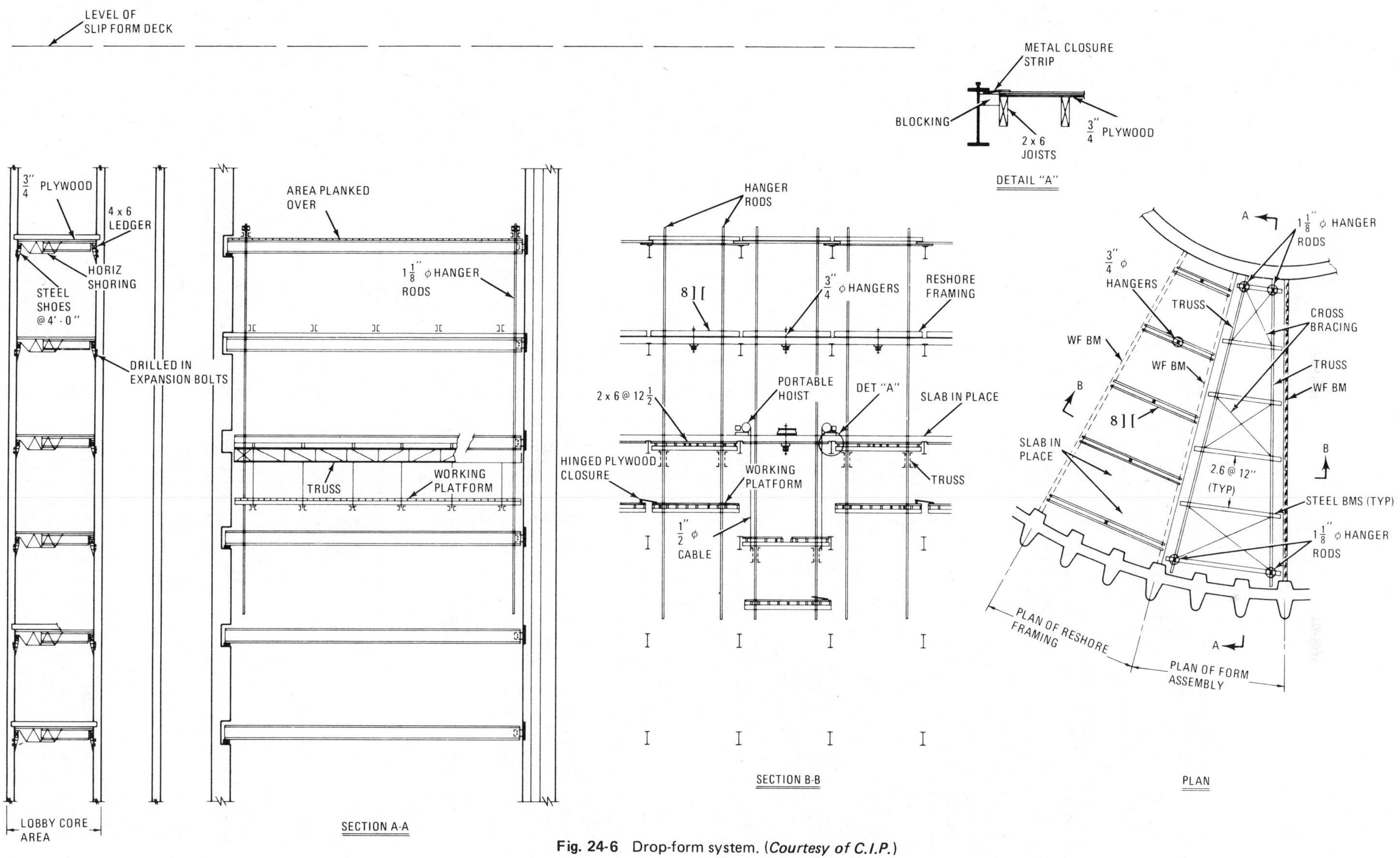

Fig. 24-6 Drop-form system. (*Courtesy of C.I.P.*)

24.3.8 Combination of Techniques

Practically all structures are built using a combination of techniques. In the planning and design of a structure, it is not a particular technique that emerges as the most desirable but rather the proper combination of techniques to obtain optimum time and cost economy. In comparing one technique against other techniques usable in the construction of a particular structure, the parameters should not be cost per yard of concrete in place. Rather, one must consider the cost of the entire structure based on combined techniques and including the values of early occupancy, tax and insurance advantages, overhead costs, financing costs, etc.

There is no value in slipforming walls very quickly if the slab-casting system is not compatible and loses the time and economy gained. On the other hand, it certainly pays to spend a few more dollars per yard for a slipformed walls if the overall floor cycle is reduced, or if significant economies are effected in the floor system. There will be no value in reducing the size of concrete members by posttensioning if the strand-tensioning cycles will result in serious extension of the construction time or inefficient discontinuous use of the labor force. On the other hand, considerable additional cost will be warranted in using prestressing techniques to reduce weight of the superstructure in areas of serious foundation problems. The entire structure must be considered as a whole, and the designer must be careful that in solving his problems in one design area, he does not create more costly ones in other areas.

Figure 24-7 shows a system study made for a high-rise building. This study included materials, forming system, equipment, hoisting and placing methods, and was in sufficient detail to yield reasonably accurate labor, schedule, and cost prediction.

Figure 24-8 shows a combination of high structural steel bents, low cast-in-place concrete bents, and precast prestressed single-tees that will be used to support a two-mile-long cross-country conveyor enclosed in a light gauge metal housing.

Figure 24-9 shows a combination of concrete construction techniques used to erect a concrete building in New York City. Included are a slipformed core, architectural precast concrete load-bearing column-spandrel units, and a cast-in-place slab-and girder floor system.

24.4 UTILIZATION OF FORMWORK

24.4.1 Planning

Formwork costs represent the major and most variable portion of the cost of concrete in place, normally 33 to 55%. As there is very little room for savings in the purchase of rigidly specified concrete and reinforcing steel, the differences in material prices are usually very small between bidding contractors and the engineer's estimate. The placing costs can vary, depending on good scheduling, proper crew planning and sequencing, efficient working spaces, and adequate materials handling. All of these items are related to and dependent on forming system selected and on how the system is implemented. The impact of the forming system is therefore greater than indicated by the 33 to 55% figure noted above.

On the concrete structures, the successful bidder is that contractor who has come up with the most economical forming system. Furthermore, the probability of the bids coming in within the engineer's estimate is predicated on a design that allows the contractor to come in with a sufficiently economical forming system. All of this emphasizes the importance of adequate study and analysis of the forming systems available so as to select the most economical system or systems for the particular project, and to do all that is possible to optimize the economy of these systems. This planning is necessary during the design, bidding, and construction phases. Table 24-3 and 24-4 show a typical comparative study.

24.4.2 Factors Affecting Economy of Formwork

There is no doubt that the one item having maximum effect on the economy of concrete formwork is the number of uses that will be made of each form section. Since this reuse item is intrinsically tied to the stripping ability of the forms, it is imperative to design and detail the forms in such a way as to make stripping quick and nondestructive. Economical reuse may also be increased by designing and sequencing structural components so that forms may be reused partially filled, bulkheaded, or slightly modified. The cheapest form obtainable will usually be more expensive per square foot of contact area when used only once, as compared to very expensive forms having many reuses.

The cost of concrete formwork in place is composed of two items. The first is the make-up item, which is the cost of the form ready for erection at the job site. This is a one-time cost; hence, the total form cost per square foot of contact area is inversely proportional to the number of uses. The second item consists of stripping, cleaning, repairing, erection, and expendable hardware. This item tends to remain constant for each reuse up to a certain point, at which the cost of repairing and cleaning starts rising rapidly. This is the point of diminishing return, at which further reuse of that form section becomes uneconomical. This point of diminishing return is of course dependent on the strength and durability of the particular form. It is also dependent on the care with which the forms are handled, cleaned, and stored. The most expensive form can be destroyed in one use by improper handling. See Tables 24-5 and 24-6, showing comparative costs for formwork.

As stated above, the forming system and reuse factors do affect the efficiency of the placing crews. Thus, the final forming system and material cost study must include full and efficient use of placing crews and the ability to meet the required construction schedule. The overall optimum reuse factor will probably be less than the diminishing-return factor discussed above. If quality requirements do not require otherwise, the forming material selected will be that which is most economical at the reuse factor selected.

Ganging reduces the stripping and erection cost much like the reuse factor reduces the make-up cost of formwork. By combining small forming sections into large assemblies, the forms can be stripped and re-erected at substantially lower cost when conditions allow (Fig. 24-10). The size of the ganged form should be the maximum size that can be lifted by the equipment available, considering accessibility to the areas involved. The ganging size will also depend on the building configuration and the reuse cycle. Ganging is possible in both vertical and horizontal casting. In order to use ganged forms, the surfaces being cast must be free of projecting dowels, brackets, and complications that will prevent easy release and lifting of the forms.

In considering the reuse factor relative to the required production, a determination will be made as to the actual number of sets of forms required. Obviously the number of sets must be kept to a minimum consistent with minimum time required before stripping. Frequently the length of time that the forms must remain in place is rather arbitrarily arrived at without the realization that overconservatism will adversely effect the cost of the structure. The use of minimum concrete strength rather than an arbitrary time

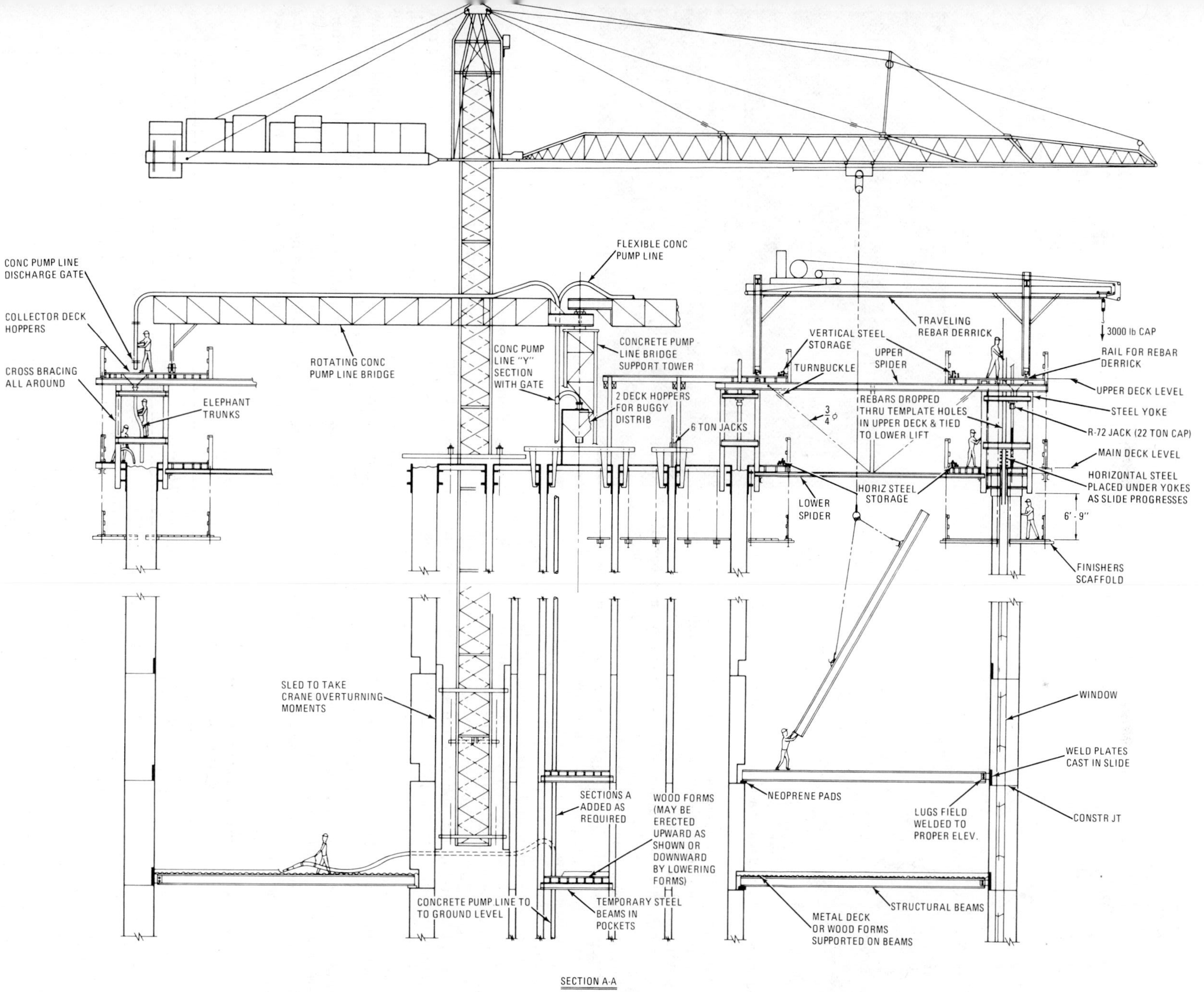

Fig. 24-7 Systems study for high-rise building. (*Courtesy of C.I.P.*)

Fig. 24-8 Combined materials in cross-country conveyor support.

requirement is recommended, as this will allow the contractor to take economic advantage of early strength and high-strength concrete. It will also furnish safety against too little strength in cold weather. When stripping time is based on concrete strength as indicated by field-cured cylinders, the use of high-strength concrete may be economically advantageous by reducing stripping time and therefore the number of form sets. It may also allow better utilization of the placing crews.

Another consideration that will promote economy in formwork is complete and detailed design of the formwork. Contractors are often overly anxious about the cost of detailed design, not realizing that this cost will be saved many times over in the field through less waste, less reworking, better production, and timely completion. In making the design, sizing of form components must be based on actual placing rates, expected temperatures, and standardized stocks on hand. If external vibration is to be used, the forms must have the necessary mass. Lifting devices and scaffold brackets should also be integrated. Every effort should be made to keep form members to a minimum consistent with strength requirements, not so much to reduce the make-up cost, but because weight is so important in the handling of the forms.

Fig. 24-9 American Bible HQ building. (*Courtesy of Brennan & Sloan.*)

24.4.3 Form Options

Forms, shores, and accessories may be rented from companies specializing in this operation. Column forms, floor domes, modular panels, shoring, etc., are available for rent at competitive rates. If the design allows, use of standard, rented form components can be the most economical solution. Conversely, whenever possible, designs should allow for use of standard form equipment. Forms may be manufactured in shops specializing in woodwork, steel, or fiberglass. The quality thus obtained is usually very high, resulting in forms capable of holding the tolerances and yielding excellent reuse. Quality required, availability and distance

TABLE 24-3 Cost Comparison by Casting Systems Type of Building and Labor Cost

Forming System	*Average Comparative Cost Factors*	*High-Rise Building*			*Medium-Rise Building*		
		Average Labor Cost	*High Labor Cost*	*Low Labor Cost*	*Average Labor Cost*	*High Labor Cost*	*Low Labor Cost*
Cast-in-Place Gang Forming	1.00	.96	1.14	.75	1.04	1.26	.82
Conventional Form + Precast Panels	1.02	.98	1.16	.79	1.06	1.26	.86
Combination of Slip and Gang Form	.98	.94	1.12	.75	1.02	1.21	.82
Slipform	.96	.86	1.02	.69	1.06	1.20	.83
Precast	.96	.95	1.04	.78	.97	1.09	.80

TABLE 24-4 System Comparison for Particular Building

Placing System	*Comparative Cost Factors*	*Finish**	*Concrete Joints*	*Construction Time Factor*	*Safety**	*Construction Tolerance*		*Special Features*
						Plumb-ness	*Wall th'k*	
Cast-in-Place Gang Forming	1.00	A–	horizontal and vertical	1.00	B	1″	−14″, +12″	No Special Skills Required
Conventional Form + Precast Panels	1.02	A+	horizontal and vertical	1.00	B+	1″	−14″, +12″	Precasting at Yard
Combination of Slip and Gang Form	.98	A–	horizontal and vertical	.75	B+	3″	−12″, +12″	Forming Operation + Conventional Slip Forming
Slipform	.97	A+	none	40	A–	2″	−14″, +12″	Special Skill and Equipment Required
Precast	.94	A+	horizontal	.50	A	12″	−14″, +12″	1. Winter Protection at Grade. 2. Inspection at Grade.

**Finish rated as follows: Uniform finish without tie holes— A+*
Uniform finish with tie holes— A–
***Safety rated as follows: Work mostly on ground or solid platform—A*
Work mostly on scaffolding— B

TABLE 24-5 Effect of Reuse on Concrete Formwork Based on One Use Equal to 1.00.

Number of Uses	*Cost Per Sq Ft CA*	
	Wood Form	*FRP*
One	1.00	1.00
Two	0.62	0.54
Three	0.50	0.38
Four	0.44	0.31
Five	0.40	0.26
Six	0.37	0.23
Seven	0.36	0.21
Eight	0.35	0.19
Nine	0.33	0.18
Ten	0.32	0.17

of shops, local labor conditions, and site congestion are factors in decision making.

On large projects, a shop may be erected on the site. If sufficient area, skill, equipment, and shelter are provided, the product will equal that of permanent shops, eliminating shipping problems and obtaining better control. The shop can be used for repair and other functions once the make-up is completed.

Last, the forms can be built in place. Some fabrication and storage facilities of a minimum standard must be provided. The quality may not be as high, but this may be the best option available in the case of highly complicated non-reusable forming or the opposite extreme of simple, undemanding formwork. Of course, combinations of rented, shop, and in-place fabrication are used as economics dictate.

24.4.4 Vertical Forming

There are basically four types of forming systems available for vertical work such as walls, columns, piers, and civil structures. They are conventional, jump, slip, and cantilever forming.

In conventional forming, an entire story or lift is constructed including the flat work. The vertical work is placed first, and after a delay of two hours or more, the floors are placed. After the concrete has attained the required strength, the forms are disassembled and passed up to the next floor from outside the building or through shafts. This entails a maximum of manual work and a minimum of equipment requirements. It is a good idea to keep form sizes down to 2 ft in width and 80 lb in weight for best efficiency. Since the floor cycles tend to be slow, a lateral subdivision is required for full continuous use of labor crews.

In jump-forming, the walls and columns usually proceed

TABLE 24-6 Comparative Cost of Forms per sq ft Contact Area for Different Materials and Reuses. (Based on field built plywood forms as base equal to 1.00)

Form Type	*Make-Up Cost*	*Assumed Reuses*	*Weight (lb/sq ft)*	*Cost (one use)*	*Cost (assumed uses)*
Plywood Forms, field built	1.00	4	10	1.50	.75
Overlaid Plywood, shop built	1.50	10	10	2.00	.65
Woodfaced, Steel Frame	2.50	10	12	3.00	.75
Epoxied Steel	3.75	40	15	4.15	.50
Fiberglass Reinforced Plastic	6.00	30	8	6.40	.60

Fig. 24-10 Ganged form being placed.

ahead of the flat work so that the wall and column forms are out of the way of the floor forming, and both systems are almost independent of each other. These forms are released and lifted straight up to the next position for reuse. They are often ganged and almost always handled by cranes (Fig. 24-11). They can be one floor high, supported on inserts set in the lift below; or, two sets, each one floor high, that leapfrog past each other, the lower one supporting the upper. Pockets, box-outs, and keys are formed for receiving the floors. Where sufficient reuse is involved with a minimum of change in wall, column, and floor dimensions, this system will have considerable merit. It should be noted here that concrete is cheaper than voids; that is, unnecessary openings and reductions in dimensions are to be avoided.

Slipforming is basically the ultimate extension of the jump-form system. The entire vertical forming system is combined into one ganged form that is raised as a unit using hydraulic jacks (Fig. 24-9). The requirements for both systems are similar but more stringent for slipforming. One or the other will be more economical based on building configuration, height, simplicity, and labor conditions. Slipforming is always faster.

Cantilevered forming lends itself to forming structures that have mild variation in sizes and slope. The forms for each lift are cantilevered by use of strongbacks from the lift below. The systems provides for relatively easy adjustment of the forms against the strong back. This system has good built-in control for holding tolerance. It is used extensively in the construction of dams and gravity cooling towers as illustrated in Figs. 24-12 and 24-13. Figure 24-14 shows a carefully designed system for forming exterior columns and core walls for a 40-story building. Note the mobile lifting device (Fig. 24-15) developed for raising the column forms. The work proceeds in a clockwise sequence of operations divided into four quadrants for a four-day cycle.

The use and sophistication of self-raising forming systems is developing rapidly. These systems may be categorized as cantilever forms or as jump-forms, but definitely not as slipforms, although they do have some of the advantages of slipforming. Climbing formwork is a name frequently applied to such forms which combine gang forms with scaffolds to form integral units equipped with pneumatic, hydraulic, or electromechanical raising devices. Several systems are marketed using different anchoring, lifting hardware, and stripping jacks. Self-raising forms have been used to construct high-rise building walls and cores, gravity cooling towers, and bridge piers.

24.4.5 Flat-Forming

There are basically six systems of floors and/or floor forming, namely, flat-slab, beam-and-slab, flying, metal-pan, leave-in-place, and cellular-steel forms.

It is basic to floor forming that the underside be perfectly level, to simplify fabrication, erection, and stripping. In flat-slab work, drop panels should be eliminated at the cost of thicker slabs and/or shear-transfer devices. In beam-and-slab work the bottom of all beams and, if possible, girders should also be kept level.

For light loadings and short to medium spans, it is hard to beat flat-slab forming for economy. Arrangement of the columns to facilitate stripping and erection or to allow the use of flying forms must be considered.

Beam-and-slab floors are used to suit structural requirements. For the sake of easy stripping, all beams should be of same depth in any one floor and of the same width in vertical projection floor to floor. Congestion at beam and girder intersections must be kept to a minimum.

Flying forms are really a system of ganging floor forms together with their supporting shores. (See Fig. 24-16). This system can be used with either flat-slab or beam-and-slab. The form can have drop panels that fold down for handling. In order to use flying forms efficiently, one side of the building must be completely open to allow removal of the flying-form sections. Adequate crane facilities are required, and column arrangements must be regularly lined up in the direction of form extraction. The economy of this system lies in designing flying sections of the maximum possible size that the cranes can handle and keeping the closure panels to a minimum. The use of flying forms has much merit and should be carefully considered.

Pan and dome floors find their area of maximum economy in forming long spans. With this system the weight of the concrete is materially decreased with little loss in strength. The forms can be steel, fiberglass reinforced plastic, hardboard, fiberboards, or even cardboard as conditions allow. To obtain maximum economy, the pan and dome sizes must all be standard and if possible the same size. As in any other form system, maximum reuse is desirable.

Leave-in-place forms derive their economy from saving the cost of stripping and cleaning. Whenever this cost exceeds the cost of the leave-in-place forms, this system becomes economical. It pays to use a slightly heavier gauge than is required structurally if a line of shoring can be eliminated. These forms can be steel, concrete, molded fiberboard, gypsum, or insulation board, depending on architectural requirements. Leave-in-place steel forms are especially economical with a steel beam system. Cellular-steel forms are a composite type of steel form that have the advantage of greater span without shoring. They are capable of longer

Fig. 24-11 Jump-forming.

spans between beams and may provide raceways for telephone and electricity.

24.4.6 Shoring

Since the load imposed by wet concrete is usually greater than the live load that will be finally placed on a hardened floor, carrying this weight to ground is an important consideration in formwork systems. The two basic shoring systems are vertical and horizontal. The vertical shores, wood, steel, etc., are in common use as shores and reshores. They have the advantages of directness, simplicity, and low cost. Their disadvantages lie in the necessity to wait until the floor below can support them, the clutter they create interfering with other operations, and the tricky reshoring and relieving operations involved.

Horizontal shores, most of them adjustable, are more expensive and have lighter capacities than vertical shores. They do, however, make each floor-forming system independent and leave the decks below free for other construction operations. The economy of the two systems can only be compared on the basis of the total construction operation.

24.5 CONCRETE PLACING EQUIPMENT

Next to formwork, concrete placing methods have the most significant effect on concrete economy and quality. It is important that the optimum system be selected that guarantees ample capacity in yards per hour, minimum labor requirements, proper handling of concrete to prevent segregation, and maximum protection from the elements. The equipment should be as nearly trouble-free as possible, and provisions should be made for spare parts and equipment and possibly for alternate handling in case of breakdown.

The method of placing must be planned in conjunction with concrete mix, admixture use, site conditions, climatic conditions, forming system, and handling of other materials. The following placing methods are in common use. None stands out under all conditions as superior to other methods; each has applications depending on specific conditions. Table 24-7 shows some characteristics of placing methods.

24.5.1 Belt Conveyors

Belt conveyors are generally divided into three types: portable, or self-contained conveyors; feeder, or series conveyors; and side discharge, or spreader conveyors. Portable conveyors are excellent for short reach and/or lift situations up to 35 ft. These conveyors require almost no set-up cost, require only one man to operate them, and are often economical in very small placements that cannot be placed directly out of a ready-mix truck but are within the 35-ft reach.

Placements beyond the reach of a portable conveyor require a number of feeder conveyors in series. These conveyors come in standard sections of 24 to 56 ft in length and can be used for lifting concrete at inclines up to 20–35°, depending on concrete mix, belt type, and capacity required (Fig. 24-17). These conveyor sections may be manhandled or self-propelled and may be brought to the job under their own power, loaded on trucks, or towed. Steel sections vary in weight from 800 to 3000 lb; aluminum sections weigh 15 to 30% less than steel sections of the same size.

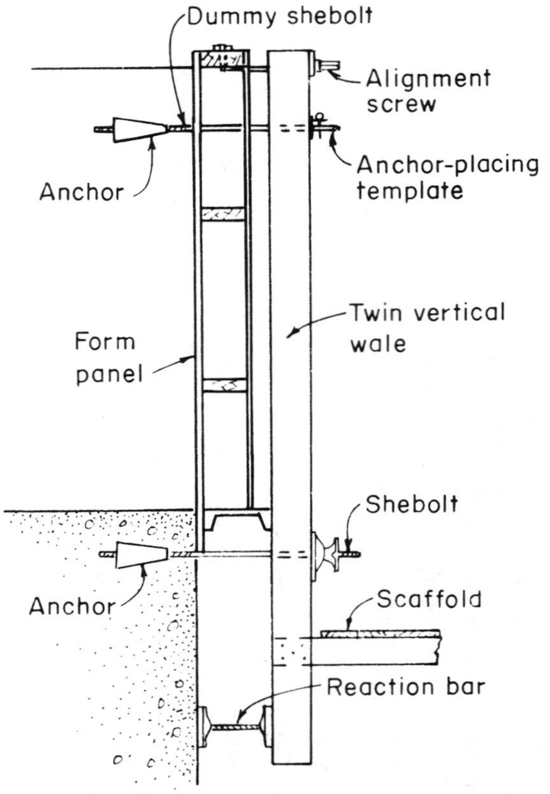

Fig. 24-12 Typical details of cantilevered form construction on a dam.

Fig. 24-14 C.B.S. HQ building. (*Courtesy of C.B.S.*)

Concrete placements often require considerable spreading, making it advantageous to use side discharge conveyors or cantilevered radial spreaders for the final depositing of concrete. Feeder and spreader conveyors are at their best

Fig. 24-13 Cantilever forms on a cooling tower.

Fig. 24-15 C.B.S. HQ Building. (*Courtesy of C.B.S.*)

TABLE 24-7 Relative Placing Costs Using Cranes, Belt Conveyors, and Pumps

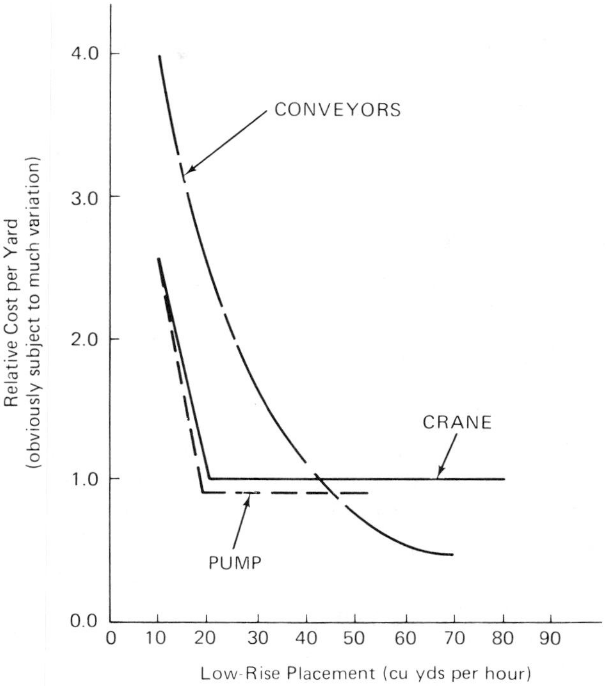

Fig. 24-16 Flying forms in action.

Fig. 24-17 A set of belt conveyors in action.

when used in large placements with fast rates of placing. Although they may be used efficiently in placements as low as 50 yd, their best efficiencies are reached in placement of 300 yd, or more, with placing rates of over 40 yds/hr. Actually, the same crews are required whether placing 30 or 100 yd. Although some conveyors are rated to 180 yd/hr, such capacities are very difficult to attain in practice. With 16-in. belts (size in general use) and belt speeds of 300 to 750 fpm, rates of 40 to 100 yd/hr are more realistic. Belts 24 in. wide make rate increases up to 67% possible.

Special concrete mixes are not required for belt handling. A slump of 2.5 to 3 in. is usually used, as this results in maximum capacity without overloading the equipment. Higher slumps are acceptable but will result in lower rates of handling. Aggregate sizes are usually limited to 2 or 3 in. for 16-in. belts (4 to 6 in. for 24-in. belts). Concrete weight will run 10 to 17 lb/ft on a 16-in. belt and 20 to 36 lb/ft on a 24-in. belt. With proper operation, the quality of the concrete will not be adversely effected by belt handling, as no significant segregation should result if concrete is discharged against or onto concrete that is already in place.

Conveyor belts are quite dependable and are not overly subject to breakdown if care and good maintenance include:

1. Flushing belts of equipment as necessary
2. Adjusting and replacing belt scrapers as required

3. Proper lubrication
4. Protection of motor and other equipment from concrete and dust
5. No continuous overloads

Rental of belt conveyors runs 60 to 70 dollars per day per 26-ft section. This will result in an average rental cost of from one to two dollars per yard. It will take a five- to eight-man crew for erection and four to six men for operation of feeder conveyors. The spreading and consolidating crew will be in addition to this.

24.5.2 Pumping Concrete

Concrete can be economically pumped at rates up to 100 yd/hr, for horizontal distances up to 2000 ft and vertical distances to 500 ft without staging. At present, no equipment is capable of accomplishing all this concurrently, and only the largest pumps and equipment can accomplish the maximum rate or distance at all. A large variety of equipment is available with capacities ranging from 15 to 200 yd/hr and rated distances varying from 300 to 3000 ft horizontally and 100 to 500 ft vertically. (Actually concrete has been pumped vertically for over 1000 ft without use of booster pumps.) Although good economy is often obtained at shorter runs, the most effective pumping distances vary from about 300 to 1000 ft horizontally and from 100 to 300 ft vertically. The pumps come skid-, trailer-, or truck-mounted, weigh 1500 to 25,000 lb, and are generally very mobile.

Hydraulic booms, as illustrated in Fig. 24-18, are available for low-rise work with reaches approaching 100 ft vertically. Concrete placing booms may also be mounted on tower crane masts. The pipe used runs in sizes from 3 to 8 in. in diameter (4 in. is most common) and includes rigid and flexible sections. Pipe that is used on booms or that must be manhandled is limited to 4 or 5 in. because of the weight. Lines over 5 in. in diameter must be a stationary installation.

The pumping of concrete is especially advantageous in placements that are physically difficult to approach with ready-mix trucks and show best economy in the range of 20 to 80 yd/hr. Equipment may be purchased for \$600 to \$2000 per yard, rated capacity. Rental, including operator, runs \$1.00 to \$5.00 per yard with a minimum yardage requirement.

Fig. 24-18 Combined vertical and horizontal pumping.

Some of the problems associated with pump concrete are pump breakdowns, line clogging and ruptures, and abrasion of equipment. The equipment must be rugged and wear-resistant. The flow of concrete must be as continuous as possible with movement at least every 15 minutes and clearing of the lines required for stops over 30 minutes to 1 hour depending on temperatures, concrete admixtures, etc. Water is required for cleaning the system, and some provision is needed to collect the resultant discharge so as not to contaminate the freshly placed work. Placements should always start at the farthest point of deposit and work toward the pump location, as this makes it possible to remove pipe sections with minimum interruption. Adding sections of pipe, on the other hand, is difficult and causes restart problems.

Concrete pumping is normally conducted at optimum line pressures of 200 psi with 400 psi as an intermittent maximum. Pressures in pump will run about three times line pressure. Flexible hose tends to increase pressure, so its use is kept to a minimum. In high-rise work the pressures tend to get very high, as the weight of the column of concrete is added to the pressure required to overcome friction. For very high lifts, extra heavy lines and connections will be required. In addition, substantial anchorages will be required to tie the pipeline to the structure to keep the pump from "walking away."

Normally two ready-mix trucks are unloaded simultaneously into the pump hopper. A minimum of five men is required, two at the truck and three at the nozzle. The spreading and consolidating crew will be in addition to this.

In order to overcome friction in the lines, a rich film of cement, sand, and water is required as a lubricant between the pipe walls and the moving slug of concrete. In order to establish this film, a grout is first pumped through the lines. Good aggregate gradation with enough cement paste and fine aggregate particles is absolutely required. Poor gradation, especially in the sand, will result in mechanical interlocking and insufficient matrix, and will not support the grout film. Excessive water can wash away this film; angular sharp aggregate can scrape it away; leaks in the line can cause loss of the film. Loss of film will cause a plug. The following minimum requirements must therefore be met in a concrete to be pumped:

1. Cement content–450 lb/yd, minimum.
2. Well-graded, fine aggregate with 10 to 30% of sand passing No. 50 sieve and 2 to 10% passing No. 100; fineness modulus of 2.40 to 3.00.
3. Maximum volume of coarse aggregate may be reduced up to 10% from ACI 211.1-70, depending on equipment and mix.
4. Slump 2 to 6 in., depending on mix, length of run, and equipment.
5. No bleeding mixtures.

Although some pump manufacturers specify minimum slumps less than 2 in., the pumping rate is slower at these low slumps, and additional slump may be required for losses during handling. It may be difficult to maintain proper moisture content with lightweight concrete. This difficulty can be solved by presoaking the aggregate; by using a vacuum- or thermal-saturation process to treat the aggregate; or by using coated pellets, adding 3 to 5 dollars per yard to the cost of the concrete. Note that the pressure in the lines tends to increase the temperature and accelerate the setting time of the concrete, which may be an advantage or disadvantage, depending on ambient temperatures.

The use of admixtures (especially those having a hydroxylated polymer base, 3 to 5%), air-entrainment, and additional paste will often prove economical by increasing production and reducing or eliminating plugs. Some difficulty has been experienced with pumping through aluminum pipe. The lime in the cement reacts with the aluminum abraded from the pipe, releasing hydrogen gas, thus resulting in lower strengths, an effect not unlike that of an excessive amount of entrained air.

24.5.3 Cranes

The use of cranes and buckets in placing concrete is general and widespread. In low-rise work, cranes find greatest usage where placements have to be made at scattered locations in the work, especially relatively small placements that do not justify installation of conveyors or pump lines (Fig. 24-19). They are especially useful in placements above ground such as columns and piers. The relatively low cost of bringing mobile cranes on the job, setting up, and demobilizing is often an important advantage.

In high-rise work, cranes are often very effective because of their ability to deliver concrete at various locations and at various levels without movement of the crane. One of the major economic advantages of crane usage is its versatility in handling other materials such as rebars, forms, inserts, and structural steel. The disadvantages of using many types of cranes lies in their inability to move to all parts of the work and in limitation in height of lift, reach, and capacity. In congested areas and on city streets, the working room they need may be a serious problem.

There are many types of cranes of widely varying characteristics and capacities. The most common type is the conventional crawler or truck-mounted crane. These cranes are either tread-mounted (must be transported over roads on trailers) or truck-mounted and self-propelled. Some smaller ones may be modified for installation on the structure. These cranes rely on a boom that may be anywhere in length from a short hydraulic telescoping arm to a 450-ft stick. The cranes revolve 360° horizontally and boom to within 15° and 20° of level and plumb. They are available in capacities from 15 to 200 tons, and more. This designation of capacity indicates the capacity when the boom is nearly vertical and quite short; a moment-capacity designation, such as used in tower cranes, is also useful. Although cranes may reach to 400 ft in height and radius, at this reach the payload is greatly reduced. For instance, a 110-ton crane with a 250-ft boom has a capacity of 1 ton at 220-ft radius. Outriggers are required to realize the capacities of many cranes.

One problem of cranes operating on the ground is that they must stay away from the building in order to accommodate the boom angle. Thus they have difficulty reaching across the face of the building to interior locations. This problem is solved by another type of crane, the tower crane. (See Figs. 24-11 and 24-20.) This crane has a vertical mast with a rotating horizontal jib at the top; the mast is stationary. The load is hoisted on a trolley that travels radially on the jib; the jib itself rotates 360°. This means that as long as the jib is high enough to clear the work and surrounding obstructions, loads can be picked up and deposited anywhere within the radius capacity of the particular crane.

These cranes take a minimum of space on the ground and can climb up with, and on, the building. They can free stand in excess of 350 ft without bracing, and can be extended to several hundred more feet when braced against the building or guyed. They are sometimes mounted on railroad carriages. In smaller capacities they are truck-mounted and have in several instances been suspended from slip-form decks. They are available in capacities up to 4500 ft tons with capacities up to 70 tons at short radii and 15 tons at 250 ft radius. They are also available with luffing booms.

Fig. 24-19 Crew placing concrete from a bucket.

TABLE "B"

MINIMUM REQUIREMENTS FOR SLABS SUPPORTING CRANES

Δ = AREA OF STEEL IN SQ IN PER LIN FT (TYP) — * = SHEAR GOVERNS

STANDARD WEIGHT CONC f'_c = 3000 p.s.i.						STANDARD WEIGHT CONC f'_c = 1500 p.s.i.					
SPAN DIM X	FREE END SLAB		FIXED END SLAB			SPAN DIM X	FREE END SLAB		FIXED END SLAB		
	SLAB THICK	BOTTOM STEEL	SLAB THICK	BOTTOM STEEL	TOP STEEL		SLAB THICK	BOTTOM STEEL	SLAB THICK	BOTTOM STEEL	TOP STEEL
1'-0"	* 19½"	0.64	* 19½"	0.113	0.55	1'-0"	*26"	0.47	*26"	0.09	0.40
2'-0"	* 19½"	1.21	* 19½"	0.26	0.98	2'-0"	*26"	0.87	*26"	0.20	0.64
3'-0"	* 19½"	1.78	* 19½"	0.44	1.38	3'-0"	*26"	1.29	*26"	0.34	0.99
4'-0"	* 19½"	2.34	* 19½"	0.65	1.74	4'-0"	*26"	1.71	*26"	0.50	1.26
5'-0"	* 19½"	2.95	* 19½"	0.89	2.13	5'-0"	*26"	2.13	*26"	0.67	1.54
7'-0"	22½"	3.55	* 19½"	1.42	2.84	7'-0"	31"	2.43	*26"	1.03	2.03
10'-0"	27½"	4.40	22½"	1.95	3.37	10'-0"	38"	2.95	31"	1.41	2.37

LIGHT WEIGHT CONC f'_c = 3000 p.s.i.						LIGHT WEIGHT CONC f'_c = 1500 p.s.i.					
SPAN DIM X	FREE END SLAB		FIXED END SLAB			SPAN DIM X	FREE END SLAB		FIXED END SLAB		
	SLAB THICK	BOTTOM STEEL	SLAB THICK	BOTTOM STEEL	TOP STEEL		SLAB THICK	BOTTOM STEEL	SLAB THICK	BOTTOM STEEL	TOP STEEL
1'-0"	*28½"	0.40	*28½"	0.08	0.34	1'-0"	*35½"	0.32	*35½"	0.07	0.2
2'-0"	*28½"	0.75	*28½"	0.17	0.60	2'-0"	*35½"	0.59	*35½"	0.14	0.4
3'-0"	*28½"	1.11	*28½"	0.28	0.86	3'-0"	*35½"	0.86	*35½"	0.23	0.
4'-0"	*28½"	1.46	*28½"	0.42	1.08	4'-0"	*35½"	1.13	*35½"	0.34	0.
5'-0"	*28½"	1.83	*28½"	0.57	1.32	5'-0"	*35½"	1.42	*35½"	0.46	1.
7'-0"	*28½"	2.58	*28½"	0.90	1.76	7'-0"	*35½"	2.10	*35½"	0.71	1.
10'-0"	*28½"	3.77	*28½"	1.44	2.46	10'-0"	36½"	2.88	*35½"	1.16	1.

WHEN TWO OR MORE SLABS ARE USED TO SUPPORT THE CRANE THE MINIMUM COMBINED SLAB THICKNESS MUST BE AS FOLLOWS;

WHEN SHEAR GOVERNS: FOR ALL OTHER CASES:

$$t\ \text{COMB} = t\ \text{REQUIRED} + n - 1 \qquad t\ \text{COMB} = \sqrt{(t_1-1)^2 + (t_2-1)^2 + (t_3-1)^2} + 1 \geq t\ \text{REQUIRED}$$

THE COMBINED STRENGTH OF THE CONCRETE SLABS SHALL BE ASSUMED AS FOLLOWS;

$$f\ \text{COMB} = \frac{t_1 f_1 + t_2 f_2 + t_3 f_3}{t_1 + t_2 + t_3}$$

THE COMBINED REINFORCING STEEL MUST BE EQUAL TO OR GREATER THAN THE REQUIRED STEEL AND IS COMPUTED AS FOLLOWS,

$$As\ \text{COMB} = \frac{As_1(t_1-1) + As_2(t_2-1) + As_3(t_3-1)}{t\ \text{COMB}}$$

n = NUMBER OF SLABS USED IN SUPPORT.
t_1, t_2, t_3, ETC ARE THE RESPECTIVE THICKNESS OF THESE SLABS
f_1, f_2, f_3, ETC. ARE THE RESPECTIVE STRENGTH OF THESE SLABS
f COMB = IS AVERAGE STRENGTH OF SLABS.
t COMB = SUM OF THICKNESSES OF SLABS USED.
t REQUIRED = THICKNESS OF ONE SLAB AS SHOWN ON TABLES.
As COMB = SUM OF STEEL AREAS OF SLABS USED.
As REQUIRED = STEEL AREAS SHOWN ON TABLES.
As_1, As_2, As_3, = STEEL AREAS OF RESPECTIVE SLABS USED.
t, n MEASURED IN INCHES.
f MEASURED IN POUNDS PER SQ IN.
As MEASURED IN SQ IN.

MAXIMUM LOADING (INCLUDING PAYLOAD and COUNTERWEIGHT)			SYMBOL	UNIT	H = 82'-0"				H = 65'-8"				H = 49'-3"			
					JIB "K"	JIB "N"	JIB "L"	JIB "LL"	JIB "K"	JIB "N"	JIB "L"	JIB "LL"	JIB "K"	JIB "N"	JIB "L"	JIB
TOTAL LOAD	VERTICAL LOAD		V	KIPS	185	190	185	191	185	190	185	191	185	190	185	19
	LONGITUDINAL OVERTURNING MOMENT		M_L	FT. KIPS	1075	1065	999	1039	1075	1065	999	1039	1075	1065	999	10
	TRANSVERSE OVERTURNING MOMENT		M_T	FT. KIPS	471	501	550	580	369	393	434	458	269	287	318	33
	SLEWING MOMENT (TOP SUPPORT ONLY)		M_S	FT. KIPS	174	225	272	335	174	225	272	335	174	225	272	33
	TRANSVERSE HORIZONTAL SHEAR (TOP SUPPORT ONLY)		S_T	KIPS	6.6	7.0	7.6	7.9	6.2	6.6	7.2	7.5	5.8	6.1	6.7	7.1
VERTICAL LOAD PER LEG			V/4	KIPS	46.2	47.5	46.2	47.8	46.2	47.5	46.2	47.8	46.2	47.5	46.2	47
SHEAR PER BOLT GROUP IN ANY DIRECTION	TOP SUPPORT	D = 21'-4"	St	KIPS	15.6	17.4	19.3	21.8	14.8	16.7	18.0	21.0	14.0	15.8	17.3	19.
		23'	"	"	14.9	16.9	18.3	21.3	13.9	15.9	17.6	20.3	13.1	15.4	16.7	19.
		25'	"	"	14.2	16.3	17.8	20.1	13.5	15.3	17.1	19.7	12.6	14.6	16.3	18.
		27'	"	"	13.5	15.8	17.5	19.9	12.9	14.8	16.5	19.3	12.2	14.2	15.9	18
		29'	"	"	12.9	15.4	16.9	19.7	12.5	14.2	16.1	18.8	11.7	13.7	15.4	18.
		31'	"	"	12.6	15.2	16.5	19.1	12.2	13.8	15.7	18.5	11.3	13.1	14.8	17.
		33'	"	"	12.2	14.7	16.3	18.8	12.0	13.5	15.5	18.0	11.0	12.8	14.7	17.
		35'	"	"	12.0	14.3	15.9	18.3	11.4	13.3	15.4	17.9	10.8	12.6	14.6	17.
	BOTTOM SUPPORT	D = 21'-4"	Sb	KIPS	12.6	12.5	11.7	12.2	12.6	12.5	11.7	12.2	12.6	12.5	11.7	12.
		23'	"	"	11.7	11.6	10.9	11.3	11.7	11.6	10.9	11.3	11.7	11.6	10.9	11.
		25'	"	"	10.8	10.7	10.0	10.4	10.8	10.7	10.0	10.4	10.8	10.7	10.0	10.
		27'	"	"	10.0	9.9	9.3	9.6	10.0	9.9	9.3	9.6	10.0	9.9	9.3	9
		29'	"	"	9.3	9.2	8.6	8.9	9.3	9.2	8.6	8.9	9.3	9.2	8.6	8.
		31'	"	"	8.7	8.6	8.1	8.4	8.7	8.6	8.1	8.4	8.7	8.6	8.1	8
		33'	"	"	8.2	8.1	7.6	7.9	8.2	8.1	7.6	7.9	8.2	8.1	7.6	7
		35'	"	"	7.7	7.6	7.2	7.4	7.7	7.6	7.2	7.4	7.7	7.6	7.2	7.

$$Sb = \frac{\sqrt{M_L^2 + M_T^2}}{5.33D}$$

BUT NOT LESS THAN $\frac{M_L}{4D}$

$$St = \frac{\sqrt{\left(\frac{M_L}{D}\right)^2 + \left(\frac{M_T}{D} + S_T + \frac{1.33M_S}{6.3}\right)^2}}{5.33}$$

BUT NOT LESS THAN

$$\frac{1}{4}\sqrt{\left(\frac{M_L}{D}\right)^2 + \left(\frac{M\ \text{SLEWING}\ ^{W}/_{O}\ \text{WIND}}{6.3}\right)^2}$$

NOTE: M_S, St, Sb — LOADS HAVE BEEN REDUCED BY 25% WHERE APPLICABLE TO ALLOW FOR 33% INCREASE IN ALLOWABLE STRESSES.

M SLEWING W/O WIND = 237[K-FT](LL), 207[K-FT](L), 168[K-FT](N), 145[K-FT](K).

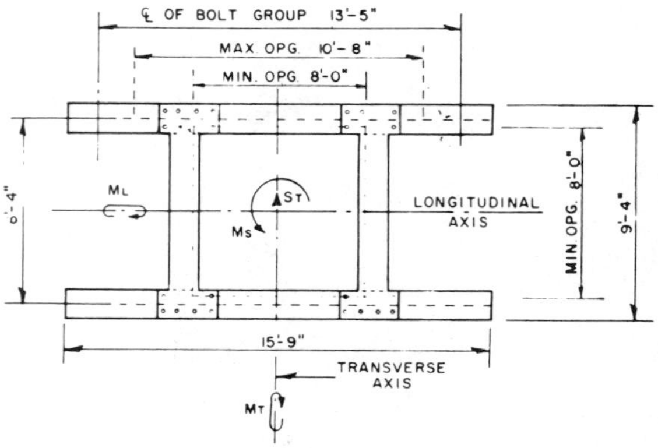

PLAN OF TOP SUPPORT
BOTTOM SUPPORT SIMILAR

Fig. 24-20 Design criteria for a climbing crane. (*Courtesy of American Pecco*).

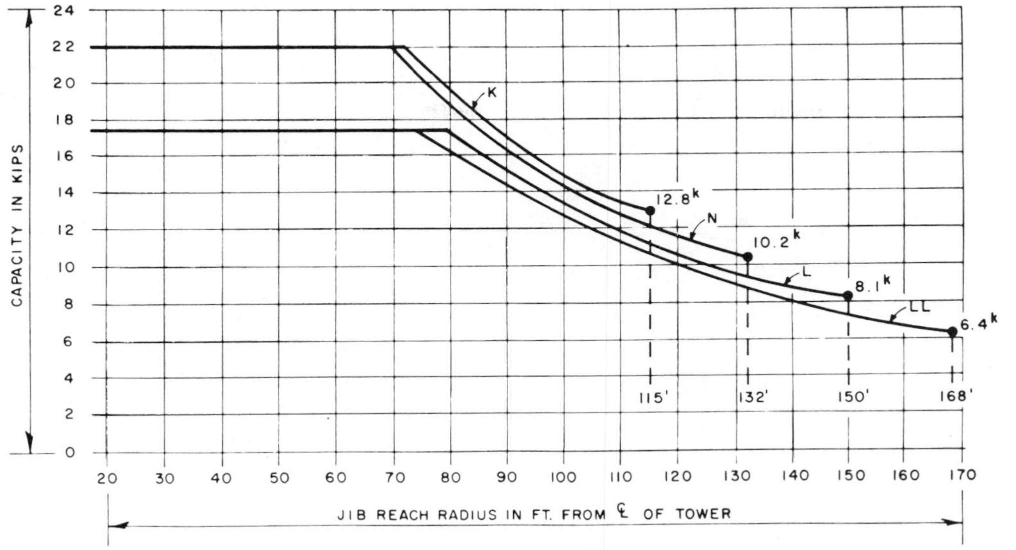

LOADING DIAGRAM

DESIGN CRITERIA

1. DESIGN BASED ON OPERATIONAL LOADS PLUS WIND LOADING FROM A 40 M.P.H. WIND.
2. WINDS OVER 40 M.P.H. ARE CONSIDERED STORM CONDITION AND NO OPERATION WILL BE PERMITTED DURING STORM CONDITIONS.
3. DURING STORM CONDITION (OVER 40 M.P.H.) THE JIB MUST BE FREE TO WEATHERVANE.
4. WHEN NOT IN USE AND DURING STORM CONDITION, THE HOIST TROLLEY MUST BE AT THE LONGEST REACH RADIUS POSITION.
5. DURING STORM CONDITIONS WITH WINDS 65 M.P.H. AND OVER A BALANCING WEIGHT (NOT INCLUDING TROLLEY) IS REQUIRED AT THE END OF THE JIB OF THE WEIGHT INDICATED IN TABLE "A". THIS LOAD SHOULD HAVE A MINIMUM PROJECTED AREA.
6. THE BALANCE LOAD SHOWN IN TABLE "D" MIGHT BE SLIGHTLY LESS DUE TO THE WIND CONDITIONS.
7. SLEWING TORQUE IS BASED ON A MAXIMUM SLEWING SPEED OF 1.0 REVOLUTIONS PER MINUTE AND A MINIMUM STOPPING TIME OF 5 SECONDS AT MAXIMUM SPEED.
8. MINIMUM SLAB THICKNESS AS GIVEN IN TABLE "B" IS BASED ON CONCRETE HAVING A MINIMUM COMPRESSIVE STRENGTH AS SHOWN BASED ON FIELD CURED CYLINDERS.
9. SLABS NOT MEETING REQUIREMENTS SHOWN MUST BE SHORED TO SLABS OR OTHER STRUCTURAL MEMBER HAVING PROPER STRENGTH.
10. IF SUPPORT CONDITIONS ARE DIFFERENT THAN THOSE SHOWN ON THIS DRAWING AN EXPERIENCED PROFESSIONAL ENGINEER SHALL BE CONSULTED.

Fig. 24-20 *(Continued)*

Tower cranes are versatile, efficient tools and are superior to conventional cranes in situations where loads must be carried up and over the construction or other obstruction, and at heights above the capacity of the conventional cranes. They are not as efficient when all points cannot be reached from one position. The main disadvantage is the cost of installing, climbing, and dismantling the cranes. Sometimes there is a clearance problem as to the jib clearing adjacent buildings. Generally speaking, cranes compete very well in economical placing of concrete when the job conditions allow free access and movement, when crane capacities are not exceeded, and when the placing area is open for vertical placement. Rentals run from $7500 to $15,000 per month depending on capacity and reach required. An operator and oiler are required in addition to rental.

Helicopters with capacities up to 9 tons have been used successfully to place concrete, to handle formwork, and to erect and dismantle cranes. Careful planning is required as the hover time is usually limited to less than one hour. In inaccessible spots they can prove very economical in terms of time and money. Figure 24-21 shows a helicopter placing the jib sections on a tower crane atop a 535-ft concrete core.

24.5.4 Materials Hoist Towers

In studying placing methods for medium- and high-rise work, the hoist tower must always be considered. It has no capability for horizontal movement of concrete either at the base or at the discharge point, but it can be used in conjunction with conveyors, pump lines, chutes, and buggies to do the job. It takes a minimum of space to operate, and has ample capacity and relatively low rental and erection cost. It is not mobile. For heavy concrete placing requirements, a hoist tower should be considered including the additional cost of required additional placing equipment and labor. Rental will run 1000 to 8000 dollars per month, depending on capacity and height including hoist, but not operator.

24.5.5 Buggies

Hand and power buggies are used for horizontal distribution of concrete. Sometimes hand buggies are lifted on materials hoist platforms to the level of discharge and then wheeled across. Hand buggies are rubber-tired and have capacities of 6 cu ft and higher. Some are made very narrow (30 in.) for tight situations. In some cases wheelbarrows

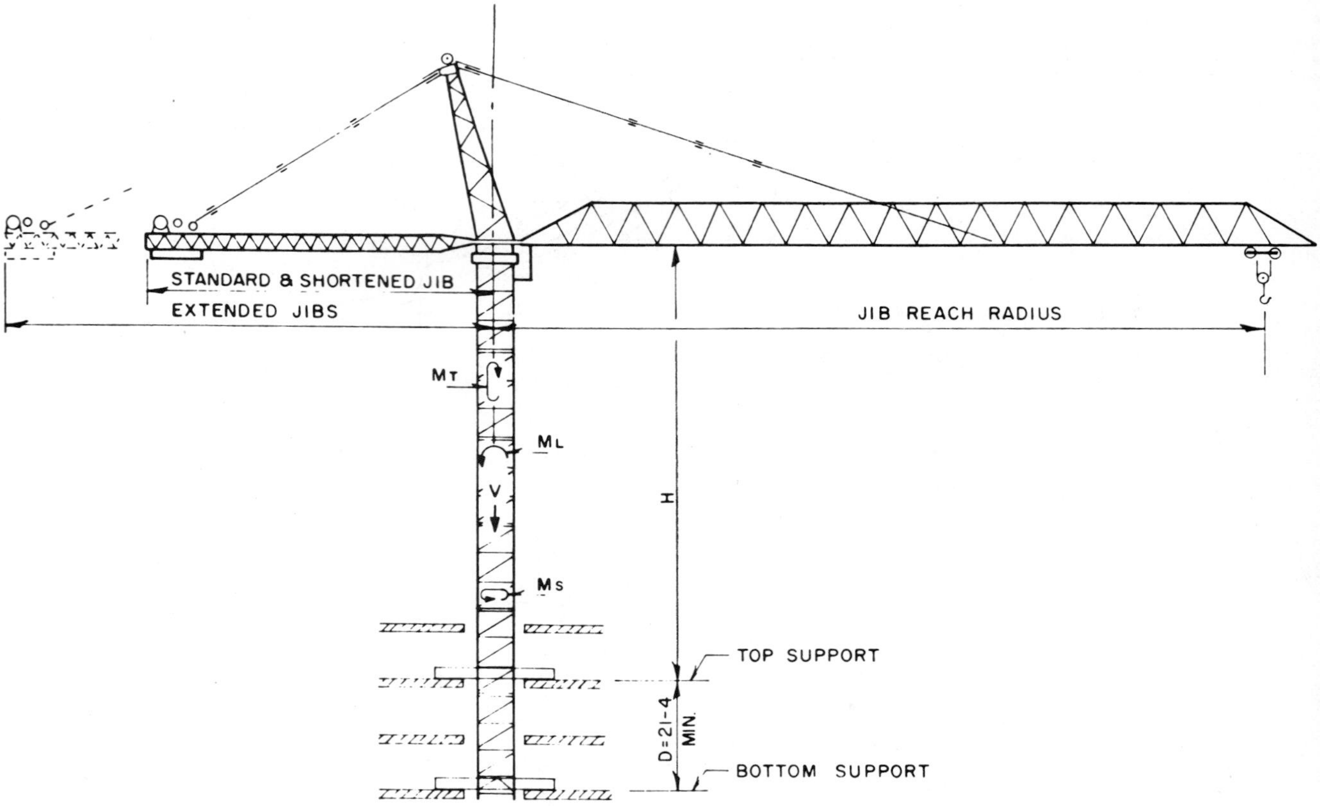

TABLE "D"			
BALANCE LOAD DURING CLIMBING			
JIB	MK	LENGTH	BALANCE LOAD
SHORT	K	115'	12,800 lbs.
NORMAL	N	132'	10,200 lbs.
EXTENDED	L	150'	8,100 lbs.
DB'LE. EXT.	LL	168'	6,400 lbs.

Fig. 24-20 *(Continued)*

may be used for the same purpose, but the capacity will be considerably less. Depending on site conditions, labor and run involved, a hand buggy can place up to 5 yd/hr.

Motorized power buggies with up to 18 cu ft capacity are available, although most units have capacities of 9 to 12 cu ft (Fig. 24-22). They are fast, very maneuverable, and capable of climbing slopes up to 35%. Power buggies will place 15 to 20 cu yd/hr. When power buggies are used, forming and shoring must be designed for at least 50% more live load than normally (75 psf) and lateral bracing must be carefully designed. Runways are required over reinforcing steel when any type buggy is used. Rental for buggies runs approximately $250 per month for plain carts to $550 for power carts.

24.5.6 Pneumatic Placing of Concrete

This method, which is illustrated in Fig. 24-23, is also known as shotcrete and gunite. Concrete is shot at high velocity against a form or other surface under pressure. Because this is a no-slump material and because of the force of the jet, the mix adheres and is firmly compacted in place. Only one-sided forms are required, and these forms will be subject to gun pressure only, which is less than the normal hydrostatic force of concrete in conventional placing. Shortcreting is advantageously used in relatively thin sections such as shell roofs, walls, tanks, chimney and hydraulic linings, and cover and repair applications for all types of structures. It is especially useful for casting curved and three-dimensional sections.

The pneumatic mortar or concrete is applied in layers 1 to 1.5 in. thick, the total thickness obtained by successive placements usually being up to 4 in. Placements considerably thicker than this are obtainable, but the cost can be high. Two to six yards of shotcrete per day per gun can be placed, depending on equipment used. For a 3-in.-thick placement the cost may run $3.50 to $7.50/sq ft subcontracted. Normal mixtures used are one part portland cement to three parts sand with a maximum aggregate size of 0.25 to 0.375 in., although mixes of one to four and aggregates to 0.75 in. are often used. Water included is usually 3 to 6 gallons per sack of cement. The shotcrete nozzle

Fig. 24-21 Erecting crane by helicopter.

Fig. 24-22 Motorized buggies in action.

Fig. 24-23 Shotcreting in action.

can be operated at distances up to 500 ft horizontally from the mix equipment and about half this distance vertically.

Care must be taken to keep "rebound" to a minimum (20% is good), blow off excess rebound material, keep form and steel surfaces clean, and prevent voids behind reinforcing bars. Definite means of checking the thickness and moist-curing are essential.

24.5.7 Other Placing Methods

Other methods of placing concrete requiring special equipment are in limited use. One such method is preplaced aggregate concrete in which the coarse aggregate is placed and compacted and then a grout of sand, cement, and water is intruded into the aggregate mass. Another placing method is dry-casting in which stone, sand, and cement are placed dry and compacted, and water is intruded at a later time.

24.5.8 Selection of Equipment

In order to determine the placing equipment to use, the various types and combinations of equipment are studied, and those that cannot be used or are obviously uneconomical are eliminated. A detailed cost study of each method is then made, including variation in concrete mix, formwork, set-up, rental, dismantling, operators, fuel, placing labor, and any other related costs.

24.6 SAFETY PAYS

The moral obligation to safeguard human life is so obvious that it will not be belabored in this chapter. As is usually the case, the moral approach is, in the long run, the best pragmatic approach. Accidents cost money; insurance against accidents costs money; court litigations cost money; job delays cost money. Workmen's compensation insurance, which amounts to 4 to 20% of payroll, is based on the manual rate established for that particular classification of work and upon the individual contractor's relative safety record within the classification. A low accident history will, for instance, tend to improve the competitive position of concrete construction versus steel erection as a whole and the competitive position of a particular contractor within his own classification. Liability insurance will likewise be based on accident records.

Anyone who has been involved in an accident investigation and/or court litigation cannot help but wonder at the extent and costs to which tests, computations, searches, and expert opinions are carried. One cannot help but wish that some of this thought and expense had been used to prevent the accident in the first place.

This chapter will not attempt to go into the organization and administration of good safety programs, or into safety codes and practices. Information on these matters is easily available from many private sources such as publishers and the Association of General Contractors and is mandated in a statutory manner by municipalities, states, government agencies at all levels, and of course by the federal government. We will, however, mention general attitude and some specific hazards that seem worthy of special note.

In the first place, safety is more a state of mind than a set of rules. Top management of the owners, designers, and builders must indicate in a way that cannot be misunderstood that they consider safety of paramount importance. They must commend those people who promote safety and discourage or remove completely from their organization those people who are careless. This attitude must filter through the whole hierarchy from president to night watchman. Above all, placing incompetent persons in safety positions just to feign safety-mindedness is criminal.

24.6.1 Good Housekeeping

One of the simplest and most basic aspects of safe construction is good housekeeping. Materials, debris, lumber scraps, empty cans, grease spots, forms and hardware lying about waiting to be tripped over, slipped on, walked into or kicked or blown off onto someone's head below, are not only a serious hazard but an indication of indifferent supervision and an encouragement to workmen to be careless and to add to the lethal litter. Any construction site that shows poor housekeeping also indicates lax supervision and a disregard for human welfare. Good housekeeping not only indicates good safe operation; it also pays dividends in better material utilization and labor efficiency.

24.6.2 Competent Personnel

Another important consideration in safety-mindedness is the use of personnel who are, in fact, competent and skilled to perform the functions assigned to them. Don't ask a carpenter who doesn't know how to place a cable clamp correctly to guy a hoist tower. Don't ask a carpenter foreman to put up shoring or bracing for a derrick or other piece of heavy equipment if the man has no knowledge of the loads and stresses involved. Don't assign a form-watcher who hasn't the slightest idea of what he is watching for. This may seem rather rudimentary, but it is very surprising how often people are assigned important tasks for which they are not qualified and for which they have not received proper instructions. If a special skill or just normal skill is required, supervisors must ensure that such skill is actually used. If skill or knowledge is missing, it is false economy to make do rather than pay the price to get proper help.

24.6.3 Scaffolding and Runways

These items tend to be very well covered by codes. Normally the sizing of the members is properly designed and furnished; it is not the member that usually fails but the connections. This is especially true of wood. Designers will very often fail to design or specify the type, number, and location of nails or bolts required in connections. Minimum and maximum laps are ignored, and cleating, bracing, and hold-downs are left to the field. Drilling holes for hanger rods will often result in very unsafe situations. Designers, supervisors, and inspectors must give their utmost attention to connections.

Another important facet often overlooked is the capacity of the scaffold or runway. This capacity, in pounds per square foot, must be carefully selected on the basis of actual field loadings expected, included storage of materials, heavy equipment handled, and moving loads such as power buggies. Once selected and designed for, this capacity must be prominently stated on the drawings and every effort taken to ensure that supervisors using the scaffolds and runways fully appreciate and understand the limitations.

24.6.4 Shoring and Formwork

Formwork and, more important, shoring are probably the major killers on construction sites. The importance of proper design, especially design for lateral loading and moving equipment, cannot be overemphasized. Deck formwork almost always fails laterally because of insufficient lateral bracing of the whole system or of the shores to each other. The lateral bracing and lacing must be completely designed including connections, and again, capacities and limitations must be prominently shown on the drawings; on wall and column pours, show placing rates as related to temperatures. During placing of concrete, forms and shoring must be carefully watched for telltale danger signals such as excessive deflection.

One serious hazard involved in placing decks is premature or accidental removal of shoring or improper reshoring procedures. A workman accidentally hitting a shore and knocking it out of position can trigger a widespread collapse. Shores therefore must be as rigidly connected as possible, and if possible protected from other operations. Education of the workers in this matter is an important safety requirement. An item that is systematically abused is the wedging at the base of shores. This wedging is required to give full bearing, as the full load of the shore must pass through the wedges. In so many instances this bearing is not provided; I have seen wedging with barely an inch of overlap and not nailed into position. *Inspectors please note.*

24.6.5 Cranes and Derricks

Cranes and derricks increasingly are being mounted on partially completed buildings, and even on formwork. As buildings get taller in height or more extensive in area, conventional boom cranes working from outside the building line can no longer reach the work. The 450-ft booms have been pushed to the limit and must be used with care.

Climbing cranes and derricks exert, in addition to the vertical loads, severe lateral loadings, that is, overturning moments. Very few failures of climbing cranes have been failures of the equipment itself; almost always the accident was caused by improper support to the building. When one considers that these cranes exert overturning moments up to 2000 ft-kips (and will get larger yet), erecting such cranes without proper engineering design is certainly courting disaster. Yet, this has been done, and serious accidents have been experienced.

Some substantial improvements are in order in the field of crane safety, for both climbing and ground-supported types. Some crane manufacturers provide users and engineers with excellent and meaningful loadings and support data. Others supply little or no information, or supply data that are incomplete or subject to interpretation. The manufacturers' representatives may be nontechnical people who do not understand the loadings themselves and who often lead undiscriminating clients into serious problems. Crane users, engineers, and approving agencies should give preference and encouragement to those manufacturers who do furnish adequate data. Figure 24-20 shows a good data sheet.

Last, the buildings must be checked by a competent structural engineer to ensure that the building members are not overloaded. This individual would preferably be the engineer of record or an engineer specializing in this field and approved by the engineer of record. The manufacturer is responsible for furnishing data but not for structural analysis of the support on the building. The cost for this engineering is properly borne by the contractor who benefits from the use of the crane.

Failures in derricks are often caused by improper guying; by damaged, rusted, or otherwise impaired parts; or by overloading. The comments made above in regard to cranes also apply here. On derricks as well as on cranes, capacities must be clearly stated and posted for the operator to see; loading must be clearly shown.

Guying failures usually result from improper use of or failure to use hardware such as clamps and thimbles, or attachment of guys to objects incapable of taking the loads. It is not uncommon to see guys subject to potentially high loads casually attached to 0.5-in. reinforcing dowels, pieces of formwork, or small footings, and at ridiculously ineffective angles, and slack, at that. When this happens, three things are indicated. First, an engineer did not locate the anchors and specify the connections. Second, an experienced, knowledgeable rigger was not used to make the installation. Third, the inspector was incompetent. Engineering design, workmanlike installation, and intelligent, alert inspection are all required.

All the design in the world will not help equipment that is not in proper working order. Rusted connections, loose bolts, cracked welds, embrittled notches, bent compression members, and damaged cables do not show in manufacturers' catalogs or on engineering drawings. You can find them with startling frequency in the field. Rigid, continuous, and periodic inspection is required by the equipment owners, users, operators, and inspection agencies. This inspection must be made by persons who know what to guard against.

One last point on cranes, derricks, and all hoisting equipment: this equipment is designed primarily to operate at normal allowable operating stresses. Capacity charts take full advantage of the stress capabilities of the equipment without considering the dynamic nature of the loading and prolonged use of the equipment; but the metal will become fatigued. At the present time there is little consideration given to this problem in the way of lower allowable stresses, accurate use records, or testing programs. Such consideration seems to be in order.

24.6.6 Fire, Smoke, and Toxic Gases

These items contribute heavily to construction injuries, property damage, and insurance costs. The codes covering these hazards are normally explicit and comprehensive. Accidents occur from failure to comply with the codes. The main requirement in this safety area seems to lie in education and frequent refreshing of the minds of workers and supervisors at all levels as to the dangers of welding sparks, salamanders, unventilated areas, improper combustion, leaky lines, damaged electrical lines or connections, and sprayed-on chemicals and materials. Exhaust from some equipment may also be dangerous.

25

Structural Analysis

KURT H. GERSTLE, Ph.D.*

25.1 INTRODUCTION

25.1.1 Purpose of Structural Analysis

Structural analysis is the process of determining the forces and deformations in structures due to specified loads so that the structure can be designed rationally, and so that the state of safety of existing structures can be checked.

In the design of structures, it is necessary to start with a concept leading to a configuration which can then be analyzed. This is done so members can be sized and the needed reinforcing determined, in order to: (a) carry the design loads without distress or excessive deformations (serviceability or working condition); and (b) prevent collapse before a specified overload has been placed on the structure (safety or ultimate condition).

Since normally elastic conditions will prevail under working loads, a structural theory based on the assumption of elastic behavior is appropriate for determining serviceability conditions. Collapse of a structure will usually occur only long after the elastic range of the materials has been exceeded at critical points, so that an ultimate strength theory based on the inelastic behavior of the materials is necessary for a rational determination of the safety of a structure against collapse. Nevertheless, an elastic theory can be used to determine a safe approximation to the strength of ductile structures (the lower bound approach of plasticity), and this approach is customarily followed in reinforced concrete practice. For this reason only the elastic theory of structures is pursued in this chapter.

*Professor of Civil Engineering, University of Colorado, Boulder, Colorado.

25.1.2 Modeling of Structures

Looked at critically, all structures are assemblies of three dimensional elements, the exact analysis of which is a forbidding task even under ideal conditions and impossible to contemplate under conditions of professional practice. For this reason, an important part of the analyst's work is the simplification of the actual structure and loading conditions to a model that is susceptible to rational analysis.

Thus, a structural framing system is decomposed into a slab and floor beams that in turn frame into girders carried by columns that transmit the loads to the foundations. Since traditional structural analysis has been unable to cope with the action of the slab, this has often been idealized into a system of strips acting as beams. Also, longhand methods have been unable to cope with three-dimensional framing systems, so that the entire structure has been modeled by a system of planar subassemblies, to be analyzed one at a time. The modern matrix-computer methods have revolutionized structural analysis by making it possible to analyze entire systems, thus leading to more reliable predictions about the behavior of structures under loads.

Actual loading conditions are also difficult both to determine and to express realistically, and must be simplified for purposes of analysis. Thus, traffic loads on a bridge structure, which are essentially both of dynamic and random nature, are usually idealized into statically moving standard trucks, or distributed loads, intended to simulate the most severe loading conditions occurring in practice.

Similarly, continuous beams are sometimes reduced to simple beams, and rigid joints to pin-joints; filler-walls are neglected; shear walls are considered as beams. In deciding how to model a structure so as to make it reasonably realistic but at the same time reasonably simple, the analyst must remember that each such idealization will make the solution more suspect. The more realistic the analysis, the greater will be the confidence it inspires, and the smaller may be the safety factor (or factor of ignorance). Thus, unless code provisions control, the engineer must evaluate the extra expense of a thorough analysis as compared to possible savings in the structure.

25.1.3 Relation of Analysis and Design

The most important use of structural analysis is as a tool in structural design. As such, it will usually be a part of a trial-and-error procedure, in which an assumed configuration with assumed dead loads is analyzed, and the members are designed in accordance with the results of the analysis. This phase is called the preliminary design; since this design is still subject to change, usually a crude, fast analysis method is adequate. At this stage, the cost of the structure is estimated, loads and member properties are revised, and the design is checked for possible improvements. The changes are now incorporated in the structure, a more refined analysis is performed, and the member design is revised. This process is carried to convergence, the rapidity of which will depend on the capability of the designer. It is clear that a variety of analysis methods, ranging from "quick and dirty" to "exact," is needed for design purposes.

An efficient analyst must thus be in command of the rigorous methods of analysis, must be able to reduce these to shortcut methods by appropriate assumptions, and must be aware of available design and analysis aids, as well as simplifications permitted by applicable building codes. An up-to-date analyst must likewise be versed in the bases of matrix structural analysis and its use in digital computers as well as in the use of available analysis programs or software.

25.1.4 Approaches in Structural Analysis

Structural analysis is based on the following building blocks:

1. *Statics.* Any portion of a structure must be in static equilibrium; in the more general case of dynamic analysis, it must satisfy Newton's laws of motion.
2. *Geometry.* The deformations of adjacent portions of a structure must be compatible with each other, so as to eliminate any jags, kinks, or other discontinuities that are inconsistent with the nature of the structure and its supports.
3. *Force-deformation relations.* In the methods presented here, elastic behavior is assumed; the classical relations of the theory of elastic bars therefore apply. In a more extended analysis that goes beyond the elastic range, inelastic force–deformation relations must be considered.

These ideas form the bases of the methods presented in this section. They are discussed further in any text on mechanics of materials, or mechanics of deformable bodies. In the presentation that follows, Section 25.2 is concerned with the statics of structures. In Section 25.3, deformations of statically determinate structures are analyzed. In Section 25.4, the force method is introduced for the analysis of statically indeterminate structures, and the displacement method is covered in Section 25.5. In Section 25.6, the analysis for moving loads is presented. Section 25.7, finally introduces the powerful finite-element method, which is an extension of the displacement method applicable to a wider range of structures than can be handled by classical means.

25.1.5 References and Design Aids

References applicable to the study of structural analysis can conveniently be split into four distinct groups:

1. Those concerned with the principles underlying the topic, in particular, books on statics, mechanics of materials, and the theories of elasticity and inelasticity:

Shames, I. H., *Engineering Mechanics*, V. 1, Prentice-Hall, Englewood Cliffs, N.J., 1980.

Merriam, J. L., *Statics*, Wiley, New York, 1978.

Popov, E. P., *Mechanics of Materials*, Prentice-Hall, Englewood Cliffs, N.J., 1976.

2. The classical texts on the analysis of civil engineering structures, stressing mainly longhand computational methods:

Sutherland, H., and Bowman, H. L., *Structural Theory*, Wiley, New York, 1950.

Norris, C. H., Wilbur, J. B., and Utku, S., *Elementary Structural Analysis*, 3rd Ed., McGraw-Hill, New York, 1976.

Timoshenko, S. P., and Young, D. H., *Theory of Structures*, McGraw-Hill, New York, 1965.

3. The modern texts on the theory of structures, stressing matrix and computer methods, but still restricting their coverage to classical civil engineering structures such as beams, trusses, arches, and frames:

Weaver, W., Jr., and Gere, J. M., *Matrix Analysis of Framed Structures*, 2nd Ed., Van Nostrand Reinhold, New York, 1980.

McGuire, W., and Gallagher, R. H., *Matrix Structural Analysis*, Wiley, New York, 1979.

Gerstle, K. H., *Basic Structural Analysis*, Prentice-Hall, Englewood Cliffs, N.J., 1974.

4. Books dealing with the extension of matrix-computer methods to more general structures such as two- and three-dimensional elements, plates, and shells by means of the finite-element method:

Zienkiewicz, O. C., *The Finite Element Method*, 3rd Ed., McGraw-Hill, London, 1977.

Cook, R. D., *Concepts and Applications of Finite Element Analysis*, Wiley, New York, 1981.

Design aids for structural analysis are often too limited to do justice to the wide variety of structural configurations faced by the designer; they may, however, be useful for conceptual purposes, or for parameter studies. Several attempts to encompass results for a wide variety of framed structures have been made:

Kleinlogel, A., *Rigid Frame Formulas*, 11th Ed., Ungar, New York, 1952.

Leontovich, V., *Frames and Arches*, McGraw-Hill, New York, 1959.

Such compilations have been rendered somewhat obsolete by advances in computational methods and particularly by the availability of efficient computer programs. Tables of beam moments, influence ordinates, flexibilities, and stiffnesses continue to be useful and will be cited at appropriate places.

Of particular use is the classic:

Roark, R. J., and Young, W. J., *Formulas for Stress and Strain*, 5th Ed., McGraw-Hill, New York, 1975.

It contains a wide assortment of useful information for the analyst.

Since the presentation of this chapter aims at a general approach valid for a wide variety of structures, it does not provide scope for specialized treatment of specific structures. Thus, basic considerations and analysis methods that may be applicable to certain classes of structures such as arches, or high-rise buildings, have not been covered, but may be studied in specialized treatises such as:

McCullough, C. B., and Thayer, E. S., *Elastic Arch Bridges*, Wiley, New York, 1931.
Amirikian, A., *Analysis of Rigid Frames*, U.S. Government Printing Office, Washington, D.C., 1942.

The Portland Cement Association has published a series of valuable pamphlets dealing with specific methods and structures:

"Continuous Concrete Bridges"
"Analysis of Rigid Frame Concrete Bridges"
"Continuity in Concrete Building Frames"
"Analysis of Small Reinforced Concrete Buildings for Earthquake Forces."

25.2 INTERNAL FORCES IN STRUCTURES

25.2.1 Introduction

The satisfaction of the equations of equilibrium is the most important aspect of structural analysis, as well as the simplest. If a structure cannot be in equilibrium with the applied loads, it cannot stand; conversely, if a sufficiently ductile structure is in static equilibrium, it will stand even though geometrical conditions may be violated. Furthermore, all of the subsequent material requires thorough familiarity with statics of structures.

25.2.2 Internal Forces

Analysis of framed structures is concerned with the determination of the internal member forces necessary for design. In the general case of three-dimensional structures, any member section may be subjected to three force components (two shear forces and one axial force), and to three moment components (two bending moments and one torsional moment), as shown in Fig. 25-1(a). All of these quantities will be designated as forces, or stress resultants.

Special cases arise in planar structures, in which all out-of-plane forces are disregarded, leaving only one axial force, one shear force, and one bending moment, as shown in Fig. 25-1(b). Other combinations are possible such as the case of combined torsion, shear, and bending moment in the bow girder of Fig. 25-1(c), or the simple case of bending moment and shear in the beam of Fig. 25-1(d). Differential relations between loads and internal forces are useful in the determination of these stress resultants.

Statically determinate structures are those in which all member forces can be found by statics alone. This requires that the number of available equilibrium equations match the number of unknown forces. For such structures, it is convenient to determine the reactions first by application of appropriate equilibrium equations to a free body of the entire structure, or, in the case of articulated structures such as three-hinged arches, to free bodies consisting of portions of the structure. Once the reactions are determined, the internal forces are found by writing the equilibrium equations of free bodies containing these desired quantities as external forces; this requires cutting the structure at the section containing these forces.

Once found, these forces are conveniently plotted along the structure to guide the designer in the sizing of members and layout of reinforcing. The most commonly used plots of this type are shear and moment diagrams, but the variation of other forces such as axial thrusts and torsional moments can be plotted similarly.

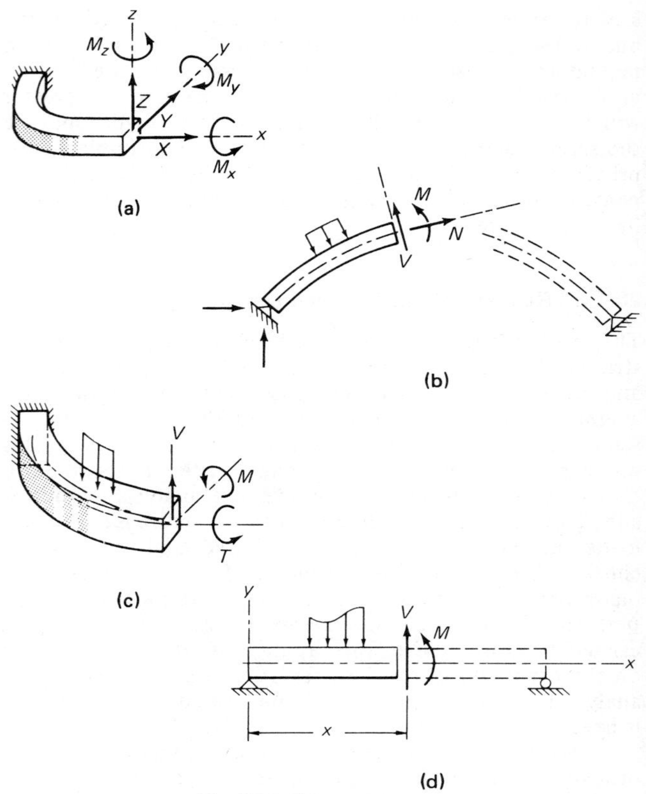

Fig. 25-1 Internal forces.

In the following examples, two different types of structures are analyzed.

EXAMPLE 25-1: Draw shear and moment diagrams for the articulated beam of Fig. 25-2(a).

SOLUTION: The reactions are found first and indicated in Fig. 25-2(a): internal shear and moment along the beam are then determined by use of appropriate free bodies and drawn as in Fig. 25-2(b) and (c). Note that at all points, the slope of the moment diagram is equal to the shear, and the extreme moment values occur at points of zero shear.

EXAMPLE 25-2: Calculate the internal forces in the horizontal can-

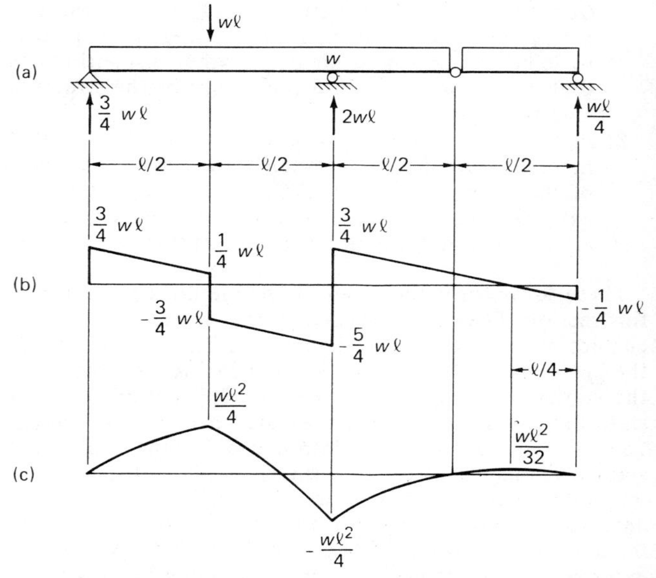

Fig. 25-2 (a) Beam and reactions; (b) shear diagram; and (c) moment diagram for Example 25-1.

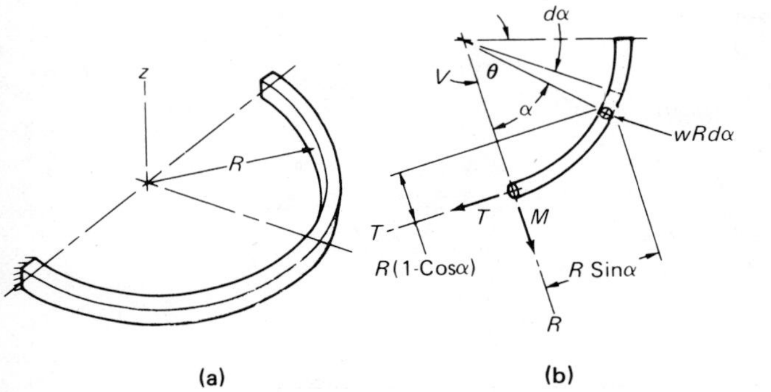

Fig. 25-3 (a) Cantilever bow girder and (b) free body diagram for Example 25-2.

tilever bow girder of Fig. 25-3(a) due to its dead load, of intensity w lb/ft.

SOLUTION: We refer to the plan view of the free body of Fig. 25-3(b), and write the equilibrium conditions

$$\Sigma F_Z = 0: -\int_{\alpha=o}^{\theta} wRd\alpha + V = 0;\ V = wR\theta$$

$$\Sigma M_R = 0: -\int_{\alpha=o}^{\theta} (wRd\alpha)(R \sin \alpha) + M = 0;$$

$$M = wR^2(1 - \cos \theta)$$

$$\Sigma M_T = 0: \int_{\alpha=o}^{\theta} (wRd\alpha)\, R\,(1 - \cos \alpha) + T = 0;$$

$$T = -wR^2(\theta - \sin \theta)$$

25.2.3 Design Approximations, Aids, and References

1. Some Aids for Calculation. The principle of superposition is a power tool in simplifying structural computations. In complicated loading systems, it often pays to consider only a part of the load at one time, compute its effects, and add to find the total effect due to all parts. In this process, each part should be so simple that its contribution can be found with minimum effort.

EXAMPLE 25-3: Compute the reactions due to the distributed load shown in Fig. 25-4(a).

SOLUTION: It is convenient here to break the load up into the two triangular distributions shown in Fig. 25-4(b), indicate their resultants, calculate their individual effects on the reactions at A, B, and add, leading to:

$$R_A = 0;\ R_B = -\tfrac{1}{2}\, w_o L.$$

Any loading applied to a symmetric structure can be decomposed into a symmetric and an antisymmetric part, and symmetry considerations applied to each portion, as shown in the next example.

EXAMPLE 25-4: Compute the reactions due to the distributed load shown in Fig. 25-5(a).

SOLUTION: The load is broken up into the symmetric part of Fig. 25-5(b), and the antisymmetric part of Fig. 25-5(c), the reactions (or any other required functions) are calculated, and added to yield:

$$R_A = (\tfrac{3}{8} + \tfrac{5}{24})\, w_o L = \tfrac{7}{12}\, w_o L;\ R_B = (\tfrac{3}{8} - \tfrac{5}{24})\, w_o L = \tfrac{1}{6}\, w_o L$$

Many types of distributed loads can be expressed in the form of polynomials. With appropriate choice of origin, such distributions can be expressed in the form:

$$p(x) = p_o \left(\frac{x}{l}\right)^n$$

where p_o represents a reference load intensity, l represents a loaded length, and n represents the order of the polynomial. In this case, the magnitude and location of the resultant, as well as the resultant internal forces, can be calculated by the formulas of Table 25-1, and used to compute reactions and forces. A more extensive tabulation of coefficients of the type shown in Table 25-1 can be found in ***Structural Analysis***, by J. J. Tuma, Schaum's Outline Series, 1969, pp. 26–33.

Another useful device for handling difficult loadings is their replacement by appropriately placed concentrated load resultants, computing their effects individually, and summing. As the number of concentrated replacement

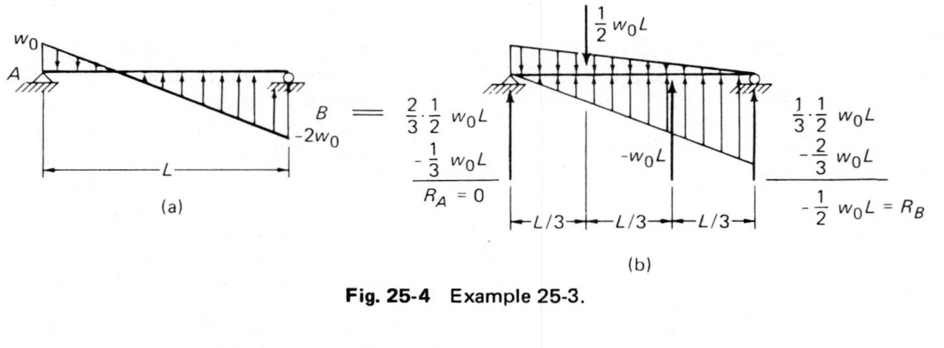

Fig. 25-4 Example 25-3.

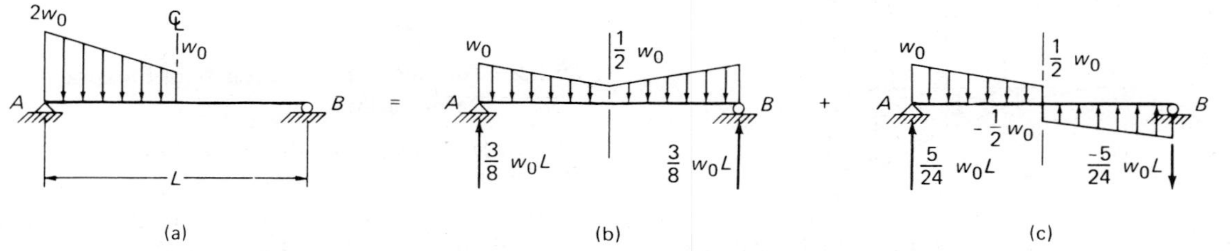

Fig. 25-5 Example 25-4.

TABLE 25-1 Beam Forces for Polynomial Loads

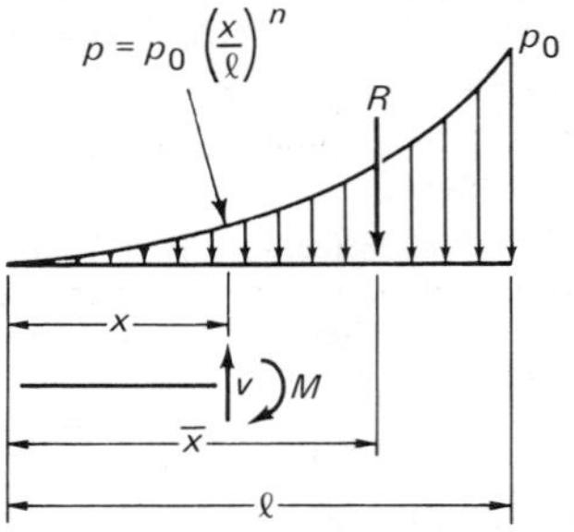

Formula \ Multiplier for	n = 0 Const. Ld.	n = 1 Linear Ld.	n = 2 Parab. Ld.	n = 3	Common Factor
$R = \frac{1}{n+1} p_o l$	1	$\frac{1}{2}$	$\frac{1}{3}$	$\frac{1}{4}$	$\times p_o l$
$\bar{x} = \frac{n+1}{n+2} l$	$\frac{1}{2}$	$\frac{2}{3}$	$\frac{3}{4}$	$\frac{4}{5}$	$\times l$
$p = \left(\frac{x}{l}\right)^n p_o$	1	(x/l)	$(x/l)^2$	$(x/l)^3$	$\times p_o$
$V = \frac{1}{n+1}\left(\frac{x}{l}\right)^{n+1} p_o l$	(x/l)	$\frac{1}{2}(x/l)^2$	$\frac{1}{3}(x/l)^3$	$\frac{1}{4}(x/l)^4$	$\times p_o l$
$M = \frac{1}{n+2}\left(\frac{x}{l}\right)^{n+2} p_o l^2$	$\frac{1}{2}(x/l)^2$	$\frac{1}{3}(x/l)^3$	$\frac{1}{4}(x/l)^4$	$\frac{1}{5}(x/l)^5$	$\times p_o l^2$

loads increases, the approximate converges toward the exact solution. Such discretizing procedures form an important part of the field of numerical analysis. The approach is indicated by Fig. 25-6, which shows a uniform load on a cantilever beam replaced by its statically equivalent quarter-point resultants, and the individual moment diagrams superposed to form the final, approximate curve.

2. References. For the principles and basic applications of the laws of statics, any one of the mechanics texts listed in Section 25.1.5 will do.

The construction of shear and moment diagrams for beams, as well as the differential relations between loads, shear, and moments, is covered in texts on mechanics of materials such as the ones listed in Section 25.1.5. The structure books listed in that section contain more advanced applications and calculations for member forces.

Design aids for the reactions and critical internal forces of statically determinate structures of standard configurations are available; for instance, see:

Manual of Steel Construction, American Institute of Steel Construction, New York.

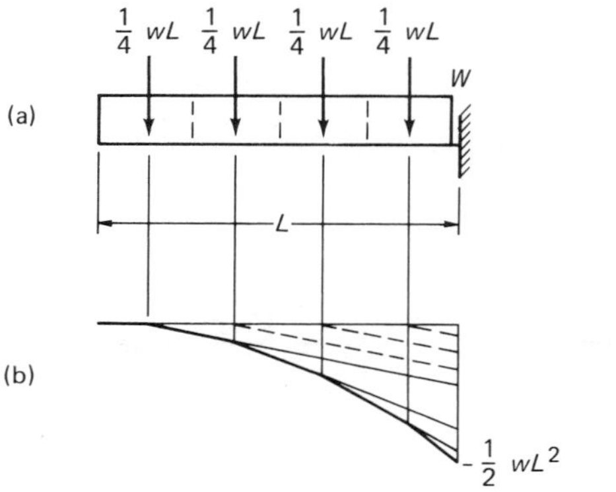

Fig. 25-6 (a) Discretized load, and (b) linearized moment diagram.

Design Handbook, Concrete Reinforcing Steel Institute, Chicago.

25.3 DEFORMATIONS OF STATICALLY DETERMINATE STRUCTURES

25.3.1 Introduction

Knowledge of the deformations of structures is necessary for two purposes: (a) to determine stiffness and deformation characteristics, and (b) to supplement the equilibrium equations with geometrical conditions for the analysis of indeterminate structures.

Two main methods are available for the calculation of deformations:

1. *The geometric methods.* These include the double-integration method, and the methods of integration of curvature, such as the curvature-area method and the conjugate-frame methods. Only the integration of curvature method will be pursued in this section.
2. *The energy methods.* These include the methods of least work, strain-energy, complementary and potential energy methods, and the method of virtual work. Only the latter will be outlined here, and extended to a matrix formulation suitable for computer work.

As presented here, all methods will be based on the assumption of small deformations, which is adequate for most purposes.

25.3.2 The Geometric Approach for Flexural Deformations

1. Angle Change and Tangential Deviation. We define the following terms:

(a) Curvature, Φ: The change of slope per unit length along the member. Fig. 25-7 shows that Φ can also be

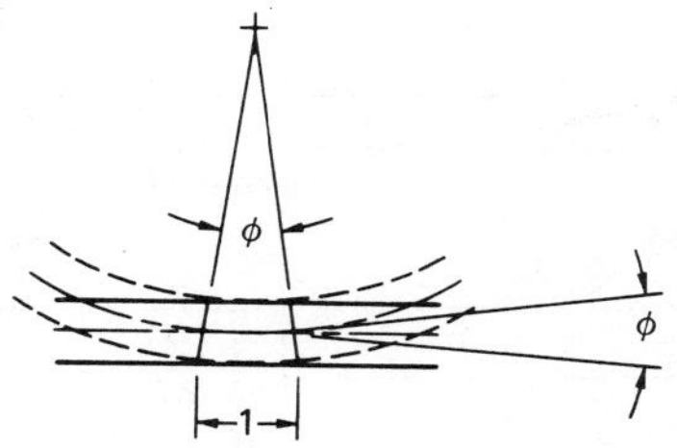

Fig. 25-7 Curvature.

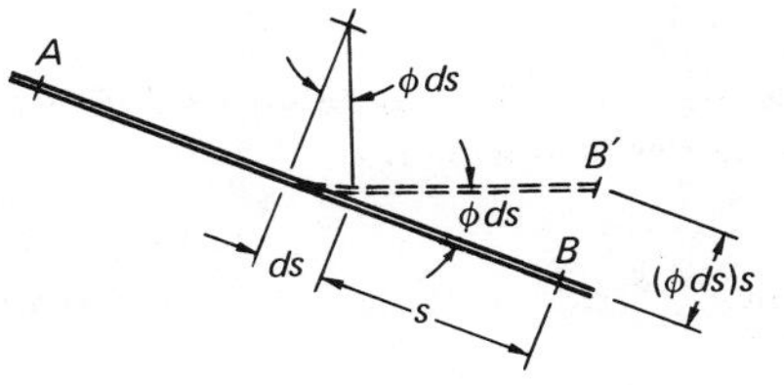

Fig. 25-8 Infinitesimal angle change and tangential deviation.

interpreted as the relative inclination of two transverse planes unity apart.

(b) Angle change, θ_{BA} : The change of slope between two points A and B of a member.

(c) Tangential deviation, δ_{BA} : The normal distance of point B on the deformed member from the tangent drawn to the deformed member at point A.

With these definitions, we consider the initially straight member AB of Fig. 25-8, subjected to a curvature Φ over an infinitesimal length ds. The geometry of this figure shows that due to this kink, of magnitude Φds:

$$d(\theta_{BA}) = \Phi\, ds$$

$$d(\delta_{BA}) = (\Phi\, ds)\, s\,.$$

Note that the origin of s is at point B.

Due to the curvature of all points between A and B:

$$\theta_{BA} = \int_A^B \Phi\, ds \qquad (25\text{-}1)$$

$$\delta_{BA} = \int_A^B \Phi\, s\, ds \qquad (25\text{-}2)$$

With these relations, all deformations of flexural structures can be determined once the curvature Φ is known; this requires visualization of the geometry of the deformed shape. Note also that only geometrical relations are involved, so

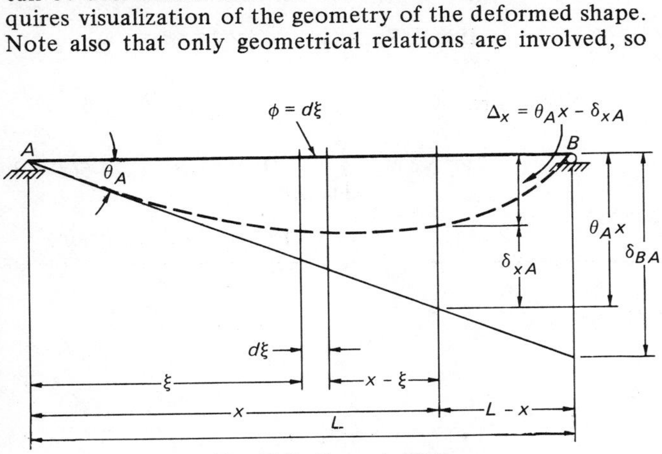

Fig. 25-9 Example 25-5.

we conclude that these deformations can be due to any effect, such as elastic or inelastic bending, temperature, or shrinkage; it is only necessary that the curvature Φ can be calculated.

EXAMPLE 25-5: The beam shown in Fig. 25-9 is subjected to a curvature $\Phi = ax$. Determine the end slope θ_A, and the equation of the elastic curve.

SOLUTION: From the geometry of the deformed shape of the beam of Fig. 25-9:

$$\theta_A = \frac{\delta_{BA}}{L} = \frac{1}{L}\int_{x=0}^{L} \Phi\,(L-x)\,dx = \frac{a}{L}\int_{x=0}^{L} (Lx - x^2)\,dx = \frac{aL^2}{6}$$

$$\Delta_x = \theta_A x - \delta_{xA}$$

$$= \frac{aL^2}{6}x - \int_{\xi=0}^{x} (a\xi)(x-\xi)\,d\xi = \frac{aL^3}{6}\left[\left(\frac{x}{L}\right) - \left(\frac{x}{L}\right)^3\right]$$

2. *The Curvature-Area Theorems.* We can interpret eqs. (25-1) and 25-2) by considering the plot of the curvature Φ along the member AB, as shown in Fig. 25-10. We observe:

$$\theta_{BA} = \int_A^B \Phi\, ds = \text{Area under curvature diagram between } A \text{ and } B$$

$$\delta_{BA} = \int_A^B \Phi\, s\, ds = \text{Static moment of area under curvature diagram between } A \text{ and } B \text{ about point } B$$

These relations are called the first and second curvature-area theorems. The integration can be performed either analytically or by subdividing the area under the curvature diagram into suitable portions and evaluating the integrals numerically or graphically.

If the curvature is due to bending of an elastic beam of stiffness EI, then it is linearly related to the moment M:

$$\Phi = \frac{M}{EI} \qquad (25\text{-}3)$$

EXAMPLE 25-6: The simple beam of flexural stiffness EI shown in Fig. 25-11(a) is prestressed by a tendon at constant eccentricity, leading to a constant curvature Φ (= Pe/EI). Determine the maximum deflection.

SOLUTION: From the geometry of the deformed shape of Fig. 25-11(b), and the curvature diagram of Fig. 25-11(b):

$$\Delta_C = \delta_{AC} = \left(\Phi \cdot \frac{L}{2}\right)\left(\frac{L}{4}\right) = \frac{\Phi L^2}{8} = \frac{PeL^2}{8EI}$$

EXAMPLE 25-7: The simple beam shown in Fig. 25-12(a) is subjected to a cranked-in end moment M at point B. Determine the end slope, θ_A, and the maximum deflection.

SOLUTION: We draw the deflected shape of the beam as in Fig. 25-12(a) and recognize the tangential deviation $\delta_{BA} = \theta_A \cdot L$, which can be computed by the second curvature-area theorem as the static moment of the area under the curvature diagram of Fig. 25-12(b)

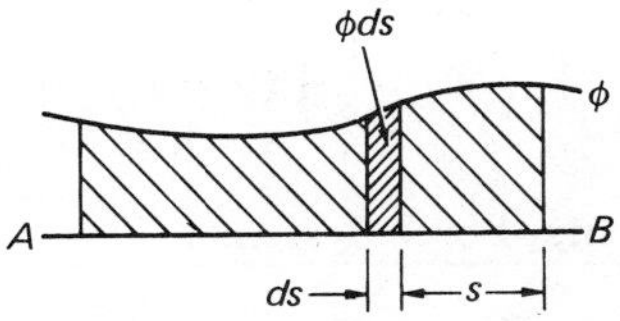

Fig. 25-10 Curvature-area concept.

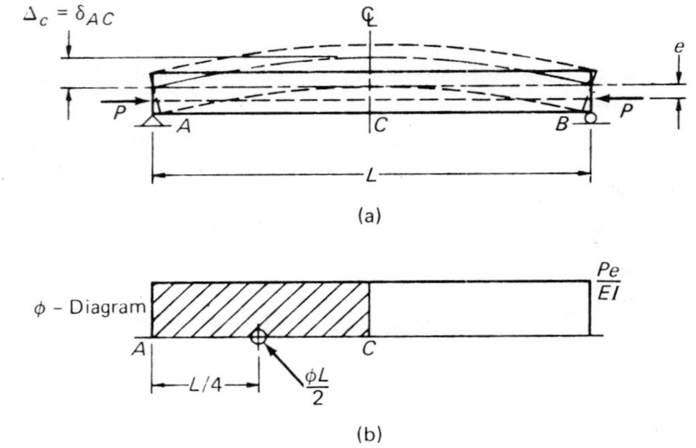

Fig. 25-11 Example 25-6. (a) Deformed shape, and (b) curvature diagram (ϕ-diagram).

about point B; therefore:

$$\theta_A = \frac{\delta_{BA}}{L} = \frac{1}{L}\left(\frac{1}{2}\frac{M}{EI}\cdot L\right)\left(\frac{L}{3}\right) = \frac{ML}{6EI}$$

(Compare this method to that used in Example 25-5.)

The maximum deflection Δ_C occurs at point C, defined by $\bar{x}$, where the slope is zero. We use the first curvature-area theorem to find this point:

$$\theta_C = 0 = \theta_A - \theta_{CA} = \frac{ML}{6EI} - \frac{1}{2}\left(\frac{\bar{x}}{L}\frac{M}{EI}\right)\bar{x} = 0;\ \bar{x} = \frac{L}{\sqrt{3}} = 0.578L$$

Having found the location $\bar{x}$ of the point of maximum deflection, we use the geometrical relations of Fig. 25-12(a) to find Δ_C:

$$\Delta_C = \theta_A\bar{x} - \delta_{CA}$$

Using again the second curvature-area theorem applied to the properties of Fig. 25-12(b), and the previously determined values of θ_A and $\bar{x}$, we set:

$$\Delta_C = \frac{ML}{6EI}\bar{x} - \left(\frac{1}{2}\frac{\bar{x}}{L}\frac{M}{EI}\cdot\bar{x}\right)\left(\frac{\bar{x}}{3}\right) = 0.0643\frac{ML^2}{EI}$$

25.3.3 Deformations by Virtual Work

1. The Theorem of Virtual Work. This theorem provides a tool for a unified formulation of both statical and geometrical equations. Before stating this theorem, we define the following:

(a) Statically compatible force system: A set of external forces P^* and corresponding internal stresses σ^* which are in equilibrium with each other.
(b) Geometrically compatible displacement system: A set of deformations Δ^{**} and corresponding internal strains ϵ^{**} which are related by the geometry of the deformed structure.

The theorem of virtual work states that for any structure in equilibrium, eq. (25-4a) holds:

$$\Sigma P^* \cdot \Delta^{**} = \int(\sigma^* dA)\cdot(\epsilon^{**}ds) = \int_{\text{vol.}} \sigma^*\epsilon^{**}\,dV \qquad (25\text{-}4a)$$

(Stat. Comp. Force System: ΣP^*, $\sigma^* dA$; Geom. Comp. Displace. System: Δ^{**}, $\epsilon^{**} ds$)

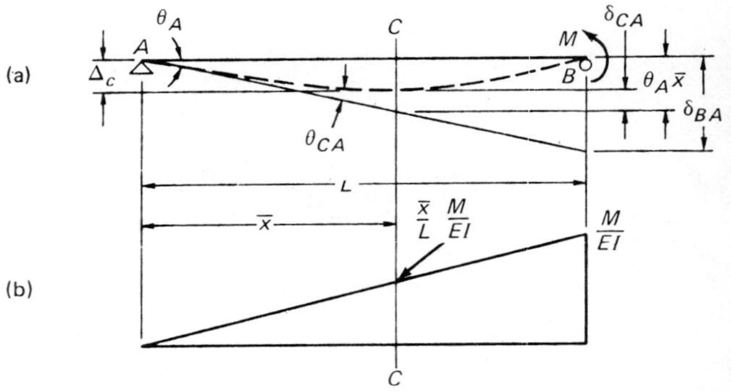

Fig. 25-12 Example 25-7. (a) Beam with EI = constant, and (b) curvature diagram (ϕ-diagram).

We define $\Sigma P^*\,\Delta^{**}$ as the external virtual work, and

$$\int_{\text{vol.}} \sigma^*\epsilon^{**}\,dV$$

as the internal virtual work, so that the theorem can also be expressed as:

$$\text{External virtual work} = \text{Internal virtual work} \qquad (25\text{-}4b)$$

There need be no relationship between the force system P^*, σ^* and the displacement system Δ^{**}, ϵ^{**}, and the theorem is therefore not restricted to elastic structures.

2. The Theorem of Virtual Forces. For the purpose of finding deformations, the displacement system Δ^{**}, ϵ^{**} is real, but the force system P^*, σ^* can be assumed for convenience; that is, it is *virtual*. In this form, eq. (25-4) is called the "Theorem of Virtual Forces." If the load P^* is assumed to be a unit load in the sense and location of the desired deformation, then this technique is called the "Unit Load Method."

The internal virtual work, defined by $\int_{\text{vol.}} \sigma^*\epsilon^{**}dV$, will now be calculated for a linear member under different loading conditions; in all cases the single-starred forces are due to the unit virtual load, the double-starred deformations are due to whatever is the real cause of the distortions.

(1) Axial load, N:

$$\text{Int. V. W.} = \int\left(\frac{N^*}{A}dA\right)(\epsilon^{**}ds) = \int_{\text{length}} N^*\epsilon^{**}ds$$

(2) Bending moment, M:

$$\text{Int. V. W.} = \int_{\text{length}} M^*\Phi^{**}ds$$

(3) Shear force, V:

$$\text{Int. V. W.} = \int_{\text{length}} V^*\gamma^{**}ds \qquad [25\text{-}5(a\text{–}d)]$$

(γ^{**} is real shear strain)

(4) Torsion moment, T:

$$\text{Int. V. W.} = \int_{\text{length}} T^{*}\theta^{**} ds$$

(θ^{**} is real unit angle of twist)

The internal virtual work is now inserted into eq. (25-4a), with the virtual external force $P^* = 1$, so that, in the absence of specified support displacements

$$\Delta^{**} = \text{Internal Virtual Work} \qquad (25\text{-}6)$$

The internal virtual work due to the different types of distortion is additive. We also note that no assumption of elasticity has been made so that at this stage the method is valid for deformations due to any cause, such as elastic or inelastic bending, shrinkage, temperature gradients, or any other effect. It is only necessary to know the value of the double-starred distortions.

3. The Theorem of Virtual Forces for Elastic Structures. The distortions due to loads applied to elastic structures are related to the real internal forces:

$$\epsilon^{**} = \frac{N^{**}}{EA}; \quad \Phi^{**} = \frac{M^{**}}{EI}$$
$$\gamma^{**} = \frac{V^{**}}{G(kA)}; \quad \theta^{**} = \frac{T^{**}}{GC} \qquad [25\text{-}7(a\text{-}d)]$$

where E and G represent the material stiffnesses, and A, I, kA, and C represent, respectively, the axial, flexural, shear, and torsional stiffness factors of the member section.

In much of the literature, the real double-starred forces in eq. (25-7) are denoted by the upper-case symbols N, M, V, and T, and the virtual single-starred forces of eq. (25-5) are denoted by the lower-case symbols n, m, v, and t. Using this notation, inserting eq. [25-7(a–d)] into eq. [25-5(a–d)] and applying eq. (25-6), we find:

$$\Delta = \int_L \frac{Nn\,ds}{EA} + \int_L \frac{Mm\,ds}{EI} + \int_L \frac{Vv\,ds}{G \cdot kA} + \int_L \frac{Tt\,ds}{GC} \qquad (25\text{-}8)$$

for an elastic structure under load causing internal axial and shear forces, and bending and twisting moments. For planar structures, no twisting is involved. In many cases, axial and shear deformations are sufficiently small that they can be neglected.

EXAMPLE 25-8: Calculate midspan deflection and end slope θ_B due to the uniform load on the restrained beam of Fig. 25-13(a) with the indicated end moments. Beam of constant stiffness EI.

SOLUTION: The real curvature diagram is drawn as in Fig. 25-13(b), and the real curvature is written conveniently by parts:

$$\Phi^{**} = \frac{M}{EI} = \frac{wL^2}{48\,EI}\left[23\left(\frac{x}{L}\right) - 24\left(\frac{x}{L}\right)^2 - 2\right]$$

For the midspan deflection, we use the virtual unit force of Fig. 25-13(c), due to which the virtual moment $m_1 = x/2$ for $0 \leqslant x \leqslant L/2$, and $m_1 = 1/2\ (1 - x)$ for $L/2 \leqslant x \leqslant L$. Use of eq. (25-9) leads to:

$$\Delta = \frac{wL^2}{48\,EI}\left\{\int_{x=0}^{L/2}\left[23\left(\frac{x}{L}\right) - 24\left(\frac{x}{L}\right)^2 - 2\right]\left[\frac{x}{2}\right]dx\right.$$
$$\left. + \int_{x=L/2}^{L}\left[23\left(\frac{x}{L}\right) - 24\left(\frac{x}{L}\right)^2 - 2\right]\cdot\frac{1}{2}\left[1 - x\right]dx\right\}$$
$$= \frac{5}{768}\frac{wL^4}{EI}$$

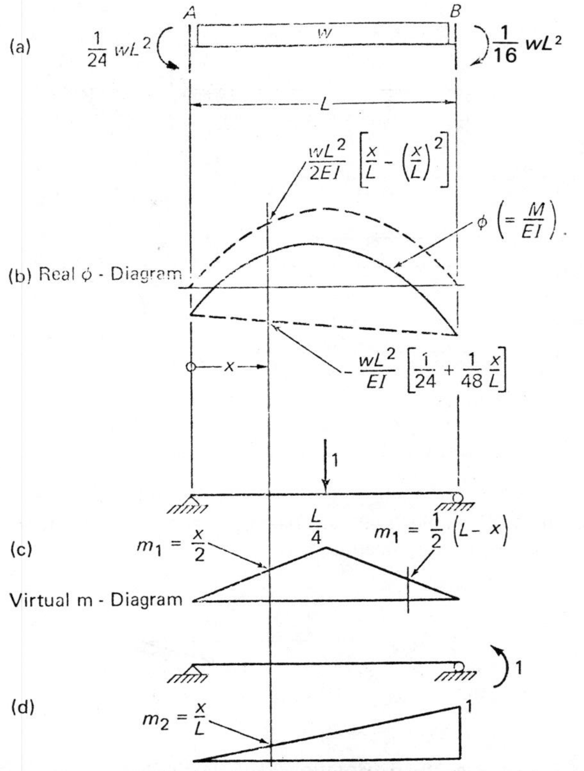

Fig. 25-13 Example 25-8. (a) Beam; (b) real curvature; (c) virtual load and moment for $\Delta_{℄}$; and (d) virtual load and moment for θ_B.

For the end slope θ_B, we use the virtual unit moment and moment diagram of Fig. 25-13(d), so that equating external and internal work leads to:

$$\theta_B = \int_{x=0}^{L}\frac{wL^2}{48\,EI}\left[23\left(\frac{x}{L}\right) - 24\left(\frac{x}{L}\right) - 2\right]\left[\frac{x}{L}\right]dx = \frac{1}{72}\frac{wL^3}{EI}$$

EXAMPLE 25-9: The simple beam shown in Fig. 25-14(a) is exposed to a transverse temperature gradient that causes increased strains in the extreme concrete surface of value +0.001 in./in., and in the tension steel of value −0.0005 in./in. Compute the maximum beam deflection due to this effect.

The real curvature Φ^{**} is computed from the strains shown in Fig. 25-14(a) as:

$$\Phi^{**} = -\frac{0.0010 + 0.0005}{24\text{ in.}} = -62.5 \times 10^{-6}\text{ rad/in.}$$

The virtual unit load P^*, and the resulting virtual moment M^* are shown in Fig. 25-8(b), from which:

$$M^* = 0.5\,x\ (0 \leqslant x \leqslant 20\text{ ft})$$

From eq. (25-4), we equate external and internal virtual work; using symmetry:

$$1 \cdot \Delta = 2\int_{x=0}^{20 \times 12}(-62.5 \times 10^{-6})(0.5\,x)\,dx = -1.8\text{ in. (upward)}$$

25.3.4 Flexibilities

1. The Member Flexibility. The flexibility of all or a portion of a structure is defined as the displacement due to a unit value of applied load.

The member flexibility, f, of a straight prismatic member

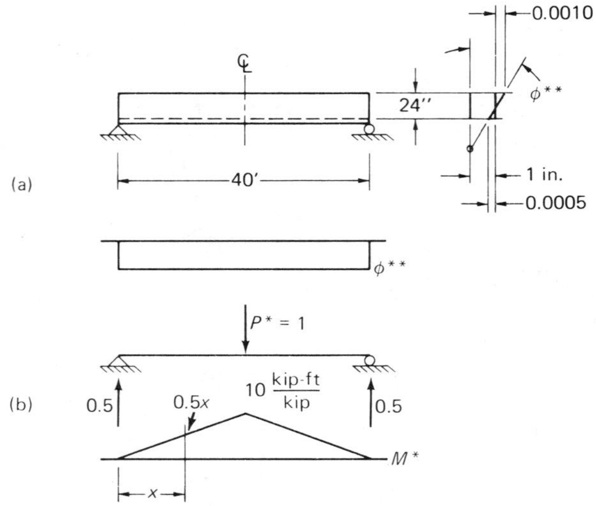

Fig. 25-14 Example 25-9. (a) Beams, strains, and curvature; and (b) virtual load and moments.

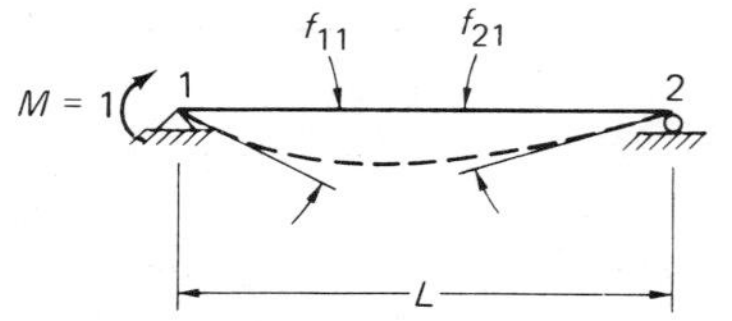

Fig. 25-15 Rotational end flexibilities.

under axial load, for instance, is its elongation due to a unit load, or $f = L/AE$.

For a beam segment, the rotational end flexibilities are of importance. We consider the simple prismatic beam shown in Fig. 25-15. We define the flexibility f_{ij} as the rotation at end i due to a unit moment at end j. There is a total of four flexibilities, the rotations at ends $i = 1, 2$, due to the unit moments at ends $j = 1, 2$. These four quantities can be found by the outlined methods of deformation analysis as:

$$f_{11} = f_{22} = \frac{L}{3EI}; \quad f_{12} = f_{21} = \frac{L}{6EI}$$

It is convenient to express the relationship between the end rotations θ and the end moments M in matrix form.

$$\begin{Bmatrix} \theta_1 \\ \theta_2 \end{Bmatrix} = \frac{L}{6EI} \begin{bmatrix} 2 & 1 \\ 1 & 2 \end{bmatrix} \begin{Bmatrix} M_1 \\ M_2 \end{Bmatrix} = [f] \{M\} \qquad (25\text{-}9)$$

in which the rotational member flexibility matrix $[f]$ is seen to contain the previously computed flexibilities.

2. *The Structure Flexibility.* The structure flexibility, δ_{ij} represents the displacement i of a structure due to an applied unit load j. To define the quantities clearly, we introduce the concept of nodes. In Fig. 25-16, for instance, node 1 represents the applied moment as well as the rotation of the left end, and node 2 stands for an applied upward load as well as a vertical upward displacement at midspan. These loads can be related to the corresponding displacements by the structure flexibility matrix $[\delta]$:

$$\begin{Bmatrix} \theta_1 \\ \Delta_2 \end{Bmatrix} = \frac{L}{48EI} \begin{bmatrix} 16 & -3L \\ -3L & L^2 \end{bmatrix} \begin{Bmatrix} M_1 \\ P_2 \end{Bmatrix}$$

or

$$\{\Delta\} = [\delta] \{P\} \qquad (25\text{-}10)$$

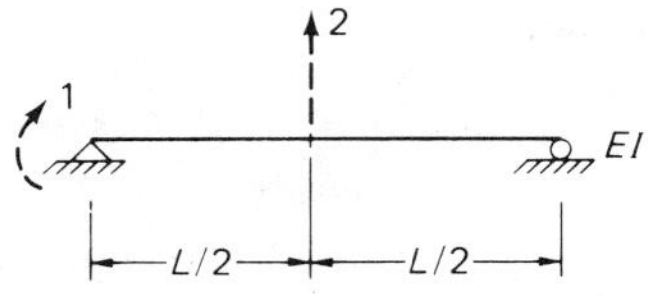

Fig. 25-16 Node designation.

where the elements of the structure flexibility method $[\delta]$ have been computed by any method. Once this matrix is known, the displacements $\{\Delta\}$ due to any combination of loads $\{P\}$ can be computed by matrix multiplication. If the displacements due to several loading conditions must be computed, the appropriate number of columns is added to the $[\Delta]$ and $[P]$ matrices.

We note that the flexibility matrices $[f]$ and $[\delta]$ are symmetric and have positive diagonal terms. The former fact is due to Maxwell's law of reciprocal deformations; the latter follows from the fact that for $i = j$, applied load and the resulting displacement at the same node must be of the same sense.

25.3.5 Beam Deformations by Matrix Method

The matrix calculation of structure deformations is suitable for computer use, and forms one of the building blocks for computer analysis of indeterminate structures. We will develop the method by use of the theorem of virtual forces, discussed earlier, in matrix form. A convenient virtual applied load will be a unit load at the point and in the sense of the desired deformation. The internal moments due to this virtual load will be called m_i, where i defines the location of the moment m. The matrix of the moments m_{ij} due to applied unit loads at j is a force transformation matrix to be determined by statics and will be called $[b] \equiv [m_{ij}]$.

The real moments, M_{ij}, are the moments at i due to the real loads $[P]$ applied at j, and can also be represented by the force transformation matrix:

$$[M_{ij}] = [b] [P]$$

EXAMPLE 25-10: For the beam of Fig. 25-17, write the $[b]$ matrix relating the moments at nodes 2 and 3 to vertical unit loads applied at these nodes. Then, evaluate the moments at these points due to $P_2 = 2$ kips, $P_3 = 3$ kips.

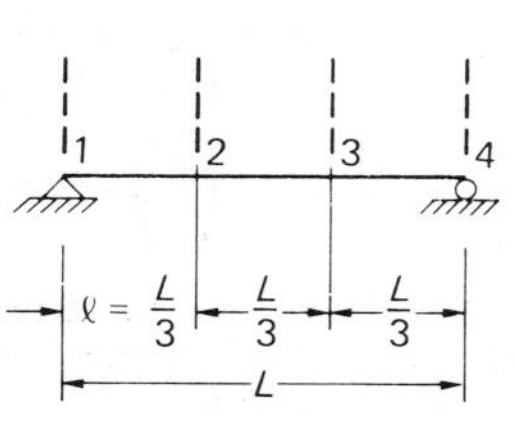

Fig. 25-17 Examples 25-10 and 25-11.

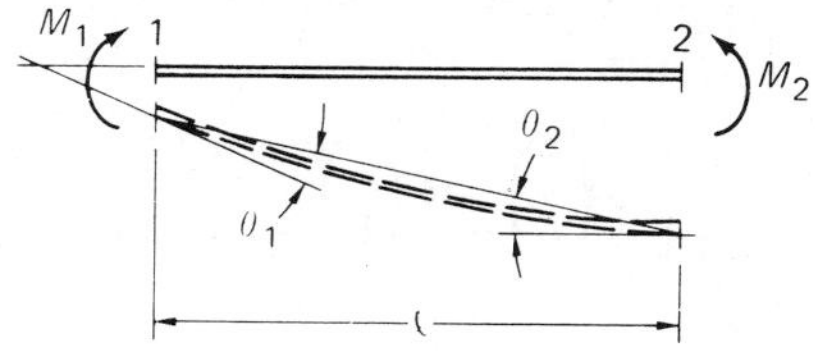

Fig. 25-18 Virtual work of beam element.

SOLUTION:

$$[b] = \begin{bmatrix} m_{22} & m_{23} \\ m_{32} & m_{33} \end{bmatrix} = \begin{bmatrix} \frac{2}{9} & \frac{1}{9} \\ \frac{1}{9} & \frac{2}{9} \end{bmatrix} L$$

$$\begin{Bmatrix} M_2 \\ M_3 \end{Bmatrix} = [b] \begin{Bmatrix} P_2 \\ P_3 \end{Bmatrix} = [b] \begin{Bmatrix} 2 \\ 3 \end{Bmatrix} = \begin{Bmatrix} \frac{7}{9} \\ \frac{8}{9} \end{Bmatrix} L \text{ (kips)}$$

We now consider the internal virtual work done in one beam element, subjected only to end moments, as shown in Fig. 25-18. This virtual work is given by the real end rotations θ_1 and θ_2 (measured with respect to the chord) multiplied by the corresponding virtual end moments m_1, m_2; this can be done in matrix form

$$\text{Internal virtual work} = [m_1 \ m_2] \begin{Bmatrix} \theta_1 \\ \theta_2 \end{Bmatrix}$$

We now find the real end rotations θ_1 and θ_2 from the real moments M_1 and M_2 by recalling the solution for a simply supported prismatic beam of length l, subject to end moments M_1 and M_2.

$$\begin{Bmatrix} \theta_1 \\ \theta_2 \end{Bmatrix} = \frac{l}{6EI} \begin{bmatrix} 2 & 1 \\ 1 & 2 \end{bmatrix} \begin{Bmatrix} M_1 \\ M_2 \end{Bmatrix} \equiv [f] \ \{M\}$$

where the element flexibility matrix $[f]$ is given by eq. (25-9).

The internal virtual work in one element is then, denoting the element by i:

$$\text{Internal V.W.} = [m_1 \quad m_2] \ [f] \begin{Bmatrix} M_1 \\ M_2 \end{Bmatrix}$$

$$= [b^i]^T \ [f^i] \ [b^i] \ \{P\}$$

For the virtual work in the entire beam, we add the contributions from all elements.

$$1 \cdot \Delta = \sum_{\text{Elm'ts}} [b^i]^T \ [f^i] \ [b^i] \ \{P\}$$

$$= [[m_1^1 \quad m_2^1][m_1^2 \quad m_2^2] \cdots] \cdot \frac{l}{6EI} \begin{bmatrix} \begin{bmatrix} 2 & 1 \\ 2 & 2 \end{bmatrix} & & \\ & \begin{bmatrix} 2 & 1 \\ 1 & 2 \end{bmatrix} & \\ & & \ddots \end{bmatrix} \begin{Bmatrix} \begin{Bmatrix} m_1^1 \\ m_2^1 \end{Bmatrix} \\ \begin{Bmatrix} m_1^2 \\ m_2^2 \end{Bmatrix} \\ \vdots \end{Bmatrix} \{P\}$$

or

$$\{\Delta\} = [b]^T \ [f] \ [b] \ \{P\}. \qquad (25\text{-}11)$$

If we consider a beam with n elements, and desire the displacements at d points due to any combination of loads of these points, then these displacements will be contained in a $[\Delta]$ matrix of order $(d \times d)$, the $[b]$ matrix will be $(2n \times d)$, the $[f]$ matrix will be $(2n \times 2n)$, and the $[P]$ matrix will be $(d \times d)$.

If we consider displacements due to unit values of applied loads, we set the $[P]$ matrix $= \begin{bmatrix} 1 & & & \\ & 1 & & \\ & & \cdot & \\ & & & \cdot \end{bmatrix} = [I]$, and recognize the resulting matrix

$$[\delta] = [b]^T \ [f] \ [b] \qquad (25\text{-}12)$$

as the structure flexibility matrix already discussed earlier.

EXAMPLE 25-11: For the beam of Fig. 25-17, calculate the structure flexibility matrix for points 2 and 3. Then calculate the displacements at nodes 2 and 3 due to $P_2 = 2$ kips, $P_3 = 3$ kips.

SOLUTION:

$$[b] = \frac{L}{9} \begin{bmatrix} \begin{bmatrix} 0 & 0 \\ 2 & 1 \end{bmatrix} \\ \begin{bmatrix} 2 & 1 \\ 1 & 2 \end{bmatrix} \\ \begin{bmatrix} 1 & 2 \\ 0 & 0 \end{bmatrix} \end{bmatrix}; \quad [b]^T = \frac{L}{9} \left[\begin{bmatrix} 0 & 2 \\ 0 & 1 \end{bmatrix} \begin{bmatrix} 2 & 1 \\ 1 & 2 \end{bmatrix} \begin{bmatrix} 1 & 0 \\ 2 & 0 \end{bmatrix} \right]$$

$$[f] = \frac{\left(\frac{L}{3}\right)}{6EI} \begin{bmatrix} \begin{bmatrix} 2 & 1 \\ 1 & 2 \end{bmatrix} & & \\ & \begin{bmatrix} 2 & 1 \\ 1 & 2 \end{bmatrix} & \\ & & \begin{bmatrix} 2 & 1 \\ 1 & 2 \end{bmatrix} \end{bmatrix};$$

$$[\delta] = [b]^T \ [f] \ [b] = \frac{L^3}{486EI} \begin{bmatrix} 8 & 7 \\ 7 & 8 \end{bmatrix}$$

$$\begin{Bmatrix} \Delta_2 \\ \Delta_1 \end{Bmatrix} = [\delta] \begin{Bmatrix} P_2 \\ P_3 \end{Bmatrix} = \frac{L^3}{486EI} \begin{bmatrix} 8 & 7 \\ 7 & 8 \end{bmatrix} \begin{Bmatrix} 2 \\ 3 \end{Bmatrix} = \frac{L^3}{486EI} \begin{Bmatrix} 37 \\ 38 \end{Bmatrix}$$

Equation (25-12) for the determination of the structure flexibility matrix is perfectly general. For instance, in the case of a member subject to bending about one axis, torsion and axial load and the corresponding displacements at each end, the $[b]$ matrix will consist of four forces for each element, and the $[f]$ matrix will consist of one (4×4) matrix for each element.

The structure flexibility matrix is an important building block for the force method of indeterminate analysis which will be discussed in the next section.

25.3.6 Design Considerations and References

1. Stiffness of Concrete Sections. In planar reinforced concrete structures, the question of appropriate choice of the flexural stiffness EI arises. While theoretically both variation of reinforcing and the effect of cracking should be considered, the *ACI Code* allows use of the gross concrete section for deformation calculations. The stiffening effect of slabs in floor systems has been discussed by:

Khan, F. R., and Sbarounis, J. A., "Interaction of Shear Walls and Frames," *ASCE Proceedings, Journal of the Structural Division*, 90 (ST3), June 1964.

When torsional deformations must be considered, the torsional stiffness factor C is needed. For rectangular cross sections, the reference below lists this factor assuming elastic behavior of the gross section. For open cross sections composed of rectangular components, such as *T*- or *L*-shaped members, the total stiffness can be obtained by adding the stiffnesses of the individual rectangular components. Further information on this topic is in the book by Roark and Young referenced in Section 25.1.5.

2. *Design Aids.* For complicated load systems, the principle of superposition is useful, especially when tabulated or graphical information for the deflections due to simple component loads is available. Numerous sources of such information are available, for instance:

Manual of Steel Construction, American Institute of Steel Construction, New York.

Design Handbook, Concrete Reinforcing Steel Institute, Chicago.

"Tables of Deflection Coefficients for Simple Beams," *Engineering Experiment Station Bulletin No. 87*, Oklahoma State University, Stillwater, Okla., 1953.

Most of this information is in the form $\Delta = C \cdot PL^3/EI$ or $C \cdot wL^4/EI$, where C is called the deflection coefficient. Most of these data apply only to prismatic beams under standard load systems and support conditions. Other quick design information is available in form of nomographs, such as:

Rose, F. O., "More on Deflections of Steel Beams," *Civil Engineering*, 39 (12), Dec. 1969, p. 49.

For other standard types of structures, some information on deformations is also available; for instance, the end deformations of circular bow girders of various types under loading normal to their plane are represented graphically in:

Hogan, M. B., "Circular Beams Loaded Normal to the Plane of Curvature," *Journal of Applied Mechanics*, 5 (2), June 1938.

Beam deflections can be controlled by specifying an allowable length-to-depth ratio, L/d, according to the following reasoning: For a balanced beam design, the maximum curvature occurs at the point of maximum moment, and is, according to standard reinforced concrete theory:

$$\Phi_{max} = \frac{1}{E_c d}\left(f_{c\,All} + \frac{f_{s\,All}}{n}\right)$$

The maximum beam deflection, Δ, can, by geometric methods, be linearly related to the maximum curvature:

$$\Delta = K\,\Phi_{max}\,L^2$$

or:

$$\left(\frac{L}{d}\right)_{All} = \frac{1}{K}\,\frac{E_c}{f_{c\,All} + \dfrac{f_{s\,All}}{n}} \cdot \frac{\Delta}{L} \qquad (25\text{-}13)$$

Values for the factor K, which depends on the loading and support conditions, are given in Table 25-2, and $E_c/[f_{c\,All} + (f_{s\,All}/n)]$ can be considered as a balanced design constant. With the allowable deflection ratio Δ/L specified, the allowable span-depth ratio of the beam can be found from eq. (25-13).

TABLE 25-2 Factors for Eq. (25-13)

Beam and Loading	*K*
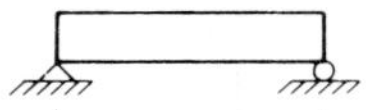	5/48
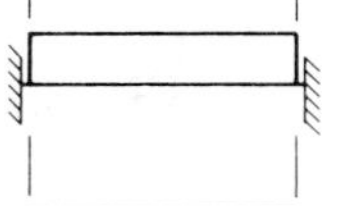	1/32
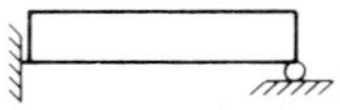	8/185
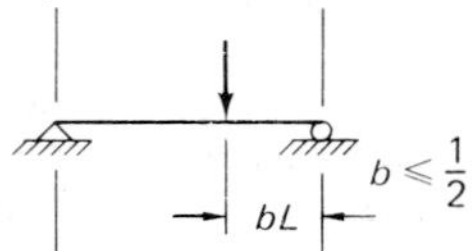	$\frac{1}{27}(1+b)\sqrt{3(1-b^2)}$
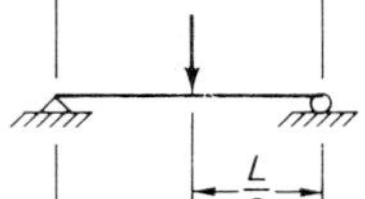	1/12
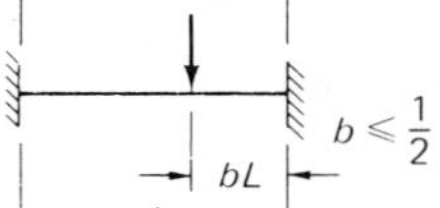	$\frac{2\,(1-b)b}{3\,(3-2b)^2}$
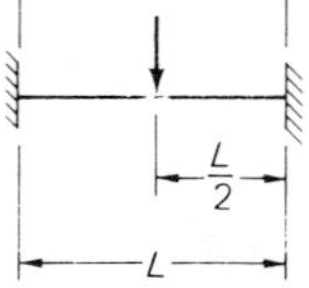	1/24

25.4 THE FORCE METHOD

25.4.1 Static Indeterminacy of Structures

Most reinforced concrete structures are statically indeterminate, that is, the number of unknown forces exceeds the number of available equilibrium equations, so that additional equations of geometry are needed for a rigorous analysis. The excess of unknown forces over the number of independent equilibrium equations is called the degree of static indeterminacy or redundancy. For an analysis by the force method, the degree of static indeterminacy must first be found.

An externally statically indeterminate structure is one that has a larger number of external reactions than equilibrium equations; for such a structure, the indeterminate analysis deals with the determination of the redundant reactions; when these are found, the remaining forces can be found by statics. Continuous beams, single-story rigid frames, and single and multiple arches fall in this category, as shown by the examples of Fig. 25-19. Under each of the structures shown in this figure, e denotes the number of equilibrium equations, f denotes the number of reactions, and $r = f - e$ denotes the degree of static indeterminacy of the structure.

Internally statically indeterminate structures have a larger number of internal forces than available independent equilibrium equations. Among this type are ring-or circuit-type structures, such as shown in Fig. 25-20. The structures of Fig. 25-19 can also be treated as internally indeterminate structures if internal forces are considered as redundants, as discussed in the next section. For the determination of the degree of indeterminacy of such structures, it is often convenient to introduce sufficient *cuts* or *releases* to reduce the structure to static determinacy, and to count up the number of redundant forces removed. In the three-story, three-bay frame of Fig. 25-21, three internal forces (shear, moment, and axial force) are removed with each of the cuts shown, so that the structure is revealed as statically indeterminate to the 27th degree.

25.4.2 The Method of Consistent Displacements

In the force method, the equations of statics are supplemented by a number of equations of geometry equal to the

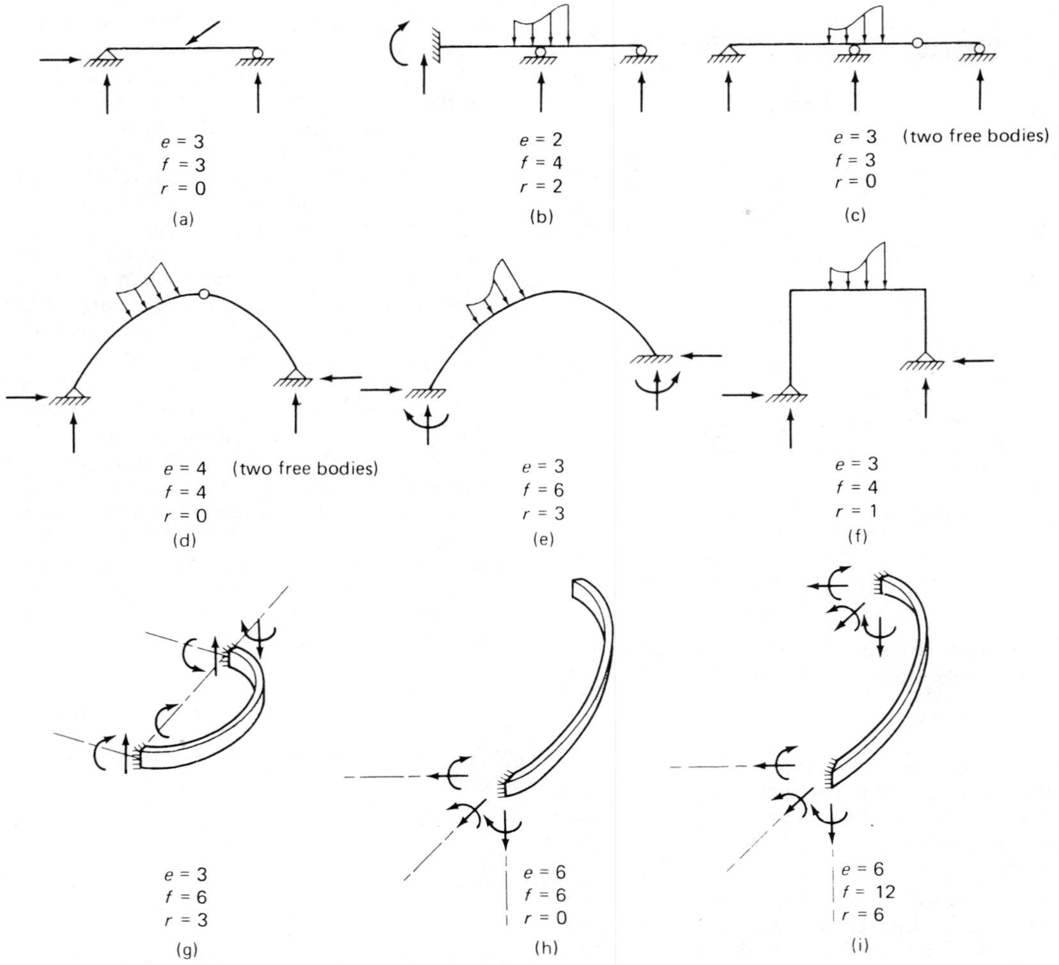

Fig. 25-19 External indeterminacy.

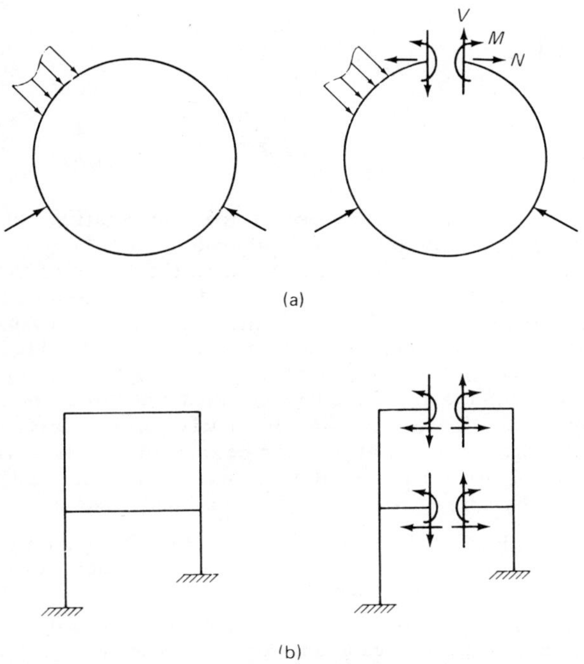

Fig. 25-20 Internal indeterminacy. (a) *left*—Ring structure; (a) *right*—cut structure; (b) *left*—two-story, single-bay frame; and (b) *right*—cut structure.

degree of indeterminacy, r, of the structure. The r unknown, or redundant, forces are the unknowns in these equations which express the consistency of the displacements with the specified support and continuity conditions. Hence the method in its classical formulation is called the method of consistent displacements.

The method will be outlined step by step in terms of the two times statically indeterminate continuous beam of Fig. 25-22(a).

Step 1. Reduce the structure to static determinacy by removing r forces (the redundants). The r points at which forces have been removed are called the cuts. The resulting structure is called the primary structure, shown in Fig. 25-22(b).

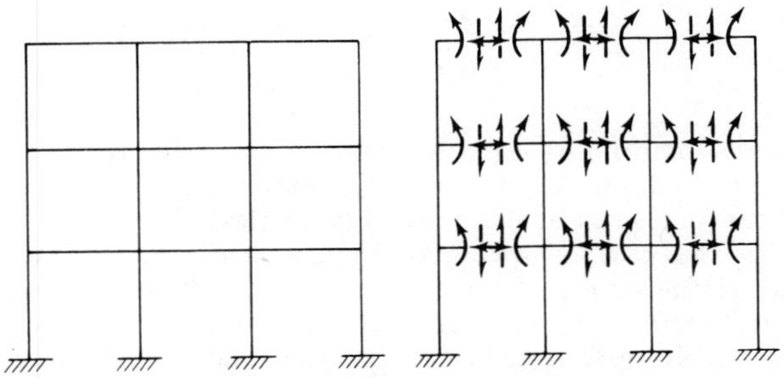

Fig. 25-21 Three-story, three-bay frame.

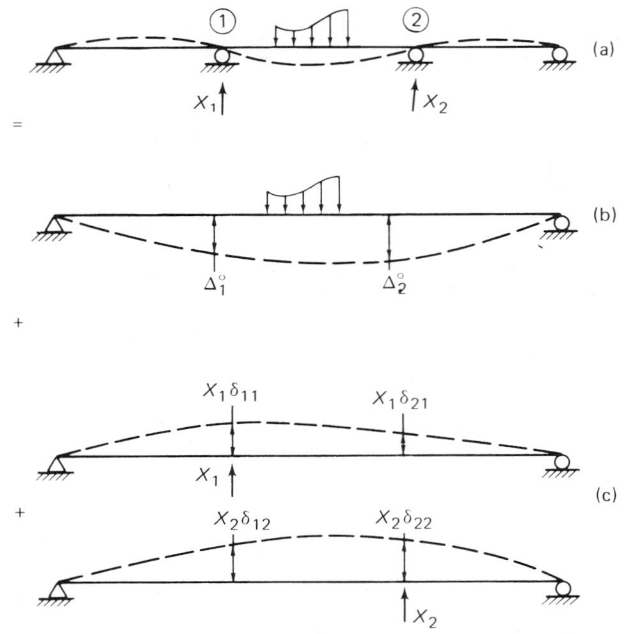

Fig. 25-22 Force method. (a) Real Beam, two times statically indeterminate; (b) primary system; and (c) effects of redundants.

Note that the conditions of geometry of the actual structure are violated at the r cuts. The aim of the subsequent calculations is to calculate those values of the r redundants necessary to restore the continuity at the cuts.

Step 2. Calculate the displacements Δ_i^0 at cuts i of the primary structure due to the applied loads. In general, r displacements Δ^0 will have to be computed for each loading conditions.

Step 3. Calculate the displacements at the cuts i of the primary structure due to unit values of the redundants at j, applied one at a time, shown in Fig. 25-22(c). These are flexibilities δ_{ij}, of total number $r \times r$.

Note that the flexibilities are properties of the structure, not of the loading, and will therefore be the same irrespective of the loading condition. By the principle of superposition, the displacement, X_j, at point i due to the actual redundant at point j, will be $X_j\,\delta_{ij}$.

Step 4. Restore continuity at the r cuts by adding the displacements for each cut i due to the applied loads, Δ_i^0, and due to all redundants, $X_j\,\delta_{ij}$, and equating to the specified displacements, which, for the problem illustrated, are zero.

For $r = 2$:

$$\sum \Delta_1 = 0:\ \Delta_1^0 + X_1\,\delta_{11} + X_2\,\delta_{12} = 0$$
$$\sum \Delta_2 = 0:\ \Delta_2^0 + X_1\,\delta_{21} + X_2\,\delta_{22} = 0 \qquad (25\text{-}14)$$

These r continuity equations have to be solved for the r unknown redundants, X_j. Once the redundants are found, all other forces can be determined by statics.

In the case of specified support displacements Δ_1 and Δ_2 at the points of redundancy, these values must be substituted for the zeros on the right-hand side of the equations. We note that in the case of statically indeterminate structures (as contrasted with determinate structures), support displacements can cause internal forces which must be considered in design.

An important consideration in solving statically indeterminate structures is the choice of the primary system. In general, the closer the structural action of the primary system resembles that of the real structure, the better conditioned will be the resulting set of simultaneous continuity equations, and the less the roundoff errors of the numerical solution. Another consideration is the ready availability of flexibilities, so that decomposition of the structure into standard components with known flexibilities (such as the rotational end flexibilities of prismatic members) should be considered. In this light, the force method of analysis of a continuous beam of $(r + 1)$ spans might be undertaken as shown in Fig. 25-23; the beam is cut over each interior support, releasing the r support moments which become the redundant forces; the requirement of slope continuity of the real beam must now be satisfied by adding the slope changes (or "kinks") over the supports due to all effects on the primary system, and equating their sum to zero.

Figure 25-23(b) through (f) shows the primary system under the applied load, and the redundant moments, applied one at a time. With the redundants and displacements as defined in these figures, the r continuity conditions become:

$$\sum\theta_1 = 0:\ X_1\,\delta_{11} + X_2\,\delta_{12} + 0 \qquad + 0 + 0 \quad + \cdots + 0 \quad =$$
$$\sum\theta_2 = 0:\ X_1\,\delta_{21} + X_2\,\delta_{22} + X_3\,\delta_{23} + 0 + 0 \quad + \cdots + 0 \quad =$$
$$\sum\theta_3 = 0:\ 0 \qquad + X_2\delta_{32} + X_3\delta_{33} + X_4\delta_{34} + 0 + \cdots + 0 =$$
$$\vdots$$
$$\sum\theta_r = 0:\ 0 \qquad + 0 \qquad + 0 \qquad + 0 + 0 + \cdots + X_r\delta_{rr} \quad =$$

In this case, the flexibilities are the differences of the appropriate end rotations due to cranked-in moments on prismatic members; that is:

$$\delta_{ij} = \left(\frac{L}{3EI}\right)_i + \left(\frac{L}{3EI}\right)_{i+1} \qquad \text{for } i = j,$$

$$\text{and } \delta_{ij} = \frac{L}{6EI} \quad \text{for } i \neq j$$

The Δ_i^0 are the differences of the end rotations of the simple spans due to the specified loads.

It is also noted that, in contrast to the action of the redundants chosen in Fig. 25-22, in this case the effect of each redundant is felt only in the adjacent beam spans, so that each continuity equation contains only three redundants, all other terms being zero. For this reason this version of the force method is called the "three-moment equation." When expressed in matrix form, it leads to a "tri-diagonal matrix" which can be solved very efficiently.

We observe that the labor of analysis can be vastly reduced by appropriate choice of the primary system.

EXAMPLE 25-12: Analyze the beam of Fig. 25-24(a) due to the following effects: (a) a uniform load w over the entire beam, and (b) a support settlement Δ of Point B.

SOLUTION: The beam is statically indeterminate to the second degree; it is decomposed into the primary system by removing the moments M_1 and M_2, which become the redundants. Figure 25-24(b), (c), and (d) shows the relative slope changes which are set

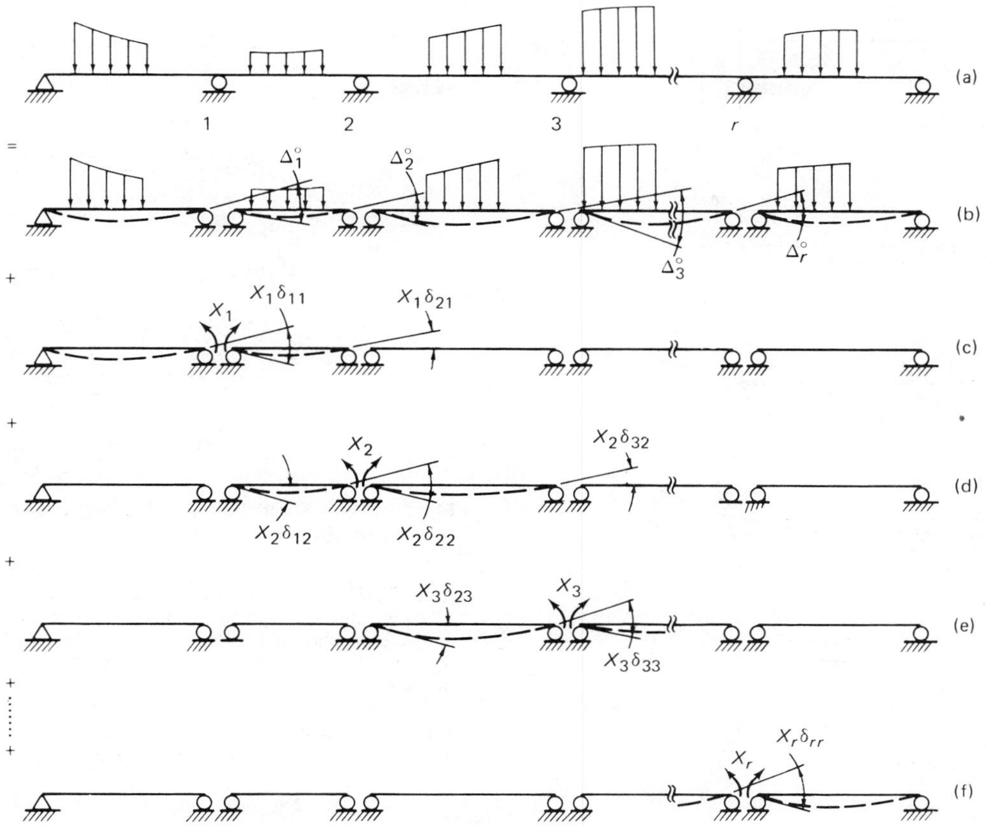

Fig. 25-23 Force method analysis of continuous beam.

equal to zero:

$$\sum \theta_1 = 0:\ \Delta_1^0 + M_1\delta_{11} + M_2\delta_{12} = 0$$

$$\sum \theta_2 = 0:\ \Delta_2^0 + M_1\delta_{21} + M_2\delta_{22} = 0$$

where the displacements are calculated for Part (a) by any method as:

$$\Delta_1^0 = 2\cdot\frac{wL^3}{24EI} = \frac{1}{12}\frac{wL^3}{EI};\ \Delta_2^0 = \frac{1}{24}\frac{wL^3}{EI}$$

$$\delta_{11} = 2\cdot\frac{L}{3EI};\ \delta_{22} = \frac{L}{3EI};\ \delta_{12} = \delta_{21} = \frac{L}{3EI}.$$

so that:

$$\frac{2}{3}M_1 + \frac{1}{6}M_2 = -\frac{1}{12}wL^2$$

$$\frac{1}{6}M_1 + \frac{1}{3}M_2 = -\frac{1}{24}wL^2$$

from which $M_1 = -0.104\ wL^2$, and $M_2 = -0.076\ wL^2$.

The moment diagram is shown in Fig. 25-24(e).

Figure 25-24(f) shows the deformed primary system due to the support settlement of part (b), from which the kinks at supports 1 and 2 are calculated as:

$$\Delta_1 = -2\frac{\Delta}{L},\quad \Delta_2 = +\frac{\Delta}{L}$$

The kinks due to the redundants are as before, so that the conditions of geometry become:

$$\frac{2}{3}M_1 + \frac{1}{6}M_2 = 2\frac{EI}{L^2}\Delta$$

$$\frac{1}{6}M_1 + \frac{1}{3}M_2 = -\frac{EI}{L^2}\Delta.$$

from which $M_1 = \dfrac{30}{7}\dfrac{EI}{L^2}\Delta;\ M_2 = -\dfrac{36}{7}\dfrac{EI}{L^2}\Delta.$

The moment diagram due to the specified support settlement Δ is shown in Fig. 25-23(g).

25.4.3 Fixed-End Moments

Fixed-end moments (FEMs) are the end moments of a structure when all ends are fixed against rotation and translation. They are calculated by the force method. Fixed-end moments are one of the building blocks used in the displacement method of analysis.

EXAMPLE 25-13: Calculate FEMs at ends A and B of the prismatic beam of Fig. 25-25 due to a load P at a from one end, b from the other.

SOLUTION: The structure is two times indeterminate; we consider the desired end moments as redundants. Calculate deformations of the primary structure by any method.

$$\Delta_A^0 = \frac{1}{6}\frac{Pab}{EIL}(a+2b);\ \Delta_B^0 = \frac{1}{6}\frac{Pab}{EIL}(2a+b)$$

$$\delta_{AA} = \frac{1}{3}\frac{L}{EI} = \delta_{BB};\ \delta_{BA} = \frac{1}{6}\frac{L}{EI} = \delta_{AB}$$

Set up equations of geometry:

$$\sum \theta_A = 0:\ M_A\frac{1}{3}\frac{L}{EI} + M_B\frac{1}{6}\frac{L}{EI} = -\frac{1}{6}\frac{Pab}{EI\cdot L}(a+2b)$$

$$\sum \theta_B = 0:\ M_A\frac{1}{6}\frac{L}{EI} + M_B\frac{1}{3}\frac{L}{EI} = -\frac{1}{6}\frac{Pab}{EI\cdot L}(2a+b)$$

Solving simultaneously:

$$M_A = -\frac{Pab^2}{L^2} = \text{FEM at } A$$

$$M_B = -\frac{Pab^2}{L^2} = \text{FEM at } B \qquad (25\text{-}15)$$

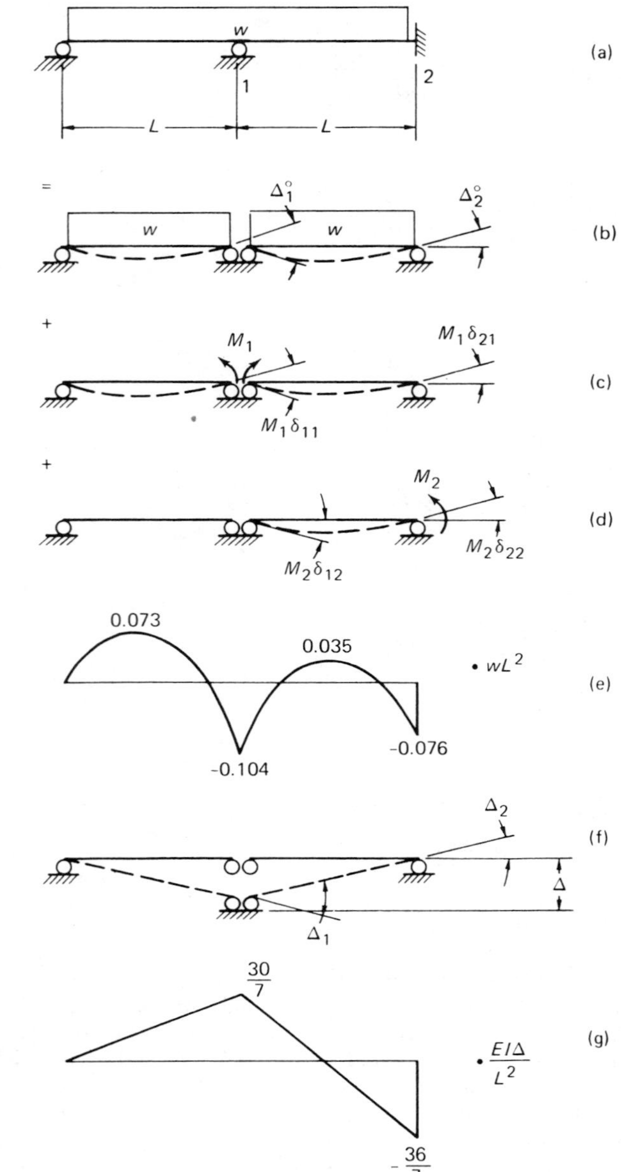

Fig. 25-24 Example 25-12.

Identical methods can be used for other structures and loads. Results of such calculations lead to the values for fixed-end moments for prismatic beams under various loading conditions shown in Table 25-3. Section 25.4.6.2 also lists references for tabulated information about fixed-end moments and stiffness factors for nonprismatic beams of interest to concrete engineers.

25.4.4 Stiffness Factors and Stiffness Matrix

Stiffness factor is the force needed to achieve a unit displacement at one node when all other nodes are fixed against displacement. Stiffness factors are one of the building blocks of the stiffness, or displacement, method for analyzing indeterminate structures.

The stiffness factor k_{ij} is the force i needed to achieve a unit displacement j, all other displacements equal to zero. Stiffness factors can be calculated by the force method.

EXAMPLE 25-14: Given a prismatic straight beam, length L, stiffness EI, shown in Fig. 25-26(a), calculate the end stiffness factors.

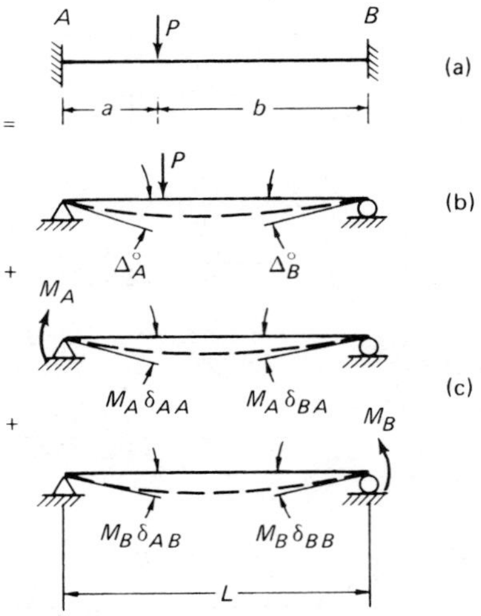

Fig. 25-25 Example 25-13. Fixed-end moments. (a) Real beam; (b) primary system; (c) flexibilities.

SOLUTION: Four displacements are considered, two end rotations θ_1 and θ_2, and two displacements, Δ_3 and Δ_4. Corresponding to each displacement, there are support forces M_1 and M_2, and V_3 and V_4. For instance, for $\theta_1 = 1$, $\theta_2 = \Delta_3 = \Delta_4 = 0$: $M_1 \equiv k_{11}$, $M_2 \equiv k_{21}$, $V_3 \equiv k_{31}$, $V_4 \equiv k_{41}$.

We will calculate these stiffnesses by the force method. The primary structure is the simple beam of Fig. 25-26(b); k_{11} and k_{21} are the redundant end moments due to $\theta_1 = 1$.

$$\Sigma\theta_A = 1: \quad k_{11}\frac{L}{3EI} - k_{21}\frac{L}{6EI} = 1$$

$$\Sigma\theta_B = 0: -k_{11}\frac{L}{6EI} + k_{21}\frac{L}{3EI} = 0$$

Solving simultaneously:

$$k_{11} = \frac{4EI}{L}; \quad k_{21} = \frac{2EI}{L}$$

The reactions, k_{31} and k_{41}, are found by statics:

$$k_{31} = -\frac{6EI}{L^2}; \quad k_{41} = +\frac{6EI}{L^2}$$

The stiffness factors k_{i3} and k_{i4} are the nodal forces corresponding to unit values of the end translations Δ_3 and Δ_4. To calculate these stiffnesses, we use the same primary structure. For k_{i4}, for instance, we refer to Fig. 25-27:

$$\Sigma\theta_A = 0: \quad k_{14}\frac{L}{3EI} - k_{24}\frac{L}{6EI} = \frac{1}{L}$$

$$\Sigma\theta_B = 0: -k_{14}\frac{L}{6EI} + k_{24}\frac{L}{3EI} = -\frac{1}{L}.$$

$$k_{14} = k_{24} = \frac{6EI}{L^2}.$$

By statics:

$$k_{34} = -k_{44} = -\frac{12EI}{L^3}.$$

It will be recognized that the stiffness factors k_{ij} can be easily assembled into a member, or element, stiffness matrix $[k]$, relating the nodal forces to the nodal displacements; for Example 25-14, for instance:

$$\begin{Bmatrix} M_1 \\ M_2 \\ Y_3 \\ Y_4 \end{Bmatrix} = \frac{EI}{L^3}\begin{bmatrix} 4L^2 & 2L^2 & -6L & 6L \\ 2L^2 & 4L^2 & -6L & 6L \\ -6L & -6L & 12 & -12 \\ 6L & 6L & -12 & 12 \end{bmatrix}\begin{Bmatrix} \theta_1 \\ \theta_2 \\ y_3 \\ y_4 \end{Bmatrix} \qquad (25\text{-}16)$$

TABLE 25-3 Fixed-End Moments for Prismatic Beams

Load Type	M_A^F	M_B^F
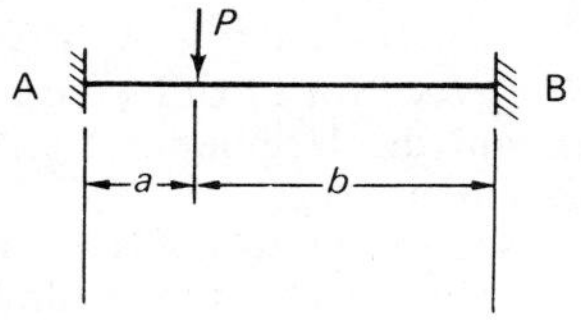	$+\dfrac{Pab^2}{L^2}$	$-\dfrac{Pa^2b}{L^2}$
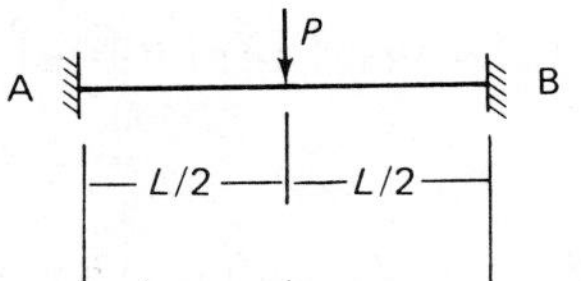	$+\dfrac{PL}{8}$	$-\dfrac{PL}{8}$
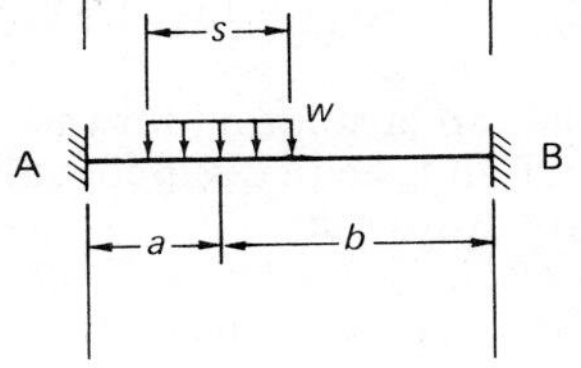	$+\dfrac{ws}{12L^2}[12ab^2 + s^2 (L - 3b)]$	$-\dfrac{ws}{12L^2}[12a^2b + s^2(L - 3a)]$
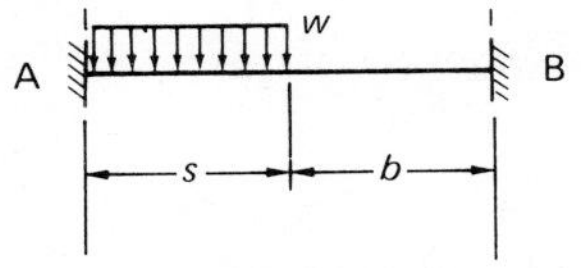	$+\dfrac{ws^2}{12L^2}[2L(3L - s) + 3s^2]$	$-\dfrac{ws^3}{12L^2}(4L - 3s)$
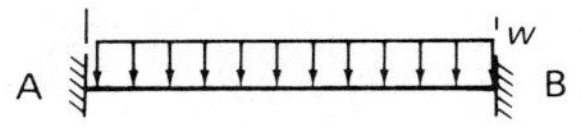	$+\dfrac{wL^2}{12}$	$-\dfrac{wL^2}{12}$
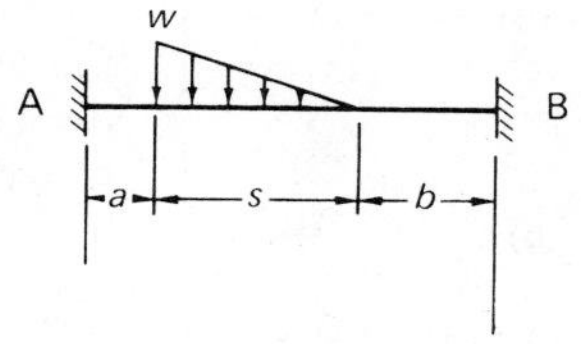	$+\dfrac{ws}{60L^2}[10b^2(3a + s)$ $+ s^2(15a + 10b + 3s) + 40abs]$	$-\dfrac{ws}{60L^2}[10a^2(3b + 2s)$ $+ s^2(10a + 5b + 2s) + 20abs]$
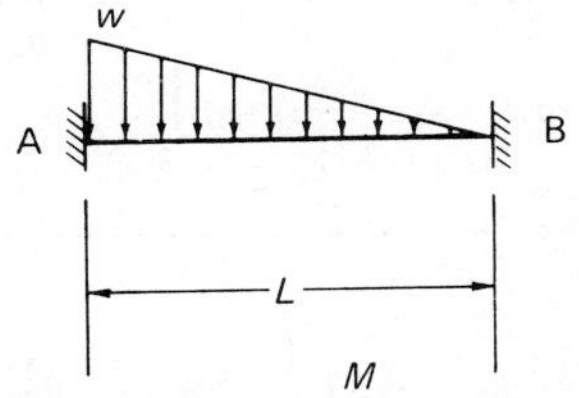	$+\dfrac{wL^2}{20}$	$-\dfrac{wL^2}{30}$
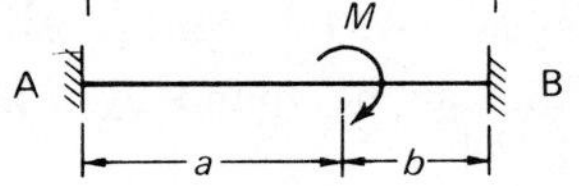	$-M\dfrac{b}{L}\left(2 - \dfrac{3b}{L}\right)$	$-M\dfrac{a}{L}\left(2 - \dfrac{3a}{L}\right)$

or:

$$\{X\} = [k]\ \{\Delta\} \tag{25-17}$$

Comparing eqs. (25-10) and (25-17), it can be deduced that if for a given structure the stiffness and flexibility matrices $[k]$ and $[\delta]$ are expressed in terms of the same nodes, then:

$$[k] = [\delta]^{-1} \tag{25-18}$$

that is, the stiffness matrix is the inverse of the flexibility matrix; it is therefore a symmetric matrix, as demonstrated by eq. (25-16).

Stiffness factors are a property of the structure, not of the load. For haunched members of interest in concrete design, values of stiffness factors are given in the references listed in Section 25.4.6.2.

25.4.5 Matrix Formulation of the Force Method

We recall that the force method of analysis of an r-times statically indeterminate structure demands satisfaction of the r conditions of geometry:

$$
\begin{aligned}
X_1\delta_{11} + \cdots + X_r\delta_{1r} + \Delta_1^0 &= \Delta_1\\
&\ \vdots\\
X_1\delta_{r1} + \cdots + X_r\delta_{rr} + \Delta_r^0 &= \Delta_r
\end{aligned}
\qquad (25\text{-}14)
$$

where δ_{ij} are the influence, or flexibility, coefficients at point i of the primary structure due to a unit redundant force at j; Δ_i^0 are the displacements of point i of the primary structure due to the applied loads; and Δ_i are the specified displacements at points of redundancy.

If we label the p points of known applied concentrated loads consecutively with the r points of redundancy, that is, $(r+1), \cdots (r+p)$, then the displacements Δ_i^0 due to applied loads X_j can be written as:

$$\Delta_i^0 = X_{(r+1)}\delta_{i(r+1)} + \cdots + X_{(r+p)}\delta_{i(r+p)}$$

Similarly, the unknown displacements Δ_i [$i = (r+1)$, etc.] at the p load points are obtained by superposition of the effects of all redundant and applied forces:

$$
\begin{aligned}
\Delta_{r+1} &= X_1\delta_{(r+1)1} + \cdots + X_r\delta_{(r+1)r} + X_{(r+1)}\delta_{(r+1)(r+1)} + \cdots + X_{(r+p)}\delta_{(r+1)(r+p)}.\\
&\ \vdots\\
\Delta_{r+p} &= X\ \delta_{(r+p)1} + \cdots \qquad\qquad \cdots + X_{(r+p)}\delta_{(r+p)(r+p)}
\end{aligned}
$$

We add these p equations to the r equations of geometry and get eq. (25-19):

r unknown forces (redundants) — p known forces (loads)

r eqs. of geometry for unknown reactions, X (known displacements of redundants):

$$
\begin{aligned}
X_1\delta_{11} + \cdots + X_r\delta_{1r} + X_{(r+1)}\delta_{1(r+1)} + \cdots + X_{(r+p)}\delta_{1(r+p)} &= \Delta_1\\
&\ \vdots\\
X_1\delta_{r1} + \cdots + X_r\delta_{rr} + X_{(r+1)}\delta_{r(r+1)} + \cdots + X_{(r+p)}\delta_{r(r+p)} &= \Delta_r
\end{aligned}
$$

p eqs. for unknown displacements, Δ (unknown displacements of loads):

$$
\begin{aligned}
X_1\delta_{(r+1)1} + \cdots + X_r\delta_{(r+1)r} + X_{(r+1)}\delta_{(r+1)(r+1)} + \cdots + X_{(r+p)}\delta_{(r+1)(r+p)} &= \Delta_{(r+1)}\\
&\ \vdots\\
X_1\delta_{(r+p)1} + \cdots + X_r\delta_{(r+p)r} + X_{(r+1)}\delta_{(r+p)(r+1)} + \cdots + X_{(r+p)}\delta_{(r+p)(r+p)} &= \Delta_{(r+p)}
\end{aligned}
\qquad (25\text{-}19)
$$

In matrix form, this set of equations can be written as:

$$
\left[\begin{array}{c|c}\delta_{\alpha\alpha} & \delta_{\alpha\beta}\\ \hline \delta_{\beta\alpha} & \delta_{\beta\beta}\end{array}\right]
\left[\begin{array}{c}X_\alpha\\ \hline X_\beta\end{array}\right]
=
\left[\begin{array}{c}\Delta_\alpha\\ \hline \Delta_\beta\end{array}\right];
\quad
\begin{array}{l}(X_\alpha \text{ unknown; } \Delta_\alpha \text{ known})\\ (X_\beta \text{ known; } \quad \Delta_\beta \text{ unknown})\end{array}
\qquad (25\text{-}20)
$$

The square flexibility matrix $[\delta]$, of order $(r+p)$, can be determined by any of the methods for finding deformations of statically determinate structures; for instance, by the matrix formulation:

$$[\delta] = [b]^T[f]\ [b] \qquad (25\text{-}12)$$

The flexibility matrix has been partitioned into $\delta_{\alpha\alpha}$ (order $r \times r$), $\delta_{\alpha\beta}$ (order $r \times p$), $\delta_{\beta\alpha}$ (order $p \times r$), $\delta_{\beta\beta}$ (order $p \times p$) in anticipation of future operations.

The matrix of forces $\{X\}$, of order $(r+p) \times 1$ for a single loading condition, or of order $(r+p) \times l$ for l different loading conditions, contains r unknown redundants, denoted by the subscript α, followed by p known loads subscripted β.

The matrix $\{\Delta\}$, of order $(r+p) \times 1$ for a single loading condition, of order $(r+p) \times l$ for l loading conditions, contains r specified redundant displacements Δ_α, followed by p unknown load displacements Δ_β.

The solution of this set of equations is accomplished in two steps. First, solve r equations for the redundants X_α:

$$[\delta_{\alpha\alpha}]\ [X_\alpha] + [\delta_{\alpha\beta}]\ [X_\beta] = [\Delta_\alpha]$$

or:

$$[X_\alpha] = [\delta_{\alpha\alpha}]^{-1}[[\Delta_\alpha] - [\delta_{\alpha\beta}]\ [X_\beta]] \qquad (25\text{-}21)$$

This corresponds to the earlier solution for the r redundants by classical methods. If all specified support displacements are zero, the Δ_α matrix drops out.

After the redundants $[X_\alpha]$ are known, the remaining p equations can be solved for the displacements at the load points:

$$[\Delta_\beta] = [\delta_{\beta\alpha}]\ [X_\alpha] + [\delta_{\beta\beta}]\ [X_\beta] \qquad (25\text{-}22)$$

The internal forces, S, can be found by using the force transformation matrix $[b]$:

$$[S] = [b]\ [X]. \qquad (25\text{-}23)$$

EXAMPLE 25-15: Find the redundant X_1, the displacements Δ_2 and Δ_3, and the moments at points 1 to 4 of the beam of Fig. 25-28(a) due to two conditions:

(a) Loading condition 1: $X_2 = 3$ kips; $X_3 = 2$ kips.
(b) Loading condition 2: A support settlement $\Delta_1 = 1$.

SOLUTION: The primary structure is as shown in Fig. 25-28(b).

Step 1. Determine flexibility matrix $[\delta] = [b]^T[f]\ [b]$.

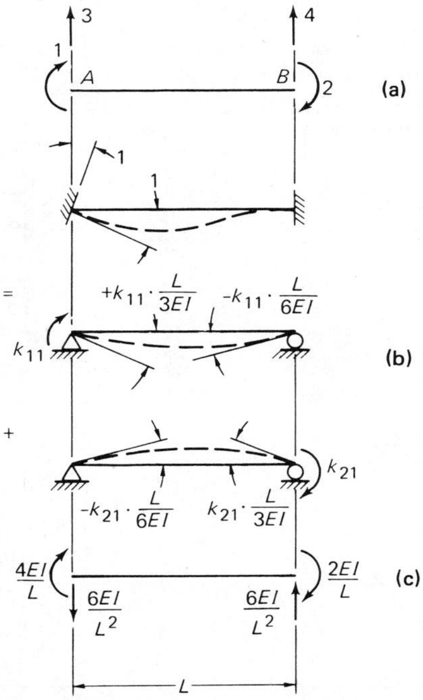

Fig. 25-26 Example 25-14. Stiffnesses (a) Real beam; (b) primary structure; and (c) final forces.

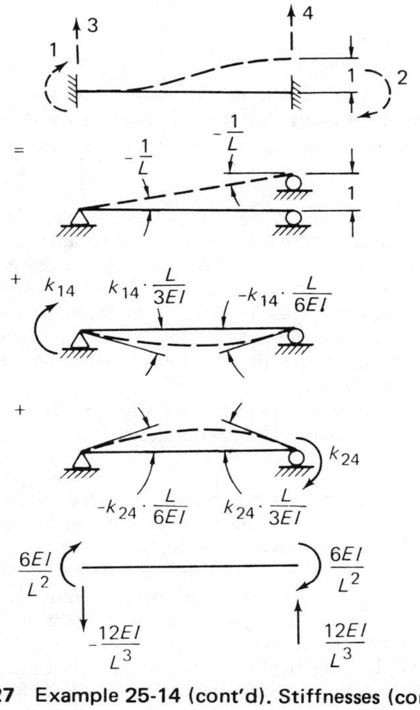

Fig. 25-27 Example 25-14 (cont'd). Stiffnesses (cont'd).

$$[f] = \frac{l}{6EI}\begin{bmatrix} 2 & 1 & & & & \\ 1 & 2 & & & & \\ & & 2 & 1 & & \\ & & 1 & 2 & & \\ & & & & 2 & 1 \\ & & & & 1 & 2 \end{bmatrix}; \quad [b] = l\begin{bmatrix} 0 & 0 & 0 \\ -1 & 0 & 0 \\ -1 & 0 & 0 \\ -2 & -1 & 0 \\ -2 & -1 & 0 \\ -3 & -2 & -1 \end{bmatrix}.$$

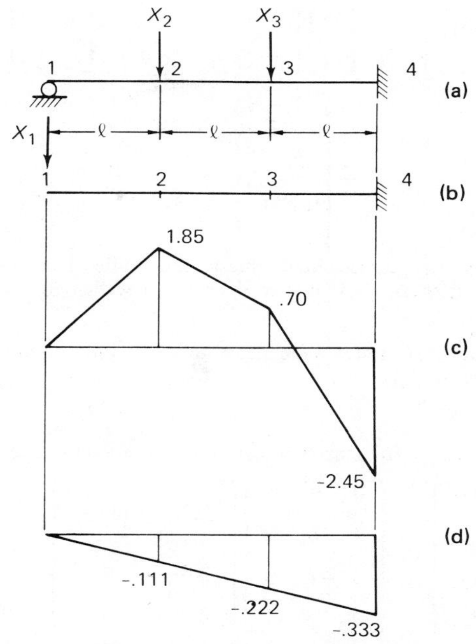

Fig. 25-28 Example 25-15. (a) Structure, constant EI; (b) primary structure; (c) moment diagram, loading condition 1 (X l); and (d) moment diagram, loading condition 2 (X EI/l^2).

$$\underset{(3 \times 3)}{[\delta]} = \underset{(3\times 6)}{[b]^T}\ \underset{(6\times 6)}{[f]}\ \underset{(6\times 3)}{[b]}$$

$$= \frac{l^3}{6EI}\begin{bmatrix} 54 & 28 & 8 \\ 28 & 16 & 5 \\ 8 & 5 & 2 \end{bmatrix} \quad \text{(note symmetry)}$$

Step 2. Set up force-displacement equations for two loading conditions.

$$\begin{bmatrix} \delta_{\alpha\alpha} & \delta_{\alpha\beta} \\ \delta_{\beta\alpha} & \delta_{\beta\beta} \end{bmatrix}\begin{bmatrix} X_\alpha \\ X_\beta \end{bmatrix} = \begin{bmatrix} \Delta_\alpha \\ \Delta_\beta \end{bmatrix}.$$

(The first column of the $[X]$ and $[\Delta]$ matrices denotes the first, the second, the second loading condition.)

$$\frac{l^3}{6EI}\begin{bmatrix} 54 & 28 & 8 \\ 28 & 16 & 5 \\ 8 & 5 & 2 \end{bmatrix}\begin{bmatrix} X_{11} & X_{12} \\ X_{21} = 3 & X_{22} = 0 \\ X_{31} = 2 & X_{32} = 0 \end{bmatrix} = \begin{bmatrix} \Delta_{11} = 0 & \Delta_{12} = 1 \\ \Delta_{21} & \Delta_{22} \\ \Delta_{31} & \Delta_{32} \end{bmatrix}$$

Step 3. Solve for redundant X_1 due to both loading conditions (α-eqs.).

$$\frac{l^3}{6EI}\left\{(54)\,[X_{11} \quad X_{12}] + [28 \quad 8]\begin{bmatrix} 3 & 0 \\ 2 & 0 \end{bmatrix}\right\} = [0 \quad 1]$$

Now, solve for $[X_{11}\ X_{12}]$.

$$[X_{11} \quad X_{12}] = \frac{1}{54}\left\{\frac{6EI}{l^3}\,[0 \quad 1] - [100 \quad 0]\right\} = \left[-1.85 \quad 0.111\,\frac{EI}{l^3}\right]$$

Note that the redundant due to the applied loads (condition 1) is 1.85 kips upward, that due to the support settlement (condition 2) depends on the beam stiffness.

Step 4. Solve for displacements Δ_2 and Δ_3 due to both loading conditions (β-eqs.).

$$\begin{bmatrix}\Delta_{21} & \Delta_{22}\\ \Delta_{31} & \Delta_{32}\end{bmatrix} = \frac{l^3}{6EI}\begin{bmatrix}28\\ 8\end{bmatrix}\begin{bmatrix}-1.85 & 0.111\dfrac{EI}{l^3}\end{bmatrix} + \begin{bmatrix}16 & 5\\ 5 & 2\end{bmatrix}\begin{bmatrix}3 & 0\\ 2 & 0\end{bmatrix}$$

$$= \begin{bmatrix}1.016\dfrac{l^3}{EI} & 0.519\\ 0.700\dfrac{l^3}{EI} & 0.148\end{bmatrix}$$

Note that the displacements due to the applied loads depend on the beam stiffness; those due to the support settlement ($\Delta_1 = 1$) do not.

Step 5. Solve for the moments at points 1 to 4 due to both loading conditions.

$$[S] = [b]\,[X]$$

(We contract the $[b]$ matrix to list only *one* moment at each point.)

$$\begin{bmatrix}M_{11} & M_{12}\\ M_{21} & M_{22}\\ M_{31} & M_{32}\\ M_{41} & M_{42}\end{bmatrix} = l\begin{bmatrix}0 & 0 & 0\\ -1 & 0 & 0\\ -2 & -1 & 0\\ -3 & -2 & -1\end{bmatrix}\begin{bmatrix}-1.85 & 0.111\dfrac{EI}{l^3}\\ 3.0 & 0\\ 2.0 & 0\end{bmatrix}$$

$$= \begin{bmatrix}0 & 0\\ 1.85l & -0.111\dfrac{EI}{l^2}\\ 0.70l & -0.222\dfrac{EI}{l^2}\\ -2.45l & -0.333\dfrac{EI}{l^2}\end{bmatrix}$$

The moment diagrams for the two loading conditions can now be plotted, as shown in Fig. 25-28(c) and (d).

25.4.6 Approximate Analysis and Design Aids

1. Preliminary Analysis. The nature of structural design as a trial-and-error process demands some facility on the part of the engineer to perform quick, approximate analyses in order to be able to estimate preliminary costs, and to determine approximate member dead loads and stiffnesses to be used in more exact analysis.

Such approximate analysis methods rely often on the ability to visualize the deformation of the structure under load; a physical feel for structural behavior is necessary for this. After the members have been sized preliminarily, an exact analysis should always follow before the design is finalized.

One approach to approximate analysis of statically indeterminate structures consists of the following steps:

Step 1. Introduce a number of assumptions regarding values of forces equal to the degree of static indeterminacy, r, of the structure. Among such assumptions might be the location of inflection points (that is, location where $M = 0$), relative distribution of shears, magnitudes of reactions as bounded by some known limiting values, etc.

Step 2. Values for r forces having been introduced, the structure is reduced to static determinacy. Thus, all other forces can be computed by statics.

The procedure is illustrated in the following example:

EXAMPLE 25-16: Perform a preliminary analysis of the two-story frame under gravity and lateral loads shown in Fig. 25-29(a).

SOLUTION: The vertical loads are treated first, as shown in Fig. 25-29(b). The extreme moment diagrams, one for the simply

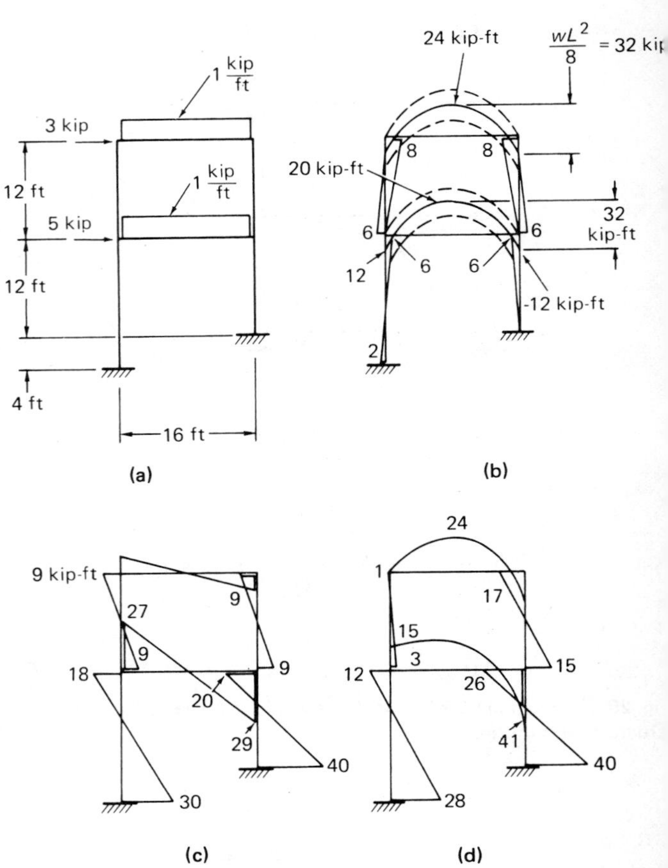

Fig. 25-29 Example 25-16. (a) Structure and loads; (b) approx analysis for vertical loads; (c) approx. analysis for lateral loads; and (d) approx. moment diagram for preliminary design.

supported case, the other for the fixed-ended case, are shown dashed for the uniform load on each of the beams. The actual end condition lies somewhere in between, leading to the beam moment diagrams shown in solid line. The beam end moments are transferred to the adjacent columns such that moment equilibrium of the joints and shear equilibrium of the frame are maintained, leading to the moments shown in solid line in Fig. 25-29(b); these moments satisfy equilibrium.

In a similar fashion, the moments of Fig. 25-29(c) are in equilibrium with the lateral loads; in drawing them, the following guidelines were followed:

1. Points of inflection were assumed at the midpoints of the second-story columns, and the column and top level beam moment diagrams drawn accordingly.
2. Of the total shear of 8 kips in the bottom level, the shorter, stiffer column was assumed to transmit the greater amount of 5 kips; the longer, more flexible column the rest. The bottom, fixed end of each column was assumed to carry more moment than the upper, elastically restrained ends, leading to the lower-story column moment diagrams shown. With the column end moments drawn, the lower beam moment diagram can be drawn.

The complete moment diagram of Fig. 25-29(d) represents an equilibrium set of forces. Whether or not these forces also satisfy geometrical compatibility can only be ascertained after member stiffnesses have been assumed. A slight lack of such compatibility can be accommodated if the members and joints possess sufficient ductility; it is therefore important to design the structure for both strength and ductility to avoid cracking.

2. Design Aids. The solutions of indeterminate structures of standard form collected in the works by Kleinlogel and Leontovich cited in Section 25.1.5 are useful for some purposes. For the design of rigid frames of rectangular, gabled, and parabolic shape, the following contains graphs, tables, and examples:

Griffiths, J. D., "Single Span Rigid Frames in Steel," American Institute of Steel Construction, New York, 1948.

Tables of fixed-end moments are available in many standard references; Table 25-3 lists them for some common loading conditions on prismatic beams. For fixed-end moments as well as stiffness factors for haunched members of the type often found in concrete structures, the first of the following provides tabular, the second graphical aids:

Portland Cement Association, *Handbook of Frame Constants*, Chicago, 1950.
Portland Cement Association, "Concrete Members with Variable Moment of Inertia," *Concrete Information ST 103*, Chicago, 1964.

By inverting the rotational end stiffnesses given in this reference in accordance with eq. (25-18), the member end flexibilities can be found for use in a force method analysis as outlined in Sections 25.4.2 or 25.4.5.

A large amount of design information for arches, including influence lines for circular and parabolic arches, is contained in:

Michalos, J., *Theory of Structural Analysis and Design*, The Ronald Press, New York, 1958.

For circular bow girders, the following references contain graphical results:

Hogan, M. B., "Circular Beams Loaded Normal to the Plane of Curvature," *ASME Transactions (Journal of Applied Mechanics)*, 5(2), June 1938, and 11(1), Mar. 1944.

25.5 THE DISPLACEMENT METHOD

25.5.1 Basic Formulation

In the force method for analysis of indeterminate structures, the redundant forces are unknowns. A set of equations of geometry, equal to the number of unknown forces (degree of static indeterminacy r), has to be solved to determine these forces.

In the displacement method (also called the stiffness method) the nodal displacements are unknowns. A set of equations of equilibrium, equal to the number of unknown displacements (degree of kinematic indeterminacy k), has to be solved to determine these displacements.

If the number of redundant forces, r, is less than the number of unknown displacements, k, the force method is advantageous. If, on the other hand, $r > k$, then the displacement method is preferred. Furthermore, the systematic formulation of the displacement method is often easier, so that for computer solutions the displacement method is commonly used irrespective of the degrees of static or kinematic indeterminacy.

Basically the displacement method is carried out in the following steps:

Step 1. Force-displacement relations (stiffness).

Step 2. Geometrical relations.

Step 3. Equilibrium relations to determine unknown displacements.

Step 4. Determination of forces by substituting displacements into stiffnesses.

These steps are illustrated by a simple example:

EXAMPLE 25-17: Calculate the moments in the structure shown in Fig. 25-30 due to the cranked-in moment M.

This structure is statically indeterminate to the fifth degree, but its kinematic degree of indeterminacy is only one, since all forces can be determined in terms of the unknown rotation θ_A. We proceed along the outlined steps:

Step 1. The end moments associated with a unit end rotation (the rotational stiffness factors) are found for all members as outlined in Section 25.4.

$$k_{AB} = \frac{M_{AB}}{\theta_{AB}} = \frac{3}{2}\frac{EI}{L};\quad k_{AC} = \frac{M_{AC}}{\theta_{AC}} = \frac{EI}{L};\quad k_{AD} = \frac{M_{AD}}{\theta_{AD}} = 3\frac{EI}{L};$$

$$k_{AE} = \frac{M_{AE}}{\theta_{AE}} = 4\frac{EI}{L}$$

Step 2. Since joint A is rigid, all member ends turn through the same angle θ_A:

$$\theta_{AB} = \theta_{AC} = \theta_{AD} = \theta_{AE} = \theta_A$$

Step 3. The single equilibrium equation needed is that of rotational equilibrium of joint A.

$$\Sigma M_A = 0:\quad M_{AB} + M_{AC} + M_{AD} + M_{AE} = M$$

Expressing the moments in terms of θ_A to obtain the equilibrium equation in terms of the displacements:

$$\frac{EI}{L}\left[\frac{3}{2} + 1 + 3 + 4\right]\theta_A = M$$

from which the unknown displacement is:

$$\theta_A = \frac{2}{19}\frac{L}{EI}M$$

Note that the total stiffness against rotation at point A of the entire structure:

$$\frac{M}{\theta_A} = \frac{19EI}{2L}$$

is obtained by adding the member stiffnesses at point A.

Step 4. To obtain the member end moments at A, we substitute the value of the unknown displacement θ_A found in Step 3 into the stiffness relations of Step 1.

$$M_{AB} = \frac{3}{19}M;\quad M_{AC} = \frac{2}{19}M;\quad M_{AD} = \frac{6}{19}M;\quad M_{AE} = \frac{8}{19}M$$

Note that the cranked-in moment M is distributed to the adjacent members in direct proportion to their stiffness. The fraction of the total moment resisted by member i is called the distribution factor and is equal to the ratio of

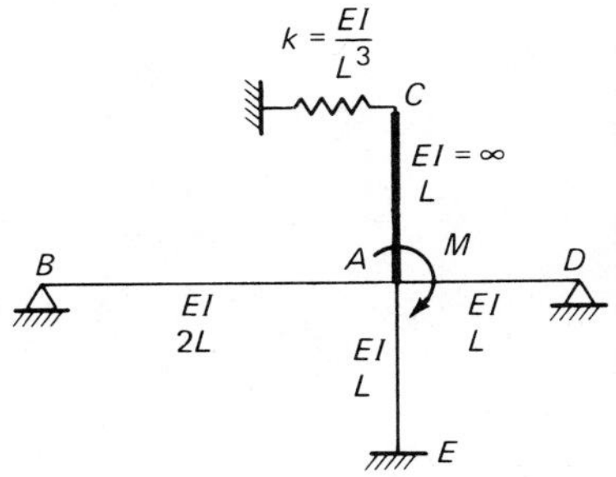

Fig. 25-30 Example 25-17. Displacement method.

the rotational stiffness of member i to the sum of the stiffnesses of all members entering the joint:

$$D.F._i = \frac{k_i}{\sum_i k} \tag{25-24}$$

25.5.2 The Element Stiffness Matrix

As already discussed in Section 25.4.4, the element stiffness matrix $[k]$ contains the stiffness k_{ij}, defined as the forces that must be applied to the structural element at node i in order to ensure a unit displacement at node j, all others zero.

The calculation of the stiffnesses for beam elements by the force method was outlined in Section 25.4.4, and eq. (25-16) gives the flexural stiffness matrix for a planar, prismatic, straight, elastic beam element. If axial deformations are to be considered as well, it is necessary to add nodes at both member ends in the axial direction, as shown in Fig. 25-31; the corresponding stiffness matrix is then given by eq. (25-25):

$$[k] = \begin{array}{c} \\ 1 \\ 2 \\ 3 \\ 4 \\ 5 \\ 6 \end{array} \begin{array}{cccccc} 1 & 2 & 3 & 4 & 5 & 6 \\ \end{array}$$

$$[k] = \begin{bmatrix} \frac{AE}{L} & & & & & \\ 0 & \frac{12EI}{L^3} & & & \text{Sym.} & \\ 0 & \frac{6EI}{L^2} & \frac{4EI}{L} & & & \\ -\frac{AE}{L} & 0 & 0 & \frac{AE}{L} & & \\ 0 & -\frac{12EI}{L^3} & -\frac{6EI}{L^2} & 0 & \frac{12EI}{L^3} & \\ 0 & \frac{6EI}{L^2} & \frac{2EI}{L} & 0 & -\frac{6EI}{L^2} & \frac{4EI}{L} \end{bmatrix} \tag{25-25}$$

(rows and columns numbered 1 to 6)

$$[k] = \begin{bmatrix} \frac{EA}{L} & & & & & & & & & & & \\ & \frac{12EI_z}{L^3} & & & & & & & & \text{Sym.} & & \\ & & \frac{12EI_y}{L^3} & & & & & & & & & \\ & & & \frac{GC}{L} & & & & & & & & \\ & & -\frac{6EI_y}{L^2} & & \frac{4EI_y}{L} & & & & & & & \\ & \frac{6EI_z}{L^2} & & & & \frac{4EI_z}{L} & & & & & & \\ -\frac{EA}{L} & & & & & & \frac{EA}{L} & & & & & \\ & -\frac{12EI_z}{L^3} & & & & -\frac{6EI_z}{L^2} & & \frac{12EI_z}{L^3} & & & & \\ & & -\frac{12EI_y}{L^3} & & \frac{6EI_y}{L^2} & & & & \frac{12EI_y}{L^3} & & & \\ & & & -\frac{GC}{L} & & & & & & \frac{GC}{L} & & \\ & & -\frac{6EI_y}{L^2} & & \frac{2EI_y}{L} & & & & \frac{6EI_y}{L^2} & & \frac{4EI_y}{L} & \\ & \frac{6EI_z}{L^2} & & & & \frac{2EI_z}{L} & & -\frac{6EI_z}{L^2} & & & & \frac{4EI_z}{L} \end{bmatrix} \tag{25-26}$$

(rows and columns numbered 1 to 12)

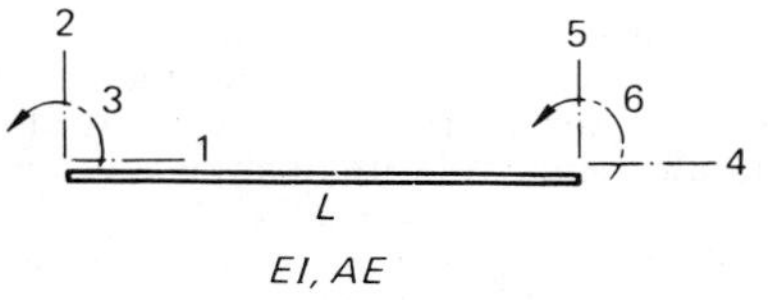

Fig. 25-31 Plane beam element.

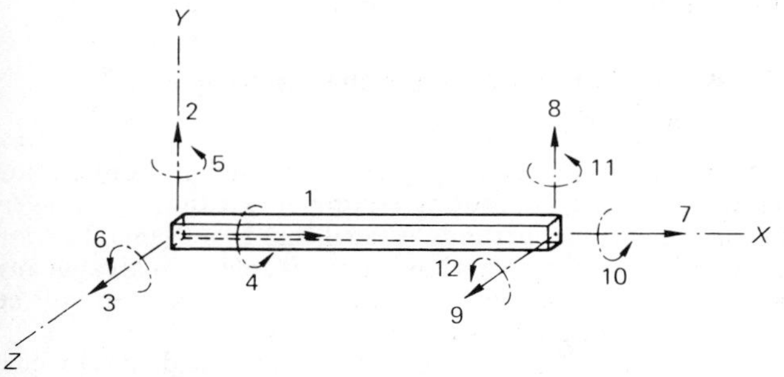

Fig. 25-32 Three-dimensional beam element.

For the more general case of a three-dimensional prismatic beam element, we must consider the 12 nodes shown in Fig. 25-32, that is, axial and transverse translatory nodes along three axes at each member end and rotatory nodes about each of these axes. With these nodes, axial forces and deformations, transverse shears and corresponding translations, twisting, and bending about two axes can be accounted for. With the numbering of Fig. 25-32, elementary beam and torsion theory leads to the stiffness matrix of eq. (25-26).

Another standard element stiffness matrix is the one for the inclined planar truss member with the nodal numbering of Fig. 25-33. Simple calculations lead to the stiffness matrix:

25.5.3 The Structure Stiffness Matrix

The structure stiffness matrix $[K]$ relates the forces and displacements of a structure composed of elements:

$$\{X\} = [K]\ \{\Delta\} \tag{25-28}$$

Each element K_{ij} of this stiffness matrix is defined as the force that must be applied to the complete structure at node i in order to ensure a unit displacement at node j, all others zero.

We recall from Section 25.5.1 that the total, or structure, stiffness at a node is obtained as the sum of the stiffnesses of all members attached to this node. This summing of element stiffness results from expressing the equilibrium equations in terms of the compatible displacements.

For instance, it may be recalled from the discussion of Section 25.5.1 that the rotational structure stiffness at Node 1 of the structure of Fig. 25-34 is obtained by summing the individual member stiffnesses:

$$K_{11} = k_{11}^1 + k_{11}^2 + k_{11}^3 = \left(\frac{4EI}{L}\right)^{①} + \left(\frac{4EI}{L}\right)^{②} + \left(\frac{4EI}{L}\right)^{③}$$

if the members are all prismatic.

Similarly, considering the translational structure stiffness along Node 2, and considering axial and flexural deformations of the prismatic members:

$$k_{22} = k_{22}^1 + k_{22}^2 + k_{22}^3 = \left(\frac{AE}{L}\right)^{①} + \left(\frac{AE}{L}\right)^{②} + \left(\frac{12EI}{L^3}\right)^{③}$$

In general it is essential that each node of the structure be carefully labeled, and that the nodal numbering of each element correspond to that of the structure. The element stiffness matrices for all members are then written, and

$$[k] = \frac{AE}{L}\begin{array}{c} \\ 1 \\ 2 \\ 3 \\ 4 \end{array}\begin{array}{cccc} 1 & 2 & 3 & 4 \\ \left[\begin{array}{c} \cos^2\alpha \\ -\sin\alpha\cos\alpha \\ -\cos^2\alpha \\ +\sin\alpha\cos\alpha \end{array}\right. & \begin{array}{c} \\ \sin^2\alpha \\ +\sin\alpha\cos\alpha \\ -\sin^2\alpha \end{array} & \begin{array}{c} \text{Sym.} \\ \\ \cos^2\alpha \\ -\sin\alpha\cos\alpha \end{array} & \left.\begin{array}{c} \\ \\ \\ \sin^2\alpha \end{array}\right] \end{array} \tag{25-27}$$

A catalogue of stiffness matrices of the type shown in eqs. (25-25) through (25-27) is necessary to perform efficient computer analysis by the displacement method. While computer programs available for production runs (see Section 25.5.7) have matrices such as eq. (25-26) built in, more specialized work may require stiffnesses of other members, such as tapering or curved members, or more general structural elements such as in the finite-element method. Otherwise, it will be necessary to model these elements as an assembly of straight prismatic members to which eq. (25-26) can be applied.

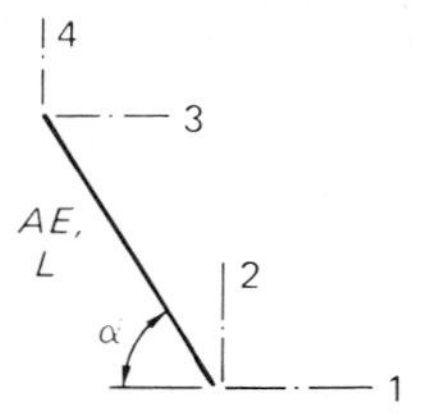

Fig. 25-33 Truss member.

superimposed, or assembled, to form the structure stiffness matrix, as shown in the following example.

EXAMPLE 25-18: Assemble the stiffness matrix for the structure and nodes shown in Fig. 25-35.

SOLUTION: We first write out the element stiffness matrices, member by member, taking care to adhere to the established numbering. These matrices follow eq. (25-26) for the two-force members 1 and 3, and eq. (25-24) for the planar beam element 2; accordingly, using the structure properties given in Fig. 25-35, we obtain:

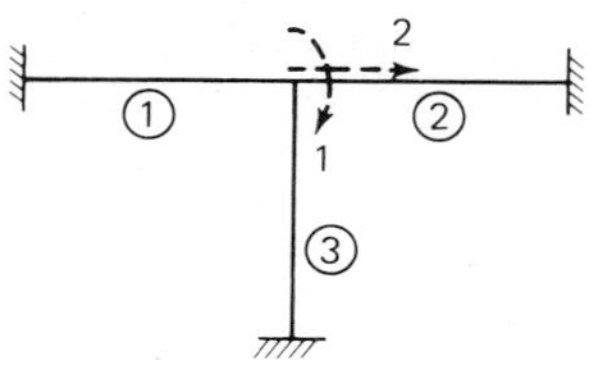

Fig. 25-34 Structure stiffness nodes.

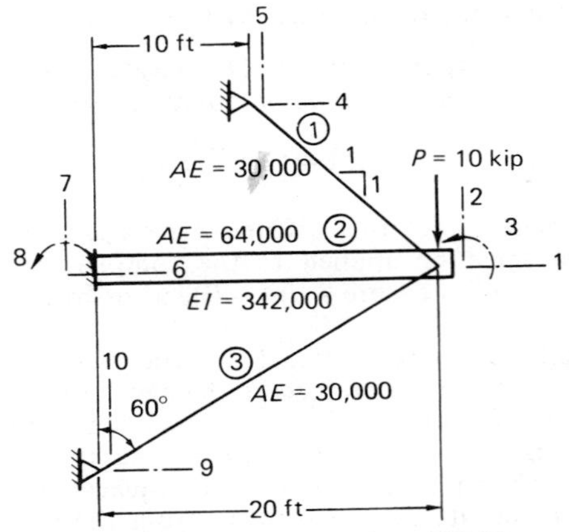

Fig. 25-35 Examples 25-18 and 25-19.

$[k]^{①} =$

	1	2	4	5
1	89		Sym.	
2	-89	89		
4	-89	89	89	
5	+89	-89	-89	89

$[k]^{②} =$

	6	7	8	1	2	3
6	267				Sym.	
7	0	0.298				
8	0	35.8	5,700			
1	-267	0	0	267		
2	0	-0.298	-35.8	0	0.298	
3	0	35.8	2,850	0	-35.8	5,700

$[k]^{③} =$

	1	2	9	10
1	81		Sym.	
2	47	27		
9	-81	-47	81	
10	-47	-27	47	27

Assembling these element stiffnesses into the locations of the 10×10 structure stiffness matrix as indicated by the numbering of the rows and columns, we obtain:

$[K] =$

	1	2	3	4	5	6	7	8	9	10
1	437									
2	-42	116.3								
3	0	-35.8	5,700							
4	-89	89	0	89				Symmetric Matrix		
5	+89	-89	0	-89	89					
6	-267	0	0	0	0	267				
7	0	-0.3	35.8	0	0	0	3			
8	0	-35.8	2,850	0	0	0	35.8	5,700		
9	-81	-47	0	0	0	0	0	0	81	
10	-47	-27	0	0	0	0	0	0	47	27

The physical meaning of these numbers should be clearly understood; element K_{11}, for instance, is about four times the value of K_{22}, indicating that the horizontal force required to stretch the structure a specified amount horizontally is four times as much as a vertical force at the same point causing the same amount of vertical displacement. The contribution of the beam member, k_{22}^{2}, is only a small fraction of the total value of the stiffness K_{22}, indicating that this stiffness is mainly due to the inclined truss members. Considerations of this type can be of great value to the designer in evaluating the effectiveness of a structural configuration.

25.5.4 Matrix Formulation of the Displacement Method

The matrix displacement method, if used in conjunction with appropriate computing equipment, is the most powerful of the various methods of analysis. In its generalization as the finite-element method, it is capable of analyzing any solid body; in this section, however, its application will be restricted to the analysis of framed structures.

We recall that the force method related the displacements $\{\Delta\}$ to the forces $\{X\}$ by means of the structure flexibility matrix $[\delta]$:

$$[\Delta] = [\delta]\,[X]. \tag{25-27}$$

Analogously, the displacement method relates the forces to the displacements by means of the structure stiffness matrix $[K]$:

$$[X] = [K]\,[\Delta]. \tag{25-28}$$

As in the force method, each node will have a force and a corresponding displacement, one of which is known, the other unknown. For a structure with k degrees of kinematic indeterminacy (or degrees of freedom), we collect the k nodes with unknown displacements and known forces (loads) at the top, the remaining f nodes corresponding to the known support displacements and unknown forces (reactions) at the bottom, call the former nodal values α, the latter β, and partition the matrix equation:

$$\begin{matrix}\text{known forces (loads)} \\ \\ \text{unknown forces (reactions)}\end{matrix} \left\{\begin{bmatrix} X_\alpha \\ \hline X_\beta \end{bmatrix}\right\} = \left[\begin{array}{c|c} K_{\alpha\alpha} & K_{\alpha\beta} \\ \hline K_{\beta\alpha} & K_{\beta\beta} \end{array}\right] \left\{\begin{bmatrix} \Delta_\alpha \\ \hline \Delta_\beta \end{bmatrix}\right\} \begin{matrix}\text{unknown displacement at load pts.} \\ \\ \text{known (specified) displacement}\end{matrix} \tag{25-29}$$

The solution is obtained in two steps: first, the first k equations are solved for the unknown displacements:

$$[X_\alpha] = [K_{\alpha\alpha}]\,[\Delta_\alpha] + [K_{\alpha\beta}]\,[\Delta_\beta]$$

from which:

$$[\Delta_\alpha] = [K_{\alpha\alpha}]^{-1}\,[[X_\alpha] - [K_{\alpha\beta}]\,[\Delta_\beta]]. \qquad (25\text{-}30)$$

Note that a $k \times k$ square matrix has to be inverted, corresponding to a solution of k simultaneous equations.

The displacements $[\Delta_\alpha]$ thus found are then substituted in the f remaining equations to solve for the unknown forces:

$$[X_\beta] = [K_{\beta\alpha}]\,[\Delta_\alpha] + [K_{\beta\beta}]\,[\Delta_\beta] \qquad (25\text{-}31)$$

With these basic concepts in mind, we outline the steps necessary for the formulation of the matrix displacement method:

Step 1. Identify the separate elements of the structure. The interconnections between these elements are called joints.

Step 2. At each joint, identify and number the nodes for which forces and corresponding displacements exist. In the most general three-dimensional case, there may be six nodes (three forces and corresponding translations, and three moments and corresponding rotations) at each joint. In other cases, there may be less. Number the nodes with unknown displacements first.

Step 3. Calculate and write the stiffness matrix for each element, adhering to the numbering established in Step 2.

Step 4. Assemble the structure stiffness matrix by superposition of the element stiffness matrices.

Step 5. Write the matrix equation $[X] = [K]\,[\Delta]$, substitute known values of forces and displacements, partition into eqs. (25-30) and (25-31), and solve for the unknown displacements Δ_α and the unknown reactions, X_β.

Step 6. To find the element (internal) forces, use the force-displacement relations for each element i:

$$[X^i] = [k^i]\,[\Delta^i] \qquad (25\text{-}32)$$

This concludes the analysis by the displacement method.

EXAMPLE 25-19: Analyze the structure shown in Fig. 25-35 by the matrix displacement method. Members 1 and 3 are two-force, while member 2 is subject to flexural and axial deformations.

SOLUTION: We identify and number the nodes as shown in Fig. 25-35, remembering to number the nodes for which the displacements are unknown first. The element siffnesses were already written and assembled into the appropriate structure stiffness matrix in Example 25-18. The given values of the applied loads, X_1 to X_3, and the specified support displacements, Δ_4 to Δ_{10} (all zero), are now inserted into the force-displacement eq. (25-29) to yield:

$$\left\{\begin{array}{c} 0 \\ -10 \\ 0 \\ \hline X_4 \\ X_5 \\ X_6 \\ X_7 \\ X_8 \\ X_9 \\ X_{10} \end{array}\right\} = \left[\begin{array}{ccc|ccccccc} 437 & & & & & & & & & \\ -42 & 116.3 & K_{\alpha\alpha} & & & & K_{\alpha\beta} & \text{Sym.} & & \\ 0 & -35.8 & 5{,}700 & & & & & & & \\ \hline -89 & 89 & & 89 & & & & & & \\ 89 & -89 & & -89 & 89 & & & & & \\ -267 & 0 & 0 & & & 267 & & & & \\ 0 & -.3 & 35.8 & & & 0 & .3 & & & \\ 0 & -35.8 & 2{,}850 & & & 0 & 35.8 & 5{,}700 & & \\ -81 & -47 & K_{\beta\alpha} & & & & K_{\beta\beta} & & 81 & \\ -47 & -21 & & & & & & & 47 & 27 \end{array}\right] \left\{\begin{array}{c} \Delta_1 \\ \Delta_2 \\ \Delta_3 \\ \hline 0 \\ 0 \\ 0 \\ 0 \\ 0 \\ 0 \\ 0 \end{array}\right\}$$

Equations (25-30) and (25-31) can now be solved to yield the matrix of unknown deformations and the matrix of unknown reactions:

$$\left\{\begin{array}{c} \Delta_1 \\ \Delta_2 \\ \Delta_3 \end{array}\right\} = \left\{\begin{array}{l} -0.0086 \text{ in.} \\ -0.0893 \text{ in.} \\ -0.0006 \text{ rad.} \end{array}\right\}$$

$$\left\{\begin{array}{c} X_4 \\ X_5 \\ X_6 \\ X_7 \\ X_8 \\ X_9 \\ X_{10} \end{array}\right\} = \left\{\begin{array}{l} -7.1802 \text{ kips} \\ 7.1802 \text{ kips} \\ 2.2904 \text{ kips} \\ 0.0067 \text{ kips} \\ 1.5977 \text{ kip-inches} \\ 4.8898 \text{ kips} \\ 2.8131 \text{ kips} \end{array}\right\}$$

Note that these reactions satisfy the equilibrium conditions. The interior element forces, that is, the end forces and moment of members 1, 2, and 3 along nodes 1, 2, and 3, can now be found by applying eq. (25-32) to each of the members. This concludes the example.

25.5.5 Moment Distribution

The method of moment distribution is a numerical application of the displacement method in which the desired quantities are determined by a method of successive approximation. For this reason it is extremely useful for longhand analysis. Because the procedure lends itself to simple physical interpretation, it can be used for quick approximate solutions, and the thought process involved can guide the designer's judgment.

Before outlining the procedure, we introduce (or review) some preliminary concepts:

(a) *Fixed-end moment:* the member end moments due to loads when the member ends are prevented from rotating, shown in Fig. 25-36(a). The calculation of fixed-end moments was discussed in Section 25.4.3.

(b) *Member end stiffness* (Fig. 25-36): the member end moment necessary to cause unit rotation of the end when all other supports are prevented from movement. Their calculation was discussed in Section 25.4.4, and it was shown there that, in particular, the rotational stiffness of a prismatic member is proportional to the member characteristic EI/L.

(c) *Distribution factor* (Fig. 25-36c): defines the member end moment at a joint as a fraction of the total unbalanced moment applied to the joint. In Section 25.5.1, this factor was shown to be equal to the ratio of the member end stiffness to the sum of stiffnesses of all members at the joint; that is, for member i:

$$\text{D.F.}_i = \frac{k_i}{\sum\limits_i k}$$

the sum of the distribution factors for any joint equals 1.

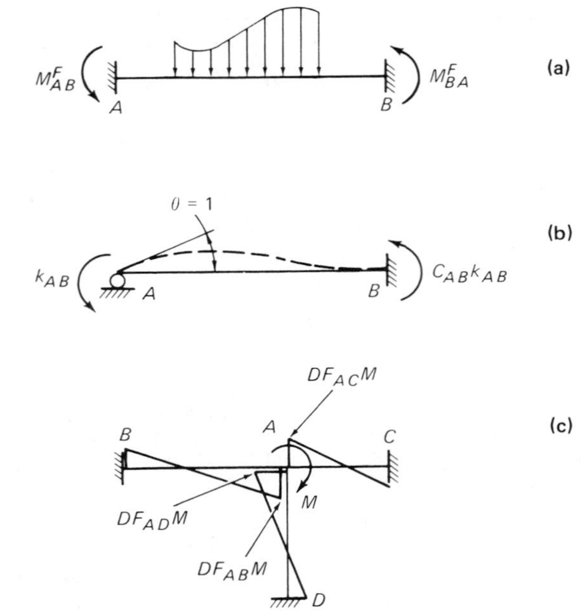

Fig. 25-36 Basic concepts for moment distribution. (a) Fixed-end moment; (b) rotational stiffness and carry-over factor; and (c) distribution factor.

(d) *Carry-over factor* C_{AB} (Fig. 25-36b): the ratio of the moment at the far, fixed end *B* of a member to the cranked-in moment at the near end *A* of the member, that is, the ratio of the rotational end stiffnesses: $C_{AB} = k_{BA}/k_{AB}$. For a prismatic beam, this value is ½; for haunched sections, the stiffnesses can be calculated by the force method, or the references listed in Section 25.4.6.2 can be consulted for values.

(e) *Sign convention.* Moments are positive when acting counter-clockwise on the member or clockwise on the joint.

The moment distribution procedure begins with the moments due to loads on the geometrically determinate structure, that is, all joints prevented from movement. These are the fixed-end moments. The structure next is gradually eased into its final deformed shape by allowing one joint at a time to rotate. Each time a joint is released in this fashion, the unbalanced moment on the joint is distributed to the adjacent members (whose far ends are fixed at this stage) in accordance with their distribution factor; a fraction of the moment thus distributed to the near member end, equal to the carry-over factor, is carried over to the far, fixed member end. This carry-over moment will in turn become an unbalanced moment to be distributed to the adjacent members upon release of that joint. As the joints are successively released, the residual unbalanced moments become smaller and smaller, allowing the total moments, obtained by addition of all incremental moments, to converge to the correct solution. In general, the procedure converges rapidly; furthermore, premature termination of the scheme will result in approximate results that may be useful for various design purposes. The successful application of the calculations depends on an efficient tabular scheme, as shown in the following example; in this scheme, all joints are released simultaneously in the "balancing cycle," following which all residual moments are carried over to the far member ends in the "carry-over cycle." This process is repeated to the required degree of convergence.

EXAMPLE 25-20: Analyze the rigid-jointed structure of Fig. 25-37 by moment distribution.

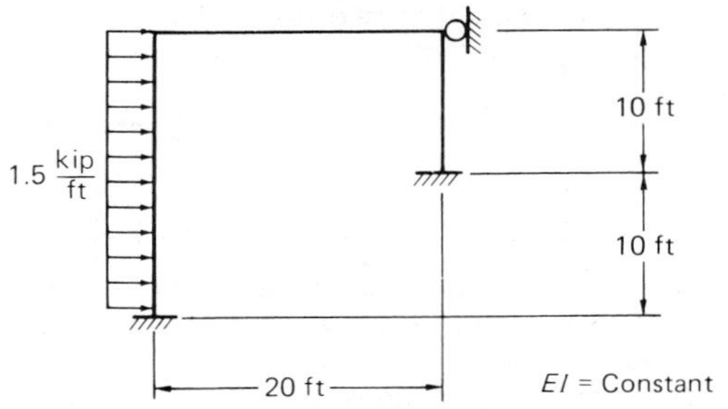

Fig. 25-37 Example 25-20.

SOLUTION: We note that the joints of this frame can only rotate but are constrained against translation. The solution is carried out in the following table (counter-clockwise moments on member end are positive):

$$\text{Fixed-End Moments: } M^F_{AB} = -M^F_{BA} = \frac{wL^2}{12} = 50 \text{ k-ft}$$

Joint	A	B		C		D	
Member	AB	BA	BC	CB	CD	DC	
k	1	1	1	1	2	4	× EI/5
D.F.	0	0.5	0.5	0.33	0.67	0	
FEM.	+50.0	−50.0					k-ft
Bal.	0	+25.0	+25.0				
C.O.	+12.5	0	0	+12.5			
Bal.	0	0	0	−6.3	−6.2		
C.O.			−3.1	0	0	−3.1	
⋮							
Final Mom.	+63	−24	+24	+8	−8	−4	k-ft

All reactions and internal forces can be obtained by statics from these end moments.

When translation of joints is possible, the moment distribution can be carried out in the following steps:

Step 1. Restrain all joints against translation, and carry out the moment distribution as before; this leads to a set of moments M_L.

Step 2. Corresponding to each translatory degree of freedom of the structure, introduce a unit displacement, calculate the resulting fixed-end moments, and distribute them in accordance with the method outlined earlier. This involves a number of separate moment distributions equal to the number of translatory degrees of freedom, each one resulting in moments M_i due to a unit displacement. The moments due to an actual (but as yet unknown) displacement Δ_i are $\Delta_i M_i$.

Step 3. Corresponding to each translatory degree of freedom, introduce a force equilibrium equation, called the shear equation. These equations are written in terms of the member end moments, and will therefore contain the unknown Δ_i. The equations are solved for the Δ_i.

Step 4. The moments M_i of Step 2 are multiplied by the appropriate translations Δ_i, and added to the moments M_L of Step 1 in order to obtain the final moments. The remaining forces can than be found by statics.

EXAMPLE 25-21: Analyze the single-story, single-bay rigid frame of Fig. 25-38(a).

SOLUTION: This structure has one degree of translatory freedom (neglecting axial deformations), the sway of the top joints. We first

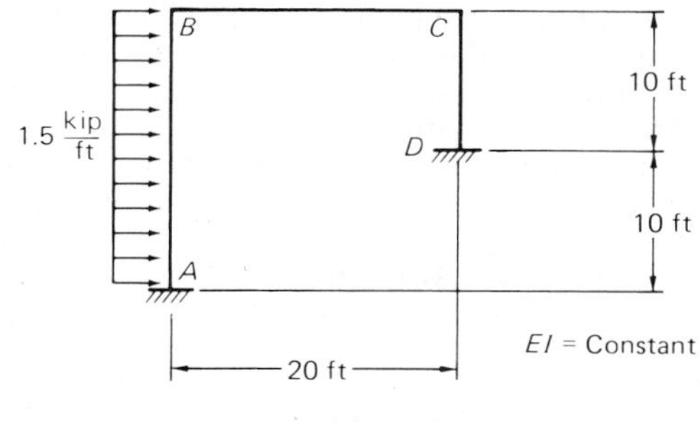

(a)

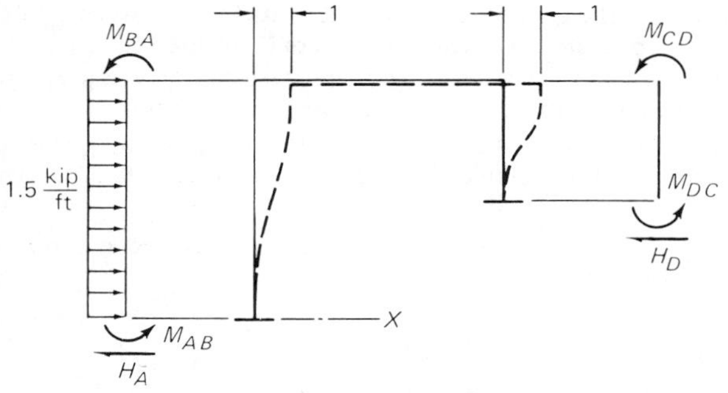

(b)

Fig. 25-38 (a) Example 25-21; and (b) effects of joint translation.

restrain the top against this sway by introducing an imaginary restraint, and perform a moment-distribution. This is identical to the preceding example, Example 25-20, and we will therefore only list here the resulting moments M_L:

Joint	*A*	*B*		*C*		*D*	
Member	*AB*	*BA*	*BC*	*CB*	*CD*	*DC*	
D.F.	0	0.50	0.50	0.33	0.78	0	
M_L	+63	−24	+24	+8	−8	−4	k-ft

We next introduce a unit sway to the right, as shown in Fig. 25-28(b), compute the fixed-end moments of the columns, which we can identify as the translatory stiffness k_{13} and k_{23} of eq. (25-16) ($= 6EI/L^2$), and perform the moment distribution; the actual moments due to the sway Δ, will be Δ times the resulting moments:

FEM	+60	+60			+240	+240	$\times \dfrac{EI}{4000}\Delta$
Bal.	0	−30	−30	−80	−160	0	
C.O.	−15	0	−40	−15	0	−80	
Bal.	0	+20	+20	+5	+10	0	
⋮							
M_{Sway}	+55	+50	−50	−80	+80	+160	$\times \dfrac{EI}{4000}\Delta \equiv \Delta'$

The actual value of the sway is determined by the shear equation; referring to Fig. 25-38(b):

$$\Sigma \overset{+}{\rightarrow} F_x = 0: (1.5\ \text{kip/ft})(20\ \text{ft}) - H_A - H_D = 0$$

where:

$$H_A = \frac{1.5 \times 20}{2} + \frac{1}{20}(M_{AB} + M_{BA})$$

$$= 15.0 + \frac{1}{20}[(63 + 55\,\Delta') + (-24 + 50\,\Delta')]$$

and:

$$H_D = \frac{1}{10}(M_{DC} + M_{CD}) = \frac{1}{10}[(-4 + 160\,\Delta') + (-8 + 80\,\Delta')]$$

so that:

$$\Sigma F_x = 0: \; 30.0 - (16.95 + 5.25\,\Delta') - (-1.20 + 25.00\,\Delta') = 0;$$

$$\Delta' = 0.487$$

We continue the tabular calculation by prorating the sway moments, and summing to determine the final moments.

M_L	+63	−24	+24	+8	−8	−4	
M_{Sway}	+27	+24	−24	−39	+39	+78	
Final Mom.	+90	0	0	−31	+31	+74	k-ft

The remaining forces can be determined by statics.

25.5.6 Design Aids and References

The presentation of the displacement method in its matrix form in Section 25.5.4 can serve only as a short introduction to the approach. For a more thorough treatment, the reader is referred to the book by Weaver and Gere, or that by McGuire and Gallagher, mentioned in Section 25.1.5.

All of these books contain some of the standard stiffness matrices for structural members, as well as discussion of important topics not covered here, such as treatment of loads applied between nodal points, and transformation of matrices from member to structure coordinates, points essential to the effective application of the method.

To incorporate the effect of member haunches in the analysis, the stiffness method permits the idealization of the nonprismatic member by a series of prismatic elements of appropriate properties in order to utilize the standard prismatic members stiffness matrices (eqs. 25-25 and 25-26), but this will lead to a vastly larger number of unknowns with consequent increase of the required computer storage and time. To avoid this, the values listed in the *Handbook of Frame Constants* referred to in Section 25.4.6.2 can be used in the required stiffness matrices.

For the application of any of the matrix methods, appropriate computing equipment, including both hardware (the machinery) and software (the programs), is essential. For routine analysis of framed elastic structures, a number of canned programs are available that contain all the calculations. The engineer supplies structure, member, and load characteristics, and the program yields all desired forces and deformations. For details regarding such a program, see:

Fenves, S. J., et al., *STRESS, User's Manual,* M.I.T. Press, Cambridge, 1964.

Of all longhand methods, the moment-distribution procedure is probably of most value for the analysis of reinforced concrete frames. A number of schemes are available to speed up the convergence, including modification of the stiffness factors to account for actual support conditions, symmetry and antisymmetry considerations, block, and overrelaxation techniques, and the like. Many of these methods are covered in:

Gere, J. M., *Moment Distribution,* Van Nostrand Reinhold, New York, 1963.

This book also contains an excellent bibliography and plotted design aids. The information contained in Portland Cement Association's *Handbook of Frame Constants* referred to in Section 25.4.6.2 is intended particularly for use in moment distribution. If it becomes necessary to evaluate fixed-end moments, stiffnesses, and carry-over factors longhand, the method of column analogy covered in textbooks such as:

Kinney, J. S., *Indeterminate Structural Analysis,* Addison-Wesley, Reading, Mass., 1957.

becomes very convenient. This method can be derived by orthogonalization of the equations of the force method discussed in Section 25.4.

Of particular importance is the application of the moment-distribution method to the analysis of lateral force effects on tall building frames with many degrees of side-sway. In this case, the shear forces associated with these side-sway effects can be solved by a method of successive corrections of unbalanced shears, analogously to the operations performed on the unbalanced moments indicated in Section 25.5.5. This very useful device for longhand lateral force analysis is described in:

Morris, C. T., Discussion of H. Cross, "Analysis of Continuous Frames by Distributing Fixed-End Moments," *Transactions ASCE,* 96 (1), 1932.

The thought process underlying the displacement method can be useful to the designer. The magnitude of relative stiffnesses of different elements can help in visualizing the flow of forces through the structure, with the stiffest elements in general the most effective.

The limiting case of the geometrically determinate (or fixed-ended) structure forms a useful counterpart to the other limiting case of the statically determinate (or pin-ended) structure and can help bracket the correct design forces. The forces on the geometrically determinate structure (the fixed-end forces) can then be relaxed, or distributed, an arbitrary number of times to arrive at improved design values. A two-cycle moment distribution based on this concept is suggested for preliminary design of building frames in:

Portland Cement Association, *Continuity in Concrete Building Frames,* 4th Ed., Chicago, 1959.

A useful concept apparent from the moment distribution process is the attenuation, or damping out, of the effects of local loadings. Thus, the *ACI Code* permits analysis of building frames for gravity loads by fixing the far ends of the columns of the story under consideration, and analyzing one story at a time.

25.6 ANALYSIS FOR MOVING LOADS

25.6.1 Introduction

In many structures a number of different loading conditions are possible. For instance, a traffic load moving across a bridge will cause different forces, depending on its location; similarly, a warehouse structure designed to carry heavy live loading may have this load distributed over different bays or different stories at any time, and it is not always obvious which of these distributions will cause the maximum stresses in a member. In such cases, the analyst must determine that load position or combination which will be critical in its effect on the forces. Influence lines are a useful tool to accomplish this, and several methods of determining these influence lines are covered in Section 25.6.2. In Section 25.6.3 these influence lines are used to determine critical member forces at given sections of the structure. Finally, Section 25.6.4 presents the concept of envelope of forces by which the designer can determine the maximum design forces at all sections of the structure due to critical positioning of the live loads.

25.6.2 Influence Lines

1. Definitions. An "influence function" (also called "Green's function" in mathematics) denotes the effect at one specified point as a function of the position of the cause, of unit value. For instance, the value of a simple beam reaction due to a unit load as a function of the position of this load would be an influence function.

An "influence line" denotes the plot of the influence function. Thus, the influence line of a specified structural effect (such as internal force, reaction, or deflection) is the value of this effect plotted as a function of the position of the unit load that causes it. For forces in statically determinate structures, influence lines for forces can be found by simple statics. In the following, several ways of performing these statical calculations will be shown.

2. Influence Lines by Free Bodies. In this basic approach, the influence line for the desired force is obtained by considering the equilibrium of appropriate free bodies to which the unit load is applied at different discrete points. More conveniently, the unit load is applied at a point defined by the variable x, and the desired force is computed as a function of x, and plotted along the structure.

EXAMPLE 25-22: Determine the influence lines for: (a) the reaction at A; (b) the moment at B: and (c) the shear at B, of the simple beam shown in Fig. 25-39(a).

(a) We consider the entire beam as free body and sum moments about B:

$$\Sigma M_B = 0: \ R_A \cdot L - 1 \cdot x = 0; R_A = \frac{x}{L}$$

The plot of this function is shown in Fig. 25-39(b).

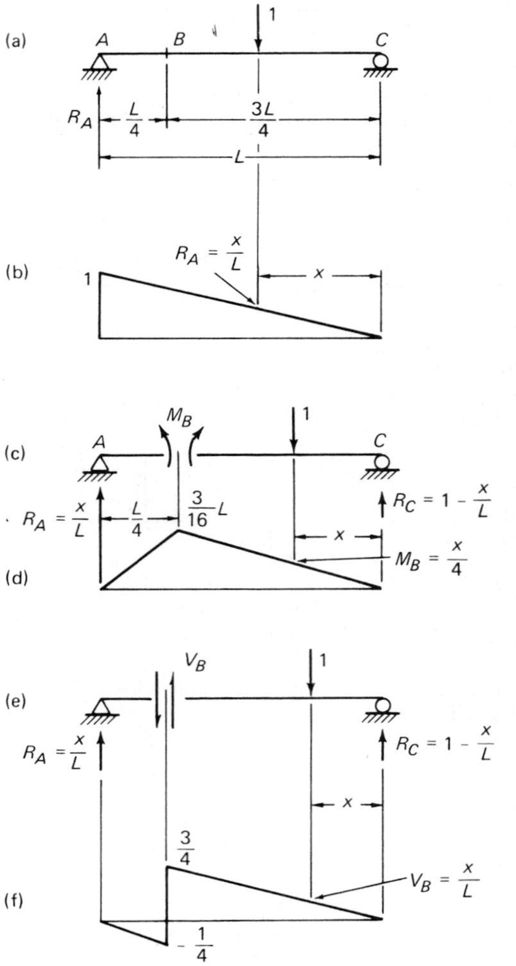

Fig. 25-39 Example 25-22. (a) Beam; (b) influence line for R_A; (c) free bodies for M_b; (d) influence line for M_B; (e) free bodies for V_B; and (f) influence line for V_B.

(b) We consider two free bodies AB and BC, shown in Fig. 25-39(c), such that the desired moment is an external force. The reactions are determined first. It is convenient to take the equilibrium of the unloaded free body; thus, for a unit load between A and B:

ΣM_B (of free body BC) = 0:

$$M_B = \frac{3}{4} L \left(1 - \frac{x}{L}\right); \quad \left(\frac{3}{4} \leqslant x \leqslant L\right)$$

and for the load between B and C:

$$\Sigma M_B \text{ (of free body } AB) = 0: \ M_B = \frac{x}{4}; \quad \left(0 \leqslant x \leqslant \frac{3}{4} L\right)$$

The influence line is shown in Fig. 25-39(d). Note that the mathematical discontinuity at B requires two separate calculations.

(c) We consider again two free bodies AB and BC, as shown in Fig. 25-39(c), and sum vertical forces of the unloaded body.

For $\frac{3}{4} L \leqslant x \leqslant L$:

$$\Sigma F_Y (\text{of } BC) = 0: \quad V_B = \left(\frac{x}{L} - 1\right)$$

For $0 \leqslant x \leqslant \frac{3}{4} L$:

$$\Sigma F_Y (\text{of } AB) = 0: \quad V_B = \frac{x}{L}$$

The influence line is plotted in Fig. 25-39(f); note that it has a unit jump in shear as the unit load crosses the critical section.

3. Alternate Methods for Determination of Influence Lines. The theorem of virtual displacements is particularly convenient for establishing influence lines. From it, the principle of Mueller-Breslau is easily established:

The influence line for any force is generated by the displaced shape of the structure resulting from a virtual unit displacement at the point and in the sense of the desired force.

The influence lines shown in Figs. 25-39 and 25-40 should be envisioned as such displaced shapes for an understanding of this concept.

The force transformation matrix $[b]$ can also serve to determine influence lines; the elements of the ith row (b_{i1}, b_{i2}, . . . , b_{ij}, . . .) represent the influence ordinates for the force i due to a unit load at points 1, 2, . . . , j, ...

4. Influence Lines for Statically Indeterminate Structures. All of the methods that have been outlined are valid for statically indeterminate structures, but the reader should be aware that a much larger amount of work is required to determine such influence lines. Here, two methods will be indicated, one suitable for longhand, the other for computer calculations.

The moment-distribution method, together with the principle of superposition, can be used for the longhand determination of influence lines. The procedure follows these steps:

Step 1. Compute the member end moments due to unit fixed-end moments, taken one at a time. This involves a number of moment-distributions equal to twice the number of loaded members.

Step 2. Compute the fixed-end moments of each member due to a moving unit load. For a straight, prismatic member, for instance, these fixed-end moments are given by the result of Example 25-13.

Step 3. Compute the actual member end moments by multiplying the end moments determined in Step 1 by the appropriate actual fixed-end moment values determined in Step 2.

Step 4. Once the member end moments have been determined in Step 3, any desired force quantity can be found by statics alone.

A well-organized tabular procedure is necessary to carry out this scheme.

The force or displacement methods in their matrix form, with appropriate computer program are very suitable for the determination of statically indeterminate influence lines. The following features are required for this use:

(a) A nodal numbering system that permits application of the unit load at all desired locations.
(b) A program that permits the solution for enough loading conditions to accommodate all load positions considered.

The output of such a computer routine should include all force quantities for which influence lines are desired.

Figure 25-40 shows two typical influence lines for a three-span continuous beam. The reader should visualize the influence lines as the deformed beam shape due to a unit kink corresponding to the moment considered, in the sense of Mueller-Breslau's theorem discussed earlier. The use of this theorem enables the analyst to visualize the shape of even very complicated influence lines without calculations, thus enabling him to anticipate critical loading positions.

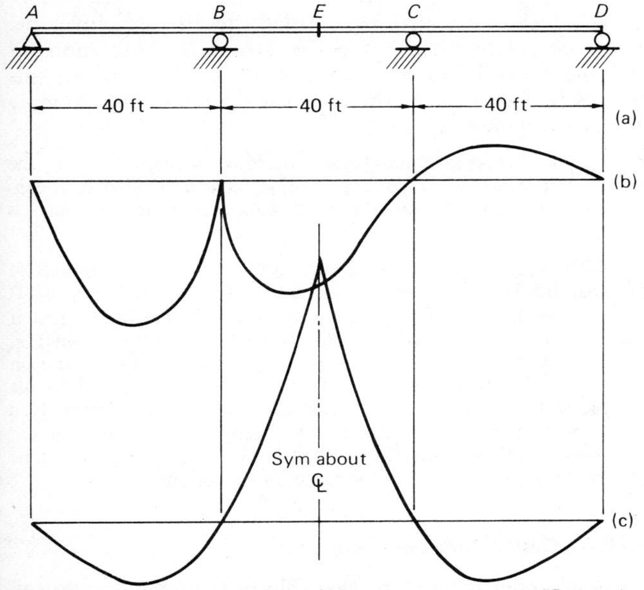

Fig. 25-40 Statically indeterminate influence lines. (a) Beam; (b) influence line for M_B; and (c) influence line for M_E.

25.6.3 Design Forces Due to Moving Loads

1. Use of Influence Lines to Compute Forces Due to Arbitrary Loads. With the influence line for a certain force, F, drawn, we can find the value of this force due to any given loading condition, such as the one shown in Fig. 25-41(a). The influence line for the force whose value is desired has been drawn as in Fig. 25-41(b). The influence ordinate at any point x is denoted by y.

The effect of the concentrated load P_1 on the desired force F equals P_1 times the effect of a unit load, y_1:

$$F = P_1 y_1$$

The resultant of the distributed load $w(s)$ acting over the element of length dx is $w \cdot dx$. The contribution of this

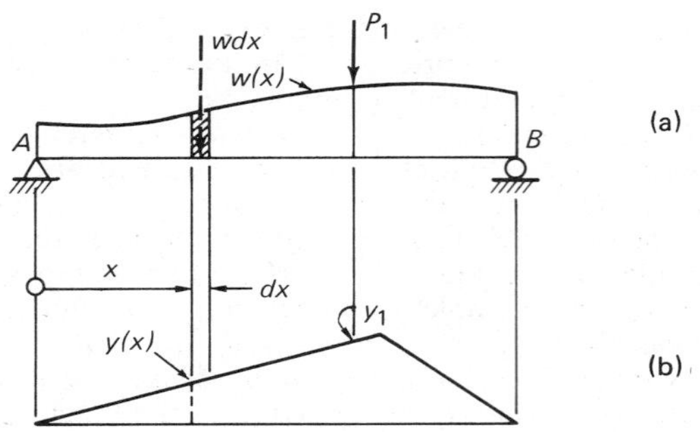

Fig. 25-41 Use of influence lines for determination of force. (a) Loads, and (b) influence line.

elementary load to the force F is $(w \cdot dx) \cdot (y)$, and that of the entire distributed load between two points A and B is:

$$F = \int_A^B w\, y\, dx$$

This integral can always be evaluated, but another, and sometimes very time-saving, method is to replace the variable $w(x)$ by its average value w_{ave}, and take it outside the integral:

$$F = w_{ave} \int_A^B y\, dx.$$

The integral $\int_A^B y\, dx$ can be interpreted geometrically as the area under the influence line between A and B, so that the value of the function due to this distributed load equals the average load intensity multiplied by the area under the portion of the influence line over which it acts. For uniform loads, this is very convenient. For linearly varying loads and influence lines, it can be shown that the average load w_{ave} is the value of the load intensity at the centroid of the area under the influence line.

The total value of the desired force due to the entire load is then:

$$F = \int w\, y\, dx + \Sigma\, P \cdot y \qquad (25\text{-}33)$$

Note that depending on the sign of the influence ordinate y, the contributions of the various forces can be either additive or canceling.

EXAMPLE 25-33: The influence line for shear at the quarter point of the simple beam of Fig. 25-42(a) has been drawn in Fig. 25-42(b); it is required to determine this shear force due to the uniform beam dead load w kip/ft and the concentrated midspan load of value wL.

SOLUTION: The net area under the influence line = $L/32 - 9L/32 = -L/4$. Using eq. (25-33), the shear force is:

$$V_B = -w \cdot \frac{L}{4} - wL \cdot \frac{1}{2} = -\frac{3}{4}\, wL$$

Note that a loading between points B and C only would lead to a larger value of shear, which indicates that partial loading can sometimes be more critical than full loading.

2. Use of Influence Lines to Compute Critical Loading Conditions. With moving or variable loads on a structure, it becomes important to determine that loading condition which causes the critical value of the forces to be used in

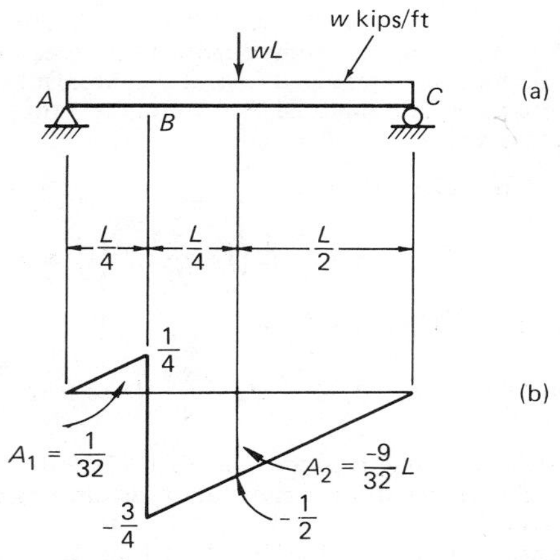

Fig. 25-42 Example 25-23. (a) Beam and load; and (b) influence line for V_B.

design. This is done by use of influence lines. A number of theoretical criteria are available for the calculation of the critical load position.

Here, however, we will only indicate some simple cases in which this determination can be done by common sense or a simple trial-and-error procedure. As a general rule, the heaviest concentrated load should be positioned at the point of maximum influence ordinate, and the remaining loads clustered around this point. Any variable distributed loads should be applied over the portion of the structure with influence ordinate of appropriate sign. In cases of sign reversal of the influence diagram, both positive and negative critical values must be determined.

The process is illustrated in Example 25-24 of the following section.

3. Envelopes of Forces. The outlined method enables the determination of the extreme values of one force, say, the moment at one section. This process now has to be repeated for every section of the structure (in practice, a finite number of sections is considered, such as $^1/_{10}$th points along each span of bridge girder), so that all sections can be designed for the critical positive and negative moment values that can possibly arise under any possible location of the design load. The curve that connects the extreme moment values at all sections is called the moment envelope. For a complete design of bridge structure, both moment and shear envelopes are necessary. Repetitive work of this type is best assigned to the computer, but a simple example will be presented here.

EXAMPLE 25-24: Compute the envelope of moments for the simple span bridge shown in Fig. 25-43(a). The design live load consists of the truck shown in Fig. 25-43(a), and the dead load is 2 kip/ft.

SOLUTION: Because of the symmetry of the bridge, only half of the span needs to be designed. The left half is split into four 10-ft segments as shown in Fig. 25-43(b), and the influence lines for their end points are calculated by any method and plotted as in Fig. 25-43(c). Next, the critical truck positions are determined for each section, as shown in Fig. 25-43(d). The critical live load moments are found by the calculations shown in Table 25-4, and the dead load moments are calculated and superposed, leading to the moment envelope is plotted as shown in Fig. 25-43(e); the span can now be designed to resist the critical moments at all sections.

25.6.4 Design Aids and References

Influence coefficients are available in tabulated or plotted form for many types of structures. For prismatic and

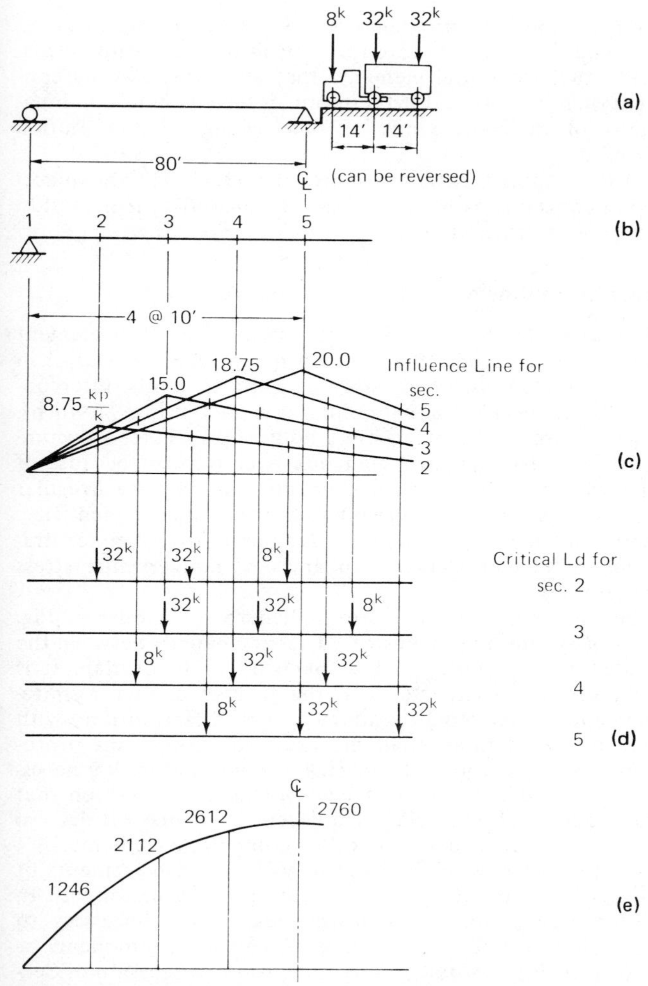

Fig. 25-43 Example 25-24. (a) Structure and load; (b) design sections; (c) influence lines; (d) critical load positions; and (e) moment envelope.

haunched continuous beams, some of the available references are:

Steel Construction Manual, American Institute of Steel Construction, 7th Ed., New York.

Griot, G., *Influence Line Tables*, Ungar, New York, 1954.

Influence lines for some types of arches are contained in the book by Michalos cited in Section 25.4.6.2.

The load systems to be considered for the design of bridges are usually contained in the appropriate building code, such as:

American Association of State Highway and Transportation Officials, *Standard Specifications for Highway Bridges,* Washington, D.C.

This set of specifications calls for consideration of two different load types, an oversize single truck at its critical location, and an equivalent uniform lane load to simulate a continuous lane of traffic. With such loading specified, it is possible to compute moment and shear envelope values for standard bridges of different spans; some of this information is also available in the above mentioned AASHTO *Specifications.*

For the design of building frames, the live load is usually applied spanwise. The critical design forces in continuous beams and frames are not due to loads over the entire structure; rather, the theorem of Mueller-Breslau cited in Section 25.6.2.3 can be used to determine the critical load conditions. Figure 25-44 shows the influence line for a typical support moment sketched quantitatively as the deformed

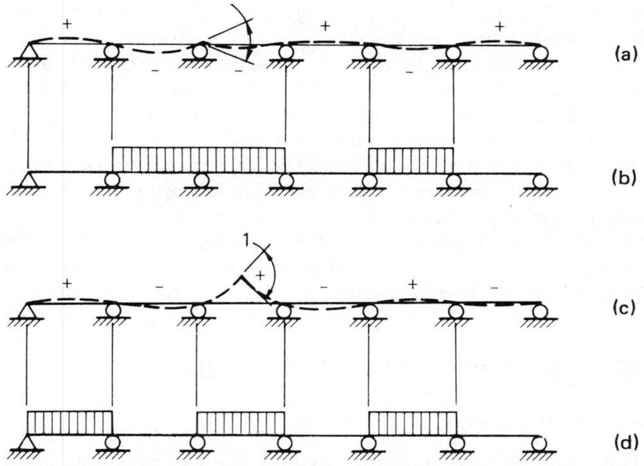

Fig. 25-44 Critical loading conditions. (a) Influence line, and (b) critical loading, for support moment; (c) influence line, and (d) critical loading, for midspan moment.

TABLE 25-4 Determination of Moment Envelope.

Section		2		3		4		5	
Axle	*Axle Ld. P*	*y*	*P · y*	*y*	*P · y*	*y*	*P · y*	*y*	*P · y*
1	8 kip	5.25	42.0	8.00	64.0	10.00	80.0	13.00	104.0
2	32 kip	7.00	224.0	11.50	368.0	18.75	600.0	20.00	640.0
3	32 kip	8.75	280.0	15.00	480.0	13.50	432.0	13.00	416.0
L. L. Total			546.0		912.0		1112.0		1160.0
D. L. Intensity									
	w	*A*	*A · w*	*A*	*A · w*	*A*	*A · w*	*A*	*A · w*
	2 k/ft	350	700.0	600	1200.0	750	1500.0	800	1600.0
D. L. + L. L. Total (in kip-ft)			1246.0		2112.0		2612.0		2760.0

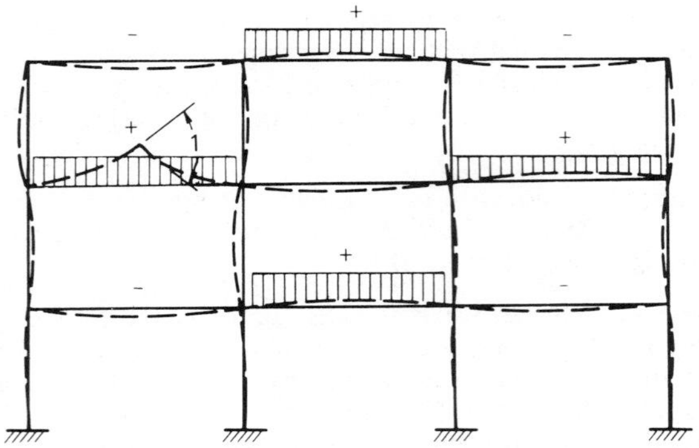

Fig. 25-45 Checkerboard loading on frame.

shape of the structure resulting from an appropriate unit kink. This influence line shows that the critical loading condition, shown in Fig. 25-44(b), consists of loads on the spans with negative influence ordinates adjacent to the support in question, and alternate spans. Similarly, Figs. 25-44(c) and (d) show that the span in question and alternate spans should be loaded for critical moments in the span. The design moments at different points of a continuous beam are thus due to different loading conditions. The moment coefficients specified in the *ACI Code* for the design of concrete beams (subject to certain restrictions) have been computed from a series of such analyses. They should thus always be shown as points on moment envelopes, with other points of these envelopes determined from the statics of the appropriate loading condition.

The same reasoning is also commonly applied to building frames. Figure 25-45 shows the influence line for a positive midspan moment resulting from an appropriate unit kink, and the corresponding "checkerboard" loading of the spans with positive influence ordinates. Similar procedures show that the critical column moments are also due to such checkerboard loadings.

25.7 THE FINITE-ELEMENT METHOD

25.7.1 Introduction

The finite-element method, as presented here and as commonly used, is an extension of the matrix displacement (or stiffness) method as covered in Section 25.5.4, applicable to the analysis of any solid; it has been used for the analysis of all kinds of bodies in their elastic and inelastic range, including ultimate strength and time-dependent behavior. Because of its versatility and ease of application when appropriate computer programs are available, it promises to become the foremost tool for analysis of slabs, walls, shells, and other structural elements that are intractable by conventional methods. Furthermore, it can handle combinations of different kinds of such elements in a unified fashion.

This chapter furnishes a short introduction to the underlying concepts, as well as some references for further study and use of the method.

25.7.2 Outline of the Finite-Element Approach

1. Basic Approach. The first step in a finite-element analysis is to model the body to be analyzed, as shown in Fig. 25-46(a), as an assembly of finite elements, interconnected at specified nodal points, as shown in Fig. 25-46(b). In the case of the plane-stress body shown here, a decomposition into triangular elements is convenient because of the ease with which such elements can simulate irregular boundaries. They also permit variation of the element size, with the smallest elements in regions of stress concentration, and larger elements in areas of more regular stress variation.

As the specified loads are applied to the model of Fig. 25-46(b), the boundaries of adjacent elements between the nodal points would tend to open up or to overlap. It is therefore to be expected that the analysis of such a finite-element model would lead to a less stiff solution, with larger deformations, than the exact solution of the prototype body. In order to avoid such discontinuities, it is necessary to model the element behavior in such a fashion that the common boundaries of adjacent elements will deform together; such elements are called compatible elements.

In the stiffness method of analysis, the displacements of the nodal points (or joints) are the primary unknowns. In order to label these unknowns clearly, it is necessary to introduce a node for each displacement component, as shown in Fig. 25-46(c). According to the discussion of Section 25.5.4, there will also be a force quantity, called nodal force, associated with each of these nodes; at each node, either the displacement or the force will be specified, the other quantity being unknown. In assigning a numbering system to the nodes, those with unknown displacement should be numbered first.

Let us now assume that it is possible to determine structural stiffness values, K_{ij}, defined as in Section 25.5.3; that is, K_{ij} is the force along node i associated with a unit displacement of node j, all other nodal displacements equal to zero. The determination of the structure stiffness matrix $[K]$ consisting of elements K_{ij} will be mentioned in Section 25.7.2.2.

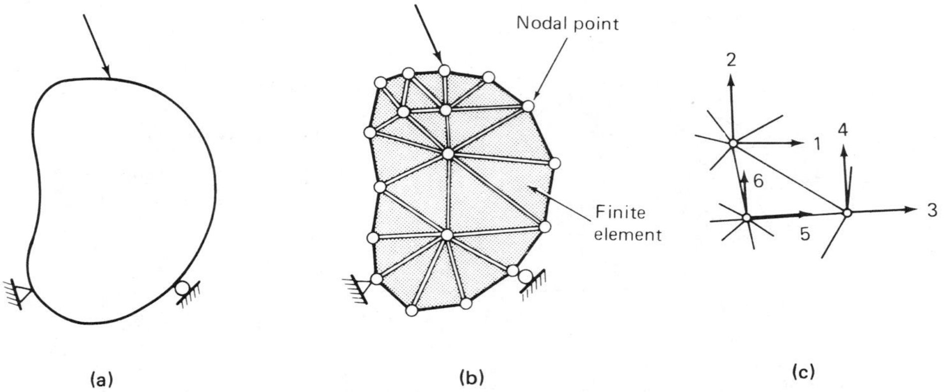

Fig. 25-46 Finite element model. (a) Prototype body; (b) finite element model; and (c) nodal numbering.

The structure stiffness matrix $[K]$ relates the nodal forces and displacements according to eq. (25-29), in which the nodal numbering, as mentioned above, has been so arranged that the unknown displacements, Δ_α, are on top, and the unknown (reactive) nodal foces, X_β, at the bottom:

$$\begin{bmatrix} K_{\alpha\alpha} & K_{\alpha\beta} \\ K_{\beta\alpha} & K_{\beta\beta} \end{bmatrix} \begin{bmatrix} \Delta_\alpha \\ \Delta_\beta \end{bmatrix} = \begin{bmatrix} X_\alpha \\ X_\beta \end{bmatrix} \tag{25-29}$$

The solution for the unknown nodal displacements and the unknown nodal forces takes place according to the two-step scheme outlined in Section 25.5.4, leading to eqs. (25-30) and (25-31):

$$[\Delta_\alpha] = [K_{\alpha\alpha}]^{-1}\,[[X_\alpha] - [K_{\alpha\beta}]\,[\Delta_\beta]] \tag{25-30}$$

$$[X_\beta] = [K_{\beta\alpha}]\,[\Delta_\alpha] + [K_{\beta\beta}]\,[\Delta_\beta] \tag{25-31}$$

The solution of eq. (25-30) involves the inversion of a matrix of order equal to the number of unknown displacements. For realistic problems, the number of unknown displacements can range from the order of tens to the order of thousands. It is apparent that problems of this type can be tackled only with the help of adequate computing equipment.

It should be clear from the above outline that the only difference between the matrix displacement method of framed structures as outlined in Section 25.5.4 and the finite element method is the choice of the stiffness matrix; this enables different types of elements to be incorporated into the analysis in a unified fashion, and contributes greatly to the usefulness of the approach.

2. Further Details. For finite-element analysis of elastic structures, stiffness matrices should be determined in accordance with the laws of the theory of elasticity. Since an exact analysis of elastic elements, such as the triangular plane-stress elements of Fig. 25-46(b), is a forbidding task, an approximate method is used that is completely general in its basic formulation, and for which results will converge to the exact solution with decreasing element size; accordingly, the larger the number of elements, the more exact will be the solution.

This method can be based on the theorem of virtual displacements, eq. (25-4), in which the virtual strains as well as the real stresses are computed by means of an assumed displacement function with unknown coefficients. These coefficients are expressed in terms of the nodal displacements; the strains are obtained by appropriate differentiation of the displacement function, and the stresses are computed from the strains by Hooke's law. The element stiffness matrix can then be calculated by insertion of these values into eq. (25-4a) in its matrix form.

Space does not permit any further details here; suffice it to say that the procedure can be expressed in standard matrix form, and is suitable for computer programming.

With the element stiffness matrices $[k]$ computed, they can be assembled into the structure stiffness matrix $[K]$ according to the procedure of Section 25.5.3. This matrix $[K]$ can now be inserted into eq. (25-29), which in turn is solved for the unknown displacements $[\Delta]$. The desired end result of a finite-element analysis is usually the distribution of stresses throughout the body; they can be obtained from the nodal displacements by manipulation of matrices already available from earlier computations. For further details, the reader is referred to one of the references cited in Section 25.7.4.

25.7.3 Linear Elastic Finite-Element Analysis

1. Example Problem. We demonstrate the effectiveness of the finite-element method by means of the deep cantilever beam shown in Fig. 25-47, loaded by a linearly varying end load. It is well known that conventional beam theory, based on the plane section assumption, is unreliable for such members.

To show the convergence of the solution with decreasing element size, we consider three solutions for different mesh sizes shown in Fig. 25-47(a) through (c). The element used in this case was a rectangular plane-stress element that admits linearly varying normal stresses but constant shear stress. The nodal numbering used is shown in the case of the 2 X 3 mesh, with the free nodes listed first (actually, the computer program used did this ordering automatically).

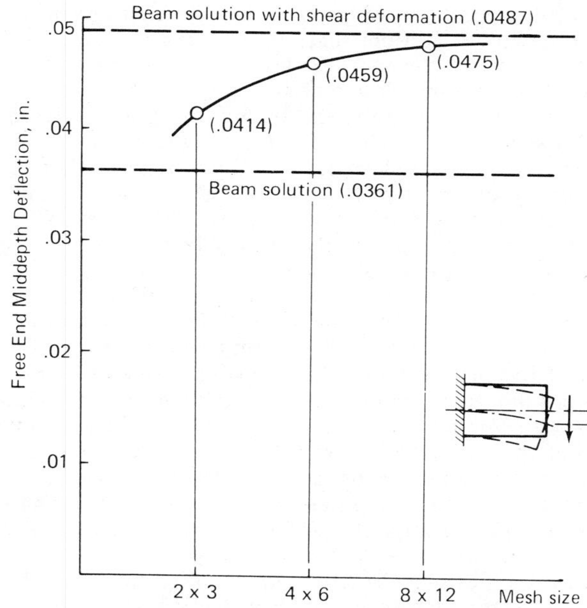

Fig. 25-48 Convergence of finite element solution.

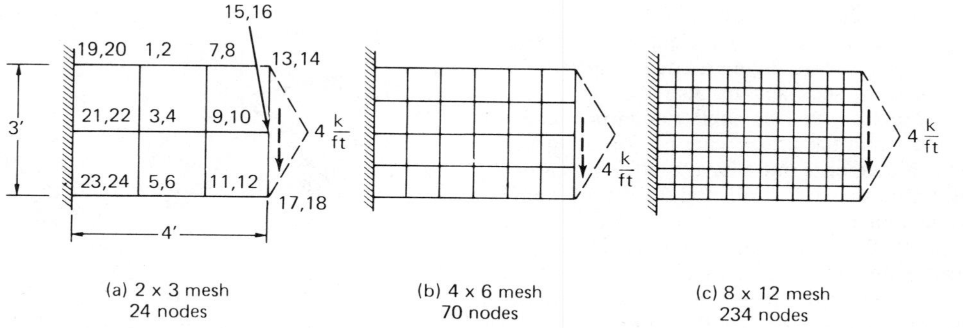

Fig. 25-47 Finite element models of deep cantilever beam. E = 3000 ksi, μ = 0.3, thickness = 12 in.

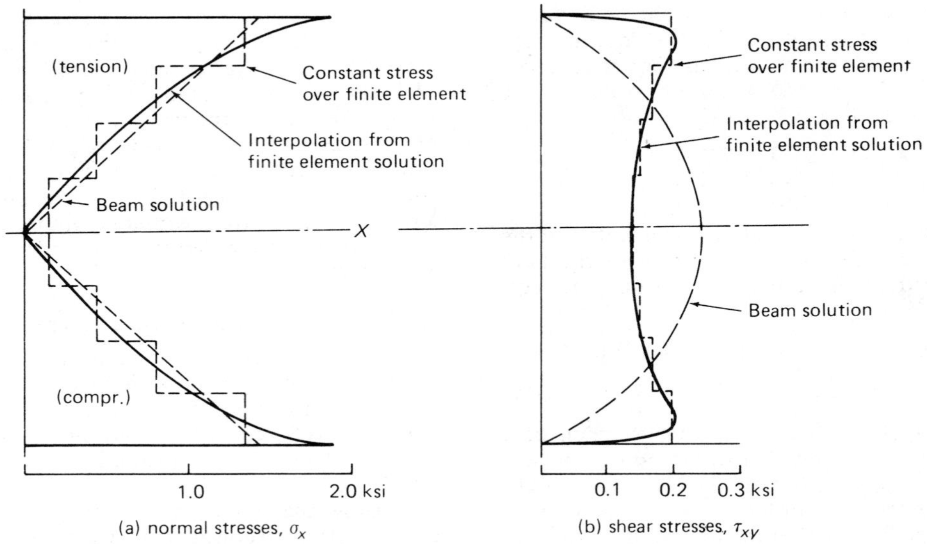

Fig. 25-49 Stress distribution near fixed end of cantilever beam.

Figure 25-48 shows the middepth deflection at the free beam end as a function of the mesh size. We note that the curve becomes asymptotic with increasing number of elements, but even for the very coarse 2 X 3 mesh the answer is sufficiently good for design purposes, though it underestimates the exact deflection (that this is always so in approximate solutions by the stiffness method can be proved by the theorem of minimum potential energy).

Figure 25-49 shows the stress distribution near the fixed end of the beam. It is clear that the strength of materials solution based on the plane section assumption would give a completely incorrect picture of the shear stress variation, and would underestimate the peak flexure stress. With stresses of this type known, the principal stress trajectories can be plotted by the usual means.

2. Different Elements. The accuracy of stiffness matrices can be improved by the introduction of additional nodes; for instance, nodal points could be added at the midpoints of the three sides of the triangular elements of Fig. 25-46, leading to six additional nodes in each element. This in turn would permit the use of more accurate displacement functions that would be better able to reproduce the actual stress variation than, for instance, the constant shear stress element used in the preceding example. Whether the improved accuracy merits the additional computations is a matter to be investigated from case to case.

Different types of structures require different elements. For framed structures, Fig. 25-32 of Section 25.5.2 shows an appropriate nodal system for a general beam element. For plane-stress structures, such as shear panels, or plane-strain structures, such as slices of a tunnel lining, triangular elements as in Fig. 25-46 or rectangular elements as in Fig. 25-47 may be used. More versatile general quadrilateral elements of various degrees of accuracy are also available.

Plate-bending elements as well as shell elements that include membrane and bending action can be found in the listed references. Such elements must have a sufficient number of degrees of freedom so that all nodal displacements and distortions to be considered can be represented. Thus, a general quadrilateral shell element, as shown in Fig. 25-50, should have at least 5 degrees of freedom at each corner in order to represent three translations and two rotations about axes within the plane of the shell, for a total of 20 degrees of freedom for the element.

Three-dimensional solids can be represented by tetrahedrons, or general "brick" elements that are generalizations of the corresponding plane elements. Except for axisymmetric cases, such three-dimensional analyses are so demanding in computer storage and time that they can be undertaken only in extreme cases.

Elements of different types can be readily combined if their nodal systems are compatible. Thus, plates or shells with edge beams, frame and shear wall combinations, and shell-type pressure vessels with ribs and solid end closures can be analyzed in a unified fashion. Thus, the finite-element method has served to bridge the traditional gap between the structural analyst and the continuum mechanicist.

3. References. For the reader with interest but little time, the structural analysis texts by McGuire and Gallagher and by Gerstle, referenced earlier, contain single chapters outlining the basics of finite-element analysis in a simple fashion.

More thorough information is contained in the books by Zienkiewicz and Cook listed in Section 25.2.5. Both of these books present an excellent overview of the state-of-the-art, including discussions of anisotropic media, material and geometrically nonlinear problems, and structural dynamics. Excellent bibliographies are included and make these books a good foundation for those willing to go further.

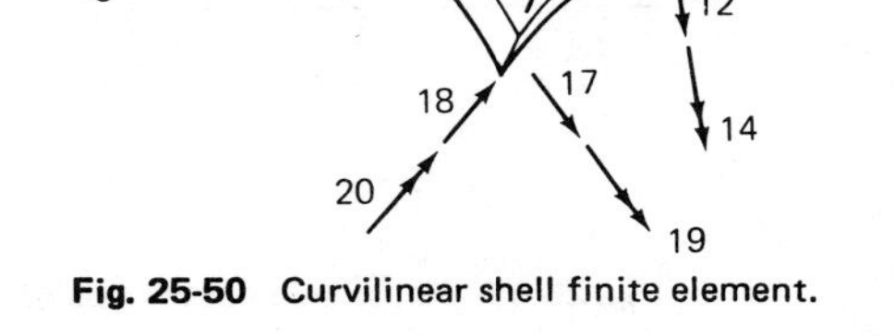

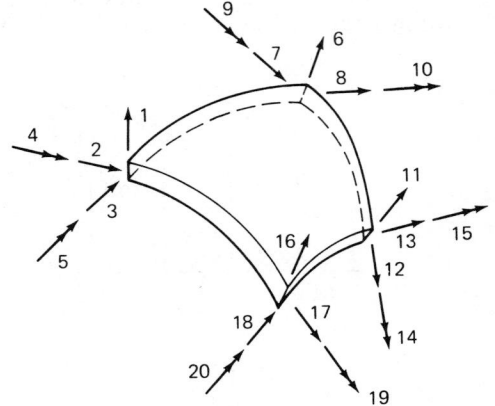

Fig. 25-50 Curvilinear shell finite element.

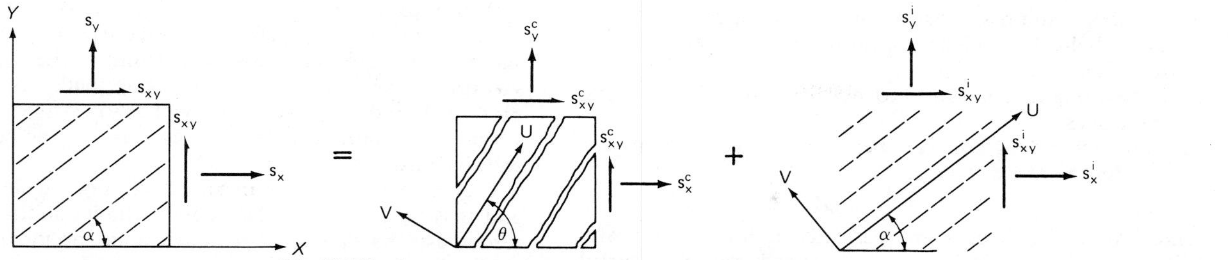

Fig. 25-51 Superpositions of reinforced concrete element.

The literature of the finite-element method has in recent years become so voluminous that any systematic listing is practically impossible. A historic milestone was provided by the following set of proceedings:

"Matrix Methods in Structural Mechanics" (*Proceedings*, Conference held at Wright-Patterson A.F.B., Ohio, 26–28 Oct. 1965), U.S. Dept. of Commerce, Nov. 1966. Available through Clearing House for Federal Scientific and Technical Information.

This volume is typical of many that have been published in recent years.

For professional use, an all-purpose, commercial computer program is probably essential. A number of such systems have been developed for use with finite-element analysis. One that has been found capable and well documented is Program SAP IV, by Bathe, Wilson, and Peterson, University of California, Berkeley. This program can analyze framed, plate, shell, or solid elastic structures in various combinations. It appears that at present such "canned" programs provide the easiest means of access to the power of finite-element analysis for structural engineers in professional practice.

25.7.4 Nonlinear Finite-Element Analysis of Reinforced Concrete Structures

Linearly elastic analysis can give only an approximation of the behavior of concrete structures under working loads because even under such conditions cracking of the plain concrete must be expected. For strength determination, some or all of the following additional aspects of structural behavior must be considered:

1. Nonlinear stress–strain relations of concrete under multiaxial stress states.
2. Actual crushing strength and failure modes in compression, and cracking in tension, of the concrete under multiaxial stresses.

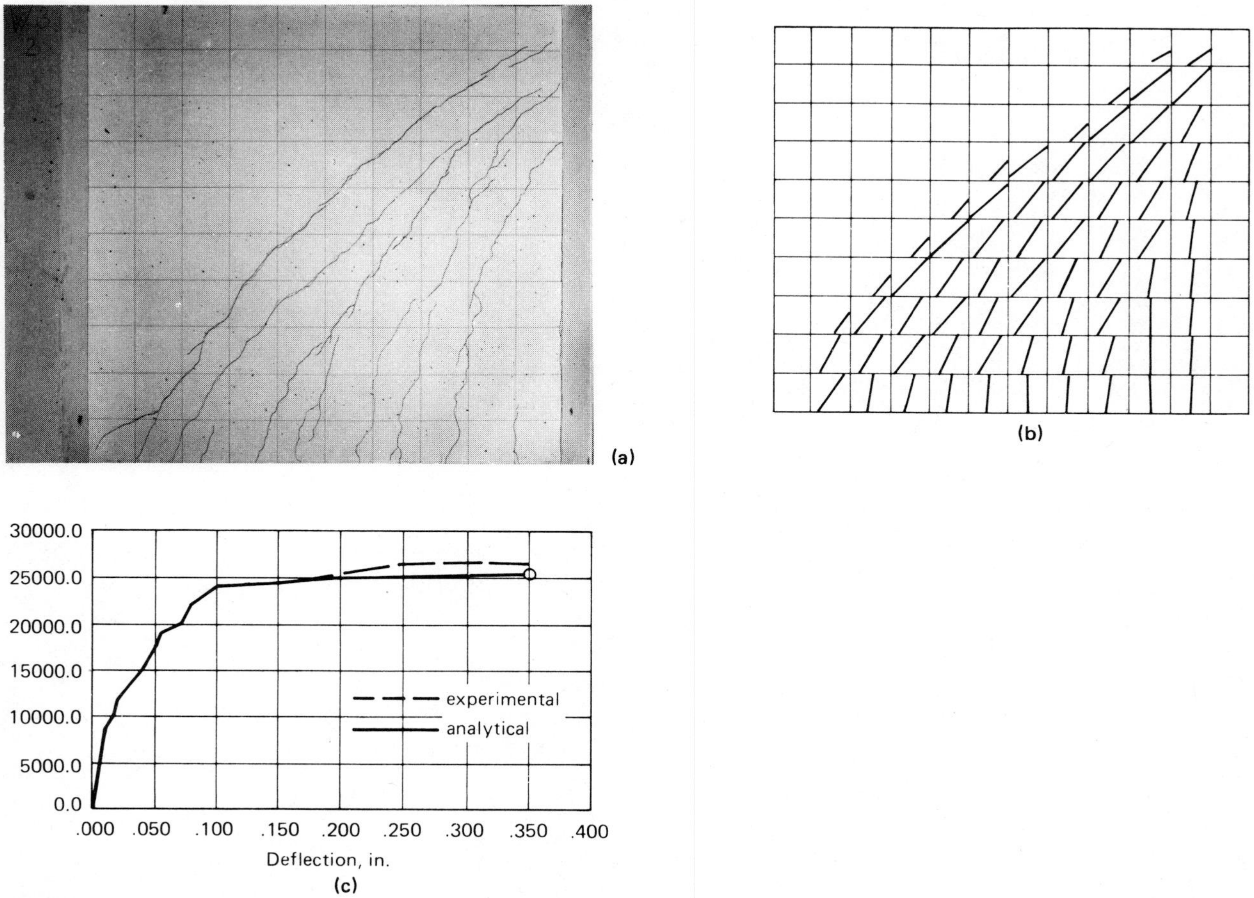

Fig. 25-52 Response of reinforced concrete panel at ultimate stage. (a) Experimental cracking; (b) analysis; and (c) load-deflection curve.

3. Yielding and possibly strain-hardening of the steel.
4. Bond behavior and deterioration of the steel–concrete interface.
5. Behavior and resistance to deformation of the crack interfaces.
6. Time-dependent concrete behavior including shrinkage and temperature effects.

Since all these effects are basically nonlinear, realistic analyses require nonlinear solution procedures. A powerful approach for such cases is a step-by-step method in which the load is applied in sufficiently small increments, all stiffnesses are adjusted during each step to account for current stress levels in each element, and a piecewise linear analysis is performed to determine stress, strain, and displacement increments, which are then added to the accumulated values from preceding load steps. Thus, the previously outlined elastic procedure must be repeated many times, leading to a lengthy sequence of operations demanding in computer capability and probably well beyond the level of routine office practice.

Reinforced concrete can be modeled for analysis purposes by superposition of plain concrete and steel elements, as shown in Fig. 25-51. Stiffnesses of concrete and steel are computed separately, taking into account the current condition of the material, and added to furnish the current reinforced concrete element stiffness, which is then assembled into the structure stiffness matrix according to the steps of Sections 25.5 and 25.7.

Figure 25-52 shows the results of such an analysis of a reinforced concrete panel, which includes the effects of tensile cracking and compressive crushing of the concrete, and yielding of the steel under increasing load. Comparison of predicted and observed behavior shows the detail with which the complete structural response can be predicted by suitable analysis.

This field is currently in an active stage of development, and it is by no means clear under which conditions the different aspects enumerated above must be included in any particular analysis. In fact, experimental evidence on plain concrete behavior, bond-slip, crack properties and propagation, and the like, is incomplete, and attempts at their analytical formulation in form suitable for inclusion in finite-element analysis must therefore be considered tentative at this time. Nevertheless, this approach has been used to good advantage in the analysis of major structures such as nuclear concrete reactor vessels and offshore structures, and its future appears promising.

Reference. An excellent and up-to-date compendium of the current state of the art of analyzing concrete structures is:

Finite-Element Analysis of Reinforced Concrete, ASCE, 1982.

It covers the range from basic analysis through detailed discussion of the different nonlinear aspects enumerated above to examples of application and a listing of available nonlinear analysis programs. A comprehensive bibliography is also included.

26

Computer Applications

STEVEN J. FENVES* and DANIEL REHAK**

26.1 INTRODUCTION

Computer applications are in daily use in essentially every branch of concrete engineering. These applications cover the principal design processes of analysis, proportioning, and detailing, auxiliary activities such as preparation of design document (specification text, bar schedules, drawings, etc.), quantity takeoff, and estimating, and many of the control functions associated with fabrication and construction. Finally, a large portion of analytical research in concrete behavior and concrete structures involves extensive use of computers.

The range of computer applications in concrete engineering is continuously expanding. New programs are being developed for problems whose solutions were inconceivable in the past, either because of the magnitude of the numerical calculations involved (e.g., the exact analysis of large, complex structures) or because of the logical complexity involved (e.g., the direct production of design drawings).

This chapter is intended to serve as an overview of computer applications for an engineer not necessarily familiar with the details of computers and programming. Throughout the chapter, emphasis is placed on the professional and managerial issues involved in computer application selection and program development, rather than on specific computers, programming languages, and applications. The objective throughout is to impress on the reader the professional responsibilities that accrue to the engineer as the computer—or more precisely, a computer program—becomes an integral part of the engineer's professional services.

*University Professor of Civil Engineering, Carnegie-Mellon University, Pittsburgh, Pennsylvania.
**Assistant Professor of Civil Engineering, Carnegie-Mellon University, Pittsburgh, Pennsylvania.

26.2 DIMENSIONS OF COMPUTER USE

Computer applications in concrete engineering can be implemented on a variety of computers, using a wide range of programs and means of man/machine interaction. The purpose of this section is to indicate the range of available tools, the principal criteria for selecting among them, and the opportunities afforded by the various choices.

26.2.1 Range of Computer Hardware

This section presents a brief description of computer hardware, and indicates the range of hardware available.

26.2.1.1 Basic Function and Components

Computers used in engineering practice can be defined as general-purpose stored-program, electronic digital computers. Digital means that, within the computer, numbers are represented by discrete symbols as in a desk calculator, in contrast to analog computers where numbers are represented by continuously varying physical quantities; digital computers can also represent and manipulate symbols other than numbers, such as alphabetic characters or geometric entities. Electronic means that the internal operations are performed by electronic devices, rather than mechanical devices such as counting wheels. Stored-program means that, for each application, the computer is provided with a sequence of steps or instructions, called the *program*, that defines the process of solution. General-purpose means that the computer is not built specifically for one type of application, so that by using different programs it is capable of solving a wide variety of problems.

A computer consists of four functional components.

The *memory unit* or storage device stores two types of information: (1) information that directs the processing, consisting of the instructions comprising the program; and (2) information that is processed, namely, the data pertaining to the particular problem being processed. The smallest unit of information directly accessible in the memory unit is called the *word-length*, measured in bits (binary digits). Depending on the size of the computer, there may be several levels of a *storage hierarchy*, of which the most important are: (1) the primary memory, directly accessible to the control unit; and (2) secondary memory, consisting of magnetic disks or tapes, for bulk storage of voluminous information. Furthermore, many computers provide *virtual memory*, where storage is organized so as to appear to the program *as if* a much larger memory were directly accessible, the information actually being moved or *swapped* between the two as needed.

The *control unit* or *arithmetic control unit* performs the actual operations, fetching instructions and data from memory as needed. In addition, the control unit can perform logical operations such as testing a number for negative sign or nonzero digits and thus make elementary decisions by selecting a path from among alternative sets of instructions.

The memory and control units are collectively referred to as the *central processing unit* (CPU). Their range is discussed in Section 26.2.1.2.

The *input unit* provides the capability for reading information (both instructions and data) from an external device and storing in memory. Similarly, the *output unit* provides for displaying output on an external device. These two units are collectively referred to as *peripherals*; their range is discussed in Section 26.2.1.3.

26.2.1.2 Range of Processors

Central processing units are available in a large range of speed and size. They are conventionally classified as:

- *Microprocessors*, usually 8-bit processors with 1000 to 16,000 words of memory.
- *Desktop* or *personal* computers, usually built around larger microprocessors.
- *Minicomputers*, with 16-bit or 32-bit processors, and up to 500,000 words of memory, operating under a multi-user operating system.
- *Superminicomputers*, with more powerful processors and sophisticated operating systems.
- *Mainframes*, with up to a million words of memory, operating under complex operating systems.
- *Supercomputers* of extremely large scale, frequently incorporating parallel processing capabilities.

26.2.1.3 Range of Peripherals

The early computers used punched cards or punched type as input media for both program and data, and usually produced punched card or printer output. Today, essentially all input is be means of a *keyboard terminal*, either directly as *interactive* input to the program running on the processor or indirectly to a *file* maintained in secondary storage, which can then be used by the program. Additional input capabilities include *graphic input* by means of digitizers or controls on a *graphic terminal*, where the position of the device is automatically translated into x–y coordinates which are transmitted to the program. The input can represent either graphic data (i.e., coordinates of points), or commands displayed on the *menu* of available commands on the terminal screen.

Output may consist of alphanumeric data on a printing or cathode ray tube (CRT) terminal screen, graphic output on the CRT or on a variety of plotters, and printed or plotted output produced on high-speed devices.

A wide range of peripherals is available for providing *secondary storage*, either to augment dynamically the processor's limited primary memory, or to provide long-term storage of programs and data. Secondary storage media range from floppy disks capable of storing from 10,000 to 40,000 characters to "hard" disks, with storage capabilities into 10^7 to 10^8 characters.

A final class of peripherals permits *remote access* or *networking* between processors and remote terminals or other processors. Remote terminals or small computers communicate over telephone lines at rates from 300 bits/sec up to 4800 bits/sec. Computer-to-computer communication usually takes place through dedicated lines or switching services at rates up to 10^6 bits/sec.

26.2.2 Range of Programs

This section reviews the range of computer programs applicable to concrete engineering along three interrelated dimensions: function, scope, and intended use.

26.2.2.1 Range of Program Functions

Program function denotes the fraction of the total design process embodied in a computer program. Four levels of program function can be distinguished.

"One-shot" programs are written to solve a specific, isolated problem occurring in the design process, such as calculating the cable profile for a particular prestressed member. They largely serve as design aids to a single individual.

Stand-alone application programs are intended to perform a specific step in the design process (e.g., the analysis of a frame). The term "stand-alone" means that input data are gathered for the one application only, and that program results are only used for subsequent manual computations or evaluations.

Design systems are typically sequences of individual programs, where the output of one program serves as input for one or more subsequent programs. Results produced by each program in the sequence are retained in secondary storage, and facilities are typically provided for the engineer to review and modify the data before initiating the next program in the sequence.

Integrated systems of programs typically extend design systems in one or more directions, as follows: (1) more general design sequences are provided, including iteration and parallel design processes; (2) applications are extended beyond design to fabrication, erection, production control, etc.; (3) general utility functions are provided for manipulating, accessing, and displaying stored data.

26.2.2.2 Range of Program Scope

Program scope defines the degree of generality of a program, that is, the range of problem types that the program can handle. It is not possible to give a completely general definition of program scope; the classification that follows is intended to be illustrative.

Unique programs are designed to handle a specific situation (e.g., prestressed girders with a specific geometry and cable pattern). Generality is sacrificed for simplicity of development and ease of use.

Special-purpose programs provide more breadth, but are still restricted to a special class of problem types. A plane frame analysis program or a flat slab design program would fall into this category: the programs can handle considerable variation in geometry, loading, etc., but cannot be used for other types of framed structures or floor systems.

General-purpose programs attempt to handle the broadest possible range of problem types. Analysis programs for framed structures or structures composed of finite elements fall into this category. Of course, in a discipline as varied as concrete engineering, there are no truly general-purpose programs—one always encounters situations that violate the built-in assumptions of the most general program. Nevertheless, the objective of general-purpose programs is to have as few limiting assumptions as is practically feasible.

26.2.2.3 Range of Program Usage

The third classification dimension deals with intended use of a program. Again, three levels can be distinguished.

Ad hoc programs are intended to be used by the program developer only, typically for a particular problem at hand. Therefore, no particular attention need to be given to input/output conventions or to documentation. Programs of this type tend to be "one-shot" and unique.

Limited production programs are intended to be used by a limited number of users other than the program developer, typically for in-house production by the organization developing the program. In addition to broader function and scope, such programs must have well-thought-out input facilities and good output capabilities, as well as complete user documentation and worked-out examples.

General production programs are intended for wide usage beyond the developer's organization. In order to accomplish this objective in a cost-effective manner, the developer must aim for breadth in function and scope, as well as provide excellent input/output facilities, complete documentation, and a wealth of tutorial material.

26.2.3 Range of User Interaction

User interaction with a program involves entering input data and receiving input results. This interaction can be implemented in a variety of ways, defined by the interaction rates, media, and levels.

26.2.3.1 Range of Interaction Rates

Interaction rate refers to the time-dimension of interaction.

In *batch programs*, all input is prepared in advance, the program is executed, and all results are output or displayed. Since the *turnaround time* between program submission and the appearance of the results can take hours or even days, this mode is frequently at odds with the normal engineering design process.

By contrast, an *interactive program* communicates directly with the user through a terminal. Input data can be requested by the program as needed, data errors can be recognized and the user requested to modify his input, output results can be displayed immediately, and the user can modify his input accordingly and request another iteration. Thus, problem solving proceeds at the user's convenience.

Time-shared computer systems typically charge a premium for interactive use. Therefore, many *hybrid* modes of interaction have been developed, typically involving an interactive *preprocessor* for the generation and checking of input data and an interactive *postprocessor* for manipulation and display of the results, the actual processing being performed in a batch mode at a reduced rate.

26.2.3.2 Range of Interaction Media

Interaction medium refers to the physical form of communication.

The "standard" medium of interaction with computer programs is through *alphanumeric symbols* representing numbers or words, entered through a keyboard or displayed on a CRT screen or printer.

Increasingly, *graphic* input and especially output are available through CRT screens or plotters. Graphic output vastly increases the usability of program results by providing visualization; it also greatly facilitates the checking and reviewing of input data.

Finally, voice input and output are becomming available through voice interpreters and synthesizers, respectively. It does not appear that this medium will become as popular and useful as the graphic medium.

26.2.3.3 Range of Interaction Levels

Interaction level is a term used to distinguish the extent to which the user's interaction with a program is "computer-like" as opposed to "engineer-like."

At the lowest level, the user enters all input in numerical form, not just for data but for options and control parameters as well. This level is frequently very constraining on the user, especially one who uses a program only infrequently, because of the necessity of either memorizing the code numbers or repeatedly consulting the user's manual.

At the next level, the user is *prompted* by the program by messages such as:

```
ENTER  1. Analysis
       2. Review of data
       3. Modification of data
       4. Quit
```

Control still resides entirely within the program, but the user's understanding and sense of control are considerably improved.

At the highest level, the user communicates his intention by *commands*, directing the program to perform a specific step or function (e.g., ANALYZE, QUIT).

On microprocessors or interactive terminals the last two levels can be effectively combined by displaying on the CRT a *menu* of available options or commands, and allowing the user to select from the menu either by number or by moving the graphic *cursor* on the screen to a specific menu area.

The concept of user commands can be extended to the point where the collection of commands constitutes a *problem-oriented language*, allowing the user to describe a problem to the computer program in essentially the same terminology he would use in instructing a colleague.

26.3 ROLE OF COMPUTERS IN THE DESIGN PROCESS

26.3.1 The Structural Design Process

This brief overview of the design process for a concrete structure serves two purposes. The primary purpose is to provide a context for classifying concrete engineering application areas presented in Section 26.3.2. A secondary purpose is to provide a comparison with the program design process discussed in Section 26.4.

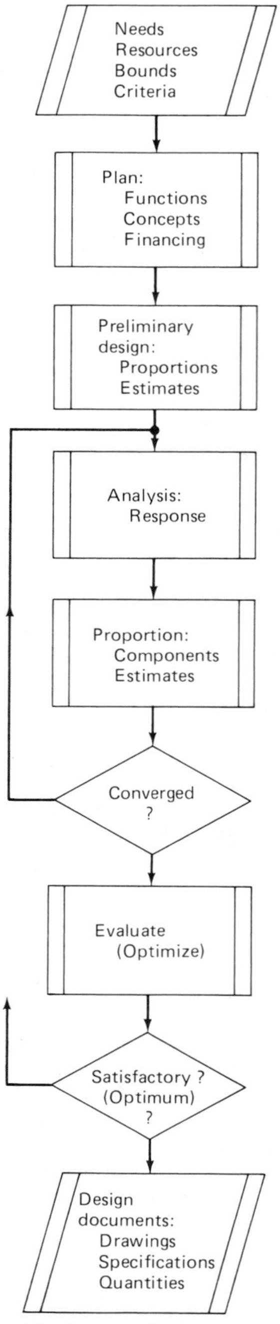

Fig. 26-1 Schematic diagram of structural design process.

26.3.1.1 Stages of Design Process

The process of designing a structure can be typically subdivided into the following stages, illustrated in Fig. 26-1.

Planning. In this first stage, the general concept of the structure is developed, based on the client's requirements, needs, and resources.

Preliminary Design. In this stage, the functional requirements and constraints are used to *synthesize* one or more possible structural concepts.

Detailed Design. The most promising structural concepts are subjected to several iterations of analysis and proportioning. *Analysis* involves determination of the response of the structure, in terms of internal forces, stresses, and deflections, to the loads imposed on it. *Proportioning* means selecting structural components (e.g., beams, columns, slabs, or their respective subcomponents) such that they satisfactorily resist the effects they are subjected to. Significant deviations between component properties assumed in the analysis and those provided by proportioning necessitate a reanalysis. The process iterates until the design is satisfactory or optimal, if an optimality criterion is available. Of course, there may be global iterations involving revisions of the overall structure, or different analytical models, proportioning rules, or both.

Preparation of Design Documents. In this stage, the design drawings and project specifications are developed.

26.3.1.2 Information Flow in the Design Process

The simplified model of the design process serves to illustrate several features that have an important bearing on computer applications.

Information Flow. The stages of the design process are tightly coupled by means of the information or data flowing between the stages. Data generated or decisions made in one stage become inputs to the succeeding stages. Thus, computer usage can be optimized if data can be "captured" when first generated, obviating the need for repeated manual entry to various programs.

Information Explosion. The volume of data representing the structure being designed grows very rapidly as design progresses. It is a major management function to track the growth of this information. Similarly, in a computer-assisted process, considerable attention must be given to the orderly management of the increasing volume of data.

Iteration. As indicated in the previous section, iteration on the analysis-proportion cycle is the norm for a concrete structure of any complexity; subiterations over portions of the design as well as global iterations, involving changes in plan or structural concepts, are also frequently required. This feature again imposes heavy data management requirements on integrated programs.

Consistency. Since information about the structure being designed is contained in different documents, it is imperative to ensure that the various types of data are consistent with each other (i.e., that they all represent the same properties of the same object). As computer programs are more closely integrated, some of the consistency problems can be eliminated by storing the information only once in a *database* and providing access to this information for all programs. However, the resolution of other consistency issues still requires engineering judgment (e.g., the decision whether the section properties provided by proportioning are consistent with those assumed in the analysis).

Constraints. Most of the design stages communicate through *constraints:* planning imposes a budget constraint on preliminary design; preliminary design imposes functional constraints on the behavior of structural components, etc. In a computer-assisted process, it is important to make clear the nature and type of constraints that each program (or segment of a larger system) is intended to satisfy, and the action to be taken if the constraints cannot be satisfied.

26.3.1.3 Global Use of Information

The objective of the structural design process is to produce the complete *specification* of the structure to be built, conventionally represented by the design documents (plans and project specifications). The computer-processible representation of the same specifications has additional uses beyond that of producing the design documents. Increasingly, this representation is shared between design and construction organizations, so that the latter can use the information for estimating, shop drawing preparation, fabrication, and construction scheduling.

26.3.2 Representative Computer Applications

The range of computer applications in concrete engineering is enormous, covering procedures from preliminary conceptual design to construction control on the site. Even though this large range exists, most computer applications intended for production use can be classified into the major categories discussed below. This classification purposely excludes the large number of "one-shot" and research programs intended to solve a specific problem and not designed for routine production use.

26.3.2.1 Preliminary Design

A number of organizations have developed programs for generating preliminary designs, that is, selecting structural configurations. Such programs tend to be of the special-purpose type, since they must incorporate assumptions about specific structural types and approximate rules derived from design experience (e.g., choice of slab type based on spans, selection of preliminary slab thicknesses, etc.). Such programs permit an engineer to quickly evaluate various structural configurations, including generating preliminary cost estimates, and to select one or more configurations for more detailed analysis.

26.3.2.2 Analysis

Structural analysis is defined as "the process of determining the forces and deformations in structures due to specified loads." Analysis is performed on a fully prescribed mathematical model of the structure, that is, one for which all relevant member properties (dimensions, stiffnesses, etc.) are assumed to be known.

Historically, analysis has been the first class of applications to be programmed, and even today it comprises the bulk of production use of computer programs. The reasons for this preponderance are: (1) analysis is a clearly identifiable, time-consuming task in the design process; (2) the increased size and complexity of structures require modeling and analytical techniques impossible to perform by manual methods; and (3) analysis of a model is based on rational principles of mathematics and behavior, and thus not subject to individual interpretation.

Methods. Analysis applications in concrete engineering are largely based on the elastic theory of structures. Furthermore, the majority of applications involve structures idealized as frames, with slabs replaced by equivalent beams and shear walls represented by equivalent columns. However, with the wide availability of programs based on the finite-element method, such gross idealization is no longer necessary, and two- and three-dimensional elements can be accurately modeled and readily incorporated into the overall model to be analyzed.

Analysis programs use, almost without exception, the deformation or stiffness method rather than the force or flexibility method commonly used for hand calculations. The reasons for this are: (1) in the stiffness method the assembly of the governing simultaneous equations requires a minimum of input data, and follows a simple, logical sequence independent of the choice of redundants, with resulting equations that can be solved quite accurately; whereas (2) in the flexibility method, the governing equations depend on the analyst's choice of redundants, their assembly is less prone to automation, the equations require considerably more input or precomputation, and they are prone to large roundoff errors if an improper primary structure is selected.

Assumptions. Many early frame analysis programs were based on the slope-deflection and moment-distribution methods, with their built-in assumption of no axial distortions. More recent programs tend to incorporate a full stiffness formulation, including axial and shearing distortions, for two reasons: (1) in tall, slender buildings, column elongations must be taken into account for realistic modeling; and (2) the efficiency and simplicity of programming the general stiffness method far outweigh the penalty of carrying along the additional degrees of freedom associated with the axial deformations. Most general-purpose analysis programs permit the input of constraints where required by the model (e.g., constraining beam elongations to be zero while still allowing column elongations).

Between plane frame programs based on the slope-deflection method and general-purpose stiffness method programs applicable to two- and three-dimensional frames (and, with finite-element capabilities, to structures comprising plate, shell, and solid elements as well), there is a host of analysis programs for special structural types, such as shear walls, shells of various types, footings, foundation mats, etc.

26.3.2.3 Proportioning

Proportioning is the process that, given the analysis results (forces and deformations) acting on a structural element, selects the size of the element and the required reinforcement, including prestressing, to ensure that the member behaves adequately under both service and ultimate conditions. Frequently, the combination of independent loading patterns to produce the critical design condition on the element being designed is considered part of proportioning, rather than analysis. On the other hand, the selection of actual bar sizes and their placement is frequently considered to be part of detailing, and will be discussed in the following section.

Iteration. If the element to be proportioned is statically determinate (e.g., a simple-span bridge girder), analysis and proportioning can be combined into a direct design procedure. If, however, the element is a component of an indeterminate structure, analysis and proportioning form an iterative loop, where initial sizes or stiffnesses are assumed, an analysis is performed, the elements are reproportioned, and the entire cycle is repeated until satisfactory results are obtained. The degree to which this cycle is automated (i.e., performed by a program without a designer's intervention) depends on both cost limitations and the design office's preferences and mode of operation. Only a limited number of fully automated design-proportioning software systems have been reported so far, due to: (1) the difficulty of specifying and programming criteria for satisfactory performance; (2) the very large number of variables that may be adjusted in order to achieve such performance; and (3) the difficulty of automatically generating initial member properties.

Optimization. Since the number of possible designs satisfying the applicable criteria is theoretically infinite, the iterative design-proportion process may be controlled so as to minimize some desired feature, such as material volume or cost. This is the motivation for the techniques of structural optimization and for programs based on such techniques. Structural optimization is still in its infancy as far as realistic reinforced concrete structures are concerned, but holds a great deal of promise.

Design Assumptions. Proportioning programs differ in several major respects from analysis programs. First, the determination of strength and serviceability of elements, and thus their proportioning, is governed to a great extent by the *Code* provisions. Unfortunately, these provisions are not presented to the profession in a format suitable for direct conversion to computer programs. The major problem is not that the formulas and methods in the *Code* are not optimal for computer formulation, but that the logic of the *Code*, such as interactions between various provisions through the text and the limitations or range of applicability of provisions, is not directly discernible. As a consequence, "programming according to the *Code*" requires a great deal of individual interpretation, and therefore limits the applicability of the program to those users who agree with the interpretations embodied in it.

It is universally accepted that the *Code* by itself is not a complete design guide, and that additional design logic, including assumptions, limitations, shortcuts, search strategies, etc., must be incorporated into any proportioning program. To cite just one example, one seldom, if ever, designs every beam of a frame separately for the most critical combination of load effects acting on it; rather, one chooses a typical, or critical, beam to be proportioned, and then replicates its design for all similar beams. The definition of what is typical, critical, or similar will generally vary widely among designers or design organizations. It is for reasons such as these that proportioning programs lack the generality of analysis programs, and tend to incorporate the assumptions and practices of the originating organization. As a consequence, programs acquired from other organizations must be carefully reviewed to ascertain whether their assumptions and limitations agree with those in use by the acquiring organizations.

26.3.2.4 Detailing

In concrete engineering applications, detailing refers primarily to the determination of the number, size, layout, and location of reinforcement, given the element dimensions and the areas of steel required. Frequently, the above calculations are combined with printing of schedules and bending diagrams.

The comments made on proportioning programs in the previous section apply equally to detailing programs. Thus, while certain details, such as lap and development lengths, hook requirements, etc., are covered by the *Code*, the logic of many other operations has to be developed individually on the basis of the practices of the particular design firm, fabricator, or client.

26.3.2.5 Preparation of Design Documents

The most time-consuming and expensive operation in a design office is the preparation of specifications, schedules, and drawings comprising the final documents leaving the office. These operations were recognized early as fruitful computer applications, and many aspects of final document preparation are today routinely performed by computer programs. At the risk of repetition, it is to be emphasized again that applications in this area contain, of necessity, assumptions and procedures that may not be applicable to every organization. However, in many instances such differences can be handled by differences in the data only, leaving the procedures essentially unchanged.

The majority of applications in the area considered may be classified into one of the three categories discussed below.

Specifications. The preparation of written job specifications has always been a time-consuming task. Most offices maintain a set of standard or *master* specifications, and then produce specifications for specific jobs by incorporating additions, modifications, and deletions. There are several computer programs in use that perform exactly the same function; the master specification is stored in computer-readable form, and the specification writer produces a set of exceptions keyed to lines or paragraphs of the master and produces the text directly on multilith of other medium for direct production. Indexing, cross-referencing, etc., of the text may also be performed.

Drawings. With the steadily decreasing cost and increasing accessibility of plotter devices, the direct production of drawings is becoming more popular. Several problems, however, have so far prevented the wide-scale use of computer graphics. Technical problems, such as the quality and reproducibility of the plots being produced, are being solved with improved hardware and materials. The programming problems, on the other hand, are quite severe, and successful applications tend to cover only: (1) schematic plots, such as erection diagrams, where little detail is to be presented; or (2) highly repetitive structural types, such as footings, beams, bridge-piers, etc., where the high cost of programming all possible cases can be economically justified. Furthermore, there is a great deal of disparity between the available plotting devices and their related systems software (plotting and control routines), so that programs designed for one type of plotter are difficult to transfer to another type of device. It is to be expected that further development and standardization will significantly reduce the cost of computer plotting.

Schedules. Between the two extremes of pure text and to-scale drawings is the large class of tables, schedules, and schematic diagrams often used in contract documents. In concrete engineering, much of the design information, such as beam-and-column schedules, bar-bending schedules, etc., has traditionally been presented in this form.

The presentation of information in such formats is a natural for computer-generated data, and much of the output of proportioning and detailing programs is presented in tabular formats ready for direct inclusion into the design documents. It is common practice in many firms to key such tables to not-to-scale generalized sketches and diagrams.

While many of the computer-produced tables and schedules are designed to duplicate the layout and contents of tables previously produced by hand, there is a definite trend, wherever contractual arrangements permit, to rearrange the contents and format to best fit the total production process. Just to cite one example, some design firms display on their beam, slab, and column schedules, in addition to the traditional data, enough placement information that the bars can be placed in the field without any detailed placement drawings. It is to be expected that this trend will continue in the future, and that much information traditionally displayed on drawings will only be tabulated on appropriate forms for direct inclusion in contract documents.

26.3.2.6 Fabrication and Construction

Contractors, concrete fabricators, and reinforcing bar suppliers are using computer programs for all phases of estimating, project management, and fabrication control. Estimating programs typically combine quantity takeoff and specification of finishes, cutouts, etc., with a "table lookup" of craft composition, labor, material, and equipment rates to generate cost estimates. Reinforcing bar suppliers' programs may combine: (1) generation of and placement shop drawings; (2) fabrication, shipping, and placement control; and (3) generation of information for numerically controlled cutting and bending machines.

At present, contractual arrangements between designers,

owners and contractors frequently require that the contractor perform a manual "quantity takeoff" from the design documents, even when those documents were generated from a computer-processable representation of the design. As discussed in Section 26.3.1.3, this transfer of information can be considerably improved if the contractual arrangements permit a more global use of the data.

26.4 PROGRAM DESIGN PROCESS

The process of designing and implementing a program is a complex task that consists of a number of steps and relies on a diverse set of tools. In many ways, it parallels the structural design process. Program design has been formalized in the last two decades and, owing to its similarity to engineering, has evolved into the discipline of *software engineering*.

The description below is oriented toward general-purpose, production programs (see Section 26.2.2). Such a program has the following attributes:

1. It is designed and written by engineers whose primary responsibility is software development.
2. It is to be used by engineers who are not necessarily familiar with many aspects of computers.
3. It is easy to use, well documented, and responsive to the needs of the users.
4. It uses computer resources efficiently.
5. It is reliable, maintainable, adaptable, and extensible.

The design of programs of more limited scope and intended usage may omit or condense some of the steps outlined below.

26.4.1 Stages of the Program Design Process

The process of designing a program is similar to other engineering design processes and consists of several steps, as shown in Fig. 26-2.

Planning. In the first stage, the general concept for the program is developed, based on requirements, needs, and available resources. The result of the planning is a document called the *functional specification* for the program. The goal of the planning stage is to determine *what* the program will do, not *how* it will be done. Topics to be covered include the features and capabilities of the program, its environment, performance goals, feasibility, resources required, costs, and user/program interactions. This last item is most important, as it provides the basis for evaluation by the users. Interaction with users during the planning stage is vital in producing a program that truly fulfills its needs.

Design. The program design stage differs the most from the traditional engineering process. Program design is still an art, and formal analysis methods and evaluation criteria have not yet evolved. Nevertheless, it consists of preliminary and detailed design stages.

The result of the preliminary design stage is a *preliminary design specification* that contains an overview of the implementation. The design process generally proceeds in a top-down manner. At the highest level, global decisions concerning data structures and program decomposition are made. Designing good data structures is fundamental because processing steps of the program center around use of the program's data. The overall processing is then subdivided into *modules,* each of which is usually designed to perform some related set of functions. Several alternative designs may be considered and an evaluation made to select the most appropriate one. Because of the "art" involved, overall design should only be performed by an experienced "program architect."

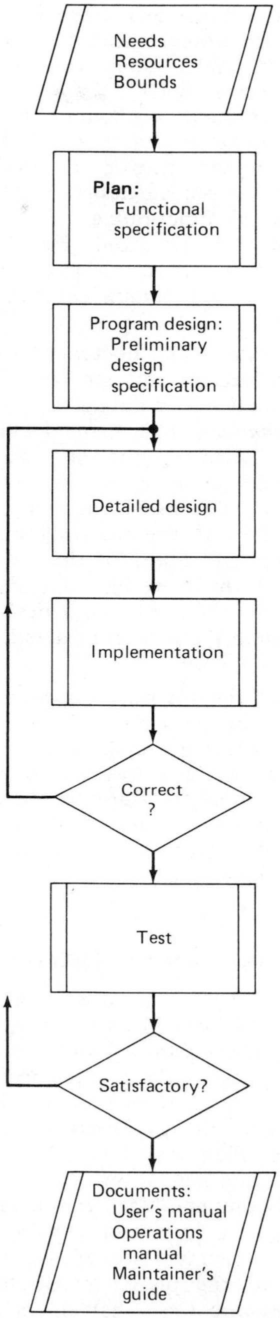

Fig. 26-2 Schematic diagram of program development process.

Once the preliminary design is completed, a detailed design is developed. The objective is a *detailed design specification*, a complete reference document from which the program can be coded. The detailed design process takes the preliminary design and fills in all details. Each of the modules is further subdivided as needed, resulting in a set of well-defined procedures for specific functions. For each component, at each level, the design document should explain the function, its design, the data used, interfaces to other components, error cases, input/output, test cases, decomposition into subparts, implementation details, etc.

Implementation. This stage is equivalent to the construction of a facility. The implementation is based solely on the design specification, and in some work may overlap with the completion of the design.

Testing. Each program is a unique, hand-crafted product, and completely general methods to ensure that it will work without extensive testing do not yet exist. Test procedures can be derived from the design specification and used to verify the program's functioning.

Program Document Preparation. Besides the documents for design and testing, a number of other documents are produced along with the program. These include:

1. *Maintainer's guide.* This document is derived from the design specification and provides all information needed to update and maintain the program.
2. *Operations manual.* This manual contains the information that the operations staff needs to install and execute the program on a particular installation.
3. *User's manual.* The user's manual is a complete description of how to use the program. It explains the basic concepts and limits of the program and gives a description of the program's operation, emphasizing commands and their actions, a description of errors and error handling, and a set of detailed examples.

The objective of the software engineering process is to provide a quality program with the features outlined above. The emphasis on applying sound engineering principles has been found to be beneficial in reducing program development costs. In general, one-third to one-half of the time should be spent on the design, no more than one-third on implementation, and the rest on testing. Emphasis on the earlier stage is fundamental.

26.4.2 Program Design and Development Tools

A variety of tools, applicable to all stages of the program development process, have been created. The objective in all cases is to simplify the process and to reduce its costs.

Design Tools. These tools are designed to provide some formal techniques to assist in program design. One early technique was the use of flow charts to represent processes. These charts have fallen from favor, as they are tedious to produce manually and have a low level of information density. A variety of other graphical techniques that show logic or data representations have been developed. A number of formal structured design methods are available. They can be applied throughout the planning and design stages, and often provide software tools that aid in the development and manipulation of the various design documents.

Programming Languages. Programs are almost exclusively coded in procedural languages such as FORTRAN. The use of machine or assembly languages is to be avoided, owing to their "low level" and machine dependence. They may be used to perform well-defined functions that require high efficiency or to access special system-dependent features not supported in the procedural language. Common procedural languages for engineering include FORTRAN, Pascal, Basic, Algol, PL/I, and APL, with FORTRAN being predominant for engineering applications. Most such languages contain similar features to support computations, input/output, control of flow, and procedures. Selection of a language is based on its applicability to the application, its availability on the host computer, and the availability of associated software (such as graphics or mathematical libraries).

Programming Techniques. Numerous programming and data structuring techniques are applicable to engineering programs. All procedural languages contain basic data structures such as vectors and arrays. From these, representations for tables, matrices, stacks, trees, graphs, queues, etc., can be developed. The data structures, along with associated processing routines for *data abstractions*, may be handled at a conceptual level, with their implementation details hidden.

The second fundamental technique for dealing with data concerns storage management. Data may exceed primary memory capacity, or it may be desired that data be saved on secondary storage independent of a program's execution. Through the use of input/output features in the procedural language, routines that support file and database structures and that provide primary and secondary storage management and data transfer may be developed.

Procedural languages are usually weak in providing features for user interaction, but improved capabilities can often be built upon the provided primitives. A basic strength of all of the procedural languages is the ability to develop individual procedures or subprograms that perform a basic function. This directly supports the program decomposition approach used in development.

Maintenance Tools. Maintaining large programs is complex, particularly if multiple implementations on different hosts exist. The maintenance tools are designed to simplify this effort.

Libraries. Numerous program libraries exist that provide many features and functions useful in engineering programs. The use of such libraries is encouraged, as the code is generally of high quality (developed by professionals), and their use reduced the total development effort. Several libraries are available that contain mathematical procedures. Typical functions of such libraries include matrix computations, equation solving, statistical computations, etc. Graphics libraries provide features to develop graphically oriented user interfaces. A wide diversity of functionality can be found in such libraries. In addition, owing to the specific features of individual graphical devices, some of these libraries are useful with only a limited number of devices. Database management is a nontrivial task. Libraries provide mechanisms to query, update, create, and maintain complex, large databases with ease.

26.4.3 Develop versus Acquire Decision

When the need for a program arises, two basic alternatives exist: develop the program using in-house personnel, or acquire an existing program that fulfills the needs. A variety of tradeoffs enter into such a decision.

Costs. Program development is costly. In the mid-1950s costs were primarily associated with hardware. Now, software development and maintenance cost accounts for over 80% of total computer costs and continues to rise. As a result, in-house development is expensive, and the resulting program may be outdated before it has recovered its development costs. The costs for acquired software are variable. Public domain programs are often sold at the cost of distribution, whereas software vendors must price their product to recover its full costs.

Personnel. In-house development requires personnel for production, maintenance, user training, and user support. Acquired software may eliminate all of these individuals with software vendors providing full maintenance, training, and consulting as part of the purchase or lease agreements, while other programs may require full or partial in-house support.

26.5 INFORMATION SOURCES

26.5.1 Sources of Programs

Computer programs, with various degrees of completeness are reliability, are available from a variety of sources. The most important sources are described below.

Public Domain. Many newer textbooks and technical articles contain source listings of programs for various applications in concrete engineering. It is to be kept in mind, however, that many such programs are essentially appendixes to technical reports, and are therefore to be classified as special-purpose programs, likely to require major modifications before being suitable for production use.

Program Libraries. Most computer manufacturers, some government agencies, and many users' groups maintain program libraries, from which copies of programs and their documentation may be obtained at nominal charges. Most libraries (except for some users' groups discussed below) maintain essentially no control over the scope, reliability, and quantity and quality of documentation of programs accepted into the libraries. Thus, testing and verification of such programs may be quite time-consuming and expensive.

Cooperative Users' Groups. Some users' groups, notably CEPA (Society for Computer Applications in Engineering, Planning and Architecture) and APEC (Automated Procedures for Engineering Consultants), maintain libraries of programs developed or commissioned by group members, with controls of the quality of programs and documentation. Many users belonging to these organizations have found programs in these libraries directly applicable to their needs, or at least as suitable bases for local modifications. Other groups in this category are HEEP (Highway Engineering Exchange Program), consisting of most state highway departments, and IUG (ICES User's Group).

Trade and Service Organizations. Some trade and service organizations and individual manufacturers provide programs under a variety of access or use mechanisms as essentially an extension of their traditional policy of providing handbooks, design charts, etc., to facilitate the use of the products or services represented by the organizations.

Proprietary Sources. There is a growing number of firms, sometimes related to traditional engineering service organizations or to computer service firms, that develop and market civil engineering application programs.

26.5.2 Sources of Other Services

Access to programs is only one type of service needed by an organization contemplating use of or currently using computers for its production work. Some of the other services available are briefly described below.

Installation Planning. The relative merits and disadvantages of an in-house computer installation versus use of outside computer services cannot be discussed here because of the host of considerations involved. For organizations owning or contemplating an in-house computer, there are consulting firms providing installation and workload analyses to determine the optimal use of existing equipment or to recommend equipment to be installed.

Computer Access. Access to computer time on a charge basis is available from a multiplicity of sources, ranging from service bureaus to other users with excess machine capacity. Time-shared and remote-batch access to computers is available on an even broader basis.

Program Access. Computer access on a charge basis usually covers use of compilers and other systems software, as well as access to "free" programs available through the supplier. In addition, many service bureaus and software firms provide access to their proprietary programs, usually on the basis of a software surcharge applied to the basic machine rate. Such arrangements are usually accompanied by some form of performance guarantee, frequently limited to reruns at no charge if software malfunctioning can be established. Many engineering firms, even those with small in-house computing facilities, find this type of program access highly advantageous for infrequent use of large-scale or highly specialized programs.

Problem Solution Services. Organizations with no in-house capability for computer use may avail themselves of complete problem solution services offered by many organizations. Under a typical arrangement, the user supplies a diagram of the structure with all the member properties and loads identified, and receives a listing of analysis results. While many engineering firms have found this approach suitable for initial familiarization or for highly unusual and infrequent problems, they eventually tend to adopt other approaches because of the lack of professional control implicit in such arrangements.

BIBLIOGRAPHY

26-1 *Computer Applications in Concrete Design and Technology*, ACI Special Publication SP-16, American Concrete Institute, Detroit, Mich., 1967.

26-2 *Impact of Computers on the Practice of Structural Engineering in Concrete*, ACI Special Publication SP-33, American Concrete Institute, Detroit, Mich., 1972.

26-3 Beaufait, F. W., Rowan, W. H. Jr., Hoadley, P. G., and Hackett, R. M., *Computer Methods of Structural Analysis*, Prentice-Hall, Englewood Cliffs, N.J., 1970.

26-4 Bell, G. and Newell, A., *Computer Structures: Readings and Examples*, McGraw-Hill, New York, 1971.

26-5 Bell, C. G., Mudge, J. C., and McNamara, J. E., *Computer Engineering*, Digital Press, Bedford, Mass., 1978.

26-6 Bowles, J. E., *Analytical and Computer Methods in Foundation Engineering*, McGraw-Hill, New York, 1975.

26-7 Brooks, F. P., *The Mythical Man-Month*, Addison-Wesley, Reading, Mass., 1979.

26-8 Date, C. J., *An Introduction to Database Systems*, Addison-Wesley, Reading, Mass., 1975.

26-9 Elson, M., *Data Structures*, SRA, Chicago, IL., 1975.

26-10 Fenves, S. J., *Computer Methods in Civil Engineering*, Prentice-Hall, Englewood Cliffs, N.J., 1967.

26-11 Gibson, J. E., *Computing in Structural Engineering*, Applied Science Publishers, London, England, 1976.

26-12 Harrison, H. B., *Computer Methods in Structural Analysis*, Prentice-Hall, Englewood Cliffs, N.J., 1973.

26-13 Jensen, R. W., and Tonies, C. C., *Software Engineering*, Prentice-Hall, Englewood Cliffs, N.J., 1979.

26-14 Kernighan, B. W., and Plauger, P. J., *The Elements of Programming Style*, McGraw-Hill, New York, 1974.

26-15 Kernighan, B. W., and Plauger, P. J., *Software Tools*, Addison-Wesley, Reading, Mass., 1976.

26-16 Knuth, D. E., *The Art of Computer Programming*,

Fundamental Algorithms, Addison-Wesley, Reading, Mass., 1969.

26-17 Knuth, D. E., *The Art of Computer Programming, Seminumerical Algorithms*, Addison-Wesley, Reading, Mass., 1969.

26-18 Knuth, D. E., *The Art of Computer Programming, Sorting and Searching*, Addison-Wesley, Reading, Mass., 1973.

26-19 Kuck, D., *The Structure of Computers and Computations*, Wiley, New York, 1978.

26-20 Litton, E., *Automatic Computational Techniques in Civil and Structural Engineering*, Halsted Press, New York, 1973.

26-21 Newman, W. M., and Sproull, R. F., *Principles of Interactive Computer Graphics*, McGraw-Hill, New York, 1979.

26-22 Rogers, B. F., and Adams, J. A., *Mathematical Elements for Computer Graphics*, McGraw-Hill, New York, 1976.

26-23 Spindel, P. D., *Computer Applications in Civil Engineering*, Van Nostrand Reinhold, New York, 1971.

26-24 Ullman, G. D., *Principles of Database Systems*, Computer Science Press, Potomac, Md., 1978.

26-25 Wright, E. W., *Structural Design by Computer*, Van Nostrand Reinhold, New York, 1976.

27

Structural Plain Concrete

FRITZ KRAMRISCH, D.Eng.*

27.1 GENERAL

27.1.1 Introduction

Many engineers look upon the use of plain concrete as structural material with suspicion and distrust. Prejudice and lack of experience seem to be the main reasons for their attitude, which is otherwise unjustified, because plain concrete has been used as structural material for many years—successfully as well as economically.

Of particular interest in this respect may be its extensive use during World War II for some of the largest industrial building complexes ever constructed. The largest of them was Chrysler's Dodge Machine and Assembly Plant in Chicago, designed by Albert Kahn Associates in Detroit, Michigan. This mainly one-story open-space building had a column spacing of 30 ft × 38 ft and covered an area of roughly 5.5 million sq ft. Other plants of similar design had even larger column spacings, but were somewhat smaller in size. Their arched roof slabs, ground floor slabs, pavements, and footings were all constructed of structural plain concrete. The remainder of the buildings was made of regular reinforced concrete, using only the minimum percentages of reinforcement permitted by the *Code* at that time. This way it was possible to reduce the amount of steel, so essential to the war effort, to the bare minimum of approximately 2.6 psf. In addition, this type of construction allowed these plants to be built at record speed.

It does not, however, require emergency conditions to realize the value of plain concrete as a structural material. Under ordinary conditions, plain concrete can just as well become structurally useful and economically advantageous, and it deserves therefore to be recognized as such.

Indiscriminate use of plain concrete for structural purposes, however, can lead to troublesome and even dangerous experiences. It is, therefore, important to know *where* it can be used and *how* to use it—designwise as well as with regard to the selection of materials and the methods of construction. And even then, in spite of the well-defined limitations and requirements in the *Building Code,* good judgment is sometimes needed on the part of the designer in order not to misuse its limited capabilities.

To clarify this matter, the American Concrete Institute formed a committee (ACI 322) with the purpose of developing maximum allowable stresses as well as minimum requirements for the desgin and construction of plain concrete to be used for structural purposes. The results of the committee's efforts were published in the *ACI Journal* in 1971. Subsequently, this report was in 1972 adopted as a standard under the name *Building Code Requirements for Structural Plain Concrete ACI 322*. From the very beginning, the intent was to consider this standard only as a temporary supplement to the General Building Code and to ACI 318, until the material could be incorporated or appended to the *Building Code Requirements for Reinforced Concrete ACI 318* proper. It took the committee quite some time to do that, partly because of the reluctance of some of its members to grant plain concrete the status of a structural material that could generally be used where applicable—this in spite of the fact that its use would be tightly controlled by limitations and requirements already clearly expressed in the standard.

*Chief Civil Engineer, ret., Albert Kahn Associates, Inc., Architects and Engineers, Detroit, Michigan; lecturer at Arizona State University, Tempe, Arizona.

The final version, as incorporated in the 1983 edition of the *Building Code Requirements for Reinforced Concrete ACI 318,* is in some respects more conservative and more restrictive than the previous standard that it now replaces, so that its economical effectiveness may be somewhat reduced. On the other hand, the fact that structural plain concrete has become a part of the main *Building Code Requirements* should help to make it find wider application in areas where it is best suited. It is also hoped that the successful structural use of plain concrete by a larger section of the engineering profession will increase their reliance on and confidence in the good aspects of this construction method, and will ultimately lead to an increase in the allowable stresses and to a relaxation of some of its restrictions.

In the author's opinion, a few items have not been dealt with sufficiently in the present *Code* presentation and deserve greater research and clarification in the future. The most important ones are discussed below because they will provide the reader with a better insight into the capabilities of this type of construction and with a better understanding of its problems.

(a) Many engineers avoid the use of plain concrete for structural purposes because of a constant fear that an unforeseen crack may develop and affect the structural integrity of the element. No guarantee, of course, can be given, that such a crack will not develop; however, in order to minimize the ill-effects of such a happening, plain concrete is only permitted to be used for structural purposes where the building elements are "continuously" supported by soil or by other structural members capable of providing continuous vertical support. It has to be admitted that a structural plain concrete element of reasonable size and good-quality material, properly mixed, placed, and cured, will not simply crack without good reason, because it requires a considerable force to split a structural element into two. The most probable conditions that are capable of producing such a force are frictional resistances and other restraints against deformations due to excessive shrinkage or temperature change. Isolated spread footings are perhaps the most commonly used structural elements made of plain concrete. Their size is usually smaller than the distance at which ordinary shrinkage cracks in plain concrete occur under consideration of the particular climatic and other conditions; their location within the soil protects them from great temperature variations; and if a rapid loss of mixing water into the subsoil can be prevented, so that no excessive drying shrinkage will occur, the chances of an unforeseen crack become, to say the least, very remote.

(b) A second item that deserves greater attention is the selection of the basis for the permissible tensile stress in flexure. The maximum permissible fiber stress in flexural tension, according to the *Building Code Requirements for Reinforced Concrete ACI 318,* is to be based on the specified compressive strength (f_c') of the concrete, which applies also to all other permissible stresses in the *Code.* Rounded aggregates, usually river gravel, are most commonly used for the making of ordinary concrete because they are economical and produce concretes of great density and high compressive strength. The "compressive" strength of the concrete is, however, rarely utilized in plain concrete design, unless we are dealing with arches or similar structural elements. It is mostly the flexural fiber stress in "tension" that governs the resistance of the cross section. It would, therefore, be much more reasonable to base the maximum permissible fiber stress in tension on the modulus of rupture f_r (obtained from a third-point loading test as commonly specified for pavement work), instead of on the compressive strength f_c'. It is known that concrete mixes using freshly crushed stone aggregates with rough fractured surfaces will develop greater tensile capacity, though smaller compressive strength, than concretes using ordinary rounded aggregates. For structures requiring large amounts of structural plain concrete, the extra cost for the special aggregate and for the flexural test may easily be offset by the permission to use a higher flexural fiber stress. This could easily be done, if the *Code* would permit utilization of the modulus of rupture for such a purpose. It may, therefore, be economically advantageous and structurally sound to permit the use of the modulus of rupture (f_r) as an alternate to the compressive strength (f_c'), in order to arrive at the maximum permissible fiber stress in flexural tension. The decision regarding which one to use could then be left to the engineer. Similar provisions were already included in *Standard ACI 322-72,* but were not incorporated in the *Building Code Requirements for Reinforced Concrete ACI 318-83.* Such matters will have to be decided upon in the future.

(c) Another item of importance concerns the use of small amounts of reinforcing steel, that is, the use of less than the minimum percentage specified in ACI 318 for each particular type of structural element. This approach is used by some engineers in the hope of acquiring greater safety and more freedom from "accidental" cracks. It is more or less accepted knowledge that a fair amount of reinforcing steel (equal to or greater than the minimum percentage specified in ACI 318) will distribute the total expected crack width over a number of hairline cracks so that it becomes practically unnoticeable. (See also under "Crack Control," Chapter 18 (Sanitary Structures) in this *Handbook.*) Real crack-free concrete, however, can only be obtained without the use of any reinforcing steel because one cannot combine a nonshrinking steel with a shrinking concrete without causing internal stresses that may ultimately result in the formation of a crack. Theoretically, an unreinforced plain concrete member, freely supported to adjust its shape to the deformations caused by shrinkage and/or temperature changes, represents the closest approach to a crack-free condition. Insufficient reinforcing may be effective in combating surface cracks, but it will not prevent brittle fracture of a flexural member; that is so because the small amount of steel will not be able to resist the impact due to a sudden stress transfer during the fracture of the plain concrete cross section.

27.1.2 Definitions, Limitations, and Requirements

Plain concrete is a concrete that either is unreinforced or contains less reinforcement than the minimum amount specified in ACI 318 for the same type of member made of reinforced concrete.

Structural is defined by the *Code* as the characteristic of a member to be used for structural purposes. This does not say very much, but it means a member that is designed either strong enough to resist calculable stresses or capable of providing stiffness and/or stability to other load carrying elements or to the whole structure.

No distinction is made by the *Code* with regard to the maximum permissible stresses, depending on whether a structural plain concrete member is cast-in-place or precast.

A basic requirement for a structural plain concrete element is that it has to be continuously supported, either by soil or by other structural members capable of providing continuous vertical support. Exempt from this requirement are only members where arch-action assures compression under all conditions of loading. Plain concrete shall not be used for structural members that have to resist forces due to earthquake, blast, impact, or other dynamic causes.

The above limitations and requirements narrow quite considerably the field where structural elements can be made of plain concrete. However, the few admissible types of members comprise sometimes the bulk, or at least a sizable portion, of all the concrete work required for certain structures. This is one of the reasons why in some cases small advantages, which appear almost insignificant if considered for the single element, may amount to appreciable savings for the whole job.

Summarizing, the most common types of structural plain concrete elements explicitly mentioned and dealt with by the *Building Code Requirements for Structural Plain Concrete ACI 318.1-83* are:

(a) All kinds of footings with the exception of those resting on piles or pier foundations.
(b) Walls that are continuously vertically supported either by footings resting on soil or by structural elements capable of supporting them, regardless of whether the supporting elements are made of steel, reinforced concrete, or any other structural material.
(c) Pedestals, but no columns or struts.

The above types of structural elements are dealt with by the *Code* in detail and will further be discussed in Section 27-3 of this chapter.

There are, however, other kinds of elements commonly made of structural plain concrete that fall within the limitations given above but are not included among the structural elements explicitly mentioned and dealt with by the *Code*.

Some types of structural elements belonging to this category are:

(d) Curved members in the form of slabs or beams, where arch action assures compression under all conditions of loading.
(e) Slabs on grade, whether used for first-floor slabs in buildings or for pavements, parking areas, or storage facilities. For additional information see "Concrete Floors on Grade," EB075, Portland Cement Assoc., 1978.
(f) Gravity walls, abutments, dead-men, various kinds of hydraulic structures, and the like. For additional information see "Small Concrete Gravity Retaining Walls," IS222, Portland Cement Assoc., 1983.
(g) Pier foundations (caissons) and the fill material for cast-in-place concrete piles. (The allowable stresses for these kinds of structural elements are usually governed by the General Building Code applicable at the location fo the construction site.) For further information see *Suggested Design and Construction Procedures for Pier Foundations, ACI 336.3R-72.*

27.1.3 Quality of Material and Joints

Concrete to be used for structural plain concrete elements shall have a minimum specified compressive strength of 2500 psi. Since the structural integrity of a plain concrete element is primarily dependent on its crack-free entity, this value was selected high enough to ensure that the material used would have sufficient homogeneity and a reliable strength quality. On the other hand, this strength value was selected low enough to make plain concrete useful for residential, commercial, and industrial purposes, economical for mass construction, and not too sensitive to shrinkage stresses.

In connection with the design of structural plain concrete elements, the provision and location of joints, in general, deserves special attention. Great care shall be exercised in their layout, which shall be decided upon during the design stage, and be strictly adhered to during construction. This is of particular importance because joints are in this type of construction equivalent to exterior edges. Every element becomes, therefore, within the confines of its exterior edges and/or joints, an independent, or as the *Code* says, a "flexurally discontinuous" unit. This is primarily due to the fact that no tension force can be transmitted across a joint without the use of steel reinforcing running perpendicular to it. This eliminates also the common "construction joint" used in ordinary reinforced concrete construction to interrupt the casting of concrete, if so desired, in areas of minimum shear.

There is, then, in the field of structural plain concrete practically only one type of joint to contend with; the *Code* calls it a "control joint" and defines it as a joint capable of separating the entire construction into individual, flexurally discontinuous units.

Structurally speaking, such joints can be located only (a) where axial or flexural tensile stresses do not exist, (b) where shear is either nonexistent or so small that it can be transmitted through a key or a greased dowel, or (c) where compression is always present to maintain bearing between the contact surfaces of adjacent elements, as in a horizontal wall joint or in a joint perpendicular to arch action. The *Code* provides no information on the maximum spacing of such control joints. It only makes reference to the various influences affecting their spacing, such as structural purpose, climate, exposure, quality of concrete, selection of materials, mixing, placing, and curing, as well as the various degrees of restraint an element may be subjected to. Similarly, the *Code* provides no information regarding the maximum size of an indiviual element; it is, however, rather obvious that the maximum individual sizes are governed by similar considerations.

As far as isolated spread footings, pedestals, and similar individual units are concerned, there will, for most ordinary cases, be no need for a joint because the elements will be smaller than indicated by an expected crack spacing. However, in the case of strip footings or other kinds of combined footings, extending over a greater length or area, the designer has to decide during the design stage where joints ought to be provided. This is important because the joints have to be structurally feasible and otherwise compatible with the expected performance of the individual element, as well as with that of the entire structure. Where such behavior cannot be attained, reinforcement has to be provided in accordance with the requirements of ACI 318.

In order to obtain a structural plain concrete element that will behave as a single independent unit, it must be constructed in one continuous casting. This means that where larger units are involved, the casting of the concrete can only be interrupted at predetermined joint locations. For this reason, no casting operation of a particular element shall be started unless a sufficient amount of concrete is available to complete the casting of the entire element.

The above considerations and construction procedures apply in principle to all structural plain concrete elements, whether included in the *ACI 318.1 Code* or not. A partial exception to these rules occurs only in connection with slabs on grade and similar structures, where a differentiation between full-depth separation joints and partial control joints is usually made. With respect to design, both kinds of joints divide the entire plain concrete structure into several independent, flexurally discontinuous units; however, in the first case a full-depth bulkhead defines the end of a pour and establishes a complete separation from the adjacent unit, whereas in the second case the joint is obtained by pressing a prefabricated sheet metal, plastic, or rubber-like insert into the just cast "green" concrete. Another way to achieve a partial control joint is by saw-cutting and tooling the setting concrete at the earliest possible time, just when it becomes capable of supporting the workman with the cutting machine. Such "control joints" should reduce the original concrete thickness by at least 25%, preferably by more, to initiate and promote the formation of a crack right beneath the insert. The roughness of the two cracked joint surfaces is expected to provide enough aggregate interlock to maintain alignment between the adjacent units. Greased, sliding steel dowels may also be used for alignment or where a more positive shear transfer is desired.

Some designers are of the opinion that plain concrete elements can be kept more crack-free, at least with respect to appearance, if small amounts of reinforcement are embedded in them; such reinforcing is usually placed close to the exposed surface. This practice was discussed in Section 27.1.1(c) of this chapter. Any such reinforcement must be in amounts smaller than the minimum specified in ACI 318 for the particular type of member. Regardless of whether it is in the form of wire mesh, fabric, or bars, it shall be stopped at least 3 in. away from the joint, in order not to interfere with its performance.

27.2 DESIGN

27.2.1 Introduction

Structural plain concrete building elements are intended to act as monolithic units made of a homogenous material. Consequently, they have not only to be constructed but also to be designed this way. Since they are in principle unreinforced, the most appropriate design approach is that of a straight-line stress and strain distribution, extending over the entire depth of the cross section. This assumption is made possible by the fact that the actual stresses are relatively small and well within the so-called elastic regions of the concrete material. Such an approach is comparable to that commonly used by the working stress or service load design method, and the maximum design stresses obtained by such an approach are equal to the stresses that we ordinarily label "allowable."

Presently, the design of ordinary reinforced concrete elements is primarily based on the strength design method. In this design method the limiting stress condition is the ultimate strength of the section, in which case the extreme fiber stresses of the concrete and of the steel reinforcement are assumed to go well beyond their elastic limits. Such a stress distribution is unthinkable for a plain concrete section because there is no means available to produce the necessary tensile resistance.

From a practical point of view it is desirable to use for structural plain concrete the same load factors as for reinforced concrete, in order not to have to change the loadings when proceeding from one material to the other. Similarly, it is desirable to arrive at a new set of limiting design stresses, called "permissible" stresses, that are in principle fictitiously increased allowable stresses. To adopt an approach comparable to that used for ordinary reinforced concrete, a strength reduction factor ϕ is also used in their evaluation. The forces, moments, and resulting stresses used in the design of a structural plain concrete member do not represent, therefore, ultimate conditions, as in ordinary reinforced concrete design, but are more comparable to multiples of calculated stress conditions obtained by a working stress design approach.

Summarizing, the design of a structural plain concrete element can be performed in two steps:

1. The service loads (dead and live) are multiplied by the same load factors used in ordinary reinforced concrete design, as prescribed in section 9 of ACI 318.
2. The resulting factored axial forces, shears, and flexural moments shall be equal to or smaller than the nominal strength capacities of the elements reduced by the strength reduction factor ϕ. The nominal strength capacities are found as the product of the area or section modulus times the applicable permissible stresses.

For reasons of simplicity a uniform strength reduction factor of $\phi = 0.65$ has been assumed to apply to *all* types of stresses in connection with structural plain concrete.

In all stress calculations involving structural plain concrete, the entire cross section shall be considered to participate in the resistance. Only where concrete is cast against soil, as in the case of footings, the overall thickness shall be reduced by 2 in. to create an allowance for unevenness of the base and contamination of the lower concrete through contact with the subsoil.

27.2.2 Flexure

The straight-line stress distribution is apparent from the equations shown in Fig. 27-1, which indicates the flexural stress relations for a T-shaped cross section, just in order to be general. In reality, the cross sections will in most cases be rectangular or consist of 12-in.-wide strips of slab sections. In this case, that is, for sections that are symmetrical about the x–x axis, $S_c = S_t = S$, and tension will always govern the design: $S = 12{,}000 M_u^{(\mathrm{kft})}/f_t^{(\mathrm{psi})}$.

27.2.3 Shear

Shear stress will seldom govern the design of a structural plain concrete section. However, where beam action is involved, and where the designer suspects it to be a significant magnitude, it shall be investigated, and the maximum shear stress shall be evaluated from the basic shear equation for a homogenous section: $v_u = V_u Q/Ib$.

The maximum shear stress due to one-way action shall be calculated at distance h from the face of the support, and sections located at a closer distance may be designed for the same shear force.

The maximum shear stress in a homogenous section occurs at the neutral axis where Q/I is a maximum. For rectangular sections, the maximum shear stress v_u is equal to $\frac{3}{2}(V_u/bh)$, and its distribution over the cross section is parabolic, with zero values at the extreme top and bottom fibers.

The maximum permissible stress for one-way shear or beam shear action is $v_{u\,\max}^{(\mathrm{psi})} = 2\phi\sqrt{f_c'}$.

Where a concentrated load, like a pier or column, rests on a slab or mat, it is resisted by two-way shear action, also called perimeter or punching shear. In this case the maximum shear stress shall be calculated along a critical section placed perpendicular to the slab or footing and located at a distance not less than $h/2$ from the concentrated load or

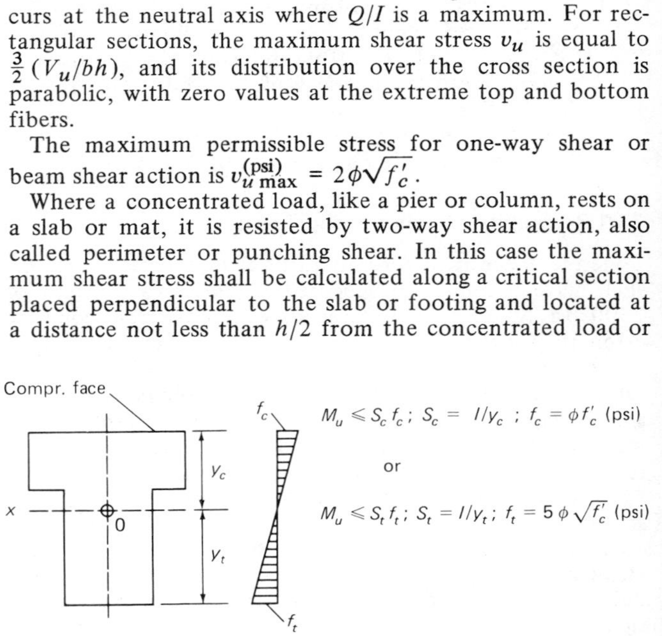

Fig. 27-1

reaction area. This is the same approach as specified for ordinary reinforced concrete in Chap. 11 of ACI 318. The maximum shear stress in this case can be evaluated as $v_{u0}^{(psi)} = 1000\, V_{u0}^{(k)}/b_0 h$.

The maximum permissible shear stress for two-way action is $v_{u0\,max}^{(psi)} = [2 + (4/\beta_c)]\ \phi\sqrt{f_c'}$, but shall never be greater than $4\phi\sqrt{f_c'}$.

In the above equation, β_c is the ratio longside/shortside of the concentrated load or reaction area. The way the equation is set up, a maximum permissible shear stress of $v_{u0} = 4\phi\sqrt{f_c'}$ is only applicable when the perimeter line is either square or has a maximum β_c-ratio of 2. For more elongated perimeters the $4/\beta_c$-value becomes a reduction factor to the maximum permissible shear stress under two-way action.

Lightweight concrete is not too often used as a material for structural plain concrete work, but it has definite advantages where precasting is employed to a considerable extent. In such cases, the following adjustments shall be made in the equations for the permissible shear stresses.

Where the quality of the lightweight concrete is specified by using a minimum splitting tensile strength f_{ct}, substitute $f_{ct}/6.7$ for $\sqrt{f_c'}$ in the permissible stress values above. In no case shall $f_{ct}/6.7$ exceed $\sqrt{f_c'}$.

Where lightweight concrete is to be used but the f_{ct} is *not* specified, the value of $\sqrt{f_c'}$ shall be multiplied by 0.75 for "all lightweight" concrete and by 0.85 for "sand-lightweight" concrete.

27.2.4 Bearing

Assuming the P_u and A_1 are the same for all three cases, the factored bearing pressure f_p will be the same in all three cases, namely, $f_p = P_u/A_1$.

This factored bearing pressure must be equal to or smaller than the maximum permissible bearing pressure of the supporting pedestal. Assuming that the pedestal is made of structural plain concrete and that the rules of ACI 318 apply, this value may be different in all three cases shown in Fig. 27-2.

Basically, the maximum permissible bearing pressure on a structural plain concrete member is $F_p = 0.85\phi f_c'$, if the contact area of the supporting part is of equal size and matching the load bearing area of the upper member as shown in Case 1. This applies not only to prismatic pedestals but also to tapered, conical-shaped, or stepped members as long as the slope of the cone envelope does not exceed an angle of 1 vertical to 2 horizontal.

Where the supporting surface of the lower member (A_2) is larger than the contact area of the upper member, as indicated in Case 2, the maximum permissible bearing pressure may be increased by the factor $\sqrt{A_2/A_1}$. In this case the area A_2 must be geometrically similar to and concentric with the upper area A_1, and the magnification factor of $\sqrt{A_2/A_1}$ shall never exceed the value of 2.

Where the shape of the supporting member is either stepped or sloped as indicated in Case 3, the area A_2 may be taken as the base area of a right cone or pyramid located entirely within the supporting structural plain concrete member and having side slopes of 1 vertical to 2 horizontal. Here as well, the magnification factor shall never exceed the value of 2.

27.2.5 Axial Compression

Columns are not permitted to be built of structural plain concrete.

The only compression members made of structural plain concrete that are permitted by the *Code* are pedestals and walls.

Arches and related structures, where the concrete is continuously in a state of compression, are permitted to be constructed of structural plain concrete, but are not specifically dealt with by the *Code*.

The design of a pedestal (i.e., the selection of its cross section) is governed by the factored load P_u and by the bearing pressure f_p it has to sustain, as discussed in Section 27.2.3 of this chapter.

The nominal compressive capacity of a wall is governed by the empirical formula given below:

$$P_{nw} = 0.45 f_c' A_g \left[1 - \left(\frac{kl_c}{32h_w}\right)^2\right]$$

$$P_u \leqslant \phi P_{nw}$$

In the above equation, the effective length factor k shall be assumed with 0.8 if at least one of its ends (top or bottom) is restrained against rotation, and with 1.0 if the wall is unrestrained against rotation at either end.

The above equation is also applicable if the resultant force P_u does not act at the centerline of the wall but at an eccentricity $e \leqslant h_w/6$, that is, within the kern area.

However, when the eccentricity, e, exceeds the kern distance, the design has to proceed in accordance with the requirements for combined stresses as described in Section 27.2.6.

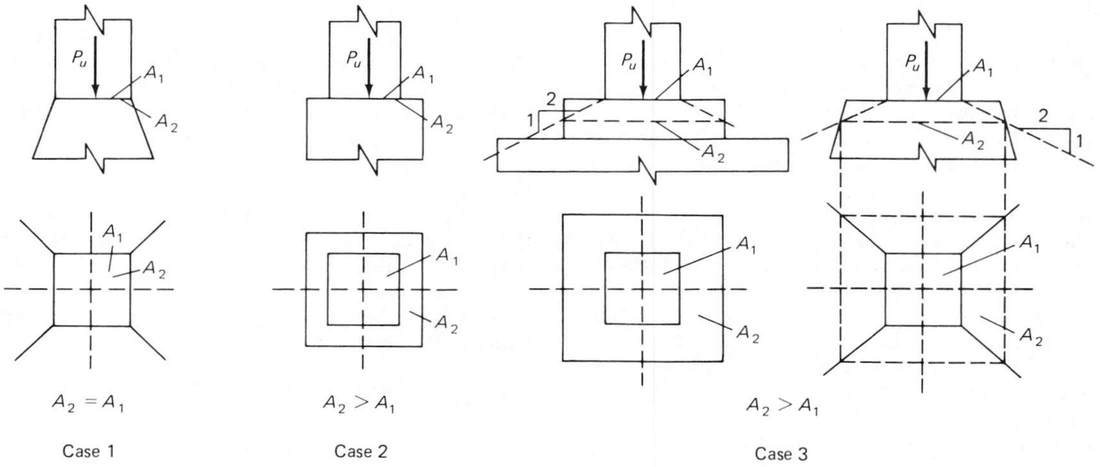

Fig. 27-2

27.2.6 Combined Stresses

Where the factord load P_u is applied at an eccentricity exceeding the kern distance $h_w/6$, or where smaller compressive loads are acting alongside larger flexural monents, so that $e = M_u/P_u > h_w/6$, then the resultant fiber stresses have to be investigated at both faces of the element. In such a case both kinds of fiber stresses, those caused by axial loads as well as those caused by flexural moments, have to be evaluated and to be combined as shown below to arrive at the critical combination.

At the compression face, the sum of the ratios of the various stress types has to be smaller than or equal to unity, and on the tension face the sum of all stresses has to be smaller than or equal to the maximum permissible tensile stress in the structural plain concrete section. Expressed in form of an equation, at the compression face $f_a/F_{ac} + f_b/F_{bc} \leqslant 1$; and at the tension face $f_b - f_a \leqslant F_t$.

In the above equations f_a = calculated axial stress from factored loading:

$$f_a = 1000P_u^{(k)}/bh;$$

F_{ac} = maximum permissible axial stress:

$$F_{ac} = 0.6\phi f_c' \left[1 - \left(\frac{kl_c}{32h_w}\right)^2\right]$$

(this stress is larger than the one used in Section 27.2.5 because it does not include the allowance for a small eccentricity); f_b = calculated flexural stress from factored loading:

$$f_b = 12{,}000 M_u^{(\text{kft})} \times 6/bh^2;$$

F_{bc} = maximum permissible flexural stress in compression:

$$F_{bc} = \phi f_c';$$

F_t = maximum permissible flexural stress in tension:

$$F_t = 5\phi\sqrt{f_c'}.$$

All stresses are in psi; b = the actual or effective width of the member, whichever is smaller.

27.3 SPECIFIC REQUIREMENTS AND EXAMPLES

27.3.1 Footings

The base area of a footing shall be determined from the total *un*factored column load and from the allowable soil pressure evaluated by principles of soil mechanics. Any eccentricities, additional bending moments, and/or horizontal shears shall be considered in the evaluation of the maximum edge pressures.

All further design procedures for structural plain concrete footings shall be based, just as for ordinary reinforced concrete footings, on factored loads, factored moments, and factored shears, and on a fictitious soil pressure evaluated from the factored loadings acting upon the previously determined base area.

The structural design of a footing made of structural plain concrete consists in principle of its investigation for flexural bending and shear. In most isolated spread footings, flexure, and in particular flexural tension, will govern the design thickness. For this reason it is advisable to start the design with this investigation and to check the selected sizes afterward for shear, if so required. In isolated spread footings, two-way shear is more critical than beam-shear; in strip footings, beam-shear is usually larger.

The critical lines about which the factored moments are to be calculated are the same as specified for reinforced concrete footings in Chap. 15 of ACI 318.

Where the cross section of a pier or pedestal is circular or polygonal, the outlines of a square, having the same cross-sectional area, may be substituted to simplify locating the critical sections for flexural moments and shears.

The least overall thickness of a footing is limited by the *Code* to 8 in. Since the lowest 2 in. of a structural element, cast directly against the soil, are to be disregarded for reasons of unevenness and soil-contamination, the least thickness to be considered in the design is only 6 in.

Footings resting on piles (pile caps) and footings supported by pier foundations (caisson grillages) shall not be made of structural plain concrete.

Isolated spread footings or sections of strip footings located between control joints must be cast in one operation, to ensure that no separation in the material (cold joint) will occur.

Where the subsoil is dry and pervious, it shall be wetted thoroughly before the start of the pouring to prevent rapid dessication, and subsequent excessive initial shrinkage of the concrete. In extreme cases vapor seals have been used with success.

On very soft subsoils, structural plain concrete (often referred to as lean concrete) is often used to keep the amount of work and traffic inside the footing excavation to a minimum.

In unstable soils with relatively high water tables, speedy, clockwork-like construction sequences of excavation, soil-mechanical inspection, and casting have often helped to solve otherwise troublesome foundation problems.

Isolated spread footings shall not be made larger than the size of a unit expected to remain crack-free under the influence of local climate, curing, shrinkage, creep, and temperature. Footings, in this respect, are less sensitive than other structural elements because they are located below grade and are therefore better protected.

Strip footings under continuous walls shall be sectioned into units of satisfactory length. The exact locations of the control joints shall be coordinated with the locations of the wall joints above.

Strip footings supporting spaced concentrated loads (columns) shall be designed as combined footings with independent units between control joints. Not less than two, better more, loads shall be supported by each unit. Depending on the stiffness relationship between footing and supporting soil, such footings can be designed either for an equally distributed soil pressure or as beams on elastic foundations. For additional information see Chapter 5 of this *Handbook*.

Where the mats of isolated spread footings become thicker than 30 to 36 in., it is advisable to substitute a sloped or stepped footing. Where a stepped footing is used, the ratio of the projection of each step to its thickness shall be selected so as to satisfy the flexural as well as the shear stresses at each level. Such footings shall preferably be cast as one monolithic unit, with the forms for the upper steps either in place from the very beginning or ready to be set immediately after casting of each lower step has been completed. At any rate, the construction procedure must ensure that all steps will act together as one monolithic unit. Where an interruption of casting or a waiting period between two successive steps is unavoidable, key-ways or bar dowels shall be provided across the plane of separation to transfer the shear forces as required.

EXAMPLE 27-1: *Structural plain concrete footing.*

column dead load:	40 kips
column live load:	60 kips
total column load: (service)	100 kips

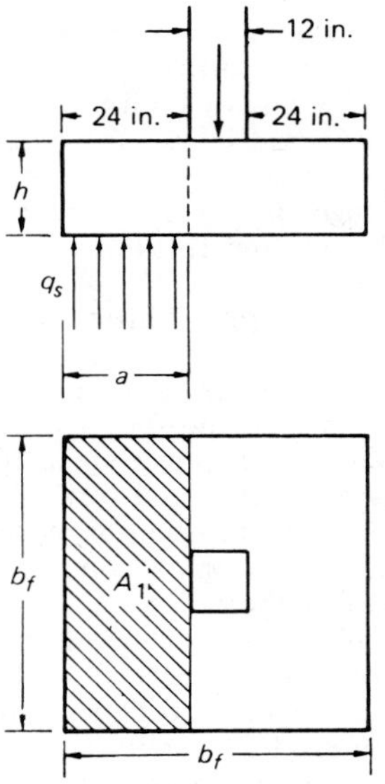

Fig. 27-3

Allowable soil pressure:

$$p_a = 4.0 \text{ kip/ft}^2$$

$$f'_c = 3000 \text{ psi}, \quad \text{pier size } 12 \times 12 \text{ in.}$$

Footing size:

$$A_F = \frac{100}{4} = 25 \text{ ft}^2, \quad \text{or} \quad 5 \times 5 \text{ ft}$$

Start with the flexural investigation because it governs the design of a square, structural plain concrete footing supporting a square column or pier. See Fig. 27-3.

$$P_u(\text{D.L.}) = 40.0 \times 1.4 = 56.0 \text{ kips}$$

$$P_u(\text{L.L.}) = 60.0 \times 1.7 = \underline{102.0} \text{ kips}$$

$$P_u = 158.0 \text{ kips}$$

$$q_s = \frac{P_u}{A_F} = \frac{158}{25} = 6.35 \text{ kips-ft}^2$$

$$M_u = b_f a q_s \frac{a}{2} = 5 \times \frac{24}{12} \times 6.35 \times \frac{24}{2 \times 12} = 63.5 \text{ kip-ft}$$

$$S_F = \frac{12 b_f \times d^2}{6} = 10d^2, \quad \text{for } b_f = 5.0$$

The permissible flexural strength is, according to Chap. 6 of ACI 318.1-83:

$$f_t = 5.0 \times \phi \times \sqrt{f'_c} = 5 \times 0.65 \times \sqrt{3000} = 179 \text{ psi}$$

$$S_{F(\text{required})} = \frac{M_u \times 12{,}000}{f_t} = \frac{63.5 \times 12{,}000}{179} = 4230$$

$$4230 = 10d^2, \quad d^2 = 423, \quad d = 20.5 \text{ in.}, \quad \text{say 21 in.}$$

Add 2 in. for unevenness and soil contamination:

$$\text{overall thickness } h = 21 + 2 = 23 \text{ in.}$$

Check footing thickness for two-way shear (for illustrative purposes only). See Fig. 27-4.

$$\Delta A_{V2} = \left(\frac{33}{12}\right)^2 = 7.55 \text{ ft}^2$$

$$\Delta V_{2u} = \Delta A_{V2} \times q_s = 7.55 \times 6.35 = 48 \text{ kips}$$

$$P_u - \Delta P_u = 158 - 48 = 110 \text{ kip}$$

$$v_{2u} = \frac{(P_u - \Delta P_u)\, 1000}{b_o d} = \frac{110 \times 1000}{4 \times 33 \times 21} = 40 \text{ psi}$$

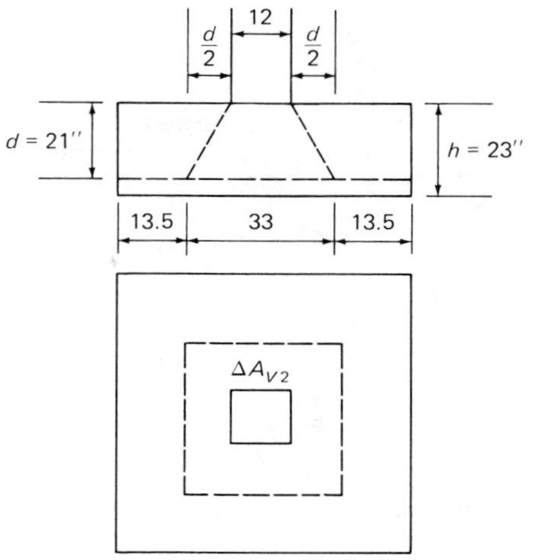

Fig. 27-4 Two-way shear in structural plain concrete footings.

Permissible:

$$v_{2u} = 4\phi \sqrt{f'_c} = 4 \times 0.65 \times \sqrt{3000} = 142 \text{ psi}$$

27.3.2 Pedestals

Structural plain concrete is often used for short intermediate-compression members, called pedestals. Such pedestals, for example, form the lowest portion of reinforced concrete columns between first-floor slab and footing mat, or they are used for the support of steel beams or under steel column base plates.

In the definition of the *ACI Code*, a pedestal is a free-standing prismatic member whose unsupported height is not greater than three time its least lateral dimension.

Their compressive strength is governed by the maximum permissible bearing capacity as calculated from the equation given in Section 27.2.4. For practical reasons, such pedestals are, wherever possible, made of a somewhat larger cross section as required by the size of the contact area of the supported member. This difference, or better the $\sqrt{A_2/A_1}$ ratio, may be utilized to magnify their permissible bearing capacity; or, vice versa, it may be used as the basis for the selection of their cross section, as long as the factor does not exceed 2.

Piers embedded in soil capable of providing lateral support may be treated like pedestals, regardless of their height-to-least-thickness ratio. Since the soil is in many cases just backfill material, it has to be well compacted to create a density equal to that of a natural soil capable of providing satisfactory lateral support.

The "allowable stresses" for the concrete used in pier foundations (caissons) or in cast-in-place concrete piles are usually governed by the General Building Code applicable for each particular site location, and do not form a part of the *Building Code Requirements for Structural Plain Concrete ACI 318.1*. These stresses are, as already indicated, "allowable" stresses and shall be used in combination with unfactored service loads.

EXAMPLE 27-2: *Pedestal.*
An 8" × 12" reinforced concrete column carries a total factored axial load $P_u = 165$ kips. Design a structural plain concrete pedestal to support the column if the top of the pedestal is flush with the top of the first floor slab and the top of the footing is 2′6" below the finished first floor elevation. See Fig. 27-5. $f'_c = 3000$ psi. For practical reasons, the pedestal is assumed to be 1" larger all around the column area, or 10" × 14". Loaded area $A_c = 8 \times 12 = 96$ in.^{2}

$$f_p = \frac{P_u}{A_1} = \frac{165}{96} = 1.72 \text{ ksi}$$

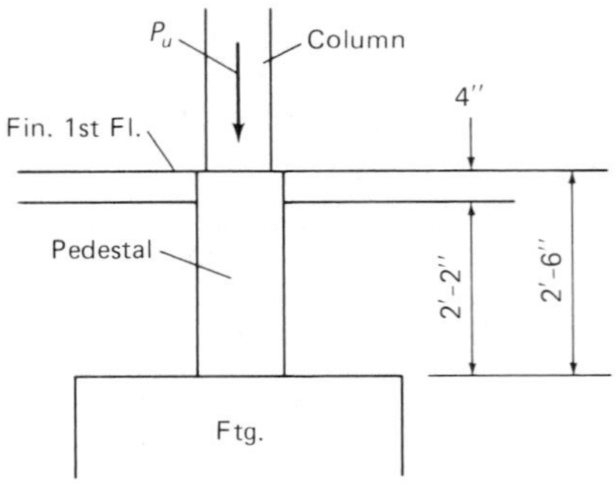

Fig. 27-5

The permissible bearing stress is, according to Section 27.2.4 of this *Handbook:*

$$F_p = 0.85\phi f_c' = 0.85 \times 0.65 \times 3.0 = 1.66 \text{ ksi}$$

Since the supporting surface is on all sides wider than the loaded area, the basic permissible F_p may be increased by the $\sqrt{A_2/A_1}$ factor, as long as it does not exceed 2. In the above factor, A_2 = larger cross-sectional area of supporting pedestal, and A_1 = smaller cross-sectional area of supported column.

$$A_2 = 10 \times 14 = 140 \text{ in.}^2$$

$$A_1 = 8 \times 12 = 96 \text{ in.}^2$$

$$\sqrt{\frac{A_2}{A_1}} = \sqrt{\frac{140}{96}} = 1.21, \ < 2.0$$

$$F_p = 1.66 \times 1.21 = 1.99 \text{ ksi} \quad 1.72 < 1.99$$

Check ratio of unsupported height to least lateral dimension of pedestal:

$$\text{Clear height} = 2'6'' - 4'' = 2'2'' = 26''$$

$$\frac{\text{Clear unsupported height}}{\text{Least lateral dimension}} = \frac{26}{10} = 2.6 < 3$$

This check can be omitted and the critical ratio of 3 can be disregarded, if the soil or backfill is well-enough compacted to provide lateral restraint for the pedestal.

27.3.3 Walls

Structural plain concrete is extensively used for the design and construction of exterior basement walls, interior bearing walls, precast wall panels, and the like.

The *Building Code Requirements for Structural Plain Concrete ACI 318.1* deals only with two basic types of wall designs:

1. Where the wall load is either applied axially or at a maximum eccentricity equal to the kern distance ($h_w/6$) of the cross section.
2. Where the eccentricity of the load exceeds the kern distance, or where small axial forces are acting alongside of larger flexural moments.

In the first case, the factored bearing capacity of the wall is determined as given in Section 27.2.5, whereas in the second case the extreme fiber stresses at both wall faces, due to factored loads and factored flexural moments, have to be determined and combined to see whether they fall within the permissible limits given in Section 27.2.6.

In addition to the evaluation of the bearing capacity or the calculation of the extreme fiber stresses, a certain number of limitations and requirements, as given below, have to be observed and followed in the design of a wall to be constructed of structural plain concrete.

Since it is rather difficult to devise a set of limitations and requirements that will control the safety of a wall design without undue curtailing of its usefulness, designers are reminded to exercise good judgment to determine where and when a wall of structural plain concrete is not only economically advantageous but also structurally safe; this applies to the wall itself and to the entire structure of which it forms a part.

In spite of all restrictions, limitations, and other requirements, there still remains quite a large field of useful applications for walls made of structural plain concrete.

The various limitations and special requirements for walls made of structural plain concrete are as follows:

1. They must be continuously vertically supported either by soil (i.e., resting on a footing) or by a structural member (like a steel beam, reinforced concrete beam, or similar) capable of supporting them.
2. They can only span in the vertical direction; that is, they must be continuously laterally supported at top and bottom. They do not have to be rotationally restrained at either end, but must be laterally supported at both ends. Retaining walls that are laterally supported only at the bottom thus cannot be built from structural plain concrete, although they are restrained at the bottom. (Gravity walls and walls used in connection with hydraulic structures do not form a part of the *Building Code Requirements for Structural Plain Concrete ACI 318.1.*)
3. They must have a minimum thickness of $\frac{1}{24}$ of their least height or width, whichever is shorter, and not be less than 6 in. thick. Exterior basement walls, foundation walls, party walls, and fire walls must have a minimum thickness of 8 in.
4. Lateral translation at top and bottom must be prevented by bracings, cross walls, piers, etc.
5. They must be surrounded by joints, that is, separated from adjoining portions of the structure, so that the size of each individual element is suitable to remain crack-free under the prevailing climatic conditions as well as under the curing, shrinkage, creep, and temperature variations particular to the site. The locations of the joints shall be carefully coordinated with the joints in the supporting footings and adjacent structure.
6. Every wall element shall be capable of resisting all forces applied to it in a safe manner.
7. Reentrant corners at windows, doors, and other openings shall be reinforced with not less than two #5 bars all around extending at least 24 in. beyond the corners of the opening.
8. The maximum horizontal length of a wall (b) considered to be effective in carrying a concentrated vertical load or reaction shall exceed neither the physical length of the wall portion, the spacing between wall joints, the center-to-center distance of the loads or reactions, nor the width of the bearing plus four times the wall thickness.
9. Like any other building element covered by this design and construction method, walls made of structural plain concrete shall not be used for resisting any forces caused by earthquake or blast. Since taller structures are more sensitive to wind and vibrations than shorter ones, designers are reminded to apply discretion and good judgment before using structural plain concrete indiscriminately for structures where sudden gusts, impact, or other shocks of related nature may cause vibrations and stresses that are not appropriate for structural plain concrete.

EXAMPLE 27-3: *Interior bearing wall.*
Design a free-standing (laterally held at top and bottom, but otherwise not restrained) interior bearing wall supporting concrete beams at 4′0″ o.c. Each beam is 10 in. wide and exerts a D.L. reaction of 20 kips and a L.L. reaction of 25 kips. The wall has a free height of

14'0" and shall be made of structural plain concrete having a specified compressive strength of $f'_c = 3000$ psi.

1. Determine min. wall thickness:

$$h_{min} = \frac{12 l_c}{24} = \frac{12 \times 14}{24} = 7 \text{ in.} \qquad \text{Use 8" wall thickness}$$

2. Determine effective wall section: The effective wall length is either $b + 4h_w$ or the spacing of the wall beams (S), whichever is smaller.

the wall is located below grade. Assume the active earth pressure with $0.03H$ in k/sq ft, where H is the depth below grade in ft. Wall is to be made of structural plain concrete having a specified compressive strength of 2500 psi.

1. Determine min. wall thickness (7.1.6.2) and (7.1.6.3):

$$h_{min} = \frac{12 l_c}{24} = \frac{12 \times 10.5}{24} = 5.25 \text{ in.} \quad h_{min} = 7\tfrac{1}{2} \text{ in.}$$

2. Find max. flexural moment per ft of wall length:

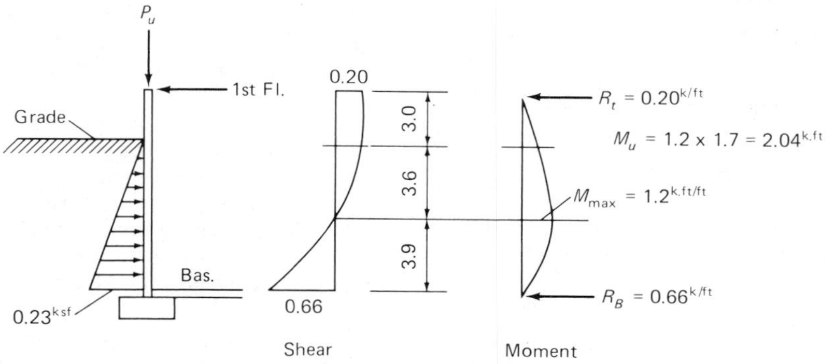

Fig. 27-6

$$b + 4h_w = 10 + 4 \times 8 = 42", \quad S = 4'0" = 48"; \quad 42" \text{ governs}$$

3. Check permissible bearing stress below beam:

$$f_p = \frac{P_u}{A_b} \qquad \begin{array}{l} P_u = \text{factored beam reaction} \\ A_b = \text{bearing area} \end{array}$$

$$P_u = 20 \times 1.4 + 25 \times 1.7 = 70.5^k, \quad f_p = \frac{70.5}{8 \times 10} = 0.88 \text{ ksi}$$

The permissible bearing strength is:

$$0.85\phi f'_c = 0.85 \times 0.65 \times 3 = 1.66^{ksi}$$

$$f_p \leqslant F_p, \qquad 0.88 < 1.66$$

4. Nominal bearing capacity of wall: Since the load is acting concentrically (or within the kern distance, i.e., $h_w/6$, the equation from Section 27.2.5 of this *Handbook* is applicable:

$$P_{nw} = 0.45 f'_c A_g \left[1 - \left(\frac{k l_c}{32 h_w}\right)^2\right]$$

A_g = effective cross-sectional area of wall; $k = 1$, because the wall is unrestrained at top and bottom.

$$P_{nw} = 0.45 \times 3 \times (8 \times 42) \left[1 - \left(\frac{1 \times (14 \times 12)}{32 \times 8}\right)^2\right] = 258^k$$

$$P_u \leqslant \phi P_{nw} \qquad \phi P_{nw} = 0.65 \times 258 = 167.7^k$$

Service D.L. of Bm:	20.0
Wt. of wall: $\frac{8 \times 42}{144}$ 0.15 × 14 =	4.9
	24.9^k
Service L.L. of Bm:	25.0^k

P_u (D.L.) = 24.9 × 1.4 = 34.9^k

P_u (L.L.) = 25.0 × 1.7 = 42.5^k

P_u = 77.4^k

$$77.4^k < 167.7^k$$

EXAMPLE 27-4: *Exterior basement wall.*
Design an exterior basement wall (laterally held at top and bottom, but otherwise not restrained), having a free height of 10'6" and supporting a D.L. of 3.0 k/ft and a L.L. of 2.5 k/ft. The lower 7'6" of

3. Investigate max. fiber stresses in wall:
   (a) At compression face (max. compression occurs at base of wall):

$$\frac{f_a}{F_{ac}} + \frac{f_b}{F_{bc}} \leqslant 1$$

f_a = actual compression stress

D.L.		3.00
Wt. of wall:	$\left(0.15 \frac{8}{12}\right) 10.5 =$	1.05
		4.05 × 1.4 = 5.65
L.L.		2.50 × 1.7 = 4.25
P_u =		9.90K

$$f_a = \frac{P_u}{A_g} = \frac{9.90}{8 \times 12} = 0.10^{ksi}$$

F_{ac} = permissible compression stress

$$F_{ac} = 0.6\phi f'_c \left[1 - \left(\frac{k l_c}{32 h_w}\right)^2\right] = 0.6 \times 0.65 \times$$

$$\times 2.5 \left[1 - \left(\frac{1 \times (10.5 \times 12)}{32 \times 8}\right)^2\right] = 0.74 \text{ ksi}$$

f_b = actual bending stress at compression face

$$f_b = -\frac{M_u 12}{S_w} = -\frac{2.04 \times 12}{128} = -0.19 \text{ ksi}$$

$$S_w = \frac{b h_w^2}{6} = \frac{12 \times 8^2}{6} = 128 \text{ in.}^3, \text{ section modulus of wall}$$

F_{bc} = permissible flexural stress in compression

$$F_{bc} = \phi f'_c = 0.65 \times 2.5 = 1.63 \text{ ksi}$$

$$\frac{f_a}{F_{ac}} + \frac{f_b}{F_{bc}} = \frac{0.10}{0.74} + \frac{0.19}{1.63} = 0.14 + 0.12 = 0.26 < 1$$

   (b) At tension face (max. flexural tension occurs at location of M_u max.)

$$f_b - f_a \leqslant F_t$$

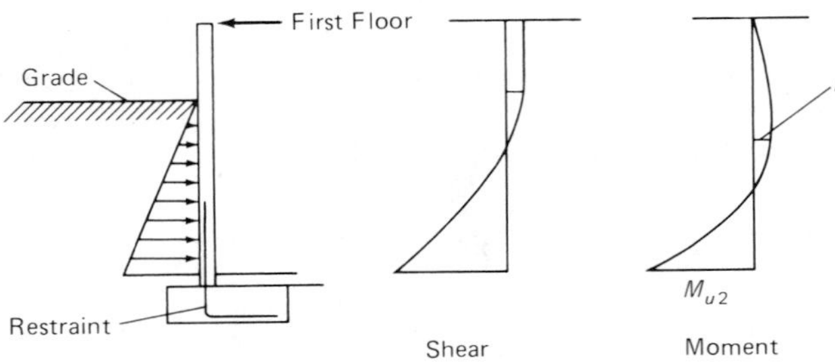

Fig. 27-7

f_b = actual flexural tension stress

$$f_b = +\frac{12M_u}{S} = \frac{12 \times 2.04}{128} = +0.19 \text{ ksi}$$

f_a = actual compression stress due to D.L. only (L.L. is disregarded to be on safe side)

D.L		3.00
Wt. of wall (at location of M_u):	$\left(0.15\frac{8}{12}\right)(3.0 + 3.6) =$	0.66
		3.66

$$P_{u(\text{D.L.})} = 3.66 \times 1.4 = 5.12^{\text{k}}$$

$$f_a = \frac{P_{u(\text{D.L.})}}{A_g} = \frac{5.12}{8 \times 12} = 0.05 \text{ ksi}$$

F_t = permissible flexural stress in tension:

$$F_t = 5\phi\sqrt{f_c'} = 5 \times 0.65 \times \sqrt{2500} = 0.16 \text{ ksi}$$

$$f_b - f_a \leqslant F_t\text{: } +0.19 - 0.05 = +0.14 \text{ ksi} \leqslant 0.16 \text{ ksi}$$

The above example shows that a considerable D.L. is required to keep the flexural tensile stress within the permissible limits, even for relatively small flexural moments. Where the D.L. is not available in sufficient magnitude, the condition can be helped by restraining the bottom end of the wall as shown in Fig. 27-7. In this case the M_{u1} is often sufficiently reduced to bring it within the permissible limits. Reinforcement, however, has to be provided to develop the M_{u2} at the bottom, and the footing has to be designed to transfer the M_{u2} into the subsoil alongside the other vertical loads. In the evaluation of the F_{ac} at the compression face, the k-factor can be reduced to 0.8, according to Section 7.1.5.2a.

Such a design approach is always recommended where groundwater is part of the lateral pressure. Provision of vertical reinforcement at the base of the wall will also help to keep the joint between the footing and wall tight.

NOTATION

A_g = gross area of section, sq in.
A_1 = loaded area, sq in.
A_2 = maximum area of the portion of the supporting surface that is geometrically similar to and concentric with the loaded area, sq in.
b = width of member, in.
b_0 = perimeter of critical section for shear in footings, in.
e = eccentricity, distance of factored load resultant from wall centroid, in.
d = effective depth of section, in.
f_a = stress due to axial compression, psi
f_b = fiber stress due to flexure, psi
f_c = fiber stress in compression, psi
f_c' = specified compressive strength of concrete, psi
$\sqrt{f_c'}$ = square root of specified compressive strength of concrete, psi
f_{ct} = average splitting tensile strength of lightweight aggregate concrete, psi
f_p = stress due to bearing, psi
f_r = modulus of rupture, psi
f_t = fiber stress in tension, psi
F_{ac} = extreme stress due to axial compression, psi
F_{bc} = extreme combined fiber stress at compression face, psi
F_p = extreme stress due to bearing, psi
F_t = extreme combined fiber stress at tension face, psi
h = overall thickness of member, in.
h_w = wall thickness, in.
I = moment of inertia of section, in^4
k = effective length factor
l_c = vertical distance between lateral supports, in.
M_u = factored moment at section, kft
P_{nw} = nominal axial load strength of wall, k
P_u = factored axial load, k
Q = static moment of cut-off area of cross section about common centroid, in^3
S = section modulus of section symmetrical about the x–x axis, in^3
S_c = section modulus of section about compression face, in^3
S_t = section modulus of section about tension face, in^3
v_u = shear stress due to factored shear force at section, psi
= shear stress due to beam action, psi
v_{uo} = shear stress due to two-way action, psi
V_u = factored shear force at section, k
V_{uo} = factored shear force along critical section for two-way action, k
y_c = distance of compression face from neutral axis, in.
y_t = distance of tension face from neutral axis, in.
β_c = ratio of long side to short side of concentrated load or reaction area
ϕ = strength reduction factor

REFERENCES

27-1 *Building Code Requirements for Structural Plain Concrete ACI 318.1-83 and Commentary ACI 318.1R-83.*

27-2 *Building Code Requirements for Structural Plain Concrete ACI 322-72* (former ACI standard).

Index